Reference use only
not for loan

Ewing's Analytical Instrumentation Handbook

Third Edition

Ewing's Analytical Instrumentation Handbook

Third Edition

Edited by Jack Cazes
Florida Atlantic University
Boca Raton, Florida, U.S.A.

MARCEL DEKKER

NEW YORK

Although great care has been taken to provide accurate and current information, neither the author(s) nor the publisher, nor anyone else associated with this publication, shall be liable for any loss, damage, or liability directly or indirectly caused or alleged to be caused by this book. The material contained herein is not intended to provide specific advice or recommendations for any specific situation.

Trademark notice: Product or corporate names may be trademarks or registered trademarks and are used only for identification and explanation without intent to infringe.

Library of Congress Cataloging-in-Publication Data
A catalog record for this book is available from the Library of Congress.

ISBN: 0-8247-5348-8

This book is printed on acid-free paper.

Headquarters
Marcel Dekker, 270 Madison Avenue, New York, NY 10016, U.S.A.
tel: 212-696-9000; fax: 212-685-4540

Distribution and Customer Service
Marcel Dekker, Inc., Cimarron Road, Monticello, New York 12701, U.S.A.
tel: 800-228-1160; fax: 845-796-1772

World Wide Web
http://www.dekker.com

Copyright © 2005 by Marcel Dekker. All Rights Reserved.

Neither this book nor any part may be reproduced or transmitted in any form or by any means, electronic or mechanical, including photocopying, microfilming, and recording, or by any information storage and retrieval system, without permission in writing from the publisher.

Current printing (last digit):

10 9 8 7 6 5 4 3 2 1

PRINTED IN THE UNITED STATES OF AMERICA

Coventry University

Preface to the Third Edition

The *Analytical Instrumentation Handbook* serves as a guide for workers in analytical chemistry, who need a starting place for information about a specific instrumental technique, either as a basic introduction to it or as a means of finding leading references dealing with theory and methodology for an instrumental technique.

The chapters which appeared in the second edition have been thoroughly expanded and updated, with concepts, applications, and key references to the recent literature. Only one chapter (Laboratory Balances) has been eliminated; eight new chapters have been added to the *Handbook*, dealing with

- Microchip technology
- Biosensor technology
- Validation of chromatographic methods
- Gel permeation and size exclusion chromatography
- Field-flow fractionation
- Countercurrent chromatography
- Hyphenated techniques in chromatography, including LC-MS, LC-NMR, and so on
- Thin-layer chromatography

The chapters have been written from the standpoint of the instrumentation as it is in use today, with an introductory description of the technique(s), and a theoretical treatment of the science and technology, wherever it is applicable or where it will facilitate an understanding of the instrumentation. However, the major emphasis is on the instrumentation. The chapters are not, simply, a "catalog" of commercially available instruments. Nevertheless, in some cases, commercially available instruments have been used as examples to illustrate design features discussed in the text.

It is sincerely intended and anticipated that this third revised and expanded edition of the *Analytical Instrumentation Handbook* will serve as a ready-reference on the desks of all practitioners of instrumental analytical chemistry.

Contents

Preface to the Third Edition .... *iii*
Contributors .... *xxi*

1. The Laboratory Use of Computers *1*
   Wes Schafer, Zhihao Lin
   I.   Introduction *1*
   II. Computer Components and Design Considerations *1*
        A. Motherboard/Central Processing Unit *2*
        B. Memory/Cache *3*
        C. Disk Storage *3*
        D. Video (Graphics Card and Monitor) *4*
        E. Other Peripherals *5*
   III. Computer Maintenance *5*
        A. Performance Monitoring *5*
        B. Virus Protection *6*
        C. Backup and Recovery *6*
   IV. Data Transfer/Instrument Interfaces *7*
        A. Transducers *7*
        B. Analog Signal Transmission *7*
        C. Analog Signal Filtering *7*
        D. Analog-to-Digital Conversion *8*
        E. Digital Signal Transmission *10*
        F. Digital Data Filtering *11*
   V.   Data Analysis/Chemometrics *12*
        A. Multivariate Calibration *12*
        B. Pattern Recognition *16*
   VI. Data Organization and Storage *18*
        A. Automated Data Storage *19*
        B. Laboratory Information Management Systems *19*
   Bibliography *20*

2. Flow Injection/Sequential Injection Analysis *21*
   Elo Harald Hansen, Jianhua Wang
   I.   Introduction *21*
   II. Flow Injection Analysis *23*

           A. Basic Principles 23
           B. Basic FIA Instrumentation 23
           C. Dispersion in FIA 24
           D. Selected Examples to Illustrate the Unique Features of FIA 25
           E. FIA Gradient Techniques—Exploiting the Physical Dispersion Process 30
           F. FIA for Process Analysis/Monitoring 32
     III. Further Developments of FIA 32
           A. Sequential Injection Analysis 33
           B. Lab-on-Valve 33
     IV. Selected Applications of FIA/SIA and their Hyphenations in On-Line Sample Pretreatments 34
           A. Separation and Preconcentration Procedures Based on Column Reactors or KRs 36
           B. FIA/SIA On-Line Solvent Extraction and Back Extraction Preconcentration 39
           C. On-Line Hydride/Vapor Generation Preconcentration Schemes 44
     V. Applications of SI-LOV for On-Line Sample Pretreatments 45
           A. The Third Generation of Flow Injection and Micro Total Analysis Systems 45
           B. Applications of μSI-LOV Systems for On-Line Sample Pretreatments with Optical Monitoring 46
           C. SI-BI-LOV: Solving the Dilemma of On-Line Column Separation/Procencentration Schemes 47
     References 52

3. Inductively Coupled Plasma Optical Emission Spectrometry 57
Tiebang Wang
     I. Introduction 57
     II. An Overview 59
           A. Principle of Atomic Emission Spectrometry 59
           B. Atomic Spectroscopic Sources 60
           C. Atomic Spectroscopic Techniques and Instruments 60
     III. Inductively Coupled Plasma 61
           A. The Discharge 61
           B. Emission Detection and Processing 62
           C. Analytical Performance and General Characteristics 62
     IV. ICP-OES Instrumentation 63
           A. Nebulizers 63
           B. Spray Chambers 65
           C. Alternative Sample Introduction Systems 65
           D. Optics and the Spectrometer 66
           E. Emission Detectors 69
           F. Radially Viewed ICP 71
     V. Applications of ICP-OES 71
           A. Geoanalysis 71
           B. Metallurgical Samples 72
           C. Agricultural and Food Samples 72
           D. Biological and Clinical Samples 72
           E. Environmental Samples 73
           F. Organics 73
           G. Nuclear Materials 73
     VI. Summary and Future Prognosis 74
     References 74

4. Atomic Absorption Spectrometry and Related Techniques 75
Bernhard Welz, Maria Goreti R. Vale
     I. Introduction 75
     II. The History of Atomic Spectroscopy 76
           A. The Early History 76
           B. Sir Alan Walsh and the Rebirth of AAS 77
           C. Boris L'vov and the Graphite Furnace 77

III. Atomic Spectra in Absorption, Emission, and Fluorescence *79*
   A. Origin and Characteristics of Atomic Spectra *79*
   B. Line Width and Line Profile *80*
   C. Measuring the Absorption—the Beer–Lambert Law *81*
   D. The Zeeman Effect *82*
IV. Spectrometers for AAS *82*
   A. Medium-Resolution LSAAS *83*
   B. High-Resolution CSAAS *92*
V. The Techniques of Atomic Spectrometry *95*
   A. The FAAS Technique *95*
   B. The GFAAS Technique *101*
   C. Chemical Vapor Generation *112*
   D. Spectrometers for AFS *117*
   E. Spectrometers for Flame OES *118*
VI. Measurement, Calibration, and Evaluation in AAS, AFS, and OES *118*
   A. Samples and Measurement Solutions *118*
   B. Calibration *119*
   C. Evaluation *121*
References *123*

5. Ultraviolet, Visible, Near-Infrared Spectrophotometers — 127
Chris W. Brown

I. Introduction *127*
   A. Beer's Law *127*
   B. Deviations from Beer's Law *128*
II. Spectrophotometer Characteristics *129*
   A. The Architecture of a Spectrophotometer *131*
III. Present and Future UV–VIS–NIR Spectrometers *137*
References *139*

6. Molecular Fluorescence and Phosphorescence — 141
Fernando M. Lanças, Emanuel Carrilho

I. Introduction *141*
II. Theory *141*
   A. Excited-State Processes *142*
   B. Excited-State Lifetimes *142*
   C. Quantum Yield *143*
   D. Quenching *143*
   E. Intensity and Concentration *143*
   F. Luminescence Spectra *144*
   G. Polarization *144*
III. Instrumentation *145*
   A. Instrument Components *145*
   B. Instrument Configurations for Conventional Fluorescence Measurements *147*
   C. Phosphorimeters *148*
   D. Instruments with Special Capabilities *150*
IV. Practical Considerations and Applications *153*
   A. Environmental Effects *155*
   B. Wavelength Calibration and Spectral Correction *156*
   C. Background and Interfering Signals *156*
   D. Direct Quantitative and Qualitative Analyses *156*
   E. Low Temperature Techniques *156*
   F. Room-Temperature Phosphorescence *157*
   G. Techniques and Applications *157*
   H. Future Outlook *158*
References *159*

7. Vibrational Spectroscopy: Instrumentation for Infrared and Raman Spectroscopy ... 163
Peter Fredericks, Llewellyn Rintoul, John Coates

    I. Introduction 163
        A. Background for Vibrational Spectroscopic Measurements 164
        B. The Basic Principles of Vibrational Spectroscopy 166
        C. Comparison of Techniques and Relative Roles 168
    II. IR Instrumentation 170
        A. Types of Instruments and Recent Trends 170
        B. Instrumentation: Design and Performance Criteria 172
        C. NIR Instrumentation 206
        D. FIR (Terahertz) Instrumentation 209
    III. Raman Instrumentation 211
        A. Types of Instrumentation and Recent Trends 211
        B. Instrumentation: Design and Performance Criteria 213
    IV. Practical Issues of Implementation of IR and Raman Instruments 230
        A. Instrument Architecture and Packaging 230
        B. Data Handling and Computer Applications 232
    V. Instrumentation Standards 233
    VI. Recommended Reference Sources 234
Suggested Reference Texts 234
General Texts and Reviews 235
Historical Reviews 235
References 235

8. X-Ray Methods ... 239
Narayan Variankaval

    I. Introduction 239
        A. Continuous and Characteristic Spectra 239
        B. Diffraction from Crystals 240
        C. Auger Effect 241
        D. Wavelength Dispersive X-Ray Spectrometry 241
        E. Energy Dispersive X-Ray Spectrometry 241
        F. X-Ray Reflectivity 241
        G. X-Ray Absorption 241
    II. Instrumentation and Methods 242
        A. Overview of Optical Components of an X-Ray Diffractometer 242
        B. X-Ray Methods 247
    III. Concluding Remarks 255
Additional Reading 255
    General 255
    X-Ray Properties of Elements 255
    X-Ray Diffraction 255
    X-Ray Detectors 255
    Particle Induced X-Ray Emission 255
    X-Ray Fluorescence 255
    X-Ray Photoelectron Spectroscopy 255
    X-Ray Reflection 255
    X-Ray Microanalysis 255
References 255

9. Photoacoustic Spectroscopy ... 257
A.K. Rai, S.N. Thakur, J.P. Singh

    I. Introduction 257
    II. Photoacoustics and PA Spectroscopy 258
        A. History 258
        B. Principle 259

Contents

   III. Relevance of PA to Plant Science *261*
   IV. Review of PA Applications in Plant Sciences *262*
   V. PA in Detection of Plant Disease *263*
   A. Experimental Setup *263*
   B. Diseases in Wheat Plants *263*
   C. Wheat Genotype EKBSN-1 AND FBPFM-2 .... *267*
   D. Virus Disease in Mung Bean Plant *267*
   E. Fungal Disease in Sugarcane *268*
   F. Virus Disease in Okra Plants *268*
   G. Seed-Borne Diseases in Wheat and Rice *268*
   VI. Conclusion *268*
   VII. Future Developments *268*
   References *269*

10. Techniques of Chiroptical Spectroscopy *271*
Harry G. Brittain, Nelu Grinberg
   I. Introduction to Chiroptical Phenomena *271*
   II. Polarization Properties of Light *272*
   III. Optical Rotation and ORD *272*
   A. Applications of Optical Rotation and ORD *275*
   IV. Circular Dichroism *276*
   A. Applications of CD *277*
   V. CPL Spectroscopy *278*
   A. Applications of CPL Spectroscopy *280*
   VI. Vibrational Optical Activity *283*
   A. Vibrational Circular Dichroism *283*
   B. Raman Optical Activity *285*
   VII. Fluorescence Detected Circular Dichroism *288*
   A. Applications of FDCD *289*
   VIII. Concluding Remarks *290*
   References *290*

11. Nuclear Magnetic Resonance *295*
Frederick G. Vogt
   I. Introduction *295*
   II. Instrument Design *296*
   A. Magnet Systems *296*
   B. NMR Probes *300*
   C. RF Generation and Signal Detection *303*
   D. Magnetic Field Gradients *306*
   E. Computer Systems *306*
   F. Accessories *307*
   III. Theoretical Background *307*
   A. Nuclear Spin Dynamics *307*
   B. External Manipulations of Spin Coherence *310*
   C. Internal Spin Interactions *311*
   D. Relaxation Phenomena and Chemical Dynamics *314*
   IV. Experimental Methods *317*
   A. Basic Pulse Methods *317*
   B. Multidimensional NMR Spectroscopy *322*
   C. Multiple Resonance and Heteronuclear Techniques *325*
   D. Diffusion, Dynamics, and Relaxation Measurements *332*
   V. Data Analysis and Interpretation *334*
   A. Spectral Processing *335*
   B. Manual Data Interpretation *337*

C. Predictive Methods *342*
D. Computer-Assisted Structure Elucidation *343*
VI. Conclusion *343*
References *344*

12. Electron Paramagnetic Resonance ... *349*
Sandra S. Eaton, Gareth R. Eaton

I. Introduction *349*
  A. EPR Experiment *350*
  B. CW Spectroscopy *352*
  C. Pulse Spectroscopy *355*
  D. Multiple Resonance Methods *355*
II. What One Can Learn From an EPR Measurement *357*
  A. Is There an EPR Signal? Under What Conditions? *357*
  B. Lineshape *358*
  C. *g*-Value *358*
  D. Spin–Spin Coupling *360*
  E. Relaxation Times *360*
  F. Saturation Behavior *361*
  G. Signal Intensity *362*
III. EPR Spectrometer *363*
  A. Microwave System *363*
  B. EPR Resonators *366*
  C. Magnet System *370*
  D. Signal Detection and Amplification *372*
  E. Data System *373*
  F. Commercial EPR Spectrometers *373*
IV. The Sample *374*
  A. Spin System *374*
  B. Temperature *374*
  C. Amount of Sample *375*
  D. Phase of Sample *376*
  E. Impurities, Overlapping Spectra *377*
  F. Intermolecular vs. Intramolecular Interactions *377*
  G. Other Environmental Effects *377*
V. Quantitative Measurements of Spin Density *378*
  A. Spectrometer Calibration *378*
  B. Calibration of Sample Tubes *379*
  C. Consideration of Cavity Q *379*
  D. Reference Samples for Quantitative EPR *380*
  E. Scaling Results for Quantitative Comparisons *380*
VI. Guidance on Experimental Technique *381*
  A. Words of Caution *381*
  B. Selection of Operating Conditions *382*
  C. Second Derivative Operation *384*
  D. CW Saturation *384*
  E. Methods of Measuring Relaxation Times *384*
  F. Measurement of B_1 .... *385*
  G. Line vs. Point Samples *386*
  H. Overcoupled Resonators *386*
  I. The ESE Flip Angle for High-Spin Systems *387*
VII. Less Common Measurements with EPR Spectrometers *387*
  A. Saturation-Transfer Spectroscopy *387*
  B. Electrical Conductivity *388*
  C. Static Magnetization *388*
  D. EPR Imaging *388*
  E. Pulsed Magnetic Field Gradients *388*

Contents

VIII. Reporting Results *388*
   A. Reporting Experimental Spectra *389*
   B. Reporting Derived Values *389*
References *390*

13. X-Ray Photoelectron and Auger Electron Spectroscopy 399
C. R. Brundle, J. F. Watts, J. Wolstenholme

I. Introduction *399*
   A. X-Ray Photoelectron Spectroscopy *400*
   B. Auger Electron Spectroscopy *401*
   C. Depth Analyzed for Solids by Electron Spectroscopy *402*
   D. Comparison of XPS and AES *403*
II. Electron Spectrometer Design *404*
   A. The Vacuum System *404*
   B. The Specimen and Its Manipulation *404*
   C. X-Ray Sources for XPS *405*
   D. Charge Compensation in XPS *406*
   E. The Electron Gun for AES *407*
   F. Electron Energy Analyzers for Electron Spectroscopy *408*
   G. Detectors *410*
   H. Spectrometer Operation for Small Area XPS *411*
   I. XPS Imaging and Mapping *411*
   J. Angle Resolved XPS *412*
III. Interpretation of Photoelectron and Auger Spectra *413*
   A. Qualitative Analysis *413*
   B. Chemical State Information: The Chemical Shift *414*
   C. Chemical State Information: Fine Structure Associated with Core Level Peaks *417*
   D. Quantitative Analysis *418*
IV. Compositional Depth Profiling *420*
   A. Angle Resolved X-Ray Photoelectron Spectroscopy *420*
   B. Variation of Electron Kinetic Energy *423*
   C. Ion Sputter Depth Profiling *424*
   D. Summary of Depth Profiling *425*
V. Areas of Application and Other Methods *426*
References *426*

14. Mass Spectrometry Instrumentation 429
Li-Rong Yu, Thomas P. Conrads, Timothy D. Veenstra

I. Introduction *429*
II. Ionization Methods *430*
   A. Matrix-Assisted Laser Desorption/Ionization *430*
   B. Electrospray Ionization *431*
III. Mass Analyzers *434*
   A. Triple Quadrupole Mass Spectrometer *434*
   B. Time-of-Flight Mass Spectrometer *435*
   C. Quadrupole Time-of-Flight Mass Spectrometer *435*
   D. Ion-Trap Mass Spectrometer *436*
   E. Fourier Transform Ion Cyclotron Resonance'Mass Spectrometry *437*
   F. Surface-Enhanced Laser Desorption/Ionization Time-of-Flight Mass Spectrometry *439*
   G. Inductively Coupled Plasma Mass Spectrometry *440*
IV. Conclusions *441*
References *441*

15. Thermoanalytical Instrumentation and Applications 445
Kenneth S. Alexander, Alan T. Riga, Peter J. Haines

I. Introduction *445*
   A. Scope of Thermal Analysis *445*

      B. Nomenclature and Definitions *446*
      C. Symbols *448*
II. Thermogravimetry *449*
      A. Development *449*
      B. Design Factors *449*
      C. The Balance *450*
      D. Furnaces and Temperature *451*
      E. Computer Workstations to Record, Process, and Control the Acquisition of Data *452*
      F. Reporting Results *453*
      G. Standards for TG *454*
III. TGA and DTG *456*
IV. DTA and DSC *457*
      A. Basic Considerations *457*
      B. Design Factors and Theory *458*
      C. Experimental Factors *463*
      D. Reporting Results *464*
      E. Calibration *465*
V. Evolved-Gas Analysis *467*
      A. Introduction *467*
      B. Instrumentation for EGA *468*
      C. The Basic Unit *469*
      D. Detector Devices *469*
      E. Precursor Considerations *470*
      F. Separation Procedures *470*
      G. Reporting Data *471*
VI. Thermomechanical Methods *471*
      A. Introduction *471*
      B. TDA and TMA *473*
      C. Dynamic Mechanical Analysis *473*
      D. Reporting Data *474*
      E. Instrumentation *474*
VII. Dielectric Thermal Analysis *476*
VIII. Calorimetry and Microcalorimetric Measurements *478*
      A. Introduction *478*
      B. Calorimeters *482*
      C. Instrumentation *482*
      D. Applications of Isothermal Calorimetry *486*
      E. Summary *490*
IX. Other Techniques *490*
X. Simultaneous Thermal Analysis Techniques *491*
      A. The Rationale Behind the Simultaneous Approach *491*
      B. Simultaneous TG–DTA *491*
      C. Simultaneous TG–DSC *492*
XI. Other Less-Common Techniques *493*
      A. Optical Techniques *493*
      B. Spectroscopic Techniques *493*
      C. X-Ray Techniques *494*
      D. DTA–Rheometry *494*
      E. Microthermal Analysis *494*
      F. Applications for TG–DTA *495*
      G. Applications of TG–DSC *496*
XII. Applications *497*
      A. Nature of the Solid State *497*
      B. Applications Based on Thermodynamic Factors *498*
      C. Applications Based on Kinetic Features *500*
      D. Applications for EGA *500*
References *503*

Contents

16. Potentiometry: pH and Ion-Selective Electrodes *509*
Ronita L. Marple, William R. LaCourse
- I. Introduction *509*
  - A. Electrochemical Cells *509*
  - B. Activity and Activity Coefficients *510*
  - C. The Nernst Equation *512*
  - D. Formal Potentials *512*
  - E. Liquid Junction Potentials *513*
  - F. Temperature Coefficients *513*
- II. Reference Electrodes *513*
  - A. The Standard Hydrogen Electrode *514*
  - B. The Calomel Electrode *514*
  - C. Silver, Silver Chloride Electrode *515*
  - D. Double Junction Reference Electrode *515*
  - E. Reference Electrode Practice *515*
- III. Indicator Electrodes *516*
  - A. Metallic Indicator Electrodes *516*
  - B. Membrane Indicator Electrodes *516*
- IV. General Instrumentation *523*
  - A. Potentiometers *523*
  - B. Direct Reading Instruments *523*
  - C. Commercial Instrumentation and Software *523*
- V. Applications of Potentiometry *524*
- VI. Current Research Activity *524*
  - A. Increasing Sensitivity, Lowering Detection Limits *524*
  - B. New Membrane Materials *525*
  - C. Biosensors *525*
  - D. Miniaturization *525*
  - E. Potentiometric Detectors in Fluidic Streams *525*
- VII. Conclusions *526*
- References *526*

17. Voltammetry *529*
Mark P. Olson, William R. LaCourse
- I. Introduction *529*
- II. General Instrumentation *529*
- III. Oxidation/Reduction *530*
- IV. Polarization and *iR* Drop *531*
- V. The Voltammogram *532*
- VI. Mass-Transport *532*
- VII. The Diffusion Layer *533*
- VIII. Faradaic Current *534*
- IX. Non-Faradaic Current *534*
- X. Voltage/Time/Current Interdependence *535*
- XI. Cyclic Voltammetry *535*
- XII. Hydrodynamic Voltammetry *536*
- XIII. Polarography *536*
- XIV. Waveforms *538*
- XV. Innovative Applications of Voltammetry *539*
  - A. Protein Film Voltammetry *539*
  - B. Voltammetry on Microfluidic Devices *540*
  - C. Fast-Scan Cyclic Voltammetry *540*
  - D. Ultramicroelectrodes *541*
  - E. Sinusoidal Voltammetry *541*

XVI. Suppliers of Analytical Instrumentation and Software *541*
XVII. Voltammetry Simulation Software *542*
References *542*

18. Electrochemical Stripping Analysis *545*
William R. LaCourse
  I. Introduction *545*
  II. Fundamentals of Stripping Analysis *546*
     A. Anodic Stripping Voltammetry *546*
     B. Cathodic Stripping Voltammetry *549*
     C. Stripping Chronopotentiometry *549*
     D. Adsorptive Stripping Voltammetry *550*
     E. Stripping Tensammetry *551*
     F. Experimental Considerations *551*
  III. Instrumentation *552*
     A. Cells *552*
     B. Electrodes *552*
     C. Stripping Analysis in Flowing Streams *554*
     D. Stripping Analyzers *555*
     E. Modeling Software and Speciation *557*
  IV. Applications *557*
  V. Conclusions *558*
  References *558*

19. Measurement of Electrolytic Conductance *561*
Stacy L. Gelhaus, William R. LaCourse
  I. Introduction *561*
     A. Principles of Conductivity *562*
     B. Strong Electrolytes *563*
     C. Weak Electrolytes *563*
     D. Ion Mobility and Transport *564*
  II. Instrumentation *566*
     A. Immersed Electrode Measurements *566*
     B. Electrodeless (Noncontacting) Measurements *574*
  III. Summary *577*
  References *578*

20. Microfluidic Lab-on-a-Chip *581*
Paul C. H. Li, Xiujun Li
  I. Introduction *581*
  II. Micromachining Methods *582*
     A. Micromachining of Silicon *582*
     B. Micromachining of Glass *583*
     C. Micromachining of Fused Silica (or Fused Quartz) *587*
     D. Micromachining of Polymeric Chips *588*
     E. Metal patterning *594*
  III. Microfluidic Flow *594*
     A. EOF and Hydrodynamic Flow *594*
     B. Surface Modifications for Flow Control *595*
     C. Fraction Collection *596*
     D. Laminar Flow for Liquid Extraction and Microfabrication *596*
     E. Concentration Gradient Generation *597*

Contents

       F. Microvalves *599*
       G. Micromixers *600*
       H. Alternative Pumping Principles *600*
       I. Microfluidic Flow Modeling Study *601*
  IV. Sample Introduction *602*
       A. Electrokinetic Injection *602*
       B. Other Sample Injection Methods *606*
  V. Sample Preconcentration *607*
       A. Sample Stacking *607*
       B. Extraction *608*
       C. Sample Volume Reduction *608*
       D. Other Preconcentration Methods *610*
  VI. Separation *610*
       A. Capillary Zone Electrophoresis *611*
       B. Capillary Gel Electrophoresis *612*
       C. Micellar Electrokinetic Capillary Chromatography *613*
       D. Derivatizations for CE for Separations *614*
       E. Isotachophoresis *615*
       F. Capillary Electrochromatography (CEC) *616*
       G. Synchronized Cyclic Capillary Electrophoresis *617*
       H. Free-Flow Electrophoresis *617*
  VII. Detection Methods *617*
       A. Optical Detection Methods *617*
       B. Electrochemical Detection *623*
       C. Mass Spectrometry *627*
  VIII. Applications to Cellular Analysis *627*
       A. Slit-Type Filters *627*
       B. Weir-Type Filters *631*
       C. Cell Adhesion *632*
       D. Studies of Cells in a Flow *633*
       E. DEP for Cell Retention *634*
  IX. Applications to DNA Analysis *635*
       A. DNA Amplification *635*
       B. DNA Hybridization *638*
       C. DNA Sequencing *642*
       D. High-Throughput DNA Analysis *642*
       E. Other DNA Applications *643*
  X. Applications to Protein Analysis *644*
       A. Immunoassay *644*
       B. Protein Separation *647*
       C. Enzymatic Assays *648*
       D. MS Analysis for Proteins and Peptides *650*
  References *659*

21. Biosensor Technology   *681*

Raluca-Ioana Stefan, Jacobus Fredeiczzk van Staden, Hassan Y. Aboul-Enein

  I. Introduction *681*
  II. Biological Materials *681*
  III. Transducers *682*
       A. Electrochemical Transducers *682*
       B. Optical Transducers *682*
       C. Screen-Printed Electrodes *683*
  IV. Immobilization Procedures of the Biological Material *683*
       A. Physical Immobilization *683*
       B. Chemical Immobilization *684*
  V. Design of a Biosensor for an Extended Use and Storage Life *684*
  VI. Array-Based Biosensors *684*

VII. Design of Flow Injection Analysis/Biosensors and Sequential Injection Analysis/Biosensors Systems 684
    A. Flow Injection/Biosensor(s) System 684
    B. Sequential Injection/Biosensor(s) System 685
References 685

22. Instrumentation for High-Performance Liquid Chromatography ... 687
Raymond P. W. Scott

I. Introduction to the Chromatographic Process 687
    A. The Plate Theory 688
    B. Molecular Interactions 690
    C. The Thermodynamic Explanation of Retention 692
    D. Control of Retention by Stationary Phase Availability 693
II. The Basic Chromatograph 696
    A. Solvent Reservoirs 696
    B. Solvent Programers 697
    C. Nonreturn Valves 699
    D. Chromatography Pumps 699
    E. Column Design and Performance 703
    F. Sample Valves 705
    G. Column Ovens 707
    H. The LC Column 708
    I. Detectors 713
III. The Modern Comprehensive LC System 724
References 725

23. Gas Chromatography ... 727
Mochammad Yuwono, Gunawan Indrayanto

I. Introduction 727
II. Basic Instrumentation 728
III. The Separation System 728
    A. Modes of GC 728
    B. Gas Chromatogram 728
IV. Gas Supply System 730
    A. Selection of Carrier Gases 730
    B. Gas Sources and Purity 730
V. Gas Chromatographic Columns 731
    A. Packed Columns 731
    B. Capillary Columns 732
    C. Stationary Phases in GC 732
    D. Optimization of the GC Column Parameters 733
    E. Connection Techniques of Capillary Column 736
VI. Sample Inlets 736
    A. Syringe Handling Techniques 736
    B. Packed Inlet System 736
    C. Split Injection 737
    D. Splitless Injectors 737
    E. Direct Injection 737
    F. Programed-Temperature Injection 738
    G. Injectors for Gas Samples 738
    H. Headspace Analysis 738
    I. Solid Phase Microextraction Headspace GC 739
    J. Purge and Trap GC Systems 739
    K. Pyrolysis GC 739
VII. Oven 740
    A. Conventional GC-Oven 740

Contents   xvii

          B. Flash GC-Oven *740*
          C. Microwave GC-Oven *740*
          D. Infrared Heated GC *740*
  VIII. Detector *741*
          A. Classification and General Properties *741*
          B. Flame Ionization Detector *742*
          C. Thermal Conductivity Detector *742*
          D. Electron-Capture Detector *742*
          E. Nitrogen/Phosphorous Detector *743*
          F. Flame Photometric Detector *744*
          G. Atomic Emission Detection *744*
          H. Gas Chromatography-Mass Spectrometry *744*
  IX. Multi Dimensional GC and Comprehensive Two-Dimensional GC *750*
          A. Multidimensional GC *750*
          B. Comprehensive Two-Dimensional GC *751*
  X. Chromatography Data System *752*
  XI. Qualitative and Quantitative Analyses *753*
          A. Qualitative Analysis *753*
          B. Quantitative Analysis *753*
  XII. Recent Developments *754*
          A. Fast GC *754*
          B. Portable GC *755*
          C. On-Line LC and GC *755*
References *755*

24. Supercritical Fluid Chromatography Instrumentation 759
Thomas L. Chester, J. David Pinkston

  I. Introduction *759*
          A. Intermolecular Forces in SFC *760*
          B. Fluids and Phase Behavior *760*
          C. Chromatography with Supercritical Fluid Mobile Phases *764*
          D. Reasons for Considering SFC *768*
  II. Instrumentation *769*
          A. Safety Considerations *769*
          B. General SFC Instrumentation Features *770*
          C. Packed-Column SFC Instrumentation *770*
          D. OT-Column SFC Instrumentation *776*
          E. Testing for Proper Operation *784*
          F. Flow Behavior with Upstream Pressure Control *784*
  III. Method Development *785*
          A. Packed-Column SFC Method Development *785*
          B. OT-SFC Method Development *789*
  IV. Supercritical Fluid Chromatography–Mass Spectrometry *791*
          A. SFC–MS Interfaces *791*
          B. Type of Mass Spectrometer *796*
  V. Solute Derivatization *796*
  VI. Chiral SFC *797*
  VII. Conclusion *798*
References *799*

25. Capillary Electrophoresis 803
Hassan Y. Aboul-Enein, Imran Ali

  I. Introduction *803*
  II. Theory of CE *804*
  III. Modes of CE *805*

IV. Instrumentation *806*
　A. Sample Injection *806*
　B. Separation Capillary *806*
　C. High-Voltage Power Supply *807*
　D. Background Electrolyte *807*
　E. Detection *807*
V. Data Integration *811*
VI. Sample Preparation *812*
VII. Method Development and Optimization *813*
VIII. Methods Validation *813*
IX. Application *813*
X. Conclusions *822*
References *822*

26. Gel Permeation and Size Exclusion Chromatography 827
Gregorio R. Meira, Jorge R. Vega, Mariana M. Yossen
I. General Considerations *827*
II. Basic Concepts *828*
　A. The Chromatographic System *828*
　B. Simple and Complex Polymers *829*
　C. Molar Mass Distribution *829*
　D. Entropy-Controlled Partitioning *832*
　E. Intrinsic Viscosity and Hydrodynamic Volume *832*
　F. SEC Resolution and Column Efficiency *833*
III. Concentration Detectors and Molar Mass Calibration *834*
　A. Differential Refractometer *834*
　B. UV Sensor *834*
　C. Molar Mass Calibrations *835*
　D. Chemical Composition *836*
IV. Molar Mass Sensitive Detectors *836*
　A. LS Detector *837*
　B. Viscosity Detector *838*
　C. Osmotic Pressure Detector *839*
V. Experimental Difficulties *839*
　A. The High Molecular Weight Saturation Problem *839*
　B. Enthalpic Interactions *840*
　C. Band Broadening *840*
　D. Polyelectrolytes *840*
VI. Instrumentation *841*
　A. High-Temperature Equipment *841*
　B. Solvent Delivery System *841*
　C. Sample Preparation and Injection *843*
　D. Fractionation Columns *844*
　E. Detectors *855*
VII. Application Examples *862*
　A. MMD of a Linear PS *862*
　B. Chain Branching in PVAc *863*
　C. Characterization of Starch *864*
　D. Sample Fractionation by Preparative SEC *865*
References *866*

27. Field-Flow Fractionation 871
Martin E. Schimpf
I. Introduction *871*
II. Principles and Theory of Retention *872*

A. The Brownian Mode of Retention *873*
B. The Steric/Hyperlayer Mode *875*
III. Experimental Procedures *876*
A. Sample Injection/Relaxation *876*
B. Selectivity *877*
C. Resolution and Fractionating Power *877*
D. Field Programing *877*
E. Ancillary Equipment *878*
F. Optimization *879*
IV. Applications *879*
A. Polymers *879*
B. Proteins and DNA *881*
C. Colloids (Suspended Particles with $d < 1$ μm) *883*
D. Starch Granules, Cells, and other Micrometer-Sized Particles ($d > 1$ μm) *885*
E. Environmental Applications *886*
References *887*

28. Instrumentation for Countercurrent Chromatography 893
Yoichiro Ito
I. Introduction *893*
II. Hydrostatic Equilibrium CCC Systems *895*
A. Development of Hydrostatic CCC Schemes *895*
B. Instrumentation of Hydrostatic CCC Schemes *895*
III. Hydrodynamic Equilibrium Systems *901*
A. Rotary-Seal-Free Centrifuge Systems *902*
B. Analysis of Acceleration on Synchronous Planetary Motions *902*
C. Instrumentation *908*
IV. Special Techniques *927*
A. pH-Zone-Refining CCC *927*
B. Affinity CCC *931*
C. CCC/MS *931*
V. Two-Phase Solvent Systems *931*
A. Retention Volume and Partition Coefficient (K) *931*
B. R_s and Retention of the Stationary Phase in CCC *932*
C. Various Parameters of High-Speed CCC Affecting Stationary Phase Retention *934*
D. Partition Coefficient of Samples *936*
VI. Future Instrumentation *937*
A. Analytical CCC *937*
B. Preparative CCC *937*
References *938*

29. HPLC-Hyphenated Techniques 945
R. A. Shalliker, M. J. Gray
I. Introduction *945*
II. Liquid Chromatography—Mass Spectroscopy *946*
A. Manual Treatment *947*
B. Mechanical or Moving Transport Interfaces *947*
C. Fast Atom Bombardment Interfaces *949*
D. Particle Beam and Monodisperse Aerosol Generation Interfaces *951*
E. MALDI and Laser-Assisted Interfaces *953*
F. Atmospheric Pressure Ionization and Nebulizer Based Interfaces *955*
G. Atmospheric Pressure Chemical Ionization Interface *955*
H. Atmospheric Sampling Glow Discharge Ionization Interface *957*
I. Thermospray Ionization Interfaces *958*
J. Atmospheric Pressure Electrospray/Ionspray Ionization Interfaces *959*
K. Sonic Spray Interface *963*

        L. Atmospheric Pressure Spray Ionization *964*
        M. Atmospheric Pressure Photoionization Interface *964*
        N. Cold Spray Ionization Interface *965*
        O. Interchangeable API Interfaces *965*
        P. High Throughput Strategies in HPLC-MS *965*
  III. Liquid Chromatography–Fourier Transform-Infrared Spectroscopy *966*
        A. Flow Cell Methods of Analysis *967*
        B. Online Removal of Water Prior to Flow Cell Detection *968*
        C. Solvent Elimination Methods *969*
        D. Deposition Substrates *969*
        E. Solvent Elimination Interfaces *970*
        F. Thermospray Nebulizer *970*
        G. Concentric Flow Nebulizer *970*
        H. Pneumatic Nebulizer *972*
        I. Ultrasonic Nebulizer *972*
        J. Electrospray Nebulizer *973*
        K. Quantitative Considerations *974*
  IV. Liquid Chromatography–Nuclear Magnetic Resonance Spectroscopy *974*
        A. Solvent Suppression *974*
        B. Sensitivity *975*
        C. Trace Enrichment *977*
  V. Liquid Chromatography-Inductively Coupled Plasma Spectroscopy *978*
  VI. Multiple Hyphenation (or Hypernation) of LC with Spectroscopy *980*
  VII. Hyphenation of Liquid Chromatography with Gas Chromatography *982*
        A. Concurrent Solvent Evaporation Techniques *982*
        B. Partially Concurrent Solvent Evaporation *983*
        C. Fully Concurrent Solvent Evaporation Techniques *985*
        D. Programmed Temperature Vaporizer/Vaporizer Interfaces *986*
        E. Miscellaneous *988*
References *988*

30. Thin Layer Chromatography 995
Joseph Sherma
  I. Introduction *995*
  II. Sample Preparation *996*
  III. Stationary Phases *997*
  IV. Mobile Phases *997*
  V. Application of Samples *998*
  VI. Chromatogram Development *999*
        A. Capillary-Flow TLC *999*
        B. Forced Flow Planar Chromatography *1001*
  VII. Zone Detection *1003*
  VIII. Documentation of Chromatograms *1004*
  IX. Zone Identification *1005*
  X. Quantitative Analysis *1006*
        A. Nondensitometric Methods *1006*
        B. Densitometric Evaluation *1007*
  XI. TLC Combined with Spectrometric Methods *1009*
        A. Mass Spectrometry *1009*
        B. Infrared Spectrometry *1009*
        C. Raman Spectrometry *1009*
  XII. Preparative Layer Chromatography *1009*
  XIII. Thin Layer Radiochromatography *1010*
  XIV. Applications of TLC *1011*
References *1012*

Contents

31. Validation of Chromatographic Methods *1015*
Margaret Wells, Mauricio Dantus
- I. Introduction *1015*
- II. Prevalidation Requirements *1017*
    - A. Instrument Qualification *1017*
    - B. System Suitability *1019*
- III. Method Validation Requirements *1020*
    - A. Precision *1020*
    - B. Accuracy *1022*
    - C. Limit of Detection *1023*
    - D. Limit of Quantitation *1023*
    - E. Range *1024*
    - F. Linearity *1024*
    - G. Ruggedness *1025*
    - H. Robustness *1026*
    - I. Solution Stability *1027*
    - J. Specificity *1027*
- IV. Method Validation Example *1028*
    - A. Purpose and Scope *1028*
    - B. Determine Method Type and Method Requirements *1028*
    - C. Set Acceptance Criteria for Method Validation Requirements *1028*
    - D. Determine System Suitability Requirements *1028*
    - E. Perform/Verify Instrument Qualification *1029*
    - F. Perform Method Validation/Evaluate Results *1029*
    - G. Method Validation Complete *1030*
- V. Concluding Remarks *1032*

References *1032*

Index *1035*

Contributors

Hassan Y. Aboul-Enein *Pharmaceutical Analysis and Drug Development Laboratory, Biological and Medical Research Department, King Faisal Specialist Hospital and Research Centre, Riyadh, Saudi Arabia*

Kenneth S. Alexander *Industrial Pharmacy Division, College of Pharmacy, The University of Toledo, Toledo, OH, USA*

Imran Ali *National Institute of Hydrology, Roorkee, India*

Harry G. Brittain *Center for Pharmaceutical Physics, Milford, NJ, USA*

Chris W. Brown *Department of Chemistry, University of Rhode Island, Kingston, RI, USA*

C.R. Brundle *C. R. Brundle & Associates, San Jose, CA, USA*

Emanuel Carrilho *Institute of Chemistry at São Carlos, University of São Paulo, São Carlos (SP), Brasil*

Thomas L. Chester *The Procter Gamble Company, Miami Valley Laboratories, Cincinnati, Ohio, USA*

John Coates *Coates Consultancy, Newtown, CT, USA*

Thomas P. Conrads *Laboratory of Proteomics and Analytical Technologies, SAIC-Frederick Inc., National Cancer Institute, Frederick, MD, USA*

Mauricio Dantus *Merck & Co. Inc, Rahway, NJ, USA*

Gareth R. Eaton *Department of Chemistry and Biochemistry, University of Denver, Denver, Colorado, USA*

Sandra S. Eaton *Department of Chemistry and Biochemistry, University of Denver, Denver, Colorado, USA*

Peter Fredericks *Queensland University of Technology, Brisbane, Queensland, Australia*

Stacy L. Gelhaus *Department of Chemistry and Biochemistry, University of Maryland, Baltimore County, Baltimore, MD, USA*

Maria Goreti R. Vale *Depto. de Quimica da UFSC, Florianopolis, S.C., Brazil*

M.J. Gray *School of Science, Food and Horticulture, University of Western Sydney, NSW, Australia*

Nelu Grinberg *Merck Research Laboratories Rahway, NJ, USA*

Peter J. Haines *Oakland Analytical Services, Farnham, Surrey, UK*

Elo Harald Hansen *Department of Chemistry, Technical University of Denmark, Lyngby, Denmark*

Gunawan Indrayanto *Faculty of Pharmacy, Airlangga University, Surabaya, Indonesia*

Yoichiro Ito *Laboratory of Biophysical Chemistry, National Heart, Lung, and Blood Institute, National Institutes of Health, Bethesda, MD, USA*

William R. LaCourse *Department of Chemistry and Biochemistry, University of Maryland, Baltimore County, Baltimore, MD, USA*

Fernando M. Lanças *Institute of Chemistry at São Carlos, University of São Paulo, são Carlos (SP), Brasil*

Paul C.H. Li *Chemistry Department, Simon Fraser University, Burnaby, British Columbia, Canada*

Xiujun Li *Chemistry Department, Simon Fraser University, Burnaby, British Columbia, Canada*

Zhihao Lin *Merck Research Laboratories, Merck & Co., Inc., Rahway, NJ, USA*

Ronita L. Marple *Department of Chemistry and Biochemistry, University of Maryland, Baltimore County, Baltimore, MD, USA*

Gregorio R. Meira *Intec (CONICET and Universidad Nacional del Litoral), Santa Fe, Argentina*

Mark P. Olson *Department of Chemistry and Biochemistry, University of Maryland, Baltimore County, Baltimore, MD, USA*

J. David Pinkston *The Procter Gamble Company, Miami Valley Laboratories, Cincinnati, Ohio, USA*

A.K. Rai *Department of Physics, G.B. Pant University of Agriculture & Technology, Pantnagar, India*

Alan T. Riga *College of Pharmacy, The University of Toledo, Toledo, OH, USA, Clinical Chemistry Department, Cleveland State University, Cleveland, OH, USA*

Llewellyn Rintoul *Queensland University of Technology, Brisbane, Queensland, Australia*

Wes Schafer *Merck & Co., Inc, Rahway, NJ, USA*

Martin E. Schimpf *Department of Chemistry, Boise State University, Boise, ID, USA*

Raymond P.W. Scott *7, Great Sanders Lane, Seddlescombe, East Sussex, UK*

R.A. Shalliker *School of Science, Food and Horticulture, University of Western Sydney, NSW, Australia*

Joseph Sherma *Department of Chemistry, Lafayette College, Easton, PA, USA*

J.P. Singh *DIAL, Diagnostic Instrumentation and Analysis Laboratory, Mississippi State University, Starkville, MS, USA*

Jacobus Fredeiczzk van Staden *Department of Chemistry, University of Pretoria Pretoria, South Africa*

Raluca-Ioana Stefan *Department of Chemistry, University of Pretoria Pretoria, South Africa*

S.N. Thakur *Department of Physics, Banaras Hindu University, Varanasi, India*

Wang Tiebang *Merck & Co., Inc, Rahway, NJ, USA*

Narayan Variankaval *Merck Research Laboratories, Rahway, NJ, USA*

Timothy D. Veenstra *Laboratory of Proteomics and Analytical Technologies, SAIC-Frederick Inc., National Cancer Institute, Frederick, MD, USA*

Jorge R. Vega *Intec (CONICET and Universidad Nacional del Litoral), Santa Fe, Argentina*

Frederick G. Vogt *GlaxoSmithKline P.L.C., King of Prussia, PA, USA*

Jianhua Wang *Research Center for Analytical Sciences, Northeastern University, Shenyang, P.R. China*

Tiebang Wang *Merck Research Laboratories, Rahway, NJ, USA*

J.F. Watts *University of Surrey, Surrey, UK*

Margaret Wells *Merck & Co. Inc, Rahway, NJ, USA*

Bernhard Welz *Depto. de Quimica da UFSC, Florianopolis, S.C., Brazil*

J. Wolstenholme *Thermo VG Scientific Corp., East Grinstead, UK*

Mariana M. Yossen *Intec (CONICET and Universidad Nacional del Litoral), Santa Fe, Argentina*

Li-Rong Yu *Laboratory of Proteomics and Analytical Technologies, SAIC-Frederick Inc., National Cancer Institute, Frederick, MD, USA*

Mochammad Yuwono *Faculty of Pharmacy, Airlangga University, Surabaya, Indonesia*

1

The Laboratory Use of Computers

WES SCHAFER, ZHIHAO LIN
Merck Research Laboratories, Merck & Co., Inc., Rahway, NJ, USA

I. INTRODUCTION

This chapter in the second edition of this work began with the statement that "processing speeds of small computers have increased by more than an order of magnitude, and the cost of memory has gone down by a comparable factor". This edition can once again truthfully begin with the same statement. The evolution of computers over the past 25 years has been truly phenomenal. Calculations that were previously only possible on the largest mainframe computers, at considerable cost, can now be performed on personal computers in a fraction of the time. This has revolutionized some aspects of analytical chemistry, especially in the field of chemometrics.

Concurrent with the increase in performance is the scientist's dependence on them. The computer is a valuable tool at almost all stages of experimentation. Tedious literature searches to determine the prior art of a subject are quickly dispatched by searching keywords against online databases. Computers are not only used to acquire data from analytical instruments but also to conveniently control them saving complex instrument parameters in method or recipe files that are easily downloaded to the instrument when needed again. Combined with automated sampling and other robotic devices, analytical instruments are often left to work late into the night and weekends, fully utilizing expensive equipment in off-hours, and freeing the scientist to concentrate on experimental design and result analysis. The computer is also extensively used in data analysis, automatically calculating results, and graphically displaying them for the scientist to best interpret. Finally, they have proven themselves invaluable in the more mundane task of storing and organizing the scientist's data for later retrieval as necessary.

This chapter is divided into two sections. The first section will briefly describe the physical components of the system and their interdependencies. The key attributes of each component as it relates to the performance of the system will be discussed. The second section will focus on the role of the computer in each stage of the experimental process: data acquisition, data analysis, and data storage.

II. COMPUTER COMPONENTS AND DESIGN CONSIDERATIONS

As one would expect, the intended use of the system should be taken into account when determining which computer should be used with a given instrument. While computers are becoming continually faster and cheaper, it is still possible to waste several thousand dollars on system components that are simply not needed.

A simple UV spectrophotometer with associated software to determine absorbance and calculate Beer's Law curves is unlikely to tax a modern personal computer. The amount of data produced by the instrument and its subsequent processing can be handled by a relatively low end computer with a relatively small hard drive. A

Michelson interferometer based infrared spectrophotometer that must perform the Fourier transforms of the data would require more resources. If it is to also perform complex chemometric calculations, processing time would probably benefit from a higher end PC. If the application is to continually compare collected spectra to a library of spectra on the hard drive, it would benefit from a system with quick disk input/output (I/O) capability.

While instrument manufacturers have little incentive to sell an underpowered computer with their instrument as it also reflects poorly on their product, marketing and testing considerations that are of little interest to the user may mean the offered PC is not the optimal one. The computers offered by instrument vendors are also often priced at a hefty premium when compared with market prices.

Still, the decision to purchase the computer from the vendor or separately must be decided on a laboratory to laboratory or even instrument to instrument basis. If the laboratory has little experience with the computer, operating system, and application software and does not have other internal organizational resources for dealing with computer issues, purchasing the computer from the vendor and including it on the vendor's service/maintenance contract is a sensible approach. This provides a single contact point for all instrument and computer issues. The instrument and software were also presumably thoroughly tested on the computer offered. If, however, a service contract will not be purchased with the instrument or if the PC is not included in the service contract, it is often advantageous to purchase the computer separately.

Frequently a laboratory's IT (information technology) organization places a number of additional requirements on the computer and the configuration of its operating system. If the instrument and its software place no special requirements on the computer, it may be best to obtain one of the computer models supported by the organization's internal IT group.

A. Motherboard/Central Processing Unit

The motherboard is the unifying component of the computer. It contains the central processing unit (CPU) as well as the system bus which is the means by which the CPU transfers data to other key components such as the memory and the hard drive. It also contains the expansion slots and communication ports that interface the computer to its peripherals such as the monitor, printer, and laboratory instrument itself. A growing number of these peripherals [e.g., video adapters and network interface cards (NICs)] are being integrated onto the motherboard as PC manufacturers search for more ways to lower the costs.

Although the motherboard and CPU are two separate components with different functions they are completely interdependent. The motherboard contains the support chips for the CPU, which dictate its architecture as well as the number of bits it processes at a time. There are four main considerations to consider when choosing the CPU and motherboard: bus speed, CPU architecture/word size (32 vs. 64 bit), and processor speed.

The motherboard contains the buses (both data and address) by which all of the system components communicate with one another. Increases in system bus speeds almost always lead to performance improvements. This is because it increases the speed by which computer components communicate with one another. Modern PCs typically split the bus into a system portion that links high-speed components such as the memory and the video card to the CPU and a slower local bus for connecting disks, modems, and printers. The speed of the system bus is especially important in terms of the CPU accessing main memory as memory access speed increases with bus speed. One should also note the number of memory slots on the motherboard as it could ultimately limit the amount of memory that can be installed on the PC.

There are two basic CPU architectures: complex instruction set computers (CISC) and reduced instruction set computers (RISC). The CISC processors are designed to complete tasks with the least amount of assembly code. Assembly code is the lowest level programing language and is used by the processor itself. The popular Intel® Pentium® family of processors are CISC based. The RISC processors are designed to complete each assembly code instruction in one clock cycle. It takes more assembly instructions to carry out tasks on the RISC based machines but each assembly code instruction is processed in one clock cycle. High performance workstations tend to be based on the RISC architecture. All of this being said, chip manufacturers have not rigidly adhered to either standard.

The processor speed is the speed at which the processor performs *internal* calculations. Applications that require intensive calculations will benefit from using the fastest CPU possible. Operations requiring the CPU to interact with other system components will still be limited by the bus speed and the capabilities of the device it is addressing. However, the relationship between processor speed and cost is far from linear and small percentage increases in speed at the high end of processor speeds can lead to large percentage increases in system cost. Therefore, one should ensure that the CPU is the bottleneck for a particular system before expending a relatively large sum of money to get the latest and fastest processor.

The word size (number of bits that the CPU and bus process per clock cycle) obviously affects the amount of data calculated per clock cycle but it also affects the maximum amount of memory the system can support. A 32 bit system is limited to addressing 4 GB (2^{32}) of

RAM. A 64 bit system can address a practically limitless amount of RAM (2^{64} = 16 exabytes or 16 billion gigabytes). The user should not, however, expect that a 64 bit system will perform twice as fast as a 32 bit system. The additional 32 bit is not necessary for every calculation and unless the operating system and application software is designed to use the 64 bit architecture it will continue to go unused.

B. Memory/Cache

In order to process instructions, the CPU must have access to the data it is to process. Data is stored in the computer's memory. Ideally the CPU should be able to access the values in memory at the same speed it processes instructions but unfortunately the cost of memory is highly dependent on its speed. The extremely expensive static random access memory (SRAM) is the only memory that can keep pace with the processor speed. In order to obtain the large amounts of memory required for the entire system, relatively inexpensive dynamic random access memory is used. As the names try to imply, the values stored in SRAM remain constant unless they are purposely changed by instructions from the CPU or the computer is powered down. Dynamic RAM must be continuously refreshed or the values stored in it fade away. This "reinforcement" of the values in DRAM is paid for as a loss in speed.

Rather than avoiding SRAM altogether because of its cost, computer manufacturers employ small amounts (1 MB or less) of it as a cache. Whenever the system must access main memory for information it copies that information into the cache as well, replacing the oldest information as necessary. Before the system turns to main memory for data, it checks the cache first. The assumption here is that if you asked for it once, you will ask for it again. As it turns out, this is a very good assumption. As you would expect, the larger the cache the more likely it is that the values being sought will be there. Caching is used extensively in computers whenever a faster device or process must turn to a slower one for information. Other examples include disk caches and web page caches.

Memory is offered in nonparity, parity, and error correction code (ECC) variations. Parity memory calculates the check sum of data as it is being written and stores the value of the checksum in the parity bit. The checksum depends on whether even or odd parity is used. With even parity the checksum is 0 if the total number of 1 values in the data byte is even and 1 if the total number is odd. Odd parity is the opposite. When the data is read from memory the checksum is calculated again and checked vs. the stored value. If they do not match a system halt is called. The ECC memory takes this process one step further. By using additional parity bits, the ECC memory is able to detect and correct single bit errors as well as detect 2 bit errors. Memory has proven itself to be very reliable and most vendors sell their systems with nonerror correction code (NECC) memory.

C. Disk Storage

As discussed, memory is used by the system to store executable statements and data that are needed on an immediate basis. It is inappropriate for long-term storage of data because there is a substantially limited amount of it and it is nondurable, that is, when the computer is shut down the information in memory is lost. The system's hard drive is used for storing software and data that are needed to boot and run the system and applications. Although there are other means of storing data, it is the first choice for storing data that are needed on a frequent basis because it is the fastest read/write durable storage media available.

The hard drive has been far from left behind in the computer's remarkable performance gains. Megabyte drives have given way to gigabyte drives; data transfer rates and sophisticated interfaces have been developed so that multiple drives can work together in concert to provide staggeringly large amounts of reliable data storage. We will consider both the hard drive itself and the interfaces that have been developed for them.

At the simplest level the hard drive consists of a platter with a thin layer of magnetic material and head mounted on a pivoting arm. The platter is rotated at speed under the head and the arm swings across the radius of the platter so that the head can access any location on the platter to read or write data.

There are three important performance attributes for the physical hard drives: data rate, seek time, and capacity. The data rate is the number of bytes per second that the drive can send to the CPU. Obviously higher data rates are better. Seek time is the time required by the drive to find a file and begin sending the information from it. Here the speed at which the disk is rotated as well as its diameter is important. The platter is rotated underneath the head, therefore the faster the disk is rotated, the faster the head will reach the desired rotational location on the disk. Since the head has to move in and out along the radius of the disk, the smaller the disk the faster it will be able to reach the desired location along the disk radius. Of course the smaller the drive, the smaller the capacity of the drive. Drives of 3.5 in. appear to have become the agreed upon compromise between speed and capacity although smaller diameter drives are built for laptops. By placing multiple platters in a single hard drive, manufacturers have developed drives with a capacity of hundreds of gigabytes.

The interface used to connect the hard drives to the system affects the number of drives and data transfer rate of the drives. The first hard drives were connected to the system via a hard drive controller that was connected to the bus. Each drive was connected separately to the motherboard via a controller, and the operating system addressed each drive individually. Later the controller was actually integrated onto the drive, integrated drive electronics (IDE), instead of being a separate card. Drives were then connected to the motherboard in a parallel master/slave arrangement that is still widely used today. The interface is called the parallel advanced technology attachment (ATA) storage interface and the drives are called ATA packet interface (ATAPI) devices. The interface is limited to two hard devices (four if two interfaces are installed on the motherboard) and they cannot be accessed concurrently. Originally the CPU was responsible for handling the data transfer process but that has been shifted to the bus master in peripheral component interconnect (PCI)-based systems.

The small computer system interface (SCSI) is a higher performance-based I/O system that handles more and faster storage devices. The SCSI bus contains a controller (host adapter) that controls the flow of data and up to 15 devices daisy-chained to it. The controller addresses each device by its unique ID (0–15) and is able to address multiple drives simultaneously as well as reorder tasks to improve throughput (asynchronous transfer). SCSI is relatively expensive and the high performance that it offers is usually not needed on a single user workstation. It is, however, a means of connecting more than four I/O devices to the computer.

D. Video (Graphics Card and Monitor)

Most PC vendors now integrate the video card onto the motherboard for cost considerations. This is generally of no consequence in the laboratory. The boom in amateur digital photography and gaming for the home PC market has greatly increased the video capabilities of the standard PC and all but the very low end of mass-produced PCs will satisfy the video requirements of most scientific applications.

There are, however, some scientific applications that place very high demands on the video subsystem of the computer. Examples include imaging applications such as digital microscopy and 3D molecular modeling applications. A basic understanding of the video subsystem is very helpful in purchasing or building the appropriate computer for the intended application as well as diagnosing video problems.

The video adapter's first and minimum required responsibility is to translate instructions from the CPU and transmit them to the monitor. The computer's graphics (video) card and monitor are codependent and they must be matched to one another in order to function optimally. Specifically they must be compatible in terms of resolution, color depth, and refresh rate.

The monitor produces video images through a grid of colored pixels. The resolution of the video card and monitor is expressed in terms of the number of horizontal pixels by the number of vertical pixels (e.g., 1024 × 768). Obviously, the higher the number of pixels, the more the information that can be displayed on the monitor, which results in sharper images.

The video card must transmit color information for each of these pixels. Any standard PC video card and monitor now support 24 bit color information which corresponds to 16,777,216 colors (2^{24}). Since the human eye can only discern about 10 million colors, this is often referred to as true color and there is no reason to increase it further. (Most video cards today are 32 bit video cards but the additional 8 bits are used to transmit translucency information for layered graphics in digital video animation and gaming.)

The refresh rate is the number of times the pixels on the monitor are redrawn per second (Hz). Generally, it will be the monitor and not the video card that will limit how fast the refresh rate may be set. Note that the maximum refresh rate is also dependent on the resolution. Since the monitor can only refresh a maximum number of pixels per second, the higher the resolution (and therefore the number of pixels), the fewer times the monitor can refresh the entire screen per second.

The conversion of operating systems and applications to graphical user interfaces (GUIs) of ever increasing complexity has required that the video card/graphics card takes on more responsibilities so as to not overtax the CPU and system memory. The better the video card the more tasks it will assume on behalf of the CPU. This added functionality can lead to dramatic performance improvements for the 3D imaging used by molecular modeling. To take on these additional tasks, video cards typically employ three major components: the graphics accelerator, the video memory, and the accelerated graphics port (AGP) connector.

The graphics accelerator is essentially a co-processor specifically designed for performing graphics calculations which are particularly intensive in 3D imaging applications. It calculates the location and color of each pixel based on more generic instructions from the system CPU to draw geometric shapes, thus freeing the system CPU from these calculations. Unfortunately, the commands used to address the graphics accelerator are specific to the particular graphic card and its drivers. This is an area where standardization has been slow in coming and many applications are optimized for a particular model or family of video cards.

Most video cards also have their own onboard memory. The more memory a video card has, the higher resolutions it will support. To support 24 bit color with a resolution of 1280×1024, 4 MB of video RAM is required. To support 1600×1200 with 24 bit color 6 MB is needed. The amount of memory does not generally increase speed unless the card sets aside a portion of its memory to assist in complex geometric calculations. It is the speed of the video RAM and the card's digital-to-analog converter that ultimately determine the maximum refresh rate supported by the card at a given resolution.

In the past, video cards were connected to the system through one of the normal expansion slots on the motherboard. Most cards now connect via a specialized AGP connector. The AGP connector interfaces the card to the motherboard and allows the graphics accelerator chip to directly access the system memory and CPU. This allows the video card to use the system memory for calculations. The speed multiplier of AGP stands for 266 Mbps, so a $2\times$ AGP will have a transfer rate of 532 Mbps.

There are two video attributes that are defined by the monitor alone: size and dot pitch. Monitor sizes are generally measured by the length of a diagonal from the upper left to the lower right corner of the screen. For cathode ray tubes (CRTs) it is important to compare the viewable size. Manufacturers do not extend the image to the edge of the CRT because it is impossible to avoid distortions near the edges. It should also be noted that higher resolutions are required on larger monitors to maintain sharp images.

The dot pitch is the distance between pixels. The larger the dot pitch, the more obvious the individual pixels will be, leading to grainy images. In general, the dot pitch should be no more than 0.28 mm, with 0.25 mm preferred.

A final word regarding monitor types is in order. Liquid crystal display (LCD) monitors have become an economical alternative to CRTs for all but the most demanding video applications. They require much less space and energy than traditional CRTs. They may also be considered safer in the laboratory environments since they operate at lower voltages, produce less heat, and there is no implosion hazard from a tube. As of yet, however, they cannot match the refresh rates of CRTs and tend to distort motion which may be an issue with 3D molecular modeling applications.

E. Other Peripherals

The system contains other peripherals that will not be discussed at any length. External storage devices such as the CD or DVD drives are widely used to read and save data not needed on a frequent basis. Most software applications have reached sizes requiring these devices to install them. The electronic revolution has also failed to wean us of our dependence/preference to paper, and the printer continues to be a key component of the computer. The serial and parallel communication ports as well as NICs will be discussed in the digital transmission section that follows.

III. COMPUTER MAINTENANCE

There are two main aspects in maintaining a computer: monitoring system performance so that corrective action can be taken when needed and protecting the system's data against hardware failure as well as malicious attacks from outside sources.

A. Performance Monitoring

The system's performance should be periodically monitored. If the computer's performance when running the instrument and/or its software leaves something to be desired, the first step is to determine what resource is being used to its maximum extent, thus limiting the speed of the entire process. Most modern operating systems provide a means of determining system resource (memory, CPU, and disk) usage. Microsoft operating systems can be monitored using its task manager facility as the "top" and "ps" commands can be used in UNIX-based systems. By monitoring system resource usage while running the instrument and its associated software, it is often a simple matter to determine what the bottleneck is.

If the system's physical memory is being completely used, increasing the system memory will alleviate the problem. Be careful not to confuse physical memory with virtual memory which includes the pagefile. Almost all systems allocate space on the system drive to the pagefile, which is used by the system as auxiliary memory. While occasional use of the page file is normal, the system should not need to use the pagefile for the active application as the data transfer rates to the hard drive are much slower than to the system memory.

If the CPU is continually running at or near capacity, you must either decrease the load on the processor by shutting down unneeded processes or by purchasing a faster processor.

If the system is spending the bulk of its time to read/write to the disk system, the first step should be to check the fragmentation status of the drive. All disks become fragmented with time and in fact a lot of systems start out that way—the installation of the operating system already fragments files on the drive. Generally, the system writes a new file to the disk in the first available open space on the drive. If that space is not large enough to contain the file, what did not fit is written to the next open space after that and if the remainder of the file still does not fit, a third open space is found and so on. A single file can be spread out over dozens of locations on the drive. The disk drive must then jump back and forth to all of these locations to read the entire file, significantly decreasing the speed at which the drive can read the file. There are several good third party disk defragmenters on the market, and some operating systems include it as part of the operating system. The defragmentation software attempts to move files around until every file is stored contiguously on the drive, thus minimizing the access time for the drive.

B. Virus Protection

There is a significant criminal element in the computer world that takes great pleasure in disrupting the computers and lives of other users. In some cases, this takes the form of a prank by merely copying itself onto as many computers as it can and in other cases it involves the very malicious destruction of data on the system's hard drive. Denial of service attacks where large numbers of computers are duped into bombarding a single website to prevent legitimate users from being able to access them have also become quite popular.

The threat posed by these "hackers" has greatly increased with high-speed connections to the internet. Computers have become very highly networked, constantly interacting with one another in order to obtain or give information requested by their users. Laboratory systems are increasingly networked to high-speed local area networks (LANs) and WANS and are susceptible as well. It is very important that the laboratory computer and its hard earned scientific data be protected. The most common threats to a computer are:

- Viruses—A virus is a small program that integrates itself with a legitimate program. Every time the program is run, the virus runs as well giving it the chance to reproduce and deliver its payload.
- Worms—A worm is a small piece of software that takes advantage of security holes in the operating systems and applications to gain access to a system. After infiltrating the system, the worm scans the network looking for systems available from the machine it just infected, which also have the security hole. It then starts replicating from there, as well.
- Trojan horses—A Trojan horse is simply a computer program that masquerades as another legitimate program. The user is duped into running the program which may cause great damage when it is run (such as erasing the hard disk).

Here are some minimal steps to protect the laboratory computer from these threats:

1. Do not open files from unknown sources.
2. Run an antivirus program and set it to scan all new files as they are written to disk. Make sure that the virus definitions used by the program are kept up to date. The software can often be set to automatically update itself at set intervals.
3. Keep the operating system current by installing security patches as they are made available.
4. Ensure that the system is protected by a firewall. By default, most networking installations allow unfettered access to the computer from outside sources. Firewalls scan incoming and outgoing data and block it from unrecognized (nontrusted) sources. Most large organizations protect their local area networks from unauthorized probing by the use of firewalls.

C. Backup and Recovery

The data on the hard drive is often a prime target for viruses and worms. The computer's hard drive is also the component most likely to fail. Indeed most people have suffered the failure of their computer's hard drive. Since the end product of all the scientists' experimentation resides on the hard drive, it is imperative that an appropriate backup and recovery strategy be implemented.

The first step in implementing a backup/recovery strategy is to define the maximum acceptable data loss and time the backups appropriately. If experiments can be reasonably repeated within a 24 h period, backing up the system daily should be sufficient.

This process should accurately define the backup and restore needs of the system. Insufficient backups will result in a needless loss of data. Strategies involving elaborate, complicated backup schemes place are difficult to implement and maintain, thus unnecessarily sapping the computer's resources and the scientist's time. They are also less likely to be faithfully executed as designed.

In cases where very little downtime due to the computer can be tolerated, disk mirroring may be an option. This functionality used to be available only on high-end servers but now redundant array of inexpensive disks (RAID) controllers can be purchased for around $100. This can be thought of as performing real-time backups of data as it is written to the disk. All data are written to both drives so that each one is a copy of the other. Should one of the drives fail, the system continues to function on the remaining drive. When the failed drive is then replaced, the controller copies all the information on the original surviving drive to the new drive to reestablish the redundancy protection. Servers often employ a second RAID technology—RAID-5. Here the RAID sets of n drives are formed (3–15 for the SCSI drives) and all

Laboratory Use of Computers

the information on each drive is copied across the remaining drives. Instead of using half the available disk capacity for redundancy, only $\frac{1}{n}$th is used for redundancy. Keep in mind that the RAID technology does not protect your system against a virus infection, as the virus will be copied to the redundant portions of the drives as well.

If the computer is on a networked enterprise-type environment such as a university or large company, backed-up file servers may be available for the storage of data. If such services exist, it can be an excellent means of decreasing the time the scientist needs to spend on nonscientific housekeeping while maintaining a high level of protection. The backup/restore procedures available on any network file server should be confirmed before entrusting your valuable data to them.

IV. DATA TRANSFER/INSTRUMENT INTERFACES

Computers dwell exclusively in the digital realm. Quantum chemistry notwithstanding, normal chemical analyses and the responses measured by them are analog phenomena. In order for computers to be able to analyze and store analog data from a laboratory instrument, the data must be converted to a digital form. This is the basic task of the instrument interface(s). In order for the computer to analyze and store the data from a given laboratory instrument, several processes must take place. In general terms they are: conversion of the physical signal to an analog signal, analog transmission, analog filtering, analog-to-digital conversion, digital transmission, and digital filtering. It may not be immediately obvious to the casual observer of the analytical instrument where these processes take place or even that they do. The transducer and the analog signal filters normally occur within the enclosure of the detector. The analog-to-digital conversion and/or the digital filtering can also occur within the detector, in the computer or in a physically separate interface. In many cases, the analog-to-digital conversion takes place in the detector to minimize the efforts required to protect the analog signal from ambient noise. Digital filtering is typically performed in the computer as it can be very computationally intensive and to simplify optimizing the digital filtering parameters by the scientist if desired.

A. Transducers

The first step is to convert the physical phenomena being studied into an electronic signal by the transducer. Typical transducers used in analytical instrumentation include: photomultiplier tube (PMT) in traditional scanning spectrophotometers using monochromators, charge-coupled devices for photodiode array instruments, particle detectors for X-ray diffraction equipment, and electrodes for measuring potentials in solutions. The transducer is obviously very specific to the instrument and will not be further discussed here except to caution that the properties of the transducer must be taken into account in the design of the analog-to-digital circuitry. For example, high impedance transducers such as electrodes likewise require that the amplifier circuitry be of high impedance.

B. Analog Signal Transmission

We will begin by examining signal transmission. Earlier this was almost always accomplished by transmitting the analog signal, usually in the form of varying voltage, from a detector such as a PMT or a transducer such as thermistor to the analog-to-digital converter installed on the PC.

Noise is everywhere but can be especially prevalent in the laboratory environment where it can weaken or worse yet alter the useful signal. Proper cable design and use is more important as the distance that the analog signal must travel increases. Voltage (as opposed to current) based signals are especially prone to electromagnetic interferences.

A simple method of reducing interference from external noise is to simply twist the signal and ground wires around one another. Any noise induced in one wire is offset by an equal but opposite noise signal in the other. Coaxial cables are another effective means of protecting data from noise. A coaxial cable consists of a central conducting (positive) wire separated from an outer conducting cylinder by an insulator. They are not affected by external electric and magnetic fields.

Cables are often shielded by surrounding them with a layer of conductive metal such as aluminum or copper to prevent them from acting as antennas for radio and other electromagnetic noise sources. The shield is then grounded only at the detector end to protect from low frequency (e.g., 60 Hz) noise or at both ends to protect high frequency signals. If the shield is grounded at both ends, care must be taken to ensure that both grounds are identical or a ground loop will form resulting in current flowing through the shield, which will affect the signal being carried by the cable. The shield should also not be left ungrounded at both ends as this will enhance the noise one is trying to eliminate.

C. Analog Signal Filtering

The most important application of analog filtering is low-pass antialiasing filters that must be applied to the analog signal when periodic signals are being measured. Unless systematic noise and harmonics at higher frequencies are removed from the analog signal, the low sampling rate in comparison with the noise can result in the noise aliasing down to the frequency of the signal itself. Aliasing will be discussed in the digital filtering section.

True analog filtering has become much less important in the digital age. Digital filters can easily simulate low-pass, band-pass, and high-pass filters and have two distinct advantages. They can be applied without permanently obscuring the original signal which is generally saved in an electronic file and they can use subsequent data points to filter the current data point such as in moving average schemes, something that is impossible in analog circuitry. All of this assumes, of course, that the analog signal was converted to a digital signal with sufficient resolution. This will be discussed in Section D.1.

D. Analog-to-Digital Conversion

There are several different methods of analog-to-digital conversion and although a comprehensive treatment of the analog-to-digital conversion is beyond the scope of this work, a basic discussion of their advantages and disadvantages is in order to ensure that the method chosen is compatible with the given application. Table 1 gives a brief summary of the various types of analog-to-digital converters along with approximate performance ranges. Most commercial devices employ some type of track and hold circuit so that the conversion can take place on a static value. Component manufacturers such as National Semiconductor, Analog Devices, and Texas Instruments among others provide detailed specification sheets for their products as well as useful application notes and guides (Fig. 1).

There are two important attributes of the analog-to-digital converters: resolution and response time. Resolution is the smallest analog change resulting in an incremental change in the digital output. The resolution of an analog-to-digital converter is determined by the number of output bits it produces. An 8 bit analog-to-digital converter is capable of separating a 1 V signal into 2^8 segments or 3.9 mV. This, for example, would be insufficient for chromatographic UV detectors that often have rated noises of 0.03 mV. A 16 bit analog-to-digital converter is capable of resolving a 1 V signal into 65,536 segments or 0.015 mV, which would be sufficient. Resolution is also referred to as the least significant bit (LSB).

Another point to consider when using analog-to-digital converters is range matching. If an instrument detector is only capable of outputting from 0 to 50 mV, using an analog-to digital converter with a 1 V range is an expensive waste of 99.5% of the ADC's range. Many commercial interfaces allow the user to adjust the range of the ADC for this reason.

1. Sampling Interval (Rise-Time)

The sampling interval is a second important consideration. If the sampling rate is too low, important "fine features" may be lost while using excessively high sampling rates wastes both computing time and disk space. Figure 2 shows the effect of sampling rate on the depiction of a modeled chromatographic peak with a partially co-eluting smaller peak with narrower peak width.

Note that the sampling interval problem also applies to experiments that at first glance do not appear to be time based. The use of the Fourier transform methodology for frequency vs. intensity experiments such as infrared and nuclear magnetic resonance spectroscopy (NMR) is an important example. Nyquist firmly established that the sampling frequency must be at least twice that of the highest frequency component in the signal. If the signal is sampled at less than the Nyquist frequency, those components above half the sampling frequency appear as components with frequencies less than half the sampling frequency (see Fig. 3).

Table 1 Analog-To-Digital Converters

Type	Description	Response time	Resolution (bit)
Parallel encoder	The input signal is simultaneously compared to a series of reference voltages and the digital output is based on the last highest voltage reference that was exceeded by the reference voltage	3–200 ns	4–10
Successive approximation	The digital voltage value is derived by comparing input voltage to sequential series of reference voltages produced by a digital-to-analog converter	3–200 μs	8–12
Voltage-to-frequency conversion	The input voltage is converted to a frequency that is proportional to the input voltage. The digital voltage value is derived from the frequency	6–33 ms	12
Single-slope integration	The input voltage is compared to an internal voltage ramp of known slope. The digital voltage value is derived from the time required for the voltage ramp to exceed the input voltage	3–40 ms	12
Dual-slope integration	The digital voltage value is derived from the time for a capacitor to discharge after being charged proportional to the input voltage	3–40 ms	10–18
Delta sigma converters	The digital voltage value is derived from successive subtractions of known voltages from the input voltage	10–100 ms	12–18

Laboratory Use of Computers

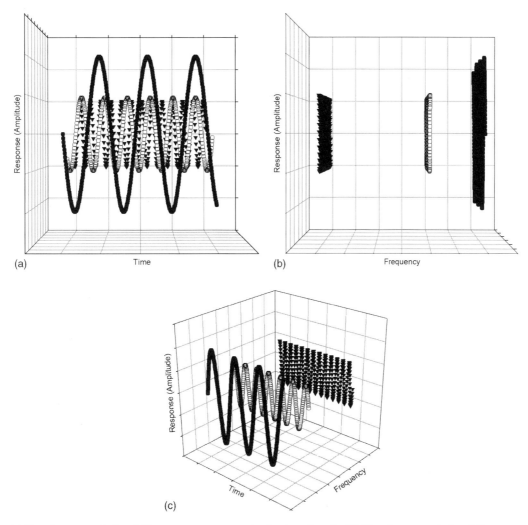

Figure 1 (a) Signal composed of multifrequency components in the time domain. (b) Signal composed of multifrequency components in the frequency domain. (c) Signal resolved into frequency components.

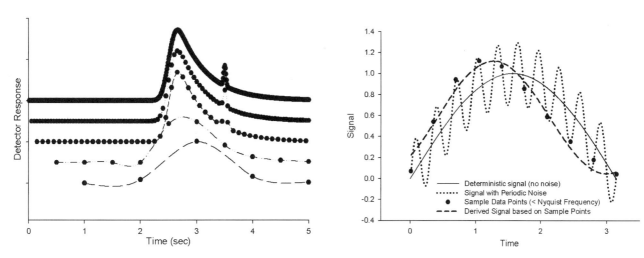

Figure 2 Effect of sampling rate on time-based measurements.

Figure 3 Effect of aliasing below Nyquist frequency.

As it is quite possible that the value of the highest frequency component in the signal is not known, an analog low-pass filter is added before the analog-to-digital converter to remove any signal with a frequency greater than half the sampling frequency. Since the filtering profile of analog filters resembles an exponential decay more than an ideal strict cut-off at the desired frequency, care must be taken to ensure filter reduces all the undesired higher frequencies to below the detection limit of the analog-to-digital converter. All of this means that analog-to-digital converter must be able to sample the signal at over twice the frequency of the highest expected component in the signal to ensure that no alias artifacts are present in the analyzed data.

E. Digital Signal Transmission

It is now much more prevalent to convert the signal to a digital one in an onboard computer on the instrument and transmit the digital signal to the computer. Digital signals are generally less susceptible to noise because they are discrete pulses, whereas analog signals are continuous. In cases where the analysis of the raw data is computationally very intensive, the workload is also split between the two computers. The detector computer can also be optimized for performing its data reduction tasks without being concerned about user interfaces.

The transmission of digital data requires that a communication protocol be defined so that the computer understands the series of discrete binary pulses being received. A brief discussion of the most popular communication methods is given later (Table 2).

1. Point-to-Point Communication (Serial, Parallel)

The simplest scenario is a direct connection between the computer and a single other device (the detector, terminal, or other computer). RS232 serial communication is one of the earliest communication protocols and is an example of such a point-to-point protocol. Both control and data signals are transmitted over eight distinct lines for hardware control or four lines when software control is used. The process of exchanging control signals to ensure proper signal transmission is also called handshaking.

As indicated by the name, serial communication serializes the native 8 bit words used by the computer in order to transmit the data bit by bit:

In order to establish communication between the two devices, they both must use the same serial port settings: transmission rate, data bits (5–8), parity (even, odd, and none), stop bits (1,2), and flow control/handshaking (hardware, XON/XOFF). The transmission rate is the number of bits per second that can be transmitted and is dependent on the actual hardware serial port. The data bit setting indicates how many data bits are sent per transmission (typically 8 bits for binary data). The parity bit is used for error checking. If set to odd, the transmitting device sets the parity bit such that the data bits plus the parity bit give an odd number of ones. If set to even, the transmitting device sets the parity bit such that the data bits plus the parity bit give an odd number of ones. For XON/XOFF (software) flow control, a device sends ASCII character 19 (Ctrl-S) in the data bits to stop transmission of data from the other device instead of using the hardware "clear to send" line. It then sends ASCII 17 (Ctrl-Q) to inform the other device that it can continue sending data.

Parallel communication is also point-to-point but sends all 8 data bits at once and is thus faster than serial communication. Parallel communication was originally unidirectional—data were only sent from the computer to the device, usually a printer. Bidirectional parallel ports are now standard.

2. Short Distance—Multiple Device Communication

Often laboratory instrumentation consists of several distinct modules that must interact with one another and the computer to function properly. The point-to-point scheme is insufficient in this case unless one dedicates one serial port for each device. Hewlett Packard developed a mulitple device interface and protocol to address this issue. The Hewlett Packard interface bus (HPIB) was adopted industry-wide and became known as the general purpose interface bus (GPIB) and the IEEE488. An active controller talks to up to 14 other devices over 8 data lines (parallel), 5 bus management lines, and 3 handshaking lines.

The SCSI described earlier to connect peripherals such as hard drives to the system bus of the computer can also be used to connect to external instruments. This is a relatively expensive solution, however, would only be used when extremely high data transfer rates are required.

The universal serial bus (USB) was developed for personal computers to simplify the connection of external

Start bit (always 0)	Data bit 1—LSB	Data bit 2	Data bit 3	Data bit 4	Data bit 5	Data bit 6	Data bit 7	Data bit 8—the most significant bit	Parity bit (odd, even, none)	Stop bit (always 1)	Stop bit (optional and always 1)

Table 2 Digital Communication Interfaces

	Maximum transmission speed	Maximum cable length	Maximum number of devices
Serial	64 kbps	10 ft	2
Parallel	50–100 kbytes/s	9–12 ft	2
USB 2.0	480 mbits/s	5 m Segments with a maximum of six segments between device and host	127
IEEE488	1 MByte/s	20 m (2 m per device)	15
Twisted pair ethernet	1000 Mbps	82 ft (329 ft at 100 Mbps)	254 per subnet using TCP/IP
SCSI	160 Mbytes/s	6 m[a]	16

[a]Single ended cable—differential on cables can be up to 25 m long.

devices. The USB cable consists of a single twisted pair for transmitting data and two wires for supplying 5 V power to low power devices (maximum 500 mA) such as mouse. The host computer automatically polls all devices on the bus and assigns them an address. Data are transmitted in 1500-byte frames every millisecond. (A frame is a communication protocol which is standardized and the user only needs to plug the device into an available USB port in order to use it.) This relatively new protocol has not been used extensively for instrumentation but will probably be adopted by instrumentation that transfers small amounts of data and currently uses the standard serial port.

3. Local Area Networks (LANs)

It is not uncommon to place instruments on an ethernet LAN with a variety of other devices and workstations all sharing the same transmission medium. A robust communication protocol is required if large numbers of devices are communicating with one another. Typically most large organizations use the TCP/IP network protocol on ethernet. Each device is assigned a unique IP address based on four 8 bit address segments corresponding to the network, subnet, and host/device, which is represented in the familiar decimal dot notation: 255.255.255.255.

A number of instrument vendors are taking advantage of the low cost of NICs and network hubs to establish small LANs to provide communications between the laboratory computer and the instrument. It is also a simple matter to configure a point-to-point network between a computer and an instrument with a cross-over cable and two NICs configured for TCP/IP.

F. Digital Data Filtering

One of the simplest noise filtering techniques is signal averaging. By repeating the experiment several times and averaging the results, the noise will tend to cancel itself out while the signal should remain constant. If the noise is truly random, the signal-to-noise ratio of the signal increases by the square root of the number of times successive signals are averaged.

Experiments need not necessarily be repeated many times in order to achieve better signal-to-noise as a result of signal averaging. In cases where there are a sufficient number of data points defining the signal curve, adjacent data points can be averaged to remove noise. This approach assumes that the analytical signal changes only slightly (and therefore linearly) within window of data points being averaged. In practice, 2–50 successive data points can be averaged in this way depending on the data being processed.

Several key analytical techniques use advanced mathematical treatment to obtain a strong signal while discriminating against noise. Examples include pulsed NMR and the Michelson interferometer-based infrared spectroscopy. In the case of NMR, the sample is irradiated with a short burst of RF radiation and then the amplitude of the decaying resonant frequencies of the excited sample is measured as a function of time. The Michelson interferometer measures the shift in frequencies caused by a fast moving mirror in the light path. These techniques have several distinct advantages over their scanning monochromator counterparts: the experiments are much quicker, thus lending themselves to signal averaging and there is no requisite loss of the source radiation from restricting the passed energy to a narrow energy band. In both cases, the signal is transformed to a response (amplitude) based on frequency instead of time by the Fourier transform method. An in-depth discussion of the Fourier transform is beyond this work, but the amplitude response S at given frequency is represented by the Fourier transform as:

$$S\hat{x}(f) = \int_{-\infty}^{\infty} x(t)e^{-i2\pi ft} \, dt \qquad (1)$$

This requires the integral to be calculated over all time in a continuous manner. Out of practical necessity, however, one can only obtain the amplitude at discrete intervals for a finite period of time. Several approximations are made in order to calculate the transform digitally, yielding the discrete fourier transform:

$$S'\hat{x}(\kappa\Delta f) \approx \frac{1}{N}\sum_{n=0}^{N-1} x(n)e^{-i(2\pi n\kappa/N)} \quad (2)$$

where N is the number of points sampled. This allows the signal to be represented in the frequency domain as in Fig. 3.

The function assumes that every component of the input signal is periodic, that is, every component of the signal has an exact whole number of periods within the timeframe being studied. If not, discontinuities at the beginning and ending border conditions develop, resulting in a distortion of the frequency response known as leakage. In this phenomena, part of the response from the true frequency band is attributed to neighboring frequency bands. This leads to artificially broaden the signal response over larger frequency bands that obscure smaller amplitude frequency signals. A technique known as windowing is used to reduce the signal amplitude to zero at the beginning and end of the time record (band-pass filter). This eliminates the discontinuities at the boundary time points, thus greatly reducing the leakage.

V. DATA ANALYSIS/CHEMOMETRICS

Besides their roles in controlling instruments, collecting and storing data, computers play a critical role in the computations and data processing needed for solving chemical problems. A good example is multivariate data analysis in analytical chemistry. The power of multivariate data analysis combined with modern analytical instruments is best demonstrated in areas where samples have to be analyzed as is and relevant information must be extracted from interference-laden data. These cases include characterization of chemical reactions, exploration of the relationship between the properties of a chemical and its structural and functional groups, fast identification of chemical and biological agents, monitoring and controling of chemical processes, and much more. The multivariate data analysis technique for chemical data, also known as chemometrics, heavily relies on the capabilities of computers because a lot of chemometrics algorithms are computationally intensive, and the data they designed to analyze are usually very large in size. Fortunately, advances in computer technology have largely eliminated the performance issues of chemometrics applications due to limitations in computer speed and memory encountered in early days.

Chemometrics is a term given to a discipline that uses mathematical, statistical, and other logic-based methods to find and validate relationships between chemical data sets, to provide maximum relevant chemical information, and to design or select optimal measurement procedures and experiments. It covers many areas, from traditional statistical data evaluation to multivariate calibration, multivariate curve resolution, pattern recognition, experiment design, signal processing, neural network, and more. As chemical problems get more complicated and more sophisticated mathematical tools are available for chemists, this list will certainly grow. It is not possible to cover all these areas in a short chapter like this. Therefore we chose to focus on core areas. The first is multivariate calibration. This is considered as the center piece of chemometrics, for its vast applications in modern analytical chemistry and great amount of research work done since the coinage of this discipline. The second area is pattern recognition. If multivariate calibration deals with predominantly quantitative problems, pattern recognition represents the other side of chemometrics—qualitative techniques that answer questions such as "Are the samples different? How are they related?", by separating, clustering, and categorizing data. We hope in this way readers can get a relatively full picture of chemometrics from such a short article.

A. Multivariate Calibration

1. General Introduction

In science and technology, establishing quantitative relationships between two or more measurement data sets is a basic activity. This activity, also know as calibration, is a process of finding the transformation to relate a data set with the other ones that have explicit information. A simple example of calibration is to calibrate a pH meter. After reading three standard solutions, the electrical voltages from the electrode are compared with the pH values of the standard solutions and a mathematical relation is defined. This mathematical relationship, often referred as a calibration curve, is used to transform the electric voltages into pH values when measuring new samples. This type of calibration is univariate in nature, which simply relates a single variable, voltage, to the pH values. One of the two conditions must be met in order to have accurate predictions with a univariate calibration curve. The measurement must be highly selective, that is, the electrode only responds to pH changes and nothing else, or interferences that can cause changes in the electrode response are removed from the sample matrix and/or the measurement process. The later method is found in chromatographic analyses where components in a sample are separated and individually detected. At each point of interest within a chromatographic analysis, the sample is pure and the detector is responding to the component of interest only. Univariate calibration works perfectly here to relate the chromatographic peak heights or areas to the concentrations of the samples. On the other side of the measurement world, measurement objects are not preprocessed or purified to eliminate things that can interfere with the measurement. Univariate calibration could suffer from erroneous data from the instrument, and worse yet there is no way for the analyst to tell if he/she is getting correct results or not. This shortcoming of univariate calibration, which is referred as zero-order calibration in tensor analysis for its scalar data nature, has seriously limited its use in modern analytical chemistry.

One way to address the problem with univariate calibration is to use more measurement information in establishing the transformation that relates measurement data with the reference data (data that have more explicit information). Modern analytical instrumentations such as optical spectroscopy, mass spectroscopy, and NMR deliver multiple outputs with a single measurement, providing an opportunity to overcome the shortcomings of univariate calibration. The calibration involving multiple variables is called multivariate calibration which has been the core of chemometrics since the very beginning. The major part of this section will be focused on the discussion of multivariate calibration techniques.

The capability of multivariate calibration in dealing with interferences lies on two bases: (1) the unique pattern in measurement data (i.e., spectra) for each component of interest and (2) independent concentration variation of the components in the calibration standard set. Let us consider the data from a multivariate instrument, say, an optical spectrometer. The spectrum from each measurement is represented by a vector $\mathbf{x} = [x_1, x_2, \ldots, x_n]$, which carries unique spectral responses for component(s) of interest and interference from sample matrix. Measurement of a calibration set with m standards generates a data matrix $\mathbf{X}$ with m rows and n columns, with each row representing a spectrum. The concentration data matrix $\mathbf{Y}$ with p components of interest will have m rows and p columns, with each row containing the concentrations of components of interest in a particular sample. The relationship between $\mathbf{X}$ (measurement) and $\mathbf{Y}$ (known values) can be described by the following equation:

$$\mathbf{Y} = \mathbf{XB} + \mathbf{E} \tag{3}$$

The purpose of calibration is to find the transform matrix $\mathbf{B}$ and evaluate the error matrix $\mathbf{E}$. Using linear regression, the transformation matrix $\mathbf{B}$, or the regression coefficient matrix as it is also called, can be found as:

$$\mathbf{B} = (\mathbf{X}'\mathbf{X})^{-1}\mathbf{X}'\mathbf{Y} \tag{4}$$

where $\mathbf{X}'$ represents the transpose of $\mathbf{X}$. The inversion of square matrix $\mathbf{X}'\mathbf{X}$ is a critical step in multivariate calibration, and the method of inversion essentially differentiates the techniques of multivariate calibration. The following sections will discuss the most commonly used methods.

2. Multiple Linear Regression

Multiple linear regression (MLR) is the simplest multivariate calibration method. In this method, the transformation matrix $\mathbf{B}$ [Eq. (4)] is calculated by direct inversion of measurement matrix $\mathbf{X}$. Doing so requires the matrix $\mathbf{X}$ to be in full rank. In other words, the variables in $\mathbf{X}$ are independent to each other. If this condition is met, then the transformation matrix $\mathbf{B}$ can be calculated using Eq. (4) without carrying over significant errors. In the prediction step, $\mathbf{B}$ is applied to the spectrum of unknown sample to calculate the properties of interest:

$$\hat{\mathbf{y}}_i = \mathbf{x}'_i(\mathbf{X}'\mathbf{X})^{-1}\mathbf{X}'\mathbf{Y} \tag{5}$$

The most important thing in MLR is to ensure the independence of the variables in $\mathbf{X}$. If the variable are linearly dependent, that is, at least one of the rows can be written as an approximate or exact linear combination of the others, matrix $\mathbf{X}$ is called collinear. In the case of collinearity some elements of $\mathbf{B}$ from least square fit have large variance and the whole transform matrix loses its stability. Therefore, for MLR collinearity in $\mathbf{X}$ is a serious problem and limits its use.

To better understand the problem, consider an example in optical spectroscopy. If one wants to measure the concentration of two species in a sample, he/she has to hope that the two species have distinct spectral features and it is possible to find a spectral peak corresponding to each of them. In some cases, one would also like to add a peak for correcting variations such as spectral baseline drift. As a result, he would have a $3 \times n\mathbf{X}$ matrix after measuring n standard samples. To ensure independence of $\mathbf{X}$ variables, the spectral peaks used for measuring the components of interest have to be reasonably separated. In some cases, such as in FTIR and Raman spectroscopy whose finger print regions are powerful in differentiating chemical compounds, it might be possible. MLR is accurate and reliable when the "no collinearity" requirement is met. In other cases, however, finding reasonably separated peaks is not possible. This is often true in near infrared (NIR) and UV/Vis spectroscopy which lack the capability in differentiating between chemicals. The NIR and UV/Vis peaks are broad, often overlapping, and the spectra of different chemicals can look very similar. This makes choosing independent variables (peaks) very difficult. In cases where the variables are collinear, using MLR could be problematic. This is probably one of the reasons that MLR is less used in the NIR and UV/Vis applications.

Due to the same collinearity reason, one should also be aware of problems brought by redundancy in selecting variables for **X** matrix. Having more variables to describe a property of interest (e.g., concentration) may generate a better fit in the calibration step. However, this can be misleading. The large variance in the transformation matrix caused by collinearity will ultimately harm the prediction performance of calibration model. It is not uncommon to see the mistake of using an excessive number of variables to establish a calibration model. This kind of calibration model can be inaccurate in predicting unknown samples and is sensitive to minor variations.

3. Factor Analysis Based Calibration

In their book, Martens and Naes listed problems encountered in dealing with complex chemical analysis data using traditional calibration methods such as univariate calibration:

1. Lack of selectivity: No single **X**-variable is sufficient to predict **Y** (property matrix). To attain selectivity one must use several **X**-variables.
2. Collinearity: There may be redundancy and hence collinearity in **X**. A method that transforms correlated variables into independent ones is needed.
3. Lack of knowledge: Our *a priori* understanding of the mechanisms behind the data may be incomplete or wrong. Calibration models will fail when new variations or constituents unaccounted by the calibration occur in samples. One wishes at least to have a method to detect outliers, and further to improve the calibration technology so that this kind of problem can be solved.

Factor analysis based calibration methods have been developed to deal with these problems. Due to space limit, we only discuss two most popular methods, principal component regression (PCR) and partial least square (PLS).

Principle Component Regression

The base of PCR is principal component analysis (PCA) which computes so-called principle components to describe variation in matrix **X**. In PCA, the main variation in $\mathbf{X} = \{\mathbf{x}_k, k = 1, 2, \ldots, K\}$ is represented by a smaller number of variables $\mathbf{T} = \{\mathbf{t}_1, \ldots, \mathbf{t}_A\}$ ($A < K$). **T** represents the principle components computed from **X**. The principle components are calculated by finding the first loading vector $\mathbf{u}_1$ that maximizes the variance of $\mathbf{u}_1 \times \mathbf{x}$ and have $\mathbf{u}_1 \times \mathbf{X} \times \mathbf{X}\mathbf{u}_1 = \mathbf{t}_1 \times \mathbf{t}_1$, where $\mathbf{t}_1$ is the corresponding score vector. The next principle component is calculated in the same way but with the restriction that $\mathbf{t}_1$ and $\mathbf{t}_2$ are orthogonal ($\mathbf{t}_1\mathbf{t}_2 = 0$). The procedure continues under this restriction until it reaches the dimension limit of matrix **X**. Figure 4 may help to understand the relationship between original variables **x** and principle component **t**.

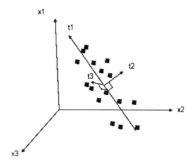

Figure 4 PCA illustrated for three **x**-variables with three principle components (factors).

Consider a sample set measured with three variables x_1, x_2, and x_3. Each sample is represented by a dot in the coordinate system formed by x_1, x_2, and x_3. What PCA does is to find the first principle component ($\mathbf{t}_1$) that points to the direction of largest variation in that data set, then the second principle component ($\mathbf{t}_2$) capturing the second largest variation and orthogonal to $\mathbf{t}_1$, and finally the third principle component ($\mathbf{t}_3$) describes the remaining variation and orthogonal to $\mathbf{t}_1$ and $\mathbf{t}_2$. From the figure it is clear that principle components replace x_1, x_2, and x_3 to form a new coordination system, these principle components are independent to each other, and they are arranged in a descending order in terms the amount of variance they describe. In any data set gathered from reasonably well-designed and measured experiments, useful information is stronger than noise. Therefore it is fair to expect that the first several principle components mainly contain the useful information, and the later ones are dominated by noise. Users can conveniently keep the first several principle components for use in calibration and discard the rest of them. Thus, useful information will be kept while noise is thrown out. Through PCA, the original matrix **X** is decomposed into three matrices. **V** consists of normalized score vectors. **U** is the loading matrix and the **S** is a diagonal matrix containing eigenvalues resulting from normalizing the score vectors.

$$\mathbf{X} = \mathbf{V}\mathbf{S}\mathbf{U}' \quad (6)$$

As just mentioned, **X** can be approximated by using the first several significant principle components,

$$\tilde{\mathbf{X}} \approx \mathbf{V}_A \mathbf{S}_A \mathbf{U}'_A \quad (7)$$

where $\mathbf{V}_A$, $\mathbf{S}_A$, and $\mathbf{U}_A$ are subsets of **V**, **S**, and **U**, respectively, formed by the first A principle components. $\tilde{\mathbf{X}}$ is a close approximation of **X**, with minor variance removed by discarding the principle components after A.

Principal components are used both in qualitative interpretation of data and in regression to establish quantitative calibration models. In qualitative data interpretation, the so-called score plot is often used. The element v_{ij} in $\mathbf{V}$ is the projection of ith sample on jth principle component. Therefore each sample will have a unique position in a space defined by score vectors, if they are unique in the measured data $\mathbf{X}$. Figure 5 illustrates use of score plot to visually identify amorphous samples from crystalline samples measured with NIR spectroscopy. The original NIR spectra show some differences between amorphous and crystalline samples, but they are subtle and complex to human eyes. PCA and score plot present the differences in a much simpler and straight forward way. In the figure, the circular dots are samples used as standards to establish the score space. New samples (triangular and star dots) are projected into the space and grouped according to their crystallinity. It is clear that the new samples are different: some are crystalline, so they fall into the crystalline circle. The amorphous samples are left outside of the circle due to the abnormalities that have shown up in NIR spectra. Based on the deviations of the crystalline sample dots on each principle component axis, it is possible to calculate statistical boundary to automatically detect amorphous samples. This kind of scheme is the bases of outlier detection by factor analysis based calibration methods such as PCA and PLS.

PCA combined with a regression step forms PCR. PCR has been widely used by chemists for its simplicity in interpretation of the data through loading matrix and score matrix. Equation (6) shows how regression is performed with the principle components.

$$\mathbf{b} = \mathbf{y}\mathbf{V}_A\mathbf{S}_A^{-1}\mathbf{U}_A' \tag{8}$$

In this equation, $\mathbf{y}$ is the property vector of the calibration standard samples and $\mathbf{b}$ is the regression coefficient vector used in predicting the measured data $\mathbf{x}$ of unknown sample.

$$\hat{y} = \mathbf{b}\mathbf{x}' \tag{9}$$

where $\hat{y}$ is the predicted property. A very important aspect of PCR is to determine the number of factors used in Eq. (6). The optimum is to use as much information in $\mathbf{X}$ as possible and keep noise out. That means one needs to decide the last factor (principle component) that has useful information and discard all factors after that one. A common mistake in multivariate calibration is to use too many factors to over fit the data. The extra factors can make a calibration curve look unrealistically good (noise also gets fitted) but unstable and inaccurate when used in prediction. A rigorous validation step is necessary to avoid these kinds of mistakes. When the calibration sample set is sufficiently large, the validation samples can be randomly selected from the sample pool. If the calibration samples are limited in number, a widely used method is cross-validation within the calibration sample set. In cross-validation, a sample (or several samples) is taken out from the sample set and predicted by the calibration built on the remaining samples. The prediction errors corresponding to the number of factors used in calibration are recorded. Then the sample is put back into the sample set and another one is taken out in order to repeat the same procedure. The process continues until each sample has been left out once and predicted. The average error is calculated as the function of the number of principle components used. The formulas for SEP (using a separate validation sample set) and for SECV cross-validation are slightly different:

$$\text{SEP} = \sqrt{\frac{\sum (y_i - \hat{y}_i)^2}{n}} \tag{10}$$

$$\text{SECV} = \sqrt{\frac{\sum_{i=1}^{n}(y_i - \hat{y}_i)^2}{n-A}} \tag{11}$$

where $\hat{y}_i$ is the model predicted value and y_i is reference value for sample i, n is the total number of samples used in calibration, and A is the number of principle components used. When the errors are plotted against the number of factors used in calibration, they typically look like the one illustrated in Fig. 6. As the principle components (factors) are added into the calibration one at a time, SEP or SECV decreases, hits a minimum, and then bounces back. The reason is that the first several principle components contain information about the samples and are needed in improving the calibration model's accuracy. The later principle components, on the other hand, are predominated by noise. Using them makes the model sensitive to irrelevant variations in data, thus being less accurate

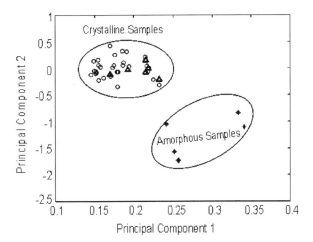

Figure 5 PCA applied to identify sample crystallinity.

and potentially more vulnerable to process variations and instrument drifts. There is clearly an optimal number of principle components for each calibration model. One of the major tasks in multivariate calibration is to find the optimum which will keep the calibration model simple (small number of principle components) and achieve the highest accuracy possible.

Partial Least Square

PLS is another multivariate calibration method which uses principle components rather than original **X**-variables. It differs from PCR by using the **y**-variables actively during the decomposition of **X**. In PLS, principle components are not calculated along the direction of largest variation in **X** at each iteration step. They are calculated by balancing the information from the **X** and **y** matrices to best describe information in **y**. The rationale behind PLS is that, in some cases, some variations in **X**, although significant, may not be related to **y** at all. Thus, it makes sense to calculate principle components more relevant to **y**, not just to **X**. Because of this, PLS may yield simpler models than PCR.

Unlike PCR, the PLS decomposition of measurement data **X** involves the property vector **y**. A loading weight vector is calculated for each loading of **X** to ensure the loadings are related to the property data **y**. Furthermore, the property data **y** is not directly used in calibration. Instead, its loadings are also calculated and used together with the **X** loadings to obtain the transformation vector **b**.

$$\mathbf{b} = \mathbf{W}(\mathbf{PW})^{-1}\mathbf{q} \quad (12)$$

where **W** is the loading weight matrix, **P** is the loading matrix for **X**, and **q** is the loading vector for **y**. Martens and Neas gave out detailed procedures of PLS in their book.

In many cases, PCR and PLS yield similar results. However, because the PLS factors are calculated utilizing both **X** and **y** data, PLS can sometimes give useful results from low precision **X** data where PCR may fail. Due to the same reason, PLS has a stronger tendency to over fit noisy data **y** than that of PCR.

B. Pattern Recognition

Modern analytical chemistry is data rich. Instruments such as mass spectroscopy, optical spectroscopy, NMR, and many hyphenated instruments generate a lot of data for each sample. However, data rich does not mean information rich. How to convert the data into useful information is the task of chemometrics. Here, pattern recognition plays an especially important role in exploring, interpreting, and understanding the complex nature of multivariate relationships. Since the tool was first used on chemical data by Jurs et al. in 1969, many new applications have been published, including several books containing articles on this subject. Based on Lavine's review, here we will briefly discuss four main subdivisions of pattern recognition methodology: (1) mapping and display, (2) clustering, (3) discriminant development, and (4) modeling.

1. Mapping and Display

When there are two to three types of samples, mapping and display is an easy way to visually inspect the relationship between the samples. For example, samples can be plotted in a 2D or 3D coordinate system formed by variables describing the samples. Each sample is represented by a dot on the plot. The distribution and grouping of the samples reveal the relationship between them. The frequently encountered problem with modern analytical data is that the number of variables needed to describe a sample is often way too large for this simple approach. An ordinal person cannot handle a coordinate system with more than three dimensions. The data size from an instrument, however, can be hundreds or even thousands of variables. To utilize all information carried by so many variables, factor analysis method can be used to compress the dimensionality of the data set and eliminate collinearity between the variables. In last section, we discussed the use of principle components in multivariate data analysis (PCA). The plot generated by principle components (factors) is exactly the same as the plots used in mapping and display method. The orthogonal nature of principle components allows convenient evaluation of

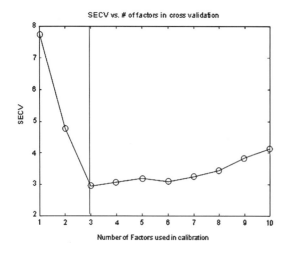

Figure 6 SECV plotted against number of principle components (factors) used in calibration.

factors affecting samples based on their positions in the principle component plot.

The distance between samples or from a sample to the centroid of a group provides a quantitative measure of the degree of similarity of the sample with others. The most frequently used ones are the Euclidean distance and the Mahalanobis distance. The Euclidean distance is expressed as:

$$D_E = \sqrt{\sum_{j=1}^{n} (x_{Kj} - x_{Lj})^2} \qquad (13)$$

where x_{Kj} and x_{Lj} are the jth coordinate of samples K and L, respectively and n is the total number of coordinates. The Mahalanobis distance is calculated by the following equation:

$$D_M^2 = (\mathbf{x}_L - \bar{\mathbf{x}}_K)' \mathbf{C}_K^{-1} (\mathbf{x}_L - \bar{\mathbf{x}}_K) \qquad (14)$$

where $\mathbf{x}_L$ and $\bar{\mathbf{x}}_K$ are, respectively, the data vector of sample L and mean data vector for class K. $\mathbf{C}_K^{-1}$ is the covariance matrix of class K. The Euclidean distance is simply the geometric distance between samples. It does not consider the collinearity between the variables that forms the coordinator system. If variables x_1 and x_2 are independent, the Euclidean distance is not affected by the position of the sample in the coordinate system and truly reflects the similarity between the samples or sample groups. When variables are correlated, this may not be true. The Mahalanobis distance measurement takes into account the problem by including a factor of correlation (or covariance).

2. Clustering

Clustering methods are based on the principle that the distance between pairs of points (i.e., samples) in the measurement space is inversely related to their degree of similarity. There are several types of clustering algorithms using distance measurement. The most popular one is called hierarchical clustering. The first step in this algorithm is to calculate the distances of all pairs of points (samples). Two points having the smallest distance will be paired and replaced by a new point located in the midway of the two original points. Then the distance calculation starts again with the new data set. Another new point is generated between the two data points having the minimal distance and replaces the original data points. This process continues until all data points have been linked.

3. Classification

Both mapping and displaying clustering belong to unsupervised pattern recognition techniques. No information about the samples other than the measured data is used in analyses. In chemistry, there are cases where a classification rule has to be developed to predict unknown samples. Development of such a rule needs training datasets whose class memberships are known. This is called supervised pattern recognition because the knowledge of class membership of the training sets is used in development of discriminant functions. The most popular methods used in solving chemistry problems include the linear learning machine and the adaptive least square (ALS) algorithm. For two classes separated in a symmetric manner, a linear line (or surface) can be found to divide the two classes (Fig. 7). Such a discriminant function can be expressed as:

$$D = \mathbf{w}\mathbf{x}' \qquad (15)$$

where $\mathbf{w}$ is called weight vector and $\mathbf{w} = \{w_1, w_2, \ldots, w_{n+1}\}$ and $\mathbf{x} = \{x_1, x_2, \ldots, x_{n+1}\}$ is pattern vector whose elements can be the measurement variables or principle component scores. Establishing the discriminant function is to determine the weight vector $\mathbf{w}$ with the restraint that it provides the best classification (most correct Ds) for two classes. The method is usually iterative in which error correction or negative feedback is used to adjust $\mathbf{w}$ until it allows the best separation for the classes. The samples in the training set are checked one at a time by the discriminant function. If classification is correct, $\mathbf{w}$ is kept unchanged and the program moves to the next sample. If classification is incorrect, $\mathbf{w}$ is altered so that correct classification is obtained. The altered $\mathbf{w}$ is then used in the subsequent steps till the program goes through all samples in the training set. The altered $\mathbf{w}$ is defined as:

$$\mathbf{w}''' = \mathbf{w} - \left\lfloor \frac{2s_i}{\mathbf{x}_i \mathbf{x}_i'} \right\rfloor \mathbf{x}_i \qquad (16)$$

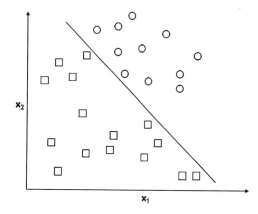

Figure 7 Example of a linear discriminate function separating two classes.

where $\mathbf{w}'''$ is the altered weight factor, s_i is the discriminant for the misclassified sample i, and $\mathbf{x}_i$ is the pattern vector of sample i. In situations where separation cannot be best achieved by a simple linear function, ALS can be used. In ALS, $\mathbf{w}$ is obtained using least squares:

$$\mathbf{w} = (\mathbf{X}'\mathbf{X})^{-1}\mathbf{X}'\mathbf{f} \qquad (17)$$

where $\mathbf{f}$ is called forcing factor containing forcing factors f_i for each sample i. When classification of sample i is correct, $f_i = s_i$, where s_i is the discriminant score for sample i. If classification is incorrect, f is modified according to the following equation:

$$f_i = s_i + \frac{0.1}{(\alpha + d_i)^2} + \beta(\alpha + d_i) \qquad (18)$$

where α and β are constants that are empirically determined and is the distance between the pattern vector and the classification surface (i.e., the discriminant score). With the corrected forcing factor, an improved weight vector $\mathbf{w}$ is calculated using Eq. (16) and used in next round of discriminant score calculation. The procedure continues until favorable classification results are obtained or preselected number of feedback has been achieved.

The nonparametric linear discriminant functions discussed earlier have limitations when dealing with classes separated in asymmetric manners. One could imagine a situation where a group of samples is surrounded by samples that do not belong to that class. There is no way for a linear classification algorithm to find a linear discriminant function that can separate the class from these samples. Apparently there is a need for algorithms that have enough flexibility in dealing with this type of situations.

4. SIMCA

SIMCA stands for soft independent modeling of class analogy. It was developed by Wold and coworkers for dealing with asymmetric separation problems. It is based on PCA. In SIMCA, PCA is performed separately on each class in the dataset. Then each class is approximated by its own principle components:

$$\mathbf{X}_i = \bar{\mathbf{X}}_i + \mathbf{T}_{iA}\mathbf{P}_{iA} + \mathbf{E}_i \qquad (19)$$

where $\mathbf{X}_i(N \times P)$ is the data matrix of class i, $\bar{\mathbf{X}}_i$ is the mean matrix of $\mathbf{X}_i$ with each row being the mean of $\mathbf{X}_i$. $\mathbf{T}_{iA}$ and $\mathbf{P}_{iA}$ are the score matrix and loading matrix, respectively, using A principle components. $\mathbf{E}_i$ is the residual matrix between the original data and the approximation by the principle component model.

The residual variance for the class is defined by:

$$S_0^2 = \sum_{i=1}^{N}\sum_{j=1}^{P} \frac{e_{ij}^2}{(P-A)(N-A-1)} \qquad (20)$$

where e_{ij} is the element of residual matrix $\mathbf{E}_i$ and S_0 is the residual variance which is a measurement for the tightness of the class. A smaller S_0 indicates a more tightly distributed class.

In classification of unknown samples, the sample data is projected on the principle component space of each class with the score vector calculated as:

$$\mathbf{t}_{ik} = \mathbf{x}_i \mathbf{P}_k^{-1} \qquad (21)$$

where $\mathbf{t}_{ik}$ is the score vector of sample i in the principle component space of class k. $\mathbf{P}_k^{-1}$ is the loading matrix of class k. With score vector $\mathbf{t}_{ik}$ and loading matrix $\mathbf{P}_k$, the residual vector of sample i fitting into class K can be calculated similarly to Eq. (19). Then the residual variance of fit for sample i is:

$$S_i^2 = \sum_{j=1}^{P} \frac{e_{ij}^2}{P-A} \qquad (22)$$

The residual variance of fit is compared with the residual variance of each class. If S_i is significantly larger than S_0, sample i does not belong to that class. If S_i is significantly larger than S_0, sample i is considered a member of that class. F-test is employed to determine if S_i is significantly larger than S_0.

The number of principle components (factors) used for each class to calculate S_i and S_0 is determined through cross-validation. Similar to what we have discussed in multivariate calibration, cross-validation in SIMCA is to take out one (or several) sample a time from a class and use the remaining samples to calculate the residual variance of fit with different number of principle components. After all samples have been taken out once, the overall residual variance of fit as the function of the number principle components used are calculated. The optimal number of principle components is the one that gives the smallest classification error.

SIMCA is a powerful method for classification of complex multivariate data. It does not require a mathematical function to define a separation line or surface. Each sample is compared with a class within the class' sub-principle component space. Therefore, it is very flexible in dealing with asymmetrically separated data and classes with different degree of complexity. Conceptually, it is also easy to understand, if one has basic knowledge in PCA. These are main reasons why SIMCA becomes very popular among chemists.

VI. DATA ORGANIZATION AND STORAGE

The laboratory and the individual scientists can easily be overwhelmed with the sheer volumes of data produced today. Very rarely can an analytical problem be answered with a single sample let alone a single analysis. Compound this by the number of problems or experiments that a scientist must address and the amount of time spent organizing and summarizing the data can eclipse the time spent acquiring it. Scientific data also tends to be spread out among several different storage systems. The scientist's conclusions based on a series of experiments are often documented in formal reports. Instrument data is typically contained on printouts or in electronic files. The results of individual experiments tend to be documented in laboratory notebooks or on official forms designed for that purpose.

It is important that all of the data relevant to an experiment be captured: the sample preparation, standard preparation, instrument parameters, as well as the significance of the sample itself. This meta data must be cross-referenced to the raw data and the final results so that they can be reproduced if necessary. It is often written in the notebook or in many cases it is captured by the analytical instrument and stored in the data file and printed on the report where it cannot be easily searched. Without this information, the actual data collected by an instrument can be useless, as this information may be crucial in its interpretation.

Scientists have taken advantage of various personal productivity tools such as electronic spreadsheets, personal databases, and file storage schemes to organize and store their data. While such tools may be adequate for a single scientist such as a graduate student working on a single project, they fail for laboratories performing large numbers of tests. It is also very difficult to use such highly configurable, nonaudited software in regulated environments. In such cases, a highly organized system of storing data that requires compliance to the established procedures by all of the scientific staff is required to ensure an efficient operation.

A. Automated Data Storage

Ideally, all of the scientific data files of a laboratory would be cataloged (indexed) and stored in a central data repository. There are several commercial data management systems that are designed to do just this. Ideally these systems will automatically catalog the files using indexing data available in the data files themselves and then upload the files without manual intervention from the scientist. In reality, this is more difficult than it would first appear. The scientist must enter the indexing data into the scientific application and the scientific application must support its entry. Another potential problem is the proprietary nature of most instrument vendor's data files. Even when the instrument vendors are willing to share their data formats, the sheer numbers of different instrument file formats make this a daunting task. Still, with some standardization, these systems can greatly decrease the time scientists spend on mundane filing-type activities and provide a reliable archive for the laboratory's data. These systems also have the added benefit of providing the file security and audit trail functionality required in regulated laboratories on an enterprise-wide scale instead of a system-by-system basis.

However, storing the data files in a database solves only part of the archiving problem. Despite the existence of a few industry standard file formats, most vendors use a proprietary file format as already discussed. If the data files are saved in their native file format, they are only useful for as long as the originating application is available or if a suitable viewer is developed. Rendering the data files in a neutral file format such as XML mitigates the obsolescence problem but once again requires that the file format be known. It will also generally preclude reanalyzing the data after the conversion.

B. Laboratory Information Management Systems

Analytical laboratories, especially quality control, clinical testing labs, and central research labs, produce large amounts of data that need to be accessed by several different groups such as the customers, submitters, analysts, managers, and quality assurance personnel. Paper files involve a necessarily manual process for searching results, requiring both personnel and significant amounts of time. Electronic databases are the obvious solution for storing the data so that it can be quickly retrieved as needed. As long as sufficient order is imposed on the storage of the data, large amounts of data can be retrieved and summarized almost instantaneously by all interested parties.

A database by itself, however, does not address the workflow issues that arise between the involved parties. Laboratories under regulatory oversight such as pharmaceutical quality control, clinical, environmental control, pathology, and forensic labs must follow strict procedures with regard to sample custody and testing reviews. Laboratory information management systems (LIMS) were developed to enforce the laboratory's workflow rules as well as store the analytical results for convenient retrieval. Everything from sample logging, workload assignments, data entry, quality assurance review, managerial approval, report generation, and invoice processing can be carefully controlled and tracked. The scope of a LIMS system can vary greatly from a simple database to store final results and print reports to a comprehensive data management system that includes raw data files, notebook-type entries, and standard operating procedures. The degree to which this can be done will be dependent upon the ability and willingness of all concerned parties to standardize their procedures. The LIMS functions are often also event and time driven. If a sample fails to meet specifications, it can be automatically programed to e-mail the supervisor or log additional samples. It can also be programed to automatically log required water monitoring samples every morning and print the corresponding labels.

It was mentioned earlier that paper-based filing systems were undesirable because of the relatively large effort required to search for and obtain data. The LIMS database addressed this issue. However, if the laboratory manually enters data via keyboard into its LIMS database, the laboratory can be paying a large up-front price in placing the data in the database so that it can be easily retrieved. Practically from the inception of the LIMS systems, direct instrument interfaces were envisioned whereby the LIMS would control the instrumentation and the instrument would automatically upload its data. Certainly this has been successfully implemented in some cases but once again the proprietary nature of instrument control codes and data file structures makes this a monumental task for laboratories. Third party parsing and interfacing software has been very useful in extracting information from instrument data files and uploading the data to the LIMS systems. Once properly programmed and validated, these systems can bring about very large productivity gains in terms of the time saved entering and reviewing the data as well as resolving issues related to incorrect data entry. Progress will undoubtedly continue to be made on this front since computers are uniquely qualified to perform such tedious, repetitive tasks, leaving the scientist to make conclusions based on the summarized data.

BIBLIOGRAPHY

Bigelow, S. J. (2003). *PC Hardware Desk Reference*. Berkeley: McGraw Hill.

Crecraft, D., Gergely, S. (2002). *Analog Electronics: Circuits, Systems and Signal Processing*. Oxford: Butterworth-Heinemann.

Horowitz, P., Hill, W. (1989). *The Art of Electronics*. 2nd. ed. New York: Cambridge University Press.

Isenhour, T. L., Jurs, P. C. (1971). *Anal. Chem.* 43: 20A.

Jurs, P. C., Kowalski, B. R., Isenhour, T. L. (1969). *Anal. Chem.* 41: 21.

Lai, E. (2004). *Practical Digital Signal Processing for Engineers and Technicians*. Oxford: Newnes.

Lavine, B. K. (1992). Signal processing and data analysis. In: Haswell, S. J., ed. *Practical Guide to Chemometrics*. New York: Marcell Dekker, Inc.

Massart, D. L., Vandeginste, B. G. M., Deming, S. N., Michotte, Y., Kaufman, L. (1988). *Chemometrics: A Textbook*. Amsterdam: Elsevier.

Materns, H., Naes, T. (1989). *Multivariate Calibration*. New York: John Wiley and Sons Ltd.

Moriguchi, I., Komatsu K., Matsushita, Y. (1980). *J. Med. Chem.* 23: 20.

Mueller, S. (2003). *Upgrading and Repairing PCs*. 15th ed. Indianapolis: Que.

Paszko, C., Turner, E. (2001). *Laboratory Information Management Systems*. 2nd ed. New York: Marcel Dekker, Inc.

Van Swaay, M. (1997). The laboratory use of computers. In: Ewing, G. W., ed. *Analytical Instrumentation Handbook*. 2nd ed. New York: Marcel Dekker, Inc.

Wold, S., Sjostrom, M. (1977). SIMCA—a method for analyzing chemical data in terms of similarity and analogy. In: Kowalski, B. R., ed. *Chemometrics: Theory and Practice*. Society Symp, Ser. No. 52. Washington D.C.: American Chemical Society.

2

Flow Injection/Sequential Injection Analysis

ELO HARALD HANSEN
Department of Chemistry, Technical University of Denmark, Kemitorvet, Kgs. Lyngby, Denmark

JIANHUA WANG
Research Center for Analytical Sciences, Northeastern University, Shenyang, P.R. China

I. INTRODUCTION

Flow injection analysis (FIA), which is method of continuous-flow analysis (CFA), was conceived in Denmark by Ruzicka and Hansen in 1974, and the first paper describing this novel and revolutionizing analytical concept for conducting chemical assays was published in early 1975 (Ruzicka and Hansen, 1975). While all the previous CFA methods had been based on complete mixing of sample and reagent(s) (physical homogenization) and awaiting all chemical reactions to proceed to equilibrium (chemical homogenization) in order to obtain so-called steady-state conditions (see Chapter 10), thus, in fact, mimicking what is done—and has been done for aeons—in batch assay, FIA broke entirely with this conceptual way of thinking. Instead, FIA is based on measuring a transient signal obtained by injecting, or inserting, a defined volume of sample solution into a continuously moving carrier stream (Fig. 1). During its transport through the system, the sample will, as a result of the axial and radial dispersion processes, result in the creation of a concentration gradient, corresponding to innumerable, sequential liquid segments representing all concentrations from 0 to C^{max}. As seen from Fig. 1, each of the concentration elements, which in turn correspond to a fixed delay time, t_i, can potentially be used for the analytical readout.

If the carrier stream in itself is a reagent [or if reagent(s) is(are) added via sidestreams], the dispersion process will consequently permit that chemical reaction can take place at the interfaces between the sample and the reagent, that is, proceeding both at the front and at the tailing part of the dispersed sample zone, giving rise to the generation of a product, a characteristic feature of which can be monitored by a suitable detector.

The fundamentals of FIA were already verbalized in the first publication (Ruzicka and Hansen, 1975), that is, injection of a well-defined volume of sample; reproducible and precise timing of the manipulations it is subjected to in the system, from the point of injection to the point of detection (so-called controlled, or rather controllable, dispersion); and the creation of a concentration gradient of the injected sample, providing a transient, but strictly reproducible readout of the recorded signal. The combination of these features, as recorded by the detector, which may continuously observe an absorbance, an electrode potential, or any other physical parameter as it changes on passage of the sample material through the flow cell, makes it unnecessary to achieve chemical equilibrium (steady-state conditions). Any point on the path toward the steady-state

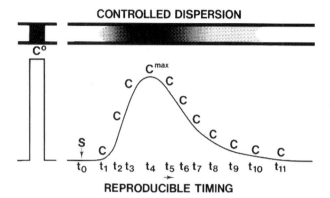

Figure 1 Dispersed sample zone, of original concentration C^0, injected at position S, and the corresponding recorder output. To each individual concentration C along the gradient corresponds a fixed delay time t_i elapsed from the moment of injection t_0. [From Ruzicka and Hansen (1988a), courtesy John Wiley and Sons.]

signal is as good a measure as the steady-state itself, provided that this point can be reproduced repeatedly, and this is certainly feasible in FIA with its inherently exact timing. This, in turn, has not only allowed to perform chemical assays much faster, and hence facilitating higher sampling rates, than in conventional procedures, but more importantly it has also permitted to implement procedures which are difficult, or, in fact, impossible, to effect by traditional means, many of which are entirely unique, as will be demonstrated in the following sections.

Since its introduction, FIA has been used extensively by researchers and analytical chemists around the world, as amply reflected in the more than 13,500 scientific papers which by 2003 have appeared world-wide (Hansen, 2003; Ruzicka and Hansen, 1988a; Ruzicka, 2003), to which should be added more than 20 monographs which have been published (Table 1). Therefore, it will be impossible in this chapter to cover all aspects of FIA, nor to pretend to make this an exhaustive review. Besides, in recent years FIA has undergone certain changes, that is, it has been supplemented by sequential injection analysis (SIA), also termed the second generation of FIA, and the lab-on-valve (LOV), the third generation of FIA. Thus, the present chapter will focus on these three generations of FIA, their characteristics, and their applications, that is, initially outlining the distinctive features of FIA when compared with conventional CFA and via selected examples demonstrate its unique capabilities, while in the following sections emphasis will be placed on the ensuing generations of FIA, detailing with their distinct advantages (and limitations when compared with FIA). Although a large part of the examples are from the authors' own laboratory, this should not be misinterpreted as lack of appropriate dues to the many persons who over the years have published ingenious and interesting procedures. Rather, it is dictated by sheer practical reasons (copyrights to individual papers). What is of importance in the present context is to demonstrate that FIA, besides allowing automation of chemical assays with high sampling frequencies and minute consumptions of sample and reagent solutions, offers potentials to implement novel and unique applications. Or as one of these authors previously wrote in characterizing FIA "the ultimate test for an analytical approach is not that it can do better what can be done by other means, but that it allows us to do something that we cannot do in any other way" (Hansen, 1989). And FIA does allow us to make unique applications. The only limitation is simply our own ingenuity.

Table 1 Significant FIA-Monographs

Ruzicka, J., Hansen, E. H. (1981). *Flow Injection Analysis*. New York: Wiley&Sons Inc.
Ruzicka, J., Hansen, E. H. (1983). *Flow Injection Analysis*. Kyoto: Kagakudonin (Japanese).
Ueno, K., Kina, K. (1983). *Introduction to Flow Injection Analysis. Experiments and Applications*. Tokyo: Kodansha Scientific (Japanese).
Valcarcel Cases, M., Luque de Castro, M. D. (1984). *Flow-Injection Analysis*. Cordoba: Imprenta San Pablo (Spanish).
Hansen, E. H. (1986). *Flow Injection Analysis*. Copenhagen: Polyteknisk Forlag.
Ruzicka, J., Hansen, E. H. (1986). *Flow Injection Analysis*. Beijing: Science Press (Chinese).
Valcarcel, M., Luque de Castro, M. D. (1987). *Flow-Injection Analysis. Principles and Applications*. Chichester: Ellis Horwood Ltd.
Ruzicka, J., Hansen, E. H. (1988a). *Flow Injection Analysis*. 2nd ed. New York: Wiley & Sons Inc.
Burguera, J. L., ed. (1989). *Flow Injection Atomic Spectroscopy*. New York: Marcel Dekker.
Karlberg, B., Pacey, G. E. (1989). *Flow Injection Analysis. A Practical Guide*. Amsterdam: Elsevier.
Ruzicka, J., Hansen, E. H. (1991). *Flow Injection Analysis*. 2nd ed. Beijing: Beijing University Press (Chinese).
Fang, Z.-L. (1993). *Flow-Injection Separation and Preconcentration*. Weinheim: VCH Verlagsgesellschaft mbh.
Frenzel, W. (1993). *Flow Injection Analysis. Principles, Techniques and Applications*. Berlin: Technical Univ.
Valcarcel, M., Luque de Castro, M. D. (1994). *Flow-Through (Bio)Chemical Sensors*. Amsterdam: Elsevier.
Fang, Z.-L. (1995). *Flow Injection Atomic Spectrometry*. Chichester: Wiley.
Calatayud, J. M., ed. (1997). *Flow Injection Analysis of Pharmaceuticals: Automation in the Laboratory*. London: Taylor&Francis.
Sanz-Medel, A., ed. (1999). *Flow Analysis with Atomic Spectrometric Detectors*. Amsterdam: Elsevier.
Trojanowicz, M. (2000). *Flow Injection Analysis. Instrumentation and Applications*. Singapore: World Scientific Ltd.

II. FLOW INJECTION ANALYSIS

A. Basic Principles

FIA has several important characteristics compared with traditional continuous-flow measurements: higher sampling rates (typically 100–300 samples/h); enhanced response times (often less than 30 s between sample injection and detector response); much more rapid start-up and shut-down times (merely a few minutes for each); and, except for the injection system, simpler and more flexible equipment. The last two advantages are of particular importance, because they make it feasible and economical to apply automated measurements to a relatively few samples of a nonroutine kind. No longer are continuous-flow methods restricted to situations where the number of samples is large and the analytical method highly routine.

As mentioned in Section I, the analytical readout in FIA can be made at any point along the concentration gradient created, but in most cases the peak height corresponding to the concentration C^{max} in Fig. 1 is used, because it is readily identified. Yet, as detailed later in Section II.E, the exploitation of several or all the concentrations along the gradient has formed the basis for a series of entirely novel analytical approaches. For the time being, we will, however, use the peak maximum as the analytical response.

The characteristics of FIA are probably best illustrated by a practical example, such as the one shown in Fig. 2, which depicts the spectrophotometric determination of chloride in a single-channel system, as based on the following sequence of reactions: (Ruzicka and Hansen, 1988b)

$$Hg(SCN)_2 + 2Cl^- \longrightarrow HgCl_2 + 2SCN^-$$
$$Fe^{3+} + SCN^- \longrightarrow Fe(SCN)^{2+}$$

in which chloride initially reacts with mercury(II) thiocyanate with the ensuing release of thiocyanate ions which subsequently react with iron(III) to form the intensely red iron(III) thiocyanate complex, the absorbance of which is measured. The samples, with chloride contents in the range of 5–75 ppm, are via a valve furnished with a 30 μL loop (S), injected into the carrier solution containing the mixed reagent, pumped at a rate of 0.8 mL/min. The iron(III) thiocyanate is formed and forwarded to the detector (D) via a mixing coil (0.5 m long, 0.5 mm ID), as the injected sample zone disperses within the reagent carrier stream. The mixing coil minimizes band broadening (of the sample zone) owing to centrifugal forces, resulting in sharper recorded peaks. The absorbance A of the carrier stream is continuously monitored at 480 nm in a microflow-through cell (volume 10 μL) and recorded [Fig. 2(b)]. To demonstrate the repeatability of the analytical readout, each sample in this experiment was injected in

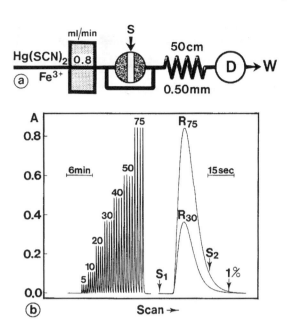

Figure 2 (a) Flow diagram (manifold) for the spectrophotometric determination of chloride: S is the point of injection, D is the detector, and W is the waste. (b) Analog output showing chloride analysis in the range of 5–75 ppm Cl with the system depicted in (a). To demonstrate the repeatability of the measurements, each sample was injected in quadruplicate. The injected volume was 30 μL, sampling rate was approximately 120 samples/h. The fast scans of the 30 ppm sample (R_{30}) and the 75 ppm sample (R_{75}) on the right show the extent of carry-over (less than 1%) if samples are injected in a span of 38 s (difference between S_1 and S_2). [From Ruzicka and Hansen (1988a), courtesy John Wiley & Sons.]

quadruplicate, so that 28 samples were analyzed at seven different chloride concentrations. As this took 14 min, the average sampling rate was 120 sample/h. The fast scans of the 75 ppm sample peaks and 30 ppm sample peaks [shown on the right in Fig. 2(b)] confirm that there was less than 1% of the solution left in the flow cell at the time when the next sample (injected at S_2) would reach it, and there was no carry-over when the samples were injected at 30 s intervals. These experiments clearly reveal one of the key features of FIA: all samples are sequentially processed in exactly the same way during passage through the analytical channel, or, in other words, what happens to one sample happens exactly the same way to any other sample.

B. Basic FIA Instrumentation

Most often the solutions in a flow injection system are propelled by a peristaltic pump, in which liquid is squeezed through plastic or rubber tubing by moving rollers. Modern pumps generally have 8–10 rollers, arranged in

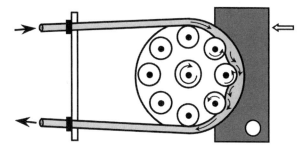

Figure 3 Schematic drawing of a peristaltic pump. While only one tube is shown, normally several tubes are accommodated. Each tube is held in place by stoppers (at left) and affixed between the rollers and the spring-loaded cam (right). [Adapted from Karlberg and Pacey (1989), courtesy Elsevier Science Publishers Company Inc.]

a circular configuration [Fig. 3 (Karlberg and Pacey, 1989; Skoog et al., 1996)], so that half the rollers are squeezing the individual tubes against a spring-loaded cam, or band, at any instant, thereby forcing a continuous, almost pulse-free, flow through the tubing. The flow is controlled partly by the speed of the motor, and partly by the internal diameter of the tubing. While all pump tubes have identical wall size, so that they fill to the same extent when completely compressed, the internal diameter of the individual tubes will, for a fixed rotational speed of the peristaltic pump, determine the flow rates. Tubes are commercially available with internal diameter ranging from 0.25 to 4 mm permitting flow rates as slow as 0.0005 mL/min and as fast as 40 mL/min. While peristaltic pumps generally suffice for most applications, allowing not only pulse-free operation, but also the operation of several tubes simultaneously, other pump devices are also applied in FIA, such as syringe pumps, but they are generally far more expensive (besides, they only permit the propagation of a single stream, which for a multiline manifold would call for several individual pumps). However, they have the advantage that one can not only freely operate them in the propulsion and aspiration mode, but also, very importantly, they allow minute volumes of liquids to be dispensed very reproducibly. For these reasons, they are ideally suited for use in sequential injection (SI) systems, as described in Section III.

The injector may consist of loop injector valves similar to those used in HPLC or in CFA, that is, furnished with an internal sample loop, or more commonly, a dedicated FIA-valve comprising a rotor and a stator furnished with four, six, or more individually accessible ports, where the injected sample volume, normally between 1 and 200 μL (typically 25 μL), is metered via an external loop of appropriate length and internal diameter. Because the injected sample volume is so small, it does not require much reagent per sample cycle, which not only makes

FIA a simple, microchemical technique, capable of providing a high sampling rate at the expense of minimum sample and reagent consumption, but also a system which generates minute amounts of waste. This is important because the disposal of chemical wastes in many instances is even more expensive than the chemicals themselves.

The conduits used in FIA-manifolds mostly consist of narrow bore plastic tubes (of materials such as PVC or PTFE), typically of inner diameter 0.5–0.8 mm. The tubes are normally coiled or knotted in order to minimize the dispersion (promote secondary flow pattern), but as a rule the tube lengths of the FIA-manifold should be made as short as possible in order to avoid adverse dilution of the injected sample solution.

The advantages of FIA are that it works continuously, virtually any number of additional lines with reagents can be added to the manifold, nearly any type of unit operation can be accommodated (e.g., heating, cooling, dialysis, or gas diffusion), and practically any type of detector can be employed. All of these features are in their own merit important, although the degree of freedom as to choice of detector is particularly significant, and undoubtedly counts for the fact that FIA so readily has been adopted as a general solution handling technique in analytical chemistry, applicable to a wide variety of tasks. This will be demonstrated in the following sections.

C. Dispersion in FIA

The degree of dispersion, or dilution, in a FIA system is characterized by the dispersion coefficient D. Let us consider a simple dispersion experiment. A sample solution, contained within the valve cavity prior to injection, is homogenous and has the original concentration C^0. When the sample zone is injected, it follows the movement of the carrier stream, forming a dispersed zone, and the shape of which depends on the geometry of the channel and the flow velocity (Fig. 1). Therefore, the response curve has the shape of a peak, reflecting a continuum of concentrations, and forms a concentration gradient, within which no single element of fluid has the same concentration of sample material as its neighbors. It is useful, however, to view this continuum of concentrations as being composed of individual fluid elements, each having a certain concentration of sample material C, since each of these elements is a potential source of readout (see also Section II.E).

In order to design a FIA system rationally, it is important to know how much the original sample solution is diluted on its way to the detector and how much time has elapsed between sample injection and readout. For this purpose, the dispersion coefficient D has been defined as the ratio of concentrations of sample material before and after the dispersion has taken place in that element of fluid that yields the analytical readout, that is (Ruzicka and Hansen, 1988b):

$$D = \frac{C^0}{C}$$

which for $C = C^{max}$, yields

$$D = \frac{C^0}{C^{max}}, \quad 0 < D < \infty$$

Hence, if the analytical readout is based on maximum peak height measurement, the concentration within that imaginary fluid element which corresponds to the maximum of the recorded curve C^{max} has to be considered. Thus, with knowledge of D, the sample (and reagent) concentrations may be estimated. The determination of D of a given FIA-manifold is readily performed. The simplest approach is to inject a well-defined volume of a dye solution into a colorless stream and to monitor the absorbance of the dispersed dye zone continuously by a spectrophotometer. To obtain the D^{max} value, the height (i.e., absorbance) of the recorded peak is measured and then compared with the distance between the baseline and the signal obtained when the cell is filled with the undiluted dye. Provided that the Lambert–Beer Law is obeyed, the ratio of respective absorbances yields a D^{max} value that describes the FIA-manifold, detector, and method of detection. Note that the definition of the dispersion coefficient considers only the physical process of dispersion and not the ensuing chemical reactions, since D refers to the concentration of sample material prior to and after the dispersion process alone has taken place. In this context, it should be emphasized that any FIA peak is a result of two kinetic processes which occur simultaneously: the *physical* process of zone dispersion and the *chemical* processes resulting from reactions between sample and reagent species. The underlying physical process is well reproduced for each individual injection cycle; yet it is not a homogenous mixing, but a dispersion, the result of which is a concentration gradient of sample within the carrier solution.

The definition of D implies that when $D = 2$, for example, the sample solution has been diluted 1:1 with the carrier stream. The parameters governing the dispersion coefficient have been the subject of detailed studies. In short, it can be stated that the most powerful means of manipulating D are the injected sample volume, the physical dimensions of the FIA system (lengths and internal diameters of the tubes), the residence time, and the flow velocity. Additional factors are the possibility of using a single line rather than a confluence manifold with several conduits, and of selecting the element of measurement on any other part of the gradient of the dispersed sample zone other than that corresponding to the peak maximum.

For convenience, sample dispersion has been defined as *limited* ($D = 1-2$), *medium* ($D = 2-10$), *large* ($D > 10$), and *reduced* ($D < 1$); the FIA systems designed accordingly have been used for a variety of analytical tasks. Limited dispersion is used when the injected sample is to be carried to a detector in undiluted form, that is, the FIA system serves as a means of rigorous and precise transport and sample presentation to the detection device [such as an ion-selective electrode (ISE) or an atomic absorption spectrometer]. Medium dispersion is employed when the analyte must mix and react with the carrier/reagent stream to form a product to be detected. Large dispersion is used only when the sample must be diluted to bring it within the measurement range. Reduced dispersion implies that the concentration of the sample detected is higher than the concentration of the sample injected, that is, on-line preconcentration is effected (e.g., by means of extraction, by incorporation of an ion-exchange column, or via coprecipitation, Section II.D.2).

The following sections give examples covering the various types of dispersion patterns. Considering the extensive literature on FIA, it is of course impossible to cover more than a few aspects, and the reader is encouraged to consult the literature for further information, which shows that FIA is established as a powerful concept in modern analytical chemistry. Originally devised as a means of providing serial analysis, FIA soon became much more than this, that is, a generally applicable solution handling technique, in the analytical laboratory as well as in industrial fields such as process control. Additionally, it was shown that it offered unique analytical possibilities, providing means for devising analytical procedures which are otherwise difficult or impossible to implement by conventional schemes, such as assays based on the generation and measurement of metastable constituents, allowing kinetic discrimination schemes, and exploiting detection principles relying on bio- and chemiluminescence. For the same reason, FIA in many contexts has been replaced by FI, emphasizing that the flow injection system is used for solution handling rather than merely for analysis.

D. Selected Examples to Illustrate the Unique Features of FIA

Although a number of applications of FIA are described in the following, they are merely a fraction of those that have appeared in the literature. However, it is hoped that they may serve to demonstrate the versatility and applicability of FIA, and possibly also as inspiration for the reader.

1. Limited Dispersion—A Means for Reproducible and Precise Sample Presentation

Because of its inherently strict timing, an obvious application of a FIA is to use it as a means for precise and reproducible presentation of a given sample to the detector, thereby ensuring that all conditions during each measuring cycle are rigorously maintained. This might for instance be advantageous if the detector is an electrode or a sensor, the response of which is diffusion controlled, or if the behavior of the detection device is affected by the time that the sample is exposed to it. Examples of the former group are ISEs and biosensors, and of the latter group the combination of FIA to atomic absorption spectrometry (AAS).

Thus, in potentiometry it is observed that many ISEs operated in the dynamic mode facilitate fast and reproducible readout. ISEs are, however, generally characterized by fairly long response times to reach steady-state conditions, and therefore it can be difficult to ascertain exactly when to make the readout by manual operation. In FIA this decision is left entirely to the system, because the sample reaches the detector after a time governed exclusively by the manifold selected. Besides, the well-known fact that many ISEs are more or less prone to interference from other ions can in many cases be eliminated, or considerably reduced, when making the measurement under FI-conditions. Thus, by taking advantage of the short residence time and exposure time of the sample in the FI system, it is often possible, via kinetic discrimination between the ion under investigation and the interfering species (which frequently exhibit longer response times), to increase the selectivity and hence the detection limit of the sensor.

The same concept of manipulating the sample exposure time can be extended to more complex sensors such as enzyme sensors, in which a membrane containing one or more immobilized enzymes is placed in front of the active surface of a detection device. The analyte is transported by diffusion into the membrane and here degraded enzymatically, forming a product which can then be sensed optically or electrochemically. A condition for obtaining a linear relationship between the concentration of analyte and the signal is, however, that pseudo-first-order reaction conditions are fulfilled, that is, the concentration of converted analyte reaching the detector surface must be much smaller than the Michaelis–Menten constant. Since this constant for most enzyme systems is of the order of 1 mmol/L, and many sample matrices (e.g., human sera or blood) contain much higher concentrations, the use of enzyme sensors in static (batch) systems often calls for complicated use of additional membrane layers aimed at restricting diffusion of analyte to the underlying enzyme layer, or, in case of electrochemical detectors, for chemically modified electrodes where the electron transfer is governed by appropriate mediators.

However, when operated in FIA, the degree of conversion can be simply adjusted by the *time* the analyte is exposed to the enzyme layer, and therefore the amount of converted analyte can be regulated directly by adjusting the flow rate of the FIA system. Thus in a FI system, used for the determinations of glucose by means of an amperometric sensor incorporating glucose oxidase, it was shown that by increasing the flow rate from 0.5 to 1.0 mL/min the linear measuring range could readily be expanded from 20 to 40 mmol/L glucose ensuring that the system could be used for physiological measurements (Petersson, 1988). Additionally, the exposure time can be exploited for kinetic discrimination, taking advantage of differences in diffusion rates of analyte and interfering species within the membrane layer of the electrode. This was very elegantly demonstrated in the mentioned system, where total spatial resolution of the signals due to glucose and paracetamol could be achieved even when paracetamol was added at excessive levels. Besides, attention should be drawn to the fact that the operation of a (bio)sensor in the FIA-mode ensures a constant monitoring of the sensor itself, that is, while the recorded signal is a measure of the concentration measured, the baseline readout indicates directly the stability of the sensor.

AAS is one of the several analytical instruments where the performance can benefit, and in some cases be significantly enhanced, when combined with FIA. Thus, by using a system consisting merely of a connecting line between the pump/injection device and a flame-AAS instrument (Fig. 4), several advantages can be gained by the possibility of repetitive and reproducible sample presentation. As seen in Fig. 4(a), the sampling frequency can be improved compared with the traditional aspiration of sample solution [Fig. 4(b)], but what is more important is that during that time normally required to aspirate one sample, it is possible to inject two individual sample aliquots, which means that the FIA approach entails not only improved precision, but also improved accuracy [Fig. 4(b)]. Furthermore, one can take advantage of the fact that the sample is exposed to the detector for only a very short period of time. The rest of the time the detector is cleansed by the carrier solution, which means that the wash-to-sample ratio is high. Therefore the chance of clogging the burner due to, for instance, a high salt content is vastly reduced or eliminated. This can be seen from Fig. 4(c) by comparing the Pb calibration runs for pure aqueous Pb standards and standards prepared in a matrix simulating seawater (3.3% NaCl), where it is obvious that the two calibration runs are identical. This feature was even more dramatically demonstrated in the literature, where some researchers, in an experiment comprising 160 repetitive injections of a 1 ppm Cu standard prepared in a 30% NaCl solution, found that hardly any deterioration of the recorded signal was observed (Schrader et al., 1989).

We now turn our attention to procedures where the dynamics of the dispersion process are superimposed by chemical reactions, and where FIA allows us to exploit the kinetic behavior of the chemistry taking place, that is, where a kinetic effect is being used either to increase

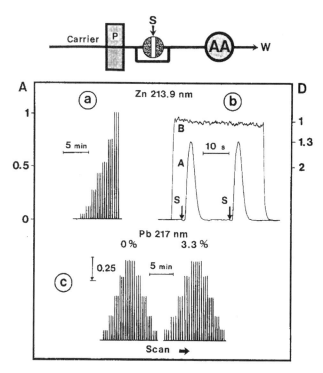

Figure 4 Single-line FIA-manifold for determination of metal ions by flame AAS (AA). Recordings obtained at a flow rate of 4.9 mL/min and an injected sample volume of 150 μL. (a) Calibration run for zinc as obtained by injection of standards in the range 0.10–2.0 ppm; (b) recorder response for the 1.5 ppm standard as obtained by (A) injection via the FIA system and (B) continuous aspiration in the conventional mode (also at 4.9 mL/min). D represents the dispersion coefficient value, which in (B) is equal to 1; and (c) calibration runs for a series of lead standards (2–20 ppm) recorded without (0%) and with (3.3%) sodium chloride added to the standards. [From Ruzicka and Hansen (1988a), courtesy John Wiley and Sons.]

the selectivity of an analytical procedure, or indeed to obtain information which is not at all accessible by conventional batchwise operation. This is well illustrated in the so-called FIA conversion techniques.

2. Medium Dispersion—FIA Conversion Techniques

The system shown in Fig. 2, where the detection of chloride obviously calls for a chemical reaction in order to sense this species, is a good example of the use of medium dispersion. In designing such systems it is important to bear in mind that the dispersion should be sufficient to allow partially mixing of sample and reagent, yet it should not be so excessive as to dilute the analyte unnecessarily, which would deteriorate the detection limit. Most FIA-procedures are based on using medium dispersion, because the analyte must be subjected to some form of intelligent "conversion". In their widest sense, the FIA conversion techniques can be defined as procedures by which a nondetectable species is converted into a detectable component through a kinetically controlled chemical reaction, with the aim of performing appropriate sample pretreatment, reagent generation, or matrix modification. In performing these tasks, *kinetic discrimination* and *kinetic enhancement* offer further advantages. In kinetic discrimination, the differences in the rates of reactions of the reagent with the analyte of interest and the interferents are exploited. In kinetic enhancement, the chemical reactions involved are judiciously driven in the direction appropriate to the analyte of interest. While, in batch chemistry, the processes are forced to equilibrium, so that subtle differences between reaction rates cannot be exploited, in the FIA-mode, small differences in reaction rates with the same reagent result in different sensitivities of measurement.

A simple demonstration of a homogeneous conversion procedure exploiting kinetic discrimination is given by the following reaction sequence, aimed at analyzing chlorate in process liquor: (Matschiner et al., 1984)

$$2ClO_3^- + 10\,Ti^{3+} + 12\,H^+ \rightarrow 10\,Ti^{4+} + Cl_2 + 6\,H_2O$$
(fast)

$$Cl_2 + LMB \rightarrow MB \qquad \text{(fast)}$$

$$MB + Ti^{3+} \rightarrow LMB + Ti^{4+} \qquad \text{(slow)}$$

The assay is performed by injecting a sample of chlorate into an acidic carrier stream of titanium(III), which is subsequently merged with a second stream of leucomethylene blue (LMB). While the first two of these reactions are very fast, the reduction of the blue species MB by the third reaction is slow. Thus, the chlorate concentration can readily be quantified via the absorbance of the MB species generated by the second reaction, while the reformation of LMB takes place after the sample plug has passed the detector and is directed to waste.

Another procedure illustrating this approach is the assay of thiocyanate, a species of interest in clinical chemistry, because it is not naturally present in humans to any significant extent, except in tobacco smokers (Bendtsen and Hansen, 1991). The half-life time of thiocyanate in the body is ~14 days, and therefore it is easy, via analysis of body fluids (saliva, blood, or urine), to distinguish smokers from nonsmokers. A fast, simple, and convenient FIA method for the determination of thiocyanate relies on the reaction of this component with 2-(5-bromo-2-pyridylazo)-5-diethylaminophenol (5-Br-PADAP) in acid media in the presence of dichromate as oxidizing reagent, resulting in the generation of an intensely colored (reddish), albeit transient, product of high molar absorptivity. The latter feature makes it ideal for quantifying low levels of thiocyanate, whereas the transient nature of the product implies that it is generated rapidly, but then fades away, having a life time of ca. 10 s. Hence, it is important that the readout is taken at the point where the color development is at its maximum. In this procedure there is an added complication: even if no thiocyanate is present, 5-Br-PADAP and dichromate will gradually react with each other, yielding a component which absorbs at the wavelength used (570 nm), that is, a passive background signal. To make things even more complicated, the signal due to the background increases with reaction time. In other words, one is faced with a situation such as that shown in Fig. 5 where the transient signal of the analyte as a function of time first increases and then decreases, whereas the background signal steadily increases.

However, by using FIA it is possible, by appropriate design of the analytical system, to adjust the sample residence time so that the detection can be effected at precisely that time where the difference between the two signals is at its maximum (indicated by the arrow and dotted vertical line).

The possibilities of achieving selectivity enhancement in FIA become even greater when we turn our attention to *heterogeneous* conversion techniques, where it is possible to incorporate into the manifold steps such as gas diffusion, dialysis, solvent extraction, or the use of packed reactors. Thus, ion-exchangers or solid reagents have been employed as column materials in order to transform a sample constituent into a detectable species, an example being the determination of various anions by means of AAS: for instance, cyanide has been assayed via interaction with a column containing CuS, resulting in the formation of soluble tetracyanocuprate, allowing the cyanide to be quantified indirectly by AAS by means of the stoichiometric amount of copper released from the solid surface (Ruzicka and Hansen, 1988c). As mentioned earlier, ion-exchangers can also be used to preconcentrate an analyte in order to accommodate it to the dynamic measuring range of a particular detection device, or simply to remove unwanted matrix components which otherwise might interfere (see also Sections IV and V). However, one of the most widely used packing materials is immobilized enzymes, as treated in detail in the following section. To complete the picture, it should be noted that reactors containing oxidants or reductants have been devised to generate reagents in *statu nascendi* [e.g., Ag(II), Cr(II), or V(II)], advantage being taken of the protective environment offered by the FIA system to form and apply reagents which, owing to their inherent instability, are impractical to handle under normal analytical conditions (Ruzicka and Hansen, 1988c).

FIA Systems with Enzymes

When applied in the FIA-mode the use of immobilized enzymes, packed into small column reactors, offers not only the selectivity and the economy and stability gained by immobilization, but also ensures that strict repetition, and hence a fixed degree of turnover from cycle to cycle, are maintained. In addition, by obtaining a high concentration of enzyme immobilized within a small volume, the ensuing high activity facilitates an extensive and rapid conversion of substrate at minimum dilution of the sample, the small dispersion coefficient in turn promoting a lower detection limit (Hansen, 1994). The enzymes needed for a particular assay can be incorporated into a single-packed reactor or into sequentially connected reactors in order to process the required sequence of events and also to afford optimal operational conditions for the reactions occurring in the individual reactors. Both optical and electrochemical detectors can be used to monitor the reaction. Thus, for oxidases, which give rise to the generation of hydrogen peroxide, detection via chemiluminescence can be employed, based on the reaction with luminol.

Bio- and chemiluminescence are particularly fascinating and attractive detection approaches, primarily because of the potentially high sensitivity and wide dynamic range of luminescent procedures, and also because the required instrumentation is fairly simple. Furthermore, luminescence has an added advantage over most optical procedures: as light is produced and measured only when sample is present, there is generally no problem with blanking. However, luminescent reactions usually generate transient emissions, because the intensity of the light emitted is proportional to the reaction rate rather than to the concentrations of the species involved (Fig. 6). Hence, the radiation is most often emitted as a flash which rapidly decreases, and for this reason the conventional approach of quantification has been to integrate the intensity over a fixed period of time and to relate this to the amount of analyte. It is obvious, however, that if the measurement of the intensity of light dE/dt can be made

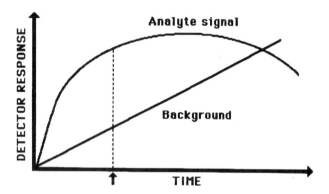

Figure 5 Exploitation of FIA for quantification of a metastable analyte species in a system which additionally entails a gradually increasing background signal response. Because of the precise and reproducible timing of FIA, it is possible to take the readout at that time which corresponds to the largest difference between the analyte and the background signals (marked by the arrow and vertical dotted line).

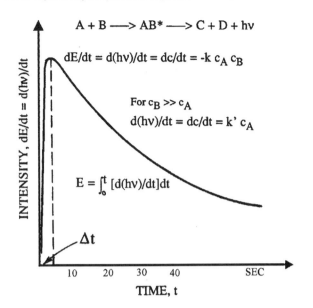

Figure 6 Typical course of light generation (hv) in a bio- or chemiluminescent reaction as a function of time. Quantification can be accomplished either by relating the amount of analyte to the energy released (i.e., integrating the area under the curve) or by using FIA and determining the intensity (dE/dt) after a fixed period of time (Δt), which, if pseudo-first-order reaction conditions are fulfilled ($c_B \gg c_A$), is directly proportional to the concentration of the analyte.

under precisely defined and reproducibly maintained conditions so that all samples are treated, physically and chemically, in exactly the same manner (i.e., the measurements can be taken repetitively at identical delay times t_i, Fig. 6), it is possible directly to relate any dE/dt-value (and preferably the one corresponding to the maximum emission, Δt) to the analyte concentration. This is feasible by means of FIA, and therefore the combination of luminescence and FIA has, in fact, revolutionized the application of bio- and chemiluminescence as analytical chemical detection procedures.

Numerous applications of enzyme assays in FIA have been reported, advantage being taken of the fact that these components constitute the selective link in the analytical chain which therefore becomes selective overall (Hansen, 1994). For the same reasons, FIA is not only a complimentary facility to (bio)sensors, but also in many cases it can be an attractive alternative.

Flow Injection Hydride Generation Schemes

Several elements (such as As, Sb, Bi, Se, Te, and Ge) can, by reaction with a strongly reducing agent, such as sodium tetrahydroborate, become chemically converted to their hydrides, as shown by Eq. (1) in Fig. 7.

Gaseous hydrides can be readily separated from the sample matrix and guided to the heated quartz flow-through cell of an AAS instrument, where they are atomised by heating and excited by radiation, so that the elements of interest can be selectively quantified [Eq. (2) in Fig. 7]. Originally, the hydride generation technique was introduced as a batch procedure, but this involved several problems, as illustrated below. The conversion of the analyte itself must necessarily take place in acidic medium. However, there are possibilities for side reactions and interferences [Eqs. (3)–(5) in Fig. 7]. The tetrahydroborate itself can react with acid and form hydrogen [Eq. (3) in Fig. 7], whereby the reagent is wasted for the hydride formation. Therefore, the tetrahydroborate must be prepared in a weakly alkaline medium and mixed with the sample and the acid precisely when it is required, under very controlled conditions. A serious possibility for interference is the presence of free metals or metal boride precipitates, particularly of Ni, Cu, and Co. Hence, if ionic species of these metal constituents are present in the sample, they become reduced by the tetrahydroborate,

Hydride generation/atomization:

$$As^{3+}, Sb^{3+}, Sn^{4+} \xrightarrow[\text{Acid (HX)}]{BH_4^-} AsH_3, SbH_3, SnH_4 \quad (1)$$

$$AsH_3, SbH_3, SnH_4 \xrightarrow{\Delta} As, Sb, Sn + nH_2 \quad (2)$$

Side reactions/interferences:

$$BH_4^- + 3HX + H^+ \longrightarrow BX_3 + 4H_2 \quad (3)$$

$$Me^{2+}(Ni, Cu, Co) \xrightarrow[\text{Acid (HX)}]{BH_4^-} Me^0 \text{ (slower)} \quad (4)$$

$$AsH_3, SbH_3, SnH_4 \xrightarrow{Me^0} As, Sb, Sn + nH_2 \quad (5)$$

Figure 7 Reactions taking place in the generation of gaseous hydrides (as exemplified for As, Sb, and Se), and possibly concurrent side reactions. The latter ones can to a large extent be suppressed, or even eliminated, by using FIA.

giving rise to the formation of colloidal free metals [Eq. (4) in Fig. 7] or metal borides, which have been shown to act as superb catalysts for degrading the hydrides before they can reach the measuring cell [Eq. (5) in Fig. 7]. However, because of the dynamic conditions prevailing in FIA, and because of the inherently short residence time of the sample in the system, these side reactions can to a large extent be eliminated or kinetically discriminated against at the expense of the main reaction. If side reactions occur, the precise timing of the FIA system ensures that they take place at exactly the same extent for all samples introduced (Fang, 1993a; Ruzicka and Hansen, 1988b). A concrete example will readily illustrate this: in his work, Åström (1982) found that it was totally impossible to determine minute quantities of Bi(III) (25 µg/L) in the presence of 100 mg/L Cu(II) in a batch system, because the hydride formed was degraded before it could reach the detector. However, when implementing the very same analytical procedure in a FI system it was perfectly feasible. In fact, it yielded close to 100% response, that is, the interference due to Cu was practically eliminated (see also Section IV.C).

3. Large Dispersion

Large dispersion is used if the sample on hand is too concentrated to be accommodated to the detection device used. This can either—within a certain range—be done by exploiting the concentration gradient formed in FIA, that is, by using a delay time corresponding to a concentration much different from that one matching C^{max} (see Section II.E). Or it may be effected by merging the sample stream with a diluent, by zone sampling, or by incorporating into the manifold a mixing chamber which, as a function of volume, reproducibly can facilitate an appropriate degree of dilution (Fang, 1989; Jørgensen et al., 1998; Ruzicka and Hansen, 1988b).

4. Reduced Dispersion—Preconcentration and On-Line Sample Preconditioning

One of the characteristics of FIA is that it allows on-line sample preconditioning or on-line preconcentration to be effected judiciously. Either in order to preconcentrate the analyte (usually a metal ion) in order to bring it into the dynamic range of the detector used, or to separate the analyte from potentially interfering matrix component, such as high salt contents which are particularly problematic when using sophisticated detection devices such as ETAAS or ICPMS. Such tasks can be accomplished in a number of avenues: (Fang, 1989; 1995a; Ruzicka and Hansen, 1988c) for instance, by incorporation of suitable column reactors (e.g., the use of an ion-exchanger to retain the ionic species, as injected by a large volume of sample solution, which later can be eluted by a small volume of injected eluent so that the recorded signals fall within the working range of the instrument). Or by forming a noncharged complex between the analyte and a suitable ligand which then can be retained on a hydrophobic column material and later eluted for quantification. Or by liquid–liquid extraction (usually via complex formation), possibly combined with back-extraction. Or simply by intelligent design of the FIA system *per se*. Several examples of such on-line procedures, which in recent years have been the subject of much interest, will be presented in Sections IV and V, and therefore are not elaborated upon at this juncture.

E. FIA Gradient Techniques—Exploiting the Physical Dispersion Process

Up to this point we have by and large confined ourselves to use the peak maximum as our analytical readout. However, any FIA peak that we create by injecting a sample necessarily contains a continuum of concentrations representing all values between 0 and C^{max} (Fig. 1), irrespective of whether we intend to use only a single element of it, such as the peak maximum, to obtain our analytical answer. However, because the carrier stream is noncompressible, and its movement can therefore be strictly controlled and reproduced from one measuring cycle to the next with high precision, each and every element of this concentration gradient is inherently characterized by two parameters: a fixed delay time t_i (Fig. 1), elapsed from the moment of injection (t_0); and a fixed dispersion value, that is, the dispersion coefficient of the injected sample is $D_S = C_S^0/C_S$, whereas that of the reagent is $D_R = C_R^0/C_R$. Therefore, it is within our power to reproducibly select and exploit any element, or elements, simply by identifying the time (t_i) associated with the particular element, its concentration being given by $C = C^0/D(t_i)$. This feature has formed the basis for the development of a number of FIA gradient techniques, opening up a series of novel analytical applications (Hansen, 1988; Ruzicka and Hansen, 1988b, c). A compilation of various gradient techniques is presented in Table 2.

Thus, it is obvious that if a sample at hand is too concentrated, one can, instead of using the C^{max}-value for evaluation, use any other element along the gradient, as identified via its delay time, t_i. Thus, at longer delay times after C^{max}, the concentration of analyte will decrease (while the concentration of reagent will increase). Also, it should be noted that the dynamic nature of the FIA gradient techniques in earnest allows to exploit the use of dynamic detectors (such as diode array instruments or fast scanning instruments), as illustrated in Fig. 8.

One application that deserves special emphasis is the stopped-flow method. As mentioned in Table 2, this

Table 2 Examples of FIA Gradient Techniques

Gradient dilution
  Selecting and using for the analytical readout specific fluid elements along the concentration gradient, the concentration being $C = C^0/D(t_i)$. To be used, for instance, to accommodate the concentration of sample to the dynamic range of a detector

Gradient calibration
  Identifying and exploiting a number of elements along the gradient, the concentrations of which are given through the dispersion coefficient of the individual elements. A multipoint calibration curve can be obtained from a single injection of a concentrated sample

Gradient scanning
  Combining the use of gradient dilution with the use of a dynamic detector which, for each concentration level, is able to continuously scan a physical parameter, such as wavelength or potential (see Fig. 8).

Stopped-flow
  Increase of sensitivity of measurement by increasing residence time, or quantifying sample concentration by measuring a reaction rate under pseudo-zero-order reaction conditions (see Fig. 9).

Titration
  Identifying elements of fluids on the ascending and descending parts of the concentration gradient where equivalence between titrand and titrant is obtained, and relating the time difference between these elements to the concentration of injected analyte.

Penetrating zones
  Exploitation of the response curves from the concentration gradients formed when two or more zones are injected simultaneously. In addition to acting as an economical way of introducing sample and reagent solutions, it can be used for measuring selectivity coefficients, and to make standard addition over a wide, controllable range of standard/analyte concentration ratios.

approach is used either to stop the flow of the sample/reagent mixture in order to obtain increased reaction time without excessive or undue dilution (as would be the case if the increased reaction time were to be obtained by pumping through a long reaction coil), or, if the stop sequence is effected within the detector itself, to monitor the reaction as it progresses, allowing us to observe it as it happens. If the conditions are manipulated to simulate pseudo-zero-order reactions (i.e., by adding excess of reagent and assuming that the sample concentration does not change during the observation period), the reaction rate, that is, the slope of the recorded stopped-flow curve becomes directly proportional to the concentration of analyte (sample), as shown in Fig. 9.

The strength of this approach stems from the high reproducibility with which various segments of the concentration gradient formed by the injected sample can be selected and arrested within the observation field of the detector, and the fact that we can change the reaction conditions to meet the pseudo-zero-order reaction criteria simply by selecting the point (or points), on the dispersion profile, which fulfills this condition, that is, the time we effect the actual stop of the zone. Reaction rate measurements, where the rate of formation (or consumption) of a certain species is measured over a larger number of data points, not only improve the reproducibility of the assay, but also ensure its reliability. This benefit occurs because interfering phenomena such as

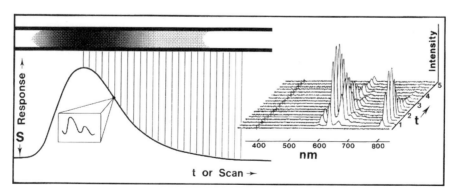

Figure 8 Gradient scanning based on selection of readouts via delay times (each corresponding to a different sample/reagent ratio) during which a detector rapidly scans a range of wavelengths (left), thus creating an additional dimension on the time–concentration matrix (right), showing a series of successive emission spectra recorded on the ascending and descending part of a dispersed zone, containing Na, K, and Ca injected into an atomic emission spectrometer furnished with a fast scanning monochromator. [From Ruzicka and Hansen (1988a), courtesy John Wiley and Sons.]

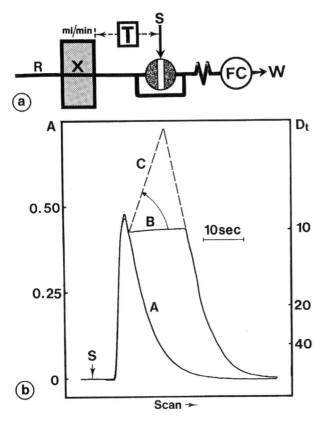

Figure 9 (a) Simple stopped-flow FIA-manifold. When the sample S is injected, the electronic timer T is activated by a microswitch on the injection valve. The time from injection to stopping of the pumping (delay time) and the length of the stop period can be both preset and controlled by a computer. (b) The principle of the stopped-flow FIA method as demonstrated by injecting a dyed sample zone into a colorless carrier stream and recording the absorbance by means of a flow-through cell: (A) Continuous pumping; (B) 9 s pumping, 14 s stop period, and continuous pumping again; (C) the dashed line indicates the curve that would have been registered if a zero- or pseudo-zero-order chemical reaction had taken place within the flow cell during the 14 s stop interval, that is, the slope would then be directly proportional to the concentration of analyte. [From Ruzicka and Hansen (1988a), courtesy John Wiley and Sons.]

blank values, existence of a lag phase, and nonlinear rate curves, may be readily identified and eliminated. Because of the inherent automatic blank control, the stopped-flow approach is an attractive option in applications such as clinical chemistry, biotechnology, and process control, where sample matrices of widely different blank values are often encountered.

F. FIA for Process Analysis/Monitoring

In emphasizing the many analytical chemical possibilities that FIA offers, it is of interest to end this section by drawing attention to an emerging application of this concept, that is, its use for monitoring purposes, which promises to become very important for future use, notably in areas such as process development and optimizations. Conventionally, most industrial processes are surveyed off-line in the laboratory after withdrawal of separate samples. However, for an efficient process monitoring and control, the time delay, the limited reliability, and the man power needed for analysis of a large number of samples are crucial parameters. Already in 1982, the first paper on utilization of FIA for process control appeared (Ranger, 1982), and since then it has (not the least supplemented by SIA) gained momentum for on- and/or at-line applications. This is not surprising, because it offers a number of advantages, the most significant ones being: (a) Because every FIA readout inherently consists of a peak that travels from the baseline, through a maximum, and back to the baseline again, one can not only monitor the analyte species via the reaction taking place in the manifold (e.g., the peak maximum), but one can also at the same time control the performance of the analytical system itself (i.e., by the baseline). (b) FIA allows occasional recalibrations or updates/checks of the calibration curve at will, that is, standards can be injected at any time desired, which is a necessity particularly if very complex samples are at hand. (c) FIA is noninvasive and very economical in sample consumption—which are important when used in pilot plant scale experiments—even if the procedure applied calls for the use of a reversed FIA assay (i.e., where the reagent is injected into sample). A recent compilation of FIA-literature for process control applications has been presented by Workman et al. (2001).

III. FURTHER DEVELOPMENTS OF FIA

In the early 1990s, a variant of FIA was introduced, that is, SIA (Ruzicka and Marshall, 1990). Termed the second generation of FIA, it was at the end of that decade supplemented by the third generation of FIA (Ruzicka, 2000), also named the lab-on-valve. Both these approaches have, in their own right, proven to entail a number of specific advantages. Thus, for instance, miniaturization of the manifolds, which, in turn, drastically reduces the consumption of sample and reagent solutions, and hence leads to generation of minute amounts of waste. Or allowing complex sample manipulations to be facilitated in simple fashions. Or readily permitting the integration of sequential unit operations. Therefore, these two analytical concepts deserve detailed, individual description and discussion. Hence, in the present section the principles of each method will be illustrated, whereas selected applications, showing their potentials and versatility, are given in Sections IV and V.

A. Sequential Injection Analysis

While most FIA procedures employ continuous, unidirectional pumping of carrier and reagent streams, SIA is based on using programable, bi-directional discontinuous flow as precisely coordinated and controlled by a computer. A sketch of a typical SIA-manifold is reproduced in Fig. 10.

The core of the system is a multiposition selection valve (here shown as a six-port valve), furnished with a central communication channel (CC) that can be made to address each of the peripheral ports (1–6), and a central communication line (CL) which, via a holding coil (HC), is connected to a syringe pump. By directing the central communication channel to the individual ports, well-defined sample and reagent zones are initially aspirated time-based sequentially into the holding coil where they are stacked one after the other. Afterwards, the selection valve is switched to the outlet port (here position 5), and the segments are propelled forward towards the detector, being dispersed on their way and thereby partial mixing with each other, and hence promoting chemical reaction, the result of which is monitored by the detector. Notable advantages of SIA are in particular that it allows the exact metering of even small volumetric volumes (of the order of a few tenths of microliters or less), and that it, thanks to the use of a syringe pump, readily and reproducibly permits flow reversals. Besides, it is extremely economical as in terms of consumption of sample and reagents, and hence in waste generation. And since all manipulations are computer controlled, it is easy and simple to reprogram the system from one application to another one. However, it is generally difficult to accommodate (stack) more than two reagents along with the sample, although additional reagents might be added further downstream, that is, by making an FIA/SIA-hybrid. And due to the use of a syringe pump, SIA has a somewhat limited operating capacity, although this in practice is rarely a constricting factor.

B. Lab-on-Valve

The LOV (Fig. 11) encompasses many of the features of SIA. However, here an integrated microconduit is placed on top of the selection valve. The microconduit, which normally is fabricated by Perspex, is potentially designed to incorporate and handle all the necessary unit operations required for a given assay, that is, it acts as a small laboratory, hence the name lab-on-valve.

Thus, it may contain facilities such as mixing points for the analyte and reagents; appropriate column reactors packed, for instance, with immobilized enzymes or small beads furnished with active groups such as ion-exchangers, which in themselves might be manipulated within the LOV in exactly the same manner as liquids; and even detection facilities. For optical assays (e.g., UV–Vis or fluorometry), this can readily be achieved by use of optical fibres, the ends of which, furthermore, can be used to define the optical path length to yield optimal measurement conditions. Thus, one of the fibres is used to direct the light from a power source into the LOV, whereas the other one serves to guide the transmitted light to an appropriate detection device [Fig. 12 (Wu et al., 2001)]. For other detectors,

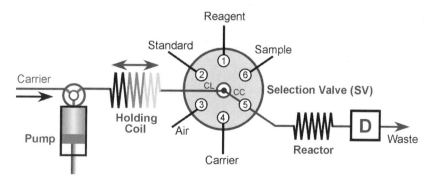

Figure 10 SIA system, as based on using a selection valve (SV), the central communication channel (CC) of which randomly can address each of the six individual ports. Initially, sample and reagent are by means of the syringe pump, and via the central communication line CL, aspirated into a holding coil and stacked there as individual zones. Thereafter the zones are through port 5 and the reaction coil forwarded to the detector D, which monitors the product formed as the result of the dispersion of the zones into each other during the transport.

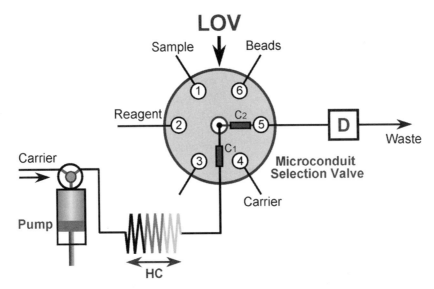

Figure 11 Schematic drawing of a LOV system, the concept of which is a microconduit placed atop a selection valve (as the one shown in Fig. 10). The microconduit should ideally contain all means for executing the sample manipulations and chemistries required plus house a detection facility, that is, it act as a small laboratory. However, when large instrumental detector devices, for example, ETAAS or ICP-MS, are to be used, it is necessary to employ external detection as shown in the figure. Besides aspirating liquids, it is also possible to handle small beads (furnished with active functional groups), which can be used to integrate small packed column reactors.

such as AAS or ICPMS, it is, of course, necessary to make use of external detection devices, as shown in Fig. 11. In SIA and LOV all flow programing is computer-controlled, which implies that it is readily possible, via random access to reagents and appropriate manipulations, to devise different assay protocols in the microsystems.

IV. SELECTED APPLICATIONS OF FIA/SIA AND THEIR HYPHENATIONS IN ON-LINE SAMPLE PRETREATMENTS

The ultimate goal of any analytical procedure is to determine the analytes of interest with satisfactory sensitivity

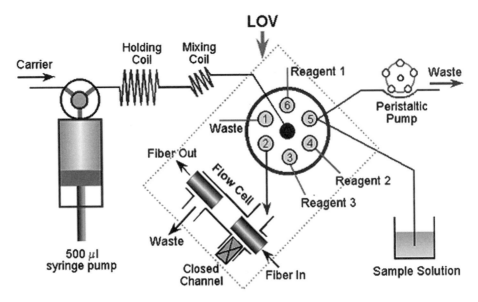

Figure 12 Schematic diagram of a μSI-LOV microsystem incorporating a multipurpose flow cell configured for real time measurement of absorbance. [Adapted from Wu et al. (2001), courtesy The Royal Society of Chemistry.]

and selectivity irrespective of the constitution of the sample matrix. In practice this is, however, frequently difficult, or rather impossible, due to the presence of various interfering effects. Moreover, in some circumstances, the analyte concentration might be too low to be analyzed directly by a given detector. In such cases, the separation of interfering components and, at the same time, the preconcentration of ultratrace levels of analytes are very often called for. Over the years a variety of separation procedures have been used based on employing solid phase extraction, solvent extraction and back-extraction, precipitation/coprecipitation, and hydride/vapor generation (Minczewski et al., 1982). However, the manual execution of these procedures are generally labor intensive and time-consuming, which tend to give rise to large sample/reagent consumption and waste production. Furthermore, sample contamination is almost inevitable in this mode of operation, which is particularly significant for samples containing ultratrace amount of analytes. Hence, appropriate sample pretreatments prior to actual analysis constitute the bottleneck of the entire analytical procedure with modern instrumentation.

In order to overcome the aforementioned drawbacks, the conventional manual procedures have to a large extent been replaced by on-line sample pretreatment methods based on the use of flow injection and SIA. Extensive studies have shown that the first and the second generations of flow injection and their hyphenations have played, and indeed continue to do so, important roles in the automation and miniaturization of on-line sample pretreatment. The miniaturized FIA/SIA on-line separation/preconcentration systems operated in an enclosed flow system offer a number of virtues when compared with their manual operational counterparts (Hansen and Wang, 2002; Wang and Hansen, 2003), including:

Reduced sample/reagent consumption and waste production
Minimized risk of sample contamination
Low limits of detection (LODs)/limits of quantification (LOQs)
Improved precision (relative standard deviation RSD)
High efficiency in terms of sample throughput/sampling frequency
Potential improvement of selectivity through kinetic discrimination under chemically nonequilibrium conditions
Ease of automation

So far, the applications of FIA/SIA on-line sample pretreatment protocols have expanded to include a variety of separation/preconcentration techniques based on the schemes summarized in Fig. 13. Some of the major techniques encompass solid phase extraction including column-based ion-exchange and adsorption, on-wall retention in a knotted reactor (KR), precipitation/co-precipitation, solvent extraction/back-extraction, and hydride/vapor generation.

Theoretically, an FIA/SIA on-line sample pretreatment system can be coupled with any kind of detector via clever design of the flow manifolds and by intelligent exploitation of the chemistries involving the analytes. A variety of on-line separation and preconcentration procedures have been extensively employed as the interface with electrochemical, spectrophotometric, chemiluminescent, fluorometric, atomic spectrometric, and mass spectrometric detectors (Hansen, 2003). The principles and practical applications of the various on-line sample pretreatment schemes are discussed separately in the following sections. For the

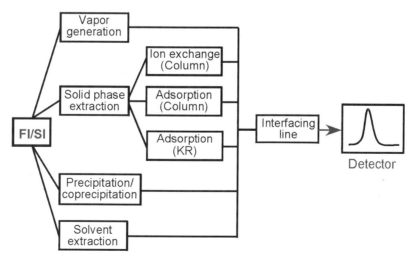

Figure 13 Illustration of frequently employed FIA/SIA on-line separation and preconcentration schemes interfaced to various analytical detectors.

reason of simplicity, the emphasis of the discussions is confined mainly on the determination of metal species.

A. Separation and Preconcentration Procedures Based on Column Reactors or KRs

1. General Requirements for Column Packing

Since the first attempt of FIA on-line ion-exchange column preconcentration by Olsen et al. (1983) column-based solid phase extraction protocols have proven to be one of the most efficient on-line separation/preconcentration approaches. As the sample solution flows through the column the analytes, or the complexes formed between the analytes and suitable chelating reagents, or the precipitates/co-precipitates, are adsorbed onto the surface of the column, which afterwards are eluted with an appropriate eluent, the eluate subsequently being introduced into the detector for quantification. Interfacing of the on-line sample pretreatment systems to most of the flow-through detectors is quite straight forward, provided that the flow rates of the on-line system and the desired sample uptake rate of the detector are compatible (Fang, 1993b). Coupling with a discrete detector, such as ETAAS, is somewhat more complicated. This is attributed to the apparent incompatibility of the continuously operating flow system with the discrete features of the ETAAS detection device. However, the major hindrances have been successfully overcome as described by Fang et al. (1990) via design of an interface for an FIA on-line column preconcentration system employing small air segments sandwiching the eluate zone.

The broad range of choice of sorbent materials, along with various chelating reagents and eluents, and the easy manipulation of the column operations make this technique most attractive for on-line sample pretreatment (Alonso et al., 2001; Burguera and Burguera, 2001). Many different kinds of sorbent materials have been used for FIA on-line column operations, including cation and anion-exchangers, chelating ion-exchangers, C_{18} octadecyl group bonded silica-gels, polymer sorbents, and activated carbon (Burguera and Burguera, 2001; Fang, 1993b; 1995b; Santelli et al., 1994). Among the most widely employed sorbents are chelating ion-exchange resins and bonded/immobilized octadecyl C_{18} silica-gel.

For the successful and practical preparation of a packed column a number of factors, which might influence the performance of an on-line column separation/preconcentration system, should be taken into account, and a balance between these factors should be carefully maintained. Some of the most important ones include:

1. The sorption (break-through) capacity of the sorbent material for both the analyte and the other constituents which might potentially act as interferents.
2. The extent of interfering effects and the level of the interferents in the sample.
3. The physical properties of the sorbent, especially its tendency to volume changes when subjected to different experimental conditions. The swelling or shrinking should be negligible during the experiments in order to minimize the creation of flow impedance.
4. The particle size of the sorbent. Although smaller particle size will result in higher break-through capacity, finer particles tend, however, to cause progressively tighter packing of the column and hence lead to build up of flow resistance.
5. The kinetics of the sorption and elution of the analytes should be favorable to fast retainment of the analytes from the sample solution and subsequently allow rapid release of it from the column.

Considering the fact that the capacities of the columns used in FIA on-line operations generally are very limited, that is, ca. 20–400 μL in volume, the performance of the columns depends not only on the parameters mentioned above, but are also significantly influenced by the geometrical designs of the column. Different kinds of column designs have been reported in the literatures, including uniform bored column and conical microcolumn. Figure 14 illustrates a commercially available conical column from Perkin–Elmer, which has proven to be sufficiently efficient for on-line solid phase extraction separation and preconcentration operations (Fang, 1993b).

2. Dispersions in FIA/SIA Column Preconcentration Systems

In the method development of FIA/SIA column pretreatment systems designed and aimed at both sensitivity and selectivity improvements, the dispersion is a critical

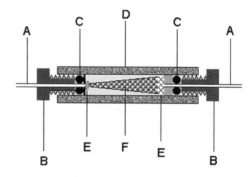

Figure 14 Schematic diagram of a Perkin–Elmer conical column. A, PTFE tubings; B, threaded fittings; C, O-rings; D, column housing; E, porous filters or glass wool; and F, column packing sorbent material. [Adapted from Fang (1993b), courtesy VCH Publishers, Inc.]

factor, a higher value of which will lead to deteriorated performance of the preconcentration process. Hence, the dispersion should be minimized by careful control of every single step of the method development process.

In the sample loading process, the use of a sample loop is a practical way to define the loaded sample volume. In such a volume-based case, the introduction of small air segments at both ends of the sample zone might help to reduce the dispersion during the sample loading process (Fang, 1995b). However, so far most of the FIA/SIA on-line separation and preconcentration systems employ time-based sample loading, which generates insignificant dispersion.

Dispersion also happens in the adsorption and elution stages, when the sorption of analytes is based on the action of functional groups or active sites on the surface of the packed column materials. The extent of dispersion in this case depends on the column geometry, the flow rate, the properties of the eluent and the sorbent, parameters which can be controlled by optimizing the system in the pertinent individual steps during the method development (Fang, 1993b).

In most of the FIA/SIA on-line separation/preconcentration procedures, the final volume of the concentrate is usually very limited, that is, at the microliter level. The dispersion of the concentrate during its direct introduction to the detector might cause loss of sensitivity. This is particularly critical for flow-through detectors such as flame-AAS, ICPAES, and ICPMS. In these cases, the introduction of small air segments at both ends of the eluate zone (Benkhedda et al., 2000a), and/or the employment of a transporting line or use of postcolumn reaction conduits made as short as possible might help to minimize the dispersion during its transport to the detector. However, this does not seem of much help when ICPAES and ICPMS are used as detection devices, because the use of a conventional cross-flow nebulization system with a spray chamber can cause large dispersion of the eluate zone. Thus, the low sample transport efficiency of a conventional cross-flow nebulizer, typically 1–2%, does not make it appropriate for transporting very limited amounts of concentrate after an FIA/SIA on-line preconcentration procedure. It is, therefore, beneficial to use a microflow high efficiency nebulizer, such as a direct injection high efficiency nebulizer (DIHEN) (Wang and Hansen, 2001a). When using ETAAS as detection device, however, the dispersion of the eluate zone during its transport to the graphite tube does not appear to impair the sensitivity. This is not merely because of the employment of air segments on both ends of the eluate zone to minimize the dispersion during the transport, but because the entire eluate zone eventually is introduced into the atomizer, and hence all the analyte contained in the eluate is quantified in the atomization process.

3. Practical Examples of On-Line Column Preconcentration

For most of the flow-through detectors, such as flame-AAS, ICPAES, or ICPMS, the interface of the on-line preconcentration system with the detectors is straightforward, because of the compatibility between the two continuous flow set-ups. In such cases, the retained analytes on the column can be eluted at a certain flow rate, and the eluate is carried into the flame or the ICP by the continuous carrier flow. Figure 15 shows a schematic flow manifold for an on-line separation/preconcentration-FAAS system using time-based sample loading and with a CPG-8Q ion-exchanger packed conical microcolumn as sorption medium (Fang and Welz, 1989). By continuous operation the microcolumn tends to create flow resistance because of the repetitive loading–elution processes and cause deterioration of the analytical performance of the entire system. In order to alleviate the adverse effect of the flow resistance, a counter-flow mode can therefore advantageously be employed, that is, after the sample solution has been loaded, the analytes retained on the column are eluted by introducing the eluent in the opposite direction, and the eluate is subsequently carried into the flame.

A SIA on-line column preconcentration system using a PTFE beads packed microcolumn as sorption medium coupled to ETAAS is illustrated in Fig. 16 (Wang and Hansen, 2002a). The entire operating procedure comprises

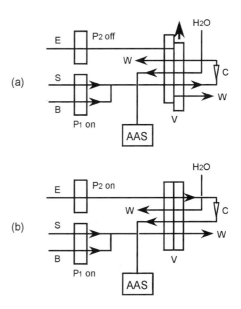

Figure 15 Flow manifold for FIA on-line column preconcentration procedure coupled to flame-AAS with countercurrent elution. (a) Loading and (b) elution P_1 and P_2, peristaltic pumps; E, eluent; S, sample solution; B, buffer; W, waste; C, column; V, injection valve. [Adapted from Fang and Welz (1989), courtesy The Royal Society of Chemistry.]

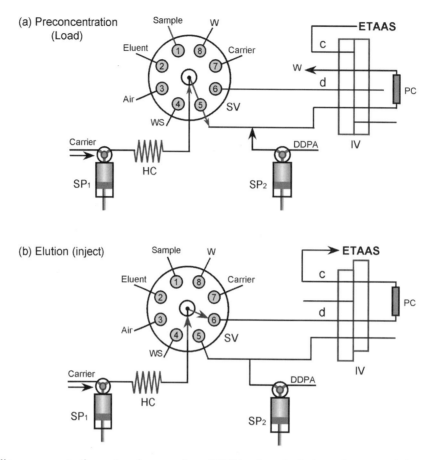

Figure 16 SIA on-line preconcentration system, incorporating a PTFE beads packed microcolumn, coupled to ETAAS: (a) preconcentration sequence and (b) elution sequence. SP_1, and SP_2, syringe pumps; SV, eight-port selection valve; IV, injection valve; HC, holding coil; PC, packed column; carrier, 0.05% HNO_3; WS, washing solution; DDPA, chelating reagent; W, waste; c and d lines, air-filled interfacing conduits between SV and ETAAS. [From Wang and Hansen (2002a), courtesy The Royal Society of Chemistry.]

two main steps:

1. Separation/preconcentration (load position): This process includes three unit operations: (a) The aspiration of sample and chelating reagent solutions into the holding coil (HC) and the syringe pump SP_2, respectively; (b) the separation and preconcentration by propelling the solutions forward followed by their confluencing and flow-through the packed microcolumn (PC), where the analyte complex is adsorbed onto the surface of the PTFE beads; and (c) the column washing with diluted chelating reagent solution in order to eliminate the loosely retained matrix components followed by the evacuation of the microcolumn and the interfacing lines "a" and "b" with air.

2. Elution of the analyte and transport of the eluate (inject position): A small amount of eluent is aspirated into the holding coil with air segments at both ends. The eluate is afterwards directed through the column to execute the elution process, whereupon the eluate is transported into the graphite tube of the ETAAS instrument for quantification.

4. Preconcentration in a KR

Since the first application of the on-line co-precipitate collection approach onto the inner wall of a PTFE tube KR (Fang et al., 1991) and the first attempt of on-line sorption of neutral metal complexes (Fang et al., 1994), KRs have been widely recognized as a trouble-free and very efficient on-line collecting medium of analytes. Made by tying the tubing into interlaced knots, the KR creates increased secondary, radial flow in the stream carrying the analytes, thus resulting in the creation of strong centrifugal forces within the stream, which, in addition to the hydrophobic nature of the PTFE surface, in most cases facilitate the retention/sorption of either the hydrophobic precipitate/co-precipitate or the neutral metal complexes onto the interior surface of the KR via molecular sorption (Fang et al., 1994). In some other cases, however, according to the nature of the retained complexes or the precipitates, the KR can also be made by using other tubing materials, such as hydrophilic microline (Nielsen et al., 1996) or nylon (Plamboeck et al., 2003).

There are two main approaches to execute on-line KR sorption/retention preconcentration with detection by various detectors. In most cases, the preconcentration is achieved through on-line merging of the sample solution and a complexing or precipitating/co-precipitating reagent, followed by the sorption/retention of the neutral metal complexes or the precipitate/co-precipitate on the interior surface of the KR. The retained analytes are, after a washing step, eluted or dissolved with an appropriate eluent, and the eluate is transported to the detector.

Another approach, which in practice only is effective for on-line sorption of metal complexes at the present, is to precoat the complexing reagent directly onto the interior surface of the KR, followed by sample loading. The analytes are thus extracted into the precoated thin layer containing complexing reagent, that is, the separation/preconcentration process takes place on the inner surface of the KR. After a washing step, the analyte is eluted and ultimately transported into the detector. This approach has so far not been much employed, but the figures of merits reported appear to indicate an improved sensitivity and overall efficiency compared with that of on-line merging of the sample and the reagent solutions (Benkhedda et al., 1999; 2000b).

The open-ended KRs entail the clear advantage of low hydrodynamic impedance, and thus allow high sample loading flow rates for obtaining better enrichment factors. However, a limiting factor in using KRs as collecting medium is its relatively low retention efficiency for most metal complexes (Yan and Yan, 2001), which is generally within the range of 30–60% and restricts significant improvement of its preconcentration capability.

The first flow manifold of on-line co-precipitate collection without filtration by using a KR as collecting medium with detection by FAAS is illustrated in Fig. 17 (Fang et al., 1991). Lead was co-precipitated with the iron(II) complex of hexamethyleneammonium and hexamethylenedithiocarbamate (HMDTC) and collected in a microline KR. The co-precipitate was afterwards dissolved by a flow of isobutyl methyl ketone (IBMK) and introduced directly into the nebulizer–burner system of a flame-AAS system. Because the normally used pump tubing cannot tolerate IBMK for extended periods of time, a displacement bottle was used for effecting the delivery of the organic eluent. This is obviously a cumbersome set-up, and it has therefore later been replaced successfully by integrated SIA-systems where inorganic as well as organic liquids readily can be manipulated (see Section IV.B).

Figure 18 illustrates another on-line preconcentration system by using a KR coupled to ICPMS for determination of rare earth elements (REE) (Yan et al., 1999). In step (a), the on-line precipitation of the REE is achieved by mixing

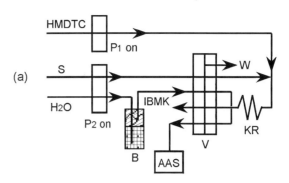

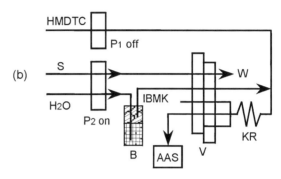

Figure 17 Flow manifold for on-line co-precipitate collection without filtration coupled to FAAS: (a) precipitation and (b) dissolution. P_1 and P_2, peristaltic pumps; B, displacement bottle; KR, microline knotted reactor; V, injection valve; S, sample; HMDTC, precipitating agent; and W, waste. [Adapted from Fang et al. (1991), courtesy The Royal Society of Chemistry.]

the sample solution with ammonia buffer solution. The precipitate is collected on the inner walls of the PTFE KR. In step (b), the collected precipitate is, after being rinsed by water, dissolved with dilute nitric acid, and the eluate is subsequently directed into the ICPMS for quantification.

B. FIA/SIA On-Line Solvent Extraction and Back Extraction Preconcentration

1. Mechanisms of Mass Transfer in an On-Line Continuous Flow System

Liquid–liquid solvent extraction is among the most effective sample pretreatment techniques in the separation of interfering matrices and preconcentration of the analytes. Operated under manual conditions, this approach has not, although extensively used, enjoyed the attraction it deserves, because of its tediousness, the high risk of contamination, and the necessity of handling large amounts of hazardous and/or toxic organic solvents (Kubáň, 1991). The employment of an FIA/SIA on-line extraction system can effectively minimize these drawbacks (Wang and Hansen, 2000a).

The process of on-line extraction of an analyte from an aqueous phase into an organic phase in a PTFE extraction

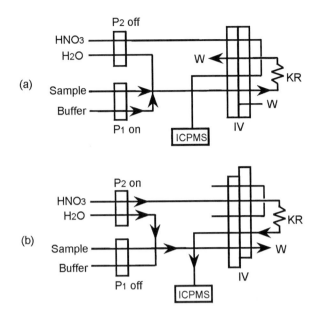

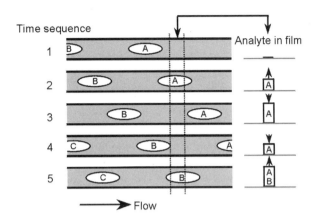

Figure 18 FIA-manifold for on-line precipitation–dissolution system coupled to ICPMS: (a) Precipitation and (b) dissolution. P_1 and P_2, peristaltic pumps; W, waste; KR, knotted reactor; and IV, injection valve. [Adapted from Yan et al. (1999), courtesy The Royal Society of Chemistry.]

Figure 19 Schematic diagram showing the mechanism of on-line solvent extraction inside a narrow bore PTFE tubing. [Adapted from Karlberg (1986), courtesy Elsevier Science Publishers.]

coil was investigated in detail by (Nord and Karlberg, 1984; Karlberg, 1986) and their suggested model being illustrated in Fig. 19 and described as follows.

In a continuous-flow system, small droplets of one phase within the other one are formed, and during this process the inner wall of the PTFE tube is coated with a very thin film of organic solvent, which substantially increases the phase contacting area. Two processes occur inside the extraction coil. One is the direct contact of aqueous and organic droplets, which explains a part of the mass transfer. Another process is illustrated in Fig. 19, that is, the analyte within the flowing aqueous droplet is first partially transferred into the organic film surrounding it, and afterwards the analyte in the film is, via molecular diffusion and convection, dispersed into the bulk of the organic droplets subsequently flowing through the extraction coil (Karlberg, 1986; Nord and Karlberg, 1984). In a continuous-flow system, the same transfer process occurs ceaselessly to the succeeding aqueous droplets, the organic film, and the following organic droplets. The analyte in the aqueous phase is, therefore, gradually transferred into the organic phase.

2. Some Practical Considerations in FIA/SIA On-Line Solvent Extractions

In order to make the extraction separation/preconcentration effective, some important factors which govern the efficiency of the on-line extraction should be considered, thus allowing the system to be operated under optimal conditions. These include the following.

The segmentation of the two immiscible phases is a prerequisite for the ensuing analyte transfer. A variety of phase segmentors have been designed, including the merging tube segmentors (Karlberg and Thelander, 1978), noncoaxial segmentors (Atallah et al., 1987), and coaxial segmentors (Kubáň et al., 1990). Although specific advantages have been claimed for some of them (Fang, 1993c), there seems no clear evidence of significant differences in the effect of their overall performance.

As expected, effective mass transfer is greatly increased if the interfacial area between the two phases is increased. This can advantageously be accomplished by decreasing the inner diameter of the extraction tubing (preferentially 0.5 mm ID). In their theoretical studies, Nord et al. (1987) furthermore showed that increased flow velocity increases the extraction rate based on residence time, which results in minimum residence time for a given degree of extraction. On the other hand, decreased flow velocity increases the extraction rate based on extraction coil length, which results in minimum extraction coil length for a given degree of extraction. However, the most efficient means of optimizing the extraction process is simply to use a KR (Wang and Hansen, 2002b). In these reactors an increased secondary, radial flow pattern is generated within the stream which causes the creation of strong centrifugal forces. This feature results in the formation of fine droplets of the two immiscible phases within each other, which thus provides a large surface contact area between them, thereby facilitating rapid and effective mass transfer.

For an on-line solvent extraction system, the phase separation unit is the most delicate part which influences the overall performance of the system. Although different phase separators (Toei, 1989) have been used, there are mainly two types of separators that have found wide applications. These are the membrane-type and gravitational phase separators.

Membrane Phase Separators

There are several different kinds of membrane-based separators, including the sandwich-type (Fang et al., 1988), the circular membrane type (Bäckström et al., 1986), the dual membrane type (Fossey and Cantwell, 1985), and the tubular membrane type (Kubáň, 1991). In these kinds of separators, the two immiscible phases are separated based on their difference of affinity to the membrane materials. One of the remarkable advantage of membrane-based separators is their high separation efficiency, even at relatively higher phase ratios (Fang, 1993c). Among the various materials employed, PTFE porous membrane is by far the most frequently used hydrophobic membrane for organic phase separations (Kubáň, 1991). A sandwich-type membrane phase separator is illustrated in Fig. 20, which was reported to have the figure of merits of high phase separation efficiency, low dispersion, high phase ratio applicable, and high total flow rates (Fang et al., 1988). The restriction is, however, its lack of robustness and stability, and thus it has limited durability. Furthermore, it is essential to wet the hydrophobic membrane with appropriate organic solvent before initiating the separation in order to prevent leakage or breakthrough of liquid through the membrane (Kubáň, 1991).

Gravitational Phase Separators

This kind of separator is operated based on the difference of the densities of the two immiscible phases (Kubáň, 1991). Until recently, however, applications of the conventional design of gravitational phase separators have been rather limited, because of their somewhat poorer performance when compared with the aforementioned separators.

A recently reported version of the gravitational phase separator, the conical gravitational phase separator (Tao and Fang, 1995), has proven to be most efficient and easy to operate. Being fabricated in two parts, with a conical cavity in the upper parts with a volume of about 100 μL, it facilitates the separation of a low-density phase. The conical gravitational phase separator allows to be operated at relatively high flow rates and large aqueous/organic phase ratios. Studies have shown that the robustness as well as the separation efficiency of the conical gravitational phase separator are quite satisfactory (Nielsen et al., 1999). This separator has so far mostly been applied for the separation of a low-density phase from the extraction mixture. Although it also has been utilized to separate a high-density phase by accommodating the conical cavity (separation chamber) in the lower part of the separator (Fang et al., 1999).

A novel designed dual-conical phase separator, schematically illustrated in Fig. 21, has proven to be very effective for separating both the low-density and high-density phases (Wang and Hansen, 2002b, c). The separator is fabricated of PEEK, with a separating chamber of approximately 140 μL being composed of two conicals of identical volumes. The extraction mixture is directed into the middle region of the chamber where the phase separation takes place. The low-density phase is directed through the upper outlet, whereas the high-density phase flows through the lower outlet. By controlling the out-flow rates, the two conicals effectively facilitate the separation of both low-density and high-density phases.

Obviously, the specific advantages of the gravitational phase separators are their easy construction and long life time when compared with their membrane-based counterparts. The restriction is that in order to ensure an efficient

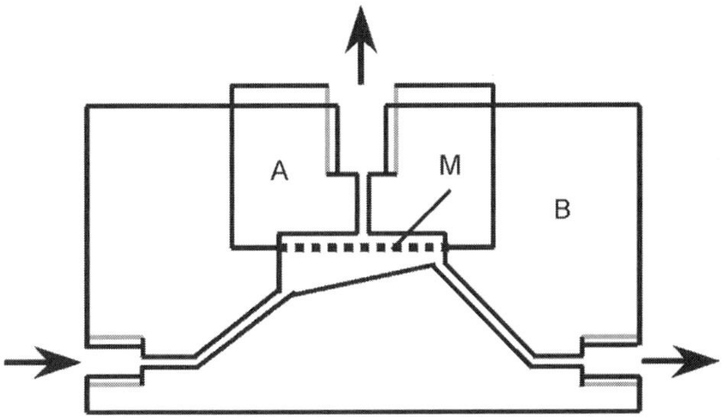

Figure 20 Schematic diagram of a sandwich-type membrane phase separator. A and B, PTFE blocks and M, microporous PTFE membrane. [Adapted from Fang et al. (1988), courtesy Elsevier Science Publishers.]

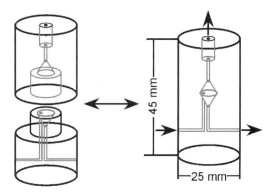

Figure 21 Schematic diagram of the dual-conical gravitational phase separator. [Adapted from Wang and Hansen (2002c), courtesy The Royal Society of Chemistry.]

separation the two immiscible phases should have an adequate difference in their densities. Fortunately, the densities of most organic solvents that are currently most often used for on-line solvent extraction are sufficiently different from that of water.

Although most of the on-line solvent extraction systems comprise a phase separation process, the use of SI makes it quite realistic to execute on-line solvent extraction without phase separation. Based on the fact that many organic solvents can form a thin film on the inner wall of a PTFE extraction coil, the film can act as a stationary phase and extract the analyte or the analyte complex with a suitable chelating reagent from the flowing aqueous phase (Lucy and Yeung, 1994; Luo et al., 1997). The analyte retained in the film can be eluted afterwards by small amount of appropriate eluent and the eluate transported to the detector.

3. Coupling On-Line Solvent Extraction to Various Detectors

The interface of on-line solvent extraction system to detectors such as a UV–Vis spectrophotometer, flame-AAS, and ICPAES generally pose no problems. But in order to achieve optimum sensitivity in the case of flame-AAS, a certain sample uptake flow rate is required, which is much higher than the flow rate of the concentrate separated and delivered from the on-line extraction system. A compromise should thus be made between these two conditions (Fang, 1993b). On the other hand, the introduction of analytes into the flame in the presence of organic solvents may create two- to threefold sensitivity enhancements for quite a few elements when compared with quantification in an aqueous phase.

The interface of on-line solvent extraction with ETAAS presents itself as a superb technique for determining ultratrace levels of metals in complicated matrices. But the continuous introduction of extracts into the graphite furnace atomizer is not feasible, because of the discrete nature of the operations of ETAAS. The extract is generally collected on-line in a sample loop, the content of which is sandwiched by air segments on both sides, and subsequently transported into the graphite tube, as illustrated in Fig. 22, where the delivery tube (DT) serves as a kind of sample loop (Nielsen et al., 1999).

ETAAS and ICPMS are among the most attractive detection devices for ultratrace metals, because of their high sensitivities, which for ICPMS is augmented by its multielement measurement capability. However, for ETAAS determinations it has been proven that quantification by injection of organic extract directly into the graphite furnace should be avoided to the extent possible. Organic solvents tend to distribute along the length of the graphite tube owing to their lower surface tension and

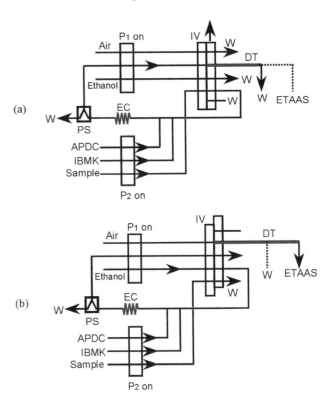

Figure 22 FIA on-line solvent extraction manifold coupled to ETAAS: (a) preconcentration sequence and (b) transportation sequence. P_1 and P_2, peristaltic pumps; W, waste; EC, extraction coil; PS, gravitational phase separator; DT, delivery tube; and IV, injection valve. [Adapted from Nielsen et al. (1999), courtesy Elsevier Science Publishers.]

good wetting ability, thus resulting in not only loss of sensitivity (Volynsky et al., 1984), but also deteriorating the reproducibility of the assay (Sturgeon et al., 1980). When coupling an on-line solvent extraction system to ICPMS, quantification by introducing the organic extracts directly into the ICP is simply prohibited as it might not only cause deposition of carbon on the sampler and skimmer cones, which results in severe loss of sensitivity, but it may also lead to an unstable plasma. Moreover, polyatomic ion interferences arising from the combination of C with O, Ar, S, and Cl might potentially cause further problems. From these points of view, the most effective approach is to employ a back-extraction step before the ETAAS and ICPMS determination in order to eliminate completely the organic solvents (Wang and Hansen, 2002b, c).

Surprisingly, preciously few attempts aimed at interfacing FIA on-line solvent extraction/back-extraction systems to ETAAS and ICPMS have been reported during the last decades (Bäckström and Danielsson, 1990; 1988; Shabani and Masuda, 1991). This might partly be attributed to the lack of robustness of the FIA systems. Or difficulties associated with manipulating the organic solvents by conventional FIA facilities, because the widely employed solvents for extraction, such as IBMK and trichlorotrifluoroethane (Freon 113), cannot be reliably delivered over long periods of time, even with solvent resistant pump tubings. The employment of a displacement bottle not only makes the manifold design more complicated, but considerable technical skills are also required for its proper manipulation (Fang and Tao, 1996). To address these problems, the SIA and the SIA/FIA hyphenated systems have shown significant advantages over conventional FIA systems in simplicity of manifold design, robustness, and versatility (Wang and Hansen, 2002b). The application of a glass syringe, a PTFE holding coil, and PTFE tubings makes it feasible to manipulate virtually any kind of organic solvents. The SIA/FIA systems are, therefore, most suitable vehicles for facilitating on-line solvent extraction/back-extraction procedures.

Their practical applicabilities can be demonstrated by a SIA/FIA hyphenated on-line solvent extraction–back extraction system interfaced to ICPMS, as is schematically shown in Fig. 23, which employs two PEEK dual-conical gravitational phase separators (Wang and Hansen, 2002c). Illustrated for the determination of copper and lead via complexation with ammonium pyrrolidinedithiocarbamate, the metal chelates are first extracted into IBMK, the organic phase is separated from the first phase separator PS_1, and stored in the holding coil (HC). The analytes are subsequently successfully back extracted into an aqueous phase by propelling the IBMK phase to meet the back extractant containing dilute nitric acid with Pd(II) as a stripping agent to accelerate the slow back-extraction process. The aqueous concentrate separated from the second phase separator PS_2 is collected in a sample loop SL, the content of which is afterwards introduced into the ICP for quantification by using a continuous infusion pump IP. In this case, analytes enrichments were

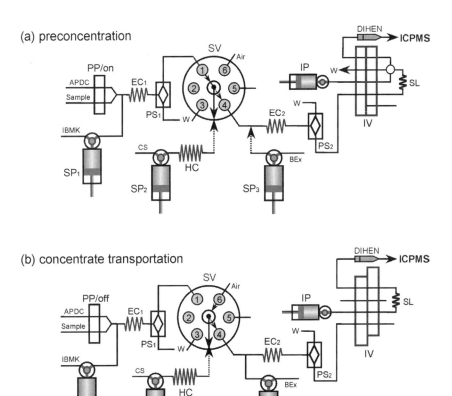

Figure 23 Hyphenated SIA/FIA-manifold for on-line solvent extraction–back-extraction system coupled to ICPMS: (a) preconcentration and (b) concentrate transportation. SP_1, SP_2, and SP_3, syringe pumps; PP, peristaltic pump; EC_1 and EC_2, extraction coils; PS_1 and PS_2, dual-conical gravitational phase separators; SV, six-port selection valve; IP, infusion pump; DIHEN, direct injection high efficiency nebulizer; HC, holding coil; SL, sample loop; IV, two-port injection valve; BEx, back extractant with stripping agent; CS, carrier; and W, waste. [From Wang and Hansen (2002c), courtesy The Royal Society of Chemistry.]

obtained in both stages of the extraction. In order to minimize dispersions during the transportation of the concentrate, a high efficiency DIHEN was used for transporting the concentrate into the ICP for quantification.

C. On-Line Hydride/Vapor Generation Preconcentration Schemes

Hydride and vapor generation is undoubtedly one of the most attractive sample pretreatment technique for the determination of hydride and vapor forming elements. In batch mode operations, the presence of transition metals, such as Ni(II), Co(II), Cu(II), and Fe(III), can severely suppress the formation and the release of the analyte hydrides, arising from the interactions of the hydrides with the reduced forms of these metals (Moor et al., 2000) (see Fig. 7). In an FIA/SIA on-line system, however, the formation rates of these free metal ions are relatively slower than the hydride generation process, thanks to the well-controlled and short reaction time as well as the rapid separation of the hydride from the reaction system. FIA/SIA on-line procedures have thus made the hydride generation operations most convenient and at the same time greatly improved the reproducibility. The majority of the applications of this protocol have so far been mostly focused on the coupling to AAS and ICPMS.

There are two main approaches for on-line obtaining the hydride and/or vapor. The most widely employed one is the chemical hydride/vapor generation scheme by using a suitable reductant (such as sodium tetrahydroborate) in acidic medium. This approach has, so far, been applied extensively for FIA on-line separation/preconcentration of hydride/vapor forming elements in various sample matrices (Menegario and Gine, 2000; Olivas et al., 1995; Tsalev et al., 2000). A second means is the FIA on-line electrochemical hydride generation technique, where the hydride is generated on-line at an appropriate electrolysis current (Bings et al., 2003). Both schemes can completely eliminate the matrix components, and they are thus most favorable for the analyses of hydride/vapor forming elements in biological samples, where serious matrix effects and consequently signal suppression are very often encountered.

Figure 24 illustrates the flow manifold of an on-line hydride generation system coupled to AAS equipped with a heated quartz tube atomizer (Yamamoto et al., 1987). Arsenic hydride is separated by employing a tubular PTFE microporous separator, and directed into the atomizer for quantification. The on-line system by using quartz tube atomizer is generally adequately robust and easy to control. Sometimes, however, the sensitivity of determination is not sufficient for ultratrace levels of hydride forming elements. It is thus most beneficial to

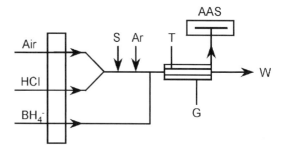

Figure 24 Flow manifold for FIA hydride generation system employing a tubular membrane gas diffusion separator coupled to AAS with a heated quartz tube atomizer. S, sample; W, waste; T, microporous PTFE tubing; G, tubular gas diffusion separator; and AAS, heated quartz tube atomizer. [Adapted from Yamamoto et al. (1987), courtesy American Chemical Society.]

introduce a preconcentration step before the hydride generation. Figure 25 depicts an FIA system demonstrating this approach as used for the determination of ultratrace levels of selenium(IV) (Nielsen et al., 1996), where the analyte is being preconcentrated via coprecipitation with lanthanum hydroxide. Thus in Fig. 25(a), which shows the system in the "fill" position, the sample is aspirated by pump P_1 and mixed on-line with buffer and coprecipitating agent, La(III). The co-precipitate generated is entrapped in the KR. In Fig. 25(b), where the system is in the "inject" position, valve V is switched over permitting the eluent, HCl, to pass through the KR to dissolve and elute the precipitate and forward it to mixing with the reductant, $NaBH_4$, leading to the formation of the hydride. The hydride is separated from the liquid matrix in the gas–liquid separator (SP) and subsequently by means of an auxiliary stream of argon gas guided to the heated quartz cell (QTA) of the AAS instrument.

The coupling of on-line hydride generation system to ETAAS presents a powerful technique for ultratrace levels of hydride forming elements. However, owing to the discontinuous operational feature of the ETAAS, continuous introduction of the hydride into the graphite atomizer in a flow-through mode is not feasible. The most widely employed approach to circumvent this problem is in-atomizer-trapping of hydrides in a graphite tube preheated to a certain temperature which is high enough for the decomposition of the hydride, but not sufficient to atomize the analyte. The successive sequestration of hydrides in the heated graphite tube provides not only an efficient means to preconcentrate the analyte, but also an elegant way to eliminate the matrix components. In addition, an appropriate coating on the interior surface of the graphite tube can offer much improved performance. Various materials have been investigated to precoat the atomizer, including Pd, Zr, Ir, Au, and Nb–Ta–W–Ir/Mg–Pd/Ir. A precoated graphite tube can be used for

Flow Injection/Sequential Injection Analysis

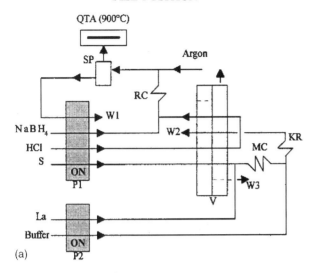

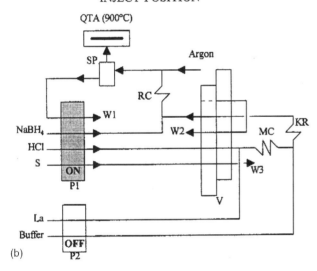

Figure 25 Schematic diagram of the flow injection-hydride generation-atomic absorption spectrometry system for on-line coprecipitation–dissolution of selenium or arsenic. (a) Coprecipitation sequence and (b) dissolution and detection sequence. [From Nielsen et al. (1996), courtesy The Royal Society of Chemistry.]

several hundreds of trapping-atomization cycles (Haug and Liao, 1996), varying according to the trapping temperature and the sequestrating time. However, this approach of repeated in-atomizer-trapping is inherently much time-consuming, and therefore it might advantageously be replaced by implementing the preconcentration on-line in FIA/SIA systems as mentioned above.

The interfacing of a FIA/SIA on-line hydride/vapor generation system to ICPMS is readily feasible. The compatibility between the two set-ups allows the hydride/vapor to be introduced directly into the ICP by using an argon flow. In many cases, however, the sensitivity by direct introduction of the hydride/vapor into the ICP is not sufficient. This can be overcome by employing electrothermal vaporization after on-line hydride/vapor generation, that is, the hydride/vapor is successively sequestrated by in-atomizer-trapping in a precoated graphite tube as in the case of ETAAS determination (Lam and Sturgeon, 1999) in order to reach a certain preconcentration level, whereupon the trapped analytes are atomized and introduced into the ICP for quantification. Or, of course, simply by facilitating the preconcentration in a FIA/SIA system coupled directly to the detection device.

V. APPLICATIONS OF SI-LOV FOR ON-LINE SAMPLE PRETREATMENTS

A. The Third Generation of Flow Injection and Micro Total Analysis Systems

As discussed in Sections II and IV, the first two generations of flow injection have proven to allow implementation of procedures which previously were difficult or, in fact, not even feasible by batch mode approaches. It is especially noteworthy that they have become the main substitutions of labor intensive manual operation of various separation/preconcentration techniques via clever on-line manipulations, which can be interfaced to virtually any kind of detectors. However, in sample/reagent limited assays, or in cases of handling highly hazardous chemicals, where waste production is a critical parameter to minimize environmental pollution, further downscaling of solution control is warranted (Ruzicka, 2000; Wu and Ruzicka, 2001; Wu et al., 2001). In this respect, the so-called micro total analysis systems (μTAS) have drawn considerable attention (Manz and Becker, 1998).

In a μTAS system, the operation of nanoliter levels of sample and reagent can be translated into ultratraces of analytes introduced into the detector, which, in turn, requires an improved sensitivity of the detection devices, posing a challenge to the miniaturization of the instruments. Therefore, the prerequisite of at least microliter levels of the final detectable volume excludes the hyphenation of a μTAS sample pretreatment system to currently available atomic spectrometric detectors. And obviously, there is still a long way to go before the atomic spectrometric detections can be performed inside a μTAS system. Besides, because of the very small channel dimensions, the chip-based μTAS systems tend to be clogged by the suspended fine particles when it is used for processing complex samples, such as those of biological origin.

The several orders of magnitude of difference for the sample/reagent consumption between the μTAS system and the conventional FIA/SIA on-line sample pretreatment systems have ostensibly revealed a gap precluding the application of the various on-line sample pretreatment protocols that so far have been successfully employed in coupling with atomic spectrometric detections. Further miniaturization of the conventional on-line sample pretreatment system is thus highly imperative in order to further reduce the consumption of expensive and rare sample and/or reagents, or to minimize the destructions of in vivo analysis, or to reduce the environmental impact.

On the other hand, in the real world, it is not always necessary to scale down the consumption of samples and reagents to nanoliter levels, as in most cases it is quite acceptable to control the consumption volume at the microliter or sub-microliter levels. In this respect, the third generation of flow injection, that is, the so-called sequential injection-bead injection-lab-on-valve (SI-BI-LOV) microsystem is not only suitable for fluidic and microcarrier beads control from the microliter to the milliliter levels, being easily performed by using conventional valves and pumps (see Section III.B). But it can also facilitate the control of virtually any kind of chemical processes in the valve, including in-LOV homogenous reaction, heterogeneous liquid–solid interaction, and in-valve real time optical monitoring of various reaction processes. This feature, therefore, makes it the most appropriate intermediate link between the conventional FIA/SIA schemes and the μTAS systems for the miniaturization of on-line sample pretreatment. Because of its much reduced operating volume compared with the conventional FIA/SIA system, the third generation of FIA is also known as the micro sequential injection-lab-on-valve (μSI-LOV) protocol (Ruzicka, 2000; Wu and Ruzicka, 2001; Wu et al., 2001).

B. Applications of μSI-LOV Systems for On-Line Sample Pretreatments with Optical Monitoring

The μSI-LOV system has been employed in a variety of microminiaturized in-valve fluidic and microcarrier beads handling procedures and appropriate in-valve interactions followed by real time detection, as well as served as a front end to capillary electrophoresis (CE). The various applications of the μSI-LOV that have been conducted so far are summarized in Table 3.

A typical configuration of a μSI-LOV system along with a close-up of the multipurpose flow cell configured for real time measurement of absorbance and fluorescence is schematically illustrated in Fig. 12, showing its working principles. The flow cell is furnished with optical fibers, communicating with an external light source and a detection device to facilitate real time monitoring of the reactions taking place inside the flow cell.

In the μSI-LOV system depicted in Fig. 12, the multipurpose flow cell is employed to accommodate the sample/reagent zone and to facilitate real time detection by recording the absorbance (facing fibers as shown in the figure) or fluorescence (at a 90° angle; not shown). In order to increase the sensitivities for kinetically slow reactions, selected segments of the sample zone can be monitored by adopting the stopped-flow technique (Section II.E). The effectiveness of arresting a preselected part of a sample zone inside the flow cell in a μSI-LOV system has been illustrated by the determination of phosphate (Ruzicka, 2000; Wu and Ruzicka, 2001), enzymatic

Table 3 The Applications of μSI-LOV Systems for Microminiaturized In-Valve Fluidic and Microcarrier Beads Handling as well as Appropriate In-Valve Interactions Followed by Real Time Monitoring and/or as a Front End to CE

LOV system	Applications	Detection	References
μSI-LOV	Phosphate assay, enzymatic activity assay of protein	A	Ruzicka, 2000; Wu and Ruzicka, 2001
	Fermentation monitoring of ammonia, glycerol, glucose, and free iron	A	Wu et al., 2001
μSI-LOV with a reduction column	Environmental monitoring of NO_3^- and NO_2^- (NO_3^- on-line reduced to NO_2^-)	A	Wu and Ruzicka, 2001
μSI-BI-LOV	Bioligand interactions assay of immunoglobulin (IgG) on protein G immobilized Sepharose beads	F/A	Ruzicka, 2000
	Lactate extrusion and glucose consumption rates by a cell culture on cytopore® beads	A	Schultz et al., 2002
μSI-LOV-CE	μSI-LOV serving as a sample pretreatment unit and sample introduction system to CE	A	Wu et al., 2002

Note: A, Absorbance and F, fluorescence.

assays (Ruzicka, 2000), and the monitoring of ammonia, glycerol, glucose, and free iron in fermentation broths (Wu et al., 2001). For carrying out on-line measurement of nitrate and nitrite, a cadmium reduction column can be incorporated into the LOV system to reduce nitrate to nitrite before confluencing this species with the chromogenic reagents (sulfanilamide and N-(1-naphthyl) ethylene-diamine) (Wu and Ruzicka, 2001). With a miniaturized reduction column, an appropriate period of stopped-flow of the sample zone in the cadmium column is employed to increase the contact time and thus the reduction efficiency.

The μSI-LOV system can also be employed as an interface for CE for anion separation. In this case, the multipurpose flow cell is reconfigured as a front end between the μSI-LOV and the CE system (Wu et al., 2002). The micro fluidic property of the μSI-LOV not only provides an efficient sample delivery conduit for the CE system with various sample injection modes, including electrokinetic, hydrodynamic, and head column field amplification sample stacking, but also at the same time serves as a versatile means of sample pretreatment to facilitate the ensuing CE separation.

In the μSI-BI-LOV mode, the flow cell is adapted to the so-called renewable column approach in order to perform bioligand interactions assays. In this case, an appropriate beads suspension is aspirated into the flow cell, whereupon the beads are trapped and ensuingly perfused with analyte solution. Afterwards, the loaded beads are exposed to various stimuli, and the (bio)chemical reactions taking place on the bead surface are recorded in real time. This has been demonstrated for the bioligand interactions assay of immunoglobulin (IgG) on protein G immobilized Sepharose beads (Ruzicka, 2000), and the monitoring of cellular lactate extrusion and glucose consumption rates by a cell culture on cytopore® beads (Schulz et al., 2002). Thus, the bioligand interaction of protein G bound on the surface of Sepharose beads with anti-mouse IgG-labeled fluorescein isothiocyanate (FITC) was monitored by fluorescence spectroscopy at 525 nm in real time (Ruzicka, 2000). The experimental protocol included four steps: (i) aspiration of bead suspension (4 μL) from port 6 into the holding coil; (ii) flow reversal to transfer a defined volume of beads into the flow cell and pack a temporary microcolumn; (iii) aspiration of 25 μL sample solution from port 5 into the holding coil; and (iv) flow reversal to bring the sample zone in contact with the protein G–Sepharose beads, resulting in the capture of IgG-FITC on the bead surface. The recorded signals for 4.0, 10.0, 25.0, 50.0, and 100.0 μg mL^{-1} antibody (IgG-FITC) and the associated calibration graph are shown in Fig. 26. At the end of the measuring period, the used beads were transferred into the holding coil and then discarded via port 1.

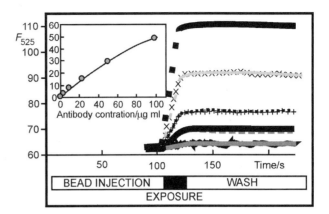

Figure 26 Fluorescence monitoring of adsorption of anti-mouse IgG-FITC molecules on the surface of Sepharose B–protein G beads. Injected bead suspension volume, 4 μL; injected sample volume, 25 μL; and sample flow rate for perfusion, 2 μL/s. Insert shows calibration graph. [From Ruzicka (2000), courtesy The Royal Society of Chemistry.]

It can be seen from Fig. 26 that during the exposure time, the fluorescence increased as the IgG-FITC molecules accumulated on the bead surface. Whereas during the washout period the IgG remains firmly retained on the bead surface, as illustrated by the constant response level, which indicates that the IgG–protein G interaction is very strong. From this point of view it is, therefore, highly beneficial to use the renewable surface approach instead of employing a permanent column.

C. SI-BI-LOV: Solving the Dilemma of On-Line Column Separation/Procencentration Schemes

1. The Current Status of On-Line Column-Based Sample Pretreatment Protocols

Column-based solid phase extraction has been extensively employed in FIA/SIA on-line separation and preconcentration assays of ultratrace levels of metals. Conventionally, as discussed in Section IV.A, the sorbent column is treated as a permanent component of the system, being used repeatedly for the sample loading/elution sequences, and it is replaced or repacked only after long-term operation. There are, therefore, some inherent limitations for the conventional mode of operation, that is, the performance of the procedures is often deteriorated by the build-up of flow resistance or back pressure caused by progressively tighter packing of the sorbent material due to repetitive operations (Fang et al., 1984; Ruzicka and Hansen, 1988c). This situation is made even worse if the sorbent beads undergo volume changes during the analytical cycle, that is, swelling or shrinking, with the change of experimental conditions (Fang et al., 1984). Moreover, the surface properties of the column material, which are associated with the retention efficiency and the kinetics during the sorption–elution process, might be irreversibly changed due to contamination, deactivation, or even loss of functional groups or active sites (Ruzicka and Scampavia, 1999; Wang and Hansen, 2000b). In addition, it is always desirable to completely elute the retained analyte from the sorbent with minimum amount of eluent in order to avoid carry-over effects. In practice, this is, however, not always feasible, which consequently leads to risks of such carry-over between sample runs.

The difficulties associated with flow resistance can be alleviated to a certain extent via various approaches, including bi-directional flows during the sample loading and elution sequences (Fang et al., 1984), or intermittent back suction of small air segments (Xu et al., 2000a, b). Nevertheless, it is difficult to completely eliminate the adverse effects of flow resistance. Furthermore, the malfunctions of the sorbent surfaces are not addressed by these means.

In this respect, a superb alternative for eliminating any problem associated with the changes of the surface properties of the sorbent materials and/or the creation of flow resistance in a column reactor is to employ a surface renewal scheme, that is, the microcolumn is simply renewed or replaced for each analytical run. The renewable surface concept has been applied to a variety of (bio)-chemical studies (Hodder and Ruzicka, 1999; Ruzicka, 1998, 2000; Ruzicka and Scampavia, 1999; Schulz et al., 2002). In fact, it is well suited for on-line solid phase extraction separation/preconcentration in a SI-LOV system incorporating a miniaturized renewable microcolumn, that is, the bead injection based lab-on-valve (SI-BI-LOV), as demonstrated by the applications illustrated in the following section.

2. Description of the SI-BI-LOV System

A SI-BI-LOV flow system with a diagram of a LOV conduit mounted atop a six-port selection valve for bead injection is illustrated in Fig. 27. Two of the channels, including the central one, are used as microcolumn positions (C_1 and C_2). In order to trap the beads—usually of a size around 100 μm—within the channel cavities serving as microcolumns, and prevent the beads from escaping during the operations, the outlets of these two channels are furnished with small pieces of PEEK tubing. The outer diameter of this tubing is slightly smaller than those of the columns and provides an internal channel which thus allows liquid to flow freely through the channel and along the walls, but entrap the beads. A close-up of the packed renewable microcolumn is also illustrated in Fig. 27.

In order to facilitate the beads handling, the sorbent beads are suspended in an appropriate amount of water or buffer solution, usually within the range of 1:10–1:20 (m/v). In operation, the suspension is first aspirated into a 1.0 mL plastic syringe which afterwards is mounted on port 6 of the LOV. The beads suspension is by means of the syringe pump (Fig. 27, left) slowly aspirated from the plastic syringe into the microcolumn position C_1. After aspiration into the LOV system, the beads can then be readily transferred back and forth between the two microcolumn positions (C_1 and C_2) as the protocol requires. These manipulations facilitate both the sample

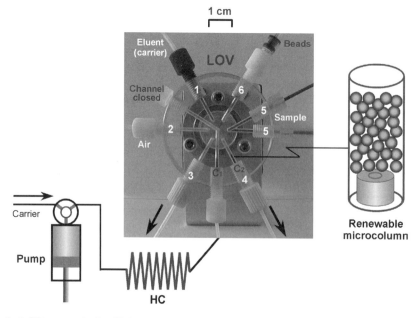

Figure 27 Diagram of a LOV system for bead injection incorporating two microcolumn positions (C_1 and C_2), along with a close-up of a packed renewable microcolumn. [From Wang and Hansen (2003), courtesy Elsevier Science Publishers.]

loading process and the treatment of the analyte-loaded beads according to which scheme is to be followed for the postanalysis, as illustrated in the following examples.

3. SI-BI-LOV Renewable Column Solid Phase Extraction of Ultratrace Levels Of Metals With Detection by ETAAS and ICPMS

Two categories of sorbent beads with appropriate diameter can be employed to pack renewable microcolumns, for example, hydrophilic SP Sephadex C-25 cation-exchanger and iminodiacetate based Muromac A-1 chelating resins, and hydrophobic materials such as PTFE and poly(styrene-divinylbenzene) copolymerized octadecyl groups (C_{18}-PS/DVB).

When using ETAAS as the detection device, two different schemes for dealing with the analyte-loaded beads can be exploited (Wang and Hansen, 2000b; 2001b, c), as depicted in Fig. 28. These include the following. (1) After the resin beads have been exposed to a certain amount of sample solution, the loaded beads are eluted with a small, well-defined volume of an appropriate eluent which ultimately is transported into the graphite tube for quantification, whereupon the used beads are discarded. New beads are aspirated and fresh columns are generated for the ensuing sample cycles. (2) The loaded beads can alternatively, as a novel and unique approach, be transported directly into the graphite tube, where advantage can be taken by the fact that the beads consist primarily of organic materials, that is, they can be pyrolyzed thereby allowing ETAAS quantification of the analytes. While the former approach can be used with ICP as well, the latter one is obviously prohibitive for this detection device. Outlines of operational sequences for these approaches are as follows.

In order to execute on-line separation/preconcentration in the renewable column, an affixed volume of sample solution is first aspirated from port 5 and stored within the HC, followed by aspiration of a small amount of beads suspension, usually 15–20 μL, which is initially captured in column position C_1. Thereafter, the central channel is directed to communicate with column position C_2, and while the syringe pump moves slowly forward the beads are hereby transferred to C_2 followed by sample solution, that is, separation/preconcentration is taking place in this column position. The analyte retained within the microcolumn can, afterwards, be dealt with according to the two schemes mentioned earlier for postanalysis.

In the first scheme, both the sample loading and the analyte elution processes take place in the same column

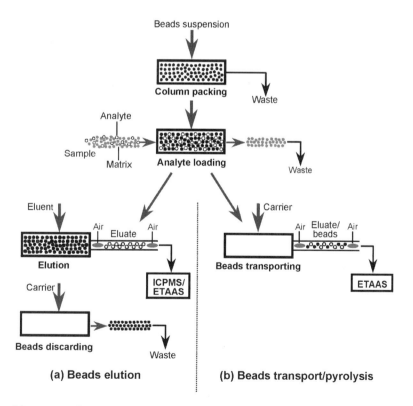

Figure 28 Illustration of the two possible schemes for dealing with analyte-loaded beads in the renewable microcolumn approach. (a) The analyte-loaded beads are eluted and the eluate is exploited for quantification by ETAAS and/or ICPMS. (b) The analyte-loaded beads are transported directly into the graphite tube of the ETAAS, where, following pyrolysis of the beads, the analyte is quantified. [From Wang and Hansen (2003), courtesy Elsevier Science Publishers.]

position C_2, the eluent having been aspirated from port 1 in an intermediate step and stored in HC, before being pumped forward and through C_2. The eluate is eventually, via air segmentation, transported into the graphite tube. Whereas in the second approach, the separation/preconcentration is carried out in column position C_2, and on completion the analyte retained beads are, along with a certain amount of carrier solution, usually 30–40 μL, transferred back to column position C_1 (this is necessary because all external communication is effected via the syringe pump, SP). The beads and the carrier zone are afterwards, via air segmentation, and through port 3 transported to the graphite tube of the ETAAS instrument, where the beads are pyrolyzed and the analytes quantified (Wang and Hansen, 2000b).

It should be emphasized that in order to pyrolyze the beads, a higher temperature than the normal pyrolysis process and a longer holding time are required. These conditions are obviously not applicable for the determination of metals with low atomization temperature, particularly for Cd, Bi, and Pb. In such cases, it is necessary to make use of the first approach comprising an elution step, since the eluate can be pyrolyzed at substantially lower temperatures when compared with the beads themselves, which helps to avoid analyte loss before the atomization process (Miró et al., 2003; Wang and Hansen, 2003).

The interface of the SI-BI-LOV on-line renewable column preconcentration system to ICPMS is, principally, quite straightforward when compared with that in the case of ETAAS. In addition, considering the fact that in many cases nitric acid potentially can be used as an eluent for the retained analytes, the protocol used for ETAAS is, after appropriate modification, readily transferable for interfacing with ICPMS, with due considerations to the continuous feature of the ICPMS system and the characteristics of the renewable microcolumn and the LOV system (Wang and Hansen, 2001a). This is illustrated in Fig. 29 which shows a SI-BI-LOV on-line renewable column preconcentration system interfaced with detection by ICPMS (Wang and Hansen, 2003).

Table 4 summarizes the characteristic performance data of a SI-BI-LOV renewable column preconcentration protocol for ultratrace level of nickel with detection by ETAAS

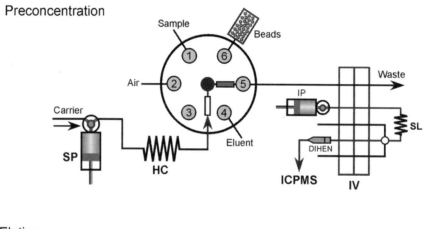

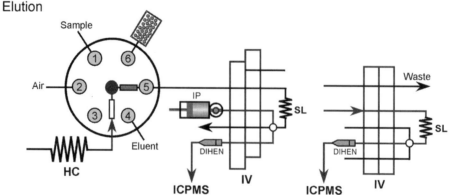

Figure 29 SI-BI-LOV on-line renewable column preconcentration system with detection by ICPMS: the beads are eluted and the eluate is introduced into the ICP via a DIHEN. SP, syringe pump; HC, holding coil; IV, two-position injection valve; SL, sample loop; IP, infusion pump; and DIHEN, direct injection high efficiency nebulizer. [From Wang and Hansen (2003), courtesy Elsevier Science Publishers.]

Table 4 Performance Data for Preconcentration of Ni(II) in the SI-BI-LOV System Using a Renewable Column Packed with SP Sephadex C-25 Ion-Exchanger with Detection by ETAAS, when Compared with a Permanent Column with Uni- and Bi-directional Repetitive Operation (Wang and Hansen, 2000b; 2001b)

	Renewable column approach		Conventional approach	
	Beads transportation	Elution	Uni-directional	Bi-directional
$RSD_H{}^a$ (%, 0.3 µg/L)		1.5	5.8	4.9
RSD_K (%, 0.3 µg/L)	3.4	1.7		
LOD_H (ng L; $n = 7$; 3σ)	9^b	10.2, 11.4^b	42	24
Enrichment factor	72.1	71.1		

[a] The subscript stands for the form of the ion exchanger.
[b] LOD value (3σ) obtained with the K^+-form of the ion exchanger.

(Wang and Hansen, 2000b; 2001b). It clearly shows that the procedures based on the employment of renewable columns give much improved precisions (RSDs) and lower LODs than those of the conventional processes, that is, by using a permanent column either with repetitive uni- or bi-directional operations. The two schemes by employing renewable columns, that is, the beads pyrolysis and column elution schemes, exhibit comparable performance, except that the precision of the elution based procedure is better than the one by pyrolyzing the beads directly, because it is easier to handle and transport homogeneous solutions than heterogeneous ones in a flow system.

It has been demonstrated that performance of a SI-BI-LOV on-line renewable column sample pretreatment system depends strongly on the morphology of the sorbent beads as well as their hydrophilicity and hydrophobicity (Miró et al., 2003).

In Fig. 30 are shown optical microscopy photographs at $100\times$ magnification of some of the commonly employed hydrophilic and hydrophobic sorbent beads, that is, Sephadex C-25, C_{18}-PS/DVB, Muromac A-1, PTFE, C_{18} covalently modified silica-gel, and controlled pore glass (CPG) (Wang et al., 2004). It is obvious that both the Sephadex C-25 and the C_{18}-PS/DVB materials are perfectly spherical and uniform in size distribution, making them ideal candidates for handling in the SI-BI-LOV format. In this sense, the hydrophilicity and hydrophobicity of the beads are not a critical factor, which was proven

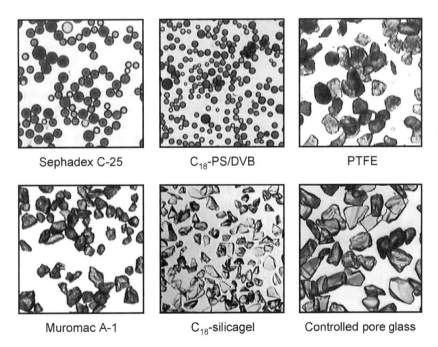

Figure 30 Optical microscopy photographs at $100\times$ magnifications of various hydrophilic (phi) and hydrophobic (pho) bead materials used in the LOV schemes, and of potential candidates. Sephadex C-25 cation-exchanger (phi); poly(styrene-divinylbenzene) copolymer alkylated with octadecyl groups (C_{18}-PS/DVB) (pho); PTFE (pho); Muromac A-1 chelating resin (phi); C_{18} covalently modified silica-gel (pho); and CPG (phi). [From Wang et al. (2003), courtesy Elsevier Science Publishers.]

experimentally (Miró et al., 2003; Wang and Hansen, 2001a). On the other hand, when looking at the nonspherical sorbents it was found that although the Muromac A-1 and PTFE beads (in fact, lumps of irregular shape rather than spherical beads) have similar morphology, they behave quite differently when their suspensions are operated in the LOV format. Thus, while the hydrophilic Muromac A-1 beads are quite easy to handle, troubles are often encountered with the hydrophobic PTFE beads, and the reproducibility of operation was not satisfactory. In this respect, it is to be expected that both C_{18} covalently modified silica-gel and CPG, therefore, will behave much alike the PTFE beads, that is, despite their inherent attractive chemical characteristics on the sorption of various metal chelates, it will prove difficult, or rather impossible, to manipulate them in the SI-BI-LOV set-up.

So far, according to the experiences gained in the exploitation of the renewable surface approach for the preconcentration of ultratrace levels of heavy metals in the LOV format prior to ETAAS and ICPMS detection, it is to be expected that this strategy is equally applicable to virtually any kind of sorbent beads, provided that suitable geometry and size-homogeneity are available. For some materials, it might be beneficial to use the renewable reactor as a semipermanent column in the LOV system, that is, the microcolumn is packed and used for an appropriate number of measuring cycles, being afterwards replaced or repacked before flow resistance or deterioration of the packing material start to pose problems.

In addition, the bead injection technique in the LOV format provides an alternative medium for precipitate/co-precipitate collection, whose operational mode also can be coupled to a hydride/vapor generation flow system in order to further improve the enrichment capability for ultratrace levels of hydride/vapor forming metals in complex matrices.

Although the third generation of flow injection is still in its infancy, the SI-BI-LOV approach has already proven to be an attractive and effective front end to various detection devices with microminiaturized sample processing along with improved efficiency and ruggedness. Considering the vast potentials, it is, therefore, to be expected that the SI-BI-LOV protocol will gain considerable momentum in the very near future (Wang et al., 2003).

REFERENCES

Alonso, E. V., Torres, A. G., Pavon, J. M. C. (2001). Flow injection on-line electrothermal atomic absorption spectrometry. *Talanta* 55(2):219–232.

Åström, O. (1982). Flow injection analysis for the determination of bismuth by atomic absorption spectrometry with hydride generation. *Anal. Chem.* 54:190–193.

Atallah, R. H., Ruzicka, J., Christian, G. D. (1987). Continuous solvent extraction in a closed-loop system. *Anal. Chem.* 59:2909–2914.

Bäckström, K., Danielsson, L. G. (1988). Design and evaluation of an interface between a continuous flow system and a graphite furnace atomic absorption spectrometer. *Anal. Chem.* 60:1354–1357.

Bäckström, K., Danielsson, L. G. (1990). Design of a continuous-flow two-step extraction sample work up system for graphite furnace atomic absorption spectrometry. *Anal. Chim. Acta* 232:301–315.

Bäckström, K., Danielsson, L.-G., Nord, L. (1986). Dispersion in phase separators for flow-injection extraction systems. *Anal. Chim. Acta* 187:255–269.

Bendtsen, A. B., Hansen, E. H. (1991). Spectrophotometric flow injection determination of trace amounts of thiocyanate based on its reaction with 2-(5-bromo-2-pyridylazo)-5-diethylaminophenol and dichromate: assay of the thiocyanate level in saliva from smokers and non-smokers. *Analyst* 116:647–651.

Benkhedda, K., Ivanova, E., Adams, F. (1999). Flow injection on-line sorption preconcentration of trace amounts of copper and manganese in a knotted reactor precoated with 1-phenyl-3-methyl-4-benzoylpyrazol-5-one coupled with electrothermal atomic absorption spectrometry. *J. Anal. Atom. Spectrom.* 14(6):957–961.

Benkhedda, K., Infante, H. G., Ivanova, E., Adams, F. (2000a). Trace metal analysis of natural waters and biological samples by axial inductively coupled plasma time of flight mass spectrometry with flow injection on-line adsorption preconcentration using a knotted reactor. *J. Anal. Atom. Spectrom.* 15(10):1349–1356.

Benkhedda, K., Infante, H. G., Ivanova, E., Adams, F. (2000b). Ultratrace determination of cobalt in natural waters by electrothermal atomic absorption spectrometry using flow injection on-line sorption preconcentration in a knotted reactor precoated with 1-phenyl-3-methyl-4-benzoyl-pyrazol-5-one. *J. Anal. Atom. Spectrom.* 15(4):429–434.

Bings, N. H., Stefánka, Z., Mallada, S. R. (2003). Flow injection electrochemical hydride generation inductively coupled plasma time of flight mass spectrometry for the simultaneous determination of hydride forming elements and its application to the analysis of fresh water samples. *Anal. Chim. Acta* 479:203–214.

Burguera, J. L., Burguera, M. (2001). Flow injection-electrothermal atomic absorption spectrometry configuration: recent developments and trends. *Spectrochim. Acta Part B* 56(10):1801–1829.

Fang, Z.-L. (1989). Analytical methods and techniques. In: Burguera, J. L., Ed. *Flow Injection Atomic Spectroscopy*. Practical Spectroscopy Series. Vol. 7. New York: Marcel Dekker Inc., pp. 103–156.

Fang, Z.-L. (1993a). Gas-liquid separation. In: *Flow Injection Separation and Preconcentration*. Weinheim: VCH, pp. 129–156.

Fang, Z.-L. (1993b). In: *Flow Injection Separation and Preconcentration*. New York: VCH Publishers, Inc., pp. 85–128.

Fang, Z.-L. (1993c). Liquid-liquid extraction. In: *Flow Injection Separation and Preconcentration*. New York: VCH Publishers, Inc, pp. 47–83.

Fang, Z.-L. (1995a). On-line preconcentration for flame and hydride generation atomic absorption spectrometry; Flow injection techniques for electrothermal atomic absorption spectrometry. In: *Flow Injection Atomic Spectrometry*. New York: John Wiley & Sons Ltd., pp. 143–202.

Fang, Z.-L. (1995b). On-line preconcentration for flame and hydride generation atomic absorption spectrometry. In: *Flow Injection Atomic Spectrometry*. New York: John Wiley & Sons Ltd., pp. 141–174.

Fang, Z.-L., Welz, B. (1989). High efficiency low sample consumption on-line ion-exchange preconcentration system for flow-injection flame atomic-absorption spectrometry. *J. Anal. Atom. Spectrom.* 4:543–546.

Fang, Z.-L., Tao, G. (1996). New developments in flow injection separation and preconcentration techniques for electrothermal atomic absorption spectrometry. *Fresenius J. Anal. Chem.* 355(5–6):576–580.

Fang, Z.-L., Ruzicka, J., Hansen, E. H. (1984). An efficient flow-injection system with on-line ion-exchange preconcentration for determination of trace amounts of heavy metals by atomic absorption spectrometry. *Anal. Chim. Acta* 164:23–39.

Fang, Z.-L., Zhu, Z.-H., Zhang, S.-C., Xu, S.-K., Guo, L., Sun, L.-J. (1988). On-line separation and preconcentration in flow injection analysis. *Anal. Chim. Acta* 214:41–55.

Fang, Z.-L., Sperling, M., Welz, M. (1990). Flow injection on-line sorbent extraction preconcentration for graphite furnace atomic absorption spectrometry. *J. Anal. Atom. Spectrom.* 5:639–646.

Fang, Z.-L., Sperling, M., Welz, B. (1991). Flame atomic absorption spectrometric determination of lead in biological samples using a flow injection system with on-line preconcentration by coprecipitation without filtration. *J. Anal. Atom. Spectrom.* 6:301–306.

Fang, Z.-L., Xu, S., Dong, L., Li, W. (1994). Determination of cadmium in biological materials by flame atomic absorption spectrometry with flow injection on-line sorption preconcentration. *Talanta* 41(12):2165–2172.

Fang, Q., Sun, Y., Fang, Z.-L. (1999). Automated continuous monitoring of drug dissolution process using a flow injection on-line dialysis sampling–solvent extraction separation spectrophotometric system. *Fresenius J. Anal. Chem.* 364(4):347–352.

Fossey, L., Cantwell, F. F. (1985). Determination of acidity constants by solvent extraction/flow injection analysis using a dual-membrane phase separator. *Anal. Chem.* 57:922–926.

Hansen, E. H. (1988) Exploitation of gradient techniques in flow injection analysis. *Fresenius' Z. Anal. Chem.* 329:656–659.

Hansen, E. H. (1989). Flow injection analysis: do we really exploit it optimally? A personal comment from a FIA-aficionado who has been in the game from the start. *Quim. Anal.* 8(2):139–150.

Hansen, E. H. (1994). Flow injection analysis: a complementary or alternative concept to biosensors. *Talanta* 41(6):939–948.

Hansen, E. H. (2003). *Flow Injection Bibliography*. http://www.flowinjection.com/

Hansen, E. H., Wang, J.-H. (2002). Implementation of suitable FI/SI sample separation and preconcentration schemes for determination of trace metal concentrations using detection by ETAAS and ICPMS. *Anal. Chim. Acta* 467:3–12.

Haug, H. O., Liao, Y.-P. (1996). Investigation of the automated determination of As, Sb and Bi by flow-injection hydride generation using in-situ trapping on stable coatings in graphite furnace atomic absorption spectrometry. *Fresenius J. Anal. Chem.* 356(7):435–444.

Hodder, P. S., Ruzicka, J. (1999). A flow injection renewable surface technique for cell-based drug discovery functional analysis. *Anal. Chem.* 71(6):1160–1166.

Jørgensen, U. V., Nielsen, S., Hansen, E. H. (1998). Dilution methods in flow injection analysis. Evaluation of different approaches as exemplified for the determination of nitrosyl in concentrated sulphuric acid. *Anal. Lett.* 31(13):2181–2194.

Karlberg, B. (1986). Flow injection analysis—or the art of controlling sample dispersion in a narrow tube. *Anal. Chim. Acta* 180:16–20.

Karlberg, B., Thelander, S. (1978). Extraction based on the flow injection principle. Part 1. Description of the extraction system. *Anal. Chim. Acta* 98:1–7.

Karlberg, B., Pacey, G. E. (1989). Components of FIA. In: *Flow Injection Analysis. A Practical Guide*. Amsterdam: Elsevier Science Publishers Company, Inc., pp. 29–65.

Kubáň, V. (1991). Liquid–liquid extraction flow injection analysis. *Crit. Rev. Anal. Chem.* 22:477–557.

Kubáň, V., Danielsson, L.-G., Ingman, M. (1990). Design of coaxial segmentors for liquid–liquid extraction/flow-injection analysis. *Anal. Chem.* 62(18):2026–2032.

Lam, J. W., Sturgeon, R. E. (1999). Determination of As and Se in seawater by flow injection vapor generation ETV-ICP-MS. *Atom. Spectrosc.* 20(3):79–85.

Lucy, C. A., Yeung, K. K.-C. (1994). Solvent extraction-flow injection without phase separation through use of the differential flow velocities within segmented flow. *Anal. Chem.* 66(14):2220–2225.

Luo, Y., Nakano, S., Holman, D. A., Ruzicka, J., Christian, G. D. (1997). Sequential injection wetting film extraction applied to the spectrophotometric determination of chromium(VI) and chromium(III) in water. *Talanta* 44(19):1563–1571.

Manz, A., Becker, A. (1998). *Microsystem Technology in Chemistry and Life Sciences*. Berlin: Springer Verlag.

Matschiner, H., Ruettinger, H. H., Sivers, P., Mann, U. (1984). Quantitative determination of chlorate ions. German (East) Patent no. 216543.

Menegario, A. A., Gine, M. F. (2000). Rapid sequential determination of arsenic and selenium in waters and plant digests by hydride generation inductively coupled plasma-mass spectrometry. *Spectrochim. Acta Part B* 55(4):355–362.

Minczewski, J., Chwastowask, J., Dybczynski, R. (1982). *Separation and Preconcentration Methods in Inorganic Trace Analysis*. Chichester: Ellis Horwood Limited.

Miró, M., Jonczyk, S., Wang, J.-H., Hansen, E. H. (2003). Exploiting the bead-injection approach in the integrated sequential injection lab-on-valve format using hydrophobic packing materials for on-line matrix removal and preconcentration for trace levels of cadmium in environmental and biological samples via formation of non-charged chelates prior to ETAAS detection. *J. Anal. Atom. Spectrom.* 18(2):89–98.

Moor, C., Lam, J. W. H., Sturgeon, R. E. (2000). A novel introduction system for hydride generation ICPMS, determination of selenium in biological materials. *J. Anal. Atom. Spectrom.* 15(2):143–149.

Nielsen, S., Sloth, J. J., Hansen, E. H. (1996). Determination of ultratrace amounts of selenium(IV) by flow injection hydride generation atomic absorption spectrometry with on-line preconcentration by co-precipitation with lanthanum hydroxide. Part II: on-line addition of coprecipitating agent. *Analyst* 121(1):31–35.

Nielsen, S. C., Sturup, S., Spliid, H., Hansen, E. H. (1999). Selective flow injection analysis of ultra-trace amounts of Cr(VI), preconcentration of it by solvent extraction, and determination by electrothermal atomic absorption spectrometry. *Talanta* 49(5):1027–1044.

Nord, L., Karlberg, B. (1984). Extraction based on the flow injection principle. Part 6. Film formation and dispersion in liquid–liquid segmented flow extraction systems. *Anal. Chim. Acta* 164:233–249.

Nord, L., Bäckström, K., Danielsson, L.-G., Ingman, M., Karlberg, B. (1987). Extraction rate in liquid–liquid segmented flow injection analysis. *Anal. Chim. Acta* 194:221–233.

Olivas, R. M., Quetel, C. R., Donard, O. F. X. (1995). Sensitive determination of selenium by inductively coupled plasma massspectrometry with flow injection and hydride generation in the presence of organic solvents. *J. Anal. Atom. Spectrom.* 10(10):865–870.

Olsen, J., Pessenda, L. C. R., Ruzicka, J., Hansen, E. H. (1983). Combination of flow injection analysis with flame atomic absorption spectrophotometry. Determination of trace amounts of heavy metals in polluted seawater. *Analyst* 108:905–917.

Petersson, B. A. (1988). Amperometric assay of glucose and lactic acid by flow injection analysis. *Anal. Chim. Acta* 209:231–237.

Plamboeck, C., Westtoft, H. C., Petersen, S. A., Andersen, J. E. T. (2003). Filterless preconcentration by coprecipitation by formation of crystalline precipitate in the analysis of barium by FIA-FAES. *J. Anal. Atom. Spectrom.* 18(1):49–53.

Ranger, C. B. (1982). Flow injection analysis. A new approach to near-real-time process monitoring. *Autom. Stream Anal. Process Control* 1:39–67.

Ruzicka, J. (1998). Bioligand interaction assay by flow injection absorptiometry. *Analyst* 123(7):1617–1623.

Ruzicka, J. (2000). Lab-on-valve: universal microflow analyzer based on sequential and bead injection. *Analyst* 125(6):1053–1060.

Ruzicka, J. (2003). *Flow Injection (CD-ROM)*. 2nd ed. Seattle: FIAlab Instruments, Inc. fialab@flowinjection.com.

Ruzicka, J., Hansen, E. H. (1975). Flow injection analysis. Part I. A new concept of fast continuous flow analysis. *Anal. Chim. Acta* 78:145–157.

Ruzicka, J., Hansen, E. H. (1988a). *Flow Injection Analysis*. 2nd ed. New York, USA: John Wiley and Sons, Inc.

Ruzicka, J., Hansen, E. H. (1988b). Principles. In: *Flow Injection Analysis*. 2nd ed. New York: John Wiley and Sons, Inc., pp. 15–85.

Ruzicka, J., Hansen, E. H. (1988c). Techniques. In: *Flow Injection Analysis*. 2nd ed. New York: John Wiley and Sons, Inc., pp. 139–256.

Ruzicka, J., Marshall, G. D. (1990). Sequential injection: a new concept for chemical sensors, process analysis and laboratory assays. *Anal. Chim. Acta* 237(2):329–343.

Ruzicka, J., Scampavia, L. (1999). From flow injection to bead injection. *Anal. Chem.* 71(7):257A–263A.

Santelli, R. E., Gallego, M., Valcarcel, M. (1994). Preconcentration and atomic absorption determination of copper traces in waters by on-line adsorption–elution on an activated carbon minicolumn. *Talanta* 41(5):817–823.

Schrader, W., Portala, F., Weber, D., Fang, Z. L. (1989). Analysis of trace elements in strong salt-containing solutions by means of FI-flame AAS. In: Welz, B., ed. *5. Colloquium Atomspektrom. Spurenanalytik*. Germany: Perkin-Elmer, pp. 375–383 (in German).

Schulz, C. M., Scampavia, L., Ruzicka, J. (2002). Real time monitoring of lactate extrusion and glucose consumption of cultured cells using a lab-on-valve system. *Analyst* 127(12):1583–1588.

Shabani, M. B., Masuda, A. (1991). Sample introduction by on-line two stage solvent extraction and back extraction to eliminate matrix interference and to enhance sensitivity in the determination of rare earth elements with inductively coupled plasma mass spectrometry. *Anal. Chem.* 63:2099–2105.

Skoog, D. A., West, D. M., Holler, F. J. (1996). Automated photometric and spectrophotometric methods. In: *Fundamentals of Analytical Chemistry*. 7th ed. New York: Saunders College Publishing International Edition, pp. 587–600.

Sturgeon, R. E., Berman, S. S., Desaulniers, A., Russel, D. S. (1980). Pre-concentration of trace-metals from sea-water for determination by graphite-furnace atomic absorption spectrometry. *Talanta* 27(2):85–94.

Tao, G., Fang, Z. (1995). On-line flow injection solvent extraction for electrothermal atomic absorption spectrometry: determination of nickel in biological samples. *Spectrochim. Acta Part B* 50(14):1747–1755.

Toei, J. (1989). Potential of a modified solvent-extraction flow-injection analysis. *Talanta* 36:691–693.

Tsalev, D. L., Sperling, M., Welz, B. (2000). Flow-injection hydride generation atomic absorption spectrometric study of the automated on-line pre-reduction of arsenate, methylarsonate and dimethylarsinate and high-performance liquid chromatographic separation of their L-cysteine complexes. *Talanta* 51(6):1059–1068.

Volynsky, A. B., Spivakov, B. Y., Zolotov, Y. A. (1984). Solvent extraction-electrothermal atomic absorption analysis. *Talanta* 31(6):449–458.

Wang, J.-H., Hansen, E. H. (2000a). Flow injection on-line two-stage solvent extraction preconcentration coupled with ETAAS for determination of bismuth in biological and environmental samples. *Anal. Lett.* 33(13):2747–2766.

Wang, J.-H., Hansen, E. H. (2000b). Coupling on-line preconcentration by ion exchange with ETAAS. A novel flow injection approach based on the use of a renewable microcolumn as demonstrated for the determination of nickel in environmental and biological samples. *Anal. Chim. Acta* 424:223–232.

Wang, J.-H., Hansen, E. H. (2001a). Interfacing sequential injection on-line preconcentration using a renewable microcolumn incorporated in a "lab-on-valve" system with direct injection nebulization inductively coupled plasma mass spectrometry. *J. Anal. Atom. Spectrom.* 16(12):1349–1355.

Wang, J.-H., Hansen, E. H. (2001b). Coupling sequential injection on-line preconcentration by means of a renewable microcolumn with ion exchange beads with detection by ETAAS. Comparing the performance of eluting the loaded beads with transporting them directly into the graphite tube, as demonstrated for the determination of nickel in environmental and biological samples. *Anal. Chim. Acta* 435:331–342.

Wang, J.-H., Hansen, E. H. (2001c). Exploiting the combination of on-line ion-exchange preconcentration in a sequential injection lab-on-valve microsystem incorporating a renewablke column with electrothermal atomic absorption spectrometry as demonstrated for the determination of trace level amounts of bismuth in urine and river sediment. *Atom. Spectrosc.* 22(3):312–318.

Wang, J.-H., Hansen, E. H. (2002a). Sequential injection on-line matrix removal and trace metal preconcentration using a PTFE beads packed column as demonstrated for the determination of cadmium by electrothermal atomic absorption spectrometry. *J. Anal. Atom. Spectrom.* 17(3):248–252.

Wang, J.-H., Hansen, E. H. (2002b). Development of an automated sequential injection on-line solvent extraction–back extraction procedure as demonstrated for the determination of cadmium with detection by electrothermal atomic absorption spectrometry. *Anal. Chim. Acta* 456:283–292.

Wang, J.-H., Hansen, E. H. (2002c). FI/SI on-line solvent extraction/back extraction preconcentration coupled to direct injection nebulization inductively coupled plasma mass spectrometry for determination of copper and lead. *J. Anal. Atom. Spectrom.* 17(10):1284–1289.

Wang, J.-H., Hansen, E. H. (2003). Sequential injection lab-on-valve: the third generation of flow injection analysis. *Trends Anal. Chem.* 22(4):225–231.

Wang, J., Hansen, E. H., Miró, M. (2003). Sequential injection/bead injection lab-on-valve schemes for on-line solid phase extraction and preconcentration of ultra-trace levels of heavy metals with determination by electrothermal atomic absorption spectrometry and inductively coupled plasma mass spectrometry. *Anal. Chim. Acta.* 499(1–2):139–147.

Workman, J. Jr., Creasy, K. E., Doherty, S., Bond, L., Koch, M., Ullman, A., Veltkamp, D. J. (2001). Process analytical chemistry. *Anal. Chem.* 73(12):2705–2718.

Wu, C.-H., Ruzicka, J. (2001). Micro sequential injection: environmental monitoring of nitrogen and phosphate in water using a lab-on-valve system furnished with a microcolumn. *Analyst* 126(11):1947–1952.

Wu, C. H., Scampavia, L., Ruzicka, J., Zamost, B. (2001). Micro sequential injection: fermentation monitoring of ammonia, glycerol, glucose, and free iron using the novel lab-on-valve system. *Analyst* 126(3):291–297.

Wu, C.-H., Scampavia, L., Ruzicka, J. (2002). Micro sequential injection: anion separation using "lab-on-valve" coupled with capillary electrophoresis. *Analyst* 127(7):898–905.

Xu, Z.-R., Xu, S.-K., Fang, Z.-L. (2000a). A sequential injection on-line column preconcentration system for electrothermal AAS determination of thallium in geochemical samples. *Atom. Spectrosc.* 21(1):17–28.

Xu, Z.-R., Pan, H.-Y., Xu, S.-K., Fang, Z.-L. (2000b). A sequential injection on-line column preconcentration system for determination of cadmium by electrothermal atomic absorption spectrometry. *Spectrochim. Acta Part B* 55(3):213–219.

Yamamoto, M., Takeda, K., Kumamaru, T., Yasuda, M., Yokoyama, S., Yamamoto, Y. (1987). Membrane gas-liquid separator for flow injection hydride-generation atomic absorption spectrometry. *Anal. Chem.* 59:2446–2448.

Yan, X.-P., Yan, X. (2001). Flow injection on-line preconcentration and separation coupled with atomic (mass) spectrometry for trace element (speciation) analysis based on sorption of organo-metallic complexes in a knotted reactor. *Trends Anal. Chem.* 20(10):552–562.

Yan, X.-Y., Kerrich, R., Hendry, M. J. (1999). Flow injection on-line group preconcentration and separation of (ultra)trace rare earth elements in environmental and biological samples by precipitation using a knotted reactor as a filterless collector for inductively coupled plasma mass spectrometric determination. *J. Anal. Atom. Spectrom.* 14(2):215–221.

3

Inductively Coupled Plasma Optical Emission Spectrometry

TIEBANG WANG

Merck Research Laboratories, Rahway, NJ, USA

There are many sources of specific, detailed information on every aspect of the inductively coupled plasma optical emission spectrometry (ICP-OES) technique. This chapter is intended to serve as an introduction to the ICP-OES technique with a few references cited. The most essential and basic information on ICP-OES is written in simple and easy to understand language for those who have some familiarity with other spectrochemical techniques such as arc and spark emission spectrometry and atomic absorption spectrometry (AAS), but want to get into the field of ICP-OES technique. It is also written for novices in the fields of spectrochemical analysis and analytical chemistry in general. For further studies and references in this fascinating field of ICP-OES, the author recommends the book by Professor A. Montaser et al. (1992), entitled "Inductively Coupled Plasmas in Analytical Atomic Spectrometry".

I. INTRODUCTION

Whenever an analyst is faced with the question "which elements are present and at what levels?", the most probable techniques the analyst resorts to are going to be ones that are based upon atomic spectrometry. As the name implies, the atomic spectrometric techniques involve detecting, measuring, and analyzing electromagnetic radiation (i.e., light) that is either absorbed or emitted from the atoms or ions of the element(s) of interest in the sample. The quantitative information (what levels) is derived from the amount of electromagnetic radiation that is either absorbed or emitted by the atoms or ions of interest, whereas the qualitative information (which elements) is obtained from the wavelengths at which the electromagnetic radiation is absorbed or emitted by the atoms or ions of interest. Although it will be dealt with in a different chapter in this handbook, the author feels obliged to mention a third technique, atomic mass spectrometry (mostly, inductively coupled plasma mass spectrometry—ICP-MS), where the quantitative information is related to the amount of ions detected by a mass spectrometer, whereas the qualitative information is related to the mass to charge ratio (m/z) at which the ions are detected.

The earliest atomic spectrometric techniques were based on flames and electrical discharges. In 1752, Thomas Melville observed a bright yellow light emitted from a flame produced by burning a mixture of alcohol and sea salt. When the sea salt was removed from the alcohol, the yellow emission was not produced.

One of the first uses of sparks for elemental analysis was reported 24 years later by Alessandra Volta. He had discovered a way to produce an electrical discharge strong enough to create sparks. Later on, Volta was successful in identifying a few gases by the colors of the light produced when he applied spark to them.

Although his work was not recognized for several decades, W.H. Talbot, in 1826, was another to report a series of experiments in which he observed different coloring of a variety of salts when they were burned in the flames.

It was not until 1859, when Kirchhoff and Bunsen speculated that the sharp line spectra from flames when burning salts were originated from atoms and not from molecules, that the nature of emission spectra began to be understood. Much of their work was made possible by Bunsen's invention of a burner (which still carries his name today), which can produce a nearly transparent and nonluminescent flame. In the following years, Kirchhoff and Bunsen developed methods based on spectroscopy that led to the eventual discovery of four elements, Cs, Rb, Tl, and In. From then on, the presence of sharp spectral lines, unobserved earlier, was the proof that scientists required for the verification of the discovery of a new element. All these facts led to these two scientists being called the pioneers of spectrochemical analysis.

During the middle of the 20th century, quantitative arc and spark spectroscopies were the best tools available that a spectroscopist could use to determine trace concentrations of a wide range of elements. Although with limited linear dynamic range (LDR), this type of analysis is still being used today in some foundries in the USA, as long as the sample can be made into conductive electrodes and a library of matched standard materials can be located.

While arc and spark emission techniques were used for the determination of metals, flame emission spectroscopy was used extensively for the determination of alkalies and other easily excited elements. The reason why the atomic spectra emitted from flames are much simpler than those from arcs and sparks is that the flames are just not hot or energetic enough to produce emissions for more elements or produce more emissions from higher energy level lines for the same element. However, many successful commercial atomic spectroscopic instrument based on flames had appeared on the markets and many manufacturers are still making those instruments even today. As these instruments are cost-effective, inexpensive to maintain, and relatively easy to operate, for laboratories where the analysis of alkalis are needed, this flame-based technique still enjoys widespread popularity.

In the 1960s and 1970s, both flame and arc/spark OES declined in application and popularity since Walsh published his first paper on AAS in 1955. Many of the analyses that had been performed using OES were increasingly being replaced by AAS. Although about half the elements in the periodic table can be determined using OES, this technique can no longer compete with AAS. The reason behind this is that the absorption of light by ground-state atoms (not excited-state atoms) is the mode for detection; the necessity for higher temperature in order to populate excited state of atoms was basically eliminated. In addition, the instabilities as well as the complex spectral interferences associated with arc/spark OES were no longer a big issue with AAS. Flame atomic absorption spectrometry (FAAS) has been used primarily in the analysis of solutions for metals. For solid samples, this technique requires that the solid samples be transformed into liquid form beforehand. FAAS does offer the analyst decent precision (<0.5% RSD) with moderate limits of detection for most elements. Graphite furnace (or electrothermal atomization) atomic absorption spectrometry (GFAAS) on the other hand, affords even greater sensitivities and lower limits of detection at the expense of poorer precision and more severe and more complexed matrix interferences than its flame-based counterpart. Fortunately, advances such as Zeeman background correction technique and stabilized temperature platform (STP) have greatly reduced or eliminated these interferences that previously plagued the GFAAS.

Both FAAS and GFAAS techniques are enjoying their widespread popularity even today as they provide fairly economical and excellent means of trace elemental analysis notwithstanding the fact that both techniques normally measure one element at a time with only extremely narrow linear calibration range. Furthermore, for both FAAS and GFAAS techniques, as each element requires its own hollow cathode lamp (HCL) and different operation conditions and parameters, these techniques will never lend themselves readily to real simultaneous multielement analysis.

It has been 39 years since Stanley Greenfield of Birmingham, England, first published his report, in 1964, on the utilization of an atmospheric pressure ICP for elemental analysis using OES (Greenfield et al., 1964). In this landmark article, Greenfield identified the following as the advantages of an ICP emission source over flames and arcs/sparks: (a) high degree of stability; (b) its high temperature that has the ability to decompose the stable compounds formed and thus overcoming their depressive interference effects; and (c) capability of exciting multiple elements and offering greatly increased sensitivities at the same time. In addition, he also noted in this article that this plasma source was far simpler to operate than the arc/spark sources especially in solution analysis. Along with Greenfield, Velmer Fassel and his colleagues at Iowa State University are generally credited with the early refinements in the ICP that make it practical in reality for the analysis of nebulized solutions by OES. The ICP source has since been refined little by little as sources of noises were tracked down and eliminated or greatly reduced, and things like gas flows, torch and nebulizer designs, and all related electronics were optimized. By the early 1970s, the low limits of detection, relative freedom from interferences, as well as the extremely wide LDR obtainable with the ICP made it an emission source indisputably far superior to any of the emission sources previously used in analytical OES.

II. AN OVERVIEW

A. Principle of Atomic Emission Spectrometry

The detection and measurement of emission of an electromagnetic radiation or light can be more easily described once the nature of atomic and ionic spectra is fully understood. A simplified Bohr model of an atom is shown in Fig. 1. An atom consists of a nucleus surrounded by one or more electrons traveling around the nucleus in discrete orbitals. Every atom has a number of orbitals in which it is possible for electrons to revolve around the nucleus. Each of the orbitals has an energy level associated with it. The further away from the nucleus an orbital is, the higher the energy level. Whenever an electron is in the orbital closest to the nucleus, that is, lowest in energy, this electron is referred to as being in *ground state*. When additional energy is absorbed by the atom in the forms of electromagnetic radiation or a collision with another entity (another electron, atom, ion, or molecule), one or more occurrences may take place. One of the most probable is excitation, that is, the electron is promoted from its ground-state orbital into an orbital further away from the nucleus with a higher energy level. This atom is referred to as being in *excited state*. An atom is unstable in its excited state and will eventually decay back to a more stable state, in other words, the electron will decay back to an orbital closer to the nucleus with lower energy. This process is associated with an energy release either through a collision with another particle or through the emission of a particle of electromagnetic radiation, also known as photon.

If the energy absorbed by the atom is high enough, an electron may be knocked off from the atom, leaving the atom with a net positive charge. This process is called ionization, and the resulting particle is called an ion. An ion also has ground and excited states through which it can absorb or emit energy by the same excitation and decay processes as an atom.

Figure 2 illustrates simplified excitation, ionization, and emission processes. The horizontal lines of this diagram represent different energy levels of an atom. The vertical arrows represent energy transitions. The energy difference between the upper and lower energy levels of a radiative transition determines the wavelength of the radiation. Mathematically, the wavelength of the emitted radiation is given by the following equation:

$$\lambda = \frac{hc}{E_2 - E_1}$$

where λ is the wavelength of the radiation in nm, h is Planck's constant (6.626196×10^{-27} erg s), c is the velocity of light (2.9979×10^{10} cm/s), and E_2 and E_1 correspond to the energies in ergs of the upper and lower energy states of the electron, respectively.

Every element has its own characteristic set of energy levels and thus its own unique set of emission spectra. It is this very property of every element in the periodic table that makes atomic emission spectroscopy extremely useful for element-specific measurements.

Generally referred to as "light", the ultraviolet/visible region (160–800 nm) of the electromagnetic spectrum is

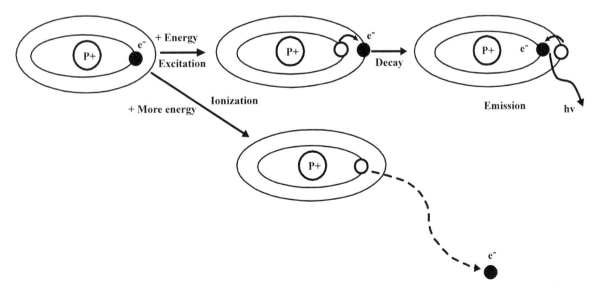

Figure 1 Simplified Bohr model of an atom.

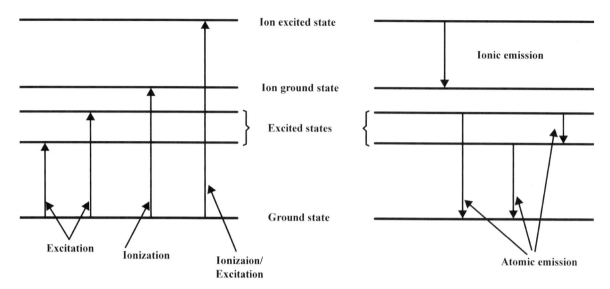

Figure 2 Simplified energy level diagrams.

the region most commonly used for analytical emission spectroscopy.

B. Atomic Spectroscopic Sources

Flames, furnaces, and electrical discharges are generally considered as the three major thermal sources used in analytical atomic spectrometry to transform sample molecules to free atoms. Flames and furnaces are mostly hot enough to thermally dissociate most types of sample molecules into free atoms with the exception of refractory compounds such as certain carbides and oxides. The breakdown of these molecules requires temperatures that far exceed the upper flame and furnace temperatures of typically 3000–4000 K. Because of the temperature limitation of the flames and furnaces, the free atoms produced in them are in their ground states; they are mostly suitable only for AAS. Alkali and alkali earth elements are mostly the only two exceptions as their lowest excited states are low enough that even the energy provided by flames and furnaces can excite these elements, thus making emission spectroscopic measurements of these elements possible.

Electrical discharges are the third type of thermal sources used in analytical OES. Before ICP source came into being, DC arcs and AC sparks dominated OES. These electrical discharges are formed by applying high electrical currents or potentials across electrodes in an inert gas atmosphere. The temperatures achievable this way are generally higher than those of traditional flame and furnace systems.

Only till more recently, other types of discharges, that is, plasmas, have been applied and used as atomization and excitation sources for OES. Technically speaking, a plasma is any form of matter (mostly gases) that is appreciably ionized, that is to say, a matter containing fraction ($>1\%$) of electrons and positively charged ions together with neutral species, radicals, and molecules. The electrical plasmas used for analytical OES are highly energetic, ionized gases. They are normally created with inert gases, such as argon or helium. These plasma discharges are considerably hotter than traditional flames and furnaces, thus are used not only to breakdown almost any type of sample molecules but also to excite and/or ionize the free atoms for atomic and ionic emission. Currently, the state-of-the-art plasma is the argon supported ICP. Other plasmas include direct current plasma (DCP) and microwave induced plasma (MIP). Only the ICP will be discussed in this chapter.

C. Atomic Spectroscopic Techniques and Instruments

Although only OES is the subject of this chapter, for the sake of completeness, all four atomic spectrometric techniques will be briefly mentioned here.

In all atomic spectroscopic techniques, the sample is decomposed by intense heat into hot gases consisting of free atoms and ions of the element of interest.

In AAS, light of wavelength characteristic of the element of interest generated by an HCL is shone through the free atomic cloud created by the thermal source. A portion of the light is then absorbed by these atoms of that element. The concentration of the element in the sample is based on the amount of light that is absorbed. If the thermal source is a flame, this technique is called flame atomic absorption spectrometry (FAAS); whereas, if the thermal source is an electrically heated graphite furnace, it is then called graphite furnace atomic absorption spectrometry (GFAAS).

In OES, the sample is exposed to a thermal source that is hot enough to induce not only dissociation into atoms (atomization) but also significant amount of excitation and ionization of the sample atoms. As mentioned previously, once the atoms or ions are in their excited states, eventually they have to decay back to lower energy states through radiative energy transitions by emitting light. The specific wavelengths at which the optical emissions are measured are used to determine the identities of the elements, and the intensities of the lights emitted are used to determine the concentrations of the elements present in the sample.

In atomic fluorescence spectrometry (AFS), free atoms are created by a thermal source as mentioned earlier, and a light source is used to excite atoms of the element of interest through radiative absorption transitions. When these selectively excited atoms decay back to lower energy levels through radiative transition (i.e., light), their emission is measured to determine concentration, similar to OES. As the excitation in AFS is selective, this technique has much less spectral interference than OES. However, as the number of excitation sources that can be used at one time is limited by the instrument, it is very difficult to analyze a large number of elements in a single run by AFS.

In atomic mass spectrometry (mostly, ICP-MS), instead of measuring absorption, emission, or fluorescence created from a thermal source, such as flame or plasma, the number of positively charged ions from the elements in the sample are measured. Very similar to a monochromator or a polychromator that separates lights according to their wavelengths, a mass spectrometer (quadrupole or magnetic sector) separates ions according to their mass to charge ratio (m/z).

III. INDUCTIVELY COUPLED PLASMA

A. The Discharge

The ICP discharge used today is not that different from that described by Fassel and his colleagues in the early 1970s. As shown in Fig. 3, argon is flown through a torch consisting of three concentric tubes made of quartz or some other material. A copper coil called load coil surrounds the top portion of the torch and is connected to a radio frequency (RF) generator. When argon is flowing tangentially through all the three tubes of the torch, a spark, normally created by Tesla coil, produces "seed electrons" which begin to ionize argon: $Ar \rightarrow Ar^+ + e^-$. The RF power applied normally at 0.5–2.0 kW and 27 or 40 MHz in the load coil causes RF electric and magnetic fields to be set up and accelerate these electrons. The way of feeding energy through coil to the electrons is known as inductive coupling in physics. These high-energy electrons in turn induce further ionization of argon by colliding with

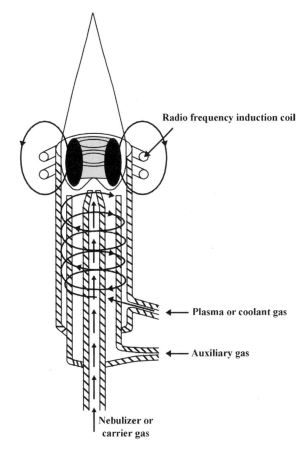

Figure 3 Inductively coupled plasma.

other argon atoms. This collisional ionization of the argon gas continues in a chain reaction, breaking down the gas into a plasma consisting of argon atoms, electrons, and argon ions, now forming the so called ICP discharge. The high temperature of this discharge ranging from 6000 to 10,000 K results from Ohmic heating (resistance to the moving of charged particles).

The formed ICP discharge is an intense, brilliant white, teardrop-shaped "fireball". A cross-sectional representation of the discharge along with the nomenclature for different regions of the plasma is given in Fig. 4. The bottom discharge is "doughnut-shaped", as the nebulizer or carrier gas flow punches a hole through it. The region is called "induction zone" (IZ) because this is the area where the inductive energy from the RF generator is transferred from the load coil to the plasma discharge. The preheating zone (PHZ) is the area where the solvent is removed from the sample aerosols (desolvation) leaving the sample as microscopic salt particles. It is also in this PHZ that the salt particles are decomposed into individual gas phase molecules (vaporization), which are subsequently broken down into single atoms (atomization).

Once the sample aerosols have been desolvated, vaporized, and atomized, the only thing left to be done by the hot

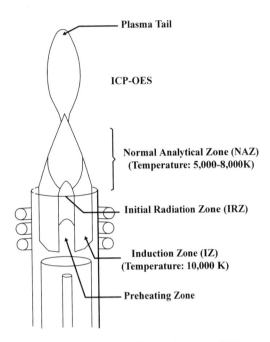

Figure 4 Cross-section and zones in ICP.

plasma discharge is to excite and ionize/excite in order for an atom or ion to emit its characteristic light. These ionization and excitation processes take place predominantly in the initial radiation zone (IRZ) and the normal analytical zone (NAZ), as shown in Fig. 4.

B. Emission Detection and Processing

Both the qualitative and quantitative information about a sample is obtained from the lights or radiations emitted by the excited atoms and ions. Because the excited species in the plasma emit light at more than one wavelength, the emission from the plasma discharge is referred to as polychromatic. The polychromatic radiation has to be sorted out into individual wavelengths so that the emission from each and every atom and ion can be identified and quantified without interferences or overlaps from emission at adjacent wavelengths. The separation of light according to their wavelengths is generally achieved using a monochromator, which is used to measure light one wavelength at a time, or a polychromator, which is used to measure light at more than one wavelength simultaneously. After the separation of the light, the actual detection of the light is done using a photosensitive device such as a photomultiplier tube (PMT) in the old days or more state-of-the-art devices such as photodiode array (PDA), charge coupled device (CCD), and charge injection device (CID) nowadays.

Qualitative information about a sample (i.e., what elements are present) is obtained by identifying the presence of emission at the wavelengths characteristic of the elements of interest. As the energetic ICP source generates large number of emission lines for each element, spectral interferences from other elements may make this task more complicated. However, the utilization of multiple lines of characteristic of each single element can ensure safe and accurate identification of the element of interest.

Quantitative information about a sample (i.e., the concentration of the elements) can be extracted from the intensities of emission lines characteristic of the elements of interest. This is accomplished by using plots of emission intensity vs. concentration, called calibration curves. Calibration curves are established by analyzing solutions with known concentrations (standard solutions) and obtaining their emission intensities for each element. These intensities can then be plotted against the concentrations of standards to establish a calibration curve for each element. When the emission intensity from an element of an unknown sample is measured, the intensity is compared to that particular element's calibration curve to determine the concentration corresponding to that intensity. Nowadays, all these can be accomplished by sophisticated computer software, and thus it is no longer necessary for the spectroscopist to manually construct these curves for quantitation of the elements in the sample.

C. Analytical Performance and General Characteristics

One of the major advantages of the ICP-OES technique is its wide element coverage. Almost all elements in the periodic table can be determined by this technique. Although most of these elements can be determined at low levels, some elements cannot usually be determined at trace levels by ICP-OES. The first group includes those elements that naturally entrained into the discharge from sources other than the samples themselves. An obvious example is the impossibility to determine argon in an argon ICP. A similar limitation would be encountered if C and H were to be determined if water or an organic solvent is used as the solvent. Entrainment of air into the plasma discharge makes H, N, O, and C determination extremely difficult, if possible at all. Halogen elements cannot generally be determined at trace levels as well because of their high excitation energies. The remaining elements that cannot be determined at trace levels or cannot be determined at all are the man-made elements, which are either too radioactive or too short-lived that gamma ray spectrometry is a better technique.

The second advantage of the ICP-OES technique is its extremely wide linear calibration range, typically four to nine orders of magnitudes (i.e., the upper limit of an element can be 10^4–10^9 times the detection limit of a particular emission line). In ICP-OES, the range of concentrations from the detection limit to its upper limit for a particular emission line is called its linear dynamic range (LDR).

Unlike AAS as well as arc and spark techniques that typically have LDRs of only one to two orders of magnitudes, an ICP spectroscopist can technically use two solutions, one blank and one high standard, to establish a calibration curve if the linear range of the emission line is known beforehand. Wide LDR also makes frequent dilution unnecessary in routine sample analysis, whereas techniques with narrower LDRs tend to require more sample dilution to keep the analyte concentrations within the linear range of their calibration curves.

In addition to wide element coverage and large LDRs, the third advantage of the ICP-OES is that many elements can be determined easily in the same analytical run (i.e., simultaneously). This characteristic of the hot plasma discharge arises from the fact that all of the emission signals needed to extract both quantitative and qualitative information are emitted from the plasma at the same time. Equipped with a polychromator, this multielement nature of the ICP can be enhanced by the simultaneous capability of the modern electronics and computer technology.

When the ICP-OES was first introduced as a technique for elemental analysis, it was thought by some optimistic experts in this field that this technique was completely free from interferences. Realistically speaking, the ICP-OES technique is not free from interferences; it does suffer the fewest and least severe interferences experienced by any of the commonly used analytical atomic spectrometry techniques. The hot plasma itself largely overcomes chemical interferences. Although spectral interferences cause the most inaccuracies in ICP-OES, the flexibility to choose from many available emission lines combined with sophisticated spectral interferences correction software make interference-free measurement of almost all elements possible. Matrix interference can be overcome by sample dilution, matrix matching, and the use of internal standards, as well as the use of method of standard additions.

IV. ICP-OES INSTRUMENTATION

A. Nebulizers

The major components and layout of a typical ICP-OES instrument is shown in Fig. 5.

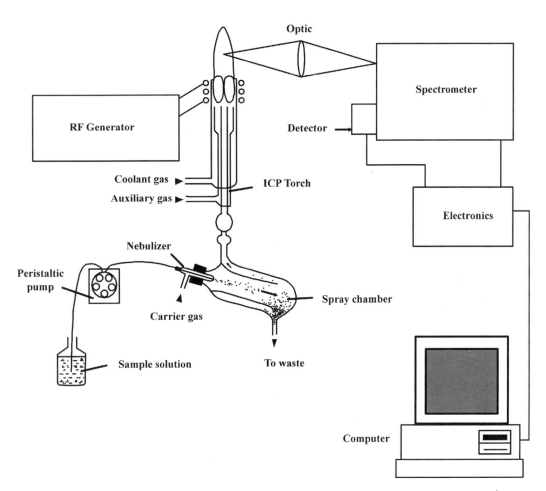

Figure 5 The major components and layout of a typical ICP-OES instrument.

In an ICP-OES, all samples are normally converted to liquid form first and then pumped by a peristaltic pump into the instrument. The liquid is converted into an aerosol or mist by a device named nebulizer. The nebulization process is one of the critical steps in ICP-OES. A perfect nebulizer should be able to convert all liquid samples into aerosol such that the plasma discharge could reproducibly desolvate, vaporize, atomize, ionize, and excite.

There are two types of nebulizers that have been successfully used in ICP-OES: pneumatic nebulizer and ultrasonic nebulizer (USN). Most commercially available ones are of pneumatic type, which uses high-speed gas flow to create aerosols. Another type is the USN, which uses an oscillating piezoelectronic transducer to make aerosols. A few of the commonly used pneumatic nebulizers and the USN will be briefly discussed here.

The most popular pneumatic nebulizer is the concentric nebulizer normally made of glass or quartz. A typical concentric nebulizer used for ICP-OES is shown in Fig. 6. In this type of nebulizer, the sample solution is sucked into a capillary tube by a low-pressure region created by fast flowing argon past the end of the capillary. The high-speed argon gas combined with the low-pressure breaks up the liquid sample into aerosol. Concentric pneumatic nebulizer is known for its excellent sensitivity and stability. However, its small capillary orifice is vulnerable to clogging, frequently by solutions containing as little as 0.1% total dissolved solids (TDS). Some new designs of concentric nebulizers have improved in this aspect, with some being able to nebulize sample solutions containing as high as 20% dissolved solids without clogging.

The second kind of pneumatic nebulizer is the cross-flow nebulizer, shown in Fig. 7. As the name implies, a high-speed stream of argon gas is directed perpendicular to the tip of the sample capillary tube (in contrast to the concentric nebulizers where the high-speed gas is parallel to the capillary). Again, the sample solution is either drawn up through the capillary tube by the low-pressure region created by the fast flowing argon or pumped into the capillary tube by a peristaltic pump. The impact of the liquid sample with the high-pressure argon gas produces the aerosol required by ICP-OES. Generally speaking, cross-flow nebulizers are not as efficient as concentric nebulizers in creating fine droplets or aerosols. However, this type of nebulizer is relatively more resistant to clogging because of the larger capillary tube used, and can be made of materials other than glass or quartz so that they are more rugged and corrosion-resistant as well compared with concentric nebulizers.

Not as popular as the first two pneumatic nebulizers, the third kind is the Babington nebulizer. It works by letting the sample liquid to flow over a smooth surface with a small orifice in it. High-speed argon gas gushing from the hole shears the sheet of sample liquid into aerosol. This kind of nebulizer is least prone to clogging because of its use of much larger sample path or hole so that sample with high TDS and sample of high viscosity can be nebulized by this type of nebulizer. It is worthwhile to mention that the variation of the Babington nebulizer is the V-groove nebulizer. In the V-groove nebulizer, instead of flowing down a smooth surface, the sample flows down a groove with a small hole in the middle for the high-speed argon gas. This nebulizer is also being used for nebulization of sample solutions containing high salt content such as seawater and urine samples.

The last type of nebulizer is the USN, in which the liquid sample is pumped into an oscillating piezoelectric transducer. The rapid oscillations of the transducer break the sample liquid into extremely fine aerosols. The efficiency of a USN is typically 10–20% or more compared with 1–2% achieved by a concentric or cross-flow nebulizer. As more sample aerosols are reaching the ICP discharge, ICP-OES equipped with a USN normally has

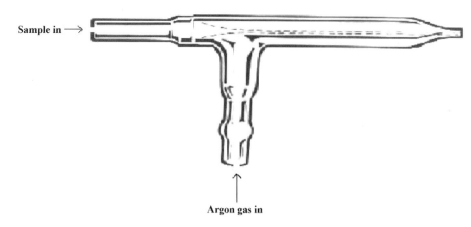

Figure 6 Concentric nebulizer for ICP-OES.

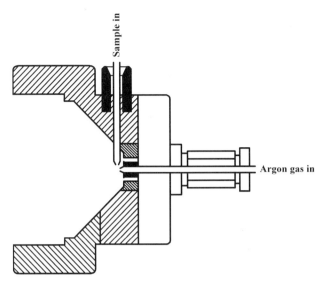

Figure 7 Cross-flow nebulizer.

sensitivities of about 10–50 times better for all elements compared with one fitted with a concentric or cross-flow nebulizer.

B. Spray Chambers

After the sample aerosol is produced by the nebulizer, it has to be carried to the torch so that it can be fed to the plasma discharge. As only very fine droplets in the aerosol can be effectively desolvated, vaporized, atomized, ionized, and excited in the plasma, large droplets in the aerosol have to be removed. To this end, a device called spray chamber is placed between the nebulizer and the torch. A typical spray chamber is shown in Fig. 8. Another purpose of the spray chamber is to minimize signal pulsation caused by the peristaltic pump.

Generally speaking, a spray chamber for ICP-OES is designed to allow droplets with diameters of roughly 10 μm or smaller to pass to the plasma. With average spray chambers and nebulizers, only about 1–5% of the sample can reach the plasma, with the remaining of the sample sent as waste to the drain. Spray chambers can be made of quartz; they can also be made of other corrosion-resistant materials so that samples containing high-concentration acids, particularly hydrofluoric acid, can be nebulized into the plasma.

C. Alternative Sample Introduction Systems

ICP discharges possess excellent characteristics that make them exceptional sources for desolvation, vaporization, atomization, ionization, and excitation. One limitation to their universal analytical application is that samples generally need to be in liquid form for analysis. The capability of introducing samples in gas or solid form would extend the universality of the technique. Considering many samples occur naturally in solid form, major efforts are required just to transfer them into liquid form before spectroscopists can introduce these samples through nebulizer–spray chamber to the plasma for ICP-OES analysis.

Several sample introduction systems have been developed to replace nebulizers and spray chambers for ICP-OES.

The most popular alternative method is hydride generation. In this technique, the samples still need to be transformed into liquid form. After being dissolved in dilute acid, the sample is mixed with a reducing agent, usually

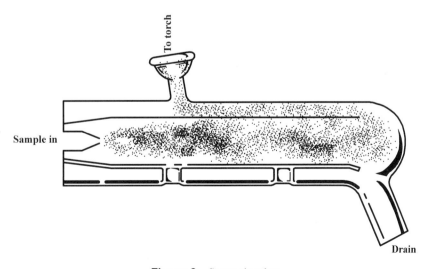

Figure 8 Spray chamber.

a solution of sodium borohydride in dilute sodium hydroxide. The reaction of sodium borohydride with the acid produces volatile hydride with a limited number of elements such as As, Se, Sb, Bi, Sn, and so on. These compounds in gaseous form are then separated from the rest of the reaction mixture and carried to plasma by a stream of argon gas.

Dramatic improvements in sensitivity and detection limit have been achieved by a factor of as high as 1000 for these limited elements by the hydride generation technique. Furthermore, various interferences have been greatly reduced as matrix components from the samples are not introduced together with the analytes in contrast to sample introduction system with nebulizers and spray chambers.

The second one is the electrothermal vaporization (ETV) or graphite furnace, basically the same set-up used in GFAAS with some modifications. ETV is generally used in research facilities to vaporize a small portion of a liquid or solid sample, and the resulting vapor is carried to the plasma discharge by a stream of argon. Although potentially higher sensitivity is possible, other aspects of this device, such as the utilization of compromised operation conditions (as more than one element is generally involved) and more severe matrix effects, have limited the number of applications of this technique to ICP-OES for sample introduction. In addition, these devices introduce sample in a noncontinuous nature, an ICP-OES spectrometer with transient signal processing capability is mandatory. In ICP-OES, ETV has not enjoyed much commercial success, whereas its counterpart ICP-MS has realized more application with ETV as an alternative sample introduction system.

The third one is the laser ablation system, which is a new and versatile solid sample introduction technique for ICP-OES. In this technique, sufficient energy in the form of a focused laser beam is shone onto a sample to vaporize it. The vaporized sample is then carried to the plasma discharge by a stream of argon. This technique has the capability of sampling a wide range of diverse materials ranging from conducting and nonconducting, inorganic and organic, to solid and powdered materials. In addition to bulk analysis, the focused laser beam permits the sampling of small areas, thus making it possible to perform *in situ* microanalysis and spatially resolved studies. The major drawback of this technique is that a perfectly matrix-matched standard is often impossible to obtain, and this makes fully quantitative determination of a lot of samples rather difficult.

There has been many other alternative sample introduction systems for ICP-OES applied or researched during the last 40 years. A few more will be mentioned briefly in what follows.

Solid samples have also been introduced into the plasma using arc and spark sources such as those first used in the early days of OES. Some researchers have inserted solid or dried liquid samples directly into the center of plasma discharge using a special computer controlled mechanical device. This technique is called direct sample insertion. In this technique, a carbon electrode with precut sample cavity is filled with solid sample or dried liquid sample and inserted into the plasma. The sample is vaporized into the plasma followed by atomization, ionization, and excitation. This technique has never been commercialized.

The study of sample introduction techniques has been an active area for ICP-related research and will continue to be so in the foreseeable future.

D. Optics and the Spectrometer

As mentioned briefly earlier, the emission is gathered for OES measurement. The radiation is usually collected by a focusing optic such as a convex lens or a concave mirror. This optic focuses the image of the plasma discharge onto the entrance slit of the wavelength-dispersing device. The capability of sorting out the components of the radiation or discriminating spectral radiation from the plasma is another critical step in OES. This section describes a few commonly used optical arrangements for wavelength discrimination.

The physical dispersion of the different wavelengths of the radiation by a diffraction grating is the most common practice of spectroscopists. Up to the early 1990s, most commercial ICP-OES spectrometers were built using grating-based dispersive devices.

A reflection diffraction grating is simply a mirror made of glass or steel with closely spaced parallel lines or grooves ruled or etched into its surface as shown in Fig. 9. Most dispersion gratings used in ICP-OES instrumentation have a groove density of about 600–4200 lines/mm. When the light strikes the grating at an incident angle of α, to the grating normal, it is then diffracted into three component beams at angles β, θ, and ω. The angle of diffraction is dependent on the wavelength and the groove density of the grating. As a general rule, the longer the wavelength and the higher the groove density, the higher the angle of diffraction will be. In an ICP instrument, the grating is incorporated in the spectrometer. The function of the spectrometer is to make the plasma radiation into a well-defined beam, disperse it according to wavelength with the grating, and focus the dispersed component light onto an exit plane or circle. In other words, the spectrometer receives white light or polychromatic radiation composed of many wavelengths and sorts it into its component or monochromatic light. Exit slits are used to allow the desired wavelengths to pass to the detector while blocking undesirable wavelengths.

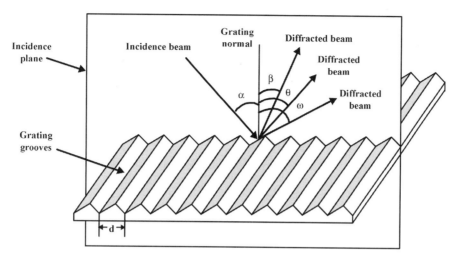

Figure 9 Light dispersion by reflection diffraction grating.

There are two ways to perform multielement analysis using conventional dispersive optics. The first is the monochromator which uses only one exit slit and one detector. Monochromators are used in multielement analysis by scanning rapidly from one wavelength to another. This is achieved by either changing the angle of the diffraction grating or moving the detector in the exit plane while leaving the grating in a fixed position. When the monochromator can scan lines with sufficient speed, fast sequential multielement analysis is possible. The most popular monochromator configuration is the Czerny–Turner type shown in Fig. 10. A polychromator, on the other hand, uses multiple exit slits (all located on the periphery of what is known as the Rowland circle) and detectors in the same spectrometer. Each exit slit in a polychromator is aligned to a specific atomic or ionic emission line for a specific element to allow simultaneous multielement analysis. The most popular design is the Paschen–Runge mount shown in Fig. 11.

In recent years, Echelle grating-based ICP-OES spectrometers have become more and more popular. It has been shown that certain advantages can be achieved by combining the characteristics of two dispersing systems such as a diffraction grating and an optical prism—the very first optical dispersing device used by spectrochemical pioneers.

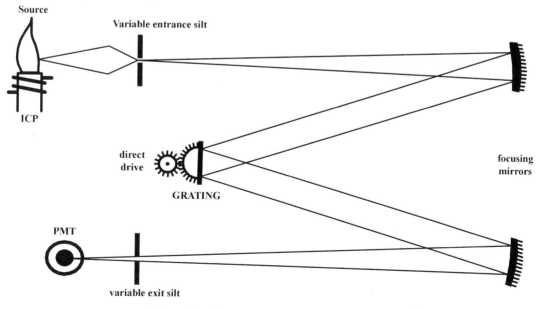

Figure 10 Czerny–Turner type monochromator.

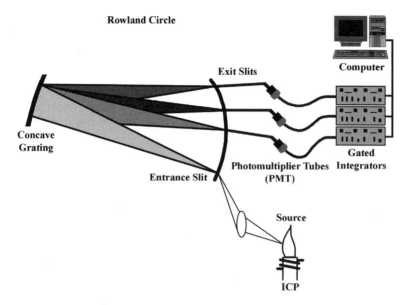

Figure 11 Paschen–Runge polychromator.

Monochromator- and polychromator-based ICP-OES spectrometers have their pros and cons. With polychromators, each emission line can be recorded during the entire sample introduction period and, technically, more samples and more elements can be analyzed in a shorter period of time. Therefore, polychromator-based instruments certainly have a higher sample throughput rate. As a conventional polychromator (also known as direct reader) uses one exit slit and one PMT for each spectral line in the system, the number of channels or lines available is generally limited to about 64 because of space requirements. Most commercial polychromators are only fitted with 20–30 spectral lines. The most important advantage of monochromator-based system is its flexibility. That is to say, the ability to access, at any time, any particular wavelength within the range of monochromator. This allows for the determination of any element whose emission line can be measured by the technique. In addition, because of their scanning capability, monochromator-based instruments are much better suited for applications that require complex background corrections often indispensable for ICP-OES analysis. Furthermore, scanning the region adjacent to the line of interest or simultaneous measurement of the immediate vicinity of the line of interest is extremely helpful in validating the analytical result. However, as elements are sequentially measured, monochromator-based ICP instruments require larger amounts of sample and have a much lower sample throughput rate than polychromator-based systems.

Echelle grating-based ICP-OES instruments have become very popular with the manufacturers and the ICP spectroscopists in recent years. It has been shown that some advantages may be taken of if the characteristics of two dispersing devices, such as that of a diffraction grating and a prism are combined. The two optical devices are placed perpendicular to each other in this configuration. One of the dispersing devices is the Echelle grating which is a coarsely ruled grating, with rulings ranging between 8 and 300 grooves/mm (compared with diffraction grating with rulings ranging from 600 to 4200 grooves/mm). The function of the Echelle grating is to separate the polychromatic radiation from the ICP by wavelengths and produce multiple, overlapping spectral orders. The second optical device, the prism, sorts or cross-disperses the overlapping orders to produce a 2D array of wavelengths on the focal plane of the spectrometer. In the 2D array, the wavelengths are in one direction, and the spectral orders are in the other. A typical optical arrangement for an Echelle grating-based ICP-OES spectrometer is illustrated in Fig. 12.

Echelle grating-based spectrometers present some unique advantages over the conventional grating-based spectrometers. They produce very high quantum efficiency in each of the spectral orders for solid-state detectors (such as PDA, CCD, and CID), thus resulting in potentially superior sensitivity. They also offer excellent spectral resolution as they are generally used in the higher spectral orders (spectral resolution is realized with increasing order). This higher spectral resolution resulting from the use of higher orders also makes a more compact instrument with smaller footprint possible.

There are many other means to disperse electromagnetic radiations such as the use of optical filters, interference filters, tunable filters, and Fourier transform

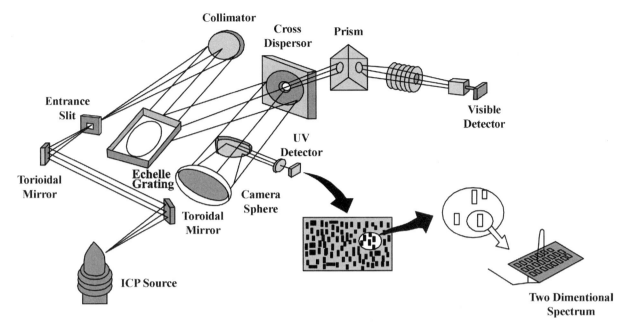

Figure 12 An Echelle grating-based ICP-OES spectrometer.

spectrometers. These have not been used extensively with ICP-OES, and have been discussed in the second edition of this handbook. Interested readers are advised to refer to the second edition of this handbook.

E. Emission Detectors

After the ICP radiation has been dispersed into its component wavelengths by the spectrometer, various detectors and related electronics are used to measure the intensity of the radiation. The most commonly used detectors are discussed in what follows.

Early ICP-OES instruments, almost all, used PMTs. The PMT is basically a vacuum tube that contains photosensitive materials such as a photocathode and a collection anode. Several electrodes, called dynodes, separate the cathode and anode, and provide electron multiplication or gain as each dynode is biased at a more positive potential. When struck by light, photoelectrons are released from the cathode, and these released electrons are accelerated toward the dynode which releases two to five secondary electrons for every single electron that strikes its surface. Those electrons then strike the second dynode and release another two to five electrons for every electron that strikes the surface of the second dynode. This process continues with each consecutive dynode causing a multiplicative effect along the way. A typical PMT contains 9–16 dynode stages. The final step is the collection of the secondary electrons from the last dynode by the anode. As many as 10^6 secondary electrons can be produced as the result of a single photon striking the photocathode of a nine-dynode PMT. The corresponding electrical current measured is a relative measure of the light intensity (hence, concentration) of the radiation reaching the PMT. Figure 13 shows schematically how a PMT amplifies the signal produced by a single photon striking the photocathode. The major advantages of the PMT over other devices are that it covers relatively wide spectral range with good sensitivity, and its range of response can be extended to over nine orders of magnitude in light intensity.

In the 1960s, solid-state devices were introduced into the electronics industry. These devices, such as transistors and diodes, were based on the properties of silicon. As their use expanded to the digital electronics industry in the form of integrated circuits (ICs), the cost of these devices as well as the cost of systems, such as digital computer using these devices were greatly reduced. It was also discovered that silicon-based sensors responded to light and were immediately integrated into linear and 2D arrays called solid-state detectors.

Consequently, three advanced solid-state detectors with superior sensitivity and resolution for spectroscopic applications have been developed. These include the PDA, the CID, and the CCD.

The detectors are typically grouped into two classes: photoconductive detectors that rely on conductivity and photodiode or junction detectors that produce a potential difference across a junction between p- and n-type semiconductors. Photodiode or junction detectors can operate in either photoconductive or photovoltaic mode. In the photovoltaic mode, the electron–hole pairs are

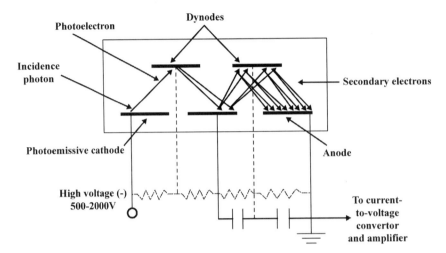

Figure 13 A PMT layout.

produced by the interaction of an incident light photon and photodiode near the p–n junction. The electrons and holes migrate to opposite sides of the junction to create a potential difference. In the photoconductive mode, a reverse bias is applied across the junction. The conductance of the junction is increased with the formation of electron–hole pairs. The resulting voltage and conductance are proportional to the intensity of the incident light.

A linear PDA consists of a series of individual silicon PDA detectors fabricated in IC form. These detectors operate in the photoconductive mode and, during operation, charge each individual capacitor on the chip. The array detector is a shift register consisting of a series of closely spaced capacitors on the chip. The capacitors store and transfer signals that are produced either optically or electronically. The transfer of stored charges in the PDA is analogous to a "bucket brigade" whereby the signals are transferred to adjacent capacitors and read sequentially at the end of the array. The measurement of optical signals and transfer of stored charges are controlled by clock circuitry. The integration time is the time between each transfer into the register. For low light level measurements, PDA detectors are available with intensifiers to enhance the light and chillers to cool the devices to reduce the dark current or electronic noise.

A CCD detector consists of a 2D array of silicon photodiode constructed with IC with over several hundreds of thousands of individual detector elements. The operation of a CCD is identical to a linear PDA detector, except that a multichannel analog to digital (A/D) convertor is required for digitizing the signal. The charge accumulated on the detector element must be read sequentially and, in the process of reading the charge, it is, in the mean time, destroyed.

A CID detector is similar to a CCD detector, except for the manner in which the individual detector is accessed and controlled. CID technology allows individual detector element to be accessed separately and the exposure time of each in the array to be controlled. In general, each individual detector element in the array may be randomly integrated to determine the amount of charge that has been accumulated during the measured time to which the device has been exposed to light. With the availability of high-speed computers, each detector element may be examined even during the integration time to determine the accumulated charge. The process of examining the charge does not destroy the charge and is known as nondestructive read-out. However, this device has an inherently higher dark current than PDA and CCD type detectors, for which effective cooling is necessary.

Later on, a new type of CCD detector was introduced—a segmented-array charge coupled device detector (SCD)—normally for Echelle ICP instrument. Instead of using a huge CCD with hundreds of thousands of contiguous detector elements, the SCD has been designed with individual collection of smaller subarrays of 20–80 detector elements. These subarrays correspond to the 200 or more of the most important ICP spectral lines of the 70 elements observed in ICP spectrometry. Most commercial CIDs and CCDs have poor sensitivity below 350 nm because of photon absorption by electrodes embedded on the surface of the devices. As the detector elements of the individual subarrays of the SCD have no embedded electrodes, the SCD has much better response to light from 160 to 782 nm.

As only an array type detector can take advantage of the 2D capability of Echelle grating based spectrometers, the ever-popular commercial ICP instruments with Echelle gratings have been always fitted with solid-state detectors such as CCDs or CIDs. With proper cooling, the detector noise of these solid-state detectors can be reduced dramatically relative to the conventional PMT detectors.

F. Radially Viewed ICP

Since the time that ICP source was used for spectrochemical analysis, efforts to improve the sensitivity and detection limit have been continuing by various groups around the world.

As early as in the mid-1970s, spectroscopists discovered that an axially viewed or "end-on" plasma discharge (as depicted in Fig. 14) presented better sensitivity and lower detection limit than the standard radially viewed or "side-on" plasma discharge (as depicted in Fig. 15). The reason for this is that by viewing the plasma in the axial direction, a longer pathlength or resident time is realized, which results in higher intensity emission. This improves sensitivity, and translates into a 5–10-fold improvement in detection limit.

Along with the sensitivity enhancement achieved, matrix interferences as well as self-absorption effects are becoming prevalent with the axially viewed plasma. The self-absorption effects were caused by including the much cooler tail plume part of the plasma discharge in the measurement. The self-absorption effects lead to much reduced LDR of the plasma. These "side effects" of the axially viewed plasma delayed the development of commercial axial plasma instrumentation by about 15 years.

A few years later after the first experiment with the axial ICP, it was found that applying a shear gas could minimize these detrimental side effects. The shear gas is a stream of air or nitrogen that flows perpendicular to the direction of the plasma discharge, which basically "pushes" the tip or the tail plume (the cooler part) of the plasma out of the optical path.

Another benefit that comes with the use of an axial ICP instrument is the laboratory productivity. Because of improved detection limits for many elements, some elements that had to be analyzed by GFAAS or hydride generation techniques to meet regulation requirements can now be analyzed by the axial ICP together with other elements. Normally, a separate sample preparation procedure is required for separate ICP, GFAAS, and hydride generation type analysis; with the axial ICP, a single sample preparation is all that is needed. This dramatically reduces total analysis time and minimizes acid usage and waste generation. In addition, as all results are generated from a single instrument, data interpretation and report generation are also simplified.

However, for real-world samples with complex matrices, there might be occasions where an axial ICP may not be the best answer for the application at hand. Many manufacturers have come up with an instrument that combines the axial and radial configurations into one single unit. With this system, the spectroscopist has the flexibility to choose and optimize the appropriate configuration for the type of samples and elements without the expense of two separate ICP units.

V. APPLICATIONS OF ICP-OES

The ICP source has had the most important impact on the field of atomic spectroscopy over the last several decades.

The versatility of the ICP-OES makes it an almost ideal analytical technique for a wide variety of applications. The versatility can be attributed not only to the large number of elements this technique can cover at trace levels at fast speed but also to the tremendous amounts of sample types that can be analyzed using this technique.

In this section, some representative applications of ICP-OES in most popular areas will be briefly described. Although non-exhaustive, the reader will get some ideas of the types of analyses where this technique has been applied successfully.

A. Geoanalysis

Multielement analysis capability coupled with high sensitivity has made the ICP-OES a versatile tool for geological

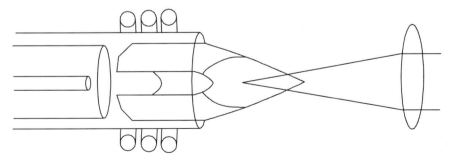

Figure 14 Axially viewed or "end-on" plasma.

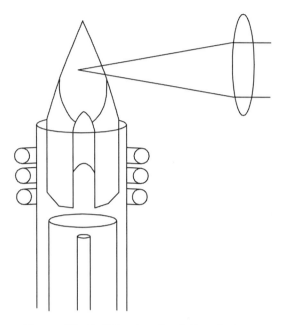

Figure 15 Radially viewed or "side-on" plasma.

applications involving the determinations of major, minor, and trace compositions of various rocks, soils, sediments, and other relevant materials. Majority of this kind of work using ICP-OES is for prospecting purposes, although it has also been used to determine the origins of rocks and other geological materials.

In addition to the acid dissolution methods commonly used with many geological samples, fusion methods using fluxes such as lithium metaborate, sodium or potassium hydroxide, and so on for high level silica containing materials are particularly important.

Typical applications of ICP-OES for analyses of geological materials include the determination of uranium in ore grade material (yellow cakes), the determination of minor and trace elements in silicate rocks, alternative procedure to the classical fire assay method for the determination of Pt and Pd in sulfide ores, and the determination of rare earth elements in rock formations.

B. Metallurgical Samples

ICP-OES is used widely together with X-ray fluorescence and other techniques for the determination of major, minor, and trace elements in metals and other metallurgical materials. This technique is used mostly for the testings of raw materials, production control, and quality control for final products. Samples are generally dissolved in or digested with acid or acid mixtures, except for certain metal oxides that often require the utilization of fusion techniques. Most solid sample introduction systems such as laser ablation, arcs and sparks, as well as glow discharge techniques can be used most appropriately in this case.

Some representative applications of the ICP-OES technique for metallurgical materials include the determination of minor and trace elements in alloys, high precision determination of Si in steels, determination of contaminants in high-purity aluminum, analysis of superconducting materials for trace contaminants, and lanthanide elements determination in manganese nodules.

C. Agricultural and Food Samples

The ICP-OES has taken root with respect to its common applications in a large variety of agricultural and food products. Types of samples include soils, fertilizers, plant materials, animal feeds, foods, and animal tissues. The purposes of these analyses are usually to determine the levels of essential nutrients as well as the levels of toxic elements present in the materials.

As most agricultural and food samples are not in liquid form, rigorous sample preparation procedures are mandatory to transform these samples into liquid form before they can be introduced into the plasma discharge for analysis. A commercially available modern microwave digestion system combined with the use of acid mixtures can make the sample preparation of these types of materials relatively easy, fast, and straightforward because of the high temperature and high pressure the microwave system can deliver nowadays.

Some typical applications of the ICP-OES technique in this field include the determination of trace metals in rice, beer, tree leaf, bark, wood samples, and wine; the measurement of 21 elements in plants and soils; analysis of infant formula for nutrients; as a finger-printing tool to determine the country of origin of orange juice through trace element analysis; and elemental concentration levels in raw agricultural crops.

D. Biological and Clinical Samples

As research reveals more information about the roles and behaviors of certain elements in biological systems, many major and trace elements are of considerable interest in biological samples such as blood, urine, animal biles, kidney stones, milk and milk powders, and serum electrolytes. Determinations by ICP-OES of essential, toxic, and therapeutic elements are extremely important in a medical research laboratory as well as in a clinical and forensic lab environments. Because of the complex nature of these types of samples, it used to be difficult to analyze these samples before the advent of the ICP-OES instrumentation. ICP-OES has made the analyses of these samples much easier and has led to much lower detection limits for many elements.

Many biological and clinical samples are not only limited in quantity but also contain elemental concentrations too low to be determined by ICP-OES using conventional pneumatic sample introduction systems. In these cases, it is frequently necessary to turn to alternative sample introduction techniques such as ETV, USN, hydride generation, and sample preconcentration procedures such as ion exchange or solvent extraction.

Examples of ICP-OES analyses of biological and clinical samples include determination of Pb, Hg, and Cd in blood; Cr, Ni, and Cu in urine; Se in liver; Cr in feces; Ni in breast milk; B, P, and S in bone; major elements in vitamin pills formulated as multimineral capsules; and of various metals in rodents.

E. Environmental Samples

Environmental samples cover too many sample types in ICP-OES applications. Many of these such as soils, river sediments, and animal and plant tissues overlap application areas discussed previously. This still leaves a number of important environmental ICP-OES applications unmentioned. The most important one is the analysis of waters from various sources.

The analysis of water can be the simplest of all ICP-OES applications, depending upon the type of water, elements to be determined, required detection limits, and protocols. Fresh or drinking waters may only need to be acidified to stabilize the elements of interest before ICP-OES analysis. Some waters may need filtering to remove large particles to avoid plugging the nebulizers. For analyses requiring extremely low detection limits, it may be necessary to resort to some kind of preconcentration procedures or the use of a USN to meet the requirement.

The analyses of soils, river sediments, various industry discharges, and coal and coal fly ash require more labor-intensive and time-consuming acid digestion and base fusion techniques for sample preparation before the samples can be introduced into the ICP-OES instrument.

Typical ICP-OES environmental applications include drinking and waste water analyses required by US EPA; determination of various metals in sea water; determination of P, Pd, Hg, Cd, and As in industry discharges, sludges, and municipal waste water; and the determination of major, minor, and trace elements in airborne particulates.

F. Organics

Analysis of organic solutions and organic compounds for elemental information by ICP-OES is indispensable for modern industrial development. Like many other applications discussed earlier, it is still necessary to perform some kind of sample preparation in which the samples are either dissolved in an appropriate organic solvent or transformed into an aqueous form before their introduction into the ICP.

If possible and economical, the transformation of organic samples into aqueous ones through direct acid dissolution and acid digestion would be the simplest and the most straightforward procedure.

Although not particularly more difficult, there are some special requirements that have to be met before the direct analysis of organics in organic solvent by ICP-OES can be carried out. Normally, the introduction of organic matrices into the plasma discharge requires that the ICP be run at higher RF power and lower nebulizer gas flow rate than that generally required for aqueous samples. Also, a small stream of oxygen is mandatory to be introduced into the plasma together with argon to avoid the carbon buildup in the ICP assembly. In most of the cases, special torches, nebulizers, and torch injectors are needed for proper and stable ICP operation in organic matrices. When peristaltic pump is used to transport the samples to the nebulizer, certain special solvent-resistant pump tubings have to be used depending on the specific organic solvent used for sample preparation.

Some of typical applications of ICP-OES to organics include the analysis of wear metals in waste lubricating oils; the analysis of solvent-extracted geological materials for trace elements; determination of Pb and S in gasoline and other petroleum products; determination of Cu, Fe, Ni, P, and so on in cooking oils; determination of major and trace elements in automobile antifreeze; the analysis of left-over catalytic elements in pharmaceutical raw materials, intermediates, and final products for process control, quality control, and trouble-shooting.

G. Nuclear Materials

The analysis of nuclear materials for major, minor, and trace elements is also one of the most important applications for ICP-OES, although these types of samples present a major difficulty because many of these materials are extremely toxic, even lethal. Therefore, extreme care must be exercised and protective gears must be in place before undertaking these types of analysis by ICP-OES.

Typical applications include the determination of plutonium to evaluate its recovery, the analysis of uranium for its purity, the determination of trace elements in uranium oxide, the determination of trace impurities in samples of liquid sodium coolant from a fast breeder reactor, and the determination of palladium and other elements in nuclear waste samples.

VI. SUMMARY AND FUTURE PROGNOSIS

ICP-OES offers multielement simultaneous capability, broad element coverage, wide LDR, and the fewest chemical, physical, and spectral interferences compared with any other sources ever used for atomic spectrochemical analysis.

It is a relatively established and matured technique that will continue to prosper and develop together with ICP-MS. There are no doubts that more automated, more powerful, cheaper, and smaller ICP-OES instruments will continue to be put on the market in the future to meet the ever-increasing demand of atomic spectroscopists in various fields around the world.

ACKNOWLEDGMENTS

The author wishes to thank Xiaodong Bu and Qiang Tu, Merck & Co., for their help in the literature search and Ivan Santos and Jean Wyvratt, Merck & Co., for their careful reviews and helpful suggestions.

REFERENCES

Greenfield, S., Jones, I. L. I., Berry, C. T. (1964). High pressure plasmas as spectroscopic emission sources. *Analyst* 89: 713–720.

Montaser, A., Golightly, D. W., Eds. (1992). *Inductively Coupled Plasmas in Analytical Atomic Spectromety*. 2nd ed. New York: VCH Publishers, Inc.

4

Atomic Absorption Spectrometry and Related Techniques

BERNHARD WELZ
Departamento de Química, Universidade Federal de Santa Catarina, Florianópolis, Santa Catarina, Brazil

MARIA GORETI R. VALE
Instituto de Química, Universidade Federal do Rio Grande do Sul, Porto Alegre, Rio Grande do Sul, Brazil

I. INTRODUCTION

Optical atomic spectral analysis is the detection (qualitative analysis) and/or determination (quantitative analysis) of chemical elements within the spectral range from 100 nm to 1 mm, using the characteristic line spectra of free atoms in the gaseous state. The term *spectroscopy* is used both for the visual observation of spectra and as a general term for all spectroscopic observation and measurement procedures. The term *spectrometry* is related to a direct photoelectric measurement of radiation using a dispersive spectral apparatus. Depending on the spectral transition used, we distinguish between *atomic absorption spectrometry* (AAS), *atomic fluorescence spectrometry* (AFS), and *optical emission spectrometry* (OES). According to IUPAC recommendation the acronym OES should be used instead of the ambiguous AES for atomic emission spectrometry in order to avoid confusion with Auger electron spectroscopy.

There are three techniques used in AAS to convert the sample to be analyzed into an atomic vapor: *flame* (F) AAS, which is widely used as a routine technique for trace element determination in the mg/L to μg/L range, and which is in essence limited to the analysis of solutions; *graphite furnace* (GF) AAS, which is one of the most sensitive and rugged techniques for trace element determination, and it can handle liquid and solid samples directly; its typical working range is μg/L to ng/L in solution, and mg/kg to μg/kg in solid samples; *chemical vapor generation* (CVG), which is widely used for such analytes that can be converted into a gaseous compound by a chemical reaction, such as hydride generation (HG).

AFS is nowadays almost exclusively used in combination with CVG, reaching limits of detection (LOD), which are at least one order of magnitude better than those obtained by CVGAAS.

There are in essence three different atomization and excitation sources in use nowadays for OES: electrical *arcs* and *sparks*, which are particularly suited for the direct analysis of conductive solid samples, and which are still widely used in the metallurgical industry. The *inductively coupled plasma* (ICP), in contrast, similar to FAAS, is predominantly used for the analysis of solutions. These two types of excitation sources will not be treated in this chapter. The third excitation source is the classical *flame*, the application of which, however, is nowadays essentially restricted to so-called *flame photometers* for the determination of sodium, potassium, and occasionally calcium and lithium, mostly in clinical laboratories.

II. THE HISTORY OF ATOMIC SPECTROSCOPY

A. The Early History

The history of atomic spectroscopy is closely connected with the observation of the sunlight. Wollaston (1802) discovered the black lines in the sun's spectrum. These were later investigated in detail by Fraunhofer (1817), who assigned letters to the strongest lines, starting on the red end of the spectrum with the letter A. Even nowadays it is common to refer to the "sodium D-line", a designation originated by Fraunhofer. In 1820, Brewster expressed the view that these "Fraunhofer lines" were caused by absorption processes in the atmosphere.

In the first half of the 19th century, several researchers started to investigate atomic spectra using electric arcs and sparks. The early use of flames as excitation sources for analytical emission spectroscopy dates back to Herschel (1823) and Talbot (1826), who identified alkali metals by flame excitation. However, one of the key contributions to atomic spectroscopy was the design of the Bunsen burner in 1856, which, for the first time, produced a nonluminous flame that made possible the undisturbed observation of atomic spectra. Soon after, Bunsen demonstrated the power of this tool by discovering and isolating two hitherto unknown elements in mineral water, cesium and rubidium. Kirchhoff and Bunsen (1861) carried out numerous experiments in which they showed that the characteristic spectral lines that are emitted by alkali and alkaline earth elements disappear from the spectrum emitted by a continuous light source, when a flame, into which these elements were introduced, is put into the light beam, as shown in Fig. 1. On the basis of these experiments, Kirchhoff formulated his general law of emission and absorption, which states that any material that can emit radiation at a given wavelength will also absorb radiation of that wavelength. Kirchhoff also showed that those lines produced with sodium in his experiments were identical to the D-lines from the sun.

In the second half of the 19th century a number of practical advances in analytical spectroscopy occurred. Beer (1852) carefully investigated the Bouguer–Lambert Law of absorption and quantified the relation between the absorption of radiation and the number of absorbing species. Lockyer (1878) stated that the brightness, thickness, and number of spectral lines were related to the quantity of the element present in the sample.

From a practical point of view, there is a distinct difference between the observation of absorption and emission spectra, particularly when smaller amounts of the element of interest are considered. In emission spectroscopy the radiation is observed in front of a nonluminous, "black" background (i.e., even a relatively weak emission can be detected quite easily). In absorption spectroscopy, however, we always have the bright emission of the primary light source as the background, and a small reduction in the emission intensity within a very limited spectral interval has to be detected, which is a much more difficult task. It is no surprise, therefore that, with the equipment available at that time, OES, using electrical arcs and sparks and chemical flames, was the preferred technique for qualitative and quantitative atomic spectrochemical analyses during the first half of the 20th century.

The work that gave principal impetus to flame excitation in terms of modern usage was that of Lundegårdh (1929). In his method the sample solution was sprayed from a nebulizer into a condensing chamber, and then into an air–acetylene flame. Since spectra produced by flames are much simpler than those produced by arc and

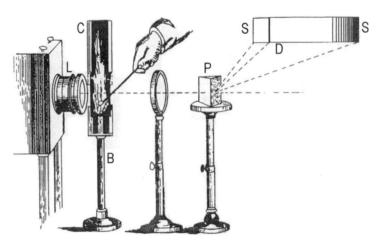

Figure 1 Schematic design of the experimental setup used by Kirchhoff and Bunsen: L, light source; B, Bunsen burner; P, prisma; S, screen; D, sodium D line, which appears as a black discontinuity in the continuous spectrum.

spark emission, simple devices for spectral isolation could be used. Griggs (1939) introduced Lundegårdh's method into the USA, and Barnes et al. (1945) reported on a simple filter photometer for alkali metal determination, which was the basis of the flame photometers introduced soon after. The burners used nowadays for FAAS, and even those for ICP-OES, are all using the principle developed by Lundegårdh.

Atomic fluorescence was studied in the early 1900s by Wood (1902), and Nichols and Howes (1924) looked at fluorescence in flames, but neither of these reports dealt with analytical applications. Winefordner and Vickers (1964) were the first to investigate the possibilities of using AFS as a practical analytical technique.

B. Sir Alan Walsh and the Rebirth of AAS

Sir Alan Walsh, after he had worked in the spectrochemical analysis of metals for seven years, and in molecular spectroscopy for another seven years at the Commonwealth Scientific and Industrial Research Organization in Australia, began to wonder in 1952 why molecular spectra were usually obtained in absorption, and atomic spectra in emission (Walsh, 1974). The result was that there appeared to be no good reason for neglecting atomic absorption spectra; in contrast, they even appeared to offer several advantages over atomic emission spectra as far as spectrochemical analysis was concerned. Among these were the facts that absorption is virtually independent of the temperature of the atomic vapor and the excitation potential, and the expectation to have much less spectral interferences due to the greater simplicity of AAS.

In his considerations about the design of an atomic absorption spectrometer, Walsh realized that using a continuum source (CS), as in the early experiments of Kirchhoff and Bunsen and others, would require a resolution of about 2 pm, which was far beyond the capabilities of the best spectrometer available in his lab at that time. He therefore concluded that the measurement of atomic absorption requires *line radiation sources* (LS) with the sharpest possible emission lines. The task of the monochromator is then merely to separate the line used for the measurement from the other lines emitted by the source. The high resolution necessary for atomic absorption measurements is in this case provided by the line source (LS).

Because of his experience with infrared spectrometers, Walsh proposed to use a *modulated radiation source*, and a synchronously tuned detection system. This way the (nonmodulated) emission spectrum of atoms in the flame produces no output signal and only the absorption spectrum is recorded. He also proposed that for analytical work the sample is dissolved and then vaporized in a Lundegårdh flame. Such flames have relatively low temperature (2500 K) and have the advantage that few atoms would be excited, the great majority being in the ground state. Thus absorption will be restricted to a small number of transitions and a simple spectrum would result. In addition, Walsh expected the method to be sensitive as transitions will be mainly from the ground level to the first excited state. One of his early reports included the diagram shown in Fig. 2, which contains all the important features of a modern atomic absorption spectrometer; the first publication of Walsh, in which he proposed AAS as an alternative to OES for spectrochemical analysis, appeared in the mid-1950s (Walsh, 1955).

Although another paper by Alkemade and Milatz (1955), in which the potential use of AAS was discussed, appeared in the same year, Alan Walsh is generally considered as the father of modern AAS. This privilege is his just due as he campaigned with untiring energy against the resistance to his new idea for more than a decade, spending much time to overcome the disinterest and misunderstanding. The first atomic absorption spectrometer built to his ideas, the Perkin Elmer Model 303, was introduced in 1963. This was the beginning of a tremendous success and a remarkable growth of this technique over the following decades.

C. Boris L'vov and the Graphite Furnace

Boris L'vov, although separated from Walsh by some 20,000 km, inspite of the very restrictive political system in Russia, was without doubt one of the earliest supporters of Walsh and his idea, and he became one of its keenest followers. L'vov remembers (L'vov, 1984), when he came across Walsh's publication in 1956, "the impression produced on me by this paper was so strong that I made up

Figure 2 The first outline by Walsh for the measurement of atomic absorption. [From Walsh (1974).]

my mind to check the validity of the author's ideas". As he did not have a flame available, he used a tubular graphite furnace, which stood unused in a corner, for his experiments. He used a sodium lamp, a prism monochromator, with the graphite furnace in between, which he heated slowly after having put some sodium chloride into the furnace. "Imagine my wonder when the bright sodium lines from the hollow cathode tube started to weaken and then disappeared completely..."

L'vov's first publication in English appeared in the early 1960s (L'vov, 1961), and a schematic design of his graphite furnace is depicted in Fig. 3. The sample was placed on a graphite electrode and atomized completely into the preheated graphite tube by a direct current arc, when the electrode was inserted into the tube. The area under the pulse-shaped absorption signal was found to be proportional to the mass of analyte introduced into the furnace. As the sample was completely atomized and the residence time of the analyte atoms in the graphite tube was much longer than in a flame, the sensitivity of the technique was typically two to three orders of magnitude better than that of FAAS.

However, similar to Walsh, L'vov also had to wait for about a decade until his ideas were accepted, and the first commercial graphite tube furnace, the Perkin Elmer HGA-70 was introduced only in 1970. However, as shown in Fig. 4, the design of this atomizer, as well as that of all other commercial graphite tube atomizers up until now, was not based on L'vov's concept of rapid atomization into a preheated tube, but on a simplified design proposed by Massmann (1967, 1968). The basic difference was that with the latter design, the sample is deposited on the wall of the graphite tube, which was heated relatively slowly because of its electrical resistance. As a result of that, considerable nonspecific

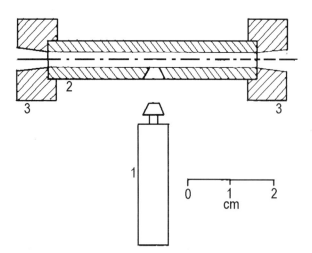

Figure 3 Graphite atomizer by L'vov: 1, graphite electrode; 2, graphite tube; 3, contacts. [From L'vov (1961).]

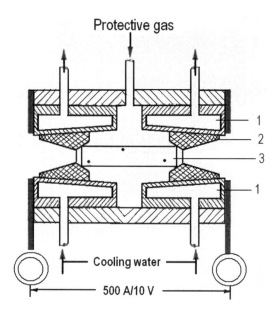

Figure 4 Perkin Elmer HGA-70: 1, cooling chambers; 2, graphite contacts; 3, graphite tube. [From Welz and Sperling (1999).]

absorption and strong matrix effects "not to be observed to this degree in the nebulizer-flame technique" were reported already in the first publications (Manning and Fernandez, 1970).

It is significant that the solution to these problems was again proposed by L'vov (1978), who, similar to Walsh, was following up his ideas all his life. In his view, the considerable difficulties in the GF technique arose mostly from the nonisothermal absorption zone in the atomizer (in terms of both time and space), from the use of peak height for signal evaluation, and from the formation of gaseous molecules of the analyte, such as monohalides. To eliminate the main problem, the temporally nonisothermal state of the absorption volume, he proposed to insert a graphite platform into the graphite tube, onto which the sample is deposited. Since the platform is largely heated by radiation from the walls of the tube it attains the atomization temperature when the tube and the gas atmosphere have already reached their final temperature. Under these conditions the formation of gaseous molecules of the analyte is greatly reduced, and variations in the release of the analyte due to matrix effects can be eliminated by integration of the peak area over time. Slavin et al. (1981) introduced the concept of the "stabilized temperature platform furnace" (STPF), in principle the translation of L'vov's ideas into practice. This concept comprised a "package" of conditions that should bring the temporally and spatially nonisothermal Massmann furnace as close as possible to the ideal of L'vov's isothermal furnace. This STPF concept is the basis of modern GFAAS and will be discussed in Section V.B.3.

III. ATOMIC SPECTRA IN ABSORPTION, EMISSION, AND FLUORESCENCE

A. Origin and Characteristics of Atomic Spectra

Atomic spectra are line spectra and are specific to the absorbing or emitting atoms (elements), that is, the spectra contain information on the atomic structure. More information about the physical principles of absorption and emission of radiation by atoms may be obtained from more detailed, specialized publications (Welz and Sperling, 1999).

Atoms consist of a nucleus surrounded by electrons. The electrons "travel" in "orbits" around the nucleus; in the classical quantum theory these orbits take the form of "orbitals". As the potential energy of the electrons in these orbitals increases with increasing distance from the nucleus, they can also be presented as energy levels. All atoms, except for the noble gases, have orbitals that are not completely filled with electrons. These electrons are called "valence" electrons, as they are involved in the formation of chemical bond; they are also termed "photo-electrons", as they are responsible for the occurrence of spectral lines. In the *ground state* the valence electron(s) is (are) in the lowest possible energy level. Next to the ground state, atoms can exist in numerous *excited states*, which can be obtained through the addition of energy, and which causes the valence electron to undergo transitions to orbitals not occupied in the ground state. These excited states are by nature unstable and the absorbed energy is quickly released. According to the law of the conservation of energy, the energy difference associated with the transition of an electron between various energy levels must be exchanged between the atom and its environment. This can take place through the exchange of kinetic energy with a collision partner or through the exchange of radiant energy. The connection between atomic structure and the interaction of atoms with radiation was established by Planck in 1900 in the quantum law of absorption and emission of radiation, according to which an atom can only absorb radiation of well-defined wavelength λ or frequency ν (i.e., it can only take up and release definite amounts of energy ε):

$$\varepsilon = h\nu = \frac{hc}{\lambda} \quad (1)$$

where h is Planck's constant and c is the speed of light. Characteristic values of ε and ν exist for each atomic species, and are reflected in their characteristic spectra.

Three forms of radiative transition are possible between the energy levels of an atom:

1. If excitation is via optical radiation, the atoms only absorb defined amounts of energy (i.e., radiation of a given frequency), and an *absorption spectrum* is observed.
2. If the energy taken up by radiation absorption according to (1) and is emitted by at least part of the atoms as radiation, a *fluorescence spectrum* is observed.
3. If excitation is via thermal or electrical energy (i.e., through collision with other particles), and at least part of the atoms emits the absorbed energy as radiation, an *emission spectrum* is observed.

Spectral transitions in absorption or emission are not possible between all the numerous energy levels of an atom, but only according to selection rules, which will not be discussed here in detail. However, these selection rules are valid only for *spectral* transitions, not for collision-induced transitions. For this reason, arbitrary energy levels can be attained through excitation with thermal energy, and the *emission spectrum* can contain all lines that are starting from these states in compliance with the selection rules.

In emission methods, we measure the excited-state population, and in atomic absorption methods we measure the ground-state population. The relative populations of the ground state (N_0) and the excited states (N_e) at a given temperature can be estimated from the Maxwell–Boltzmann expression:

$$\frac{N_e}{N_0} = \left(\frac{g_e}{g_0}\right) e^{-(E_e - E_0)/kT} \quad (2)$$

where g_e and g_0 are the statistical weights of the excited and ground states, respectively; E_e and E_0 are the energies of the two states; k is the Boltzmann constant; and T is the absolute temperature. Table 1 summarizes the relative population ratios for a few elements at 2000, 3000 and 10,000 K. We see that even for a relatively easily excited element such as sodium, the excited-state population is small, except at 10,000 K, as obtained in plasmas. Short-wavelength elements require much more energy for excitation and exhibit poor sensitivity by flame OES, where temperatures rarely exceed 3000 K. Those with long-wavelength emissions will exhibit better sensitivity. Measurement of ground-state atoms, as is done in AAS,

Table 1 Values of N_e/N_0 for Selected Resonance Lines and Temperatures

Element/	N_e/N_0		
line (nm)	2,000 K	3,000 K	10,000 K
Na 589.0	9.9×10^{-6}	5.9×10^{-4}	2.6×10^{-1}
Ca 422.7	1.2×10^{-7}	3.7×10^{-5}	1.0×10^{-2}
Zn 213.8	7.3×10^{-15}	5.4×10^{-10}	3.6×10^{-3}

will be less dependent on the wavelength. We also see from Table 1 that the fraction of excited-state atoms is small and temperature-dependent, whereas the fraction of ground-state atoms is virtually constant and nearly 100%. Atomic emission is still sensitive, as we have to measure only the emitted radiation and not a small decrease in a high signal, which has some noise, as in absorption. As, to a first approximation, only transitions from the ground state can be observed in absorption, *absorption spectra* are always much simpler and contain significantly fewer lines than emission spectra.

As atomic fluorescence is the reversal of absorption of energy by radiation, only a limited number of excited states can be reached, and the *fluorescence spectrum* should be as simple as the absorption spectrum. However, firstly, the excited electron need not return to the ground state directly, but can do so in different steps, and secondly, there are several side reactions possible that lift the electron to a level that cannot be reached by absorption of radiation (Sychra et al. 1995). Hence, fluorescence spectra are usually somewhat more complex than absorption spectra, but still much simpler than emission spectra.

B. Line Width and Line Profile

Up until now we have been talking about exactly defined energies that can be absorbed or emitted by the atoms, which would correspond to exactly defined frequencies of the corresponding spectral lines. This is in fact not true as every atomic line, and assuming that the phenomena yet to be discussed are absent, has a line profile as depicted in Fig. 5. This profile is characterized by the central frequency, ν_0, the peak amplitude, I_p, and the frequency distribution (line profile) with a width of $\Delta\nu_{\text{eff}}$. The latter is generally quoted as the width of the profile at half maximum (FWHM = full width at half maximum), $I_{p/2}$, or *half-width* for brevity. The spectral range within the half-width is the *line core* and the ranges to either side are the *line wings*.

The minimum possible half-width is termed the *natural line width*. An excited atom remains only for a very short period of typically 10^{-9}–10^{-8} s in the excited state, before it releases the energy of excitation, for example, as a photon. According to the Heisenberg uncertainty principle the energy levels of the transition can only be determined with an uncertainty of ΔE over the observation time Δt. Because of this uncertainty, the natural width of atomic lines is of the order of 0.01–0.2 pm, which is negligible compared to the broadening mechanisms discussed subsequently.

Firstly, spectral lines undergo broadening because of the random thermal movement of the atoms, which can be described by the Maxwell distribution. This *Doppler broadening* leads (e.g., for the sodium D line at a temperature of 2500 K and at atmospheric pressure) to a broadening of 4.5 pm, which is two orders of magnitude greater than the natural line width. Secondly, at atmospheric pressure and temperatures of 2500–3000 K, which are typical for flames and graphite furnaces, an atom undergoes in the order of 10 collisions/ns with other particles. The lifetime of a collision is typically less than a few picoseconds. Thus in normal flames an excited atom undergoes numerous collisions with other particles during its natural lifetime of a few nanoseconds, leading to a change in its velocity. This *collisional broadening* obviously depends on the pressure and is less in hollow cathode lamps (HCLs) than in a flame or furnace at atmospheric pressure. Without going into further details of these and other broadening mechanisms, the half-widths of emission and absorption lines under atmospheric pressure and temperatures around 2500 K are typically of the order of 1.5–5 pm, depending on the wavelength, as shown in Fig. 6.

Besides the earlier-discussed line broadening mechanisms there are two more phenomena that have to be mentioned here, *self-absorption* and *self-reversal*. A photon emitted by an atom can be absorbed by another atom of the same species within the emission source when the emitted radiation emanates from a transition that ends at the ground state (*resonance line*). This self-absorption leads to the attenuation of the emitted radiation mainly of the line core. Particularly, for very high atom concentrations and hot emission sources the emission line is also significantly broadened. In the case of a pronounced temperature gradient from the center to the outer zones, the absorption profile of the atoms in the ground state can be significantly narrower than the emission profile, leading, in extreme cases, to a complete absorption of the line core, called self-reversal. This is

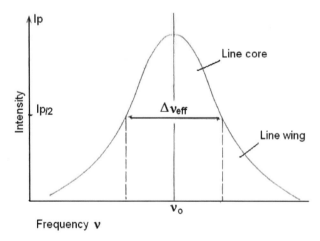

Figure 5 Schematic presentation of line profile and half-width of an atomic spectral line; $\Delta\nu_{\text{eff}}$ = FWHM; ν_0 = line center.

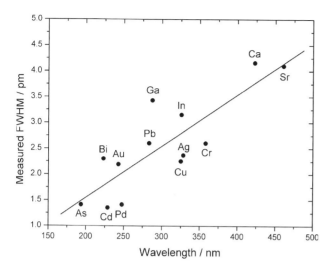

Figure 6 Correlation between FWHM, measured for a number of elements, and the wavelength of their analytical lines. [from Welz et al. (2003).]

one of the reasons why HCLs should not be operated at a very high current. It is also the reason why in emission spectrometry the most intense resonance lines frequently cannot be used and less-intense nonresonance lines must be employed. On the other hand, this phenomenon has been employed in AAS for background correction using high-current pulsing, as will be discussed in Section IV.A.6.

C. Measuring the Absorption—the Beer–Lambert Law

Free atoms in the ground state are able to absorb radiant energy of exactly defined frequency (light quantum $h\nu$) with concomitant transformation in an excited state. The amount of energy absorbed, E_{abs}, per unit time and volume is proportional to the number N of free atoms per unit volume, the radiant energy $h\nu_{jk}$, and the spectral radiant intensity S_ν at the resonance frequency:

$$E_{abs} = B_{jk} N S_\nu h \nu_{jk} \qquad (3)$$

The proportionality factor B_{jk} is the Einstein probability coefficient of absorption of the transition $j \rightarrow k$. The product $B_{jk}S_\nu$ is an expression for the fraction of all atoms present in the ground state that can absorb a photon of the energy $h\nu_{jk}$ per unit time. In unit time a radiation unit of cS_ν ($c =$ speed of light), or $cS_\nu/h\nu$ photons, passes through a unit volume. The fraction of the photons that is absorbed by atoms in the ground state is proportional to the total number N of free atoms and to the "effective cross-section" of an atom, the so-called *absorption coefficient* κ_{jk}. The total amount of energy absorbed per unit volume can then be expressed as the product of the number of absorbed photons and their energy:

$$E_{abs} = \kappa_{jk} N c S_\nu \qquad (4)$$

By equating the energies in Eqs. (3) and (4), we can express the absorption coefficient as:

$$\kappa_{jk} = \frac{h\nu B_{jk}}{c} \qquad (5)$$

Till now we have considered only the absorption of radiation per unit volume, a quantity that is difficult to measure in practice. If we use the quantity normally employed in absorption measurements, the radiant power Φ, we obtain the *Beer–Lambert Law* in the form:

$$\frac{\Phi_{tr}(\lambda)}{\Phi_0(\lambda)} = e^{-N\ell\kappa(\lambda)} \qquad (6)$$

where $\Phi_{tr}(\lambda)$ is the radiant power leaving the absorption volume, $\Phi_0(\lambda)$ is the incident radiant power entering the absorption volume, ℓ is the length of the absorbing layer, $\kappa(\lambda)$ is the spectral atomic absorption coefficient, and N is the total number of free atoms.

From this it is clear why the transmittance $\Phi_{tr}(\lambda)/\Phi_0(\lambda)$ does not decrease linearly with the concentration, as would be the case if the total absorption cross-section was the sum of all individual absorption cross-sections. Because a photon is absorbed as soon as an absorbing species is in the path, it is irrelevant whether or not other absorbing species are in the "shadow" of the first one. The effective absorption cross-section of the individual absorbing species thus decreases with increasing number of these species. By logarithmizing Eq. (6) we obtain the expression:

$$A \equiv \log \frac{\Phi_0(\lambda)}{\Phi_{tr}(\lambda)} = 0.43 N\ell\kappa(\lambda) \qquad (7)$$

which states that the *absorbance A* is proportional to the total number of free atoms N and to the length ℓ of the absorbing layer. If the absorption coefficient is known, it is possible using the preceding equation to perform absolute measurements of N. However, in the majority of cases the absorption coefficient is not known with sufficient accuracy for this purpose. However, this is not important for routine measurements as the analyst is not interested in the absolute concentration of atoms in the absorption volume, but in the concentration or mass of the analyte in the sample.

Nevertheless, the number of absorbing atoms does not stand in simple relation to the concentration of the analyte in the sample, as a number of phase changes must take place on the way to the formation of gaseous atoms, as will be discussed in Section V.A.3. Under the assumption that these parameters remain constant under constant experimental conditions, they can be described by an "effective factor" which can be determined via calibration samples. Like most other spectrometric techniques (including AFS and OES), AAS is thus a *relative technique* in which the relationship between quantity or mass of the analyte and the measured value is determined by *calibration*, for example, by using calibration samples or calibration solutions.

D. The Zeeman Effect

The Dutch physicist Pieter Zeeman discovered the phenomenon that emission lines of atoms split under the influence of a magnetic field (Zeeman, 1897). This effect, named Zeeman effect, arises from the interaction of the external magnetic field with the magnetic moment of the emitting or absorbing atoms. It will be discussed here only to the extent that is necessary to understand its application for background correction purposes in AAS, which will be treated in Section IV.A.6.

The magnetic moments of atoms stem from the movement of the electrons in the orbitals and the spin of the electrons. As a result of the interaction with the external magnetic field the terms of the atom depend on its orientation in the field. The relative orientations of the atoms in a magnetic field obey selection rules and are quantized. The splitting of the levels increases proportionally to the magnetic field strength, and the spaces are equidistant as the energy levels can only take values symmetrically distributed about the degenerate level without magnetic field. Electronic transitions between these energy levels are only permitted according to the selection rule $\Delta M = 0, \pm 1$. As depicted in Fig. 7, the resulting multiplet pattern comprises an inner group of lines with $\Delta M = 0$, termed the π components, and two symmetrical outer groups with $\Delta M = +1$ and $\Delta M = -1$, termed the σ^+ and σ^- components, respectively.

A special case arises when the energy splitting of both levels is the same, or if one energy level does not exhibit splitting. In this case there is only *one* π component and only *one* σ^+ and σ^- component, respectively. This situation is termed the *normal Zeeman effect*. The general situation of multiple splitting is termed the *anomalous Zeeman effect*. Further distinction must be made between splitting into an *uneven* number of π components, in which the original wavelength is retained in at least *one* component, and splitting into an *even* number of components, in which the original wavelength *disappears* completely from the spectrum.

In addition to the splitting of the spectral lines into π and σ components, the radiation is also *polarized*. The π components are linearly polarized in a direction *parallel* to the direction of the magnetic field, whereas the σ components are *circularly* polarized in the direction perpendicular to the magnetic field. The visible intensities of the individual components depend on the orientation

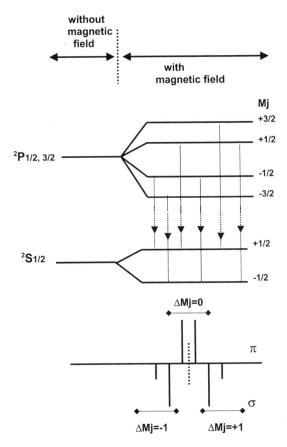

Figure 7 Splitting of the energy levels in a magnetic field with the resulting splitting pattern for the transition $^2S_{1/2} \leftrightarrow {}^2P_{1/2,3/2}$. [From Welz and Sperling (1999).]

of the direction of view with respect to the magnetic field and obey quantum selection rules, which describe the transitional probabilities. The two most important ones are that the intensities of the components are distributed symmetrically about the position of the noninfluenced line, and that the sum of the intensities of all π components must equal the sum of the intensities of all σ components.

IV. SPECTROMETERS FOR AAS

The combination of all optical and mechanical assemblies required for the generation, conductance, dispersion, isolation, and detection of radiant energy is termed a spectrometer. The quality of a spectrometer is determined largely by the *signal-to-noise* (S/N) *ratio*, which in itself derives mainly from the radiation (optical) conductance (i.e., which percentage of the radiation emitted by the radiation source finally arrives at the detector). The usable wavelength range of an atomic absorption spectrometer depends on the radiation source, the optical components used in the radiation train, and the detector. In practice this is usually from 193.7 nm, the most widely used wavelength of arsenic, to 852.1 nm, the most sensitive wavelength of cesium.

A. Medium-Resolution LSAAS

1. LSs and Continuum Sources

In conventional AAS, LSs are used that emit the spectral lines of one or a few elements. HCLs and electrodeless discharge lamps (EDLs) are the main types of lamp employed. Lamps that emit a spectral continuum are used for background correction purposes only.

LSs are spectral radiation sources in which the analyte element is volatilized and excited so that it emits its spectrum. Excitation can be caused by a low-pressure electrical (glow) discharge, by microwaves or radiowaves, or by thermal energy. By using LSs in AAS it is possible to do without high-resolution monochromators, as the concomitant elements cannot, in principle, absorb radiation from the element-specific radiation source. As the "resolution" in conventional AAS is largely determined by the width of the emission lines from the radiation source, the line width has a major influence on the absence of spectral interferences caused by line overlapping. The width of the emission lines also has an influence on the linearity of the calibration function.

The emission intensity from the radiation source does not have a direct influence on the sensitivity in AAS, as the absorbance depends on the *ratio* of the radiant power entering the absorption volume to that leaving it [Eq. (7)]. Nevertheless, the emission intensity has a marked influence on the S/N ratio and thus on the precision of a measurement, as well as on the detection limit. Generally, for all sources, the emission intensity increases with increasing energy input. At the same time, however, higher energy input leads to higher temperatures within the source, and thus to line broadening and self-absorption.

Hollow Cathode Lamps

The HCL has been the source type used most frequently and for the longest period in AAS. It is a spectral lamp with a hollow, usually cylindrical, cathode containing one or few analyte elements. The anode is mostly made of tungsten or nickel. These are enclosed in a glass cylinder usually with a quartz window. The lamp is filled with an inert gas, usually neon or argon, at a reduced pressure of about 1 kPa. If a voltage of 100–200 V is applied across the electrodes, a glow discharge takes place in the reduced gas-pressure atmosphere. In a very simplified way the complex processes can be described as follows: electrons emitted from the cathode are accelerated by the strong electrical field and undergo inelastic collisions with the fill gas atoms, resulting in their ionization. The ions are attracted by the cathode and accelerated in the electrical field, and when they impinge upon the cathode they eject metal atoms from the surface, which are excited to radiation in the intense discharge.

Figure 8 shows a typical design of an HCL as it is used for AAS. Single-element lamps typically have the highest energy output and provide the best protection against spectral interferences due to line overlap. *Multielement lamps*, which have been designed to facilitate the change between frequently determined elements, facilitate routine analysis, but have to be chosen with care. Lamps with two or three elements can typically be used without problems, but lamps with more elements cannot be recommended for all applications.

Several attempts have been undertaken over the years to increase the radiation output of HCL without the usually associated disadvantages of line broadening and self-absorption. *Boosted discharge lamps*, which have an additional electrical discharge in order to excite neutral atoms that have been sputtered from the cathode, exhibit higher radiation intensities and narrower line widths than conventional HCL, but they can only be manufactured for relatively volatile elements and require an additional power supply.

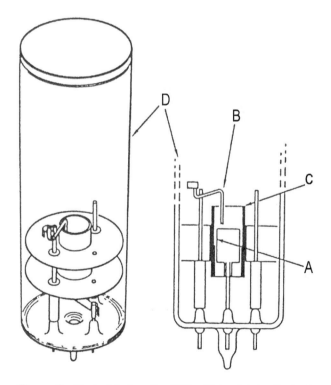

Figure 8 Typical design of an HCL with insulated cathode: A, hollow cathode; B, anode; C, ceramic shield; D, glass cylinder. [From Welz and Sperling (1999).]

Electrodeless Discharge Lamps

EDLs are among the radiation sources exhibiting the highest radiant intensity and the narrowest line widths. The actual lamp, comprising a sealed quartz bulb containing the analyte element and a noble gas under reduced pressure, is either permanently mounted in a radiofrequency coil or inserted into the coil. The radiation from an EDL is based on an inductively coupled discharge that is generated in an electromagnetic radiofrequency field (e.g., 27.12 MHz). The physical processes involved in the emission of radiation from an EDL have been described in detail by Gilmutdinov et al. (1996). Similarly to boosted discharge lamps, EDL can only be manufactured successfully for relatively volatile elements, and they also require an additional power supply.

Continuum Sources

In a CS the radiation is distributed continuously over a greater wavelength range. In conventional AAS, a CS is used exclusively for the sequential or quasi-simultaneous background measurement and correction (Section IV.A.6). Deuterium, hydrogen, and halogen lamps are mostly used for this purpose. The *deuterium lamp* is a spectral lamp with deuterium as the discharge gas in a quartz bulb. The deuterium lamp emits a sufficiently high radiant power in the short wavelength range from about 190 nm to around 330 nm. The *hydrogen lamp* is typically used in the form of an HCL with hydrogen as the fill gas. The characteristics of this lamp and the wavelength range are similar to that of a deuterium lamp, with the advantage that the spatial intensity distribution emitted by this lamp is more similar to that of the primary radiation source. A *halogen lamp* is an electrically heated metal band or coil, usually of tungsten, in a quartz bulb. The blackening of the bulb by metallic deposits is prevented by gaseous halogenated additives. The halogen lamp emits a sufficiently high radiant power in the spectral range above 300 nm, thus complementing the working range of the former lamps.

2. *The Radiation Train*

In AAS it is normal to distinguish between single-beam and double-beam spectrometers. In a single-beam spectrometer the primary radiation is conducted through the absorption volume without geometric beam splitting, whereas in a double-beam spectrometer the beam is divided into two sections. A portion of the radiation, the *sampling radiation* (or sample beam), is passed through the absorption volume (flame, furnace, etc.), whereas the other portion, the *reference radiation* (or reference beam), bypasses the absorption volume. The geometric splitting and recombining of the beam can be effected by means of a rotating chopper made from a partially mirrored rotating quartz disk, by means of semitransparent mirrors, or by combinations of these. Typical switching (modulation) frequencies are between 50 and 100 Hz.

Single-beam spectrometers have the advantage that they contain fewer optical components and thus radiation losses are lower and the optical conductance higher. The major advantage claimed for *double-beam spectrometers* is better long-term stability as they compensate for changes in the intensity of the source and the sensitivity of the detector. Nevertheless, this advantage is frequently overvalued, as during the warm-up phase not only does the radiant intensity of the source change, but also the line profile, and thereby the sensitivity. Also, the double-beam system is capable of neither recognizing nor compensating for changes in the atomizer, such as flame drift during the warm-up phase of the burner. Nowadays it has been accepted that a double-beam spectrometer has clear advantages for routine FAAS in order to avoid frequent checks for baseline drift and recalibration. For GFAAS, where usually the baseline is reset before each atomization stage, long-term stability is not of interest, so the better optical conductance of a single-beam spectrometer is of advantage. Figure 9 shows the radiation train of an instrument that combines the design of a single- and a double-beam spectrometer and allows choosing the preferred configuration according to the specific application.

3. *Dispersion and Separation of Radiation*

When LSs are used the analytical lines are separated from other emission lines from the source and from broadband emission from the atomizer by monochromators. For dispersion of the radiation, diffraction gratings are mostly used in AAS. Prisms are less suitable as the optical conductance of prism monochromators deteriorates by orders of magnitude toward longer wavelengths. Because of the importance of the UV range in AAS, it is advantageous to use gratings that have blaze wavelengths as far as possible in the UV. As a relatively wide wavelength range has to be covered by an atomic absorption spectrometer, configurations employing two gratings or gratings with several blaze wavelengths are frequently employed to improve the efficiency. The gratings used in AAS usually have a ruling density of 1200–2800 lines/mm, so that a good grating can have up to 100,000 lines.

The most frequently used monochromator configurations in AAS are the Littrow and the Czerny–Turner. For a point radiation source the symmetrical reflection on two identical concave spherical mirrors in the Czerny–Turner mounting produces a strongly astigmatic image for complete compensation of the coma error. In comparison with most other monochromator mountings,

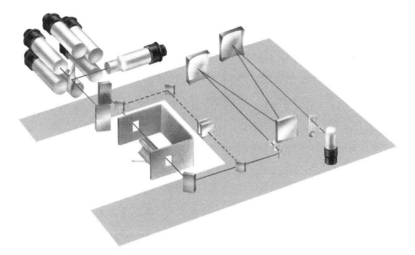

Figure 9 Radiation train of an atomic absorption spectrometer which combines single- and double-beam design; solid line: sample beam for single-beam and double-beam operation; broken line: reference beam for double-beam operation only. (By kind permission of Analytik Jena AG.)

in a Czerny–Turner monochromator the agreement between the optical system and the detector is almost perfect.

Through the use of element-specific LSs and modulation of the radiation, because the selectivity and specificity of LSAAS depend only on the half-width of the emission and absorption lines, the monochromator has the sole task of separating the analytical line from other emission lines of the source. Spectral slit widths of 0.2–2 nm are generally adequate for this purpose.

The *spectral slit width* $\Delta \lambda$ is the product of the geometric width s of the slit and the *reciprocal linear dispersion* $d\lambda/dx$ in the plane of the slit:

$$\Delta \lambda = s \left(\frac{d\lambda}{dx} \right) \tag{8}$$

As the maximum usable spectral slit width is determined by the spectrum emitted by the primary source, the reciprocal linear dispersion determines the geometric slit width. A reciprocal linear dispersion of 2 nm/mm means that at a geometric slit width of 1 mm the spectral slit width is 2 nm, or a geometric slit width of 0.1 mm is necessary to obtain a spectral slit width of 0.2 nm. In an atomic absorption spectrometer because the image of the radiation source—a radiation beam of several millimeters diameter—is formed on the entrance slit, the geometric width of the entrance slit determines the amount of radiation falling on the dispersing element (grating), and subsequently on the detector. Therefore, with a wide entrance slit, a relatively large amount of radiant energy falls on the detector; this means that the noise always present in the signal is relatively small when compared with the signal (high S/N ratio). It is thus desirable to have as small a numerical reciprocal linear dispersion as possible (i.e., to use a strongly dispersive element). The significance of the geometric slit width, on the other hand, means that the largest spectral slit width just meeting the requirement for the isolation of the analytical line should always be chosen.

If a larger slit width is used (i.e., when more than one emission line from the radiation source is allowed to pass) AAS does not lose any of its specificity and selectivity, provided that the analytical lines of two elements do not fall on the detector when multielement lamps are being used. The disadvantages brought by a slit width that is too large are reduction in the sensitivity and an increasing nonlinearity of the calibration curve.

4. Detection and Modulation of Radiation

Detectors operating on the photoelectric principle are used exclusively for the detection of radiation in AAS. The requirements of AAS are best met by broadband photomultipliers (multialkali types). The photomultiplier tube (PMT) is a radiation detector in which the incident radiation falling on a photocathode causes the emission of primary electrons (outer photoelectric effect), which are released into the surrounding vacuum. Resulting from the applied dynode voltage each primary electron is accelerated so rapidly that when it strikes a dynode 2–10 secondary electrons are emitted, leading to a cascade effect as shown schematically in Fig. 10. Nevertheless, there are limits to the voltage that can be applied, as this leads to a higher dark current and thus to increased noise. PMTs with 9–12 dynodes are mostly used in AAS so that the anode output varies as the sixth to tenth power of changes in the applied voltage. As the output signal is

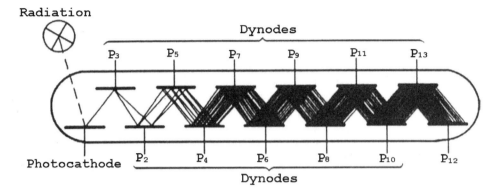

Figure 10 Schematic of a PMT with 11 dynodes. [From Welz and Sperling (1999).]

so extremely sensitive to changes in the applied voltage, the stability and freedom from noise of this source are especially important. In modern AAS instruments microprocessors are mostly used to automatically regulate and control the applied voltage.

In principle, semiconductor barrier layer detectors can also be used in AAS. Here the incident radiation produces electron–cavity pairs (inner photoelectric effect), which are separated by the electric field at the barrier layer. Recombination takes place via the electric current (photocurrent) through the external circuit. Photoelements, photodiodes, and phototransistors belong to this category. The most important characteristics of all photoelectric detectors are the spectral sensitivity, the quantum efficiency, the usable wavelength range, the linear range, the S/N ratio, the response time, and the dark current (Welz and Sperling, 1999).

Using electronic measuring techniques, modulation of the radiation allows the absorption at the analytical line to be discriminated from other radiation. The periodic change in the radiant power is usually achieved by either modulation of the discharge current of the radiation source or by rotating choppers. By the use of a *selective amplifier* that is tuned to the modulation frequency, only the radiation modulated at that frequency is processed. Other radiation, especially the nonmodulated emission from the atomizer, is received continuously, and is thus subtracted by the electrical measuring system in all measurement phases. The modulation of the radiation thus leads to the high selectivity of AAS; it is also the reason why relatively large spectral slit widths can be used in AAS.

5. Data Acquisition and Output

The incident radiant power entering the absorption volume, Φ_0, and the radiant power leaving the absorption volume, Φ_{tr}, are converted, after spectral dispersion and separation, by the detector into electrical signals and amplified. Ratioing both signals leads to the calculation of the transmittance (spectral transmission factor), $\tau_i(\lambda)$, as a measure of the transmission:

$$\tau_i(\lambda) = \frac{\Phi_{tr}(\lambda)}{\Phi_0(\lambda)} \qquad (9)$$

and of the absorptance (spectral absorption factor), $\alpha_i(\lambda)$, as the measure of the absorption of the radiant energy in the absorption volume:

$$\alpha_i(\lambda) = \frac{\Delta\Phi(\lambda)}{\Phi_0(\lambda)} = \frac{\Phi_0(\lambda) - \Phi_{tr}(\lambda)}{\Phi_0(\lambda)} = 1 - \tau_i(\lambda) \qquad (10)$$

The correlation between these quantities and the concentration or mass of the analyte is given by the Beer–Lambert Law (refer to Section III.C), which states that for an ideal dilution the absorbance A (or the integrated absorbance A_{int}), given as the negative decadic logarithm of the transmittance, is proportional to the product of the length of the absorbing layer ℓ and the concentration c or mass m of the analyte:

$$A = \varepsilon(\lambda)c(i)\ell \qquad (11)$$
$$A_{int} = \varepsilon'(\lambda)m(i)\ell \qquad (12)$$

The specific absorbance coefficients ε and ε', as the proportionality factors, are a function of the wavelength λ and in practice are influenced by a number of marginal conditions such as dissociation in the atomizer and the physical properties of the measurement solution. The pathlength ℓ is the distance the absorption beam passes through the absorption volume, which is determined by the geometry of the atomizer. For following *steady-state signals*, as they are produced in FAAS, it is of advantage to use absorbance values, preferably in the form of mean values, which are formed from a series of momentary values. For atomization techniques with discrete sample dispensing, such as GFAAS, all the analyte species or a constant proportion thereof contained in the sample are atomized over a defined time period. The output of the signal as the (*time-*) *integrated absorbance* is of advantage in this case. As the absorbance is a dimensionless quantity, the integrated absorbance has the dimension s (second).

Atomic absorption spectrometers contain a logarithmic amplifier to convert the experimentally observed transmittance into absorbance. As long as a steady-state signal is generated by the atomizer the conversion is relatively easy. It is only necessary to follow the steady-state transmittance for as long as necessary for an adequate approximation. For a transient atomization peak, however, the conversion to absorbance is complicated by two contradictory requirements: (i) the measurement of the transmission requires an instrument with a very short response time of typically 10–20 ms (modulation frequency 50–100 Hz) and (ii) for every measurement the shot noise should be as low as possible, and this requires long response times, leading to a falsification of the transmission values. The way out of the dilemma can be found in the radiant power. The more energy that falls on the detector, the better will be the S/N ratio, under otherwise unchanged conditions, and the noise will be low in comparison to the signal. This means that the entire atomic absorption spectrometer must be optimized for measuring fast, transient signals, and a high optical conductance is one of the key requirements. A further means of optimization is in the *effective measurement time* during a measurement cycle. The longer the measurement time and the shorter the dark period, the better will be the S/N ratio.

During a measurement cycle the spectrometer generates several measures, which derive, for example, from the radiant power of the sample beam and the reference beam. During signal processing the measures for the various phases are used for calculation so that the resulting signal corresponds to the concentration or mass of the analyte atoms in the atomizer. However, owing to drift in the electronics, radiation losses at optical components, and so on, the absorbance resulting from both signals will never be $A = 0$, even when there are no analyte atoms in the atomizer. It is thus necessary to set the readout of the instrument to zero prior to the start of a series of measurements and at regular intervals during the measurements.

One of the major reasons for constructing double-beam spectrometers was to compensate for drift phenomena and thus to avoid the necessity of having to reset the baseline frequently. However, not all the phenomena contributing to baseline drift in AAS can be compensated by a double-beam system. In addition, a double-beam spectrometer always requires a larger number of optical components, resulting in lower optical conductance, which is in contrast to the requirements for measuring fast transient signals. The technique that has proven to be the most reliable for that purpose is to use a single-beam spectrometer and automatic baseline offset correction (Auto Zero) immediately prior to atomization of the analyte.

6. Measurement and Correction of Background Attenuation

As mentioned earlier, the high selectivity and specificity of AAS derive from the use of element-specific LS, modulation of this radiation, and selective amplifiers. As a result, far fewer spectral interferences occur than with OES. Nevertheless, the radiation passing through the atomizer can be attenuated not only by atomic absorption but also by a number of other effects that are frequently termed collectively as "nonspecific" or "background" absorption. Typically these can be absorption by gaseous molecules or radiation scattering on particles.

Till date as there is no technique that allows the atomic absorption of the analyte to be measured exclusively, it is necessary to measure the total absorption (specific and nonspecific), then the background attenuation, and then to subtract this from the total absorption. On the basis of theoretical considerations it is possible to place a number of principal requirements on an ideal system for the measurement and correction of background attenuation:

1. The background attenuation should be measured exactly at ("under") the analytical wavelength and with the same line profile as the total absorption.
2. The background attenuation should be measured at exactly the same location in the atomizer as the total absorption.
3. If the absorption changes with time (transient signals), the total absorption and the background attenuation should be measured simultaneously.
4. Background correction should not cause a worsening of the S/N ratio.
5. The technique employed for background measurement and correction should be applicable for all elements (over the entire spectral range).

None of the techniques for the measurement and correction of background discussed subsequently meets all these requirements. None of the techniques, for example, is capable of measuring the total and background absorption simultaneously, but only in rapid sequence. Nevertheless this "catalog of wishes" is a useful means of evaluating the individual techniques.

CS Background Correction

Background correction (BC) with CS is based on the assumption that background attenuation, in contrast to atomic absorption, is a broadband phenomenon, which does not change within the range of the selected spectral slit width. The mode of function is depicted schematically in Fig. 11. The exit slit of the monochromator isolated the analytical line from the spectrum emitted by the LS, whereas a band of radiation, corresponding to the selected spectral slit width (e.g., 0.2 nm) is isolated from the

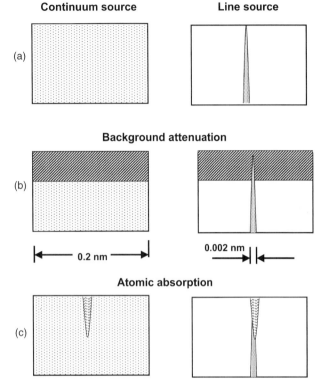

Figure 11 Mode of function of BC with CS: (a) the radiant intensity, represented schematically by dots, for the CS is distributed over the entire width of the spectral band isolated by the slit, whereas for the LS it is limited to a few picometers; (b) background absorption attenuates the radiation emitted by both sources to the same extent; and (c) atomic absorption in the first approximation attenuates only the radiation from the LS. [From Welz and Sperling (1999).]

continuum emitted by the CS. Analyte atoms absorb radiation from the LS at the analytical line proportional to their content. Radiation from the CS, on the other hand, is only attenuated in the very narrow wavelength range of a few picometers, in which the analyte atoms absorb. Depending on the selected slit width, this amounts to 1–2% maximum of the continuous radiation reaching the detector and is negligible. Radiation scattering or broadband molecular absorption attenuates the radiation from the LS and the CS to equal degrees, so that the ratio of the radiant power of each beam does not change.

BC using CS deviates in virtually all aspects from the requirements stated earlier for an ideal system:

1. The background is measured as a broadband to either side of the analytical line. If BC is to be correct under the line, the mean attenuation over the observed spectral range must be the same as that at the center of the analytical line.
2. Even if it is possible to align the LS and the CS exactly along the same optical axis, nevertheless, differences in the geometry and distribution of the radiant power of each source mean that different absorption volumes are measured.
3. The measurement is not simultaneous; in the case of transient signals the quality of BC will depend on the measurement frequency.
4. Through the use of two radiation sources and additional optical components for beam splitting and recombination the S/N ratio is significantly worsened.
5. There is no CS available for BC that covers the entire spectral range of AAS.

Despite these weaknesses, BC using CS can still be successfully utilized, provided that the requirements are not too high. This is generally the case for FAAS and CVGAAS, whereas it can be employed for GFAAS only after thorough investigation.

BC Using High-Current Pulsing

Smith and Hieftje (1983) took up the proposal of earlier researchers for BC using high-current pulsing. The technique is based on the strong line broadening and self-reversal of resonance lines observed in HCL at high operating currents (refer to Section III.B.). The total absorption is measured at normal operating currents and thus with normal line profiles, whereas the background absorption is measured next to the analytical line with the strongly broadened profile caused by the self-reversal due to high-current pulsing.

The advantages of this technique when compared with BC with CS are that only one radiation source is used and that the background absorption is measured close to the analytical line and not integrated over a wide spectral range. Among the most obvious disadvantages is the fact that this technique can be used successfully only for those elements that exhibit almost total line reversal under the conditions used (i.e., for the most volatile elements). For elements of medium volatility, such as Al, the loss of sensitivity with this technique is already 75%, and for less volatile elements this technique cannot be used at all.

Compared with the requirements for an ideal system for BC, this technique does not meet most of the aspects. Although only one radiation source is used there are differences in the beam geometry, and the background attenuation is measured next to the analytical line with a significantly changed profile. The S/N ratio for many elements is worsened because of the loss of sensitivity associated with this technique. Obviously, it can be used only with volatile elements. Besides being a sequential correction technique, the slow dissipation of the line broadening after a high current pulse of some 40 ms means that the system can be operated only at a frequency of about 20 Hz, which is insufficient for the rapid signals of GFAAS. This means that it brings about improvements when compared with BC with CS only in a few isolated cases.

Zeeman-Effect BC

The Zeeman effect has been described briefly in Section III.C. Zeeman effect BC (ZBC) is nowadays practically only used in combination with GFAAS. There are various configurations possible to apply the ZBC in AAS: The magnet can be mounted at the radiation source (direct) or at the atomizer (inverse), the magnetic field can be parallel (longitudinal) or perpendicular (transverse) to the radiation beam, and the magnet can produce a constant (DC) or an alternating (AC) magnetic field. This results in eight theoretically possible configurations. However, owing to the fact that the π components are polarized parallel to the radiation beam, they disappear permanently when a DC field is applied longitudinally, hence this configuration cannot be used at all in AAS, as there is no component available to measure the total absorption. Because of other problems associated with a DC magnetic field, this approach has been abandoned in the meantime. Another problem is that typical LS for AAS cannot be operated in a strong magnetic field, requiring the development of special lamp types that can tolerate a magnetic field. The direct Zeeman configuration with the magnet at the radiation source has therefore been abandoned as well, leaving only the two possibilities with an AC magnet either longitudinally or transversely orientated at the atomizer.

This inverse configuration has the advantages that conventional radiation sources, such as HCL or EDL can be used without problems and that all measurements (i.e., of total and of background absorption) are carried out at the analytical line with the same line profile emitted by the same radiation source. In the phase "magnet off" conventional atomic absorption (i.e., total absorption) is measured. In the phase "magnet on" the absorption lines of the analyte atoms are split into π and $\sigma^{+/-}$ components, and with a sufficiently strong magnetic field the $\sigma^{+/-}$ components are completely moved out of the emission profile of the radiation source, as shown in Fig. 12.

Obviously, the π component has to be removed as well in order to measure background absorption only, because otherwise the π component will cause atomic absorption. The *longitudinal orientation* of the magnet at the graphite furnace has the advantage that the π component disappears during the magnet on phase, because the π component is polarized parallel to the radiation (i.e., in the direction of observation), and hence "invisible". This means that no additional optical components are required for the measurement of the background absorption only. The

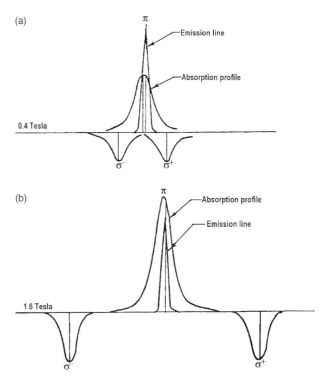

Figure 12 Line profiles for cadmium at 228.8 nm with magnetic fields of (a) 0.4 T and (b) 1.6 T at the atomizer. [From Welz and Sperling (1999).]

disadvantage of the longitudinal configuration is that, in order to maintain a sufficiently strong magnetic field over the length of the absorption volume, the graphite tube has to be as short as possible. This results in a loss of sensitivity and the risk of increasing imprecision. The *transverse configuration* has the disadvantage that a polarizer (i.e., an additional optical component) is required to remove the π component from the spectrum. This inevitably results in 50% loss of radiant power from the radiation source, and hence in a slight deterioration of the S/N ratio. This loss, however, is usually more than compensated by the use of significantly longer graphite tubes, which results in a corresponding increase in sensitivity, and by a stronger and more homogeneous magnetic field within the atomizer (Gleisner et al., 2003).

The situation depicted in Fig. 12 that the $\sigma^{+/-}$ components are completely moved out of the emission profile of the radiation source in the presence of a sufficiently strong magnetic field (typically 0.8–1 T), unfortunately is only valid for analytical lines that exhibit a normal Zeeman splitting. For the majority of lines and elements the situation is more like that shown in Fig. 13 for chromium. In the case of the anomalous Zeeman effect the π and $\sigma^{+/-}$ components split into more than one line, which results in a significant broadening of all components, and often in a partial overlap of the wings

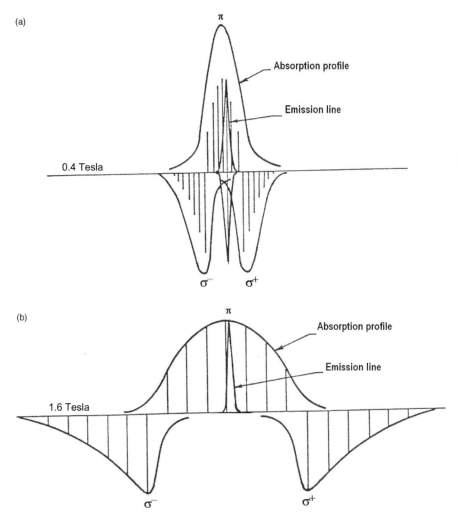

Figure 13 Line profiles for chromium at 357.9 nm with magnetic fields of (a) 0.4 T and (b) 1.6 T at the atomizer. [From Welz and Sperling (1999).]

of the $\sigma^{+/-}$ components with the wings of the emission line, even in strong magnetic fields. This means that, depending on the degree of overlap, some atomic absorption is measured in the magnet on phase in addition to the background absorption, which can be easily detected by measuring the "background absorption" of a matrix-free solution of the analyte.

In ZBC as the absorbance measured during the magnet on phase is subtracted from that measured during the magnet off phase, the earlier effect has two consequences: firstly, as some atomic absorption is also subtracted from the total absorption, there is a proportional loss in sensitivity associated with ZBC, which is called *Zeeman factor* (sensitivity ratio ZBC/without BC). In the case of no overlap this factor is 1.0, but in extreme cases this factor can assume values around 0.5, corresponding to a 50% loss of sensitivity. Secondly, as the absorbance measured owing to the overlap with the $\sigma^{+/-}$ components is true atomic absorption, it increases linearly with the analyte mass or concentration (i.e., a regular calibration can be established). As the linear working range in AAS is rather limited and deviation from the linear relationship between absorbance and analyte mass or concentration typically starts between 0.5 and 1 absorbance, the calibration curve measured in the magnet off phase starts to bend toward the concentration axis at much lower analyte concentrations than the (less-sensitive) calibration curve measured in the magnet on phase, as shown in Fig. 14. As in ZBC the two curves are subtracted from each other, the resulting calibration curve has a reduced slope (corresponding to the Zeeman factor), and it exhibits a maximum at the point, where the slope of the magnet off calibration curve and that of the magnet on curve ($\sigma^{+/-}$ absorption) become equal. Beyond that point the resulting curve assumes a negative slope, as the slope of the σ absorption curve becomes greater than that for total absorption. This

AAS and Related Techniques

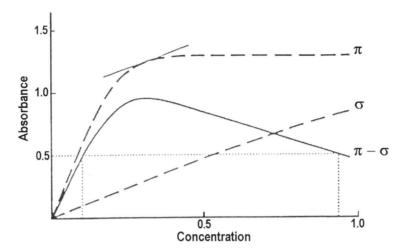

Figure 14 Schematic presentation of the calibration curves for the π and the $\sigma^{+/-}$ components with insufficient removal of the $\sigma^{+/-}$ components from the emission profile, and the resulting calibration curve with rollover. [From Welz and Sperling (1999).]

phenomenon is termed *rollover*, and it means that calibration curves with ZBC have an even smaller linear range than those of conventional AAS and they become ambiguous as two concentration or mass values can be assigned to each absorbance reading, as shown in Fig. 14.

Fortunately, in GFAAS this rollover of the calibration curve can be recognized from the signal shape as shown in Fig. 15. In GFAAS as we measure a transient signal, we start from an analyte atom concentration of zero, reach a maximum, and return to zero. This means we are moving with time along the concentration axis in Fig. 14 from left to right and back to left. In terms of absorbance, for high analyte concentrations we move along the resulting calibration curve through the maximum, down the negative slope, and for decreasing atom concentrations again up the negative slope to the maximum and finally down the positive slope. This easily explains the "dip" in the signal for high analyte concentrations. It has to be mentioned that this rollover phenomenon is essentially only observed in absorbance, but not when the signal is integrated over time (i.e., when peak area is measured instead of peak height, as is common practice now in GFAAS).

One of the limitations of ZBC, as mentioned earlier, is the limited linear range, which can be significantly inferior to conventional AAS, depending on the splitting pattern of the respective line. De Loos-Vollebregt et al. (1986) have proposed a solution to that problem, the so-called "three-field mode"; however, this option has been included in commercial instruments only recently (Gleisner et al., 2003). The term *three-field mode* denotes a switching cycle consisting of a sequence of three magnetic field phases; in addition to the usual phases of magnet on and magnet off, a third phase is introduced with an intermediate magnetic field. When this intermediate field is used, the $\sigma^{+/-}$ components are less separated from the emission line (i.e., cause more atomic absorption), and the difference between the curves with maximum magnetic field and reduced magnetic field becomes less, resulting in a calibration curve of reduced sensitivity. By proper selection of the two magnetic field strengths, a second calibration curve can be obtained with a slope, reduced according to the specific requirements. Figure 16 shows a typical example for the application of the three-field mode for the determination of high as well as low analyte concentrations.

ZBC with the magnet at the atomizer fulfills almost all the requirements on an ideal system outlined in

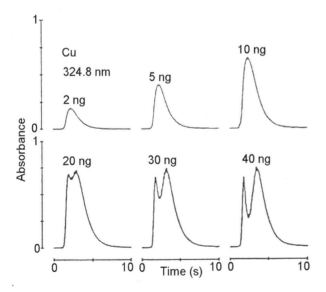

Figure 15 Effect of rollover of the calibration curve on the signal shape for increasing masses of copper in a magnetic field of 0.8 T at the atomizer. [From Welz and Sperling (1999).]

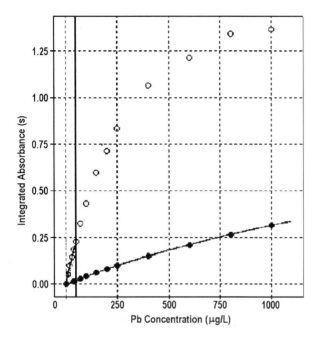

Figure 16 High- (○) and low-sensitivity (●) analytical curves for lead using the three-field technique with 0, 0.6, and 0.8 T magnetic field strength at the atomizer. [From Gleisner et al. (2003).]

Section IV.A.6. As the same radiation source is used for atomic and background measurements, background absorption is measured "under" the line with the same line profile and at exactly the same location. The technique can be employed for all elements, and it causes for most elements only a slight reduction in the S/N ratio. The only problem, which is common to all BC techniques in LSAAS, is that total and background absorptions are not measured simultaneously. The modulation frequency hence becomes very important for an accurate correction of background absorption. However, modulation frequencies of 200 Hz have already been realized in modern instrumentation (Gleisner et al., 2003), reducing this problem to a minimum.

B. High-Resolution CSAAS

Substitution of a CS for LS, without changing the rest of the instrument, is not a reasonable approach. The instability of the most intense CS, xenon arc lamps, gives noisy baselines and poor detection limits. Medium-resolution monochromators that are ideal for isolating LS emission lines provide a spectral bandwidth that is too large for use with a CS, resulting in poor sensitivity and specificity, nonlinear calibration curves, and greater susceptibility to spectral interferences. In addition, the intensity of most common CS decreases dramatically below 280 nm. Consequently, the use of a CS for AAS requires the redesign of the whole instrument (Harnly, 1999). An overview of some historic approaches to CSAAS can be found in a review article by Harnly (1999). The breakthrough in this technique has been achieved by Becker-Ross and coworkers (1996) with an instrumental concept that has been continuously improved over the past years, and which is outperforming conventional LSAAS in many respects. It consists of a high-intensity xenon short arc lamp, a high-resolution double Echelle monochromator (DEMON), and a linear charge-coupled device (CCD) array detector as shown in Fig. 17.

1. Instrumental Concept

The *xenon short arc lamp* with a nominal power of 300 W, which is used as the primary radiation source, has an electrode distance of ≤1 mm. Owing to its specific electrode design and a high gas pressure of about 17 atm in cold

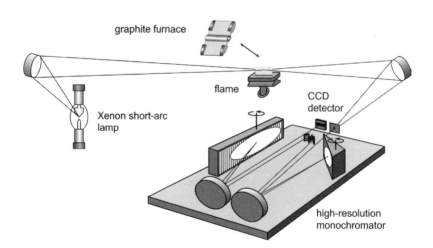

Figure 17 Instrumental concept for high-resolution CSAAS. [From Welz et al. (2003).]

condition, the lamp operates in a hot spot mode, which leads to an increase in radiation intensity, especially in the UV range. The spectral radiance of this lamp exceeds that of conventional LS by at least one to two orders of magnitude over the entire spectral range covered by AAS. Special hardware and software have been developed to compensate for the "migration" of the arc (i.e., its local instability) using a "center of gravity" correction of its position within the radiation train. The fluctuations over time, which are typical for xenon arc lamps, are controlled using selectable correction pixels of the CCD array detector, as will be discussed later.

The radiation is focused by an off-axis ellipsoidal mirror into the absorption volume of an atomizer, and subsequently onto the variable entrance slit of the DEMON. The DEMON consists of a prism premonochromator for order separation and an Echelle monochromator for simultaneous recording of small sections of the high-resolved spectrum. Both units are in Littrow-mounting with focal lengths of 300 and 400 mm, respectively, resulting in a total spectral resolution of $\lambda/\Delta\lambda \approx 140{,}000$ combined with a stable and compact design. For wavelength selection both components (prism and grating) are rotated by means of stepping motors. Moreover, the system includes the possibility of active wavelength stabilization via spectral lines from an internal Ne lamp. A spectral range of 190–900 nm is covered, and the instrumental bandwidth (with a spectral slit width of 24 μm) was determined to be 1.6 pm at 200 nm and 8.8 pm at 900 nm.

A UV sensitive linear array (CCD) detector with 512×58 pixels, size 24 μm × 24 μm, records the spectral radiation distribution. About 200 pixels are typically used in the UV to record the absorbance at the analytical line and in a spectral environment of ca. ± 0.2–0.3 nm around the analytical line. All the photosensitive pixels convert the incident photons independently and simultaneously into photoelectrons and store them within the irradiation time of typically 10 ms. The stored charge pattern is transferred for all pixels simultaneously into the readout register, and subsequently converted into charge-proportional voltage impulses in the on-chip amplifier, where they are amplified and digitalized. The next irradiation of the photosensitive pixels is already going on during this readout. The CCD controller permits the recording of up to 60 subsequent scans per second of the complete spectrum covered by the CCD array detector.

2. Readout and BC Capabilities

Figure 18 shows a typical readout of the integrated absorbance measured at each pixel over a spectral range of about 0.2 nm. This figure also exhibits the resolution of the absorption line; the center pixel measures the absorbance only in the line core, whereas the full width of the absorption line extends over about five pixels. The greatest linear working range is obtained when only the center pixel is used for measurement. However, using the

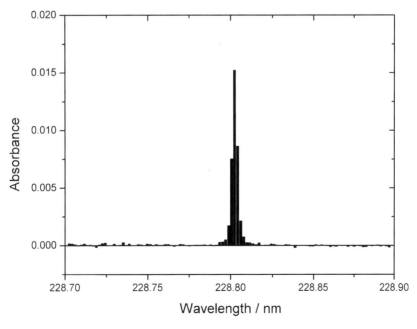

Figure 18 Integrated absorbance spectrum for 40 μg/L Cd recorded in the environment of the resonance line at 228.802 ± 0.1 nm. [From Welz et al. (2003).]

center ±1 pixel results in higher sensitivity and a better S/N ratio. The absorbance is measured over time with each individual pixel, resulting in a 3D output, as shown in Fig. 19, which provides a much greater amount of information when compared with LSAAS, particularly in the case of transient absorption pulses.

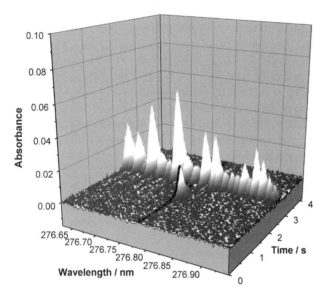

Figure 19 Atomic and molecular absorbance signals recorded in the vicinity of the thallium resonance line at 276.787 nm during the atomization of marine sediment. [From Welz et al. (2002).]

One of the important features of the software is the automatic correction for all "events" that are "continuous" within the observed spectral range (i.e., that influence all pixels of the CCD array in the same way). The most important assumption for this kind of correction is that variations in the intensity of the CS as well as continuous background absorption are perfectly correlated in time within the small spectral range of 0.3–0.6 nm that is recorded. This is guaranteed by the fact that all pixels simultaneously convert the incident photons into photoelectrons and are read out simultaneously, so that proportional variations in the intensity are precisely converted into proportional changes in the digitalized signals for each individual pixel. Hence any pixel or set of pixels can be selected to correct for changes in the lamp intensity and/or for background absorption. And as the measurement is truly simultaneous, even rapid changes in intensity can be corrected without problems.

Obviously, this procedure cannot correct for any absorption due to atoms other than the analyte or to gaseous molecules that exhibit a fine structure at the position of the analytical line. Fortunately direct overlap of two atomic lines within the small spectral range of a few picometers (i.e., within the three pixels that are typically used for measurement) is extremely rare in atomic absorption. Hence the high resolution of the spectrometer avoids these problems *a priori*, and even in the case of a molecular absorption with fine structure it is often possible to separate the atomic from the molecular absorption, as is shown in Fig. 20. However, in the case that neither a spectral separation nor

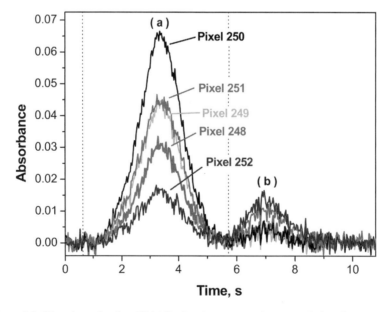

Figure 20 Determination of thallium in coal using CSAAS: absorbance over time recorded at the center pixel (250) and at the two neighbor pixels at both sides of the thallium line at 276.787 nm: (a) atomic absorption and (b) molecular absorption; dotted lines are the integration limits for measuring atomic absorption. [From Silva et al. (2004).]

a separation in time between the absorption pulse of the analyte and the background absorption is possible, there is yet another option to correct for spectral interferences "under the line". The software offers the possibility to measure and store reference spectra of atoms and molecules, which may later be subtracted from the spectrum measured for an actual sample, using a least squares algorithm (Becker-Ross et al., 2000). The mathematical procedure used in this case is a linear fit of the reference spectrum to every single sample spectrum. The reference spectrum will be increased or decreased by multiplication with a magnification factor. The differences between the reference spectrum and the sample spectrum as well as their squares will be calculated pixel by pixel, and the sum of the square values over all pixels will be added up. After that the mentioned magnification factor will be varied in order to minimize the sum of the squares or, in other words, to find the "least squares". Using this procedure, specifically that part of the structured background will be eliminated that corresponds to the fine structure of the reference spectrum. A linear combination of more than one reference spectrum can be used with the same target. Obviously, using this option not only makes it possible to correct for any type of structured background absorption, even under the line, but also presents a valuable tool for identifying the source of spectral interferences.

V. THE TECHNIQUES OF ATOMIC SPECTROMETRY

The flame, graphite furnace, and chemical vapor generation techniques are discussed in some detail in this section, especially with respect to their particular characteristics. The typical processes taking place in each type of atomizer are treated. Emphasis is placed on practical analytical aspects, such as, for example, mechanisms of atomization and interferences characteristic for each technique and their avoidance or elimination.

The *atomizer* is the "place" in which the analyte is atomized (i.e., the flame, the graphite tube, or the quartz tube). The *atomizer unit* encompasses, in addition to the atomizer, all assemblies required for operation (e.g., a burner with nebulizer and gas supply or a graphite furnace with power supply). The portion of the atomizer through which the measurement radiation beam passes is termed the *absorption volume*. The task of the atomizer is to generate as many free atoms in the ground state as possible and to maintain them in the absorption volume as long as possible in order to obtain optimum sensitivity. The most important criteria for the selection of a suitable atomizer for a given analytical task are the concentration of the analyte in the sample, the amount of sample available, and the state of the sample (solid, liquid, and gas).

In OES the analyte not only has to be atomized, but also has to be excited, and the atomizer is in this case termed the *radiation source*, and the part of the flame that is used for measuring the emission intensity is termed the *observation zone*.

A. The FAAS Technique

The flame technique is the oldest of the AAS techniques. For many years it was the "work horse" for the determination of secondary and trace elements, and even nowadays it is difficult to imagine a routine analytical laboratory without this technique. In flame atomization, either an indeterminate volume or a fixed aliquot of the measurement solution is converted into an aerosol in a nebulizer and transported into the flame. The flame must possess enough energy not only to vaporize but also to atomize the sample rapidly and quantitatively. The chemical composition of the flame can have a major influence on these processes.

1. Spectroscopic Flames

The task of the flame is to vaporize and convert the entire sample as far as possible into gaseous atoms (i.e., not only the analyte but also the concomitants). At the same time the flame is also the absorption volume, a fact that places a number of prerequisites on an ideal flame:

1. The flame must supply sufficient thermal energy to rapidly atomize the sample independent of the nature and quantity of the concomitants, but without causing noticeable ionization of the analyte.
2. The flame must provide a suitable chemical environment that is advantageous for atomization.
3. The flame should be transparent to the absorption radiation and should not emit too strongly.
4. The flame should allow low gas flow rates (burning velocity) so that the atoms remain in the absorption volume for as long as possible.
5. The flame should be as long as possible to provide high sensitivity.
6. Flame operation should be safe, and both the flame gases and the combustion products should not pose health and safety risks.

Nowadays, two flame types are used almost exclusively in AAS, the air–acetylene flame and the nitrous oxide–acetylene flame. The two flame types complement each other in an ideal manner and come amazingly close to the requirements of an ideal flame. The fuel-to-oxidant ratio should be optimized for each analyte with the S/N ratio (not only the sensitivity) as the main criterion.

For some 35 elements the air–acetylene flame, with a maximum temperature of 2250°C and a burning velocity of 158 cm/s, offers an environment and a temperature suitable for atomization; only the alkali elements are noticeably ionized. The flame is completely transparent over a wide spectral range and only starts to absorb radiation below 230 nm. The emission of the air–acetylene flame is very low as well so that ideal conditions are given for many elements. Normally this flame is operated stoichiometrically or slightly oxidizing; however, the ratio of the flame gases is variable over a wide range, thus further increasing its applicability.

There are about 30 elements that in effect cannot be determined satisfactorily in the air–acetylene flame as they form refractory oxides, and the partial pressure of oxygen is rather high in this flame. For these elements the nitrous oxide–acetylene flame with a temperature of 2700°C, a burning velocity of 160 cm/s, and a much more reducing atmosphere offers a highly suitable thermal and chemical environment for high degrees of atomization. Nevertheless, the nitrous oxide–acetylene flame has two disadvantages that must be taken into consideration. Firstly, many analytes, particularly alkaline earth and rare-earth elements are more or less strongly ionized in the hot flame and thereby show reduced sensitivity. Secondly, the flame exhibits a relatively strong emission that can on occasions significantly contribute to the noise level.

2. Nebulizer–Burner Systems

Nowadays, so-called premix burners, as shown schematically in Fig. 21, are used exclusively in FAAS because the laminar flame offers excellent conditions for performing determinations with minimum interference. In systems of this type the measurement solution is aspirated by a pneumatic nebulizer and sprayed into a spray chamber. The sample aerosol is mixed thoroughly with the fuel gas and the auxiliary oxidant in the chamber before leaving the burner slot, above which the flame is burning. Depending on the burner head, the flame is generally 5–10 cm long and a few millimeters wide. Normally, the radiation beam passes through the entire length of the flame. Although the flame is supported by a homogeneous gas mixture, both the temperature distribution and chemical composition of the flame are not homogeneous, mostly because of ambient air mixing with the outer flame zones. This results in the formation of a concentration gradient for the analyte atoms over the flame height and perpendicular to the radiation beam, which makes it necessary to adjust the burner so that the radiation beam passes through the zone of highest atom concentration.

Concentric nebulizers are used almost exclusively in FAAS to aspirate the sample solution. Their popularity

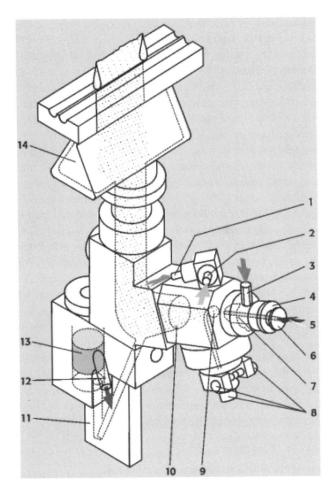

Figure 21 Typical premix burner with nebulizer and burner head: 1, oxidant supply; 2, fuel supply; 3, nebulizer oxidant supply; 4, nebulizer adjustment screw; 5, nebulizer capillary (measurement solution supply); 6, locking nut; 7, nebulizer; 8, impact bead adjustment; 9, impact bead; 10, mixing chamber; 11, siphon; 12, siphon outlet; 13, float; 14, burner head. (By kind permission of Analytik Jena AG.)

stems from their relatively simple design, which allows the manufacture of nebulizers with reproducible properties, their mechanical stability (they are made of corrosion-resistant metals, such as Pt, Pt/Ir, Ti, or of polymers), and the fact that they can aspirate the measurement solution unaided. The aspiration rate of concentric nebulizers is determined by the pressure drop at the liquid capillary, and is described by the Hagen–Poiseuille equation:

$$Q = \frac{P r^4 \pi}{800 \eta l} \qquad (13)$$

where Q is the aspiration rate in mL/s, P is the pressure in Pa, r is the radius of the capillary in mm, η is the viscosity of the aspirated liquid in poise, and l is the length of the capillary in mm. The only parameter that depends on the sample is the viscosity, which must be maintained constant in order to avoid problems, as will be shown in Section V.A.4. The aspiration rate of nebulizers used in FAAS is typically between 5 and 10 mL/min.

Besides aspirating the measurement solution, the nebulizer also converts the aspirated solution into an aerosol, which is a particularly suitable form of presentation for the measurement solution as the drying and volatilization processes can be favorably influenced because of the very large surface-to-mass ratio. In a pneumatic nebulizer the nebulizing gas provides the energy that is required to divide a quantity of liquid into fine droplets in order to generate the *primary aerosol*. In FAAS the nebulizing gas is mostly the oxidant or part of it.

The primary aerosol generated by the nebulizer generally has a wide range of droplet sizes and in the time available the larger droplets cannot be completely vaporized and atomized in the flame. Furthermore, in pneumatic nebulizers the aerosol is generated with a large excess of energy that must be reduced by calming the turbulent gas streams if negative influences on the flame stability are to be avoided. This "conditioning" of the aerosol is performed in the spray chamber with its inserts or baffles; postnebulization and/or separation of large droplets takes place, the aerosol is thoroughly mixed with the flame gases, and partial vaporization of the solvent occurs, resulting in the *secondary aerosol*. Frequently, *impact beads* are used in spray chambers to fragment larger droplets through the high-velocity impact with the surface of the bead, and also to separate droplets from the aerosol stream that were not sufficiently reduced in size through postnebulization. The larger droplets can also be removed from the aerosol stream by deflecting the gas stream in the spray chamber by means of *flow spoilers* because, as a result of their inertia, the larger droplets cannot follow rapid changes in direction of the gas stream. In addition, flow spoilers enhance mixing of the aerosol with the flame gases, leading to an improvement of the flame stability. The processes of sedimentation, vaporization, and turbulent and centrifugal separation form the *tertiary aerosol*, which is finally transported into the flame. Owing to all the separation processes in the spray chamber, where larger droplets are eliminated, only about 5% of the originally aspirated sample volume eventually reaches the flame. Most of the attempts to increase this figure were not very successful, as the gain in sensitivity is in most cases accompanied by a proportional increase in interferences, unless almost matrix-free solutions are investigated.

The task of the *burner head* is to conduct the premixed gases to the combustion reaction in such a way that a stable flame is produced. In order to obtain high sensitivity, the burner slot should be as long as possible. Nevertheless, the geometry of the burner head must be suitable for the oxidant and fuel gas to be used; otherwise compromises in the performance must be taken into account. The dimensions of the burner slot must be such that the exit flow velocity of the flame gases is always greater than the burning velocity of the flame. If this condition is not met the flame can flash back into the spray chamber and cause explosive combustion of the gas mixture. Another important feature of the burner head is a low tendency to form encrustations from the fuel gas or sample constituents, even with high dissolved solids content in the solution.

3. Atomization in Flames

The processes that lead from aspiration of the measurement solution to the formation of free atoms in the vapor phase are depicted schematically in Fig. 22; nebulization and the formation of the aerosol have been described in detail in Section V.A.2. The tertiary aerosol enters the flame where it is dried (i.e., the remaining solvent is vaporized); the particles generated this way are volatilized and the gaseous molecules dissociate into free atoms.

We shall not consider the notoriously poor efficiency of the pneumatic nebulizer in our discussion on atomization. As our starting point we shall consider the aerosol that effectively reaches the flame so that we can distinguish between improvements in the degree of nebulization and improvements in the degree of atomization. Moreover, particularly for matrix-free solutions, we can assume that with a premix burner and a flame of sufficient temperature the analyte is transferred completely to the vapor phase (i.e., all particles are volatilized). The validity of this assumption obviously depends on the maximum particle size in the tertiary aerosol, which should be ≤ 10 μm. As the burning velocity of the flames used in FAAS is about 160 cm/s, the time available to complete all the processes shown in Fig. 22 is only a few milliseconds. We must therefore pose the question whether the available time is sufficient for equilibrium to be reached. Deviations from equilibrium can occur when the reaction speed of a process is too slow. In this case, concomitants can have a catalytic effect on atomization or a competitive process, and hence cause signal enhancement or depression.

In the *vapor phase* the analyte can exist as atoms or in molecular form; both forms can be ionized or excited to varying degrees. The sum of the partial pressures of all these forms of the analyte, assuming complete volatilization, is of the order of 10^{-1}–10 Pa. This is a negligibly small value in comparison to the partial pressures of the flame gas components. Therefore, if a matrix-free analyte solution is nebulized, the composition of the components in the flame can change as a result of the solvent, but not as a result of the analyte.

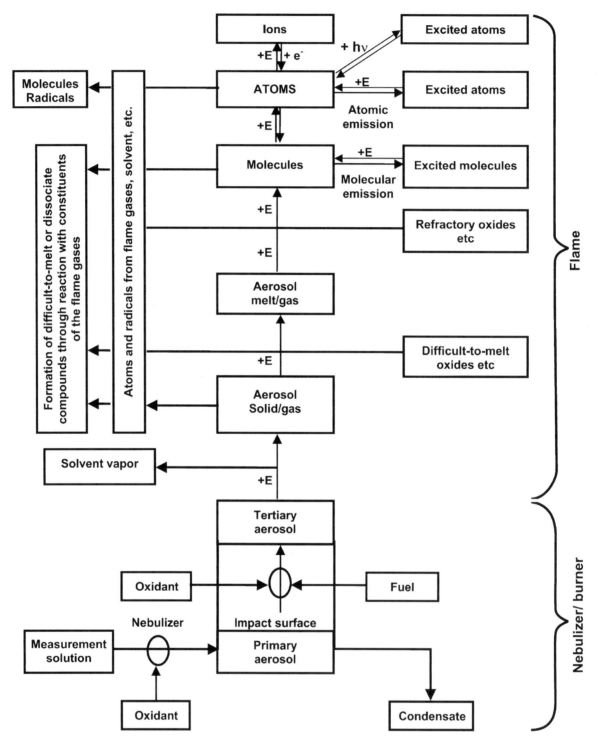

Figure 22 Schematic representation of the most important processes possible in a flame. In addition to the introduction of thermal energy E the chemical processes taking place in the flame also have a major influence. [From Welz and Sperling (1999).]

Diatomic molecules predominate in the gas phase; triatomic species are limited to the monohydroxides of alkali and alkaline earth elements, to monocyanides, and to suboxides such as Cu_2O. The majority of other polyatomic compounds dissociate very rapidly at temperatures well below those prevailing in analytical flames. The dissociation energies E_D of diatomic molecules are generally in the order of 3–7 eV. In the usual analytical flames, molecules are mostly atomized completely at $E_D < 3.5$ eV; compounds with $E_D > 6$ eV are classed as difficult-to-dissociate. The *degree of dissociation* can change significantly with the flame temperature and composition; the solvent can also have an influence. The description of equilibria by means of dissociation constants is valid, independent of the mechanism; in other words, provided that the system is in equilibrium, it is immaterial whether it is a true thermal dissociation or whether chemical reactions are involved. Merely the *velocity* with which equilibrium is attained depends on the reaction mechanism.

Dissociation equilibria involving *components of the flame gases* must be accorded special attention. Typical examples are the dissociation of oxides, hydroxides, cyanides, and hydrides. The partial pressures of O, OH, CN, and H in a flame are determined by reactions between the natural components of the flame. The influence of any sample constituent on these components is negligible, since every reaction that leads to a reduction in the concentration of a species is immediately counteracted by infinitesimal shifts in the equilibria of the main components. This *buffer effect of the flame gases* with respect to the concentrations of O, OH, CN, and H still takes place even when the components of the flame gases are not in equilibrium. For this reason mechanisms of interference cannot be explained by "competition in the flame for available oxygen", and so on. Apparent discrepancies between calculated and experimentally determined values for the degree of atomization of monoxides can be explained, at least in part, by the participation of free radicals in the reduction process.

Organic solvents were used already in the early years of flame emission spectrometry to increase the sensitivity (Bode and Fabian, 1958a, b) and have been described as early as 1961 for FAAS (Allan, 1961). Most organic solvents have a lower viscosity and surface tension than water and are thus more easily aspirated and more finely nebulized. Organic solvents generally shift the range of droplet sizes toward smaller droplets (Farino and Browner, 1984), which results in higher nebulization efficiency, so that a higher proportion of the measurement solution reaches the flame. The dissociation of water is a strongly endothermic reaction that noticeably reduces the flame temperature, whereas the combustion of organic solvents (except for highly halogenated) is generally an exothermic reaction that increases the flame temperature. This, together with the lower thermal stability of organic molecules can lead to a significant enhancement of the degree of atomization. An important factor, however, which has to be considered, is the water content of the organic solvent. Solutions saturated with water, such as occur with solvent extractions, frequently enhance the signal far less than water-free solutions.

4. Interferences in Flames

Spectral Interferences

Spectral interferences caused by direct overlapping of the analytical line emitted by the radiation source and the absorption line of a concomitant element have not been reported for the main resonance lines, and are limited to a few individual cases, when less sensitive alternate lines are used. The background attenuation observed in FAAS is more or less exclusively caused by molecular absorption; nevertheless, it rarely reaches such a magnitude that it causes real problems. This kind of interference occurs most frequently when an easily atomized element is determined in the presence of a concomitant element that, for example, forms a difficult-to-dissociate oxide or hydroxide in the air–acetylene flame. The alkali halides are further examples of molecular absorption spectra that have been carefully investigated (Fry and Denton, 1979). The background absorption caused by real matrices, such as urine or seawater, could in all cases be easily controlled using BC with a CS.

The only exceptions to this are the few cases where the molecular absorption spectrum exhibits a *fine structure*, as has been reported for the InCl absorption in the environment of the 267.6 nm line for gold (Höhn and Jackwerth, 1976), and for the OH molecular absorption around the 306.8 nm bismuth line (Massmann, 1982). The latter case is shown in Fig. 23, using high-resolution CSAAS. Obviously, in such situations conventional BC with a CS is not only unable to remove the interference, but can even make it worse. Becker-Ross et al. (2002) have recently also shown the appearance of broad-range, narrow-structure background absorption of high concentrations of phosphoric acid, iron, and copper in an air–acetylene flame using a high-resolution spectrometer.

Nonspectral Interferences

Transport interferences occur when the mass flow of the analyte that finally arrives in the flame is changed by the presence of concomitants in the measurement solution in comparison to the calibration solution. The only parameter that influences the aspiration rate of the nebulizer is the *viscosity* of the measurement solution (refer to Section V.A.2). In addition to the aspiration rate, the range of droplet sizes, and hence the portion of the measurement solution that reaches the flame is also influenced by the viscosity and the *surface tension* of the solution. If it is not possible to keep these parameters constant in all measurement solutions, the differences have to be considered by the use of an appropriate calibration technique, such as the analyte addition technique (refer to Section VI.B.2) to avoid errors.

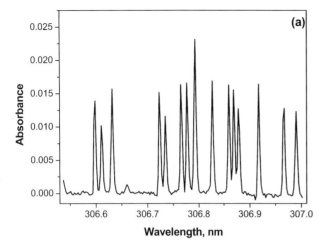

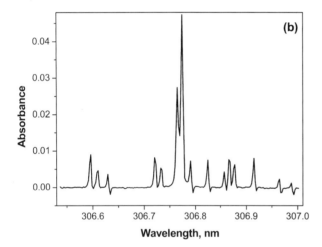

Figure 23 Molecular absorption of the OH molecule in the vicinity of the bismuth line at 306.77 nm in an air–acetylene flame recorded with HR-CSAAS: (a) OH absorption spectrum for water only and (b) 10 mg/L Bi solution; the two doublet lines of Bi coincide directly with two of the OH absorption lines.

Solute volatilization interferences are based on a change in the rate of volatilization of the aerosol particles in the presence of a concomitant. The volatilization of solid particles in hot gases is one of the least researched processes in flame spectrometry and thus based largely on conjecture. For this reason this kind of interference will be treated only in a very qualitative manner; for more details see Welz and Sperling (1999). Firstly, this interference depends to a great extent on the range of droplet sizes and particularly on the maximum droplet size that is reaching the flame (i.e., from the efficiency of the spray chamber and the quality of the tertiary aerosol produced there). Obviously, solute volatilization interferences are caused predominantly by concomitants that form difficult-to-volatilize compounds and that are present at high concentration. A typical example is the interference of aluminium on the determination of calcium in an air–acetylene flame (Welz and Luecke, 1993). Similar interferences are caused by phosphates and other refractory oxides. There are in essence two ways to deal with this kind of interference: firstly, there is the use of the hotter nitrous oxide–acetylene flame that provides a much more reducing atmosphere and prevents the formation of big clusters of refractory oxides that can trap the analyte; secondly, there is the use of releasers such as lanthanum chloride or ammonium salts that break the refractory oxide clusters at elevated temperatures.

Vapor phase interferences are all those effects that are caused by concomitants in the gas phase after complete volatilization of the analyte. A flame is a dynamic system, but we can nevertheless assume that the residence time of the sample in the absorption volume is long enough in the first approximation to permit equilibrium between free atoms and compounds (dissociation equilibrium), ions (ionization equilibrium), and excited atoms (excitation equilibrium) to be reached. These equilibria can be described by the law of mass action, the Saha equation, and the Boltzmann distribution law. All processes are temperature dependent, and as temperature varies within the flame, the concentrations of individual components change with flame height. As AAS is a relative technique, we cannot consider the incomplete conversion of the analyte into gaseous atoms as a result of the preceding equilibria to be an interference. Merely the *change in the equilibrium* brought about by concomitants constitutes an interference. A vapor phase interference presupposes that the analyte and the interferent are simultaneously present in the vapor phase and that they have a *common third partner*, such as a common anion or a free electron, which can influence the equilibrium.

As the excitation equilibrium is shifted completely in the direction to the non-excited state under normal conditions used for AAS, we do not observe *excitation interferences*. The only possible interferences in the vapor phase are thus dissociation and ionization interferences.

Dissociation interferences can be expected when equilibria contain components that are not part of the flame gases, such as halides. If the analyte is present as the halide, it is usually completely volatilized, but dissociation into atoms does not by any means have to be complete. The existence of nondissociated halides has been demonstrated repeatedly through the molecular spectra of monohalides. As the dissociation constants of halides are temperature dependent, the strongest dissociation interferences are found in flames of low temperature. In general we can say that when the acid concentration is >1 mol/L this determines the anion in the environment around the analyte.

The dissociation energies of the halides increase with decreasing atomic mass of the halogens. The determination of indium and copper are examples in which the increase in the dissociation energies in the sequence HI < HBr < HCl agrees with the level of depression. Often, though, the interferences caused by halogens are much more complex as they can also (positively) influence volatilization.

Thermal ionization is unlikely at the temperatures prevailing in flames normally used in AAS. On the other hand, ionization occurs in the primary reaction zone because of charge transfer from molecular ions such as $C_2H_3^+$ or H_3O^+ and collisional ionization with relatively long times of relaxation takes place in higher reaction zones. This can lead to a shift in the Saha equilibrium. All ionization reactions involving species produced in the primary reaction zone come to equilibrium slowly, so that the concentration of neutral analyte atoms increases with increasing observation height.

The question as to whether noticeable ionization will occur and whether an ionization interference is to be expected depends on the ionization energy of the analyte, on possible interfering concomitants, and on the flame used. In the hot nitrous oxide–acetylene flame, ionization normally occurs for elements having ionization energies of <7.5 eV. In flames of lower temperature, ionization is practically limited to the alkali metals. Moreover, the extent of ionization for a given element and temperature is dependent on the concentration of the analyte. As shown in Fig. 24 for the example of barium, ionization is stronger at lower concentrations than at higher. This results in a concave calibration curve, which can be explained by the more efficient recombination of ions and electrons at higher concentrations.

It should be pointed out that the ionization of an analyte under given conditions is not an interference, but only the alteration of the degree of ionization by a concomitant element. Nevertheless, the nonlinearity of the calibration curve caused by ionization has two major disadvantages for practical analysis: firstly, numerous calibration solutions are required to establish the curve. Secondly, the loss of sensitivity in the lower range precludes trace analysis. For these reasons elimination of ionization is very desirable.

In principle there are two ways of suppressing ionization. Firstly, we can determine the analyte in a flame of lower temperature. This is possible for the alkali elements, but it is not practicable for the majority of elements. The second and usually more amenable way to suppress ionization is to add a large excess of a very easily ionized element (usually potassium or cesium), which acts as an ionization buffer by producing a large excess of electrons in the flame. The corresponding increase in the analyte signal (see Fig. 24) can be explained by the fact that the partial pressure of electrons generated by the ionization buffer shifts the ionization equilibrium of

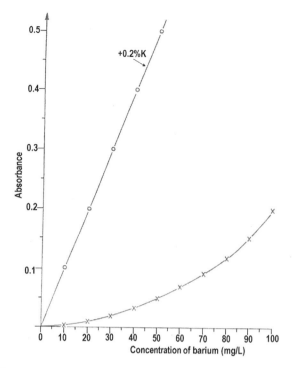

Figure 24 Effect caused by ionization in the determination of barium in a nitrous oxide–acetylene flame. The lower curve (×) was obtained for measurement of aqueous solutions, while the upper curve (○) was obtained after addition of 2 g/L K to suppress ionization. (From Welz and Sperling (1999).]

the analyte in favor of the uncharged atoms. The ease with which the ionization equilibrium can be shifted even by relatively low concentrations of concomitants makes ionization a potentially real interference and is an additional good reason why it should be suppressed.

As solute volatilization interferences and dissociation interferences decrease with increasing flame temperature, but ionization interferences increase, we must often judge which one is the more serious. The choice often falls to the flame of higher temperature since ionization is in general much easier to control.

B. The GFAAS Technique

In GFAAS a measured volume of the sample solution, usually 10–50 μL, is dispensed into the atomizer and the temperature is increased stepwise to remove the solvent and concomitants as completely as possible before atomization. As the entire aliquot introduced into the graphite tube is atomized within a short time, typically 1–3 s, a peak-shaped, time-dependent signal is generated whose area (integrated absorbance) is proportional to the mass of the analyte in the measurement solution. As the entire sample is atomized and because of the much longer residence time of the atoms in the absorption volume when compared with FAAS, the sensitivity of GFAAS is some two to three orders of magnitude higher than that of FAAS.

1. Graphite and Graphite Tube Atomizers

Graphite is an allotropic form of carbon formed by sp^2 hybridization of the orbitals. The σ electrons form strong covalent bonds in the basal plane, whereas the π electrons form weak van der Waals bonds between the planes. Atomizers for GFAAS are nowadays made largely of polycrystalline electrographite (EG), which is particularly suited for this purpose because of its mechanical and electrical properties. Polycrystalline EG consists of graphite grains held together by a binder of amorphous carbon. Micropores with dimensions less than 2 nm are found to be within the grains, whereas macropores in the micrometer range can be found in the carbon binder bridges. Due to its "open" structure that makes polycrystalline EG rather reactive, particularly at higher temperatures, it is not well suited from a chemical point of view and as "atomization surface". Hence, the surface of polycrystalline EG has to be sealed, and an $\sim$50 μm thick coating with pyrolytic graphite (PG) is typically used for that purpose, as shown in Fig. 25. A coating of PG, due to the extremely low permeability of this material, markedly reduces the reactivity of the atomizer surface. All open pores of the EG substrate are sealed, and the susceptibility of PG to oxidation vertically to the planes is low because of the high crystalline order.

The *dimensions and shape* of graphite tubes used as atomizers in GFAAS are determined by the conflicting conditions for high sensitivity, good S/N ratio, and low electrical power consumption. The mean residence time of the atoms in the absorption volume, and hence also the sensitivity, is directly proportional to the square of the tube length. In addition, because the volume increases as the square of the diameter, the cloud of atoms will be correspondingly diluted (i.e., the absolute sensitivity in a graphite tube is inversely proportional to the tube diameter). From this point of view, in order to obtain the maximum sensitivity, graphite atomizers should be as long and as thin as possible. However, the energy requirement for heating also increases with the length of the tube, and the diameter has to be large enough to allow the radiation beam to pass the tube without obstruction. Unless there are other limiting conditions, such as a longitudinal magnetic field (refer to Section IV.A.6), graphite tubes are typically 30 mm long with an internal diameter of 6–7 mm.

Graphite atomizers are heated because of their ohmic resistance by passing a high current at low voltage along or across the tube, which means that the tubes have to be electrically contacted. In order to avoid contamination

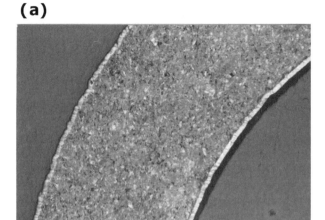

Figure 25 Polished section through an EG tube with a ca. 50 μm coating of PG viewed in polarized light: (a) 50× magnification and (b) 500× magnification. [From Welz and Sperling (1999).]

from metal parts, these contacts have to be made of graphite as well, and they have to be cooled in order to avoid arcing and sparking between contact and tube. There are two fundamentally different design approaches for graphite tubes, longitudinally and transversely heated atomizers. *Longitudinally heated graphite tubes* are simple in design, but as they are contacted at their ends, they exhibit a significant temperature gradient over their length. *Transversely heated graphite tubes* are more complex in design, as the contacts have to be integrated in the tube (i.e., tube and contact are made of one piece of graphite, as shown in Fig. 26); however, they offer the great advantage of a nearly isothermal temperature distribution over the tube length (i.e., throughout the entire absorption volume).

Since the pioneering research of L'vov (1978) and the transfer of his ideas into practice by Slavin et al. (1981) with the STPF concept (refer to Section V.B.4), it is common knowledge that the sample should be deposited

Figure 26 Graphite tube with PIN platform and integrated contacts for transverse heating. (By kind permission of Analytik Jena AG.)

onto and atomized from a *platform* in the graphite tube, and not from the tube wall. As the purpose of the platform is to delay atomization of the analyte until the atomizer and the gas atmosphere have stabilized in temperature, it has to be heated by radiation and not by the electrical current. This means that the platform should have *minimal contact* with the tube, although it has to be fixed in a stable position in the tube in order to guarantee reproducible conditions. These conditions are met best by so-called "pin" or "integrated" platforms that are connected with the tube in a single "point" (see Fig. 26).

In order to allow safe operation the graphite tube has to be installed in a housing, the *furnace* that protects the operator from excessive thermal and optical radiation, and the graphite tube from contact with ambient air. The graphite tube is usually surrounded by a pair of *graphite contacts* that transfer the electrical energy to the atomizer and keep the radiation losses, and hence the atomizer temperature, within narrow limits. These contacts are mounted in a *water-cooled metal housing* that keeps the furnace at a safe temperature and cools the graphite tube after the atomization stage to ambient temperature quickly in order to enable a high sampling frequency. The furnace is flushed by a stream of argon, and we have to distinguish between the protective gas and the purge gas flow. The *protective gas flow* is conducted around the graphite tube and protects the hot graphite tube from contact with ambient air (i.e., from being burnt); this protective gas flow is on all the time during the use of the graphite furnace. The *purge gas flow* is conducted through the graphite tube, entering from both ends, in order to quickly remove sample constituents that are volatilized during the drying and pyrolysis stages (refer to Section V.B.2); this purge gas flow is turned off during atomization in order to increase the residence time of the analyte atoms in the absorption volume. The furnace is usually closed at its ends with removable *quartz windows* to further protect the graphite atomizer from entraining air, and to allow better control of the inert gas flows in the furnace. Another feature that is usually included in the furnace compartment is a device for measuring and controlling the *atomizer temperature* in order to have reproducible atomization conditions over the lifetime of the tube. On the other hand, there is usually no need for *furnace adjustment*, as the radiation beam, which has to pass through the center of the tube without being obstructed, determines the position of the graphite tube.

2. Temperature Program and Heating Rate

As already mentioned, in GFAAS a known volume of the measurement solution is introduced into the graphite tube and subject to a *temperature program* in order to remove as much of the sample constituents as possible before atomization. Typical stages of this temperature program are drying, pyrolysis, atomization, cleaning, and cooling, and the individual stages may include a temperature *ramp* (i.e., a controlled increase of temperature over time) and isothermal *hold* times.

For *drying* usually a relatively fast ramp is used to a temperature some 10°C below the boiling point of the solvent (i.e., 90°C for water), followed by a short hold time (e.g., 10 s) and a slow ramp to a temperature about 20°C above the boiling point, followed by a longer hold time of about 20–40 s, depending on the injected volume. Complex matrices such as blood or urine obviously might need a more sophisticated drying program. It is strongly recommended to watch the drying process with a mirror during method development.

Pyrolysis and atomization curves are usually established in order to find the optimum conditions for these stages. In the *pyrolysis curve*, an example of which is shown in Fig. 27(a), the integrated absorbance measured for atomization at the optimum temperature is plotted against the pyrolysis temperature as the variable. In the *atomization curve*, which is shown in Fig. 27(b), at optimum pyrolysis temperature the atomization temperature is varied and the integrated absorbance is measured in each case. The pyrolysis curve usually consists of a "flat" part, where the pyrolysis temperature has no influence on the integrated absorbance measured in the atomization stage, indicating that no analyte is volatilized (and lost) under these pyrolysis conditions. Above a certain maximum pyrolysis temperature, however, the signal measured during atomization begins to drop significantly, indicating analyte losses in the pyrolysis stage. Obviously the pyrolysis temperature chosen should be one at which no analyte losses occur, but high enough to remove as much of the concomitants as possible prior to

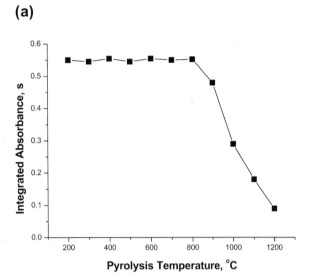

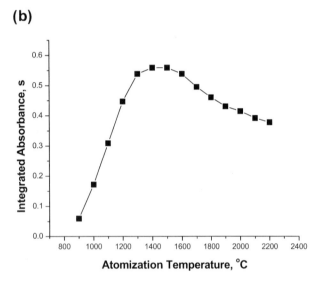

Figure 27 Typical pyrolysis and atomization curves for a volatile element using STPF conditions: (a) pyrolysis curve and (b) atomization curve.

atomization. The atomization curve, at least for volatile and medium volatile elements, typically exhibits a maximum in integrated absorbance, followed by a gradual decrease in sensitivity. At temperatures lower than the maximum the analyte is incompletely atomized; the decrease in sensitivity at higher temperatures is caused by the shorter residence time of the analyte atoms in the absorption volume due to the faster diffusion at higher temperatures. Usually temperatures 100–200°C higher than the maximum are optimum, as the transient signal often is very broad at the maximum, calling for long integration times that result in a deterioration of the S/N ratio. Low volatile elements usually exhibit a maximum, followed by a plateau (in transversely heated atomizers) or do not even reach a maximum (in longitudinally heated atomizers). The ramp and hold times required for pyrolysis depend largely on the matrix, and relatively slow ramp rates and longer hold times are required in the presence of complex matrices. In contrast, the highest available heating rate is typically used for atomization in order to increase the "platform effect" (i.e., the time delay for analyte atomization). The time for atomization should be selected as short as possible, but it should obviously allow the atomization pulse to return to baseline.

The *cleaning stage* has the sole purpose of removing potential residues of the matrix from the graphite tube and avoiding memory effects or reducing them to a minimum. Usually temperatures of 2600–2650°C are applied for the shortest time necessary so as to not affect tube lifetime. There is a significant difference in the necessity of a cleaning stage between transversely and longitudinally heated graphite tube atomizers. As the latter ones exhibit a pronounced temperature gradient over the tube length, there is a high risk that low-volatile analytes and/or matrix constituents condense at the cooler ends, from where they are difficult to remove. Figure 28 shows the memory effects for molybdenum in the two types of atomizers, which are negligible in the transversely heated tube, even using a lower atomization temperature, when compared with the longitudinally heated furnace, which does not allow a reasonable determination of this element.

The *cooling stage* after atomization and cleaning has to be long enough to allow safe injection of the next measurement solution. In many instruments this time is predetermined by the software. Some authors have recommended introducing a cooling stage prior to atomization in order to increase the platform effect (i.e., the delay of analyte atomization); this practice, although widely used, has not found general acceptance.

3. Chemical Modification

In GFAAS we use a temperature program to separate the analyte and concomitants *in situ* prior to the atomization stage. The highest possible pyrolysis temperature is required to effectively separate the concomitants. However, as the analyte must not be volatilized during the pyrolysis stage, there are limitations to the maximum temperature. To determine this maximum temperature we establish pyrolysis curves (refer to Section V.B.2).

However, every element can occur in a large number of chemical compounds, which often differ substantially in their physical properties and thus in their volatilities. Hence the pyrolysis curve depends on the analyte species, which is present in a test sample, or which is formed during pyrolysis. Obviously, this compound depends very strongly on the concomitants and is very often not known. If a pyrolysis curve is established using

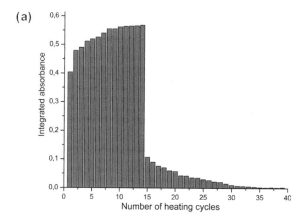

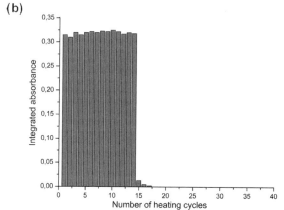

Figure 28 Memory effects for molybdenum atomized in different furnaces (signals 1–14), followed by a series of blank measurements (signals 15–40); (a) longitudinally heated atomizer, atomization from the wall at 2650°C and (b) transversely heated atomizer, atomization from a platform at 2500°C. [From Welz and Sperling (1999).]

matrix-free, aqueous calibration solutions, we have absolutely no guarantee that the analyte species present in the test sample will behave in the same manner. To eliminate this uncertainty and to gain control over the form in which the analyte is present, we use chemical additives that bring the physical and chemical properties of the sample and calibration solutions as close together as possible. This procedure is termed *chemical modification*. In most cases the aim is to form the most stable compound or phase of the analyte. In the ideal case the thermal stability of the concomitants is at the same time lowered. Under no circumstances, however, should the thermal stability of concomitants be increased as the aim of chemical modification is to achieve separation of the analyte and the matrix during the pyrolysis stage.

Schlemmer and Welz (1986) have made a list of selection criteria for an ideal modifier:

1. It should be possible to pretreat the analyte to the highest feasible temperature; a pyrolysis temperature of $\geq 1000°C$ is often required to reduce the bulk of the concomitants significantly.
2. The modifier should stabilize as many elements as possible.
3. The reagent should be available in high purity to prevent the introduction of blank values.
4. The modifier, which is normally added in large excess, should not contain any element that might later have to be determined in the trace range.
5. The modifier should not shorten the lifetime of the graphite tube and platform.
6. The modifier should only make the lowest possible contribution to the background attenuation.

The mixed palladium and magnesium nitrates modifier (Pd–Mg modifier), proposed by the same authors, does not meet these criteria in all points, but it meets them better than any other modifier that has been described to date. Welz et al. (1992) investigated 21 elements which could be stabilized by this modifier. The attainable maximum pyrolysis temperatures and optimum atomization temperatures are presented in Table 2. We must naturally point out that the maximum pyrolysis temperature

Table 2 Maximum Usable Pyrolysis Temperatures Without Modifier and With the Pd–Mg Modifier, and the Optimum Atomization Temperatures with the Pd–Mg Modifier

Element	Maximum pyrolysis temperature (°C)		Optimum atomization temperature (°C)
	Without modifier	Pd–Mg modifier	
Ag	650	1000	1600
Al	1400	1700	2350
As	300	1400	2200
Au	700	1000	1800
Bi	600	1200	1900
Cd	300	900	1700
Cu	1100	1100	2600
Ga	800	1300	2200
Ge	800	1500	2550
Hg	<100	250	1000
In	700	1500	2300
Mn	1100	1400	2300
P	200	1350	2600
Pb	600	1200	2000
Sb	900	1200	1900
Se	200	1000	2100
Si	1100	1200	2500
Sn	800	1200	2400
Te	500	1200	2250
Tl	600	1000	1650
Zn	600	1000	1900

Source: From Welz et al., 1992.

determined without modifier was measured for matrix-free calibration solutions and may be considerably lower in the presence of concomitants. On the other hand, the maximum pyrolysis temperature determined with modifier does not change noticeably even in the presence of higher concentrations of concomitants (Welz et al., 1992).

The stabilization mechanism of the palladium modifier has been investigated in detail by Ortner et al. (2002). Palladium penetrates into the pyrolytic graphite subsurface to a depth of about 10 μm already during the drying stage; the analytes also penetrate into the subsurface, the essential place for graphite-modifier–analyte interaction. During the pyrolysis stage palladium is intercalated into the graphite lattice which results in an activation of the intercalated metal atoms, This activated palladium atoms are able to form strong covalent bonds with the analyte, which explains why many volatile analytes are stabilized to similar temperatures by this modifier (Table 2).

Besides the usual application of a modifier in solution, added together with each measurement solution, the use of *permanent chemical modifiers* is finding increasing acceptance. Ortner and Kantuscher (1975) were the first to propose the impregnating of graphite tubes with metal salts, which resulted in higher sensitivity and longer tube lifetime. Nowadays permanent modifiers, usually platinum-group metals with high melting points, such as Ir, Rh, or Ru; carbide-forming elements, such as W or Zr; or combinations of them, are usually applied by "thermal deposition". A concentrated solution of the modifier is deposited on the platform and a temperature program is applied to dry the solution, reduce the modifier to the metal, and to form the metal carbide. This process is repeated several times until a modifier mass of typically 200–500 μg is deposited on the platform. This "coating" may then be active for several hundred up to more than 1000 consecutive atomization cycles, depending on the analyte and the matrix. The major advantages that have been claimed for permanent modifiers over modifiers applied with the measurement solution are:

1. A simplified operation and a shorter analysis time, as the modifier does not need to be added to each solution of measurement.
2. Longer tube lifetime as the "coating" protects the graphite surface from oxidative attack.
3. Lower blank values due to an *in situ* purification of the modifier during the conditioning program.

Although significant improvement in tube lifetime has also been reported for liquid samples in the presence of corrosive matrices (Silva et al., 1999) the greatest advantages in the use of permanent modifiers was found with the direct analysis of solid samples, and for sequestration of analyte hydrides in the graphite tube (refer to Sections V.B.5 and V.C.5).

4. The STPF Concept

The STPF concept, introduced by Slavin et al. (1981), is a "package" of measures to achieve three aims, or to put it another way, to achieve interference-free GFAAS determinations in three steps, (i) to separate concomitants as far as possible by volatilization prior to atomization, (ii) to control the reactions in the condensed phase to prevent analyte losses and to better separate the concomitants, and (iii) to minimize the influence of nonseparated concomitants on the analyte in the gas phase by as complete atomization as possible.

The STPF concept includes four major measures to achieve these aims: well-controlled, largely inert graphite surfaces, chemical modification, atomization under isothermal conditions, and effective BC. The topic *graphite surface* has been discussed in Section V.B.1, and will not be addressed again, particularly as the use of graphite tubes with a dense layer of pyrolytic graphite is common practice nowadays. *Chemical modifiers* are added to the measurement solution in order to control reactions in the condensed phase with the goal of stabilizing the analyte to high temperatures and to separate concomitants as far as possible (refer to Section V.B.3). The *atomization under isothermal conditions* has been discussed in Sections V.B.1 and V.B.2, and is best realized by atomization from a platform in a transversely heated graphite tube using a high heating rate. This is demonstrated in Fig. 29, which shows the gas phase temperature close to the

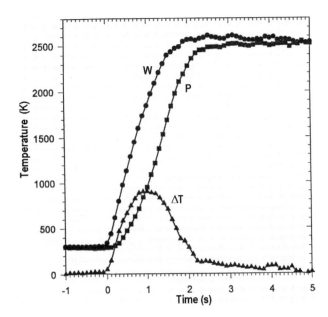

Figure 29 Temperature increase for the gas phase close to the platform (P) and close to the wall (W) of a transversely heated graphite tube, and the temperature difference (ΔT) between the platform and the tube wall. [From Sperling et al. (1996).]

platform and close to the tube wall of a transversely heated graphite tube as well as the temperature difference between platform and wall during the rapid heating cycle of the atomization stage (Sperling et al., 1996). The temperature difference reaches maximum values of almost 1000°C, but disappears just after the tube wall has attained the maximum temperature. Using these conditions and using *integrated absorbance* for signal evaluation, the vast majority of matrix effects can be eliminated in GFAAS. The aspect of *effective background correction* will be discussed in detail in Section V.B.7. The only point that has to be stressed here is that the STPF concept is a package of measures, and it is essential to follow all of the conditions. The risk of interferences increases substantially when only parts of the concept are used, as all of its components are interrelated.

5. Analysis of Solid Samples

Of all the atomization techniques discussed in this chapter, the graphite furnace technique is the only one that is really suitable for the direct analysis of solid samples. In fact, graphite atomizers have been used for this purpose since their introduction. The main reasons for performing the direct analysis of solid samples without prior digestion or dissolution can be summarized as follows:

1. Higher sensitivity for the determination of trace elements, since every solubilization procedure causes substantial dilution.
2. Significantly less time is required for sample preparation when dealing with samples that cannot be dissolved or digested easily, and thus the total analysis time is reduced.
3. The risk of contamination from reagents, apparatus, and the laboratory atmosphere is much lower; this is a particularly important factor for trace analysis.
4. The risk of loss of the analyte during sample preparation due to volatilization, sorption on the apparatus, or incomplete solubilization is much lower.
5. The use of toxic or corrosive reagents is avoided.
6. A microdistribution analysis or homogeneity test can be performed.

Nevertheless, the pros and cons have to be evaluated for each type of application for direct analysis of solid samples. One of the main arguments against the direct analysis of solid samples, the difficulty of *introducing the samples* into the atomizer is no longer valid, as nowadays both manual and automatic accessories for sample introduction are commercially available. Figure 30 shows an automatic system, where a pair of tweezers transfers a graphite boat with the sample to a microbalance and then into the graphite tube for atomization. The sample

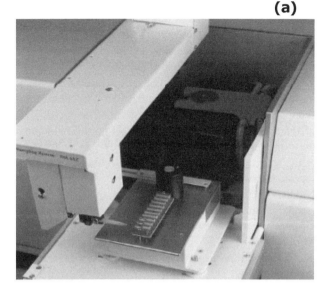

Figure 30 Solid sampling accessory for GFAAS: (a) automatic introduction system for 10 solid sampling platforms and (b) solid sampling platform. (By kind permission of Analytik Jena AG.)

weight is communicated from the balance to the software of the instrument, where the measured absorbance is converted into the absorbance normalized for 1 mg of sample. Obviously, a new sample weighing, typically in the range of 0.1–10 mg, must be performed for every replicate measurement, which is more demanding than the injection of a liquid.

Another disadvantage, which is actually typical for all kinds of solid analysis techniques, including, for example, OES with arcs or sparks, laser ablation, or X-ray fluorescence, is the relatively poor *repeatability precision* when compared with solution analysis, which is largely due to the inhomogeneity of natural samples. For this reason it is typically necessary to grind solid samples to a particle size ≤50 μm for GFAAS analysis, which is more than what is usually required when the sample is digested.

However, the higher imprecision, which is typically around 5–10% RDS, is usually no problem in trace analysis, and has to be seen in relation to the improved sensitivity and accuracy as sample preparation and the associated risk of contamination is reduced to a minimum.

Calibration problems are frequently mentioned as one of the disadvantages of solid sampling GFAAS. Obviously, reference materials with a certified value for the analyte of interest are expensive, on occasions not available, and they introduce an additional uncertainty into calibration, as the analyte content is only certified within a certain confidence interval. However, it has been shown that even with complex matrices, such as coal, ash, or sediment aqueous standards might be used for calibration if the STPF concept is applied rigorously (Silva et al., 2002, 2004). This is necessary as appearance time and shape of the peaks often differ significantly for the various matrices and particularly in comparison with matrix-free standards.

It has also been argued that *chemical modification* is less effective for solid samples as the modifier does not come sufficiently into contact with the analyte species that are enclosed within sample particles. Firstly, it has been shown that many refractory matrices themselves act as kind of a modifier, releasing the analyte only at significantly higher temperatures so that the requirement for modifiers might be less than for solution analysis. Secondly, it has further been shown that an intimate contact between analyte and modifier is not a prerequirement for analyte stabilization, as the analyte becomes mobile already at relatively low temperatures and is attracted by the modifier (Maia et al., 2002). This also explains why even permanent modifiers that are applied to the graphite boat as kind of a surface coating, and which are not in contact with the majority of the sample at all, can be successfully applied in solid sampling GFAAS (Silva et al., 2002; Vale et al., 2001).

The analysis of *suspensions* (*slurry technique*) has often been recommended as an alternative, as it combines the advantages of solid sampling with solution analysis, and permits, for example, the use of conventional liquid sample handling apparatus, such as autosamplers (Bendicho and de Loos-Vollebregt, 1991). Slurries may also be diluted similar to solutions, although precision is degrading with increasing dilution. In principle, the precision for the analysis of suspensions cannot be higher than for the analysis of solids when the same number of particles is dispensed into the atomizer. The precision can, however, be considerably better when a substantial portion of the analyte is leaching out into the liquid phase, which is frequently the case. This leaching process could, however, also cause problems, as the analyte is distributed between two phases, and might behave in an entirely different way in these two forms (Maia et al., 2001). A problem that is additionally new for slurry analysis is the production of a homogeneous and stable suspension at the time of dispensing. This might include the necessity for additional grinding to a particle size smaller than that usually necessary for direct solid sampling and/or the addition of surfactants or reagents that increase viscosity. Another possibility, which has actually been commercially available, is to mix the slurry in the autosampler cup using an ultrasonic probe immediately before dispensing it into the graphite tube (Miller-Ihli, 1989).

The final decision on which technique should be applied, direct solid sampling, slurry sampling, or sample digestion, depends on the type of sample to be investigated, the analyte and its concentration, the number of elements that have to be determined in a sample, and various other criteria. Hence the decision has to be made individually, but it should be kept in mind that GFAAS is offering all these alternatives for sample introduction.

6. Atomization in Graphite Furnaces

Despite the enormous advances that have been made in GFAAS since its introduction by L'vov at the end of the 1950s, there are still a number of fundamental mechanisms that have not been fully explained. These include the actual process of atomization and the transport of the atoms into and from the absorption volume; in other words, the processes that determine the *shape of the atomization signal*. Without going into the process of atomization itself, L'vov proposed a simple mathematical model that described the time dependence of the atom population in the absorption volume. The change in the number N of atoms in the absorption volume over the time t is given by:

$$\frac{dN}{dt} = n_1(t) - n_2(t) \qquad (14)$$

where $n_1(t)$ is the number of atoms entering the absorption volume per time unit and $n_2(t)$ is the number leaving it per time unit.

The two processes have often been treated separately in order to facilitate their description, although in practice they are naturally inseparable. Obviously, the atomization signal is determined by a supply function and a loss function; however, no model has been proposed until now that could predict these processes exactly. One of the reasons is obviously an interaction of the analyte atoms with the graphite surface, which is very difficult to predict, and in addition is influenced by the presence of modifiers and concomitants. This topic will therefore not be further discussed here.

Similarly complex are the atomization mechanisms for the various elements, which are, in part, still discussed controversially. Therefore they can only be treated here in a very general and simplified manner. There appears to be consensus nowadays that the interaction of analyte atoms with the graphite surface plays an important role in the atomization of many elements. The macroscopic interactions between the gas phase and the surface can influence the diffusion of gaseous species from the surface into the absorption volume. Such interactions can lead to the formation of a *Langmuir film* (i.e., a more dense gas layer above the surface), within which equilibrium gas pressure can form. This film is formed when equilibrium between the pressure and the gas velocity is attained, and the diffusion of particles out of this film is determined by the concentration gradient.

The main problem with all methods to investigate mechanisms is that only the solid or gaseous *end products* are visible, *not* the processes of their generation. Therefore only a combination of several methods allows an extrapolation to be made. According to current understanding, there have at least five different atomization mechanisms to be considered for the different classes of analytes.

For copper, silver, gold, and the noble metals a *reduction of the oxide* (generated, e.g., by decomposition of the nitrate) to the element by graphite in the condensed phase has been proposed according to:

$$CuO_{(s)} + C_{(s)} \longrightarrow Cu_{(s,l,ad)} + CO_{(g)} \longrightarrow Cu_{(g)} \quad (15)$$

The actual release mechanism apparently depends on the analyte mass: for an analyte mass <3 ng the mechanism appears to be *desorption* from a submonolayer on the graphite, whereas *volatilization* from microdroplets appears to prevail for analyte masses >4 ng. Similar reduction–desorption mechanisms have also been proposed for cobalt and iron.

The atomization mechanism for manganese and nickel also appears to depend on the analyte mass: for masses <3 ng the mechanism appears to be a *gas phase dissociation of the oxide*, desorbed from a submonolayer on the graphite surface:

$$MnO_{(s)} \longrightarrow MnO_{(g)} \longrightarrow Mn_{(g)} + O_{(g)} \quad (16)$$

For an analyte mass >4 ng a vaporization from microdroplets with *concurrent dissociation* appears to prevail:

$$MnO_{(s)} \longrightarrow Mn_{(g)} + O_{(g)} \quad (17)$$

The first step in the atomization of gallium, indium, and thallium is a reduction of the oxide to a suboxide, which is released into the gas atmosphere, readsorbed at the graphite tube surface, where it is reduced to gaseous analyte atoms:

$$Ga_2O_{3(s)} + C_{(s)} \longrightarrow Ga_2O_{(g)} + CO_2$$
$$\longrightarrow Ga_2O_{(ad)} \longrightarrow Ga_{(g)} + CO \quad (18)$$

Gilmutdinov et al. (1991, 1992) have shown this two-step process impressively using spectral shadow filming. At first the molecular absorption of Ga_2O can be observed, starting from the platform and expanding into the gas atmosphere, and only later in the atomization stage could atomic absorption, originating from the tube wall, be measured. Several authors have pointed to the risk of analyte losses in the form of the gaseous suboxide for these elements, particularly under nonisothermal conditions.

Similar mechanisms with readsorption of the volatilized oxides have also been proposed for some of the alkaline earth and rare earth elements. For most of these elements, however, formation of gaseous carbides is competing with atom formation, and the extent to which one or the other reaction prevails, is strongly dependant on analytical conditions.

The complex atomization mechanisms that have been proposed for arsenic, selenium, and other volatile elements without the addition of a modifier will not be discussed here, as these elements can only be determined reasonably in the presence of a modifier such as palladium. The stabilization mechanism of this modifier has already been discussed in Section V.B.3.

7. Interferences in GFAAS

Virtually no *transport interferences* occur in GFAAS as the measurement solution is normally dispensed as a predefined volume into the atomizer, usually with an autosampler, and not with a nebulizer. Similarly, no transport interferences are to be expected when solid samples are introduced into the graphite tube on a graphite boat. Also, almost no references to *ionization interferences* are to be found in the literature, as the temperatures used in AAS are too low for thermal ionization (see Section V.A.3) and neither charge-transfer nor collisional ionization is likely to occur in the inert gas atmosphere of the graphite atomizer. In contrast, however, there appears to be an almost endless number of publications on solute volatilization and gas phase interferences in GFAAS. Many of these interferences are caused by the incorrect use of this technique and are greatly reduced or eliminated when the STPF concept is applied consistently (refer to Section V.B.4). We do not consider it to be of any great use to make a detailed presentation of these avoidable interferences. Hence, in the following we shall only discuss those interferences and their elimination that persist even under STPF conditions.

Spectral Interferences

GFAAS is a technique used in extreme trace analysis. In practice this means that a concentration difference of six to seven orders of magnitude between the analyte and concomitants is no rarity, especially for the analysis of solids. This vast excess of concomitants automatically leads to an increased risk of interferences. This problem can be reduced only by separating the concomitants as much as possible prior to the atomization stage. The completeness of this separation depends in the first place on the volatility of the analyte and the concomitants. It is not always possible to alter these parameters by chemical modification to such an extent that largely complete separation can be achieved in the pyrolysis stage.

Even if a large fraction of the concomitants can be removed, the remaining quantity still makes BC necessary for almost any kind of sample, including natural water. CS BC, which was developed for FAAS, and is perfectly adequate for this technique, is frequently overtaxed in GFAAS. This is, for example, when the background corrector cannot follow rapid changes in the background signal at the start of atomization, or when the background attenuation exceeds the correctable range with absorbance values greater than about 0.5–1, or when the background exhibits fine structure, which cannot be corrected at all by the use of CS in conventional AAS. The inability to correct for background attenuation properly can be recognized either by an excessively noisy signal or by a deflection of the signal to negative absorbance values, as shown in Fig. 31(a). As there cannot be less than zero absorbance, the latter is a clear sign for a BC problem.

After the introduction of ZBC in the 1980s, numerous papers were published demonstrating the superiority of this technique in comparison with BC with CS. Among the best-known examples is the interference caused by iron and phosphate on the determination of selenium (Welz et al., 1983), the elimination of which is shown in Fig. 31(b). These interferences make the determination of selenium by GFAAS almost impossible without using ZBC, not only in biological materials, but also in environmental and metallurgical samples. This is obviously only one example out of the large number that has been published; overall, ZBC is without doubt the system of choice for GFAAS.

However, the impression should not be given that ZBC is completely without interferences; nevertheless, any interferences are powers of magnitude lower than with CS BC, and most of them can be avoided or eliminated by relatively simple means. In principle there are two possible sources of spectral interferences with ZBC: *atomic lines* that are less than 10 pm from the analytical line can cause overlapping because of their own Zeeman splitting, and *molecular bands* with a rotational fine structure which exhibit the Zeeman effect. The first category was investigated thoroughly in particular by Wibetoe and Langmyhr (1984, 1985, 1986). All interferences caused by atomic lines in ZBC published in the literature are presented in Table 3. Of the 22 cases of line overlap, merely seven are recommended primary analytical lines. Whereas the most sensitive analytical lines are used almost exclusively in the analysis of solutions by GFAAS, suitable

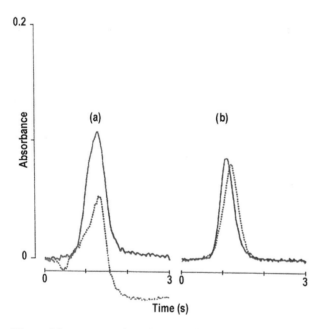

Figure 31 Determination of selenium in blood serum: (a) CS BC and (b) ZBC; solid signal: aqueous calibration solution; dotted signal: serum sample. [From Gilmutdinov et al. (1992).]

Table 3 Spectral Interferences in ZBC due to Direct Overlap with σ-Components of Other Atomic Lines

Analyte	Wavelength (nm)	Nature of wavelength[a]	Interferent
Ag	328.1	P	Rh
Al	308.2	s 1.5	V
Au	267.6	s 2	Co
B	249.7	P	Co
Bi	227.6	s 15	Co
Co	243.6	s 10	Pt
Cr	360.5	s 2	Co
Eu	459.4	P	V, Cs
Fe	271.9	s 3	Pt
Ga	287.4	P	Fe
Hg	253.7	P	Co
Ni	341.5	s 3	Co
	305.1	s 4	V
Pb	261.4	s 25	Co
Pd	247.6	P	Pb
Pt	265.9	P	Eu
	273.4	s 3	Fe
	306.5	s 1.5	Ni
Si	250.5	s 3	Co, V
Sn	300.9	s 3	Ca
	303.4	s 2	Cr
Zn	213.9	P	Fe

[a]P, primary resonance line; s, secondary analytical line; the figure indicates the factor by which the sensitivity is less than at the primary resonance line.

care is recommended for the analysis of solid samples, where less sensitive lines are employed more frequently.

Massmann (1982) was the first to point out the possibility of interferences in ZBC by molecular spectra; the excitation spectra with a large number of lines can undergo splitting in a magnetic field, and this risk is mostly present for light, diatomic molecules, such as PO. When the splitting occurs within the emission profile of the radiation source, the background measured with the magnetic field on is different from that without magnetic field (i.e., it cannot be corrected properly). All spectral interferences by the PO molecule in ZBC mentioned in the literature are compiled in Table 4. Although about half of the interferences occur at secondary lines or are not very pronounced, they have nevertheless to be taken into account to avoid potential errors.

The ultimate solution for the correction of these residual BC problems can only be expected from high-resolution CSAAS (see Section IV.B.2) as with this technique any atomic or molecular absorption that does not directly overlap with the analytical line is not detected at all by the selected pixel(s). Moreover, least squares BC can be applied to correct for direct line coincidence within the small spectral range selected for analyte measurement. Last but not the least, even the fastest changes in background absorption (e.g., at the beginning of the atomization stage) can be corrected without problems in high-resolution CSAAS, as background measurement and correction are truly simultaneous with the measurement of atomic absorption. Although high-resolution CSAAS was not yet commercially available at the point when this chapter was written, the expectations in this technique are fully justified.

Nonspectral Interferences

Customarily we classify nonspectral interferences in GFAAS into solute volatilization interferences and gas phase interferences. The majority of solute volatilization interferences are caused by the formation of volatile compounds of the analyte in the condensed phase, which are then vaporized during the pyrolysis stage. This kind of interference is also called *preatomization losses*. Gas phase interferences are caused by incomplete dissociation of gaseous molecules of the analyte in the atomization stage. Nevertheless, it is not always easy to clearly assign an interference to one or the other category, and a solute volatilization interference might turn into a gas phase interference by using a lower pyrolysis temperature, when the same molecule that has previously been lost in the pyrolysis stage is now incompletely dissociated in the atomization stage. It has to be stressed, however, that preatomization losses of the analyte can get almost completely excluded by the use of a proper chemical modifier and by establishing the pyrolysis curve, and hence the optimum pyrolysis temperature with the sample to be analyzed and not with a matrix-free calibration solution. Similarly, gas phase interferences can be reduced to a minimum by atomization under isothermal conditions (i.e., from a platform in a transversely heated graphite atomizer using the highest possible heating rate).

The most notorious and best-investigated interferents in GFAAS are the *halogens*, in particular the chlorides. *Sulfates* follow in second place. The magnitude of the interference as well as its mechanism depends on the cation to which the interferent is bond after the drying stage, as this determines the point in time and the temperature at which the interferent is released into the gas phase. Obviously, a gas phase interference can be observed only when the analyte and the interferent are in the gas phase at the same time. Hence, to avoid gas phase interferences, the best way is to separate the appearance of analyte and interferent in time; the best separation is obviously obtained when the interferent is already volatilized in the pyrolysis stage. One way to attain this, at least when the matrix is sodium chloride, is through the addition of *ammonium nitrate* according to (Ediger et al., 1974):

$$NaCl + NH_4NO_3 \longrightarrow NaNO_3 + NH_4Cl \qquad (19)$$

where low-volatile sodium chloride (melting point 801°C; boiling point 1413°C) is converted to ammonium chloride that sublimes at 335°C. Frech and Cedergren (1976) found that *hydrogen* was suitable to eliminate the interference of chloride on the determination of lead in steel; the chloride was removed as HCl in this case. A number of organic modifiers such as ascorbic acid or oxalic acid appear to have the same effect as they form hydrogen on decomposition.

Although hydrochloric acid itself is quite volatile and probably completely removed in the drying stage, a significant part of the chloride might be left in the graphite tube

Table 4 Spectral Interferences Caused by Phosphate (Molecular Spectrum of PO) in ZBC and Means of Avoiding Them

Analyte	Wavelength (nm)	Avoidance of interference
Ag	328.1	Minimal interference
Cd	326.1	Primary analytical line at 228.8 nm
Cu	244.2/247.3	Primary analytical line at 324.7 nm
Fe	246.3	Primary analytical line at 248.3 nm
Hg	253.7	Temperature program
In	325.8	Analytical line at 304.0 nm
Pb	217.0	Analytical line at 283.3 nm
Pd	247.6/244.8	Analytical line at 276.3 nm

after drying, bound to a matrix cation or intercalated in the graphite structure (graphite chlorides), and cause the earlier-described interferences. The use of hydrochloric acid and other chloride containing acids should hence be limited to the necessary minimum during sample preparation for GFAAS; if large quantities of hydrochloric acid cannot be avoided, it might be advantageous to fume off the excess by boiling with nitric acid prior to GFAAS analysis.

C. Chemical Vapor Generation

In this section we shall discuss systems and atomizers used to determine those analytes that can be vaporized in the atomic state or as molecules, such as hydrides, by *chemical reaction at ambient temperature*. These analytes are essentially mercury (by cold vapor generation), and antimony, arsenic, bismuth, germanium, lead, selenium, tellurium, tin, and also more recently cadmium and thallium (by HG). Sodium tetrahydroborate is the preferred reductant nowadays, although tin(II) chloride is used occasionally for the reduction of mercury. As the same or very similar apparatus is used for both techniques, they will be treated together in this section. This similarity is not, however, valid for atomization as mercury in the course of CVG is released as the atomic vapor, so that an atomizer is not required for this element.

1. Systems for CVG

The early systems for cold vapor (CV) and HGAAS consisted of normal laboratory apparatus, such as flasks, dropping funnels, gas inlet tubes, and so on, assembled to meet the requirements. A system of this type is depicted schematically in Fig. 32. The measurement solution is placed into the flask, the air is driven out by an inert gas, the reductant is added, and the gaseous analyte species is transferred in the inert gas stream to the atomizer or absorption cell. A time-dependent signal is generated in this technique and the profile is largely determined by the kinetics of the release of the gaseous analyte from the solution.

In modern apparatus we have to distinguish between batch systems and flow systems. All *batch systems* essentially operate on the earlier-described principle; improvements are reflected in the ease of operation, the degree of automation, and the efficiency with which the analyte is driven out of solution. With batch systems the

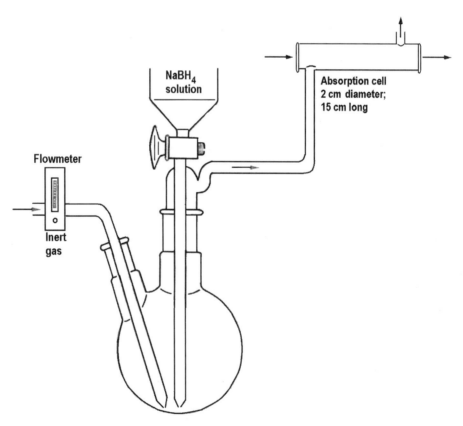

Figure 32 Laboratory-made accessory for HGAAS or CVAAS using sodium tetrahydroborate or tin(II) chloride as reductant. [From Welz and Sperling (1999).]

measurement signal is proportional to the mass of the analyte in the measurement solution and not to its concentration. Thus, as in GFAAS, it should be possible to compensate for influences on the release velocity of the analyte by measuring the integrated absorbance. However, when NaBH$_4$ is used as reductant, the pH value of the solution changes during the reaction and thus also the volume of hydrogen generated per time unit. The resulting change in the gas stream during the measurement time makes integration difficult for an online determination so that peak height measurement is usually preferred. The greatest advantage of batch systems is the large volumes that can be handled, and which lead to a high *relative sensitivity* (in concentration). Another advantage is that slurries can be used in batch systems (Flores et al., 2001; Vieira et al., 2002), which reduces significantly the risk of contamination during sample preparation, and which is of particular advantage for mercury determination. The biggest disadvantage of these systems is the large dead volume that leads to a relatively poor *absolute sensitivity*.

Flow systems for CVG can be of the continuous-flow (CF) or of the flow-injection (FI) type. Although there are a few basic differences between the two approaches, they have a number of common features. A common feature of both systems is that the reaction takes place within tubes and that a separator is thus required to separate the gaseous analyte from the liquid phase. These gas–liquid separators (GLS) can be either vessels made of glass, silica, or plastic, or membranes that are permeable for gases only. The major requirements for that device are that the GLS is large enough to prevent liquid from foaming over and solution droplets reaching the atomizer; on the other hand, the dead volume of the GLS should be as small as possible to guarantee good sensitivity. A further common feature is that both systems can be easily automated and that the sample preparation can be entirely or partially incorporated into the automatic analytical sequence. In *CF systems* the measurement solution, the reductant, and further reagents as required, are transported continuously in tubes in which they are mixed and, after passing through a reaction zone, the phases are separated in a GLS.

CF systems generate time-independent signals in which the absorbance is proportional to the *concentration* of the analyte in the measurement solution. In *FI systems*, instead of the sample solution a carrier solution, such as a dilute acid, is pumped continuously and a small volume of sample solution (e.g., 100–500 µL) is injected into the carrier stream at regular intervals. A time-dependent signal is hence generated whose form depends on the dispersion of the measurement solution in the carrier solution. Nevertheless, the release of the gaseous analyte species influences the height of the signal, as with CF systems.

2. The Generation of Hydrides

The net reaction of sodium tetrahydroborate in acidic solution and the simultaneous reduction of the hydride-forming element can be described quite simply as follows:

$$BH_4^- + H_3O^+ + 2H_2O \rightarrow H_3BO_3 + 4H_2 \quad (20)$$

and

$$3BH_4^- + 3H^+ + 4H_2SeO_3 \rightarrow 4SeH_2 + 3H_3BO_3 + 3H_2O \quad (21)$$

where Eq. (21) merely serves as an example for all the hydride-forming elements. Nevertheless, Eqs. (20) and (21) are coarse simplifications of the true processes, which to date are still unknown in detail. We only have to mention the term "nascent hydrogen", which is still not explained, and a number of other observations in connection with the decomposition of sodium tetrahydroborate in acidic solution, which cannot be discussed in detail. The finding that under optimum conditions the generation of the hydride and the transport to the atomizer are virtually quantitative for all hydride-forming elements is of major significance in practice (Dědina, 1988). The HG itself appears to be fast for most elements, and the release of the hydride from the solution is usually the rate-determining step.

As the analyte hydride is generated in a chemical reaction, the *chemical form*, in which the analyte is present, such as its *oxidation state*, plays an important role. The situation is most pronounced for the elements of Group VI, selenium and tellurium, as the hexavalent oxidation state does not generate a measurable signal in any situation. Reduction to the tetravalent state prior to determination by HGAAS is thus mandatory for these two elements. For this purpose boiling hydrochloric acid in the concentration range 5–6 mol/L is used almost exclusively. For the elements in Group V, mainly arsenic and antimony, the difference in sensitivity between the trivalent and pentavalent oxidation states depends strongly on the system used and the experimental conditions. In batch systems and at a pH $\leq$ 1 the hydride is formed more slowly from the pentavalent than from the trivalent oxidation states. This results in a peak height that is 20–30% lower, but the peak areas are essentially the same. The situation is significantly different in flow systems, where the difference in sensitivity depends very much on the length of the reaction coil, and can amount to more than an order of magnitude with short reaction coils. Potassium iodide (30–100 g/L KI), with the addition of ascorbic acid in acid solution (5–10 mol/L HCl) has frequently been used for the prereduction of As(V) and Sb(V). Brindle et al. (1992) proposed L-cysteine

(5–10 g/L in 0.1–1 mol/L HCl) as an alternative reductant for avoiding the high acid and reagent concentrations.

Although the differing behavior of the individual oxidation states is troublesome for the determination of the total concentration, it can be utilized for *speciation analysis*. This is particularly simple for selenium and tellurium as without prereduction only the tetravalent oxidation state can be selectively determined, whereas after prereduction the total concentration of ionic selenium or tellurium is determined. For arsenic and antimony we can make use of the fact that at pH values of 5–7, instead of the usual pH of ≤1, the trivalent state is reduced to the hydride, but not the pentavalent state.

With the exception of purely inorganic samples, the hydride-forming elements can be present not only in various oxidation states but also in a multitude of *organic compounds*. A number of these compounds also form volatile hydrides, but their behavior is often significantly different during atomization and typically exhibit differing sensitivity. Numerous organic compounds of the hydride-forming elements, however, do not react at all with sodium tetrahydroborate and cannot thus be determined directly by HGAAS. In the HG technique *digestion of the samples* is thus mandatory and should be part of the method, not only for the analysis of biological materials and environmental materials, but also for the analysis of water. As some of these organic compounds are very resistant to digestion, particularly harsh conditions are often required (Welz and Sperling, 1999).

3. The CV Technique for Mercury

Mercury is the only metallic element that has a vapor pressure as high as 0.0016 mbar at 20°C, which corresponds to a concentration of approximately 14 mg/m^3 of atomic mercury in the vapor phase. The possibility thus exists of determining mercury directly by AAS without an atomizer; it must merely be reduced to the element from its compounds and transferred to the vapor phase. This procedure is termed the CVAAS. Usually the names of Hatch and Ott (1968) are associated with this technique, although they were not the first ones to describe the reduction of mercury salts to metallic mercury with tin(II) chloride for the determination by AAS. With this reagent it is necessary to drive the reduced mercury from the liquid phase by a gas stream, and there have been various designs proposed to speed up this process and make it less dependent on sample properties. The most successful approach was a collection of the volatilized mercury vapor on gold gauze (Welz et al., 1984) or a similar material by amalgamation. The mercury can be rapidly liberated from the collector by heating, with the advantage of eliminating all kinetic effects of mercury release from solution, accompanied by a significant increase in sensitivity. Later, when sodium tetrahydroborate became available for analytical purposes, this reagent was also used for the reduction of mercury, with the advantage of an easier release of the mercury vapor from solution due to the hydrogen gas developed during the reaction, but at the expense of increased interferences due to transition metals (refer to Section V.C.6).

Usually, the same or similar apparatus is used for HGAAS and CVAAS, with the only difference that the absorption cell is not heated or only heated to 50–90°C in order to prevent condensation of water vapor, as mercury need not to be atomized. Alternatively, mercury can also be collected in a graphite tube treated with a permanent modifier, such as gold or palladium (Flores et al., 2001) and then determined by GFAAS. The high toxicity of mercury and many of its compounds, and the necessity to determine very low concentrations of this element in widely varying types of sample have led to the construction of various special apparatus just for the determination of mercury. This kind of instrument typically is nondispersive (i.e., has no monochromator), and works with a low-pressure mercury lamp, an absorption cell of 20–30 cm length and small inner diameter, and a UV-sensitive detector.

Very similar to the discussion in the previous section, mercury has to be present as the inorganic ion Hg^{2+} in order to be reduced to the element; most organic mercury compounds are not reduced, particularly not when tin(II) chloride is used as the reducing agent. This means that a sample digestion is mandatory for CVAAS, not only when solid samples have to be analyzed, but also whenever organic mercury compounds might be present. This digestion, like any sample treatment for mercury determination, has to be carried out with utmost care, as errors because of analyte loss and/or contamination are particularly notorious in the case of this element (Welz and Sperling, 1999).

4. Atomization Units for HGAAS

In essence, there are two different atomization units in use nowadays for gaseous hydrides; externally heated *quartz tube atomizers* (QTA), which are typically used *online* with HG, and *graphite tube atomizers*, which are exclusively used with a *sequestration* (i.e., an *in situ* collection and preconcentration of the analyte in the graphite tube prior to its atomization). For externally heated QTA a distinction is made between electrical and flame heating. Flame heated QTA are cheaper to acquire but more expensive to run; another problem is the rapid devitrification of the silica tube. Electrically heated QTA are easier to operate and usually offer good temperature control. As shown in Fig. 33 the analyte hydrides are normally introduced at the center of the heated silica tube and

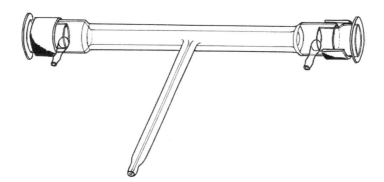

Figure 33 QTA for HGAAS with an inlet tube of 1 mm internal diameter. [From Welz et al. (1984).]

leave it close to the ends, which are often closed with silica windows for better control of the absorption volume.

5. Atomization of Hydrides

Atomization in Heated Silica Tubes

In the 1970s and even in the 1980s, the prevailing opinion was that atomization of hydrides in heated quartz tubes took place via a simple thermal dissociation. However, according to thermodynamic equilibria calculations, neither arsenic nor antimony or the other hydride-forming analytes can exist as gaseous atoms at temperatures below 1000°C (i.e., temperatures that are typically used in these atomizers). Fundamental conclusions about the atomization in the HG technique were elucidated by Dědina and Rubeška (1980) using a QTA in the inlet of which a small fuel-rich oxygen–hydrogen flame was burning. The authors concluded that the atomization of hydrides must be due to collision with *free hydrogen radicals* generated in this flame. Welz and Melcher (1983) found later that a small amount of oxygen is required for the atomization of arsine, and that the oxygen requirement is higher at lower temperatures. From these and other observations they concluded that even without a flame burning in the QTA, atomization is through H radicals that are produced by the reaction of hydrogen with oxygen, and proposed a three-step mechanism for the atomization of arsenic according to:

$$AsH_3 + H \longrightarrow AsH_2 + H_2 \quad (\Delta H = -196 \, kJ/mol) \tag{22}$$

$$AsH_2 + H \longrightarrow AsH + H_2 \quad (\Delta H = -163 \, kJ/mol) \tag{23}$$

$$AsH + H \longrightarrow As + H_2 \quad (\Delta H = -163 \, kJ/mol) \tag{24}$$

This mechanism was confirmed in the following years by a large number of further experiments (Dědina, 1992, 1993; Dědina and Welz, 1992; Stauss, 1993; Welz and Schubert-Jacobs, 1986; Welz et al., 1990), and we can obtain the following impression of the processes occurring in the QTA: a cloud of H radicals is formed in the hot zone of the QTA through the reaction of hydrogen and oxygen. The exact location of this cloud of radicals is determined by the temperature profile inside the atomizer, the flow rate and composition of the purge gas, and the design of the atomizer. The number of H radicals is largely determined by the oxygen supply to the atomizer; it is *not the number* of H radicals that is decisive for effective atomization but rather the *cross-sectional density* of the radical cloud. Hence, the gas flow rate is selected such that the radical cloud is located in the *side inlet tube* of the atomizer, the inner diameter of which should be as small as possible.

Thermodynamically, free analyte atoms are forbidden outside the radical cloud, as already mentioned. However, the atmosphere inside the QTA is far from thermodynamic equilibrium. It has been shown, for example, that the prevailing H radical concentration exceeds the equilibrium concentration by at least six to seven orders of magnitude (Stauss, 1993). Analyte atoms are removed from the absorption volume by forced convection and/or decay, probably a first-order reaction, which takes place at the silica surface of the atomizer. This assumption is supported by the observation that the surface in HGAAS has a major influence on the sensitivity. Summarizing we can say that in the QTA analyte atoms can be observed merely because the decay of both the H radicals and the analyte atoms is kinetically hampered (for details see Welz and Sperling, 1999).

Sequestration in Graphite Tube Atomizers

The direct introduction and atomization of hydrides in heated graphite tubes brings about two major disadvantages: firstly, a significantly reduced sensitivity when compared with QTA, mostly because of the much shorter length of the absorption tube, and secondly, a significantly shortened tube lifetime, as hydrogen reacts with carbon at the temperatures necessary for atomization. Both problems can be eliminated by trapping and preconcentrating the hydride in a graphite tube, treated with a noble metal, such as palladium or iridium, at a temperature of typically 400°C, and separating the atomization from the HG and collection step (Sturgen et al., 1989; Zhang et al., 1989). Further advantages of this technique, besides the preconcentration, are the elimination of all kinetic effects that are associated with an online determination, and a significant reduction of interferences in the atomizer, as will be discussed in Section V.C.6. Bulska et al. (2002) have shown that the trapping mechanism, at least for antimony, arsenic, and selenium is very similar to the stabilization mechanism of analytes added in aqueous solution. Using palladium, iridium, and rhodium as modifiers, the authors found that all analytes, together with the modifiers, had partially penetrated into the graphite subsurface to varying degrees. It can therefore be assumed that the atomization mechanisms for the trapped hydrides are identical to those for aqueous solutions in GFAAS (see Section V.B.3).

6. Interferences in CVGAAS

The interferences in the HG technique can be of widely varying nature and often depend very strongly on the apparatus used. *Spectral interferences* caused by background absorption or line overlap are essentially absent in CVGAAS as the concomitants are normally retained in the condensed phase and only small amounts of gaseous products are formed. This statement holds for QTA as well as for atomization in graphite furnaces after sequestration. *Nonspectral interferences* can be classified mainly into interferences that occur during CVG and release (i.e., interferences in the condensed phase) and those in the atomizer (i.e., in the gas phase). Whereas the former are mostly the same, independent of the atomization system used, the latter are strongly related to the atomizer, and essentially no interferences of this type are observed when the hydrides are sequestrated and atomized in a graphite tube. Obviously no gas phase (atomization) interferences are observed in the determination of mercury.

Condensed-Phase Interferences

In contrast to FAAS and GFAAS, in CVGAAS there is a chemical reaction prior to the actual determination step, namely, the generation of elemental mercury (CVAAS) or hydride (HGAAS). If the analyte in the sample is not exactly in the same form as the analyte in the calibration solution, the generation of the gaseous analyte may not be with either the same speed or the same yield. This interference cannot be eliminated by the analyte addition technique as the added analyte species might behave "normally" but the species present in the sample may not (refer to Section VI.B.2). The influence of the oxidation stage on the determination by CVGAAS, which requires a prereduction for the determination of the total analyte content, as well as the need for a sample digestion under harsh conditions in case organic analyte species are present, have been discussed in Section V.C.2. Compounds such as arsenobetaine, arseno sugars, selenomethionine, or the trimethyl selenonium ion cannot be mineralized with nitric acid alone, not even under pressure and with microwave assistance.

Besides these "avoidable" interferences, which are in essence due to an insufficient sample preparation, there are interferences caused by concomitants, especially by transition metals of the iron and copper groups and also by the platinum group metals, which influence primarily the *release of the hydride* or the mercury vapor from solution. As the interferents are usually present in large excess when compared with the analyte, the interference is a first-order process. This means that the interference does not depend on the analyte-to-interferent ratio, but only on the *concentration of the interferent*, and that it can be reduced significantly by simple dilution. This type of interference is caused by a reduced form of the interferent, usually the metal, and is a two-step process. In a first step the interfering ion is reduced by sodium tetrahydroborate, and a black precipitate of the finely dispersed metal can be observed in most cases. This highly reactive form of the interferent then adsorbs and traps or decomposes the analyte hydride that has not yet been released into the gas phase. Mercury might be trapped in an analogous manner by amalgamation at the precipitated metal.

Knowledge of this mechanism makes it possible to take several measures in order to avoid or correct for these interferences. The first one, which has been applied empirically long before the mechanism was known, is the use of *masking agents* such as ethylendiamine tetraacetic acid, thiourea, citric acid, pyridine-2-aldoxime, L-cysteine, thiosemicarbazide, and so on, which form complexes with the concomitant metal ions and hence prevents their reduction to the interfering species. Another possibility is an *increase of the acid concentration* of the measurement solution and/or a *decrease of the sodium tetrahydroborate concentration*. The former increases the solubility of the metal; the latter avoids or delays the reduction of the metal ion and its precipitation. As the reduction of the analyte to the hydride is always the fastest process (maybe except for the pentavalent forms of antimony and arsenic), it is often sufficient to retard the reduction of the transition metal ion until the hydride is released from the solution. This technique, called *kinetic discrimination* has been most successfully used in flow systems, where the reaction time can be precisely controlled via the length of the reaction coil. A third possibility is the addition of an *electrochemical buffer*, which is reduced preferentially, and the reduced form of which does not interfere with the hydride. Iron(III) has been proposed for this purpose to eliminate the interference of nickel on the determination of several hydride-forming elements in steel (Welz and Melcher, 1984). The potentials of possible reduction reactions that can take place when sodium tetrahydroborate is added to a solution containing Fe(III) and Ni(II) are listed in Table 5. As a result of the electrochemical potentials Fe(III) is initially reduced to Fe(II) before Ni(II) is reduced to the metal and precipitated. The same buffer has also been used for the suppression of the interference of copper on the determination of selenium (Bye, 1987).

Interferences in the QTA

It is logical that interferences in the atomizer can only be caused by concomitants that are vaporized during the chemical reaction and also reach the atomizer together with the analyte hydride. This means that this kind of interference under normal analytical conditions is limited to mutual interactions of the hydride-forming elements. If we consider the atomization mechanism and the transport of free atoms in the absorption volume, then there are two possibilities for interferences in the atomizer. Either the interferent reduces the *concentration of H radicals* to such an extent that the analyte can no longer be completely atomized, or it accelerates the *decay of free atoms*. In the first case the interference should be reduced when more radicals are generated (e.g., by increasing the rate of supply of oxygen). In the second case the interference should decrease when the probability of contact of the analyte atoms with the quartz surface (if the decay takes place there) is reduced or if the probability of contact with the interferent in the gas phase (if the interference is based on a reaction analyte-interferent in the gas phase) is reduced.

Selenium is the element that has been investigated in most detail in the QTA. The interferences caused by the other hydride-forming elements on the determination of selenium published in the literature are very inconsistent, but it could be shown that most of the substantial differences are based on the variety of systems used for HG and the varying experimental conditions (Welz and Stauss, 1993). Large-volume batch systems require relatively high sodium tetrahydroborate concentrations and a purge gas flow rate of several hundred mL/min to attain optimum sensitivity. These conditions demand an inlet tube for the QTA with relatively wide inner diameter in which the cross-sectional density of the H radicals is low. A shortage of radicals is thus the most common problem in such systems, and the tolerance limit to other hydride-forming elements is correspondingly low. In properly designed systems, however, radical deficiency appears to be no problem, and the interference of other hydride-forming elements is due to heterogeneous gas–solid reactions between free (analyte) atoms in the gaseous phase and finely dispersed particles, formed by free (interfering) atom recombination (D'Ulivo and Dědina, 2002).

In order to eliminate or reduce interferences in the QTA, depending on the system used, it might be worthwhile to investigate if the addition of a small amount of oxygen (e.g., 1% v/v) to the purge gas has an effect. Alternatively these interferences could in some cases also be significantly reduced by chemical means (i.e., by inhibiting the volatilization of the interfering hydride). In the case of selenium and tellurium this can be reached by converting them into the hexavalent oxidation state, which does not form a hydride, or by the addition of potassium iodide, which reduces arsenic and antimony to the trivalent state, but selenium and tellurium to the elemental state, so that HG is also prevented (Barth et al., 1992).

D. Spectrometers for AFS

As the fluorescence intensity is directly proportional to the quantity of absorbed radiant energy, and hence the intensity of the incident radiation, the investigation and development of high-intensity *radiation sources* is closely related with AFS. Whereas Winefordner and Vickers (1964) used metal vapor discharge tubes (i.e., LS in their early work), Veillon et al. (1966) were able to demonstrate the usefulness of a CS for AFS. In general, as the incident radiation does not reach the detector with this technique, a radiation source producing narrow spectral lines is not required. Nowadays EDL and boosted HCL for the hydride-forming elements and metal vapor discharge tubes for mercury are used almost exclusively in AFS.

The general design of an AFS instrument is depicted in Fig. 34. Although a variety of other designs have been investigated (Sychra et al., 1975) the fluorescence signal usually is observed at a right angle to the exciting radiation, decreasing the possibility of scattered radiation

Table 5 Electrochemical Potentials for the Reduction of Fe(III), Fe(II), and Ni(II) by Sodium Tetrahydroborate

Reaction	Electrochemical potential (V)
$Fe^{3+} + e^- \leftrightarrows Fe^{2+}$	+0.77
$Fe^{2+} + 2e^- \leftrightarrows Fe^0$	−0.41
$Ni^{2+} + 2e^- \leftrightarrows Ni^0$	−0.23

Source: From Dědina and Welz, 1993.

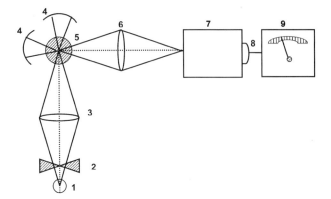

Figure 34 Basic layout of an AFS instrument: 1, primary radiation source; 2, chopper; 3, illuminating optics; 4, reflecting mirrors; 5, atomizer; 6, entrance optics; 7, monochromator; 8, detector; 9, amplifier and readout system. [From Sychra et al. (1975).]

from the excitation source entering the monochromator. Mirrors and lenses can be used at appropriate positions along the optical path to collect the radiant energy and thereby increase the intensity. Use of a chopper or other means of modulating the exciting radiation and of an amplifier that will respond to the frequency of the modulated signal are highly desirable. If the amplifier does not respond to a DC signal, *thermal emission* from the sample cell will not be detected. This is most important, as thermal emission frequently is more intense than the fluorescence signal.

Atomic fluorescence spectra are simple, with relatively few spectral lines; therefore, similar to AAS, high-resolution monochromators are not required. Moreover, as the exciting radiation, in contrast to AAS, is not reaching the detector, other lines emitted by the radiation source, including fill gas lines, do not have to be separated. For this reason it is possible to even use *nondispersive* systems (i.e., instruments without monochromator), in which a relatively wide spectral range is detected using broadband optical filters and/or so-called solar-blind PMT, which only detect radiation in the UV range. Such systems often provide the greatest signal intensity because of their large radiation-gathering ability.

As the atomic fluorescence originates from the absorption of electromagnetic radiation by gas phase atoms, the analyte has to be vaporized and atomized in the same way as for AAS determination. In order to reduce radiationless transitions from the excited state ("quenching"), which increase significantly with temperature, the atomization cell should be at the lowest possible temperature. Obviously, as in AAS, matrix interferences increase in low-temperature atomizers. For this reason, commercial AFS instruments are nowadays only available for CVG (i.e., in combination with HG and CV for the determination of mercury), as the matrix does not reach the atomizer in these cases. The atomization can then be in heated quartz tubes, low-temperature flames, or heat pipes. Obviously, all comments made about CVG in AAS (refer to Section V.C) are equally valid for CVG AFS. As mentioned earlier, the main advantage of AFS is its higher sensitivity, and the disadvantage is that a separate instrument is required for the determination of mercury (which is not unusual) and/or the hydride-forming elements.

E. Spectrometers for Flame OES

In the 1960s and the early 1970s there was still a lot of discussion about the relative merits of FAAS and FOES, particularly in the USA. In order to stay out of this discussion, most instrument manufacturers at that time were offering the possibility to also carry out FOES with their AAS instruments. This situation changed with the introduction of ICP-OES in the mid-1970s, which has been advertised as being "simpler than AAS", which obviously was true only in part. But there was one thing that ICP-OES clearly succeeded to do, and that was to terminate the use of FOES, except for a very few elements in a restricted area of application, which is nowadays served by the so-called *flame photometers* [i.e., instruments that are based on the concept of Barnes et al. (1945)].

Flame photometers are low-resolution single- or dual-channel spectrometers that use interference filters for spectral resolution for the determination of mainly the alkali metals Na, K, and Li, and occasionally the alkaline earth elements Ca and Ba. They are predominantly used in clinical laboratories to determine these elements, after a 100-fold dilution, routinely in blood serum and urine samples. These instruments typically use low-temperature flames with propane, butane, or natural gas as the fuel and air as the oxidant in order to avoid ionization of the alkali elements and to keep background emission to a minimum. These flames, however, are far from optimum for the determination of the alkaline earth elements, as they cannot dissociate the strong metal–oxygen bonds. The determination of Ca and Ba is therefore two to three orders of magnitude less sensitive than that of Na and K, and in addition prone to interferences, unless a strict control of the matrix is possible.

VI. MEASUREMENT, CALIBRATION, AND EVALUATION IN AAS, AFS, AND OES

A. Samples and Measurement Solutions

Initially a portion is removed from the material to be examined and is delivered to the laboratory; this is the *laboratory sample*. The material obtained by suitable treatment or preparation of the laboratory sample is the *test sample*. A *test portion* is removed from this for analysis. Under given prerequisites it is possible to analyze solid samples directly by GFAAS (refer to Section V.B.5), but in most cases the analytical portion must be taken into solution prior to further examination. The *test sample solution* is produced by treating the test portion with solvents, acids, and so on, or by digestion. This solution can be used directly or after further pretreatment steps, such as dilution, addition of buffers, and so on, for the measurement.

One or more *calibration samples* are required for the quantitative determination of an element by any atomic spectrometric technique, usually in the form of *calibration solutions*. For the preparation of solutions only laboratoryware that meet the requirements of titrimetric analysis should be used. Prior to use, all laboratoryware should be cleaned in warm, dilute nitric acid, thoroughly rinsed with water, and checked for contamination. Only doubly distilled water or water of similar quality should be used. All chemicals should be of analytical reagent or a higher grade of purity. The concentration of the analyte in the chemicals used should be negligible in comparison to the lowest concentration expected in the measurement solutions.

Calibration solutions are normally prepared from a *stock solution* that contains the analyte in an appropriately high, known concentration, frequently 1000 mg/L. Such a stock solution can be prepared by dissolving 1 g of the ultrapure metal or a corresponding weight of an ultrapure salt in a suitable acid, and making up to 1 L. For the preparation of stock solutions, only metals and salts are suitable, where the content of the analyte is known accurately to within at least 0.1%. The content of other elements should also be known and should as far as possible be negligible. Depending on the atomization or excitation technique used, certain acids or salts may be unsuitable for the preparation of stock solutions (e.g., sulfates for flame techniques or chlorides for GFAAS). Calibration solutions are prepared from the stock solution by serial dilution.

In AAS a *zeroing solution* is required to set the null signal (baseline) of the spectrometer; usually this is the pure solvent (e.g., deionized water). No such solution is required in AFS and OES, where zero emission is used as the reference point.

As for every type of sample pretreatment procedure the risk exists of contaminating the sample with the analyte through reagents, laboratoryware, and the ambient air, suitable *reagent blank solutions* must be prepared, treated, and measured in the same way as the test sample solution. The analyte content determined in such blank solutions must be taken into consideration during calibration and when evaluating the results.

Reference materials are of considerable importance for the accuracy of an analysis. These are materials of which one or more characteristics are known so accurately that they can be used for calibration or for analytical control within a quality assurance program. Reference materials are either synthesized from high-purity materials or selected from preanalyzed samples. If the characteristics of a reference material have been determined, documented, and guaranteed by a number of independent and recognized procedures, it is called a *certified reference material* (CRM).

B. Calibration

The relationship between the concentration c or the mass m of the analyte in the measurement solution (or in the test portion in the case of solid sample analysis) as the desired quantity and the measured quantity—the absorbance A or integrated absorbance A_{int} in AAS, or the emission intensity I in AFS and OES—is established and described mathematically by the use of calibration samples, usually calibration solutions. The relationship among the absorbance, the integrated absorbance, or the emission intensity and the concentration or mass of the analyte is given by the *calibration function*:

$$A = f(c) \tag{25}$$
$$A_{int} = f'(m) \tag{26}$$
$$I = f''(c) \tag{27}$$

The graphical presentation of the calibration function is the *calibration curve* (or analytical curve). In the optimum working range there is a linear relationship between the absorbance A, the integrated absorbance A_{int}, or the emission intensity I and the concentration c or mass m of the analyte in the measurement solution. This is an advantage for the evaluation of the measurement as the linear equation represents the simplest of all possibilities. In AAS this *linear range* typically only extends over one to two orders of magnitude, whereas in AFS and OES it covers typically three to five orders of magnitude.

1. Standard Calibration Technique

The standard calibration technique is the simplest, fastest, and most commonly used calibration technique in atomic spectrometry. In this technique the measured quantity (A, A_{int}, or I) of the measurement solution prepared from the test sample is compared directly with the measured quantity of calibration solutions. The concentration or mass of the analyte in the test sample solution is determined by interpolation. The calibration solutions must cover the entire expected concentration range of the test sample solutions. The evaluation can be performed graphically, in particular for linear calibration functions, as depicted in Fig. 35. In general though the calibration function is determined by *least mean squares* calculation.

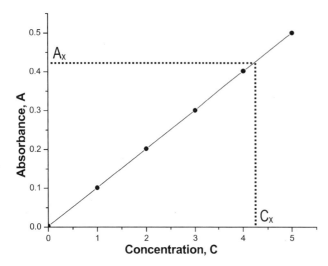

Figure 35 Graphical evaluation according to the standard calibration technique. [From Welz and Sperling (1999).]

The necessary algorithms are included in the software of all modern instruments. The use of least mean squares calculations is nevertheless coupled to a number of prerequirements:

1. The independent variable (the analyte content) is without error; more realistically, the error in preparing the calibration solutions is significantly smaller than the error in measuring the dependent variables (A, A_{int}, or I).
2. The experimental uncertainty for the determination of the dependent variables follows a normal distribution.

Resulting from the possible sources of error in the preparation of calibration solutions, the first condition must be put in question as soon as the repeatability standard deviation of the measurements is very good (e.g., <1%). The second condition is usually met when solutions are analyzed, but not necessarily for solid sample analysis. Evaluation in the *nonlinear range* can also be performed with suitable methods of calculation. Nevertheless, the precision and trueness deteriorate with increasing nonlinearity, but in fact this deterioration is far less dramatic than is often assumed.

A prerequirement for the applicability of the standard calibration technique is that the calibration and test sample solutions exhibit *absolutely identical behavior* in the atomizer (in AAS and AFS), or the radiation source (in OES) employed. A different behavior need not necessarily be caused by interferences due to concomitants, it might also be due to a different chemical form or a different oxidation state of the same analyte, as has been discussed for the HG technique (refer to Section V.C.2). If concomitants in the sample cause interferences using this calibration technique, and a large number of similar samples have to be analyzed, this can be compensated by mixing the interferent(s) with the calibration solutions. This is referred to as *matrix matching*.

2. Analyte Addition Technique

The analyte addition technique is often used when the composition of the test sample is unknown and interferences from the concomitants can be expected. In this technique the test sample itself is used to match the matrix. The test sample solution is divided into a number of aliquots; one aliquot remains unspiked, whereas the other aliquots are spiked with increasing masses of the analyte at equidistant intervals. All aliquots are then diluted to the same volume. Evaluation is performed by extrapolating the calibration function to zero (i.e., $A = 0$, $A_{int} = 0$ s, or $I = 0$), as shown schematically in Fig. 36(a). The intercept c_x

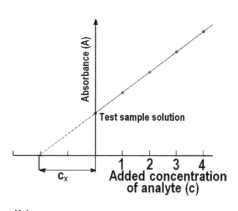

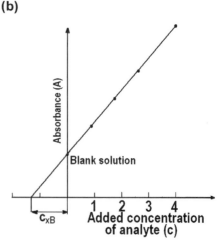

Figure 36 Graphical evaluation according to the analyte addition technique: (a) for the test sample solution and (b) for the blank solution. [From Welz and Sperling (1999).]

corresponds to the analyte content in the unspiked measurement solution.

The gradient of the calibration function determined by the analyte addition technique is specific for the analyte in the unknown test sample solution. The content of the analyte in the *blank solution* c_{xB} must always be determined by a *separate addition* and taken into account in the evaluation, as the gradient of the calibration function mostly differs in this case from that of the test sample solutions, as indicated in Fig. 36(b). If the compositions of the test samples differ significantly, it is necessary to apply the analyte addition technique separately to every test sample. If a number of samples of similar composition are to be examined, the calibration function determined by the analyte addition technique for a single representative test sample might be used for the other test samples. We refer to this procedure as the *addition calibration technique*.

The analyte addition technique was originally used in flame techniques, such as FAAS, to eliminate transport interferences, caused by differing viscosities of the measurement solutions (refer to Section V.A.4), an application for which it is ideally suited. This calibration technique, however, is completely overtaxed when it is used routinely to eliminate interferences in GFAAS (Welz, 1986). The analyte addition technique has the distinct advantage that every test sample is calibrated individually, so that the influence of concomitants can be eliminated. The analyte addition technique compensates *nonspecific, multiplicative interferences* (i.e., those interferences that alter the gradient of the calibration function). On the other hand it *cannot* eliminate any *additive interferences*. These include, in addition to problems associated with contamination or loss of the analyte, all *spectral interferences*.

A prerequirement for the applicability of the analyte addition technique is *identical behavior* of the analyte contained in the test sample solution and the added analyte. For certain techniques, such as GFAAS or HGAAS, it is essential for the accuracy of the determination that the analyte be present in the identical form (e.g., oxidation state or species). To obtain identical behavior, it may be necessary, for example, to add the analyte prior to the digestion of the test sample, so that it is taken through the entire sample preparation procedure. Proper application of the STPF concept (refer to Section V.B.4), and in particular of chemical modifiers, can also make a significant contribution to identical behavior.

A general condition for the application of the analyte addition technique is that all measurement values, including the highest addition must be in the *linear range* of the calibration function, as otherwise too high values for the analyte content in the test samples are determined with a linear regression of the measurement values. This restriction also means that the working range is limited to about one-third or one-quarter of the range that can be attained with direct calibration. Also, the *precision is generally poorer* as the test sample solution is always measured in the lower section of the available range, and as the zero point is not fixed but extrapolated. Finally, there is no doubt that the analyte addition technique requires *much more time and effort* for calibration and should thus be avoided for reasons of economy when it does not bring any major advantages.

3. Reference Element Technique

In this technique a reference element is added to all calibration and test sample solutions in known concentration (this technique is also referred to as the *internal standard* technique, a term deprecated by IUPAC). The measurement values for the analyte and the reference element are then placed in relation. This technique is based on the prerequirement that in the event of an interference the reference element behaves in an identical manner to the analyte (i.e., it undergoes the same signal enhancement or depression). This prerequirement is only met in principle for nonspecific interferences, such as transport interferences in flame techniques or dilution errors during sample preparation.

A practical limitation of the reference element technique is that it can only be performed meaningfully with a two- or multichannel spectrometer. For this reason it is used frequently in OES, such as in flame photometers with a reference channel, but only very rarely in AAS. Further, its application is basically limited to flame techniques as transport interferences do not occur with the other techniques, and an identical behavior of two elements is very unlikely in the case of GFAAS or HGAAS.

C. Evaluation

The performance characteristics of an analytical method are given by a set of quantitative, experimentally determinable *statistical quantities*. Such characteristic data can be used to compare various analytical methods or techniques, to choose a suitable method for a given task, for quality control, or to judge analytical results. It is essential that the analyst is aware of the statistical quantities for an analytical method if a given analytical task is to be performed successfully. They are the prerequisites for the correct choice of method and for effective quality assurance (Christian, 2003; Welz and Sperling, 1999).

The evaluation of the analytical results is performed via the *evaluation function*, which is nothing but than the inverse of the calibration function [Eqs. (25)–(27)]

$$c = g(A) \tag{28}$$

$$m = g'(A_{\text{int}}) \tag{29}$$

$$c = g''(I) \tag{30}$$

The graphical representation of the evaluation function is the evaluation curve. The slope S of the calibration curve is termed *sensitivity*:

$$S = \frac{dA}{dc} \tag{31}$$

$$S = \frac{dA_{int}}{dm} \tag{32}$$

$$S'' = \frac{dI}{dc} \tag{33}$$

It is very important to realize that sensitivity, although its definition is the same, has different dimensions, depending on the technique that is used, as shown in Table 6. Hence, although statements such as "technique A is more sensitive than technique B" are possible in obvious cases, the sensitivity is *not* suitable for a quantitative comparison of different techniques. In AAS the absorbance (for time-independent signals) and the integrated absorbance (for transient signals) are directly related to the analyte concentration and mass, respectively, through the Beer–Lambert Law, which makes them relatively stable figures under given conditions, as will be discussed later. In AFS and OES, however, the emission intensity depends on a large number of factors and is usually given in arbitrary numbers, which allows only a day-to-day comparison of the results obtained with the same instrument under the same conditions, but not a comparison between instruments or techniques. The decisive criterion for the performance capabilities of any spectrometer, and for the comparison of instruments or techniques is the S/N ratio, or the *limit of detection* (LOD) as a derived quantity.

To provide a measure for the sensitivity of the analyte under given conditions in AAS we use the terms *characteristic concentration*, c_0, and *characteristic mass*, m_0. This is the concentration or mass of the analyte corresponding to an absorbance $A = 0.0044$ (1% absorption) or an integrated absorbance $A_{int} = 0.0044$ s. The sensitivity in AAS is thus given by physical quantities such as the absorption coefficient of the analytical line and by characteristics of the atomizer unit. The characteristic concentration and especially the characteristic mass are important quantities in quality assurance. Virtually all manufacturers of atomic absorption spectrometers provide tables of reference values or the like so that the current performance of instruments can be checked.

Under *precision* we understand in general the closeness of agreement between independent test results obtained under prescribed conditions. As these conditions can be very different from case to case, two extreme situations should be observed, termed the repeatability and the reproducibility conditions, respectively. *Repeatability* conditions are those conditions under which the *same operator* using the *same equipment* within short intervals of time obtains independent test results with the same method on identical test objects in the *same laboratory*. In contrast, the *reproducibility* conditions are those conditions under which *different operators* using *different equipment* obtain independent test results with the same method on identical test objects in *different laboratories*.

Generally, the reliability of a measurement result increases with the number of measurements. For this reason measurement results should not be based on single values, but rather on *mean values* obtained from a number of measurements. If all the single values only deviate from each other randomly (i.e., they are distributed normally about a single central value in a Gaussian distribution), we can calculate the mean $\bar{x}$ from n single values x_i according to:

$$\bar{x} = \frac{\sum_{i=1}^{N} x_i}{n} \tag{34}$$

If merely the instrumental measurement is repeated and an aliquot of the same measurement solution is used, we speak of *replicate measurements*. The scatter of the single values x_i about the mean $\bar{x}$ is a measure of the *repeatability*; the precision of the repeated analytical step can be determined by variance analysis in the form of the *repeatability standard deviation*, s_r:

$$s_r = \sqrt{\frac{\sum_{i=1}^{N}(\bar{x} - x_i)^2}{n-1}} \tag{35}$$

Frequently the *relative standard deviation* (RSD) is used, which is given by:

$$s_{rel} = \frac{s}{\bar{x}} \tag{36}$$

An important characteristic of an analytical technique or method is the LOD, which is a measure for that

Table 6 Examples for the Dimensions of Sensitivity, Depending on the Measured Quantity

Measured quantity	Symbol	Dimension	Concentration or mass	Sensitivity
Absorbance	A	–	mg/L	L/mg
Integrated absorbance	A_{int}	s	ng	s/ng
Intensity	I	s^{-1}	mg/L	L/(mg s)

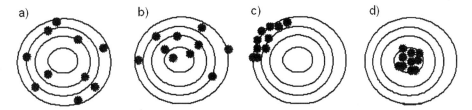

Figure 37 Schematic representation of various types of analytical error in the form of hits on a target. a) and b) poor precision; c) and d) good premises; b) and c) systematic error; a) and d) good trueness.

concentration or mass of the analyte which, when exceeded, allows recognition with a given statistical certainty that the analyte content in the test sample (test solution) is larger than that in the blank test sample (blank test solution). As the measured signal must be distinguished with a given certainty from the signal for the blank test sample (the blank value), it stands in close relationship to the precision of the determination of the blank test sample. The blank test sample is a sample that contains a very low content of the analyte and comes close to the remaining composition of the test sample.

As signals are not particularly suitable for interlaboratory comparison, the value of the signal is converted via the calibration function into a concentration or mass of the analyte. The concentration that gives a signal equal to $3\times$ the standard deviation of the blank test sample, s_{blank}, is generally taken as the LOD, and the standard deviation should be determined from at least 10, preferably 20 readings of the blank value:

$$\chi_{LOD} = \frac{3 s_{blank}}{S} \qquad (37)$$

The precision at the LOD is by definition about 33%. As the uncertainty of a measurement sets the *number of significant figures* in an answer, the LOD must in no case be reported with more than two significant figures. In addition, according to definition because the standard deviation of analytical signals close to the LOD is of the same magnitude as their mean, they cannot be applied for the quantification of the analyte concentration or mass in test samples. The smallest concentration or mass of an analyte that can be determined quantitatively from a single analysis with a risk of error $<5\%$ with the required statistical certainty $P = 95\%$ is termed *limit of quantitation* (LOQ), and is usually defined as:

$$\chi_{LOQ} = \frac{9 s_{blank}}{S} \qquad (38)$$

Although detection is positive in the range between the LOD and the LOQ, quantitative statements are inadmissible. For *quantitative measurements*, concentrations should be at least $10\times$ the LOD.

The most important goal of every quantitative analysis is to obtain the correct analytical result; this goal can only be achieved when no systematic errors are made. The *trueness* (often also termed *accuracy*) is the degree of agreement between the measured value and the true value. As an absolute true value is seldom known, a more realistic definition of trueness then would assume it to be the agreement between a measured value and the *accepted* true value. We can, by good analytical technique, such as making comparison against a known reference material of similar composition, arrive at a reasonable assumption about the trueness of a method, within the limitations of the knowledge of the "known" sample. Clear distinction must be made between precision and trueness, and also between *random errors* and *systematic errors* (bias). These relationships are depicted schematically in Fig. 37 in the form of hits on a target. It is nearly impossible to have trueness without good precision, and the term *accuracy* actually includes both, trueness and precision.

REFERENCES

Alkemade, C. Th. J., Milatz, J. M. W. (1955). A double-beam method of spectral selection with flames. *Appl. Sci. Res., Sect. B* 4:289.

Allan, J. E. (1961). The use of organic solvents in atomic absorption spectrophotometry. *Spectrochim. Acta* 17:467.

Barnes, R. B., Richardson, D., Berry, J. W., Hood, R. L. (1945). Flame photometry—a rapid analytical procedure. *Ind. Eng. Chem., Anal. Ed.* 17:605.

Barth, P., Krivan, V., Hausbeck, R. (1992). Cross-interferences of hybride-forming elements in hybride-generation atomic absorption spectrometry. *Anal. Chim. Acta* 263:111.

Becker-Ross, H., Florek, S., Heitmann, U. (2000). Observation, identification and correction of structured molecular background by means of continuum source AAS—determination of selemium and arsenic in human urine. *J. Anal. Atom. Spectrom.* 15:137.

Becker-Ross, H., Okruss, M., Florek, S., Heitmann, U., Huang, M. D. (2002). Echelle-spectrograph as a tool for studies of structured background in flame atomic absorption spectrometry. *Spectrochim. Acta Part B* 57:1493.

Beer, A. (1852). *Ann. Phys.* 86:78.

Bendicho, C., de Loos-Vollebregt, M. T. C. (1991). Solid sampling in electrothermal atomic absorption spectrometry using commercial atomizers. A review. *J. Anal. Atom. Spectrom.* 6:353.

Bode, H., Fabian, H. (1958a). Organische, mit Wasser nicht mischbare Lösungsmittel in der Flammenphotometrie—die Bestimmung des Kupfers nach Extraktion als Innerkomplexverbindung. *Fresenius Z. Anal. Chem.* 162:328.

Bode, H., Fabian, H. (1958b). Zur flammenphotometrischen Bestimmung des Kupfers. *Fresenius Z. Anal. Chem.* 163:187.

Brindle, I. D., Alarabi, H., Karshman, S., Le, X. C., Zheng, S. G. (1992). Combined generator/separator for continuous hydride generation: application to on-line pre-reduction of arsenic(V) and determination of arsenic in water by atomic emission spectrometry. *Analyst* 117:407.

Bulska, E., Jędral, W., Kopyść, E., Ortner, H. M., Flege, S. (2002). Secondary ion mass spectrometry for characterizing antimony, arsenic and selenium on graphite surfaces modified with noble metals and used for hydride generation atomic absorption spectrometry. *Spectrochim. Acta Part B* 57:2017.

Bye, R. (1987). Iron(III) as releasing agent for copper interference in the determination of selenium by hydride-generation atomic absorption spectrometry. *Anal. Chim. Acta* 192:115.

Christian, G. D. (2003). *Analytical Chemistry*. 6th ed. Hoboken, NJ: WILEY.

D'Ulivo, A., Dědina, J. (2002). The relation of double peaks, observed in quartz hydride atomizers, to the fate of free analyte atoms in the determination of arsenic and selenium by atomic absorption spectrometry. *Spectrochim. Acta Part B* 57:2069.

Dědina, J. (1988). Evaluation of hydride generation and atomization for AAS. *Prog. Anal. Spectrosc.* 11:251.

Dědina, J. (1992). Quartz tube atomizers for hydride generation atomic absorption spectrometry: mechanism of selenium hydride atomization and fate of free atoms. *Spectrochim. Acta Part B* 47:689.

Dědina, J., Rubeška, I. (1980). Hydride atomization in a cool hydrogen—oxygen flame burning in a quartz tube atomizer. (1980). *Spectrochim. Acta Part B* 35:119.

Dědina, J., Welz, B. (1992). Quartz tube atomizers for hydride generation atomic absorption spectrometry: mechanism for atomization of arsine. Invited lecture. *J. Anal. Atom. Spectrom.* 7:307.

Dědina, J., Welz, B. (1993). Quartz tube atomizers for hydride generation atomic absorption spectrometry: fate of free arsenic atoms. *Spectrochim. Acta Part B* 48:301.

De Loos-Vollebregt, M. T. C., De Galan, L., Van Uffelen, J. W. M. (1986). Extended range Zeeman atomic absorption spectrometry based on a 3-field a.c. magnet. *Spectrochim. Acta Part B* 41:825.

Ediger, R. D., Peterson, G. E., Kerber, J. D. (1974). Application of the graphite furnace to saline water analysis. *Atom. Absorpt. Newslett.* 13:61.

Farino, J., Browner, R. F. (1984). Surface tension effects on aerosol properties in atomic spectrometry. *Anal. Chem.* 56:2709.

Flores, É. M. M., Welz, B., Curtius, A. J. (2001). Determination of mercury in mineral coal using cold vapor generation directly from slurries, trapping in a graphite tube, and electrothermal atomization. *Spectrochim. Acta Part B* 56:1605.

Fraunhofer, J. (1817). *Ann. Phys.* 56:264.

Frech, W., Cedergren, A. (1976). Investigations of reactions involved in flameless atomic absorption procedures : Part II. An experimental study of the role of hydrogen in eliminating the interference from chlorine in the determination of lead in steel. *Anal. Chim. Acta* 82:93.

Fry, R. C., Denton, M. B. (1979). Molecular absorption spectra of complex matrices in premixed flames. *Anal. Chem.* 51:266.

Gilmutdinov, A. Kh., Zakharov, Yu. A., Ivanov, V. P., Voloshin, A. V. (1991). Shadow spectral filming: a method of investigating electrothermal atomization. Part 1. Dynamics of formation and structure of the absorption layer of thallium, indium, gallium and aluminium atoms. *J. Anal. Atom. Spectrom.* 6:505.

Gilmutdinov, A. Kh., Zakharov, Yu. A., Ivanov, V. P., Voloshin, A. V., Dittrich, K. (1992). Shadow spectral fiming: a method of investigating electrothermal atomization. Part 2. Dynamics of formation and structure of the absorption layer of aluminium, indium and gallium molecules. *J. Anal. Atom. Spectrom.* 7:675.

Gilmutdinov, A. Kh., Radziuk, B., Sperling, M., Welz, B., Nagulin, K. Yu. (1996). Spatial distribution of radiant intensity from primary sources for atomic absorption spectrometry .2. Electrodeless discharge lamps. *Appl. Spectrosc.* 50:483.

Gleisner, H., Eichardt, K., Welz, B. (2003). Optimization of analytical performance of a graphite furnace atomic absorption spectrometer with Zeeman-effect background correction using variable magnetic field strength. *Spectrochim. Acta Part B* 58:1663.

Griggs, M. A. (1939). *Science* 89:134.

Höhn, R., Jackwerth, E. (1976). Nicht kompensierbarer Untergrund als systematischer Fehler bei der Atomabsorptionsspektrometrie. *Anal. Chim. Acta* 85:407.

Harnly, J. M. (1999). The future of atomic absorption spectrometry: a continuum source with a charge coupled array detector. *J. Anal. Atom. Spectrom.* 14:137.

Hatch, W. R., Ott, W. L. (1968). Determination of submicrogram quantities of mercury by atomic absorption spectrophotometry. *Anal. Chem.* 40:2085.

Heitmann, U., Schütz, M., Becker-Ross, H., Florek, S. (1996). Measurements on the Zeeman-splitting of analytical lines by means of a continuum source graphite furnace atomic absorption spectrometer with a linear charge coupled device array. *Spectrochim. Acta Part B* 51:1095.

Herschel, J. F. W. (1823). *Trans. R. Soc. Edin.* 9:445.

Kirchhoff, G., Bunsen, R. (1861). Ueber das Verhältnis zwischen dem Emissionsvermögen und dem Absorptionsvermögen der Körper für Wärme und Licht. *Philos. Mag.* 22:329.

L'vov, B. V. (1961). The analytical use of atomic absorption spectra. *Spectrochim. Acta* 17:761.

L'vov, B. V. (1978). Electrothermal atomization—the way toward absolute methods of atomic absorption analysis. *Spectrochim. Acta Part B* 33:153.

L'vov, B. V. (1984). Twenty-five years of furnace atomic absorption spectroscopy. *Spectrochim. Acta Part B* 39:149.

Lockyer, J. N. (1878). *Studies in Spectrum Analysis*. London: Appleton.

Lundegårdh, H. (1929). *Die quantitative Spektralanalyse der Elemente*. Jena: Gustav Fischer.

Maia, S. M., Vale, M. G. R., Welz, B., Curtius, A. J. (2001). Feasibility of isotope dilution calibration for the determination of thallium in sediment using slurry sampling electrothermal vaporization inductively coupled plasma mass spectrometry. *Spectrochim. Acta Part B* 56:1263.

Maia, S. M., Welz, B., Ganzarolli, E., Curtius, A. J. (2002). Feasibility of eliminating interferences in graphite furnace atomic absorption spectrometry using analyte transfer to the permanently modified graphite tube surface. *Spectrochim. Acta Part B* 57:473.

Manning, D. C., Fernandez, F. J. (1970). Atomization for atomic absorption using a heated graphite tube. *Atom. Absorpt. Newslett.* 9:65.

Massmann, H. (1967). Zweikanalspektrometer zur Korrektur von Untergrundabsorption beim Atomisieren in der Graphitküvette. *Fresenius Z. Anal. Chem.* 225:203.

Massmann, H. (1968). Vergleich von Atomabsorption und Atomfluoreszenz in der Graphitküvette. *Spectrochim. Acta Part B* 23:215.

Massmann, H. (1982). The origin of systematic errors in background measurements in Zeeman atomic-absorption spectrometry. *Talanta* 29:1051.

Miller-Ihli, N. J. (1989). Communications. Automated ultrasonic mixing accessory for slurry sampling into a graphite furnace atomic absorption spectrometer. *J. Anal. Atom. Spectrom.* 4:295.

Nichols, E. L., Howes, H. L. (1924). *Phys. Rev.* 23:472.

Ortner, H. M., Kantuscher, E. (1975). Metallsalzimprägnierung von Graphitrohren zur verbesserten AAS-Siliziumbestimmung. *Talanta* 22:581.

Ortner, H. M., Bulska, E., Rohr, U., Schlemmer, G., Weinbruch, S., Welz, B. (2002). Modifiers and coatings in graphite furnace atomic absorption spectrometry—mechanisms of action (A tutorial review). *Spectrochim. Acta Part B* 57:1835.

Schlemmer, G., Welz, B. (1986). Palladium and magnesium nitrates, a more universal modifier for graphite furnace atomic absorption spectrometry. *Spectrochim Acta Part B* 41:1157.

Silva, J. B. B., Silva, M. A. M., Curtius, A. J., Welz, B. (1999). Determination of Ag, Pb and Sn in aqua regia extracts from sediments by electrothermal atomic absorption spectrometry using Ru as a permanent modifier. *J. Anal. Atom. Spectrom.* 14:1737.

Silva, A. F., Welz, B., Curtius, A. J. (2002). Noble metals as permanent chemical modifiers for the determination of mercury in environmental reference materials using solid sampling graphite furnace atomic absorption spectrometry and calibration against aqueous standards. *Spectrochim. Acta Part B* 57:2031.

Silva, A. F., Borges, D. L. G., Welz, B., Vale, M. G. R., Silva, M. M., Klassen, A., Heitmann, U. (2004). Method development for the determination of thallium in coal using solid sampling graphite furnace atomic absorption spectrometry with continuum source, high-resolution monochromator and CCD array detector. *Spectrochim. Acta Part B* 59:841.

Slavin, W., Manning, D. C., Carnrick, G. R. (1981). The stabilized temperature platform furnace. *Atom. Spectrosc.* 2:137.

Smith, S. B., Hieftje, G. M. (1983). A new background-correction method for atomic-absorption spectrometry. *Appl. Spectrosc.* 37:419.

Sperling, M., Welz, B., Hertzberg, J., Rieck, C., Marowsky, G. (1996). Temporal and spatial temperature distributions in transversely heated graphite tube atomizers and their analytical characteristics for atomic absorption spectrometry. *Spectrochim. Acta Part B* 51:897.

Stauss, P. (1993). *Dissertation*, Universität Konstanz, Germany.

Sturgeon, R. E., Willie, S. N., Sproule, G. I., Robinson, P. T., Berman, S. S. (1989). Sequestration of volatile element hydrides by platinum group elements for graphite furnace atomic absorption. *Spectrochim. Acta Part B* 44:667.

Sychra, V., Svoboda, V., Rubeška, I. (1975). *Atomic Fluorescence Spectroscopy*. London: Van Nostrand Reinold Company.

Talbot, W. H. F. (1826). *Brewster's J. Sci.* 5:77.

Vale, M. G. R., Silva, M. M., Welz, B., Lima, E. C. (2001). Determination of cadmium, copper and lead in mineral coal using solid sampling graphite furnace atomic absorption spectrometry. *Spectrochim. Acta Part B* 56:1859.

Veillon, C., Mansfield, J. M., Parsons, M. L., Winefordner, J. D. (1966). Use of a Continuous Source in Flame Fluorescence Spectrometry. *Anal. Chem.* 38:204.

Vieira, M. A., Welz, B., Curtius, A. J. (2002). Determination of arsenic in sediments, coal and fly ash slurries after ultrasonic treatment by hydride generation atomic absorption spectrometry and trapping in an iridium-treated graphite tube. *Spectrochim. Acta Part B* 57:2057.

Walsh, A. (1955). The application of atomic absorption spectra to chemical analysis. *Spectrochim. Acta* 7:108.

Walsh, A. (1974). Atomic absorption spectrometry – stagnant or pregnant? *Anal. Chem.* 46:698A.

Welz, B. (1986). Abuse of the analyte addition technique in atomic-absorption spectrometry. *Fresenius Z. Anal. Chem.* 325:95.

Welz, B., Melcher, M. (1983). Investigations on atomisation mechanisms of volatile hydride-forming elements in a heated quartz cell. Part 1. Gas-phase and surface effects; decomposition and atomisation of arsine. *Analyst* 108:213.

Welz, B., Melcher, M. (1984). Mechanisms of transition metal interferences in hydride generation atomic-absorption spectrometry. Part 3. Releasing effect of iron(III) on nickel interference on arsenic and selenium. *Analyst* 109:577.

Welz, B., Schubert-Jacobs, M. (1986). Investigations on atomization mechanisms in hydride-generation atomic-absorption spectrometry. *Fresenius Z. Anal. Chem.* 324:832.

Welz, B., Luecke, W. (1993). Interference of aluminum chloride and nitrate on alkaline earth elements in an air-acetylene flame. *Spectrochim. Acta Part B* 48:1703.

Welz, B., Stauss, P. (1993). Interferences from hydride-forming elements on selenium in hydride-generation atomic-absorption spectrometry with a heated quartz tube atomizer. *Spectrochim. Acta Part B* 48:951.

Welz, B., Sperling, M. (1999). *Atomic Absorption Spectrometry*. 3rd ed. Weinheim, New York: Wiley-VCH.

Welz, B., Melcher, M., Schlemmer, G. (1983). Determination of selenium in human-blood serum – comparison of 2 atomic-absorption spectrometric procedures. *Fresenius Z. Anal. Chem.* 316:271.

Welz, B., Melcher, M., Sinemus, H. W., Maier, D. (1984). Picotrace determination of mercury using the amalgamation technique. *Atom. Spectrosc.* 5:37.

Welz, B., Schubert-Jacobs, M., Sperling, M., Styris, D. L., Redfield, D. A. (1990). Investigation of reactions and atomization of arsine in a heated quartz tube using atomic absorption and mass spectrometry. *Spectrochim. Acta Part B* 45:1235.

Welz, B., Schlemmer, G., Mudakavi, J. R. (1992). Palladium nitrate–magnesium nitrate modifier for electrothermal atomic absorption spectrometry. Part 5. Performance for the determination of 21 elements. (1992). *J. Anal. Atom. Spectrom.* 7:1257.

Welz, B., Vale, M. G. R., Silva, M. M., Becker-Ross, H., Huang, M. D., Florek, S., Heitmann, U. (2002). Investigation of interferences in the determination of thallium in marine sediment reference materials using high-resolution continuum-source atomic absorption spectrometry and electrothermal atomization. *Spectrochim. Acta Part B* 57:1043.

Welz, B., Becker-Ross, H., Florek, S., Heitmann, U., Vale, M. G. R. (2003). High-resolution continuum-source atomic absorption spectrometry – What can we expect? *J. Braz. Chem. Soc.* 14:220–229.

Wibetoe, G., Langmyhr, F. J. (1984). Spectral interferences and background overcompensation in zeeman-corrected atomic absorption spectrometry : Part 1. The effect of iron on 30 elements and 49 element lines. *Anal. Chim. Acta* 165:87.

Wibetoe, G., Langmyhr, F. J. (1985). Spectral interferences and background overcompensation in inverse zeeman-corrected atomic absorption spectrometry : Part 2. The effects of cobalt, manganese and nickel on 30 elements and 53 elements lines. *Anal. Chim. Acta* 176:33.

Wibetoe, G., Langmyhr, F. J. (1986). Spectral interferences and background overcompensation in inverse zeeman-corrected atomic absorption spectrometry Part 3. Study of eighteen cases of spectral interference. *Anal. Chim. Acta* 186:155.

Winefordner, J. D., Vickers, T. J. (1964). Atomic fluorescence spectroscopy as a means of chemical analysis. *Anal. Chem.* 36:161.

Wollaston, W. H. (1802). *Philos. Trans.* 92:365.

Wood, R. W. (1902). *Philos. Mag.* 3:128.

Zeeman, P. (1897). *Philos. Mag.* 5:226.

Zhang, L., Ni, Z. M., Shan, X. Q. (1989). *In situ* concentration of metallic hydride in a graphite furnace coated with palladium—determination of bismuth, germanium and tellurium. *Spectrochim. Acta Part B* 44:751.

5

Ultraviolet, Visible, Near-Infrared Spectrophotometers

CHRIS W. BROWN
Department of Chemistry, University of Rhode Island, Kingston, RI, USA

I. INTRODUCTION

Spectrophotometers for the ultraviolet (UV), visible (VIS), and near-infrared (NIR) regions will be discussed in this chapter. Typically, the UV region is considered to extend from 190 to 350 nm, the visible region from 350 to 800 nm, and the NIR from 800 to 2500 nm. However, considerable latitude can be found in the definitions of these regions; the UV might extend up to 400 nm and the short wavelength NIR (SWNIR) from 600 to 1100 nm. The total wavelength range from 190 to 2500 nm is equivalent to a frequency range of about 10^{15}–10^{14} Hz, or a wavenumber range of 50,500–4000 cm^{-1}. In this discussion, we will specifically define the abscissa in terms of wavelength (λ) in nm, although it should be pointed out that wavenumbers are often used in the NIR.

Absorptions in the UV–VIS regions are due to electronic transitions. Most UV–VIS absorptions by organic molecules are attributed to transitions involving nonbonding (n) electrons or electrons in molecular orbitals found in unsaturated molecules. Generally, the absorptions are due to n $\rightarrow$ π^* and $\pi \rightarrow \pi^*$ transitions; thus, molecules with double bonds and especially conjugated double bonds such as aromatic compounds are stronger absorbers. UV–VIS spectra of some typical aromatic molecules are shown in Fig. 1 and spectra of several organic dyes in Fig. 2. Most visible dyes consist of polynuclear aromatic ring systems with low energy transitions in the visible region. All aromatic molecules have UV absorptions in the 200–250 nm region. Most NIR absorptions are due to overtone and combinations of fundamentals normally observed in the mid-IR region. Generally, the NIR bands are at least one order of magnitude weaker than the parent absorptions in the mid-IR. One advantage of the NIR is that the decreases in intensities when compared with the mid-IR bands are not the same for all molecules. For example, water is virtually opaque throughout much of the mid-IR region at most pathlengths, but it is much more transparent in the NIR. Thus, it is often possible to observe NIR spectra of solutes dissolved in water. Spectra of three different organic molecules in the NIR are shown in Fig. 3. The patterns in the longer wavelength regions of 2000–2500 nm form a "fingerprint" similar to that observed in the mid-IR region of 600–1800 cm^{-1}.

A. Beer's Law

Absorbance in any spectral region depends on the Beer–Lambert–Bouguer expression commonly known as Beer's Law:

$$A = abc \tag{1}$$

where A is the absorbance, a is the absorptivity or molar extinction coefficient, b is the pathlength, and c is the concentration of the absorbing substance. Both absorbance and absorptivity are wavelength dependent. Actually, a spectrophotometer does not measure the absorbance

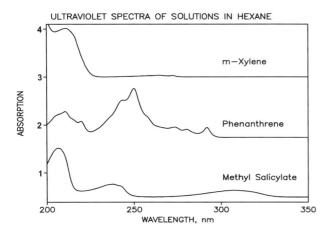

Figure 1 UV spectra of *m*-xylene, phenanthrene, and methyl salicylate.

directly, but rather it must be calculated by taking the negative logarithm of the fraction of light transmitted through a sample. If P is the power of the light passing through a sample and P_0 is the power of the light detected when the concentration of the absorbing material is zero, the fraction of light transmitted is given by

$$T = \frac{P}{P_0} \qquad (2)$$

This is often written as percent transmittance or $\%T$. Absorbance is given by

$$A = -\log T = \log \frac{P_0}{P} \qquad (3)$$

The quantities P and P_0 can be measured sequentially on a sample and a blank, or simultaneously with a sample and a blank in two identical cuvets. It is important to keep in mind that A is calculated from T, so that, though

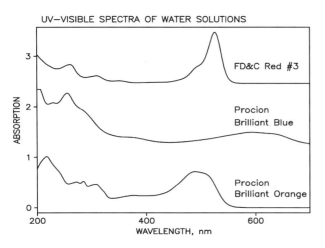

Figure 2 UV–VIS spectra of three dyes.

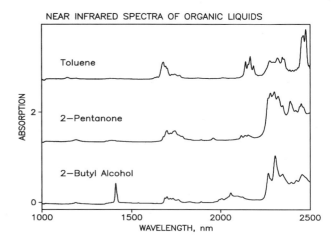

Figure 3 NIR spectra of toluene, 2-pentanone, and 2-butyl alcohol.

noise in the measurements arises in T, its effect on absorbance and concentration is a logarithmic relationship. (The symbol P for the power carried by a light beam is sometimes replaced by I for intensity, but P is the correct term.)

B. Deviations from Beer's Law

Both chemical and instrumental deviations from Beer's Law may occur. Chemical deviations are caused by reactions such as molecular dissociation or the formation of a new molecule. The most prevalent cause is some type of association due to hydrogen bonding. For example, at very low concentrations in an inert solvent, alcohols exist as monomers but, as the concentration is increased the monomers combine to form dimers, trimers, and so on. Thus, the concentrations and absorbances due to the monomers do not increase linearly with the total concentration of the species. An example of this effect is given in Fig. 4, which shows the NIR spectra of methanol in CCl_4 as a function of concentration. The first overtone of the O—H stretching vibrations of the monomer at 1404 nm does not increase linearly with the total concentration. Moreover, as the total concentration increases, bands due to dimers and oligomers start to appear at slightly longer wavelengths (1450–1650 nm).

There are two primary instrumental deviations from Beer's Law. The first has to do with the fact that the law was defined for monochromatic radiation. Should a sample have entirely different absorptivities at two wavelengths, both falling on the detector at the same time, Eq. (3) would have to be rewritten as

$$A = \log\left(\frac{P_0' + P_0''}{P_0' 10^{-a'bc} + P_0'' 10^{-a''bc}}\right) \qquad (4)$$

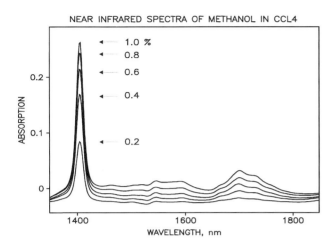

Figure 4 NIR spectra of methanol in CCl$_4$ as a function of concentration.

where the single and double primes refer to the two wavelengths. Should a' be identical to a'', Eq. (4) reverts to Eq. (3); however, nonlinearity can occur at higher concentrations if there is a significant difference between a' and a''.

The second source of instrumental error is stray light. This problem can be inherent in the spectrometer, but it can also be caused by the operator. Stray light is any light reaching the detector without passing through the sample. Thus, if a sample were completely opaque at a certain wavelength, any photons that were detected would be due to stray light. These photons could be passing around the sample, through holes in the sample as might be caused by air bubbles in a liquid, or they might be the result of poor shielding, permitting room light to reach the detector from some external light source, for example, room lights. The effect of stray light is to add a constant power (intensity) of light, P_s, to both the numerator and denominator in the absorbance expression:

$$A = \log\left(\frac{P_0 + P_s}{P + P_s}\right) \quad (5)$$

As the concentration increases, P approaches zero, and A asymptotically approaches a maximum level given by $\log[(P_0 + P_s)/P_s]$. For P_s equal to 10% of P_0, the maximum value of A is 1.04. As this is a log relation, A approaches 1.04 asymptotically with the concentration. Thus, stray light should be suspected any time when nonlinear data is encountered.

Stray light in a spectrophotometer can be measured by inserting an opaque blocking filter into the optical path. A signal observed by the detector under these conditions is due solely to stray radiation. A 10 g/L solution of potassium iodide does not transmit appreciably below 259 nm, but is essentially completely transparent above 290 nm when observed in a 10 mm cuvet (Poulson, 1964). If a spectrophotometer is set at a lower wavelength, say 240 nm, any signal that is observed must originate from stray radiation. In determining the stray light below 259 nm, the cuvet holding the solution should be placed in the spectrophotometer and scanned from the longer wavelength, transparent region to the shorter, opaque region. In this way the detector signal will be gradually decreased to its lowest level. False readings can sometimes be obtained by inserting a cuvet containing the sample in the spectrophotometer at an opaque wavelength, as the abrupt decrease in signal may not register correctly. In addition, there are other more elaborate procedures given in the literature (Kaye, 1981; 1983) for measuring stray light.

II. SPECTROPHOTOMETER CHARACTERISTICS

The components of a spectrophotometer depend on the region of the electromagnetic spectrum. Historically, a specific region of the spectrum has been defined by the availability of components, sources, and detectors for that region. This is especially true of the UV region in which transmission optics must be made of very pure silica, reflection optics must have a special coating, the source is generally a deuterium lamp, and the detector must be UV-enhanced. In fact, the cross-over from the visible to the UV occurs at about 350 nm, where the visible tungsten-halogen lamp stops emitting and the deuterium lamp is required. At very short wavelengths, ultrapure silica is needed for adequate transmission.

Ideally, a UV–VIS–NIR monochromator should be able to produce either a light beam of nearly monochromatic radiation at the exit slits, or focus nearly monochromatic radiation on individual pixels of an array detector for any wavelength in the range from 190 to 2500 nm. The sensitivity in any part of the region depends on the light source, the composition of the optical components, and the response of the detector. The sensitivity should be such that accurate measurements of transmission can be made from nearly 100% to as low as 0.01% corresponding to between 0 and 4 absorbance units (AU).

The limits of detection are determined by the signal-to-noise (S/N) ratio. At low absorbances (or high transmission), the values of P and P_0 are nearly equal to that their ratio may be lost in the instrument noise. The noise at that level is primarily due to Johnson (thermal) noise in the input resistors and shot noise from detectors, especially, photomultipliers (PMTs) and electrical component connections. At high absorbances (low levels of transmission), the principle source of noise is stray light. In single monochromators with conventional diffraction gratings, stray light is typically about 0.1% of the total radiant energy at the detector. This can be reduced to about 0.001% by passing the light twice through the grating by modulating the optical beam, or by using a double monochromator. Stray light can be further decreased through the use of holographic gratings to about 0.0005%.

It is necessary to distinguish between slit-width and bandwidth. The term "slit width" refers to the physical separation (in mm) between the slit jaws, whereas "bandwidth" denotes the range of wavelengths passed through the physical slit. For a grating spectrometer, the bandwidth at any wavelength setting is determined by the angular dispersion of the grating and the width of the entrance and exit slits. This is demonstrated for a Czerny–Turner monochromator in Fig. 5. White light enters the entrance slit in the upper left of the figure. The light rays are collimated by a concave mirror so that parallel rays fall on the grating. The light dispersed by the grating is then focused by the second mirror (top right of figure) onto the exit slits. As shown in the figure, the white light is separated into three different wavelengths. Two of these, λ_1 and λ_3, fall on the slit jaws and do not reach the detector, whereas λ_2 falls on the slit opening and then onto the detector. The linear dispersion D (in mm/nm) is given by the relation

$$D = \frac{dy}{d\lambda} = F \frac{dr}{d\lambda} \quad (6)$$

where $dy/d\lambda$ is the wavelength derivative of linear distance across the exit slit, F is the focal length, and $dr/d\lambda$ is the wavelength derivative of the angle of dispersion (the angle between the normal to the grating and the diffracted rays for a particular wavelength). For a given angle of incidence, the angular dispersion of a grating can be expressed as

$$\frac{dr}{d\lambda} = \frac{n}{d \cos r} \quad (7)$$

where n is the order of the grating and d is the distance between grooves. At small angles of dispersion, the $\cos r$ approximates unity and

$$D = \frac{nF}{d} \quad (8)$$

Thus, at small angles the linear dispersion of a grating monochromator is a constant, which depends upon the focal length of the monochromator, the distance between the grating grooves, and the grating order.

The resolving power of a monochromator is given as

$$R = \frac{\lambda}{\Delta \lambda} \quad (9)$$

This represents the ability of a monochromator to distinguish between images having slightly different wavelengths. It can be shown (Skoog et al., 1998) that the resolving power is equal to the grating order times the number of grooves across the face of the grating. Thus, the resolving power increases with the order and the physical width of the grating.

The "f-number" of a grating monochromator is defined as the focal length divided by the diameter of the collimating mirror. The light gathering power of the monochromator increases as the inverse square of the f-number. Thus, an $f/2$ monochromator gathers light $4\times$ as much as an $f/4$ monochromator.

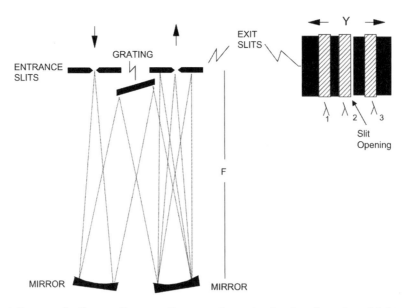

Figure 5 Optical diagram of a Czerny–Turner grating monochromator showing dispersion of light across the exit slits.

A. The Architecture of a Spectrophotometer

There are four basic parts to a spectrophotometer. The source provides radiation over the wavelength range of interest. White light from the source is passed through a wavelength selector that provides a limited band of wavelengths. The radiation exiting the wavelength selector is focused onto a detector which converts the radiation into electrical signals. Finally, the selected signal is amplified and processed as either an analog or a digital signal. We will consider each of the four major components separately.

1. Sources

A deuterium lamp is the primary source of UV radiation. An electric discharge passing through the gas excites the deuterium atoms which then lose their excess energy as the characteristic radiation. High-pressure gas-filled arc lamps containing argon, xenon, or mercury can be used to provide a particularly intense source for specific UV regions. Tungsten-halogen lamps, such as those intended for use in 35 mm projectors, are used for the visible and NIR regions.

2. Wavelength Selectors

The most commonly used wavelength selectors for UV–VIS–NIR spectrophotometers are grating monochromators. However, filter instruments, interferometers, and prism monochromators are sometimes used, and will be discussed here.

Filter Photometers

Absorption filters consist of colored glass or dyed gelatin sandwiched between glass plates. Effective bandwidths range from 30 to 250 nm. Narrow bandwidth filters absorb most of the radiant energy striking them, allowing only a small fraction to pass. Thus, the filters may require air cooling to prevent overheating.

Interference filters are more efficient and more expensive. These filters rely on optical interference to pass a narrow band of wavelengths. The filters consist of two parallel transparent plates coated on their insides with semitransparent metallic films. The space between the two films is filled with a dielectric such as calcium fluoride. Light passing through the first film is partially reflected at the second film. In turn, this reflected beam is partially reflected by the first film, producing interference effects. Those wavelengths that are equal to $2t\eta/n$, where t is the distance between plates, η is the refractive index of the dielectric, and n is the order of the filter (an integer 1, 2, etc.), will be in phase. The bandwidth of these filters is about 1.5% of the wavelength at peak transmittance, where they have a maximum transmittance of about 10%.

Prism Monochromators

The use of prism monochromators has decreased significantly in recent years, primarily because of the fact that the spectrum is a nonlinear function of wavelength and the resolution is not as good as with gratings. However, many prism spectrometers are still in use. The easiest design to visualize is the Bunsen monochromator shown in Fig. 6. Light focused onto the entrance slit is collimated by the first lens, and the parallel rays directed to a glass prism. Light passing through the prism is refracted. Shorter wavelengths are refracted more strongly than longer wavelengths (i.e., blue light is bent more than red). The dispersed light is then focused by the second lens onto the exit slits. Different wavelengths can be focused through the exit slit by either moving the exit slits and lens or by rotating the prism.

Grating Monochromators

Diffraction gratings have become much less expensive and of improved quality in recent years. Gratings can be fabricated for either transmission application or reflection application, but the latter is far more common and will be discussed here. Early gratings were made with elaborate and expensive ruling machines which used a diamond point to scribe each groove in the grating face, one at a time. At present, replica gratings are made from a master that was scribed with the ruling engine on a very hard, optically flat, polished surface. The replicas are made

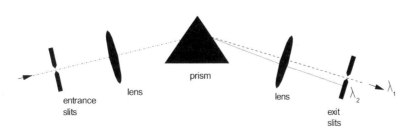

Figure 6 Optical diagram of a Bunsen prism monochromator.

by casting a liquid resin onto the surface of the master. The surface of the replica is made reflecting with a thin coating of aluminum or gold.

Replica gratings are now rapidly being replaced by holographic gratings. These are made using a pair of mutually interfering laser beams to develop a film of photo-resist on a glass surface. The sensitized regions of the photo-resist are etched away leaving a grooved structure that can be coated with a reflective metallic film. The resulting gratings produce more perfect line shapes and provide spectra with reduced stray light, virtually free from the images or ghosts often found in their predecessors.

There are a number of different classic designs for grating monochromators. Probably the most frequently encountered is the Czerny–Turner design shown in Fig. 5. This is a symmetric design with the entrance slit and concave collimating mirror on one side of the normal to the grating and the focusing mirror (identical to the collimating mirror) and exit slit on the other side. The Ebert design shown in Fig. 7(a) is similar, but the two collimating mirrors in the Czerny–Turner design are replaced by a single, large mirror. The Littrow mount shown in Fig. 7(b) is a somewhat more cumbersome, but more compact design. The big advantage of this design is that the size of the monochromator is reduced by using a single mirror to collimate and focus the radiation. Commercial Littrow instruments often place the entrance and exit slits at 90° to each other on the same side of the grating and use a small plane mirror to direct the dispersed light onto the exit slit, as shown in the figure.

Recent emphasis on miniaturization and the increased use of fiber optics have supported the development of miniature spectrophotometers, small enough to fit on a computer card. These monochromators incorporate very small gratings and, in some cases, eliminate the need for mirrors by using a concave grating, which disperses and focuses the light. Exit slits may be used in the focal plane with a single detector, or the slits can be replaced by an array detector for measuring the entire spectrum simultaneously. More will be said about miniature spectrometers and array detectors later.

Interferometers

Michelson interferometers have become the mainstay of mid-IR spectroscopy, to the extent that nearly all grating or prism spectrophotometers for use in the IR have been replaced by Fourier-transform instruments (FTIR), which make use of interferometers. We are now seeing increasing use of interferometers in the NIR region. They have not been as popular in the UV and visible, because the major advantages of the FT approach are not as effective at shorter wavelengths as they are in the mid-IR. The use of interferometers in the NIR warrants a short discussion of their general design.

An optical diagram of the Michelson interferometer is given in Fig. 8. The design is simple and symmetrical.

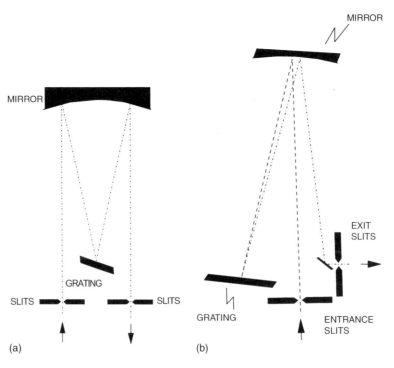

Figure 7 Optical diagram of (a) Ebert grating monochromator and (b) Littrow grating monochromator.

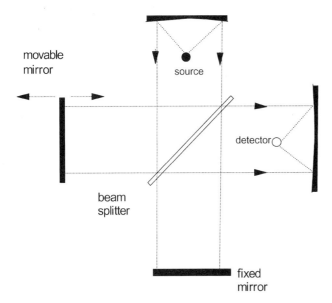

Figure 8 Optical diagram of a Michelson interferometer.

Light rays from a source are collected by a collimating mirror. The collimated rays are split by a beam splitter and directed into the two legs of the interferometer, one half of the light going to a fixed mirror and the other half to a movable mirror. The two reflected beams are recombined at the beam splitter. The phases of the two reflected rays at the beam splitter will depend upon the location of the movable mirror. At any instant in time, some wavelengths will be in-phase and others out-of-phase; the maximum in-phase conditions will occur when the distance of the movable mirror from the beam splitter is exactly equal to the distance of the fixed mirror from the beam splitter. One half of the recombined light will return to the source, and the other half will be transmitted to the detector after passing through the sample. The resulting signal at the detector as a function of the location of the movable mirror is called *interferogram*. The interferogram can be mathematically converted into a spectrum by taking its Fourier transform. Each pass of the movable mirror produces a single interferogram. Successive interferograms can be averaged to improve the S/N ratio.

3. Detectors

Certainly, the greatest recent improvements in the spectrophotometers discussed in this chapter have to be in the area of detectors. The mainstay detectors have been phototubes and PMTs; these still have a sizable share of the detector market, but the numerous advances in solid-state detectors have greatly increased their popularity. Improvements and developments in solid-state detectors, especially in the array type, are driven by their use in the consumer market. In the following sections, we will address single and multichannel detectors.

Single-Channel Detectors—Phototubes and PMTs

A vacuum phototube consists of a concave cathode made from (or coated with) a photoemissive material that emits electrons when irradiated with photons. The electrons from the cathode negatively charged by the photons, cross the vacuum to the positive anode and into the amplifier. The number of electrons ejected is directly proportional to the power of the incident radiation. The sensitivity of a photoemissive cathode to various wavelengths depends upon the composition of the emitting surface. For example, a red-sensitive surface is coated with a sequence of metals typically Na/K/Cs/Sb, whereas NIR sensitivity is obtained from coatings made with Ga/In/As. Generally, phototubes have a small dark current resulting from thermally induced electron emission.

PMTs are ideal devices for measuring low-power radiation. In addition to a cathode and an anode, the PMT is provided with a number of other specially shaped electrodes called dynodes. The cathode is the same as that for the photodiode, but is at a much higher negative voltage with respect to ground (e.g., -900 V). The electrons released from the cathode are attracted to the first dynode, which is maintained at a more positive potential (e.g., -810 V) than the cathode. When the electrons from the cathode strike this dynode, they release even more electrons. These electrons are then accelerated to the second dynode, which is at a still more positive potential of -720 V. This multiplication process is repeated over and over until the electrons reach the anode. The collision of each electron with a dynode causes several secondary electrons to be released. In this way, a single photon can produce as many as 10^6–10^7 electrons, with a series of 9 or 10 dynodes. Dark current caused by thermal emission of electrons can be a major problem with PMTs; however, cooling the PMT to $-30°C$ can reduce the thermal causes of dark current to a negligible amount. It should also be mentioned that PMTs are designed for measuring very low levels of light; hence they can be damaged by more intense illumination.

Photovoltaic Detectors

A photovoltaic detector is fundamentally a solid-state DC generator depending on the presence of light to produce a potential. A semiconducting material such as silicon or selenium is deposited as a thin film on a flat piece of steel. The semiconductor is coated with a transparent film of gold or silver to serve as the collector electrode. When radiation of high enough energy reaches the semiconductor, it causes covalent bonds to break producing electrons and holes. The electrons migrate to the metallic film, and the holes to the steel plate. The electrons produce a current in the external circuit, which is proportional to the radiation. Photovoltaic detectors are used primarily in the visible region, but lack sensitivity at low light levels.

Photodiodes

The photodiode detector consists of a p–n junction diode formed on a silicon chip. A reverse bias is applied to the diode to deplete the electrons and positive holes at the p–n junction as shown in Fig. 9. These are similar to Zener diodes, which pass current when a threshold voltage is reached. In this case, the reverse bias potential is first below the threshold voltage. Photons cause electrons and holes to form in the depletion layer, and these provide a current proportional to the incident radiant power. These devices are more sensitive than phototubes, but less sensitive than PMTs.

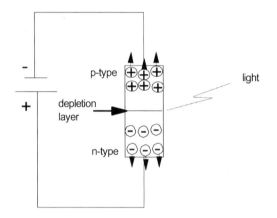

Figure 9 Diagram of a photodiode showing light incident on the depletion layer.

Multichannel Array Detectors

Multichannel detectors consist of 1D or 2D arrays of detector elements. They are all solid-state devices, basically multiple adaptations of the photodiodes discussed earlier. The main advantage of array detectors is speed, as they can measure a complete range of wavelengths simultaneously (Sedlmair et al., 1986).

A 1D array can be made by imbedding small bars of p-type elements in a substrate of n-type silicon on a single chip as shown in Fig. 10. Each p-type element would be connected to a negative potential and the n-type substrate to a positive potential to form the reverse bias. Generally, a 10 μF capacitor is connected in parallel to each p–n pixel. The capacitors are selected sequentially and charged under computer control. Light entering the chip forms electrons and holes at the depletion boundaries of each junction. The amount of current required to recharge the capacitor is proportional to the power of the light falling on that pixel. If this array was used to replace the exit slits of a conventional spectrophotometer, the entrance slit width would be adjusted to match the width of the p-type element, so that the resolution is determined by the size of the p-type elements.

Charge-Transfer Detectors

It is desirable to replace the single-channel PMT tube with a multichannel array for measuring very weak radiations. Attempts have been made to do this with the photodiode arrays, the vidicon tube, and a number of intensified adaptations of these devices, but they do not have sufficient sensitivity, dynamic range, and noise performance to compete with the PMT. These devices, nevertheless, have limited use for multichannel detection where noise and dynamic range are not so important.

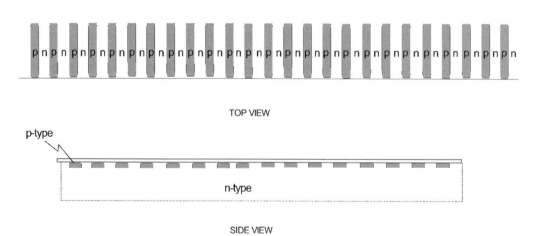

Figure 10 Photodiode array. The top view shows the face that the light would fall upon. The side view shows that the p-type elements are embedded in a continuous layer of n-type material.

Analytical instrumentation has again benefited from developments in the consumer markets. The design and commercialization of video and digital cameras have provided the ideal multichannel array detector for spectroscopic instrumentation. The detectors used in these cameras are a form of charge-transfer detectors (CTD). These detectors (Sweedler et al., 1988) come in two basic types: charge-injection devices (CID) and charge-coupled devices (CCD). Both of these are based on p- and n-doped silicon, but differ in their construction and their methods for handling the charge caused by photons impinging on the surface.

An example of a photon being detected and read from a single-element CID is shown in Fig. 11. Each pixel is connected to two electrodes, one for collecting the charges and one for sensing the total charge. These electrodes are attached across a silica insulator, and each forms a capacitor with the n-doped region. When a photon enters this region, it causes the formation of an electron and a positive hole. The electron goes to the p-doped region and the positive hole is collected near the capacitor formed by the more negative electrode. To read the charge that is built up at this point, the other electrode is disconnected from the applied potential and the voltage (V_1) between it and ground is measured. Next, the voltage on the collector electrode is reversed to positive, which repels all the positive holes. These transfer to the other electrode and are measured as V_2; the difference ($V_2 - V_1$) is proportional to the number of holes, hence to the number of photons. In the last frame of Fig. 11, the charge is removed by making the sensing electrode positive, and the procedure is started all over again.

A diagram of a CCD detector is shown in Fig. 12. For this device, each pixel element consists of a photodiode and a metal-oxide-semiconductor capacitor, which can accumulate and store charge as discussed for the single-element CID. However, the detector/capacitor elements in each row are coupled with the elements in the next lower row. After the charges accumulate in the elements of a row, they are transferred to the next lower row. This transfer is controlled by the parallel clock. Basically, the transfer is performed from the bottom row upward, that is, the bottom is emptied and the charges in the next higher row are transferred downward. Then the charges in the third row are transferred to the second row, and so on. The charges in the elements of the bottom row are transferred to the right side of the sensor array element to element; this transfer is controlled by the serial clock.

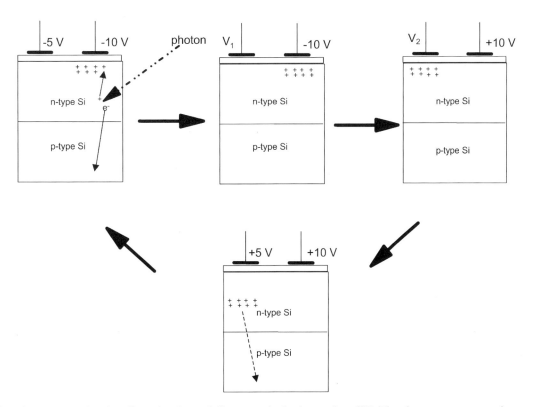

Figure 11 Diagrams showing the effect of a photon falling on a single element in a CID. The photon generates an electron and a positive hole. The measurement is based on counting the number of positive holes accumulated at the more negative electrode. The voltage generated by the positive holes is given by the difference ($V_2 - V_1$), that is, the voltages measured at the left electrode during the second and third frames.

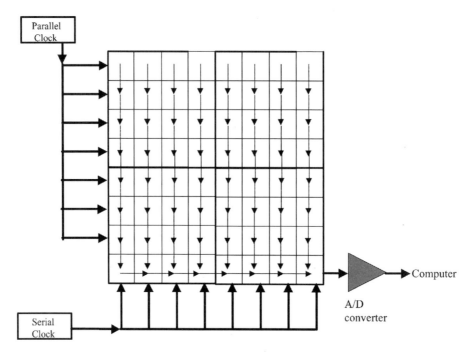

Figure 12 Schematic of a 2D CCD. Each rectangle represents a detector element. The accumulated charges are transferred down the array one row at a time until they reach the registers at the bottom. The charges are transferred serially from left to right across the registers to the A/D converter.

At the right end of the bottom row, the charges are converted to digital signals by an analog-to-digital (A/D) converter. The pixels in a CCD detector have to be read sequentially, whereas those in the CID can be read randomly as each individual element of the CID is connected to an A/D multiplexer.

Basically, the CTDs integrate the charge at each pixel in order to improve the S/N ratio compared with photodiode arrays and vidicons. The CCDs used in video cameras can be incorporated in miniature spectrometers; thus, opening up entirely new uses for analytical instruments.

4. Signal Processing

Currently, most spectrophotometers come either with built-in processors or with provision for interfacing to a personal computer. The computer controls the operating settings, adjusts the slits, sets the scan times, and turns on and turns off the light sources. Probably, the more important job for the computer is to acquire and store the spectral data and to convert the data from transmission to absorbance. It may also subtract a background spectrum from the raw data. In addition, the computer may have access to a spectral library for matching an unknown spectrum to a library spectrum, which helps in identification of unknown samples. In any case, it should be possible to display the spectrum on a monitor. It is also possible to smooth the spectrum to reduce the noise, as well as to subtract one spectrum from another, to name a few of the manipulation procedures. Several spectrometer packages provide the user with software for performing multicomponent analyses.

5. Calibrations

Some instruments are provided with internal calibration ability, such as testing for photometric accuracy. However, external calibration is often desirable. This is especially true for wavelength calibrations. For complete details on calibrations, the reader should refer to two reviews on the calibration of instruments (Mark, 1992; Mark and Workman, 1992).

Wavelength Calibration

Generally, a glass filter containing didymium or holmium oxide is used for wavelength calibration (Venable and Eckerle, 1979). Both of these can be obtained from the National Institute of Standards and Technology (NIST). Recently, strategies have been described for evaluating wavelength and photometric accuracies, spectral bandwidths, and S/N ratios (Ebel, 1992). Moreover, a wavelength calibration procedure using two well-separated spectral lines has been developed for low-resolution photodiode array spectrophotometers (Brownrig, 1993).

Table 1 Absorptivities in kg/(g cm) of $K_2Cr_2O_7$ in 0.001 M Perchloric Acid at 23.5°C (Brownrig, 1993)

$K_2Cr_2O_7$ (g/kg)	235(1.2)[a] min.	257(0.8) max.	313(0.8) min.	350(0.8) max.	Uncertainty[b]
0.020	12.243	14.248	4.797	10.661	0.034
0.040	12.291	14.308	4.804	10.674	0.022[c]
0.060	12.340	14.369	4.811	10.687	0.020[c]
0.080	12.388	14.430	4.818	10.701	0.020[c]
0.100	12.436	14.491	4.825	10.714	0.019[c]

[a]Wavelength (in nm); spectral bandwidths inside parentheses.
[b]Includes estimated systematic errors and the 95% confidence interval for the mean.
[c]The uncertainty for the 313 nm wavelength is reduced to half for the concentrations marked.

Photometric Accuracy

Solutions of potassium chromate and potassium dichromate have been widely used for checking photometric accuracy. A listing of absorptivities of $K_2Cr_2O_7$ in $HClO_4$ solutions at four wavelengths is given in Table 1 (Mielenz et al., 1977). It should be noted that absorptivities change slightly as a function of concentration. Recently, standard reference materials from NIST for monitoring stability and accuracy of absorbance or transmittance scales have been described (Messman and Smith, 1991). A listing of some standard materials available from NIST is given in Table 2.

6. *Sampling Devices*

Common to most spectrophotometers for the UV–VIS–NIR regions are sampling cuvets with 10.0 mm pathlength. These are available in vitreous silica ("quartz"), UV-transmitting glass, borosilicate glass, and various plastics. Cuvets are also made with pathlengths from 0.1 to 100 mm.

Recent developments in fiber optics technology have greatly enhanced their potential as interfaces between spectrophotometers and samples (Brown et al., 1992).

Table 2 NIST Standard Reference Materials for Spectrophotometers (Steward, 1986)

Property tested	Material	Wavelength (nm)
Transmittance	Glass filters	440–635
Absorbance	Liquid filters	302–678
Pathlength	Quartz cuvet	—
UV absorbance	Potassium dichromate	235–350
Fluorescence	Quinine sulfate dehydrate	375–675
Wavelength	Didymium-oxide glass	400–760
Transmittance	Metal-on-quartz filters	250–635
Stray light	Potassium iodide	240–280
Wavelength	Holmium-oxide solution	240–650

Care must be taken in selecting fibers for the desired optical region. Low-hydroxy silica fibers for the NIR have improved greatly in the past few years, and transmit with very little O–H absorption out to 2000 nm. In the UV region, pure synthetic silica is required, but there is still some loss at the very short wavelengths. In the visible region, silica or glass fibers or even plastic fibers can be used. Fiber optic interfaces are commercially available and a number of instruments are on the market designed for fiber optics, such that the fibers can be fitted directly into the casing of the spectrometer. Although fiber optics offer solutions to many problems, care has to be taken in designing the fiber–sample interface. The cone of light exiting from a fiber can have a relatively large angle and light can be lost, especially if there are changes in refractive index in going from air to sample container to sample. On the whole, fiber optics greatly increases the potential applications for spectroscopic instruments by making it possible to use laboratory instruments for process control and environmental monitoring.

III. PRESENT AND FUTURE UV–VIS–NIR SPECTROMETERS

In the early 1970s, UV–VIS–NIR spectrometers were large, stand-alone instruments, which produced a spectrum on a strip-chart recorder. The footprints of the instruments were on the order of several square feet. A spectrum was scanned by moving a grating or a prism and the amplified detector signal was used to move a pen on the strip-chart recorder. By the end of the 1970s, the spectrometers were still the scanning type, but microprocessors and/or computer interfaces had been added to provide spectra in digital format. Array detectors were added to the spectrometers in the 1980s; these eliminated the need for scanning and greatly reduced the time required to obtain a single spectrum. However, the footprints of the instruments were still a few square feet in size.

Two major goals of the developers of chemical and biological instrumentation are miniaturization and spatial image information. These two goals were exemplified by the developments in UV–VIS–NIR instrumentation during the 1990s. By the start of the 21st century, the footprints of many optical spectrometers had been reduced to a few square inches and, in some cases, as small as an integrated circuit chip. Moreover, 2D array type detectors were being used to provide spatial images as a function of wavelength.

Two current, bench top spectrometers, one for the UV–VIS and the other for the NIR region, are shown in Fig. 13. Both of these instruments require a fraction of the space required by their predecessors from the 1980s. Both instruments are interfaced to computers for acquiring, processing, storing, and displaying the spectra. The UV–VIS spectrometer uses diode-array detection, whereas the NIR spectrometer is a rapid-scanning grating instrument.

Miniature spectrometers were marketed in the early 1990s by several companies. The sizes of these spectrometers were on the order of 10×10 cm^2 and could be fitted to a circuit board for plugging directly into a personal computer slot. A popular example of such an instrument is shown in Fig. 14. The footprint of the Ocean Optics package is $\sim 15 \times 15$ cm^2, but the actual spectrometer part of the pack is small. The spectrometers can be fitted with several different gratings and CCD

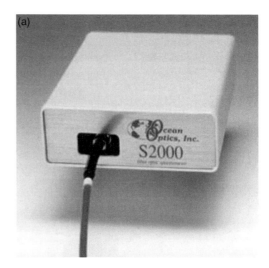

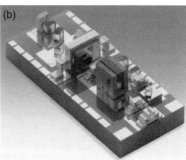

Figure 14 Miniaturized spectrometers. (a) Ocean Optics Model Mode S2000. This spectrometer can be purchased optimized for various spectral regions from 200 to 1100 nm. The miniature spectrometer inside the package is the size of a small circuit board and the footprint of the package is $\sim 15 \times 15$ cm^2. (b) AXSUN Technologies microspectrometer showing components on aluminum nitride optical baseplate. The bench is 14 mm long.

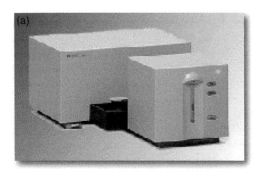

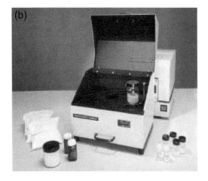

Figure 13 Two 21st Century bench top spectrometers. (a) Agilent Technologies UV–VIS spectrometer with a wavelength range of 190–1100 nm. (b) Foss NIRSystems NIR Rapid Content Sampler with an wavelength range of 1100–2500 nm.

array detectors, which are optimized for the spectral region of choice. These spectrometers are designed to be interfaced to samples and sources through a fiber.

The next level of miniaturization of spectrometers has just appeared on the market and is also shown in Fig. 14. All the components of the AXSUN Technologies spectrometer including the source, wavelength selector, and detector are on a chip, which is 14 mm long. This spectrometer is operated in the NIR and is interfaced to a sample by means of a fiber. The wavelength selector is a Fabry–Perot filter, which is simply an interference filter with controllable distances between the parallel reflecting surfaces. The very small size of the Fabry–Perot makes it possible to vary the distance between the reflecting surfaces rapidly, and high-resolution spectra can be acquired in a short time period. This is certainly the first step in the development of microspectrometers, an area that will blossom in the coming years.

Figure 15 Spectral Dimensions Model Sapphire, NIR Chemical Imaging system for measuring NIR microscopic images.

The other important area of 21st century spectroscopy development is in spectral imaging. An example of such a spectrometer for the NIR is shown in Fig. 15. This instrument appears more like a microscope, and it is a microscope that can collect microscopic images over a wavelength range of 1100–2450 nm. An InSb camera with 81,920 pixels (320 × 256 array) is used to measure the spatial image at each wavelength. The wavelength selector for this spectrometer is a liquid crystal tunable filter, which can be rapidly scanned. Imaging spectrometers for the UV–VIS, NIR, and mid-IR came into the market within the last decade. The technology is still very sophisticated and expensive, but major developments have already been made and we will see simpler, less expensive instruments over the next decade.

UV–VIS–NIR spectroscopy will continue to grow and find many uses in this 21st century. More consumer applications of miniature and imaging spectrometers will be found in the medical, transportation, pharmaceutical, and agriculture industries, to name just a few. Technologies developed in laboratories will eventually be used by the layman.

REFERENCES

Brown, C. W., Donahue, S. M., Lo, S-C. (1992). Remote monitoring with near infrared fiber optics. In: Patonay, G., ed. *Advances in Near-IR Measurements*. Vol. 1. JAI Press, p. 1.

Brownrig, J. T. (1993). Wavelength calibration method for low resolution photodiode array spectrometers. *Appl. Spectrosc.* 47:1007.

Ebel, S. F. (1992). *J. Anal. Chem.* 342:1007.

Kaye, W. (1981). Stray light ratio measurements. *Anal. Chem.* 53:2201.

Kaye, W. (1983). *Am. Lab.* 15(1):18.

Mark, H. (1992). *Pract. Spectrosc.* 13:107.

Mark, H., Workman, J. (1992). Statistics in spectroscopy: developing the calibration model. *Spectrosc.* 7:14.

Messman, J. D., Smith, M. V. (1991). *Spectrochim. Acta, Part B* 46B:1653.

Mielenz, K. U., Velapoldi, R. A., Mavrodineanu, R. (1977). *Standardization in Spectrophotometry and Luminescence Measurements*. Special Publication. Washington: National Institute for Standards and Technology, p. 466.

Poulson, R. F. (1964). *Appl. Optics* 3:99.

Sedlmair, J., Ballard, S. G., Mauzrail, D. C. (1986). *Rev. Sci., Instrum.* 57:2995.

Skoog, D. A., Holler, J. F., Nieman, T. A. (1998). *Principles of Instrumental Analysis*. 5th ed. Philadelphia: Harcourt Brace & Company.

Steward, R. W., ed. (1986). *NBS Standard Reference Materials Catalog 1986–87*. Washington: National Institute for Standards and Technology, p. 100.

Sweedler, J. V., Bilhorn, R. B., Epperson, P. M., Sims, C. R., Denton, M. B. (1988). High performance charge transfer device detectors. *Anal. Chem.* 60:282A.

Venable, W. FL., Jr., Eckerle, K. L. (1979). *Didymium Glass Filters for Calibrating the Wavelength Scale of Spectrophotometers*. Special Publication. Washington: National Institute for Standards and Technology, pp. 260–266.

6

Molecular Fluorescence and Phosphorescence

FERNANDO M. LANÇAS, EMANUEL CARRILHO
Institute of Chemistry at São Carlos, University of São Paulo, São Carlos (SP), Brasil

I. INTRODUCTION

Molecular luminescence spectroscopy can be used for fundamental studies of molecular excited states as well as for selective and sensitive analysis of luminescent samples. Luminescence processes, such as fluorescence and phosphorescence, are emission processes and, as a result, luminescence techniques have dynamic ranges and detection limits that are several orders of magnitude greater than molecular ultraviolet–visible (UV–Vis) absorption techniques for highly luminescent compounds. Selectivity can be derived from several sources. Firstly, only certain groups of compounds that absorb UV–Vis radiation are likely to undergo de-excitation by luminescence. Secondly, selectivity among luminescent molecules can often be accomplished on the basis of excitation and emission spectral characteristics and excited-state lifetimes.

Polarization and selective quenching techniques can be extended to nonluminescent compounds by the use of indirect methods in which the analyte is involved in a chemical reaction or physical interaction with luminescent reagents.

There are numerous sources of information in the literature on luminescence spectroscopy. Most instrumental analysis textbooks provide descriptions of components and configurations of luminescence instruments, as well as some background theory and applications. A number of books and book chapters on fluorescence and phosphorescence analysis have been written that include discussions of theory, instrumentation, practical considerations, and applications (Guilbault, 1990; Lakowicz, 1999; McGown, 1966; Rendell, 1987; Schulman, 1977). Several recent series and reviews have focused on state-of-the-art developments and current trends in luminescence instrumentation, techniques, and applications (Baeyens et al., 1991; Cline Love and Eastwood, 1985; Dewey, 1991; Eastwood, 1983; Eastwood and Cline Love, 1988; Ichinose et al., 1991; Lakowicz and Thompson, 2001; Schulman, 1985; Warner and McGown, 1991; Wehry, 1975; Wolfbeis, 1993). The biennial Fundamental Reviews issue of Analytical Chemistry contains a review of molecular luminescence spectrometry that focuses on recent developments in theory, instrumentation, and applications (Agbaria et al., 2002).

II. THEORY

The fundamental aspects of luminescence phenomena have been discussed in a number of books (McCarrol et al., 2000; Mycek and Pogue, 2003; Rendell and Mowthorpe, 1987; Rigler and Elson, 2001; Valeur, 2001). Berlman (1971) has provided a compilation of fluorescence spectra that also includes values for quantum yields, excited-state lifetimes, and other fundamental quantities. An introduction to the theory of luminescence spectroscopy is presented in this section, including descriptions of fundamental processes and characteristics, spectra, and quantitative relationship between intensity and concentration.

A. Excited-State Processes

The processes involved in molecular photoluminescence are shown in Fig. 1. Molecular absorption of photons in the UV–Vis range causes the electronic transition from a lower energy level [typically the lowest vibrational level v_0, of the ground electronic singlet state (S_0)] to an excited singlet state S^*. The molecule rapidly undergoes vibrational relaxation to the lowest vibrational level of the excited state from which it may be de-excited by one of several competitive pathways. De-excitation directly back to the ground state can occur either by nonradiative processes or by emission of photons. The latter process is referred to as *fluorescence*. Alternatively, under favorable conditions, the molecule may undergo a "forbidden" transition to an overlapping triplet state (T^*—a metastable excited electronic state) by a process known as *intersystem crossing*. Once in the triplet state, the molecule may undergo vibrational relaxation to the lowest vibrational level of T^* or return to S^* by intersystem crossing. In the latter case, de-excitation by photon emission results in *delayed fluorescence*, which is spectrally identical to prompt *fluorescence*, but it exhibits the longer decay associated with phosphorescence (see later). In the former case, de-excitation may occur by photon emission from the triplet state to the ground state, known as *phosphorescence*, or by nonradiative processes. Phosphorescence generally occurs at longer wavelengths than fluorescence because the triplet state presents a lower energy level than the overlapping excited singlet state.

B. Excited-State Lifetimes

Phosphorescence can be distinguished from fluorescence on the basis of the kinetics of their decay, which is a reflection of the lifetimes of their respective excited states. Both processes follow first-order kinetics:

$$I(t) = I_0 \, e^{-kt} \tag{1}$$

where $I(t)$ is the intensity at time t and I_0 is the initial intensity [$I(t)$ at $t = 0$]. The rate constant k for de-excitation can be expressed in terms of radiative (k_R) and nonradiative (k_{NR}) de-excitation rate constants:

$$k = k_R + k_{NR} \tag{2}$$

The luminescence lifetime τ is the time required for the intensity to decay to $1/e$ of the initial intensity [$t = \tau$ when $I(t) = I_0/e$] and is inversely proportional to k. Fluorescence lifetimes, τ_F, are on the order of nanoseconds to microseconds. Phosphorescence lifetimes, τ_P, are much longer because of the forbidden singlet–triplet–singlet transitions, and range from milliseconds to seconds.

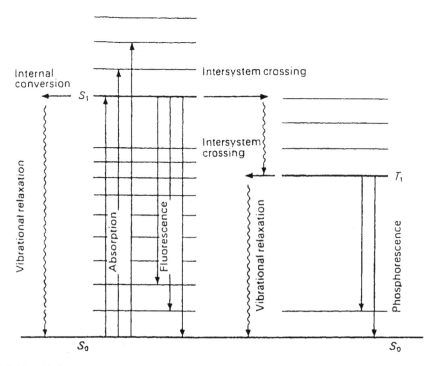

Figure 1 A Jablonski diagram showing electronic processes. [Adapted from John and Soutar (1981) with permission.]

C. Quantum Yield

The quantum yields or efficiencies of the fluorescence and phosphorescence processes can be expressed as

$$\phi_F = \frac{\text{Number of photons emitted as fluorescence}}{\text{Total number of photons absorbed}} \quad (3)$$

and

$$\phi_P = \frac{\text{Number of photons emitted as phosphorescence}}{\text{Total number of photons absorbed}} \quad (4)$$

respectively. The quantum yield is related to the radiative rate constant k_R and the observed excited-state lifetime:

$$\phi = k_R \tau \quad (5)$$

If the quantum yield is unity, the observed lifetime τ is equal to $1/k_R$ and is called the *intrinsic lifetime* (also referred to as the natural or radiative lifetime).

D. Quenching

Quenching refers to any process that causes a reduction in the quantum yield of a given luminescence process. Quenching can be categorized as either collisional or static. Collisional quenching is described by the Stern-Volmer equation

$$\frac{I_0}{I} = 1 + k_q[Q]\tau \quad (6)$$

where I_0 is the luminescence intensity in the absence of quencher, I is the intensity in the presence of quencher at concentration $[Q]$, k_q is the rate of collisional quenching, and τ is the observed lifetime. Collisional quenching is evidenced by a linear decrease in the observed luminescence lifetime with increasing quencher concentration. In contrast, static quenching is due to the formation of a ground state complex between the luminescent molecule and the quencher, with formation constant K_c, described by:

$$\frac{I_0}{I} = 1 + K_c[Q] \quad (7)$$

The observed lifetime does not appear in this equation and is independent of quencher concentration in static quenching.

E. Intensity and Concentration

The fluorescence intensity I_F of a simple solution of a fluorescent compound is directly proportional to the intensity of the excitation beam (I_{ex}) and the absorbance (εbc, where ε is the molar absorptivity, b is the pathlength, and c is the concentration) and quantum yield (ϕ_F) of the compound:

$$I_F = K_F I_{ex} \phi_F \varepsilon bc \quad (8)$$

The proportionality constant K_F is a function of instrumental response factors and geometry.

As I_F is proportional to I_{ex}, sensitivity and detection limits can be improved by increasing the intensity of the source [up to limits imposed by photodecomposition (photobleach) or nonlinear response]. This is in contrast to absorption spectroscopy in which an increase in source intensity does not directly affect sensitivity and detection limits. In addition, the directly measured quantity I_F is linearly proportional to the concentration of the fluorescent compound, unlike absorptiometric experiments in which concentration is linearly related to the log of the ratio of two measured intensities. The combination of these proportionality features results in sensitivities and detection limits for fluorescence measurements of strongly fluorescent molecules that are several orders of magnitude better than those for absorptiometric determinations of the same molecules. Fluorescence spectroscopy is also more selective than absorption spectroscopy, because not all molecules that absorb UV–Vis radiation are fluorescent; fluorescence analysis can be extended to nonfluorescent molecules through derivatization or indirect detection methods.

An equation analogous to Eq. (8) relates phosphorescence intensity to incident intensity, quantum yield, and absorbance:

$$I_P = K_P I_{ex} \phi_P \varepsilon bc \quad (9)$$

The subset of absorbing molecules that exhibit phosphorescence is smaller than the group exhibiting fluorescence, for several reasons, including the forbidden nature of the transitions involved and the longer excited-state lifetimes that make phosphorescence much more susceptible to quenching. Phosphorescence is generally observed only at low temperatures (liquid nitrogen) or with special techniques such as deposition of the sample in a solid matrix or dissolution in micellar solutions to eliminate the quenching processes that occur in solution and reduce the rates of competitive de-excitation processes. Additionally, the addition of atoms or molecules, such as the "heavy atoms" (Br, I, Tl, Ag, etc.), which increase the probability of intersystem crossing, can be used for further enhancement of phosphorescence.

Dissolved molecular oxygen in solution is probably the most troublesome and ubiquitous quencher of both fluorescence and phosphorescence. The effect of oxygen is especially serious in the case of phosphorescence, which is generally not observed in solution even in the presence of only small concentrations of dissolved oxygen.

Solutions are often deoxygenated prior to measurement to eliminate this source of quenching.

F. Luminescence Spectra

Luminescence excitation spectra are acquired by scanning excitation wavelengths at a constant emission wavelength. Emission spectra are collected by scanning emission wavelengths at a constant excitation wavelength. The excitation spectrum of a molecular species is analogous to its absorption spectrum, and is generally independent of the emission wavelength due to vibrational relaxation to the lowest vibrational level of S* prior to photon emission (Fig. 1). For the same reason, the emission spectrum is generally independent of the excitation wavelength. The two independent dimensions of spectral information can be more fully explored by means of total luminescence spectroscopy in which fluorescence intensity is plotted as a function of excitation wavelength on one axis and emission wavelength on the other (Fig. 2).

Fluorescent molecules that have similar vibrational structures in the ground and excited singlet states will have excitation and emission spectra that are roughly mirror images of each other (Fig. 3). Spectral resolution and detailing increase as temperature decreases and the rigidity of the sample matrix increases; these effects are exploited in the low temperature techniques such as matrix isolation and fluorescence line-narrowing (Section IV.E).

G. Polarization

A ray of light consists of a magnetic field and an electric field that are perpendicular to each other and to the direction of propagation of the ray. The electric field vector has a particular angular orientation with respect to a given coordinate system, and a beam composed of many rays can be characterized in terms of the overall angular distribution of the electric field vectors. The angular distribution of the rays may be random, resulting in a nonpolarized beam, or the rays may have parallel vectors and be completely polarized. If a molecule is excited with completely polarized light, the emission may be depolarized because of the effects from the absorption process, difference in orientation between the absorption and emission dipoles of the molecule, Brownian rotation of the molecule, and absorption of emitted light by a second molecule, which then emits light with different orientation. Other effects may be observed because of scattered light and energy transfer. The polarization P is generally determined by exciting the sample with vertically polarized light and measuring the intensities of the vertical and horizontal

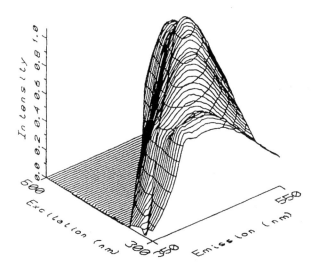

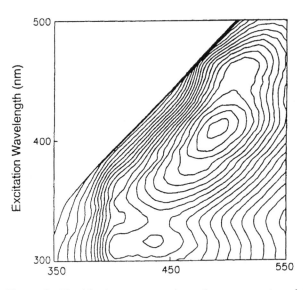

Figure 2 Total luminescence spectrum, shown as a contour plot and a surface plot of an aqueous solution of a humic substance.

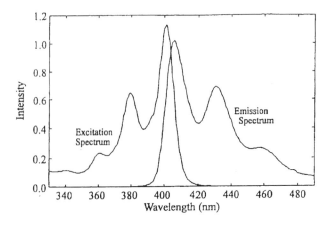

Figure 3 Fluorescence excitation and emission spectra of benzo(k)fluoranthene in ethanol.

components of the emitted light (I_V and I_H, respectively):

$$P = \frac{I_V - I_H}{I_V + I_H} \quad (10)$$

Fluorescence anisotropy r, which is largely measured in protein biophysics,

$$r = \frac{I_V - I_H}{I_V + 2I_H} \quad (11)$$

is closely related to polarization and may be calculated from the same experimental data.

III. INSTRUMENTATION

A. Instrument Components

A general schematic diagram of a luminescence spectrometer is shown in Fig. 4. Emission is usually measured at 90° to the excitation beam to avoid background from nonabsorbed radiation, although angles other than 90° are sometimes used for specific applications. The instrumental components include a light source with its own power supply, an excitation wavelength selector, a sample chamber, an emission wavelength selector, a detector, an output device, and in most modern instruments a computer controller for data acquisition and analysis.

1. Light Sources

Luminescence intensity is directly proportional to the intensity of the light source [Eqs. (8) and (9)], and high intensity sources can therefore be used to increase sensitivity and to lower detection limits in luminescence analysis. The xenon arc lamp is a commonly used source. The output is continuous over a broad wavelength range that extends from the near-UV to the near-IR (NIR) and is therefore well suited to spectral scanning. Another common source is the high pressure mercury arc lamp. The output is a continuum with a line spectrum superimposed on it, making the mercury lamp better suited to nonscanning filter instruments. Other sources include halogen lamps and combination xenon–mercury lamps. The outputs of the sources used for luminescence measurements are shown in Fig. 5. The intensity of the excitation beam can be increased by the use of a mirrored backing in the source compartment to redirect emitted light in the direction of excitation.

Lasers are often used in luminescence experiments in which continuous scanning of excitation wavelength is not required. The properties and principles of lasers and laser light have been described (Vo-Dinh and Eastwood, 1990; Wilson, 1982). Tunable lasers (Latz, 1976) can provide multiwavelength excitation. The properties of a wide variety of dyes used in tunable dye lasers have been surveyed (Maeda, 1984).

Improvements in detection limits can be obtained by using lasers, particularly with respect to absolute amounts, as laser beams can be focused to irradiate very small volumes (pL range). Improvements in detectability may be limited, however, by an increase in the background signals and localized heating effects. Also, the excitation beam must often be greatly attenuated to avoid photodecomposition of the sample constituents. Specialized applications of lasers include time-resolved fluorimetry, detection in very small sample volumes such as in HPLC or capillary electrophoresis, and spatial imaging. The use of lasers for chemical analysis has been discussed in recent books and articles (Richardson, 1981; Wright, 1984; Zare, 1984).

Pulsed sources, including both lamps and lasers, are used for special applications such as for dynamic measurements of luminescence lifetimes and for time-resolved elimination of background signals (Sections III.C, III.D.3, and IV.C).

2. Wavelength Selectors

Either filters or monochromators can be used for wavelength selection. Filters offer better detection limits and are therefore well suited to quantitative analysis, but do not provide wavelength scanning capabilities desirable for acquisition of spectra. Often, a filter is used in the excitation beam along with a monochromator in the emission beam to collect emission spectra. Full emission and excitation spectral information can be acquired only if monochromators or polychromators (dispersive devices without slits for use with array detectors, Section III.D.1) can be

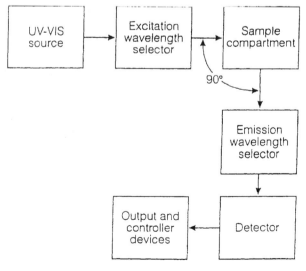

Figure 4 General schematic diagram of a single beam fluorometer.

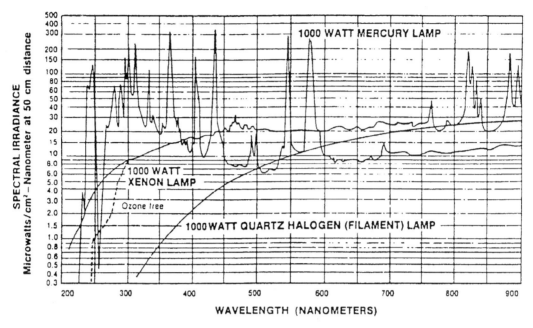

Figure 5 Output of several light sources commonly used for fluorescence excitation. (From the literature of Oriel Corp., with permission.)

used in both the excitation and the emission beams. This is necessary for techniques such as synchronous excitation and total luminescence spectroscopy. Lasers provide monochromatic light and do not require wavelength selectors except to isolate one of several lines if there is a multi-line output; this can be achieved using an interference filter. Lasers cannot be used for collection of excitation spectra.

3. Sample Compartments

Cuvettes for fluorescence measurements of solutions are usually rectangular with at least two adjacent faces that are transparent. The remaining two faces may also be transparent or they may have reflective inner surfaces to direct more of the fluorescence emission to the detector. Quartz cuvettes are used for measurements in the UV–Vis region, and less expensive glass cuvettes can be used for measurements in the visible region only. Inexpensive disposable cuvettes made of polystyrene or polyethylene can also be used for work in the visible range with certain solvents. Disposable acrylic cuvettes can be used for the 275–350 nm wavelength range and are useful for measurements of native protein fluorescence. Special cuvettes are available for microvolume work and flow systems, such as flow injection analysis, fluorimetric detection for HPLC, and stopped flow measurements. Mirrored backings on these cells are especially useful to maximize the detected emission intensities.

Low temperature phosphorescence measurements are generally made on samples that are contained in special dewar cells. The sample is contained in a quartz tube that is surrounded by liquid nitrogen in the dewar container. Samples must be cooled to a solid state very carefully to avoid cracking and "snow" formation. For room-temperature phosphorescence, samples may be deposited onto a strip of filter paper or other support material. A dry, inert gas is continuously passed over the sample to eliminate oxygen and moisture. Phosphorescence measurements of liquid samples and solutions may be made using conventional fluorescence cuvettes. Special cuvettes designed to facilitate oxygen removal can also be used.

4. Detectors

Photomultiplier tubes (PMTs) are the most common detectors and various types are available for different applications. In general, they are sensitive in the range of 200–600 nm, with maximum sensitivity obtained in the 300–500 nm range (Fig. 6). Red-sensitive PMTs are also available for detection above 600 nm. The PMT housings are sometimes cooled to temperatures as low as $-40°C$ to minimize temperature-dependent noise.

Photomultipliers can be operated either as analog detectors in which the generated current is proportional to the incident radiative intensity or as photon counters in which the current pulses generated by individual incident photons are counted. A preset threshold value allows noise and other low energy events to be ignored. Single-photon detection offers greatly improved detection limits and is useful for the measurement of very low intensity

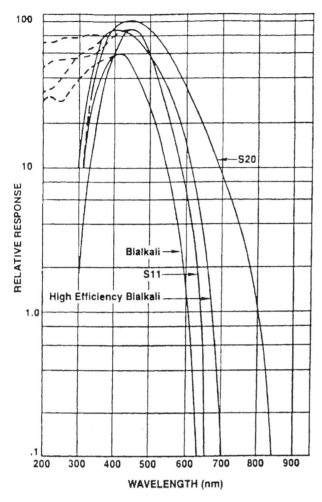

Figure 6 Spectral response curves for some common PMT detectors. (———) glass window, (– – –) quartz window. (From Hamamatsu, Ltd., with permission.)

signals. The linear response range is much narrower than that of the analog detector, being limited on the low end by signal-to-noise deterioration and on the high end by the response time of the counter.

The use of multichannel detectors (Christian et al., 1981; Ingle and Ryan, 1983; Johnson et al., 1979; Talmi, 1975; Warner et al., 1975) for the simultaneous acquisition of emission or total luminescence spectra has increased the range of applications of luminescence experiments to include real-time experiments, kinetic measurements, and on-line detection for chromatography and other flow systems. The ability to acquire complete spectral information instantaneously has also greatly facilitated qualitative analysis by dramatically reducing the time required per analysis. Among the more commonly used detectors are diode arrays, vidicons, silicon-intensified target vidicons, charge-coupled and charge-injection devices, and numerous other devices made possible by recent technological advances.

5. Output Devices and Computers

Simple instruments such as those with filters for wavelength selection often have simple analog or digital readout devices. X–Y recorders can be used to record spectra with a spectrofluorometer. Computers are frequently interfaced to more sophisticated spectrofluorometers to control data acquisition and for storage, manipulation, and analysis of spectral data. For large data arrays, such as those acquired in total luminescence spectroscopy, and complex experiments such as those involving dynamic measurements of luminescence lifetimes, computerized data collection and analysis are indispensable.

B. Instrument Configurations for Conventional Fluorescence Measurements

Instruments designed for luminescence spectroscopy can usually be used for either fluorescence or phosphorescence measurements with a few instrumental modifications generally required for the latter. In this section, we will discuss the instrumental configurations used for fluorescence instruments. Subsequent sections will address the specific instrumental requirements and some of the special techniques required for phosphorescence measurements.

1. Filter Fluorometers

Instruments in which wavelength selection is accomplished with filters are generally referred to as fluorometers, whereas monochromator-based instruments capable of spectral scanning are called spectrofluorometers. The simplest, least expensive, fluorometers are single beam filter instruments with halogen lamp sources, phototube or PMT detectors, and simple output devices. Ratiometric fluorometers are also available in which the ratio of the sample signal to a reference signal is used in order to compensate for fluctuations in line voltage, drift, and so on. In some instruments, such as the one shown in Fig. 7, the sample and reference light paths are taken from the same source and alternately measured by the same detector through the use of a mechanical cam. A similar configuration in which two detectors are used, one for the sample beam and one for the reference beam, will correct for source output fluctuations only. Single detector configurations are also capable of minimizing effects of detector variability.

Filter fluorometers are suitable for quantitative analysis applications in which spectral scanning and high resolution are not required. Filters transmit more light and cost less than monochromators, thereby providing better detection limits with less expensive instrumentation.

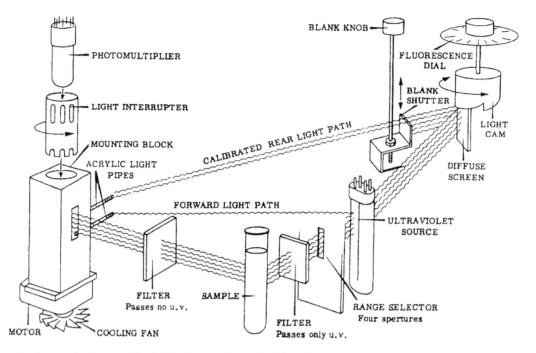

Figure 7 Schematic diagram of a double-beam (ratiometric) filter fluorometer. (From Sequoia-Turner, with permission.)

2. Spectrofluorometers

Spectrofluorometers are also available in both single beam and ratiometric configurations. A single beam instrument is shown in Fig. 8. A ratiometric spectrofluorometer configuration in which two separate detectors are employed is depicted in Fig. 9. The reference PMT monitors the excitation beam after it has passed through the monochromator so that corrections are based on fluctuations in the wavelength of interest only. The reference PMT is placed after the sample to monitor the transmitted beam, thereby allowing absorption measurements to be made on the sample.

Single-photon counting instruments are used to measure very low light levels and are designed for maximum signal-to-noise performance. Figure 10 illustrates the single-photon counting detection scheme. Double monochromator arrangements are often used in the excitation or emission beams of single-photon counting instruments to minimize stray light. Cooled PMT housings may be useful to minimize detector noise. Effects of fluctuations in source output can be minimized by ratiometric detection in which a reference PMT detects a portion of the excitation beam that is diverted to the detector by a beam splitter placed between the excitation monochromator and the sample. If an appropriate blank solution is placed in the reference beam, absorption measurements can also be made.

Spectrofluorometers offer the most flexibility for quantitative and qualitative applications. Many instruments are modular so that monochromators can be replaced with filters in the emission or excitation beams if desired.

C. Phosphorimeters

Phosphorescence can be measured with a conventional, continuous source fluorescence instrument that is equipped with a phosphoroscope attachment. The phosphoroscope is used to eliminate fluorescence signals on the basis of the large difference in lifetimes between fluorescence and phosphorescence, which allows phosphorescence to be measured after the fluorescence signal has completely decayed. Early phosphoroscopes used the rotating can assembly (Fig. 11), but modern instruments most commonly employ such optical choppers as the rotating mirror (Fig. 12) (Vo-Dinh et al., 1977). Pulsed-source instruments can be used in time-resolved spectroscopy to minimize rapidly decaying background signals (Hamilton and Naqvi, 1973; Vo-Dinh et al., 1972) or to recover the phosphorescence decay curve. The time-resolved approach has the advantage of allowing the resolution of multiexponential phosphorescence decay signals in order to perform multicomponent determinations (Boutillier and Winefordner, 1979; Goeringer and Pardue, 1979).

Room-temperature phosphorescence (RTP) spectroscopy (Vo-Dinh, 1984) can be performed on conventional fluorescence instruments. In solid surface techniques (Hurtubise, 1981), modifications are necessary to measure the luminescence that is either reflected or transmitted

Molecular Fluorescence and Phosphorescence

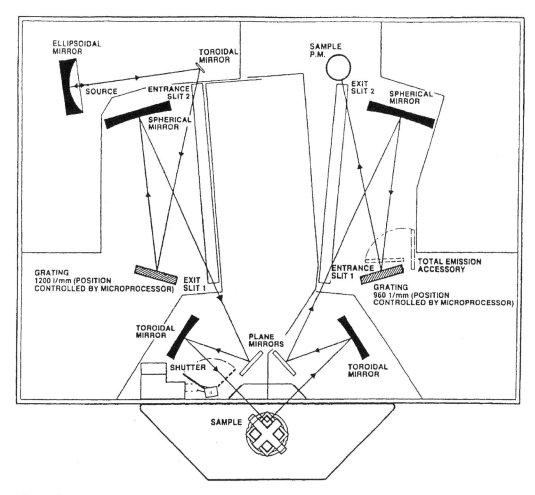

Figure 8 Schematic diagram of a single-beam spectrofluorometer. (From Perkin–Elmer, with permission.)

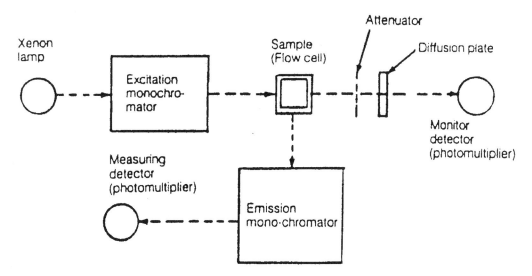

Figure 9 Schematic diagram of a ratiometric spectrofluorometer. (From Perkin–Elmer, with permission.)

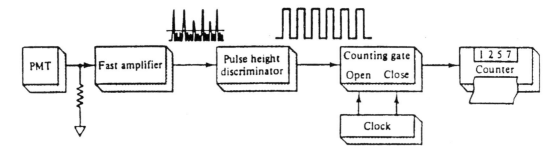

Figure 10 Depiction of single photon counting. The signal is detected by the PMT, emerging as a pulse that is then amplified. If the magnitude of the amplified pulse exceeds a certain threshold, it is passed through the discriminator and counted as a signal. Smaller pulses due to noise and dark current are much more likely to be rejected, thereby increasing signal-to-noise. [From Ingle and Crouch (1988), with permission.]

from the solid surface. Alternatively, commercial luminescence instruments with spectrodensitometers are available for RTP as well as fluorescence measurements on solid surfaces.

D. Instruments with Special Capabilities

1. Spectrofluorometers for Extended Spectral Acquisition

Spectrofluorometers can be used to generate excitation spectra, emission spectra, synchronous excitation spectra, and total luminescence spectra. Conventional emission and excitation spectra simply require emission and excitation monochromators. Acquisition of synchronous excitation spectra (Lloyd, 1971; Vo-Dinh, 1978, 1981) requires that the emission and excitation monochromators can be simultaneously and synchronously scanned. Total luminescence spectra (Christian et al., 1981; Giering and Hornig, 1977; Suter et al., 1983; Warner et al., 1975, 1976; Weber, 1961) can be acquired on any instrument that is equipped with both emission and excitation monochromators. As the name implies, total luminescence spectra may contain phosphorescence as well as fluorescence signals, especially for solid surfaces or solutions that contain micelles or other organized media to enhance phosphorescence.

Array detection can greatly reduce the time required for data acquisition by allowing simultaneous measurement of intensity at many wavelengths. The instrument shown in Fig. 13 has array detection of the emission beam so that the emission spectrum at a given excitation wavelength can be acquired in essentially the same time that it would take to measure the intensity at a single emission wavelength in a scanning instrument. A spectrograph is used instead of an emission monochromator in order to disperse the emission beam and simultaneously detect all of the wavelengths in a given range. Similarly, total luminescence spectra can be acquired in much less time relative to conventional spectrofluorometers because emission spectra can be simultaneously collected at the sequentially scanned excitation wavelengths.

The use of two-dimensional array detectors can further facilitate the collection of total luminescence spectra. A rapid-scanning spectrofluorometer is shown in Fig. 14 (Johnson et al., 1979). The exit slits are removed from the excitation and emission monochromators to convert

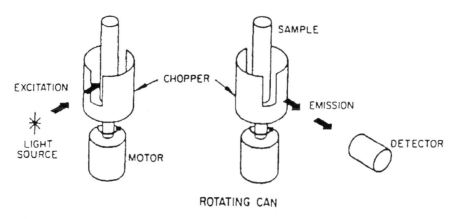

Figure 11 A rotating cylinder phosphoroscope. [From Vo-Dinh (1984), with permission.]

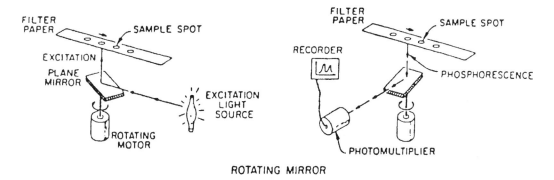

Figure 12 A rotating mirror phosphoroscope. [From Vo-Dinh (1984), with permission.]

them to polychromators. The excitation wavelengths are spatially resolved along the vertical axis of the sample cuvette and therefore of the detector, and the emission wavelengths are spatially resolved along the horizontal axis of the detector (Fig. 15). A silicon-intensified target vidicon serves as the two-dimensional detector. Other imaging devices, including those listed in Section III.A.4, can also be used. In addition to the reduction in analysis time and concomitant potential for improvements in the signal-to-noise ratio, advantages of array detection include the ability to perform multiwavelength kinetic studies that would be difficult or impossible with sequential scanning and minimization of sample decomposition due to inherent instability or photodegradation.

2. Polarization Measurements

Fluorescence polarization can be measured with either a fluorometer or a spectrofluorometer. Polarizers are placed in both the excitation and the emission beams, each between the corresponding monochromator or filter and the sample. Two different formats can be used (Lakowicz, 1999). The L-format (Fig. 16) uses the conventional fluorometer configuration with a single excitation channel and a single emission channel. A correction factor must be applied to account for partial polarization that occurs in the monochromators and for the dependence of monochromator transmission efficiency on the polarizer angle. The T-format (Fig. 17) uses a single excitation

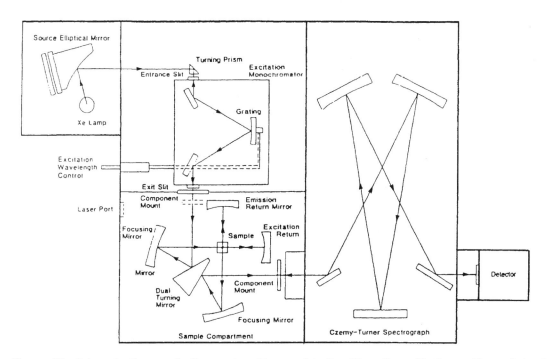

Figure 13 Schematic diagram of a fluorometer with array detection. [From Tracor-Northern, with permission.]

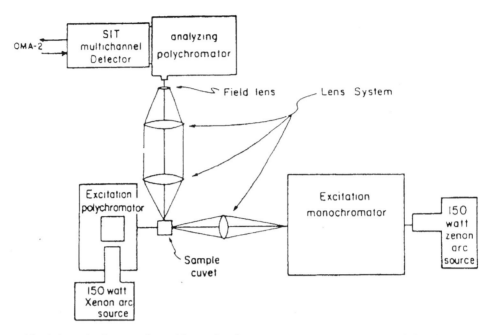

Figure 14 Schematic diagram of a rapid-scanning fluorometer. [From Warner et al. (1979), with permission.]

channel and two emission channels, one for the vertically polarized component and the other for the horizontal component. The two emission signals are measured ratiometrically to eliminate errors due to polarization that occur in the excitation monochromator. A correction factor must still be used to account for the different sensitivities and polarizer settings of the two emission channels.

3. Luminescence Lifetime Measurements and Related Techniques

Numerous books, chapters, and monographs have discussed techniques and instrumentation used for the determination of excited-state lifetimes and related time-resolved techniques (Cundall and Dale, 1983; Demas, 1983; Hieftje and Vogelstein, 1981; Lakowicz, 1999; Visser, 1985; Ware, 1971; Yguerabide and Yguerabide, 1984). The most conceptually direct approach is the use of pulsed-source excitation and direct acquisition of the decay curve that can be analyzed to yield the exponential decay characteristics of the emitting components. The range and performance of this technique depends on the width and reproducibility of the excitation pulses relative to the lifetime range under investigation, as well as on the detector response, and is generally limited to lifetimes of several nanoseconds or longer. Measurements of subnanosecond fluorescence lifetimes with pulsed-source excitation may require more sophisticated technology. The instrument shown in Fig. 18 uses a flashlamp source with a high repetition rate and time-correlated single-photon counting (TCSPC) (O'Conner and Phillips, 1984) to extend the range of lifetimes that can be determined

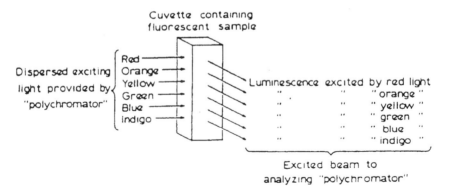

Figure 15 Illumination of a sample with polychromatic light. [From Warner et al. (1976), with permission.]

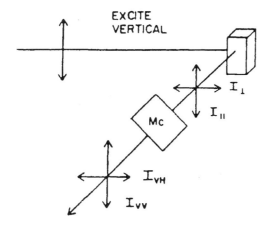
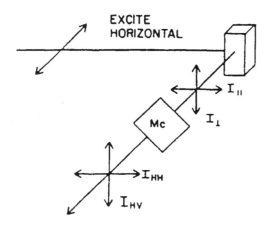

Figure 16 L-format for polarization measurements. MC, monochromator. Subscripts refer to horizontal (H), vertical (V), perpendicular, and parallel. [From Lakowicz (1999), with permission.]

into the subnanosecond range. Each pulse generates a start signal at a PMT that triggers the voltage ramp of the time-to-amplitude converter (TAC, Fig. 19). The ramp is terminated when a photon emitted from the sample is detected. Therefore, the amplitudes of the TAC output pulses are proportional to the time between the start and stop signals, which in turn are statistically related to the luminescence lifetime. Lasers with high repetition rates can be used with fast PMT detection to extend the use of TCSPC well into the subnanosecond region. As with steady-state single-photon counting measurements, TCSPC is designed for the detection of very low level photon signals.

The long lifetimes associated with phosphorescence emission are readily measured in the time domain by simple pulsed-source systems with direct acquisition of the decay curves. A delay time is imposed to allow fluorescence emission to completely decay before measurements are begun. Alternatively, a simple chopper device may be used to eliminate the fluorescence signal.

Phase-modulation luminescence spectroscopy (Birks and Dyson, 1961; Gaviola, 1927; Lakowicz, 1999; Spencer and Weber, 1961) is a frequency-domain alternative to pulsed methods for luminescence lifetime determinations. The excitation beam from a continuous source (lamp or laser) is modulated at an appropriate frequency (megahertz range for fluorescence lifetimes, hertz to millihertz range for phosphorescence lifetimes). As shown in Fig. 20, the luminescence emission will be phase-shifted by angle ϕ and demodulated by a factor m relative to the excitation beam. The values of ϕ and m depend on the modulation frequency and the luminescence lifetime of the sample, which can be independently calculated from ϕ and m. A phase-modulation instrument for fluorescence lifetime determinations is shown in Fig. 21.

Time-resolved methods (Cundall and Dale, 1983; Ware, 1971), based on the pulsed-excitation approach, and phase-resolved methods (Lakowicz and Cherek, 1981; Mattheis et al., 1983; McGown and Bright, 1984, 1987; Mousa and Winefordner, 1974; Veselova et al., 1970) based on the phase-modulation technique, can provide selectivity for chemical analysis based on luminescence lifetime differences, as well as providing a means for eliminating background interference (Section IV.C).

4. Fiber Optics

Optical fibers can be used to transport light to and from luminescent samples. Recent reviews have described the principles and use of fiber optic technology (Chabay, 1983; Rabek, 1982), which is a rapidly growing area of interest with a wide range of applications. Wavelengths ranging from 250 to 1300 nm can be transmitted, depending on the fiber material. Advantages of optical fibers include their low cost, the simplicity and flexibility of experimental configurations, and the ability to take *in situ* measurements. Relatively inaccessible locations can be probed and analytes can be directly measured in the field (Jones, 1985), in process streams, and *in vivo* for biological studies (Peterson and Vurek, 1985). Optical fiber fluoroprobes (Wolfbeis, 1985), which have a reagent phase immobilized on the optical probe, can be used as chemical sensors, as an alternative to electrochemical devices.

IV. PRACTICAL CONSIDERATIONS AND APPLICATIONS

Discussions of practical considerations, such as calibration, spectral correction, data handling, and other topics, can be found in various general texts, chapters, and monographs (Guilbault, 1990; Lakowicz, 1999; McGown, 1966; Mielenz, 1982; Rendell, 1987; Schulman, 1977). Some of these considerations, as well as areas and examples of applications, are discussed in this section.

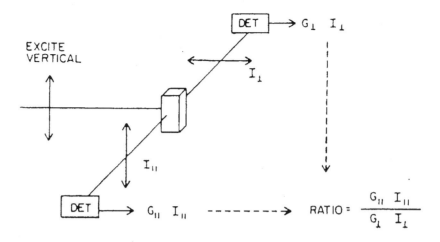
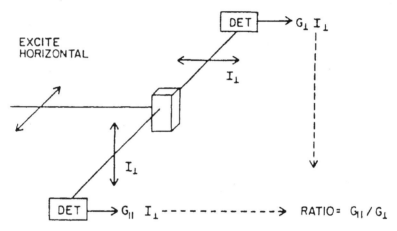

Figure 17 T-format for polarization measurements. DET, detector; G, gain. Subscripts as in Fig. 16. [From Lakowicz (1999), with permission.]

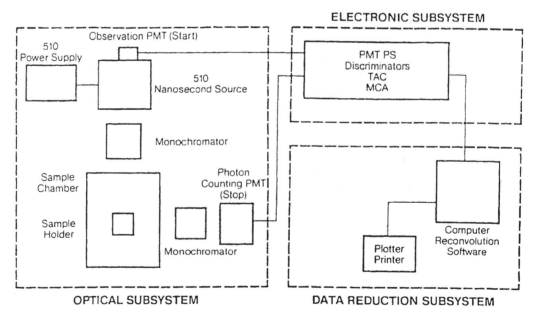

Figure 18 TCSPC fluorescence lifetime instrument. (From PRA, with permission.)

Molecular Fluorescence and Phosphorescence

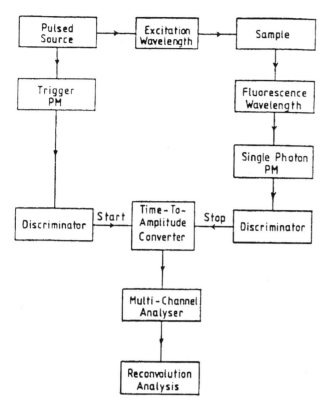

Figure 19 The single-photon correlation technique. [From Birch and Imhof (1991), with permission.]

A. Environmental Effects

Luminescence measurements are very sensitive to experimental conditions owing to the competition of the fluorescence process with other de-excitation processes.

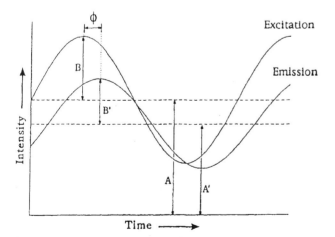

Figure 20 Depiction of the phase-modulation fluorescence lifetime technique, showing intensity-modulated excitation and emission beams with a.c. amplitudes B and B', and d.c. intensities A and A', respectively. Lifetimes are determined from phase shift f and demodulation m, where $m = (B'/A')/(B/A)$.

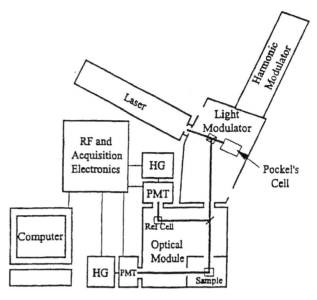

Figure 21 Frequency-domain fluorescence lifetime instrument. A multiharmonic Fourier transform technique is used to acquire phase and modulation data at multiple frequencies simultaneously. A base frequency generated by radio frequency (RF) electronics is converted into a phase-coherent electrical pulse at the harmonic modulator, which drives the Pockel's cell to convert the output of a CW laser into a pulsed signal containing the base frequency and harmonics. Harmonic generators (HG) modulate the gain of the PMTs at a base frequency that is phase-locked to the base excitation frequency and slightly offset from it, for phase-sensitive detection of the multiharmonic emission signal. (From SLM Instruments, with permission.)

Sensitivity of molecular luminescence to the chemical microenvironments of the emitting molecules is an important tool for probing macromolecular structures (Lakowicz, 1999), such as micelles (Gratzel and Thomas, 1976), proteins (Churchich, 1976), biological membranes (Badley, 1976) and cells, and for studying solvadon, TM excited-state processes (Froehlich and Wehry, 1976; Schulman, 1976), molecular and macromolecular rotations (Lakowicz, 1999), and many other processes. On the other hand, experimental conditions such as solvent composition and temperature must be very carefully controlled for qualitative and quantitative analyses in order to obtain accurate and reproducible results. For aqueous solutions, effects of pH and ionic strength must also be considered.

As discussed in Section II.E, dissolved oxygen quenches luminescence and decreases fluorescence lifetime. Deoxygenation is desirable and often essential for many luminescence experiments, and can be accomplished by passing high purity helium or nitrogen through the sample or by alternate methods such as rapid freeze–thaw cycles or molecular dialysis (Rollie et al., 1983, 1987). Other quenchers besides oxygen in a sample matrix, as well as inner-filter effects, self-quenching, and excited-state quenching processes, must also be considered.

B. Wavelength Calibration and Spectral Correction

Emission monochromators are commonly calibrated with a standard source such as a mercury arc lamp, which has a number of accurately known emission lines ranging from 253.65 to 809.32 nm. Excitation monochromators can then be calibrated by using a scattering solution to direct the calibrated excitation line through the emission monochromator to the detector.

Acquired spectra can be corrected for the wavelength dependencies of source output, monochromator efficiency, and detector response by various methods that include the use of thermopiles, quantum counters, calibrated phototubes with known spectral response curves, and solutions of standard fluorescent compounds.

C. Background and Interfering Signals

A common source of background intensity is scattered light, which may result from particulate, Rayleigh, or Raman scattering phenomena. Fluorescence and phosphorescence may also be mutually interfering. Scattered light is usually easily identified in spectra and can often be avoided by appropriate choices of excitation and emission wavelength. Additionally, as scattered light has an effective lifetime of zero on the time scale of luminescence decay, time-resolved or phase-resolved approaches can be used to minimize the contribution of scattered light to an emission signal (Mousa and Winefordner, 1974). Phase-resolved techniques can also be used directly to suppress longer-lived emission in order to measure shorter-lived emission, such as measurements of Raman scattered light in the presence of fluorescence (Demas and Keller, 1985; Nithipatikom and McGown, 1986).

Using blank subtraction background signals can be corrected for, provided that the background signal is the same in the blank and the sample, and that the background and sample signals are additive. Derivative spectroscopy, wavelength modulation techniques (O'Haver, 1976), and time-resolved or phase-resolved techniques can be used to improve selectivity and reduce background interferences.

D. Direct Quantitative and Qualitative Analyses

Luminescent molecules can be directly determined from measurements of luminescence intensity [Eqs. (8) and (9), Section II.E]. Molecules that exhibit native luminescence include organic molecules with conjugated z-bond systems, such as polycyclic and polynuclear aromatic hydrocarbons, fluorescent metal chelates, certain rare earth elements, and inorganic lanthanide and uranyl compounds. Simultaneous multicomponent determinations of luminescent compounds in mixtures can be performed if sufficient selectivity is available and the compounds absorb and emit independently in the absence of synergistic effects. Selectivity can be achieved through the exploitation of emission and excitation spectral characteristics, luminescence lifetimes, and selective enhancement or quenching (Warner et al., 1985). Qualitative analyses are also based on one or more of these dimensions of information. For complex mixtures, techniques such as synchronous excitation and total luminescence spectroscopy (Section III.D.1) can be used for multicomponent determinations and qualitative analysis (Giering and Hornig, 1977; Lloyd, 1971; Suter et al., 1983; Vo-Dinh, 1978, 1981; Warner et al., 1976; Weber, 1961).

E. Low Temperature Techniques

As discussed in Sections II.E and III.A.3, phosphorescence measurements are often made on solid materials at low temperatures. Low temperature fluorescence techniques are also sometimes appropriate (Jankowiak et al., 2000; Metzger et al., 1969; Shik and Bass, 1969; Shpol'skii, 1959; Shpol'skii et al., 1959; Stroupe et al., 1977; Wehry and Mamantov, 1981). In matrix isolation spectroscopy (Metzger et al., 1969; Shik and Bass, 1969; Stroupe et al., 1977), the sample is vaporized, mixed with a large excess of a diluent gas, and deposited on a solid surface at a low temperature (10–15 K or lower) in order to minimize aggregation and quenching effects. Another technique employs the *Shpol'skii effect* (Shpol'skii, 1959; Shpol'skii et al., 1959) to obtain high resolution spectra in *n*-alkane or other solvents at low temperatures (77 K). Highresolution Shpol'skii spectra can be used for fingerprinting of fossil fuels and other organic samples. Resolution can be increased by using Shpol'skii media with the matrix isolation technique. These techniques can provide lower detection limits, longer linear response ranges, and better resolution and selectivity for many samples. However, they have not been used widely on a routine basis for chemical analysis because of the care and skill required for sample preparation.

Even lower temperatures are needed for fluorescence line-narrowing spectroscopy (FLNS). FLNS is a low temperature ($\sim$4.2 K) technique used to excite a narrow region in a broadened absorption band of an analyte. The line-narrowing effect typically reduces inhomogeneous broadening from 100–300 cm^{-1} down to $\sim$5 cm^{-1} (Jankowiak et al., 2000).

F. Room-Temperature Phosphorescence

The application of phosphorescence spectroscopy to chemical analysis may be significantly increased by the use of RTP (Hurtubise, 1981; Vo-Dinh, 1984) to avoid the need for low temperatures and frozen samples. Techniques include solid surface RTP, micelle-stabilized RTP, and solution sensitized RTP. Solid surfaces that have been used include cellulose, alumina, silica gel, sodium acetate, sucrose, chalk, inorganic mixtures, and polymer salt mixtures. Micelle stabilization involves the entrapment of the phosphorescent molecule within a micelle in order to enhance its phosphorescence by increasing its rigidity and minimizing solution quenching interactions. Heavy atoms are often used in both solid surface and micelle RTP techniques to increase the rate of intersystem crossing from excited singlet to overlapping triplet states. Solution sensitized RTP can be used indirectly to determine molecules that emit little or no phosphorescence. The analyte absorbs photons and transfers the energy via nonradiative excitation transfer to a phosphorescent acceptor molecule. The phosphorescence of the acceptor is measured and used to determine the analyte.

G. Techniques and Applications

1. Chemical Analysis and Characterization

A wide variety of approaches have been taken for the application of luminescence spectroscopy to chemical analysis and characterization. Indirect luminescence methods in which a nonluminescent analyte is coupled to a luminescent reagent, as exemplified by solution sensitized RTP in the previous section, can be used to extend the applicability of luminescence analysis. Kinetic methods (Ingle and Ryan, 1981) for both direct and indirect luminescence analysis have been described for a variety of analytes, including determinations of enzymes and their substrates. Immunochemical techniques (Karnes et al., 1985; Smith et al., 1981) involving luminescent-labeled antigens, haptens, or antibodies have proven to be an important alternative to methods that employ radioactive-labeled reagents. Spectral fingerprinting by techniques that combine the multidimensional nature of luminescence, such as total luminescence spectroscopy, synchronous excitation spectroscopy, and lifetime-based methods, have been used for the characterization of complex samples such as fossil fuels (Goldberg and Negomir, 1989; John and Soutar, 1976; von der Dick and Kalkreuth, 1984; Siegel, 1985), humic substances (Goldberg and Negomir, 1989), bacteria (Baek et al., 1988; Rossi and Warner, 1985), and human serum (Stevens et al., 1991; Wolfbeis and Leiner, 1985).

Spectroscopic characterization of biomolecules such as nucleic acids and proteins can be achieved with a different luminescent process in which the excitation radiation is internally transferred from another fluorescent moiety in the same molecule. This is the principle of fluorescence resonance energy transfer (FRET), which helps to excite a large number of different fluorophores with only one excitation wavelength. The molecular design consists of a single donor moiety, which is excited by the excitation wavelength, and internally transferring the energy to an acceptor moiety. Depending on the experimental setup sample multiplexing is possible using different acceptors (Klostermeier and Millar, 2001; Periasamy, 2001).

2. Luminescence Detection in Chemical Separations

Luminescence detection for HPLC has become increasingly popular and includes the use of conventional fluorescence detection (Froehlich and Wehry, 1981; Shelly and Warner, 1983), fluorescence lifetime detection (Desilets et al., 1987; Smalley et al., 1993), and micelle-stabilized RTP (Cline Love et al., 1984; Weinberger et al., 1982) and sensitized RTP (Donkerbroek et al., 1982) techniques.

Capillary electrophoresis has become a very popular and powerful technique for chemical separations because of its high speed, high resolution, applicability to a large range of solute sizes, and its ability to handle very small sample volumes, on the order of pico- to femtoliters, and even single blood cells (Albin et al., 1993; Jadamec et al., 1977; Kuhr and Monnig, 1992; Li, 1992). Detection can be provided by radiometry, electrochemistry, mass spectrometry, or absorption or fluorescence spectroscopy. Fluorescence detection generally provides the best sensitivity, linearity, and selectivity in capillary electrophoresis. In particular, laser excited fluorescence (Gassman et al., 1985; Guthrie et al., 1984; Lee et al., 1992, 1993; Liu et al., 1991; Yeung et al., 1992) has provided some of the lowest detection limits for commercial instruments, as low as 10^{-11} M or the equivalent 10^{-20} mol in a nanoliter volume. Sensitivity is, in many instances, the main issue in important applications such as in DNA adducts studies. It is important to be able to detect one modified DNA base (adduct) in $10^8 - 10^{11}$ base pairs from only 100 μg of DNA. Several fluorescent detection schemes have been successfully applied to adducts detection: native fluorescence (Formenton-Catai and Carrilho, 2003), internal labeling (Schmitz et al., 2002), and FLNS (Jankowiak et al., 2000).

The combination of multicolor fluorescence detection followed by capillary electrophoresis separation is the fundamental tool that helped to finish the sequencing of the human genome ahead of its schedule. Several issues were fundamental to its success such as the optics design, FRET dyes, enzymes, and capillary electrophoresis conditions (Franca et al., 2002). Improvements in

DNA sequencing is still undergoing rapid improvement as new approaches become more robust and reliable (Marziali and Akeson, 2001).

Another separation platform is changing the face of how analytical chemistry will be made in the future. Miniaturized separation channels fabricated in microdevices (Reyes, et al., 2002) (microchips) are one of the most exciting developments in instrumentation development in analytical chemistry recently and one of the best-suited detection system is based on laser-induced fluorescence detection (Auroux et al., 2002).

3. Biological Applications

Measurements of fluorescence spectra, lifetime, anisotropy, and dynamic anisotropic decay, have been widely used for the study of biological systems (Section IV.A) (Lakowicz, 1999) Native fluorescence has been utilized extensively to study proteins. Extrinsic fluorescent probes that are added to a sample to report on various structures and processes in biological systems are also very useful in studies of proteins, nucleic acids, membranes, and other macromolecules and molecular assemblies. A variety of fluorescent probes have been described for these purposes as well as for studies of micelles, nucleic acids, and vesicles (Haugland, 1994; Lakowicz, 1999).

An interesting development in the area of biological applications is the technique of fluorescence-detected circular dichroism (FDCD). In FDCD, the optical activity of ground state molecules is measured through fluorescence detection (Tinoco and Turner, 1976; Turner et al., 1974) rather than absorption as in conventional circular dichroism spectroscopy (see Chapter 13). FDCD offers higher sensitivity and detectability than absorption circular dichroism due to lower background signals and the direct dependence of fluorescence emission on the intensity of the excitation beam. Moreover, FDCD is highly selective because nonfluorescent chirophores do not contribute to the signal. The FDCD technique has been used to study proteins, nucleic acids, and molecular interactions (Dahl et al., 1982; Egusa et al., 1985; Lamos et al., 1986a, b; Lobenstine et al., 1981). New developments in this field include FDCD in total luminescence spectroscopy (Thomas et al., 1986) and the incorporation of lifetime resolution in FDCD for multicomponent analysis (Geng and McGown, 1992; Wu et al., 1993).

Molecular diagnostics is one of the main applications of separation microchip technologies using fluorescence detection. The combination of several analytical steps in one single microdevice enables fluorescence detection of DNA starting from a few microliters of blood for the detection and diagnostics of mutations, point mutations, polymorphisms, and even DNA sequencing (Landers, 2002).

4. Fluorescence Measurements in NIR

Fluorescent dyes that absorb and emit in the NIR region (650–1000 nm) have been gaining popularity recently as alternatives to UV–Vis dyes as labels and probes (Akiyama, 1993; Casay et al., 1993; McClure, 1994; Miller et al., 1993; Patonay, 1993; Patonay and Antoine, 1991). Interest in these compounds arises from several very important advantages of working in the NIR region. NIR dyes have high molar absorptivities and quantum yields approaching 70% when the dyes are adsorbed on solid surfaces such as gels. Background noise due to native sample fluorescence and scattered light are much lower in the NIR region so that lower detection limits can be achieved. A major attraction of NIR spectroscopy is the availability of small, inexpensive devices for excitation and detection. Solid-state diode lasers are rugged, long-lived sources of relatively high excitation power in the NIR region. Silicon photodiode detectors provide highly sensitive detection with faster response than PMTs and very low noise levels.

Several families of NIR dyes have been identified (Akiyama, 1993; Casay et al., 1993; Miller et al., 1993; Patonay, 1993). Among these, some of the most fruitful are the polymethines. These dyes include the carbocyanines, the merocyanines, and metal-containing phthalocyanines and naphthalocyanines. In the xanthene family that includes the visible fluorescein and rhodamine dyes, several derivatives exhibit excitation and emission approaching the NIR. The rapidly growing interest in NIR spectroscopy has stimulated efforts to identify and synthesize new NIR dyes for applications in fluorescence analysis.

H. Future Outlook

The excellent sensitivity and relative simplicity of luminescence analysis will continue to stimulate the development of new technology and applications for chemical measurements. Instrumentation to facilitate rapid multiwavelength detection, lifetime-based measurements, FDCD, and other advanced techniques will increase the accessibility of multidimensional approaches for routine applications. Explorations will continue for new luminescent probes and tags, in both the UV–Vis and the NIR spectral regions, to be used in both fundamental research and analytical applications. Lasers will continue to play a critical role in the development of new luminescence approaches in areas such as chromatographic detection, spatial imaging, and fiber optic sensing. Methods based on single molecule fluorescence detection will become routine for studies involving biomolecules (Ishijima and Yanagida, 2001; Nie and Zare, 1997).

REFERENCES

Agbaria, R. A., Oldham, P. B., McCarroll, M., McGown, L. B., Warner, I. M. (2002). *Anal. Chem.* 74:3952.

Akiyama, S. (1993). In: Schulman, S. G., ed. *Molecular Luminescence Spectroscopy. Part 3*. New York: Wiley, pp. 229–251.

Albin, M., Grossman, P. D., Moring, S. E. (1993). *Anal. Chem.* 65:489A.

Auroux, P.-A., Iossifidis, D., Reyes, D. R., Manz, A. (2002). *Anal. Chem.* 74:2637.

Badley, R. A. (1976). In: Wehry, E. L., ed. *Modern Fluorescence Spectroscopy*. Vol. 2. New York: Plenum, pp. 91–168.

Baek, M., Nelson, W. H., Hargraves, P. E., Tanguay, J. F., Suib, S. L. (1988). *Appl. Spectrosc.* 42:1405.

Baeyens, W. R. G., De Keukeleire, D., Korkidis, K., eds. (1991). *Luminescence Techniques in Chemical and Biochemical Analysis*. New York: Dekker.

Berlman, I. B. (1971). *Handbook of Fluorescence Spectra of Aromatic Molecules*. 2nd ed. New York: Academic.

Birch, D. J. S., Imhof, R. E. (1991). In: Lakowicz, J. R., ed. *Topics in Fluorescence Spectroscopy, Vol. 1: Techniques*. New York: Plenum.

Birks, J. B., Dyson, D. J. (1961). *J. Sci. Instrum.* 38:282.

Boutillier, G. D., Winefordner, J. D. (1979). *Anal. Chem.* 51:1384.

Casay, G. A., Czuppon, T., Lipowski, J., Patonay, G. (1993). *SPIE Proceedings* 1885:324.

Chabay, J. (1983). *Anal. Chem.* 54:1071A.

Christian, G. D., Callis, J. B., Davidson, E. R. (1981). In: Wehry, E. L., ed. *Modern Fluorescence Spectroscopy*. Vol. 4. New York: Plenum, pp. 111–165.

Churchich, J. E. (1976). In: Wehry, E. L., ed. *Modern Fluorescence Spectroscopy*. Vol. 2. New York: Plenum, pp. 217–237.

Cline Love, L. J., Eastwood, D., eds. (1985). *Advances in Luminescence Spectroscopy*. STP 863. Philadelphia: ASTM.

Cline Love, L. J., Habarta, J. G., Dorsey, J. G. (1984). *Anal. Chem.* 56:1132A.

Cundall, R. B., Dale, R. E., eds. (1983). *Time-Resolved Fluorescence Spectroscopy in Biochemistry and Biology*. NATO ASI Series A: Life Sciences. Vol. 69. New York: Plenum.

Dahl, K. S., Apardi, Tinoco, I. (1982). *Biochemistry* 21:2730.

Demas, J. M. (1983). *Excited State Lifetime Measurements*. New York: Academic.

Demas, J. N., Keller, R. A. (1985). *Anal. Chem.* 57:538.

Desilets, D. J., Kissinger, P. T., Lytle, F. E. (1987). *Anal. Chem.*, 59:1830.

Dewey, T. G., ed. (1991). *Biophysical and Biochemical Aspects of Fluorescence Spectroscopy*. New York: Plenum.

von der Dick, H., Kalkreuth, W. (1984). *Fuel* 63:1636.

Donkerbroek, J. J., Van Eikema Hommes, N. J. R., Gooijer, C., Velthorst, N. H., Frei, R. W. (1982). *Chromatographia* 15:218.

Eastwood, D., ed. (1983). *New Directions in Molecular Luminescence*. STP 822. Philadelphia: ASTM.

Eastwood, D., Cline Love, L. J., eds. (1988). *Progress in Analytical Luminescence*. STP 1009. Philadelphia: ASTM.

Egusa, S., Sisido, M., Imanishi, Y. (1985). *Macromolecules*. 18:882.

Formenton-Catai, A. P., Carrilho, E. (2003). *Anal. Bioanal. Chem.* 376:138.

Franca, L. T. C., Carrilho, E., Kist, T. B. L. (2002). *Q. Rev. Biophys.* 35:169.

Froehlich, P., Wehry, E. L. (1976). In: Wehry, E. L., ed. *Modern Fluorescence Spectroscopy*. Vol. 2. New York: Plenum, pp. 319–438.

Froehlich, P., Wehry, E. L. (1981). In: Wehry, E. L., ed. *Modern Fluorescence Spectroscopy*. Vol. 3. New York: Plenum, pp. 35–94.

Gassman, E., Kuo, J., Zare, R. N. (1985). *Science* 230:813.

Gaviola, Z. (1927). *Z. Phys.* 42:852.

Geng, L., McGown, L. B. (1992). *Anal. Chem.* 64:68.

Giering, L. P., Hornig, A. W. (1977). *Am. Lab.* 9:113.

Goeringer, D. E., Pardue, H. L. (1979). *Anal. Chem.* 51:1054.

Goldberg, M. C., Negomir, P. M. (1989). In: Goldberg, M. C., ed. *Luminescence Applications in Biological, Chemical, Environmental, and Hydrological Sciences*. ACS Symposium Series 383. Washington, DC: American Chemical Society, pp. 180–205.

Gratzel, M., Thomas, J. K. (1976). In: Wehry, E. L., ed. *Modern Fluorescence Spectroscopy*. Vol. 2. New York: Plenum, pp. 169–216.

Guilbault, G. G., ed. (1990). *Practical Fluorescence*. 2nd ed. New York: Dekker.

Guthrie, E., Jorgenson, J., Dluzneski, P. (1984). *J. Chromatogr.* 22:171.

Hamilton, T. D. S., Naqvi, K. R. (1973). *Anal. Chem.* 45:1581.

Haugland, R. P. (1994). *Handbook of Fluorescent Probes and Research Chemicals 1992–1994*. Eugene, Oregon: Molecular Probes, Inc.

Hieftje, G. M., Vogelstein, E. E. (1981). In: Wehry, E. L., ed. *Modern Fluorescence Spectroscopy*. Vol. 4. New York: Plenum, pp. 25–50.

Hurtubise, R. J. (1981). *Solid Surface Luminescence Analysis: Theory, Instrumentation, Applications*. New York: Dekker.

Ichinose, N., Schwedt, G., Schnepel, F. M., Adachi, K. (1991). *Fluorometric Analysis in Biomedical Chemistry*. New York: Wiley-Interscience.

Ingle, J. D. Jr., Ryan, M. A. (1981). In: Wehry, E. L., ed. *Modern Fluorescence Spectroscopy*. Vol. 3. New York: Plenum, pp. 95–142.

Ingle, J. D. Jr., Ryan, M. A. (1983). In: Talmi, Y., ed. *Multichannel Image Detectors*. ACS Symposium Series 102. Vol. II. Washington, DC: American Chemical Society, pp. 155–170.

Ingle, J. D. Jr., Crouch, S. R., eds. (1988). *Spectrochemical Analysis*. Prentice Hall.

Ishijima, A., Yanagida, T. (2001). *Trends Biochem. Sci.* 26:438.

Jadamec, J. R., Saner, W. A., Talmi, Y. (1977). *Anal Chem.* 49:1316.

Jankowiak, R., Roberts, K. P., Small, G. J. (2000). *Electrophoresis* 21:1251.

John, P., Soutar, I. (1976). *Anal. Chem.* 48:520.
John, P., Soutar, I. (1981). *Chem. Br.* 17:278.
Johnson, D. W., Callis, J. B., Christian, G. D. (1979). In: Talmi, Y., ed. *Multichannel Image Detectors*. ACS Symposium Series 102. Washington, DC: American Chemical Society, pp. 97–114.
Johnson, D. W., Gladden, J. A., Callis, J. B., Christian, G. D. (1979). *Rev. Sci. Instrum.* 50:118.
Jones, B. E. (1985). *J. Phys. E.* 18:770.
Karnes, H. T., O'Neal, J. S., Schulman, S. G. (1985). In: Schulman, S. G., ed. *Molecular Luminescence Spectroscopy: Methods and Applications: Part 1*. New York: Wiley-Interscience, pp. 717–779.
Klostermeier, D., Millar, D. P. (2001). *Biopolymers* 61:159.
Kuhr, W., Monnig, C. (1992). *Anal. Chem.* 64:389R.
Lakowicz, J. R. (1999). *Principles of Fluorescence Spectroscopy*. 2nd ed. New York: Plenum.
Lakowicz, J. R., Cherek, H. (1981). *J. Biochem. Biophys. Methods* 5:19.
Lakowicz, J. R.; Thompson, R. B. (2001). *Advances in Fluorescence Sensing Technology V*. Proc. SPIE-Int. Soc. Opt. Eng., 2001; 4252. Bellingham: SPIE.
Lamos, M. I., Lobenstine, E. W., Turner, D. H. (1986a). *J. Am. Chem. Soc.* 108:4278.
Lamos, M. L., Walker, G. T., Krugh, T. R., Turner, D. H. (1986b). *Biochemistry* 25:687.
Landers, J. P. (2002). *Anal. Chem.* 75:2919.
Latz, H. W. (1976). In: Wehry, E. L., ed. *Modern Fluorescence Spectroscopy*. Vol. 1. New York: Plenum.
Lee, T. T., Lillard, S. J., Yeung, E. S. (1992). *J. Chromatogr.* 595:319.
Lee, T. T., Lillard, S. J., Yeung, E. S. (1993). *Electrophoresis* 14:429.
Li, S. F. Y. (1992). *Capillary Electrophoresis: Principles, Practice and Applications*. New York: Elsevier.
Liu, J., Shirotu, O., Novotny, M. (1991). *Anal. Chem.* 63:413.
Lloyd, J. B. F. (1971). *Nat. Phys. Sci.* 231:64.
Lobenstine, E. W., Schaefer, W. C., Turner, D. H. (1981). *J. Am. Chem. Soc.* 103:4936.
Maeda, M. (1984). *Laser Dyes: Properties of Organic Compounds for Dye Lasers*. New York: Academic.
Marziali, A., Akeson, M. (2001). *Annu. Rev. Biomed. Eng.* 3:195.
Mattheis, J. R., Mitchell, G. W., Spencer, R. D. (1983). In: Eastwood, D., ed. *New Directions in Molecular Luminescence*. ASTM STP 822. Baltimore: ASTM, pp. 50–64.
McCarrol, M. E., Warner, I. M., Agbaria, R. A. (2000). *Encyclopedia of Analytical Chemistry*. John Wiley & Sons, pp. 10251–10280.
McClure, W. F. (1994). *Anal. Chem.* 66:43A.
McGown, L.B. (1966). Metals handbook. *Materials Characterization*. 9th ed. Vol. 10. Metals Park, Ohio: ASM Handbook Committee, American Society for Metals, pp. 72–81.
McGown, L. B., Bright, F. V. (1984). *Anal. Chem.* 56:1400A.
McGown, L. B., Bright, F. V. (1987). *Crit. Rev. Anal. Chem.* 18:245.
Metzger, J. L., Smith, B. E., Meyer, B. (1969). *Spectrochim. Acta*, 25A:1177.
Mielenz, K. D. (1982). *Measurement of Photoluminescence*. New York: Academic.
Miller, J. N., Brown, N. B., Seare, N. J., Summerfield, S. (1993). In: Wolfbeis, O. S., ed. *Fluorescence Spectroscopy: New Methods and Applications*. Berlin: Springer-Verlag, pp. 189–196.
Mousa, J. J., Winefordner, J. D. (1974). *Anal. Chem.* 46:1195.
Mycek, M.-A., Pogue, B. W., eds. (2003). *Handbook of Biomedical Fluorescence*. New York: Marcel Dekker.
Nie, S. M., Zare, R. N. (1997). *Annu. Rev. Biophys. Biomol. Struct.* 26:567.
Nithipatikom, K., McGown, L. B. (1986). *Anal. Chem.* 58:3145.
O'Conner, D. V., Phillips, D. (1984). *Time-Correlated Single Photon Counting*. London: Academic.
O'Haver, T. C. (1976). In: Wehry, E. L., ed. *Modern Fluorescence Spectroscopy*. Vol. 1. New York: Plenum, pp. 65–81.
Patonay, G. (1993). In: Patonay, G., ed. *Advances in Near-Infrared Measurements*. Vol. I. New York: JAI Press, pp. 113–138.
Patonay, G., Antoine, M. D. (1991). *Anal. Chem* 63:321A.
Periasamy, A. (2001). *J. Biomed. Opt.* 6:287.
Peterson, J. I., Vurek, G. G. (1985). *Science* 224:123.
Rabek, J. F. (1982). *Experimental Methods in Photophysics, Part 1*. New York: Wiley-Interscience, pp. 272–286.
Rendell, D. (1987). *Fluorescence and Phosphorescence Spectroscopy*. New York: Wiley.
Rendell, D., Mowthorpe, D. (1987). *Fluorescence and Phosphorescence Spectroscopy: Analytical Chemistry by Open Learning*. John Wiley & Sons.
Reyes, D. R., Iossifidis, D., Auroux, P.-A., Manz, A. (2002). *Anal. Chem.* 74:2623.
Richardson, J. H. (1981). In: Wehry, E. L., ed. *Modern Fluorescence Spectroscopy*, Vol. 4. New York: Plenum, pp. 1–24.
Rigler, R., Elson, E. S. (2001). *Fluorescence Correlation Spectroscopy: Theory and Applications*. Berlin: Springer.
Rollie, M. E., Ho, C.-N., Warner, I. M. (1983). *Anal. Chem.* 55:2445.
Rollie, M. E., Patonay, G., Warner, I. M. (1987). *Anal. Chem.* 59:180.
Rossi, T. M., Warner, I. M. (1985). In: Nelson, W. H., ed. *Instrumental Methods for Rapid Microbial Analysis*. New York: VCH, pp. 1–50.
Schimtz, O. J., Wörth, C. C. T., Stach, D., Wießler, M. (2002). *Angew. Chem. Int. Ed.* 41:445.
Schulman, S. G. (1976). In: Wehry, E. L., ed. *Modern Fluorescence Spectroscopy*. Vol. 2. New York: Plenum, pp. 239–275.
Schulman, S. G. (1977). *Fluorescence and Phosphorescence Spectroscopy: Physicochemical Principles and Practice*. London: Pergamon.
Schulman, S. G., ed. (1985). *Molecular Luminescence Spectroscopy: Methods and Applications*, Part 1. New York: Wiley-Interscience.
Shelly, D. C., Warner, I. M. (1983). *Chromatogr. Sci.* 23:87.
Shik, J. S., Bass, A. M. (1969). *Anal. Chem.* 41:103A.
Shpol'skii, E. V. (1959). *Sovt. Phys. Uspekhi* 2:378.
Shpol'skii, E. V., Il'ina, A. A., Klimova, L. A. (1959). *Dokl. Akad. Nauk. SSSR* 87:935.

Siegel, J. A. (1985). *Anal. Chem.* 57:934A.

Smalley, M. B., Shaver, J. M., McGown, L. B. (1993). *Anal. Chem.* 65:3466.

Smith, D. S., Hassan, M., Nargessi, R. D. (1981). In: Wehry, E. L., ed. *Modern Fluorescence Spectroscopy*. Vol. 3. New York: Plenum, pp. 143–191.

Spencer, R. D., Weber, G. (1961). *Ann. NY Acad. Sci.* 158:361.

Stevens, R. D., Cooter, M. S., McGown, L. B. (1991). *J. Fluores.* 1:235.

Stroupe, R. C., Tokousbalides, P., Dickinson, R. B. Jr., Wehry, E. L., Mamantov, G. (1977). *Anal. Chem.* 49:701.

Suter, G. W., Kallir, A. J., Wild, U. P. (1983). *Chimia* 37:413.

Talmi, Y. (1975). *Anal. Chem.* 47:658A, 699A.

Thomas, M. P., Patonay, G., Warner, I. M. (1986). *Rev. Sci. Instrum.* 57:1308.

Tinoco, I., Turner, D. H. (1976). *J. Am. Chem. Soc.* 98:6453.

Turner, D. H., Tinoco, I., Maestre, M. F. (1974). *J. Am. Chem. Soc.* 96:4340.

Valeur, B. (2001). *Molecular Fluorescence: Principles and Applications*. Weinhein: Wiley-VCH.

Veselova, T. V., Cherkasov, A. S., Shirokov, V. I. (1970). *Opt. Spectrosc.* 29:617.

Visser, A. J. W. G., ed. *Anal. Instrum.* Vols. 3 and 4. pp. 193–566, 1985.

Vo-Dinh, T. (1978). *Anal. Chem.* 50:396.

Vo-Dinh, T. (1981). In: Wehry, E. L., ed. *Modern Fluorescence Spectroscopy*. Vol. 4. New York: Plenum, pp. 167–192.

Vo-Dinh, T. (1984). *Room Temperature Phosphorimetry for Chemical Analysis*. New York: Wiley-Interscience.

Vo-Dinh, T., Eastwood, D. (1990). *Laser Techniques in Luminescence Spectroscopy*. STP 1066. Philadelphia: ASTM.

Vo-Dinh, T., Paltzold, R., Wild, U. P. (1972). *Z. Phys. Chem.* 251:395.

Vo-Dinh, T., Walden, G. L., Winefordner, J. D. (1977). *Anal. Chem.* 49:1126.

Ware, W. R. In: Lomola, A. A., ed. (1971). *Creation and Detection of the Excited State*. Vol. IA. New York: Dekker, pp. 213–302.

Warner, I. M., McGown, L. B., eds. (1991). *Advances in Multidimensional Luminescence*. Vol. 1. New York: JAI Press (see also Vol. 2, 1993).

Warner, I. M., Callis, J. B., Davidson, E. R., Gouterman, M., Christian, G. D. (1975). *Anal. Lett.* 8:665.

Warner, I. M., Callis, J. B., Davidson, E. R., Christian, G. D. (1976). *Clin. Chem.* 22:1483.

Warner, I. M., Fogarty, M. P., Shelly, D. C. (1979). *Anal. Chim. Acta* 109:361.

Warner, I. M., Patonay, G., Thomas, M. P. (1985). *Anal. Chem.* 57:463A.

Weber, G. (1961). *Nature* 190:27.

Wehry, E. L., ed. (1975). *Modern Fluorescence Spectroscopy*. Vols. 1 and 2. New York: Plenum (see also Vols. 3 and 4, 1981).

Wehry, E. L., Mamantov, G. (1981). In: Wehry, E. L., ed. *Modern Fluorescence Spectroscopy*. Vol. 4. New York: Plenum, pp. 193–250.

Weinberger, R., Yarmchuk, P., Cline Love, L. J. (1982). *Anal. Chem.* 54:1552.

Werner, T. C. (1976). In: Wehry, E. L., ed. *Modern Fluorescence Spectroscopy*. Vol. 2. New York: Plenum, pp. 277–317.

Wilson, J. (1982). In: Evans, T. R., ed. *Applications of Lasers to Chemical Problems*. New York: Wiley, pp. 1–34.

Wolfbeis, O. S. (1985). *Trends Anal. Chem.* 4:184.

Wolfbeis, O. S., ed. (1993). *Fluorescence Spectroscopy; New Methods and Applications*. Berlin: Springer-Verlag.

Wolfbeis, O. S., Leiner, M. (1985). *Anal. Chim. Acta* 167:203.

Wright, J. C. (1984). In: Kompa, K., Wanner, J., eds. *Laser Applications in Chemistry*. New York: Plenum, pp. 67–73.

Wu, E., Geng, L., Joseph, M. J., McGown, L. B. (1993). *Anal. Chem.* 65:2339.

Yeung, E. S., Wang, P., Li, W., Giese, R. W. (1992). *J. Chromatogr.* 608:73.

Yguerabide, J., Yguerabide, E. E. (1984). In: Rousseau, D. L., ed. *Optical Techniques in Biological Research*. New York: Academic, pp. 181–290.

Zare, R. N. (1984). *Science* 226:298.

7

Vibrational Spectroscopy: Instrumentation for Infrared and Raman Spectroscopy

PETER FREDERICKS, LLEWELLYN RINTOUL
Queensland University of Technology, Brisbane, Queensland, Australia

JOHN COATES
Coates Consultancy, Newtown, CT, USA

I. INTRODUCTION

In reviewing the contents of this *Handbook*, a decision had to be made regarding the presentation of the techniques for infrared (IR) and Raman spectroscopy. The problem lies in the complementary nature of these two techniques and the significant amount of overlap in both the instrumentation and the applications. It was decided that the readers would be served better if the two techniques were presented side-by-side within the same chapter. In this way, the similarities and dissimilarities can be assimilated conveniently, and for a person new to vibrational spectroscopy, it provides a more even-handed approach to these two powerful methods of analysis. However, having made the case for combining the techniques into a single chapter, the integrity of the techniques will be retained where necessary, and the individual techniques will be discussed separately, if appropriate.

In the most basic terms, IR and Raman spectroscopies are actually dissimilar in terms of the origins of the phenomena and how they are observed and measured. However, both come together when the final analytical data are presented, that is, the net effect of both techniques is to provide a spectrum produced from vibrational transitions within a molecular species. The fundamental theory and underlying concepts involved in the measurement of vibrational spectra are summarized in the next section.

The IR region of the electromagnetic spectrum is divided into three subregions, which are quite different from theoretical and application standpoints, and also require significantly different instrumentation. The most important region by far is the mid-IR, which contains the fundamental absorption bands, and has been used for qualitative and quantitative analysis for more than 50 years. It would be quite unusual for an analytical laboratory not to possess a mid-IR spectrometer.

Near-infrared (NIR) spectroscopy has gained popularity since the mid-1980s. It is now an important branch of IR spectroscopy, covering a broad range of industrial qualitative and quantitative methods of analysis. Today, its prime role is for material screening, quality control, and process analytical applications, and it has little to offer as an analytical problem-solving tool. NIR is included in this chapter, and is discussed in terms of basic theory and instrumentation technology.

Far-infrared (FIR) has been mainly of academic interest in the past. However, new instrumentation has made access to the FIR region, and the adjacent region towards microwave wavelengths, much easier. This part of the spectrum (approximately 300–3 cm^{-1}) is known as the "terahertz" region. Considerable effort is being expended to define applications for this lately rediscovered part of the electromagnetic spectrum.

As noted, this chapter covers four main branches of spectroscopy: traditional IR (mid-IR), NIR, FIR, and Raman spectroscopies. It has been integrated to provide a consistent and balanced approach to these interrelated techniques. The result is a long text, and to help the reader, each major section is self-contained. This is intended to reduce the need to cross-reference facts from other parts of the chapter. As a consequence, there has been some deliberate repetition of the key points.

Strictly speaking, the word *light* defines only the visible components of the spectrum, from 380 to 780 nm. However, because of the breadth of this text, where the subject matter covers radiation from the ultraviolet to the IR, the term light will be used at times in its generic form as an alternative wording for electromagnetic radiation.

A. Background for Vibrational Spectroscopic Measurements

IR spectroscopy is undoubtedly one of the most versatile techniques available for the measurement of molecular species in the analytical laboratory today. Its role has now been extended way beyond the laboratory and is being considered for applications ranging from a diagnostic probe for medical research to the controlling technology for modern chemical plants and refineries. A major benefit of the technique is that it may be used to study materials in all three physical states—solids, liquids, and gases.

The classical view of IR spectroscopy was based on what is termed the mid-IR. This region covers the fundamental vibrations of most of the common chemical bonds featuring the light- to medium-weight elements. In particular, organic compounds are well represented in this spectral region. Today, the mid-IR region is normally defined as the frequency range of 4000–400 cm^{-1}. The upper limit is more or less arbitrary, and was originally chosen as a practical limit based on the performance characteristics of early instruments. The lower limit, in many cases, is defined by a specific optical component, such as, a beamsplitter with a potassium bromide (KBr) substrate which has a natural transmission cut-off just below 400 cm^{-1}. Raman spectroscopy has somewhat different frequency or wavelength limits imposed on it because it is measured as a wavelength/frequency shift from a well-defined, fixed wavelength. Theoretically, one can extend Raman spectroscopy down to a region close to zero wavenumbers and to an upper limit which may be as high as 4000 cm^{-1}, but is typically less—both are dependent on the optical system used.

Note that today, for most chemical-based applications, it is customary to use the frequency term *wavenumber* with the units of cm^{-1}. Some of the early instruments, which utilized prism-based dispersion optics, fabricated from sodium chloride, provided a scale linear in wavelength (microns or more correctly, micrometers, μm). These instruments provided a spectral range of 2.5–16 μm (4000–625 cm^{-1}). With the wavelength (λ) described in microns, the relationship to the frequency (ν) in wavenumbers is provided by:

$$\nu = \frac{10,000}{\lambda} \quad (1)$$

The terms *frequency* and *wavelength* may be used interchangeably, and will be used accordingly, throughout this text. Wavelength is still used today by the optical physics, electro-optics, and astronomy communities. However, in general, chemists prefer to use the frequency term because of its fundamental relationship to the energy of vibrational transitions, which in turn form the basis of the characteristic vibrational group frequencies. Group frequencies are the key to spectral interpretation for chemical structure and functionality.

The region below 400 cm^{-1}, generally classified as the FIR or more recently *terahertz region*, is characterized by low-frequency vibrations typically assigned to low-energy deformation vibrations and the fundamental stretching modes of heavy atoms. There is only one IR-active fundamental vibration that extends beyond 4000 cm^{-1}, and that is the H–F stretching mode of hydrogen fluoride. Generally, the spectral region above 4000 cm^{-1} is known as NIR. The classical definition of NIR is the region between the IR and the visible part of the spectrum, which is nominally from 750 to 3300 nm (0.75–3.3 μm, approximately 13,300–3000 cm^{-1}).

Originally, the NIR spectrum was serviced by extensions to ultraviolet–visible (UV–Vis) instrumentation. In the late 1970s, NIR was recognized to be useful for quantitative measurements for a wide variety of solid and liquid materials, and as a result a new class of instruments, dedicated to this spectral region, was developed.

In the mid-IR region, the main class of commercial instruments are Fourier transform infrared (FTIR) spectrometers. The original FTIR spectrometers, available in the 1960s, were produced for the FIR measurements. However, in general, today it is possible to extend the technology to include all measurements from the UV–Vis region, to the FIR. Extending the technology above the mid-IR region has required improved optical and mechanical designs to meet the more stringent tolerances necessary for good performance. FTIR spectrometers configured for operation in the NIR region are now readily available. Initially, the role of FTIR for the NIR was considered to be of marginal importance, because of dynamic range considerations and the perceived loss of the multiplex advantage. However, these objections have in part been overcome by improved signal handling electronics and also by extended sampling capabilities—such as the use of optical fibers for sample interfacing.

The entry of FTIR spectrometers into the NIR market has brought into question the "uniqueness" of this technique. The promotion of NIR as a unique "science" beyond the mid-IR is really a commercial issue, and the boundaries between the techniques are quite artificial. In reality, both the near- and mid-IR should be considered as the same subject: the NIR being derived from the first, second, and third overtones of the fundamental region, and combination (summation or difference) bands, all attributed to information from the mid-IR. However, NIR spectroscopy is used in a very different manner to mid-IR spectroscopy because its strengths lie in quantitative rather than qualitative analysis. Hence, the authors have decided to discuss it within a separate section.

One of the main reasons for the rise in popularity of NIR, over mid-IR, has been the issue of sampling. The overtone and combination bands measured in the NIR are at least one to two orders of magnitude weaker than the fundamental absorptions. It will be seen later that sample thickness is generally measured in micrometers, or tenths of millimeters in the mid-IR, whereas millimeters and centimeters are appropriate in the NIR to measure comparable levels of light absorption. This simplifies sampling requirements for all condensed-phase materials, and in particular for powdered or granular solids. Also, common silicate materials such as quartz and glass are transparent throughout the NIR, in contrast to the mid-IR. This eliminates the necessity for hygroscopic or exotic materials, for sampling and instrument optics, that are commonplace in the mid-IR region.

For most common IR applications (NIR, mid-IR, or FIR) the calculations are made from data presented in a light absorption format. In practice, this involves comparing light transmission through the sample with the background transmission of the instrument. The output, normally expressed as percent transmittance, is I/I_0, where I is the intensity (or power) of light transmitted through the sample, and I_0 is the measured intensity of the source radiation (the background) illuminating the sample. The actual absorption of IR radiation is governed by the fundamental molecular property, absorbance, which in turn is linked to molar absorptivity for a light transmission measurement by Beer's Law:

$$A = \alpha c \quad (2)$$

Or more generally, by the Beer–Lambert–Bouguer Law:

$$A = \alpha b c \quad (3)$$

where α is the *molar absorptivity*, b is the sample thickness or optical pathlength, and c is the analyte concentration.

It must be appreciated that each absorption observed within a spectrum can be assigned to a vibrational mode within the molecule, and each has its own contribution to the absorptivity term, α. Absorbance in turn is determined from the primary measurement as:

$$A = \log_{10}\left(\frac{1}{T}\right) \quad \text{or} \quad \log_{10}\left(\frac{I_0}{I}\right) \quad (4)$$

Not all IR spectral measurements involve the transmission of radiation. Such measurements include external reflectance (front surface, "mirror"-style or specular reflectance), bulk diffuse reflectance, and photoacoustic determinations. In all cases more complex expressions are required to describe the measured spectrum in terms of IR absorption. In the case of photoacoustic detection, the spectrum produced is a direct measurement of IR absorption. While most IR spectra are either directly or indirectly correlated to absorption, it is also possible to acquire IR spectra from emission measurements. This is particularly useful for remote spectral measurements, especially for astronomical, geological, and environmental applications. For such applications, the main requirement is to have a detection system that is cooler than the measurement environment.

In the Raman studies, light is scattered from a sample after irradiation from a high intensity monochromatic source—typically a laser. Most of the radiation collected in a Raman measurement results from elastic scattering, and this remains unchanged with respect to frequency or wavelength compared with the original incident beam. This is known as Rayleigh scattering. An extremely small fraction of the scattered radiation results from inelastic scattering, where the energy from the source is modified by vibrational transitions that occur during the energy transfers of the scattering process. The changes, known as the *Raman effect*, are observed as shifts to both lower and higher frequencies—a phenomenon first predicted by Smekel, and later demonstrated by C.V. Raman (Raman, 1928; Raman and Krishnan, 1928). These shifts, known as Stokes and anti-Stokes shifts, respectively, are small compared with the original frequency (or wavelength); therefore, a spectrometric analyzer with good resolving power is required to provide a useful, analytical spectrum. As implied, the Raman effect is an extremely weak phenomenon, with observed intensities several orders of magnitude less than the intensity of the illuminating monochromatic light source. Similarly, the intensity of the anti-Stokes lines are weaker than the Stokes lines, in this case, by many orders of magnitude, dependent on the temperature of the sample.

As noted above, the Raman effect is the result of the molecule undergoing vibrational transitions, usually from the ground state ($v = 0$), to the first vibrational energy level ($v = 1$), giving rise to the Stokes lines, observed as spectral lines occurring at lower frequency (longer wavelength) than the incident beam. Because the effect involves a vibrational transition, with a net gain in energy, it is comparable to the absorption of photon energy experienced in the generation of the IR vibrational spectrum. Therefore both techniques, IR and Raman spectroscopies, are derived from similar energy transitions, and the information content of the spectra will have some commonality. However, the spectra are not identical because not all energy transitions are allowed by quantum theory, and different selection rules govern which transitions are allowed. These are usually discussed in terms of IR-active and Raman-active vibrations, and are defined as follows:

1. For an IR vibration to be active, that is, for it to absorb IR energy, the molecular vibration must induce a net change in dipole moment during the vibration.
2. For a molecular vibration to be Raman-active vibration, there must be a net change in the bond polarizability during the vibration.

The result of these different rules is that the two sets of spectral data are complementary. In general, molecules that have polar functional groups with low symmetry tend to exhibit strong IR spectra and weak Raman spectra, and molecules with polarizable functional groups with high symmetry provide strong Raman spectra. In practice, for most molecular compounds reality is somewhere in between these two extremes. As a consequence, most compounds produce unique IR and Raman spectra that feature a mixture of strong, medium, and weak spectral bands—typically, a strong feature in one spectrum will either not show or may be weaker in the other spectrum, and vice versa.

Like the IR spectrum, the Raman spectrum is also presented in wavenumbers, cm^{-1}, as a displacement from the main excitation, or the Rayleigh line. In concept, the spectral range starts from zero, and extends to the full range of the fundamental ($v = 0$ to $v = 1$) molecular vibrations, out to a nominal 4000 cm^{-1}. In practice, the extreme intensity of the main excitation line limits how close one may get to zero wavenumbers, and dependent on the properties of the instrument optics used, this may be anywhere from 15 to 200 cm^{-1}. Likewise, the upper end of the spectrum may or may not extend out to 4000 cm^{-1}, and is often limited by detector response. In reality, there are few significant Raman-active vibrations above 3500 cm^{-1}, and so this is seldom an issue.

Unlike the traditional IR measurement, the Raman effect is a scattering phenomenon, and is not constrained by the laws of absorption. The intensity of a recorded spectral feature is a linear function of the contribution of the Raman scattering center, and the intensity of the incident light source. The measured intensity function is not constant across a spectrum, and is constrained by the response of the detector at the absolute frequency (wavelength) of the point of measurement. Also, the Raman scattering is not a linear effect, and varies as a function of λ^4, where λ is the absolute wavelength of the scattered radiation. While the Raman intensity may be used analytically to measure the concentration of an analyte, it is necessary to standardize the output in terms of the incident light intensity, the detector response, and the Raman scattering term, as a function of the absolute measured wavelength.

This outline of IR and Raman spectroscopies has been included to provide a fundamental view of the two techniques, with a broad assessment of their similarities and differences. These are relevant in terms of how instrumentation is implemented, and how it is used. A detailed discussion of the underlying principles of vibrational spectroscopy is outside of the scope or the requirements of this *Handbook*. However, an appreciation of the basics is helpful to supplement the outline provided earlier, and for a more in-depth discussion, the text by Colthup et al. (1990) is recommended.

B. The Basic Principles of Vibrational Spectroscopy

The basic concept of vibrational spectroscopy is first considered in terms of a simple diatomic molecule, to simplify the discussion to the vibration of a single, isolated bond. The model of a simple harmonic oscillator is based on Hooke's law where the vibration frequency ($\bar{v}$) in wavenumbers is defined as:

$$\bar{v} = \frac{1}{2\pi c} \sqrt{\frac{\kappa}{\mu}} \quad (\text{cm}^{-1}) \tag{5}$$

where c is the velocity of light, κ is a force constant, μ is the reduced mass ($m_1 m_2 / (m_1 + m_2)$), and m_1 and m_2 are the masses of the atoms forming the molecular bond.

From quantum theory, the vibrational energy levels E_{vib} are defined as:

$$\frac{E_{\text{vib}}}{hc} = \left(v + \frac{1}{2}\right)\bar{v} \quad (\text{cm}^{-1}) \tag{6}$$

where h is the Plank's constant, v is the vibrational quantum number (0, 1, 2, 3,...), and $\bar{v}$ is the intrinsic vibrational frequency of the bond (cm^{-1}).

In practice, the quantized energy levels are not equally spaced because the molecule deviates from ideality, and it behaves as an anharmonic oscillator, requiring the expression (6) to be modified by an anharmonicity term:

$$\frac{E_{vib}}{hc} = \left(v+\frac{1}{2}\right)\bar{v}_e - \left(v+\frac{1}{2}\right)^2 \bar{v}_e \chi_e \quad (\text{cm}^{-1}) \quad (7)$$

where $\bar{v}_e$ is the equilibrium vibrational frequency (cm^{-1}) and χ_e is the anharmonicity constant, the value of which is small and positive.

For the fundamental vibration, that is, from $v=0$ to $v=1$, the net energy of the transition is given by:

$$\frac{E_{vib}}{hc} = \bar{v}_e(1 - 2\chi_e) \quad (\text{cm}^{-1}) \quad (8)$$

Transitions from the ground state vibrational quantum level, $v=0$, to the higher vibrational levels, $v=2$, $v=3$,..., are weaker, and correspond to the first and second overtones, respectively. The energy for these transitions is provided by the expressions:

$$\frac{E_{vib}}{hc} = 2\bar{v}_e(1-3\chi_e) \quad (\text{cm}^{-1}, \text{ first overtone}) \quad (9)$$

$$\frac{E_{vib}}{hc} = 3\bar{v}_e(1-4\chi_e) \quad (\text{cm}^{-1}, \text{ second overtone}) \quad (10)$$

At room temperature, ~293 K, most transitions occur from the ground state. Transitions from the first vibration level, $v=1$, to the next level, $v=2$, increase with increase in temperature. These very weak "hot bands" can be observed with slightly lower frequency than the fundamental vibration, as indicated (in cm^{-1}) by:

$$\frac{E_{vib}}{hc} = \bar{v}_e(1-4\chi_e) \quad (\text{cm}^{-1}) \quad (11)$$

The overall energy possessed by a molecule can be approximately defined as the sum of four energy terms, originating from translation (displacement), and changes in energy states resulting from the interaction with electromagnetic radiation—electronic, vibrational, and rotational:

$$E_{total} = E_{trans} + E_{elect} + E_{vib} + E_{rot} \quad (12)$$

In magnitude, the rotational energy term is significantly smaller than the vibrational energy term. While this term contributes to the overall spectrum, the evidence of the rotational transitions is only recorded in the spectra of light molecules, normally in the gas phase. For very light molecules, these may be observed as fine structure absorptions, centered about the mean vibrational absorption, assuming the spectrum is recorded with adequate resolution. Dependent on the molecule, most of these lines have natural widths at half height of a few tenths of a wavenumber, or less.

At the beginning of this section, the use of the harmonic oscillator model was first applied to a diatomic molecule. For most practical applications, with the exception of the diatomic gas molecules, it is necessary to adjust the model to accommodate polyatomic molecules. The fundamental theory can still be applied for all bonded atoms within a molecule. This is discussed in terms of normal modes of vibration, based on the number of degrees of freedom in three-dimensional space, and are determined as follows:

$3N-5$ normal modes, for a linear molecule
$3N-6$ normal modes, for a nonlinear molecule

Here N is the number of atoms in the molecule.

If the $3N-6$ rule is applied to a moderately complex molecule containing 50 atoms, it can be seen that 144 normal modes of vibration are predicted. This might imply that with an increase in the number of atoms in a molecule, there is a corresponding increase in the complexity of the spectrum. In reality, this is seldom observed because many of the vibrations are degenerated (or nearly so), and consequently the number of observed vibrations is much fewer than the predicted. Most of the fundamental vibrations or modes are described in terms of stretching or bending of chemical bonds within the molecule. In the IR spectrum, added complexity is observed, as indicated earlier, by the presence of overtones of the fundamental frequencies. A further complication arises from the occurrence of combination and/or difference bands, which occur when the fundamental modes are combined. These are generally weak features in the spectrum, but they occur in the spectra of most complex molecules.

Not all the fundamental modes of vibration are necessarily IR-active, and likewise for the Raman-active vibration. As noted earlier, the Raman spectrum is an indirect measurement of vibrational transition based on inelastic scattering of monochromatic radiation. A general expression describing the Raman effect is:

$$\bar{v}_s = \bar{v}_0 \pm \frac{\Delta E_{vib}}{hc} \quad (\text{cm}^{-1}) \quad (13)$$

where $\bar{v}_s$ is the frequency of the Raman scattered radiation and $\bar{v}_0$ is the frequency of exciting line (usually a laser line).

As noted earlier, and as indicated by the $\pm \Delta E_{vib}/hc$ term, the Raman effect produces two sets of lines—Stokes ($\bar{v}_s = \bar{v}_0 - (\Delta E_{vib}/hc)$) and anti-Stokes ($\bar{v}_s = \bar{v}_0 + (\Delta E_{vib}/hc)$). The abundance of the anti-Stokes lines is a function of how many molecules exist in the first vibrational energy state, $v=1$, compared with the ground state. The ratio is dependent on the measurement temperature (T, expressed in K), and is determined from the Boltzmann distribution:

$$\frac{n_{v=1}}{n_{v=0}} = \exp\left(\frac{\Delta \bar{E}_{\text{vib}}}{kT}\right) \tag{14}$$

where $\Delta \bar{E}_{\text{vib}}$ is the vibrational transition energy expressed in J/molecule.

The observed differences in intensity of the anti-Stokes vs. the Stokes compared with the value obtained from Eq. (14) are attributed to the λ^4 law for the scattered radiation. In this case, the higher frequency anti-Stokes lines are more efficiently scattered than the lower frequency Stokes lines. An adjustment may be made for scattering efficiency as follows:

$$\frac{I_{\text{anti-Stokes}}}{I_{\text{Stokes}}} = \frac{(\bar{\nu}_0 + \bar{\nu})^4}{(\bar{\nu}_0 - \bar{\nu})^4} \frac{n_{v=1}}{n_{v=0}} \tag{15}$$

A summary of the energy transition processes for IR, NIR, and Raman spectroscopies, including the Stokes and anti-Stokes transitions, are provided in Fig. 1.

One final factor to consider for the measured Raman spectrum is the effect of polarizability, and the ability to distinguish purely symmetric vibrations from the interaction with plane-polarized light. The depolarization ratio ρ is determined from the ratio of the spectral intensity for perpendicularly polarized radiation ($I_\updownarrow$) to the intensity of the spectrum with parallel polarized radiation (I). A value of 3/4 for this ratio indicates a depolarized band, and a value between 0 and 3/4 indicates a polarized band, from a symmetric vibration.

C. Comparison of Techniques and Relative Roles

The previous two sections have provided the fundamental basis for IR, NIR, and Raman spectroscopies. They also have established the ground rules for the information content of data acquired from these techniques. It is informative, from a practical point of view, to review the respective advantages and disadvantages they offer to the analytical chemist. Table 1 helps to summarize, on a comparative basis, the pros and cons of each technique. All three techniques are mature, in terms of the understanding of the techniques, and in terms of implementation for standard analytical laboratory applications.

IR (primarily the mid-IR) is still the largest single market, and the majority of instruments are now used today for routine analysis, environmental analysis, and quality control applications in the manufacturing industries. It is still the technique of choice for investigatory applications, and in this area, microscope-based methods have become increasingly important. NIR is a growing area of application, in particular for quantitative measurements for manufacturing and process control. Although the ancestry of Raman spectroscopy is the longest, it has had several rebirths, and today the technique is enjoying modest growth, with new interest in industrial areas of application. Like IR, there is a growing interest in microscope-based applications, and the focus for Raman spectroscopy is on materials science and medical research.

From this point on, the discussion relative to the specific aspects of instrumentation design will be dealt with separately for the individual techniques of IR and Raman spectroscopies, except where obvious areas of overlap occur. The lion's share goes to IR (both

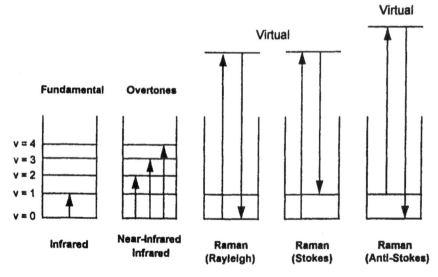

Figure 1 Summary of the vibrational transition process.

Vibrational Spectroscopy

Table 1 A Comparative Study of Mid-IR, NIR, FIR, and Raman Spectroscopies for Analytical Applications

	Mid-IR spectroscopy	NIR spectroscopy	FIR spectroscopy	Raman spectroscopy
Spectrum appearance (average)	Detailed, often highly structured	Broad, diffuses absorptions, typically lacks structure	Detailed, often highly structured	Detailed, often highly structured
Nature of phenomenon	Strong	Weak	Strong	Weak
Information content	Spectrum reflects properties and composition	Spectrum reflects properties and composition	Spectrum reflects properties and composition	Spectrum reflects properties and composition
	Characteristic property of material	Characteristic property of material	Characteristic property of material	Characteristic property of material
	Good for first-order interpretation for chemical functionality	Limited for first-order interpretation for chemical functionality	Poor for first-order interpretation for chemical functionality except for heavy atom containing compounds	Moderate-to-weak for first-order interpretation for chemical functionality
	Moderate for skeletal/backbone analysis	Moderate-to-weak for skeletal/backbone analysis	Good for skeletal/backbone analysis	Good for skeletal/backbone analysis
	Quantitative—normally follows Beer–Lambert law	Quantitative—normally follows Beer–Lambert law	Quantitative	Quantitative—emission technique, difficult to standardize
Sample handling matrix				
Solid	Yes—sampling may be difficult	Yes—sampling easy	Yes—sampling easy	Yes—sampling easy
Liquid	Yes—sampling usually easy	Yes—sampling easy	Yes—sampling easy	Yes—sampling easy
Gas	Yes—sampling usually easy	Unsuitable—sampling easy, but limited application	Yes—vibrational and rotational spectra can be measured	Possible—limited application
Optics				
General/instrument	Special IR transmitting—KBr, ZnSe, and so on. Main instrument optics feature front surface mirrors	Typically low OH quartz. Refractive optics used in instrumentation	Mylar or wire grid beamsplitters, polyethene windows. Vacuum bench to remove water vapor	Typically quartz or glass. Refractive optics used in instrumentation
Sampling	IR transmitting or mirrors. Exotic material, such as diamond used for difficult applications	Quartz or sapphire windows	Polyethene, CsI, and diamond	Quartz or glass
Special	Catadioptic lenses for remote sensing, and Cassegrain optics for microscope applications	Optical fibers commonly used for sample interfacing	Optical fibers not used	Optical fibers may be used for sample interfacing
Typical/common analytical applications	Materials characterization	Compositional analysis	Few analytical applications	Confirmational analysis for certain functional groups—complement to IR
	Compound identification	Property measurements	Characterisation of organometallics	Structural analysis
	Quality control and process analysis	Quality control testing and screening	Skeletal modes of polymers, lattice modes of crystals	Process analysis and some quality control
	Microsampling	Process analysis		Can be good for aqueous media
	Hyphenated techniques—GC-IR, TGA-IR, and so on			
	Gas mixture analysis			
Common interferences	Water vapor and liquid state	Water vapor and liquid state	Water vapor	Sample broad-band fluorescence

mid- and NIR) because of its maturity and its broader acceptance.

II. IR INSTRUMENTATION

A. Types of Instruments and Recent Trends

In common with most instrumental techniques, instrumentation for IR spectroscopy has evolved into various categories. Furthermore, a subclass of instruments for specific analyses in the NIR spectral region has grown out of the 1980s. For the most part, the focus of the discussion will be on traditional IR (mid-IR) instruments. However, there is a large degree of overlap, and where appropriate, NIR instruments, their designs, and their applications, will be included in this discussion.

As a result of the evolution of IR instruments, we have seen four main categories emerge:

1. Research (high-end) instruments: typified by high performance, with spectral resolution of $0.1\,cm^{-1}$ or better, very flexible, and with a large number of options and configurations.
2. Advanced analytical (mid-range) instruments: similar to research instruments, but with less resolution (cut-off around $0.1\,cm^{-1}$, with $0.5\,cm^{-1}$ or lower resolution being adequate for most applications), comparable flexibility but more emphasis on problem-solving options—high throughput is often traded for resolution in these instruments (consistent with the problem-solving role).
3. Routine analytical (low-end) instruments: compact instruments, simple to use, with practical sampling options, moderate performance (average resolution around $2-4\,cm^{-1}$) but reduced configuration flexibility compared with advanced analytical instruments.
4. Dedicated instruments: defined as an instrument or analyzer that is configured to perform a limited set of functions, and is optimized for a specific analysis or set of analyses. This may be a modified instrument from one of the above categories, or it may be uniquely designed for optimum performance—most NIR instruments fall into this category.

The number of high-end instruments sold continues to decline. Although not as drastic, there has been a slowdown in the growth of the number of advanced analytical instruments. The main areas for growth have been in the last two categories—the routine and the dedicated instruments, with the lion's share going to the routine analytical instruments. In common with the personal computer market, which in itself has helped to fuel the low-cost instrument market, the number of options and the performance of low-end instruments have grown. For many analytical applications, it is possible to purchase an instrument today for less than half of the price paid 10 years ago, with comparable or improved performance.

All current commercial mid-IR instruments, sold for analytical applications, are based on an interferometric measurement, that is, they are classified as FTIR instruments. The Fourier transform relates to the mathematical conversion of the data from its primary interferometric format (time or spatially based), to the traditional wavelength of spectral frequency-based format. Before 15 years, there was a 1:1 mix of dispersive and FTIR instruments. However, today there are probably not more than 5% of dispersive instruments still in regular service for mid-IR analyses.

For applications outside of the mid-IR region, such as in the NIR, or for applications outside of the laboratory, such as dedicated IR analyzers, other technologies are available. In the NIR region, a significant number of commercial instruments use optical dispersion techniques featuring either rapid-scanning monochromators or fixed grating spectrographs combined with multielement detector arrays. One novel technology that has been successfully applied to NIR measurements is the acousto-optical tunable filter (AOTF) (Wang et al., 1993). While this device may also be applied to mid-IR measurements, no commercial products have yet been produced. Array detector devices have been developed to operate in all the IR spectral regions, however, based on cost, quality, and availability, the most widely used region for these detectors is the silicon photodiode array (PDA), applicable in the short-wave NIR. This is definitely one area for future technology expansion.

Today, in the mid-2000s, instrumentation hardware and measurement technology continue to be refined in certain areas for traditional laboratory instruments, but there are currently no immediate changes anticipated. The most significant changes are now occurring in sampling technology and software.

Sampling is the key because it involves the optical interface between the sample and the IR radiation from the instrument. A complete industry has been formed around sampling, with many companies now specializing in general-purpose and dedicated sampling accessories for virtually all commercial IR instruments. Users of these instruments have a broad selection of sampling techniques available to handle most sample types. An important area of development is remote sampling, where the sampling point is separated from the main instrument. This has taken several forms: from open path measurements of ambient air, to the use of sample probes coupled to the instrument via optical light guides or optical fibers (Ganz and Coates, 1996). This has become an important area for NIR, but is insignificant for mid-IR because of the cost and performance of mid-IR fibers.

A second, very significant area has been the use of high-quality IR microscopes (Katon and Sommer, 1992; Katon et al., 1986), which have been applied to a vast array of applications, including forensics, advanced materials research (such as high-tensile fibers, ceramics, and semiconductors), and biological and biomedical applications. An important extension has been the combined use of computer software and video imaging, with the spectroscopic microscopes—both IR and Raman spectroscopies. In this case, a powerful imaging technology has been developed where the compound or functional group-specific IR maps of a material may be combined with the confocal visual graphics image. So important has this become that new sections on the IR imaging and the Raman imaging have been added to this chapter for the current edition.

Proprietary computer systems, developed by the instrument manufacturers and common in the 1980s, have been completely replaced by standard personal computers usually running Microsoft Windows®. As a result, software tools, for general spectral manipulation, have become more or less standardized into a common set, available on most current IR spectrometers. Functions, such as multiple-order derivatives, spectral deconvolution, Kubelka–Munk conversion (providing photometric correction for diffuse reflectance data), and Kramers–Kronig transformation (for removing the refractive index related components from reflectance spectra), have become commonplace on most instruments.

In many cases, there is a desire to automate procedures, and so recent emphasis has been on the development of method generation software, making full use of the graphical user interface of Microsoft Windows®. An important aspect that has been developed as an off-shoot of good laboratory practice (GLP) regimes has been the need to document and maintain audit trails for analytical procedures, including method, software, and instrument performance validation. For certain applications, such as those required for pharmaceutical quality control and environmental measurements, this is a government-regulated requirement in many countries.

Methods for the qualitative and quantitative assessment of spectral data have continued to advance. The traditional methods of library searching—by the use of matching, either partially resolved spectra or peak parameters—have remained essentially the same for 20 years. The most significant developments in this area have been in the use of interactive graphics, in particular animated graphics and multimedia tools, as well as taking advantage of the newer high capacity storage media, in particular gigabyte (or greater) hard disks and optical, or magneto-optical, disk drives. Comparative methods which are in essence both qualitative and quantitative in nature, based on vector metrics or statistical evaluation methods, are gaining popularity. These are used primarily for the automated verification of compound identification—both mid-IR and NIR are natural candidates for this application, especially in the pharmaceutical industry.

Standardization of both instrument output and data formats is an important issue. This is particularly the case with the increased use of the internet as a communication medium. The internet could become a major international repository for spectral data, and already a number of databases, both free and commercial, are available. Standardized data formats have allowed such concepts to become practical. The JCAMP-DX format (McDonald and Wilks, 1988) has been widely adopted. Furthermore it has become more common for software packages from a particular manufacturer to be able to import from, or export to, the software packages of other manufacturers. It is also possible to use third party software such as GRAMS^{TM}, from Thermo Galactic, which will accept data from almost all instruments and therefore will allow a laboratory with several different IR and Raman instruments to have consistency of data handling and manipulation.

Simplifying data transfer by standardized formats is only one side of the problem. Instrument output standardization, a localized issue, is equally important, especially with the increased use of multivariate methods for quantitative analysis. It has long been recognized that a commercial IR instrument, while conforming to a general specification, does have unique characteristics that are convoluted with the spectral output. This means that a calibration for a quantitative measurement is often a unique feature of an individual instrument—meaning that every instrument has to be individually calibrated. Work continues to be directed towards the standardization of data and instrument to remove or reduce this instrument-specific dependency (Workman and Coates, 1993).

Multivariate methods for quantitative analysis have been used for IR analyses since the mid-1970s (Workman et al., 1996). The use of statistical evaluation tools to establish correlations between measured spectral output and material composition or physico-chemical characteristics or properties has resulted in a hybrid science, known as *chemometrics*. Chemometric methods are used widely for a range of different near- and mid-IR applications. New methods for data evaluation, regression, and analysis validation algorithms are constantly being proposed and implemented. Concerns aired relative to the proliferation of chemometric methods, in part, have been addressed by the publication of a standard recommended practice for multivariate IR quantitative analysis (ASTM, 1996a). A new area of application for chemometrics is modeling the spectral characteristics of an instrument and any associated sampling method. The objective is to provide a self-learning system tool for the on-line diagnostics of any instrument while in use. This is linked to the validation issue raised earlier, for routine instruments, and to the need to provide intelligent instruments for dedicated and/or automated analyses.

Today, there is a close relationship between traditional laboratory instruments, which tend to be multifaceted and flexible, and IR analyzers. Analyzers are often dedicated versions of traditional instruments, in which a customized system is produced from a combination of the base instrument, a specific sampling accessory, and software—the latter featuring some or all of the characteristics described above. In this case, the software could include a method, a calibration matrix, and the necessary output-formatting functions. Extensions of this concept can lead to the development of a tailor-made, turnkey analyzers, possibly in one of the following formats:

Laboratory-based, bench-top analyzers
Plant and process analyzers
Portable and transportable instruments

A reasonable number of dedicated IR commercial analyzers are available for a broad range of applications, which include petroleum, petrochemical, chemical, and pharmaceutical analysis, as well as food and agricultural product analysis, semiconductor analysis, and environmental gas analysis. This is a growing segment of the commercial instrument market, which now also includes the Raman instrumentation.

B. Instrumentation: Design and Performance Criteria

In its basic form, IR spectroscopy is a technique for measuring radiation absorption. The measurement is normally made on a relative basis, where the absorption is derived indirectly by comparing the radiation transmitted throughout the sample from a source with the unattenuated radiation from the same source. For most measurements, the signal obtained is expressed as a percentage of the radiation transmitted through either the sample or reflected from the sample. In its simplest form, the measurement may be made at a nominal single wavelength. However, most laboratory measurements are made from a multiwavelength, scanning spectrometer. The main components of such an instrument are summarized in the block diagram in Fig. 2. In essence, this is a generic rendering of almost any analytical instrument that features an energy source, an energy analyzer, a sampling point, a detection device, and electronics for control, signal processing, and data presentation.

Although the traditional IR measurement is based on an absorption measurement, the technique is not limited to a configuration of absorption photometry. If the IR source is replaced by energy emitted from a sample, the measurement becomes one of the *IR emission* spectrometry. A special case of emission spectrometry are the Raman measurements, where the scattered radiation from the sample takes the place of the source. Modern FTIR instruments, which are capable of supporting multiple optical configurations, can provide near/mid-IR and Raman spectroscopies from a single instrument.

The four main instrument components—source, radiation or energy analyzer, detector, and electronics for signal processing—can vary, dependent on the instrument type and application. Each of these principal components will now be discussed in terms of application, available technology, and performance. In many cases there is a choice, and the final selection must be based on performance for a particular application, or simply on cost considerations. Information on the technologies is available in a broad range of publications, such as *Photonics Spectra*, *Laser Focus World*, and specialist *SPIE* publications. Basic descriptions can be found in various texts (Hudson, 1969; Wolfe and Zissis, 1985), but this may be dated in terms of performance and technology.

The key to performance of an IR instrument is the ability to convert efficiently the source energy into useable information at the detector. This relies on maintaining a high

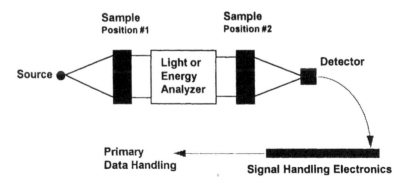

Figure 2 Schematic layout of a generic optical spectrometer.

optical throughput in the energy analyzer (interferometer, monochromator, etc.), an efficient optical coupling with the sample (before and after interaction), and eventually efficient collection and imaging on the detector. Dependent on the design of the instrument, and the nature of the measurement technology, a number of factors can effect the success of the optimal passage of light through the instrument. The sample is the most important of these factors (Messerschmidt, 1986) and is probably the least controllable: it is a key optical element and its mode of interaction with radiation can have a profound impact on the overall instrument performance.

Placement of the sample is important. A choice for sample position is indicated in Fig. 2. In light dispersion instruments, the sample is normally placed before the monochromator or polychromator (spectrograph). This is essentially to maximize the collection of the low levels of radiation that results after dispersion, and to minimize interference from unwanted, stray sources of radiation. Stray light is one of the main sources of photometric error in this type of instrumentation. One negative influence of placing the sample between the source and the analyzer is that the full impact of the radiation from the source is focused on the sample. This may cause localized heating if the sample is kept in the beam for an extended period. Not only might this cause unwanted thermally induced changes in the sample (decomposition, evaporation, or changes in morphology), but it can also lead to photometric errors, in particular at longer wavelengths, as a result of energy re-radiation effects. This error manifests itself on strong absorption bands, where the sample emits radiation at exactly the same wavelengths as the absorptions, which is observed as a reduction of the measured band intensity.

Placing the sample after the light analyzer and in front of the detector will eliminate the potential for this problem to occur, assuming that the radiation reaching the detector is modulated. However, in certain experiments, if the sample is hot, and is placed in front of the detector without effective shielding, the thermal radiation from the sample can cause a DC offset on the detector signal. Examples where this situation can occur are with applications such as GC-IR, where the sample is monitored in a cell heated to between 200°C and 250°C. In such cases, the DC offset signal can be sufficiently high that it results in a noticeable loss in detector sensitivity.

In FTIR instruments, where the sample is normally placed after the interferometer, the energy re-radiation problem does not occur. The radiation from the interferometer is uniquely modulated, and the light originating from the source can be differentiated from any thermal radiation originating from the sample. However, one sample-related artifact that can have a comparable impact on the instrument's photometric accuracy is double modulation. This occurs when source radiation is reflected from the front surface of the sample back into the interferometer cavity. The effect may be reduced by paying attention to the geometry of the interferometer relative to the sample, and it may be successfully eliminated by the use of an intermediate aperture (Jacquinot stop or J-stop). In this case, the use of a semicircular (instead of the traditional circular) aperture takes advantage of the image inversion of the reflected beam. This inverted image is effectively blocked by the nontransmitting portion of the semicircular aperture.

1. Mid-IR Sources

In theory, the ideal source for any type of optical spectroscopy would be a point source with a stable output and a flat emission response across the entire spectral range. A tunable laser operating in the IR region might be perceived to come close to this ideal. A considerable amount of work has been carried out on tunable diode lasers, and it is anticipated that a practical solution will be available for commercial instrumentation within the near future. Work has been reported on the use of tunable laser sources for several specialist applications (Schlossberg and Kelley, 1981). Commercial instruments based on older tunable laser technology were introduced in the 1980s, but the performance was limited to a very narrow spectral range, and only a few were installed. Laser sources are used for special, dedicated applications, in particular for specific gases or vapors, where the laser tuning envelope overlaps with vibrational–rotational fine structure of the analyte—an example of application is isotope analysis (Sauke et al., 1994). Pulsed light emitting diodes are used for some NIR applications for dedicated measurements. In this case, the light source is finely tuned with a narrow bandpass filter, limiting the spectral measurement to a single nominal wavelength.

Internal Sources

The traditional IR source is, in fact, an extended source with a continuous output providing an energy distribution close to that of a blackbody radiator. Figure 3 provides typical blackbody radiation emission profiles as a function of temperature. The normal operating range for a mid-IR source is between 1000 and 1500 K. It can be seen from these blackbody profiles that there is little to be gained by increasing the source temperature when operating only in the mid-IR region. Raising the temperature increases the NIR and visible components in the spectral output of the source, but has a minimal impact on the output in the mid-IR fingerprint region of the spectrum. While raising the source temperature may provide an incremental increase in the amount of useful energy reaching the detector, it may have less impact

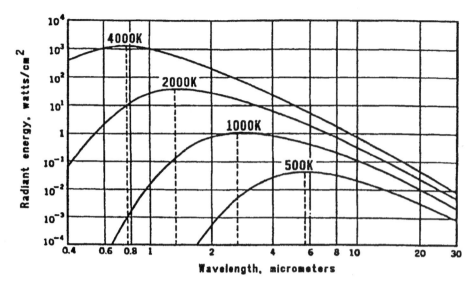

Figure 3 Radiant energy distribution curves for a black body source operating at various temperatures.

than paying attention to efficiency of the overall system optics.

There are two popular styles of IR source based on the nature of construction: (1) metal oxide, silicon carbide, or ceramic rigid sources and (2) metal filament (coilform) sources. One of the original sources used was the *Nernst glower*, which was composed of a mixture of rare earth metal oxides (mainly zirconium and yttrium). This source has a negative coefficient of electrical resistance such that the mixture becomes a conductor at elevated temperatures. It can only be "ignited" after preheating, and becomes self-sustaining as the temperature increases—note that ballast resistors are required to regulate the output. The emission profile is close to a black body, with a nominal operating temperature of 1800 K. Although energy-wise the source is ideal, there are practical difficulties: the source suffers "runaways" (due to the negative coefficient of resistance), the source is mechanically fragile, and there is a need to preheat the source before it ignites.

An important variant to the Nernst glower is the electrically heated ceramic source. One of these sources, the *Opperman*, was originally implemented by Perkin–Elmer, and features a mixture of rare metal oxides contained within a ceramic sheath, with a platinum or nichrome wire located coaxially down the center. The electric current is passed through the wire, which in turn heats the oxide mixture. In this form, characteristics of the Nernst were obtained without its drawbacks. Both the Opperman and the Nernst were designed as extended (cylindrical) sources, which were a good match for the mechanical slit geometry of the early dispersive instrument designs. With modern instruments, this is not necessarily a requirement, and more compact designs are possible. Commercial ceramic heaters, which may be used as IR sources, are available in a variety of shapes and sizes.

A robust source designed for operation at higher temperatures and for high output is the *Globar*—an electrically heated rod made from silicon carbide. For some implementations, the high power output associated with this source results in an excessive amount of heat being dissipated within the instrument. In such circumstances, supplemental water-cooling is applied, to remove excess heat, and to help reduce thermal/oxidative degradation of the electrical contacts.

Typical early metal filament sources used in the mid-IR were in essence an extension of the traditional incandescent light bulb—they were comprised of a nichrome wire coil. The only reason that a light bulb is not used in conventional mid-IR instruments is that the quartz or glass envelope has an optical cut-off over most of the mid-IR spectral region, below 2500 cm^{-1}. The coiled filament is heated electrically in air at temperatures around 1000–1100°C. The main failure mode of this style of source is aging, caused by thermal stressing and oxidation of the filament—resulting in some modification to the output characteristics and an embrittlement of the metal, which usually results in fracture of the source. If the filament is used essentially unsupported, with time a sagging or deformation of the coiled source can be experienced. This results in a long-term shifting in the position of the source image, and a consequent need for source realignment. A variant in design, which helps to support the source coilform, and to provide a more uniform light output, involves a filament loosely coiled around a ceramic rod.

Interestingly, in recent years, many of the sources considered or actually used for commercial mid-IR instruments are modifications or adaptations of commercially available heating elements. The point of interest being that the original component was not necessarily intended to function as an IR spectral source. Examples are gas igniters used in gas heaters, these include coilform filaments constructed from Kanthal and silicon carbide tipped elements. The key issues here are that the devices are designed to be robust, and are expected to operate without failure. Another example is the use of a silicon carbide tipped *glo-plug* used for preheating the combustion chamber in a diesel engine. These devices operate quite well as IR sources, requiring relatively low voltages and current, especially if they are well insulated. Again, the ability to operate without failure under enduring conditions makes this type of component ideal as a long-life mid-IR source, particularly for nonlaboratory applications.

The stabilization of source output can be a major issue, especially for sources in which the heating element is exposed. One source of instability is air turbulence in front of the source, which produces some localized cooling, as well as generating source noise caused by refractive index gradients between the hot air adjacent to the source, and the nearby cooler air. Some of these effects can be reduced by surrounding the source element by a thermally insulated enclosure. The insulation helps to stabilize the output, plus it permits the source to be operated at a lower voltage. Certain silicate structures, similar to the thermal tiles used on the space shuttle, provide effective insulation for this application.

Other source-related instabilities are attributed to short- and long-term changes in output. Short-term output variations are typically linked to voltage fluctuations, and the longer term effects are often related to actual material changes in the characteristics of the source element—such as oxidation or other forms of thermal degradation. Nonhomogeneity in output can also be associated with material changes, manifested as localized hot spots, thermal gradients, and physical deformation of the element. Short-term fluctuations can be addressed either by the use of some form of optical feedback, or by the use of a stabilized DC power supply.

Most sources have a finite lifetime, and replacement will usually be required at some point in time. Like traditional light bulbs, it is difficult to guarantee a definite lifetime. On average, for a mid-IR source, a lifetime of 2–5 years may be achieved, but in certain unusual situations, this could be as short as 6 months. Common failure modes are fracturing (as noted earlier), the formation of hot spots or cold spots, or possibly electrical connection problems, often associated with oxidation at the physical connection points. Symptoms of these failures range from a change in color temperature over the spectral range, observed as a change in output at certain wavelengths, to an obvious total loss of energy. The color temperature changes can often be diagnosed by unexpected or inconsistent distortions to the spectral background, which in turn may be observed as unstable baselines, especially in the short wavelength (high wavenumber) region of the spectrum. Under these circumstances the replacement of the source, and possibly the entire source assembly, is often the only remedy. In some modern instruments, the use of prealigned source assemblies may simplify this procedure.

The Synchrotron Source

In recent years a novel source for IR radiation called a synchrotron has become available in certain parts of the world. In this device a beam of electrons generated in a linear accelerator (linac) is accelerated in an approximately circular path, called a booster ring, to close to the speed of light. They are then injected into a larger ring, the storage ring, where the electrons continue to circulate at relativistic velocity. Under these conditions the electrons, and indeed any charged particle undergoing a curved motion, emit intense broad-band electromagnetic radiation across the whole spectrum from X-rays to the far-IR. The latest generation of synchrotrons utilizes magnetic devices called "wigglers" and "undulators" around the ring to increase the intensity of the radiation. The radiation is emitted tangentially to the path of the electrons so that if a port is incorporated into the ring an intense beam can be obtained, which then passes through various optics into a small laboratory for use in experiments. This is called a beamline. A synchrotron facility may incorporate many beamlines (the Advanced Light Source currently has 27), all of which can be used simultaneously. While most beamlines are used for X-ray experiments, a small number are designed for IR radiation. The intensity of mid- and far-IR radiation obtained from a synchrotron is orders of magnitude greater than that of a conventional source. Moreover, the effective source size can be as little as 100 µm. A schematic of a synchrotron showing the electron source, the storage ring, and the beamlines is given in Fig. 4. Synchrotrons are very expensive, and available time for using them is limited. In a typical situation the scientist books a beamline well in advance and travels to the synchrotron facility with the samples and any special equipment required to carry out a predetermined program of experiments.

While many IR experiments would not benefit greatly from a large increase in source intensity, it has been found that IR microscopy in particular can be significantly enhanced (Carr, 1999). The weak source in a typical FTIR spectrometer leads to low light levels at the detector in a microspectroscopy experiment, particularly when a small

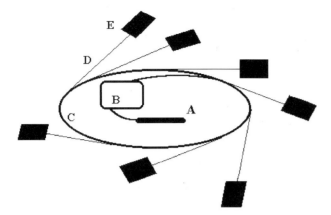

Figure 4 Schematic of a sychrotron facility: (A) linac (linear accelerator); (B) booster ring; (C) storage ring; (D) beamline; (E) beamline laboratory.

aperture is in use to achieve a high spatial resolution. Synchrotron radiation is so intense that very small apertures can be set in the IR microscope, possibly even less than the wavelength of the radiation. The spatial resolution is then limited by diffraction. Despite the intensity of synchrotron radiation, it has been shown that it does not significantly heat the sample (Martin et al., 2001). This is very important for biological samples, such as tissues, which are likely to be one of the most significant applications of synchrotron radiation.

Despite their high cost, many countries around the world now have a synchrotron facility and countries with a strong research culture, such as the United States, have several. The storage ring circumference, which correlates with the light intensity, varies widely from about 70 m for the smaller facilities, up to more than 1000 m for the largest. While there are now many synchrotron facilities around the world, some of the better known are: the Advanced Light Source at the Lawrence Berkely National Laboratory, Berkeley, CA; the National Synchrotron Light Source at Brookhaven National Laboratory, Brookhaven, NY; the Advanced Photon Source at the Argonne National Laboratory, Chicago, IL, Daresbury Synchrotron Radiation Source, Cheshire, UK; the European Synchrotron Radiation Facility, Grenoble, France; and the Photon Factory, Tsukuba, Japan.

A recent variant of the synchrotron is the free electron laser (FEL) in which a relativistic beam of electrons is passed through a periodic magnetic field produced by an undulator or wiggler to generate tunable, coherent, high-power radiation, currently spanning wavelengths from millimeter to visible and potentially to shorter wavelengths. The FEL is much smaller and less expensive than a synchrotron and may become the method of choice for producing very bright light of IR wavelengths.

2. The Light or Energy Analyzer

Current IR instruments can be categorized according to the application and the technology used:

FTIR instruments, based on an interferometric measurement of radiation
Dispersive instruments, based on scanning and nonscanning systems
Filter instruments, single and multiple wavelengths
Novel technologies—examples being optoacoustic tunable filters, Hadamard transform devices, and laser sources.

For the entire IR, the FTIR instrumentation is the most flexible and the most popular in the analytical laboratory for both routine and research applications. A major benefit offered by the modern FTIR instruments is a flexible configuration, where the interferometer, source, and detector may be selected to provide optimum performance in the selected spectral range. This includes the Raman experiment, where the scattered radiation from the sample becomes the source input to the interferometer. The overall flexibility makes the FTIR (interferometry) the most important class of instrument technology, and will be given the greatest focus in this section.

The only area where the older dispersive technology has been retained is in the NIR region (as well as the UV–Vis spectral regions), where the performance benefits of the interferometric system have less overall impact. Current dispersive NIR instruments feature both scanning—typically rapid scanning—and nonscanning configurations. The nonscanning systems are normally integrated with a detector array, where the spectrum is acquired from the signal provided by each pixel in the array.

Filter instruments, for the most part, are used for single function, dedicated measurements. This restriction is dictated by the need for filter elements with fixed spectral characteristics, selected according to the absorption profile of a specific analyte. However, filter instruments are not limited to fixed frequency measurements, and scanning systems have been devised, based on the use of multiple wavelength, variable filters—in the form of either a linear variable filter (LVF) or a tilting wedge interference filter. Although filter instruments may seem limited, either to specific measurements, or by their performance, they constitute an important class of IR instrumentation. Outside of the laboratory, they form the largest class of instrumentation in current use, mainly as dedicated IR analyzers. Furthermore, because of their mechanical simplicity, they are typically more reliable and cost-effective than traditional laboratory-style instruments.

As with any technology-based application there is always pressure for performance improvement, typically in the areas of scan speed, optical throughput, measurement stability, and overall output reproducibility. Mechanical stability is a key issue and this can be achieved with optical designs that require no critical moving parts. Detector arrays are an important technology in this regard, however, by themselves, they do not constitute a spectral measurement system. They are spatial sensing devices that require a dispersing element to be practical. For this reason, detector arrays, although perceived as novel technology for IR measurements, are covered by the section dealing with dispersive instrumentation.

AOTF technology has been successfully applied to commercial instruments, and this also provides the benefits of no moving parts, high scan speed, and a reasonably high optical throughput. Hadamard transform instruments have been discussed for many years (Decker, 1977; Marshall and Comisarow, 1975; Treado and Morris, 1989), but difficulties in constructing the critical encoding masks have limited the implementation of this technology, mainly to the visible and NIR spectral regions. The dream of being able to use laser-based sources, either as monochromatic point sources or as tunable devices, has also existed for many years (Wright and Wirth, 1980). The benefits of their use in a photometric system are linked to the properties of laser light, namely its monochromaticity, its coherence, its high intensity, and its minimal beam divergence. All of which can potentially simplify the design of a laser-based spectral system. To date, their use is limited to some single analyte applications, plus some routine NIR applications for multivariate analysis. However, this is expected to be one of the future growth areas for IR spectroscopy.

Fourier Transform Instruments

The light analyzer at the heart of a Fourier instrument is an interferometric modulator, and for most implementations this is based on the *Michelson* interferometer. The Michelson interferometer has a simple design. In its most basic form, it is made up of three main elements: a fixed mirror, a moving mirror, and an optical beam splitter. This combination, which is often referred to as the modulator, is the component that generates the interferogram, and this is illustrated schematically in Fig. 5.

The key component in the modulator is the beam splitter. Referring to Fig. 4, this optical element splits the beam into two paths, a reflected path and a transmitted path. An ideal beam splitter divides the beam equally at all wavelengths. Both light beams in the two separated paths are reflected back from the fixed and moving mirrors to the beam splitter, where they recombine. The combined beam is reflected towards the sample, and then towards the detector. In practical implementations of the Michelson interferometer, the basic architecture may appear differently. It will be seen later that the key

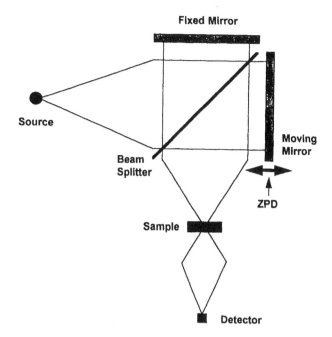

Figure 5 Schematic layout for a Michelson interferometer.

attribute of the moving mirror component is to generate an optical path difference (OPD) between the two paths separated by the beam splitter. The same net effect can be generated if moving components are featured in both paths, and this is featured in several commercial designs, including the use of moving refractive optics to produce a variable OPD.

In the schematic illustrations the intervening optics are not taken into account. For an ideal optical system, the radiation from the source is collimated (made parallel) before interacting with the beam splitter. After the beam splitter, the combined beam remains collimated, and is normally re-imaged to a focus at the sampling point. Finally, the IR energy from the sample is collected and further re-imaged on the detector. In a mid-IR instrument, the imaging optics are normally front surface polished curved mirrors constructed from aluminum. Gold-coated surfaces may be used to improve reflectivity and to maximize corrosion resistance. A typical set of mirrors for a basic interferometer system would include a pair of parabolic mirrors, to collimate the source image, and to re-focus the combined beam on the sample. An off-axis ellipsoidal mirror is commonly used for collecting the image from the sample, and refocusing the image, providing as much as a 6:1 image reduction.

Refractive imaging optics (lenses) are unusual in mid-IR instruments, primarily because of the limited availability of IR transmitting materials capable of being ground and polished into a lens form. Also, the image quality from appropriate refractive optics is poorer in the mid-IR region than that obtained from equivalent reflective optics. Note, however, for some specialist applications, lens systems constructed from materials such as potassium bromide and zinc selenide, have been used. In the NIR, the situation is less acute, and for some applications quartz lens optics are utilized.

The interferometer geometry, as illustrated (Fig. 5), features 45° placement of the beam splitter relative to the beam. Although the 45° geometry is traditional, it is not essential, and other designs based on angles as high as 60° are featured in some commercial instruments. A higher angle helps to reduce polarization effects; and with attention to other aspects of the optical design, it may also reduce the incidence of double modulation, caused by back reflections from the sample and potentially the source optics.

Beamsplitter Materials

Beamsplitter fabrication is a critical issue, in terms of both the optical tolerances involved and the composition. An ideal beamsplitter has a 1:1 transmission to reflection ratio at all wavelengths. In practice, a real beam splitter deviates from this ideal, and as a consequence it performs with reduced efficiency. The choice of material(s) is dependent on the wavelength range selected for the instrument. For applications throughout the IR spectral region, materials with a moderately high refractive index, such as silicon and germanium, have been used. When implemented in the mid-IR (germanium) and NIR (germanium or silicon) IR, the material is applied as a thin film deposited on an IR transmissive substrate. In common terminology, reference is sometimes made to the use of a potassium bromide (KBr) beamsplitter: this refers to the use of a KBr substrate used in the construction, not to the actual material used for the beamsplitter.

A very flat surface finish on the beam splitter and its substrate is essential to meet the criterion for interferometric measurement, with typical tolerances of around $1/8\lambda$. In the case of more common coated beamsplitters, the substrate is polished to the required degree of flatness, and the beam splitter coating is vapor deposited on the polished surface. As noted, germanium on KBr is the most common combination for the mid-IR, and for some of the longer wavelength NIR. The normal range of this beam splitter is between 6500 and 400 cm^{-1}; the upper limit defined by the transparency of the germanium, and the lower limit by the optical cut-off of potassium bromide. In practice, the overall transmission characteristics of the beam splitter, including the upper transmission limit, can be modified by the use of multiple coatings, involving other vapor-deposited materials, such as zinc selenide and KRS-5 (thallium bromide/iodide eutectic).

The choice of substrate can be as equally important as the beamsplitter material. As noted, although not ideal because of its hygroscopic nature, KBr tends to be the most common material of choice for laboratory-based analytical applications in the mid-IR. The capability to obtain a high-quality polish on KBr with the desired degree of flatness, plus its overall optical transmission characteristics tends to favor its use. Additional coatings for moisture protection may be needed, and these have included barium fluoride and certain organic polymeric materials. In this case, the coating will provide modest protection against moisture, while having a minimum impact on the transmission characteristics of the beam splitter elements. For specialist applications, typically outside of the laboratory, where a lower spectral range is less important than resistance to atmospheric moisture, materials such as calcium fluoride ($\sim$1100 cm^{-1} lower limit) and zinc selenide ($\sim$800 cm^{-1}) may be used.

The Operation of an FTIR Interferometer

RAPID SCAN. In order to appreciate the principle of the interferometric measurement, it helps to start with a simple optical system composed of a basic two mirrors and beam splitter combination, as shown in Fig. 6, with a monochromatic source, of wavelength λ. We start with the situation where the two mirrors are equidistant from the beam splitter. The light striking the beam splitter is sent on the two paths to the mirrors—in the ideal case, 50% is reflected to the fixed mirror, and 50% is transmitted to the moving mirror. The two beams are reflected back along the same paths, where they are recombined back at the beam splitter.

With both mirrors equidistant, the two beams travel identical distances, and so when they recombine they are mutually in phase. Based on the concepts of the wave theory of light, a condition for constructive interference is established. The combined beam is passed to the sample, and eventually reaches the detector, where an electrical signal is generated. This signal has a maximum value, which is designated by the value of unity in Fig. 6. In this unique situation, where both paths are equal, we have the moving mirror located at a position known as the zero path difference (ZPD).

Now we consider a situation where the mirror has been physically displaced by a very small distance, equivalent to a quarter of a wavelength, $\lambda/4$ (a movement to the right, as shown in Fig. 6). Obviously, the paths are no longer equal, and the OPD, in terms of the distance that the light travels (to and from the mirror, $2 \times \lambda/4$), is one half of a wavelength, $\lambda/4$. We now have two light waves that are fully out-of-phase, and here, upon recombination, destructive interference of the two beams occurs. In the ideal situation, the light beams cancel, and no energy is transferred to the detector, and the signal falls to zero.

At the next point, where the mirror is moved to one half of a wavelength, $\lambda/2$ from ZPD, the beam reflected from

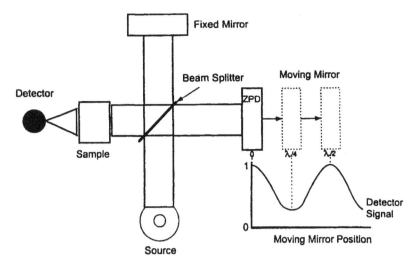

Figure 6 A simplified schematic illustrating the formation of an interferogram from single wavelength source.

the moving mirror has one full wavelength difference in optical path. At the point of recombination, both beams are back in phase, and there is a return to the situation of full constructive interference, with the maximum signal re-established at the detector. In a practical optical system the mirror moves continuously, as a result, the signal measured at the detector varies sinusoidally between the two extremes where the light beams go from full constructive to full destructive interference. As the mirror continues to move, we experience an oscillating signal at the detector, which has a cosine relationship (more accurately a raised cosine) with the mirror displacement, where maxima are observed at integral numbers of $\lambda/4$ and $\lambda/2$ in terms of the distance traveled by the mirror ($\lambda/2$ and λ in terms of OPD). The movement of the mirror with the subsequent generation of an OPD is sometimes known as optical retardation, where the distance (x) traveled by the mirror is called the *retardation*. The term *interferometric signal* is used to define the modulated output at the detector.

When a second wavelength, λ_2, is selected, a similar waveform is generated with an initial maximum signal at the ZPD, and subsequent maxima are observed at integral wavelength values (OPD) from ZPD. If the cosine waveforms from λ and λ_2 are compared, it is noted that both start with a maximum at the ZPD, and that the successive maxima occur at different intervals, corresponding to λ and λ_2, respectively. As third and fourth wavelengths are evaluated, comparable waveforms are generated, each with a common maximum at the ZPD, and each with maxima at distances corresponding to the individual wavelengths (λ_3 and λ_4).

Extending this discussion to a polychromatic source, as we have in a *real* instrument, we see all wavelengths (available within the range of the source), and therefore, conceptually, we would generate unique cosine waves for each component wavelength, where each waveform is uniquely encoded as a function of its wavelength. In practice, the signal collected at the detector is the result of the summation of all of the waveforms. The fact that each waveform has a maximum at ZPD results in a large characteristic signal where all of the waveforms coincide. Beyond this central point, which is known as the center burst, the result of adding all the waveforms together is to produce a summation signal that rapidly decays with increasing distance from ZPD. The additive effect of the component waveforms for a simple system composed of five overlapped cosine waves is shown in Fig. 7.

Until now, the discussion is that with a waveform originating from ZPD, the scan of the moving mirror starts at ZPD. In practice, this is not the case, and most modern interferometers used in FTIR instruments scan more or less symmetrically about the ZPD. This produces an interferogram with the centerburst located in the center, as seen in Fig. 8, resulting in a double-sided interferogram. In theory, both sides are symmetrical, and both sides represent the same spectral information.

Obviously, if each of the component wavelength cosine signatures can be isolated from the interferogram, then it would be possible to determine the contribution of each individual wavelength. The summation of each individual wavelengths would then yield a reconstruction of the spectral output of the source in wavelength (or frequency) space. This function is performed mathematically by the Fourier transform, a numerical method in common use for frequency analysis. The Fourier transform can be expressed as a pair of integrals for the inter-relationship between the intensity in the interferogram, as a function

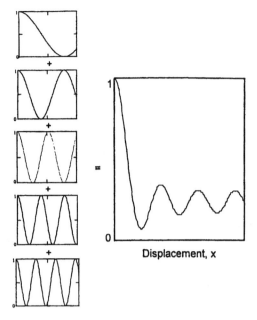

Figure 7 The summation of cosine waveforms from a common origin (zpd).

of retardation $I(x)$, and the intensity in spectral space, as a function of frequency or wavenumber $I(\nu)$, as:

$$I(x) = \int_{-\infty}^{+\infty} I(\nu) \cos 2\pi\nu x \, d\nu \qquad (16)$$

and

$$I(\nu) = \int_{-\infty}^{+\infty} I(x) \cos 2\pi\nu x \, dx \qquad (17)$$

Normally, x would be expressed in cm and ν would be expressed in cm^{-1} for the generation of the IR spectrum.

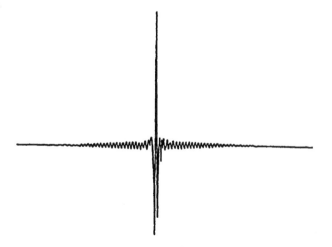

Figure 8 An idealized interferogram.

Returning to the interferogram, as noted earlier, in theory the two sides of the interferogram are symmetrical. In practice, the interferogram is seldom symmetrical, and in some cases, dependent on the actual configuration of the interferometer, it may be very asymmetric. Typically, the interferogram is preprocessed by additional numerical methods prior to the computation of the spectrum. These can include phase correction, dispersion corrections, and apodization. Phase correction numerically adjusts for certain optical and electronic effects that impact the measurement of the recorded signal. Dispersion corrections (linearizations) are applied in systems that feature refractively scanning optics, and where the optical retardation is generated via a varying pathlength of a moving refractive element. Finally, while apodization does not affect the symmetry of the interferogram, it is a method of preprocessing that is applied to the raw interferogram to change the amplitude distribution across the full width of the interferogram. The role of apodization will be discussed subsequently.

We have established how an interferogram is formed, and the relationship of the recorded signal and the optical retardation, or the displacement of the moving component. Now, it is worthwhile to discuss the relationship of the actual spectrum to the actual data collection and to the interferogram. An important aspect to consider is spectral resolution. There are several methods used to define resolution, based on linewidth, or the ability to separate two adjacent lines. In practical terms, the resolution of an instrument can be considered as the ability of an instrument to separate two spectral lines that are very close in wavelength or frequency. Normally, the closer the two lines, the higher the resolving power of the instrument that is required to separate (to observe) the two lines. Translating this to an interferometer, and to the acquisition of an interferogram, we consider the relationship between the two spectral lines and their representative waveforms in the interferogram. In effect, it is necessary to consider what is required to distinguish and to separate two cosine frequencies originating from the same fixed point (ZPD).

In a low-resolution experiment, the two cosine waves can be differentiated after a relatively short distance (retardation) from the ZPD. At higher resolution, the differences between the frequencies are smaller, and so a much larger retardation is required before the two individual cosine functions can be defined uniquely. While this is a simplistic explanation, it helps one to appreciate the relationship between the collection of the interferogram and spectral resolution: that is, the higher the resolution required to represent the final spectrum, the larger the interferogram must be, in terms of distance from the ZPD, to achieve that resolution. In order for the two individual cosine waves to be differentiated, it is necessary for them to go out-of-phase, and back in phase with each other, at least once.

This requires that the interferogram is collected out to a retardation of $1/\Delta\nu$ cm, where: $\Delta\nu = \nu_2 - \nu_1$ (the separation between the two spectral bands at ν_1 and ν_2, expressed in cm^{-1}).

Normally, an instrument is designed to provide a maximum, constant retardation, and therefore the maximum achievable resolution of the spectrometer can be expressed as $1/\Delta\nu_{max}$, where $1/\Delta\nu_{max}$ is the maximum distance that the moving mirror travels from ZPD. Lower resolution spectra can be obtained on the same instrument by recording the interferogram with a shorter retardation. Alternatively, if a spectrum is recorded at the maximum resolution, the resolution may be degraded by truncating the data, thereby numerically reducing the length of the interferogram.

Looking at the fundamental equations to express the Fourier transform, it can be seen that the integration limits extend from $-\infty$ to $+\infty$. Because the integration to determine $I(\nu)$, Eq. (17), is made in respect to the retardation x, it implies that there is an infinite length to the interferogram. This is obviously impractical because this implies an infinite distance traveled by the moving mirror. As noted above, it is normal to record the interferogram to a distance consistent with the resolution, user-defined for the particular measurement, or defaulted to the maximum resolution of the instrument. In consequence, the interferogram is truncated relative to the fundamental equation. The truncation, as noted earlier, limits the resolution of the instrument to $1/\Delta\nu_{max}$, and as a consequence imposes a spectral line width consistent with this resolution. When dealing with the issue of resolution and IR spectral data, there are two factors to consider: the natural line width of the spectral line, dictated by the energy processes involved in the transition that is occurring, and the limiting resolution of the measurement device, the spectrometer.

The result is a convolution of the natural spectral line width and the limiting resolution of the spectrometer. If the natural line width is narrower than the limiting resolution of the instrument, then the truncation results in a distortion of the measured line width. This distortion appears as artifacts at the base of the measured spectral line, in the form of negative lobes, and baseline oscillations (Fig. 9), which are defined in the form of a sinc function. This distortion can be reduced, or removed by the use of a process called apodization. In practice, a weighting function is applied to the interferogram to reduce the impact of truncation by de-emphasizing the contribution of the individual frequencies at the truncation limits of the data. This weighting or apodization function can take many forms. In the simplest case—known as triangular apodization—the function takes the form of a linear ramp with full contribution at the centerburst (ZPD) extending to zero contribution at the truncation limit, as illustrated in Fig. 9. This change in information distribution in the interferogram has impact on the final recorded linewidth: it can reduce the truncation artifacts to acceptable limits, but with a trade-off in line width, the line becomes broader, and there is a corresponding decrease in line intensity. An inspection of the information distribution across the interferogram reveals that the contribution of noise relative to information content, increases at the extremes of the interferogram. Therefore, another side-effect of apodization is a reduction in the noise content, which is also related to the de-emphasizing of the data at the extremes.

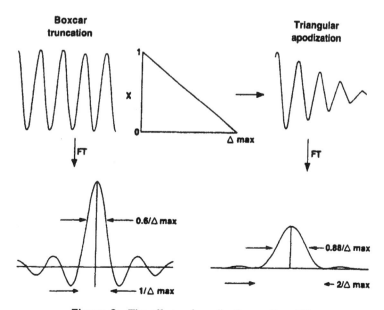

Figure 9 The effects of apodization on linewidth.

Triangular apodization is one of the most common implementations, in part because of its simplicity. However, the selection of a linear apodization function is arbitrary, and does not take into account the actual distribution of information. Alternative numerical functions are used to optimize the apodization as a function of line width and line shape vs. artifact level, and these include bell-shaped curves, such as a raised cosine (examples being Happ-Genzel and Hanning apodizations), and other nonlinear convolutions, such as the Beer–Norton series. Most IR instrument manufacturers provide a choice which typically includes: no apodization (normal truncation—originally known as boxcar), triangular apodization, and one or more of the nonlinear functions. Typically, the default is set to triangular or one of the nonlinear functions. Note, however, that apodization only really needs to be applied if the natural spectral line width is less than the instrument's line width (resolution). If the measured spectral lines are always wider, then apodization may not be needed, although some benefits can be gained in terms of noise reduction, especially if the line broadening effect of the apodization has no significant impact on the measured spectrum. Note that if numerically comparing two spectra, such as in absorbance subtraction, it is essential that the same apodization is used for both spectra.

While on the subject of factors that influence the quality of the final spectral data, it is important to revisit the issue of phase correction. The signal produced from the interferometer is a phase-dependent phenomenon. Key optical components, such as the beamsplitter, and electronic components, including the detector, can introduce phase shifts and phase-related errors. In the case of the detector, internal time constants can impose a lag in the output signal relative to the measured event. As noted earlier, these phase-related phenomena are addressed or compensated by numerical phase correction. Several schemes have been proposed for this correction, the most popular being the *Mertz* correction. Failure to apply adequate phase correction will lead to photometric errors in the final spectrum. The Mertz method has some limitations, and alternative methods, such as the Forman (Chase, 1982) convolution approach, have been suggested as better solutions. However, this latter method is computationally intensive, and is not implemented for routine applications.

STEP SCAN. In continuously scanned (most) FT-IRs, the mirror of one arm of the interferometer moves with a linear (constant) velocity, and this feature confers a time dependence on the interferogram. The moving mirror modulates the intensity of the beam with a modulation frequency dependent on the mirror velocity and the IR wavelength. If time-dependent processes are studied then the time dependencies of the interferogram and the process mix and the scans are no longer useful unless the frequencies involved in the process are well outside the range of the lowest and highest modulation (Fourier) frequencies of the IR beam. Step scan is a method of eliminating the time dependence of the scan by constructing the interferogram from a series of measurements at discrete mirror positions as the mirror is stepped, rather than moved continuously, through the range from zero to maximum OPD. After each step the measurement is made either while the mirror is static or while the mirror is "dithered" sinusoidally about its set position.

A static mirror measurement is useful for time-resolved experiments (see later section), particularly for reversible or repeatable processes. A typical measurement sequence is as follows. The mirror is moved to the required position, and allowed to stabilize. A trigger pulse initiates the time-dependent process under study, and the detector output is recorded. The mirror is stepped to the next position, and the process is repeated. At the end of the experiment, after all positions of the OPD have been measured, the data from each mirror position are apportioned to an interferogram according to the time separation from the trigger. Inverse FT produces a series of spectra for each time interval.

Dithering the mirror, phase modulation (PM) of the beam is achieved on some step-scan spectrometers by applying an AC voltage to piezoelectric transducers actuating on the fixed mirror. Two important parameters under user control are the modulation frequency and the magnitude of the mirror displacement (the amplitude of the modulation). The depth of the phase modulation depends on the mirror displacement and IR wavelength. Typical PM frequencies range from 5 Hz to 1 kHz and amplitudes from 0.5 to 3.5 wavelengths of HeNe laser line (300–900 nm). PM offers several advantages, such as noise reduction through synchronous signal detection, and is particularly advantageous in photoacoustic spectroscopy (PAS) where the depth of penetration is controlled by the PM frequency, and remains constant across the whole spectrum, unlike the case in rapid-scan PAS.

PM step-scan has been applied to polymer stretching experiments in which polymers are subjected to repeated cycles of stretching and contracting to study molecular reorientation rates. Another important area of application is the study of thin films with PM-IR reflection absorption spectroscopy. This is a double modulation experiment where an IR beam near grazing angle is rapidly switched between p and s polarization by a ZnSe photoelastic modulator (PEM) inserted between the interferometer and the sample. The PEM superimposes a second modulation frequency on the beam but only for those wavelengths that are absorbed by sample. Simultaneous detection of the differential signal and "DC" background signal and subsequent ratioing of these signals to obtain the spectrum dramatically improves sensitivity by reducing artifacts due to water absorption and instrumental drift.

Until recently a step-scan FTIR bench was also a necessary component of IR imaging systems. Because there is a finite readout time involved in the use of detector arrays, it was not possible to read the whole array before the mirror had moved beyond the next zero-crossing position. With step scan, a reading from all pixel positions could be obtained from each OPD position before the mirror was stepped to the next position. The latest generation of detector arrays are fast enough such that a step-scan bench is no longer a requirement for IR imaging systems.

Practical Interferogram Acquisition Issues

As noted earlier, many instruments operate by the collection of the full double-sided interferogram. This will depend on the design of a specific instrument, and it may depend on the resolution required for the final spectrum. At high resolution, a large number of data points is required, and dependent on the data handling capacity of the instrument, it may only be practical to acquire data from one side of the interferogram. In such cases, the full centerburst region is recorded, plus some data from the second part of the interferogram. This additional information is required for the phase correction calculations (typically the Mertz correction). In the actual data collection mode, the data can be acquired in both directions, as well as from both sides of the interferogram. This is termed bidirectional data acquisition, and instruments that provide both functions are said to provide double-sided, bidirectional data acquisition.

There are pros and cons to be considered, relative to the method of data acquisition. Instruments that scan in one direction only have a fast-flyback mode where the scanning element, usually the moving mirror, returns to the start of scan position, at the end of the scan, in a fast, nondata acquisition sequence. This, time, albeit short, can be perceived as lost time. Conversely, instruments that are bi-directional have a turn-around phase, where the scanning device has to be decelerated to a stand-still, and then accelerated back up to the nominal scan speed in the opposite direction. This turn-around time can also be perceived as lost time. In the end, it depends on the type of experiment, and its performance requirements, the scan speed, the time to stabilize the scanning device at turn-around. High-performance instruments often provide the user with the flexibility to select the preferred acquisition sequence.

One final issue in the actual mode of data acquisition is how the data is to be used. In most FTIR measurements it is normal to co-add scans (spectra), to produce a final signal-averaged spectrum with an improved signal-to-noise ratio (SNR). The debate can be where to perform the co-addition—with the raw data (interferograms) or the transformed spectra. Now consider the option of double-sided data acquisition. Asymmetry of the interferogram dictates that the collected data must be co-added in different registers (storage areas). Similar considerations will apply for bi-directional acquisition, where subtle differences may exist between the forward and reverse scans. In such systems, the incompatibilities between the different data sets are accommodated by signal averaging into different registers. The collected interferograms are individually transformed, and then the final spectra may be co-added to produce a spectrum from the full set of scan cycles. Therefore, in an instrument that provides double-sided, bi-directional acquisition, a single scan cycle may actually be made up from four individual interferogram segments.

The layout of a traditional FTIR instrument is shown in Fig. 10. Relating to this figure, the modulator (interferometer), which is composed of the IR beamsplitter, the fixed mirror, and the moving mirror, has been covered. A further important item, which is integral in the operation of the modulator, is the helium-neon (HeNe) reference laser. Most commercial FTIR instruments incorporate a HeNe, which serves a critical function in the data acquisition, providing a very accurate trigger or clocking signal for the digitization of the interferogram. The critical issue in the acquisition is the accurate location of all acquired data points, in terms of time or distance from ZPD. A high level of signal repeatability is necessary to gain full advantage of noise reduction by signal averaging. In a perfect spectrometer the random noise in the spectrum is reduced by the square root of the number of co-added scans. The perfect situation only occurs for highly reproducible spectral data, where there is exact registration between interferograms—for most signal averaging, it is normal to co-add interferograms, but co-addition of transformed transmission spectra can yield a comparable result.

The collection of the interferometric data is event-based, where the acquisition of a data point is triggered by a laser-based interferometer that is synchronized with the main IR interferometer. In the system illustrated (Fig. 10), the laser follows the same effective optical path as the main IR beam. At the beamsplitter the laser light interacts with a small region that has an alternative coating to provide a visible interferometer at the laser wavelength. The monochromaticity of the laser line (632.8, 15,804 cm^{-1}) results in the production of a single sinusoidal signature directly linked to the wavelength of the laser line and the position of the moving mirror. The zero-crossings, the point where the sinusoidal wave changes phase (passes through zero), are used as the

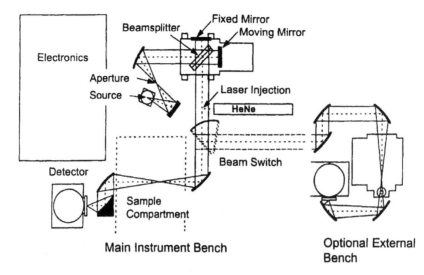

Figure 10 An example layout of a commercial FTIR spectrometer. (Courtesy of Nicolet Instrument Corporation.)

trigger point to access the data from the IR interferogram, which is intimately coupled to the laser signature. The points are not only collected at well-defined and highly reproducible intervals (defined by the laser wavelength), but also their position from the centerburst can be accurately assigned.

In order to fix a data point accurately, it is necessary to know exactly when data acquisition started, and when a data point is acquired. In early instruments, and in some current instruments, a secondary white light interferometer, coupled to the main IR interferometer, is used as a trigger point to initiate data acquisition from the IR interferogram. For early instruments, the "white" light was provided by an independent tungsten lamp. In modern implementations, this is often the visible component from a small portion of the source radiation, extracted from the main IR beam via a small pick-off mirror. However, in many modern instruments, white light is not used to trigger data acquisition, instead, some system simply acquires all of the interferogram, locates accurately the position of the centerburst (by a correlation method), and uses this as an equivalent of a trigger point to signify the start of data acquisition. The only negative of this approach is that when analyzing samples that only transmit a low level of light, it is difficult to accurately locate and correlate the position of the centerburst. In an extreme case, where no light is transmitted, the interferometer servo drive can lose itself and go out-of-control, unless an override is implemented. An alternative approach, often used on more expensive instruments, is *fringe counting*, where the absolute position of individual reference laser fringes is located. This approach uses a measurement principle known as *laser quadrature*, which is based on the polarization of the laser light, to provide the mechanism for counting the fringes as they change phase.

The number of points acquired is linked to the length of the interferogram and spectral range. As noted above, it is normal to collect the IR data at the zero-crossings of the reference laser interferogram. The frequency of data collection is determined by the Nyquist criterion—that is, at minimum, two points are required to represent and/or reconstruct an individual cosine wave. For any spectrum, fixed interferogram points may be taken at every zero-crossing, which provides a certain size of a digital data array. However, it is possible to take a smaller number of data points, at say every alternate zero-crossing or every third zero-crossing, thereby reducing the size of the final data array. There is a trade-off, in terms of the final spectral range that can be represented, relative to the highest wavenumber (smallest wavelength) that can be represented, again determined by the Nyquist criterion. The common zero-crossing selections and their corresponding upper wavenumber limits are presented in Table 2.

For low-cost instruments that are only intended for operation in the mid-IR region (below $4000\,cm^{-1}$), the option to use every third zero-crossing reduces the memory requirements from processing, and speeds up

Table 2 Maximum Wavenumber Limit as a Function of the Number of Zero Crossings

Zero crossings	Maximum wavenumber limit (cm^{-1})
1	15,804
2	7,902
3	5,268

the processing of the interferogram. If this option is taken, it is necessary to apply optical and/or numerical filtering to remove higher frequencies that can be folded back into the spectrum array. Note that every third zero-crossing (or every second crossing) yields data point frequencies that also uniquely represent higher frequencies (wavelengths). This folding of data, which would be observed as spurious peaks or as an attenuation of spectral features within the measurement range, is called *aliasing*.

For the final spectrum, the number of data points is normally a function of the spectral resolution, where the Nyquist criterion of two points per half bandwidth is upheld. For many instruments, the data increment between points is a noninteger value, directly related to the laser frequency. There are some exceptions, with certain commercial instruments, where the raw data are interpolated to provide integer data increments, in such cases, more points may exist in the spectrum than dictated by the Nyquist criterion.

One comment regarding spectral resolution: as noted in the physical description of the interferometer, the highest resolution available from an interferometer is defined by the maximum OPD achievable. One issue not discussed was the use of apertures to ensure that the desired resolution is achieved over the entire measurement range. In a discussion of throughput, it is implied that all of the source radiation is available to be transferred to the sample and onto the detector. In reality this is not necessarily the case. To achieve maximum resolution, it is necessary to maintain optimal light transfer through the interferometer. This requires a collimated beam, with minimal divergence, and in practice this does not occur at full aperture through the beam splitter. To minimize the impact, the beam size is reduced by the use of an aperture. This aperture is often referred to as the J-stop. For most instruments that have adjustable J-stops, located at an intermediate focal point either before or after the interferometer, optimal apertures are defined for a given spectral resolution. Most high-performance instruments have this feature, and many provide the option for automatic selection when the resolution is defined by the operator. If attention is not paid to the system apertures, then wavelength shifts and variable resolution may be experienced across the measurement range. Note also that sampling methods that reduce the sampling area and become the effective aperture will also exhibit these undesirable effects.

Mechanical and optical stability is of prime importance in the design of an interferometer. The encoded spectral data, within the interferogram, has to be recorded with a precision comparable or better than the wavelengths of light being measured. This requires high precision drive mechanisms, especially for high resolution systems that feature larger retardations. In systems that incorporate moving mirror optics, the mounting of the fixed mirror (and its flatness), the flatness, uniformity and parallelism of the beamsplitter, and its mounting are of equal importance to the precision of motion of the moving mirror. As the moving mirror scans, there is no tolerance for beam disturbances caused by mirror wobble or tilt. In some commercial instruments, corner cube optics are incorporated in place of planar mirrors to help combat scan errors caused by irregularities in the drive system. High-pressure gas bearings have been used in the past, for high-performance instruments, to produce a smoother running and frictionless drive system. In recent years, however, the trend has been towards stabilized or compensating mechanical drive systems, and the use of corrective optics, such as full or partial corner cubes.

Other factors that impact operation in a particular operating environment are temperature and vibration. Temperature changes within the interferometer, unless compensated in the design, will lead to dimensional changes, resulting in a loss of optical alignment and an associated loss in light throughput. This is often observed in the final spectral data as baseline instability, especially at the high frequency (short wavelength) end of the spectrum. Thermal problems can result from aspects of the internal design, such as improper placement of the source relative to the interferometer, or externally, from the operating environment.

Vibration is another undesirable factor, which again may originate from internal sources, such as the drive system, cooling fans, transformers, or from external sources. Internal sources of vibration are often linked to the electrical line frequency, of 50 or 60 Hz and higher harmonics. These will be observed as line spikes, at the equivalent frequencies, in the final transformed spectrum. External vibration sources can be more difficult to determine, especially if they are intermittent, or a mixture of very low frequencies. Extreme vibrations can result in uncorrelatable scans, which ideally are rejected in the interferogram co-addition process. Instrument operation outside of a normal laboratory can lead to exposure to both variations (and/or extremes) of temperature, and mechanical vibrations from heavy equipment, reciprocating engines, pumps, and so on. If this situation is anticipated, then it is possible to compensate for both temperature and vibration in the design and mounting of the interferometer within the instrument.

The interferometer is a single-channel device, and consequently operates in a single-beam mode. In order to produce the spectrum of a sample, with the instrument response function, or background characteristic removed, it is necessary to collect the transformed sample and background spectra independently, and to calculate the ratio of the two. The instrument background is the result of a convolution of many parameters: the source emission characteristics, the beamsplitter design construction and materials, the detector characteristics, and the signal processing are some of the most important. The background is acquired with either an open beam or a sampling accessory in place (if the accessory contributes to the recorded background), and typically at a point in time close to the acquisition of the sample spectrum—an example background is provided in Fig. 11.

If the acquisitions are separated by a long period of time, this may result in some artifacts in the final spectrum. Thermal drift changes, which show up as changes in baseline position and slope, and atmospheric water vapor and carbon dioxide are the most common. Atmospheric water vapor typically shows up as a series of narrow bands above 3000 cm^{-1} and between 1650 and 1400 cm^{-1}, carbon dioxide occurs as a broader feature centered around 2250 cm^{-1} in the final spectrum as well as a small spike at 668 cm^{-1}, as seen in the single-beam spectrum (Fig. 11). The direction of the bands can be positive or negative in the ratioed spectrum, dependent on the relative concentrations, represented individually, in the spectra. One method to reduce the spectral interference from these two components is to purge the instrument optics—interferometer and sampling area—with dry air or nitrogen. Care must be taken in the design of the purge system not to generate turbulence or thermal gradients in the light path, especially in front of the source or inside the interferometer cavity. Either can result in the production of extraneous noise and optical instability, in particular if high flow-rates are used. An alternative approach is to seal and desiccate the spectrometer optics, leaving only a small area exposed to the atmosphere. This is a common feature on lower cost instruments where the need to purge is considered undesirable.

Dispersive or Monochromator-Based Instruments

The use of dispersive instruments in the mid-IR essentially ceased more than 10 years ago because of the benefits of the interferometric approach which gave FTIR a much higher performance. Once FTIR instruments became available they rapidly replaced dispersive instruments except for certain applications where photometric accuracy is important. For many years, the optics industry has made critical measurements close to zero transmission on optical materials by IR spectroscopy. Straylight performance is obviously an issue, but generally the more advanced, ratio recording dispersive instruments were able to provide the necessary performance. Attempts to reproduce the performance on equivalent FTIR instruments were generally unsuccessful, primarily due to double modulation effects, defined earlier. However, the application of a half (semicircular) aperture, also described earlier, provides an effective block to back reflected radiation. This modification has enabled FTIR instruments to be applied to critical optical measurements.

Filter Instruments

Filter IR instruments are very popular as IR analyzers, especially for the determination of the carbon dioxide and of unburned hydrocarbons (C–H) in combustion gases. In this case, a narrow bandpass filter is used with a wavelength that corresponds to absorption of the particular gas. The instrument, strictly a photometer or a filtometer (Wilks, 1992), is very simple in construction, where the illustration used in Fig. 2 is simplified by replacing the "light analyzer" component by an optical filter. For the best measurements, it is normal to acquire data equivalent to the analyte peak (filter 1) and a clear portion of baseline (filter 2). Various designs have been proposed to generate these two measurement wavelengths in a simple filter instrument, from the relatively complex, two filter wheel design, shown in Fig. 12, to a simple two-channel system, where both channels illuminate a single detector, and a physical chopping device (a rotating sector wheel) selects the two beams independently.

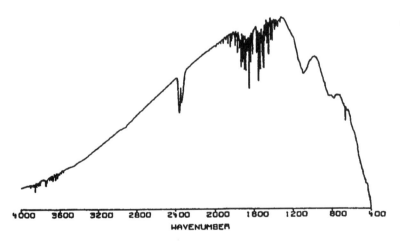

Figure 11 An example of an FTIR instrument background.

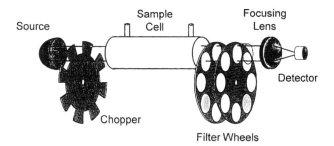

Figure 12 A schematic of a two-filter wheel, multiwavelength filter photometer.

An interesting variant of this normal configuration, which greatly simplifies the instrument design, is the use of two or more detectors, with bandpass filters located just in front of the detector, or bonded to the detector casing as a window material. Figure 13 shows such a configuration, with an integrated system—source, sample cell, filters, and detectors—intended for gas analysis. One innovative commercial design features an integrated filter-detector module with two filters mounted as windows on a single detector canister (two detector elements inside), providing a peak and baseline measurement from a single physical assembly. The concept of dedicating a filter to a detector helps to simplify the design, and to provide a compact construction. In such systems, the light path is so short that interferences from atmospheric carbon dioxide and water vapor are insignificant

The instrument design shown in Fig. 12 is for a multicomponent gas analyzer, with the opportunity to mount up to 16 filters in two-filter wheels (eight per wheel)—providing peak and baseline measurements for up to eight analytes. This particular analyzer can be configured for the continuous measurement of a liquid or gas stream. For certain gas analyses, where a high degree of specificity and/or sensitivity is required, it is possible to utilize the analyte itself as the filter medium. This technique is known as gas filter correlation analysis.

While the use of fixed, discrete filters is the traditional approach to filter instruments, instruments have been designed around a variable filter. It is possible to construct a interference filter that has a variable bandpass across the filter. Such LVFs have been used in scanning instruments, where the filter is moved laterally across a slit aperture. The benefit here is that the design can be simple and compact, and this approach has been used for portable instrumentation. Although these scanning filter instruments typically provide low spectral resolution (16 cm^{-1}, or lower), the performance has been shown to be adequate for portable gas analyzers (Foxboro Miran instruments). With the availability of suitable detector arrays, it becomes unnecessary to mechanically scan the LVF.

Other IR Measurement Technologies

HADAMARD TRANSFORM DEVICES. For many years, Hadamard transform devices were proposed as alternative approaches for collecting spectral data from a dispersive instrument, without one of the major limitations imposed by traditional scanning instruments—that is, the need to detect one wavelength at a time (Decker, 1977; Marshall and Comisarow, 1975). It will be appreciated from earlier discussions that in a scanning dispersive instrument, the dispersed light image passes over a single mechanical slit, where "individual" wavelengths sequentially pass through the slit and onto the detector. This means that for a given wavelength the measurement is made only once and the accuracy of the measured signal is impacted by the presence of noise, which often originates from the detector in the mid-IR. This limitation is countered in the FTIR measurement by both the throughput and multiplex advantages, and the use of signal averaging. The latter providing the $\sqrt{n}$ improvement in noise content as a function of n co-added scans.

It was proposed that an alternative approach, using the Hadamard transform instead of the Fourier transform, could impart the $\sqrt{n}$ advantages for a dispersive-based instrument. In a Fourier transform instrument, the IR frequencies are encoded into a series of unique cosine

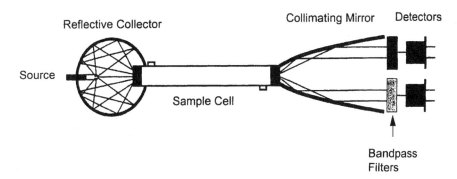

Figure 13 Schematic of a simple filter-based gas analyzer. (Courtesy of Janos Technology Inc.)

waves, whereas on a Hadamard transform device, the information is encoded by a series of square wave functions. The principle is to use a special mask that generates a multislit image from the dispersion device on the detector (Decker, 1977; Marshall and Comisarow, 1975). The mask is cut to provide a unique combination of open and closed aperture situations that can be represented by a binary sequence (such as 100101101...), where 1 is open and 0 is closed. A series of such, uniquely different masks are generated where the total number in the series is equal to the number of resolution elements to be measured. Also, each unique mask must have more than half the number of resolution elements as "slit" openings. This approach generates as many unique equations as the number of resolution elements (and unique mask combinations), which in turn can be solved and can be used to generate spectral information via the transform.

The problems with the method are obviously how to generate and use such a large number of digital masks. Various approaches have been suggested, including a combination of sliding masks (Marshall and Comisarow, 1975). The only practical implementations in recent years have been in the visible and NIR, and in these spectral regions, the use of programmable LCD arrays, for electronically generating the digital masks, have been proposed. In general the technique is little used, most likely because little performance gain is achieved in the NIR (not detector noise limited) and because detector arrays probably offer a more practical solution.

LASER-BASED SYSTEMS. To date, most laser-based systems are limited to the measurement of specific analytes using solid-state or gas laser line sources. One specific use is for open path environmental measurements, where a laser with a frequency (or a tunable frequency range) that coincides with one or more of the fine structured lines from the vibrational–rotational envelope of a particular gas species is selected. In most cases, the measurement may be made on an overtone rather than the fundamental. Examples are the measurement of hydrogen sulfide (1.57–1.59 μm) and hydrogen fluoride (1.312 μm). While traditional tunable laser systems are limited currently to experimental research systems, because of their high cost, systems based on multiple, less expensive NIR diode lasers, have been developed for multivariate NIR analyses. Commercial analyzer systems have been developed on this principle, and have been applied to food and fuel analyses. Also, work continues on tunable diode lasers that operate both in the NIR and the mid-IR, although no commercial instruments have been produced yet.

3. Detectors

There are two basic types of detector used for IR instrumentation: thermal detectors and quantum or photon-sensitive detectors (Lerner, 1996). Thermal detectors measure the total energy absorbed from the IR beam by the corresponding change in temperature. Quantum detectors rely on the interaction of individual photons with electrons in the detector substrate, causing the electrons to become excited into a high-energy state.

Thermal Detectors

Thermal detectors tend to be slow, being influenced by the thermal characteristics (heat capacity, thermal transfer rate, etc.) of the detector material. Photon detectors are much faster being constrained only by the speed of movement of electrons through the substrate. Thermal detectors are generally inexpensive, they do not require any significant cooling to operate (if any), and they have a flat response over a wide range of wavelengths, that is, their response is essentially wavelength independent.

There are at least three different types of thermal detector: thermocouples, pyroelectric detectors, and bolometers. In the past, there was also a pneumatic detector, known as the Golay detector, where the thermal expansion of a nonabsorbing gas was used to monitor the small temperature changes. This was a sensitive detector, but it was also delicate, and susceptible to mechanical failure. An interesting variant on the Golay detector is the photoacoustic detector, which operates as a thermal detector, but is also a flexible sample handling accessory. This will be described in more detail later.

As noted, the temperature changes experienced by a thermal detector are very small, often of the order of 10^{-3}–10^{-2} K. In all measurements the IR radiation is modulated, such that the signal is constantly compared to a zero value. This means that the detector sees a large thermal differential rather than a small incremental change, on an already small thermal signal. The thermocouple and Golay detectors were used on early dispersive instruments, the Golay detector being particularly useful for slow-scan FIR instruments.

THERMOCOUPLES. A thermocouple is produced by the formation of a junction between two dissimilar metals. The temperature differential between the two junctions generates a voltage proportional to the amount of heat energy falling on one of the junctions. The operating thermal function is known as the Seebeck coefficient, which can be as high as 100 μV/°C for metal combinations such as bismuth and antimony. The junction is often blackened to improve the response, which is nearly constant over most of the operating wavelength range. At longer wavelengths, typically below 25 μm (400 cm^{-1}) the efficiency falls off. The sensitivity is inversely proportional to the target area of the detector, and so this must be kept to a minimum. A further benefit of keeping the dimensions small is that the thermal mass of the detector element is reduced, which reduces thermal lag, and improves the response. Sensitivity is also improved by thermally isolating the junction, and by housing it in an evacuated enclosure, thereby reducing environmental thermal noise.

Finally, a lens is normally mounted in front of the detector element with approximately 5–6× magnification. This is made from an IR transmitting material, such as KRS-5 (thalium bromide/iodide eutectic) or cesium iodide. In the case of the latter material, it is usual to coat the lens with a thin film of a moisture protecting polymer. The lens is typically bonded to the housing with a low-volatility epoxy-style adhesive. Care has to be taken to make a nonporous seal, and that the bonding material does not out-gas—in both cases, the vacuum within the detector cavity would be compromised.

As noted, the traditional thermocouple is slow in response, and limits the speed performance of an IR instrument. The modulation (chopping) rate of older dispersive instruments was typically not greater than 17 Hz, primarily because of this speed limitation. In order to collect a good, high signal-to-noise spectrum at a nominal 4 cm^{-1} resolution, over the full spectral range (4000–400 cm^{-1}), it was necessary to scan for 10–15 min. In later dispersive instruments, pyroelectric detectors were used. While these are also thermal detectors, they are considerably faster in response than the thermocouple. However, the pyroelectric devices provide adequate performance at moderate scan rates (such as 1 scan/s, 4 cm^{-1} resolution) when used on a modern FTIR instrument.

PYROELECTRIC DETECTORS. Pyroelectric detectors are ferroelectric devices, electrical conductors, or semiconductors, which change electric polarization as a function of temperature: the degree of polarization decreases with increase in temperature. A signal is produced at electrodes which are placed across the surface of the detector material as small polarization changes occur. Several materials are available as pyroelectrics, the most popular for IR instrumentation being deuterated triglycine sulfate (DTGS). Other materials, such as strontium barium niobate, and especially lithium tantalate, which is used commercially in fire detectors, also function as suitable detectors for IR instruments. Lithium tantalate is relatively inexpensive, and is often featured in lower cost instruments. While its detectivity is about an order of magnitude less than DTGS, its lower cost and its thermal characteristics make it still a viable option for routine instruments.

The response of the TGS (DTGS) detector is inversely proportional to the modulation frequency (rate) of the incident beam. The modulation rates encountered in an interferometer scanning at 0.2 cm/s are 1600 Hz at 4000 cm^{-1} and 160 Hz at 400 cm^{-1}. This implies that the detectivity is up to an order of magnitude less at the high frequency end of the spectrum compared with the low frequency end. Another consequence of this is that if the scan rate of the interferometer is slowed down by a factor of 10, a corresponding increase in performance is experienced. Slowing the scan rate down can expose the system to potential scan errors, and pick-up of interferences from power-line frequencies (50/60 Hz, and harmonics). Such interferences can be addressed in an instrument's design, with the electronics (using filtration) and by vibrational isolation of components, such as transformers, that respond to AC line frequency.

One important performance issue which impacts the DTGS detector is that the temperature–electric polarizability phenomenon of the detector occurs below the Curie point. At the Curie point, the detector may depolarize, either temporarily or permanently. In the case of the DTGS detector, this occurs at a relatively low temperature, around 40°C. If the ambient temperature of the detector exceeds this value, the detector will stop functioning, possibly permanently (dependent on design). One method for preventing this situation occurring with the DTGS detector is to provide thermoelectric (Peltier) cooling of the detector element. A study of detector performance vs. operating temperature indicates that for the DTGS detector an optimum range exists between 20°C and 30°C. If the detector element temperature is stabilized within this temperature range, then the detector will operate with optimum sensitivity, and without the risk of depolarization. It should be noted that the Curie point of lithium tantalate is significantly higher, and therefore, this detector is well suited to high temperature applications.

Photon-Sensitive or Quantum Detectors

There is a relatively broad range of detectors that can be characterized as IR quantum detectors. Much of the technology for these detectors has originated from government research for the aerospace and military defence programs, where high sensitivity and fast response are important criteria. Four types of IR detector, based on semiconductor materials, are in use for these applications: intrinsic and extrinsic semiconductors, photoemissive detectors, and quantum well detectors (Lerner, 1996).

The underlying principle of semiconductor photodetectors is that the absorbed photon raises a valence electron across an energy gap to the conduction band of the semiconductor. As the energy of the photon gets lower (longer wavelengths), the bandgap also must become narrower in order to detect the photon. As a result, the bandgap of the detector material defines the lower wavenumber (or longest wavelength) limit, or cut-off, of the system. Silicon used in normal photodiodes has its cut-off around 1100 nm (9090 cm^{-1}) but there are several other intrinsic semiconductors that cover the longer wavelengths. These include lead sulfide (PbS) and lead selenide (PbSe), indium antiminide (InSb) and indium arsenide (InAs), as well as the "ternary compounds or alloys", such as indium gallium arsenide (InGaAs) and mercury cadmium telluride (HgCdTe) (Wolfe and Zissis, 1985). The optimum frequency response of many of these detector materials ranges from 10^2 to 10^4 Hz, which is a good match to the normal modulation rates encountered in the FTIR instruments. Of these materials, the HgCdTe (MCT) offers the best coverage of the mid-IR range—typically 10,000–750 cm^{-1}, and is the most widely used. InSb detectors (10,000–1850 cm^{-1}) cover the high frequency region at higher detectivity than the MCT, and often find application in gas phase spectroscopy and other specialist areas of research where the limited range is not a disadvantage.

Conventional single point MCT FTIR detectors are photoconductive devices, comprising a slab of MCT semiconductor material between two contacts. In a typical configuration a small current is passed between contacts. When light of sufficient energy is absorbed, a hole and a free electron are generated, which increases the current in the device. The change in current is AC-coupled to an amplifier, and subsequently measured. It is also possible to suitably dope MCT to create a photovoltaic device in which the photo-generated electron hole pair is separated across the internal potential barrier of the p–n junction. The choice between photovoltaic vs. photoconductive MCT detectors has several important consequences for the spectroscopist. The smaller junction capacitance of photovoltaic MCTs leads to a faster time response, an important consideration in time-resolved FTIR. However, this comes at the price of increased noise and a decreased range at the low wavenumber end.

MCT detectors are normally operated at liquid nitrogen temperatures, and for most laboratory applications, cryogenic cooling is necessary. However, with an increased interest in field-portable instrumentation, and in instruments for nonlaboratory applications, it is desirable to consider alternative methods of cooling. Commercial detectors featuring triple stage thermoelectric cooling are available. While these are convenient to use, they provide a limited cooling range, and as a consequence, both the measurement range (typically down to 1300 cm^{-1}) and the detector response are sacrificed. Mechanical or pneumatic cooling devices are also used, such as Sterling cycle coolers, Joule–Thompson compressed gas coolers, and closed-cycle refrigerators. Sterling cycle coolers have been used by the military for field-portable IR sensing systems; however, they are expensive and they have a limited operating lifetime (often 5000–10,000 h or less), which makes them less desirable for instrumentation applications. A new generation of closed-cycle refrigerators is becoming available, and these look attractive for nonportable applications.

MCT is defined as a ternary compound or alloy, that is, typically used in a nonstoichiometric combination (Hg$_{1-x}$Cd$_x$Te). Varying the mercury:cadmium ratios influences the bandwidth, which, as noted earlier, impacts the operating spectral range (Wolfe and Zissis, 1985 p. 11.86) as indicated below:

$x = 0.21$, lower limit $= \sim 830$ cm^{-1} (12 μm)

$x = 0.02$, lower limit $= \sim 710$ cm^{-1} (14 μm)

$x = 0.17$, lower limit $= \sim 600$ cm^{-1} (17 μm)

Commercially, these detectors are sold as narrow (type A), medium, and broad bandwidth devices with an integrated Dewar for liquid nitrogen. The broad bandwidth devices cover most of the spectral range down to 400 cm^{-1}, but at the sacrifice of sensitivity. While the broad-band detectors have the high-speed response, characteristic of photon detectors, their relative detectivity may only be marginally better than a top-end DTGS detector.

When operating in the mid-IR region with a FTIR instrument, the normal choice for detector selection is between MCT and DTGS. For most routine applications, DTGS is the most common and appropriate choice, for the following reasons:

1. It is less expensive.
2. It has an excellent linear dynamic range (up to three orders of magnitude, or better).
3. It has moderate to high sensitivity (dependent on application and implementation).
4. it does not require cooling (with the exception of possible thermal stabilization).

When used at or around optimal conditions, the MCT detector offers a significant gain in sensitivity and/or speed compared with a DTGS detector. It saturates readily, which means that it is unsuitable at high light levels, but conversely it is ideal at low light level measurements. Typical applications that benefit from this characteristic are highly absorbing samples, microsampling, especially with IR microscopes, and other low-throughput applications, such as on-line GC-IR, where sampling requires a small cross-section light pipe. In this latter application, the speed advantage of this detector is important for capturing the rapidly changing information across the peaks of a highly resolved chromatogram. Scan rates as high as 60–100 scans per second are practical, dependent on the spectral resolution selected. The main constraints to higher speeds are the speed and width of the analog-to-digital (A/D) converter and the data transfer rates, between the instrument and the external computer, and within the computer. For applications that involve a reduced sample image, such as the IR microscope or the GC-IR light pipe, optimum performance can be gained by matching the detector element size to the size of the image. Normal instrument detector elements are between 1 and 2 mm^2. Reducing this dimension down to between 250 and 100 μm provides a better match for microapplications, with a resultant signal-to-noise improvement.

Two important issues with most of the photon detectors are linearity and detector saturation. For most spectrometric applications it is desirable to have linearity out to at least between two and three absorbance units, dependent on application. This is particularly the case in the mid-IR where a wide dynamic range is required, even for qualitative measurements—remembering that the normal spectral presentation is from 0%T to 100%T (%T indicates % transmittance), and that absorbances of 2 and 3 correspond to 1 and 0.1%T, respectively. Dependent on the detector electronics and the preamplifier gain settings, a detector may only be linear to an absorbance of one (or maybe less). Correct preamplifier gain set-up, for the full aperture, unattenuated light levels, and for normal operation is essential. Further linearization of the signal is possible by the detector electronics or by postprocessing of the signal, based on assumptions of the detector response characteristics as a function of frequency and light level. A nonzero transmission value (as high as 2–3%T, or more), in regions of total absorbance, such as a transmission cut-off, is an indicator of nonlinear response by the detector.

Detector saturation is also an undesirable situation, which can lead to nonlinear response, and in extreme circumstances, to severe spectral distortion. This situation can occur when too much light is focused onto the detector (as a function of time), and this can occur in any type of instrument. If the instrument is set-up for reduced aperture or low-energy applications, then it may be necessary to add an additional aperture or a neutral density filter (or mesh) when the system is applied to full aperture applications. Note that detector saturation can also occur when the light modulation frequency of an instrument is reduced—such as scanning the interferometer more slowly.

Detector Arrays

MCT-FOCAL PLANE ARRAYS. The development of mid-IR detector arrays has been driven mainly by the military for missile guidance and surveillance purposes. The detector that showed greatest promise for FTIR applications was the MCT focal plane array (FPA). MCT is not a suitable semiconductor for use in normal electronic circuitry, and so it is not possible to construct "monolithic" MCT IR detector arrays in the style of the silicon charge-coupled device (CCD). This problem was overcome through the development of a hybrid device comprising an array of MCT detectors "bump-bonded" with indium connections (bumps) to a CCD style amplifier and readout chip. Each detector element and its corresponding readout device denote a pixel.

First generation FTIR FPA devices were military surplus and were found to be unreliable and suffered poor sensitivity and limited wavelength range. For missile use, the detector only requires cooling once and was not designed to withstand the cycling to liq. N$_2$ temperatures that daily use in an FT-IR entailed. As a result, a common failure mode was delamination of the two layers of the hybrid chip. Pixel dropouts and variations in pixel sensitivity across the detector were other adverse features of the early FPA detectors. Signal readout was slow, necessitating the use of step-scan bench—further increasing the cost and complexity of use. With further development, these problems were largely overcome and now so-called second generation "built-for-purpose" FTIR FPAs have been released. Current MCT-FPAs are mechanically stable and the readout time is sufficient that stepscan benches are no longer a requirement, at least for the smaller arrays.

In FPAs the photodetector elements have been implemented in a photovoltaic configuration because photoconductive detectors are unsuited to CCD type readout and its relatively high steady-state current results in too great a thermal load for ready heat dissipation across the array. An important implication for the spectroscopist is that the low wavenumber range is limited to ~900 cm^{-1} in photovoltaic MCTs as opposed to 750 cm^{-1} or less for photoconductive MCTs.

An IR microscope may, depending on the available funds, be equipped with a 128 × 128 element or larger FPA. When used with a 15× microscope objective this will cover ~700 μm^2 with an effective pixel separation of 5.5 μm, which is closely matched to the diffraction-limited spatial resolution in the mid-IR.

MCT LINEAR ARRAYS. A slightly different approach has led to the development of a small linear MCT array for use with FTIR microscopy. Currently available is a 16 × 1 photoconductive array with a larger single MCT detector on the same chip. By restricting the array to a single line of detectors, connections can be made conventionally, from the side, rather than from the rear as in the FPA. This feature neatly circumvents the problems inherent in the hybrid construction of the latter. Conventional connectivity also means that the MCT can be used in the photoconductive configuration, resulting in a lower wavenumber limit than is currently achieved with the photovoltaic FPA. The disadvantage is, of course, that coverage is only 16 pixels at a time rather than, say, 16,384 pixels at a time for a 128 × 128 FPA. On the other hand, the 16×1 linear array is a fraction of the price of a 128 × 128 array.

4. Electronics and Primary Signal Handling

Several different measurement technologies have been highlighted, from the sequential wavelength-based scanning of a monochromator-based instrument, to the near-instantaneous and highly encoded output of an interferometer. In all cases, it is necessary to capture the analog signal generated by the detector, to preamplify the signal with closely coupled electronics (sometimes integrated into the detector circuitary), and to convert the signal into a readable format for the eventual photometric or spectral measurement. For a spectrometer, this signal (the y-axis information) must be synchronized to the scanned wavelengths or frequencies (the x-axis information). In an interferometer, this signal is sampled and encoded by the internal laser frequency counter. In essence, there are three types of signal that are collected:

1. *The comparative signal obtained from a simple photometer, such as a filter or a monochromator instrument (single- or double-beam).* In these instruments, a modulated signal is generated, typically by a mechanical beam "chopping" device, where a zero (dark) signal is measured and interspersed with the photometric signal from the sample. An additional measured signal can be used as a reference, which may be obtained as a baseline point (in a filter instrument) or as an open beam measurement in a dispersive instrument. In the latter case a traditional double-beam spectrum is generated, where the signals at each measurement point (as a function of wavelength) are ratioed against each other as the instrument is scanned.
2. *A full spectral signal generated from an array-based instrument, or a high-speed scanning device such as an AOTF.* In the case of the array-based instrument, a dark current, measured in the absence of light, is necessary for the highest readout accuracy.
3. *An interferometric signal*, which normally exhibits an wide dynamic range, from a maximum signal, at the centerburst, that rapidly decays, to a near zero value at the extremes (the wings) of the measured interferogram.

In all three cases, it is necessary to convert the amplified analog signal from the detector preamplifier into a digital numeric format. This requires an A/D converter on the primary output. The number of bits supported by the A/D converter defines the useable dynamic range of the data, the representation of the noise, and the vertical resolution of the data—ultimately, this defines the quality of the final spectral data. Low-cost detector systems typically feature anything from a 12-bit to a 16-bit A/D converter. A 12-bit converter will provide a digital resolution of 1 part in 4096, of which 1 or 2 bits are required to represent the noise. Such a system will provide a basic signal-to-noise performance of around 1000:1. Some of the low-cost array-based devices may use such an inexpensive A/D converter. It is more common to use a 16 bit device (digital resolution of 1:65,536), which is readily available at low cost. For an interferometric instrument, there are a few other considerations to bear in mind:

1. *The speed of the A/D converter.* Especially at high scan rates (as used with the MCT detector, for GC-IR and high-speed kinetics measurements) where the modulation rates of the optical signal are high. Also, the rate of change of the signal around the centerburst is very high. It is essential that the interferometric signal is recorded accurately. Failure to do this will result in distortions to the final Fourier transformed spectral data.
2. *The dynamic range of the signal.* The interferometric signal is very wide, and it is essential to record all parts of the interferogram accurately, representing the data over the full height of the centerburst signature, as well as the noise that is distributed across entire interferogram. The centerburst contains much of the information that defines the broad-band background features of the final spectrum. Any numerical truncation of the maxima or minima, or irregularity in recording the contours of the signature will result in distortions to the spectral background.
3. *Digitization of the noise* is also an important issue. It is normal to fill the A/D converter with the signal to maximize the definition of the noise. Normally, it is accepted that at least 2 bits are required to represent the noise. Therefore, with a fixed width, the level of noise that can be represented before truncation is defined by the number of bits within the A/D converter (allowing 2 bits for the noise). Consequently, the width of the A/D converter defines the limiting signal-to-noise performance of the instrument.

In low-cost instrumentation 16-bit A/D converters are in common use because they can be obtained with high conversion rates at relatively low price. However, this width is not adequate for high-performance FTIR instruments that are capable of providing high signal-to-noise levels, especially for high sensitivity measurements with low noise DTGS, MCT, or many of the NIR detectors.

With such instruments, the trend is towards 18 or 20 bit A/D converters. These come at much higher costs: typically increasing the bit width and/or speed have a large impact on cost, at one time an additional 1–2 bits increased the cost by up to a factor of 10.

One practical alternative is to use gain ranging, a technique for boosting the gain by multiples of two to emulate a broader bit width in areas of low signal. Note that the gain switching must be applied in exactly the same position in all collected interferograms if, as usual, signal averaging is to be performed. Also, the gain expansion multiplier must be an exact multiple of two. Any imprecision here will result in distortion to the final Fourier transformed spectrum. Some modern instruments use a combination of a wider A/D (such as 18 bits) combined with two (or more) levels of gain ranging to provide 20 effective bits (or greater).

Beyond the A/D converter, the digital signal is passed to an accumulator, normally a series of registers, located either on a dedicated on-board processor (within the instrument) or on an external processor (within a PC workstation). The registers are typically set to be 32, 64, or 128 bits wide, and the numbers are stored and processed as scaled integers. After accumulation and scaling or normalization the numbers are handled as either scaled integers or floating point numbers, dependent on the internal architecture of the data handling and manipulation software, and the numerical function being used.

The final spectral data can be represented in several forms, dependent on the nature of the data acquisition, and the measurement experiment. Obviously, for simple filter-based instruments, the data output is a series of scalar values: typically at least a pair of values (often more), representing a sample measurement data point and a reference or baseline data point. With a spectrometric instrument, which provides a full spectrum output, the final data may be presented in the most basic form, which could be a single-beam spectrum or an interferogram (representing a single-beam spectrum). In such cases, the ordinate axis of the single-beam spectrum is some form of intensity value, one exception being photoacoustic measurements, where the raw signal is actually related directly to sample absorption.

For most practical applications, the single-beam, which is a convolution of the instrument background spectrum and the sample spectrum, is ratioed against a background spectrum, obtained in the absence of a sample. For many measurements, this ratioed spectrum can be presented in either a transmittance format (T or % transmittance, %T) or an absorbance format (A, or $\log 1/T$). With the exception of reflectance and emission experiments, transmittance is the natural format of the data, and absorbance is the form used when correlation with material composition (concentrations) is required, as defined by the Beer–Lambert–Bouguer relationship.

5. The Sample and Sample Handling

The range of applications based on IR spectral measurement are generally broader than any other instrumental method of analysis. As already noted, essentially any type of matrix can be considered as a potential candidate for IR analysis. Originally, the technology was limited in its application, and at one time, IR spectroscopy was branded as an energy limited technique. Newer instrument designs, improved optics and electronics, more sensitive detectors, and significantly better methods of sampling and sample handling have all contributed to the overall gains that IR spectroscopy has made in the past two decades (Porro and Pattacini, 1993a, b).

For the most part, the technology side of the instrumentation is well understood, and the basic hardware provided by the commercial instrument vendors is more or less at the state-of-the-art level, in terms of performance. The critical issue is always the sample, and how it should be presented to the instrument. It is both an issue of the form of the sample matrix and the optical characteristics of the sample–instrument interface. This section deals with critical aspects associated with the sample, and it provides an overview of the most popular methods of sampling, with a focus on the techniques that are generally available, and are serviced by instrument or specialist accessory vendors.

First, it is necessary to evaluate the constraints imposed by the instrumentation itself. Traditionally, the instrument had a very monolithic design, where the sampling point was within a sample compartment. The benefit of the sample compartment approach was that the sample could be handled in a controlled environment, with a stable optical arrangement, and within a purged environment, if required. However, too often, the sample compartment was an afterthought, and it imposed too many constraints on the measurement in terms of the space available, and the size and type of sample to be handled. Today, such constraints may continue to exist, dependent on the design of the basic instrument, but there are usually options for alternative sampling positions, such as, the availability of external beams, both for input (emission ports) and output. Also, with certain instruments and sampling accessories, it is possible to take the sampling point away from the instrument to a remote sampling location. In these cases, the instrument–sample interface may involve close-coupled, dedicated optics, flexible optics (such as optical waveguides or lightpipes) or optical fibers. The final selection here depends on the application, the performance requirements, and the cost of implementation.

Before addressing specific issues of sampling, it is worthwhile to discuss the design issues of the traditional sampling compartment. In general, the need to use a sampling compartment tends to be limited to traditional laboratory use of instrumentation. Outside of the laboratory, where applications tend to be dedicated or at least more specialized, the use of external sampling, with sampling stations or remote probes tends to be the normal choice. In general more attention is paid today on the dimensions and the design of the sampling compartment. For most instruments, the beam is brought to a focus, either in the center or close to the side of the compartment. The argument for side focus has been that larger sampling accessories can be accommodated within a given size of compartment. However, most accessory manufacturers prefer the center-focus configuration; typically the optical design is simpler (less re-imaging required), and for most applications, the design is symmetrical, again making the overall design simpler.

Imaging at a focal point is important. On slit-dispersive instruments the normal image is rectangular. The image size, in this case, at the focus may be a few millimeters (2–4 mm) wide by 8–15 mm high—dependent on the instrument and/or the intended application. Originally, this type of imaging was most popular, and it imposed some geometrical constraints on the design of sampling accessories, including simple light transmission cells. While some instruments may still utilize this geometry, the most common imagery is circular or elliptical. Another feature of a modern IR instrument is that most of the common measurement technologies use a single beam architecture. Original dispersive, mid-IR instruments provided two beams with real-time compensation for the instrument's background (imposed by the optics and electronics). A single-beam design provides more flexibility and space within the sample compartment. Some manufacturers have provided a simulation of the double-beam (dual-beam) operation by the use of a sample shuttle, where the sample is temporarily moved out of the path of the IR beam to enable a background to be recorded within the same timeframe as the sample. Obviously, such accessories are constrained by the space available, which in turn limits their use to the simplest of sampling techniques.

As noted, the location of the focal point, and image dimensions and shape are important design issues. Another important consideration, especially when specialized sampling accessories are concerned, is how the image is formed, and its integrity and homogeneity. First, the choices are focused beam, or collimated or parallel beam. Some instrument designs provide both options, which can be important when using certain accessories. If a focused beam is used, the choice may be between a field stop image or a pupil image—a pupil image being comparable to the image used in a camera or the eye, where changing a limiting aperture does not change the size of the beam image at the sample. For the most part, most instruments are designed to utilize a field stop image, where the field stop is usually the limiting aperture of the instrument—typically a slit or aperture stop, but sometimes the sample or the sampling device.

In the case of an FTIR instrument, where the reference laser image is often placed confocally, in the center of the main IR beam, the impact of this image must be considered. With field stop imaging, the obscuration imposed by the laser optics has minimal impact, however, with a pupil image, where the sample focus is effectively an image of the beam splitter, this does generate an effective hole at some point in the beam. This may produce problems when working with microsampling applications, where only a part of the beam is effective in the measurement, and where the hole generates a dead spot.

Accessories are an important part of day-to-day operation of laboratory instruments. In most cases, the manufacturer has optimized the design, but there is usually a need to fine tune the accessory to the "unique" optics of the instrument. Dependent on the specific design of the sample compartment, this alignment, when made, may not be retained. In some cases, minor adjustments may be required each time the accessory is used. To remove this inconvenience, some instrument (or accessory) vendors provide special accessory mounting plates that attach to fixed locations in the base plate of the sample compartment. This enables accessories to be prealigned and removed from an instrument at anytime. This feature is particularly important for larger accessories, such as the multipass gass cells, or microaccessories, where even minor changes in alignment can result in a significant loss in optical throughput.

An important variant to this approach is the use of a separate external sample compartment, where accessories may be permanently mounted for dedicated measurements. Most of the major IR spectrometer vendors provide external sampling beams, which can be interfaced to a separate sample compartment or sampling station. This enables a user to configure a single instrument to perform several optimized experiments. Usually under software control, a simple solenoid controlled mirror is used to redirect the IR beam. Different approaches have been used, often allowing several external sampling benches to be connected, either in series or as branches from a central optical unit. Some of the more versatile FTIR instruments are designed to allow users to configure the systems with advanced sampling technologies, such as GC-IR, IR microscopy, and even a Raman attachment, all mounted on a single central optical measurement module.

The external bench is only one of several approaches for interfacing a sample outside of the main sample compartment. This approach normally requires a permanent, direct optical coupling between the instrument and the external bench, which obviously imposes limits on how the systems are configured. Also, it requires additional bench-top space, a factor that can be a problem in overcrowded laboratories. In part, this is an issue of the conventional design, which still tends to be based on a monolithic style of instrumentation.

A more flexible approach to external sampling involves the use of optical light guides, these can be in the form of lightpipe-based technology or optical fibers. The lightpipe technology provides a semirigid structure, constructed from relatively large diameter (2–3 cm) pipes that channel the light from the spectrometer to the sampling point, which may be up to several meters from the instrument. The lightpipes, which feature all reflective internal optics, are articulated with special joints to function like limbs, providing some degree of directional flexibility. The radiation is normally transferred from point to point in the form of a collimated beam, and is re-imaged at the connecting joints to reduce beam divergence. Light losses to the walls of the lightpipes from extreme rays and at each reflecting surface (re-imaging) can be significant. But, with an efficient sample–optical interface, and with a sensitive detector, this approach to external sampling is practical for some applications. These include reaction monitoring (laboratory- and plant-based) and remote batch sampling in drums or blending vessels.

An alternative, and sometimes more attractive light guide approach, is to use optical fibers. In this case, the light is transferred between the instrument and the sampling point along an IR transmitting fiber, typically a few hundred microns in diameter. Dependent on the application and the type of instrument used, the fibers are either single filaments or bundles. Generally for the mid-IR fiber bundles are required to obtain sufficient light. For operation into the mid-IR (below 2500 nm, 4000 cm^{-1}) very few materials are available which will transmit the radiation. The commonest materials are polycrystalline silver halides, and chalcogenide glasses, based on a mixture of germanium, selenium, and arsenic. Both of these materials allow a substantial amount of the mid-IR spectrum to be collected, but they are expensive and high attenuation restricts the fiber length to a matter of a few meters, often less. Approximate transmission ranges are 2500–600 cm^{-1} for silver halide, and 4000–900 cm^{-1} for chalcogenide glass. The probe end of the fiber bundle may be a simple diffuse reflectance tip, or it may be interfaced to an attenuated total reflectance (ATR) element such as zinc selenide. Heavy metal fluoride fibers incorporating, for example, zirconium or barium may also be used but have a limited spectral range to only about 2100 cm^{-1}. Fluorozirconates are available with high optical transmission, and have the potential to provide useable spectral performance over several hundred meters.

The optical impact of the sample and/or the sampling accessory on the imagery of the instrument are seldom considered. Both the sample and any sampling accessory act as important optical element within the system, and they can distort, modify, or defocus the image that reaches the detector. This is an important factor to consider in the design of an instrument. A lot of effort may be expended in the development of an instrument with the highest optical throughput in the absence of a sample, but simply adding a sample may be all that is required to disturb optimal throughput, with the result of only mediocre performance. Likewise, when a sampling accessory is designed it is important to minimize optical distortion, and to attempt to maintain the image quality of the overall instrument. Too often, using commercial instrumentation and off-the-shelf accessories, it is necessary to accept compromised performance.

It is important that detector is always uniformly illuminated, and that the image does not move around on the detector from sample to sample. Some detector devices, such as MCTs, may have regions of inhomogeneity, and nonlinearities associated with an image moving around the surface of the detector element may result in serious photometric problems or nonreproducibilities. To minimize these errors, it is important to pay attention to the sampling process as a whole, to make all manual procedures as reproducible as possible, and to pay attention to the manufacturer's recommendations for accessory mounting and alignment. When aligning an accessory, it is always helpful to see the location of the IR beam. For many systems, where there is no visible component in the beam, this can pose a problem. Most FTIR instruments do allow a component of the laser image to enter the sampling area. This may serve as a guide, but it must be realized that this does not necessarily track the center of the beam, or the region of the beam that produces the highest throughput. Typically throughput monitoring programs are available on most commercial instruments that allow the user to monitor the level of optical throughput when aligning an accessory.

Practical Methods of Sample Measurement

The innovations in sampling handling and software, which have been developed over the past decade and a half, have elevated IR spectroscopy to its current level of importance and versatility. Originally, routine IR sampling was limited to optical transmission measurements through materials in cells (gases and liquids) and in the form of films (liquids and solids), and simple reflection experiments. This restricted the scope of useful applications of the technique. Until alternative technologies, such as

Table 3 Coordinates of the IR Sampling Matrix

Index	1 Phase	2 Light measurement	3 Sample size
(1)	Solid	Transmission	Macro-bulk
(2)	Liquid	Reflectance	Macro-trace
(3)	Gas	Absorption	Micro
(4)		Emission	

Note: $1(1) + 2(2) + 3(1)$ = Solid + reflectance + bulk sample; $1(2) + 2(1) + 3(3)$ = liquid + transmission + microsample; and $1(3) + 2(4) + 3(2)$ = gas + emission + macrotrace analysis.

FTIR, and advanced data processing methods became available, the technique was considered useful only for simple qualitative and quantitative measurements on a limited range of materials. Today, the options for sampling can be expanded into a three-dimensional matrix, with the coordinates shown in Table 3.

In essence, any combination of the above sampling coordinates can be assembled to provide a sampling scenario for IR analysis, and for most, commercial accessories exist to make the measurement practical. For convenience, the most popular techniques will be reviewed on the basis of the light measurement. Several specialized methods exist, in particular the IR-hybrid (or hyphenated) techniques, such as IR microscopy and the gas/liquid/solid time-dependent measurements (GC-IR, LC-IR, TG-IR, step-scan, etc.), which warrant separate mention. Each one of these provide unique measurements, and can be considered as a full discussion topic, beyond the scope of this chapter. The reader is directed to "General Text and Reviews" supplied with this chapter. Subject areas are discussed in greater detail in many of the references (Bourne et al., 1984; Chalmers et al., 2002; Compton et al., 1991; Cook, 1991; Culler, 1993; Fabian et al., 2003; Griffiths et al., 1983; Hammiche et al., 1999; Harrick, 1987; Hirschfeld, 1980; Jones and McClelland, 1990; McClelland, 1983; McClelland et al., 1993; Millon and Julian, 1992; Mirabella, 1993; Nafie, 1988; Ohta and Ishida, 1988; Porro and Pattacini, 1993b; Shafer et al., 1988; Suzuki and Gresham, 1986; Vassallo et al., 1992; Wehry and Mamantov, 1979; White, 1991; Wilkins, 1994).

Transmission Measurements

Light transmission was, for many years, the standard approach for mid-IR measurements for liquids, solids, and gases. When measuring liquids, it is necessary to provide a means for retaining the liquids. There are several practical options, dependent on the type of analysis required—qualitative or quantitative—and the nature of the sample. For mobile liquids, cells are available with fixed or variable pathlengths.

Pathlength is a critical issue with liquids, because most compounds in the mid-IR region have relatively high absorptivities, and normal practical pathlengths range from around 10–200 μm. Another issue is the construction of the cell—in terms of IR transmissive window materials and the method of sealing/defining the pathlength. A problem in the IR is finding a material that offers transmission throughout the desired measurement range, that is, robust, sample resistant, and relatively inexpensive. Table 4 provides a list of the most common IR transmissive materials, mostly available as windows. The transmission range quoted is based on an average window thickness of a few millimeters. The alkali metal halides, in particular potassium bromide (KBr) and cesium iodide (CsI), are completely compatible with the desired spectral ranges for the mid-IR. However, there are obvious mechanical and physical difficulties with these materials. Both are hygroscopic (especially CsI), KBr readily cracks along its main crystal planes, and CsI is essentially plastic, readily deforming under slight mechanical pressure. Windows made from materials such as calcium and barium fluoride, and zinc selenide have reasonable mechanical strength, and are essentially water insoluble, but they have limited spectral ranges, and are much more expensive than comparable windows made from the alkali metal salts.

Fixed pathlength cells are available in two forms: permanently sealed cells and demountable cells. Fixed cells are traditionally fabricated with lead or tin foil spacers, with a permanent seal produced by amalgamation with mercury. Such sealing procedures, while being efficient, tend to be unpopular in the 1990s because of environmental and worker safety regulations. Alternative sealing methods include the use of alternative bonding agents, such as metal-filled epoxy resins or ceramic-based adhesives. Also, stainless steel spacers have been used in place of lead. In the case of demountable cells, teflon is used both as a spacer and as a seal.

For simple qualitative measurements, other approaches can be used for low-volatility liquids with medium to high viscosity. First there is the capillary film formed as a sandwich with one or two drops of liquid between two IR windows. If the sample has high viscosity or is a semisolid, it may be sampled as a thin film smear over a single window. A variant of this approach is to evaporate a film of material, from solution. Two recent commercial enhancements to the single window approach is the use of 3M sampling cards (disposable IR Cards), or the Janos Screen cards. The 3M cards feature a thin polymeric film substrate made from polyethylene (PE) or PTFE. In both cases, the card-mounted films are sufficiently thin that the spectral interference from the polymeric material is minimal, and may be removed by spectral subtraction. The Janos product is a cardmounted fine wire mesh

Table 4 Optical and Physical Properties of Common IR Window Materials

Material	Useful range (cm^{-1}, transmission)	Refractive index at 1000 cm^{-1}	Water solubility (g/100 mL, H$_2$O)
Sodium chloride (NaCl)	40,000–590	1.49	35.7
Potassium bromide (KBr)	40,000–340	1.52	65.2
Cesium iodide (CsI)	40,000–200	1.74	88.4
Calcium fluoride (CaF$_2$)	50,000–1,140	1.39	Insoluble
Barium fluoride (BaF$_2$)	50,000–840	1.42	Insoluble
Silver bromide (AgBr)	20,000–300	2.2	Insoluble
Zinc sulfide (ZnS)	17,000–833	2.2	Insoluble
Zinc selenide (ZnSe)	20,000–460	2.4	Insoluble
Cadmium telluride (CdTe)	20,000–320	2.67	Insoluble
AMTIR[a]	11,000–625	2.5	Insoluble
KRS-5[b]	20,000–250	2.37	0.05
Germanium (Ge)	5,500–600	4.0	Insoluble
Silicon (Si)	8,300–660	3.4	Insoluble
Cubic zirconia (ZrO$_2$)	25,000–~1,600	2.15	Insoluble
Diamond (C)	45,000–2,500, 1,650–<200	2.4	Insoluble
Sapphire	55,000–~1,800	1.74	Insoluble

[a]Infrared glass made from germanium, arsenic, and selenium.
[b]Eutectic mixture of thalium iodide/bromide.

screen that retains a stable thin liquid film for medium viscosity, nonvolatile liquids.

The success of sampling solids for transmission measurements is dependent on the sample and the quality of spectral data required. Moldable materials, such as polymer films, may be hot-pressed or cast into thin, self-supporting films, which are ideal for transmission. Other forms of solid require a more rigorous approach. If the material is amorphous, and dissolves in an inert, volatile solvent, then a cast film (on an IR window) may be adopted.

Beyond these simple, direct methods, the traditional procedure is to grind the sample to a submicron particle size, and to disperse it in an IR transmitting medium of comparable refractive index. Usually, the ground sample is mixed with potassium bromide (or similar alkali halide) and is compressed into a pellet—typically requiring 10 tons or greater pressure for a 13 mm diameter pellet. If well prepared, the final pellet should be clear, or at least translucent, with a uniform appearance. A second, less popular approach is to grind the sample into a thick paste with a refined white mineral oil and to present the sample as a thin film of the paste between a pair of windows. This method in the past has been referred to as the "Nujol mull method". In both the pellet method and the mull method it is undesirable for the sample to be fully opaque, producing a high degree of light scattering. If this occurs, a significant amount of light transmission is lost, and spectral distortion can result. In such cases, either the sample preparation procedure may be repeated, or if the sample is soluble in a suitably IR transmissive solvent, a solution approach may be used with a standard liquid cell.

A relatively new approach of handling solids is to use diamond as a sampling aid. There are two types of accessory, the diamond anvil cell and diamond-based internal reflection (ATR) accessories (this latter technique will be discussed later, in "Reflectance Measurements"). In the case of the diamond anvil cell, the solid is compressed between two parallel diamond faces to produce a thin film of the material. The spectrum is measured by transmission through the diamond anvils. Diamond transmits throughout most of mid-IR, with the exception of a region centered around 2000 cm^{-1}, where a single broad absorption occurs. A beam condenser (or a microscope) is normally required to improve throughput, because of the small size of the diamonds, and resultant small aperture. One advantage of this approach is that the high mechanical strength of the diamonds provides a means for sampling hard materials—a high pressure can be generated between the two diamond faces, enabling most materials to be compressed.

GAS CELLS. Gases typically lend themselves to the simplest form of sample handling. For routine sampling, they are retained in a short pathlength cell between

1 and 10 cm in pathlength, dependent on application and concentration range of the analyte. The most basic cells are glass or metal tubes with circular windows mounted or bonded at both ends. For convenience of filling, usually two sampling ports are attached, equipped with stopcock valves. If a trace analysis is to be performed with the analyte concentration in the ppm region, it is necessary to move to an extended pathlength cell. Normally these cells contain additional internal optics that permit the light path to be folded (multiple bounces) within the cell. Such cells are normally configured with effective pathlengths between 2 and 20 m. When working with gases and/or vapors, it is necessary to maintain good control of both temperature and pressure, especially for quantitative measurements.

CELLS FOR LIQUIDS OR SOLIDS. One of the main limitations of the transmission method for most condensed-phase materials is that it is constrained to submillimeter pathlengths by the naturally high levels of IR absorptivity. Also, the light transmission ranges of sampling substrates and window materials impose practical restrictions on certain measurements. Many of these problems can be reduced by switching to the NIR for analyses, especially for quantitative or comparative (quality testing) purposes. Because of the lower absorptivities in the NIR overtone and combination band regions, pathlengths greater than 1 mm (first overtone) to 10 cm (third overtone) can be employed. Transmission sampling can be implemented with fiber-optics, and often with a single-ended transflectance dip probe. In principle, this can make the sampling as simple as using a pH electrode. However, while transmission is obviously easier in the NIR, it is not always a popular method, and is normally limited to clear liquids. Turbidity or suspended particulates can cause a high degree of scattering, which is a more efficient phenomenon in the short wavelength regions. The consequent high-energy losses make transmission often less appealing for routine or process-oriented measurements.

Reflectance Measurements

There are three methods of reflectance that have practical application in IR spectroscopy:

External or specular reflectance
Diffuse reflectance
Internal reflectance

External reflectance is essentially a mirror-style reflection, based on the simple laws of reflection with the angles of incidence and reflection being equal. By definition, this approach requires a smooth polished surface. The sample beam is deflected onto the sample surface by one mirror, and is intercepted, and redirected towards the detector by a second mirror. Although the technique may in principle be used for any type of reflective surface, most analytical applications involve the study of films on metal surfaces. In such cases, the experiment is essentially a double transmission measurement, twice through the coating: once to the metal substrate, and then from the metal with a mirror reflection. This type of measurement is sometimes called a *transflectance* or an absorption–reflection measurement. It is ideal for investigations of both organic and inorganic coatings on metal surfaces, because the coating is examined intact, without removal (Millon and Julian, 1992).

The coating may be anything from a 100 μm thick to submicrometer thicknesses, dependent on the sample and the method of measurement. For thick films, a relatively low angle of incidence is used, typically between near normal and 30° to the normal. This is suitable for film thicknesses of a few microns to 100 microns or more. For thinner films, it is necessary to increase the angle of incidence to what is described as a grazing angle of incidence, typically between 75° and 85° from the normal. Further increases in sensitivity may be gained by the use of a polarizer, set for perpendicular polarization. Simple, inexpensive accessories exist for both low and grazing angles of incidence. Accessories designed for grazing angle measurements normally feature gold-coated mirrors to eliminate interference from surface contaminants, such as the oxide layer that forms on polished aluminum surfaces.

Not all external reflectance measurements feature a transflectance mode, and reflection can occur from the surface (or close to the surface) of the sample itself. Interaction still occurs with the sample, and the resultant mode is dependent on its optical properties. The reflectivity from the surface of a sample is governed by the refractive index, and this property changes rapidly in the region of an absorption band. The resultant spectral data collected from such a reflective surface is typically highly distorted, composed of a mixed mode spectrum, containing contributions from both the absorption spectrum and the refractive index spectrum. These spectra are not directly useful; however, it is possible to perform a data conversion, similar to a Fourier transform, known as the Kramers–Kronig transform (Ohta and Ishida, 1988), to separate the absorption component from the refractive index component. The measured spectrum n' is defined by the complex function:

$$n' = n - ik \qquad (18)$$

where the absorbance function (pseudo absorbance) is obtained as the imaginary term, ik, and the refractive index is represented by the *real* term, n.

Vibrational Spectroscopy

Diffuse reflectance is obtained from a roughened surface, which may be continuous or may be a powder. It is not strictly a surface phenomenon, because it requires absorption to occur by interaction between the sample matrix and the incident IR beam. For the ideal measurement, the beam penetrates the surface, interacts with the sample, up to a depth of 1–3 mm, and eventually exits the surface at any angle, as indicated in Fig. 14. For efficient collection of the scattered radiation it is necessary to use a large collecting mirror above the sample. Most commercial accessories use a pair of off-axis ellipsoid mirrors, typically with a 6:1 magnification ratio. The IR beam is focused on one side of the sample with one mirror, and with a symmetrical arrangement, the other mirror is used to collect the scattered, diffusely reflected radiation. For NIR measurements, an integrating sphere is often used to provide a more effective diffuse reflectance measurement, especially for irregular samples.

Diffuse reflectance can be difficult to optimize in the mid-IR region, and it is often necessary to perform some degree of sample preparation (Culler, 1993; Suzuki and Gresham, 1986). Most applications are with powders, and here it is necessary to obtain a sample with a uniform particle size, if practical. One of the first difficulties is that the diffusely reflected component is not the only reflected beam from most samples. It is common to experience a reasonably high degree of front surface reflection from the sample, which may exist in a couple of forms. In one case, there may be no interaction at all between the light and the sample. When this occurs, the light collected, which is unaltered, acts as a straylight component—equivalent to having a hole in the sample for transmission measurements. In the second situation, the beam interacts with just the surface of the sample, and the equivalent of an external reflectance occurs. The resultant reflected radiation, which typically contains a major refractive index component, produces a distorted mixed mode spectrum, as described earlier.

The mix of reflectance effects, in the measurement of a diffuse reflectance spectrum, is influenced by the nature of the sample, and the way that the sample and the sample surface is prepared. Highly absorbing samples, in particular inorganic compounds, tend to restrict the propagation of the beam through the sample, and as a result, most of the reflections originate from close to the surface. In such cases, the spectra tend to be highly distorted. Conversely, many organic compounds, which are weak or moderate absorbers, often do not present such a problem. In order to minimize front surface reflection-related distortions, and to promote propagation of the beam, up to depths of 2–3 mm through the sample, it is recommended that the sample be ground, and mixed with a nonabsorbing matrix, such as KCl or KBr. As a rule of thumb, dilutions around 10% work well for organics, and dilutions between 1% and 5% work best for inorganics. Once diluted, the sample spectra are generally free from refractive index related distortions.

As described, the diffuse reflectance measurement involves penetration of the IR radiation into the sample. However, the pathlength is not defined, and varies as a function of sample absorptivity. That is, in regions of low absorption the penetration, and hence the effective pathlength, is greater than in regions of high absorption. As a result, the overall spectrum of a material is distorted, and unlike absorbance, the signal produced is a nonlinear function relative to concentration. Various approaches to linearization have been proposed, with one of the most popular being the Kubelka–Munk relationship:

$$f(R_\infty) = \frac{(1 - R_\infty)^2}{2R_\infty} = K_2 C \tag{19}$$

where R_∞ is the sample reflectance spectrum at infinite sample depth ratioed against a nonabsorbing sample matrix (usually KBr or KCl), also at infinite sample depth; K_2 is the proportionality constant, equivalent to the product of absorptivity and pathlength in normal transmission measurements; and C is the concentration of absorbing species.

Compare with Beer's Law for transmission:

$$A = -\log_{10} T = K_1 C \tag{20}$$

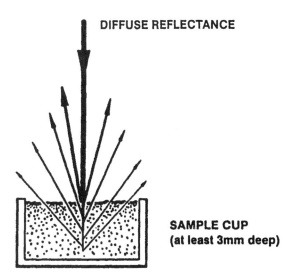

Figure 14 Sample depth of penetration in diffuse reflectance measurements.

The Kubelka–Munk modification is typically applied to undiluted or concentrated mixtures (in KBr) for quantitative measurements, and for the presentation of undistorted spectral data. Note that the $1 - R_\infty$ term inverts the spectrum, which appears in an absorbance-like format.

Internal reflectance or ATR operates in a very different manner compared with the previous two reflectance methods (Harrick, 1987; Mirabella, 1993). It involves the measurement of the internal reflectance from an optical element (or IRE—internal reflectance element) with a high refractive index. Total internal reflection occurs at angles above the critical angle, defined from the ratio of refractive indices of the optical element and its external interface. If a sample is in intimate contact with the surface, and the critical angle criterion is maintained, then interaction will occur between the internally reflected radiation and the sample. The penetration depth is typically no more than a micrometer. The following relationship defines the depth of penetration into the sample in terms of the wavelength of the radiation, the angle of the incident IR beam, and refractive indices of the element and the sample:

$$D_p = \frac{\lambda}{2\pi n_1 \sqrt{\sin^2\theta - n_{21}^2}} \tag{21}$$

where D_p is the depth of penetration, λ is the measurement wavelength, n_1 and n_2 are the refractive indices of the IRE and the sample, n_{21} is the ratio of n_2/n_1; and θ is the angle of incidence.

The interaction with the sample results in the generation of an absorption-like spectrum, whose band intensities increase as a function of increasing wavelength (decreasing wavenumber). In practice, for conventional accessories, it is uncommon that a single reflection is used. Most provide an IRE geometry that permits multiple internal reflection (multiple bounces) to occur. If the sample is in contact with the IRE at the point of each reflection, then the effect is additive, and the effective pathlength is the product of the number of reflections and the depth of penetration.

The original ATR accessories were designed for extended solids, primarily self-supporting film materials and compliant polymers. The sample was clamped to the surface of the IRE, and the combined assembly was located into a set of transfer optics. Modern designs of the accessories take various forms, with the horizontally mounted and the cylindrical versions being the most practical. These not only handle extended solids, but are also used for liquids, pastes, and soft powders. Hard and noncompliant materials are not well suited for ATR—they can cause damage to the IRE, and also they produce poor quality spectra because of nonintimate contact (remembering that a micrometer or less of the sample surface is measured). One exception is a new generation of diamond-based ATR sampling accessories, where samples are brought into intimate contact with the aid of a high-pressure clamping device. In this case, most forms of materials, including hard cured resins, rigid fibers, and even rocks and minerals, may be sampled.

A range of materials is available as IREs, and the final choice is usually dependent on effective spectral range, refractive index, robustness, and price. Early accessories featured a material called KRS-5, at eutectic mixture of thalium iodide/bromide. Although this material provided an excellent spectral range, it was generally undesirable because of material toxicity, partial water solubility, and surface softness (plasticity). Even after only a few uses, an IRE may need repolishing because of low throughput caused by surface scratches. Modern accessories usually use IREs made from zinc selenide, germanium, or AMTIR (a glass based on germanium, selenium, and arsenic). These materials are reasonably hard, and can withstand average usage in a laboratory. The surfaces are attacked, however, by strong acids (except AMTIR) or alkalis. Germanium, which has a relatively high refractive index of 4.0 (compared with 2.4–2.5 for the other materials), is used where a reduced effective pathlength and/or a lesser depth of penetration is required. Materials such as cubic zirconia, sapphire, and diamond are starting to become available for special applications. Diamond is the most ideal of these materials, providing transmission for most of the mid-IR region—the other two materials are limited by a cut-off of around 2000 cm^{-1}. Note that several diamond-based accessories are now available, including the ATR devices mentioned earlier, and diamond-tipped immersion ATR probes for process monitoring applications.

Most ATR measurements are made in the mid-IR spectral region. However, in principle, the ATR effect may be observed in any wavelength region, as long as the critical angle criterion is met. In practice, with traditional accessories, the depth of penetration, and the absorption coefficients are so low that the technique is less useful in the NIR. However, work has been reported on the use of optical fibers with their cladding layers removed. If the fiber is then placed in intimate contact with a material, such as a liquid, then the fiber acts as an ATR element, with an extended contact region, where up to 100 reflections or more may be achieved. This approach has been alternatively named *evanescent wave spectroscopy*.

Photothermal Measurements

True absorption measurements are seldom made in optical spectroscopy. Generally, absorption is measured indirectly from a transmission measurement, and is calculated by logarithmic conversion. Photothermal techniques do, however, provide a direct measurement of absorption. There are several photothermal techniques but the most important by far is PAS (McClelland, 1983; McClelland et al., 1993). Practically, PAS is implemented by an accessory that serves as both a sampling device and as a detector.

Sampling takes place within a chamber, which may be purged for sample conditioning, and sealed for the actual measurement—as illustrated in Fig. 15. The sample, is placed in a small cup, typically a few millimeters in diameter, under an IR transmitting window. It is preferable to replace the atmosphere above the sample with dry nitrogen, argon, or helium. The modulated IR beam from the instrument is focused on the sample, where absorption occurs at the characteristic IR frequencies. Absorbed IR energy is converted into heat energy, and a thermal wave is formed. Dependent on the overall absorptivity of the sample, and its thermal conductivity, the thermal wave diffuses to the surface of the sample. At the surface, the thermal energy is transferred to the gas molecules immediately above the sample. As the result of this transfer of energy, a pressure wave is formed in the gas at the same characteristic modulation frequencies as the absorption bands of the sample. These vibrations, which result in the generation of a high frequency acoustic signal, are picked up with a high sensitivity microphone. This signal, in turn is amplified, and used by the FTIR instrument in exactly the same way as the signal generated by a conventional thermal detector, such as a DTGS detector.

The signal produced can be defined in terms of the intensity of the incident beam, I_0, and the modulation produced by the interferometer, which in turn is defined by the wavenumber and the scan rate. This can be expressed as follows:

$$P_n = \frac{KI_0}{\sqrt{\omega}} \qquad (22)$$

where P_n is the photoacoustic signal normalized against a carbon black reference, K is the proportionality constant, I_0 is the intensity of modulated incident beam, and ω is the modulation frequency (wavenumber × scan velocity).

This is a generalized expression for the photoacoustic signal, where K combines several variables related to the diffusion of the thermal wave and the transfer of energy, that impacts the measurement. These include the absorption and the thermal conductivity of the matrix, and the thermal conductivity of the gas above the sample.

The photoacoustic signal can be enhanced by increasing the magnitude of K. One way to do this is to increase the thermal conductivity of the gas above the sample. Changing from nitrogen to helium will give this effect, with approximately a fourfold increase in sensitivity. It should be appreciated that the photoacoustic effect is inversely proportional to the absorption frequency, and so at lower wavenumbers (longer wavelengths) the signal intensity increases because of the associated decrease in modulation rate. Similarly, the signal intensity increases with a decrease in modulation rate. This can be achieved by reducing the scan speed of the interferometer. The signal is presented as a ratio against the background signal obtained from carbon black, a "universal" IR absorber.

The photoacoustic technique is sometimes termed the method of last resort. In part, this is due to the nature of the measurement, and the fact that there can be some skill required to obtain the best quality results. Two important issues are the impact of environmental sound/vibrations, and the influence of water vapor. Vibrations can generally be reduced or eliminated by careful mounting within the sample compartment, and by ensuring vibrational isolation of the accessory. Problems associated with water vapor can be more troublesome. The incoming radiation interacts readily with any water vapor above the sample prior to interacting with the sample. Because water is a strong IR absorber, its spectrum tends to be dominant

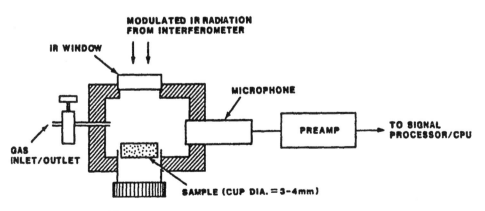

Figure 15 Schematic of an IR photoacoustic detector.

over the spectrum of the sample. The impact of water can be reduced or eliminated by effective purging with dry gas prior to the analysis. Note that solid surfaces, especially powders, can retain surface moisture, which can degas during a measurement. Surface moisture can only be removed by extensive purging.

While PAS may produce an acceptable spectrum of a small sample, such as a piece of fiber, it cannot be used for microscopic measurements. However, recently a new type of photothermal measurement was described (Hammiche et al., 1999), which used a thin 5 μm diameter wire (Pt with 10% Rh) bent to a sharp V-shape with the point placed in contact with the sample. As modulated IR radiation in an FTIR spectrometer was incident on the sample the resistance of the wire was used to measure the temperature and hence the IR absorption. In other words the variation of the resistance of the wire during the scan constituted an interferogram which was transformed to a spectrum of that small part of the sample in contact with the wire. Currently, a spatial resolution of a few μm is achievable by this technique, but the authors believe that resolution in the hundreds of nm is possible in the future. The ultimate aim is to combine the IR method with an atomic force microscope.

Emission Measurements

In principle, IR emission measurements can be made on any sample, where the sample temperature is elevated above the temperature of the detector element. The sample may be placed either in the instrument or it may be located at a remote point. In the latter case, the sample may be a flame or a hot gaseous emission from a smoke stack. For this type of experiment, a telescope optic is used to image the IR emission from the hot sample into the spectrometer, in place of the source. Alternatively, for a laboratory measurement, the sample is placed in a suitable accessory, usually equipped with electrical heating. The accessory can be substituted for the source at the source location or at an alternative source position within the instrument such as the second position in a dual source turret. Another method is to replace the source with a detector and install the emission accessory in the sample compartment of the spectrometer. Most of the current "advanced" instruments offer an external emission sampling port and so perhaps the easiest system is to use an off-axis parabolic mirror to direct the light emitted from the heating accessory into the external port. A flexible accessory with a temperature range from ambient to 1500°C based on a modified graphite atomic absorption furnace has been described in the literature (Vassallo et al., 1992). A commercial accessory (Harrick, 1987) that features a tilting sample stage and a microscopic eyepiece to assist sample alignment is also available.

Normally, the emission profile of a sample is compared with a characteristic emission curve of a black body radiator under the same experimental conditions. For laboratory-based measurements, a blackened block of copper, or a graphite stub may be used in place of the sample to serve as a suitable reference. To obtain the true emission spectrum of a sample, free from background contributions, it is necessary to record the spectrum of the sample, the instrument's emission background (minus sample), and the spectrum of the blackbody radiator. The absolute emissivity (the spectrum) is determined from the following expression (Compton et al., 1991).

$$\varepsilon = \frac{E_s - E_{bg}}{E_{bb} - E_{bg}} \quad (23a)$$

where ε is the absolute emissivity; E_s is the emission from sample; E_{bg} is the background emission—instrument, surroundings, and so on; and E_{bb} is the emission from black body.

Equation (23a) does not account for black body radiation from the detector. Further complications occur if the sample is placed on a support such as a heater, in which case emission from the support and subsequent sample absorption must also be considered. Emittance is not linear with analyte concentration, and so for analytical purposes the effective absorbance should be calculated by taking the logarithm.

$$A = -\log[1 - \varepsilon] \quad (23b)$$

where A is the effective absorbance or linear emittance scale.

Most successful emission measurements are made from thin films of materials deposited or coated on a nonabsorbing (usually metallic) substrate. Materials such as polymer films and laminates will also produce meaningful data. However, problems of reabsorption can be encountered with thick samples, if the surface of the sample is at a lower temperature than the bulk. In these cases, the emission wavelengths naturally coincide with the absorption wavelengths of the sample, and as a result there is an attenuation of the signal, with possible band distortions, or even band inversions (where the net absorption is greater than the emission). Benefits associated with emission experiments are that the measurement is effectively non-contact (the sample may even be remote), and there is normally no atmospheric interference—emission is being measured, not transmission. Emission has great potential for the real-time monitoring of materials at elevated temperature such as thermal oxidation, polymer curing kinetics, catalysis, and process measurements and as an adjunct to other thermal analysis techniques.

One technique, known as TIRS—transient IR spectroscopy, seems particularly suited to process measurements

(Jones and McClelland, 1990). In TIRS, a heated sample stream, for example, a polymer melt stream, flows past the field of view of an FTIR, where a temperature gradient is created in the upper layer by a jet of gas directed at the surface. The depth of the temperature gradient depends on the flow speed of process stream. With sufficient flow the layer under analysis remains thin as it continuously generated but then passes out of the field of view before thermal diffusion to the underlying material occurs. If a jet of hot gas is used, the surface layer is heated and a net emission gain is measured. If a jet of cold gas is used, the cooler surface layer absorbs the emission from the underlying bulk material, which in this case acts as a source, and a transmittance spectrum may be obtained.

Microsampling and Microscopy

Microsampling includes all forms of microspectroscopy, it is IR microscopy, with its combined imaging capabilities, which has gained most importance. Microsampling has been performed with IR spectroscopy for decades, and in itself is not unique. In principle, it is based on Beer's Law, where the units of measurement can be expressed as weight per unit area. That is, for a given thickness of sample, if you reduce the measurement area, you reduce the sample weight requirements proportionately. Experiments with original dispersive instruments were extremely limiting because of low throughput. Sample sizes were constrained to around 1 mm diameter, and this corresponded to a minimum sample weight of a few micrograms (based on the use of a beam condenser, and for an acceptable SNR).

With the introduction of FTIR instruments, and especially the use of MCT detectors, smaller sample sizes became practical, and detection limits were reduced by one to two orders of magnitude, dependent on the actual sampling procedure. One accessory that has gained popularity is the diamond anvil cell, in particular the microversion, which enables sampling on a broad range of materials (including hard plastics and minerals). The aperture of this cell is typically around 100 μm in diameter. One problem with this accessory, and most microaccessories, is the difficulty associated with correct placement of the cell in the beam, with respect to the sample. The IR microscope overcomes these problems by providing a magnified visual image of the sample at exactly the same point as the passage of the IR beam (Katon and Sommer, 1992; Katon et al., 1986). This is made practical by the use of confocal imaging where both the visible and the IR beams follow identical paths through the microscope imaging optics to the sample—as illustrated in Fig. 16. This is not a new concept, and IR microscopes have been available since the 1950s, but it was not until the last decade that the performance of the instrumentation matched the demands of the application.

One of the benefits of the microscope is that it is possible to view the sample with a relatively large field of view—up to about 1 mm. High quality imaging, for both the IR and the visible illumination, is made possible by the use of all reflecting optics, usually with an on-axis

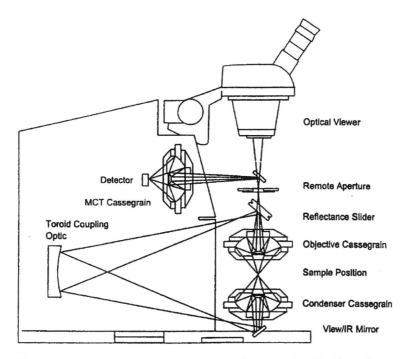

Figure 16 Schematic of an IR microscope. (Courtesy of Perkin Elmer Corp.)

Cassegrain lens design. A specific area of the sample can be centered and focused by the use of normal x, y, z translation controls. By the use of aperture controls, located at an intermediate focus, it is possible to select regions of the sample as small as 5–10 μm across. Once the exact area for study is located, the visible illumination is switched out, and the IR beam is switched in. The IR beam, which is normally supplied from a standard instrument bench, is passed to a dedicated MCT detector once it interacts with the sample. This detector is optimized for small sampling with an element size between 250 μm and 1 mm.

Most microscope accessories provide the opportunity to observe and analyze samples in both transmission and reflectance modes. Also, special objective lenses are available to permit grazing angle and ATR experiments to be performed on the microscope. A wide range of samples can be studied, including fibers, powders, laminates, electronic components, biological specimens, and so on. Modern systems are equipped with video imaging which can be connected directly to the instrument computer; this in combination with a computer-controlled microscope stage enables a user to produce two-dimensional video graphic and spectroscopic images from the sample.

The key parameters for IR microscopy are SNR and spatial resolution. Obviously these parameters are related because as one closes the aperture of the microscope to view a smaller sample, the amount of light available is reduced. This means that it is difficult and time-consuming to measure samples of 15 μm or below, at a reasonable SNR. Furthermore the minimum sample size is also determined by diffraction which limits it to about the wavelength of the radiation. For mid-IR, with wavelengths in the range 2.5–25 μm, an estimate of the practical diffraction limit is about 10 μm, which therefore defines the practical spatial resolution of the technique. This is a very significant drawback of IR microscopy, especially when compared with Raman microscopy where, because excitation is generally in the visible region, a spatial resolution of around 1 μm can be achieved routinely. Optimum (diffraction limited) spatial resolution in IR microscopy can be achieved by the use of a synchrotron source, which is sufficiently bright that apertures can be regularly set at the diffraction limit (Carr, 1999), or by using an imaging microscope equipped with a FPA detector (see below) for which an aperture is not required. Applications utilizing near field light also have the potential to improve spatial resolution.

Imaging IR Spectroscopy

The original use of the IR microscope was to take spectra of small samples or of small parts of larger samples. However, it soon became apparent that collecting multiple spectra from a heterogeneous surface allowed the construction of images showing the distribution of components or properties across the surface. This approach is sometimes termed "chemical imaging" because it relies on molecular structural changes derived from IR (or Raman spectra) spectra. There are two experimental approaches for obtaining the IR images. The first method, usually termed mapping, uses an IR microscope equipped with a computer-controlled stage to move the sample so that many independent spectra are taken in a grid pattern over the sample. For example, one might take spectra of a 1 mm^2 section of the sample by setting the aperture to 20 μm and the step-size to 20 μm. This will lead to a grid pattern of 50 spectra by 50 spectra, or 2500 spectra in total. Images can then be constructed by plotting the intensity of a band (e.g., the carbonyl band) after normalization, or a band ratio. While the images generated by this method are excellent, the collection of so many independent spectra consecutively is very time-consuming, and if high spatial resolution is required the SNR may be poor.

An alternative approach is to utilize a FPA detector. This is an MCT detector which consists of a square array of independent detector elements. The number of elements may be as many as 128 × 128, though 64 × 64 is more common. A section of the sample is directly imaged onto the detector and all the spectra are collected simultaneously. No aperture is required, so the method optimizes both SNR and spatial resolution. A diffraction-limited spatial resolution of about 5.5 μm has been claimed, and this is likely to be the case particularly if images are generated from the C–H stretching region of the spectrum. When FPA based systems were introduced they required step-scan FTIR spectrometers, and the interferograms were collected one point at a time. However, improvements in signal processing and computer speed now allow the use of a rapid-scan spectrometer, which helps to reduce the cost somewhat. Currently, for the larger arrays (128 × 128) so much data are generated that step-scanning is still required.

FPA-based systems are now available commercially from several manufacturers (for example, Bruker, Digilab, and Thermo-Nicolet), but are very expensive mainly because of the cost of the detectors which have generally been developed for military use. Somewhat less expensive systems are also available with smaller arrays of 32 × 32 and even 16 × 16. If images of larger samples are required from these smaller arrays then a process called "mosaicing" is used to "stitch together" several images to make one large image.

The IR imaging has many uses, but one of the most important is in the biological and biomedical areas where images of tissues are studied in detail in order to detect and understand disease processes, particularly cancer. In recent years there have been many publications in this area (see, e.g., Fabian et al., 2003). A second important area is polymers because of the heterogeneous nature of many polymer surfaces (Chalmers et al., 2002).

Specialized Measurement Techniques

During the past two decades, FTIR has gained popularity because it has enabled users to perform experiments that were generally impractical or impossible with earlier dispersive instruments. Two important characteristics in favor of FTIR are speed and high optical throughput. The following are examples of some of the techniques that have benefited and/or have been made possible by these features of FTIR spectrometry.

HYPHENATED TECHNIQUES. Hybrid or hyphenated techniques (Hirschfeld, 1980) include chromatographic sampling (GC-IR, LC-IR, SFC-IR, etc.) and thermal analysis methods (such as TG-IR) interfaced on-line to a spectroscopic instrument, in this case to a FTIR. Although attempts have been made in the past to combine chromatographic techniques and IR, most of the successful experiments were performed off-line with trapped samples (condensed or adsorbed). Of these techniques, the GC/IR has undoubtedly received the most attention, to the extent that there are texts dedicated to the subject (Bourne, 1984; Cook, 1991; Griffiths et al., 1983; Hirschfeld, 1980; Wehry and Mamantov, 1979; White, 1991; Wilkins, 1994), and protocols exist for routine analyses (ASTM method E1642-94). Two widely different FTIR based methods for GC/IR have been promoted. The traditional approach is to use a low-volume, gold-coated, heated lightpipe interfaced directly to the chromatographic column via heated transfer lines. This method is a real-time, on-line measurement, where the light from the interferometer is focused on the lightpipe, and the transmitted light is captured by a dedicated, small-target high-to medium sensitivity MCT detector.

The second approach is to make an on-line sample collection and measurement, where the effluent from the column is trapped on a moving cooled surface. The original versions featured matrix isolation (Wehry and Mamantov, 1979), where the collection surface was cryogenically cooled to liquid helium temperatures, and the eluted components were trapped in localized zones within a solid argon matrix. The matrix was evaluated *in situ* by a reflected microfocused beam, as the surface moved. Eluted components were measured as long as necessary, to produce high-quality, high-resolution matrix isolated spectra. A commercial system was developed by Mattson Instruments, known as the *Cryolect* (Bourne, 1984). More recent methods have featured subambient trapping, where the eluted components are trapped on a moving cooled surface, but without an argon matrix. The hardware in this case is simpler, with the collection substrate being a zinc selenide window, and the cooling being performed by a Peltier (thermoelectric) device. The eluants are evaluated by microscope-style optics, in transmission (Katon et al., 1986). In this case, a commercial system called the *Tracer* was developed by BioRad-Digilab.

In LC/IR and TLC/IR, the largest constraint is the sampling matrix—typically an strongly absorbing mobile phase, in the case of LC, and an intensely absorbing solid substrate, such as silica gel or alumina, in the case of TLC. Many attempts, most featuring trapping methods, have been considered for solvent elimination in LC applications. These include methods similar to the Tracer-style method mentioned above, such as the system offered by Lab Connections, known as the *LC-Transform*, where the sample is deposited onto the surface of a position-encoded, slowly rotating germanium disc. After the chromatographic run, the disc is transferred to a stepper motor-driven reflectance sampling stage, where the disc is rotated at a rate comparable to the speed used at the time of sample collection. Spectra are recorded sequentially from the disc surface, and are used for the reconstruction of the chromatogram, as well as obtaining individual spectra from the separated components.

SFC/IR is an interesting alternative, when appropriate, because certain mobile phases, such as super-critical carbon dioxide, are reasonable IR solvents, providing regions of IR transparency. However, applications of SFC are limited, and so the technique has not gained significant popularity. TLC/IR has had limited success, in some cases by direct measurement using diffuse reflectance measurements, and others involving some form of extraction of the separated components from the TLC plate (Shafer et al., 1988). Some work has also been reported on direct Raman studies, in this case the interference from the TLC solid substrate is less of a problem because silica and alumina are relatively weak Raman scatterers. Also, some of the enhanced methods have been used, such as the resonance Raman spectroscopy, where samples are colored and absorbed close to the excitation wavelength, or surface-enhanced Raman spectroscopy (SERS) by including silver particles within the chromatography medium or spraying silver colloid on the developed TLC plate.

TIME-RESOLVED FTIR. FTIR is a valuable tool for the study of reaction kinetics, providing both molecular and kinetic information on the various reactive species. The measurement speed of a rapid-scan FTIR is, however, limited by the cycle time of the moving mirror, currently of the order of 20–50 ms and so, for the study of fast kinetics, alternative measurement techniques are required. Step-scan provides one solution; at least for reactions that are reversible or that can be repeatedly, reproducibly activated, such as polymer stretching experiments, as an example of the former, and photochemical reactions in a flowing gas cell, as an example of the latter. There is no inherent limit to the time-resolution attainable in step-scan measurements, but in practice, the rise time of the detector, amplifiers, or digitizers is the "rate-determining step". For standard IR detectors and internal digitizers 10 μs is typically achieved, but, with fast external photovoltaic MCTs and built-for-purpose electronics, speeds up to 10 ns are possible.

For time-resolution shorter than the ns timescale, noninterferometric methods such as pump-probe techniques may be employed. In a pump-probe experiment the sample is excited or a reaction is initiated with a short laser pulse (the pump), prior to interrogation after a controlled (short) time interval with a pulse from a second probe laser. The probe laser is tuned to an appropriate absorption band to gain spectroscopic information of the transient state or species. Pump-probe methods avoid the poor sensitivity of fast MCTs and gated electronics but at the expense of complicated laser systems, narrow spectroscopic windows, and the other advantages that FTIR confers. Pump-probe techniques may gain popularity with the development of FEL sources at synchrotron installations.

DICHROISM AND OPTICAL ACTIVITY. One of the few methods available for determining the absolute configuration of chiral molecules is a specialist IR spectroscopic technique known as vibrational circular dichroism (VCD). VCD is analogous to traditional CD in that it involves the measurement of the degree of dichroism in the absorption of left and right circularly polarized light. However, VCD operates in the IR region, and involves vibrational transitions; whereas traditional CD operates in the UV–Vis, and involves electronic transitions. The strength and sign of the VCD signal are given by the relative direction (dot product) of the electric and magnetic dipole transition moments, which are not orthogonal for chiral molecules. Among the advantages of VCD is that the absolute configuration of a molecule may be obtained by comparing the measured VCD spectrum with a calculated spectrum of the molecule of known configuration. If the sign and the spectra match then the configuration is the same as used in the calculation. If the spectra match but are of opposite sign then the molecule has the opposite configuration to that used in the calculation. In recent years VCD has matured as a technology to the point where commercial instruments are available either as stand-alone units or as accessory on an external bench. With the availability of accurate molecular orbital calculation programs, such as Gaussian 98/03, VCD in combination with *ab initio* calculations is increasingly adopted as standard practice particularly for small molecules of biological significance. Other applications of VCD include determination of the relative enantiometric purity, and in the study of proteins and nucleic acids. The basic components of a VCD-IR spectrometer are outlined in Fig. 17.

C. NIR Instrumentation

NIR instrumentation can be broadly separated into two types, process and laboratory instruments. Process instruments need to be simple, robust, and reliable, and are generally filter-based. Process NIR is mainly used for repetitive measurements for qualitative or quantitative analysis in conjunction with multivariate software, and there is not usually a requirement for a full spectrum. The filters are chosen by using a laboratory-based scanning instrument and applying chemometric techniques for finding the optimum wavelengths for the measurement and the mathematical model which gives the most accurate results. Simple filter instruments can then be used to carry out the analysis in the process situation.

For laboratory-based scanning instruments, two possibilities exist: dispersive or the Fourier transform. Both instruments have a similar level of performance. We have seen that the Fourier transform methods outperform dispersive, where the predominant noise source is thermal noise in the detector. However, where the predominant noise source is shot noise the multiplex advantage is lost, and the dispersive approach is more effective. The cross-over point for the two approaches is around the middle of the NIR region and hence both methods are used in NIR instrumentation, whereas only FT is now used for mid-IR.

1. Sources

For the NIR operation, typically above 4000 cm^{-1}, it is possible to use the same source as used in the mid-IR region. However, it will be appreciated from the black body curves in Fig. 3 that output of this type of source is expected to fall off from around 5000 cm^{-1} (2 μm). In this case, there is a benefit in boosting the temperature in excess of

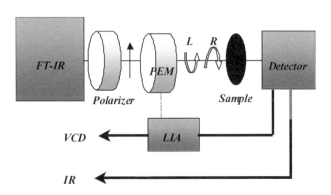

Figure 17 Schematic of a VCD-IR.

2000 K. However, this may not be practical as it may cause localized heating problems and it will significantly shorten the lifetime of the source. For this reason it is common to use a quartz halogen lamp, which operates at temperatures around 3000 K. The quartz envelope of the bulb does not cause an IR transmission problem in this spectral region.

The high output of the quartz halogen lamp is linked to the halogen cycle generated within the envelope between the tungsten filament and the iodine vapor (the halogen). This regenerative process is influenced by the external temperature of the quartz envelope. It is essential that no cold spots are allowed to develop, otherwise a deposition of metal will be experienced, with a consequent darkening in the cooler regions of the envelope. This not only reduces the output efficiency of the source, but it also has a negative impact on the lifetime. The lifetimes of quartz halogen lamps are typically shorter than the other styles of source mentioned earlier. With continuous operation, typical lifetimes are in the range of 1000–5000 h.

2. Light or Energy Analyzer

Filter Instruments

FIXED FILTER INSTRUMENTS. As mentioned above, many NIR analyses are developed in laboratories using scanning instruments, but are implemented with small, rugged instruments containing up to 20, but generally fewer, discrete, fixed interference filters, the wavelength characteristics of which have been chosen to optimize the particular analysis being carried out. The filters are contained in a wheel controlled by a stepper motor. Usually a chemometric model has been developed to relate the filter measurements to the significant properties of the sample. Such analyzers are used in plant situations, often by minimally trained operators. They are very important in the agricultural area, where analyses of, for example, protein, oil, and moisture in grain are a common application.

ACOUSTO-OPTICAL TUNABLE FILTERS. An AOTF functions as an electronically tunable bandpass filter. It consists of a birefringent crystal material with a RF transducer bonded to one face. An acoustic wave is generated across the crystal, which in turn produces a modulation of the index of refraction. Under the right conditions, if light is passed though the crystal, a portion of the beam will be diffracted. For a fixed RF frequency, only a narrow band of optical radiation satisfies the condition, and only those frequencies are diffracted. If the RF frequency is changed, the centroid of the optical frequencies will change correspondingly. Hence, it is possible to scan a spectral region with a particular optical bandpass by scanning a range of RF frequencies (Wang, 1992).

From a schematic of an AOTF device, shown in Fig. 18, it can be seen that there are three emergent beams from the crystal. Two beams are diffracted and one beam passes undiffracted, the latter being the main, zero-order beam. The other two beams are orthogonally polarized, and are effectively "monochromatic". The beams are designated the + and − beams, or the ordinary and extraordinary beams. Normally, as illustrated, two of the beams are block, and one is used for the analytical measurements. The spectral range is dictated by the crystal material and the range of the applied RF frequencies. The most common material used is tellurium dioxide, TeO_2, which covers the visible, the NIR, and some of the mid-IR down to 4.5 μm. Although, in the past, some reference has been made to materials providing transmission further into the mid-IR, problems of crystal fabrication and material purity have limited their availability. Several commercial NIR instruments have been fabricated around TeO_2 crystal devices.

The bandpass of the crystal is a function of the length of the lightpath through the crystal. The shorter the lightpath

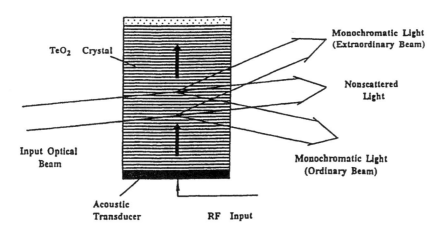

Figure 18 Principle of operation of an AOTF device (Burns and Ciurczak, 1992).

(crystal length) the wider the bandwidth. There is typically a compromise between the size of the device and the desired performance. The main issues are often associated with crystal fabrication—reproducibility of cutting dimensions, crystal finish, and material homogeneity being the most important. During operation, maintaining crystal dimensions is important for reliable performance. This requires rigid temperature control; note that the RF transducers dissipate a large amount of heat.

The main benefits of the AOTF device are no physical moving parts and the speed of scan. It is typically possible to sweep the frequencies for the spectral measurement range within 200 ms. Another attractive feature is that the electronic wavelength selection permits random access to specific wavelengths at very high speed.

Dispersive Instruments

In recent years the number of grating based NIR instruments has declined in favor of FT instruments. Many of the established manufacturers of FTIR instruments such as Bruker, Thermo-Nicolet, and Perkin Elmer have produced dedicated FT-NIR instruments for the QC/QA market. While these instruments are highly suitable for many applications involving powdered solids and liquids, they are less suitable for the important application of agricultural products. This is because the relatively small beam size of about 1 cm does not allow a sufficiently large sample for routine analysis of materials such as grain or soybeans. Hence there remains a market for dispersive NIR instruments where a beam size of 2.5 cm can be achieved.

An important class of dispersive instrumentation that is gaining interest for NIR use is detector array-based instruments, also described as multichannel spectrometers. In their simplest form these consist of a polychromator or spectrograph constructed from a dispersing element, such as a fixed concave holographic grating and a suitable detector array. It was once suggested that it would be inconceivable to consider a multichannel analyzer for IR applications, based on resolution elements and cost (Marshall and Comisarow, 1975). In reality, the basis of the argument was hypothetical, and had no bearing on modern IR applications. Most array detector applications are performed on low-resolution spectra, and over a limited spectral range. Under such conditions it is practical to measure spectra with array densities ranging from 256 to 1024. In fact, for a dedicated application, between 25 and 100 detector elements (pixels) can be adequate. Military projects requiring IR night imaging and sensing have helped to provide cost-effective array devices. Also, the extension of the silicon detector into the short-wave NIR (down to 1080 nm) has provided an additional source of reliable, cost-effective arrays (often less than $1000) for certain applications. In total 2048 element silicon arrays have become available, and these provide an interesting opportunity to combine the NIR and visible spectral regions with little or no loss in spectral resolution (dependent on grating selection).

These instruments are typically used for dedicated applications, rather than for routine IR measurements. For some applications in the NIR this technology is sometimes combined with optical fibers to simplify the sample–instrument interface. In most cases, external optical filters—short wavelength cut-off and/or bandpass filters—are used to reduce the impact of other grating orders and straylight. Fabrication of the spectrograph raises important issues relative to the reduction (or elimination) of straylight, the geometric stability of the array, and spectral calibration. The array is a physical element that can change position and dimension with variations in temperature. While these may seem small, they are of the same order of the wavelengths being measured. For best performance, care must be taken in the mounting of the array, and where necessary, suitable thermal control may be applied. The dispersion is typically nonlinear, and so some form of internal or external calibration has to be applied to linearize the output (wavelength scale) from the diode array. It should be noted that most practical systems use arrays that function in the NIR. However, materials are becoming available for some limited mid-IR applications. Some additional discussion on arrays and detector materials will be covered under the section on detectors.

Fourier Transform Instruments

In order to use an FTIR spectrometer for NIR measurements one must have the correct system of source, beamsplitter, and detector. The source has been dealt with earlier and is generally a quartz halogen lamp. There are two common choices for the beam splitter substrate: quartz and calcium fluoride. For most applications in the NIR, a quartz substrate is used, coated with germanium, silicon, or a mixture of coatings, dependent on the spectral range and performance required. Quartz is a much more robust material than the KBr used for mid-IR beamsplitters, and is not hygroscopic. Calcium fluoride is a more expensive material, but has slightly better NIR transmission and is useable, though not very efficient, well into the mid-IR to about 1200 cm^{-1}, whereas the cut-off for quartz is around 3300 cm^{-1}. Both quartz and CaF_2 have an upper wavenumber limit of about $15,000 \text{ cm}^{-1}$. Detectors for the NIR use are discussed as follows.

3. Detectors

The silicon used in normal photodiodes has its cut-off around 1100 nm (9090 cm^{-1}). While this makes silicon detectors unsuitable for most IR applications, they are used in instrumentation that operates in the short-wave NIR (750–1100 nm). This spectral region can be used to measure third overtone and combination band information, and this is the basis of some commercial IR analyzers. Typical applications include food and agricultural products (for moisture, fat, and protein) and hydrocarbons (such as refined petroleum).

For the longer wavelength NIR region, several intrinsic semiconductor materials are available as photon detectors. These include lead sulfide (PbS) and lead selenide (PbSe), indium antiminide (InSb) and indium arsenide (InAs), as well as the "ternary compounds or alloys", such as indium gallium arsenide (InGaAs) (Wolfe and Zissis, 1985). These intrinsic semiconductors provide varied response characteristics and spectral detection ranges. The characteristics can be modified and/or improved by cooling down as far as liquid nitrogen temperatures (77 K). Note that optimum response of many of these detector materials requires modulation of the radiation in ranges from 10^2 to 10^4 Hz. This is a good match to the normal modulation rates encountered in FTIR instruments. Of these materials, the most widely used for NIR applications with FTIR spectrometers is the InSb detector, which operates well in the range from 10,000 to about 2000 cm^{-1}. InGaAs may be used with or without cooling. Cryogenic cooling is not required, and thermoelectric (two-stage Peltier device) cooling down to $-30°C$ is adequate. Operating the detector without cooling provides a slightly better spectral range (lower limit) but at the sacrifice of signal-to-noise (S/N) performance.

4. Sampling Techniques for NIR Spectroscopy

The standard techniques of transmission, reflection, and transflection are still in common use. The low absorptivities in the NIR region allow transmission samples to be much thicker, by up to two orders of magnitude compared with the mid-IR samples. This certainly makes the sampling of powdered and granular solids much easier. For most NIR work, reflectance methods are used (Wetzel, 1983), and instruments sometimes incorporate an integrating sphere, which makes the optics very simple. The wavelength of the light means that the measurement is less affected by particle size than a similar measurement in the mid-IR. Almost any solid material that can fill a 2.5 cm sampling cup with a quartz window can be measured. Generally, NIR reflection measurements do not exhibit the distortions commonly encountered in the mid-IR because of the considerably lower absorptivities of overtone and combination bands. As a result more complete penetration occurs and very little interference from front surface reflections is experienced. The ATR method is also possible for NIR spectroscopy, but is much less commonly used compared with the mid-IR region.

There have been a number of advances in the NIR sampling techniques in recent years. The first of these is a redesign of the instrument to make the measurement even simpler by having the light exit through a quartz window on the top of the instrument. When a sample in a glass container, perhaps even a standard sample bottle, is placed on the window the reflected light returns into the instrument and reaches the detector which measures the NIR spectrum. This approach minimizes sample handling leading to faster measurements with less possibility of contamination or exposure to toxic materials. In some instruments the sample container is rotated to obtain a more representative sample. The container must be included in any chemometric model since it will make a small contribution to the spectrum.

The second advance in sampling has been the widespread adoption of fiber-optic sampling probes to simply measure spectra of liquids and fine powders by immersion, and other solids merely by pointing the probe at the sample. Typically the fiber bundle is attached to the front of the instrument, and has a length of about 1 m. Alternatively, if the spectrometer has a standard design then an accessory can be purchased which fits into the sample compartment and provides a fiber bundle with delivery and return fibers. The probe end often has a gun-shaped handle. There are many designs for fiber-optic probes to suit a wide variety of samples. Examples include transmission, transflection, ATR, and reflection probes. Quartz fibers are the most cost-effective, and the most readily available. They are typically used as single filaments (200–600 μm diameter) or as bundles. Bundles are expensive, but are necessary for certain applications where light has to be collected over a relatively large area—such as in diffuse reflectance measurements. Ultralow hydroxyl content fibers are essential for use over most of the NIR spectral region, especially for measurements at the longer wavelengths. Although quartz transmits into the mid-IR, normal quartz (low OH) fibers have a practical cut-off around 2560 nm ($\sim 3900\ cm^{-1}$), or even lower, due to the long absorption pathlength and the resultant high signal attenuation at the longer wavelengths. Fiber-optic probes are particularly important in quality control applications where they minimize sample handling and provide more reproducible spectra.

D. FIR (Terahertz) Instrumentation

FIR spectroscopy is not common and is relatively difficult and expensive to do well. It is mainly used for the study of molecules containing heavy atoms, such as organometallics, leading to low-energy fundamental modes. It is also used in the study of the phonon modes of solids as well as molecular deformation modes. Water vapor can be a problem in this region of the spectrum, and although purging with a suitable gas can help, the best FIR spectrometers have vacuum benches. The FIR region has been variously defined in the past, but is now considered to cover the region $400–10\ cm^{-1}$. Recently, a new description for light of these frequencies has appeared: the terahertz region, which covers light of the approximate frequency range $0.1 \times 10^{12}–10 \times 10^{12}$ Hz (about $3–330\ cm^{-1}$) and therefore overlaps with what was traditionally known as the FIR. Terahertz spectrometers have become commercially available only very recently (2003) and therefore, for most applications, the FIR region is accessed using an FTIR spectrometer equipped with the relevant source, beamsplitter, and detector as described below.

1. Source

From Fig. 3, it is obvious that the traditional blackbody sources become progressively inefficient in the FIR region, with little benefit from an increase in operating temperature. A practical alternative is the high-pressure mercury discharge lamp, which in the FIR provides a continuum—in contrast to its line source characteristics in the UV and visible spectral regions. The radiation, which extends from approximately 100 μm to 2500 μm, is believed to be thermal in nature, originating from the hot mercury arc plasma, and from the hot quartz envelope of the discharge lamp (Vasko, 1968).

For terahertz spectroscopy the source is completely different. A terahertz emitter is used consisting of a material, such as zinc telluride, which emits a pulse of broad-band terahertz radiation when it is excited by a femtosecond pulsed laser with a power around 200 μJ, and, for example, an 800 nm wavelength. The spectral range is around $1–133\ cm^{-1}$.

2. Light or Energy Analyzer

The significant part of the interferometer is of course the beamsplitter. In the past, cesium iodide has provided beamsplitters with an extended range below $400\ cm^{-1}$, down to $200\ cm^{-1}$, thus giving access to part of the FIR. However, this material is both highly hygroscopic and is also plastic in nature. This latter characteristic is highly undesirable for an optical substrate—it is difficult to polish, and to retain its flatness (the material may distort with time owing to cold-flow within its mechanical mounting).

The most common FIR beamsplitters over many years have been Mylar® (polyethyleneterephthalate) films which give acceptable performance but tend to vibrate and hence generate additional noise. Mylar beamsplitters are available in different thickness which are effective for different spectral ranges. The very thin films (3–6 μm) are effective over the whole FIR range, but are the most difficult to use. Slightly thicker Mylar films, 12.5 and 25 μm, are effective over approximate ranges of $300–30\ cm^{-1}$ and $120–10\ cm^{-1}$, respectively.

In recent years manufacturers have developed specialized FIR beamsplitters based, for example, on a block of silicon with a thickness of several mm. These have fair performance over the range $400–50\ cm^{-1}$, which covers most of the FIR region and do not suffer from vibration.

3. Detectors

Bolometers

A bolometer is essentially a tiny resistor that changes its resistance with temperature. In order to provide adequate sensitivity as an IR detector, the device must be very small. Today, microdevices are manufactured based on silicon micromachining technology from the semiconductor industry, in which the sensing element is only a few micrometers across. For measurement, the element is typically placed within one arm of a Wheatstone bridge with a second thermally shielded device in the balancing arm. Originally, bolometers offered comparable response problems to thermocouple detectors, however, these problems no longer exist with the modern micro versions.

Other Detectors

Thermal devices are usually well suited for FIR detection, especially with interferometric instruments because the lower light modulation rates, which result at longer wavelengths, are matched to the slower response of these detectors. A DTGS detector, equipped with a PE window (for transmission below $400\ cm^{-1}$) will suffice for routine applications. As mentioned earlier, the photoacoustic detector operates as a thermal detector, and this also provides good performance in the FIR.

The situation is different for the MCT detector (an intrinsic semiconductor) where the bandgap is too small for use at long wavelengths, especially below $400\ cm^{-1}$ (25 μm). An alternative is to use an extrinsic semiconductor, based on a traditional semiconductor material, such as silicon or germanium, which is suitably doped with impurities. The impurities provide a source of electrons below the normal cut-off of the detector material. The spectral detection range of the material is defined by the dopant. Note that liquid helium cooling is required for this type of detector.

Terahertz detection relies on the fact that the presence the terahertz pulse causes an optical birefringence in a second crystal of zinc telluride, which means that the polarization of light, other than the terahertz light, traveling through the material changes. A very short pulse of

Vibrational Spectroscopy

light with a wavelength of 800 nm is used as a *gate* pulse, and the polarization change of this gate pulse is measured. Terahertz spectrometers can be made relatively simple because the same laser light used to generate the terahertz radiation is also used as the detection gate pulse.

4. Sampling Techniques for Far-IR Spectroscopy

The standard sampling method in the FIR for solids is to prepare a PE pellet from PE powder in the same manner as a KBr pellet is made in the mid-IR. Many of the other sampling methods used for mid-IR will also work for FIR, but if, as is typical, a globar source and DTGS detector are used then SNR can be a significant problem.

III. RAMAN INSTRUMENTATION

A. Types of Instrumentation and Recent Trends

Raman instrumentation has an interesting heritage, and in some ways the technique has seen more changes and diversity than any other analytical technique. This may be due to two aspects of its evolution—one technical and the other perception vs. acceptance. The Raman effect is extremely weak, and Raman spectroscopy has obviously gained over the years by technology that improves signal and detectability. Not all these technological "breakthroughs" follow the same developmental trail. Second, the technique is in some ways too closely related to IR spectroscopy, and is treated as a poor relative.

Raman spectroscopy, in practical terms and for specific applications, can be demonstrated to have considerable advantages over IR spectroscopy. However, as a general analytical technique, it is difficult to demonstrate that it offers any net advantages. And, it has been seen to have some clear disadvantages, in particular the common interference from broad-band sample fluorescence, which can totally mask the spectrum. Consequently, when people justify the purchase of new laboratory instrumentation, the safe decision tends to be IR spectroscopy, which is well established, and has evolved consistently over the past 40+ years. With Raman spectroscopy, there is less practical history, and the instrument platforms are constantly changing. Today, there are a couple of technology choices which are not necessarily mutually exclusive—there are pros and cons to the selection. When it comes to a specific application, however, it can be easy to demonstrate whether the Raman spectroscopy or IR spectroscopy is the better choice.

There are currently two main technological approaches to the design of Raman instrumentation—a monochromator or spectrograph-based CCD system and FT-Raman. As mentioned above, because of practical and technical constraints, these two approaches do not necessarily produce exactly the same end result. Qualitatively, both generate the same fundamental spectrum for a given material, however, the overall appearance of the spectrum or the impact of the sample, may be different. This section will provide an overview of instrumentation as it exists currently, and it will provide a general discussion of the trends and applications.

The greatest technological difficulty for Raman spectroscopy has been the weakness of the Raman spectral signal, compared with the magnitude of the main excitation wavelength. The intensity of a particular Raman line can be in the range 10^{-6}–10^{-10} of the main excitation line (possibly even as low as 10^{-12}). The issues are: how to measure such a low signal in the presence of a dominant signal, how to remove effectively the interference from the dominant signal, and, if possible how to enhance the weaker signal. One approach to increase the absolute intensity of the signal is to increase the power of the source—the laser power. While this may be possible, it does not remove the fundamental dynamic range problem, or the interference problem, which only becomes worse with increased laser power. For years, the traditional Raman instruments featured scanning monochromators. The important criteria for the monochromator design were to minimize straylight, to enable the very weak signals to be measured, and to design for maximum Rayleigh line rejection. The original solution was to use more than one monochromator, with commercial systems being based on double and triple monochromators. While these had the desired optical properties, the use of up to three monochromators made these instruments mechanically complex, and severely limited the optical throughput of the instrument (down to 5% or less overall efficiency). With the introduction of laser line rejection filters (Carrabba et al., 1990; Schulte, 1992; Tedesco et al., 1993), and in particular the holographic notch filters, the need for the second and third monochromators was effectively eliminated. A side benefit of the use of these rejection filters and the elimination of the additional monochromators was the associated reduction in instrument size (Chase, 1994; Messenger, 1991).

The next significant technology boost was provided by the move towards detector arrays—initially with intensified PDAs, and more recently with the cooled CCD arrays. With such devices, the need to scan the monochromator mechanically is removed. The result is a high-efficiency spectrographic system with no moving parts. Today, the limitations of the technology are the cost of high-performance spectroscopic CCD array cameras, and the overhead associated with cooling. However, with major advances in imaging technologies in the 1990s there has been a corresponding expansion in CCD technology. As a result, it is anticipated that CCDs with improved performance and lower costs will continue to become available. The trades that have to be made with a CCD are spectral range vs. spectral bandwidth, and the impact of the signal cut-off between 1000 and 1100 nm for silicon. These issues will be discussed in greater detail later.

With the gain in performance experienced with FTIR instrumentation compared with dispersive IR instruments, there has been a natural desire to determine whether or not the same level of performance can be achieved for Raman spectroscopy. Originally, this experiment was considered to be impractical (Hirschfeld and Schildkraut, 1974) because of noise considerations and the extraordinarily large dynamic range involved between the excitation (Rayleigh) line and the Raman signals. However, with the advent of the laser line rejection filters mentioned above, Hirschfeld and Chase (1986) demonstrated the feasibility of FT-Raman spectroscopy. Original experiments were performed with the 647.1 nm line of a krypton gas laser, and with a silicon detector. However, the real justification for the move to FT-Raman spectroscopy was the ability to use NIR lasers. Moving from visible to NIR excitation helped to remove one of the major interferences encountered with Raman spectroscopy—the occurrence of broad-band fluorescence. A second, practical advantage for a user is that Raman spectroscopy can be performed on an existing FTIR instrument, without significant redesign. In fact, most of the major FTIR vendors offered Raman spectroscopy as an accessory for their high-end instruments. In most cases, the 1064 nm line of the Nd:YAG laser, operating with powers of up to 4 W, is used for excitation. Following the success of the FT-Raman accessories, some dedicated FT-Raman instruments were produced, with notable gain in performance linked to the optimization of the optical system. In particular, gains were experienced by the use of high reflectivity optics, the reduction in the number of optical elements, and the use of high sensitivity detectors, matched to the laser image.

In recent years the popularity of FT-Raman has declined significantly. Many users found that FT-Raman as an accessory to an FTIR spectrometer was not sustainable, and that a dedicated and expensive FT-Raman instrument was required to achieve reasonable outputs. It was also found that the sensitivity of FT-Raman instruments was limited. The multiplex advantage is seen for this type of instrument only when the system is detector noise limited (Chase, 2002), and this, together with the v^4 disadvantage of the long-wavelength excitation, leads to long measurement times. Micro-FT-Raman, while commercially available had only limited success. Another, and major, reason for the decline of FT-Raman has been the upsurge in interest in bench-top dispersive micro-Raman systems. The use of CCD detectors, together with holographic filters to remove the Rayleigh line, confers excellent sensitivity on these small systems. Furthermore, the microscopic nature of the system requires laser powers of up to about 10 mW at the sample and hence reliable, stable, and relatively inexpensive laser sources of around 20 mW can be used compared with the 1–2 W laser systems required for FT-Raman. The *raison d'etre* of FT-Raman is fluorescence suppression, however, many users of micro-Raman systems have found that fluorescence is not as big a problem as they expected. One reason for this is the widespread use of diode lasers with a wavelength around 785 nm, which is long enough to reduce fluorescence for many samples. The second reason is that in some instances it is possible that fluorescence is quickly "burned out" in micro-Raman instruments because of the very high power density of the focused light. Whatever the reason, most users agree that micro-Raman instruments suffer somewhat less from fluorescence interference than might be expected from the previous Raman work. However, there will always be some samples where the move to near-IR excitation is beneficial, and so access to an FT-Raman instrument will be useful. In fact, the use of an even longer wavelength around 1.3 μm for a few highly fluorescing samples has been demonstrated to be worthwhile (Asselin and Chase, 1994).

One of the major applications of Raman spectroscopy has been microscopy, with the benefits of the spatial resolution of the laser light source. Raman microscopy gained popularity in the mid-1970s from the pioneering work of Delhaye and Dhamelincourt (1975), and by the introduction of the MOLE (Dhamelincourt et al., 1979) by Instruments SA. Later, following the gain in popularity of IR microscopy, with microscope accessories optimized for commercial FTIR instruments, a parallel implementation was made for FT-Raman (Messerschmidt and Chase, 1989). Likewise, commercial Raman microscopes are offered either as accessories or as dedicated systems for use with CCD-based technology.

An important recent development is in the area of Raman spectroscopic imaging microscopy, a two-dimensional experiment, where the Raman spectrum is scanned with a liquid crystal tunable filter (LCTF) device, and the main image is generated by a CCD array (Treado et al., 1992a, b). This technology is expected to have significant impact on studies in materials science, in polymer chemistry, and in the area of biological and medical research.

One of the major benefits of Raman spectroscopy is the fact that the primary measurement involves visible or NIR radiation. This permits the use of conventional glass or quartz optics for imaging. Furthermore, it opens up the opportunity to use optical fibers for remote sampling. In such an arrangement, a single fiber is used for the transmission of laser radiation to the sample, and second fiber, or a series of fibers (Fig. 19), transmits the Raman

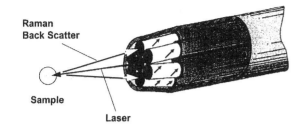

Figure 19 Schematic diagram of a Raman fiber-optic sampling probe featuring a fiber-optic bundle.

scattered radiation back to the spectrometer. The construction of the sample–light interface in this case is very important to maximize the coupling between the laser and the sample, and the subsequent collection of the scattered radiation. Silica fibers are normally used, which can transmit visible or NIR radiation over relatively long distances without significant light loss. Usually, the Raman spectrum from silica is very weak, however, over the distance covered by optical fibers, the contribution can be significant. To overcome this problem, sampling probes featuring optical filtering in the measurement head (Owen et al., 1994) are utilized. An example of such a probe head is shown in Fig. 20.

As noted earlier, there are several important areas of application where Raman spectroscopy excels over IR. Often, these are based on practical issues, such as the ability to use glass in the optical system, the lower Raman scattering of water, which permits the study of aqueous media, and the opportunity to have noncontact sampling. A secondary issue is that unlike mid-IR spectroscopy, there is no interference from atmospheric water vapor or carbon dioxide. This, coupled to the scaling down in instrument size, the ability to perform remote measurements, and the availability of mechanically simple instruments, has made Raman spectroscopy a practical tool for process applications. Applications have ranged from raw material screening to reaction monitoring, with a major focus on the analysis of polymeric products.

The inherent stability associated with modern CCD-based instruments was discussed earlier: achieved by the use of fixed monochromators, temperature insensitive optics, and stabilized laser sources. In combination with the ease-of-sampling, the use of glass optical components, and the convenience of optical fibers, Raman spectroscopy has found a new range of applications in process monitoring. With easy sampling, if not easier than NIR, and with species sensitivity and selectivity comparable to mid-IR, Raman spectroscopy offers distinct advantages for this area of application. It is expected that this new focus will help to promote Raman spectroscopy to its rightful status as a routine analytical technique alongside IR spectroscopy. In keeping with this trend, it is anticipated that new, lower-cost instruments will become available, and that there will be further competition from companies that offer unique packaging technologies, in particular in the area of instrument miniaturization. Likewise, FT-Raman is hoped to strengthen its position by providing a practical resolution to the fluorescence issue, especially as an analytical problem-solving tool. It provides a commercial analytical solution that takes advantage of the complementary nature of IR and Raman spectroscopies, with both techniques residing in the same instrument. In essence, if the instrument is sufficiently flexible, Raman spectroscopy, NIR, and mid-IR can be performed on the same system.

B. Instrumentation: Design and Performance Criteria

Virtually all of the discussion of IR instrumentation has centered around the fact that the technique is fundamentally a light absorption-based method of measurement. The main exceptions being specific cases where IR

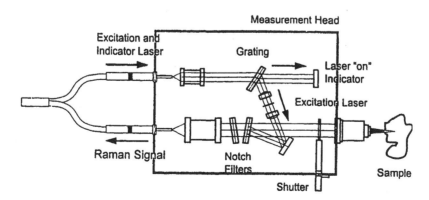

Figure 20 Schematic diagram of a Raman fiber-optic sampling probe featuring filter elements in the measurement head. (Courtesy of Kaiser Optical Systems Inc.)

emission measurements are made of a hot source or heated sample. As noted, Raman spectroscopy is an especially weak phenomenon, and that reasonably high resolution is required to resolve the spectral lines (Stokes and anti-Stokes) from the intense Rayleigh line. There are a number of practical factors to take into account when selecting the measurement technology for a Raman spectroscopy application. It is not as simple as IR, where the main issues are typically spectral performance and system flexibility.

In itself, the selection of a laser source might seem simple. Normally the choice might be perceived as being simply wavelength selection, and nominal output power for the selected line. However, there are other key issues that impact not only the selection of laser line, but also the laser type. For most Raman spectroscopy applications, continuous (cw) lasers are used, although Raman measurements may be made with pulsed lasers. However, a very different style of instrumentation is required for the signal handling from a pulsed laser source.

Laser stability is one important issue. There are at least two factors to consider here—output stability and wavelength stability. The measured signal from the scattered radiation is directly related to the output or power of the laser. Increasing the laser power will increase proportionately the intensity of the Raman lines, as well as a corresponding increase in the intensity of the Rayleigh line. Hence, fluctuations in laser output will be reflected as unstable Raman band intensities. Qualitatively, this may not cause a problem, but quantitatively it can be very serious. The application of Raman spectroscopy to continuous on-line process measurements is becoming popular, and for this application, stable peak intensities are essential. In the event that drift or fluctuations occur, intensity corrections, possibly based on either the monitoring of the laser output or the intensity of the Rayleigh line, can provide a solution.

Laser wavelength instability is potentially a worse situation. Slight shifts in wavelength will cause corresponding shifts to the entire Raman spectrum. If this occurs within the timeframe of the spectrum acquisition, then the entire spectrum will be degraded. Mode hops and laser line frequency drift will cause such shifts to occur with a resultant loss in spectral definition. This can be particularly prevalent with diode lasers. In order to maintain an acceptable level of performance with this type of laser it is usually necessary to thermally stabilize the laser with the aid of a thermoelectric (Peltier) device.

Having stated that the scattered Raman band intensities are a function of the laser power, it must be realized that the Rayleigh line intensity is impacted by an equivalent amount if the power level is raised. This often means that little is gained, in terms of enhancing the Raman signal, by increasing the laser power beyond a certain limit. Each sample responds differently, and the "limit" is not clearly defined. However, increasing the power will increase straylight problems, especially in the region of the Rayleigh line. Also, if the sample is potentially either thermally or photolytically unstable, then increasing laser power may result in the destruction of the sample. And so, the "limit" is best assessed on a sample-to-sample basis.

One of the main obstacles to the successful implementation of Raman spectroscopy is sample fluorescence. Fluorescence is often observed as a broad-band background that superimposes all or part of the measured Raman spectrum. For some samples, this fluorescence can be one or more orders of magnitude greater than the intensity of the Raman lines. As a result, the entire spectrum may be dominated by the fluorescence with little or no evidence of the Raman spectrum. The source of fluorescence is not always well characterized. Obviously, compounds that fluoresce naturally when excited by shorter wavelengths will produce an intense interference as the wavelength of the laser approaches or interacts with the electronic transition envelope of the sample. In such cases the only solution is to select a longer excitation wavelength, outside the range of the absorption envelope. Sometimes, simply moving from a green laser line to a red laser line will suffice.

On other occasions, the source of the fluorescence is less obvious. Often samples that are not normally considered to be fluorescent will exhibit fluorescence, which is believed to be associated with the presence of impurities. A sample clean-up procedure, through an absorbent, such as activated carbon, alumina, or silica will sometimes remove the problem. Alternatively, a burn-out process, where the sample is exposed to the laser for some period before the measurement, may result in a reduction of the fluorescence to an acceptable level. In this case, it is assumed that there is a selective quenching of the fluorescent sites within the sample. However, it is important that the sample is not moved during this process (or prior to measurement). Also, dependent on the exact nature of the fluorescent component, the process can take quite a long time—anywhere between 5 min to as much as 1 h—for reduction to useable levels. There are occasions where none of these methods works or they are inappropriate, and the only practical solution is to use a longer wavelength laser line. Also, when using the burn-out method, there is a natural assumption that the sample itself remains intact during the prolonged exposure to the laser.

Other regimes for removing or reducing the impact of fluorescence have been considered, and these include moving to shorter wavelength (as mentioned), the use of quenching agents and time-based discrimination methods. The use of quenching agents may have limited applicability, dependent on the type of compound being investigated. Examples are compounds that form charge-transfer complexes with materials such as butan-2,3-dione and tetracyanoethylene. The types of compounds that interact in this manner with these reagents are aromatic and polycyclic hydrocarbons. Note that relatively high levels of quenching agents may be required to reduce the fluorescence, and their presence may cause a spectral interference.

The use of time-based discrimination techniques is particularly interesting because it takes advantage of the fact that fluorescence, as a temporal event, is relatively slow compared with the virtually instantaneous transition producing the Raman signal. If in the collection of the Raman spectroscopy data, the time taken to acquire the signal can be reduced, only a small portion of the fluorescence signal is collected. Relative to the Raman signal, the contribution of the fluorescence will be significantly reduced. One approach is to consider using a pulsed laser system, but typically the repetition rates are too slow to provide adequate temporal discrimination. However, the use of a gated PDA, where the signal is collected for only a few nanoseconds, will provide good fluorescence rejection as long as the fluorescence build-up is slow in comparison with the acquisition period.

It must be realized that switching to a longer wavelength may not always be practical. First, as noted earlier, the Raman scattering efficiency is a function of $1/\lambda^4$ (v_0^4)—this results in a significant reduction in the observed line intensities at longer wavelengths. In order to maintain line intensities it is necessary to increase the laser power accordingly, if practical—laser design or sample stability may limit this option. Also, the older instruments featured photomultiplier tube (PMT) detection, and these components typically had poorer performance at the red end of the spectrum, although versions with improved red performance were available. It must be realized, however, that the spectral information is observed in a region shifted to longer wavelengths. For example, if the 632.8 nm HeNe line is used, the Stokes Raman lines may extend as far as 850 nm, well within the short-wave NIR region. The net effect of the λ^4 losses and detector insensitivity can make changes to longer wavelength impractical, unless the choice of detector is flexible or a deterioration in performance can be tolerated. Newer detection technology, using silicon-based detectors (linear and CCD arrays), which are sensitive to just beyond 1050 nm, or InGaAs detectors, which are sensitive to $\sim$1700 nm (FT-Raman instruments), have provided the ability to record good quality spectra with longer wavelength lasers.

Up to this point, only traditional Raman spectroscopy has been addressed, and the emphasis has been on the interference from fluorescence, the relatively poor overall efficiency of the Raman scattering process, and the dramatic loss of Raman intensity at longer laser excitation wavelengths. An obvious conclusion might be that the technique is very sample dependent, and that there is not a general optimum solution—the trades often being wavelength (sensitivity) vs. fluorescence. There are some enhanced Raman methods, which for some compounds produce significantly intensified Raman spectra, and which overcome some aspects of this dichotomy. Two such techniques are resonance Raman (Asher, 1993a, b; Asher et al., 1993; Chen et al., 1994) and SERS (Fleischmann et al., 1974; Garrell, 1989). One other Raman-based technique worthy of mention is coherent anti-Stokes Raman spectroscopy (Valentini, 1985)—further discussion of this technique is beyond the scope of this book.

Resonance Raman is particularly interesting because it can turn Raman spectroscopy into a highly specific probe for certain functional groups or chemical sites. Resonance Raman occurs when the laser excitation frequency coincides with an electronic absorption band. In this case, the vibrations associated with the absorbing chromophore are enhanced by as much as 10^3–10^6 times the normal Raman intensity. These intensified Raman lines are linked to the specific chromophore site and functional groups or sites, within the molecule, that interact with the chromophoric group. The early resonance Raman experiments were with the visible lines of the argon ion laser, and this obviously constrained the technique to a limited set of colored compounds. Of these, the work with the heme chromophore of the hemeglobin molecule was the most significant. It is possible to observe the influence of external molecular ligands, such as oxygen, carbon monoxide, and cyanide, on the critical heme site, free of interference from the remaining of the protein structures.

Moving to shorter wavelengths from the visible towards, and into the ultraviolet regions might, at first site, seem to be impractical because one normally equates high levels of native fluorescence with the use of UV excitation. However, many compounds absorb in the ultraviolet spectral region, and so there is a high probability for the resonance Raman effect to occur. Also, it has been observed that below 260 nm excitation that there is virtually no interference from fluorescence (Asher, 1993a). One of the main issues here has been the appropriate selection of a laser operating in the UV range. One approach is to use a dye laser pumped by a Nd:YAG or a XeCl excimer laser, coupled to frequency doubling and tripling crystals to provide a wavelength selectable range of 200–750 nm (Nd:YAG) and 206–950 nm (excimer). Both these laser systems provide a pulsed laser output. A practical alternative, where continuous wavelength tuning is not a requirement, is an intracavity frequency doubled argon ion laser, which provides continuous output of five excitation lines in the range 230–260 nm.

As noted, another technique that provides an enhanced Raman spectral output is SERS. Unlike resonance Raman, the laser wavelength is not important, unless aqueous-based measurements are made with NIR excitation. As the name implies, the measurement is specific in nature, and somewhat limited in application to surfaces or interfaces. Most studies of SERS have been performed on electrode surfaces. An signal enhancement, in the order of 10^6 is observed for adsorbed species on certain metallic electrode surfaces, notably metals such as gold, silver, platinum, and to some extent copper. The original experiments (Fleischmann et al., 1974) involved calomel (Hg_2Cl_2) on a mercury surface, and later work involved organics, such as pyridine adsorbed on roughened silver electrode surfaces. The enhancement phenomenon is not restricted to electrodes, and similar effects have been observed for other substrates involving metals, such as colloidal suspensions of metals and metals deposited or embedded in oxides. A critical factor in all the experiments is the surface roughness, which is nominally at the atomic scale. The coinage metals, copper, silver, and gold, appear to exhibit the most consistent SERS effects. The origin of the enhancement is believed to be associated with two different mechanisms: (1) an increased electric field in the region of the roughened surface (electromagnetic enhancement) and (2) a chemical enhancement caused by a charge-transfer state between the adsorbate and the metal. The electromagnetic enhancement is thought to be a much larger effect than chemical enhancement. Even greater enhancement is available by utilizing a combination of SERS and surface-enhanced resonance Raman spectroscopy. Although experiments involving metal surfaces or colloids might appear to be limited, the phenomenon does open up interesting applications in the area of analytical chemistry because of the potential selectivity and sensitivity which might be achieved. In recent years there has been a considerable effort to obtain spectroscopic information from a single molecule by SERS techniques (Kneipp and Kneipp, 2003).

One final measurement-related issue for Raman spectroscopy is the influence of water and atmospheric absorptions. Unlike IR measurements, Raman measurements are not impacted by the presence of atmospheric water vapor or carbon dioxide. Although carbon dioxide does possess a strong symmetrical stretching vibration, which is Raman-active vibration, the Raman cross-section of gases is extremely low, and the interference in the optical path of a Raman instrument at the sampling point is infinitesimally small, and has no significance relative to the measurement. Likewise, with water, which is also a poor Raman scatterer, the presence of water vapor is inconsequential.

Normally, for most measurements, water is considered to be an appropriate solvent for the Raman spectroscopy because of its weak Raman spectrum. However, this is not a general statement. In the case of NIR illumination, a problem can arise from the fact that absorption can occur in the regions of the overtones of water. The Stokes Raman lines from a sample excited by the Nd:YAG laser can fall in the region of these water absorptions, which can be quite intense at long pathlengths. The observed result is a loss of signal intensity in the regions of water absorption. A similar problem can be encountered with optical fibers. The choice of fibers is dependent on the laser wavelength selected for a specific measurement. It is normal to use silica or quartz fibers—these are relatively inexpensive, and normally transmit both visible and NIR radiation. Silica fibers are available with different levels of hydroxyl groups—the ultralow OH content fibers are required for NIR based Raman measurements. However, for measurements involving visible lasers, with wavelengths in the red (such as a HeNe) or that involve scattering into the red region, it is necessary to use high OH content fibers. In this case, metallic dopants used in the cladding of ultralow OH fibers produce a strong absorption around 640 nm.

1. Base Measurement Technologies

Like IR spectrometric instruments, Raman instruments today fall into two main categories—dispersive instruments and interferometric (FT-Raman) instruments. Many relevant aspects of the available technologies for these two methods of measurement have already been covered in the section dealing with IR instruments. The main issues are: handling low light levels as efficiently as possible, providing adequate resolution to measure the scattered spectral lines, which are characteristically narrow, and removing the interference and associated straylight from the main laser excitation line (Rayleigh scattering). The Raman bands are particularly narrow relative to the normal spectral line widths in the measurement region covered—the ultraviolet through to the NIR, dependent on laser excitation selected. For traditional dispersive instruments, these are demanding criteria. However, as noted previously, the advent of high-performance laser line rejection filters has helped to eliminate a major practical instrument design issue.

The issue of optical efficiency in terms of light gathering from the sample and the reduction of light losses at all reflective or transmissive optical surfaces has to be addressed irrespective of the "light" analyzer technology used. The first issue relates to sampling. Dependent on the type of sample—liquid, powder, single crystal, and so on—an optimum geometry for sample presentation, illumination, and scattered light collection must be selected. Various illumination and collection regimes have been proposed, and used, from near 360° backscatter, to 90° or 180° straight-through collection geometry. Typically, for sample compartment experiments, some form of curved mirror, such as a parabolic reflector, is used to collect the scattered radiation, which is then re-imaged onto the entrance slits or aperture of the main spectrometer or analyzer.

One of the main benefits of Raman spectroscopy is that the excitation energy, and most of the scattered light energy, falls into the transmission range of glass or quartz optics. Commercial camera lenses, which are manufactured to high precision, corrected for minimal image distortion and chromatic aberrations, as well as anti-reflection coated for maximum image brightness, are often used as imaging optics between the sample and the slit/aperture. Ideally, all other transmissive or reflective optical surfaces should also be appropriately coated to reduce reflection losses. In older style monochromator-based instruments, the attention to such coatings was essential to ensure adequate light throughput, especially when double or triple monochromators were required. This is one of the reasons why concave (focusing) holographic gratings were favored, because their use reduced the number of optical surfaces required within the monochromator.

Dispersive Instrumentation

As noted, the trade-off among light throughput, resolution, and straylight rejection dictates the style of monochromator used for a particular Raman instrument. At the time when laser Raman started to gain popularity, from 1970, the instruments of that period were based on either double or triple monochromator designs. Typically with a Czerny–Turner style configuration. The first monochromator of the double monochromator (the first two, in the case of the triple monochromator) design was used for straylight rejection. While the triple monochromator excelled in straylight rejection, the throughput was very low, with a resultant poor signal-to-noise performance. In most cases, the use of this design was only justified for studies of spectral lines within a few wavenumbers of the exciting line. Single monochromator instruments were generally considered unacceptable because of straylight consideration, and so the double monochromator systems became the most popular design.

An interesting variant for the triple monochromator design was the combination of a double monochromator spectrometer (a scanning instrument) with a fixed grating spectrograph as the main spectral analyzing component. In this case, a linear array was used as the detector. Instruments with this configuration could be used as either a double monochromator with a traditional PMT detector, or as described with the array—providing optimum performance and flexibility.

A second consideration, relative to the use of a monochromator, is resolution. This is dependent on the wavelength region selected and the dispersion of the grating within that region—defined by the ruling density. The higher the groove or ruling density, in terms of lines per mm, the greater the resolution, but the narrower the spectral range covered by the grating. The selection of a suitable grating would be based on the required nominal resolution, and the Raman shifted spectral range (normally in cm^{-1}). Because the actual measurement range for the Stokes scattered lines is in either the visible or the NIR, it is important to appreciate the relationship between the spectrum range and the actual measured wavelength. Table 5 is provided to give an appreciation of the absolute wavelength range covered for a Raman shift spectral range of $0-4000$ cm^{-1} (comparable to the mid-IR spectral range) with the common laser excitation lines.

A typical grating density for operation in the visible range is usually between 1000 and 2000 lines/mm for a scanning grating instrument. As noted, if higher resolution is required, a higher ruling density must be selected, with a corresponding reduction in spectral range. Some instrument designs feature more than one grating on a rotatable turret. This permits coverage of a full spectral range at higher resolution, by the use of more than one grating, or it permits an optimum combination of low-resolution and high-resolution capability within a single instrument. Such instruments are often customized for a specific end-users requirements.

Scanning monochromators are perceived to be mechanically complex, and indeed when three monochromators have to be driven in synchronization then this is indeed a fact. Older systems often involved mechanical clutches and linkages between monochromators, and the monochromator drives themselves were cam driven. This often resulted in problems associated with "back-lash" and hysteresis within the drive mechanisms—generating the potential for frequency scale errors from scan-to-scan. More modern instruments feature direct drive

Table 5 Absolute Wavelength for a $0-4000$ cm^{-1} Raman Shift for Common Laser Lines

Laser	Excitation (λ, nm)	Absolute range[a] (nm)
Ar$^+$	Blue (488)	488–606
Ar$^+$	Green (514.5)	514–647
Nd:YAGx2	Green (532)	532–676
HeNe	Red (632.8)	633–847
Diode	sw-NIR (785)	785–1144
Nd:YAG	NIR (1064)	1064–1852

[a] Defines the Stokes Raman scattering range from the Rayleigh line.

monochromators where the grating is mounted on the shaft of a digitally controlled stepper motor. This removes most of the error sources described, and it allows easy synchronization between monochromators.

With the advent of laser rejection filter, such as the holographic notch filter (Tedesco et al., 1993), the monochromator designs were simplified. The need for the premonochromators (of a double or triple monochromator system) for the Rayleigh line rejection and the straylight reduction was removed. Based on the comments above, this also helped to remove some of the complexity of the optomechanical design. The next step is the use of array detectors, which may be used in both scanning and static grating systems, dependent on the wavelength range and resolution requirements. Initially linear silicon diode arrays were used, but today, the two-dimensional CCD array, in the form of a self-contained camera, is the preferred technology.

One of the benefits of using an array detector is that the dispersion element, usually a grating, can be static. That is the spectrometer, or more accurately, the spectrograph, has no critical moving parts associated with the frequency/wavelength measuring scale. This results in a very stable and reproducible optical system. One major benefit here is that signal averaging may be performed easily, and without the risk of signal degradation caused by nonreproducibility of the scanning system. Several spectrograph configurations are offered, some based on traditional monochromator designs, others feature alternative grating technologies such as the use of holographic transmission gratings (see Fig. 21).

Most of the latest instrument designs are built around a commercial CCD camera. In essence, the camera shutter is the only mechanical component, and this is not critical to the wavelength scale measurement. In the set-up of a fixed monochromator system, there is still the requirement to linearize the dispersion. In the scanning instruments, this linearization is normally performed by either a cam drive on the grating, or via programed control of the direct grating drive. In the case of a static system, the spectrum from a standard material or a gas emission line source is recorded, where the spectral wavelengths are accurately established (usually NIST traceable). From these data, an error correction table is established. All spectra subsequently recorded are then corrected by interpolation, or via a convolution with an error correction function.

An important factor to appreciate with an array-based instrument is that there is a fixed spectral range and resolution, defined by the grating and the system aperture/slit, and a fixed spectral digitization, defined by the array size. It is important to appreciate that a trade has to be made between spectral range and spectral resolution. In addition to the grating considerations discussed earlier, it is important to realize that the number of resolution elements is also defined by the array size. A common dimension used for the array is 1024 elements (or pixels) wide. Assuming that the detector is sensitive over the entire measured spectral range, and if there is the need to record a spectrum over the full $0-4000\ cm^{-1}$ wavelength range (in common with IR instruments), then the spectral resolution has to be limited to at least $8\ cm^{-1}$ with a 1024 element array. This provides a minimum digitization of one point per $4\ cm^{-1}$, which only just meets the criterion of two points per half bandwidth. On a dispersive instrument it is debatable whether this level of digitization is adequate. Especially because the noise, which has a much higher frequency, is not well represented at this digital resolution.

In practice, it is normal for many applications to record the spectrum at higher resolution, with at least $4\ cm^{-1}$

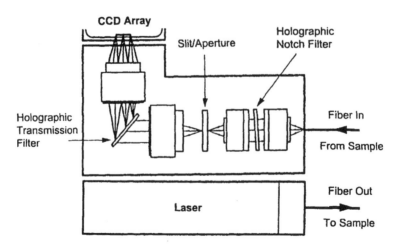

Figure 21 Schematic diagram of a fiber-optic Raman CCD-based spectrograph featuring a transmission grating. (Courtesy of Kaiser Optical Systems Inc.)

bandwidth at half height. This means that a decision has to be made about which portion of the spectrum is to be covered, remembering now that the digitization criterion now limits the spectral range to 2000 cm^{-1}. One option for many organic applications would be to ignore the lower frequency regions, and to consider a spectral range of say 2500–500 cm^{-1}. The only negative of this is that important hydride-related vibrations, such as the C–H and N–H stretch, are not covered by this spectral range. Note also that if higher spectral resolution, say 1 or 2 cm^{-1}, or a denser data point resolution (better than two points per bandwidth) is required, then correspondingly smaller spectral ranges are obtained (500 and 1000 cm^{-1}, respectively), and further applications-based decisions must be made.

Note that some of the lower cost CCD array camera offers less pixel density, and so the spectral range can be further constrained. New higher resolution array chips are becoming available, with improved quality and performance, and offering more than 1024 elements at reasonable cost—driven by the information machine market (copiers, telefaxes, scanners, etc.). A 2048 array has recently become available, and this would obviously have a favorable impact on the Raman instrumentation.

The Raman instrument manufacturers in part handle the current resolution/spectral range dilemma by providing mechanisms for exchanging gratings. This is achieved either by providing a mechanical system for changing the grating, such as a rotatable turret with multiple gratings, or by special grating mounts that permit easy grating exchange, while maintaining optical alignment. A minor complication here is that each time a grating is changed, the dispersion linearization mechanism must be changed accordingly. This may be automatic in some systems, but others may require operator intervention.

One advantage of the use of a CCD array detector is that typically only one dimension of the array is fully utilized. It is common for the dispersed image from the grating to illuminate several rows of an array. However, this area is typically only a small percentage of the available number of rows. An interesting option is to use more than one grating, located on the same mount, one above the other, featuring different groove densities. This projects a two or more dispersion patterns, one above the other, on the same array. This is an interesting configuration because if correctly set-up, the two separate spectra produced can be combined in software to produce a larger spectrum than produced by single illumination of a 1024 element array. Instruments are available with this feature, based on both reflective and transmissive gratings. In this way, a reasonable compromise can be established for a full spectrum at 4 cm^{-1} resolution—typically with some truncation of the extremes of the range to provide adequate spectral overlap of the two recorded segments.

Interferometric or FT-Based Instruments

By now, the reader must be aware that in the selection of a Raman instrument certain decisions must be made based on the application, and also that some compromises may have to be accepted. FT-Raman, featuring an interferometric-based measurement, is another choice, offering different performance characteristics, but again, not necessarily a clear-cut solution. For some time it has been recognized that scanning monochromators tend to be mechanically complex, and are inefficient in terms of their data collection and optical throughput. With IR instruments, issues of wavelength accuracy (Connes advantage), optical throughput (Jacquinot advantage), and signal multiplexing (Fellgett advantage) are considered net advantages for FTIR instruments over their dispersive counterparts.

Today, wavelength accuracy can often be handled by software. The throughput advantage is still an issue with a scanning monochromator instrument, but the multiplex advantage may be less of an issue outside of the IR region. It is only valid if the system is detector noise limited. Measurements with Raman spectroscopy are normally in the visible or the short-wave NIR region. In both cases, high sensitivity photon detectors are used, and these are usually shot noise limited, where the noise level is dependent on the signal level. This is counter to the situation where a measurement is detector noise limited, and so the multiplex advantage is not normally realized. At first glance, based on these considerations, the benefits of using an interferometer to measure the Raman spectrum might seem questionable. Furthermore, with the move towards fixed grating spectrographs, with CCD array detection, one clearly overcomes most of the original disadvantages cited for dispersive instruments, as well as a multiplex issue.

It is necessary to go back to one of the major obstacles of recording Raman spectra of materials in general to appreciate why there is strong interest in FT-Raman, and that is the dominance of broad-band fluorescence in many recorded spectra. Fluorescence occurs with most commercial materials and a wide range of organic compounds. This is one of the main reasons why Raman spectroscopy has been shown to be accepted for general analysis. While it is often possible to quench fluorescence and/or clean-up the sample, it is an extra and unwelcome step in a routine analytical laboratory. Furthermore, when fluorescence occurs, it may be so dominant that it is impossible to obtain a useable Raman spectrum. This is one of the big differences with IR spectroscopy, where it is almost always possible to record a useful spectrum.

For this reason, for the past twenty years, IR has been the technique of choice for routine vibrational spectral analysis. This has been a big disappointment for proponents of Raman spectroscopy, which on the surface is more attractive as a technique, in terms of implementation and sampling, than IR. FT-Raman has provided the opportunity to narrow the gap, and to make Raman spectroscopy nearly as easy to use as FTIR.

The key issue with FT-Raman is that the system can be configured to work in the NIR, and can be used with a Nd:YAG laser, operating at 1064 nm. In this spectral region there is virtually no fluorescence. It is probably incorrect to say that no fluorescence occurs. For example, there has been some evidence that asphaltenes, highly conjugated ring compounds found in petroleum products, do have electronic transition envelopes that extend into the NIR, and that these may exhibit weak fluorescence with the 1064 nm excitation. However, the occurrence of such material is not widespread, and their presence does not preclude the acquisition of good quality Raman spectra. Normally, with the 1064 nm illumination, it is not possible to use a CCD-based system because the detector cut-off starts around this wavelength. However, as a point of interest, it is possible to measure the anti-Stokes Raman lines which extend to shorter wavelengths, down to around 750 nm (from 1064 nm), for a full range (or near full range) spectrum. A commercial instrument, based on this mode of operation, is available from Kaiser Optical Systems.

The key to successful implementation of FT-Raman, like the recent dispersive instrument technology, has been the availability of high-performance laser line rejection holographic notch filters. Without such a filter, the laser line would swamp the A/D converter, and the shot noise generated in the detector would be distributed across the entire spectrum, negating the multiplex advantage. However, with the filter in place, and with the low intensity of the Raman scatter lines, the multiplex advantage is, in part, regained.

The implementation of FT-Raman is straightforward, and in many cases is treated simply an add-on to an existing FTIR instrument. A typical optical arrangement is presented in Fig. 22, where the output from the Raman accessory is coupled to the emission port of the interferometer. This arrangement offers significant flexibility and benefits for the analytical chemist, especially with a modern interferometer that features interchangeable optics. In principle, the requirements to adapt an FTIR instrument to Raman spectroscopy, apart from the Raman specific components, are a NIR transmitting beam splitter, typically a quartz substrate element, and a NIR sensitive detector, such as a germanium or InGaAs detector. The same instrument can be readily reconfigured for mid-IR work, by returning the beamsplitter to a KBr substrate, and changing the detector to a DTGS or a MCT.

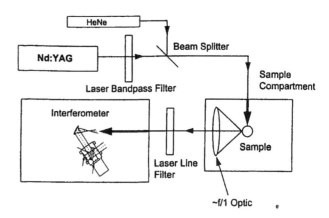

Figure 22 Schematic diagram of an FT-Raman spectrometer.

One implementation factor to consider is the power requirements for the laser. Although fluorescence, an important obstacle, is removed, the move to longer wavelength excitation has a few downsides. First, and most importantly, is the loss of scattering efficiency, which arises from the $1/\lambda^4$ relationship. With 1064 nm illumination, the Raman scattering efficiency is $8\times$ less than HeNe (632.8 nm) and $18\times$ less than the argon ion laser (514.5 nm). In order to retain comparable line intensities to normal visible lasers, which are normally operated with power between 5 and 200 mW, it is necessary to use a laser with power output as high as 1–4 W. This higher power requirement has at least three negative attributes: the lasers are more expensive, higher heat dissipation in the sample, with a potential for thermal damage, and the safety factor. This latter point is important, a full powered Nd:YAG beam can cause physical damage, and more importantly, it is invisible, leading to a potential safety hazard.

All instruments provide safety interlocks to help prevent operator exposure to the beam when working with the sample. A question arises, however, with optical fiber sampling where it may be difficult to obtain feedback of a "laser-on" situation, especially in a situation where a broken fiber is experienced. The secondary HeNe laser shown in Fig. 22 is placed with its beam confocal with the main Nd:YAG. When the safety interlock is in operation, the Nd:YAG beam is blocked and the HeNe is focused on the sampling point, allowing the operator to optimize the position of the sample. As soon as the sample compartment cover is closed, the interlock is deactivated, and the HeNe switches out and the Nd:YAG beam is unblocked.

Filter-Based Instruments

LIQUID CRYSTAL TUNABLE FILTERS. An LCTF also functions as an electronically tunable filter, but with

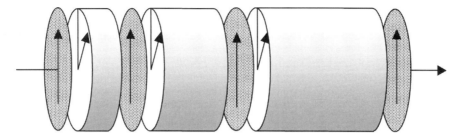

Figure 23 Schematic diagram of a three-stage fixed Lyot filter.

considerably less wavefront distortion than the AOTF. LCTFs are polarization devices based on the principles of the Lyot filter. A three-stage Lyot filter is shown in Fig. 23. Each stage consists of a birefringent waveplate oriented at 45° to, and sandwiched between, two parallel linear polarizers. The polarizers convert incoming light into plane-polarized light aligned at 45° to the fast axis of the waveplate, wherein it is decomposed equally into an ordinary wave and an extraordinary wave. The two waves travel with different velocities in the waveplate and recombine upon exit to produce a beam with modified polarization characteristics. The accumulated phase separation of the two beams is called the retardance, and is a product of the birefringence, Δn, and the thickness of the waveplate. Transmittance through the second polarizer only occurs when the retardance of the waveplate is a multiple of the wavelength. When this condition is met, the two beams will be in phase to combine to form plane-polarized light parallel to the exit polarizer.

The throughput of a single stage is given by

$$T(\lambda) = \cos^2\left(\frac{\pi R}{\lambda}\right)$$

When the retardance is high, the transmittance reaches a maximum over several orders depending on the wavelength of interest (see Fig. 24). A thick waveplate (high retardance) gives a narrow bandpass but the close proximity of the adjacent transmitted peak limits the useful wavelength range. These undesirable higher orders are suppressed in a multistage Lyot filter by cascading a number of stages together, typically in an arrangement whereby the retardance of each subsequent stage is twice the size of the previous stage. This simple relationship ensures that the wavelength of peak transmittance is the same for each stage, but at the same time the unwanted higher orders of the thickest stage are suppressed by the thinner stages. The total transmittance of the filter is the product of the transmittance of each stage (see Fig. 24). The overall bandpass (spectral resolution) of the multistage filter is governed by the largest stage (highest

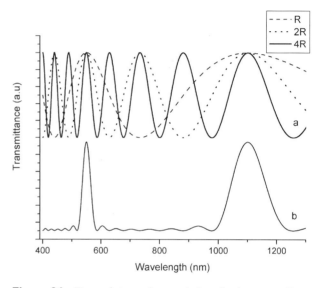

Figure 24 Transmittance characteristics of a three-stage Lyot filter.

retardance), and the maximum wavelength range is governed by the shortest stage (lowest retardance).

In an LCTF one or more liquid crystal waveplates are inserted into each stage of the Lyot filter. In liquid crystal waveplates, a liquid crystal material is held between two parallel glass windows that are coated with a conducting film such as indium-tin oxide. The windows are further treated so the molecules of the liquid crystal adopt an initial, or at rest, orientation. When a voltage is applied to the windows, the induced electric field causes the molecules to re-align according to the strength of the field. The re-orientation affects the birefringence of the device, and provides a means of changing the retardance by varying the applied voltage. The retardance and therefore the wavelength of peak transmission can be continuously tuned across the wavelength range of the LCTF. To maintain the same principles of operation as the fixed Lyot filter, the variable portion of the retardance must also cascade in multiples of two in each subsequent stage. This can be achieved either by placing double the number of liquid crystal waveplates in each subsequent stage, which has the advantage that the wavelength can be tuned by applying the same voltage to each liquid crystal element, or by increasing the applied voltage to double the retardance in each subsequent stage.

High spectral resolution and large spectral range are achieved by cascading over many stages. However, because of the significant losses that occur at each of the polarizers, these desirable traits are achieved at the price of throughput. Many clever alternative configurations have been developed involving split and interleaved waveplates that reduce the number of stages. With these recent developments the throughput and bandpass capabilities of LCTFs have surpassed that of AOTFs but the best feature of the LCTF arises from the linear optical path through the device and low wavefront distortion. Thus, LCTFs are ideal devices for spectral imaging. In the case of Raman spectral imaging, near diffraction limit image quality has been demonstrated using an LCTF with a spectral resolution of up to 7.6 cm^{-1} and transmittance between 6.7% and 16.3% across the spectral range (Morris et al., 1996). On the downside LCTFs are very expensive. Stand-alone units that operate in the visible region or out to 2000 nm are available from Cambridge Research and Instrumentation, Inc (Woburn MA).

2. Excitation Sources

Visible Lasers

Since the 1970s, lasers are the accepted source for Raman measurements. Early work was performed on conventional gas lasers, with the most common choices being the HeNe (632.8 nm), and the argon ion (488.0 and 514.5 nm) and krypton ion (530.9, 568.2, and 647.1 nm) lasers. Both the Ar^+ and Kr^+ lasers or sometimes a mixed Ar^+/Kr^+ laser was favored because they offered relatively high power up to a Watt or more with a selection of laser lines throughout the entire visible region. The choice would be based on the applications considered, and the cost factor. The higher powered Ar^+ and/or Kr^+ lasers have low wall-plug conversion efficiency, and require water-cooling, at relatively high flow-rates and with filtered (clean) water. The mixed gas lasers were the most expensive, and often the most difficult to maintain. The HeNe offered the benefit of portability, but was generally low powered (up to 50 mW), and provided less performance with the earlier instruments because of the lower Raman scattering $(1/\lambda^4)$ efficiency and lack of detector sensitivity in the red.

With the improved efficiency of modern instruments, lower powered lasers may be used, and both argon ion and HeNe are common, cost-effective choices. With the lower power, water-cooling is no longer a requirement for Ar^+ laser, the only negative issue is that they lose output with time due to the progressive loss of gas pressure, erosion of the electrodes, and deterioration of the windows of the laser tube.

A more recent development is the continuous (cw) frequency doubled Nd:YAG laser (532 nm), and this has become a serious alternative to the Ar^+ laser. Stability, size, and high wall-plug efficiency makes this laser attractive particularly for process-oriented Raman applications, where gas lasers are less practical.

Dye lasers offer a tunable source of visible laser excitation, and were once considered to be a potential solution for the Raman fluorescence problem. The concept was to tune the laser until the fluorescence was minimized relative to the Raman signal.

In a dye laser a jet of dye solution is continuously cycled through an optical cavity. An Ar^+ laser is used to "pump" the dye, to produce a broad-band coherent beam which is "tuned" with a suitable dispersion optic. Different laser/dye combinations may be used to provide specific wavelength ranges. With an Ar^+ laser pump and a suitable selection of dyes, it is possible to cover most of the visible region. The advent of higher efficiency instruments has enabled greater use of the "red"-end lasers (HeNe NIR diode lasers) to reduce the effects of fluorescence. Dye lasers are now mainly used for special applications, such as resonance Raman where the tunability confers considerable advantage in the study of excitation profiles.

NIR Lasers

By moving the laser excitation further into the NIR the incidence of fluorescence is dramatically reduced, and for a short period in the early 1990s FT-Raman using NIR excitation in the form of a Nd:YAG laser operating at 1064 nm became the centre of much research. In a Nd:YAG laser (or just YAG laser for short), the active medium is a yttrium aluminum garnet glass rod that has been doped with about 1% Nd^{3+}. In early models the population inversion was maintained by xenon lamp discharge but these designs possessed inherent noise characteristics, associated with the use of flashlight pumping, which made a significant noise contribution to the FT-Raman spectrum. This problem was reduced by the move to a laser diode-pumped system, which resulted in lower noise, and a smaller, more efficient, and more convenient package. While the most common implementation of the Nd:YAG is with the 1064 nm line, there is the opportunity, with modification, to use a longer wavelength laser line (actually a doublet) at 1339 nm (Asselin and Chase, 1994). At this wavelength, the interference from fluorescence is further reduced. The only practical problem is detection, where the common detectors are unable to handle the full range of scattered wavelengths, which extend into the top-end of the mid-IR. With a standard detector (InGaAs or Ge), it is possible to operate with a Raman shift as high as 1800 cm^{-1}.

YAG laser excitation at 1064 nm precludes the use of high detectivity devices such as silicon CCDs which cut-off around 1000 nm. However, many of the advantages in terms of reduced fluorescence can still be attained by using laser diodes, with typical wavelengths of 720, 785, and 830 nm with the added bonus of increased detectivity afforded by CCD detection. Originally, the use of diode lasers, although attractive because of their size and lower cost, was considered to be impractical because of wavelength stability issues—primarily due to mode hopping and thermal drift. Both of these issues have been addressed by modifications to the basic design, with an external cavity, and by the use of thermoelectric cooling/temperature stabilization. Commercial laser diodes suitable for Raman spectroscopy are now available at output powers of 30 mW in addition to higher power models of up to 300 mW that are suitable for Raman imaging applications.

Another important solid-state laser, which provides tunability from the visible (red) to NIR spectral region, is the Ti:sapphire, providing laser output wavelengths in the range of 670–1000 nm. Ti:sapphire lasers require powerful visible laser pumps, and are large and expensive systems.

UV Lasers

A recent, interesting extension to the Ar$^+$ laser involves frequency doubling by the use of an intracavity β-barium borate crystal. This generates a series of laser lines from the ultraviolet and into the visible (228.9–457.9 nm). An important attribute of this laser system is its production of a continuous (cw) output of lines in the ultraviolet, which is required for UV-resonance Raman applications (Asher, 1993a, b; Asher et al., 1993). Apart from resonance Raman studies, UV lasers have been exploited as a method to circumvent fluorescence. It seems counterintuitive to use UV laser to reduce the problem of fluorescence, but the high excitation energy allows the possibility of nonradiative pathways of relaxation prior to fluorescence transitions and this opens up a window between the excitation wavelength and the fluorescence band. So, in some cases it is possible to obtain useful portions of the spectrum in this window.

The He/Cd laser is another readily available source in the gas laser family, and has lines at 441.6 and 325 nm. He/Cd lasers are of similar construction to, and fall somewhere between, the Ar$^+$ and He/Ne lasers in terms of efficiency, size, and lifetime. Small air-cooled units are available giving around 50 mW of light. Unfortunately, the 325 nm line is not far enough into the UV to confer the fluorescence advantage outlined above.

Pulsed Lasers

The use of pulsed lasers was mentioned regarding the issue of fluorescence, and the use of time-based discrimination. Repetition rates of YAG and excimer pulsed lasers are around 20 Hz and 200–500 Hz, respectively, but the actual pulse duration may be only ~10 ns (5–15 ns). A gated detector is required to gain the temporal benefit of such a short pulse duration, otherwise the noise collected between pulses becomes a negating factor. One of the main problems associated with pulsed laser systems is that at the peak power of the pulse there is the possibility of sample damage, and it is necessary to defocus the beam to minimize the damage. This defocusing results in loss of the Raman sensitivity.

3. *Detectors*

Today, the detector technology is more or less predefined by the type of instrument that is selected. Dispersive instruments are mostly CCD array-based (multichannel), and FT-Raman instruments use singlechannel detectors. With Raman spectroscopy being a weak emission-based measurement, there are some important issues to consider when evaluating detectors. First are detector noise and dark current. Detector cooling is often an requirement to reduce these to acceptable levels when recording weak Raman signals. Second is the issue of detector range and the overall detector response profile. Earlier, when discussing laser selection, the limitations imposed by the detector were also raised. Dependent on the detector type selected, it may not be possible to record an entire Raman spectrum. The flatness of the detector response curve may also be an issue, because this will dictate the relative intensities of the measured spectrum. In this case, the output of the system may be normalized against a "white light" source of known output, taking into account the $1/\lambda^4$ reduction in the Raman scattered line intensities.

Photomultiplier Tubes and Other Single-Channel Detectors

Until the mid-1980s, the only significant detector for Raman spectroscopy was the PMT. These were installed in the standard, scanning grating monochromator instruments. The useable wavelength range of a PMT varies, dependent on the type of tube that is used. Typically, they are most sensitive in the ultraviolet and visible regions. Extended wavelength range tubes can be obtained with enhanced red sensitivity, extending the normal working range from about 400 to 800 nm, to out beyond 900 nm with a cooled photocathode. It must be remembered that if a red line excitation is used, such as the 632.8 nm HeNe, it is necessary to have such extended performance (see Table 5). One additional factor to consider is that the dark current of the extended range detectors tends to be higher than standard PMTs. Because of the high sensitivity, a PMT can readily saturate from over-exposure to high levels of laser radiation. This can cause a short-term memory effect, but normally, the tube will recover without permanent damage.

In FT-Raman, the instrument still features a single-channel detector. However, as discussed previously, the throughput and some of the multiplex advantage can be realized with the interferometric method of measurement, and so the negative aspects of the dispersive scanning instrument are not at issue. Although the early FT-Raman experiments were performed with visible lasers, notably the HeNe, the main attraction of the technique is the use of NIR sample illumination. With the Nd:YAG laser as the main excitation source at 1064 nm, the detector requirements are sensitivity in the mid-range NIR, to at least $\sim$1700 nm ($\sim$3500 cm^{-1}).

The original work on FT-Raman featured lead sulfide and germanium detectors, but these were generally inadequate in terms of signal-to-noise performance. High-purity germanium and InGaAs detectors are currently used, and of these InGaAs tends to be the most popular. The long-wavelength enhanced germanium detector is typically cryogenically cooled (77 K, LN$_2$). InGaAs is used with or without cooling (cryogenic or thermoelectric)—the trades being range vs. sensitivity. Cooling reduces detector noise, but also reduces the wavelength range covered by the detector. The consequence is that with cooling the Raman spectrum can be recorded out to 3000 cm^{-1}, and without cooling, this extends to around 3500 cm^{-1}. For routine analytical applications, the latter is obviously more desirable.

Earlier, it was mentioned that the Nd:YAG laser can be configured to operate at 1339 nm, which can significantly reduce the impact of fluorescence (Asselin and Chase, 1994). However, operating at such wavelengths can create a detector problem if a full range spectrum is required. With a Raman shift out to 3500 cm^{-1} (actual wavelength $\sim$2500 nm), a detector sensitive in the top-end of the mid-IR is required. With a standard InGaAs detector the spectral range only extends to 1800 cm^{-1}. If data is required at higher frequencies, it is possible to switch to an indium arsenide (InAs) detector, but with some loss in sensitivity.

Multichannel Detectors Including CCD Arrays

A major gain in the Raman spectroscopy performance from dispersive instruments has been achieved by switching from singlechannel to multichannel detection. The first systems featured the optical multichannel analyzer, which were offered as cooled PDAs with 512 or 1024 detector elements (pixels). Originally, the cost of these devices was perceived to be high, but with time, the benefits gained in terms of speed and design simplicity outweighed the negatives. In practice, the detector becomes the main element of the instrument, which contrasts with the FT-Raman instrument, where the interferometer tends to be the central component. Today, the same situation applies, where the CCD camera, the CCD cooler, and all associated electronics tend to dictate the instrument form and dimensions (as well as price).

As previously noted, there is always a trade in terms of the number of detector elements in the array, and spectral range, and that this varies as a function of wavelength and spectral bandwidth (resolution). A relatively wide range of CCD chips exist, and these vary in terms of quality, pixel dimension, the number of pixels, the array size, and so on, and all these factors impact the performance and the price of the CCD camera. The availability of higher quality chips with higher pixel densities is increasing, with a general decline in prices, fueled by the video imaging market. This has a net gain for the CCD Raman spectroscopy, where the camera price tends to define the instrument price, which in relative terms is high compared with FTIR.

Issues of CCD performance are tied to dark current, which is reduced by cooling. In most cases, multistage thermoelectric (Peltier) cooling is applied, and here the issue is how efficiently the heat dissipated from the device is removed. Both liquid and air cooling are used, and liquid cooling (from a closed-cycle cooler), which is more efficient, is preferred to acquire a lower dark current. Other issues of performance enhancement are related to the fabrication and implementation of the CCD chip. Today, the best performance is achieved from a thinned, back-illuminated device. In such an arrangement, a high quantum efficiency is achieved, and a spectral range to around 1000 nm or better can be obtained. Again, this detection limit constrains the choice of lasers. If the 785 nm diode laser is used, a popular choice for fluorescence reduction and cost and size reasons, the spectral performance declines when approaching 3000 cm^{-1}.

A final and important aspect to consider with the CCD array is that it is indeed a two-dimensional sensing device, and at least two factors can be considered. The typical image from the spectrograph extends over several pixels in the vertical direction on the array—horizontal is defined by the spectral dispersion. For a given column of illuminated pixels, the signal essentially comes from a single wavelength (dependent on spectral bandwidth). The output from these pixels can be combined to improve the overall signal level: a process known as binning. Binning can be performed on-chip, with the integrated electronic components, or in software. Here the selection depends on time constraints and signal-to-noise considerations.

The second factor is that because of the two-dimensional nature of the CCD, it is possible to place more than one image across the array at one time. The opportunity to utilize more than one grating by placing one image above the other was discussed in the section dealing with fixed grating spectrographs. A second option is to use a single grating, but to use images from different sampling points. In this manner, it is possible to do optical multiplexing of more than one sample at a time, when optical fibers are used. Dependent on the sampling probe geometry, more than one image is produced from a single sample when a bundle of fibers is used for collection of the scattered radiation. At the spectrograph end, the input fiber bundle is normally mounted to provide a vertical displacement of each image. These images in turn are focused one above the other on the CCD array. Even with this configuration, there is normally a large unexposed area on the array, which permits interfacing with more than one fiber-optic sampling probe.

To this point, all discussion has centered around silicon-based arrays. While these are extremely sensitive and practical detectors for Raman spectroscopy, they are limited by their wavelength range. Arrays based on germanium and InGaAs can be used but they are less sensitive than the silicon. Also the cost is high and the quality is generally not so good; there is a higher percentage of dead pixels. These arrays have been produced for military applications, and they are available commercially, but with lower availability than CCD devices.

4. Electronics and Primary Signal Handling

The selection of detector and the overall instrument technology tend to dictate the requirements for the overall signal handling. Some of the key issues pertaining to the way that signals are handled, and the importance of the A/D converter were covered in the IR instrumentation technology section. Where appropriate, some additional commentary will be provided here relative to the differing needs of dispersive and interferometric instruments, and the types of detector technology.

Dispersive Instruments

For the most part, the only important detector technology today is the silicon-based CCD array. The concept of binning, a factor in signal acquisition, has already been discussed. Beyond this, important signal acquisition factors that relate to the device are read noise and the digitization (number of bits provided). The lower cost CCDs typically have more read noise and have lower digitization levels—often around 11 or 12 bits. As discussed in the section on FTIR, the digitization level can define the ultimate sensitivity of a spectral measurement.

In the case of a CCD there is an interesting trade related to binning. On-chip binning will reduce the impact of read noise because the signal is only read once from the chip, but the digitization will be constrained to the bit level of the chip. Binning in software will result in a higher read noise, but also because of noise there is the opportunity to increase the bit level. The preferred method of binning tends to be application dependent, and is dictated by the required noise levels and/or measurement discrimination for low-level signals.

FT-Raman Instruments

As noted throughout this section on the Raman instruments, the laser line rejection filter plays a critical role in all modern Raman spectrometers. This is especially the case with FT-Raman. The critical issue being the signal acquisition. In the absence of adequate rejection of the Rayleigh line, the A/D converter would be filled with the radiation from the laser line, and consequently, the frequency components from the Raman lines would be too weak to be above the bit level. Also, as noted earlier, with a high level of laser light, the system would become detector shot noise limited, and there would be no opportunity to realize the multiplex advantage, in fact, because of the source of the noise, it would be turned into a disadvantage. However, modern holographic notch filters provide excellent laser line rejection, and the Raman signal is well differentiated from any residual laser light. Also, the A/D converter range of most FTIR instruments, which is normally around 18 effective bits, is more than adequate at the measured light levels.

Virtually all the discussion so far has implied the use of continuous laser output (cw lasers). Reference was made to pulsed lasers and the use of gated electronics to help discriminate the Raman signal from broad-band fluorescence. Pulsed systems are used, but the electronics for the data acquisition are relatively sophisticated, and most systems in use are custom-designed by the end-user. It is not usual to consider the implementation of a pulsed system based on a FT-Raman instrument. However, it has been demonstrated that it is possible to trigger the laser pulses from the HeNe zero-crossing signals of the interferometer. In this way the pulses and the data acquisition remain synchronized (Chase, 1994).

5. Factors that Influence the Recorded Spectrum

Raman spectroscopy is an emission-based technique, and there are potential difficulties in generating "standard" spectra that are consistent from instrument to instrument. With absorption spectroscopy, spectra are corrected for the instrument response function either by a double-beam instrument geometry or by a numeric ratio of the sample spectrum with the instrument background. Also, the peak intensities are uniquely defined in terms of the Beer–Lambert law, where the measured peak intensity is dependent on the concentration of the absorbing species, the optical thickness or pathlength of the sample, and a unique molecular property of the absorbing species, known as absorptivity.

Factors that govern the intensity of the Raman signal are the Raman scattering cross-section of the sample (in part, an equivalent to absorptivity), the polarization of the laser beam, and the intensity or power at the sample of the laser (dependent on the degree of focus). This may be expressed quantitatively in a similar form to the Beer–Lambert equation:

$$I \approx NI_0 \left(\frac{d\sigma}{d\Omega}\right) \quad (24)$$

where I is the intensity of the Raman band, N is the number of scattering molecules per unit volume, I_0 is the intensity of laser beam, and $(d\sigma/d\Omega)$ is the differential scattering cross-section.

There is no direct equivalent of the pathlength term in Raman spectroscopy. However, the position of the sample is critical, and for maximum signal, a representative portion of the sample must be at the focus of the laser. Most sampling devices permit fine adjustment of the sample position which may be optimized to the spectrum output. In cases where invisible NIR lasers are used, especially in FT-Raman, a secondary visible (HeNe) is used to help to visibly locate the sample, and typically motorized positioners are used to optimize the sample position when the main laser beam is switched in.

Because the laser power is variable, and is typically varied from sample-to-sample, there is no way to be sure of 100% reproducibility for a given sample. It is possible to place a power meter at the sampling position, but this only provides an approximate reading. Laser stability, both short-term (during spectrum acquisition) and long-term (sample-to-sample), is therefore an important issue. Also, in the data acquisition, the detector preamplifier gain is also a variable. Normally, this is set to fill the A/D converter, but variations in the overall spectral background will contribute to the energy falling on the detector, and this means that the gain setting is based on light level, not necessarily peak intensity.

The comments so far have concentrated on factors that influence the overall spectrum intensity, and does not address the issue of relative peak intensities. At least two factors influence the relative intensities of a recorded spectrum—the $1/\lambda^4$ factor, which is going to be constant for a given laser wavelength, and the instrument response function, which is constant for a given instrument, in a given configuration. If corrections are not made for these two factors, then the spectrum of given sample can vary significantly from instrument to instrument. Of these, the instrument response function is the most critical, with the detector response often having the greatest influence, especially at the higher Raman shifts. In some cases, this can be the difference between seeing a band in one spectrum and not in another for the same sample. This is common when attempting to observe the C–H stretching frequencies, where many detectors exhibit a fall off in response towards the red end of the spectrum. Another variable across the spectrum, when comparing two instruments will be the dispersion efficiency (for a grating instrument) or the beam splitter transmission efficiency (for a FT-Raman instrument). When measured, this function will include the attenuation effects of all filters, including the laser line rejection filter.

Various approaches are available for correcting Raman spectra to provide a more consistent spectral output between instruments. The $1/\lambda^4$ correction was discussed earlier, and is easy to implement. The user must attempt to be consistent with the laser output and the detector gain settings. In such cases, the use of a "standard" reference material can be beneficial to monitor consistency. With regard to the unique instrument response function, in particular for FT-Raman instruments, where the contribution can be significant, a blackbody emitter or a sample that provides a well-defined broad-band fluorescence in the spectral region of interest, can be used to characterize the function. In the case of a blackbody-based correction, it is possible to estimate a correction function from the temperature, and to fine tune this based on the actual output of the blackbody source emission curve.

6. Sample Handling and Accessories

Sample handling in Raman spectroscopy is relatively easy compared with sampling in the mid-IR, and to some extent easier than sampling for the NIR. There is no requirement to prepare the sample in a special way, unless it is necessary to spatially differentiate a region of the sample, such as in Raman microscopy. For most applications, the analysis is performed on liquids or solids. Gases can be studied, but normally a special cell with integrated optics is required to optimize the interaction between the laser radiation and the sample.

A major benefit of sampling with Raman spectroscopy is that glass or quartz can be used without significant spectral interference. One standard approach for handling liquids is to draw the sample into a small capillary tube, and to place the tube at the focus of the laser beam, within the sample compartment of the instrument. Alternatively, the sample may be retained in a sealed cuvette, or even within a glass bottle. In the latter case, external sample probes are sometimes used to simplify the sampling. Although glass may be freely used, it is important to realize that when working with NIR laser excitation, the presence of OH in the glass might cause attenuation of certain spectral regions because of some absorption at the longer scattered wavelengths. The same can apply when working with optical fibers, where low OH fibers are required for NIR based measurements.

There are few restrictions on the types of liquid samples that can be studied. The ability to use glass is particularly important because it provides the opportunity to study corrosive materials, such as concentrated acids. The narrow range of acceptable IR transmissive windows limits studies on these materials in the mid-IR to exotic materials such as sapphire, cubic zirconia, and diamond, and only diamond provides the opportunity to study the fingerprint region. Aqueous solutions and wet samples are generally not a problem, because the spectral interference from water is small. However, again when working with NIR laser excitation, it is important to realize that the Raman bands in some areas of the spectrum will be attenuated by water (OH-based) absorption.

Solids can be studied as powders, as continuous sheets or fibers, and as single crystals. Like liquids, powdered samples can be handled conveniently in capillary tubes or directly in sample vials. Alternatively, accessories exist where the sample is packed into a cup, similar to the approach used for IR diffuse reflectance. In some cases, an option is provided to spin the cup to continuously change the position of the focused laser beam on the sample to reduce the possibility of thermal degradation. Single crystals, and other crystalline or structurally oriented materials can be studied on goniometer equipped accessories, where the orientation needs to be accurately defined.

The role of optical fibers is becoming increasingly important in Raman spectroscopy for sample handling, and it is in this area where an increased use of Raman spectroscopy is anticipated. The opportunity to sample away from the instrument without the physical constraints of a sample compartment is very attractive. Sample interfacing with optical fibers is already well established with NIR spectroscopy (Ganz and Coates, 1996), but is somewhat limited in the mid-IR. In Raman spectroscopy, the measurement is relatively simple, and the materials— glass and quartz—are inexpensive to implement.

The style of probe that can be used typically features a single fiber for the laser illumination, and one or more collection fibers for return to the spectrometer. Several designs are commercially available, and two examples are provided in Figs. 19 and 20. Figure 20 is just one variation of a design theme that is provided by several of the instrument manufacturers. In this case, the probe contains certain active optical elements that remove unwanted or stray radiation. One of the problems with the use of long lengths of quartz optical fibers is that although the Raman spectrum of quartz is weak, over the length of the fiber, the spectral contribution can become a significant interference. This is removed at the probe head by a laser wavelength bandpass filter—effectively this rejects any weak Raman scattered lines from the quartz. At the collection side of the probe there is a laser line rejection filter. The net result is that only the weak Raman scattered radiation returns to the spectrograph. All spurious Raman lines and the main laser radiation are rejected at the probe head. The design of the focusing optics at the tip of the probe head can be varied to set different focal points from the probe. This is convenient for a permanent set-up where the probe is located outside of a glass port into a reactor, or simply mounted next to a glass reaction flask in a fume cupboard.

The probe shown in Fig. 19 is a simple insertion-style probe that features a bundle of fibers for the collection of the scattered radiation. The geometry at the head of this probe is important, relative to the angles of the fibers and the angles of the polished faces at the end of the fibers. This is important for locating a focal position of the laser beam, and for reducing the impact of the silica Raman scattering. This particular probe can be machined in materials such as Hastaloy or Monel to minimize corrosion, and the tip is sealed to permit operation under moderate pressures. This type of probe can be used for insertion into pipelines or reactors, for process monitoring, as a dip probe in a laboratory, and as a hand-held probe for the quality testing of materials in drums or vials.

From the nature of the sampling probes described above, it is obvious that Raman spectroscopy opens up new areas of application beyond the traditional laboratory. As instrumentation becomes more reliable, more compact, and more self-contained, it is becoming practical to consider Raman spectroscopy as a tool for quality control and process monitoring applications. The CCD array-based instruments are becoming particularly important in the area. One side advantage, mentioned earlier, is that it is possible to optically multiplex several images on the CCD array from different sampling points, again made easier by the use of optical fibers.

7. Raman Microscopy

The concept of combining a microscope with a Raman spectrometer, to produce an instrument often known as a Raman microprobe, was developed simultaneously in France and in the United States in the mid-1970s. For many years the instruments were large, expensive, and limited to research applications. Recently, less expensive, bench-top microprobe spectrometers, based on a single monochromator and a CCD detector, have become available. At the same time microscope accessories have become available for FT-Raman spectrometers utilizing near-IR excitation. Hence, application of Raman microspectroscopy has expanded rapidly in research, and also in industrial, analytical, and forensic laboratories. Recent publications have discussed developments in the field (Sommer, 1998; Turrell and Corset, 1996). Without doubt microscopy has become the most important sampling method for Raman spectroscopy. For some time it has been appreciated that Raman spectroscopy should offer advantages over IR microscopy because of the use of shorter wavelengths, where the lower diffraction limit enables the technique to examine smaller sample areas. In the mid-IR, this limit starts to take effect around 10 μm, whereas in Raman spectroscopy, dependent on the excitation wavelength, this can be below 1 μm. Spectral and spatial discrimination at the one micron level is particularly important for microelectronic, biological, biomedical, and forensic applications.

One important issue to be aware of during the sampling is the destructive effect of the laser. At the high magnification of the microscope there is a resultant high laser light flux at the focal point at the sample. This can cause materials to evaporate or to thermally degrade if attention is not paid to the problem. However, it is not difficult to see when this is happening either by visual inspection of the sample after the measurement, or by noting changes in the spectrum during data collection.

Microscopes that have the capability to limit the volume viewed to a thin layer of sample around the focal plane are known as confocal microscopes. For samples with some transparency it is possible to move the focal plane progressively into the sample and therefore to selectively view horizontal "slices" of the sample. For optical microscopes, confocal capability is achieved by incorporating a pinhole in the back image plane of the microscope. This enhances spatial resolution slightly in the focal plane (the xy-directions) and, more importantly, leads to a dramatic improvement in resolution along the optical axis (the z-direction). The physical limit applied in confocal microscopy leads to a minimum depth of focus (Δz) that has been estimated by Juang et al. (1988) to depend on the numerical aperture of the objective lens (NA), on the refractive index of the immersion medium (n), and on the wavelength of the light (λ) according to the expression:

$$\Delta z = \pm \frac{4.4\, n\lambda}{2\pi(\mathrm{NA})^2}$$

Application of this expression with light of wavelength 514.5 nm and an objective lens with NA = 0.92 gives a maximum achievable depth resolution of 0.6 μm (Barbillat et al., 1994).

The technique of confocal microscopy has been applied to Raman microspectroscopy where resolution along the optical axis is the major benefit (Puppels et al., 1990; Tabaksblat et al., 1992; Turrell et al., 1996). Figure 25 shows a schematic of the principle of confocal Raman microspectroscopy. The laser spot on the sample forms an image at the image back plane, which is mainly blocked by the small pinhole placed on the optical axis. The size of the pinhole is in the range 100–500 μm, and it has the effect of eliminating light originating from the out-of-focus regions of the sample, both in the focal plane, but more importantly from above and below the focal plane. Performance is further improved by also incorporating an illumination pinhole in the incident laser beam optics to remove speckle and diffracted light so that a clean focus is achieved. Figure 26 shows in diagrammatic form the effect of the pinholes. If the Raman microprobe is focused on a sample consisting of a thin slice, light will reach the detector. However, if the focal plane is slightly above or below the sample, then the greater proportion of the Raman scattered light will be eliminated by the pinhole and will not reach the detector. While the pinhole approach has been the most common method of achieving a confocal arrangement, the theory allows for restriction of the light at any image plane of

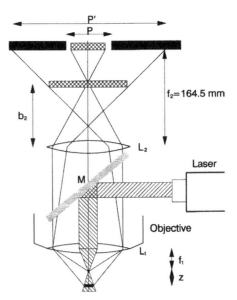

Figure 25 Diagram illustrating the principle of confocal Raman microspectroscopy (Tabaksblat et al., 1992). A laser spot on the focal plane is imaged onto the pinhole P and the light is able to pass through. A laser spot at a distance z below the focal plane is imaged to size P′ and is largely blocked by the pinhole P. L, lens; M, beamsplitter; f_1 and f_2, focal lengths of lenses L_1 and L_2, respectively; b_2, image distance of out-of-focus laser spot. (Reproduced by permission of the Society for Applied Spectroscopy.)

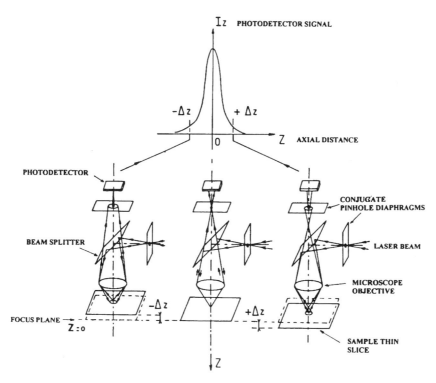

Figure 26 Principle of confocal microscopy and depth discrimination (Barbillat et al., 1994). (Reproduced by permission of John Wiley & Sons.)

the sample within the optical system. Renishaw used this fact to great advantage by developing a bench-top Raman microprobe spectrometer, in which part of the confocal capability is achieved by restriction of the light by control of the slit width at the entrance of the monochromator. Further discrimination of the light from the focal plane is then achieved by limiting the active area of the CCD detector. This detector is an array of approximately 400×600 pixels, the long axis of which is used to the describe the spectrum, and the shorter axis to describe the image height. In a typical CCD camera the pixel size is around $20\ \mu m \times 20\ \mu m$. For normal Raman mode, an image height of about 20 pixels is recommended. When confocal capability is required the image height is restricted to only 4 pixels, so that the active area becomes 4×600 pixels. A resolution of about $2.5\ \mu m$ in the z-direction has been claimed.

The major benefit of confocal operation is that "optical slicing" becomes possible so that spectra can be obtained from beneath the surface of the sample merely by moving the sample a known distance towards the objective of the microscope, allowing samples to be mapped from the surface into the bulk material. This generated some excitement within the spectroscopic community until a warning was sounded by Everall (2000) that the depth discrimination of Raman microprobe spectrometers may be much worse than previously thought when probing depths more than a few micrometres from the surface. The reason for this is the refraction at the air–sample interface which occurs when using a "dry" metallurgical objective, as typically used in a Raman microprobe. Everall found that as the depth in a depth profiling experiment was increased, the position of the centre of the focal plane in the z-direction increased dramatically, and also the depth of focus became progressively larger. The use of an immersion objective is recommended to overcome refraction at the sample surface so that movement of the microscope stage in the z-direction, and the position of the focal plane within the sample are more closely related. The confocal aspects of Raman microprobes are very important in terms of understanding where the Raman signal originates. In confocal mode, the signal is restricted to a few μm at the surface, but if the system is operated nonconfocally then the Raman scattered light from deeper in the sample may reach the detector. For a homogeneous sample this would not matter, but for a sample where the surface layer has different characteristics, it could be significant.

8. Raman Imaging and Mapping Spectroscopy

Raman imaging is the process of studying an area of the surface of a sample (Barbillat, 1996; Garton et al., 1993). Imaging refers to the experiment that utilizes global illumination of the area under study, which is then imaged directly onto a CCD detector. In the past, imaging has only been possible at a single Raman wavelength, or a narrow band of wavelengths (Williams et al., 1994). Systems of filters have been used to provide the narrow band of wavelengths, and in fact by using a set of up to five filters and changing the angle of each of the filters a reasonable part of the Raman spectrum can be obtained, albeit at rather low-resolution, $15-20\,cm^{-1}$. Images can then be generated by selecting a particular filter and an angle that corresponds to a certain Raman band and imaging the sample through the filter onto the CCD camera. Clearly it is a slow process to generate images on a number of Raman bands. Recently instrumentation has become available which allows the complete Raman spectrum to be collected for each pixel. Such instrumentation is based on an LCTF, and is the major product of the company ChemImage Inc. The LCTF is tuned electrically, and has no moving parts. When a portion of the sample is globally illuminated by the laser source, the scattered light passes through the LCTF before being imaged onto the CCD camera. The LCTF then allows spectroscopic analysis of the light to build up a Raman spectrum at each pixel of the CCD, which corresponds to a particular pixel of the sample surface. The method produces many thousands of spectra simultaneously and the images are created by software manipulation of the data. The LCTF approach has the advantage that the data are measured quickly, and that once collected the data can be investigated at leisure to obtain the maximum information. A further advantage is that the spatial resolution is essentially diffraction limited and, for example, for an argon ion laser with a wavelength of 514 nm would be at least 0.5 μm. However, there are several drawbacks. The system is expensive, not only because the LCTF is expensive, but also because a high output laser around 2 W is required to illuminate a section of the sample, say 30 μm × 30 μm, at a reasonable power density. Simple arithmetic shows that the source would need to be 900× more powerful to achieve a similar measurement to the standard single point micro-Raman measurement using a 10 mW source. The expense of the LCTF and the laser sources also precludes having several laser lines available because a different LCTF is required for each excitation wavelength. It is now common practice in Raman spectroscopy to have at least two, and possibly more, lasers available so that problems of fluorescence or sample degradation might be avoided by moving to a different source.

Raman mapping is an equivalent procedure which uses point illumination to obtain the spectrum at a single point on the sample (Appel et al., 2000; Keen et al., 2001). The sample is then moved under computer control so that spectra are gathered in a grid pattern over an area of the sample. The spectra are saved as a multifile, and may be displayed and interpreted using commercially available software. Raman mapping requires the collection of large numbers of spectra and therefore may be very time-consuming. For example, Keen et al. (2001, 2002) describe Raman maps of a heterogeneous polymer surface for which a 50 μm × 50 μm section of the surface was studied in detail by collecting spectra with a laser spot size of about 1 μm and moving the sample 1 μm between steps. Hence a total of 2601 spectra were collected. Each spectrum took around 20 sec, and additional time was required to move the sample. The whole data collection process took in excess of 16 h. However, a large region of the Raman spectrum or even the entire spectrum may be collected at each point, providing a large amount of data for further interpretation.

Raman mapping may be expedited by a line scanning process. There are two approaches: the first is to use a small scanning mirror to cause the laser beam to scan over a small distance of about 20–50 μm. The line is imaged onto the CCD camera, and independent spectra can be collected at a spatial resolution of about 1 μm. In order to collect mapping information the sample needs to be moved in one dimension only, perpendicular to the direction of the scanned line. Thus, for a 50 μm × 50 μm map, the sample needs to be moved only 50 times. There is little benefit in line scanning unless the laser power is increased substantially, since the scanned line would require 50× longer to collect data equivalent to 50 individually collected spectra. To obtain the benefit the laser power is usually increased from 10–15 mW, typically used for single point measurements, to 200–300 mW. The fact that the laser spot is in continuous motion helps to prevent sample degradation with this much higher laser power. The second approach is similar but utilizes a line focus objective. This method has the benefit of fewer moving parts, but the confocal aspects of the measurement are more difficult to define.

IV. PRACTICAL ISSUES OF IMPLEMENTATION OF IR AND RAMAN INSTRUMENTS

In various sections of this chapter there have been discussions relating to the changing role of all forms of vibrational spectroscopy. Today, vibrational spectroscopy is no longer just a tool for the specialist spectroscopist. For the three main areas covered in this chapter—mid-IR, NIR, and Raman spectroscopies—there now exist basic simple instruments that can be used for routine analysis. Also, in all cases, special customized versions of these instruments are available for dedicated applications—these extend from sophisticated microscope instruments for technology and medical applications, to rugged, compact instruments for manufacturing applications—near-line and on-line. Considering the range of instrument types and their targeted roles, it is important to discuss briefly issues that relate to system architecture and packaging.

A. Instrument Architecture and Packaging

Originally, spectroscopic instruments were designed as large, monolithic structures. This type of design was considered to be necessary to gain performance and stability. Also, it was always envisaged that the instrument would be located on a bench in a laboratory (probably air conditioned), and with a specialist operator. Also, early instruments were mechanically intensive, which to some extent defined their size and form, and added complexity, which in turn translated to reliability problems. Today, with modern microelectronics, with high-performance electrooptic components, and with high-powered compact computers, these problems and constraints no longer exist.

When defining an instrument architecture it is essential to evaluate the needs of the end-user, the planned role for the instrument, and the location of the instrument. The starting point for spectroscopic instruments is the research laboratory. Today, research instruments are still relatively large. This is primarily associated with the need for high-performance and high-resolution optics. This tends to apply whether dealing with an IR instrument or a Raman instrument. Most research instruments offer maximum flexibility in terms of the use of sources, the main optical analyzer component (dispersive or interferometric), detectors, and the use of a wide array of accessories. Nearly all of the high-performance FTIR instruments offer multiple beam geometries which provide this flexibility. Similarly, the research grade Raman instruments, whether dispersive or interferometric, offer a similar degree of flexibility.

Having stated the needs, the architecture for a research instrument requires the ability to provide multiple, switched beams, usually collimated; the ability to attach optical assemblies to the main optical bench, and maintain alignment; and the ability to access specific electric signals from attached devices (such as detector inputs). The computer system should not only provide all standard instrument control, data acquisition, and data manipulation functions, but it should enable the use to perform low-level functions, such as accessing control signals, and modifying control and data acquisition functions. In summary, the system is designed with a totally open architecture, and the end-user is provided with access to all operating levels within the instrument.

The next grade of instrument to consider is the nonroutine analytical instrument. In this case, there may be a single operator, or the instrument may be available for open access as a service instrument. Much of the flexibility of the research instrument is still required. Generally, the performance and/or the spectral resolution requirements may be lower. For many applications, optical throughput is important, and this is often traded for resolution. The architecture is still required to be somewhat open, providing the option to couple different accessories. The ability to make the instrument multifunctional is often one of the purchasing justifications. However, the end-user typically wants easy operation, and there is seldom the requirement to access low-level functions within the instrument. Software tools are important, and these include data manipulation functions, interactive graphics, and search and match algorithms for materials characterization. Important accessories for this area of application include microscopes and chromatographic interfaces.

The routine instrument is typically smaller, lower performance, and less flexible. Often, the main requirements of such instruments are to produce a spectrum from a liquid or solid sample (sometimes a gas with FTIR). On standard laboratory instruments, the sampling is normally performed within the sample compartment. If special accessories are used, these are often prealigned, and these either "plug-in" or are located on some form of kinematic mount. Alignment of such accessories must be minimal. Relative to the use of a computer, the user interface must be simple, and in most cases, the most important operations are simply obtaining a spectrum, displaying a spectrum, and storing the data (on disk). Important add-on software would include spectral searching and quantitative analysis programs. Some routine instruments provide the option for a single external optical beam, and this is typically dedicated to a specialized sampling accessory or a basic microscope. The most desirable features of this style of instrument, which impacts the architecture, is a compact design and ease-of-use. The user interface can be from a custom keyboard and display, or from a personal computer—sometimes an integrated lap-top computer.

Up to this point, most of the instrument platforms described are normally designed to function within a laboratory environment, and with some form of trained operator. For these instruments, it is normally adequate to use a molded external cover which bolts to the baseplate of the spectrometer. The optical bench of the spectrometer is usually vibrationally isolated from the external baseplate, and this ensures that vibrations transmitted from the external cover do not influence the optical performance. The main optical components are normally housed within a sealed compartment. In the case of a dispersive instrument, this is normally light-tight, with optically blackened internal surfaces. Light exclusion may not be required for interferometer-based IR instruments because only light originating from the source, which is uniquely modulated, is detected. However, most IR instruments are sealed and desiccated to protect sensitive optics and to remove the spectral interference of water vapor. Thermal compensation, or even thermal control may be needed in some instruments, especially where thermal expansion/contraction interferes with optical alignment and/or the accuracy of a measurement.

The electronics of most modern instruments are located inside the instrument. This includes the power supplies, which in lower cost instruments are usually of the switching variety. In higher cost instruments, where stability and attention to noise sources is important, linear power supplies may be used. The main control and data acquisition functions are typically located on different circuit boards. For these functions, there is often a move towards the use of DSP chips and/or ASICs for speed, ease-of-implementation, and reliability. Often, these provide important processing functions, such as on-board fast Fourier transforms (for FT instruments), and linearization and calibration functions (for dispersive instruments). Some instruments include additional on-board intelligence, in terms of additional computer functions and even some limited end-user programmability. In terms of architecture, circuit boards are normally either interfaced via some form of internal bus structure (wire harness or a passive backplane), or they plug into a mother board (usually a control board).

With the move towards instrumentation for operation outside of the laboratory, it is also necessary to consider both architecture and packaging. Most instruments used outside of the laboratory environment are used for dedicated applications—including raw material screening in a warehouse, near-line monitoring of batch-produced products, on-line monitoring of a process stream, field monitoring for an environmental application, and so on. Today, most dedicated applications for vibrations spectroscopy tend to be industrially oriented. However, in the future it is reasonable to expect them to extend to some consumer or medical-based applications. These different areas have been highlighted to help to define various scenarios involving instruments, and to indicate the different environmental operating issues that impact design, architecture, and packaging.

Three basic types of overall instrument platforms emerge from these application areas—bench-top instruments, permanently installed, industrially hardened instruments (on-line analyzers), and field-portable instruments. If suitably designed, bench-top instruments may also serve as transportable instruments. Ideally, the bench-top and the industrially hardened instruments feature common measurement technology for a given area of implementation. The reason being that the bench-top unit serves as a plant support instrument to the on-line analyzer. If both instrument platforms are served, a common architecture is usually desirable—it simplifies implementation, it reduces manufacturing costs, and it minimizes maintenance overhead.

Most instruments used outside of a laboratory must be sealed and protected against moisture and dust. They require a robust design, and where practical the entire instrument should be housed within a single package. It should be expected that there will be extremes of temperature and vibration—both are normally detrimental to the operation of optical instruments. This requires that all sensitive optical elements are adequately protected and isolated from the operating environment. When working outside of the protected laboratory environment, it is usually necessary to conform to specified electrical and fire safety codes. This may require the instrument to be housed within a purged safety enclosure or an explosion-proof housing to eliminate potential fire-hazard problems. In practice, most forms of vibrational spectroscopy instrumentation can be adapted to work in such environments, if attention is paid at the beginning to the system architecture and the packaging requirements. However, it is seldom practical to work backwards.

As we progress in the 21st century, it is predicted that more instrumentation will move out of the laboratory. Also, it is expected that instrumentation will get smaller and less expensive. Field-portable instruments are becoming more popular, and these are expected to move towards simple hand-held devices, even for relatively advanced spectroscopic measurement techniques, including IR and Raman spectroscopies. Obviously, this will require miniaturized spectrometers, battery-powered operation, and remote communications, possibly via cellular phone, satellite, or radio. Already, micromachined components that will provide a spectrometer on a chip are being offered (Zuska, 1996).

B. Data Handling and Computer Applications

Almost without exception, all modern instruments feature some form of communication to an external computer. In a few cases, computers are built into the instrument, but even these instruments are usually required to provide an external output of data. Today, the personal computer, usually an IBM-compatible, is the data system of choice and has replaced workstations manufactured by individual spectrometer makers. For most laboratory applications a Pentium-based processor is usually sufficient. Microsoft Windows is the standard operating environment. One Windows-based package worthy of note is *Lab-View* (National Instruments), which provides easy instrument interfacing and an ideal environment for prototyping and designing the instrument user interface. Computers from

Apple have been used for certain applications, especially where the graphics interface provides an advantage, however, in general their use has not been widespread.

With the transition to a standard, generic computer platform, and away from proprietary computers, the issue of software for spectroscopic applications is relatively easy to address. Most instrument manufacturers offer their own basic software for spectral data acquisition, spectral display, data manipulation, and storage/archiving. In addition, a full software package is also available from Thermo-Galactic Industries, known as *GRAMS*, which is offered with software drivers for virtually every commercial spectrometer. In addition, this software will read spectral data files from the instrument manufacturers' software. Some manufacturers offer a customized version of *GRAMS* for their instruments, in place of their own software, or as a supplement.

All the spectroscopic software packages offer a standard suite of data manipulation and numerical processing functions. They also provide graphical display functions and the ability to produce high-quality hard copy spectral output to a laser printer. In terms of data storage, most packages offer the option for alternative data formats, including the JCAMP-DX format, and a simple ASCII, comma separated PRN file format. The latter is particularly useful because it enables one to import data into other, nonspectroscopic software products, such as Microsoft *Excel*; mathematical/numerical packages, such as *Matlab*, *Mathematica*, and *MathCad*; and graphical analysis and presentation packages, such as Jandel's *SigmaPlot*, Advanced Graphics Software's *Slidewrite Plus*, and SoftShell's *IRKeeper*.

Spectral library searching is still an important software function for IR spectral data. With an increase in the usage of both the NIR and Raman instruments, there is now also a requirement for spectral search data bases to include these forms of spectral data. While some of the instrument manufacturers offer their own search packages and data bases, the major part of the data base business is covered by two noninstrument manufacturers—Bio-Rad, Sadtler division, and Aldrich Chemicals. Stand-alone software is provided from both of these companies, and in the case of Sadtler, a program known as *IR Mentor* provides assistance with spectral interpretation. Similar software interpretation programs are also offered by a couple of the instrument manufacturers.

Quantitative analysis is one of the most important areas of application in all three of the vibrational spectroscopy techniques. Although quantitative analysis started with mid-IR, it was NIR that increased its popularity for multi-component analysis. Chemometrics became established in the mid-1980s as a legitimate branch of analytical chemistry, and NIR spectroscopy was one of the reasons for its acceptance. It is usually necessary to use one of the multivariate chemometric methods of processing to extract the quantitative data from raw NIR spectra. However, these techniques are not limited to just NIR, and they may be used to advantage for the other vibrational spectroscopy techniques. In the cases of mid-IR and Raman spectroscopies, it is not always necessary to adopt a multivariate approach. If the spectral data is adequately resolved, then a traditional Beer–Lambert style of approach may be used, where the analyte peaks are identified, and correlated with concentration.

Again, many of the instrument manufacturers offer quantitative analysis software packages. These typically feature both simple univariate methods, and more comprehensive multivariate methods—with partial least squares and principal component regression currently being the most popular. In addition to the instrument manufacturers' software, several other companies offer multivariate packages, including Thermo-Galactic, Camo (*Unscrambler*), and InfoMetrix (*Pirouette*). *Matlab* from Math Works, Inc. has become an important development environment for chemometrics tools, and add-on packages and toolboxes are available that run under *Matlab*. These include locally weighted regression and neural network tools—both chemometrics methods, being considered for nonlinear modeling.

With a move towards instruments for dedicated applications, there is a need to develop tools to allow users to construct customized programs for specific analyses. These might include data acquisition, spectral data preprocessing, qualitative or quantitative analysis, reporting of the results, and communications with a host computer—as well as customization of the user interface. In the case of the user interface, both *Visual Basic* (Microsoft) and *LabView* (National Instruments) are important design and prototyping tools. In the past, most software packages offered a macroprogramming language to provide many of these functions. However, now there are software packages that allow a user to build up a method or a procedure using a graphical user interface based on Windows. The advantage of these packages are that they can help generate audit trails, which are essential for modern day GLP requirements. With applications in government-regulated industries, such as pharmaceutical and environmental, the need exists for software validation. Many software manufacturers are going through the certification requirements needed for validation, and some packages are already available.

V. INSTRUMENTATION STANDARDS

The issue of standardization was raised earlier in terms of instrument performance, validation, and standardized spectral data formats. This subject is important for all forms of instrumentation, and applies equally to both the IR and Raman instruments. It is a sufficiently important topic that it is worthy of some further mention. First, there is the issue of the instrumentation itself. All manufacturers strive to make a product that meets specifications. These specifications include a definition of performance in terms of wavelength/wavenumber accuracy, repeatability, and scan-to-scan precision. These may be referenced to standards available from NIST and IUPAC standards (Cole, 1977; IUPAC, 1961). With FTIR instruments, in principle, the wavenumber scale is controlled by the laser reference frequency. Its accuracy holds true in the absence of a sample or any constraining optics, such as an accessory. However, whenever attenuating optics are used that redefine the aperture of the instrument, this accuracy statement no longer holds true.

It is understood that every spectrometer, independent of measurement technique, has unique output characteristics that originate from variations in the source, the light analyzer optics, and the detector. In the case of IR spectroscopy, which is traditionally an absorption-based technique, it is customary to ratio the spectrum of the sample to the spectrum of the instrument background. This removes the convolved instrument spectral components and produces a first-order spectrum, which may be compared to spectra of the same sample produced on similar instruments. The term first-order is used here because the ratio does not take into account optical distortions caused by the sample or the sampling accessory. However, if an accessory has a unique background this can be included when the instrument background is recorded. Also, it must be realized that some unique characteristics remain in the spectrum relative to the instrument lineshape function, and any photometric errors in recording the band intensities.

There is now a movement towards providing methods for characterizing the performance of instruments. The ASTM provides some standard practices (ASTM, 1996b) for the different types of spectrometer. Once suitable standards are identified, it is possible to make all instruments of a given type conform to the standard. At the most basic level, this involves some form of correction to the frequency or wavelength scale. Beyond this, if the full instrument function can be characterized, it is possible to make software corrections that generate a standardized data format. Using such an approach, it is possible to transfer spectra between different instruments (of the same brand). In the future, it is expected that chemometrics will play a greater role in characterizing an instruments performance, and this will include methods for differentiating the differing contributions of the instrument and the sample to the final spectrum.

VI. RECOMMENDED REFERENCE SOURCES

Analytical IR spectroscopy has been practiced since the mid-1950s and as a consequence, there is a plethora of published literature on this technique. Much has been recorded in journals in terms of basic theory, instrumentation concepts and designs, and applications. Unfortunately, with the move towards digital cataloging of literature, some of the early work, which is still very appropriate today, is no longer readily available. The early work is particularly useful for the understanding of the basic concepts of measurement, and provides helpful practical hints and tips relative to sample preparation. Unfortunately, without access to this past literature we see reinvention of techniques, once regarded as second nature.

References to analytical Raman spectroscopy have been less abundant, but can still be traced back over a similar timeframe. Also, the interest in this technique has been somewhat cyclic, a result of several "renaissances" linked to major changes in the available technology. Because of the long timescale involved in both techniques, and because of the large amount and diversity of reference material, most of the sources quoted here are limited to key texts, published as books, and journal references that are more review or technique-oriented.

SUGGESTED REFERENCE TEXTS

Burns, D. A., Ciurczak, E. W., eds. (1992). *Handbook of Near-Infrared Analysis. Practical Spectroscopy Series*. Vol. 13. New York: Marcel Dekker.

Chalmers, J. M., Griffiths, P. R., eds. (2002). *Handbook of Vibrational Spectroscopy*. Chichester: John Wiley & Sons Ltd. (This is a superb set of five reference volumes covering the theory, instrumentation, and applications of vibrational spectroscopy. It is absolutely the most up to date, authoritative, and comprehensive source for all aspects of the field.)

Chase, D. B., Rabolt, J. F., eds. (1994). *Fourier Transform Raman Spectroscopy: From Concept to Experiment*. San Diego: Academic Press.

Chia, L., Ricketts, S. (1988). *Basic Techniques and Experiments in Infrared and FT-IR Spectroscopy*. Norwalk, CT: Perkin-Elmer Corporation.

Coleman, P. B., ed. (1993). *Practical Sampling Techniques for Infrared Analysis*. Boca Raton: CRC Press.

Colthup, N. B., Daly, L. H., Wiberley, S. E. (1990). *Introduction to Infrared and Raman Spectroscopy*. 3rd ed. New York: Academic Press.

Durig, J. R. (1985). *Chemical, Biological and Industrial Applications of Infrared Spectroscopy*. New York: Wiley.

Ferraro, J. R., Basile, L. J. (1978). *Fourier Transform Infrared Spectroscopy*. Vol. 1. Orlando: Academic Press. (See also Vols. 2–4, 1979, 1982, 1985.)

Ferraro, J. R., Krishnan, K., eds. (1989). *Practical Fourier Transform Infrared Spectroscopy: Industrial and Chemical Analysis*. San Diego: Academic Press.

Grasselli, J. G., Snavely, M. K., Bulkin, B. J. (1981). *Chemical Applications of Raman Spectroscopy*. New York: Wiley.

Gremlich, H.-U., Yan, B., eds. (2001). *Infrared and Raman Spectroscopy of Biological Materials*. New York: Marcel Dekker.

Griffiths, P. R., de Haseth, J. A. (1986). *Fourier Transform Infrared Spectrometry*. New York: Wiley-Interscience (Vol. 83 of Chemical Analysis).

Günzler, H., Gremlich, H.-U. (2002). *IR Spectroscopy. An Introduction*. Weinheim: Wiley-VCH.

Harrick, N. J. (1987). *Internal Reflection Spectroscopy*. Ossining, New York: Harrick Scientific Corporation (Original publication -John Wiley & Sons, New York, 1967).

Harthcock, M. A., Messerschmidt, R. G., eds. (1988). *Infrared Microspectroscopy Theory and Applications*. New York: Marcel Dekker.

Hendra, P., Jones, C., Warnes, G. (1991). *Fourier Transform Raman Spectroscopy: Instrumentation and Chemical Applications*. Ellis Horwood Series in Analytical Chemistry. Chichester, UK: Ellis Horwood Limited.

Herres, W. (1987). *Capillary Gas Chromatography-Fourier Transform Infrared Spectroscopy, Theory and Applications*. New York: Huthig.

Humecki, H. J., ed. (1995). *Practical Guide to Infrared Microspectroscopy, Practical Spectroscopy Series*. Vol. 19. New York: Marcel Dekker.

Katsuyama, T., Matsumura, H. (1989). *Infrared Optical Fibers*. Philadelphia: Adam Hilger.

Kerker, M. (1990). *Selected Papers on Surface-Enhanced Raman Scattering*. Bellingham, WA: SPIE-ISOE.

Lewis, I. R., Edwards, H. G. M. (2001). *Handbook of Raman Spectroscopy: From the Research Laboratory to the Process Line*. New York: Marcel Dekker.

Mackenzie, M. W. (1988). *Advances in Applied Fourier Transform Infrared Spectroscopy*. New York: John Wiley & Sons.

McCreery, R. L. (2000). *Raman Spectroscopy for Chemical Analysis*. New York: Wiley-Interscience.

Mirabella, F. M. Jr., ed. (1993). *Internal Reflection Spectroscopy: Theory and Applications, Practical Spectroscopy Series*. Vol. 15. New York: Marcel Dekker.

Parker, F. S. (1983). *Applications of Infrared, Raman and Resonance Raman Spectroscopy in Biochemistry*. New York: Plenum Press.

Pelletier, M. J., ed. (1999). *Analytical Applications of Raman Spectroscopy*. Abingdon: Blackwell Science.

Schrader, B., ed. (1995). *Infrared and Raman Spectroscopy: Methods and Applications*. NY: VCH Publishers, Inc.

Smith, B. C. (1999). *Infrared Spectral Interpretation: A Systematic Approach*. Boca Raton: CRC Press.

Strommen, D. P., Nakamoto, K. (1984). *Laboratory Raman Spectroscopy*. New York: Wiley.

Suetaka, W., Yates, J. T. (1995). *Surface Infrared and Raman Spectroscopy: Methods and Applications*. New York: Plenum Press.

Turrell, G., Corset, J. (1996). *Raman Microscopy: Developments and Applications*. New York: Academic Press.

Weber, W. H., Merlin, R., eds. (2000). *Raman Scattering in Materials Science*. Berlin: Springer.

White, R. (1991). *Chromatography/Fourier Transform Infrared Spectroscopy and its Applications*. New York: Marcel Dekker.

GENERAL TEXTS AND REVIEWS

Cooke, P. M. (1996). Chemical microscopy (IR, UV and Raman microscopy). *Anal. Chem.* 68(12):339R–340R.

Covert, G. L. (1966). Infrared spectrometry. In: Snell, F. D., Hilton, C. L., eds. *Encyclopedia of Industrial Chemical Analysis*. General Techniques. Vol. 2. New York: John Wiley & Sons, pp. 253–304.

Jones, R. N. (1985). Analytical applications of vibrational spectroscopy—a historical review. In: Durig, J. R., ed. *Chemical, Biological and Industrial Applications of Infrared Spectroscopy*. Chichester, UK: John Wiley & Sons, Ltd.

Kagel, R. O. (1991). Raman spectroscopy. In: Robinson, J. W., ed. *Practical Handbook of Spectroscopy*. Boca Raton: CRC Press Inc., pp. 539–562.

Smith, A. L. (1991). Infrared spectroscopy. In: Robinson, J. W., ed. *Practical Handbook of Spectroscopy*. Boca Raton: CRC Press Inc., pp. 481–535.

Wright, J. C., Wirth, M. J. (1980). Principles of lasers. *Anal. Chem.* 52(9):1087A–1095A.

HISTORICAL REVIEWS

Ferraro, J. R. (1996). A history of Raman spectroscopy. *Spectroscopy* 11(3):18–25.

Griffiths, P. R. (1992). Strong-men, connes-men, and blockbusters or how mertz raised the hertz. *Anal. Chem.* 64(18):868A–875A.

Jones, R. N. (1992). The UK's contributions to IR spectroscopic instrumentation: from wartime fuel research to a major technique for chemical analysis. *Anal. Chem.* 64(18):877A–883A.

Miller, F. A. (1992). Reminiscences of pioneers and early commercial IR instruments. *Anal. Chem.* 64(17):824A–831A.

Wilks, J. A. Jr. (1992). The evolution of commercial IR spectrometers and the people who made it happen. *Anal. Chem.* 64(17):833A–838A.

REFERENCES

Appel, R., Zerda, T. W., Waddell, W. H. (2000). Raman microimaging of polymer blends. *Appl. Spectrosc.* 54(11):1559–1566.

Asher, S. A. (1993a). UV resonance Raman spectroscopy for analytical, physical and biophysical chemistry, Pt I. *Anal. Chem.* 65(2):59A–66A.

Asher, S. A. (1993b). UV resonance Raman spectroscopy for analytical, physical and biophysical chemistry, Pt 2. *Anal. Chem.* 65(2):20IA–210A.

Asher, S. A., Bormett, R. W., Chen, X. G., Lemmon, D. W., Cho, N., Peterson, P., Arrigoni, M., Spinelli, L., Cannon, J. (1993). UV resonance Raman spectroscopy using a new cw laser source: convenience and experimental simplicity. *Appl. Spectrosc.* 47(5):628–633.

Asselin, K., Chase, B. (1994). FT-Raman spectroscopy at 1.339 micrometers. *Appl. Spectrosc.* 48(6):699–701.

ASTM (The American Society for Testing and Materials) (1996a). Standard practice for infrared, multivariate quantitative analysis. *ASTM Annual Book of Standards*. Vol 03.06. West Conshohocken, PA 19428-2959, USA: ASTM, pp. 755–779 (Practice E1655-94 (1995)).

ASTM (1996b). Molecular spectroscopy. *ASTM Annual Book of Standards*. Vol 03.06. West Conshohocken, PA 19428-2959, USA: ASTM, pp. 589–785.

Barbillat, J. (1996). Raman imaging. In: Turrell, G., Corset, J., eds. *Raman Microscopy: Developments and Applications*. London: Academic Press, p. 379.

Barbillat, J., Dhamelincourt, P., Delhaye, M., Da Silva, E. (1994). Raman confocal microprobing, imaging and fiber-optic remote sensing: a further step in molecular analysis. *J. Raman Spectrosc.* 25(1):3–11.

Bourne, S., Reedy, G. T., Coffey, P. I., Mattson, D. (1984). *Am. Lab.* 16(6):90.

Carr, G. L. (1999). High-resolution microspectroscopy and sub-nanosecond time-resolved spectroscopy with the synchrotron infrared source. *Vib. Spectrosc.* 19(1):53–60.

Carrabba, M. M., Spencer, K. M., Rich, C., Rauh, D. (1990). The utilization of a holographic Bragg diffraction filter for Rayleigh line rejection in Raman spectroscopy. *Appl. Spectrosc.* 44(9):1558–1561.

Chalmers, J. M., Everall, N. J., Schaeberle, M. D., Levin, I. W., Lewis, E. N., Kidder, L. H., Wilson, J., Crocombe, R. (2002). FT-IR imaging of polymers: an industrial appraisal. *Vib. Spectrosc.* 30(1):43–52.

Chase, D. B. (1982). Phase correction in FT-IR. *Appl. Spectrosc.* 36:240–244.

Chase, B. (1994). A new generation of Raman instrumentation. *Appl. Spectrosc.* 48(7):14A–19A.

Chase, B. (2002). Fourier Transform near-infrared Raman spectroscopy. In: Chalmers, J. M., Griffith, P. R., eds. *Handbook of Vibrational Spectroscopy*. Vol. 1. Chichester: John Wiley & Sons, pp. 522–533.

Chen, C., Smith, B. W., Winefordner, J. D., Pelletier, M. J. (1994). Near-ultraviolet resonance Raman spectroscopy using incoherent excitation. *Appl. Spectrosc.* 48(7):894–896.

Cole, A. R. H. (1977). *Tables of Wavenumbers for the Calibration of Infrared Spectrometers*, 2nd ed. Oxford, England: International Union of Pure and Applied Chemistry (IUPAC), Pergamon Press.

Colthup, N. B., Daly, L. H., Wiberley, S. E. (1990). Vibrational and rotational spectra. *Introduction to Infrared and Raman Spectroscopy*. Chap. 1. 3rd ed. New York: Academic Press.

Compton, S. V., Compton, D. A. C., Messerschmidt, R. G. (1991). Analysis of samples using infrared emission spectroscopy. *Spectroscopy* 6(6):35–39.

Cook, B. W. (1991). The molecular analysis of organic compounds using hyphenated techniques. *Spectroscopy* 6(6):22–26.

Culler, S. R. (1993). Diffuse reflectance spectroscopy: sampling techniques for qualitative/quantitative analysis of solids. In: Coleman, P. B., ed. *Practical Sampling Techniques for Infrared Analysis*. Boca Raton: CRC Press, pp. 93–105.

Decker, J. A. (1977). Hadamard transform spectroscopy. In: Vanasse, G. A., ed. *Spectrometric Techniques*. Vol. I. New York: Academic Press, pp. 189–227.

Delhaye, M., Dhamelincourt, P. (1975). Raman microprobe and microscope with laser excitation. *J. Raman Spectrosc.* 3:33–43.

Dhamelincourt, P., Wallart, F., Leclercq, M., N'Guyen, A. T., Landon, D. O. (1979). Laser Raman molecular microprobe (MOLE). *Anal. Chem.* 51(3):414A–42IA.

Everall, N. J. (2000). Modelling and measuring the effect of refraction on the depth resolution of confocal Raman microscopy. *Appl. Spectrosc.* 54(6):773–782.

Fabian, H., Lasch, P., Boese, M., Haensch, W. (2003). Infrared microspectroscopic imaging of benign breast tumor tissue sections. *J. Mol. Struct.* 661–662 (see also 411–417).

Fleischmann, M., Hendra, P. J., McQuillan, A. J. (1974). Raman spectra of pyridine adsorbed at a silver electrode. *Chem. Phys. Lett.* 26(2):163–166.

Ganz, A., Coates, J. P. (1996). Optical fibers for on-line spectroscopy. *Spectroscopy (Eugene, Oregon, USA)* 11(1):32–38.

Garrell, R. L. (1989). Surface-enhanced Raman spectroscopy. *Anal. Chem.* 61(6):401A–411A.

Garton, A., Batchelder, D. N., Cheng, C. (1993). Raman microscopy of polymer blends. *Appl. Spectrosc.* 47(7):922–927.

Griffiths, P. R., de Haseth, J. A., Azarraga, L. V. (1983). Capillary gas chromatography-Fourier transform IR spectrometry. *Anal. Chem.* 55(13):1361A–1387A.

Hammiche, A., Pollock, H. M., Reading, M., Claybourn, M., Turner, P. H., Jewkes, K. (1999). Photothermal FT-IR spectroscopy: a step towards FT-IR microscopy at a resolution better than the diffraction limit. *Appl. Spectrosc.* 53(7):810–815.

Harrick, N. J. (1987). *Internal Reflection Spectroscopy*. Ossining, New York: Harrick Scientific Corporation (Original publication -John Wiley & Sons, New York, 1967).

Hirschfeld, T. (1980). The hyphenated methods. *Anal. Chem.* 52(2):297A–312A.

Hirschfeld, T., Schildkraut, E. R. (1974). In: Lapp, M., ed. *Laser Raman Gas Diagnostics*. New York: Plenum, pp. 379–388.

Hirschfeld, T., Chase, B. (1986). FT-Raman spectroscopy: development and justification. *Appl. Spectrosc.* 40(2):133–137.

Hudson, R. D. Jr. (1969). *Infrared System Engineering*. Wiley Series in Pure and Applied Optics. New York: Wiley-Interscience.

International Union of Pure and Applied Chemistry (IUPAC). (1961). *Tables of Wavenumbers for the Calibration of Infrared Spectrometers*. Washington, DC: Butterworths.

Jones, R. W., McClelland, J. F. (1990). Quantitative analysis of solids in motion by transient infrared emission spectroscopy using hot-gas jet excitation. *Anal. Chem.* 62:2074–2079.

Juang, C.-B., Finzi, L., Bustamante, C. J. (1988). Design and application of a computer-controlled confocal scanning differential polarization microscope. *Rev. Sci. Instrum.*, 59(11):2399–2408.

Katon, J. E., Sommer, A. J. (1992). IR microspectroscopy: routine IR sampling methods extended to the microscopic domain. *Anal. Chem.* 64(19):93IA–940A.

Katon, J. E., Pacey, G. E., O'Keefe, J. F. (1986). Vibrational molecular micro-spectroscopy. *Anal. Chem.* 58(3):465A–481A.

Keen, I., Rintoul, L., Fredericks, P. M. (2001). Raman and infrared microspectroscopic mapping of plasma-treated and grafted polymer surfaces. *Appl. Spectrosc.* 55(8):984–991.

Keen, I., Rintoul, L., Fredericks, P. M. (2002). Raman microscopic mapping: a tool for the characterisation of polymer surfaces. *Macromol. Symp.* 184:287–298.

Kneipp, K., Kneipp, H. (2003). Surface enhanced Raman scattering—a tool for ultrasensitive trace analysis. *Can. J. Anal. Sci. Spectrosc.* 48(2):125–131.

Lerner, E. J. (1996). Infrared detectors offer high sensitivity. *Laser Focus World* 32(6):155–164.

Marcott, C., Dowrey, A. E., Noda, I. (1994). Dynamic two-dimensional IR spectroscopy. *Anal. Chem.* 66(21):1065A–1075A.

Marshall, A. G., Comisarow, M. B. (1975). Fourier and Hadamard transform methods in spectroscopy. *Anal. Chem.* 47(4):491A–504A.

Martin, M. C., Tsvetkova, N. M., Crowe, J. H., McKinney, W. R. (2001). Negligible sample heating from synchrotron infrared beam. *Appl. Spectrosc.* 55(2):111–113.

McClelland, J. F. (1983). Photoacoustic spectroscopy. *Anal. Chem.* 55(1):89A–105A.

McClelland, J. F., Jones, R. W., Luo, S., Seaverson, L. M. (1993). A practical guide to FTIR photoacoustic spectroscopy. In: Coleman, P. B., ed. *Practical Sampling Techniques for Infrared Analysis*. Boca Raton: CRC Press, pp. 107–144.

McDonald, R. S., Wilks, P. A. Jr. (1988). J-CAMP-DX: a standard form for exchange of infrared spectra in computer readable form. *Appl. Spectrosc.* 42(I):151–162.

Messenger, H. W. (1991). Researchers reduce size of Raman spectrometers. *Laser Focus World* July:145–150.

Messerschmidt, R. G. (1986). Optical properties of the sample/apparatus interface in FTIR spectrometry. *Spectroscopy (Eugene, Oregon, USA)* 1(2):16–20.

Messerschmidt, R. G., Chase, D. B. (1989). FT-Raman microspectroscopy: discussion and preliminary results. *Appl. Spectrosc.* 43(1):11–15.

Millon, A. M., Julian, J. M. (1992). FTIR techniques for the analysis of coating problems: solid sampling accessories. *ASTM STP1119* 173–195.

Mirabella, F. M. Jr., ed. (1993). *Internal Reflection Spectroscopy: Theory and Applications*. Practical Spectroscopy Series. Vol. 15. New York: Marcel Dekker.

Morris, H. R., Hoyt, C. C., Miller, P., Treado, P. J. (1996). Liquid crystal tunable filter Raman chemical imaging. *Appl. Spectrosc.* 50(6):805–811.

Nafie, L. A. (1988). Polarization modulation FTIR spectroscopy. In: Mackenzie, M. W., ed. *Advances in Applied Fourier Transform Infrared Spectroscopy*. New York: John Wiley & Sons, pp. 68–104.

Nafie, L. A. (1996). Vibrational optical activity. *Appl. Spectrosc.* 50(5):14A–26A.

Ohta, K., Ishida, H. (1988). Comparison among several numerical integration methods for Kramers-Kronig transformation. *Appl. Spectrosc.* 42:952–957.

Owen, H., Tedesco, J. M., Slater, J. B. (1994). Remote Optical Measurement Probe. United States Patent no. 5,377,004.

Porro, T. J., Pattacini, S. C. (1993a). Sample handling for mid-infrared spectroscopy, part I: solid and liquid sampling. *Spectroscopy (Eugene, Oregon, USA)* 8(7):40–47.

Porro, T. J., Pattacini, S. C. (1993b). Sample handling for mid-infrared spectroscopy, part II: specialized techniques. *Spectroscopy (Eugene, Oregon, USA)* 8(8):39–44.

Puppels, G. J., de Mul, F. F. M., Otto, C., Greve, J., Robert-Nicoud, M., Arndt-Jovin, D. J., Jovin, T. M. (1990). Studying single living cells and chromosomes by confocal Raman microspectroscopy. *Nature* 347(6290):301–303.

Raman, C. V. (1928). A new radiation. *Indian J. Phys.* 2:387–398.

Raman, C. V., Krishnan, K. S. (1928). A new type of secondary radiation. *Nature* 121:501–502.

Sauke, T. B., Becker, J. F., Loewenstein, M., Gutierrez, T. D., Bratton, C. G. (1994). An overview of isotope analysis using tunable diode laser spectrometry. *Spectroscopy* 9(5):34–39.

Schlossberg, H. R., Kelley, P. L. (1981). Infrared spectroscopy using tunable lasers. In: Vanasse, G. A., ed. *Spectrometric Techniques*, Vol. II. New York: Academic Press, pp. 161–238.

Schulte, A. (1992). Near-infrared Raman spectroscopy using CCD detection and a semiconductor bandgap filter for Rayleigh line rejection. *Appl. Spectrosc.* 46(6):891–893.

Shafer, K. H., Herman, J. A., Bui, H. (1988). The use of TLC for sample preparation in FTIR. *Am. Lab.* February:142–147.

Sommer, A. J. (1998). Raman microspectroscopy. In: Mirabella, F. M., ed. *Modern Techniques in Applied Molecular Spectroscopy*. Chapter 7. New York: John Wiley & Sons, pp. 291–322.

Suzuki, E. M., Gresham, W. R. (1986). Forensic science applications of diffuse reflectance infrared Fourier transform spectroscopy (DRIFTS): I. Principles, sampling methods and advantages. *Forensic Sci.* 31(3):931–952.

Tabaksblat, R., Meier, R. J., Kip, B. J. (1992). Confocal Raman spectroscopy: theory and application to thin polymer samples. *Appl. Spectrosc.* 46(1):60–68.

Tedesco, J. M., Owen, H., Pallister, D. M., Morris, M. D. (1993). Principles and spectroscopic applications of volume holographic optics. *Anal. Chem.* 65:441A–449A.

Treado, P. J., Morris, M. D. (1989). A thousand points of light: The Hadamard transform in chemical analysis and instrumentation. *Anal. Chem.* 61(11):723A–734A.

Treado, P. J., Levin, I. W., Lewis, E. N. (1992a). Near-infrared acousto-optic filtered spectroscopic microscopy: a solid-state approach to chemical imaging. *Appl. Spectrosc.* 46(4):553–559.

Treado, P. J., Levin, I. W., Lewis, E. N. (1992b). High-fidelity Raman imaging: a rapid method using an acousto-optic tunable filter. *Appl. Spectrosc.* 46(8):1211–1216.

Turrell, G., Corset, J., eds. (1996). *Raman Microscopy: Developments and Applications*. London: Academic Press.

Turrell, G., Delhaye, M., Dhamelincourt, P. (1996). Characteristics of Raman microscopy. In: Turrell, G., Corset, J., eds.

Raman Microscopy: Developments and Applications. Chapter 2. London: Academic Press, pp. 27–49.

Valentini, J. J. (1985). Coherent anti-stokes Raman spectroscopy. In: Vanasse, G. A., ed. *Spectrometric Techniques.* Vol. IV. Orlando: Academic Press, pp. 1–62.

Vasko, A. (1968). *Infra-red Radiation.* Cleveland, OH: CRC Press, p. 65

Vassallo, A. M., Cole-Clarke, P. A., Pang, L. S. K., Palmisano, A. J. (1992). Infrared emission spectroscopy of coal minerals and their thermal transformations. *Appl. Spectrosc.* 46(1):73–78.

Wang, X. (1992). Acousto-optic tunable filters spectrally-modulate light. *Laser Focus World* 28(5):173–180.

Wang, X., Soos, J., Li, Q., Crystal, J. (1993). An acousto-optical tunable filter (AOTF) NIR spectrometer for on-line process control. *Process Control Qual.* 5:9–16.

Wehry, E. L., Mamantov, G. (1979). Matrix isolation spectroscopy. *Anal. Chem.* 51(6):643A–656A.

Wetzel, D. L. (1983). Near-IR reflectance analysis: sleeper among spectroscopic techniques. *Anal. Chem.* 55(12):1165A–1176A.

White, R. (1991). *Chromatography/Fourier Transform Infrared Spectroscopy and Its Applications.* New York: Marcel Dekker.

Wilkins, C. L. (1994). Multidimensional GC for qualitative IR and MS of mixtures. *Anal. Chem.* 66(5):295A–301A.

Wilks, P. A. (1992). High performance infrared filtometers for continuous liquid and gas analysis. *Process Control Qual.* 3:283–293.

Williams, K. P. J., Pitt, G. D., Smith, B. J. E., Whitley, A., Batchelder, D. N., Hayward, I. P. (1994). Use of a rapid scanning stigmatic Raman imaging spectrograph in the industrial environment. *J. Raman Spectrosc.* 25(1):131–138.

Wolfe, W. L., Zissis, G. J. (1985). *The Infrared Handbook.* Revised Edition, SPIE. (Prepared by The Infrared Information and Analysis (IRIA) Center, Environmental Research Institute of Michigan, Office of Naval Research, Dept. of the Navy, Washington, DC).

Workman, J. Jr., Coates, J. (1993). Multivariate calibration transfer: the importance of standardizing instrumentation. *Spectroscopy (Eugene Oregon, USA)* 8(9):36–42.

Workman, J. J. Jr., Mobley, P. R., Kowalski, B. R., Bro, R. (1996). Review of chemometrics applied to spectroscopy: 1985–1995, Part 1. *Appl. Spectrosc. Rev.* 31(1&2):73–124.

Wright, J. C., Wirth, M. J. (1980). Lasers and spectroscopy. *Anal. Chem.* 52(9):988A–996A.

Zuska, P. (1996). Spectrometers shrinking to fit new uses. *Photon. Spectra* 30(7):111–114.

8

X-Ray Methods

NARAYAN VARIANKAVAL
Merck Research Laboratories, Rahway, NJ, USA

I. INTRODUCTION

Analytical methods using X-rays are some of the most powerful techniques for materials characterization. The purpose of this chapter is to review various techniques that utilize X-rays for probing molecular and supramolecular structure, in addition to discussing the instrumental aspects that are characteristic for each method. When matter is bombarded with high-energy electrons, protons, deuterons, α-particles, or heavier ions, X-rays are emitted. X-rays usually have a wavelength between 0.05 and 100 Å. They propagate at the velocity of light and travel in straight lines, undeflected by electric or magnetic fields. When X-rays encounter matter, a variety of processes may take place and each of these processes can be utilized to study particular properties of materials. Some of these processes include reflection, refraction, diffraction, scatter, absorption, fluorescence, and polarization.

It should first be noted that the overall choice of the analytical method in materials characterization depends on the physicochemical characteristics of the material to be investigated and the length scale to be probed. As a general rule, the length scale that can be probed in a scattering experiment is on the order of the wavelength of the scattered radiation. Also, the depth of penetration of a particular type of radiation in a sample depends on the energy of the radiation—for example, a wavelength of 0.1 nm corresponds to $\sim$100 eV for electrons, 10 keV for X-rays, and 25 meV for neutrons. Thus, neutrons can penetrate much farther into the sample when compared with electrons or X-rays. X-rays are not suited to studying light elements and cannot distinguish isotopes of the same element.

A. Continuous and Characteristic Spectra

X-rays can be produced by deceleration of electrons and/or transitions in the inner orbital electrons of atoms. Consequently, X-rays can be emitted as a continuous spectrum or as characteristic spectra. *Continuous radiation* is also called *bremsstrahlung* or braking radiation because it is primarily produced by the deceleration of the electrons by the atom. However, when an atom is bombarded by electrons or photons of sufficient energy it is also possible to remove an electron from its inner shell. Another electron from an outer shell may fall into this vacancy in the inner shell radiating energy in the process. This radiation is characteristic of the element and hence is referred to as *characteristic radiation*, and the resulting spectrum is called the characteristic or line spectrum (cf. Fig. 1). The process of characteristic X-ray emission is termed *X-ray fluorescence* (XRF). The characteristic radiations are designated by a combination of Roman and Greek letters referring to the classical designations of electron energy levels. Thus, a spectral peak labeled Kα refers to

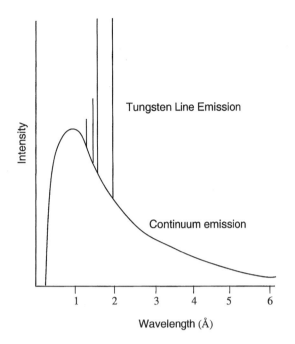

Figure 1 X-ray spectral output from a tungsten X-ray tube showing the continuum or bremsstrahlung background and the characteristic line emission of tungsten. [Re-created from Birks (1978).]

an electron transition from the L to the K shell; Kα1 and Kα2 correspond to electrons originating from different suborbitals of the L shell, and so on, as indicated in Fig. 2.

In many cases, the photons emitted in an X-ray transition can be absorbed by another electron within the atom that is ejected as a result of an internal photoelectric effect. This conversion of X-rays into photoelectrons is called the Auger effect and the resulting photoelectrons are referred to as Auger electrons.

B. Diffraction from Crystals

The diffraction of X-rays by matter is a consequence of two different phenomena: (a) scattering by each individual atom and (b) interference of the scattered waves. This interference occurs because the waves scattered by the atoms are coherent with the incident wave, and therefore, coherent with each other. When the difference between the incident and diffracted waves is an integral multiple of the wavelength of the X-rays, constructive interference occurs and a beam of enhanced intensity is seen. This is Bragg's Law and is illustrated in 2D representation in Fig. 3. From the geometry of the system it can be seen that $AB + BC = 2d \sin \theta$. The lines in Fig. 3 represent planes of

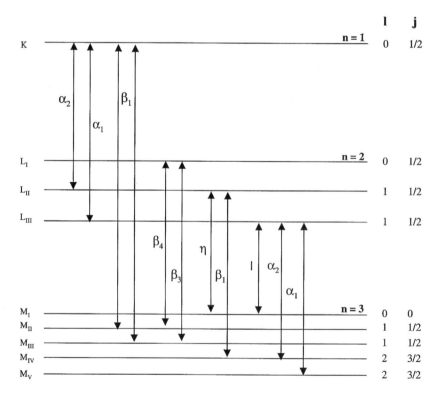

Figure 2 Energy level symbols used in association with X-ray levels. [Re-created from Markowicz (1992).]

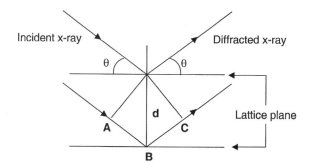

Figure 3 The Bragg diffraction condition. The difference in path lengths between the two X-ray beams is $AB + BC$. From the geometry of the system it can be seen that $AB + BC = 2d \sin \theta$. The Bragg Law states that constructive interference occurs when this difference is equal to an integer multiple of X-ray wavelengths.

atoms within the crystal. As the maximum value of $\sin \theta$ is 1, the minimum d-spacing that can be observed is one-half the X-ray wavelength. n Represents the order of reflection from a particular plane, and d is fixed for a given crystal plane and is determined by the size of the unit cell of the crystal. Thus, by varying the angle θ between the crystal and the source, it is possible to sample different d-spacings in the crystal. The basic step in X-ray crystallography involves measuring the intensity of the diffracted X-rays at various angles. This diffracted intensity depends on the intensity of the incident beam and the concentration of electrons along a given plane in the crystal. From a typical diffraction experiment, the intensity of diffracted X-rays is obtained as a function of diffraction angle. It is possible to determine the structure of crystals from this information.

C. Auger Effect

The emission of an electron from an atom accompanying the filling of a vacancy in an inner-electron shell is termed the Auger effect. The vacancy in the inner shell may be produced as a result of X-ray photon or electron bombardment. The energy of the incident photon or electron should be in the 3–25 keV range for the creation of the vacancy in the inner shells of atoms. The emitted electrons are characteristic of the element and/or bonding-state of atoms of the element in the material. Thus, the Auger effect can be used as a sensitive probe of elemental composition on surfaces and for depth-profiling. As this effect is a secondary process with regard to X-rays, it will not be considered in detail in this chapter.

D. Wavelength Dispersive X-Ray Spectrometry

Wavelength dispersive X-ray spectrometry involves measuring X-ray intensity as a function of wavelength.

In this method, the radiation emanating from a specimen is analyzed by a diffraction crystal mounted on a 2θ goniometer at specific angles. The wavelength of the radiation is obtained from the angle by the application of Bragg's Law. Here, d represents the lattice spacing in the analyzing crystal. The X-rays are usually collimated before being detected by a suitable scintillation or flow proportional detector. The variation of X-ray intensity as a function of diffraction angle is obtained by this technique.

E. Energy Dispersive X-Ray Spectrometry

When X-rays strike a sample, they can be absorbed and/or scattered through the material. Sometimes the innermost electrons can absorb all the energy and be ejected from the inner shells creating vacancies. To gain stability, electrons from outer shells can be transferred to the inner shells. This process results in the emission of X-rays that are characteristic of the material. The energy of these emitted X-rays corresponds to the difference in binding energies of the corresponding shells. As each element has a unique set of energy levels, these characteristic X-rays can be used to identify the particular element. The process of emission of characteristic X-rays from an element is termed XRF. The study of XRF is also referred to as energy dispersive X-ray spectrometry (EDX or EDS).

F. X-Ray Reflectivity

The techniques that are encompassed by this phenomenon are total X-ray reflection fluorescence (TXRF), X-ray specular reflectivity, and X-ray diffuse scattering. X-ray reflectivity takes advantage of the fact that when X-rays are directed at a sample at an angle lower than the critical angle, virtually all photons will be reflected at an equally small angle. The few X-rays directed at the surface will excite atoms close to the surface, which in turn will emit their characteristic radiation in all directions with virtually no backscatter.

G. X-Ray Absorption

X-rays can be absorbed by atoms and molecules. Typically, the proportion of X-rays absorbed (absorption coefficient) increases as the energy decreases. At certain specific values of energy that are characteristic of an element, known as absorption edges, there is an abrupt increase in the amount of energy absorbed. This energy corresponds to the ejection of an electron from the element. As the absorption edge is characteristic of the element, the absorption energy can be used as an identification tool for different elements.

II. INSTRUMENTATION AND METHODS

A. Overview of Optical Components of an X-Ray Diffractometer

The function of the X-ray optics is to condition the primary X-ray beam into the required wavelength, beam focus size, beam profile, and divergency. The optimum combination of X-ray optics usually depends on the specific application. For example, high-resolution X-ray diffraction for solving crystal structures from powder diffraction data usually requires a high-energy source of X-rays such as a synchrotron, a monochromator that permits only a single wavelength to interact with the sample, and a high-sensitivity detector. A description of the various optical components is given in the following section.

1. X-Ray Tubes

The four main components of an X-ray tube (cf. Fig. 4) are the cathode, which generates electrons; the anode target, which generates X-rays when impacted by the electrons; a vacuum system for maintaining very low pressure in the tube generator; and a beryllium exit window, which is essentially transparent to X-rays. The radiation produced by the tube can be analyzed in a number of ways and will be discussed in the later sections of this chapter.

2. Beam Attenuators, Filters, Slits, Masks, and Antiscatter Devices

The beam attenuator is an absorber that is placed in the X-ray beam to reduce the intensity by a specific factor. Specific beam attenuators are used with X-ray mirrors and beam monochromators.

Divergence slits are fitted in the incident beam path to control the equatorial divergence of the incident beam. The length of the area on the sample that is irradiated by the incident beam is dependent on the divergence of the X-ray beam and the position of the sample with respect to the beam. It can be calculated as follows:

$$L = \frac{R(\sin \omega \times \sin \delta)}{\sin^2 \omega - \sin^2(\delta/2)} \quad (1)$$

where L is the irradiated length of the sample, R is the radius of the goniometer, δ is the divergence angle, and ω is the angle between the incident beam and the sample surface.

Beam masks are fitted in the incident beam path to control the axial width of the incident beam. The size of the beam mask opening must be such that the sample encounters the incident X-ray beam completely during the entire measurement. The total width of the area on the sample irradiated by the incident beam is dependent on the size of the X-ray beam and the location of the sample with respect to the beam.

3. Collimators, Mirrors, and Monochromators

In most X-ray techniques, it is necessary to produce a finely collimated beam. The main function of the collimator is to control the shape of the X-ray beam hitting the sample. It usually consists of closely spaced parallel metal plates that are placed between the source and the sample. Metal tubes are usually used in single crystal experiments. The inner radius of the collimator is slightly bigger than the sample, so that the X-ray beam can irradiate the entire sample. Incident beam collimators usually have two narrow regions. The region closest to the X-ray source functions as the collimator. The second narrow region has a larger diameter and is used to remove stray radiation that takes a bent path due to interaction with the edge of the first narrow region of the collimator. Collimators can also be used in the diffracted beam side, but in this position they function only to prevent stray radiation from hitting the detector.

When an intense, highly focused point source is required, X-ray mirrors can also be used. These mirrors shape the X-ray beam and can also act as filters for the radiation used. Such mirrors usually find application in high-intensity sources such as synchrotron or rotating anode generators. Mirrors are also used to produce a parallel beam of X-rays from a divergent source and can be fabricated in the same module as a monochromator to produce a focused single-wavelength X-ray beam. For example, Göbel mirrors are a new device based on a layered crystal, which transforms the primary divergent X-ray beam into a highly brilliant, parallel beam. These mirrors can be planar (divergent-beam optics), elliptical (focusing optics), or parabolic (parallel-beam optics). Parabolically curved mirrors can reduce a divergent beam from a source with a 1° opening to an intense focused beam with a divergence of less than a few tenths of an arc-minute. Using such an intense parallel incident

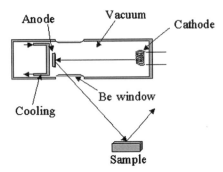

Figure 4 Schematic of a typical X-ray tube.

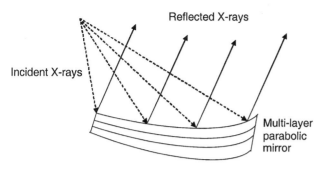

Figure 5 Schematic of multilayer mirror used for parallel focusing.

beam from a laboratory X-ray source in powder diffraction eliminates instrumental broadening and specimen aberrations caused by surface displacement, flat specimen, and specimen transparency. A schematic of a curved multilayer mirror is shown in Fig. 5. In addition to providing a parallel beam, the mirror increases the beam intensity. The d-spacing of the multilayer can have a gradient along the mirror in order to take into account the changing angle of incidence. The principle use of multilayer dispersion elements is in instruments for XRF analysis.

Monochromators have a critical role to play in high-resolution X-ray diffraction. Two different technologies are employed to produce a beam at the sample that is convergent both vertically and horizontally. Graphite-based focusing monochromators provide versatile and economical focusing in applications where resolution requirements are not extreme (unit cells less than 100 Å with Cu radiation). It can be advantageous to use graded d-spacing multilayer monochromators for those applications requiring higher resolution. Common crystals used as dispersion elements in a monochromator along with the typical d-spacing and applicable range are listed in Table 1.

4. Detectors

The most common type of detectors are X-ray film, imaging plate, counting, charge coupled device (CCD), multiwire, scintillation, and energy resolving semiconductor. The choice of detector depends on the specific application. The critical characteristics of detectors include speed, signal-to-noise ratio, active surface, resolution, and energy range. A comparison among different detectors is given in Table 2.

The X-ray film is the simplest and oldest type of detector but is now rarely used. The film offers very high spatial resolution and is particularly useful in studying large viruses. Counting detectors have the advantage of very low noise, particularly when measuring weak signals but suffer from poor spatial resolution. A brief description of other detectors used in X-ray analysis is provided below.

Imaging Plate Detectors

This detector works by the photo-stimulation of phosphor in an organic binder that is deposited onto a polymer film as shown in Fig. 6. The impinging X-rays are stored as an image in the plate. The image is read by scanning with a suitable laser and exciting luminescence, which is read by a photomultiplier tube. Some advantages of the imaging plate detector are high sensitivity, low noise, wide dynamic range, good linearity, good-to-high spatial resolution (50–200 μm), lack of image distortion, and finally, a large X-ray aperture, which enables collection of several orders of reflections. A limitation of the imaging plate is that it cannot be used as an energy detector.

Imaging plates have been used for solving structures of proteins and for fiber and polymer diffraction studies. They can be used for long exposure studies as well as time-resolved investigations by rapid displacement of the imaging plate as a function of time.

CCD Detectors

The CCD detector differs from the imaging plate detector in that the X-ray photons are detected by a CCD. CCDs are usually made of silicon doped with impurities to create sites that differ in conductivities (cf. Figs. 7 and 8). When optical photons strike the doped Si surface, photoelectrons are emitted. An electrical potential applied across the surface of the doped Si then propels these photoelectrons to a CCD chip. The energy of the photon is determined from the charge of the photoelectron.

CCD detectors combine very good spatial resolution with a moderate energy discrimination that is significantly better than gas-proportional detectors. Spatial resolution down to 1 μm can be obtained using these detectors. Some applications include wide-angle and small-angle X-ray crystallography, protein structure solution, X-ray astronomy, and medical and nondestructive testing.

Gas-Proportional Detectors

The operating principle of this type of detector is the ionization of a gas (usually argon) by incident X-ray photons. This ionization produces electrons that travel to a wire anode and positive gas ions that are collected at the cathode under the influence of an electric field. The collected ions generate a small current through a resistor. The magnitude of this current is proportional to the intensity of the incident X-rays. The response obtained from the detector is improved by increasing the voltage such that

Table 1 Crystals used as Dispersive Elements in X-Ray Spectrometers and Monochromators

Crystal	Miller indices	$2d$ (Å)	Useful wavelength region (Å)	Applications
α-Quartz	50$\bar{5}$2	1.624	0.142–1.55	Shortest $2d$ of any practical crystal; good for high-Z K-lines excited by 100 kV generators
Lithium fluoride (LiF)	422	1.652	0.144–1.58	Better than quartz (50$\bar{5}$2) for the same applications
Corundum, aluminum oxide	146	1.660	0.145–1.58	Same applications as quartz 50$\bar{5}$2
LiF	420	1.801	0.157–1.72	Similar to LiF (422)
Calcite	633	2.02	0.176–1.950	
α-Quartz	22$\bar{4}$3	2.024	0.177–1.96	
α-Quartz	31$\bar{4}$0	2.360	0.205–2.25	Transmission crystal optics
α-Quartz	22$\bar{4}$0	2.451	0.213–2.37	
Corundum, aluminum oxide	030	2.748	0.240–2.62	Diffracted intensity ~2–4× quartz (203) with same or better resolution
α-Quartz	20$\bar{2}$3	2.749	0.240–2.62	Improves dispersion for V–Ni K-lines and rare earth L-lines
Topaz	006	2.795	0.244–2.67	
LiF	220	2.848	0.248–2.72	Same applications as quartz (20$\bar{2}$3), with 2–4× their diffracted intensity
Mica	331	3.00	0.262–2.86	Transmission crystal optics
Calcite	422	3.034	0.264–2.93	
α-Quartz	21$\bar{3}$1	3.082	0.269–2.94	
α-Quartz	11$\bar{2}$2	3.636	0.317–3.47	
Si	220	3.840	0.335–3.66	Lattice period known to high accuracy
Fluorite	220	3.862	0.337–3.68	
Germanium (Ge)	220	4.00	0.349–3.82	
LiF	200	4.027	0.351–3.84	Best general crystal for K K- to Lr L-lines; highest intensity for largest number of elements of any crystal; combines high intensity and high dispersion.
Aluminum	200	4.048	0.353–3.86	Curved, especially doubly curved optics
α-Quartz	20$\bar{2}$0	4.246	0.370–4.11	"Prism" cut
α-Quartz	10$\bar{1}$2	4.564	0.398–4.35	Used in prototype Laue multichannel spectrometer
Topaz	200	4.638	0.405–4.43	
Aluminum	111	4.676	0.408–4.46	Curved, especially doubly curved optics
α-Quartz	11$\bar{2}$0	4.912	0.428–4.75	
Gypsum	002	4.990	0.435–4.76	Efflorescent
Rock salt	200	5.641	0.492–5.38	S Kα and Cl Kα in light matrices; good general crystal from S K to Lr L
Calcite	200	6.071	0.529–5.79	Very precise wavelength measurements; extremely high degree of crystal perfection with resultant sharp lines
Ammonium dihydrogen phosphate (ADP)	112	6.14	0.535–5.86	
Si	111	6.2712	0.547–5.98	Very rugged and stable general purpose crystal; high degree of perfection obtainable
Sylvite, potassium chloride	200	6.292	0.549–6.00	
Fluorite	111	6.306	0.550–6.02	Very weak second order, strong third order
Ge	111	6.532	0.570–6.23	Eliminates second order; useful for intermediate- to low-Z elements where Ge Kα emission is eliminated by pulse height selection
Potassium bromide	200	6.584	0.574–6.28	
α-Quartz	10$\bar{1}$0	6.687	0.583–6.38	P Kα in low Z-matrices, especially in calcium; intensity for P–K K lines greater than EDDT, but less than PET
Graphite	002	6.708	0.585–6.4	P, S, Cl K-lines, P Kα intensity >5× EDDT; high integrated reflectivity but relatively poor resolution

(continued)

Table 1 *Continued*

Crystal	Miller indices	$2d$ (Å)	Useful wavelength region (Å)	Applications
Indium antimonide	111	7.4806	0.652–7.23	Important for K-edge of Si
ADP	200	7.5	0.654–7.16	Higher intensity than EDDT
Topaz	002	8.374	0.730–7.99	
α-Quartz	10$\bar{1}$0	8.512	0.742–8.12	
Pentaerythritol (PET)	002	8.742	0.762–8.34	Al, Si, P, S, Cl Kα; good general crystal for Al–Sc Kα; soft deteriorates with age and exposure to X-rays
Ethylenediamine D-tartrate (EDDT)	020	8.808	0.768–8.4	Rugged and stable
Na β-alumina	0004	11.24	0.980–10.87	
Oxalic acid dihydrate	001	11.92	1.04–11.37	
Sorbitol hexa-acetate	110	13.98	1.22–13.34	High resolution; stable in vacuum; available in small pieces only
Rock sugar, sucrose	001	15.12	1.32–14.42	
Beryl	10$\bar{1}$0	15.954	1.39–15.22	Difficult to obtain; good specimens have $\lambda/\delta\lambda \sim 2500$–3000 at 12 Å
Bismuth titanate	040	16.40	1.43–15.65	
Silver acetate	001	20.0	1.74–19.08	
Rock sugar, sucrose	100	20.12	1.75–19.19	
Na β-alumina	0002	22.49	1.96–21.74	
Thallium hydrogen phthalate	100	25.9	2.26–24.7	
Rubidium hydrogen phthalate	100	26.121	2.28–24.92	Diffracted intensity $\sim 3\times$ KHP for Na, Mg, Al Kα, and Cu Lα; $\sim 4\times$ KHP for F Kα, $\sim 8\times$ KHP for O Kα
Potassium hydrogen phthalate (KHP)	100	26.632	2.35–25.41	Good general crystal for all low-Z elements down to O

Note: The Miller indices are given for the diffracting planes parallel to the surface of the dispersive element. The indicated useful wavelength region lies in the 2θ interval between $10°$ and $140°$.
Source: Adapted from Underwood (2001).

additional molecules are ionized by the electrons liberated in the initial ionization event. If the voltage is raised even higher, saturation occurs and all pulses are of equal magnitude, regardless of the incident photon. This is the principle of operation of the *Geiger–Müller counter*.

Scintillation Detector

Scintillation detectors rely on the ability of X-rays to cause certain substances to fluoresce. Typically, NaI crystals are used as scintillation sources. A photomultiplier

Table 2 Properties of Common X-Ray Detectors

Detector	Energy range (keV)	$\Delta E/E$ at 5.9 keV (%)	Maximum count rate (s^{-1})
Gas ionization (current mode)	0.2–50	n/a	10^{11}[a]
Gas proportional	0.2–50	15	106
Multiwire and microstrip proportional	3–50	20	10^6/mm^2
Scintillation [NaI(Tl)]	3–10,000	40	2×10^6
Energy-resolving semiconductor	1–10,000	3	2×10^5
Surface-barrier (current mode)	0.1–20	n/a	10^8
Avalanche photodiode	0.1–50	20	10^8
CCD	0.1–70	n/a	n/a
Superconducting	0.1–4	<0.5	5×10^3
Image plate	4–80	n/a	n/a

[a]Maximum count rate is limited by space-charge effects to around 10^{11} photons/s per cm^3.
Source: Thompson (2001).

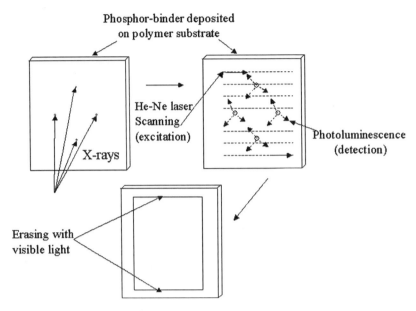

Figure 6 Schematic of an imaging plate detector.

tube is employed, as in the imaging plate, to convert optical photons to an electrical current that is amplified and measured. Advantages of the scintillation detector include good angular resolution (which is independent of beam dimensions), fast readout, and inexpensive optics. Major limitations are that parallel measurements cannot be implemented and there is a limit to the count rate set by the photomultiplier pulse width.

Multiwire Detectors

Proportional wire detectors have been used for several decades in high-energy physics. They are well-developed and reliable instruments. A grid of perpendicular anode and cathode wires forms a pixel array. Xenon is used as the counter gas and the chamber is covered by a large Beryllium window. The readout electronics are based on time delay circuitry and are extremely reliable.

Broadband High-Resolution Microcalorimeter Detector

The microcalorimeter detector works by measuring the rise in temperature upon absorption of an X-ray photon. The X-ray photon is captured by an absorber and the temperature rise measured by a thermistor. If E is the energy of an X-ray photon then the rise in temperature dT is given by

$$dT = \frac{E(\text{photon})}{C} \qquad (2)$$

where C is the specific heat capacity of the absorber. A schematic of a microcalorimeter detector is shown in Fig. 9. The X-ray absorber is usually mercury telluride. Further details of the mechanism of operation of this detector can be found in the article by Labov et al. (1996).

The energy resolution of the microcalorimeter detector is at least an order of magnitude higher than the most

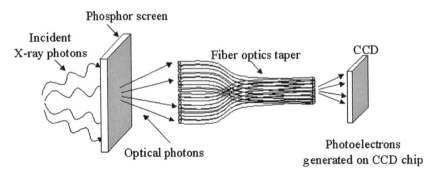

Figure 7 Schematic of a CCD detector.

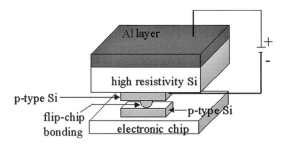

Figure 8 Schematic of a semiconductor pixel detector (CCD chip). (Re-created from http://www.nikhef.nl/~giggio/pixdet.html.)

recent semiconductor detectors. The quantum efficiency of such a detector is 95% at 6 eV and the energy resolution is 7 eV in the range of 0.2–20 eV. This detector couples the energy resolution of a wavelength dispersive spectrometer with the quantum efficiency of an EDS spectrometer. The disadvantage of such a detector is the enormous degree of cooling required in maintaining the device at <1 K.

B. X-Ray Methods

1. Conventional X-Ray Diffractometry

In conventional X-ray diffractometry, the intensity of X-rays diffracted from a sample is measured as a function of diffraction angle. The separation of the technique into wide- and small-angle methods is based on the size scale probed by the X-ray beam.

Wide-Angle X-Ray Diffraction

The size range probed by wide-angle X-ray diffraction (WAXD) is in the 1.5–10 Å range. The corresponding angular range that can be investigated by WAXD is 0.5–170°. Powders, single crystals, and films can be studied using WAXD. The diffraction pattern obtained for a material is almost a fingerprint of the crystalline nature of the material and can subsequently be used in qualitative and quantitative analyses. By far the most popular use is in single crystal structure solution and identification and quantification of polycrystalline samples.

INSTRUMENTATION. A typical instrumental setup for performing a WAXD experiment consists of an X-ray tube for generation of X-rays, monochromator, beam masks, divergence and antiscatter slits to focus the X-ray beam on the sample, Ni filters to remove the Kβ line, and finally a sensitive detector. Sample holders are usually metal disks, capillary tubes, or clamps for holding films/fibers depending on whether reflective or transmission geometry is used. The specifics of the instrument configuration and optics depend on the application. The two most common diffractometer geometries are Debye–Scherrer (cf. Fig. 10) and Bragg–Brentano (cf. Fig. 11). The former is typically used in transmission experiments, whereas the Bragg–Brentano setup is a reflection configuration. In this configuration, the detector is always 2θ and the angle between the incident beam and the sample is θ. Hence, this geometry is also referred to as θ–2θ geometry. The source and the detector along with the slits move as a unit. The locus of this motion is a circle centered at the sample.

In powder diffraction the sample is usually prepared in the form of a fine homogeneous powder and a thin layer is inserted in the path of the X-ray beam, though theoretically, this technique can also equally well analyze slurry samples or gels. If necessary, a finely cut Si disk can be used in the sample holder to reduce unnecessary background scattering. The result of the diffraction experiment is a pattern which represents the intensity of diffracted X-rays as function of scattering angle. This pattern is characterized by features that could potentially depend on the morphology of the particles, particle size, and residual stresses and strains in the sample. These

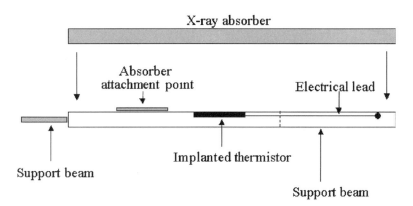

Figure 9 Schematic of a microcalorimeter detector. (Re-created from http://lheawww.gsfc.nasa.gov/docs/xray/astroe/ae/calorim-details.html#picture.)

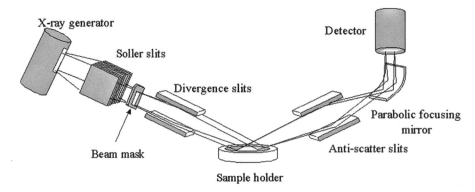

Figure 10 Schematic of Debye–Scherrer geometry.

properties affect the resolution and broadening of peaks. It follows that any quantitative analysis that relies on deconvoluting peak areas must necessarily take into account the differences in these properties between samples.

The position of diffraction peaks in a pattern is solely determined by the unit cell of the crystal, whereas the peak intensity is primarily a function of the positions of the atoms in the crystal. Although peak positions are invariant with respect to bulk powder properties, the observed peak intensity can deviate significantly from the true scattered intensity as a result of the orientation of the crystallites in a polycrystalline sample (cf. Fig. 12). In the figure on the left, the presence of acicular or needle-like crystals causes an artificial alignment of the crystals on the sample holder. This will result in overrepresentation of certain planes with the crystallite, and hence the intensities corresponding to these planes will be higher than usual. This problem is not present in a crystal with an equant morphology (Fig. 12) where an approximately equal representation of all planes will be provided to the incoming X-ray beam.

Accurate intensities that are devoid of preferred orientation effects can be obtained by collecting the diffraction pattern from a rotating sample. Theoretically, rotation of a sample about every possible axis effectively randomizes the orientation of individual crystallites, thereby producing peaks, the intensities of which solely depend on atomic positions. The best method to implement this is by using a four-circle goniometer. A goniometer is a mechanical assembly consisting of the sample holder, detector housing, and other gear components. By rotating a crystal/powder sample through three Euler angles—ϕ, χ, and ω (cf. Fig. 13), appropriate orientational averaging can be achieved. The resulting peak intensity is thus dependent only on atomic positions.

Conversely, the same setup can also be used to investigate crystal orientation in a polycrystalline sample by measuring the intensity at a particular scattering angle as a function of

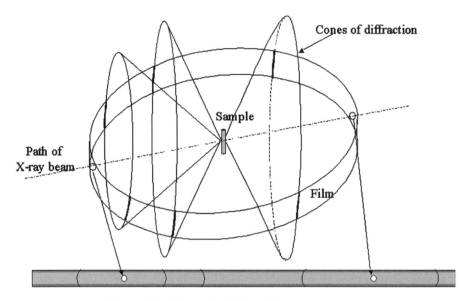

Figure 11 Schematic of Bragg–Brentano geometry.

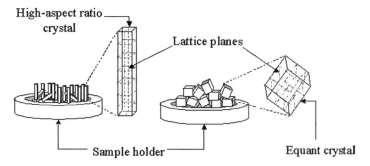

Figure 12 Alignment of crystals on a sample holder. High aspect ratio crystals align preferentially in one direction, whereas equant crystals have a high probability of random packing.

various sample orientations. This is termed *texture analysis*. By collecting intensity data for several angles the complete orientation distribution can be mapped out.

APPLICATIONS. WAXD is typically used to study the crystal structure and other interatomic level structural characteristics of materials such as organic and inorganic crystals, polymer films, and powders. Typically, a plot of intensity as a function of diffraction angle is utilized to determine the unit cell parameters of crystalline materials. When the sample is a single crystal grown under suitable conditions, the diffraction pattern can be used to solve the crystal structure of materials, which involves determining the atomic positions in the crystal lattice. On a macroscopic level, WAXD patterns can be used routinely for the identification of crystalline phases in mixtures, determination of amorphous content in organic and inorganic materials, estimation of texture in polycrystalline or oriented materials such as polymer films and fibers, determination of residual stresses and strains in mechanically stressed systems, and also identification of impurities.

Small-Angle X-Ray Scattering

Small-angle X-ray scattering (SAXS) is primarily used for investigating long-range periodicity or order in materials. The length scales probed by SAXS is 10–1000 Å corresponding to an angular range of 0°–0.5°. The inverse relationship between scattering angle and particle size is utilized to glean information on the structure of solids and liquids.

INSTRUMENTATION. The experiment is performed by observing a coherent scattering pattern from a sample bombarded with a collimated, monochromatic X-ray beam with small cross-section. The scattering pattern (intensity as a function of scattering angle) is the result of heterogeneities in electron density in the sample. As the angles observed in the SAXS experiment are usually less than $5°\text{-}2\theta$, the sample-to-detector distance is much greater than in a WAXD setup. Transmission mode is the preferred mode for soft materials but in the case of opaque samples, a combination of grazing incidence diffraction geometry and SAXS can be employed.

Interpreting the results of a SAXS experiment is usually nontrivial. In a WAXD experiment, Bragg peaks are observed at well-defined angles that are representative of the lattice spacing in the crystals. As the length scale probed by SAXS is significantly larger, the scattering function is usually featureless and the data must usually be reduced to an intensity vs. angle plot prior to analysis.

APPLICATIONS. An important feature of SAXS is that it can be used to study liquids or solutions. The advent of fast 2D detectors has greatly enlarged the scope of SAXS to include time-resolved structural studies. At very small angles, the SAXS curve can be used to provide information on the radius of gyration of macromolecular structures. At slightly higher angles the surface-to-volume ratio of scattering particles can be obtained. These two regimes are referred to as the Guinier and Porod regions, respectively. SAXS has been used to study protein residue conformations (overall size and shape, and not molecular

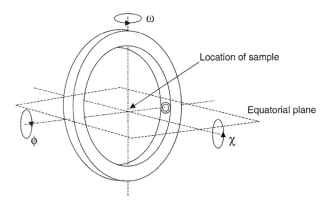

Figure 13 Schematic of a four-circle goniometer. The sample is located at the intersection of the three axes.

details) and time-resolved SAXS has been utilized in the study of folding kinetics (Russell et al., 2000). For most high-resolution work, a synchrotron source is essential.

As the length scales probed by WAXD and SAXS are very different, the results are complementary, and often data from one technique is used in conjunction with that from the other to solve problems such as protein conformation, protein–ligand interactions, and protein–DNA/RNA interactions under both static and dynamic conditions.

2. X-Ray Fluorescence

As mentioned before, the emission of characteristic X-rays is called XRF. There are two types of XRF studies—wavelength dispersive (WDXRF) and energy dispersive (EDXRF). In WDXRF, the intensity of X-rays is measured as a function of wavelength, whereas in EDXRF it is measured as a function of energy of the X-rays. The main principle of XRF is the creation of a vacancy in the inner orbitals. This can be achieved by bombarding the sample with high-energy X-rays, electrons, or protons. In X-ray bombardment the phenomenon responsible for fluorescence is photon absorption, whereas in proton and electron bombardment the incident particles undergo a Coulombic interaction with the sample. The exciting sources could be X-ray tubes, radioactive, synchrotron radiation, particle beam, and electron. WDXRF uses *analyzing* or *Bragg* crystals for wavelength selection. The wavelength will depend on the d-spacing of the Bragg crystal. EDXRF uses solid-state detectors and multichannel analyzers to measure X-ray intensity as a function of energy.

X-Ray Tube Excitation

It is convenient to use X-ray emission tubes to excite fluorescence in samples. The radiation from the tube is used directly (primary flux) or after passing through a filter to reduce background scattering from the primary X-ray wavelengths. Alternatively, the primary X-ray beam can be used to excite secondary X-rays from an auxiliary target. These secondary radiations are then used to excite fluorescence from the sample. Schematics of both configurations are presented in Fig. 14. The breadth of the excitation energy can be increased by using the continuum radiation rather than the characteristic radiation. The main drawback of this method is a loss in sensitivity. Similar to WAXD, collimators and filters are usually employed to remove continuum background features because of emissions from the target element to the detector.

Radioisotope Excitation

Although X-ray tube systems can achieve better sensitivities than radioisotopes because of greater X-ray flux,

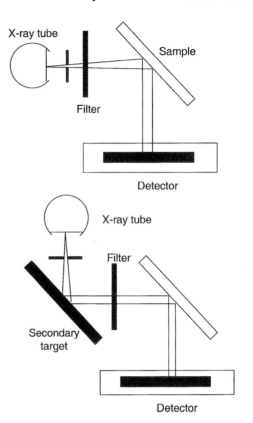

Figure 14 Schematic configuration of X-ray tube excitation. [Re-created from Jaklevic and Goulding (1978).]

simpler, less expensive, lighter, more portable, and rugged systems can be built using gamma radiation emitting radioisotopes (cf. Fig. 15). The energy range of gamma radiation emitted by radioisotopes is in the same range required to excite XRF and hence, gamma rays can be directed at a sample and the resulting fluorescence can be detected by an EDS system. The elements that can be analyzed are source-specific, that is, the technique is most sensitive for those elements that have absorption energies similar to, but less than, the energy of the exciting

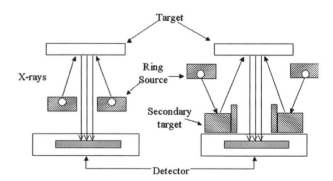

Figure 15 Schematic configuration of radioisotope excitation. [Re-created from Jaklevic and Goulding (1978).]

source. Some of the common radioisotope sources are Fe-55, Cd-109, Gd-153, and Am-241. Multiple sources can be incorporated in a single instrument to enable the analysis of a wide range of elements. Gas-filled or solid-state detectors such as mercuric iodide semiconductor detectors can be used with radioisotope sources. Some common applications of radioisotope-excited XRF include analysis of lead-based paint, detecting elements such as arsenic and selenium in soil sediments, metal analysis in water, and other types of field analyses of contaminants at hazardous sites. Radioisotope excitation is most commonly used for portable applications. The disadvantages of this type of excitation are the necessity of additional safety precautions required when operating a radioactive source and periodic replacement of source material because of radioactive decay.

Electron Beam Excitation

XRF can be excited by electrons. This type of excitation can be used to obtain elemental analysis of samples and is also referred to as EDX microanalysis. The combination of high-resolution microscopy and XRF is a valuable tool in developing industrial processes, failure analysis of solid-state electronic components, and single particle analysis of air pollutants. The critical components of such systems are high-throughput acquisition electronics and high-resolution EDS detector such as a Li-doped silicon detector. This detector works by monitoring the change in conductivity that occurs when the semiconductor absorbs energy from X-ray radiation. The increase in conductivity is directly proportional to the energy of absorbed X-rays.

Particle-Induced X-Ray Emission

Particle-induced X-ray emission (PIXE) is a more recent method of X-ray excitation in observing XRF. The particles most commonly used are protons although heavier ions such as Li, C, and O have also been used. The energy of the particles is usually of the order of 1–4 MeV. This method is usually associated with high precision in the quantitative analysis of samples ($\pm 3\%$) and high sensitivity (sub-ppm level). The advantages of PIXE over other methods such as electron microprobe microanalysis are reduced background radiation, increased signal-to-noise ratio, and good resolution even for thicker samples. The latter is due to the fact that protons are not multiply scattered as much as electrons in a sample.

Experimental Considerations

XRF measurements are similar to atomic and molecular fluorescence measurements. However, the sampling depth of the X-ray beam in the material varies significantly with the wavelength and therefore, the energy of the incident radiation. The extent of penetration can range from a few micrometers to several centimeters. This has implications in estimations of the analyte concentration primarily owing to interelement effects. In theory, all elements that can absorb the incident X-ray energy will contribute either absorption or emission effects to the ultimate emission signal at a particular wavelength or energy. In addition, the chemical composition and morphology of the sample also affect the scattering of the X-ray beam consequently increasing the complexity of the calculations. However, algorithms that can reliably perform elemental analysis with an accuracy of a few percent exist.

The analytical capability of an XRF instrument depends on the excitation source, source-to-sample geometry, instrument stability, counting time, and sample matrix. The detection limit, accuracy, and precision of the measurement are directly determined by the magnitude of the total counts and resolution width of the peak.

Counting Statistics

In all XRF measurements, the ultimate signal is the result of counting the pulses that arrive at the readout after detection and pulse height discrimination. The precision of the measurement is a function of the number of counts, and hence related to the overall time of measurement, t. There are two ways of controlling the precision or relative standard deviation (RSD) of the results generated by an XRF instrument—fixed count time or fixed count. Usually, fixed count time is preferred over fixed count because of the long counting times required by fixed count. The fixed count time allows a known RSD to be calculated and sample turn-around time to be managed effectively. The RSD of the net signal is given by:

$$\sigma_n = \sqrt{\sigma_t^2 + \sigma_b^2} = \sqrt{\frac{R_T}{t_T} + \frac{R_b}{t_b}} \quad (3)$$

where the subscripts n, t, and b stand for net, total, and background, respectively, σ is the standard deviation, R is the counting rate, and t is the count time.

The overall *accuracy* of the XRF measurement will depend on the precision of the measurement, the reproducibility of the sample preparation, and the similarity between the characteristics of the standards used for reference.

Applications

XRF spectroscopy is primarily used for determining bulk inorganic composition. It is especially useful in analyzing elements with an atomic number greater than that of aluminum. Specific applications include quantification of inorganic fillers in plastic materials, chemical analysis of glass and glazed materials, metal speciation in soils, and pigment analysis in paints and sculptures.

3. Surface X-Ray Methods

X-Ray Photoelectron Spectroscopy

X-ray photoelectron spectroscopy (XPS) is also known as electron spectroscopy for chemical analysis (ESCA). It was developed in the mid-1960s by K. Siegbahn and his research group. In this method, X-rays are used to eject electrons from the inner-shell orbital. The kinetic energy of the ejected electrons depends on the energy of the incident X-rays and the binding energy of the electrons in the orbital. As this binding energy depends on the chemical environment (bonding) of the atom, XPS can be used to identify the ligands and oxidation-state of an atom. Spectral information is usually collected from 2 to 20 atomic layers, depending on the material studied. Thus, XPS is primarily a surface technique.

XPS instruments typically consist of a monochromatic X-ray source, a photoelectron detector, and an electron energy analyzer. The sample is usually placed in a high vacuum chamber. The detector is usually an electron multiplier tube or a multichannel detector such as a microchannel plate and the energy is measured by an electrostatic analyzer.

APPLICATIONS. The primary applications of ESCA are chemical analyses of surfaces of organic and inorganic materials. Any element with an atomic number greater than that of helium can be investigated using XPS. Both qualitative and quantitative depth-profiling can be carried out. Some specific applications include the investigations of catalysts, adsorption, electronic surface states, oxidation of semiconductors and alloys, minerals, glasses, radiation damage, and surface diffusion.

X-Ray Specular Reflectivity

Specular reflectivity is also referred to as grazing incidence X-ray reflectivity (GIXR). The length scale that can be investigated by this technique is 10–10,000 Å. Consider the scattering geometry shown in Fig. 16. Here k_i and k_r are the incident and reflected wave vectors, respectively. The angles of incidence and reflection are θ_i and θ_r, respectively, and are fixed in a surface scattering experiment. When $\theta_i = \theta_r$, the resultant wave vector q is perpendicular to the surface. This defines the condition of specular reflectivity. The resultant wave vector, q ($= k_r - k_i$), is a unique function of the density profile $\rho(z)$ in the plane perpendicular to the surface and averaged over the xy plane. The specularly reflected intensity is

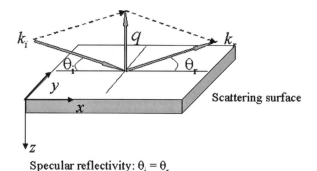

Specular reflectivity: $\theta_i = \theta_r$

Figure 16 Geometry of the X-ray specular reflection process.

given by:

$$I(q_z) \propto \frac{1}{q_z^4} \left| \int \frac{d\rho(z)}{dz} \exp(-iq_z z) \, dz \right|^2 \qquad (4)$$

The critical parameters affecting the specular intensity are surface roughness, density variations, and film thickness.

This X-ray method requires a high intensity, monochromatic, and collimated X-ray beam with the sample aligned in the grazing incidence configuration. A specular reflectivity experiment is carried out by measuring the reflected intensity of the X-ray beam as a function of the scattering vector or exchanged momentum, $Q = 4\pi/\lambda \sin \theta$. Similar to XRF, this experiment can be implemented either in wavelength dispersive or energy dispersive mode. In the former, monochromatic radiation is used as a function of the incident angle, θ, whereas in the latter a white incident beam is used with an energy dispersive X-ray detector.

Specific instances of the applications of X-ray specular reflectivity are well documented (Evmenenko et al., 2003; Méchin et al., 1998; Weissbuch et al., 2002; Xu et al., 2003).

X-Ray Total Reflection

BASIC CONCEPTS. The refractive index (RI) in the X-ray region is a complex quantity given by:

$$n = 1 - (\delta + i\beta) \qquad (5)$$

where δ is the dispersion part and β is the absorption part. These two parameters are very small quantities and depend on the atomic scattering factors. The real part of the RI is smaller than unity, which means that X-rays can experience total external reflection at the vacuum–sample interface. A critical angle, $\theta_c = \sqrt{2\delta}$ can be defined as the angle below which total reflection of X-rays can occur. However, for certain combinations of elements and energies of X-ray radiation, the dispersion part is negative and this prevents total reflection from being observed. A schematic of the processes in X-ray total reflection is shown in Fig. 17.

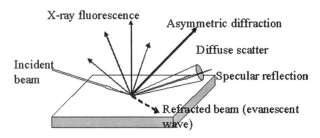

Figure 17 Processes in X-ray total reflection. In specular reflection the angle of incidence is equal to the angle of reflection. [From Stoev and Sakurai (????).].

INSTRUMENTATION. The basic requirements for performing grazing angle X-ray reflection measurements are a monochromatic, well-collimated, narrow incident beam, an accurate setting of the incident and reflected beam angles, a calibrated zero-angle position, and a very accurate and precise goniometer for sample positioning. The detector should possess a high dynamic range. Typical requirements and parameter ranges are listed in Table 3. It should be noted that at low scattering angles, Compton scattering produces photons that are as energetic as elastically scattered photons. Because of this consideration, for angles above the critical angle, energy dispersive detectors with high dynamic range may be required to study scattered photons.

LIMITATIONS. The instrumentation required for X-ray total reflection can be quite expensive. This arises owing to the need for—(a) a high-power X-ray source—usually hundreds of kilowatts, (b) small spot size on the anode, (c) high-precision optics and need for frequent alignment, and (d) long collimator or waveguide, as very low angles are involved. The sample needs to be finely polished and the angle and the height, carefully calibrated.

Grazing Emission X-Ray Fluorescence

The geometry of scattering in grazing emission XRF (GEXRF) is the inverse of TXRF. In this technique the sample is irradiated at ~90° with an uncollimated polychromatic source, and only that part of the fluorescence emitted at grazing angles (or exit angles close to the critical angle for total reflection) is detected by a wavelength dispersive detector. Both TXRF and GEXRF rely on scatter properties close to or below the Bragg angle to reduce background intensities and to improve detection limits. A detector placed at angles lower than the Bragg angle will not detect background scatter.

INSTRUMENTATION. A GEXRF is usually an accessory to a conventional WDXRF instrument. Because a wavelength dispersive detector is used, this technique is conducive for light-element analysis ($5 < Z < 14$). This is an advantage over conventional TXRF. However, the linear response in GEXRF is insufficient when compared with TXRF, as soft characteristic radiation is more strongly absorbed in its longer path through the matrix than in TXRF. The capabilities of GEXRF can, however, be improved using synchrotron exciting radiation. The advantage of GEXRF over TXRF is that X-rays can be directed at the sample with scant regard to the spot size or angle.

APPLICATIONS. The primary application is in depth-profiling of materials containing light elements, though it can also be applied to heavy atoms. Surface mapping and microanalysis are possible with this technique. This technique has been applied in water analysis and in investigating scale on oxidized alloys (Koshelev et al., 2001). However, quantitative analysis in such cases is not yet practical.

Extended X-Ray Absorption Fine Structure

PRINCIPLE. For an isolated atom, the absorption coefficient decreases monotonically beyond the absorption edge. However, for an ensemble of atoms such as in a molecule, the environment of the atom produces fluctuations in the absorption coefficient. These fluctuations in the post-edge region arising from the backscattering off neighboring atoms of the emitted electron wave, are related to the radial distribution function of atoms in the sample and is referred to as extended X-ray absorption fine

Table 3 Some Instrumental Requirements for X-Ray Total Reflection

Parameter	Values
Angular divergence	0.01°
Line focus	0.3–0.5 mm
Beam width (take-off angle)	30–50 mm (6°)
Detector range	$0–10^6$ cps

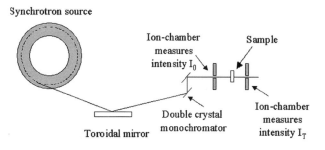

Figure 18 Schematic of EXAFS instrumental setup.

structure (EXAFS). The chemical environment of the atoms in a material can thus be analyzed by measuring the amplitude and frequency of these oscillations.

INSTRUMENTATION AND METHODOLOGY. An important requirement for observing EXAFS is an intense X-ray source such as a synchrotron. The experiment is performed by directing high-energy X-ray photons at a sample and by varying the incident energy around the absorption edge of the element of interest (Fig. 18). The X-ray absorption as a function of incident energy is then measured typically in transmission mode. In this mode, the intensity of the X-rays is measured before (I_0) and after (I_T) passing through the sample using an ion-chamber gas detector and the ratio (I_0/I_T) is plotted against energy (eV). A theoretical model is derived which describes this variation, and experimental data is fit to the model. One of the main limitations of EXAFS is that complex models may be required to adequately interpret the results of the experiment.

APPLICATIONS. Specific applications of EXAFS include, but are not limited to, the structure elucidation of oxidation catalysts (Barker et al., 2002), amorphous

Table 4 Synchrotron Sources Around the World

Location	Institution
USA	Advanced Light Source (ALS), Berkeley, CA
	Advanced Photon Source (APS), Argonne, IL
	Center for Advanced Microstructure and Devices (CAMD), Baton Rouge, LA
	Duke Free Electron Laser Laboratory (DFELL), Durham, NC
	Cornell High Energy Synchrotron Source (CHESS), Ithaca, NY
	National Synchrotron Light Source (NSLS), Upton, NY
	Synchrotron Radiation Center (SRC), Madison, WI
	Stanford Synchrotron Radiation Laboratory (SSRL), Stanford, CA
	National Institute of Standards and Technology—SURF II, Gaithersburg, MD
Brazil	Laborotório Nacional de Luz Sincrotron (LNLS), Campinas, SP
Canada	Canadian Light Source (CLS), Saskatoon
Denmark	Institute for Storage Ring Facilities, University of Aarhus, Denmark
France	European Synchrotron Radiation Facility (ESRF), Grenoble
	Laboratoire pour l'Utilisation du Rayonnement Electromagnétique (LURE), Orsay
	Soleil, Orsay
Germany	Angströmquelle Karlsruhe GmbH (ANKA), Karlsruhe
	Berliner Elektronenspeicherring, Gesellschaft für Synchrotronstrahlung m. b. H. (BESSY), Berlin
	Dortmund Electron Test Accelerator (DELTA), Dortmund
	Elektronen Stretcher-Anlage (ELSA), Bonn
	Hamburger Synchrotronstrahlungslabor (HASYLAB), Hamburg
Italy	Elettra, Trieste
United Kingdom	Diamond, Didcot Synchrotron Radiation Source (SRS), Daresbury
Spain	Laboratori de Llum de Sincrotró (LLS), Barcelona
Sweden	Max-Lab, Lund University, Lund
Switzerland	Swiss Light Source at the Paul Scherrer Institute, Villigen
China	Beijing Synchrotron Radiation Facility (BSRF), Beijing
India	Center for Advanced Technology (INDUS-I and INDUS-II), Indore
Japan	Nano-Hana, Ichihara
	Photon Factory, Tsukuba
	SPring-8, Japan Synchrotron Radiation Research Institute, Nishi Harima
	VSX Light Source, Kashiwa
	UVSOR facility, Okazaki
Russia	Siberian Synchrotron Radiation Center (SSRC), Novosibirsk
South Korea	Pohang Accelerator Laboratory, Pohang
Thailand	National Synchrotron Research Center, Nakhon Ratchasima
Singapore	Singapore Synchrotron Light Source, Singapore
Taiwan	Synchrotron Radiation Research Center, Hsinchu
Australia	Australian Synchrotron, Melbourne

Ge_xSe_{1-x}, and $Ge_xSe_yZn_z$ thin films (Choi et al., 2002); investigations of arsenic uptake from aqueous solutions by minerals (Farquhar et al., 2002); and speciation of metals such as Pd (Hamill et al., 2002).

III. CONCLUDING REMARKS

Although most X-ray instruments require similar optical components—X-ray source, collimator, monochromator, slits, sample holder, filters, detector, and so on—it should be emphasized that the fundamentals of each X-ray process and the geometry of the source–sample–detector configuration are unique to each technique. In many instances, the electronic components used in each technique are also specialized to maximize sensitivity and response. Some methods such as *reflectivity* and *absorption* demand intense X-ray sources, whereas others such as WAXD and SAXD can be successfully implemented with a laboratory source. However, with the burgeoning field of crystal structure determination from powder diffraction data and the need for real-time monitoring of processes such as nucleation, even diffraction studies may require a powerful X-ray source such as the synchrotron. Table 4 lists some of the synchrotron sources available worldwide. These sources are dedicated to both basic and applied research. It is common to find various experiments such as diffraction, fluorescence, reflectivity, and absorption being performed using several beamlines feeding off a single synchrotron source.

ADDITIONAL READING
General

Compton, A. H., Allison, S. K. (1935). *X-rays in Theory and Experiment*. 2nd ed. Princeton: Van Nostrand.

X-Ray Properties of Elements

http://www-cxro.lbl.gov/data_booklet/ (accessed August 13, 2003).

X-Ray Diffraction

Bish, D. L. (1989). *Modern Powder Diffraction*. Washington, D.C.: Mineralogical Society of America.
Cullity, B. D. (1978). *Elements of X-ray Diffraction*. 2nd ed. Reading, MA: Addison-Wesley Pub. Co.
Jenkins, R., Snyder, R. L. (1996). *Introduction to X-ray Powder Diffractometry*. New York: John Wiley & Sons.

X-Ray Detectors

Debertin, K., Helmer, R. G. (1988). *Gamma and X-ray Spectrometry with Semiconductor Detectors*. Amsterdam: North-Holland.
http://imagine.gsfc.nasa.gov/docs/science/how_l2/xray_detectors.html (accessed August 13, 2003).

Particle Induced X-Ray Emission

Johansson, S. A. E., Campbell, J. L., Malmqvist, K. G., eds. (1996). *Particle Induced X-ray Emission Spectrometry (PIXE)*. New York: John Wiley and Sons.

X-Ray Fluorescence

Jenkins, R. (1999). *X-ray Fluorescence Spectrometry*. 2nd ed. New York: John Wiley and Sons.
Jenkins, R., Gould, R. W., Gedcke, D. (1995). *Quantitative X-ray Spectrometry*. 2nd ed. New York: Marcel Dekker.

X-Ray Photoelectron Spectroscopy

Hufner, S. (2003). *Photoelectron Spectroscopy*. 3rd ed. Heidelberg, Germany: Springer Verlag.
McIntyre, N. S., Davidson, R. D., Kim, G., Francis, J. T. (2002). New frontiers in X-ray photoelectron spectroscopy. *Vacuum* 69(1–3):63–71.
Venezia, A. M. (2003). X-ray photoelectron spectroscopy (XPS) for catalysts characterization. *Catalysis Today* 77(4):359–370.

X-Ray Reflection

Klockenkämper, R. (1996). Chemical analysis: a series of monographs on analytical chemistry and its applications. *Total Reflection X-ray Fluorescence Analysis*. 1st ed. New York: John Wiley & Sons.

X-Ray Microanalysis

Hayat, M. A. (1980). *X-ray Microanalysis in Biology*. Baltimore: University Park Press.

REFERENCES

Barker, C. M., Gleeson, D., Kaltsoyannis, N., Catlow, C. R. A., Sankar, G., Thomas, J. M. (2002). On the structure and coordination of the oxygen-donating species in up arrow MCM-41/TBHP oxidation catalysts: a density functional theory and EXAFS study. *Phys. Chem. Chem. Phys.* 4:1228–1240.
Birks, L. S. (1978). Wavelength dispersion. In: Herglotz, H. K., Birks, L. S., eds. *X-Ray Spectrometry*. New York: Marcel Dekker, Inc., p. 7.

Choi, J., Gurman, S. J., Davis, E. A. (2002). Structure of amorphous Ge_xSe_{1-x} and $Ge_xSe_yZn_z$ thin films: an EXAFS study. *J. Non-Cryst. Solids* 297:156–172.

Evmenenko, G., van der Boom, M. E., Yua, C.-J., Kmetkoa, J., Dutta, P. (2003). Specular X-ray reflectivity analysis of adhesion interface-dependent density profiles in nanometer-scale siloxane-based liquid films. *Polymer* 44(4):1051–1056.

Farquhar, M. L., Charnock, J. M., Livens, F. R., Vaughan, D. J. (2002). Mechanisms of arsenic uptake from aqueous solution by interaction with goethite, lepidocrocite, mackinawite, and pyrite: an X-ray absorption spectroscopy study. *Environ. Sci. Technol.* 36:1757–1762.

Hamill, N. A., Hardacre, C., McMath, S. E. J. (2002). In situ XAFS investigation of palladium species present during the Heck reaction in room temperature ionic liquids. *Green Chem.* 4:139–142.

Jaklevic, J. M., Goulding, F. S. (1978). Energy dispersion. In: Herglotz, H. K., Birks, L. S., eds. *X-Ray Spectrometry*. New York: Marcel Dekker, Inc., p. 32.

Koshelev, I., Paulikas, A. P., Beno, M., Jennings, G., Linton, J., Uran, S., Veal, B. W. (2001). Characterization of the scale on oxidized Fe–Ni–Cr alloys using grazing emission X-ray fluorescence. *Phys. B: Condens. Matter* 304(1–4):256–266.

Labov, S. E., Mears, C. A.; Frank, M., Hiller, L. J., Netel, H., Lindeman, M. A. (1996). Assessment of low temperature X-ray detectors. *Nucl. Instrum. Methods Phys. Res. Sect. A* 370(1):65–68.

Markowicz, A. A. (1992). X-ray physics. In: Van Grieken, R. E., Markowicz, A. A., eds. *Handbook of X-ray Spectrometry*. New York: Dekker, p. 10.

Méchin, L., Chabli, A., Bertin, F., Burdin, M., Rolland, G., Vannuffel, C., Villégier, J.-C. (1998). A combined x-ray specular reflectivity and spectroscopic ellipsometry study of CeO_2/yttria-stabilized-zirconia bilayers on Si(100) substrates. *J. Appl. Phys.* 84(9):4935–4940.

Russell, R., Millett, I. S., Doniach, S., Herschlag, D. (2000). Small angle x-ray scattering reveals a compact intermediate in folding of the tetrahemena group I RNA enzyme. *Nat. Struct. Biol.* 7(5):367–370.

Stoev, K. N., Sakurai, K. (1999). Review on grazing incidence X-ray spectrometry and reflectometry. *Spectrochim. Acta Part B* 54:41.

Thompson, A. C. (2001). X-ray detectors. *X-ray Data Booklet*. University of California, Berkeley: Lawrence Berkeley National Laboratory.

Underwood, J. (2001). Multilayers and crystals. *X-ray Data Booklet*. University of California, Berkeley: Lawrence Berkeley National Laboratory.

Weissbuch, I., Buller, R., Kjaer, K., Als-Nielsen, J., Leiserowitz, L., Lahav, M. (2002). Crystalline self-assembly of organic molecules with metal ions at the air-aqueous solution interface. A grazing incidence X-ray scattering study. *Colloids Surf. A-Physicochem. Eng. Aspects* 208(1–3):3–27.

Xu, C., Kaminorz, Y., Reiche, J., Schulz, B., Brehmer, L. (2003). Supramolecular structures of VD films based on 2,5-diphenyl-1,3,4-oxadiazole derivatives. *Synth. Met.* 137(1–3):963–964.

http://www.nikhef.nl/~giggio/pixdet.html (accessed August 13, 2003).

http://lheawww.gsfc.nasa.gov/docs/xray/astroe/ae/calorimdetails.html#picture (accessed August 20, 2003).

9

Photoacoustic Spectroscopy

A.K. RAI
Department of Physics, University of Allahabad, Allahabad, India

S.N. THAKUR
Department of Physics, Banaras Hindu University, Varanasi, India

J.P. SINGH
DIAL, Diagnostic Instrumentation and Analysis Laboratory, Mississippi State University, Starkville, MS, USA

I. INTRODUCTION

The last decade has witnessed a rapid expansion of the science of photoacoustics (PA), photothermal (PT), and related phenomena (Bicanic, 1998). These processes have emerged as valuable tools for optical and thermal characterization of a wide range of samples, offering significant improvements (high sensitivity and precision) over traditional methods. This rapidly growing multidisciplinary field of research (Scudieri and Bertolotti, 1999) brings together physicists; chemists; biologists, agricultural, food, environmental, and medical scientists; and so on. Basically, PA and PT involve studies of the heat produced in an absorbing sample when exposed to modulated or pulsed optical radiation. The effects of optically induced heating are observed either in the absorbing sample itself or in the adjacent medium.

PT phenomena cause a *photoinduced* change in the *thermal* state of the absorbing sample. Light energy absorbed by the sample, followed by nonradiative transitions, results in sample heating. This heating leads to a temperature change, as well as to changes in thermodynamic parameters of the sample, which are related to temperature. Measurements of the temperature, pressure, or density changes that occur because of optical absorption are ultimately the basis for PT detection. There are a variety of detection techniques used to monitor the thermal state of the analytical sample (Bell, 1880; Dovichi, 1987; Tam, 1986, 1989) as summarized in Table 1.

Temperature can be directly measured using thermocouples, thermistors, or pyroelectric devices in techniques called *photothermal calorimetry*. Temperature changes can also be indirectly measured using methods that monitor infrared (IR) emissions. As the thermal IR emission is related to sample temperature, this method is called *photothermal radiometry*.

Two other temperature-dependent thermodynamic parameters that are commonly exploited in PT detection are pressure and density. The pressure changes that occur upon periodic or pulsed sample heating can be detected by using a microphone or other pressure transducers to monitor the resulting acoustic wave. The method is called *optoacoustics* or *photoacoustics*.

Table 1 Detection Techniques Related to Thermodynamic Parameters of the Sample

Thermodynamic parameter	Measured property	Detection technique
Temperature	Temperature	Calorimetry
	Infrared emission	Photothermal radiometry
Pressure	Acoustic waves	Photoacoustic spectroscopy
Density	Refraction index	Photothermal interferometry
		Photothermal lens
		Photothermal deflection
		Photothermal refraction
	Surface deformation	Photothermal diffraction
		Surface deflection

PT spectroscopy refers to methods that monitor the temperature-dependent refractive index (density) changes, usually with a probe laser. There is a wide range of nomenclature used to describe methods for refractive index change detection in the PT spectroscopy literature, but all of them rely on a few basic principles of light propagation, namely, optical pathlength changes, diffraction, and refraction. The optical pathlength changes that occur because of the PT-induced refractive index change can be measured by interferometry. The phase of probe laser light passing through the heated sample, relative to the phase passing through the reference arm, results in a change in light intensity at a photoelectric detector. This method is known as *photothermal interferometry*. Spatial gradients in the refractive index result in a direction change in the propagation of the probe laser beam. Thus, light will exit a medium with a refractive index gradient at an angle relative to the incident ray. This bending of light path is commonly called *photothermal deflection*. Spatially dependent refractive index profiles can also result in focusing or defocusing of the probe beam of laser light. This occurs when the refractive index profiles are curved. Thus, the thermally perturbed sample can act as a lens. Light transmitted through an aperture placed beyond such a PT lens will vary with the strength of the lens. PT methods based on measurement of the strength of this lens are called *photothermal lensing spectroscopy*. Some experimental apparatuses measure a signal that is due to the combined effects of deflection and lensing. These may be generally classified as *photothermal refraction spectroscopy* methods. Lastly, a periodic spatial refractive index modulation results in a volume phase diffraction grating, which diffracts light at an angle that meets requirements from Bragg's Law. The amount of diffracted light is proportional to the refractive index change and it is measured with a photoelectric detector. Methods used to measure spectroscopic signals on the basis of volume phase diffraction gratings formed by the PT effects are called *photothermal diffraction spectroscopy*.

It is beyond the scope of the present article to describe all the work related to PT detection. We will describe here only the measurements using transducers to monitor the pressure changes that occur upon periodic or pulse sample heating.

II. PHOTOACOUSTICS AND PA SPECTROSCOPY

A. History

The PA effect was discovered in the 19th century and was first reported by Alexander Graham Bell (Bell, 1880). He said, "... thin disks of very many substances emitted sound when exposed to the action of a rapidly interrupted beam of sunlight ... ", while giving an account to the American Association for the Advancement of Science of his work on the photophone. Bell's photophone consisted of a voice-activated mirror, a selenium cell, and an electrical telephone receiver. A beam of sunlight was intensity modulated at a particular point by means of the voice-activated mirror. This intensity-modulated beam was then focused onto a selenium cell incorporated in a conventional electrical telephone circuit. As the electrical resistance of the selenium depends on the intensity of light falling on it, the voice-modulated beam of sunlight resulted in electrically reproduced telephone speech. While experimenting with the photophone, Bell observed that it was possible to obtain, at times, an audible signal directly in a nonelectrical fashion. This phenomenon occurred if the beam of light was rapidly interrupted with a rotating slotted disk, and then focused onto solids that were kept in a tube closed at one end, with a flexible diaphragm at the open end in contact with the ear of the observer. Bell described in considerable detail his investigations of this new effect (Bell, 1881). He found that audible sound could be heard from an airtight transparent tube filled with various materials when the light shining on the transparent tube was modulated (Fig. 1). The sound was loud when the tube was filled with materials having large absorption coefficients. The sound also depended on the intensity of the light, it increased with the increasing intensity of light. The operational principles are now well understood. Modulation of the light impinging on an absorbing substance produces a similar modulation in temperature through the PT effect. In a gas of restricted volume, temperature modulation produces a pressure modulation. The periodic pressure modulation is an acoustic signal. The PA technique remained almost dormant for

Photoacoustic Spectroscopy

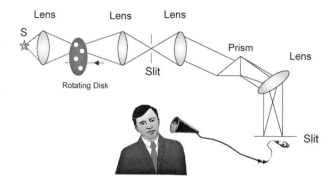

Figure 1 Schematic diagram of the spectrophone.

90 years after its discovery because of unavailability of a powerful source of light and a sensitive detector.

Viengerov (1938) used the PA effect to study light absorption in gases and obtained quantitative estimates of concentration in gas mixtures on the basis of signal magnitudes. This may have been the first use of photoacoustic spectroscopy (PAS). Sensitive chemical measurement application followed the work of Kerr and Atwood (1968) who used a laser to excite the samples. PAS attracted many workers when Kreuzer (1971) demonstrated parts per billion (ppb) detection sensitivities of methane in nitrogen using a 3.39 μm helium–neon laser excitation source. Later Kreuzer et al. (1972) demonstrated sub-ppb detection of ammonia and other gases using IR CO and CO_2 lasers. Rosencwaig (1973) made an extensive attempt to develop PAS and was responsible for the worldwide rebirth of interest in the technique. Rosencwaig (1973), Maugh (1975), Krikbright (1978), Prasad et al. (2002a, b), and Kapil et al. (2003) have described multifarious applications of PAS in a variety of samples.

B. Principle

Adams et al. (1976), Rosencweig (1978), and Rosencweig and Gersho (1976) presented a simple theoretical explanation for the generation of a PA signal in an airtight PA cell containing a condensed sample filled with a coupling gas. The process involved is given here in detail. Light absorbed by condensed matter is converted, partially or totally, into heat by nonradiative transitions. The heat produced in the interior of the sample is periodically diffused across the sample boundary. The acoustic signal in the gas microphone cell is due to this periodic heat flow from the solid sample to the surrounding gas. Acoustic signal generation involves four steps as shown in Figure 2. Only a relatively thin layer of air (∼0.2 cm for chopping frequency of 100 Hz) adjacent to the surface of a solid responds thermally to the periodic heat flow from the solid to the surrounding air. This boundary layer of the air can then be regarded as a vibratory piston, creating the acoustic signal detected in the cell. The acoustic signal is proportional to the product of following four terms:

$$IP_\alpha(\lambda)f_\delta(\lambda)f_\gamma(\lambda)$$

where I is the intensity of the incident radiation, $P_\alpha(\lambda)$ is the probability of the incident photon being absorbed, $f_\delta(\lambda)$ is the fraction of the absorbed energy that is converted to heat energy by nonradiative transitions, $f_\gamma(\lambda)$ is the fraction of this heat energy which reaches the surface of the sample and heats the thin layer of the gas/air. If the incident light is modulated at an angular frequency ω, then the intensity of the incident radiation on the surface of the sample is assumed to have the following sinusoidal form

$$\frac{1}{2}I_0(1 - \cos \omega t)$$

The heat density at any point in the solid sample is proportional to the absorption in accordance with the Beer–Lambert Law and may be written as:

$$\frac{1}{2}\beta I_0 \exp(-\beta x)(1 - \cos \omega t)$$

where β is the absorption coefficient of the material.

The thermal diffusion equation in the sample (Fig. 2), taking account of the heat sources distributed inside the sample, can be written as

$$\frac{\partial^2 \phi}{\partial x^2} = \frac{1}{\alpha_s}\frac{\partial \phi}{\partial t} + A \exp(-\beta x)(1 - \cos \omega t)$$

$$\text{for } -1 \leq x \leq 0 \quad (1)$$

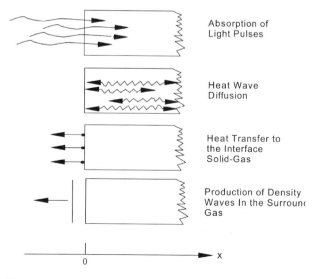

Figure 2 Four-step process from light absorption to pressure variation.

where, ϕ is temperature at position x and α_s is thermal diffusivity of the sample given by

$$\alpha_s = k_s/\rho_s C_s$$

where ρ_s is density, C_s is specific heat, and k_s is thermal conductivity of the material and, $A = \beta I_0 f_\delta/2k_s$. Here, f_δ is the efficiency of conversion of absorbed light energy into heat energy by nonradiative transition at a particular wavelength of the incident radiation.

Similarly, the diffusion equation for the backing material and the gas is given by

$$\frac{\partial^2 \phi}{\partial x^2} = \frac{1}{\alpha_b}\frac{\partial \phi}{\partial t} \quad \text{for } -1-1_b \leq x \leq -1 \quad (2)$$

$$\frac{\partial^2 \phi}{\partial x^2} = \frac{1}{\alpha_g}\frac{\partial \phi}{\partial t} \quad \text{for } 0 \leq x \leq 1_g \quad (3)$$

The solution of Eqs. (1)–(3) may be obtained by using the following boundary conditions:

$$\phi_g(0, t) = \phi_s(0, t)$$

$$\phi_b(-l, t) = \phi_s(-l, t)$$

Now it is clear that the main source of the acoustic signal is the periodic heat flow from the solid to the surrounding gas. The diffusion of heat from the sample produces a periodic temperature variation in the gas and this time-dependent component of temperature in the gas attenuates rapidly to zero with increasing distance from the surface of the sample. At a distance of $2\pi\mu_g$ [where μ_g is the thermal diffusion length of the gas(air)], the periodic temperature can be taken as fully damped, so it is only the gas within this layer which is heated periodically and thus expands and contracts periodically. This layer, therefore, acts as a piston on the rest of the gas column. Considering that the piston acts adiabatically, the acoustic pressure in a cell resulting from the displacement of this gas piston can be derived as

$$\delta P(t) = Q \exp\left[j\left(\omega t - \frac{\pi}{4}\right)\right]$$

where Q is a very complex quantity given by

$$Q = \frac{\beta I_0 f_\delta P_0}{2\sqrt{2} k_s l_g a_g T_0 (\beta^2 - \sigma_s^2)}$$

$$\times \frac{(r-1)(b+1)e^{\sigma sl} - (r+1)(b+1)e^{-\sigma sl} + 2(b-r)e^{-\beta l}}{(g+1)(b+1)e^{\sigma sl} - (g-1)(b+1)e^{-\sigma sl}}$$

where

$$b = \frac{ka_g}{k_s a_s}$$

Here, a represents thermal diffusion coefficient of the material.

$$g = \frac{k_g a_g}{k_s a_s}$$

$$r = \frac{(1-j)\beta}{2a_s}$$

$$\sigma_i = (1+j)a_i$$

Expression for $\delta P(t)$ is very complicated because of the complex nature of Q, but this may be simplified by considering special cases.

Case 1 Optically transparent solid sample ($\mu_\beta > 1$)
In this case, the incident light will be able to penetrate the whole of the sample and absorption takes place throughout the length of the sample.

Case 1a Thermally thin solid ($\mu_s \gg 1; \mu_s > \mu_\beta$)
For this case the complex equation for Q is

$$Q = \frac{(1-j)\beta l \mu_b}{2 a_g k_b} Y$$

where

$$Y = \frac{f_\delta P_0 I_0}{2\sqrt{2} l_g T_0}$$

and a, the thermal diffusion coefficient $= (\omega/2\alpha)^{1/2}$, ω is the modulation frequency of light, μ is thermal diffusion length $= 1/a = (2\alpha/\omega)^{1/2}$.

This shows that the acoustic signal has a ω^{-1} dependence and is proportional to βl. It also depends on the thermal properties of the backing materials. This is easily understandable because in this case a part of the heat generated in the sample also goes to the backing plate.

Case 1b Thermally thick solid ($\mu_s < 1; \mu_s \ll \mu_\beta$)
The heat generated in the deep interior of the sample is not able to reach the front surface in this case and the expression for Q is given by

$$Q = -j\frac{\beta \mu_s \mu_s}{2 a_g k_s} Y$$

It shows that the signal depends only on thermal behavior, and the absorption coefficient of the sample and its frequency dependence is of the type $\omega^{-3/2}$.

Case 2 Optically opaque solid ($\mu_\beta \ll 1$)
In this case most of the light is absorbed within a very small thickness of the sample near the front surface.

Case 2a Thermally thin solid sample ($\mu_s \gg 1; \mu_s \gg \mu_\beta$)
The simplified equation for Q is given by

$$Q = \frac{(1-j)\mu_b}{2 a_g k_b} Y$$

In this case also, the frequency dependence is of the type ω^{-1}.

Case 2b Thermally thick solid sample ($\mu_s < 1$; $\mu_s > \mu_\beta$)

The thermal diffusion length is smaller than the length of the sample, so that the heat generated in the deep interior (near the back plate) does not reach the front surface and Q is given by

$$Q = \frac{(1-j)\mu_s}{2a_g k_s}$$

This expression is similar to the result in Case 2a, but differs in that the signal depends on the thermal properties of the sample rather than on that of the backing plate.

Case 2c Thermally thick solids ($\mu_s \ll 1$; $\mu_s < \mu_\beta$)

$$Q = -j\frac{\beta\mu_s(\mu_s)}{2a_g k_s}Y$$

Frequency dependence is $\omega^{-3/2}$.

III. RELEVANCE OF PA TO PLANT SCIENCE

Many quick and reliable physical methods are available to diagnose human diseases, and advance research is also going on in this area. However, in plant sciences, as a review of the literature suggests, there is almost no application of physical methods for quick, reliable, and cost-effective detection/diagnosis of plant diseases. Work has been in progress for the last three decades to evolve, screen, and develop disease resistant varieties of plants through conventional methods, but plant diseases continue to cause heavy losses through reduction in quality and quantity of plant products. Therefore, the development of cost-effective methods for quick and reliable screening of plant materials is of great significance as it would reduce the time taken in releasing plant varieties resistant to different diseases for common use. This will simultaneously reduce the losses by diseases, as it will enable faster protection of plants from diseases by application of remedial measures immediately after detection of the disease. The PAS technique has several advantages, as summarized below:

1. For PA studies, no special sample preparation is required; powders, gels, liquids, fibrous material, and so on, can be easily studied.
2. Unlike conventional spectroscopy, scattering does not pose any problem in PAS as scattering and reflection losses do not produce a PA signal. This aspect makes it particularly attractive for application to biological samples, which generally correspond to scattering media.
3. The PA technique can be used to study the nonradiative de-excitations, which is a major pathway of energy transfer in plants after absorption of electromagnetic radiation. This is not possible with conventional optical techniques.
4. Some of the chemicals used as drugs and pesticides either do not dissolve easily to form solutions or get dissociated in the solution phase (Prasad and Thakur, 2002). The technique of PAS is ideally suited for them to be studied in solid or powder phase.
5. The potential merit of the PAS is the dependence of the PA signal on modulation frequency (i.e., depth profile analysis). Thus PA spectra of layered plant material like whole leaf can be studied simply by changing the chopping frequency.
6. *In vivo* and *in situ* studies of the intact green leaf can be done. The advantage of PA measurements with intact leaves lies in the fact that (a) absorption properties can be measured, without the typical problems associated with optical techniques resulting from strong light scattering properties of leaves; (b) the measurement of absorption properties at different depths can be carried out ("depth profiling") by simply changing the chopping frequency of the incident light; (c) the measurement of nonradiative transitions helps in the understanding of the energy balance of photosynthetically active samples. Light energy absorbed by chlorophyll, carotenoid, and other pigments (i.e., flavonoids), can be transferred into heat or emitted as fluorescence. In its natural condition, a photosynthetically active leaf evolves light-induced O_2 during photochemistry of carbohydrate syntheses, and this O_2 evolution can be measured by PAS (Butler, 1978; Malkin and Cahen, 1979). Thus, PAS is capable of measuring directly the photosynthetic O_2 evolution. This is not necessarily observable in growth rate and biomass formation, especially in short-term experiments.
7. Standard PA spectra for healthy and different types of diseased plant materials can be obtained in a short time, thus allowing quick and efficient detection/diagnosis of diseases that may help in an early application of control measures or selection of disease-free planting materials for further use and rejection of susceptible ones.
8. This technique, apart from being quick and having the potential of detecting diseases at an early stage, will have an added advantage of allowing the use of the remaining part of the sample plant material, especially in vegetatively propagated crops, if it is found to be disease free.

Thus, PA measurements open up new unique possibilities for both applied and basic research in plant sciences (Buschmann, 1990; Buschmann and Kocsanyl, 1989). In plant pathology, this technique is gaining importance as evidenced by comparison studies of PA spectra of normal and diseased plants (Plaria et al., 1998a).

IV. REVIEW OF PA APPLICATIONS IN PLANT SCIENCES

Realizing the importance of the PA technique, to both applied and basic research in agricultural science (Bicanic et al., 1989a; Buschmann and Prehn, 1990; Plaria et al., 1998a; Rai and Mathur, 1998), several research groups have started work in this area. Scientists at Botanical Institute, University of Karlsrulhe, Karlsrulhe, Germany are extensively using PAS to study the photosynthetic activity in leaves and have published many papers (Buschmann, 1989, 1990, 1999; Buschmann and Kocsanyl, 1989; Malkin and Cahen, 1979; Szigeti et al., 1989). They have also studied the PA spectra of herbicide-treated bean leaves and shown that PA spectra can be applied to detect herbicide effects which are related to changes in photosynthetic activity (Fuks et al., 1992; Szigeti et al., 1989). These workers have demonstrated that PA spectra are influenced by the composition of pigments and also by the photosynthetic activity of leaves. Their depth profile analyses in leaves have demonstrated that at higher frequencies only the peaks of anthocyanins contained in the epidermis are detected. By lowering the modulation frequency, a chlorophyll peak in the red region appears, which is caused by the chlorophyll contained in the mesophyll layer below the epidermis. The location of the depth in the leaf from where the PA signal is detected is determined by selecting the appropriate modulation frequency. This enables the measurement of absorption spectra of inner layers of whole leaf, which otherwise would only be possible by cutting horizontal sections from it. Cahen and coworkers at the Weizmann Institute of Sciences, Rehovot, Israel, have given extensive explanation of the generation of PA in whole leaves (Bults et al., 1982; Cahen et al., 1980; Flaisher et al., 1989; Poulet et al., 1983). They have shown that for modulation frequencies up to 150 Hz, the PA signal is due to superposition of heat emission and O_2 evolution during photosynthesis, whereas for chopping frequency greater than 150 Hz, the PA signal is only due to nonradiative transitions (heat emission) (Bukhov and Carpentier, 1996; Bults et al., 1982). Recently, Bajra and Mansanares (1998) have demonstrated the usefulness of the open PA cell in the determination of photosynthetic parameters such as photochemical loss and O_2 evolution.

The department of Molecular and Laser Physics, University of Nijmegen, The Netherlands, is engaged in research on exchanged gases from biological samples utilizing the PA effect (Bicanic et al., 1989a; Haiept et al., 1997; Harren, 2000; Harren and Reuss, 1997; Vries et al., 1996). They have developed a facility for the detection of biologically interesting molecular gases like ethylene, ethane, methane, water vapor, acetaldehyde, ethanol, and other small hydrocarbons and nitrogen- and sulfur-containing compounds at and below the ppb level. They have used the PA technique for detection of C_2H_4 produced by specific plant species during senescence and inundation. They have also used this technique to measure the level of orthophosphate in water and soil (Bicanic et al., 1989b; Doka et al., 1998). Greene et al. (1992), applied the PAS technique in the study of fungal infections in seeds. Another active group using PAS in agricultural science headed by Professor Doka and his associates at Department of Physics, Pannom Agricultural University, Hungary, have discussed the possibility of measuring the Fe content (Doka et al., 1991) of milk protein concentrates and red lead (Pb_3O_4) in the ground sweet red paprika (Doka et al., 1997a). They have also monitored the radiation-induced changes in egg powders (Doka et al., 1997b). These scientists have used PA for routine investigation of foodstuff (Dadarlat et al., 1996; Favier et al., 1997; Frandas and Bicanic, 1999).

The PA spectra of Pb and toxic Al-containing plants were compared with those of normal plant (Marquezini et al., 1990; Pal et al., 1989). PAS was used to study the effect of SO_3 and SO_2 on isolated corn mesophyll chloroplasts by monitoring the photochemical energy storage (Veeranjaneyulu et al., 1990). The measurement of photosynthetic activities in intact leaves under Cu stress was performed by Ouzounidou et al. (1993). IR laser PAS was used for detecting gases of agricultural, biological and environmental interest (i.e., C_2H_4, CH_4, C_6H_6, NO_2, HCHO, etc.) (Buscher et al., 1994; Prasad and Thakur, 2003). PAS in the UV, visible, and IR regions was employed for quantitative analysis and semiquantitative estimates in food additives (Haas and Vinha, 1995). PAS has also been used for the measurement of the water vapor diffusion coefficient in wood (Balderas-Opez et al., 1996).

The chromophore–protein interactions in biological systems, with particular emphasis on photochromes, were studied by Braslavsky and her coworkers at Max-Planck-Institute of Strahlenchemie, Malheim, Germany. They used laser-induced PA spectroscopy involving time intervals in the range of hundreds of microseconds to milliseconds (Habib Jiwan et al., 1995a, b). Muratikov et al. used PT and PA measurements for thermal and thermoelectric characterization of ceramics with residual stress (Muratikov, 1998; Muratikov et al., 1997).

By using Fourier transformed IR PAS, Jones and his coworkers (Letzelter et al., 1995) developed a method to enable samples to be taken from single seeds without affecting the ability to grow plants from analyzed seeds, thus allowing the genetic study of biosynthetic pathways.

At Laser Physics Center of the Hungarian Academy of Sciences, a facility using an external cavity diode laser light source in combination with a PA detector has been developed for high-sensitivity gas detection, and several papers have been published (Bozoki et al., 1999; Sneider et al., 1997).

The Scientists at Katholieke University at Leuven, Belgium, are using a photopyroelectric setup for the study of specific heat capacity and thermal conductivity of liquid samples (Caerels et al., 1998; Glorieu et al., 1999). Gordon et al. (1998) have used PA Fourier transformed IR spectra for the detection of mycotoxigenic fungi in food grains. Ageev et al. (1998) have shown that laser PA can be used to determine the kinetics of plant CO_2 evolution under exposure to hypobaria and elevated concentration of pollutants. Bergevin et al. (1995) have used PAS to assess the maturity of strawberries. Pulsed laser PAS was used for depth-resolved and laterally resolved analysis of artificial tissue models (Beenen et al., 1997). Foster et al. (1999) have demonstrated the use of pulsed laser PAS to nondestructively monitor aqueous Cr(VI) concentrations at trace levels. Boucher and Carpentier (1999) have been using PA to study the effect of Hg^+, Cu^+, and Pb^+ on photosystem ll. They also used this technique to see the changes in nonradiative dissipation and photosynthesis in barley leaves after mild heating of the leaves (Bukhov et al., 1998a, b). Dahnke et al. (2000) have used PAS for online monitoring of biogenic isoprene emission.

To the best of our knowledge, the use of the PAS technique in disease diagnosis of plants is not even in the contemplative stages at any university/center. No published work has come from any of the laboratories either. For the first time, at Pantnagar, India, PAS experimental facilities have been set up to carry out diagnosis and studies of plant diseases. In the following, sections we describe the use of PAS technique in studying diseases related to plants.

V. PA IN DETECTION OF PLANT DISEASE

Two different modes have been adopted for recording the PA spectra of diseased and healthy plants: (i) by varying the wavelength of the exciting radiation and keeping modulation frequency constant and (ii) by varying the modulation frequency and keeping the wavelength of exciting radiation constant. The first approach is called the wavelength scanning mode and is expected to give characteristic peaks (bands) due to synthesis of new molecules or particular nutritional deficiency/disorders like zinc, sulfur, and so on, in diseased plants. It is, thus, possible to identify a disease by its characteristic peaks at different wavelengths. The second technique is called depth-profiling mode and was used for detection of damage due to disease in leaf tissues at various depths. In this way, the intensity of the vertical spread of the disease, which is directly proportional to the strength of the PA signal, was estimated at various depths in the leaf tissue.

A. Experimental Setup

In the first approach, the PA spectra were recorded using a single-beam photoacoustic spectrometer (Fig. 3) constructed in our laboratory (Joshi and Rai, 1995). A 300 W xenon arc lamp (Oriel Corporation 66083) served as the excitation source. Before entering the monochromator, the incident light was modulated by a mechanical chopper (Stanford SR 540). The monochromator (CEL, India, HM 104) with a 1200 grooves/mm grating blazed at 500 nm, has a spectral dispersion of 3.3 nm/mm. All spectra were manually recorded between 300 and 720 nm in steps of 5 nm. The modulated radiation (17 Hz) was focused onto the sample compartment of an indigenous PA cell (2.0 cm diameter, 2 mm depth) equipped with a sensitive microphone (Fig. 4). The PA signal from the cell was amplified and fed to the lock-in amplifier (Stanford SR 530) for synchronous phase sensitive detection using a time constant of 30 s for all measurements. The spectra were normalized by dividing (at each wavelength) the PA signal from the sample and that obtained from a carbon black standard.

In the depth-profiling mode, the PA signals were recorded using 10 mW 632.8 nm radiation from a linearly polarized helium–neon laser (510 P, Aerotech USA). The laser beam intensity was modulated by a mechanical chopper (Stanford SR 540) before entering the PA cell equipped with a sensitive microphone. The PA signals were recorded by varying the chopping frequency from 5 to 400 Hz. The PA signal was processed by a lock-in-amplifier (time constant: 30 s, type Stanford SR 530).

About 1.0 cm^2 of the infected leaf area was taken (in four replicates) for the analysis; the healthy leaves of corresponding plants served as control samples. Great care was exercised to minimize the damage and dehydration of leaves used, and each measurement is an average of four independent experiments.

B. Diseases in Wheat Plants

We have studied three types of diseases in wheat plants, namely, loose smut, brown rust, and leaf blight. The leaves infected with different diseases and corresponding healthy leaves were collected in a wet polythene bag from the disease nursery raised at Crop Research Center,

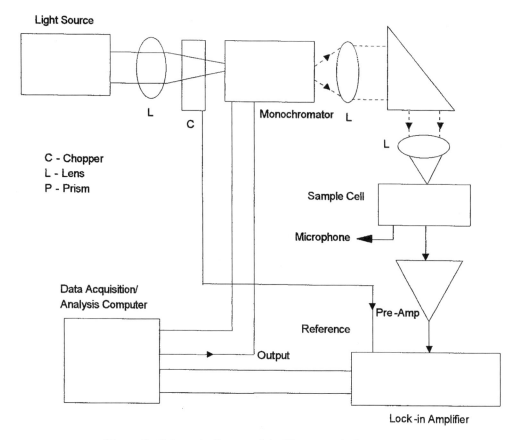

Figure 3 Schematic diagram of the Photoacoustic Spectrometer.

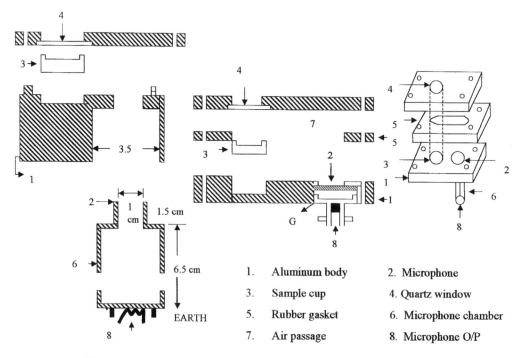

Figure 4 Cross-sectional view of a sample cylindrical photoacoustic cell.

Photoacoustic Spectroscopy

G.B. Pant University of Agriculture and Technology, Pantnagar, India. About 1 cm^2 of each leaf tissue was enclosed in the PA cell (Joshi and Rai, 1995) to record their spectra.

1. Loose Smut

Loose smut is one of the common diseases of wheat, which is present in all the wheat-growing areas of India. The characteristic symptom of the disease is production of black powdery spore of casual fungus, in place of the grain. The disease originates from an infected seed where the fungus is present in the form of dormant mycelicum (fungus growth). As the seed germinates after sowing, the fungus in the seed also becomes active and moves along with the growing points of plants and finally produces an infected earhead filled with spores. The PA spectra of wheat leaves infected with loose smut and those of healthy leaves, were recorded by varying the chopping frequency from 5 to 400 Hz, both in abaxial (upper surface of the leaf facing the incidents radiation) and adaxial (lower portion of the leaf facing the incident radiation) modes (Rai and Mathur, 1998). For the abaxial mode, the PA signal strength of leaves infected with the loose smut is higher than that of healthy leaves in the frequency range from 5 to 200 Hz (see Fig. 5). The PA signal strength of both leaves (infected and healthy) was the same in the frequency range of 200 to 400 Hz. The vertical length of the sample, which contributes the PA signal strength, depends on modulation frequency, as shown in Table 2. It is clear from this table that up to modulation frequency of 200 Hz the whole width (0.1–0.5 mm) of the leaf is contributing to the PA signal. Beyond 200 Hz, the PA signal is contributed from the layers close to the surface of the leaf facing the incident radiation. When the leaf is exposed to photons having appropriate wavelength, the leaf pigments capture the photons and absorbed energy is utilized in three competitive processes: (i) fluorescence, (ii) photosynthesis reaction center, and (iii) nonradiative transition (heat). The last two parts of the absorbed energy are measured as a PA signal. Morphologically, the leaf infected with disease is photosynthetically inactive; thus, primary photosynthetic pigments do not capture the incident photons. On the other hand, these photons are absorbed by other biomolecules developed in the diseased leaves and these biomolecules lose energy nonradiatively, resulting in an enhanced PA signal in the frequency range of 5–200 Hz. Beyond 200 Hz, the PA signal starts coming from the

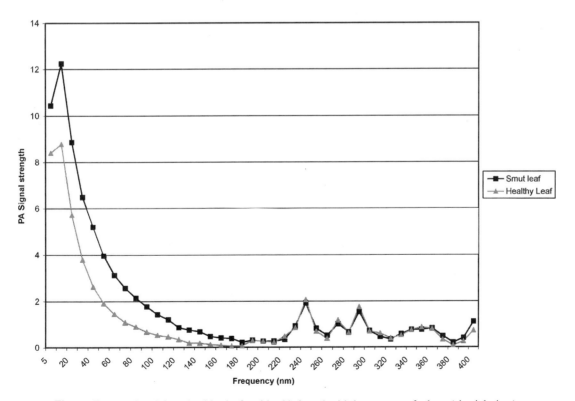

Figure 5 PA signal from healthy leaf and leaf infected with loose smut of wheat (abaxial view).

Table 2 Influence of Chopping Frequency on Thermally Active Layer

Chopping frequency v (Hz)	Chopping frequency ω (rad/s)	Thermal diffusion length μ_s^a (μm)	Thermally active layer L_s^b (μm)
5	31.4	96	602.8
10	62.8	67	420.8
17	106.8	52	326.6
50	314.0	30	188.0
100	628.0	20	125.6
200	1256.0	15	94.2
300	1884.0	12	75.4
400	2512.0	11	69.0

$^a\mu_s = (2k_s/\omega\rho_s C_s)^{1/2}$ where k_s is conductivity of sample, ρ_s is density of sample, and C_s is specific heat of sample.
$^bL_s = 2\pi\mu_s$.

upper layers of the leaf. In this region, the magnitude of the signal for both infected and healthy leaves is the same, showing that the photosynthetic pigment composition is the same for healthy and infected leaves in layers between 0 and 15 μm depth from the upper surface. Thus, our result indicates that infection in loose smut is entering from the lower (adaxial) surface of the leaf and depth of penetration is equal to the thickness of 15 μm from the adaxial surface. This is further supported by the fact that in the the adaxial mode, the PA signal strength of the infected leaf is higher than the healthy one at all chopping frequencies (Fig. 6). This is because, in the adaxial mode, the infected portion is common for all frequencies (0–400 Hz) giving an enhanced signal in the case of the infected leaf.

2. Brown Rust

Brown rust, or leaf rust, is the most serious of the three rusts (brown, yellow, and stem rust) affecting wheat. It produces small brown-colored scattered pustules on the leaf surface, which causes reduced photosynthesis and premature defoliation of leaves, affecting the yield severely. Primary induction of the brown rust comes from hills

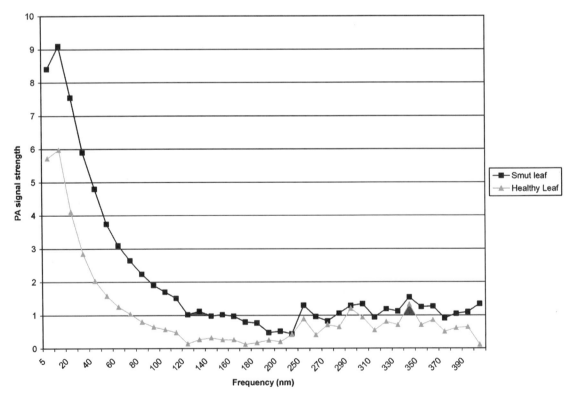

Figure 6 PA Signal from healthy leaf and leaf infected with loose smut of wheat (adaxial view).

through air currents. The PA signals of leaves moderately and severely infected by brown rust caused by (*Puccinia recondita* L.), along with the healthy leaf of wheat (*Tritcum aestivum*), were recorded in both abaxial and adaxial modes in order to ascertain the depth of pathogen penetration. Both the vertical and horizontal penetrations of the pathogen were analyzed to predict the infection patterns. For the abaxial mode, the PA signal from leaves severely infected with brown rust disease is higher than that obtained from the healthy leaf (Rai et al., 2001). This is in marked contrast to a situation observed when the adaxial surface is facing the incident radiation; the PA signal from severely infected leaves is lower than that from the healthy leaf (Rai et al., 2001). In the case of a horizontal spread of the disease in leaves, the tendency toward stem side was observed (Rai et al., 2001). Our result also reveals the progress of disease vertically toward the leaf. In contrast to the loose smut, the infection in brown rust and leaf blight (Singhal et al., 2002) penetrates from the upper surface of the leaf. This observation is convincing because of the fact that leaf blight and brown rust disease is an air-borne disease, so infection comes from the air and affects the upper surface of the leaf; whereas, loose smut is a seed-borne disease.

3. Leaf Blight

Among the fungal diseases of wheat, leaf blight caused by *Alternaria* and *Helminthosporium* is one of the major diseases causing heavy losses ranging between 16.5% and 28.3% (Singh et al., 1996). Leaf blight pathogen was isolated from the infected wheat varieties/materials grown in experimental plots at the Crop Research Center. Cultures isolated on potato dextrose agar medium were purified by single spore isolation. The single spore culture was used for proving pathogenicity on susceptible varieties. The pathogenic isolate was used for measuring the PA spectra at different intervals (i.e., after 2, 4, 8, and 10 days of inoculations), and compared with the measurements on healthy leaves and that of the older leaf spots already present on other infected plants on the date of inoculation. Similarly, PA spectra of healthy, inoculated, and sprayed (with fungicides) leaves after different dates of inoculation were also measured to know the effect of fungicidal application on disease/pathogen development in the tissues. Our result (Singhal et al., 2002) shows that in the case of leaf blight also, the disease is penetrating from the upper portion of the leaf.

C. Wheat Genotype EKBSN-1 AND FBPFM-2

We have also tried to differentiate two wheat genotypes EKBSN-1 and FBPFM-2, in terms of productivity of the grain of wheat by PAS technique (Singhal et al., 1998).

The leaves of the two genotypes, EKBSN-1 and FBPFM-2, were also taken from the Crop Research Center of G.B. Pant University of Agriculture and Technology, Pantnagar, India.

It is a well-known fact that in healthy leaves, the PA signal is a combination of PT (nonradiative de-excitation) and photobaric (oxygen evolution) contribution. Therefore, if in a particular plant the photosynthetic rate is large, the oxygen evolution will also be large, giving a strong PA signal. The PA signal of wheat genotype EKBSN-1 is large when compared with the PA signal strength of wheat genotype FBPFM-2 in the frequency rage of 5–160 Hz (Singhal et al., 1998). Beyond this frequency range (from 160 to 400 Hz), the two signals have the same strength. This observation shows that the rate of O_2 evolution in the healthy leaf of wheat genotype EKBSN-1 is higher, giving a stronger PA signal than that of a healthy leaf of wheat genotype FBPFM-2 in the frequency range of 5–160 Hz. Our experimental result also supports the conclusion drawn by Bults et al. (1982) that at higher modulation frequencies (above 160 Hz) O_2 evolution becomes homogeneous and the microphone of the PA cell then only detects the pulses of heat emission induced by the light pulses of the excitation light. The PA signal strength of the two genotypes is nearly equal above 160 Hz, because in this frequency range (160–400 Hz) the PA signal is arising only because of the contribution of heat emission (nonradiative deexcitation), which is equal in both wheat genotypes. In the case of wheat genotype EKBSN-1, greener leaves indicate the presence of higher amounts of chlorophyll, resulting in greater photosynthesis, which is likely to produce bolder grains, giving higher productivity when compared with FBPFM-2. Our results showing a higher PA signal for wheat genotype EKBSN-1 also indicates the presence of higher amounts of chlorophyll, contributing to more O_2 evolution or more photosynthesis, giving higher strengths of the photoacoustic signals.

D. Virus Disease in Mung Bean Plant

The leaves of a mung bean (*Vigna radiata* L.), severely infected by the virus exhibited a mosaic of yellow and green colouration on the leaf blade. The PA spectra of healthy and infected leaves were recorded in wavelength scanning mode using a modulation frequency of 17 Hz (Plaria et al., 1998a). The leaves of a mung bean (*V. radiata* L.) infected with yellow mosaic virus exhibits a higher PA signal than the untreated, healthy leaves of same cultivars at all wavelengths examined. The ratio of PA signals obtained from diseased and healthy leaves in the UV region was lower when compared with the corresponding ratio in the visible region. PA spectra from viral-infected leaves were influenced by both the evolutions of O_2 and heat emission.

E. Fungal Disease in Sugarcane

The leaf sheaths of a field-grown sugarcane (*Saccharum officinarum* L.), infected with a sheath blight disease caused by *Rhizoctonia* sp., a fungi, were collected and brought to the laboratory in a wet polythene bag. The leaves of sugarcane (*S. officinarum* L.), severely infected with sheath blight exhibit a higher PA signal than the untreated, healthy leaves at all wavelengths examined (Plaria et al., 1998a). Further, these leaf sheaths exhibited the strongest PA signal in the UV region; at longer wavelengths, the PA signal decreases consistently. In fungal infections, the higher PA signal was attributed to nonradiative de-excitation (heat emission).

F. Virus Disease in Okra Plants

The PA spectra of okra (*Abelmoscus* sp.) leaves, infected with virus, and those of healthy leaves were recorded in the wavelength range of 200–800 nm at a modulation frequency of 18 Hz. The PA signal strength was higher in the UV region for the infected leaves, whereas no appreciable difference could be detected in the visible part of the spectrum (Plaria et al., 1998b). The higher PA signal in the UV region is due to the virus-induced proteins and nucleic acids.

G. Seed-Borne Diseases in Wheat and Rice

Wheat and rice are two of the major food crops of the world. We have successfully used the PAS technique to differentiate the dry spores of K.B. (*Tilletia indica*) with other seed-borne pathogens (Gupta et al., 2001). Spores of *Ustiloginoidea virens* (false smut) and *Tilletia barclayana* were extracted from the seeds of infected rice with the help of needles, forceps, and surgical blades. Similarly, teliospores of *T. indica* and spores of *Ustilago tritici* were extracted from the infected wheat seeds of cultivar HD 2328. Cultures of *Helminthosporium sativum* and *Alternaria triticina* were grown, and after sporulation, spores were collected. The PA spectra of these six seed-borne pathogens of wheat and rice (viz., *T. indica*, *T. barclayana*, *U. tritici*, *U. virens*, *H. sativum*, and *A. triticina*) were recorded using the wavelength scanning mode (Gupta et al., 2001) in the wavelength range of 200–800 nm at a modulation frequency of 18 Hz. The number of peaks and the intensity of the signals in PA spectra of all pathogens were compared. The differences were observed in the number of peaks and intensity of the signals for each pathogen.

Our results show that *U. tritici* is a different group of pathogen when compared with the other five pathogens and may be distinguished from the other five pathogens by the presence of a weak band at 272 nm. The band at 232 nm in the PA spectra of *T. barclayana* is broad when compared with that in the other five pathogens, indicating the presence of different functional moities in the molecule of *T. barclayana*, and it may be distinguished by the presence of this band from the other pathogens. On the basis of the intensity of bands at 292 and 232 nm, one can clearly distinguish the pathogens, which is not possible by using conventional methods. The intensity ratio of the bands (intensity of bands at 292 nm/intensity of bands at 232 nm) in pathogens *T. indica*, *T. barclayana*, *U. virens*, *U. tritici*, *A. triticina*, and *H. sativum* is 5.3, 2.5, 0.3, 0.00, 1.8, and 2.8, respectively. This result shows that *T. indica*, *U. virens*, and *U. tritici* are easily identified by high, very low, and zero intensity ratios, respectively.

Helminthosporium may be distinguished from other pathogens by its well-defined characteristic bands in the UV–visible region, at 352, 412, and 752 nm.

In contrast to PAS, conventional absorption spectroscopy cannot make the differential diagnosis of seed-borne pathogens. As observed (Gupta, 1999), the absorption spectra of these pathogens showed nearly similar band patterns, and the intensities of the bands were almost similar. In addition, being a destructive technique, the amount of spore samples required is higher (i.e., preparation of aqueous sonicated extracts of spores) for recording the absorption spectra; whereas the PAS technique is a nondestructive technique, and no sample preparation is required.

VI. CONCLUSION

Our results showed that PAS is a suitable, nondestructive technique for distinguishing different plant diseases and plant products. This technique has proven to be useful for differential diagnosis of various seed-borne pathogens of wheat and rice. PAS can detect the early development of disease caused by a virus in a leaf tissue. This is based on the virus-induced proteins and nucleic acids, which contribute to a higher signal in the UV region. The depth profile analysis using PAS is vital in diagnosing the extent of leaf infection spread both vertically and horizontally. The extent of the disease spreading in the vertical direction is directly proportional to the strength of the PA signal.

VII. FUTURE DEVELOPMENTS

The development of the PAS technique in agricultural science has progressed rapidly since the 1980s, with considerable amount work on the study of photosynthesis activity, trace gases, and different food stuffs.

The PAS application in the disease diagnosis of plants needs more detailed study, specially by separating the PA signal coming from the O_2 evolution and heat emission (nonradiative transition in biomolecules synthesized because of disease developments). It requires more extensive work to find out the characteristics, peaks/bands/lines, expected to arise because of synthesis and/or destruction of different biomolecules during the course of disease development in plants.

Further work is also required to diagnose the seed-borne disease at the seed level so that this technique can be extended for use as a rapid, sensitive, and economical detection method in seed certification standards and plant quarantine regulation.

REFERENCES

Adams, M. J., Beadle, B. C., Krikbright, G. F. (1976). *Analyst* 101:553.

Ageev, B. G., Ponomarev, Yu. N., Sapozhnikova, V. A. (1998). *Appl. Phys.* B67:467.

Bajra, P. R., Mansanares, A. M. (1998). *Instrum. Sci. Technol.* 26:209.

Balderas-Opez, A., Thomas, S. A., Vargas, H., Olaid-Portugal, V., Baquero, R. (1996). *Forest Prod. J.* 46:84.

Beenen, A., Spanner, G., Niessner, R. (1997). *Appl. Spectrosc.* 51:51.

Bell, A. G. (1880). *Am. J. Sci.* 20:305.

Bell, A. G. (1881). *Philos. Mag.* 11:510.

Bergevin, M., N = Soukpoekossi, C. N., Charlebois, D., Leblanc, R. M., Willemot, C. (1995). *J. Appl. Spectrosc.* 49:397.

Bicanic, D. ed. (1998). Photoacoustic, photothermal and related phenomena: their recent developments and applications in the field of agriculture, biological and environmental sciences. *Special Issue of Instrumentation Science and Technology*. New York: Marcel Dekker, Publisher.

Bicanic, D., Harren, F., Reuss, J., Woltering, E., Snel, J., Voesenek, L. A. C. J., Zuidberg, B., Jalink, H., Bijnen, F., Blom, C. W. P. M., Sauren, H., Kooiljman, M., Van Hove, L., Tonk, W. (1989a). In: Hess, P., ed. *Phtoacoustic, Photothermal and Photochemical Processes in Gases*. Vol. 46. Springer Verlag, 213.

Bicanic, D., Kunze, W. D., Sauren, H., Jalink, H., Lubbers, M., Strauss, E. (1989b). *Water, Air Soil Pollut.* 45:115.

Boucher R, N., Carpentier, R. (1999). *Photosynth. Res.*, 59:167.

Bozoki, Z., Sneider, J., Gingl, Z., Mohacsi, A., Szakall, M., Bor, Zs., Szabo, G. (1999). *Meas. Sci. Technol.* 10:999.

Bukhov, N. G., Carpentier, R. (1996). *Photosynth. Res.* 47:13.

Bukhov, N. G., Boucher, N., Carpentier, R. (1998a). *Physiol. Plant.* 104:563.

Bukhov, N. G., Boucher, N., Carpentier, R. (1998b). *Can J. Bot.* 75:1399.

Bults, G., Horwitz, B. A., Malkin, S., Cahen, D. (1982). *Biochim. Biophys. Acta* 679:452.

Buscher, S., Fink, T., Dax, A., Yu, Q., Urban, W. (1994). *Int. Agrophys.* 8:547.

Buschmann, C. (1989). *Philos. Trans. R. Soc. Lond.* B323:423.

Buschmann, C. (1990). *Bot. Acta* 103:9.

Buschmann, C. (1999). *Photosynthetica* 36:149.

Buschmann, C., Kocsanyl, L. (1989). *Photosynth. Res.* 21:129.

Buschmann, C., Prehn, H. (1990). In Linskens, H. F., Jackson, J. F., ed. *Modern Methods of Plant Analysis*. Springer-Verlag, 148.

Butler, W. L. (1978). *A Rev. Plant Physiol.* 29:345.

Caerels, J., Glorieu, C., ThoeN, J. (1998). *Rev. Sci. Instrum.* 69:452.

Cahen, D., Bults, G., Garty, H., Malkin, S. (1980). *J. Biochem, Biophys. Methods* 3:293.

Dadarlat, D., Bicanie, D., Gibkes, J., Pasca, A. (1996). *J. Food Eng.* 30:155.

Dahnke, H., Kahl, J., Schuler, G., Boland, W., Urban, W., Kuhnemann, F. (2000). *Appl. Phys. B Lasers Optics*, 70(2):275.

Doka, O., Kispeter, J., Lorincz, A. (1991). *J. Dairy Res.* 58:453.

Doka, O., Bicanic, D., Szollosy, L. (1997a). *Instrum. Sci. Technol.* 26:203.

Doka, O., Kispeter, J., Bicanic, D. (1997b). *Instrum. Sci. Technol.* 25:297.

Doka, O., Bicanic, D., Szucs, M., Lubbers, M. (1998). *Appl. Spectrosc.* 52:1526.

Dovichi, N. J. (1987). *CRC Crit. Rev. Anal. Chem.* 17:357.

Favier, J. P., Bicanic, D., Doka, O., Chirtoc, M., Helander, P. (1997). *J. Agric. Food Chem.* 45:777.

Flaisher, H., Wolf, M., Cahen, D. (1989). *J. Appl. Phys.* 66:1832.

Foster, N. S., Amonette, J. E., Autrey, S. T. (1999). *Appl. Spectrosc.* 53:735.

Frandas, A., Bicanic, D. (1999). *J. Sci. Food Agric.* 79:1381.

Fuks, B., Homble, F., Figeys, H. P., Lannoye, R., Eyekcn, F. V. (1992). *Weed Science* 40:371.

Glorieu, C., Li. Voti, R., Theon, J., Bertolotti, M., Sibilia, C. (1999). *J. Appl. Phys.* 85:7059.

Gordon, S. H., Wheeler, B. C., Schudy, R. B., Wicklow, D. T., Greene, R. V. (1998). *J. Food Protection* 61:221.

Greene, R. V., Gordon, S. H., Jackson, M. A., Bennett, G. A., McClelland, J. F., Jones, R. W. (1992). *J. Agric. Food Chem.* 40:1144.

Gupta, V. (1999). M.Sc. Thesis, G. B. Pant University of Ag. & Tech., Pantnagar, India.

Gupta, V., Kumar, A., Garg, G. K., Rai, A. K. (2001). *Instrum. Sci. Technol.* 29:283.

Haas, U., Vinha, C. A. (1995). *Analyst* 120:351.

Habib Jiwan, J. L., Chibisov, A. K., Braslavsky, S. E. (1995a). *J. Phys. Chem.* 99:9617.

Habib Jiwan, J. L., Wegewijs, B., Indelli, M. T., Scandola, F., Braslavsky, S. E. (1995b). *Rec. Trav. Chim. Pays - Bas* 114:542.

Haiept, K., Bicanic, D., Gerkema, E., Frandas, A. (1997). *Biosci. Biotechnol. Biochem.* 59:1044.

Harren, F. J. M., Cotti, G., Oomens, J., te Lintel Hekkert, S. (2000). In: Meyers, R. A., ed. *Encyclopedia of Analytical Chemistry*. John Wiley Ltd, Chichester, 2203–2226.

Harren, F., Reuss, J. In: Trigg, G. L. (1997). *Encyclopedia of Applied Physics*. Vol. 19. Weinheim, VCM, pp. 413–435.

Joshi, S., Rai, A. K. (1995). *Asian J. Phys.* 4:265.
Kapil, J. C., Joshi, S. K., Rai, A. K. (2003). *Rev. Sci. Instrum.* 74:3536–3543.
Kerr, E. L., Atwood, J. G. (1968). *Appl. Opt.* 7:915.
Kreuzer, L. B. (1971). *J. Appl. Phys.* 42:2934.
Kreuzer, L. B., Kenyon, N. D., Patel, C. K. N. (1972). *Science* 177:367.
Krikbright, C. F. (1978). *Opt. Pure Appl.* 11:25.
Letzelter, N., Wilson, R., Jones, D. A., Sinnaeve, G. (1995). *J. Food Sci. Agric.* 67:239.
Malkin, S., Cahen, D. (1979). *Photochem. Photobiol.* 29:803.
Marquezini, M. V., Cella, N., Silva, E. C., Serra, D. B., Lima, C. A. S., Vargas, H. (1990). *Analyst* 115:341.
Maugh, T. H. (1975). *Science* 188:38.
Muratikov, K. L. (1998). *Tech. Phys. Lett.* 23:536.
Muratikov, K. L., Glazov, A. L., Rose, D. N., Dumor, J. E., Quay, G. H. (1997). *Tech. Phys. Lett.* 23:188.
Ouzounidou, G., Lannoye, R., Karataglis, S. (1993). *Plant Sci.* 89:221.
Pal, S., Vidyasagar, P. B., Gunale, V. R. (1989). *Curr. Sci.* 58:1096.
Plaria, P., Rai, A. K., Mathur, D. (1998a). *Instrum. Sci. Technol.* 26:221.
Plaria, P., Mathur, D., Rai, A. K. (1998b). *J. Sci. Res.* 48:33.
Poulet, P., Cahen, D., Malkin, S. (1983). *Biochim. Biophys. Acta* 724:433.
Prasad, R. L., Thakur, S. N. (2002). *Spectrochim. Acta Part A* 58:441.
Prasad, R. L., Thakur, S. N. (2003). *Ind. J. Chem. Soc.* 80:341.
Prasad, R. L., Prasad, R., Bhar, G. C., Thakur, S. N. (2002a). *Spectrochim. Acta Part A* 58:3093.
Prasad, R. L., Thakur, S. N., Bhar, G. C. (2002b). *Pramana J. Phys.* 59:487.
Rai, A. K., Mathur, D. (1998). Proc. of the National Symposium on Recent Advances in Laser Molecular Spectroscopy at DDU. Gorakhpur, 64.
Rai, A. K., Mathur, D., Singh, J. P. (2001). *Instrum. Sci. Technol.* 29:355.
Rosencwaig, A. (1973). *Science* 181:657.
Rosencwaig, A. (1978). *Adv. Electr. Elec. Phys.* 45:207.
Rosencwaig, A., Gersho, A. (1976). *Appl. Phys.* 47:64.
Scudieri, F., Bertolotti, M. eds. (1999). A. I. P. Conference Proceeding 463 on 10th International Conference on Photoacoustic and Photothermal Phenomena di March.
Singh, A. K., Singh, K. P., Tiwari, A. N. (1996). Wheat pathology. In: *Wheat Research at Pantnagar*. Research Bulletin No. 128. Directorate of Experiment Station, G.B.P.U.A.& T. Pantnagar, India, 25.
Singhal, S. K., Rai, A. K., Singh, K. P. (1998). Proceedings of National Laser Symposium at I.I.T. Kanpur, India.
Singhal, S. K., Singh, K. P., Joshi, S. K., Rai, A. K. (2002). *Curr. Sci.* 82:172.
Sneider, J., Bozoki, Z., Szabo, G., Bor, Zs. (1997). *Opt. Eng.* 36:482.
Szigeti, Z., Nagel, E. M., Buschmann, C., Lichtenthaler, H. K. (1989). *J. Pant Physiol.* 134:104.
Tam, A. C. (1986). *Rev. Mod. Phys.* 58:381.
Tam, A. C. (1989). In: J.A., ed. *Photothermal Investigation in Solid and Fluids Sell*. Academic Press Inc., New York.
Veeranjaneyulu, K., Charlebois, D., N' soukpo'e-Kossi, C. N., Leblanc, R. M. (1990). *Environ. Pollut.* 65:127.
Viengerov, M. L. (1938). *Dokl. Akad. Nauk SSSR* 19:687.
Vries, H. S. M., Wasono, M. A. S., Harren, F. J. M., Woltering, E. J. (1996). *Postharvest Biol. Technol.* 8:1.

10

Techniques of Chiroptical Spectroscopy

HARRY G. BRITTAIN
Center for Pharmaceutical Physics Milford, NJ, USA

NELU GRINBERG
Merck Research Laboratories Rahway, NJ, USA

I. INTRODUCTION TO CHIROPTICAL PHENOMENA

Most forms of optical spectroscopy are usually concerned with measurement of the absorption or emission of electromagnetic radiation, the energy of which lies between 10 and 50,000 wavenumbers. Any possible effects associated with the polarization of the electric vectors of the radiant energy are not normally considered during the course of commonly-performed experiments. For molecules lacking certain types of molecular symmetry, interactions with polarized radiation are important and can be utilized to study a wide variety of phenomena. Materials are, therefore, classified as being either isotropic (incapable of influencing the polarization state of light) or anisotropic (having the ability to affect the polarization properties of transmitted light).

Molecules for which the mirror images cannot be superimposed are denoted as being dissymmetric or chiral, and these enantiomer structures are capable of being resolved. The fundamental requirement for the existence of molecular dissymmetry is that the molecule cannot possess any improper axis of rotation, the minimal interpretation of which implies additional interaction with light whose electric vectors are circularly polarized. This property manifests itself in an apparent rotation of the plane of linearly polarized light (polarimetry) or in a preferential absorption of either left- or right-circularly polarized light [circular dichroism (CD)]. CD can be observed in either electronic or vibrational bands. The Raman scattering of a chiral compound can also reflect the optical activity of the molecule. If the excited state of a compound is both luminescent and chiral, then the property of circularly polarized luminescence (CPL) can be observed. CD of an optically active molecule can be used in conjunction with fluorescence monitoring to provide a differential excitation spectrum [fluorescence detected circular dichroism (FDCD)].

A variety of chiroptical techniques are available, and these can be useful in analytical work. The foremost problem is a determination of the absolute configuration of all dissymmetric atoms within a molecule. Any chiroptical technique can be used to obtain the enantiomeric purity of a given sample, although certain methods have been found to be more useful than others. Questions of optical activity are of extreme importance to the pharmaceutical industry, where an ever-increasing number of synthesized drug substances contain chiral substituents. Documentation of the properties of any and all dissymmetric centers is essential for the successful registration of such substances.

The entire field of asymmetric chemical synthesis is supported by studies of molecular optical activity.

Each chiroptical technique is somewhat different in its instrumentation, its experimental design, and in what information is most readily deduced from the measurements. These various methods will be discussed, in turn, within the scope of the present review, and representative examples will be provided to illustrate the power associated with each particular technique. A general introduction to optical activity has been provided by Charney (1) who has provided the most readable summary of the history and practice associated with chiroptical spectroscopy. A number of other monographs have been written, which concern various applications of chiroptical spectroscopy, ranging from the very theoretical to the very practical (2–9) and these texts contain numerous references suitable for entrance into the field. The areas of CD and optical rotatory dispersion (ORD) have received the most extensive coverage, and these have been exhaustively studied since the effects were discovered. The newer methods are covered primarily in review articles, which will be cited when appropriate.

II. POLARIZATION PROPERTIES OF LIGHT

An understanding of the polarization properties of light is essential to any discussion of chiroptical measurements. It is the usual practice to consider only the behavior of the electric vector in describing the properties of polarized light, even though it is possible to use the magnetic vector equally well. Unpolarized light propagating along the Z-axis will contain electric vectors whose directions span all possible angles within the X–Y plane. Linearly polarized light represents the situation where all the transverse electric vectors are constrained to vibrate in a single plane. The simplest way to produce linearly polarized light is by dichroism, which would be the passage of the incident light beam through a material that totally absorbs all electric vectors not lying along a particular plane. The other elements suitable for the production of linearly polarized light are crystalline materials that exhibit optical double refraction, and these are the Glan, Glan-Thompson, and Nicol prisms.

As with any vector quantity, the electric vector describing the polarization condition can be resolved into projections along the X and Y axes. For linearly polarized light, these will always remain in-phase during the propagation process unless passage through another anisotropic element takes place. Attempted passage of linearly polarized light through another polarized (referred to as the analyzer) results in the transmission of only the vector component that lies along the axis of the second polarizer. If the incident angle of polarization is orthogonal to the axis of transmission of the analyzer, then no light will be transmitted.

Certain crystalline optical elements have the property of being able to alter the phase relationship existing between the electric vector projections. When the vector projections are rendered 90° out of phase, the electric vector executes a helical motion as it passes through space, and the light is now denoted as being circularly polarized. As the helix can be either left- or right-handed, the light is referred to as being either left- or right-circularly polarized. It is preferable to consider linearly polarized light as being the resultant formed by combining equal amounts of left- and right-handed circularly polarized components, the electric vectors of which are always exactly in phase. When the phase angle is between the two vector components in any light beam lies between 0° and 90°, the light is denoted as being elliptically polarized. The production of a 90° phase shift is termed as quarter-wave retardation, and an optical element that effects such a change is a quarter-wave plate. Passage of linearly polarized light through a quarter-wave plate produces a beam of circularly polarized light, the sense of which depends on whether the phase angle has been either advanced or retarded by 90°. The passage of circularly polarized light through a quarter-wave plate will produce linearly polarized light, whose angle is rotated by 90° with respect to the original plane of linear polarization. Circularly polarized light will pass through a linear polarizer without effect.

The chiroptical spectroscopic methods that have been developed take advantage of the fact that anisotropic materials are capable of producing the same effects in polarized light, as do anisotropic crystalline optical elements. As these effects are determined by the molecular stereochemistry, the utility of chiroptical methods to the study of molecular properties is self-evident.

III. OPTICAL ROTATION AND ORD

Charney (1) has provided an excellent summary of the history associated with chiroptical methods of study. The study of molecular optical activity can be considered as beginning with the work of Biot, and with the publication of his notes (10). Biot demonstrated that the plane of linearly polarized light would be rotated upon passage through an optically active medium, and designed a working polarimeter capable of quantitatively measuring the effect. The use of calcite prisms was introduced by Mitscherlich in 1844 (11), and the double-field method of detection was devised by Soleil in 1845 (12–14). Because of these early developments, many advances in polarimetry have been made, and a large number of detection schemes are now possible. An extensive summary of methods has been provided by Heller (15).

In principle, the measurement of optical rotation is extremely simple, and a suitable apparatus is shown schematically in Fig. 1. The incident light is collimated and plane-polarized, and allowed to pass through the medium of interest. In most common measurements, the medium consists of the analyte dissolved in an appropriate solvent. The polarization plane of the incident light is set by the orientation of the initial polarizer, and then the angle of rotation is defined with respect to this original plane. This is carried out by first determining the orientation of the polarizer and the analyzer for which no light can be transmitted (the null position). The null position is properly determined by inserting the sample cell (filled with solvent) between the polarizer and the analyzer. The cell is then emptied, filled with the medium containing the optically active material, and then the cell is introduced between the prisms. The analyzer is rotated until a null position is again detected, and the observed angle of rotation is taken as the difference between the two null angles.

The measurement of the null angle as a minimum in the transmitted light, when conducted visually, is experimentally difficult to observe. For this reason, a more efficient detection mechanism was developed, which is commonly known as the "half-shade" technique. The half-shade is a device that transforms the total extinction point into an equal illumination of two adjacent fields. This mode of detection was found to be superior for manual polarimeters, as the human eye is much more adept at balancing fields of transmitted light, rather than at detecting a simple minimum in light intensity. In most cases, setting an appropriate device in front of a simple analyzing prism, such as the Lippich prism, brings about the half-shade effect.

With the development of photoelectric devices, the manual detection of null positions in polarimetry became superseded by instrumental measurements of the endpoint. As would be anticipated, measurements can be made far more easily and accurately using a photoelectric detection of the null position. Early versions of automated polarimeters used the half-shade method, but instead the two light intensities were measured using photomultiplier tubes. The position of the analyzer was rotated until the difference in signals detected by the two detectors reached a minimum. Other methods have made use of modulated light beams and variations on the method of symmetric angles. A wide variety of detection methods have been developed, enabling accurate measurements of optical rotation to be made on a routine basis. The influence of polarimeter design on the observed signal-to-noise characteristics has been discussed by a number of investigators (16–18).

The velocity of light (v) passing through a medium is determined by the index of refraction (n) of that medium:

$$v = \frac{c}{n} \qquad (1)$$

where c is the velocity of light in vacuum. For a nonchiral medium, the refractive index will not exhibit a dependence on the sense of the polarization state of the light. For a chiral medium, the refractive index associated with left-circularly polarized light will not normally equal the refractive index associated with right-circularly polarized light. It follows that the velocities of left- and right-circularly polarized light will differ on passage through a chiral medium. As linearly polarized light can be resolved into two in-phase, oppositely signed, circularly polarized components, then the components will no longer be in phase once they pass through the chiral medium. Upon leaving the chiral medium, the components are recombined, and linearly polarized light is obtained whose plane is rotated (relative to the original plane) by an angle equal to half the phase angle difference of the circular components. Charney (1) has shown that this phase angle difference is given by β:

$$\beta = \left\{ \frac{2\pi b'}{\lambda_0} \right\} (n_L - n_R) \qquad (2)$$

In Eq. (2), β is the phase difference, b' is the medium path (in cm), λ_0 is the vacuum wavelength of the light used, and n_L and n_R are the refractive indices for left- and right-circularly polarized light, respectively. The quantity $(n_L - n_R)$ defines the circular birefringence of the chiral medium, and this quantity is the origin of what is commonly referred to as optical rotation. The observed rotation of the plane-polarized light is given in radians by:

$$\varphi = \frac{\beta}{2} \qquad (3)$$

The usual practice to express rotation is in terms of degrees, and in that case Eq. (2) becomes:

$$\alpha = \left\{ \frac{1800 b}{\lambda_0} \right\} (n_L - n_R) \qquad (4)$$

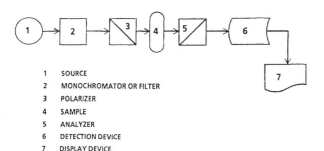

1 SOURCE
2 MONOCHROMATOR OR FILTER
3 POLARIZER
4 SAMPLE
5 ANALYZER
6 DETECTION DEVICE
7 DISPLAY DEVICE

Figure 1 Block diagram of a simple polarimeter suitable for the measurement of optical rotation. For use in ORD, a tunable wavelength source would be used.

In Eq. (4), b represents the medium path length in decimeters, which is the conventional unit. The optical rotation (α) of a chiral medium can be either positive or negative depending on the sign of the circular birefringence. It is not practical to directly measure the circular birefringence, as the magnitude of ($n_L - n_R$) is typically in the range of (15) 10^{-8}–10^{-9}. The optical rotation (α) exhibited by a chiral medium depends on the optical path length, the wavelength of the light used, on the temperature of the system, and on the concentration of dissymmetric analyte molecules. If the solute concentration (c) is given in terms of grams per 100 mL of solution, then the observed rotation (an extrinsic quantity) can be converted into the specific rotation (an intrinsic quantity) using:

$$[\alpha] = \frac{100\alpha}{bc} \quad (5)$$

The molar rotation, $[M]$, is defined as

$$[M] = \frac{(FW)\alpha}{bc} \quad (6)$$

$$= \frac{(FW)[\alpha]}{100} \quad (7)$$

where FW is the formula weight of the dissymmetric solute. When the solute concentration is given in units of molarity (M), Eq. (6) becomes

$$[M] = \frac{10\alpha}{bM} \quad (8)$$

The temperature associated with a measurement of the specific or molar rotation of a given substance must be specified. Thermal volume changes or alterations in molecular structure (as induced by a temperature change) are capable of producing detectable changes in the observed rotations. In the situation where solute–solute interactions become important at high concentrations, it may be observed that the specific rotation of a solute is not independent of concentration. Therefore, it is an acceptable practice to obtain polarimetry data at a variety of concentration values to verify that a true molecular parameter has been measured.

The verification of polarimeter accuracy is often not addressed on a routine basis. The easiest method to verify the performance of a polarimeter is to use quartz plates that have been cut to known degrees of retardation. These plates are normally certified to yield a specified optical rotation at a specific wavelength. A reasonable criterion for acceptability is that the observed optical rotation should be within $\pm 0.5\%$ of the certified value. The quartz plates are available from a variety of sources, including manufacturers of commercial polarimeters or houses specializing in optical components.

Another possibility to verify the accuracy of polarimetry measurements is to measure the optical rotation of a known compound. Owing to the stability of their rotatory strengths once dissolved in a fluid medium, steroids are probably most suitable for this purpose. Chafetz (19) has provided a compilation of the specific rotation values obtained for a very extensive list of steroids. When possible, data have been reported for alternate solvents, as well as for wavelengths other than 589 nm.

The most important parameter in determining the magnitude of optical rotation is the wavelength of the light used for the determination. Generally, the magnitude of the circular birefringence increases as the wavelength becomes shorter, and hence specific rotations increase in a regular manner at lower wavelengths. This behavior persists until the light is capable of being absorbed by the chiral substance, whereupon the refractive index exhibits anomalous behavior. The variation of specific or molar rotation with wavelength is termed as ORD.

The anomalous dispersion observed in ORD spectra arises as the refractive index of a material is actually the sum of real and imaginary parts:

$$n = n_0 + ik \quad (9)$$

where n is the observed refractive index at some wavelength, n_0 is the refractive index at infinite wavelength, and k is the absorption coefficient of the substance. It is evident that if ($n_L - n_R$) does not equal 0, then ($k_L - k_R$) will not equal 0 either. The relation between the various quantities was first conceived by Cotton (20), and is illustrated schematically in Fig. 2.

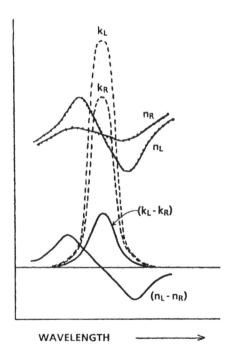

Figure 2 Schematic diagram of the Cotton effect, illustrating the effects of circular birefringence and CD within an isolated absorption band.

A. Applications of Optical Rotation and ORD

Since the discovery of the optical activity of asymmetric molecules by Biot (10) in the early 19th century, the optical rotation of light has become an important tool in the analysis of chiral compounds. The earliest work involving chiral organic molecules was entirely based on ORD methods, since little else was available at that time. One of the largest data sets collected to date concerns the chirality of ketone and aldehyde groups (21), which eventually resulted in the deduction of the octant rule (22). The octant rule was an attempt to relate the absolute stereochemistry within the immediate environment of the chromophore with the sign and intensity of the ORD Cotton effects. To apply the rule, the CD within the $n-\pi^*$ transition around 300 nm is obtained, and its sign and intensity are noted. The rule developed by Djerassi and coworkers states that the three nodal planes of the n- and π^*-orbitals of the carbonyl group divide the molecular environment into four front octants and four back octants. A group or atom situated in the upper-left or lower-right rear octant (relative to an observer looking at the molecule parallel to the C=O axis) induces a positive Cotton effect in the $n-\pi^*$ band. A negative Cotton effect would be produced by substitution within the upper-right or lower-left back octant. Although exceptions to the octant rule have been shown, the wide applicability of the octant rule has remained established (23). The ability to deduce molecular conformations in solution on the basis of ORD spectra data has proven to be extremely valuable to synthetic and physical organic chemists, and enabled investigators of the time to develop their work without requiring the use of more heroic methods.

Polarimetry is a well-established technique in the pharmaceutical industry, where the synthesis of chiral drugs has become a well-recognized necessity, and the enantiomeric purity of drugs is a critical issue (24–26). In the organic chemistry laboratory, specific rotation is used to ascertain the optical purity of certain synthesized compounds (27–35).

Other applications are concerned with the correlation between the absolute configuration of optically active compounds and their optical rotation. Thus, Zhang et al. (36) studied the effect of urea and sodium hydroxide on the molecular weight and conformation of α-(1 → 3) D-glucan from *Lentinus edodes* in aqueous solutions. The results showed a conformational change of α-glucan that occurs when the compound is dissolved in aqueous 0.5 M NaOH containing 0.4–0.6 M urea. These conformational changes correlated with the specific rotation of the compounds.

Ayscough et al. (37) studied the epimerization of triphenyl phosphine metal complexes coordinated with stereogenic metal centers, and their use as molecular optical switches. The specific rotation was found to undergo an inversion in sign upon epimerization, suggesting that the inversion of the propeller configuration of the coordinated PPh_3 ligand is a major contributor to the switch. Azobenzene-modified polyaramides containing atropisomeric 2,2'-binaphthyl linkages were found to exhibit thermo- and photo-responsive chiroptical behavior when dissolved in dilute solutions. The magnitude of the specific rotation at the sodium D-line ranged into hundreds of degrees, and was dependent on the extent of binaphtyl loading along the polymer chain. The irradiation of the polymer samples to induce the trans → cis isomerization process resulted in an immediate chiroptical response, with the CD band intensities and optical rotation becoming significantly diminished. These effects were fully reversible and were attributed to the presence of one-handed helical conformations in the trans azobenzene-modified polymers that were severely disrupted following the trans → cis isomerization reaction (38).

There has been a special interest in the use of polarimetry as a means of detection for the chromatographic analysis of chiral compounds. Thus, Yeung et al. built a laser-based polarimeter for detecting optically active compounds during the conduct of high performance liquid chromatography (HPLC) (39).

The most recent advance in optical rotation measurement is cavity ring-down polarimetry (40). The principle underlying this technique consists of an optical cavity having two mirrors. When a short laser pulse is injected into the cavity, it is reflected back and forth within the cavity, but its intensity will decrease over time due to leakage through the mirrors. The intensity of the laser pulse within the cavity follows the exponential decay pattern, which is referred to as ring-down. The change in the pattern due to rotation of plane of polarization within the cavity by an optically active sample is the observable, measured in cavity ring-down polarimetry. As light is reflected back and forth within the cavity, it is necessary to maintain the additivity of optical rotation during numerous reflections. For this purpose, the laser pulse is first circularly polarized before being allowed to enter the cavity. Two quarter-wave retardation plates are placed within the cavity, and another quarter-wave plate is placed at the exit port of the cavity. The estimated sensitivity of this technique for the measurement of optical rotation is 2.5×10^{-7} degrees/cm. As a consequence of such high sensitivity, the optical rotation of low vapor pressure samples could be measured in the vapor phase.

There are many other applications of optical rotation in the literature, the coverage of which is beyond the scope of this Chapter. The paper of Polavarapu contains a comprehensive review on advances in optical rotation (41).

IV. CIRCULAR DICHROISM

Optical rotation and ORD phenomena are easily interpretable outside regions of electronic absorption, but exhibit anomalous properties when measured within absorption bands. This effect arises as the complex refractive index also contains a contribution related to molecular absorptivity, as described in Eq. (9). Not only the phase angle between the projections of the two circularly polarized components will be altered by passage through the chiral medium, but also their amplitudes will be modified by the degree of absorption experienced by each component. This differential absorption of left- and right-circularly polarized light is termed as CD, and is given by $(k_L - k_R)$.

The effect of one circularly polarized component being more strongly absorbed than the other is that when the projections are recombined after leaving the chiral medium, they no longer produce plane-polarized light. Instead, the resulting components describe an ellipse, whose major axis lies along the angle of rotation. The measure of the eccentricity of the ellipse that results from the differential absorption is termed as ellipticity (φ). It is not difficult to show that (1):

$$\varphi = \left\{\frac{\pi z}{\lambda}\right\}(k_L - k_R) \tag{10}$$

where z is the path length in cm, and λ is the wavelength of the light. If C is the concentration of absorbing chiral solute in moles/liter, then the mean molar absorptivity, a, is derived from the absorption index by:

$$a = \frac{4\pi k}{2.303 \lambda C} \tag{11}$$

In that case, the ellipticity in radians becomes:

$$\varphi = 2.303 \frac{Cz}{4} \tag{12}$$

The expression of ellipticity in radians is cumbersome, and consequently it is a usual practice to convert this quantity into units of degrees using the relation:

$$\theta = \varphi\left(\frac{360}{2\pi}\right) \tag{13}$$

so that

$$\theta = (a_L - a_R)\, Cz(32.90) \tag{14}$$

The molar ellipticity is an intensive quantity, and is calculated from:

$$[\theta] = \frac{\theta(\text{FW})}{LC'(100)} \tag{15}$$

where FW is the formula weight of the solute in question, L is the medium path length in decimeters, and C' is the solute concentration in grams/mL. The molar ellipticity is related to the differential absorption by:

$$(a_L - a_R) = \frac{[\theta]}{3298} \tag{16}$$

Most instrumentation suitable for measurement of CD is based on the design of Grosjean and Legrand (9), and a block diagram of this basic design is shown in Fig. 3. Linearly polarized light is passed through a dynamic quarter-wave plate that modulates the beam alternately into left- and right-circularly polarized light. The dynamic quarter-wave plate is a piece of isotropic material, which is rendered anisotropic through the external application of stress. The device can be a Pockels cell (in which stress is created in a crystal of ammonium dideuterium phosphate through the application of high-voltage AC) or a photoelastic modulator (in which the stress is induced by the piezoelectric effect). The light leaving the sample cell is detected by a photomultiplier tube, whose current output is converted to voltage, and split. One signal consists of an alternating current signal proportional to the CD, and is due to the differential absorption of one component over the other. This signal is amplified by means of phase-sensitive detection. The other signal is averaged, and is related to the mean light absorption. The ratio of these signals varies linearly as a function of the CD amplitude, and is the recorded signal of interest.

The calibration of CD spectrometers has been approached most often by the use of reference standard materials, with various salts of d-10-camphorsulfonic acid having received the highest degree of attention. Highly accurate CD spectra of the free acid (42,43), the n-propylammonium and n-butylammonium salts (44), and the *tris*(hydroxymethyl)aminomethane salt (45) have been published, and the relative virtues of each material are discussed in detail. Instrumental techniques, based purely on the use of suitable optical elements, have also been proposed as methods for the calibration of CD spectrometers (46,47).

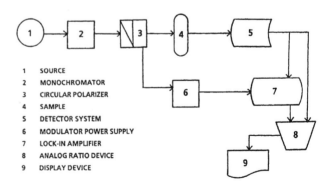

Figure 3 Block diagram of an analog apparatus suitable for the measurement of CD.

A. Applications of CD

It would be impossible to summarize the utility of CD spectroscopy for the study of molecular stereochemistry, as entire monographs have been written on the subject (48–55). As a comprehensive review is beyond the scope of this chapter, only several examples of applications of CD spectroscopy will be presented here to illustrate the power of the technique.

The oxime chromophore is formed upon reaction of a carbonyl group with hydroxylamine, and that group is particularly well suited for CD spectroscopic studies. The UV absorption and CD spectra simplify so that the sign and intensity of the Cotton effect accurately reflects the stereochemistry within its immediate vicinity (56). The CD spectra of most oximes consist of one major peak (without any accompanying fine structure) around 240 nm. The CD spectrum of an oxime usually yields more reliable stereochemical information than that of the parent ketone, as the single-signed CD peak has been shown to be a reliable indicator of the absolute stereochemistry adjacent to the chromophore. Although such information would be available from considerations of the octant rule, confirmation is possible through formation and characterization of the oxime.

As another example, β-lactam compounds represent a versatile and commercially significant class of antibiotics. The resistance of some bacteria to common antibiotics prompted researchers to investigate other classes of antibiotics, such as the O-analogs of penicillin. However, understanding their structure can lead to an understanding of their mechanism of interactions. Thus, Frelek et al. (57) studied the relationship between the molecular structure of 5-dethia-5-oxacephams and clavams and their chiroptical properties using X-ray diffraction analysis, molecular modeling calculations, and CD spectroscopy. The helicity of the β-lactam chromophore in both groups of compounds is caused by the nonplanarity of the system, and depends upon the absolute configuration of the bridgehead carbon atom. This indicates that the sense of chirality is controlled by the absolute configuration at the C-6 or C-5 groups in the oxacephams and clavams, respectively. The helicity appears to be independent of the type and the position of other substituents present in the oxacepham and clavam moieties, and is also the sign-determining factor for the $n \rightarrow \pi^*$ CD band. The results correlate the negative/positive torsional angle of the β-lactam subunit O=C-N-C with the negative/positive sign of the $n \rightarrow \pi^*$ Cotton effect for both classes of compounds.

Solid-state CD spectroscopy is another type of application. Such experiments can provide valuable information on conformation and intermolecular interactions, including chirality induction (58). For example, solid-state CD can provide information on solute–solvent interactions in solution, as well as intermolecular interactions in the solid state. The effects of solvents on optical activity are caused by the formation of co-ordination compounds between the optically active molecules concerned and the solvent molecules such as counter ions in solution. This may affect optical activity by alteration of conformation in the case of flexible compounds, or through vicinal effects. In the solid state, molecules are densely packed and under the much stronger influence of the neighboring molecules. Such a situation can be regarded as an extreme case of solvent effect. Thus, an unusual conformation of a chiral molecule, which is unstable in solution, may be studied in the solid state. Nonchiral molecules may become optically active when the degree of freedom of molecules, for example, rotation and orientation about chemical bonds, is severely restricted by the formation of a crystal lattice. For instance, 2-thioaryloxy-3-methylcyclohexen-1-one is a nonchiral compound, and its solution does not show any CD peaks in the 450–250 nm region. When the compound was trapped in a 1:1 inclusion crystal with a chiral host compound, $(-)$-(R, R)-$(-)$-trans-2,3-bis(hydroxydiphenylmethyl)-1,4-dioxaspiro-[4, 4] nonane, it exhibited substantial CD peaks in nujol mull or a KBr matrix. The two solid-state spectra were similar. The solution CD of the inclusion compound did not exhibit CD peaks in the 400–280 nm region. Hubal and Hofer (59) studied chirality and CD of oriented molecules and anisotropic phases. Thus, the results of the CD spectra along different viewing directions within molecules and phases—the anisotropy of CD—can give information, suitable for checking helicity rules or analyzing the suprastructural chirality of films of organic materials. The authors' results showed that the anisotropic CD (ACD)—the CD of anisotropic phases and oriented molecules—in an oriented state would result in different information about chirality being gained from different viewing directions.

Assignment of absolute configurations of molecules requires the knowledge of the molecular structure and an observable signal that can be assigned to one of two possible enantiomers. More recently, chiroptical spectroscopic techniques have been used to assign absolute configuration in system where the molecular structure is known. Among the most useful techniques, because of its nonempirical nature, is exciton coupled circular dichroism (ECCD) (60). This method requires spatial proximity of two chromophores in a chiral environment. Dipole–dipole interaction between the electric transition moments of chromophores can lead to splitting of the excited states (exciton coupling), which can generate Cotton effects of mutually opposite signs. Several reports present the correlation between the CD spectra and the absolute configuration of compounds (60–63).

Studies of the complexation of chiral crown hosts with different chiral analytes were described by Farkas et al. (64). The authors used CD spectroscopy to explore the discriminating efficiency of pyridino- and thiopyridino-, phenazino- and acridino-18-crown-6 hosts, as well as pyridino- and phenazino-18-crown-6 hosts with allylic moieties attached either to the macrocyclic ring or to the heterocyclic subunit using enantiomers of α-(1-naphtyl)-ethylamine hydrogen perchlorate. CD spectroscopy showed that the relative stability of heterochiral [(R, R)-host/(S) guest or (S, S)-host/R-guest] generally exceeds homochiral complexes [(R, R)/(R) or (S, S)/(S)]. Increasing the bulk of the substituents decreases the complex stability, but increases the discriminating power of the host.

Molecular interactions between chiral and achiral compounds can give rise to induced circular dichroism (ICD) of the achiral counterpart, with the requirement that the achiral component of the system absorbs in the UV or visible region. The CD spectra obtained in this way is an indicator of the absolute configuration of the chiral component, as well as the orientation of the molecules in the complex, relative to each other. The process can be regarded as a salvation of the achiral partner of the complex by the chiral one. This salvation can give rise to an asymmetric perturbation of the chromophoric substrate, which can be traced as a CD signal in the UV region (65). Coupling between identical or near identical chromophores is regarded as exciton coupling, resulting in bisignate CD curves due to a spilt Cotton effect (65). In this way, the interaction of cyclodextrins with aromatic guests has been studied (66–68). Harata et al. showed that the ICD of a chromophore located inside the cyclodextrin host would be positive when the electric transition dipole moment is parallel to the principal axis of the cyclodextrin. If the alignment is perpendicular to the principal axis of the cyclodextrin, the resulting ICD will be negative (69,70). However, Kodaka showed that the situation is reversed if the chromophoric guest is located outside the cyclodextrin (71,72). Many applications of ICD are presented in Ref. (65). Porphyrins and metalloporphyrins are a very interesting class of compounds capable of inducing CD. Porphyrins have powerful CD chromophores that are characterized by their intense and red-shifted Soret band, propensity to undergo $\pi-\pi$ stacking, ease of incorporating metals, and ease in varying solubility. They offer the possibility of studying the stereochemistry of chiral porphyrins assemblies, large organic molecules, biopolymers, and compounds available in minute quantities. The tendency of porphyrins to undergo $\pi-\pi$ stacking, and of zinc porphyrins to coordinate with amines enables the CD exciton chirality method to be extended to configurational assignment of flexible compounds containing only one stereogenic center. The ICD of the host porphyrin's Soret band reflects the identity of guests and binding modes of host/guest complexation with high sensitivity (73–77).

The CD spectra are used extensively for studying the conformation of proteins and polypeptides. The correlation of certain spectral futures with well-defined peptide conformation has been used to develop a computational procedure for conformational analysis (78). These results yield reasonable estimates of α helix and β strand and sheets, as well as various types of bends present in a polypeptide or protein (79), and information on quaternary structure (80). Similarly, CD spectroscopy can also give an important information on the conformation of nucleic acids (81) along with the interactions between the proteins and nucleic acids (82,83).

The use of CD as an extension of UV detection in HPLC brought a new dimension to the separation of enantiomers (84,85). The CD coupled with HPLC has been used to determine the stereochemistry of a pure eluting enantiomer using single wavelength monitoring (86,87). The combination of a CD detector and nonchiral liquid chromatography was used for the chiral analysis of unresolved enantiomers to determine the enantioselectivity of a molybdenum-catalyzed asymmetric allylic alkylation reaction. The CD/UV peak area ratio of the unresolved enantiomers was calculated and compared with that of a reference standard to determine the enantiomeric purity. The limit of quantitation obtained was 20 ng for the chiral ligand and 1 mg for the branched product (88).

V. CPL SPECTROSCOPY

The optical rotation, ORD, and CD methods yield information relating to the ground electronic state of the chiral molecule. CPL spectroscopy involves a measurement of the spontaneous differential emission of left- or right-circularly polarized light by an optically active species. As with CD spectroscopy, the chirality can either be natural or induced by a magnetic field. CPL spectroscopy differs from CD spectroscopy in that it is a measure of the chirality of a luminescent excited state. If the geometry of the molecule remains unchanged during the excitation process, then the same chiroptical information could be obtained by using either the CD or the CPL method. In the situations where geometrical changes are associated with excitation into higher electronic states, then comparison of CD and CPL results can be used to deduce the nature of these structural modifications. Several reviews covering the applications to which CPL spectroscopy has been put have been written by several authors (89–92).

The major limitation associated with CPL spectroscopy is that it is confined to luminescent molecules only. However, this restriction can be used to advantage in that it imparts selectivity to the technique. The CPL technique has been particularly useful in the study of chiral lanthanide complexes (93,94), as the absorptivity of these compounds is sufficiently low so that high-quality CD spectra can only be obtained under conditions for which the chemistry of the analytes is incompletely known. It is appropriate to consider CPL spectroscopy as a technique that combines the selectivity of CD spectroscopy with the sensitivity of luminescence spectroscopy.

Even though CPL spectroscopy has been known for some time, no commercial instrumentation has yet become available for its measurement. Instruments have been described that are based on either an analog design following the CD principles of Grosjean and Legrand (4), or on digital methods that employ photon counting electronics (95). The production of artifact-free data has been discussed in great detail, both from the viewpoint of spurious signals that would be introduced by imperfect optical components (96,97), and from photoselection effects (98).

A block diagram describing the basic design of an analog CPL spectrometer is shown in Fig. 4. The excitation source can be either a laser or an arc lamp, but it is important that the source of excitation is unpolarized to avoid possible photoselection artifacts. It is best to collect the emitted light at 0° to the excitation beam (often denoted as the "head-on" geometry) so that spurious effects due to linear polarization in the emission will not find their way through the imperfect electronics used for the phase-sensitive detection. Extensive discussions of these effects have been provided by a number of investigators (89,91). Any unabsorbed excitation energy is purged from the optical train through the use of a long-pass filter.

The circular polarization in the emitted light is detected by the circular analyzer, as the insensitivity of photomultiplier tubes to light polarization states requires the use of a transducer. Through the use of a dynamically operated quarter-wave retardation element (a photoelastic modulator or a Pockels cell), the circularly polarized component of the emission is transformed into periodically interconverting orthogonal planes of linearly polarized light. This effect is accomplished by advancing and retarding the phase angle difference between the electric vectors by 90°. The linear polarizer following the modulator extinguishes one of the linear components, producing a modulated light intensity proportional to the CPL intensity in the emission.

The wavelength dependence of the emission spectrum is analyzed in the usual manner by an emission monochromator, and detected by a photomultiplier tube. The current output of the tube is converted to a voltage signal, and divided. One analog output can be sent either to a chart recorder, or digitized for input into a laboratory computer. The other signal is fed into a lock-in amplifier, where the AC ripple present in the large DC background is amplified by means of phase-sensitive detection. The rectified AC output of the phase-sensitive detector is proportional to the CPL intensity, and is the other signal to be processed. It is useful to ratio the differential (CPL) and total luminescence (TL) signals, and to display this quantity as well.

The CPL experiment thus produces two measurable quantities that are obtained in arbitrary units and are directly related to the circular polarization condition of the luminescence. The TL intensity is defined as:

$$I = I_L + I_R \quad (17)$$

and the CPL intensity is defined as:

$$\Delta I = I_L - I_R \quad (18)$$

In Eqs. (17) and (18), I_L and I_R represent the emitted intensities of left- and right-circularly polarized light, respectively. The ratio of these quantities must be dimensionless, and therefore free of unit dependence. Analogous to the Kuhn anisotropy factor defined for CD spectroscopy, the luminescence dissymmetry factor of CPL spectroscopy is defined as:

$$g_{lum} = \frac{2\Delta I}{I} \quad (19)$$

In the digital measurement method, the number of left-circularly polarized photons is counted separately from the number of right-circularly polarized photons (95). However, the definitions for the various CPL quantities are still the same.

The calibration of CPL spectra is an important question as long as all CPL spectrometers are laboratory constructed. The best method reported to date is still that originally reported by Steinberg and Gafni (99), who used the birefringence of a quartz plate to transform the empirically observed dissymmetry factors into absolute quantities.

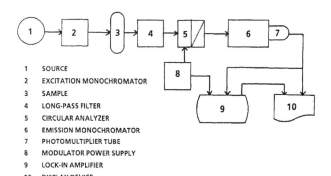

1 SOURCE
2 EXCITATION MONOCHROMATOR
3 SAMPLE
4 LONG-PASS FILTER
5 CIRCULAR ANALYZER
6 EMISSION MONOCHROMATOR
7 PHOTOMULTIPLIER TUBE
8 MODULATOR POWER SUPPLY
9 LOCK-IN AMPLIFIER
10 DISPLAY DEVICE

Figure 4 Block diagram of an analog apparatus suitable for the measurement of CPL.

A. Applications of CPL Spectroscopy

CPL spectroscopy is a powerful tool for studying chiral structures in the excited state either in the solution phase or in optically active crystals (100). In fact, CPL should be envisioned as the emission analog of CD. In CD, one measures the differential adsorption coefficients of left (L)- and right (R)-circularly polarized light, $\Delta\varepsilon = \varepsilon_L - \varepsilon_R$. In the acquisition of the CPL effect, one measures the differential emission of left (L)- and right (R)-circularly polarized photons in the molecular luminescence (being either fluorescence or phosphorescence), $\Delta I = I_L - I_R$. The type of information obtained from the two techniques is not only similar, but also complementary (101).

When CD and CPL effects are combined, there is a possibility of probing the optical activity of a racemic mixture without the need of chemical resolution. In such work, the CD effect is used to create an excess of one enantiomer in the excited state by irradiating the racemic sample with circularly polarized light. Consequently, the enantiomeric excess in the luminescent state is observable by the emission of circularly polarized light. The technique can be used for direct determination of enantiomeric purity (102), with applications to the study of the dynamics of fast racemization reactions.

Dekkers and Moraal studied the racemization of cis-3,4-dimethylcyclopentanone (DMCP) using CPL (101). DMCP exists as two half-chair conformers, and the conversion between the two occurs as a very fast dynamical equilibrium. The two conformers are enantiomers, and their interconversion is a racemization process (Fig. 5). In this work, a 1:1 mixture of methyl cyclopentane and methylcyclohexane was used to dissolve the DMCP. The solution was introduced into a cryostat and was excited with circularly polarized light. The circular polarization was effected by passing the light through a Glan polarizer, followed by a quartz quarter-wave plate appropriate to the excitation wavelength. The incident polarization was modulated between left- and right-handedness by rotating the quarter-wave plate over 90°. The circular polarization

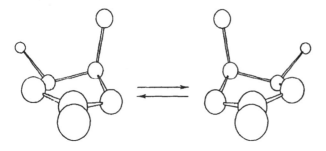

Figure 5 Half-chair forms of DMCP [Reprinted from Dekkers and Moraal (1993)].

of the DMCP fluorescence was followed at 420 nm, at a bandwidth of 20 nm, and as a function of temperature. At high temperatures, the racemization rate in the excited state (k^*) was large when compared with the reciprocal lifetime (τ_F^{-1}) of the fluorescence state. As a consequence, the effect of the chiral photoselection vanishes before the emission event occurs. Lowering the temperature does not affect τ_F^{-1}, but it decreases k^*. In the medium temperature range, the interconversion process became sufficiently slow so as to enable the observation of some optical activity. The maximum CPL effect was observed near 125 K.

Schauerte et al. designed an instrument for the determination of time-resolved CPL (TR-CPL) with subnanosecond resolution, and studied its application to several chiral systems (103). CPL spectra are very sensitive to the environmental medium in which the lumiphores are located. Such situations are limiting factors, especially for biomolecules, which contain several lumiphores in different environments (such as exist for tryptophan residues in proteins). At the same time, such biomolecules can interact with each other in solution, forming ensembles of molecules, which can exist in equilibrium with each other, as well as in equilibrium with different conformational states. As a consequence, an observed CPL spectrum must be the result of superposition of several spectra overlapping each other. One possibility of resolving such spectra is to follow the time evolution of the CPL signal and the luminescence decay after electronic excitation of the lumiphore, and to correlate the CPL components to those obtained from analysis of the luminescence decay. Individual terms in the CPL may then be assigned to distinct decay components. Time dependent CPL may also arise from excited state interactions of the lumiphore that affect the local chiral properties during the excited state lifetime. Each lumiphore may possess a time dependent CPL, and the TR-CPL method can be used to obtain information on the dynamics of excited state interaction.

Schauerte et al. used their instrument to determine anisotropy factor values as small as 10^{-4} with subnanosecond resolution, enabling the measurement of changes in CPL on the short time scale that is typical for protein fluorescence decay. The capabilities of their instrument were demonstrated using a model system, where they recorded the TR-CPL signal generated by $(+)$- and $(-)$-camphorequinones placed in the two compartments of a tandem cuvette. To create the time dependent optical activity, the fluorescence decay time of one enantiomer was shortened by quenching. In subsequent experiments, the authors time-resolved the circularly polarized emission of reduced nicotinamide adenine dinucleotide bound to horse liver alcohol dehydrogenas in the binary complex, as well as in the tight ternary complex formed in the presence of the substrate analog isobutyramide. The results showed the existence of a time dependent CPL, while the optical activity of the ternary complex was found to be essentially time independent.

The excited state properties of synthetic polymers can also be studied using CPL spectroscopy. Thus, Peeters and co-workers (104) studied the excited state properties of a series of α,ω-dimethyl-oligo{2,5-bis[2-(S)-methylbutoxy]-p-phenylene vinylene}$_s$ (OPV)$_{ns}$, where n is the number of phenyl rings in the oligomeric molecule, $n = 2-7$. The authors performed their experiments in solution at ambient temperature, under matrix isolated conditions at low temperature, and as nanoaggregates using absorption (time-resolved), photoluminescence, photoinduced absorption, CD, and CPL spectroscopies. It was found that the singlet $S_1 \leftarrow S_0$ and triplet $T_n \leftarrow T_1$ transition energies decreased with the length of conjugation. For the S_1 state of OPV$_n$, the lifetime strongly decreased with chain length due to enhanced nonradiative decay and radiative decay. The increase in the nonradiative decay rate constant was much more pronounced, and as a result, the photoluminescence quantum yield was less for longer oligomers. Low temperature studies yielded spectra characterized by well-resolved vibronic fine structure.

Lanthanides and actinides are groups of elements that form dissymmetric complexes with a number of organic ligands. Solutions of the Gd(III) complexes are becoming increasingly important due to their potential use as contrast agents in NMR imaging (105) and as a result the luminescent analogs have been extensively studied using CPL spectroscopy. As an example of a situation where CPL spectroscopy was used as unique advantage, the formation of lanthanide (Ln) ternary complexes is offered. It had been known that lanthanide complexes of ethylenediamine tetraacetate (EDTA) were co-ordinatively unsaturated, and that Ln(EDTA) complexes could be used as aqueous shift reagents for the simplification of nuclear magnetic resonance spectra (106). This property was based on the ability of these compounds to form ternary complexes:

$$\text{Ln(EDTA)} + S \Longleftrightarrow \text{Ln(EDTA)(S)} \quad (20)$$

When the Ln ion was paramagnetic, then the induced shifts in the resonance lines of S could be used to deduce molecular conformations. The substitution of one methyl group on the ethylenediamine backbone of EDTA yields the PDTA ligand, but this ligand is dissymmetric and can be resolved into its enantiomers. The CPL measured within the Tb(III) luminescence bands of the Tb(PDTA) complex upon formation of ternary complexes with resolved ligands enables one to study the possible stereochemical changes that accompany the formation of the ternary complexes.

It was found that Tb(R,R-PDTA) could indeed form ternary complexes with a large variety of achiral substrate ligands (107). In many cases, it was found that the CPL spectra of the Tb(R,R-PDTA)(S) ternary complexes were quite different from the CPL spectrum of the parent Tb(R,R-PDTA) complex. These findings indicate that the ternary ligands were capable of deforming the PDTA ring structure. When extremely flexible ternary ligands were studied (such as succinic acid), it was found that the CPL spectra of the ternary complexes were not altered relative to the spectra of the parent complex. However, for this particular ring system, another study (108) demonstrated that the succinate-type ring structure was forced to adopt a new conformation upon complexation with a Tb(EDTA) complex. These works indicate that, although the lanthanide complexes with EDTA are capable of functioning as aqueous shift reagents, steric interactions between the EDTA ligand and the substrate can be significant. The NMR data may be used to deduce solution-phase conformations of the analyzed substrate, but this stereochemical information represents only the situation for the perturbed ligand. It is probably not possible to deduce the stereochemistry of the free ligand in solution from data obtained using Ln(EDTA) complexes as shift reagents. This information could only have been obtained from a chiroptical study, and the CD spectroscopy would not have been the method of choice for such work.

Abdollahi et al. used CPL and TL to study the Tb(III) and Eu(III) complexes with transferins (109), with the lanthanide ion serving as a substitutional replacement for iron in a series of iron-binding transferins. The spectral region used corresponded to $^5D_4 \to {}^7F_5$ transition of Tb(III), as this band is the most intense emissive transition for this ion. Direct excitation of Tb(III) via its $^5D_4 \to {}^7F_6$ transition was accomplished using the 448-nm line of an Ar ion laser. This mode of excitation was used to avoid problems associated with sample photodegradation that occurred when using UV radiation to excite the Tb(III) ions via intermolecular energy transfer from neighboring aromatic molecules. The total emission showed some structure due to the complex crystal field splitting expected for $J = 4$ and $J = 5$ states in a low symmetry site. The authors concluded that the total emission spectra associated with the $^5D_4 \to {}^7F_5$ transition of Tb(III), when substituted for Fe(III) in transferins, were not a very sensitive measure of minor structural differences, although the emission intensity may be used as a measure of metal binding. As the total emission from Tb(III) complexes in aqueous and nonaqueous media often shows considerably well-resolved crystal field splitting, the lack of observable splittings in the protein systems to some extent reflects the absence of rigidity in the lanthanide co-ordination sites.

In an interesting study, Mondry et al. (110) reported the observation of CPL from solutions of Eu(III) complexes with triethylenetetra-aminehexaacetic acid (TTHA) [Eu(TTHA)$^{3-}$] following circularly polarized excitation. This is the first example of CPL from a racemic lanthanide complex that does not have approximate D_3 symmetry. The results showed that the CPL had very little temperature dependence from 10°C to 80°C, indicating that the complex does not undergo significant rearrangement that would lead to racemization in the temperature range studied. Experiments involving quenching of Tb(TTHA)$^{2-}$ luminescence by resolved Ru(phen)$_3^{2+}$ indicated that no enantioselective quenching was observed for the Tb(TTHA)$^{2-}$ complex. Such results shed light on the lack of diastereomeric discriminating interactions between the oppositely charged ions in this system.

The addition of a chiral environment compound to a kinetically labile racemic mixture of metal complexes may result in an equilibrium shift such that the concentrations of the two enantiomeric metal complexes are no longer equal. This equilibrium shift may lead to a measurable optical rotation, or to the appearance of CD or CPL. This phenomenon is often referred to as the "Pfeiffer Effect", and such perturbation effects can be used to study specific diastereomeric interactions. Thus, Huskowska and Riehl (111) used CPL spectroscopy to study the Pfeiffer perturbation of the racemic equilibrium of the D_3 complexes of lanthanide(III) ions with 2,6-pyridinecarboxylate by a number of sugars. However, various spectroscopic probes show that the addition of the chiral sugar has no observable effect on the structure of the lanthanide complex. The dependence of the enantiomeric excess on the concentration of added sugar was shown to be linear for all the sugars studied. Such dependence was shown to be in agreement with an equilibrium model, in which the optically active lanthanide complexes form weakly bound diastereomeric outer sphere associated complexes with various sugars. The authors did not find any correlation between the configuration and conformation of the sugar studied and the sign of the enantiomeric excess.

Synthesis of enantiopure ligands for lanthanide complexes requires care, as small changes in reaction conditions can lead to addition or loss of co-ordinated solvent molecules, leading to possible racemization. Chiral luminescent lanthanide complexes can be discriminating probes of local environments. They should be highly emissive and be conformationally rigid on the time scale of the luminescence emission. Muller et al. (112) reported the first selective formation of a single stereoisomer in a mononuclear lanthanide coordination compound resulting from the interplay of three independent terdentate chiral C_2 symmetrical ligands. The use of a pinene fragment in the structure of the ligand introduces steric bulk that in turn leads to the preferential formation of a limited number of stereoisomers. The authors used an enantiopure ligand of the type shown in Fig. 6. The studies indicated

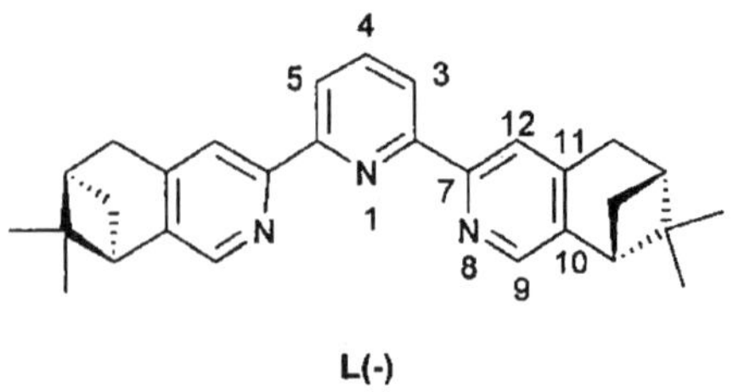

Figure 6 Structure of the pinene-based ligand [Reprinted from Muller et al. (2002)].

that the presence of the pinene group in positions 9 and 10 led to the formation of only LnL$_2$ complexes (Ln = La, Eu, Lu), due to steric interactions. However, when the pinene moiety was in positions 10 and 11 (see Fig. 6), and facing away from metal center complexes, Ln[L(−)$_3$](ClO$_4$)$_3$ complexes were isolated with in 80–90% yields. In addition, the introduction of the pinene moiety was found to be important in controlling the stereochemistry of the luminescent lanthanide coordination compound. The resultant emitted light was polarized in a direction determined by the helicity of the metal complex, which in turn was controlled by the absolute configuration at the remote carbon center in the pinene group.

Macrocycles are compounds that form complexes with a number of metals of the periodic table, as well as with the lanthanides (113). The complexes of lanthanides with macrocycles containing 18-member macrocyclic ligands that contain pyridine rings have several advantages. They are relatively easy to prepare, and have a more symmetric and rigid structure. Similarly, chiral macrocyclic complexes with lanthanides are very suitable compounds for studying their chiroptical properties in solution, and are perfect candidates for CPL spectroscopy. Tsubomura et al. (114) synthesized a new chiral 18-membered lanthanide complex from lanthanide nitrates, 2,6-pyridinecarboxaldehyde, and chiral 1,2-diaminocyclohexane. The structure of the complex is presented in Fig. 7. The structures of these

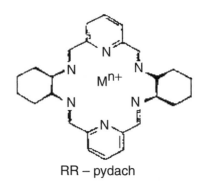

Figure 7 Structure of macrocycle complexes, M = La^{3+}, Eu^{3+}, and Tb^{3+}. [Reprinted from Huskowska and Riehl (1995)].

complexes were shown by NMR and luminescence spectroscopy to have D_2 molecular symmetry. Strong CPL was also detected for the Eu(III) and Tb(III) complexes owing to the twisted conformation. Large values for the dissymmetry factor were found, indicating a strong chiral environment at the Tb(III) ion as well as for the Eu(III) ion, and the absolute chiral structures were determined from CD data. The intramolecular energy transfer processes were found to occur between the π^* electronic state of the ligand and the 4f level of the central lanthanide ion in Eu(III) and Tb(III) complexes.

VI. VIBRATIONAL OPTICAL ACTIVITY

As all chiral molecules exhibit strong absorption in the infrared region of the spectrum, extensive investigation into the optical activity of vibrational transitions has been carried out. The IR bands of a small molecule can easily be assigned with the performance of a normal coordinate analysis, and these can usually be well resolved. The optical activity observed within a vibrational transition is determined by the symmetric properties of its group vibration, and consequently, can represent a probe of local chirality. When coupled with other methods capable of yielding information on solution-phase structure, vibrational optical activity can be an extremely useful method.

One of the problems associated with vibrational optical activity is the weakness of the effect. The rotational strengths of vibrational bands are much smaller than those of electronic bands, as the magnetic dipole contribution is significantly smaller. In addition, instrumental limitations of infrared sources and detectors create additional experimental constraints on the signal-to-noise ratios. In spite of these problems, two methods suitable for the study of vibrational optical activity have been developed. One of these is vibrational circular dichroism (VCD), in which vibrational optical activity is observed through the use of the classic method of Grosjean and Legrand. The other method is Raman optical activity (ROA), in which chirality is studied through the scattering effects of Raman spectroscopy.

A. Vibrational Circular Dichroism

VCD was first measured in the liquid phase by Holzwarth and coworkers in 1974 (115,116) and shortly confirmed by Nafie et al. (117). These and other works carried out during this period demonstrated that VCD could be measured at good signal-to-noise levels, and that the effect was quite sensitive to details of molecular stereochemistry and conformation (118).

An infrared CD spectrometer can be constructed using the same detection scheme as described for a UV/VIS CD spectrometer, using optical components suitable for infrared work. Reflective optics optimized for infrared reflection are normally used to eliminate problems associated with the transmission of infrared energy by lenses. The light source can be a tungsten lamp, glowbar, or carbon rod source, depending on the particular spectral range to be covered. The monochromator used to select the analyzing wavelengths should contain a grating blazed for first-order work in the infrared.

The circular polarization in the infrared source is obtained first by linearly polarizing the light, and then by passing it through a photoelastic modulator. The linear polarizer is either a CaF_2 crystal overlaid by a wire grid (producing the linear polarization through dichroism), or by a $LiIO_3$ crystal polarizer. The optical element within the photoelastic modulator is made up of either CaF_2 or ZnSe, chosen on the basis of the desired spectral range. The detector is usually a photovoltaic device capable of responding to infrared energy. The AC component within the detector output is discriminated and amplified by phase-sensitive detection, and the mean detector output is obtained to normalize the VCD signal. The ratio of the differential absorbance to the total absorbance:

$$g_{abs} = \frac{2(a_L - a_R)}{(a_L + a_R)} \quad (21)$$

where a_L and a_R represent the absorptivities for left- and right-circularly polarized light, respectively.

Early VCD measurements were complicated by artifacts arising from imperfections in the optical components (Fig. 8). These effects were detected by running the VCD spectrum of the racemic compound, and corrected spectra were obtained by subtracting the two spectra. In most instances, careful alignment of the optical components is able to eliminate most effects.

An alternate method for the detection of VCD was introduced by Nafie, who was able to show that a modified

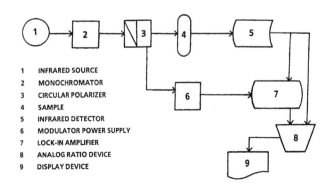

Figure 8 Block diagram of an analog apparatus suitable for measurement of VCD.

Fourier transform infrared (FTIR) transform spectrometer could be used (119). The instrumental advantages of the FTIR method have provided significant enhancements in signal-to-noise ratios, and thus reduce the time required to obtain a VCD spectrum. The origin of an artifact peculiar to FTIR-VCD has been identified, and a proposal advanced for its elimination (120).

Interestingly, it appears that use of the FT method does not bring about a significant improvement in terms of data acquisition time relative to that associated with the dispersive technique, and may, in fact, be less advantageous (121). Lee and Diem have vigorously pursued the development of their VCD instrumental design, and have built an extremely sensitive spectrometer designed specifically to operate around a wavelength of 6 μm, or between 1600 and 2000 cm^{-1} (122). Working within this spectral region permits study of the carbonyl chromophore, and this particular group has been found to be a useful probe of the structure existing in a wide variety of biological systems. It has been determined that the use of high-throughput optical elements, digital data manipulation, and high-quality lock-in amplifiers permit VCD to be measured at superior signal-to-noise levels, even in the presence of high background absorption.

1. Applications of the Vibrational Optical Activity

Initially, VCD was very difficult to measure because the intensities of VCD bands, relative to their parent unpolarized infrared absorption bands, are typically an order of magnitude smaller than the intensities found for electronic CD. However, instrumental advances have enabled the maturing of VCD as an area of molecular spectroscopy with vastly improved methods for routine measurement and interpretation of spectra. Dedicated VCD instrumentation has become commercially available, enabling the application of VCD to a variety of stereochemical problems (123,124).

For example, the VCD within the carbon–hydrogen stretching region of 21 amino acids, five *bis*(amino acid)Cu(H) complexes, and the *tris*(alaninato)Co(HI) complex was obtained in deuterium oxide (125). For free amino acids, a positive VCD band at 2970 cm^{-1} was noted, corresponding to the methine–CH stretching mode. The intensity of this VCD was associated with the presence of an intramolecular hydrogen bond existing between the ND^{3+} and COO^- functionalities. Upon complexation with transition metal ions, the VCD intensity of this band increased even further due to increased bonding strength between the amine and carboxylate groups through the bonds of the Cu(II) or Co(III) ion. It was theorized that the mechanism of this effect originated in a ring current that was closed, either by the intramolecular hydrogen bond or by the metal–ligand bonds.

VCD spectroscopy has become a widely used technique for the determination of the absolute configuration of small organic molecules (126–139). For a more comprehensive review on the applications of VCD, the reader is advised to consult the works of Nafie and Freedman (140) and Keiderling (141).

Aamouche et al. (142) developed a new methodology for predicting the VCD spectra of chiral molecules using density functional theory (DFT) and gauge-invariant atomic orbitals (GIAO). The method permits the direct determination of the absolute configuration of organic molecules in solution. The application of the method to Tröger's base leads to the absolute configuration opposite of that deduced from electronic CD, confirming the conclusion arrived from X-ray analysis of a diastereomeric salt. Using the same method, Devlin et al. (143) determined the absolute configuration of 1-thiochromanone S-oxide (**1**). Analysis of the VCD spectrum using DFT and GIAO predicts the existence of two stable conformations, separated by less than 1 kcal/mol. The VCD spectrum predicted for the equilibrium mixture of the two conformations of (S)-**1** was in good agreement with the experimental spectrum of (+)-**1**. The absolute configuration of **1** was assigned as R(−)/S(+).

Similarly, Stephens et al. (144) reported the absolute configuration of 1-(2-methyl-naphtyl) methyl sulfoxide. The DFT calculations predicted the existence of two stable conformations of the compound, **E** and **Z**, with **Z** having a lower energy than **E** by less than 1 kcal/mol. In both conformations, the S–O bond is rotated from coplanarity with the naphtyl moiety by 30–40°. The predicted unpolarized absorption IR spectrum of the equilibrium mixture of the two conformations permitted assignment of the experimental IR spectrum in the mid-IR spectral region. The VCD spectrum predicted for the compound is in good agreement with the experimental spectrum of the (−)-enantiomer, defining the absolute configuration of the compound as R(+)/S(−) (145).

Devlin et al. (146,147) and Drabowicz et al. (148) each studied the absolute configuration of organic sulfoxides in detail using VCD spectroscopy. Very good correlation was observed between the predicted and the experimental values. Wang et al. described the assignment of the absolute configuration of *tert*-butyl-1-(2-methylnaphtyl)phosphine oxide and *tert*-butylphosphinothioic acid using VCD spectroscopy (129,149). The absolute configuration assignment of the compounds was supported by *ab initio* simulation predictions.

Pieraccini et al. (150) reported a very interesting application of VCD spectroscopy. For the first time, these authors showed a systematic experimental and theoretical analysis of cholesteric induction due to solutes whose chirality stemed only from a single stereogenic center. The twisting power of a series of alkyl aryl sulfoxides has been determined in several nematic solvents. The sign of the twist reflects the handedness of the induced helical arrangement of the solvent molecules, and correlates well with the configuration of the stereogenic sulfur in the nematic solvents. It was found that (S)-configured dopants induced (M)-chiral nematics. Instead, (S)-configured cyclic sulfoxides, which are forced to adopt a different conformation with respect to the parent acyclic compounds, induce right-handed nematics. The calculations reliably reproduced the experimentally observed behavior. The more flexible, open-chain compounds induced chiral nematics of opposite handedness.

He et al. (151) reported the first *ab initio* VCD calculation for a molecule containing a transition metal, namely a closed-shell Zn(II) complex. At the same time, the authors presented a new VCD enhancement mechanism in the corresponding open-shell Co(II) and Ni(II) complexes, which increased the VCD intensity by an order of magnitude, with changes in sign for many transitions, leaving the unpolarized IR unchanged. Intensity-enhanced VCD shows promise as a sensitive new spectroscopic probe for stereochemistry evaluation.

VCD spectroscopy has been used to measure the percent enantiomeric excess (%ee), which is defined as the excess of the number of moles of one enantiomer over that of the opposite enantiomer as a percentage of the total number of moles of both enantiomers. For a sample consisting of a mixture of enantiomers, the VCD intensity is directly proportional to the %ee. Urbanova et al. (152) used VCD spectroscopy in the mid-infrared region to measure the enantiomeric purity of terpene solutions in carbon tetrachloride. The concentration dependence of the VCD was statistically analyzed with the aim of obtaining a reliable correlation between the VCD band area and the concentration of individual enantiomers. It was found that statistical methods, such as partial least squares, were optimal for the analysis of spectra in the determination of %ee. The best fit for the training spectra has to be obtained, and then the resulting database is used to make predictions of an unknown (140). Analysis of enantiomeric purity in a number of compounds has led to the conclusion that as long as one obtains spectra of sufficient quality, an accuracy of prediction at less than 1%ee can be achieved (153).

VCD spectroscopy can be used not only to study the absolute configuration of small molecules, but also to study the conformational changes of large molecules such as peptides and proteins. Thus, VCD provides alternative views of protein and peptide conformations, and exhibit advantages over electronic (UV) CD or IR spectroscopy. VCD is sensitive to short-range order, allowing it to discriminate among β-sheet, various helices, and disordered structures. The theoretical analysis of the VCD of peptides yields a detailed prediction of helix and hairpin spectra, and site-specific applications of isotopic substitution for structure and folding (154–158).

Measurements of the VCD of RNA and DNA in aqueous solutions are characterized by large signals (in terms of $\Delta A/A$) for a variety of modes. VCD enables access to the in-plane base deformation modes, facilitating study of interbase relative disposition and stacking interactions. In addition, the phosphate P–O stretching modes are VCD active, enabling one to sense backbone stereochemistry. Finally, the VCD of coupled C–H or C–O motions permits one to monitor the ribose conformation. Single-stranded RNA gives rise to a positive VCD couplet, located in the range 1600–1700 cm^{-1}, which is typically centered over the most intense in-plane base deformation band. This band is due to a C=O stretching mode of the planar base.

In general, the VCD of DNA in the base stretching modes is very similar to that of RNA, allowing for helical conformation differences. While RNA molecules are mostly in an A form, DNA molecules are mostly found in the B form in aqueous solution, which can be transformed to the A form upon dehydration. The left-handed Z-form can, in some cases, be formed in solutions of high ionic strength. The VCD spectra of poly(dG–dC) and related DNA oligomers in the B-form and Z-form have different band shapes for their base stretching modes (159). The B–Z conformational transition can be induced by ions such as Mn^{2+}, and monitored by VCD. The transition is most pronounced in the region 1600–1770 cm^{-1} (corresponding to the C=O stretching modes), where the absorption maxima are shifted by up to 19 cm^{-1} and the negative/positive VCD couplet of the B-form (at 1691/1678 cm^{-1}) changes sign to positive/negative, and its frequencies follow the absorption maxima to lower wavenumbers of 16771/1656 cm^{-1} (160).

Andrushchenko et al. studied the interaction of other transitional metals such as Cu^{2+} with DNA using VCD (161). The Cu^{2+} ions bind to phosphate groups at low concentrations. Upon increasing the concentration, chelates are formed in which Cu^{2+} binds to the N7 group of guanidine (G) and a phosphate (P) group. Detectable only by VCD, significant distortion of most guanine–cytosine (GC) base pairs occurs at a [Cu/P] ratio of 0.5, with a minor effect on the adenine–thymine (AT) base pairs. Such a configuration favors a sandwich complex with the Cu^{2+} ion inserted between two adjacent guanines in a GpG sequence. The AT base pairs become significantly distorted when the metal concentration is increased to 0.7 [Cu/P].

B. Raman Optical Activity

A method complementary to VCD spectroscopy is the ROA. The ROA effect is the differential scattering of left- or right-circularly polarized light by a chiral substrate. The ROA is particularly useful within the 50–1600 cm^{-1} spectral region, and remains as the preferred method for obtaining optical activity within vibrational bands having energies less than 600 cm^{-1}. This situation has arisen because the spectral limitations of VCD optical components provide substantial barriers to the performance of work below about 600 cm^{-1}. The ROA effect was first demonstrated in 1973 by Barron and Buckingham (162,163), who observed differential scattering effects in the Raman spectra of α-phenethylamine and α-pinene. Development of the technique required substantial experimental effort, as small imperfections in the optical elements could yield artifacts substantially larger in magnitude than a genuine ROA effect. Barron has provided both general introductions (164) and detailed review (165) of the field. With the ability of ROA data to be obtained in aqueous solutions, it appears that the technique would be extraordinarily useful in the characterization of biological systems (166).

The basic design of functional ROA spectrometers has been described by McCaffrey (167,168). With the introduction of a special optical system, Hug was able to drastically reduce ROA artifacts (169), and this design is summarized in the block diagram of Fig. 9.

The laser output is passed through a small monochromator to remove unwanted plasma lines, and totally polarized by a linear polarizer. The beam is then circularly polarized by a quarter-wave plate, and allowed to impinge on the sample cell. The scattered light is passed through a photoelastic modulator, which is set to advance and retard phase angles by 90°. As the ROA effect employs visible light scattering, no special modulator materials are required. The modulated light beam is passed through a linear polarizer, thus providing an AC signal in the scattered light that is proportional to the ROA intensity. The scattered light is collected over a 20° cone with respect to the laser (167) analyzed by a high-resolution double grating monochromator, and detected by a photomultiplier. The PMT output is split, with one signal being amplified by a DC device and the other discriminated through phase-sensitive detection. The two quantities can either be displayed directly, or ratioed before being displayed. The possible artifacts associated with deflection of the incident laser beam as it passes through an electro-optic modulator have been analyzed, and a procedure for the minimization of these effects has been presented (170).

In a very detailed article, Hecht and Barron have used the Stokes–Mueller calculus to extensively analyze the various experimental methods for obtaining ROA spectra (171). In their analysis, they considered the dual-beam modulation approach, scattered beam modulation, and incident beam modulation. After considering the type of artifacts that would be associated with each method, they concluded that only the incident beam modulation approach, combined with backscattering geometry, was suitable for the study of chiral molecules randomly oriented in an optically anisotropic phase. The power of this experimental approach has been amply demonstrated in a detailed study of the vibrational chirality of monosaccharides (172).

Several proposals have been advanced regarding the design of ROA spectrometers, whose method of detection is based on multichannel detectors. Diem and coworkers have described a diode array detector equipped spectrometer, which made use of a dual lens light collection scheme to reduce artifacts (173). On the other hand, Barron was able to develop a conventional single lens system in a spectrometer equipped with an intensified diode array detector, and obtained perfectly acceptable spectra (174). A charge-coupled-device detector has also been used in conjunction with a scattered circular polarization ROA spectrometer (175).

In the conventional spectrometer design based on the method of Grosjean and Legrand, the amplified DC signal is proportional to the mean scattered light intensity:

$$I = I_L + I_R \quad (22)$$

whereas the AC signal (after amplification by means of phase-sensitive detection) is proportional to the differential scattering intensity:

$$\Delta I = I_L - I_R \quad (23)$$

Taking the ratio of these quantities removes any unit dependence, producing a quantity which is termed as the circular intensity differential (CID):

$$\text{CID} = \frac{\Delta I}{I} \quad (24)$$

As with VCD spectroscopy, these CID magnitudes are quite small, with most values ranging between 10^{-3} and 10^{-4}.

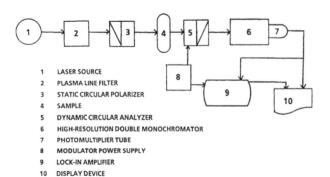

1 LASER SOURCE
2 PLASMA LINE FILTER
3 STATIC CIRCULAR POLARIZER
4 SAMPLE
5 DYNAMIC CIRCULAR ANALYZER
6 HIGH-RESOLUTION DOUBLE MONOCHROMATOR
7 PHOTOMULTIPLIER TUBE
8 MODULATOR POWER SUPPLY
9 LOCK-IN AMPLIFIER
10 DISPLAY DEVICE

Figure 9 Block diagram of an analog apparatus suitable for the measurement of ROA.

1. Applications of ROA

Vibration motion associated with chiral molecules produce chiroptical spectra that are dependent on the circular polarization direction of the excitation light and the detection optics. As would be the case for VCD, ROA is a form of optical activity sensitive to the chirality associated with all $3N-6$ fundamental vibrational transitions, where N is the number of atoms (176). In many instances, ROA offers some practical advantages over VCD. One of the advantages of ROA is that it is able to cut through the complexity of the corresponding vibrational co-ordinates that sample the most rigid and chiral parts of the structure (177) Another advantage is that the inefficient Raman scattering of water enables one to perform Raman experiments in aqueous solutions. At the same time, the low attenuation of the excitation light removes the requirement of preparing ultra thin samples, as is the case with infrared absorption (178).

One extremely detailed ROA investigation into vibrational chirality was performed by Nafie and coworkers, who performed detailed studies on $(+)$-$(3R)$-methyl-cyclohexanone (179). The Raman scattering observed between 100 and 600 cm^{-1} was found to contain strong ROA in all the major bands. The assignment of these bands to 11 skeletonal motions was made on the basis of a full normal coordinate analysis and through studies on several deuterated compounds. The ROA bands were found to occur as couplets, an effect that was explained in terms of chiral vibrational perturbations due to the presence of the methyl group mixing the A' and A'' skeletonal modes of cyclohexanone.

Terpenes are a class of substances that have strong VCD and ROA signals. This property makes them useful as standard calibration compounds for VCD and ROA spectrometers. Thus, Bour and et al. (180) measured and calculated the ROA of α-pinene and $trans$-pinene. The experiments were performed on a computer-controlled, home built spectrophotometer with backscattering incident circular polarization ROA. The spectra were obtained and interpreted on the basis of ab $initio$ quantum chemical calculations. Good agreement between the observed and the calculated spectra was observed in the mid-IR region.

Crassous and Collet (181) reported on the synthesis of enantiomerically enriched samples of bromochlorofluoromethane (CHFClBr), obtained using fractional crystallization of the strychnine salts of bromochlorofluoroacetic acid (FClBrCCO$_2$H). The absolute configuration of the acid was established by X-ray crystallography, and the absolute configuration of the haloform was determined using ROA, and molecular modeling of the enantioselective molecular recognition process of CHFClBr by the chiral cryptophane-C. From these stereochemical assignments, it was observed that the decarboxylation used to obtain the S-$(-)$ and R-$(-)$-CHFClBr from S-$(+)$ and R-$(-)$-FClBrCO$_2$H, respectively, occurred with retention of configuration.

ROA is an excellent technique for studying polypeptide and protein structures in aqueous solution, as their ROA spectra are dominated by bands originating in the peptide backbone that directly reflect the solution conformation. Furthermore, the special sensitivity of ROA to dynamic aspects of structure makes it a new source of information on order–disorder transitions (176). Vibrations of the backbone in polypeptide and proteins are usually associated with three main regions of Raman spectrum. First, there is the backbone skeletal stretch region (~ 870–1150 cm^{-1}), originating mainly in the C$_\alpha$–C, C$_\alpha$–C$_\beta$, and C$_\alpha$–N stretching modes. Second, there is the amide III region (~ 1230–1310 cm^{-1}), which is often thought to involve mainly the in-phase combination of the N–H in-plane deformation with the C$_\alpha$–N stretch. Third, there is the amide I region (~ 1630–1700 cm^{-1}), which arises mostly from the C=O stretch (182,183). It has been shown that in small peptides, the amide III regions involve mixing between the N–H and C$_\alpha$–H deformations, and should be extended to 1340 cm^{-1} (184). The extended amide region is of interest for ROA studies because the coupling between the N–H and C$_\alpha$–H deformations is sensitive to geometry and generates ROA band structure information.

As the time scale of the Raman scattering event (2.2×10^{-14} s for a vibration shift 100 cm^{-1} excited in the visible region) is much shorter than that of the fastest conformational fluctuation in the biomolecules, an observable ROA spectrum is a superposition of snapshot spectra from all the distinct chiral conformers present in the sample at equilibrium. For example, for unfolded lysozime and ribonuclease A, there are a number of differences in the ROA spectra, as compared with those of the native state. ROA in the backbone skeletal stretch region is less intense in the unfolded state than in the native state, and more intense in ribonuclease A than in lysozime, suggesting that more secondary structure has been lost. Much of the ROA band structure in the extended amide III region has disappeared, being replaced by a large broad couplet, negative at low wavenumber, and positive at high wavenumber (185).

ROA of the native and A states of α-lactalbumin (partially denatured at low pH) is very indicative to the sensitivity of ROA for small changes in the molecule, as compared with the parent Raman spectra. Thus, much of the ROA band structure in the extended amide III region of the A-state spectrum has disappeared, and is replaced by a large couplet consisting of a single negative band observed at ~ 1236 cm^{-1} and two distinct positive bands at ~ 1297 and 1312 cm^{-1}. The negative ~ 1236 cm^{-1} signal may originate in residues clustering around an average of the conformations (corresponding to the two negative signals in the native spectrum) observed at ~1222 and 1246 cm^{-1}, and assigned to β-structure. The positive band at ~1340 cm^{-1} may arise from the same α-helical sequences in a hydrophobic environment. The positive ~1340 cm^{-1} band (assigned to the α-helix in a hydrophobic state) that dominates the ROA spectrum of the native state is completely lost in the A-state, with new positive band appearing at ~1312 cm^{-1} that may originate in reduced hydration in some α-helical sequences. In the amide I region, the positive signal of the couplet has shifted by ~10 cm^{-1} to higher wavenumber, as compared with that for the native state. The signal originating from the W3 vibrational mode of tryptophan residues at ~1551 cm^{-1} almost completely disappeared in the ROA spectrum of the A-state, and replaced by a weak positive band at 1560 cm^{-1}. This indicates the existence of conformational heterogeneity among these side chains, associated with a loss of the characteristic tertiary interactions found in the native state (184).

These studies demonstrate that ROA is an important tool for assessing the conformational changes of peptides and proteins (186–193), as well as studying the molecular structure of viruses and conformation of their constituent proteins (194–197).

VII. FLUORESCENCE DETECTED CIRCULAR DICHROISM

Fluorescence detected circular dichroism (FDCD) is a chiroptical technique first developed by Tinoco and coworkers. A CD spectrum is obtained by measuring the difference in TL obtained after the sample is excited by left- and right-circularly polarized light. For the FDCD spectrum of a given molecular species to match its CD spectrum, it must happen that the luminescence excitation spectrum be identical to the absorption spectrum.

A block diagram of a typical FDCD spectrometer is shown in Fig. 10. A tunable UV source is required, for which the output must be relatively independent of wavelength. High-powered xenon lamps (whose output is selected by a suitable grating monochromator) are appropriate for this purpose. The spectral output is linearly polarized by a Glan or Glan-Thompson polarizer, selected for its UV transmission. A modulated circular polarization is induced in the beam through the use of a dynamic quarter-wave plate, in exactly the same way as would be used to obtain a conventional CD spectrum. This modulated circularly polarized excitation beam is allowed to enter the sample cell and to produce the required luminescence. Up to this point, the optical train is identical to that of a conventional CD spectrometer.

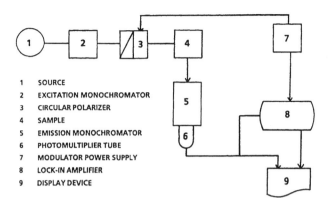

Figure 10 Block diagram of an analog apparatus suitable for the measurement of FDCD.

In the FDCD experiment, any fluorescence is detected at right angles to the excitation beam by a photomultiplier tube, and this emitted light contains the chiroptical information. As was described for CPL spectroscopy, the tube output will contain an AC signal equal to $(I_L - I_R)$, which is proportional to the differential absorption spectrum. A measurement of the mean DC output of the tube is equal to $(I_L + I_R)$, and is proportional to the mean absorption spectrum. The ratio of these two signals yields the FDCD quantity:

$$\text{FDCD} = \frac{I_L - I_R}{I_L + I_R} \quad (25)$$

This FDCD signal can either be displayed on a chart recorder, or digitized for introduction into a computer system. As in the case of the CPL experiment, accurate results can only be obtained if the two measurements are made simultaneously.

A general theory relating to the FDCD of isolated chiral fluorescent compounds has been presented, and the effect of solute concentration considered (198). The FDCD is potentially of great use in the study of solute–solute interactions, as the added selectivity of fluorescence in either species can be very useful. Three such situations has been identified and analyzed for the possible effects on the FDCD spectra (199), namely, (a) when the optically active material is nonfluorescent, but can be coupled with an optically inactive fluorescent species, (b) when a fluorescent optically active material is coupled to a nonfluorescent achiral compound, and (c) when an optically active fluorescent species is coupled to a chiral fluorescent compound.

The FDCD experiment can easily yield instrumental artifacts when photoselection effects are possible (200,201). When the fluorescent molecule is situated in a medium where molecular motion is restricted (as would exist for viscous or rigid phases), then the excitation energy may not be absorbed equally by all molecular orientations. These effects would be most pronounced for planar aromatic hydrocarbons, whose transition dipoles exhibit a strong dependence on the molecular orientation with respect to the incident energy. When these photoselection effects occur, the excitation spectrum can be quite different from the absorption spectrum and the FDCD spectrum will not match the CD spectrum. Turner and Lobenstein have described a dual photomultiplier tube detection system in which the effects due to photoselection can be minimized (202). The use of ellipsoidal mirrors to collect the fluorescence has also been found to reduce the magnitude of instrumental artifacts (203).

Warner and coworkers have described a multichannel spectrometer for the measurement of FDCD (204). Through the use of a polychromator and photodiode array detection, a system was produced, which could be used to produce rapid scanning of FDCD. As their device could obtain spectra in exceedingly short periods of time, these workers were able to produce contour plots of chirality as a function of both excitation and emission wavelengths.

More recently, a phase-modulation spectrofluorometer has been modified so as to permit the measurement of time-resolved FDCD data (205). This particular methodology would find application in the study of dissymmetric microenvironments. The signals produced by such instrumentation have been analyzed on a theoretical basis to demonstrate how to evaluate the Kuhn and luminescence dissymmetry factors.

A. Applications of FDCD

Determination of the absolute configuration of an organic compound is an important application of FDCD. Thus, Sugimoto et al. (206) determined the absolute configuration of the native pteridine isolated from *Tetrahymena pyriformis*. The compound was determined as (6R)-5,6,7,8-tetrahydro-D-monapterin (=(6R)-2-amino-5,6,7,8-tetrahydro-6-[(1R,2R)-1,2,3-trihydroxypropyl]pteridin-4(3H)-one). To accomplish this task, the authors determined the configuration of the 1,2,3-trihydroxypropyl side chain as D-threo by measuring the FDCD spectrum of the aromatic pterin derivative obtained by I_2 oxidation. The authenticity of the finding was proven by comparison with an authentic sample of D-monapterin at wavelengths below 300 nm, where the first (−)-Cotton effect was detected at 270 nm, and the second (+)-Cotton effect at 230 nm. L-Monapterin exhibited a mirror-image FDCD spectrum.

Similarly, Chen et al. (207) used FDCD for the determination of the absolute configuration of naturally occurring pteridines. The method proved to be ten times more sensitive than CD spectroscopy.

The determination of enantiomeric excess represents another area where FDCD spectroscopy has the potential of practical applications. Geng and Mcgown (208) demonstrated the feasibility of FDCD for mixtures of binaphthyl compounds [(S)- and (R)-1,1′-binaphthyl-2,2′-diyl hydrogen phosphate (BNPA)]. The FDCD measurement yielded good accuracy in the determination of enantiomeric excess. In addition, the FDCD method provided good detectability of the enantiomeric compound, with a detection limit of 20 mM, corresponding to a CD signal $\Delta A = 3.4 \times 10^{-7}$ absorbance units, being observed for pure (S)-BNPA. The enantiomeric excess was shown to be equal to the FDCD signal divided by the Kuhn anisotropy factor of the enantiomeric compound. As the anisotropy factor is independent of the concentration of the compound, the enantiomeric purity can be anisotropy determined for any total concentration of the compound within the linear dynamic range, once the anisotropy factor has been determined.

The power of FDCD spectroscopy has also been illustrated through studies of the CD and FDCD associated with poly-L-tryptophan (209). It had been inferred from ORD and CD studies that this polypeptide adopted a right-handed α-helical conformation in 2-methoxyethanol, but owing to the effects associated with the indole side chains the actual conformation of the peptide backbone could not be studied directly (210,211). While the CD spectrum would reflect the optical activity of both the backbone and side chains, the FDCD spectrum would only reflect the chirality experienced by the indole side chains. In studies of the CD and FDCD of this polymer system, it was noted that both similarities and differences were observed. Through a series of assumptions regarding how the CD and the FDCD effects were observed, these workers were able to use the chiroptical data to deduce separate CD spectra corresponding to the side chains and the backbone. The backbone CD spectrum was found to closely resemble the genuine CD spectra, normally observed for α-helical polymers (212). This result was taken as confirmation that poly-L-tryptophan indeed adopt the right-handed α-helical conformation in 2-methoxyethanol.

The conformational analysis of poly (L-tyrosine) (PLT) and tyrosyl oligomers, based on their backbone CD spectra, was reported by Watanabe et al. (213). The separation of the backbone CD component of the tyrosyl compounds from the natural CD was achieved using FDCD spectroscopy. The backbone of PLT in methanol led to the observation of two negative CD maxima at 213 and 222 nm, enabling a deduction that the solution conformation of PLT was the α-helix conformation. The conformational transition of PLT from α-helix to the β structure was caused by the addition of an aqueous solution of sodium hydroxide to the methanolic PLT solution.

The origin of the positive CD band of the tyrosyl compound near the amide $n-\pi^*$ was experimentally proven by applying the FDCD method to N-acetyl-(L)-tyrosinamide (which is a monomer model compound for PLT). The observed CD of N-acetyl-(L)-tyrosinamide was attributed to the optical rotation originating from the L_a transition of the phenolic ring on the side chain. N-acetyl-(L)-tyrosinamide did not exhibit a backbone CD band in the amide $n-\pi^*$ transition region, whereas the tyrosine dimmer, trimer, and hexamer showed a positive dichroic band around 230 nm derived from the optical activity of their backbone amide chromophores.

ECCD characterized by split Cotton effects, is a microscale chiroptical method that can be used to determine the absolute configurations or conformations of the dissymmetric atoms in a variety of compounds. The exciton split CD of molecules consisting of multiple interacting chromophores, identical or different, can be reproduced by pairwise summation of interacting chromophores. The sensitivity of conventional CD remains limited because it depends on the dichroic absorption. In contrast, the sensitivity of FDCD is generally enhanced, because it is based on direct measurement of emitted radiation against a zero background (214). In addition, in FDCD, only CD active and fluorescent transitions give rise to a signal, enabling the selective measurement of multichromophoric molecules.

Dong et al. (214) reported the extension of FDCD to an exciton coupled system consisting of identical chromophores with well-defined and intense transition moments, such as 1(S),2(S)-trans-cyclohexanediol bis(2-naphtoate), the steroid 3β,6α-bis-(2-anthroate), and ouabegenin 1,3,19-tris-(2-naphthoate). The results showed that FDCD is 50–100 fold more sensitive than conventional CD. The increased sensitivity represents a significant asset of FDCD, but the selectivity of fluorescence detection holds greater possibilities in terms of practical applications of FDCD. The main issue in term of fluorescence detected exciton chirality is finding the criterion that constitutes a reasonable fluorophore, which produces an agreement between spectra obtained from FDCD and conventional CD. Thus, the fluorophore should have

1. A large molar absorptivity and a known direction of the electric transition moment.
2. High fluorescence quantum yield.
3. Chemical and photochemical stability.
4. An appropriate substituent (e.g., carboxyl) suitable for derivatizion with common functional groups such as hydroxyl or amines.
5. No, or very limited, effects due to photoselection (215).

An interesting study was reported by Castaguetto and Canary (216), who designed a quinoline-based ligand, able to complex with metals such as Zn(II), Cu(II), Cd(II), and Fe(II). After formation of the metal/ligand complexes, the resulting products yielded both fluorescence and (ECCD). This fluorescence and the ECCD properties of the complexes result in the interesting situation where the ligand not only signals the presence of the metal ion, but also, evaluation of both properties, allows identification of the metal ion.

FDCD can be also used to study protein–dye associations, with the aim of establishing the possible existence of subtle conformational changes in the protein structure. As with ordinary fluorescence measurements, increases in FDCD signals are indicative of an increased number of bound fluorophores. Interactions of achiral dyes with proteins can cause disssymmetric perturbations of fluorophores, creating an induced CD. A decrease in FDCD is indicative of quenching, disassociating dyes, and/or protein denaturation, all of which results in fluorophores without induced chirality. In instances where optical impurities exist, as with some sample cells or lenses, FDCD signals can indicate optical activity in an inoptically active sample.

Meadows et al. (217) used FDCD to study the binding of squarylium dyes with BSA. The authors found that upon denaturation of the protein with urea, the magnitude of the FDCD signal decreased with an increase in the concentration of urea. This phenomenon indicated that the observed FDCD signal occurs as a result of the dyes interacting with the three-dimensional arrangement of the protein.

VIII. CONCLUDING REMARKS

The range of techniques suitable for the study of molecular optical activity is extensive, and sophisticated instrumentation for each method continues to evolve. Each technique is associated with its own particular selectivity, and each is characterized by its advantages and disadvantages. The chirality within electronic transitions can be studied by ORD, CD, CPL, or by FDCD. Chirality within vibrational bands can be studied either by VCD or by ROA. Optical activity within rotational bands has been shown to be theoretically feasible (218), and undoubtedly, a method will eventually be developed to permit measurement of this predicted effect. As the development of dissymmetric chemistry is currently an extremely active research area, it is anticipated that the development of chiroptical methods and applications will proceed at an equivalent pace.

REFERENCES

1 Charney E. The Molecular Basis of Optical Activity. New York: Wiley, 1979.

2. Djerassi C. Optical Rotatory Dispersion: Application to Organic Chemistry. New York: McGraw-Hill, 1960.
3. Crabbe P. ORD and CD in Chemistry and Biochemistry. New York: Academic, 1972.
4. Velluz L, Legrand M, Grosjean, M. Optical Circular Dichrois: Principles, Measurements, and Applications. Weinheim: Verlag Chemie, 1965a.
5. Snatzke G, ed. Optical Rotatory Dispersion and Circular Dichroism in Organic Chemistry. London: Heyden, 1967a.
6. Nakanishi K. Circular Dichroism Spectroscopy: Exciton Coupling in Organic Stereochemistry. Mill Valley, CA: University Science Books, 1983a.
7. Caldwell DJ, Eyring H. The Theory of Optical Activity. New York: Wiley-Interscience, 1971a.
8. Barron L. Molecular Light Scattering and Optical Activity. Cambridge: Cambridge University Press, 1982a.
9. Mason, S. Molecular Optical Activity and Chiral Discrimination. Cambridge: Cambridge University Press, 1982a.
10. Biot JB. Mem. Inst. de France 1812; 50:1.
11. Mitscherlich R. Lehrb. der Chem. 4th ed. 1844.
12. Soleil H. Compt Rend 1845; 20:1805.
13. Soleil H. Compt Rend 1847; 24:973.
14. Soleil H. Compt Rend 1848; 26:162.
15. Heller W. Optical rotation — experimental techniques and physical optics. In: Weissberg A, Rossiter BW, eds. Physical Methods of Chemistry. Chapter 2. New York: Wiley, 1972.
16. Yeung ES. Talanta 1985; 33:1097.
17. Kankare J, Stephens R. Talanta 1986; 33:571.
18. Voigtman E. Anal Chem 1992; 64:2590.
19. Chafetz L. Pharmacopeial Forum 1992; 19:6159.
20. Cotton A. Compt Rend 1895; 120(989):1044.
21. Kirk DN. Tetrahedron 1986; 42:777.
22. Moffitt W, Woodward RB, Moscowitz A, Klyne W, Djerassi C. J Am Chem Soc 1961; 83:4013.
23. Juaristi E. Introduction to Stereochemistry & Conformational Analysis. New York: John Wiley & Sons, Inc., 1991:58.
24. Brittain HG. Microchem J 1997; 57:137.
25. Brittain HG. J Pharm Biomed Anal 1998; 17:933.
26. Zhao C, Polavarapu PL. Appl Spectrosc 2001; 55:913.
27. Ensch C, Hesse M. Helv Chim Acta 2002; 85:1659.
28. Tararov VT, Kadyrov R, Kadyrova Z, Dubrovina N, Borner A. Tetrahedron Asymm 2002; 13:25.
29. Tori M, Ohara Y, Nakashima K, Sono M. J Nat Prod 2001; 64:1048.
30. Tao T, Parry RJ. Org Lett 2001; 3(3):3045.
31. Wu Y, Cui X, Zhou N, Maoping S, Yun H, Du C, Zhu Y. Tetrahedron: Asymmetry 2000; 11:4877.
32. Pache S, Botuha C, Franz R, Kundig PE. Helv Chim Acta 2000; 83:2436.
33. Ohba M, Izuta R, Shimizu E. Tetrahedron Lett 2000; 41:10251.
34. Kakuchi T, Narumi A, Kaga H, Ishibashi T, Obata M, Yokota K. Macromolecules 2000; 33:3964.
35. Di Nunno L, Franchini C, Nacci A, Scilimati A, Sinicropi MS. Tetrahedron: Asymmetry 1999; 10:1913.
36. Zhang P, Zhang L, Cheng S. Carbohydr Res 2000; 327:431.
37. Ayscough AP, Costello J, Davies SG. Tetrahedron Asymmetry 2001; 12:1621.
38. Lustig SR, Everlof GJ, Jaycox GD. Macromolecules 2001; 34:2364.
39. Yeung ES, Steenhoek LE, Woodruff SD, Kuo JC. Anal Chem 1980; 52:1399.
40. Muller T, Wilberg KB, Vaccaro PH. J Phys Chem 2000; 104:5959.
41. Polavarapu PL. Chirality, 2002; 14:768.
42. De Tar DF. Anal Chem 1969; 41:1406.
43. Chen GC, Yang JT. Anal Lett 1977; 10:1195.
44. Gillen MF, Williams RE. Can J Chem 1975; 53:2351.
45. Pearson KH, Zadnik VC, Scott JL. Anal Lett 1979; 53:1049.
46. Davidson A, Norden B. Spectrochim Acta 1976; 32A:717.
47. Norden B. Appl Spectrosc 1985; 39:647.
48. Barron L. Molecular Light Scattering and Optical Activity. Cambridge: Cambridge University Press, 1982b.
49. Berova N, Nakanishi K, Woody RW, eds. Circular Dichroism Principles and Applications. New York: John Wiley & Sons, Inc., 2000.
50. Caldwell DJ, Eyring, H. The Theory of Optical Activity. New York: Wiley-Interscience, 1971b.
51. Fasman GD, ed. Circular Dichroism and Conformational Analysis of Biomolecules. New York: Plenum Press, 1996.
52. Mason S. Molecular Optical Activity and Chiral Discrimination. Cambridge: Cambridge University Press, 1982b.
53. Nakanishi K. Circular Dichroic Spectroscopy. Exciton Coupling in Organic Stereochemistry. Mill Valley, CA: University Science Books, 1983b.
54. Snatzke G, ed. Optical Rotatory Dispersion and Circular Dichroism in Organic Chemistry. London: Heyden, 1967b.
55. Veluz L, Legrand M, Grosjean M. Optical Circular Dichroism: Principles, Measurements, and Applications. Weinheim: Verlag Chemie, 1965b.
56. Crabbe P, Pinelo L. Chem Ind 1966; 158.
57. Frelek J, Lysek R, Borsuk K, Jagodzinski J, Furman B, Klilmek A, Chmielewski M. Enantiomer 2002; 7:107.
58. Kuroda R, Honma T. Chirality 2000; 12:269.
59. Hubal H-G, Hofer T. Chirality 2000; 12:278.
60. Zhang J, Holmes A, Sharma A, Brooks NR, Rarig RS, Zubieta J, Canary JW. Chirality 2003; 15:180.
61. Chisholm J, Golik DJ, Krishnan BM, Watson JA, Van Vranken DL. J Am Chem Soc 1999; 121:3801.
62. Gawronski J, Brzostowska M, Kacprzak K, Kolbon H, Skowronek P. Chirality 2000; 12:263.
63. Rosini C, Superchi S, Bianco G, Mecca T. Chirality 2000; 12:256.
64. Farkas V, Szalay L, Vass E, Hollosi M, Horvath G, Huszthy P. Chirality 2003; 15:S65.
65. Allenmark S. Chirality 2003; 15:409.
66. Takenaka S, Kondo KTN. J Chem Soc Perkin Trans 2 1975; 1520.
67. Takenaka S, Matsura NTN. Tetrahedron Lett 1974a; 2325.
68. Takenaka S, Kondo KTN. J Chem Soc Perkin Trans 2 1974b; 1749.
69. Harata K, Uedaira H. Bull Chem Soc Japan 1975; 48:375.
70. Harata K. Bioorg Chem 1981; 10:255.

71. Kodaka M. J Phys Chem 1991; 95:2110.
72. Kodaka M. J Am Chem Soc 1993; 115:3702.
73. Huang X, Nakanishi K, Berova N. Chirality 2000; 12:237.
74. Huang X, Fujioka N, Pescitelli G, Koehn FE, Williamson RT, Nakanishi K, Berova N. J Am Chem Soc 2002; 124:10320.
75. Matile S, Berova N, Nakanishi K, Fleischauer J, Woody RW. J Am Chem Soc 1996; 118:5198.
76. Kurtan T, Nesnas N, Li Y-Q, Huang X, Nakanishi K, Berova N. J Am Chem Soc 2001; 123:5962.
77. Rickman BH, Matile S, Nakanishi K, Berova N. Tetrahedron, 1998; 5041.
78. Bhatnagar SR, Gough CA. In: Fasman GD, ed. Circular Dichroism and the Conformational Analysis of Biomolecules. New York: Plenum Press, 1996:183.
79. Yang TJ, Wu C-S, Martinez MH. In: Timasheff SN, Hirs CHW, eds. Methods in Enzymology. 1986; 130:208.
80. Li R, Nagai Y, Nagai M. Chirality 2000; 12:216.
81. Jonson WC Jr. In: Fasman GD, ed. Circular Dichroism and the Conformational Analysis of Biomolecules. New York: Plenum Press, 1996:433.
82. Gray DM, Gray CW, Mou, T-C, Wen J-D. Enantiomer 2002; 7:49.
83. Gray DM. In: Fasman GD, ed. Circular Dichroism and the Conformational Analysis of Biomolecules. New York: Plenum Press, 1996:469.
84. Edkins TJ, Bobbitt DR. Anal Chem 2001; 73:488A.
85. Bobbitt DR, Linder SW. Trends Anal Chem 2001; 20(3):111.
86. Salvadori P, Bertucci C, Rosini C. Chirality 1991; 3:376.
87. Salvadori P, Bertucci C, Rosini C. In: Woody RW, ed. Circular Dichroism: Application and Interpretation. New York: VCH, 1994:541.
88. Miller MT, Zhihong G, Mao B. Chirality 2002; 14:659.
89. Richardson FS, Riehl JP. Chem. Rev. 1986; 86:1.
90. Brittain HG. Excited state optical activity. In: Molecular Luminescence Spectroscopy: Methods and Applications, Part I. Chapter 6. New York: Wiley Interscience, 1985.
91. Richardson FS, Riehl JP. Chem Rev 1977; 77:773.
92. Brittain HG. Photochem Photobiol 1987; 46:1027.
93. Brittain HG. J Coord Chem 1989; 20:331.
94. Brittain HG. Coord Chem Rev 1983a; 48:243.
95. Schippers PH, Van Den Beukel A, Dekkers HPJM. J Phys E 1982; 15:945.
96. Shindo Y, Nagakawa M. Appl Spectrosc 1985; 39:32.
97. Dekkers HPJM, Moraal PF, Timper JM, Riehl JP. Appl Spectrosc 1985; 39:818.
98. Blok PML, Dekkers HPJM. Appl Spectrosc 1990; 44:305.
99. Steinberg IZ, Gafni A Rev Sci Instrum 1972; 43:409.
100. Herren M, Morita M. J Lumin 1996; 66&67:268.
101. Dekkers HPJM, Moraal PF. Tetrahedron: Assymmetry 1993; 4(3):473.
102. Schippers PH, Dekkers HPJM. Tetrahedron 1982; 38:2089.
103. Schauerte JA, Schkyer BD, Steel DG, Gafni A. Proc Natl Acad Sci USA 1995; 92:569.
104. Peeters E, Ramos AM, Meskers SCJ, Janssen RAJ. J Chem Phys 2000; 112:9445.
105. Lauffer RB. Chem. Rev 1987; 87:901.
106. Elgavish GA, Reuben J. J Am Chem Soc 1976; 98:4755.
107. Spaulding L, Brittain HG, O'connor LH, Pearson KH. Inorg Chem 1986; 25:188.
108. Brittain HG. Inorg Chim Acta 1983b; 70:91.
109. Abdollahi S, Harris WR, Riehl JP. J Phys Chem 1996; 100:1950.
110. Mondry A, Meskers SCJ, Riehl JP. J Lumin 1994; 51:17
111. Huskowska E, Riehl JP. Inorg Chem 1995; 34:5615.
112. Muller G, Bunzli J-C, Riehl JP, Suhr D, Von Zelewski A, Murner H. Chem Commun 2002; 14:1522.
113. Zolotov YA, ed. Macrocyclic Compounds in Analytical Chemistry. New York: John Wiley & Sons, Inc, 1997.
114. Tsubomura T, Yasaku K, Sato T, Morita M. Inorg Chem 1992; 31:447.
115. Hsu EC, Holzwarth J. J Chem Phys 1973, 59:4678.
116. Holzwarth J, Hsu EC, Mosher HS, Faulkner TR, Moscowitz A. J Am Chem Soc 1974; 96:251.
117. Nafie LA, Cheng JC, Stephens PJ. J Am Chem Soc 1975; 97:3842.
118. Keiderling TA. Appl Spectrosc Rev 1981; 17:189.
119. Nafie LA, Diem M, Vidrine DW. J Am Chem Soc 1979; 101:196.
120. Malon P, Keiderling TA. Appl Spectrosc 1988; 42:32.
121. Diem M, Roberts G, Lee MO, Barlow A. Appl Spectrosc 1988; 42:20.
122. Lee O, Diem M. Anal Instrum 1992; 20:23.
123. Nafie LA, Freedman TB. Enantiomer 1997; 3:283.
124. Freedman TB, Nafie LA. Biopolymers 1995; 37:265.
125. Oboodi MR, Lal BB, Young DA, Freedman TB, Nafie LA. J Am Chem Soc 1985a; 107:1547.
126. Solladie-Cavallo A, Balaz M, Salisova M, Suteu C, Nafie LA, Cao X, Freedman TB. Tetrahedron Asymmetry 2001a; 12:2605.
127. Solladie-Cavallo A, Sedy O, Salisova M, Biba M, Welch CJ, Nafie LA, Freedman T. Tetrahedron Asymmetry 2001b; 12:2703.
128. Freedman TB, Dukor RK, Van Hoof PJCM, Kellenbach ER, Nafie LA. Helv Chim Acta 2002; 85:1160.
129. Wang F, Wang Y, Polavarapu PL, Li T, Drabowicz J, Pietrusievicz MK, Zygo K. J Org Chem 2002a; 67:6539.
130. Bour P, Navratilova H, Setnicka V, Urbanova M, Volka K. J Org Chem 2002; 67:161.
131. Dyatkin AB, Freedman TB, Cao X, Dukor RK, Maryanoff B, Maryanoff CA, Mathews JM, Shah RD, Nafie LA. Chirality 2002; 14:215.
132. Solladie-Cavallo A, Marsol C, Pescitelli G, Bari LD, Salvadori P, Huang X, Fujioka N, Berova N, Cao X, Freedman TB, Nafie LA. Eur. J Org Chem 2002; 1788.
133. Wang F, Polavarapu PL, Lebon F, Longhi G, Abbate S, Catellani M. J Phys Chem 2002b, 106:5918.
134. Kuppens T, Langenaeker W, Tollenaere Bultinck, P. J Phys Chem 2003a; 107:542.
135. Mazaleyrat J-P, Wright K, Gaucher A, Wakselman M, Oancea S, Formaggio F, Toniolo C, Setnicka V, Kapitan J, Keiderling, TA. Tetrahedron Asymmetry 2003; 14:1879.
136. Monde K, Taniguchi T, Miura N, Nishimura S-I, Harada N, Dukor RK, Nafie LA. Tetrahedron Lett 2003; 44:6017.

137 Holmen A, Oxelbark J, Allenmark S. Tetrahedron Asymmetry 2003; 14:2267.
138 Kuppens T, Langenaeker W, Bultinck TJPP. J Phys Chem 2003b; 107:542.
139 Freedman TB, Cao X, Oliviera RV, Cass QB, Nafie LA. Chirality 2003; 15:196.
140 Nafie LA, Freedman TA. Pract Spectrosc 2001; 24:15.
141 Keiderling TA. Pract Spectrosc 2001; 24:55.
142 Aamouche A, Devlin FJ, Stephens PJ. Chem Commun 1999; 361.
143 Devlin FJ, Stephens PJ, Superchi SPS, Rosini C. Chirality 2002a; 14:400.
144 Stephens PJ, Aamouche A, Devlin FJ, Superchi S, Donnoli MI, Rosini C. J Org Chem 2001a; 66:3671.
145 Stephens PJ, Aamouche A, Devlin FJ, Superchi S, Donnoli MI, Rosini C. J Org Chem 2001b; 66:3671.
146 Devlin FJ, Stephens PJ, Scafato P, Superchi S, Rosini C. Chirality 2002b; 14:400.
147 Devlin FJ, Stephens PJ, Scafato P, Superchi S, Rosini C. Tetrahedron Asymmetry 2001; 12:1551.
148 Drabowicz J, Dudzinski B, Mikolajczyk M, Wang, F, Dehlavi A, Goring J, Park M, Rizzo C, Polavarapu PL, Biscarini P. Wieczorek M, Majzner WR. J Org Chem 2001; 66:1122.
149 Wang F, Polavarapu PL. J Org Chem 200; 66:9015.
150 Pieraccini S, Donnoli MI, Ferrarini A, Gottarelli G, Licini G, Rosini C, Superchi S, Spada GP. J Org Chem 2003; 68:519.
151 He Y, Cao X, Nafie LA, Freedman TB. J Am Chem Soc 2001; 123:11320.
152 Urbanova M, Setnicka V, Volka K. Chirality 2000; 12:199.
153 Nafie LA, Freedman TB. Enantiomer 1997; 3:283.
154 Keiderling TA, Silva RAGD, Yoder G, Dukor RK. Bioorg. Med. Chem. 1999; 7:133.
155 Vass E, Hollosi M, Besson F, Buchet R. Chem Rev 2003a; 103:1917.
156 Keiderling TA. Curr Opin Chem Biol 2002; 6:682.
157 Kuznetsov SV, Hilario J, Keiderling TA, Ansari A. Biochemistry 2003; 42:4321.
158 Borics A, Murphy RF, Lovas S. Biopolymers 2003; 72.
159 Keiderling TA. In: Fasman GD, ed. Circular Dichroism and the Conformational Analysis of Biomolecules. New York: Plenum Press, 1996:587.
160 Andrushchenko V, Wieser H, Bour P. J Phys Chem 2002; 106:12623.
161 Andrushchenko V, Van De Sande H, Wieser H. Biopolymers 2003; 72:374.
162 Barron LD, Buckingham AD. J Chem Soc Chem Comm 1973; 152.
163 Barron LD, Boggard MP, Buckingham AD. J Am Chem Soc 1973; 95:603.
164 Barron LD. Am Lab 1980a; 12:64.
165 Barron LD. Acc Chem Res 1980b; 13:90.
166 Barron LD, Hecht L. In: Clark RJH, Hester RE, eds. Biomolecular conformational studies with vibrational optical activity. Biomolecular Spectroscopy, Part B. New York: Wiley, 1993.
167 McCaffery AJ, Shatwell RA. Rev Sci Instrum 1976; 47:247.
168 Horvath LI, McCaffery AJ. J Chem Soc Far II 1977; 73:562.
169 Hug W. Appl Spectrosc 1981; 35:562.
170 Gunia U, Diem M. J Raman Spectrosc 1987; 18:399.
171 Hecht L, Barron LD. Appl Spectrosc 1990; 44:483.
172 Wes ZQ, Barron LD, Hecht L. J Am Chem Soc 1993; 115:285.
173 Oboodi MR, Davies MA, Gunia U, Blackburn MB, Diem M. J Raman Spectrosc 1985b; 16:366.
174 Barron LD, Torrance LF, Cutler DJ. J Raman Spectrosc 1987; 18:281.
175 Hecht L, Che D, Nafie LA. Appl Spectrosc 1991; 45:18.
176 Barron LD, Blanch EW, Hecht L. Adv Protein Chem 2002; 62:51.
177 Barron LD, Hecht L, Blanch EW, Bell AF. Prog Biophys Mol Biol 2000; 73:1.
178 Nafie LA. Raman Rev 2000; 3.
179 Freedman TB, Kallmerten J, Zimba CG, Zuk WM, Nafie LA. J Am Chem Soc 1984; 106:1244.
180 Bour P, Baumouk V, Hazlikova J. Collect Czech Chem Commun 1997; 62:1384.
181 Crassous J, Collet A. Enantiomer 2000; 5:429.
182 Tu AT. Adv Spectrosc 1986; 13:47.
183 Miura T, Thomas GJ Jr. In: Roy S, ed. Subcellular Biochemistry. New York: Plenum Press, 1995:55.
184 Blanch EW, Hecht L, Barron LD. Methods 2003; 29:196.
185 Smyth E, Syme CD, Blanch EW, Hecht L, Vasak L, Barron LD. Biopolymers 2001; 58:138.
186 Blanch EW, Hecht L, Barron LD. Protein Sci. 1999; 8:1362.
187 Blanch EW, Morzova-Roch A, Duncan A. Cochran E, Doig AJ, Hecht L, Barron LD. J Mol Biol 2000a, 301:553.
188 Pappu VR, Rose GD. Protein Sci 2002; 11:2437.
189 Hecht L, Barron LD, Blanch EW, Bell AF, Day LA. J Raman Spectrosc 1999; 30:815.
190 Bochicchio B, Tamburro AM. Chirality 2002; 14:782.
191 Mccoll IH, Blanch EW, Gill AC, Rhie, AGO, Ritchie MA, Hecht L. Nielsen L, Barron LD. J Am Chem Soc 2003; 125:10019.
192 Blanch EW, Morzova-Roch A, Hecht L, Noppe W, Barron LD. Biopolymers 2000b; 57:235.
193 Vass E, Besson F, Buchet R. Chem Rev 2003b; 103:1917.
194 Blanch EW, Hecht L, Day LA, Pederson DM, Barron LD. J Am Chem Soc 2001a; 123:4863.
195 Blanch EW, Hecht L, Syme CD, Volpetti V, Lomonossoff, GP, Nielsen L, Barron LD. J Gen Virolog 2002a; 83:2593.
196 Blanch EW, Robinson DJ, Hecht L, Syme CD, Nielsen L, Barron LD. J Gen Virolog 2002b; 83:241.
197 Blanch EW, Robinson DJ, Hecht L, Barron LD. J Gen Virolog 2001b; 82:1499.
198 Tinocco I, Turner DH. J Am Chem Soc 1976; 98:6453.
199 White TG, Pao Y-H, Tang MM. J Am Chem Soc 1975; 79:4751.
200 Ehrenberg B, Steinberg IZ. J Am Chem Soc 1976; 98:1293.
201 Tinocco I, Ehrenberg B, Steinberg IZ. J Chem Phys 1977; 66:916.

202. Lobenstein EW, Turner, DH. J Am Chem Soc 1980; 102:7786.
203. Bicknesse SE, Maestre MF. Rev Sci Instrum 1987; 58:2060.
204. Thomas M, Patonay G, Warner I. Rev Sci Instrum 1986, 57:1308.
205. Wu K, McGown LB. Appl Spectrosc 1991; 45:1.
206. Sugimoto T, Ikemoto K, Murata S, Tazawa M, Nomura T, Hagino Y, Ichinose H, Nagatsu T. Helv Chim Acta 2001; 84:918.
207. Chen N, Ikemoto K, Sugimoto T, Murata S, Ichinose H, Nagatsu T. Heterocycles 2002; 56:387.
208. Geng L, McGown LB. Anal Chem 1994; 66:3243.
209. Muto K, Mochizuki H, Yoshida R, Ishii T, Handa T. J Am Chem Soc 1986; 108:6416.
210. Cosani A, Peggion E, Verdini AS, Terbojevich M. Biopolymers 1968; 6:963.
211. Peggion E, Cosani A, Verdini AS, Del Pra D, Mammi M. Biopolymers 1968; 6:1477.
212. Woody RJ. J Poly Sci Macromol Rev 1977; 12:181.
213. Watanabe K, Muto K, Ishii T. Biospectroscopy 1997; 3:103.
214. Dong J-G, Wada A, Takakuwa T, Nakanishi K, Berova N. J Am Chem Soc 1997; 119:12024.
215. Nehira T, Parish CA, Jockusch S, Turro NJ, Nakanishi K, Berova N. J Am Chem Soc 1999, 121:8681.
216. Castaguetto JM, Canary JW. Chem Commun 1998; 203.
217. Meadows F, Narayanan N, Patonay G. Talanta 2000; 50:1149.
218. Polavarapu PL. J Chem Phys 1987; 86:1136.

11

Nuclear Magnetic Resonance

FREDERICK G. VOGT

GlaxoSmithKline P.L.C., King of Prussia, PA, USA

I. INTRODUCTION

Nuclear magnetic resonance (NMR) spectroscopy is a powerful method of elucidating the structure and dynamics of a wide range of materials. Most NMR analyses are conducted on condensed phases, such as organic, organometallic, and biological molecules in solution, as well as solid-state materials such as glasses and polymers. The capabilities of NMR have been extended constantly over the past 60 years, and it has become one of the most powerful and widely used spectroscopic techniques available. NMR spectroscopy was first developed by the research groups of Bloch and Purcell in the 1940s (Bloch, 1946; Purcell, 1946). This work was recognized by the 1952 Nobel Prize in Physics. The initial chemical applications of NMR were to ^{1}H and ^{19}F nuclei in solution, and it soon became apparent that the highly resolved spectra of organic molecules could be used to determine chemical structure (Arnold et al., 1951; Dickenson, 1950). New developments in theory and experiment also followed quickly, starting with Hahn's work on spin echoes (Kaplan and Hahn, 1957), the advent of double-resonance experiments (Emshwiller et al., 1960; Hartmann and Hahn, 1962), and magic angle spinning (MAS) for solid-state samples (Andrew et al., 1959; Lowe, 1959). Sensitivity was the major limitation of NMR at the start of the 1960s, with the only available instrumentation being of the continuous-wave (CW) variety. The development of Fourier transform (FT)-NMR using pulsed excitation, first studied by Lowe and Norberg, and later perfected by Anderson and Ernst, led to a great increase in achievable signal-to-noise ratio (Ernst and Anderson, 1966; Lowe and Norberg, 1957). The appearance of algorithms and computers to rapidly perform FTs, coupled with greatly improved magnet technology, resulted in routine NMR of "sparse" spin-$\frac{1}{2}$ nuclei (those that are 0.1–5% naturally abundant), such as ^{13}C, ^{15}N, and ^{29}Si. These events quickly brought multinuclear NMR spectroscopy into the mainstream of analytical chemistry. The next major milestone, multidimensional NMR, first appeared in 1973, and within another decade had been thoroughly explored (Ernst et al., 1987). R.R. Ernst would receive the 1991 Nobel Prize in Chemistry for the development of multidimensional NMR, and for previous work on the development of FT-NMR. Multiple pulse techniques for 1D, 2D, and 3D NMR became widespread during the 1980s. All forms of NMR spectroscopy were greatly affected by these new developments, including the structural analysis of proteins in solution (Cavanagh et al., 1996; Wuthrich, 1986). Successful research in protein structural analysis led to the awarding of part of the 2003 Nobel Prize in Chemistry to K. Wuthrich. NMR has impacted virtually every aspect of organic, inorganic, and organometallic chemistry, along with biochemistry and materials sciences. Even computational chemistry has been involved, by playing a key role in predicting NMR properties and interpreting NMR data.

The versatility of NMR derives from the detailed information about chemical structure that can be obtained from the spectra. Compared with other analytical and spectroscopic methods, NMR is unsurpassed in its information content, but suffers from relatively poor sensitivity. Virtually all modern NMR experiments are run in the FT mode, which relies on pulsed coherent radiation to excite a broad range of spectral frequencies. Two other modes are available for generation of NMR signals: swept CW excitation and stochastic excitation (Blümich, 1996; Ernst et al., 1987). However, these techniques are only used selectively at the present time. Therefore, development of NMR instrumentation has focused on the delivery of pulsed, coherent radiofrequency (RF) radiation to the sample area, where it is harnessed by special coils to produce a phase coherence in nuclear spins polarized by a large magnet. The subsequent motion of these spins produces a signal that is usually processed using an FT, resulting in an interpretable NMR spectrum. Modern NMR spectroscopy is thus greatly dependent on high-quality, low-noise RF electronics, pulse programming systems, stable magnetic fields, and fast computers. The focus of this chapter is on the current practice of NMR spectroscopy as an analytical technique; however, specific coverage of liquid chromatography interfaces (LC-NMR) or magnetic resonance imaging (MRI) is not presented, although the instrumentation and theory are closely related.

II. INSTRUMENT DESIGN

Technological and engineering advances of the past decade have lead to great improvements in NMR capabilities. Available NMR magnetic field strengths have increased beyond 20 T, while the advent of shielded magnet systems has greatly reduced the space requirements of new NMR installations. The electronics of signal generation and detection have improved along with computer hardware. The widespread use of pulsed field gradients in high-resolution NMR has both made routine 2D NMR more reliable and has pushed the limits of detection for more advanced techniques. Despite these advances, the basic design of an FT-NMR spectrometer has not changed. Each of the components in a modern NMR instrument (shown in the block diagram of Fig. 1) is covered in the following sections. Major technological developments are also summarized along with considerations for the purchase of an instrument. Additional information is available in general references on NMR instrumentation (Ellet et al., 1971; Redfield and Gupta, 1971).

A. Magnet Systems

The vast majority of magnets currently used in NMR are superconducting solenoids or *cryostats*, with the basic

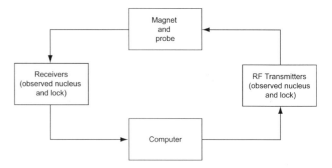

Figure 1 Basic block diagram showing the major components of a modern NMR spectrometer, which essentially consists of computer-controlled transceivers, a delivery and detection system for monochromatic pulsed RF, and a magnet system.

arrangement depicted in Fig. 2 (Laukien and Tschopp, 1993). Although electromagnets and permanent magnets are still in limited use, and have new roles as small-scale NMR sensors, they cannot achieve the stable, strong fields in excess of 3 T needed in most NMR applications.

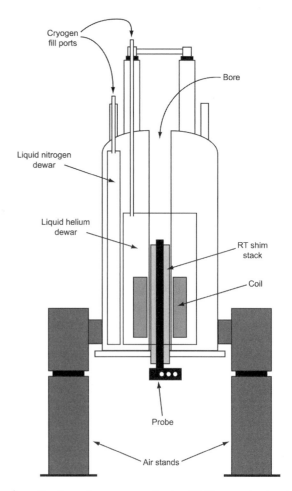

Figure 2 A modern superconducting NMR magnet, showing the coil, dewars, and surrounding assembly, including the probe.

Superconducting cryostats contain a large solenoid immersed in liquid helium at a temperature of 4.2 K or below. The superconducting part of the coil is usually made from NbTi, Nb_3Sn, or $(NbTa)_3Sn$ alloys, often in multifilament configurations inside a supporting copper sheath (Laukien and Tschopp, 1993). The coil resides inside a liquid helium dewar at the center of the magnet (through which the bore passes). This dewar is suspended inside an evacuated chamber at 10^{-6} torr. The vacuum provides excellent insulation, but thermal contact occurs with the suspension joints and guide rods that center the dewar. To improve the situation and reduce the boil-off of the expensive helium, the inner dewar is surrounded by a second suspended dewar containing liquid nitrogen at 77 K. The internal design of the magnet is optimized so that as He and N_2 gas boils away, the cold gas flows past joints and connecting rods and keeps them cold, to lessen thermal conduction. A radiation shield is then placed between the two dewars for additional thermal insulation. All of this engineering has resulted in modern systems in which helium is typically replenished after several months of use, whereas nitrogen refills usually follow a weekly or bi-weekly schedule.

At the center of the cryostat is a room temperature (RT), open-air bore in which the sample is placed. The magnetic field produced by the solenoid is strongest and most homogeneous near the center of the bore, and the probe is therefore designed to reach this area. The sample is placed into this area either manually, by way of a compressed air lift, or through a pump (in the case of a flow probe). Two vertical bore sizes are commonly encountered in solution-state NMR (54 and 89 mm) and dictate the probe and sample sizes that can be used. Horizontal bore magnets of larger diameters are used in MRI.

Despite their advantages, superconducting magnets also have their limitations. They are vulnerable to mechanical impact or other outside stress. The coil cable must balance several competing factors, and can only act as a superconductor if its critical temperature (T_c), field strength, and current density are not exceeded. Above these limits, the superconductor can become resistive, and the resulting undesired vaporization of cryogens is known as a *quench*. Higher-temperature superconductors are currently being explored for both the primary NMR magnet coil and for the secondary shielding and shim coils discussed subsequently, and may help overcome some of these limitations in the future (Hara et al., 1997).

The past decade has seen the introduction of new magnet technologies that have significantly impacted NMR installations. Although the large size and immobility of NMR magnets is still a fixture in most laboratories, mobile magnets have recently become available and are finding applications in a variety of areas (Eidmann et al., 1996; Rokitta et al., 2000). As the core NMR applications in chemistry, biology, and, physics still make use of stationary magnets, another important advance in magnet engineering has been particularly welcome, namely, the introduction of *actively shielded* magnet systems (Smith et al., 1986). These systems include a second superconducting coil outside of the main solenoid, to cancel out the effects of the field in the surrounding room. The 5 G line, typically used as a safety guideline for magnet installations, has been reduced in some cases to within the magnet itself. (However, strong fields still remain above and below the magnet in most cases.) Many laboratories have benefited from the alleviation of space requirements made possible by shielded magnets. Finally, sophisticated *supercooled* (or *subcooled*) magnet systems have now become available for obtaining field strengths as high as 21 T (900 MHz). These systems rely on Joule–Thomson cooling to reduce the liquid helium temperature and improve the performance of the superconducting cable (Laukien and Tschopp, 1993; Wu et al., 1984). To achieve this, the lower part of the magnet (containing the coil) is actively refrigerated to 2 K, whereas the upper section of the helium dewar remains at the usual 4.2 K and ambient pressure. At 2 K, most superconductors have higher critical current densities, and magnetic fields in excess of 18 T are accessible with $(NbTa)_3Sn$ cable. The higher static fields possible with super-cooled magnets are useful in ultra-high-field applications.

The choice of magnetic field strength for a particular application is dependent on the desired resolution and sensitivity, the cost of the instrument, and the space requirements of the installation. The use of higher fields generally results in improved spectra, as the resolution of chemical shifts in an NMR spectrum increases linearly with B_0 and the overall sensitivity generally increases as the 7/4 power of B_0 (depending on the probe design). The effects of field strength are shown in Fig. 3, where the ^1H spectra of cholesteryl acetate at 7.05 and 16.4 T are compared. As frequency in NMR is given by:

$$\nu_0 = \frac{\gamma B_0}{2\pi} \quad (1)$$

where ν_0 is given in Hz and is known as the Larmor frequency and γ is the gyromagnetic ratio (see Section III.A), these two field strengths correspond to instrument frequencies of 300 and 700 MHz. The higher field spectrum has 2.33× the spectral dispersion of that at 300 MHz, which can be seen in the better-resolved peak shapes. Like many NMR spectra, these are plotted on the parts per million (ppm) scale. A ppm is defined by:

$$\delta(\text{ppm}) = 10^6 \frac{\nu_x - \nu_{\text{ref}}}{\nu_{\text{ref}}} \quad (2)$$

where ν_x is the frequency (in Hz) of a signal of interest and ν_{ref} is the frequency of a reference compound.

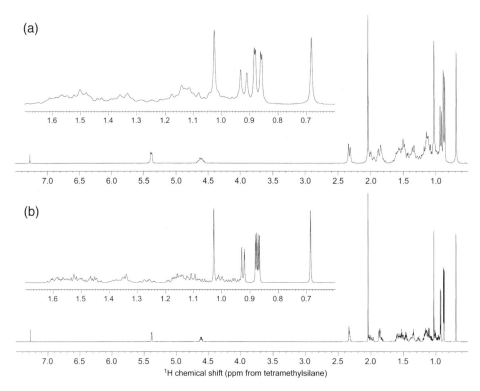

Figure 3 ^{1}H NMR spectra of cholesteryl acetate in CDCl$_3$ at 300 MHz (a) and 700 MHz (b). The results illustrate the general effects of increasing magnetic field strength on NMR resolution.

The change in appearance of the spectra in Fig. 3 is caused by the linear increase in chemical shifts, whereas the *J*-coupling and to some extent the linewidth remains constant. However, at higher fields, relaxation effects caused by chemical shift anisotropy can broaden lines, and stronger and more sophisticated RF pulses are required to excite wider spectral ranges. (These effects are discussed further in Section III). Still, higher fields remain extremely useful for many common analytical problems. As a rough guide, typical field strengths for small molecule solution-state analysis and many solid-state applications are 7.0–11.7 T (300–500 MHz), with higher fields of 14.1–18.8 T (600–800 MHz) usually reserved for macromolecules, flow NMR (including hyphenated techniques), and specialized solid-state experiments.

The static magnetic fields of superconductors are generally stable, but can drift several microteslas per hour, which leads to an undesirable change in resonance frequencies during the course of an NMR experiment. One way to avoid this problem is to continuously observe an selected NMR resonance and compare its frequency to a very stable reference frequency. The field can then be rapidly adjusted in real-time using an additional B_0 coil in the shim stack (Fig. 2). This type of arrangement is known as a *lock system*. As deuterated solvents are widely used in NMR to avoid strong ^{1}H solvent signals, their ^{2}H resonance(s) can be easily locked in this manner to maintain a constant field. In principle any nucleus can be used as a lock signal, although ^{19}F is the only other used in common practice (especially if ^{2}H is to be observed in the experiment). Most modern spectrometers use a graphical display to report the status of the lock system. A CW sweep across the ^{2}H spectrum is observed, the frequency is centered, and the phase can be adjusted. Once the signal is on resonance, an integrator mode can be switched on, and the response of the lock resonance as a function of time is monitored. On newer systems a digital lock system with higher stability and faster response is available.

High-resolution NMR requires a field homogeneity on the order of 10^{-10}, which cannot be achieved by the large coil alone over even a small sample volume at the center of the field. Additional field gradient coils are therefore used to null out the "defects" in the field. These coils are built directly into the superconducting magnet and are known as *shims* or *cryoshims* (if they are maintained at low temperature inside the magnet dewar). Cryoshims are normally only adjusted during the installation of the magnet, whereas RT shims are routinely optimized for maximum field homogeneity. The RT shims are installed in a stack that inserts into the magnet bore and surrounds the probe (Fig. 2). Modern RT shim systems contain coils that can generate a variety of orthogonal spherical field gradients (Table 1). The profiles of the common axial gradients are shown in Fig. 4.

Table 1 Common RT Shims

Shim name	Function	Gradient order	Interaction order
Z0	1	0	0
Z1	z	1	0
Z2	$2z^2 - (x^2 - y^2)$	2	1
Z3	$z[2z^2 - 3(x^2 + y^2)]$	3	2
Z4	$8z[2z^2 - 3(x^2 + y^2)] + 3(x^2 + y^2)^2$	4	2
Z5	$48z^3[2z^2 - 5(x^2 + y^2)] + 90z(x^2 + y^2)^2$	5	2
X	x	1	0
Y	y	1	0
ZX	zx	2	2
ZY	zy	2	2
XY	xy	2	1
$X^2 - Y^2$	$x^2 - y^2$	2	1
Z^2X	$x[4z^2 - (x^2 + y^2)]$	3	2
Z^2Y	$y[4z^2 - (x^2 + y^2)]$	3	2
ZXY	zxy	3	2
$Z(X^2 - Y^2)$	$z(x^2 - y^2)$	3	2
X^3	$z(x^2 - 3y^2)$	3	1
Y^3	$z(3x^2 - y^2)$	3	1

High-quality shims are important as they can greatly improve the achievable resolution and information content of NMR spectra. Suggested procedures for manual shimming are described in the literature, and typically involve inspection of the lock level, the free-induction decay (FID) from a pulsed NMR experiment, and the lineshape of selected signals in the NMR spectrum (Chmurny and Hoult, 1990; Miner and Conover, 1996). An iterative adjustment procedure is then followed. For example, the Z1 shim may be adjusted first, followed by the Z2 shim and other "axial" shims. For the shims with higher interaction orders, a multistep adjustment process may need to be used (Chmurny and Hoult, 1990; Miner and Conover, 1996). As minor changes in sample composition (e.g., solvent changes) usually result in negligible magnetic susceptibility changes, day-to-day NMR operations only require small changes to shim settings. The shim values need significant adjustments when the probe is changed or modified, because of alterations in the magnetic susceptibility of the construction materials or their spatial distribution. The shimming of solid-state NMR probes (such as MAS probes) is less demanding but is still necessary (Sodickson and Cory, 1997).

Shimming can be time consuming and can sometimes test the patience of even the most diligent NMR spectroscopist, and automated procedures are therefore of great interest as NMR sample throughput increases. A common form of automatic shimming system, which has been in use for some time, allows for computer-controlled modification of shim values using algorithms to maximize the lock signal. This is time consuming but can yield good results if properly used (Chmurny and Hoult, 1990). However, a new technique has begun to revolutionize the process of achieving high-quality shims, by using the pulsed field gradients and shim coils already built into most probes and shim stacks. These techniques are collectively known as *gradient shimming* methods (Barjat et al., 1997; Hu et al., 1995; van Zijl et al., 1994). Briefly, a 1D or

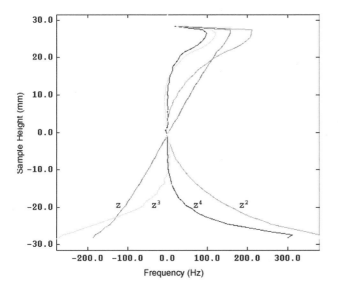

Figure 4 Shim profile plots for common z-shims (z–z^4). The curves represent the response of the field to the shim current, and are used to calibrate gradient shimming software routines. (Figure produced from Bruker XWINNMR Software Version 3.5, Bruker-Biospin.)

3D image of a sample containing a strong resonance is acquired using linear field gradient pulses along the x, y, and z directions. The response of the probe sample area to changes in shim currents is observed (Fig. 4) and a field map is created. (Generally, a field map is created for each probe used on the system.) This map is used to iteratively adjust the actual shims using repeated image acquisitions until the best homogeneity is achieved. The process is rapid and only a few iterations are required, and is especially useful for biological samples dissolved in water, or for probes containing flow cells. The most common nuclei employed in gradient shimming are ^{1}H and ^{2}H.

Besides shimming the sample, the homogeneity of the static magnetic field can be improved by mechanically rotating the sample around the z-axis of the static magnetic field (i.e., parallel to the magnet bore). Spinning at 20 Hz is sufficient to average out minor field inhomogeneities, and is often used during the acquisition of basic 1D spectra. However, spinning is not recommended for most 2D experiments and some 1D experiments (especially those involving spectral subtraction), as the vibrations imposed on the sample lead to serious artifacts. In the case of many 2D experiments, an deleterious effect known as "t_1 noise" is observed in the second dimension when spinning is used.

Finally, it should be noted that most modern magnet systems are supported on stands seated on compressed air cushions for insulation against mechanical building vibration. Magnet systems are preferably situated on the base floor of a building on a large concrete slab. Systems that are on higher floors are especially vulnerable to vibration from the building, although the air-supported float systems greatly reduce this problem. The effects of vibration can often be observed in spectra as small distortions around the edges of narrow NMR signals.

B. NMR Probes

The sample compartment in an NMR spectrometer is referred to as the *probe* or *probehead*. The quality of the probe strongly influences the capability of an instrument, and probe design remains a significant area of ongoing research and development for NMR manufacturers. A simplified picture of the internal layout of a typical solution-state NMR probe is shown in Fig. 5. (Note that the position of the probe in the magnet is depicted in Fig. 2). At the top of the probe is the coil (or coils), surrounded by the necessary electronics and tuning components. The coil is designed to transfer energy to the sample via a pulsed alternating electromagnetic signal in the RF frequency range. It delivers a short-range local magnetic field B_1 to the sample via the principle of induction. The field must to be perpendicular to the large B_0 field. The coil also serves a second purpose, detecting

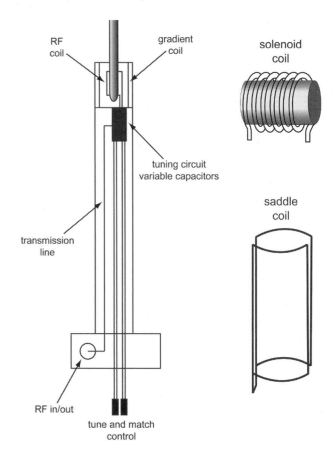

Figure 5 Diagram of a typical solution-state NMR probe, highlighting the coils and RF circuitry. A single-resonance design is shown for simplicity, with a sample tube inserted from the top into a saddle-shaped coil. On the right, solenoidal and saddle-shaped coil designs are depicted. Solenoidal coils are more commonly used in flow probes and in solid-state probes.

the inductance that results in the sample after the initial pulse. The changing magnetic field in the sample induces an electromotive force in the loops of the coil, which is carried to the receiver (see Section II.E). The RF circuitry and the coil represent two major components of the probe discussed here; other components (such as pulsed field gradients) are also discussed in later sections. General reviews of probe design have appeared in the literature and should be consulted for more details (Chen and Hoult, 1989; Doty, 1996a; Hill, 1996; Hoult, 1978).

1. RF Circuitry

The probe circuit is responsible for efficiently transmitting the RF power from the RF source to the sample. This is accomplished using a resonant circuit that includes the coil inductor and other tuning components. Variable capacitors and plug-in fixed capacitors are used to adjust the resonance frequency of a coil to the desired value, which can range over hundreds of MHz in the case of broadband probe channels. Most probes are capable of tuning multiple frequencies (in both single- and double-coil arrangements) for heteronuclear work. Probe circuits are characterized by the complex impedance Z and the quality factor Q, the former of which is given by:

$$Z = R + i\left(\omega L - \frac{1}{\omega C}\right) \quad (3)$$

Here R, C, and L are the resistance, capacitance, and inductance of the coil circuit, respectively, and ω is the resonance frequency, which is given by:

$$\omega = \frac{1}{\sqrt{LC}} \quad (4)$$

The Q of the circuit includes the inductor(s) and is a measure of the circuit's energy storage efficiency:

$$Q = \frac{\text{max. energy stored}}{\text{avg. energy dissipated per radian}} = \frac{\omega}{\Delta\omega}$$
$$= \frac{1}{\omega RC} = \frac{\omega L}{R} \quad (5)$$

The Q factor also specifies the effective rate of energy dissipated, which is also referred to as the ring-down time. The design of probe circuits is strongly affected by the inductor (discussed in Section II.B.2). Other important factors in circuit design include impedance transform (tuning and matching) considerations and circuit isolation. Note the Q of the probe, or the ring-down time of a pulse, is not necessarily as important in terms of sensitivity as other factors, namely, the amount of power placed into the probe circuit, and the efficiency at which this power is converted into current and subsequently into the transverse RF field. More information about the circuit is available in the literature (Cho and Sullivan, 1992a, b; Fukushima and Roeder, 1981; Traficante, 1989, 1993).

2. RF Coil Design and Sensitivity

The heart of the probe circuit is the inductor. The signal-to-noise ratio of an NMR experiment is strongly affected by the design and sensitivity of the RF coil. Most coils used in modern NMR probes act as both transmitters and receivers, as efficient decoupling of the two can be achieved in the preamplifier (see Section II.C). (This is possible even though the power of the pulse is nine orders of magnitude larger than the NMR signal.) However, *crossed-coil* configurations have been used in the past with separate RF transmit and receive coils. These are avoided because of losses caused by the nesting of coils. Note that this is the same rationale behind the design of inverse and standard configuration probeheads. Double-tuned coils can also be employed. The signal-to-noise ratio measured at the NMR receiver is given by (Hoult and Richards, 1976):

$$\frac{S}{N} = \frac{\text{peak signal}}{\text{RMS noise}}$$
$$= \frac{k_0(B_1/i)v_s N_s \gamma(h/2\pi)^2 I(I+1)(\omega_0^2)}{3\sqrt{2}V_{\text{noise}} k_B T}$$
$$\propto \frac{\omega_0^2 (B_1/i) v_s}{V_{\text{noise}}} \quad (6)$$

The expression predicts a ω_0^2 dependence of the signal on the field (i.e., that the EMF induced in the coil is proportional to the square of the Larmor frequency). This dependence can be counteracted by the noise term, as shown subsequently. As Eq. (6) also contains the initial magnetization of the sample, it is meant to be a total expression for the sensitivity of the NMR experiment [as long as the assumptions are fulfilled; see Hoult and Richards (1976) for more details].

The NMR signal-to-noise ratio is proportional to (B_1/i), which is the magnitude of the transverse magnetic field induced in the coil per unit current. The S/N is also proportional to the square of the Larmor precession frequency (ω_0), the sample volume v_s, the number of spins N_s, the spin quantum number I, and the gyromagnetic ratio γ. The gyromagnetic ratio and spin quantum number are properties of a given nucleus discussed in Section III.A. The scaling constant k_0 accounts for RF homogeneity (the deviation in effective RF field across the sample or coil volume) (Hoult and Richards, 1976). NMR experiments are primarily subject to thermal (or Johnson) noise, given by (Doty et al., 1988):

$$V_{\text{noise}} = \sqrt{4 k_B T_c R_{\text{noise}} \Delta f} \quad (7)$$

$$R_{\text{noise}} = \frac{l}{p}\sqrt{\frac{\mu\mu_0 \omega_0 \rho(T_c)}{2}} \quad (8)$$

where V_{noise} is the noise and the resistance R_{noise} refers to that of the entire probe circuit, including the coil and its surroundings (Hoult and Lauterbur, 1977). Equation (6) also predicts that lowering the temperature of this resistor (T_c) will reduce the noise, an effect exploited in the cryogenic probes discussed in Section II.B.4. (Note that a common coil and sample temperature is assumed in this equation.) The noise is measured over a spectral bandwidth defined by Δf. For $R_{\text{noise}} = 50\,\Omega$, a noise value of -168 dB m Hz$^{-1/2}$ is obtained (Doty, 1996a).

The term B_1/i in Eq. (5) is a direct measure of probe sensitivity and depends on the actual coil in the probe. The two coil designs in widespread use for NMR are the solenoid and the saddle coils depicted in Fig. 5. Probes for solid-state NMR often make use of solenoids wrapped at the magic angle or perpendicular to the field. The saddle type of design is important for solution-state because of the need for top-loading pneumatic sample changing with most instruments, as these coils generate a transverse B_1 field while still allowing for axial sample access. Examining the expressions for B_1/i for these two coil types is informative. For a solenoidal coil (Hoult and Richards, 1976; Peck et al., 1995):

$$\frac{B_1}{i} = \frac{\mu_0 n}{d\sqrt{1+(h/d)^2}} \quad (9)$$

where n is the number of turns, d is the diameter of the coil, and h is its length. For a saddle coil:

$$\frac{B_1}{i} = \frac{n\mu_0 \sqrt{3}}{\pi}\left[\frac{2dh}{(d^2+h^2)^{3/2}} + \frac{2h}{d\sqrt{d^2+h^2}}\right] \quad (10)$$

Saddle coils are found to be inherently less sensitive than solenoids (Hoult and Richards, 1976). (A number of other factors also affect the S/N of the NMR experiment, including the circuit quality factor, which degrades with increasing field.) Other special coil designs are prevalent throughout the field of MRI and may eventually impact certain NMR applications. In addition, the advent of flow probes allows for a more flexible choice of coil arrangement as the pneumatic lift is no longer a concern. One final useful proportionality that can be derived from Eq. (6) is (Doty, 1996a):

$$\frac{S}{N} \propto \frac{N_s}{\tau_{\pi/2}\sqrt{P}} \quad (11)$$

where $\tau_{\pi/2}$ is the average $\pi/2$ pulse width obtained with power P.

There are a number of tube and coil sizes in common use in NMR. These include coils for accommodating sample tubes with diameters of 10, 5, 3, 2.5 and 1.7 mm, and flow cells with diameters of 4 mm down to several micrometers. The most popular sample tube size is the 5 mm tube, owing both to its ease of use and readily available, inexpensive disposable tubes. The volumes contained by these tubes are typically in the range of 0.75–1 mL. For many mass-limited samples, this volume results in very dilute solutions. To improve the concentration of these samples, sample tubes and flow cells have been designed to contain volumes of 1 nL–100 μL. Generally, NMR probes which operate on these volumes are referred to as *microprobes*. The rationale for developing microprobes and microcoils is twofold: many samples of interest are already in solution and are of limited volume, and the sensitivity per unit volume is better for smaller coil sizes. This can be seen by insertion of Eq. (9) into Eq. (6) (Lacey et al., 1999; Peck et al., 1995):

$$\frac{S}{N} = \frac{\omega_0^2 [n/d\sqrt{1+(h/d)^2}]}{\sqrt{n^2 d \omega_0^{1/2}/h}} \propto \frac{\omega_0^{7/4}}{d} \quad (12)$$

The S/N therefore improves for a microcoil as the coil diameter decreases, as long as a fixed length-to-diameter ratio is maintained. It should also be noted that although the field dependence of the S/N expected from Eq. (6) is ω_0^2, factoring in the noise in this manner predicts a scaling of $\omega_0^{7/4}$. There are other well-known equations that predict a 3/2 dependence of S/N on magnetic field strength (Hoult and Richards, 1976); however, these are based on other assumptions and it is generally accepted that Eq. (13) provides a better description of the NMR system for a solenoid. A recent review covers solenoidal micro- and nanocoils and other aspects of NMR on small sample volumes (Lacey et al., 1999).

3. Flowprobes

These have become widespread in recent years, with the introduction of liquid sample handling systems and the popularity of hyphenated liquid chromatography and NMR (LC-NMR). Flow cells in probes are generally of the order of 50–300 μL and can be oriented in a number of ways with respect to the magnetic field. Samples are introduced via a flowing stream of solvent, and are stopped in the cell for analysis. The use of flowprobes and microcoils has also led to the development of more efficient multiplex NMR systems, which can analyze several samples at the same time by time sharing of the spectrometer electronics, although this has not yet found widespread use (Doty et al., 1988).

4. Cryogenic Probes

One of the most important new probe designs of the last decade is the cryogenically cooled probe (Jerosch-Herold and Kirschman, 1989; Styles et al., 1984). This complex but effective innovation greatly improves the sensitivity of the NMR probe by cooling its electronics to about 25 K. For many years, NMR research at cold temperatures has benefited from this improved sensitivity, as Eq. (6) predicts that the noise level will fall with decreasing temperature T. As most coil materials show decreased resistance R at lower temperatures, an even greater decrease in noise is predicted (Jerosch-Herold and Kirschman, 1989; Styles et al., 1984). However, the required thermal insulation deteriorates the coil filling factor, so that in practice the signal-to-noise gain for a given sample when compared with a similar conventional probe is usually a factor of 3 or 4. Cryogenic NMR probes have recently become commercially available in both sample tube and flow models, and newer probes can be changed between these two popular modes. Cryogenic probes have already found applications in drug discovery and natural product analysis (Martin, 2002; Russell et al., 2000). It should be noted that lowering the temperature of the sample also increases the signal-to-noise ratio by increasing the magnetization, but this is not feasible for most compounds of interest in solution-phase, and use of low-temperature samples (<25 K) is generally confined to the area of solid-state physics.

5. Solid-State Probes

Most solid-state probes are designed for MAS, where the sample is rotated rapidly (1–35 kHz) at an angle of 54°44′ relative to the static magnetic field (Doty, 1996b; McKay, 1996). The double-bearing rotor/stator design is the most commonly employed scheme, as it offers stability combined with high spinning rates. This design uses two air collars to suspend the rotor (usually a zirconia or boron nitride cylinder) on a plastic tip, where an additional amount of air pressure is then applied to vanes situated at one of the rotor ends. A variety of rotors sizes are available, from 7.5 mm rotors holding 500 mg of material and spinning to ∼6 kHz all the way up to 2.5 mm and smaller designs holding 10 mg of material and spinning at 35 kHz. The higher powers needed in solid-state NMR are handled by special transmission line-based circuitry (McKay, 1996). Other solid-state probes, including static probes and probes for single-crystal studies, are available as specialty items or are sometimes constructed by researchers.

6. Sample Temperature Control

Most NMR probes are also designed for variable temperature (VT) work, by piping gas from an external supply across a resistive heater and then onto the sample. Both cold and hot gas can be used, and dry nitrogen is usually preferred. (Cold nitrogen can be generated directly from a boil-off system or by passing RT gas through a cold heat exchanger.) The temperature is usually monitored electronically via a thermocouple placed near the sample in the top of the probe, which can be calibrated by observing the NMR spectra of temperature-sensitive compounds (see Section II.G). The construction materials used in the probe dictate the temperature range. Typical commercial probes have ranges from $-150°C$ to $+200°C$. Probes for solid-state NMR have similar ranges; this includes MAS probes when dry nitrogen is used as a spinning and bearing gas.

7. Guidelines for Probe Selection

The earlier discussion highlights a number of issues with NMR probes that can impact their selection for a particular experiment. To summarize, the factors that generally need to be kept in mind include:

1. sample volume within the coil(s), NMR tube size, or flow cell volume;
2. RF power levels, efficiency, and sensitivity to inductance in the sample;
3. homogeneity of the RF field delivered during pulses;
4. decay time and Q of the resonant circuit;
5. tunable range over multiple resonances of interest.

This range of considerations obviously complicates the decision of a potential probe buyer. Some simplified guidelines can be offered, however. It is generally best to have two or three general-purpose probes per instrument, plus additional specialty probes. Most commercial probes have very good tolerance for high RF powers, excellent RF homogeneity, and very broadband tuning range from ^{15}N to ^{31}P frequencies. Probe vendors have developed automatic tuning probes, which are now readily available for a reasonable price, and are very useful for automated multinuclear work and even for maintaining proper tune after solvent changes without user intervention. Specialty probes are often needed for ^{19}F and triple or quadruple resonance work (e.g., $^1H/^{13}C/^{15}N$ probes for protein NMR). It should be mentioned that many commercial broadband probes include a stop filter for ^2H frequencies to suppress noise from the lock channel, which must be removed if the user is planning ^2H NMR experiments.

C. RF Generation and Signal Detection

1. Transmitter and Pulsed RF Generation

NMR spectroscopy is performed in the RF regime, although it is the magnetic component of the field oscillating at these frequencies that is used to induce and detect resonance. Modern NMR spectrometers can deliver pulses of coherent RF radiation ranging from a few megahertz up to nearly 1 GHz. The pulses can range from the microsecond time regime up to milliseconds and seconds in duration, and can consist of complex amplitude, phase, and frequency shifting steps. The pulsed RF can be delivered in synchrony across several carrier frequencies and in time with other clocked events (such as gradient pulses). All of this functionality needs to be programmable and highly flexible. RF generation for an NMR spectrometer is therefore a formidable task. At the present time, both *analog* and *digital* schemes are used at various stages in modern spectrometers for RF generation and signal detection. The basic arrangements will be reviewed here, although different manufacturers employ their own unique proprietary enhancements. Block diagrams of analog and digital NMR transceivers are shown in Fig. 6. These two basic architectures are encountered in the majority of systems in use, including both manufactured and custom-built installations. Analog systems date back to the earliest days of NMR and have evolved considerably in the interim (Ellet et al., 1971; Redfield and Gupta, 1971; Wittebort and Adams, 1992). Digital transceivers

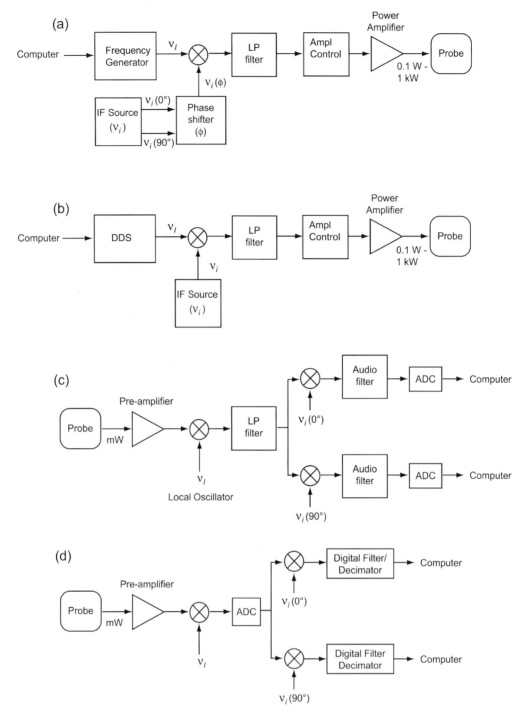

Figure 6 Block diagrams of analog (a) and digital (b) NMR transmitters. Block diagrams of analog (c) and digital (d) NMR receivers. Analog-to-digital converters are abbreviated as ADC.

are a fairly recent innovation, having been developed in the past decade (Kasal et al., 1994; Momo et al., 1994; Redfield and Kunz, 1994; Sternin, 1995; Villa et al., 1996; Yun et al., 2002).

In analog systems, a heterodyne mixing scheme is often used. This involves mixing a variable generated frequency (known as the local oscillator or LO frequency) with phase shifted and amplitude modulated IF frequencies to produce RF, which is then delivered to the sample. The choice of IF must take into account possible interference with NMR signals of interest and the capabilities of the RF circuit components. Upon reception, the signal is mixed with the LO frequency to again obtain a signal near the IF frequency (with deviations from IF caused by resonance shifts in the spectrum). The signal is then mixed with the real IF frequency to produce audio frequency signals. The resonance shifts then are seen as deviations around 0 Hz. The role of the IF in signal generation and reception is to allow for easier phase shifting (using cables cut to fractions of the IF wavelength), filtering, and frequency switching. The phase stability of the analog systems is generally quite good, with typical deviations of $0.2°/h$. In addition, analog systems are generally driven by digital components. This is because the timing of NMR experiments requires sub-microsecond precision, which can be obtained by locking the transmitter gates to a master RF clock (10 or 100 MHz for many systems). Pulse programmers for analog systems contain digital high-power pulse forming gates and low-power shaped pulse generation circuits.

The most recent significant advance in the field of NMR signal detection has been the introduction of digital transceivers. The transmitter portions are based on direct digital synthesis (DDS), a technology which creates RF from digitized waveform tables stored in memory. The system is essentially a digital-to-analog converter (DAC), which can produce very rapid frequency and phase shifts. This enables a variety of special NMR experiments that would otherwise be difficult to perform with conventional analog signal generation. For example, fast frequency shifts on the order of ± 100 kHz (or even greater) are possible. Fast amplitude control can be implemented with traditional fast gates or directly in the DDS. A mixer-driven shift (or "bump") to IF frequency is still often employed, so that the output from the DDS (often limited to 10–100 MHz) can be brought up to RF frequencies of interest in NMR (which span from 10 to 900 MHz). Note that the receivers in such systems may also be digital or analog, depending on the design of the system.

2. Power Amplifiers

The final stage of RF pulse generation is amplification. A number of configurations are in common use, and can involve multiple amplifiers. Many modern amplifiers are *linear*, in that they respond linearly to a wide range of input amplitudes. This is useful in many NMR pulse sequences that switch from high power to low power (sometimes as much as 50 dB difference) in a matter of microseconds. (Examples include pulse sequences that contain nonselective and selective pulses). As the strength of the RF field is given by $\omega_1 = \gamma B_1$, higher amplifier outputs are needed for nuclei with lower gyromagnetic ratios to achieve comparable RF field strengths and pulse widths. Power amplifier outputs vary from the tens of watts up to kilowatts for solid-state and imaging applications.

3. Preamplifier

The role of the preamplifier is simply to amplify the tiny detected NMR signal returned from the probe. Since this circuit must be sensitive, fast switching diodes, transmission line tees, and quarter-wavelength cables are used to protect it from the input power of the pulse (Cho and Sullivan, 1992a, b; Traficante, 1989, 1993). The use of these circuits causes the preamplifier to look like an open circuit while the RF pulse is going in, then "reconnecting" it when the detection phase of the experiment begins. More than 60 dB of isolation is possible between the circuits using the properties of crossed-diodes.

4. Receiver

The NMR signal is obtained as an analog AC signal at RF frequencies, and needs to be processed prior to further analysis. NMR signals are measured as offsets from a reference frequency. The reference frequency is usually placed in the center of the spectrum to allow for efficient RF excitation across the spectrum. In order to discriminate between the signs of resonances above and below the reference (transmitter) frequency, a phase-sensitive detector must be employed. In NMR, the most common detector of this type is referred to as *quadrature* detection. The basic scheme of analog quadrature detection is shown in Fig. 6(c). As noted, the signal is mixed with the LO frequency to recover a signal near IF but shifted by NMR effects. However, the second mixing stage, in which this signal is converted to audio frequencies, actually occurs after the IF signal is split into two components. The first of these two components is mixed with the true IF frequency at its original phase, and the second is mixed with the true IF frequency shifted in phase by $90°$. The two separate signals are passed through audio filters, and sent to the analog-to-digital converter (ADC). The ADC is the final step in an analog receiver, and digitally samples the signal at a rate usually chosen based on the Nyquist relation. This indicates that required spectral widths for NMR experiments should be of the order of the expected range of chemical shifts. For example, at 400 MHz, a ^{1}H spectrum might require a 7500 Hz bandwidth to encompass signals from -1 to 12 ppm. Imbalance between the real and imaginary sections of the analog detector causes artifacts to appear at multiples of the frequency of strong peaks, which can be folded into the spectrum. (These can be suppressed by phase cycling.) The audio filters used are typically high-quality Butterworth or Bessel filters that have a sharp cutoff combined with limited phase effects on the spectrum.

Digital quadrature detection (DQD) has been introduced for NMR spectroscopy in the past decade. DQD has found immediate acceptance and is now available on most commercial systems. The detection process is carried out using *oversampling* in the time domain, which refers to sampling at an order of magnitude rate higher than that required by the Nyquist condition. Digital filters can be used, which offer extremely sharp frequency cutoff points, and other beneficial characteristics when compared with analog filters. Signals outside the spectral window are fully suppressed and are not folded. This allows for expanded spectral regions to be studied using narrow spectral widths, without interference from signals which would normally leak through analog filters. Other advantages of DQD include better dynamic range, better signal-to-noise, reduction of first-order phase distortions, and improved spectral baselines. The improvement in dynamic ranges is effectively achieved by oversampling, and a 16-bit digitizer can sometimes effectively achieve a few more bits using this scheme. Finally, the receiver phase shifts obtained in DQD are exceptional, so that the balance between the real and imaginary does not need to be adjusted and artifacts are minimized to the point that phase cycling is not required to suppress quadrature images. A block diagram of a DQD is pictured in Fig. 6(d). The signal from the preamplifier is mixed with LO to obtain IF, filtered, and sampled at a higher rate than required by the Nyquist relation. A digital signal processor stage then achieves quadrature detection by digital mixing with the appropriate signals from the DDS. The number of points is then reduced by decimation, which amounts to averaging n points, causing the dwell to increase by a factor of n and reducing the spectral width by $1/n$. In many cases, the DSP output is the convolution of a square wave filter and the signal, so that the FID has a disturbing build-up for the first points prior to the expected long exponential decay.

D. Magnetic Field Gradients

In addition to the B_0 field, it is often useful at times in NMR to have magnetic field gradients across the sample. These differ from the gradients produced by poor shimming in high-resolution work, as they are actually desired by the experimentalist. The gradients can either be time-dependent pulsed gradients (the usual case) or static constant gradients. Unlike shim gradients, these gradients are generally linear so that the observed frequency spectrum corresponds directly to a spatial position via the Larmor relationship [Eq. (2)]. The technology for implementing pulsed field gradients was initially developed for imaging applications, and NMR implementations have only advanced recently because of important applications in coherence pathway selection (see Section III).

The gradient coils used in NMR are usually integrated directly into the probe. (Wide-bore imaging systems usually have gradient coils separate from the probe, but the basic principles are the same.) In particular, the pulsed gradient coils used for coherence pathway selection are wrapped around the upper section of the probe, around the RF coil(s) and sample chamber. One important consideration in selecting or designing a probe is whether to use a single-axis gradient (usually aligned along the z-axis) or triple-axis gradients. Triple axis gradients can be useful for certain pulsed experiments in which no correlation between gradients is desired. Pictorially speaking, it is sometimes desirable to "wind" the magnetization along two orthogonal axes, so that there is no chance that subsequent gradients will "undo" the encoding and expose other coherence orders until the desired time. There are also a few experiments which make use of magic angle gradients. One disadvantage of triple axis gradients is the 10–20% reduction typically seen in probe sensitivity.

Axial (z) coils are usually created by solenoids, often in a paired configuration to cancel higher-order nonlinearity (especially if imaging is to be conducted). Several methods are available for producing transverse linear field gradients. One common design uses arrays of circular arcs similar to a saddle coil. Coil designs are further complicated by the need for active shielding in modern gradient coil systems. This is necessary to avoid the deleterious effects of eddy currents produced by the gradients. Rectangular-shaped gradient pulses were used in the early days of gradient-enhanced NMR, but these suffered from problems with amplifier rise times and eddy currents. The eddy currents are produced by the pulsed gradient coils in the surrounding conducting material, and forces the use of a recovery period (dead time) after the gradient pulse. These problems can be greatly reduced by the use of sine-shaped gradients, which are now standard in most NMR pulse sequences. Thus, the electronics that drive the gradients in modern NMR systems include complete digital control of the current being sent to the gradient coil(s). A DAC unit generates analog voltages to produce the desired current waveforms. Gradient amplifiers are then used to produce the final current, with audio frequency amplifiers used to generate time-dependent or modulated gradients (especially for imaging).

E. Computer Systems

Modern NMR spectrometers usually come equipped with at least one computer workstation. A variety of operating systems are in current use by spectrometer vendors. These include Unix and its variations (Solaris®, IRIX®, LINUX), Microsoft Windows® and the Apple Macintosh® OS, running on a variety of processors. The selection of the computer system for a particular NMR installation is commonly driven by the offerings of the manufacturer and the hardware capabilities and compatibilities needed by the purchaser. Offline systems can usually be chosen more freely as modern networking software can accommodate and share data over different operating platforms. For example, a Windows PC can be used to process data acquired on a Silicon Graphics IRIX computer via network file sharing. In addition, the offline data systems can run third-party simulation and modeling software (see Section V for more information). As a guide to those purchasing instruments, NMR systems typically benefit from fast computer architectures with as large an amount of main memory and disk space as available, with processors running in the range of 1–3 gigaflops, and memory and disk space in the 256–1024 megabyte and 10–200 gigabyte ranges, respectively. In addition, high-quality video displays are helpful for viewing detailed spectral features, pulse programs, acquisition macros, and a multitude of other information. Finally, the latest printing and output systems are usually available for use with NMR spectrometers, and electronic formats for data sharing are in common use.

F. Accessories

Although NMR probes often contain heaters for VT work, the rest of the VT system is usually externally located. Modern systems now include refrigerated air chillers that supply the probe with air or nitrogen in the −20°C to +5°C range. This is useful for maintaining stable temperatures in the probe, as the heater can adjust the cold gas to the desired temperature easily. In addition, long-term studies upto several days can be easily performed at lower temperatures. For example, a molecule that degrades rapidly at RT can often be more readily analyzed at −5°C in appropriate solvents (such as $CDCl_3$). To achieve colder temperatures, nitrogen boil-off or heat exchanger systems are used. Computerized units are used to control the VT accessories, usually by adjusting the current delivered to the probe heater and the nitrogen boil-off heater. As VT systems obtain the temperature via a thermocouple near the sample, the actual sample temperature may differ and temperature calibration standards must be used. For high-temperature solution-state work (300–420 K), the difference in ^1H chemical shift between the hydroxyl and methylene protons in ethylene glycol (1,2-ethanediol) can be used as a temperature calibration, whereas methanol serves a similar purpose for lower temperatures (170–300 K) (van Geet, 1970; Martin et al., 1980). A similar procedure is possible in solid-state NMR, for MAS and other solid-state applications, using the sensitivity of the ^{207}Pb nucleus in lead nitrate (Bielecki and Burum, 1995).

Another common accessory is the automatic sample changer, or *autosampler*. There are a variety of units in service and under development. Units which change samples contained in 5 mm glass tubes (or other sizes) usually employ a carousel and sometimes a robotic arm for transferring the sample. New units utilize a sliding rail system for faster sample changing. Flowprobes allow for convenient and quick sample changing, without preparation of the traditional NMR glass tube. Samples can be taken from small vials (resembling those used for chromatography applications) or directly from multiwell plates. Automatic sample handling systems then retrieve the sample through a syringe, transfer it to the probe using a pump, and stop it in the probe for analysis. Bruker's BEST™ and Varian's VAST™ systems are two well-known examples of automatic flow-based sample handlers.

III. THEORETICAL BACKGROUND

A variety of interactions among nuclei, surrounding electrons, and external fields are observed in NMR spectra, and are fundamentally responsible for the power of NMR as an analytical technique. External RF fields are used to manipulate the nuclear spin-state. Because quantum mechanics is able to describe NMR phenomena in great detail, the technique can be exploited and manipulated to obtain a remarkable amount of information about atomic and molecular structures. The theory presented here is intended to serve as an cursory introduction, and the major texts in the field should be consulted for more information (Abragam, 1961; Ernst et al., 1987; Gerstein and Dybowski, 1985; Slichter, 1996). The monographs by Abragam (1961) and Slichter (1996) are especially notable for their rigorous treatment of NMR theory, whereas the text of Ernst et al. (1987) is especially suited to understanding FT-NMR and multidimensional experiments.

A. Nuclear Spin Dynamics

Many atomic nuclei have a special property known as spin. The presence of nuclear spin implies angular momentum, and nuclei with spin act as dipoles (or small magnets). NMR has its roots in the interaction of spin with strong static magnetic fields, otherwise known as the *Zeeman effect*. Nuclei have high-energy excited-states characterized by quantum numbers, but in the ground-state available to NMR, only a single nuclear spin quantum number I is necessary to define the possible states. The nuclear spin quantum number is a positive multiple of $1/2$, and can be predicted from the shell model of the nucleus, by way of the order of filling of nuclear energy levels and pairing-type interactions. The presence of one or more "unpaired" nucleons will give rise to an overall nuclear spin. The predictions for nuclear spin are based on A, the mass number of a nucleus, N, the number of neutrons, and Z, the atomic number or number of protons (note that $N = A - Z$). Nuclei with an even A, N, or Z have a spin of zero. If A is even and Z is odd, the nucleus has an integral spin and is known as a boson. If A and Z are both odd, the nucleus has a half-integral spin $>1/2$ and is known as a fermion. And finally, if A is odd and Z is even, the nucleus has a spin of $1/2$ and is again known as a fermion. Some specific examples are the ^{1}H and ^{13}C nuclei with $I = 1/2$, whereas the ^{139}La nucleus has $I = 7/2$, and NMR-inactive nuclei like ^{12}C have $I = 0$.

The angular momentum of a nucleus with spin I is related to the magnetic dipole moment by a scaling constant γ (the gyromagnetic ratio):

$$\boldsymbol{\mu} = \gamma \mathbf{I} \tag{13}$$

where $\boldsymbol{\mu}$ and $\mathbf{I}$ are collinear vector quantities. In the presence of an external magnetic field $\mathbf{B}_0$, the energies of the nuclear spin-states will be:

$$E = -\boldsymbol{\mu} \cdot \mathbf{B}_0 \tag{14}$$

The angular momentum has eigenvalues $I(I + 1)$ where I is an integer or a half-integer. Quantum mechanics allows for one of the three components of the angular momentum to be specified simultaneously with the magnitude of $\mathbf{I}$, and the z-component (I_z) is chosen by convention. The eigenvalues of Eq. (14) are then multiples of the angular momentum. Choosing the variable m as the magnetic spin quantum number, with possible values of $(-I, -I + 1, \ldots, I - 1, I)$, and noting the axial symmetry of the fields used in NMR leads to an interaction energy with the B_0 field:

$$E_m = -\frac{\gamma h m B_0}{2\pi} \tag{15}$$

where h is Planck's constant. The resonance condition for a bare nucleus ($\omega_0 = \gamma B_0$), also given in Eq. (2), defines the general region in which RF irradiation is needed for NMR spectroscopy of a particular nucleus, to excite transitions between the levels in Eq. (15).

From Eq. (15), the relative populations of the m spin-states in the field can be determined from a Boltzmann distribution:

$$\frac{N_m}{N} = \frac{\exp(-E_m/k_B T)}{\sum_{m=-I}^{I} \exp(-E_m/k_B T)} \tag{16}$$

where N_m is the number of nuclear spins with magnetic quantum number m, N is the total number of spins, T is the temperature in K, and k_B is the Boltzmann constant. At normal temperatures ($\sim$273 K), the terms inside the exponential functions are small. The exponentials can be expanded in a Taylor series and truncated at the first term to obtain:

$$\frac{N_m}{N} = \frac{1 - (E_m/k_B T)}{2I + 1} \tag{17}$$

(This truncation is known as the high-temperature approximation.) For a spin-$\frac{1}{2}$ nucleus the spin-states are commonly referred to as α and β, with α being the slightly more populated lower-energy state. The population excess is small at RT for example, the population difference for ^{1}H nuclei in a field of 11.7 T is 1 spin out of every 10^5. Thus the sensitivity of NMR is inherently less than that of other common spectroscopic techniques, such as mass spectrometry and ultraviolet spectroscopy.

Nearly every element of interest on the periodic table has a stable isotope with a nonzero spin quantum number (Table 2) (Harris, 1996; Harris and Mann, 1978). There are also a number of radioactive isotopes, some with long half-lives, that are amenable to NMR spectroscopy. Nuclei with $I > 1/2$ have a nonspherical distribution of charge and are known as quadrupolar nuclei, as they have a quadrupole moment (Table 2). The gyromagnetic ratio is negative in some cases, which can affect referencing and the general sense of phase when working with these nuclei.

Only the microscopic magnetization has been considered so far; however, it is the bulk effect of the microscopic magnetic moments which is detected in NMR. The bulk magnetization is given by the vector sum of the nuclear moments:

$$M = \gamma \hbar \sum_{m=-I}^{I} m N_m \tag{18a}$$

$$M \approx \frac{N \gamma^2 \hbar^2 B_0 I(I + 1)}{3 k_B T} \tag{18b}$$

where Eq. (18b) is obtained by substitution of Eq. (17). The vector sum of the microscopic magnetization is the experimental observable in NMR. The motion of the bulk magnetic moment in a magnetic field is given by:

$$\frac{dM}{dt} = M(t) \times \gamma B(t) \tag{19}$$

The bulk magnetization undergoes precession at the Larmor frequency around the axis of the static field. The classical view of the situation is sketched in Fig. 7 for both a spin-$\frac{1}{2}$ and a spin-1 nucleus, with the excess population in the lower-energy state shown for the spin-$\frac{1}{2}$ nucleus only.

The theory of magnetic resonance, while remarkably accurate, can be difficult to absorb for the student. As a guide, we note that there are essentially three "levels" of theory in widespread use: the semiclassical vector model, formalisms based on operator algebra (Sorensen et al., 1984), and the full density matrix approach (Abragam, 1961; Ernst et al., 1987; Gerstein and Dybowski, 1985;

Nuclear Magnetic Resonance

Table 2 Properties of Selected NMR-Active Nuclei

Nucleus	Spin (I)	Magnetic moment (μ)	Gyromagnetic ratio (γ)[a]	Frequency in MHz at 2.344 T	Natural abundance (%)	Receptivity[b]	Quadrupole moment (Q)[c]
^{1}H	1/2	4.837354	26.7522128	100.000	99.985	1.0000	–
^{2}H	1	1.212601	4.10662791	15.350	0.015	0.00000145	0.002860
^{6}Li	1	1.162556	3.9371709	14.716	7.5	0.000638	−0.00082
^{7}Li	3/2	4.204075	10.3977013	38.864	92.5	0.272	−0.041
^{11}B	3/2	3.471031	8.5847044	32.084	80.1	0.132	0.0022
^{13}C	1/2	1.216613	6.728284	25.145	1.10	0.000175	–
^{14}N	1	0.571004	1.9337792	7.226	99.634	0.00100	0.02044
^{15}N	1/2	−2.712618	−2.71261804	10.137	0.366	0.00000382	–
^{17}O	5/2	−2.24077	−3.62808	13.556	0.038	0.0000111	−0.02558
^{19}F	1/2	4.553333	25.18148	94.094	100.0	0.834	–
^{23}Na	3/2	2.862981	7.0808493	26.451	100.0	0.0927	0.1089
^{27}Al	5/2	4.308687	6.9762715	26.056	100.0	0.207	0.1403
^{29}Si	1/2	−0.96179	−5.3190	19.867	4.67	0.000367	–
^{31}P	1/2	1.95999	10.8394	40.480	100.0	0.0665	–
^{93}Nb	9/2	6.8217	6.5674	24.476	100	0.488	−0.32
^{139}La	7/2	3.155677	3.8083318	14.126	99.9	0.0605	0.20
^{207}Pb	1/2	1.00906	5.58046	20.921	22.1	0.00201	–

Source: Values are from Harris and Mann (1978) and Harris (1978).
[a] Reported in units of 10^7 rad/T/s
[b] Relative to ^{1}H at 1.0000
[c] Reported in barn (10^{-28} m^2)

Slichter, 1996). The simplest approach is the semiclassical vector model, used to this point, which can describe and is the primary theory discussed here. Mathematically, this phenomenon is described by the Bloch equations, given at the end of this section. The limitations of the Bloch equations are seen when coupling interactions are considered, and quantum mechanics (either operator algebra or density matrix theory) is necessary to explain the effects. Relaxation effects can be included in any of the models to varying degrees of accuracy.

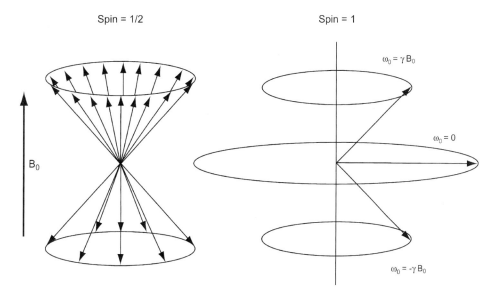

Figure 7 Larmor precession for spin-½ and spin-1 nuclei. For the spin-½ case, a range of vectors are shown, with slightly more vectors in the lower-energy state (α). Only a single vector is shown for each state in the spin-1 situation. When viewed from a quantum mechanical perspective, these diagrams represent angular momentum, and only the z-component can be specified with certainty.

B. External Manipulations of Spin Coherence

The primary externally induced interaction of interest in NMR, besides the Zeeman interaction, is the applied RF fields used to generate and manipulate spin coherences. The irradiating field, known as B_1, is applied at the resonance frequency of the nucleus of interest, as determined by the gyromagnetic ratio of the nucleus and the static field strength. The effects of RF are usually visualized by transforming the classical precession of the nucleus into a frame rotating at the Larmor frequency. In this frame, the magnetization is static and is initially aligned along the z-axis, as shown in Fig. 8. Upon application of a pulse, the magnetization is rotated around through the x–y plane. A pulse can be calibrated to place the magnetization directly into this plane; this is referred to as a $\pi/2$ pulse. Once the magnetization is in this plane it can induce a detectable signal in the coil. A π pulse places the magnetization along the $-z$-axis (essentially inverting the spin populations). The relative phase of the pulse in NMR experiments places the magnetization vector along different axes in the x–y plane. Most phase shifts are confined to the $+x, +y, -x,$ and $-y$ directions (corresponding, respectively, to 0°, 90°, 180° and 270° degree phase shifts of the RF waveform). However, arbitrary phase shifts are needed in a number of experiments. Using the convention shown here, a pulse of $+x$ phase rotates the magnetization into the $-y$ direction, a pulse of $+y$ phase rotates the magnetization into the x direction, and so on.

The rotation induced by pulses is only ideal if the pulse is perfectly calibrated and exactly on resonance. If the pulse is calibrated, but the spectrum contains peaks spaced by a frequency difference that is of the order of the B_1 field (as do most spectra), then the excitation for these peaks will be noticeably affected by resonance offset. The transmitter is placed at the center of the spectrum for optimal excitation, as the rectangular $\pi/2$ pulse

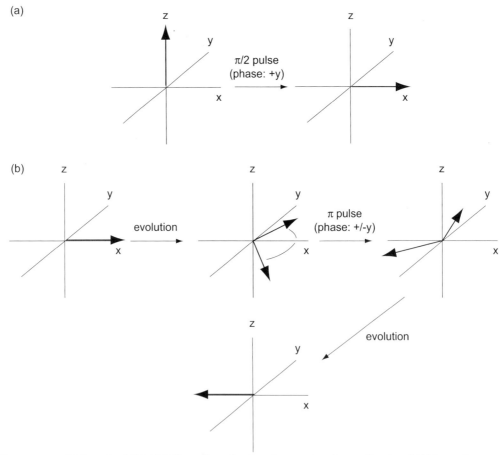

Figure 8 The vector model for pulsed RF. (a) Effects of rotation into the x–y plane by a $\pi/2$ pulse. (b) Effects of a two-pulse spin echo sequence, where magnetization placed into the x–y plane by a $\pi/2$ pulse evolves under the influence of an interaction (such as the chemical shift). Two vectors are shown accumulating opposite phases, as would be the case for two nuclei resonating above and below the transmitter frequency. After a fixed time t, these vectors are rotated around the $+/-y$ axis by a π pulse, and continue to evolve at their frequency during a second period t, after which an echo appears. The effects of relaxation are neglected here.

of duration t_p yields a power distribution given by:

$$P(\nu) = \left(\frac{\sin(2\pi\nu t_p)}{2\pi\nu t_p}\right)^2 \quad (20)$$

Equation (20) predicts that the excitation power falls off as the transmitter frequency moves away from the resonance for the rectangular $\pi/2$ pulse.

More sophisticated pulse sequences make use of multiple pulses, and periods of evolution between the pulses are also used to allow the effects of internal interactions (discussed later) to accumulate. One common element in many pulse sequences is the spin echo, shown in Fig. 8(b). An initial coherence is created in the transverse plane (along the $-y$-axis) by an initial $\pi/2$ pulse as in Fig. 8(a), and is then allowed to evolve for a set period of time. Two vectors are shown, representing two resonances under the influence of chemical shifts. The vectors process in difference directions, relative to the transmitter in the rotating frame. A π pulse is then applied to interchange the x components of the two vectors while leaving the y component alone.

The standard rectangular RF pulses discussed to this point are the simplest employed in NMR. More flexibility in manipulating spin coherence can be achieved by use of composite pulses, which are made up of a series of pulses of differing phase. These can have superior excitation, refocusing, and inversion properties (Ernst et al., 1987; Levitt, 1986). Pulses may also be shaped, to tailor their excitation or inversion properties (see Section IV.A). Finally, pulses may be employed to "decouple" (or remove) the effects of another interaction, or to transfer polarization through an interaction from one spin species to another (both homonuclear and heteronuclear). The coherences generated by pulses are often selected to allow for evolution through desired quantum-states. The concept of *coherence order* has been introduced to help visualize the spin dynamics. In the absence of a coupling interaction, rectangular RF pulses generate single-quantum coherence of order ± 1. Multiple quantum coherences (of order $\pm 2, 3, \ldots$) and zero quantum coherences can be generated by RF pulses when the effects of scalar, dipolar, or quadrupolar interactions are active. It is often desirable to select certain orders of coherence while avoiding others. Two methods of coherence pathway selection are used in modern NMR experiments. *Phase cycling*, which is also used to remove artifacts, is used to select desired coherence orders by constructive and destructive interferences between FIDs acquired during a cycle of pulsed RF phase shifts (Ernst et al., 1987). *Pulsed field gradients* are also used for the same task, but can accomplish the selection in a single transient (by gradient encoding and decoding) and offer better performance for many NMR pulse sequences of interest (Cavanagh et al., 1996; Hurd, 1990). Phase cycling and gradient-selection are discussed further in the literature and in the context of actual pulse sequences in Section IV (Cavanagh et al., 1996; Ernst et al., 1987; Hurd 1990).

C. Internal Spin Interactions

There are several interactions inherent to a molecule or chemical system that are detectable in NMR experiments. These (in contrast to interactions with externally applied magnetic fields) interactions are generally broken into the following categories: chemical shielding, magnetic dipolar coupling, J-coupling, and quadrupolar interactions.

1. Chemical Shift or Shielding

Magnetic chemical shielding of the nucleus by surrounding electrons depends primarily on the electronic environment around a nucleus (Jameson, 1987). The effect is caused by magnetic fields produced by the motion of the electrons in the presence of the magnetic field. These fields alter the fundamental precession frequency of the nucleus. Chemical shifts are usually reported in units of parts per million (ppm) relative to a internal or external reference standard, as given in Eq. (2). The chemical shielding is defined as:

$$B = B_0(1 - \sigma) \quad (21)$$

The chemical shielding tensor σ (a 3×3 real second rank tensor) contains the orientational dependence of the shielding with respect to the B_0 field (Grant, 1996). The tensor has six independent components, which can be observed in the NMR of single-solid crystals. In solid powders, unique spectra containing the three principle components (or eigenvalues) of the tensor can be observed, as shown in Fig. 9. These three components (σ_{11}, σ_{22}, and σ_{33}) are also defined in terms of an anisotropy $\Delta\sigma = \sigma_{33} - (\sigma_{22} + \sigma_{11})/2$ and an asymmetry $\eta = (\sigma_{22} - \sigma_{11})/(\sigma_{33} - \sigma_{iso})$. In solution, the rapid reorientational motion of molecules causes the tensor to average to its *isotropic* value:

$$\sigma_{iso} = \frac{1}{3}(\sigma_{11} + \sigma_{22} + \sigma_{33}) \quad (22)$$

Because of this, solution-state chemical shielding is reported as a single number, relative to an accepted standard reference compound. The effects of isotropic chemical shift on spin dynamics are shown in Fig. 9(b). The shift causes magnetization in the transverse plane to gain or lose phase unless the transmitter is exactly in resonance.

The chemical shielding tensor can be calculated from quantum mechanical models. Theoretical treatments of σ divide it into two components

$$\sigma = \sigma_{dia} + \sigma_{para} \quad (23)$$

where σ_{dia} and σ_{para} represent diamagnetic and paramagnetic shielding terms, respectively. The diamagnetic term

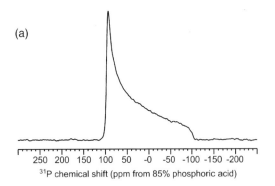

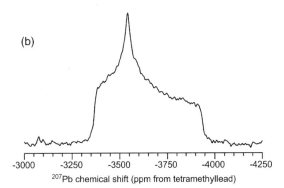

Figure 9 Chemical shift tensors in solid-state NMR. (a) ^{31}P chemical shift tensor of triphenylphosphine oxide is shown. Two of the principal components are equal ($\delta_{11} = \delta_{22} = +90$ ppm, $\delta_{33} = -100$ ppm), and the tensor is referred to as being axially symmetric. In (b) the asymmetric ^{207}Pb chemical shift tensor for lead sulfate is shown ($\delta_{11} = -3371$, $\delta_{22} = -3532$ ppm, $\delta_{33} = -3935$ ppm). These tensors collapse to a single isotropic peak in solution.

describes undistorted spherically symmetric motion of the electrons. The paramagnetic term is a correction for distorted electronic motions and nonspherical charge distribution, and is usually negative with respect to the diamagnetic term. Calculation of σ is possible using modern *ab initio* quantum mechanics, especially Hartree–Fock, density functional theory, and higher correlational methods (Webb, 1996; Helgaker et al., 1999).

2. Magnetic Dipolar Coupling

Magnetic dipolar coupling between nuclei occurs directly through the intervening space, and perturbs the spin-states of the engaged nuclei. The coupling depends on the orientation of the internuclear dipolar vector with respect to the B_0 field. The dipolar interaction energy between any two nuclear moments $\boldsymbol{\mu}_1$ and $\boldsymbol{\mu}_2$ separated by a vector $\mathbf{r}$ (where r is the magnitude of $\mathbf{r}$, or the internuclear distance) is:

$$E = \frac{(\boldsymbol{\mu}_1 \cdot \boldsymbol{\mu}_2)}{r^3} - \frac{3(\boldsymbol{\mu}_1 \cdot \mathbf{r})(\boldsymbol{\mu}_2 \cdot \mathbf{r})}{r^5} \quad (24)$$

From this interaction six types of terms arise (Slichter, 1996). At high fields, only two of these contribute, both of which are scaled by an angular term $(1 - 3\cos^2 \theta)$, where θ is the angle between the internuclear vector and the magnetic field. In the case of heteronuclei, where the gyromagnetic ratios differ, only a single term survives (Slichter, 1996). In any case, the interaction is also scaled by the dipolar coupling, given by ω_D:

$$\omega_D = \frac{\gamma_I \gamma_S \hbar \mu_0}{r_{IS}^3} \quad (25)$$

The strength of the dipolar coupling depends on the inverse cube of the internuclear distance.

Since most molecules in solution tumble isotropically, with no preferred direction, the angular term $(1 - 3\cos^2 \theta)$ causes dipolar coupling to average to zero. In solution-state, the effects of dipolar couplings can only be observed via relaxation phenomena [such as the nuclear Overhauser effect (NOE)]. Dipolar coupling can be observed in molecules dissolved in liquid crystals, and greatly affects NMR spectra in the solid-state. Solution order often makes use of field oriented liquid crystalline phases (Emsley and Lindon, 1975), but it is possible to observe dipolar effects at high field when molecules with anisotropic magnetic susceptibilities show a measurable deviation from isotropic tumbling in the B_0 field (Lohman and MacLean, 1981). Similarly, applied electric fields have been used to partially orient molecules in solution (Buckingham and Pople, 1963; Hilbers and MacLean, 1969). Recently, weakly oriented phases have become popular for measuring residual dipolar couplings (RDCs) (Simon and Sattler, 2002). In these phases, the orientation is accomplished by a variety of means, including lipid bicelles, rod-shaped viruses, and dilute liquid crystal media, and the dipolar coupling is scaled to a manageable level of 10–20 Hz while high resolution is still maintained. In the solid-state, dipolar couplings are generally overlapped with other interactions of similar magnitude, but can be averaged using sample spinning methods, and selectively reintroduced for observation (Gerstein and Dybowski, 1985; Slichter, 1996). The interest in direct measurement of dipolar couplings in solutions and solids stems from the availability of internuclear distances through the direct relationship in Eq. (25).

3. Scalar or J-Coupling

J-coupling, also known as the *indirect coupling*, is a nuclear magnetic dipolar coupling effect mediated by the surrounding electrons. It is best known for causing splitting of resonance lines in high-resolution solution-state NMR, especially of the ^{1}H nucleus (Coiro and Smith, 1996). The *J*-coupling possesses anisotropy, which can be quite large in solids and in liquid crystalline phases.

Only the isotropic portion of the *J*-coupling is generally of interest in solution; these are independent of field strength, and the values of *J*-coupling constants are usually reported in Hertz. A reduced coupling constant K is also used, especially when comparing couplings between isotopes (such as ^{1}H–^{13}C and ^{2}H–^{13}C couplings). K is defined as:

$$K_{IS} = \frac{4\pi^2}{h\gamma_I \gamma_S} J_{IS} \tag{26}$$

Solution-state NMR at high fields is dominated by chemical shifts and *J*-coupling interactions. The relative energies of these two major effects helps delineate how they interact. Two distinct situations are encountered. The first, known as the *weak coupling* limit, occurs when the chemical shift difference between two mutually coupled resonances is significantly greater than the size of the coupling interaction between them. This can be the case for heteronuclei (which obviously have large chemical shift differences) or for well-separated nuclei of the same isotope, such as two ^{1}H resonances separated by several ppm. If all of the coupling nuclei are spin-$\frac{1}{2}$, then the maximum number of peaks for a given resonance is $n + 1$, where n is the number of magnetically distinct coupled nuclei. The intensity ratio of the split lineshapes follows "Pascal's triangle", or a binomial distribution, leading to the well-known 1:2:1 triplet and 1:3:3:1 quartet for coupling to a spin-$\frac{1}{2}$ nucleus. Nuclei with spin $>1/2$ cause additional splitting of the resonance; for coupling of a spin-$\frac{1}{2}$ to a single spin-$\frac{3}{2}$, for example, results in a 1:1:1:1 pattern. When two resonances are separated by chemical shifts that are smaller or are equal to the size of the *J*-coupling, the weak coupling limit no longer applies, as the *J*-coupling cannot be treated as a perturbation of the chemical shift interaction. This case is known as the *strong coupling* limit, and second-order effects dominate (Corio and Smith, 1996). The simple splitting patterns cannot be rationalized and a more complete quantum mechanical calculation is needed to analyze the patterns. The effects of weak and strong *J*-coupling are illustrated in the ^{1}H and ^{19}F spectra of Fig. 10.

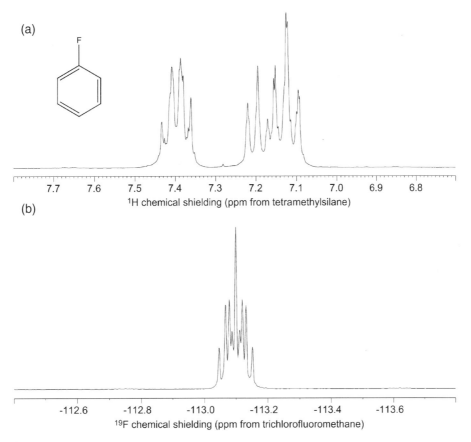

Figure 10 Homonuclear and heteronuclear *J*-coupling in the CDCl$_3$ solution-state spectra of monofluorobenzene obtained at ^{1}H frequency of 300 MHZ. (a) ^{1}H spectrum showing the effects of ^{19}F heteronuclear coupling in addition to the ^{1}H homonuclear coupling. (b) ^{19}F NMR spectrum showing a complex multiplet caused by heteronuclear coupling to ^{1}H nuclei.

The *J*-coupling is made up of contributions from several distinct terms:

$$J = J^{FC} + J^{DD} + J^{DSO} + J^{PSO} \quad (27)$$

where J^{FC} is the Fermi contact interaction between the electron spin and the nucleus, J^{DD} is the dipolar interaction between the magnetic moments of the nucleus and the electrons, and J^{PSO} and J^{DSO} are, respectively, the paramagnetic and diamagnetic spin orbit contributions, which arise because of interactions between the orbital motions of the electrons and the nuclear magnetic moment. The FC contribution can be calculated by the finite perturbation theory (FPT) method, using semiempirical approaches (Facelli, 1996). The latest density functional methods include all of the terms in Eq. (27), and have greatly improved the predictability of *J*-coupling constants (Helgaker et al., 2000).

4. Electric Quadrupolar Coupling

Electric quadrupolar coupling arises because of coupling between the quadrupole moment of a nucleus with the electric field gradient (EFG) at the nuclear site. The EFG is caused by the surrounding nuclei and electrons, and can vanish for sites of high symmetry. When an EFG is present, it can interact with a spin $>1/2$ nucleus and its nonspherical electrostatic charge distribution. Quadrupolar coupling causes additional splitting of lineshapes in oriented phases, but is not directly observed in isotropic solution-state spectra and is only manifested through relaxation phenomena. As an example, the static (nonspinning) ^{2}H solid-state NMR spectrum of d_{18}-hexamethylbenzene is shown in Fig. 11. This interaction is a first-order interaction that causes a splitting of the resonance into $2I$ lines, which for spin-1 is given by (Freude and Haase, 1993):

$$\Delta \nu_Q = (3e^2qQ/4)(3\cos^2\theta - 1) \quad (28)$$

where Q is the quadrupole moment, e is the elementary charge, and $eq = V_{zz}$ (the largest component of the traceless EFG tensor). The strength of the quadrupolar interaction can be quite large, on the order of several MHz, and can rival the Zeeman interaction in size. Because of this, second-order effects also appear in solid-state and oriented-phase spectra (Freude and Haase, 1993).

D. Relaxation Phenomena and Chemical Dynamics

1. Relaxation Phenomena

The effects of nuclear spin relaxation affect all NMR spectra. Two primary types of macroscopic relaxation are distinguished in NMR: *longitudinal relaxation* (characterized by a relaxation time T_1) and *transverse relaxation* (characterized by T_2). Longitudinal relaxation, also known as *spin–lattice relaxation*, is the process by which longitudinal magnetization reaches its equilibrium value. Longitudinal magnetization is static and aligned with B_0, and is created by the difference in populations between the nuclear spin-states [as in described in Eq. (17)]. For example, the first time a sample is placed into a magnet, the duration of time it takes to reach the maximum population difference can be determined from T_1. In addition, T_1 determines the frequency at which NMR transients can be collected in FT mode, as the system needs to relax to a certain level to ensure detectable populations. The magnetization present after a time t is given by:

$$(M_z(t) - M_z^0) = (M_z^{\text{initial}} - M_z^0)\exp\left(\frac{-t}{T_1}\right) \quad (29)$$

Here, M_z^0 is the equilibrium value of the magnetization, given by Eq. (18b), and M_z^{initial} is the initial magnetization present at time zero.

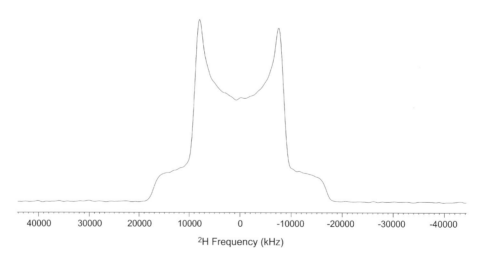

Figure 11 ^{2}H quadrupolar coupling in solid d_{18}-hexamethylbenzene, observed at 55.2 MHz using the quadrupolar echo pulse sequence. The methyl groups rotate rapidly at room temperature (the energy barrier to rotation is 1–2 kcal/mol), causing averaging of the quadrupolar interaction. The observed splitting between the sharp singularities is ~15.6 kHz.

Transverse magnetization, also known as *spin–spin relaxation*, is created by the RF fields discussed in Section III.B. It is perpendicular to the B_0 field and has an equilibrium value of zero, as it represents spin coherence precessing at the Larmor frequency. It is also independent of longitudinal relaxation. RF pulses create a spin coherence of individual nuclear magnetic moments, so that they are all precessing in concert. These moments eventually lose phase coherence and fan out into a random distribution. Note that if M_z has relaxed to the equilibrium value M_z^0, there can no longer be a transverse magnetization, so that $T_2 \leq T_1$. T_2 is given by:

$$T_2 = \frac{1}{\pi(\Delta\nu_{1/2})} \tag{30}$$

where $\Delta\nu_{1/2}$ is the observed full linewidth of a Lorentzian peak at half-maximum amplitude, which should be corrected for nonrelaxation effects (i.e., from poor field homogeneity).

Addition of these two relaxation time effects into Eq. (19) allows for a new equation that describes the motion of the magnetization under the influence of the static B_0 field, the transverse B_1 field, and the two relaxation times:

$$\frac{d\mathbf{M}}{dt} = \mathbf{M} \times \gamma\mathbf{B} - \frac{M_x\mathbf{i} + M_y\mathbf{j}}{T_2} - \frac{(M_z - M_z^0)}{T_1} \tag{31}$$

Equation (31) represents a set of equations, for each component of the magnetization, that are collectively known as the Bloch equations.

2. Spectral Density and Motional Regimes

The correlation time τ_c varies with the size of a molecule, the viscosity of the solvent, and the temperature. Generally, it is on the order of picoseconds for small molecules and nanoseconds for larger molecules. The relaxation analysis given here is only valid for spherical rotational motion, also known as isotropic tumbling. When the frequency of this motion matches a transition frequency, exchange of energy occurs and relaxation ensues. Information about molecular motion frequency is contained in spectral density functions:

$$J(0) = 2\tau_c \tag{32a}$$

$$J(\omega_I) = \frac{2\tau_c}{1 + (\omega_I\tau_c)^2} \tag{32b}$$

$$J(2\omega_I) = \frac{2\tau_c}{1 + (2\omega_I\tau_c)^2} \tag{32c}$$

These expressions can be thought of as the probabilities that the molecular reorientation velocity matches with zero, single-, and double-quantum transitions. (The zero quantum term only affects T_2 and arises because of nonresonant dephasing effects on the magnetization). These terms will be employed in the relaxation expressions given subsequently. In the analysis of relaxation, simplifications can be obtained in two commonly encountered conditions known as the extreme narrowing limit ($\omega_0\tau_c \ll 1$) and the spin diffusion limit ($\omega_0\tau_c \gg 1$). The extreme narrowing limit is often applicable in studies of small molecules in nonviscous solvents; at this limit the values of T_1 and T_2 are usually equal.

Both T_1 and T_2 can be made up of several components with different origins, although a single component dominates in many cases of interest. Relaxation of a spin-½ nucleus is usually governed by dipolar effects if other spin-½ nuclei are situated within a few angstroms. This can lead to complex models if many nuclei are dipolar-coupled to the nucleus of interest. The experimentally measured rate can be the sum of all the possible relaxation mechanisms:

$$\frac{1}{T_1^{\exp}} = \frac{1}{T_1^{CSA}} + \frac{1}{T_1^{Q}} + \frac{1}{T_1^{SR}} + \frac{1}{T_1^{P}} + \sum_j \frac{1}{T_1^{DD}} + \sum_k \frac{1}{T_1^{J}} \tag{33}$$

where the following abbreviations are used for the individual mechanisms: CSA is chemical shift anisotropy, Q is quadrupolar, DD is dipole–dipole, P is paramagnetic, SR is spin–rotation, and J is J-coupling. A similar equation can be written for T_2. The individual mechanisms for this relaxation are summarized here, with more detail available in the literature (Abragam, 1961, Slichter, 1996; Sudmeier et al., 1990).

CSA relaxation is an especially important mechanism for the relaxation of nuclei with significant anisotropic chemical shielding. The effect occurs because of interaction between the nuclear magnetic dipole and the surrounding electrons in a magnetic field, and grows stronger as the static field increases, and as the nuclear shielding increases (i.e., with increasing atomic number). The relaxation rates are for a spin I are given by:

$$\frac{1}{T_1^{CSA}} = \frac{[\Delta\sigma\gamma B_0]^2}{40}\left(1 + \frac{\eta^2}{3}\right)6J(\omega_1) \tag{33a}$$

$$\frac{1}{T_2^{CSA}} = \frac{[\Delta\sigma\gamma B_0]^2}{40}\left(1 + \frac{\eta^2}{3}\right)[3J(\omega_1) + 4J(0)] \tag{33b}$$

where $\Delta\sigma$ and η are defined in Section III.C. Simplifications occur in the extreme narrowing limit, so that $T_2^{CSA} = (6/7)T_1^{CSA}$ (Abragam, 1961; Sudmeier et al., 1990). This relaxation mechanism has become more important with the increasing fields available in NMR laboratories, as Eq. (33) depends on the square of B_0. In solution-state NMR, relaxation by chemical shielding anisotropy is typically important for spin-1/2 nuclei without nearby dipolar coupling partners (i.e., surrounded by spin-0 nuclei). CSA relaxation is also important for heavier nuclei, and for lighter nuclei in cases of large anisotropy, such as for ^{13}C nuclei in carbonyl and acetylene groups.

The contributions of dipolar and CSA relaxation effects can be separated by determining relaxation times at different field strengths.

The *dipolar relaxation* for two isolated homonuclear dipolar coupled nuclei of the same isotope is given by:

$$\frac{1}{T_1^D} = \frac{2\gamma^4 h^2}{40\pi^2 r^6} I(I+1)[J(\omega_I) + 4J(2\omega_I)] \quad (34a)$$

$$\frac{1}{T_2^D} = \frac{2\gamma^4 h^2}{40\pi^2 r^6} I(I+1)[3J(0) + 5J(\omega_I) + 2J(2\omega_I)] \quad (34b)$$

The situation for unlike spins I and S (analogous to the weak coupling limit for coherent interactions) is given by:

$$\frac{1}{T_1^D} = \frac{\gamma_I^2 \gamma_S^2 h^2}{30\pi^2 r^6} S(S+1)[J(\omega_I - \omega_S) + 3J(\omega_I)$$
$$+ 6J(\omega_I + \omega_S)] \quad (35a)$$

$$\frac{1}{T_2^D} = \frac{\gamma_I^2 \gamma_S^2 h^2}{120\pi^2 r^6} S(S+1)[4J(0) + J(\omega_I - \omega_S)$$
$$+ 3J(\omega_I) + 6J(\omega_S) + 6J(\omega_I + \omega_S)] \quad (35b)$$

where the heteronuclear spectral densities are given by:

$$J(\omega_I - \omega_S) = \frac{2\tau_c}{1 + [(\omega_I - \omega_S)\tau_c]^2} \quad (36a)$$

$$J(\omega_I + \omega_S) = \frac{2\tau_c}{1 + [(\omega_I + \omega_S)\tau_c]^2} \quad (36b)$$

The relaxation of the I spin will be exponential with these decay times if the S-spin population is near equilibrium; if not, more complex coupled relaxation theory is needed. When multiple nuclei are coupled, these equations can be used to yield sums over the pairs of interactions for analysis.

Electric *quadrupolar relaxation* typically dominates for nuclei with $I > 1/2$, unless the quadrupolar coupling tensor around the nucleus is small or spherically symmetric. As a quadrupolar nucleus tumbles through the field, this interaction generates torque on the nuclear magnetic dipole, and can change its spin-states when the velocity of this motion matches the frequencies of allowed transitions. This relaxation mechanism is similar to the CSA mechanism in its effect (Abragam, 1961; Sudmeier et al., 1990).

Scalar relaxation, caused by the J-coupling interaction, differs from dipolar and chemical shift relaxation in that the small J-coupling anisotropy does not allow for motionally induced relaxation. However, other processes can still produce relaxation. *Scalar relaxation of the first kind* occurs when one of the J-coupling nuclei participates in a chemical exchange in which the lifetimes of the states are less than the inverse of the coupling constants (Abragam, 1961; Sudmeier et al., 1990). *Scalar relaxation of the second kind*, the more commonly encountered form, occurs when one of the coupled nuclei has a T_1 relaxation time shorter than the inverse of the J-coupling constant (Abragam, 1961; Sudmeier et al., 1990). In solvents with no exchangeable protons (such as d_6-DMSO), this mechanism is responsible for the broadening of signals from protons directly attached to ^{14}N nuclei, which make up >99% of the natural abundance of this element. The rapid quadrupolar relaxation of this nucleus induces relaxation of the J-coupled ^1H partner.

Paramagnetic relaxation occurs through coupling between nuclear magnetic moments and the magnetic moments of unpaired electrons. The effect can be mediated through direct dipolar coupling (pseudocontact) or through scalar coupling (FC). More details are available in the literature (Abragam, 1961; La Mar et al., 1973); from the cursory point of view taken here, paramagnetic relaxation in NMR has its widest application in the use of relaxation agents. Paramagnetic relaxation agents are used to reduce the T_1 relaxation time of nuclei, and can improve sensitivity by allowing for faster repetition rates in pulsed NMR. For example, chromium acetylacetonate is widely used to improve the sensitivity and quantitative accuracy of ^{13}C and ^{29}Si NMR in organic solutions. For aqueous solutions, copper sulfate is a popular choice.

The *NOE* is a consequence of cross-relaxation between dipolar-coupled spins (Derome, 1987; Neuhaus and Williamson, 2000). The steady-state NOE enhancement obtained on saturation of S is given by:

$$\eta_I(S) = \frac{\sigma_{IS}^{NOE} \gamma_S}{\rho_I \gamma_I} \quad (37)$$

where σ_{IS}^{NOE} defines the cross-relaxation rate between spins I and S, and ρ_I is the spin–lattice relaxation rate for spin I. An analysis of the relaxation rates for a dipole-coupled two-spin systems results in a relatively simple result; in the homonuclear case ($\gamma_I = \gamma_S$), the maximum NOE is simply $\eta_{IS} = 0.5$, or a 50% enhancement, in the extreme narrowing limit. In the spin diffusion limit, the maximum NOE enhancement is -100% and effects can disappear altogether. In heteronuclear cases, the maximum NOE enhancement is given by one-half of the gyromagnetic ratios of the involved nuclei. For ^1H and ^{13}C, the maximum NOE is about 200%, whereas for nuclei with negative gyromagnetic ratios (^{15}N and ^{29}Si) the enhancement is negative, so that peaks may invert or disappear in the spectrum.

The steady-state NOE is extremely useful as a qualitative tool for detecting whether two spins can participate in mutual dipolar relaxation (Derome, 1987; Neuhaus and Williamson, 2000). Quantitative interpretations of the steady-state NOE are difficult, as the relaxation contributions of other nearby spins may be important, and J-coupling can also confuse the issue. While the steady-state NOE is best interpreted in a qualitative sense, the kinetics of the NOE (or the build-up) can be used to quantitatively extract out distance information. The most common way of measuring kinetics is via *transient NOE*. Although both the steady-state and transient NOE depend on a number of competing relaxation pathways, the initial rate of dipolar cross-relaxation depends on the inverse sixth power of the distance. The growth of the NOE could be measured in a steady-state experiment, but the fact that saturation is not instantaneous makes it simpler to use the transient NOE. Actual pulse sequences used to study the NOE are discussed subsequently, but in general the steady-state NOE is analyzed in the NOE difference experiment, and the transient NOE is studied using the 2D NOESY (and related 1D experiments) (Neuhaus and Williamson, 2000).

3. Chemical Dynamics

Hindered bond rotation, ring inversions, tautomerism, proton exchange, and many other dynamic effects are commonly studied by NMR (Oki, 1985; Kaplan and Fraenkel, 1980). Not only can the processes be observed (in a variety of solvent media), but thermodynamic parameters can be extracted from NMR spectra run as a function of temperature. The energy barrier of processes observed in typical NMR experiments is usually in the 20–100 kJ/mol range, as the ability of NMR to observe such processes is affected by the so-called *NMR timescale*. Dynamics can be studied in detail by NMR if the frequency of the process is of the same order as the frequency differences seen the spectrum. For example, a two-site exchange between sites A and B can be observed (through separate signals) only if k is much less than the difference in resonance frequencies between the sites. Modified versions of the Bloch equations, known as the McConnell equations, are often used to describe these phenomena (Cavanagh et al., 1996; Ernst et al., 1987). VT NMR is commonly used to detect the presence of rotational isomers (or rotamers) in molecules with hindered rotation or partial double bond character. Finally, chemically induced dynamical nuclear polarization (CIDNP) is observed in reactions in which radicals combine, so that the unpaired electrons can exchange angular momentum with coupled nuclei (Closs, 1974). The NMR spectrum of the products can show extremely large changes in the amplitude of resonances because of this effect.

IV. EXPERIMENTAL METHODS

This section contains a brief survey of the currently available experimental methods. There is an enormous flexibility in the design of experiments for NMR spectroscopy, which can include a variety of manipulations using RF and gradient pulses, timed delays, and mechanical sample motion, all using multiple acquisition dimensions. Only the basic and most widely used experimental methods are covered, and the interested reader is referred to reviews of NMR experimental design for more information (Braun et al., 1998; Ernst et al., 1987; Freeman, 1998; Levitt, 2001; Stejskal and Memory, 1994).

A. Basic Pulse Methods

The simplest NMR experiment is a single $\pi/2$ pulse on or near the resonance of interest, followed by an acquisition period in which the resulting FID is detected. This sequence has already been discussed with reference to the vector model in the rotating frame. The single-pulse sequence is shown in Fig. 12(a), and is still the most widely used experiment, because it is all that is needed to obtain a solution-state ^{1}H NMR spectrum. The basic two-pulse spin echo, also discussed in Section III.B, is depicted in Fig. 12(b) (Kaplan and Hahn, 1957). A delay follows the initial excitation pulse, during which magnetization dephases under the influence of an interaction (such as the chemical shift). The second pulse then refocuses the magnetization. The spin echo often serves as a basic component of more complex pulse sequences, but is useful by itself for time-shifting the FID signal, perhaps to avoid initial time-series distortions from pulse breakthrough, or to allow time for the electronics to acquire a fast-decaying (broad) signal. A different form of echo, commonly referred to as the quadrupolar echo, is used for spin-1 nuclei and related coupled-spin systems and is widely used in ^{2}H NMR (Slichter, 1996). It consists of two $\pi/2$ pulses and is run in a manner similar to the spin echo.

Simple phase cycles are also employed with these basic experiments. In the single-pulse experiment, the CYCLOPS phase cycle is used to suppress artifacts (Braun et al., 1998; Ernst et al., 1987). This involves shifting the transmitter phase in 90° increments while simultaneously shifting the receiver phase. Phase cycles are also employed in the basic spin echo experiment so that only signal refocused by the echo is acquired; signal which has been excited by the initial pulse but not by the second pulse (e.g. because of resonance offset effects) is canceled out over multiple scans (Braun et al., 1998; Ernst et al., 1987).

The basic pulse sequences shown here are also useful for setting up and adjusting experimental parameters. These include setting the pulse widths and amplitudes, which for abundant nuclei can be done with the pulse sequence of Fig. 12(a). The pulse width or power is varied, while the response of the signal is monitored (Braun et al., 1998). The relaxation (or recovery) delay between pulse sequences must be chosen appropriately for quantitative work, or can be made rapid if this is not

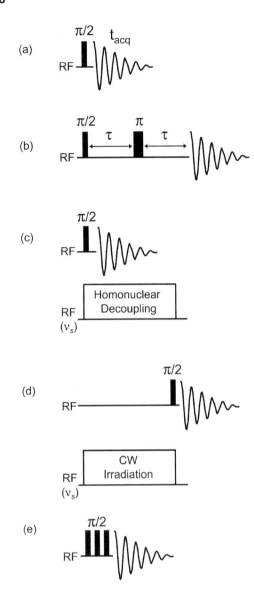

Figure 12 Basic single-resonance 1D pulse sequences. (a) The basic single-pulse sequence, used for ^{1}H solution-state spectroscopy. A selective RF pulse can be substituted for the hard RF pulse shown. (b) The two-pulse spin echo, used to acquire broad spectra and as a basic element of more sophisticated pulse sequences. (c) Single-pulse with homonuclear decoupling, primarily used to study J-coupling connectivity in ^{1}H NMR. (d) NOE difference experiment, primarily used to study dipolar connectivity via NOE effects in ^{1}H NMR. (e) ARING ring-down elimination experiment, used for obtaining 1D spectra of low-frequency nuclei.

a concern. Other considerations include setting the receiver (or audio) gain, usually by trial and error. The number of points and the acquisition time must be set to obtain the appropriate spectral width and resolution, while avoiding truncation of the time-domain signal. One final consideration is the desired S/N, which can be improved by acquiring more transients. However, as in other areas of Fourier spectroscopy, the S/N only increases as the square root of the number of scans, so a point of diminishing returns can be rapidly reached.

1. Shaped, Composite, and Adiabatic Pulses

The RF pulses discussed to this point have been rectangular in shape, in that the transmitter is gated on and off so that the pulse reaches full RF intensity quickly and shuts off immediately at the end of the desired pulse width. Shorter rectangular pulses on the order of 2–30 μs are the most commonly used in NMR, and are generally nonselective. However, as excitation efficiency falls off as a function of resonance offset [see Eq. (20)], signals far from the transmitter frequency will not be efficiently perturbed when a longer pulse is applied to a sample. Therefore, longer square pulses can be used as *selective pulses*, for applications in which a single-resonance (or a group of resonances) with a given bandwidth are perturbed, whereas the rest of the spectrum is left alone (Braun et al., 1998; Kessler et al., 1991; McDonald and Warren, 1991). In addition, the pulses may also be shaped in amplitude patterns other than the rectangle, to further tailor their excitation, refocusing, and inversion bandwidths and even their phase properties. Soft shaped pulses can be used very effectively to excite a single-resonance in a spectrum, as illustrated in Fig. 13. In Fig. 13(a) the assigned ^{1}H spectrum of a pyridyl sulfonamide is shown. A 20 ms Gaussian-shaped $\pi/2$ pulse was applied to the 2-position resonance to yield the spectrum in Fig. 13(b). Only this resonance is significantly excited by the RF irradiation; the rest of the spectrum is virtually unperturbed. (Note that there is some minor phase dispersion; this is a property of the type of pulse used.) Shaped selective pulses are used in a variety of pulse sequences for various purposes (Braun et al., 1998; Kessler et al., 1991; McDonald and Warren, 1991). Phase modulation and frequency modulation are also possible. *Composite pulses*, already mentioned in Section III.B, are made up of a series of pulses of differing phase. The excitation, refocusing, and inversion capabilities of these pulses have made them very popular for both wideband and band-selective applications (Levitt, 1986). Composite pulses consist of discrete and "instantaneous" phase shifts (within the limits of the electronics). Continuous phase shifts and frequency shifts, which are closely related, are the basis of *adiabatic pulses* (Abragam, 1961; Tannus and Garwood, 1997). These pulses rotate the magnetization by sweeping through the frequency during the course of a generally longer pulse. One of their main advantages is insensitivity to RF (B_1) inhomogeneity, a problem which affects all NMR probes, providing that the "adiabatic condition" is satisfied (i.e., that the

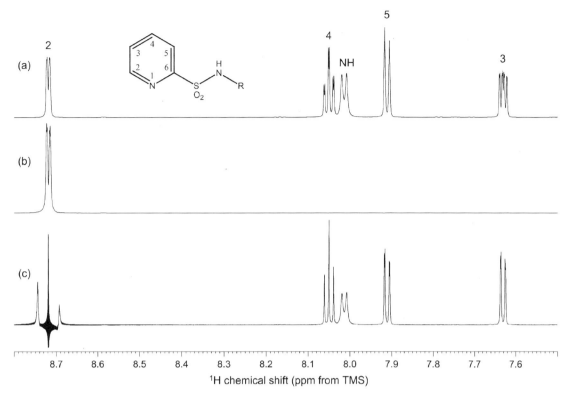

Figure 13 Applications of basic ^{1}H NMR pulse sequences. The aromatic region of the 700 MHz ^{1}H spectrum of a substituted pyridyl sulfonamide in d_6-DMSO solution is shown in (a), obtained using a single hard RF pulse. A 20 ms Gaussian-shaped pulse was applied at the frequency of the 2-position signal in spectrum (b), resulting in selective excitation of this resonance. In spectrum (c), homonuclear CW decoupling was applied at the same frequency during acquisition to collapse the $^3J_{H2,H3}$ and $^4J_{H2,H4}$ couplings (4.8 and 1.7 Hz, respectively).

effective magnetic field changes slowly so that the magnetization can follow it). They also can be made extremely broadband without the excessive power input characteristic of a rectangular pulse, and have found application at higher fields for replacement of rectangular pulses and in decoupling. Instead of a calibrated pulse width, $\pi/2$ and π rotations are achieved by adiabatic half-passage and full-passage pulses, respectively, and arbitrary rotations are possible using multistep phase-shifted adiabatic pulses (Tannus and Garwood, 1997).

2. Homonuclear Decoupling

One of the simplest and still very useful applications of the selective pulse has been in homonuclear decoupling (Jesson et al., 1973). A weak RF field of constant phase and amplitude (known as a CW pulse) is applied on resonance to peak of interest after the usual $\pi/2$ pulse, as shown in Fig. 12(d). The field must be gated to protect the preamplifier when the receiver needs to require a data point. This is not a problem for ^{1}H NMR, where homonuclear decoupling is most often employed, as the dwell times between points is on the order of 10–50 μs. (The field only needs to be gated off for 1–2 μs to acquire the point.) An example is shown in Fig. 13(c), with the 2-position resonance decoupled by a weak RF field (∼500 Hz) applied continuously during the 2 s acquisition period. If only a few couplings are of interest to the spectroscopist, then homonuclear decoupling remains one of the fastest ways to obtain the relevant information, more quickly than the 2D techniques discussed subsequently. More recently, band selective homonuclear decoupling has become available for use in simplifying ^{1}H NMR spectra, especially of biological macromolecules.

3. NOE Difference Experiments

The steady-state NOE difference experiment is another widely used 1D single-resonance pulse sequence (Braun et al., 1998; Derome, 1987; Neuhaus and Williamson, 2000). It is shown in Fig. 12(d), and consists of soft, selective homonuclear CW irradiation prior to the $\pi/2$ RF pulse. (The sequence is similar to homodecoupling, but there is no need to gate the transmitter off during small detection windows.) This pulse saturates a dipolar transition as noted in the theory discussed in

Section III.D, resulting in a strong difference signal for any resonance in close spatial proximity (2–4 Å) of the irradiated site. An example from small molecule ^{1}H NMR is shown in Fig. 14. Although quantitative use of the steady-state NOE is difficult, qualitative interpretations have been used effectively to answer stereochemical and regiochemical questions for many years. The major drawback of the technique is its reliance on a difference spectrum, which requires the subtraction of two large signals from each other. This magnifies slight differences in phase and random variations between experiments, and can cause significant artifacts that can obscure valuable information. To circumvent these problems, new 1D experiments that utilize the transient NOE instead of the steady-state NOE have become available. Presently, the most popular of these are the GOESY and DPFGSE-NOE experiments (Neuhaus and Williamson, 2000).

They offer substantial advantages despite their reliance on the weaker transient effect.

4. Solvent Suppression

The NOE difference experiment shown in Fig. 12(d) can also be used to suppress strong solvent signals if a protic solvent is present in a sample of interest. When applied in this manner, the long CW pulse in the sequence is often referred to as a *presaturation* period. The weak RF field is applied for sufficient time to saturate the solvent resonance, so that the population difference is equalized and the signal is eliminated (until it has recovered via longitudinal relaxation). Several methods are available for solvent suppression in NMR (Braun et al., 1998; Gueron et al., 1991). Although simple presaturation is still used in many laboratories, multiple-pulse solvent

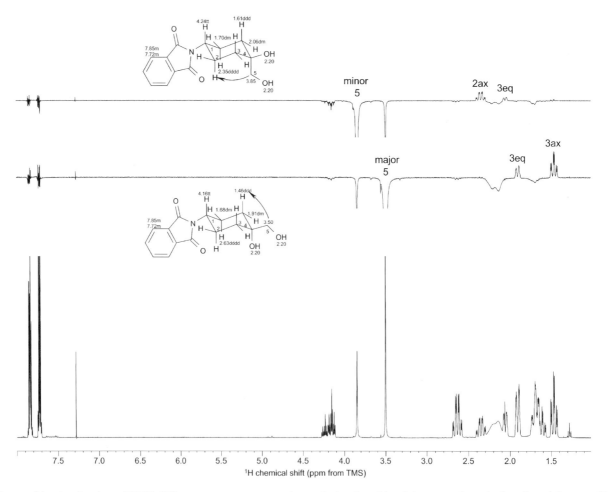

Figure 14 Application of NOE difference spectroscopy to the analysis of positional isomers of a substituted cyclohexane. Upon irradiation of the 5-position in both isomers (referred to as major and minor, based on their relative concentration), NOE effects are observed as positive signals to the 2- and 3-positions. In the minor isomer, the 5-position is not sufficiently close to the axial 3-position for an NOE to occur. Instead the axial 2-position is affected. This is contrasted by the major isomer, in which the 5-position shows NOE contacts with both 3-position protons. A full ^{1}H and ^{13}C assignment of these compounds was performed prior to interpretation of the NOE difference data.

suppression offers tangible advantages and has received a great deal of attention. Some of the more popular multiple-pulse experiments in current use include the WET sequence and other sequences based on gradient-encoded spin echoes and selective pulses (Braun et al., 1998). Solvent suppression is most useful for allowing the dynamic range of the ADC to be applied to signals of interest, while also having the beneficial effect of suppression of distortions caused by radiation damping.

5. Basic Pulse Sequences for Other Nuclei

Single-resonance 1D experiments are not limited to ^{1}H NMR. A variety of other nuclei including ^{19}F and ^{31}P can be studied using these simple pulse sequences. However, coupling to protons often necessitates the use the double-resonance experiments discussed in Section IV.C, to either decouple the heteronuclei or to use them as an additional source of information. However, there are many cases in which nuclei with unresolvable couplings, such as quadrupolar nuclei in solution, are often studied using basic single-resonance methods. For example, the ^{17}O spectrum of a 10% solution of isopropyl acetate in CDCl$_3$, as shown in Fig. 15(a), is readily obtained using a single pulse. At lower frequencies, and with rapidly decaying signals, the problem of *acoustic ringing* can become a major issue (Braun et al., 1998). Spurious signals are observed in the NMR if acoustic ringing is present. The effect is caused by the electromagnetic generation of ultrasonic waves in metallic probe materials. The strength of the ringing can be estimated from (Fukushima and Roeder, 1981).

$$\text{Acoustic ringing} \propto \frac{B_1(B_0^2)}{mv_s[1 + 2.5 \times 10^{13}(\rho^4/v_s^4)v_0^2]} \tag{38}$$

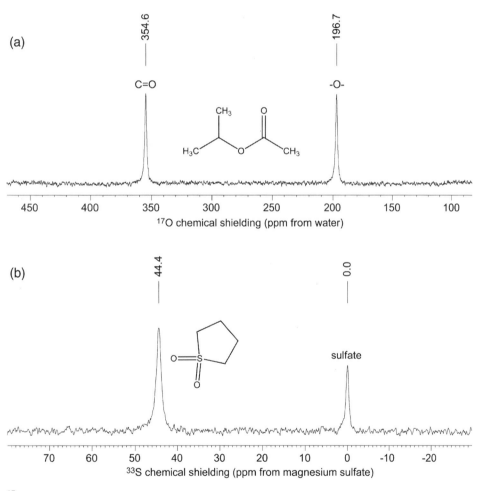

Figure 15 (a) ^{17}O NMR spectrum of isopropyl acetate, measured at 94.9 MHz with a single-pulse sequence. Linear prediction (discussed subsequently) was used to correct the first few points of the time-domain signal, which were corrupted by acoustic ringing. (b) ^{33}S spectrum of a 3:1 mixture of tetramethylene sulfone and magnesium sulfate (85.0 mg MgSO$_4$, 252 mg tetramethylsulfone). The spectrum was obtained at 53.7 MHz using the three pulse ARING sequence shown in Fig. 11.12(e) to suppress acoustic ringing. The sulfone signal is arbitrarily referenced to the magnesium sulfate, which is generally within 2 ppm of the recommended ^{33}S reference (cesium sulfate), because of concentration and counterion effects. No linear prediction was used during data processing.

where m and v_s are the mass density and the acoustic shear velocity of the material, respectively, and ρ is its resistivity. Therefore, in the solution-state analysis of quadrupolar nuclei, it is useful to have a pulse sequence that eliminates the effects of acoustic ringing. Several experiments are available for this task (Braun et al., 1998). The ARING pulse sequence, shown in Fig. 12(e), is an easily implemented example. Its use in ^{33}S NMR is demonstrated on a 3:1 molar mixture of tetramethylene sulfone and magnesium sulfate, as shown in Fig. 15(b).

B. Multidimensional NMR Spectroscopy

Multidimensional NMR spectroscopy makes use of two or more time dimensions (Ernst et al., 1987; Kessler et al., 1988). Whereas the first dimension is *directly* detected by the NMR receiver, the second and higher dimensions are *indirectly* detected by inserting variable delay(s) into the pulse sequence, automatically incrementing these delays, and storing the data for a given delay separately. Here we concentrate primarily on 2D experiments, which are the most useful for the greatest variety of problems, although several 3D experiments are considered. Generalized reference is often made to experiments with more than one dimension; these are referred to as "nD" experiments for brevity.

The actual acquisition of 2D spectra resembles the acquisition of a series of 1D spectra. The first 2D experiment to be proposed was referred to as "correlation spectroscopy" or COSY (Ernst et al., 1987; Kessler et al., 1988). In this experiment, the J-coupling is used to mix the spin states of protons with different chemical shifts. A variable t_1 delay in this experiment is stepped through a range of equally spaced values, serving as a second time dimension. During this time evolution period, the spins are frequency-labeled by their chemical shift, and develop multiple quantum coherences from their mutual J-couplings. Upon application of the second pulse, mixing of the spin-states occurs. The result is a connectivity map that identifies ^{1}H nuclei engaged in significant $^{2-5}J_{\mathrm{HH}}$ couplings (usually from 2 to 15 Hz, depending on the setup of the experiment) by a cross-peak appearing at the frequency of the first proton in one dimension, and at the frequency of the second proton in the second dimension. The two dimensions are generally labeled F_1 and F_2 to correspond, respectively, to the two time periods t_1 and t_2. By convention, t_1 and F_1 are always used to label the indirectly detected dimension. Modern incarnations of the COSY experiment are heavily used in analytical chemistry laboratories; two popular versions, which make use of pulsed field gradient-selection, are shown in Fig. 16 (Ancian et al., 1997; Shaw et al., 1996). A demonstration of a gradient-selected double-quantum filtered (DQF) COSY experiment applied to sucrose in deuterated water is given in Fig. 17.

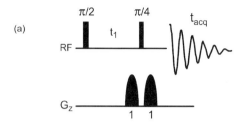

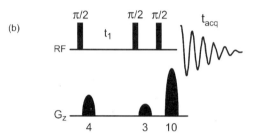

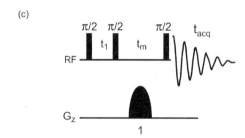

Figure 16 Commonly used homonuclear 2D pulse sequences. (a) The magnitude-mode COSY-45 experiment with pulse field gradient-selection. (b) The magnitude-mode DQF-COSY experiment, again with gradient-selection. This experiment filters out the signals from uncoupled protons along the diagonal. (c) A phase sensitive gradient-enhanced NOESY experiment, for detecting the transient NOE in solution.

All 1D NMR spectra are normally taken with quadrature phase detection, as discussed in Sections II.C and III.B. This allows for sign discrimination when the transmitter is placed in the center of the spectrum. As 2D spectroscopy is indirectly detected, the effects of a quadrature detector must be simulated in many cases to allow for proper sign discrimination. This may be achieved directly in the pulse sequence; for example, the COSY-class of experiments achieves phase discrimination in its second dimension as a by-product of coherence pathway selection. The 2D spectrum can then be processed in the magnitude mode to produce positive signs for all peaks (see Section V.A). In addition to phase discrimination, phase sensitivity in both dimensions is often desired for enhanced resolution. This requires separation of true real and imaginary components for both dimensions, and is

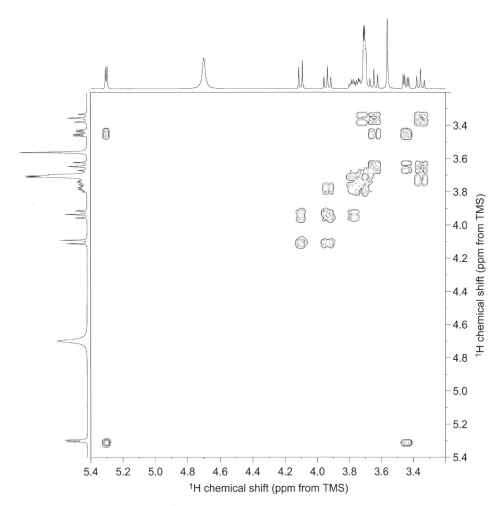

Figure 17 Gradient-selected magnitude-mode ^{1}H DQF-COSY spectrum of sucrose in D$_2$O, obtained at 700 MHz. All contours are positive. The ^{1}H spectrum is shown along the two axes for reference. Correlations are observed off the diagonal between ^{1}H spins with mutual coupling constants in the approximate range of 3–15 Hz. Several of the more obvious assignments are noted on the spectrum in reference to the numbered structure. These results can be used to study ^{1}H spin networks and hence elucidate certain aspects of molecular structure.

presently achieved by either the time-proportional phase incrementation (TPPI) method, the States (hypercomplex) method, or by the echo–antiecho method (in gradient-enhanced spectroscopy) (Braun et al., 1998; Cavanagh et al., 1996; Ernst et al., 1987; Neuhaus and Williamson, 2000). These methods can in principle produce pure-absorption lineshapes requiring complex Fourier transformation in both dimensions and phase correction (see Section V.A.). Both magnitude-mode and phase-sensitive 2D experiments are discussed here.

Another class of experiment makes use of RF mixing schemes to achieve isotropic (strong) coupling amongst all of the spins in a network (by effectively removing the influence of chemical shifts). This is achieved by spin locking the magnetization, or more commonly by broadband sequences which eliminate the chemical shift while retaining strong *J*-coupling terms. Total correlation spectroscopy (TOCSY) and homonuclear Hartman-Hahn (HOHAHA) experiments are examples of isotropic mixing, in which ^{1}H chemical shifts are correlated (Braun et al., 1998; Cavanagh et al., 1996; Ernst et al., 1987). A number of mixing schemes have been derived using analyses similar to those used for optimizing decoupling schemes (Shaka and Keeler, 1987). Some of the more capable variants include the MLEV, WALTZ, and DIPSI pulse trains. In addition, highly effective phase-sensitive gradient-enhanced versions of the TOCSY experiment for ^{1}H spins are available (Kover et al., 1998). They are useful for identifying spin networks in small molecules and have been extensively used in the study of macromolecules.

The COSY and TOCSY families of experiments provide correlation through *J*-coupling, which normally means through covalent bonds. A third class of experiment achieves chemical shift correlation via through-space dipolar cross-relaxation. This NOE-based experiment is known as NOESY (Cavanagh et al., 1996; Ernst et al., 1987). The NOESY sequence produces negative peaks for small molecules in the extreme narrowing limit, and it is useful to run the experiment in the phase-sensitive mode to confirm this sign. A modern gradient-enhanced variant of NOESY sequence is given in Fig. 16(c) and demonstrated in the application shown in Fig. 18 (Wagner and Berger, 1996). Many molecules in the intermediate size range of 500–2500 Da have rotational correlation times on the order of several tens of nanoseconds, which lead to weakly observed NOEs and potentially confusing conformational or configurational information. In these cases, a rotating-frame analog of the NOESY experiment known as ROESY (or CAMELSPIN) can be extremely useful (Neuhaus and Williamson, 2000). The most recent versions of the ROESY experiment suppress undesired TOCSY correlations and have contributed to the increased usage of the experiment (Braun et al., 1998; Hwang and Shaka, 1992).

It is not necessary for both dimensions in a homonuclear 2D experiment to contain information about the

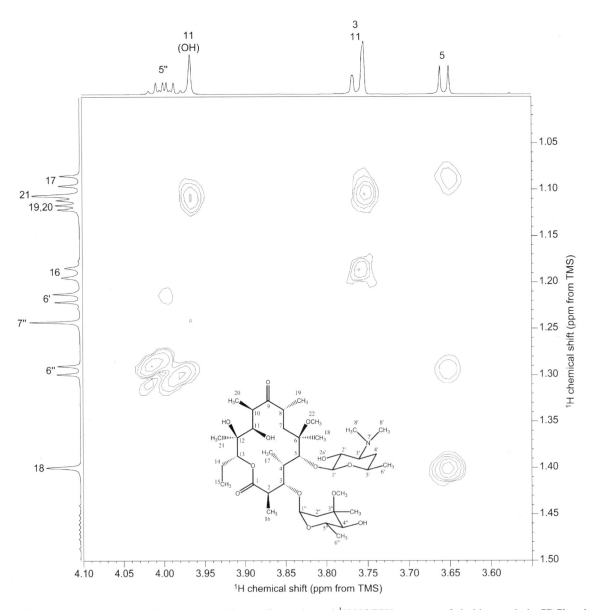

Figure 18 An expanded portion of the phase-sensitive gradient-enhanced ^{1}H NOESY spectrum of clarithromycin in CDCl$_3$ solution, obtained at 700 MHz. The assignments were obtained from a set of 1D and 2D ^{1}H and ^{13}C experiments. All of the contours shown except the 5″–6″ correlation are negative, indicating NOE contacts between the involved ^{1}H nuclei. The 5″–6″ correlation contains positive components from zero-quantum J-coupling mixing between these sites.

same interaction. Ernst et al. (1987) have reviewed the concept of *separation of interactions* by 2D NMR. For example, instead of a correlation plot of chemical shifts, another common ^{1}H experiment correlates chemical shift in the F_2 dimension with the magnitude of J-couplings in the F_1 dimension. This experiment is known as *J-resolved spectroscopy*, and actually predates most of the major 2D correlation experiments. When applied to protons, it consists of a spin echo pulse sequence in which the dephasing and refocusing τ delays are simultaneously incremented to create a t_1/F_1 dimension in which only J-coupling is active. However, while the refocusing period achieves a pure J-coupling interaction in F_1, both the J-coupling and chemical shift are active during F_2. As the separation of interactions is not perfect, a shearing transformation is normally applied to the data to compensate (Ernst et al., 1987). The J-resolved experiment is still widely employed for homonuclear J-coupling measurement, as the spin echo refocuses inhomogeneity in the static field and usually allows for increased accuracy (Ernst et al., 1987).

C. Multiple Resonance and Heteronuclear Techniques

The irradiation of multiple homonuclear resonances has been demonstrated to this point by the homonuclear decoupling and NOE difference pulse sequences. These homonuclear resonances differ in frequency by at most several kilohertz, and the experiments may be preformed using a single-output channel of a modern NMR spectrometer. However, most currently available probes have several channels that are designed to irradiate heteronuclear resonances that differ in frequency by tens or hundreds of megahertz (see Section II.B). Experiments which make use of these channels are historically labeled *double-resonance* (or *triple-resonance*, in the case of three channels). Multiple resonance experiments can have a single-acquisition dimension, or may incorporate the concepts of multidimensional spectroscopy discussed in the preceding section.

1. Heteronuclear Decoupling and NOE Enhancement

One of the simplest forms of double-resonance is heteronuclear decoupling. For example, ^{13}C spectra are often obtained with ^{1}H decoupling using the pulse sequence shown in Fig. 19(a). Decoupling can consist of a single long RF pulse of constant phase, frequency, and amplitude, as was the case in the homonuclear decoupling experiment discussed previously. However, CW pulses of this sort are selective in their decoupling ability, and are not particularly useful for fully decoupling an entire spectral range, such as a ^{1}H spectrum covering $\sim$10 ppm. [In the earlier days of double-resonance NMR, selective decoupling was used to identify directly attached heteronuclei, a role that is nowadays normally filled by the 2D heteronuclear correlation (HETCOR) experiments]. *Broadband decoupling* sequences have been developed to more effectively remove the effects of heteronuclei (Shaka and Keeler, 1987). For the common case of decoupling ^{1}H from a heteronucleus, the WALTZ family of sequences are extremely popular, with an excellent combination of effective field strength combined with reasonable bandwidth (Cavanagh et al., 1996; Ernst et al., 1987; Shaka and Keeler, 1987). For the extremely wide bandwidths needed to decouple ^{13}C nuclei from ^{1}H, the less-efficient but broadband GARP sequence has been introduced (Shaka and Keeler, 1987). The heteronuclear decoupling period can be extended to cover the entire relaxation delay, to build-up an NOE enhancement for superior signal-to-noise ratio in ^{13}C spectroscopy (but at the loss of quantitative accuracy). Examples of heteronuclear decoupling are shown in Figs. 19 and 20.

2. Spectral Editing

In heteronuclear coupled spin systems, the absence of the decoupler RF field produces lineshapes that are split by the heteronuclear coupling constant(s). Given that single-bond couplings for many nuclei fall into a narrow range (see Section V.B), observation of these splittings can be used to determine the number of attached protons. In ^{13}C spectroscopy, a quaternary carbon would result in a singlet, and methine (CH), methylene (CH$_2$), and methyl (CH$_3$) signals would appear as doublets, triplets, and quartets, respectively, in the ^{1}H-coupled spectrum. For most compounds, these multiplets show coupling constants in the vicinity of 150 Hz. Unfortunately, the splitting of the ^{13}C resonances reduces the sensitivity of the experiment and increases spectral overlap. Multiplicity-editing experiments have been developed to avoid this problem, primarily for ^{13}C NMR and also for ^{15}N, ^{29}Si, and other relevant nuclei. One the earliest, and still most useful spectral editing methods is the gated spin echo (GASPE), also known as the spin echo FT (SEFT) and J-modulated spin echo experiment, a close cousin of the more sophisticated attached proton test (APT) pulse sequence (Braun et al., 1998; Turner, 1996). The GASPE experiment is depicted in Fig. 21(b). Experiments based on pulsed polarization transfer have also been developed; these are exemplified by the DEPT and INEPT family of pulse sequences (Braun et al., 1998; Doddrell et al., 1982; Ernst et al., 1987; Turner, 1996). The oft-used DEPT-135 pulse sequence is depicted in Fig. 21(c). Both the GASPE and DEPT-135 experiments are demonstrated for ^{13}C spectroscopy, along with inverse-gated decoupling and NOE-enhanced experiments, in Fig. 20. Although the multiplicity is not known absolutely (especially for

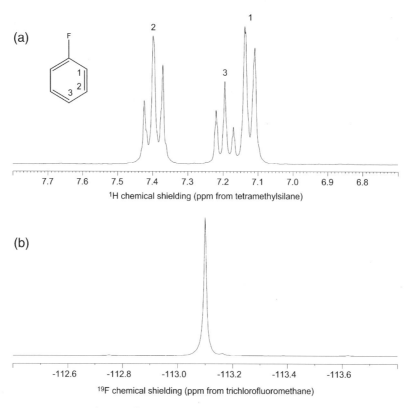

Figure 19 Heteronuclear decoupling in ^{1}H and ^{19}F NMR. (a) ^{1}H spectrum of monofluorobenzene acquired with CW decoupling applied directly to the ^{19}F resonance. With the effects of J-coupling to ^{19}F of the fluorine removed, the multiplicity of the resonances is clear. (b) ^{19}F spectrum of the same molecule acquired with ^{1}H broadband decoupling (using the WALTZ-16 sequence). The coupled ^{1}H and ^{19}F spectra of this molecule are shown in Fig. 10.

CH and CH$_3$ resonances with similar chemical shifts), the GASPE and DEPT-135 results show that editing based on proton multiplicity serves as a useful confirmation of assignments. In recent years, other spectral editing techniques have been developed for enhanced spectral editing. These include the SEMUT method of generating edited subspectra (Bildsoe et al., 1983; Turner, 1996) and the newer polarization-transfer PENDANT experiment (Homer and Perry, 1995), which (unlike DEPT) detects quaternary carbons as well.

3. Inverse-Detected 2D HETCOR

HETCOR refers to pulse sequences which correlate heteronuclear chemical shifts (e.g., ^{1}H and ^{13}C) via J-coupling or dipolar cross-relaxation. Although spectral editing and NOE enhancement experiments give some indication as to the nature of attached protons, HETCOR experiments make more explicit use of relevant interactions to obtain direct connections between heteronuclei. These methods are tremendously useful in determining molecular structures; hence, a great variety of HETCOR experiments have been developed (Braun et al., 1998; Cavanagh et al. 1996; Ernst et al., 1987). Both phase cycled and gradient-selected experiments have been designed. Because of their improved characteristics and great popularity, only gradient selected experiments are discussed in detail here. [Phase cycled sequences differ in that they often make use of the BIRD subsequence for suppressing ^{1}H–^{12}C signals (Braun et al., 1998)]. In addition, *inverse-detected* HETCOR experiments are given priority in this section. Inverse detection refers to detection of the nucleus with the higher gyromagnetic ratio, which offers a sensitivity advantage proportional to $\gamma_e \gamma_d^{3/2}$, where γ_e is the excited nucleus and γ_d is the detected nucleus. It is optimal to select the nucleus with the highest gyromagnetic ratio for detection (usually ^{1}H), and design the probe to have maximum sensitivity for this nucleus. This is the common practice in most NMR laboratories, although X-detected experiments (where X is ^{13}C, ^{19}F, ^{31}P, ...) are still used in some cases.

4. Heteronuclear Multiple-Quantum and Single-Quantum Coherence

The heteronuclear multiple-quantum coherence (HMQC) and the heteronuclear single-quantum coherence (HSQC) experiments are currently the most frequently used

Nuclear Magnetic Resonance

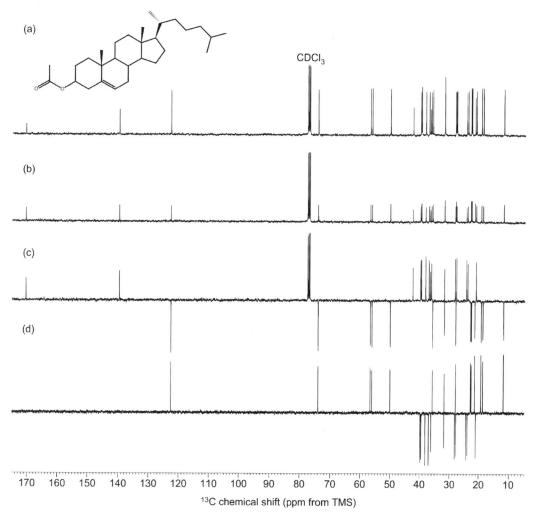

Figure 20 Spectral editing and hetero-NOE enhancements in the 100 MHz ^{13}C NMR spectra of cholesteryl acetate. In (a), the ^{13}C single-pulse spectrum with continuous ^{1}H decoupling (during both acquisition and relaxation delays) is shown. For the sake of comparison, (b) shows the spectrum obtained with decoupling only during the acquisition period. The full heteronuclear NOE is observed in (a) but not in (b). The results of two widely used editing techniques on this sample are also shown. The GASPE experiment in (c) yields quaternary and CH$_2$ carbons phased up, whereas CH and CH$_3$ carbons are phased down. In (d), a DEPT-135 experiment shows CH and CH$_3$ carbons phased up, with CH$_2$ carbons phased down and quaternary carbons suppressed. A relaxation time of 10 s and 2048 scans were used for all experiments.

HETCOR experiments (Braun et al., 1998; Cavanagh et al., 1996; Ernst et al., 1987). A magnitude-mode version of the popular gradient-enhanced HMQC experiment is depicted in Fig. 22(a) (Hurd and John, 1991). This experiment transfers ^{1}H magnetization via the $^1J_{CH}$ coupling to ^{13}C (or another heteronucleus), where it is allow to experience the ^{13}C chemical shift for the t_1 period. The magnetization is then returned to the ^{1}H nuclei for detection. The narrow range of $^1J_{CH}$ is used to set delays so that the magnetization is transferred efficiently. A phase-sensitive version of the gradient-enhanced HSQC pulse sequence is shown in Fig. 22(b) (Kontaxis et al., 1994). Its primary advantage over HMQC is suppression of ^{1}H–^{1}H couplings during the t_1 evolution period, which leads to improved resolution. Sensitivity-enhanced HSQC pulse sequences are available as well, which work by reintroducing components of the signal normally lost during gradient selection (at the expense of more RF pulses) (Cavanagh et al., 1996; Kay et al., 1992). Composite pulses, shaped pulses and adiabatic pulses, can be substituted into these sequences to improve performance in some cases. The results of the phase-sensitive, gradient-enhanced HSQC experiment in Fig. 22(b) as applied to clarithromycin are shown in Fig. 23. For reference, 1D ^{1}H and ^{13}C GASPE spectra are shown along the F_2 and F_1 axes, respectively. Furthermore, to illustrate the process of spectral assignment, all of

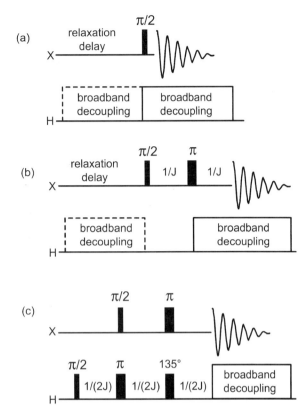

Figure 21 Basic heteronuclear pulse sequences. (a) Single-pulse experiment using broadband decoupling during acquisition. The decoupler may be left on continuously (as denoted by the dashed region) if NOE enhancements are desired. If the decoupler is not left on, the experiment is often referred to as the inverse-gated experiment. (b) The gated spin echo (GASPE) experiment. Primarily used for ^{1}H and ^{13}C double-resonance work, it encodes the number of attached protons for a given ^{13}C signal into its phase in a proton decoupled spectrum. Quaternary carbons and methylene carbons acquire an opposite phase to methine and methylene carbons. The signal may be enhanced by NOE effects by leaving the decoupler on during relaxation. The two delays denoted by $1/J$ are set based on a $^1J_{CH}$ of 140–150 Hz (typically ~7 ms). (c) The DEPT-135 pulse sequence, also used for multiplicity editing but with pulsed polarization-transfer signal enhancement. The delays are set in the same manner as the GASPE, except to $1/(2J)$ or ~3.5 ms.

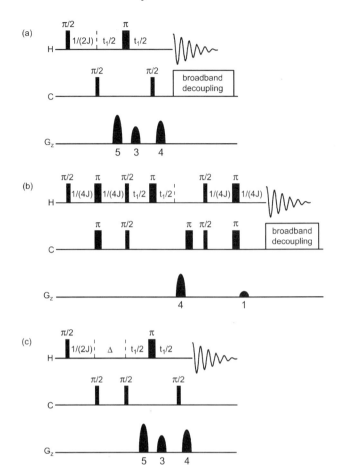

Figure 22 ^{1}H-detected gradient-enhanced pulse sequences for solution-state HETCOR. (a) The basic magnitude-mode HMQC pulse sequence. This sequence is used for ^{1}H–^{13}C correlation via $^1J_{CH}$, or for longer range correlations by setting the initial delay to several milliseconds, with suitable adjustment of the gradients. The gradient ratio is 11:9:4 for ^{1}H–^{15}N. (b) Phase-sensitive HSQC pulse sequence used to acquire the data in Fig. 23. For ^{1}H–^{15}N work, the gradient ratio is ~10:1. The sign of the first gradient in this sequence is negated to perform echo–antiecho selection [see Kontaxis et al. (1994)]. (c) Magnitude-mode HMBC pulse sequence, used to acquire the data in Fig. 25. Note that all gradient ratios are approximate and should be adjusted on the actual instrument, and that phase cycles for these sequences are given in the references.

the correlations have been annotated, as determined from a full ^{1}H and ^{13}C analysis (including other supporting 2D experiments).

The *heteronuclear multiple-bond correlation (HMBC)* experiment, shown in Fig. 22(c), is an extension of the basic HMQC experiment designed to detect long-range coupling (Braun et al., 1998; Cavanagh et al., 1996; Ernst et al., 1987). Again, the narrow range of $^1J_{CH}$ values encountered in organic compounds is exploited. However, unlike HMQC and HSQC, the HMBC experiment suppresses this strong coupling so that weaker longer-range couplings can be more easily analyzed. This is accomplished by first pulse on the ^{13}C channel, which acts as a "low-pass filter", by forcing magnetization transferred by $^1J_{CH}$ into unobservable double- and zero-quantum states. The longer evolution delay (Δ) then transfers magnetization through the two, three, and four bond J-couplings. The experiment is normally run without decoupling, as the detected couplings are usually small "antiphase" couplings that would collapse upon decoupling. The lack of decoupling also helps in locating residual unsuppressed $^1J_{CH}$ couplings. The

Nuclear Magnetic Resonance

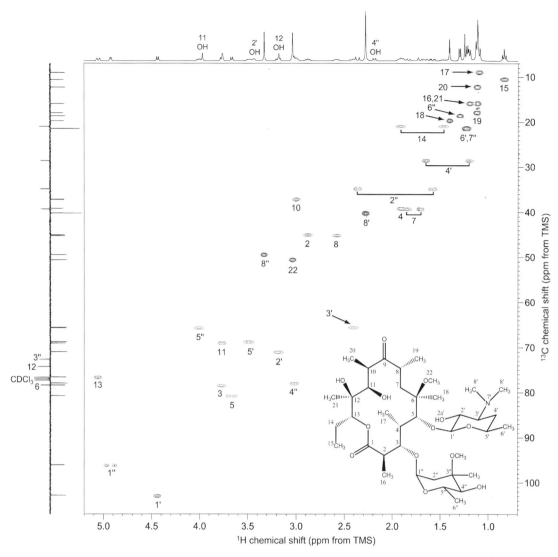

Figure 23 ^{1}H–^{13}C gradient-selected phase-sensitive HSQC spectrum of clarithromycin in CDCl$_3$, obtained at a ^{1}H frequency of 400 MHz. A single-pulse ^{1}H spectrum is shown at the top, and a ^{13}C GASPE spectrum is shown on the left side. The spectrum shows correlations between ^{1}H and ^{13}C nuclei that were transferred by $^1J_{CH}$ couplings on the order of 150 Hz. The annotated correlations show the assignments for methine, methylene, and methyl carbons (which were obtained by a series of ^{1}H and ^{13}C 1D and 2D experiments, including this experiment). Assignments for quaternary carbons and for hydroxyl protons are shown on the directly detected ^{1}H and ^{13}C spectra.

gradient-enhanced version shown in Fig. 22(c) is run in magnitude-mode and produces spectra with very few artifacts (Willker et al., 1993). An example of an HMBC experiment applied to carvedilol is shown in Fig. 24.

^{1}H–^{15}N spectroscopy of small molecules in natural abundance has seen a resurgence in the past few years, primarily because of the development of gradient-enhanced inverse-detected experiments (Martin and Hadden, 2000). An example of this powerful analytical method is shown for the antibiotic telithromycin in Fig. 25, using the simple magnitude-mode HMQC experiment given in Fig. 22(a). Resonance assignments for the ^{15}N and correlated ^{1}H nuclei are shown, and the nitrogen environments in the molecule can be easily studied. (^{15}N spectral interpretation is discussed in Section V.B.)

5. Other HETCOR Experiments

"Hybrid" heteronuclear experiments combine features of other experiments and are particularly useful in studies of biological macromolecules (see subsequently) (Braun

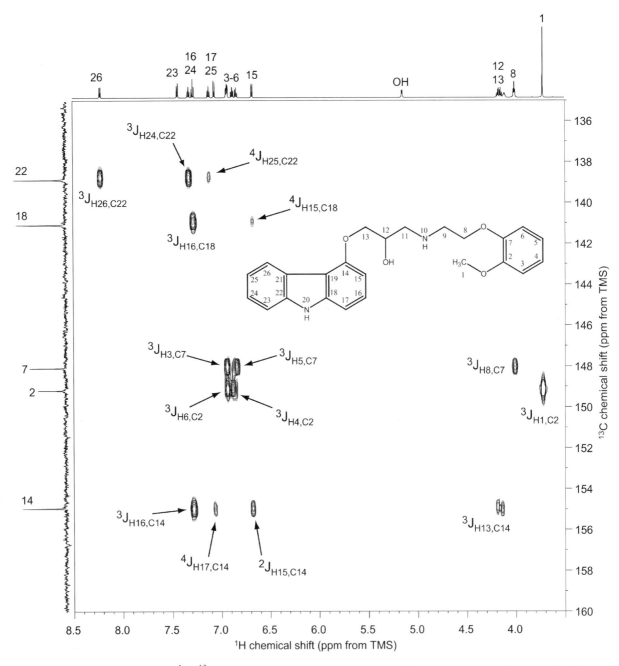

Figure 24 Expanded region of the ^{1}H–^{13}C gradient-selected magnitude-mode HMBC spectrum of carvedilol in d_6-DMSO solution, obtained at a ^{1}H frequency of 700 MHz. Only long-range two, three, and four-bond correlations are detected. Residual ^1J signals are often seen at low levels, but the contours have been adjusted so that these are not visible. The long-range data shown here plays a critical role in the full suite of 2D experiments needed for total assignment of the ^{1}H and ^{13}C spectra of this compound.

et al., 1998; Cavanagh et al., 1996). One example often applied in small molecule work is the HMQC-TOCSY experiment, which correlates ^{1}H spin networks with directly attached carbons via a TOCSY mixing period performed after the magnetization has been transferred back to the protons (Braun et al., 1998). Although inverse detection has supplanted other methods for the HETCOR spectroscopy of the common ^{1}H–^{13}C and ^{1}H–^{15}N spin systems, directly detected correlation techniques are still applicable, especially in sequences designed for less-common spin pairs (Berger et al., 1996). Although J-coupling is the transfer mechanism in the majority of these pulse sequences, the heteronuclear NOE can be important for ^{1}H–^{19}F, ^{1}H–^{31}P, and ^{1}H–^{7}Li

Nuclear Magnetic Resonance

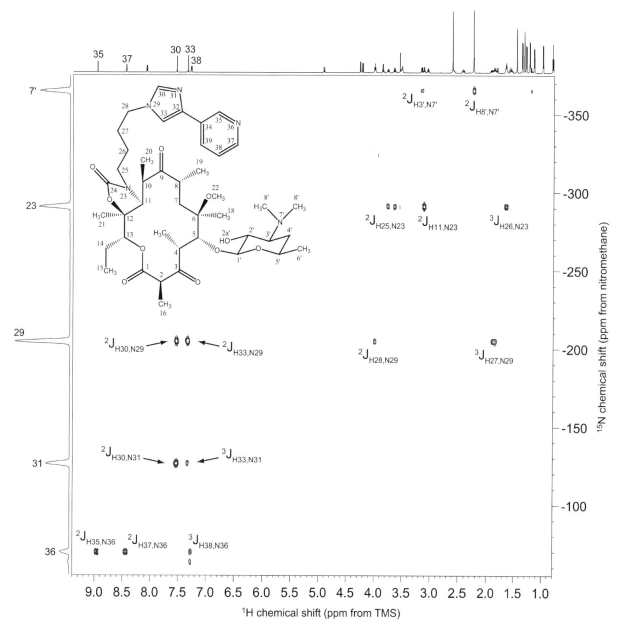

Figure 25 ^{1}H–^{15}N gradient-selected magnitude-mode HMQC spectrum of telithromycin in CDCl$_3$, obtained at a ^{1}H frequency of 700 MHz. The polarization transfer delay was adjusted for optimal detection of ~10 Hz couplings. A single-pulse ^{1}H spectrum is shown at the top, and a positive projection of the columns is on the left side. The spectrum shows correlations between ^{1}H and nitrogen that involve a ^1J on the order of 5–90 Hz. (Assignments and data interpretation are also discussed in Section V.B.)

spectroscopy. A heteronuclear analog of the NOESY experiment (known as HOESY) is available in 2D and selective 1D versions, although routine steady-state NOE experiments are still widely employed (Braun et al., 1998; Neuhaus and Williamson, 2000). Likewise, a heteronuclear version of isotropic mixing known as hetero-TOCSY has also appeared in both 1D and 2D guises, and has been applied in cases where inverse ^{1}H detection is undesirable (Braun et al., 1998; Brown and Sanctuary, 1991). Experiments for the precise measurement of heteronuclear couplings have also seen extensive development (Eberstadt et al., 1995; Marquez et al., 2001). The latest pulse sequences are gradient-enhanced inverse-detected methods that have been designed to detect the *J*-couplings of interest by scaled (or unscaled) splitting in one or more dimensions, by quantitative intensity variations, or by extraction of lineshapes for special fitting procedures.

Table 3 Selected NMR Experiments for Macromolecules

Experiment	Information content
3D NOESY HSQC	Combines NOESY and HSQC into a 3D experiment for the simultaneous analysis of ^{1}H–^{13}C direct connectivity and ^{1}H–^{1}H dipolar connectivity, for protein backbone assignments
3D TOCSY HSQC	As above, but with isotropic J-coupling mixing for ^{1}H–^{1}H correlation
HCCH-COSY, HCCH-TOCSY	Used for the assignment of aliphatic ^{1}H and ^{13}C resonances of ^{13}C-labeled proteins, via a three step magnetization transfer: ^{1}H → ^{13}C, ^{13}C → ^{13}C, and ^{13}C → ^{1}H. Constant time versions are also available
3D CBCANH	Establishes connectivity between the carbon of a peptide residue and its amide proton and ^{15}N, and subsequently to neighboring residues
HN(CO)CA	Supplies sequential correlations between peptide residues. The amide ^{1}H and ^{15}N chemical shifts are correlated with the α-carbon shift via a transfer through the carbonyl ^{13}C
HCNA-J	Used for measurement of J-coupling constants in labeled proteins

6. HETCOR in the Study of Macromolecules

A large class of HETCOR experiments have been developed for the specific purpose of studying biological macromolecules in solution. The experiments are typically designed to detect special features in heavily overlapped spin systems of macromolecules (especially proteins). Both isotopically labeled and natural abundance proteins are the subjects of investigation, for purposes of determining structure, conformation, and mobility. Some selected experiments are listed in Table 3 (Cavanagh et al., 1996; Wuthrich, 1986). Multiple coherence transfer steps using heteronuclear and homonuclear J-coupling and NOE/ROE mixing are often imbedded in these sequences. The sequential assignment of a ^{13}C- and ^{15}N-labeled protein, for example, can involve a series of nD experiments to determine backbone connectivity followed by sidechain structure and conformation and even tertiary structure.

7. Solid-State NMR Experiments

One final class of double-resonance experiment is the popular cross-polarization MAS (CP-MAS) method used in solid-state NMR (Gerstein and Dybowski, 1985; Slichter, 1996). The spinning sidebands produced by MAS can be a nuisance to interpretation. They can be suppressed using spinning at a faster rate (i.e., exceeding the magnitude of the chemical shift interaction), or by making use of pulse sequences designed for the suppression of spinning sidebands. The total suppression of sidebands (TOSS) experiment is such an experiment, and is based on a series of π pulses applied after the CP period that cancel out the sideband intensity (Antzutkin, 1999). The combined CP-TOSS experiment is demonstrated in the example shown in Fig. 26(a), where the ^{13}C CP-TOSS spectrum of a solid form of carvedilol is shown. The spectrum was taken at 5 kHz, and the residual sidebands obtained at this speed are suppressed by the TOSS sequence. The spectrum is not quantitative because of the suppression of sidebands from higher-CSA sites and because of variable CP rates (Gerstein and Dybowski, 1985; Slichter, 1996). In Fig. 26(b), a form of editing used in solid-state NMR is demonstrated. This experiment is known as nonquaternary suppression (NQS), and involves turning off the ^{1}H decoupling field during the TOSS cycle, which allows for strong dipolar dephasing of methine and methylene carbons. (Methyl and quaternary carbons are unaffected because of weaker dipolar coupling; in the case of methyl carbons this is the result of rapid rotation.) The NQS experiment is a reliable tool for assigning crowded spectral regions in solid-state NMR.

D. Diffusion, Dynamics, and Relaxation Measurements

Molecular rotational correlation times and local flexibility information can be obtained from NMR relaxation times. There are several basic methods for measuring longitudinal relaxation (T_1), the most prevalent of which is the inversion recovery pulse sequence. This is the only sequence discussed here; other alternatives can be found in the literature (Braun et al., 1998; Derome, 1987; Neuhaus and Williamson, 2000). In the inversion recovery sequence, a pseudo-2D experiment depicted in Fig. 27(a), the magnetization is inverted from its equilibrium-state (along the $+z$-axis) to the $-z$-axis by a π pulse. This effectively inverts the populations of the states from their Boltzmann values [Eq. (16)]. The return of the magnetization to its equilibrium-state is monitored by a second $\pi/2$ pulse, as shown, to "readout" the state of the recovery. An example is shown for a ^{1}H T_1 measurement for carvone in CDCl$_3$ in Fig. 28. The time delay (τ) is incremented through a series of chosen values, which need not be linear. (The example in Fig. 28 uses linearly spaced delays from 0 to 15 s). The spectra taken with small τ values show the inverted magnetization, which passes through 0 and then recovers to its equilibrium value as τ nears and then exceeds T_1.

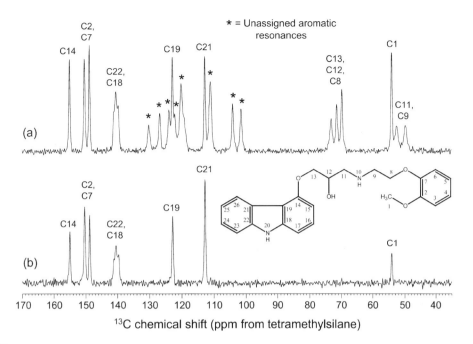

Figure 26 (a) ^{13}C CP-TOSS and (b) CP-TOSS with NQS spectra of a solid form of carvedilol at obtained at 90 MHz. The spinning rate was 5 kHz and a CP period of 3 ms was used.

Transverse relaxation rates can be estimated from the peak shape as in Eq. (30). However, this estimate includes broadening effects from static field inhomogeneities. For a true measurement of T_2, the Carr–Purcell sequence with the Meiboom–Gill modification (the CPMG sequence) is used (Braun et al., 1998; Slichter, 1996). This sequence consists of a series of n spin echoes, as shown in Fig. 27(b), which are phase-cycled so as to avoid deleterious effects from pulse calibration errors, RF inhomogeneity, and molecular diffusion during the pulse sequence. The sequence is run in with increasing values of n to produce an exponential decay from unity to zero, as the T_2 relaxation mechanism(s) irreversibly dephase the magnetization.

The analysis of relaxation data usually involves nonlinear fitting of the exponential decay to obtain T_1 or T_2 values of interest. For the inversion recovery experiment, the relaxation delay between transients must be long enough to allow for full relaxation (which obviously requires a prior estimate of the longest T_1 in the system). In addition, it should be noted that the results shown in Fig. 28 are nonquantitative, because a common paramagnetic impurity, oxygen (O_2), has not been excluded from the sample. In order to perform a quantitative inversion recovery measurement, the sample is usually degassed by the freeze–pump–thaw technique, or the oxygen is purged from the sample by bubbling nitrogen or argon through it (Braun et al., 1998; Neuhaus and Williamson, 2000). The success of the degassing procedure can be checked by repeated measurements until a constant T_1 or T_2 is achieved. In general, ^{13}C relaxation times are less sensitive to proper degassing than those for ^{1}H, which tend to be on the exterior of molecules and closer to dissolved paramagnetic impurities.

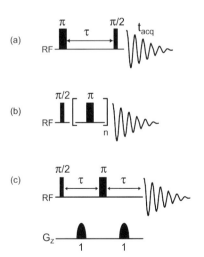

Figure 27 Pulse sequences used for relaxation and diffusion measurements. (a) Inversion recovery pulse sequence for T_1 measurement. (b) CPMG sequence for T_2 measurement. (c) Basic pulsed-field gradient spin echo experiment for diffusion measurements.

The analysis of dynamics by NMR forms a large subdiscipline, and can include the study of any solution or solid-state process that must overcome an energy barrier (Perrin and Dwyer, 1990). Tautomerization, conformational

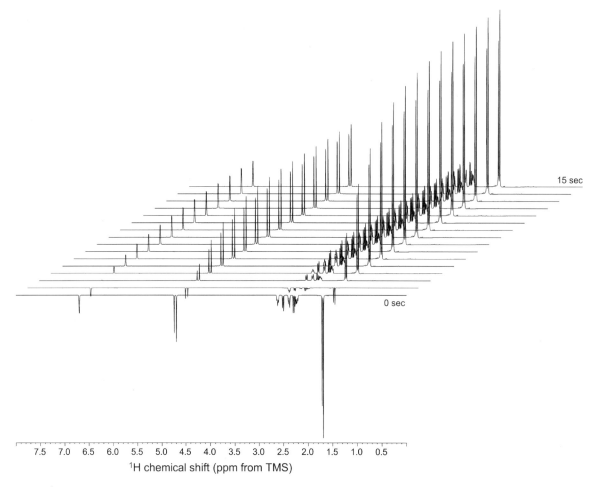

Figure 28 ^{1}H T_1 relaxation time measurement for carvone in CDCl$_3$ using the inversion recovery pulse sequence of Fig. 27(a). Sixteen linearly spaced delay values were chosen for this demonstration. The peak areas as a function of time are fitted to exponentials to extract out the T_1 value for each ^{1}H site. For relaxation time analysis, the sample is degassed to remove the effects of dissolved paramagnetic dioxygen.

exchange, rearrangements, and solid-state dynamics of polymers are but a few examples of such processes. The basic experiments used to study dynamics have already been introduced under different names. The NOE difference experiment of Fig. 12(d) can be used to monitor the transfer of saturation to an exchanging spin species; in this role it is referred to as the *saturation-transfer* experiment (Braun et al., 1998). Likewise, the NOESY experiment of Fig. 16(c) is relabeled the EXSY experiment for the study of exchange processes (Perrin and Dwyer, 1990). VTs and different mixing times are used to enhance the information content of these experiments.

Molecular diffusion can also be measured by specially designed NMR techniques (Johnson, 1996). The first of these, which serves as a basic building block of many more complex sequences, is the pulsed field gradient spin echo (PFG-SE) sequence shown in Fig. 27(c) (Braun et al., 1998). This experiment is normally run by increasing the gradient strength from zero up toward values in excess of 0.5 T/m (50 G/cm), depending on the instrumentation available and the size of the diffusion constant of interest. More sophisticated sequences have replaced the PFG-SE experiment for routine diffusion studies, with full 2D versions generally falling under the title of *diffusion-ordered spectroscopy* (DOSY) (Johnson, 1996). Analysis of the data generally involves fitting or special processing techniques along with knowledge of the diffusion model (Johnson, 1996).

V. DATA ANALYSIS AND INTERPRETATION

NMR spectra can vary greatly in their complexity and information content. However, all spectra share the need for analysis and interpretation, so that the underlying biological, chemical, or physical process can be understood. Processing of NMR data has grown into its own field, and can greatly affect the final outcome of an analysis. After processing is completed, spectral interpretation is still necessary. This is often a detailed undertaking, and relies on empirical or tabulated evidence in conjunction with theoretical knowledge. The basic procedures of spectral processing and data interpretation are reviewed here, with special emphasis on the current state-of-the-art. The literature contains numerous reviews on the subject of NMR data processing (Ernst et al., 1987; Harris, 1978; Higinbotham and Marshall, 2000; Hoch and Stern, 1996; Stephenson, 1988).

A. Spectral Processing

NMR spectroscopy in the FT mode normally requires processing of time-domain responses to RF irradiation if the desired data is to be extracted. In most cases this spectral processing is usually centered around the FT. Signals are characterized in the time-domain by their complex sinusoidal frequency and their decay rate. Overlapping signals are separated in the frequency domain using the FT. Noise in NMR is usually present as a statistically random series of numbers with a normal (or Gaussian) distribution around an average value of zero. Random noise is uncorrelated, meaning that different members of the series of numbers. Systematic noise may also be present. The general processing procedure for a 1D NMR spectrum is to apply a window function and perform zero-filling before the discrete Fourier transform (DFT). Linear prediction (LP) may also be applied at this stage. Following the transform, phase correction and baseline adjustments may be made and other postprocessing steps may be employed. For 2D and higher-dimensional spectra, the procedure can involve applying window functions, zero-filling, and LP. The procedure is normally carried out sequentially in multiple dimensions, with the directly detected dimension processed first. Fourier transformation is then applied to each of the dimensions, possibly including data shuffling to provide the complex values needed for a phase-sensitive transform. (The shuffling scheme varies; accommodations for TPPI, echo–antiecho, or hypercomplex-States methods are commonly available in processing software.)

1. Apodization and Window Functions

Apodization is commonly applied to weight time-domain data for the purpose of resolution enhancement or cosmetic noise "suppression", and is used in vibrational and other forms of FT spectroscopy, besides NMR. One primary criteria for apodization functions is the "leakage" into surrounding frequencies caused by their application. The most commonly used apodization functions in NMR have been well documented (Ernst et al., 1998; Hoch and Stern, 1996); one dedicated review lists over 20 different functions (Harris, 1978). Exponential and Gaussian window functions are two of the most useful, as are sine-bell and shifted sine-bell functions used in 2D work. The exponential function leads to Lorentzian lineshapes upon Fourier transformation:

$$a_k = e^{-\pi W_L k \Delta t} \tag{39a}$$

where a_k is the kth point in the time-series (i.e., the FID), W_L is the "line-broadening" constant (in Hz), and Δt is the sampling time between points (or dwell time). The Gaussian function is given by:

$$a_k = e^{-W_G(k\Delta t)^2} \tag{39b}$$

where W_G is the Gaussian broadening constant. An interesting property of this function is that its FT is another Gaussian function. These functions are then multiplied by the kth point in the data set to apodize the data. More complex apodization often consists of multiple functions. One example is the Lorentz-to-Gauss transformation, which combines Eqs. (39a) and (39b). This function is used to enhance resolution in spectra with good S/N, for identifying overlapping resonances or for the purpose of extracting J-coupling values (which can be distorted by Lorentzian overlap) (Marquez et al., 2001). The Lorentz-to-Gauss transformation is given by (Hoch and Stern, 1996):

$$a_k = e^{+\pi W k \Delta t} e^{-W_G(k\Delta t)^2} \tag{39c}$$

2. Fourier Transformation

The FID acquired in the NMR experiment contains both signal and noise, and must be used to estimate the spectrum. The straightforward method for accomplishing this task is the DFT (Ernst et al., 1987; Harris, 1978; Hoch and Stern, 1996)

$$f_n = \frac{1}{\sqrt{N}} \sum_{k=0}^{N-1} d_k e^{-2\pi i k n/N} \tag{40}$$

where N is the number of complex data points, and the transformation changes a time series (or vector) **d** into a frequency series **f**. The DFT is based on the continuous integral form of the FT, which can be used to transform analytical functions between domains. As mentioned in Section II.C, the sampling rate or dwell time (dw) determines the spectral width (sw = 1/dw) that will be obtained in the DFT. The *sampling* (or *Nyquist*) *theorem* states that the sampling rate of the digitizer must be at least twice as fast as the highest frequency in the signal, or the spectrum will contain "folded" lineshapes at incorrect (but usually predictable) frequencies. Without other means of extending the time-domain data set, the sampling time (or number of points acquired) determines the digital resolution.

The time series acquired in NMR can be extended by a padding end of the data set with zeros, in a process known as *zero-filling*. Zero-filling may be used to extend the data set to a size more amenable to fast Fourier transform (FFT) techniques (e.g., to a power of two), and also serves as a means of increasing the number of points in the final frequency domain spectrum. In this role, zero-filling essentially interpolates between data points, and can be useful to improve the shapes of peaks and facilitate further analysis. Zero-filling at the end of the data set can cause problems if the NMR signal has not decayed to zero, as this convolutes the signal with a box function and causes severe baseline distortions known as "sinc-wiggles".

LP is time-series analysis method that is superior to zero-filling for signals which have not decayed to zero (Hoch and Stern, 1996; Stephenson, 1988). In linear prediction, an extrapolated data point is determined by a weighted linear combination of the previous data points. A number of LP algorithms are in use (Hoch and Stern, 1996; Stephenson, 1988). It is common practice to use forward LP in the indirectly detected dimension(s) of nD experiments, where it is very useful in enhancing the resolution and in saving time in long nD acquisitions. Backward LP can be used to correct the first few points of a data set in case of corruption by acoustic ringing or pulse breakthrough, as was done for the ^{17}O data set shown in Fig. 15.

Maximum entropy spectral reconstruction is an alternative method to Fourier transformation (Hoch and Stern, 1996; Stephenson, 1988; Vogt and Mueller, 2002). It operates by using the time-domain series to produce a spectrum that is maximized in its entropy (or in other words, the most likely spectrum). Other closely related special processing methods are available for nonsinusoidal time domain signals (Vogt and Mueller, 2002). These include matrix inversion methods and regularization methods (for computing maximum entropy spectra or spectra with suppressed noise features) that are also useful for the analysis of decaying data such as that produced in DOSY experiments, via the Laplace transform (Johnson, 1996).

3. Magnitude Spectra and Phase Correction

Phase correction is a procedure for obtaining frequency domain spectra with absorptive phase peaks. This is not the same as computing spectra with positive peaks, which can be produced from a calculated *magnitude spectrum*. This is calculated by taking the absolute value of the sum of the real and imaginary parts for each data point. However, magnitude spectra are broadened by the mixing of absorptive and dispersive lineshapes, and cannot be used when phase sensitivity is needed. (Two 1D experiments that need phase sensitivity, e.g., are NOE difference spectroscopy and the GASPE/APT spectral editing methods.) In these cases, phase-sensitive spectra are obtained by complex Fourier transformation and phase correction of the separate real and imaginary data arrays. The need for *zeroth-order phase correction* results from variations in the sample-dependent settings of the transmitter and receiver phases. This type of phase correction is not frequency dependent, and does not affect the experiment as long as the transmitter and receiver phases still have the same relative relationship. In practice, the zeroth-order phase shift is numerically applied to phase 1D and 2D spectra to absorption mode as follows:

$$x_j^{\text{cor}} = x_j^{\text{real}} \cos(\phi_0) - x_j^{\text{imag}} \sin(\phi_0) \quad (41\text{a})$$

where j is the index of the data point x being processed and ϕ_0 is the zeroth-order phase angle. Equation (41a) is thus applied to each data point before or after Fourier transformation (i.e., in the time or the frequency domains) and corresponds to a "balancing" of the real and imaginary signal portions. An identical effect can also be accomplished by manual or software adjustment of the receiver phase, so that the acquired FID will already have the desired phase. The most accurate adjustment of ϕ_0 is generally obtained in the frequency domain, with the spectrum at a high vertical expansion, so that the base of the peaks can be equalized. *First-order phase correction* refers to a linear frequency-dependent contribution to the phase, and is usually digitally corrected by the following expression

$$x_j^{\text{cor}} = x_j^{\text{real}} \cos\left(\frac{j\phi_1}{N}\right) + x_j^{\text{imag}} \sin\left(\frac{j\phi_1}{N}\right) \quad (41\text{b})$$

where N is the number of points and ϕ_1 is the first-order phase angle. The two most common cause of frequency-dependent phase shifts are the time delay before acquisition and the effects of analog filters on the spectrum. The delay before acquisition, usually a few microseconds, allows for frequency-dependent evolution of the transverse magnetization. Analog filters cause a similar effect, slowing the wider frequencies of signal as they pass through the electronics. In general, adjustment of ϕ_1 is also done in the frequency domain simultaneously with ϕ_0. A point on the spectrum (often the maximum of a strong peak) is selected as a "pivot point" and is phased to absorption using ϕ_0. The other peaks are then adjusted to absorption using ϕ_1. This adjustment is often done manually by the spectroscopist, as proper phasing can have a positive impact on integral accuracy. Alternatively, automatic methods are also available for phase correction and have become popular on high-throughput instruments (Heuer, 1991). The phase correction of nD spectra is generally more complex than that of 1D spectra, although the procedure for manually phasing the data is essentially the same. The rows and columns of the data matrix are generally phased independently. The phase corrections for indirectly detected dimensions of nD experiments can generally be calculated; this is of great assistance in the otherwise complicated procedure of phasing 3D and higher-dimensional spectra.

4. Baseline Correction

Even with the advent of digital quadrature detectors, rolling baselines are still encountered in NMR. Correction of the spectral baseline can be very important, especially for quantitative work. One of the most commonly used methods for baseline correction is to automatically fit the offending baseline to a third- or fourth-order polynomial, and then subtract the fitted function from the actual data. Other methods allow the user to manually select points in the spectrum to be used as the baseline, and construct a piecewise linear model between these points for subtraction (Hoch and Stern, 1996). In many cases, baseline distortions in NMR can be corrected by identifying the fundamental problem with the FID. Echo pulse sequences can be used to shift the signal to a time that is less affected by instrumental affects (such as pulse breakthrough). Baseline distortion from acoustic ringing can be avoided by use of special pulse sequences such as the ARING experiment discussed in Section I.A, or by applying backward LP to correct the initial FID points responsible for the distortion. A problem with the treatment of the first data point in FFT algorithms has historically caused baseline problems, but is easily avoided by appropriate scaling of the first point. New algorithms have become available in recent years for efficient baseline correction of 1D and 2D spectra (Brown, 1995). Most commercial spectrometers come equipped with software routines for automatic and manual baseline correction.

Two other important postprocessing tasks are *peak picking* and *integration*. More sophisticated postprocessing is covered in Section V.D as it is more closely allied with computer-assisted structure determination, although the distinction is somewhat arbitrary. Peak picking refers to the process of finding the centroid and possibly the multiplicity of an NMR peak. It is useful for generating concise tabulations of results from lengthy data sets. Integration of peaks, both 1D and 2D, is especially useful for quantification of the number of spins contributing to the peak in ^{1}H spectroscopy. For NMR studies of other nuclei, polarization transfer or NOE-enhanced methods are generally used, so that integration is generally restricted to comparisons of similar sites, unless integral responses are quantified. Volume integrals are used in 2D NOE and exchange spectroscopy to assess relaxation and dynamic rates.

Symmetrization is a postprocessing method primarily used to clarify 2D homonuclear correlation spectra (Ernst et al., 1987; Hoch and Stern, 1996). It is especially popular for processing routine ^{1}H COSY data sets, as it clarifies the spectrum by artificially suppressing t_1 noise. As there is reflection symmetry about the diagonal in COSY spectra, and across the center of the F_1 dimension in *J*-resolved spectra, symmetrization is accomplished by the rapid procedure of replacing the two symmetry-related values with their average. For example, two points in a COSY spectrum with intensities given by $f(\omega_1, \omega_2)$ and $f(\omega_2, \omega_1)$ would be replaced by $f(\omega_1, \omega_2)/2 + f(\omega_2, \omega_1)/2$. Although the use of symmetrization is accepted practice for magnitude-mode COSY and *J*-resolved spectra, it is generally not recommended for processing phase-sensitive NOESY and ROESY data sets, or in any case where real spectral information could be substantially altered or destroyed by the averaging procedure. When symmetrization is employed, it is advisable to compare the results with the unsymmetrized data to verify its integrity.

Background subtraction is also employed in some NMR applications. It can be used for subtraction of large resonances in 1D spectra, but this is not as common as in other forms of spectroscopy, as the results are generally poor. Background subtraction is used to help ameliorate the effects of t_1 noise in 2D spectra (a more serious problem prior to the advent of pulsed field gradients). A noise region can be defined along several rows of a 2D spectrum, which is then averaged and numerically subtracted from all of the F_1 rows (Klevit, 1985).

B. Manual Data Interpretation

The wide variety of experimental NMR techniques discussed earlier hints at the ever-present need to untangle spectral parameters. Current state-of-the-art NMR methods are quite efficient at extracting pertinent information. The experimentalist wishes to obtain as many needed chemical shift and *J*-coupling parameters as possible, along with NOE contacts, relaxation rates, and other parameters detectable in oriented phases. This can include quantitative integration and connectivity information. In addition, stereochemistry, conformation, and/or chemical dynamics can be studied using the experiments discussed in Section IV. The analysis of chemical shifts observed in both 1D and 2D NMR spectra is generally considered to be the most important stage in the interpretation process, followed by *J*-coupling and dipolar cross-relaxation analysis. Both manual and computer-assisted interpretation methods are used for small molecules, but the latter finds more favor in the analysis of larger molecules (see Sections V.C and V.D). To some extent, the level of understanding of the NMR parameters for a given nucleus is a function of the chemical importance of the corresponding atom, and the amount of work done by researchers in the field. As hydrogen and carbon make up the backbones of organic molecules, spectroscopy of the ^{1}H and ^{13}C nuclei has been of utmost importance and has received the most attention. Many other nuclei, such as ^{15}N, ^{19}F, ^{29}Si, and ^{31}P have found widespread applications in diverse areas of biochemistry, materials science, and polymer chemistry. A number of quadrupolar nuclei have also been the subject of extensive analyses. This chapter can only present a cursory overview of the information available from these nuclei. However, the general references supplied in the following should serve as a guide to the interested reader.

1. Proton NMR

Because of the sensitivity of the ^{1}H nucleus, its incorporation in so many compounds of chemical interest, and the wealth of information available from the observed chemical shift and J-coupling constants, ^{1}H has traditionally served as the cornerstone of NMR. Interpretation of proton NMR spectra is a fundamental skill taught to most undergraduate chemistry students. A great number of chemical shifts have been tabulated (Günther, 1998; Pretsch et al., 2000; Silverstein and Webster 1998) ranges for common organic functional groups are listed in Table 4. Proton chemical shifts are usually referenced to tetramethylsilane (TMS) at 0.0 ppm, often indirectly by tabulated values for residual ^{1}H in deuterated solvents (Silverstein and Webster 1998).

Several important factors affect ^{1}H shielding. The first of these is the electronegativity of substituents on the attached carbon or heteronucleus. In general (neglecting other effects), more electronegative substituents lead to more deshielding. For example, as carbon is more electronegative than hydrogen, the series CH_4, CH_3R, CH_2R_2, and CHR_3 trends from shielded (~0.2 ppm) to deshielded (~2.0 ppm). ppm). Electronegative elements like oxygen, nitrogen, and fluorine attached to a carbon will tend to deshield ^{1}H nuclei attached to the same carbon. Longer-range effects of electronegativity are also frequently observed. Diamagnetic anisotropy effects can alter the shielding of protons in many cases, such as the somewhat abnormal situation for acetylene. Similarly, ring current effects are seen with aromatic systems and resonance effects can occur in amides and other molecules.

The interpretation of coupling constants in ^{1}H NMR spectra is also an important process (Schaefer, 1996; Thomas, 1997). Despite the complexity of their fundamental parameters, ^{1}H–^{1}H two-bond (^2J), three-bond (^3J) and four-bond (^4J) coupling constants can be readily interpreted in a large number of cases. Empirical relationships have been derived from theoretical considerations and from model compounds. For example, the well-known Karplus relationship describes the torsional (or dihedral)

Table 4 Typical Chemical Shift Ranges for ^{1}H, ^{13}C, and ^{15}N Nuclei for Selected Organic Functional Groups (Berger et al., 1997; Kalinowski et al., 1988; Martin et al., 1981; Mason, 1996; Pretsch et al., 2000; Silverstein and Webster, 1998)

Functional group	^{1}H chemical shift range (ppm relative to TMS)	^{13}C chemical shift range (ppm relative to TMS)	^{15}N chemical shift range (ppm relative to CH_3NO_2)
Cyclopropyl	0.3–0.8	0–10	N/A
Alkyne	2.0–3.0	70–90	N/A
Aliphatics	1.0–3.5	10–40	N/A
—S—CH$_3$ and >N—CH$_3$	2.0–3.0	10–20 (—S—CH$_3$) 25–55 (>N—CH$_3$)	−380 to −300 (>N—CH$_3$)
Ethers	3.0–5.5	40–70	N/A
Amines (NH$_2$, NHR, NR$_2$)	3.6–4.6 (NH)	N/A	−380 to −300
Alcohols	(OH)	40–70	N/A
Phenols	4.0–10.0 (OH)	115–160	N/A
Alkenes	4.5–8.0	105–145	N/A
Aromatic (benzene)	7.0–8.5	110 to 160	N/A
Heteroaromatic (pyridine)	7.4–8.6 (CH)	125–150	−75
Heteroaromatic (pyrazine)	8.6 (CH)	145 (CH)	−46
Pyrroles	4.0–6.0 (NH) 6.0–7.0 (CH)	100–120 (CH)	−230
Imines	N/A	155–175	0 to −80
Furan/pyran	(CH)	110–160	N/A
Nitro	N/A	N/A	+20 to −50
Amides	5.0–9.0 (NH)	160–172	−260 to −290
Esters	N/A	162–175	N/A
Aldehydes	9.5–10.5	185–205	N/A
Ketones	N/A	195–225	N/A
Carboxylic acids	9.0–12.0	172–184	N/A

angle dependence for *vicinal proton couplings* ($^3J_{HH}$) (Thomas, 1997):

$$^3J_{HH} = A\cos\phi + B\cos^2\phi + C \qquad (42)$$

This equation was developed from electronic valence bond theory and is generally calibrated from known structural parameters (from crystallography and other methods). The relationship between torsional angle and $^3J_{HH}$ holds for a variety of compounds. For the common H—C—C—H system, the constants A, B, and C depend in a complex way on the electronegativity of the substituents, and also on the H—C—C bond angles. Fortunately, these constants have been derived empirically from studies on a great number of model compounds. One common choice is $A = 7.76$, $B = -1.10$, and $C = 1.40$ (Haasnoot et al., 1980). The empirical Karplus correlation is just one method for analysis of $^3J_{HH}$ values; more sophisticated equations with improved reliability have been derived (Haasnoot et al., 1980; Schaefer, 1996; Thomas, 1997). It is important to note that empirical correlations of this sort are only as good as their parameterization and should be used with caution in unknown or poorly characterized systems. Other empirical relationships have been studied for *geminal proton couplings* ($^2J_{HH}$) as a function of bond angle. For example, in aliphatic systems, $^2J_{HH}$ generally increases from 5 Hz toward 40 Hz as the H—C—H bond angle changes from 120° to 90° (Silverstein and Webster, 1998; Thomas, 1997). *Long-range couplings* ($^4J_{HH}$ and $^5J_{HH}$) are also important, and can either be detected in the 1D spectrum, in a 2D COSY spectrum, or in a 2D J-resolved experiment. For example, the well-known W-type ^4J coupling is widely used to analyze the stereochemistry of cyclic aliphatic organic compounds (Schaefer, 1996). $^4J_{HH}$ couplings as large as 25 Hz can be encountered in cases of strong π-orbital overlap (Thomas, 1997). Long-range couplings are also useful in everyday structural work; a routinely encountered example is $^4J_{HH}$ and $^5J_{HH}$ couplings (also known as meta- and para-couplings) in aromatic systems. With the advent of sophisticated *ab initio* coupling constant predictions, most of the empirical relationships noted here can also be theoretically studied and predicted, even for unknown systems (see Section V.C).

The most widely used NOE experiments are performed solely on ^{1}H spins, so interpretation of this data forms a major area of ^{1}H NMR spectroscopy. NOE effects are generally observed between nuclei within 3.0 Å of each other, but multiple-spin effects on cross-relaxation can confuse the issue. In addition, interpretation of NOE build-up rates and extraction of interatomic distance information can also be a complex process, but is generally achieved by assuming a total correlation time for a molecule, finding known groups to use as a calibration, and then using the r^{-6} dependence to find unknown distances. The topic is covered in dedicated reviews and will not be discussed in detail here.

2. Carbon-13

The ^{13}C nucleus serves as the second cornerstone of structural NMR, despite its fairly low sensitivity (about 0.02% that of ^{1}H) (Kalinowski et al., 1988; Stothers, 1972; Wehrli et al., 1988). Its enormous importance in organic chemistry and biochemistry stems from its role in the structural skeletons of so many molecules. Interpretation of ^{13}C spectra differs from that of ^{1}H spectra, primarily because of the lack of homonuclear coupling in the spectra. ^{13}C spectra are actually a sum of the subspectra of isotopomers, as the chance of encountering two ^{13}C nuclei is quite low ($\sim 1/10,000$). (Although ^{13}C–^{13}C J-coupling information can be recovered in INADEQUATE-type experiments, this is not a common occurrence in general ^{13}C work.) However, from the point of view of the chemist, interpretation of ^{13}C spectra is aided by parallels with ^{1}H NMR, in that the chemical shielding follows similar trends (albeit over a much larger 220 ppm range). The combined use of ^{13}C and ^{1}H NMR with correlation methods is particularly powerful, and can often circumvent spectral overlap in the individual spectra. Full ^{1}H and ^{13}C assignments are generally possible for most molecules up to larger biomolecules. Likewise, structural unknowns can be elucidated from the combination of chemical shift and J-coupling information.

Solution-state substituent effects on ^{13}C shifts are generally more predictable than those for ^{1}H, because of lessened dependence on conformation and dynamics. In addition, as the spectra cover a greater range, characteristic shifts for functional groups can often be unequivocally defined and used as "anchor points" for full assignments. A great number of shifts have been tabulated and large databases of known compounds are available. In general, ^{13}C sites are deshielded (to higher ppm) by electron withdrawing groups. Typical ^{13}C chemical shift values are given for many functional groups in Table 4. As in ^{1}H NMR, ^{13}C spectra are referenced to TMS at 0.0 ppm. The deuterated solvent resonance can be used for approximate referencing (Silverstein and Webster, 1998), or TMS, DSS (2,2-dimethyl-2-silapentane-5-sulfonic acid), or TSP (3-trimethylsilyl propionate, sodium salt) can also be employed for more exact work. In solid-state ^{13}C NMR, hexamethylbenzene and adamantane are widely employed as secondary chemical shift references.

The magnitude of one-bond J-couplings between ^{13}C and ^{1}H can be interpreted in terms of the amount of s-character in the bond. This can be useful for identification of three-membered rings (such as epoxides and cyclopropyl groups) directly from the ^{13}C satellites in the 1D ^{1}H spectrum. $^1J_{CH}$ values are positive and range between 120 and 280 Hz. The strong dependence on the degree of hybridization of the carbon is empirically given by (Kalinowski et al., 1988):

$$^1J_{CH} = 570\left(\frac{\%s}{100}\right) - 18.4 \qquad (43)$$

where %s represents the percentage of orbital s-character on the carbon in the C—H bond. Three-member rings exhibit unusually high $^1J_{CH}$ because of ring strain; for example, in the series cyclopropane, cyclobutane, cyclopentane, and cyclohexane, $^1J_{CH}$ values of, respectively, 160, 134, 128, and 125 Hz are found. For acetylene, $^1J_{CH}$ ranges as high as 249 Hz. However, most $^1J_{CH}$ values for organic molecules cluster around 150 Hz, so that this average is typically used to set delays in a variety of spectral editing and HETCOR pulse sequences. Whereas $^1J_{CH}$ is relatively easy to measure, $^1J_{CC}$ presents a challenge, as it only occurs in isotopomers containing two ^{13}C nuclei. $^1J_{CC}$ can be observed in natural abundance on concentrated solutions (Braun et al., 1998), or when isotopic labeled material is available. ($^1J_{CC}$ values are also obtained in the INADEQUATE series of experiments.) Interpretation of these couplings can be very useful when values are available (see Table 5 for typical values) (Kalinowski et al., 1988).

The longer-range nature of two- and three-bond couplings ($^{2/3}J_{CH}$) is extremely useful for structural analysis via the HMBC experiment, in addition to conformational and stereochemical (or configurational) analysis. The sign of longer range $^nJ_{CH}$ couplings is usually determined relative to the sign of a well-known coupling (such as a $^1J_{CH}$). From the stereochemical point of view, the vicinal ($^3J_{CH}$) coupling is generally the most important. A study of a number of rigid hydrocarbons yielded the following useful empirical relationship (Aydin and Günther, 1990):

$$^3J_{CH} = 4.50 - 0.87\cos\phi + 4.03\cos 2\phi \qquad (44)$$

The effects of α and β substitution on this equation can be very significant. Carbonyl groups, hydroxyl groups, and many other substituents can alter the predicted values by several hertz. These deviations can be taken into account in some cases, if model studies have been conducted for closely related substituents. However, as this is not always feasible, another possibility is the prediction of J-coupling constants by the latest theoretical methods (see Section V.C). Interpretation of J-couplings also plays an important role in *planar* structural analysis. One of the best-known examples is the relationships between $^2J_{CH}$, $^3J_{CH}$, and $^4J_{CH}$ on an aromatic ring (into the same carbon). Typically, the $^3J_{CH}$ value is in the neighborhood of 7.5 Hz, whereas the $^2J_{CH}$ and $^4J_{CH}$ are much smaller (1–2 Hz). However, these trends can be altered significantly by substituents and by inclusion of heteroatoms in the ring. Empirical knowledge of this sort is very useful for interpreting HMBC data sets and in making correct assignments. ^{13}C–^{19}F coupling can often be used in a similar manner, but are usually observed in ^{13}C-detected spectra. Other trends in longer-range couplings between ^{13}C and other nuclei (including ^{13}C) are not

Table 5 One-Bond J-Coupling Ranges for a Variety of Spin Pairs

Nuclei	^1J (Hz)	Comments
^1H–^1H	287[a]	H_2
^1H–^{11}B	100–190[a]	Terminal H
	<80[a]	H-bridge
^1H–^{13}C	96–320	
	125[a]	Ethane
	156.2[a]	Ethylene
	248.7[a]	Acetylene
^1H–^{15}N	−61.2[b]	Ammonia (NH_3)
	−73.3[b]	Ammonium (NH_4Cl in H_2O)
	−90.3[b]	Urea (in D_2O)
^1H–^{17}O	79	H_2O (gas)
^1H–^{29}Si	−147 to −382	
^1H–^{31}P	−200	P^{III} Compounds
	400–1100	P^{IV} Compounds
^6Li–^{13}C	6–17[c]	$[LiCH_3]_4$
^{11}B–^{11}B	14–28[a]	Difficult to resolve
^{11}B–^{13}C	18–76[a]	
^{11}B–^{19}F	1.4–189[a]	
^{11}B–^{31}P	30–100[a]	
^{13}C–^{13}C	34.6[d]	Ethane
	67.2[d]	Ethylene
	170.6[d]	Acetylene
^{13}C–^{15}N	5.8[d]	Tetramethylammonium
^{13}C–^{19}F	158[d]	CFH_3
	372[d]	$CFBr_3$
^{13}C–^{29}Si	−37 to −113[b]	
^{13}C–^{31}P	0 to −25	P^{III} Compounds
	45 to 300[a]	P^{IV} Compounds
^{15}N–^{15}N	19[a]	
^{15}N–^{29}Si	6–12[b]	
^{17}O–^{31}P	90–203[a]	Substituted phosphine oxides
^{19}F–^{17}O	424[c]	F_2O_2
^{19}F–^{29}Si	167–488[b]	
^{19}F–^{31}P	800–1500[a]	
^{29}Si–^{29}Si	53–186[b]	
^{29}Si–^{31}P	7–50[b]	
^{29}Si–^{195}Pt	−1600[b]	*trans*-$PtCl(SiCH_2Cl)(PEt_3)_2$
^{31}P–^{31}P	−620–800[b]	

[a]In Harris and Mann (1978).
[b]In Berger et al. (1997).
[c]In Günther (1996).
[d]In Kalinowski et al. (1988).

discussed here but can be found in the references (Kalinowski et al., 1988; Stothers, 1972; Wehrli et al., 1988).

3. Nitrogen-15

Nitrogen plays a critical role in many areas of chemistry, and has two stable isotopes available for study by magnetic resonance techniques (^{14}N and ^{15}N). Although ^{14}N has a natural abundance of >99%, as a quadrupolar nucleus it generally has broad lines in solution and is not as amenable to high-resolution work as the spin-$\frac{1}{2}$ ^{15}N isotope. (It has been used extensively in nuclear quadrupole resonance, and its effects are routinely observed in NMR via relaxation.) As previously noted, gradient-selected inverse detected ^{1}H–^{15}N correlation has led to a renewed interest in ^{15}N spectroscopy as a tool in structural characterization, especially since this nucleus plays an important role in pharmaceutical and natural product chemistry. Even before this renaissance, ^{15}N chemical shielding and J-coupling parameters were exhaustively determined by direct methods for a tremendous number of compounds (Berger et al., 1997; Martin et al., 1981; Mason, 1996). Nitrogen can participate in bonding with many different elements, and thus a great variety of trends have been discovered. Chemical shielding spans a large range of 1350 ppm (Mason, 1996), but most organic compounds of interest are confined to a smaller 400 ppm region of this area. Two reference compounds are used in ^{15}N NMR (nitromethane and ammonia); nitromethane is at 380.4 ppm on the ammonia scale. Amines, amides, imines, imides, nitriles, azo compounds, pyridine-like nitrogens, and many other common functional groups all resonate at well-spaced and fairly characteristic frequencies. Some representative ^{15}N shifts are listed in Table 4. Nitrogen chemical shielding can be very sensitive to medium effects, especially when hydrogen bonding to a lone pair on a nitrogen is possible. For example, amino ^{15}N sites are generally deshielded by increasing involvement in hydrogen bonding or by protonation. Pyridine and nitrile nitrogen groups are shielded under the same circumstances (Berger et al., 1997).

The magnitude of $^{1}J_{NH}$ is generally around 75–90 Hz (see Table 5 for examples) and signs are always negative, because of the negative gyromagnetic ratio of ^{15}N. The couplings can be correlated with nitrogen hybridization via (Berger et al., 1997):

$$\%s = 0.43|^{1}J_{NH}| - 6 \qquad (45)$$

However, this equation must be calibrated for specific compound classes, as surrounding bond angles can influence this value. This coupling finds use in HETCOR experiments as a transfer mechanism for directly attached protons, and its magnitude can be of use in a wide range of roles, including dynamics studies (Berger et al., 1997).

Values of $^{2}J_{NH}$ are generally ~2 Hz for aliphatic systems, but can be larger for aromatic systems. The $^{2}J_{NH}$ coupling found for pyridine is -10.8 Hz (Berger et al., 1997). (All of these couplings are observable in properly configured long-range correlation experiments) (Martin and Hadden, 2000). Three-bond Karplus-type relationships of the form of Eq. (42) also exist for nitrogen–proton couplings, and are of great use in studies of peptides and proteins using parameters are available in the literature (Berger et al., 1997; Wang and Bax, 1996). Homonuclear J-coupling involving nitrogen, and couplings with other heteronuclei, are also useful in some cases, but are usually difficult to detect without spin labeling (Berger et al., 1997). Exceptions occur for ^{15}N coupling to the abundant nuclei ^{19}F and ^{31}P.

4. Phosphorous-31

The critical role of phosphorous in many biological systems and materials, combined with its high sensitivity, has made it an important area for NMR applications (Berger et al., 1997; Gorenstein, 1983). ^{31}P is a spin-$\frac{1}{2}$ nucleus and is the only naturally occurring isotope of phosphorous. Inorganic phosphorus compounds and organic phosphines, phosphites, phosphonium salts, and a host of other compounds, all have proven amenable to ^{31}P NMR analysis. Overall, ^{31}P chemical shifts span a range of almost 700 ppm. The generally useful region of the ^{31}P spectrum is restricted to the area of about ±50 ppm around the reference at 0.0 ppm (85% H_3PO_4). Phosphorus chemical shift ranges are notoriously difficult to predict. Even different valences of phosphorus do not fall into discernable shielding patterns.

One-bond phosphorus homonuclear J-couplings cover a substantial range (Table 5) (Berger et al., 1997). Longer-range Karplus-type relationships exist for ^{31}P as well. For example, three H—C—O—P couplings are needed to determine torsional angles around the backbone subunits of deoxyribonucleic acid (DNA) and ribonucleic acid (RNA). Measurement of these $^{3}J_{HP}$ values helps determine the backbone conformation in solution using the following relationship (Berger et al., 1997; Thomas, 1997):

$$^{3}J_{\underline{H}CO\underline{P}} = 15.3\cos^2\phi - 6.2\cos\phi + 1.5 \qquad (46)$$

5. Other Important Nuclei

There are significant applications of NMR to other nuclei besides those already discussed. One of the first nuclei studied by NMR was ^{19}F, which is a spin-$\frac{1}{2}$ nucleus that is nearly as sensitive as ^{1}H. The range of ^{19}F chemical shielding covers ppm and includes many characteristic shifts. A great variety of ^{19}F chemical shifts have been tabulated, usually relative to $CFCl_3$ (Berger et al., 1997; Emsley and Phillips, 1971). Although rarely encountered in nature, fluorine is a terminal group in many synthetic organic molecules (as opposed to backbone elements like C and N), and can report on its local region of a molecule. This can be an advantage, as fluorine can function as a spin label or a tag, with particularly simple and background-free spectra. In the analysis of organic molecules, the relative magnitudes of fluorine J-couplings can be useful, especially for determining fluorination sites in conjunction with ^{1}H and ^{13}C spectra. However, Karplus-type relationships for $^3J_{FX}$ are complicated by through space effects (which have also limited theoretical methods; see Section V.C). Another important element with an NMR-active nucleus is silicon, which, like carbon, is tetravalent and serves as a backbone element in many polymers (usually in conjunction with oxygen). Silicon plays an increasingly important role in synthetic organic chemistry, in organometallics, reagents, and silyl tags and protecting groups. The ^{29}Si nucleus has a natural abundance of 4.7%, and a similar receptivity to that of ^{13}C. It also has a rich range of chemical shifts and J-couplings, which have been thoroughly reviewed (Marsmann, 1981). Applications of ^{29}Si NMR to the solid-state analysis of silicates and a wide variety of related materials have also been presented (Engelhardt and Michel, 1987). Finally, several quadrupolar nuclei with applications in organic and organometallic chemistry also deserve mention, namely, ^{17}O, ^{6}Li/^{7}Li, and ^{11}B (Berger et al., 1997; Boykin, 1991; Günther, 1996; Kennedy, 1987). Another quadrupolar nucleus with diverse applications throughout inorganic chemistry, environmental science, and electrochemistry is ^{27}Al (Akitt, 1989; Öhman and Edlund, 1996). This relatively sensitive nucleus has also been studied extensively in the solid-state, because of its fundamental role in zeolite and ceramic frameworks (Engelhardt and Michel, 1987). Although this list must necessarily be concise, there are many other NMR-active nuclei of great importance in various areas of science, and relevant reviews should be consulted for more information (Harris and Mann, 1978).

C. Predictive Methods

Methods for predicting NMR parameters such as chemical shifts and coupling constants take advantage of the voluminous empirical data available in the literature, or the great strides that have been made in quantum chemistry in recent years. For convenience, the methods can be grouped into several categories. *Additivity rules* make use of the bonding connectivity of the molecule, using tabulated empirical substituent effects and additivity rules, and are primarily limited to the prediction of chemical shifts. These have seen wide application to both ^{1}H and ^{13}C spectra through published tables and automated programs (Kalinowski et al., 1988; Pretsch et al., 1992; Schaller and Pretsch, 1994; Schaller et al., 1995; Tusar et al., 1992). For example, in fluorine-substituted toluenes, the carbon bearing the CH_3 (the *ipso* carbon) is predicted to have a shift of 124.7 ppm if the fluorine is meta to the CH_3, and 133.3 ppm if the fluorine is para to the CH_3 (Kalinowski et al., 1988). (The actual experimental values are 122.8 and 132.2 ppm, respectively.) This sort of additivity effect takes advantage of electric field effects, steric effects, resonance effects, inductive effects, and anisotropy considerations and has been tabulated for a large number of organic compounds. Additivity rules of this sort have also been obtained for other nuclei besides ^{1}H and ^{13}C (Berger et al., 1997; Marsmann, 1981).

Another class of predictive method are referred to here as *database-driven predictive methods*. These can range from basic search algorithms up to sophisticated learning algorithms, but have the common feature of using tabulated data as a source for predicting unknown parameters (as opposed to the additivity methods, which reduce the tabulated data into rules). Large databases have been used for this task for many years (Small, 1992). The algorithms used to pool the database values and predict chemical shifts include regression analysis techniques and neural networks (Jurs et al., 1992; Meiler et al., 2000).

From a quantum theoretical standpoint, the most advanced methods are those based directly on solving the electronic structure of a molecule given the 3D atomic coordinates. *Ab initio* methods (which begin from basic principles) are the most complex and computationally demanding, but can be highly accurate. Approaches to chemical shift and J-coupling prediction require accurate 3D structures. These structures can be obtained by a number of means, including single-crystal X-ray crystallography, NMR restraints, or computed structures. Even conformationally averaging structures can be studied if the dynamics are characterized and the structure of the conformers is known (or is at least predictable). The growth of density functional theory has enabled better predictions of NMR parameters than traditional wavefunction-based Hartree–Fock methods (Koch and Holthausen, 2001). Density functional theory has a balance of speed and accuracy that has outperformed other *ab initio* methods in most cases. Density functional theory methods can even predict J-coupling values that were previously unavailable (because of triplet instability problems in Hartree–Fock-type calculations). In addition, EFG tensors can be predicted for solid-state NMR work on quadrupolar nuclei. *Semiempirical* methods are similar to *ab initio* calculations, but are parameterized with experimental data and are less computationally demanding. Methods based on semiempirical calculations combined with a 3D view of substituent effects have had success in the prediction of ^{1}H chemical shifts (Abraham, 1999).

D. Computer-Assisted Structure Elucidation

All modern NMR data interpretation is computer-assisted to some extent; in this section we refer only to methods in which computer algorithms are used to identify the structure of a molecule either directly from the raw spectral data or indirectly from peak tables and user-supplied connectivity information. The empirical and theoretical considerations used in manual data interpretation can also be applied to computer-assisted interpretation. The difference is that a computer algorithm is used to match peaks and correlations in 1D and 2D data sets to a chemical structure (or structures). As it has been shown that it is possible in many cases to reliably predict spectral parameters given a chemical structure, there is ample reason to employ these methods in conjunction with computer-assisted interpretation.

1. Automatic Analysis of 1D Spectra

Solution-state ^{1}H NMR has received a great deal of attention from researchers because of the sensitivity of the nucleus and its facile application to high-throughput NMR. One of the first necessities of *computer-assisted structure interpretation* (CASE) is efficient identification and tabulation of useful signals. The most basic task is simply peak peaking, which often can be manually (and tediously) guided by the spectroscopist. As ^{1}H NMR signals are complicated by field-dependent homonuclear *J*-coupling, labile (exchangeable) protons, and dynamic phenomena, more sophisticated methods have been developed for automatic extraction of integrals and *J*-coupling constants, and subsequent conversion into numbers of protons and coupling constant value. There are several recent approaches to this problem (Golotvin et al., 2002; Griffiths, 2000; Griffiths and Bright, 2001, 2002; Laatikainen et al., 1996; Strokov and Lebedev, 1999). Once the parameters have been extracted from the 1D data set they can be compared with predicted values and ranked. The situation is similar for ^{13}C spectra, but is generally simplified as both resonance position and multiplicities are easily extracted from 1D spectra (including edited methods such as APT and DEPT).

2. Automatic Analysis of 2D Spectra

Pattern matching algorithms and other procedures can be applied to 2D spectra as well. In combination with the 1D data, molecular fragments may be identified using functional group trends and direct connectivity data (e.g., 1D, COSY, HMQC, and HSQC) and then assembled into molecules using longer range COSY or HMBC data. A recent example of a structural elucidation using CASE tools serves to illustrate the capabilities of these methods as applied to ^{1}H and ^{13}C NMR with a variety of 2D methods (Blinov et al., 2003). Commercial software available in this field includes NMR-SAMS (Spectrum Research LLC, Madison, WI) and ACD Structure Elucidator (ACD, Toronto, Canada).

3. Macromolecules

Macromolecules, especially biomolecules with detailed structures, are obvious candidates for CASE. The sheer complexity of the resonance assignment process for a single protein, for example, can require many weeks or months of human analysis time. The software tools used in this field are often closely integrated with molecular modeling and dynamics routines (Cavanagh et al., 1996; Neuhaus and Williamson, 2000). Higher-dimensional spectra are handled by the programs, which are specifically designed to locate structural motifs amongst the linked amino acid residues in proteins (in contrast to small molecule tools, which must handle arbitrary organic structures).

When used in conjunction with the predictive methods discussed above, CASE tools are a powerful assistant for structural determinations. It remains to be seen whether there will be general acceptance of these tools into NMR laboratories, as the currently available software has a number of drawbacks. These include proprietary architectures, high costs for purchasing and licensing software, and difficult user interfaces. Spectrometer vendors have recently begun development of CASE tools as a logical progression of the data processing software used on modern NMR spectrometers. With the continuing push toward greater sample throughput, often driven by parallel synthesis and combinatorial chemistry, it seems likely that CASE tools will eventually see widespread use for many applications.

VI. CONCLUSION

Modern NMR spectroscopy is an analytical technique with much to offer to the practicing scientist. As field strengths push higher and accessible sample quantities get lower, NMR cannot help but find even more applications. Some of the areas for future expansion include the use of RDCs to supplement and assist NOE work in biomolecules. Advances in solid-state NMR, including full assignments and structural analysis in the solid-state, are now becoming widely available (Lesage and Emsley, 2001; Potrzebowski, 2003). New 2D experiments have been developed that obtain the entire spectrum in a single scan, using gradient encoding (Shrot and Frydman, 2003). New NMR techniques for studying protein–protein and protein–ligand interactions in solution have begun to shed light on these binding events, and provide details on drug–receptor behavior (Meyer and Peters, 2003). The size of macromolecules which can be analyzed by NMR is still increasing (>25 kDa), with advances in isotopic labeling, and pulse sequences which interfere with dipole–dipole and chemical shift relaxation mechanisms to obtain higher resolution (the TROSY class of sequences) (Keniry and Carver, 2002). Finally, continued advances in hardware and software, greater access to cryogenic probes, and further sensitivity improvements are expected to advance the detection limits and power of NMR.

ACKNOWLEDGMENTS

The author thanks his colleagues and mentors for helpful discussions, especially Dr. Charles W. DeBrosse, Dr. Alan J. Freyer, Dr. Alan J. Benesi, Dr. Gerald Terfloth, and Prof. Karl T. Mueller. In addition, Dr. Alan J. Freyer is acknowledged for providing the data in Fig. 14, and Ms. Amanda L. Brown is thanked for help with preparation of the figures.

REFERENCES

Abragam, A. (1961). *The Principles of Nuclear Magnetism.* Oxford: Clarendon Press.

Abraham, R. J. (1999). A model for the calculation of proton chemical shifts in non-conjugated organic compounds. *Prog. NMR Spectrosc.* 35:85–152.

Akitt, J. W. (1989). Multinuclear studies of aluminum compounds. *Prog. NMR Spectrosc.* 21:1–149.

Ancian, B., Bourgeois, I., Dauphin, J. F., Shaw, A. A. (1997). Artifact-free pure absorption PFG-enhanced DQF-COSY spectra including a gradient pulse in the evolution period. *J. Magn. Reson.* 125:348–354.

Andrew, E. R., Bradbury, A., Eades, R. G. (1959). Removal of dipolar broadening of nuclear magnetic resonance spectra of solids by specimen rotation. *Nature* 183:1802.

Antzutkin, O. N. (1999). Sideband manipulation in magic-angle spinning NMR. *Prog. NMR Spectrosc.* 35:203–266.

Arnold, J. T., Dharmatti, S. S., Packard, M. E. (1951). *J. Chem. Phys.* 19:507.

Aydin, R., Günther, H. (1990). Carbon-13, proton spin-spin coupling. X. Norbornane: a reinvestigation of the Karplus curve for $^3J(^{13}C,^1H)$. *Magn. Reson. Chem.* 28:448–457.

Barjat, H., Chilvers, P. B., Fetler, B. K., Horne, T. J., Morris, G. A. (1997). A practical method for automated shimming with normal spectrometer hardware. *J. Magn. Reson.* 125:197–201.

Berger, S., Fäcke, T., Wagner, R. (1996). Two-dimensional correlation spectroscopy by scalar couplings: a walk through the periodic table. *Magn. Reson. Chem.* 34:4–13.

Berger, S., Braun, S., Kalinowski, H. O. (1997). *NMR Spectroscopy of the Non-Metallic Elements.* New York: Wiley.

Bielecki, A., Burum, D. P. (1995). Temperature dependence of ^{207}Pb MAS spectra of solid lead nitrate. An accurate sensitive thermometer for variable-temperature MAS. *J. Magn. Reson. A* 116:215.

Bildsoe, H., Donstrup, S., Jakobsen, H. J., Sorenson, O. W. (1983). Sub-spectral editing using a multiple quantum trap: analysis of J-cross talk. *J. Magn. Reson.* 53:154–162.

Blinov, K., Elyashberg, M., Martirosian, E. R., Molodtsov, S. G., Williams, A. J., Tackie, A. N., Sharaf, M. M. H., Schiff, P. L., Crouch, R. C., Martin, G. E., Hadden, C. E., Guido, J. E., Mills, K. A. (2003). Quindolinocryptotackieine: the elucidation of a novel indoloquinoline alkaloid structure through the use of computer-assisted structure elucidation and 2D NMR. *Magn. Reson. Chem.* 41:577–584.

Bloch, F. (1946). *Phys. Rev.* 70:460–474.

Blümich, B. (1996). Stochastic excitation. In: Grant, D. M., Harris, R. K., eds. *Encyclopedia of Nuclear Magnetic Resonance.* New York: Wiley, pp. 4581–4591.

Boykin, D. W. (1991). *^{17}O NMR Spectroscopy in Organic Chemistry.* Boca Raton: CRC Press.

Braun, S., Kalinowski, H. O., Berger, S. (1998). *150 and More Basic NMR Experiments.* 2nd ed. New York: Wiley-VCH.

Brown, D. E. (1995). Fully automated baseline correction of 1D and 2D NMR spectra using Bernstein polynomials. *J. Magn. Reson. A.* 114:268–270 (and references therein).

Brown, L. R., Sanctuary, B. C. (1991). Hetero-TOCSY experiments with WALTZ and DIPSI mixing sequences. *J. Magn. Reson.* 91:413–421.

Buckingham, A. D., Pople, J. A. (1963). High-resolution nuclear magnetic resonance spectra in electric fields. *Trans. Faraday Soc.* 59:2321–2330.

Cavanagh, J., Fairbrother, W. J., Palmer, A. G., Skelton, N. J. (1996). *Protein NMR Spectroscopy: Principles and Practice.* San Diego: Academic Press.

Chen, C. N., Hoult, D. I. (1989). *Biomedical Magnetic Resonance Technology.* New York: Adam Hilgar.

Chmurny, G. N., Hoult, D. I. (1990). The ancient and honorable art of shimming. *Concepts Magn. Reson.* 2:131–149.

Cho, S. I., Sullivan, N. S. (1992a). Parameterized description of NMR signal reception systems—Part I: impedance of transmission line and probe. *Concepts Magn. Reson.* 4:227–243.

Cho, S. I., Sullivan, N. S. (1992b). Parameterized description of NMR signal reception systems — Part II: S/N of NMR signal reception systems. *Concepts Magn. Reson.* 4:293–306.

Closs, G. L. (1974). Chemically induced dynamic nuclear polarization. *Adv. Magn. Reson.* 7:157–229.

Corio, P. L., Smith, S. L. (1996). Analysis of high-resolution solution-state spectra. In: Grant, D. M., Harris, R. K., eds. *Encyclopedia of Nuclear Magnetic Resonance.* New York: Wiley, pp. 2516–2520.

Derome, A. (1987). *Modern NMR Techniques for Chemistry Research.* Oxford: Pergamon.

Dickenson, W. C. (1950). *Phys. Rev.* 77:736.

Doddrell, D. M., Pegg, D. T., Bendall, M. R. (1982). Distortionless enhancement of NMR signals by polarization transfer. *J. Magn. Reson.* 48:323–327.

Doty, F. D. (1996a). Probe design and construction. In: Grant, D. M., Harris, R. K., eds. *Encyclopedia of Nuclear Magnetic Resonance.* New York: Wiley, pp. 3753–3761.

Doty, F. D. (1996b). Solid-state probe design. In: Grant, D. M., Harris, R. K., eds. *Encyclopedia of Nuclear Magnetic Resonance*. New York: Wiley, pp. 4475–4485.

Doty, F. D., Connick, T. J., Ni, X. Z., Clingan, M. N. (1988). Noise in high-power, high-frequency double-tuned probes. *J. Magn. Reson.* 77:536–549.

Eberstadt, M., Gemmecker, G., Mierke, D. F., Kessler, H. (1995). Scalar coupling constants—their analysis and their application for the elucidation of structures. *Angew. Chem. Int. Ed. Engl.* 34:1671–1695.

Eidmann, G., Savelsburg, R., Blümler, P., Blümich, B. (1996). The NMR MOUSE, a mobile universal surface explorer. *J. Magn. Reson. A* 122:104–109.

Ellet, J. D., Gibby, M. G., Haeberlen, U., Huber, L. M., Mehring, M., Pines, A., Waugh, J. S. (1971). Spectrometers for multi-pulse NMR. *Adv. Magn. Reson.* 5:117–176.

Emshwiller, M., Hahn, E. L., Kaplan, D. (1960). *Phys. Rev.* 118:414–424.

Emsley, J. W., Phillips, L. (1971). Fluorine chemical shifts. *Prog. NMR Spectrosc.* 7:1–520.

Emsley, J. W., Lindon, J. C. (1975). *NMR Spectroscopy Using Liquid Crystal Solvents*. New York: Pergamon.

Engelhardt, G., Michel, D. (1987). *High-Resolution Solid-State NMR of Silicates and Zeolites*. New York: Wiley.

Ernst, R. R., Anderson, W. A. (1966). *Rev. Sci. Instrum.* 37:93.

Ernst, R. R., Bodenhausen, G., Wokaun, A. (1987). *Principles of Nuclear Magnetic Resonance in One and Two Dimensions*. New York: Oxford University Press.

Facelli, J. C. (1996). Indirect coupling: semiempirical calculations. In: Grant, D. M., Harris, R. K., eds. *Encyclopedia of Nuclear Magnetic Resonance*. New York: Wiley, pp. 797–816.

Freeman, R. (1998). *Spin Choreography: Basic Steps in High Resolution NMR*. New York: Oxford University Press.

Freude, D., Haase, J. (1993). Quadrupole effects in solid-state nuclear magnetic resonance. *NMR Basic Prin. Prog.* 29:1–90.

Fukushima, E., Roeder, S. B. W. (1981). *Experimental Pulse NMR: A Nuts and Bolts Approach*. Reading, MA: Addison-Wesley.

Gerstein, B. C., Dybowski, C. R. (1985). *Transient Techniques in the NMR of Solids*. San Diego: Academic Press.

Golotvin, S., Vodopianov, E., Williams, A. (2002). A new approach to automated first-order multiplet analysis. *Magn. Reson. Chem.* 40:331–336.

Gorenstein, D. G. (1983). Non-biological aspects of phosphorus-31 NMR spectroscopy. *Prog. NMR Spectrosc.* 16:1–98.

Grant, D. M. (1996). Chemical shift tensors. In: Grant, D. M., Harris, R. K., eds. *Encyclopedia of Nuclear Magnetic Resonance*. New York: Wiley, pp. 1298–1321.

Griffiths, L. (2000). Towards the automatic analysis of ^{1}H NMR spectra. *Magn. Reson. Chem.* 38:444–451.

Griffiths, L., Bright, J. D. (2001). Towards the automatic analysis of ^{1}H NMR spectra: Part 2. Accurate integrals and stoichiometry. *Magn. Reson. Chem.* 39:194–202.

Griffiths, L., Bright, J. D. (2002). Towards the automatic analysis of ^{1}H NMR spectra: Part 3. Confirmation of postulated chemical structure. *Magn. Reson. Chem.* 40:623–634.

Gueron, M., Plateau, P., Decorps, M. (1991). Solvent signal suppression in NMR. *Prog. NMR Spectrosc.* 23:135–209.

Günther, H. (1996). Lithium NMR. In: Grant, D. M., Harris, R. K., eds. *Encyclopedia of Nuclear Magnetic Resonance*. New York: Wiley, pp. 2807–2828.

Günther, H. (1998). *NMR Spectroscopy*. New York: Wiley.

Haasnoot, C. A. G., DeLeeuw, F. A. A. M., Altona, C. (1980). The relationship between proton-proton NMR coupling constants and substituent electronegativities—I. *Tetrahedron* 36:2783–2792.

Hara, T., Iwata, Y., Ishii, H. (1997). Compact high-T_c superconducting power cable: summary and propects. In: *Advances in Superconductivity IX*. Proceedings of the International Symposium on Superconductivity. Springer, Tokyo.

Harris, F. J. (1978). On the use of windows for harmonic analysis with the discrete Fourier transform. *Proc. IEEE* 66:51–83.

Harris, R. K. (1996). Nuclear spin properties and notation. In: Grant, D. M., Harris, R. K., eds. *Encyclopedia of Nuclear Magnetic Resonance*. Wiley: New York, pp. 3301–3314.

Harris, R. K., Mann, B. E. (1978). *NMR and the Periodic Table*. London: Academic.

Hartmann, S. R., Hahn, E. L. (1962). *Phys. Rev.* 128:2042–2053.

Helgaker, T., Jaszunski, M., Ruud, K. (1999). Ab initio methods for the calculation of NMR shielding and indirect spin-spin coupling constants. *Chem. Rev.* 99:293–352.

Helgaker, T., Watson, M., Handy, N. C. (2000). Analytical calculation of nuclear magnetic resonance indirect spin-spin coupling constants and the generalized gradient approximation and hybrid levels of density functional theory. *J. Chem. Phys.* 113:9402–9409.

Heuer, A. (1991). A new algorithm for automatic phase correction by symmetrizing lines. *J. Magn. Reson.* 91:241–253 (and references therein).

Higinbotham, J., Marshall, I. (2000). NMR lineshapes and lineshape fitting procedures. *Ann. Rep. NMR Spectrosc.* 43:59–120.

Hilbers, C. W., MacLean, C. (1969). Electric field effects in nuclear magnetic resonance. *Mol. Phys.* 16:275–284.

Hill, H. D. W. (1996). Probes for high resolution. In: Grant, D. M., Harris, R. K., eds. *Encyclopedia of Nuclear Magnetic Resonance*. New York: Wiley, pp. 3762–3767.

Hoch, J. C., Stern, A. S. (1996). *NMR Data Processing*. New York: Wiley-Liss.

Homer, J., Perry, M. C. (1995). Enhancement of the NMR spectra of insensitive nuclei using PENDANT with long-range coupling constants. *J. Chem. Soc. Perkin Trans. 2*: 533–536.

Hoult, D. I. (1978). The NMR receiver: a description and analysis of design. *Prog. NMR Spectrosc.* 12:41–77.

Hoult, D. I., Richards, R. E. (1976). The signal-to-noise ratio of the nuclear magnetic resonance experiment. *J. Magn. Reson.* 24:71–85.

Hoult, D. I., Lauterbur, P. C. (1979). The sensitivity of the zuegmatographic experiment involving human samples. *J. Magn. Reson.* 34:425–433.

Hu, J., Javaid, T., Arias-Mendoza, F., Liu, Z., McNamara, R., Brown, T. R. (1995). A fast, reliable, automatic shimming procedure using ^{1}H chemical shift imaging spectroscopy. *J. Magn. Reson. B* 108:213–219.

Hurd, R. E. (1990). Gradient-enhanced spectroscopy. *J. Magn. Reson.* 87:422–428.

Hurd, R. E., John, B. K. (1991). Gradient-enhanced proton-detected heteronuclear multiple-quantum coherence spectroscopy. *J. Magn. Reson.* 91:648–653.

Hwang, T. L., Shaka, A. J. (1992). Cross-relaxation without TOCSY: transverse rotating-frame Overhauser effect spectroscopy. *J. Am. Chem. Soc.* 114:3157–3159.

Jameson, C. J. (1987). Chemical shielding. In: Mason, J., ed. *Multinuclear NMR*. New York: Plenum.

Jerosch-Herold, M., Kirschman, R. K. (1989). Potential benefits of a cryogenically cooled NMR probe for room-temperature samples. *J. Magn. Reson.* 85:141–146.

Jesson, J. P., Meakin, P., Kneissel, G. (1973). Homonuclear decoupling and peak elimination in Fourier transform magnetic resonance. *J. Am. Chem. Soc.* 95:618–620.

Johnson, C. S. (1996). Diffusion measurements by magnetic field gradient methods. In: Grant, D. M., Harris, R. K., eds. *Encyclopedia of Nuclear Magnetic Resonance*. New York: Wiley, pp. 1626–1644.

Jurs, P. C., Ball, J. W., Anker, L. S., Friedman, T. L. (1992). Carbon-13 number magnetic resonance spectrum simulation. *J. Chem. Inf. Comput. Sci.* 32:272–278.

Kalinowski, H. O., Berger, S., Braun, S. (1988). *Carbon-13 NMR Spectroscopy*. New York: Wiley.

Kaplan, D., Hahn, E. L. (1957). *Bull. Am. Phys. Soc.* 2:384.

Kaplan, J. I., Fraenkel, G. (1980). *NMR of Chemically Exchanging Systems*. New York: Academic.

Kasal, M., Halamek, J., Husek, V., Villa, M., Ruffina, U., Confrancesco, P. (1994). Signal processing in transceivers for nuclear magnetic resonance and imaging. *Rev. Sci. Instrum.* 65:1897–1902.

Kay, L. E., Keifer, P., Saarinen, T. (1992). Pure absorption gradient enhanced heteronuclear single quantum correlation spectroscopy with improved sensitivity. *J. Am. Chem. Soc.* 114:10663–10665.

Keniry, M. A., Carver, J. A. (2002). NMR spectroscopy of large proteins. *Ann. Rep. NMR Spectrosc.* 48:31–69.

Kennedy, J. D. (1987). Boron. In: Mason, J., ed. *Multinuclear NMR*. New York: Plenum.

Kessler, H., Mronga, S., Gemmecker, G. (1991). Multidimensional NMR experiments using selective pulses. *Magn. Reson. Chem.* 29:527–557.

Kessler, H., Gehrke, M., Griesinger, C. (1988). *Angew. Chem. Int. Ed. Engl.* 27:490–536.

Klevit, R. E. (1985). Improving two-dimensional NMR spectra by t_1 ridge subtraction. *J. Magn. Reson.* 62:551–555.

Koch, W., Holthausen, M. C. (2001). *A Chemist's Guide to Density Functional Theory*. New York: Wiley-VCH.

Kontaxis, G., Stonehouse, J., Laue, E. D., Keeler, J. (1994). The sensitivity of experiments which use gradient pulses for coherence-pathway selection. *J. Magn. Reson. A* 111:70–76.

Kover, K. E., Uhrin, D., Hruby, V. J. (1998). Gradient and sensitivity enhanced TOCSY experiments. *J. Magn. Reson.* 130:162–168.

La Mar, G. N., Horrocks, W. D., Holm, R. H., eds. (1973). *NMR of Paramagnetic Molecules*. New York: Academic Press.

Laatikainen, R., Niemetz, M., Weber, U., Sundelin, J., Hassinen, T., Vepsalainen, J. (1996). General strategies for total-lineshape-type spectral analysis of NMR spectra using integral-transform iterator. *J. Magn. Reson. A* 120:1–10.

Lacey, M. E., Subramanian, R., Olson, D. L., Webb, A. G., Sweedler, J. V. (1999). High-resolution NMR spectroscopy of sample volumes from 1 nL to 10 µL. *Chem. Rev.* 99:3133–3152.

Laukien, D. D., Tschopp, W. H. (1993). Superconducting NMR magnet design. *Concepts Magn. Reson.* 6:255–273.

Lesage, A., Emsley, L. (2001). Through-bond heteronuclear single-quantum correlation spectroscopy in solid-state NMR, and comparison to other through-bond and through-space experiments. *J. Magn. Reson.* 148:449–454.

Levitt, M. H. (1986). Composite pulses. *Prog. NMR Spectrosc.* 18:61–122.

Levitt, M. H. (2001). *Spin Dynamics: Basics of Nuclear Magnetic Resonance*. New York: John Wiley and Sons.

Lohman, J. A. B, MacLean, C. (1981). The determination of magnetic susceptibility anisotropies from quadrupolar interactions in NMR. *J. Magn. Reson.* 42:5–13.

Lowe, I. J. (1959). *Phys. Rev. Lett.* 2:285.

Lowe, I. J., Norberg, R. E. (1957). *Phys. Rev.* 107:46.

Marquez, B. L., Gerwick, W. H., Williamson, R. T. (2001). Survey of NMR experiments for the determination of $^nJ(C,H)$ heteronuclear coupling constants in small molecules. *Magn. Reson. Chem.* 39:499–530.

Marsmann, H. (1981). Silicon-29 spectroscopic results. *NMR Basic Principles and Progress.* 17:66–235.

Martin, G. E. (2002). Cryogenic NMR probes: applications. In: Grant, D. M., Harris, R. K., eds. *Encyclopedia of Nuclear Magnetic Resonance*. New York: Wiley, pp. 3762–3767.

Martin, M. L., Delpuech, J. J., Martin, G. J. (1980). *Practical NMR Spectroscopy*. London: Heyden.

Martin, G. E., Hadden, C. E. (2000). Long-range ^1H-^{15}N heteronuclear shift correlation at natural abundance. *J. Nat. Prod.* 63:543–585.

Martin, G. J., Martin, M. L., Gouesnard, J. P. (1981). ^{15}N NMR Spectroscopy. *NMR Basic Principles and Progress* 18:1–359.

Mason, J. (1996). Nitrogen NMR. In: Grant, D. M., Harris, R. K., eds. *Encyclopedia of Nuclear Magnetic Resonance*. New York: Wiley, pp. 3222–3251.

McDonald, S., Warren, W. S. (1991). Uses of shaped pulses in NMR: a primer. *Concepts Magn. Reson.* 3:55–81.

McKay, R. A. (1996). Probes for special purposes. In: Grant, D. M., Harris, R. K., eds. *Encyclopedia of Nuclear Magnetic Resonance*. New York: Wiley, pp. 3768–3771.

Meiler, J., Meusinger, R., Will, M. (2000). Fast determination of ^{13}C NMR chemical shifts using artificial neural networks. *J. Chem. Inf. Comput. Sci.* 40:1169–1176.

Meyer, B., Peters, T. (2003). NMR spectroscopy techniques for screening and identifying ligand binding to protein receptors. *Angew. Chem. Int. Ed. Engl.* 42:864–890.

Miner, V. W., Conover, W. W. (1996). Shimming of superconducting magnets. In: Grant, D. M., Harris, R. K., eds. *Encyclopedia of Nuclear Magnetic Resonance*. New York: Wiley, pp. 2585–2603.

Momo, F., Sotgiu, A., Testa, L., Zanin, A. (1994). Digital frequency synthesizers for nuclear magnetic resonance spectroscopy. *Rev. Sci. Instrum.* 65:3291–3292.

Neuhaus, D., Williamson, M. P. (2000). *The Nuclear Overhauser Effect in Structural and Conformational Analysis*. 2nd ed. New York: Wiley-VCH.

Öhman, L. O., Edlund, U. (1996). Aluminum-27 NMR of solutions. In: Grant, D. M., Harris, R. K., eds. *Encyclopedia of Nuclear Magnetic Resonance*. New York: Wiley, pp. 742–751.

Oki, M. (1985). *Applications of Dynamic NMR Spectroscopy to Organic Chemistry*. New York: Academic Press.

Peck, T. L., Magin, R. L., Lauterbur, P. C. (1995). Design and analysis of microcoils for NMR microscopy. *J. Magn. Reson. B* 108:114–124.

Perrin, C. L., Dwyer, T. J. (1990). Application of two-dimensional NMR to kinetics of chemical exchange. *Chem. Rev.* 90:935–967.

Potrzebowski, M. J. (2003). What high-resolution solid-state NMR spectroscopy can offer to organic chemists. *Eur. J. Org. Chem.* 8:1367–1376.

Pretsch, E., Fürst, A., Badertscher, M., Bürgin, R., Munk, M. E. (1992). C13Shift: a computer program for the prediction of ^{13}C NMR spectra based on an open set of additivity rules. *J. Chem. Inf. Comput. Sci.* 32:291–295.

Pretsch, E., Bühlmann, P., Affolter, C. (2000). *Structure Determination of Organic Compounds*. New York: Springer.

Purcell, E. M. (1946). *Phys. Rev.* 69:681.

Redfield, A. G., Gupta, R. K. (1971). Pulsed Fourier transform nuclear magnetic resonance spectrometer. *Adv. Magn. Reson.* 5:81–115.

Redfield, A., Kunz, S. D. (1994). Simple NMR input system using a digital signal processor. *J. Magn. Reson. A* 108:234–237.

Rokitta, M., Rommel, E., Zimmermann, U., Haase, A. (2000). Portable nuclear magnetic resonance imaging system. *Rev. Sci. Instr.* 71:4257–4262.

Russell, D. J., Hadden, C. E., Martin, G. E., Gibson, A. A., Zens, A. P., Carolan, J. L. (2000). A comparison of inverse-detected heteronuclear NMR performance: conventional vs. cryogenic microprobe performance. *J. Nat. Prod.* 63:1047–1049.

Schaefer, T. (1996). Stereochemistry and long range coupling constants. In: Grant, D. M., Harris, R. K., eds. *Encyclopedia of Nuclear Magnetic Resonance*. New York: Wiley, pp. 4571–4581.

Schaller, R. B., Arnold, C., Pretsch, E. (1995). New parameters for predicting ^{1}H NMR chemical shifts of protons attached to carbon atoms. *Anal. Chim. Acta* 312:95–105.

Schaller, R. B., Pretsch, E. (1994). A computer program for the estimation of ^{1}H NMR chemical shifts. *Anal. Chim. Acta* 290:295–302.

Shaka, A. J., Keeler, J. (1987). Broadband spin decoupling in isotropic liquids. *Prog. NMR Spectrosc.* 19:47–129.

Shaw, A. A., Salaun, C., Dauphin, J. F., Ancian, B. (1996). Artifact-free PFG-enhanced double-quantum-filtered COSY experiments. *J. Magn. Reson. A* 120:110–115.

Shrot, Y., Frydman, L. (2003). Single-scan NMR spectroscopy at arbitrary dimensions. *J. Am. Chem. Soc.* 125:11385–11396.

Silverstein, R. M., Webster, F. X. (1998). *Spectrometric Identification of Organic Compounds*. New York: Wiley.

Simon, B., Sattler, M. (2002). De novo structure determination from residual dipolar couplings by NMR spectroscopy. *Angew. Chem. Int. Ed. Engl.* 41:437–440.

Slichter, C. P. (1996). *Principles of Magnetic Resonance*. 3rd ed. Heidelburg: Springer-Verlag.

Small, G. W. (1992). Database retrieval techniques for carbon-13 nuclear magnetic resonance spectrum simulation. *J. Chem. Inf. Comput. Sci.* 32:279–285.

Smith, T. I., McAshan, M. S., Fairbank, W. (1986). Eur. Pat. Appl. EP1985-303259, 19850508, US Pat. Appl 1984-628452.

Sodickson, A., Cory, D. G. (1997). Shimming a high-resolution MAS probe. *J. Magn. Reson.* 128:87–91.

Sorensen, O. W., Eich, G. W., Levitt, M. H., Bodenhausen, G., Ernst, R. R. (1984). *Prog. NMR Spectrosc.* 16:163–192

Stejskal, E. O., Memory, J. D. (1994). *High Resolution NMR in the Solid State*. New York: Oxford University Press.

Stephenson, D. S. (1988). Linear prediction and maximum entropy methods in NMR spectroscopy. *Prog. NMR Spectrosc.* 20:515–626.

Sternin, E. (1995). Radio frequency phase shifting at the source simplifies NMR spectrometer design. *Rev. Sci. Instrum.* 66:3144–3145.

Stothers, J. B. (1972). *Carbon-13 NMR Spectroscopy*. New York: Academic Press.

Strokov, I. I., Lebedev, K. S. (1999). Computer aided method for chemical structure elucidation using spectral databases and ^{13}C NMR correlation tables. *J. Chem. Inf. Comput. Sci.* 39:659–665.

Styles, P., Soffe, N. F., Scott, C. A., Cragg, D. A., Row, F., White, D. J., White, P. C. J. (1984). A high-resolution NMR probe in which the coil and preamplifier are cooled with liquid helium. *J. Magn. Reson.* 60:397–404.

Sudmeier, J., Anderson, S. E., Frye, J. S. (1990). Calculation of nuclear spin relaxation times. *Concepts Magn. Reson.* 2:197–212.

Tannús, A., Garwood, M. (1997). Adibatic pulses. *NMR Biomed.* 10:423–434.

Thomas, W. A. (1997). Unraveling molecular structure and conformation—the modern role of coupling constants. *Prog. NMR Spectrosc.* 30:183–207.

Traficante, D. D. (1989). Impedance: what it is, and why it must be matched. *Concepts Magn. Reson.* 1:73–92.

Traficante, D. D. (1993). Introduction to transmission lines: basic principles and applications of quarter-wavelength cables and impedance matching. *Concepts Magn. Reson.* 5:57–86.

Turner, C. J. (1996). Heteronuclear assignment techniques. In: Grant, D. M., Harris, R. K., eds. *Encyclopedia of Nuclear Magnetic Resonance*. New York: Wiley, pp. 2335–2343.

Tusar, M., Tusar, L., Bohanec, S., Zupan, J. (1992). ^{1}H and ^{13}C NMR spectra simulation. *J. Chem. Inf. Comput. Sci.* 32:299–303.

van Geet, A. L. (1970). Calibration of methanol nuclear magnetic resonance thermometer at low temperature. *Anal. Chem.* 42:679–680.

van Zijl, P. C. M., Sukumar, S., O'Neil-Johnson, M., Webb, P., Hurd, R. E. (1994). Optimized shimming for high-resolution NMR using three-dimensional image-based field mapping. *J. Magn. Reson. A* 111:203–207.

Villa, M., Tian, F., Confrancesco, P., Halamek, J., Kasal, M. (1996). High-resolution digital quadrature detection. *Rev. Sci. Instrum.* 67:2123–2129.

Vogt, F. G., Mueller, K. T. (2002). Orientationally-broadened NMR spectra: analysis by special processing methods. In: Grant, D. M., Harris, R. K., eds. *Encyclopedia of Nuclear Magnetic Resonance*. Vol. 9. New York: Wiley, pp. 112–125.

Wagner, R., Berger, S. (1996). Gradient-selected NOESY—a fourfold reduction of the measurement time for the NOESY experiment. *J. Magn. Reson. A* 123:119–121.

Wang, A. C., Bax, A. (1996). Determination of backbone dihedral angles ϕ in human ubiquitin from reparameterized empirical Karplus equations. *J. Am. Chem. Soc.* 118:2483–2494.

Webb, G. A. (1996). Shielding: overview of theoretical methods. In: Grant, D. M., Harris, R. K., eds. *Encyclopedia of Nuclear Magnetic Resonance*. New York: Wiley, pp. 4307–4318.

Wehrli, F. W., Marchand, A. P., Wehrli, S. (1988). *Interpretetation of Carbon-13 NMR Spectra*. New York: Wiley.

Willker, W., Leibfritz, D., Kerssebaum, R., Bermel, W. (1993). Gradient selection in inverse heteronuclear correlation spectroscopy. *Magn. Reson. Chem.* 31:287–292.

Wittebort, R. J., Adams, J. (1992). High-speed phase and amplitude modulator. *J. Magn. Reson.* 96:624–630.

Wu, K. C., Brown, D. P., Schlafke, A. P., Sondericker, J. H. (1984). Helium refrigerator with features for operation at supercritical pressure. *Adv. Cryogenic Eng.* 29:495–501.

Wuthrich, K. (1986). *NMR of Proteins and Nucleic Acids*. New York: Wiley.

Yun, J., Yu, J., Hongyan, T., Gengying, L. (2002). A complete digital radio-frequency source for nuclear magnetic resonance spectroscopy. *Rev. Sci. Instrum.* 73:3329–3331.

12

Electron Paramagnetic Resonance

SANDRA S. EATON, GARETH R. EATON
Department of Chemistry and Biochemistry, University of Denver, Denver, Colorado, USA

I. INTRODUCTION

Electron paramagnetic resonance (EPR), also sometimes called electron spin resonance (ESR) or electron magnetic resonance (EMR), is a powerful technique for examining the environment of unpaired electrons in organic free radicals or transition metal complexes, present in biological systems or materials. To perform these measurements meaningfully and quantitatively requires knowledge of the nature of the analytical instrument. Consequently, this chapter focuses on the instrumental and experimental aspects of EPR. Throughout the chapter, maximal use will be made of references to other readily available sources for theoretical background, tabulations of spectral parameters, alternative spectrometer configurations, and so on. The chapter attempts to be a fairly comprehensive guide to most of what the novice will need to learn eventually, but wherever good discussions already exist reference will be made to them rather than repeating the details here. For the mathematical background and general theory of continuous wave (CW) EPR, the reader is guided to the books by Weil et al. (1994) and by Atherton (1973). Introductory treatments are given in Carrington and McLachlan (1967), Drago (1992a, b), Pake and Estle (1973), and Slichter (1990). The principles of pulsed EPR are given in the book by Schweiger and Jeschke (2001). The books by Alger (1968), Poole (1967, 1983), and Wilmshurst (1968, Chapter 4) provide information on EPR instrumentation. Comments in this chapter concerning commercial instruments are based primarily on Bruker spectrometers because these are widely available in the USA. Information concerning the earlier Varian spectrometers, which are no longer sold, can be found in the 1st and 2nd editions of this *Handbook*. After the sample has been properly prepared, and the spectra have been competently obtained, the data have to be interpreted. For this aspect of EPR, in addition to the texts cited earlier, see the following references for spectra of organic radical (Gerson, 1970), for transition metal ions (Abragam and Bleaney, 1970; Pilbrow, 1990), for applications to biological systems (Berliner, 1976, 1979, 1998; Berliner and Reuben, 1989, 1993; Berliner et al., 2000; Dalton, 1985; Likhtenshtein, 1976; Swartz et al., 1972), and for general discussions (Gordy, 1980; Harriman, 1978; Poole and Farach, 1971, 1972, 1994, 1999). References to examples of EPR spectra for many types of paramagnetic species were given in the first edition of this *Handbook*.

EPR is a technique for measuring the absorption of electromagnetic radiation by a sample with unpaired electron spins. Usually the sample is placed in a magnetic field, and the transitions that are monitored are between Zeeman levels that differ by ± 1 in m_S, where m_S is the quantum number for the projection of the electron spin along the direction of the external magnetic field, which is labeled as the z-axis. The physical phenomenon was discovered by Zavoisky in Kazan, Russia, in 1944, building upon prior studies of nonresonant absorption and relaxation, especially by Gorter in the Netherlands. Much of the early development of EPR occurred in the laboratory of Bleaney at Oxford. The history of the first 50 years of EPR is described in the book, "Foundations of Modern EPR" (Eaton et al., 1998b).

The most common experiments use ~9.0–9.6 GHz microwaves (in the X-band), for which the free-electron resonance occurs at ~3200–3400 G (0.32–0.34 T).[1] However, an increasingly wide range of resonant frequencies are being explored (Eaton and Eaton, 1999a, 2004).

A variety of paramagnetic centers can be studied by EPR, including organic free radicals, radiation-induced radicals, paramagnetic transition metal complexes, triplet species, defect centers in solids, and conduction electrons in metals. The information that can be obtained from EPR spectra is discussed in Section II. The samples can be in the solid, liquid, or gas phase; a wide range of examples is described in "EPR in the 21st century" (Kawamori et al., 2002). The sample is discussed in Section IV. Common applications of EPR include studies of electronic structure, reaction kinetics and mechanism, radiation damage, defect centers of semiconductors, organic triplet states, transition metal ions, and spin labels or spin probes attached to biological molecules.

There are fundamental similarities in the physics of NMR (nuclear magnetic resonance) and EPR. However, several characteristics of EPR are different from NMR:

1. The sign of the electron moment is the opposite of that of the proton moment.
2. EPR spectra at a fixed frequency commonly extend over a wider range of magnetic fields than do NMR spectra.
3. EPR relaxation times are much shorter than NMR relaxation times, which restrict the use of multiple-pulse and Fourier-transform (FT) techniques in EPR at room temperatures to organic radicals or requires the use of cryogenic temperatures.
4. The wavelengths used in EPR are much shorter than those used in NMR. For example, the X-band wavelength is ~3 cm, whereas the wavelength for 200 MHz NMR is 150 cm. Some lumped circuit designs commonly used in NMR cannot be readily carried over into EPR.

5. The EPR sample interacts with the EPR spectrometer to a greater extent than samples generally interact with most spectrometers, including NMR.
6. EPR has enjoyed less widespread use than NMR. The manufacturers of EPR spectrometers have commonly assumed that the user was more sophisticated about the use of the spectrometers than the users of NMR spectrometers. Consequently, the EPR user has to know more about the way the spectrometer works than do users of other types of commercial magnetic resonance spectrometers.

It is the interaction of the sample with the spectrometer that results in the special focus of this chapter. The emphasis throughout this text on resonator Q, microwave B_1, and modulation amplitude reflect a need for the user of EPR to have at least a phenomenological understanding of the EPR instrument (magnet, microwave, and signal detection systems) at the level provided by this chapter. There are many aspects of these systems that are important to the design of the spectrometer, but the readers of this book will be able to assume that these aspects are fully engineered and are "transparent to the user". Those who seek to build their own instruments will need a different type and level of knowledge of the individual microwave components. For this information refer to EPR books by Alger (1968), Poole (1967, 1983), Wilmshurst (1968), microwave books (Baden Fuller, 1990; Blackburn, 1949; Laverghetta, 1976, 1981; Marcuvitz, 1951; Montgomery, 1947; Montgomery et al., 1948; Ragan, 1948; Reich et al., 1953; Smullin and Montgomery, 1948), and papers describing EPR spectrometers (Alecci et al., 1992; Halpern et al., 1989; Huisjen and Hyde, 1974a; Mailer et al., 1985; Norris et al., 1980; Percival and Hyde, 1975; Quine et al., 1987, 1992, 1996, 2002).

A. EPR Experiment

An electron has quantized angular momentum. Although both orbital angular moment, L, and spin angular momentum, S, contribute, the fundamentals of EPR are best illustrated by first considering just the spin angular momentum for a single unpaired electron, $S = 1/2$. When an electron is placed in a magnetic field, the projection of the magnetic moment on the axis defined by the magnetic field (conventionally called the z-axis) takes two discrete values, $m_S = +1/2$ and $m_S = -1/2$. The energy for the $m_S = +1/2$ level is greater than the energy for the $m_S = -1/2$. The separation between these levels is linearly proportional to magnetic field strength, B (Fig. 1). When the energy of the electromagnetic radiation is equal to the separation between the spin energy levels, transitions between the spin states occur. The transitions are magnetic-dipole allowed, and are effected by the component of the

---

[1] Adoption of SI units is occurring slowly in the magnetic resonance literature. 10,000 Gauss (G) = 1 Tesla (T), 1 ampere-turn/meter = $4\pi \times 10^{-3}$ oersteds. In most cases, the assumption is made that the magnetic field inside the sample is the same as the external magnetic field; that is, gauss and oersted are not usually distinguished. Early literature used units of gauss.

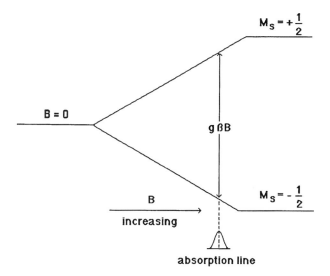

Figure 1 Zeeman energy level splitting for an electron in a magnetic field. The energy separation is linearly proportional to magnetic field strength, B. Transitions between the two electron energy levels are stimulated by microwave radiation when $h\nu = g\beta B$. If the lineshape is due to relaxation, it is Lorentzian. The customary display in EPR spectroscopy is the derivative of the absorption line that is shown here.

microwave magnetic field, B_1, perpendicular to the magnetic field created by the external electromagnet. The sample is placed in a resonator that maximizes the perpendicular B_1 field at the sample. The match of $h\nu$ with $g\beta B$ is called resonance; hence the name electron paramagnetic resonance. β is a fundamental constant and g is a characteristic value for a particular sample. The following calculation illustrates the numerical values associated with the EPR experiment at X-band at 3400 G and room temperature.

$$\Delta E = g\beta B$$

$$\Delta E = (2.0023)(0.92731 \times 10^{-20}\,\text{erg}\,\text{G}^{-1})(3400\,\text{G})$$

$$= 6.3129 \times 10^{-17}\,\text{erg}$$

Then $\Delta E/kT = 1.52 \times 10^{-3}$ and $e^{\Delta E/kT} = 0.998477$. Thus, the energy difference between the two energy levels is a very small fraction of thermal energies at room temperature and the population difference between the levels is only $\sim 0.2\%$. Conversion of ΔE to frequency units (1 erg = $1.5094 \times 10^{26}\,\text{s}^{-1}$) yields 9.53 GHz; in wave numbers, this is 0.3 cm^{-1}. The g-value usually is dependent on the orientation of the molecule with respect to the external magnetic field, which is called anisotropy.

The magnetic field of neighboring nuclei or electrons can add to or subtract from the external magnetic field, resulting in spin–spin splitting analogous to the nuclear spin–spin splitting in NMR. This splitting for coupling to a single nucleus with $I = 1/2$ is illustrated in Fig. 2.

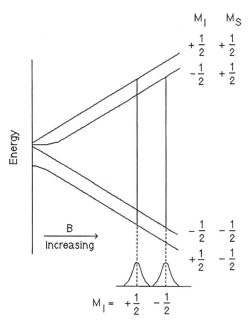

Figure 2 The coupling of nuclear spins to the electron spin splits the energy levels shown in Fig. 1 into additional levels. The nuclear hyperfine coupling, and the selection rule that nuclear spins do not flip when electron spins flip, results in $2nI+1$ absorption lines in the EPR spectrum, where n is the number of equivalent nuclei and I is the nuclear spin quantum number. In the usual EPR experiment the microwave frequency is held constant and the magnetic field is swept. Resonance absorption of microwave energy occurs at each magnetic field for which the nuclear-hyperfine-shifted energy levels are separated by the energy equivalent of the microwave quantum, $h\nu$, and the selection rules are $\Delta m_S = \pm 1$ and $\Delta m_I = 0$.

If hyperfine couplings are less than linewidths, the couplings are not resolved. These smaller couplings can be measured by ENDOR (Section I.D) or echo envelope modulation (Section I.C.2).

EPR experiments fall into two general categories. In a CW experiment the microwaves are continuous, the magnetic field is scanned to achieve resonance, and the signal is reported as a function of magnetic field. In a pulse experiment high-power microwaves are applied to the sample, often at constant magnetic field, and the response of the spin system is monitored. CW experiments are discussed first, followed by pulsed experiments, including Fourier-transform EPR (FT-EPR), electron spin echo (ESE), and saturation recovery (SR), which are discussed in Section I.C.

The instrumentation needed for a CW EPR experiment consists of four partially separate subsystems: (a) microwave system including source, cavity or other resonator, and detector; (b) magnet system including power supply and field controller; (c) modulation and phase-sensitive signal detection system; and (d) data display and manipulation system, including oscilloscope or computer display,

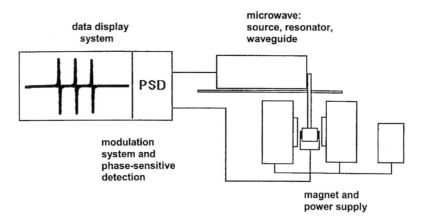

Figure 3 The fundamental modules of a CW EPR spectrometer include (a) the microwave system including source, cavity or other resonator, and detector; (b) the magnet system including power supply and field controller; (c) the modulation and phase-sensitive detection system; and (d) the data display and manipulation system. Each is described in detail in the text and may be largely independent of other modules. The user interaction with the modules is typically via a computer.

analog to digital converter (A/D), and computer (Fig. 3). The subsystems are discussed in detail in Section III.

B. CW Spectroscopy

1. Use of Magnetic Field Modulation

Because an understanding of the process by which the signal is detected in CW experiments is fundamental to the rest of the information in this chapter, the topic is discussed here, rather than in Section III. As a first step it is necessary to recognize that the spectrometer measures absorption of microwave energy by the sample. Absorption of energy by the sample changes the microwave energy in the standing wave pattern in the cavity (i.e., it changes the cavity Q, see Section III.A), which modifies the match between the cavity and the waveguide, thereby changing the amount of microwave energy reflected back from the cavity to a detector. If the magnetic field is scanned slowly through resonance, the absorption of energy for a nitroxyl radical looks like Fig. 4. There are three lines because of the coupling to the ^{14}N nucleus ($I = 1$, 99.63% abundance). The three lines have equal intensity because the three nitrogen m_I values (1, 0, −1) have equal probability. There are small lines on each side of these three lines that are the contribution to the spectrum from molecules that contain a ^{13}C ($I = 1/2$, 1.11% abundance), which splits the signal into two lines.

The problem with direct-detection of the EPR signal is that the signal-to-noise (S/N) is poor. Throughout science, a common method of improving S/N is the use of phase-sensitive detection. Commercial CW EPR spectrometers use magnetic field modulation and phase-sensitive detection at the modulation frequency. The physical principles

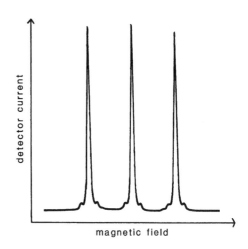

Figure 4 The spectrum shown here is the absorption spectrum (integral of the usual first-derivative spectrum) of a nitroxyl radical. At each magnetic field for which there is resonant microwave energy absorption, there is a change in cavity Q, which results in a reflection of microwave power from the cavity, and hence an increase in detector crystal current. Resonance occurs at a different magnetic field for radicals in which the nitrogen nuclear spin state, m_I, is 1, 0, or −1.

are, briefly:

1. *Modulation.* A lower frequency variation (called modulation) is imposed on a high-frequency signal. In EPR the high frequency is the microwaves, and the lower frequency is magnetic field modulation, usually at 100 kHz.
2. *Phase-sensitive detection.* The low-frequency signal imposed on the detected microwave signal is compared with the wave form that generated the low-frequency modulation. Only the components of the signal that have a particular phase

relative to that of the original generating signal pass through the circuit. As the noise is random, most of the noise is rejected and the S/N is improved. The output of a modulation and phase-sensitive detection system is a signal that is the first derivative of the absorption signal.

Although the typical 100 kHz modulation is a low frequency relative to the microwave frequency (10^5 vs. 10^{10} Hz), it is a very high frequency relative to the magnetic field scan rate (e.g., 0.1 G modulation at 100 kHz results in a maximum field change of $\pi \times 10^4$ G/s and a 4 min scan of 100 G is $\sim$0.4 G/s). The modulation is achieved by applying a sinusoidal voltage, varying at 100 kHz, to coils on the walls of the cavity, which creates an oscillating magnetic field. The magnetic field generated in the coils adds to or subtracts from the magnetic field produced by the large electromagnet. As the magnetic field of the electromagnet is swept slowly, you can consider it to be constant during a few cycles of the modulation field. As illustrated in Fig. 5, the 100 kHz modulation field causes the magnetic field seen by the sample to vary between B_{m1} and B_{m2} 100,000 times per second. The difference $B_{m2} - B_{m1}$ is equal to the "modulation amplitude" setting. If the modulation amplitude is set to 2 G, the magnetic field seen by the sample oscillates from 1 G below the magnetic field established by the electromagnet to 1 G above the field established by the electromagnet. When the 100 kHz modulation causes the magnetic field to be at B_{m1}, the detector current is i_1. As the field increases sinusoidally from B_{m1} to B_{m2}, the detector current increases sinusoidally from i_1 to i_2. The amplitude of the 100 kHz modulation of the detector output is proportional to the slope of the absorption signal between B_{m1} and B_{m2}. Note that the phase of the detected signal for the rising portion of the absorption signal is the inverse of that for the falling portion of the absorption signal. Thus the output of the phase-sensitive detector gives the value of the derivative (slope) of the absorption signal. However, the signal will be a true derivative only if the absorption curve is a straight line between B_{m1} and B_{m2}. This is never exactly true, but an adequately close approximation is achieved if $B_{m2} - B_{m1}$ is less than 1/10 of the distance between the inflection points of the absorption curve. These inflection points in the absorption spectrum correspond to the maximum and minimum in the first-derivative curve. The separation between the maximum and minimum in the first derivative spectrum is denoted as ΔB_{pp}. For a more detailed picture, see Poole (1967, Fig. 10-3).

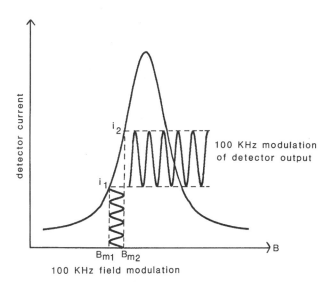

Figure 5 This figure displays how the detector current changes as a function of magnetic field modulation (here assumed to be at 100 kHz). Relative to the modulation frequency, the sweep of the magnetic field generated by the electromagnet can be considered to be stationary. The 100 kHz modulation of the current in the coils associated with the cavity results in a small magnetic field that oscillates at 100 kHz between values that symmetrically add to and subtract from the main field. As a result, the magnetic field oscillates at 100 kHz between B_{m1} and B_{m2}. As the absorption of microwave energy depends on magnetic field, the detector crystal current oscillates at 100 kHz also, between i_1 at B_{m1} and i_2 at B_{m2}. If $B_{m1} - B_{m2}$ is small enough relative to the linewidth of the EPR signal, $i_2 - i_1$ is proportional to the slope of the absorption signal. The output of the detector crystal is amplified and compared with the output of the 100 kHz oscillator in the phase-sensitive detector, resulting in a first-derivative display.

2. Dispersion Effects on EPR Spectra

The EPR signal from a sample is proportional to the microwave magnetic susceptibility (Rinard et al., 1999a), which has real and imaginary components. EPR spectrometers usually are tuned to detect the absorption signal, which is the imaginary part of the microwave susceptibility. In any region of the electromagnetic spectrum, including EPR, absorption is accompanied by dispersion, which is the real part of the microwave magnetic susceptibility. Spectrometers can be designed to detect the dispersion signal by using two detector crystals. There are experiments for which this is important (Poole, 1983). The fundamental paper describing dispersion lineshapes is Pake and Purcell (1948). In some cases the fact that the dispersion signal saturates much less readily than the absorption signal can be used to advantage (DeRose and Hoffman, 1995; Gaffney and Silverstone, 1993; Gaffney et al., 1993; Gulin et al., 1992; Hales et al., 1989; Harbridge et al., 2002b; Hyde et al., 1982; Un et al., 1994).

Even though the spectrometer is designed to detect the absorption signal, the observed lineshape may be a mixture of absorption and dispersion. At the peak of the absorption signal, the dispersion component has a zero-crossing, so mixing of a dispersion component into the detected EPR signal causes asymmetry in the lineshape. For example, the conductivity and dielectric absorption of some solutions can provide sufficient mixing of the real and imaginary components of the microwave magnetic susceptibility to cause asymmetry in the lines in the spectrum. Published examples of such effects include spectra of 75 mM aqueous vanadyl solutions (Rogers and Pake, 1960) and 0.05–0.6 M aqueous solutions of Cr^{3+} and Fe^{3+} (Levanon et al., 1970). Related work on the EPR of paramagnetic ions in metal was reported (Peter et al., 1962). These papers and that of Poole (1983, p. 490) contain several diagrams illustrating mixtures of absorption and dispersion lineshapes. The mixture of absorption and dispersion is easy to observe on a routine spectrometer. The standard irradiated fused silica sample (Eaton and Eaton, 1993a) sold by Wilmad as WGSR-01-4 has a long electron spin relaxation time T_1 and spin–spin relaxation time (see later), and the CW absorption spectrum mixes with the dispersion spectrum unless the frequency is very carefully set and controlled. Hence, this sample is a good monitor of the functioning of the automatic frequency control (AFC) circuit of the spectrometer (Ludowise et al., 1991).

3. Passage Criteria

The usual treatments of EPR spectroscopy assume slow-passage conditions, unless stated otherwise. In practice, routine experiments may not actually satisfy slow-passage criteria. Hence, it is important to understand passage criteria at least to the point of knowing whether failure to achieve slow-passage conditions could be affecting the experimental spectrum. Physically, the terms "slow", "rapid", and "fast" passage describe the relationship of the spectral scan rate ("passage" through the spectrum) to relaxation times (Weger, 1960) (also, see Alger, 1968, Table 2-1).

1. *Slow passage.* The spin system is always in thermal equilibrium.
2. *Rapid passage.* The modulation field sweeps through a spin packet in a time shorter than the relaxation time of the spin packet.
3. *Fast passage.* The electron spins do not have time to relax between modulation sweep cycles.

Although the criteria can become quite complicated when one considers the wide range of relaxation times and experimental conditions that might be realized in various experiments, the essential features are that one should avoid modulation frequencies that are faster than the electron spin relaxation rate, and the microwave power should be kept well below saturation. If it is not possible to satisfy these criteria, the power and modulation amplitude should be kept as low as possible because the key parameters are the degree of saturation and the magnetic field sweep rate in Gauss per second. In the extreme case of partial saturation and modulation frequency that is fast relative to relaxation, the EPR spectrum can look like an absorption rather than a derivative, and invert when the field is scanned in the reverse direction (Fig. 6). When in doubt about the conditions of saturation and modulation, comparing spectra obtained with reverse scans and different scan rates is good practice. With organic radicals, or with some metals at low temperature, it is difficult to achieve slow-passage conditions. Often,

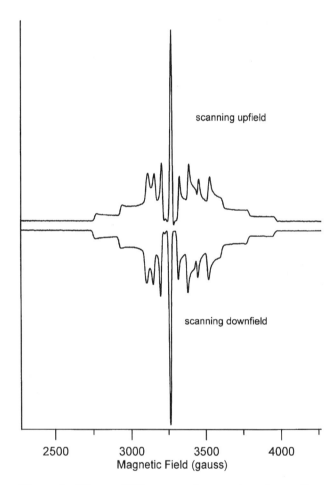

Figure 6 When an EPR spectrum is scanned too fast to allow the spins to be at thermal equilibrium, passage effects occur. In extreme cases the first-derivative spectrum looks like absorption spectrum, and the spectrum turns upside down when the field-scan direction is reversed. This figure illustrates passage effects for a vanadyl porphyrin complex at ~5 K. For less extreme cases of passage effects, distorted spectra are observed, which are best identified by recording spectra with both increasing and decreasing magnetic field scan rates. In the absence of passage effects the scans should be superimposable, assuming that the correct filter time constant has been selected.

the use of modulation frequencies less than 100 kHz is needed. Failure to achieve slow-passage conditions invalidates estimates of intensities and the use of CW saturation methods for estimating relaxation times.

C. Pulse Spectroscopy

A pulsed EPR experiment measures the response of the electron spin after a sequence of one or more pulses of microwaves. Some of the more commonly used experiments are described briefly here. Details can be found in Schweiger and Jeschke (2001). The simplest pulsed EPR experiment is FT-EPR, which is the analog of FT-NMR (Bowman, 1990). A single pulse is applied and the ensuing free induction decay is recorded. Fourier transformation gives the corresponding frequency-domain spectrum. As discussed later, the largest microwave magnetic field, B_1, that is available on most spectrometers can excite spins over only a few gauss, which defines the widest signal that can be characterized in a single FT-EPR experiment.

In a SR experiment a long microwave pulse is applied to saturate the spin system. The return to equilibrium is then monitored with low-power CW microwaves and the recovery time constant can be related to electron spin relaxation processes (Harbridge et al., 2003; Huisjen and Hyde, 1974b; Hyde, 1979).

1. Electron Spin Echo

The basic principle involved in ESE spectroscopy is the formation of an echo (Berliner et al., 2000; Mims, 1972; Schweiger and Jeschke, 2001). The simplest spin echo experiment uses two microwave pulses to form an echo, which is also known as a Hahn echo (Slichter, 1990). The experiment can be described in terms of a vector model in the rotating reference frame. At equilibrium, in the presence of an external magnetic field, the net magnetization can be viewed as a vector precessing about the direction of the external magnetic field. The Larmor precession frequency is characteristic of the environment of the unpaired electron. When a short (tens of ns) pulse of microwaves is applied along an axis perpendicular to the external field (the z-axis), the microwave magnetic field exerts a torque on the net magnetization. The length of the pulse and magnitude of the microwave magnetic field are selected to cause the magnetization to rotate into the x–y plane. This is called a 90° pulse. Owing to small differences in the Larmor frequency for different spins in the sample, the spins precess at slightly different frequencies and diverge from the average. After a variable time interval, τ, a 180° pulse is applied that flips the average magnetization vector from, for example, the $+x$ direction to the $-x$ direction and interchanges the roles of faster and slower precessing spins. After a second interval τ, the vectors reconverge to form the echo. In a two-pulse electron spin echo envelope modulation (ESEEM) experiment the time interval τ is varied and the echo intensity is recorded as a function of τ. In the absence of specific electron-nuclear hyperfine coupling the echo decays monotonically with a time constant T_m, the phase memory time constant. T_m reflects the effects of stochastic processes that occur on the time scale of the experiment and are not refocused by the 180° pulse (Eaton and Eaton, 2000a). More complicated pulse sequences are used to obtain specific information about the spin system (Schweiger and Jeschke, 2001).

2. ESE Envelope Modulation

When ESE data are obtained in frozen solution, ESE decay curves are almost always complicated by the presence of modulation on the exponential decay curve. In the present context it is important to know that the modulation is due to coupling to nuclei in the vicinity of the unpaired electron, and occurs at times after the pulse that relate to the Larmor frequencies of the nuclei (modified by dipole and exchange coupling to the electron, and, in some important cases, the quadrupole couplings). The modulation is called ESEEM, and abbreviated ESEEM or ESEM. The ESEEM pattern reveals the nuclear environment of the unpaired electron. It is much deeper (better defined) at lower microwave frequency and at lower magnetic field at the same frequency. There are some special frequencies at which some of the terms cancel and the ESEEM frequencies are almost pure quadrupole frequencies. ESEEM is one of the very important EPR observables, especially for metal ions. The only requirement for observing the ESEEM is that the B_1 of the microwave pulse must be large enough to encompass energy levels split by a nuclear interaction that is significantly anisotropic. ESEEM provides a powerful method of characterizing the nuclear spin (Chasteen and Snetsinger, 2000; Dikanov and Tsvetkov, 1992; Mims and Peisach, 1981; Narayana and Kevan, 1983).

D. Multiple Resonance Methods

Irradiation with a second frequency during an EPR experiment can provide much more detailed information about the spin system than can the usual single-frequency measurement. The most common of these multiple resonance methods are ENDOR, ELDOR, and TRIPLE. Figure 7 shows the types of transitions observed in these experiments.

In CW ENDOR one adjusts the microwave frequency and magnetic field to a resonance of interest, and then sweeps a radiofrequency signal through the range of nuclear resonant frequencies. It is necessary to operate at

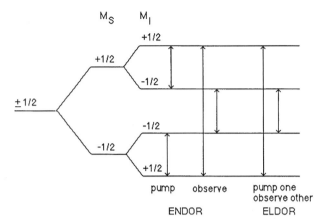

Figure 7 ENDOR and ELDOR transitions. This energy level diagram is sketched for the case of $S = 1/2$, $I = 1/2$, $A > 0$, $g_n > 0$, and $A/2 > v_n$. The allowed EPR transitions have $\Delta m_S = \pm 1$, and $\Delta m_I = 0$. Forbidden EPR transitions with $\Delta m_I = \pm 1$ sometimes have observable intensities. NMR transitions have $\Delta m_I = \pm 1$ and $\Delta m_S = 0$. Observation of a change in the intensity of one of the EPR transitions when one of the NMR transitions is irradiated is the ENDOR experiment. Observation of a change in the intensity of one of the EPR transitions when another EPR transition is irradiated is the ELDOR experiment.

a microwave power level that partially saturates the EPR transition (i.e., at microwave power greater than the linear portion of the curve in Fig. 8), and one must be able to saturate the NMR transition. The radiofrequency (RF) field at the nuclear Larmor frequency induces transitions between states such that the overall effect is to relieve the degree of saturation of the electron spin system. This causes an increase in the amplitude of the EPR signal. When the RF is swept through the frequency of a nuclear transition of a nucleus whose small coupling to the electron is within the EPR linewidth, there are changes in the intensity of the EPR transition. Thus, there is a "peak" corresponding to each nuclear transition that is coupled to the electron spin. It is the change in the EPR intensity at the nuclear resonance frequency that is the ENDOR effect. Electron–nuclear couplings are observed via ENDOR with an effective resolution that is much higher than in conventional EPR. In addition, quadrupolar couplings can be observed by ENDOR, whereas the quadrupole couplings are not observable, to first-order, in EPR transitions. The ENDOR experiment can be extended to the use of two simultaneous radiofrequency fields (TRIPLE). Whereas the ENDOR experiment provides only the magnitudes of the nuclear coupling constant, the TRIPLE experiment also provides the signs of the coupling constants. ENDOR can be performed in either CW or pulsed modes. ENDOR accessories and bridges for pulsed EPR are commercially available.

The ELDOR experiment involves two microwave frequencies. One EPR transition is saturated, and a

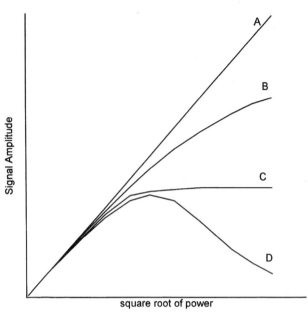

Figure 8 A plot of signal amplitude as a function of the square root of the microwave power incident on the cavity is called a power saturation curve. The plot would be linear if no saturation of the electron spin system occurred. To do quantitative EPR, it is necessary to operate at power levels below the power at which the curve becomes nonlinear. The figure presents CW EPR power saturation curves for four cases: (A) no power saturation within the available microwave power of the spectrometer—short relaxation times; (B) a small degree of homogeneous saturation at the highest available power levels—intermediate relaxation times; (C) extensive inhomogeneous saturation, exhibiting a "leveling" of the saturation curve at a relatively low power, but no subsequent decrease in amplitude of the EPR signal; (D) extensive homogeneous saturation, exhibiting a maximum in the saturation curve, due to long relaxation times.

second EPR transition is observed with a nonsaturating microwave field from a second source. For experiments performed with a single-mode resonator, the difference between the two frequencies must be less than the bandwidth of the resonator. When one transition is saturated, various relaxation mechanisms transfer magnetization between the energy levels of the transition that is saturated and the energy levels of the transition that is observed. The result is a reduction or enhancement in the intensity of the second transition (Harbridge et al., 2003).

Note that in both the ENDOR and the ELDOR experiments the experimental observable is a change in the intensity of the EPR line upon irradiation with a second frequency. Thus, the S/N is poorer than for the conventional EPR experiment for the same species. Each of the measurements depends on having relaxation times in an appropriate range, so not every compound can be studied by these techniques over the range of concentrations, temperatures, and so on, that might be of interest. This strong dependence on relaxation times translates into a powerful tool for studying relaxation behavior.

Introductions to ENDOR, TRIPLE, and ELDOR are provided in Atherton (1973), Chasteen and Snetsinger (2000), Dorio and Freed (1979), Kevan and Kispert (1976), Box (1977), Schweiger (1982), and Eachus and Olm (1985). A convenient table comparing the techniques is in Poole (1983, p. 650). The conditions needed to observe ENDOR of various nuclei in organic radicals in solution are discussed in detail in Plato et al. (1981). An extensive review of ENDOR covering the period 1978–1989 is in Piekara-Sady and Kispert (1994) and Goslar et al. (1994). Illustrations of TRIPLE provide leading references to the literature (Kirste et al., 1985; Kurreck et al., 1984). A wide range of variations on the basic theme of multiple resonance has been developed, with CW and pulsed methods, field sweep and frequency sweep, and various combinations. Poole (1983) also describes double resonance experiments in which in addition to the microwave field, there is optical irradiation (optically detected magnetic resonance, ODMR), pulse radiolysis (dynamic electron polarization, DEP), an electric field (Mims, 1976), and acoustic or ultrasonic paramagnetic resonance (UPR) (Devine and Robinson, 1982).

An alternative approach to EPR, called multiple-quantum EPR (MQEPR) has been developed in the laboratory of J.S. Hyde (Hyde et al., 1995; Mchaourab and Hyde, 1993a, b; Mchaourab et al., 1993; Sczaniecki et al., 1990, 1991; Strangeway et al., 1995). The EPR spectrometer bridge for MQEPR uses two microwave frequencies locked a specific frequency apart. Numerous combinations of relative microwave powers and sweeps of the magnetic field or one or both microwave frequencies are conceptually possible. Irradiation with two microwave frequencies is equivalent to irradiation with a single frequency that has been sinusoidally modulated. Nonlinear response of the spin system can result in intermodulation sidebands, which can be detected. The outputs are the multiquantum transitions, which can be combined in various ways to get useful displays. No magnetic field modulation is used in MQEPR.

II. WHAT ONE CAN LEARN FROM AN EPR MEASUREMENT

A. Is there an EPR Signal? Under What Conditions?

The first thing we learn via EPR is whether the sample exhibits a signal at the selected frequency, magnetic field, temperature, microwave power, and/or time after pulse(s). This is a nontrivial observation. It must be emphasized that a spin system may give an EPR signal (CW or pulsed) under some conditions and not under other conditions. The failure to observe an EPR signal does not mean that the system does not have unpaired electrons. For example, many paramagnetic transition metals have relaxation times so short that the lines are too broad to observe at room temperature. In general, relaxation times get longer as the temperature is decreased, so some systems with $S > 1/2$ (or even with $S = 1/2$ and low-lying excited states) can only be observed at subambient, and often cryogenic, temperatures. Similarly, some triplet state organic radicals have large anisotropic interactions that, when incompletely averaged, cause the spectra to be too broad to observe in solution. They can be seen, however, when they are immobilized in a rigid lattice (van der Waals, 1998). Thermally populated levels may have different spin states than the ground state, so changes of observability of species as a function of temperature can be due to both changes in relaxation times and changes in population. For example, the ground state of the antiferromagnetically coupled copper acetate dimer is a singlet, so the EPR signal disappears as the sample is cooled to liquid helium temperature. CW EPR is usually more sensitive than pulsed EPR, because of the use of a high-Q cavity and modulation/phase-sensitive detection. Hence, one would usually search for an "unknown" EPR signal first with CW techniques. Then, more detailed information about the spin system could be obtained with one of the pulsed and/or multiple resonance techniques.

1. How Many Species are Present in the Sample?

The presence of multiple peaks in an EPR spectrum could be due to two species or anisotropy or hyperfine coupling within one species. One should always suspect multiple chemical species and even multiple isomers or conformers of a single chemical species. However, there are in fact very few chemical species whose EPR spectra consist of only a single line because most species have nuclei with spins that are in the vicinity of the unpaired electron. The splittings due to these nuclear spins are discussed in the paragraph "electron–nuclear coupling". Usually, the best approach is the traditional chemical equilibrium and chemical purity tests—do the relative intensities change with sample preparation? Multiple species likely will show up as multiple time constants in pulsed EPR measurements such as SR and ESE. However, multiple time constants in these measurements do not necessarily imply the presence of multiple species, as there are various relaxation mechanisms that may be simultaneously effective resulting in multiple time constants (Eaton and Eaton, 1993a; Harbridge et al., 2003; Mazur et al., 1997a, b, 2000; Nagy, 1994, 1997). Furthermore, if the relaxation times of multiple species differ sufficiently, some species may not be seen in the observation time window. The studies of the species produced by reduction of C_{60} provides an interesting case study because of the presence in the sample of signals with very different linewidths. Early studies focused only on the sharpest signal that dominated the first-derivative spectra and missed the broad signal. Those studies call attention to the importance of reproducibility of sample preparation and careful signal quantitation (Eaton and Eaton, 1996a; Eaton et al., 1996). To first-order, hyperfine splittings are independent of magnetic field, whereas differences in resonant field that arise from g-value differences increase linearly with magnetic field. Therefore comparison of spectra at two microwave frequencies is a definitive approach to distinguish between splittings due to hyperfine and multiple species.

2. How Long Does it Take to Obtain an EPR Spectrum?

Often the question is asked "how long does it take to obtain an EPR spectrum?" Even ignoring S/N questions (and attendant signal averaging time requirements), the time required depends on what one wants to learn. The most comprehensive definition of "EPR spectrum" conceptually includes all CW and time-domain information available from the ESR phenomenon. Such a comprehensive multiperspective study is almost never performed for a single chemical species. For example, to answer the question of whether a particular column chromatography fraction contains a nitroxyl radical would take only a few minutes, but to determine the relaxation characteristics of a spin system as a function of temperature might take several days of experiments. In addition, it is sometimes necessary to perform the series of measurements at more than one microwave frequency to analyze the spin system. The importance of knowing what one wants to learn from an EPR measurement cannot be stressed too strongly. The care with which various parameters are adjusted in obtaining a spectrum also is a function of the information that one wishes to obtain from the spectrum.

3. Dependence of EPR Spectra on Chemical Structure and Physical Environment

The resonant field for the signal and the splittings of the signal provide information about the environment of unpaired electrons in the material under study. Some of the parameters that are commonly examined are enumerated in what follows. Any of the parameters may be studied as a function of a variety of factors, including solvent, temperature, sample phase (solid solution, liquid solution, glassy solution, single crystal, powder, etc.), other species in the sample, sample size, sample orientation, or irradiation (frequencies from acoustic to ionizing, including resonant frequencies for electrons or nuclei). Each such variable studied will provide additional physical/chemical/biological information about the sample. A key question to address at the outset of any EPR study is what measurement is most likely to provide the desired information most efficiently and unambiguously.

B. Lineshape

If the linewidth is due entirely to relaxation phenomena, it is Lorentzian, and the electron spin relaxation time will be characterized by a simple exponential. The lines in the EPR spectrum of Fremy's salt (peroxylamine disulfonate) in fluid solution are almost perfectly Lorentzian. However, most samples exhibit some unresolved hyperfine structure, which makes the lineshape more complicated. If there is extensive unresolved hyperfine structure, the lineshape will approach Gaussian. [See Weil et al. (1994) and Poole (1983) for extensive discussions of the details of Lorentzian and Gaussian lines.] Note that the Lorentzian line has a large fraction of its area far from the center of the line. Hence, it is necessary to integrate rather wide scans to obtain proper integrations of EPR spectra (Eaton et al., 1979; Poole, 1983). Only if the EPR lineshape is determined by relaxation and not by unresolved hyperfine or unaveraged g- and a-anisotropy is it proper to use the linewidth to estimate the electron–electron spin–spin relaxation time, T_2. This physical condition is less common than are reports of T_2 based on linewidths. However, even when there is a substantial unresolved hyperfine, the temperature-dependent contribution to the linewidth can be analyzed to determine T_2 (Rakowsky et al., 1998). The EPR lineshape can reveal kinetic information such as the slow tumbling of nitroxyl spin labels (Berliner, 1976, 1979, 1998; Berliner and Reuben, 1989; Jost and Griffith, 1972; McConnell and McFarland, 1970) and rates of chemical reactions in equilibrium mixtures (Sullivan and Bolton, 1970).

C. g-Value

The g-value reflects the mixing of orbital angular momentum (L) with the spin angular momentum (S). Hence, it is a characteristic of the chemical species. For the free electron $g = 2.0023$. Many organic radicals and defect centers in solids have g-values close to 2. Transition metal ions have a wider range of g-values. The signal from immobilized samples exhibit g-anisotropy. For a species that is tumbling rapidly in solution, an averaged g-value is observed. Incomplete motional averaging results in broadening of the lines. The most complete information concerning g-anisotropy is obtained by the study of an oriented single crystal. In frozen solution or in powdered solids most of the detailed (off-diagonal) g-anisotropy information is lost, and at best one can determine the principal g-values (the diagonal elements) from powder spectra. Note that the term "powder" spectrum is used to describe any system with random orientation of the molecular species, whether this was obtained from a true powdered crystal or from a glassy solution. Usually one would have to perform a computer simulation of the powder spectrum to obtain reliable values.

The full g-matrix can be written as:

g_{xx} g_{xy} g_{xz}
g_{yx} g_{yy} g_{yz}
g_{zx} g_{zy} g_{zz}

For example, g_{yx} is the g-value along the y-axis when the magnetic field is applied along x. The mathematics of g-values is discussed in Abragam and Bleaney (1970, Section 15-8). The matrix product of g with its transpose is the g^2 tensor, which can be diagonalized. Some of the literature is a bit careless about the nomenclature with regard to g-values, which are incorrectly called elements of the g tensor. However, for practical purposes in EPR of fluid and frozen solutions, very little intellectual integrity is lost in treating the g-values as numbers characteristic of the spectrum. Determination of all components of the matrix requires orientation-dependent studies of a single crystal. Diagonalization then gives the principal values, that is, the values along the magnetic axes of the paramagnetic center (Weil et al., 1994).

Incomplete resolution of spin–spin splitting (see later) and incomplete averaging of anisotropies in g and in hyperfine couplings, along with second-order effects, may make it difficult to measure g directly from the spectrum. Computer simulation should be used. Fortunately, precise values of g are most useful for interpretation for just those species for which it is easiest to make precise measurements—well-resolved spectra of organic radicals. As discussed in the section on the instrumental aspects of EPR, accurate g-value measurements are not trivial.

The information summarized by the g-value may be sampled differently by CW and pulsed EPR spectroscopies. For example, different features are observed in the magnetic-field-swept ESE-detected and CW-detected spectra of high-spin Co(II) complexes in frozen solution (Kang et al., 1994), so different g-values might be reported in the different experiments if the data are not analyzed with the selectivity in mind. The signals missing from the ESE-detected spectrum have very short relaxation times and are not detected by the ESE experiment but are detected by the CW experiment.

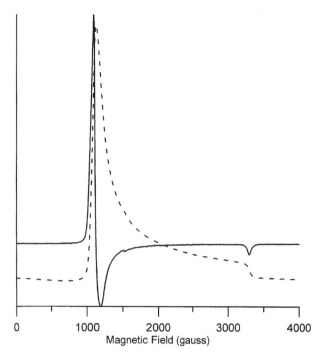

Figure 9 CW EPR spectra are normally recorded as the derivative of the absorption spectrum. Consequently, in a region of the absorption spectrum where the amplitude is changing only very slowly with change in magnetic field, the derivative is nearly zero, that is, it looks like there is no spectrum present. This is illustrated here for the case of high-spin Fe(III), whose derivative EPR spectrum is commonly described as having peaks at $g = 6$ and $g = 2$. In fact, the spectrum extends over this entire region, as shown by the absorption spectrum.

1. The Meaning of "Peaks" in a Derivative Spectrum

It is important to recognize that the first-derivative display emphasizes sharp features at the expense of broad features. This is the reason that derivative presentations are being used increasingly in other types of spectroscopy. A simple example illustrates the relation between broad absorption spectra and signals in the derivative curve. The EPR spectrum of high-spin Fe(III) porphyrins in frozen solution is commonly described as having two EPR lines, at $g = 6$ and at $g = 2$ (Fig. 9).[2] In fact, there is signal extending all the way from $g = 6$ to $g = 2$, but

---

[2]For paramagnetic systems with more than one unpaired electron, the g-value calculated from the resonant field via $h\nu = g\beta B$ is called an "effective" g-value. The effective g-value differs from the g that would be obtained by analyzing the spectrum in terms of a spin Hamiltonian. For example, for high-spin Fe(III) $g = 2$ in the spin Hamiltonian, even though the field values for the extrema yield "g effective" values of 2 and 6. This is discussed in Weil et al. (1994).

because the derivative display is used, the small slope, except at the turning points of the powder pattern, causes the first derivative to be approximately zero over a large portion of the spectrum. It is worth studying carefully the relation between the absorption spectrum distribution and the derivative spectrum (which is the slope of the absorption spectrum). Note, for example, how the slope of the high-field extremum leads to a negative-going peak in the derivative curve. The broadness of this spectrum is due to the anisotropy of high-spin Fe(III). Much of the area of the EPR spectrum occurs in the region of field sweep where the derivative EPR signal is nearly zero. This is a phenomenon largely of rigid lattice spectra. In fluid solution, where anisotropies are largely averaged away, a portion of the derivative spectrum that looks like baseline usually is devoid of absorption signal. The most common exception would be cases of intermediate chemical exchange. In contrast, the echo-detected field-swept spectrum gives the absorption signal, not the first derivative, which can be advantageous in dealing with very broad lines (Gaffney et al., 1998).

D. Spin–Spin Coupling

1. Electron–Nuclear Coupling

The electron–nuclear coupling constant, generally denoted by a_n, is independent of magnetic field. The observed splittings, measured on a magnetic field plot, may depend on m_I and on magnetic field, due to second-order corrections that are called Breit–Rabi corrections [see Weil et al. (1994) and Carrington and McLachlan (1967) for further discussion]. However, most electron–nuclear couplings are small enough that the second-order corrections are sufficiently small at X-band that quartets, sextets, and so on are clearly recognizable. Many nuclear couplings are observable in the standard CW EPR spectrum. Some of those that are too small to resolve in the standard CW spectrum can be obtained from electron nuclear double resonance (ENDOR), electron nuclear nuclear triple resonance (TRIPLE), or ESEEM spectra (see Section I.D).

For example, nitrogen-14 ($I = 1$) hyperfine coupling causes the nitroxyl free radical EPR spectrum to exhibit splitting into three major lines (Fig. 4). Sometimes these three lines are further split into multiplets. The ENDOR spectrum reveals the "unresolved" hyperfine splitting within one of the lines of the three-line nitroxyl CW EPR spectrum due to coupling of the electron to hydrogens in the molecule. These couplings are small enough that they merely broaden the CW EPR spectrum in most nitroxyl radicals. This is a specific example of the general phenomenon called "inhomogeneous broadening".

2. Electron–Electron Coupling

Electron–electron spin–spin coupling can lead to resolved splitting in EPR spectra just as can electron–nuclear coupling in EPR and nuclear–nuclear coupling in NMR (Eaton and Eaton, 1978, 1988, 1989). Dipolar interaction between paramagnetic centers is the basis for distance measurements in proteins (Berliner et al., 2000; Persson et al., 2001) and polymers (Jeschke, 2002).

E. Relaxation Times

The following is a brief overview of electron spin relaxation. Detailed discussions are given in Bowman (1990), Eaton and Eaton (2000a), Standley and Vaughan (1969), and Bertini et al. (1994a, b). A physical system that is moved away from equilibrium by a change in one or more variables of state will return to equilibrium by relaxation processes. The study of relaxation is a more general view of kinetics than the usual reaction rate measurement. Relaxation times are, effectively, a different dimension to view chemistry. Relaxation times are parameters characteristic of an ensemble of spins, and have no direct meaning for isolated chemical species. This is in contrast to g-values and spin–spin splittings, which are characteristic of the isolated molecule (though they change depending on the environment). The measurement of relaxation times is important because the relaxation times (a) affect the choice of experimental parameters (see Section VI.B) and (b) provide chemical and physical information that is not available by other techniques.

Saturation and relaxation measurements are more important in magnetic resonance than in higher frequency spectroscopies because rates of spontaneous emission from an excited state are proportional to the square of the frequency. Thus relaxation rates are slower for EPR than for electronic energy levels, but faster than that for NMR. Saturation is also a major concern in EPR because the population differences between the energy levels involved in the transitions are so small (see Section I.A). There are numerous contributions to the relaxation processes. The measured relaxation times are interpreted in terms of all of the possible contributions, so a list of these contributions serves to guide the experimental design. Relaxation times may depend on the species being studied, intramolecular vibrations, rotations, intermolecular interactions, including solvent and collisions with like molecules or other paramagnetic species, magnetic field, temperature, diffusion, spin diffusion, viscosity, cross relaxation, lattice phonons, phonon bottleneck, and electric field. The interpretation of relaxation times in terms of various contributions can represent a major effort, and is beyond the scope of this chapter. Discussion of several of these contributions can be found in Poole and Farach (1971), Eaton and Eaton (2000a), Standley and Vaughan (1969), Kevan and Schwartz (1979), and Muus and Atkins (1972). The contributions to spin lattice relaxation can be obtained by analysis of the temperature dependence of T_1 (Zhou et al., 1999).

If an electron spin system is displaced from equilibrium by the application of a microwave magnetic field, B_1, perpendicular to the magnetic field of the electromagnet, the spin system will attain a new macroscopic state that has (a) higher energy and (b) lower entropy (nonzero magnetization in the xy plane, M_x, M_y, requires some degree of phase coherence, and hence has lower entropy). Removal of B_1 allows the spin system to relax toward equilibrium. Achieving thermal equilibrium may be slower than the loss of phase coherence, as there can be absence of phase coherence without thermal equilibrium, but there cannot be thermal equilibrium while there is still phase coherence (Reisse, 1983).

The relaxation time for attaining thermal equilibrium of the M_z magnetization is called T_1, the longitudinal or spin–lattice relaxation time. T_1 processes change populations of electron spin states, which means that the total energy of the spin system is changed by a T_1 process. T_1 can contribute to the linewidth because of the finite lifetime of the spin states. A short T_1 leads to an uncertainty in energies and therefore a broadening of the EPR lines. The relaxation time for loss of phase coherence (i.e., achieving $M_x = M_y = 0$) is T_2, the transverse or spin–spin relaxation time. T_2 processes are adiabatic, that is, the total energy of the spin system is not changed by T_2 processes. For example, a mutual spin flip

↑↓   ↔   ↓↑

between one spin in the upper state and one in the lower state does not change the total number of spins in each state. Relaxation time values are surveyed in Eaton and Eaton (2000a), Standley and Vaughan (1969), Bertini et al. (1994a), Bowman and Kevan (1979), Brown (1979), and Al'tshuler and Kozyrev (1974).

One important application of relaxation data is the selection of spectrometer operating conditions. For example, if the electron spin relaxation time is longer than the reciprocal of the modulation frequency, a distorted spectrum will result. If the spectrum is obtained at saturating microwave power levels, the quantitation of spin concentration will not be correct.

F. Saturation Behavior

In the absence of spectral diffusion processes, CW power saturation curves can be analyzed to determine the product of the relaxation times, T_1T_2 (Poole, 1983; More et al., 1984, 1986). A convenient way to display power saturation behavior is to plot spectral amplitude vs. the square root of microwave power (Fig. 8). The spectral amplitude can be defined by the difference between the signal intensities at the magnetic fields that correspond to the maxima and minima of the unsaturated signal (a plot at constant field) or by the difference between signal maxima and minima at each power. Because the lines broaden as the signal is saturated, these two definitions will give different plots. A plot at constant magnetic field is more sensitive to changes in relaxation times and easier to simulate using a computer. A linear plot indicates no saturation. The lower the microwave power at which the plot deviates from linearity, the more readily the signal saturates. Alternative displays of the CW power saturation information are favored by some biochemists (Beinert and Orme-Johnson, 1967; Rupp and Moore, 1979). To obtain EPR intensities, spin–spin splittings, and so on, one normally seeks to record an unsaturated EPR spectrum. That is, one operates in the linear region of Fig. 8. Practice, as reported in the literature, is not always consistent with this guidance. Some authors use the term "saturation" to mean operation at a power higher than the value that gives the maximum signal intensity in Fig. 8.

The shape of the power saturation curve reveals whether the spin system is homogeneously or inhomogeneously broadened (Poole and Farach, 1971). Pure homogeneous broadening will occur only if the line is Lorentzian. Any degree of unresolved hyperfine splitting or other broadening of the relaxation-determined linewidth will result in a power saturation curve that does not decrease as fast (or at all) when the spin packet saturation factor S becomes less than 1.

$$S = \frac{1}{1 + \gamma^2 B_1^2 T_1 T_2}$$

There is a large literature related to attempts to obtain T_1 and T_2 from power saturation curves of inhomogeneously broadened lines. Entry to this literature can be obtained through the references cited in Poole and Farach (1971), More et al. (1985), and Eaton and Eaton (1985). Even when carefully performed, the resulting relaxation times may be strongly impacted by spectral diffusion processes that move spins off resonance (Harbridge et al., 2002a, 2003). With the increasing availability of pulse spectrometers, it is now preferable to measure relaxation times directly by SR or ESE. Nevertheless, for a closely related series of systems, comparisons of power saturation curves reliably provide useful relative values of an effective T_1T_2 product. It is quite common, for example, to characterize a spin system by measuring $P_{1/2}$, the microwave power at which the EPR signal is 1/2 the amplitude it would have been if there had been no saturation. The special significance of this value is that in the limit of homogeneous broadening (Lorentzian line) $P_{1/2}$ is the power at which the saturation factor $S = 1/2$. This parameter is, for example, a useful monitor of spin label accessibility to paramagnetic reagents (Mchaourab and Perozo, 2000).

The microwave power saturation behavior of a sample reveals much about the nature of the spin system. The ease of saturation could reveal, for example, how well the sample is degassed, whether aggregation is occurring, whether a good glass formed upon freezing, whether unsuspected paramagnetic impurities are present, or whether the species being studied is in close proximity to another paramagnetic species.

G. Signal Intensity

EPR is a quantitative analytical method. The intensity of the EPR signal, whether CW, FT, SR, or ESE, is a measure of the number of species present, so long as the nature of the spin system is known and appropriate instrument settings are selected. In this regard, nomenclature is poor. Many reports do not distinguish between the amplitude of the spectrum in the usual derivative display and the intensity of the signal as revealed by the double integration of the derivative signal. Whenever EPR is used to demonstrate the presence of radicals in a system, the concentration should be determined by comparison of the integrated EPR signal with that of an appropriately chosen standard (see Section V). In addition, because the derivative mode of spectral presentation that is almost always used in CW EPR selects for sharp spectra makes broad spectra look relatively less important, care must be taken to recognize superimposed sharp and broad spectra. Note that if the lineshapes are the same, regardless of the details of the shape (Chesnut, 1977), the amplitude of the signal (height on the spectral display) is inversely proportional to the square of the width of the peak. For two signals with the same area, if one is twice as wide as the second, the amplitude of the first-derivative spectrum will be only 1/4 that of the second. Section V provides a detailed discussion of aspects of quantitative EPR. With the sensitivity of modern EPR spectrometers, it is possible to get a "full scale" almost noise-free spectrum of something that is less than a few percent of the stoichiometric concentration. There are many such examples in the literature. To avoid errors of assignment or interpretation, it is important to perform quantitative EPR measurements.

Very few reports even attempt to quantitate time-domain signals. The amplitude of an electron spin FID or echo signal is proportional to the z-axis spin magnetization, M_z, at the time of the microwave pulse. Any spectrometer has a "dead-time", a time after a pulse during which it is not possible to record accurate signals because of cavity ringing (a Q effect), response times of

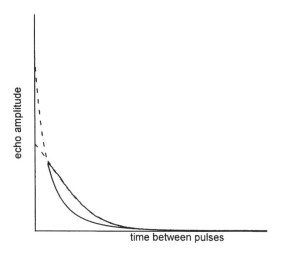

Figure 10 Time-domain EPR spectra cannot be obtained starting at the true zero of time (except in very special cases) because of the instrument response to the microwave pulses. To compare time-domain EPR spectra quantitatively, the amplitudes have to be extrapolated back through the instrument dead-time to the true zero of time, as shown in this figure, unless the relaxation times are identical. The solid lines in this figure are the ESE decay curves for two different species. The dotted lines extrapolate the decay curves through the instrument dead-time to time = 0. Note that two species with different relaxation times could have different concentrations even though they yield echoes of the same amplitude at a given interpulse spacing, τ.

switches, recovery of circuits from overload, and so on. Hence, the data recorded in an ESE or SR experiment do not start at time = 0. Consider the simple case of purely exponential decay (Fig. 10). Extrapolation of an exponential back to time = 0 should give a signal amplitude proportional to the magnetization that existed at the time of the pulse, and hence to the spin concentration. Hence, if two samples with the same spin concentrations are compared, the time = 0 signal amplitudes should be the same, even if they have different relaxation times, so long as the relaxation behaves in accordance with a single exponential. If the signal amplitudes do not behave this way, then not all of the spins in the sample are being observed. This could be due to some of the spins having a much faster spin relaxation, as might be caused, for example, by aggregation, varying degrees of spin–spin interactions or chemical heterogeneity (e.g., different ligation at the metal site). Comparison with a standard of known concentration and known relaxation time will reveal whether all of the spins are being observed. In the more complicated case of multiple exponential behavior, it is still possible to make such quantitative comparisons, but both the time constants and the relative contributions of each of the component exponentials would have to be known. For example, quantitation of spin echo data has been used to confirm the predicted dependence of signal intensity on microwave frequency (Rinard et al., 1999b).

Similar arguments apply to the SR experiment. In the ideal case, with full saturation of the spin system, there is zero signal at time = 0. However, there is a dead-time after the saturating pulse, so there will always be some degree of recovery before signal recording can start. The infinite-time SR signal is just the steady-state CW signal under direct-detection (i.e., no magnetic field modulation) conditions. If the recovery is exponential (i.e., a true measure of T_1) one should be able to extrapolate back to time = 0 and compare the difference in signal amplitude from time = 0 to infinite time. We will term this the amplitude of the SR signal. If two samples have the same spin concentration (spins per gauss of spectrum), they should give the same amplitude of the SR signal, even if they have different T_1s. If this is not true, then not all of the spins are being observed. Note the important distinction concerning spins per gauss. For the same spin concentration in moles per liter, and the same T_1, a broad EPR spectrum will yield a proportionately weaker SR signal relative to a narrow EPR spectrum, as the signal depends on the number of spins in the observation window.

III. EPR SPECTROMETER

The user, or potential user, of EPR spectroscopy should learn about the instrumentation because:

1. Commercial spectrometers are so modular that one has to be well informed about the function of each component to be an intelligent customer.
2. The sample interacts with the spectrometer to a greater extent than in most techniques.
3. Some important experiments require home-built equipment.

There are two general types of EPR spectrometers. CW spectrometers are the most commonly used spectrometers. These can have accessories for multiple resonance, including ENDOR, TRIPLE, and ELDOR (see Section I.D). The second type is pulsed spectrometers, including ones designed for FT, SR, and spin echo experiments. Both types of instrumentation are commercially available. The following paragraphs discuss features and components that are common to most EPR spectrometers.

A. Microwave System

Diagrams of the microwave circuits are given in Figs. 11, 12, and 13 for CW, spin echo (Norris et al., 1980; Quine et al., 1987), and SR (Hyde, 1979; Mailer et al., 1985) EPR spectrometers, respectively. This part of the spectrometer is called the microwave bridge.

1. Reference Arm

All modern CW EPR spectrometers use two microwave paths that are phase-coherent: an observe arm that includes the resonator and sample and a reference arm (Fig. 11). ESE spectrometers also have two arms, a reference arm and a pump and observe arm (Fig. 12). An FT EPR spectrometer bridge is similar to an ESE spectrometer bridge. SR spectrometers require three arms—pump, observe, and reference (Fig. 13). There is another fundamental difference between the various spectrometers: microwave power is incident on the sample during the observation period in CW and SR spectrometers, but not in ESE or FT spectrometers. This difference is very fundamental, not only to the spin physics but also to the design philosophy of the spectrometer. Multiple resonance spectrometers have another layer of complexity to permit more than one microwave frequency to be used at once (or in quick succession) or to permit both microwave and RF (radiofrequency, NMR frequencies) to be used simultaneously.

The spectrometer schematics in Figs. 10, 11, and 12 are simplified to show only the major functional features. In addition there will be, for example, many microwave isolators to prevent signals from interacting in bizarre ways. Many research spectrometers also have elaborate switching among components needed for different types of experiments (Mailer et al., 1985; Norris et al., 1980; Percival and Hyde, 1975; Quine et al., 1987, 1992, 2002). In the authors' laboratory, for example, there are spectrometers designed to perform CW, ESE, and SR EPR with the same bridge and resonator without changing the sample. The circuits needed to achieve this flexibility are much more complicated than shown in the figures in this chapter.

In each case there are attenuators to set power levels, phase shifters to establish the correct phase relationships for the two or three arms of the bridge, circulators to direct the flow of the microwaves, and isolators to prevent unwanted interaction of components. In the pulsed spectrometers one also uses PIN diode switches to shape the pulses and protect the detector. Limiters are also used to protect sensitive components. Information about microwave components is provided in the first edition of Poole's treatise (1983) and in Baden Fuller (1990) and Laverghetta (1981). The important advance of the Varian E-Line and Bruker ER420 and ER200 spectrometers over the earlier (e.g., Varian V4500) systems was the introduction of the reference arm and the use of a circulator in place of the magic T. The reference arm permits biasing the crystal at critical coupling of the cavity, so that signal amplitudes as a function of incident power truly reflect the power response of the spin system. The microwave power incident on a magic T is

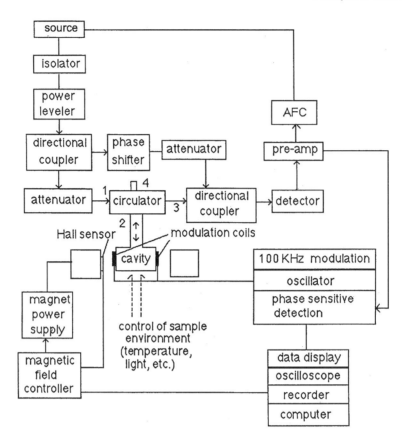

Figure 11 CW EPR spectrometer block diagram. This figure extends the information in Fig. 3 to the component level, with particular emphasis on the microwave system. Pictured here is the fundamental reference-arm-bridge, with four-port circulator, which is the basis for all modern CW EPR spectrometers. The key features are the use of directional couplers (unequal power dividers) to pick out part of the source output for the reference arm, where its phase relative to the sample arm is established, and then to recombine the two arms prior to the detector crystal. Hence, the detector crystal can be biased by using power from the reference arm, and the cavity can be used critically coupled. The use of a circulator instead of a magic T as in earlier designs uses power and signal more efficiently, increasing S/N. Additional components are added to this basic system for dispersion operation (two detector crystals are needed), bimodal cavity operation (two reference arms are needed), and so on.

divided equally with half going to each of two arms of the T. Thus, half of the source power goes to the cavity, and half of the reflected power goes to the detector. The circulator directs the microwave power without dividing it, and hence gives a factor of 4 greater signal power at the detector relative to the magic T, if the signal does not saturate. Almost all modern EPR spectrometers use a circulator. The exceptions are some special purpose spectrometers that use directional couplers in place of the circulator or that use transmission resonators.

2. Choice of Microwave Frequency

Considerable thought should be given with respect to selection of microwave frequency. Although most EPR spectrometers operate at X-band (between ~9 and 10 GHz), commercial spectrometers are becoming available over a widening range of frequencies: currently ~1 GHz (L-band) to 95 GHz (W-band). It now becomes important to consider what EPR frequency is optimum to answer a particular question for a particular sample. X-band has proven to be a convenient compromise, and remains the frequency of choice if one is to have only one spectrometer. Increased spectral dispersion and greater sensitivity to rotational motion is achieved at higher microwave frequency (Budil et al., 1989; Eaton and Eaton, 1993b, 1999a; Lebedev, 1990). For example, the 10-fold increase in microwave frequency from 9.5 to 95 GHz results in a 10-fold increase in separations of features in the spectra that arise from g-anisotropy. Resolution of these features may be key to determine, for example, whether the effective symmetry at a metal site is axial or rhombic. When the zero-field splitting for metal ions with $S > 1/2$ is greater than the EPR quantum ($h\nu$), the energies for some transitions may be too large to detect. By increasing the microwave

Electron Paramagnetic Resonance

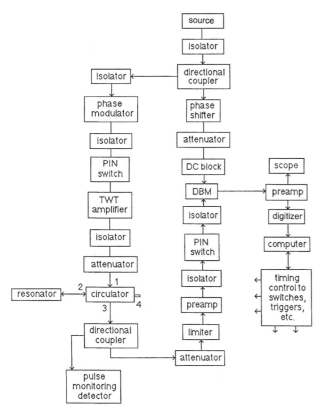

Figure 12 Spin-echo spectrometer block diagram. The spin-echo spectrometer requires only two arms in the microwave bridge, but requires sophisticated timing control for pulse shaping and signal detection. Usually the ESE spectrometer uses a higher power pulse than does the SR spectrometer, so a TWT amplifier is used to amplify the klystron output. Because of the high pulse power, special care is needed in the microwave echo detection components. GaAsFET amplifiers are used to amplify the echo prior to the DBM. GaAsFET amplifiers have low noise figures but they need to be protected with limiters on their input. The limiter is a device that clips high-level signals and only passes low-level signals.

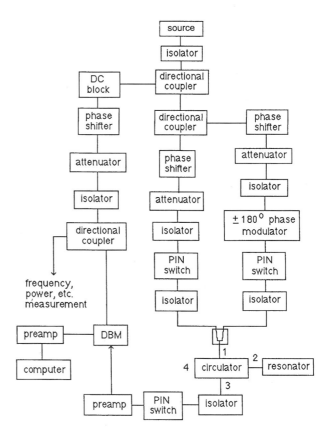

Figure 13 SR spectrometer block diagram. Only the microwave aspects of the spectrometer are shown. The magnet and data display aspects are similar to the CW spectrometer in Fig. 11, except that the analog-to-digital converter has to be faster, and the computer does more control of the system timing. To achieve pulsed SR (or stepped power EPR) operation, a third arm is added to the microwave bridge. In the third arm are components to establish the phase and the power level of the saturating pulse. PIN diode switches and phase shift modulators are bounded by isolators (three-port circulators with one port terminated in a load, so that microwaves pass in only one direction) to prevent switching transients from getting back to the source or into the signal. It is common in pulsed spectrometers to compare the reference arm and the signal arm with a DBM, which acts in this application as a phase-sensitive detector of microwaves. The DC block is a device that does not have electrical continuity for DC signals, but does for microwave signals. It prevents ground loops in the microwave system. The microwave system shown here for use with a reflection cavity (or other reflection resonator) can be modified to use a bimodal cavity.

frequency, higher energy transitions become accessible, so higher microwave frequencies and the corresponding high magnetic field strengths are particularly useful for metal ions with $S > 1/2$ and large zero-field splittings (Wood et al., 1999). Sometimes higher magnetic fields (and the higher accompanying microwave frequencies) make it possible to obtain EPR signals for complexes with an even number of unpaired electrons such as Ni(II) ($S = 1$) (van Dam et al., 1998) and Mn(III) ($S = 2$) (Barra et al., 1997). Home-built spectrometers at 150 GHz (Krinichnyi, 1995; Lebedev, 1990), 250 GHz (Budil et al., 1989), and higher frequencies (Brunel, 1996; Earle et al., 1996; Eaton and Eaton, 1999b; Hassan et al., 2000; Prisner, 1997) have provided very useful insights, but more labs have selected ~94–95 GHz (Allgeier et al., 1990), and this is the frequency range chosen by Bruker for their commercial high-field accessory to the ESP300 series spectrometer. For some questions, the clearest answers are obtained by comparing spectra as a function of microwave frequency (Hustedt et al., 1997).

However, bigger is not always better. Higher microwave frequency yields higher sensitivity (for a fixed sample size), but increasingly inconveniently small sample sizes. There is also increasing problem with sample lossiness at higher frequencies. Recently it has been recognized that some important characterization information is more readily obtained at lower frequencies than at higher frequencies (Hyde and Froncisz, 1982). A particularly dramatic example is the improved resolution of nitrogen hyperfine structure in Cu(II) complexes that can be achieved with spectra at ~2 GHz (S-band). Consequently, there has been intense activity in developing the S-band (2–4 GHz) region for EPR. The effort to obtain EPR spectra in living organisms has driven spectroscopy to regions of the spectrum in which water is less lossy than at X-band, so there has been some activity at even lower frequencies, 1 GHz and below (Eaton and Eaton, 2004; He et al., 1999; Koscielniak et al., 2000; Quine et al., 2002; Swartz and Halpern, 1998; Ueda et al., 1998).

3. Microwave Source

Until recently klystrons were the most common microwave source in EPR spectrometers, and have dominated the X- and Q-band systems. Many spectrometers still in operation use klystrons as the microwave source. Recent improvements in solid state microwave sources, as well as their longer life, smaller size, and lower power supply requirements, have caused them to be used in recent EPR systems. Adequate power (~200 mW) with low noise (amplitude and phase noise) is readily available. Many X-band pulsed EPR systems, and all recent X-band and lower frequency CW EPR spectrometers, use solid state microwave sources. The spectrometer manufacturers are making significant improvements in lowering the noise of solid state sources. Bruker and JEOL use solid state sources at X-band and at lower frequencies.

For g-value calculations and for subtraction of spectra it is important to know the field/frequency relations of the spectra. Hence, a microwave frequency counter is an almost essential tool for quantitative EPR spectroscopy. The Bruker X-band spectrometers have an option for a built-in frequency counter, which is very useful. If a spectrometer does not have a built-in counter, an external one should be used to monitor the operating frequency. A sample of the microwave source output is available at an SMA connector on the back of newer commercial bridges to facilitate such measurements. At lower frequencies the actual frequency may be known because the frequency source could be a synthesizer that is as accurate as a counter. Frequency measurements are more difficult at higher frequencies and it may be necessary to calibrate the field/frequency relationships with a standard sample of known g-value.

4. Waveguide vs. Coaxial Cable

Wherever it is mechanically convenient, waveguide is used because of its lower loss and higher shielding properties compared with coaxial cable. However, waveguide becomes unmanageably large at frequencies below X-band, and even at X-band it is awkward to build a microwave bridge assembly using waveguide components. In addition, many of the newer high-performance microwave components are becoming available only in coaxial units. Most microwave bridges are assembled using semirigid coaxial cable. The use of flexible coaxial cable should be avoided except for prototype modules because in some respects it has poorer performance than semirigid coaxial cable.

New fabrication techniques make it possible to design most of the microwave system of an EPR spectrometer on a printed circuit board. Thus, there is a trend toward nonwaveguide components in EPR spectrometers. In assembling systems, and especially in attaching the resonator, coupling schemes that are mechanically suitable for waveguide have to be replaced by coupling schemes suitable for coax, stripline, and so on. The component for which it is still essential to use waveguide is the rotary vane attenuator, which is a very useful component because it does not introduce phase shifts.

B. EPR Resonators

In the fifty-plus-year history of EPR spectroscopy, the majority of EPR measurements have been done using commercial spectrometers and TE_{102} rectangular cavities (Wilmhurst, 1968). Many articles have been published that provide detailed characterization of the performance of the Varian E-231 TE_{102} X-band cavity (Dalal et al., 1981; Eaton and Eaton, 1980; Mailer et al., 1977) and references therein. Analogous characterization of the Bruker ER4102ST rectangular cavity, which is similar to the Varian E-231 cavity, are being performed (Mazur et al., 1997a, 2000). JEOL spectrometers have used a cylindrical TE_{011} cavity. Cylindrical cavities inherently have a higher Q. Other things being equal, signal intensity increases with Q, but microwave source noise is more of a problem, the higher the Q. If the microwave source noise is low enough, and if the sample geometry is appropriately chosen, a cylindrical resonator can give a higher S/N than a rectangular resonator. For general laboratory use, a rectangular resonator has been found more versatile and convenient to use by many experimentalists.

An EPR cavity is a rectangular or right-circular cylindrical "box" with conducting walls, an opening for inserting the sample, and a small opening to a piece of microwave waveguide through which microwave energy gets to the sample. Near this opening is an adjustable device that has the effect of varying the "electrical size" of the opening, and hence is called an "iris". Electronically it is a transformer. To "match" the cavity to the waveguide one adjusts the iris. The term match in this use means to make the impedance of the iris and cavity combination the same as the impedance of the waveguide at the resonant frequency. Since a sample absorbs microwave energy because of its dielectric (and sometimes resistive) properties, the effective resistive losses in the cavity change when the sample is inserted. Hence, the match has to be adjusted for each sample or change in temperature of a sample. Sometimes experimentalists refer to this step as "tuning" the system. From a microwave viewpoint, the concept is "coupling" of the (almost enclosed) sample cavity to the transmission line (wave guide from the bridge to the cavity). When the impedances are exactly matched, the cavity is said to be critically coupled. There are no EPR experiments in which it is useful to have the cavity "undercoupled", but as outlined in what follows, in ESE experiments it is often useful to have the cavity overcoupled. An overcoupled resonator appears to have an impedance that is lower than the waveguide impedance.

Resonator is a more general term than cavity. The most important features of a resonator are

1. *Background signal*. This can be due to contamination, or it can be due to impurities in the materials from which the resonator is fabricated. For example, see the discussion by Buckmaster et al. (1981) on impurities in brass cavities.
2. *Mechanical stability*. Microphonics in the resonator may be the limiting feature in the S/N of home-built systems and even for commercial systems at high modulation amplitude.
3. *Filling factor*. The filling factor is the fraction of the B_1^2 in the resonator that is filled with the sample to be studied. The filling factor of cavity resonators is usually very small. One of the main advantages of some of the newer types of resonators is their much larger filling factor relative to a cavity. Signal amplitude is proportional to filling factor.
4. *Resonator efficiency*, Λ. This is the B_1 generated at the sample per square root of watt incident on the resonator.
5. *Q*. The signal amplitude is proportional to resonator Q. The Q is strongly influenced by the sample and must be known or controlled to do quantitative EPR.

Several factors have spurred the interest in improved resonators. (1) The very small filling factor of the standard sample geometry in a resonant cavity is a major limitation in the application of EPR spectroscopy. (2) The large size of cavities at low frequencies is inconvenient. (3) The urgency to obtain larger B_1 at the sample per watt of incident microwave power in time-domain (pulsed) EPR has also driven development of alternative resonator designs. The larger B_1 and the lower Q, the better for pulsed EPR. Low Q with high B_1 is also desirable for ELDOR (electron double resonance) and dispersion EPR studies.

Researchers have explored alternative resonator designs, including slotted-tube resonators, loop-gap resonators (LGRs) (Eaton and Eaton, 2004; Froncisz et al., 1986; Hornak and Freed, 1985; Hyde and Froncisz, 1986, 1989; Hyde et al., 1985; Mehdizadeh et al., 1983), split-ring resonators (Hardy and Whitehead, 1981), slotted-tube (Mehring and Freysoldt, 1980; Schneider and Dullenkopf, 1977), folded half-wave resonators (Lin et al., 1985), dielectric resonators (Walsh and Rupp, 1986), and ferroelectric resonators (Dykstra and Markham, 1986). Each has special features that make it particularly useful for certain experiments. The figure of merit for a new resonator is the increase in B_1 at the sample relative to the TE_{102} cavity, and the increased S/N obtainable with the new resonant structure relative to a cavity. The fundamental problem for the design and use of these new resonant structures is to achieve the high B_1 fields simultaneous with a mechanically convenient coupling design that is consistent with the experimental realities of the types of samples that chemists, biologists, and materials scientists want to study. Some resonators described in the literature are fine for some types of sample, but not practical for other types of samples. For example, the resonator developed by Mims (1974, 1976) was optimized for using liquid samples that were then to be frozen and studied at liquid He temperature. It has not been used for the more common sample in dilute liquid solution at room temperature, and probably would not function well under those conditions. The greatest versatility has been found with the LGR (Eaton and Eaton, 2004; Froncisz et al., 1986; Hornak and Freed, 1985; Hyde and Froncisz, 1986, 1989; Hyde et al., 1985; Mehdizadeh et al., 1983). The LGR design is sufficiently flexible that the resonator can be designed to fit an experiment, rather than designing an experiment to fit the constraints of the resonator. The Bruker ER4118XMS "split ring" resonator has the three-loop-two-gap resonator topology.

Aqueous samples are particularly challenging because of the lossiness (high nonresonant absorption of microwaves) of water. The X-band LGR reported by Hyde (Froncisz et al., 1985) is designed to give improved S/N for sample-limited aqueous biological samples. The sample holder intended for use with this resonator is now available from Wilmad. It is made of a plastic that is highly permeable to gases, so that a sample can be deoxygenated by passing nitrogen over the outer surface. Bruker markets a cavity (ER4103TM) designed specifically to improve the S/N of lossy samples at or near room temperature. Recently Bruker introduced the AquaX sample assembly in which the sample is distributed between multiple small-diameter capillaries to maximize the volume of sample that can be positioned in the resonator without adversely lowering resonator Q. This design is based on the principles demonstrated in Hyde (1972) and Eaton and Eaton (1977).

At Q-band, cylindrical resonators are common, although there are examples of use of LGRs and other resonators. Required sample sizes become very small at Q-band: even a 1 mm o.d. quartz tube is "large" in two ways. First, unless the sample is very nonlossy, it will decrease the cavity Q too much to make it possible to obtain spectra. For example, toluene can be used in tubes with a few tenths mm i.d., but with 10% chloroform in toluene and the same tube it may not be possible to match the resonator. Aqueous samples have to be put in very thin capillary tubes. The other feature is that the amount of dielectric in the sample tube can cause leakage of microwaves out the hole through which the sample is inserted into the cavity. To compensate for the dielectric when using the Varian Q-band cavity, for example, it is common to insert a conductor, such as a strip of aluminum, alongside the sample tube, and vary the position of the metallic strip until match is achieved. The maximum sample size at W-band is even smaller than at Q-band. Modern Q-band resonators include some that use a dielectric cylinder (synthetic sapphire), analogous to the X-band dielectric resonator. The dielectric resonator can accommodate larger samples than the cylindrical resonator.

In addition to resonator Q, EPR signal intensity is proportional to the "filling factor". The filling factor is the fraction of the microwave power in the resonator that is effective in EPR transitions. The geometrical factor of the ratio of sample volume to resonator volume is weighted by the microwave power (B_1^2). Thus, a sample in a region of high B_1 has a higher filling factor than the same sample in a lower B_1 region of the same resonator. The formulae are given in Poole (1983). As a rough approximation, a sample in a standard 4 mm o.d. quartz sample tube in an X-band rectangular cavity resonator has a filling factor of $\sim 1\%$. In an LGR designed for 4 mm tubes, the same sample has a filling factor of $\sim 10\%$. The wall thickness of the tube, and the fact that some of the microwave B_1 is outside the sample region of the LGR, between it and the shield (which is necessary to have a return path), keep the filling factor from approaching 100%. A thinner-walled tube (fragile) would give a substantially increased EPR signal in the same LGR. LGRs have a lower Q than cavity resonators, somewhat compensating for the increased filling factor. The lower Q does, however, mean that microwave source noise is less of a problem. LGRs are an attractive alternative to cavity resonators.

Combining the concepts sketched earlier, one might seek a high-filling-factor, high-Q, cylindrical resonator. This is possible with a dielectric resonator. A material of very high dielectric constant can concentrate microwaves (by changing the wavelength) to the point that small devices can become resonant. The dimensions are appropriate for commonly used frequencies. A right-circular cylinder with a hole down the center for a sample tube becomes attractive as an EPR resonator. The use of dielectric resonators is becoming increasingly common, and Bruker markets X-band and Q-band ones in the Flexline series. The main problem with dielectric resonators is that it is extremely difficult to obtain materials with appropriate dielectric properties that do not have sufficiently high levels of paramagnetic ions to give a large "background" signal in EPR measurements. Synthetic sapphire (a form of aluminum oxide) is used in the Bruker ER4118DR resonator. There are signals observable in a wide CW spectrum, especially at low temperature. Some of the signals have long enough relaxation times that they can be seen in pulsed EPR measurements.

Bruker sells a special "dual mode" resonator (ER4116DM). This cavity is built to support two microwave modes at slightly different frequencies. One mode has the usual arrangement with B_1 perpendicular to the external magnetic field B_0, which is needed for observing the normal "allowed" EPR transitions. The second mode has B_1 polarized parallel to B_0, which enhances "forbidden" transitions and suppresses allowed transitions. Such resonators are useful for observing, for example, $\Delta m_S = \pm 2$ transitions and for studying integer spin systems (Hendrich and Debrunner, 1998).

There is no "best" resonator for all experiments, because one needs to weigh many factors. The optimum depends on sample and experiment. Among the tradeoffs to keep in mind are that for large B_1 per $\sqrt{\text{watt}}$ we need high Q, small resonator volume, and critical coupling. For constant flip angle we need a sample that is small relative to nonuniformities in B_1. For maximum spectral bandwidth in pulsed EPR we need low Q. However, for high S/N we need high Q and large sample volume, and for short relaxation times we need low Q and short t_p.

1. Q of a Resonator

Q is sometimes called the quality factor of a resonator. In this chapter Q is used to refer to the loaded Q of the cavity, which is sometimes designated as Q_L. The word "loaded" means that the resonator is coupled to a waveguide, independent of whether a sample is present in the resonator. One has to read the EPR literature carefully because of ambiguities about whether the loaded or unloaded Q is meant.

There are several ways to define Q. Each definition focuses on a different feature of these resonant structures.

$$Q = \frac{2\pi(\text{frequency})(\text{energy stored})}{\text{power dissipated}}$$

$$Q = \frac{\nu}{\Delta \nu}$$

where $\Delta \nu$ is the difference in frequency at the half-power points, that is, the bandwidth. The Q determines the time constant for the resonator to respond to a perturbation and therefore is related to the ring-down time after a pulse (Geschwind, 1972). The various contributions to Q are discussed in Dalal et al. (1981), Rinard et al. (1994), and Eaton et al. (1998a).

Significance of Q

Quantitative EPR, whether CW or pulsed, requires either keeping resonator Q constant or measuring Q and accounting for different values of Q in data analysis. The microwave magnetic field, B_1, for a particular incident microwave power increases as the square root of Q. Consequently, it is important to know Q for quantitative EPR and measurement of relaxation times. The AFC stability depends on Q. If the Q is too small, the frequency dependence of the cavity response is too shallow and the AFC cannot maintain constant frequency. If the Q is too large, the AFC circuit will not be stable enough to hold the frequency. Thus, it is important to adjust the gain of the AFC circuit to match the Q of the resonator. The highest S/N commercial spectrometers individually match source, resonator Q, and overall bridge performance to optimize these tradeoffs.

Note that the Q depends on the surface conductivity of the resonator, which increases rapidly with decreasing temperature. Thus, if the variable temperature system cools the entire resonator and not just the sample, the Q will change with the temperature of the sample. At liquid He temperatures, a copper cavity may have too high a conductivity to have a usable Q. The Q of a dielectric resonator also increases as the temperature decreases. The minimum length microwave pulse that can be introduced into a resonator depends on the Q. Mims (1965) showed that in a pulsed EPR experiment the minimum pulse time, t_p, assuming a Gaussian shape, whose energy can be transmitted to the resonator is related to the resonator Q by: $t_p = 4Q/\omega$. Furthermore, the Q-determined ringing time after a pulse limits how soon an echo, FID, or recovery signal can be measured following a pulse. Thus, Q importantly impacts both CW and pulsed EPR experiments.

Measurement of Q

There are many ways of describing resonator Q in terms of experimental observables. On a test bench, for example, with the appropriate equipment one can measure the frequency bandwidth by measuring the points at which there is a 45° phase shift or 3 dB[3] change in reflected power or use a sweeper with a Smith chart display. The following notes specifically focus on measurements that can be made in EPR spectrometers using the displays available in the spectrometer (i.e., without using additional test equipment). As these measurements depend on the detector used, we review the characteristics of detectors in detail later in this chapter.

CW MEASUREMENT OF Q. On a CW spectrometer the most convenient method of measuring Q is the measurement of the resonator band pass, $\Delta \nu$ (Fig. 14). Details for performing this measurement on the Varian E-line spectrometer are presented in Dalal et al. (1981). With nonlossy solvents the value of the loaded Q is ~ 3500 for Varian or Bruker rectangular X-band resonators. Lossy solvents cause a reduction in Q. The software on modern Bruker spectrometers includes a readout that estimates resonator Q from the mode pattern in "tune" mode. With a modern Gunn diode source the reflected power is nearly constant, except for the resonator dip, and there would be an approximately horizontal line at the power level of the dashed line of Fig. 14.

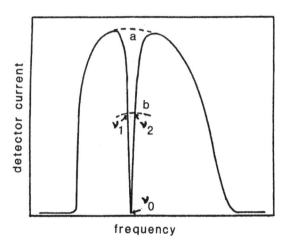

Figure 14 Klystron power mode, showing the "dip" due to cavity absorption. The vertical axis is power (detector current), and the horizontal axis is frequency. a is the detector current off resonance, b is the detector current at half-power, ν_0 is the resonant frequency of the cavity, and $\nu_1 - \nu_2$ is the half-power bandwidth of the cavity. The width of the dip is a measure of cavity Q. The dip should be in the middle of the power mode to get maximum power from the spectrometer. When the reference arm power is turned on, the display increases in amplitude and the dip does not touch the baseline because in this case the power from the reference arm adds to the power reflected from the cavity. When the reference arm is on, the symmetry of the power mode with respect to the dip is a measure of the relative phases of the reference arm and the signal arm.

---

[3] A power ratio in dB is defined by dB = 10 log $(P/P_{\text{reference}})$.

CAVITY RING-DOWN METHOD. Following a step change in incident microwave power, the power in the cavity reaches the new power level exponentially. The higher the Q of the resonator, the longer the response time constant. The following discussion is in terms of response of the resonator to an input pulse of microwave power. The exponential response applies both to the increase in stored power at the beginning of the pulse and to the decrease (ring-down) at the end of the pulse. The measurements of Q are based on the ring-down at the end of a pulse. In texts this approach to measuring Q is sometimes called the "transient decay" or "decrement method".

If a crystal detector is used and it is operating in the square-law region, then the detected voltage is proportional to power.

$$P(t) = P_0 e^{-t/\tau} = P_0 e^{-\omega t/Q}$$

where $P(t)$ is the power emitted from the resonator and detected, t is the time after the pulse, and τ is the time constant for the ring-down (Gallay and van der Klink, 1986). Typically, this voltage output from the crystal is recorded with a digital oscilloscope and the decay is analyzed with a computer to obtain the time constant τ for the ring-down. If $\tau = Q/\omega$ and $\omega = 2\pi\nu$, then $Q = 2\pi\nu\tau$. The Bruker E580 "receiver monitor" signal is the output from a double-balance mixer (DBM) that is biased into the linear region. The ring-down time measured with this detector is proportional to $\sqrt{P}$. Thus, if the ring-down of this signal is used to calculate Q, the equation that should be used is $Q = \pi\nu\tau$. Note the factor of 2 difference, depending on the type of detector used.

In practice in pulsed EPR τ is in ns and ν is in GHz, so the exponents cancel and the calculation involves simple small numbers. At X-band, since $2\pi\nu \approx 60$ and τ is the $1/e$ ($=0.37$) point on the oscilloscope display, a "one significant figure" measure of Q can be estimated from the scope display of the ring-down. This approach is useful for tuning the spectrometer, even though a more accurate measurement is required after tuning is complete, if signal quantitation is desired.

Because of the importance of these measurements, we include numerical examples. If the time constant for the decay is 60 ns and the frequency is 9 GHz,

$$Q = 2\pi(9 \times 10^9 \, s^{-1})(60 \times 10^{-9} \, s) = 3400$$

This is roughly what one would expect for a critically coupled standard rectangular (TE_{102}) resonator.

Relationship Between Cavity Q and Instrument Dead-time

Consider the use of such a rectangular resonator in an SR spectrometer. If we were to pump at, for example, 600 mW, and observe at 600 nW, the pump power from the resonator must decay by a factor of 10^6 to become roughly comparable to the observe power. This would require $\sim$14 time constants, as $e^{14} = 1.2 \times 10^6$. The ring-down time constant for a resonator with $Q \sim 3000$ is 60 ns, and 14×60 ns = 840 ns, so the ring-down of pump power is significant for $0.5-1$ μs.

For the ESE experiment the pump power is larger relative to the observed signal, so the power following the incident pulse must decay for even more time constants. Two things help in the ESE experiment. First, there is no observe power during the data collection, so a higher-gain microwave preamplifier can be used. Second, one usually uses over-coupling (Rinard et al., 1994) to achieve a lower resonator Q, so the ring-down time constant is shorter. The time constant of the power ring-down equals $Q_L/(2\pi\nu)$. The dead-time may be as long as $\sim$20 time constants. As the ring-down time is linearly dependent on Q, it is necessary to decrease the Q to very low values to measure short relaxation times. Note also that the frequency is in the denominator of the expression for τ. Thus, the lower the frequency, the longer the ring-down time for the same Q. To get equally short dead-times on a lower frequency spectrometer requires Q values to be decreased by the ratio of the frequencies.

C. Magnet System

Up through $\sim$70 GHz, EPR spectrometers usually use electromagnets. An X-band spectrometer designed for a narrow spectral range near $g = 2$ (such as the Bruker e-scan) can use a permanent magnet with sweep coils. The Resonance Instruments (formerly Micro-Now) 8400 table-top EPR uses a narrow-gap electromagnet that does not require water cooling. At low frequencies, air-core coils, such as Helmholtz coils, may be more practical than an iron-core magnet (Eaton and Eaton, 2004; Rinard et al., 2002c). At frequencies above 70 GHz, and importantly at 95 GHz, it is necessary to use superconducting magnets (Eaton and Eaton, 1999a). The main problem with superconducting magnets is that sweeping the main magnetic field boils off a lot of helium, and room-temperature sweep coils that will fit within the bore of the supercon magnet sweep only a limited range—currently $\sim$1000 G. Special supercon magnet systems for 250 GHz EPR were reported by Freed (Earle et al., 1996). High-field magnets for ESR have been described by workers at the National High Magnetic Field Laboratory (Brunel, 1996; Hassan et al., 2000).

1. Magnet Homogeneity

The magnetic field homogeneity requirements are not as demanding for EPR as for NMR because EPR linewidths are usually rather large in comparison with NMR linewidths. Because EPR systems usually require a wider magnet gap than NMR systems, and because a wide field sweep is required for most EPR systems, it is a challenge to make a good EPR magnet. The commercial systems have magnetic field homogeneity specifications of about ± 15 mG over a volume with 2.5 cm diameter and 1.3 cm length. It is important to confirm the homogeneity of the field, especially after a magnet has been moved. The easiest way to do this is to use an oscilloscope display of an NMR signal. Note that this homogeneity is specified at a particular magnetic field, usually at ~ 3400 G. Away from this field the homogeneity degrades. However, EPR linewidths are usually greater for transitions far from $g = 2$, so there is usually no problem with lower magnetic field homogeneity far from $g = 2$.

Inherent in the use of 100 kHz magnetic field modulation is the assumption that approximately ± 35 mG sidebands thereby created will contribute negligibly to the overall linewidths. Important cases with narrower lines are known [e.g., the samples used for oximetry: Ardejaer-Larsen et al. (1998) and Yang et al. (2001)], and for these samples it is necessary to use lower frequency modulation.

2. Size of Magnet

For many EPR experiments, a 3 or 4 inch diameter pole face is adequate. However, a larger magnet makes possible a wider variety of studies. There are three factors that should be considered in selection of the size of the magnet—maximum field required, volume of region with homogeneous field, and gap size. For X-band studies the maximum magnetic field needed exceeds $\sim 10,000$ G only for some transition metal compounds, so for most studies the magnetic field strength needed does not usually dictate the size of the magnet. For Q-band (35 GHz) the center field needed for $g = 2$ is 12,500 G, so the maximum field available should be a few thousand gauss larger than this. Higher-field EPR becomes sufficiently demanding of the magnet, so that the primary focus shifts from the microwave system to the magnet (Eaton and Eaton, 1999a).

The volume of space needed for the experiment places the most stringent requirements on the size of magnet. The magnet pole face diameter needed is determined primarily by the homogeneous volume needed for use of a dual cavity. For example, with two TE_{102} cavities combined, as in the Bruker ER4105DR dual cavity assembly, the two samples are 43 mm apart. This cavity would most commonly be used for comparative g-value measurements with highly resolved spectra, so the magnet homogeneity needed is maximum for this case. The larger the pole face diameter, the easier it is to produce homogeneous fields over a given diameter. Usually, ~ 9–10 inch diameter is adequate. The magnet gap (the distance between the pole faces) has to be at least 35 mm (Bruker) to accommodate the width of the standard X-band EPR rectangular cavities. Less space could be needed for some of the newer resonant structures, but even here a shield of these dimensions is commonly used. Some accessories require a minimum gap, which for standard magnets requires a minimum pole diameter. A 10 inch diameter is needed for the Bruker variable temperature Q-band accessory or the Oxford 935 cryostat used with the Bruker Flexline resonators. Care needs to be taken to avoid size incompatibilities. For example, some of the closed cycle cryogenic systems will not reach to the center of a 12 inch magnet unless the gap is very large. Experiments involving external coils, such as imaging studies, require additional gap space. If resources are available, a wider gap is preferable, as it facilitates future experiments. The tradeoffs are that magnet weight and cost increase with pole face diameter and gap size for a given field strength. Also the magnet power supply size needed increases with gap size for a given field strength.

3. Magnetic Field Measurement and Control

The accuracy of magnetic field measurement and control determines the accuracy with which one can measure g- and a-values, compare areas under spectra, and subtract background spectra. The most common field measurement devices incorporated in EPR spectrometers for feedback control are Hall probes and, occasionally, rotating coil gaussmeters. Commercial magnetic field controllers have gain and offset controls that permit calibration of the field scan. The portable Hall-probe gaussmeters that have recently become available are very handy, but are not sufficiently accurate for calibration of EPR spectrometers. The temperature compensated Hall probes built into the commercial EPR spectrometers are more accurate than the test equipment. The Bruker Hall probe has especially impressive specifications: absolute field accuracy better than 500 mG over the full 20 kG range, 10 mG resetability, sweep linearity ± 5 mG or $\pm 1 \times 10^{-3}$%. For calibration of the magnetic field it is necessary to use an NMR gaussmeter or a sample whose spectral features are well documented. The Bruker NMR teslameter (gaussmeter, ER035M) is an important spectrometer component for users who are doing careful g-value measurements. If an NMR gaussmeter is not available, secondary standards can be used with accuracy adequate for most studies. Near $g = 2$, Fremy's salt is useful (see Poole, 1983, p. 445). Over wider fields some Mn(II) and Fe(III) samples can be used (Eaton and Eaton, 1980).

D. Signal Detection and Amplification

1. Background Information on Crystal Detectors

Many types of crystal detectors have been used in EPR spectrometers for various purposes. Bruker has used a zero-bias Schottky diode for CW operation. This type of diode generally gives the best sensitivity, and is fairly robust, but does not give fast pulse response and is very temperature-sensitive. For the most careful quantitative work the temperature of the diode has to be controlled. As mentioned earlier, the output of a diode in response to microwave input can be simplified as consisting of a "square-law" region at low-incident microwave power and a "linear" region at high power, followed by a "saturated" region at still higher power. The terms are a bit awkward, and historically relate to a tradition of plotting output current vs. microwave voltage. Such a plot is linear at high microwave power and parabolic (square-law) at low power. The transition from square-law to linear regions is gradual. In the square-law region the voltage output of the crystal is proportional to the input microwave power. For most crystals this region extends from approximately -50 dBm[4] to approximately -15 dBm, at which power the output voltage is ~ 10 mV. Above ~ 0 to $+5$ dBm most crystals are in the linear region. In this region the output voltage is linear in the square root of the microwave power incident on the crystal. These are very rough numbers and vary more than a factor of 2 for various types of crystals.

A good rule of thumb seems to be that if the crystal output is less than ~ 10 mV the crystal is in the square-law region. For critical applications this should be verified by individual test. Background information is in the Radiation Lab Series (Montgomery, 1947, Montgomery et al., 1948) and in Poole (1983). Some typical response curves are in the literature of various manufacturers. Conversion loss and noise are related in complicated ways, so the optimization of a crystal detector in a spectrometer system requires careful engineering. The S/N sometimes is best in the transition region between square-law and linear. The reference-arm bias is designed to put the crystal in the power range for optimum performance.

A crystal detector rectifies the microwaves, producing a voltage that is subsequently amplified. This voltage carries the various modulation information, such as the 70 kHz (Varian) or 76 kHz (Bruker) AFC and 100 kHz (or lower) magnetic field modulation, used in the modulation phase-sensitive detectors in later stages of the detection system. The oscilloscope display during the "tune" mode of the spectrometer is a plot of microwave power reflected from the cavity (vertical) vs. microwave frequency (horizontal). The power output from a klystron is a nonlinear function of frequency, so the display is a broad hump. The power output from a solid state Gunn diode is more uniform over frequency. When the source frequency is tuned to the resonant frequency of the resonator, the energy incident on the cavity is absorbed in the cavity, and there is a "dip" in the power reflected from the cavity. The relative frequency width of this dip reflects the Q of the cavity (see Section III.B.1 for details). The dip touches the baseline (zero reflected power) when the cavity is critically coupled to the source.

DBMs are useful detectors in EPR spectrometers, but they are less common in commercial instruments (except for the Bruker pulse and transient bridges) because it is harder for the casual user to interpret the output. The reason is that DBMs reveal both phase and amplitude information, so the output has a sinusoidal response superimposed on the resonator response if the electrical path lengths are not equal. A biased crystal detector also reveals the phase information in the detected signal relative to the phase of the microwaves used to bias the crystal. This is why one has to select the reference-arm phase. Most pulsed EPR spectrometers use DBMs, with the reference arm to the local oscillator (LO) side and the detected signal to the RF side. As these inputs have the same frequency, the DBM works as a microwave phase-sensitive detector, and the output is a voltage (FID, echo, and SR). In CW operation with a DBM, the output voltage carries the modulation information. In most applications, DBMs are operated with a LO input of approximately $+5$ to $+10$ dBm. This biases the crystals that make up the mixer circuit into their linear region. Thus, the output from a DBM detector is almost always in the linear region (i.e., the voltage output is linear in $\sqrt{P}$).

As the first amplifier usually determines the noise level of the detection system, and EPR gives a very low-level signal, the use of a low-noise (e.g., NF $\ll 2$) microwave amplifier is an attractive concept for improved S/N in EPR spectroscopy. For ESE systems, a gallium arsenide field-effect transistor (GaAsFET) amplifier is a standard component of a detection system. In CW EPR spectrometers, the use of such an amplifier must be considered carefully. In the ideal case, with a perfectly critically coupled resonator, a microwave preamplifier would provide a significant S/N advantage over the usual detectors. To take optimum advantage of a high-gain low-noise preamplifier, very good coupling to the resonant structure is essential. For example, if the coupling is only 40 dB (1 μW reflected when the incident power is 10 mW) a preamplifier gain $>20-30$ dB will result in saturation of the output of the preamplifier (this usually occurs at about $+10$ dBm). Coupling of 40 dB is the best that is usually

---
[4] The notation dBm denotes power relative to a reference power of 1 mW. Thus, 10 dBm is 10 mW and -50 dBm is 10 nW.

achieved in waveguide systems with iris and tuning screw coupling. GaAsFET preamplifiers are useful for very low-level signals, such as for easily saturable CW signals, and for SR, FT, and ESE experiments, but will not soon be useful for high-power CW studies, such as for metals. When installed in older EPR spectrometers (such as the Varian E-line series) a factor of ~2–3 improvement in S/N at low operating powers is attained at relatively low cost. The improved detector systems in modern EPR spectrometers make the S/N improvement achieved with a GaAsFET preamplifier less.

2. Noise

The limiting noise of a modern EPR spectrometer varies with power incident on the resonator. At the lowest incident powers the detection system may limit the noise performance, but at high incident powers the microwave source is the limiting noise source in almost all cases. This assumes, of course, that other obvious noise sources, such as power supplies, switches in the microwave system, ground loops, and so on, have been properly engineered. There are special exceptions to this generalization: homemade resonators may be microphonic or may radiate (hand-waving effects); thermal drift of resonator or detector may be caused by environment changes or modulation or power changes; flow of gas or boiling of cryogens may introduce sample motion or changes in Q, and so on. The standard "weak pitch" S/N test is particularly challenging because it tests so many features of the spectrometer simultaneously (Eaton and Eaton, 1992). The limit to S/N often could be very low-frequency noise, even at the level of building vibrations.

E. Data System

Early use of computers in EPR was reviewed comprehensively by VanCamp and Heiss (1981), and there is an update by Kirste (1994). Operation of a modern EPR spectrometer is largely via a computer that controls data acquisition, manipulation, and display. The real-time data-acquisition function is handled by a computer that is largely invisible to the user. The user interacts with high-level software on an industry-standard hardware platform.

A key step in computer acquisition of EPR spectra is A/D conversion. Relative to the state of the art in A/D converters, CW EPR requires relatively slow A/D. In the Bruker console there is an A/D whose resolution and sampling rate depends on the magnetic field-scan rate in order to optimize digitization. Substantial additional information about the spin response in a CW spectrometer can be obtained by digitizing the complete response, prior to phase-sensitive detection. One way to do this uses what Hyde calls time-locked subsampling (Hyde et al., 1998). Time-domain EPR experiments make greater demands on the digitizer than phase-sensitive detected CW spectroscopy. This is a rapidly changing technology, so the digitizer is usually a separate module that can be replaced easily as better capability becomes available. The Bruker pulse spectrometers are based on a proprietary chip in their SpecJet module (Eaton and Quine, 2000). Components are not yet available to directly digitize signals at the X-band microwave frequency.

F. Commercial EPR Spectrometers

There is very little competition in the commercial EPR market. One must either accept the features of the spectrometers available, modify one to fit one's own needs, or build one's own. As a practical matter, this means that there is not much room for negotiation of price. However, there is more than an order of magnitude range in capability, complexity, and cost of commercial EPR spectrometers, and the range of capabilities of commercial EPR spectrometers encompasses most of what might be required for most analysis, instruction, and main-stream research.

Early commercial EPR spectrometers are listed in Alger (1968, Table 3-4). In the early years of EPR Varian was a major vendor, but they no longer sell EPR spectrometers. Today JEOL and Bruker are the major commercial suppliers of EPR spectrometers. Resonance Instruments also has made a significant contribution. The total market is fairly small, relative to the market for other types of analytical instrumentation, so marketing decisions by large corporations have caused major fluctuations in availability of various models. The routine CW JEOL spectrometers have been sold in the USA, but the pulsed spectrometers have not been. The authors' limited experience with JEOL spectrometers is the reason for not describing them in detail in this chapter.

Bruker sells CW spectrometers at frequencies between 1 and 95 GHz and pulse spectrometers at 9, 34, and 95 GHz. Accessories are available for ENDOR and ELDOR. The Bruker e-scan system is designed for quantitative EPR, with a special focus on routine monitoring (Maier and Schmalbein, 1993). It is designed for radiation dosimetry monotoring based on the radical formed in irradiated alanine, and includes an automatic sample changer. Alanine dosimetry is an increasingly important application of EPR spectroscopy (Nagy et al., 2000).

The Resonance Instruments 8400 is a routine research EPR spectrometer. This is a table-top unit light enough for one person to lift. Its 6000 G scan range (without water cooling) and 5×10^{10} spins/G sensitivity make it a worthy successor to the E-4. It is a very good instrument for student use. Resonance Instruments also has a system that is designed for analysis of gemstones.

Oxford Instruments provides cryogenic accessories specifically designed to interface with Bruker EPR spectrometers. Wilmad Glass provides EPR tubes and all of the glassware and quartzware needed for JEOL, Varian, and Bruker spectrometers.

IV. THE SAMPLE

The chemical and physical nature of the sample, as well as expected magnetic behavior, relates to the requirements for and limitations placed on the spectrometer. For example, the effect of the loss factor of the solvent, whether the sample is limited in size, whether it is stable, and so on, determines the type of EPR experiment that can be performed. Some other features of the interaction of instrumentation with the information you want to get from a sample were mentioned in Section III. Special resonators, such as loops, cut torroids, and so on, can be used on the surface of large samples, avoiding the need to have a sample small enough to put inside a resonator (Eaton and Eaton, 2004; Hirata and Ono, 1997; Hirata et al., 2000, 2001). In such a case the limit becomes the size of access to the magnetic field. As there are no commercial spectrometers with these capabilities, this section will focus on the more common situation in which a sample is put into a loop-gap, dielectric, or cavity resonator and placed in a homogeneous magnetic field. For these cases some of the sample conditions to be considered in the design of EPR experiments include (a) amount of sample; (b) spin concentration in the sample; (c) phase of the sample (solid, liquid, and gas); (d) temperature at which the sample is to be studied; (e) uniformity of the sample (aggregation and heterogeneity); and (f) environmental conditions of interest (oxygen concentration and hydrostatic pressure).

A. Spin System

EPR spectra are more readily observed for systems with an odd number of unpaired electrons (Weil et al., 1994). For systems with even numbers of unpaired electrons EPR spectra frequently cannot be observed. See Section II.D for a discussion on interacting spins and Section VI.H for a discussion on pulsed EPR flip angle for $S > 1/2$.

B. Temperature

1. Selection of Temperature for Operation

The first question to consider is the temperature range in which prior work would lead one to expect to be able to observe the spectrum of interest. It is important to know how sensitive T_1 and T_2 might be to temperature. In general, the lower the temperature the more likely it is that an EPR signal can be observed, both because of relaxation times becoming longer and because the Boltzmann population difference becomes more favorable. One should be careful to avoid misinterpretation due to the change of relaxation times with temperature. The relief of saturation due to shortening of relaxation times as temperature is increased can lead to a stronger EPR signal at higher temperature, and thus an apparent increase in unpaired spin population with increasing temperature, the opposite of what one expects from the Boltzmann population changes. If the relaxation time becomes too long relative to the modulation frequency in CW EPR, passage effects can affect the signal shape (see Section I.D). This can occur at room temperature for samples with long relaxation times, but more commonly occurs at low temperature where relaxation times for many samples become longer.

2. Temperature Control

Temperature control in magnetic resonance has not been engineered as carefully as the microwave and electronics aspects. Usually, the nature of the sample and the interaction of the cavity or resonator with anything put into it makes it most convenient to change the sample temperature by passing heated or cooled gas over the sample. If the resonator is small, as in Q-band measurements, or some of the newer S- and X-band resonators such as loop-gap or dielectric resonators, it is mechanically more convenient to cool the resonator and the sample together. This changes the mechanical dimensions of the resonator, and its electrical conductivity, so frequency and Q change with temperature. Mechanical instabilities can also be a problem. In gas flow systems there is usually a temperature gradient over the sample. This gradient might be more than a degree Kelvin. Consequently, at very low temperatures, where the spectrum changes rapidly with temperature, or at room temperature for some biological samples (such as membrane melting or protein unfolding studies), the gradient might be larger than the temperature range of interest. Special efforts should be taken to map the temperature gradient if temperature is a critical variable. For precise temperature control near room temperature, a recirculating liquid (remember it must be nonlossy, such as a silicone fluid, or a hydrocarbon or fluorocarbon) system is probably better than a gas stream. This would have to be locally constructed because no commercial unit is available. Berlinger (1985) reviews both high- and low-temperature experimental techniques.

A sample sealed into a quartz tube may not be in thermal equilibrium with the heat-transfer gas. This is a practical problem at temperatures below 77 K (in the liquid helium range), especially if the sample is an "insulating" material such as an organic solvent or a biological membrane. If the sample was carefully evacuated to remove oxygen, and the sample contracts more with decreasing temperature than does the quartz tube, the sample can become very effectively isolated from the quartz tube and hence from the cooled gas stream. Samples can remain for very long times (minutes to hours) at temperatures much higher than that of the cryogen. To prevent this, it is convenient to back-fill the EPR tube with a low pressure of He gas as an internal heat-transfer agent before sealing the tube. If the top of the tube is at room temperature, as in the Oxford ESR900 system, then it is necessary to put a baffle (e.g., a plug of quartz wool) in the cooled portion of the tube to prevent convective heat transfer. The Bruker cryostat based on the Oxford ESR935 puts the entire tube in the cooled He gas. If the tube is not sealed (tube caps will not fit into the apparatus) the sample comes to thermal equilibrium very quickly. If it is sealed, the earlier cautions concerning heat transfer apply. Note that it is difficult to avoid oxygen contamination (and EPR signals of O_2) with the Bruker/Oxford ESR935 system unless the tube is sealed or purged with He gas during the transfer to the cryostat.

C. Amount of Sample

The amount of sample to be used in an experiment depends on the information sought. In normal CW EPR experiments there is a limit on the strength of the EPR signal. In general, the smaller the amount of sample the better, so long as adequate S/N can be obtained. Too large a sample can cause problems. For the spectrometer response to be linear the Q change due to the microwave resonant energy absorption by the sample must be a small perturbation of the cavity Q. Goldberg has described the limits in detail (Goldberg and Crowe, 1977). With the standard TE_{102} cavity at X-band the number of spins must be kept below $\sim 10^{18}$, corresponding to less than about a milligram of paramagnetic compound, in order to avoid excessive change in Q at resonance (Goldberg and Crowe, 1977). Commercial X-band EPR spectrometers can detect roughly 10^9 spins of a material having a 1 G wide line. In practice, with saturable organic radicals in solution, one would use ~ 100–200 μL of solution, and expect to be able to observe radicals at micromolar concentration with reasonable S/N. Concentrations between 10^{-4} and 10^{-3} M are common practice, yielding good S/N and avoiding most problems of high concentrations unless lines are particularly narrow or the sample tends to aggregate. Note that the S/N is lower for SR and ESE measurements than for CW measurements, so in general one needs higher concentrations for these time-domain measurements than for CW. The wider bandwidth required in the detection path for the pulse signals than for CW signals is a major contributor to this difference because noise increases with the square root of bandwidth. Poorer S/N for pulse measurements than for CW is a problem, because the time-domain measurements are more sensitive monitors of concentration-dependent phenomena than are CW experiments.

For perspective on signal intensities, consider that the Varian strong pitch sample had 3×10^{15} ΔH spins/G/cm of sample length. The Bruker pitch samples are comparable to the Varian samples, but they have different spectral widths and each sample is marked with relative concentration. If the spectrum is ~ 2.8 G wide, then this is $\sim 8.3 \times 10^{15}$ spins/cm. Such a sample yields an almost noise-free spectrum. The weak pitch sample is ~ 0.003 times the concentration of strong pitch, and hence $\sim 2.5 \times 10^{13}$ spins/cm. The values in this paragraph are given only to illustrate practical S/N values for various numbers of spins. The S/N specification of the Bruker E580 spectrometer for the weak pitch sample is 3000:1.

When positioning the sample in the cavity, one must be aware of the B_1 distribution in the cavity and be aware of the effect of sample on B_1. See Section V.A for a discussion of B_1 and modulation distribution over the sample. Position is more important for small samples than for samples that extend the full length of the cavity. This is especially important in efforts to compare two samples, as in spin concentration determinations, and in efforts to measure relaxation properties.

It is common to use 4 mm o.d. quartz sample tubes for X-band EPR in rectangular resonators, unless the sample is lossy, in which case a smaller diameter capillary or a flat cell is used. A commonly used LGR (from Medical Advances) uses 1 mm o.d. tubes at X-band. Bruker produces both 3 and 5 mm split-ring resonators at X-band. Even at low frequencies, commercial LGRs are commonly designed for 4 mm samples. At Q-band roughly an order of magnitude less sample is used relative to X-band.

The fundamental article on EPR sensitivity is by Feher (1957). Most textbook treatments are based on Feher's article. A summary of Feher's results is provided in the Varian E-Line Century Series EPR instruction manual. Reading these various sources can be confusing, even after differences in notation are sorted out, because so many of the variables are functionally related to other variables, permitting many quite different looking "correct" expressions. Furthermore, many treatments seek expressions for the dependence of minimum detectable number of spin on, for example, frequency. These calculations can be done for many different situations. For example, the Varian manual summarizes the dependence of signal amplitude on frequency for eight combinations of whether the signal can be saturated, whether it is limited in size, and whether it is lossy. Several articles present relevant material (Goldberg and Crowe, 1977; Randolph, 1972; Warren and Fitzgerald, 1977a, b). An emphasis on minimum detectable number of spins does not provide formulae directly applicable to the usual case of a sample in a lossy solvent contained in a cylindrical sample tube. A practical summary of the effects of lossy solvents is in Dalal et al. (1981).

A related issue is the dependence of sensitivity on microwave frequency. Predictions for nonlossy samples with varying assumptions concerning the scaling of resonator and sample dimensions are presented in Rinard et al. (1999a, c, 2004) and Eaton et al. (1998a), and experimental verification is presented in Rinard et al. (1999b, 2002a, b).

D. Phase of Sample

Intermolecular electron–electron spin interaction can cause loss of resolution of EPR spectra. Thus spectra are usually obtained on samples that are magnetically dilute. Magnetic dilution can be achieved by dissolving a sample in a diamagnetic solvent or doping into a diamagnetic solid. It is much harder to interpret EPR spectra of magnetically concentrated samples. Unless the goal is to characterize the magnetically concentrated material, it is better to dissolve the solid (if it is stable to the solvation). One exception to this generalization is the use of pure solid diphenylpicryl hydrazyl (DPPH) as a g-value standard (Poole, 1983). In this case the exchange interaction between the paramagnetic centers averages the hyperfine interaction and produces a single relatively narrow line. Similarly exchange-narrowed lines in lithium phthalocyanine (LiPc) have been found to be especially sensitive to oxygen and are being used to measure local oxygen concentration, which is denoted as oximetry (Ilangovan et al., 2000a, b; Liu et al., 1993).

1. Solid State

Solid state samples can be single crystals, powders, or glassy solutions. If frozen solutions are used, solvents that give glasses should be selected. Solvents that crystallize can cause locally high concentrations of radicals, resulting in distorted CW lineshapes and shortened relaxation times.

Techniques for single crystal EPR are discussed in Morton and Preston (1983) and Chien and Dickenson (1981). Note that molecules are not "rigid" in solids or in glassy solutions, and motion of the molecule may still be significant even in a superficially "solid" matrix (Barbon et al., 1999; Du et al., 1992).

2. Fluid Solution

Concentration

In NMR of diamagnetic species one usually strives for the highest sample concentration that does not lead to viscosity broadening of the spectrum. In contrast, in EPR one strives for the lowest concentration for which adequate S/N can be obtained. The difference is the effect of intermolecular collisions in fluid solution on spin relaxation, and hence on broadening of the spectrum. In general known, low, concentrations should be used. For some purposes it will be necessary to extrapolate to infinite dilution, and to check for solvent effects, including aggregation. In ESE measurements high concentrations of spins can cause multiple echoes, at τ, 2τ, and so on because of radiation damping effects. Instantaneous diffusion can result in shortened T_m if local spin concentrations are high (Eaton and Eaton, 1993a, 2000a).

Solvent

Lossy solvents make it more difficult to obtain quantitative EPR spectra. Note, in this regard, that the solvent loss tangent can be very temperature dependent and frequency dependent. One dramatic case is CH_2Cl_2 at X-band—small changes in temperature near 0°C can dramatically change the apparent concentration of species dissolved in CH_2Cl_2 unless care is taken to account for changes in solvent lossiness (and hence in Q). The mere fact that a solute dissolves in a solvent means that there is significant solvent–solute interactions. This chemistry can affect spectra. For example, the effect of various solvents on hyperfine splitting of nitroxyl radicals has been examined and has been used to estimate the polarity of the environment (Knauer and Napier, 1973). In frozen solution a_{zz} appears to reflect mobility, and in some solvents it continues to increase as the solvent is cooled to well below liquid nitrogen temperatures (Du et al., 1995; Freed, 1976). For many samples spin echo dephasing is dominated by nuclear spins and becomes substantially longer in deuterated solvents (Zecevic et al., 1998). However, ESEEM due to deuterium is very deep, so in some measurements ESEEM due to solvent deuterons may interfere with measurements of echo modulation by other nuclei or with spin echo meausurements of T_m.

Oxygen

Most fluid solution EPR spectra are sensitive to the presence of oxygen or other paramagnetic gases. The longer the relaxation time, the more sensitive the spectrum is to the presence of oxygen. Degassing usually means getting rid of oxygen. Bubbling argon or nitrogen through the sample, or evacuating with a few cycles of freeze–pump–thaw, will reduce the oxygen concentration to the point that it will not broaden the CW EPR spectrum. However, for relaxation time measurements or for measurements such as ENDOR and ELDOR, which are very sensitive to relaxation times, one has to continually improve degassing techniques. The longest relaxation time measured is the closest to the correct value. NMR relaxation times are particularly sensitive to removal of air, and the lack of effectiveness of some of the common oxygen-removal techniques has been documented (Fukushima and Roeder, 1981, p. 161; Homer et al., 1973). The best advice in this aspect of experimental technique is never trust your result, and keep trying. However, because at low concentrations of O_2 the effects on relaxation times occur primarily via collisions (Heisenberg exchange effects) (Hausser, 1998; Hyde and Subczynski, 1989) freezing the sample reduces the effect of O_2 on relaxation times.

Oxygen in the normal atmosphere is used as a S/N standard at Q-band. Because the EPR signal is dependent on O_2 concentration, the S/N test was confounded when used in our lab at $\sim$1600 m elevation. At cryogenic temperatures O_2 trapped in condensed phases (including the ice on the outside of EPR tubes that have been stored in liquid nitrogen!) can yield strong EPR signals that interfere with the spectrum being studied. Low concentrations of gaseous O_2 in a tube cooled to near the O_2 boiling point can give very strong sharp EPR spectra superimposed on the spectrum of interest.

Other Paramagnetic Impurities

Other paramagnetic impurities can affect EPR spectra and relaxation times. Beyond the presence of unsuspected components of the sample (e.g., it was not known that it contained iron), the most likely problem is dirty sample tubes. Also be alert that attempts to clean tubes may introduce paramagnetic impurities, ranging from rust particles to Cr(III) from cleaning solutions.

3. Gaseous Samples

EPR has only limited applicability to gaseous samples. Most of the species that have been studied are very small molecules. See Westenberg (1975) for a discussion of gas-phase EPR.

E. Impurities, Overlapping Spectra

Always check the "background" signal, whether the measurement is CW, ESE, or SR. There can be instrumental artifacts such as transient signals from switches, sloping baselines due to interaction of modulation with the sweeping magnetic field, or thermal responses. To the extent that these are demonstrated to be reproducible, they can be subtracted from spectra of the sample. Background signals are present in most EPR resonators, whether metallic or dielectric. Even high-quality synthetic quartz sample tubes and variable temperature quartzware yield EPR signals. For critical experiments it may be necessary to select the quartzware for minimum background signal. Pyrex or other borosilicate glasses, commonly used because they are so much cheaper and are easier to fabricate into special apparatus, yield strong EPR signals. The signal due to the tube has been reported in the literature and attributed to the sample. Just as stopcock grease keeps showing up in NMR spectra, iron and copper ions recur in EPR spectra and often are incorrectly attributed to other species.

The "impurities" might be inherent in the sample because of incomplete reactions, side products, and so on. (*Note*: You might be looking at only a small part of the sample.) EPR spectra can provide detailed information about chemical equilibria and purity. Conversely, if the chemistry of the sample is not taken into consideration, superposition of spectra due to chemical equilibria, decomposition products, or impurities in the sample may be misinterpreted as features of the spectra of the species the experimenter thinks (or hopes) is present. Remember that the special benefits of derivative spectra for highlighting sharp spectra also can be a liability when sharp and broad spectra are superimposed. Even <1% of a sharp spectrum, when recorded "full scale" in the usual way can mislead the unwary into ignoring the rest of the signal, which may be the signal of the species of interest. Many examples of this problem can be found in the literature.

F. Intermolecular vs. Intramolecular Interactions

The best way to distinguish these interactions is to examine spectra as a function of concentration. For example, EPR can be used to determine whether a species is a doublet or triplet. Often the observation of a half-field ($\Delta m_S = \pm 2$) transition is used to argue for the presence of a triplet species. These measurements have to be extrapolated to infinite dilution, as intermolecular as well as intramolecular electron–electron dipole–dipole interactions result in such transitions. For example, half-field transitions can be observed in simple frozen solutions of monomeric nitroxyl radicals (Eaton and Eaton, 1989; Eaton et al., 1983). Pulse turning angles also can be used to determine S for a species, see Section VI.I.

G. Other Environmental Effects

Varying other environmental conditions, such as pressure, controlled atmosphere, light or other radiation, may be central to the information one wants from the EPR spectrum. The standard commercial X-band cavities are designed with slots in the end of the cavity for irradiation of the sample. Irradiation is harder to do with the Q-band resonator and some of the newer resonators. However, the loop-gap and slotted-tube resonators can be designed with adequate slots for irradiation. Cryostats are available with windows so that samples can be irradiated at low temperature. Special sample arrangements for high-pressure studies have been described (Goldberg and McKinney, 1984; Grandy and Pretakis, 1980). The paramagnetic species might be generated by almost the entire range of ways of introducing energy into a sample: light, ultrasonics, electrochemistry, γ-irradiation, heat, mechanical stress, and so on.

V. QUANTITATIVE MEASUREMENTS OF SPIN DENSITY

The older literature (e.g., Alger, 1968) suggests that EPR is inherently nonquantitative. This impression is erroneous. Unfortunately, many practitioners do not take advantage of the information that can be obtained from quantitative EPR spectra, whether CW or time-domain. Both improvements in instrumentation and improvements in awareness of relevant experimental parameters make it reasonable to aspire to $\sim 1\%$ accuracy in signal area measurements. However, it is necessary to be careful about many facets of the experiment to obtain reliable quantitative EPR spectra. This section provides guidance. Several reviews provide details and full documentation (Eaton and Eaton, 1980; Goldberg and Bard, 1983; Nagy, 1997; Randolph, 1972).

The key to quantitative EPR is attention to detail, and awareness both of the sample and of the spectrometer. Special attention must be paid to the effect of solvent either by measurement of cavity Q or by keeping all sample parameters constant, and the selection of the reference material (including knowing its behavior as a function of temperature). The general principles in the following discussions apply to most spectrometers that use a reflection cavity and magnetic field modulation, but the details relate specifically to the Varian E-Line spectrometers and the modern Bruker spectrometers, as most relevant data in the literature are for these spectrometers. The reference-arm bridge was a major contribution towards making routine quantitative measurements of signal area possible.

A. Spectrometer Calibration

Goldberg (1978) provided detailed consideration of linearity in Varian V-4500, E-3, and E-line spectrometers.

The important message is that each investigator should check the linearity of modulation amplitude, microwave power, receiver amplification, and magnetic field scan of their spectrometer. The Bruker EMS104 and e-scan were designed with special attention to the demands of quantitative EPR.

In addition to checking linearity of modulation amplitude settings, it is important to calibrate the absolute magnitude of the modulation amplitude. This is best done with a small sample that has a narrow line. A speck of DPPH is adequate. One measures the broadening of the line due to excessive modulation amplitude. Reference to Fig. 5, which shows how the derivative lineshape is obtained, reveals that when the modulation amplitude exceeds the linewidth, a signal such as that given in Fig. 15 will be obtained. The splitting between the positive and negative excursions, corrected for the original nonbroadened linewidth, gives the magnitude of the modulation amplitude. Details of the procedure and tables necessary for the task are given in Poole (1983, p. 242). The modulation amplitude at the sample depends on the modulation frequency, and has to be separately calibrated and adjusted for each frequency and for each cavity. The current Bruker

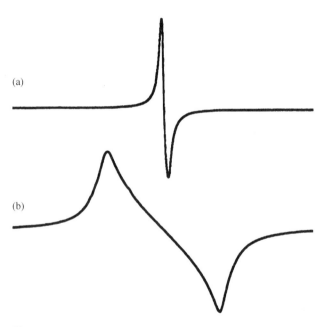

Figure 15 (a) A spectrum of solid DPPH, obtained with modulation amplitude $\ll$ linewidth. (b) Spectrum of the same sample obtained with modulation amplitude $\gg$ linewidth, which demonstrates the distortion of the spectrum due to over-modulation. The distortion follows directly from the picture in Fig. 5. For example, if the field is centered on resonance, and the modulation amplitude is large relative to linewidth, i_1 and i_2 would both be ~ 0. The separation between the two "peaks" in this display depends on the modulation amplitude and on the linewidth of the undistorted EPR signal.

software includes an automated modulation calibration routine.

Note also that the shape of the modulation field depends on the cavity design. The modulation field in the Bruker ER4102ST cavity and the Varian E-231 cavity has a roughly cosine-squared shape relative to the center of the cavity, decreasing toward the top and bottom. This must be accounted for if samples of different length are compared. In the Varian E266 Q-band cavity assembly the modulation coils are mounted on the outside of the variable temperature dewar, and are larger than the cavity, so the modulation is nearly constant over the cavity. The details of these considerations are discussed in Mailer et al. (1977). For many LGR and other small resonators the modulation coils can be larger relative to the active sample volume than in the X-band rectangular cavity.

The intensity of a CW EPR signal for a sample at a given position in a resonator depends linearly on B_1 (is proportional to the square root of microwave power). However, for identical samples at two different positions in the resonator (or the same point sample moved to a new position in the resonator) the EPR signal intensity is proportional to the ratio of B_1^2 at the two locations (Sueki et al., 1996). B_1 has an approximately cosine dependence relative to the center of the TE_{102} cavity resonator (it would be perfect cosine dependence except for the effect of the sample entry port). At X-band, using a standard rectangular resonator, the combined effect of B_1 distribution and magnetic field modulation distribution produces a $\cos^4$ dependence on position relative to the center of the cavity for the CW EPR signal intensity. In the case of a line sample extending the height of the cavity, most of the signal amplitude comes from the center ~1 cm of sample. The position-dependence is different for pulsed EPR signals. Because a modulation field is not used for SR or ESE, the $\cos^2$ dependence of the modulation amplitude is irrelevant. In the SR experiment, the recovery signal has the B_1^2 dependence on position described earlier for CW spectra. In the ESE experiment, however, the dependence on position is linear in B_1 if the power is adjusted to produce the same turning angle (e.g., 90° and 180° pulses) at the various positions. Special attention has to be paid to the B_1 distribution in large samples at low frequency. When samples with dimensions of the order of a significant fraction of wavelength are used there will be phase changes and concentration of B_1 that can make the EPR signal stronger in some positions and null, or even inverted phase, in other positions (Sueki et al., 1996).

B. Calibration of Sample Tubes

Unless one buys the most expensive EPR tubes, there will be a range of sizes (often 3.8–4.2 mm for nominal 4.0 mm o.d. tubes) in each batch. Therefore, one should select similar tubes (with a micrometer) to minimize tube corrections. Internal diameters should be calibrated gravimetrically using standard procedures for volumetric glassware (e.g., weighing water in tubes). The simplest way to perform routine measurements of signal area is to calibrate a few tubes with aliquants of the same sample. Such tubes, used with the same type of sample and the same solvent, temperature, and so on, carefully positioned in the cavity, can yield results accurate to within 5%, and within 2–3% with care. In favorable cases, with good S/N, one may reasonably aspire to 1% accuracy.

If a small amount of solid sample is used it must be positioned carefully at the center of the cavity. For aqueous samples, such as protein and other biological preparations with low-spin concentrations, it is important to get as much sample into the active volume of the resonator as possible, but without lowering resonator Q too much. Commercial flat cell assemblies are optimized for this purpose. However, they are expensive and fragile and difficult to position properly. An alternative experimental arrangement, found convenient in our lab, which also has the advantage of being disposable so that cross-contamination is avoided, is the use of "microslides" (Eaton and Eaton, 1977). These are flat tubes intended for providing a flat surface for optical microscopic examination of liquids. These microslides can be positioned in the EPR cavity fairly reproducibly by inserting them into Teflon tubing and then inserting this assembly into the quartz dewar insert, which provides support. Removing and reinserting the sample yields reproducibility of the EPR signal intensity to within ~5%. One striking observation using these sample holders is that best results are obtained in the Bruker 4102ST rectangular resonator if the flat cell is perpendicular to the nodal plane of the microwave distribution in the cavity, rather than parallel to the nodal plane as would be expected from elementary considerations of field distributions. The positioning of the sample is done by observing the width of the dip in the klystron mode pattern while rotating the sample to obtain the maximum Q.

C. Consideration of Cavity Q

At constant incident power the signal amplitude is linearly proportional to cavity Q. Insertion of any material into the EPR cavity changes the resonant frequency of the cavity, the Q, and the distribution of electric and magnetic fields in the cavity. This effect is presented pictorially in Randolph (1972, Figs. 3–8) and in Casteleijn et al. (1968). Particularly important manifestations of this effect include the distortions caused by the quartz variable temperature dewar commonly used in EPR, and the distortions caused by the sample tube and solvent themselves.

A sample with a high dielectric constant, such as aqueous samples, or samples with high electrical conductivity have strong effects on resonator Q. The presence of dielectric in the cavity shifts the resonance frequency to lower frequency (a conductor has the opposite effect). The range of sample tubes and solvents commonly used in chemical laboratories can affect Q by much more than a factor of two even when sizes, sample amounts, and so on are chosen to make it easy to tune the spectrometer. The most detailed examination of this matter is given in Casteleijn et al. (1968), to which reference should be made for derivations of the relevant formulae. In view of the importance of resonator Q, a discussion of the measurement of Q is included in Section III.A. Even a nonlossy dielectric sample, such as the irradiated fused silica standard sample, causes a large reduction in Q of a dielectric resonator if the filling factor is large.

Several papers in the analytical chemistry literature have described sample tube/holder assemblies intended to keep the experimental conditions (sample tube diameter and positioning) as constant as possible (Chang, 1974; Mazur et al., 1997a, 2000; Nakano et al., 1982). There are many experimental situations in which such an approach is not very convenient—for example, samples that have to be prepared on a vacuum line.

D. Reference Samples for Quantitative EPR

One always does relative measurements of EPR spectral areas (Eaton and Eaton, 1980; Goldberg and Bard, 1983). Absolute measurements are so difficult that in practice one would seldom have reason to attempt such measurements (Rinard et al., 1999b, c; 2002a). The more similar the reference standard is to the unknown, the more accurate the quantitation of the EPR spectrum will be. Some especially egregious errors have been made in sincere attempts to achieve as accurate a result as possible. For example, researchers have gone to great effort to obtain carefully hydrated single crystals of copper sulfate to use as a reference for the quantitation of the EPR signal of an aqueous solution of an enzyme. The difference in the effects on cavity Q of a small single crystal and of an aqueous solution presents one of the worst possible cases for accurate comparison. Wherever possible, use a sample of the same spin state, and hence the same spectral extent, at about the same concentration in the same solvent, and in the same tube, or one that has been calibrated relative to the tube used for the unknown. A variety of samples that have proven useful as reference samples in EPR are listed in Nagy (1997), Eaton and Eaton (1980), and Yordanov (1994).

E. Scaling Results for Quantitative Comparisons

Unless the spectra to be compared are obtained under exactly the same conditions, it will be necessary to scale the area (double integral of the first-derivative signal) obtained for the unknown to compare it with the standard. This involves:

1. Subtract the background spectrum.
2. Correct for modulation amplitude. Area scales linearly with modulation amplitude, even when spectra are overmodulated.
3. Correct for gain. Area scales linearly with gain settings of the detector amplifier. However, note that in the current Bruker software gain is reported in units of decibels with the relationship gain = 20 log(signal amplitude).
4. Correct for microwave power. Be sure to obtain spectra at power levels below saturation. Under these conditions, area scales as the square root of the incident microwave power.
5. Correct for Q differences. Area scales linearly with Q.
6. Correct for temperature differences. Note that EPR is a measurement of bulk spin magnetic susceptibility. Thus, the area scales with the population difference of the ground and excited states. For isolated $S = 1/2$, the temperature dependence typically obeys a simple Boltzmann behavior. Interacting spin systems, and cases of $S > 1/2$, can result in complicated temperature dependences. In this regard, note that most solid reference materials are not magnetically dilute, so their temperature dependence does not obey a simple Curie law (Goldberg and Bard, 1983; Molin et al., 1966; Slangen, 1970). A paper on chromic oxide is illustrative of the efforts needed in these types of cases (Goldberg et al., 1977). Also note that the density of solutions, and hence the number of spins at a given position in the B_1 field varies with temperature. In addition, solvent loss tangent varies with temperature. The temperature dependence of the sample and reference have to be the same or known and the differences mathematically compensated for.
7. Correct for spin differences. The transition probability, and hence the area, scales as $S(S+1)$. For example, comparing $S = 1/2$ and $S = 3/2$, assuming all transitions are observed, the relative areas are in the ratio 3/4 to 15/4 from this effect. If one observes only part of the transitions, which is common for $S > 1/2$, it is also necessary to correct for the relative multiplicities, which go as

(2S + 1). In some cases, simulation of the spectra may be required.

8. Correct for g-value differences. Area scales as g. Note that papers and books published prior to Aasa and Vanngard (1975) state incorrectly that area scales as g^2.
9. Correct for field-scan width. For first-derivative spectra, the scan width correction factor is (1/sweep width)^2.
10. When an integrating digitizer is used, as in the Bruker E500 series spectrometers, the recorded signal area varies with scan time. This is taken into account in the "normalized acquisition" option of the latest (2003) Linux version of the Bruker software, but not in earlier versions.

If the software includes a "normalization" option it is important to know which parameters are included. It is also important to ensure that the signal plus noise nearly fills the full scale of the digitizer as much as possible.

All of these general considerations (except modulation amplitude) apply to both CW and time-domain spectra. In the case of time-domain spectra there are two other major considerations. First, the spectrum can depend very sensitively and very strongly on the microwave pulse conditions and the observation conditions. For example, a magnetic field-swept, spin echo-detected spectrum depends on the interpulse time τ at which the echo is recorded. Most ESE spectrometers use boxcar detection of the echo amplitude. The shape of the field-swept spectrum depends on the portion of the echo that is sampled by the boxcar averager and the digitizer. Refer also to Section II.F where the effects of dead-time and multiple exponential decays are discussed.

VI. GUIDANCE ON EXPERIMENTAL TECHNIQUE

A. Words of Caution

Unlike many modern instruments, EPR spectrometers, having been built in a tradition of use primarily by specialists, are not designed to be fool-proof. Considerable care is required in using an EPR spectrometer to (a) prevent costly damage and (b) obtain useful results.

1. Cooling Water

Water cooling is needed for the microwave source and for the electromagnet. The recent Bruker spectrometers have interlocks to prevent damage to the source if you forget to turn the cooling water on.

2. Cleanliness

Most dirt and tobacco smoke have an EPR signal. Some materials used for suspended ceilings contain significant amounts of Mn(II). The cavity must be protected from dirt. It is good practice to prohibit smoking in the spectrometer room, and to clean the outside of all sample tubes just prior to inserting them into the cavity. A tissue such as "Kimwipe" and a cleaner such as a residue-free electronic contact spray cleaner is handy for this purpose. Fingerprints will transfer dirt to the cavity.

Inevitably, the cavity will become contaminated, possibly from a broken sample tube, causing significant background signals. The mildest cleaning that will remove the known or suspected contaminant should be used. Check the manufacturer's guidance on solvent compatibility. For their rectangular resonator Bruker suggests ethanol, hexanes, toluene, and/or 0.1 M EDTA followed by methanol. Some people use an ultrasonic cleaner, but this is not wise—it may loosen critically torqued parts of the cavity. After cleaning, purge with clean dry nitrogen gas to remove the last traces of solvent. Other resonators may need to be returned to the vendor for cleaning.

3. Changing Samples

Always attenuate the power to ~40 dB below full power before changing samples when doing CW EPR, to avoid unnecessary spike of power to the detector crystal. In pulse systems, turn high-power amplifiers such as traveling wave tube (TWT) amplifiers to standby when changing samples. A single mistimed or misdirected pulse from a TWT can destroy components of the detection system.

4. Detector Current

The output of the crystal detector depends on the magnitude of the bias current to the crystal detector. Each spectrometer should be checked to determine the range of detector current values within which the signal amplitude is independent of detector current. If the detector current drifts, as can happen with lossy solvents, or when the temperature is changed, significant errors in signal amplitude can result, S/N is degraded, and quantitative measurements are prevented. The output of the detector crystal is dependent on temperature. As the bridge warms up during the first hour or so after power is turned on, the accuracy of quantitative spectra may change during this period. Similarly, the detector crystal itself changes temperature as a result of changes in incident power, so spectra run immediately after large power changes may not be equilibrium responses.

Pulsed EPR spectrometers typically use a double-balanced mixer or a quadrature mixer (the latter is used in the Bruker E580). The output of the mixer depends on the power applied to the reference side of the mixer. Reproducible setting of the reference power as well as the reference phase is important. In pulse mode on the E580 one simply uses the maximum available reference power to bias the mixer.

B. Selection of Operating Conditions

As CW is the most common EPR experiment, we emphasize parameters for CW spectroscopy, but also point out some of the considerations in making spin echo measurements. Even if spin echo capability is not available to the reader, this section should help to understand reports in the literature.

While initially searching for a signal in a sample whose spectroscopic properties are not known, one can use relatively high spectrometer settings, such as 10 mW power, 1 G modulation amplitude, a fast scan, and short filter time constant. This is likely to be adequate to at least detect a signal, if it is present, with reasonable S/N. Recognize, though, that such a cursory scan could miss samples at two extremes: (a) a signal with such long relaxation time and narrow line that it is saturated or filtered out or (b) a broad signal in the presence of a more obvious sharp signal. Always look for spectra you do not expect. To obtain a quantitatively correct spectrum requires adjustment of microwave power, phase, modulation amplitude, gain, scan rate, and filter time constant. The criteria for selection of these settings is discussed in the following paragraphs.

1. Microwave Power

To obtain quantitative CW EPR spectra and to obtain undistorted EPR lineshapes, it is necessary to obtain spectra at microwave powers below those that cause significant saturation of the EPR spectrum. Most EPR samples can be saturated with the power levels available in commercial spectrometers. Thus, it is always important to check for saturation.

The incident microwave power should be set to a value below that at which the power saturation curve deviates from linearity (see Section II.G and Fig. 8). A quick way to check that you are operating in a range in which the signal intensity varies linearly with the square root of microwave power is to decrease the attenuation by 6 dB (a factor of 4 increase in power). The spectral amplitude should increase by a factor of 2; if it does not, reduce the power and try again.

If there is no unresolved hyperfine splitting the relationship between linewidth, B_1, and relaxation times is

$$(\Delta B_{pp})^2 = \frac{4}{3\gamma^2 T_2^2}(1 + \gamma^2 B_1^2 T_1 T_2)$$

where B_{pp} is the peak-to-peak linewidth and γ is the electron magnetogyric ratio (Eastman et al., 1969; Poole and Farach, 1971; Schreurs and Fraenkel, 1961). If B_1 is small enough that the $\gamma^2 B_1^2 T_1 T_2$ product term is $\ll 1$, the signal is unsaturated. However, consider typical X-band relaxation times for a nitroxyl radical at liquid nitrogen: $T_1 = 200$ μs, $T_2 = 2$ μs. With an attenuation of 40 dB below 200 mW, $P = 0.02$ mW. For an X-band rectangular resonator $B_1 \sim 2 \times 10^{-2} \sqrt{(QP)}$ with P in watts (Bales and Kevan, 1970; More et al., 1984) and $Q \sim 3500$, so 40 dB produces $B_1 \sim 4.2 \times 10^{-3}$ G and $\gamma^2 B_1^2 T_1 T_2 \sim 1$, which indicates severe saturation. In practice, this power does not cause severe saturation because spectral diffusion processes cause the effective relaxation times to be much shorter than the values measured by pulse methods (Eaton and Eaton, 2000a).

The selection of microwave power to be used for time-domain EPR is different for each type of measurement. For example, for SR it is important to use an observe power low enough that it does not saturate the spectrum and shorten the apparent recovery time (Huisjen and Hyde, 1974b; Mailer et al., 1985). For two-pulse ESE one normally seeks to maximize the echo intensity (90°, 180° or 120°, 120° pulses). However, the pulse power would be changed for selective excitation or for checks of instantaneous diffusion.

2. Phase

There are two phase settings in the normal CW spectrometer, the reference-arm phase and the phase of the phase-sensitive detector operating at the field modulation frequency (often 100 kHz). If the reference-arm phase is wrong the signal amplitude will be low, and dispersion will be mixed with absorption. This phase setting is adjusted to maximize the detector current at high microwave power. If the 100 kHz detector phase is wrong the signal amplitude will be low. Because a null is easier to see than a maximum, the best approach is to set the phase for null signal, and then change phase by 90°. For saturation-transfer spectroscopy this setting is very critical, and a substantial literature has been devoted to it (Beth et al., 1983; Hyde, 1978; Hyde and Thomas, 1980; Watanabe et al., 1982). In pulsed EPR, where a DBM or quadrature detector is used there is similar need to establish the reference phase for the mixer. It is also important to adjust the phase of the AFC system (normally 70 or 77 kHz), but there is no user control of this phase in commercial instruments.

3. Modulation Amplitude

The rigorous experimental criterion is that the modulation amplitude ($B_{m2} - B_{m1}$ in Fig. 5) should be kept less than 1/10 of the derivative peak-to-peak linewidth, ΔB_{pp}.

However, modulation amplitudes up to about $1/3$ of ΔB_{pp} cause relatively little distortion. If the modulation amplitude is too large, the EPR signal will be distorted. In fact, it is possible to obscure hyperfine splitting when the modulation amplitude is too large. For the most accurate lineshapes you should always scan the spectrum with a very small modulation amplitude, determine the narrowest linewidth, and set the modulation amplitude to $1/10$ of that linewidth. If you were using too large a modulation amplitude during the initial scan, repeat the measurement of the linewidth, and adjust the modulation amplitude again if necessary. For small modulation amplitudes the S/N increases linearly with modulation amplitude. It may happen that the S/N is too low to obtain a spectrum with the modulation amplitude meeting the earlier criteria. In this case, one has to go back to the basic question—what information do you want from the sample? If lineshape is the crucial information, then signal averaging will be needed to improve the S/N. If subtleties of lineshape are of less significance, it may be acceptable to increase the modulation amplitude up to roughly $1/2$ of the linewidth [see Poole (1983, Section 6H), for details on degree of line distortion]. If the area under the peak is the information desired, overmodulation is acceptable, as the area is linearly proportional to modulation amplitude, even when the modulation is so large as to cause distortion of the lineshape. Computational approaches have been developed to correct for the lineshape distortion due to overmodulation (Robinson et al., 1999a, b).

4. Gain

The gain is adjusted to give the desired size of display and where possible should be increased to use the full range of the digitizer. Note, however, that the gain should not be so high as to cause the amplifier to saturate. On the Varian E-Line the PSD output becomes nonlinear when the "receiver level" meter is at the high end of the scale. These problems seem to be designed out of the recent Bruker CW systems, but there is no indicator to the operator when the Bruker pulse system amplifiers are saturating and becoming nonlinear. The operator must always check for linearity if blessed with a signal strong enough to saturate the amplifier.

5. Scan Rate and Filter Time Constant

The S/N in a spectrum depends strongly on the bandwidth of the detection system. CW EPR typically uses a very narrow bandwidth via the 100 kHz phase-sensitive detector, and subsequently removes $1/f$ noise via a user-selectable filter time constant. Pulsed EPR has to use a much larger bandwidth to record the rapidly changing signals, so there is inherently poorer S/N in the recorded spectra. White noise increases linearly with the square root of the bandwidth.

Scan rate and filter time constant are related to each other and to the CW linewidth in the following formula. The inequality must be satisfied to obtain undistorted lines.

$$\frac{\text{Spectrum width (G)}}{\text{Linewidth (G)}} \times \frac{\text{Time constant (s)}}{\text{Sweep time (s)}} < 0.1$$

A faster sweep or longer time constant does not give the system enough time to respond to changes in signal amplitude as the line is traversed.

For example, if the narrowest line in the spectrum has $\Delta B_{pp} = 1$ G, and a 200 G sweep is to be conducted in 100 s, then

$$\text{Time constant} < \frac{(0.1)(100 \text{ s})(1 \text{ G})}{(200 \text{ G})} < 0.05 \text{ s}$$

If the time constant is greater than this, the line will be distorted.

In the Bruker software one sets time constants in the way discussed earlier, but conversion times for the digitizer and number of points in the spectrum determine the scan time. The slower the scan the higher the resolution of the digitizer. The ranges are 0.33 s to 45 min and 9–22 bits.

S/N can be improved by using a longer time constant and a slower scan. Why then, would one want to use an expensive computer system for S/N improvement, when the spectrometer is designed for extensive analog filtering? The answer is that with a perfectly stable sample and stable instrument, roughly equal time is involved in either method of S/N improvement. The problem is that perfect stability is not achieved, and the filtering discussion focuses on high-frequency noise. Long-term spectrometer drift due to air temperature changes, drafts, vibration, line voltage fluctuations, and so on, limit the practical lengths of a scan. Signal averaging will tend to average out baseline drift problems along with high-frequency noise. Drifts in the magnetic field magnitude are not averaged out by filtering or slow scans, and always increase apparent linewidth. Ultimately, the resultant line broadening limits the spectral improvement possible with any averaging or filtering technique. In addition, computer collected spectra can subsequently be digitally filtered without changing the original data, whereas analog-filtered data is irreversibly modified.

If the sample decays with time, a separate set of problems emerges. Assume, for example, that you want to compare lineshapes of two peaks in a noisy nitroxyl EPR spectrum, and that the amplitude of the spectrum is changing with time because of chemical reaction (shifting equilibria, decay, oxygen consumption or diffusion, etc.). In this case one wants to minimize the time spent scanning between points of interest. It would be wise to scan the narrowest portion of the spectrum that will give the information of interest. Then a numerical correction for the measured rate of change in the spectrum is the best way to handle the problem. The impact of the time dependence can be minimized more effectively by averaging rapid scans than by filtering a slow scan.

6. Choice of Modulation Frequency

As a tradeoff between low-frequency noise and distortion by modulation sidebands, most CW EPR spectra are obtained using 100 kHz modulation. However, for samples with narrow-line spectra such as deuterated triaryl methyl radicals, lower modulation frequencies are required to avoid broadening of ~30 mG linewidths (Yong et al., 2001). ST-EPR spectra are usually obtained with modulation at a lower frequency and detection at the second harmonic, for example, 50 and 100 kHz. For slow-passage EPR, it is necessary to have the reciprocal of the modulation frequency much greater than T_1.

$$\nu_m^{-1} \gg T_1$$

This criterion is not met as often as it is assumed to be. Some samples at liquid nitrogen temperature, and many samples at liquid helium temperature, have T_1 that is too long to permit use of 100 kHz modulation. Passage effects (see Section I.D), recognizable as distortions of lineshapes, or even inversion of signals upon reversal of the field-scan direction, alert you to the need to use a lower modulation frequency. On older instruments, such as the Varian E-line and Century series spectrometers, there was a large degradation in S/N at lower modulation frequencies. In the current Bruker spectrometers there is very little increase in noise at low-modulation frequencies. The increase in electron spin relaxation times with decrease in temperature often forces one to use the lowest modulation frequency available on the spectrometer. Bruker spectrometers have modulation frequencies between 1 and 100 kHz and calibration of the modulation amplitude and phase is performed by the software using a DPPH sample.

C. Second Derivative Operation

With magnetic field modulation and phase-sensitive detection, the voltage at the detector can be expressed in a Fourier series (Noble and Markham, 1962; Poole, 1983; Russell and Torchia, 1962; Wilson, 1963). The ν_m term is the first derivative, and the component detected at $2\nu_m$ is the second derivative. Note that whereas the amplitude of the first derivative EPR signal is proportional to the modulation amplitude, A, the amplitude of the second derivative EPR signal is proportional to the square of the modulation amplitude. If you integrate a first-derivative spectrum twice (I_1) and a second derivative spectrum three times (I_2) the results are related by $I_2 = I_1 A/4$, if the actual absorption signal areas are identical (Wilson, 1963).

It is also possible to obtain a second derivative of the EPR signal by using two modulation frequencies and two phase detectors. This follows directly from the discussion of how the first-derivative lineshape is obtained (Section I.B), as everything stated there would remain true if the original signal were the first derivative. Thus, modulation at 100 and 1 kHz, followed by phase detection at 100 kHz, yields a first-derivative spectrum with a 1 kHz modulation signal on it, and then phase detection at 1 kHz yields the second derivative spectrum. If you go one step further and use the second harmonic of the high-frequency modulation plus a low-frequency modulation, you get the third derivative display.

Derivatives can also be generated with computer manipulation of digitally stored data. However, straightforward numerical derivative computation causes such a noisy looking derivative (due to the discrete nature of the data array) that it is not very useful unless the original data were virtually noise-free or unless extensive multipoint averaging is used. Much more useful is the pseudo-modulation technique developed by Hyde et al. (1992). The software is available from the National Biomedical ESR Center, Milwaukee.

D. CW Saturation

Even as pulsed EPR spectrometers become more common, CW continuous saturation characterization of spin relaxation will remain important. Faster relaxation times can be better characterized by CW methods than by pulse methods. In addition, the methods are complementary in the information they provide. See Section II.F for ways to plot CW saturation data. In the literature $P_{1/2}$ is commonly used as a saturation parameter (Mailer et al., 1977). $P_{1/2}$ is the incident microwave power at which the EPR signal has half the amplitude it would have in the absence of saturation.

E. Methods of Measuring Relaxation Times

Both CW and pulsed EPR methods can be used to measure relaxation times (Poole, 1971, Chapters 3 and 4). For each method, it is necessary to know the microwave magnetic field at the sample. Hence, it is necessary to know how to measure cavity Q (see Section III.A) and B_1 (see Section VI.F). Each method of measuring relaxation times gives different results, because it measures different contributions to the relaxation. See, for example, the comparison of CW progressive saturation and SR

(Hyde and Sarna, 1978) and of various pulse methods (Harbridge et al., 2003). Because pulse methods offer the possibility to distinguish between various contributions, we focus on those methods here. A more extensive discussion of relaxation times is given in Eaton and Eaton (2000).

1. Saturation Recovery

Conceptually, one of the simplest relaxation time measurements is the SR method (Huisjen and Hyde, 1974b). This is sometimes heuristically called "stepped EPR". In the SR method one saturates the spin system with a high microwave power, then reduces the power and observes the recovery of the spin system toward equilibrium by performing a CW EPR measurement at very low power level. In the absence of other processes that take saturated spins off resonance, this is a measure of T_1. Increasing the length of the saturating pulse often can mitigate the effects of spectral diffusion (Harbridge et al., 2003). The capabilities of current instrumentation limit these measurements to values of T_1 longer than about a few hundred nanoseconds. The S/N is low in an SR measurement because it is a "direct detection" method, that is, it does not use magnetic field modulation and phase-sensitive detection at this modulation frequency. In addition, to obtain a correct recovery time constant it is necessary to use observe powers that are low enough that the spectrum is not saturated. This results in very weak signals, which have to be averaged tens to hundreds of thousands of times.

2. Spin Echo

The Hahn spin echo technique, first developed for NMR, also can be used in EPR. Any two microwave pulses generate an echo. Detailed diagrams of the magnetization following the microwave pulses can be found in Kevan and Schwartz (1979), Levitt (2002), Mims and Peisach (1981), Poole (1983), Schweiger and Jeschke (2001), and Thomann et al. (1984). The amplitude of the echo following a two-pulse sequence in which the time (τ) between pulses is increased, decays with a time constant that is usually denoted as T_m, the phase memory decay time. Under certain circumstances T_m can be identified as the transverse relaxation time T_2 (Eaton and Eaton, 2000a). The value does not depend on magnetic field inhomogeneity, or on hyperfine coupling (in the absence of spin diffusion), so it is T_2 not T_2^*. This is a better measure of relaxation times than is linewidth, except for the case of the purely homogeneously broadened line, for which T_2 can be obtained accurately from linewidth. With modern instrumentation such as the Bruker E580 it is possible to measure T_m times as short as ~50 ns.

Estimates of T_1 can be made with various pulse sequences on a spin echo spectrometer, just as in NMR. If the two-pulse sequence that gives a spin echo is repeated faster than about once every five times the T_1 relaxation time, the spin system will not return to equilibrium between pulse sequences. In this case, the z-magnetization is decreased, less magnetization is available to project into the xy plane, and the echo amplitude is decreased. Thus, T_1 can be determined from the dependence of echo amplitude on pulse repetition rate. If spectral diffusion makes a constant contribution over the range of repetition times used, the measured T_1 approximates the actual T_1.

Another spin echo approach is to first invert the spin system with a 180° pulse, and then sample this inverted magnetization with a normal two-pulse spin echo (180-τ-90-T-180-T-echo, varying τ). During the time between the inverting pulse and the sampling pulse, the spin system relaxes by T_1 processes. Hence, a plot of echo amplitude vs. delay time after the inverting pulse yields T_1. This method, which is called inversion recovery, may also yield a decay time faster than T_1 because of spectral diffusion. One can also perform a stimulated echo experiment (90-τ-90-T-90-τ-echo, varying T). This is the pulse sequence commonly used for ESEEM studies. The decay of the echo is ideally T_1, but in reality the decay time is shorter than T_1 because of spectral diffusion. Comparison of the relaxation times measured by SR and various ESE methods reveals the kinetics of spectral diffusion (Harbridge et al., 2003). Many pulse sequences are described in texts on NMR (Akitt, 1983; Ernst et al., 1987; Farrar and Becker, 1971; Freeman, 1997; Fukushima and Roeder, 1981; Harris, 1983) and details of the application to EPR are discussed in Schweiger and Jeschke (2001), Dalton (1985), and Kevan and Schwartz (1979).

F. Measurement of B_1

One of the most uncertain parameters in EPR measurements is the magnitude of the microwave magnetic field, B_1, at the sample. Because all materials in the cavity, including the sample tube and the sample itself, affect the distribution of the microwave magnetic and electric fields, it is virtually impossible to know the value of B_1 at the sample exactly, except by a pulse method that measures B_1 directly. An extensive discussion of the measurement of B_1 is given by Bales and Kevan (1970). Some of the more practical ways to estimate B_1 are summarized in what follows.

B_1 can be calculated from the dimensions of the resonator and the Q (Poole, 1983; Rataiczak and Jones, 1972). For a rectangular X-band resonator, such as the Varian E-231 cavity or the Bruker ER4102ST cavity, one finds that $2B_1 \sim 4 \times 10^{-2} (QP)^{1/2}$, where P is the power in watts. The value of B_1 calculated with this formula is not valid if there is substantial dielectric in the cavity that distorts the B_1 distribution. The quartz dewar used for variable temperature studies can increase the B_1 at the sample by nearly a factor of two, depending on the thickness of the quartz (Wardman and Seddon, 1969; Wyard and Cook, 1969). Each dewar has to be calibrated. Similarly, the lens effect of the sample tube and solvent itself needs to be calibrated: it has been found to increase the integrated signal area by up to 55% (Dalal et al., 1981).

The interaction of a conductor with the microwave field is a measure of B_1. On the basis of this phenomenon, the "method of perturbing spheres", which is described in Ginzton (1957), has been applied to EPR resonators (Freed et al., 1967). One measures the frequency change due to the presence of a conducting sphere.

If the relaxation times are known for a sample with a homogeneously broadened line, it is possible to use that sample as a standard to determine B_1. Note that the relaxation times used in these measurements will be effective values under the conditions of the CW measurements and may include significant contributions from spectral diffusion (Harbridge et al., 1998, 2003). Fremy's salt has been advocated for this purpose (Beth et al., 1983). Other well-characterized samples, such as irradiated sugar and irradiated glycylglycine (Copeland, 1973; Mottley et al., 1976), also can be used. The linewidth of a homogeneously broadened line (ΔB) is related to the linewidth in the absence of saturation, ΔB_0, and the relaxation times by the following equation (Eastman et al., 1969; Schreurs and Fraenkel, 1961):

$$(\Delta B)^2 = (\Delta B_0)^2 + \frac{4T_1 B_1^2}{3T_2}$$

As $B_1^2 = KP$, where K is a proportionality constant and P is the microwave power, a plot of $(\Delta B)^2$ vs. P has a slope of $(4KT_1)/(3T_2)$ and the intercept $= (\Delta B_0)^2$. The value of K is then determined from the slope of the line and the known values of T_1 and T_2. This technique also can be used to measure K for various experimental arrangements such as dewar inserts and flat cells in the cavity.

In pulsed EPR, B_1 for an $S = \frac{1}{2}$ sample is related to the flip angle in radians, Θ, and the length of the pulse in s, t_p, by the equation (Mims, 1972, Chapter 2)

$$B_1 = \frac{\Theta}{(1.76 \times 10^7 \text{ rad}/(\text{sG}))(t_p)}$$

In practice, one uses the flip angle to estimate B_1. Precise setting of the pulse flip angle is difficult, except in special cases. If the spectrum is narrow relative to the B_1 that can be achieved in a pulse, a very convenient technique uses the "third echo" (Perman et al., 1989). In a three-pulse sequence $\pi/2 - \tau - \pi/2 - T - \pi/2$ there are echoes at τ, $T - \tau$, T, and $T + \tau$. By selecting $\tau < T - \tau$, the third stimulated echo in the sequence is at time T [this is one of the echoes that Mims (1972) described as "unwanted"]. If the pulses are all exactly $\pi/2$, the T echo will be nulled. This is a very sharp null, and permits the pulse power to be set to within a few tenths of a decibel. A short sample that exhibits a narrow line spectrum, such as the standard irradiated fused silica sample, can be used to give a precise calibration of B_1. More commonly, the spectrum of interest is wider than the B_1 achievable, and the T echo null technique described earlier does not work. In this case a close approximation to the $\pi/2$ pulse time can be obtained by adjusting the microwave pulse power to achieve a maximum echo for either of two cases: (a) If the second pulse is twice as long as the first, the maximum echo occurs when the pulses are $\pi/2$, π; (b) If the pulses are of equal length, the maximum echo occurs when the pulses are $2\pi/3$ (120°).

The measurement of B_1 is so fundamental that many alternate ways of estimating B_1 have been published. For example, Peric et al. (1985) reported a method for obtaining B_1 by measuring the splittings between sidebands as a function of modulation frequency, using modulation frequencies from 100 kHz to 1.5 MHz. Other reports discuss the utility of a spec of DPPH as a secondary standard whose absolute signal intensity is measured (Hemminga et al., 1984b), and the use of the magnetization hysteresis spectrum and power-dependent linewidth of a small crystal of the TCNQ salt of methyl phenazine (Vistnes and Dalton, 1983).

G. Line vs. Point Samples

Because both modulation amplitude and microwave B_1 vary over the height of the standard TE_{102} cavity (Kooser et al., 1969; Mailer et al., 1977; Schreurs et al., 1960), the experimentalist has a difficult tradeoff. A stronger signal will be obtained, for a constant sample concentration, if a "line" sample is used that extends the entire height of the cavity. However, for a constant number of spins, without saturation, a point sample at the center of a TE_{102} rectangular cavity gives a stronger EPR spectrum than the same number of spins distributed along a line sample that exactly matches the cavity length. Interpretation of saturation behavior and relaxation times is more difficult for a line sample than for a point sample because each portion of the line sample experiences a different B_1. If a point sample is used, the placement of the sample is critical. Several papers have discussed these problems in detail (Dalal et al., 1981; Fajer and Marsh, 1982; Mailer et al., 1977).

H. Overcoupled Resonators

Long ago Mims stated that it was better to use an overcoupled resonator for ESE than to purposefully lower the

Q by adding lossy materials to the resonator. As lossy materials convert microwave energy to heat, this is intuitively obvious. However, recently this principle has been given rigorous expression (Rinard et al., 1994). The echo signal from an overcoupled resonator is larger, by as much as 3 dB, than the echo from the same sample in the same resonator with the Q reduced to the same value by adding loss rather than by overcoupling. However, overcoupling causes power to be reflected from the resonator because of impedance mismatch. Thus, significantly less B_1 is achievable with the same incident power using an overcoupled resonator than with a critically coupled resonator. Using a low-Q resonator requires 3 dB more incident microwave power as the same experiment in a resonator with high Q lowered by overcoupling to the same Q as the low-Q resonator. The decision whether to use a critically coupled low-Q resonator or an overcoupled high-Q resonator depends on the available microwave power, and the ability of the detection system to tolerate the reflected power that occurs because of the impedance mismatch created when one overcouples a resonator.

I. The ESE Flip Angle for High-Spin Systems

Next consider the relation of spin flip angle to the spin state of the chemical species being studied. The equation is commonly given (see Section VI.F) as

$$\Theta = \gamma B_1 t_p$$

but this is actually for $S = 1/2$. For $S > 1/2$ the fundamental paper is Sloop et al. (1981). The background quantum mechanics is presented in Weil et al. (1994, Appendix B). Qualitatively, because the physical model is a coupling of angular momenta, the B_1 required for a $\pi/2$ pulse of a given length t_p is smaller for a system with larger spin angular momentum. Hence, less incident microwave power is required to produce a $\pi/2$ pulse for larger electron spin S. Note that this is true for, for example, $S = 3/2$, even if the transition being observed is $+1/2 \leftrightarrow -1/2$. To include the dependence on S, the flip angle equation can be written

$$\Theta = c\gamma B_1 t_p \quad \text{where} \quad c = [S(S+1) - m_S(m_S + 1)]^{1/2}$$

for the transition between m_S and $m_S + 1$ for a system with spin S.

The B_1 required to achieve a $\pi/2$ pulse can distinguish between values of S, if the spectrometer is calibrated. A very common question is whether a species is a triplet, or whether a doublet impurity is being observed. This equation shows that only half the power is required for the triplet as for the doublet. This is 3 dB in power, which should be enough for an unambiguous determination. Thus, the power required for a $\pi/2$ pulse will tell you whether the species is a triplet.

Another application of this relationship is that a lower power spectrometer can be used to study high-spin systems. The amount of power required for a $\pi/2$ pulse becomes very much smaller for a high-spin system. For example, it is fairly common to study Mn^{2+} in biological systems. The $-1/2 \leftrightarrow 1/2$ transition is almost always observable, at any temperature, even if the other transitions are unobservable because of large ZFS. Only 1/3 as much B_1, or 1/9 the power, is required to study this Mn^{2+} transition as is needed for, for example, a nitroxyl spin label or a tyrosyl radical (which might be present in the same system).

Similarly, different transitions for the same spin system have maximum intensity at different powers. A field-swept echo-detected EPR spectrum will selectively enhance one transition over another for the same spin system. This will result in differences between linear CW spectra and ESE spectra. This effect also permits one to selectively enhance a transition of interest, such as aiding in the identification of transitions in coupled spin systems. For example, note that the $-1/2 \leftrightarrow 1/2$ transition for $S = 3/2$ (such as Cr^{3+}) requires only 1/2 the B_1 that is required for an $S = 1/2$ system such as a nitroxyl spin label. That is, for these transitions, a $\pi/2$ pulse for the Cr^{3+} transition is a π pulse for the nitroxyl. Were all pulses ideal and if B_1 did not vary over the dimensions of the sample, there would be no echo from the nitroxyl under conditions that optimized the echo from the Cr^{3+}.

VII. LESS COMMON MEASUREMENTS WITH EPR SPECTROMETERS

Less common means simply that. It does not mean less important, nor does it mean that they will continue to be less common. For example, in the years since the first edition of this book there has been explosive growth in EPR imaging and *in vivo* EPR, and in the types of pulsed EPR techniques. ENDOR is becoming so common that it has moved from this section in the first edition to an earlier section of this chapter. High-field EPR is also becoming accessible. Many combinations of heretofore separate methodologies are yielding new insights, for example, via various multidimensional spectroscopies.

A. Saturation-Transfer Spectroscopy

The technique of saturation-transfer spectroscopy can be used to measure rates of molecular motion that are a little faster than electron spin relaxation rates (Beth and Hustedt, 2004; Dalton et al., 1976; Hyde and Thomas,

1980; Marsh et al., 2004). This time-scale is particularly applicable to the study of biological systems labeled with nitroxyl spin labels. Standard conditions for the measurement of saturation-transfer EPR spectra have been delineated (Hemminga et al., 1984a). For the most common ST-EPR technique, the spectrometer must be capable of phase-sensitive detection at twice the magnetic field modulation frequency. The Varian Century Series and Bruker ER200 and later spectrometers have this capability.

B. Electrical Conductivity

Because electrical conductivity of a sample affects the magnitude of the EPR signal observed from the sample, it is possible to measure the microwave electrical conductivity of a sample with an EPR spectrometer (Setaka et al., 1970). The conductivity of the walls of a resonant cavity affect the Q of the cavity. This effect has been used to measure surface resistance of nonferromagnetic metals at X-band (Hernandez et al., 1986). Electron spin diffusion rates in conducting crystals have been measured with ESEs in a magnetic field gradient (Alexandrowicz et al., 2000; Callaghan et al., 1994; Maresch et al., 1984; Wokrina et al., 1996). EPR can be used to monitor aspects of superconductivity in new materials (Emge et al., 1985). A novel combination of EPR and the ac Josephson effect provided a new type of spectrometer (Baberschke et al., 1984).

C. Static Magnetization

The dc magnetization of a sample can be measured by using the EPR of a standard sample attached to the sample whose magnetization is to be measured. The EPR probes the magnetic field outside the sample (Schultz and Gullikson, 1983).

D. EPR Imaging

With suitable magnetic field gradients, EPR can provide pictures of spin concentration as a function of the three spatial dimensions x, y, and z. Most of the early work in EPR imaging has emphasized making pictures of objects. In this effort there has been an implied concept that the dimensions of interest were the three Cartesian dimensions of the laboratory coordinate system (Eaton et al., 1991; Ohno, 1986, 1987). However, the properties of the spin systems provide access to several additional dimensions, which may, in some cases, provide more insight into the nature of a sample than the spatial dimensions alone. We urge an expanded view that considers as dimensions of the imaging problem several additional features of EPR spectroscopy, including g-values, hyperfine splitting, relaxation times T_1 and T_2, microwave field distribution, spin flip angles, and chemical kinetics (e.g., formation and decay of radicals or diffusion). Combinations of these dimensions yield numerous possible multidimensional imaging experiments. Subsequent papers emphasized relaxation times and spin flip angles as imaging dimensions (Eaton and Eaton, 1986a, 1987a, b). There is a mathematical isomorphism between a two-dimensional spatial imaging experiment and the spectral-spatial problem (Bernardo et al., 1985; Eaton and Eaton, 1986b; Lauterbur et al., 1984; Maltempo, 1986, Maltempo et al., 1987, 1991; Stillman et al., 1986). Reviews of EPR imaging include Ohno (1987), Eaton et al. (1991), Eaton and Eaton (1986b, 1993c, 1996b, 1999c, 2000b). For a tutorial introduction to EPR imaging see Carlin et al. (1994). Recent examples include *in vivo* imaging (Kuppusamy et al., 1998; McCallum et al., 1996; Williams et al., 2002).

E. Pulsed Magnetic Field Gradients

The discussion of passage effects noted that spectra could be distorted if the magnetic field were changed rapidly relative to electron spin relaxation times. To understand electron spin relaxation one often wants to change the magnetic field rapidly. For certain experiments, the magnetic field should be pulsed, that is, changed in a stepwise manner analogous to changes in the microwave power. For example, one might want to change the magnetic field by 100 G in 100 ns. No commercial EPR spectrometer has this capability, but pulsed magnetic fields and field gradients are being explored in several research labs (Callaghan et al., 1994; Dzuba et al., 1984; Ewert et al., 1991; Forrer et al., 1990).

VIII. REPORTING RESULTS

In order to communicate to others the results of the measurements made, it is important to report a subset of the experimental parameters that most strongly affect the spectral results. Minimally, one needs to describe the sample preparation (whether it was degassed, what the spin concentration was, etc.) and the magnetic field, microwave frequency and power, modulation frequency and amplitude, resonator type, and field-scan rate. For pulsed measurements the time and power of the pulses must be stated.

This section is a summary guide to nomenclature, and to what parameters should be reported to communicate the results of the measurement. It is assumed that the chemistry of the problem is adequately described, including source of the sample, concentration, and temperature, so only the spectroscopic aspects of the results are covered in this section.

A. Reporting Experimental Spectra

Spectra should be reported with magnetic field increasing to the right. First-derivative spectra should be reported such that the low field rise of the first line has a positive excursion.

The figure or its caption or the "methods" section of a paper should include the following information: (a) magnetic field at some point in the spectrum, or g-factor marker; (b) magnetic field scale, and scan direction; (c) microwave frequency; (d) microwave power level; (e) magnetic field-scan rate; (f) modulation frequency and amplitude; (g) whether a dewar or other cavity insert was used; (h) whether these parameters were calibrated or "read from the spectrometer"; and (i) what standards were used. If multiple derivatives, saturation-transfer EPR, or other such techniques were used, the earlier list should be expanded accordingly. For commercial spectrometers the manufacturer and model number should be stated. For experiments performed on locally constructed spectrometers or specialty accessories item such as new resonators, details of design and construction should be included or referenced.

B. Reporting Derived Values

The EPR field uses a chaotic mix of units. Some effort is going on within the International EPR Society to establish a consistent set of conventions. Until such time, the following notation and units are recommended. For the details of spin Hamiltonian notation for multielectron systems the notation can become obscure, and care must be taken to define all terms.

EPR	Call the spectroscopy EPR, not ESR
g	Unless particular attention is paid in the paper to the mathematical properties of g, use a term such as g-factor, g-value, or element of g-matrix. Thus, one would report, for example, $g = 2.0031$ as the isotropic g-factor, and $g_Z = 2.31$ as a component of the g-factor.
B	The magnetic field generated by the electromagnet should be denoted B, not H, and the units should be gauss or tesla.
B_1	The microwave magnetic field should be denoted B_1. The units should be gauss or tesla. It should be clearly stated whether the B_1 reported is the maximum value or the r.m.s. value, and whether it is the amplitude of the linearly polarized microwave magnetic field or the circularly rotating microwave component.
a, A	All electron–nuclear interactions should be called hyperfine interactions. The term superhyperfine interactions does not serve a theoretically useful role and should be discarded. When a peak splitting is measured on a magnetic field scan it should be denoted a, given in units of gauss or tesla, and called a "hyperfine splitting". When the peak splitting is obtained by computer fitting of the spectrum including second-order corrections, it should be denoted A, given in energy units (cm^{-1} or MHz), and called a hyperfine coupling.
T_1, T_2, T_m	The electron spin relaxation times should be given in units appropriate to their experimental definition. In most cases this will be seconds per radian, not seconds as is usually reported. However, the convention in SI units is to omit the rad.
τ_c, τ_r	The correlation time for molecular motion of the species whose EPR signal is being studied should be given in units appropriate to the experimental definition. When molecular rotation is the motion being considered the units are seconds per radian, not seconds as is usually reported.
Exchange/dipolar interactions	This is one of the most chaotic nomenclature areas in magnetic resonance—for a detailed discussion see Eaton and Eaton (1989). With careful attention to the mathematics, one of the following terms should be used: isotropic exchange, antisymmetric exchange (no unambiguous examples are known), symmetric anisotropic exchange (do not call it pseudodipolar), isotropic dipolar contribution that arises from g-anisotropy, antisymmetric dipolar contribution that arises from g-anisotropy, and dipolar (the major portion of the dipolar interaction). Use of other terms should be discontinued. Terms such as "fast" or "slow" should not be used to qualitatively describe magnitudes of J. The isotropic exchange should be given with sign and magnitude appropriate for an energy separation of $-2 J h$ between the ground and first excited states.

ACKNOWLEDGMENTS

This chapter builds on the contributions of many EPR spectroscopists, who have, via their papers or conversations, taught us what we have attempted to communicate here. We are especially grateful to James S. Hyde, Michael K. Bowman, Richard W. Quine, and George A. Rinard for patient tutorials over several years. We also thank vendors for providing unpublished information about their products, with the understanding that by the time this is read the products may have evolved to the point that information in this chapter is out of date. Our research on EPR has been supported in part by National Science Foundation grants CHEM9103262 and BIR9316827 and National Institutes of Health NIBIB grant EB002807.

REFERENCES

Aasa, R., Vanngard, T. (1975). EPR signal intensity and powder shapes. Reexamination. *J. Magn. Reson.* 19:308–315.

Abragam, A., Bleaney, B. (1970). *Electron Paramagnetic Resonance of Transition Ions*. Oxford: Oxford University Press.

Akitt, J. W. (1983). *NMR and Chemistry: An Introduction to the Fourier Transform—Multinuclear Era*. 2nd ed. Chapman and Hall.

Al'tshuler, S. A., Kozyrev, B. M. (1974). *Electron Paramagnetic Resonance in Compounds of Transition Elements*. 2nd ed. Halsted-Wiley: Jerusalem.

Alecci, M., Penna, S. D., Sotgiu, A., Testa, L., Vannucci, I. (1992). Electron paramagnetic resonance spectrometer for three-dimensional *in vivo* imaging at very low frequency. *Rev. Sci. Instrum.* 63:4263–4270.

Alexandrowicz, G., Tashma, T., Feintuch, A., Grayevsky, A., Dormann, E., Kaplan, N. (2000). Spatial mapping of mobiity and density of the conduction electrons in $(FA)_2PF_6$. *Phys. Rev. Lett.* 84:2973–2976.

Alger, R. S. (1968). *Electron Paramagnetic Resonance: Techniques and Applications*. New York: Wiley-Interscience.

Allgeier, J., Disselhorst, A. J. M., Weber, R. T., Wenckebach, W. T., Schmidt, J. (1990). High-frequency pulsed electron spin resonance. In: Kevan, L., Bowman, M. K., eds. *Modern Pulsed and Continuous-Wave Electron Spin Resonance*. New York: John Wiley, pp. 267–283.

Ardenjaer-Larsen, J. H., Laursen, I., Leunbach, I., Ehnholm, G., Wistrand, L.-G., Petersson, J. S., Golman, K. (1998). EPR and DNP properties of certain novel single electron contrast agents intended for oximetric imaging. *J. Magn. Reson.* 133:1–12.

Atherton, N. M. (1973). *Electron Spin Resonance: Theory and Applications*. Wiley.

Baberschke, K., Bures, K. D., Barnes, S. E. (1984). ESR *in situ* with a Josephson tunnel junction. *Phys. Rev. Lett.* 53:98–101.

Baden Fuller, A. J. (1990). *An Introduction to Microwave Theory and Techniques*. 3rd ed. Oxford: Pergamon Press.

Bales, B. L., Kevan, L. (1970). Paramagnetic relaxation of silver species in γ-irradiated frozen aqueous solutions. *J. Chem. Phys.* 52:4644–4653.

Barbon, A., Brustolon, M., Maniero, A. L., Romanelli, M., Brunel, L. C. (1999). Dynamics and spin relaxation of tempone in a host crystal. An ENDOR, high field EPR and electron spin echo study. *Phys. Chem. Chem. Phys.* 1:4015–4023.

Barra, A.-L., Gatteschi, D., Sessoli, R., Abbati, G. L., Cornia, A., Fabretti, A. C., Uyttcrhoeven, M. G. (1997). Electronic structure of manganese(III) compounds from high-frequency EPR spectra. *Angew. Chem. Int. Ed. Engl.* 36:3239–2331.

Beinert, H., Orme-Johnson, W. H. (1967). Electron spin relaxation as a probe for active centers of paramagnetic enzyme species. In: Ehrenberg, A., et al., ed. *Magnetic Resonance in Biological Systems*. Pergamon Press, pp. 221–257.

Berliner, L. J., ed. (1976). *Spin Labeling: Theory and Applications*. New York: Academic Press.

Berliner, L. J., ed. (1979). *Spin Labeling II*. New York: Academic Press.

Berliner, L. J., ed. (1998). *Spin Labeling: The Next Millenium*. New York: Plenum.

Berliner, L. J., Reuben, J., eds. (1989). *Spin Labeling Theory and Applications*. Biological Magnetic Resonance. Vol. 8. New York: Plenum Press.

Berliner, L. J., Reuben, J., eds. (1993). *EMR of Paramagnetic Molecules*. Biological Magnetic Resonance. Vol. 13. New York: Plenum.

Berliner, L. J., Eaton, G. R., Eaton, S. S., eds. (2000). *Distance Measurements in Biological Systems by EPR*. Biological Magnetic Resonance, Vol. 19. New York: Kluwer.

Berlinger, W. (1985). Variable temperature EPR in solid state physics. *Magn. Reson. Rev.* 10:45–80.

Bernardo, J. M. L., Lauterbur, P. C., Hedges, L. K. (1985). Experimental example of NMR spectroscopic imaging by projection reconstruction involving an intrinsic frequency dimension. *J. Magn. Reson.* 61:168–174.

Bertini, I., Martini, G., Luchinat, C. (1994a). Relaxation data tabulation. In: Poole, J. C. P., Farach, H., eds. *Handbook of Electron Spin Resonance: Data Sources, Computer Technology, Relaxation, and ENDOR*. New York: American Institute of Physics, pp. 79–310.

Bertini, I., Martini, G., Luchinat, C. (1994b). Relaxation, background, and theory. In: Poole, J. C. P., ed. *Handbook of Electron Spin Resonance*. New York: American Institute of Physics, pp. 51–77.

Beth, A., Hustedt, E. J. (2004). Saturation transfer EPR: rotational dynamics of membrane proteins. *Biol. Magn. Reson.* 24, 369–407.

Beth, A., Balasubramanian, K., Robinson, B. H., Dalton, L. R., Venkataram, S. K., Park, J. H. (1983). Sensitivity of V_2' saturation transfer electron paramagnetic resonance signals to anisotropic rotational diffusion with [15N]nitroxide spinlabels. Effects of noncoincident magnetic and diffusion tensor principal axes. *J. Chem. Phys.* 87:359–367.

Blackburn, J. F. (1949). *Components Handbook*. MIT Radiation Laboratory Series. Vol. 17. McGraw Hill.

Bowman, M. K. (1990). Fourier transform electron spin resonance. In: Kevan, L., Bowman, M. K., eds *Modern Pulsed and Continuous-Wave Electron Spin Resonance.* New York: John Wiley, pp. 1–42.

Bowman, M. K., Kevan, L. (1979). Electron spin-lattice relaxation in non-ionic solids. In: Kevan, L., Schwartz, R. N., eds. *Time Domain Electron Spin Resonance.* New York: John Wiley, pp. 68–105.

Box, H. C. (1977). *Radiation Effects: ESR and ENDOR Analysis.* New York: Academic Press.

Brown, I. M. (1979). Electron spin echo studies of relaxation processes in molecular solids. In: Kevan, L., Schwartz, R. N., eds. *Time Domain Electron Spin Resonance.* New York: John Wiley, pp. 195–229.

Brunel, L. C. (1996). Recent developments in high frequency/ high magnetic field CW EPR. Applications in chemistry and biology. *Appl. Magn. Reson.* 11:417–423.

Buckmaster, H. A., Hansen, C., Malhotra, V. M., Dering, J. C., Gray, A. L., Shing, Y. H. (1981). Baseline offset in EPR spectrometers. *J. Magn. Reson.* 42:322–323.

Budil, D. E., Earle, K. A., Lynch, W. B., Freed, J. H. (1989). Electron paramagnetic resonance at 1 millimeter wavelength. In: Hoff, A. J., ed. *Advanced EPR: Applications in Biology and Biochemistry.* Amsterdam: Elsevier, pp. 307–340.

Callaghan, P. T., Coy, A., Dormann, E., Ruf, R., Kaplan, N. (1994). Pulsed-gradient spin-echo ESR. *J. Magn. Reson. A* 111:127–131.

Carlin, R. T., Trulove, P. C., Eaton, G. R., Eaton, S. S. (1994). Electrochemistry and EPR spectroscopy of C_{60}^{n-} in the presence of water, OH^- and H^+. Proceedings—Electrochemical Society (Recent Advances in the Chemistry and Physics of Fullerenes and Related Materials). Vol. 94–24, pp. 986–994.

Carrington, A., McLachlan, A. D. (1967). *Introduction to Magnetic Resonance.* Harper and Row.

Casteleijn, G., TenBosch, J. J., Smidt, J. (1968). Error analysis of the determination of spin concentration with the electron spin resonance method. *J. Appl. Phys.* 39:4375–4380.

Chang, R. (1974). Simple setup for quantitative electron paramagnetic resonance. *Anal. Chem.* 46:1360.

Chasteen, N. D., Snetsinger, P. A. (2000). ESEEM and ENDOR spectroscopy. In: Que, L. J., ed. *Physical Methods in Bioinorganic Chemistry: Spectroscopy and Magnetism.* Chapter 4. Sausalito, CA: University Science Books.

Chesnut, D. B. (1977). On the use of AW2 method for integrated line intensities from first-derivative presentations. *J. Magn. Reson.* 25:373–374.

Chien, J. C. W., Dickenson, L. C. (1981). EPR crystallography of metalloproteins and spin-labeled enzymes. *Biol. Magn. Reson.* 3:155–211.

Copeland, E. S. (1973). Simple method for estimating H1 [microwave magnetic field strength] in ESR experiments. Microwave power saturation of γ-irradiation induced glycylglycine radicals. *Rev. Sci. Instrum.* 44:437–442.

Dalal, D. P., Eaton, S. S., Eaton, G. R. (1981). The effects of lossy solvents on quantitative EPR studies. *J. Magn. Reson.* 44(3):415–428.

Dalton, L. R. (1985). *EPR and Advanced EPR Studies of Biological Systems.* Boca Raton, FL: CRC Press.

Dalton, L. R., Robinson, B. H., Dalton, L. A., Coffey, P. (1976). Saturation transfer spectroscopy. *Adv. Magn. Reson.* 8:149–259.

van Dam, P. J., Klaasen, A. A. K., Reijerse, E. J., Hagen, W. R. (1998). Application of high frequency EPR to integer spin systems: unusual behavior of the double-quantum line. *J. Magn. Reson.* 130:140–144.

DeRose, V. J., Hoffman, B. M. (1995). Protein structure and mechanism studied by electron nuclear double resonance spectroscopy. *Methods Enzymol.* 246:554–589.

Devine, S. D., Robinson, W. H. (1982). Ultrasonically modulated paramagnetic resonance. *Adv. Magn. Reson.* 10:53–117.

Dikanov, S. A., Tsvetkov, Y. D. (1992). *Electron Spin Echo Envelope Modulation Spectroscopy.* Boca Raton, FL: CRC Press.

Dorio, M. M., Freed, J. H., eds. (1979). *Multiple Electron Resonance Spectroscopy.* Plenum Press.

Drago, R. S. (1992a). Electron paramagnetic resonance spectroscopy. *Physical Methods for Chemists.* Ft. Worth: Saunders College Publishing, pp. 360–401.

Drago, R. S. (1992b). Electron paramagnetic resonance spectra of transition metal ions. *Physical Methods for Chemists.* Ft. Worth: Saunders College Publishing, pp. 559–603.

Du, J. L., More, K. M., Eaton, S. S., Eaton, G. R. (1992). Orientation dependence of electron spin phase memory relaxation times in copper(II) and vanadyl complexes in frozen solution. *Israel J. Chem.* 32(2–3):351–355.

Du, J.-L., Eaton, G. R., Eaton, S. S. (1995). Temperature, orientation, and solvent dependence of electron spin-lattice relaxation rates for nitroxyl radicals in glassy solvents and doped solids. *J. Magn. Reson. A* 115(2):213–221.

Dykstra, R. W., Markham, G. D. (1986). A dielectric sample resonator design for enhanced sensitivity of EPR spectroscopy. *J. Magn. Reson.* 69:350–355.

Dzuba, S. A., Maryasov, A. G., Salikhov, A. K., Tsvetkov, Y. D. (1984). Superslow rotations of nitroxide radicals studied by pulse EPR spectroscopy. *J. Magn. Reson.* 58:95–117.

Eachus, R. S., Olm, M. T. (1985). Electron nuclear double resonance spectroscopy. *Science* 230:268.

Earle, K. A., Budil, D. E., Freed, J. H. (1996). Millimeter wave electron spin resonance using quasioptical techniques. *Adv. Magn. Reson. Opt. Reson.* 19:253–323.

Eastman, M. P., Kooser, R. G., Das, M. R., Freed, J. H. (1969). Heisenberg spin exchange in E.S.R. spectra. I. Linewidth and saturation effects. *J. Chem. Phys.* 51:2690–2709.

Eaton, S. S., Eaton, G. R. (1977). Electron paramagnetic resonance sample cell for lossy samples. *Anal. Chem.* 49(8):1277–1278.

Eaton, S. S., Eaton, G. R. (1978). Metal-nitroxyl interactions. Part 5. Interaction of spin labels with transition metals. *Coord. Chem. Rev.* 26(3):207–262.

Eaton, S. S., Eaton, G. R. (1980). Signal area measurements in EPR. *Bull. Magn. Reson.* 1(3):130–138.

Eaton, G. R., Eaton, S. S. (1985). Relaxation times for the organic radical signal in the EPR spectra of oil shale, shale oil, and spent shale. *J. Magn. Reson.* 61(1):81–89.

Eaton, G. R., Eaton, S. S. (1986a). Electron spin-echo-detected EPR imaging. *J. Magn. Reson.* 67(1):73–77.

Eaton, S. S., Eaton, G. R. (1986b). EPR imaging. *Spectroscopy (Duluth, MN, United States)* 1(1):32–35.

Eaton, G. R., Eaton, S. S. (1987a). EPR imaging using T_1 selectivity. *J. Magn. Reson.* 71(2):271–275.

Eaton, G. R., Eaton, S. S. (1987b). Dimensions in EPR imaging. In: Weil, J., ed. *Electronic Magnetic Resonance in the Solid State*. Ottawa: Canadian Institute of Chemistry, pp. 639–650.

Eaton, S. S., Eaton, G. R. (1988). Interaction of spin labels with transition metals. Part 2. *Coord. Chem. Rev.* 83:29–72.

Eaton, G. R., Eaton, S. S. (1989). Resolved electron–electron spin–spin splittings in EPR spectra. *Biol. Magn. Reson.* 8:339–397 (Spin labeling).

Eaton, S. S., Eaton, G. R. (1992). Quality assurance in EPR. *Bull. Magn. Reson.* 13(3–4):83–89.

Eaton, S. S., Eaton, G. R. (1993a). Irradiated fused-quartz standard sample for time-domain EPR. *J. Magn. Reson.* 102(3):354–356.

Eaton, S. S., Eaton, G. R. (1993b). Applications of high magnetic fields in EPR spectroscopy. *Magn. Reson. Rev.* 16:157–181.

Eaton, G. R., Eaton, S. S. (1993c). Electron paramagnetic resonance imaging. In: Morris, M. D., ed. *Microscopic and Spectroscopic Imaging of the Chemical State*. New York: Marcel Dekker, pp. 395–419.

Eaton, S. S., Eaton, G. R. (1996a). EPR spectra of C_{60} anions. *Appl. Magn. Reson.* 11(2):155–170.

Eaton, S. S., Eaton, G. R. (1996b). EPR imaging. *Electron Spin Resonance* 15:169–185.

Eaton, G. R., Eaton, S. S. (1999a). High-field and high-frequency EPR. *Appl. Magn. Reson.* 16(2):161–166.

Eaton, S. S., Eaton, G. R. (1999b). High magnetic fields and high frequencies in ESR spectroscopy. *Handbook of Electron Spin Resonance*. Vol. 2. pp. 345–370.

Eaton, G. R., Eaton, S. S. (1999c). ESR imaging. *Handbook of Electron Spin Resonance* 2:327–343.

Eaton, S. S., Eaton, G. R. (2000a). Relaxation times of organic radicals and transition metal ions. *Biol. Magn. Reson.* 19:29–154 (Distance Measurements in Biological Systems by EPR).

Eaton, S. S., Eaton, G. R. (2000b). EPR imaging. *Electron Paramagnetic Resonance* 17:109–129.

Eaton, S. S., Eaton, G. R. (2004). EPR at frequencies below X-band. *Biol. Magn. Reson.* 21, 59–114.

Eaton, G. R., Quine, R. W. (2000). Comparison of four digitizers for time-domain EPR. *Appl. Spectrosc.* 54(10):1543–1545.

Eaton, S. S., Law, M. L., Peterson, J., Eaton, G. R., Greenslade, D. J. (1979). Metal-nitroxyl interactions. 7. Quantitative aspects of EPR spectra resulting from dipolar interactions. *J. Magn. Reson.* 33(1):135–141.

Eaton, S. S., More, K. M., Sawant, B. M., Eaton, G. R. (1983). Use of the ESR half-field transition to determine the interspin distance and the orientation of the interspin vector in systems with two unpaired electrons. *J. Am. Chem. Soc.* 105(22):6560–6567.

Eaton, G. R., Eaton, S. S., Ohno, K., eds. (1991). *EPR Imaging and in vivo EPR*. Boca Raton, FL: CRC Press.

Eaton, S. S., Kee, A., Konda, R., Eaton, G. R., Trulove, P. C., Carlin, R. T. (1996). Comparison of electron paramagnetic resonance line shapes and electron spin relaxation rates for C60- and C603- in 4:1 toluene: acetonitrile and dimethyl sulfoxide. *J. Phys. Chem.* 100(17):6910–6919.

Eaton, G. R., Eaton, S. S., Rinard, G. A. (1998a). Frequency dependence of EPR sensitivity. In *Spatially resolved magnetic resonance: methods, materials, medicine, biology, rheology, geology, ecology, hardware*. Based on Lectures Presented at the 4th International Conference on Magnetic Resonance Microscopy, Albuquerque, Oct. 1997, pp. 65–74.

Eaton, G. R., Eaton, S. S., Salikhov, K. M., eds. (1998b). *Foundations of Modern EPR*. Singapore: World Scientific.

Emge, T. J., Wang, H. H., Beno, M. A., Leung, P. C. W., Firestone, M. A., Jenkins, H. C., Cook, J. D., Carlson, K. D., Williams, J. M., Venturini, E. L., Azevedo, J., Schirber, J. E. (1985). A test of superconductivity vs. molecular disorder in (BEDT-TTF)$_2$X synthetic metals: synthesis, structure (298, 120 K), and microwave/ESR conductivity of (BEDT-TTF)$_2$I$_2$Br. *Inorg. Chem.* 24:1736–1738.

Ernst, R. E., Bodenhausen, G., Wokaun, A. (1987). *Principles of Nuclear Magnetic Resonance in One and Two Dimensions*. New York: Oxford University Press.

Ewert, U., Crepeau, R. H., Dunnam, C. R., Xu, D. J., Lee, S. Y., Freed, J. H. (1991). Fourier transform electron spin resonance imaging. *Chem. Phys. Lett.* 184:25–33.

Fajer, P., Marsh, D. (1982). Microwave and modulation field inhomogeneities and the effect of cavity Q in saturation transfer ESR spectra. Dependence on sample size. *J. Magn. Reson.* 49:212–224.

Farrar, T. C., Becker, E. D. (1971). *Pulse and Fourier Transform NMR*. New York: Academic Press.

Feher, G. (1957). Sensitivity considerations in microwave paramagnetic resonance absorption techniques. *Bell Syst. Tech. J.* 36:449–484.

Forrer, J., Pfenninger, S., Eisenegger, J., Schweiger, A. (1990). A pulsed ENDOR probehead with the bridged loop-gap resonator: construction and performance. *Rev. Sci. Instrum.* 61:3360–3367.

Freed, J. H. (1976). Theory of slow tumbling ESR spectra of nitroxides. In: Berliner, L. J., ed. *Spin Labeling: Theory and Applications*. New York: Academic Press, pp. 53–132.

Freed, J. H., Leniart, D., Hyde, J. S. (1967). Theory of saturation and double resonance effects in electron spin resonance spectra. III. Radio frequency coherence and line shapes. *J. Chem. Phys.* 47:2762–2773.

Freeman, R. (1997). *Spin Choreography: Basic Steps in High Resolution NMR*. Sausalito, CA: University Science Books.

Froncisz, W., Lai, C. S., Hyde, J. S. (1985). Spin-label oximetry: kinetic study of cell respiration using a rapid-passage T_1-sensitive electron spin resonance display. *Proc. Natl. Acad. Sci. USA* 82(2):411–415.

Froncisz, W., Oles, T., Hyde, J. S. (1986). Q-band loop-gap resonator. *Rev. Sci. Instrum.* 57(6):1095–1099.

Fukushima, E., Roeder, S. B. W. (1981). *Experimental Pulse NMR: A Nuts and Bolts Approach*. Addison-Wesley.

Gaffney, B. J., Silverstone, H. J. (1993). Simulation of the EMR spectra of high-spin iron in proteins. *Biol. Magn. Reson.* 13:1–57.

Gaffney, B. J., Mavrophilipos, D. V., Doctor, K. S. (1993). Access of ligands to the ferric center in lipoxygenase-1. *Biophys. J.* 64:773–783.

Gaffney, B. J., Eaton, G. R., Eaton, S. S. (1998). Electron spin relaxation rates for high-spin Fe(III) in iron transferrin carbonate and iron transferrin oxalate. *J. Phys. Chem. B* 102(28):5536–5541.

Gallay, R., van der Klink, J. J. (1986). Resonator and coupling structure for spin-echo ESR. *J. Phys. E Sci. Instrum.* 19:226–230.

Gerson, F. (1970). *High Resolution E.S.R. Spectroscopy.* New York: Wiley.

Geschwind, S., ed. (1972). *Electron Paramagnetic Resonance.* New York: Plenum Press.

Ginzton, E. L. (1957). *Microwave Measurements.* New York: McGraw-Hill.

Goldberg, I. B. (1978). Improving the analytical accuracy of electron paramagnetic resonance spectroscopy. *J. Magn. Reson.* 32:233–242.

Goldberg, I. B., Bard, A. J. (1983). Electron-spin-resonance spectroscopy. In: Kolthoff, I. M., Elving, P. J., eds. *Treatise on Analytical Chemistry.* Part I, Vol. 10. New York: Wiley-Interscience, pp. 226–289.

Goldberg, I. B., Crowe, H. R. (1977). Effect of cavity loading on analytical electron spin resonance spectrometry. *Anal. Chem.* 49:1353–1357.

Goldberg, I. B., McKinney, T. M. (1984). High-performance coaxial EPR cavity for investigations at elevated temperatures and pressures. *Rev. Sci. Instrum.* 55:1104–1110.

Goldberg, I. B., Crowe, H. R., Robertson, W. M. (1977). Determination of the composition of mixtures of sodium chromite and chromic oxide by electron spin resonance spectrometry. *Anal. Chem.* 49:962–966.

Gordy, W. (1980). *Theory and Applications of Electron Spin Resonance.* Techniques of Chemistry. Vol. 15. New York: Wiley.

Goslar, J., Piekara-Sady, L., Kispert, L. D. (1994). ENDOR data tabulations. In: Poole, J. C. P., Farach, H. A., eds. *Handbook of Electron Spin Resonance.* New York: American Institute of Physics Press, pp. 360–629.

Grandy, D. W., Pretakis, L. (1980). A high-pressure, high-temperature electron paramagnetic resonance cavity. *J. Magn. Reson.* 41:367–373.

Gulin, V. I., Dikanov, S. A., Tsvetkov, Y. D., Evelo, R. G., Hoff, A. J. (1992). Very high frequency (135 GHz) EPR of the oxidized primary donor of the photosynthetic bacteria Rb. sphaeroides R-26 and Rps. viridis and of YD (signal II) of plant photosystem II. *Pure Appl. Chem.* 64:903–906.

Hales, B. J., True, A. E., Hoffman, B. M. (1989). Detection of a new signal in the ESR spectrum of vanadium nitrogenase from *Azotobacter vinelandii*. *J. Am. Chem. Soc.* 111:8519–8520.

Halpern, H. J., Spencer, D. P., van Polen, J., Bowman, M. K., Nelson, A. C., Dowey, E. M., Teicher, B. A. (1989). Imaging radio frequency electron-spin-resonance spectrometer with high resolution and sensitivity for *in vivo* measurements. *Rev. Sci. Instrum.* 60:1040–1050.

Harbridge, J. R., Eaton, G. R., Eaton, S. S. (1998). Impact of spectral diffusion on apparent relaxation times for the stable radical in irradiated glycylglycine. In *Modern Applications of EPR/ESR: from biophysics to materials science. 1st Proceedings of the Asia-Pacific EPR/ESR Symposium*, Kowloon, Hong Kong, Jan. 20–24, 1997, pp. 220–225.

Harbridge, J. R., Eaton, S. S., Eaton, G. R. (2002a). Electron spin–lattice relaxation in radicals containing two methyl groups, generated by γ-irradiation of polycrystalline solids. *J. Magn. Reson.* 159(2):195–206.

Harbridge, J. R., Rinard, G. A., Quine, R. W., Eaton, S. S., Eaton, G. R. (2002b). Enhanced signal intensities obtained by out-of-phase rapid-passage EPR for samples with long electron spin relaxation times. *J. Magn. Reson.* 156(1):41–51.

Harbridge, J. R., Eaton, S. S., Eaton, G. R. (2003). Electron spin-lattice relaxation processes of radicals in irradiated crystalline organic compounds. *J. Phys. Chem. A* 107(5):598–610.

Hardy, W. N., Whitehead, L. A. (1981). A split-ring resonator for use in magnetic resonance from 200–2000 MHz. *Rev. Sci. Instrum.* 52:213–216.

Harriman, J. E. (1978). *Theoretical Foundations of Electron Spin Resonance.* New York: Academic Press.

Harris, R. K. (1983). *Nuclear Magnetic Resonance Spectroscopy.* Pitman Books.

Hassan, A. K., Pardi, L. A., Krzystek, J., Sienkiewicz, A., Goy, P., Rohrer, M., Brunel, L.-C. (2000). Untrawide band multifrequency high-field EMR technique: a methodology for increasing spectroscopic information. *J. Magn. Reson.* 142:300–312.

Hausser, K. H. (1998). The effect of concentration and oxygen in EPR. In: Eaton, G. R., Eaton, S. S., Salikhov, A. K., eds. *Foundations of Modern EPR.* Singapore: World Scientific, pp. 469–481.

He, G., Shankar, R. A., Chzhan, M., Samoulinlov, A., Kuppusamy, P., Zweier, J. L. (1999). Noninvasive measurement of anatomical structure and intraluminal oxygenation in the gastrointestinal tract of living mice with spatial and spectral imaging. *Proc. Natl. Acad. Sci. USA* 96:4586–4591.

Hemminga, M. A., deJager, P. A., Marsh, D., Fajer, P. (1984a). Standard conditions for the measurement of saturation-transfer ESR spectra. *J. Magn. Reson.* 59:160.

Hemminga, M. A., Leermakers, F. A. M., de Jager, P. A. (1984b). Quantitative measurements of B1 in ESR and saturation-transfer ESR spectroscopy. *J. Magn. Reson.* 59:137–140.

Hendrich, M. P., Debrunner, P. G. (1998). EPR of non-Kramers systems in biology. In: Eaton, G. R., Eaton, S. S., Salikhov, A. K., eds. *Foundations of Modern EPR.* Singapore: World Scientific, pp. 530–547.

Hernandez, A., Martin, E., Margineda, J., Zamarro, J. M. (1986). Resonant cavities for measuring the surface resistivities of metals at X-band frequencies. *J. Phys. E.* 19:222–225.

Hirata, H., Ono, M. (1997). A flexible surface-coil-type resonator using triaxial cable. *Rev. Sci. Instrum.* 68:1–2.

Hirata, H., Walczak, T., Swartz, H. M. (2000). Electronically tunable surface-coil-type resonator for L-band EPR spectroscopy. *J. Magn. Reson.* 142(1):159–167.

Hirata, H., Walczak, T., Swartz, H. M. (2001). Characteristics of an electronically tunable surface-coil-type resonator for L-band electron paramagnetic resonance spectroscopy. *Rev. Sci. Instrum.* 72:2839–2841.

Homer, J., Dudley, A. R., McWhinnie, W. R. (1973). Removal of oxygen from samples used in nuclear magnetic resonance studies of spin–lattice relaxation times. *J. Chem. Soc. Chem. Commun.* 893–894.

Hornak, J. P., Freed, J. H. (1985). Electron spin echoes with a loop-gap resonator. *J. Magn. Reson.* 62:311–313.

Huisjen, J., Hyde, J. S. (1974a). A pulsed EPR spectrometer. *Rev. Sci. Instrum.* 45:669–675.

Huisjen, J., Hyde, J. S. (1974b). Saturation recovery measurement of electron spin-lattice relaxation times. *J. Chem. Phys.* 60:1682–1683.

Hustedt, E. J., Smirnov, A. I., Laub, C. F., Cobb, C. E., Beth, A. H. (1997). Molecular distances from dipolar coupled spin-labels: the global analysis of multifrequency continuous wave electron paramagnetic resonance data. *Biophys. J.* 72:1861–1877.

Hyde, J. S. (1972). A new principle for aqueous sample cells for EPR. *Rev. Sci. Instrum.* 43:629–631.

Hyde, J. S. (1978). Saturation-transfer spectroscopy. *Methods Enzymol.* XLIX:480–511.

Hyde, J. S. (1979). Saturation recovery methodology. In: Kevan, L., Schwartz, R. N., eds. *Time Domain Electron Spin Resonance*. New York: John Wiley: pp. 1–30.

Hyde, J. S., Froncisz, W. (1982). The role of microwave frequency in EPR spectroscopy of copper complexes. *Annu. Rev. Biophys. Bioeng.* 11:391–417.

Hyde, J. S., Froncisz, W. (1986). Loop gap resonators. *Specialist Periodical Reports on Electron Spin Resonance*. Royal Society of Chemistry, pp. 175–184.

Hyde, J. S., Froncisz, W. (1989). Loop gap resonators. In: Hoff, A. J., ed. *Advanced EPR: Applications in Biology and Biochemistry*. Amsterdam: Elsevier, pp. 277–306.

Hyde, J. S., Sarna, T. (1978). Magnetic interactions between nitroxide free radicals and lanthanides or Cu2+ in liquids. *J. Chem. Phys.* 68:4439–4447.

Hyde, J. S., Thomas, D. D. (1980). Saturation-transfer spectroscopy. *Annu. Rev. Phys. Chem.* 31:293–317.

Hyde, J. S., Subczynski, W. K. (1989). Spin-label oximetry. *Biol. Magn. Reson.* 8:399–425 (Spin labeling).

Hyde, J. S., Froncisz, W., Kusumi, A. (1982). Dispersion electron spin resonance with the loop gap resonator. *Rev. Sci. Instrum.* 53:1934–1937.

Hyde, J. S., Yin, J.-J., Froncisz, W., Feix, J. B. (1985). Electron–electron double resonance (ELDOR) with a loop-gap resonator. *J. Magn. Reson.* 63:142–150.

Hyde, J. S., Jesmanowicz, A., Ratke, J. J., Antholine, W. E. (1992). Pseudomodulation: a computer-based strategy for resolution enhancement. *J. Magn. Reson.* 96(1):1–13.

Hyde, J. S., McHaourab, H. S., Strangeway, R. A., Luglio, J. R. (1995). Multiquantum ESR: physics, technology and applications to bioradicals. In: Ohya-Nishiguchi, H., Packer, L., eds. *Bioradicals Detected by ESR Spectroscopy*. Basel, Switzerland: Birkhaeuser, pp. 31–47.

Hyde, J. S., McHaourab, H. S., Camenisch, T. G., Ratke, J. J., Cox, R. W., Froncisz, W. (1998). Electron paramagnetic resonance detection by time-locked subsampling. *Rev. Sci. Instrum.* 69(7):2622–2628.

Ilangovan, G., Zweier, J. L., Kuppusamy, P. (2000a). Electrochemical preparation and EPR studies of lithium phthalocyanine: evaluation of the nucleation and growth mechanism and evidence for potential-dependent phase formation. *J. Phys. Chem. B* 104:4047–4059.

Ilangovan, G., Zweier, J. L., Kuppusamy, P. (2000b). Electrochemical preparation and EPR studies of lithium phthalocyanine. Part 2. Particle-size dependent line broadening by molecular oxygen and its implication as an oximetry probe. *J. Phys. Chem. B* 104:9404–9410.

Jeschke, G. (2002). Determination of the nanostructure of polymer materials by electron paramagnetic resonance spectroscopy. *Macromolecular Rapid Communications* 23(4):227–246.

Jost, P., Griffith, O. H. (1972). Electron spin resonance and the spin labeling method. *Methods Pharmacol.* 2:223–276.

Kang, P. C., Eaton, G. R., Eaton, S. S. (1994). Pulsed electron paramagnetic resonance of high-spin cobalt(II) complexes. *Inorg. Chem.* 33(17):3660–3665.

Kawamori, A., Yamauchi, J., Ohta, H., eds. (2002). *EPR in the 21st Century: Basics and Applications to Material, Life, and Earth Sciences*. Amsterdam: Elsevier.

Kevan, L., Kispert, L. D. (1976). *Electron Spin Double Resonance Spectroscopy*. New York: Wiley.

Kevan, L., Schwartz, R. N., eds. (1979). *Time Domain Electron Spin Resonance*. New York: Wiley.

Kirste, B. (1994). Computer techniques. *Handbook of Electron Spin Resonance: Data Sources, Computer Technology, Relaxation, and ENDOR*. New York: American Institute of Physics, pp. 27–50.

Kirste, B., Harrer, W., Kurreck, H. (1985). ENDOR studies of novel di- and triphenylmethyl radicals generated from galvinols. *J. Am. Chem. Soc.* 107:20–28.

Knauer, B. R., Napier, J. J. (1973). The nitrogen hyperfine splitting constant of the nitroxide functional group as a solvent polarity parameter. The relative importance for a solvent polarity parameter of its being a cybotactic probe vs. its being a model process. *J. Am. Chem. Soc.* 98:4395–4400.

Kooser, R. G., Volland, W. V., Freed, J. H. (1969). E.S.R. relaxation studies on orbitally degenerate free radicals. I. Benzene anion and tropenyl. *J. Chem. Phys.* 50:5243–5257.

Koscielniak, J., Devasahayam, N., Moni, M. S., Kuppusamy, P., Mitchell, J. B., Krishna, M. C., Subramanian, S. (2000). 300 MHz continuous wave electron paramagnetic resonance spectrometer for small animal *in vivo* imaging. *Rev. Sci. Instrum.* 71:4273–4281.

Krinichnyi, V. I. (1995). *2-mm Wave Band EPR Spectroscopy of Condense Systems*. CRC Press: Boca Raton, FL.

Kuppusamy, P., Afeworki, M., Shankar, R. A., Coffin, D., Krishna, M. C., Hahn, S. M., Mitchell, J. B., Zweier, J. L. (1998). *In vivo* electron paramagnetic resonance imaging of tumor heterogeneity and oxygenation in a murine model. *Cancer Res.* 58:1562–1568.

Kurreck, H., Bock, M., Bretz, N., Elsner, M., Lubitz, W., Muller, F., Geissler, J., Kroneck, P. M. H. (1984). Fluid solution and solid-state electron nuclear double resonance of flavin model compounds and flavoenzymes. *J. Am. Chem. Soc.* 106:737–746.

Lauterbur, P. C., Levin, D. N., Marr, R. B. (1984). Theory and simulation of NMR spectroscopic imaging and field plotting by projection reconstruction involving an intrinsic frequency dimension. *J. Magn. Reson.* 59:536–541.

Laverghetta, T. S. (1976). *Microwave Measurements and Techniques*. Artech House.

Laverghetta, T. S. (1981). *Handbook of Microwave Testing*. Artech House.

Lebedev, Y. S. (1990). High-frequency continuous-wave electron spin resonance. In: Kevan, L., Bowman, M. K., eds. *Modern Pulsed and Continuous-Wave Electron Spin Resonance*. New York: John Wiley, pp. 365–404.

Levanon, H., Charbinsky, S., Luz, Z. (1970). ESR and electronic relaxation of Cr^{3+} and Fe^{3+} in water and water–glycerol mixtures. *J. Chem. Phys.* 53:3056–3064.

Levitt, M. H. (2002). *Spin Dynamics: Basics of Nuclear Magnetic Resonance*. Chichester: John Wiley.

Likhtenshtein, G. I. (1976). *Spin Labeling Methods in Molecular Biology*. New York: Wiley.

Lin, C. P., Bowman, M. K., Norris, J. R. (1985). A folded half-wave resonator for ESR spectroscopy. *J. Magn. Reson.* 65:369–374.

Liu, K. J., Gast, P., Moussavi, M., Norby, S. W., Vahidi, N., Walczak, T., Wu, M., Swartz, H. M. (1993). Lithium phthalocyanine: a probe for electron paramagnetic resonance oximetry in viable biological systems. *Proc. Natl. Acad. Sci. USA* 90:5438–5442.

Ludowise, P., Eaton, S. S., Eaton, G. R. (1991). A convenient monitor of EPR automatic frequency control function. *J. Magn. Reson.* 93(2):410–412.

Maier, D., Schmalbein, D. (1993). A dedicated EPR analyzer for dosimetry. *Appl. Radiat. Isotop.* 44:345–350.

Mailer, C., Sarna, T., Swartz, H. M., Hyde, J. S. (1977). Quantitative studies of free radicals in biology: corrections to ESR saturation data. *J. Magn. Reson.* 25:205–210.

Mailer, C., Danielson, J. D. S., Robinson, B. H. (1985). Computer-controlled pulsed electron-paramagnetic-resonance spectrometer. *Rev. Sci. Instrum.* 56:1917–1925.

Maltempo, M. M. (1986). Differentiation of spectral and spatial components in EPR imaging using 2-D image reconstruction algorithms. *J. Magn. Reson.* 69:156–161.

Maltempo, M. M., Eaton, S. S., Eaton, G. R. (1987). Spectral-spatial two-dimensional EPR imaging. *J. Magn. Reson.* 72(3):449–455.

Maltempo, M., Eaton, S. S., Eaton, G. R. (1991). Spectral-spatial imaging. In: Eaton, G. R., Ohno, K., Eaton, S. S., eds. *EPR Imaging and In Vivo Spectroscopy*. Boca Raton, FL: CRC Press, pp. 135–144.

Marcuvitz, N. (1951). *Waveguide Handbook*. MIT Radiation Laboratory Series. Vol. 10. New York: McGraw Hill.

Maresch, G. G., Mehring, M., von Schutz, J. U., Wolf, H. C. (1984). Time-resolved ESR investigation of electron-spin diffusion in the radical cation salt $(fluoranthenyl)^{2+}AsF_6^-$. *Chem. Phys. Lett.* 85:333–340.

Marsh, D., Horvath, L. I., Pali, T., Livshits, V. A. (2004). Saturation transfer studies of biological membranes. *Biol. Magn. Reson.* 24, 309–376.

Mazur, M., Morris, H., Valko, M. (1997a). Analysis of the movement of line-like samples of variable length along the x-axis of a double TE_{104} and a single TE_{102} rectangular resonator. *J. Magn. Reson.* 129:188–200.

Mazur, M., Valko, M., Pelikan, P. (1997b). Quantitative EPR spectroscopy in solid state chemistry. *Chem. Pap.* 51:134–136.

Mazur, M., Valko, M., Morris, H. (2000). Analysis of the radial and longitudinal effect in a double TE_{104} and a single TE_{102} rectangular cavity. *J. Magn. Reson.* 142:37–56.

McCallum, S. J., Alecci, M., Lurie, D. J. (1996). Modification of a whole-body NMR imager with a radio frequency EPR spectrometer suitable for *in vivo* measurements. *Meas. Sci. Tech.* 7:1012–1018.

McConnell, H. M., McFarland, B. G. (1970). Physics and chemistry of spin labels. *Quart. Rev. Biophys.* 3:91–136.

Mchaourab, H. S., Hyde, J. S. (1993a). Dependence of the multiple-quantum EPR signal on the spin-lattice relaxation time. Effect of oxygen in spin-labeled membranes. *J. Magn. Reson. B* 101(2):178–184.

Mchaourab, H. S., Hyde, J. S. (1993b). Continuous-wave multiquantum electron paramagnetic resonance spectroscopy. III. Theory of intermodulation sidebands. *J. Chem. Phys.* 98(3):1786–1796.

Mchaourab, H. S., Perozo, E. (2000). Determination of protein folds and conformational dynamics using spin-labeling EPR spectroscopy. *Biol. Magn. Reson.* 19:185–247.

Mchaourab, H. S., Christidis, T. C., Hyde, J. S. (1993). Continuous wave multiquantum electron paramagnetic resonance spectroscopy. IV. Multiquantum electron-nuclear double resonance. *J. Chem. Phys.* 99(7):4975–4985.

Mehdizadeh, M., Ushii, T. K., Hyde, J. S., Froncisz, W. (1983). Loop-gap resonator: a lumped mode microwave resonant structure. *IEEE Trans. Microwave Theory Tech.* 31:1059–1064.

Mehring, M., Freysoldt, F. (1980). A slotted tube resonator for pulsed ESR and ODMR experiments. *J. Phys. E.* 13:894–895.

Mims, W. B. (1965). Electron echo methods in spin resonance spectrometry. *Rev. Sci. Instrum.* 36:1472–1479.

Mims, W. B. (1972). Electron spin echoes. In: Geschwind, S., ed. *Electron Paramagnetic Resonance*. New York: Plenum Press, pp. 263–351.

Mims, W. B. (1974). Measurement of the linear electric field effect in EPR using the spin echo method. *Rev. Sci. Instrum.* 45:1583–1591.

Mims, W. B. (1976). *The Linear Electric Field Effect in Paramagnetic Resonance*. Oxford.

Mims, W. B., Peisach, J. (1976). Assignment of a ligand in stellacyanin by a pulsed electron paramagnetic resonance method. *Biochemistry* 15:3863–3869.

Mims, W. B., Peisach, J. (1981). Electron spin echo spectroscopy and the study of metalloproteins. *Biol. Magn. Reson.* 3:213–263.

Molin, Y. N., Chibrikin, V. M., Shabalkin, V. A., Shuvalov, V. F. (1966). *Zavod. Lab.* 32:933.

Montgomery, C. G. (1947). *Technique of Microwave Measurements*. MIT Radiation Laboratory Series. Vol. 11. New York: McGraw Hill.

Montgomery, C. G., Dicke, R. H., Purcell, E. M. (1948). *Principles of Microwave Circuits*. MIT Radiation Laboratory Series. Vol. 8. New York: McGraw Hill.

More, K. M., Eaton, G. R., Eaton, S. S. (1984). Determination of T_1 and T_2 by simulation of EPR power saturation curves and saturated spectra. Application to spin-labeled iron porphyrins. *J. Magn. Reson.* 60(1):54–65.

More, K. M., Eaton, G. R., Eaton, S. S. (1985). Metal-nitroxyl interactions. 45. Increase in nitroxyl relaxation rates in fluid solution due to intramolecular interaction with iron(III) and manganese(III). *Inorg. Chem.* 24(23):3820–3823.

More, K. M., Eaton, G. R., Eaton, S. S. (1986). Metal-nitroxyl interactions. 47. EPR spectra of two-spin-labeled derivatives of EDTA coordinated to paramagnetic metal ions. *Inorg. Chem.* 25(15):2638–2646.

Morton, J. R., Preston, K. F. (1983). EPR spectroscopy of single crystals using a two-circle goniometer. *J. Magn. Reson.* 52:457–474.

Mottley, C., Kispert, L. D., Wang, P. S. (1976). An ELDOR study of methyl radical production at 77 K in irradiated acetate powders as a function of metal cation. *J. Phys. Chem.* 80:1885–1891.

Muus, L. T., Atkins, P. W., eds. (1972). *Electron Spin Relaxation in Liquids*. Plenum Press.

Nagy, V. (1994). Quantitative EPR: some of the most difficult problems. *Appl. Magn. Reson.* 6:259–285.

Nagy, V. Y. (1997). Choosing reference samples for electronic paramagnetic resonance (EPR) spectrometric concentration measurements. Part 1. General introduction and systems of S = 1/2. *Anal. Chim. Acta* 339:1–11.

Nagy, V., Sleptchonok, O. F., Desrosiers, M. F., Weber, R. T., Heiss, A. H. (2000). Advancements in accuracy of the alanine EPR dosimetry system. Part III: usefulness of an adjacent reference sample. *Radiat. Phys. Chem.* 59:429–441.

Nakano, K., Tadano, H., Takahashi, S. (1982). Quantitative multimode cavity electron spin resonance spectrometry by two-step spectral integration. *Anal. Chem.* 54:1850.

Narayana, P. A., Kevan, L. (1983). Fourier transform analysis of electron spin echo modulation spectrometry. *Magn. Reson. Rev.* 7:239–274.

Noble, G. A., Markham, J. J. (1962). Analysis of paramagnetic resonance properties of pure and bleached F-centers in potassium chloride. *J. Chem. Phys.* 36:1340–1353.

Norris, J. R., Thurnauer, M. C., Bowman, M. K. (1980). Electron spin echo spectroscopy and the study of biological structure and function. *Adv. Biol. Med. Phys.* 17:365–416.

Ohno, K. (1986). ESR imaging and its applications. *Appl. Spectros. Rev.* 22:1–56.

Ohno, K. (1987). ESR imaging. *Magn. Reson. Rev.* 11:275–310.

Pake, G. E., Purcell, E. M. (1948). Lineshapes in nuclear paramagnetism. *Phys. Rev.* 74:1184–1188.

Pake, G. E., Estle, T. L. (1973). *The Physical Principles of Electron Paramagnetic Resonance*. 2nd ed. Reading, Mass.: W. A. Benjamin, Inc.

Percival, P. W., Hyde, J. S. (1975). Pulsed EPR spectrometer II. *Rev. Sci. Instrum.* 46:1522–1529.

Peric, M., Rakvin, B., Dulcic, A. (1985). Measurement of microwave field strength in ESR by a pulsed modulation technique. *J. Magn. Reson.* 65:215–221.

Perman, W. H., Bernstein, M. A., Sandstrom, J. C. (1989). A method for correctly setting the rf flip angle. *Magn. Reson. Med.* 9:16–24.

Persson, M., Harbridge, J. R., Hammarstrom, P., Mitri, R., Martensson, L.-G., Carlsson, U., Eaton, G. R., Eaton, S. S. (2001). Comparison of electron paramagnetic resonance methods to determine distances between spin labels on human carbonic anhydrase II. *Biophys. J.* 80(6):2886–2897.

Peter, M., Shaltiel, D., Wernick, J. H., Williams, H. J., Mock, J. B., Sherwood, R. C. (1962). Paramagnetic resonance of S-state ions in metals. *Phys. Rev.* 126:1395–1402.

Piekara-Sady, L., Kispert, L. D. (1994). ENDOR spectroscopy. In: Poole, C. P. J., Farach, H. A., eds. *Handbook of Electron Spin Resonance*. Chapter V. New York: American Institute of Physics.

Pilbrow, J. R. (1990). *Transition Ion Electron Paramagnetic Resonance*. London: Oxford.

Plato, M., Lubitz, W., Mobius, K. (1981). A solution ENDOR sensitivity study of various nuclei in organic radicals. *J. Phys. Chem.* 85:1202–1219.

Poole, C. P. Jr. (1967). *Electron Spin Resonance: A Comprehensive Treatise on Experimental Techniques*. New York: Interscience Publishers.

Poole, C. P. (1983). *Electron Spin Resonance: A Comprehensive Treatise on Experimental Techniques*. 2nd ed. New York: Wiley.

Poole, C. P., Farach, H. (1971). *Relaxation in Magnetic Resonance*. New York: Academic Press.

Poole, J. C. P., Farach, H. A. (1972). *The Theory of Magnetic Resonance*. New York: Wiley.

Poole, J. C. P., Farach, H., eds. (1994). *Handbook of Electron Spin Resonance*. Vol. 1. New York: AIP Press.

Poole, J. C. P., Farach, H., eds. (1999). *Handbook of Electron Spin Resonance*. Vol. 2. New York: Springer-Verlag.

Prisner, T. (1997). Pulse high-frequency/high-field EPR. *Adv. Magn. Opt. Reson.* 20:245–299.

Quine, R. W., Eaton, G. R., Eaton, S. S. (1987). Pulsed EPR spectrometer. *Rev. Sci. Instrum.* 58(9):1709–1723.

Quine, R. W., Eaton, S. S., Eaton, G. R. (1992). Saturation recovery electron paramagnetic resonance spectrometer. *Rev. Sci. Instrum.* 63(10):4251–4262.

Quine, R. W., Rinard, G. A., Ghim, B. T., Eaton, S. S., Eaton, G. R. (1996). A 1–2 GHz pulsed and continuous wave electron paramagnetic resonance spectrometer. *Rev. Sci. Instrum.* 67(7):2514–2527.

Quine, R. W., Rinard, G. A., Eaton, S. S., Eaton, G. R. (2002). A pulsed and continuous wave 250 MHz electron paramagnetic resonance spectrometer. *Magn. Reson. Eng.* 15:59–91.

Ragan, G. L. (1948). *Microwave Transmission Circuits*. MIT Radiation Laboratory Series. Vol. 9. New York: McGraw Hill.

Rakowsky, M. H., Eaton, G. R., Eaton, S. S. (1998). Comparison of the effect of high-spin and low-spin Fe(III) on nitroxyl T1 in a spin-labeled porphyrin. In *Modern Applications of EPR/ESR: from biophysics to materials science. 1st Proceedings of the Asia-Pacific EPR/ESR Symposium*, Kowloon, Hong Kong, Jan. 20–24, 1997, pp. 19–24.

Randolph, M. L. (1972). Quantitative considerations in electron spin resonance studies of biological materials. In: Swartz, H. M., Bolton, J. R., Borg, D. C., eds. *Biological Applications of Electron Spin Resonance*. Chapter 3. New York: Wiley.

Rataiczak, R. D., Jones, M. T. (1972). Investigation of the cw [continuous wave] saturation technique for measurement of electron spin–lattice relaxation. Application to the benzene anion radical. *J. Chem. Phys.* 56:3898–3911.

Reich, H. J., Ordung, P. F., Krauss, H. L., Skalnik, J. G. (1953). *Microwave Theory and Techniques*. D. Van Nostrand.

Reisse, J. (1983). In: Lambert, J. B., Riddell, F. G., eds. *The Multinuclear Approach to NMR Spectroscopy*. Boston: Reidel, p. 63.

Rinard, G. A., Quine, R. W., Eaton, S. S., Eaton, G. R., Froncisz, W. (1994). Relative benefits of overcoupled resonators vs. inherently low-Q resonators for pulsed magnetic resonance. *J. Magn. Reson. A* 108:71–81.

Rinard, G. A., Eaton, S. S., Eaton, G. R., Poole, C. P. Jr., Farach, H. A. (1999a). Sensitivity in ESR measurements. *Handbook of Electron Spin Resonance* 2:1–23.

Rinard, G. A., Quine, R. W., Song, R., Eaton, G. R., Eaton, S. S. (1999b). Absolute EPR spin echo and noise intensities. *J. Magn. Reson.* 140(1):69–83.

Rinard, G. A., Quine, R. W., Harbridge, J. R., Song, R., Eaton, G. R., Eaton, S. S. (1999c). Frequency dependence of EPR signal-to-noise. *J. Magn. Reson.* 140(1):218–227.

Rinard, G. A., Quine, R. W., Eaton, S. S., Eaton, G. R. (2002a). Frequency dependence of EPR signal intensity, 250 MHz to 9.1 GHz. *J. Magn. Reson.* 156(1):113–121.

Rinard, G. A., Quine, R. W., Eaton, S. S., Eaton, G. R. (2002b). Frequency dependence of EPR signal intensity, 248 MHz to 1.4 GHz. *J. Magn. Reson.* 154(1):80–84.

Rinard, G. A., Quine, R. W., Eaton, S. S., Eaton, G. R., Barth, E. D., Pelizzari, C. A., Halpern, H. J. (2002c). Magnet and gradient coil system for low-field EPR imaging. *Magn. Reson. Eng.* 15:51–58.

Rinard, G. A., Quine, R. W., Eaton, S. S., Eaton, G. R. (2004). Frequency dependence of EPR sensitivity. *Biol. Magn. Reson.* 21, 115–154.

Robinson, B. H., Mailer, C., Reese, A. W. (1999a). Linewidth analysis of spin labels in liquids. I. Theory and data analysis. *J. Magn. Reson.* 138:199–209.

Robinson, B. H., Mailer, C., Reese, A. W. (1999b). Linewidth analysis of spin labels in liquids. II. Experimental. *J. Magn. Reson.* 138:210–219.

Rogers, R. N., Pake, G. E. (1960). Paramagnetic relaxation in solutions of VO^{++}. *J. Chem. Phys.* 33:1107–1101.

Rupp, H., Moore, A. L. (1979). Characterization of iron–sulfur centers of plant mitochondria by microwave power saturation. *Biochim. Biophys. Acta* 548:16–29.

Russell, A. M., Torchia, D. A. (1962). Harmonic analysis in systems using phase sensitive detection. *Rev. Sci. Instrum.* 33:442–444.

Schneider, H. J., Dullenkopf, P. (1977). Crossed slotted tube resonator (CSTR)—a new double resonance NMR probehead. *Rev. Sci. Instrum.* 48:832–834.

Schreurs, J. W. H., Fraenkel, G. K. (1961). Anomalous relaxation of hyperfine components in electron spin resonance. II. *J. Chem. Phys.* 34:756–758.

Schreurs, J. W. H., Blomgren, G. E., Fraenkel, G. K. (1960). Anomalous relaxation of hyperfine components in electron spin resonance. *J. Chem. Phys.* 32:1861–1869.

Schultz, S., Gullikson, E. M. (1983). Measurement of static magnetization using electron spin resonance. *Rev. Sci. Instrum.* 54:1383–1385.

Schweiger, A. (1982). *Electron Nuclear Double Resonance of Transition Metal Complexes with Organic Ligands*. Structure and Bonding. Vol. 51. Springer-Verlag.

Schweiger, A., Jeschke, G. (2001). *Principles of Pulse Electron Paramagnetic Resonance*. Oxford: Oxford University Press.

Sczaniecki, P. B., Hyde, J. S., Froncisz, W. (1990). Continuous wave multi-quantum electron paramagnetic resonance spectroscopy. *J. Chem. Phys.* 93(6):3891–3898.

Sczaniecki, P. B., Hyde, J. S., Froncisz, W. (1991). Continuous wave multiquantum electron paramagnetic resonance spectroscopy. II. Spin-system generated intermodulation sidebands. *J. Chem. Phys.* 94(9):5907–5916.

Setaka, M., Sancier, K. M., Kwan, T. (1970). Electron spin resonance investigation of electrical conductivity parameters of zinc oxide during surface reactions. *J. Catal.* 16:44–52.

Slangen, H. J. M. (1970). Determination of the spin concentration by electron spin resonance. *J. Phys. E. Sci. Instrum.* 3:775–778.

Slichter, C. P. (1990). *Principles of Magnetic Resonance*. 3rd ed. Berlin: Springer-Verlag.

Sloop, D. J., Yu, H.-L., Lin, T.-S., Weissman, S. I. (1981). Electron spin echoes of a photoexcited triplet: pentacene in p-terphenyl crystals. *J. Chem. Phys.* 75:3746–3757.

Smullin, L. D., Montgomery, C. G. (1948). *Microwave Duplexers*. MIT Radiation Laboratory Series. Vol. 14. New York: McGraw Hill.

Standley, K. J., Vaughan, R. A. (1969). *Electron Spin Relaxation Phenomena in Solids*. New York: Plenum Press.

Stillman, A. E., Levin, D. N., Yang, D. B., Marr, R. B., Lauterbur, P. C. (1986). Back projection reconstruction of spectroscopic NMR images from incomplete sets of projections. *J. Magn. Reson.* 69:168–175.

Strangeway, R. A., McHaourab, H. S., Luglio, J. R., Froncisz, W., Hyde, J. S. (1995). A general purpose multiquantum electron paramagnetic resonance spectrometer. *Rev. Sci. Instrum.* 66(9):4516–4528.

Sueki, M., Rinard, G. A., Eaton, S. S., Eaton, G. R. (1996). Impact of high-dielectric-loss materials on the microwave field in EPR experiments. *J. Magn. Reson. A* 118(2):173–188.

Sullivan, P. D., Bolton, J. R. (1970). The alternating linewidth effect. *Adv. Magn. Reson.* 4:39–85.

Swartz, H. M., Halpern, H. (1998). EPR studies of living animals and related model systems (*in vivo* EPR). *Biol. Magn. Reson.* 14:367–404 (Spin labeling).

Swartz, H. M., Bolton, J. R., Borg, D. C., eds. (1972). *Biological Applications of Electron Spin Resonance*. New York: Wiley.

Thomann, H., Dalton, L. R., Dalton, L. R. (1984). Biological applications of time domain ESR. *Biol. Magn. Reson.* 6:143–186.

Ueda, Y., Yokoyama, H., Ohya-Nishiguchi, H., Kamada, H. (1998). ESR spectroscopy for analysis of hippocampal elimination of nitroxide radical during kainic acid-induced seizure in rats. *Magn. Reson. Med.* 40:491–493.

Un, S., Brunel, L.-C., Brill, T. M., Zimmermann, J.-L., Rutherford, A. W. (1994). Angular orientation of the stable tyrosyl radical within photosystem II by high-field 245-GHz electron paramagnetic resonance. *Proc. Natl. Acad. Sci. USA* 91:5262–5266.

Van Camp, H. L., Heiss, A. H. (1981). Computer applications in electron paramagnetic resonance. *Magn. Reson. Rev.* 7:1–40.

Vistnes, A. I., Dalton, L. R. (1983). Experimental methods to determine the microwave field strength in electron spin resonance. *J. Magn. Reson.* 54:78–88.

van der Waals, J. H. (1998). The path that led to the study of luminescent triplet states by EPR with optical detection in zero field. *Foundations of Modern EPR*. Singapore: World Scientific, pp. 496–529.

Walsh, J. W. M., Rupp, J. L. W. (1986). Enhanced ESR sensitivity using a dielectric resonator. *Rev. Sci. Instrum.* 57:2278–2279.

Wardman, P., Seddon, W. A. (1969). Electron spin resonance studies of radicals condensed from irradiated water vapor: paramagnetic relaxation of trapped electrons in ice. *Can. J. Chem.* 47:2155–2160.

Warren, D. C., Fitzgerald, J. M. (1977a). Precision of quantitative electron spin resonance using copper(II) bound to ion-exchange resins. *Anal. Chem.* 49:250–255.

Warren, D. C., Fitzgerald, J. M. (1977b). Parameters influencing the electron spin resonance signal intensity of metal ion complex-exchanged resins. *Anal. Chem.* 49:1840–1842.

Watanabe, T., Sasaki, T., Fujiwara, S. (1982). Phase dependence of saturation transfer EPR signals and estimated rotational correlation times. *Appl. Spectrosc.* 36:174.

Weger, M. (1960). Passage effects in paramagnetic resonance experiments. *Bell Syst. Tech. J.* 39:1013–1112.

Weil, J. A., Bolton, J. R., Wertz, J. E. (1994). *Electron Paramagnetic Resonance: Elementary Theory and Practical Applications*. New York: John Wiley & Sons, Inc.

Westenberg, A. A. (1975). Use of ESR for the quantitive determination of gas phase atom and radical concentrations. *Prog. React. Kinet.* 7:73–82.

Williams, B. B., Al Hallaq, H., Chandramouli, G. V. R., Barth, E. D., Rivers, J. N., Lewis, M., Galtsev, V. E., Karczmar, G. S., Halpern, H. J. (2002). Imaging spin probe distribution in the tumor of a living mouse with 250 MHz EPR: correlation with BOLD MRI. *Magn. Reson. Med.* 47(4):634–638.

Wilmshurst, T. H. (1968). Spectrometer cavities. *Electron Spin Resonance Spectometers*. Chapter 4. Plenum.

Wilson, G. V. H. (1963). Modulation broadening of nuclear magnetic resonance and electron spin resonance line shapes. *J. Appl. Phys.* 34:3276–3285.

Wokrina, T., Dormann, E., Kaplan, N. (1996). Conduction-electron spin relaxation and diffusion in radical cation salt diperylene hexafluorophosphate. *Phys. Rev. B* 54:10492–10501.

Wood, R. M., Stucker, D. M., Jones, L. M., Lynch, W. B., Misra, S. K., Freed, J. H. (1999). An EPR study of some highly distorted tetrahedral manganese(II) complexes at high magnetic fields. *Inorg. Chem.* 38:5384–5388.

Wyard, S. J., Cook, J. B. (1969). Electron spin resonance studies of radiation damage in biological materials. In: Wyard, S. J., ed. *Solid State Biophysics*. New York: McGraw-Hill, pp. 61–103.

Yong, L., Harbridge, J., Quine, R. W., Rinard, G. A., Eaton, S. S., Eaton, G. R., Mailer, C., Barth, E., Halpern, H. J. (2001). Electron spin relaxation of triarylmethyl radicals in fluid solution. *J. Magn. Reson.* 152(1):156–161.

Yordanov, N. D. (1994). Quantitative EPR spectroscopy – "state of the art". *Appl. Magn. Reson.* 6:241–257.

Zecevic, A., Eaton, G. R., Eaton, S. S., Lindgren, M. (1998). Dephasing of electron spin echoes for nitroxyl radicals in glassy solvents by non-methyl and methyl protons. *Mol. Phys.* 95(6):1255–1263.

Zhou, Y., Bowler, B. E., Eaton, G. R., Eaton, S. S. (1999). Electron spin lattice relaxation rates for $S = 1/2$ molecular species in glassy matrices or magnetically dilute solids at temperatures between 10 and 300 K. *J. Magn. Reson.* 139(1):165–174.

13

X-Ray Photoelectron and Auger Electron Spectroscopy

C. R. BRUNDLE
C. R. Brundle & Associates, Soquel, CA, USA

J. F. WATTS
University of Surrey, Surrey, UK

J. WOLSTENHOLME
Thermo Electron Corp., East Grinstead, UK

I. INTRODUCTION

This chapter is concerned with X-ray photoelectron spectroscopy (XPS) and Auger electron spectroscopy (AES) as analytical techniques. As will be shown, both of these techniques, when applied to analyze solids, are sensitive to the outermost atomic layers of material. They can answer, to varying degrees, the following important questions:

1. Which elements are present at or near the surface?
2. What chemical states of these elements are present?
3. How much of each chemical state of each element is present?
4. What is the spatial distribution of the materials in 3D?
5. If material is present as a thin film at the surface:
   (a) How thick is the film?
   (b) How uniform is the thickness?
   (c) How uniform is the chemical composition of the film?

In electron spectroscopy, we are concerned with the emission and energy analysis of electrons (generally in the range of 20–2000 eV). These electrons are liberated from the specimen being examined as a result of the photoemission process (in XPS) or the radiationless de-excitation of an ionized atom by the Auger emission process in AES.

In the simplest terms, an electron spectrometer consists of the specimen under investigation, a source of primary radiation (X-rays for XPS; electrons for AES), and an electron energy analyzer, all contained within a vacuum chamber operating in a vacuum good enough to allow transmission of the ejected electrons around the analyzer without collisions with residual gaseous molecules (about 10^{-4} torr) (Fig. 1). Because these same residual gaseous atoms could also contaminate the surfaces of the materials being analyzed (if the surface is clean and reactive) modern spectrometers are usually capable of reaching much better vacuum (10^{-9} or 10^{-10} torr) to reduce or eliminate this problem. Even if surface contamination were not a problem the pressure should be below about 10^{-6} torr to prevent damage to the various detectors and hot filaments in the instrument.

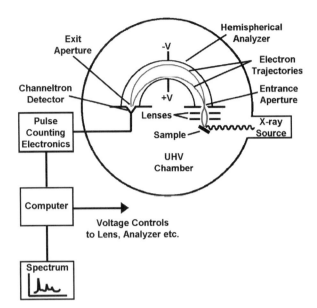

Figure 1 Schematic diagram of a typical electron spectrometer showing the necessary components. A hemispherical electrostatic electron energy analyzer is depicted.

There will usually be a vacuum load lock arrangement of some sort to allow sample entry without major compromise of the vacuum in the analysis chamber, and an ion gun fitted to clean surfaces and sputter into the subsurface *in situ* in order to produce depth profiles. There may also be a secondary chamber fitted with various specimen preparation facilities and perhaps ancillary analytical facilities. A data system will be used for data acquisition and subsequent processing.

The source of the primary radiation for the two methods is different; XPS makes use of soft X-rays, whereas AES relies on the use of an electron gun. In principle, the same energy analyzer may be used for both XPS and AES; consequently, the two techniques are sometimes to be found in the same analytical instrument, particularly in academic environments. However, the different strengths of the two techniques, high spatial resolution in AES and better chemical state analysis in XPS, have resulted in different practical applications and commercial equipment being designed primarily for either technique.

A brief review of the basics of the two processes is given in this section, before moving on to more detailed descriptions of the instrumentation and the spectroscopic interpretations.

For further reading we recommend the edited volumes by Brundle and Baker (1977 to 1982), the textbook by Carlson (1976), the book by Briggs and Seah (1990) on surface analysis and the very recent volume on surface analysis by Watts and Wolstenholme (2003).

A. X-Ray Photoelectron Spectroscopy

In XPS, we are concerned with the ejection of an electron from one of the tightly bound core or weakly bound valence shells of an atom or molecule (usually a core level for analytical purposes, as we will see later) under the influence of a monochromatic source of X-rays. This is schematically illustrated in Fig. 2(a), showing the ejection of a 1s electron (K level) from a carbon atom, which has the atomic electronic structure, $1s^2, 2s^2, 2p^2$.

The kinetic energy, KE, of the ejected electron (the photoelectron) depends on the energy of the photon, $h\nu$, (given as 1486.6 eV in Fig. 2(a), as that is the most usual X-ray energy used in XPS), and the binding energy, BE, of the electron involved (the 1s electron, here). It is given by the Einstein photoelectric law:

$$\mathrm{KE} = h\nu - \mathrm{BE} \qquad (1)$$

In XPS we measure the energy of the ejected electron. As $h\nu$ is a known monochromatic value, measurement of KE leads directly to determination of the BE. The usefulness of determining the BE is obvious when we remember that the energy of an electron shell in a particular atom is characteristic of that atom. Thus, if we determine BE through measuring KE, we identify which atom is involved in the process. This is the basis of the analytical technique. Let us consider an ensemble of C atoms. In Fig. 2(a), we have indicated that the C 1s electron is being ejected under the influence of an X-ray photon. Alternatively, for any individual C atom in the ensemble, the interaction may lead to the ejection of a 2s or a 2p electron. So, for the ensemble, all three processes will occur, and three groups of photoelectrons with three different KEs will be ejected. This will lead to a photoelectron spectrum (the distribution of KEs of all the ejected electrons) as schematically shown in Fig. 2(c). Using Eq. (1), the KE scale can be replaced by the BE scale and a direct determination of the electronic energy levels for the C atom is provided. Note that in Fig. 2(c) the peak intensities are not identical. This is because of two factors. The number of electrons available for removal in each energy level is not the same and the probability for ejection of an electron from the different levels can be very different. For C atoms, using an X-ray energy of 1486.6 eV energy, the probability (called the photoionization cross-section) is far higher for the C 1s level.

Thus, so far, we have implied that the number of peaks in a photoelectron spectrum corresponds to the number of occupied energy levels in the atoms present which are lower than the X-ray energy used; the position of the peaks directly measures the BEs and identifies the atoms present; and the intensities of the peaks depend on the number of atoms present, the number of electrons

XPS and AES

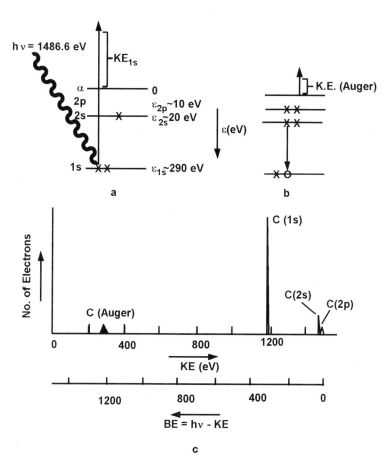

Figure 2 (a) Schematic representations of the electronic energy levels of a carbon atom and the photoionization of a C 1s electron. (b) Auger emission relaxation process for the C 1s hole-state produced in (a). (c) Schematic of the kinetic energy distribution of photoelectrons ejected from an ensemble of carbon atoms subjected to 1486.6 eV X-rays.

in the energy level concerned, and the photoionization probability for electrons in the level concerned. In practice we know sufficient about cross-section values to be able to turn peak intensities observed into atomic compositional information (Brundle and Baker, 1977–1982).

The premise that the number of peaks in a spectrum corresponds to the number of occupied energy levels is, however, only an approximation. It depends on the idea that electrons behave independently of each other. When this approximation breaks down, additional features appear in the spectrum owing to the involvement of electrons in addition to the ejected electron. This will become obvious later when we consider actual spectra in more detail.

The final important features in a photoelectron spectrum have to do with the ability to distinguish atoms of a given element which are in different chemically bonded states. This comes about because though, on a gross scale, electron BEs are characteristic of the atom concerned, at a finer level there are slight changes in the BE depending on how the atom is chemically bonded. In addition, the additional features referred to in the previous paragraph are often strongly dependent on chemistry. All of these, can be used to help distinguish the same atom in different chemical (i.e., bonding) environments, if the photoelectron spectrum is analyzed in detail at high enough spectral resolution to detect these effects (Siegbahn et al., 1967).

B. Auger Electron Spectroscopy

Once a core level hole has been created, as in Fig. 2(a) by X-ray impact, the ionized atom must relax in some way. This is achieved by an electron from a less tightly bound shell dropping into the core hole; the energy released in the process being dissipated by either the emission of an X-ray photon (X-ray fluorescence) or ejection of another electron, called the Auger electron (Auger, 1925), as indicated in Fig. 2(b). Thus, Auger features will appear in photoelectron spectra as additional peaks. However, owing to the ability to finely focus electron beams, and therefore get high spatial resolution for analytical purposes, AES is usually performed using electron beam columns, not X-rays, to create the core holes.

When a specimen is irradiated by high-energy electrons, core electrons are ejected in the same way that an X-ray beam will cause core electrons to be ejected in XPS. However, the ejected electrons in this case do not provide direct measurement of the BE because the excess energy of the process does not all get transferred to this electron, but is shared with the scattered primary electron. The purpose of the electron beam is simply to create the core hole. Once that is created, the subsequent decay process leading to the ejection of the Auger electron depends only on the relative energies of the energy levels involved. This is illustrated in Fig. 3, where a $KL_{2,3}L_{2,3}$ Auger process is shown occurring. It has this notation because it involves an electron relaxing from the $L_{2,3}$ level into the K shell core hole, which releases sufficient energy to eject another electron (the Auger electron) from the $L_{2,3}$ level. The Auger electron has a discrete KE, which enables identification of the atom concerned as follows.

The kinetic energy of a $KL_{2,3}L_{2,3}$ Auger electron is approximately equal to the difference between the energy of the core hole and the energy levels of the two outer electrons, $E_{L_{2,3}}$ (the term $L_{2,3}$ is used in this case because, for light elements, L_2 and L_3 cannot be resolved):

$$E_{KL_{2,3}L_{2,3}} \approx E_K - E_{L_{2,3}} - E_{L_{2,3}} \quad (2)$$

This equation does not take into account the interaction energies between the core holes ($L_{2,3}$ and $L_{2,3}$) in the final atomic state nor the inter- and extra-relaxation energies which come about as a result of the additional core screening needed. The calculation of the energy of Auger electron transitions is much more complex than the simple model outlined earlier, but there is a satisfactory empirical approach which considers the energies of the atomic levels involved and those of the next element in the periodic table.

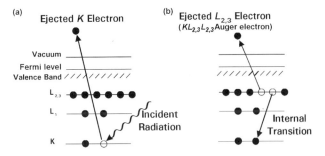

Figure 3 (a) The incident high-energy electron causes the ejection of a core electron (a K electron in this case). (b) The relaxation process involves an internal transition from a higher level and the ejection of an Auger electron. In this case, both of these are $L_{2,3}$ electrons.

Following this empirical approach, the Auger electron energy of transition $KL_1L_{2,3}$ for an atom of atomic number Z is written:

$$E_{KL_1L_{2,3}}(Z) = E_K(Z) - \frac{1}{2}[E_{L_1}(Z) + E_{L_1}(Z+1)]$$
$$- \frac{1}{2}[E_{L_{2,3}}(Z) + E_{L_{2,3}}(Z+1)] \quad (3)$$

For the $KL_{2,3}L_{2,3}$ transition the second and third terms of Eq. (3) are identical and the expression is simplified to:

$$E_{KL_{2,3}L_{2,3}}(Z) = E_K(Z) - [E_{L_{2,3}}(Z) + E_{L_{2,3}}(Z+1)] \quad (4)$$

As the energy levels involved in the Auger process described earlier are all core levels, they are all characteristic of the atom involved, and therefore so is energy of the Auger electron, $E_{KL_{2,3}L_2}$. Thus, we can get atomic identification in a manner parallel to that of XPS. In principle we can also get chemical state information as in XPS. Finally, the use of a finely focused electron beam for AES enables us to achieve surface analysis at a high spatial resolution, in a manner analogous to EPMA in the scanning electron microscope (Brookes and Castle, 1986).

C. Depth Analyzed for Solids by Electron Spectroscopy

Electrons of the energies analyzed in XPS and AES can travel only short distances in solids before being inelastically scattered, losing energy in the process (Seah and Dench, 1979). This process, shown schematically in Fig. 4(a), is the reason the techniques are surface sensitive. In XPS the soft X-rays used travel far into the solid in atomic layer terms. Photoelectrons originating near the surface have a high probability of escaping through the surface without being inelastically scattered and therefore appearing in the photoelectron spectrum at the "correct" KE corresponding to the BE. Those originating much deeper have little chance of escaping without losing energy in scattering processes and so do not appear in the photoelectron peak, but somewhere in the scattered electron background at lower KE (apparent higher BE), as in Fig. 4(b). Thus, the peaks in a spectrum originate mostly from atoms nearer the surface and the background comes mostly from the bulk. AES is very similar as the Auger electrons coming out have energies in a similar range to those in XPS.

Having said the techniques are surface sensitive, the actual depth analysis in any particular situation in both XPS and AES varies with the KE of the electrons under consideration, and with the material being analyzed. It is determined by a quantity known as the attenuation length (λ) of the electrons, which is closely related to the inelastic mean free path (IMFP). It varies as $E^{0.5}$ in the energy range of interest in electron spectroscopy, and various relationships have been suggested which relate λ

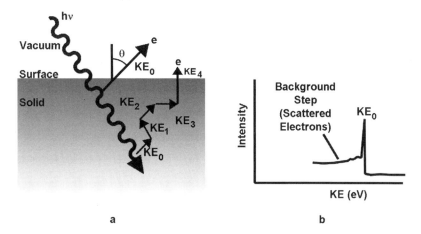

Figure 4 (a) Schematic of inelastic electron scattering occurring as a photoelectron, initial energy KE_0, tries to escape from the solid, starting at two different depths. $KE_4 < KE_3 < KE_2 < KE_1 < KE_0$. (b) KE distribution (i.e., electron spectrum) obtained because of the inelastic scattering in (a). Note that the peak at E_0 must come mainly from the surface region, and the background step, consisting of the lower energy scattered electrons, from the bulk.

to both electron energy and material properties. One such equation proposed by Seah and Dench (1979) of the National Physical Laboratory is given as follows:

$$\lambda = \frac{538 a_A}{E_A^2} + 0.41 a_A (a_A E_A)^{0.5} \quad (5)$$

where E_A is the energy of the electron in eV, a_A^3 is the volume of the atom in nm^3, and λ is in nm.

Values for λ and IMFP have been derived from optical spectroscopy, as well as from XPS and AES measurements made on layers of known thickness. Various databases exist from which values of IMFP and attenuation length can be obtained, and the extensive work done over the years in this area allows us to know the values to within a factor of two, and in many cases to within 10%.

The intensity of electrons, I, escaping the surface without scattering from all depths greater than d in a direction whose angle is θ with respect to the surface normal is given by the Beer–Lambert relationship:

$$I = I_0 \exp\left(-\frac{d}{\lambda} \cos \theta\right) \quad (6)$$

where I_0 is the intensity from an infinitely thick, uniform substrate. The Beer–Lambert equation can be manipulated in a variety of ways to provide information about overlayer thickness and, if there is the possibility to vary θ, to provide a nondestructive depth profile (i.e., without removing material by mechanical, chemical, or ion-milling methods). Using the appropriate analysis of Eq. (6), it can be shown that if you collect electrons emerging at 90° to the specimen surface, some 65% of the signal in electron spectroscopy will emanate from a depth of λ, 85% from a depth of 2λ, and 95% from a depth of 3λ. At angles of less than 90° the depth analyzed is reduced by the $\cos \theta$ term. The actual values of the IMFP vary between less than one and a few nanometers for different materials and over the range of energies involved.

D. Comparison of XPS and AES

Although it is difficult to make a comparison of the techniques before they are described and discussed in detail, it is pertinent at this point to outline the strengths and weaknesses of each to provide background information.

XPS is also known by the acronym ESCA (electron spectroscopy for chemical analysis) (Siegbahn et al., 1967). It is this *chemical* specificity which is the major strength of XPS as an analytical technique, for which it has become deservedly popular. By this we mean the ability to define not only the elements present in the analysis but also the chemical state. In the case of iron, for instance, the spectra of Fe^0, Fe^{2+}, and Fe^{3+} are all slightly different and to the expert eye, are easily distinguishable (Brundle and Chuang, 1977). AES, on the other hand, is only occasionally used for chemical state differentiation. The practical reason is that AES experiments are usually designed to maximize the signal strength by using lower spectral resolution analyzers; so chemical state information is often not resolvable.

Small area XPS is available on most modern instruments. A spectroscopic spatial resolution of about 15 μm is possible. As signal is being traded for spatial resolution, analysis times can become long. Set beside a spatial resolution of 10–15 nm, which can be achieved on the latest commercial Auger microprobes, it becomes clear that the XPS is not the way to proceed for surface analysis at high spatial resolution. On the other hand, high spatial resolution AES is often the major goal of Auger analysis.

In this case as much excitation beam current as possible is focused onto the smallest spot. There are limitations to this, however, and the Auger intensity generated from a 10 nm area is very small. This is why electron energy analyzers for AES are often designed to collect the maximum signal at the expense of other desirables, such as spectral resolution, or the capability of varying θ.

In addition to the chemical state information referred to earlier, XPS spectra can be quantified in a very straightforward manner and meaningful comparisons can be made between specimens of a similar type. Quantification of Auger data is rather complex and the accuracy obtained is generally not so good. Because of the complementary nature of the two methods, they have come to be regarded as the most important general methods (i.e., applicable to a wide range of materials and situations) of surface analysis in the context of materials science and technology.

II. ELECTRON SPECTROMETER DESIGN

The design and construction of electron spectrometers is a very complex undertaking and will usually be left to the handful of specialist manufactures worldwide, although many users specify minor modifications to suit their own requirements. An example of an XPS spectrometer is shown in Fig. 5.

A. The Vacuum System

All commercial spectrometers are now based on vacuum systems designed to operate in the ultrahigh vacuum (UHV) range of 10^{-8}–10^{-10} torr, and it is now generally accepted that XPS and AES experiments should be carried out in this pressure range. The reason a UHV environment is necessary is the surface sensitivity of the techniques themselves. At 10^{-6} mbar, the equivalent of about a monolayer of gas is delivered to the surface every second. If the surface is passivated (e.g., by prior air exposure) this will not matter too much. At the other extreme of a clean, reactive surface, this means that the surface will be contaminated by a monolayer in about 1 s. This time period is short compared with that required for a typical spectral acquisition, clearly establishing the need for a UHV environment during analysis.

UHV conditions are usually obtained in a modern electron spectrometer by use of ion pumps or turbomolecular pumps. Diffusion pumps, which were popular some time ago (Brundle et al., 1974), have now largely disappeared from modern commercial instruments. Whichever type of pump is chosen, it is common to use a titanium sublimation pump to assist the prime pumping and to achieve the desired vacuum level. All UHV systems need baking from time to time to remove adsorbed layers from the chamber walls.

The trajectory of electrons can be strongly influenced by the Earth's magnetic field, severely compromising the resolution capability of the electron energy analyzer. Consequently, some form of magnetic screening is required around the specimen and electron analyzer. The methodology depends on the manufacturer. Sometimes, it results in the whole vacuum chamber being made out of mu-metal.

B. The Specimen and its Manipulation

Although electron spectroscopy can be carried out successfully on gases and liquids as well as solids, the vast majority of practical work done is on solids. The criteria for practical analyses of solids by AES and XPS are not the same, the requirements for specimens for AES being somewhat more stringent (Powell and Czanderna, 1998).

Specimens for both XPS and AES must be stable within the UHV chamber of the spectrometer. Very porous materials (such as some ceramic and polymeric materials) can pose problems, as can those which either have a high vapor pressure or have a component which has a high vapor pressure (such as a solvent residue). In this context, 10^{-7} torr is considered a high vapor pressure.

As far as XPS is concerned, once these requirements have been fulfilled, the specimen is amenable to analysis. For AES, however, the use of an electron beam dictates that for routine analysis the specimen should be conducting and effectively grounded. As a guide, if a specimen can be imaged (in an uncoated condition) in an SEM without any charging problems, a specimen of similar type can be analyzed by AES. The analysis of insulators such as polymers and ceramics by AES is quite feasible but its success relies heavily on the skill and experience of the instrument operator. Such analysis is achieved by ensuring that the incoming beam current is exactly balanced by the combined current of emitted electrons (all secondaries including Auger electrons, backscattered

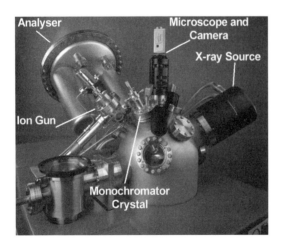

Figure 5 An example of an XPS spectrometer.

electrons, elastically scattered electrons, etc.), by adjusting the beam energy (3–5 keV), specimen current (very low, probably <10 nA), and electron take-off angle. With the recent introduction of ion guns capable of beam energies below ~50 eV, it is now possible to obtain high-quality Auger data from some insulators. The positive ions neutralize the surface and their energies are too low to cause atoms to be sputtered from the surface. In the special case of a thin insulating layer on a conducting or semiconducting specimen, the use of a high primary beam energy can induce a conducting track within the insulating layer and excess charge can be dissipated.

For mounting of conducting specimens, a fine strip of conducting paint, in addition to the adhesive tape, is all that is necessary to prevent specimen charging. Solvents in the conductive paint can cause the pump down-time to be extended as they evaporate into the vacuum. Alternatively, metal tape with a metal-loaded (conducting) adhesive may be used. Most laboratories have a selection of specimen holders, usually fabricated in-house, to accommodate large and awkwardly shaped specimens. Discontinuous specimens present rather special problems. In the case of powders, the best method is embedding in indium foil, but if this is not feasible, dusting onto double-sided adhesive tape can be a very satisfactory alternative. Fibers and ribbons can be mounted across a gap in a specimen holder, ensuring that no signal from the mount is detected in the analysis.

The type of specimen mount varies with instrument design and most modern spectrometers use a specimen stub similar to the type employed in scanning electron microscopy (SEM). For analysis, the specimen is held in a high-resolution manipulator with x, y, and z translations, tilt, and rotation about the z-axis (azimuthal rotation). For scanning Auger microscopy, where the time taken to acquire high-resolution maps may be several hours, the stability of the stage is critical, as any drift during analysis will degrade the resolution of the images. Image registration software, used during acquisition, can mitigate the effects of a small amount of drift.

Once mounted for analysis, heating or cooling of the specimen can sometimes be carried out *in vacuo*. Cooling is generally restricted to liquid nitrogen temperatures although liquid helium stages are available. Heating may be achieved by direct (contact) heating using a small resistance heater or by electron bombardment for higher temperatures. Such heating and cooling will either be a preliminary to analysis or be carried out during the analysis itself (with the obvious exception of electron bombardment heating). Heating, in particular, will often be carried out in an auxiliary preparation chamber because of the possibility of severe outgassing encountered at higher temperatures.

The routine analysis of multiple similar specimens by AES and, in particular, XPS can be a time-consuming business, and some form of automation is desirable. This is available from several manufacturer's in the form of a computer-driven carousel or table, which enables a batch of specimens to be analyzed when a machine is left unattended, typically overnight.

C. X-Ray Sources for XPS

X-rays are generated by bombarding an anode material with high-energy electrons. The electrons are emitted from a thermal source, usually in the form of an electrically heated tungsten filament, but, in some focusing X-ray monochromators, a lanthanum hexaboride emitter is used because of its higher current density (brightness). The efficiency of X-ray emission from the anode is determined by the electron energy, relative to the X-ray photon energy. For example, the Al Kα (energy 1486.6 eV) photon flux from an aluminum anode increases by a factor of more than five if the electron energy is increased from 4 to 10 keV. At a given energy, the photon flux from an X-ray anode is proportional to the electron current striking the anode. The maximum anode current is determined by the efficiency with which the heat, generated at the anode, can be dissipated. For this reason, X-ray anodes are usually water-cooled.

The choice of anode material for XPS determines the energy of the X-ray transition generated. It must be of high enough photon energy to excite a photoelectron peak from all elements of the periodic table; it must also possess a natural X-ray line width that will not broaden the resultant spectrum excessively. The most popular anode materials are aluminum and magnesium. These are usually supplied in a single X-ray gun with a twin anode configuration providing Al Kα or Mg Kα photons of energy 1486.6 and 1253.6 eV respectively (Barrie, 1977). This is possible because it is the anode and not the filament which is at a high potential (10–15 kV).

Such twin anode assemblies provide a modest depth profiling capability, as with 233 eV difference in the X-ray energies, the KEs of all peaks in a spectrum will be lower by 233 eV using the Mg source, reducing the λ value. This also provides the ability to differentiate between Auger and photoelectron transitions and separate them when the two overlap in the spectrum using one X-ray energy. XPS peaks will change to a position 233 eV higher on the KE scale on switching from Mg Kα to Al Kα, whereas the energy of Auger transitions remains constant. Converted to the BE scale, of course, the true photoelectron peaks remain unshifted and the "apparent BE" of the Auger peaks change by 233 eV, as shown in Fig. 6.

Higher energy anodes, such as Ti (4510.9 eV) and Cr (5417.0 eV), have been used in XPS (Diplas et al., 2001), but this is a very specialized capability. The advantages of using one of these elements in a twin anode is that

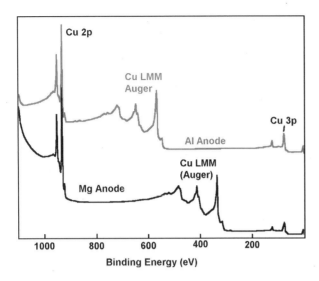

Figure 6 Comparison of XPS spectra recorded from copper using Al Kα (upper) and Mg Kα (lower) radiation. Note that on the BE scale the XPS peaks remain at constant values but the X-AES transitions move by 233 eV on switching between the two sources.

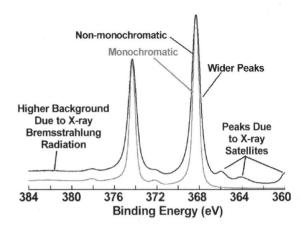

Figure 7 A comparison of the Ag 3d spectra acquired with monochromatic and nonmonochromatic X-rays. (The spectra are normalized to the maximum peak intensity.)

higher energy core levels become accessible and, as the KEs of the ejected electrons can be much higher than with Al or Mg sources, the analysis depth can be increased. The disadvantages are that spectral resolution is poorer because the X-ray line width contribution is greater and it is much harder to maintain stability of these anodes. Very recently examples of Cr 1s and Fe 1s core levels have been obtained using a monochromated Cu Kα source, $h\nu = 8048$ eV (Beamson et al., 2004).

Modern XPS instruments are often fitted with an X-ray monochromator, which narrows the line width of the X-ray coming from the anode (typically, from 1 down to 0.25 eV). At present, all commercially available X-ray monochromators used for XPS employ a quartz crystal [usually the $(10\bar{1}0)$ crystal face] as the diffraction lattice for the monochromator. Quartz is a convenient material because it is inert and compatible with UHV, and it can be bent or ground into the correct shape. Usually, the monochromator on an XPS instrument is used for Al Kα radiation, because the spacing in the quartz crystal lattice means that first order reflection occurs at a convenient angle. However, other materials and other diffraction orders have been used.

The resolution advantage is illustrated in Fig. 7. This shows the 3d region of the XPS spectrum of Ag acquired using monochromatic and nonmonochromatic X-rays. In addition to the improved resolution, the background is lower using the monochromatic X-rays, and the X-ray satellites, visible when the nonmonochromatic source is used, are removed.

By the use of a focusing X-ray monochromator, illustrated in Fig. 8, it is possible to produce a small area analysis, and this now forms the basis of commercial instruments with a spatial resolution of 20 μm or less. This is one route to small area XPS. The other commercially available method, electron optical aperturing, is discussed later in this chapter. In Fig. 8, the quartz crystal is curved in such a way that it focuses the X-ray beam as well as causes it to be diffracted. By this means, the size of the X-ray spot on the surface of the specimen is approximately equal to that of the electron spot on the anode. Thus, by varying the focusing of electron source, the analyst can vary the analysis area.

D. Charge Compensation in XPS

Photoemission from an insulating specimen may cause electrostatic charging to occur, resulting in a shift in the

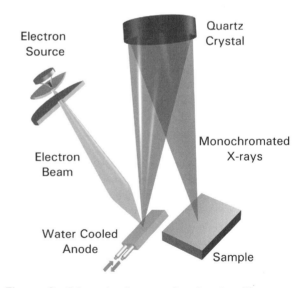

Figure 8 Schematic diagram of a focusing X-ray monochromator.

peak position in the direction of higher BE. When the photoemission is excited by a nonmonochromatic X-ray source, there are usually a sufficient number of low-energy electrons available in the neighborhood of the specimen to effectively neutralize the specimen and allow high-quality XPS spectra to be obtained.

When monochromatic X-ray sources are used, these low-energy electrons are not produced in such large numbers near the specimen and so effective neutralization does not always take place (it should also be uniform over the analysis area to prevent broadening of the XPS peaks). When charge compensation is necessary, it is a normal practice to flood the specimen with low-energy electrons. This technique minimizes the risk of differential or non-uniform charging. Any overall negative shift is corrected by calibration during data processing. Typically, the energy of the electrons is less than 5 eV. There has been extensive work dealing with the problem of charging in XPS.

E. The Electron Gun for AES

Since 1969 (the year in which AES became commercially available), AES has become a well-accepted analytical method for the provision of surface analyses at high spatial resolution. The intervening years have seen the electron guns improve from 500 μm beam spot size of the converted oscilloscope gun of the late 1960s to 10 nm, which is typical of top-of-the range Auger microprobes of today. In between these two extremes are the 5 μm and 100 nm guns, which are typical of electron guns in use with instruments on which high spatial resolution Auger is not the prime technique (multitechnique instruments or flat film depth profiling AES).

The critical components of the electron gun are the electron source and the lens assemblies for beam focusing, shaping, and scanning. Electron sources may be either thermionic emitters or field emitters (cold or Schottky) and lenses for the electron gun may be electrostatic or, for high-resolution applications, electromagnetic. If we consider the lenses first, the criterion, which distinguishes an electron gun for AES from one conventional in electron microscopy, is the need to operate in a UHV environment. In the early days of AES, this effectively precluded the use of electromagnetic lenses, as the coils were not able to withstand UHV bakeout temperatures. Consequently, electrostatic lenses were much favored and by gradual design improvement the stage has been reached where electron guns with electrostatic lenses can achieve a spatial resolution of <100 nm. The major step forward as far as quality scanning Auger microscopy was concerned, was the development of a gun with bakeable electromagnetic lenses; by the late 1970s, such lenses, with a spatial resolution of 50 nm and a thermionic emitter became routinely available.

Nowadays, the combination of electromagnetic lenses and a field emission source of electrons results in Auger instruments whose electron spot size can be less than 10 nm.

To be useful as an electron source for AES a source should have the following properties in addition to just being able to generate a small beam spot:

1. *Stability*: The current emitted from the source should be highly stable over long periods.
2. *Brightness*: High emission currents from a small emitted area are required if the eventual spot size at the specimen is to be small.
3. *Monoenergetic*: The focal length of electromagnetic and electrostatic lenses is dependent upon the energy of the electrons. This means that the optimum focusing conditions for a small spot can only occur for electrons having a very small range of KE.
4. *Longevity*: Under normal operating conditions the emitter must last for many hundreds of hours. Replacement of emitters requires that the vacuum be broken and the instrument be baked before it can be used.

Table 1 compares the important characteristics of the types of emitter considered here.

The stringent vacuum requirements and the relatively poor stability mean that the cold field emitter is rarely used for AES. In commercially available medium and high performance Auger instruments, either LaB_6 or Schottky field emitters are generally used.

The spot size attainable with a particular electron gun is a function of the primary beam current. For example, the smallest spot size obtainable on a scanning Auger microscope with electromagnetic lenses and LaB_6 filament is about 20 nm at 0.1 nA, but increases to 100 nm at 10 nA. The intensity of Auger electrons emitted depends on the specimen current and, at 0.1 nA, spectrum acquisition will be a lengthy process, but at 10 nA the current of Auger electrons will be increased to the point where analysis becomes practical. A satisfactory compromise must be reached between spatial resolution and spectral intensity. A field emission source in a similar electron beam column will give somewhat better resolution at an improved current owing to its superior brightness. They give usable signal intensities down to 10–15 nm. Such electron guns are bulky and are the preserve of specialized high-resolution scanning Auger microscopes (SAM). When SAM is required on a multitechnique XPS instrument, the best configuration probably includes a high-brightness, hot field emission gun with electrostatic lenses, which is a much more compact design. Such an assembly will provide reliable routine operation with the minimum of attention and an analytical spot size of

Table 1 Comparison of Electron Emitters

	Thermionic	LaB$_6$	Cold field emitter	Schottky emitter
Work function (eV)	4.5	2.7	4.5	2.95
Brightness (A/cm^2/srad)	<10^5	~10^6	10^7–10^9	>10^8
Current into a 10 nm spot	1 pA	10 pA	10 nA	5 nA
Maximum beam current	1 µA	1 µA	20 nA	200 nA
Minimum energy spread (eV)	1.5	0.8	0.3	0.6
Operating temperature (K)	2700	2000	300	1800
Short term stability (%)	<1	<1	>5	<1
Long term stability	High	High	>10%/h	<1%/h
Vacuum required (mbar)	<10^{-4}	<10^{-6}	<10^{-10}	<10^{-8}
Typical lifetime (h)	<200	~1000	>2000	>2000
Relative cost	Low	Medium	High	High

<100 nm. The lack of magnetic objective lenses also removes the problem of magnetic shielding when trying to run high-resolution XPS with such lenses located close by.

F. Electron Energy Analyzers for Electron Spectroscopy

There are two types of electron energy analyzer in general use for XPS and AES: the cylindrical mirror analyzer (CMA) and the hemispherical sector analyzer (HSA).

The CMA is used when it is not important that the highest resolution is achieved and when the highest throughput is desirable. For example, it is routinely used for AES if chemical state information is not required. Typically, commercial CMA analyzers can provide an energy resolution of about 0.4–0.6% of the energy to which they are tuned. HSAs can be operated in such a way that they have a similar resolution but can also operate with more than a factor of 10 better resolutions, at the cost of lower throughput.

The development of these two analyzers is a reflection of the requirements of AES and XPS at the inception of the techniques. The primary requirement for AES was high sensitivity (analyzer transmission), the instrument resolution (the contribution of analyzer broadening to the resultant spectrum) being of lesser importance. The need for high sensitivity led to the development of the CMA. For XPS, on the other hand, it is spectral resolution that is the cornerstone of the technique and this led to the development of the HSA as a design of analyzer with sufficiently good resolution. The addition of a transfer lens to the HSA and multichannel detection increase its sensitivity to the point where both high transmission and high resolution are possible, and this type of analyzer may now be used with acceptable results for both XPS and AES. The nature of the CMA does not permit the addition of a transfer lens. It also does not lend itself to performing angle resolved measurements (variation of θ for nondestructive depth profiling).

The CMA consists of two concentric cylinders as illustrated in Fig. 9, the inner cylinder is held at earth potential, whereas the outer is ramped at a negative potential. An electron gun is often mounted coaxially within the analyzer. A certain proportion of the Auger electrons emitted will pass through the defining aperture in the inner cylinder, and, depending on the potential applied to the outer cylinder, electrons of the desired energy will pass through the detector aperture and be re-focused at the electron detector. Thus, an energy spectrum can be built up by merely scanning the potential on the outer cylinder to produce a spectrum of intensity (in counts per second) vs. electron KE.

This spectrum will contain, not only the Auger electrons, but also all the other emitted electrons, Auger

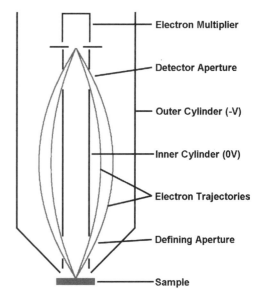

Figure 9 Schematic diagram of the CMA.

peaks being superimposed, as weak features, on an intense background. For this reason the differential spectrum is often recorded rather than the direct energy spectrum. This used to be achieved by applying a small a.c. modulation to the analyzer and comparing the output at the detector with this standard a.c. signal by means of a phase sensitive detector (lock-in-amplifier) Seah et al. (1983), give a discussion on the use of differentiation in Auger spectra. The resultant signal is displayed as the differential spectrum. In modern Auger spectrometers, however, the differential spectrum, if required, is calculated within the data system from the digitally collected data. A comparison of direct (pulse counted) and differential Auger electron spectra, from a nickel foil, is presented in Fig. 10.

Though the CMA can provide good sensitivity for AES, it suffers from a number of disadvantages, making it totally unsuitable for XPS. In addition to the poor resolution already mentioned, the energy calibration of the analyzer is strongly dependant on the distance of the surface of the sample from the analyzer, and also the area from which electrons can be collected is very small. In an attempt to overcome these disadvantages, a double pass CMA was developed with limited success but all modern commercial XPS instruments are now equipped with an HSA.

An HSA, otherwise known as a concentric hemispherical analyzer (CHA) or a spherical sector analyzer (SSA), consists of a pair of concentric hemispherical electrodes between which there is a gap for the electrons to pass. Between the specimen and the analyzer there is usually a lens system, which serves several purposes, one of which is to reduce the KE of the electrons before entering the analyzer, thereby improving the absolute resolution of the analyzer.

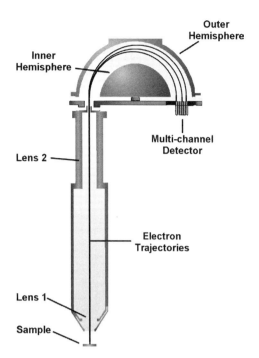

Figure 11 Schematic diagram of a modern HSA and transfer lens. Showing also the trajectories of three electrons each having a slightly different KE.

The schematic diagram of Fig. 11 shows a typical HSA configuration for XPS.

A potential difference is applied across the two hemispheres with the outer hemisphere being more negative than the inner one. Electrons injected tangentially at the input to the analyzer will only reach the detector if their energies are given by

$$E = e\Delta V \left(\frac{R_1 R_2}{R_2^2 - R_1^2} \right) \qquad (7)$$

where the KE of the electrons is given by E, e is the charge on the electron, ΔV is the potential difference between the hemispheres, and R_1 and R_2 are the radii of the inner and outer hemispheres, respectively. The radii of the hemispheres are constant and so Eq. (7) can be expressed as

$$E = ke\Delta V \qquad (8)$$

in which k is known as the spectrometer constant and depends upon the design of the analyzer. An HSA also acts as a lens and so electrons entering the analyzer on the mean radius will reach the exit slit even if they enter the analyzer at some angle with respect to the tangent to the sphere given by the mean radius.

Electrons whose energy is higher than that given by Eq. (8) will follow a path whose radius is larger than the mean radius of the analyzer and those with a lower KE will follow a path with smaller radius. Provided that the energy of these electrons does not differ too greatly from

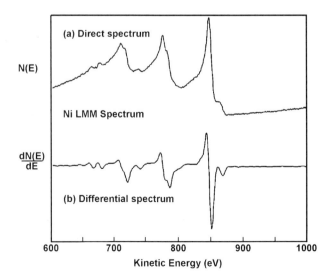

Figure 10 Comparison of (a) direct and (b) differential Auger spectra for Ni LMM.

that given by the equation, these electrons will also reach the output plane of the analyzer. It is possible, therefore, to provide a number of detectors at the output plane, increasing the sensitivity by parallel detection. Instruments with up to nine discrete channel electron multipliers (channeltrons) are commercially available. Some instruments are fitted with 2D detectors (channel plates), which allow the user to select the number of channels collected at any one time. Up to 112 channels can be defined on some instruments; the width of each channel is small, so this does not imply that the sensitivity is 112× that of an equivalent instrument fitted with a single channeltron. The advantage of having such a large number of channels is that it allows high-quality spectra to be recorded without scanning the analyzer.

The HSA is typically operated in one of two modes: constant analyzer energy (CAE), sometimes known as fixed analyzer transmission (FAT), and constant retard ratio (CRR), also known as fixed retard ratio (FRR).

In the CAE mode electrons are accelerated or retarded by the lens system. The analyzer is set to pass a user-defined fixed electron energy. The selected pass energy affects both the transmission of the analyzer and its resolution. Selecting a low pass energy will result in high resolution, whereas a high pass energy will provide higher transmission but poorer resolution. As the pass energy remains constant throughout the electron kinetic energy range, the resolution (in electron volts) is constant across the entire width of the spectrum.

Figure 12 shows part of the XPS spectrum of silver recorded at a series of pass energies, showing the effect of pass energy on resolution. The spectra have been normalized but the increasing noise with decreasing pass energy is a result of the decreasing sensitivity. In a typical XPS experiment, the user will select a pass energy in the region greater than 100 eV for survey or wide scans and in the region of 20 eV for higher resolution spectra of individual core levels. These narrow scans are used to establish the chemical states of the elements present and for quantification purposes. It is normal practice to collect XPS spectra in the CAE mode, so that the energy resolution (in electron volts) remains constant across the spectrum.

In the CRR mode electrons are retarded to some user-defined *fraction* of their original KE as they pass through the analyzer (known as the retard ratio). In order to achieve analysis in the CRR mode, the voltages on the hemispheres are scanned with a varying ΔV. In this mode the pass energy is proportional to the KE. The percentage resolution in this mode of operation is constant (not the absolute resolution in electron volts) and inversely proportional to the retard ratio. The constant of proportionality depends upon the design of the instrument. Typically, the resolution available from a good commercial HSA can be selected from the range 0.02–2.0%. (A CMA is likely to be around 4%.) It is normal practice to collect e-impact Auger electron spectra in the CRR mode.

The performance of the HSA is strongly dependent upon the nature and quality of the transfer lens or lenses between the specimen and the entrance to the analyzer. They move the analyzer away from the analysis position allowing other components of the spectrometer to be placed closer to the specimen; maximize the collection angle to ensure high transmission and sensitivity; retard the electrons prior to their injection into the analyzer; control the area of the specimen from which electrons are collected, allowing small area XPS measurements to be made; and control the acceptance angle.

G. Detectors

Electron multipliers are used to count the individual electrons arriving at the detector. Although there are many types of electron multiplier, only two types are commonly used in electron spectrometers: channeltrons and channel plates. Channeltrons have been in use for over 30 years, so will not be described here. Suffice it to say their job is to provide a signal amplification of the order of 10^7 and are capable of handling up to about 1×10^6 counts/s before severe nonlinearity sets in.

A channel plate is a disc having an array of small holes. Each of these holes behaves as a small channeltron. The gain of an individual channel is much lower than that of a single channeltron; so it is common to use a pair of channel plates in tandem. The maximum count rate using channel plates is about 3×10^5 counts/s for 2D detection. Channel plates are used when it is necessary

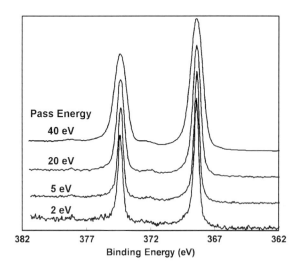

Figure 12 XPS spectrum of Ag showing the effect of pass energy upon the Ag 3d part of the spectrum. The spectra have been normalized for clarity.

to detect data in 2D. Spectrometers have been designed using channel plates to measure signals in an $X-Y$ array for parallel acquisition of photoelectron images in an X-energy array for parallel acquisition of XPS line scans; and in an energy–angle array for the parallel acquisition of angle resolved XPS (ARXPS) spectra.

H. Spectrometer Operation for Small Area XPS

It is often necessary to analyze a small feature or imperfection on the surface of a specimen. For the analysis to be effective, as much as possible of the signal from the surrounding area should be excluded. This is usually done by either flooding the analysis area with X-rays but limiting the area from which the photoelectrons are collected using the transfer lens (lens-defined small area analysis), or focusing a monochromated beam of X-rays into a small spot on the specimen (source-defined small area analysis).

In most spectrometers the transfer lens fitted to the analyzer are operated in such a way that they produce a photoelectron image at some point in the electron optical column. If a small aperture is placed at this point, then only the electrons emitted from a defined area of the specimen can pass through the aperture and reach the analyzer. If the magnification of the lens is M and the diameter of the aperture is d, then the diameter of the analyzed area is d/M. In some instruments, an aperture can be selected from a number of fixed apertures, whereas in other instruments an iris is used to provide a continuous range of analysis areas. Spherical aberrations, which occur in any electron optical lens system, mean that the acceptance angle of the lens has to be limited to provide good edge resolution in the analysis area. This is achieved using either another set of fixed apertures or another iris placed at some point remote from the image position of the lens. Using this technique, commercial instruments can provide small area analysis down to about 15 μm.

This is an effective method for producing high-quality, small area XPS data but it suffers from a disadvantage. Reducing the angular acceptance of the lens reduces the detected flux per unit area of the specimen and the analysis times can become very long.

In the second method a quartz crystal is bent so that it can focus a beam of X-rays and provide monochromatic X-rays by diffraction. In this respect, it behaves rather like a concave mirror. The focusing is usually achieved using a magnification of unity, which means that the size of the X-ray spot on the specimen is approximately equal to the size of the electron spot on the X-ray anode. Analysis areas down to about 10 μm can be achieved in commercial instruments using this method.

Because the source of X-rays is defining the analysis area, aberrations in the transfer lens will not affect the analysis area and so the lens can be operated at its maximum transmission, regardless of how small the analysis area becomes. The sensitivity of a spectrometer operating in this mode is therefore much higher than that of an equivalent instrument operating in the lens-defined mode.

I. XPS Imaging and Mapping

A logical extension to small area XPS is to produce an image or map of the surface. Such an image or map shows the distribution of an element or a chemical state on the surface of the specimen. There are two distinct approaches, used by manufacturers, to obtain XPS maps: serial acquisition, in which each pixel of the image is collected in turn (mapping mode) and parallel acquisition, in which data from the whole of the analysis area is collected simultaneously (true imaging, as in an optical microscope or a TEM).

Serial acquisition of images is based on a 2D, rectangular array of small area XPS analyses. The ultimate spatial resolution in the image is determined by the size of the smallest analysis area (this depends upon the instrument, but 10 μm is possible using a source-defined approach with a high-quality, modern spectrometer). Serial acquisition is generally slower than parallel acquisition but has the advantage that one can collect a range of energies (i.e., a portion of the spectrum) at each pixel, whereas in parallel acquisition, only a single energy can be collected during each acquisition. There are several methods by which the analysis area can be stepped over the field of view of the image, each having advantages and disadvantages. They are scanning the specimen stage, the lenses, and the X-ray spot.

In parallel acquisition of photoelectron images, the whole of the field of view is imaged simultaneously without scanning voltages applied to any component of the spectrometer (imaging mode). To obtain images via this route, additional lenses are required in the spectrometer and a 2D detector must be used at the image plane.

The spatial resolution of parallel imaging is dependent upon the spherical aberrations in the lens. Limiting the angular acceptance of the lens can reduce the effect of the aberrations and so resolution can be improved at the expense of sensitivity. The use of a magnetic immersion lens in the specimen region also reduces aberrations and therefore, allows higher sensitivity at a given resolution. This method of imaging is relatively fast and commercial instruments can produce images with an image resolution of down to 3 μm.

Parallel imaging clearly provides the best image resolution, and is faster than the serial methods but it only collects an image at a single energy. It is customary to make a measurement at a photoelectron peak energy and a second measurement at some energy remote from the peak where the signal intensity is approximately equal to the estimated background signal under the peak maximum. By subtracting the background signal from the signal at the peak maximum, a more accurate measurement can be made. This is in contrast with the use of a serial mapping method in conjunction with a multichannel detector, to produce a "snapshot" spectrum at each pixel of the map. Such spectra can then be treated with advanced data processing techniques in order to extract the maximum chemical information from the image.

Figure 13 shows examples of XPS images from parallel acquisition using monochromated Al Kα X-rays. This specimen was a glass substrate with gold features. Images from Au 4f and Si 2p are shown along with an intensity line scan measured from the Au 4f image. The line scan indicates that the spatial resolution in this image is about 3 μm. These images can be used to define areas for spectroscopy. A 25 μm^2 area was selected from the gold region of the specimen and from the glass; the respective spectra are shown in Fig. 13. Images such as these can be acquired in just a few minutes; those in Fig. 13 were collected in 4 min (2 min for the peak image and 2 min for the background) and the subsequent spectra were collected in less than 2 min each. The specimen was an insulator and so required the use of an electron flood gun to control the specimen charging.

To estimate the lateral resolution in small area XPS, a knife-edge specimen (often silver) is translated through the analysis area while measuring the XPS signal from the Ag 3d$_{5/3}$ peak. The signal is initially zero until the knife-edge begins to intercept the analysis area when it begins to rise. The intensity of the signal continues to rise until the silver completely fills the whole of the analysis area. The distance through which the knife-edge has to be translated for the signal to change between two prescribed percentages of the total signal change is then determined. This distance is described as the lateral resolution for spectroscopy. The percentages used vary with the instrument manufacturer. The range 20–80% is used by some because the reported spatial resolution is approximately equal to the radius of the analysis area (within <2%). Others use the range 16–84%. Measurement of spatial resolution should be made in two orthogonal directions in case the analysis area is not circular.

J. Angle Resolved XPS

As mentioned earlier, the small mean free path of electrons within a solid means that the information depth in XPS analysis is of the order of a few nanometers if the electrons are detected at a direction normal to the specimen surface. If electrons are detected at some angle to the normal, the information depth is reduced by an amount equal to the cosine of the angle between the surface normal and the analysis direction. This is the basis for the powerful

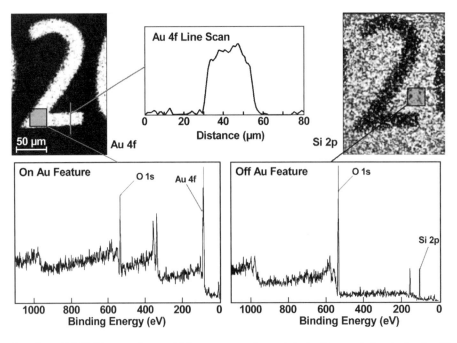

Figure 13 Examples of parallel XPS images from gold features on a glass substrate. The resolution in the Au 4f and Si 2p images can be measured from the line scan to be about 3 μm. The images can be used to define small areas from which spectra can be acquired.

analysis technique, ARXPS. One of the reasons for the usefulness of the method is that it can be applied to films which are too thin to be effectively analyzed by conventional sputter depth profiling techniques. There are also sputter-induced artifacts, such as preferential sputtering of one component and changes in chemical state that can compromise this method. ARXPS is a nondestructive technique which can provide chemical state information.

To obtain ARXPS data the angular acceptance of the transfer lens is set to provide good angular resolution, usually the half angle is set to be in the region of 1–3°. A series of spectra is then acquired as the specimen surface is tilted with respect to the lens axis. Figure 14 illustrates how the spectra might appear at each end of the angular range if the specimen consists of a thin oxide layer on a metal substrate. Note that the relative intensity of the oxide peak is larger at the higher grazing emission angle. ARXPS measurements such as these provide information on the thickness and chemical composition of very thin (a few nanometers) surface layers, as will be illustrated later.

There is one commercial instrument which produces ARXPS spectra without tilting the specimen. It is capable of parallel collection of angle resolved data. The 2D detector at the output plane has photoelectron energy dispersed in one direction (as with a conventional lens analyzer arrangement) and the angular distribution dispersed in the other direction, Fig. 15. Such an arrangement can provide an angular range of 60° with a resolution close to 1°. This has a number of advantages over the conventional method:

1. ARXPS measurements can be taken from very large specimens, such as complete semiconductor wafers. Such specimens are too large to be tilted inside an XPS spectrometer.

2. The analysis position remains constant throughout the angular range. When combining small area XPS and ARXPS it is difficult to ensure that the analysis point remains fixed while tilting the sample.

3. The analysis area remains constant during the analysis. If lens-defined small area XPS is combined with ARXPS then the analysis area would increase by a large factor as the specimen is tilted away from its normal position. Using a combination of source-defined small area analysis and parallel collection, the analysis area becomes independent of angle.

III. INTERPRETATION OF PHOTOELECTRON AND AUGER SPECTRA

Photoelectron and Auger spectra are amenable to many levels of interpretation, ranging from a simple qualitative assessment of the elements present to a full-blown quantitative analysis complete with assignments of chemical states, and determination of the phase distribution for each element. The most common interpretation is an estimation of the relative amounts of each element present. As the techniques are closely related it is not surprising that there are similarities in the way that AES and XPS spectra are treated.

A. Qualitative Analysis

The first step to be taken in characterizing the surface chemistry of the specimen under investigation is the identification of the elements present. To achieve this it is usual to record a survey, or wide scan, spectrum over a region that will provide fairly strong peaks for all elements in the periodic table. In the case of both XPS and AES, a range of 0–1000 eV is often sufficient, though the current IUVSTA[1] recommendations for the acquisition of XPS survey spectra extend this range.

An example Auger spectrum is shown in Fig. 16(a). The spectra in Fig. 16 are from a 1 nm layer of silicon dioxide on elemental silicon. The peaks produced by the elements present, in this case Si, O, C, are observed superimposed on a high background typical of Auger spectra. This is because the Auger intensities are very small compared with the scattered electron background from the primary excitation beam. The spectrum is cut off deliberately at the low KE end, where there would be a intense background of low-energy secondary electrons (these are the electrons used in SEM for imaging). At the upper

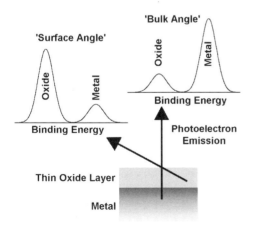

Figure 14 Illustration of XPS spectra taken from a thin oxide film on a metal at near normal collection angle (bulk angle) and near grazing collection angle (surface angle).

[1] IUVSTA is the International Union for Vacuum Science, Technique, and Applications.

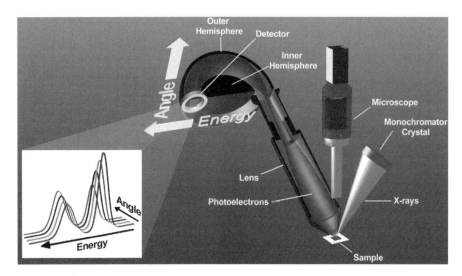

Figure 15 The arrangement of a spectrometer capable of collecting angle resolved spectra in parallel.

end the elastic reflected peak of the excitation beam will show up as the next most intense feature. Again it is deliberately cut off here. Auger spectra may be recorded in either the direct or the differential mode [Fig. 16(b)]. The latter has the effect of flattening out the background, but is otherwise cosmetic. Nowadays, the direct mode is becoming more popular, as detailed interpretation is more straightforward. However, much published data is in the differential mode (in analog form), so it is often necessary to use the differential mode subsequently simply for comparison purposes. In the modern world of digitally acquired data this can always be done after acquisition, during the processing step. Figure 16(c) shows the Si KLL region of the direct Auger spectrum, recorded at higher resolution and showing both the chemical shift associated with oxidation of silicon and a plasmon loss feature.

The equivalent photoelectron spectrum from the same specimen, Fig. 16(d), is composed of the individual photoelectron peaks and the associated Auger lines resulting from the de-excitation process following photoemission. Unlike the Auger spectrum of Fig. 16(a) the electron background is relatively small and increases in a step-like manner after each spectral feature. This is a result of the scattering of the characteristic photoelectrons within the matrix, bringing about a loss of KE, as discussed earlier (Fig. 4). The shape of this background itself contains valuable information that, to the experienced electron spectroscopist, provides a means of assessing the way in which near-surface layers are arranged. In the case of a perfectly clean bulk homogeneous material, the photoelectron peaks will have a horizontal background after each one (see Fig. 4), or one with a slightly negative slope. If the surface is covered with a thin overlayer, the peaks from the buried phase will have a positive slope and if the film is thick when compared with the probing depth, the peak itself will be absent and the only indication will be a change in background slope at the appropriate energy.

The survey spectrum in XPS or AES will generally be followed up by the acquisition of more detailed, higher resolution spectra around the elemental peaks of interest. Both XPS and AES spectra contain valuable chemical state information. That XPS usually contains more chemical state information than AES is illustrated by comparing Fig. 16(c) with Fig. 16(e). The Auger spectrum can show the presence of both oxidized and elemental silicon but the XPS spectrum can reveal the presence of the suboxides of silicon that occur at the oxide/element interface.

The rapid increase in the use of monochromatic sources, and high-resolution analyzers, in the early 1990s led to a step change in the level of chemical state information attainable from core level spectra in XPS, particularly for the analysis of polymers. The complexity of core level XPS spectra recorded at high resolution has seen a parallel growth in the level of sophistication of peak fitting routines to enable the analyst to assign the various components of a convoluted spectrum with a degree of confidence. Important parameters in such treatments include the number of components, the shape of the peak (usually, a Voight function), the degree of asymmetry (important in metals which are asymmetric as a result of core hole lifetime effects), the width of the components (constrained together or allowed to vary), and the shape of the background which will be influenced by both elastic and inelastic scattering of the electrons.

B. Chemical State Information: The Chemical Shift

The XPS chemical shift is the cornerstone of the technique and the reason why high-resolution analyzers and accurate

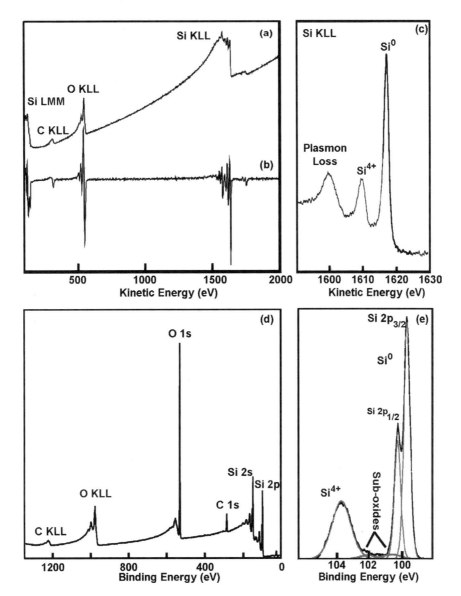

Figure 16 Electron spectra from a thin layer (∼1 nm) of SiO_2 on Si. (a) Direct Auger spectrum. (b) Differential Auger spectrum. (c) A direct Auger spectrum of the Si KLL region using high-energy resolution so that the elemental peak, oxide peak, and plasmon peak can be resolved. (d) The XPS survey spectrum (recorded with monochromatic Al Kα radiation) of the same sample. Note the C 1s peak resulting from the deposition of adventitious carbon from the atmosphere. (e) The Si 2p region recorded at higher resolution showing the elemental (Si^0) and oxide (Si^{4+}) components along with contributions from the sub-oxides found at the SiO_2/Si interface.

calibration of energy scales were seen in XPS long before in AES. All elements with core levels in the periodic table exhibit a chemical shift, which can vary from a small fraction of an electron volt up to several electron volts. Set alongside the line width of the X-ray sources used in XPS (0.25–1.0 eV) it is clear that using chemical shifts is not a high-resolution spectroscopy. This is why some form of data processing is often required to extract the maximum level of information from a spectrum. The computer curve fitting of high-resolution XPS spectra is now a routine undertaking, as indicated earlier, and international standards are now being drafted to provide a unified framework within which such procedures can be undertaken.

The shifts observed in XPS have their origin in initial state and/or final state effects (Bagus et al., 1999). In the case of initial state effects, it is the charge on the atom prior to photoemission that plays the major role in the determination of the magnitude of the chemical shift. For example, the C—O bond in an organic polymer has a C 1s shift of 1.6 eV relative to the unfunctionalized (methylene) carbon, whereas both C=O and O—C—O are shifted by 2.9 eV. In essence, the more electronegative the groups bonded to the C atom, the greater the polarization of electrons away from it, and the greater the positive XPS chemical shift of the C 1s peak. This is illustrated in a striking manner for fluorocarbon species, the C 1s chemical shifts being larger than those of carbon–oxygen compounds as fluorine is a more electronegative element. The C—F group is shifted by 2.9 eV, whereas CF_2 and CF_3 functionalities are shifted by 5.9 and 7.7 eV, respectively. Unfortunately, such examples of the chemical shift are unusually large and, in general, values of 1–3 eV are encountered. An example of the manner in which the peak fitting of a complex C 1s spectrum is achieved is shown in Fig. 17. The specimen is an organic molecule, diglycidyl ether of bisphenol A, which is a precursor to many thermosetting paints, adhesives, and matrices for composite materials. By the consideration of the structure of the molecule it is possible to build up a synthesized spectrum, the relative intensities of the individual components reflecting the stoichiometry of the sample. For polymer XPS, at this level of sophistication, the resolution attainable with a monochromatic source is absolutely essential. Even then it may not be possible, using chemical shifts, to distinguish, in a polymer, between a C atom bonded to four other C atoms and one bonded to one C atom and three H atoms. This is because the electronegativity of C and H are nearly the same; so there is no charge transfer and very nearly no chemical shift.

Final state effects, such as core hole screening, that occur following photoelectron emission, relaxation of electron orbitals, and the polarization of surrounding ions are often dominant in influencing the magnitude of the chemical shift. They tend to counter initial state shifts. However, in most metals there is still a positive shift between the elemental form and mono-, di-, or trivalent ions, but in the case of cerium the very large final state effects give rise to a negative chemical shift of ca. 2 eV between Ce and CeO_2. There are several other anomalies like this, but in general, most elements behave in a predictable manner: the more positive charge on the atom in the initial state, the higher the BE as measured in XPS. Table 2 lists some typical chemical shifts observed between neutral and positively charged states (oxidized) of some important industrial elements. A great deal of theoretical work has been devoted to understanding XPS chemical shifts and being able to calculate them reliably. These range from simple point charge electrostatic models to full-blown molecular orbital quantum chemistry or many body calculations (Fadley et al., 1969).

Despite the availability of theory, databases of known chemical shifts in known chemical state environments still provide the best way of pursuing chemical speciation identification in XPS. There are various compilations of binding energies and the most extensive is that promulgated by NIST, which is available free of charge over the internet (http://srdata.nist.gov/xps/index.htm). This provides a ready source of standard data with which the individual components of a spectrum can be assigned with a high degree of confidence. Another ready laboratory source is the handbook of spectra that some manufacturers provide with their instruments (Moulder et al., 1992; Childs et al., 1995).

The thrust in the early development of AES was the use of analyzers, such as the retarding field or the CMA, which provided a high level of transmission but at the expense of spectral resolution. Thus, the peaks from early Auger spectrometers were very broad and superimposed on the intense electron background from the inelastic scattered primary beam. Small signals on a high background, coupled with the fact that early analyzers were analog and noisy that led to the practice of using phase sensitive

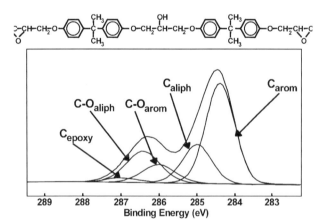

Figure 17 C 1s spectrum of the basic building block of epoxy product, the diglycidyl ether of bisphenol A, the structure of which is shown above the spectrum. This spectrum was recorded using monochromatic Al radiation.

Table 2 Examples of Chemical Shifts Between Elemental and Oxidized States of a Number of Metallic Elements

Element	Oxidation state	Chemical shift from zero-valent state (eV)
Ni	Ni^{2+}	~2.2
Fe	Fe^{2+}	~3.0
	Fe^{3+}	~4.1
Ti	Ti^{4+}	~6.0
Si	Si^{4+}	~4.0
Al	Al^{3+}	~2.0
Cu	Cu^{+}	~0.0
	Cu^{2+}	~1.5
Zn	Zn^{2+}	~0.1
W	W^{4+}	~2.0
	W^{6+}	~4.0

detection to acquire differential spectra, both to be able to more easily see the wanted features, and to reduce the instrument noise by the phase sensitive detection. The combination of poor resolution and differentiation meant that even if there were well-defined chemical information in the spectra, the practices used for spectral acquisition often effectively obliterated it! Frankly, it was also not of much interest to most of the community using it at that time for looking at solids. They were concentrating on achieving the greatest sensitivity to the presence of low-concentration elements. Another reason was the superposition of the degenerate band structure onto the shape of the Auger peak in the case of transitions involving valence levels. This may lead to changes in the shape of Auger transitions from different chemical environments but, generally, not the discrete chemical shift that is observed in XPS core levels. However, if the two outer electrons are not valence electrons (i.e., CCC Auger transitions) a sharp peak may result as observed, for example, in the KLL series of peaks of aluminum and silicon, and the LMM series of copper, zinc, gallium, germanium, and arsenic. The Ge LMM Auger spectra of Fig. 18 show components attributable to Ge^0 and Ge^{4+} separated by over 8 eV.

The cornerstone of any spectral analysis which relies on peak position to provide information presupposes the ability to determine such values with the necessary accuracy, ±0.1 eV in electron spectroscopy. The two possible sources of error are those owing to spectrometer calibration and those resulting from electrostatic charging of the specimen. The former is easily overcome by accurate calibration of the spectrometer against known (standard) values for copper and gold. The latter is resolved for metallic specimens by proper mounting procedures, but for insulators and semiconductors a slight shift as a result of charging is always a possibility. Although it is possible to use an internal standard, such as the adventitious carbon 1s position in XPS, this is not foolproof, as there may be differential charging.

A more attractive method, if there are both photoelectron and the Auger peaks in an XPS spectrum, is to make use of the chemical shift on both the Auger and photoelectron peak, and to record the separation of the two lines. An example is given in Fig. 19 for the XPS of PTFE, where both XPS and Auger peaks are observable for F and C atoms. This quantity, the energy difference between the XPS and Auger peaks, is known as the Auger parameter, α, (Wagner and Joshi, 1988) and is numerically defined as the sum of the peaks

$$\alpha = E_B + E_K \qquad (9)$$

where E_B is the BE of the photoelectron emission peak and E_K is the KE of the Auger transition. The measured value will thus be independent of any electrostatic charging of the specimen. The reason this provides chemical state information is that whereas the photoemission process of XPS is a one-electron process, leaving a single core hole, the Auger process is a three-electron process, starting with one hole and leaving three holes. This leads, in general, to much larger chemical shifts on Auger peaks than on photoelectron peaks. The elements that yield useful Auger parameters in conventional (Al Kα) XPS include F, Na, Cu, Zn, As, Ag, Cd, In, and Te.

C. Chemical State Information: Fine Structure Associated with Core Level Peaks

Within the high-resolution spectra of individual core levels there may exist fine structure (additional peaks or broadenings) that gives the electron spectroscopist additional information concerning the chemical environment

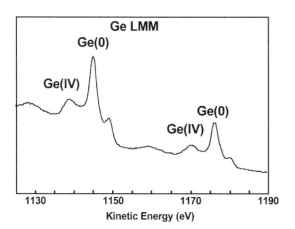

Figure 18 Auger chemical state information for a germanium single crystal with a thin layer of oxide.

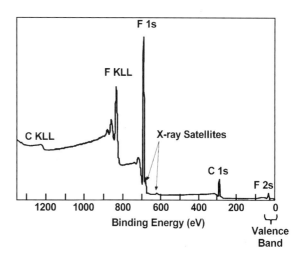

Figure 19 Survey XPS spectrum of PTFE showing KLL Auger peaks of fluorine and carbon.

of an atom. The major features in this category are "shake-up" satellites, multiplet splitting, and plasmon losses (Fadley, 1978).

Shake-up satellites occur when there is a probability for the outgoing photoelectron to simultaneously interact with a valence electron and excite it (shakes it up) to a higher energy level. If this happens, the KE of the ejected core electron is reduced slightly giving a satellite structure a few electron volts below (but above on the BE scale) the core level position without the shake-up process. In any ensemble of identical atoms, some may undergo this process, but most will not (it is an example of a deviation from the one-electron approximation). Thus, such shake-up features are generally weak or even nonobservable for many elements. They do appear quite strongly in some of the important industrial elements, however, and can be extremely useful in identifying bonding situations (Brundle et al., 1976); for example, in the 2p spectra of the d-band metals and the bonding to antibonding transition of the π molecular orbital ($\pi \to \pi^*$ transition) brought about by C 1s electrons in aromatic organics. The former is illustrated by the spectrum of NiO compared with Ni metal, Fig. 20. A strong shake-up satellite is observed for Ni^{2+}, but not for Ni^0 in the Ni 2p spectrum. For elements like Ni, Cu, Fe, Co, it is easily possible for the expert eye to assign an oxidation state by simply recognizing the pattern of the whole 2p region (fingerprinting). Again, this avoids issues of determining absolute BEs accurately and relying entirely on chemical shifts.

Multiplet splitting of a photoelectron peak may occur in a compound that has unpaired electrons in the valence band (Fadley, 1978). It arises from the different possible spin distributions of the electrons of the band structure. This results in a doublet of the core level peak being considered. Usable multiplet splitting effects are observed for Mn, Cr (3s levels), Co, Ni ($2p_{3/2}$ levels), and the 4s levels of the rare earths. The $2p_{3/2}$ spectrum of nickel shows multiplet splitting for NiO, as shown in Fig. 20, but not for $Ni(OH)_2$—a feature that has proved very useful in the examination of passive films on nickel.

Plasmon losses occur in both Auger and XPS spectra and are specific to materials which have a free-electron-like band structure. They arise when the outgoing electron excites collective oscillations in the conduction band electrons and thus suffer a discrete energy loss. The characteristic plasmon loss peaks for clean aluminum metal are shown in Fig. 21. It shows several losses in multiples of the characteristic bulk plasmon energy, about 15 eV. In addition, it also shows a weaker surface plasmon loss at about 11 eV. If the top layer of Al atoms were covered with oxygen, the surface plasmon would disappear, but some of the bulk plasmon intensity would remain until the Al was oxidized beyond the probing depth. Thus, plasmon features can provide useful analytical information. For instance, in both XPS and Auger spectra, the KLL Auger of Al and Si in their elemental form show strong plasmon losses, which are absent for Al_2O_3 and SiO_2, allowing a simple distinction of the element from the oxide. Of course, the two forms can easily be distinguished also by chemical shifts, provided there is no charging problem for the insulating oxide. This is likely the case for XPS, but, depending on the exact circumstances, may not be so in Auger. So, like shake-up structure, plasmon losses provide "charging resistant" information.

D. Quantitative Analysis

In order to quantify spectra from either XPS or AES, one must convert peak intensities (peak areas, or peak-to-peak heights in Auger spectra displayed as differentials) to atomic concentrations. In this section, quantification will be primarily concerned with homogeneous material; the situation is more complicated for specimens which have films at their surface that are either thinner than the information depth of the technique, or discontinuous.

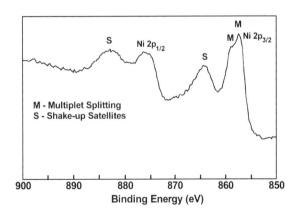

Figure 20 Multiplet splitting for the Ni 2p spectrum of NiO.

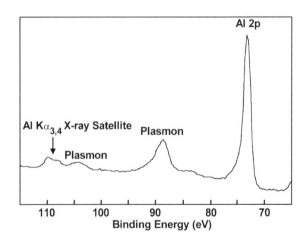

Figure 21 Plasmon loss features from clean aluminum.

There are many factors which must be considered when attempting to quantify electron spectra. These are either sample-related factors or spectrometer-related factors. Sample-related factors include:

1. The cross-section for emission, which is the probability of the emission of an electron owing to the effect of the incoming radiation (X-ray photon in XPS or electron in AES). The cross-section depends upon a number of factors such as the element under investigation, the orbital from which the electron is ejected, and the energy of the exciting radiation.
2. The escape depth of the electron emitted from the atom, which depends upon its KE (the escape depth passes through a minimum with increasing KE. The minimum occurs in the region of 20–50 eV) and the nature of the specimen.
3. The angle between the incoming X-ray beam and the emitted photoelectron. This affects the sensitivity to electrons in different orbitals in different ways via the angular asymmetry factor. It can be shown that if the angle is 55.7° (the so-called "magic angle"), this factor becomes unity and does not have to be taken into account.

Spectrometer-related factors include:

1. The transmission function of the spectrometer (the proportion of the electrons transmitted through the spectrometer as a function of their KE).
2. The efficiency of the detector.
3. Stray magnetic fields which affect the transmission of low-energy electrons to a greater extent than high-energy electrons and so must be taken into account in the quantification.

A quantitative evaluation of surface composition may be made on the basis of first principles, or an empirical relationship, together with cross-sections or sensitivity factors, which may be published or determined in-house. In practice, only the latter is used, as it factors out difficult to determine instrument factors. In the following we give only the simplest approach. There are many published works on quantification procedures (e.g. Seah, 1980) and commercial software can be obtained. In addition, commercial instruments usually come with software incorporating their own approaches to quantification.

In XPS, the intensity (I_A) of a photoelectron peak from element A in a solid is given in a simplified form, by:

$$I_A = K(E) J \sigma(h\nu) \int_{z=0}^{\infty} N_A(z) \exp\left(\frac{-z}{\lambda(E)\cos(\theta)}\right) dz \quad (10)$$

where $K(E)$ is a term which covers all of instrumental and geometrical factors described earlier and is a function of electron KE, J is the X-ray photon flux falling in the area of the sample that can be viewed by the analyzer, $N_A(z)$ is the distribution of atoms A with depth (it is assumed here that the distribution is uniform in x and y), $\sigma(h\nu)$ is the cross-section for photoelectron production (which depends on the element, the orbital from which the photoelectron is ejected, and the energy of the X-ray photons), $\lambda(E)$ is the electron attenuation length which is dependent upon the energy of the photoelectron and the material from which it is ejected, z is the distance into the solid in a direction normal to its surface, and θ is the angle between the sample normal and the direction in which the photoelectron is emitted from the material.

The intensity referred to will usually be taken as the integrated area under the peak following the subtraction of a linear or S-shaped background. Background subtraction is a research area in itself! There are many ways to do it, and some are more theoretically acceptable than others. Usually, because one is looking for changes from one situation to another, it is more important to be completely consistent in how it is done, rather than theoretically correct.

Equation (10) can be used for direct quantification (first principles approach) but, more usually, experimentally determined sensitivity factors (F) are employed. The parameter F includes the terms σ, K, and λ, in the standard equation, as well as additional features of the photoelectron spectrum such as characteristic loss features. Once a set of peak areas has been calculated for the elements detected, I in Eq. (10) can be determined. The terms σ, K, and λ are incorporated into a set of sensitivity factors appropriate for the spectrometer used, or explicitly incorporated into the algorithm used for quantification. We can determine the atomic percentage of the elements concerned, by dividing the peak area by the sensitivity factor and expressing as a fraction of the summation of all normalized intensities:

$$[A] \text{ atomic\%} = \frac{I_A/F_A}{\sum I/F} \times 100\% \quad (11)$$

The calculation of surface composition by this method assumes that the specimen is homogeneous within the volume sampled by XPS. This is rarely the case but, even so, the previously mentioned method provides a valuable means of comparing similar specimens. For a more rigorous analysis, angular dependent XPS may be employed to determine whether the material is homogeneous in depth, and to elucidate the hierarchy of layers if it is not. Lateral homogeneity is dealt with by the small spot approach, which is obviously much superior in Auger than in XPS. If there is lateral inhomogeneity on a scale smaller than the lateral resolution of the instrument, it cannot be dealt with by XPS.

The quantitative interpretation of Auger spectra is not as straightforward as in XPS. The first problem encountered is the form of the spectrum. In the differential mode the intensity measurement is the peak-to-peak height. This is a measure of the gradient of the leading or trailing edge of a peak and is only proportional to the area of the peak, provided all the peaks have the same shape (and do not change with chemistry!). For low-resolution spectrometers, this is approximately achieved by the instrumental broadening of all peaks. For high-resolution studies, the peak shape changes and the fine structure, which becomes apparent in the spectrum, reduces apparent peak-to-peak height and all relationship to area is lost. It is for this reason that the integrated peak area of a direct-energy spectrum is preferable for quantitative AES. The relative peak areas in a spectrum will also depend on the primary beam energy used for the analysis. Owing to these issues, it is necessary to fabricate binary or ternary alloys and compounds of the type under investigation to provide calibration by means of a similar Auger spectrum; however, the sensitivity factors produced have a narrow range of applicability. In this manner, it is possible to determine the concentration of an element of interest (N_A) as follows:

$$N_A = \frac{I_A}{I_A + F_{AB}I_B + F_{AC}I_C + \cdots} \quad (12)$$

where I is the measured intensity of the element represented by the subscript, and F is the sensitivity factor determined from the binary standard such that:

$$F_{AB} = \frac{I_A/N_A}{I_B/N_B} \quad (13)$$

Various semiquantitative methods are employed by laboratories throughout the world which relate a measured Auger electron intensity to that of a standard material under the same experimental conditions, and this seems to be a fairly satisfactory approach where the time and expense of producing the relevant standard specimens are not warranted.

IV. COMPOSITIONAL DEPTH PROFILING

Although both XPS and AES are essentially methods of *surface* analysis, it is possible to use them to provide compositional information as a function of depth. This can be achieved in four ways:

1. ARXPS. By manipulating the Beer–Lambert equation either to increase or to decrease the integral depth of analysis by changing the geometry of the experiment, and hence the information depth. This is a nondestructive method, but it is restricted in the depth it can probe by the λ value involved. The approach works best for the top 4 nm or so.
2. By using more than one electron peak, of differing KEs, for the analysis and achieving as mentioned in 1 above.
3. By removing material from the surface of the specimen *in situ* by ion sputtering (Hoffman, 1983; Stevie, 1992). Analysis is then alternated with material removal and a compositional depth profile gradually built up. This method is typically used over depths of tens to hundreds of nanometers.
4. By removing material mechanically and examining the freshly exposed surface (Lea and Seah, 1981). Common methods for doing this are angle lapping and ball cratering. This is more appropriate to get information over micrometers.

There are also chemical (wet) methods for removing layers, but these are very specific to the material involved and will not be discussed here.

A. Angle Resolved X-Ray Photoelectron Spectroscopy

Angle resolved measurements in electron spectroscopy are used almost exclusively in ARXPS. From the Beer–Lambert equation, it is clear that the depth of analysis is dependent on the angle of electron emission, θ. By recording spectra with good angular resolution at a high value of θ, say 75° (relative to the specimen normal), an analysis is recorded which is extremely surface sensitive. As normal electron emission is approached ($\theta = 0°$) the analysis depth moves towards the limiting value of $\sim 3\lambda$ (containing 95% of the total signal). A thin uniform overlayer will give a characteristic angular distribution predicted by the Beer–Lambert expression, as shown in Fig. 22 An island-like distribution will show a weaker angular dependence.

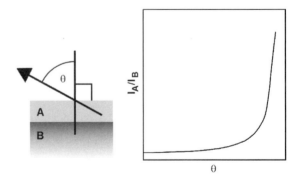

Figure 22 The ratio of signal intensities from the overlay and substrate as a function of emission angle.

This manner of depth profiling is invaluable for compositional changes that occur very close to the surface (first few nanometers) and has been employed for studies of thin passive films on metals (Brundle and Madix, 1979) and surface segregation in polymers (Perruchot et al., 2003). In conventional ARXPS, the angular acceptance range of the spectrometer is reduced by the user to provide the required angular resolution. Clearly, there must be a compromise between angular resolution and sensitivity (acquisition time). Spectra are then collected at each number of take-off angles by tilting the specimen. Figure 23 presents an ARXPS data set for a specimen of Ga As with a thin oxide layer at its surface.

It is clear from the montage of As 3d spectra that the oxide peak is dominant at the surface and falls off as the probing depth increases. This phenomenon is repeated in the gallium part of the spectrum (not shown). Thus, we can conclude straight away that the bulk Ga As is covered by a layer of material in which the Ga and the As are oxidized. The data was collected serially, by tilting and measuring at each angle. An alternative method would have been to use parallel detection over the angular range, such as is possible in the Theta Probe from Thermo Electron Corporation, eliminating the need for tilting the sample.

The angular data provide a useful guide to the relevant abundance of each element in the near-surface layers, but there is often a need to provide the thickness of an individual layer and to attempt to extract a depth profile. This can be approached using the Beer–Lambert equation and an ARXPS data set such as that of Fig. 23, which can be processed to provide such information. Consider a thin layer of thickness d, of material A on a substrate B. To obtain an expression for the signal from A, the Beer–Lambert equation must be integrated between 0 and d and becomes:

$$I_A = I_A^\infty \left[1 - \exp\left(-\frac{d}{\lambda_{A,A} \cos \theta}\right)\right] \quad (14)$$

The signal from B arriving at the B–A interface is I_B^∞, assuming layer B is thick. This signal is then attenuated by passing through layer A. The signal emerging is therefore given by:

$$I_B = I_B^\infty \exp\left(-\frac{d}{\lambda_{B,A} \cos \theta}\right) \quad (15)$$

Note that the term $\lambda_{B,A}$ is the attenuation length in layer A for electrons emitted from layer B.

Taking the ratio of these signals:

$$\frac{I_A}{I_B} = R = R^\infty \frac{[1 - \exp(-d/\lambda_{A,A} \cos \theta)]}{\exp(-d/\lambda_{B,A} \cos \theta)} \quad (16)$$

where $R^\infty = I_A^\infty / I_B^\infty$

$$R = R^\infty \left[\exp\left(\frac{d}{\lambda_{B,A} \cos \theta}\right) - \exp\left(\frac{d}{\cos \theta}\left[\frac{1}{\lambda_{B,A}} - \frac{1}{\lambda_{A,A}}\right]\right)\right] \quad (17)$$

If $\lambda_{A,A} = \lambda_{B,A} = \lambda_A$, which will be approximately true if measurements are being taken from a thin layer of oxide on its own metal, then

$$R = R^\infty \left[\exp\left(\frac{d}{\lambda_A \cos \theta}\right) - 1\right] \quad (18)$$

Rearranging and taking the natural logarithm

$$\ln\left[1 + \frac{R}{R^\infty}\right] = \frac{d}{\lambda_A \cos \theta} \quad (19)$$

Plotting the left-hand side of this equation against $1/\cos \theta$ will then produce a straight line whose gradient is equal to d/λ_A.

The simple form of this equation can be used only if $\lambda_A \approx \lambda_B$. This is true if the electrons detected from layers A and B have approximately the same energy (e.g., both are emitted from Si 2p). If this is not the case, then the more rigorous Eq. (16) must be used. The value for R^∞ is the ratio of the intensities of the appropriate peaks from bulk specimens of the materials. In this context, bulk means well beyond the probing depth. The values for the individual intensities will depend upon X-ray flux density, sensitivity factors, atom densities, and so on. For a thin layer of SiO_2 on Si, for instance, most of these factors cancel (assuming the Si 2p peaks are used for both materials) except for the atom densities and the attenuation lengths in the two materials. This

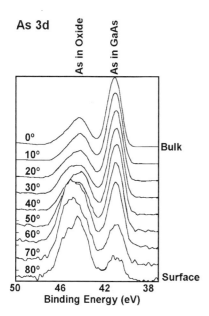

Figure 23 ARXPS data acquired by tilting the specimen. In this case, the specimen is gallium arsenide with a thin layer of oxide at the surface.

means that

$$R^\infty = \frac{\sigma_{Si,SiO_2} \lambda_{Si,SiO_2}}{\sigma_{Si,Si} \lambda_{Si,Si}} \quad (20)$$

where $\sigma_{x,y}$ is the atom number density (atoms per unit volume) of the element x in the material y. Note that R^∞ contains the term $\lambda_{Si,Si}$. The ratio of the atom number densities is given by:

$$\frac{\sigma_{Si,SiO_2}}{\sigma_{Si,Si}} = \frac{D_{SiO_2} F_{Si}}{D_{Si} F_{SiO_2}} \quad (21)$$

where D_x is the density (mass per unit volume) of material x and F_x is the formula weight of x. When there is more than one atom of an element represented by the formula, then the formula weight should be multiplied by the number of atoms present. For example, if the number density of oxygen atoms in silicon dioxide is required then F_{SiO_2} should be multiplied by two.

Figure 24a shows a set of data, plotted as in Eq. (19) from a number of samples where the oxide thickness was also measured using ellipsometry. Using a value for λ of 3.4 nm, the thickness of the oxide layers can be calculated from the ARXPS data. This is shown in Fig. 24b. Note that in this figure the line does not pass through the origin; this is because a thin layer of contamination is present on the sample surface. This layer cannot be distinguished from the oxide layer using ellipsometry and its thickness is included with that of the oxide. Using ARXPS, however, that layer is not included in the measured thickness.

There are, of course, limits in terms of the thickness of an overlayer that can be resolved using ARXPS analysis method. At the thick end of the depth scale, the signal generated by the substrate becomes too weak to use. In general, the thickest layer, which can be analyzed using this method, is about 3λ which, for silicon dioxide, is ~9 nm. One should also be careful of extending data to too high emission angles. Above 60–70° elastic scattering of electrons causes a significant deviation from linearity for plots as shown in Fig. 24 even when the layer is uniform. Experiments and theory show the effect is negligible at lower angles (Powell and Jablonski, 2002).

Polymers present some special problems in surface analysis in that they are not generally amenable to analysis by AES or ion beam compositional depth profiling because of specimen charging and degradation problems. Consequently, ARXPS is one of the few ways of probing near-surface compositional gradients. There are many approaches that can be taken to model such profiles but work from the National Physical Laboratory in the UK has produced validated software in the form of a spreadsheet which incorporates several ways of handling data; the choice depends on the type of information that is desired from the ARXPS data set. Named ARCtick, the

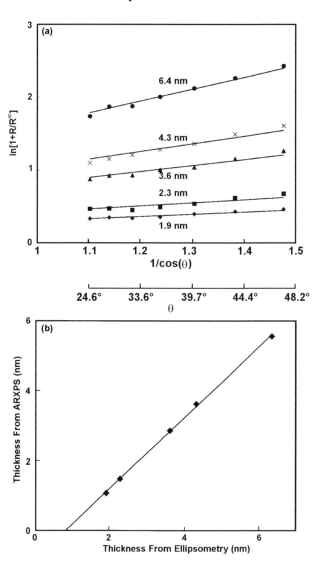

Figure 24 (a) Plot of $\ln[1 + R/R^\infty]$ vs. $1/\cos\theta$ for a series of silicon oxide on silicon samples. The gradient of these plots provide a value for the thickness of the layer after multiplication by λ, the attenuation length. (b) Comparison of the ARXPS data with ellipsometry for the same samples. It should be noted that the linearity of this plot is excellent but there is an intercept because ellipsometry measurements include the thin layer of carbonaceous contamination at the surface.

routine is fully described at http://www.npl.co.uk/npl/cmmt/sis/arctick.html.

No unique transformation from angle dependent intensities to depth dependent concentration exists. This implies that a least squares fit of trial profiles to experimental data is not sufficient to determine accurate concentration profiles. The concept of maximum entropy has therefore been introduced to produce a smooth profile, avoiding the "over fitting" that a method based on least squares fitting would produce (Livesey and Smith, 1994; Chang et al., 2000; Opila, 2002).

In outline, the procedure is as follows. A random depth profile is generated and the ARXPS intensities expected from such a profile are calculated. The calculated ARXPS data are compared with the experimental data and the error is calculated:

$$\chi^2 = \sum \frac{(I_k^{calc} - I_k^{obs})^2}{\sigma_k^2} \quad (22)$$

where σ is the standard deviation. The entropy term is then calculated from the trial profile:

$$S = \sum_j \sum_i c_{j,i} - c_{j,i}^0 - c_{j,i} \log\left(\frac{c_{j,i}}{c_{j,i}^0}\right) \quad (23)$$

The quantity $c_{j,i}$ is the concentration of element i in layer j. The maximum entropy solution is derived by minimizing χ^2 while maximizing the entropy. This can be achieved by maximizing the joint probability function, Q:

$$Q = \alpha S - 0.5\chi^2 \quad (24)$$

where α is a regularizing constant, providing a suitable balance between the least squares term and the entropy term. Concentration space is then searched to find the optimum value of Q. This method has been applied to the reconstruction of depth profiles from layers whose thickness is less than about twice the attenuation length of electrons within the layer. An example of the use of this method is shown in Fig. 25 for a layer of silicon oxynitride on silicon (acquired using parallel detection over a 60° range in a Theta Probe without tilting the sample). In this example the distribution of oxygen and nitrogen within the ultrathin layer is clearly seen. Using the maximum entropy approach reliably requires expertise and knowledge; however, as both the α value and the weighting factors for the χ^2 of each element involved are operator chosen. Over emphasizing α will result in an over-smoothed profile. Over emphasizing χ^2 will result in unrealistic spikes. In the case shown it is known that there is a sharp interface where Si concentration goes to 100% at the substrate end of the film of interest, and there is no extra contaminant layer on the surface. If there were not these constraints, it would be harder to get a reliable fit. Small variations in the parameters chosen might make large differences. Because of these issues the modeling of ARXPS data is most appropriate when looking for changes in well-characterized layer structures for which maximum entropy "recipes" have been developed and least appropriate when the depth structure of the sample is completely unknown in advance.

B. Variation of Electron Kinetic Energy

An alternative way of obtaining in-depth information in a nondestructive manner is by examining electrons from different energy levels of the same atom. The IMFP varies with KE, and by selecting a pair of electron transitions that are both accessible in XPS but have widely separated energies, it is possible to obtain a degree of depth selectivity. The Ge 3d spectrum (KE = 1450 eV, $\lambda \sim 2.8$ nm) of Fig. 26(a) shows Ge^0 and Ge^{4+}

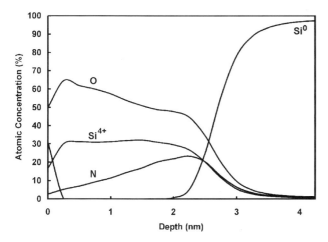

Figure 25 Depth profile through an ultrathin layer of silicon oxynitride reconstructed from ARXPS data using the maximum entropy method.

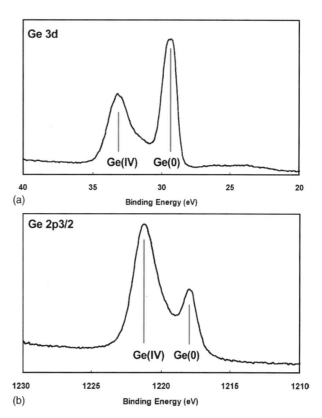

Figure 26 Ge 3d and Ge $2p_{3/2}$ spectra showing variation of sampling depth with electron KE.

components with the oxide component being about 80% of the elemental Ge. If we also record the spectrum of the Ge $2p_{3/2}$ region (KE = 260 eV, $\lambda \sim 0.8$ nm) we see the elemental component is much smaller compared with the Ge^{4+} peak, Fig. 26(b), thus confirming the presence of the oxide layer as a surface phase.

It is possible to obtain a similar effect by using the same electron energy level but exciting the photoelectrons with a series of different X-ray energies. For instance, C 1s electrons with a BE of 285 eV, which corresponds to a KE of 969 eV in Mg Kα radiation, 1202 eV in Al Kα, and 2700 eV in Ag Lα radiation. These KEs yield analysis depths of approximately 6, 7, and 10 nm for a typical polymer. Although high-energy X-ray sources for XPS are still rare, the conventional Al/Mg twin anode fitted to most spectrometers does provide a modest depth profiling capability which is often sufficient to distinguish between a surface layer and an island-like distribution.

C. Ion Sputter Depth Profiling

Although the nondestructive methods described earlier are extremely useful for assessing compositional changes in the outer 1–10 nm of material, it is necessary to remove material to go deeper. For depths down to hundreds of nanometers this is usually done by ion bombardment, using the inert gases Ar or Xe (usually Ar because of cost), within the spectrometer. This is, of course, not a method limited to XPS or AES and is, perhaps, best known in the SIMS approach to depth profiling.

The literature available on the subject of ion beam–solid interactions is enormous. All that can be achieved here is to make the reader aware of general principles and the possible causes of profile distortion (Stevie, 1992). The primary process is that of sputtering surface atoms to expose underlying atomic layers. At the same time, some of the primary ions are implanted into the substrate and will appear in subsequent spectra. Atomic (cascade) mixing results from the interaction of the primary ion beam with the specimen and leads to a degradation of depth resolution. Enhanced diffusion and segregation may also occur and will have the same effect. The sputtering process itself is not straightforward; there may be preferential sputtering of a particular type of atom. Ion-induced reactions may occur; for instance, copper (II) is reduced to copper (I) after exposure to a low-energy low-dose ion beam. As more and more material is removed the base of the etch crater increases in roughness and eventually, interface definition may become very poor indeed.

A high-quality vacuum is essential if a good depth profile is to be measured. If there are high partial pressures of reactive impurities present, the surface may not reflect the true material composition. This arises because the act of sputtering produces a highly reactive surface, which can getter residual gases from the vacuum. Oxygen, water, and carbonaceous materials are common contaminants. For this reason also, the gas feed to the ion gun must be free from impurities.

The current density profile of an ion beam is generally not uniform; many approximate to a Gaussian cross-section. Such a beam cross-section would produce a crater in the specimen which does not have a flat bottom. Poor depth resolution would result from extracting data from such a crater. To overcome this difficulty, the ion beam is usually scanned or rastered over an area which is large with respect to the diameter of the beam. Rastering produces a crater, which has a flat area at the center from which compositional data can be obtained.

When collecting the spectroscopic data from the crater it is important to limit the data acquisition to the appropriate area within the crater, avoiding the area close to the walls where the crater bottom is not flat. This is usually a simple matter in AES because the electron beam used for the analysis usually has a much smaller diameter than the ion beam used for etching the specimen. The electron beam can therefore be operated in point analysis mode in the center of the crater or rastered over a small area. For XPS, the methods of small area analysis are generally used, as described earlier. Figure 27 shows an example of a simple XPS depth profile through a tantalum oxide layer grown on tantalum metal.

There are many points to bear in mind when selecting the conditions for a depth profile (Stevie, 1992; Lea and Seah, 1981). Most of these are concerned with either the speed of acquisition or the depth resolution; the requirements for speed are usually opposite to those for

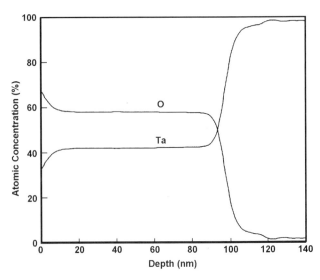

Figure 27 XPS depth profile through a layer of tantalum oxide on tantalum metal.

resolution. It should be possible, in principle, to increase the etch rate by using the maximum beam current available. However, in a normal ion gun the spot size of the beam increases with the increasing beam current and so the rastered area must be increased to ensure that the crater bottom remains flat. So increasing the beam current will not necessarily increase the etch rate. It is the ion beam current density, which is the important parameter. In the energy range normally employed in XPS and Auger profiling, the sputter yield increases with ion energy. Higher energies also mean smaller spot sizes at a given beam current and so will lead to better crater quality. The higher etch rate will, however, be accompanied by poorer depth resolution because the ions can penetrate deeper into the material causing atomic mixing.

Depth resolution in a depth profile is a measure of the broadening of an abrupt interface brought about by physical or instrumental effects. A commonly accepted method for measuring depth resolution is to measure the depth range, Δz, over which the measured concentration changes from 16% to 84% of its total change while profiling through an abrupt concentration change. Many of the factors, which affect the sputter rate, also determine the depth resolution available. Some of these factors, discussed later, relate to the characteristics of the sample, some to the instrument and some the physical process of sputtering.

The extent to which the ion beam characteristics affect depth resolution generally relate to the depth range of the ions after striking the specimen. This is because the passage of an energetic ion through a solid causes atomic mixing along the whole trajectory of the ion. Hence low ion energy, grazing incidence angles, and heavy ions lead to the best depth resolution because these minimize the depth range over which mixing can occur.

The crater must be as flat as possible over the analysis area. If this is not the case then information is being collected from a range of depths and the resolution suffers. Generally, the raster dimensions should be at least 5 ion beam diameters to get good flatness over a reasonable distance within the crater.

The sputtering process can cause the surface to become rough during the profiling experiment, degrading the resolution as a function of depth. This problem can be significantly reduced or eliminated by rotating the specimen beneath the ion beam (azimuthal rotation) during the sputtering cycles. Many commercial instruments now offer specimen stages, which incorporate azimuthal rotation. In a multicomponent sample the sputter yields from different elements can be different. Under these conditions there will be roughening, which may not be controllable by azimuthal rotation.

To analyze to greater depths than is practical using sputtering, it is necessary to resort to an *ex-situ*, mechanical process for removing material. Two related methods are briefly mentioned here: angle lapping and ball cratering.

When angle lapping is employed, the material is removed by polishing the specimen at a very shallow angle (<3°) and then introducing the specimen, with any buried interface now exposed, into the spectrometer. A brief ion etch to remove contamination is all that is needed prior to analysis. By carrying out Auger point analyses in a stepwise manner across the taper, the variation of concentration with depth is established and it is a matter of simple geometry to convert position of the analysis in the $x-y$ plane to the distance from the original surface, the z plane.

The problem of cutting the taper is overcome, to a large extent, by using an allied process known as ball cratering. In this process, mechanical sectioning of the specimen is carried out by a rotating steel ball of known diameter (usually about 30 mm) coated with fine (~1 μm) diamond paste which rotates against the specimen and produces a shallow, saucer-like crater. The ball can be removed from time to time to assess the progress of the lapping and, on replacement, automatically "self-centers" in the crater.

By recording Auger point analyses along the surface of the crater, a compositional depth profile can be determined. If there are buried interfaces of special interest, ion sputtering may be used at a point on the crater close to the interface to obtain better depth resolution. Ball cratering devices are available commercially. They consist of a horizontal shaft with a reduced diameter in the form of a "V". The specimen is mounted near the drive shaft and the hardened steel ball rests on it, driven by the horizontal shaft. Diamond paste is commonly used as an abrasive. Inspection of the crater, using the integral microscope, is carried out during the erosion process after removing the ball. The ball can be readily relocated in the crater if further material removal is required.

Ball cratering works well for metals and oxides but there are problems with both soft and certain brittle materials. Polymers are extremely difficult to handle but some success has been obtained by using a ball cratering machine equipped with a cryo-stage.

An ultra-low angle microtomy approach can be used on polymers, with depth resolutions of 10s of microns being achieved many 10s of microns below the original surface (Watts, 2004).

D. Summary of Depth Profiling

Sputter depth profiling is by far the most popular means of producing a compositional depth profile in surface analysis. Although the earlier discussions suggest that it is fraught with difficulties, it is fair to say that the majority of the problems can be circumvented or reduced to an acceptable level by careful experimental techniques. It is for this reason that, as a method for depth profiling, it is widely used in studies of metals, oxides, ceramics, and semiconductors. The analysis depth that is feasible varies with the sample and the system employed but the upper limit is of the order of a few microns, and more usually profiles are done this way only to hundreds of nanometers. The main reasons for this are the time for the experiment and the depth resolution degradation with depth. When the thickness is so great that the time for the experiment would be excessive, either angle lapping or ball cratering should be considered.

For very thin layers there may be problems associated with the attainment of the so-called sputter equilibrium leading to uncertainties in the sputter yield. Furthermore, the process of sputtering can lead to changes in the chemical composition of the material. Under these circumstances the use of ARXPS has many advantages, but it is only appropriate for layers up to about 4 nm usually.

V. AREAS OF APPLICATION AND OTHER METHODS

The main concern of this chapter has been the practice of electron spectroscopy, the instrumentation, how to use it, and how to interpret the data. It is not appropriate in this volume to discuss in any detail how it is applied in any particular discipline; so we restrict this short final section to some general comments.

XPS and AES are widely used in materials science (Holloway and Vaidynathan, 1993), polymer science (Beamson and Briggs, 1992; Watts, 2001), electronic devices and -chemistry, biological science, catalysis, electrochemistry; in fact wherever surface, near surface, or thin film elemental and/or chemical composition are needed and the sample can withstand a vacuum. They are used in industry in the earlier mentioned areas during development programs, for trouble shooting, quality assurance, and most recently for metrology. If there is a choice to be made between using one or the other technique, that choice is usually simple. If one does not need spatial resolution beyond the capability of XPS and one is not limited by the time available for analysis, XPS is the superior technique. It has better quantitative analysis capability, more developed chemical state identification capability, a nondestructive approach to ultrathin film depth profiling, reduced difficulty in dealing with insulating materials, and reduced probability of causing sample damage. However, that said, many, many applications require far better lateral resolution than XPS can achieve, but are trivial for AES or SAM to achieve. For this reason alone AES is used just as much in industry (particularly in the semiconductor and disk drive industries, where, when lateral resolution is needed it is nearly always beyond the reach of XPS).

XPS and AES are not the only choices for surface analysis. They are two of the major, generally applicable, ones, but there are others, such as SIMS, which are equally important. SIMS's strength is its large dynamic range, which allows it to deal with trace concentrations down to ppm, or even ppb, whereas XPS or AES cannot handle below about 0.1–1%, depending on circumstances. As depth profiling trace dopants is critical in semiconductors, SIMS has long dominated that area. Ion scattering is another major technique. In its common form, Rutherford backscattering, where high-energy ions (1–3 MeV typically) are used, is a standardless quantitative method for atomic composition (no chemistry), but only for films nowadays regarded as quite thick (tens of nanometers to a micron), because of its poor depth resolution. It also is incapable, in its usual form, of distinguishing between some important elements (like Ni and Fe). In the medium energy ion scattering form, both these drawbacks are overcome, but this type of instrumentation is rarely available.

REFERENCES

Auger, P. (1925). *J. Phys. Radium.* 6:205.
Baer, D. R. (1984). *Appl. Surf. Sci.* 19:382.
Bagus, P. S., Illas, F., Pachioni, G., Parmagiani, F. (1999). *J. Electron Spectrosc.* 100:215.
Barrie, A. (1977). In: Briggs, D., ed. *Handbook of Ultraviolet and X-Ray Photoelectron Spectroscopy.* London, UK: Heyden & Sons, Ltd.
Beamson, G., Briggs, D. (1992). *High Resolution XPS of Organic Polymers.* Chichester, UK: John Wiley & Sons, Ltd.
Beamson, G., Haines, S., Moslemzadeh, N., Tsakiropoulos, P., Weightman, P., Watts, J. F. (2004). *Surf. Interface Anal.* 36:275.
Briggs, D., Seah, M. P. (1990). *Practical Surface Analysis by Auger and X-Ray Photoelectron Spectroscopy.* Chichester, UK: John Wiley & Sons, Ltd.
Brooker, A. D., Castle, A. D. (1086). *Surf. Interface Anal.* 8:113.
Brundle, C. R., Baker, A. D. (1977). *Electron Spectroscopy: Theory, Techniques and Applications.* Vol. 1. New York: Academic Press (See also Vols. 2–5, 1978–1982).
Brundle, C. R., Chuang, T. J. (1977). *Surf. Sci.* 68:459.
Brundle, C. R., Madix, R. J. (1979). *J. Vac. Sci. Technol.* 16:474.
Brundle, C. R. et al. (1974). *J. Electron Spectroc.* 3:241.
Brundle, C. R., Chuang, T. S., Rice, D. W. (1976). *Surf. Sci.* 59:413.
Brundle, C. R., Evans, C. A., Wilson, S. (1992). *Encyclopedia of Materials Characterization.* Stoneham, MA, USA: Butterworth-Heinemann.
Carlson, T. A. (1976). *Photoelectron and Auger Spectroscopy.* New York: Plenum.
Castle, J. E. (1986). *Surf. Interface Anal.* 9:345.

Chang, J. P., Green, M. L., Donnelly, V. M., Opila, R. L., Eng, J., Sapjeta, J., Silverman, P. J., Weir, B., Lu, H. C., Gustafsson, T., Garfunkel, E. (2000). *J. Appl. Phys.* 87:4449.

Childs, K. D., Carlson, B. D., La Vanier, L. A., Moulder, J. F., Paul, D. F., Stickle, W. F., Watson, D. G. (1995). *Handbook of Auger Electron Spectroscopy*. Eden Prairie, USA: Physical Electronics, Inc.

Diplas, S., Watts, J. F., Morton, S. A., Beamson, G., Tsakiropoulos, P., Clark, D. J., Castle, J. E. (2001). *J. Electron. Spec.* 113:153.

Fadley, C. S. (1978). In: Brundle, C. R., Baker, A. D., eds. *Electron Spectroscopy, Theory, Techniques, and Applications*. Vol 2.

Fadley, C. S., Shirley, D. A., Freeman, A. J., Bagus, P. S., Mallow, J. V. (1969). *Phys. Rev. Lett.* 23:1397.

Heckingbottom, R. (1986). *Surf. Interface Anal.* 9:265.

Hofmann, S. (1983). In: Briggs, D., Seah, M. P., eds. *Practical Surface Analysis by Auger and X-Ray Photoelectron Spectroscopy*. Chichester, UK: John Wiley & Sons, Ltd.

Holloway, P. H., Vaidyanathan, P. N. (1993). *Characterization of Metals and Alloys*. Stoneham, MA, USA: Butterworth-Heinemann.

Lea, C., Seah, M. P. (1981). *Thin Solid Films* 81:279.

Livesey, A. K., Smith, G. C. (1994). *J. Electron Spectrosc.* 67:439.

Moulder, J., Stickle, W. F., Sobol, P. E., Bomben, K. D. (1992). *Handbook of Photoelectron Spectroscopy*. Eden Prairie, USA: Perkin-Elmer Corp.

Opila, R. L., Eng, J. (2002). *Prog. Surf. Sci.* 69:125.

Paynter, R. W., Ratner, B. D. (1985). In: Andrade, J., ed. *Surface and Interfacial Aspects of Biomedical Polymers*. New York, USA: Plenum Press.

Perruchot, P., Watts, J. F., Lowe, C., Beamson, G. (2003). *Int. J. Adhes.* 23:101.

Powell, C. J., Czanderna, A., eds. (1998). *Methods of Surface Characterization*. Vol 5.

Powell, C. J., Jablonski, A. (2002). *Surf. Interface Anal.* 33:211.

Seah, M. P. (1980). *Surf. Interface Anal.* 2:222.

Seah, M. P., Dench, W. A. (1979). *Surf. Interface Anal.* 1:2.

Seah, M. P., Anthony, M. T., Dench, W. A. (1983). *J. Phys. E* 16:848.

Siegbahn, K. (1967). *ESCA: Atomic, Molecular, and Solid State Structure Studied by Means of Electron Spectroscopy*. Uppsala, Sweden: Almqvist & Wiksells.

Stevie, F. A. (1992). In: Brundle, C. R., Evans, C. A., Wilson, S., eds. *Encyclopedia of Materials Characterization*. Stoneham, MA, USA: Butterworth-Heinemann.

Wachs, I. E. (1992). *Characterization of Catalytic Materials*. Stoneham, MA, USA: Butterworth-Heinemann.

Wagner, C. D., Joshi, A. (1988). *J. Electron Spectrosc.* 47:283.

Wagner, C. D. et al. (1981). *Surf. Interface Anal.* 3:211.

Watts, J. F. (2001). *J. Electron Spectrosc.* 121:233.

Watts, J. F. (2004). *Surf. Interface Anal.* 37.

Watts, J. F., Wolstenholme, J. (2003). *An Introduction to Surface Analysis by XPS and AES*. Chichester, UK: John Wiley & Sons, Ltd.

ARXPS of Polymers

14

Mass Spectrometry Instrumentation

LI-RONG YU, THOMAS P. CONRADS, TIMOTHY D. VEENSTRA
Laboratory of Proteomics and Analytical Technologies, SAIC-Frederick Inc., National Cancer Institute, Frederick, MD, USA

I. INTRODUCTION

The momentum produced by the sequencing of the human genome has rapidly spilled over into the study of gene transcripts, proteins, and small molecule metabolites. The current trend in biology is no longer to study individual components, but to study the cell as a whole. This trend has not only spawned several new technologies (i.e., cDNA and protein arrays), but also has brought about the realization of unrecognized utilities possessed by existing technologies (i.e., metabonomics analysis using nuclear magnetic resonance spectroscopy). One of the largest trends is the development of methods to analyze large numbers of proteins in single experiments. These so-called "proteomics" approaches provide a powerful means to examine global changes in protein levels and expression under changing environmental conditions.

Although mass spectrometry (MS) instrumentation has been available for decades, it recently has undergone a dramatic increase in popularity with the recognition that MS is an invaluable tool in biological analysis. This resurgence has been fueled by the interest in the development of proteomic technologies, and MS is the leading technology in this field today. There are several attributes that have led to its position as the premier analytical technology in this field. The sensitivity of MS instrumentation allows for the routine identification of molecules in the femtomole (fmol, 10^{-15} mol) and even in the high attomole (amol, 10^{-18} mol) range (Miyashita et al., 2001). The mass measurement accuracy afforded with current MS technology is routinely less than 50 parts-per-million (ppm), and can be less than 5 ppm, providing more confident protein identifications (Chaurand et al., 1999). Probably, most importantly, the partial sequence information provided by tandem MS (MS/MS) approaches enables the confident identification of mixtures of proteins based on the primary sequence information obtainable (Griffiths et al., 2001). In addition to all of these attributes, the throughput by which proteins can be identified by MS is unparalleled by any other biophysical technique.

MS is arguably the most important technology in proteomics today because of its ability to identify proteins. While proteins exist *in vivo* as biopolymers of amino acids ranging in size from a few hundred daltons (Da) up to complexes of greater than one million daltons, typically the first step in the analysis of a protein by MS is to digest it into peptides (Rappsilber and Mann, 2002). There are two main reasons for doing this digestion. The primary reason for this is that it is difficult to identify a protein based solely on its molecular weight. Though the identification of a purified protein can be confirmed on the basis of a measured molecular weight, trying to identify an intact protein *de novo* on the basis of this measurement is extremely difficult. A much more reliable method is to acquire a spectrum of the peptides resulting from the digestion of a protein (Fenyo, 2000). This so-called "peptide map" or "peptide fingerprint" is then compared to theoretical peptide maps derived from a protein or genomic database to identify the correct protein (Fig. 1). Two peptide maps generated by matrix-assisted laser desorption ionization time-of-flight mass spectrometry (MALDI-TOF-MS) analysis of isoforms of the protein 14-3-3 are shown in Fig. 2. As the homology of these two proteins is quite high, many of the peptides within each of the spectra match to peptides within both of the proteins; however, the isoforms can be differentiated on the basis of unique masses (in bold italics) map to either the β or ε isoform. Though peptide mapping works very well for a single isolated protein or a simple mixture of proteins, it will not work well for complex mixtures. In these instances, individual peptide ions can be automatically isolated and fragmented by (collision-induced) dissociation (CID) within the mass spectrometer in a process known as tandem MS or MS/MS (Martin et al., 1987). The masses of the fragment ions are measured to obtain partial or complete sequence information, as shown in Fig. 3. If the appropriate collisional energy is used, the resulting MS/MS spectrum will contain an ensemble of masses representing various lengths of the peptide. This experimental information is then compared to theoretical MS/MS spectra by various software programs to identify the peptide on the basis of partial sequence information (Eng et al., 1994).

II. IONIZATION METHODS

The mass spectrometer can be thought of as two distinct components: the ionization source and the mass analyzer. The ionization source is the region of the instrument in which the sample of interest is ionized, with a positive or negative charge, and then desorbed into the gas phase. The two most common methods to do this are MALDI (Karas et al., 1987) and electrospray ionization (ESI) (Dole et al., 1968). The development of ESI and MALDI has enabled significantly larger proteins (i.e., greater than several hundred thousand daltons) to be analyzed. Moreover, the interface of MS with separation methods, such as liquid chromatography (LC), became routine with the implementation of ESI. This combination of LC with MS made the characterization of complex mixtures, such as entire cell lysates, practical.

A. Matrix-Assisted Laser Desorption/Ionization

During the same time that ESI was being developed, significant advancements were being made in the laser desorption of biological molecules. Though laser desorption was effective at desorbing and ionizing small molecules and peptides, it was not until the proper organic matrices were identified that laser desorption could be used to analyze proteins with masses of >10,000 Da (Karas et al., 1987; Tanaka et al., 1988). MALDI generates high mass ions by using a pulsed laser beam to irradiate a solid mixture of an analyte dissolved in a suitable matrix compound (Ryzhov et al., 2000).

To prepare the sample for MALDI analysis, it is mixed with an equal volume of a saturated solution of matrix prepared in a solvent such as water, acetonitrile, acetone, or tetrahydrofuran. The matrix is a small, highly conjugated organic molecule that strongly absorbs energy in the

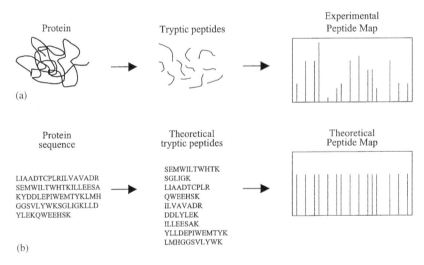

Figure 1 Identification of a protein via peptide mapping. (a) The protein of interest is enzymatically or chemically digested and the masses of the resulting peptides are measured using MS. (b) Each protein sequence in the database is digested *in silico* according to the specificity of the digestion scheme used. From these masses, a theoretical mass spectrum is constructed. The experimental mass spectrum is then compared with the theoretical mass spectra to identify the most probable protein from which the peptide fragments originated.

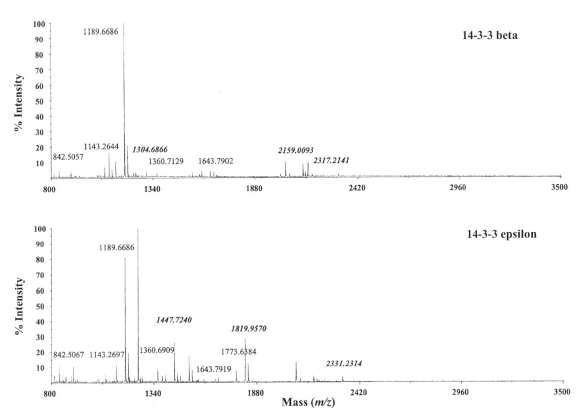

Figure 2 MALDI-TOF-MS and identification of a tryptic digest of two isoforms of the protein 14-3-3. The protein is identified by comparing the observed peptide masses with the theoretical mass spectra generated from a virtual digestion of the proteins within the appropriate protein database, as described in Fig. 1. Though 14-3-3 β and ε are highly homologous, the two isoforms can be unambiguously assigned by peptide masses that are unique to each protein (bold italics).

UV region. The most commonly used matrices for proteins and peptides include α-cyano-4-hydroxycinnamic acid (CHCA), 2,5-dihydroxybenzoic acid (DHB), and 3,5-dimethoxy-4-hydroxycinnamic acid (sinapinic acid). A few microliters of this mixture is deposited onto a MALDI target plate and dried, resulting in the integration of the peptides into a crystal lattice. The MALDI target plate is then inserted into the source region of the mass spectrometer and irradiated with a laser pulse, as shown in Fig. 4. A nitrogen (N_2) laser (337 nm) operating at 2–20 Hz is commonly used to irradiate the sample because of its size, cost, and ease of operation; however, other lasers operating at different wavelengths can be used. Higher repetition rate lasers operating at 200–1000 Hz are also becoming popular as a means to decrease the spectral acquisition time. The matrix molecules upon irradiation, because of the high matrix-to-analyte concentration ratio, absorb most of the photon energy provided by the laser. The energy is then transferred to the analytes (i.e., peptides) in the sample, which are subsequently ejected from the target surface into the gas phase. The MALDI source region of most spectrometers is not at atmospheric pressure; however, it is maintained at a higher pressure relative to that of the mass analyzer region enabling ions to be drawn into this region of the instrument. Recently, however, MALDI sources operating at atmospheric pressure have demonstrated relatively high sensitivity for peptide mass fingerprinting (Dainese et al., 1997). By operating at atmospheric pressure, MALDI can be interfaced to analyzers, such as ion traps and quadrupole TOF analyzers, which have historically been reserved for ESI applications. Unlike ESI, MALDI typically produces singly charged ions regardless of whether the analyte is a peptide or a large protein. This propensity to produce species with high mass-to-charge (m/z) values has made the coupling of MALDI with large m/z range mass analyzers, such as TOF spectrometers, a popular choice.

B. Electrospray Ionization

About the same time that effective methods to analyze large biomolecules by MALDI were being developed, the ability to characterize proteins and peptides by MS was being greatly enhanced by the development of ESI (Aleksandrov et al., 1984; Dole et al., 1968; Fenn et al., 1989). The mechanism describing how ESI works is

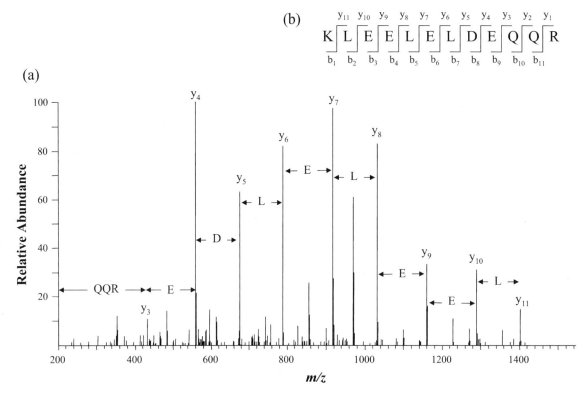

Figure 3 Protein identification using MS/MS. In MS/MS, a parent peptide ion of interest is selected by the mass analyzer and subjected to CID to produce fragment ions. After refocusing, the fragment ions are guided to the detector. (a) ESI-MS/MS mass spectrum of a peptide observed from a tryptic digest of mitogen activate protein kinase kinase (MAPKK). Primary sequence information is determined by comparing the mass differences between major peaks in the spectrum with the calculated molecular masses of the amino acid monomers within the peptide. (b) When fragmentation occurs across the peptide bond, a series of y and b ions (as numbered in the sequence and on the spectra) are formed.

shown in Fig. 5. ESI requires the sample to be in solution and flow into the ionization (or source) region of the spectrometer. As the sample flows toward the source region, a high voltage is applied to a stainless steel or other conductively coated needle that makes contact with the solution. This applied voltage results in a charge being applied to the sample prior to desolvation and its introduction to the mass analyzer. Upon exiting the spray tip, the solution produces submicrometer-sized droplets that contain ions produced from the sample of interest as well as the solvent. Prior to entering the mass analyzer region of the mass spectrometer the sample still needs to be desolvated. To desolvate the sample, the droplets pass through either a heated capillary or a curtain of nitrogen gas within the source region of the instrument, which is maintained at atmospheric pressure. As the mass analyzer region of the spectrometer is maintained at low pressure, the ions are drawn into the spectrometer where their masses are recorded.

One of the attributes of ESI that distinguishes it from other ionization methods is its ability to produce multiply charged ions from large biological molecules. For example, a singly protonated 25,000 Da protein has an m/z ratio of 25,001, making it only within the detectable range of higher m/z range instruments such as TOF-MS. A protein of this size, however, is capable of accepting 10–30 protons depending on the protein and the solution conditions. Therefore, there will be protein populations that contain between 10 and 30 protons having m/z values ranging from 2501 (25,010/10) to 834.3 (2503/30). All of the various charge states of the protein will be observed within the mass spectrum. For example, the ESI-MS spectrum of the calcium-binding protein calbindin shown in Fig. 6, is populated by signals representing populations of the protein with a variety of different charges attached. The molecular mass of this protein is 29,866 Da; however, the m/z ratio of the signal representing the protein with 10 charges, for example, is only 2987.6.

As mentioned earlier, MS most commonly measures proteins after they have been proteolytically digested, usually with trypsin. Peptides typically exist as singly, doubly, or triply charged ions, depending on their size and number of basic residues present. For a peptide of mass 2000 Da, its singly charged species (i.e., the $[M + H]^+$ ion) will have an m/z value of 2001, whereas

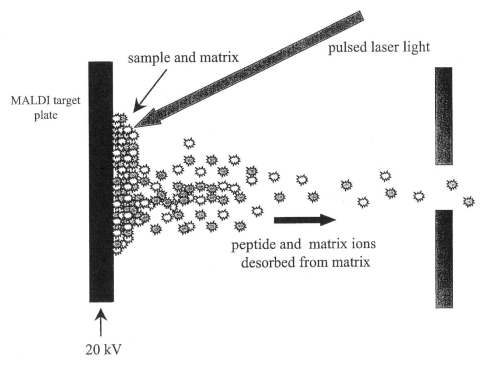

Figure 4 Principles of MALDI. The sample is cocrystallized with a large excess of matrix, and short pulses of laser light are focused on to the sample spot. The matrix absorbs the laser energy and dissipates it into the sample, causing part of the illuminated substrate to vaporize. The rapidly expanding plume of matrix and sample ions are then drawn into the mass analyzer via a pressure differential between the analyzer and the source region.

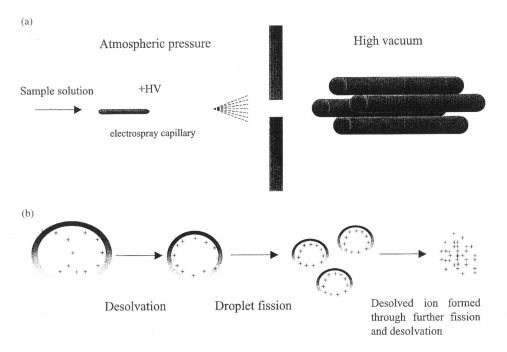

Figure 5 Principles of ESI. (a) The sample solution is passed through a stainless steel or other conductively coated needle, and a high positive potential is applied to the capillary (cathode) causing positive ions to drift toward the tip. (b) The droplets travel toward the mass spectrometer orifice during which they evaporate and eject charged analyte ions. The desolvated ions are drawn into the mass analyzer by the relative low pressure maintained behind the orifice.

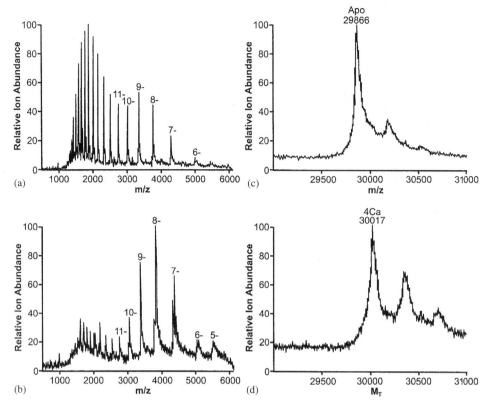

Figure 6 (a) Multiply charged ESI mass spectra of apocalbindin and (b) the same protein in the presence of calcium acetate. (c) Deconvoluted spectra of apocalbindin and (d) in the presence of calcium acetate.

its doubly ($[M + 2H]^{2+}$) and triply ($[M + 3H]^{3+}$) charged ions will have m/z values of 1001 and 667.7, respectively. It is common to observe both 1+ and 2+-charged species of peptides that have been produced via a tryptic digestion because of the basic sites on the N-terminal and the C-terminal lysine or arginine residue.

III. MASS ANALYZERS

A. Triple Quadrupole Mass Spectrometer

The quadrupole mass spectrometer has historically been the most commonly used combination of mass analyzer with ESI (Yost and Boyd, 1990). A quadrupole consists of four metal rods arranged in parallel to which direct current and RF voltages are applied. The two most common types of quadrupole mass spectrometers are single-stage and triple-stage quadrupoles. Though the use of single quadrupole mass spectrometers has been limited in proteomics, they have been readily adopted to the study of small molecules in pharmaceutical laboratory settings. The triple quadrupole is an extremely versatile instrument capable of product ion, precursor ion, and neutral loss scanning. These instruments have been used to identify proteins extracted from 2D-PAGE gels (Dainese et al., 1997), phosphopeptide characterization (Steen et al., 2001), and glycopeptide identification (Carr et al., 1993). The triple quadrupole instrument is composed of three quadrupole regions, Q1, Q2, and Q3. Quadrupoles Q1 and Q3 are used to guide and select ions through the analyzer region and onto the detector. Q1 and Q3 are separated by Q2, which is an RF-only quadrupole and acts as a collision cell to dissociate ions selected using Q1. Collisions using neutral gases, such as N_2 and argon, are used in Q2 to fragment the ions of interest. The mass measurement accuracy of triple quadrupole analyzers is at least 0.5 amu for the fragment ions produced by MS/MS (Shevchenko et al., 1997).

To perform CID of molecules using a triple quadrupole mass analyzer, the instrument is alternatively switched between two different scan modes. In the first mode, called a "full-scan" analysis, a broad m/z-range of ions generated from the source region is allowed to pass through Q1. The ions that pass through Q1 also pass freely through Q2 and Q3 onto the detector. Essentially, all of the ions produced in the source are measured. In the second scan mode, Q1 is used as a mass filter, by setting the RF voltage to allow a specific ion to pass through. The ion that passes through Q1 is then subjected to fragmentation within Q2 by filling this quadrupole with an inert gas. The resulting fragment ions then pass through Q3 onto the detector.

For proteomic applications, quadrupole mass spectrometers are most typically coupled with ESI sources; however, chemical ionization sources are also a popular choice in the analysis of nonpolar molecules. A major step in the use of these types of spectrometers was their coupling to LC separations. This coupling allows very complex mixtures to be fractionated and the components analyzed directly by the spectrometer as they elute from the chromatographic column. It is rare to see a quadrupole mass spectrometer that is not directly coupled to an LC system. Indeed, all of the mass analyzers that can utilize ESI sources can be directly coupled to LC fractionation.

B. Time-of-Flight Mass Spectrometer

An extremely popular choice in proteomic applications is the TOF-MS. The principal factors that make TOF-MS so popular are their high speed, sensitivity, and resolution. The method by which TOF spectrometers measure the m/z ratios is based on the time it takes for the ions generated in the source to strike the detector (i.e., its time of flight). Larger ions travel slower than smaller ions and thus take a longer time to reach the detector (Cotter, 1997). To move ions through the analyzer, a potential, V_s, is applied across the source to extract and accelerate the ions from the source into the field-free "drift" zone, or the tube, of the instrument.

The initial TOF analyzers operated in a linear mode, in which ions were continually extracted from the source region and sent through the flight tube to the detector (Chernushevich et al., 2001; Shevchenko et al., 2000). This mode does not provide the highest resolution as ions with the same m/z will have varying velocities because they have different initial energies and positions as they move from the source to the analyzer region. This deficiency has been solved by two developments. To correct these deficiencies, reflectron TOF (Cornish and Cotter, 1993), which focuses ions with the same m/z values and allows them to strike the detector, and time pulsed-laser ionization with delayed extraction, in which there is a slight delay between the ionization of the sample and the extraction of the ions into the flight tube, were developed (Brown and Lennon, 1995; Juhasz et al., 1996). Delayed extraction allows all the ions to get an equal start time, enabling ions of equal m/z to reach the detector simultaneously.

The primary use of TOF analyzers has been to generate peptide fingerprints to identify proteins. TOF analyzers have typically been suited with MALDI sources for this application because of the simplicity in both sample preparation and operation of the instrument. Though TOF instruments have generally not contained a true collision cell to provide MS/MS sequence data, instruments equipped with a reflectron, however, can measure fragmentation products using a process called "post-source decay" (PSD) (Kaufmann, 1995). In PSD, the reflectron voltage is adjusted so that fragment ions generated during the ionization and acceleration of the species of interest are focused and detected (Kaufmann et al., 1996). Though PSD analysis can be relatively slow, it does provide useful complementary information to substantiate the identification of a molecule such as a peptide. An exciting development in the field of TOF analyzers is the recent release of MALDI-TOF/TOF instruments that contain a true collision cell separated by two TOF tubes. Ions generated in the source region are accelerated through the first drift tube and can be dissociated through collisions with an inert gas in the collision cell. The resulting fragment ions are subsequently accelerated through a second TOF tube and detected. Proteins can thus be identified through high mass accuracy peptide fingerprinting as well as MS/MS.

C. Quadrupole Time-of-Flight Mass Spectrometer

The quadrupole time-of-flight mass spectrometer (QqTOF) MS can be regarded either as the addition of a mass-resolving quadrupole and collision cell to a TOF, or as the replacement of the third quadrupole (Q3) in a triple quadrupole by a TOF, as shown in Fig. 7 (Chernushevich et al., 2001). The usual QqTOF configuration comprises three quadrupoles, whereas one of the quadrupoles, Q0, is an RF-only quadrupole that provides collisional damping for ions moving from the source region into the mass analyzer. Q1 and Q2 act as described earlier for the standard triple quadrupole, whereas Q3 is replaced by a reflecting TOF mass analyzer. For MS measurements, Q1 is operated in the RF-only mode, allowing all of the ions to pass directly through onto the TOF tube. When operating in MS/MS mode, Q1 is used as a mass filter to allow only the ion(s) of interest to pass onto the collision cell, Q2. The ion(s) is (are) then accelerated to energies between 20 and 200 eV before it enters the collision cell, where it undergoes collisions with a neutral gas such as Ar or N_2, resulting in fragmentation. Prior to entering the TOF analyzer, the ions are collisionally cooled, re-accelerated, and focused. The result is a parallel beam of ions that continuously enters the TOF region of the mass analyzer. The ions are then pushed in an orthogonal direction to their original trajectory by applying an electric field that is pulsed at a frequency of several kilohertz (kHz). The ions thus enter the field-free drift space of the TOF analyzer, and are separated on the basis of their m/z.

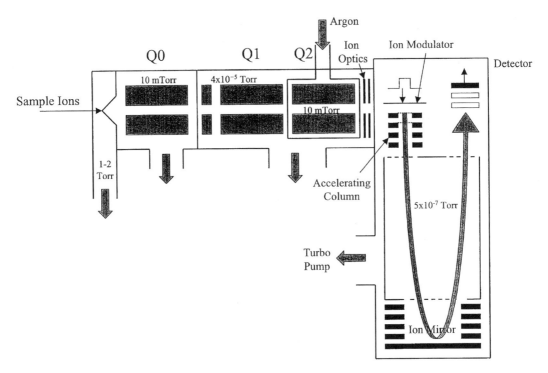

Figure 7 Schematic of a QqTOF mass spectrometer. In this instrument the Q3 region of a triple quadrupole mass spectrometer has been replaced with a TOF tube. This combination provides the instrument with ion selection, MS/MS capabilities of a triple quadrupole mass analyzer, and the high mass accuracy and resolution capabilities of a TOF analyzer.

This instrument combines the benefits of ion selectivity and sensitivity (quadrupole) with high mass resolution and mass accuracy (TOF) in both the MS and MS/MS modes, and as such is a popular choice of mass analyzer in proteomics. The QqTOF has a resolution in the range of 13,000–20,000 and a sensitivity level in the low femtomole range (Shevchenko et al., 2000). The instrument can be equipped with both ESI and MALDI sources and can be readily coupled with LC for the analysis of complex mixtures. The combination of high sensitivity and high mass accuracy for both precursor and product ions, and also the simplicity of operation for those already familiar with LC/MS analysis on quadrupole and triple quadrupole instruments, have made this type of mass analyzer a popular choice for proteomic applications.

D. Ion-Trap Mass Spectrometer

Probably the most popular mass analyzer in use in the field of proteomics today is the quadrupole ion trap. The popularity of the quadrupole ion trap has its roots in the discovery and development of the mass-selective axial instability scan (Stafford et al., 1984). The first development was the mass-selective *instability* mode of operation in which the ions created over a given time period could be trapped and then sequentially ejected into a conventional electron multiplier detector. Unlike the mass-selective *stability* mode of operation where only one m/z value could be stored, ions with a wide range of m/z values could be stored. The next major development showed that the mass resolution of this mass analyzer could be improved by adding about 1 mtorr of helium gas to the ion trap. This increase in resolution is because of a reduction of the kinetic energy of the ions and the contraction of their trajectories to the center of the trap (Stafford et al., 1984). This contraction allows packets of ions of a given m/z to form, enabling them to be ejected faster and more efficiently than a diffuse cloud of ions, thereby improving resolution and sensitivity.

An ion-trap mass analyzer works quite differently than a quadrupole or TOF (Fig. 8). Quadrupole mass analyzers essentially measure ions as they pass through onto the detector, whereas an ion trap collects and stores ions. Once trapped, manipulations such as MS/MS can be performed on the ions. Once the required ion manipulations are completed, the stored ions are then scanned out of the trap and detected. This entire sequence of trapping, storing, manipulating, and detecting the ions is performed in a continuous cycle. MS/MS analysis is conducted by filling the trap with all of the available ions and ejected all but a particular ion of interest. The energy of trapped ions is then increased and He_2 is introduced into the trap, resulting in collisions with the trapped ions causing them to fragment. These fragments are retained within

MS Instrumentation

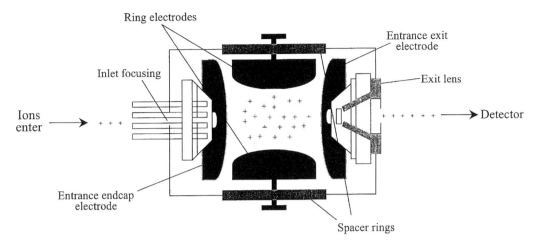

Figure 8 Schematic of an ion-trap mass analyzer. The ion trap consists of the ring electrode and the entrance and exit endcap electrodes. These electrodes are placed in a configuration to form a "trap" in which ions can be captured, manipulated, and analyzed. Ions produced from the source are focused through a small orifice in the entrance endcap electrode resulting in their capture within the trap. Various voltages are applied to the electrodes to trap and eject ions according to their m/z ratios. Applying a potential to the ring electrode causes the ions within the trap to oscillate in a stable trajectory. The electrode system potentials are altered to produce instabilities in the ion's oscillation trajectories causing them to be ejected in the axial direction and then detected.

the trap, scanned out according to their m/z, and detected. Fragment ions can also be retained within the trap and subjected to further rounds of MS/MS (i.e., MS/MS/MS or MS^n); however, such fragmentation is rarely used in proteomic studies, but is particularly useful in the characterization of small molecules.

The ion-trap mass spectrometer is a very popular choice for a wide variety of diverse applications in biological, pharmaceutical, environmental, and industrial laboratories. The ion trap can be readily interfaced to LC and ion sources such as ESI (Moyer, et al., 2002), and more recently MALDI (Krutchinsky et al., 2001). This mass analyzer is the most widely used instrument for global proteomic studies designed to characterize complex mixtures of proteins. This wide use is because of its inherent sensitivity and its ability to operate in a data-dependent MS/MS mode. In this mode, each full MS scan is followed by a specific number (usually three to five) of MS/MS scans where the most abundant peptide molecular ions are dynamically selected for fragmentation. Tryptic peptides resulting from the proteolytic digestion of a complex protein extract are injected onto a reversed phase column and separated prior to MS analysis. The column outlet is placed near the orifice of the mass analyzer, enabling peptides to be drawn into the instrument as they elute from the chromatographic column and be analyzed by MS and MS/MS. Such a configuration is routinely capable of identifying over 500 peptides in a single LC/MS/MS experiment.

This type of analysis forms the basis of several studies in which large numbers of proteins have been identified from a specific cell type or tissue. Among these is the analysis of a mouse cortical neuron cell lysate digested with trypsin and analyzed by μLC/MS/MS using an ion-trap mass spectrometer operating in a data-dependent MS/MS mode, as shown in Fig. 9. Over the course of this analysis, the mass spectrometer acquired approximately 3500 MS/MS, which resulted in the identification of 715 unique peptides that mapped to over 300 proteins.

E. Fourier Transform Ion Cyclotron Resonance Mass Spectrometer

Though it was developed almost 30 years ago (Comisarow and Marshall, 1974), Fourier transform ion cyclotron resonance (FTICR) MS has only recently generated great enthusiasm within the proteomics community. FTICR-MS functions much like an ion-trap analyzer; however, the trap positioned within a high magnetic field, typically ranges in field strength from 3 to 12 tesla, as shown in Fig. 10 (Marshall et al., 1998). Working at higher magnetic fields benefits at least eight parameters related to FTICR performance, the two most critical being resolution and mass accuracy. FTICR instruments have provided the highest resolution (Solouki et al., 1997), mass accuracy (Bruce et al., 1999), and sensitivity (Hakansson et al., 2001) for peptide and protein measurements so far achieved. The magnetic field causes ions captured within the trap to resonate at their cyclotron frequency. A uniform electric field that oscillates at or near the cyclotron frequency of the trap ions is then applied to excite the ions into a larger orbit that can be measured as they pass by detector plates on opposite sides of the trap. The ions within the trap can also be dissociated or ejected depending on the amount of

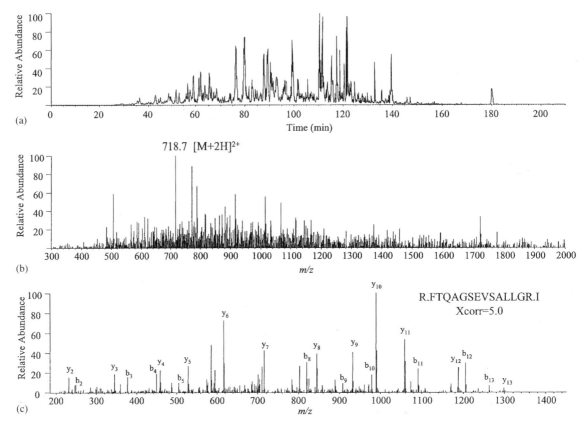

Figure 9 Data-dependent tandem mass spectrometry (MS/MS) identification of peptides in a complex proteome mixture. (a) The complexity of the base peak chromatogram of a proteolytically digested proteome sample is reflective of the complexity of the mixture being analyzed by μLC/MS/MS. In the data-dependent MS/MS mode, the most intense peptide detected in the MS scan (b) is isolated and fragmented by CID to obtain the MS/MS spectrum through which the peptide is identified (c).

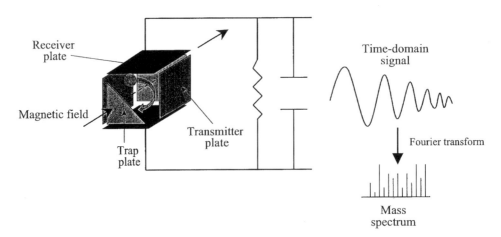

Figure 10 Principles and schematic of FTICR. Ions caught within the trap resonate at their cyclotron frequency because of the presence of the high magnetic field. The ions are excited into a larger orbit by the application of the appropriate electric field energy, and their resonance frequency is then measured as they pass by detector plates on opposite sides of the trap. Ions can also be dissociated through application of energy or introduction of a gas into the trap and then ejected from the trap by accelerating them to a cyclotron radius larger than the radius of the trap. The measured cyclotron frequency of all the ions in the trap is converted into m/z values using a Fourier transform.

energy applied. The cyclotron frequencies of all the ions in the trap are then recorded and the frequency values are converted into m/z values using a Fourier transform.

The performance capabilities of FTICR make it a potentially powerful tool in the characterization of global proteome mixtures. One of its greatest advantages compared with other types of mass analyzers is its wide dynamic range (i.e., $>10^3$). This wide dynamic range enables the identification of low abundance species in the presence of higher abundance components. The high mass accuracy of FTICR (potentially <1 ppm) also enables it to conduct multiplexed MS/MS studies in which multiple ions are accumulated and fragmented in a single scan and the product ions correctly assigned to their parent ion (Li et al., 2001). This multiplexed capability has the potential to significantly increase the number of species that can be identified within a single LC/MS/MS analysis compared with the existing conventional technologies. FTICR can be equipped with both MALDI and ESI sources, enabling it to be coupled with online LC and electrophoretic separations. Though still a very expensive and somewhat technically challenging technology, these scenarios are rapidly changing with some of the new Fourier transform mass spectrometry (FT-MS) instruments becoming commercially available. It is likely that FTICR will make an ever-increasing impact on proteomics research in the near future.

F. Surface-Enhanced Laser Desorption/Ionization Time-of-Flight Mass Spectrometry

Surface-enhanced laser desorption/ionization time-of-flight (SELDI-TOF) MS is a novel approach, whose novelty is not based on the mass analyzer used but on the surface from which protein and peptides are ionized and desorbed (Hutchens and Yip, 1993). The mass analyzer is a very simple linear TOF; however, the SELDI-TOF-MS system has provided researchers opportunities to analyze proteomic samples with simplicity that was not previously available to them. The principle of this approach is very simple; the proteins of interest are captured by adsorption, partition, electrostatic interaction, or affinity chromatography on a solid-phase protein chip surface (Issaq et al., 2003).

The SELDI-TOF-MS system is made up of two major components: the protein chip arrays and the mass analyzer. The mass analyzer is a relatively simple TOF-MS equipped with a pulsed UV nitrogen laser (Fig. 11). Though the mass analyzer is relatively sensitive, it lacks high resolution and has poor mass accuracy for a TOF-MS. Even with the use of time lag focusing, the achievable mass accuracy is much less than that provided by the more conventional, high resolution TOF-MS instruments described previously. What distinguish SELDI-TOF-MS from the other MS-based systems are the protein chip arrays. The arrays

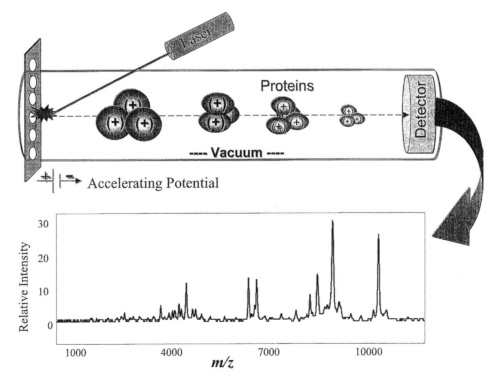

Figure 11 Schematic diagram of the SELDI-TOF mass spectrometer. After preparation, the samples are analyzed by a laser desorption ionization (LDI) TOF-MS resulting in a profile of the m/z ratios of the various proteins that are retained in the array.

are composed of spots of different chromatographic surfaces designed to retain specific proteins on the basis of the character of the surface present on the chip. The spots are made up of either a chemically (anionic, cationic, hydrophobic, hydrophilic, metal ion, etc.) or a biochemically (immobilized antibody, receptor, DNA, enzyme, etc.) active surface. The chemically active chromatographic surfaces retain whole classes of proteins, whereas the biochemically active surfaces are designed to couple with an affinity agent, such as an antibody, that interacts specifically with a single target molecule.

The single unique capability that has made SELDI-TOF-MS popular is its ability to analyze very crude samples with minimal cleanup prior to MS analysis. Biological fluids such as serum, urine, and plasma can be spotted directly on the protein chip arrays surfaces. After the protein chip is processed using a few simple processing steps, a matrix is applied and the spectrum of the sample can be acquired. A common misnomer is that SELDI is a novel method to ionize and desorb molecules for MS analysis; however, it is really just MALDI for a unique type of chip surface. In SELDI method the surface is enhanced; not the surface results in an enhanced desorption and ionization.

SELDI-TOF-MS analysis results in a low resolution spectrum of species that are bound to the protein chip surface. The instrument does not have MS/MS capabilities, and the poor resolution and mass accuracy of the analyzer make absolute protein identification impossible, unless a protein of interest is selectively targeted using an affinity-based surface. SELDI-TOF-MS does, however, allow an investigator to obtain and compare spectra from a significant number of samples in a relatively short time period. For example, a single operator can acquire mass spectra of more than 150 different samples in a single day. The analysis of a large number of samples will ideally reveal a protein signal that is unique, or over-expressed, in one sample set when compared with a different sample set. The net result is the molecular mass of a protein(s) that is (are) differentially expressed in different samples.

Much of the excitement generated by the use of SELDI-TOF-MS is in the use of the spectral patterns that it acquires and its potential ability as a diagnosis of disease states such as cancer. In the seminal study describing this application, mass spectra of serum samples from healthy and ovarian cancer-affected women were acquired (Petricoin et al., 2002a). Though visual inspection of the mass spectra did not reveal any obvious feature that would allow the origin of the samples to be ascertained, application of an artificial intelligence program was able to decipher diagnostic "patterns" within the profiles and determine that the serum had been obtained from a healthy or a disease-affected individual. Since this original study, several laboratories have confirmed the potential diagnostic ability of combining serum proteomic pattern analysis with sophisticated bioinformatic tools for breast (Li et al., 2002) and prostate (Adam et al., 2002; Petricoin et al., 2002b; Qu et al., 2002) cancer.

G. Inductively Coupled Plasma Mass Spectrometry

Inductively coupled plasma (ICP)-MS is similar to all the other technologies described earlier in that the instrument is made up of a source and a mass analyzer. ICP-MS is widely used in determining the element composition of different types of samples, and is an excellent choice in analyzing samples such as water for common elements (magnesium, sodium, iron, calcium, etc.) and a number of trace elements (zinc, lead, selenium, manganese, etc.) that have specific biological functions, but can be hazardous if present in high concentrations. The schematic of an ICP-MS instrument is shown in Fig. 12. In this instrument, the ICP is the ion source. This ion source operates at atmospheric pressure and at very high temperatures

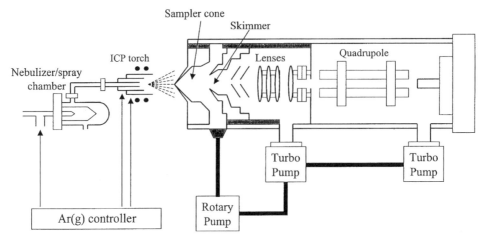

Figure 12 Schematic of an ICP-MS.

(5000–10,000 K). An argon plasma is generated in a quartz torch within the source (O'Connor and Evans, 1999) in the presence of a frequency electromagnetic field operating at a power of 600–1800 W and a frequency between 27 and 40 MHz (Fisher and Hill, 1999). As the argon flows through the inner tube of the torch, it is seeded with free electrons from a discharge coil. As the charged particles are forced to flow in a closed annular path, an eddy current of electrons and cations is formed. Further ionization is produced as these rapid moving ions and electrons collide with other argon atoms, leading to high thermal energy as they meet resistance to their flow. A second stream of argon gas passes through the outer tubes of the torch to keep the torch cool, as well as to provide a gas flow to center and stabilize the plasma. This technique, with the high efficiency of atomization and ion formation of the ICP is complemented by the specific and sensitive multielement detection capability offered by atomic mass spectrometry.

For the analysis of a sample, it is passed into the plasma stream in a solution usually via an HPLC column or capillary effluent, CE. The sample is introduced as an aerosol through the center tube of the torch into the plasma by means of a nebulizer connected to a spray chamber, which separates and removes the larger droplets of the aerosol. As the analyzer region of an ICP-MS is maintained at low pressure, the ions produced in the plasma are drawn into the mass analyzer via a pressure differential. The ions enter through the sampler and skimmer cones and are focused into the mass analyzer via a series of lenses. As in the other types of MS analyzers described earlier, the ions being analyzed by ICP-MS are separated on the basis of their m/z ratio and then are detected (Houk, 1986; O'Connor and Evans, 1999). The response for the majority of commercially available instruments is linear over 4–11 orders of magnitude. Another attribute of ICP-MS is that it can be directly coupled with many different chromatographic and electrophoretic separation methods including reversed-phase LC, partition, micellar, ion exchange, and size exclusion chromatography. This ability provides a great deal of flexibility when analyzing for a particular element within different sample matrices.

As mentioned earlier, ICP-MS is used primarily for elemental analysis and is not generally thought of as a biological or proteomics analytical technology. A method that combines ESI and ICP-MS for the identification and quantitation of phosphopeptides, however, has been developed by Wind et al. (2001). In this strategy, the eluent from an LC separation of a tryptic digest of a phosphoprotein is interfaced alternatively to ICP-MS and ESI-MS. The ICP-MS is used to monitor for the presence of ^{31}P, and ESI-MS measures the molecular masses of the corresponding peptides. Phosphopeptides can be identified by aligning the two separate LC runs and determining the peptides that produce a ^{31}P signal. The two advantages of this strategy are its high selectivity and the fact that the signal intensity of the ^{31}P detection is directly proportional to the molar amount of ^{31}P in the LC eluate, unlike ESI or MALDI-MS in which the phosphopeptide ionization efficiency (and hence its quantitation) is compound-dependent. In addition, the detection limit is approximately 1 pmol of phosphopeptide injected. Though this method has not been widely used, promising results have been demonstrated for β-casein, activated human MAP kinase ERK1, and the catalytic subunit of protein kinase A (Wind et al., 2001). Though the combined ESI/ICP-MS has relied on peptide mapping to identify the phosphopeptide, it is obvious that the strategy is amenable to tandem MS identification as well.

IV. CONCLUSIONS

Though the concept of MS is over a hundred years old, it has really been in the last decade that MS has taken its place as one of the most powerful analytical tools in science. Purchasing these types of instruments is almost analogous to buying a computer: shortly after you buy it, another more powerful system is being introduced. With the drive for ever-increasing sensitivity it is unlikely that the development of more powerful MS technologies will soon end. With the rising interest in proteomics, the market for MS instruments as well as trained operators has never been greater. Though this chapter attempts to provide a description of some of the available MS tools, each type by itself requires a whole book for its in-depth description. Indeed, as with the MS technologies themselves, any chapter of this topic can become quickly out-of-date.

ACKNOWLEDGMENTS

This project has been funded in whole or in part with Federal funds from the National Cancer Institute, National Institutes of Health, under Contract No. NO1-CO-12400. By acceptance of this article, the publisher or recipient acknowledges the right of the US Government to retain a non-exclusive, royalty-free license and to any copyright covering the article. The content of this publication does not necessarily reflect the views or policies of the Department of Health and Human Services; nor does the mention of trade names, commercial products, or organizations imply endorsement by the US Government.

REFERENCES

Adam, B. L., Qu, Y., Davis, J. W., Ward, M. D., Clements, M. A., Cazares, L. H., Semmes, O. J., Schellhammer, P. F., Yasui, Y., Feng, Z., Wright, G. L. Jr. (2002). Serum protein fingerprinting coupled with a pattern-matching algorithm distinguishes prostate cancer from benign prostate hyperplasia and healthy men. *Cancer Res.* 62(13):3609–3614.

Aleksandrov, M. L., Gall, L. N., Krasnov, V. B., Nikolaev, V. I., Pavlenko, V. A., Shkurov, V. A. (1984). Ion extraction from solutions at atmospheric pressures: a mass spectrometric method of analysis of bioorganic compounds. *Dokl. Akad. Nauk. SSSR* 277:379–383.

Brown, R. S., Lennon, J. J. (1995). Sequence-specific fragmentation of matrix-assisted laser-desorbed protein/peptide ions. *Anal. Chem.* 67:3990–3999.

Bruce, J. E., Anderson, G. A., Wen, J., Harkewicz, R., Smith, R. D. (1999). High-mass-measurement accuracy and 100% sequence coverage of enzymatically digested bovine serum albumin from an ESI-FTICR mass spectrum. *Anal. Chem.* 71:2595–2599.

Carr, S. A., Huddleston, M. J., Bean, M. F. (1993). Selective identification and differentiation of N- and O-linked oligosaccharides in glycoproteins by liquid chromatography–mass spectrometry. *Protein Sci.* 2:183–196.

Chaurand, P., Luetzenkirchen, F., Spengler, B. (1999). Peptide and protein identification by matrix-assisted laser desorption ionization (MALDI) and MALDI-post-source decay time-of-flight mass spectrometry. *J. Am. Soc. Mass Spectrom.* 10:91–103.

Chernushevich, I. V., Loboda, A. V., Thomson, B. A. (2001). An introduction to quadrupole-time-of-flight mass spectrometry. *J. Mass Spectrom.* 36:849–865.

Comisarow, M. B., Marshall, A. G. (1974). Fourier transform ion cyclotron resonance spectroscopy. *Chem. Phys. Lett.* 25:282–283.

Cornish, T. J., Cotter, R. J. (1993). A curved-field reflectron for improved energy focusing of product ions in time-of-flight mass spectrometry. *Rapid Commun. Mass Spectrom.* 7:1037–1040.

Cotter, R. J. (1997). *Time-of-Flight Mass Spectrometry: Instrumentation and Applications in Biological Research.* Oxford: Oxford University Press.

Dainese, P., Staudenmann, W., Quadroni, M., Korostensky, C., Gonnet, G., Kertesz, M., James, P. (1997). Probing protein function using a combination of gene knockout and proteome analysis by mass spectrometry. *Electrophoresis* 18:432–442.

Dole, M., Mack, L. L., Hines, R. L., Mobley, R. C., Ferguson, L. D., Alice, M. B. (1968). Molecular beams of macroions. *J. Chem. Phys.* 49:2240–2249.

Eng, J. K., McCormack, A. L., Yates, J. R. (1994). An approach to correlate tandem mass spectral data of peptides with amino acid sequences in a protein database. *J. Am. Soc. Mass Spectrom.* 5:976.

Fenn, J. B., Mann, M., Meng, C. K., Wong, S. F. (1989). Electrospray ionization for mass spectrometry of large biomolecules. *Science* 246:64–71.

Fenyo, D. (2000). Identifying the proteome: software tools. *Curr. Opin. Biotechnol.* 11:391–395.

Fisher, A., Hill, S. J. (1999). In: Hill, S. J., ed. *Inductively Coupled Plasma Spectrometry and its Applications.* Sheffield, UK: Sheffield Academic Press.

Griffiths, W. J., Jonsson, A. P., Liu, S., Rai, D. K., Wang, Y. (2001). Electrospray and tandem mass spectrometry in biochemistry. *Biochem. J.* 355:545–561.

Hakansson, K., Emmett, M. R., Hendrickson, C. L., Marshall, A. G. (2001). High-sensitivity electron capture dissociation tandem FTICR mass spectrometry of microelectrosprayed peptides. *Anal. Chem.* 73:3605–3610.

Houk, R. S. (1986). Mass-spectrometry of inductively coupled plasmas. *Anal. Chem.* 58:A97.

Hutchens, T. W., Yip, T. T. (1993). New desorption strategies for the mass spectrometric analysis of macromolecules. *Rapid. Commun. Mass Spectrom.* 7:576–580.

Issaq, H. J., Conrads, T. P., Prieto, D. A., Tirumalai, R., Veenstra, T. D. (2003). SELDI-TOF MS for diagnostic proteomics. *Anal. Chem.* 75:148A–155A.

Juhasz, P., Roskey, M. T., Smirnov, I. P., Haff, L. A., Vestal, M. L., Martin, S. A. (1996). Applications of delayed extraction matrix-assisted laser desorption ionization time-of-flight mass spectrometry to oligonucleotide analysis. *Anal. Chem.* 68:941–946.

Karas, M., Bachmann, D., Bahr, U., Hillenkamp, F. (1987). Matrix-assisted ultraviolet laser desorption of non-volatile compounds. *Int. J. Mass Spectrom. Ion Processes* 78:53–68.

Kaufmann, R. (1995). Matrix-assisted laser desorption ionization (MALDI) mass spectrometry: a novel analytical tool in molecular biology and biotechnology. *J. Biotechnol.* 41:155–175.

Kaufmann, R., Chaurand, P., Kirsch, D., Spengler, B. (1996). Post-source decay and delayed extraction in matrix-assisted laser desorption/ionization-reflectron time-of-flight mass spectrometry. Are there trade-offs? *Rapid Commun. Mass Spectrom.* 10:1199–1208.

Krutchinsky, A. N., Kalkum, M., Chait, B. T. (2001). Automatic identification of proteins with a MALDI-quadrupole ion trap mass spectrometer. *Anal. Chem.* 73:5066–5077.

Li, J., Zhang, Z., Rosenzweig, J., Wang, Y. Y., Chan, D. W. (2002). Proteomics and bioinformatics approaches for identification of serum biomarkers to detect breast cancer. *Clin. Chem.* 48(8):1296–1304.

Li, L., Masselon, C. D., Anderson, G. A., Pasa-Tolic, L., Lee, S. W., Shen, Y., Zhao, R., Lipton, M. S., Conrads, T. P., Tolic, N., Smith, R. D. (2001). High-throughput peptide identification from protein digests using data-dependent multiplexed tandem FTICR mass spectrometry coupled with capillary liquid chromatography. *Anal. Chem.* 73:3312–3322.

Marshall, A. G., Hendrickson, C. L., Jackson, G. S. (1998). Fourier transform ion cyclotron resonance mass spectrometry: a primer. *Mass Spectrom. Rev.* 17:1–35.

Martin, S. A., Rosenthal, R. S., Biemann, K. (1987). Fast atom bombardment mass spectrometry and tandem mass spectrometry of biologically active peptidoglycan monomers from *Neisseria gonorrhoeae*. *J. Biochem.* 262:7514–7522.

Miyashita, M., Presley, J. M., Buchholz, B. A., Lam, K. S., Lee, Y. M., Vogel, J. S., Hammock, B. D. (2001). Attomole level protein sequencing by Edman degradation coupled with accelerator mass spectrometry. *Proc. Natl. Acad. Sci. USA* 98:4403–4408.

Moyer, S. C., Cotter, R. J. (2002). Atmospheric pressure MALDI. *Anal. Chem.* 74:468A–476A.

Moyer, S. C., Cotter, R. J., Woods, A. S. (2002). Fragmentation of phosphopeptides by atmospheric pressure MALDI and ESI/Ion trap mass spectrometry. *J. Am. Soc. Mass Spectrom.* 13(3):274–283.

O'Connor, G., Evans, E. H. (1999). In: Hill, S. J., ed. *Inductively Coupled Plasma Spectrometry and its Applications*. Sheffield, UK: Sheffield Academic Press.

Petricoin, E. F., Ardekani, A. M., Hitt, B. A., Levine, P. J., Fusaro, V. A., Steinberg, S. M., Mills, G. B., Simone, C., Fishman, D. A., Kohn, E. C., Liotta, L. A. (2002a). Use of proteomic patterns in serum to identify ovarian cancer. *Lancet* 359(9306):572–577.

Petricoin, E. F. III, Ornstein, D. K., Paweletz, C. P., Ardekani, A., Hackett, P. S., Hitt, B. A., Velassco, A., Trucco, C., Wiegand, L., Wood, K., Simone, C. B., Levine, P. J., Linehan, W. M., Emmert-Buck, M. R., Steinberg, S. M., Kohn, E. C., Liotta, L. A. (2002b). Serum proteomic patterns for detection of prostate cancer. *J. Natl. Cancer Inst.* 94(20):1576–1578.

Qu, Y., Adam, B. L., Yasui, Y., Ward, M. D., Cazares, L. H., Schellhammer, P. F., Feng, Z., Semmes, O. J., Wright, G. L. Jr. (2002). Boosted decision tree analysis of surface-enhanced laser desorption/ionization mass spectral serum profiles discriminates prostate cancer from noncancer patients. *Clin. Chem.* 48(10):1835–1843.

Rappsilber, J., Mann, M. (2002). What does it mean to identify a protein in proteomics? *Trends Biochem. Sci.* 27:74–78.

Ryzhov, V., Bundy, J. L., Fenselau, C., Taranenko, N., Doroshenko, V., Prasad, C. R. (2000). Matrix-assisted laser desorption/ionization time-of-flight analysis of Bacillus spores using a 2.94 microm infrared laser. *Rapid Commun. Mass Spectrom.* 14:1701–1706.

Shevchenko, A., Chernushevich, I., Ens, W., Standing, K. G., Thomson, B., Wilm, M., Mann, M. (1997). Rapid 'de novo' peptide sequencing by a combination of nanoelectrospray, isotopic labeling and a quadrupole/time-of-flight mass spectrometer. *Rapid Commun. Mass Spectrom.* 11:1015–1024.

Shevchenko, A., Loboda, A., Shevchenko, A., Ens, W., Standing, K. G. (2000). MALDI quadrupole time-of-flight mass spectrometry: a powerful tool for proteomic research. *Anal. Chem.* 72:2132–2141.

Solouki, T., Emmett, M. R., Guan, S., Marshall, A. G. (1997). Detection, number, and sequence location of sulfur-containing amino acids and disulfide bridges in peptides by ultrahigh-resolution MALDI FTICR mass spectrometry. *Anal. Chem.* 69:1163–1168.

Stafford, G. C., Kelley, P. E., Syka, J. E. P., Reynolds, W. E., Todd, J. F. J. (1984). Recent improvements in and analytical applications of advanced ion-trap technology. *Int. J. Mass Spectrom. Ion Processes* 60:85–98.

Steen, H., Kuster, B., Mann, M. (2001). Quadrupole time-of-flight versus triple-quadrupole mass spectrometry for the determination of phosphopeptides by precursor ion scanning. *J. Mass Spectrom.* 36:782–790.

Tanaka et al. (1988). Protein and polymer analyses up to m/z 100,000 by laser ionization time-of-flight mass spectrometry. *Rapid Commun. Mass. Spectrom.* 2:151–153.

Wind, M., Edler, M., Jakubowski, N., Linscheid, M., Wesch, H., Lehmann, W. D. (2001). Analysis of protein phosphorylation by capillary liquid chromatography coupled to element mass spectrometry with ^{31}P detection and to electrospray mass spectrometry. *Anal. Chem.* 73:29–35.

Yost, R. A., Boyd, R. K. (1990). Tandem mass spectrometry: quadrupole and hybrid instruments. *Methods Enzymol.* 193:154–200.

15

Thermoanalytical Instrumentation and Applications

KENNETH S. ALEXANDER
Industrial Pharmacy Division, College of Pharmacy, The University of Toledo, Toledo, OH, USA

ALAN T. RIGA
College of Pharmacy, The University of Toledo, Toledo, OH, USA
Clinical Chemistry Department, Cleveland State University, Cleveland, OH, USA

PETER J. HAINES
Oakland Analytical Services, Farnham, Surrey, UK

I. INTRODUCTION

The term "thermal analysis" is commonly applied to a variety of techniques in which a property of a system is recorded as a function of temperature while the system is driven by a controlled temperature program (Hill, 1991). The name has been approved by IUPAC but many physical chemistry textbook authors apply the term specifically to cooling curves used to construct condensed phase diagrams (Adamson, 1986; Atkins, 1994). In this chapter, the IUPAC definition is used.

A. Scope of Thermal Analysis

As noted, thermal analysis implies a determination of a specified physical property as a function of temperature or time. The plot of the property of the system plotted against temperature is said to be a *thermal analysis curve*.

In practice, there are other conditions which have to be satisfied in modern thermal analysis instruments, namely:

1. The physical property and the sample temperature should be measured continuously.
2. Both the property and the temperature should be recorded automatically.
3. The temperature of the sample should be altered in a predetermined manner.
4. The processing of the temperature program and the measurement of the property should be accomplished using a dedicated computer workstation. The workstation should be capable of recording data and processing as well as controlling the instrumentation.

The purpose of making thermal analysis measurements is to study the physical and chemical changes that occur in the sample, or system, subjected to the temperature program. Therefore, one has to interpret a thermal analysis curve by relating the features of the property against temperature with possible chemical or physical events that have taken place in the system under observation. The most common property measured is that of mass, but calorimetry experiments predate this technique and provide information primarily concerning the enthalpy changes that take place. The analysis and detection of evolved gas is also an important subject. Another group of studies is called *thermomechanical mechanical analyses*. These deal with dimensional changes and changes properties connected with the strength of materials when subjected to temperature changes.

Generally, thermal analysis techniques may be classified into three groups, depending on the manner in which the physical property is recorded:

1. The absolute value of the property itself can be measured, for example, the sample mass.
2. The differential method measures the difference between some property of the sample and that of a standard material, such as their temperature difference.
3. The rate at which the property is changing with temperature can be measured; this forms the basis of derivative measurements and often may be interpreted on a kinetic basis.

These fundamental techniques are used for the characterization of drugs and drug products while processing or under aging conditions, which may be simulated, and the method also gives access to thermodynamic data. Owing to the different information delivered, thermal analysis methods are concurrent or complementary to other analytical techniques such as spectroscopy, chromatography, melting, loss on drying, and assay, for identification, purity, and quantitation. They are basic methods in the field of polymer analysis and in physical and chemical characterization of pure substances as well as mixtures. They find good applications for preformulation, processing, and control of the drug product. The introduction of automation considerably increases the advantages of these methods. New horizons are opening with the availability of combined techniques and microthermal analysis techniques.

Measurements by thermal analysis are conducted for the purpose of evaluating the physical and chemical changes that may take place in a heated sample. This requires that the operator interpret the observed events in a thermogram in terms of plausible reaction processes. The reactions normally monitored can be endothermic (melting, boiling, sublimation, vaporization, desolvation, solid–solid phase transitions, chemical degradation, etc.) or exothermic (crystallization, oxidative decomposition, etc.) in nature.

Thermal methods can be extremely useful during the course of preformulation studies, as carefully planned work can be used to indicate the existence of possible drug-excipient interactions in a prototype formulation (Haines, 1995). During the course of this aspect of drug development, thermal methods can be used to evaluate compound purity, polymorphism, solvation, degradation, drug–excipient compatibility, and a wide variety of other desirable characteristics. Several recent reviews have been written on such investigations (Ford and Timmins, 1989; Giron-Forest, 1983; Haines, 1995; Turl, 1997; Wunderlich, 1990).

B. Nomenclature and Definitions

Considering the number of physical parameters of a substance, which may be measured, the number of techniques derived is very large. Details on most techniques are well described by Wendlandt (Wendlandt, 1986). For pharmaceutical applications, the methods generally used are differential scanning calorimetry (DSC), thermogravimetry (TG) (or thermogravimetric analysis, TGA), and to a lesser extent thermomechanical analysis (TMA). All techniques are automated and have data acquisition. Hyphenated techniques and modulated DSC are evolving as state-of-the-art techniques of the 21st century. There are excellent books or review articles dealing with the principle instrumentation and applications of thermal analysis methods for pharmaceuticals available (Brennan, 1997; Cheng et al., 2000; Ford and Timmins, 1989; Giron, 1986, 1990, 1995, 1998, 1999; Giron-Forest, 1983; Haines, 1995; Thompson, 2000; Turl, 1997; Wendlandt, 1986; Wunderlich, 1990). As emphasized by Cheng et al. (2000), the tendency in the next two decades will be more toward precise and meaningful measurements in these techniques, which will provide new developments in obtaining the temperature dependence of a material's structure and dynamics.

In the field of thermal analysis, national and international organizations have collectively made certain recommendations concerning nomenclature, abbreviations, definitions, and standards. The recommended nomenclature has been formulated by an international committee that is part of the International Confederation for Thermal Analysis and Calorimetry (ICTAC). The recommendations have been widely circulated in the booklet by Hill (1991). The recommended names and abbreviations for the most commonly used techniques are listed in Table 1. The definitions of these techniques, the property measured, and other details follow.

TG or TGA is defined as "a technique in which the mass of the sample is monitored against time (t) or temperature (T) while the temperature of the sample in a specified atmosphere is programed". The name of the instrument is a *thermobalance* or *thermogravimetric* analyzer. As in all other thermal analysis techniques, a thermocouple or any other convenient temperature measurement device may be used to record the temperature automatically. The data is initially presented as mass (m) against temperature (T) but again as in all the other techniques the

Table 1 Nomenclature in Thermal Analysis

Type	Name of method	Abbreviation
General	Thermal analysis	TA
Static	Methods associated with mass change	
	Isobaric mass change determination	
	Isothermal mass change determination	
Dynamic	Thermogravimetry	TG
	Derivative thermogravimetry	DTG
Evolved volatiles	Evolved gas detection	EGD
	Evolved gas analysis	EGA
Temperature changes[a]	Heating curve determinations	
	Heating rate curves	
	Inverse heating rate curve	
	Differential thermal analysis	DTA
	Derivative differential thermal analysis	
Enthalpy changes[a]	Differential scanning calorimetry	DSC
Dimensional changes	Thermodilatometry	TMA
	Derivative thermodilatometry	
	Thermomechanical analysis	
	Dynamic thermomechanometry	

[a]As classical calorimetry relies initially on noting a temperature change as the basis of the calculation of enthalpy, the distinction between these two categories is often blurred.

temperature program and the environmental atmosphere should be noted. In decomposition reactions, the reactant material degrades, often to be replaced by the solid product. An example of this is the decomposition of calcium carbonate to form calcium oxide; from a record of the mass of the system plotted against the temperature the decomposition of the material can easily be followed. This is often plotted as the percentage mass loss or the fractional decomposition (α) against temperature.

Derivative thermogravimetry (DTG) is not a separate technique from TG. The same equipment is used but instead of plotting mass (m) against time (t), the computer will plot the rate of change of mass against time (dm/dt) against time (t).

Isobaric mass change determinations refer to the equilibrium mass of a substance at a constant partial pressure of the volatile product(s) measured as a function of temperature while the substance is subjected to a controlled temperature program. The isobaric mass change curve is plotted with mass as the ordinate decreasing downward and temperature on the abscissa increasing from left to right.

Evolved-gas analysis (EGA) is a technique in which the nature and/or the amount of gas or vapor evolved from the sample, is monitored against time or temperature while the temperature of the sample, in a specified atmosphere, is programed. There is no single type of instrument for EGA, but the technique will always involve a furnace and a gas analyzer or detector. Most commonly, the gas is analyzed by a mass spectrometer (MS) or Fourier transform infrared spectrograph (FT-IR). However, other gas analysis techniques can be used. If the gas is merely detected instead of being analyzed then the technique is termed gas analysis detection (EGD). Some specific forms of EGA have found increasing applications for studying aspects of catalysis, such as reduction, oxidation, or desorption. In this context, EGA in a hydrogen atmosphere is known as temperature-programed reduction (TPR) and EGA in an oxygen atmosphere is temperature-programed oxidation (TPO). EGA in the absence of decomposition, in an inert atmosphere or vacuum, is temperature-programed desorption (TPD). The method of analysis should always be clearly stated in describing these methods.

There are other techniques that relate to mass in thermal analysis. Thus emanation thermal analysis (ETA) is a technique in which the release of trapped (usually radioactive) gas from the sample is monitored against time or temperature in a specified atmosphere while the temperature of the sample is programed. The gas is trapped during a pretreatment step and is released during the experiment as a result of structural or morphological changes resulting from chemical or physical processes.

Methods listed in Table 1 under the heading of "temperature change" or "enthalpy change" tend to overlap, as the basic measurement of enthalpy in the ultimate evaluation is often a change of temperature.

Differential thermal analysis (DTA) is a technique in which the temperature difference between a substance and a reference material is measured as a function of temperature, while the sample and the reference material are subjected to a controlled temperature program. The record is a DTA curve. The temperature difference ΔT should be plotted on the ordinate with endothermic processes downward and exothermic processes upward, and temperature or time on the abscissa increasing from left to right. The instrument manufacturers provide DTA instruments that can respond quantitatively to enthalpy changes, and almost without exception these are referred to as differential scanning calorimeters.

There are a variety of techniques grouped under the heading of TMA. Thus, *thermodilatometry* is defined as a technique in which the dimensions of a substance under negligible load is measured as a function of temperature while the substance is subjected to a controlled temperature program. The plot in the thermodilatometric curve is then the dimension plotted on the ordinate increasing upward with temperature, T, or time, t, on the abscissa increasing from left to right. TMA is specifically a technique in which the deformation of a substance under a nonoscillatory load is measured as a function of temperature while the substance is subjected to a controlled temperature program. It finds increasing use in polymer technology. The mode, as determined by the type of stress applied (compression, torsion, flexure, or torsion), should always be stated.

Dynamic thermomechanometry on the other hand, is a technique in which the dynamic modulus and/or clamping of a substance under oscillatory load is measured as a function of temperature while the substance is subjected to a controlled temperature program. A related technique is *torsional braid analysis*. in which the substance is prepared by impregnating a glass braid or thread substrate with the sample. The braid is then subjected to free torsional oscillations. Other thermal analytical techniques are found in the literature but not always in the instrument manufacturers' catalogs.

One such technique is *thermosonometry* (Lonvik, 1978) in which the sound emitted by a substance is measured as a function of temperature while the substance is subjected to a controlled temperature program. Another technique is *thermoacoustimetry*, in which the characteristics of the imposed acoustic waves are measured as a function of temperature while the substance is subjected to a controlled temperature program. In *thermo-optometry* the optical characteristic of a substance is measured as the sample is subjected to a controlled temperature regime. *Thermophotometry, thermospectrometry, thermofractometry, and thermoluminescence* involve the measurement of total light, light of specific wavelengths, refractive index, and luminescence, respectively, as the system is subjected to a controlled temperature program. *Thermomicroscopy* refers to observations under a microscope while the sample is taken through a range of temperatures. Two other techniques may be mentioned, *thermoelectrometry*, in which an electrical characteristic of a substance is measured as a function of temperature and *thermomagnetometry*, in which the magnetic susceptibility of a substance is measured as a function of temperature.

The term "simultaneous" is used to describe the operation of two or more techniques upon the same sample subjected to the same controlled temperature regime. This overcomes the disadvantages of using the various techniques of thermal analysis separately. These disadvantages may be listed as follows:

1. Samples may not be the same (always a problem with such materials as coals, lime stones, cements, etc.).
2. The thermal environment and sample temperatures may not be the same in the two sets of equipment.
3. The ambient atmosphere and the flow rate may not be the same in both techniques and even when these are formally the same, atmospheric environment immediately in contact with the sample may differ because of differences in the geometry of the container and furnace around the sample.

The disadvantage of simultaneous application of the techniques may be that the environment necessary for, say, TG may not be the optimum for DTA or DSC. It is not strictly exact to describe EGA as simultaneous. The reason is that although the sampling probe may be located immediately above the sample, the measurement may be made at a distance. In traversing this distance there is always a possibility of reaction among the gaseous products.

C. Symbols

The abbreviations for the various techniques have already been introduced. The confusion that can arise, however, between the use of, for example, TG and T_g, the latter representing the glass transition temperature, has caused a number of investigators and instrument manufacturers to use TGA for TG, therefore avoiding confusion. The recommendations regarding symbols to be used in thermal analysis have been listed (Hill, 1991) and are reproduced here:

1. The International System of Units (SI) should be used wherever possible.
2. The use of symbols with superscripts should be avoided if possible.
3. The use of double subscripts should be avoided.
4. The symbol T should be used for temperature whether expressed in degrees Celsius (°C) or in Kelvins (K). For temperature intervals, either symbol (K or °C) can be used.

5. The symbol t should be used for time, whether expressed in seconds (s), minutes (min), or hours (h).
6. The heating rate can be expressed either as dT/dt, when a true derivative is intended, or as K/min or °C/min. The heating rate so expressed need not be constant and can be positive or negative.

The symbols m for mass and W for weight are recommended. The symbol α is recommended for the fraction reacted or changed.

The following rules are recommended for subscripts:

Where the subscript relates to an object, it should be a capital letter:

m_S represents the mass of sample
T_R represents the temperature of the reference material

Where the subscript relates to a phenomenon occurring it should be in lower case:

T_g represents glass transition temperature
T_c represents the temperature of crystallization
T_m represents the temperature of melting
T_σ represents the temperature of a solid–state transition

Where the subscript relates to a specific point in time or to a point on the curve, it should be in lower case or in figures:

T_i represents the initial temperature
T_f represents the final temperature
$T_{0.5}$ represents the time at which the fraction reacted is 0.5
$T_{0.3}$ represents the temperature at which the fraction reacted is 0.3
T_p represents the temperature of the peak
T_e represents the temperature of the extrapolated onset

II. THERMOGRAVIMETRY

A. Development

The general production of commercial thermobalances did not occur until the mid-1950s. Reliable instruments were not available until the introduction of the Chévenard thermobalance in 1943. Even this apparatus left much to be desired. It is mainly because of (Duvai, 1963) that a more reliable modification was produced in 1947. Figure 1 shows in schematic form, the basic units present in contemporary TGA equipment. It should be noted that most modern units are constructed on a modular basis. One workstation can collect process and control data from more than one thermal analysis unit. The early history of thermobalance design is to be found in several publications (Keattch and Dollimore, 1975; Mackenzie, 1984), and will not be discussed here.

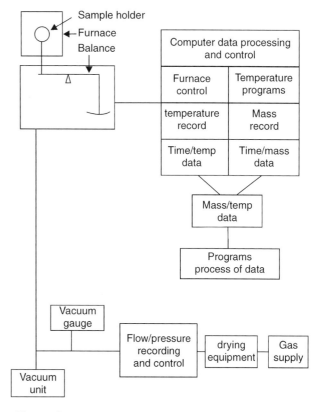

Figure 1 A schematic diagram of contemporary TGA units.

B. Design Factors

The basic instrumental requirements for TGA include a precision balance and a furnace capable of being programed for a linear rise of temperature with time. If a computer is interfaced with the equipment then it should be capable of controlling the temperature program, data acquisition, data programing, and data presentation. Most instrument manufacturers present equipment that has access to this kind of dedicated computer. These requirements can be incorporated into the equipment in a variety of ways, and with a vast range of purchase price. Choice of equipment must be based on specific requirements but must also take note of the features for good thermobalance design and requirements of the operator. These are the points to be considered.

1. Decide on the maximum working temperature required and choose a model that can reach around 150°C in excess of this.
2. Check out the hot zone of the furnace. It should be uniform and of reasonable length. The word uniform is an elastic term, but probably for most purposes ±2°C would be satisfactory. The uniform hot zone should be of a size such that a crucible can be easily located within it, together

with the temperature-measuring sensors. If the balance is not a null deflection type and the crucible moves, then the uniform hot zone must be large enough to accommodate this movement.

3. The shape (or shapes) of the available crucibles must be considered. Some crucibles are designed to prevent the easy loss of volatiles. The rate of loss of volatile material from the crucible can be heavily dependent on the crucible geometry.

4. The winding of the furnace must be noninductive to avoid anomalous effects on mass with magnetic or conducting samples. This has been utilized as a method of calibration.

5. The heating rate should be reproducible. A variety of heating programs should be available and the user must consider in advance the requirements imposed by the field of study. Temperature-jump programs, isothermal heating modes, holding a rising temperature experiment for required periods of time at a predetermined temperature, should all be available. In kinetic determinations it is advantageous to be able to impose parabolic or logarithmic heating rates on the samples under investigation. Finally, in considering the heating rate, the user should decide if a controlled cooling rate is also required.

6. The sensitivity of the balance must be in accordance with the mass of sample being used and the alteration in mass expected. There is considerable concern in some areas—coal testing, cement hydration, clay identification—that the choice of a suitable representative sample is jeopardized by the limited load capabilities of the balance.

7. The positioning of the balance with respect to the furnace is important in order to prevent radiation and convection currents from the furnace from affecting the recorded mass.

8. The recorded temperature should ideally be that of the sample. If possible this should also be the sensor that controls the temperature program.

9. Chemical attack by volatiles liberated in the decomposition process should be avoided. This is often most easily achieved by passing a carrier gas over the balance system, thus venting the products of decomposition away from the balance system.

10. Provision should be made for measuring the rate of flow of gas through the system and over the sample.

C. The Balance

The requirements for a good automatic, continuously recording balance are essentially those of any analytical balance plus the ability to continuously record the data. The specifications must include accuracy, sensitivity, reproducibility, and maximum load. The recording balance should have the capability of covering an adequate range of mass adjustment, and it should show a high degree of mechanical and electronic stability. There should be rapid response to mass change, and the equipment should be mounted so that it is unaffected by vibration. The balance should also be unaffected by ambient temperature changes. Various weighing systems can be discussed but deflection and null-point balances are most often found in practice. The principle of the deflection balances is simply that the deflection or movement can be calibrated to read in terms of mass. Several different systems of deflection balances are shown in Fig. 2. These may be discussed separately. The beam balance utilizes conversion of the beam deflection about the fulcrum into a form suitable for reproduction. In early units a photographically recorded trace was used, but signals generated by suitable displacement— measuring transducers may be used or curves drawn electromechanically.

In the helical spring balance, Hooke's law is utilized to convert an elongation or contraction of the spring into a record of mass change. The spring material is important. Usually, a quartz fiber is employed, as this avoids anomalous results associated with temperature change and fatigue problems. The method has been used extensively in adsorption studies of gases and vapors on solids (Gregg and Sing, 1982). A balance of this kind was used by Loners (Loners, 1952) for the study of metal oxidation. The Aminco Thermo-Grav unit employed a precision spring that could be selected to give the desired sensitivity and damped magnetically or with a fluid dashpot to reduce oscillations (Daniels, 1973). The cantilevered beam has one end fixed, with the other end, on which the sample is placed, free to undergo deflections. The deflection can be translated into mass loss by methods utilized for

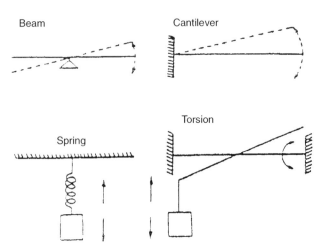

Figure 2 Schematic representation of the different types of deflection balances.

beam type balances. In the torsion-wire balance the beam is attached to a taut wire, which acts as the fulcrum. Deflections are proportional to changes in mass and the torsional characteristics of the wire. In all these deflection-type balances various techniques may be used for the measurement and recording of the deflection which, by calibration, can be converted to mass change. An optical lever arrangement can be used, involving measuring the deflections by means of a light beam reflected from a mirror mounted on the balance beam. These deflections can be recorded photographically or measured electronically by means of a shutter attached to the balance beam, to intercept a light beam impinging on a phototube. The light intensity is a measure of the beam deflection and is thus related to mass change. A further method of measuring deflection employs a linear variable differential transformer (LVDT) using an armature freely suspended from the balance beam into the coil of a differential transformer. Strain gauges can also be used to measure beam deflections.

The principle employed in the null-point balance is shown in Fig. 3. This system is used widely in TGA. A sensor on the balance detects the deviation from the null position, and this operates a servomechanism to restore the deviation to the null position. A review by Gordon and Campbell (1960) summarizes the various methods that have been employed to detect the deviation of a balance beam, the methods of restoring the beam to the null position, and the manner of recording the mass change. All the signals employed to detect and restore deviations in null-type balances can be collected by a dedicated computer suitably interfaced. Electronic laboratory balances are the subject of a review by Ewing (1976), and are discussed in detail in Chapter 2 of this Handbook.

D. Furnaces and Temperature

The furnace and control system should be capable of producing a wide range of temperature programs accurately. These should include isothermal runs, rising or falling temperature, temperature-jump, and parabolic and logarithmic temperature rise programs. The position of the furnace with respect to the balance is important. Figure 4 indicates the possible positions for a furnace in a TGA apparatus. The unit with two furnaces is designed to eliminate convection currents and other errors in

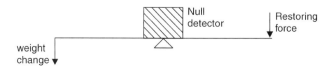

Figure 3 Schematic representation of the null-point balance.

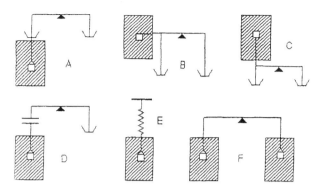

Figure 4 Schematic of the position of the furnace with respect to the balance. A, Below the balance beam. B, Extended balance beam. C, Above the balance beam. D, Remote coupling. E, Spring balance. F, Two furnaces.

weight associated with temperature changes, but it is essential to have both furnaces at exactly the same temperature. The prime considerations are ease of operation, minimal convection effects, and rapid cooling. A non-inductively wound furnace should be standard and is usually provided in commercial equipment.

Two common methods of making a furnace involve (a) winding a suitable metallic wire or ribbon on a ceramic form and then covering the wound wire with a ceramic paste or (b) coating a form with a thin layer of ceramic paste, winding the furnace wire or ribbon on top of this layer, and then covering the wire with a thicker layer of paste, and withdrawing the form. The winding process is followed by careful baking, surrounding the baked furnace with insulating material, then incorporating the assembly in a suitable housing, and finally providing the necessary electrical connections.

The temperature required governs the choice of material. However, the atmosphere is an additional factor that must be considered. It is possible to construct furnaces going up to 2800°C, but about 1650°C represents an easily attainable maximum temperature using commercial equipment. Table 2 indicates the maximum temperatures obtainable for furnace resistance elements for use in thermobalance construction. From this table it can be seen that up to 1100°C, Kanthal or Nichrome wire or ribbon can be used. There are special winding techniques, such as "coiled-coil", to accommodate differential expansion problems as the temperature is altered. Various platinum alloys, especially with high rhodium content allow temperatures to be reached as high as 1750°C. Higher temperatures are rarely required in TGA, but specialty furnaces employing tungsten, molybdenum, or graphite can be used for temperatures beyond 1750°C. These high temperatures require furnace operation in an inert atmosphere or even a reducing atmosphere.

Table 2 Maximum Temperature Limits in Furnace Construction for Thermobalances and Other Thermal Analysis Furnaces

Furnace element	Maximum temperature (°C)
Nichrome	1000
Chromel A	1100
Tantalum	1330
Kanthal or Nichrome	1350
Platinum	1400
Platinum—10% rhodium	1500
Platinum—20% rhodium	1600
Silicon carbide	1600
Rhodium	1800

Not all furnaces used in thermobalances are wire-wound. Sestak (1969) has, for example, constructed an infrared heater, its advantage being that it enables studying decompositions under vacuum conditions. It is not possible to construct one furnace suitable for all purposes. In a given commercial thermobalance assembly there may be as many as four interchangeable furnaces used to cover the ranges (a) −150–500°C, (b) ambient to 1400°C, (c) ambient to 1600°C, and (d) 400–1800°C. The low temperatures are achieved by a combined heating and cooling unit, usually with the furnace working against liquid nitrogen to allow accurate temperature programing. Furnaces have a finite lifetime, but without knowledge of the workload it is difficult to put this in terms of absolute figures. The higher the temperature reached by the furnace, the greater its use with an oxidizing atmosphere, the shorter the lifetime expectancy.

Some units sold commercially have large, massive furnaces, whereas in other units the furnaces are miniature. Furnaces of low mass cool quickly but hold little heat, whereas the converse holds true for high-mass furnaces. In a low-mass furnace a linear temperature rise is relatively more difficult to control. A high-mass furnace may hold an isothermal temperature but takes considerable time to achieve the required temperature. However, in everyday use, cooling requirements are paramount and the time required to reach a desired isothermal temperature must be as short as possible, and this usually means that the choice would be a system of low heat capacity. If a uniform hot zone is required, this demands some skill in furnace winding, and is invariably based to some extent on trial and error. The high-mass furnaces should provide a larger, uniform, hot zone. This is important, and with smaller furnaces the positioning of the sample holder must be considered very carefully, especially if a deflection balance system is used, when movement of the sample holder might bring the sample out of the hot zone.

Temperature measurement is most commonly accomplished by use of thermocouples. Chromel/alumel thermocouples can be used up to 1100°C, whereas platinum metal alloys can be used up to 1750°C. Beyond this temperature, tungsten/rhenium thermocouples can be used. One must consider the possibility of the thermocouple reacting with the sample, reactant, or decomposition products. The highest signal output is achieved using a base/metal thermocouple with the additional advantage that the chromel/alumel thermocouples' response to temperature is approximately linear. The platinum/platinum alloy thermocouples have lower sensitivities, and in the higher temperature ranges a nonlinear response. Another temperature sensor that is sometimes employed is the resistance thermometer. The resistance of the furnace winding can be used if it has a high temperature coefficient. Optical and radiation pyrometers are rarely used to measure sample temperature (Terry, 1965).

It is important to position the temperature-measuring device as close to the sample as possible. With the temperature sensor in other positions there can be a significant thermal lag between the furnace temperature and that of the sample. If the thermocouple is placed within the sample, the leads from the thermocouple can affect the sensitivity of the balance system. This was prevented by Dial and Knapp (1969) who used inductive coupling. The mechanical control of the furnace temperature has never really been satisfactory. The most often used technique is to actuate temperature control from temperature sensor measurements of the furnace temperature.

E. Computer Workstations to Record, Process, and Control the Acquisition of Data

The material discussed in this section applies to all thermal analysis techniques. A computer workstation is commonly used to control, collect, and process data for any thermal analysis technique. All the large instrument manufacturers supply such computer interfacing for TGA units. There are specialist computer firms that will interface any unit as required and allow the capability of processing thermal analysis data from commercially available programs or from programs devised by the investigators.

One problem that the user should keep in mind is that the computer data acquired from the temperature sensors is not always linear and the computer acquisition system must accurately convert the data signal to a temperature. Most computer users will require a "change of scale" potential from their equipment and also a program to record percent mass change against temperature. They will also require the derivative (DTG) plot. These are minimum requirements, and most interfaced computers can process the data in many more ways, as required by the objectives of the particular experiment.

In most commercial instruments, the experimental data is recorded either on the hard drive or on diskettes and a printer or plotter provides a printout of the data for the material with scales as required and all experimental conditions noted. In the case of TGA data, the initial and final temperatures can be presented on the plot. The TG data can be processed to produce the DTG curve. The weight change can be expressed as weight or percentage scales. Each stage, in a multiple stage decomposition, can be plotted as a fraction decomposed. This last method of expressing the data is especially useful in the first stage of computer calculations for kinetic parameters based on rising temperature data instead of the more classical isothermal kinetic runs at various temperatures (Chen et al., 1992; Dollimore et al., 1991). The computer workstation usually provides a real-time presentation of the experimental data so that the experiment may be stopped or altered in design as the data is unfolded on the monitor screen.

F. Reporting Results

In reporting TG results (or for that matter, any other thermal method) the recommendations of the standardization committee of ICTA must be considered (Dunn, 1993). There is a wide diversity of equipment and no single instrument represents the optimum design for all possible studies. The TG technique is dynamic and the results often involve kinetic factors, producing data that are highly dependent on procedure.

The use of the word "sample" refers to the actual material investigated. The use of this word is common to all thermal analysis techniques. In TG equipment, the balance assembly and furnaces are referred to as a *thermobalance*. The sample in TG is rarely diluted, but it might be diluted in simultaneous TG–DTA. The sample holder in TG is then the container or support for the sample.

In thermal analysis techniques a thermocouple or some other temperature sensor is used to measure the temperature. The position of this thermocouple with reference to the sample should be recorded. Again, in all techniques, the heating rate should be noted as the rate of temperature increase (degrees per minute). The use of the computer workstation enables more sophisticated temperature programs to be employed, but it is imperative that such programs should be accurately noted. If the equipment is undergoing cooling, it is termed the "cooling rate." The "heating rate" is said to be constant when the temperature/time curve is linear, but this should be checked. The plot of mass against temperature from the TG equipment is called the *thermogravimetric curve* or TG curve,

and if differentiated it is called the *derivative thermogravimetric curve* or DTG curve.

A single stage TG curve, shown in Fig. 5, serves to explain the various terms used in connection with TGA. The points made from this figure can readily be extended to a multistage process, which can simply be considered as a series of interconnected single-stage processes. Here a plateau (AB in Fig. 5) is that part of the TG curve where the mass remains essentially constant. The *initial temperature*, T_i (B in Fig. 5), is that temperature at which the cumulative mass change can first be detected on the thermobalance curve. Likewise, the *final temperature*, T_f, (C in Fig. 5), is that temperature at which the cumulative mass change reaches a maximum. The *reaction interval* is sometimes quoted representing the difference between T_f and T_i. The curve is often represented in differential form (i.e., rate of weight loss plotted against temperature or time). The following recommendations on reporting data apply not only to TG but also to other thermal methods. The following information should be provided:

1. Identification of all substances (sample, reference, and diluent) by a definitive name, an empirical formula, or equivalent compositional data.
2. A statement of the source of all substances, details of their histories, pretreatments, and chemical purities, so far as these are known.
3. Measurement of the average rate of linear temperature change over the range involving the phenomena of interest. Nonlinear temperature programing should be described in detail.
4. Identification of the sample atmosphere in terms of pressure, composition and purity; whether the atmosphere is static, self-generated, or dynamic, pressure and humidity should be specified. If the pressure is other than atmospheric, full details of the method of control should be given. If the

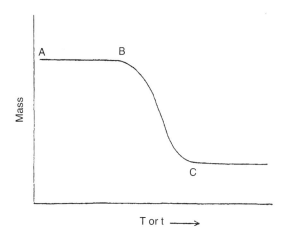

Figure 5 Schematic TG curve.

system involves a flowing gas supply, then the flow rate should be specified. Where purification of the flowing gas supply is practiced, the purification methods should be stipulated.

5. A statement of the dimensions, geometry, and materials of the sample holder.
6. Sample mass should be recorded.
7. Identification of the abscissa scale in terms of time or of temperature at a specified location. Time should be plotted to increase from left to right.
8. A statement of the method used to identify intermediates or final products.
9. Faithful reproduction of all original records.
10. Identification of the apparatus used, by type and/or commercial name, together with details of the location of the temperature-measuring thermocouple.

In reporting TG data the following additional information should be recorded:

1. A statement of the sample mass and mass scale for the ordinate. Mass loss should be plotted downward, but this follows automatically if the ordinate scale is simply that of mass of sample. Additional scales such as fraction decomposed, percent mass loss, or molecular composition are very useful and may be included as desired.
2. If DTG is employed then the method of obtaining the derivative should be indicated and the units of the ordinate noted.

G. Standards for TG

In TG two parameters need verification, namely, the recorded mass and the temperature. The mass scale can be checked by use of standardized masses. Often the changes in mass can be checked by checking performance against a multistage degradation of a compound for which these parameters are well established.

Calcium oxalate monohydrate comes into this category and Fig. 6 provides an example of such a plot. It goes through three separate decomposition changes:

$$CaC_2O_4 \cdot H_2O \longrightarrow CaC_2O_4 + H_2O$$
$$CaC_2O_4 \longrightarrow CaCO_3 + CO$$
$$CaCO_3 \longrightarrow CaO + CO_2$$

The calibration of temperature has proved difficult, as the temperature sensor may not be at the temperature of the sample because of temperature gradients existing between the furnace and the sample. Two methods have been proved useful. The use of materials with solid–liquid phase changes has been developed in such a way that on changing to a liquid the material is lost from the balance sample assembly and so there is a dramatic mass loss. McGhie and coworkers (McGhie, 1983; McGhie et al., 1983) have described this method, termed "fusible-link temperature calibration". The equipment is shown in Fig. 7. In this a fusible metal standard is suspended from a coil of platinum wire. All this is held in position in the sample container of the DuPont balance. A hole is positioned in the bottom of the sample container so that when the fusible metal standard melts, it and the platinum

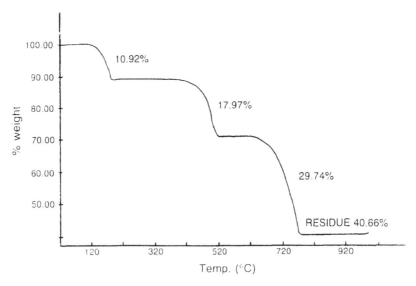

Figure 6 The TG plot for calcium oxalate monohydrate in air. Plateau A: 182.4–336°C, 89.1% residue; plateau B: 503–538°C, 70.5% residue; plateau C: 759–1020°C, 40.7% residue.

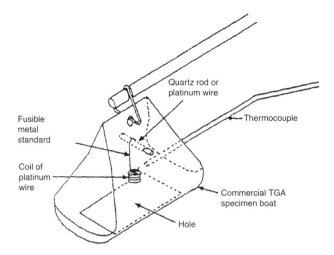

Figure 7 Fusible link unit on the DuPont TG balance.

Table 3 Fusible-Link Temperature (°C) Calibration Data

Material	Observed corrected	Literature	Deviation
Indium	154.20	156.63	−2.43
Lead	331.05	327.50	3.55
Zinc	419.68	419.58	0.10
Aluminum	659.09	660.37	−1.25
Silver	960.25	961.93	−1.68
Gold	1065.67	1064.43	1.24

Source: McGhie (1983).

coil are lost from the sample container, resulting in an immediate change of mass. A typical result using zinc is shown in Fig. 8. Suitable materials and relevant data are shown in Table 3.

The use of ferromagnetic standards requires a small modification to the thermobalance, in that the ferromagnetic material is placed in the sample container and suspended within a magnetic field. At the temperature of the material's Curie point the magnetic mass diminishes and a mass loss is recorded by the thermobalance. Norem et al. (1970) indicate the following criteria for the effective operation of this calibration:

1. The transition must be sharp.
2. The energy required should be small (to ensure its production under dynamic scanning conditions as a sharp transition).
3. The transition temperature should be unaffected by the chemical composition of the atmosphere and be independent of pressure.
4. The transition should be reversible so that the sample can be used repeatedly.
5. The transition should be unaffected by the presence of other standards so that a single experiment will determine several temperatures.
6. The transition should be observable when using milligram quantities.

The data in Fig. 9 shows a typical run. An ICTA-certified Magnetic Reference Materials Set GM761 was reported by Blame and Fair (1983) to give the results shown in Table 4. There is a dependence of the initial transition temperature on the rate of heating. This is small over the heating range from 5°C/min to 20°C/min but appreciable for fast heating rates (EIder, 1982; Gallagher et al., 1980).

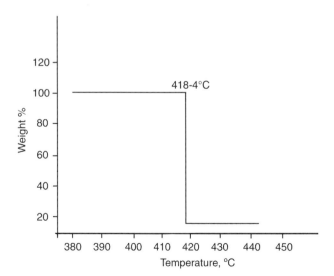

Figure 8 Fusible link mass-drop curve for zinc.

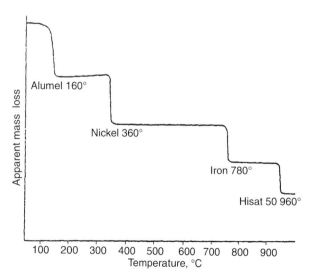

Figure 9 The Curie point for the ferromagnetic standards alumel, nickel, iron, and HISAT-50.

Table 4 Results Obtained by Blaine and Fair (1983) for Magnetic Transition Temperatures (°C) of Magnetic Standards GM761

Material	Experimental	Literature	Deviation
Permanorm 3	259.6 ± 3.7	266.4 ± 6.2	−6.8
Nickel	261.2 ± 1.3	351.4 ± 5.4	6.8
Mumetal	403.0 ± 2.5	385.9 ± 7.2	17.1
Permanorm 5	431.3 ± 3.7	459.3 ± 7.3	−28.0
Trafoperm	756.0 ± 1.9	754.3 ± 11.0	2.2

III. TGA AND DTG

In TG or TGA the change in sample mass is determined as a function of temperature and/or time. The instrument is a thermobalance that permits the continuous weighing of a sample as a function of time. The sample holder and a reference holder are bound to each side of a microbalance. The sample holder is in a furnace, without direct contact with the sample, the temperature of which is controlled by a temperature programer. The balance part is maintained at a constant temperature. The instrument is able to record the mass loss or gain of the sample as a function of temperature and time $[m = f(T)]$. Most instruments also record the DTG curve which is the rate of the mass change $dm/dt = f(T)$.

The DTG curves allow a better distinction of overlapping steps as demonstrated in Fig. 10 for $CuSO_4 \cdot 5H_2O$. The area under the DTG curve is proportional to the mass change and the height of the DTG peak at any temperature gives the rate of mass change. The real advantage of DTG is that it permits an accurate location of the end of a desolvatation process if decomposition follows desolvatation by use of the minima in the DTG curve.

The instrument used in TGA is a thermobalance (balance controller, sample chamber, furnace, and furnace controller) with data processing. In order to check the stability of the system a baseline at the highest sensitivity has to be made for all heating rates in the temperature range of analysis: The highest deviation will be observed at the highest heating rate. The thermobalance may have a vertical or horizontal construction. The sensitivity of new thermobalances attains 0.1 μg. Some manufacturers offer combined DSC/TG instruments.

The mass accuracy is generally not a problem in modern TG. Calibration of the mass with certified mass can be used as for all other balances. Electrostatics, temperature fluctuation, sensitivity of the sensor, and thermal lags have to be known, which is best done with regular calibration. For automatic TG, the pans have to be tightly closed and pierced just before the measurement; therefore, the TG curves for desolvatation may be different

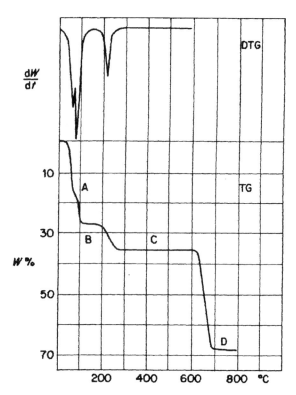

Figure 10 Overlapping steps of DTG curve compared with the TG curve for $CuSO_4 \cdot 5H_2O$ heated in air at 3°C/min in a platinum crucible. The loss of three water molecules is seen in the DTG plot whereas the fifth water has not been driven off even at 500°C in the TG plot.

for open pans. The use of a protective gas and its flow, as well as the sample mass and the heating rate play a role in the comparison of the temperature of thermal events. The influence of heating rate is exemplified in Fig. 11 with $CuSO_4 \cdot 5H_2O$. The limit of detection can be calculated by determining the maximum deviation of the baseline in the temperature range of interest. Depending upon the material being investigated the TG curve can occur in a number of recognizable shapes as seen in Fig. 12 and demonstrated by the metallic oxalates given in Table 5.

Table 6 shows an example of calibration performed with hydrates in which, over the starting of dehydration, temperatures range from 50°C to 120°C and, at the end of dehydration, from approximately 150°C until 270°C with different heating rates.

As there is no contact between the pan and furnace the "thermal lag" is higher than that found in DSC. The standards recommended by ICTA and distributed by NBS are ferromagnetic standards (discussed in Section II) exhibiting loss of ferromagnetism at their curie point temperature within a magnetic field: nickel (354°C), Permanorm 3 (266°C), Numetal (386°C), Permanorm 5 (459°C), Trafoperm (754°C). The method does not permit the temperature measurement with high precision.

Thermoanalytical Instrumentation

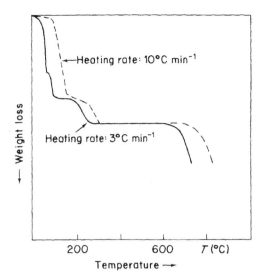

Figure 11 Influence of heating rate for CuSO$_4\cdot$5H$_2$O heated in air in a platinum crucible.

Table 5 Classification of TGA Results into Types of Shape of the Curve

Oxalate	Decomposition	
	In air	In N$_2$
Copper	D	E
Zinc	A	A
Cadmium	E	E
Aluminum	B	B
Tin	E	E
Lead	D	E
Thorium	A	A
Antimony	D	E
Bismuth	D	E
Chromium	B	B
Manganese	A	A
Ferrous	C	C
Ferric	C	—
Cobalt	A[a]	A
Nickel	A[a]	A

[a]Rapid heating in a limited supply of air can result in a curve of type D.

These standards have been studied by several authors (Charsley et al., 1987a). The ICTA temperatures are within 5–10°C. McGhie et al. (1983) proposed a calibration technique which provides for a small inert platinum weight suspended by a fusible link composed of a calibration standard which releases the platinum weight at the temperature of melting. The Mettler instrument TGA 850 is constructed so that the melting curve for standards can be measured and used as a calibration as demonstrated in Table 6. Table 7 provides a TG experiment using ferromagnetic materials in a magnetic field.

TG can be used with different atmospheres and under vacuum. TG has a huge number of pharmaceutical applications. Automated TG is extremely efficient in replacing the loss of drying assay in drug substances, being able to separate loss of solvent from decomposition by using very small amounts of substance. Solvent entrapped or bound as solvate is easily determined (Giron, 1986; Giron et al., 1999; Komatsu et al., 1994). A comprehensive chapter on TG has been recently written by Dunn and Sharp (1993). Ozawa proposes the use of modulated TG for kinetic analysis (Ozawa, 2000).

Water sorption–desorption isotherms can be carried out by using thermobalances. Newer specific instruments allow measuring water sorption–desorption isotherms at different constant temperatures and include the following [e.g., dynamic vapour sorption instrument (DVS) and Surface Measurement Systems Ltd., Monarch Beach, USA].

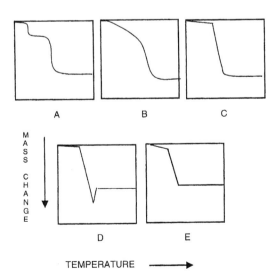

Figure 12 Shapes of TG curves.

IV. DTA AND DSC

A. Basic Considerations

DTA must be regarded as being a qualitative technique in the sense that the response, through lack of calibration or because it does not respond in a quantitative manner, cannot be interpreted in terms of enthalpy change. The temperature at which changes are observed to take place should, however, be an accurate response. Once properly calibrated as a calorimeter, the equipment is termed a *differential scanning calorimeter*.

Two modes of DSC are currently available by instrumentation manufacturers, namely, power-compensation differential calorimetry (power-compensation DSC) and

Table 6 Four Sets of NBS-ICTA-Certified Reference Materials Available for Temperature Calibration of DTA Apparatus

GM-757[a] (−83°C to −58°C)	GM-758 (125–435°C)	GM-759 (295–575°C)	GM-750 (570–940°C)
1.2-Dichloroethane	Potassium nitrate	Potassium perchlorate	Quartz
Cyclohexylchloride	Indium	Silver Sulfate	Potassium Sulfate
Phenyl ether	Tin	Quartz	Potassium chromate
o-Terphenyl	Potassium perchlorate	Potassium sulfate	Barium carbonate
	Silver sulfate	Potassium chromate	Strontium carbonate

[a]Melting point.

Table 7 TG Experiment Using Ferromagnetic Materials in a Magnetic Field (°C)

Substance	Observed temperature	Actual temperature
Alumel	160	163
Nickel	360	354
Iron	780	780
Hisat	960	1000

heat-flux differential scanning calorimetry (heat-flux DSC). The two methods need to be distinguished from each other, as power-compensation DSC was for a long time a copyrighted term employed by one of the instrument manufacturers and therefore the term DSC is often found in the literature applying to power-compensation equipment only (Watson et al., 1964).

Typically a heat-flux unit would employ multiple sensors, as in the Calvet-type arrangement (Calvet and Prat, 1963) or a controlled heat leak, as in the Boesma arrangement (Boersma, 1955). It should be noted that because DSC is considered by some to measure thermodynamic properties, the DSC plots are often found with the endothermic peak plotted in an upward direction and the exothermic peak in the downward direction. This is to conform with IUPAC requirements on the presentation of thermodynamic parameters.

When a material is heated or cooled, there is a change in its structure or composition. These transformations are connected with heat exchange. DSC is used for measuring the heat flow into and out of the sample as well as for determining the temperature of the thermal phenomenon during a controlled change of temperature. The first method developed by Le Chatelier in 1887 was DTA where only the temperature induced in the sample was measured.

B. Design Factors and Theory

In classical DTA (heat-flux DSC), a sample cell and a reference cell are subjected to the same temperature program. The names of the cells indicate their use. Thermocouples placed in each cell work against one another and record an AT signal against the temperature of the system. This is illustrated in Fig. 13. As already mentioned, if the DTA can be calibrated to produce enthalpy data, then most commercial instrument manufacturers would call the technique DSC. There are, however, two clearly recognized assemblies of cells and heaters in DSC. The first is the power-compensation equipment (Watson et al., 1964). This is schematically shown in Fig. 14. It can be seen that two heaters are employed. Then in any heating program heat energy absorbed or evolved by the sample is compensated for by adding or removing an equivalent amount of energy in the heater located in the sample holder. Hence, the power needed to maintain the sample holder at the same temperature as the reference holder provides an electrical signal opposite but equivalent to the varying thermal behavior of the sample. Because the contents of the two cells are

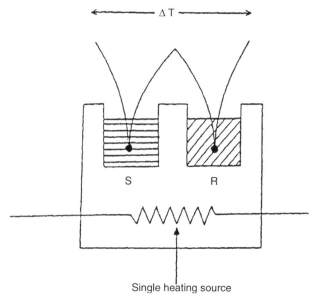

Figure 13 Classical DTA (heat flux DSC) (S, sample cell; R, reference cell).

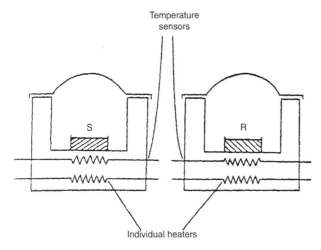

Figure 14 Power compensation DSC (S, sample cell; R, reference cell).

maintained at an identical temperature by electronically controlling the rate at which heat is supplied to the sample and the reference materials, the ordinate of the DSC curve represents the rate of energy absorption by the test sample relative to that of the reference material, the rate depending on the heat capacity of the sample.

The "Boersma" cell also gives a good calorimetric response. This DSC assembly is shown in Fig. 15.

A simplified heat transfer theory of DTA/DSC is based on the method developed by Vold (1949). The equation of heat balance for the cell containing the reactant is:

$$C \, dT_1 = dH + K(T_3 - T_1) \, dt \quad (1)$$

and for the cell containing the inert material:

$$C \, dT_2 = K(T_3 - T_2) \, dt \quad (2)$$

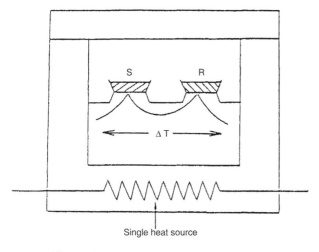

Figure 15 Schematic of a "Boersma" DTA.

where C is the heat capacity for each cell. T_1, T_2, and T_3 are the temperatures of the reactant, reference material, and block, respectively, K is the heat transfer coefficient between the block and the cell, and dH is the heat evolved by the reaction in time, dt.

Assumptions which make it possible to write the preceding two equations are:

1. The heat capacities of the two cells are the same and do not change during the reaction.
2. There is a uniform temperature throughout the sample at any instant.

Subtraction then gives

$$C(dT_1 - dT_2) = dH + K(T_2 - T_1) \, dt \quad (3)$$

But $T_2 - T_1 = \Delta T$, the difference signal recorded in DTA/DSC. Thus,

$$C \, d\Delta T = dH - K\Delta T \, dt \quad (4)$$

The total heat of reaction is then found by integration from $t = 0$ to $t = \infty$

$$\therefore \quad \Delta H = C(\Delta T - \Delta T_0) + \int_0^\infty \alpha_0 \quad (5)$$

When $t = 0$, $\Delta T = 0$ and when $t = \infty$, $\Delta T = 0$

$$\therefore \quad \Delta H = KS \quad (6)$$

where S is the area beneath the DTA peak.

The influence of physical properties on the baseline can be considered more realistically by assigning values of C and K to each cell, and considering the simple case where there is no reaction (i.e., $dH = 0$), then

$$C_1 \, dT_1 = K_1(T_3 - T_1) \, dt \quad (7)$$

and

$$C_2 \, dT_2 = K_2(T_3 - T_2) \, dt \quad (8)$$

C_1 and K_1 refer to the reactant cell and C_2 and K_2 to the reference cell. On rearrangement this gives

$$T_1 = T_3 - \frac{C_1}{K_1} \frac{dT_1}{dt} \quad (9)$$

and

$$T_2 = T_3 - \frac{C_2}{K_2} \frac{dT_2}{dt} \quad (10)$$

The values of dT_1/dt and dT_2/dt represent heating rates and should be identical, that is:

$$\frac{dT_1}{dt} = \frac{dT_2}{dt} = \frac{dT}{dt} \quad (11)$$

then

$$\Delta T = T_1 - T_2 = \frac{dT}{dt}\left(\frac{C_2}{K_2} - \frac{C_1}{K_1}\right) \quad (12)$$

We now have three cases:

$$\frac{C_2}{K_2} = \frac{C_1}{K_1} \quad \text{and} \quad C \neq f(T) \quad (13)$$

which is demonstrated by a zero value of ΔT (Fig. 16).

$$\frac{C_2}{K_2} > \frac{C_1}{K_1} \quad \text{and} \quad CK \neq f(T) \quad (14)$$

which gives a positive value of ΔT (see Fig. 16) and

$$\frac{C}{K_2} < \frac{C_1}{K_1} \quad \text{and} \quad CK \neq f(T) \quad (15)$$

which gives a negative constant value of ΔT (Fig. 16).
If

$$\frac{C_2}{K_2} < \frac{C_1}{K_1} \quad (16)$$

but CK is a linear function, $f(t)$, then the linear plot shown in Fig. 17 results,
If, however,

$$\frac{C_2}{K_2} < \frac{C_1}{K_1} \quad (17)$$

but cK is a quadratic juction, $f(I)$, then the curved plot of Fig. 16 results.
If, however,

$$\frac{C_2}{K_2} > \frac{C_1}{K_1} \quad (18)$$

then the slopes are in the opposite direction.

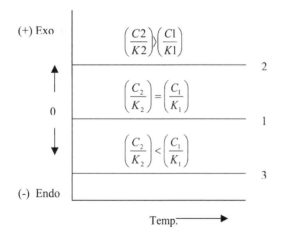

Figure 16 DTA baseline. In all cases $(CK) \neq f(T)$.

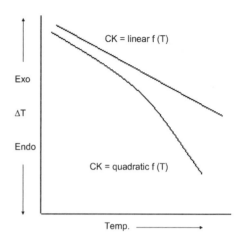

Figure 17 DTA baseline, where $C_2/K_2 < C_1/K_1$.

For a more comprehensive treatment, the publications of Wilburn and his co-workers should be consulted (Wilburn et al., 1969, 1991).

A block diagram of a DTA unit is shown in Fig. 18. The sample holder measuring system comprises the thermocouples, sample and reference containers, and associated equipment. The furnace system varies in size from one instrument to another but must have a uniform temperature zone around the sample and reference cell.

A linear response with time was previously regarded as the major requirement in temperature programing. Now, however, the computer workstation will include

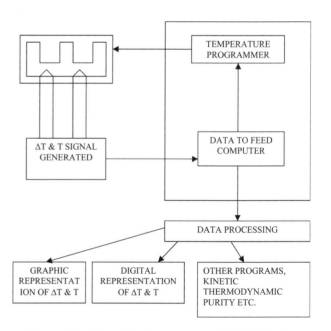

Figure 18 Schematic of contemporary DTA/DSC units.

programs that allow a whole variety of optional temperature responses against time to be used. These include stepped temperature jumps and, for example, logarithmic temperature–time control. The range of temperatures available (although not in a single unit) is from around −200°C to around +2000°C. Generally, equipment is commercially available as a single unit for the temperature ranges from −200 to +500°C and for temperatures from ambient to around 1000°C. Units have also been introduced operating up to 2000°C. DSC units normally operate up to a maximum temperature of about 700°C. Beyond this temperature other heat flux conditions begin to operate, which complicates calibration, but this has not prevented units being developed up to about 1000°C. High-pressure systems have also become available. This variety of operating conditions leads to a correspondingly greater range of cell designs. They all give responses which are dictated by experimental conditions. This is true for both low- and high-temperature systems.

The principle of DSC is as follows: two ovens are linearly heated; one oven contains the sample in a pan, the other contains an empty pan as a reference pan. If no change occurs in the sample during heating, the sample pan and the reference pan are at the same temperature. If a change, such as melting, occurs in the sample, energy is used by the sample and the temperature remains constant in the sample pan, whereas the temperature of the reference pan continues to increase. Therefore, a difference of temperature occurs between the sample pan and reference pan.

In power-compensated DSC, two individual heaters are used to monitor the individual heating rates of the two individual ovens. A system controls the temperature difference between the sample and reference. If any temperature difference is detected, the individual heatings are corrected in such a way that the temperature is kept the same in both pans. That is, when an endothermic or exothermic process occurs, the instrument delivers the compensation energy in order to maintain equal temperature in both pans.

In heat-flux DSC the temperature is primarily measured; in power-compensated DSC energy is primarily measured. The differentiation of measuring principles with modern instrumentation is not very significant under normal applications. Owing to calibration and integrated data handling the instruments produce similar qualities of reported results. Each instrument can deliver the same information, namely, heat flow as a function of temperature (or time). The peak shape, resolution, and sensitivity depend upon the principle of measurement and the specification of the instrument.

For first-order transitions such as melting, crystallization, sublimation, boiling, and so on, the integration of the curve gives the energy involved in the transition. For second-order transitions, the signal gives the change in the specific heat (i.e., glass transitions).

Figure 19 shows typical transitions. Melting and crystallization are first-order transitions. The extrapolated onset temperature (T_e) is the melting or boiling point. The peak temperature T_p is dependent on instrument and measurement parameters. The glass point is determined as an inflexion point. Manufacturers represent the heat flow in a variety of ways, the endotherms on the positive side for power-compensation DSC and, on the negative side for heat-flux DSC. Melting, boiling, and sublimation are endothermic, which means they need energy. Crystallization is exothermic, which means it supplies energy. Desolvatations without melting are generally endothermic. Solid-state phase transition and decomposition may be endothermic or exothermic. Modern instruments provide heating, cooling, and isotherms between subambient temperatures (with a cooling device) and higher temperatures in the range of 1200–1500°C. In order to avoid reactions with the atmosphere the measurements are carried out under nitrogen. The major components of the systems include the DSC sensors, the furnace, the program, and data handling. The temperature plotted on the abscissa is the programed temperature, not the temperature of the sample. The difference between the programed and the actual temperature of the sample is called "thermal lag". It depends on the thermal resistance of the instrument and the heating rate. Modern instruments are dedicated to accurate analytical measurements for pharmaceuticals, the sensors are in direct contact with the bottom of the pans and the sample size is in the milligram range or less. Therefore, any correction is not very high, but it has to be taken into consideration. Generally, pure indium (>99.9999%) is used for the correction of the thermal lag.

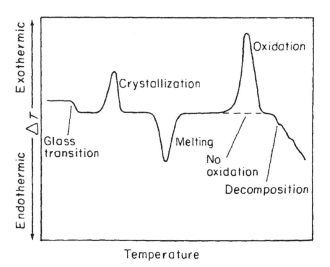

Figure 19 Melting and crystallizations transitions schematic showing the type of changes encountered with polymeric materials using a DTA/DSC recording.

Extreme efforts have been made in recent years to validate the different instruments, not only in comparing principle and results, but also in determining the critical parameters such as heating and cooling rates, particle size, weight, resolution, atmosphere, and type of pans (crimped pan, sealed pan, open pan, etc.).

The instruments are automated and data acquisition is instantaneous via computer. The calibration of the instrument should be done on a defined periodic basis not just yearly. This includes the measurement of temperature and enthalpy. Most certified standards are highly purified metals. Indium is the preferred reference standard, but it covers only one temperature. It is recommended to include several organic substances for which the melting point or the melting enthalpy has been accurately determined. Sarge and coworkers (Sarge and Gmelin, 2000; Sarge et al., 2000) proposed several organic substances and metals. The heat determination of quartz was also recommended. Sabbah (1999) published a broad review of data for organic substances. It is usually suitable to have several certified materials covering a broad range of temperatures corresponding to the thermal events of interest (Giron et al., 1989).

Tables 8 and 9 are, respectively, examples of temperature calibration and the calorimetric response for a PE-DSC-7 instrument using different materials (Giron, 2000). It is very important that the confidence of the laboratory which delivers the reference be maintained. As the heating rate may have an influence on the data, it is recommended to compare the melting point and the melting enthalpy of organic standards in addition to indium, at different heating rates covering the measurement range. For very accurate determinations it is recommended to use standards with a melting in the range of the considered temperature in a serial of measurements.

In pressure DSC (PDSC), the sample can be submitted to different pressures, which allows the characterization of substances at the pressures of processes or to distinguish between overlapping peaks observed, for example, by desolvatation (Han et al., 1998).

Modulated DSC (MDSC) is a new technique introduced in 1993 (Gill et al., 1993), which has been thoroughly examined and discussed (Hohne, 1999). The main advantage of this technique is the separation of overlapping events in the DSC scans. In conventional DSC, a constant linear heating or cooling rate is applied. In MDSC the normally linear heating ramp is overlaid with a sinusoidal function (MDSC) defined by a frequency and amplitude to produce a sine wave-shaped temperature vs. time function. Using Fourier mathematics, the DSC signal is split into two components: reflecting nonreversible events (kinetic) and reversible events.

$$T = T_0 = bt + B\sin(\omega t) \quad (19)$$

$$\frac{dq}{dt} = C(b + B\omega\cos(\omega t)) + f(t, T) + K\sin(\omega t) \quad (20)$$

where T is temperature, C is the specific heat, t is time, ω is frequency, $f(t,T)$ is the average underlying kinetic

Table 8 Example of Calibration of Perkin–Elmer DSC-7 Instruments with Melting Standards at 10 K/min Under Nitrogen

Certified substances	Onset T (°C) certificate	Instrument 1 onset T (°C)	ΔT (°C)	Instrument 2 with intracooler onset T (°C)	ΔT (°C)
Iodobenzene	−31.3			−32.2	0.9
H$_2$O	0.0			0.1	0.1
4-Nitrotoluene	51.5	50.4	1.1	51.2	0.3
Biphenyl	69.3	68.2	1.1	68.6	0.7
Naphthalene	80.2	79.4	0.8	80.1	0.1
Benzil	94.7	94.21	0.6	94.5	0.2
Acetanilide	114.0	113.9	0.1	113.6	0.4
Benzoic acid	122.1	122.0	0.1	121.8	0.4
Diphenylacetic acid	146.5	146.9	0.4	146.9	0.4
Indium	156.6	156.8	0.2	156.5	0.1
Anisic acid	183.1	183.6	0.5	183.2	0.1
2-Chloro-anthraquinone	210.0	210.1	0.1	210.1	0.1
Tin	231.9	232.7	0.8	232.7	0.8
Anthraquinone	284.5	285.2	0.7	284.8	0.7
Lead	327.5	328.6	1.1	—	—
Zinc	418.9	420.3	1.4	—	—

Table 9 Example of Calorimetric Measurements of Standards with two Different DSC-7 Instruments and Measurement Cell at Different time at 10 K/min Under Nitrogen

Standard substance	ΔH (J/g) Theory	A ΔH (J/g)	% Deviation	B ΔH (J/g)	% Deviation	C ΔH (J/g)	% Deviation
Naphthalene (80.2°C)	148.6	147.1	1.0	148.6	0.0	—	—
Benzil (94.7°C)	112.0	110.1	1.7	112.8	0.7	—	—
Benzoic acid (80.2°C)	147.2	—	—	—	—	14.66	0.4
Biphenyl (69.3°C)	120.4	120.0	0.6	—	—	120.5	0.1
Diphenyl acetic Acid (146.5°C)	146.9	—	—	146.8	0.1	—	—
Indium (156.6°C)	28.7	28.63	0.2	28.8	0.35	28.7	0.1
Tin (231.9°C)	60.2	60.0	0.3	—	—	60.8	1.0

function once the effect of the sine wave modulation has been subtracted. The value K is the amplitude of the kinetic response to the sine wave modulation, and $(b + B\omega \cos(\omega t))$ is the measured quantity dT/dt or "reversing" curve.

The total DSC curve, the reversing curve giving reversible transitions, and the nonreversing curve giving irreversible transitions (e.g., the glass transitions) are obtained. MDSC is a valuable extension of conventional DSC. Its applicability (Coleman et al., 1996) is recognized for precise determination of the temperature of glass transitions and for the study of the energy of relaxation and depends on the number of important parameters to be studied. It has been recently applied for the determination of glass transitions of hydroxypropylmethylcellulose films (McPhillips et al., 1999) and for the study of amorphous lactose (Craig et al., 2000) as well as for the study of some glassy drugs (Bottom, 1999).

Microwave thermal analysis (MWTA) is also a newer technique (Parkes et al., 2000), in which microwaves are used both to heat a material and as a means of detecting thermal transitions.

Micro-DSC is becoming very popular. The instruments for conventional DSC allows one to measure very small amounts of material. It was possible to characterize the melting peak of indium with 0.032 mg using any of today's modern DSC instruments. Newer instrument generations will permit increased sensitivity and a reduction in the amount of material studied to the nanorange (Olson et al., 2000). This technique will be discussed in detail later.

C. Experimental Factors

DTA and DSC units are all responsive to changes in experimental conditions. These conditions include the packing of the sample, the container and thermocouple design, the ambient pressure, and the actual temperature program. Disc-shaped thermocouples are utilized in some low-temperature DTA/DSC units. A major problem in their operation is transporting heat uniformly away from the sample, which often precludes the use of ceramic blocks. The use of the disc-shaped thermocouples (usually either chromel/alumel or iron/constantan) ensures optimum thermal contact, provided the sample container is similarly flat-bottomed. The early design by Jensen and Beevers (1938) for low-temperature DTA/DSC units used a block of metal in contact with a liquid nitrogen reservoir to transmit the low ambient temperature to the region around the sample and reference containers. The temperature program is then achieved by use of a furnace operating against the low ambient temperature achieved initially. In other designs the furnace assembly is initially cooled by a liquid nitrogen pumping system. Thermal control is then achieved by keeping the nitrogen flow constant and programing the furnace heat input.

The use of high-pressure DSC cells allows phase transitions and chemical dissociations to be examined as a function of the pressure of the ambient atmosphere. However, in such cases it is important to have accurate pressure measuring devices attached to the equipment. These high-pressure cells are also extremely useful in operating the DTA/DSC system at high vacuum or controlled pressures, which may be less than one atmosphere. There is a significant dependence of the DTA/DSC results upon the rate of heating. The effect on kaolin of lowering the heating rate has been demonstrated (Spiel et al., 1945). The effect of the heating rate is often reported in terms of the initial deviation temperature (T_j), the magnitude of the ΔT signal at peak maximum (ΔT_p), and the actual temperature measured. The increase in heating rate increases T_j, ΔT, and T_f. The choice of the temperature measurement is all-important in considering peak area. Wilburn et al. (1991) concluded that if ΔT [given by $(T_s - T_r)$ where T_s is the temperature of the sample and T_r the temperature of the reference material] is plotted against T_s then the peak area is proportional to the heating rate. If $(T_s - T_r)$ is measured against (t) then it is independent of the heating rate. Provided the peak area is defined by the latter method, then, on a simple theoretical basis, it must also be independent of the rate at which the reaction takes place and the specific heat of the sample, provided it remains constant. It does, however, depend on the heat of reaction per unit volume of the sample, on the thermal conductivity of the materials in the furnace and the sample, and on the heat transfer coefficient between the holder and the furnace wall (Cunningham and Wiburn, 1971). Furthermore, provided that ΔT is less than 10°C, then the addition of terms in refined treatments is not likely to seriously affect the conversion of peak area to reliable calorimetric data. It can be shown in addition that ΔT_p is dependent on heating rate.

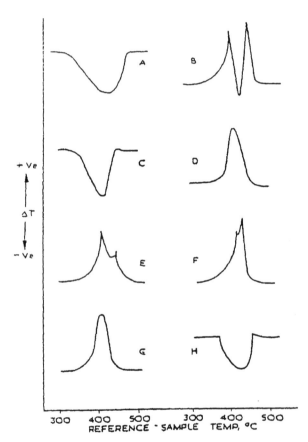

Figure 20 DTA of zinc oxalate dehydrate under various experimental conditions. (The dehydration peak is endothermic in all cases: this section of the curves has been omitted.) (A) sample weight 1.00 g, heating rate 5°C/min, static air; (B) sample weight 95 mg, heating rate 10°C/min, static air; (C) sample weight 82 mg, heating rate 10°C/min, nitrogen, 1000 cm³/min; (D) sample weight 82 mg, heating rate 10°C/min, oxygen, 1000 cm³/min; (E) sample weight 97 mg, heating rate 5°C/min, static air; (F) sample weight 80 mg, heating rate 5°C/min, static air; (G) sample weight 66 mg, heating rate 5°C/min, static air; and (H) sample weight 206 mg, heating rate 5°C/min, static air.

Figure 20 shows data obtained by Judd and Pope (1971) for the decomposition of hydrated zinc oxalate. The dehydration of the zinc oxalate dihydrate ($ZnC_2O_4 \cdot 2H_2O$) is always endothermic:

$$ZnC_2O_4 \cdot 2H_2O(s) \longrightarrow ZnC_2O_4(s) + 2H_2O(g) \qquad (21)$$

and occurs as a separate stage. The thermal decomposition is represented by

$$ZnC_2O_4(s) \longrightarrow ZnO(s) + CO(g) + CO_2(g) \qquad (22)$$

and is endothermic. However, in air, the zinc oxide provides a catalytic surface for the reaction

$$2CO(g) + O_2(g) \longrightarrow 2CO_2(g) \qquad (23)$$

and this reaction is exothermic. Experimental conditions are such, however, that the various cells and different experimental conditions can make the overall reaction in air exothermic or, by virtue of the two processes not matching, much more complicated. It is at once apparent that the actual experimental conditions pertinent to the type of cell, heating rate, sample mass, and environmental gas should be considered in any DTA/DSC study that is proposed.

D. Reporting Results

The word "sample" has already been defined as the material being investigated, and for DTA the sample may be used diluted or undiluted (this should be noted). A *reference material* is often required in DTA; it is "a known substance, inactive thermally over the temperature range of interest" The term *specimens* applies to both the sample and reference materials. The *sample holder* is the container or support for the sample, whereas the *reference holder* is the container or support for the reference material. These should be identical in size and shape in any single unit. The specimen holder assembly is the complete assembly in which the specimens are housed. Sometimes, the source of heating or cooling is part of the same unit with the containers or supports for the sample and reference material, so this must then be considered as part of the specimen holder assembly. In certain units the specimen holder assemblies are in the form of a relatively large mass of material in intimate contact with the specimen or specimen holder, and this is termed a *block*. The *differential thermocouples* or ΔT *thermocouples* are the thermocouples used to measure the temperature differences.

The results from the DTA units are produced as plots of ΔT vs. T, termed the DTA curve. A complication is that the ordinate in DTA is labeled ΔT but the output from the ΔT thermocouple system will normally be in volts. If the emf signal is represented by E, it must be noted that though

$\Delta T = \beta E$, it is also true that $\beta = f(T)$ (i.e., β varies with temperature). A similar situation arises with other temperature sensor systems.

A set of definitions is recommended for DTA. All of these definitions refer to a single peak, as shown in Fig. 21. These definitions and their usage can be extended logically to multiple peak systems, showing shoulders or more than one maximum or minimum.

The baselines on the DTA curve, AE and CD, represent regions where lT does not change significantly. An endothermic or exothermic peak is a portion of the DTA curve that has departed from and subsequently returned to the baseline (EHC in Fig. 21). In an endothermic peak or "endotherm" the temperature of the sample falls below that of the reference material (i.e., T is negative). In an exothermic peak or "exotherm" the temperature of the sample rises above that of the reference material (i.e., ΔT is positive). The peak width (EC in Fig. 21) is defined as the time or temperature interval between the points of departure from and return to the baseline. Peak height (HB in the figure) is the vertical distance between the interpolated baseline and the peak tip (H in Fig. 21). It is usually quoted in degrees Kelvin or Celsius. The peak area (EHCE) is the area enclosed between the peak and the interpolated baseline. This area in a properly calibrated unit will be proportional to the enthalpy, and hence can be used to estimate the amounts of material involved. Finally, with regard to DTA the extrapolated onset (B in Fig. 21) is the point of intersection of the tangent (HJ in Fig. 21) drawn at the point of greatest slope on the leading edge of the peak with the extrapolated baseline (EJ in Fig. 21). Specifically, in the actual reporting of experimental data the general items mentioned under TG should be followed. In addition, for DTA and DSC additional details should be provided.

1. Wherever possible each thermal effect should be identified and supporting evidence stated.
2. If the sample mass is diluted, then not only should the mass of the sample be noted, but also the nature and amount of the diluent.
3. If the reference cell is filled with diluent then the amount and nature of the inert material should be specified.
4. The nature and type of crucible used in DTA or DSC should be specified. If the reference cell is used empty then this should be stated.
5. The geometry and type of thermocouples should be described.
6. The ordinate scale should indicate the deflection per degree Celsius at a specified temperature. The preferred plotting should indicate an upward deflection as a positive temperature differential and a downward deflection as a negative temperature differential with respect to the reference.

E. Calibration

There are four sets of reference materials available for temperature calibration of DTA units (Table 5). The sets include materials exhibiting solid phase transitions and solid–liquid transitions. All materials are intended to be used in the heating mode as temperature standards. The materials are intended for use in DTA units under normal operating conditions. Consequently, the temperature values quoted are generally higher than true equilibrium values given in the literature. A certificate is provided with each set defining the reference points on the DTA curve and giving the mean temperature values together with standard deviations. The certificates also include a discussion of the reasons for variation of the temperatures recorded in different instrument types. The substances used in the calibration sets can easily be obtained in a high state of purity. Table 10 lists details for highly pure inorganic materials (Pope and Judd,

Table 10 Accepted Equilibrium Transition Temperatures (°C) and Extrapolated DTA onset, as Recommended by ICTA

Material	Transition type	Equilibrium transition temperature	Extrapolated onset Temperature (ICTA standards)
KNO_3	Solid–solid	127.7	128 ± 5
In	Solid–liquid	156.6	154 ± 6
Sn	Solid–liquid	231.9	230 ± 5
$KClO_4$	Solid–solid	299.5	199 ± 6
Ag_2SO_4	Solid–solid	424	242 ± 7
SiO_2	Solid–solid	573	571 ± 5
K_2SO_4	Solid–solid	583	582 ± 7
K_2CrO_4	Solid–solid	665	665 ± 7
$BaCO_3$	Solid–solid	810	808 ± 8
$SrCO_3$	Solid–solid	925	928 ± 7

Source: From Pope and Judd (1977) and Wunderlich and Bepp (1974).

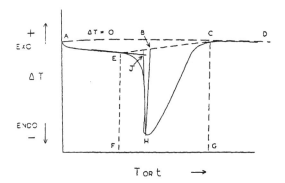

Figure 21 Schematic DTA peak, demonstrating definitions mentioned in text.

Table 11 Transition Temperatures (°C) Shown by Selected Organic Compounds

Compound	Transition type	Transition temperature	Extrapolated onset temperature
p-Nitrotoluene	Solid–liquid	51.5	51.5
Hexachloroethane	Solid–solid	71.4	—
Naphthalene	Solid–liquid	80.3	80.4
Hexamethyl benzene	Solid–solid	110.4	—
Benzoic acid	Solid–liquid	122.4	122.1
Adipic acid	Solid–liquid	151.4	151.0
Anisic acid	Solid–liquid	183.0	183.1
2-Chloroanthraquinone	Solid–liquid	209.1	209.4
Carbazole	Solid–liquid	245.3	245.2
Anthraquinone	Solid–liquid	284.6	283.9

Source: From Kambe et al. (1972), Pope and Judd (1977), and Wunderlich and Bepp (1974).

1977; Wunderlich, and Bopp, 1974). However, for DSC in particular, not only the temperature of the transition, but also the enthalpy change associated with the transition is important. Table 11 gives the transition temperatures for selected organic compounds (Kambe et al., 1972; Pope and Judd, 1977; Wunderlich, and Bopp, 1974). Table 12 gives enthalpy for certain materials covering a range of 80–1065°C (David, 1964; Kambe et al., 1972; Pope and Judd, 1977; Wunderlich, and Bopp, 1974).

The choice of temperature program may often involve compromises. The desire for a large signal by using large samples or high heating (or cooling) rates may lead to an unacceptable thermal lag. The thermal lag is less of a problem when using high-sensitivity modern equipment which then allows the use of small samples. A heating rate of 10°C/min is commonly used as the starting point in a preliminary investigation. Most experiments should be started some 30°C below the temperature of interest in order that a quasi-steady-state be established prior to obtaining the data.

Table 12 Enthalpies of Fusion for Selected Materials

Material	Melting point (°C)	Enthalpy of fusion (J/g)
Naphthalene	80.3	149
Benzoic acid	122.4	148
Indium	156.6	28.5
Tin	231.9	60.7
Lead	327.5	22.6
Zinc	419.6	113
Aluminum	660.4	393
Silver	961.9	105
Gold	1064.4	62.8

Source: From Charsley et al. (1987a), Gallagher et al. (1980), and McGhie et al. (1983).

Figure 22 depicts the effects of heating rate on the fusion peak for indium which is displayed against temperature and time. The curves illustrate the efficiency of extrapolated onset temperature when compared with the peak temperatures.

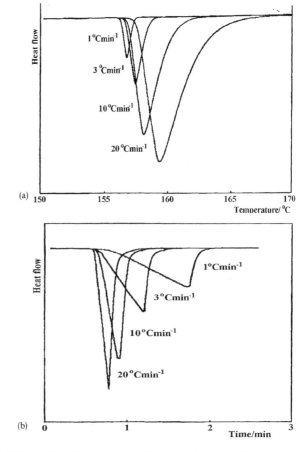

Figure 22 The influence of heating rate on the thermal analysis curves for the fusion of indium: (a) plotted against temperature and (b) plotted against time.

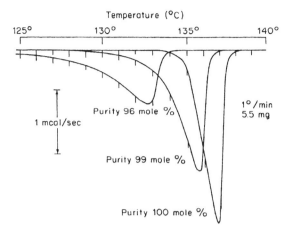

Figure 23 A DSC trace for the drug phenacetin.

The nature and flow rate of the atmosphere in the DTA/DSC cell can have a significant effect on the thermal analysis curve. Samples in an open crucible may be more reactive in air or oxygen. It is more common to use inert N_2 and argon to blanket the sample and sweep out evolved gaseous products. The atmosphere is also known to play an important role in heat exchange. The control of the flow rate is even important in experiments where the sample is in a sealed crucible. A dry atmosphere must be used for measurements below room temperature, otherwise condensation will occur in the sample holder and give use to spurious signals.

A further available standard is a polymer of polystyrene with a highly reproducible glass-transition temperature at around 100°C. It is for use on DTA units, and the certificate provided gives the mean temperatures for the various initial points on the curve and the variations introduced by different specimen-holder assemblies.

The purity of the sample can have a distinct effect on the shape of the curve as well as the onset temperature. A DSC trace for the drug phenacetin is given as a function of mole % purity in Fig. 23. It should be pointed out that quality control units of the pharmaceutical/chemical plant should provide standardized curves for the chemicals that they use as well as mixtures with their most common adjuvant. Materials that can be studied by DTA and DSC are given in Table 13.

Table 13 Materials Studied by DTA and DSC

Polymers, glasses, and ceramics	Pharmaceuticals
Oils, fats, and waxes	Biological materials
Clays and minerals	Metals and alloys
Coal, lignite, and wood	Natural products
Liquid crystals	Catalysts
Explosives, propellants, and pyrotechnics	Polymorphs

V. EVOLVED-GAS ANALYSIS

A. Introduction

The ICTAC definition of EGA of "a technique in which the nature and/or amount of volatile products released by a substance are measured as a function of temperature as the substance is subjected to a controlled temperature program" is unsatisfactory in describing the current practice of EGA as a thermal analysis technique. In many ways it is too broad, as it includes other techniques such as pyrolysis-GC, and in other ways too narrow, as it would exclude studying the changing nature of the gas stream as it is passed over the sample, as in catalytic studies. The definition does not require an EGA analyzer to be coupled to another technique, but in practice this is nearly always the case. This section will deal with EGA, and it is usually carried out simultaneously with another technique. The commonest combinations are those with TG and TG–DTA/DSC, where EGA assists in interpreting the chemistry of the events leading to weight losses.

There have been few reviews of the topic as a whole (Warrington, 1992) and one book dedicated to it (Lodding, 1967). The technique, is however, widely used, though the subject matter is dispersed through the literature, and not readily found. There seems to be a problem in the correct choice of keywords: the official *Chemical Abstracts* term, "thermal analysis—evolved gas", is rarely used, and searches using this are disappointing.

The forerunner of EGA, evolved gas detection (EGD) aimed to determine whether a DTA peak was connected with gas loss. EGA implies either specific identification of the product and/or quantification. The value of this is easily appreciated: the main TA techniques furnish physical data, and chemical interpretation is usually obtained by inference or analysis of the sample before and after a thermal event. This approach can be insecure, because transient intermediates may exist that cannot be isolated for analysis, or the product may transform on exposure to the atmosphere. EGA offers the opportunity of obtaining specific chemical information during the thermal experiment.

There has been a multitude of approaches to EGA, ranging from simple methods for answering a particular need to sophisticated research tools. Nearly all thermal analysis instruments use a flowing purge gas, which contains the chemical information that is sought. At the simplest level, holding a piece of wet litmus paper or another indicator in the effluent stream from the instrument, may answer the question at hand. Beyond this, almost every type of gas detector/analyzer has been linked to every type of thermal analysis equipment, and most manufacturers offer one or more methods for EGA, usually in combination with TG or TG–DTA/DSC. The approach taken in any instance depends on the information required, and there is no truly universal solution.

B. Instrumentation for EGA

EGA has employed almost every known type of gas detector. These may be specific, capable of responding to a single gas, or be general-purpose detectors, capable in principle of responding to all, or at least many gases. Specific detectors include hygrometers (Warrington and Barnes, 1981), nondispersive infrared cells (Morgan, 1977), paramagnetic oxygen detectors, chemiluminescense sensors for nitrogen oxides, fuel cell-based devices for hydrocarbons, and many others. Generally, the specific detectors are capable of quantitative performance. A typical EGA system is given in Fig. 24.

A useful approach that has appeared sporadically is to pass the evolved-gas/purge gas mixture through an absorbing solution, and then follow the changing concentration of the product of interest by titrimetry (Hoppler, 1991; Paulik and Paulik, 1972), ion-selective electrodes (Fennell et al., 1971), conductimetry (Brinkworth et al., 1981), colorimetry (Chiu, 1986), and so on. These methods are relatively easy to set up, inexpensive, and capable of excellent quantitative performance. In general, linking to the thermal analysis unit will be a simple matter, though bubbling of the effluent gas through the absorbing solution may cause disturbances to a thermobalance.

Gas chromatography (GC) has been used, when the main advantage is its ability to separate mixtures. Conclusive identification of the components would be possible if a mass spectrometer (MS) was connected to the GC detector (Barnes et al., 1981). GC is necessarily a batch technique: a portion of the gas products is collected over a chosen time span, and the ability to build up a profile of the evolution of a given species is limited by how many samples can be taken in the course of the experiment. Connecting a GC to a thermobalance requires a carefully designed interface if the TG data is to remain unaffected. A simple and convenient technique has recently been described (Lever et al., 2000) in which organic vapors are trapped in an absorbent tube for subsequent "off-line" analysis by desorption GC-MS.

The method of linking the TA unit to the gas detector requires careful design to achieve good performance. Ideally, the quality of the information from the thermobalance, for example, should not be compromised. This implies that the pressure inside it should not be changed significantly by the connection to the EGA unit. The aim then is to transfer evolved gases from the vicinity of the sample to the detector as quickly as possible, so that the EGA data is effectively simultaneous with, for example, the TG curve. Although transfer delays can be allowed for, apparatus with a long transfer time will result in smearing of the EGA curve due to diffusion broadening. A long residence time in the transfer system might also lead to unwanted secondary reactions. The transfer path should be inert with respect to the products of interest; certain species can degrade in contact with metal tubing, for instance, especially when it is heated, as is normally the case to avoid condensation of heavier products. Pipework of vitreous silica is the usual choice. Joining to an FT-IR is relatively simple. The cells used create only a slight resistance to the effluent gases from the unit, and a short length of small-bore tubing is used to carry all the gases to the cell. Joining to an MS is more difficult, as these operate under vacuum, typically ca. 10^{-6} mbar. To maintain this pressure, only $1-2$ cm^3 of the effluent gas is passed to the ion source. Some form of split is therefore needed to remove the bulk of the purge gas/product mixture. The most common solution is to use a capillary-and-bypass arrangement. Other approaches are a "skimmer" coupling (Kaiserberger et al., 1977), or separator (Charsley et al., 1987b), which has to be used with a helium purge gas, but gives dramatic increase in sensitivity.

Both FT-IR and MS require a powerful data system which, in addition to controlling the equipment, and displaying data, may also have libraries of standard spectra to assist identification. A complete integration of thermal analysis and EGA software is unfortunately rarely achieved. The literature on the techniques and applications of TG-IR till 1997 has been summarized by Materazzi (1997) and excellent review of the use of MS for EGA are available (Dollimore et al., 1984; Korobeinochev, 1987; Raemaekers and Bart, 1997).

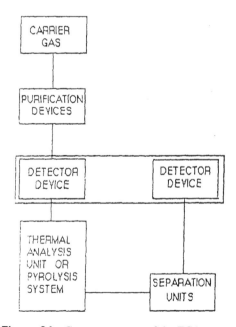

Figure 24 Component parts of the EGA system.

C. The Basic Unit

EGA refers to the quantitative analysis of gaseous materials, whereas EGD refers to the detection of an evolved gaseous species without its actual measurement. Logically, the terms bear no reference to thermal analysis and in fact the use of the methods described here has evolved merely with respect to the identification and analysis of gases evolved in any chemical process. Logically, too, the word "detection" could be held to refer to the identification of gaseous species evolved without quantitative analysis. However, in the following discussion the emphasis will be on EGA of volatile product species from the thermal decomposition of materials subjected to programed temperatures. It will be further seen that in this context EGA is a technique added to an already existing thermal analysis facility and is rarely used alone. The detectors used for gas analysis include:

Mass spectrometer
Titration cell
Infrared cell
Various chemical detectors
Flame-ionization detector
Thermal conductivity detector
Gas density detector

A precursor to the EGA unit is the appropriate thermal analysis system, including a pyrolysis unit or some other thermal "breakdown" device. Normally, a carrier gas is used; this should be purified as required. Pyrolysis separator devices are often employed before the analysis stage. Strictly, GC should be regarded as a separator and not a detector device, although often it is quoted as an analyzer or detector. The gas detector devices are often matched in pairs, one located in the inlet system and the other on the outlet. It is then the difference signal that is measured; Fig. 25 indicates the scheme outlined earlier. Certain gas detector devices, such as the mass spectrometer, and FT-IR spectroscopy units are not used in the "paired" manner; the devices most often used in this way are the flame-ionization and thermal conductivity detectors.

D. Detector Devices

Detector units such as the flame-ionization detector or thermal conductivity detector have found extensive use in chromatography and offer no problems in the context of EGA. They are almost always used in pairs—one on the inlet system and the other in the outlet stream. Gas density detectors can be used in the same manner.

In dynamic atmospheres the production of volatiles on decomposition causes a rise in pressure proportional to the total amount of volatiles produced. Various sensitive manometers can be used for the purpose of measuring the amount or rate of volatilization—the Prout and Tompkins (1946) experiments used in the study of the decomposition of potassium permanganate serve as a classic example.

The use of chemical detectors is often important when the evolved species is difficult to detect by other methods. The Dupont moisture evolution analyzer utilizes an electrochemical cell to determine water coulometrically. Another commercial unit, made by Panametrics, comprises a thin aluminum–alumina–aluminum sandwich, and relies on alteration in the electrical impedance as water is adsorbed in the pores. Gallagher and Gyorgy (1980) used this technique and reported excellent agreement with mass-loss techniques.

FT-IR spectroscopy can be used for EGA. At least three commercial units are available having this capability associated with a thermal analysis unit, either a DSC/DTA or TG. A well-established infrared spectroscopic method is the use of an infrared cell set to cover a particular absorption band. This is particularly useful for carbon monoxide, carbon dioxide, and water, which are difficult to analyze on an MS. A review by Low (1967) covering nondispersive analyzers, dispersion spectrometers, bandpass filter instruments, and interference spectrometers, is especially useful. One method of improving sensitivity is to convert the volatile product to another material that has a better response to infrared spectroscopy. A typical example is the use of calcium carbide to convert water to acetylene (Kiss, 1969).

Mass spectroscopy enjoys wide application in EGA. There is first the exciting prospect of placing the thermal analysis furnace directly within the MS. This avoids interfacing problems, but the conditions of thermal decomposition under a programed temperature regime are rather unique and results, although very basic, must be viewed cautiously.

A particularly useful design has been described by Gallagher (1978) and applied by Kinsbrow et al. (1979). Another design, employing a time-of-flight unit with a furnace incorporated within the unit, is described by Price et al. (1980) who used the equipment to study kinetics of decomposition. The equipment described by Lurn uses laser heating together with differential pumping to produce a molecular beam of volatile products (Lurn, 1977). The earliest report of the use of mass spectroscopy in this manner is probably the paper by Gotilke and Langer (1966). These instruments really combine the detector device and the thermal analysis equipment in a single unit.

The more usual application of mass spectroscopy is as a detector device combined with either TG or DTA/DSC. Commercial units are available but there are many instances of equipment being interfaced in the laboratory. Zitomer (1968) made early use of this technique, and a schematic of his equipment is shown in Fig. 25. The important requirement is an efficient interface to reduce

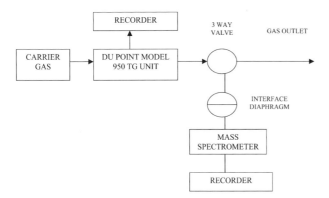

Figure 25 TG-MS apparatus described by Zitomer (Charsley et al., 1993).

the pressure from atmospheric (or whatever pressure the TG or DTA/DSC unit uses) to the low pressure of the MS. This can be a capillary tube or a diaphragm. In a typical unit, gas is withdrawn at about 15 cm away from the sample chamber through a flexible heated silica-lined steel capillary, some 1.2 m long. This is usually just the first stage in pressure reduction. The final pressure reduction can be made by means of a molecular leak such as a silicon carbide frit (Charsley et al., 1993a). The aim should be to reduce the chance of secondary reactions and obtain the EGA data as soon after the event as possible. Commercial units are now available.

Dyszel (1983) has used a combined TO atmospheric pressure chemical ionization MS system with a Sciex TAGA 300 MS. Initially ionization takes place with electron bombardment of nitrogen, and the N_2^+ ion so produced undergoes a charge transfer reaction with O_2 forming O^+. This then forms a cluster with moisture in the air to eventually produce a hydrated proton H_3O^+. The volatile products entering the ion source from the TG unit then react by an addition mechanism with the H_3O^+ and the ionized molecules pass to the quadrupole analyzer. Of course, the evolved species can be trapped and transferred later to the MS (Chiu and Beattie, 1980).

E. Precursor Considerations

Precursor units for the EGA are the thermal analysis devices employed, or a pyrolysis chamber, together with the carrier gas supply and a purification train. Pyrolysis techniques are now an essential part of organic characterization. A simple pyrolysis unit is described by Rogers et al. (1960) The equipment described by Rogers et al. (1960) and Vassallo (1961) was applied in studies on the pyrolysis of various polymeric materials. The aim of the Pyrochrom device is elemental analysis (Liebman et al., 1973). Automated pyrolysis equipment has been described by Nesbitt and Wendlandt (1974). Irwin (1993) provides the most in-depth description of analytical pyrolysis techniques. He lists various types of pyrolysis units which includes heated filament pyrolyzers, Curie-point pyrolyzers (induction heating), furnace pyrolyzers, and pyrolysis by laser beam. In the last process a laser beam is focused onto the surface of an analytical sample to produce a temperature increase. In the case of a ruby laser (694.3 nm) or neodymium (1040 nm) source, a burst of energy (1 ms) is focused on a spot 1 pm in diameter. Energy dissipation is sufficient to increase the temperature of the sample in excess of 10,000°C thus providing pyrolysis fragments (Kimelt and Speiser, 1977). Alternatively, low-power lasers in the infrared region may be utilized, for example CO_2 at 9.1–11.1 μm, which avoids some of the problems found with the higher energy laser systems (Coloff and Vanderborgh, 1973; McClennen et al., 1985).

Temperature-programed desorption is conveniently discussed in this section. A typical use of this technique is the temperature programed desorption of methyl naphthalene from alkali-metal faujasite (Ferino et al., 1993) in which a single peak is observed which changes with the nature and amount of the alkali-metal cation and the Si/Al ratio of the faujasite.

Temperature-programed reduction is a modification of the earlier-mentioned technique. In typical application of this modification, 5% hydrogen in nitrogen is passed over a catalyst sample and the change in hydrogen content of the effluent gas is monitored as the temperature is continuously increased (Robertson et al., 1975; Tsuchiya et al., 1971; Tsuchiya and Nakamura, 1977).

F. Separation Procedures

It is often found convenient to effect a separation or partial separation of volatile products from EGA prior to the analysis process. The most convenient tool for this is often GC. This separator is used with pyrolysis equipment or a thermal analysis unit. The following points may be made:

1. The analysis is not continuous.
2. The analysis is rarely in real-time.
3. Production of secondary pyrolysis products is lower in effluent gas from thermal analysis equipment than from some pyrolysis units.
4. The loss of volatiles recorded on the GC can be matched with the process noted in the thermal analysis unit.
5. The temperature at which a particular volatile fraction or component is released can be noted on the thermal analysis unit.

Equipment using TG–GC has been described by Chiu (1968, 1970). Special attention has been devoted to obtaining an efficient and reliable coupling system. Cukor and Persiani coupled a DuPont Model 950 thermobalance to a Perkin-Elmer Model 900 GC (Cukor and Persiani, 1974). Other descriptions of TG–GC have been given by Wiedermann (1969) and Uden et al. (1976). The equipment described by Uden coupled the thermal analyzer and a pyrolysis reaction system with an online vapor-phase infrared spectroscopic elemental and functional analysis, together with mass spectroscopy. The partial separation of evolved gas using GC followed by analysis of separated fractions using MS has become common practice. Coupling of GC with DTA is also possible, but sampling time is greater than with DTA–MS (Yamada et al., 1975). A thermal analysis separator device has been constructed by McNeill (1970) which proved to be very useful. The thermal volatilization unit has a pyrolysis chamber containing the sample (usually a polymer), which is subjected to a predetermined temperature program while pumping with a vacuum pump. The effluent-gas stream is passed through four isothermal traps (at 0°C, −45°C, −75°C, and −100°C), each with a pressure gauge before reaching the main trap at −196°C. This effectively allows the collection of fractions of different volatility. Each of the fractions can then be investigated separately. Any orthodox high-vacuum fractionation unit would afford a similar separation.

G. Reporting Data

In reporting EGA or EGD data the general recommendations already reported should be followed. The following additional information is recommended:

1. A statement of the temperature environment of the sample during the reaction.
2. Identification of the ordinate scale, in specific terms if possible. In general, an increasing concentration of evolved gas should be plotted upward. In the case of gas density detectors, the increasing gas density should be plotted upward.
3. The flow rate, total volume, construction, and temperature of the system between the sample and detector should be given, together with an estimate of the time delay within the system.
4. Location of the interface between the systems for sample heating and detecting or measuring evolved gases.
5. In the case of EGA when exact units are not used, the relationship between signal magnitude and concentration of species measured should be stated.

VI. THERMOMECHANICAL METHODS
A. Introduction

The term thermomechanical is somewhat misleading. It is defined as covering thermal analysis techniques that measure changes in volume, shape, and lengths, and other properties relating to the physical shape of a solid substance subjected to a programed temperature regime. However, logically the word covers other mechanical effects, such as the viscoelastic response of a sample. The word dynamic has come to mean oscillatory. The techniques measuring viscoelastic responses are becoming increasingly important.

Techniques most often found are: (1) thermodilatometry (TDA) or dilatometry, (2) TMA, and (3) dynamic thermomechanometry, where the dimensional changes under zero or minimal load are measured. Technically, changes in density, measured by all the traditional devices such as a pycnometer, should be included under this heading.

In TMA a nonoscillatory stress is applied to the solid sample and deformation under load is measured. DMA refers to techniques in which oscillatory stress is applied to the sample and the dynamic modulus or the mechanical damping of the sample is measured. There is probably as much application of these techniques isothermally as under rising temperature conditions. The techniques are increasingly used to characterize the behavior of "finished products", and this is especially true of DMA.

Many materials will deform under the applied stress at a particular temperature which is often connected with the material melting or undergoing a glass–rubber transition. Alternatively, the specimen may possess residual stresses which have been "locked-in" during preparation. Dimensional changes will occur on heating as a consequence of the relaxation of these stresses. Stress (σ) may be defined as the ratio of the mechanical force applied (F) divided by the area over which it acts (A):

$$\sigma = \frac{F}{A} \tag{24}$$

The stress is usually applied in compression or tension mode, as shown in Fig. 26. The units of stress are either Newton/meter square (N m^2) or Pascals (Pa).

If the applied stress is negligible then the technique becomes that of TDA. This technique is used to determine the coefficient of thermal expansion of the material from the relationship:

$$\alpha l_o = \frac{\mathrm{d}l}{\mathrm{d}T} \tag{25}$$

where α is the coefficient of thermal expansion (ppm °C or µm/m/°C); l_o is the original sample length (m); and dl/dT is the rate of change of the sample length with temperature (µm/°C).

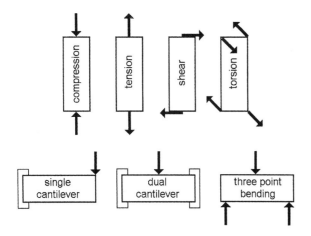

Figure 26 Common mechanical deformation modes: compression, tension, shear, to bending (single cantilever, dual cantilever, three-point bending).

Dynamic mechanical analysis DMA is concerned with the measurement of the mechanical properties (mechanical modulus or stiffness and damping) of a specimen as a function of temperature. DMA is a sensitive probe for molecular mobility within materials and is most commonly used to measure the glass transition temperature and other transitions in macromolecules, or to follow changes in mechanical properties brought about by chemical reactions.

For this type of measurement the specimen is subjected to an oscillating stress, usually following a sinusoidal waveform:

$$\sigma(t) = \alpha_{max} \sin \omega t \quad (26)$$

where $\sigma(t)$ is the stress at time t, σ_{max} is the maximum stress, and ω is the angular frequency of oscillation. Please note that $\omega = 2\pi f$ where f is the frequency in Hertz. The applied stress produces a corresponding deformation or strain (ε) defined by:

$$\varepsilon = \frac{\text{change in dimension}}{\text{original dimension}} = \frac{\Delta l}{l_o} \quad (27)$$

The strain is measured according to how the stress is applied (e.g., compression, tension, bending, shear, etc.). Strain is dimensionless, but often expressed as a percentage.

For an elastic material, Hooke's law applies and the strain is proportional to the applied stress according to the relationship.

$$E = \frac{d\sigma}{d\varepsilon} \quad (28)$$

Where E is the elastic, or Young's, modulus with units of N/m^2 or Pa. Such measurements are normally carried out in tension or bending, when the sample is a soft material or a liquid then measurements are normally carried out in shear mode, thus a shear modulus (G) is measured. The two moduli are related to one another by:

$$G = \frac{E}{2 + 2\nu} \quad (29)$$

where ν is known as Poisson's ratio of the material. This normally lies between 0 and 0.5 for most materials and represents a measure of the distortion which occurs (i.e., the reduction in breadth accompanying an increase in length) during testing.

If the material is viscous, Newton's law holds. The specimen possesses a resistance to deformation or viscosity, η, proportional to the rate of application of strain, such that:

$$\eta = \frac{d\sigma}{d\varepsilon/dt} \quad (30)$$

The units of viscosity are Pa.

A coil spring is an example of a perfectly elastic material in which all of the energy of deformation is stored and can be recovered by releasing the stress. Conversely, a perfectly viscous material is exemplified by a dashpot, which resists extension with a force proportional to the strain rate but affords no restoring force once extended, all of the deformation energy being dissipated as heat during the loading process. In reality, most materials exhibit behavior intermediate between springs and dashpots—viscoelasticity.

If, as in the case of DMA, a sinusoidal oscillating stress is applied to a specimen, a corresponding oscillating strain will be produced. Unless the material is perfectly elastic, the measured strain will lag behind the applied stress by a phase difference (δ) shown in Fig. 27. The ratio of peak stress to peak strain gives the complex modulus (E^*) which consists of an in-phase component or storage modulus (E') and a 90° out-of-phase (quadrature) component or loss modulus (E'').

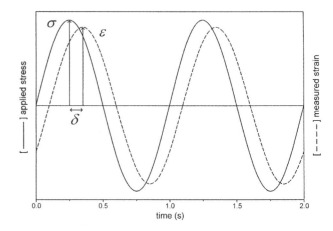

Figure 27 Relationship between stress (σ) and strain (ε) during a dynamic mechanical test.

The storage modulus, being in-phase with the applied stress, represents the elastic component of the material's behavior, whereas the loss modulus, deriving from the condition at which $d\varepsilon/dt$ is a maximum, corresponds to the viscous nature of the material. The ratio between the loss and storage moduli (E''/E') gives the useful quantity known as the mechanical damping factor (tan δ) which is a measure of the amount of deformational energy that is dissipated as heat during each cycle. The relationship between these quantities can be illustrated by means of an Argand diagram, commonly used to visualize complex numbers, which shows that the complex modulus is a vector quantity characterized by magnitude (E^*) and angle (δ) as shown in Fig. 28 where E' and E'' represent the real and imaginary components of this vector thus:

$$E^* = E' + iE'' = \sqrt{(E'^2 + E''^2)} \quad (31)$$

So that:

$$E' = E^* \cos \delta \quad (32)$$

and

$$E'' = E^* \sin \delta \quad (33)$$

B. TDA and TMA

A typical TDA unit is shown in Fig. 29. Instrumentation problems in TDA and TMA have been discussed by Paulik and Paulik (1979), Gill (1984), and Riesen and Sommerauer (1983). In the figure, an LVDT is shown as the instrument by which dimensional changes of the sample may be measured. The movement is transferred into the LVDT via a rod, usually of quartz or some other material not subject to appreciable dimensional change on heat treatment. A mechanical guide is then necessary to position this

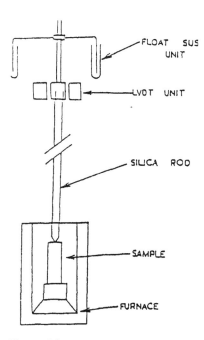

Figure 29 Basic design of a TDA unit.

rod correctly. The samples must be machined to convenient shapes. The testing of powders is difficult but not impossible. The achievement of zero weight on the sample is also difficult to achieve but is lessened by some system such as the float system depicted. Provision for expansion against a defined load is generally provided. Equipment is available enabling measurements to be made over a temperature range from below ambient to around 1200°C, although not in the same unit.

Equipment and methodology is discussed in depth by Ish-Shalom (1993). The technique can be used with variable temperature programs. These include isothermal runs and heating runs at predetermined temperature rates. A technique, which predates controlled rate thermogravimetry, has been employed in TDA, where the rate of shrinkage or expansion of the sample is kept constant (Palmour and Johnson, 1967).

C. Dynamic Mechanical Analysis

The word dynamic here refers to the fact that the instrument is used to study the viscoelastic response of a sample under oscillatory load. Commercial instruments use different techniques:

1. The sample is forced into oscillation or made to take on its natural frequency.
2. Stress can be applied in a variety of ways, for example, in flexure, tension, compression, or torsion.
3. The sample has a load applied continuously and measurement is made of the modified oscillatory response.

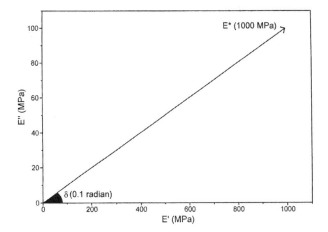

Figure 28 Argand diagram to illustrate the relationship between complex modulus (E^*) and its components.

The application of the instrument is to shape samples, generally polymeric end products (Lofthouse and Burroughs, 1978; Wetton, 1981). In experiments where the sample is in free vibration; the dynamic modulus is related to the natural frequency (maximum amplitude).

In many of today's DMA units, Young's modulus is also written as:

$$E = \frac{4\pi^2 f^2 J - K}{2w(L/2 + D)^2} \left(\frac{L}{T}\right)^3 \quad (34)$$

where f is the DMA frequency, J is the moment of inertia of the sample arms (which are fixed to a rigid block via low-friction flexure pivots), K is the spring constant of the pivots, D is the clamping distance, w is the sample width, T is the sample thickness, and L is the sample length. In polymer laboratory units a bar sample is used, clamped rigidly at both ends and subjected at its center point to a sinusoidal vibration using a drive shaft attached to the sample by a clamp. The stress on the sample is proportional to the current supplied to the vibrator, whereas the strain on the sample is proportional to the sample displacement. Special clamps allow the testing of soft materials (rubbers, adhesives, fats, etc). Films and fibers can also be tested, whereas liquid polymers are tested on a support. Thus, in cured epoxy systems, the DMA technique can easily observe β and δ transitions, which are often beyond the capability of DSC techniques.

Torsional braid analysis (TBA) is another technique employing oscillation imposed on the sample. It was introduced by Gillham (1996) and Lewis and Gillham (1962). In this technique the sample is prepared in solution form and impregnated onto glass braid or thread, followed by evaporation of the solvent. The sample is heated while subjected to free torsional oscillations. The relative rigidity parameter is noted (defined as $1/p^2$, where p is the period of oscillation) and used as a measure of the shear modulus. A measure of the logarithmic decrement, termed the *mechanical damping index*, $1/n$ (where n is the number of oscillations between two arbitrary but fixed boundary conditions in a series of waves), is also noted. It is used extensively in polymer systems to reveal glass transitions, melting, and the effect of chemical reactions.

D. Reporting Data

In reporting TMA data, besides the general points made regarding reporting, the following additional details should be given:

1. A clear statement of the temperature environment of the sample.
2. The type of deformation (tensile, torsional, bending, etc.) and the dimensions, geometry, and materials of the loading elements.
3. Identification of the ordinate scale in specific terms where possible.
4. For static procedures, increasing expansion, elongation, or extension, and torsional displacement should be plotted upward. Increased penetration or deformation in flexure should be plotted downward. For dynamic mechanical procedures, the relative modular anchor mechanical loss should be plotted upward.

E. Instrumentation

A schematic diagram of a typical TMA instrument is shown in Fig. 30. The sample is placed in a temperature-controlled environment with a thermocouple or other temperature-sensing devices, such as a platinum resistance thermometer, placed in close proximity. The facility to circulate a cryogenic coolant such as cold nitrogen gas from a dewar vessel of liquid nitrogen is useful for subambient measurements. The atmosphere around the sample is usually controlled by purging the oven with air or nitrogen from a cylinder. Because of the much larger thermal mass of the sample and oven when compared with a DSC or a thermobalance, the heating and cooling rates employed are usually much slower for TMA. A rate of 5°C/min is usually the maximum recommended value for good temperature equilibration across the specimen. Even this rate can be a problem for some specimens where appreciable temperature gradients can exist between the middle and ends of the sample, particularly around the test fixtures—which can represent a significant heat sink.

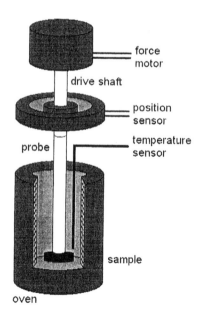

Figure 30 Schematic diagram of a thermomechanical analyzer.

For compression measurements (as illustrated) a flat-ended probe is rested on the top surface of the sample and a static force is applied by means of a weight or (more commonly in the case of modern instrumentation) an electromagnetic motor similar in principle to the coil of a loudspeaker. Some form of proximity sensor measures the movement of the probe. This is usually achieved by using a LVDT which consists of two coils of wire which form an electrical transformer when fed by an AC current. The core of the transformer is attached to the probe assembly and the coupling between the windings of the transformer is dependent upon the displacement of the probe. Other transducers such as capacitance sensors (which depend on the proximity of two plates—one fixed, the other moving) or optical encoders are used in certain instruments.

Most commercial instruments are supplied with a variety of probes for different applications (Fig. 31). A probe with a flat contact area is commonly used for thermal expansion measurements where it is important to distribute the applied load over a wide area. Probes with sharp points or round-ended probes are employed for penetration measurements so as to determine the sample's softening temperature. Films and fibers, which are not self-supporting, can be measured in extension by clamping their free ends between two grips and applying sufficient tension to the specimen to prevent the sample buckling. Volumetric expansion can be determined using a piston and cylinder arrangement with the sample surrounded by an inert packing material such as alumina powder or silicone oil.

The equipment must be calibrated before use. The manufacturers, as well as various standardization agencies, usually provide recommended procedures. Temperature calibration is usually carried out by preparing a sample that consists of a number of metal melting point standards, such as those used for DSC, sandwiched between steel or ceramic discs. The melting of each standard causes a change in height of the stack as each metal melts and flows. Force calibration is often performed by balancing the force generated by the electromagnetic motor against a certified weight added to the drive train. Length calibration can be more difficult to carry out. A common check on the performance of the instrument is to measure the thermal expansion of a material whose values are accurately known (such as aluminum or copper).

The distinction between a TMA analyzer and a DMA analyzer is blurred nowadays as many instruments can perform TMA-type experiments. The configuration of a DMA is essentially the same as the TMA shown in Fig. 30 with the addition of extra electronics which can apply an oscillating load and the ability to resolve the resulting specimen deformation into in-phase and out-of-phase components so as to determine E', E'', and tan δ. The facility for sub ambient operation is more common on a DMA than a TMA. The same recommendations about modest rates of temperature change are even more important for the larger samples used in DMA. Stepwise isothermal measurements are often carried out for multiple frequency operation. In this experiment the oven temperature is changed in small increments and the sample allowed to come to thermal equilibrium before the measurements are made. The frequency range over which the mechanical stress can be applied commonly covers 0.01–100 Hz. The lower limit is determined by the amount of time that it takes to cover enough cycles to attain reasonable resolution of tan δ (10 s for one measurement of 0.01 Hz—though normally some form of data averaging is applied meaning that a measurement at this frequency can take a minute or more). The upper limit is usually determined by the mechanical properties of the drive system and clamps.

Different clamping geometries are used to accommodate particular specimens (Fig. 32). Single or dual cantilever bending modes are the most common for materials which can be formed into bars. Shear measurements are used for soft, thick samples. Films and fibers are usually mounted in tension with loading arranged so that the sample is always in tension. Torsion measurements are normally done with a special design of instrument since most DMAs can only exert a linear rather than a rotational force.

The effect of temperature on the mechanical properties of a liquid can be investigated using a special type of DMA analyzer called an oscillatory rheometer. In this instrument the sample is contained as a thin film between two parallel plates. One of the plates is fixed, whereas the other rotates back and forth so as to subject the liquid to a shearing

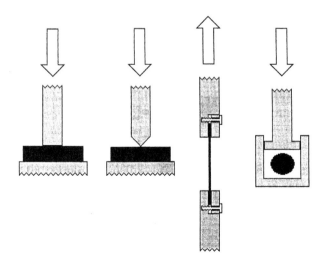

Figure 31 TMA probe types (left–right): compression, penetration, tension, volumetric.

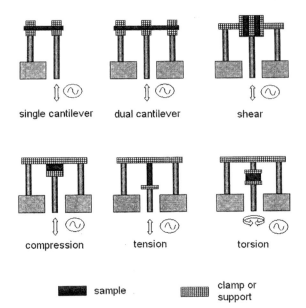

Figure 32 Common clamping geometries for dynamic mechanical analysis (cf. Fig. 1).

motion. It is possible to calculate the shear modulus from the amplitude of the rotation and the resistance of the sample to deformation. Because the test is performed in oscillation, it is possible to separate the shear modulus (G) into storage (G') and loss modulus (G'') by measuring the phase lag between the applied strain and measured stress. Other geometries such as concentric cylinders or cone and plate are often used, depending on the viscosity of the sample.

An alternative method of examining the dynamic mechanical properties of liquids is to coat them onto an inert support (typically a glass fiber braid). This measure is termed TBA and does not provide quantitative modulus measurements as it is difficult to decouple the response of the substrate from that of the sample.

The method of calibration of DMA analyzers varies from instrument to instrument and it is essential to follow manufacturer's recommendations. Temperature calibration can sometimes be done as for TMA analyzers since many instruments can operate in this mode. Load or force calibration is often carried out using weights. It is difficult to achieve the same degree of accuracy and precision in modulus measurements from a DMA as might be obtained by using an extensometer without taking great care to eliminate clamping effects and the influence of instrument compliance (which can be estimated by measuring the stiffness of a steel beam). Extensometers are much bigger instruments and the size of test specimens is correspondingly larger. Additionally, they often only operate at room temperature. For many applications the user is, however, mainly interested in the temperature at which changes in mechanical properties occur and the relative value of a material's properties over a broad range of temperatures.

VII. DIELECTRIC THERMAL ANALYSIS

A sample can be subjected to either a constant or oscillating electric field rather than some other parameter. Dipoles in the material will attempt to orient with the electric field, whereas ions, often present as impurities, will move toward the electrode of opposite polarity. The resulting current flow is similar in nature to the deformation brought about by mechanical tests. It represents a measure of the freedom of charge carriers capable of responding to the applied field. The sample may be present as a thin film between two metal electrodes so as to form a parallel plate capacitor. Two types of tests can be performed. The first, thermally stimulated current analysis (TSCA), subjects a sample to a constant electric field. The current, which flows through the sample, is measured as a function of temperature. Often, the sample is heated to a high temperature under the applied field and then quenched to a low temperature. This process aligns dipoles within the specimen similarly to drawing a material under a mechanical stress brings about orientation of the molecules in the sample. The polarization field is then switched off, and the sample is reheated while the current flow, resulting from the relaxation of the induced dipoles to the unordered state, is monitored.

The second, dielectric thermal analysis (DETA), subjects a sample to an oscillating sinusoidal electric field. The applied voltage produces a polarization within the sample and causes a small current to flow which leads the electric field by a phase difference (δ) (Fig. 33). Two fundamental electrical characteristics, conductance and capacitance, are determined from measurements of the amplitude of the voltage (V), current (I), and δ. These are used to determine the admittance of the sample (Y) given by:

$$Y = \frac{1}{V} \quad (35)$$

where Y is a vector quantity which is characterized by its magnitude [V] and direction δ.

The capacitance C is the ability to store electrical charge and is given by:

$$C = [Y]\frac{\sin \delta}{\omega} \quad (36)$$

The conductance G_c is the ability to transfer electric charge and is given by:

$$G = [Y]\cos \delta \quad (37)$$

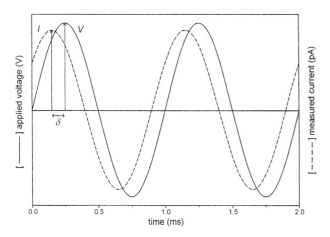

Figure 33 Relationship between voltage and current in a capacitor (cf. Fig. 2).

It is more common to present data in terms of the relative permittivity ε' and dielectric loss factor ε''—which are related to capacitance and conductance by:

$$\varepsilon' = \frac{C}{(\varepsilon_o A/D)} \quad (38)$$

and

$$\varepsilon'' = \frac{G_e}{\omega \varepsilon_o A/D} \quad (39)$$

where ε_o is the permittivity of free space (8.86 × 10^{-12} F/m) and A/D, in m, is the ratio of electrode area (A) to plate separation of sample thickness (D) for a parallel plate capacitor. More generally, A/D is a geometric factor which is found by determining the properties of the measuring cell in the absence of a sample. Both ε' and ε'' are dimensionless quantities.

The ratio $\varepsilon''/\varepsilon'$ is the amount of energy dissipated per cycle divided by the amount of energy stored per cycle and known as the dielectric loss tangent or dissipation factor (tan δ).

A schematic diagram of a typical instrument is given in Fig. 34. The sample is presented as a thin film, typically no more than 1 or 2 mm thick. It is placed between two parallel plates so as to form a simple electrical capacitor. A grounded electrode surrounding one plate, known as a guard ring, is sometimes incorporated so as to improve performance by minimizing stray electric fields. A thermocouple or platinum resistance thermometer is placed in contact with one of the plates (sometimes one on each plate) so as to measure the specimen's temperature. For specialized applications, such as remote sensing of large components, an interdigitated electrode is used (shown in the inset in Fig. 34). These employ a pair of interlocking comb-like electrodes and often incorporate a temperature sensor (resistance thermometer). They can be embedded in structures such as a thermosetting polymer composite and the dielectric properties of the material monitored while it is cured in an autoclave.

A usual part of the calibration protocol for DETA is to measure the dielectric properties of the empty dielectric cell so as to account for stray capacitances arising from the leads which must be of coaxial construction. Temperature calibration can be done by measuring the melting transition of a crystalline low molecular weight organic crystal such as benzoic acid placed between the electrodes.

DETA involves monitoring the viscosity of a system via its ability to store or transport electrical charge. Changes in the degree of alignment of dipoles and the ion mobility provide information pertaining to physical transitions in the material and to material properties such as viscosity, rigidity, reaction rate, and cure state. By use of remote dielectric sensors, the measurements can be made in actual processing environments such as presses, autoclaves, and ovens.

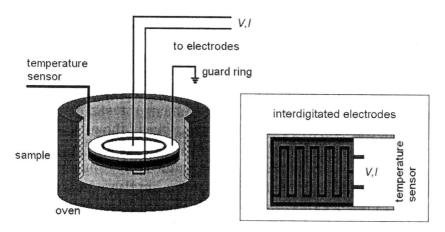

Figure 34 Schematic diagram of a dielectric, thermal analysis instrument. Inset shows a single-surface interdigitated electrode.

In dielectric cure monitoring, the ion mobility (electrical conductivity) of the material is of greatest interest. Almost all materials contain current carriers which are charged atoms or charged molecular complexes. The application of a voltage between a set of electrodes will create an electric field that forces those ions to move from one electrode toward the other. Ions encounter something analogous to viscous drag as they flow through a medium filled with molecules, and their mobility through this medium determines the conductivity. Conductivity is inversely proportional to viscosity. Ions moving through very fluid, watery materials have high mobility and conductivity, resulting in low resistivity that correlates with low viscosity. Conversely ions moving through very rubbery materials have low mobility and conductivity corresponding to the high viscosity. It is important to note that beyond some point in the cure the physical viscosity will climb so high that it is no longer measurable. This occurs even though the cross-linking reaction has not reached completion. Increasing polymerization continues to affect ionic motion. Dielectric measurements retain sensitivity past the time when ionic and physical viscosity deviate. Consequently, with proper interpretation, dielectric measurements are useful throughout the entire curing process for determining changes in viscosity and rigidity, and are extremely sensitive in determining the end of cure.

One can demonstrate the conductivity of a glass fiber-reinforced epoxy resin composite during curing in a heated press. An interdigitated electrode can be embedded in the sample during preparation of the specimen. A resistance thermometer in the electrode monitors the temperature of the sample directly during processing. Initially, the conductivity of the specimen increases as it is heated to the cure temperature and the resin becomes less viscous. Cross-linking causes the viscosity to increase and, as a consequence, the conductivity decreases. Toward the end of the reaction the material forms a highly cross-linked network and should show a strong frequency dependence in conductivity. Dielectric measurements readily lend themselves to being carried out simultaneously with DMA when experiments are performed in compression or torsion. The sample is usually mounted between parallel plates which are used to apply the mechanical stress—electrical connections can be established to these and used to make a dielectric measuring cell (http://www.ifpina.org/lchl.html).

A second example of using DETA is the practice of placing tinted self-adhesive plastic films as a means of limiting the effects of sunlight on the interiors of buildings and vehicles. Stuck to the inside surfaces of windows, they are used to filter infrared and ultraviolet (UV) radiation, thereby avoiding heat build-up in offices and vehicles, and fading of interior upholstery. After the glass itself, the adhesive is first to receive solar radiation and must be stabilized against photodegradation by an appropriate choice of polymer and stabilizer package. This not only protects the adhesive but also blocks short wavelength radiation, which suppresses fading of the dyes used to color the film.

The performance of a candidate system may be assessed by exposure to intense radiation equivalent to the solar spectrum in a hot, humid environment during accelerated aging. Measurement of the UV transmission of the film is made at regular intervals in addition to critical evaluation of the optical appearance of the plate by eye. As these products are to be used in vehicles, any blemishes in the film brought about by degradation of the adhesive are extremely undesirable.

In such a study (Trombly and Shepard, 1994), test panels of window film were mounted on glass plates for accelerated aging. A small hole was cut in the plastic film and an interdigitated single surface dielectric sensor applied to the exposed surface of the adhesive after chilling the plate in a freezer to aid removal of the backing. Measurement of the dielectric loss and permitivity were carried out from 0.1 to 100 kHz in decade steps at ambient temperature before and after 600 and 1200 h exposure.

Values of dielectric loss (ε'') at 0.1 Hz for the same adhesive containing three candidate stabilizer packages are given in Table 14 at different stages of weathering. The data indicates that, of the three stabilizer formulations tested, the "poor" package showed the largest increase in ε''. UV transmission measurements showed that, although starting with the same value (ca. 0.4%), the transmission of the "standard" package rises to 5.7% when compared with around 2.5% for both adhesives containing the "good" and "poor" formulation after 1200 h of weathering. The "poor" sample exhibited severe optical distortion of the film, but remained light fast, the "good" sample (the same formulation as the "poor" sample but with an additional stabilizer) maintained good UV absorption and showed no blemishes. Even after 600 h exposure it was apparent that changes in adhesive behavior could be detected by this technique before deterioration in appearance could be seen by eye (Price, 1997).

VIII. CALORIMETRY AND MICROCALORIMETRIC MEASUREMENTS

A. Introduction

The Second Law of Thermodynamics states that: "everything naturally tends to its most stable state". Some materials benefit from change and become more useful; however, for most materials, change represents a reduction

Table 14 Processes Amenable to Study by TG and DTA/DSC

Process	TG effect		DTA/DSC effect	
	Gain	Loss	Exotherm	Endotherm
Adsorption	×		×	
Desorption		×		×
Dehydration/desolvation		×		×
Sublimation		×		×
Vaporisation		×		×
Decomposition		×	×	×
Solid–solid transition			×	×
Solid–gas reaction	×	×	×	×
Solid–solid reaction	Maybe	Maybe	×	×
Crystallisation			×	
Melting				×
Polymerisation		Maybe	×	
Catalytic reactions	Maybe	Maybe	×	

in useful properties and loss in quality. In some cases, change can even result in a danger to health. For most materials around us, change can be obvious; discoloring of paint, rusting of metal, decay or a raging brush or forest fire. Other changes can be subtle, such as a polymorphic change of a crystal, fatigue in a metal, or the surface sorption of vapor. For the majority of cases, change can have a considerable influence on the useful properties of a material. What appears to be the same substance by casual observation may have very different properties if change has actually occurred.

Industries, which rely on materials, have responsibility to ensure constant quality. Reliability comes from this knowledge of the process of change and the predictions about this change which results from environment conditions. Most materials are at their highest quality at the time of manufacture but depreciate with time. As it is important to understand change, effort and money are therefore invested in an attempt to ensure quantification of the changes which occur. This provides the basis for a shelf life specification and recommendation for the conditions under which a material can be used.

There are many recognized techniques for evaluating the structure and consistency of material, for example, spectroscopic analysis. These are often used to ensure batch-to-batch uniformity. Calorimetriy recognizes subtle differences in materials that are not apparent using any other technique. In addition, calorimetry can quantify change in terms of rate, how much, and the probability for a change to occur.

Change can be promoted by a variety of means. Chemical modifications observed from oxidation and reduction, hydrolysis and photolytic reactions, reactions with excipients and with solvents, and autocatalytic reactions. Change also comes about from modifications to the solid state. Crystalline materials undergo polymorphic changes, changes in hydrate or solvate forms, phase changes, and change in surface area. Change is inevitable if change is possible. The rate of change may be dependent on environmental conditions such as temperature, pressure, humidity, and mechanical action.

Calorimetry is defined as the measurement of heat. It has been used to study reactive systems since 1780 when Lavoisier and De Laplace first studied the respiration of a guinea pig in an ice calorimeter (Lavoisier and De Laplace, 1780). The quantity of water collected and the rate of melting gives a thermodynamic and kinetic evaluation for respiration. Since that time, considerable progress has been made in the technology of calorimeters. Lavoisier was restricted to measuring exothermic reactions at 273.15 K and the sensitivity of the instrument was dependent on the accuracy of weighing the melted ice. Modern calorimeters can directly record, exo or endothermic reactions with signals as low as 5×10^{-8} W. Such sensitivity permits greater specificity in the interpretation of calorimetric data than what the ice calorimeter of Lavoisier could achieve.

Instruments have been designed to measure energy changes from temperatures a little above absolute zero up to temperatures in excess of 2000 or 3000 K. Homemade instruments were the rule until 1960; however, high-quality measurements were very common. Calorimetry was not a routine tool for use in the laboratory. Advances in electronics and the development of new designs have changed this situation so dramatically that calorimetry is now widely used for the day-to-day physical and chemical characterization of materials within a wide range of research applications. Physical and chemical processes for solids, liquids, and gases have been investigated. Each application addresses specific problems and descriptions of many different types of calorimeters may be found in the literature (Kemp, 1988; Pugh, 1966; Rossini, 1956; Sturtevant, 1959; Skinner, 1962). Modern calorimetry has found many applications in material development. The high sensitivity and signal stability over long time periods make modern calorimeters very suitable for one study of change. Most instruments require small amounts of material, are very sensitive and, at least in principle, can record all changes that occur. Many texts (Atkins, 1978; Price, 1998; Smith, 1990) provide reviews of the laws of physics, physical and chemical changes, as well as the development and mathematics of thermodynamics under which calorimetric instruments operate.

1. *The First Law of Thermodynamics: The Energy of an Isolated System is Constant*

Another expression of this law considers the internal energy (U) of a system and how it is effected by the work done on it (w) and heat energy added to it (q). The change in internal energy of a system (ΔU) may be written as

$$\Delta U = q + w \tag{40}$$

If heat is added to a system, the temperature rises. If the system is at constant volume, no work is done, so for an infinitesimal change:

$$dU = dq_v = C_v \, dT \tag{41}$$

where C_v is the heat capacity at constant volume. This applies when combustion takes place in a closed bomb calorimeter. At constant pressure, enthalpy (H) is defined to take into account the work done on the system by the pressure:

$$H = U + PV \tag{42}$$

Heat added to raise the temperature at constant pressure is given as

$$dH - dq_p = C_p \, dT \tag{43}$$

where C_p is the heat capacity at constant pressure. The heat capacities may not be constant with temperature. Applying an electrical power for a measured time allows the instrument to be calibrated. The temperature rise is a consequence of the applied power and can be determined from the current (I) passed through a resistance for a time (t) as follows:

$$Pt = I^2 Rt \tag{44}$$

When a change occurs with the evolution of heat to the surroundings at constant pressure it is referred as exothermic and the enthalpy of the system decreases (ΔU is negative). If the change takes in heat from the surroundings it is endothermic (ΔU is positive). Many reactions have been studied and the value of the enthalpy of reaction determined accurately. Tables of standard molar enthalpies of formation ΔH_8 at 25°C can be found in several sources (Lang, 1956; Weast and Lide, 1989) and the standard enthalpy change for another reaction may be calculated from them using Hess's Law. This law states that "the enthalpy change of a reaction may be expressed as the sum of the enthalpy changes into which the overall reaction may be divided". Hess's law allows the calculation of enthalpy changes in reactions which cannot be measured directly, either because they occur too slowly for the instrument to follow, or because they do not occur naturally.

2. *The Second Law of Thermodynamics: In an Isolated System Spontaneous Processes Occur in the Direction for Increasing Entropy*

The entropy (S) of a system is a thermodynamic function related to the statistical distribution of energy within that system. The Second Law means that the system and surroundings will change spontaneously from a state of low probability to a state of maximum probability, which will be the equilibrium state. An example is the mixing of two ideal gases at atmospheric pressure. This will occur spontaneously if a barrier between them is removed. No heat or pressure change is involved, but the change clearly involves an increase in the randomness of the system as the gases become mixed.

3. *The Third Law of Thermodynamics: All Perfect Crystals Have the Same Entropy at Absolute Zero*

To use entropy as a criterion for spontaneous change, it is necessary to investigate both the system and surroundings. For that reason, a further thermodynamic function, the free energy, or Gibbs function (G), is used. This combines the enthalpy (H) and the entropy (S) and allows the combination of the effects of both H and S on the system only, generally at functions, as they depend only on the state of the system.

$$G = H - TS \tag{45}$$

The free energy change (ΔG) serves as a good indicator for the potential for a reaction to take place and how far a reaction will go toward completion. Under standard conditions, ΔG is related to the equilibrium constant (K_{eq}) of a reaction at a given temperature:

$$\Delta G^\theta = -RT \ln(K_{eq}) \tag{46}$$

All reactions may be written as equilibrium processes. Some equilibrium lies far to the right, indicating that the majority of the reactants form products. Some lie far to the left, indicating the majority of reactants remain as reactants. Similarly, some lay between these extremes. The International Conference on Harmonisation (ICH) (http://www.ifpma.org/ich1.html) of technical requirements for registration of pharmaceuticals publish guidelines recommending the reporting threshold for degradation products in pharmaceutical materials. The amount of degradation product allowed before the requirement of degradation identification is currently 0.1% for a high-dose drug product. A reactant product quantity of 0.1%, at equilibrium, relates to an equilibrium constant (K_{eq}) of 1×10^{-3}. Beezer et al. (2001) have shown that the calculation of (K_{eq}) is possible from isothermal calorimetric data where enthalpy changes for a reaction are determined as a function of temperature. From Eq. (46), an equilibrium constant of 1×10^{-3} corresponds to a ΔG of $+17.12$ kJ/mol, at 298.15 K. Most spontaneous reactions (most natural reactions) will result in a decrease in Gibbs energy and an increase in the overall entropy.

Crystallization from a saturated solution is a favorable process with a negative ΔG. This occurs because the process is exothermic (ΔH negative) and entropically favorable (ΔS positive). The removal of solute from the solvent has an associated entropy gain greater than the decrease in entropy from building the crystal lattice. The calculation of the equilibrium constant can provide a powerful investigative tool for comparing materials. The equilibrium solubility of polymorphs can be studied as a function of temperature. A phase diagram can then be made (e.g., a plot of ΔG against temperature). This can give considerable insight into how a material would behave under different environmental conditions and monotonic and enantiotropy behavior identified.

Measuring the equilibrium constant at several temperatures allows a further equation to be derived that permits the enthalpy change to be calculated for the reaction from the van't Hoff equation:

$$\frac{d(\ln K)}{dT} = \frac{\Delta H^\theta}{RT^2} \quad (47)$$

Another relationship which may be used for reactions that take place in an electrochemical cell is that between the free energy and the electromotive force (E) of the cell:

$$\Delta G = -nFE \quad (48)$$

Measuring the emf for the cell at several different temperatures allows the calculation of the enthalpy change, giving another indirect measurement technique. Calorimetric measurements also rely upon the laws of heat transfer. Transfer of heat by radiation is most important at high temperatures, and is minimized by using bright clean metal surfaces. Conduction will occur through the material of the calorimeter, but not through a vacuum. The conduction law for heat transfer across a sample of area (A) thickness dx is:

$$\frac{dq}{dt} = -kA\left(\frac{dT}{dx}\right) \quad (49)$$

Here, dq/dt is the heat transferred per unit time, dT/dx is the temperature gradient in the $+x$ direction, and k is the thermal conductivity of the sample. On occasion, the geometric constants and thermal conductivity may be combined into a heat transfer coefficient (h), giving the heat conduction from the sample to environment as:

$$\frac{dq}{dt} = -h(T_S - T_E) \quad (50)$$

where T_S is the temperature of the sample and T_E is the temperature of the environment. Newton's law can usually be used to approximate the cooling of a vessel occurring by conduction or by convection in a draft, if the temperature difference is small:

$$\frac{dT}{dt} = K'(T_S - T_E) \quad (51)$$

where K' is a constant for the system and conditions. This law also generally demonstrates that if a system is perturbed at time $t = 0$, such that a temperature difference (ΔT) is reached, then the system will return to equilibrium exponentially according to the following expression:

$$\Delta T_t = \Delta T_0 \exp\left(\frac{-t}{\tau}\right) \quad (52)$$

where τ is the time constant for the system. The use of single or multiple thermocouples in calorimetry is extremely important. If the difference between the temperatures of the sample and reference thermocouple junctions is ΔT, then an emf (E) is produced which depends on the thermoelectric constant (ε) and the number of thermocouples (N). Therefore:

$$E = N\varepsilon\Delta T \quad (53)$$

The conduction of heat through the material and holding chamber produces a power output that changes with time, as demonstrated by the Tian equation:

$$P = \frac{k}{N\varepsilon}E + \tau\frac{dE}{dt} \quad (54)$$

The exponential decay of the power with time may be integrated to measure the heat produced. Cooling in the material and holding chamber may be induced by passing a current in the reverse direction, allowing a more efficient means of temperature control which can reduce experimental time. Such principles are used in Calvet or thermoelectric heat pump calorimeters.

B. Calorimeters

Because of the profusion of calorimeter designs there is no agreed system of classification. Hemminger and Höhne (1984) have suggested a method based on three criteria:

1. The measuring principle
2. The mode of operation
3. The principle of construction

Any calorimeter has basically two regions—the sample and the surroundings. The "sample" with a temperature (T_S) refers not only to the process under investigation (e.g., a phase change or a reaction) but also the associated containers, heaters, and thermometers. The "surroundings" also refers to the controlled region around the sample with a temperature (T_E). The temperature control of the surroundings may be active, as in the case of a Peltier unit; or passive, as in the case of a heat sink. The term (T_E) does not refer to the general laboratory conditions, which may need to be controlled to minimize unwanted effects. A crucial element of calorimetry is the measurement of (T_S) and (T_E) and their difference (ΔT) as a function for time (t).

$$\Delta T = T_S - T_E \tag{55}$$

1. Measuring Principles

1. Heat conduction calorimeters operate at constant temperature. Heat liberated from a reaction is, to a good approximation, entirely diluted within a heat sink. As seen in the case of Lavoisier's ice calorimeter the heat of a reaction results in the melting of ice, allowing the reaction to be followed. In addition, the endothermic phase change associated with the melting of ice maintains the system at constant temperature. Modern isothermal calorimeters measure the conduction of heat as it travels between the reaction ampoule and the surroundings and often have a very high degree of sensitivity.
2. Heat accumulation calorimeters allow a rise in temperature reaction system for exothermic reactions or a decrease in temperature for endothermic reactions. A reaction is followed by measurement of a temperature change as a function for time, although modern calorimeters allow the signal to be converted into power. An adiabatic solution calorimeter is typical of this class.
3. Heat exchange calorimeters actively exchange heat between the sample and surroundings often during a temperature scanning experiment. The heat flow rate is determined by the temperature difference along the thermal resistance between the sample and the surroundings. Heat-flux DSC uses this principle.

2. Modes of Operations

Three modes of operation are important.

1. *Isothermal* where sample and surroundings are held at a constant temperature ($\Delta t = 0$, $T_S =$ constant).
2. *Isoperibol*, or constant temperature jacket, where the surroundings stay at a constant temperature, whereas the sample temperature may alter ($\Delta T \neq 0$, $T_E =$ constant).
3. *Adiabatic*, where ideally, no heat exchange takes place between the sample and surroundings because they are both maintained at the same temperature, which may increase during the reaction ($\Delta T = 0$, $T_S \neq$ constant).

3. Construction

The construction of a calorimeter may have a single measuring system, or a twin or differential measuring system. Simple solution calorimeters have a single cell, whereas a DSC has twin cells and operates in the scanning mode. The use of twin cells reduces the effects of internal and external noise and transient fluctuations. Although calorimetry is intimately associated with thermodynamics, isothermal or adiabatic conditions are never exactly achieved. Allowances or corrections are made for the slight differences between theoretical and actual behavior. Systematic errors, which may be unsuspected, can cause problems.

A specific instrument must always be calibrated in some way. An electrical method may be available as part of the overall instrument package supplied by the manufacturer and is always a good start. It does not, however, eliminate unsuspected systematic errors and results should always be checked by measurements of a known "standard" system that is similar to the system under investigation. This is particularly relevant for modern microcalorimeters where a signal, often of the order of a few microwatts, can be much influenced by systematic errors. The method of calibration should be noted for each type of calorimetric experiment.

C. Instrumentation

1. Solution Calorimeters

Solution calorimeters are usually adiabatic calorimeters. They are mainly used for the study of rapid reactions, for example, heats of solution, heat capacity of liquids, heat capacity of solids by a method mixture, or the enthalpy change of rapid reactions in solution. A schematic diagram is given in Fig. 35. The temperature sensor, plus a means of electrical calibration and a device for mixing reactants are all enclosed within a dewar flask, or other adiabatic assembly.

Calibration may be performed by electrical heating or by using a reaction of known enthalpy change, such as

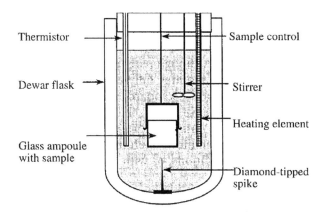

Figure 35 Schematic of an adiabatic solution calorimeter. The assembly may be kept in a thermostated bath. Activation is done by impaling the glass ampoule onto the spike, releasing the sample into the contents of the dewar flask.

the neutralization of a strong acid by a strong base. The whole system is allowed to equilibrate at some initial temperature T_0, and power P is supplied to the calorimeter for a time t raising the temperature to T_1. The total energy supplied raises the temperature of the system according to the following equation:

$$P_c = C_{p,\text{total}}(T_1 - T_2) \quad (56)$$

where, $C_{p,\text{total}}$ is the heat capacity of the whole system. A measurement of temperature change $(T_1 - T_0)$ is thus proportional to the reaction heat flow (power). The total heat capacity of the system is the sole unknown in [Eq. (46)] and successive experiments lead to a curve of $C_{p,\text{total}}$ as a function of temperature. The term $C_{p,\text{total}}$ is equal to $(C_{p,0} + m_s C_{p,s})$ where $C_{p,0}$ refers to the empty calorimeter assembly. Two sets of measurements are needed so that the "empty" data can be subtracted from those for the loaded system to give the specific heat capacity ($C_{p,s}$) of a sample of mass m_s. Subtraction of data for full and empty sample holders is a common requirement in calorimetry. An example may help to explain this process.

Passing a current of 0.15 A through a resistor of 1000 Ω for 60 s results in a temperature rise of 3.0°C (from 22°C to 25°C) in a dewar flask containing 100 cm³ of distilled water. The heat supplied, calculated from Eq. (5) is 1350 J, so that $C_{p,\text{total}}$, the average heat capacity of the system, is 450 J/K. As the heat capacity of the water is about 4.18 J/K, $C_{p,s} = 418$ J/K and $C_{p,o} = 32$ J/K.

If the temperature interval includes a phase transition, the final *apparent* sample ($C_{p,s}$) may appear exceptionally large because it includes a contribution from $\Delta H_{\text{tr}}(T_{\text{tr}})$ where tr = fusion or transition at some intermediate temperature T_1, as well as from low and high temperature phases (subscripts 1 and 2 respectively):

$$C_{p,s}(T_1 - T_{t0}) = C_{p,1}(T_{\text{tr}} - T_0) + \Delta H_{\text{tr}}(T_{\text{tr}}) + C_{p,2}(T_t - T_{\text{tr}}) \quad (57)$$

Equation (57) shows that, by adjusting the start and end temperatures and thus the total energy supplied, adiabatic calorimeters may be used to measure both C_p and ΔH_{tr}.

2. Combustion Calorimeters

The enthalpy change that occurs on the combustion of a material in a reactive atmosphere, usually oxygen under pressure, is the most important method of obtaining thermochemical data, both for the stability of materials and for the characterization of the energy content of fossil fuels. Complete combustion of organic compounds to $CO_2(g)$ and $H_2O(l)$ is essential, and corrections may be needed for elements such as nitrogen and sulfur.

In an adiabatic flame calorimeter, the fuel (e.g., natural gas or oil), is burned completely in an atmosphere of oxygen at 1 atmosphere pressure, and the heat is transferred into a known mass of liquid (usually water) and the temperature rise (ΔT) is measured. Provided that heat losses are minimized, the enthalpy change (ΔH) per mole of fuel is given by the equation:

$$\Delta H = C_s \frac{\Delta T}{n} \quad (58)$$

where C_s is the heat capacity of the system, obtained by calibration and n is the moles of fuel consumed during the experiment, found from measurements of the volume or mass of fuel consumed. In flow calorimetry, a steady state is set up where a constant input of power, either by means of electrical heating or by burning a fuel supplied at a constant rate, is balanced by efficient heat transfer to a heat sink.

A diagram of a flame calorimeter for combustion is seen in Fig. 36. This type of calorimeter is used in several ASTM methods to determine the calorific value of gases or other fuels [e.g., D 1826 (1994)].

For example, if one considers the apparatus in Fig. 30, where a gas is supplied at 9×10^{-6} m³/s and water flows at 90 g/s, and a temperature rise of 0.9°C, at steady state, was measured. Assuming the heat capacity of water to be 4.18 J/g/K, then the heat produced per second is:

$$\frac{dq}{dt} = 90 \times 4.18 \times 0.9 = 338.6 \text{ W} \quad (59)$$

The calorific value of the gas is 37.6 MJ/m³, assuming an ideal gas behavior is ΔH equal to -843 kJ/mol.

More widely used is the adiabatic bomb calorimeter seen in Fig. 37. The sample is placed in the crucible, in contact with an ignition wire. The crucible is placed in a

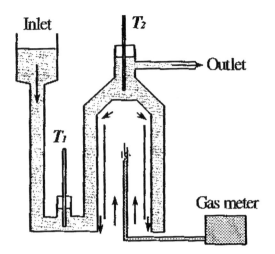

Figure 36 Schematic of a flame calorimeter for gases. The combustion of the regular flow raises the temperature of the water from T_1 at the inlet to T_2 at the outlet.

strong bomb of stainless steel, which is firmly sealed and pressurized to about 25–30 atm with oxygen. The bomb is placed inside a container with a measured amount of water. This is carefully supported within a shielding container with an outer jacket. The shield contains an electrically conducting solution with a heater, or some other means of temperature control. Thermistor sensors measure the temperature of the inner container, and of the outer jacket. The system is allowed to equilibrate until there is no difference in temperature between the sensors, the bomb is "fired" by passing a current through the ignition wire. This burns in the oxygen oven and ignites the sample which combusts completely. The temperature of the outer shield is raised to match the temperature within the bomb, maintaining the system under adiabatic conditions.

The calorimeter is calibrated using a substance whose internal energy change (ΔU) of combustion is accurately known. Benzoic acid is often used, having

$$\Delta U = -3215 \, \text{kJ/mol}^{-1} = -26.35 \, \text{kJ/g} \tag{60}$$

Corrections for the burning of the ignition wire, and for the conversion of any nitrogen in the sample into nitric acid should be applied. It must be noted that the bomb is a constant volume system, and therefore the calorimetry determines ΔU. Equation (61) can be used to convert ΔU into ΔH:

$$\Delta H = \Delta U + \Delta n RT \tag{61}$$

where Δn is the increase in the number of moles of gas during combustion.

Applications for the combustion calorimetry include the determination of the calorific value of fuels, and the energy content of foods, as well as obtaining primary thermochemical data, such as the resonance energy of aromatic compounds (Ingold, 1953; Vollhardt, 1987). Bomb calorimetry is used in several ASTM methods, for example, D 2382 (1988) and D 4809 (1995).

A further example of a combustion calorimeter, although of an oxygen specific nature, is the cone calorimeter (Babrauskas, 1981). This operates using the oxygen consumption principle, which states that for most combustible materials there is a unique relation between the heat released during a combustion reaction and the amount of oxygen consumed from the atmosphere as:

$$\Delta H = -13.1 \, \text{MJ/kg O}_2 \tag{62}$$

Using this relationship, the measurement of the concentration of oxygen in the combustion product stream, together with the flow rate provides a measure of the rate of heat release (RHR). The equipment may be applied to measure the flammability characteristics of materials such as plastics and fabrics by heating a sample measuring approximately ($10 \times 10 \times 1$ cm) using an electric heater to provide controlled irradiation. Any electric spark may provide ignition and, in addition to the measurement of heat, the mass loss of the sample, the intensity of smoke produced, and the concentration of product gases can all be measured during the course of the combustion. The heats of combustion and smoke parameters compare well with those determined by other methods and the method is now recognized as ASTM E 1354 (1994).

3. Reaction Hazard Calorimeters

The avoidance of hazards is of prime importance at industrial sites and in laboratory testing. The determination of the conditions under which rapid exothermic reactions, explosions and unwanted side reactions can occur is vital, so that the process may be carried out with the least amount of risk and the most profitable production of the required material. A small-scale laboratory reactor with facilities for stirring, heating, cooling addition of reagents, sampling or recycling of products, plus measurement of the system temperature and pressure may be employed to model chemical reactions and simulate larger scale processes. The fundamental characteristics of the process depend upon the total heat of reaction, the heat capacity of the system, and the effectiveness of heat transfer. These are measured by monitoring the product and jacket temperatures, enabling the calculation of heat flow at any time. Especially important are the possibilities of runaway heating and of explosive build up of pressure or adverse reactions, The possibility of online monitoring of chemical composition by spectroscopic techniques such as FT-IR spectrometry allows the study of the chemical nature of the processes involved and adds to the usefulness of the instrument (Schneider and Behl, 1997; Singh, 1993).

The measurement of temperature and pressure is a very important aspect of process safety. Knowing the enthalpy change ($\Delta_r H$) of a reaction and the system heat capacity then, under adiabatic conditions, the temperature rise $\Delta T_{\text{adiabatic}}$ can be determined by:

$$\Delta T_{\text{adiabatic}} = \frac{-\Delta_r H}{C_p} \qquad (63)$$

Increasing the cooling rate may contain a thermal runaway reaction. However, if the cooling fails, switching off the feed may stop the reaction. Sometimes even this will not counteract the runaway heating or pressure rise. Measurement in a calorimeter which is designed so that the heat capacity of the container (mC_{cont}) is much lower than that of the chemical reactants (mC_{chem}), which simulates the design in industrial plant and gives a better indication of likely runaway (Technical Information Sheet #4, Thermal Hazard Technology, Bletchley, UK). This is sometimes referred to as having a phi-factor close to 1.0:

$$\phi = 1 + \frac{mC_{\text{cont}}}{mC_{\text{chem}}} \qquad (64)$$

One system, which combines calorimetric measurements with the monitoring of temperature and pressure, encloses the sample in an inert cell, to which pressure and temperature sensors are attached. The cell is then enclosed in a bomb-like container and heated at a slow rate, between 0.5°C/min and 2°C/min, by the integral furnace. Any rapid rise in pressure in temperature during the heating indicates that the sample is undergoing a degradation, which may pose a risk. A schematic of an adiabatic bomb calorimeter is given in Fig. 37.

4. Isothermal Calorimeters

Isothermal calorimeters, referred to as heat conduction or as heat flow calorimeters, maintain a reaction system close to constant temperature. Necessary for the measurement of heat conduction, a temperature gradient is propagated from the reaction amp to or from a surrounding heat sink. Correction is made for this small imbalance of temperature. Peltier units are located within the temperature gradient of a reacting system, between the reaction ampule and surrounding heat sink, so that the transmission of energy can be measured as in Fig. 32. Various designs of heat sinks in heat conduction calorimeters have been used. These include a thermostatted water/oil bath, as in the case of the Thermal Activity Monitor (Isothermal microcalorimeter, Thermometric, Jafalla, Sweden) and the Calorimetry Science Corporation series of instruments (Calorimetry Science Corporation, Provo, UT), thermostatted air box used for the LKB batch calorimeters (LKB-Produkter AB, Bromma, Sweden), and temperature-regulated Peltier units as used in the (Setaram S.A. France, Caluire, France). Modern instruments have a twin ampule configuration incorporating a reaction ampule linked to a reference ampoule. The observed calorimetric signal is the differential of the two signals eliminating a considerable amount of external noise. Such calorimeters are highly sensitive and have exceptional long-term base-line stability. A schematic is given in Fig. 38.

Calibration of isothermal calorimeters is made using an electrical heater. A known amount of power is supplied

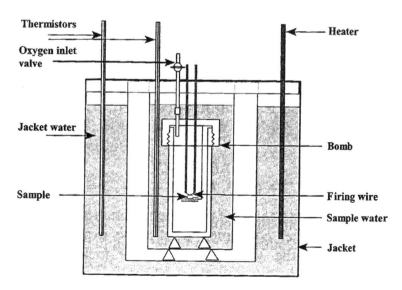

Figure 37 Schematic of an adiabatic bomb calorimeter.

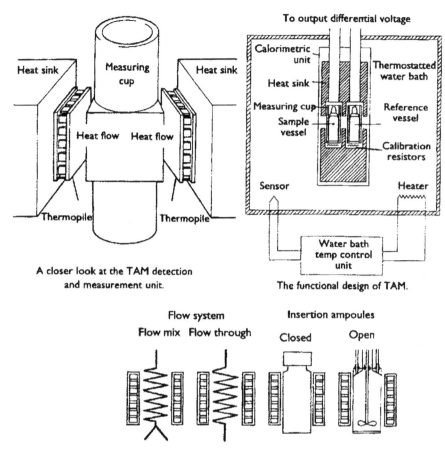

Figure 38 Schematic of an isothermal calorimeter unit showing the position of the reaction ampoule, Peltier units, and heat sink. The instrument has different combination of application, including flow through or ampoule insertion. (Courtesy Thermometric Ltd. Jafalla, Sweden.)

over a given time period to which the calorimetric signal is set. Some advantage is gained using chemical calibration and there have been several approaches for such purposes (Guan and Kemp, 2000; Willson et al. 1999). Briggner and Wadso (1992) have given a comprehensive description of the fundamental workings of heat conduction calorimeters.

D. Applications of Isothermal Calorimetry

Calorimetry has found many uses within many different disciplines in the scientific community. There are active areas of research in such diverse industries as metals, foods, agriculture, textiles, explosives, ceramics, chemicals, biological systems, pharmaceuticals, polymers, and the nuclear industry.

In the metal industry studies have been made of the phase transitions of metal alloys (Dantzer and Millet, 2001), modeling capacitors (Louzguine and Inoue, 1991), metal hydration kinetics (Seguin et al., 1999), precipitation of solutionized aluminum (Smith, 1997), and mechanisms of solid state transitions (Smith, 1988).

Food applications include the shelf life of goods (Riva et al., 2001), cooking of rice (Riva et al., 1994), oxidation of rapeseed oil (Kowalski et al., 1997), photocalorimetric study of plants (Johansson and Wadso, 1997), properties of amorphous sugar (Lelay and Delmas, 1998), and metabolism of dormant fruit buds (Gardea et al., 2000).

The study of mineral systems and ceramics has given valuable information for the construction and materials industry (Bonneau et al., 2000; Clark and Brown, 2000; Salonen et al., 1999, 2000; Stassi and Schiraldi, 1994), and the chemical industry has also benefited from studies of reactions and physical processes (Lehto and Laine, 1997; Machado et al., 1999; Stassi and Schiraldi, 1994; Ulbig et al., 1998a, b).

Calorimetric studies relating to polymers include the curing of resins (Smirnov and D'yachkova, 2000), aggregation of PEO–PPO–PEO [PEO = poly(ethylene oxide) PPO = poly(propylene oxide)] polymers (Gaisford et al., 1998), behavior of block copolymers (Beezer et al., 1994). Investigation of biological systems, particularly of ligand protein interactions, have been carried out using calorimetry (Guan et al., 1998; Hegde et al., 1998; Hoffmann and van Mil, 1999; Morgon et al., 2001; O'Keefe et al., 2000; Kemp, 2000a, b; Kerns et al., 1999; Wadso, 1995), often complementing DSC studies in some cases.

Calorimetric study of slow reactions (Montanari et al., 2000), polymorphism (Gallis et al., 2000; Hanson, 2000), and degradation (Durucan and Brown, 2000; Koenigbauer et al., 1992; Starink and Zahra, 1999), as well as many other applications, show the need for these techniques in the pharmaceutical laboratory.

In addition, several review articles have been published listing recent developments and current activities in isothermal calorimetry (Beezer, 2000; Beezer et al., 1995; Buckton and Darcy, 1999; Kemp, 2000; Kemp and Lamprecht, 2000; Koenigbauer, 1994; Otsuka et al., 1994; Wadso, 1997a, b, 2001).

1. Chemical Reactions

An arbitrary distinction can be made between chemical reaction and physical changes. Both sorts of change can be equally harmful quality of material.

Solution-Phase Reactions

In general, they are less complex than solid-state reactions to study by calorimetry. Reactions are often comparable with reasonable straightforward reaction mechanisms. There are many references to solution calorimetric studies in the literature (Arnett, 1999; Lis et al., 2000; Pikal and Dellerman, 1989; Willson et al., 1996).

The hydrolysis of ethyl ethanoate (Fairclough and Hinshelwood, 1937) is often provided as a practical analysis because of its apparent simplicity. The study of this reaction by isothermal calorimetry shows that a change in reaction mechanism occurs during the course of the reaction. A test reaction was made by mixing 2 cm^3 of a 0.011 M ethyl ethanoate solution with 0.2 cm^3 of a 1 M solution of sodium hydroxide. The resulting calorimetric signal is shown in Fig. 39. The analysis of the data indicates there are three regions along the calorimetric signal, consistent with:

1. an initial second-order process,
2. a period where the analysis of the calorimetric data gave constantly changing results as a function of time,
3. a final region where the reaction became first-order and the reaction went to completion.

A rationale of these observations is that the reaction is initially first-order with respect to ethyl ethanoate and first-order with respect to hydroxy ions, hence second-order overall. As the reaction proceeds and the concentration of reactants declines, the order with respect to hydroxy ions decreases, and finally the reaction becomes the first-order ethyl ethanoate alone.

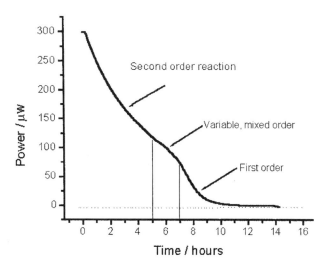

Figure 39 Isothermal calorimetric study of the solution phase reaction of ethyl ethanoate.

Solid-State Reactions

The majority of solid-state reactions studied by calorimetry are morphological changes rather than chemical changes. This is a reflection of the complexity of chemical reactions and the difficulty of extracting meaningful information from such calorimetric data. Personal experience of physical property characterization within the pharmaceutical industry provides assurance that solid-state morphological stability issues are common organic solids. About 80% of new chemical entities entering early development, in the pharmaceutical industry, have a propensity for change during the rigors of normal manufacturing processes. As a point of interest, most of the chemical entities were not initially recognized as having a potential morphological problem using conventional microscopic type analytical techniques commonly used for such purposes.

The morphological diversity of compounds is clearly recognized (Florence and Atwood, 1998). Processing may induce morphological change owing to stress. Storage conditions may also provide an opportunity for change, especially when storage extends over long periods. The consequences of poor control during processing and storage are, at times, forgotten. The mechanisms of solid-state reactions are complex. Molecular restriction within the structure hinders chemical reaction. There are numerous examples in the literature giving interpretations of solid-state reactions (Brown et al., 1997; Byrn, 1982). However, a general method for interpretation has yet to gain popularity. When reaction rates are relatively slow, mathematics can be used to extrapolate from the observation period (days) into the future (years) if one wishes to determine the progress of a reaction (Willson et al., 1995).

Compatibility Studies

Isothermal calorimetry has been used as a rapid screen for pharmaceutical materials. The main advantage is that a large number of materials, for example, a number of excipients, can be tested against a drug substance in a relatively short period. There have been some publications describing quantitative methods of analysis (Selzer et al., 1998). More commonly, compatibility studies are qualitative, utilizing the rapid determination of the presence of a reaction using binary drug–excipient mixtures.

2. Physical Changes

Physical changes tend to be relatively fast and often recorded within the observation period of a study. Physical changes are typically performed by stressing a material and following any resulting change. Such applications include crystallinity, morphological change, hygroscopicity, and vapor sorption.

Crystallinity

Of all solid-state transitions, the crystallization of amorphous materials provides very abundant literature (Buckton and Darcy, 1999; Larsen et al., 1997; Schmitt et al., 1999). A useful method for analysis is to couple a heat conduction calorimeter with a vapor perfusion device. The perfusion device allows control over the reaction environment, including vapor partial pressure. Crystallization can be induced in an amorphous sample by sequentially increasing the partial pressure of a vapor in contact with the sample, effectively lowering the glass transition temperature. On reduction of the T_g, spontaneous recrystallization can occur. This is typically observed in three characteristic steps: first is vapor sorption and is usually enhanced by the high surface area of amorphous materials; second is a crystallization exotherm; and the third step is endothermic desorption of vapor as a consequence of surface area reduction.

Morphological Change

The investigation of polymorphs by practical application of calorimetry cited in the literature employs almost every type of isothermal or isoperibolic calorimetry. The number of polymorphic forms of crystalline materials is dependent on how many different orientations the molecules can be arranged to form a crystal lattice. There is, at present, no way to predict the number of polymorphic forms a material may have. The more effort that is put into searching, the more

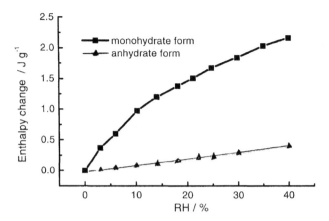

Figure 40 Two different polymorphic forms of a drug substance showing different vapor sorption enthalpy changes. Exploitation of different thermodynamic properties of material allows the identification of a form or mixture of forms (RH = relative humidity).

forms may be found. The molecular arrangement of molecules in a lattice can have a significant effect on the physicochemical properties of a material. An anecdotal example dates back to the Napoleonic wars. Tin buttons on the tunics of solders became brittle and broke off when marching in the cold climates of northern Europe. Below 13.2°C tin undergoes a polymorphic conversion from an α form that is strong and malleable into a β form that is fragile and brittle. Calorimetry can be used as a method for identification of polymorphic forms by exploitation of differences in physical properties, such as transition temperatures, solubility, vapor interactions crystal lattice energy, and so on. Figure 40 provides an example of this.

Hygroscopicity

Hygroscopicity (Campben et al., 1983) can only be properly defined if stated in conjunction with surface area. The units of hygroscopicity should be (moles of water)/m^2/mol of material. As a general guide, a material with a surface area of 3000 m^{-2}/mol, which is typical of a micronized drug substance, will adsorb 0.5% by weight for a monolayer surface coverage. Sorption in excess of 0.5% corresponds to ever-increasing degrees of hygroscopicity typically involve RH perfusion coupled with calorimetry. A material is subjected to increasing levels of water vapor and the mass increase followed by measuring the associated enthalpy change. Quantification of water sorption can be made using the enthalpy change for vapor condensation of 2.44 kJ/g water. An example of hygroscopicity of a pharmaceutical drug substance can be seen in Fig. 41.

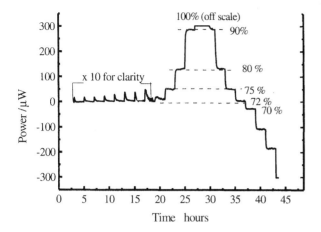

Figure 41 Graph illustrating the application of vapor perfusion to study hygroscopicity. Here, sodium chloride was used as an example. Starting at a relative humidity of 50%, the RH was stepped in increments of 2.5% to 70% RH. The resulting signal was seen to be adsorption type peaks. Above 72% RH the signal became stepped, indicating the onset of deliquescence.

Vapor Sorption

Vapor sorption studies are performed extensively to characterize surface properties, such as surface energy, surface area, and affinity for a given vapor (Buckton et al., 1999; Campben et al., 1983; Silvestre-Albero et al., 2001). Surface properties can be characterized by vapor sorption where the sorption data are attributed to a sorption model that conforms to basic sorption type. Brunauer et al. (1938a) have described five such types. From the adsorption model, thermodynamic information such as enthalpy change for vapor interaction and surface area can be determined. The Brunauer–Emmett–Teller (BET) type equations are typically applied to sorption data up to monolayer vapor coverage (Brunauer et al., 1938b; Pudipeddi et al., 1996). An extension of this equation is the Guggenheim–Andersson–DeBoer model (GAB) (Menkov and Dinkov, 1999; Veltchev and Menkov, 2000) and has been successfully applied to data where sorption levels exceed the formation of a monolayer. Calorimetric vapor perfusion allows good control of such experiments (Bakri). A plot of the enthalpy change for sorption against the vapor partial pressure allows BET type equations to be applied to the date, see Fig. 42.

The mathematical form of the BET equation can be transformed into an equation that can be applied to calorimetric data,

$$q = \frac{CV_m[\Delta H_1 x + \Delta H_1 - \Delta H_1 x^2]}{(1-x)(1-x+Cx)} \quad \text{where}$$

$$C = \exp\left(\frac{\Delta H_1 - \Delta H_L}{RT}\right) \quad (65)$$

Here V_m is the moles of water adsorbed per gram of solid for the formation of a monolayer, ΔH_1 is the enthalpy change for surface adsorption of water vapor, ΔH_L is the enthalpy change for water condensation, x is the partial pressure of vapor (p/p^0), and q is the measured enthalpy change for the sorption of x.

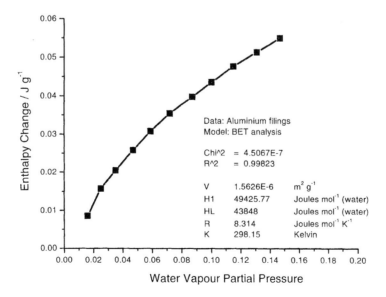

Figure 42 Water vapor adsorption isotherm for aluminum foil. The analysis of the data using a type II isotherm model reveals a surface area of 0.04 m^2/g/aluminium and an enthalpy change of 49.4 kJ/(mol water)

E. Summary

Several different types of calorimeters have been described, also the extraordinary diverse selection of applications associate with them. It should be recognized that there is a wide variety of other calorimeters designed for a specific use. For example, the US nuclear industry makes a disposable instrument for monitoring radioactivity in nuclear power stations. Others include a calorimeter that can be placed over growing plants, in particular to measure thermoregulation. Higher technological equipment includes disposable solid-state calorimeters printed on a circuit board requiring a few milligrams of sample, capable of temperature scanning at near to one million degrees per second.

The versatile nature of calorimeters, both commercial and homemade, provide instruments which allow direct access to the thermodynamic properties of materials being studied. Calorimetry is unintrusive in the way information is extracted during a study and highly versatile. They are capable of measuring from nanowatts to megawatts, from near absolute zero to several thousand Kelvin. The sample studied can be in any phase or mixtures of phases. Calorimetry can, in principle, be used to obtain all the thermodynamic and kinetic parameters resulting in a reaction, and is limited only by the sensitivity of the instrument to detect a change.

Calorimeters faithfully record all the changes that occur in a sample. A temptation is to mold the resulting calorimetric signal into a preconceived idea. Although getting a result from a calorimeter is relatively easy, a correct result takes a lot of time, effort, and cogitation.

IX. OTHER TECHNIQUES

Other thermal analysis techniques have been reported. These include those methods grouped under *thermoelectrometry*, covering the variation in the electrical characteristics of a substance when subjected to a controlled temperature program. Paulik and Paulik (1978) have reviewed this topic. Optical microscopes can also have the sample subjected to temperature programing (Vaughan and Burroughs, 1978). There are thermal analysis techniques based on sound emitted by the sample during heat treatment (Lonvik, 1978). The simultaneous use of small-angle X-ray scattering and wide-angle X-ray diffraction during the programed heating of various polymeric materials have been described (Ryan, 1993).

One technique that has emerged is thermally stimulated current depolarization (TSC) which is available commercially. The principle of TSC is to orient polar molecules or pendant polar groups of macromolecules, by applying a high voltage field at a high temperature and then quenching the material to a much lower temperature where molecular motion occurs. After this polarization, the material is heated at a constant rate causing it to depolarize, thereby creating a depolarizing current. This thermally stimulated current can be related directly to molecular mobility, indicating the physical and morphological structure of materials. Ibar (1993) has produced a book on the subject. The technique has found a wide application to polymers (Bernes et al., 1991, 1992; Vanderschueren et al., 1991).

Some techniques are really modifications in the use of existing equipment. Controlled rate thermal analysis is reviewed by Reading (1992). In this, the rate of reaction is kept constant or controlled, achieved by changing the sample temperature. Basically, this is programing the reaction rate and noting the temperature as the resultant parameter. The technique was introduced by Rouquerol (1970) and by Paulik and Paulik (1971). It has been utilized in various publications to determine kinetic parameters (Reading et al., 1984, 1992). A related technique is termed *high-resolution thermal analysis*. In this the heating rate on the TG equipment is decreased whenever a mass loss is detected, establishing a minimum or almost zero heating rate, and the heating rate is then reintroduced when the rate of mass loss lends to zero. This introduces a very sharp loss of mass for a transition over a small range of temperature (Gill et al., 1992).

Another technique which is also an adaptation of existing equipment is MDSC. In this the temperature is subjected to a sinusoidal ripple (modulated) while maintaining an overall heating rate. Now, if an irreversible effect is present on the heating part of the ripple then by definition it will be absent on the cooling part of the ripple. The word "reversible" is not used here in the thermodynamic sense but in the practical sense. Thus dissociation or boiling are thermodynamically reversible, but in a DSC system the loss of material incurred in the transition renders it experimentally irreversible. The conventional DSC signal can then be resolved into a MDSC signal for the nonreversing component and an MDSC reversing component, Examples are given by Gill et al. (1993) and the theory is outlined by Reading et al. (1993). It is particularly useful in the determination of the heat capacity and the glass transition point in polymers.

The possibility exists with modern equipment of obtaining DTA type data from a TG balance or single pan unit. This relies on the fact that the heating rate applied to the sample is in fact perturbed by the reaction heat resulting from the transition. By plotting the perturbation (ΔT) against the projected heating curve (temperature T) a curve is obtained which is identical with the DTA/DSC signal.

X. SIMULTANEOUS THERMAL ANALYSIS TECHNIQUES

A. The Rationale Behind the Simultaneous Approach

There is a large family of simultaneous thermal analysis (STA) techniques, in which two or more types of measurements are made at the same time on a single sample. The methodology entails using a more complex instrument, often specially built, which is essential to a variety of thermal studies. Instruments for simultaneous measurement have been constructed for more than 50 years since the benefits of this approach were rapidly appreciated.

During the investigation of a new material it is unlikely that any single thermal analysis technique will provide all the information required to understand its behavior. Complementary information may be needed, from another thermal technique and another form of analysis. The challenge is to correlate the data from the individual analyses. Information obtained from separate TG and DSC instruments cannot be expected to correlate precisely when the sample is experiencing different conditions during thermal treatment. There are difficulties even in comparing data from different instruments of the same type, such as DSC instruments from different manufacturers, as is born out by the experiences of standardization bodies trying to develop measurement procedures that are independent of the make of instrument. This is true even for "well-behaved" systems where there is only one possibility for the type of event occurring on heating. When complex systems are studied, and alternative reaction schemes are possible, or the nature of the transformation is markedly influenced by the size, form, or environment of the sample, discrepancies between the sets of data are to be expected. There are some materials, even though they may be chemically pure, that do not give precisely repeatable behavior. The overall reaction may result in the same products each time, but the course of the transformation can differ in their details. In such cases it is advantageous to determine the two or more measurements on the same sample. Simultaneous measurements are also valuable when dealing with nonhomogeneous samples. Homogenization (e.g., size reduction) in order to obtain a more representative sample, sometimes cannot be carried out without risk of altering the system under study.

In terms of an overall understanding of the system, the availability of two or more independent measurements that correlate precisely leads to a synergistic effect. Therefore, the total value of the information is greater than merely the sum of the parts. Paulik and Paulik in their review of their own STA studies (Paulik and Paulik, 1981), speak of this as a "multiplying" effect.

At first sight, a simultaneous instrument may seem attractive as an alternative to buying separate units. A case has been made for this on economic grounds (initial purchase, running costs, etc.). However, the many advantages are technical, not economic. Modern simultaneous devices are capable of excellent performance, but they will not perform the individual measurement tasks as well as dedicated separate units. Their advantage lies in the unambiguous correlation of two or more sets of information.

B. Simultaneous TG–DTA

TG and DTA were the two most common techniques for many years. DSC has now taken over from DTA in recent times. Given the essentially complementary nature of TG–DTA/DSC it is not surprising that their combination was the first to be realized. TG is inherently quantitative, after appropriate calibration and corrections, but responds only to reactions accompanied by a mass change. DTA is capable in principle of detecting any reaction or transition that entails a change in enthalpy or heat capacity, but requires a good deal of effort before it is truly quantitative. The same is still true, though to a lesser extent, with DSC. For reactions involving a mass change, DTA and DSC can never be quantitative, as the material lost carries heat from the system, and the changing thermal properties of the remaining material affect the heat transfer to it. The complementary nature of the techniques can be appreciated from the processes amenable to a given study by each of them (Table 14).

Early attempts to couple the two techniques led to devices with separate TG and DTA sensors heated close together in the same furnace. This approach has been termed "concurrent" TA, and though an improvement over separate instruments, it is not truly STA. The main problem in mounting a DTA head in a thermobalance lies in the means of taking the DTA signal from the thermocouples without interfering with the operation of the balance. In general, this is done using very fine flexible wires or ribbons.

In one of the few reviews dedicated to STA (Paulik and Paulik, 1978), the Paulik brothers claim to have been the first to build a truly simultaneous TG–DTA, in 1955. Over many years they developed their instrument, the "derivatograph" to incorporate dilatometry and EGA, and used it to pioneer the development of controlled rate methods. By modern standards, the instrument performance was pedestrian, and used samples of up to about 1 g. The next step forward was the versatile Mettler TA-1 described by Wiedemann (1964), which had a higher sensitivity and could be operated under vacuum or in reactive atmospheres. Scantor Redcroft (now Rheometric Scientific and part of TA Instruments), Netzsch and several other manufacturers soon followed with instruments all mounting a DTA head on to the thermobalance suspension or rise rod. A simplified view of a typical sensor arrangement is seen in Fig. 43.

A different approach has recently been taken by TA Instruments, in using two horizontal balances. The balance arms, each with a thermocouple attached to a pan carrier holds the sample and reference adjacent to each other in the furnace. In this instrument there is no heat flow path between the sample and reference, other than through the surrounding atmosphere which limits the quality of the DTA data. Mettler-Toledo now offers a technique in which a form of DTA is obtained by comparing the sample temperature with calculated reference temperature profile. Modern TG-DTA instruments are capable in general of a TG resolution around 1 μg, use samples typically from 5 to 100 mg, and can give sensitive and quantitative DTA performance when the head is of the appropriate type (i.e., there is a heat flow link between sample and reference).

C. Simultaneous TG–DSC

The recent trend has been to incorporate heat-flux DSC sensors rather than DTA sensors into the thermobalance. The greater complexity of power-compensated sensors, and the fact that they are restricted to lower temperatures, has precluded their use in STA. In general, the DSC heat-flux sensors are merely modified quantitative DTA plates. The relative position of the sample and reference being more precisely defined and with a more reproducible heat flow link between them. The implied advantage is that real calorimetric measurements are possible. This is true to some extent. The linearization of the sensor output and conversion of the DTA signal and heat flow units in software or hardware reduces the effort required. Careful calibration is again vital. One problem that may arise is that the position of the DSC sensor may alter slightly with different sample weights. This alters the heat transfer characteristics and therefore the baseline of the DSC curve, and possibly the sensitivity. Some manufacturers, therefore, offer the option to clamp the TG–DSC head rigidly for best DSC performance, and thereby lose the STA ability. The advantages of TG–DSC include all of those discussed for TG-DTA previously, with the possibility of improved quantification of the heat flow data. Figure 44 illustrates the measurement of the heat of vaporization by TG–DSC.

A TG–DSC instrument is capable of subambient temperature operation and has been available for a number of years. It is especially valuable in the study of systems containing moisture, or other volatiles (Charsley et al., 1980). In this case, a DSC heat-flux plate was adapted to allow it to be suspended from the thermobalance beam, instead of the pair of thermocouples. A completely different approach to TG–DSC is taken by SETARAM, in their TG-DSC 111 instrument. The sample and reference in this case are suspended in tubes surrounded by Calvet-type heat flux transducers which do not need to contact them. The symmetrical chemobalance automatically compensates for the significant buoyancy corrections that would arise with the large samples that this instrument can accommodate.

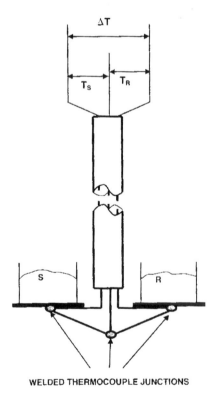

Figure 43 Schematic diagram of DTA head mounted on a thermobalance suspension.

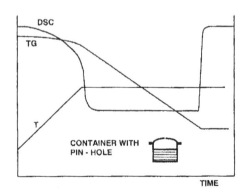

Figure 44 Illustration of the method for measuring the heat of vaporization by TG–DSC.

XI. OTHER LESS-COMMON TECHNIQUES

A. Optical Techniques

The value of using an extremely sophisticated sensor (the human eye) in the study of heated materials has been appreciated for some time. In the context of thermal analysis, the monitoring of the optical properties during a thermal program can be of immense assistance. The term thermoptometry is used to cover observations, or measurements, of an optical property as a function of temperature. A useful review of the field is provided by Haines (1995). Samples can be viewed by reflectance, transmission, and light intensity. In each case light can be measured using a photocell, thereby making the technique into a measurement instead of an observation. Temperature-controlled stages for samples being viewed through a microscope (hot-stage microscopy) have been available for many years. Modern versions are available from, for example, Mettler and Linkam. The observations are often helpful in the interpretation of other thermal data revealing the nature of the reaction or transition taking place. With known materials, a preliminary screening can avoid damage to TG or DSC equipment by showing whether the sample melts, bubbles, creeps, inks, swells, decrepitates, or damages the crucible. Solid–solid transitions may be detectable by a color change, or movement due to a change in volume. Microscopic examination of reaction interfaces may suggest an appropriate choice of reaction model. Several systems have been described for combining DSC or DTA with microscopy (Forslund, 1984; Perron et al., 1980). A more unusual combination is that of TG with the ability to view the sample shown in Fig. 45, which was used to study the pyrolysis and combustion of coal particles (Matzakos et al., 1993). The video camera was able to produce 640 × 480 pixel images of an area approximating 2 × 2 mm, with the help of a microscope with an extra-long working length. Partial fusion, swelling, and changes in structure could be related to the progress of the TG curve, whereas particles of around 0.5 mm were heated upon a platinum mesh support. These workers made a special furnace, for mounting in a Perkin–Elmer thermobalance that allowed heating rates of up to 100°C/s, similar to those experienced by pulverized coal in power plants.

B. Spectroscopic Techniques

Microscope hot-stages designed for transmitted light work, and special heating accessories, have been used in conjunction with the FT-IR spectrophotometer to study physical and chemical changes in a heated sample. By using a hot-stage with an incorporated DSC cell simultaneous DSC–FT-IR can be achieved (Mirabella, 1986). The range of experiments possible in this case was restricted by the need to use thin films of sample, which resulted in poor DSC curves. More flexibility was obtained by using an IR microscope accessory to collect reflectance spectra from a sample in a DSC cell (Johnson et al., 1992). This equipment was used to follow the curing reaction in an amine-cured epoxy material, and to study structural changes in PET through its glass transition and melting regions. Figures 46 and 47 provide the evidence for the use of simultaneous TGA–DTA studies combined with hot-stage microscopy to confirm the thermal events seen in the thermogram.

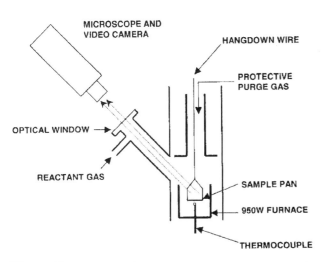

Figure 45 Equipment for simultaneous TG and microscopic observation.

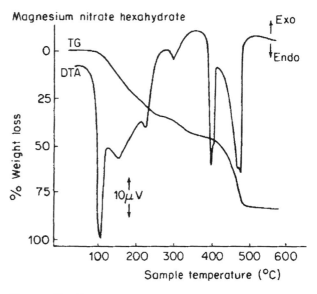

Figure 46 Simultaneous TG–DTA studies on $Mg(NO_3)_2 \cdot 6H_2O$.

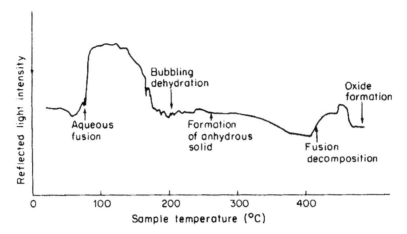

Figure 47 Hot stage microscopy studies on $Mg(NO_3)_2 \cdot 6_2O$.

C. X-Ray Techniques

Whereas EGA is useful in obtaining chemical information in cases where gases are lost during a reaction, it gives no direct data on the condition of the solid phases. The technique of X-ray diffraction (XRD) has been used extensively to identify reaction intermediates and products, and can be applied to small samples generally used in thermal studies. Extracting samples from, for example, a thermobalance for analysis might give misleading results if the samples are susceptible to a reaction with atmospheric gases (primarily oxygen, water, and carbon dioxide) or the solid was liable to undergo a phase transition. The ability to apply XRD simultaneously with other methods has therefore been exploited in several instances.

XRD was coupled with TG (Gerard, 1974) by building a top-loading thermobalance onto the diffractometer in such a way as to place the sample layer in the X-ray beam. Where the heating rate would be limited by the scanning rate of the diffractometer, it is possible to scan only small angular regions characteristic of the products of interest, thus saving time. This type of apparatus was used to investigate a complex series of overlapping stages in the reduction, by hydrogen, of tungstic oxide, which results in a fairly featureless TG curve, but was shown by the XRD spectra to proceed via the production of many phases in the tungsten–oxygen system, and two modifications of tungsten. XRD has also been coupled with DSC. An instrument based on a conventional X-ray source was described by Newman et al. (1987) that was used to study catalyst regeneration. This instrument incorporated an MS for simultaneous EGA.

A difficulty with the DSC–XRD coupling is that the XRD requires low-temperature scans to obtain sufficient sensitivity, but this results in lower DSC sensitivity. The problem has been solved by several workers through the use of a high-intensity synchrotron X-ray source. Caffrey used such a combination to study biological liquid crystals (Caffrey, 1991). The need for higher intensity X-rays is greater when carrying out small-angle (SAXS) and wide-angle scattering experiments (WAXS), and when synchrotron radiation has been used, as typified by the work of Rayan and coworkers on polymer structure determinations (Ryan et al., 1994). Another unusual simultaneous coupling was that of temperature-programed reduction with XRD and Mossbauer spectroscopy, which was applied to Fe–Mo–O catalysts (Zhang et al., 1995).

D. DTA–Rheometry

Rheometry is not usually classified as a thermal analysis technique, but when data are collected as a function of temperature it must be considered as such. Correlation of temperature-resolved rheological measurements with a separate DSC experiment would present problems owing to the very different sample sizes and environments employed in each technique. A recently described device (Application Note, Rheometric Scientific, Piscataway, NJ) combines a dynamic stress rheometer with a simple DTA sensor, both heated and cooled by a Peltier element. An example of the information available is seen in Fig. 48 where both the DTA and modulus are shown when a margarine sample is cooled, then heated. The rheological properties of the material are not recovered after cooling, suggesting that the structure of the emulsion has been altered irreversibly. The melting and crystallization are revealed clearly on the thermal signal.

E. Microthermal Analysis

The recent technique of microthermal analysis (μ-TA), which now has a variety of measurement modes, is included because usually two or more measurements are made simultaneously. μ-TA combines the imaging

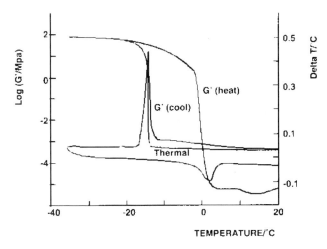

Figure 48 Simultaneous DTA and rheology curves for a low fat margarine.

capabilities of atomic force microscopy (AFM) with a form of localized thermal analysis, and is capable of measuring thermal transitions on an area of a few microns. A good introduction to the whole family of these methods is available on the internet, from which application studies can be downloaded (www.anasys.co.uk). In brief, the method entails using a very fine platinum wire loop as the AFM tip, which acts as a heater, and a resistance thermometer. This tip is scanned over the sample surface to obtain the AFM topographic image. By measuring the power required to keep the tip at a fixed temperature, an image is produced on the basis of variations in thermal conductivity of the region below the tip. From the scanned area, points can be selected for localized thermal analysis. When the tip is applied to the surface with a constant small force, the tip temperature can be raised linearly. Thermal transitions are detected by measuring the power applied to the tip and its position, which shows expansion or indentation during the event, or by a change in the rate of these processes.

The technique has been extended to give direct chemical information. The heated tip can be used to pyrolyze a small area of interest by rapidly heating to about 800°C. The resulting plume of gas is drawn into a small sampling tube containing approximately 0.1 mL of Tenax and Porapak, where organic volatiles are adsorbed. The tube is then transferred to an automatic thermal desorption unit, from which the products are passed to a GC-MS system for separation and identification (Price et al., 1999). Applications include the identification of polymers in multilayer films, and the analysis of layers of emulsion paint. A recent addition to the capabilities of μ-TA is localized FT-IR analysis. The area selected for examination is irradiated with an intense focused IR beam. The tip in this case acts as a bolometer and records the interferogram, which is them deconvoluted using the FT-IR software to produce an IR spectrum of a small area. The entire field of μ-TA is covered in a comprehensive recent review by (Pollock and Hammiche, 2001).

F. Applications for TG–DTA

An example of the complementary nature of TG and DTA curves obtained simultaneously is shown in Fig. 49, which represents the partial reduction of a superconducting ceramic from the "YBCO" or "123" family. The stoichiometric composition is $Yba_2Cu_3O_7$; however, processing usually results in a substoichiometric phase conducting between 6.5 and 7 oxygen atoms. Much effort went into establishing the correct tempering conditions (time, temperature, and oxygen pressure) to produce the desired composition. Various approaches were tried as a means of determining the oxygen content, including reduction

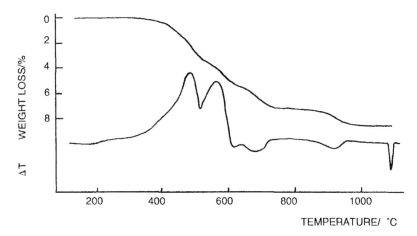

Figure 49 TG and DTA curves for the reduction of $Yba_2Cu_3O_7$ superconductor: 50 mg sample, 15°C/min, 5% H_2 in N_2.

using a $H_2(4\%)/N_2(96\%)$ purge gas mixture. The TG curve in Fig. 48 shows a multistage weight loss over a broad temperature range leading to a plateau of 1000°C. The overall weight loss corresponds to the loss of about 3 oxygen atoms, but give no clues as to the specific reduction process(es) taking place. The DTA curve shows that the first two stages are exothermic, and the second two endothermic. It then crucially shows a sharp peak with no weight loss at 1083°C, due to the melting of copper. The area of the peak confirms that only oxygen initially associated with copper is reduced.

Accurate temperature calibration was necessary for the earlier example, which leads to another advantage of TG–DTA instruments. The temperature calibration can be properly carried out as the sample temperature is measured directly, which is not the case in conventional thermobalances in which the temperature sensor is not in contact with the sample. A TG–DTA instrument was used to provide accurate transition temperatures for the ICTAC Curie point standards for TG calibration, after calibration with the melting temperatures of pure metals (Charsley et al., 1980). Precise temperature calibration of TG instruments is important because of the current stress not only on quality assurance, but also in the performance of kinetic studies. Imprecision in temperature measurement in TG work renders many kinetic studies meaningless, once an error of a few degrees (and this is easily achieved) has a marked influence on the derived parameters, particularly when the studies are often restricted for practical reasons to a small overall temperature range.

The fact that the sample mass is continuously monitored means that DTA peak measurements can be referred back to a true sample weight after allowing for prior lower temperature weight losses such as drying. Confidence in the quality of peak area measurements, especially at high temperatures, is increased when the TG data show that no sublimation decomposition is taking place. Glass transition measurements can be markedly influenced by the moisture content in the case of many important polymers. With STA, the moisture content at the time of measurement is known exactly.

Figure 50 shows the results from an experiment on a coal sample. Results for combustion reactions are highly sensitive to experimental conditions, and often difficult to reproduce, making the STA approach valuable. In this case the curving of the samples is seen before ca. 200°C on the TG curve, and the large exothermic effect due to oxidation can be assigned to the dry weight.

Comparison of the curves shows that a substantial portion of the heat output occurs during a period of weight gain, information that would be difficult to obtain reliably by any other means. Interestingly, EGA reveals also that, during this period of a weight gain, large amounts of water and CO_2 are concurrently evolved.

DTA data can help to explain unusual features of the TG curve. Abrupt changes in the rate of weight loss can occur when the sample melts, or at least partial fusion occurs, and this may be detectable by DTA. During the loss of material by sublimation, or solid decomposition, if the experiment extends through the melting point, the rate of weight loss will show a slight arrest, owing to the loss in surface area of the sample. The DTA curve might indicate a fusion peak at this point. The rate of reaction in an initially solid-state system, or the rate of a solid–gas reaction, may be influenced by the establishment of a fluid phase. Irregular sharp features on a TG curve due to bubbles bursting in a viscous melt are usually seen clearly as sharp endothermic spikes on the DTA trace.

G. Applications of TG–DSC

Figure 51 shows TG and DSC curves obtained from the low-temperature instrument referred to earlier, where the

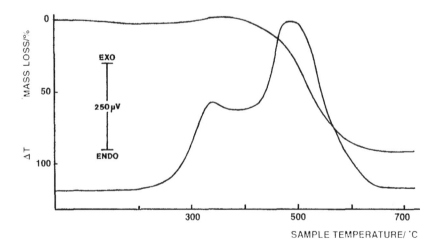

Figure 50 TG and DTA curves for a bituminous coal: 18 mg sample, 10°C/min, air.

multiple solid-state phase transitions in ammonium nitrate are examined. The curves show immediately that partial sublimation, and then vaporization, of the material occurs before and after the melting point. More accurate estimates of the heats transition and melting are thus available on the basis of the correct sample weight. Furthermore, the literature describing the solid phase transitions is confusing, as it has been shown that traces of moisture can affect the exact course of the transitions on heating and cooling. Using STA, moisture content is known precisely at each stage.

The sample is sealed in a container with a small pinhole, equilibrated at the chosen temperature, and loses weight at a constant rate until it is exhausted. At the same time, the DSC curve shows a deflection from the baseline equivalent to the heat of vaporization, which can be directly referred to the rate of weight loss. Calibration with materials of known vapor pressure would also allow this arrangement to derive absolute vapor pressures with good accuracy.

XII. APPLICATIONS

A. Nature of the Solid State

Applications in thermal analysis are often governed by the fact that the material being investigated is in the solid state and a condensed phase (solid or liquid) remains throughout any particular investigation (Sharp et al., 1966). Solids have bulk and surface properties and their prehistory is frozen into the structure which may contain impurities and defect structures that affect the reactivity. It is possible to prepare solids that have the same composition but exhibit very different properties. It is necessary then, when describing a solid sample, to include as many of the following points as possible, namely, sample size, particle size, methods of sizing (e.g., the use of crushing and grinding apparatus), size fractionation methods, thermal and mechanical history, method of preparation of the solid (including industrial and process conditions), surface texture (e.g., the porosity of the surface), surface area, any form of surface modification, which may be physical (e.g., compaction) or chemical, compositional, and phase analyses, the isotropic and anisotropic nature of the solid (e.g., direction of the fiber in composite materials) (Fenner, 1913).

The application of thermal analysis is often to elucidate this prehistory of the solid sample. However, it is possible to remove the prehistory by a preheat stage prior to the thermal analysis experiment, though such pretreatment may cause sintering with loss of surface area and increase in particle size, and loss of rigidity. In many cases industrial laboratories require rising temperature experiments that give information about the prehistory of the sample. On the other hand, measurement of thermodynamic properties will require removal of prehistory, for example, by melting, followed by measurement on a cooling curve. The question of the elimination of prehistory is dependent on what data are required and what can be done to the sample without loss of important structural features.

Many samples can be stored without alteration over time without special precautions, whereas others cannot. Thus, calcium hydroxide will often contain calcium carbonate due to recarbonation. Storage conditions are important and should be noted. In particular reactive solids, whether minerals are freshly removed from an anaerobic environment or are prepared with a very high surface area, need to be noted. The difference in reactivity of finely divided nickel powder produced from nickel formate and a nickel bar is due largely to the high

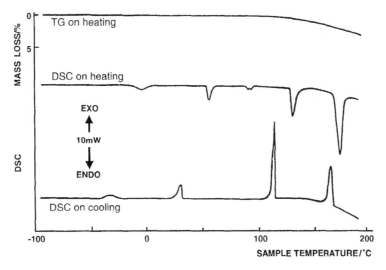

Figure 51 TG and DSC curves for the multiple solid-state phase transitions of ammonium nitrate utilizing a 5 mg sample reheating rate of 10°C/min in nitrogen using a low-temperature TG–DSC.

surface area of the powder. In the powder it is often difficult to maintain the material in a pure state, as the surface is liable to undergo rapid change.

B. Applications Based on Thermodynamic Factors

It must be noted that as far as thermodynamic factors are concerned with materials in the liquid state a conventional textbook approach is possible. The difficulties in dealing with the solid state have already been discussed. As noted, the solid phase is dominated by its prehistory which presents special features in the application of thermodynamics (Dollimore, 1992). The rigid nature of the solid phase forms means that solid phases can exist in a lower temperature region where, from purely thermodynamic considerations, a transformation might be expected. The original work of Fenner (1913) on silica is summarized in Table 15. This concept forms the basis of very rapid quenching systems from a high temperature to a lower temperature in order to ascertain the high-temperature form. However, the product solid in a solid-state decomposition is often formed in an amorphous or energy-rich state owing to the inability of the solid phase to retain the stresses involved in rearrangement and possible volume changes. It is best to retain the term "amorphous" for this state of affairs and to use the term "vitrious" to apply to a super-cooled liquid. In both cases only short-range order exists and long-range order is lost.

These "metastable" states persisting at temperatures lower than usual owe their existence to a kinetic factor. The rate of change is lowered with a drop in temperature and if the "metastable" solid phase is brought successfully to a temperature where the rate of change is negligibly slow, then to all intents and purposes it has to be regarded as stable.

First we have to recognize that the forms of the equations relating to phase transitions, equilibrium reaction systems, or rates of kinetic change are similar. For a phase change (e.g., liquid to vapor)

$$\ln p = \frac{-\Delta H}{RT} + \text{constant} \quad (66)$$

where p is the equilibrium vapor pressure above the liquid, T is the temperature in Kelvin, and $-\Delta H$ is the latent heat of vaporization. A similar equation holds for the equilibrium between a solid and a vapor. For the equilibrium condition, for solid phase decomposition, calcium carbonate is often quoted in textbooks:

$$\ln pCO_2 = \frac{-\Delta H}{RT} + \text{constant} \quad (67)$$

where for the reaction

$$CaCO_3(s) \xrightarrow{\Delta} CaO(s) + CO_2(g) \quad (68)$$

pCO_2 is the equilibrium pressure of CO_2 at temperature T. Calcium carbonate decomposition is often quoted as being reversible, but certainly with various limestone some different results will be obtained, although the validity of the form of the equation can be shown. In many dehydrations, reversibility may be demonstrated (Brown et al., 1980) and in some carbonates the oxides may be recarbonated (Beruto and Searcy, 1974), but in other systems reversibility may be absent. Taking carbonates, then, two systems can be cited: reversible systems

$$MCO_3(s) \rightarrow MO(s) + CO_2(g) \quad (69)$$
$$MO(s) + CO_2(g) \rightarrow MCO_3(s) \quad (70)$$

and nonreversible systems where carbonization of the oxide does not take place.

The reason for reversibility or nonreversibility in the carbonation reaction must lie in the nature of the carbonate layer initially produced; if it prevents further access of CO to the oxide layer, then realistic progress of the carbonization is impeded and the process eventually becomes nonreversible.

Figure 52 shows schematically some of these points using DTA traces. The figure shows a schematic illustration of a reversible phase change, typically a solid-to-liquid change. Such a change would be endothermic on the heating curve and exothermic on the cooling curve. Runs of this kind distinguish the amorphous to crystalline transition, which is exothermic and irreversible. In studying single-stage decompositions, consider calcium carbonate dissociation as an example. Similar results would be seen for oxide dissociation in the presence of an increasing pressure of oxygen,

However, for such simple systems and phase changes, a controlled atmosphere manometer experiment would probably produce the same results more cheaply and simply, but the endothermic (or exothermic) character of the

Table 15 Phase Transitions in Silica (°C)

Temperature	Transition
117	$\alpha T \leftrightarrow \beta_1 T$
163	$\beta_1 T \leftrightarrow \beta_2 T$
198–241	$\beta C \rightarrow \alpha C$
220–275	$\alpha C \rightarrow \beta C$
570	$\beta Q \rightarrow \alpha Q$
575	$\alpha Q \rightarrow \beta Q$
850 ± 10	$Q \leftrightarrow T$
1470 ± 10	$T \leftrightarrow C$
1625	$\beta C \rightarrow$ silica glass

Note: T, tridymite; C, cristobalite; Q, quartz; α, polymorph stable at low temperatures; β, polymorph stable at high temperatures.

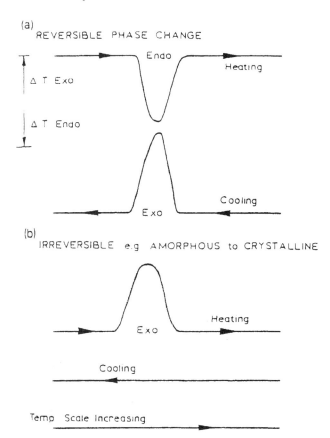

Figure 52 Schematic DTA showing phase changes: (a) reversible phase changes and (b) irreversible amorphoris to crystalline phase changes.

transition would not be immediately apparent. In two-stage decompositions, the DTA is superior to manometer systems. This behavior is shown in outline in Fig. 53. This would hold for oxide dissociation or dehydration sequences. As the pressure of product gas is increased, both stages of dissociation are moved to higher temperatures. In manganese dioxide dissociation there are several steps involved (Dollimore and Tonge, 1964):

$$MnO_2 \rightarrow Mn_2O_3 \quad (71)$$
$$Mn_2O_3 \rightarrow Mn_3O_4 \quad (72)$$
$$Mn_3O_4 \rightarrow MnO \quad (73)$$

All these dissociations are pressure-dependent. Phase changes of the solid–solid type occur which are well characterized in the Mn_2O_3 form. These transitions are almost invariant with respect to pressure. All the phases characterized here by formulas are in fact capable of sustaining a high degree of nonstoichiometry (Fig. 54). The variability found in these types of measurements is basically related to a kinetic ingredient which cannot be separated from the equilibrium conditions.

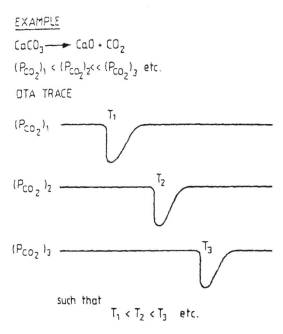

Figure 53 Schematic DTA plots for single-stage decompositions carried out with increasing pressure of product gas.

There is a continuing change in heat capacity of the sample, which determines the position of the baseline in DTA (Flynn, 1993). This forms the basis of a method of determining heat capacity. A typical schematic of phase transitions in polymers is shown in Fig. 55. A transition not discussed so far is the second-order transition (marked A) involving only a change in the baseline. The scale is exaggerated in Fig. 55. For polystyrene and

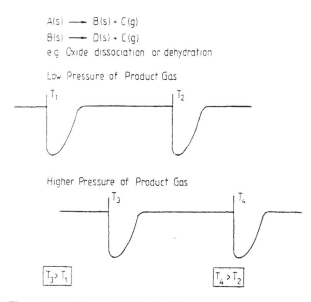

Figure 54 Schematic DTA plots for two-stage decompositions carried out with increasing pressure of product gas.

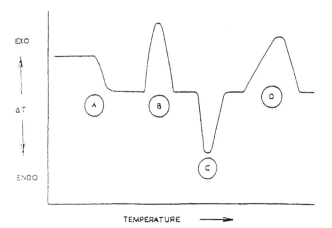

Figure 55 Hypothetical DTA trace for a polymer heated in air.

polyvinyl chloride, if the ΔT value associated with this change is 0.1–0.2°C, then a typical first-order change would be 10–50× as great as shown (Keavney and Eberlin, 1963). These second-order transitions are termed glass transitions. On the high-temperature side the material has lost its rigidity and is plastic, whereas on the low temperature side it is rigid. There is a degree of dependence of the T_g point in polymers on experimental conditions and on the method of determination. Obviously, TMA methods, DMA methods, and DTA/DSC can all be used in its evaluation.

C. Applications Based on Kinetic Features

It should be noted that unlike thermodynamics, kinetic features are pathway dependent. The kinetic parameters associated with the Arrhenius equation are

$$k = Ae^{-E/RT} \tag{74}$$

where k is the specific reaction rate constant and varies with temperature, A is the pre-exponential term, E is the energy of activation, R is the gas constant, and T is the absolute temperature. These are classically determined by running a series of kinetic experiments at various temperatures. In solid-state reactions we then have:

$$d\alpha = k(T)f(\alpha) \tag{75}$$

where α is the fraction decomposed at time t, $d\alpha/dt$ is the rate of reaction, $k(T)$ is specific reaction rate at temperature T, and $f(\alpha)$ is some function of α describing the progress of the reaction. There is no reason why this classical method cannot be investigated by thermal methods, as almost all commercial thermal instruments can be operated under constant-temperature conditions. The time taken to reach the isothermal temperature in such instruments, however, poses problems. It is possible in theory to obtain the Arrhenius parameters in a rising-temperature experiment. If the temperature regime imposed on the system is represented by $T = T_0 + bt$ where T is the temperature at time t, T_0 is the starting temperature, and b is the heating rate, then combination of the above equations gives

$$k(T) = \frac{(d\alpha/dT)b}{f(\alpha)} \tag{76}$$

$$\ln k = \ln A - \frac{E}{RT} \tag{77}$$

where k is defined as indicated. Instead of the traditional Arrhenius plot of $\ln k$ against $1/T$ this expression is usually given as

$$\ln\left(\frac{d\alpha/dT}{f(\alpha)}\right) = \ln\frac{A}{B} - \frac{R}{RT} \tag{78}$$

in this trivial change the plot is of $\ln[(d\alpha/dT)/f(\alpha)]$ against $1/T$. Either plot would lead to the evaluation of E and A. The differential signal $d\alpha/dt$ is, however, attended with difficulties, and most investigators would prefer an integral method. Yet, the use of an integral method involves the evaluation of $\int e^{-E/RT} dT$ and this proves impossible analytically.

Resort must be made to numerical methods or alternative similar expressions that can be analytically evaluated. The whole problem is reviewed in detail by Brown et al. (1980). The sample history is important in any solid-state kinetic problem, and there are articles dealing with the complications of thermal history. The article by Flynn (1981) is particularly useful. There are many other applications in which the kinetics are employed more empirically or in such a way that they serve to characterize the system under study, Typical of these studies are those dealing with the stability (shelf life) of pharmaceuticals (Radecki and Wesolowski, 1979; Rosenvold et al., 1982) and the proximate analysis of coal (Morgan, 1977). These examples are selective, and there are many other applications to be found in the literature.

D. Applications for EGA

A good example of what can be achieved by combining "off-the-shelf" equipment is the DTA–EGA apparatus used by Morgan for a number of years. A DTA with a large bore (37 mm) tube furnace was connected to a series of nondispersive infrared detectors, specific to water CO_2, SO_2, and NH_3. A high (300 mL/min) purge gas rate gave a response time, whereas the detectors had adequate sensitivity to cope with the dilution of the products in such conditions (Milodowski and Morgan,

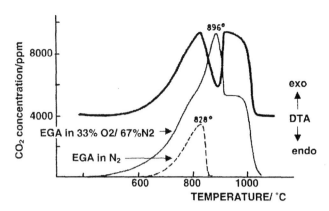

Figure 56 DTA and EGA curves for a graphite–calcite schist: 100 mg sample, 10°C/min, N_2 (67%), O_2 (33%).

1980). The apparatus was used in the assessment of the graphite content of a schist, from Burma, that also contained calcite (Morgan, 1977). Figure 56 shows the DTA curve recorded in oxidizing conditions, and the EGA curves for CO_2 in oxidizing and inert conditions.

The large exothermic peak shows a superimposed endothermic peak due to calcite dissociation. Subtracting the EGA peak recorded in nitrogen, arising from the calcite dissociation, from that in oxidizing gas gives the amount of CO_2 coming only from the graphite oxidation, which then allows a quantitative estimate of the graphite content to be made.

A TG–FT-IR combination has been used to study the degradation of recyclable car body underseal materials containing PVC (Post, 1995). In this case the small TG furnace, FT-IR gas cell and transfer line, each of a few cubic centimeters in volume, needed only ca. 20 cm^3/min of purge gas, which resulted in high sensitivity and speedy response. A stacked plot of partial spectra collected at intervals during heating is shown in Fig. 57.

The spectra show the rotational lines due to HCl released on heating. Strongly polar molecules like HCl are ideal for study by IR, as their absorbance is high, and this experiment showed clearly that the gas was released at temperatures as low as ca. 150°C. The instrument was used to estimate the PVC content of the materials by comparing the EGA peak areas with those for pure PVC.

A TG–DTA–MS instrument was developed (Charsley et al., 1987c, 1993b) and used by Haines (1995) in a wide range of projects, covering most classes of material. A simple capillary coupling was shown to operate with all common purge gases, up to 1500°C (Fig. 58). A special feature of the interface is the option to select a jet separator, which is mounted in parallel to the commoner bypass arrangement, in a temperature-controlled enclosure. With this and the capillary heated to around 180°C, nearly all products are transferred efficiently through the 0.3 mm ID silica connecting tube. The interface can isolate the MS from the atmosphere, in contrast to some systems that sample all the time. This has advantages in being

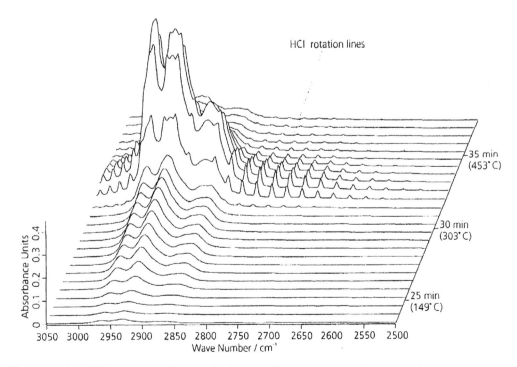

Figure 57 Changes in the FT-IR spectrum of the evolved gases with temperature during the heating of a PVC-containing material.

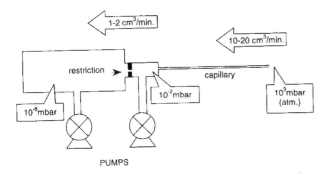

Figure 58 Schematic diagram of a capillary-and-bypass inlet for MS sampling from atmospheric pressure.

able to bake out the analyzer to reduce background signals due to contamination, and in carrying out MS diagnosis. Over time, it has become clear that the best results are obtained only with a carefully designed all-metal plumbing system. All polymeric tubing is permeable to water, oxygen, or both, which are detectable by the MS. A great virtue of EGA is its ability to detect leaks in the gas path, and monitor purging procedures which are essential to obtain a truly inert atmosphere. It is revealing to find what effort has to be expended to reduce oxygen levels in standard TA equipment to a level where they react to only an insignificant degree with sensitive materials at high

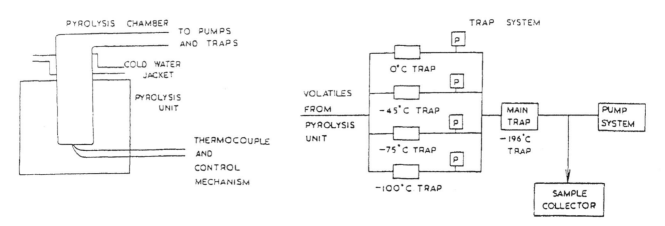

Figure 59 Thermal volitization equipment.

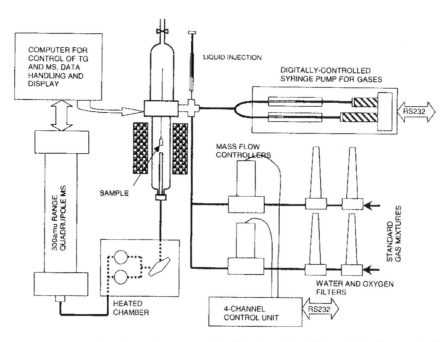

Figure 60 TG–MS equipment showing various means of calibration of the MS sensitivity for quantitative EGA studies.

temperature. Therma volitilization equipment is shown in Fig. 59.

This instrument was used in a study of a pyrotechnic initiator composition (Berger et al., 1995), based on fine (1.7 μm) zirconium powder, potassium perchlorate, and a small amount of nitrocellulose as a binder. Mixtures were studied in helium (a) to take advantage of the higher sensitivity of the separator, (b) to reduce the likelihood of ignition of these sensitive mixtures by using a purge gas of high thermal conductivity, and (c) to use the ancient purging properties of helium in sweeping traces of oxygen from internal spaces of the equipment.

First, the nitrocellulose decomposes around 200°C, leaving a carbon rich residue of only ca. 10% of the weight of nitrocellulose. Following the sharp orthorhombic–cubic phase transition in potassium perchlorate at 300°C, an exothermic reaction takes place between the zirconium and perchlorate while, at the same time, the carbon residue is oxidized by the perchlorate, as shown by the CO_2 curve. The persistence of CO_2 evolution up to ca. 600°C is a real and repeatable phenomenon, and remains unexplained. Oxygen evolution from excess perchlorate starts around 400°C, where it is more sensitively detected by the MS than the TG curve. The O_2 curve dips temporarily during the fusion of first a eutectic mixture of the perchlorate and its chloride reaction product, and then the remaining perchlorate, as shown by the DTA peaks in this region. Excess perchlorate then decomposes exothermically. The sensitivity of the EGA measurements can be judged from the CO_2 peak around 400°C, which represents ca. 30 μg of the gas. The detailed interpretation of the course of events undergone by this mixture was possible only by using the additional information given by EGA.

The same MS instrument was also linked to a TA Instruments model 951 thermobalance, which is ideal for EGA work, but is sadly no longer available (Charsley et al., 1995). This combination was used to investigate the problem of calibration of a TG–MS for quantitative EGA. A schematic diagram of the layout is shown in Fig. 60.

Calibrations could be made on the basis of the decomposition of solid standards, injections of volatile liquids, standard gas mixtures, injection of pulses or gas, or steady injection rates of gases into the purge gas stream. A good linear relationship between EGA peak area and amount of product was found for between a few and a few hundred micrograms of gas when using a jet separator.

REFERENCES

Adamson, A. W. (1986). *A Textbook of Physical Chemistry*. Orlando, Florida: Academic Press, p. 401.

Arnett, E. M. (1999). *J. Chem. Thermodyn.* 31:711.

Atkins, P. W. (1978). *Physical Chemistry*. Oxford: Oxford University Press.

Atkins, P. W. (1994). *Physical Chemistry*. 5th ed. San Francisco: Freeman, p. 244.

Babrauskas, V. (1981). *J. Fire Flammabil.* 12:51.

Bakri, A. *Thermometric Application Note*, TM, AB, Jafalla, Sweden, No 22021.

Barnes, P. et al. (1981). *J. Thermal Anal.* 25:299.

Beezer, A. E. (2000). *Thermochim. Acta* 349:1.

Beezer, A. E., Loh, W., Mitchell, J. C., Royall, P. G., Smith, D. O., Tute, M. S., Armstrong, J. K., Chowdhry, B. Z., Leharne, S. A., England, D., Crowther, N. J. (1994). *Langmuir* 10:4001.

Beezer, A., Mitchell, J. C., Colgate, R. M., Scally, D. J., Twyman, L. J., Willson, R. J. (1995). *Thermochim. Acta* 250:277.

Beezer, A. E., Morris, A. C., O'Neil, A. A., Willson, R. J., Hills, A. K., Mitchell, J. C., Connor, J. A. (2001). *J. Phys. Chem. B* 105:1212.

Berger, B., Charsley, E. L., Warrington, S. B. (1995). *Propellants, Explosives Pyrotechnics*. 20:266.

Bernes, A., Chatain, D., Lacabanne, C. (1991). *Calorim. Anal. Therm.* 22:lii.

Bernes, A., Chatain, D., Lacabanne, C. (1992). *Thermochim. Acta* 204:69.

Beruto, D., Searcy, A. W. (1974). *J. Chem. Soc., Faraday Trans.* 70:949.

Blame, R. L., Fair, P. O. (1983). *Thermochim. Acta* 67:233.

Boersma, S. L. (1955). *J. Am. Ceram. Soc.* 38:281.

Bonneau, O., Vernet, C., Moranville, M., Aitcin, P. C. (2000). *Chem. Concr. Res.* 340:1861 (Special issue SI).

Bottom, R. (1999). The role of MTDSC in the characterization of a drug molecule exhibiting polymorphic and glass forming tendencies. *Int. J. Pharm.* 192(1):47–53.

Brennan, W. P. (1997). Some applications of thermal analysis as a supplement to or replacement for ASTM testing standards. *Thermochim. Acta* 18:10–13 (See also 101–111).

Briggner, L., Wadso, I. (1992). *J. Biochem. Biophys. Methods* 22:101.

Brinkworth, S. et al. (1981). In: Dollimore, D., ed. *Proceedings 2nd ESTA Conference*. London: Heyden.

Brown, M., Dollimore, D., Galwey, A. K. (1980). In: Bamford, S. H., Tipper, C. F. H., eds. *Comprehensive Chemical Kinetics*. Vol. 22. Amsterdam: Elsevier, p. 117.

Brown, M. E., Galwey, A. K., Guarini, G. G. T. (1997). *J. Thermal. Anal.* 49:1135.

Brunauer, S., Emmett, P. H., Teller, E. (1938a). *J. Am. Chem. Soc.* 60:309.

Brunauer, S., Emmett, P. H., Teller, E. (1938b). *Bureau of Chemistry and Solids*. George Washington University, February, p.309.

Buckton, G., Darcy, P. (1999). *Int. J. Pharm.* 179:141.

Buckton, G., Dove, J. W., Davies, P. (1999). *Int. J. Pharm.* 193:13.

Byrn, S. R. (1982). *Solid State Chemistry of Drugs*. New York: Academic Press.

Caffrey, M. (1991). *Trends Anal. Chem.* 10:156.

Calvet, E., Prat, H. (1963). In: Skinner, H. A., ed. *Recent Progress in Microcalorimetry*. New York: Pergamon, p. 177.

Campben, L., Amidon, G. L., Zografi, G. (1983). *J. Pharm. Sci.* 72:1381.

Charsley, E. L. et al. (1980). In: Wiedemann, H. G., ed. *Thermal Analysis*. Vol. 1. Basel: Birkhauser, p. 237.

Charsley, E. L., Warne, S.St., Warrington, S. B. (1987a). Thermogravimetric apparatus temperature calibration using melting point standards. *Thermochim. Acta* 114:53–61.

Charsley, E. L., Manning, N. J., Warrington, S. B. (1987b). *Thermochim. Acta* 114:47.

Charsley, E. L., Manning, N. J., Warrington, S. B. (1987c). *Thermochim. Acta* 114:147.

Charsley, E. L., Warrington, S. B., McGhie, A. R., Jones, G. K. (1990). *Am. Lab.* 22:21.

Charsley, E. L., Walker, C., Warrington, S. B. (1993a). *J. Therm. Anal.* 40:983.

Charsley, E. L., Walker, C., Warrington, S. B. (1993b). *J. Therm. Anal.* 40:983.

Chen, I. D., Gao, X., Dollimore, D. (1992). *Anal. Instrum.* 20:137.

Cheng, S. Z. D., Li, C. Y., Calhoun, B. H., Zhu, L., Zhou, W. W. (2000). Thermal analysis: the next two decades. *Thermochim. Acta* 355:59–68.

Chiu, J. (1968). *Anal. Chem.* 40:1516.

Chiu, J. (1970). *Thermochim. Acta* 1:231.

Chiu, J. (1986). *Thermochim. Acta* 101:231.

Chiu, J., Beattie, A. J. (1977). *Thermochim. Acta* 21:263 (See also **1980**, 40, 251; **1981**, 50, 49).

Clark, B. A., Brown, P. W. (2000). *Adv. Chem. Res.* 12:137.

Coleman, N. J., Craig, D. Q. M. (1996). Modulated DSC: a novel approach to pharmaceutical thermal analysis. *Int. J. Pharm.* 135:13–29.

Coloff, S. C., Vanderborgh, N. E. (1973). *Anal. Chem.* 45:1507.

Craig, D. Q. M., Barsnes, M., Royall, P. G., Kett, V. L. (2000). An evaluation of the use of modulated temperature DSC as a means of assessing the relaxation behavior of amorphous lactose. *Pharm. Res.* 17(6):696–700.

Cukor, P., Persiani, C. (1974). In: Chiu, J., ed. *Polymer Characterization by Thermal Methods of Analysis*. New York: Marcel Dekker, p. 107.

Cunningham, A. D., Wiburn, F. W. (1971). In: MacKenzie, R. C., ed. *Differential Thermoanalysis*, Vol. 1. New York: Academic, p. 31.

Daniels, T. (1973). *Thermal Analysis*. London: Kogan Page, p. 49.

Dantzer, P., Millet, P. (2001). *Thermochim. Acta* 370:1.

David, D. (1964). *J. Anal. Chem.* 36:2162.

Dial, H. W., Knapp, G. S. (1969). *Rev. Sci., Instrum.* 40:1086.

Dollimore, D. (1992). *Thermochim. Acta* 203:7.

Dollimore, D., Tonge, K. H. (1964). In: Schwab, G. M., ed. *Proc. 5th Int. Symp., React. Solids*. Amsterdam: Elsevier, p. 497.

Dollimore, D. et al. (1984). *Thermochim. Acta* 75:59.

Dollimore, D., Evans, T. A., Lee, Y. F., Wilburri, F. W. (1991). *Thermochim. Acta* 188:77.

Dunn, I. G. (1993). *Therm. Anal.* 40:1431.

Dunn, J. G., Sharp, J. H. (1993). In: Kolthoff, I. M., ed. *Treatise on Analytical Chemistry*. Part 1. 2nd ed. Vol. 13. New York: John Wiley, pp. 127–267.

Durucan, C., Brown, P. W. (2000). *Mater. Res.* 5:717.

Duvai, C. (1963). *Inorganic Thermogravimetric Analysis*. 2nd revised ed. Amsterdam: Elsevier, p. 772.

Dyszel, S. M. (1983). *Thermochim. Acta* 61:169.

Elder, J. P. (1982). *Thermochim. Acta* 52:235.

Ewing, O. W. (1976). *J. Chem. Educ.* 53:A251, A291.

Fairclough, R. A., Hinshelwood, C. N. (1937). *J. Chem. Soc.* 538.

Fennell, T. et al. (1971). In: Wiedenmann, H. G., ed. *Thermal Analysis*. Vol. 1. Basel: Birkhauser, p. 245.

Fenner, C. N. (1913). *Am. J. Sci.* 36:331.

Ferino, I., Monaci, R., Rombu, E., Solinas, V. (1993). *J. Thermal Anal.* 40:1233.

Florence, A. T., Atwood, D. (1998). *Physiochemical Principles of Pharmacy*. London: MacMillan Press.

Flynn, J. H. (1981). Thermal analysis. In: Turi, E. A., ed. *Polymer Characterization*. Philadelphia: Heyden, p. 43.

Flynn, J. H. (1993). *Thermochim. Acta.* 217:129.

Ford, J. L., Timmins, P. (1989). Pharmaceutical thermal analysis techniques and applications. In: Rubinstein, M. H., ed. *Series in Pharmaceutical Technology, Ellis Horwood Books in Biological Sciences*. New York: John Wiley & Sons.

Forslund, B. (1984). *Chem. Scr.* 24:107.

Gaisford, S., Beezer, A. E., Mitchell, J. C., Bell, P. C., Fakorede, F., Finnie, J. K., Williams, S. J. (1998). *Int. J. Pharm.* 174:39.

Gallagher, P. K. (1978). *Thermochim. Acta* 26:175.

Gallagher, P. K., Gyorgy, B. M. (1980). In: Weidemann, H. O., ed. *Thermal Analysis*. Vol. 1. Basel: Birkhauser, p. 113.

Gallagher, P. K., Coleman, E., Sherwood, R. C. (1980). *Thermochim. Acta* 37:291.

Gallis, H. E., van Miltenburg, J. C., Oonk, H. A. H. (2000). *Phys. Chem. Phys.* 2:5619.

Gardea, A. A., Carvajal-Millan, E., Orozco, J. A., Guerrero, V. M., Llamas, J. (2000). *Thermochim. Acta* 349:89.

Gerard, N. (1974). *J. Phys. E* 7:509.

Gill, P. S. (1984). *Am. Lab.* 16:39.

Gill, P. S., Sauerbrunn, S. R., Reading, M. (1993). Modulated differential scanning calorimetry. *J. Thermal. Anal.* 40:931–939.

Gill, P. S., Sauerbrunn, S. R., Crowe, B. S. (1992). *J. Therm. Anal.* 38:255.

Gill, P. S., Sauerbrunn, S. R., Reading, M. J. (1993). *J. Therm. Anal.* 40:931.

Gillham, J. K. (1996). *Appl. Polym. Symp.* 2:45.

Giron, D. (1986). Applications of thermal analysis in the pharmaceutical industry. *J. Pharm. Biochem. Anal.* 40:755–770.

Giron, D. (1990). Thermal analysis in pharmaceutical routine analysis. *Acta Pharm. Jugosl.* 40:95–157.

Giron, D. (1995). Thermal analysis of drugs and drug products. In: Swarbrick, J., Boylan, J. C., eds. *Encyclopedia of Pharmaceutical Technology*. Vol. 15. New York: Marcel Dekker, Inc., pp. 1–79.

Giron, D. (1998). Contribution of thermal methods and related techniques for rational development of pharmaceuticals. *P.S.S.T.* 1:191–262.

Giron, D. (1999). Thermal analysis, microcalorimetry and combined techniques for the study of pharmaceuticals. *J. Therm. Anal. Calorim.* 56:191–262.

Giron, D. (2000). Characterization of pharmaceuticals by thermal analysis. *Am. Pharm. Rev.* 3(2):53–61, 3(3):43–53.

Giron, D., Goldbronn, C., Piechon, P. (1989). Thermal analysis methods for pharmacopea materials. *J. Pharm. Biochem. Anal.* 7:1421–1430.

Giron, D., Goldbronn, C., Pfeffer, S. (1999). Automation in thermogravimetry: application in pharmaceutical industry. Poster Presented at the 4th Symposium on Pharmacy and Thermal Analysis, Karsruhe.

Giron-Forest, D. (1983). Thermoanalytische Verfahren. In: Feltkamp, H., Fuchs, P., Sucker, H., eds. *Pharmazeutischer Qualitätskontrolle*. Georg Thieme Verlag: Stutgart, Germany, pp. 298–310.

Gohlke, R. S., Langer, H. C. (1966). *Anal. Chim. Acta* 36:530.

Gordon, S., Campbell, C. (1960). *Anal. Chem.* 32:271R.

Gregg, S. J., Sing, K. S. W. (1982). *Adsorption Surface Area and Porosity*. 2nd ed. New York: Academic, p. 303.

Guan, Y. H., Kemp, R. B. (2000). *Thermochim. Acta* 349:163, 176.

Guan, Y., Evans, P. M., Kemp, R. B. (1998). *Biotechnol. Bioeng.* 58:464.

Haines, P. J. (1995). *Thermal Methods of Analysis—Principles, Applications and Problems*. London: Blackie Academic Professional.

Han, J., Gupte, S., Suryanarayanan, R. (1998). Applications of pressure differential scanning calorimetry in the study of pharmaceutical hydrates. II. Ampicillin trihydrate. *Int. J. Pharm.* 170:63–72.

Hanson, L. D. (2000). *Ind. Eng. Chem. Res.* 39:3541.

Hegde, S. S., Kumar, A. R., Ganesh, K. N., Swaninathan, C. P., Khan, I. (1998). *Biochem. Biophys. Acta* 1388:93.

Hemminger, W., Höhne, G. W. (1984). *Calorimetry—Fundamental Practice*. Weinheim: Verlag-Chemie.

Hill, J. O. (1991). *For Better Thermal Analysis and Calorimetry*. 3rd ed. ICTA, p. 7.

Hoffmann, M. A. M., van Mil, P. J. J. M. (1999). *J. Agric. Food Chem.* 47:1898.

Hohne, G. W. H. (1999). Modulated DSC. *Thermochim. Acta* 330:45–51.

Hoppler, H. U. (1991). *Labor Praxis* 15:763.

Ibar, J. P. (1993). *Fundamentals of Thermally Stimulated Current and Relaxation Mop Analysis*. New Canaan, Connecticut: SLP Press, p. 667.

Ingold, C. K. (1953). *Structure and Mechanism in Organic Chemistry*. London: Bell.

Irwin, W. J. (1993). In: Wineforder, J. D., Dollimore, D., Dunn, J., eds. *Treatise on Analytical Chemistry*, Pt. I, Thermal Methods. 2nd ed. New York: John Wiley, p. 309.

Ish-Shalom, M. (1993). In: Wineforder, J. D., Dollimore, D., Dunn, J., eds. *Treatise on Analytical Chemistry*. Pt. I. Thermal Methods. 2nd ed. New York: John Wiley, p. 309.

Jensen, E. T., Beevers, C. A. (1938). *Trans. Faraday Soc.* 34:1478.

Johansson, P., Wadso, I. (1997). *J. Biochem. Biophys. Methods* 35:103.

Johnson, D. J. et al. (1992). *Thermochim. Acta* 195:5.

Judd, M. D., Pope, M. I. (1971). *J. Inorg. Nucl., Chem.* 33:365.

Kaiserberger, E. et al. (1977). *Thermochim. Acta* 295:73.

Kambe, A., Hone, K., Suski, T. (1972). *J. Thermal Anal.* 4:461.

Keattch, C. J., Dollimore, D. (1975). *An Introduction to Thermogravimetry*, 2nd ed. London: Heyden, p. 164.

Keavney, J. J., Eberlin, E. C. (1963). *J. Appl. Polym. Sci.* 3:394.

Kemp, R. B. (1988). In: Brown, E., ed. *Handbook of Thermal Analysis and Calorimetry*. Vol. L. Chapter 14. Amsterdam: Elsevier.

Kemp, R. B. (2000a). *J. Therm. Anal. Cal.* 60:831.

Kemp, R. B. (2000b). *Thermochim. Acta* 355:115.

Kemp, R. B. (2000c). *Thermochim. Acta* 349:XI–XII.

Kemp, R. B., Lamprecht, I. (2000). *Thermochim. Acta* 348:1.

Kerns, R. T., Kini, R. M., Stefansson, S., Evans, H. J. (1999). *Arch. Biochem. Biophys.* 369:107.

Kimelt, S., Speiser, S. (1977). *Chem. Rev.* 77:437.

Kinsbrow, F., Gallagher, P. K., English, A. T. (1979). *Solid State Electron.* 22:517.

Kiss, A. B. (1969). *Acta Chim. Sci. Hung.* 61:207.

Koenigbauer, M. J. (1994). *Pharm. Res.* 11:777.

Koenigbauer, M. J., Brooks, S. H., Rullo, C. G. (1992). *Pharm. Res.* 939.

Komatsu, H., Yoshii, K., Okada, S. (1994). *Chem. Pharm. Bull.* 42:1631–1635.

Korobeinochev, O. P. (1987). *Russ. Chem. Rev.* Dec.:957.

Kowalski, B., Ratusz, K., Miciula, A., Krygier, K. (1997). *Thermochim. Acta* 307:117.

Lang, N. A. (1956). *Handbook of Chemistry*. Sandusky, OH: Handbook Publishers.

Larsen, M. J., Hemming, D. J. B., Bergstrom, R. G., Wood, R. W., Hansen, L. D. (1997). *Int. J. Pharm.* 154:103.

Lavoisier, A. L., De Laplace, P. S. (1780). *Hist. Acad. R. Sci.* 1784:355.

Lehto, V. P., Laine, E. (1997). *Pharm. Res.* 14:899.

Lelay, P., Delmas, G. (1998). *Carbohydr. Polym.* 37:49.

Lever, T. et al. (2000). Proc. 28th NATAS Conf. Savannah, GA, Oct. 2000, p. 720.

Lewis, A. F., Giliham, J. K. (1962). *J. Appl. Polym. Sci.* 6:422.

Liebman, S. A., Ahlstrom, D. H., Crichton, T. G., Prudor, G. D., Levy, E. I. (1973). *Thermochim. Acta* 5:403.

Lis, G. Q., Qu, S. S., Zhou, C. P., Liu, Y. (2000). *Chem. J. Chin. Univ.-Chin.* 21:791.

Lodding, W., ed. (1967). *Gas Effluent Analysis*. New York: Marcel Dekker.

Lofthouse, M. O., Burroughs, P. J. (1978). *Therm. Anal* 13:439.

Loners, I. (1952). *Rev. Met.* 49:807.

Lonvik, K. (1978). *Thermochim. Acta* 27:27.

Louzguine, D. V., Inoue, A. (1991). *Mater. Res. Bull.* 34(12):1991.

Low, M. J. D. (1967). In: Lodding, W., ed. *Gas Effluent Analysis*. Chapter 6. New York: Dekker.

Lurn, R. M. (1977). *Thermochim. Acta* 18:73.

Machado, C., Mascimento, M. D., Rezende, M. C., Beezer, A. E. (1999). *Thermochim. Acta* 328:155.

Mackenzie, R. C. (1984). *Thermochim. Acta* 73:307.

Materazzi, S. (1997). *Appl. Spectrosc. Rev.* 32:385.

Matzakos, A. N. et al. (1993). *Rev. Sci. Instrum.* 64:1541.

McClennen, W. H., Richards, J. M., Mauzlaar, H. L. C., Paysch, J. B., Lattimer, R. P. (1985). *Polym. Mater. Sci. Eng.* 53:203.

McGhie, A. R. (1983). *Anal. Chem.* 55:987.

McGhie, A. R., Chiu, J., Fair, P. G., Blaine, R. L. (1983). Studies on ICTA reference materials using simultaneous TG-DTA. *Thermochim. Acta* 67:241–250.

McNeill, I. C. (1970). *Eur. Polym. J.* 6:373.

McPhillips, H., Craig, D. Q. M., Royall, P. G., Hill, V. L. (1999). Characterization of the glass transition of HPC using modulated DSC. *Int. J. Pharm.* 180:83–90.

Menkov, N. D., Dinkov, K. T. (1999). *J. Agr. Eng. Res.* 74:261.

Milodowski, A. E., Morgan, D. J. (1980). *Nature* 286:248.

Mirabella, F. M. (1986). *J. Appl. Spectrosc.* 40:417.

Montanari, M., Beezer, A. E., Montanari, C. A., Pilo-Valoso, D. (2000). *J. Med. Chem.* 43:3448.

Morgan, D. J. (1977). *J. Thermal Anal.* 12:245.

Morgon, T. D., Beezer, A. E., Mitchell, J. C., Bunch, A. W. (2001). *J. Applied Microbiol.* 90:53.

Nesbitt, L. E., Wendlandt, W. W. (1974). *Thermochim. Acta* 10:85.

Newman, R. A. et al. (1987). *Adv. X-Ray Anal.* 30:493.

Norem, S. D., O'Neill, M. J., Gray, A. P. (1970). *Thermochim. Acta* 1:29.

O'Keefe, B. R., Shenoy, S. R., Xie, D., Zhang, W. T., Muschik, J. M., Currens, M. J., Chaiken, I., Boyd, M. R. (2000). *Mol. Pharmacol.* 58:982.

Olson, E. A., Efremov, M. Y., Kwan, A. T., Lai, S., Petrova, V. (2000). Scanning calorimeter for nanoliter-scale liquid samples. *Appl. Phys. Lett.* 77(17):2671–2673.

Otsuka, T., Yoshioka, S., Aso, Y., Terao, T. (1994). *Chem. Pharm. Bull.* 42:130.

Ozawa, T. (2000). Kinetic analysis by repeated temperature scanning. Part 1. Theory and methods. *Thermochim. Acta* 356:173–180.

Palmour, H., Johnson, P. O. (1967). In: Kuczynski, O. C., Hooten, N. A., Gibbon, C. F., eds. *Sintering and Related Phenomena*. New York: Gordon and Breach, p. 779.

Parkes, G. M. B., Barnes, P. A., Bond, G., Charsley, E. L. (2000). Qualitative and quantitative aspects of microwave thermal analysis. *Thermochim. Acta* 356:85–96.

Paulik, J., Paulik, F. (1971). *Anal. Chim. Acta* 56:328.

Paulik, J., Paulik, F. (1972). *Thermochim. Acta* 4:189.

Paulik, F., Paulik, J. (1978). *Analyst* 103:417.

Paulik, F., Paulik, J. (1979). *J. Therm. Anal.* 16:399.

Paulik, F., Paulik, J. (1981). In: Svehla, G., ed. *Comprehensive Analytical Chemistry*. Vol. XII(A). Amsterdam: Elsevier.

Perron, W. et al. (1980). In: Wiedemann, H. G., ed. *Thermal Analysis*. Vol. 1. Basel: Birkhauser.

Pikal, M. J., Dellerman, K. M. (1989). *Int. J. Pharm.* 50:233.

Pollock, H. M., Hammiche, A. (2001). *J. Phys. D: Appl. Phys.* 34:R23.

Pope, M. L., Judd, M. D. (1977). *Introduction to Differential Thermal Analysis*. London: Heyden, p. 37.

Post, E. et al. (1995). *Thermochim. Acta* 263:1.

Price, D. M. (1997). *J. Therm. Anal. Catal.* 49:953.

Price, G. (1998). *Thermodynamics of Chemical Processes*. Oxford: Oxford University Press.

Price, D. M. et al. (1999). *Int. J. Pharm.* 192:85.

Price, D., Dollimore, D., Fatemi, N. J., Whitehead, R. (1980). *Thermochim. Acta* 42:323.

Price, D. M. et al. (1999). Proc. 28th ATAS Conf. Orlando, FL, p. 705.

Prout, E. G., Tomkins, F. C. (1946). *Trans. Faraday Soc.* 42:482.

Pudipeddi, M., Sokoloski, T. D., Duddu, S. P., Carstensen, J. T. (1996). *J. Pharm. Sci.* 85:381.

Pugh, B. (1966). *Fuel Calorimetry*. London: Butterworth.

Radecki, A., Wesolowski, M. (1979). *J. Therm. Anal.* 17:73.

Raemaekers, K. G. H., Bart, J. C. J. (1997). *Thermochim. Acta* 295:1.

Reading, M. (1992). In: Charsiey, F. L., Warrington, S. B., eds. *ThermalAnalysis—Techniques and Applications*. Royal Society of Chemistry, p. 1126.

Reading, M., Dollirmore, D., Rouquerol, J., Rouquerol, F. (1984). *J. Therm. Anal.* 29:775.

Reading, M., Dollimore, D., Whitehead, R. (1992). *J. Therm. Anal.* 37:2165.

Reading, M., Elliot, D., Hill, V. L. (1993). *J. Therm. Anal.* 40:9949.

Riesen, R., Sommerauer, H. (1983). *Am. Lab.* 15:30.

Riva, M., Schiraldi, A., Piazza, L. (1994). *Thermochim. Acta* 246:317.

Riva, M., Fessas, D., Schiraldi, A. (2001). *Thermochim. Acta* 370:73.

Robertson, S. D., McNicol, B. D., de Baas, J. H., Kloet, S. C., Jenkins, W. (1975). *J. Catal.* 37:424.

Rogers, R. N. S., Yasuda, S. K., Zinn, J. (1960). *Anal. Chem.* 32:672.

Rosenvold, R. J., Dubow, J. B., Rajeshwar, K. (1982). *Thermochim. Acta* 53:321.

Rossini, F. D., ed. (1956). *Experimental Thermochemistry I*. New York: Wiley Interscience.

Rouquerol, J. (1970). *J. Therm. Anal.* 2:123.

Ryan, A. J. (1993). *Therm. Anal.* 40:887.

Ryan, A. J. et al. (1994). *ACS Symp. Ser.* 581:162.

Sabbah, R., Xu-wu, A., Chickos, J. S., Leita, M. L., Roux, M. V., Torres, L. A. (1999). Reference materials for calorimetry and differential thermal analysis. *Thermochim. Acta* 331:93–204.

Salonen, J., Lehto, V. P., Laine, E. (1999). *J. Appl. Phys.* 86:5888.

Salonen, J., Lehto, V. P., Laine, E. (2000). *J. Porous Mater.* 7:335.

Sarge, S. M., Gmelin, E. (2000a). Temperature, heat and heat flow rate calibration of differential scanning calorimeters. *Thermochim. Acta* 347:9–13.

Sarge, S. M., Hohne, G. W. H., Cammenga, H. K., Eysel, W., Gmelin, E. (2000b). Temperature, heat and heat flow rate calibration of scanning calorimeters in the cooling mode. *Thermochim. Acta* 361:1–20.

Schmitt, E. A., Law, D., Zhang, G. G. (1999). *J. Pharm. Sci.* 88:291.

Schneider, K., Behl, H. (1997). *Mettler-Toledo Mag.* (2).

Seguin, B., Gosse, J., Ferrieux, J. P. (1999). *Eur. Phys. J. Appl Phys.* 8(3):275.

Selzer, T., Radau, M., Kreutr, J. (1998). *Int. J. Pharm.* 171:227.

Sestak, I. (1969). In: Schwenker, R. F., Cam, P. D., eds. *Thermal Analysis*. Vol. 2. New York: Academic, p. 1985.

Sharp, J. H., Brindley, G. H., Narahari Achar, B. N. (1966). *J. Am. Ceram. Soc.* 49:379.

Silvestre-Albero, J., deSalazar, C. G., Sepulveda-Escribano, A., Rodriguez-Reinoso, F. (2001). *Colloid Surf. A Physicochem. Eng. Asp.* 187:151.

Singh, J. (1993). *Thermochim. Acta* 226:211.

Skinner, H. A., ed. (1962). *Experimental Thermochemistry II.* New York: Wiley Interscience.

Smirnov, Y. N., D'yachkova, S. L. (2000). *Russ. J. Appl. Chem.* 73:870.

Smith, E. B. (1990). *Basic Chemical Thermodynamics.* Oxford: Oxford Science.

Smith, G. W. (1988). *Thermochim. Acta* 323:123.

Smith, G. W. (1997). *Thermochim. Acta* 291:59.

Spiel, S., Berkeihamer, L. H., Pask, J. A., Davis, B. (1945). *U.S. Bureau of Mines*, Tech. Papers. p. 664.

Starink, M. J., Zahra, A. M. (1999). *J. Mater. Sci.* 34:1117.

Stassi, A., Schiraldi, A. (1994). *Thermochim. Acta* 246:417.

Sturtevant, J. M. (1959). In: Weissberger, A., ed. *Techniques of Organic Chemistry Methods.* Chapter XIV. New York: Interscience.

Technical Information Sheet #4. Bletchley, UK: Thermal Hazard Technology.

Terry, D. R. (1965). *J. Sci. Instrum.* 42:507.

Thompson, K. C. (2000). Pharmaceutical applications of calorimetric measurements in the new millenium. *Thermochim. Acta* 366:83–87.

Trombly, B., Shepard, D. D. (1994). *Instrum. Sci. Technol.* 22:259.

Tsuchiya, S., Amenomiya, Y., Cvetanovic, R. J. (1971). *J. Catal.* 20:1.

Tsuchiya, S., Nakamura, M. (1977). *J. Catal.* 50:1.

Turl, E. A. (1997). *Thermal Characterization on Polymeric Materials.* 2nd ed. New York: Academic Press.

Uden, P. C., Henderson, D. E., Lloyd, R. J. (1976). In: Dollimore, D., ed. *First European Symposium on Thermal Analysis.* London: Heyden, p. 1029.

Ulbig, P., Surya, L., Schulz, S., Seippel, J. (1998a). *J. Therm. Anal. Cal.* 54:33.

Ulbig, P., Friese, T., Schulz, S., Seippel, J. (1998b). *Thermochim. Acta* 310:217.

Vanderschueren, J., Niezette, J., Yuianakopoulos, G., Thielen, A. (1991). *Thermochim. Acta* 192:287.

Vaughan, H. P., Burroughs, P. J. (1978). *Thermal Anal.* 13:439.

Vassalio, O. A. (1961). *Anal. Chem.* 33:1823.

Veltchev, Z. N., Menkov, N. D. (2000). *Drying Technol.* 18:1127.

Vold, M. J. (1949). *Anal. Chem.* 21:683.

Vollhardt, K. P. C. (1987). *Organic Chemistry.* New York: Freeman.

Wadso, I. (1995). *Thermochim. Acta* 267:45.

Wadso, I. (1997a). *Thermochim. Acta* 294:1.

Wadso, I. (1997b). *Chem. Soc. Rev.* 26:79.

Wadso, I. (2001). *J. Therm. Anal. Cal.* 64:75.

Warrington, S. B. (1992). In: Charsley, E. L., Warrington, S. B., eds. *Thermal Analysis—Techniques and Applications.* Special Publication 117. Cambridge: Royal Society of Chemistry.

Warrington, S. B., Barnes, P. A. (1981). In: Dollimore, D., ed. *Proceedings 2nd ESTA Conference.* London: Heyden.

Watson, E. S., O'Neill, M. J., Justin, J., Brenner, N. (1964). *Anal. Chem.* 36:1233.

Weast, R. C., Lide, D. R., eds. (1989). *CRC Handbook of Chemistry and Physics.* Boca Raton, FL: CRC Press.

Wendlandt, W. W. (1986). Thermal analysis. In: Elving, P. J., Winefordner, J. D., Kolthoff, I. M., eds. *Chemical Analysis.* 3rd ed. New York: John Wiley, p. 19.

Wetton, R. E. (1981). *Anal. Proc.* 416.

Wiedemann, H. G. (1964). *Chem. Ing.-Tech.* 36:1105.

Wiedemann, H. G. (1969). In: Schwenker, R. F., Garn, P., eds. *Thermal Analysis*, Vol. 1. New York: Academic, p. 229.

Wilburn, F. W., Melling, R., Mcintosh, R. M. (1969). *Anal. Chem.* 41:1275.

Wilburn, F. W., Dollimore, D., Crighton, J. S. (1991). *Thermochim. Acta* 173(18):191.

Willson, R. J., Beezer, A. E., Mitchell, J. C., Loh, W. (1995). *J. Phys. Chem.* 99:7108.

Willson, R. J., Beezer, A. E., Mitchell, J. C. (1996). *Int. J. Pharm.* 132:45.

Willson, R. J., Beezer, A. E., Hills, A. K., Mitchell, J. C. (1999). *Thermochim. Acta* 325:125.

Wunderlich, B. (1990). *Thermal Analysis.* New York: Academic Press.

Wunderlich, B., Bopp, R. C. (1974). *Thermochim. Acta* 6:335.

Yamada, K., Orra, S., Haruki, T. (1975). In: Buzas, I., ed. *Proc. 4th ICTA Conf.* Vol. 3. London: Heyden, p. 1029.

Zhang, H. et al. (1995). *J. Solid-State Chem.* 117:127.

Zitomer, F. (1968). *Anal. Chem.* 40:1091.

Web address for ICH guidelines http://www.ifpina.org/lchl.html

www.anasys.co.uk

16

Potentiometry: pH and Ion-Selective Electrodes

RONITA L. MARPLE, WILLIAM R. LACOURSE
Department of Chemistry and Biochemistry, University of Maryland, Baltimore County, Baltimore, MD, USA

I. INTRODUCTION

The technique based on measuring the potential of electrochemical cells in the absence of appreciable current is known as *potentiometry*. Potentiometric measurements require a reference electrode, an indicator electrode, and a reliable potential-measuring instrument, such as a voltmeter. The test solution must be in direct contact with the indicator electrode. The reference electrode may be placed in the test solution, or it can be brought into contact with the test solution through a salt bridge.

Owing to its simplicity and versatility, potentiometry is perhaps the most widely used analytical technique. It is most commonly used for selective determination of analyte concentrations in titrations and in a wide variety of test solutions. There are many types of electrodes for many different applications, and the number of applications grows everyday.

Modification of electrode sizes and surfaces is a very fruitful and interdisciplinary area of research in analytical chemistry. Current research in potentiometric measurements and indicator electrodes focuses on increasing sensitivity, decreasing size, and using new materials to increase stability and reproducibility. Integration into microfluidic platforms is the latest area of interest for micro- and nanoelectrodes. Also, there is now great emphasis on biosensors, including enzyme electrodes, immunosensors, and DNA sensors. These improvements will greatly advance research in many fields, including that of biology and bioanalytical chemistry.

The fundamentals of potentiometry constitute the first part of this chapter. The areas of pH and ion-selective electrodes are included in this chapter, as well some applications of potentiometry. Finally, available instrumentation and current research activity are briefly summarized.

A. Electrochemical Cells

An electrochemical cell is the combination of two *half-cell(s)*, a half-cell being an electrical conductor, or *electrode*, immersed in a suitable electrolyte solution. Half-reaction(s) are used to describe the chemical reactions occurring in the half-cell(s), and each has a characteristic potential, E, associated with it. When all the components of the half-reaction are at standard-state conditions, this potential is the *standard electrode potential*, E^0:

$$Cu^{2+} + 2e^- \rightleftarrows Cu \quad E^0 = +0.337 \text{ V} \quad (1)$$

$$Zn^{2+} + 2e^- \rightleftarrows Zn \quad E^0 = -0.763 \text{ V} \quad (2)$$

Table 1 contains a selective list of E^0 values for half-cell(s) vs. the *standard hydrogen electrode* (SHE) (Dean, 1999a). A more complete list can be found in Dean (1999a). By convention, the half-reaction(s) are written as reductions. As the standard potential becomes more positive, the tendency for the reaction to occur in the forward direction

Table 1 Standard Electrode Potentials of Selected Half-Reactions at 298 K vs. SHE

Reaction	E^0 (V)
$F_2(g) + 2H^+ + 2e^- \rightleftarrows 2HF$	+3.053
$O_3 + 2H^+ + 2e^- \rightleftarrows O_2 + H_2O$	+2.075
$Ag^{2+} + e^- \rightleftarrows Ag^+$	+1.980
$H_2O_2 + 2H^+ + 2e^- \rightleftarrows 2H_2O$	+1.763
$Ce^{4+} + e^- \rightleftarrows Ce^{3+}$	+1.72
$MnO_4^- + 4H^+ + 3e^- \rightleftarrows MnO_2(s) + 2H_2O$	+1.70
$H_5IO_6 + H^+ + 2e^- \rightleftarrows IO_3^- + 3H_2O$	+1.603
$MnO_4^- + 8H^+ + 5e^- \rightleftarrows Mn^{2+} + 4H_2O$	+1.51
$Cr_2O_7^{2-} + 14H^+ + 6e^- \rightleftarrows 2Cr^{3+} + 7H_2O$	+1.36
$Cl_2 + 2e^- \rightleftarrows 2Cl^-$	+1.3583
$O_2 + 4H^+ + 4e^- \rightleftarrows 2H_2O$	+1.229
$Br_2(lq) + 2e^- \rightleftarrows 2Br^-$	+1.065
$Ag^+ + e^- \rightleftarrows Ag$	+0.7991
$Hg_2^{2+} + 2e^- \rightleftarrows 2Hg$	+0.7960
$Fe^{3+} + e^- \rightleftarrows Fe^{2+}$	+0.771
$O_2 + 2H^+ + 2e^- \rightleftarrows H_2O_2$	+0.695
$2HgCl_2 + 2e^- \rightleftarrows Hg_2Cl_2(s) + 2Cl^-$	+0.63
$I_3^- + 2e^- \rightleftarrows 3I^-$	+0.536
$Cu^{2+} + 2e^- \rightleftarrows Cu$	+0.340
$AgCl + e^- \rightleftarrows Ag + Cl^-$	+0.2223
$2H^+ + 2e^- \rightleftarrows H_2$	+0.0000
$Ni^{2+} + 2e^- \rightleftarrows Ni$	−0.257
$Ag(CN)_2^- + e^- \rightleftarrows Ag + 2CN^-$	−0.31
$PbSO_4 + 2e^- \rightleftarrows Pb + SO_4^{2-}$	−0.3505
$Cd^{2+} + 2e^- \rightleftarrows Cd$	−0.4025
$Fe^{2+} + 2e^- \rightleftarrows Fe$	−0.44
$Zn^{2+} + 2e^- \rightleftarrows Zn$	−0.7626
$Na^+ + e^- \rightleftarrows Na$	−2.714
$Li^+ + e^- \rightleftarrows Li$	−3.045

Source: Dean (1999a).

is favored. Conversely, as the standard potential becomes more negative, the reaction is favored in the opposite direction (i.e., oxidation is favored).

To obtain the net cell reaction, Eq. (2) is subtracted from Eq. (1), canceling out the electrons and giving the overall reaction:

$$Cu^{2+} + Zn \rightleftarrows Cu + Zn^{2+} \quad E^0_{cell} = +1.100\,V \quad (3)$$

This electrochemical cell's potential, +1.100 V, is a measure of its *electromotive force* (emf). The emf is a numerical representation of the driving force behind the reaction, or its tendency to move toward equilibrium. Here the emf is positive, so this cell will proceed spontaneously in the direction it is written, producing energy.

A cell that produces energy is known as a *galvanic cell*. If the emf of an electrochemical cell is negative, it requires energy in order to move toward equilibrium, and it is classified as an *electrolytic cell*.

A shorthand notation can be used to simplify the description of electrochemical cells. The cell described by Eq. (3) is written as:

$$Zn\,|\,Zn^{2+}(\alpha=1)\,\|\,Cu^{2+}(\alpha=1)\,|\,Cu \quad (4)$$

By convention, the *anode*, or the electrode at which oxidation occurs, is written on the left, whereas the *cathode*, the electrode at which reduction occurs, is written on the right. Single vertical lines represent phase boundaries across which a potential difference develops. The double vertical line corresponds to the liquid junction region (i.e., a salt bridge) that isolates the contents of the two half-cell(s) but maintains electrical contact between them. Activity, α, (described in the following section) or concentration data of liquid phases are indicated in parentheses. A schematic of the electrochemical cell described by Eq. (4) is shown in Fig. 1.

The emf of the cell represented in Eq. (4) can now be written as:

$$\begin{aligned}E_{cell} &= E_{right} - E_{left} + E_j \quad \text{or} \\ E_{cell} &= E_{anode} - E_{cathode} + E_j\end{aligned} \quad (5)$$

where E_{anode} is the potential of the anode, $E_{cathode}$ is the potential of the cathode, and E_j is the liquid junction potential at the salt bridge.

B. Activity and Activity Coefficients

Owing to interionic effects, described by the solution's *ionic strength*, the effective concentration of the analyte

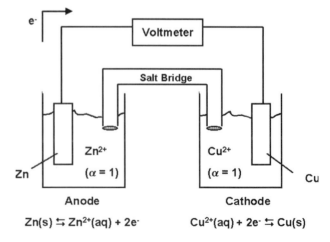

Figure 1 Simple galvanic electrochemical cell.

Potentiometry

of interest is oftentimes less than the actual concentration. This effective concentration is known as the *activity*. At dilute concentrations, it can be assumed that the activity of all reactants and products is unity, and Eq. (3) holds true. However, the activity of an ion i is related to its molar concentration C_i by:

$$\alpha_i = \gamma_i C_i \qquad (6)$$

where γ_i is the *activity coefficient* of ion i. Also, the molar concentration is related to the ionic strength by:

$$\mu = \frac{1}{2}\sum C_i Z_i^2 \qquad (7)$$

where Z_i is the charge of ion i. Therefore, activities other than unity (or different concentrations) would produce a different cell emf.

The theoretical calculation of activity coefficients is performed using the Debye–Hückel equation. For dilute solutions in which the ionic strength is <0.01:

$$-\log \gamma_i = 0.512 Z_i^2 \mu^{1/2} \quad \text{(at 25°C)} \qquad (8)$$

For solutions in which the ionic strength is up to 10-fold higher:

$$-\log \gamma_i = \frac{0.512 Z_i^2 \mu^{1/2}}{1 + \mu^{1/2}} \quad \text{(at 25°C)} \qquad (9)$$

At higher ionic strengths, the Debye–Hückel equation fails, and activity coefficients must be determined experimentally by physicochemical methods.

There are several factors to consider when calculating potentials using the Nernst equation, which relates the activity to cell potential (discussed in the following section). First, solids such as zinc and copper metal always have unit activity, so they cancel out of the Nernst equation. Second, the activity coefficient of uncharged species is unity, so partial pressures of gases and molar concentrations of uncharged species are substituted in the Nernst equation directly. Finally, in dilute solutions of univalent ions, the concentration can be substituted for activity as an approximation.

Kielland (1937) has calculated approximate activity coefficients for many ions in aqueous solution at concentrations up to 0.2 M based on experimental data, taking into account both ionic charge and size. Some of these values are summarized in Table 2; a more comprehensive list can be found in Kielland (1937).

Calculating the pH of a solution containing 0.01 N HCl and 0.09 N KCl can be used to illustrate the principles outlined earlier:

1. The simplest calculation considers HCl as the only source of acidity. Activity coefficients are ignored:

$$pC_H = -\log C_{H^+} = -\log(0.01) = 2.00$$

2. Calculation of the ionic strength of the solution using Eq. (7) followed by calculation of the activity coefficient from Eq. (9) (pαH is the pH as measured by a standardized pH meter) is an even

Table 2 Activity Coefficients of Single Ions in Water Calculated from the Debye–Hückel Equation

Ion	Ionic concentration (M)			
	0.005	0.01	0.05	0.1
H^+	0.950	0.993	0.88	0.83
Li^+	0.948	0.929	0.87	0.835
$K^+, Cl^-, Br^-, I^-, CN^-, NO_2^-, NO_3^-, OH^-, F^-, ClO_4^-, MnO_4^-$	0.946	0.926	0.855	0.81
$Na^+, CdCl^+, ClO_2^-, IO_3^-, HCO_3^-, H_2PO_4^-, HSO_3^-$	0.947	0.928	0.86	0.82
$Ca^{2+}, Cu^{2+}, Zn^{2+}, Sn^{2+}, Mn^{2+}, Fe^{2+}, Ni^{2+}, Co^{2+}$	0.809	0.749	0.57	0.485
$PO_4^{3-}, [Fe(CN)_6]^{3-}, [Cr(NH_3)_6]^{3+}, [Co(NH_3)_5H_2O]^{3+}$	0.612	0.505	0.25	0.16
$Al^{3+}, Fe^{3+}, Cr^{3+}, Sc^{3+}, La^{3+}, In^{3+}, Ce^{3+}, Nd^{3+}$	0.632	0.54	0.325	0.245
$HCOO^-, H_2\ citrate^-, CH_3NH_3^+, (CH_3)_2NH_2^+$	0.946	0.926	0.855	0.81
$NH_3^+CH_2COOH, (CH_3)_3NH^+, C_2H_5NH_3^+$	0.947	0.927	0.855	0.815
$[OC_6H_2(NO_3)_3]^-, (C_3H_7)_3NH^+, CH_3OC_6H_4COO^-$	0.948	0.930	0.875	0.845
$(C_6H_5)_2CHCOO^-, (C_3H_7)_4N^+$	0.949	0.931	0.880	0.85
$(COO)_2^{2-}, H\ citrate^{2-}$	0.804	0.741	0.55	0.45
$Citrate^{3-}$	0.616	0.51	0.27	0.18

Source: Kielland (1937).

better approximation:

$$\mu = \frac{1}{2}\sum c_i z_i^2$$
$$= \frac{1}{2}[(0.01 \times 1) + (0.01 \times 1)$$
$$+ (0.09 \times 1) + (0.09 \times 1)]$$
$$= 0.10\,N$$

$$-\log \gamma_{H^+} = \frac{0.512 z_i^2 \mu^{1/2}}{1 + \mu^{1/2}}$$

$$\log \gamma_{H^+} = -\frac{[(0.512)(1)(0.10)]^{1/2}}{1 + (0.1)^{1/2}}$$

$$\gamma_{H^+} = 0.753$$
$$p\alpha H = -\log \alpha_{H^+} = -\log(\gamma_{H^+})(C_{H^+})$$
$$= -\log(0.753)(0.01)$$
$$p\alpha H = 2.12$$

3. The best approximation uses the activity coefficient given in Table 2 for the calculation:

$$\mu = 0.10$$
$$\gamma_{H^+} = 0.83$$
$$p\alpha H = -\log \alpha_{H^+} = -\log(\gamma_{H^+})(C_{H^+})$$
$$= -\log(0.83)(0.01)$$
$$p\alpha H = 2.08$$

A value of 2.08 is best for this solution and agrees with the measurement obtained with a pH meter.

C. The Nernst Equation

The Nernst equation is an expression of the activity (or concentration) dependence of emf in an electrochemical cell. It can be used to calculate cell potentials at conditions other than standard-state or unit activity.

For the general half-reaction:

$$\text{Ox} + ne^- \rightleftarrows \text{Red} \quad (10)$$

where Ox is the species being reduced to Red and n is the number of electrons transferred in the reaction, the potential of the half-cell at something other than standard-state conditions, E, is given by the Nernst equation:

$$E = E^0 + \frac{RT}{nF} \ln \frac{\alpha_{Ox}}{\alpha_{Red}}$$
$$= E^0 + \frac{2.303 RT}{nF} \log \frac{\alpha_{Ox}}{\alpha_{Red}} \quad (11)$$

Here, E^0 is the standard electrode potential, R is the molar gas constant (8.31441 J mol/K), T is the absolute temperature (in Kelvin), F is the Faraday constant (96,487 coulombs per equivalent), n is the number of electrons transferred in the reaction, α_{Ox} is the activity of the oxidized form of the analyte, and α_{Red} is the activity of the reduced form. The *Nernst coefficient* ($2.303RT/F$) has the value 59.16 mV at 25°C. Table 3 lists values of the Nernst coefficient at various temperatures (Dean, 1999b).

D. Formal Potentials

Standard electrode potentials, E^0, were introduced in Section I.A and Table 1. As mentioned earlier, E^0 describes the half-cell's potential when all of its components are at standard-state conditions, or unit activity. The *formal potential*, $E^{0'}$, is the potential of the half-cell when the concentrations of reactants and products are 1 M and when the analytical concentrations of all other cell components are carefully noted. Formal potentials are used to partially compensate for activity effects and errors due to side reactions, such as dissociation, association, and complex-formations. Here, activity is not used, but concentration is used directly. The notation "1 F HCl" stands for one *formula weight* of HCl per liter rather than one *molecular weight* per liter. For example, consider the ferrous–ferric couple below.

For the reaction at unit activities:

$$Fe^{3+} + e^- \rightleftarrows Fe^{2+} \quad E^0 = +0.771$$

For the reaction in 1 F HCl:

$$Fe^{3+} + e^- \rightleftarrows Fe^{2+} \quad (1\,F\,HCl) \quad E^{0'} = +0.70\,V$$

For the reaction in 1 F H_2SO_4:

$$Fe^{3+} + e^- \rightleftarrows Fe^{2+} \quad (1\,F\,H_2SO_4) \quad E^{0'} = +0.68\,V$$

Table 3 Values of the Nernst Coefficient ($2.3026RT/F$) at Various Temperatures

Temperature (°C)	Nernst coefficient (mV)
0	54.197
10	56.181
20	58.165
30	60.149
40	62.133
50	64.118
60	66.102
70	68.086
80	70.070
90	72.054
100	74.038

Source: Dean (1999b).

Potentiometry

The Nernst equation [Eq. (11)] can also be rewritten in terms of the formal potential and concentrations:

$$E = E^{0'} + \frac{RT}{nF} \ln \frac{[\text{Ox}]}{[\text{Red}]} \quad (12)$$

where $E^{0'} = E^0 + (RT/nF) \ln(\alpha_{\text{Ox}})/(\alpha_{\text{Red}})$.

A list of formal potentials for various half-cell(s) has been compiled by Swift (1939).

E. Liquid Junction Potentials

In many electrochemical cells, solutions of different composition are brought into contact with each other. Hence, boundaries, or *liquid junctions*, are formed at the interface of these solutions. For example, liquid junctions occur at the ends of the salt bridge in Fig. 1. Differences in diffusion rates of ions migrating across the liquid junction (i.e., H^+ will migrate faster than Cl^- owing to differences in size, weight, shape, etc.) result in a potential difference, or *liquid junction potential*, E_j. Junction potentials occur at all solution interfaces and are dependent on the mobilities of the ions and the magnitude of the concentration gradient.

It is nearly impossible to eliminate liquid junction potentials. However, it has been determined experimentally that this potential can be greatly reduced by using a concentrated electrolyte solution, or *salt bridge*, between the two solutions of the electrochemical cell. As the concentration of salt in the bridge increases and as the mobilities of the ions in the bridge approach each other in magnitude, the effectiveness of the salt bridge increases. The best example is a saturated potassium chloride solution, greater than 4 M at room temperature and having only a 4% difference in the mobilities of the potassium and chloride ions (Skoog et al., 1998a). If chloride ion interferes in a particular measurement, less ideal salts can be substituted, such as nitrate or sulfate, with the creation of a somewhat larger liquid junction potential. Values of liquid junction potentials based on theoretical estimations are listed in Table 4 (Bates, 1973a).

F. Temperature Coefficients

The *temperature coefficient* of an electrochemical cell is the rate of change of cell emf with temperature. Specifically, it describes the magnitude of the change in the equilibrium potential of the electrode when the temperature of the system is changed by 1°C (Serjeant, 1984a). The temperature coefficient can be determined experimentally in two ways. First, the two electrodes can be operated at the same temperature, which is known as the *isothermal* operation of the electrodes, and the difference between the coefficients of the individual electrodes is the temperature coefficient of the system. Secondly, the two electrodes are at different temperatures, and the system is under *thermal* operation. In this case, the temperature coefficient is calculated in a different manner.

Table 4 Liquid Junction Potentials (E_j) at 25°C Between Listed Solutions and a Saturated Solution of KCl

Solution, concentration	E_j in saturated KCl (mV)
HCl, 1 M	14.1
HCl, 0.1 M	4.6
HCl, 0.01 M	3.0
HCl, 0.01 M; NaCl, 0.09 M	1.9
KCl, 0.1 M	1.8
KH phthalate, 0.05 M	2.6
KH_2PO_4, 0.025 M; Na_2HPO_4, 0.025 M	1.9
$NaHCO_3$, 0.025 M; Na_2CO_3, 0.025 M	1.8
NaOH, 0.01 M	2.3
NaOH, 0.1 M	-0.4
NaOH, 1 M	-8.6

Source: Bates (1973a).

The Nernst equation can be rewritten, emphasizing the temperature term:

$$E = E^0 + \frac{0.19841T}{n} \log \alpha_i \quad (13)$$

and the cell temperature coefficient is given by:

$$\frac{dE}{dT} = \frac{dE^0}{dt} + \frac{0.19841}{n} \log \alpha_i + \frac{0.19841T}{n} \frac{d \log \alpha_i}{dT} \quad (14)$$

Although most potentiometric instruments allow for automatic temperature compensation, this only corrects for the term ($0.19841T$) from Eq. (13) and not for the temperature coefficients of the electrodes or solutions in the measurement. Therefore, it is important that standardization and measurement of the test solution be done at the same temperature. Table 5 lists the thermal and isothermal temperature coefficients for common reference electrodes (Hampel, 1964).

II. REFERENCE ELECTRODES

Unfortunately, the potential of a single half-cell cannot be measured directly. Therefore, the potential of a lone indicator electrode must be measured against a *reference electrode*, an electrode that does not respond to the test solution and whose potential is known and constant. There are several criteria that an ideal reference electrode should meet: (1) it should be reversible and follow the Nernst equation; (2) it should maintain a constant potential with time; (3) it should maintain its potential after being subjected to finite currents; and (4) it should not exhibit

Table 5 Calculated Temperature Coefficients of Common Reference Electrodes at 25°C

Reference electrode	Calculated thermal temperature coefficient (mV/°C)	Isothermal temperature coefficient[a]
SHE/ sat KCl	+0.871	0.000
Hg/Hg$_2$Cl$_2$/1.0 M KCl	+0.573	−0.298
Hg/Hg$_2$Cl$_2$/sat KCl (SCE)	+0.165	−0.706
Ag/AgCl/1.0 M KCl	+0.235	−0.636
Ag/AgCl/($\alpha = 1$) KCl	+0.213	−0.658

[a]Isothermal temperature coefficients are calculated by subtracting 0.871 mV/°C from the thermal temperature coefficient.
Source: Hampel (1964).

thermal hysteresis (Skoog et al., 1998b). There are only a few reference electrodes commonly used today.

A. The Standard Hydrogen Electrode

As mentioned earlier in Section I.A, all half-reaction potentials are measured against the SHE, shown in Fig. 2. This electrode, Pt | H$_2$(g) | H$^+$(aq), has an effective hydrogen gas pressure of 1 bar and a hydrogen ion activity of unity or a concentration of approximately 1 M and is defined by:

$$2H^+ + 2e^- \rightleftarrows H_2 \quad E^0 = 0.0000 \text{ V}$$
$$\text{(at all temperatures)} \quad (15)$$

However, the SHE is inconvenient to construct and very fragile, so other reference electrodes are more commonly used.

B. The Calomel Electrode

The *calomel electrode* shown in Fig. 3 consists of mercury, mercury I chloride (calomel), and a solution of potassium chloride of known concentration. It is represented by:

$$\text{Hg} \mid \text{Hg}_2\text{Cl}_2 \text{ (saturated), KCl } (x\text{M}) \quad (16)$$

The half-cell reaction for this electrode is:

$$\text{Hg}_2\text{Cl}_2 + 2e^- \rightleftarrows 2\text{Hg} + 2\text{Cl}^- \quad (17)$$

and its potential is dependent on the concentration of chloride present in the electrode. Therefore, the chloride ion concentration must always be specified when describing this electrode. For example, the *saturated calomel electrode* (SCE) contains a saturated solution of potassium chloride that is also saturated with calomel, so this electrode's standard potential vs. the SHE at 25°C is +0.244 V. Table 6 lists the common reference electrodes and their potentials at varying temperatures and concentrations of calomel (Bates, 1973b) (Ag/AgCl electrodes are discussed below). The calomel electrode cannot be used at high temperatures (>80°C) owing to disproportionation of Hg(I) into Hg and Hg(II) ion, and it equilibrates slowly to temperature changes (LaCourse, 1997).

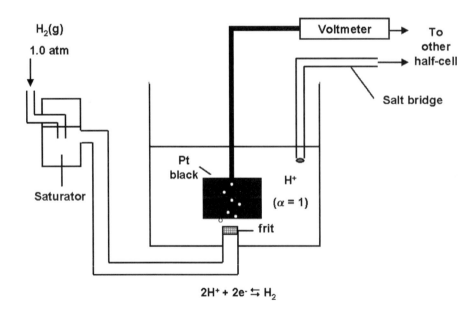

Figure 2 Schematic of a typical SHE.

Potentiometry

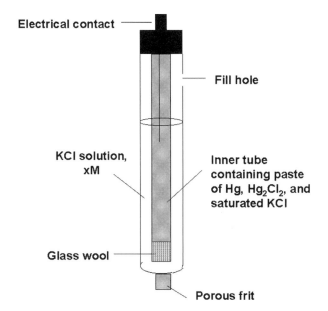

Figure 3 Calomel reference electrode.

C. Silver, Silver Chloride Electrode

The most common and widely marketed reference electrode is the *silver, silver chloride electrode* (Ag/AgCl) depicted in Fig. 4. Here, a silver wire coated with silver chloride is immersed in a solution of potassium chloride saturated with silver chloride. The half-cell is given by:

$$\text{Ag} \mid \text{AgCl (saturated), KCl } (x\text{M}) \tag{18}$$

The half-reaction is:

$$\text{AgCl} + e^- \rightleftarrows \text{Ag} + \text{Cl}^- \tag{19}$$

It is important to note that the redox reaction occurring at the electrode is actually $\text{Ag}^+ + e^- \rightleftarrows \text{Ag}^0$, and the potential is dependent on the Ag^+ concentration. However, the Ag^+ concentration is dependent on the solubility product equilibrium of AgCl, so the potential then depends on the chloride ion concentration. Therefore, the standard potential of the Ag/AgCl electrode at 25°C is +0.2223 V. Potentials for Ag/AgCl electrodes at varying temperatures and concentrations of KCl are listed in Table 6 (Bates, 1973b, p. 335).

The Ag/AgCl electrode has several advantages over the calomel electrode. This electrode is easily constructed, it can be operated over a wider range of temperatures than the calomel electrode, and it can be used in nonaqueous solutions (LaCourse, 1997, p. 30).

D. Double Junction Reference Electrode

A double junction reference electrode has two junctions. The first junction separates the reference electrode from an intermediate solution, and the second isolates the intermediate solution from the test solution. This reference electrode can be used to minimize contamination of the test solution.

E. Reference Electrode Practice

Erratic behavior or drifting response in potentiometric measurements often signifies a problem with the reference electrode. Problems arise when the junction becomes partially plugged or when wires are broken inside the electrode. There are several guidelines to follow when caring for reference electrodes to minimize the occurrence of problems.

The internal solution should always be filled to a level above that of the test solution. This will prevent contamination of the internal solution by plugging of the junction due to a reaction between the components of the test solution and the ions of the internal solution. If contamination is occurring, a common way to minimize it is to use a second salt bridge between the reference electrode and the test solution. Also, the electrode can be soaked in order to attempt restoration of the junction (Fisher, 1984).

The correct filling solution concentration should always be used to replenish the electrode. Manufacturer's filling instructions should be strictly followed.

Table 6 Standard Potentials (mV) of Common Reference Electrodes at Varying Temperatures

Temperature, °C	Calomel, 0.1 M KCl	Calomel, sat KCl	Ag/AgCl, 3.5 M	Ag/AgCl, sat KCl
10	0.3362	0.2543	0.2152	0.2138
15	0.3362	0.2511	0.2117	0.2089
20	0.3359	0.2479	0.2082	0.2040
25	0.3356	0.2444	0.2046	0.1989
30	0.3351	0.2411	0.2009	0.1939
35	0.3344	0.2376	0.1971	0.1887
40	0.3336	0.2340	0.1933	0.1835

Source: Bates (1973b).

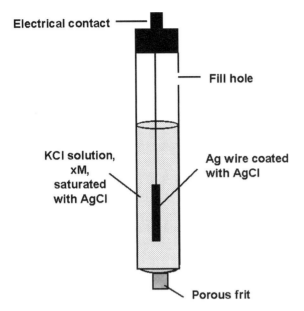

Figure 4 Ag/AgCl reference electrode.

One should keep in mind that any ion-specific electrode can be used as a reference electrode, as long as the ion is constant in solution. This type of electrode has no liquid junction and eliminates problems that may arise from the liquid junction. Fisher (1984) has published an article detailing the care and maintenance of reference and pH electrodes.

III. INDICATOR ELECTRODES

Ideally, an indicator or working electrode responds rapidly and reproducibly to changes in activity of the analyte. There are two types of indicator electrodes, metallic and membrane.

A. Metallic Indicator Electrodes

There are four types of metallic indicator electrodes, including electrodes of the first kind, electrodes of the second kind, electrodes of the third kind, and redox indicators.

Metallic electrodes of the first kind are represented by, for example:

$$Cu^{2+} + 2e^- \leftrightarrows Cu(s)$$

where the electrode is in equilibrium with the cation derived from the electrode itself. These electrodes are not typically used for potentiometric measurements because they are not very selective, responding to other, more easily reduced cations. Some metals dissolve in acid, so they can only be used in basic or neutral solutions. Also, some metals used are so easily oxidized that the test solutions must be degassed prior to measurements, and harder metals (iron, nickel, etc.) do not give reproducible potentials.

Electrodes of the second kind are responsive to an anion with which its cation forms a precipitate or stable complex ion. For example, silver cations from silver electrodes respond to halide and halide-like anions:

$$AgCl(s) + e^- \leftrightarrows Ag(s) + Cl^- \quad E^0 = 0.222\,V$$

Therefore, the potential of the indicator electrode depends on the activity of the chloride ion in solution.

An electrode of the third kind is made to respond to a different cation other than its own. An example of this electrode is mercury which can be used to determine the activity of the calcium ion in a calcium-containing solution.

Electrodes of the fourth kind are inert electrodes, such as gold or platinum, immersed in a solution of a redox couple. These electrodes do not take part in the redox reaction, but act as a source of electrons for the reaction.

B. Membrane Indicator Electrodes

Membrane electrodes are also called *ion-selective electrodes* owing to their high selectivity toward particular ions of interest. This section discusses pH measuring electrodes, the most common membrane electrode, and other ion-selective electrodes.

1. pH Measuring Electrodes

The determination of pH, or hydrogen ion activity, is one of the most important and practical applications of potentiometry. The pH of a solution is typically measured using a *glass pH electrode*. In the glass membrane pH electrode, the membrane separates the test solution from an internal reference solution of known acidity, and the potential difference across this membrane is measured to determine the pH. This electrode is unique in that its mechanism of response involves an ion-exchange process and not an electron-transfer reaction, therefore making it impervious to interferences by oxidizing and/or reducing agents present in the test solution. In addition, its rapid and accurate response to sudden changes in pH makes it a logical choice for potentiometric measurements (Peters et al., 1974).

Corning 015 glass, consisting of approximately 22% Na_2O, 6% CaO, and 72% SiO_2, is typically used to manufacture glass pH electrodes. Each silicon atom is bonded to four oxygen atoms, thus creating a 3D network of SiO_4^{4-} groups. The negative charges are balanced by cations located within the interstices of the structure. These small, singly charged cations, such as sodium, can move throughout this lattice and exchange with protons from

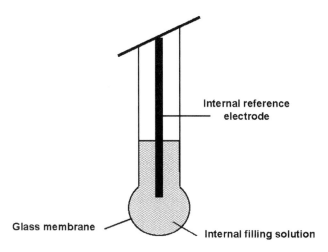

Figure 5 Glass pH electrode.

the test solution, causing the ion-exchange phenomenon mentioned earlier (Skoog et al., 1998b, pp. 597–598).

A schematic of a typical glass pH electrode is depicted in Fig. 5. It consists of a glass membrane (described in the preceding paragraph) that is shaped into a bulb and is highly sensitive to hydrogen ions in solution. Inside the glass bulb is an internal filling solution, usually a KCl solution saturated with AgCl or Hg_2Cl_2, and an internal reference electrode, usually silver or calomel (Ag | AgCl or Hg | Hg_2Cl_2), extended upward through the glass tube to provide electrical contact to the external circuit. If the internal reference electrode (inside the pH electrode) and the external reference electrode (of the overall cell) are chemically matched, the temperature coefficient for this cell is minimized. An example of a cell for measurement of pH is described by:

Ag | AgCl (sat), [xCl$^-$] ∥ test solution | glass membrane
(external reference electrode)

| pH 7 buffer [xCl$^-$], AgCl (sat) | Ag
(internal reference electrode) (20)

If this electrode was immersed in a test solution of pH = 7 at 25°C, the potential should theoretically be 0.0 V, aside from a few millivolts owing to small liquid junction potentials and such.

From the Nernst equation, the potential for the preceding cell is given by:

$$E = E^0 - (0.19841T)(\text{pH}) \quad \text{or} \quad \text{pH} = \frac{E^0 - E}{0.19841T} \quad (21)$$

where E is the cell emf when the electrode is immersed in the pH 7 buffer. If the emf is 0.0 V, as theoretically it should be, then:

$$E^0 = 414.1 \text{ mV} \quad (\text{at } 25°\text{C}) \quad (22)$$

As discussed earlier, the standard electrode potential depends on temperature, so a different temperature (other than 25°C) would result in a different E^0 value.

Glass membrane electrodes for measuring pH have a few limitations. At high pH values (>9) a negative error occurs, or the pH values measured are lower than the true values. This is due to the fact that the electrode is responding to alkali metal ions as well as hydrogen ions. This effect is known as the *alkaline error*, and the extent of this error with different glass electrodes is shown in Fig. 6 (Bates, 1973c).

Additionally, glass membrane electrodes suffer from the *acid error* which is also illustrated in Fig. 6. In this case (pH < 0.5), pH readings tend to be higher than the true values.

Commercial glass electrodes are available in many forms, depending on the application. Some of these include immersion electrodes for general laboratory use, the hypodermic microtype which fit into needles, miniature glass and calomel electrodes, and many others.

Although the glass membrane electrode is the best-known electrode for measuring pH, there are others that can be used. The *quinhydrone* and the *antimony* electrodes are two examples (Bates, 1973d, pp. 295–304).

Quinhydrone (QH) is an equimolecular compound of benzoquinone (OC_6H_4O) and hydroquinone (HOC_6H_4OH) and undergoes a reversible oxidation–reduction reaction with hydrogen ions:

$$\text{Quinone} + 2H^+ + 2e^- \rightleftharpoons \text{Hydroquinone} \quad (23a)$$

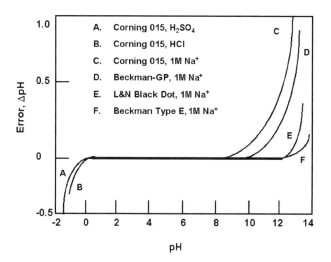

Figure 6 Glass electrode corrections in both acidic and alkaline solutions at 25°C.

This electrode uses a platinum wire indicator electrode, a Ag/AgCl saturated KCl reference electrode, and a small amount of quinhydrone mixed into the test solution. It is known from Table 3 that $E_{QH} = 699.4$ mV and from Table 6 that $E_{Ag/AgCl, \text{ sat'd KCl}} = 198.8$ mV. So, from Eq. (21):

$$E_{cell} = E_{QH} - E_{ref} + 59.16 \log[H^+]$$
$$= 699.4 - 198.8 + 59.16 \log[H^+] \quad (23b)$$
$$pH = -\log[H^+] = \frac{500.6 - E_{cell}}{59.16} \quad (23c)$$

The quinhydrone electrode is easy to construct, comes to equilibrium quickly, and is applicable to use in nonaqueous solutions. However, this electrode is unstable at pH >8, and it is susceptible to interference because of the presence of proteins, certain oxidizing agents, and high concentrations of salt. Also, it is not stable for long periods of time, especially at temperatures above 30°C.

The antimony electrode consists of a stick of antimony cast in air, and its potential is a function of the pH of the test solution in which the stick is immersed (Britton and Robinson, 1931). The antimony is covered with an oxide layer (Sb_2O_3) involved with the oxidation–reduction reaction producing this electrode's potential:

$$Sb_2O_3 + 6H^+ + 6e^- \leftrightarrows 2Sb + 3H_2O \quad (24)$$

This electrode can be used at elevated temperatures and in alkaline solutions. It can be used in solutions containing cyanide and sulfite (glass membrane and quinhydrone electrodes cannot). However, this electrode is not as accurate as the glass membrane electrode, and it is disturbed by the presence of oxidizing and reducing agents present in the test solution (Bates, 1973b, pp. 302–304).

There are other hydrogen ion electrodes available, but they are discussed elsewhere (Bates, 1973b, pp. 304–306).

Definitions of pH

The dissociation of water is written as:

$$H_2O \leftrightarrows H^+ + OH^- \quad (25)$$

and its *ionization constant* is given by:

$$K_W = (\alpha_{H^+})(\alpha_{OH^-}) \quad (26)$$

For pure water and very dilute solutions, it is assumed that activity coefficients are unity, and Eq. (26) becomes:

$$K_W = [H^+][OH^-] \quad (27)$$

$K_W = 1.012 \times 10^{-14}$ at 25°C, and $pK_W = 13.995$. The ionization constant of water is dependent on temperature, as shown in Table 7 (Light and Licht, 1987).

Table 7 Ionization Product of Water (K_w), and the pH of Neutral Water at Varying Temperatures

Temperature (°C)	K_w ($\times 10^{14}$)	pH of neutral water
0	0.1153	7.469
10	0.2965	7.264
20	0.6871	7.081
25	1.012	6.998
30	1.459	6.918
40	2.871	6.771
50	5.309	6.638
60	9.247	6.517
70	15.35	6.407
80	24.38	6.307
90	37.07	6.215
100	54.45	6.132

Source: Light and Licht (1987).

A solution in which $[H^+] = [OH^-]$ is said to be *neutral*:

$$[H^+]_{neutral} = [OH^-] = (K_W)^{1/2} = 1.006 \times 10^{-7} \quad (28)$$
$$pH_{neutral} = -\log[H^+]_{neutral} = (pK_W)/2 = 6.997 \quad (29)$$

As is the case with the ionization constant, the pH of a neutral solution varies with temperature (see Table 7).

The pH of a solution is defined by:

$$pH = -\log[H^+] \quad (30)$$

Note that a change in $[H^+]$ by a factor of 10 results in a unit change in pH. Because $[H^+]$ and $[OH^-]$ are not usually greater than 10 M, the practical range of the pH scale is -1 to $+15$.

A more modern definition of pH is:

$$pH = p\alpha H = -\log \alpha_{H^+} = \log[H^+]\alpha_{H^+} \quad (31)$$

In this definition, the brackets denote molar concentration. However, molality, a temperature-independent concentration unit, can be used instead in this definition, as long as the concentration units are specified.

Operational Definition of pH

The *operational definition* of pH is given by:

$$pH_X = pH_S - \frac{(E_X - E_S)F}{RT \ln 10} \quad (32)$$

This is an experimental definition of pH, based on potentiometric measurements with glass electrodes and standard solutions. Here, pH_X is the pH of an unknown solution, pH_S is the pH of the standard solution, E_X is the emf of the cell with the unknown pH solution, E_S is the emf of the cell with the standard pH solution, and F, R, and T

Potentiometry

have the usual meanings. The electrochemical cell is described by:

reference electrode | salt bridge || solution $pH_{X \text{ or } S}$
| glass electrode   (33)

Table 8 lists the seven primary pH standard solutions over a range of temperatures, developed by the US National Institute for Standards and Technology (NIST) (Bates, 1973d).

The pH Meter

The type of equipment used for measuring pH is usually determined by the internal resistance of the cell. For cells with low internal resistance ($10^5 \, \Omega$ or less), one uses the *potentiometer*, an instrument that compares an unknown emf directly to that of a known cell. However, because of the high resistance of the glass electrode, high impedance voltmeter circuits are required for measuring pH. Typically, a *direct reading pH meter* is employed for such measurements.

A pH meter should be able to measure pH accurately, even with high resistance from the cell, and it should be able to compensate for temperature fluctuations, preferably automatically. Common, simplified meters measure pH reproducibly with an accuracy of ± 0.1 pH unit. An excellent, more expensive pH meter is able to read the pH of a solution reproducibly to a few thousandths of a unit. Some meters have memory containing the pH values of standard buffers. Others have the capacity of determining concentration from the Nernst equation.

Bates outlines several guidelines to consider when measuring pH (Bates, 1973e):

1. The pH meter measures the difference between the pH values of two solutions, a standard and an unknown, and both solutions should be at the same temperature.
2. When standardizing the electrode, one should use two buffer solutions to bracket, if possible, the unknown.
3. Errors due to temperature fluctuations and liquid junction potentials can be minimized by standardizing the meter at a pH value close to that of the unknown.
4. The most suitable electrode should be chosen for the measurement, considering the temperature of the solution and the expected pH value. One should keep in mind that some electrodes display major errors at elevated temperatures and at high or low pH values (alkaline and acid errors).

2. *Other Ion-Selective Electrodes*

Other than the glass membrane electrode for measuring pH, there are membrane electrodes available for the determination of many cations and anions by direct potentiometric measurement. These membrane electrodes are termed *ion-selective electrodes* (ISEs) because of their high selectivity for particular ions in the presence of interfering ions. ISEs are also called *pIon electrodes* due to the fact that their output is usually measured as a p-function, such as pH, pNa, and pCl. Just like the pH electrode, all ISEs respond to changes in the activity of the ion of interest, the potential difference across the membrane is measured with respect to a reference electrode, and the response follows the Nernst equation.

Table 8 pHs of NIST Primary Buffer Solutions at Various Temperatures

Temperature (°C)	KH tartate (sat at 25°C)	KH_2 citrate (0.05 M)	KH phthalate (0.05 M)	KH_2PO_4 (0.025 M) Na_2HPO_4 (0.025 M)	KH_2PO_4 (0.008695 M) Na_2HPO_4 (0.03043 M)	Borax (0.01 M)	$NaHCO_3$ (0.025 M) Na_2CO_3 (0.025 M)
0	—	3.863	4.003	6.984	7.534	9.464	10.317
10	—	3.820	3.998	6.923	7.472	9.332	10.179
20	—	3.788	4.002	6.881	7.429	9.225	10.062
25	3.557	3.776	4.008	6.865	7.413	9.180	10.012
30	3.552	3.766	4.015	6.853	7.400	9.139	9.966
40	3.547	3.753	4.035	6.838	7.380	9.068	9.889
50	3.549	3.749	4.060	6.833	7.367	9.011	9.828
60	3.560	—	4.091	6.836	—	8.962	—
70	3.580	—	4.126	6.845	—	8.921	—
80	3.609	—	4.164	6.589	—	8.885	—
90	3.650	—	4.205	6.877	—	8.850	—

Source: Bates (1973d).

In general, a cell containing an ISE can be represented as:

reference electrode ‖ test solution ‖ ISE (34)

and the cell's emf is given by:

$$E = E^0 + \frac{2.303RT}{nF} \log \alpha_i \quad (35)$$

for ion i with charge n. Because the n term includes charge, the second term in Eq. (35) will be positive for cations and negative for anions.

Properties of ISEs

All ISEs share common properties, making them both sensitive and selective toward particular ions. These include (Skoog et al., 1998b, pp. 596–597):

1. The solubility of the electrode in the analyte solution of interest approaches zero. Therefore, all membranes are constructed with large molecules or molecular aggregates like silica glasses and polymeric resins. Even ionic inorganic compounds with low solubility have been used.
2. All membranes used in ISEs exhibit some small amount of electrical conductivity.
3. All membranes in ISEs are capable of selectively binding the ion of interest.

Selectivity of Electrodes

There is no ISE that responds only to the ion of interest, although it is typically more responsive to this primary ion than others. If an interfering ion is present in a large enough concentration, the electrode response is due to contributions from both the primary and interfering ions. The selectivity of the electrode for primary ion i with respect to interfering ion j is dependent on the *selectivity coefficient*, k_{ij}, of the ISE. Equation (35) can be written as:

$$E = E^0 + \frac{2.303RT}{nF} \log\left(\alpha_i + \sum k_{ij}\alpha_j^{n/z} + \cdots\right) \quad (36)$$

where α_i is the activity of primary ion i with charge n, α_j is the activity of interfering ion j with charge z, and k_{ij} is the ISE's selectivity coefficient. When $k_{ij} \ll 1$, there is little interference from ion j. However, when $k_{ij} \gg 1$, the ISE will respond better to interfering ion j than to primary ion i. When $k_{ij} = 1$, then the ISE will respond with equal sensitivity to both ions i and j.

Classification of Ion-Selective Membrane Electrodes

There are two main classes of ion-selective membrane electrodes: crystalline and noncrystalline (LaCourse, 1997, p. 31; Skoog et al., 1998b, pp. 596–606). Crystalline membrane electrodes are further classified as single crystal and polycrystalline (or mixed crystal). Noncrystalline membrane electrodes are broken down into the categories of glass (discussed earlier), liquid, and immobilized liquid in a rigid polymer.

Crystalline membrane electrodes are made from an ionic compound (single crystal) or a homogeneous mixture of ionic compounds (polycrystalline). Most of these ionic compounds are insulators, having little or no conductivity at room temperature. Those that are conductive have small, singly charged ions that are mobile in the solid phase.

An example of a single crystal electrode is the lanthanum fluoride, LaF_3, electrode for determining F^-. At the membrane–solution interface, ionization creates charge on the membrane surface:

$$LaF_3 \rightleftharpoons LaF_2^+ + F^- \quad (37)$$

The magnitude of this charge depends on the fluoride ion concentration in solution, and the side of the membrane with the higher fluoride ion concentration becomes more negative with respect to the other side. This charge difference is used to determine the difference in fluoride ion concentration on either side of the membrane.

The silver sulfide, Ag_2S, electrode, for example, is a mixed crystal electrode used for determining S^{2-} and Ag^+. Here, the electrode is immersed into a test solution, and a small amount of silver sulfide dissolves saturating the film of liquid directly surrounding the electrode. The solubility (and the silver ion concentration) depends upon the sulfide concentration of the analyte.

Liquid membranes, made from immiscible liquids, work by selectively binding and directly determining the activities of particular ions. The earliest liquid membranes were formed using liquid ion-exchangers immobilized on an inert porous solid support, as shown in Fig. 7. The electrode is made from two concentric plastic or glass tubes, and the porous support is placed in contact with both inner and outer volumes of the tubes. An organic liquid ion-exchanger fills the outer tube and soaks the support, whereas an aqueous solution of KCl and a salt of the ion to which the electrode responds is contained in the inner tube. The reference electrode is also contained inside the inner tube. An electrode potential is generated when the activities of the inner and outer solutions surrounding the membrane differ. These electrodes can be operated over a temperature range of 0–50°C, a smaller range than that of crystalline and glass electrodes, and they are restricted to use in aqueous solutions.

Rather than using a porous solid support, a newer type of liquid membrane is constituted from a liquid ion-exchanger immobilized in a tough polyvinyl chloride membrane. This particular membrane is made by dissolving the polyvinyl chloride and liquid ion-exchanger in a solvent (i.e., tetrahydrofuran) and then allowing the

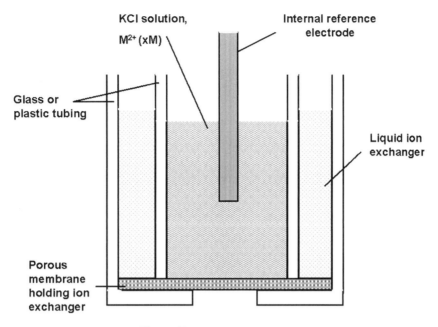

Figure 7 Liquid membrane ISE.

solvent to evaporate, creating a flexible membrane that can be cut and shaped to fit the end of a tube. This membrane works in the same manner as those in which the ion-exchanger is an actual liquid in the pores of the support, as described in the previous paragraph. Today, most liquid membranes are of this newer type.

Gas-Sensing Probes

Gas-sensing probes are often referred to as "gas-sensing electrodes", although these sensors are not electrodes but electrochemical cells made up of an ISE, a reference electrode, a suitable ionic solution, and a gas permeable membrane (Bailey, 1980a). The membrane separates the test solution and the indicator electrode. These probes are very sensitive and selective for determining dissolved gases such as CO_2 and NH_3. Figure 8 depicts a gas-sensing probe.

An example of a gas-sensing probe would be that for determining CO_2. In this case, the internal filling solution is sodium bicarbonate and sodium chloride. The indicator electrode is the glass pH electrode, and the Ag/AgCl electrode serves as the reference. When the probe is placed in a solution containing CO_2, the gas diffuses through the membrane, and equilibrium is quickly established on either side of the membrane (between the test solution and the thin film of internal solution adjacent to the membrane). Then, another equilibrium is established, causing the pH of the internal surface film to change:

$$CO_2(aq) + H_2O \rightleftharpoons H^+ + HCO_3^- \quad (38)$$

It is this pH change that is used to measure the carbon dioxide content of the test solution.

Enzyme ISEs

An *enzyme ISE* is a type of biosensor, comprising an enzyme coating the membrane of an electrochemical transducer, used for both biological and biochemical analyses (Hibbert and James, 1984). Here, the sample of interest undergoes an enzyme-catalyzed reaction, producing a product whose concentration is proportional to the amount of reactant that is determined by the ISE.

An example of such a biosensor is the urea $(NH_2)_2CO$ electrode:

$$(NH_2)_2CO + H_2O \rightarrow NH_4^+ + HCO_3^- \quad (39)$$

Here, urea is hydrolyzed in the presence of urease, and the ISE responds to the ammonium ion formed by this reaction.

Enzyme ISEs have both advantages and disadvantages. They are easy to use and respond quickly. The combination of the selectivity of the enzyme-catalyzed reaction and the ISE produces responses that are virtually free from interferences. However, the steady decrease of enzyme activity with use produces a drifting response, and the membrane layer must be periodically renewed. There are many enzyme ISEs described for different analytes including urea, creatinine, amygdalin, cholesterol, penicillin, amino acids, and glucose (Bailey, 1980b).

Ion-Selective Field-Effect Transistors

Ion-selective field-effect transistors (ISFETs) are semiconductor devices used for measuring ionic species in solution. The device consists of a gate insulator, a drain, a source, a substrate, and a reference electrode.

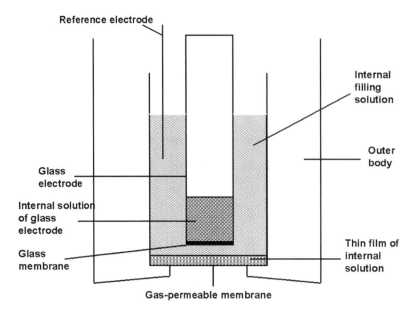

Figure 8 Gas-sensing ISE.

The gate is covered with an insulating layer of silicon nitride (Si_3N_4), and it is in contact with the test solution, which is also in contact with the reference electrode. The silicon nitride adsorbs ions of interest in its available sites. A variation in the concentration of the ions of interest in the test solution will change the concentration of adsorbed ions, varying the voltage between the gate and the source, and thus changing the conductivity of the channel of the ISFET. This conductivity is monitored electronically, providing a signal that is proportional to the logarithm of the concentration of ions of interest in solution (Skoog et al., 1998b, pp. 606–607). Figure 9 depicts an ISFET sensitive to hydronium ions in solution.

ISFETs have the advantages of ruggedness, faster response time than membrane electrodes, small size, inertness in harsh environments, and low electrical impedance. Additionally, ISFETs do not require hydration and can be

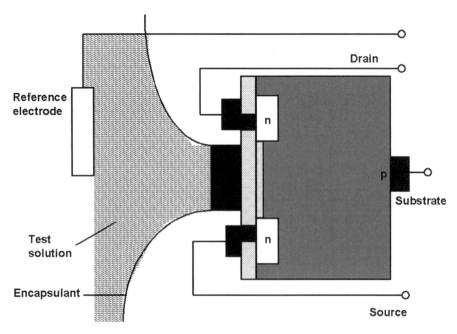

Figure 9 An ISFET for measuring pH.

stored dry. ISFETs have been reviewed in the literature (Zemel, 1975).

IV. GENERAL INSTRUMENTATION

There are two types of instruments used in potentiometry, the potentiometer and the direct reading electronic voltmeter. Both instruments are used to measure the emf of electrochemical cells. Both types are often referred to as pH meters, which is true when their internal resistances are high enough to be used with membrane electrodes (as discussed in Section III.C). Both instruments are briefly described here.

A. Potentiometers

The *potentiometer* is an instrument used to measure cell emf that relies on a compensation method. Here, an unknown emf is compared directly to that of a known, standard cell. When the emfs of both the unknown and standard cells are in "voltage balance", no current flows, and the "open-circuit voltage" of the cell is obtained (Bates, 1973e, pp. 391–392).

The potentiometer tends to be more accurate than the direct reading type instrument. However, it is less useful than the direct reading instrument for following pH changes that occur during a reaction or titration. Most modern instruments are of the direct reading type.

B. Direct Reading Instruments

Typically, a direct reading instrument consists of an ISE whose output is connected to a high-resistance field-effect transistor or a voltage follower. Direct potentiometric measurements are made by comparing the potential developed by the ISE in the test solution to its potential in a standard solution of the analyte of interest. Preliminary separation or clean-up steps are seldom required owing to the selectivity of most indicator electrodes, making direct potentiometric measurements rapid and easily adapted to continuous monitoring of activities of ions in solution.

C. Commercial Instrumentation and Software

The pH meters described earlier are available in a variety of configurations to meet various measurement needs. They are available as both bench top and portable systems in models suited for basic to advanced measurements. Many meters allow for direct temperature measurement and compensation without a separate

Table 9 Manufacturers, Internet Addresses, and Commercially Available Instrumentation

Manufacturer	Internet address	Instrumentation
Analytical Instrument Systems, Inc.	www.aishome.com	Portable analytical potentiostat
BVT Technologies	www.bvt.cz	Miniature electrochemical and biosensor substrates
Cole-Parmer Instrument Company	www.coleparmer.com	Handheld meters, disposable pH electrodes, industrial pH controllers, water testing instrumentation
Corning	www.corning.com	Ion-selective, pH, reference electrodes (liquid membrane, solid-state, glass), portable meters, gas-sensing probes
DeltaTRAK	www.deltatrak.com	Pocket ISFET pH meter
Digimed	www.digimed.ind.br	Field pH meters, gas-sensing probes, online potentiometric analyzer, combination pH electrodes
EXTECH INSTRUMENTS	www.extech.com	Handheld pH meters, field meters
Hanna Instruments, Inc.	www.hannainst.com	High-pressure pH electrodes, pocket meters
HORIBA Instruments, Inc.	www.horibalab.com	Field pH/ion meters, pocket and handheld meters
MAN-TECH Associates, Inc.	www.titrate.com	Titration/ion analysis systems (with autosampler), ion-selective, pH, and reference electrodes, field meters
Sentron	www.sentron.nl	Non-glass ISFET pH/ion meters
WTW Measurement Systems, Inc.	www.wtw-inc.com	Polymer and glass electrodes, field instruments, handheld and pocket meters
LTSN Physical Sciences	http://dbweb.liv.ac.uk/ltsnpsc/swrevs/1electLAB.htm	electrochemLAB software
Metrohm	www.metrohm.com	822 Titration Curve Simulator

temperature probe. Dual- and multichannel meters allow the use of multiple electrodes on one meter.

As seen subsequently, many companies manufacture handheld and pocket-size meters that can be used in the field as well as in the laboratory. Some companies make micro- and full-size electrodes and tailor them to the needs of the customer (they add substrates). Others have modified electrodes and meters to withstand high temperatures and pressures. Still others manufacture waterproof meters to allow under water measurements.

Some companies sell software that can be used to simulate curves obtained from potentiometric titrations, compute the potential of a given cell, determine activities of components, and illustrate temperature dependence of a cell. This type of software is available for academic use as well as for professional scientists.

Table 9 lists a few companies and some of their major products. This is by no means a list of all the companies that sell electroanalytical instrumentation. Also, the summary of the products sold at these companies is not all-inclusive. This table merely serves as a small sample of what types of products are available and where one may look to purchase this instrumentation.

V. APPLICATIONS OF POTENTIOMETRY

The most common use of potentiometry is to measure the activity of an ion of interest in solution and to measure the cell emf. However, the potential of an indicator electrode can also be used to establish the equivalence point in a titration experiment, a method known as a *potentiometric titration*.

Potentiometric titrations are very useful, providing more accurate data than the analogous method using indicators. They can be used to determine ionization constants, acid–base equilibria, and stability constants. Other potentiometric titrations include complexometric, oxidation–reduction, and precipitation titrations. Serjeant thoroughly outlines potentiometric applications to the determination of solution equilibrium data and to titrimetric analysis (Serjeant, 1984b).

Modern pH meters and electrodes are used for a variety of applications. Often, they are used in the laboratory for routine measurements and research. However, they are finding much use in many other fields. They can be used for environmental samples, measuring analytes in drinking and wastewater, air, and soil. They are used to determine heavy metals in various matrices. They are used for monitoring various entities in both physiological and biological samples. They measure food quality, monitoring fruits, cheeses, and meats. They are very important instruments in the pharmaceutical and chemical industries. They can be applied in a variety of matrices, including aqueous and nonaqueous mediums, viscous samples, slurries, emulsions and oils, and paints and inks. Overall, pH meters and electrodes are very diverse and flexible in their applications, bringing simplicity, ruggedness, and selectivity to analytical measurements. With continuing research on the horizon, the number and types of applications will keep growing.

VI. CURRENT RESEARCH ACTIVITY

There are some thousands of references in the literature on electrodes and potentiometry, as it is a very diverse and fruitful area of research. Although there are many directions in which the research diverges, there are several common trends that emerge. Here is an attempt to summarize these trends, although it is by no means a critical review, and only a very limited amount of papers are cited as examples in each area.

A. Increasing Sensitivity, Lowering Detection Limits

It has now been shown that even under the zero-current conditions of potentiometry, there are ion fluxes and concentration polarizations across ion-selective membranes (Gyurcsanyi et al., 2001). It is these ion fluxes that can impair analytical measurements at trace levels. By optimizing the concentration of the ionophore in the membrane and inner electrolyte compositions, these ion fluxes can be decreased, thus vastly improving detection limits. Bakker and Pretsch have complied a review on this topic, detailing how these ion fluxes can be minimized in various ways (Bakker and Pretsch, 2001). By lowering these concentration polarizations, it is possible to achieve detection limits as low as 10^{-7}–10^{-12} M for various ions (Peper et al., 2001). This concept has also been recently applied to potentiometric titrations, achieving the same kind of sensitivities (Peper et al., 2001).

In other cases, very high sensitivities have been achieved by applying small currents across the ion-selective membrane (Michalska et al., 2003; Pergel et al., 2001). Detection limits in the range of 10^{-12} M have been reported using this method.

Another area that shows promise is the use of rotating electrodes for potentiometric measurements. Here, the detection limits are shifted to lower concentrations with increasing rotation speed, typically one order of magnitude lower than that achieved using a stir bar (Ye and Meyerhoff, 2001).

It is likely that research in lowering the detection limits of potentiometric measurements will continue. Any advances in this area would be of great consequence in many fields, allowing trace measurements in environmental, clinical, and biological samples, to name a few.

B. New Membrane Materials

Ion-selective membranes are usually made of plasticized poly(vinyl chloride). However, this material has several distinct disadvantages such as poor chemical resistance to extreme pH conditions, poor heat resistance, and poor biocompatibility. As a result, a search for alternative materials has ensued. Various materials have been used, proving to be fully functional as ion-selective membranes, such as methyl–acrylic copolymers (Heng and Hall, 2000; Malinowska et al., 2000; Qin et al., 2003) which require no plasticizers and allow the covalent immobilization of many ionophores.

Others have manufactured ion-selective membranes based on sol–gel modification (Kimura et al., 2001). Modifying glass electrodes with sol–gel allows the design of glass-based sensors with covalently attached ionophores, capable of determining anions as well as cations. Because glass electrodes are not toxic, these types of membranes will allow measurements of various ions in biological systems.

Creating alternate membrane materials will aid in continuing research. New materials make it possible to increase selectivity and ruggedness while improving basic characteristics (i.e., biocompatibility).

C. Biosensors

Active research in the development of biosensors is prevalent in the current literature. A biosensor is an analytical device, that is, an electrode that incorporates biological materials, such as enzymes, microbes, antibodies, receptors, plant or animal cells, organelles, or tissues, for highly selective molecular recognition (Nakamura and Karube, 2003). Biosensors include enzyme-modified electrodes, immunosensors, and, more recently, DNA sensors (Bakker and Telting-Diaz, 2002). Owing to their high specificity, fast response times, low cost, portability, ease of use, and continuous real-time signal, biosensors are in constant use. They have a wide variety of applications, including but not limited to monitoring wastewater quality, acid rain sensing, monitoring toxins such as cyanide, pesticide sensing, meat quality sensing, blood glucose monitoring, cell process monitoring, and bacteria and virus sensing. Advances in biosensors will have a tremendous impact on many fields of study, including clinical and diagnostic biomedicine, farm, garden and veterinary analysis, microbiology, pharmaceuticals and drug analysis, environmental analysis, fermentation control and analysis, and food and drink production and analysis. Nakamura and Karube (2003) have compiled a good review on current research activity and future trends in biosensors.

D. Miniaturization

Microelectrodes have been around since the 1950s (Hartree, 1952; Joels and Macnaughton, 1957; Thompson and Brudevold, 1954). Today, miniaturized electrodes are routinely applied in physiological studies (Nagels and Poels, 2000). However, it was not until recently that nanoscale fabrication techniques have become significant for miniaturization of biosensing devices and chips (Nakamura and Karube, 2003). The development of microelectrodes on microfluidic chips and chip arrays has advanced rapidly in recent years, offering the advantages of increased speed of analysis, a decreased cost of mass production, and reduction of required sample and solvent amount. Chips have been developed for many areas, including capillary electrophoresis, proteomics, DNA sequencing, and chemiluminescence and photochemical reaction detection (Nakamura and Karube, 2003). A μTAS (total analysis system) including a flow cell having an enzyme reactor, a mixing cell, and a spiral capillary on the flow cell has also been constructed by micromachining techniques (Nakamura and Karube, 2003). In the near future, nanotechnology will provide nanobiosensing devices as part of new total analysis systems on chips or arrays, allowing ultrasensitive devices for single molecule detection (Nakamura and Karube, 2003).

E. Potentiometric Detectors in Fluidic Streams

The first potentiometric detectors used for liquid chromatography were developed in the late 1980s by Manz and Simon (1987) and Trojanowicz and Meyerhoff (1989). Since then, the trend of integrating electrochemical sensors into fluidic systems has continued. Solid-state electrodes have been used in liquid chromatography to detect a variety of analytes, including carboxylic acids and organic acids (Nagels and Poels, 2000) and amino acids and glucose (Bakker and Telting-Diaz, 2002). Since then, ISEs have been used as detectors in microchips (Tantra and Manz, 2000) with both capillary electrophoresis (Nagels and Poels, 2000; Tanyanyiwa et al., 2002) and micrototal analysis systems (Nagels and Poels, 2000). Although potentiometric detection is not firmly established (Tanyanyiwa et al., 2002), there is enormous opportunity to develop sensitive and selective detectors on the basis of the fact that there are so many ways that potential can develop (Nagels and Poels, 2000). Because of its simplicity, ruggedness, and ease of miniaturization, it is likely

VII. CONCLUSIONS

Potentiometry is a very simple yet robust electrochemical technique used in a number of analytical measurements. Since its inception, it has changed and broadened to encompass a vast array of applications for many fields. Current research allows for increased sensitivity and selectivity for potentiometric measurements, continually introducing new and modified instruments to meet the needs of today's research scientist. Potentiometric detectors bring the method's simplicity, selectivity, and ruggedness to chromatographic separations. Miniaturization allows for incorporation of ISEs on microchips, and the near future will allow for single molecule detection on a chip. Overall, potentiometry is an interdisciplinary field, and advancements here will benefit all researchers alike.

REFERENCES

Bailey, P. L. (1980a). Gas sensing probes. In: *Analysis with Ion-selective Electrodes*. 2nd ed. Philadelphia: Heyden & Son Ltd., p. 158.

Bailey, P. L. (1980b). Miscellaneous sensors. In: *Analysis with Ion-selective Electrodes*. 2nd ed. Philadelphia: Heyden & Son Ltd., p. 188.

Bakker, E., Pretsch, E. (2001). Potentiometry at trace levels. *Trends Anal. Chem.* 20(1):11–19.

Bakker, E., Telting-Diaz, M. (2002). Electrochemical sensors. *Anal. Chem.* 74(12):2781–2800.

Bates, R. G. (1973a). Liquid junction potentials and ionic activities. In: *Determination of pH, Theory and Practice*. 2nd ed. New York: John Wiley & Sons, p. 38.

Bates, R. G. (1973b). Cells, electrodes, and techniques. In: *Determination of pH, Theory and Practice*. 2nd ed. New York: John Wiley & Sons, pp. 327–335.

Bates, R. G. (1973c). Glass electrodes. In: *Determination of pH, Theory and Practice*. 2nd ed. New York: John Wiley & Sons, p. 365.

Bates, R. G. (1973d). pH Standards. In: *Determination of pH, Theory and Practice*. 2nd ed. New York: John Wiley & Sons, p. 73.

Bates, R. G. (1973e). Measurement of electromotive force. The pH meter. In: *Determination of pH, Theory and Practice*. 2nd ed. New York: John Wiley & Sons, p. 422.

Britton, H. T. S., Robinson, R. A. (1931). Analytical chemistry. *J. Chem. Soc.* 458:185.

Dean, J. A. (1999a). *Lange's Handbook of Chemistry*. 15th ed. New York: McGraw-Hill, Inc., pp. 8.137–8.139.

Dean, J. A. (1999b). *Lange's Handbook of Chemistry*. 15th ed. New York: McGraw-Hill, Inc., p. 8.115.

Fisher, J. E. (1984). Measurement of pH. *Am. Lab.* (6):54–60.

Gyurcsanyi, R. E., Pergel, E., Nagy, R., Kapui, I., Lan, B. T. T., Toth, K., Bitter, I., Linder, E. (2001). Direct evidence of ionic fluxes across ion-selective membranes: a scanning electrochemical microscopic and potentiometric study. *Anal. Chem.* 73(9):2104–2111.

Hampel, C. A., ed. (1964). Electrode potentials, temperature coefficients. In: *Encyclopedia of Electrochemistry*. New York: Reinhold Publishing Corporation, pp. 432–434.

Hartree, E. F. (1952). A micro glass electrode. *Biochem. J.* 52(4):619–621.

Heng, L. H., Hall, E. A. H. (2000). Producing "self-plasticizing" ion-selective membranes. *Anal. Chem.* 72(1):42–51.

Hibbert, D. B., James, A. M. (1984). Ion-selective electrode. In: *Dictionary of Electrochemistry*. 2nd ed. New York: John Wiley & Sons, Inc., pp. 172–173.

Joels, N., Macnaughton, J. I. (1957). A micro-pH electrode system suitable for routine laboratory use. *J. Physiol.* 135(1):1–2P.

Kielland, J. (1937). Individual activity coefficients of ions in aqueous solutions. *J. Am. Chem. Soc.* 59:1675–1678.

Kimura, K., Yajima, S., Takase, H., Yokoyama, M., Sakurai, Y. (2001). Sol–gel modification of pH electrode glass membranes for sensing anions and metal ions. *Anal. Chem.* 73(7):1605–1609.

LaCourse, W. R. (1997). Electrochemical fundamentals. In: *Pulsed Electrochemical Detection in High-Performance Liquid Chromatography*. Techniques in Analytical Chemistry. New York: John-Wiley & Sons, pp. 28–30.

Light, T. S., Licht, S. L. (1987). Conductivity and resistivity of water from the melting to critical points. *Anal. Chem.* 59:2327–2330.

Malinowska, E., Gawart, L., Parzuchowski, P., Rokicki, G., Brzozka, Z. (2000). Novel approach of immobilization of calix[4]arene type ionophore in 'self-plasticized' polymeric membrane. *Anal. Chim. Acta.* 421(1):93–101.

Manz, A., Simon, W. (1987). Potentiometric detector for fast high-performance open-tubular column liquid chromatography. *Anal. Chem.* 59(1):75–79.

Michalska, A., Dumanska, J., Maksymiuk, K. (2003). Lowering the detection limit of ion-selective plastic membrane electrodes with conducting polymer solid contact and conducting polymer potentiometric sensors. *Anal. Chem.* 75(19):4964–4974.

Nagels, L. J., Poels, I. (2000). Solid state potentiometric detection systems for LC, CE and μTAS methods. *Trends Anal. Chem.* 19(7):410–417.

Nakamura, H., Karube, I. (2003). Current research activity in biosensors. *Anal. Bioanal. Chem.* 377(3):446–468.

Peper, S., Ceresa, A., Bakker, E., Pretsch, E. (2001). Improved detection limits and sensitivities of potentiometric titrations. *Anal. Chem.* 73(15):3768–3775.

Pergel, E., Gyurcsanyi, R. E., Toth, K., Lindner, E. (2001). Picomolar detection limits with current-polarized Pb^{2+} ion-selective membranes. *Anal. Chem.* 73(17):4249–4253.

Peters, D. G., Hayes, J. M., Hieftje, G. M. (1974). Direct potentiometry and potentiometric titrations. In: *Chemical Separations and Measurements: Theory and Practice of Analytical Chemistry*. Philadelphia: W.B. Saunders Company, pp. 360–361.

Qin, Y., Peper, S., Radu, A., Ceresa, A., Bakker, E. (2003). Plasticizer-free polymer containing a covalently immobilized Ca2+-selective ionophore for potentiometric and optical sensors. *Anal. Chem.* 75(13):3038–3045.

Serjeant, E. P. (1984a). Procedures of analytical potentiometry. In: Elving, P. J., Winefordner, J. D., Kolthoff, I. M., eds. *Potentiometry and Potentiometric Titrations*. Chemical Analysis. Vol. 69. New York: John Wiley & Sons, p. 247.

Serjeant, E. P. (1984b). In: Elving, P. J., Winefordner, J. D., Kolthoff, I. M., eds. *Potentiometry and Potentiometric Titrations*. Chemical Analysis. Vol. 69. New York: John Wiley & Sons, pp. 305–667.

Skoog, D. A., Holler, F. J., Nieman, T. A. (1998a). An introduction to electroanalytical chemistry. In: *Principles of Instrumental Analysis*. 5th ed. Philadelphia: Harcourt Brace & Company, p. 571.

Skoog, D. A., Holler, F. J., Nieman, T. A. (1998b). Potentiometry. In: *Principles of Instrumental Analysis*. 5th ed. Philadelphia: Harcourt Brace & Company, p. 591.

Swift, E. H. (1939). In: Latimer, W. M., ed. *A System of Chemical Analysis*. New York: Prentice-Hall, Inc., pp. 540–543.

Tantra, R., Manz, A. (2000). Integrated potentiometric detector for use in chip-based flow cells. *Anal. Chem.* 72(13):2875–2878.

Tanyanyiwa, J., Leuthardt, S., Hauser, P. C. (2002). Conductimetric and potentiometric detection in conventional and microchip capillary electrophoresis. *Electrophoresis* 23:3659–3666.

Thompson, F. C., Brudevold, F. (1954). A micro-antimony electrode designed for intraoral pH measurements in man and small experimental animals. *J. Dent. Res.* 33(6):849–853.

Trojanowicz, M., Meyerhoff, M. E. (1989). Potentiometric pH detection in suppressed ion chromatography. *Anal. Chem.* 61(7):787–789.

Ye, Q., Meyerhoff, M. (2001). Rotating electrode potentiometry: lowering the detection limits of nonequilibrium polyion-sensitive membrane electrodes. *Anal. Chem.* 73(2):332–336.

Zemel, J. N. (1975). Ion-sensitive field effect transistors and related devices. *Anal. Chem.* 47:255A–268A.

17

Voltammetry

MARK P. OLSON, WILLIAM R. LACOURSE
Department of Chemistry and Biochemistry, University of Maryland, Baltimore County, Baltimore, MD, USA

I. INTRODUCTION

Voltammetry is an electroanalytical technique that measures current as a function of potential. The term *voltammetry* encompasses a broad area of electroanalytical chemistry that includes polarography, linear scan voltammetry (LSV), cyclic voltammetry (CV), pulsed voltammetry (PV), and stripping voltammetry. An entire chapter has been devoted to stripping voltammetry, and therefore it will not be discussed here. In general, voltammetry is an analytical tool that can be used for the quantitative analysis of various redox-active compounds and is very useful in the analysis of analytes that lack a chromophore or fluorophore. In recent years, an excellent example of the power of voltammetry has been with the use of fast-scan cyclic voltammetry (FSCV) to study the release and uptake of neurotransmitters in the brain *in vivo* (Adams, 1976; Stamford, 1986) and in real-time. The electrodes used in conjunction with this technique can be made sufficiently small so that they can be inserted directly into a portion of the brain to measure neurotransmitter concentrations. One specific example of the application of this technique has been demonstrated recently and shows that dopamine release in the ventral tegmental area of the brain coincides with the initiation of drug-seeking behaviors in rats (Phillips et al., 2003). Findings such as these may eventually provide researchers with some answers on how to treat drug and alcohol addiction and may also help in evaluating general neuronal processes. Overall, voltammetry is very versatile, and can be used for the analysis of many redox-active species.

Voltammetric currents are generated by the oxidation or reduction of an analyte at an electrode surface by applying a potential between two electrodes. The current generated in voltammetric measurements is analogous to the ultraviolet (UV) absorbance observed when obtaining a simple UV scan. With voltammetry, potentials are scanned to give a current response and with UV, wavelengths are scanned to give an absorbance reponse. Whereas UV measurements require that an analyte absorbs a photon at a particular wavelength, voltammetry requires that a sufficient potential is applied to the electrode to force a nonspontaneous redox reaction to occur. The process of forcing nonspontaneous reactions to occur is known as *electrolysis*, and allows direct control of the reaction under study. That is, electrochemical reactions can be accelerated, stopped, or reversed simply by varying electrochemical potential.

II. GENERAL INSTRUMENTATION

The electrochemical cell used in voltammetric measurements usually consists of three electrodes. A typical electrochemical cell is shown in Fig. 1. The electrode at

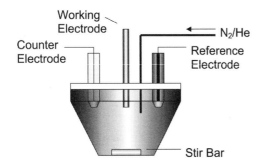

Figure 1 A typical electrochemical cell used in voltammetry consists of a working electrode, a counter (auxiliary) electrode, and a reference electrode. Hydrodynamic conditions require either a stir bar or an RDE. The cell may be sparged with an inert gas to remove dissolved O_2.

which an analyte is oxidized or reduced is known as the *working electrode*. The remaining two electrodes are the *counter* and *reference* electrodes. The counterelectrode serves to create a closed circuit. The reference electrode, whose potential remains constant throughout any given experiment, is necessary so that the potential applied at the working electrode is known. That is, the potential applied at the working electrode is measured against the standard potential of the reference electrode as absolute potentials cannot be measured. Because reference potentials vary from one reference electrode to another, it is important to report the applied potential in any given experiment as "vs." the type of reference electrode used. Owing to their ruggedness, the most widely used reference electrodes are the Ag/AgCl and saturated calomel electrode (SCE).

The electrochemical cell is connected to a potentiostat, which controls the potential applied at the working electrode. A generic illustration of a potentiostat is given in Fig. 2. It consists of a waveform generator (waveforms will be discussed later), a potentiostatic control circuit, a current-to-voltage (*i*-to-*E*) converter, and an output device. Modern instrumentation can be complicated and

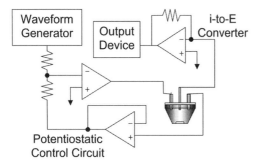

Figure 2 A potentiostat used in voltammetry consists of a waveform generator, a potentiostatic control circuit, a current-to-voltage (*i*-to-*E*) converter, and an output device.

the electronics needed for voltammetric measurements may differ depending on the application. Here, Fig. 2 is shown only to illustrate the basic requirements for voltammetric measurements.

Although it is not required, an electrochemical cell may also be fitted with a *sparging* line and/or stirring apparatus. *Sparging* is a process where an inert gas such as N_2 or He is bubbled into the solution in the electrochemical cell to remove dissolved O_2. This is done because dissolved O_2 can be reduced and sometimes interferes with analyte response. The solution may also be stirred to study the response of analyte under forced convection. This is discussed in Section XII. Hydrodynamic conditions may be used to simulate the response of an analyte in flowing systems. Two common methods may be used to generate hydrodynamic conditions. The first is simply to stir externally by adding a magnetic stir bar to the solution and stir at a fixed rate. The second is to use a rotating disk electrode (RDE). The use of an RDE is advantageous over external stirring as it allows more precise control over hydrodynamics. Hydrodynamic voltammetry will be discussed in greater detail in a later section.

III. OXIDATION/REDUCTION

Whether oxidation or reduction of analyte occurs depends on the potential applied between the working electrode and counterelectrode. Consider a reaction:

$$a\text{A} + n\text{e}^- \longleftrightarrow b\text{B} \tag{1}$$

where A is a reactant, B is a product from the reduction of A, and a and b are their respective reaction coefficients. The reaction quotient, Q, is then defined as:

$$Q = \frac{[\text{B}]^b}{[\text{A}]^a} \tag{2}$$

and the Nernst equation is given by:

$$E_\text{w} = E_\text{A}^0 - \frac{0.0592}{n \log Q} - E_\text{ref} \tag{3}$$

where E_w is the potential applied between the working electrode and the counterelectrode, E_A^0 is the standard potential for the half-reaction [Eq. (1)], and E_ref is the standard potential of the reference electrode. When the system is at equilibrium, no reaction occurs and Q is equal to unity. This is the equilibrium potential (E_eq) where:

$$E_\text{eq} = E_\text{A}^0 - E_\text{ref} \tag{4}$$

Here, there are no redox reactions taking place. Hence, there is no current flow resulting from redox processes. When potentials positive of the equilibrium potential are applied to the working electrode, analyte oxidation occurs (ignoring polarization effects and iR drop). When analyte oxidation occurs, the working electrode is considered to be the *anode* of the electrochemical cell and an *anodic* current response is therefore observed. When potentials negative of the equilibrium potential are applied to the working electrode, analyte reduction occurs (ignoring polarization effects and iR drop). When analyte reduction occurs, the working electrode is considered to be the *cathode* of the electrochemical cell and a *cathodic* current response is therefore observed. From Eq. (2), the magnitude of Q is indicative of the quantity of analyte oxidized or reduced and is therefore directly related to the current passed in the cell. Unfortunately, the oxidation or reduction of analytes at the solid/liquid interface is not so straightforward and this explanation has been purposely oversimplified. There are many other factors to consider when discussing the relationship between current and potential. The effects of the solution on the applied potential should be considered.

IV. POLARIZATION AND iR DROP

An electrode is *polarized* when the potential that is required to maintain a given current is different from the value predicted by the Nernst equation. The degree to which the electrode is polarized is known as the *overpotential* (η) of the electrode. Overpotentials are facilitated through three different types of electrode polarization, including concentration polarization (η_{conc}), kinetic polarization ($\eta_{kinetic}$), and charge-transfer polarization (η_e). Overpotentials require more energy to be added to the system in order to pass an equal amount of current (designated as i_1) predicted by the Nernst equation. This means that the effects of electrode polarization (η_{total}) must be compensated for and are added to Eq. (3) to give a new potential (E_η) that passes current i_1 such that:

$$E_\eta = E_w + \eta_{total} \tag{5}$$

where:

$$\eta_{total} = \eta_{conc} + \eta_{kinetic} + \eta_e \tag{6}$$

Polarization is not the only parameter that alters the potential needed to pass the current i_1. Another factor is the drop in potential that results from the electrical resistance of the solution in the cell. This is known as *ohmic potential loss* or *iR Drop*. The ohmic potential loss (E_{Ohmic}) obeys Ohm's law such that:

$$E_{Ohmic} = i_1 R_{cell} \tag{7}$$

where R_{cell} is the electrical resistance of the cell. This adds another factor to Eq. (5) and the total applied potential (E_{app}) that is needed to pass the current i_1 is therefore given as:

$$E_{app} = E_w + \eta_{total} + E_{Ohmic} \tag{8}$$

In terms of energy, this shows that more energy (in the form of electrode potential) is needed to overcome the activation energy barrier for reduction when taking η_{total} and E_{Ohmic} into account [Fig. 3(a)]. The effects of iR drop and polarization on the current–potential response [Figure 3(b)] for the reduction of A to B is directly related to the energy diagram. The potential required to pass the current i_1 is shifted to more negative values when the iR drop and overpotential are accounted for. This demonstrates that redox potentials for a given analyte are not absolute and may be shifted depending

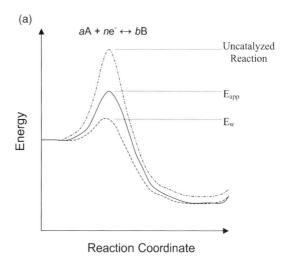

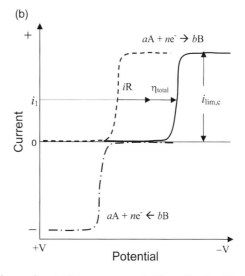

Figure 3 (a) The energy needed for reduction (E_{app}) is higher than that predicted by the Nernst equation (E_w) due to iR drop and polarization effects. This is reflected in the voltammogram (b) where the potential needed for the reduction of A to B is shifted to more negative potentials than predicted by the Nernst equation.

on solution conditions that affect the total energy needed for the redox reaction to occur. This includes variables such as supporting electrolyte concentration (lower concentration = higher resistance), temperature, and solution viscosity.

V. THE VOLTAMMOGRAM

When a linear voltage scan is applied to the system, current passes through the cell as a function of E_{app}. For reaction [Eq. (1)], scanning from positive to negative potentials (and assuming that no B is initially present) results in the reduction of reactant A to product B [Fig. 3(b)] and gives a cathodic current response that is positive by convention. Scanning from negative to positive potentials (assuming that the reduction is reversible) results in the oxidation of B to A and gives an anodic current response that is negative by convention. Anodic and cathodic currents are always opposite in sign as the electron flow in oxidation reactions is opposite of the electron flow in reduction reactions. Each plot is an example of a *voltammetric wave*, and a plot of one or more voltammetric waves is referred to as a *voltammogram*.

The sigmoidal-shaped voltammetric waves of Fig. 3(b) are a result of *mass-transport* limited redox reactions. As E_{app} is scanned to increasingly negative potentials, Q increases in magnitude to balance Eq. (3). As Q increases, the degree of analyte reduction also increases [see Eq. (2)] and the magnitude of the cathodic current therefore increases. This explains the rise in current as E_{app} is made more negative. At extreme negative values, the current plateaus and a limited current ($i_{lim,c}$) is observed. The reason for this is that the current is limited by the rate at which the analyte (A) can be transported to the working electrode surface. Redox reactions take place *at the surface* of the electrode and the current will level off to a value that is determined by the rate of mass-transport if the analyte cannot be transported to the electrode surface at a fast enough rate to keep up with reaction kinetics.

VI. MASS-TRANSPORT

Mass-transport is the process by which an analyte is transported in solution. In electrochemistry, this generally refers to the mechanism of analyte transport to or from an electrode surface. Mass-transport can occur through three separate processes. They are *convection*, *migration*, and *diffusion*.

Natural convection is always present in the electrolytic cell because of thermal gradients or density gradients in the cell and does not usually contribute greatly to mass-transport processes. However, voltammetric experiments in a quiescent solution are usually limited to ca. 5 min because of the effects of natural convection. Forced convection (hydrodynamic conditions) can increase the rate of mass-transport to the electrode surface. Forced convection is usually applied to the study of the redox behavior of an analyte in flowing systems as the rate of mass-transport in hydrodynamic systems is greater than in quiescent systems.

Migration is a result of the influence of an electric field and becomes negligible as the total concentration of electrolytes in solution increases. Efforts are made to minimize the effects of analyte migration by introducing an inactive supporting electrolyte whose concentration is 50–100 times that of the analyte. As analyte migration is under the influence of a potential gradient, the use of large concentrations of supporting electrolyte makes mass-transport independent of E_{app}.

Diffusion contributes to mass-transport processes through concentration gradients and can be explained by LeChatlier's Principle. It states that "If a system at equilibrium is disturbed the system will, if possible, shift to partially counteract the change". Therefore, when reactants are consumed at the electrode surface, their concentration at the surface decreases and reactants from bulk solution are transported to the surface by diffusion to counteract the consumption of reactants.

Mathematically, the effect of mass-transport on the current is given by:

$$i = nF\frac{dC}{dt} \tag{9}$$

where n is the number of electrons (moles of electrons per mole of analyte) involved in the half-reaction, F is the Faraday constant (96,485 C/mol), and dC/dt is the rate of concentration change (mol/s) at the electrode surface. The rate of concentration change per unit area (A, in cm^2) of the electrode is a measure of mass-transport and is defined as the concentration flux (J) where:

$$\frac{dC}{dt} = AJ \tag{10}$$

and J is in units of mol/s/cm^2. The total flux J is a combination of the flux due to diffusion ($J_{diffusion}$), migration ($J_{migration}$), and convection ($J_{convection}$) such that:

$$J = J_{diffusion} + J_{migration} + J_{convection} \tag{11}$$

Combining Eqs. (9) and (10) results in the expression:

$$i = nFAJ \tag{12}$$

so that current is proportional to the product of the total flux and the electrode surface area. Equation (12) demonstrates that current is a quantitative measure of the rate at which the analyte is brought to the surface of the electrode so that the greater is the flux, the larger is the current.

VII. THE DIFFUSION LAYER

In an unstirred cell and with high concentrations of electrolyte, the flux due to convection and migration may (for the most part) be ignored. Under quiescent conditions, only diffusion will be considered. As stated earlier, the process of diffusion generates a concentration gradient. When a potential is applied that is sufficient to oxidize or reduce an analyte at the electrode surface such that the redox reaction is fast in comparison with mass-transfer, the analyte is quickly consumed at the electrode surface so that its concentration becomes very small with respect to its value in bulk solution. This effect is known as *concentration polarization* and results in a concentration gradient where the amount of analyte at the electrode surface is much lower than in bulk solution. The concentration gradient extends from the electrode surface into bulk solution and is linear with distance from the electrode surface (X_0) to the diffusion layer boundary (X_δ), at which it becomes nonlinear. This is illustrated in Fig. 4, where the concentration (mol/cm^3) of analyte A at the electrode surface has been defined as C_A^0, the concentration of analyte A at the diffusion layer boundary has been defined as C_A^δ, and the linear portion of the concentration gradient from the electrode surface to the diffusion layer boundary has been defined as δ. The region δ is known as the *Nernst diffusion layer* and is defined as:

$$\delta = X_\delta - X_0 \tag{13}$$

for planar electrodes where X is distance (cm) from the electrode surface. As the concentration gradient is defined as dC/dx, the concentration gradient over the diffusion layer is given by the slope of the straight line of Fig. 4 such that:

$$\frac{dC}{dx} = \frac{C_A^\delta - C_A^0}{X_\delta - X_0} = \frac{C_A^\delta}{\delta} \tag{14}$$

The flux from diffusion can be related to the diffusion coefficient (D) by:

$$J_{\text{diffusion}} = -D\frac{dC}{dx} \tag{15}$$

where D is given units of cm^2/s. Combining Eqs. (12), (14), and (15) (assuming no flux from convection or migration) gives the limiting current (i_{lim}) that is a function of the diffusion layer thickness and the diffusion coefficient where:

$$i_{\text{lim}} = nFAD\frac{C_A^\delta}{\delta} \tag{16}$$

This shows that the *diffusion-limited current* passed in the cell increases linearly with increasing concentrations of analyte at the diffusion layer boundary (assuming all other parameters remain constant). As C_A^δ is essentially equal to the concentration of analyte in bulk solution (C_A^b), i_{lim} increases linearly with analyte concentration.

Equation (16) also shows that the diffusion-limited current is inversely proportional to the thickness of the diffusion layer. Under quiescent conditions, the diffusion layer grows over time so that the value of i_{lim} decays according to the *Cottrell equation*:

$$i_{\text{lim}}(t) = nFAD^{1/2}\frac{C_A^\delta}{(\pi t)^{1/2}} \tag{17}$$

where t is the time (in s). This means that if an oxidizing or reducing potential is applied (and not varied) in unstirred solutions, the current will eventually drop to zero. Therefore, the practice of obtaining limiting currents under quiescent conditions (at planar electrodes) is not practical. In contrast, the thickness of the diffusion layer remains constant over time when applying forced convection. This is the basis for electroanalytical techniques such as DC amperometry, where a single potential is maintained throughout any given experiment for analyte oxidation or reduction. DC amperometric determinations are easily coupled to separations by high performance liquid chromatography (HPLC) as the flow of solution aids in maintaining *Faradaic* currents at a single potential.

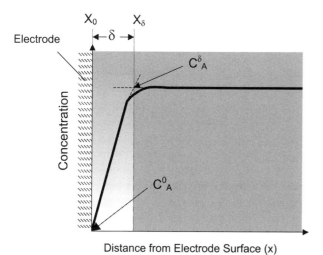

Figure 4 Concentration polarization results from the oxidation or reduction of analyte at the electrode surface, where analyte concentration (C_A^0) is lowest at the electrode surface (x_0) and increases linearly with distance from the electrode surface. The distance from x_0 to the point where the concentration gradient ceases to be linear (x_δ) is known as the Nernst diffusion layer and is given as δ. The concentration of analyte at x_δ (C_A^δ) is essentially equal to the concentration of analyte in bulk solution.

VIII. FARADAIC CURRENT

Faradaic currents are the currents that result from the direct transfer of electrons via redox reactions at the heterogeneous electrode–solution interface. The charge-transfer overpotential (η_e), mentioned earlier, is the potential that promotes a Faradaic response. The Faradaic response of analytes in unstirred cells is not constant and decays over time according to the Cottrell equation. When an E_{app} is applied such that the charge-transfer overpotential is sufficient to oxidize or reduce an analyte, its concentration at the electrode surface decreases over time as the reaction consumes the analyte. As the analyte is consumed over time, Faradaic current decreases as a result of concentration polarization (Fig. 5). Mass-transport then brings more analyte to the electrode surface as the analyte is consumed. In unstirred solutions, the Faradaic current eventually decays to zero. If the solution is stirred, the current will reach a steady-state that is proportional to the analyte concentration in bulk solution and the thickness of the diffusion layer. The steady-state current [i_{lim} from Eq. (16)] is dependent on the bulk analyte concentration.

The overall (observed) current response in voltammetric techniques is not simply a function of analyte oxidation or reduction. *Non-Faradaic* currents are the currents that result from all other processes where direct transfer of electrons to or from the electrode surface does not occur.

IX. NON-FARADAIC CURRENT

Non-Faradaic currents can contribute significantly to the overall (observed) current response in voltammetric systems. In voltammetry, one of the major contributors of non-Faradaic current is known as *charging current*.

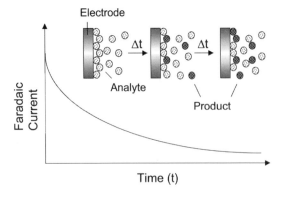

Figure 5 When an electrode is immersed in solution and a potential is applied, Faradaic current decreases over time (Δt) as analyte is consumed (oxidized or reduced) at the electrode surface.

When the working electrode is immersed in a solution and a potential is applied, ions in solution that are adjacent to the electrode surface will rearrange themselves to compensate for the charge developed at the electrode surface. This process results in a brief surge of current that is not derived from any redox processes. The charging current decays over time as a result of the charge at the electrode surface being fully compensated for. Once the charge is fully compensated for, it is considered to be polarized. This is known as *charge polarization*. Charge polarization is very similar to the charging of a capacitor in electronic circuits. Working electrodes in solution therefore act as capacitors and have a natural capacitance associated with them.

Working electrodes act as capacitors in two ways. First, when a potential is applied between the working and counterelectrodes, there is an initial surge in current that decays to zero over time. Second, any instantaneous change in potential will ultimately alter the charge at the electrode surface so that a surge in current is observed. This means that if potentials are scanned linearly with time, charging currents will be observed.

Mathematically, the charge on the electrode (Q) is a function of the electrode area (A), the specific capacitance (capacitance per unit area, C), and potential (E), and is given by:

$$Q = EAC \qquad (18)$$

and the charging current can be expressed as:

$$i_{charge} = \frac{dQ}{dt} \qquad (19)$$

Combining Eqs. (18) and (19) gives:

$$i_{charge} = \frac{d(EAC)}{dt} = AC\frac{dE}{dt} + EC\frac{dA}{dt} + EA\frac{dC}{dt} \qquad (20)$$

so that the charging current is a function of the change in potential, electrode area, and specific capacitance with time. The largest contribution to charging currents is the change in potential (E) with time for most voltammetric techniques. For polarographic determinations, the change in electrode area over time is also significant as the dropping Hg electrodes used in polarography become larger over time (until they are released) so that their surface area is always changing (this excludes stationary Hg drop electrodes). Changes in capacitance may be observed over time and also contribute to charging current in voltammetric measurements, but the most significant portion of charging current is usually a result of changes in potential.

Unfortunately, charging currents result in high background noise in voltammetric analyses. This ultimately compromises signal-to-noise ratios and thus limits of detection for low analyte concentrations. Fortunately,

Voltammetry

charging currents decay to zero so that unnecessary noise can be avoided in some applications by delaying the time at which current is sampled. In order to maximize signal-to-noise, linear time–potential scans have been supplemented by more complex *waveforms* where time and potential are manipulated to maximize the ratio of Faradaic to non-Faradaic current. This is typically done by sampling current at a specific time interval where the Faradaic response is large and the non-Faradaic charging current is small.

X. VOLTAGE/TIME/CURRENT INTERDEPENDENCE

When scanning linearly across a series of potentials, the observed current is a function of potential and time. The rate at which the potentials are scanned is known as the *scan rate* (v), usually expressed as mV/s. It is very important to specify the scan rate since the current output can change drastically with changes in the scan rate. For example, consider the case for reaction [Eq. (1)] where reaction kinetics are very slow when compared with the scan rate. Oxidation/reduction would occur only to a small degree as the reaction kinetics cannot keep up with the scan rate, and the voltammogram would display only a small current output (assuming no *non-Faradaic* currents). If the scan rate is slowed so that the reaction kinetics are fast when compared with the scan rate, a larger Faradaic current is passed. One of the implications of this is that variations in scan rate allow the study of kinetic parameters for oxidation or reduction. This is usually done using *cyclic voltammetry* (CV), where potentials are scanned between two limits, V_{anodic} and $V_{cathodic}$, at a specific scan rate. Either single cycling or continuous cycling between these two limits, known as the *switching potentials*, can be done.

XI. CYCLIC VOLTAMMETRY

CV is the most fundamental voltammetric technique and is important in the initial stages of developing and optimizing electroanalytical techniques (Adams, 1969; Bard and Faulkner, 1980). CV is also instrumental in deducing reaction kinetics. In studying the mechanism of electrode reactions, the use of stationary electrodes with a cyclic potential scan makes it possible to investigate the products of the electrode reaction and detect electroactive intermediates (Galus and Adams, 1963; Galus et al., 1963). Furthermore, the time scale for the method can be varied over an extremely wide range, and both relatively slow and fairly rapid reactions can thus be studied with a single technique (Nicholson and Shain, 1964).

CV measures Faradaic current as a function of time and thus the scan rate (v) by (Nicholson and Shain, 1964):

$$i = kn^{2/3}AD^{1/2}Cv^{1/2} \qquad (21)$$

where k is a constant, v is the scan rate, C is the concentration of analyte in bulk solution, and all other terms have been defined earlier. Consider the general case:

$$R + ne^- \longleftrightarrow [R] \longrightarrow P \qquad (22)$$

where reactant R is electrolytically reduced to an intermediate [R] and Q is rapidly converted to P in solution. It will be assumed that the conversion of [R] to product P is irreversible and that P is electroinactive (it cannot be oxidized or reduced at the electrode surface). If a fast forward scan rate (from negative to positive potentials) is applied to the system, a cathodic wave (i_c) will be observed for the reduction of R (Fig. 6). This is also true of a slow scan rate. The magnitude of the Faradaic response will increase with the square root of the scan rate according to Eq. (21) so that a larger response is observed for the fast scan rate than with the slow scan rate. The reason for this is that, according to the Cottrell equation, the current decreases over time as the thickness of the diffusion layer increases.

When applying the reverse scan, an anodic wave will be observed with the fast scan rate but not with the slow scan rate (after subtracting out non-Faradaic currents). With the fast scan rate, the intermediate [R] does not have enough time to be completely converted to P and is oxidized back to R on the reverse scan. With the slow scan, a

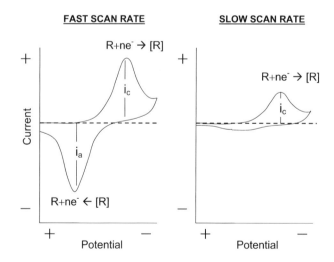

Figure 6 (a) The reduction of R to the intermediate [R] is observed on the forward scan and the oxidation of [R] to R is observed on the reverse scan for fast CV scan rates. (b) The reduction of R to [R] is observed on the forward scan, but the oxidation of [R] to R is not observed on the reverse scan since the intermediate [R] has been almost entirely converted to P by the time the reverse scan is applied.

sufficient amount of time has passed so that [R] is converted completely to P and, since the conversion of [R] to P is irreversible, P cannot be oxidized back to R. Voltammetry can therefore be used to determine the rate at which an intermediate [R] is converted to P. It can also be used to deduce whether the system in question is reversible. In reversible systems, as in the case of R ↔ [R], the cathodic (i_c) and anodic (i_a) peak currents are approximately equal in value (not sign) and are separated by a value of $0.0592/n$ V (where n is the number of electrons involved in the half-reaction).

Although reaction (22) represents a rather generic case for the study of reaction kinetics, CV can be applied to the study of the heterogeneous (electrode–analyte) and homogeneous kinetics of various systems. Owing to its relative simplicity, CV has become almost a standard technique in the study of electrode kinetics (Schmitz and Van der Linden, 1982). More recently, FSCV has been used for the determination of heterogeneous rate constants, rates of chemical reaction following electron transfer, and redox potentials of short-lived chemical species (Wipf et al., 1988). Newer applications of FSCV will be discussed along with other innovative voltammetric techniques in a later section. For a more in-depth discussion of CV and the mathematics that couple potential, time, current, and so on, see Nicholson and Shain (1964).

XII. HYDRODYNAMIC VOLTAMMETRY

As discussed earlier, hydrodynamic voltammetry is the application of voltammetry under forced convection. When the electrode is under forced convection, the concentration gradient (δ) remains constant and i_{lim} does not decay over time. To understand this, it seems necessary to develop a picture of the electrode in solution under hydrodynamic conditions (Fig. 7). Just adjacent to the surface of the electrode is the Nernst diffusion layer. Under hydrodynamic conditions, the Nernst diffusion layer remains stagnant owing to the friction between the solution and the electrode surface. Just beyond the Nernst diffusion layer is the *laminar flow region* where the flow of the solution is parallel to the electrode surface. Any solution beyond the laminar flow region is called the *turbulent flow region* as there is no specific direction of flow at this point. The diffusion layer remains narrow and fixed under hydrodynamic conditions since convection maintains the analyte concentration throughout the laminar flow and turbulent flow regions. A constant current can therefore be maintained under hydrodynamic conditions. This is fundamental to electroanalytical techniques such as DC amperometry, where analytes are oxidized or reduced at a constant potential under hydrodynamic conditions.

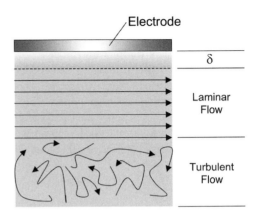

Figure 7 Under hydrodynamic conditions, the solution beyond the electrode surface consists of the Nernst diffusion layer (δ) just adjacent to the electrode surface, the laminar flow region, and the turbulent flow region.

Hydrodynamics aid in analyte detection as the increase in total flux results in larger Faradaic currents. The detection of an analyte following its introduction into a flowing system benefits from the added flux according to Eq. (12). A magnetic stir bar may be used to mechanically agitate the solution. Alternatively, a mechanical rotator equipped with an RDE may be used for more precise control over forced convection.

XIII. POLAROGRAPHY

Invented in 1922 by Jaroslav Heyrovsky (Heyrovsky, 1992), polarography was the first voltammetric technique to be used. Polarography is simply a special case of voltammetry and is distinct in two ways. The first is that polarography makes use of a dropping mercury electrode (DME) that hangs from a glass capillary. The second is that forced convection (stirring) is avoided. Polarography has been used for many years and continues to be used today (Barek and Zima, 2003; Barek et al., 2001; Zima et al., 2001) because of certain advantages.

One of the advantages of polarography is that Hg can tolerate largely negative potentials for the reduction of analytes that cannot be reduced at other electrode surfaces. The negative potential range is limited by the potential at which water is reduced to hydrogen gas:

$$2H_2O + 2e^- \rightarrow H_2 + 2OH^- \quad (23)$$

This happens at largely negative potentials for Hg when compared with other electrode materials such as Au, Pt, or carbon (Fig. 8). This makes it possible to obtain current–potential curves for alkali metal ions, aluminum ion, manganous ion, and other analytes whose current–potential curves are otherwise inaccessible (Lingane, 1958).

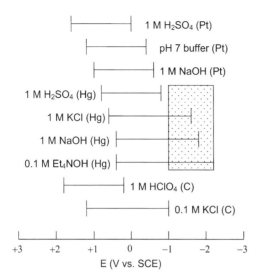

Figure 8 The range of electrochemical potentials for any given voltammetric measurement is limited by the working electrode material. Hg can be used in negative potential regions where materials such as carbon (C) and platinum (Pt) cannot be used. Hg cannot be used in very positive potentials regions. [Adapted from Skoog et al. (1998).]

convection is necessary. It is also a disadvantage because a lack of convection results in a smaller amperometric response [Eq. (12)] when compared with systems where convection occurs. Polarographic limiting currents are therefore one or more orders of magnitude smaller than hydrodynamic limiting currents (Skoog et al., 1998).

A fourth disadvantage of polarography is that a large charging current is observed as each new drop of Hg is formed. This is due to the large increase in surface area as each drop of Hg is formed [see Eq. (20)]. In *DC polarography*, which incorporates a linear potential scan, limits-of-detection are compromised by polarograms (the polarographic equivalent of a voltammogram) that exhibit rising baselines owing to charging current.

An example of a typical polarogram (simulation) is given and shows (a) the Faradaic background, (b) Faradaic plus charging background, (c) a complete polarogram where an analyte has been reduced, and (d) the sampled portion of the polarogram (Fig. 9). The Faradaic response

A second advantage of polarography is that a new electrode surface is generated with each drop of Hg. A mechanical drop-knocker forces the initial drop of Hg to fall and a new one is then formed from the continuous flow of Hg. This happens repeatedly, and a clean electrode surface is maintained over time so that analyte response is not compromised. In comparison, voltammetry at noble metal electrodes requires pulsed potentials to maintain a clean electrode surface (LaCourse, 1997). Likewise, voltammetric measurements at other electrode materials such as glassy carbon typically requires daily mechanical polishing prior to use.

There are disadvantages to polarography as well as advantages. The first disadvantage to polarographic analysis is that Hg is easily oxidized so that the potential range for anodic polarography is limited (Fig. 8). If largely positive potentials are applied to Hg, analyte response will be hindered by a large anodic response from Hg oxidation. Anodic polarography is therefore restricted to analytes that are relatively easily oxidized.

A second disadvantage of polarography is that large quantities of Hg may be necessary for analyses. Hg toxicity becomes a large factor if the technique is to be used in laboratories that require many analyses on a daily or weekly basis.

The third disadvantage of polarography is that forced convection must be avoided. Stirring may result in an ill-defined electrode surface or may even unpredictably dislodge the Hg drop from the glass capillary. This is a disadvantage as it is difficult to use in situations where

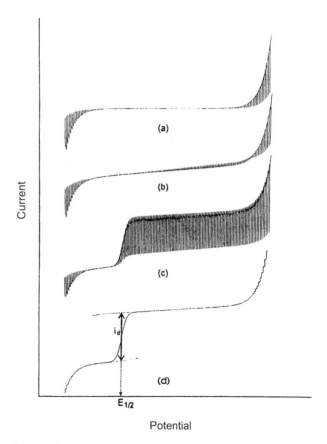

Figure 9 A simulation of DC polarograms shows (a) Faradaic background current; (b) FARAdaic plus charging background current; (c) a complete DC polarogram that includes Faradaic background current, charging background current, and an analyte reduction wave; (d) sampled current from the same polarogram showing the half-wave potential ($E_{1/2}$) and diffusion limited current (i_d).

without analyte is near zero except at the extremes where the oxidation of Hg occurs at extreme positive potentials and the reduction of water to H_2 occurs at extreme negative potentials [Fig. 9(a)]. The charging current results in a baseline that shifts upward slightly with increasingly negative potentials [Fig. 9(b)]. The reduction of analyte then results in a cathodic wave where the observed response is dependent on the frequency of the Hg drop. When a new Hg drop is formed, the total current observed is initially near zero. As the drop increases in size, Faradaic and non-Faradaic currents increase and a current maximum is reached when the Hg drop is near its largest size. When an analyte is reduced at a DME, the overall effect is then to generate a polarogram such as that given by Fig. 9(c). In modern DC polarography, current is sampled at a specified time near the end of the lifetime of each drop of Hg so that the large fluctuations in current are not observed. The net result is a polarogram that is very similar in appearance to a voltammogram [Fig. 9(d)]. This is known as current-sampled (TAST) polarography.

Improvements in polarographic analyses have been made because of the inherent disadvantages of classical DC polarography. One method is to use a stationary mercury drop electrode (SMDE) instead of a DME. Here, the drop of Hg is allowed to grow to a specified size and is then stopped. Once the growth is stopped, current is sampled. This results in much smaller charging currents and improves limits of detection. One problem with this is that the lifetime of a drop is limited to only a few seconds to prevent surface contamination. Reproducibility is also compromised as it is difficult to duplicate the size of the drop. Another method that improves upon classical DC polarography is *pulsed polarography*.

Pulsed polarography takes advantage of the fact that a large potential pulse rapidly converts analyte to product to promote a large Faradaic response. The Faradaic current is proportional to the rate of concentration change of analyte at the electrode surface (dC/dt) according to Eq. (9). Comparably, a linear scan (voltage ramp) does not provide such a dramatic change in potential and the analyte is not consumed as rapidly at the electrode surface. This results in a smaller Faradaic response when compared with pulsed techniques. One of the difficulties in pulsed techniques is that the application of potential pulses results in large charging currents. However, these can be alleviated by allowing a sufficient delay time (t_{del}) before sampling the current so that the charging current can decay to values near zero. In polarography, the Faradaic current is sampled near the end of the lifetime of the Hg drop to further minimize charging current. The overall result of applying pulsed potentials is to increase signal-to-noise and lower limits of detection by maximizing Faradaic currents while minimizing non-Faradaic currents.

Pulsed techniques are important to voltammetry as well as polarography. They are very similar with the exception that pulsed polarography uses a DME and therefore the rate of change in the surface area of Hg is a factor that is not present in pulsed voltammetric methods. The application of pulsed waveforms to polarography and voltammetry will be considered in greater detail.

XIV. WAVEFORMS

Waveforms are simply the variation in potential with time. The simplest waveform is a linear scan (voltage ramp) where potential is scanned linearly with time [Fig. 10(a)]. The waveform used for CV is simply a linear scan that is scanned forward, switched, and then scanned backward [Fig. 10(b)]. The CV waveform may be repeated over many cycles (three cycles are shown here). Pulsed waveforms are more complex and these can be divided primarily into normal pulse voltammetry (NPV), differential pulse voltammetry (DPV), and square wave voltammetry (SWV).

NPV [Fig. 10(c)] consists of a series of pulses of increasing amplitude, with the potential returning to the initial value after each pulse. The current is measured at some sampling interval (S) near the end of each pulse maximum. The Faradaic response is maximized by the potential pulse (when compared with linear scans) and a significant amount of charging current is avoided in the measurement since the current is sampled near the end of the pulse.

DPV [Fig. 10(d)] differs from NPV in that it uses a waveform that consists of small pulses (of constant amplitude) superimposed on a staircase wave form. Like NPV, the Faradaic response is maximized and charging current is minimized by sampling the current (S_2) near the end of each pulse maximum. Again, this is very advantageous over linear-scan techniques since signal-to-noise is maximized. DPV differs from NPV in that current is also sampled at a second interval (S_1). By subtracting S_1 from S_2, a voltammogram is obtained that is a measure of the *change in current* with potential. This will appear very different from the shape of the NPV voltammogram. If the NPV voltammogram gives a classical, sigmoidal response (i vs. E) for the oxidation or reduction of analyte, the equivalent response using DPV will be the derivative of the NPV response (Δi vs. E) and will result in a peak instead of a wave. This can be attributed to the fact that Δi becomes small and approaches zero as the limiting current is approached. This differential output allows the observation of individual peak maxima that differ by as

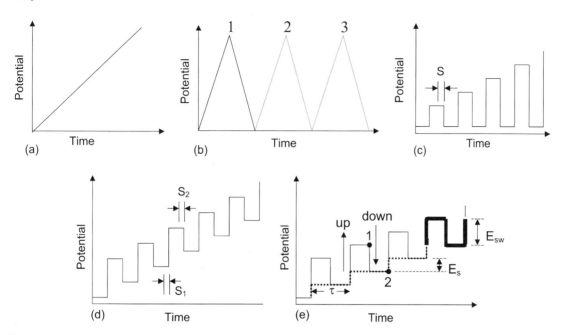

Figure 10 (a) LSV uses the simplest waveform where potential varies linearly with time. (b) CV uses a waveform where potential varies linearly with time up to some switching potential where the potential then varies linearly with time back to its original potential. (c) NPV uses a potential pulse to increase the Faradaic response of analyte. Current is sampled for some time (S) at the end of the pulse. (d) DPV uses a potential pulse to increase Faradaic response where the current measured at S_2 is subtracted from S_1 to give a differential current measurement (Δi vs. E). (e) SWV uses a square wave pulse (pulses of magnitude E_{sw}) incorporated onto a staircase waveform (pulses of magnitude E_s) for a differential current measurement ($i_1 - i_2$). The frequency of one cycle of the waveform is given by τ.

little as 50 mV. In comparison, NPV (as well as classical techniques) requires approximately 200 mV between waves to resolve them.

SWV is similar to DPV in that current is sampled at two different times in the waveform and results in a differential output (Δi vs. E). The SWV waveform [Fig. 10(e)] is essentially the combination of a square wave (bold, solid line) of period τ with a staircase voltage function (dotted lines), also of period τ. The "forward" current is measured at t_1 just before the "down" pulse is applied. The "reverse" current is measured at t_2 (O'Dea et al., 1981). The net current is defined as the current measured at t_1 minus the current measured at t_2, or $\Delta i = i_1 - i_2$. The SWV waveform is advantageous for various reasons.

One advantage of the application of an SWV waveform is that the detrimental effects of charging current are reduced (Barker and Jenkins, 1952; Barker, 1958), allowing very fast potential scans. The charging current developed on the forward scan (from the "up" pulse) is essentially subtracted out when the reverse scan (the "down" pulse) is applied. Typical SWV measurements therefore take only 1–5 s whereas DPV requires much longer analysis times (2–4 min). One example of the speed of SWV (square wave polarography in this case) is that a complete potential scan can be performed on one drop of Hg (using a DME). Also, because of the rapid analysis time, repetitive scanning can be used with SWV to increase signal-to-noise by signal-averaging (Anon, 1980).

A second advantage of SWV is that analyte response can theoretically be increased for reversible systems. For example, if an analyte is reduced on the up pulse and can be reversibly oxidized back to its original state on the down pulse, analyte response is actually being measured twice in one cycle of the waveform. This increases the magnitude of the measured response. Despite this, limits of detection for SWV are similar to those for DPV (10^{-7}–10^{-8} M). The true advantage of SWV is therefore analysis speed.

XV. INNOVATIVE APPLICATIONS OF VOLTAMMETRY

A. Protein Film Voltammetry

The modification of electrodes with enzymes (proteins) to increase analyte specificity in electroanalytical techniques has been done since the 1960s (Guilbault and Montalvo,

1969; Guilbault et al., 1969; Montalvo, 1969). Electrodes that have been modified with enzymes have been used as biosensors and are specific toward certain substrates. However, only recently have enzyme-modified electrodes been used to investigate the role of electrochemical potential in enzyme function. This is known as protein film voltammetry (PFV) (Armstrong et al., 1997). PFV has been applied to the investigation of the mechanisms of enzymes that exhibit redox behavior (Anderson et al., 2001; Armstrong, 2002a, b; Camba and Armstrong, 2000; Hirst et al., 1997; Jeuken et al., 2002; Jones et al., 2000; Leger et al., 2003; Zu et al., 2002). By using protein film voltammetry, it has been determined that certain enzymes display optimum activity at a particular potential. This can be both mechanistically informative and physiologically relevant (Leger et al., 2003). This also adds a level of control over an enzyme's function. Imagine the ability to turn an enzyme on or off simply by varying electrical potential. This could be of great consequence in the fields of chemistry, biology, and medicine as it implies that scientists or engineers can attain a certain level of control over biological systems by simply turning a knob or flipping a switch.

B. Voltammetry on Microfluidic Devices

Microfluidic devices are essentially analytical systems that have been shrunk to a very small scale. As such, they are generally referred to as "Lab-on-a-Chip" devices. These small analytical systems are advantageous in terms of speed, minimal sample/reagent consumption, miniaturization, efficiency, and automation (Hadd et al., 1997; Jakeway et al., 2000). Voltammetric measurements on such a device were shown only very recently as a reality for the first time (Wang et al., 2000), but it is very likely that interest in such devices will continue to grow dramatically owing to their simple design and low costs. The coupling of voltammetry with these devices enhances their power as it adds a new dimension of information (based on redox activity) and opens up highly sensitive detection schemes (Wang, et al., 2000). Others have already begun to apply different types of voltammetry to such systems. Hebert, Kuhr, and Brazill recently used sinusoidal voltammetry (SV) on microfluidic devices to enhance the detection of neurotransmitters (Hebert et al., 2003). In SV, the raw time domain is collected from the electrochemical cell and converted into the frequency domain in order to better decouple the Faradaic signal from the background components, as well as to generate a unique "fingerprint" frequency spectrum to aid in identification and isolation of the chemical species (Hebert et al., 2003). SV was shown to be an order of magnitude more sensitive than DC amperometry (which has not been considered here as a voltammetric technique as it involves current measurement at a constant potential) and limits of detection in the low attomole range were observed for the neurotransmitters dopamine and isoproterenol.

C. Fast-Scan Cyclic Voltammetry

The ability to make fast electrochemical measurements is desirable so that rapid heterogeneous or homogeneous reaction rates can be measured directly (Wipf, et al., 1988). FSCV allows fast measurements and has become almost standard in studying neurotransmitters as the small size of the electrodes (ultramicroelectrodes, <50 μm radius) used in conjunction with FSCV allows the probing of various regions of the brain (in animal studies). Although one may initially think that CV and FSCV are not very different, there are a few variables to consider when applying very fast scan rates to extremely small electrodes. First, the application of very fast scan rates will result in a large charging current that greatly increases background noise. Second, ohmic potential loss (E_{Ohmic}) can distort the measured current because E_{Ohmic} varies as the potential (and thus current) are rapidly changing. From Eq. (8), it is easy to see that this complicates things greatly as the applied voltage (E_{app}) is coupled with E_{Ohmic} and with current.

Charging currents increase with scan rate for all sizes of disk electrodes. This is unfortunate as signal-to-noise can be severely compromised by charging current. The charging current increases linearly with scan rate whereas Faradaic currents increase with the square root of the scan rate. Therefore the ratio of Faradaic current to charging current decreases as scan rates increase. For low concentrations of analyte at very fast scan rates, response may then be overwhelmed by background noise. Luckily, many reactions do not occur so quickly that they require extremely fast scan rates and FSCV can then be used to determine reaction rates of various analytes in solution. But, for applications such as *in vivo* studies where concentrations of analyte are usually small, background noise can be very detrimental. In neurochemical studies, minimized noise is necessary to observe the low concentrations of neurotransmitters that are physiologically important (Michael et al., 1999). Work has therefore continued to move in the direction of increasing signal-to-noise for FSCV (Michael et al., 1999).

Fortunately, the use of ultramicroelectrodes results in smaller values of E_{Ohmic} so that fast scan rates can be employed with less distortion. The reason for this is that with a smaller electrode surface area, both Faradaic [Eq. (16)] and charging [Eq. (20)] currents are greatly reduced. From Ohm's law, E_{Ohmic} is linearly dependent on the total current passed in the cell so that smaller currents result in smaller values for E_{Ohmic} and thus minimal distortion occurs. This allows scan rates of up to ~ 20 kV/s whereas practical CV with electrodes of conventional size is restricted to an upper limit of ~ 100 V/s (Howell and Wightman, 1984).

D. Ultramicroelectrodes

Ultramicroelectrodes are electrodes that are typically less than 50 μm in radius. As current decreases with electrode surface area (A), only very small currents are passed with ultramicroelectrodes. This allows voltammetry in highly resistive solutions (Bond et al., 1984; Ciszkowska and Stojek, 2000; Howell and Wightman, et al., 1984; Hyk and Stojek, 2002). Voltammetry with little or no supporting electrolyte has become an area of interest to electroanalytical chemists. For many years (and even to this date for most studies), the first requirement of any voltammetric experiment was to add excess of supporting electrolyte to decrease solution resistivity and subsequently decrease E_{Ohmic}. With ultramicroelectrodes, this is unnecessary as the small currents passed in the electrochemical cell give small values of E_{Ohmic}.

Another aspect of ultramicroelectrodes is that the cell time constants (τ_C) are very small. The cell time constant (in s) is equivalent to the time constant of a capacitor:

$$\tau_C = RC \tag{24}$$

where R is resistance in Ω and C is capacitance in F. As the surface area of an electrode decreases, its capacitance decreases. Therefore, ultramicroelectrodes exhibit smaller time constants than conventional electrodes. Less time is therefore needed for charging currents to decay to zero. This can be beneficial to applications where long delay times are necessary to reduce background noise from charging currents.

One research area where lower time constants may have great impact is in pulsed electrochemical detection (PED). PED following HPLC has been used successfully for many years primarily for the determination of carbohydrates, amines, amino acids, and sulfur compounds. Long delay times are necessary prior to current sampling in PED to reduce the effects of charging current. This has limited PED to waveform frequencies of near 2 Hz and is not of great consequence in typical HPLC separations. However, in microfluidic systems, separations are typically much faster and the low frequency of waveform cycles can compromise chromatographic integrity. This would make PED on microfluidic devices virtually impossible if conventional electrodes were used. However, with the small time constants observed with ultramicroelectrodes, PED could be used for the determination of various analytes following separations of microfluidic devices. This has been demonstrated very recently, where carbohydrates, amino acids, and antibiotics were separated (less than 2 min per separation) on a microfluidic device that utilized a 25 μm Au electrode for detection (Garcia and Henry, 2003).

E. Sinusoidal Voltammetry

Sinusoidal voltammetry (SV) is a type of voltammetry that utilizes a sinusoidal waveform such that the electrochemical current response may be examined in the frequency domain rather than the time domain. Examination in the frequency domain can lead to sensitivity and selectivity advantages not found in the time domain (Cullison and Kuhr, 1996). For example, when using a sinusoidal waveform, the major component of charging current is a phase-shifted sine wave at the fundamental frequency of the waveform (Bauer, 1964). In contrast, the Faradaic current response from oxidation or reduction may have components in the higher harmonics (Oldham, 1960). These differences in the harmonic frequency composition of the Faradaic and charging current allow the discrimination of Faradaic current from charging current (Singhal et al., 1997a). With this advantage, SV has been used for the sensitive detection of carbohydrates (Singhal et al., 1997a), amino acids (Brazill et al., 2000), nucleotides (Singhal and Kuhr, 1997a), DNA (Singhal and Kuhr, 1997a), and neurotransmitters (Hebert et al., 2003).

SV also allows the discrimination of some analytes that are typically oxidized or reduced at similar potentials. This selectivity stems from the kinetics of the redox process. Recently, sinusoidal voltammetry was used to selectively detect glucose (a monosaccharide) over maltose (a disaccharide) (Singhal et al., 1997a). It was reasoned that the slower kinetics of oxidation of the disaccharide when compared with the monosaccharide allowed its selective detection. The current response at higher harmonics is much larger for a faster reacting species, thus allowing selectivity based on harmonic frequencies. With this selectivity, SV can easily compete with other voltammetric methods such as FSCV and may find more widespread use in the future.

XVI. SUPPLIERS OF ANALYTICAL INSTRUMENTATION AND SOFTWARE

The following is a list of some suppliers for voltammetric instrumentation and software for those who may be interested in obtaining such equipment. It is not meant to be all-inclusive and websites are subject to change.

Bioanalytical Systems (BAS)
2701 Kent Avenue, West Lafayette, IN 47906, USA
www.bioanalytical.com

Princeton Applied Research
Multiple distributors
www.princetonappliedresearch.com

Metrohm
Multiple distributors
www.metrohm.com

Topac
101 Derby Street, Hingham, MA 02043, USA
www.topac.com

Trace Detect
Seattle, WA 98105, USA
www.tracedetect.com

Gamry Instruments
734 Louis Drive, Warminster, PA 18974, USA
www.gamry.com

Cypress Systems
2300 West 31st St., Lawrence, KS 66047, USA
www.cypresssystems.com

Pine Instrument Company
101 Industrial Dr., Grove City, PA 16127, USA
www.pineinst.com

XVII. VOLTAMMETRY SIMULATION SOFTWARE

Because of the various parameters (electrode surface area, diffusion, rates of electrode rotation, temperature, solution resistance, etc.) that affect the outcome of voltammetric experiments, simulation software have been developed to aid in understanding the response of the system when changing one or more of these variables. An excellent tutorial (and free!!) program for CV is named Cyclic VoltSim and can be found at the website listed in the following table. More advanced simulation software includes DigiSim and POLAROGRAPH.com 5.3. These are not free downloads. However, there is a free downloadable version of POLAROGRAPH.com 5.3 that allows restricted access to the program. This list of programs is not meant to be all-inclusive and websites are subject to change.

Company (or website)	Simulation software
BAS	DigiSim
www.electrochemistrysoftware.com	POLAROGRAPH.com 5.3
colossus.chem.umass.edu/bvining/free.htm	Cyclic VoltSim (Free!!)

REFERENCES

Adams, R. N. (1969). *Electrochemistry at Solid Electrodes*. New York, Marcel Dekker, Inc.

Adams, R. N. (1976). Probing brain chemistry with electroanalytical techniques. *Anal. Chem.* 48:1128 A.

Anderson, L. J., Richardson, D. J., Butt, J. N. (2001). Catalytic protein film voltammetry from a respiratory nitrate reductase provides evidence for complex electrochemical modulation of enzyme activity. *Biochemistry* 40(38): 11294.

Anon. (1980). Square wave voltammetry. A challenger to differential pulse polarography emerges. *Anal. Chem.* 52(2): 229A.

Armstrong, F. A. (2002a). Protein film voltammetry: revealing the mechanisms of biological oxidation and reduction. *Russ. J. Electrochem. (Translation of Elektrokhimiya* 38:49.

Armstrong, F. A. (2002b). Insights from protein film voltammetry into mechanisms of complex biological electron-transfer reactions. *Dalton Trans.* 661.

Armstrong, F. A., Heering, H. A., Hirst, J. (1997). Reaction of complex metalloproteins studied by protein-film voltammetry. *Chem. Soc. Rev.* 26:169.

Bard, A. J., Faulkner, L. R. (1980). *Electrochemical Methods: Fundamentals and Applications*. New York, John Wiley & Sons.

Barek, J., Zima, J. (2003). Eighty years of polarography—history and future. *Electroanalysis* 15(5): 467.

Barek, J., Cvacka, J., Muck, A., Quaiserova, V., Zima, J. (2001). Electrochemical methods for monitoring of environmental carcinogens. *Fresenius J. Anal. Chem.* 369(7–8): 556.

Barker, G. C. (1958). Square wave polarography and some related techniques. *Anal. Chim. Acta.* 18:118.

Barker, G. C., Jenkins, I. L. (1952). Square-wave polarography. *Analyst* 77:685.

Bauer, H. H. (1964). *Aust. J. Chem.* 17:715.

Bond, A. M., Fleischmann, M., Robinson, J. (1984). Electrochemistry in organic solvents without supporting electrolyte using platinum microelectrodes. *J. Electroanal. Chem.* 168:299.

Brazill, S. A., Singhal, P., Kuhr, W. G. (2000). Detection of native amino acids and peptides utilizing sinusoidal voltammetry. *Anal. Chem.* 72:5542.

Camba, R., Armstrong, F. A. (2000). Investigations of the oxidative disassembly of Fe–S clusters in *Clostridium pasteurianum* 8Fe ferredoxin using pulsed-protein-film voltammetry. *Biochemistry* 39:10587.

Ciszkowska, M., Stojek, Z. (2000). Voltammetric and amperometric detection without added electrolyte. *Anal. Chem.* 72:755A.

Cullison, J. K., Kuhr, W. G. (1996). *Electroanalysis* 8:314.

Galus, Z., Adams, R. N. (1963). Anodic oxidation of N-methylaniline and N,N-dimethyl-p-toluidine. *J. Phys. Chem.* 67:862.

Galus, Z., Lee, H. Y., Adams, R. N. (1963). Triangular wave cyclic voltammetry. *J. Electroanal. Chem.* 5:17.

Garcia, C. D., Henry, C. S. (2003). Direct determination of carbohydrates, amino acids, and antibiotics by microchip electrophoresis with pulsed amperometric detection. *Anal. Chem.* 75(18): 4778.

Guilbault, G. G., Montalvo, J. G., Jr. (1969). Urea-specific enzyme electrode. *J. Am. Chem. Soc.* 91(8): 2164.

Guilbault, G. G., Smith, R. K., Montalvo, J. G., Jr. (1969). Use of ion selective electrodes in enzymic analysis. Cation electrodes for deaminase enzyme systems. *Anal. Chem.* 41(4): 600.

Hadd, A., Raymond, D., Halliwell, J., Jacobson, S., Ramsey, M. (1997). Microchip device for performing enzyme assays. *Anal. Chem.* 69:3407.

Hebert, N. E., Kuhr, W. G., Brazill, S. A. (2003). A microchip electrophoresis device with integrated electrochemical detection: a direct comparison of constant potential amperometry and sinusoidal voltammetry. *Anal. Chem.* 75(14): 3301.

Heyrovsky, J. (1922). *Chem. Listy* 16:256.

Hirst, J., Ackrell, B. A. C., Armstrong, F. A. (1997). Global observation of hydrogen/deuterium isotope effects on bidirectional catalytic electron transport in an enzyme: direct measurement by protein-film voltammetry. *J. Am. Chem. Soc.* 119(32): 7434.

Howell, J. O., Wightman, R. M. (1984). Ultrafast voltammetry and voltammetry in highly resistive solutions with microvoltammetric electrodes. *Anal. Chem.* 56:524.

Hyk, W., Stojek, Z. (2002). Generalized theory of steady-state voltammetry without a supporting electrolyte. Effect of product and substrate diffusion coefficient diversity. *Anal. Chem.* 74:4805.

Jakeway, S., de Mallo, A. J., Russell, E. L. (2000). Miniaturized total analysis systems for biological analysis. *Fresenius J. Anal. Chem.* 366:525.

Jeuken, L. C., Camba, R., Armstrong, F. A., Canters, G. W. (2002). The pH-dependent redox inactivation of amicyanin from *Paracoccus versutus* as studied by rapid protein-film voltammetry. *J. Biol. Inorg. Chem.* 7:94.

Jones, A. K., Camba, R., Reid, G. A., Chapman, S. K., Armstrong, F. A. (2000). Interruption and time-resolution of catalysis by a flavoenzyme using fast scan protein film voltammetry. *J. Am. Chem. Soc.* 122:6494.

LaCourse, W. R. (1997). *Pulsed Electrochemical Detection in High-Performance Liquid Chromatography*. New York, John Wiley & Sons, Inc.

Leger, C., Elliot, S. J., Hoke, K. R., Jeuken, L. J., Jones, A. K., Armstrong, F. A. (2003). Enzyme electrokinetics: using protein film voltammetry to investigate redox enzymes and their mechanisms. *Biochemistry* 42(29): 8653.

Lingane, J. J. (1958). *Electroanalytical Chemistry*, 2nd ed. New York, Interscience Publishers, Inc.

Michael, D. J., Joseph, J. D., Kilpatrick, M. R., Travis, E. R., Wightman, R. M. (1999). Improving data acquisition for fast-scan cyclic voltammetry. *Anal. Chem.* 71(18): 3941.

Montalvo, J. G., Jr. (1969). Electrode for measuring urease enzyme activity. *Anal. Chem.* 41(14): 2093.

Nicholson, R. S., Shain, I. (1964). Theory of stationary electrode polarography. single scan and cyclic methods applied to reversible, irreversible, and kinetic systems. *Anal. Chem.* 36(4): 706.

O'Dea, J. J., Osteryoung, J., Osteryoung, R. A. (1981). Theory of square wave voltammetry for kinetic systems. *Anal. Chem.* 53(4): 695.

Oldham, K. B. Theory of Faradaic distortion. *J. Electrochem. Soc.* 1960, *107*, 766.

Phillips, P. E. M., Stuber, G. D., Heien, M. L. A. V., Wightman, R. M., Carelli, R. M. (2003). Subsecond dopamine release promotes cocaine seeking. *Nature*, 422:614.

Schmitz, J. E. J., Van der Linden, J. G. M. (1982). Temperature-dependent determination of the standard heterogeneous rate constant with cyclic voltammetry. *Anal. Chem.* 54(11): 1879.

Singhal, P., Kuhr, W. G. (1997a). Direct electrochemical detection of purine- and pyrimidine-based nucleotides with sinusoidal voltammetry. *Anal. Chem.* 69:3552.

Singhal, P., Kuhr, W. G. (1997b). Ultrasensitive voltammetric detection of underivatized oligonucleotides and DNA. *Anal. Chem.* 69:4828.

Singhal, P., Kawagoe, K. T., Christian, C. N., Kuhr, W. G. (1997a). Sinusoidal voltammetry for the analysis of carbohydrates at copper electrodes. *Anal. Chem.* 69(8): 1662.

Skoog, D. A., Holler, J. F., Nieman, T. A. (1998). *Principles of Instrumental Analysis*, 5th ed. Philadelphia, Harcourt Brace & Company.

Stamford, J. A. (1986). Effect of electrocatalytic and nucleophilic reactions on fast voltammetric measurements of dopamine at carbon fiber microelectrodes. *Anal. Chem.* 58(6): 1033.

Wang, J., Polsky, R., Tian, B., Chatrathi, M. P. (2000). Voltammetry on microfluidic chip platforms. *Anal. Chem.* 72(21):5285.

Wipf, D. O., Kristensen, E. W., Deakin, M. R., Wightman, R. M. (1988). Fast-scan cyclic voltammetry as a method to measure rapid heterogeneous electron-transfer kinetics. *Anal. Chem.* 60(4): 306.

Zima, J., Barek, J., Moreira, J. C., Mejstrik, V., Fogg, A. G. (2001). Electrochemical determination of trace amounts of environmentally important dyes. *Fresenius J. Anal. Chem.* 369(7–8): 567.

Zu, Y., Fee, J. A., Hirst, J. (2002). Breaking and re-forming the disulfide bond at the high-potential, respiratory-type rieske [2Fe–2S] center of *Thermus thermophilus*: characterization of the sulfhydryl state by protein-film voltammetry. *Biochemistry* 41(47): 14054.

18

Electrochemical Stripping Analysis

WILLIAM R. LACOURSE
Department of Chemistry and Biochemistry, University of Maryland, Baltimore County, Baltimore, MD, USA

I. INTRODUCTION

The determination of trace levels of an analyte or analytes in complex matrices continues to be an important challenge in modern analytical chemistry. In order to detect target compounds at or below the limit of detection of an instrumental technique, a preconcentration or enrichment step is required prior to quantitation. Aside from enrichment, the preconcentration step may serve to isolate the analyte from the matrix, and, as a consequence, selectivity and stability can be improved.

Often, the solution to the isolation of trace components in complex matrices is accomplished via chromatography. The target analyte(s) is (are) enriched on the head of a miniature column, washed, and eluted for eventual analysis either in a batch (beaker) or flowthrough cell. Preconcentration by chromatography involves (i) additional equipment or instrumentation, (ii) additional reagents and sample handling, and (iii) development of an entire procedure, which may result in a new sample matrix that is incompatible with detection protocols. All of these may lead to increased cost, time, and risk of contamination.

Electrochemical preconcentration represents an efficient alternative to chromatography for the isolation and enrichment of trace components (Sioda et al., 1986; Wang, 1996) in that the preconcentration step and subsequent "stripping" and quantitation protocol all occur at a single working electrode in the same cell using the same equipment. Hence, *rapid, sensitive, economical*, and *simple* trace analysis can be performed. In addition, the risk of contamination is minimized because very little or no reagent addition is needed. The preconcentration step can also be combined with a wide range of instrumental techniques to achieve unique advantages (i.e., selectivity, reduced sensitivity to dissolved oxygen, and overcoming capacitance effects).

This chapter will briefly present the fundamentals of various stripping procedures. Then, the instrumentation requirements, including the selection of the voltammetric analyzer, working electrode, cell and other variables, will be discussed. It is recommended that the readers familiarize themselves with the chapters on conductance, potentiometry, and, especially, voltammetry. The literature of stripping analysis is reviewed briefly to help potential users in understanding and choosing among several variants of the technique. Review articles (Copeland and Skogerboe, 1974; Esteban and Casassas, 1994; Florence, 1984; Wang, 1982a, 1990) and books (Brainina and Neyman, 1993; Vydra et al., 1976; Wang, 1985, 1989) give a comprehensive outline of modern stripping analysis, embracing new strategies, associated problems, and numerous applications.

II. FUNDAMENTALS OF STRIPPING ANALYSIS

Owing to the possibility of preconcentrating the analyte onto the working electrode (by factors of 100 to more the 1000), stripping analysis has the lowest limits of detection among all electroanalytical methods (i.e., 10^{-11} M). Of no less importance are its wide linear dynamic range, its ability to analyze more than one (four to six) analyte at a time, and its relatively low cost. Hence, stripping analysis is ideally suited for:

1. Determination of chemical species at trace levels in complex samples of environmental, clinical, food, pharmaceutical, or industrial origin.
2. Trace metal speciation in natural waters or *in situ* field screening.
3. Online, high-throughput screening applications.

Overall, it is now possible to use stripping analysis for detecting more than 40 trace elements and a plethora of organic compounds of critical chemical and biochemical interest.

Stripping analysis consists of two or sometimes three steps:

1. *Preconcentration*: Deposition or adsorption of the analyte or analytes onto or into the working electrode during time *t*. This step often occurs under potential control or, possibly, at open circuit.
2. *Medium-exchange*: Exchange sample matrix with inert supporting electrolyte. This step is often unnecessary, but it is an option that can improve the sensitivity and stability of stripping analysis. Since medium-exchange is not always used, stripping analysis is commonly described as only a two-step process of preconcentration and stripping.
3. *Stripping*: The accumulated species is oxidized or reduced back into the solution. This can be achieved by varying the applied potential over time, applying a fixed current, or by inducing oxidation/reduction by another chemical species in solution. In all cases, the resulting response is proportional to the concentration of that analyte in or on the electrode, and thus, the sample solution. Of particular note, stripping voltammograms or chronopotentiograms can yield qualitative as well as quantitative information.

There are many manifestations of stripping analysis, with each dependent on the nature of preconcentration and stripping steps. With the long history (Esteban and Casassas, 1994; Wang, 1990) and recent advances, some confusion has arisen in the selection of effective nomenclature for stripping analysis. In this chapter, the terms recommended by Fogg and Wang (1999) will be adopted throughout. The nomenclature in stripping analysis is as follows:

1. *Stripping voltammetry (SV)*: Determination of an accumulated analyte by monitoring the faradaic current during a potential scan.
2. *Stripping chronopotentiometry (SCP)*: Determination of an accumulated analyte by observing the change of electrode potential with time during the stripping (at a constant rate) by either chemical or electrochemical means. In this case, the time between two significant changes in electrode potential is related to the analyte concentration.
3. *Stripping tensammetry (ST)*: Determination of an accumulated analyte by monitoring the capacitance current produced by a desorption process during a potential scan.

A. Anodic Stripping Voltammetry

Anodic stripping voltammetry (ASV) involves the electrochemical oxidation of a preconcentrated analyte. The term ASV should only be used when the analyte(s) is accumulated either by reduction (e.g., a metal ion) or by direct adsorption (e.g., organic compounds) and determined by its subsequent oxidation (Fogg and Wang, 1999).

The first step involves the reduction of a metal ion to the metal, which usually forms an amalgam with the mercury electrode [Eq. (1)]:

$$M^{n+} + ne^- + Hg \longrightarrow M(Hg) \qquad (1)$$

The applied potential on the working electrode should be maintained at least 0.4 V more negative than the standard potential of the most difficult to reduce ion (see Fig. 1). Under these conditions the deposition step is controlled by mass-transport. Thus, hydrodynamic control via rotation, stirring, or flow is typically used to facilitate the deposition step. Increased mass transfer due to curvilinear diffusion often mitigates the need for hydrodynamic control for microelectrodes. The concentration of the reduced metal in the mercury, C_{Hg}, is given by Faraday's law:

$$C_{Hg} = \frac{i_L t_d}{nFV_{Hg}} \qquad (2)$$

where i_L is the limiting current for the deposition of the metal, t_d is the length of the deposition period, n is the number of electrons transferred, F is the Faraday constant, and V_{Hg} the volume of the mercury electrode. The deposition time is typically 1–10 min for measurements in the range of 10^{-7}–10^{-9} M analyte concentrations (Wang, 1996). It should be noted that the metal concentration in the amalgam is not uniformly distributed (Shain and Lewinson, 1961).

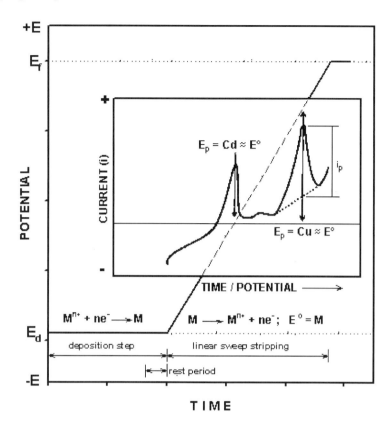

Figure 1 The potential–time waveform used in linear-sweep ASV with delineation of the deposition potential (E_d) and final potential (E_f). The resulting stripping voltammograms (inset) is positioned to show the relationship between the stripping potential and the potential–time waveform. The peak current (i_p) is a function of the concentration of the species being stripped from the electrode, which is in direct correlation to the solution concentration.

The preconcentration step can be followed by a medium-exchange step, where the sample solution is replaced with more ideal supporting electrolyte in which stripping is carried out. Medium-exchange can minimize interferences caused by sample components, and the result may be improved sensitivity and enhanced resolution between neighboring stripping peaks. In any case, the stripping step is typically performed under quiescent conditions, and any stirring or flow is stopped, followed by a rest period (ca. 10–15 s) to allow the system to equilibrate.

At the end of the rest period, which is incorporated into the preconcentration step, the stripping step begins by the application of a potential–time waveform going in the positive direction. Figure 1 shows the application of a linear-sweep, potential–time waveform, which is the simplest waveform. When the potential reaches the standard potential of the metal–metal ion redox couple, that particular amalgamated metal is oxidized, or stripped (dissolved), from the mercury electrode [Eq. (3)]:

$$M(Hg) \rightarrow M^{n+} + ne^- + Hg \qquad (3)$$

Anodic peaks are observed for the analytes of interest in the current–potential voltammogram, which is recorded during the entirety of the stripping step (also shown in Fig. 1). The peak potential (position), E_p, of each metal is characteristic of that metal and is related to the standard potential of its redox couple. The E_p can be used to identify the metal. The peak current (height), i_p, is proportional to the concentration of the corresponding metal ion in the test solution. The concentration is determined by a standard addition or a calibration curve. Hence, both qualitative and quantitative information can be obtained in SV.

For reversible stripping reactions, the peak current for linear-sweep voltammetry at thin mercury film electrodes (MFEs) (rapid depletion of all metal from the amalgam) is given by:

$$i_p = \frac{n^2 F^2 v A l C_{Hg}}{2.7 RT} \qquad (4)$$

where v is the scan rate, A is the film area, l is the film thickness, and R and T have the usual meaning (Wang, 1996). As the mercury film becomes thicker, the peak current is proportional to the square root of the scan rate.

Detailed theories and peak current equations for stripping peaks produced at the mercury film and hanging drop electrodes using a variety of waveforms have been published (Lund and Onshus, 1976; Wang, 1985).

The instrumental simplicity and speed of linear-sweep stripping voltammetry (LSSV) have resulted in its use in numerous stripping studies (particularly those aimed at the 10^{-6}–10^{-8} M level of the analyte). Unfortunately, the continuous change in potential generates a relatively large charging-background current. Linear scan rates of typically 50–100 mV/s are employed and offer the desired compromise between the relative magnitudes of the analytical and charging currents, resolution of adjacent peaks, and speed. Using a well-prepared MFE, LSSV is only slightly (three- to fivefold) less sensitive than the more complex and slower differential pulse waveform (Florence, 1980).

Other potential–time waveforms used during the anodic scan to oxidize the metals out of the electrode are designed mainly to discriminate against the charging current. Hence, increased signal-to-background characteristics are observed, particularly when using the hanging mercury drop electrode (HMDE). At present, the most commonly used excitation waveforms in stripping analysis are differential pulse and square wave (Fig. 2). The deposition step for these is no different from that employed in LSSV and most other stripping strategies.

In addition to the effective compensation of the charging current, differential pulse stripping voltammetry (DPSV) often produces narrower peaks, which are desirable when mixtures are analyzed. Another feature of DPSV is that some of the material stripped from the electrode during the potential pulse is redeposited into the electrode during the waiting period between pulses. Therefore, the same material is "seen" many times, resulting in enhanced sensitivity. A disadvantage of anodic DPSV is that the potential scan must be quite slow (5 mV/s), hence lengthening the analysis time. Pulse amplitudes of 25 or 50 mV are commonly employed. An increase in the scan rate or pulse amplitude will result in peak broadening and impaired selectivity.

Square wave stripping voltammetry (SWSV) has the advantage over the differential pulse mode in that much faster scan rates (up to 1 V/s) can be used. Hence, a complete stripping voltammogram can be obtained in 1 s. Rapid scanning offers an additional advantage—the oxygen that is depleted from the electrode surface during the preconcentration step cannot be replenished during fast SWSV. Hence, the analytical time can be further reduced, as removal of oxygen is not as critical to

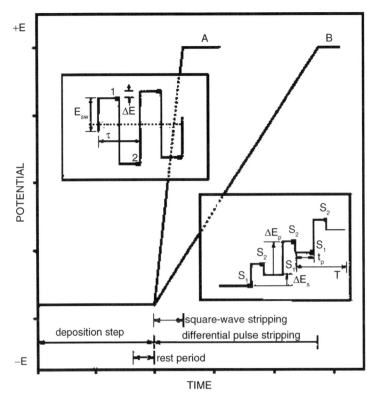

Figure 2 Potential–time waveforms for (A) SWSV and (B) DPSV showing that SWSV can be performed at much faster rates than DPSV techniques. Insets show details of the specific waveforms used for the stripping step; where, potential (E), time (T, t, or τ), and current sampling (■) parameters are denoted.

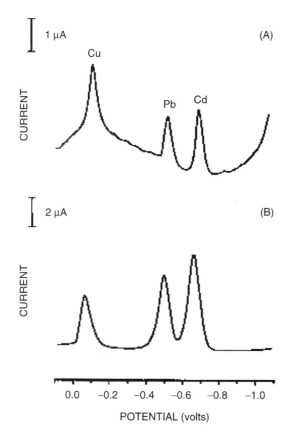

Figure 3 Stripping voltammograms comparing (A) DPSV and (B) SWSV approaches for a nondeaerated solution containing 50 ppb Cu, Pb, and Cd ions. SWSV typically shows improved sensitivity and stability. Deposition at 2 min at an MFE held at -1.1 V in 0.1 M acetate buffer, pH 4.5. [Adapted from Wang (1996), reprinted with permission.]

obtain analytically useful stripping voltammograms (Wojciechowski et al., 1985). Figure 3 compares differential pulse and square wave stripping voltammograms for a non-deaerated solution containing 50 ppb of Cd, Pb, and Cu ions.

In addition to DPSV and SWSV, other voltammetric modes capable of minimizing various background-current contributions during the stripping step have been attempted (Wang, 1985). These approaches include modulated waveforms, such as alternating current or staircase, as well as twin electrode strategies like subtractive SV or anodic stripping with collection. An early study (Turner et al., 1984), critically compared the performance of these stripping procedures, and it was shown that they offer detection limits comparable to those obtained using the differential pulse mode. Although modern instrumentation with computer control has alleviated the complexity of performing the experiment, the performance issue may not justify the effort.

B. Cathodic Stripping Voltammetry

Cathodic stripping voltammetry (CSV) is the "mirror image" of ASV (Florence, 1979). The term CSV should only be used when the analyte(s) is (are) accumulated either by oxidation (e.g., a metal ion) or by direct adsorption (e.g., organic compounds) and determined by its subsequent reduction (Fogg and Wang, 1999). Quantitation is accomplished by measuring the height of the resulting reduction peak. In general, CSV applications are characterized by the anodic deposition of sparingly soluble mercury compounds on the mercury electrode surface [Eqs. (5) and (6)]:

$$A^{n-} + Hg \rightarrow HgA + ne^- \quad \text{Deposition} \quad (5)$$
$$HgA + ne^- \rightarrow A^{n-} + Hg \quad \text{Stripping} \quad (6)$$

As a result, it is not unusual for calibration curves to display nonlinearity at high concentrations. Most of the reported applications of CSV have been carried out at HMDE or mercury pool electrodes. A rotating silver disk electrode offers various advantages for the determination of anions that form insoluble silver salts. In this case, the preconcentration [Eq. (7)] and stripping [Eq. (8)] steps involve the following reactions:

$$Ag + X^- \rightarrow AgX + ne^- \quad \text{Deposition} \quad (7)$$
$$AgX + ne^- \rightarrow Ag + X^- \quad \text{Stripping} \quad (8)$$

C. Stripping Chronopotentiometry

SCP encompasses the techniques presently known as potentiometric stripping analysis (PSA) and stripping potentiometry (SP). SCP resembles ASV as the preconcentration step is similar in that the analyte is reduced and concertedly deposited onto a mercury electrode surface. Rather than scanning a positive-going potential–time waveform and monitoring the current, the metal amalgam is oxidized ("stripped") chemically from the mercury electrode surface [Eq. (9)]:

$$M(Hg) + \text{oxidant} \rightarrow M^{n+} + Hg \quad \text{Stripping} \quad (9)$$

Typical oxidants used for the oxidation are O_2, Hg(II) and Cr(VI). In another approach, the amalgam can be stripped off by applying a constant anodic current to the electrode. By either means, the potential of the working electrode is recorded as a function of time, and a stripping curve, like the one shown in Fig. 4 (Jagner and Graneli, 1976), is obtained. A sudden change in potential occurs when all the metal deposited in the electrode has been oxidized from the surface. The transition time needed for the oxidation of a given metal, t_M, is a quantitative measure of the metal concentration

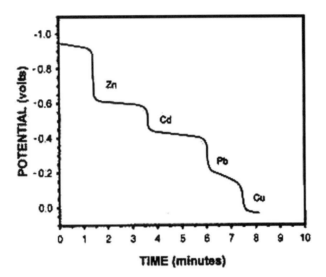

Figure 4 Representative potentiogram of stripping chronopotentiometric analysis of a solution containing Zn(II), Cd(II), Pb(II), and Cu(II) ions.

in the sample:

$$t_M \approx [M^{n+}]t_{dep} \qquad (10)$$

For constant current SCP, the stripping time is inversely proportional to the applied stripping current. As predicted by the Nernst equation, the potential at which the reoxidation takes place serves as a quantitative identification of the different metals. The sigmoidal-shaped curves are easily converted to peaks by taking the first derivative of the analytical signal (i.e., dE/dt vs. E), which is easily accomplished with modern computer-controlled systems. Hence, both qualitative and quantitative information is provided by SCP.

While the sensitivity of SCP approaches that of SV, it requires simpler instrumentation, can accommodate very tiny electrodes, and is less prone to interferences (e.g., dissolved oxygen or organic surfactants). Indeed, the use of nondeaerated samples is the main reason for the growing interest in this approach. Hence, SCP has proven useful for the rapid and accurate determination of several trace metals in a wide variety of media and materials. These applications, as well as the theory of SCP techniques, have been reviewed (Jagner, 1982; Ostapczuk, 1993).

D. Adsorptive Stripping Voltammetry

Adsorptive stripping voltammetry (AdSV) has extended the scope of stripping analysis toward numerous analytes. It is similar to conventional stripping analysis in that the analyte is concentrated onto the working electrode prior to its voltammetric measurement, but it differs from the conventional scheme in the way that the preconcentration is accomplished. Rather than using electrolytic deposition, AdSV utilizes controlled adsorptive accumulation for preconcentration. Hence, for a wide range of surface-active organic and inorganic species that cannot be preconcentrated electrolytically, the adsorption approach serves as an effective alternative. AdSV exploits the formation of a solution phase metal chelate that is allowed to accumulate onto the working electrode for a specific length of time, under conditions of maximum adsorption (e.g., electrolyte, pH, potential, and mass-transport). The accumulated material is determined by applying a negative- or positive-going scan for reducible or oxidizable species, respectively. The adsorptive stripping response reflects the corresponding adsorption isotherm, as the amount of analyte (Z) adsorbed is proportional to the bulk concentration of the analyte. On the basis of the Langmuir adsorption isotherm:

$$Z = Z_m \frac{BC_A}{1 + BC_A} \qquad (11)$$

where Z_m is the surface concentration corresponding to monolayer coverage, B is the adsorption coefficient, and C_A is the bulk concentration. Hence, a linear response is expected (i.e., typically $<10^{-7}$ M), on the basis of the Langmuir adsorption isotherm, provided that full coverage of the electrode is avoided (Wang, 1996). In addition to voltammetric quantification of the surface-bound chelate, an SCP approach on the basis of applying a constant reducing current and monitoring the potential–time behavior, has been utilized (Eskilsson et al., 1985).

Typically the HMDE is used for measuring reducible species, whereas carbon paste electrodes are employed for oxidizable ones. The other instrumental requirements (voltammetric analyzer, cell, etc.) are the same as those used in conventional stripping analysis. With relatively short preconcentration times, very low concentrations (10^{-9}–10^{-11} M) can be determined. Figure 5 shows voltammograms for solutions of increasing levels of the anticancer agent mitomycin C (Wang et al., 1986).

Because of the nature of the preconcentration step, interfering surfactants (that compete for adsorption sites) should be destroyed. Additional improvements in the sensitivity (and hence pM detection limits) can be achieved by coupling the adsorptive accumulation with catalytic effects (to regenerate the oxidized form of the chelate) (Wang et al., 1987; Yokoi and van de Berg, 1991; Wang et al., 1992a). For more details of AdSV the reader is referred to several reviews (van de Berg, 1991; Wang, 1983; Paneli and Voulgaropoulos, 1993).

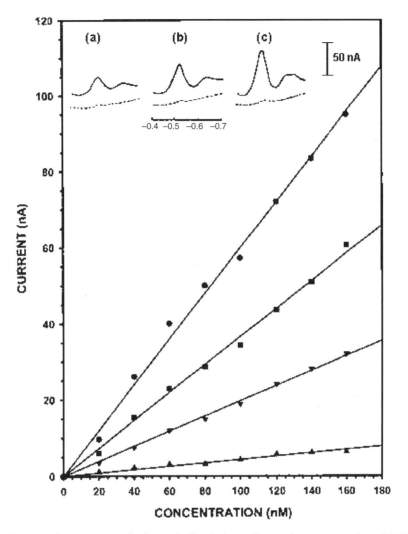

Figure 5 Adsorptive stripping voltammograms of mitomycin C solutions of increasing concentrations: (a) 40, (b) 80, and (c) 120 nM. (---) Responses for each of the experiments without the 2 min accumulation step. The inset shows the calibration plots for (▲) 9, (▼) 30, (■) 60, and (●) 120 s preconcentration steps.

E. Stripping Tensammetry

Electroinactive organic compounds can also be determined using ST. In ST, the analytes that adsorb onto the electrode surface are subsequently desorbed by the application of a potential scan. The capacitance current is monitored as a function of voltage (Kalvoda, 1984). Applications have focused on non-electroactive surfactants (e.g., detergents).

F. Experimental Considerations

Table 1 lists detection limits reported for the various stripping strategies discussed in the previous sections. Clearly, stripping analysis offers extremely low detection limits that compare favorably with those obtained by other (non-electrochemical) "trace" techniques. It is emphasized that these limits can be obtained by operators with expertise in stripping analysis and trace chemistry. Experimentally, because only a small fraction of the analyte is deposited, it is essential that all experimental parameters (e.g., deposition time, forced convection rate, and temperature) be reproducible during a series of measurements. In particular, the ability to obtain such low values strongly depends on the degree to which contamination can be minimized, especially from reagents and laboratory water. Hence, all principles of good laboratory practice such as glassware cleanliness and proper sample collection and storage must be stringently adhered to in order to obtain high accuracy and low limits of detection.

Most stripping measurements require the addition of appropriate supporting electrolyte and removal of dissolved oxygen. The former is needed to decrease resistance of the solution and to ensure that the metal ions of

Table 1 Relative Sensitivity of Various Stripping Strategies

Stripping technique	Limit of detection (M)	Analyte	Working electrode	Reference
Differential pulse	2×10^{-10}	Lead	Mercury drop	Florence (1986)
Square wave	1×10^{-10}	Lead	Mercury drop	Florence (1986)
Linear scan	5×10^{-11}	Lead	Mercury film	Florence (1986)
Differential pulse	1×10^{-11}	Lead	Mercury film	Florence (1986)
Square ware	5×10^{-12}	Lead	Mercury film	Florence (1986)
PSA	1.5×10^{-10}	Lead	Mercury film	Graneli et al. (1980)
Chelate adsorption	1×10^{-10}	Nickel	Mercury drop	Torrance and Gatford (1985)
Analyte adsorption	2.5×10^{-11}	Riboflavin	Mercury drop	Forsman (1983)
CSV	2×10^{-10}	Penicillin	Mercury drop	Forsman (1983)
Catalytic adsorption	1×10^{-12}	Titanium	Mercury drop	Yokoi and van de Berg (1992)

interest are transported toward the electrode by diffusion and not electrical migration. Contamination of the sample by metal impurities in the reagents used for preparing the supporting electrolyte is a serious problem. Dissolved oxygen severely hampers the quantitation (except in SCP and SWSV) and must be removed, usually by purging the sample with high-purity nitrogen.

The main types of interferences in stripping analysis are overlapping stripping signals, the adsorption of organic surfactants on the electrode surface, and the formation of intermetallic compounds between metals (e.g., Cu–Zn and Cu–Cd) codeposited in the mercury electrode. Overlapping signals cause problems in the simultaneous determination of metals with similar redox potentials (e.g., Pb/Sn, Bi/Cu, and Tl/Cd). Intermetallic compound formation and surfactant adsorption cause a depression of the stripping response, shifts of the signal may also be observed. Various chemical, instrumental, or mathematical strategies available for the elimination or minimization of these interfering effects have been described (Copeland and Skogerboe, 1974; Vydra et al., 1976; Wang, 1985).

III. INSTRUMENTATION

A. Cells

Many designs of electrochemical cells for stripping measurements have been reported in the literature or made available commercially. Typically, the cell contains three electrodes (working, reference, and auxiliary), which are immersed in the sample solution. The electrochemical cell can be as simple as a beaker (i.e., 10–100 mL volume). The specific shape depends primarily on the working electrode used. For stripping work, aimed at measurements at the 10^{-7}–10^{-9} M level, precleaned glass or quartz cells can be used. For quantitation of lower concentration levels (10^{-10}–10^{-12} M) Teflon™ cells are desirable, because of the possible release of traces of heavy metals from glass or quartz cells. The electrodes, as well as a tube for the purging gas, are supported in four holes in the cover. The cover is usually made of Teflon or Plexiglas™, and offers good sealing against dust particles. In most cases, the performance requirements for the cell are minimal, so that it can be designed for convenience of use (e.g., cleaning, purging, stirring, size, etc.).

Most manufacturers of voltammetric analyzers make electrochemical cells suitable for stripping measurements. Other special-purpose cell configurations (micro, flow) are also available for work in small volumes or for online measurements. These include miniaturized cells with stationary (DeAngelis et al., 1977) or rotating (Eggli, 1977) MFEs, or cells in which the solution flows through a thin-layer channel (Anderson et al., 1982), onto a stationary disk in a wall-jet design (Jagner, 1983) or through an open-tubular electrode (Lieberman and Zirino, 1974). In addition, extremely inexpensive (disposable) cells, with screen-printed planar electrodes on a plastic or ceramic support, have been introduced to meet the growing needs for single-use decentralized metal testings (Wang and Baomin, 1992). Specially designed cells, aimed at minimizing or eliminating the time for oxygen removal, including a rotating cell assembly, a large-volume wall-jet, or a multicell system (with separated deaeration and measurement racks) have also been described (Clem et al., 1973; Wang and Freiha, 1985).

B. Electrodes

Although the performance requirements of the cell design are not critical, the working electrode is at the heart of SV. The working electrode in stripping analysis must be stationary, which obviates the dropping mercury electrode that is so popular in polarography. In addition, the working electrode must have a redox behavior favorable to the analyte, a reproducible area, and a low background current over a wide potential range. Stripping analysis is typically performed at a micromercury electrode, because reducing the volume of the mercury enhances the concentration of deposited metals [Eq. (2)].

The most commonly used electrodes, which fulfill the earlier requirements, are the MFE and the HMDE. Thin MFEs offer improved sensitivity (larger area-to-volume ratio) and excellent resolution of neighboring peaks, when compared with the HMDE (Fig. 6) (Mart et al., 1980). HMDE offers reproducible renewal of the surface, reduced intermetallic interferences, and high hydrogen overvoltage. The HMDE is preferred for AdSV and CSV, whereas MFE is preferred for SCP. The MFE is also more compatible with field work owing to its rugged nature.

1. Mercury Film Electrodes

The MFE consists of a thin (1–100 μm) film of mercury on a solid support. The mercury film can be applied to the substrate prior to the voltammetric measurement (prepared MFE) or during the measurement (*in situ* plated MFE). *In situ* plating requires adding a spike of Hg^{2+} ions to the sample (ca. 5×10^{-5} M); mercury and analytes are codeposited (Florence, 1970). Carbon, especially glassy carbon, is well suited as a support for mercury film. Other carbon materials have been used successfully as substrates, for MFE. Among these are wax-impregnated graphite (Magjer and Branica, 1977), Kelgraf (Anderson and Tallman, 1976), and graphite-epoxy (Wang, 1981). Best results are obtained when a rotating glassy carbon electrode is coupled with an *in situ* plated mercury film. Aside from high plating efficiency, the mercury-coated glassy carbon has a wide accessible potential range and low background current. A simpler approach to achieving hydrodynamic control is to use a stationary MFE with stirring of the solution (Wang, 1982b). Coverage of MFEs with permselective coatings (e.g., Nafion™ or Kodak AQ™ ionomers) has been extremely useful in protecting the electrode from surfactant poisoning (Hoyer et al., 1987; Wang and Taha, 1990).

Whereas most stripping applications of MFEs are based on a disk-shaped geometry, other electrode configurations may be employed for specific applications. These include MFE-based flow cells for online monitoring; dual-electrode designs (e.g., ring-disk) for anodic stripping with collection (Laser and Ariel, 1974), and split-disk electrode for subtractive SV (Sipos, et al., 1977).

2. Hanging Mercury Drop Electrode

The HMDE is widely used in stripping analysis. With the popular screw-type (Kemula-type) HMDE the mercury drop is formed by extrusion of mercury from a capillary, fed from a reservoir by a micrometer-driven syringe. The calibrated micrometer permits reproducible delivery of drops of known area. To ensure proper operation, the capillary should be cleaned, and there must be no entrapped air in the mercury reservoir or the capillary.

The use of a static mercury drop electrode (SMDE), developed by EG&G PARC, instead of the Kemula-type HMDE, improves the reliability and convenience of stripping measurements. This is due to the automatic production of highly reproducible drops that hang at the capillary tip. A built-in valve is utilized for this purpose, allowing the mercury flow to be stopped to produce the stationary electrode. The SMDE also improves the ability to hold the drop, over long deposition periods, in a stirred solution. SMDEs that offer a renewable mercury microelectrode with drops of about 50 μm diameter are also available (Novotny, 1990).

The controlled growth mercury drop electrode (CGME) offered by Bioanalytical Systems (West Lafayette, IN) is the latest innovation in mercury electrode technology. In CGME, the drop growth and size are controlled by a fast response valve, which gives the user total control via computer programing.

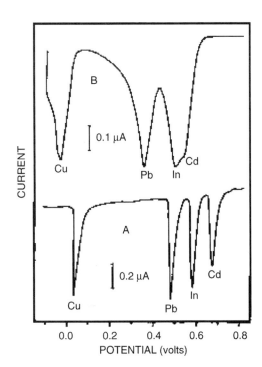

Figure 6 Anodic stripping voltammograms for a mixture of Cd(II), In(III), Pb(II), and Cu(II) ions at 200 nM each using (A) MFE and (B) HMDE. Deposition times of (A) 5 and (B) 30 min were used. [Adapted from Florence (1970), reprinted with permission.]

3. Solid Electrodes

The determination of metals, such as Au, Ag, Se, As, Re, Te, and Pt, possessing either oxidation potentials more positive than that of mercury or low solubility in mercury, requires the use of bare solid electrodes. Gold and glassy carbon electrodes are especially suited for this purpose, with the metal of interest being determined following its preconcentration as a metallic layer. These electrodes have also been used with AdSV for oxidizable organic compounds and mercury determinations.

Such measurements yield low sensitivity and reproducibility when compared with analogous measurements at mercury electrodes. In addition, solid-electrode stripping measurements may suffer from difficulties such as the appearance of multiple peaks (and hence a nonlinear response) or incomplete stripping. Nevertheless, reliable procedures for trace measurements of the earlier-mentioned elements in various real samples have been reported (Gil and Ostapczuk, 1994; Wang, 1985). These methods often require precise electrode cleaning, polishing, and pretreatment procedures; the nature of these steps depends on the materials involved. Substantial improvements in the quantification of selenium, arsenic, and tellurium are observed after codeposition with either gold or copper (Andrews, 1980). Most stripping applications of bare solid electrodes have utilized a rotating-disk or flowthrough electrode configurations.

4. Chemically Modified Electrodes

Chemically modified electrodes (CMEs) can add a new dimension to stripping analysis. In particular, modification of electrode surfaces can provide alternative approaches for accumulation of surface-bound species, thus extending the scope of stripping measurements. Most popular schemes used to trap metal species on the electrode surface are based on complexation and electrostatic attraction. The modifying agent (e.g., ligand or ion-exchanger) can be introduced to the surface by functionalizing an appropriate polymer coating or directly into the matrix of a carbon paste electrode. Hence, preconcentration is accomplished by a purely non-electrolytic deposition step; the collected analyte is subsequently measured during a voltammetric scan. Major requirements for CMEs include selective accumulation, prevention of surface saturation, and a convenient regeneration of a "fresh", or analyte-free, surface (Wang, 1996).

Baldwin et al. (1986) reported on the determination of traces of nickel based on dimethylglyoxime-containing carbon paste electrode. A poly(vinylpyridine)-coated platinum electrode can be used for measuring low levels of Cr(VI) (Cox and Kulesza, 1983). Trace measurements of uranium can be obtained at a trioctylphosphine oxide-coated glassy carbon electrode (Izutsu et al., 1983).

Nafion- (Szentirmay and Martin, 1984) or amine-modified (Price and Baldwin, 1980) electrodes can be used to preconcentrate organic analytes; in the latter case, the preconcentration is based on covalent attachment. Bioaccumulation (Gardea-Torresdey et al., 1988) and biocatalytic processes (Kulys et al., 1982) can also be exploited to obtain desired sensitivity and selectivity enhancements.

5. Micro- and Array Electrodes

As evinced by Eq. (2), the concentration of deposited material at a mercury electrode can be dramatically increased by reducing the volume of the mercury, which leads to lower limits of detection. The physical properties and voltammetric behavior of carbon fiber microelectrodes (Edmonds, 1985) have been reviewed, and they have been used as a substrate to make MFEs (Baranski and Quon, 1986). These electrodes exhibit high plating efficiency in the absence of forced convection, reduced ohmic drop or sample volume, and low cost. These characteristics result in major simplification of the instrumentation and operation of stripping analysis, as stirring or rotating devices are eliminated and potentiostatic two-electrode systems can be used. Impressively, stripping analysis can be performed in volumes as small as 0.5 μL (Wang et al., 1994). In addition to carbon, it is possible to use iridium wire for supporting the tiny MFEs (DeVitre et al., 1991). Single-use, disposable, screen-printed electrodes have been described (Wang and Baomin, 1992).

C. Stripping Analysis in Flowing Streams

Highly sensitive and versatile automated methods for stripping analysis can be developed by coupling it with flow systems, particularly those based on flow injection. The interest in stripping analysis/flow systems has increased considerably in recent years (Fogg and Wang, 1999; Gunasingham and Fleet, 1989; Jagner, 1983; Luque de Castro and Izquierdo, 1991; Wang, 1983, 1996). Besides the obvious advantages of online monitoring and possible automation, such coupling offers several advantages over stripping analysis performed in batch (beaker) systems. These include reduced analysis time, analysis of small volume (100–500 μL) samples, improved reproducibility, and minimization of errors due to contamination or adsorption losses. Flow systems offer also a simple and convenient way to perform the medium-exchange procedure, where the sample solution—after preconcentration—is replaced with more ideal supporting electrolyte in which stripping is carried out. The manifold of flow injection systems is particularly useful for this purpose; after the sample plug passes through the detector, stripping can occur in the carrier environment. As a result, stripping analysis can be

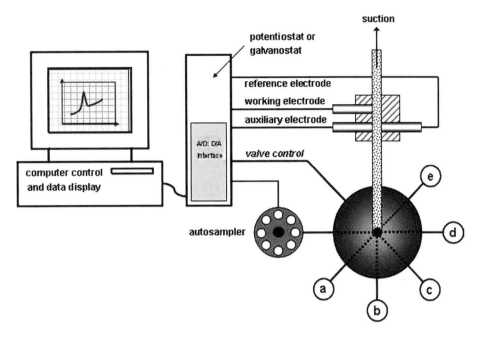

Figure 7 Generalized diagram of an automated flow system for stripping voltammetry; where, valve ports are typically (a) Hg plating solution, (b) stripping solution, (c) Hg removing solution, (d) rinsing solution, and (e) an open channel for additional reagent solutions.

substantially improved with respect to resolution between overlapping peaks or sample deoxygenation. Improved sensitivity can be obtained using subtractive stripping measurements in which the carrier (blank) response is subtracted from that of the sample plug (Wang and Dewald, 1984). Figure 7 shows a schematic of a computerized flow system for automated potentiometric stripping analysis possessing several of the above advantages. Besides flow injection/ASV systems, automated flow systems have been applied to shipboard stripping analysis of natural waters for trace metals (Zirino et al., 1978). Such a system is controlled by a specifically designed program which controls the potentiostat, pumps, valves, and recorder. Specially designed cells (e.g., dual coulometric–voltammetric cells) can offer additional advantages such as reduced intermetallic or oxygen interferences (Wang and Dewald, 1983a).

D. Stripping Analyzers

The popularity of stripping analysis depends on the availability of suitable instrumentation. The basic instrumentation required for stripping analysis has been described in detail in Chapter 17. Typically, a modern microprocessor-based voltammetric analyzer, with an electronic potentiostat and associated signal generator, provides the desired potential control (during preconcentration) and potential programing (during stripping). Most systems use computers to output the data and perform data manipulation. Some stripping experiments require simultaneous control of two working electrodes. Bipotentiostats for such dual-electrode operation are available commercially from Pine Instrument Company (Pine Grove, PA). Table 2 is a partial list of suppliers of stripping analysis systems.

Figure 8 shows the basic instrumentation required for potentiometric stripping analysis. This includes potentiostatic circuitry for controlling the working-electrode potential during the deposition, a timing switch control, and input impedance for measuring the potential during the stripping period; the latter must be larger than 10^{12} Ω, to prevent electrochemical oxidation of the amalgamated metals. A microcomputer is desirable for rapid data acquisition, experimental control, and background subtraction. For example, the microcomputer can transform the basic SCP/PSA response to a differential peak-shaped potentiogram (of dt/dE vs. E), in addition to being used for registering rapid stripping events. For example, the stripping step for concentrations of metals below the 10^{-8} M level takes only 50–200 ms.

Computer-based technology has offered other advances in instrumentation and methodology for stripping analysis. For example, Bond et al. (1984) described a compact microprocessor-based voltammetric analyzer for stripping measurements in harsh environments (e.g., hazardous or radiation laboratories) as well as clean laboratories for ultratrace measurements. The microprocessor systems were interfaced to a microcomputer system external to the laboratories. The same group also described a

Table 2 List of Some Companies and Their Products for Stripping Analysis

Supplier	Email address	Product description
Amel	www.amelchem.com	High sensitivity potentiometric stripping analyzers
Bioanalytical Systems, Inc.	www.bioanalytical.com	Complete line of electrochemical instrumentation, including potentiostats, CGME technology, cells, and so on
Cypress	www.cypresssystems.com	Stripping analysis systems and custom potentiostats
ECO Chemie	www.ecochemie.nl	Stripping analysis systems including a variety of cells
ESA	www.esainc.com	Electrochemical blood analyzers with disposable electrodes
Maran & Company	www.maran.co.uk	Dual cell ASV systems
Gamma Analysen Technik GMBH	www.gatgmbh.com	Automatic polarography/voltammetry systems, field powered units
Gamry Instruments	www.gamry.com	Pulse voltammetry software and electrochemical systems
IVA Co., Ltd.	—	Batch and flowthrough stripping analysis systems
Metrohm	www.brinkman.com	Complete line of electrochemical/polarography instrumentation
Pine Instrument Company	www.pineinst.com	Dual-electrode potentiostats and rotating electrode technology
Princeton Applied Research	www.princetonapplied research.com	Complete line of electroanalytical instrumentation and related software
Trace|Detect	www.tracedetect.com	Trace metals analysis system, nanoband™ electrode technology

Note: Websites are subject to change.

computerized multi-time-domain method for DPSV or anodic SWSV (Bond et al., 1986). The system can detect possible matrix effects (e.g., organic surfactants) and accordingly "make decisions" regarding the need for a pretreatment procedure or the method of quantification. Valenta and coworkers (1982, 1984) described a fully automated system for anodic DPSV that greatly simplifies routine trace metal analysis. Apart from controlling the

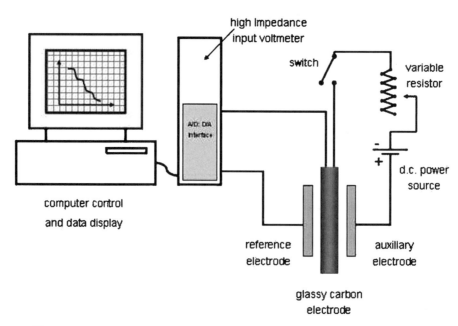

Figure 8 Schematic diagram illustrating the main components of an SCP analyzer.

Electrochemical Stripping Analysis

voltammetric measurement and processing the data, all stages of the sample pretreatment (filtration, acidification, and UV-irradiation) can be automated. Computer control also greatly simplifies the use of stripping flow systems.

E. Modeling Software and Speciation

Stripping analysis has the potential to perform speciation of metals. Careful control of solution chemistry and deposition parameters can preferentially preconcentrate free metal ions and labile complexes. Computer modeling programs for generating pseudopolarograms are commonly used (in speciation analysis) for studying trace metal complexation parameters (Florence, 1986), such as coordination number and dissociation constant, and for determining the complexation capacity of natural waters (Brown and Kowlski, 1979). Several instrument manufacturers also supply simulation software.

IV. APPLICATIONS

The availability of reliable, software-controlled pulse polarographs capable of performing a variety of stripping techniques has resulted in a plethora of applications. Hence, Table 3 lists only a sampling of those applications.

Electrochemical stripping analysis in all its forms is well established for monitoring trace analytes in environmental, clinical, food, pharmaceutical, or industrial samples. For example, stripping analysis appears to be the best analytical tool for the direct, and simultaneous, determination of four metals of prime environmental concern (i.e., Pb, Cd, Zn, and Cu) in seawater. Other metals conveniently measured by stripping analysis include Bi, In, Sn, Th, Ga, Hg, Ag, Se, Mn, As, and Au. Adsorptive collection of metals as their surface-active complexes on the HMDE has been shown to be extremely useful for ultratrace measurements of metals (e.g., Ti, Al, Cr, Ni, Mo, V, U, Fe, or Th) that cannot be conveniently determined by conventional stripping analysis. This strategy also offers effective alternative procedures for metals such as Cu, Ga, Mn, and Sn, which are measurable only with some difficulty by conventional ASV. Overall, it is now possible to use stripping analysis for detecting more than 40 trace elements, including applications for the analysis of metals in soils (Lukaszewski and Zembrzuski, 1992), blood and urine (Christensen et al., 1982; Jagner et al., 1981; Nygren et al., 1990), human hair (Liu and Jiao, 1990), wine (Wang and Mannino, 1989), fruit juices (Mannino, 1989), fish (Golimowski and Gustavsson, 1984), and gunshot residue (Bobrowski and Bond, 1991; Konaunur and van Loon, 1977).

In addition, low levels of various inorganic and organic compounds can be detected using AdSV or CSV. Among these are ions such as halide, cyanide, sulfide, or selenide, as well as various thiols, peptides, or penicillins. Stripping analysis has been applied for measuring trace levels of biological compounds and pharmaceutical significance. These compounds include folic acid (Luo, 1986), nitro-containing pesticides (Benadikova and Kalvoda, 1984), progesterone, streptomycin, riboflavin (Wang et al., 1985a), digoxin (Wang et al., 1985b), codeine (Kalvoda, 1983), bilirubin (Wang et al., 1985c), diazepam (Kalvoda, 1984), and adriamycin (Chaney and Baldwin, 1982).

Table 3 Representative Applications of Stripping Analysis

Analyte	Matrix	Stripping strategy	Working electrode	Reference
Pb, Cd, Se, Co, Ni, and Mn	Rain water	Differential pulse, chelate adsorption, CSV	Mercury drop	Vos et al. (1986)
Bi, Cd, Cu, Pb, Sb, and Zn	Sea water and marine samples	Linear scan	Mercury film	Florence (1972)
Cu and Zn	Pharmaceutical tablets	Differential pulse	Mercury film	Florence (1972)
Ni	Nail	Chelate adsorption	Mercury drop	Gammelgoard and Andersen (1985)
Cd, Pb, and Tl	Fly-ash	PSA	Mercury drop	Christensen et al. (1982)
Hg	Fish	ASV	Gold disk	Mannino and Granata (1990)
U	Groundwater	Chelate adsorption	Mercury drop	Wang et al. (1992b)
Co	Seawater	Chelate adsorption	Mercury drop	Donat and Bruland (1988)
Sb, Cu, and Pb	Gunshot residue	Linear scan	Mercury film	Brineer et al. (1985)
Pb and Sn	Fruit juices	PSA	Mercury film	Mannino (1982)
Cd, Cu, and Pb	Urine	Differential pulse	Mercury drop	Lund and Eriksen (1979)
Chlorpromazine	Urine	Analyte adsorption	Carbon paste	Wang and Freiha (1983c)
As	Urine	Staircase	Mercury film	Davis et al. (1978)
Thioamide	Plasma and urine	CSV	Mercury drop	Davison and Smyth (1979)

On the forefront of stripping analysis research is its application to the determination of nuclei acids such as DNA by stripping techniques (Wang et al., 2000). Ultratrace levels of manganese (i.e., limit of detection is 2.5×10^{-10} M) have been determined by combining cathodic SWSV with ultrasonic irradiation (Jin et al., 2000). Katano and Senda used SV to study the behavior of nonionic surfactants and anionic surfactants (Katano and Senda, 2001). They were able to determine these compounds at nanomolar levels. The utility and versatility of stripping analysis is shown in recent efforts of Brainina and coworkers (Brainina et al., 1999, 2000, 2001a, b; Stojko et al., 1998; Zaharchuk et al., 1999).

V. CONCLUSIONS

Stripping analysis has been shown to be a very sensitive, rapid, reproducible, and economical technique for measurements at trace and ultratrace levels of metals, inorganic ions, and organic compounds. Because the preconcentration is done in the same cell as the final electrochemical measurement, contamination risks are greatly reduced. Recent developments, particularly the introduction of new procedures based on chelate adsorption or chemically modified sensors for measuring a great number of metals, the introduction of low-cost sophisticated ("push-button, do-all") instrumentation and of disposable electrodes coupled to compact instruments, or the use of microelectrodes, have stimulated further interest in the field. In particular, stripping analysis plays a major role in oceanographic studies owing to its ability to achieve low detection limits in saline. Also, its small size, low power needs, and versatility make it ideal for field applications. A review of its application to natural water has been published (Tercier and Buffle, 1993). Automation and flow-systems have increased its presence as an online monitoring technique of industrial processes and for environmental monitoring. Not to be ignored is its analytical utility for samples of critical biochemical significance, such as Pb in children's blood and sensitive DNA detection. Certainly, stripping analysis is a technique that supplements and complements the analytical capabilities of any lab.

REFERENCES

Anderson, J., Tallman, D. (1976). *Anal. Chem.* 48:209.
Anderson, L., Jagner, D., Josefson, M. (1982). *Anal. Chem.* 54:1371.
Andrews, R. W. (1980). *Anal. Chim. Acta* 119:47.
Baldwin, R. P., Christensen, J. K., Kryger, L. (1986). *Anal. Chem.* 58:1790.
Baranski, A., Quon, H. (1986). *Anal. Chem.* 58:407.
Benadikova, H., Kalvoda, R. (1984). *Anal. Lett.* 171:195.
van de Berg, C. M. (1991). *Anal. Chim. Acta* 250:265.
Bobrowski, A., Bond, A. M. (1991). *Electroanalysis* 2:157.
Bond, A. M., Greenhill, H. B., Heritage, I. D., Reust, J. B. (1984). *Anal. Chim. Acta* 165:209.
Bond, A. M., Heritage, I. D., Thormann, W. (1986). *Anal. Chem.* 58:1063.
Brainina, Kh., Neyman, E. (1993). *Electrochemical Stripping Methods*. Washington, D.C.: American Chemical Society.
Brainina, Kh., Henze, G., Stojko, N., Malakhova, N., Faller, K. (1999). Thick film graphite electrodes in stripping voltammetry. *Fresenius J. Anal. Chem.* 364:285–295.
Brainina, Kh., Malakhova, N. A., Stojko, N. (2000). Stripping voltammetry in environmental and food analysis. *Fresenius J. Anal. Chem.* 368:307–325.
Brainina, Kh., Ivanova, A. V., Khanina, R. M. (2001a). Long-lived sensors with replaceable surface for stripping voltammetric analysis: Part I. *Anal. Chim. Acta* 436:129–137.
Brainina, Kh., Kubysheva, E., Miroshnikoa, E., Parshakov, S., Maksimov, Y., Volkonsky, A. (2001b). Small-size sensors for the in-field stripping voltammetric analysis of water. *J. Field Anal. Chem. Technol.* 5:260–271.
Brineer, R., Chouchoiy, S., Webster, R., Popham, R. (1985). *Anal. Chim. Acta* 172:31.
Brown, S., Kowlski, B. (1979). *Anal. Chem.* 51:2133.
Chaney, E. N., Baldwin, R. P. (1982). *Anal. Chem.* 54:2556.
Christensen, J., Kryger, L., Pind, N. (1982). *Anal. Chim. Acta* 136:39.
Clem, R., Litton, G., Ornelas, L. (1973). *Anal. Chem.* 45:1306.
Copeland, T. R., Skogerboe, R. K. (1974). *Anal. Chem.* 46:1257A.
Cox, J., Kulesza, P. (1983). *Anal. Chim. Acta* 154:71.
Davis, P., Berlandi, F., Dulude, G., Griffin, R., Matson, W. (1978). *Am. Ind. Hyd. Assoc. J.* 6:480.
Davison, I., Smyth, F. (1979). *Anal. Chem.* 51:2127.
DeAngelis, T. P., Bond, R. E., Brooks, E. D., Heineman, W. R. (1977). *Anal. Chem.* 49:1792.
DeVitre, R., Tercier, M., Tsacopoulos, M., Buffle, J. (1991). *Anal. Chim. Acta* 249:419.
Donat, J., Bruland, K. (1988). *Anal. Chem.* 60:240.
Dorten, W., Valenta, P., Nurnberg, H. W. (1984). *Fresenius J. Anal. Chem.* 317:264.
Edmonds, T. (1985). *Anal. Chim. Acta* 175:1.
Eggli, R. (1977). *Anal. Chim. Acta* 91:129.
Eskilsson, H., Haraldsson, C., Jagner, D. (1985). *Anal. Chim. Acta* 175:79.
Esteban, M., Casassas, E. (1994). *Trends Anal. Chem.* 13:110.
Florence, T. M. (1970). *J. Electroanal. Chem.* 27:273.
Florence, T. M. (1972). *J. Electroanal. Chem.* 35:237.
Florence, T. M. (1979). *J. Electroanal. Chem.* 97:219.
Florence, T. M. (1980). *Anal. Chim. Acta* 119:217.
Florence, T. M. (1984). *J. Electroanal. Chem.* 169:207.
Florence, T. M. (1986). *Analyst* 11:489.
Fogg, A. G., Wang, J. (1999). Terminology and convention for electrochemical stripping analysis. *Pure Appl. Chem.* 71:891–897.
Forsman, U. (1983). *Anal. Chim. Acta* 146:71.
Gammelgoard, B., Andersen, J. (1985). *Analyst* 110:1197.
Gardea-Torresdey, J., Darnall, D., Wang, J. (1988). *Anal. Chem.* 60:72.

Gil, E., Ostapczuk, P. (1994). *Anal. Chim. Acta* 293:55.

Golimowski, J., Gustavsson, I. (1984). *Fresenius J. Anal. Chem.* 317:481.

Graneli, A., Jagner, D., Josefson, M. (1980). *Anal. Chem.* 52:2220.

Gunasingham, H., Fleet, B. (1989). In: Bard, A. J., ed. *Electroanalytical Chemistry.* Vol. 16. New York: Marcel Dekker.

Hoyer, B., Florence, T. M., Blately, G. (1987). *Anal. Chem.* 59:1609.

Izutsu, K., Nakamura, T., Takizawa, R., Hanawa, H. (1983). *Anal. Chim. Acta* 149:14.

Jagner, D. (1982). *Analyst* 107:593.

Jagner, D. (1983). *Trends Anal. Chem.* 2:53.

Jagner, D., Graneli, A. (1976). *Anal. Chim. Acta* 83:19.

Jagner, D., Josef, M., Westerlund, S., Aren, K. (1981). *Anal. Chem.* 53:1406.

Jin, J.-Y., Xu, F., Miwa, T. (2000). Square-wave Cathodic stripping voltammetry of ultratrace manganese in the presence of ultrasound irradiation. *Anal. Sci.* 16:317–319.

Kalvoda, R. (1983). *Anal. Chim. Acta* 138:11.

Kalvoda, R. (1984). *Anal. Chim. Acta* 162:197.

Katano, H., Senda, M. (2001). Ion-transfer stripping voltammetry of nonionic and ionic surfactants and its application to trace analysis. *Anal. Sci.* 17:i337–i340.

Konaunur, N., van Loon, G. (1977). *Talanta* 24:184.

Kulys, J., Cenas, N., Svirmicka, G., Svirmickiene, V. (1982). *Anal. Chem. Acta* 138:19.

Laser, D., Ariel, M. (1974). *J. Electroanal. Chem.* 49:123.

Lieberman, S., Zirino, A. (1974). *Anal. Chem.* 46:20.

Liu, C., Jiao, K. (1990). *Anal. Chim. Acta* 238:367.

Lukaszewski, Z., Zembrzuski, W. (1992). *Talanta* 39:221.

Lund, W., Onshus, D. (1976). *Anal. Chim. Acta* 86:207.

Lund, W., Eriksen, R. (1979). *Anal. Chim. Acta* 107:37.

Luo, D. B. (1986). *Anal. Chim. Acta* 189:277.

Luque de Castro, M., Izquierdo, A. (1991). *Electroanalysis* 3:457.

Magjer, T., Branica, M. (1977). *Croat. Chem. Acta* 49:L1.

Mannino, S. (1982). *Analyst* 107:1466.

Mannino, S. (1989). *Analyst* 108:1257.

Mannino, S., Granata, G. (1990). *Italian J. Food Sci. Technol.* 2.

Mart, L., Nurnberg, H., Valenta, P. (1980). *Fresenius J. Anal. Chem.* 300:350.

Novotny, L. (1990). *Electroanalysis* 2:257.

Nygren, O., Vaughan, G., Florence, T. M., Morrison, G., Warner, I., Dale, L. (1990). *Anal. Chem.* 62:1637.

Ostapczuk, P. (1993). *Anal. Chim. Acta* 273:35.

Paneli, M., Voulgaropoulos, A. (1993). *Electroanalysis* 5:355.

Price, J. F., Baldwin, R. P. (1980). *Anal. Chim.* 52:1940.

Shain, I., Lewinson, J. (1961). *Anal. Chem.* 33:187.

Sioda, R., Bailey, G., Lund, W., Wang, J., Leach, S. (1986). *Talanta* 33:421.

Sipos, L., Valenta, P., Nurnberg, H., Branica, M. (1977). *J. Electroanal. Chem.* 77:263.

Stojko, N., Brainina, Kh., Faller, C., Henze, G. (1998). Stripping voltammetric determination of Hg at modified solid electrode. I. Development of modified electrodes. *Anal. Chim. Acta* 371:145–153.

Szentirmay, M., Martin, C. (1984). *Anal. Chem.* 56:1898.

Tercier, M., Buffle, J. (1993). *Electroanalysis* 5:187.

Torrance, K., Gatford, C. (1985). *Talanta* 32:273.

Turner, D., Robinson, S., Whitfield, M. (1984). *Anal. Chem.* 56:2387.

Valenta, P., Sipos, L., Kramer, I., Krumpen, P., Rutzel, H. (1982). *Fresenius J. Anal. Chem.* 312:101.

Vos, L., Komy, Z., Reggers, G., Roekens, E., Van Grieken, R. (1986). *Anal. Chim. Acta* 184:271.

Vydra, F., Stulik, K., Julakova, E. (1976). *Electrochemical Stripping Analysis.* New York: Halsted Press.

Wang, J. (1981). *Anal. Chem.* 53:2280.

Wang, J. (1982a). *Environ. Sci. Technol.* 16:104.

Wang, J. (1982b). *Talanta* 29:125.

Wang, J. (1983). *Am. Lab.* 15(7):14.

Wang, J. (1985). *Stripping Analysis: Principles, Instrumentation and Applications.* Deerfield Beach, Florida: VCH Publishers.

Wang, J. (1989). In: Bard, A. J., ed. *Electroanalytical Chemistry.* Vol. 16. New York: Marcel Dekker, pp. 1–87.

Wang, J. (1990). *Fresnius. J. Anal. Chem.* 337:508.

Wang, J. (1996). Electrochemical preconcentration. In: Kissinger, P. T., Heineman, W. R., eds. *Laboratory Techniques in Electroanalytical Chemistry.* New York: Marcel Dekker, Inc., pp. 719–737.

Wang, J., Dewald, H. D. (1983a). *Anal. Chem.* 55:933.

Wang, J., Dewald, H. (1983b). *Anal. Lett.* 16:925.

Wang, J. and Freiha, B. (1983c). *Anal. Chem.* 55:1285.

Wang, J., Dewald, H. D. (1984). *Anal. Chem.* 56:156.

Wang, J., Freiha, B. (1985). *Anal. Chem.* 57:1776.

Wang, J., Mannino, S. (1989). *Analyst* 114:643.

Wang, J., Taha, Z. (1990). *Electroanalysis* 2:383.

Wang, J., Baomin, T. (1992). *Anal. Chem.* 64:1706.

Wang, J., Luo, D. B., Farias, P. A. M., Mahmoud, J. S. (1985a). *Anal. Chem.* 57:158.

Wang, J., Lou, D. B., Farias, P. A. M. (1985b). *Analyst* 110:885.

Wang, J., Luo, D. B., Farias, P. A. M. (1985c). *J. Electroanal. Chem.* 110:885.

Wang, J., Lin, M. S., Villa, V. (1986). *Anal. Lett.* 19:2293.

Wang, J., Zadeii, J., Lin, M. S. (1987). *J. Electroanal. Chem.* 237:281.

Wang, J., Lu, J., Taha, Z. (1992a). *Analyst* 117:35.

Wang, J., Setiadji, R., Chen, L., Lu, J., Morton, S. (1992b). *Electroanalysis* 4:161.

Wang, J., Rongrong, X., Baomin, T., Wang, J., Renschler, C., White, C. (1994). *Anal. Chim. Acta* 233:43.

Wang, J., Grundler, P., Flechsig, G.-U., Jasinski, M., Rivas, G., Sahlin, E., Paz, J. L. L. (2000). Stripping analysis at a heated carbon paste electrode. *Anal. Chem.* 72:3752–3756.

Wojciechowski, M., Winston, G., Osteryoung, J. (1985). *Anal. Chem.* 57:155.

Yokoi, K., van de Berg, C. M. G. (1991). *Anal. Chim. Acta* 245:167.

Yokoi, K., van de Berg, C. M. G. (1992). *Anal. Chim. Acta* 257:293.

Zaharchuk, N. F., Yu, S., Borisova, N. S., Brainina, Kh. (1999). Modified thick-film graphite electrodes: morphology and stripping voltammetry. *Electroanalysis* 9:614–622.

Zirino, A., Lieberman, S. H., Clavell, C. (1978). *Environ. Sci. Technol.* 12:73.

19

Measurement of Electrolytic Conductance

STACY L. GELHAUS, WILLIAM R. LACOURSE
Department of Chemistry and Biochemistry, University of Maryland, Baltimore County, Baltimore, MD, USA

I. INTRODUCTION

The foundations of conductivity were established throughout the 19th century by four scientist in particular, Alessandro Volta, Georg Ohm, Michael Faraday, and Friedrich Kolrausch. In 1800, Volta invented the first electric battery. The invention of a battery that supplied a continuous electric current paved the way for Georg Ohm who began measuring electric current through metals and in 1837, established Ohm's law. Around the same time Ohm established this law, Faraday was working on the law of electrolysis and determined that aside from electric current being carried by metals, ions in solution could carry a current as well. However, it was not until 1869, that Kolrausch began measuring the conductivity of electrolytes, giving electrolytic conductivity a physical and mathematical basis. All solutions possess some degree of conductivity. The conductance of a solution is an important characteristic for a variety of fields and applications to industry, from measuring the purity of water to monitoring the conditions of processes in the dairy and brewing industries. All along, instrumentation has developed to meet the needs of the new and changing applications. Bench top conductivity meters have been replaced by handheld field devices; nonintrusive electrodeless sensors have been developed, and conductivity cells have been reduced so much in size that they are now able to fit inside microchips or even slide along the outside of capillary tubing.

The two main type of non-faradaic electrochemical analysis can be categorized as potentiometric, discussed in a previous chapter, and conductometric. This chapter will focus on conductometric measurement techniques, specifically those which provide information about the total ionic content of an aqueous solution—the electrolytic conductance. The motion of ions in a solution can be altered by the application of a potential difference between two electrodes immersed in the solution. When a voltage is applied across two platinum electrodes (usually platinized) placed in an electrolytic solution, an electric current will be transferred to an extent that is in accordance with the amount and mobility of free positive and negative ions present in the solution. This characteristic of the solution is nonspecific, a bulk property, and this can be perceived as a drawback in some detection scenarios because measuring conductivity will give the overall solution conductance, not the conductance of each individual analyte. However, there are many situations where the bulk property of conductivity can be useful. One obvious case is in characterizing the purity of potable waters and monitoring the effectiveness of water demineralizers. Conductance can be used effectively to measure the concentration of acids and bases, as it gives a monotonic, though not always linear, conductance–concentration plot over wide concentration ranges. As electrolytes vary with respect to their ability to conduct current, conductance measurements provide a valuable method of analyzing binary mixtures of electrolytes. By measuring conductance, ionic mobilities, diffusion coefficients, and transport numbers can also be determined. Conductometric titrimetry is a useful method for following the course of reactions involving electrolytes. Along with titrimetry, electrode cells and sensors have also been developed both in lab and commercially for the measurement of conductance. Conductance detectors with flowthrough cells are commonly used for the detection of inorganic ions and organic acids following separation by high performance liquid chromatography (HPLC). There are many different types of instrumentation available for conductivity measurement. It is through the principles and theory of conductivity that the reader will obtain an adequate understanding of the available instrumentation and applications of conductivity.

A. Principles of Conductivity

By applying a rapidly oscillating field (>1 kHz) in an electrochemical cell, the mobility of ions in solution, or conductivity can be measured (see Fig. 1). No oxidation or reduction takes place in conductivity detection, but charging and discharging of cell electrodes do occur (LaCourse, 1997). The fundamental measurement used to study conductivity is resistance, R, of the solution. Under the precautions required the earlier experimental description should follow Ohm's law in which

$$R = \frac{\rho l}{A} \qquad (1)$$

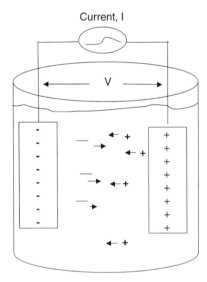

Figure 1 The movement of ions in solution under an applied potential. Anions move to the cathode and cations move to the anode.

where resistance, R, equals the resistivity (ρ) multiplied by the length (l) between electrodes divided by the area (A) of that geometry between electrodes. The resistivity, also called specific resistance, has the units of ohms/centimeters. For solutions, the area is usually taken as the area of each parallel electrode and the length becomes the distance between them. The conductance of a solution, G, is the inverse of the resistance, R, of the solution; therefore, $G = 1/R$. As resistance is expressed in ohms (Ω), conductance is expressed in Ω^{-1}. The reciprocal ohm used to be called mho, but it is now officially called Siemens, S, and $1\,S = 1\Omega^{-1}$.

The conductance of a sample is highly dependent on the cross-sectional area and the length between electrodes. The conductance will decrease as l increases and increases with cross-sectional area (Atkins, 1998).

$$G = \frac{\kappa A}{l} \qquad (2)$$

In the equation above, κ, is the conductivity. The units of κ are S/m. This equation is also the inverse of the resistivity where $\kappa = 1/\rho = (1/R)(d/A) = G(d/A) = G\theta$. In this instance, θ, is equal to d/A in cm^{-1}. As θ is a function of the geometry of the electrode and is a cell constant; it is useful in the characterization of cells. When electrode geometry diverges from plane and parallel, the cell constant can be best determined by measuring solutions of known specific conductance ($\kappa = G\theta$). Reference solutions for this purpose have been characterized in cells of known geometry. A number of such solutions are listed in Table 1.

As conductance is determined by the totality of ions present in a solution, largely acting independently of each other, it can be expressed as a summation. The κ can be thought of as the conductance of one cubic centimeter of solution. Suppose 1 cm^3 contains 1 gram-equivalent

Table 1 Reference Solutions for Calibration of Cell Constants

Approx. molarity	Method of preparation	Temp (°C)	G (μS/cm)
1.0	74.2460 g KCl per L of solution at 20°C	0	65,176
		18	97,838
		25	111,342
0.1	7.4365 g KCl per L of solution at 20°C	0	7,138
		18	11,167
		25	12,856
0.01	0.7440 g KCl per L of solution at 20°C	0	773.6
		18	1,220.5
		25	1,408.8
0.001	Dilute 100 mL of 0.01 M to 1 L at 20°C	25	146.93

Source: ASTM (1992).

of electrolyte. The equivalent conductivity Λ can be written in the following terms:

$$\Lambda = \frac{1000}{c}\kappa \quad (3)$$

The molar concentration is c and the SI units of equivalent conductivity is Siemens meter-squared per mol (S m^2/mol). The summation covers all ions of both signs. For some purposes it is desirable to define a molar conductivity, for which the usual symbol is λ_m. Typical values for this are 10 mS m^2/mol where 1 mS = 10^{-3} S. The λ_m is given by the relation

$$\lambda_m = \lambda^+ + \lambda^- = \frac{\kappa}{c} \quad (4)$$

The molar conductivity, λ_m, is a property of ions (either positive or negative) that gives quantitative information about their relative contributions to the conductance of the solution (Atkins, 1998; Dahmen, 1986). Arrhenius postulated in 1887, that an appreciable amount of electrolyte will dissociate into free ions in solution. This value is to some extent dependent on the total ionic concentration, increasing with increasing dilution. The molar conductivity of an electrolyte would be independent of concentration if κ were proportional to the concentration of the electrolyte (Dahmen, 1986). Unfortunately, the molar conductivity is found to vary with concentration experimentally. A possible reason for this is that the number of ions in the solution might not be proportional to the concentration of the electrolyte. The concentration of ions in a weak acid solution depends on the concentration of the acid in a complicated way and doubling the concentration of acid does not double the number of ions. Secondly, because ions act strongly with one another, the conductivity of the solution is not exactly proportional to the number of ions present. This concentration dependence indicates that there are two possible classes—strong electrolytes and weak electrolytes. The classification of electrolytes into either category depends not only on the solute, but also the solvent (Atkins, 1998).

B. Strong Electrolytes

Strong electrolytes are substances that completely ionize in solution and include ionic solids (NaCl) and strong acids (HCl). As they completely ionize, the concentration of ions in solution is proportional to the concentration of electrolyte added (Atkins, 1998). In an extensive series of experimentation during the 19th century, Friedrich Kohlrausch showed that at low concentrations the molar conductivities of strong electrolytes vary linearly with the square root of the concentration:

$$\lambda_m = \lambda_m^0 - Kc^{1/2} \quad (5)$$

This variation is called Kohlrausch's law (Atkins, 1998). The constant λ_m^0 is the limiting molar conductivity, or in other words, the molar conductivity in the limit of zero concentration (when ions are effectively far apart and do not interact with one another). The constant K is found to depend more on the stoichiometry of the electrolyte (MA or M$_2$A) rather than on its specific identity (Atkins, 1998). He was also able to show that λ_m^0 can be expressed as the sum of contributions from its individual ions. If the limiting molar conductivity of cations is denoted λ_+ and the anions λ_-, then his law of the independent migration of ions states that

$$\lambda_m^0 = \nu_+ \lambda_+ + \nu_- \lambda_- \quad (6)$$

where ν_+ and ν_- are the numbers of cations and anions per formula unit of electrolyte.

C. Weak Electrolytes

Weak electrolytes do not fully ionize in solution and they include Brønsted acids and bases, such as CH$_3$OOH and NH$_3$. The marked concentration dependence of their molar conductivities arises from the displacement of the equilibrium toward products at low molar concentrations.

$$HA(aq) + H_2O(l) H_3O^+(aq) + A^-(aq) K_a$$

$$= \frac{a(H_3O^+)a(A^-)}{a(HA)}$$

A weak electrolyte has a molar conductivity, which is normal at concentrations close to zero, but then falls drastically to low values as the concentration increases (Atkins, 1998). When Kolrausch plotted weak electrolyte conductivity vs. the square root of concentration graph was tangential. The conductivity depends on the number of ions in the solution, and therefore, on the degree of ionization, α, of the electrolyte. Equivalent conductivity reaches a limiting value for infinite dilution (see Table 2).

$$\lambda_0 = \lambda_0^+ + \lambda_0^- \quad (7)$$

The degree of dissociation from conductivity, α, the activity.

$$\alpha = \frac{\lambda}{\lambda_0} \quad (8)$$

so that

$$\lambda = \alpha \lambda_0 = \alpha(\lambda_0^+ + \lambda_0^-)$$

This was also derived by Arrhenius and is similar to Ostwald's dissociation of a weak acid (Dahmen, 1986).

$$K = \frac{[H+][A-]}{[HA]} = \frac{\alpha^2 c}{1-\alpha} \text{ or } K = \frac{\alpha^2}{(1-\alpha)V}$$

For 1 equivalent of HA in $c = 1/V$.

Table 2 Equivalent Ionic Conductivity of Selected Ions at Infinite Dilution in S*cm^2*mol at 25°C

Cations[a]	Λ_0	Temp. coeff.[b]	Anions (a)	Λ_0	Temp. coeff.[b]
H^+	349.8	0.0139	OH^-	109.6	0.018
$Co(NH_3)_6^{3+}$	102.3	—	$Fe(CN)_6^{4-}$	110.5	0.02
K^+	73.5	0.0193	$Fe(CN)_6^{3-}$	101.0	—
NH_4^+	73.5	0.019	$Co(CN)_6^{3-}$	98.9	—
Pb_2^+	69.46	0.02	SO_4^{2-}	80.0	0.022
La_3^+	69.6	0.023	Br^-	78.14	0.0198
Fe_3^+	68.0	—	I^-	76.8	0.197
Ba_2^+	63.64	0.023	Cl^-	76.4	0.0202
Ag^+	61.9	0.021	$C_2O_4^{2-}$	74.2	0.02
Ca_2^+	59.5	0.0230	NO_3^-	71.42	0.020
Cu_2^+	53.6	0.02	CO_3^{2-}	69.3	0.02
Fe_2^+	54.0	—	ClO_4^-	67.3	0.020
Mg_2^+	53.06	0.022	HCO_3^-	44.5	—
Zn_2^+	52.8	0.02	$CH_3CO_2^-$	40.9	0.022
Na^+	50.11	0.0220	$HC_2O_4^-$	40.2	—
Li^+	33.69	0.0235	$C_6H_5CO_2^-$	32.4	0.023
$(n\text{-Bu})_4N^+$	19.5	0.02	$Picrate^-$	30.4	0.025

[a]For ions of charge z, the figures given are on an equivalent basis, so that they apply to the fraction $(1/z)$ of a mole.
[b]The temperature coefficient, when known, is given as $(1/\lambda_0)(d\lambda_0/dT)$ with units of K^{-1}.
Source: Frankenthal (1963).

The degree of ionization is defined so that, for the acid HA at molar concentration, c, at equilibrium

$$[H_3O^+] = \alpha c \quad [A^-] = \alpha c \quad [HA] = (1-\alpha)c$$

If we ignore activity coefficients, the acidity constant, K_a, is approximately

$$K_a = \frac{\alpha^2 c}{1-\alpha} \quad (9)$$

from which follows that

$$\alpha = K_a \left[\frac{(1+4c)^{1/2}}{K_a} - 1 \right] \quad (10)$$

The electrolyte is fully ionized at infinite dilution, and its molar conductivity is then λ_m^0 (Dahmen, 1986). As only a fraction of α is actually present as ions in the actual solution, the measured molar conductivity λ_m is given by Robinson and Stokes (1959) to Table 3:

$$\lambda_m = \alpha \lambda_m^0 \quad (11)$$

D. Ion Mobility and Transport

To interpret conductivity measurements, it is useful to know why ions move at different rates, why they have different molar conductivities, and why the molar conductivities of strong electrolytes decrease with the square root of the molar concentration. The motion of ions in solution is largely random; however, the presence of an electric field does bias this movement causing the ions to undergo net migration through solution (Brett and Brett, 1993). The current, I, that passes between two parallel electrodes is related to the flux or charge, j, and to the potential difference between them, $\Delta\emptyset$, by

$$I = jA = \kappa \frac{\Delta\emptyset A}{l} = \kappa E A \quad (12)$$

Table 3 Mobilities of Some Ions in Water at Infinite Dilution

Cations	10^{-4} μ/cm^2 s V	Anions	10^{-4} μ/cm^2 s V
H^+	36.2	OH^-	20.6
Li^+	4.0	F^-	5.7
Na^+	5.2	Cl^-	7.9
K^+	7.6	Br^-	8.1
Rb^+	8.1	I^-	8.0
Cs^+	8.0	NO_3^-	7.4
NH_4^+	7.6	ClO_4^-	7.0
Mg_2^+	11.0	SO_4^{2-}	8.3
Ca_2^+	6.2	CO_3^{2-}	7.5
Cu_2^+	5.6		
Zn_2^+	5.5		

Source: Brett and Brett (1993).

The ions in the solution between them experience a uniform electric field of magnitude

$$E = \frac{\Delta\emptyset}{l} \quad (13)$$

For each ion

$$\kappa_i = z_i c_i u_i F \quad (14)$$

In such a field, an ion of charge ze experiences a force of magnitude

$$F = zeE = ze\frac{\Delta\emptyset}{l} \quad (15)$$

A cation responds to the application of the field by accelerating toward the negative electrode and an anion responds by accelerating toward the positive electrode. As the ion moves through, the solvent experiences a frictional retarding force, F_{fric}, proportional to its speed (Brett and Brett, 1993). The two forces act in opposite directions, and the ions reach a terminal speed, the drift speed (s), when accelerating force is balanced by the viscous drag. The net force is zero when

$$s = \frac{zeE}{f} \quad (16)$$

Because the drift speed governs the rate at which charge is transported, we might expect the conductivity to decrease with increasing solution viscosity and ion size.

Molar conductivity

$$\lambda_{\text{mi}} = \frac{\kappa_i}{c_i} = z_i u_i F \quad (17)$$

and the equivalent conductivity

$$\Lambda_m = \sum \lambda_i = \sum \frac{\kappa_i}{c_i} \quad (18)$$

Unfortunately this measurement is not species selective and individual ionic conductance can only be calculated if conductance or mobility of one ion is known.

In electric fields of high intensity (order of 100 kV/cm) the conductivity increases with field strength. This is experimentally true for large bulky ions, but not for small ones. Strong electrolytes move without a solvent sheath as relaxation time for the ionic atmosphere becomes too large; a limiting current is reached as field strength is increased. The molar conductivities for alkali metals increase from Li$^+$ to Cs$^+$ even though atomic radii increase. This is due to the fact that smaller ions have stronger electric fields; therefore, they are able to solvate more extensively, giving them a larger hydrodynamic radius. This increase is inversely proportional to the atomic number of alkali metals. Stoke's law states that there is a decrease of size in that order for entire ionic moiety as it is moved by electric driving force because of the decreasing degree of hydration of alkali metal ions (Dahmen, 1986). However, weak electrolytes in the electric field interact with dipoles of undissociated molecules, increasing the dissociation constant. H$_3$O$^+$ and $^-$OH have exceptionally high mobilities. This is caused by the proton transfer between neighboring H$_2$O molecules and is confirmed by the special properties of acids and bases. The mobility of Cl$^-$ and NO$_3^-$ is nearly equal to that of K$^+$. Diffusion potentials are avoided in cells when these combinations of salt bridges are employed (Dahmen, 1986). Relaxation mobility and the diffusion coefficient are related by the chemical potential.

$$\mu_i = \mu_i^\theta + RT \ln c_i \quad (19)$$

One can also differentiate with respect to distance

$$\left(\frac{\partial \mu_i}{\partial x}\right)_{P,T} = \frac{RT}{c_i}\left(\frac{\partial c_i}{\partial x}\right)_{P,T} \quad (20)$$

and the diffuse force felt by the particle is

$$F = -\left(\frac{\partial \mu_i}{\partial x}\right)_{P,T} = \frac{RT}{c_i}\left(\frac{\partial c_i}{\partial x}\right)_{P,T} \quad (21)$$

The number of flux ions, i, J_i, is

$$J_i = \frac{j_i}{z_i e} = c_i \mu_i E \quad (22)$$

Substituting for electric field intensity, E

$$J_i = \frac{\mu_i RT}{z_i F}\left(\frac{\partial \mu_i}{\partial x}\right)_{P,T} \quad (23)$$

A comparison with Fick's law shows that

$$D_i = \frac{\mu_i RT}{z_i F} \quad (24)$$

This is the Einstein relation and shows the direct proportionality between the diffusion coefficient and mobility. The relation between conductivity and diffusion coefficient can be seen in the Nernst–Einstein relation and is easily derived from

$$\lambda_i = \frac{z_i^2 F^2 D^2}{RT} \quad (25)$$

This permits the estimation of diffusion coefficients from the conductivity measurements (Brett and Brett, 1993).

The transport number $t_\pm$ is defined as the fraction of total current carried by the ions of a specified type. For a solution with two kinds of ion, the transport numbers of the cations (t_+) and anions (t_-) are

$$t_\pm = \frac{I_\pm}{I} \qquad (26)$$

where $I_\pm$ is the current carried by the cation (I_+) or anion (I_-) and I is the total current through the solution. It follows that the sum of total current of anions and cations must be equal to 1 (Atkins, 1998). The limiting transport number $t_\pm^0$ is defined in the same way, but for the limit of zero concentration of the electrolyte solution. The current that can be ascribed to each type of ion is related to the mobility of the ion by the following equation:

$$t_\pm^0 = \frac{z_+ v_+ u_+}{z_+ v_+ u_+ + z_- v_- u_-} \qquad (27)$$

Because the ionic conductivities are related to the mobilities, it follows that

$$t_0^+ = \frac{\lambda_0^+}{\lambda_0} \quad \text{and} \quad t_0^- = \frac{\lambda^-}{\lambda_0} \qquad (28)$$

The transport number varies with ionic constitution of solution and is another way of expressing conductivities or mobilities. The transport number for each ion at infinite solution was determined by Hittorf (1854–1859). Another method for determining mobilities from $\Lambda+/\Lambda-$ ratio can be calculated on the basis of absolute velocities of ions under the influence of a potential gradient. The idea was postulated by Lodge in 1886, and the experiments were performed by Masson in 1899 (Dahmen, 1986).

Much information pertaining to ionic equilibria can be obtained from conductometric data, particularly in situations where ions tend to be removed from solution by an equilibrium process. This applies, for example, to the combination of anions with hydrogen cations to form partially dissociated molecular acids, to the formation of complexes between metallic cations and various ligands, and to the formation of sparingly soluble salts. Thus the measurement of conductance can lead to the establishment of acidic and basic dissociation constants, stability constants, and solubility product constants. Further details of theory can be found in any modern text on physical chemistry or in a textbook on electrochemistry.

II. INSTRUMENTATION

There are two general types of devices for measuring conductance. The first, and most widely used, employs a pair of contacting electrodes, frequently platinum, immersed in the test liquid. The second type of instrumentation is noncontacting or "electrodeless" and depends on inductive or capacitive effects to measure conductance. The remainder of this chapter is divided into two parts, which will include the description of each type of instrumentation, commercial examples of each, and examples of applications.

A. Immersed Electrode Measurements

1. Conductivity Cells

The conductivity cell can be represented schematically by the equivalent circuit shown in Fig. 2. Most measurements are made using two electrodes with the same geometric surface area, hence the same cell constant, θ. The distance between the electrodes, the area, is also a known dimension.

The capacitance and resistance of the cell connectors and the contacts they make are shown as C_C and R_C, respectively. As the electrodes are assumed to be identical, the double-layer capacitances, C_d, of the two electrodes are assumed to be equal. The ohmic resistance of the solution between the electrode surfaces is R_S. An interelectrode capacitance term, C_I, is included to account for the dielectric properties of the bulk solvent. And lastly, a frequency dependent faradaic impedance, Z_f, which includes both charge-transfer resistance and Warburg impedance, is shown for each electrode (Coury, 1999).

When making low impedance connections to the cell, it is usually valid to ignore the contribution of C_C and R_C. Also, the complications imposed by Faradaic impedance can be minimized through experimental design. Conductance measurements are often made by applying an AC potential and then measuring the current. At high frequencies, the branch of the circuit containing Z_f can be neglected as it varies as the reciprocal of the square root of frequency. The circuit in Fig. 2 can be simplified to the circuit seen below in Fig. 3. In this diagram, C_p and C_S represent the combined parallel and series capacitances, respectively. R_s is the ohmic resistance of the solution between the electrodes (Coury, 1999). If an AC potential is applied, alternating current will flow through R_s and at the same time through C_p. If C_p can be kept small, the effect of R_s can be studied by itself. In practice,

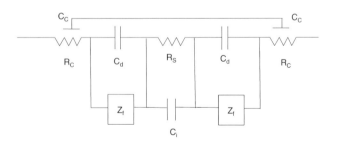

Figure 2 The electronic equivalent to a conductivity cell circuit. [Redrawn from Coury (1999).]

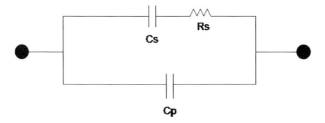

Figure 3 The simplified version of the conductivity cell circuit shown in Fig. 2. [Redrawn from Coury (1999).]

Table 4 Recommended Cell Constants for Various Conductance Ranges

Conductance range (μS/cm)	Cell constant (cm^{-1})
0.05–20	0.01
1–200	0.1
10–2000	1.0
100–20,000	10.0
1000–200,000	50.0

Source: ASTM (1992).

the double-layer capacitances can be increased substantially by platinization, which greatly increases the effective surface area of the electrodes. Then C_p becomes significant only with solutions of high resistance (low conductance), when large electrodes close together must be selected in order to keep the measurement within range. Hence, most commercial instruments for measuring electrolytic conductance operate on AC. Some use the 60 or 50 Hz power-line frequency for convenience; however, higher frequencies favor low impedance for double-layer capacitances; therefore, many instruments include built-in oscillators to provide excitation (typically) 1 or 2 kHz at 5 V (RMS). The most common contacting electrodes used to measure conductivity have two- or four-electrode cells.

Two-Electrode Cells

Figure 4 shows a simplified drawing of a two-electrode conductivity cell. Certain cells are primarily intended for precision physicochemical measurements, whereas others are more convenient for routine use or for titrimetry. For precise work, cells should be held at a constant temperature, as the conductance of most electrolytic solutions increases at the rate of about 2% per Kelvin (see Table 2). Cells are made with various cell constants, of which 1.0 and 0.1 are the most widely useful. Table 4 indicates the cell constants appropriate for various ranges of conductance (ASTM, 1992).

The electrodes are usually fabricated from platinum, though graphite, titanium, and tungsten are also occasionally used. Platinum electrodes are best coated with a finely divided form of the metal, known as platinum black. This coating can be produced *in situ* by a few minutes of electrolysis in a solution of chloroplatinic acid containing a small amount of lead acetate. The electrolysis should be repeated with reversed polarity. After platinizing, the electrodes should be stored in distilled water.

Four-Electrode Cells

The innovative design of the four-electrode for contacting conductivity cells reduces the problem of polarization error and fouling. The technique using these cells, illustrated in Fig. 5, minimizes the effect of resistive electrode coatings caused by solution contaminants.

The current-carrying outer electrodes function similarly to the two-electrode cell in which the measured resistance depends not only on the solution but also the resistance of any coatings that might develop on the electrodes. Ohm's law is given for this situation by

$$V = I(R_{\text{solution}} + R_{\text{coating}}) \tag{29}$$

where V is the voltage across and I the current through the electrodes. Thus, with the two-electrode circuit, the measured resistance (or its reciprocal, conductance) is the sum of the solution and coating resistances. The alternating current is only applied to the outer rings. The

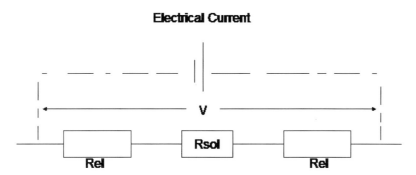

Figure 4 A schematic of a two-electrode conductivity cell (Redrawn from www.radiometer-analytical.com.)

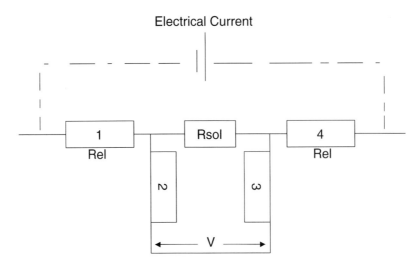

Figure 5 A schematic of a four-electrode conductivity cell (Redrawn from www.radiometer-analytical.com.)

resistance of the electrode coating does not affect E_{VME}, the measured voltage between the inner electrodes, because no current is drawn through them. With an independently measured value of the current, I, through the current-carrying electrodes, the resistance of the solution can be calculated from Eq. (29) and the potentiometrically measured voltage:

$$R_{solution} = \frac{E_{VME}}{I} \quad (30)$$

It is essential for both two- and four-electrode cells, that the source voltage must be from an alternating-current supply, as direct current could cause unwanted electrochemical reactions and polarization at the electrodes. Table 5 lists the advantages and disadvantages of the two- and four-electrode cells (www.radiometer-analytical.com).

2. Circuitry

The traditional circuit for measuring resistance is the Wheatstone bridge (Fig. 6). The bridge circuit consists of four resistors, an AC voltage source, and a detector. This circuit is well adapted to static measurements, but as the bridge must be balanced to give a reading, it is not readily applicable to continuous measurements. The Wheatstone bridge circuit is used to measure medium resistance values (1 Ω to 1 MΩ). However, the two-ganged multiple-point switch shown selects between several ranges (only two are shown, but there may be more).

Table 5 Advantages and Disadvantages of Two- and Four-Electrode Cells

Advantages	Disadvantages
Two-Electrode Cell	
Easy to maintain	Field effects cell must be positioned in the center of the measuring vessel
Use with sample changer (no carryover)	Only cells with no bridge between the plates
Economical	Polarization in high conductivity samples
Recommended for viscous media or samples with suspension	Calibrate using a standard with a value close to the measuring value, measurement accurate over two decades
Four-Electrode Cell	
Linear over a very large conductivity range	Unsuitable for micro samples, depth of immersion 3–4 cm
Calibration and measurement in different ranges	Unsuitable for use with a sample changer
Flowthrough or immersion type cells	
Ideal for high conductivity measurements	
Can be used for low conductivity measurements if cell capacitance is compensated	

Source: Coury (1999).

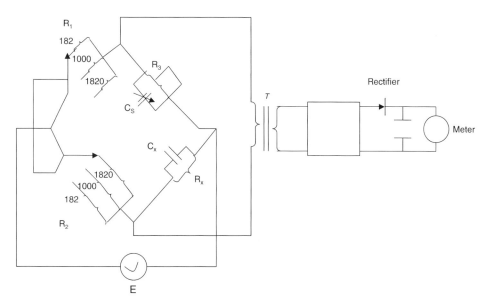

Figure 6 An AC Wheatstone bridge for measuring conductance. The two-pole and three-pole position switch permits choice of 0.1, 1.0, or 10 as ratios multiplying the readings of the variable resistor, R_3. [From Braunstein and Robbins (1971).]

In commercial models, accuracies on the order of $\pm 0.1\%$ are possible (Nilsson and Riedel, 2001). At balance, when the meter shows a null, it is easily demonstrated that the resistance of the test cell is given by

$$R_x = \frac{R_1 R_3}{R_2} \quad (31)$$

or the conductance by

$$L_x = \frac{R_2}{R_1 R_3} \quad (32)$$

Note that R_3 is a calibrated variable resistor or a bank of decade resistors, so that its reading a balance, multiplied by the R_1/R_2 ratio, gives the resistance of the test cell directly. In some bridges a switch is included to interchange the positions of R_x and R_2, which permits a dial reading (on R_3) directly proportional to conductance rather than resistance:

$$L_x = \frac{R_3}{R_1 R_2} \quad (33)$$

For precise results, it is necessary to include a small variable capacitor to cancel the effects of the cell capacitance, as without this the balance point of the bridge would not be sufficiently sharp. A detailed discussion of the capacitive effect has been published (Braunstein and Robbins, 1971). A number of electronic circuits have been developed to give an output voltage proportional to the conductance of a sample without the need for balancing a bridge. Such an instrument is needed for continuous monitoring of flow streams; such an example would be the output of a detector for liquid chromatography. If is also useful for conductometric titrimetry with constant inflow or reagent, the voltage signal being displayed against time on chart recorder or computer (Ahmon, 1977; Daum and Nelson, 1973; Muha, 1977). A self-balancing bridge controlled by a microcomputer has been reported (Kiggen and Laumen, 1981).

3. Commercial Instruments

Commercial instruments for the measurement of conductance come in the form of meters, probes, and sensors. The majority of instruments in today's market for the measurement of electrolytic conductance are self-contained electronic units provided with digital readout. Top-of-the-line instruments include a temperature-sensing probe and automatic temperature compensation based on a range of value of the temperature coefficient. They may also have an option of displaying the temperature of the solution. Provision is usually made for entering the cell constant pertaining to the cell in use, so that the displayed readings are directly expressible in terms of Siemens per centimeter. Some instruments offer a selection of operating frequencies, with lower frequencies for solutions of low conductivity, where cell constants are low ($0.1\,\mathrm{cm}^{-1}$, e.g.) requiring higher interelectrode capacitance.

4. Applications

One of the most common applications of conductivity is the measurement of ions in water sources. Figure 7 shows specific conductances for a number of materials.

Water itself is a very poor conductor; its specific conductance due to dissociation into H_3O^+ and OH^- ions is

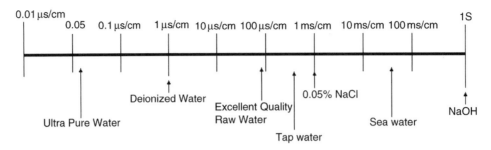

Figure 7 Specific conductance ranges for some typical solutions. [Compiled from Ewing (1985) and www.analyzer.com.]

0.055 μS/cm at 25°C (Light, 1997; Morash, et al., 1994; Thornton and Light, 1989) Water of this theoretical purity can be produced using commercially available nuclear-grade ion-exchange resins. It is used extensively in the semiconductor, power, and pharmaceutical industries. The conductivity measurement is extremely sensitive to traces of ionic impurities. The presence of 1 ppb (1 μg/L) of sodium chloride will increase the conductivity by 4% from 0.0550 to 0.0571 μS/cm at 25°C. The resistivity, which is the unit more commonly used for measurement of ultrapure water, will correspondingly decrease from 18.2 to 17.5 MΩ/cm at 25°C. Ordinary distilled or deionized water with a conductivity of about 1 μS/cm falls far short of this purity. The conductivity and resistivity of ultrapure water over the range of 0–100°C are shown in Table 6 (Light, 1997; Morash, et al., 1994; Thornton and Light, 1989). Table 7 gives the conductivity of different types of water at 25°C (www.topac.com/conductivityprobes.html).

Water purification equipment is often provided with conductance monitors that can be configured to shut down a still or initiate regeneration of the ion-exchange bed of a demineralizer if the conductance becomes too high.

Quantitative analysis for many common acids, bases, and salts may be carried out rapidly by conductivity measurements as well. Table 8 is an extensive compilation of equivalent conductivities over the range of commonly useful concentrations (MacInnes, 1951).

This table used in conjunction with $\kappa = G\theta$ and Eq. (3), permits accurate calculation of concentration of many of the most common binary electrolyte solutions. Solutions of strong electrolytes show a nearly linear increase of conductance with concentration up to about 10% or 20% by weight. At higher concentrations the conductance decreases again, due to such interactions as complexation reactions, formation of dimers or higher polymers, or increased viscosity. Figure 8 shows the relation between conductance and concentration for a few representative solutes.

Conductance measurements can be useful in following the kinetics of reactions that involve a change in ionic content or mobility. For example, see a report by Queen and Shabaga (1973) on the kinetics of a series of solvolytic reactions. Kiggen and Laumen (1981) have used similar measurements to observe diffusion processes in solution.

Conductometric Titrimetry

Conductometric titrimetry is widely applicable for titration reactions involving ions. Figure 9 shows an example of the type of curve that result in the titration of a strong acid with a strong base.

Any type of titration can be carried out conductometrically if a substantial change in conductance takes place before and/or after the equivalence point. Conductometric titration has been used in acid–base, precipitation and complex formation titrations, and displacement titrations.

Table 6 Conductivity and Resistivity of Theoretically Pure Water

Temperature (°C)	Conductivity (μS/cm)	Resistivity (MΩ/cm)
0	0.01162	86.09
25	0.0550	18.18
50	0.1708	5.855
75	0.4010	2.494
100	0.7768	1.287

Source: Light (1997), Morash et al. (1994), Thornton and Light (1989).

Table 7 Conductivity of Different Types of Water at 25°C

Water	μS/cm
Ultrapure water	0.055
Distilled water	0.5–5
Rain water	20–100
Mineral water	50–200
River water	250–800
Tap water	100–1500

Source: www.topac.com/conductivityprobes.html.

Electrolytic Conductance

Table 8 Equivalent Conductances of Some Electrolytes at 25°C

Electrolyte	Concentration (M)							
	0.0000	0.0005	0.001	0.005	0.01	0.02	0.05	0.10
NaCl	126.45	124.50	123.73	120.65	118.51	115.76	111.06	106.74
KCl	149.86	147.81	146.95	143.55	141.27	138.34	133.37	128.96
LiCl	115.03	113.15	112.4	109.4	107.32	104.65	100.11	95.86
HCl	426.16	422.74	421.36	415.80	412.00	407.24	399.09	391.32
NH_4Cl	149.7	—	—	—	141.28	138.33	133.29	128.75
KBr	151.9	—	—	146.09	143.43	140.48	135.68	131.39
KI	150.3	—	—	144.37	142.18	139.45	134.97	131.11
NaI	126.94	125.36	124.25	121.25	119.24	11.70	112.79	108.78
NaO_2CCH_3	91.0	89.2	88.5	85.72	83.76	81.24	76.92	72.80
$NaO_2CCH_2CH_3$	85.92	84.24	83.54	80.90	79.05	76.63	—	—
$NaO_2C(CH_3)_2CH_3$	82.70	81.04	80.31	77.58	75.76	73.39	69.32	65.27
KNO_3	144.96	142.77	141.84	138.48	132.82	132.41	126.31	120.40
$KHCO_3$	118.00	116.10	115.34	112.24	110.08	107.22	—	—
$AgNO_3$	133.36	131.36	130.51	127.20	124.76	121.41	115.24	109.14
NaOH	248	246	245	240	237	233	227	221

Source: MacInnes (1951).

Conductometric titrimetry is rarely used for redox titrations because the needed conditions cannot be met.

The concentration of electrolytes not participating in the titration reaction should be kept small in order to keep the background conductance low and improve sensitivity. Conductance ($G = \kappa/\theta$) is usually plotted against the titration parameter λ, the degree of conversion. For the case of strong acids and bases, the plot of G vs. λ shows a plot yielding straight lines of positive and negative slopes. It is where these slopes intersect that is the equivalence point. A different type of plot is obtained when titrating a weak acid with a strong base. The equivalence point on these graphs is sometimes hard to distinguish, especially at low acid concentrations with high pKa values. In these cases the curves lie lower and start to rise at smaller λ value making the inflection point harder to discern (see Fig. 10) (Dahmen, 1986).

Ion Chromatography

Conductance detectors are routinely used with ion-exchange and similar types of chromatography, including ion chromatography (IC). This latter analytical tool has been developed during the last several decades, and has rapidly established for itself a major position in analytical instrumentation. IC can be used for the analysis of

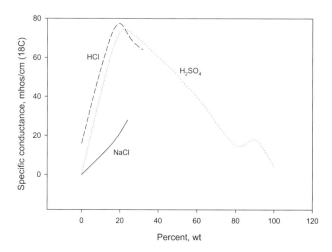

Figure 8 Conductance vs. concentration for selected electrolytes. [Adapted from Rosenthal (1957).]

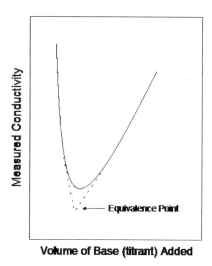

Figure 9 Titration of a strong acid with a strong base. [Adapted from Shugar and Dean (1990).]

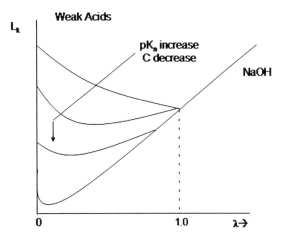

Figure 10 Conductometric titration of weak acids. [Adapted from Dahman (1986).]

chloride, nitrate, sulfate, and hydrogen carbonate in drinking and boiler water; nitrate in food products; fluoride in toothpaste; ammonium, potassium, nitrate, and phosphate in soil and fertilizers; bromide, sulphate, and thiosulphate in fixing baths, and sodium and potassium in body fluids and infusion solutions (Franklin, 1985; Meyer, 1997). Dionex corporation has recently been able to detect perchlorate ions in water as well. As one can see, subsequent development has broadened the application of IC beyond simple anion analysis of inorganic ions. Organic ions, such as acids found in fruit juices (i.e., citric acid) can also be detected (Meyer, 1997). Table 9 lists the four major classes of compounds amenable to IC analysis (Franklin, 1985; Meyer, 1997).

IC is a special form of liquid chromatography (covered in another chapter) and so modern HPLC equipment, with proper accessories, can be used for IC. IC was first reported, by Small et al. (1975). Figure 11 shows, schematically, the arrangement of parts in a typical ion chromatograph.

The unique contribution of these authors was the use of a "suppressor column", which is a second ion-exchange column to react with the ions of the eluent. They also used a sensitive small-volume conductivity cell as a detector to measure only the conducting analyte ions in the sample. Conductivity is a nonspecific measurement and cannot distinguish between the eluent and sample ions. Sensitivities in the μg/L (ppb) range may be achieved. Detection of inorganic ions can be accomplished with either chemical or electronic suppression.

Detection using a chemical suppressor has also been called dual-column chromatography. Figure 12 shows the principle of chemical suppression of background conductivity, which is due to the buffers and other salt solutions used as the mobile phase. Chemical suppression is not always necessary. If the background is low, nonsuppressed IC can be used; however, the limits of detection are usually higher.

The column used for separation is packed with far fewer ionic groups than would be found in an ion-exchange column, allowing for the use of a mobile phase of low ionic strength. Mobile phase conductivity is reduced by the suppressor column, which contains oppositely charged ions. The suppressor column must exchange large amounts of weakly concentrated mobile phase; therefore, a high exchange capacity resin is essential to its function. Because the suppressor column needs to be regenerated from time to time, it is easier to use a hollow-fibre or membrane suppressor. Both allow for suppression, without the need for regeneration (Haddad et al., 2003). A packed fibre may be used to reduce extra column volume. Using these fibers and membranes, gradient elution is possible. Typical ion chromatograms are shown in Fig. 13.

IC with chemical suppression is more sensitive than electronic suppression, but requires more sophisticated equipment.

Conductivity detection with electronic suppression can also be termed single-column IC. A suppressor is not essential if the eluents are carefully chosen to have a low equivalent conductance and sufficient chromatographic strength (i.e., phthalate solutions) and the background

Table 9 Compounds Amenable to Chromatographic Analysis

Major classes	Examples	Primary detectors
Inorganics	Sulfate, sulfite, thiosulfate, sulfide, nitrate, nitrite, borate, phosphate, hypophosphite, pyrophosphate, tripolysulfate, trimetaphosphate, selenate, selenite, arsenate, arsenite, chlorite, perchlorate, chlorate, hypochlorite, carbonate, cyanide, fluoride, bromide, other halides, potassium, sodium	Conductivity, amperometry
Metals	Alkalis, alkaline earths, gold (I and II), platinum, silver, palladium, iron (II and III), copper, nickel, tin, lead, cobalt, manganese, cadmium, chromium, aluminum	Conductivity, colorimetry
Organics	Organic acids (citric), amines, amino acids, carbohydrates, alcohols, saccharin, sugars, surfactants (anionic or cationic)	Conductivity, amperometry, fluorescence

Source: Franklin (1985) and Meyer (1997).

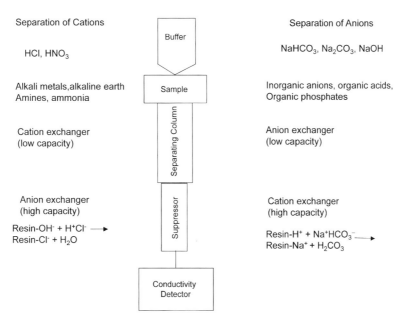

Figure 11 Typical ion chromatograph setup. [Adapted from Meyer (1997).]

current is electronically compensated (Haddad et al., 2003). However, the noise is higher than with chemical suppression. Temperature stability also requires attention so as not to change ion mobility. This method is preferred when sensitivity is not an issue because working with the one column simplifies the method (Meyer, 1997). Cation determinations are carried out in a similar manner by using a cation-exchange resin in the separator column. The sample cations are eluted with a strong acid, which in turn is eliminated in the suppressor column by using an anion-exchange resin in the hydroxide form.

Conductivity is the most used detector for IC. A host of improvements have been made in cell design and detector circuitry, driven by the needs of this new analytical method. These include miniaturization of cell volume, improved electronic suppression and stabilization techniques, and better temperature control and regulation. One of the latest improvements to the ion chromatograph has been the development of reagent-free IC. Solvents used for the mobile phase no longer have to be made because the system runs on deionized water alone. The solvents needed for separation are made by the eluent generator which is coupled to the self-regenerating suppressor. What eluent is generated is dependent upon whether the separation is for anions or cations, KOH and MSA respectively. Along with the reagent-free system, Dionex is continually developing new forms of suppressors such

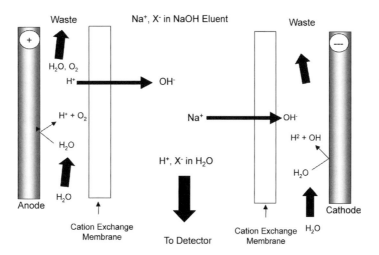

Figure 12 Principles of chemical suppression. [Redrawn from Haddad et al. (2003).]

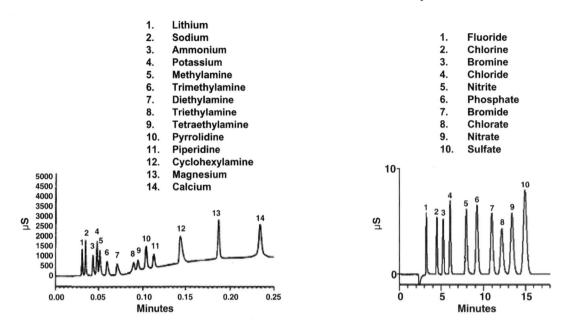

Figure 13 Chromatograms of cation and anion inorganic standards.

as the Atlas Electrolytic Suppressor, The SRS—Ultra Self-Regenerating Suppressor, the MMS III or micromembrane suppressor, and the AMMS-ICE, the anion ion exclusion suppressor. Discussion of these improvements (suppressors) and others pertaining to IC has been presented in several review papers (Haddad et al., 2003; Hatsis and Lucy, 2003; Lopez-Ruiz, 2000; Rabin et al., 1993).

B. Electrodeless (Noncontacting) Measurements

Conductance measurements can be without physical contact between the solution and any metallic conductors. Information is transferred from the solution to the electronic sensing circuits by electromagnetic induction. Two techniques are available: high frequency AC (radiofrequency) or low frequency.

In the first of these two techniques, the sample cell, made of glass or plastic, is surrounded by two metallic bands cemented onto the glass and separated from each other by a few millimeters. The two bands constitute the two electrodes of a capacitor, with a dielectric of glass. This capacitor is small in terms of microfarads, but offers very low impedance at radiofrequency. This is symbolized (in Fig. 2) by the sum of the double-layer capacitances. The high-frequency current passes easily through these impedances, but at the same time, both series resistances, R_C, become essentially infinite (i.e., open circuit). In this way the resistance of the cell becomes the desired R_s, paralleled by the very small value C_p. This approach, sometimes called oscillometry,

has been treated extensively by Pungor and several instruments have been on the market (Pungor, 1965).

1. Cell Design

The low-frequency method (20–50 kHz) of electrodeless conductivity, often called "inductive conductivity" has become popular, especially in chemical processes and industrial solution applications. This system utilizes a probe consisting of two encapsulated toroids in close proximity to each other, as shown in Fig. 14.

One toroid generates an alternating electric field in the solution, whereas the other acts as a receiver to pick up a signal from the field. The transformer core consists of the solution itself. The efficiency with which alternating current is transferred from the primary to the secondary depends on the conductivity of the solution. Several configurations are possible. One form is designed for immersion. The toroids are covered with a chemically resistant fluorocarbon or other high-temperature thermoplastic material. Any precipitates or coatings adhering to this probe generally have little or no effect on the measured conductance. A configuration in which the probe does not come in contact with the solution illustrated in Fig. 15. The unit is installed around a section of nonconducting pipe, such as a capillary in capillary electrophoresis, which contains a solution being separated. A complete liquid loop must exist for this arrangement to work.

In either case, the generating toroid is energized from a stable audiofrequency source, typically 20–50 kHz. The pick-up toroid is connected to a receiver that measures

Electrolytic Conductance

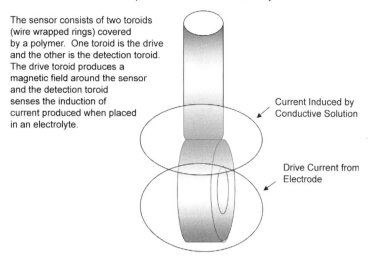

Figure 14 Simplified representation of an electrodeless conducutivity measuring circuit. [From Light (1997).]

the current through this secondary winding. The current is then amplified and sent to an analog recorder or to a computer. This current is a direct function of the conductance of the solution in the loop, in a manner completely analogous to the traditional measurement with contacting electrodes (Light and Licht, 1987).

The useful range of commercially available instruments extends from 0–100 μS/cm to 0–2 S/cm, with relative accuracy of a few tenths of a percent of full-scale, after temperature compensation. A temperature sensor is incorporated into the toroid probe, and a compensation circuit corrects all reading to the standard reference temperature of 25°C.

2. Circuitry

The circuitry of a contactless detection system varies from detector to detector. Several different types include the

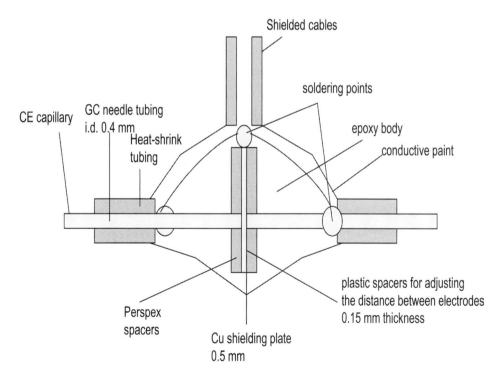

Figure 15 Electrodeless capacitative cell on a CE capillary. [From Macka et al. (2003).]

oscillator-based detector, bridge-based detector, and the single-electrode detector. In the oscillator-based detector, the electrodes are driven by an oscillator with a capacitive coupling to the measurement electrodes. The conductance between the electrodes determines the frequency of oscillation, which is then amplified before the signal is sent out to the processor (Alder et al., 1984; Vairneau et al., 2000). The second circuit design is relatively novel and is being investigated by Treves Brown et al. (2000). Three capacitors are used to isolate two arms of an AC bridge. This circuit for a conductivity cell would be applicable for capillary zone electrophoresis (CZE), isoelectric focusing, miniature flow injection analysis, and also stacking methods such as isotachophoresis (ITP). The last type of circuit, the single-electrode detector, was reported in 1999 (Prest et al., 1999). An electrode measures the potential at a single point in the separation forming a potential divider. Preparation of a device for use with the single-electrode detector was reported previously (Prest et al., 1999).

3. Commercial Instruments

The electrodeless conductivity technique using low-frequency cells has been known since 1951 (Relis, 1951). Such instruments are manufactured commercially for analysis and control in the chemical process industries such as dairy and fermentation and in other continuous monitoring applications. Contactless detectors have several advantages over contacting such as improved stability, accuracy, and freedom from maintenance. Advancements in design and accuracy have become very important in the development of detectors for capillary electrophoresis (CE) and chip-based separations.

4. Applications

Numerous applications of electrodeless conductivity have been published for the chemical, pulp and paper, aluminum, mining, and food industries (Calvert et al., 1958; Fulford, 1985; Gow et al., 1966; Muscow, 1968; Muscow and Ballard, 1984; Ormod, 1978; Queeney and Downey, 1986; Timm et al., 1978). Similar instrumentation has been used for *in situ* measurements of the salinity of seawater (Hinkelmann, 1957). Another application in which contactless conductivity measurements has made a debut is CE.

Capillary Electrophoresis

Conductivity detection following separation found a foothold in CE. The traditional contacting conductivity detectors were used originally by direct galvanic contact of the run buffers and sensing electrodes on column or at the end of the capillary. However, the development of more creative contactless detection techniques was to follow. The contactless detector has several advantages over the contacting mode. There is the absence of passivation and bubble formation associated with the electrode–solution contact, effective isolation from high voltages, a simplified construction and alignment of the detector (even at different locations), and the use of narrower capillaries (Pumera et al., 2002). Contactless detection for conventional CE systems is based on a two-electrode cell with tubular or semitubular electrodes placed over the capillary. Zemann et al. (1998) describe a capacitively coupled conductivity detection system (CCCD). In their design, two cylindrical conducting surfaces are placed around the polyimide coated capillary. The electrodes can be made of conducting silver varnish painted on the capillary or of two syringe cannulas. Figure 16 illustrates the ion separations using CCCD.

The detector can be placed anywhere along the length of the capillary and it is not necessary to burn a window in the polyimide coating, thus weakening the strength of the capillary (Zemann et al., 1998). Moving away from inorganic ions, several papers have been published recently using contactless conductivity to detect mono- and disaccharides and high voltage contactless conductivity detection of amino acids following separation by CE (Carvalho et al., 2003; Tanyanyiwa et al., 2003). Detection of the carbohydrates was based upon indirect principles; however, the amino acids could be detected directly or indirectly.

Microchips

The idea of separations and lab on a chip has become very popular. It minimizes costs of solvents, systems, and system components, as well as minimizes time. At

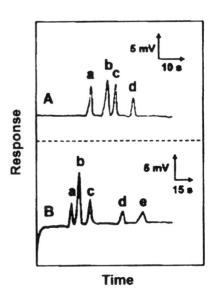

Figure 16 Electropherograms of cation and anion inorganic standards. [From Macka et al. (2003).]

Electrolytic Conductance

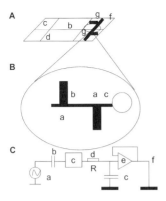

Figure 17 Microchip device used for ion separation with contactless conductivity detection. [From Pumera et al. (2002).]

one time, conductivity detection for microchips also relied on a galvanic contact of the run buffer and electrode; however, this technique has also turned toward contactless detection. One type of detector (see Fig. 17) is comprised of two external metallic film electrodes placed on the cover of a plastic such as polymethyl methacrylate.

The low-cost plastic microchips and easily constructed detectors allow for the possibility of designing disposable CE-conductivity microchips (Pumera et al., 2002). Figure 18 displays a diagram of the chip-based separation and detection of inorganic ions.

Wang et al. (2002a, b) has also separated and detected low-explosive ionic components and degradation products from chemical warfare agents using microchip CE-contactless conductivity.

Theoprax has software which uses conductivity measurements to create a pH-spectra. Another company, JAHM, has developed a database of 2200 materials and 16,000 sets of temperature-dependent data for elastic modulus, including thermal expansion and thermal conductivity to name a few.

III. SUMMARY

The theory behind conductivity is well studied and measurements such as resistance, ionic mobility, concentration, and diffusion can be calculated through measuring the conductivity of a solution. Measurement takes place using contacting cells, such as the two- or four-electrode cells, and contactless measurements using toroids. Contacting and contactless cells can take the form of meters, probes, sensors, or flowthrough cell detectors. Table 10 is a limited list of some of the companies and the products and instrumentation they sell for conductivity measurements.

Conductivity measurements are not only important in the lab, but in various industries as well. Applications range from water distillation, titrimetry, dairy and brewing industries, electrochemical reactions, IC, and CE to name a few. Aside from software designed to run the instrumentation for these various applications there is also a small amount of conductivity modeling software available. Most of this modeling software focuses on the assessment of soil and ground water conductivity estimates. Theoprax-Research has designed software that takes analytical conductivity measurements and converts this data into pH-spectra to study the interaction between different molecules in dependence on the degree of dissociation of their functional groups. All of

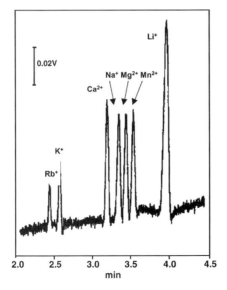
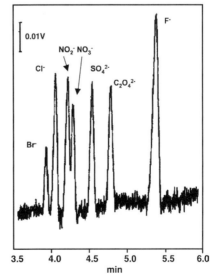

Figure 18 Example of cation and anion chip-based separation. [From Pumera et al. (2002).]

Table 10 List of Some Companies and Their Products for Conductivity Measurement

Company and location	Website	Product description
Aanderaa, Norway	www.aanderaa.com	Submersible conductivity/temperature sensors
Applied Microsystems, Canada	www.appliedmicrosystems.com	Various conductivity/temperature sensors for underwater measurements
Topac, USA	www.topac.com	Conductivity, salinity, resistivity, portable and handheld meters, probes, cells, kits
QA Supplies, USA	www.qasupplies.com	Conductivity testers and meters
LTH Electronics Ltd, UK	www.lth.co.uk	Contacting and electrodeless conductivity instruments and sensors for measuring food quality in dairy, brewing, and so on
ESA, USA	www.esainc.com	HPLC and CE instruments and supplies
Dionex, USA	www.dionex.com	Conductivity detectors, suppressors, ion chromatography systems
Honeywell, USA	http://content.honeywell.com	Toriodal conductivity cells, analyzers, transmitters for measuring conductivity in electrical utilities, water, pulp and paper, pharmaceuticals, and metal finishing
Aquarius Tech, Australia	www.aquariustech.com	Conductivity controller, monitor, and data log for automatic dosing and control for commercial and process water treatment
GLI International, USA	www.gliint.com	Contacting and electrodeless analyzers, transmitters, and sensors
JUMO, USA	www.jumoprocesscontrol.com	Analytical instruments, transmitters, electrodes, handheld meters, inductive conductivity transmitters, conductivity cells for measuring electrolytic conductivity
ICM, USA	www.icmmeters.com	Bench top and field conductivity meters
OI Analytical, USA	www.oico.com	Ion analyzers for flow injection analysis, and segmented flow analysis, electrolytic conductivity detector for GC
Quadrex Corp., USA	www.quadrexcorp.com	SRI portable GC detectors such as FID/DELCD, NP/DELCD, HID, and TCD
Foxboro Co., USA	www.foxboro.com	Contacting and electrodeless conductivity analyzers, transmitters and sensors
Analytical Sensors, Inc. USA	ww.asi-sensors.com	Glass body and epoxy body conductivity sensors, ion selective electrodes
Unidata America, USA		Precision water conductivity, electrode conductivity
Yellow Springs Instrument Co., USA		Handheld and bench top instruments

the instrumentation, software, and applications have one common interest—measurement of the total ionic content of an aqueous solution; hence, the measurement of electrolytic conductance.

REFERENCES

Ahmon, M. (1977). One-chip conductivity meter monitors salt concentration. *Electronics* 15:132–133.

Alder, J. F., Fielden, P. R., Clark, A. (1984). Simultaneous conductivity and permittivity detection with a single cell for liquid-chromatography. *Anal. Chem.* 56:985–988.

ASTM D1125-91. (1992). Standard test methods for electrical conductivity and resistivity of water. *Annual Book of ASTM Standards.* Vol. 11.01. American Society for Testing and Materials.

Atkins, P. W. (1998). Molecules in motion. *Physical Chemistry.* 6th ed. New York: W.H. Freeman and Company, pp. 737–744.

Braunstein, J., Robbins, J. D. (1971). Electrolytic conductance measurements and capacitive balance. *J. Chem. Educ.* 48:52–59.

Brett, C. M. A., Brett, A. M. O. (1993). *Electrochemistry Principles, Methods, and Applications.* New York: Oxford Science Publishing, pp. 26–31.

Calvert, R., Cornelius, J. A., Griffiths, V. S., Stock, D. I. (1958). The determination of the electrical conductivities of some concentrated electrolyte solutions using a transformer bridge. *J. Phys. Chem.* 62:47–53.

Carvalho, A. Z., da Silva, J. A. F., do Lago, C. (2003). Determination of mono- and disaccharides by capillary electrophoresis with contactless conductivity detection. *Electrophoresis* 24:2138–2143.

Coury, L. (1999). Conductance measurements part 1: theory. *Curr. Separations* 18(3):91–96.

Dahmen, E. A. M. F. (1986). Non-faradaic methods of electrochemical analysis. *Electroanalysis Theory and Applications in Aqueous and Non-aqueous Media and in Automated Chemical Control*. Techniques and Instrumentation in Analytical Chemistry. Vol. 7. New York: Elsevier, pp. 11–25.

Daum, P. H., and Nelson, D. F. (1973). Bipolar current method for the determination of solution resistance. *Anal. Chem.* 45:463–470.

Frankenthal, R. P. (1963). In: Meites, L., ed. *Handbook of Analytical Chemistry*. New York: McGraw-Hill, pp. 5–30.

Franklin, G. O. (1985). Development an applications of ion chromatography. *Am. Lab.* 17(6):65–80.

Fulford, G. D. (1985). Use of conductivity techniques to follow Al_2O_3 extraction at short digestion times. In: Bohner, H. O., ed. *Light Metals*. Warrendale, Pennsylvania: The Metallurgical Society of America, AIME, p. 265.

Gow, W. A., McCreedy, H. H., Kelly, F. J. (1966). *Can. Mining, Metallurgy Bull.* July.

Haddad, P. R., Jackson, P. E., Shaw, M. J. (2003). Developments in suppressor technology for inorganic ion analysis by ion chromatography using conductivity detection. *J. Chromatogr. A* 1000:725–742.

Hatsis, P., Lucy, C. A. (2003). Improved sensitivity and characterization of high-speed ion chromatography of inorganic ions. *Anal. Chem.* 75:995–1001.

Hinkelmann, H. (1957). Ein verfahren zur elektrodenlosen messung der elektrischen leitfähigkeit von elektrolyten. *Z. Agnew. Phys., Einschl. Nukl.* 9:505.

Kiggen, H. J., Laumen, H. (1981). Self-balancing high precision ac wheatstone bridge for observing diffusion processes in electrolyte solutions. *Rev. Sci. Instrum.* 52:1761–1764.

LaCourse, W. R. (1997). Amperometric detection in HPLC. *Pulsed Electrochemical Detection in High Performance Liquid Chromatography*. Techniques in Analytical Chemistry. New York: John Wiley and Sons, pp. 60–63.

Light, T. S. (1997). Measurement of electrolytic conductance. *Ewing's Analytical Instrumentation Handbook*. 2nd ed. Galen Wood Ewing. New York: Marcel Dekker Inc., pp. 1099–1122.

Light, T. S., Licht, S. L. (1987). Conductivity and resistivity of water from the melting to the critical point. *Anal. Chem.* 59(19):2327–2330.

Lopez-Ruiz, B. (2000). Advances in the determination of inorganic anions by ion chromatography. *J. Chromatogr. A* 881(1–2):607–621.

MacInnes, D. A. (1951). *The Principles of Electrochemistry*. New York: Dover, p. 339.

Macka, M., Hutchinson, J., Zemann, A., Shusheng, Z., Haddad, P. (2003). Miniaturized movable contactless conductivity detection cell for capillary electrophoresis. *Electrophoresis* 24:2144–2149.

Meyer, V. (1997). Ion chromatography. *Practical High-Performance Liquid Chromatography*. 2nd ed. New York: John Wiley, pp. 190–196.

Morash, K. R., Thornton, R. D., Saunders, C. C., Bevilacqua, A. C., Light, T. S. (1994). Measurement of the resistivity of high-purity water at elevated temperatures. *Ultrapure Water* 11(9):18.

Muha, G. M. (1977). A simple conductivity bridge for student use. *J. Chem. Educ.* 54:677.

Musow, W. (1968). On-line causticity sensor and programmable monitor applied to slaked lime addition and control. Canadian Pulp and Paper Association, Montreal.

Musow, W., Ballard, A. (1984). Toroidal conductivity sensor technology applied to cyanidation of flotation tailings circuits. Instrument Society of American, 12th Annual Mining and Metallurgy Industries Symposium, Vancouver.

Nilsson, J. W.; Riedel, S. A. (2001). Simple resistive circuits. *Electric Circuits*. 6th ed. New Jersey: Prentice Hall, pp. 77–78.

Ormod, G. T. W. (1978). Electrodeless conductivity meters in the measurement and control of the amount of lime in alkaline slurries. NIM-SAIMC Symposium in Metallurgical Process Instrumentation, Nat. Inst. Metallurg., Johannesburg.

Prest, J. E., Baldock, S. J., Bektas, N., Fielden, P. R., Treves Brown, B. J. (1999). Single electrode conductivity detection for electrophoretic separation systems. *J. Chromatogr. A* 836:59–65.

Pumera, M., Wang, J., Opekar, F., Jelinek, I., Feldman, J., Lowe, H., Hardt, S. (2002). Contactless conductivity detector for microchip capillary electrophoresis. *Anal. Chem.* 74:1968–1971.

Pungor, E. (1965). *Oscillometry and Conductometry*. Oxford: Pergamon.

Queen, A., Shabaga, R. (1973). A simple automatic conductance bridge for measuring the rates of chemical reactions in solution. *Rev. Sci. Instrum.* 44:494–496.

Queeney, K. M., Downey, J. E. (1986). Applications of a microprocessor-based electrodeless conductivity monitor. *Adv. Instrum.* 41(Pt 1):339.

Rabin, S., Stillian, J., Barreto, V., Friedman, K., Toofan, M. (1993). New membrane-based electrolytic suppressor device for suppressed conductivity detection in ion chromatography. *J. Chromatogr.* 640:97–109.

Relis, M. J. (1951). Method and apparatus for measuring the electrical conductivity of a solution. U.S. Patent 2,542,057.

Robinson, R. A., Stokes, R. H. (1959). Electrolyte solutions. London: Butterworth.

Rosenthal, R. (1957). Electrical conductivity measurements. In: Considine, D. M., ed. *Process Instruments and Controls Handbook*. New York: McGraw-Hill, pp. 6–159.

Shugar, G. J., Dean, J. A. (1990). *The Chemist's Ready Reference Handbook*. New York: McGraw-Hill, pp. 20.10–20.17.

Small, H., Stevens, T. S., Bauman, W. C. (1975). Novel ion exchange chromatographic method using conductimetric detection. *Anal. Chem.* 47:1801.

Tanyanyiwa, J., Schweizer, K., Hauser, P. C. (2003). High-voltage contactless conductivity detection of underivitized amino acids in capillary electrophoresis. *Electrophoresis* 24:2119–2124.

Thornton, R. D., Light, T. S. (1989). *Ultrapure Water* 6(5):14.

Timm, A. R., Liebenberg, E. M., Ormrod, G. T. W., Lombard, S. C. (1978). Nat. Inst. Metallurg., Report 2003, Johannesburg.

Treves B., B. J., Vairieanu, D.-I., Fielden, P. R. (2000). Conductivity detectors for microscale analytical chemistry. MicroTec Conference, Hanover, Germany, Sept 25–27.

Vairneau, D.-I., Fielden, P. R., Treves Brown, B. J. (2000). A capacitively coupled conductivity detector for electroseparations. *Meas. Sci. Technol.* 11:244–251.

Wang, J., Pumera, M., Collins, G. E., Mulchandani, A. (2002a). Measurements of chemical warfare agent degredation products using an electrophoresis microchip with contactless conductivity detector. *Anal. Chem.* 74(23):6121–6125.

Wang, J., Pumera, M., Collins, G. E., Opekar, F., Jelinek, I. (2002b). A chip based capillary electrophoresis-contactless conductivity microsystem for fast measurements of low explosive ionic components. *Analyst* 127(6):719–723.

www.analyzer.com (accessed August, 2003)

www.radiometer-analytical.com (accessed August 2003)

www.theoprax-research.com (accessed November, 2003).

www.topac.com/conductivityprobes.html (accessed August 2003)

Zemann, A. J., Schnell, E., Volgger, D., Bonn, G. K. (1998). Contactless conductivity detection for capillary electrophoresis. *Anal. Chem.* 70:563–567.

20

Microfluidic Lab-on-a-Chip

PAUL C. H. LI, XIUJUN LI
Chemistry Department, Simon Fraser University, Burnaby, British Columbia, Canada

I. INTRODUCTION

Miniaturized analysis has various advantages such as fast analysis time, small reagent consumption, and less waste generation. Moreover, it has the capability of integration, coupling to sample preparation, and further analysis. A miniaturized gas chromatography (GC) column with a thermal conductivity detector (TCD) on silicon (Si) was first constructed in 1979 (Terry et al., 1979), and a high-performance liquid chromatography (HPLC) column with a conductometric detector constructed on Si-Pyrex in 1990 (Manz et al., 1990a). The first demonstration of liquid-based miniaturized chemical analysis system was based on capillary electrophoresis (CE), and this appeared in 1992 (Manz et al., 1992). In this work, capillary zone electrophoresis (CZE) separation of calcein and fluorescein as detected by laser induced fluorescence (LIF) was achieved on a glass chip with 10 μm deep and 30 μm wide channels in 6 min. The success is attributed to the use of electroosmotic flow (EOF) to pump reagents inside small capillaries which could develop high pressure, preventing the use of HPLC. Since then, different CE modes have been demonstrated and different analyses (i.e., cellular, oligonucleotide, and protein analyses) have been achieved by numerous research groups, as mentioned in subsequent sections. In this book chapter, we focus on the microfluidic lab-on-a-chip (coined in 1992) (Harrison et al., 1992) which consists of the micromachined channels and chambers. Other important chip-based technology such as microarray or microwells is beyond the scope of this chapter. Applications other than analysis, such as chemical synthesis, have been reviewed recently (Hodge et al., 2001), is also not covered here.

Since 1990, there are numerous review articles summarizing research performed in the area of the microfluidic chip (Auroux et al., 2002; Reyes et al., 2002). Reviews on specific topics have also been published (see Table 1). Whereas these reviews summarize exciting research and proposing new directions, this chapter is focused on an overview of the available technology, its limitations, and breakthrough over the years. Beginners in the field should find this chapter useful to navigate the vast literature available on the technology; and experienced researchers will find useful information compiled (see Tables 1–17) for easy comparison and references. Although some comparisons among different approaches are made here, the readers should judge on the suitability of a certain technology for their needs. Moreover, references will mostly be made on complete studies. As the field is fast expanding, the proceedings of dedicated conferences, such as Transducers, and micro total analysis system (μTAS), will also be referred to include the latest findings.

The chapter is subdivided into several sections which include micromachining methods, microfluidic

Table 1 Reviews on the Microfluidic Technology

Review topics	References
History	Becker and Gärtner (2000), Effenhauser et al. (1997a), Haswell (1997), Polson and Haves (2001) and Reyes et al. (2002)
Miniaturization	Burbaum (1998), Lindern (1987), Manz et al. (1993), Manz et al. (1991a), Ramsey et al. (1995), Reyes et al. (2002), Service (1995) and Service (1996)
μTAS	Haswell (1997), Manz et al. (1990a, 1991b, 1993, 1995), Reyes et al. (2002), Shoji (1998) and van de Berg and Lammerink (1998)
Separation	Auroux et al. (2002), Bruin (2000), Campaña et al. (1998), Colyer et al. (1997a), Dolnik et al. (2000), Effenhauser et al. (1997a), Effenhauser (1998), Jacobson and Ramsey (1997, 1998), Khandurina and Guttman (2002), Kutter (2000), Manz et al. (1993), Manz et al. (1993, 1994), Ramsey et al. (1995) and Rossier et al. (1999a)
Clinical analysis	Auroux et al. (2002), Kricka and Wilding (1996) and Landers (2003)
Cellular assay	Auroux et al. (2002), Fuhr and Shirley (1998) and Mcdonald et al. (2000)
Protein assay	Auroux et al. (2002), Colyer et al. (1997a), Dolnik et al. (2000), Hodge et al. (2001), Khandurina and Guttman (2002) and Sanders and Manz (2000)
DNA analysis	Auroux et al. (2002), Bruin (2000), Colyer et al. (1997a), Dolnik et al. (2000), Figeys and Pinto (2000), Khandurina and Guttman (2002), Landers (2003), Mathies et al. (1998), O'Donnell et al. (1996) and Sanders and Manz (2000)
Detector	Auroux et al. (2002), Bruin (2000), Dolnik et al. (2000), Haswell (1997), Henry (1999), Khandurina and Guttman (2002) and Wang (2002)
Injection	Alarie et al. (2000), Auroux et al. (2002), Haswell (1997), Jacobson and Ramsey (1998) and Polson and Haves (2001)
Micromachining	Campaña et al. (1998), Haswell (1997) and Reyes et al. (2002)
Chip bonding methods	Reyes et al. (2002) and Shoji and Esashi (1995)
Polymeric chips	Becker and Gärtner (2000), Dolnik et al. (2000), Mcdonald et al. (2000), Qin et al. (1998), Reyes et al. (2002) and Rossier et al. (1999a)
Microfluidic flow	Bruin (2000), Figeys and Pinto (2000), Polson and Haves (2001), Reyes et al. (2002) and van de Berg and Lammerink (1998)
Micropumps	Cheng et al. (1998a), Elwenspoek et al. (1994), Gravesen et al. (1993), Haswell (1997), Manz et al. (1993), Reyes et al. (2002), Shoji (1998) and van de Berg and Lammerink (1998)
Microvalves	Cheng et al. (1998a), Elwenspoek et al. (1994), Gravesen et al. (1993), Manz et al. (1993), Reyes et al. (2002), Shoji (1998) and van de Berg and Lammerink (1998)
Micromixers	Auroux et al. (2002) and Elwenspoek et al. (1994)

operations (liquid flow, sample introduction, and preconcentration), chemical separations, detection technology, and various chemical and biochemical analyses (applications on cellular, oligonucleotide, and protein analyses). Emphasis will be placed on analytical applications although the basic principles about micromachining and fluid flow and control will also be covered only to the extent that their understanding will assist the exploitation of the microfluidic technology on analytical applications.

Many of the principles of "operations" on the chip are adapted from conventional wisdom. However, the original citations of these theories or principles, which have already been given in the original research articles, will not be repeated here to avoid increasing the number of microchip references in the chapter. However, the readers are encouraged to consult the original citations from the references.

II. MICROMACHINING METHODS

Micromachining methods, which include film deposition, photolithography, etching, access-hole formation, and bonding of microchip, have first been achieved on Si (Manz et al., 1992). Thereafter, glass (Pyrex, soda-lime, and fused silica) was used as the micromachining substrate. Recently, polymeric materials are being widely used as substrates. The micromachining methods of these substrates differ and are discussed subsequently.

A. Micromachining of Silicon

The Si substrate was micromachined using a photolithography process (Ko and Suminoto, 1999). After photolithography, the exposed portions of photoresist were dissolved by a developer, and the remaining unexposed areas were hardened by heating (baking). The exposed areas of the Si substrate were etched subsequently. Meanwhile, a Pyrex glass plate was patterned with metal electrodes. Then, the Pyrex plate was bonded to the Si wafer using the anodic bonding process. The etching, drilling, and bonding processes are described in more detail.

1. Si Etching

Isotropic wet-etch using $HF-HNO_3$, whose etch rate is independent of the etching direction, has been employed to produce approximately rectangular grooves, oriented in any direction on the Si $\langle 100 \rangle$ wafer, using thermal SiO_2 (~1 μm thick) as the etch mask (Terry et al., 1979).

Anisotropic etch, whose etch rate depends on the etching direction, can be achieved using KOH on $\langle 100 \rangle$ Si or $\langle 110 \rangle$ Si. This method allows V-grooves or vertical walls to be formed on the substrate. Apparently, the anisotropic etch rate of Si at the $\langle 111 \rangle$ crystal plane was very small in comparison with the $\langle 100 \rangle$ plane (Stemme and Kittilsland, 1998). Therefore, the etch depth will be dependent on the opening width, rather than on the etch time. However, this method will produce the desired groove profiles only if the groove lies along specific crystallographic axes on the wafer, and certain shapes (square corner and circle) cannot easily be realized (Terry et al., 1979), except by corner compensation (Murakami et al., 1993).

To achieve anisotropic etch on Si, dry (plasma) etch was also used. The different etch methods used to produce microfluidic chips on Si are tabulated in Table 2.

As Si is a semiconductor, it is not electrically insulating. However, electrical insulation (<500 V) of Si can be provided by a SiO_2/Si_3N_4 film (~200 nm) (Harrison et al., 1991) or low-temperature oxide (LTO) (Nieuwenhuis et al., 2001). It was found that SiO_2 insulation was better achieved by plasma deposition rather than by thermal growth; a 13 μm thick plasma-deposited SiO_2 film would withstand an operation voltage of 10 kV (Mogensen et al., 2001a). For better insulating properties, glass (e.g., Pyrex, and fused silica), which sustains at least an electric field of 10^5 V/cm without dielectric breakdown, is used.

After etching, access holes can be created on the Si substrate by etch-through (Jeong et al., 2000; Murakami et al., 1993; Raymond et al., 1994; Terry et al., 1979) or by drilling (Bökenkamp et al., 1998; Kamholz et al., 1999).

2. Bonding of Si Chips

Anodic bonding of Si to Pyrex was mostly used to create a sealed Si chip (Brivio et al., 2002; Brody et al., 1996; Kamholz et al., 1999; Murakami et al., 1993; Terry et al., 1979). Various bonding conditions are shown in Table 3.

Bonding of Si to glass was also achieved using UV-curable optical adhesives (Burns et al., 1998).

On the other hand, bonding of Si to Si was achieved by a low-temperature curing polyimide film (Hsueh et al., 1998) or by using an intermediate deposited layer of borophosphosilicate glass and subsequent anodic bonding (300 V, 350°C) (Sobek et al., 1993).

B. Micromachinig of Glass

Similar to the procedure for Si micromachining, micromachining for glass also includes thin-film deposition, photolithography, etching, and bonding (see Fig. 1) (Fan and Harrison, 1994).

Table 2 Etching of Si Substrates

Etch conditions	Etch rates	References
Isotopic wet-etch		
$HF-HNO_3$		Terry et al. (1979)
Anisotropic wet-etch		
Tetramethylammonium hydroxide (TMAH) and water (1:4), 90°C	28 μm/30 min	Martinoia et al. (1999)
Ethylenediamine/pyrocatechol (EDP)	100 μm/h	Brody et al. (1996)
40% KOH/10% isopropyl alcohol, 76°C	~0.4 μm/min	Ocvirk et al. (2000)
KOH		Harrison et al. (1991), Terry et al. (1979) and Wilding et al. (1994a)
Anisotropic dry etch		
Reactive ion etching (RIE)		Hua et al. (2002), Jeong et al. (2000), Juncker et al. (2002) and Wilding et al. (1994a)
$SF_6-C_2ClF_5$ plasma etch		Parce et al. (1989)
SF_6 deep plasma etch (with C_4F_4 sidewall passivation to produce vertical walls)	5 μm/min	Hediger et al. (2000)

Table 3 Various Anodic Bonding Conditions to Fabricate Si–Pyrex Chips

Temperature (°C)	Voltage (V)	References
400	600	Terry et al. (1979)
400	800	Brivio et al. (2002)
450	1000 (positive on Si, negative on Pyrex)	Murakami et al. (1993)
430	1600	Jeong et al. (2000)
400	400	Brody et al. (1996)

1. Etching of Glass

Etching of glass is mostly achieved by isotropic wet-etch. Various etching conditions based on HF are tabulated in Table 4. A typical glass channel created by wet-etch is shown in Fig. 2.

For mechanically polished glass plates, thermal annealing should be performed before etching to avoid poor wet-etch quality (Fan and Harrison, 1994; Fu et al., 2002a; Hibara et al., 2002a).

In the photolithographic process, a photomask is needed. The photomask is usually generated by laser or e-beam ablation. It is found that e-beam ablation has produced smoother edges (10× better than the UV laser-ablated mask) in the photomask, thus leading to smoother channel walls after etching, and higher efficiency in CE separations (Jacobson et al., 1994b). Uniformity in channel depth and width can be experimentally verified by examining the linearity of a plot of separation efficiency (N) vs. channel length (L) (Moore et al., 1995).

In one report, it was mentioned that wet HF etch was performed in an ultrasonic bath (Fu et al., 2002a), and this step may also be generally adopted in HF-based etching. Channel etching using HF was usually performed before bonding, but etching could also be achieved after glass bonding (for enlarging some channels) (Liang et al., 1996).

Because wet chemical etch on glass results in an isotropic etch, the resulting channel cross-section is trapezoidal or semicircular (Jacobson et al., 1994c, d, e). In order to form a circular channel, two glass plates were first etched with a mirror image pattern of semicircular cross-section channels. Then, the two plates were aligned and thermally bonded to form a circular channel (Hasselbrink et al., 2002; Liu et al., 2001a; Zhang et al., 1999, 2000).

Normally, a one-mask process was employed, but for specific applications, a two-mask process was used to create the micropump channel (1–6 μm deep) and the sample channel (20–22 μm deep) (Lazar and Karger, 2002).

There are other dry micromachining processes for glass, which includes powder blasting (Brivio et al., 2001, 2002; Hediger et al., 2000; Schasfoort and van Duijn et al., 2000; Wensink and Elwenspoek, 2001) and mechanical machining by a wafer saw (Liu et al., 2001d; Ueno et al., 2001).

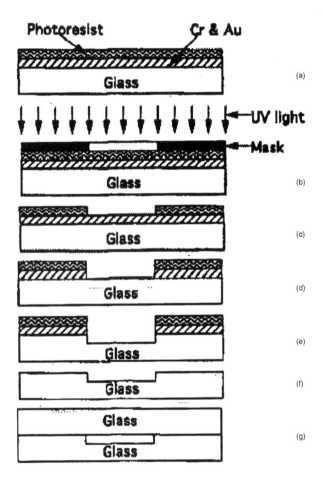

Figure 1 Sequence for fabrication of the glass microfluidic chip. (a) Cr and Au masked glass plate coated with photoresist, (b) sample exposed to UV light through a photomask, (c) photoresist developed, (d) exposed metal mask etched, (e) exposed glass etched, (f) resist and metal stripped, and (g) glass cover plate bonded to form sealed capillary. [(Fan and Harrison, 1994) Courtesy of American Chemical Society.]

2. Drilling of Glass for Access-Hole Formation

In order to provide liquid access to the microfluidic chips, holes are usually created on the cover plate. These access holes are usually created on glass by drilling using diamond drill bits. However, other methods were also used (see Table 5).

Table 4 HF-Based Wet Etching of Glass

Etch conditions	Etch rates	References
Buffered oxide etch (BOE, 10:1)		Fan and Harrison (1994)
Dilute, stirred NH$_4$F/HF	5.2 μm in 20 min	Jacobson and Ramsey (1996) and Jacobson et al. (1994a, 1999b)
BOE: 40% NH$_4$F/49% HF (10:1)	~10 nm/min	Ocvirk et al. (2000)
BOE: 40%/49% (7:1)	0.03 μm/min	Weiller et al. (2002)
BOE	8 μm in 15 min	Woolley and Mathies (1994)
BOE (6:1) or NH$_4$F/HF		Fu et al. (2002a) and Mao et al. (2002a)
	0.66 μm/min	Baldwin et al. (2002)
HF/water (1:4)	30 μm for 15 min	Keir et al. (2002)
Conc. HF	30 μm for 4 min	Emrich et al. (2002)
49% HF	7 μm/min	Liu et al. (1999), Paegel et al. (2000) and Simpson et al. (1998)
	9 μm/min	Weiller et al. (2002)
50% HF	~10 μm/min	Dodge et al. (2001)
	13 μm/min	Hibara et al. (2002a)
HF/HNO$_3$ (7:3)	6.5 μm/min	Handique et al. (2000), Handique et al. (2001) and Namasivayam et al. (2002)
HF/HNO$_3$		Chiem and Harrison (1997), Chiem et al. (2000) and Harrison et al. (1993a)
HF/HNO$_3$/H$_2$O (20:14:66)	25 μm in 6 min	Xue et al. (1997)
		Fan and Harrison (1994b)
HF/HCl/H$_2$O (1:2:4)	16 μm/2.5 min	Haab and Mathies (1999)
BOE (6:1)/HCl/H$_2$O (1:2:4) stirred at 250 rpm		Lapos et al. (2002)

For protection against chipping during drilling by diamond bits, the cover plate was sandwiched between two protecting glass slides using Crystal Bond (Chiem et al., 2000). After hole drilling, the hole could be etched to provide a smooth conical channel exit (Lapos et al., 2002).

Drilling has also been performed on the bonded plate, rather than on the cover plate before bonding. To avoid plugging the channel in the bonded plate with glass particles during drilling, the channel was filled with Crystal Bond (Bings et al., 1999). In some earlier devices, no drilling was performed, and a circular cover slip was used with the end of channel protruding out to reach the solution reservoir (Jacobson et al., 1994d, e).

3. Glass Bonding

After etching and access-hole formation, the channel plate and cover plate are bonded. Thermal bonding is the major method (see Table 6), though alternative schemes exist.

Bonding can be achieved in two ways: (1) between an etched glass plate (containing patterned electrodes) and a drilled cover plate or (2) between an etched glass plate and a drilled cover plate (containing patterned electrodes) (Woolley et al., 1998).

Normally, the glass plates are cleaned and dried before thermal bonding. In one report, the two glass plates were brought into contact while wet. With an applied pressure of 50 g/cm^2, the plates were bonded thermally (Baldwin et al., 2002).

Low-temperature bonding of glass plates was accomplished using potassium silicate with the following thermal treatment: 90°C (1 h), 0.3°C/min to 200°C (12 h) (Khandurina et al., 1999, 2000).

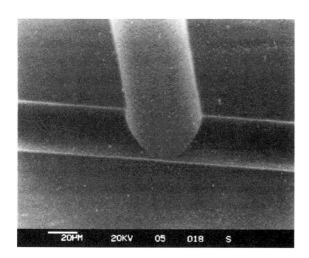

Figure 2 Electron micrograph of a T-intersection of a Pyrex glass microchip. [(Fan and Harrison, 1994) Courtesy of American Chemical Society.]

Table 5 Access-Hole Formation on Glass Chips

Methods	References
Diamond-bit drilling	Baldwin et al. (2002), Chiem and Harrison (1997), Liu et al. (1999), Woolley and Mathies (1994) and Woolley et al. (1997)
HF-etching (4 min, 1 mm dia., 145 μm deep)	Haab and Mathies (1999)
CO_2 laser drilling	Cheng et al. (1998c)
Ultrasonic abrasion	Eijkel et al. (2000b), Fan and Harrison (1994), Gottschlich et al. (2001) and Hibara et al. (2002a)
Electrochemical discharge drilling	Arora et al. (2001), Shoji et al. (1992) and Walker et al. (1998)
Powder blasting	Brivio et al. (2002), Chmela et al. (2002) and Timmer et al. (2001)

Besides bonding Si to Pyrex glass, anodic bonding can be used to bond Pyrex to Pyrex. In this case, a silicon nitride film (200 nm) was used as an adhesive layer, with the following anodic bonding conditions: 1500 V, 450°C, 10 min (Berthold et al., 1999).

There are other nonthermal methods for bonding. For instance, bonding of glass was carried out by 1% HF (Fu et al., 2002a). Bonding of glass plates at room temperature (RT) (20°C) has been achieved only after rigorous cleaning. This allows bonding of different types of glass substrates having different thermal expansion coefficients (Chiem et al., 2000).

Optical UV adhesive was also used to bond glass to glass (Namasivayam et al., 2002), glass to silicon (Burns et al., 1996; Handique et al., 2000, 2001), or glass to quartz (Burns et al., 1996). In one report, such bonded device can be separated and rebonded after heating up to 500°C (Divakar et al., 2001).

In one report, pressure sealing of glass plates by clamping was used as a nonthermal temporary bonding method to avoid thermal degradation of a chemically modified (ODS) layer (Hibara et al., 2002a).

After glass bonding, solution reservoirs were created over access holes to hold reagents. The reservoirs were formed by various methods. Most commonly, short plastic or glass tubings were glued to the access holes using epoxy resin. Septa have also been used to define reservoirs around drilled holes. The septa were compressed on the

Table 6 Thermal Bonding of Glass Chips

Bonding conditions	References
Up to 440°C	Effenhauser et al. (1993)
500°C	Culbertson et al. (2000a) and Jacobson et al. (1994d, 1999a, 1999b)
560°C (3 h)	Haab and Mathies (1999)
580°C (20 min)	Fu et al. (2002a)
595°C (6 h)	Liang et al. (1996)
600°C	Arora et al. (2001)
Up to 600°C using a program	Woolley and Mathies (1994)
620°C (3 h), 10°C/min	Xue et al. (1997)
620°C (4 h)	Effenhauser et al. (1993)
650–660°C (4–6 h) (Pyrex)	Harrison et al. (1993)
680°C (borofloat glass)	Emrich et al. (2002)
Up to 620°C (3°C/min), hold for 30 min, to RT (3°C/min)	Baldwin et al. (2002)
Up to 500°C (from 60°C) for 1 h, hold for 1 h, increased to 570°C and hold for 5 h, cooled to 60°C and left for 12 h	Keir et al. (2002)
Equilibrated at 500°C, 550°C, 600°C then annealed at 625°C for 2 h (Pyrex)	Manz et al. (1991b)
440°C (0.5 h), 473°C (0.5 h), 605°C (6 h), 473°C (0.5 h), cooled overnight	Li and Harrison (1997)
500°C (1 h), 550°C (0.5 h), 620°C (2 h), 550°C (1 h), cooled overnight	Harrison et al. (1992)
Up to 550°C (40°C/min) (0.5 h), 20°C/min to 610°C (0.5 h), 20°C/min to 635°C (0.5 h), 10°C/min to 650°C (6 h), cooled to RT	Fan and Harrison (1994)
Up to 400°C (20°C/min), 10°C/min to 650°C (4 h), 10°C/min to 450°C, cooled to RT (borofloat glass)	Lapos and Ewing (2000)
550°C (1 h), 580°C (5 h), 555°C (1 h), cool for at least 8 h	Sirichai and deMello (2000)
Up to 400°C (280°C/h) (4 h), 280°C/h to 588°C (6 h), cooled to RT	Mao et al. (2002a)

Table 7 Fused Quartz Etching

Etch conditions	Etch depth (μm)	References
Wet HF etch		
Dilute, stirred NH_4F/HF at 50°C	8	Jacobson and Ramsey (1995)
BOE (NH_4F/HF 6:1)		Chen et al. (2001b)
NH_4F/HF (1:1)	28	Koutny et al. (1996)
HF	19	Yang et al. (2002b)
Dry etch		
Fast atom beam	10	Sato et al. (2000, 2001a)
CHF_3-based plasma (DRIE)	10	He et al. (1998)
C_3F_8/70% CF_4 plasma	<10	Oki et al. (2001)
SF_6	10	Sato et al. (2000)
Excimer laser		Hisamoto et al. (2001) and Tokeshi et al. (2000a)
CO_2 laser	100	Sato et al. (2000, 2001a)

chip using Plexiglas support plates (Deng and Henion, 2001; Deng et al., 2001; Li et al., 1999, 2000).

To avoid any change in the buffer concentration due to solution evaporation, the solution reservoirs were sealed with thin rubber septa (with a small hole for Pt wire electrode insertion) held in place with Parafilm (Kutter et al., 1997; Moore et al., 1995). With this strategy, good reproducibility (RSD of peak area and migration time are less than 1.3, and 1.2%, respectively) can be maintained (Moore et al., 1995). Moreover, an overflowing structure was constructed to maintain the solution level (Fang et al., 2001). Change of buffer concentration due to electrolysis was alleviated by using a porous bridge which separated the electrode from the buffer solution (Wallenborg et al., 2000).

On the other hand, thick poly(dimethylsiloxane) (PDMS) slabs (with punched holes) have been directly placed on the chip and aligned with the access holes to create solution reservoirs (Dodge et al., 2001; Haab and Mathies, 1999; Simpson et al., 1998; Woolley et al., 1998). Usually, no sealant is needed. But in one report, a silicone sealant was used to attach the PDMS slab (Zhang et al., 2000). When high voltages were used, the solution reservoirs should be sufficiently apart to avoid electric arcing (Moore et al., 1995); PDMS layers have also been employed to prevent arcing between adjacent electrodes inserted in the access holes (Cheng et al., 2001).

C. Micromachining of Fused Silica (or Fused Quartz)

1. Fused Quartz Etching

Wet and dry etch methods have been employed to micromachine fused silica chips (see Table 7). An HF-etched channel in quartz was shown in Fig. 3. In order to provide access holes, they can be created on fused quartz cover chips by laser drilling (He et al., 1999; Koutny et al., 1996; Swinney et al., 2000) or ultrasonic drilling (He et al., 1998).

2. Bonding of fused quartz chips

The etched and cover quartz chips are bonded using various methods (see Table 8).

Sealed quartz channels could also be fabricated by first depositing a layer of SiO_2 on the quartz substrate which was then wet-etched through a tiny hole to create a quartz channel and a thin SiO_2 roof. Therefore, in this case, no bonding plate was needed (Ericson et al., 2000).

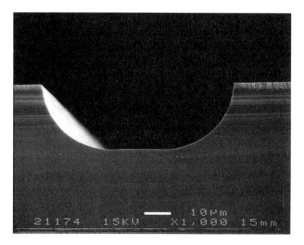

Figure 3 Electron micrograph of the cleaved edge of a quartz chip showing the channel cross section. [(Koutny et al., 1996) Courtesy of American Chemical Society.]

Table 8 Bonding of Fused Quartz Chips

Bonding conditions	References
Thermal bonding	
1110°C	Jacobson and Ramsey (1995)
1110°C, 5 h	Jacobson et al. (1995)
200°C (2 h) then	Koutny et al. (1996)
1000°C (overnight)	
1150°C	Hisamoto et al. (2001), Sato et al. (2000, 2001a) and Tokeshi et al. (2000a)
Other bonding methods	
Sealed to a PDMS layer	Yang et al. (2002b)
1% HF with 1.3 MPa pressure	Hosokawa et al. (1998) and Ichiki et al. (2001)

D. Micromachining of Polymeric Chips

Polymeric chips are fabricated with various methods, as developed from the plastic industry. These processes include casting, injection molding, laser-ablation, imprinting, and compression molding.

1. Casting

A procedure for micromolding in capillaries (MIMIC) for fabricating microstructures of polymeric materials has been proposed (Kim et al., 1995).

The casting method has mostly been employed to fabricate PDMS chips. A procedure for casting of PDMS replicas from an etched Si master with positive relief structures (in Si) has been developed (see Fig. 4) (Effenhauser et al., 1997b). The Si were silanized [in 3% (v/v) dimethyl-octadecylchlorosilane/0.025% H_2O in toluene] for 2 h to facilitate peeling off the PDMS replicas. The PDMS replica is a negative image (see Fig. 5) of the positive relief structure. A 10:1 mixture of Sylgard 184 and its curing agent was poured over the master mounted within a mold. After curing, the PDMS replica could be easily peeled off the master. Holes were punched through the replica. Then it was placed on another thin slab of PDMS for sealing. Because an anisotropic KOH-based etching method was employed to etch the Si master without corner compensation, the channel intersection in the replica were limited in shape by the $\langle 111 \rangle$ plane of the Si master (Effenhauser et al., 1997b).

Although the surface free energy of PDMS is not particularly high ($\sim$22 mN/m), the smoothness of the casted surface of PDMS in combination with its elastomeric nature ensures good adhesion of the PDMS layer on both planar or curved substrates. These properties of the PDMS elastomer simplify the sealing method (i.e., no thermal bonding needed). Pressure (up to 1 bar) could

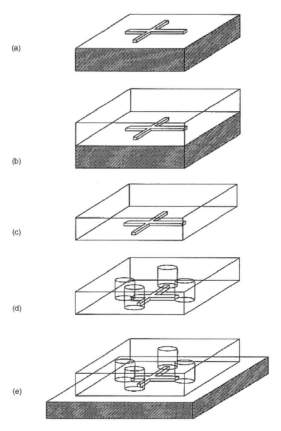

Figure 4 Fabrication procedure of a PDMS chip: (a) silicon master wafer with positive surface relief, (b) premixed solution of Sylgard 184 and its curing agent poured over the master, (c) cured PDMS slab peeled from the master wafer, (d) PDMS slab punched with reservoir holes, and (e) ready-to-use device placed on another slab of PDMS. [(Effenhauser et al., 1997b) Courtesy of American Chemical Society.]

be applied to the channels without the use of clamping. Moreover, since the sealing is not permanent, the two slabs can be peeled off for more thorough cleaning, if needed (Effenhauser et al., 1997b).

Rapid prototyping (<24 h) was suggested to produce PDMS chips. A transparency photomask was used to create a master consisting of positive relief structures (made of photoresist) on a Si wafer. PDMS was then casted against the master to yield numerous polymeric replicas (>30 replicas per master) (see Fig. 6). This method allows the channel width of >20 μm to be made. Channel depth of 1–200 μm can be casted by using different thickness of photoresist (Duffy et al., 1998). PDMS chips have also be casted from a glass (not Si) master containing photoresist (SU-8) as the positive relief structure (Seong et al., 2002; Zhang and Manz, 2001).

PDMS can be casted from two commercially available kits: (1) General Electric-RTV 615 (Chou et al., 1999;

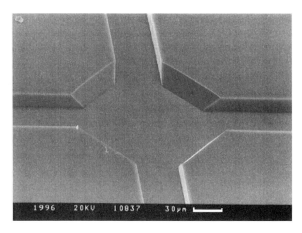

Figure 5 Electron micrograph showing the sample injection region of the PDMS microdevice. Channel cross-section, 50 μm (width) × 20 μm (height). The geometrical features of the channel junction result from the anisotropy of the wet chemical etching process of the Si master. [(Effenhauser et al., 1997b) Courtesy of American Chemical Society.]

Fu et al., 2002b; Unger et al., 2000) or (2) Dow Corning-Sylgard 184 (mostly used)

Both kits contain the PDMS elastomers which are cured by cross-linking the silicone hydrides using a proprietary platinum-based catalyst. Components A and B from either source are mixed in a definite ratio (10:1). The mixture is either spin-coated to obtain a thin film (10–12 μm) or poured in a mold to obtain a slab of greater thickness (2 mm). The mixture is then incubated for either curing or bonding. RTV 615 needs higher temperature to cure, whereas Sylgard 184 is stiffer and more chemically inert to acids and bases (see Table 9 for details on various curing conditions) (Fu et al., 2002b).

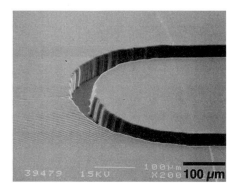

Figure 6 Electron micrograph of a turn in the channel fabricated in a PDMS chip, created by casting the polymer against a positive relief, which is made of photoresist patterned on a glass substrate. The roughness in the side wall arises from the limited resolution of the transparency used as a photomask in photolithography. [(Duffy et al., 1998) Courtesy of American Chemical Society.]

Table 9 Various Curing Conditions of PDMS Elastomers

Curing conditions	References
Sylard 184	
60°C, 60 min	Esch et al. (2001)
60°C, 3 h	Wu et al. (2002)
65°C, 1 h	Duffy et al. (1998), Hosokawa et al. (1999) and Seong et al. (2002)
65°C, >3 h	Bodor et al. (2001) and Liu et al. (2000a)
70°C, 12 h	Tokano et al. (2002)
75°C, 1 h	Grzybowski et al. (1998)
75°C, 3 h	Ocvirk et al. (2000)
70°C, at least 72 h	Chan et al. (1999b)
80°C, 25 min	Fu et al. (2002b)
200°C, 2 h	Gao et al. (2001) and Jiang et al. (2001)
RTV615	
80°C, 1 h	Fu et al. (2002b)
80°C, 1.5 h	Unger et al. (2000)
90°C, 2 h	Chou et al. (1999)
Not specified	Liu et al. (2000d)

Thermal aging of PDMS was also carried out at 115°C for 12 h in order to homogenize the elastomer surface properties (Linder et al., 2001).

A two-mask process has been used to create a glass master containing positive relief structures with depressions to cast the channel/chamber (25–30 μm deep) and weir (7–12 μm clearance) on a PDMS chip (Seong et al., 2002). Another two-mask process was used to create a Si molding master consisting of 3 μm-high Si relief structures and 25 μm-high photoresist relief structures (Hosokawa et al., 1999).

Bonding between a PDMS channel slab to a glass plate can be achieved by reversible (Martin et al., 2000; Takayama et al., 1999) or irreversible bonding (Duffy et al., 1998; Linder et al., 2001; Mao et al., 2002b; Seong et al., 2002). In the latter case, the PDMS surfaces are oxidized in an O_2 plasma. Then, PDMS seals irreversibly to various substrates: PDMS, glass, Si, SiO_2, quartz, Si_3N_4, polyethene, glassy carbon, and oxidized polystyrene (PS). Within 1 min of oxidation, the two bonding surfaces must be brought into conformal contact, otherwise bonding will fail. But this method of irreversible sealing did not work well with PMMA, polyimide, and polycarbonate (Duffy et al., 1998).

The oxidized PDMS slab also yields negatively charged channels which then supports EOF (Duffy et al., 1998). However, in another report, it was mentioned that native PDMS supported EOF, which was stable on the basis of EOF and contact angle measurements, probably because the surface charge of PDMS was dictated by the amount of silica fillers present in the PDMS formulations (Ocvirk et al., 2000).

Permanent bonding of two PDMS layers was also achieved by having the bottom layer (with an excess component of the elastomer base A in the mixture, i.e., 30A:1B) and the upper layer (with excess curing agent B, i.e., 3A:1B). On further curing, the excess components allowed new PDMS to form at the interface for bonding (Unger et al., 2000).

In reversible bonding, no oxidation was done. When compared with irreversible bonding, channels formed by reversible bonding are more stable (in EOF), and easy to clean (after chip removal). However, reversible bonding is prone to leakage and cannot withstand pressure greater than 5 psi (Martin et al., 2000). Reversible sealing was more consistent after rinsing the PDMS and glass plates with methanol and then drying in an oven at 65°C for 10 min (Liu et al., 2000a). Nonpermanent bonding was also made between a PDMS layer and a gold-coated glass slide which has been deposited with thiolated DNA capture probes (Esch et al., 2001).

Reversible bonding of PDMS to PMMA was also achieved (Hosokawa et al., 1999; Xu et al., 2000). A PDMS replica containing microchannels ($<100\ \mu m$ deep) were sealed against a PMMA plate (or a PDMS replica of it) that had deep (300–900 μm) reservoirs machined in it (Duffy et al., 1999). PDMS was also sealed against a patterned hydrophobic fluorocarbon film (Lee and Kim, 2001). Moreover, PDMS (after O_2 plasma treatment) has been reversibly bonded to Au, glass, and $Si-SiO_2$ substrates (Delamarche et al., 1997).

Deeper ($>1.5\ \mu m$) channels in PDMS were prone to collapse, either spontaneously (because of gravity) or during suction (Delamarche et al., 1997). Therefore, support pillars were put within channels to avoid bowing of the high-aspect-ratio PDMS channel (100 μm wide and 3 μm deep) during sealing it to glass (Chou et al., 1999).

3D microfluidic channel system could also be fabricated in PDMS. The system was created using a "membrane sandwich" method. In a three-layer PDMS device, the middle layer has channel structures molded on both faces (and with connections between them), and this middle layer was sandwiched between two thicker PDMS slabs to provide channel sealing and structural support (Anderson et al., 2000a). A three-level structure was also created with a basketweave pattern of crossing but nonintersecting channels. A five-level channel system was fabricated with a straight channel surrounded by a coiled channel (Anderson et al., 2000a). A simple stacking of two PDMS channels also realized a Z-shaped 3D interface (Ocvirk et al., 2000).

A solid-object printer, using a melted thermal plastic build material (mp 80–90°C) as the "ink", was used to produce a molding mask for casting PDMS chips (McDonald et al., 2002). This printer provides an alternative to photolithography but only creates features of $>250\ \mu m$. As the mold is easily damaged by deformation, only about 10 replica can be made from one master. The roughness of the master is $\sim 8\ \mu m$ which can be reduced by thermally annealing it against a smooth surface. In this case, a negative master is first made and annealed, and a positive master is created from the smoothed negative master (McDonald et al., 2002).

Reduction photolithography using arrays of microlenses (40 μm) was demonstrated, and this achieved a lateral size reduction of 10^3. This method generated a feature size down to 2 μm over large areas ($2 \times 2\ cm^2$) (Wu et al., 2002). Reduction can be achieved using 35 mm film photography ($8\times$) or microfiche ($25\times$) (Deng et al., 2000). On the other hand, photolithography using gray scale masks further enables the generation of micropatterns that have multilevel and curve features on photoresist upon a single exposure (Wu et al., 2002).

Instead of using a planar molding master, a fused-silica capillary (50 μm i.d. and 192 μm o.d.) was used as a template for casting PDMS channels, as well as the fluid inlet/outlet capillary. After curing of PDMS, the middle prescored section (4 cm) of the capillary was removed to reveal the PDMS channel (192 μm wide and deep) (Gao et al., 2001). A capillary was also used, as an electrospray emitter, to mold a PDMS channel. In this case, after curing of PDMS, the last 0.5 cm section of the capillary was removed to create a channel (Jiang et al., 2001).

Besides using PDMS, 3D structures for mass spectrometry (MS) analysis have been casted in epoxy (Epofix) using two negative masters (in PDMS), which in turn were created from Lucite positive masters. Epofix was selected as the chip substrate, among acrylic–polyester resin (Casolite) and epoxy resin (Araldite), because of the best mechanical properties and the least chemical interference needed for fabricating the MS chip (Liu et al., 2000d). In another report, PDMS was chosen over epoxy to fabricate MS chips because of less chemical noises (interferents) in MS, and over polyurethane because of good adhesion properties. Even so, in the use of PDMS, its curing (at 70°C) should be carried out for at least 72 h to further reduce the chemical noise (Chan et al., 1999b).

2. Injection Molding

Injection molding is a common procedure in the plastic industry. Therefore, this method has been used to fabricate acrylic chips (McCormick et al., 1997). The channel pattern was first fabricated by wet-etching a Si master. Then Ni electroforms made out of the Si master were used to create acrylic chips (500 per electroform) via injection molding. A negative image was formed on the Si master because it is more difficult to etch a 45 μm-deep channel than to create a 45 μm-high ridge pattern in Si. Access holes (3 mm) were drilled through the acrylic chip and it was sealed with a cover film (2 mm thick, Mylar Sheet) using a thermally activated adhesive (McCormick et al., 1997).

Furthermore, three PS plastic plates were molded to create the microfluidic channels, which were employed for biomolecular interaction study using surface plasmon resonance (SPR) measurement. The upper plastic plate with ridge patterns was ultrasonically welded to the middle plate. The middle plate was sealed to the bottom plate with molded silicone rubber layers (Sjoelander and Urbaniczky, 1991).

3. Laser Ablation

Photoablation using lasers of various wavelengths (IR, visible, UV, or X-ray) has been employed to fabricate plastic chips using various polymeric substrates.

For instance, pulsed UV laser ablation (193 nm) has been used to make plastic chips out of polyethylene terephthalate (PET, 100 μm thick sheet) (Bianchi et al., 2001; Rossier et al., 1999a, b; Schwarz et al., 1998) and polycarbonate (PC, 125 μm thick) (Bianchi et al., 2001; Rossier et al., 1999a). Channels of size as narrow as 30 μm and as deep as 100 μm can be made (Bianchi et al., 2001; Rossier et al., 1999b).

A UV excimer laser (193 nm) was used to ablate channels in polystyrene (1.2 mm), polycarbonate (1 mm), cellulose acetate (100 μm), and PET substrates (100 μm) (Roberts et al., 1997). Low-temperature (125°C) lamination was used for fast bonding (<3 s) the ablated substrates to PET films (35 μm thick) with PE adhesive (5 μm thick) (Roberts et al., 1997; Rossier et al., 1999a). Channels (37 μm deep) of 1 mm × 0.04 mm were ablated through a copper foil mask. Relative to the original polymer, the photoablated surface is rougher and has increased hydrophilicity. The EOF increases in the following order: PC < PS < cellulose acetate < PET (Roberts et al., 1997). The excimer-laser ablation can also increase the surface charge of the PMMA channel at a 90° turn, reducing band broadening (plate height) by 40% (Tohnson et al., 2001). An excimer laser was also used to machine a PC chip (6 mm thick) to create 160 μm-wide channels (60 μm deep) (Xu et al., 1998), or on a polyimide sheet (Giordano et al., 2001a; Xu et al., 1998).

Laser ablation (diode-pumped Nd:YVO$_4$ laser, $\lambda = 532$ nm) was used to create a master on the PMMA layer (doped with Rhodamine B to facilitate the absorption of laser radiation) coated on a Si wafer [see Fig. 7(a)]. The width of the ablated features depended on the diameter of the laser beam in the focal point and the position of the focal point. The best aspect ratios were obtained with the

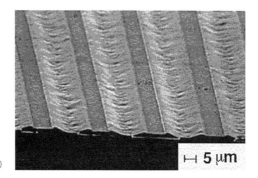

Figure 7 Electron micrograph of (a) lines generated by laser-ablation on a PMMA master and (b) the PDMS replica casted from the PMMA master. [(Grzybowski et al., 1998) Courtesy of American Chemical Society.]

laser beam focussed 3–4 μm into the PMMA film. The ablated PMMA–Si master was used to cast a PDMS layer, which had smoother surfaces than the PMMA master [see Fig. 7(b)] (Grzybowski et al., 1998).

X-ray micromachining (7–9 Å synchrotron radiation) was used to create PMMA channel structures (20 μm wide and 50 μm deep) (Ford et al., 1998). Sealing was achieved by heating both PMMA plates on a hot plate (150°C) for 5–10 min. Then they were aligned, pressed together, and the two-plate assembly was allowed to cool. This surface heating prevents outgassing and bubble inclusions which occur in bulk heating (Ford et al., 1998). In contrast, conventional (rather than X-ray) machining on PMMA can only create channels of large depth (from 125 μm to 3 mm) and widths (>125 μm) (Duffy et al., 1999).

In another report, an X-ray mask (using 10 μm-thick Au as the X-ray absorber on a Kapton film) was employed to ablate a PMMA wafer. For sealing, a PMMA cover was first spin-coated with a 2 μm layer of poly(butyl methacrylate-co-methyl methacrylate) and then bonded to the developed PMMA plate thermally (120°C for 1 h) (Meng et al., 2001).

A CO$_2$ laser (1060 nm) was used to cut through a polycarbonate PC (black) (carbon-coated) wafer of 250 μm thickness to create microfluidic channels. The laser-machined black PC wafer was then thermally bonded between two transparent PC wafers at 139°C under 2 tons of pressure for 45 min (Liu et al., 2002). CO_2 laser was also used to ablate Mylar sheets (with adhesive). Then, the machined Mylar sheets were laminated together (Schilling et al., 2002). Moreover, a PMMA substrate was machined by an infrared CO_2 laser. However, the microchannel (~200 μm deep) has a Gaussian-like cross-section and a certain degree of surface roughness (Goranovic et al., 2001).

4. Wire Imprinting

Plastic channels could also be fabricated by the hot wire imprinting method. A channel was imprinted on a piece of Vivak (copolyester) by a nichrome wire (50 μm dia.) clamped between the plastic and a glass slide (80°C for 15 min). Then, a perpendicular channel was similarly imprinted on a second piece of Vivak. The two pieces were then sealed at 75°C for 25 min. (Note that the 2 channels are not coplanar!) The channels were filled with water for storage (Wang and Morris, 2000; Wang et al., 2000a).

Microchannels were wire-imprinted on Plexiglas using a 90 μm-dia. tungsten wire at 175°C. The channels were triangular-like, typically 200 μm wide and 75 μm deep (Chen et al., 2001a).

A chromel wire (13 or 25 μm dia.) was used to hot-imprint microchannels on PMMA or Plexiglas (1.6 mm thick) substrates. The plastic was heated for 10 min at 105°C (softening temperature of PMMA). Although another imprinting procedure can be carried out using a wire perpendicular to the first imprinted channel, the plastic substrate appeared to be more rigid following the first heating cycle. Therefore, the second channel was imprinted on a second PMMA plate. Then the two plates were bonded together at 108°C for 10 min using a press with the application of a uniform pressure. Sometimes, the PMMA plate was not well-sealed owing to bubble entrapment. These failed devices could generally be salvaged by reheating the device for a longer time at the same temperature (Martynova et al., 1997).

To fabricate more complex structures, a Si master with positive relief structures was first created; then, the master was used to imprint the channel pattern on a PMMA plate (135°C for 5 min) (Martynova et al., 1997).

Room-temperature (RT) imprinting on PMMA (Lucite) and copolyester (Vivak) plates was achieved using a Si master. A hydraulic press was employed to apply the pressure (4500–2700 psi) at room temperature. Such an RT operation prevents breakage of the Si master owing to the differences in thermal expansion coefficients of Si and PMMA. This procedure improves the lifetime of the master, so that 100, instead of 10, devices per master can be imprinted (Xu et al., 2000). Wire imprinting was also achieved on PS, PMMA, and a copolyester material (Vivak) (Locascio et al., 1999). Low-temperature bonding was applied for channel sealing for 10 min. The EOF was found to be the highest in the copolyester chip and the lowest in the PS chip. The chromel wire was stretched taut and placed on a plastic plate (for 7 min) heated to a temperature higher than the softening point (Locascio et al., 1999). Copolyester sheets were also imprinted using a Si master (RT, 1600 psi, 5 min) to produce microchannels (30 μm deep and 100 μm wide) (Jiang et al., 2001).

A PS substrate was imprinted by a Si template with channels at RT using pressure (890 psi or 6.1×10^6 Pa) applied by a hydraulic press for 3 min. The imprinted PS chip was then sealed with a PDMS lid (Barker et al., 2000a).

When high-temperature bonding is used to fabricate PMMA chips, hot imprinting must be used to create channels on the PMMA substrate (110°C, 5.1×10^6 Pa or 740 psi for 1 h). Then the chip was bonded to another PMMA cover at 103°C for 12 min. RT-imprinted PMMA channel will be distorted during the thermal bonding process (Barker et al., 2000a; Ross et al., 2001).

5. Compression Molding

Another common procedure, compression molding, was also used to fabricate microstructures on PC chips (1 mm thick) using high temperature (188°C) and pressure (11 metric ton pressure applied by a hydraulic press). Before bonding, the hydrophobic channel surface was irradiated with UV (220 nm) to assist aqueous solution transport. The molded chip was thermally bonded to another PC wafer. The bonded chip did not yield to a pressure up to 150 psi (134°C, 4 metric ton, 10 min) (Liu et al., 2001e).

Hot-embossing was also used to make polymer chips on PMMA (Becker et al., 1998) or PC (Becker et al., 1998; Soper et al., 2001).

Moreover, a Zeonor plastic plate was micromachined with microchannels (60 μm wide and 20 μm deep) by the hot-embossing technique (130°C, 250 psi) using a Si master. The embossed chip was thermally bonded to another Zeonor plate (85°C, 200 psi, 10–15 min). Zeonor, normally used to manufacture CD and DVD, has lower water absorption (<0.01%) than PC (0.25%) and PMMA (0.3%). In addition, Zeonor has strong chemical resistance to alcohols, ketones, and acids (Kameoka et al., 2001). In another report, a 2-mm thick cyclo-olefin (Zeonor 1020 R) substrate was embossed using a Si master to create 20 μm-wide and 10 μm-deep channels (Kameoka et al., 2002).

6. Other Plastic Micromachining Processes

Microchannels have also been directly patterned on SU-8 photoresist (Lao et al., 2002; Madou et al., 2000; Renaud et al., 1998). Multilevel structures were fabricated using the SU-8 photoresist (Renaud et al., 1998). In another report, the photoresist was spun (1250 rpm for 30 s) on an ITO-coated glass plate, which was first treated by an O_2 plasma to increase the adhesion of SU-8 on ITO. The photoresist channel was of ribbon-like structure with triangular ends (40 μm height, 10 mm breadth, and 90 mm length). The channel was sealed by hot-pressing another ITO-glass on top of the photoresist structures (Lao et al., 2002). SU-8 photoresists have also been used to create multilayered structures (i.e., as the channel wall materials), which were sandwiched between Si and Pyrex or between quartz and quartz (Jackman et al., 2000).

Microcavities (53 μm-dia. and 8 μm-deep) have been fabricated using polyimide layers (two 4 μm layers sandwiched by Au/Cr electrodes) (Henry et al., 2001).

Liquid-phase photopolymerization was used to fabricate plastic chips. To create a microchannel, a UV-mask was used to prevent polymerization at the channel area, while the unexposed areas were photopolymerized. Subsequent suction and flushing removed the unexposed monomer mixtures (Beebe et al., 2001).

Fabrication of hydrogel structures in microchannels was first achieved by functionalizing the glass–PDMS channel with 3-(trichlorosilyl)propyl methacrylate (TPM). Then, the channel was filled with a solution consisting of 1% (v/v) 2-hydroxy-2-methylpropiophenone (HMPP, the photoinitiator) and 50% poly(ethyleneglycol)-diacrylate (PEG-DA) (Seong et al., 2002; Zhan et al., 2002). PEG-DA was first purified using an alumina column to remove the stabilizer and impurities. Free-radical polymerization was initiated by UV (365 nm) through a photomask to dissociate HMPP into methyl radicals which attacked the acrylate functionalities in both PEG-DA in the solution and TPM at the channel wall, allowing formation and attachment of the hydrogel structure, respectively (Seong et al., 2002). In another report, a transparent channel (of width 500–200 μm and of depth 50–180 μm) was filled with a photopolymerizable liquid mixture consisting of acrylic acid and 2-hydroxyl methacrylate (1:4 molar ratio), ethylene glycol dimethacrylate (1 wt%) and a photoinitiator (3 wt% Irgacure 651 or 2,2-dimethoxy-2-phenylacetophenone). Polymerization was completed in less than 20 s to produce the hydrogel structures (Beebe et al., 2000).

Fluoropolymers (PTFE, FEP) were etched by IBE (Ar ion beam etching) to create microstructures (Lee et al., 1998).

Plastic microstructures were formed with Parlyene C by an additive process. The structures were supported on a PC substrate. As PC was dissolved in acetone used to remove the sacrificial photoresist layer, Parlyene C was first coated on the PC substrate to protect it (Webster et al., 1998).

7. Problems Encountered in the Use of Polymeric Chips

Polymeric materials may have some unfavorable properties for microfluidic operations.

For instance, there is autofluorescence emitted from the plastic substrates and sealing materials. However, the use of a longer excitation wavelength (530 nm) is one way to alleviate the problem (McCormick et al., 1997). In addition, as the UV transparency of olefin polymers is down to 300 nm, these materials may be better substrates than PMMA or PC (Chi et al., 2001). Moreover, commercial PDMS is optically transparent down to ~230 nm. Its relatively low refractive index ($n = 1.430$) reduces the amount of reflected excitation light and hence lowers the background (Effenhauser et al., 1997b).

Polymeric materials usually have low dielectric breakdown voltages. But the use of a lower E field (~1100 V/cm) helps alleviate this problem (Zhang and Manz, 2001). Fortunately, the electrical insulation property of PDMS is high ($R > 10^{15}$ Ω/cm) (Effenhauser et al., 1997b).

The thermal conductivity of plastics is typically a factor of 3–5× poorer than silica, but because the channels in the plastic chip are usually narrow (i.e., with high SVR), the heat dissipation properties of the plastic (e.g., acrylic) channel compared favourably with that of a fused silica capillary (75 μm i.d.) (McCormick et al., 1997). Furthermore, the thermal conductivity (in W/m/K) of PDMS (0.15) is sufficient, although it is lower than glass (0.7–1.0) and fused silica (1.38). The EOF in PDMS is 10-fold less than those found in glass or fused silica (Effenhauser et al., 1997b).

There is a need to prime the hydrophobic channels, especially in PDMS, before filling. But the addition of ethanol (up to 5%) assisted in filling hydrophobic (e.g., PDMS) channels (Delamarche et al., 1997). The PDMS channel can be rendered hydrophilic by oxidation (O_2 plasma treatment) (Delamarche et al., 1997; Duffy et al., 1998). But, PDMS treated with O_2 plasma had a short useful lifetime (~15 min) of hydrophilic surface in air before the surface became hydrophobic again. Storing the treated PDMS under water maintained the channel hydrophilicity for more than 1 week (Delamarche et al., 1997). PDMS channels can also be rendered hydrophilic by acid treatments: (1) 0.01% HCl, pH 2.7, 43°C for 40 min (Chou et al., 1999); (2) 5% HCl, 5 min (Linder et al., 2001); or (3) 9% HCl for 2 h (Bernard et al., 2001). The flow of liquid in thin PDMS channels is also facilitated by a large filling pad plus a second large flow-promoting pad (Delamarche et al., 1997).

Although PDMS is swollen by many organic solvents, it is unaffected by water, polar solvents (e.g., ethylene glycol), and perfluorinated compounds (Grzybowski et al., 1998). Compatibility of PDMS to other organic solvents can be improved by using a coating of sodium silicate (Jo et al., 2000).

In composite plastic microchannels, an additional problem of extra dispersion (Taylor dispersion) arose from the difference in zeta potentials of the different materials forming the channels (Bianchi et al., 2001). Caged fluorescent dye [fluorescein *bis*(5-carboxymethyoxy-2-nitrobenzyl) ether dipotassium salt] was used to visualize the greater dispersion in EOF obtained in acrylic or composite channels due to nonuniformity in the surface charge density (Ross et al., 2001).

E. Metal patterning

Metal layers were deposited on microchip substrates for use as an etch mask in micromachining or as electrodes for electrochemical detection. Various metals have been used as the overlayer and adhesion layer (see Table 10).

Thermal degradation of the Au/Cr metal layers has been studied using Auger electron spectroscopy (AES). It was found that Cr diffused easily into the upper Au layer and got oxidized at the surface when the glass plate was heated at 200–800°C (Murakami et al., 1993).

In addition to metal, other materials, such as ITO (Fiedlerr et al., 1998; Hsueh et al., 1996) were used as electrodes on microfluidic chips. To facilitate fusion bonding of glass chips, electrode materials could be deposited in pre-etched recesses on glass chips (Jeong et al., 2001; Lichtenberg et al., 2001).

III. MICROFLUIDIC FLOW

The liquid flow in the microfluidic chip is mostly achieved by using EOF (Harrison et al., 1993). Other mechanisms have also been employed for microfluidic flow. Surface modification is essential to manipulate the direction and magnitude of EOF. Flow has been employed for fraction collection, and generation of concentration gradient. Laminar flow in microfluidic channel allows liquid–liquid extraction and microfabrication to occur within the channels. Moreover, valving and mixing are needed in order to achieve a better flow control. All these microfluidic flow operations are further described in subsequent sections.

A. EOF and Hydrodynamic Flow

The EOF in a network of channel has been modeled by the flow of electric current in a network of resistors using Kirchoff's rules. On the basis of this study, a better control of liquid flow can be achieved by designing channels of

Table 10 Metal Patterning on Microchips

Metal	References
Au	
Au (10 nm)/Cr (5 nm) on PMMA	Ford et al. (1998)
Au (45 nm)/Cr (1.5 nm)	Esch et al. (2001)
Au (200 nm)/Cr (20 nm) on glass	Murakami et al. (1993)
Au (250 nm)/Cr (50 nm) on glass	Eijkel et al. (2000b)
Au (1 μm)/Cr (50 nm) on Si	Hsueh et al. (1996)
Au (12 nm)/Ti (1.5 nm) on polyurethane	Mrksich et al. (1996)
Au (38 nm)/Ti (5 nm) on glass	Grzybowski et al. (1998)
Au (50 nm)/Ti (10 nm) on Si	Juncker et al. (2002)
Au (50 nm)/Ti (5 nm) on glass	Gallardo et al. (1999)
Au (100 nm)/Ti (5 nm) on glass	Liu et al. (2000a)
Au on SU-8 vertical wall	Lu et al. (2001)
Pt	
Pt (200–300 nm)/Cr (40–60 nm) on PMMA	Grass et al. (2001)
Pt (200 nm)/Cr (50 nm) on PMMA	Vogt et al. (2001)
Pt (∼150 nm)/Ti (5 nm) on Pyrex	Ueno et al. (2001)
Pt (200 nm)/Ti (30 nm) on glass	Namasivayam et al. (2002)
Pt (200 nm)/Ti (20 nm) on glass	Lapos et al. (2002)
Pt (260 nm)/Ti (20 nm) on glass	Woolley et al. (1998)
Pt (300 nm) on Si	Hua et al. (2002)
Pt (150 nm)/Ti (10 nm) on Si	Hsueh et al. (1998)

different solution resistances and by applying different voltages (Harrison et al., 1993). With voltage-controlled EOF, dilution between two liquid streams is feasible by controlling their relative flow speeds (see Fig. 8) (Harrison et al., 1993; Seiler et al., 1994).

For a better flow control using only EOF, secondary hydrodynamic flow (HDF) should be avoided. This could be achieved by ensuring that all solution reservoirs were filled to the same liquid level to avoid HDF (Effenhauser et al., 1997b). In addition, HDF could be prevented by closing the inlet reservoir to the atmosphere using a valve. In this way, better EOF control and CE separation (RSD of migration time decreased by 10–30×) can be achieved (Kaniansky et al., 2000). On its own right, HDF is a major pumping mechanism and is used in many applications. HDF is mainly achieved using a syringe pump or peristaltic pump.

B. Surface Modifications for Flow Control

In some applications, surface modification is needed to suppress or modify EOF. EOF can be suppressed by coating the channel surface with methylhydroxyethylcellulose (MHEC) (Kaniansky et al., 2000), or photopolymerized polyacrylamide (Kirby et al., 2001). Reversal of EOF can be also achieved (e.g., the PMMA surface was modified with N-lithioethylenediamine or N-lithiodiaminopropane (Henry et al., 2000) or the PDMS surface was coated by hydrophobic interaction with TBA^+) (Ocvirk et al., 2000).

In some applications, an increase in the surface charge is needed in order to increase or support EOF. For instance, after O_2 plasma oxidation, PDMS supports EOF because of the increase in surface negative charges by oxidation. However, because of the instability of the charge created on the polymer surface, EOF was unstable. Better stability can be achieved by immediately filling the PDMS channel with liquids, rather than letting it expose to air. The useful lifetime of these devices for quantitative analysis requiring stability of EOF is probably 3 h (Duffy et al., 1998).

UV irradiation can increase surface charge (and hence hydrophilicity) in PC chips (Liu et al., 2001b). The surface charge on the glass channel can also be modified by UV irradiation when the surface is coated with a TiO_2 film (Khoury et al., 2000).

UV polymerization was used to graft acrylic acid, acrylamide, dimethylacrylamide, 2-hydroxylethyl acrylate, or poly(ethyleneglycol)monomethoxyl acrylate on the PDMS channel in order to create a hydrophilic surface. The grafted PDMS channels become more readily filled with liquids when compared with native PDMS, and show reduced adsorption of solutes (e.g., peptides) when compared with oxidized PDMS. The magnitude of the EOF of the grafted PDMS chip is intermediate between that of native PDMS and oxidized PDMS. Unlike oxidized PDMS, the EOF of grafted PDMS remains stable upon exposure to air (Hu et al., 2002).

For Vivak polyester channels, alkaline hydrolysis of surface groups (e.g., ester) to ionizable group (e.g., carboxylate) has produced a more reproducible EOF. In combination with a dynamic coating (e.g., CTAB), EOF may be eliminated or reversed in direction depending on the CTAB concentration (Wang et al., 2000a).

Polyelectrolyte multilayers (PEMs) have been used to alter surface charges and control the direction of EOF in chips made of PS (Barker et al., 2000a, b; Kim et al., 1995), PMMA (Barker et al., 2000a), and PETG (Barker et al., 2000b). PEM deposition was carried out by exposing the microchannel to alternating solutions of positively charged PEM [poly(allylamine hydrochloride), polycallyamise hydrochloride (PAAH)] and negatively charged PEM [poly(styrene sulfonate), PSS] (Barker et al., 2000c). Moreover, by depositing the PEM at different parts of the PMMA channels, special flow control patterns can be achieved (Barker et al., 2000a). PDMS channels

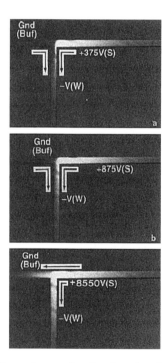

Figure 8 Photomicrographs showing the direction of flow during the controlled mixing of a buffer solution and a 100 μM fluorescein solution. The glass channels were 30 μm wide. The buffer reservoir (Buf) was at ground, and the sample reservoir (S) had the indicated positive voltages applied, whereas the waste reservoir (W) was at −3750 V. Increase in the voltage applied at S indicated a higher degree in mixing. [(Seiler et al., 1994) Courtesy of American Chemical Society.]

were also coated by PEM using alternating cationic layers (polybrene, PB) and anionic layers (dextran sulfate, DS). Stable EOF can be achieved for up to 100 runs (Liu et al., 2000a). PB and DS are able to stabilize EOF because these polyelectrolytes have high and low pK_a values, respectively. So in the pH 3–10 range, PB is essentially positive and DS is essentially negative. For comparison, in glass channels, the silanol groups have intermediate pK_a values, and so the surface charge strongly depends on pH (Liu et al., 2000a).

C. Fraction Collection

Accurate flow switching has been employed for fraction collection of single or a group of oligonucleotides (on a glass chip) (Effenhauser et al., 1995).

After CGE separation of a mixture of fluorescently labelled oligonucleotides [p(dT)$_{10-25}$] for 170 s, preselected components were withdrawn by applying a brief 5 s fraction collection pulse (see Fig. 9). Since the EOF has been suppressed, the oligonucleotide migrated from the negative voltage to the GND (i.e., positive). In a typical procedure, a CGE run occurs for 140 s (phase I), fraction selection for 5 s (phase II), removal of other components for 40 s (phase III), and fraction collection for 65 s (phase IV), and all these steps lead to a total operation time of 250 s. Such a microchip fraction collection system alleviates the problem of dead volume in a conventional system. However, a dilution factor of three is still involved based on the calculation of peak band broadening (Effenhauser et al., 1995).

Collection of fragments from a 100 bp DNA ladder was also achieved in a PDMS chip (Hong et al., 2001) or a glass chip (Khandurina et al., 2002). In the latter example, a small reversed field was maintained in the separation column to halt or slow down later migrating DNA in order to assist collection of a DNA fraction.

D. Laminar Flow for Liquid Extraction and Microfabrication

Laminar flow is prevalent in microfluidic channels, and this phenomenon is exploited for various applications. For instance, a two-phase flow between an organic and an aqueous stream has been achieved. This allows the ionophore-based ion-pair extraction to be carried out for the detection of potassium and sodium (Hisamoto et al., 2001) and iron (Tokeshi et al., 2000a) in microchannels. In such a system, degradation in the signal response caused by leaching of ion-sensing components, as found in conventional ion-selective optodes, is not an issue, because fresh organic phase can be used in every measurement. In addition, a low-viscosity organic solvent can be used to increase solute diffusion which is usually slow in

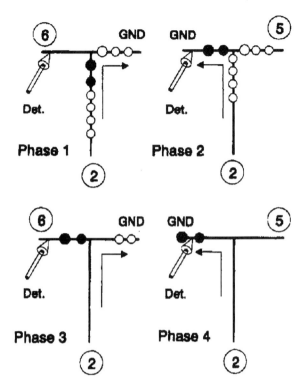

Figure 9 Various phases implemented for fraction collection at reservoir 6. During the whole sequence, reservoir 2 is kept at a negative high voltage. Phase 1: Application of ground at reservoir 5 initiates the separation of the injected sample plug toward reservoir 5. Phase 2: When the sample zones of interest (filled circles) arrive at the intersection, reservoir 6 is grounded for a few seconds and reservoir 5 left floating. Phase 3: The remaining zones are removed from the system by switching back to the phase 1 conditions. Phase 4: The extraction fractions are moved past the detector by grounding reservoir 6 again while leaving reservoir 5 flowing. [(Effenhauser et al., 1995) Courtesy of American Chemical Society.]

polymeric membrane-based ion sensors (Hisamoto et al., 2001).

To further stabilize multiphase flow, the liquid–liquid interface can be stabilized with the boundary formed at a constricted opening (Shaw et al., 2000). Moreover, guide structures (5 μm high) were fabricated in a microchannel (20 μm deep) which was etched from three closely spaced (35 μm apart) parallel lines (10 μm wide) on a photomask (Tokeshi et al., 2002). The multiphase flow (aqueous/aqueous, aqueous/organic, and aqueous/organic/aqueous) became more stable in such microchannels, allowing extraction reactions to proceed over a long distance along the channel.

The conventional procedure (with 40 steps!) for the determination of Co became simpler using the microfluidic chip with guide structures to stabilize multiphase flow (see Fig. 10). Co^{2+} in the sample (aqueous phase 1)

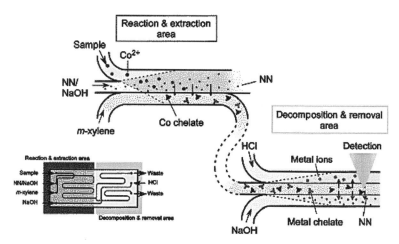

Figure 10 Schematic illustration of Co (II) determination in a quartz chip fabricated with guided structures in the channel for multiphase flow. For details in operation, see text. [(Tokeshi et al., 2002) Courtesy of American Chemical Society.]

was first mixed with a chelating agent [2-nitroso-1-naphthol (NN)] (aqueous 2) to form a complex which was then extracted into *m*-xylene (organic 1) along a microchannel. The Co-chelates were then separated from the chelates of other metal interferences (Cu^{2+}, Ni^{2+}, and Fe^{2+}) which were decomposed by HCl and were then extracted into the aqueous 3 phase. The decomposed NN was extracted into aqueous 4 (NaOH). Finally, the undecomposed Co-chelates (still in organic 1) were detected by TLM (Tokeshi et al., 2002).

A two-phase liquid–liquid (water–nitrobenzene) crossing flow has been demonstrated in glass chips to mimic counter-current flow for liquid–liquid extraction. The flow was stabilized by an octadecylsilane (ODS)-modified channel (Hibara et al., 2002a).

A two-phase flow was utilized to perform liquid ion-exchange applied for conductivity suppression in ion chromatography. Tetraoctylammonium hydroxide (TOAOH) or Amberlite LA-2 (a secondary amine) was dissolved in an organic solvent (butanol) to exchange away H^+ for conductivity detection of heavy metal ions (Ni, Zn, Co, Fe, Cu, and Ag) (Kuban et al., 2002).

Laminar flow, which is prevalent in the microchannel, has also been utilized for microfabrication. For instance, a gold-coated (250 Å thick) glass–PDMS channel was patterned with a zig–zag Au wire using a laminar flow of Au etchant flowing parallel to a water flow. In addition, a silver wire was formed chemically at the interface of two parallel laminar flows of silver salt and reductant (Kenis et al., 1999).

Glass etching was also achieved at the interface of two parallel laminar flows of HCl and KF (i.e., creating HF at the interface). Moreover, a polymeric structure was precipitated at the interface of two flows of oppositely charged polymers. The classic example was the fabrication of a three-electrode system inside a channel. First, Au etchant was flowed to separate a gold patch, creating two isolated Au electrodes. Subsequently, a silver wire was formed in the middle, and 1% HCl then converted silver on its surface to AgCl, creating a silver/AgCl reference electrode (see Fig. 11) (Kenis et al., 1999).

The self-assembled monolayer (SAM), together with multistream laminar flows, was used to pattern flow streams in glass chips. For instance, a stream was confined by virtual walls to the central hydrophilic region, which was adjacent to two streams flowing in hydrophobic regions patterned by two organosilanes, OTS and HFTS (see Fig. 12). An increase in the liquid pressure allowed the central liquid stream to burst into the adjacent hydrophobic regions with successively lower surface free energy. This behavior is like a pressure-sensitive valving system. Moreover, using photocleavable SAM (with the 2-nitrobenzyl photosensitive group), the SAM can be patterned through a UV mask, giving rise to complex flow patterns. The virtual wall created can be employed to carry out gas–liquid reactions. For instance, acetic acid vapor was allowed to react with a pH indicator solution at the virtual wall, allowing kinetic studies to be carried out (Zhao et al., 2001).

E. Concentration Gradient Generation

Gradients of different concentrations were generated by parallel and serial mixing (Cheng et al., 1998b; Dertinger et al., 2001; Jacobson et al., 1999a; Kamholz et al., 1999; Murakami et al., 2001; Weigl and Yager, 1999; Weigl et al., 1998; Yang et al., 2002a, b). In parallel mixing, seven different concentrations, including zero concentration, were generated by mixing three different sample concentrations with three buffer streams in parallel

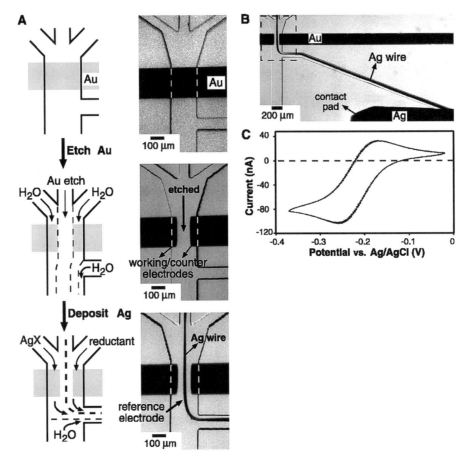

Figure 11 (a) Optical micrographs of the stepwise fabrication of a three-electrode system inside a 200 μm wide PDMS channel. Two gold electrodes (counter and working) are formed by selectively etching the gold stripe with a three-phase laminar flow system. A silver reference electrode is fabricated at the interface of a two-phase laminar flow. (b) Overview picture of the three-electrode system including the silver contact pad. The dashed box corresponds to the last picture shown in (a). The silver wire widens to a large silver contact pad. (c) Cyclic voltammogram of 2 mM $Ru(NH_3)_6Cl_3$ in a 0.1 M NaCl solution (scan rate: 100 mV/s) [(Kenis et al., 1999) Courtesy of American Association for the Advancement of Science.]

(Jacobson et al., 1999a). In serial mixing, five concentrations were generated by mixing one sample concentration with one buffer stream (Jacobson et al., 1999a). More complex channel networks for parallel mixing even produced various concentration gradients (e.g., linear, parabolic, or periodic) on a PDMS chip. In this manner, the gradient can be maintained over a long period of time (Dertinger et al., 2001).

Concentration gradients were generated along the interface of two parallel HDF streams (Kamholz et al., 1999; Weigl and Yager, 1999; Weigl et al., 1998). Moreover, a concentration gradient was generated along a long parallel dam at the border of two parallel channels, where the sample and buffer flowed separately. Diffusion across the dam created a concentration gradient in the same direction of the flow (Yang et al., 2002a, b).

A concentration gradient of solvent B (65% acetonitrile/10 mM acetic acid) and solvent A (10 mM acetic acid/10% methanol) was generated on-chip for gradient frontal separation of peptides in a C18 cartridge (Figeys and Aebersold, 1998).

A concentration gradient was created by mixing two streams of liquid at a Y-junction and allowing diffusional mixing to occur downstream for a fixed distance (i.e., 2 cm). Thereafter, these laminar flow streams of various concentrations were separated by dividing the main channel into a series of small tributaries (see Fig. 13). In the same study, in addition to the concentration gradient, a temperature gradient has been created on the chip. When the temperature gradient is perpendicular to a series of channels, each channel is held at a discrete temperature. On the other hand, if it is parallel, the temperature varies as the liquid flows downstream. The hot end is realized by a square brass tube heated by a cartridge heater; whereas the cold end is constructed by another square brass tube with cooling water flowing inside the tube.

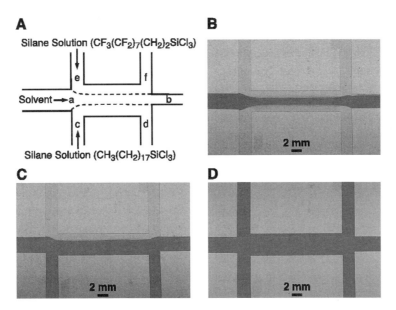

Figure 12 Pressure-sensitive valves. The laminar flow scheme for patterning surface free energies inside channels with two different trichlorosilanes, OTS and HFTS, is shown in (a). Images of flow patterns of an Rhodamine B solution obtained in increasing pressures from (B), (C), to (D). [(Beebe et al., 2000) Courtesy of Nature Publishing Group.]

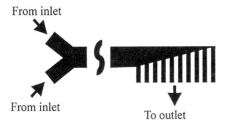

Figure 13 Twelve channels emanating downstream of a Y-mixing junction, creating a gradient of 10 different concentrations. [(Mao et al., 2002a) Courtesy of American Chemical Society.]

The temperature gradient spanned from 36°C to 77°C by the cold and hot brass tubes separated by 3.4 mm. By combining the concentration and temperature gradients, the fluorescent intensity of a dye (carboxyfluorescein) was used to illustrate the capability of the device to obtain 3D information (Mao et al., 2002a).

F. Microvalves

For better flow control, microfluidic valves are designed. For instance, a latex membrane (150 μm thick) has been used to construct a valve. A sample is loaded by liquid pressure (10–12 psi) through the microfluidic valve opened by applying vacuum (30 mm Hg) on the opposite side of the valve diaphragm. To facilitate sample filling, while the valve is open, air is simultaneously evacuated through a hydrophobic membrane vent (1.0 μm pore size). Both the valve and vent have dead volumes of about 50 nL (Lagally et al., 2001).

Multiple valves have been constructed on a PDMS valve control layer. Multilayer (up to seven) soft lithography has been used to generate the valve control layer (4 mm thick for strength) plus a fluid layer (40 μm thick) on PDMS. The small Young modulus (~750 kPa) of PDMS allows a large deflection (1 mm) using a small actuation force (~100 kPa on a 100 μm × 100 μm valve area). The response time is on the order of 1 ms. Round channels were more easily sealed (~40 kPa) than trapezoidal channels were (>200 kPa). Round channels were casted from rounded photoresist relief (after baking it at 200°C for 30 min). By pneumatic actuation, the valving rate was attained at ~75 Hz to produce a pumping flow rate of 2.35 nL/s (Unger et al., 2000).

Valving on a PDMS sheet has also been achieved after an elliptical hole was punched in the PDMS, and an elliptical needle was inserted in the hole. Then, valve actuation was achieved by having an elliptical needle rotated for 90°, thus compressing the fluidic channel below for valve closure (Mao et al., 2002b).

Pluronics F127 has been exploited to construct microfluidic valves. This material, which is an uncharged tri-block copolymers [$(EO)_{106}(PO)_{70}(EO)_{106}$], is a commercially available surfactant. But within a certain concentration (18–30%), the material, which is of low viscosity (<2 cP) at low temperature (0–5°C), will form a self-supporting gel (cubic liquid crystalline) at a higher temperature (e.g., RT). Therefore, this material has been used as a one-shot, phase-change valve for polymerase chain reaction (PCR) when activated by raising the material to room temperature. The Pluronics valve can hold up to 20 psi pressure, which is above the holding pressure (6.8 psi) required for PCR (up to 94°C) (Liu et al., 2002).

Hydrogel valves were also created within microfluidic channels to provide fluid control. To increase the mechanical stability, the hydrogel was formed around prefabricated posts in the channel. Because of the reduced thickness of hydrogel, the response time to chemical stimuli was much reduced to 8 s (from 130 s). The swelling and contraction provide valve-close and valve-open actuation, respectively, due to chemical stimuli (e.g., pH change). When acrylic acid (A) was used in the hydrogel, high pH caused swelling and low pH caused contraction. On the other hand, when 2-(dimethylamino)ethyl methacrylate (B) was used instead, high pH caused contraction of the hydrogel. This property was exploited to produce two hydrogel gating valves. In this case, only one gate (A) will contract and open at pH 4.7 (lower pH). On the other hand, at pH 6.7 (higher pH), only gate B will contract and open (Beebe et al., 2000).

Passive and capillary burst valves were fabricated on a plastic disk. A wider channel would require less force to burst through (into a bigger chamber) and hence a lower rotation rate or a lower centrifugal force was used for valve to burst open (Duffy et al., 1999).

Nonstick photopolymer formed inside a glass microfluidic channel was used as a mobile piston for flow control. Sealing against high pressure (>30 MPa or 4500 psi) could be achieved by the piston. The polymer was compatible with organic solvents. Actuation time less than 33 ms was observed. The photopolymer was formed by irradiating an unmasked section filled with the monomer (trifluoroethyl acrylate/1,3-butanediol diacrylate in a 1:1 ratio) and photoinitiator (2,2'-azobisisobutyronitrile) using UV (355 nm). Using this method, a check valve (one-way flow), a diverter valve (like an exclusive OR logic gate), and a 10 nL pipette with a check valve were fabricated (Hasselbrink et al., 2002; Rehm et al., 2001).

Electrochemically generated bubbles were also used as microfluidic valves. The valves closed when the bubbles inflated, and vice versa (Hua et al., 2002). A capillary retention valve was achieved at a constriction which had the highest capillary pressure and hence pinned the interfacial meniscus of the liquid, thus confining the liquid (Juncker et al., 2002). Thermal gelation of methyl cellulose has been used for valving in a Y-channel to sort fluorescent beads (Tashiro et al., 2001). A valve was created using a cold finger. An ice plug formed in the channel created the off-state. When a heater was turned on to melt the ice, the channels was opened, creating the on-state (Jannasch et al., 2001). Movement of ferrofluid (controlled by a magnet) was also employed as a valve control (Hartshorne et al., 1998). Field-effect control was achieved for valve control by applying an electric field perpendicular to the channel in order to alter the zeta potential for EOF (Buch et al., 2001; Schasfoort et al., 1999).

G. Micromixers

Because the laminar flow is prevalent in the microfluidic channel, solution mixing, which has mainly been accomplished by diffusion, is slow. Laminar flow in microchannels did not lead to fast mixing even in the presence of pillar structures (Keir et al., 2002).

Therefore, various strategies have been employed to facilitate mixing. A PDMS fluidic mixer was created, and it comprised one large channel with smaller, chevron-shaped indentations which were not centered in the channel. Such an arrangement forced the fluid to recirculate and achieved a mixing effect (McDonald et al., 2002).

Mixing could also be achieved in a PDMS microchannel using square grooves in the channel bottom (Stroock et al., 2002) or in a PMMA T-junction with laser-ablated slanted grooves (Tohnson et al., 2001). Two droplets (600 pL) were merged and mixed by a push–pull (shuttling) method in a PDMS device consisting of a hydrophobic microcapillary vent (HMCV) (Hosokawa et al., 1999). Bidirectional flow, which occurs at a funnel-shaped channel because of the interaction of EOF and HDF, also creates a mixing effect (Boer et al., 1998). Other micromixers based on various principles have been constructed. These principles include vortex (Bohm et al., 2001), eddy diffusion (Bökenkamp et al., 1998; Esch et al., 2001; Fujii et al., 1998; He et al., 2001; Larsen et al., 1996; Manz et al., 1998; Mensinger et al., 1995; Ro et al., 2001; Rohr et al., 2001; Svasek et al., 1996) rotary stirring (Lu et al., 2001b), turbulence (Bökenkamp et al., 1998; Hong et al., 2001b), EK instability (Choi et al., 2001; Oddy et al., 2001), chaotic advection (Liu et al., 2001c), and piezoelectric actuation (Woias et al., 2000; Yang et al., 1998).

H. Alternative Pumping Principles

Besides EOF and HDF, other pumping mechanisms are employed to generate the microfluidic flow. For instance, centrifugal force was first used to pump fluids through microchannels in a rotating plastic disk. The fluids are loaded at the center of the disk. Various flow rates (5 nL/s to >0.1 mL/s) were achieved in channels of different dimensions and at different rotation speeds (60–3000 rpm). This pumping method provides a wider range of flow rates when compared with EOF (10 nL/s–0.1 μL/s) or HDF (10 nL/s–10 μL/s). In addition, the centrifugal flow is insensitive to various physiochemical properties (e.g., pH and ionic strength) of the liquids and works in different conditions of the channel (e.g., wall adsorption and trapped air bubbles). Therefore, this method can be used to pump aqueous, biological, and organic liquids. However, one limitation of the centrifugal flow is that the flow direction cannot be reversed (Duffy et al., 1999).

Moreover, the EOF can induce a HD flow, and this is termed as the EOF-induced flow. This flow can be generated at a T-intersection (Ramsey and Ramsey, 1997) or near a thin gap (Alarie et al., 2001; Guijt et al., 2001a). At a T-intersection, EOF was first initiated to a sidearm. However, when the sidearm was coated to reduce EOF, a flow was induced in the field-free main channel, resulting in an EOF-induced flow. This strategy has been used to maintain a stable electrospray for MS analysis (Ramsey and Ramsey, 1997).

The EOF-induced flow has also caused preferential mass transport of anions (relative to cations and neutrals) into the field-free main channel (Culbertson et al., 2000b). This is because the EOF (after its reduction) can no longer overcome the negative electrophoretic mobility of the anions in the side channel. Therefore, the anions can only flow in the field-free channel (Culbertson et al., 2000b). Selective ion extraction was extended by using an additional HDF applied to the field-free channel. An increase in the pressure applied at the field-free channel increased the amount of extraction of negatively charged ions into this channel (Kerby et al., 2002).

Further exploitation of the EOF-induced flow in multiple capillary channels (of width 1–6 μm) combined the multiple flow to produce adequate hydraulic pressure for liquid pumping (Lazar and Karger, 2002). The multiple channels (100) ensure the generation of sufficient flow rate (10–400 nL/min), while the small dimensions (of depth 1–6 μm) result in the necessary hydraulic pressure to prevent pressurized back-flow leakage (up to 80 psi) (Lazar and Karger, 2002). Another narrow-gap EOF pump was constructed to produce ~400 Pa pressure with 850 nm-deep channels cascaded in three stages to produce a 200 μm/s flow velocity (Takamura et al., 2001).

Moreover, the thermal effect is exploited to produce liquid pumping. For instance, a thermocapillary pump was constructed. The pumping method is based on surface-tension change due to local heating created by heaters fabricated on a Si–Pyrex chip. This method was used to pump reagents to perform successive PCR, gel electrophoresis, and detection (Burns et al., 1996; Darhuber et al., 2001).

Thermopneumetic pressure was generated on a Si–glass chip to pump discrete droplets. Trapped air (100 nL) was heated by a resistive metal heater (by tens of degree Celsius) to generate an air pressure of about 7.5×10^{-3} Pa. A flow rate of 20 nL/s could be obtained on channels using a heating rate of ~6°C/s. Hydrophobic patches were used to define the discrete droplet and to prevent the liquid from entering the zones of the trapped air and the vent (Handique et al., 2001).

Capillary forces were used to pump reagents through Si microchannels (Juncker et al., 2002; Sohn et al., 2001). Additional gradients in surface pressure, which could be created by electrochemically generating and consuming surface-active species at the two ends of a channel, has been used for liquid pumping (Gallardo et al., 1999).

Reduced pressure was employed to create liquid flow so as to facilitate filling of PDMS channels with aqueous solutions. Better results, without trapped air bubbles, can be achieved, when compared with the methods by capillary force or pressure gradient. This technique also allowed the filling of a 3D microchip when a single solution entry was used and reduced pressure was simultaneously applied to all 11 reservoirs. This was achieved by placing the whole chip under reduced pressure (Monahan et al., 2001).

Hydrostatic flow (based on liquid level difference) has been used to introduce beads into microchannels. This method is found to be superior than the use of HDF. Pneumatic (N_2) pumping is also used to introduce probe DNA samples in microchannels via a 10-port gas manifold. This method appears to be more effective than the use of EOF (Fan et al., 1999).

Additional pumping principles based on magnetohydrodynamics (Eijkel et al., 2001; Lemoff and Lee, 2000; Schasfoort et al., 2001) and electrodynamics (Green, 2001) have also been exploited for microfluidic pumping.

I. Microfluidic Flow Modeling Study

Mathematical modeling and simulation have been applied to various flow studies (see Table 11).

An important consequence of the modeling study is to reduce dispersion at the channel turns (*vide infra*). Resolution enhancement for separation requires longer serpentine channels to be fabricated. This creates dispersion problems due to the turn geometry. Various approaches have been employed to reduce the dispersion. For instance, compensating turns, which have constrictions to even out the differences in the path lengths and electric field strengths of the inner and outer tracks at the turns, have been proposed (Molho et al., 2001; Paegel et al., 2002). Simulation and experiment (using photobleached or caged-fluorescence visualization) were employed to illustrate the improvement obtained by using compensating turns (Molho et al., 2001). The advantage of the use of tapered turns over noncompensating (90° or 180°) turn was nicely illustrated by CGE separation of DNA samples (Paegel et al., 2000).

Table 11 Various Mathematical Modeling and Simulation to Study Microfluidic Flow

Topics of study	References
Electrokinetic transport and diffusion for electrokinetic focussing	Ermakov et al. (1998)
pH gradient formed by transverse IEF in pressure-driven flow	Cabrera et al. (2001)
Pressure-driven flow over grooved channel surfaces	Stroock et al. (2002)
EOF in composite microchannels (PET-PE and PC-PE) for dispersion reduction	Bianchi et al. (2001)
Pinched pressure-driven flow injection	Bai et al. (2002)
Electrokinetic flow at turns for dispersion reduction	Griffiths and Nilson (2002), Griffiths and Nilson (2000) and Molho et al. (2001)
Dispersion at turns	Culbertson et al. (1998), Dutta and Leighton (2002) and Griffiths and Nilson (2001)
Hydrodynamic flow and diffusion for generation of a concentration gradient	Yang et al. (2002a)
Field-amplified sample stacking	Bharadwaj and Santiago (2001)
Flow parameters determined in a PDMS channel	Dittrich and Schwille (2002)
EOF in the presence of finite inertial and pressure forces	Santiago (2001)
Simulation and experiment for injection	Fu et al. (2002a)
EOF in 2D rectangular channel	Theemsche et al. (2002)
EOF visualized by particle tracking technique	Devasenathipathy et al. (2002)
HDF and diffusion: concentration gradient generated in a T-sensor	Kamholz et al. (1999) and Weigl et al. (1996)
EOF in channels with nonuniform zeta potential distribution (a chaotic EOF stirrer)	Qian and Bau (2002)

IV. SAMPLE INTRODUCTION

In most cases, sample introduction on-chip is achieved using electrokinetic (EK) flow (Manz et al., 1992). Two important EK injection modes, namely, pinched injection and gated injection have been developed. Furthermore, some alternative injection methods are described.

A. Electrokinetic Injection

In EK injection, sample introduction is usually biased. However, the sample can be introduced in a nonbiased manner, as long as the electric voltage is applied for a sufficient time so that the slowest migrating component has passed the intersection and entered the analyte waste channel. This method resembles frontal chromatography. Accordingly, the sample composition at the intersection is representative of that in the original sample, though it is not the case somewhere downstream in the waste channel (Effenhauser et al., 1993; Moore et al., 1995).

A cross injector and a double-T injector have been constructed. The injection and separation modes for a cross-injector vs. double-T injectors (with the 100 and 250 μm "sampling loop") were visualized and shown in Fig. 14 (Wallenborg and Bailey, 2000).

The aforementioned injection of a solution plug is termed as plug injection. Another sample introduction mode is stack injection. Stack and plug injections have been compared (see Fig. 15). Under similar conditions, the plug injection produced a better resolution, whereas the stack injection produced a higher sensitivity (Woolley and Mathies, 1994).

With EK injection, reasonable reproducibility (RSD in migration time and peak area are 0.1% and 2%, respectively) in 11 successive injection/separation has been reported using a double-T injector consisting of a sampling loop 150 μm long (Effenhauser et al., 1993). In another report, repetitive injections and separations also resulted in good migration time RSD (0.06%) and peak height RSD (1.7%) (Effenhauser et al., 1994).

It was found that a 250 μm double-T injector increased in the peak signal when compared with a straight-cross injector (Liu et al., 1999; Wallenborg and Bailey, 2000). Although a fivefold increase was expected on the basis of the intersection volume calculations of a 250 μm injector vs. a cross injector, the discrepancy was likely to be caused by the back flow of analyte upon applying the push-back voltage (for preventing leakage) (Wallenborg and Bailey, 2000).

In a T-injector, which eliminates the need for a waste channel and reservoir, biased EK injection may result in

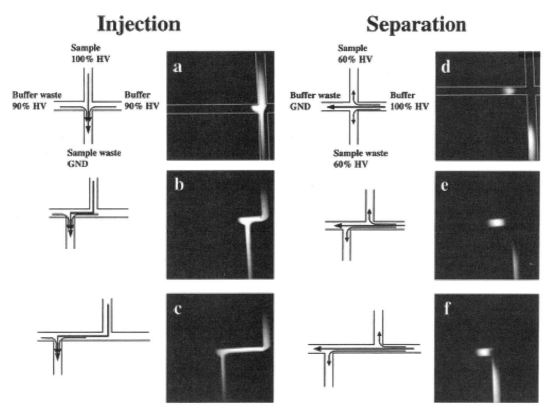

Figure 14 Fluorescence micrographs of injection and separation of a marker (0.4 mM fluorescein) performed on a glass microchip, using different injector configurations. Injections and separations are shown in (a–c) and (d–f), respectively. Injector configurations: straight cross (a, d), 100 μm offset double-T (b, e), and 250 μm offset double-T (c, f). The illustrations to the left of the photographs show the injector configurations and applied voltages during injection and separation modes. [(Wallenborg et al., 2000) Courtesy of American Chemical Society.]

loading higher mobility analytes at the expense of the lower mobility ones. A delayed (25 s) back biasing was useful, which allowed sufficient loading of higher-MW (low mobility in gel) DNA fragments, before a push-back voltage was applied (Emrich et al., 2002).

1. Pinched Injection

In plug injection, there is leakage of the sample plug around the intersection. The leakage has been attributed to diffusive and convective phenomena (Seiler et al., 1993). This leakage can be reduced by using the pinched mode in which the buffers from two adjacent channels are flowed in to shape the plug (see Fig. 16) (Jacobson et al., 1994d). For clarity, the plug injection without using pinching voltage is thus called floating injection. It was found that an increase in the pinching voltage resulted in less injected materials, whereas a decrease in the pinching voltage caused less peak symmetry (von Heeren et al., 1996a).

The temporal stability of the injection plug is excellent in the pinched mode (RSD 1%) regardless of the loading time. But in the floating mode, the plug size continually increased with the loading time. Reproducibility for five successive injections gives 1.7% RSD in the peak area for the pinched injection, when compared with 2.7% RSD for the floating injection (Jacobson et al., 1994d).

Sample leakage can cause peak tailing (Fan and Harrison, 1994). To avoid this, the buffer is pushed back into the analyte channel and analyte waste channel by applying a push-back voltage (Harrison et al., 1993; Jacobson et al., 1994d). This can be applied for a short duration ($\sim$120 ms), rather than continuously, to provide a clean-cut of the injection plug (von Heeren et al., 1996a). Sample leakage during preinjection and separation can also be avoided by putting at least 30 MΩ of resistors between the sample waste and ground (Liu et al., 2000a). In the case of repetitive injection/separation, the loading time for a subsequent injection should be long enough to compensate for the pushed-back sample front created in the previous separation (Effenhauser et al., 1994).

Various injector geometries (simple-cross, double-T, and triple-T) were investigated on a glass microchip. The triple-T injector allowed for a selection of three

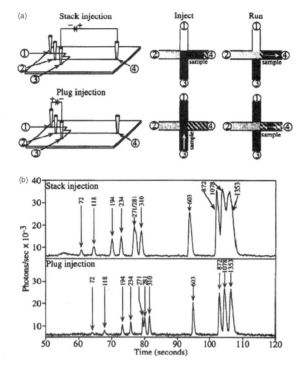

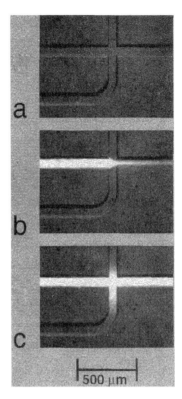

Figure 15 (a) Schematic diagram of the stack injection and plug injection method. The hatched regions indicate the sieving medium in the separation channel. (b) Electropherogram comparing the stack and plug injection methods. A sample containing φX174 Hae III fragments at 10 ng/μL was injected for 1 s in each experiment. The buffer consisted of the standard TAE/HEC sieving medium with 1 μM of the TO dye. A signal of 8000 photons/s over background corresponds to DNA levels of 100 pg/μL in the separation channel. [(Woolley and Mathies, 1994) Courtesy of National Academy of Sciences.]

Figure 16 Images of sample injection at a cross-injector in a glass microchip; (a) no fluorescent analyte, (b) pinched injection of Rhodamine B, and (c) floating injection of Rhodamine B. [(Jacobson et al., 1994d) Courtesy of American Chemical Society.]

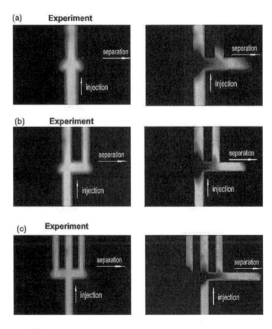

Figure 17 Sample loading and injection at a cross, double-T, and triple-T injector fabricated on a glass microchip. [(Fu et al., 2002a) Courtesy of American Chemical Society.]

injection volumes depending on whether a cross, double T, or triple-T injector was used (see Fig. 17) (Fu et al., 2002a). In order to create sample plugs of improved plug width and symmetry, a six-port injector with two additional channels were also designed (Bousse et al., 2000; Deshpande et al., 2000).

It is found that if the sample channel width is narrower than the separation channel width (by fivefold), floating injection did not result in sample leakage, and no pinched injection was necessary. Moreover, a push-back voltage was not necessary during separation. In addition, a T injector, rather than a cross injector, was sufficient (only when the sample channel width was one-tenth of the separation channel width, but the resolution was inferior to the cross injector). Accordingly, the number of reservoirs in a multichannel (S) system is reduced to $S + 2$, which is the real theoretical limit (Zhang and Manz, 2001), rather than $S + 3$ (Simpson et al., 1998).

The floating injection was successful without sample leakage in some DNA separations which used viscous

Microfluidic Lab-on-a-Chip

solutions, possibly because of slow diffusion of the DNA molecules in these solutions (Khandurina et al., 2000; Woolley and Mathies, 1995).

2. Gated Injection

Gated injection was adopted in which the sample was continually loaded in parallel with a separation buffer to a waste port by EK flow (see Fig. 18). Injection of the sample was achieved by interrupting the flow of the separation buffer for a short time (known as the injection time) so that the sample stream was injected (Jacobson et al., 1994a). Gated injection has also been achieved using one power supply and three solution reservoirs (Jacobson et al., 1999b).

This gated injection method has allowed the sample loading to be achieved in a continuous manner, whereas a pinched injection mode cannot (Jacobson et al., 1994a). Therefore, the gated injection could also be employed for 2D separation: OCEC/CE (see Fig. 19). The initial injection (a → b) was performed for OCEC for 0.5 s. Subsequently, serial sampled injection (a → c) for the second dimensional CE (for 0.2 s every 3.2 s) was achieved at a faster speed using HV relays (see Table 12) (Gottschlich et al., 2001).

The relay-controlled gated injection has RSD values of 1.9% in migration time and of 5.5% in peak area for Rhodamine B (10 μM) during 200 serial injections. The plate height remained below 2 μm as long as the concentration of Rhodamine B was below 100 μM (Gottschlich et al., 2001).

However, the gated mode has two problems: (1) the injection plug length increases with the injection time, and (2) the plug length is longer for faster migrating species, leading to a biased injection (Jacobson et al., 1994a). Moreover, because of the turn at the injector, there was a second level of sampling bias due to transradial electrokinetic selection (TREKS) (i.e., the faster moving

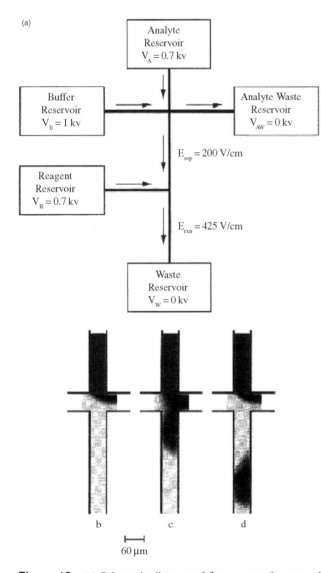

Figure 18 (a) Schematic diagram of flow pattern for a gated injector. CCD images of the gated injection using Rhodamine B (b) prior to injection, (c) during injection, and (d) after injection into separation column with $E = 200$ V/cm. [(Jacobson et al., 1994a) Courtesy of American Chemical Society.]

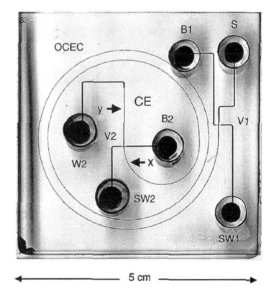

Figure 19 Image of the glass microchip used for 2D chemical separations. The separation channel for the OCEC (first dimension) extends from the first valve V1 to the second valve V2. The CE (second dimension) extends from the second valve V2 to the detection point y. Reservoirs for sample (S), buffer 1 and 2 (B1, B2), sample wastes 1 and 2 (SW1, SW2), and waste (W) are positioned at the terminals of each channel. The arrows indicate the detection points in the OCEC channel (x) and CE channel (y). [(Gottschlich et al., 2001) Courtesy of American Chemical Society.]

Table 12 Sample Injection Scheme for OCEC/EC in Fig. 19

	Electric voltages (kV)					
	S	B1	SW1	B2	SW2	W2
a	9.5	10.0	5.0	3.0	2.5	0.0
b (0.5 s)	9.5	8	5.0	3.0	2.5	0.0
c (0.2 s)	9.5	10.0	5.0	Open	Open	0.0

molecule traced out a larger turning radius than the slower moving molecules) (Slentz et al., 2002).

To avoid sampling bias, hydrodynamic injection (due to liquid pressure difference) could be implemented after EK loading (see Fig. 20). As soon as the sample was loaded up to the intersection junction, the voltage was turned off for hydrodynamic injection to occur. The sample plug was truncated when voltage was turned on again. If a positive voltage was applied, the injection procedure worked only for cations and slowly migrating anions (because of the need of EK loading). For the injection of other anions (e.g., Cl^-, and $Cr_2O_7^{2-}$), negative voltages should be applied (Backofen et al., 2002).

Similar injection procedure was also reported for CEC of three FITC-labeled peptides leading to less biased amounts of the faster moving components (see Fig. 21) (Slentz et al., 2002).

Another variation is using a continuous HD flow for loading and injection, gated using an EK flow. As long as the gated voltage is greater than a critical value, the HD flow is prevented from injecting downstream. Injection is implemented by interrupting the EK flow for a short time (see Fig. 22) (Chen et al., 2002).

B. Other Sample Injection Methods

Pure HDF injection can be used. Pressure sample introduction was achieved using vacuum suction. The sample was sucked through an inserted capillary into a short section of microchannels between three ports. Different amounts could be selected by filling different sections of the microchip (Zhang et al., 2000).

An optically gated injection was demonstrated for the CZE separation of four amino acids labeled with 4-chloro-7-nitrobenzofurazan (NBD) in one-channel (Lapos and Ewing, 2000), or four-channel chips (Xu et al., 2002). The gating beam was used to continuously photobleach the sample, except for a short time during injection by interrupting the beam (100–600 ms) using an electronic shutter. With only a sample reservoir and a waste reservoir, the sample continuously flowed electrokinetically. Six consecutive separations of the same sample mixture have been accomplished in under 30 s (Lapos and Ewing, 2000; Xu et al., 2002).

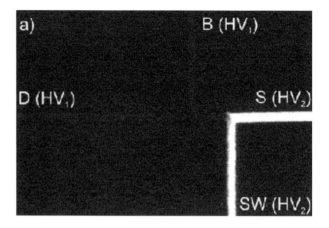
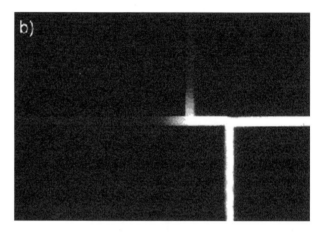
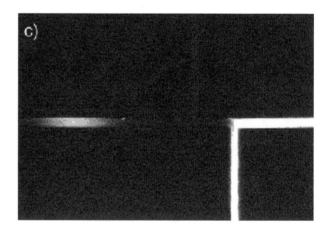

Figure 20 Fluorescent micrographs of the injection process. (a) The high-voltage sources HV1 and HV2 are switched on. The sample is transported from the sample reservoir (S) to the sample waste (SW), and run buffer from the reservoir (B) to the detector (D). (b) For injection, both HV sources are switched off for ~3 s. Sample flows in due to hydrodynamic flow. (c) The sample plug is introduced into the separation channel by switching on the HV sources and at the same time the two separate streams of electrolytes are re-established. [(Backofen et al., 2002) Courtesy of American Chemical Society.]

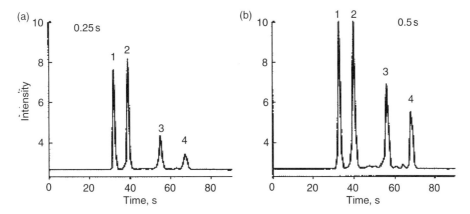

Figure 21 Separation of an FITC-labeled synthetic peptide mixture following (a) electrokinetic and (b) diffusion-based injection on a microchip. Mobile phase: 1 mM carbonate buffer (pH = 9.0); separation E field: 300 V/cm; 1, FITC–Gly-Phe-Glu-Lys-OH; 2, FITC–Gly-Phe-Glu-Lys(FITC)-OH; 3, FITC; 4, FITC–Gly-Tyr-OH. Analyte concentrations: $1 + 2 = 4 = 50\ \mu M$. [(Slentz et al., 2002) Courtesy of American Chemical Society.]

V. SAMPLE PRECONCENTRATION

Several preconcentration strategies have been devised to enrich dilute samples. The microfluidic chip has the advantage to integrate various on-chip structures to achieve preconcentration by stacking, extraction, or other methods.

A. Sample Stacking

Field-amplified stacking has been achieved on-chip for sample preconcentration. The stacking is based on a lower electrical conductivity in the sample buffer relative to the run buffer (Bharadwaj and Santiago, 2001; Lichtenberg et al., 2000, 2001b; Liu et al., 2000).

Although stack injection has been employed previously (Harrison et al., 1992; Woolley and Mathies, 1994), the benefit of stacking for sample preconcentration was only studied in detail later (Jacobson and Ramsey, 1995). With the sample buffer (0.5 mM) at a 10-fold lower conductivity than the separation buffer (5 mM), simple EK stacking using the gated injection was observed for the separation of dansylated amino acids (dansyl-lysine, didansyl-lysine, dansyl-isoleucine, and didansyl-isoleucine) (Jacobson and Ramsey, 1995).

Owing to the EK bias in favor of faster migrating species in stack injection, the signal enhancement ranged from 31 to 8. However, stack injection produced less plate numbers (N) when compared with those obtained in pinched injection. In addition, RSD ($n = 6$) in peak areas for stack, nonstacked, and pinched injections are 2.1%, 1.4%, and 0.75%, respectively (Jacobson and Ramsey, 1995).

Stacking of a neutral analyte can also be achieved, but only when a high-salt (NaCl) sample buffer is employed. The co-ion (Cl^-) should be present in the sample buffer at a concentration sufficiently higher than that of the electrokinetic vector ion (cholate). Therefore, the velocity of Cl^- is less than that of cholate. This causes the formation of a pseudo-steady-state co-ion boundary that forces an

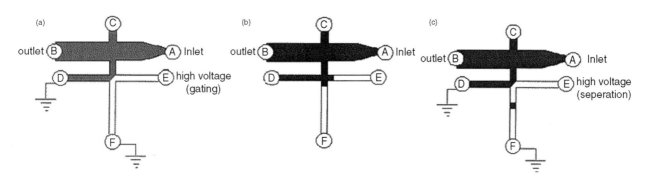

Figure 22 Voltage switching scheme for sample introduction. [(Chen et al., 2002) Courtesy of American Chemical Society.]

increased concentration of cholate near the boundary, leading to the stacking effect (Palmer et al., 2001).

B. Extraction

An OTS-coated glass channel was employed for enrichment of samples (the neutral coumarin dye C460). Acetonitrile (15%) was used for the loading and enrichment in the liquid-phase extraction procedure. Subsequently, 60% acetonitrile was used for elution. It was found that the dye (8.7 nM) was enriched by 80-fold (in 160 s) (Kutter et al., 2000).

Sample preconcentration was also achieved based on solid-phase extraction. This was carried out on ODS-coated silica beads (of diameter 1.5–4 μm) trapped in a cavity (10 μm deep) bound by two weir-type structures (9 μm high) (see Fig. 23). Concentration enhancement

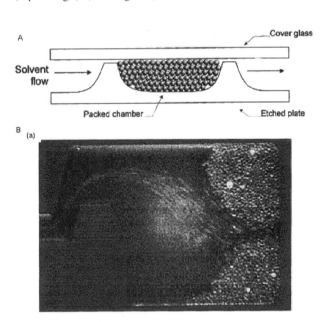

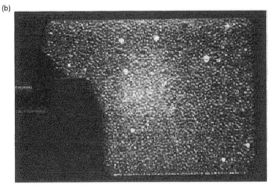

Figure 23 (A) Drawing of cross-section of a packed chamber, showing weir heights in relation to channel depth and particle size. (B) Images of the chamber at (a) an intial stage of EK packing and (b) after it is completely filled with beads. [(Oleschuk et al., 2000) Courtesy of American Chemical Society.]

up to 500-fold was demonstrated using a nonpolar analyte (BODIPY) eluted by acetonitrile (Oleschuk et al., 2000).

A 2 μL unpurified sample is concentrated into a smaller volume (∼10 nL) in an on-chip chamber containing a capture matrix, yielding a volumetric concentration factor of ∼200. Longer residence time can be achieved by using a doubly tapered chamber. Here, the electric field is 10-fold lower within the chamber than within the tapered channel. So, a high-field (fast) sample introduction is followed by a low-field (slow) sample flow past the capture matrix, increasing the residence time so that the binding kinetics is dominant over electromigration (Paegel et al., 2002).

Sample preconcentration was also achieved for gaseous samples.

A flow-through cell (Pyrex) consisting of the silica-based solid absorbent was used for preconcentration of the BTX gaseous mixture. A thin-film heater was used to desorb the adsorbed gas molecules, which flowed downstream for UV absorbance detection. LOD of 1 ppm (toluene) was achieved when compared with 100 ppm without preconcentration (Ueno et al., 2001, 2002b). With an additional air-cooled cold-trap channel located after the adsorbent region, the desorbed molecules were prevented from being diluted before reaching the detection cell, and a further improvement of LOD of toluene to 0.05 ppm was achieved (Ueno et al., 2002a).

C. Sample Volume Reduction

A porous membrane structure (made of sodium silicate) was fabricated to preconcentrate large molecules such as DNA. This structure (of width 3–12 μm) allows electric current to flow, yet preventing large molecules from passing through [see Fig. 24(A)]. This is a physical stacking method based on a reduction in the sample volume by removing small solvent molecules (Khandurina et al., 1999). With this method, preconcentration of PCR products (199 bp) has been achieved with starting template copy number of 15 in an injection volume of 80 pL [see Fig. 24(B)]. Therefore, as few as 10 cycles, with a short analysis time of less than 20 min can be achieved (Khandurina et al., 2000).

Another preconcentration technique is based on sample volume reduction due to flow confinement. A sample flow through a channel (20 μm deep) can be confined by a perpendicular makeup flow into a thin layer (of thickness 2–6 μm) above the sensing area (see Fig. 25). This 3D microfluidic confinement reduces the sample volume, increases the flow velocity, and benefits any mass transport limited processes such as heterogeneous immunoassays. The degree of confinement, and hence preconcentration,

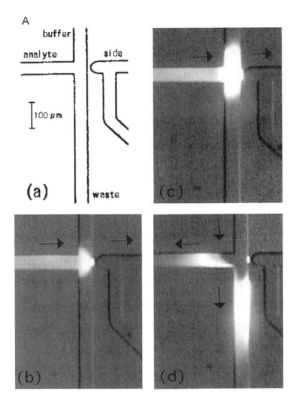

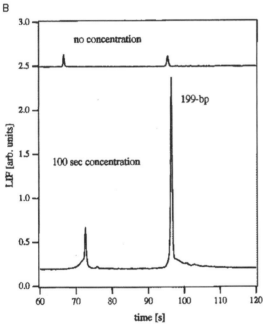

Figure 24 (A) Schematic of the injection tee and porous membrane is shown in (a). CCD images of analyte concentrated for (b) 2 min and (c) 3 min. Injection of concentrated analyte plug is depicted in (d). Porous membrane region width is 7 μm. All channels are filled with 3% LPA in 1× TBE buffer. DNA sample: 25 μg/mL φX174 HaeIII digest with 6.0 μM TO-PRO added. (B) Electrophoretic analysis with and without preconcentration of PCR product after 20 cycles. [(Khandurina et al., 1999, 2000) Courtesy of American Chemical Society.]

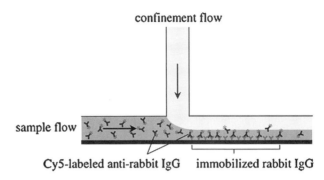

Figure 25 Illustration of the flow confinement concept. In microchannels, a sample flow is joined with a confinement flow (e.g., water or sample medium) in a perpendicular orientation. Under laminar flow conditions, no mixing occurs. The sample flow is confined into a thin layer of higher velocity. For immunoassay application, rabbit IgG is immobilized on a planar waveguide and Cy5-labeled anti-rabbit IgG is introduced as analyte in the sample flow. [(Hofmann et al., 2002) Courtesy of American Chemical Society.]

can be adjusted through the volume flow rate of the confining flow (Hofmann et al., 2002; Slyadnev et al., 2001).

Conventional dialysis methods are actually based on reduction of the sample volume by the selective removal of small molecules. Affinity dialysis and preconcentration of aflatoxins were achieved in a copolyester chip (see Fig. 26). After affinity binding to the aflatoxin B_1 antibody, various aflatoxins (B_1, B_2, G_1, G_2, and G_{2a}) were retained, whereas the other small molecules passed through a PVDF dialysis membrane. Thereafter, the solution was exposed to a countercurrent flow of dry air, leading to water evaporation and analyte concentration (Jiang et al., 2001).

Countercurrent dialysis was achieved through a dialysis membrane (MW cutoff of 8000) sandwiched between a laser-machined PC sheet with a 160 μm-wide channel

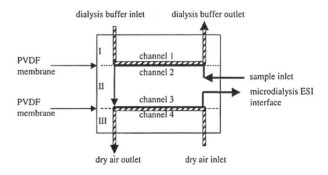

Figure 26 Side view schematic of miniaturized affinity dialysis and concentration system. I, II, and III indicate top, middle, and bottom imprinted copolyester pieces, respectively. Piece II is imprinted on both sides. Two PVDF membranes separate the copolyester channels. [(Jiang et al., 2001) Courtesy of American Chemical Society.]

for sample flow (0.5 μL/min) and a laser-machined polyimide sheet with a 500 μm-wide channel for buffer flow (100 μL/min) (see Fig. 27). This device produced a desalted sample (horse heart cytochrome C) for subsequent MS analysis (Xu et al., 1998).

D. Other Preconcentration Methods

Common preconcentration strategies suffer from various disadvantages such as the requirement of multiple buffers (for stacking), need of reproducible surface modification (for extraction), applicability only to large molecules or particles (for dialysis), inaccessibility of pI values or low solubility of most proteins at their pI values (for IEF), and generation of electrode products (for stacking) (Ross and Locascio, 2002).

A new preconcentration concept temperature gradient focusing (TGF) is devised. In this method, balancing of the electrophoretic flow against the bulk flow (EOF) has been achieved in the presence of a temperature gradient (cf. pH gradient in IEF) using an appropriate buffer in polycarbonate chips. This also introduces the focusing effect (cf. IEF), leading to a preconcentration factor of 10,000 or greater (from 8 nM to 90 μM for Oregon Green). Although TGF can be applied to a capillary, this method has not been first demonstrated in the capillary format, but in the microfluidic format. It is mainly because of the ease of temperature gradient implementation on the planar chip surface and the ease of visualization through the optically transparent planar glass substrate. TGF has been demonstrated with various ionic species (Oregon Green 488 carboxylic acid, Cascade Blue hydrazide, FQ-labeled aspartic acid, CBQCA-labeled serine and tyrosine, GFP, TAMRA-labeled 20-mer oligonucleotide, and fluorescently labeled PS particles) (Ross and Locascio, 2002).

VI. SEPARATION

The first miniaturized chip was fabricated on a Si wafer for GC (Terry et al., 1979). Isotropic etch on a 200 μm thick Si ⟨100⟩ wafer (5 cm diameter) has produced a spiral GC channel (200 μm wide, 30 μm deep, and 1.5 m long) incorporated with a thermal conductivity detector (TCD) (Terry et al., 1979). An etched Si miniature valve with a Ni diaphragm (activated by a solenoid plunger) was used for injections. GC analysis of hydrocarbons using an OV-101 stationary phase was achieved. Since then, GC of alkanes (Hannoe et al., 1998; Frye-Mason et al., 1998; Lehmann et al., 2000; Naji et al., 2001) and methyl esters (Frye-Mason et al., 2001) were performed on chip.

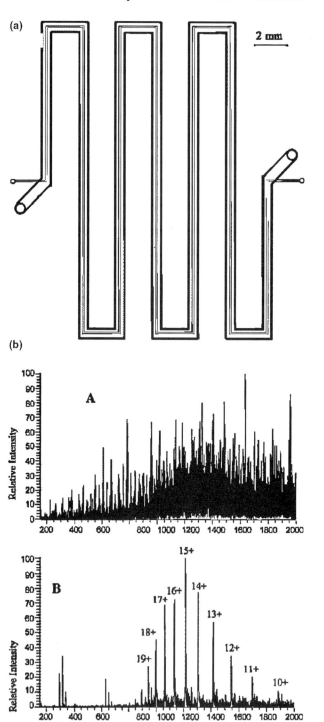

Figure 27 (a) Schematic representation of the image of overlapping sample and buffer channels after alignment of the microchannel device. (b) (A) ESI-mass spectrum of 5 μM horse heart myoglobin in 500 mM NaCl, 100 mM Tris, and 10 mM EDTA by direct infusion; (B) ESI-mass spectrum of previous myoglobin sample after on-line microdialysis using 10 mM NH$_4$OAc and 1% acetic acid as dialysis buffer. [(Xu et al., 1998) Courtesy of American Chemical Society.]

A. Capillary Zone Electrophoresis

Six fluorescein-labeled amino acids are separated by CZE and detected by laser-induced fluorescence (LIF) on a Pyrex-glass chip (10 μm-deep and 30 μm-wide channel), all achieved in a very short time of about 15 s (see Fig. 28) (Harrison et al., 1993). Calcein and fluorescein were also separated by CZE (Harrison et al., 1992). Separation of a binary mixture of Rhodamine B and dichlorofluorescein was even achieved in only 0.8 ms using a short separation length of 200 μm (Jacobson et al., 1998).

An index of separation speed, N/t, was defined, where N is the number of theoretical plates and t is the migration time (Effenhauser et al., 1993; Jacobson et al., 1994e). The theoretical upper limit of $8500 \, s^{-1}$ for Gln–FITC ($H = 0.3$ μm) is close to the experimental value of $8300 \, s^{-1}$.

The plate height, H, is given by Eq. (1) (Effenhauser et al., 1993, 1994; Harrison et al., 1992; Jacobson et al., 1994d),

$$H = H_{\text{diff}} + H_{\text{inj}} + H_{\text{det}} \quad (1)$$

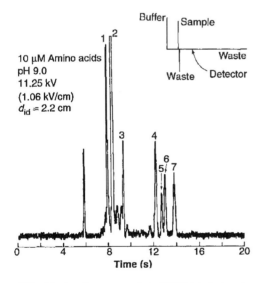

Figure 28 Electropherogram of six FITC-labeled amino acids in pH 9.0 buffer with 2330 V applied between the injection and detection points and a potential applied to the side channels to reduce leakage of the sample. The peaks were: 1, Arg; 2, FITC hydrolysis product; 3, Gln; 4, Phe; 5, Asn; 6, Ser; 7, Gly. The inset shows the approximate layout of device, with a buffer to separation channel–waste distance of 10.6 cm. [(Harrison et al., 1993) Courtesy of American Association for the Advancement of Science.]

The parameters, H_{diff}, H_{inj}, and H_{det} are given in Eq. (2)–(4) as follows (Effenhauser et al., 1993, 1994; Fan and Harrison, 1994; Seiler et al., 1993),

$$H_{\text{diff}} = \frac{2Dt_m}{L} + \frac{2D(t_{\text{inj}} + t_{\text{dl}})}{L} \quad (2)$$

$$H_{\text{inj}} = \frac{L_{\text{inj}}^2}{12} + \frac{2D(t_{\text{inj}} + t_{\text{dl}})}{L} \quad (3)$$

$$H_{\text{det}} = \frac{L_{\text{det}}^2}{12} + \frac{(\mu E \tau)^2}{L} \quad (4)$$

where H_{diff}, H_{inj}, and H_{det} are the plate heights due to longitudinal or axial diffusion, injector length, and detector length, respectively; t_m is the migration time; D is the diffusion coefficient; t_{dl} is the delay time between loading and separation; t_{inj} is the injection time; L is the distance from the injection to detection points; L_{inj} is the injector length; L_{det} is the detector window = pinhole size divided by microscope magnification; and τ is the time constant of the detection system.

H_{inj} usually accounts for about 50% of the total H (Effenhauser et al., 1994).

In order to increase the separation channel length, a serpentine channel (165 mm long) within an area of 10 mm × 10 mm was designed (see Fig. 29). Band distortion around the turns (of radius 0.16 mm) is substantial owing to longer migration distance and lower electric field strength of a molecule traveling on the outer track of the channel when compared with the inner track (see Fig. 30). For one 90° turn and four 180° turns, the additional contribution to band broadening, H_{geo}, is 0.85 μm [see Eq. (5)]. With this additional term, Eq. (1) is modified to give Eq. (6). It is found that H_{geo} is a significant portion of the solute-independent plate heights (sum of first three terms) [see Eq. (6)] (Jacobson et al., 1994d).

$$H_{\text{geo}} = \frac{n(\omega\theta)^2}{12L} \quad (5)$$

$$H = \underset{(0.23\,\mu m)}{H_{\text{inj}}} + \underset{(0.025\,\mu m)}{H_{\text{det}}} + \underset{(0.85\,\mu m)}{H_{\text{geo}}} + H_{\text{diff}} \quad (6)$$

The Joule heating effect was first studied by examining the nonlinearity in a plot of current vs. E and plot of linear velocity vs. E (Jacobson et al., 1995; Moore et al., 1995). Deviation from linearity appeared when E was greater than 720 V/cm. The power was around 2.3–3.8 W/m which was a twofold improvement over 1 W/m typically used in fused silica capillary (Fan and Harrison, 1994).

The additional plate height contribution due to Joule heating is given by Eq. (7) (Jacobson et al., 1995):

$$H_{\text{Joule}} = \frac{7 \times 10^{-9} \mu E^5 d^6 \lambda^2 C^2}{D \kappa^2} \quad (7)$$

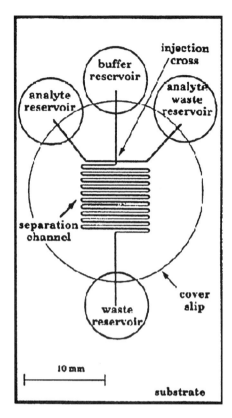

Figure 29 Schematic of serpentine channel geometry for the microchip with the large circle representing the cover slip and the smaller circles the reservoirs. [(Jacobson et al., 1994d) Courtesy of American Chemical Society.]

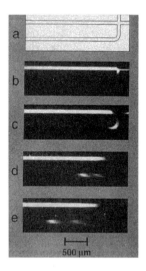

Figure 30 (a) Schematic of region imaged (injection cross); CCD images of (b) a pinched sample loading, and of (c, d, and e) separations of rhodamine B (less retained) and sulforhodamine (more retained) at 1, 2, and 3 s, respectively, after switching to the separation mode using a separation field strength of 150 V/cm. [(Jacobson et al., 1994d) Courtesy of American Chemical Society.]

However, the calculated value is insignificant when compared with the observed deviations (Jacobson et al., 1995). Although the power dissipated in a separation channel may be less than 1 W/m, caution must be taken not to overlook possible higher values in narrower adjacent sample channel causing the flow of heated fluid into the separation channel (Culbertson et al., 2000a).

CE separation usually requires high voltage. But, application of voltage >30 kV was not feasible because of arcing between the Pt electrodes even though they were wrapped in Tygon tubings (Culbertson et al., 2000a). Lower voltage can be achieved using a moving electric field achieved on a series of electrodes constructed along the separation channel (Lin and Wu, 2001).

In one example, open access channel electrophoresis was achieved on a glass channel plate, with no cover plate (Liu et al., 2001d). Other CZE separations were tabulated in Table 17 (Badal et al., 2001; Duffy et al., 1998; Effenhauser et al., 1993; Fruetel et al., 2001; Hu et al., 2002; Hutt et al., 1999).

B. Capillary Gel Electrophoresis

Capillary gel electrophoresis (CGE) has mostly been associated with DNA analysis, which is also covered in a separate section.

The first miniaturized CGE separation of single-stranded antisense oligonucleotides (10–25 b) was achieved in a 12 μm deep and 50 μm wide channel, using a non-cross-linked sieving medium (10% T 0% C polyacrylamide). Separation was achieved at a separation speed of 5000 plates/s and plate height of 0.2 μm within 45 s! Separation efficiency of 2/3 of the theoretical limit ($N = 330{,}000$) was achieved, suggesting band broadening caused by Joule heating was likely to be small (Effenhauser et al., 1994).

CGE separation for oligonucleotide DNA fragments (ΦX174 Hae III) has been achieved in about 120 s in a microfabricated glass channel of 8 μm depth and 50 μm width using a separation sieving matrix of TAE buffer and 0.75% (w/v) HEC, and an intercalating dye (1 μM TO or 0.1 μM TO6) for LIF detection ($\lambda_{ex} = 488$ nm and $\lambda_{em} = 530$ nm). An increase in the electric field strength from 100 to 180 V/cm shortens the separation time from 400 to 200 s, with only a slight loss in resolution (Woolley and Mathies, 1994).

Another CGE separation was also performed in a 50 μm wide and 20 μm deep channel casted on PDMS. The heat dissipation of PDMS (∼0.3 W/m) is less efficient than fused silica (1 W/m), but it is sufficient for most applications in E of about 100–1000 V/cm (Effenhauser et al., 1997b).

A cross-linked photopolymerized polyacrylamide gel was formed within microchannels. Unlike in non-cross-linked gel systems, high E field or long separation length were not required. The use of cross-linked gel also allowed sample compaction and injection. Sample compaction was accomplished using electrophoretic injection at 12 V/cm through the cross-linked gel. CGE was performed at 20 V/cm over a separation length of 0.18 cm (see Fig. 31) (Brahmasandra et al., 2001)!

Other CGE applications are included in the DNA application section and Table 17 (Chen et al., 1999; Duffy et al., 1998; Effenhauser et al., 1997b; McCormick et al., 1997; Munro et al., 1999).

C. Micellar Electrokinetic Capillary Chromatography

Micellar electrokinetic capillary chromatography (MECC) separation of three neutral coumarin dyes was achieved on a soda-lime glass chip. Detection was achieved using the 350 nm excited light (Moore et al., 1995).

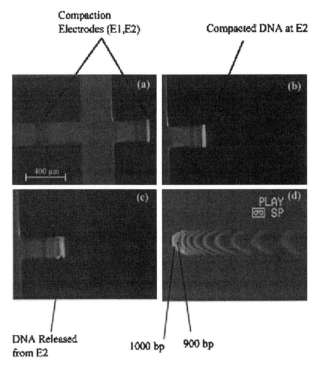

Figure 31 Video sequence depicting the feasibility of microseparations using electrode-defined sample compaction/injection and photopolymerized polyacrylamide gel sieving matrix. In (a) and (b), a fluorescently labeled 100 bp ladder DNA was compacted at a 50 mm electrode by applying E of $\sim$12 V/cm. (c) Compacted sample was released by switching the electric field to two electrodes spanning the gel matrix. (d) Separation was initiated from left to right at $E = 20$ V/cm. Complete resolution of all fragments was achieved in a separation length of 1.8 mm in less than 15 min. [(Brahmasandra et al., 2001) Courtesy of Wiley-VCH.]

The elution range (or window) as determined by t_0/t_m in Eq. (8) was found to be 0.43, which is similar to the literature value (Moore et al., 1995).

$$k' = \frac{kV_s}{V_m} = \frac{t_R - t_0}{t_0[1 - (t_R/t_m)]} \quad (8)$$

In this mode of chromatography, sorption/desorption kinetics and polydispersity of micelles contribute to the decrease in efficiency [see Eq. (9)] (Moore et al., 1995)

$$H = H_{ec} + H_{diff} + H_{mc} + H_{ep} \quad (9)$$

where H_{ec} is the extra-column broadening due to finite size of injection and detection—still the dominant factor; H_{mc} is the sorption/desorption kinetics—small for fast hydrophobic interactions between neutral solute and micelles but large for slow ionic interactions between charged solutes and micelles; and H_{ep} is the electrophoretic dispersion due to polydispersity in micelle sizes (as large as 20% RSD)—reduced if the micelle concentration is larger, but will cause excessive Joule heating. Both H_{mc} and H_{ep} are larger for more retained solute with high k' (Moore et al., 1995).

The separation resolution of the three neutral coumarin dyes was found to be better in MECC (Moore et al., 1995) than in CEC (see a later section) (Jacobson et al., 1994c), even when high E was used in MECC (Moore et al., 1995).

Direct comparison of MECC performed on-chip and in fused silica capillary has been made (von Heeren et al., 1996a). It was found that separation efficiency for FITC–Ser (10 μM) was higher in chip than in capillary because of two reasons: (1) the channel cross-section is smaller in chip (40 μm × 10 μm) than in capillary (75 μm id) and (2) the E value is higher in chip (1175 V/cm) than in capillary (215 V/cm).

MECC (SDS system (von Heeren et al., 1996a), Tween 20 system (Chiem and Harrison, 1997)) was employed in uncoated glass chips for determination of theophylline in human serum without the severe adsorption problem of the protein (anti-theophylline antibody). The separation was achieved in 50-fold shorter analysis time when compared with a competitive immunoassay first achieved on a fused silica capillary (von Heeren et al., 1996a).

The MECC separation of explosives was achieved, except that three isomers of nitrotoluenes cannot be resolved (Wallenborg and Bailey, 2000). A peak height RSD was 1.7–3.8% for TNB, DNB, TNT, tetryl, 2,4-DNT, 2,6-DNT, and 2-amine-4,6-DNT. But the linear ranges for TNB, DNB, TNT, and tetryl were only 1–5 ppm! This narrow linear range, which is caused by the indirect LIF detection based on fluorescent quenching, may be sufficient for screening, but is certainly not useful for quantitation (Wallenborg and Bailey, 2000).

The MECC separation of 19 naturally occurring amino acids has been achieved with an impressive average N of 280,000, mainly because of the use of a 25 cm long channel. The dispersion introduced by the turn geometry was reduced using a spiral, instead of a serpentine channel with turns of much larger radius of curvature, r_c (Culbertson et al., 2000a), as first introduced in a GC chip (Terry et al., 1979).

The separation was achieved in only 165 s with a resolution greater than 1.2 when 10% (v/v) propanol, instead of 20% (v/v) methanol, was used as the organic modifier (Culbertson et al., 2000a).

The MECC separation of amino acids using a more conducting buffer (2370 μS) generated a less N (average 280,000) than a CZE separation of DCF and contaminants ($N = 1,100,000$) using a less conducting buffer (438 μS) because of the generation of less Joule heating (Culbertson et al., 2000a).

The MECC separation of organophosphate nerve agents was achieved using amperometric detection. Studies show that the use of MES buffer (pH 5.0) and 7.5 mM SDS provides the best separation performance (short analysis time and adequate resolution) (Wang et al., 2001a). MECC with gradient elution using different compositions of MeOH or CH_3CN was accomplished on a glass chip (Hofmann et al., 1999).

Later, 2D liquid phase separation was conducted on-chip. The first MECC was followed by a second CZE to separate a peptide mixture or the tryptic digests of cytochrome c, α-lactalbumin, and ribonuclease A. However, the CZE and MECC separations are not completely orthogonal (or uncorrelated), limiting the resolving power of the 2D method (Rocklin et al., 2000). There are other MECC applications (Wheeler et al., 2001), which are given in Table 17.

D. Derivatizations for CE for Separations

Derivatizations have been employed to enhance the detection for CE separations. Derivatization can be achieved either by postcolumn or precolumn means.

1. Postcolumn Derivatization

Postcolumn derivatization was investigated to detect the separated amino acids (Jacobson et al., 1994a) or proteins (see Fig. 32) (Colyer et al., 1997b). For instance, after gated injection, separation of arginine (2 mM) and glycine (2 mM) in a buffer (20 mM sodium borate, pH 9.2, 2% v/v methanol, 0.5% v/v β-mercaptoethanol) was detected after the individual components mixed and

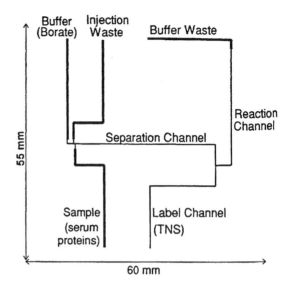

Figure 32 Schematic diagram of the post-separation labeling design of the chip. Thick lines were 256 μm wide, thin lines were 66 μm side. Channels were 14 μm deep. [(Colyer et al., 1997b) Courtesy of Elsevier Science.]

reacted with a stream of a derivatizing agent consisting of 3.7 mM OPA. Turbulence or band broadening was found in the channel after mixing unless E was above a threshold value (840 V/cm). The plate height was found to decrease monotonically with increasing E. As OPA had a typical half-time of reaction with amino acids of 4 s, the residence time should be long enough, and this was achieved by adjusting E for maximal product formation. For detection purpose, sufficient product detection occurred with E at 1700 V/cm (Jacobson et al., 1994a).

Postcolumn derivatization by OPA was also achieved to analyze amino acids. With the use of fused silica with substantially lower fluorescence background ($\lambda_{ex} = 325$ nm), a two-fold improvement in S/N for phenylalanine detection was achieved (Fluri et al., 1996).

Postseparation labeling was also achieved for separation of four human serum proteins (IgG, transferrin, α-1-antitrypsin, and albumin) using 0.2 mM of 2-toluidinonaphthalene-6-sulfonate (TNS). TNS is a virtually nonfluorescent reagent which, upon noncovalent association with proteins, produces a fluorescent complex ($\lambda_{ex} = 325$ nm and $\lambda_{em} = 450$ nm) for detection. TNS was selected among the commonly used labels: OPA (for amine-containing protein) or ANS (for protein). This postseparation labeling avoids the problems of multiple-site labeling, which is manifested as multiple peaks after separation (Colyer et al., 1997b). Postcolumn noncovalent labeling of protein was also achieved using Nano Orange as the labeling reagent (Liu et al., 2000c).

2. Precolumn Derivatization

CZE with LIF detection ($\lambda_{ex} = 351.1$ nm and $\lambda_{em} = 440$ nm) of two amino acids (0.58 mM glycine and 0.48 mM arginine) was also achieved by precolumn derivatization with 5.1 mM OPA (Jacobson et al., 1994b). This method may be more advantageous than postcolumn derivatization (Jacobson et al., 1994a), when the analysis time is faster than the product reaction time ($t_{1/2}$ for OPA ~4 s) (Jacobson et al., 1994b). A reaction chamber of 96 μm wide (at half-depth) and 6.2 μm deep was constructed "before" the injection cross. The chamber is wider than the separation channel of width 31 μm to allow for lower electric field and hence longer residence time for the derivatization reaction (see Fig. 33) (Jacobson et al., 1994b).

Precolumn OPA derivatization were also employed on a PDMS chip for MECC separation of biogenic amines (Ro et al., 2001a). Precolumn OPA derivatization and MECC of amino acids were performed on a glass chip using amperometric detection (the OPA derivatives were also electroactive). Voltage (for separation) programing was used to decrease the migration time of late migrating species (Wang et al., 2000b).

E. Isotachophoresis

Isotachophoresis (ITP) alone (Masar et al., 2001) and ITP preconcentration before CZE (Kaniansky et al., 2000; Masar et al., 2001) have been achieved on a PMMA chip (see Fig. 34). With valves, appropriate solutions are

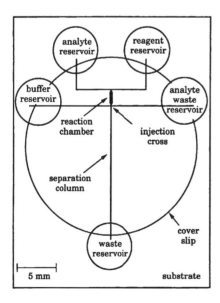

Figure 33 Schematic of the microchip with integrated precolumn reactor. The reaction chamber is 2 mm long, and the separation column is 15.4 mm long. [(Jacobson et al., 1994b) Courtesy of American Chemical Society.]

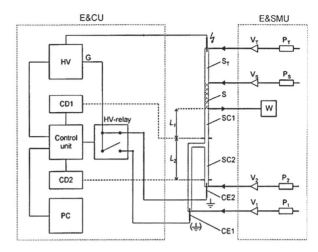

Figure 34 Block scheme of the ITP. E&CU, electronic and control unit; HV, high-voltage power supply; CD1, and CD2, platinum conductivity detectors for the first and second separation channels, respectively; HV-relay, a high-voltage relay switching the direction of the driving current in the separation compartment; S_T, a high-voltage (terminating) channel; S, sample injection channel; SC1, the first separation; SC2, the second separation; CE1, and CE2, counter electrodes for the first and second separation channels, respectively; L_1, and L_2, separation paths for the ITP measurements of the migration velocities on the chip; E&SMU, electrolyte and sample management unit; V_1, V_2, V_T, needle valves for the inlets of the separation and terminating channels; V_S, a pinch valve for the inlet of the sample injection channel; W, waste container; P_1/P_2, P_S, P_T, syringes for filling the separation, sample injection, and terminating channels, respectively. [(Kaniansky et al., 2000) Courtesy of American Chemical Society.]

filled via FEP (fluorinated ethylene propylene copolymer) tubings (V_1, V_2, V_3, and V_4) to four consecutive columns on-chip in the following order: (1) P_2 to W for filling LE to SC2 (and SC1), (2) P_1 to W for filling SC1, (3) P_T to W for filling TE to S_T (and S), (4) P_s to W for filling S (Kaniansky et al., 2000).

When a second ITP was cascaded with first ITP, known as concentration-cascade ITP, SC2 was filled with LE (as in SC1). When ITP was followed by CZE (ITP–CZE), SC2 was filled with a background electrolyte (not necessarily with the greatest mobility). A concentration-cascade ITP was achieved for 14 anions (200 μM) within 600 s (see Fig. 35). ITP preconcentration and subsequent CZE were achieved for nitrate, fluoride, and phosphate (10 μM) in the presence of high contents of sulfate (800 μM) and chloride (600 μM) (Kaniansky et al., 2000).

Bidirectional ITP with a common terminating electrolyte (TE) was performed to achieve simultaneous cationic and anionic separations on a PMMA chip. Without a complex injector design, sample introduction was achieved hydrodynamically for ITP separation (Prest et al., 2002).

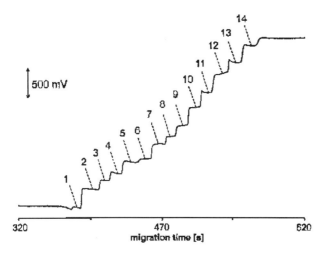

Figure 35 An isotachopherogram from the separation of a 14-component model mixture of anions using concentration-cascade ITP. CD2 was used to monitor the separation. The injected sample contained the anions at 200 μM concentrations. Zone assignments: LE, leading anion (chloride); 1, chlorate; 2, methanesulfonate; 3, dichloroacetate; 4, phosphate; 5, citrate; 6, isocitrate; 7, glucuronate; 8, β-bromopropionate; 9, succinate; 10, glutarate; 11, acetate; 12, suberate; 13, propionate; 14, valerate; TE, terminating anion (capronate). [(Kaniansky et al., 2000) Courtesy of American Chemical Society.]

F. Capillary Electrochromatography (CEC)

Electrochromatography was first demonstrated in the open microfluidic channel coated with octadecylsilane (ODS) as the chromatography stationary phase. EOF is used for the pumping system within the 5.6 μm deep and 66 μm wide open channels (Jacobson et al., 1994c). Plate heights of 5.0–44.8 μm were achieved for three coumarin dyes. Detailed considerations in band broadening (Van Deemter equation) due to extra column effects (injection plug length, detector observation length, and column geometry), axial diffusion, and mass transfer for mobile phase have been provided [see Eq. (11)],

$$H = \underset{(0.13\,\mu m)}{H_{inj}} + \underset{(0.014\,\mu m)}{H_{det}} + \underset{(0.53\,\mu m)}{H_{geo}} + H_{diff} + H_m \quad (11)$$

H_{geo}, which has the greatest contribution among the first three factors, can be reduced by decreasing the channel width (at the expense of detector path length), the number of turns, and the turn angle. The plate heights as contributed by Joule heating and mass transfer for the thin stationary phase film were neglected. Other phenomena, such as electric field effects or eddy flow, though conceivable, were claimed to be not observed (Jacobson et al., 1994c).

Upon curve fitting to the Van Deemter equation $(A + B/v + C_m v)$, the fitted C_m values are two orders of magnitude higher than the calculated C_m values (on the basis of channel height and diffusion coefficient of the analyte in aqueous solution). This discrepancy has been attributed to the trapezoidal geometry of the channel cross-section, leading to more broadening near the triangular cross-section than the middle rectangular cross-section (Jacobson et al., 1994c).

Open channel CEC (with coupled solid-phase extraction) was performed for PAHs (Broyles et al., 2001). CEC with ODS beads was used to separate five bioactive peptides (papain inhibitors, proctobin, opiod fragment 90–95, ileangiotensin III, and angiotensin III). CEC of a mixture of BODIPY and fluorescein was achieved in less than 20 s on a reverse-phase chromatography bed (a 200 μm bed packed with ODS-coated silica beads) (see Fig. 36) (Oleschuk et al., 2000). CEC of two coumarin dyes was also performed using ODS beads (Finot et al., 2001). CEC of four FITC-labeled synthetic peptides was achieved on a PDMS chip (Penner et al., 2001; Slentz et al., 2002). A polystyrene-sulfonic acid stationary phase immobilized on a collocated monolith support structure (COMOSS) was used in CEC (He et al., 1998). An OCEC study with gradient elution of the mobile phase (H_2O/CH_3CN) was also reported (Kutter et al., 1998).

2D Open-channel electrochromatography (OCEC)/CE has been achieved on a glass chip consisting of a spiral column (see Fig. 19). The separation of a tryptic digest of β-caesin was achieved using a mobile phase of 10 mM sodium borate (for CE) and 30% v/v acetonitrile (for OCEC) (Gottschlich et al., 2001). A 2D contour plot was generated by plotting each electropherogram at the corresponding time on the OCEC axis (see Fig. 37) (Gottschlich et al., 2001). Here, 17 peaks from OCEC generated 26 spots in the 2D plot (Gottschlich et al., 2001).

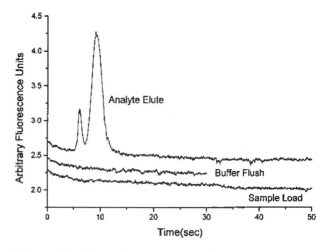

Figure 36 Electrochromatogram of fluorescein (first) and BODIPY (second), showing different steps of the separation: step 1, 100 s loading; step 2, 30 s buffer flush; step 3, an isocratic elution from the 200 μm long column with a 30% acetonitrile/70% 50 mM ammonium acetate mobile phase. [(Oleschuk et al., 2000) Courtesy of American Chemical Society.]

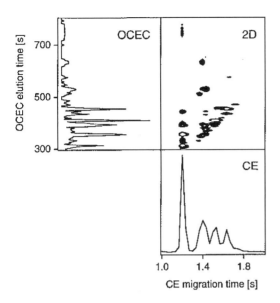

Figure 37 2D Separation of TRITC-labeled tryptic peptides of β-casein. The projections of the 2D separation into the first dimension (OCEC) and second dimension (CE) are shown to the left and below the 2D contour plot, respectively. The field strengths were 220 V/cm in the OCEC channel and 1890 V/cm in the CE channel. The buffer was 10 mM sodium borate with 30% (v/v) acetonitrile. The detection point y in Fig. 19 was 0.8 cm past valve V2 in the CE channel. [(Gottschlich et al., 2001) Courtesy of American Chemical Society.]

G. Synchronized Cyclic Capillary Electrophoresis

Column switching to increase column length and hence N has been demonstrated in HPLC. However, because of instrumentation limitations, this advantage has not been exploited in CE. With micromachining on glass, a complex channel system can then be constructed on-chip, leading to synchronized cyclic capillary electrophoresis (SCCE) (Burggraf et al., 1994). With the use of four HV power supplies and 11 HV relays, switching between four columns for CE has been achieved. An enhancement of R_s from 1.7 to 3.0 has been achieved for the CZE separation between two major peaks in a FITC sample (Burggraf et al., 1994).

Further development of SCCE was demonstrated in the separations of amino acids (von Heeren et al., 1996a, b) and oligonucleotides (von Heeren et al., 1996b). Separation of dsDNA was achieved on a Si chip (coated with thermal SiO_2 for insulation) (Jeong et al., 2000). Although SCCE has the advantage of increasing column length by cycling, it suffers from some drawbacks (Burggraf et al., 1994; Culbertson et al., 2000a; Jeong et al., 2001; von Heeren et al., 1996a).

H. Free-Flow Electrophoresis

Free-flow electrophoresis (FFE) of various Rhodamine-B labeled amino acids was achieved on a Si–Pyrex chip. To avoid electrical breakdown of Si, it has been deposited with a composite SiO_2 and Si_3N_4 layer (Raymond et al., 1994). Separation depends on the deflection angle of a sample flow stream of flow speed of v due to an applied perpendicular E field. The deflection angle is greater if v decreases, E increases, and μ (electrophoretic mobility) increases. This gives rise to possible fraction collection at the outlet array (70 μm wide, 50 μm deep, and 5 mm long) (Raymond et al., 1994).

Isolation of the separation bed from the two electrode-containing beds was achieved by two arrays of narrow grooves (12 μm wide, 10 μm deep, and 1 mm long). Whereas the side-bed flow should be high enough (>15 μL/min) for effective removal of electrophoretic gas bubbles, the flow should not be too high (<50 μL/min), to avoid increasing the separation-bed flow rate and reducing the residence time. It is because the residence time should be maximized to increase sample deflection and to enhance separation resolution (Raymond et al., 1994). Greater resolution was obtained when a greater potential difference across the separation bed was obtained by increasing voltage and increasing side-bed buffer conductivity. Separation of a neutral, monanion, and dianion amino acid (labeled) was demonstrated (Raymond et al., 1994). FFE was also conducted on a Si chip (with a 25 μL bed) for separation of various proteins (HSA, bradykinin, and ribonuclease A) or tryptic digests of mellitin and cytochrome c (Raymond et al., 1996).

Other chromatographic methods were also performed on-chip. Hydrodynamic chromatography of fluorescent nanospheres (polystyrene) and macromolecules (dextran) has been achieved on a Si–Pyrex chip. Separation was based on faster movement of larger particle or molecules because they follow the faster fluid density near the center of a channel (Chmela et al., 2002).

Open tubular LC using an ODS-coated channel was attempted on a Pyrex chip (Cowen and Craston, 1995). Separation of two proteins (BSA and IgG) by anion-exchange chromatography was demonstrated on a PDMS chip packed with beads (Seki et al., 2001).

VII. DETECTION METHODS

A. Optical Detection Methods

1. Fluorescence Detection

The most commonly used detection system for microchip applications has been LIF (see Fig. 38) (Effenhauser et al., 1997b; Lapos et al., 2002; Seiler et al., 1993; Woolley and Mathies, 1994).

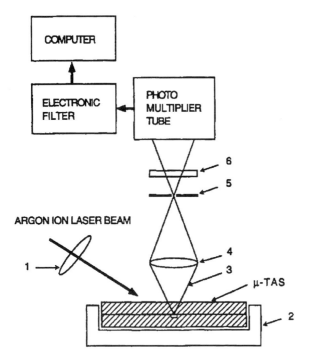

Figure 38 Schematic of the laser-induced fluorescence detection system. Argon ion laser light was focused with a lens (1) onto the separation channel, which was held in place with a Plexiglas holder (2). Fluorescence emission (3) was collected with a microscope objective (4), focused onto a spatial filter (5), emission filter (6), and then detected with a photomultiplier tube. [(Seiler et al., 1993) Courtesy of American Chemical Society.]

In LIF detection, band-pass filters are usually used to help reduce background. In one report, even two 530 nm band-pass filters have been employed (Simpson et al., 1998). The detection limit can be improved by reducing background fluorescence using the fused silica substrate (Jacobson et al., 1994b).

Scanning LIF was performed using a galvanoscanner which probed 48 channels sequentially (Simpson et al., 1998). Scanning LIF detection based on the acoustic–optical effect offered a much faster scan rate (200 Hz), than a translating-stage-based scanner (3.3 Hz scan rate). Three scanning modes—raster (uni- or bi-directional), step, and random addressing—are available. The last one is difficult to achieve with a translating-stage-or galvanometric-based scanner (Huang et al., 1999).

With the use of an optical fiber, fluorescence was measured in which the excitation laser beam (~18 μm wide) was directed along the channel (longitudinal excitation). This results in a 20-fold improvement in S/N when compared with conventional (transverse) excitation with fiber, leading to a detection limit of fluorescein of 3 nM (20,000 molecules). This is because the excitation beam is narrower than the channel wall and there is less light scattering from the solution sheath (Liang et al., 1996).

Further reduction in the background could be achieved by optimizing the chip bonding procedure to reduce light scattering centers, by avoiding the curved channel wall in excitation, and by adjusting of incidence angle. With these improvements, an LOD of 30 pM (fluorescence) was achieved (Chiem and Harrison, 1997).

Another strategy of enhancing the measured fluorescence intensity (in the use of inverted microscopy) is to use a greater channel depth. This is achieved with a channel geometry with a depth (20 μm) that is twice as large as the width (10 μm) on a PDMS replica, made possible only by using anisotropic etch for creating the Si $\langle 100 \rangle$ master (Effenhauser et al., 1997b).

Another way to reduce the fluorescence background, especially from plastic chips, is to modulate the velocity of an analyte moving pass a spatial filter (by modulation of the separation voltage in 7–20 Hz). Noise rejection with a decrease in LOD by one order of magnitude was achieved using a lock-in amplifier (synchronous decrease demodulation), because the fluorescence background (from the chip substrate) was not modulated (Wang and Morris, 2000).

Fluorescence detection of DNA was enhanced by focusing the sample stream so that it was within the illumination region of the laser probe beam (Haab and Mathies, 1999).

Photobleaching, as estimated by the photochemical lifetime (τ), can be obtained by measuring the fluorescence signal (F) of OPA-glycine as a function of residence time (Δt) in a frontal electropherogram. A value of τ (51 ms for OPA–arginine and 58 ms for OPA–glycine) can be obtained from the slope of a plot $\ln F$ vs. Δt (Jacobson et al., 1994b).

An array of circular or elliptical microlenses (of 175 μm in diameter) made of photoresist was fabricated on both the bottom glass plate (for focusing the excitation beam) and the top glass plate (for collecting emission). An array of chromium aperture array (3000 Å thick) was also formed around the focusing microlenses to block off the beam (larger than the lens diameter) and around the collecting microlenses to block off unabsorbed excitation beam from being scattered into the detector. Detection limit of Cy5 was found to be 3.3 nM (Roulet et al., 2002).

LIF has been carried out to detect proteins on PDMS chips with the detection optical fiber (coupled to a blue LED light source) and the microavalanche photodiode both embedded in the PDMS chip (Chabinyc et al., 2001). Integrated LIF detection was also achieved by a Si photodiode interference filter fabricated on a Si substrate on which a parylene channel was built (Webster et al., 2001).

Time-resolved fluorescence was used to detect Rhodamine 6G (R6G), sulforhodamine 101 (SR101), and Rhodamine B (RB) using a Ti-sapphire laser (800 nm, 50 fs) converted to 400 nm (for R6G) or 532 nm (SR101 or RB) using an optical parametric amplifier or second harmonic generation (Hibara et al., 2002b). Single chromophore molecules could be detected on the basis of fluorescence burst detection (Fister et al., 1998).

Evanescent field-based fluorescence detection has been used to detect Cy5-labeled anti-rabbit IgG which binds to rabbit IgG immobilized on a planar optical waveguide within a PDMS chip. The waveguide consists of a 150 nm thick silicon nitride layer deposited on a 2.1 μm thick SiO_2 buffer layer on a Si substrate. This format allows excitation of the fluorescent molecules present within ~200 nm of the waveguide surface (Hofmann et al., 2002).

Dual LIF and amperometric detection were used to detect a five-component mixture (see Fig. 39). Each fluorescence detection peak (i.e., NBD–Arg, NBD–Phe, and NBD–Glu) is normalized to an internal standard, CAT, in order to reduce the RSD of migration time (from 2.7% to 0.8%). Here, the EOF marker (DA) and internal standard, catechol (CAT) are detected in a second channel, avoiding overlapping with any analyte peak (Lapos et al., 2002).

2. Indirect Fluorescent Detection

Besides the commonly used direct LIF detection, indirect LIF detection has also been reported. Indirect LIF using 5 μM Cy7 was employed to detect 1 ppm of explosives in spiked soil samples. In contrast to fluorophore displacement (from the SDS micelle interior), it was the quenching of the fluorophore that led to negative peaks (Wallenborg and Bailey, 2000).

In contrast to a capillary-based system, an increase in E from 185 to 370 V/cm did not result in an unstable background fluorescence due to excessive Joule heating, probably because of the effective heat dissipation in the glass chip. Upon multiple injection, it was found that the detection sensitivity decreased, most likely caused by the degradation of the visualizing dye (Cy7) (Wallenborg and Bailey, 2000). For other applications using indirect detection (Munro et al., 2002; Sirichai and deMello, 2000), see Table 17.

3. Multipoint Fluorescent Detection

Multiplepoint (Shah function) detected, time-domain detector signals were converted into frequency-domain plots by Fourier transformation (in the forward direction). This technique was dubbed Shah convolution Fourier transform detection (SCOFT) (Crabtree et al., 1999; Kwok and Manz, 2001a, c). In comparison, single-point detection time-domain response is commonly known as the electropherogram.

The Shah function is a detector function that can be realized using multiple slits. With microfabrication, the separation channel and detection slits (55) can be fabricated and aligned to each other (Crabtree et al., 1999).

The 488 nm Ar^+ laser line was expanded to produce a parallel beam illuminating along the 4.5 cm separation channel (15 μm deep, 50 μm wide at the top, and 20 μm wide at the bottom) consisting of 55 slits (300 μm wide and 700 μm spaced center to center). The time-domain signal for a single component (fluorescein) being detected via 55 slits (the Shah detector function) is shown together with the frequency-domain signal (indicating the migration time) obtained after Fourier transform (see Fig. 40). Using this technique, the baseline drift can also be eliminated (Crabtree et al., 1999).

Multipoint detections of a two-component sample (Kwok and Manz, 2001b, c) or four-component sample (Kwok and Manz, 2001a) were also achieved, but the separation resolution was not as good as that obtained from single-point detection. The use of multiple sample injections (up to a maximum of three) was found to enhance S/N, that is, the S/N is slightly higher than the square root of the number of injected sample plugs (Kwok and Manz, 2001c).

Instead of applying a physical mask, the Shah (or comb) function was applied after collecting the signal over an unmasked channel (9 mm long) using a cooled CCD detector. As the detector function was applied after data collection, greater flexibility can occur in the choice of the detector function (Shah, modified-Shah, or sine). Reprocessing of data can always be performed to achieve the best results (McReynolds et al., 2002). When the Shah function was used, although the second harmonic was not seen, new artifact peaks in the frequency domain appeared. However, when the sine function was used before FT, there was no artifact peak, and this provided an advantage over the SCOFT method (McReynolds et al., 2002).

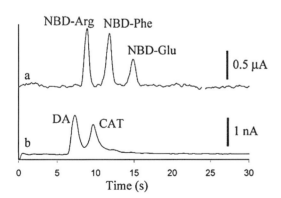

Figure 39 Simultaneous detection of a five-component sample in TES buffer. Electropherograms obtained using (a) LIF and (b) EC detection. Peaks are identified as DA, CAT, and NBD-labeled Arg, Phe, and Glu. [(Lapos et al., 2002) Courtesy of American Chemical Society.]

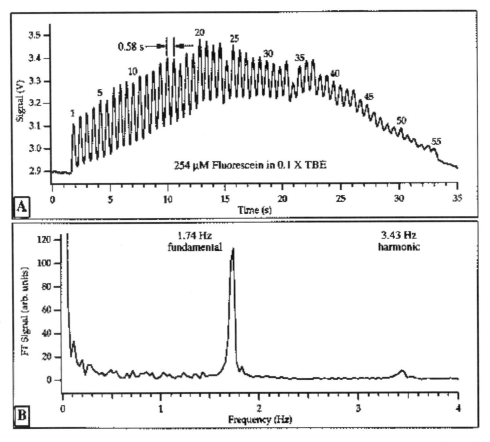

Figure 40 One-component injection detected by SCOFT. (A) Time-domain data produced when a fluorescein sample is injected down the separation channel: 55 peaks, spaced at 0.58 s, superimposed on the Gaussian distribution of the expanded laser beam. (B) Fourier transformation of part A, displayed in magnitude formats: peaks for 1.74 Hz (~1/0.58 s) fundamental and 3.43 Hz harmonic can be seen. [(Crabtree et al., 1999) Courtesy of American Chemical Society.]

Multiple injections, in which injections were performed in a continuous but random sequence, were also used in cross-correlation chromatography. A single-point detection output was correlated with the injection profile, leading to the detection sensitivity enhancement due to the multiplex advantage (see Fig. 41) (Fister et al., 1999).

4. Absorbance Detection

A UV absorbance detector has been achieved on-chip by fabricating a U-cell which provides a longitudinal optical path length of 140 μm. This provides a 11-fold increase in absorbance, giving a detection limit of 6 μM for hydrolyzed FITC. This is higher than a sevenfold increase based on a sevenfold increase in the path length. Insertion of a single-mode optical fiber (3.1–50 μm dia. silica core, 125 μm dia. silica cladding, and 250 μm dia. jacket polyimide coating) into a 230 μm wide and 150 μm deep channel was achieved by first stripping the polyimide coating and then etching the silica cladding. A 3.1 μm fiber was used for launching and a 7.9 μm fiber was used for collecting light (see Fig. 42). An index matching fluid (1, 4-dibromobutane, $n = 1.519$) was used to fill the channel (Liang et al., 1996). It was also reported that the use of the collimating lens and detection slit increased the absorbance detection sensitivity in a PDMS chip (Ro et al., 2001b).

UV absorbance detection was used to determine a gaseous mixture of BTX using optical fibers along a 20 mm long channel (Ueno et al., 2001, 2002b). Fabrication of a monolithic optical waveguide along a U-shaped detection cell was used to detect various compounds (see Fig. 43), using absorbance detection at 254 nm (Mogensen et al., 2001a) or 488 nm (Mogensen et al., 2001b).

Optical absorbance detection (converted from reflectance measurement) was achieved on a plastic disk for *p*-nitrophenol detection (430 nm). The disk contained 1 wt% of titanium white pigment so that the disk was reflective to white light. Instead of using a scanning absorbance detector for 48-channel enzyme measurements, a stationary detector was used with the disk spinning at high speed (60–300 rpm) needed to generate the centrifugal force for fluid flow (Duffy et al., 1999).

Microfluidic Lab-on-a-Chip

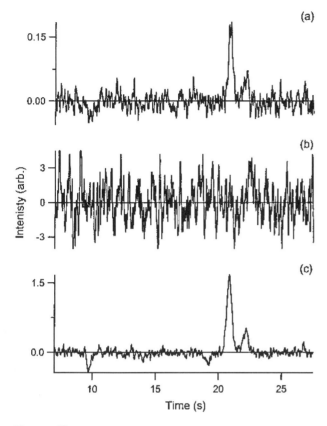

Figure 41 Electropherogram and correlograms of dichlorofluorescein (70 pM) and fluorescein (63 pM): (a) an individual 7 bit correlogram, (b) electropherogram from a single 0.25 s gated injection, (c) the average of 12 bit correlogram. [(Fister et al., 1999) Courtesy of American Chemical Society.]

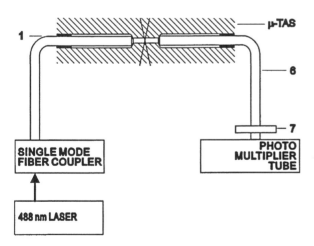

Figure 42 Schematic of absorbance detection on a glass microchip. 1, Launch fiber; 6, longitudinal absorbance collection fiber; 7, filter. [(Liang et al., 1996) Courtesy of American Chemical Society.]

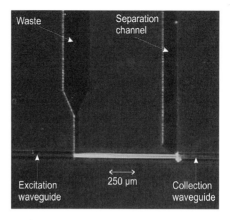

Figure 43 Micrograph of a U-shaped absorption cell with an optical path length of 1000 μm on a silica glass microchip. A 250 nM fluorescein solution was used for visualization. [(Mogensen et al., 2001b) Courtesy of Wiley-VCH.]

Increased optical path length (720 μm) for absorbance detection was also achieved by constructing a 3D fluid path (Wolk et al., 2001). A multireflection cell was designed to enhance optical absorbance detection on a glass chip. Unlike in a Si chip where the crystal plane $\langle 111 \rangle$ could be used as the reflective surface (Verpoorte et al., 1992), a reflective metal film was needed on a glass chip. It was found that an Al film (80-nm thick) provided the best reflectance. Using the bromothymol blue (633 nm) absorbance, 5- to 10-fold enhancement in absorbance was achieved, corresponding to an increased effective path length of 50–272 μm, as obtained from the channel depth of 10–30 μm (Salimi-Moosavi et al., 2000).

5. Optical Emission Detection

An optical emission detector was constructed on-chip using a plasma chamber in which an atmospheric pressure DC glow discharge was generated in helium (99.995% pure) (see Fig. 44) (Eijkel et al., 1999, 2000a). A number of carbon-containing compounds (hexane, methanol, ethanol, 1-propanol, 1-butanol, and 1-pentanol) were detected in a GC (conventional) effluent by recording the emission at 519 nm. For hexane, the detection limit was 10^{-12} g/s (or 800 ppb) with only 2 decades of dynamic range (Eijkel et al., 2000a).

Although microwave-induced plasma (MIP) is the most popular plasma used for GC-OES, the DC glow discharge has recently received more attention because it can be operated at low temperature (but at a low pressure of 1–30 torr). But in a miniaturized device, the pressure does not have to be very low (e.g., 860 torr or atmospheric) so as to avoid excessive gas heating and arcing because of a decrease in the device dimension. The device has an improved lifetime, probably because of the

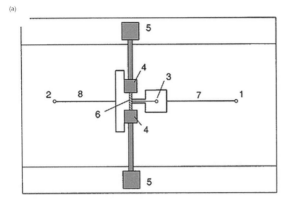

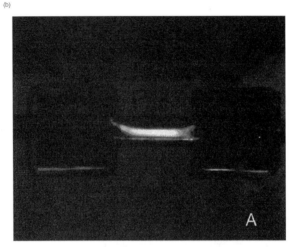

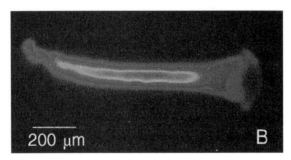

Figure 44 (a) Schematic of a plasma chip. Features of the $20 \times 30 \times 0.5$ mm bottom plate are: 1, gas inlet; 2, gas outlet; 3, pressure sensor connection; 4, electrodes; 5, electrode connection pads. Etched in the $14 \times 30 \times 0.5$ mm top plate are: 6, plasma chamber; 7, inlet channel; 8, outlet channel. (b) An image of a plasma in a plasma chamber ($1000 \times 350 \times 150$ μm) at 750 torr, 500 V, and 60 μA is given in (A). False-colour image of the same plasma is given in (B). [(Eijkel et al., 2000b) Reproduced by permission of Royal Society of Chemistry.]

low-temperature operation at atmospheric pressure, minimizing cathodic sputtering (Eijkel et al., 2000a).

But the detector signal shows a marked peak broadening and tailing when compared with the FID signal. This is mainly attributed to the dead volumes of the connection between the conventional GC column and the plasma chip detector, thus necessitating the on-chip integration of GC and plasma detector (Eijkel et al., 2000a). The LOD of methane was reported to be 10^{-14} g/s (7 ppm) with over 2 decades of linear range (Eijkel et al., 2000b). This compares well with the LOD (for carbon) obtained from an FID and from a microwave-induced plasma atomic emission detector (10^{-12} g/s) (Eijkel et al., 2000a).

6. Chemiluminescence Detector

Chemiluminescence (CL) detection of an HRP-goat antimouse IgG (HRP-Ab) was based on the HRP/luminol/H_2O_2 system. An aluminum mirror was deposited on the back side of the microchip to enhance light collection. The amount of decrease in HRP-Ab with an increasing amount of mouse IgG (10–60 μm/mL) was quantified using an internal standard (microperoxidase) after CZE separation (Mangru and Harrison, 1998).

CL for codeine determination using the $Ru(bpy)^{2+}$ (or TBR) system was performed on a glass chip (Greenway et al., 2000a). Oxidation of TBR was achieved using cerium (IV) sulfate or lead oxide as the oxidizing agent. The addition of a nonionic surfactant (Triton X-45) strongly enhanced the CL emission.

CL can also be achieved using the electrode for oxidation, and this mode is called electrochemiluminescence (ECL). Such a detection based on the $Ru(bpy)^{2+}$ (TBR) and $Ru(phen)^{2+}$ (TPR) ($\lambda_{em} = 610$ nm) systems has been achieved on-chip (Arora et al., 2001). ECL can occur at room temperature in aqueous buffered solutions and in the presence of dissolved O_2 and other impurities. To provide the electric field for ECL to occur at the detector region, a pair of connecting floating Pt electrodes (50 μm wide, 50 μm apart, and 100 nm thick) was used with the use of a Cr underlayer. Unfortunately, corrosion of Cr was observed in the presence of Cl^- (Arora et al., 2001).

$Ru(bpy)^{2+}$ and TPA are oxidized electrochemically to form $Ru(bpy)^{3+}$ and TPA^+, respectively. Then, TPA^+ becomes deprotonated and almost immediately forms TPA–H which subsequently transfers an electron to $Ru(bpy)^{3+}$ to form an excited state of the reduced form, $Ru(bpy)^{2+*}$. This species will emit and relax back to $Ru(bpy)^{2+}$ (Arora et al., 2001). The other ion, $Ru(phen)^{2+}$ has a similar ECL property (Arora et al., 2001; Hsueh et al., 1996).

Separation of $Ru(bpy)^{2+}$ and $Ru(phen)^{2+}$ was barely observed using MECC (SDS system). With the floating electrodes, the emission intensity, which increased with increasing voltage across the electrodes, could only be modified by increasing the separation voltage. Indirect ECL detection was demonstrated for separation of 300 μM of three amino acids (L-valine, L-alanine, and L-aspartic acid) using 0.5 mM $Ru(bpy)^{2+}$ in the run buffer (Arora et al., 2001).

ECL of Ru(bpy)$^{2+}$ (or TBR) was conducted on a Si–glass chip with an ITO anode and Au cathode (Hsueh et al., 1996). Through the transparent ITO anode, orange light (620 nm) was observed and recorded by a detector. It was found by CV that the oxidation potential was more positive, and the peak current density was less on an ITO anode when compared with a Pt anode. In this work, Au cannot be used as anode, presumably because of polymerization of TPA at the gold surface (Hsueh et al., 1996). Other CL and ECL applications are given in Table 17 (Hsueh et al., 1998; Yakovleva et al., 2002).

7. Other Optical Detection Methods

There are alternative optical detection techniques such as refractive index (RI) measurement, thermal lens microscopy (TLM), and Raman scattering.

(a) Refractive Index (RI)

On-chip RI detector based on backscatter interferometry was achieved to detect glycerol (743 μM). Improved sensitivity was achieved in the backscatter format, rather than the forward scatter one, because the probe beam passed the detector channel more than once in the backscatter format. This was enhanced in the unique hemicylindrical shape of the channel. Upon a change from water to glycerol, there was a change in RI, and the portion of the backscatter fringes shifted (191 μm/mRIU), upon irradiation by a He/Ne laser of a beam diameter of 0.8 mm (Swinney et al., 2000).

An holographic-based RI detection was proposed. When a collimated laser beam passed through an holographic optical element, it was divided into two coherent beams. One beam (probe beam) was directed through the channel and the second beam (reference beam) passed through the glass and acted as a control (Burggraf et al., 1998). The two beams subsequently diverged in the far field and interfered, which generated an interference pattern to be detected by a photodiode array. A change in the RI of the solution in the channel resulted in a lateral shift in the fringe pattern (Burggraf et al., 1998).

(b) Thermal Lens Microscope

TLM, which is a type of photothermal spectroscopy, depends on the coaxial focusing of the excitation and probe laser beams using the chromatic aberration of a microscopic objective lens. A YAG laser (532 nm) (Sato et al., 2000, 2001a) or an Ar ion laser (514.5 nm (Sato et al., 2000) or 488 nm (Tokeshi et al., 2001)) was used for excitation. A He/Ne (632.8 nm) was used as a probe beam (Sato et al., 2000, 2001a). The probe beam was detected by a photodiode and the output was recorded as a TLM signal. This signal depends on the radiationless relaxation process after optical excitation of the analytes (Tokeshi et al., 2001).

TLM has a detection limit at the single-molecule level (Hisamoto et al., 2001). Spatial resolution of 1 μm can be attained. Moreover, TLM is not strongly affected by scattering of the laser beam (e.g., by the cell membranes) (Tamaki et al., 2002).

(c) Raman Scattering

Raman scattering was measured (in the 700–1600 cm^{-1} range) using a NaYVO$_4$ laser (532 nm) for the detection of herbicides (diquat or paraquat) in a glass chip (Walker et al., 1998).

Surface-enhanced resonance Raman scattering (SERRS) has been employed to detect an azo dye, 5-(2′-methyl-3′,5′-dinitrophenylazo) quinolin-8-ol, which is a derivative of trinitrotoluene (TNT), using silver colloid aggregates produced *in situ* in the chip. It was possible to detect 10 μL of 10^{-9} M dye (or 10 fmol) using SERRS, representing a 20-fold increase in sensitivity over that achieved using a macroflow cell (Keir et al., 2002).

B. Electrochemical Detection

1. Amperometric Detection

The first electrochemical (EC) detection was carried out on a glass chip for analysis of DNA restriction fragments and for PCR product sizing (see Fig. 45) (Woolley et al., 1998). The working and counterelectrodes were RF-plasma-sputtered (2600 Å Pt with 200 Å Ti) and the reference electrode was a silver/AgCl wire. The integrated microelectrodes allowed the facile and stable location of the working electrode at the exit of the separation channel (Woolley et al., 1998).

Fe(phen)$_3^{2+}$ was used as the electrochemically active intercalation reagent. The constant background current from free Fe(phen)$_3^{2+}$ decreased in the presence of the DNA–Fe(phen)$_3^{2+}$ complexes. Therefore, this is an indirect amperometric detection method. It was found that a distance of 300 μm, instead of 600 μm, between the working electrode and reference electrode has produced less interference (in the form of a sloping baseline), allowing the use of a separation voltage up to 1200 V (240 V/cm) (Woolley et al., 1998).

Gated injection is essential for applications using EC detection, because it will avoid applying a finite voltage to the buffer waste vial, because it must be at ground for EC detection (Martin et al., 2000).

There are numerous advantages of EC detection on-chip (Backofen et al., 2002), which include (1) miniaturization of the electrodes without compromising LOD; (2) the short response time of the detector; (3) the minimal

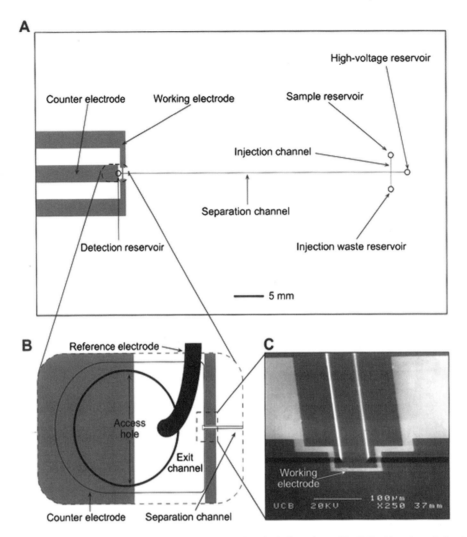

Figure 45 Capillary electrophoresis chip with integrated electrochemical detection. (A) Full chip view. Injection channel length, 5 mm; separation channel length, 50 mm; channel full width at half-depth, 46 μm; channel depth, 8 μm. (B) Expanded view of the integrated electrochemical detector. (C) Scanning electron micrograph (140×) of the detection region showing the location of the working electrode in the exit channel 30 μm beyond the end of the separation channel. Image has been rotated 90° for viewing clarity. [(Woolley et al., 1998) Courtesy of American Chemical Society.]

dead volume of the detector; and (4) preparation of electrodes (working) compatible with the planar technology.

Amperometric detection of catecholamines has been achieved using a Pd film decoupler. The Pd film has been thermally evaporated onto a plastic chip (without the use of the Cr or Ti adhesion underlayers). Owing to the fast diffusion of H_2 on a Pd surface, gas bubbles will not form. This reduces one of the interferences to the EC signal (Chen et al., 2001a).

Amperometric detection of glucose was achieved using a gold (200 nm Au/20 nm Cr) electrode patterned on glass (Pyrex) which was bonded to a Si chip. The surface of the Si chip has been oxidized for insulation of the electrode. Glucose oxidase was covalently immobilized on the inner surface of the channel (Murakami et al., 1993).

Unlike CE/EC devices previously reported, the working electrode as well as the reference and counter-electrodes were all patterned (10 nm Ti/300 nm Pt) on the glass chip. No external wire electrodes were used. LOD of dopamine and catechol were in the 4–5 μM range (Baldwin et al., 2002). [These electrodes were deposited via a lift-off process in trenches (0.3 μm deep) previously etched into the glass substrate, ensuring leak-free bonding. The lift-off process, as opposed to the wet-etch process, allowed the dual-composition electrode to be patterned in a single step.] The electrodes were situated under a "shelf" (owing to the cover plate) so that the detection volume was restricted and dispersion at the column exit was minimized. The stability of the chip was found to be more than 2 months (Baldwin et al., 2002).

Better isolation of the high separation field (similar baseline noise at various separation voltages) was achieved using a sputtered Au film formed at the exit of the channel, which was perpendicular to the axis of the channel (Wang et al., 1999a).

In-channel amperometric detection was possible without using a decoupler. This was achieved using an electrically isolated (nongrounded, floating) potentiostat (Martin et al., 2002).

Metal electrodes cannot easily be made on PDMS because of its pliable nature. Therefore, the Au/Cr microband electrodes have been formed on a glass plate to be aligned with the PDMS channel plate. Even so, room-temperature bonding is essential to avoid the increase in the grain-boundary diffusion rate (from the Cr adhesion layer to the Au overlayer). To avoid working electrode fouling, it is cleaned by applying a bipolar square wave voltage to the electrode after 25 injections; however, more than 10 repeated applications will still destroy the electrode (Martin et al., 2000). The LOD of catechol (single-electrode detection) was found to be 4 μM (Martin et al., 2000), which is 3× lower than a previous report (Woolley et al., 1998), possibly because of the application of the bipolar square wave voltage between injections to clean the electrode (Martin et al., 2000).

Other than metal, carbon electrodes were also used for EC detection. Amperometric detection (end-column) was achieved at the end side of the chip using the amperometric current–time (i–t) curve mode. The detector, which was situated at the waste reservoir near the channel outlet side, consisted of the silver/AgCl wire (reference), platinum wire (counter), and screen-printed carbon (working) electrodes (Wang et al., 2001a).

A relatively flat baseline with low noise was observed despite the use of a high negative detection potential (-0.5 V vs. silver/AgCl), the use of nondeaerated run buffer along a channel of 72 mm long, and the absence of a decoupling mechanism. However, the use of higher separation voltage (>2000 V) will increase background noise (see Fig. 46) (Martin et al., 2000).

It was reported that when the carbon-fiber electrode was used in a PDMS chip, the in-channel format gave better peak symmetry than the end-channel format (Martin et al., 2001a).

Conducting carbon polymer ink, which filled a UV-ablated microchannel, was used to construct the integrated microelectrode on a plastic chip. Both CV and chronoamperometry were employed to detect a model compound (ferrocenecarboxylic acid) to 3 μM, corresponding to 0.4 fmol within a volume 120 pL (Rossier et al., 1999b).

A replaceable C disk electrode (two-electrode system) was used for end-column amperometric detection of biogenic amines. The electrode was inserted through a guide tube and situated at a central position to maximize

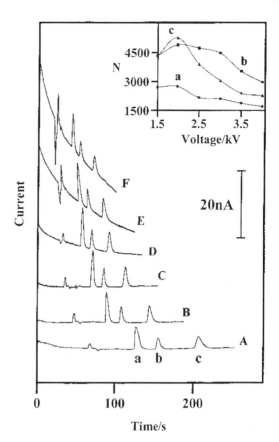

Figure 46 Influence of the separation voltage upon the response for a mixture containing 3.0×10^{-5} M paraoxon (a), 3.0×10^{-5} M methyl parathion (b), and 6.0×10^{-5} M fenitrothion (c). Separation performed using (A) +1500, (B) +2000, (C) +2500, (D) +3000, (E) +3500, and (F) +4000 V. Also shown (inset) are the resulting plots of plate number (N) vs. separation voltage. Separation buffer, 20 mM MES (pH 5.0) containing 7.5 mM SDS; injection voltage, +1500 V; injection time, 3 s; detection potential, -0.5 V (vs. silver/AgCl wire) at bare carbon screen-printed electrode. [(Wang et al., 2001a) Courtesy of American Chemical Society.]

Coulombic efficiency with respect to the axis of the capillary and was 30 ± 5 μm away from the capillary exit. This configuration also allowed easy replacement of the electrode especially after biofouling (Zeng et al., 2002).

Microdisk electrodes (30 μm dia. C disk) were used for amperometric detection of ascorbic acid (LOD = 5 μM) in a PDMS–glass chip (Hinkers et al., 1996).

The first dual-electrode (Au/Cr) amperometric detection was achieved on a PDMS–glass device to positively identify catechol (100 μM) among a complex mixture containing ascorbic acid (see Fig. 47) (Martin et al., 2000).

To avoid deterioration of the Au/Cr electrode, dual C-fiber electrodes were constructed on PDMS. The LOD of catechol was found to be 500 nM. Two C fibers (33 μm dia.) were inserted into the PDMS channels (35 μm wide and 35 μm deep) fabricated on a second

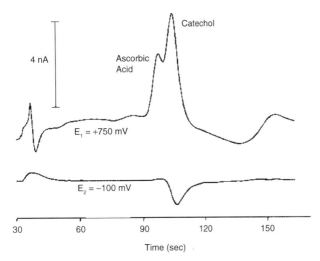

Figure 47 Dual-electrode detection of an unresolved mixture of 100 μM ascorbic acid and catechol. Separation conditions: 25 mM boric acid, pH 9.2; applied potential 870 V (250 V/cm). Injection: 1 s (S to SW) at 870 V. $E_1 = +750$ mV; $E_2 = -100$ mV. [(Martin et al., 2000) Courtesy of American Chemical Society.]

plate (bottom). The separation channel (25 μm wide and 50 μm deep) was fabricated on the top plate. Consecutive injections (up to 41) could be performed before the electrode was cleaned with a bipolar square wave voltage. Because of the use of the C electrode, detection of peptides (e.g., Des-Tyr-Leu-enkephalin), which formed stable Cu (II) and Cu (III) complexes, could be made by the dual-electrode detector. Nevertheless, amino acids would not be detected by this method (Gawron et al., 2001). Dual-electrode amperometric detection also allowed the positive detection of two peptides (Martin et al., 2001b).

Thick-film C electrode (by screen printing) was constructed. Carbon ink (10 μm thick) was first printed on an alumina plate, and cured thermally. Then, the silver ink (28 μm thick) was printed and cured to partially overlap with and hence connected to the C layer. An insulating ink layer (70 μm thick) was finally printed to cover the C–silver junction. The thick-film C electrode was found to enhance the detection sensitivity, as compared to the thin-film amperometric detector (Wang et al., 1999b).

A carbon-paste electrode was constructed by filling a laser-ablated (PET or PC) channel with C-ink. The whole structure was then cured at 70°C for 2 h (Rossier et al., 1999a).

2. Potentiometry

An on-chip potentiometric detector or ion-selective electrode (ISE) has been used to detect Ba^{2+} in a flow stream (1 μL/min) based on a liquid polymeric membrane containing a barium ionophore (Vogtle): N,N,N',N'-tetracyclohexylbis(o-phenyleneoxyldiacetamide) (Tantra and Manz, 2000). The ISE chip comprises two distinct channels (10 μm deep): (a) a sample channel holding the flowing sample solution (Ba^{2+}) and (b) a U-channel entrapping the solvent polymeric membrane and the barium ionophore (see Fig. 48). The two channels are connected by a junction structure (point D) of 20 μm in width for the sample to come into contact with the membrane. To prevent potential drift, reservoir A has been filled with an internal filling solution (0.1 M $BaCl_2$), which is usually missing in other types of miniaturized ISE (Tantra and Manz, 2000). The responses to changes of Ba^{2+} are rapid within seconds.

Potassium and nitrite ionophores have also been incorporated into an optode membrane to detect K^+ and NO_2^- potentiometrically on a plastic disk (Badr et al., 2002).

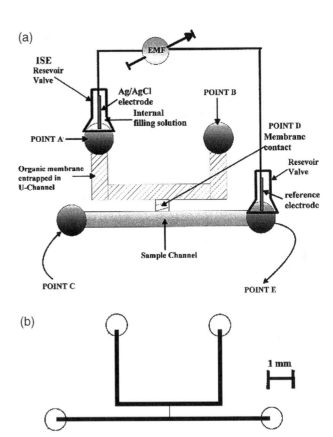

Figure 48 Micromachined ISE chip. (a) Schematic drawing of a sensor chip design with channels and reservoirs: point A, inlet for U-channel; point B, outlet for U-channel; point C, inlet for sample channel; point D, junction structure, where membrane contacts sample solution; and point E, outlet for sample channel. The diagram illustrates the complete filling of the silanized U-channel with organic membrane cocktail. Circles denote drilled holes. The diagram also illustrates the measuring system. (b) Scaled diagram of the 12 mm × 6 mm chip. [(Tantra and Manz, 2000) Courtesy of American Chemical Society.]

3. Conductivity Detection

Conductivity detection of NaCl was performed on a Si–Pyrex chip with Pt electrodes (Darling et al., 1998). Conductivity detection was made possible on a Pt electrode sputtered on a PMMA chip (Kaniansky et al., 2000).

Contactless conductivity detection which was capacitively coupled to the electrolyte (but not in direct contact) in the channel, was used to detect inorganic ions (Lichtenberg et al., 2001; Prest et al., 2002).

Voltage of several hundred volts was used to achieve low detection limits. The electrodes were placed in ultrasonically ablated wells so that they were very close (0.2 mm) to the channel. In addition, a Faraday shield was placed between the two electrodes to avoid direct coupling between them (Tanyanyiwa and Hauser, 2002). Anodic stripping voltammetry (ASV) was performed on a Si–Pyrex chip with Pt electrodes to detect Pb^{2+} (Darling et al., 1998).

C. Mass Spectrometry

The microfluidic chip has been coupled to different MS interfaces (ESI, MALDI) and mass-analyzed (by single quadrupole, triple-Q, ion-trap, MS/MS, and FTICR). The chip can be single or multichannel integrated with various sample handling processes such as preconcentration, digestion, and separation.

Other than glass, polymeric materials have been used to fabricate the MS chip. These materials, including PMMA (Meng et al., 2001), Epofix (Liu et al., 2000d), and PDMS (Chan et al., 1999a), have been evaluated for possible organic contaminants. No sample cross-contamination has been found for 10 μM cytochrome c vs. 10 μM ubiquitin (Albin et al., 1996), and tryptic digests of βlac, CA, and BSA (Figeys et al., 1998). Good stability (for 4 h) was also obtained for the PMMA chip (Meng et al., 2001).

For electrospray generation, the emitter used is usually in the form of a tip, though electrospray from a flat edge has also been demonstrated (Ramsey and Ramsey, 1997). To reduce the dead volume at the electrosprayed tip, it was inserted (then glued) into a specially drilled flat-bottom hole, as opposed to a conical-bottom hole (Bings et al., 1999).

Various methods have been reported to fabricate monolithic needle tips in polymeric materials (Guber et al., 2001), parylene (Licklider et al., 2000), or PDMS (Kim and Knapp, 2001a). Moreover, monolithic Si nozzles were fabricated by DRIE. This approach showed greater signal stability and intensity (Schultz et al., 2000).

An array of triangular-shaped tips was fabricated from a 5 μm polymeric sheet (parylene C) by O_2 plasma etching. The sheet was sandwiched between two plastic plates (Kameoka et al., 2002). The wicking tip protruding from the end a microchannel aided in forming a triangular shaped droplet (0.06 nL in volume) and assisted in forming a stable Taylor electrospray cone.

An open-access channel (i.e., no cover plate) has been used for CE separation before the MALDI MS analysis (Liu et al., 2001d).

The research in the area of on-chip MS analysis is very active and the important differences are summarized in Table 13.

VIII. APPLICATIONS TO CELLULAR ANALYSIS

Cellular studies are facilitated in the microfluidic chips because of their small dimensions together with excellent optical properties (for observation) and fluidic control capabilities (for reagent delivery).

Cell retention, manipulation, and subsequent cellular analysis on-chip can be achieved by using slit- or weir-type filters, or by cell adhesion, fluid flow, and dielectrophoresis (DEP). To demonstrate successful cell retention or manipulation on-chip, particles are sometimes used to carry out modeled studies.

A. Slit-Type Filters

Slit-type filters have been fabricated for retention of suspension cells (e.g., blood cells) or micrometer-sized particles.

A filter consisting slits of 5 μm spacings was fabricated on a 400 μm thick Si substrate using RIE. After sealing with a Pyrex glass plate, this filter was found to be able to retain latex beads of 5.78 μm in diameter within a 500 μm channel (see Fig. 49). However, in the study of red blood cells (RBC), the cells tended to deform and passed through the filter (Sutton et al., 1997; Wilding et al., 1994a). A fluid filter was also constructed, using two silicon membranes (containing numerous 10 μm diameter holes), which displaced laterally relative to each other and separated by a distance with silicon dioxide spacers of submicron thickness (Stemme and Kittilsland, 1998). Retention of oligonucleotides-immobilized beads was also achieved on a slit-type filter chamber fabricated on Si (Andersson et al., 2000b; Russom et al., 2001).

Slit-filters are usually fabricated on Si or quartz, but they can also be fabricated on glass (but with greater gap spacings of 11–20 μm). A glass filter was fabricated, essentially to prevent particles from clogging channels downstream (Lazar and Karger, 2002).

A lateral percolation filter was fabricated near the entry port of a channel on quartz. Filter elements having 1.5 μm channel width and 10 μm depth were anisotropically etched with an aspect ratio greater than 30:1 (He et al., 1999). In contrast to the usual axial slit filter (in which

Table 13 Summary of MS Analysis Using Various Techniques of Chip-MS

Chip design	Flow	MS mode	Analytes	Sample treatment	Separation	References
Multichannel, glued capillary tip, glass chip		ESI	Melittin	Trypsin		Ekström et al. (2000)
Si chip		MALDI	Lysozyme 1 µM	Trypsin on porous Si		Ekström et al. (2000)
Glass chip		ESI	Cytochrome c	Trypsin on beads	CE	Wang et al. (2000d)
PDMS chip		ESI, single Q	Cytochrome c	Trypsin on PVDF membrane	CITP/CZE	Gao et al. (2001)
Nine-channel/hydrophobically coated flat edge, glass chip	Pressure flow 100–200 nL/min	ESI-triple-Q	Myoglobin (60 nM)	Nil	Nil	Xue et al. (1997)
Flat edge emitter, glass chip	EOF induced flow 1.5 nL/s	ESI-ITMS	Tetrabutylammonium iodide 10 µM	N/A	Nil	Ramsey and Ramsey (1997)
Coupled micro sprayer, glass chip	4 µL/min pressure flow	ESI-triple-Q	Carnitine and acylcarnitines in fortified human urine 35–124 µM (10–20 µg/mL)	N/A	CE	Deng and Henion (2001)
Inserted/glued capillary tip, glass chip	1 µL/min pressure flow	QTOF	17–62 µM QTOF carnitine and acylcarnitines in fortified human urine	N/A	CE	Deng and Henion (2001)
Transfer capillary, PMMA chip	Pressure flow 20–200 nL/min	ESI-FTICR	Cytochrome c 500 nM	Nil	Nil	Meng et al. (2001)
Inserted glued tip, glass chip	EOF 100 nL/min, pressure flow 4 µL/min	ESI-triple-Q-SIM	Carnitine or acetylcarnitine (1–500 µg/mL) imipramine or desipramine (5–500 µg/mL) in fortified human plasma	N/A	CE	Deng et al. (2001)
Inserted glued tip, glass chip	EOF 20–30 nL/min	ESI-TOF	Tryptic digest of bovine hemoglobin (4 µM) cytochrome c (0.8 µM) human hemoglobin (normal and sickle cell 0.24 µM)	Mixed off-chip/on-chip tryptic digestion	Nil	Lazar et al. (2001)
Glass chip	N/A	MALDI-FTMS	[Lys¹]-bradykinin (0.5 mg/mL)	Nil	CE	Liu et al. (2001d)
Inserted glued tip, Zeonor chip	Pressure flow 6 µL/min	ESI-triple Q	Carnitine acylcarnitines, butyl carnitine	N/A	CE	Kameoka et al. (2001)

Embedded tip, PDMS chip	Pressure flow 0.4 µL/min	ESI-single Q	0.4 µM Phenobarbital	Dialysis	Jiang et al. (2001)
Nine-channel/one transfer cap sheath, glass chip	EOF flow	ESI-ITMS	Tryptic digest of BSA (182 nM), Horse myoglobin (237 nM), Human haptoglobin (222 nM), 2D gel yeast proteins (40 µg)	Trypsin	Figeys et al. (1998)
96-Channel 96 tip, epoxy resin chip	N_2 pressure 200 nL/min	ESI-ITMS	8-mer peptide for HIV-1 protease (with inhibitors; e.g., Pepstatin A)	N/A	Liu et al. (2000d)
90-Channel Si–glass chip	Vacuum 10^{-6} Pa		41-base deoxyoligonucleotide	Digested by snake venom phosphodiesterase	Brivio et al. (2002)
			Adrenocorticotropin (ACTH)	Digested by carboxypeptidase	
Inserted glued cap tip, glass chip with an internal standard side channel	EOF 200 nL/min, pressure flow 50 nL/min	ESI-Q TOF	1D gel of tryptic digest of membrane protein of *H. influenzae* (nonpathogenic Rd and pathogenic Eagan strains)	Trypsin	Li et al. (2000)
		ESI-triple-Q	Mixture of leu-enkephalin, somatostatin, angiotensin II bradykinin, LHRH		
		ESI-triple-Q	Tryptic digest of seed lectin of *P. vulgaris* L.		
Orthogonal Inserted glued tip, glass chip	EOF 20–30 nL/min	ESI-TOF	Gramicidin S (0.1–10 µM), cytochrome *c* (0.1 µM), bradykinin, leu-enkephalin, methionine enkephelin	Trypsin	Lazar et al. (1999)
Transfer capillary PDMS chip	EOF plus pressure 2–3 psi	ESI-ITMS	Rat serum albumin separated by 2D gel, angiotensin I (100 nM)	Trypsin	Chan et al. (1999)
Inserted tip, PC chip	0.3 µL/min pressure flow	ESI-ITMS	Horse heart cytochrome *c* (5 µM), 30-mer oligonucleotide (10 µM)	Trypsin	Xu et al. (1998)

(*continued*)

Table 13 *Continued*

Chip design	Flow	MS mode	Analytes	Sample treatment	Separation	References
Two samples/1 transfer capillary glass chip	EOF 200–300 nL/min	ESI-ITMS	Fibrinopeptide A (33 nM) tryptic digest of CA (290 nM), BSA (130 nM)	Trypsin	Nil	Figeys et al. (1997)
Transfer capillary integrated nebulizer, glass chip	Pneumatic nebulization 150 nL/min	ESI-ITMS	Angiotensin peptides (20 μg/mL) cytochrome *c* tryptic digest	Trypsin	ITP-CE	Zhang et al. (1999)
Inserted glued capillary, glass chip Microsprayer, glass chip	Pressure flow 50 nL/min Pressure flow 1.5 μL/min	ESI-triple-Q	Angiotensin I, leu-enkephalin vascoactive intestinal peptide, Glu-fibrinopeptide B, tryptic digest from *D. biflorus, P. sativum* lectins substance P, brakykinin (20 μg/mL) oxytoccin, met-enkephalin, Leu-enkephalin, bombesin, LHRH, Arg^8-vasopressin, bradykinin	Trypsin	CE	Li et al. (1999)
Transfer capillary, glass chip	EOF	ESI-ITMS	Tryptic digest of horse apomyoglobin (7.4 nM), yeast protein	Trypsin	Nonchip gradient frontal CEC separation C18 column	Figeys et al. (1998)
Inserted cap tip (no glue), glass chip	Pressure flow 100 nL/min	ESI-ITMS (LCQ)	Tryptic digest of cytochrome *c* myoglobin, β-lactoglobulin A and B, BSA	Trypsin	CE	Zhang et al. (2000)
Triangular-shaped tip on parylene C sheet, cyclo-olefin chip	Pressure flow 300 nL/min	ESI-TOF	1 μM desipiramine and 1 μM imipramine, chicken cytochrome *c*	N/A	Nil	Kameoka et al. (2002)
Inserted tip, glass chip	EOF-hydraulic pump 200 nL/min	ESI-TOF	Bovine hemoglobin tryptic digest	Mixed off-chip, on-chip tryptic digestion	Nil	Lazar and Karger (2002)

Inserted tip, PC chip	ESI-ITMS	200 nL/min	BSA (30 μM), cytochrome *c* (8 μM), ubiquitin (2.4 μM), *E. coli* lysate	Dialysis	Xiang et al. (1999)
Monolithic nozzle	ESI-TOF	0–5 psi N₂, 100 nL/min	Cytochrome *c* 10 nM, nine-peptide mix (160 pg/μL each)	Trypsin	Schultz et al. (2000)
Micromachined emitter, parylene chip	ESI-ITMS	2–4 psi gas, 35–77 nL/min	Gramicidin S, cytochrome *c*	Trypsin	Licklider et al. (2000)
Inserted emitter, silver glue coated, PMMA chip	ESI-ITMS		BSA, fibrinopeptide A, osteocalcin fragment 7–19, bradykinin (all 10 pmol/μL) off-chip (ACTH 1–17)	Trypsin	Chen et al. (2001b)
16-Channel, monolithic PDMS tip, PDMS		300 nL/min	Adrenocorticotropic hormone fragment 1–17 angiotensin I and III (Arg I and Arg III) (all 1 μM)	Trypsin	Kim and Knapp (2001a)

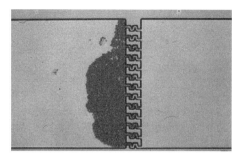

Figure 49 Filtrations of 5.78 μm diameter latex beads in a 500 μm wide Si channel by a slit-type filter (5 μm channels between 73 μm wide posts set across a 500 μm wide, 5.7 μm deep channel). [(Wilding et al., 1998) Courtesy of Elsevier Science.]

the filter area is dictated by the channel area), the fluid flow in the lateral percolation filter is at right angle to the filtering direction. This filter provides a much larger filter area than can be offered by the channel width. Because of the channel depth of 10 μm, even though the channel entrance is blocked by a particle, the liquid is still capable of flowing under the particle. Filtering of 5 μm silica particles, soybean cells (*Glycine max* var. *Kent*), human KB cancer cells, and *Escherichia coli* cells has been achieved with the lateral percolation filter (He et al., 1999).

B. Weir-Type Filters

Other than slit-type filters, weir-type filters have been fabricated to improve cell retention. It was found that a weir-type Si microfilter (3.5 μm gap), which was covered by glass, performed better than the slit-type filter (7 μm gap) to retain white blood cells (WBC) from whole blood. Although the slit-type filter can retain nondeformable latex microspheres, deformable objects such as blood cells squeeze through the gaps (Wilding et al., 1998). Microbeads are shown to be retained by the weirs (dams) around the chambers fabricated in a Si chip (Chiem et al., 2000), a glass chip (Oleschuk et al., 2000), or a PDMS chip (Volkmuth and Austin, 1992).

V-shaped grooves of 9 μm top opening and 14.4 μm in length were fabricated on a Si substrate, which was mechanically pressed to an optically flat Pyrex glass plate for sealing. The channels were used to study rheology (flow dynamics) of WBC. Apparently, the WBCs, after activation by a chemotactic peptide (formyl-methionyl-leucyl-phenylalanine, FMLP), showed a greater resistance to channel passage (Kikuchi et al., 1992).

A barrier (or weir) with a gap of 0.1 μm was fabricated on a Si wafer. After bonding with Pyrex glass, the chip was used as a fluid filter. The barrier was along a V-shaped fluid channel where particle-laden fluid passed from the entry to the exit ports, and the filtered fluid plasma passed over the barrier to a side channel (Brody et al., 1996). The V-shaped geometry of the channel was critical to prevent the surface tension lock. A vertical wall (i.e., 90°), instead of the V-shaped wall (i.e., 55°), would require a much higher pressure (1 atm) to overcome the surface tension present in a 0.1 μm gap (Brody et al., 1996). A similar device was fabricated to retain microspheres (Sohn et al., 2001).

Immunosorbent assay of human secretory immunoglobulin A (s-IgA) (∼200 μg/mL in human saliva) was achieved on polystyrene beads (45 μm dia.), which were introduced in a microchannel (100 μm deep) consisting of a weir (or shallow channel) of 10 μm clearance (see Fig. 50) (Sato et al., 2000). Detection of colloidal gold, which has been conjugated to anti-s-IgA (goat antiserum), was achieved by TLM. The total assay time was shortened from 24 h to less than 1 h. Antigen–antibody reaction was completed in 10 min, which was comparable to liquid-phase (homogenous) immunoassay at room temperature (Sato et al., 2000). A similar strategy has also been used to detect carcino embryogenic antigen (CEA) (Sato et al., 2001a).

HL-60 cells have been docked along a parallel dam fabricated in a quartz chip for cell immobilization. In contrast to a usual perpendicular (or axial) dam structure, the cells docked along a parallel dam suffered less shear stress because the main fluid flow was not reduced (Yang et al., 2002b). According to a mathematical model, at least 40-fold increase in liquid pressure occurred across the particles when immobilized at a perpendicular dam, leading to a large stress. The threshold ATP concentration for Ca uptake (as measured by intracellular fluo 3) was estimated to be within a range of 0.16 ± 0.02 μM. The same batch of docked cells (∼80) was used to respond to four consecutive ATP concentrations (0, 2, 5, and 10 μM). Although the ATP-stimulated Ca uptake reaction was reversible, desensitization of the calcium ion channel occurred after multiple ATP simulations (Yang et al., 2002b).

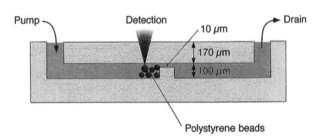

Figure 50 Cross-section of the silica glass microchip for immunosorbent assay. [(Sato et al., 2000) Courtesy of American Chemical Society.]

C. Cell Adhesion

For adherent cells, the retention of cells for measurements is straightforward. For instance, a microphysiometer was constructed to detect the change in acidity (due to lactate and CO_2 formation) of mammalian cells using the light-addressable potentiometer sensor (LAPS) fabricated on a Si chip (Parce et al., 1989).

Various types of adherent cells were grown on a cover slip, which was then laid on top of the LAPS chip. Measurements were made for acidification of (1) normal human epidermal keratinocytes due to epidermal growth factor or organic chemicals and (2) human uterine sacroma cells due to doxorubicin and vincristine (chemotherapeutic drugs). In addition, the inhibition (by ribavirin) of the viral infection of murine fibroblastic L cells by vesicular stomatitis virus could be investigated by following the acidification rate. A limitation of these studies is the requirement for a low-buffered medium (low bicarbonate content) to achieve maximum sensitivity (Parce et al., 1989).

The response of CHO-K1 cells (adherent), which expressed the m1-muscarinic acetylcholine receptors, due to the agonist carbachol was measured using the LAPS chip with eight fluidic channels (see Fig. 51). A cover slip with the adherent cells grown was put on the channels for sealing (Bousse and McReynolds, 1994; Bousse et al., 1993).

Microcontact printing using a PDMS stamp was employed to pattern a contoured (with ridges and grooves) polyurethane film deposited on a Au-coated glass slide. Then, the ridges or plateaus were patterned with a methyl-terminated SAM (hexadecanethiol), where fibronectin would adsorb; whereas the grooves were patterned with triethyleneglycol-terminated SAM, which resisted the protein. Thereafter, bovine capillary endothelial (BCE) cells, which attached only to the fibronectin surfaces, were used for apoptosis studies (Chen et al., 1998; Mrksich et al., 1996).

On the other hand, multiple laminar flow was used to pattern cells and their environments in PDMS chips (Takayama et al., 1999). For instance, two cell types (chicken erythrocyte and *E. coli*) have been shown to deposit next to each other on fibronectin-treated surfaces (see Fig. 52). Moreover, cell retention could occur only at the patterned regions [e.g., cell detachment occurred only to the cells (BCE) that was passed with a patterned stream containing trypsin/EDTA] (Takayama et al., 1999).

Chicken embryo spinal cord neurons have been deposited in a micromachined Si chip coated with a synthetic additive protein (poly-lysine). It was found that groups of neurites have grown toward the channel (50 μm wide) connecting between pits where the neurons were deposited (Martinoia et al., 1999).

Microfluidic Lab-on-a-Chip

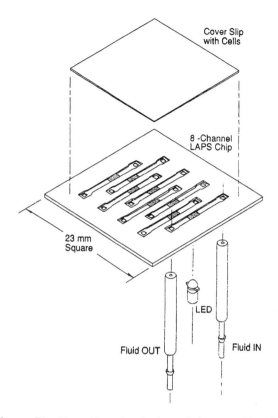

Figure 51 Three-dimensional view of the assembly of an eight-channel LAPS chip with a cover slip, LEDs, and fluid connections. [(Bousse et al., 1993) Courtesy of Institute of Electrical Engineers of Japan.]

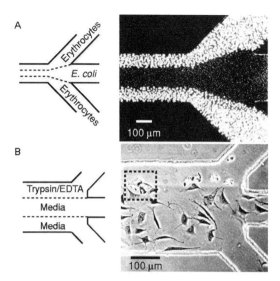

Figure 52 Examples of types of patterns that can be formed by laminar flow patterning in PDMS channels sealed to a polystyrene Petri dish. (A) Two different cell types patterned next to each other. A suspension of chicken erythrocytes was placed in the top and bottom inlets, and PBS in the middle inlet and allowed to flow by gravitational force for 5 min followed by a 3 min PBS wash; this flow formed the pattern of bigger cells (outer lanes). Next, a suspension of E. coli (RB 128) was placed in the middle inlet and PBS in inlets 1 and 3 and allowed to flow by gravitational forces for 10 min followed by a 3 min PBS wash; this flow created the pattern of smaller cells (middle lane). Both cell types adhered to the Petri dish by nonspecific adsorption. Cell were visualized with Syto 9 (15 μM in PBS). Fluorescence micrographs taken from the top through PDMS. (B) Patterned detachment of BCE cells by treatment with trypsin/EDTA. Cells were allowed to adhere and spread in a fibronectin-treated capillary network for 6 h and nonadherent cells removed by washing. Trypsin/EDTA and media were allowed to flow from the designated inlet for 12 min by gravity. Phase-contrast image observed by an inverted microscope looking through the polystyrene Petri dish. [(Takayama et al., 1999) Courtesy of National Academy of Sciences.]

For transport studies, epithelial cells (Madin-Darby canine kidney, MDCK) have been grown into tight monolayers on a permeable polycarbonate membrane (0.4 μm pores, 1–4 mm^2 area) glued in between two micromachined silicon or glass wafers (Hediger et al., 2000). The glue (about 1 μm thick) was applied using a paraffin foil and a rolling procedure to ensure sufficient glue for bonding, yet not too much to block the nanopores on the PC membrane (Hediger et al., 2000).

In another study, mast cells (RBL-2H3), which were grown in a PDMS chip, released histamine upon stimulation by an allergen. The concurrent release of a fluorescent dye (quinacrine) was detected (Matsubara et al., 2001). Moreover, cytochrome C released from mitochondria to cytosol in neuroblastoma–glioma hybrid cells was measured by scanning TLM on a quartz chip (Tamaki et al., 2002).

By using a PDMS channel layer sealed against a glass coverslip, cell adhesive (poly-L-lysine) and nonadhesive (agarose 1%) microdomains can be created on the coverslip (Tokano et al., 2002). For instance, endothelia or astrocytes–neuron cocultures were plated only on the poly-L-lysine microdomains of the treated coverslip.

Subsequently, calcium wave measurements (using Fluo 3) was made to study calcium signaling of cells within confluent cell microdomains and across neighboring, but spatially disconnected, microdomains (Tokano et al., 2002).

D. Studies of Cells in a Flow

Sometimes, cellular analysis was performed with cells in a flow stream. Human blood cells (WBC, RBC) rheology was studied in channels fabricated on the Si–Pyrex substrates. The channels were either uncoated or coated with albumin (Wilding et al., 1994a).

Mouse sperm cells were allowed to swim through a meander channel in a glass (Pyrex or soda lime glass) chip to fertilize an egg to demonstrate (*in vitro*) fertilization (IVF) (Kicka et al., 1995).

Saccharomyces cerevisiae (yeast) cells were mobilized in a microfluidic glass chip using electrokinetic (EK) flow (see Fig. 53) (Li and Harrision, 1997). In addition, erythrocytes were lysed using a detergent (3 mM SDS), then light scattering and video imaging were used to monitor the lysing reaction (Li and Harrision, 1997; McClain et al., 2001a).

However, the use of EK flow to manipulate lymphocytes in microchannels may inactivate the cells because of the pH change occurring at the cell reservoir. This problem can be alleviated by introducing a salt-bridge between the cell reservoir and the electrode (Oki et al., 2001).

Flow cytometry of fluorescent (0.972 μm dia.) and non-fluorescent (1.94 μm dia.) latex particles was achieved on a glass chip using dual-channel laser light scattering and fluorescent measurements (Schrum et al., 1999). A maximum sample throughput of 34 particles was obtained using pinched injection which was essentially the use of electrokinetic focusing (cf. hydrodynamic focusing used in conventional flow cytometry) (Schrum et al., 1999).

Flow cytometry of *E. coli* has been demonstrated on a glass chip. The membrane-permeable nuclei acid stain (Syto15, $\lambda_{em} = 546$ nm) and membrane-impermeable nuclei acid stain (propidium iodide, PI, $\lambda_{em} = 617$ nm) were used to detect viable and nonviable cells, respectively (Culbertson et al., 2001; McClain et al., 2001b).

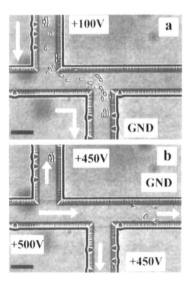

Figure 53 Photomicrographs showing yeast cell transport in a glass chip. White arrows show the direction of solvent flow, and the black bar shows the scale (40 μm). (a) Cell loading vertically downwards. (b) Cell selection to the right. [(Li and Harrision, 1997) Courtesy of American Chemical Society.]

In addition, the fluorescein-labeled ($\lambda_{em} = 520$ nm) polyclonal antibody to *E. coli* was used. Besides using a PDMA surface coating, BSA (1 mg/mL) was added to prevent cell adhesion and cell clumping. Light-scattering (forward) was used to provide information about the cell size (of length 0.7–1.5 μm) of *E. coli*. EK focusing was used to confine the cells. Cell counting rates of 30–85 Hz have been achieved (McClain et al., 2001b).

The measurements of capacitance changes due to varied amounts of polarizable cellular DNA has also been used for flow cytometry studies (Sohn et al., 2000).

An integrated microfabricated cell sorter has been constructed using a control layer and a fluidic layer fabricated on PDMS (Fu et al., 2002b). The control layer consists of valves that will be pneumatically controlled by pressurized N_2. This device is superior to an electrokinetic sorter which suffers from (1) buffer incompatibilities and (2) frequent voltage adjustments because of ion depletion or pressure imbalance due to evaporation.

E. coli cells expressing EGFP were sorted and selected out from a mixture which also contained *E. coli* cells expressing the *p*-nitrobenzyl (PNB) esterase. About 480,000 cells have been sorted at a rate of 44 cells/s in 3 h. The recovery yield was 40% and the enrichment ratio was about 83-fold (Fu et al., 2002b).

Inhibition of rolling of cells (neutrophils) by various anti-cell-adhesion molecules (anti-E-selectin, anti-P-selectin, sialyl Lewis X, Furoidan) was studied in a Y-channel. Comparison can be easily be made between cells with rolling and no-rolling (control) in the two arms of the Y-channel (Salimi-Moosavi et al., 1998).

Cell retention can be achieved when cell movement is balanced by an opposing force. Well-defined vortices formed by opposing EOF and HDF have served to trap beads without the use of a physical barrier (Lettieri et al., 2001).

E. DEP for Cell Retention

DEP has been employed to trap or manipulate cells such as HL-60 (Voldman et al., 2002), Jurkat cells (Grad et al., 2000), and mouse fibroblasts (3T3 cells) (Fuhr and Wagner, 1994).

DEP refers to the force acting on induced polarizations or charge dipoles in a nonuniform (1–20 MHz) electric field of low voltage (1–3 V). Usually, in highly conductive aqueous media, the cells will be less polarizable than the media at all frequencies. The heating generated in this operation can be reduced in the microscale. In addition, an high-frequency (>100 kHz) AC field: (1) eliminates any electrophoretic movement of the cell due to its charged membrane, (2) eliminates the electrochemical reactions (gas formation and electrode corrosion), and (3) mimimizes the alternating voltage imposed upon the

Figure 54 Trapped latex particles, 15 μm diameter, have been confined from a particle jet by DEP (the drive is 1 MHz). Streaming from left to right. [(Ficdlcrr et al., 1998) Courtesy of American Chemical Society.]

resting (static) transmembrane voltage (at 2 V and 20 MHz, the alternating voltage was 12 mV which was 20×-less than the value at DC) (Voldman et al., 2002).

Negative DEP, in which the particles moved toward the field minima, has been employed to trap and sort latex particles (of diameter 3.4–15 μm) and live mammalian cells (mouse L929) (see Fig. 54) (Fiedlerr et al., 1998). Due to the high permittivity of water, polymer particles and living cells show negative DEP at high-field frequencies (10 V across a 20 μm electrode gap or 0.5 MV/m). The repelling force, which is about 0.2–15 pN for latex spheres (diameter, 3.4–15 μm), is sufficient to hold particles at a flow rate up to 10 mm/s (Fiedlerr et al., 1998). DEP (1 MHz) has also been used to stretch λ DNA molecules (48 kbp) (Namasivayam et al., 2002).

The planar quadrupoles, which have been used so far in DEP work, are inadequate to trap the cells against the fluid flow (e.g., 12 μL/min or ~50 pN drag force). So four 50 μm high cylindrical Au electrodes were used which could confine particles over 100× more strongly than planar quadrupole electrodes. A surrogate assay of calcium loading to HL-60 cells was employed to demonstrate the observation of the luminescent dynamics in trapped cells, followed by cell sorting (Voldman et al., 2002). A similar DEP method was also used to study the uptake and cleavage of calcein-AM in a Jurkat cell (Grad et al., 2000).

With regard to the use of an electric field, the microfluidic chip has also been employed for the electrofusion of liposomes (5 μm dia.) and RBC (Strömberg et al., 2001), and for electric field-flow fractionation (EFFF) to separate nanoparticles (Lao et al., 2002).

IX. APPLICATIONS TO DNA ANALYSIS

DNA amplification (mostly by polymerase chain reaction, PCR) and other subsequent DNA analysis (including hybridization, sequencing, and genotyping) have been facilitated by the use of the microfluidic chip. These applications are then described in details in subsequent sections.

A. DNA Amplification

Much effort has been made to integrate PCR chambers to carry out amplifications of DNA molecules prior to their analysis. PCR has been achieved on a Si-based reaction chamber (25 or 50 μL) integrated with a polysilicon thin-film (2500 Å thick) heater for the amplification of the GAG gene sequence (142 bp) of HIV (cloned in bacteriophage M13) (Northrup et al., 1993).

PCR of bacteriophage λ DNA (500 bp) was performed in a 10 μL chamber fabricated in a Si–Pyrex PCR chip. Subsequently, off-chip agarose gel electrophoresis was performed and it indicated that PCR on chip was not as efficient as conventional PCR (Wilding et al., 1994b).

Surface passivation on the PCR chip was needed to avoid inhibition of the enzyme (polymerase) and adsorption of DNA (template and product). It was found that native silicon is an inhibitor of PCR and an oxidized silicon (SiO_2) surface was required to provide the best passivation (Albin et al., 1996).

On-chip PCR benefits from the high surface-to-volume ratio (SVR) of the PCR chamber. For instance, SVR of Si chambers are 10 (Wilding et al., 1994b) and 17.5 mm²/μL (Cheng et al., 1996a), compared with 1.5 mm²/μL in conventional plastic reaction tubes and 8 mm²/μL in glass capillary reactive tubes (Cheng et al., 1996a; Shoffner et al., 1996).

PCR of β-globin target (cloned in M13) (268 bp) and of genomic *Salmonella* DNA (159 bp) as well as their CGE analysis were integrated. First, the target DNA was amplified in a microfabricated Si PCR reactor (~20 μL) heated by a polysilicon heater (3000 Å Si doped with boron). Then the PCR products were directly injected into a glass CE chip for CGE separations. The heating and cooling rates are 10°C/s and 2.5°C/s, respectively, compared with typical rates of 1°C/s in conventional thermal cyclers (Woolley et al., 1996).

Anti-*Taq* DNA polymerase antibody (Hot-start PCR), which inhibited the *Taq* polymerase before PCR reagents attained a high temperature, was employed to reduce the loss of *Taq* polymerase due to nonspecific binding. This strategy produced a more consistent and higher yield than that obtained in the conventional PCR reaction tube and that obtained in the PCR chip (without Hot-start) (Cheng et al., 1996a).

PCR results of the human CFTR (~100 bp) that was amplified directly from isolated human lymphocytes were compared with those from human genomic DNA extracted from the cells. The results indicated that tedious DNA extraction was not necessary (Cheng et al., 1996a).

PCR and subsequent CGE separation of λ phage DNA were integrated on a glass microchip (see Fig. 55). An on-chip DNA concentration device was included to reduce the

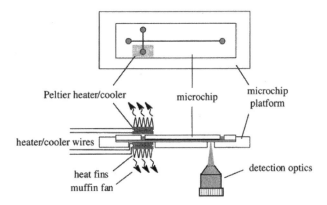

Figure 55 Schematic of the dual Peltier assembly for rapid thermal cycling of PCR followed by CE analysis. [(Khandurina et al., 2000) Courtesy of American Chemical Society.]

analysis time to 20 min (by decreasing the number of thermal cycles required to 10 cycles), and the starting DNA copy number to 15 (0.3 pM) (Khandurina et al., 2000).

A polymeric coating [poly(vinylpyrrolidone)] has been formed in glass microchips for robust CGE analysis of PCR products containing a high salt content (Munro et al., 2001). Among other coatings tested, this coating, which was applied after silianization (using trimethylsilane, TMS) of the channel wall, has produced the least degradation after 635 analysis (Munro et al., 2001).

The use of a thin-film heater permitted the PCR cycle time as fast as 30 s in a 280 nL PCR chamber. Stochastic amplification has been demonstrated using one to three templates. The PCR mixture was prevented from flowing into the CE channel owing to the passive barrier formed by the HEC sieving medium (Lagally et al., 2001).

PCR amplification and subsequent hybridization of a DNA sequence in E. coli (221 bp) and the *Enterococcus faecalis DNA E* gene (195 bp) have been achieved in a disposable polycarbonate (PC) chip (Liu et al., 2002). The higher transition temperature of PC (145°C) allows the thermal cycling at high temperature possible (Liu et al., 2001e). The chip has a well-positioned window (1.7 cm^2) to provide access for surface treatment and oligonucleotide probe attachment. Afterwards, the window was enclosed with a second PC piece using adhesive tape and epoxy (Liu et al., 2002). In order to bypass a post-amplification step for denaturing of dsDNA, asymmetrical PCR was employed. This approach utilized strand-biasing to preferentially amplify the target strand of choice. Hybridization was performed on a microarray spotted onto a bybridization channel (7 µL) downstream of the PCR chamber. To evaluate the probe attachment and hybridization chemistry, the probes were Cy3-labeled and the targets were Cy5-labeled. While the normal approach is not to label the probe, this strategy allows the visualization of probes even though no hybridization has occurred (Liu et al., 2002).

Alternative methods have been reported for thermal cycling in PCR. PCR, which depends on thermal cycling, has been achieved by moving the PCR mixture in a glass chip through three separate heating zones for melting, annealing, and extension (see Fig. 56) (Chou et al., 2001; Kopp et al., 1998a, b; Soper et al., 2001). A 20-cycle PCR amplification of the gyrase gene (176 bp) of *Neisseria gonorrhoeae* was performed at flow rates of 5.8–15.9 nL/s, corresponding to cycling times of 18.8–1.5 min for 20 cycles (Kopp et al., 1998a, b).

IR-mediated temperature control has been employed for PCR. Because the chip material (polyimide) does not absorb IR, the low thermal mass of the solution allows for fast thermal cycling, and 15 cycles have been achieved in 240 s! (Giordano et al., 2001a)

Furthermore, integrated PCR devices have been constructed. An integrated DNA analysis device has been constructed on a Si–glass chip to perform various tasks such as measuring of nanoliter-size reagents and DNA samples, solution mixing, DNA amplification, DNA digestion, CGE separation, and fluorescent detection. Both the heater (boron-doped) and temperature sensor (diode photodetector with the TiO_2/SiO_2-interference filter) are also integrated. Hydrophobic regions are present for fluid control. The only external electronic component not integrated with the chip is a blue-light emitting diode (Burns et al., 1998).

In a dual-function Si–glass microchip, the isolation of WBC from whole blood (3.5 µL) using weir-type (3.5-µm gap) filters was followed by PCR of the exon 6 region of the dystrophin gene (202 bp) (Wilding et al., 1998). After removal of RBC (hemoglobin in RBC is a PCR inhibitor), the DNA from the retained WBC was released by the initial high-temperature denaturation step (94°C) (Wilding et al., 1998).

Real-time PCR for DNA amplification and detection was facilitated using the microfluidic chip. PCR of β-actin DNA (294 bp) has been integrated with real-time detection at 518 nm using a fluorescent reporter probe (Albin et al., 1996). Real-time monitoring of PCR amplification was achieved by sequential CGE after 15, 20, 25, and 30 s (see Fig. 57) (Woolley et al., 1996).

Bacillus subtilis spore samples (10^4 mL^{-1}) were sampled, filtered, and sonicated in the presence of 6 µm glass beads for spore disruption, and the released DNA underwent real-time PCR in a plastic microfluidic cassette (Taylor et al., 2001). Real-time PCR was performed by on-chip thermal cycling for the detection of Hantavirus (430 bp), *Borrelia burgdorferi*, human β-actin (294 bp), HIV, orthopoxviruses (266–281 bp), and the human complement C6 gene (73b) (Ibrahim et al., 1998).

For complex samples, multiplex PCR has been carried out. On-chip multiplex PCR was achieved on four regions

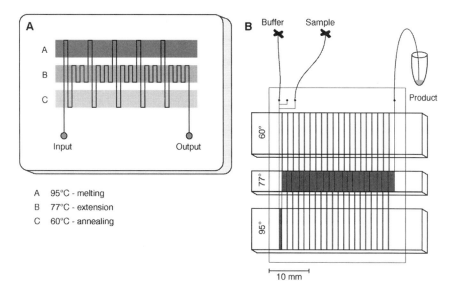

Figure 56 (a) Schematic of a chip for flow-through PCR. Three well-defined zones are kept at 95°C, 77°C, and 60°C by means of thermostated copper blocks. (b) Layout of the device used in this study. The glass chip incorporates 20 identical cycles, except that the first one includes a threefold increase in DNA melting time. [(Kopp et al., 1998a) Courtesy of American Association for the Advancement of Science.]

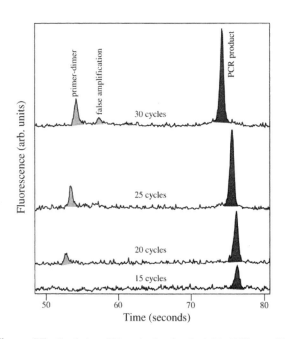

Figure 57 Real-time CE analysis of a β-globin PCR amplification using an integrated PCR-CE microdevice. Chip CE separations of the same sample were performed sequentially in the integrated chip after 15, 20, 25, and 30 cycles at 96°C for 30 s and 60°C for 30 s. The PCR product peak is shaded with dark gray; the false amplification and primer-dimer peaks are shaded with light gray. [(Woolley et al., 1996) Courtesy of American Chemical Society.]

in the bacteriophage λ DNA (199 bp and 500 bp), *E. coli* genomic DNA (346 bp), and *E. coli* plasmid DNA (410 bp). Subsequent CGE separation was performed downstream after the dye (TO-PRO) was added to the PCR reservoir (Waters et al., 1998a).

Lysis of *E. coli*, multiplex PCR amplification and electrophoretic sizing were executed sequentially on a glass chip (Waters et al., 1998b). Multiplex PCR was achieved for five regions of the genomic and plasmid DNA of *E. coli*, and four regions in the λ DNA (Waters et al., 1998a). Subsequent CGE separation was performed, and the 3% PDMA sieving medium appeared to be stable during thermal cycling. The PCR reaction mixture did not appear to adversely affect the CGE analysis of the PCR products (Waters et al., 1998b).

Multiplex PCR of the muscular dystrophin gene was performed on-chip after using the degenerate-oligonucleotide primed PCR (DOP-PCR) for increasing the number of DNA templates. Therefore, DOP-PCR can provide the template DNA from the whole human genome. But the procedure is feasible only when the amplicon size is less than 250 bp (Cheng et al., 1998c).

Research on the integration of PCR and DNA analysis has been very active. For comparison, the chip designs, PCR conditions, and PCR mixtures are tabulated in Tables 14, 15 and 16, respectively.

Besides PCR, there are other DNA amplification techniques employed on-chip. An isothermal cycling probe

Table 14 Fabrication of On-Chip PCR Chambers

Volume	Depth	Material	Bonding	Surface treatment	References
50 μL	500 μm	Si–glass	Silicone rubber	Silianization	Northrup et al. (1993)
10 μL	115 μm	Si–Pyrex	Anodic	Thermal oxide 1000 Å	Cheng et al. (1996a) and Shoffner et al. (1996)
5 μL		Si–Pyrex	Anodic	Thermal oxide 4000 Å	Albin et al. (1996)
20 μL		Si	Polyimide	Thin-walled polypropylene liner	Woolley et al. (1996)
10–20 μL	400 μm	Glass	Potassium silicate	BSA (2.5 mg/mL) treatment	Khandurina et al. (2000)
280 nL	42 μm	Glass	Thermal	coated	Lagally et al. (2001)
38 μL	250 μm	PC	Thermal (139°C), pressure (2 ton)		Liu et al. (2002)
20 μL	1 mm	PC	Thermal (134°C), pressure (4 ton)	UV	Liu et al. (2001e)
		plastic	Resin cured at 105°C for 50 min		Yu et al. (2000)
12 μL		Si–glass	Anodic	Thermal oxide 2000 Å	Cheng et al. (1998c)
1.7 μL		polyimide			Giordano et al. (2001a)

technology (CPT) reaction has been achieved on-chip to perform DNA amplification without thermal cycling (Tang et al., 1997). This technique, which gives linear amplification, works by producing an increased amount of the cleaved chimeric probe for detection. However, the template DNA does not increase in amount. Detection of the 391 bp DNA used for diagnosis of the Newcastle disease has been achieved by measuring the amount of the labeled 12-mer fragment (dA_{10}-rA_2) resulting from the cleavage of the chimeric 24-mer probe (dA_{10}-rA_4-dA_{10}) which binds to the 391-mer target (Tang et al., 1997).

The DNA amplification based on strand displacement amplification (SDA) for the 106-bp product from *Mycobacterium tuberculosis* was reported (Burns et al., 1998).

Ligase chain reaction (LCR) was achieved on a Si–glass chip (containing a chamber of 10 μL) for detecting known point mutation in the *LacI* gene sequence (Cheng et al., 1996b). The yield from the chip was found to be 44% of that obtained from using the conventional reaction tube, possibly because of nonspecific binding of both the template DNA and ligase enzyme to the chip surface (Cheng et al., 1996b). The LCR method was also used to detect K-ras mutations (Soper et al., 2001).

Amplifications involving mRNA were also carried out. Viable *Cryptosporidium parvum* in water treatment plants was detected after amplification of the heat-shock mRNA (103-mer) by nuclei-acid-sequence-based amplification (NASBA). A sandwich hybridization assay was subsequently conducted within a PDMS channel (Esch et al., 2001). Purification of mRNA with subsequent single strand cDNA synthesis by reverse transcription were all achieved on the same chip (Harrison et al., 2001; Jiang et al., 2000)

B. DNA Hybridization

Hybridization between complementary DNA strands is generally adopted for the detection of specific DNA sequences.

Dynamic DNA hybridization was performed by pumping fluorescently labeled DNA probes through microchannels containing target-bearing paramagnetic polystyrene beads. Simultaneous interrogations of four DNA targets (in duplicates) consecutively by five probes was achieved in eight microchannels within a microfluidic chip. A second hybridization has been performed (at 37°C) by a new probe DNA, after the thermal denaturation (87°C) of the duplex formed by the old probe DNA. Four DNA targets [from mouse m-actin (50-mer), *Bacillus subtilis* (30-mer), universal bacterial probe (27-mer), *E. coli* (25-mer), *Staphylococcus aureus* (25-mer)] were immobilized on paramagnetic beads. These were achieved either by interaction between the streptavidin-coated beads and biotinylated DNA targets (25–30-mer) or by base-pairing between $(dT)_{25}$ oligonucleotide-beads and poly-A-tailed target DNA (Fan et al., 1999).

Hybridization of the DNA target to the biotinylated DNA probe retained on streptavidin-coated microbeads (15.5 μm dia.) was achieved in three serial microchambers. The chamber array was defined by weirs and by hydrogel [poly(ethyleneglycol) diacrylate] plugs (Seong et al., 2002). Subsequently, fluorescently labeled targets were electrophoretically transported to the chamber

Table 15 Conditions for On-Chip PCR

Cycling temperatures and times	Cycle time	Heating rate (°C/s)	Cooling rate (°C/s)	Heating methods	References
Initial 50 min at 95°C; 95°C (15 s), 55°C (15 s), 72°C (60 s); final 72°C for 5 min	3 min/cycle	7	6	Peltier heater/cooler	Wilding et al. (1994b)
96°C (30 s) 54°C (30 s) 72°C (120 s) or 94°C (60 s); then 94°C (15 s), 60°C (15 s), 72°C (60 s), 72°C for 5 min		5	5	Peltier	Albin et al. (1996)
96°C (2 s), 55°C (5 s), 72°C (2 s); final 72°C for 30 s for β-globin DNA	15 min	10	2.5	Polysilicon heater 3000 Å doped with boron	Woolley et al. (1996)
95°C (10 s), 56°C (15 s), 72°C (20 s) for *Salmonella* DNA	39 min				
Initial 94°C for 1 min; 94°C (15 s), 55°C (60 s); 72°C (60 s); final 72°C for 10 min for *C. jejuni* bacterial DNA	1 min/cycle 28 cycles	13	35	Polysilicon (2500 Å) heater Peltier heater/cooler	Northrup et al. (1993) Cheng et al. (1996a)
Initial 94°C for 6 min; 94°C (30 s), 53°C (30 s), 65°C (120 s); final 65°C for 5 min for human genomic DNA	35 cycles				
94°C for 2 min to induce cell lysis, 94°C (2 min), 50°C (3 min), 72°C (4 min); final 72°C for 7 min	25 cycles			Commercial thermal cycler	Waters et al. (1998b)

(*continued*)

Table 15 *Continued*

Cycling temperatures and times	Cycle time	Heating rate (°C/s)	Cooling rate (°C/s)	Heating methods	References
94°C (30 s), 50°C (20 s), 72°C (25 s) (25 cycles)	1.25 min/cycle	2	3–4	Peltier heater/cooler	Khandurina et al. (2000)
95°C (60 s); 94°C (5 s), 64°C, touch-down (in 2°C increments) to 50°C (15 s), 72°C (10 s)	15 min/30 cycles	10	10	1 cm² resistive heater N₂ cooling	Lagally et al. (2001)
94°C (70 s); 36 cycles of 94°C (20 s), 55°C (20 s); 72°C (20 s)	19 s/cycle	7.9	~4.6	Peltier	Liu et al. (2002)
94°C (2 min), 37°C (3 min), 72°C (4 min), 72°C for 7 min	24 cycles			Commercial thermal cycler	Liu et al. (2001e)
94°C (5 min); 30 cycles of 94°C (10 s), 63°C (60 s), 65°C (60 s); 65°C (10 min)				Peltier heater/cooler	Wilding et al. (1998)
94°C (4 min); 94°C (2 min) 50°C (3 min) 72°C (4 min), 24 cycles; 72°C (7 min)				Commercial thermal cycler	Waters et al. (1998a)
94°C (60 s), 50°C (30 s), 72°C (30 s)		2.4	2	Resistive heater plus Peltier heater/cooler	Yu et al. (2000)
94°C (10 s); 15 cycles of 94°C (2 s), 68°C (2 s), 72°C (2 s); 72°C (10 s)		10	10	IR heating	Giordano et al. (2001a)

Table 16 Composition of PCR Mixtures

Primer conc.	Nucleotide conc.	Polymerase	Amount of DNA template	Additives for surface modifications[a]	Volume of mixture	References
400 μM	200 μM	Taq 1.25 U	10^8 copy		50 μL	Northrup et al. (1993)
	200 μM	Taq 1.0 U	0.2 ng		10 μL	Wilding et al. (1994b)
0.3 μM	200 μM	Taq 2.5 U	1 ng			Shoffner et al. (1996)
0.5 μM	400 μM	Taq 2.5 U	500 ng (4×10^5 copies)/Salmonella DNA		50 μL	Woolley et al. (1996)
0.6 μM	200 μM	Taq 0.6 U	1.2 ng C. jejuni DNA		12 μL	Cheng et al. (1996a)
0.6 μM	200 μM	Taq 0.4 U	125 ng Human genomic DNA			
0.6 μM	200 μM	Taq 0.6 U	1.2 ng C. jejuni DNA			
0.2 μM	200 μM	(Taq Start antibody 132 ng) Taq 0.4 U	Human lymphocytes (1500 and 3000 cells)			
1 μM	200 μM	Taq 2.5 U	10^8 copies	Zwitterion buffer (tricine) and nonionic surfactant (Tween 20) as dynamic coating	10 μL	Kopp et al. (1998a)
1 μM	200 μM	Taq 25 U/mL	E. coli cells	0.001% Gelatin, 250 μg/mL BSA	10–25 μL	Waters et al. (1998b)
0.1–1 μM	200 μM	Taq 25 U/mL	10 ng/mL λ Phage DNA	0.001% Gelatin, 250 μg/mL BSA	6 μL	Khandurina et al. (2000)
0.2 μM	200 μM	Taq 2.5 U/50 μL	200 Copies (M13/pUC, 136 bp)/50 μL 1000 Copies (S65C, 231bp)/50 μL	1.5 μM BSA	280 nL	Lagally et al. (2001)
0.25 μM						
1.0 μM	200 μM	Taq 25 U/mL	10 ng/mL	0.001% Gelatin, 250 μg/mL BSA	20 μL	Liu et al. (2001e)
reverse primer, 1.2 μM forward primer, 12 nM	125 μM	Taq 25 U/mL	50 pg/μL	0.001% Gelatin, 250 μg/mL BSA	38 μL	Liu et al. (2002)
0.5 μM	2 mM	Taq 10 U 2.2 μg Taq start antibody	White blood cells	Possibly proteins from white blood	50 μL	Wilding et al. (1998)
1.0 μM	200 μM	Taq 25 U/mL	10 ng/mL λ DNA white E. coli cells	0.001% Gelatin, 250 μg/mL BSA	12 μL	Waters et al. (1998a)
1.0 μM	200 μM	Taq 25 U/mL	E. coli cells (346 bp product)	0.001% Gelatin, 250 μg/mL BSA	1.5 μL	Yu et al. (2000)
80 μM	2.5 mM	Taq 1.2 U Taq Start antibody 132 ng	9.6 ng Human genomic DNA	0.1% Triton X-100, 0.01% gelatin	12 μL	Cheng et al. (1998c)
0.2 μM	200 μM	Taq 10 U	0.1 ng λ Phage DNA (500 bp)	0.001% gelatin, 0.75% (w/v) PEG	50 μL	Giordano et al. (2001a)

[a] To prevent adsorption of PCR polymerase or nuclei acids, surface treatment is necessary.

consecutively through the hydrogel plugs. Detection was achieved by fluorescent imaging of the fluorescein-labeled DNA targets which were hybridized to the probe immobilized on the beads. Hybridization was 90% complete within 1 min. After hybridization, the targets in a particular chamber could be released for subsequent analysis after denaturation using a pump-driven flow of 0.1 N NaOH into that chamber. Therefore, multiple analysis could be performed using the same bead set. However, there was degradation of performance because extended exposure to high-pH solution will lead to (1) shrinkage of the hydrogel plugs and hence leakage of the targets and (2) deterioration of streptavidin and hence loss of the probes (Seong et al., 2002).

Moreover, DNA probes were directly immobilized onto hydrogel plugs formed in a PC chip. Acrylamide-modified 20-mer oligomers were incorporated in the hydrogel during its photopolymerization. The hydrogel was porous to fluorescently tagged DNA target for hybridization under an electrophoretic flow (Olsen et al., 2002).

C. DNA Sequencing

Because of the need of fast and high-throughput DNA sequencing in the human genome project, research in these areas using the microfluidic chip has been very active.

DNA sequencing of M13mp18A has been achieved on a glass chip using a denaturing sieving medium (9% T, 0% C polyacrylamide). Using the one-color detection, separation to 433 bases was achieved in 10 min. Using the four-color detection, separation to 150 bases was achieved in 540 s. In contrast, the sequencing rates by slab-gel and CE are 50–60 and 250–500 bases/h, respectively. However, 12× more primers and templates are needed for the sequencing reaction performed in the CE chip compared with conventional CE (Woolley and Mathies, 1995).

The separation matrix [denaturing 3–4% linear polyacrylamide (LPA)], separation temperature (35–40°C), channel length (7.0 cm), channel depth (50 μm), and injector parameters (100 or 250 μm double-T injector, 60 s loading time) have been optimized for DNA sequencing on-chip. This facilitates the one-color detection of separation of 500 bases of M13mp18 ssDNA in 9.2 min and four-color detection of 500 bases in 20 min (Liu et al., 1999).

Further optimization of DNA sequencing of M13mp18 DNA (up to 550 bases) on a microfabricated glass chip was achieved on various parameters, such as separation buffer composition (no borate), sieving polymer concentration (>2% LPA), device temperature (≤50°C), electric field strength (lower E for longer bases), and channel length (>6 cm) (Salas-Solano et al., 2000).

With a longer CE channel (40 cm), the average DNA read length (four-color-detection) has been successfully increased to 800 bases in 80 min (98% accuracy) for M13mp18 DNA (or the DNA with a 2 kb human chromosome 17 insert). For comparison, a commercial capillary-based DNA sequencer took 10 h to complete the same 800 bases analysis (Koutny et al., 2000) DNA sequencing up to 700 bp (with 98% accuracy) has also been achieved in 40 min using a 18 cm plastic microchannel (Boone et al., 2000).

The dependence of various parameters, such as selectivity, diffusion, injector size, device length, and channel folding, on the resolution of the LPA sieving medium for ssDNA separation was investigated. Separations of 400 bases in under 14 min at 200 V/cm, and of 350 bases in under 7 min at 400 V/cm ($R \geq 0.5$) were achieved (Schmalzing et al., 1998).

A low-viscosity gel capture matrix, containing an acrylamide-copolymerized oligonucleotide, was loaded into a 60 nL capture chamber to purify DNA sequencing products prior to DNA sequencing on a chip. For comparison, conventional sample purification or cleanup has been performed by ethanol precipitation of DNA (Paegel et al., 2002). The capture oligonucleotide, which was a 20 base complementary sequence, bound to the sequencing products, whereas chloride, excess primer, and excess DNA template were unretained and were washed off. The process (at 50°C) took only 120 s to complete. The sequencing products were then released at 67°C and subsequently introduced into a 15.9 cm CE microchannel. CGE separation completed in 32 min to produce the sequencing information up to 560 bp (Paegel et al., 2002).

D. High-Throughput DNA Analysis

The microfluidic chip uniquely facilitates high-throughput DNA sequencing or analysis. For instance, 12 different samples were analyzed in parallel on a capillary array electrophoresis (CAE) chip. The two-color method was used for the multiplex fluorescent detection of the HLA-H DNA (on-column labeled with thiazole orange, emitting in green) and the sizing ladder, pBR332 Msp1 (prelabeled with butyl TOTIN, emitting in both green and red) (Woolley et al., 1997). A reciprocating confocal fluorescence scanner was used to detect 12 channels at a rate of 0.3 s per scan (Woolley et al., 1997). Thereafter, the use of the 13-channel (Fan et al., 2001) or 16-channel chips (Liu et al., 2000e) for DNA sequencing were also reported.

The research on high-throughput genetic analysis moved on, and a 10 cm glass wafer with 48 separation channels (but with 96 samples) was used to perform DNA sequencing in less than 8 min, which translated to <5 s per sample (see Fig. 58). The variants of the *HFE* gene, which was correlated with hereditary hemochromatosis (HHC, iron storage disorder), were detected using

Microfluidic Lab-on-a-Chip

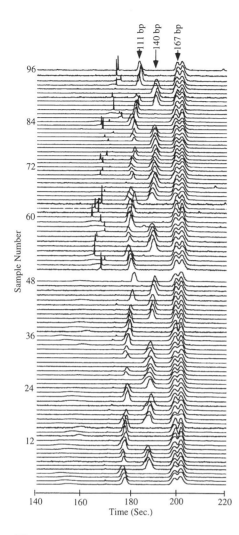

Figure 58 Electropherograms generated from 96 concurrent separations obtained in glass chip with 48 channels. The various samples show the genotyping of the *HFE* gene. The 845G type show a single peak at 140 bp; the 845A type shows a single peak at 111 bp; the heterozygote type exhibits both the 140 and 111 bp peaks. [(Simpson et al., 1998) Courtesy of National Academy of Sciences.]

DNA which was isolated from peripheral blood leukocytes (Simpson et al., 1998).

Thereafter, 96 channel CAE was achieved on a microchip by adopting a radial design in the channel arrangement. The channels shared a common anode reservoir at the center of a 10 cm glass microplate (see Fig. 59). Grouping the anode, cathode, and waste reservoirs reduced the number of reservoirs per plate to 193. A four-color rotary confocal fluorescence scanner was used for detection. Loading of 96 samples from a 96-well microplate was achieved using a capillary array loader. CGE separation of the pBR322 restriction digest samples was achieved in 120 s (Shi et al., 1999).

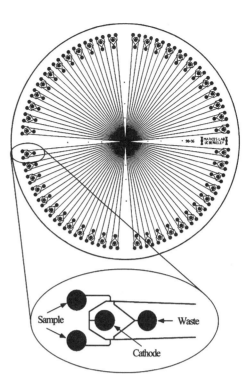

Figure 59 Mask pattern for the 96-channel radial capillary array electrophoresis microplate (10 cm in diameter). Separation channels with 200 μm double-T injectors were masked to 10 μm width and then etched to form 110 μm wide by ∼50 μm deep channels. The diameter of the reservoir holes is 1.2 mm. The distance from the injector to the detection point is 33 mm. [(Shi et al., 1999) Courtesy of American Chemical Society.]

To date, 384 capillary lanes have been incorporated into a 20 cm diameter glass wafer to perform simultaneous genotyping. CGE of DNA samples from 384 individuals was performed on a 8 cm long channel in only 325 s (Emrich et al., 2002). Another 384 lane plate for DNA sequencing has also been reported (Ehrlich et al., 2001).

E. Other DNA Applications

There are other DNA applications. For instance, DNA analysis after restriction enzymatic digestion was performed. Digestion of pBR322 (125 ng/μL) by *HinfI* (4 U/μL) and subsequent CE analysis were integrated in a chip at room temperature (20°C). Fluorescent detection of the restriction fragments was achieved using an intercalating dye TOTO-1 (1 μM) (Jacobson and Ramsey, 1996).

Plasmid DNA (supercoiled Bluescript SK) was digested by *TaqI* restriction enzyme in a Si–glass device with two channels. The flow streams of DNA and enzymes were first introduced in individual channels, and then moved to mix and stopped to react (at 65°C for 10 min) (Burns et al., 1996).

Moreover, single-molecule DNA analysis was performed. Single-molecule DNA fragments (2–200 kbp) were sized on a PDMS chip based on the differences in the molecule size (length), but not in the electrical mobility (Chou et al., 1999). The method (3000 molecules analyzed in 10 min) is 100× faster than pulsed-field gel electrophoresis. DNA fragments obtained from λ-phage DNA were digested with *HindIII* or ligated with the T4 ligase. Fragment lengths were measured by the amount of intercalating dye, YOYO-1 (1 molecule per 4 bp), using quantitative fluorescence. For instance, 28 fg of DNA (about 3000 molecules) were detected in a volume of 375 fL (Chou et al., 1999).

Single λ-DNA molecules (41 pM) were detected as bursts on-chip using the intercalating dye YOYO-1 (Effenhauser et al., 1997b). Single molecule detection of λ-DNA (310 fM) and other small DNA molecules was also performed on PC or PMMA chips, using TOPRO-5 (Wabuyele et al., 2001).

With micromachining, a physical sieving matrix was designed for the analysis of long DNA molecules. An array of microfabricated microposts facilitated the study of the electrical mobility of long DNA molecules under an electrical field in a well-controlled sieving matrix (unlike the polymeric gel matrix) (Volkmuth and Austin, 1992). An array of 0.15 μm high posts (of 1.0 μm diameter and 2.0 μm spacing) was microfabricated on a Si wafer bonded to Pyrex glass. The effective pore size of 1.0 μm corresponds roughly to a 0.05% agarose gel which is never realized because it is physically unstable. The DNA used in the study was a highly purified 100 kb DNA (∼30 μm contour length) from *Micrococcus luteus* (Volkmuth and Austin, 1992). Tribranched DNA molecules (from bacteriophage λ DNA) were also studied in a microfabricated micropost array (Volkmuth et al., 1995). A similar array was also fabricated on Si (using an excimer laser) to sort DNA on the basis of diffusion (Cabodi et al., 2001).

Moreover, a microfabricated entropic trap array was constructed for separating long DNA molecules. The channel comprised narrow constrictions (of width 75–100 nm) and wider regions (of width 1.5–3 μm) that allowed size-dependent trapping and separation of two DNA molecules (37.9 and 164 kbp) to occur in 15 min, compared with 12–24 h achieved in pulse field slab-gel electrophoresis. Surprisingly, it was found that a longer DNA molecule had a higher probability to escape the constriction and hence had a greater mobility (Han and Craighead, 2000).

X. APPLICATIONS TO PROTEIN ANALYSIS

In proteomics studies, research on protein analysis is very important for cellular protein functional assay and clinical diagnostics. Once again, the microfluidic technology plays an essential role in protein assays. Immunoassay, protein separation, enzymatic assay, and mass spectrometric analysis will be described in detail in subsequent sections.

A. Immunoassay

Immunoassay is generally classified into the homogeneous and the heterogeneous formats.

1. Homogeneous Immunoassay

Homogeneous immunoassay is based on the CE separation of the labeled antibody (Ab) and the antibody–antigen (Ab–Ag) complex.

In competitive homogeneous immunoassay, separation (in 30 s) and quantitation (1–60 μg/dL) of free and bound labeled antigen (cortisol) were carried out in a fused silica chip. As the Ab–Ag complex was not detected, an internal standard (fluorescein) was added to aid quantitation. In addition, as most of the total cortisol in serum was bound, a releasing agent: 8-anilino-1 naphthalenesulfonic acid (ANS) should be added (Koutny et al., 1996).

A robust assay for theophylline was established within the clinical range of 10–20 μg/mL (Chiem and Harrison, 1997). As the Ab–Ag complex can be mobilized, no additional internal standard was needed (Chiem and Harrison, 1997), as in an immunoassay study for cortisol (Koutny et al., 1996).

Separation of Cy5-BSA from unreacted Cy5 and the complex that was formed from Cy5-BSA and anti-BSA was achieved in a flow-through sampling chip (Chen et al., 2002).

Immunoassay of goat IgG was also achieved on a PMMA chip (Martynova et al., 1997). The PMMA chip should be placed on a piece of electrically grounded aluminum foil to reduce the electrostatic charges acquired by the plastic surface, otherwise the fluid flow by voltage control could not be performed. PMMA was selected because it was reported to be the least hydrophobic of the commonly available plastic materials (Martynova et al., 1997). Competitive immunoassay for BSA was demonstrated by performing a CE separation on-chip (Harrison et al., 1996).

Separation of antigen and Ab–Ag complex was achieved by dialysis with a polymeric microfluidic chip containing a PVDF dialysis membrane (Jiang et al., 2001). Simultaneous immunoassays for ovalbumin and anti-estradiol were performed on a six-channel microfluidic device within 60 s. The limit of detection of anti-estradiol ranged from 4.3 to 6.4 nM (see Fig. 60) (Cheng et al., 2001).

A human serum albumin (HSA) assay was performed on a T-sensor (on a Si–Pyrex chip) in a flow stream

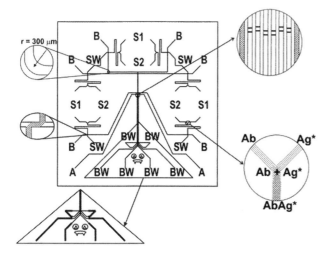

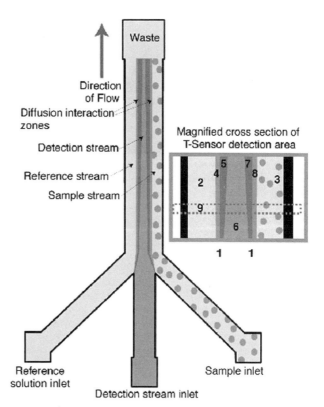

Figure 60 Illustration of the overall layout of the flow channel manifolds in the glass SPIDY device. Each reaction cell has reservoirs for sample (S1), antibody (S2), sample injection waste (SW), and running buffer (B). Six reaction cells are replicated around the outside of the wafer. Then expansion of the left illustrates the double-T injector design for loading sample from the reactor into the separation column. In the bottom right expansion, the antibody (Ab) and antigen (Ag) mixer in the reaction cell is illustrated. An expansion in the bottom depicts the pattern of channels at the buffer waste (BW) reservoirs used during separations. The thicker lines identify channel segments that are 300 μm wide, while the thinner lines represent 50 μm wide segments. The expansion in the upper right of the detection zone, across which a laser beam is swept, illustrates a separation of two components in the six separation channels. Two optical alignment channels filled with fluorescent dye sandwich the separation channels. [(Cheng et al., 2001) Courtesy of American Chemical Society.]

Figure 61 Schematic of flow and diffusion within the T-sensor at a 1:1:1 flow ratio. A reference solution enters the device from the left, a detection solution from the middle, and a sample stream enters from the right. The magnified schematic shows (1) original flow boundaries, (2) reference stream, (3) particle-laden sample stream, (4) diffusion of detector substance into reference stream, (5) diffusion of reference analyte into detector stream, (6) detection stream, (7) diffusion of sample anlyte into detection stream, (8) diffusion of detector substance into sample, and (9) detector cross-section. [(Weigl and Yager, 1999) Courtesy of American Association for the Advancement of Science.]

with an internal control (see Fig. 61) (Kamholz et al., 1999; Weigl and Yager, 1999). The left reference solution (1 μM HSA), center indicator solution (albumin Blue, AB580: $\lambda_{ex} = 580$ nm, $\lambda_{em} = 606$ nm), and right sample solution (HAS, 0, 2, 4, 6, 8 μM) interacted based on diffusion. Simultaneous measurements of the reference and sample solutions were feasible (Weigl and Yager, 1999; Weigl et al., 1998; Yager et al., 1998). The indicator, AB580, which has a high affinity for serum albumin, but low for other proteins, produced an enhanced fluorescence upon binding to HSA (Kamholz et al., 1999).

2. *Heterogeneous Immunoassay*

In a heterogeneous immunoassay format, a dinitrophenyl (DNP)-conjugated lipid bilayer was coated on a 12-channel glass–PDMS chip, and 12 concentrations (0.15–13.2 μM) of anti-DNP IgG antibodies (labeled with Alexa fluor 594) were analyzed (Yang et al., 2001). This enabled an entire binding curve to be obtained in a single experiment. Total internal reflection fluorescence microscopy (TIRFM) was used for detection (see Fig. 62) (Yang et al., 2001).

For the immunoassay of IgG, physisorption of goat anti-mouse IgG (biotin-conjugated) was first made to the PDMS surface. Then neutravidin was introduced to bind to the biotin. Immobilization of biotinylated goat anti-human IgG on the neutravidin-coated channel was first achieved by confining the reagent flow at a T-intersection (see Fig. 63). The antigen (Cy5-human IgG) was applied via a perpendicular channel. Fluorescence measurement allowed the Ab–Ag binding, background, and nonspecific binding to be accomplished simultaneously (Linder et al., 2001).

Polystyrene beads coated with a capture (mouse) antibody against the carcinoembryonic antigen (CEA) were introduced in a microchannel consisting of a weir structure

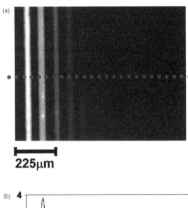

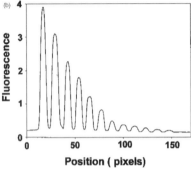

Figure 62 (a) Bulk-phase epifluorescence image of Alexa 594 dye-labeled anti-DNP inside bilayer-coated microchannels. Starting from the left-hand side, the antibody concentrations are 13.2, 8.80, 5.87, 3.91, 2.61, 1.74, 1.16, 0.77, 0.52, 0.34, 0.23, and 0.15 mM, respectively. A line scan of fluorescence intensity across the microchannels is plotted in (b). [(Yang et al., 2001) Courtesy of American Chemical Society.]

(Fig. 50) (Sato et al., 2001a). Thereafter, a serum sample containing CEA (from human colon adenocarcinoma) was introduced. Then the first detection antibody (rabbit anti-human) was allowed to bind with the bound CEA. Finally, a second detection antibody (goat anti-rabbit) conjugated to colloidal gold was used for detection by TLM. The assay time was shortened from 45 h (conventional ELISA) to 35 min. A detection limit of ∼0.03 ng/mL CEA was achieved, which was adequate for normal serodiagnosis of colon cancer (5 ng/mL) (Sato et al., 2001a). A similar method was used to assay interferon gamma, IFNγ (Sato et al., 2001b).

Anti-atrazine antibodies were coupled to a Si microchip comprising 42 coated porous flow channels. The porous Si layer was achieved by anodizing the channel in 40% HF/ 96% ethanol (1:1). Detection of atrazine was achieved in a direct competitive format in which the decrease in a known amount of horseradish peroxidase-labeled atrazine was determined by chemiluminescence detection (Yakovleva et al., 2002).

Heterogeneous immunoassay was performed in a glass microchip with Protein A (PA) (0.1 mg/mL) first immobilized on the microchannel wall (Dodge et al., 2000,

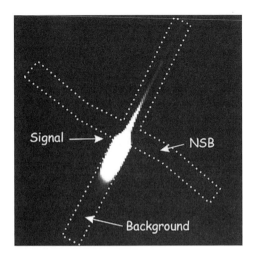

Figure 63 Bound antigen image at the cross area obtained with the fluorescent scanner. Signal is observed at the intersection where Cy5–human IgG binds to the immobilized biotin-conjugated goat anti-human IgG pattern. Information on background levels and nonspecific binding (NSB) can be extracted from the same experiment. Quantitation was performed by averaging the fluorescence intensity in a small area in the region of interest. [(Linder et al., 2001) Courtesy of American Chemical Society.]

2001; Fluri et al., 1998). Then, rabbit IgG (the analyte) was introduced using the EK flow and incubated in the channel. After washing, Cy5-labeled IgG was admitted and competed with the unlabeled IgG for the Protein A sites. After a second washing, the amount of Cy5-IgG was eluted (using a chaotropic buffer containing glycine at pH 2.0) and detected fluorescently. The whole assay procedure (incubation, washing, and elution) was completed within 5 min using the EK flow. A concentration down to 50 nM for rabbit IgG has been achieved (Dodge et al., 2001).

After patterning antibodies as an array using the PDMS microchannel, TNT was detected using the heterogeneous immunoassay format (Sapsford et al., 2002). Different modes (direct, competitive, displacement, and sandwich) produced various LODs (5–20 ng/mL) and dynamic ranges (20–200 ng/mL) (Sapsford et al., 2002).

An array of antigens patterned on a substrate was constructed for detection of IgG antibodies from various species. This method, termed as miniaturized mosaic immunoassay, is rendered possible by using the PDMS channel for patterning materials (Bernard et al., 2001). Various antibodies were patterned via the PDMS channel on a substrate for detection of several antigens [Staphylococcal enterotoxin B, F1 antigen from *Yersinia pestis*, and D-dimer (DDi), which is a marker of sepsis and thrombotic disorder] (Rowe et al., 1999).

Anti-DDi antibodies were also adsorbed on a laser-ablated PET channel for detection of DDi based on a competitive assay with DDi-ALP (DDi-alkaline phosphastase) (Rossier et al., 1999a; Schwarz et al., 1998).

Microfluidic Lab-on-a-Chip

An assay of a cardiac marker (human C-reactive protein, CRP) based on a solid-phase sandwich immunoassay was achieved on a Si–PDMS chip (Juncker et al., 2002).

PDMS microfluidic networks were used to pattern biomolecules (mouse IgG and chicken IgG) in low volume (μL) and on small areas (mm^2). These channels are 3 μm wide and 1.5 μm deep (Delamarche et al., 1997). Subsequently, the PDMS layer was peeled off and the underivatized areas were blocked by BSA. Then, the patterned surface was exposed to three fluorescein-labeled antibodies: tetramethylrhodamine-conjugated anti-chicken IgG (red), fluorescein-labeled anti-mouse IgG (green), and R-phycoerythrin-labeled anti-goat IgG (orange). Only the binding for chicken IgG and mouse IgG were observed through the red and green channels; the nonbinding for goat IgG (orange) was not observed (Delamarche et al., 1997).

PDMS channels were used to pattern mouse IgG (in every other channel) in a 12-channel chip which conformed to the standard liquid introduction format using a 12-channel pipettor (McDonald et al., 2002).

B. Protein Separation

In the separation of samples containing protein, protein adsorption to the channel wall is always a problem. For instance, adsorption of green fluorescent protein (GFP) to the walls of plastic polycarbonate microchannels is problematic (Ross and Locascio, 2002). To avoid protein adsorption, the microchannels normally need to be coated. Coatings, such as (acryloylaminopropyl)trimethylammonium chloride (BCQ) (Deng et al., 2001; Li et al., 1999, 2000), or methacryloyloxyethylphosphorylcholine (MPC) (Oki et al., 2000), were used. In homogeneous immunoassay achieved on-chip, a zwitterionic buffer (tricine) combined with a neutral surfactant (Tween 20) were used additionally to produce a dynamic coating (Chiem and Harrison, 1997). A three-layer coating was applied to a PDMS channel to reduce analyte and reagent adsorption while maintaining a modest cathodic EOF (Linder et al., 2001).

Fluorescently labeled peptides have been separated on a PDMS chip (Effenhauser et al., 1997b). Separations of the charge ladders of various proteins, such as bovine carbonic anhydrase II (CA), insulin, and lysozme, have been achieved on an oxidized PDMS chip. Since lysozme is positively charged, the oxidized PDMS channel has been coated with a cationic polymer, polybrene, to prevent wall adsorption of protein (Duffy et al., 1998).

A protein sizing assay was performed on-chip, using a universal noncovalent fluorescent labeling method, which involved the loading of a fluorescent dye to the SDS–protein complex (Bousse et al., 2001; Giordano et al., 2001). The problem of high fluorescent background, which was caused by the binding of the dye to free SDS in the solution, was alleviated by a post-column dilution step (see Fig. 64). Eleven samples were applied to reservoirs A1–A3, B1–B3, C1–C3, and D1 and D2. As dilution from D4 was needed to remove the SDS–dye complexes in the solution after protein separation (at junction T), the 11 samples could only be analyzed sequentially. On-chip dilution helped break up the SDS micelles, thereby allowing more dye molecules to bind to the protein, and this is a faster process (0.3 s) than the conventional destaining process (1 h) based on diffusion. This represents a much greater speed increase for fast separation plus fast destaining. With this method, a sizing range of 9–200 kDa can be achieved, with good accuracy (5%) and sensitivity (30 nM for carbonic anhydrase) (Bousse et al., 2001).

Moreover, protein separation has commonly been achieved by isoelectric focusing (IEF). IEF of BSA was achieved on-chip using a neutral pH gradient developed under EK flow conditions. The flow cell was constructed by sandwiching two gold electrodes by Mylar, which were held together by a pressure-sensitive adhesive (Cabrera et al., 2001).

Transverse IEF in a pressure-driven flow has been demonstrated using BSA and soybean lectin on -chip (Macounová et al., 2001). Pd electrodes were used (in preference to Au) because of the non-gassing character of Pd. The protein sample was sandwiched between two buffer streams and was prevented from direct contact with the channel wall (and hence the electrode), a process akin to hydrodynamic focusing (Macounová et al., 2001).

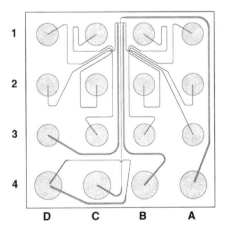

Figure 64 Chip design for protein separation. The wells are shown in light gray. Well D4 is the SDS dilution well and is connected to both sides of the dilution intersection. Wells A4 and C4 are the separation buffer and waste wells, and B4 and D3 are used as load wells. All other wells contain samples. [(Bousse et al., 2001) Courtesy of American Chemical Society.]

IEF of BSA (Macounova et al., 2000), bacteria (Yager et al., 2000), bovine hemoglobin (Macounova et al., 2000), and protein markers (Fan et al., 2001; Van der Noot et al., 2001) were also reported. IEF of cytochrome c and β-lactoglobulin was achieved on a PET chip (Rossier et al., 1999a).

IEF of EGFP was achieved in a PMMA chip sealed with PDMS. Observation of fluorescence was achieved through the optically transparent PDMS layer (Xu et al., 2000). One-step IEF of several Cy5-labeled peptides was achieved on a glass chip without simultaneous focusing and mobilization (by EOF) (Hofmann et al., 1999). IEF of pI markers was demonstrated on a quartz chip using whole-column UV (280 nm) absorption imaging (Mao and Pawliszyn, 1999).

C. Enzymatic Assays

A huge protein application area involves the use of enzymes. For instance, the enzyme, β-galactosidase (β-Gal) was assayed on a chip in 20 min. β-Gal will react on a substrate, resorufin β-D-galactopyranoside (RBG), to form fluorescent resorufin (Hadd et al., 1997). Michaelis–Menten constants were determined by varying the substrate concentrations and monitoring the amount of resorufin by LIF. In addition, the inhibition constants of phenylethyl β-D-thiogalactoside, lactose, and p-hydroxymercuribenzoic acid to β-Gal were determined (Hadd et al., 1997).

Multiple enzymatic assays were achieved on a PDMS–PMMA disk (Duffy et al., 1999). In particular, alkaline phosphatase (ALP) (0.1 mg/mL) and p-nitrophenol phosphate (5 mM) were mixed with 15 concentrations (0.01–75 mM) of theophylline (inhibitor), which were carried out in triplicate. This summed up to 45 assays; this together with three calibrations (for the product p-nitrophenol) were simultaneously accomplished in a single 48-channel disk (see Fig. 65). A complete isotherm for the inhibition of alkaline phosphatase by theophylline was obtained in one single experiment, giving rise to $K_i = 9.7 \pm 0.9$ mM. In addition, 15 (in triplicate) substrate concentrations ($K_m/2–8\ K_m$) were used to generate the necessary data to establish the Michaelis–Menten kinetic parameters (K_m and V_m) in one single experiment (Duffy et al., 1999).

Another enzyme inhibition assay using a peptide substrate was performed on a PMMA chip (Boone and Hooper, 1998). In another report, several acetylcholinesterase (AChE) inhibitors, including tacrine, edrophonium, tetramethyl- and tetraethyl-ammonium chloride, carbofuran, and eserine were assayed on a chip (Hadd et al., 1999). AChE converted the substrate, acetylthiocholine, to thiocholine, which, after reacting with coumarinylphenyl-maleimide (CPM), formed a thioether (thiocholine-CPM) for LIF detection. As the acetonitrile solvent used to

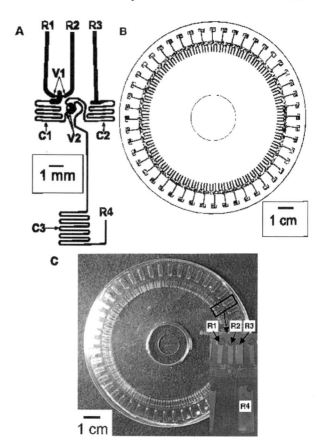

Figure 65 (a) Design of the fluidic structure used to perform an assay composed of mixing an enzyme with an inhibitor, followed by mixing with a substrate, and detection. The solutions of enzyme, inhibitor, and substrate were loaded in reservoirs that were connected to channels labeled R1, R2, and R3 respectively. Enzyme and inhibitor combined after being released by capillary burst valves, V1, and mixed in a meandering 100 μm wide channel, C1. The enzyme–inhibitor mixture was combined with the substrate in a chamber after being released by burst valve, V2. These solutions mixed in a meandering channel, C3, and emptied into a cuvette (not shown) from a section of channel labeled R4. (b) The complete PDMS replica sealed to a PMMA layer comprising 48 of the structures shown in (a). (c) Photograph of the disk. The inset shows the magnified detail of one of the fluidic structures of the disk. [(Duffy et al., 1999) Courtesy of American Chemical Society.]

dissolve CPM inhibited AChE activity, CPM was added postreaction (Hadd et al., 1999).

Dual precolumn enzymatic reactions, followed by CE separation and amperometric detection, were achieved on-chip (Wang et al., 2001b). A sample containing glucose and ethanol was placed in the sample reservoir, and an enzyme mixture containing GOx, alcohol dehydrogenase (ADH) and NAD^+ was placed in the reagent reservoir. Glucose was converted to hydrogen peroxide by glucose oxidase (GOx), whereas ethanol converted

NAD$^+$ to NADH by ADH. After separation of H$_2$O$_2$ and NADH, they were detected amperometrically at the anode (Wang et al., 2001b).

Enzymatic microreactors have been prepared in the microfluidic chip by immobilizing trypsin on porous polymer monoliths. The polymers were made (in situ) by reacting 2-vinyl-4,4-dimethylazlactone, ethylene dimethacrylate, and acrylamide (or 2-hydroxyethyl methacrylate). The azlactone functionality reacts readily with the amine and thiol groups of the enzyme (trypsin) to form stable covalent bonds (Peterson et al., 2002). The proteolytic activity of the microreactor (7.5 nL) was demonstrated by the cleavage of equine (horse) myoglobin (14.2 pmol/μL). After collection of the trypsin digest, MALDI-TOF was performed off-line (Peterson et al., 2002).

There are other reports on the use of immobilized enzymes. HaeIII restriction endonuclease enzyme was immobilized on an amine-modified PMMA surface (Henry et al., 2000). Trypsin and chymotrypsin were immobilized on an enzyme reactor fabricated on Si (Ekström et al. 2000).

Glucose oxidase and HRP were physically entrapped during polymerization of poly(ethyleneglycol) (PEG) hydrogel-based micropatches for the detection of glucose (Zhan et al., 2002). The micropatches were first molded within 17 μm high and 135 μm wide PDMS channels. Then the PDMS mold was removed and the micropatches were enclosed within a larger (32 μm high and 200 μm wide) PDMS channel. After GOx converted glucose to gluconolactone and H$_2$O$_2$, in the presence of HRP, H$_2$O$_2$ reacted with a dye (Amplex Red) to form a fluorescent product: resorufin. GOx and HRP were entrapped in the hydrogel; whereas glucose (the analyte), O$_2$, and Amplex Red were able to diffuse into the hydrogel within a reasonable time scale. As a control, a hydrogel micropatch containing only HRP, but no GOx, was also formed inside the channel (Zhan et al., 2002).

A microchip-based enzyme assay for protein kinase A (PKA) was achieved by performing on-chip electrophoretic separation of the fluorescently labeled substrate (kemptide: (Leu-Arg-Arg-Ala-Ser-Leu-Gly)) and its product. The effect of a protein kinase A inhibitor [H89: N-2-[(p-bromocinnamyl)aminoethyl]-5-isoquinoline sulfonamide hydrogenchloride] was then studied on-chip (Cohen et al., 1999).

In another report, streptavidin-conjugated alkaline phosphatase was linked to biotinylated phospholipid bilayers coated inside PDMS microchannels (Mao et al., 2002b). The surface-bound enzyme was found to have a lower (sixfold) turnover rate than the free enzyme in solutions. One-shot Lineweaver–Burk (reciprocal) plot can be achieved by 12 independent concentrations of the substrate (4-methylumbelliferyl phosphate) created on-chip to obtain the Michaelis–Menten parameters after measuring

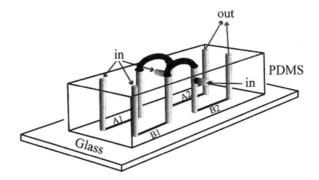

Figure 66 Schematic representation of the microfluidic device used for the two-step glucose detection reaction. [(Mao et al., 2002b) Courtesy of American Chemical Society.]

the blue fluorescence of the product: 7-hydroxy-4-methyl courmarin. After diffusion mixing between two streams (substrate and buffer) for a fixed distance (2 cm), the main channel was partitioned into a series of 12 smaller channels that emanated from the main flow stream (see Fig. 13). This method created a dilution series that ranged over a factor of 33 (i.e., 0.103–3.41 mM) of the substrate (Mao et al., 2002b).

The one-shot experiment was done in one order of magnitude less time to establish Michaelis–Menten kinetics (needs 12 concentrations) and with one-third less error in the kinetic raw data (Mao et al., 2002b).

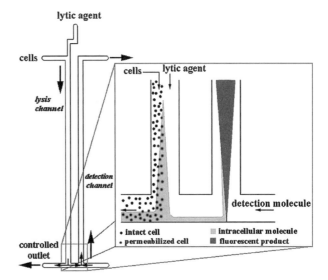

Figure 67 Schematic of microfluidic device for cell lysis and fractionation/detection of intracellular components. Pump rates are controlled at all inlets and one outlet. Lytic agent diffuses into the cell suspension, and lysing the cell. Intracellular components then diffuse away from the cell stream and some are brought around the corner into the detection channel, where their presence can be detected fluorescently. [(Schilling et al., 2002) Courtesy of American Chemical Society.]

A two-step enzymatic detection of glucose can also be achieved using two immobilized enzymes. Avidin-conjugated glucose oxidase was immobilized in A1 and B1, and streptavidin-conjugated HRP was immobilized in A2, B2 (see Fig. 66). After incubation, A1 and A2 (also B1 and B2) were connected with a reversibly attachable U-shaped plastic tube (700 μm o.d.). After admitting a O_2-saturated glucose-containing sample into A1 (only buffer into B1 as a control channel), H_2O_2 was formed. The H_2O_2 formed (only in A1) was subsequently flowed to A2 to convert Amplex Red (in the presence of HRP) to resorufin for fluorescent detection (Mao et al., 2002b).

Lysis of *E. coli*, extraction of a large intracellular enzyme (β-galactosidase), and its detection (using RBG as the substrate) were achieved on-chip using the pressure-driven flow (Schilling et al., 2001, 2002). The first two steps were achieved in a H-filter, which was coupled to a T-sensor to complete the third step (see Fig. 67). As the cells were large, they remained in the left half of the channel. Only the enzyme, after meeting with a lytic agent (a proprietary mild detergent), by diffusion moved to the right channel (Schilling et al., 2002).

D. MS Analysis for Proteins and Peptides

Tremendous effort has been taken to accelerate proteomics research using on-chip mass spectrometric analysis of proteins and peptides.

A nine-channel glass microchip was interfaced to MS to detect myoglobin (see Fig. 68). Stable electrospray could be achieved at the flat edge of the channel exit with a pressure flow (100–200 nL/min). Detection of recombinant human growth hormone (2 μM), ubiquitin (10 μM), endorphin (30 μM) have also been demonstrated (Xue et al., 1997).

A PDMS microchannel has been coupled to a porous PVDF membrane (0.45 μm pore) with adsorbed trypsin

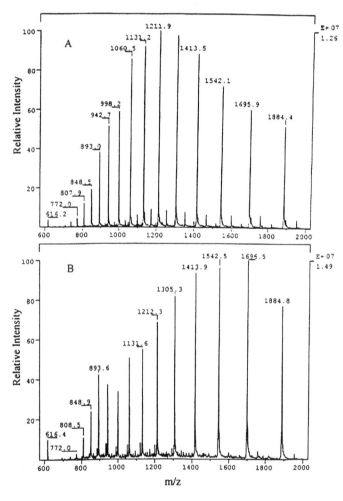

Figure 68 Comparison of ESI-mass spectra of myoglobin (6 μM in 75% methanol, 0.1% acetic acid) obtained from (a) HF-etched capillary (50 μm i.d.), 800 nL/min and (b) microchip with 60 μm wide and 25 μm deep channels, 200 nL/min. [(Xue et al., 1997) Courtesy of American Chemical Society.]

for on-line protein digestion of horse heart cytochrome c and subsequent MS analysis (Gao et al., 2001).

A microchip comprising 96 channels with 96 imbedded capillary tips has been constructed for high-throughput MS analysis of peptides (see Fig. 69) (Liu et al., 2000d).

An 8-mer peptide substrate was used for the study of HIV-1 protease activity in the presence of various inhibitors (e.g., Pepstatin A). With a tripeptide as an internal standard, inhibitors of various concentrations were placed in the 96-channel plate to establish the inhibition constant (Liu et al., 2000d).

A pulled capillary tip inserted and glued to the end of a microchannel was employed as a disposable nanoelectrospray emitter. Membrane proteins from nonpathogenic (Rd) and pathogenic (Eagan) strains of H. influenzae were isolated (in 1D SDS–PAGE) and in-gel digested with trypsin. The digests were subsequently placed on the chip to carry out on-chip separation for clean-up and partial separation and subsequent ESI-QTOF detection and peptide mass-fingerprint database search (Li et al., 2000). Various proteins and peptides have also been studied by ESI-TOF (Lazar et al., 1999).

Tryptic digest of BSA, horse myoglobin, human haptoglobin, and 2D-gel yeast proteins were sequentially analyzed in a nine-channel chip with a single transfer capillary coupled to an ESI-ITMS instrument (Figeys et al., 1998). Sequential infusion of tryptic digests of CA (290 nM) and BSA (130 nM) into ESI-ITMS was achieved using EOF without cross-contamination when a central flow of buffer confined the other sample in reservoir and channel by precise voltage control (Figeys et al., 1997).

A disposable nanoelectrospray emitter was inserted and glued into a drilled hole with low dead volume for analysis of various peptide standards and tryptic digests of lectins from D. biflorus and P. sativum (Li et al., 1999).

In addition, a sheath flow microionsprayer was used, and CE was used to remove spectral interferences (Li et al., 1999).

Gradient frontal separation was performed on a C18 cartridge before MS analysis of tryptic digest of horse apo-myoglobin and yeast proteins (Figeys and Aebersold, 1998).

When trypsin was mixed off-chip and digestion carried out on-chip, MS analysis of cytochrome c and hemoglobin was reported (Lazar et al., 2001). Instead of ammonium acetate (pH 6.5), which was commonly used for ESI-MS, ammonium bicarbonate (pH 8) was employed so that the enzyme trypsin was most active. Paradoxically, it is possible to detect positively charged ions from high pH solutions (wrong-way-round electrospray) (Lazar et al., 2001).

Adrenocorticotropin (ACTH), which is a peptide hormone, has been digested by carboxypeptidase Y in a 90-well plate constructed in a MALDI plate format. Peptide digestion was initiated in the MALDI interface where mixing of peptide and enzyme were self-activated in the vacuum conditions. Subsequent TOF-MS analysis produced kinetic information of the peptide digestion reaction (Brivio et al., 2002).

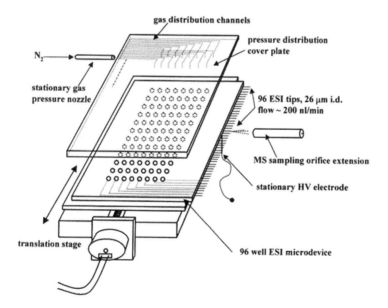

Figure 69 Exploded view of a 96-channel MS chip design. The plate with individual channels and electrospray tips was positioned on a translation stage in front of the extension of the MS sampling orifice. The electrospray analysis of individual samples was activated by sequential pressurization of the sample wells through the pressure distribution cover plate and connection of the ESI high voltage through the stationary HV electrode positioned under the ESI device. The silicone rubber sealing gasket placed between the ESI device and the pressure distribution cover plate. [(Liu et al., 2000d) Courtesy of American Chemical Society.]

Table 17 Analytical Applications Using the Microfluidic Technology

Class	Compound	Sample matrix	Concentration	Chip	Separation mode	Detection	References
Amino acid	Cysteine		200 µM	PDMS–PDMS	CZE	Dual amp (C)	Martin et al. (2001b)
	Glycine		150 µM	PDMS–PDMS	CZE	Dual amp (C)	Martin et al. (2001b)
	19 Amino acids		10 mM	Glass	CZE	LIF (TRITC)	Culbertson et al. (2000a)
	19 Amino acids	In urine	32.9 µM LOD (average)	Glass	CZE	Indirect LIF (fluo 0.5 mM)	Munro et al. (2002)
	Amino acids		140–220 µM	Glass	CZE	LIF (NBD)	Weiller et al. (2002)
	D/L Val, Ala, Glu and Asp		1 mM	Glass	CD-MECC	LIF (FITC)	Hutt et al. (1999)
	Glycine			PDMS–PDMS	CZE	Amp (C fiber)	Martin et al. (2002)
	Glycine, tryptophan, histidine		1 mM	Glass	CZE	Amp (Cu/Pt)	Schwarz et al. (2001a)
	Phenylalanine		2 µM	Fused silica	CZE	LIF (OPA)	Fluri et al. (1996)
	Valine, alanine, aspartic acid		300 µM	Si–Pyrex	MECC	ECL (TBR)	Hsueh et al. (1996)
	Valine, leucine			Glass	MECC (SDS)	Amp (Au on C)	Wang et al. (2000b)
Biogenic amine	5-Hydroxyindole-3-acetic acid (metabolite of serotinin)		2.5 µM LOD	PDMS–PDMS	CZE	Amp	Gawron et al. (2001)
	Antidepressant (nortriptyline, amidrityline)		0.10–0.15 mg/mL	Quartz	CEC	UV (239 nm)	Ericson et al. (2000)
	Catechol			PDMS		Amp/LIF	Manica et al. (2001)
	Catechol		0.47 µM LOD	Plastic (Plexiglas)	CZE	Amp (C)	Martin et al. (2001a)
	Catechol		100 µM	PDMS–glass		Amp (Pd)	Chen et al. (2001a)
	Catechol		4 µM	PDMS–glass		Dual amp (Au)	Martin et al. (2000)
	Catechol		4–5 µM	Glass		1st stage amp (Au)	Martin et al. (2000)
	Catechol	Spiked cerebral spinal fluid	110 µM	Glass	CZE	Amp (Pt)	Baldwin et al. (2002)
	Catechol					Amp	Lapos et al. (2002)
	Catechol		4 µM LOD	PDMS–PDMS	CZE	Amp (C fiber)	Martin et al. (2002)
	Catechol		0.78 µM LOD	Glass	CZE	Amp (C)	Wang et al. (1999b)
	Catechol		0.5 µM	PDMS–PDMS	CZE	Amp (C fiber)	Gawron et al. (2001)
	D/L-Metanephrine				CE	Amp	Schwarz and Hauser (2001b)
	Dopamine		3.7 µM LOD	Glass	CZE	Amp (Pt)	Woolley et al. (1998)
	Dopamine		0.29 µM LOD	Plastic (Plexiglass)	CZE	Amp (Pd)	Chen et al. (2001a)
	Dopamine		0.24 µM	Glass	CZE	Amp (C)	Zeng et al. (2002)
	Dopamine		4–5 µM	Glass		Amp (Pt)	Baldwin et al. (2002)
	Dopamine	Spiked cerebral spinal fluid	50 µM	Glass	CZE	Amp	Lapos et al. (2002)
	Dopamine		0.38 µM LOD	Glass	CZE	Amp (C)	Wang et al. (1999b)
	Dopamine		20 µM	PDMS	CZE	Amp	Liu et al. (2000a)
	Dopamine		20–200 µM (1 µM LOD)	Glass	CZE	Amp (Au)	Wang et al. (1999a)
	Dopamine		10 µM	Glass	CZE	Amp (Au/Pt)	Schwarz et al. (2001a)
	Dopamine		1–165 µM	PDMS–glass	CZE	Amp (Pt)	Fanguy and Henry (2002)
	Epinephrine or adrenaline		6.5 µM LOD	Glass	CZE	Amp (Pt)	Woolley et al. (1998)
	Epinephrine or adrenaline		0.1 µM	Glass	CZE	Amp (C)	Zeng et al. (2002)
	Epinephrine or adrenaline		10 µM	Glass	CZE	Amp (Au/Pt)	Schwarz et al. (2001a)
	Epinephrine or adrenaline			PDMS–PDMS	CZE	Amp (C)	Gawron et al. (2001)

Microfluidic Lab-on-a-Chip

Category	Analyte	Sample	Concentration/LOD	Chip material	Mode	Detection	Reference
	Isoproterenol		1.3 μM	Glass	CZE	Amp (Au)	Wang et al. (1999a)
	L-DOPA		100 mM	PDMS-glass	CZE	Amp (C)	Backofen et al. (2002)
	L-DOPA		100 μM	Glass	CZE	Amp (Au)	Wang et al. (1999a)
	Noradrenaline		100 mM	PDMS-Glass	CZE	Amp (C)	Backofen et al. (2002)
	Noradrenaline		10 μM	Glass	CZE	Amp (Au/Pt)	Schwarz et al. (2001a)
	Serotinin			PDMS–PDMS	CZE	Amp (C)	Gawron et al. (2001)
	Serotinin		0.1 mM	Glass			Zeng et al. (2002)
	Catechol		12 μM LOD	Glass	CZE	Amp (Pt)	Woolley et al. (1998)
	Catechol		5–200 μM (1 μM LOD)	PDMS–PDMS	CZE	Dual amp (C)	Martin et al. (2001b)
DNA	DNA			Si–Pyrex	Nil	ECL (TBR)	Hsueh et al. (1998)
	HaeIII digests of pBR322		40 fmol LOD	Glass	CGE(HEC)	LIF (YO-PRO-1)	Huang et al. (2001)
	HaeIII digests of pBR322			Glass	CGE(HEC)	LIF (YO-PRO-1)	Huang et al. (1999)
	HaeIII digests of phiX174		75 fg of 72 bp fragment	Parylene C	CGE (HEC)	LIF (SYBR Green I)	Webster et al. (2001)
	HaeIII digests of phiX174			Glass	CGE (cross-linked PA)	LIF (YOYO-1)	Brahmasandra et al. (2001)
	HLA-DQα			Glass	CGE	LIF (TO6)	Woolley and Mathies (1994)
	LDL-receptor gene			Glass	CGE	LIF	Cantafora et al. (2001)
	Oligo p(dT)10–25			Glass	CGE (LPA)	LIF	Effenhauser et al. (1995)
	pBluescript (100–1000 bp)			Glass	CGE	Fluoro burst counting	Haab and Mathies (1999)
	RT–PCR product of hsp72 mRNA			Glass	CGE	LIF	Gottwald et al. (2001)
	Salmonella DNA			Glass	CGE (HEC)	Indirect amp (TPR)	Woolley et al. (1998)
	Serotonin Transporter gene (5-HTTLPR)			Glass	CGE (agarose-LPA)	LIF (EtBr)	Nemoda et al. (2001)
Drug	Cortisol	In serum	10–600 μg/L	Fused silica	CZE	LIF (FITC)	Koutny et al. (1996)
	Theophylline	In serum	10–20 μg/mL	Glass	CZE	LIF	Harrison and Chiem (1996)
	Theophylline	In serum	7–20 μg/mL	Glass	CZE/MECC	LIF	von Heeren et al. (1996a)
	Theophylline		2.5–40 μg/mL	Glass	CZE	LIF	Chiem and Harrison (1997)
	4-Acetamidophenol (or acetaminophen)		100 μM	Glass	CZE	Cond (contactless)	Tanyanyiwa and Hauser (2002)
	4-Acetamidophenol (or acetaminophen)			Glass	CZE	Amp (Au/Pt)	Wang et al. (2000e)
	Atropine (alkaloid)					CL	Greenwood and Greenway (2001)
	Caffeine		12.9 mM	Si monolithic		UV (254 nm)	Mogensen et al. (2001a)
	Codeine					CL	Nelstrop and Greenway (1998)
	Codeine		0.5–2 μM 0.83 μM LOD	Glass		CL (TBR, 620 nm)	Greenway et al. (2000a)
	Hydroquinone		40 μM	PDMS	CZE	Amp	Liu et al. (2000a)
	Ibuprofen		100 μM	Glass	CZE	Cond (contactless)	Tanyanyiwa and Hauser (2002)
	Ketoprofen (anti-inflammatory drug)		9.8 mM	Si		UV (254 nm)	Mogensen et al. (2001a)

(continued)

Table 17 *Continued*

Class	Compound	Sample matrix	Concentration	Chip	Separation mode	Detection	References
	Paracetamol (or acetaminophen)		16.5 mM	Si		UV (254 nm)	Mogensen et al. (2001a)
	Pethidine(derived from morphine)					CL	Greenwood and Greenway (2001)
Enzymes	Alkaline phosphatase		0.1 mg/mL	Plastic	Nil	UV (430 nm for pNP)	Duffy et al. (1999)
	GPT, GOT	Serum	10–300 U/l	Glass	Nil	Amp	Jobst et al. (1996)
Inorganics	Al^{3+}		30 ppb LOD	Quartz	CZE	LIF (HQS)	Jacobson et al. (1995)
	Ba^{2+}		10 μM–0.1 M	Glass	Nil	Potent-ISE	Tantra and Manz (2000)
	Ca^{2+}	Muscle cell		Glass–quartz	Nil	TSM Acoustic wave	Li et al. (2001)
	Cd^{2+}					UV	Chudy et al. (2001)
	Cd^{2+}		57 ppb LOD	Quartz	CZE	LIF (HQS)	Jacobson et al. (1995)
	Cl^-					Cond	Weber et al. (2000)
	Cl^-		20–30 μM	PMMA	ITP	Cond	Bodor et al. (2001)
	CO_2		30–60 mmHg	Si–Pyrex	Nil	Smp	Shoji and Esashi (1992)
	Co^{2+}					CL	Greenway et al. (2000b)
	Co^{2+}					Cond	Fielden et al. (1998)
	Co^{2+}	As Co-complex in water	0.1–1 μM	Quartz	Toluene extract	TLM	Tokeshi et al. (2000b)
	Co^{2+}		18 nM LOD	Quartz	Nil	TLM	Tokeshi et al. (2002)
	Cr (VI)					CL	Kim et al. (2001b)
	Cr^{3+}		6.8 mM LOD	Glass	CZE	Cond (contactless)	Tanyanyiwa and Hauser (2002)
	Cu^{2+}					Optical em	Enkins and Manz (2001)
	Cu^{2+}					UV	Chudy et al. (2001)
	F^-					Cond	Weber et al. (2000)
	F^-					Cond	Baldock et al. (1998)
	F^-		10 μM	PMMA	ITP-CZE	Cond	Kaniansky et al. (2000)
	F^-	Tap, mineral, river water	0.5–0.7 μM	PMMA	ITP-CZE	Cond	Bodor et al. (2001)
	Fe^{2+}		1–21 μM	Quartz	Nil	TLM	Tokeshi et al. (2000a)
	H^+ (pH)			Si	Nil	Potent (ISFET)	Fiehn et al. (1995)
	H^+ (pH)		1–9	Si–Pyrex	Nil	Amp	Shoji and Esashi (1992)
	Hg^{2+}					UV	Chudy et al. (2001)
	K^+					Cond	Schasfoort et al. (2000)
	K^+		100 μM–1 M	Quartz	Nil	TLM	Hisamoto et al. (2001)
	K^+		2.52 mM	PDMS–PDMS	ITP	Cond	Prest et al. (1999)
	K^+			Plastic disk	Nil	Potent	Badr et al. (2002)
	K^+		0.49 mM LOD	Glass	CZE	Cond (contactless)	Tanyanyiwa and Hauser (2002)
	K^+		500 μM LOD	Glass	CZE	Cond	Guijt et al. (2001b)
	Li^+			Glass		Cond	Schasfoort et al. (2000)
	Li^+		500 μM LOD	Glass	CZE	Cond	Guijt et al. (2001b)

Analyte	Notes	Concentration	Substrate	Separation	Detection	Reference
Mg^{2+}		0.35 mM LOD	Glass	CZE	Cond (contactless)	Tanyanyiwa and Hauser (2002)
Mn^{2+}					Cond	Fielden et al. (1998)
Mn^{2+}		2.1 mM LOD	Glass	CZE	Cond (contactless)	Tanyanyiwa and Hauser (2002)
Na^+			Si–Pyrex		Cond	Weber et al. (2000)
Na^+		100 μM–1 M	Quartz	Nil	Cond (contactless)	Darling et al. (1998)
Na^+		2.17 mM	PDMS–PDMS	ITP	TLM	Hisamoto et al. (2001)
Na^+		0.41 mM LOD	Glass	CZE	Cond	Prest et al. (1999)
Na^+					Cond (contactless)	Tanyanyiwa and Hauser (2002)
Na^+		1–50 mM (500 μM LOD)	Glass	CZE	Cond	Guijt et al. (2001b)
$Na^+, NH_4^+, Li^+, SO_4^{2-}, NO_3^-, F^-$		0.01 M	PMMA	ITP (bidirectional)	Cond (contactless)	Prest et al. (2002)
NH_4^+		10 μM–0.1 M	Si	Nil	Potent (ISFET)	Fiehn et al. (1995)
Ni^{2+}					Cond	Fielden et al. (1998)
Nitrate					Cond	Schasfoort et al. (2000)
Nitrate		20–30 μM	PMMA	ITP	Cond	Bodor et al. (2001)
Nitrite		10 μM	PMMA		Amp (C)	Martin et al. (2001a)
Nitrite	Tap, mineral, river water		PMMA	ITP-CZE	Cond	Kaniansky et al. (2000)
Nitrite		0.5–0.7 μM	PMMA	ITP-CZE	Cond	Bodor et al. (2001)
Nitrite		10 μM	PDMS–PDMS	CZE	Amp (C fiber)	Martin et al. (2002)
Nitrite			Plastic disk	Nil	Potent	Badr et al. (2002)
Nitrite	For NO-releasing compound SIN-1	10–250 μM 1 μM LOD	PDMS–PDMS	CZE	Amp (C)	Kikura-Hanajiri et al. (2002)
NO_3^-		10 μM–0.1 M	Si	Nil	Potent (ISFET)	Fiehn et al. (1995)
O_2		20–100 %	Si–Pyrex	Nil	Amp	Shoji and Esashi (1992)
Pb^{2+}		100 ppm			UV	Chudy et al. (2001)
Pb^{2+}	As OEP complex in benzene	97–780 pM or 0.4–3.4 molecules	Si–Pyrex		ASV	Darling et al. (1998)
Pb^{2+}			Quartz	Nil	TLM	Tokeshi et al. (2001)
Phosphate		9.67–96.7 μM	Si	Nil	Absorbance	Schoot et al. (1994)
Phosphate		0.01–1.0 mM	Si	Nil	Amp	Hinkers et al. (1996)
Phosphate					Cond	Schasfoort et al. (2000)
Phosphate		10 μM	PMMA	ITP-CZE	Cond	Kaniansky et al. (2000)
Phosphate	White wine	12–42 mg/L	PMMA	ITP	Cond	Masar et al. (2001)
Phosphate	Tap, mineral, river water	0.5–0.7 μM	PMMA	ITP-CZE	Cond	Bodor et al. (2001)
Phosphate		38.7–96.7 μM	Si	Nil	Absorbance (660 nm)	Verpoorte et al. (1993)
Sulfate		20–30 μM	PMMA	ITP	Cond	Bodor et al. (2001)
Zn^{2+}					UV	Chudy et al. (2001)

(continued)

Table 17 Continued

Class	Compound	Sample matrix	Concentration	Chip	Separation mode	Detection	References
	Zn^{2+}		46 ppb LOD	Quartz	CZE	LIF (HQS)	Jacobson et al. (1995)
	Zn^{2+}		2.8 mM LOD	Glass	CZE	Cond (contactless)	Tanyanyiwa and Hauser (2002)
Organic acid	Ascorbic acid		14.2 mM	Si monolithic		UV (254 nm)	Mogensen et al. (2001a)
	Ascorbic acid	Only 1st stage oxid		PDMS–glass		Dual amp (Au)	Martin et al. (2000)
	Ascorbic acid		5 µM LOD	PDMS–glass	CZE	Amp (C)	Backofen et al. (2002)
	Ascorbic acid		5 µM LOD	Glass	CZE	Amp (Au/C)	Schwarz et al. (2001a)
	Ascorbic acid		100 µM	Glass	CZE	Amp (Au/Pt)	Wang et al. (2000e)
	Citric acid	White wine	12–42 mg/L	PMMA	ITP	Cond	Masar et al. (2001)
	Citric acid		100 µM	Glass	CZE	Cond (contactless)	Tanyanyiwa and Hauser (2002)
	Citric acid		10 µM LOD	Glass	CZE	Cond	Guijt et al. (2001b)
	Citric acid	Red wine		PMMA	ITP	Cond	Grass et al. (2001)
	Fumaric acid		5 µM LOD	Glass	CZE	Cond	Guijt et al. (2001b)
	Glutamic acid	Red wine		PMMA	ITP	Cond	Grass et al. (2001)
	Lactic acid	White wine	12–42 mg/L	PMMA	ITP	Cond	Masar et al. (2001)
	Lactic acid		100 µM	Glass	CZE	Cond (contactless)	Tanyanyiwa and Hauser (2002)
	Lactic acid	Red wine		PMMA	ITP	Cond	Grass et al. (2001)
	Malic acid	White wine	12–42 mg/L	PMMA	ITP	Cond	Masar et al. (2001)
	Malic acid		10 µM LOD	Glass	CZE	Cond	Guijt et al. (2001b)
	Malic acid	Red wine		PMMA	ITP	Cond	Grass et al. (2001)
	Oxalic acid	Red wine		PMMA	ITP	Cond	Grass et al. (2001)
	Salicylic acid		100 µM	Glass	CZE	Cond (contactless)	Tanyanyiwa and Hauser (2002)
	Succinic acid	Red wine		PMMA	ITP	Cond	Grass et al. (2001)
	Tartaric acid	White wine	12–42 mg/L	PMMA	ITP	Cond	Masar et al. (2001)
	Tartaric acid		5 µM LOD	Glass	CZE	Cond	Guijt et al. (2001b)
	Tartaric acid	Red wine		PMMA	ITP	Cond	Grass et al. (2001)
	Uric acid		5 µM LOD	Glass	CZE	Amp (Au/C)	Wang et al. (2000e)
	Acetate	Red wine		PMMA	ITP	Cond	Grass et al. (2001)
	Uric acid	Urine	15–110 µM	PDMS–glass	CZE	Amp (Pt)	Fanguy and Henry (2002)
Organics	Alkylphenones		0.3–0.6 mg/mL	Quartz	CEC	UV (240 nm)	Ericson et al. (2000)
	Amino sugars	From BSM	12.5 µM	Quartz	CZE	LIF (NBD-F)	Suzuki et al. (2001)
	ATP	Jurkat cells	0.16 µM	PDMS-quartz	Nil	LIF (fluo 3)	Yang et al. (2002b)
	Atrazine		3.7–209 pM	Si	Nil	CL (HRP/luminol)	Yakovleva et al. (2002)
	Azo dyes		1 nM or 10 fmol LOD			SERRS	Keir et al. (2002)
	Benzene		10 ppm	Pyrex	Nil	UV	Ueno et al. (2002b)
	CD-3	Fresh or seasoned developer solutions	0.17 mM LOD	Glass	CZE	Indir LIF (5 mM fluorescein)	Sirichai and De Mello (2001)

656 Analytical Instrumentation Handbook

Analyte	Sample	LOD/Range	Substrate	Mode	Detection	Reference
CD-3	Photographic developer solution	5–20 mg/L	Glass	CZE	Indir LIF (2 mM fluorescein)	Sirichai and deMello (2000)
CD-4	Fresh or seasoned developer solutions	0.39 mM LOD	Glass	CZE	Indir LIF (5 mM fluorescein)	Sirichai and De Mello (2001)
Chlorophenols		100 µM	Glass	CZE	Amp (Au/Pt)	Schwarz et al. (2001a)
Citrulline		150 µM	PDMS–PDMS	CZE	Dual amp (C)	Martin et al. (2001b)
Cy5		3.3 nM	Glass		LIF	Roulet et al. (2002)
Dimethyl methyl phosphate					SAW	Frye-Mason et al. (1998)
Ethanol					Diffraction	Schumacher et al. (1998)
Ethanol	Red wine	200 mM	Glass	CZE	Amp (Au/C)	Wang et al. (2001b)
Ethyl acetate					IR	Jackman et al. (2001)
Explosives					Indir LIF	Wallenborg et al. (2000)
Explosives	Spiked soil	1 ppm	Glass (borofloat)	MECC	Indir LIF (5 uM Cy7)	Wallenborg and Bailey (2000)
Explosives		2–15 – mM	Glass	MECC	Cond-amp	Wang and Pumera (2002)
Explosives (2,4-DNT and 2,6-DNT)		20 mg/L	Glass	MECC	Amp (C)	Wang et al. (1999b)
Explosives (4-NT)		2 mg/L	Glass	MECC	Amp (C)	Wang et al. (1999b)
Explosives (DNB, TNT)		0.6 mg/L	Glass	MECC	Amp (C)	Wang et al. (1999b)
Ferrocenecarboxylic acid		3 µM	Plastic		CV (C) and chronoamp (C)	Rossier et al. (1999b)
FITC		6 µM	Glass		UV/LIF	Liang et al. (1996)
Fluorescein		300 fM	Glass	CZE	LIF	Ocvirk et al. (1998)
Fluorescein		25 nM	PDMS–PDMS	CZE	LIF	Chabinyc et al. (2001)
Fluorescein		10 µM	Si–Pyrex	CZE	UV absorbance (488 nm)	Mogensen et al. (2001b)
Fluorescein		10 µM	Glass	CZE	LIF	Han and Craighead (2000)
Fluorescein		63 pM	Glass	CZE	LIF	Sirichai and De Mello (2001)
Fluorescein		0.5 mM	Glass	CZE	Indirect LIF (fluo 0.5 mM)	Munro et al. (2002)
Fructose and galactose		1 mM	Glass	CZE	Amp (Cu/Pt)	Schwarz et al. (2001a)
Glucose		1–80 mM	Quartz	Nil	CL	Rong and Kutter (2001)
Glucose		0.01–10 mM	Si–Pyrex	Nil	Amp (Au)	Forssen et al. (1994)
Glucose		1–25 mM	Si–Pyrex		CL	Blankenstein et al. (1996)
Glucose		0.7 mM	Glass		Amp (Au)	Murakami et al. (1993)
Glucose	Red wine	6 µM LOD	Glass	CZE	Amp (Au/C)	Wang et al. (2001b)
Glucose		742 µM	Silica	CZE	Amp (Au/C)	Wang et al. (2000e)
Glycerol			Glass	Nil	RI	Swinney et al. (2000)
Hexane	10–12 g/s or 800 ppb		Glass	Conv GC	Op em (519 nm)	Eijkel et al. (2000a)
Hydrocarbons			Si	GC	TCD	Terry et al. (1979)
LDL & HDL			Pyrex	CZE	LIF	Weiller et al. (2002)

(continued)

Table 17 Continued

Class	Compound	Sample matrix	Concentration	Chip	Separation mode	Detection	References
	Methane		10–13 g/s or 400 ppb	Glass	Nil	Op em	Eijkel et al. (2000b)
	Methane		10e–12 g/s or 600 ppm LOD + D195	Glass	Nil	Op em (CH)	Eijkel et al. (1999)
	Methyl formate					IR	Floyd et al. (2001)
	N-Acetylglucosamine		33 mM	Glass (soda lime)	CZE	RI	Burggraf et al. (1998)
	Organophosphate nerve agents		48–86 µg/L LOD	Glass	CZE	Cond	Wang et al. (2002)
	O-Xylene		10 ppm	Pyrex	Nil	UV	Ueno et al. (2002b)
	Paraoxon, methylparaoxon, parathions, fernitrothion	Spiked river water	30–60 µM	Glass	MECC	Amp (C)	Wang et al. (2001a)
	Paraquat, diquat			Glass	Raman		Reshni et al. (1998)
	Paraquat, diquat		0.23 µM	Glass	ITP	Raman	Walker et al. (1998)
	Penicillamine		200 µM	PDMS–PDMS	CZE	Dual amp (C)	Martin et al. (2001b)
	Propane					Photoacoustic	Firebaugh et al. (2000)
	Raffinose		33 mM	Glass (soda lime)	CZE	RI	Burggraf et al. (1998)
	Sucrose					RI	Jakeway and DeMello (2001)
	Sucrose		33 mM	Glass (soda lime)	CZE	RI	Burggraf et al. (1998)
	Sucrose		10–100 mM	Si	nil	FTIR	Lendl et al. (1997)
	Sucrose		1 mM	Glass	CZE	Amp (Cu/Pt)	Schwarz et al. (2001a)
	Toluene		10 ppm	Pyrex	Nil	UV	Ueno et al. (2002b)
	Uracil			Quartz	CEC	UV (254 nm)	Ericson et al. (2000)
	Urasil (or uracil)					UV	Nishimoto et al. (2000)
	VOC					UV	Horiuchi et al. (2001)
	Water		25–70% RH	Si	Nil	Microcantilever, optical deflection	Berger et al. (1996)
Peptide	Des-Tyr-Leu-enkephalin		450 µM	PDMS–PDMS	CZE	Dual amp (C)	Martin et al. (2001b)
	Des-Tyr-Leu-enkephalin		PDMS–PDMS	CZE	Amp (C)		Gawron et al. (2001)
	tri-, tetra-, penta-, octa-peptides		300 nM	Glass	IEF	LIF (Cy5)	Hofmann et al. (1999)
	Tyr-Gly-Gly		340 µM	PDMS–PDMS	CZE	Dual amp (C)	Martin et al. (2001b)
Protein	alpha-Lactalbumin		5 µM	PDMS–PDMS	CZE	LIF	Chabinyc et al. (2001)
	Anti-estradiol		4.3 nM LOD	Glass (soda lime)	CZE	LIF	Cheng et al. (2001)
	Anti-rabbit IgG			PDMS–Si		LIF(Cy5)	Hofmann et al. (2002)
	Bovine alpha-lactalbumin		85 nM LOD	Glass	CZE	LIF (NanoOrange)	Liu et al. (2000c)
	Bovine beta-lactoglobulin A and B		70 nM LOD	Glass	CZE	LIF (NanoOrange)	Liu et al. (2000c)
	BSA		0.1 mg/mL	Fused silica	CZE	LIF (OPA)	Fluri et al. (1996)
	CA		30 nM	Plastic	CZE	LIF	Bousse et al. (2001)
	Carbonic anhydrase		5 µM	PDMS–PDMS	CZE	LIF	Chabinyc et al. (2001)

Analyte	Sample	LOD	Material	Separation	Detection	Reference
CEA	In human serum		Quartz	Nil	TLM	Sato et al. (2001a)
Cyano-methemoglobin				Nil	UV	Takao et al. (2001)
DD			PDMS	Nil	LIF	Rowe et al. (1999)
Human C-reactive protein (CRP)			Glass	CGE	LIF	Juncker et al. (2002)
Human serum albumin (HSA)			Glass	Nil	LIF	Weigl and Yager (1999)
HSA			Glass	CGE	LIF	Kamholz et al. (1999)
Human T-cell protein tyrosine phosphatase			Glass (soda lime)	Pressure flow	LIF (DiFMU)	Kerby and Chien (2001)
Kinase			Plastic	CZE	LIF (FITC)	Xue et al. (2001)
Monoclonal mouse IgG (anti-BSA)	Mouse ascites fluid	34–134 µg/mL	Glass	CZE	LIF	Chiem and Harrison (1997)
Mouse IgG		10–60 µg/ml	Glass	CZE	CL (HRP/luminol)	Mangru and Harrison (1998)
Mouse IgG (anti-BSA)	Mouse ascites fluid	46.8 µg/ml	Glass	CZE	LIF	Qiu and Harrison (2001)
Myoglobin		30 µg/mL or 2.4 ng LOD	Quartz	IEF	UV absorb imaging	Mao and Pawliszyn (1999)
pI marker 6.6		0.3 µg/mL or 24 pg	Quartz	IEF	UV absorb imaging	Mao and Pawliszyn (1999)
s-IgA	Human saliva	200 µg/mL	Glass	Nil	TLM	Sato et al. (2000)
Thrombin		0.5 5 µg/mL	Au-Cu channel	Nil	Optical	Schmitt et al. (1996)

REFERENCES

Alarie, J. P., Jacobson, S. C., Culbertson, C. T., Ramsey, J. M. (2000). Effects of the electric field distribution on microchip valving performance. *Electrophoresis* 21(1):100–106.

Alarie, J. P., Jacobson, S. C., Broyles, B. S., Mcknight, T. E., Culbertson, C. T., Ramsey, J. M. (2001). Electroosmotically induced hydraulic pumping on microchips. Micro Total Analysis Systems 2001. Proceedings µTAS 2001 Symposium. 5th ed. Monterey, CA, Oct. 21–25, 2001, pp. 131–132.

Albin, M., Kowallis, R., Picozza, E., Raysberg, Y., Sloan, C., Winn-Deen, E., Woudenberg, T., Zupfer, J. (1996). Micromachining and microgeneticsw: what are they and where do they work together? *Transducer* Jun 2–6:253–257.

Anderson, J. R., Chiu, D. T., Jackman, R. J., Cherniavskaya, O., McDonald, J. C., Wu, H., Whitesides, S. H., Whitesides, G. M. (2000a). Fabrication of topologically complex three-dimensional microfluidic systems in PDMS by rapid prototyping. *Anal. Chem.* 72(14):3158–3164.

Andersson, H., Ahmadian, A., van der Wijngaart, W., Nilsson, P., Enoksson, P., Uhlen, M., Stemme, G. (2000b). Micromachined flow-through filter-chamber for solid phase DNA analysis. Micro Total Analysis Systems 2000. Proceedings of the µTAS Symposium. 4th ed. Enschede, The Netherlands, May 14–18, 2000, pp. 473–476.

Arora, A., Eijkel, J. C. T., Morf, W. E., Manz, A. (2001). A wireless electrochemiluminescence detector applied to direct and indirect detection for electrophoresis on a microfabricated glass device. *Anal. Chem.* 73:3282–3288.

Auroux, P.-A., Iossifidis, D., Reyes, D. R., Manz, A. (2002). Micro total analysis systems. 2. Analytical standard operations and applications. *Anal. Chem.* 74(12):2637–2652.

Backofen, U. B., Matysik, F., Lunte, C. E. (2002). A chip-based electrophoresis system with electrochemical detectin and hydrodynamic injection. *Anal. Chem.* 74:4054–4059.

Badal, M. Y., Wong, M., Chiem, N., Salimi-Moosavi, H., Harrison, D. J. (2001). Developing a routine coating method for multichannel flow networks on a chip using pyrolyzed poly(dimethylsiloxane). Micro Total Analysis Systems 2001. Proceedings µTAS 2001 Symposium. 5th ed. Monterey, CA, Oct. 21–25, 2001, pp. 535–536.

Badr, I. H. A., Johnson, R. D., Madou, M. J., Bachas, L. G. (2002). Fluorescent ion-selective optode membranes incorporated onto a centrifugal microfluidics platform. *Anal. Chem.* 74(21):5569–5575.

Bai, X., Josserand, J., Jensen, H., Rossier, J. S., Girault, H. H. (2002). Finite element simulation of pinched pressure-driven flow injection in microchannels. *Anal. Chem.* 74(24):6205–6215.

Baldock, S. J., Bektas, N., Fielden, P. R., Goddard, N. J., Pickering, L. W., Prest, J. E., Snook, R. D., Brown, B. J. T., Vaireanu, D. I. (1998). Isotachophoresis on planar polymeric substrates. Micro Total Analysis Systems '98. Proceedings µTAS '98 Workshop. Banff, Canada, 13–16 Oct. 1998, pp. 359–362.

Baldwin, R. P., Roussel, T. J. Jr., Crain, M. M., Bathlagunda, V., Jackson, D. J., Gullapalli, J., Conklin, J. A., Pai, R., N, J. F., Walsh, K. M., Keynton, R. S. (2002). Fully integrated on-chip electrochemical detection for capillary electrophoresis in a microfabricated device. *Anal. Chem.* 74:3690–3697.

Barker, S. L. R., Ross, D., Tarlov, M. J., Gaitan, M., Locascio, L. E. (2000a). Control of flow direction in microfluidic devices with polyelectrolyte multilayers. *Anal. Chem.* 72(24):5925–5929.

Barker, S. L. R., Tarlov, M. J., Branham, M., Xu, J., Maccrehan, W., Gaitan, M., Locascio, L. E. (2000b). Derivatization of plastic microfluidic device with polyelectrolyte multilayers. Micro Total Analysis Systems 2000. Proceedings of the μTAS Symposium. 4th ed. Enschede, The Netherlands, May 14–18, 2000, pp. 67–70.

Barker, S. L. R., Tarlov, M. J., Canavan, H., Hickman, J. J., Locascio, L. E. (2000c). Plastic microfluidic devices modified with polyelectrolyte multilayers. *Anal. Chem.* 72(20):4899–4903.

Becker, H., Gärtner, C. (2000). Polymer microfabrication methods for microfluidic analytical applications. *Electrophoresis* 21:12–26.

Becker, H., Dietz, W., Dannberg, P. (1998). Microfluidic manifolds by polymer hot embossing for μ-TAS applications. Micro Total Analysis Systems '98. Proceedings μTAS '98 Workshop. Banff, Canada, 13–16 Oct. 1998, pp. 253–256.

Beebe, D. J., Moore, J. S., Bauer, J. M., Yu, Q., Liu, R. H., Devadoss, C., Jo, B. (2000). Functional hydrogel structures for autonomous flow control inside microfluidic channels. *Nature* 404:588–590.

Beebe, D. J., Mensing, G., Moorthy, J., Khoury, C. M., Pearce, T. M. (2001). Alternative approaches to microfluidic systems design, construction and operation. Micro Total Analysis Systems 2001. Proceedings μTAS 2001 Symposium. 5th ed. Monterey, CA, Oct. 21–25, 2001, pp. 453–455.

van de Berg, A., Lammerink, T. S. J. (1998). Micro total analysis system: microfluidic aspect, integration concept and applications. In: *Microsystem Technology in Chemistry and Life Science*. Berlin Heidelberg: Springer-Verlag, pp. 21–50.

Berger, R., Gerber, C., Gimzewski, J. K. (1996). Nanometers, picowatts, femtojoules: thermal analysis and optical spectroscopy using micromechanics. Micro Total Analysis Systems '96. Proceedings of 2nd International Symposium on μTAS '96. Basel, 19–22 Nov. 1996, pp. 74–77.

Bernard, A., Michel, B., Delamarche, E. (2001). Micromosaic immunoassays. *Anal. Chem.*, 73(1):8–12.

Berthold, A., Nicola, L., Sarro, P. M., Vellekoop, M. J. (1999). Microfluidic device for airborne BTEX detection. *Transducers* 99:1324–1327.

Bharadwaj, R., Santiago, J. G. (2001). Optimization of field amplified sample stacking on a microchip. Micro Total Analysis Systems 2001. Proceedings μTAS 2001 Symposium, 5th ed. Monterey, CA, Oct. 21–25, 2001, pp. 613–614.

Bianchi, F., Wagner, F., Hoffmann, P., Girault, H. H. (2001). Electroosmotic flow in composite microchannels and implications in microcapillary electrophoresis systems. *Anal. Chem.* 73(4):829–836.

Bings, N. H., Wang, C., Skinner, C. D., Colyer, C. L., Thibault, P., Harrison, D. J. (1999). Microfluidic devices connected to fused-silica capillaries with minimal dead volume. *Anal. Chem.* 71(15):3292–3296.

Blankenstein, G., Scampavia, L., Branebjerg, J., Larsen, U. D., Ruzicka, J. (1996). Flow switch for analyte injection and cell/particle sorting. Micro Total Analysis Systems '96. Proceedings of 2nd International Symposium on μTAS '96. Basel, 19–22 Nov. 1996, pp. 82–84.

Bodor, R., Madajova, V., Kaniansky, D., Masar, M., Johnck, M., Stanislawski, B. (2001). Isotachophoresis and isotachophoresis—zone electrophoresis separations of inorganic anions present in water samples on a planar chip with column-coupling separation channels and conductivity detection. *J. Chromatogr. A* 916(1–2):155–165.

Boer, G., Dodge, A., Fluri, K., Schoot van der, B. H., Verpoorte, E., de Rooij, N. F. (1998). Studies of hydrostatic pressure effects in electrokinetically driven μTAS. Micro Total Analysis Systems '98. Proceedings μTAS '98 Workshop. Banff, Canada, 13–16 Oct. 1998, pp. 53–56.

Bökenkamp, D., Desai, A., Yang, X., Tai, Y.-C., Marzluff, E. M., Mayo, S. L. (1998). Microfabricated silicon mixers for submillisecond quench-flow analysis. *Anal. Chem.* 70(2):232–236.

Bohm, S., Greiner, K., Schlautmann, S., De Vries, S., Van Den Berg, A. (2001). A rapid vortex micromixer for studying high-speed chemical reactions. Micro Total Analysis Systems 2001. Proceedings μTAS 2001 Symposium. 5th ed. Monterey, CA, Oct. 21–25, 2001, pp. 25–27.

Boone, T. D., Hooper, H. H. (1998). Multiplexed, disposable, plastic microfluidic systems for high-throughput applications. Micro Total Analysis Systems '98. Proceedings μTAS '98 Workshop. Banff, Canada, 13–16 Oct. 1998, pp. 257–260.

Boone, T. D., Ricco, A. J., Gooding, P., Bjornson, T. O., Singh, S., Xiao, V., Gibbons, I., Williams, S. J., Tan, H. (2000). Sub-microliter assays and DNA analysis on plastic microfluidics. Micro Total Analysis Systems 2000. Proceedings of the μTAS Symposium. 4th ed. Enschede, The Netherlands, May 14–18, 2000, pp. 541–544.

Bousse, L., McReynolds, R. (1995). Micromachined flow-through measurement chambers using LAPS chemical sensors. Micro Total Anal. Syst. Proc. μTAS '94 Workshop. 1st ed. Meeting Date 1994, pp. 127–138.

Bousse, L., McReynolds, R. J., Kirk, G., Dawes, T., Lam, P., Bemiss, W. R., Parce, J. W. (1993). Micromachined multichannel systems for the measurement of cellular metabolism. *Transducer* 916–920.

Bousse, L., Minalla, A., West, J. (2000). Novel injection schemes for ultra-high speed DNA separations. Micro Total Analysis Systems 2000. Proceedings of the μTAS Symposium. 4th ed. Enschede, The Netherlands, May 14–18, 2000, pp. 415–418.

Bousse, L., Mouradian, S., Minalla, A., Yee, H., Williams, K., Dubrow, R. (2001). Protein sizing on a chip. *Anal. Chem.* 73(6):1207–1212.

Brahmasandra, S. N., Ugaz, V. M., Burke, D. T., Mastrangelo, C. H., Burns, M. A. (2001). Electrophoresis in microfabricated devices using photopolymerized polyacrylamide gels and electrode-defined sample injection. Electrophoresis 22(2):300–311.

Brivio, M., Tas, N. R., Fokkens, R. H., Sanders, R. G. P., Verboom, W., Reindhoudt, D. N., van de Berg, A. (2001).

Chemical microreactors in combination with mass spectrometry. Micro Total Analysis Systems 2001. Proceedings μTAS 2001 Symposium. 5th ed. Monterey, CA, Oct. 21–25, 2001, pp. 329–330.

Brivio, M., Fokkens, R. H., Verboom, W., Reinhoudt, D. N., Tas, N. R., Goedbloed, M., van den Berg, A. (2002). Integrated microfluidic system enabling (bio)chemical reactions with on-line MALDI-TOF mass spectrometry. *Anal. Chem.* 74:3972–3976.

Brody, J. P., Osborn, T. D., Forster, F. K., Yager, P. (1996). A planar microfabricated fluid filter. *Sensors Actuators A* 54:704–708.

Broyles, B. S., Jacobson, S. C., Ramsey, J. M. (2001). Sample concentration and separation on microchips. Micro Total Analysis Systems 2001. Proceedings μTAS 2001 Symposium. 5th ed. Monterey, CA, Oct. 21–25, 2001, pp. 537–538.

Bruin, G. J. M. (2000). Recent development in electrokinetically driven analysis on microfabricated devices. *Electrophoresis* 21:3931–3951.

Buch, J. S., Wang, P.-C., DeVoe, D. L., Lee, C. S. (2001). Field-effect flow control in a polydimethylsiloxane-based microfluidic system. *Electrophoresis* 22(18):3902–3907.

Burbaum, J. (1998). Engines of discovery. *Chem. Britain* June:38–41.

Burggraf, N., Manz, A., Verpoorte, E., Effenhauser, C. S., Widmer, H. M., de Rooij, N. F. (1994). A novel approach to ion separation in solution: synchronized cyclic capillary electrophoresis (SCCE). *Sensors Actuators B* 20:103–110.

Burggraf, N., Krattiger, B., de Rooij, N. F., Manz, A., de Mello, A. J. (1998). Holographic refractive index detector for application in microchip-based separation systems. *Analyst* 123(7):1443–1447.

Burns, M. A., Mastrangelo, C. H., Sammarco, T. S., Man, F. P., Webster, J. R., Johnson, B. N., Foerster, B., Jones, D., Fields, Y., Kaiser, A. R., Burke, D. T. (1996). Microfabricated structures for integrated DNA analysis. *Proc. Natl. Acad. Sci. USA* 93:5556–5561.

Burns, M. A., Johnson, B. N., Brahmasandra, S. N., Handique, K., Webster, J., Krishnan, M., Sammarco, T. S., Man, P. M., Jones, D., Heldsinger, D., Mastrangelo, C. H., Burke, D. T. (1998). An integrated nanoliter DNA analysis device. *Science* 282:484–487.

Cabodi, M., Chen, Yi-F., Turner, S., Craighead, H. (2001). Laterally asymmetric diffusion array with out-of-plane sample injection for continuous sorting of DNA molecules. Micro Total Analysis Systems 2001. Proceedings μTAS 2001 Symposium. 5th ed. Monterey, CA, Oct. 21–25, 2001, pp. 103–104.

Cabrera, C. R., Finlayson, B., Yager, P. (2001). Formation of natural ph gradients in a microfluidic device under flow conditions: model and experimental validation. *Anal. Chem.* 73(3):658–666.

Campaña, A. M. G., Baeyens, W. R. G., Aboul-Enein, H. Y., Zhang, X. (1998). Miniaturization of capillary electrophoresis system using micromachining techniques. *J. Microcolumn Sep.* 10(4):339–355.

Cantafora, A., Blotta, I., Bruzzese, N., Calandra, S., Bertolini, S. (2001). Rapid sizing of microsatellite alleles by gel electrophoresis on microfabricated channels: application to the D19S394 tetranucleotide repeat for cosegregation study of familial hypercholesterolemia. *Electrophoresis* 22(18):4012–4015.

Chabinyc, M. L., Chiu, D. T., McDonald, J. C., Stroock, A. D., Christian, J. F., Karger, A. M., Whitesides, G. M. (2001). An integrated fluorescence detection system in poly (dimethylsiloxane) for microfluidic applications. *Anal. Chem.* 73(18):4491–4498.

Chan, J. H., Timperman, A. T., Qin, D., Aebersold, R. (1999b). Microfabricated polymer devices for automated sample delivery of peptides for analysis by electrospray ionization tandem mass spectrometry. *Anal. Chem.* 71:4437–4444.

Chen, C. S., Mrksich, M., Huang, S., Whitesides, G. M., Ingber, D. E. (1998). Micropatterned surfaces for control of cell shape, position, and function. *Biotechnol. Prog.* 14(3):356–363.

Chen, Y.-H., Wang, W.-C., Young, K.-C., Chang, T.-Y., Chen, S.-H. (1999). Plastic microchip electrophoresis for analysis of PCR products of hepatitis C virus. *Clin. Chem. (Washington, D.C.)* 45(11):1938–1943.

Chen, D., Hsu, F., Zhan, D., Chen, C. (2001a). Palladium film decoupler for amperometric detection in electrophoresis chips. *Anal. Chem.* 73:758–762.

Chen, S.-H., Sung, W.-C., Lee, G.-B., Lin, Z.-Y., Chen, P.-W., Liao, P.-C. (2001b). A disposable poly(methylmethacrylate)-based microfluidic module for protein identification by nanoelectrospray ionization–tandem mass spectrometry. *Electrophoresis* 22(18):3972–3977.

Chen, S., Lin, Y., Wang, L., Lin, C., Lee, G. (2002). Flow-through sampling for electrophoresis-based microchips and their applications for protein analysis. *Anal. Chem.* 74:5146–5153.

Cheng, J., Shoffner, M. A., Hvichia, G. E., Kricka, L. J., Wilding, P. (1996a). Chip PCR. II. Investigation of different PCR amplification systems in microfabricated silicon-glass chips. *Nucl. Acids Res.* 24:380–385.

Cheng, J., Shoffner, M. A., Mitchelson, K. R., Kricka, L. J., Wilding, P. (1996b). Analysis of ligase chain reaction products amplified in a silicon-glass chip using capillary electrophoresis. *J. Chromatogr. A* 732:151–158.

Cheng, J., Kricka, L. J., Sheldon, E. L., Wilding, P. (1998a). Sample preparation in microstructure device. In: *Microsystem Technology in Chemistry and Life Science*. Berlin, Heidelberg: Springer-Verlag, pp. 215–232.

Cheng, S. B., Skinner, C. D., Harrison, D. J. (1998b). Integrated serial dilution on a microchip for immunoassay sample treatment and flow injection analysis. Micro Total Analysis Systems '98. Proceedings μTAS '98 Workshop. Banff, Canada, 13–16 Oct. 1998, pp. 157–160.

Cheng, J., Waters, L. C., Fortina, P., Hvichia, G., Jacobson, S. C., Ramsey, J. M., Kricka, L. J., Wilding, P. (1998c). Degenerate oligonucleotide primed-polymerase chain reaction and capillary electrophoretic analysis of human DNA on microchip-based devices. *Anal. Biochem.* 257(2):101–106.

Cheng, S. B., Skinner, C. D., Taylor, J., Attiya, S., Lee, W. E., Picelli, G., Harrison, D. J. (2001). Development of a multichannel microfluidic analysis system employing affinity capillary electrophoresis for immunoassay. *Anal. Chem.* 73(7):1472–1479.

Chi, J., Kim, S., Trichur, R., Cho, H. J., Puntambekar, A., Cole, R. L., Simkins, J., Murugesan, S., Kim, K., Lee, J., Beaucage, G., Nevin, J. H., Ahn, C. H. (2001). A plastic micro injection molding technique using replaceable mold-disks for disposable microfluidic system and biochips. Micro Total Analysis Systems 2001. Proceedings μTAS 2001 Symposium. 5th ed. Monterey, CA, Oct. 21–25, 2001, pp. 411–412.

Chiem, N., Harrison, D. J. (1997). Microchip-based capillary electrophoresis for immunoassays: analysis of monoclonal antibodies and theophylline. *Anal. Chem.* 69:373–378.

Chiem, N., Lockyear-Shultz, L., Andersson, P., Skinner, C., Harrision, D. J. (2000). Room temperature bonding of micromachined glass devices for capillary electrophoresis. *Sensors Actuators B* 63:147–152.

Chmela, E., Tijssen, R., Blom, M. T., Gardeniers, H. J. G. E., van den Berg, A. (2002). A chip system for size separation of macromolecules and particles by hydrodynamic chromatography. *Anal. Chem.* 74:3470–3475.

Choi, J.-W., Hong, C.-C., Ahn, C. H. (2001). An electrokinetic active micromixer. Micro Total Analysis Systems 2001. Proceedings μTAS 2001 Symposium. 5th ed. Monterey, CA, Oct. 21–25, 2001, pp. 621–622.

Chou, H., Spence, C., Scherer, A., Quake, S. (1999). A microfabricated device for sizing and sorting DNA molecules. *Proc. Natl. Acad. Sci. USA* 96:11–13.

Chou, C. F., Changrani, R., Roberts, P., Sadler, D., Lin, S., Mulholland, A., Swami, N., Terbrueggen, R., Zenhausern, F. (2001). A miniaturized cyclic PCR device. Micro Total Analysis Systems 2001. Proceedings μTAS 2001 Symposium. 5th ed. Monterey, CA, Oct. 21–25, 2001, pp. 151–152.

Chudy, M., Wróblewaski, W., Dybko, A., Brzózka, Z. (2001). PMMA/PDMS based microfluidic system with optical detection for total heavey metals concencentration assessment. Micro Total Analysis Systems 2001. Proceedings μTAS 2001 Symposium. 5th ed. Monterey, CA, Oct. 21–25, 2001, pp. 521–522.

Cohen, C. B., Chin-dixon, E., Jeong, S., Nikiforov, T. T. (1999). A microchip-based enzyme assay for protein kinase A. *Anal. Biochem.* 273:89–97.

Colyer, C. L., Tang, T., Chiem, N., Harrison, D. J. (1997a). Clinical potential of microchip capillary electrophoresis systems. *Electrophoresis* 18:1733–1741.

Colyer, C. L., Mangru, S. D., Harrison, D. J. (1997b). Microchip-based capillary electrophoresis of human serum proteins. *J. Chromatog. A* 781(1 + 2):271–276.

Cowen, S., Craston, D. H. (1995). An on-chip miniature liquid chromatography system: design, construction and characterization. Micro Total Anal. Syst. Proc. μTAS '94 Workshop. 1st ed. Meeting Date 1994, pp. 295–298.

Crabtree, H. J., Kopp, M. U., Manz, A. (1999). Shah convolution Fourier Transform detection. *Anal. Chem.* 71:2130–2138.

Culbertson, C. T., Jacobson, S. C., Ramsey, J. M. (1998). Dispersion sources for compact geometries on microchips. *Anal. Chem.* 70(18):3781–3789.

Culbertson, C. T., Jacobson, S. C., Ramsey, J. M. (2000a). Microchip devices for high-efficiency separations. *Anal. Chem.* 72(23):5814–5819.

Culbertson, C. T., Ramsey, R. S., Ramsey, J. M. (2000b). Electroosmotically induced hydraulic pumping on microchips: differential ion transport. *Anal. Chem.* 72(10):2285–2291.

Culbertson, C. T., Alarie, J. P., Mcclain, M. A., Jacobson, S. C., Ramsery, J. M. (2001). Rapid cellular assays on microfabricated fluidic device. Micro Total Analysis Systems 2001. Proceedings μTAS 2001 Symposium. 5th ed. Monterey, CA, Oct. 21–25, 2001, pp. 285–286.

Darhuber, A. A., Davis, J. M., Reisner, W. W., Troian, S. M. (2001). Thermocapillary migration of liquids on patterned surfaces: design concept for microfluidic delivery. Micro Total Analysis Systems 2001. Proceedings μTAS 2001 Symposium. 5th ed. Monterey, CA, Oct. 21–25, 2001, pp. 244–246.

Darling, R. B., Yager, P., Weigl, B., Kriebel, J., Mayes, K. (1998). Integration of microelectrodes with etched microchannels for in-stream electrochemical analysis. Micro Total Analysis Systems '98. Proceedings μTAS '98 Workshop. Banff, Canada, 13–16 Oct. 1998, pp. 105–108.

Delamarche, E., Bernard, A., Schmid, H., Michel, B., Biebuyck, H. (1997). Patterned delivery of immunoglobulins to surfaces using microfluidic networks. *Science* 276:779–781.

Deng, Y., Henion, J. (2001). Chip-based capillary electrophoresis/mass spectrometry determination of carnitines in human urine. *Anal. Chem.* 73:639–646.

Deng, T., Wu, H., Brittain, S. T., Whitesides, G. M. (2000). Prototyping of masks, masters, and stamps/molds for soft lithography using an office printer and photographic reduction. *Anal. Chem.* 72(14):3176–3180.

Deng, Y., Zhang, H., Henion, J. (2001). Chip-based quantitative capillary electrophoresis/mass spectrometry determination of drugs in human plasma. *Anal. Chem.* 73:1432–1439.

Dertinger, S. K. W., Chiu, D. T., Jeon, N. L., Whitesides, G. M. (2001). Generation of gradients having complex shapes using microfluidic networks. *Anal. Chem.* 73(6):1240–1246.

Deshpande, M., Greiner, K. B., West, J., Gilbert, J. R., Bousse, L., Minalla, A. (2000). Novel designs for electrokinetic injection in μTAS. Micro Total Analysis Systems 2000. Proceedings of the μTAS Symposium. 4th ed. Enschede, The Netherlands, May 14–18, 2000, pp. 339–342.

Devasenathipathy, S., Santiago, J. G., Takehara, K. (2002). Particle tracking techniques for electrokinetic microchannel flows. *Anal. Chem.* 74(15):3704–3713.

Dittrich, P. S., Schwille, P. (2002). Spatial two-photon fluorescence cross-correlation spectroscopy for controlling molecular transport in microfluidic structures. *Anal. Chem.* 74(17):4472–4479.

Divakar, R., Butler, D., Papautsky, I. (2001). Room temperature low-cost UV-cured adhesive bonding for microfluidic biochips. Micro Total Analysis Systems 2001. Proceedings μTAS 2001 Symposium. 5th ed. Monterey, CA, Oct. 21–25, 2001, pp. 85–386.

Dodge, A., Fluri, K., Linder, V., Lettieri, G., Linchtenberg, J., Verpoorte, E., de Rooij, N. F. (2000). Valveless, sealed microfluidic device for automated heterogeneous immunoassay: design and operational consideratios. Micro Total Analysis Systems 2000. Proceedings of the μTAS Symposium. 4th ed. Enschede, The Netherlands, May 14–18, 2000, pp. 407–410

Dodge, A., Fluri, K., Verpoorte, E., de Rooij, N. F. (2001). Electrokinetically driven microfluidic chips with surface-modified chambers for heterogeneous immunoassays. *Anal. Chem.* 73(14):3400–3409

Dolnik, V., Liu, S., Jovanovich, S. (2000). Capillary electrophoresis on microchip. *Electrophoresis* 21:41–54.

Duffy, D. C., McDonald, J. C., Schueller, O. J. A., Whitesides, G. M. (1998). Rapid prototyping of microfluidic systems in poly(dimethylsiloxane). *Anal. Chem.* 70(23):4974–4984.

Duffy, D. C., Gillis, H. L., Lin, J., Sheppard, N. F. Jr., Kellogg, G. J. (1999). Microfabricated centrifugal microfluidic systems: characterization and multiple enzymatic assays. *Anal. Chem.* 71(20):4669–4678.

Dutta, D., Leighton, D. T. Jr. (2002). A low dispersion geometry for microchip separation devices. *Anal. Chem.* 74:1007–1016.

Effenhauser, C. S. (1998). Integrated chip-based microcolumn separation systems. In: *Microsystem Technology in Chemistry and Life Science*. Berlin, Heidelberg: Springer-Verlag, pp. 51–82.

Effenhauser, C. S., Manzz, A., Widmer, H. M. (1993). Glass chips for high-speed capillary electrophoresis separation with submicrometer plate heights. *Anal. Chem.* 65:2637–2642.

Effenhauser, C. S., Paulus, A., Manz, A., Widmer, H. M. (1994). High-speed separation of antisense oligonucleotides on a micromachined capillary electrophoresis device. *Anal. Chem.* 66:2949–2953.

Effenhauser, C. S., Manz, A., Widmer, H. M. (1995). Manipulation of sample fractions on a capillary electrophoresis chip. *Anal. Chem.* 67:2284–2287.

Effenhauser, C. S., Bruin, G. J. M., Paulus, A. (1997a). Integrated chip-based capillary electrophoresis. *Electrophoresis* 18:2203–2213.

Effenhauser, C. S., Bruin, G. J. M., Paulus, A., Ehrat, M. (1997b). Integrated capillary electrophoresis on flexible silicone microdevices: analysis of DNA restriction fragments and detection of single DNA molecules on microchips. *Anal. Chem.* 69:3451–3457.

Ehrlich, D., Adourian, A., Barr, C., Breslau, D., Buonocore, S., Burger, R., Carey, L., Carson, S., Chiou, J., Dee, R., Desmarais, S., El-Difrawy, S., King, R., Koutny, L., Lam, R., Matsudaira, P., Mitnik-Gankin, L., O'Neil, T., Novotny, M., Saber, G., Salas-Solano, O., Schmalzing, D., Srivastava, A., Vazquez, M. (2001). BioMEMS-768 DNA sequencer. Micro Total Analysis Systems 2001. Proceedings μTAS 2001 Symposium. 5th ed. Monterey, CA, Oct. 21–25, 2001, pp. 6–18.

Eijkel, J. C. T., Stoeri, H., Manz, A. (1999). A molecular emission detector on a chip employing a direct current microplasma. *Anal. Chem.* 71(14):2600–2606.

Eijkel, J. C. T., Stoeri, H., Manz, A. (2000a). A dc microplasma on a chip employed as an optical emission detector for gas chromatography. *Anal. Chem.* 72:2547–2552.

Eijkel, J. C. T., Stoeri, H., Manz, A J. (2000b). An atmospheric pressure dc glow discharge on a microchip and its application as a molecular emission detector. *J. Anal. Atomic. Spectrom.* 15:297–300.

Eijkel, J. C. T., Dalton, C., Hayden, C. J., Drysdale, J. A., Kwok, Y. C., Manz, A. (2001). Development of a micro system for circular chromatography using wavelet transform detection. Micro Total Analysis Systems 2001. Proceedings μTAS 2001 Symposium. 5th ed. Monterey, CA, Oct. 21–25, 2001, pp. 541–542.

Ekström, S., Önnerfjord, P., Nilsson, J., Bengtsson, M., Laurell, T., Marko-Varga, G. (2000). Integrated microanalytical technology enabling rapid and automated protein identification. *Anal. Chem.* 72:286–293.

Elwenspoek, M., Lammerink, T. S. J., Miyake, R., Fluitman, J. H. (1994). Towards integrated microliquid handling systems. *J. Micromech. Microeng.* 4:227–245.

Emrich, C. A., Tian, H., Medintz, I. L., Mathies, R. A. (2002). Microfabricated 384-lane capillary array electrophoresis bioanalyzer for ultrahigh-throughput genetic analysis. *Anal. Chem.* 74:5076–5083.

Enkins, G., Manz, A. (2001). Optical emission detection of liquid analytes using a micro-machined d.c. glow-discharge device at atmospheric pressure. Micro Total Analysis Systems 2001. Proceedings μTAS 2001 Symposium. 5th ed. Monterey, CA, Oct. 21–25, 2001, pp. 349–350.

Ericson, C., Holm, J., Ericson, T., Hjerten, S. (2000). Electroosmosis- and pressure-driven chromatography in chips using continuous beds. *Anal. Chem.* 72(1):81–87.

Ermakov, S. V., Jacobson, S. C., Ramsey, J. M. (1998). Computer simulations of electrokinetic transport in microfabricated channel structures. *Anal. Chem.* 70(21):4494–4504.

Esch, M. B., Locascio, L. E., Tarlov, M. J., Durst, R. A. (2001). Detection of visible *Cryptosporidium parvum* using DNA-modified liposomes in a microfluidic chip. *Anal. Chem.* 73(13):2952–2958.

Fan, Z. H., Harrison, D. J. (1994). Micromachining of capillary electrophoresis injectors and separators on glass chips and evaluation of flow at capillary intersections. *Anal. Chem.* 66(1):177–184.

Fan, Z. H., Mangru, S., Granzow, R., Heaney, P., Ho, W., Dong, Q., Kumar, R. (1999). Dynamic DNA hybridization on a chip using paramagnetic beads. *Anal. Chem.* 71:4851–4859.

Fan, Z. H., Tan, W., Tan, H., Qiu, X. C., Boone, T. D., Kao, P., Ricco, A. J., Desmond, M., Bay, S., Hennessy, K. (2001). Plastic microfluidic devices for DNA sequencing and protein separations. Micro Total Analysis Systems 2001. Proceedings μTAS 2001 Symposium. 5th ed. Monterey, CA, Oct. 21–25, 2001, pp. 19–21.

Fang, Q., Xu, G.-M., Fang, Z.-L. (2001). High throughput continuous sample introduction interfacing for microfluidic chip-based capillary electrophoresis systems. Micro Total Analysis Systems 2001. Proceedings μTAS 2001 Symposium. 5th ed. Monterey, CA, Oct. 21–25, 2001, pp. 373–374.

Fanguy, J. C., Henry, C. S. (2002). The analysis of uric acid in urine using microchip capillary electrophoresis with electrochemical detection. *Electrophoresis* 23(5):767–773.

Fiedlerr, S., Shirley, S. G., Schnelle, T., Fuhr, G. (1998). Dielectrophoretic sorting of particles and cells in a microsystem. *Anal. Chem.* 70:1909–1915.

Fiehn, H., Howitz, S., Pham, M. T., Vopel, T., Bürger, M., Wegner, T. (1995). Components and technology for a fluidic-isfet-microsystem. Micro Total Analysis Systems '94. Proceedings of the μTAS '94 Workshop. University of Twente, The Netherlands, 21–22 Nov. 1994, pp. 289–293.

Fielden, P. R., Baldock, S. J., Goddard, N. J., Pickering, L. W., Prest, J. E., Snook, R. D., Brown, B. J. T., Vaireanu, D. I. (1998). A miniaturized planar isotachophoresis separation device for transition metals with integrated conductivity detection. Micro Total Analysis Systems '98. Proceedings μTAS '98 Workshop. Banff, Canada, 13–16 Oct. 1998, pp. 323–326.

Figeys, D., Aebersold, R. (1998). Nanoflow solvent gradient delivery from a microfabricated device for protein identifications by electrospray inonization mass spectrometry. *Anal. Chem.* 70:3721–3727.

Figeys, D., Pinto, D. (2000). Lab-on-a-chip: a revolution in biological and medical sciences. *Anal. Chem.* May 1: 330a–335A.

Figeys, D., Ning, Y., Aebersold, R. (1997). A microfabricated device for rapid protein identification by microelectrospray ion trap mass spectrometry. *Anal. Chem.* 69:3153–3160.

Figeys, D., Gypi, S. P., Mckinnon, G., Aebersold, R. (1998). An integrated microfluidics-tandem mass spectrometry system for automated protein analysis. *Anal. Chem.* 70:3728–3734.

Finot, M., Jemere, A. B., Oleschuk, R. D., Takahashi, L., Harrison, D. J. (2001). High throughput pharmaceutical formulation evaluation and analysis using capillary electrochromatography on a microfluidic chip. Micro Total Analysis Systems 2001. Proceedings μTAS 2001 Symposium. 5th ed. Monterey, CA, Oct. 21–25, 2001, pp. 480–482.

Firebaugh, S. L., Jensen, K. F., Schmidt, M. A. (2000). Miniaturization and integration of photoacoustic detection with a microfabricated chemical reactor system. Micro Total Analysis Systems 2000. Proceedings of the μTAS Symposium. 4th ed. Enschede, The Netherlands, May 14–18, 2000, pp. 49–52.

Fister, J. C. III, Jacobson, S. C., Davis, L. M., Ramsey, J. M. (1998). Counting single chromophore molecules for ultrasensitive analysis and separations on microchip devices. *Anal. Chem.* 70(3):431–437.

Fister, J. C., III Jacobson, S. C., Ramsey, J. M. (1999). Ultrasensitive cross-correlation electrophoresis on microchip devices. *Anal. Chem.* 71(20):4460–4464.

Floyd, T. M., Schmidt, M. A., Jensen, K. F. (2001). A silicon microchip for infrared transmission kinetics studies of rapid homogeneous liquid reactions. Micro Total Analysis Systems 2001. Proceedings μTAS 2001 Symposium. 5th ed. Monterey, CA, Oct. 21–25, 2001, pp. 277–279.

Fluri, K., Fitzpatrick, G., Chiem, N., Harrison, D. J. (1996). Integrated capillary electrophoresis devices with an efficient postcolumn reactor in planar quartz and glass chips. *Anal. Chem.* 68(23):4285–4290.

Fluri, K., Lettieri, G. L., Schoot, B. H. V. D., Verpoorte, E., de Rooij, N. F. (1998). Chip-based heterogeneous immunoassay for clinical diagnostic applications. Micro Total Analysis Systems '98. Proceedings μTAS '98 Workshop. Banff, Canada, 13–16 Oct. 1998, pp. 347–350.

Ford, S. M., Kar, B., Mcwhorter, S., Davies, J., Soper, S. A., Klopf, M., Calderon, G., Saile, V. (1998). Microcapillary electrophoresis device fabricated using polymeric substrates and X-ray lithography. *J. Microcol. Separations.* 10(5):413–422.

Forssen, L., Elderstig, H., Eng, L., Nordling, M. (1995). Integration of an amperometric glucose sensor in a μ-TAS. Micro Total Anal. Syst. Proc. μTAS '94 Workshop. 1st ed. Meeting Date 1994, pp. 203–207.

Fruetel, J., Renzi, R., Crocker, R., VanderNoot, V., Stamps, J., Shokair, I., Yee, D. (2001). Application of microseparation arrays to the detection of biotoxins in aerosol backgrounds. Micro Total Analysis Systems 2001. Proceedings μTAS 2001 Symposium. 5th ed. Monterey, CA, Oct. 21–25, 2001, pp. 523–524.

Frye-Mason, G., Kottenstette, R. J., Heller, E. J., Matzke, C. M., Casalnuovo, S. A., Lewis, P. R., Manginell, R. P., Schubert, W. K., Hietala, V. M., Shul, R. J. (1998). Integrated chemical analysis systems for gas phase CW agent detection. Micro Total Analysis Systems '98. Proceedings μTAS '98 Workshop. Banff, Canada, 13–16 Oct. 1998, pp. 477–481.

Frye-Mason, G., Kottenstette, R., Mowry, C., Morgan, C., Manginell, R., Lewis, P., Matzke, C., Dulleck, G., Anderson, L., Adkins, (2001). Expanding the capabilities and applications of gas phase miniature chemical analysis systems (μChemLab). Micro Total Analysis Systems 2001. Proceedings μTAS 2001 Symposium. 5th ed. Monterey, CA, Oct. 21–25, 2001, pp. 658–660.

Fu, L.-M., Yang, R.-J., Lee, G.-B., Liu, H.-H. (2002a). Electrokinetic injection techniques in microfluidic chips. *Anal. Chem.* 74(19):5084–5091.

Fu, A. Y., Chou, H., Spence, C., Arnold, F. H., Quake, S. R. (2002b). An integrated microfabricated cell sorter. *Anal. Chem.* 74:2451–2457.

Fuhr, G., Shirley, S. G. (1998). Biological application of microstructures. In: *Microsystem Technology in Chemistry and Life Science*. Berlin, Heidelberg: Springer-Verlag, pp. 83–116.

Fuhr, G., Wagner, B. (1995). Electric field mediated cell manipulation, characterization and cultivation in highly conductive media. Micro Total Anal. Syst. Proc. μTAS '94 Workshop. 1st ed. Meeting Date 1994, pp. 209–214.

Fujii, T., Hosokawa, K., Shoji, S., Yotsumoto, A., Nojima, T., Endo, I. (1998). Development of a microfabricated biochemical workbench-improving the mixing efficiency. Micro Total Analysis Systems '98. Proceedings μTAS '98 Workshop. Banff, Canada, 13–16 Oct. 1998, pp. 193–176.

Gallardo, B. S., Gupta, V. K., Eagerton, F. D., Jong, L. I., Craig, T. V. S., Shah, R. R., Abbott, N. L. (1999). Electrochemical principles for active control of liquids on sub-millimeter scales. *Science* 283:57–61.

Gao, J., Xu, J., Locascio, L. E., Lee, C. S. (2001). Integrated microfluidic system enabling protein digestion, p. separation, and protein identification. *Anal. Chem.* 73(11):2648–2655.

Gawron, A. J., Martin, R. S., Lunte, S. M. (2001). Fabrication and evaluation of a carbon-based dual-electrode detector for poly(dimethylsiloxane) electrophoresis chips. *Electrophoresis* 22(2):242–248.

Giordano, B. C., Ferrance, J., Swedberg, S., Huhmer, A. F. R., Landers, J. P. (2001a). Polymerase chain reaction in polymeric microchips: DNA amplification in less than 240 seconds. *Anal. Biochem.* 291(1):124–132.

Giordano, B. C., Couch, A. J., Ahmadzadeh, H., Jin, L. J., Landers, J. P. (2001b). Dynamic labeling of protein-sodium dodecyl sulfate (SDS) complexes for laser induced fluorescence (LIF) detection on microchips. Micro Total Analysis Systems 2001. Proceedings μTAS 2001 Symposium. 5th ed. Monterey, CA, Oct. 21–25, 2001, pp. 109–110.

Goranovic, G., Klank, H., Westergaard, C., Geschke, O., Telleman, P., Kutter, J. P. (2001). Characterization of flow in laser-machined polymeric microchannels. Micro Total Analysis Systems 2001. Proceedings μTAS 2001 Symposium. 5th ed. Monterey, CA, Oct. 21–25, 2001, pp. 623–624.

Gottschlich, N., Jacobson, S. C., Culbertson, C. T., Ramsey, J. M. (2001). Two-demensional electrochromatography/capillary electrophoresis on a microchip. *Anal. Chem.* 73:2669–2674.

Gottwald, E., Muller, O., Polten, A. (2001). Semiquantitative reverse transcription-polymerase chain reaction with the Agilent 2100 Bioanalyzer. *Electrophoresis* 22(18):4016–4022.

Grad, G., Müller, T., Pfennig, A., Shirley, S., Schnelle, T., Fuhr, G. (2000). New micro devices for single cell analysis, cell sorting and cloning-on-a0chip: the cytocon instrument. Micro Total Analysis Systems 2000. Proceedings of the μTAS Symposium. 4th ed. Enschede, The Netherlands, May 14–18, 2000, pp. 443–446.

Grass, B., Neyer, A., Johnck, M., Siepe, D., Eisenbeiss, F., Weber, G., Hergenroder, R. (2001). A new PMMA-microchip device for isotachophoresis with integrated conductivity detector. *Sensors Actuators B: Chem.* B72(3):249–258.

Gravesen, P., Branebjerg, J., Jensen, O. S. (1993). Microfluidics—a review. *J. Micromech. Microeng.* 3:168–182.

Green, N. G. (2001). Integration of a solid state micropump and a sub-micrometre particle analyser/separator. Micro Total Analysis Systems 2001. Proceedings μTAS 2001 Symposium. 5th ed. Monterey, CA, Oct. 21–25, 2001, pp. 545–546.

Greenway, G. M., Nelstrop, L. J., Port, S. N. (2000a). Tris(2,2-bipyridyl)ruthenium (II) chemiluminescence in a microflow injection system for codeine determination. *Anal. Chim. Acta* 405(1–2):43–50.

Greenway, G. M., Nelstrop, L. J., McCreedy, T., Greenwood, P. (2000b). Luminol chemiluminescence systems for metal analysis by μTAS. Micro Total Analysis Systems 2000. Proceedings of the μTAS Symposium. 4th ed. Enschede, The Netherlands, May 14–18, 2000, pp. 363–366.

Greenwood, P. A., Greenway, G. M. (2001). Development of a μTAS screening device for drug analysis by chemiluminescence. Micro Total Analysis Systems 2001. Proceedings μTAS 2001 Symposium. 5th ed. Monterey, CA, Oct. 21–25, 2001, pp. 343–344.

Griffiths, S. K., Nilson, R. H. (2000). Band spreading in two-dimensional microchannel turns for electrokinetic species transport. *Anal. Chem.* 72(21):5473–5482.

Griffiths, S. K., Nilson, R. H. (2001). Modeling electrokinetic transport for the design and optimization of microchannel systems. Micro Total Analysis Systems 2001. Proceedings μTAS 2001 Symposium. 5th ed. Monterey, CA, Oct. 21–25, 2001, pp. 456–458.

Griffiths, S. K., Nilson, R. H. (2002). Design and analysis of folded channels for chip-based separations. *Anal. Chem.* 74(13):2960–2967.

Grzybowski, B. A., Haag, R., Bowden, N., Whitesides, G. M. (1998). Generation of micrometer-sized patterns for microanalytical applications using a laser direct-write method and microcontact printing. *Anal. Chem.* 70(22):4645–4652.

Guber, A. E., Dittrich, H., Heckele, M., Herrmann, D., Musliga, A., Pfleging, W., Schaller, Th. (2001). Polymer micro needles with through-going capillaries. Micro Total Analysis Systems 2001. Proceedings μTAS 2001 Symposium. 5th ed. Monterey, CA, Oct. 21–25, 2001, pp. 155–156.

Guijt, R. M., Lichtenberg, J., Baltussen, E., Verpoorte, E., de Rooij, N. F., van Dedem, G. W. K. (2001a). Indirect electro-osmotic pumping for direct sampling from bioreactors. Micro Total Analysis Systems 2001. Proceedings μTAS 2001 Symposium. 5th ed. Monterey, CA, Oct. 21–25, 2001, pp. 399–400.

Guijt, R. M., Baltussen, E., Van der Steen, G., Schasfoort, R. B. M., Schlautmann, S., Billiet, H. A. H., Frank, J., Van Dedem, G. W. K., Van den Berg, A. (2001b). New approaches for fabrication of microfluidic capillary electrophoresis devices with on-chip conductivity detection. *Electrophoresis* 22(2):235–241.

Haab, B. B., Mathies, R. A. (1999). Single-molecule detection of DNA separations in microfabricated capillary electrophoresis chips employing focused molecular streams. *Anal. Chem.*, 71(22):5137–5145.

Hadd, A. G., Jacobson, S. C., Ramsey, J. M. (1999). Microfluidic assays of acetylcholinesterase inhibitors. *Anal. Chem.* 71(22):5206–5212.

Hadd, A. G., Raymond, D. E., Halliwell, J. W., Jacobson, S. C., Ramsey, J. M. (1997). Microchip device for performing enzyme assays. *Anal. Chem.* 69(17):3407–3412.

Han, J., Craighead, H. G. (2000). Separation of long DNA molecules in a microfabricated entropic trap array. *Science* 288:1026–1029.

Handique, K., Burke, D. T., Mastrangelo, C. H., Burns, M. A. (2000). Nanoliter liquid metering in microchannels using hydrophobic patterns. *Anal. Chem.* 72(17):4100–4109.

Handique, K., Burke, D. T., Mastrangelo, C. H., Burns, M. A. (2001). On-chip thermopneumatic pressure for discrete drop pumping. *Anal. Chem.* 73(8):1831–1838.

Hannoe, S., Sugimoto, I., Katoh, T. (1998). Silicon-micromachined separation columns coated with amino acid films for an integrated on-chip gas chromatograph. Micro Total Analysis Systems '98. Proceedings μTAS '98 Workshop. Banff, Canada, 13–16 Oct. 1998, pp. 145–148.

Harrison, D. J., Chiem, N. (1996). Microchip lab for biochemical analysis. Micro Total Analysis Systems '96. Proceedings of 2nd International Symposium on μTAS '96. Basel, 19–22 Nov. 1996, pp. 31–33.

Harrison, D. J., Manz, A., Glavina, P. G. Electroosmotic pumping within a chemical sensor system integrated on Silicon. *Transducer* 91:792–795.

Harrison, D. J., Manz, A., Fan, Z., Lüdi, H., Wildmer, H. M. (1992). Capillary electrophoresis and sample injection systems integrated on a planar glass chip. *Anal. Chem.* 64:1926–1932.

Harrison, D. J., Fluri, K., Seiler, K., Fan, Z., Effenhauser, C. S., Manz, A. (1993). Micromachining a miniaturized capillary electrophoresis-based chemical analysis system on a chip. *Science* 261:895–897.

Harrison, D. J., Fluri, K., Chiem, N., Tang, T., Fan, Z. (1996). Micromachining chemical and biochemical analysis and reaction systems on glass substrates. *Sensors Actuators B* 33:105–109.

Harrison, D. J., Majid, E., Attiya, S., Jiang, G. (2001). Enhancing the microfluidic toolbox for functional genomics and recombinant DNA methods. Micro Total Analysis Systems 2001. Proceedings μTAS 2001 Symposium. 5th ed. Monterey, CA, Oct. 21–25, 2001, pp. 10–12.

Hartshorne, H., Ning, Y., Lee, W. E., Backhouse, C. (1998). Development of microfabricated valves for μTAS. Micro Total Analysis Systems '98. Proceedings μTAS '98 Workshop. Banff, Canada, 13–16 Oct. 1998, pp. 379–381.

Hasselbrink, E. F. Jr., Shepodd, T. J., Rehm, J. E. (2002). High-pressure microfluidic control in lab-on-a-chip devices using mobile polymer monoliths. *Anal. Chem.* 74(19):4913–4918.

Haswell, S. J. (1997). Development and operating characteristics of micro flow injection analysis systems based on electroosmotic flow. *Analyst* 122:1R–10R.

He, B., Tait, N., Regnier, F. (1998). Fabrication of nanocolumns for liquid chromatography. *Anal. Chem.* 70(18):3790–3797.

He, B., Tan, L., Regnier, F. (1999). Microfabricated filters for microfluidic analytical systems. *Anal. Chem.* 71:1464–1468.

He, B., Burke, B. J., Zhang, X., Zhang, R., Regnier, F. E. (2001). A picoliter-volume mixer for microfluidic analytical systems. *Anal. Chem.* 73(9):1942–1947.

Hediger, S., Fontannaz, J., Sayah, A., Hunziker, W., Gijs, M. A. M. (2000). *Sens. Actuators B* 63:63–73.

von Heeren, F., Verpoorte, E., Manz, A., Thormann, W. (1996a). Micellar electrokinetic chromatography separations and analyses of biological samples on a cyclic planar microstructure. *Anal. Chem.* 68(13):2044–2053.

von Heeren, F., Verpoorte, E., Manz, A., Thormann, W. (1996b). Characterization of electrophoretic sample injection and separation in a gel-filled cyclic planar microstructure. *J. Microcol. Separations* 8(6):373–381.

Henry, C. M. (1999). The incredible shrinking mass spectrometer, miniaturization is on tract to take MS into space and the doctor's office. *Anal. Chem.* Apr. 1:264A–268A.

Henry, A. C., Tutt, T. J., Galloway, M., Davidson, Y. Y., McWhorter, C. S., Soper, S. A., McCarley, R. L. (2000). Surface modification of poly(methyl methacrylate) used in the fabrication of microanalytical devices. *Anal. Chem.* 72(21):5331–5337.

Henry, C. S., Vandaveer, W. R. IV., Mubarak, I., Gray, S. R., Fritsch, I. (2001). Self-contained microelectrochemical detectors for analysis in small volumes of static and flowing fluids. Micro Total Analysis Systems 2001. Proceedings μTAS 2001 Symposium. 5th ed. Monterey, CA, Oct. 21–25, 2001, pp. 321–322.

Hibara, A., Nonaka, M., Hisamoto, H., Uchiyama, K., Kikutani, Y., Tokeshi, M., Kitamori, T. (2002a). Stabilization of liquid interface and control of two-phase confluence and separation in glass microchips by utilizing octadecylsilane modification of microchannels. *Anal. Chem.* 74(7):1724–1728.

Hibara, A., Saito, T., Kim, H.-B., Tokeshi, M., Ooi, T., Nakao, M., Kitamori, T. (2002b). Nanochannels on a fused-silica microchip and liquid properties investigation by time-resolved fluorescence measurements. *Anal. Chem.* 74(24):6170–6176.

Hinkers, H., Conrath, N., Czupor, N., Frebel, H., Hüwel, S., Köckemann, K., Trau, D., Wittkampf, M., Chemnitius, G., Haalck, L., Meusel, M., Cammann, K., Knoll, M., Spener, F., Rospert, M., Kakerow, R., Köster, O., Lerch, T., Mokwa, W., Woias, P., Richter, M., Abel, T., Mexner, L. (1996). Results of the development of sensors and μTAS-modules. Micro Total Analysis Systems '96. Proceedings of 2nd International Symposium on μTAS '96. Basel, 19–22 Nov. 1996, pp. 110–112.

Hisamoto, H., Horiuchi, T., Tokeshi, M., Hibara, A., Kitamori, T. (2001). On-chip integration of neutral ionophore-based ion pair extraction reaction. *Anal. Chem.* 73(6):1382–1386.

Hodge, C. N., Bousse, L., Knapp, M. R. (2001). Microfluidic analysis, screening, and synthesis. In: *High-Throughout Synthesis Principles and Practice*. New York: Marcel Dekker Inc., pp. 303–330.

Hofmann, O., Che, D., Cruickshank, K. A., Mueller, U. R. (1999). Adaptation of capillary isoelectric focusing to microchannels on a glass chip. *Anal. Chem.* 71(3):678–686.

Hofmann, O., Viorin, G., Niedermann, P., Manz, A. (2002). Three-dimensional microfuidic confinement for efficient sample delivery to biosensor surfaces. Application to immunoassays on planar optical waveguides. *Anal. Chem.* 74:5243–5250.

Hong, J. W., Hagiwara, H., Fujii, T., Machida, H., Inoue, M., Seki, M., Endo, I. (2001a). Separation and collection of a specified DNA fragment by chip-based CE system. Micro Total Analysis Systems 2001. Proceedings μTAS 2001 Symposium. 5th ed. Monterey, CA, Oct. 21–25, 2001, pp. 113–114.

Hong, C.-C., Choi, J.-W., Ahn, C. H. (2001b). A novel in-plane passive micromixer using Coanda effect. Micro Total Analysis Systems 2001. Proceedings μTAS 2001 Symposium. 5th ed. Monterey, CA, Oct. 21–25, 2001, pp. 31–33.

Horiuchi, T., Ueno, Y., Niwa, O. (2001). Micro-fluidic device for detection and identification of aromatic VOCs by optical method. Micro Total Analysis Systems 2001. Proceedings μTAS 2001 Symposium. 5th ed. Monterey, CA, Oct. 21–25, 2001, pp. 527–528.

Hosokawa, K., Fujii, T., Endo, I. (1998). Hydrophobic microcapillary vent for pneumatic manipulation of liquid in μTAS. Micro Total Analysis Systems '98. Proceedings μTAS '98 Workshop. Banff, Canada, 13–16 Oct. 1998, pp. 307–310.

Hosokawa, K., Fujii, T., Endo, I. (1999). Handling of picoliter liquid samples in a poly(dimethylsiloxane)-based microfluidic device. *Anal. Chem.* 71(20):4781–4785.

Hsueh, Y.-T., Smith, R. L., Northrup, M. A. (1996). A microfabricated, electrochemiluminescence cell for the detection of amplified DNA. *Sensors Actuators B* 33:110–114.

Hsueh, Y., Collins, S. D., Smith, R. L. (1998). DNA quantification with an electrochemiluminescence microcell. *Sensors Actuators B* 49:1–4.

Hu, S., Ren, X., Bachman, M., Sims, C. E., Li, G. P., Allbritton, N. (2002). Surface modification of poly(dimethylsiloxane) microfluidic devices by ultraviolet polymer grafting. *Anal. Chem.* 74(16):4117–4123.

Hua, S. Z., Sachs, F., Yang, D. X., Chopra, H. D. (2002). Microfluidic actuation using electrochemically generated bubbles. *Anal. Chem.* 74(24):6392–6396.

Huang, Z., Munro, N., Huehmer, A. F. R., Landers, J. P. (1999). Acousto-optical deflection-based laser beam scanning for fluorescence detection on multichannel electrophoretic microchips. *Anal. Chem.* 71(23):5309–5314.

Huang, Z., Sanders, J. C., Dunsmor, C., Ahmadzadeh, H., Landers, J. P. (2001). A method for UV-bonding in the fabrication of glass electrophoretic microchips. *Electrophoresis* 22(18):3924–3929.

Hutt, L. D., Glavin, D. P., Bada, J. L., Mathies, R. A. (1999). Microfabricated capillary electrophoresis amino acid chirality analyzer for extraterrestrial exploration. *Anal. Chem.* 71(18):4000–4006.

Ibrahim, M. S., Lofts, R. S., Jahrling, P. B., Henchal, E. A., Weedn, V. W., Northrup, M. A., Belgrader, P. (1998). Real-time microchip PCR for detecting single-base differences in viral and human DNA. *Anal. Chem.* 70(9):2013–2017.

Ichiki, T., Ujiie, T., Hara, T., Horiike, Y., Yasuda, K. (2001). On-chip cell sorter for single cell expression analysis. Micro Total Analysis Systems 2001. Proceedings μTAS 2001 Symposium. 5th ed. Monterey, CA, Oct. 21–25, 2001, pp. 271–273.

Jackman, R. J., Floyd, T. M., Schmidt, M. A., Fensen, K. F. (2000). Development of methods for on-line chemical detection with liquid-phase microchemical reactors using conventional and unconventional techniques. Micro Total Analysis Systems 2000. Proceedings of the μTAS Symposium. 4th ed. Enschede, The Netherlands, May 14–18, 2000, pp. 155–158.

Jackman, R. J., Queeney, K. T., Herzig-Marx, R., Schmidt, M. A., Jensen, K. F. (2001). Integration of multiple internal reflection (MIR) infrared spectroscopy with silicon-based chemical microreactors. Micro Total Analysis Systems 2001, Proceedings μTAS 2001 Symposium. 5th ed. Monterey, CA, Oct. 21–25, 2001, pp. 345–346.

Jacobson, S. C., Ramsey, J. M. (1995). Microchip electrophoresis with sample stacking. *Electrophoresis* 16:481–486.

Jacobson, S. C., Ramsey, J. M. (1996). Integrated microdevice for DNA restriction fragment analysis. *Anal. Chem.* 68(5):720–723.

Jacobson, S. C., Ramsey, J. M. (1997). Microfabricated devices for performing capillary electrophoresis. In: *Handbook of Capillary Electrophoresis*. 2nd ed. Boca Raton: CRC Press, pp. 827–839.

Jacobson, S. C., Ramsey, J. M. (1998). Microfabricated chemical separation devices. In: *HPLC: theory Techniques and Applications*. New York: Wiley, pp. 613–633.

Jacobson, S. C., Koutny, L. B., Hergenröder, R., Moore, A. W., Ramsey, J. M. (1994a). Microchip capillary electrophoresis with an integrated postcolumn reactor. *Anal. Chem.* 66:3472–3476.

Jacobson, S. C., Hergenröder, R., Moore, A. W., Ramsey, J. M. (1994b). Precolumn reactions with electrophoretic analysis integrated on a microchip. *Anal. Chem.* 66:4127–4132.

Jacobson, S. C., Hergenröder, R., Koutny, L. B., Ramsey, J. M. (1994c). Open channel electrochromatography on a microchip. *Anal. Chem.* 66:2369–2373.

Jacobson, S. C., Hergenröder, R., Koutny, L. B., Warmack, R. J., Ramsey, J. M. (1994d). Effects of injection schemes and column geometry on the performance of microchip electrophoresis devices. *Anal. Chem.* 66:1107–1113.

Jacobson, S. C., Hergenröder, R., Koutny, L. B., Ramsey, J. M. (1994e). High-speed separation on a microchip. *Anal. Chem.* 66:1114–1118.

Jacobson, S. C., Moore, A. W., Ramsey, J. M. (1995). Fused quartz substrates for microchip electrophoresis. *Anal. Chem.* 67:2059–2063.

Jacobson, S. C., Culbertson, C. T., Daler, J. E., Ramsey, J. M. (1998). Microchip structures for submillisecond electrophoresis. *Anal. Chem.* 70(16):3476–3480.

Jacobson, S. C., McKnight, T. E., Ramsey, J. M. (1999a). Microfluidic devices for electrokinetically driven parallel and serial mixing. *Anal. Chem.* 71(20):4455–4459.

Jacobson, S. C., Ermakov, S. V., Ramsey, J. M. (1999b). Minimizing the number of voltage sources and fluid reservoirs for electrokinetic valving in microfluidic devices. *Anal. Chem.* 71:3273–3276.

Jakeway, S. C., De Mello, A. J. (2001). A single point evanescent wave probe for on-chip refractive index detection. Micro Total Analysis Systems 2001. Proceedings μTAS 2001 Symposium. 5th ed. Monterey, CA, Oct. 21–25, 2001, pp. 347–348.

Jannasch, H. W., Mcgill, P. R., Zdeblick, M., Erickson, J. (2001). Integrated micro-analyzers with frozen plug valves. Micro Total Analysis Systems 2001. Proceedings μTAS 2001 Symposium. 5th ed. Monterey, CA, Oct. 21–25, 2001, pp. 529–530.

Jatisai, T., Peter, C. H. (2002). High-voltage capacitively coupled contactless conductivity detection for microchip capillary electrophoresis. *Anal. Chem.*, 74(24):6378–6382.

Jeong, Y. W., Kim, B. H., Lee, J. Y., Park, S. S., Chun, M. S., Chun, K., Kim, B. G., Chung, D. S. (2000). A cyclic capillary electrophoresis separator on silicon substrate with synchronized-switching. Micro Total Analysis Systems 2000. Proceedings of the μTAS Symposium. 4th ed. Enschede, The Netherlands: May 14–18, 2000, pp. 375–378.

Jeong, Y. W., Kim, S. Y., Chung, S., Paik, S. J., Han, Y. S., Chang, J. K., Cho, D. D., Chuang, D. S., Chun, K. (2001). Methodology for junction dilution compensation pattern and embedded electrode in CE separator. Micro Total Analysis Systems 2001. Proceedings μTAS 2001 Symposium. 5th ed. Monterey, CA, Oct. 21–25, 2001, pp. 159–160.

Jiang, G., Harrison, D. J. (2000). mRNA isolation for cDNA library construction on a chip. Micro Total Analysis Systems 2000. Proceedings of the μTAS Symposium. 4th ed. Enschede, The Netherlands, May 14–18, 2000, pp. 537–540.

Jiang, Y., Wang, P.-C., Locascio, L. E., Lee, C. S. (2001). Integrated plastic microfluidic devices with esi-ms for drug screening and residue analysis. *Anal. Chem.*, 73(9): 2048–2053.

Jo, B., Moorthy, J., Beebe, D. (2000). Polymer microfluidic valves, membranes and coatings. Micro Total Analysis Systems 2000. Proceedings of the μTAS Symposium. 4th ed. Enschede, The Netherlands, May 14–18, 2000, pp. 335–338.

Jobst, G., Moser, I., Svasek, P., Svasek, E., Varahram, M., Urban, G. (1996). Rapid liver enzyme assay with low cost μTAS. Micro Total Analysis Systems '96. Proceedings of 2nd International Symposium on μTAS '96. Basel, 19–22 Nov. 1996, p. 221.

Juncker, D., Schmid, H., Drechsler, U., Wolf, H., Wolf, M., Michel, B., de Rooij, N., Delamarche, E. (2002). Autonomous microfluidic capillary system. *Anal. Chem.* 74(24):6139–6144.

Kameoka, J., Craighead, H. G., Zhang, H., Henion, J. (2001). A polymeric microfluidic chip for CE/MS determination of small molecules. *Anal. Chem.* 73:1935–1941.

Kameoka, J., Orth, R., Ilic, B., Czaplewski, D., Wachs, T., Craighead, H. G. (2002). An electrospray ionization source for integration with microfluidics. *Anal. Chem.* 74: 5897–5901.

Kamholz, A. E., Weigl, B. H., Finlayson, B. A., Yager, P. (1999). Quantitative analysis of molecular interaction in a microfluidic channel: the T-sensor. *Anal. Chem.* 71(23): 5340–5347.

Kaniansky, D., Masár, M., Bielčková, J., Iványi, F., Eisenbeiss, J., Stanislawski, B., Grass, B., Neyer, A., Jöhnck, M. (2000). Capillary electrophoresis separations on a planar chip with the column-coupling configuration of the separation channels. *Anal. Chem.* 72:3596–3604.

Keir, R., Igata, E., Arundell, M., Smith, W. E., Graham, D., Mchugh, C., Cooper, J. M. (2002). SERRS. *In situ* substrate formation and improved detection using microfluidics. *Anal. Chem.* 74:1503–1508.

Kenis, P. J. A., Ismagilov, R. F., Whitesides, G. M. (1999). Microfabrication inside capillaries using multiphase laminar flow patterning. *Science* 285:83–85.

Kerby, M., Chien, R.-L. (2001). A fluorogenic assay using pressure-driven flow on a microchip. *Electrophoresis* 22(18):3916–3923.

Kerby, M. B., Spaid, M., Wu, S., Parce, J. W., Chien, R.-L. (2002). Selective ion extraction: a separation method for microfluidic devices. *Anal. Chem.* 74(20):5175–5183.

Khandurina, J., Guttman, A. (2002). Bioanalysis in microfluidic devices. *J. Chromatogr. A* 943:159–183.

Khandurina, J., Jacobson, S. C., Waters, L. C., Foote, R. S., Ramsey, J. M. (1999). Microfabricated porous membrane structure for sample concentration and electrophoretic Analysis. *Anal. Chem.* 71(9):1815–1819.

Khandurina, J., Mcknight, T. E., Jacobson, S. C., Waters, L. C., Foote, R. S., Ramsey, J. M. (2000). Integrated system for rapid PCR-based DNA analysis in microfluidic devices. *Anal. Chem.* 72:2995–3000.

Khandurina, J., Chován, T., Guttman, A. (2002). Micropreparative fraction collection in microfluidic devices. *Anal. Chem.* 74(7):1737–1740.

Khoury, C., Moorthy, J., Stremler, M. A., Moore, J. S., Beebe, D. J. (2000). TiO2 surface modifications for light modulated control of flow velocity. Micro Total Analysis Systems 2000. Proceedings of the μTAS Symposium. 4th ed. Enschede, The Netherlands, May 14–18, 2000, pp. 331–334.

Kicka, L. J., Faro, I., Heyner, S., Garside, W. T., Fitzpatrick, G., Wilding, P. (1995). Micromachined glass-glass microchips for *in vitro* fertilization. *Clin. Chem.* 41:1358–1359.

Kikuchi, Y., Sato, K., Ohki, H., Kaneko, T. (1992). Optically accessible microchannels formed in a single-crystal silicon substrate for studies of blood rheology. *Microvasc. Res.* 44:226–240.

Kikura-Hanajiri, R., Martin, R. S., Lunte, S. M. (2002). Indirect measurement of nitric oxide production by monitoring nitrate and nitrite using microchip electrophoresis with electrochemical detection. *Anal. Chem.* 74(24):6370–6377.

Kim, E., Xia, Y., Whitesides, G. M. (1995). Polymer microstructures formed by moulding in capillaries. *Nature* 376:581–584.

Kim, J.-S., Knapp, D. R. (2001a). Miniaturized multichannel electrospray ionization emitters on poly(dimethylsiloxane) microfluidic devices. *Electrophoresis* 22(18): 3993–3999.

Kim, D. J., Cho, W. H., Ro, K. W., Hahn, J. H. (2001b). Microchip-based simultaneous on-line monitoring for CR(III) and CR(VI) using highly efficient chemiluminescence detection. Micro Total Analysis Systems 2001. Proceedings μTAS 2001 Symposium. 5th ed. Monterey, CA, Oct. 21–25, 2001, pp. 525–526.

Kirby, B. J., Wheeler, A. R., Shepodd, T. J., Fruetel, J. A., Hasselbrink, E. F., Zare, R. N. (2001). A laser-polymerized thin film silica surface modification for suppression of cell adhesion and electroosmotic flow in microchannels. Micro Total Analysis Systems 2001. Proceedings μTAS 2001 Symposium. 5th ed. Monterey, CA, Oct. 21–25, 2001, pp. 605–606.

Ko, W. H., Suminto, J. T. (1989). In: Göpel, W., Hesse, J., Zemel, J. N. eds. *Sensors, a Comprehensive Survey*. Vol. 1. Weinheim: VCH, pp. 107–168.

Kopp, M. U., de Mello, A. J., Manz, A. (1998a). Chemical amplification: continuous-flow PCR on a chip. *Science* 280:1046–1048.

Kopp, M. U., Luechinger, M. B., Manz, A. (1998b). Continuous flow PCR on a chip. Micro Total Analysis Systems '98. Proceedings μTAS '98 Workshop. Banff, Canada, 13–16 Oct. 1998, pp. 7–10.

Koutny, L. B., Schmalzing, D., Taylor, T. A., Fuchs, M. (1996). Microchip electrophoretic immunoassay for serum cortisol. *Anal. Chem.* 68(1):18–22.

Koutny, L., Schmalzing, D., Salas-solano, O., El-Difrawy, S., Adourian, A., Buonocore, S., Abbey, K., McEwan, P., Matsudaira, P., Ehrlich, D. (2000). Eight hundred-base sequencing in a microfabricated electrophoretic device. *Anal. Chem.* 72:3388–3391.

Kricka, L. J., Wilding, P. (1996). Micromachining: a new direction for clinical analyzers. *Pure Appl. Chem.* 68(10):1831–1836.

Kuban, P., Dasgupta, P. K., Morris, K. A. (2002). Microscale continuous ion exchanger. *Anal. Chem.* 74(21):5667–5675.

Kutter, J. P. (2000). Current developments in electrophoretic and chromatographic separation methods on microfabricated devices. *Trends Anal. Chem.* 19:352–363.

Kutter, J. P., Jacobson, S. C., Ramsey, J. M. (1997). Integrated microchip device with electrokinetically controlled solvent mixing for isocratic and gradient elution in micellar electrokinetic chromatography. *Anal. Chem.* 69(24):5165–5171.

Kutter, J. P., Jacobson, S. C., Matsubara, N., Ramsey, J. M. (1998). Solvent-programmed microchip open-channel electrochromatography. *Anal. Chem.* 70(15):3291–3297.

Kutter, J. P., Jacobson, S. C., Ramsey, J. M. (2000). Solid phase extraction on microfluidic device. *J. Microcol. Separations* 12(2):93–97.

Kwok, Y. C., Manz, A. (2001a). Characterization of Shah convolution Fourier transform detection. *Analyst* 126(10):1640–1644.

Kwok, Y. C., Manz, A. (2001b). Shah convolution differentiation Fourier transform for rear analysis in microchip capillary electrophoresis. *J. Chromatogr. A* 924(1–2):177–186.

Kwok, Y. C., Manz, A. (2001c). Shah convolution Fourier transform dtection: multiple-sample injection technique. *Electrophoresis* 22:222–229.

Lagally, E. T., Medintz, I., Mathies, R. A. (2001). Single-molecule DNA amplification and analysis in an integrated microfluidic device. *Anal. Chem.* 73:565–570.

Landers, J. P. (2003). Molecular diagnostics on electrophoretic microchips. *Anal. Chem.* 75:2919.

Lao, A. I. K., Trau, D., Hsing, I. (2002). Miniaturized flow fractionation device assisted by a pulsed electric field for nanoparticle separation. *Anal. Chem.* 74:5364–5369.

Lapos, J. A., Ewing, A. G. (2000). Injection of fluorescently labeled analytes into microfabricated chips using optically gated electrophoresis. *Anal. Chem.* 72:4598–4602.

Lapos, J. A., Manica, D. P., Ewing, A. G. (2002). Dual fluorescence and electrochemical detection on an electrophoresis microchip. *Anal. Chem.* 74:3348–3353.

Larsen, U. D., Branebjerg, J., Blankenstein, G. (1996). Fast mixing by parallel multilayer lamination. Micro Total Analysis Systems '96. Proceedings of 2nd International Symposium on µTAS '96. Basel, 19–22 Nov. 1996, pp. 228–230.

Lazar, I. M., Karger, B. L. (2002). Multiple open-channel electroosmotic pumping system for microfluidic sample handling. *Anal. Chem.*, 74(24):6259–6268.

Lazar, L. M., Ramsey, R. S., Sundberg, S., Ramsey, J. M. (1999). Subattomole-sensitivity microchip nanoelectrospray source with time-of-flight mass spectrometry detection. *Anal. Chem.* 71:3627–3631.

Lazar, L. M., Ramsey, R. S., Ramsey, J. M. (2001). On-chip proteolytic digestion and analysis using "Wrong-Way-Round" electrospray time-of-flight mass spectrometry. *Anal. Chem.* 73:1733–1739.

Lee, S., Kim, Y. (2001). Metering and mixing of nanoliter liquid in the microchannel networks driven by fluorocarbon surfaces and pneumatic control. Micro Total Analysis Systems 2001. Proceedings µTAS 2001 Symposium. 5th ed. Monterey, CA, Oct. 21–25, 2001, pp. 205–206.

Lee, L. P., Berger, S. A., Pruitt, L., Liepmann, D. (1998). Key elements of a tranperent Teflon microfluidic system. Micro Total Analysis Systems '98. Proceedings µTAS '98 Workshop. Banff, Canada, 13–16 Oct. 1998, pp. 245–248.

Lehmann, U., Krusemark, O., Muller, J. (2000). Micro machined gas chromatograph based on a plasma polymerised stationary phase. Micro Total Analysis Systems 2000. Proceedings of the µTAS Symposium. 4th ed. Enschede, The Netherlands, May 14–18, 2000, pp. 167–170.

Lemoff, A. V., Lee, A. P. (2000). An AC magnetohydrodynamic microfluidic switch. Micro Total Analysis Systems 2000. Proceedings of the µTAS Symposium. 4th ed. Enschede, The Netherlands, May 14–18, 2000, pp. 571–574.

Lendl, B., Schindler, R., Frank, J., Kellner, R., Drott, J., Laurell, T. (1997). Fourier transform infrared detection in miniaturized total analysis systems for sucrose analysis. *Anal. Chem.* 69(15):2877–2881.

Lettieri, G.-L., Verpoorte, E., de Rooij, N. F. (2001). Affinity-based bioanalysis using freely moving beads as matrices for heterogeneous assays. Micro Total Analysis Systems 2001. Proceedings µTAS 2001 Symposium. 5th ed. Monterey, CA, Oct. 21–25, 2001, pp. 503–504.

Li, P. C. H., Harrison, D. J. (1997). Transport, manipulation, and reaction of biological cells on-chip using electrokinetic effects. *Anal. Chem.* 69:1564–1568.

Li, J., Thibault, P., Bings, N. H., Skinner, C. D., Wang, C., Colyer, C., Harrison, D. J. (1999). Integration of microfabricated devices to capillary electrophoresis–electrospray mass spectrometry using a low dead volume connection: application to rapid analyses of proteolytic digests. *Anal. Chem.* 71:3036–3045.

Li, J., Kelly, J. F., Chernuschevich, I., Harrison, D. J., Thibault, P. (2000). Separation and identification of peptides from gel-isolated membrane proteins using a microfabricated device for combined capillary electrophoresis/nanoelectrospray mass spectrometry. *Anal. Chem.* 72:599–609.

Li, P. C. H., Wang, W., Parameswaran, A. M. (2001). Microfluidic biochip integrated with acoustic wave sensor for single heart muscle cell analysis. Micro Total Analysis Systems 2001. Proceedings µTAS 2001 Symposium. 5th ed. Monterey, CA, Oct. 21–25, 2001, pp. 295–296.

Liang, Z., Chiem, N., Ocvirk, G., Tang, T., Fluri, K., Harrison, D. J. (1996). Microfabrication of a planar absorbance and fluorescence cell for integrated capillary electrophoresis devices. *Anal. Chem.* 68:1040–1046.

Lichtenberg, J., Daridon, A., Verpoorte, E., de Rooij, N. F. (2000). Combination of sample pre-concentration and capillary electrophoresis on-chip. Micro Total Analysis Systems 2000. Proceedings of the μTAS Symposium. 4th ed. Enschede, The Netherlands, May 14–18, 2000, pp. 307–310.

Lichtenberg, J., Verpoorte, E., De Rooij, N. F. (2001a). Operating parameters for an in-plane, contactless conductivity detector for microchip-based separation methods. Micro Total Analysis Systems 2001. Proceedings μTAS 2001 Symposium. 5th ed. Monterey, CA, Oct. 21–25, 2001, pp. 323–324.

Lichtenberg, J., Verpoorte, E., de Rooij, N. F. (2001b). Sample preconcentration by field amplification stacking for microchip-based capillary electrophoresis. *Electrophoresis* 22(2):258–271.

Lin, Y.-C., Wu, C.-Y. (2001). Design of moving electric field driven capillary electrophoresis chips. Micro Total Analysis Systems 2001. Proceedings μTAS 2001 Symposium. 5th ed. Monterey, CA, Oct. 21–25, 2001, pp. 553–554.

Licklider, L., Wang, X.-Q., Desai, A., Tai, Y.-C., Lee, T. D. (2000). A micromachined chip-based electrospray source for mass spectrometry. *Anal. Chem.* 72(2):367–375.

Linder, V., Verpoorte, E., Thormann, W., de Rooij, N. F., Sigrist, H. (2001). Surface biopassivation of replicated poly(dimethylsiloxane) microfluidic channels and application to heterogeneous immunoreaction with on-chip fluorescence detection. *Anal. Chem.* 73(17):4181–4189.

Lindern, W. E. van der. (1987). Minaturisation in flow injection analysis practical limitations from a theoretical point of view. *Trends Anal. Chem.* 6:37–40.

Liu, S., Shi, Y., Ja, W. W., Mathies, R. A. (1999). Optimization of high-speed DNA sequencing on microfabricated capillary electrophoresis channels. *Anal. Chem.* 71:566–573.

Liu, Y., Fanguy, J. C., Bledsoe, J. M., Henry, C. S. (2000a). Dynamic coating using polyelectrolyte multilayers for chemical control of electroosmotic flow in capillary electrophoresis microchips. *Anal. Chem.* 72(24):5939–5944.

Liu, Y., Foote, R. S., Jacobson, S. C., Ramsey, R. S., Ramsey, J. M. (2000b). Transport number mismatch induced stacking of swept sample zones for microchip-based sample concentration. Micro Total Analysis Systems 2000. Proceedings of the μTAS Symposium. 4th ed. Enschede, The Netherlands, May 14–18, 2000, pp. 295–298.

Liu, Y., Foote, R. S., Jacobson, S. C., Ramsey, R. S., Ramsey, J. M. (2000c). Electrophoretic separation of proteins on a microchip with noncovalent, Postcolumn Labeling. *Anal. Chem.* 72(19):4608–4613.

Liu, H., Felten, C., Xue, Q., Zhang, B., Jedrzejewski, P., Karger, B. L., Foret, F. (2000d). Development of multichannel devices with an array of electrospray tips for high-throughput mass spectrometry. *Anal. Chem.* 72:3303–3310.

Liu, S., Ren, H., Gao, Q., Roach, D. J., Loder, R. T. Jr., Armstrong, T. M., Mao, Q., Blaga, L., Barker, D. L., Jovanovich, S. B. (2000e). Parallel DNA sequencing on microfabricated electrophoresis chips. Micro Total Analysis Systems 2000. Proceedings of the μTAS Symposium. 4th ed. Enschede, The Netherlands, May 14–18, 2000, pp. 477–480.

Liu, S., Zhang, J., Ren, H., Zheng, J., Liu, H. (2001a). A microfabricated hybrid device for DNA sequencing. Micro Total Analysis Systems 2001. Proceedings μTAS 2001 Symposium. 5th ed. Monterey, CA, Oct. 21–25, 2001, pp. 99–100.

Liu, Y., Ganser, D., Schneider, A., Liu, R., Grodzinski, P., Kroutchinina, N. (2001b). Microfabricated polycarbonate CE devices for DNA analysis. Micro Total Analysis Systems 2001. Proceedings μTAS 2001 Symposium. 5th ed. Monterey, CA, Oct. 21–25, 2001, pp. 119–120.

Liu, R. H., Ward, M., Bonanno, J., Ganser, D., Athavale, M., Grodzinski, P. (2001c). Plastic in-line chaotic micromixer for bilogical applications. Micro Total Analysis Systems 2001. Proceedings μTAS 2001 Symposium. 5th ed. Monterey, CA, Oct. 21–25, 2001, pp. 163–164.

Liu, J., Tseng, K., Garcia, B., Lebrilla, C. B., Mukerjee, E., Collins, S., Smith, R. (2001d). Electrophoresis separation in open microchannels. A method for coupling electrophoresis with MALDI-MS. *Anal. Chem.* 73:2147–2151.

Liu, Y., Ganser, D., Schneider, A., Liu, R., Grodzinski, P., Kroutchinina, N. (2001e). Microfabricated polycarbonate CE devices for DNA analysis. *Anal. Chem.* 73:4196–4201.

Liu, Y., Rauch, C. B., Stevens, R. L., Lenigk, R., Yang, J., Rhine, D. B., Grodzinski, P. (2002). DNA amplification and hybridization assays in integrated plastic monolithic devices. *Anal. Chem.* 74:3063–3070.

Locascio, L. E., Perso, C. E., Lee, C. S. (1999). Measurement of electroosmotic flow in plastic imprinted microfluid devices and the effect of protein adsorption on flow rate. *J. Chromatogr. A* 857:275–284.

Lu, H., Jackman, R. J., Gaudet, S., Cardone, M., Schmidt, M. A., Jensen, K. F. (2001a). Microfluidic devices for cell lysis and isolation of organelles. Micro Total Analysis Systems 2001. Proceedings μTAS 2001 Symposium. 5th ed. Monterey, CA, Oct. 21–25, 2001, pp. 297–298.

Lu, L., Ryu, K. S., Liu, C. (2001b). A novel microstirrer and arrays for microfluidic mixing. Micro Total Analysis Systems 2001. Proceedings μTAS 2001 Symposium. 5th ed. Monterey, CA, Oct. 21–25, 2001, pp. 28–30.

Macounova, K., Cabrera, C. R., Holl, M. R., Yager, P. (2000). Generation of natural pH gradients in microfluidic channels for use in isoelectric focusing. *Anal. Chem.* 72(16):3745–3751.

Macounová, K., Cabrera, C. R., Yager, P. (2001). Concentration and separation of proteins in microfluidic channels on the basis of transverse IEF. *Anal. Chem.* 73(7):1627–1633.

Madou, M. J., Lu, Y., Lai, S., Lee, J., Daunert, S. (2000). A centrifugal microfluidic platform-a comparison. Micro Total Analysis Systems 2000. Proceedings of the μTAS Symposium. 4th ed. Enschede, The Netherlands, May 14–18, 2000, pp. 565–570.

Man, A., Harrison, D. J., Verpoorte, E., Widmer, H. M. (1993). Planar chips technology for miniaturization of separation systems: a developing perspective in chemical monitoring. In: *Advances in Chromatography*. Vol. 33. New York-Basel-Kong Kong: Marcel Dekker Inc., pp. 2–66.

Mangru, S. D., Harrison, D. J. (1998). Chemiluminescence detection in integrated post-separation reactors for microchip-based capillary electrophoresis and affinity electrophoresis. *Electrophoresis* 19(13):2301–2307.

Manica, D. P., Lapos, J. A., Jones, A. D., Ewing, A. G. (2001). Dual electrochemical and optical detection on a microfabricated electrophoresis chip. Micro Total Analysis Systems 2001. Proceedings µTAS 2001 Symposium. 5th ed. Monterey, CA, Oct. 21–25, 2001, pp. 262–264.

Manz, A., Miyahara, Y., Miura, J., Watanabe, Y., Miyagi, H., Sato, K. (1990a). Design of an open-tubular column liquid chromatograph using silicon chip technology. *Sensors Actuators* B1:249–255.

Manz, A., Graber, N., Widmer, H. M. (1990b). Miniaturized total chemical analysis system: a novel concept for chemical sensing. *Sensors Actuators* B1:244–248.

Manz, A., Fettinger, J. C., Verpoorte, E., Lüdi, H., Widmer, H. M., Harrison, D. J. (1991a). Micromachining of monocrystalline silicon and glass for chemical analysis systems. *Trends Anal. Chem.* 10:144–149.

Manz, A., Harrison, D. J., Fettinger, J. C., Verpoorte, E., Lüdi, H., Widmer, H. M. (1991b). Integrated electroosmotic pumps and flow manifolds for total chemica analysis systems. *Transducer* 939–941.

Manz, A., Harrison, D. J., Verpoorte, E. M. J., Fettinger, J. C., Paulus, A., Lüdi, H., Widmer, H. M. (1992). Planar chips technology for miniaturization and integration of separation techniques into monitoring system. *J. Chromatogr.* 593:253–258.

Manz, A., Verpoorte, E., Effenhauser, C. S., Burggraf, N., Raymond, D. E., Harrison, D. J., Widmer, H. M. (1993). Miniaturization of separation techniques using planar chip technology. *J. High Res. Chromatogr.* 16:433–436.

Manz, A., Effenhauser, C. S., Burggraf, N., Harrison, D. J., Seiler, K., Fluri, K. (1994). Electroosmotic pumping and electrophoretic separations for miniaturized chemical analysis systems. *J. Micromech. Microeng.* 4(4):257–265.

Manz, A., Verpoorte, E., Raymond, D. E., Effenhauser, C. S., Burggraf, N., Widmer, H. M. (1995). µTAS-miniaturized total chemical analysis system. Micro Total Analysis Systems '94. Proceedings of the µTAS '94 Workshop. The Netherlands: University of Twente, 21–22 Nov. 1994, pp. 5–27.

Manz, A., Bessoth, F., Kopp, M. U. (1998). Continuous flow versus batch processing- a few examples. Micro Total Analysis Systems '98. Proceedings µTAS '98 Workshop. Banff, Canada, 13–16 Oct. 1998, pp. 235–240.

Mao, Q., Pawliszyn, J. (1999). Demonstration of isoelectric focusing on an etched quartz chip with UV absorption imaging detection. *Analyst* 124(5):637–641.

Mao, H., Holden, M. A., You, M., Cremer, P. S. (2002a). Reusable platforms for high-throughput on-chip temperature gradient assays. *Anal. Chem.* 74(19):5071–5075.

Mao, H., Yang, T., Cremer, P. S. (2002b). Design and characterization of immobilized enzymes in microfluidic systems. *Anal. Chem.* 74(2):379–385.

Martin, R. S., Gawron, A. J., Lunte, S. M. (2000). Dual-electrode electrochemical detection for poly(dimethylsiloxane)-fabricated capillary electrophoresis microchips. *Anal. Chem.* 72:3196–3202.

Martin, R. S., Kikura-Hanajiri, R., Lacher, N. A., Lunte, S. M. (2001a). Studies to improve the performance of electrochemical detection for microchip electrophoresis. Micro Total Analysis Systems 2001. Proceedings µTAS 2001 Symposium. 5th ed. Monterey, CA, Oct. 21–25, 2001, pp. 325–326.

Martin, R. S., Gawron, A. J., Fogarty, B. A., Regan, F. B., Dempsey, E., Lunte, S. M. (2001b). Carbon paste-based electrochemical detectors for microchip capillary electrophoresis/electrochemistry. *Analyst* 126(3):277–280.

Martin, R. S., Ratzlaff, K. L., Huynh, B. H., Lunte, S. M. (2002). In-channel electrochemical detection for microchip capillary electrophoresis using an electrically isolated potentiostat. *Anal. Chem.* 74(5):1136–1143.

Martinoia, S., Bove, M., Tedeso, M., Margesin, B., Grattarola, M. (1999). A simple microfluidic system for patterning populations of neurons on silicon micromachined substrates. *J. Neuro. Methods* 87:35–44.

Martynova, L., Locascio, L. E., Gaitain, M., Kramer, G. W., Christensen, R. G., Maccrehan, W. A. (1997). Fabrication of plastic microfluid channels by imprinting methods. *Anal. Chem.* 69:4783–4789.

Masar, M., Kaniansky, D., Bodor, R., Johnck, M., Stanislawski, B. (2001). Determination of organic acids and inorganic anions in wine by isotachophoresis on a planar chip. *J. Chromatogr. A* 916(1–2):167–174.

Mathies, R. A., Simpson, P. C., Woolley, A. T. (1998). DNA analysis with capillary array electrophoresis microplates. Micro Total Analysis Systems '98. Proceedings µTAS '98 Workshop. Banff, Canada: 13–16 Oct. 1998, pp. 1–6.

Matsubara, Y., Murakami, Y., Kinpara, T., Morita, Y., Yokoyama, K., Tamiya, E. (2001). Allergy sensor using animal cells with microfuidics. Micro Total Analysis Systems 2001. Proceedings µTAS 2001 Symposium. 5th ed. Monterey, CA, Oct. 21–25, 2001, pp. 299–300.

McClain, M. A., Culbertson, C. T., Jacobson, S. C., Ramsey, J. M. (2001a). Single cell lysis on microfluidic devices. Micro Total Analysis Systems 2001. Proceedings µTAS 2001 Symposium. 5th ed. Monterey, CA, Oct. 21–25, 2001, pp. 301–302.

McClain, M. A., Culbertson, C. T., Jacobson, S. C., Ramsey, J. M. (2001b). Flow cytometry of escherichia coli on microfluidic devices. *Anal. Chem.* 73:5334–5338.

McCormick, R. M., Nelson, R. J., Alonso-Amigo, M. G., Benvegnu, D. J., Hooper, H. H. (1997). Microchannel electrophoretic separation of DNA in injection-molded plastic substrates. *Anal. Chem.* 69:2626–2630.

Mcdonald, J. C., Duffy, D. C., Anderson, J. R., Chiu, D. T., Wu, H., Schueller, O. J. A., Whitesides, G. M. (2000). Fabrication of microfluidic systems in poly(dimethylsiloxane). *Electrophoresis* 21:27–40.

McDonald, J. C., Chabinyc, M. L., Metallo, S. J., Anderson, J. R., Stroock, A. D., Whitesides, G. M. (2002). Prototyping of microfluidic devices in poly(dimethylsiloxane) using solid-object printing. *Anal. Chem.* 74(7):1537–1545.

McReynolds, J. A., Edirisinghe, P., Shippy, S. A. (2002). Shah and sine convolution Fourier transform detection for microchannel electrophoresis with a charge coupled device. *Anal. Chem.* 74:5063–5070.

Meng, Z., Qi, S., Soper, S. A., Limbach, P. A. (2001). Interfacing a polymer-based micromachined device to a nanoelectrospray ionization Fourier Transform ion cyclotron resonance mass spectrometer. *Anal. Chem.* 73:1286–1291.

Mensinger, H., Richter, Th., Hessel, V., Döpper, J., Ehrfeld, W. (1995). Microreactor with integrated static mixer and analysis system. Micro Total Analysis Systems '94. Proceedings of the μTAS '94 Workshop. University of Twente, The Netherlands, 21–22 Nov. 1994, pp. 237–243.

Mogensen, K. B., Petersen, N., Hübner, J., Kutter, J. (2001a). In-plane UV absorbance detection in silicon-based electrophoresis devices using monolithically integrated optical waveguides. Micro Total Analysis Systems 2001. Proceedings μTAS 2001 Symposium. 5th ed. Monterey, CA, Oct. 21–25, 2001, pp. 280–282.

Mogensen, K. B., Petersen, N. J., Hubner, J., Kutter, J. P. (2001b). Monolithic integration of optical waveguides for absorbance detection in microfabricated electrophoresis devices. *Electrophoresis* 22(18):3930–3938.

Molho, J. I., Herr, A. E., Mosier, B. P., Santiago, J. G., Kenny, T. W., Brennen, R. A., Gordon, G. B., Mohammadi, B. (2001). Optimization of turn geometries for microchip electrophoresis. *Anal. Chem.* 73(6):1350–1360.

Monahan, J., Gewirth, A. A., Nuzzo, R. G. (2001). A method for filling complex polymeric microfluidic devices and arrays. *Anal. Chem.* 73(13):3193–3197.

Moore, A. W. Jr., Jacobson, S. C., Ramsey, J. M. (1995). Microchip separation of neutral species via micellar electrokinetic capillary charomatography. *Anal. Chem.* 67:4184–4189.

Mrksich, M., Chen, C. S., Xia, Y., Dike, L. E., Ingber, D. E., Whitesides, G. M. (1996). Controlling cell attachment on contoured surfaces with self-assembled monolayers of alkanethiolates on gold. *Proc. Natl. Acad. Sci. USA* 93(20):10775–10778.

Munro, N. J., Snow, K., Kant, J. A., Landers, J. P. (1999). Molecular diagnostics on microfabricated electrophoretic devices: from slab gel- to capillary- to microchip-based assays for T- and B-cell lymphoproliferative disorders. *Clin. Chem.* 45(11):1906–1917.

Munro, N. J., Hühmer, A. F. R., Landers, J. P. (2001). Robust polymeric microchannel coating for microchip-based analysis of neat PCR products. *Anal. Chem.* 73:1784–1794.

Munro, N. J., Huang, Z., Finegold, D. N., Landers, J. P. (2002). Indirect fluorescence detection of amino acids on electrophoretic microchips. *Anal. Chem.* 72:2765–2773.

Murakami, Y., Takeuchi, T., Yokoyama, K., Tomiya, E., Karube, I. (1993). Integration of enzyme-immobilized column with electrochemical flow cell using micromachining techniques for a glucose detection system. *Anal. Chem.* 65:2731–2735.

Murakami, Y., Kanekiyo, T., Kinpara, T., Tamiya, E. (2001). On-chip bypass structure for sample segment division and dilution. Micro Total Analysis Systems 2001. Proceedings μTAS 2001 Symposium. 5th ed. Monterey, CA, Oct. 21–25, 2001, pp. 175–176.

Naji, O. P., Bessoth, F. G., Manz, A. (2001). Novel injection methods for miniaturised gas chromatography. Micro Total Analysis Systems 2001. Proceedings μTAS 2001 Symposium. 5th ed. Monterey, CA, Oct. 21–25, 2001, pp. 655–657.

Namasivayam, V., Larson, R. G., Burke, D. T., Burns, M. A. (2002). Electrostretching DNA molecules using polymer-enhanced media within microfabricated devices. *Anal. Chem.* 74(14):3378–3385.

Nelstrop, L. J., Greenway, G. M. Investigation of chemiluminescent microanalytical systems. *YAS* 98:355–358.

Nemoda, Z., Ronai, Z., Szekely, A., Kovacs, E., Shandrick, S., Guttman, A., Sasvari-Szekely, M. (2001). High-throughput genotyping of repeat polymorphism in the regulatory region of serotonin transporter gene by gel microchip electrophoresis. *Electrophoresis* 22(18):4008–4011.

Nieuwenhuis, J. H., Lee, S. S., Bastemeijer, J., Vellekoop, M. J. (2001). Particle-shape sensing-elements for integrated flow cytometer. Micro Total Analysis Systems 2001, Proceedings μTAS 2001 Symposium. 5th ed. Monterey, CA, Oct. 21–25, 2001, pp. 357–358.

Nishimoto, T., Fujiyama, Y., Abe, H., Kanai, M., Nakanishi, H., Arai, A. (2000). Microfabricated CE chips with optical slit for UV absorption detection. Micro Total Analysis Systems 2000. Proceedings of the μTAS Symposium. 4th ed. Enschede, The Netherlands, May 14–18, 2000, pp. 395–398.

Northrup, M. A., Ching, M. T., White, R. M., Watson, R. T. (1993). DNA amplification with a microfabricated reaction chamber. *Transducer* 924–926.

Ocvirk, G., Tang, T., Jed Harrison, D. (1998). Optimization of confocal epifluorescence microscopy for microchip-based miniaturized total analysis systems. *Analyst* 123(7):1429–1434.

Ocvirk, G., Munroe, M., Tang, T., Oleschuk, R., Westra, K., Harrison, D. J. (2000). Electrokinetic control of fluid flow in native poly (dimethylsiloxane) capillary electrophoresis device. *Electrophoresis* 21:107–115.

Oddy, M. H., Santiago, J. G., Mikkelsen, J. C. (2001). Electrokinetic instability micromixers. Micro Total Analysis Systems 2001. Proceedings μTAS 2001 Symposium. 5th ed. Monterey, CA, Oct. 21–25, 2001, pp. 34–36.

O'Donnell, M., Maryanne, J., Little, D. P. (1996). Microfabrication and array technologies for DNA sequencing and diagnostics. *Genet. Anal. Biomol. Eng.* 13(6):151–157.

Oki, A., Adachi, S., Takamura, Y., Ishihara, K., Kataoka, K., Ichiki, T., Honike, Y. (2000). Glucose measurement in blood serum injected by electroosmosis into phospholipid polymer coated microcapillary. Micro Total Analysis Systems 2000. Proceedings of the μTAS Symposium. 4th ed. Enschede, The Netherlands, May 14–18, 2000, pp. 403–406.

Oki, A., Adachi, S., Takamura, Y., Onoda, H., Ito, Y., Horiike, Y. (2001). Electrophoresis velocity measurement of lymphocytes under suppression of pH change of PBS in microcapillary. Micro Total Analysis Systems 2001. Proceedings μTAS 2001 Symposium. 5th ed. Monterey, CA, Oct. 21–25, 2001, pp. 505–506.

Oleschuk, R. D., Shultz-Lockyear, L. L., Ning, Y., Harrison, D. J. (2000). Trapping of bead-based reagents within microfluidic systems. On-chip solid-phase extraction and electrochromatography. *Anal. Chem.* 72(3):585–590.

Olsen, K. G., Ross, D. J., Tarlov, M. J. (2002). Immobilization of DNA hydrogel plugs in microfluidic channels. *Anal. Chem.* 74(6):1436–1441.

Paegel, B. M., Hutt, L. D., Simpson, P. C., Mathies, R. A. (2000). Turn geometry for minimizing band broadening in microfabricated capillary electrophoresis channels. *Anal. Chem.* 72(14):3030–3037.

Paegel, B. M., Yeung, S. H., Mathies, R. A. (2002). Microchip bioprocessor for integrated nanovolume sample purification and DNA sequencing. *Anal. Chem.* 74:5092–5098.

Palmer, J., Burgi, D. S., Munro, N. J., Landers, J. P. (2001). Electrokinetic injection for stacking neutral analytes in capillary and microchip electrophoresis. *Anal. Chem.* 73:725–731.

Parce, J. W., Owicki, J. C., Kercso, K. M., Sigal, G. B., Wada, H. G., Muir, V. C., Bousse, L. J., Ross, K. L., Sikic, B. I., McConnell, H. M. (1989). Detection of cell-affecting agents with a silicon biosensor. *Science* 246:243–247.

Penner, N. A., Slentz, B. E., Regnier, F. (2001). Capillary electrochromatography on microchips with in situ fabricated particles. Micro Total Analysis Systems 2001. Proceedings μTAS 2001 Symposium. 5th ed. Monterey, CA, Oct. 21–25, 2001, 559–560.

Peterson, D. S., Rohr, T., Svec, F., Fréchet, J. M. J. (2002). Enzymatic microreactor-on-a-chip: protein mapping using trypsin immobilized on porous polymer monoliths molded in channels of microfluidic devices. *Anal. Chem.* 74(16):4081–4088.

Polson, N. A., Haves, M. A. (2001). Microfluidics controlling fluids in small places. *Anal. Chem.* June 1:312A–319A.

Prest, J. E., Baldock, S. J., Bektas, N., Fielden, P. R., Treves B. B. J. (1999). Single electrode conductivity detection for electrophoretic separation systems. *J. Chromatogr. A* 836(1):59–65.

Prest, J. E., Baldock, S. J., Fielden, P. R., Goddard, N. J., Brown, B. J. T. (2002). Bidirectional isotachophoresis on a planar chip with integrated conductivity detection. *Analyst (Cambridge)* 127(11):1413–1419.

Qian, S., Bau, H. H. (2002). A chaotic electroosmotic stirrer. *Anal. Chem.* 74(15):3616–3625.

Qin, D., Xia, Y., Rogers, J. A., Jackman, R. J., Zhao, X., Whitesides, G. M. (1998). Microfabrication, microstructures and microsystems. In: *Microsystem Technology in Chemistry and Life Science*. Berlin, Heidelberg: Springer-Verlag, pp. 1–20.

Qiu, C. X., Harrison, D. J. (2001). Integrated self-calibration via electrokinetic solvent proportioning for microfluidic immunoassays. *Electrophoresis* 22(18):3949–3958.

Ramsey, R. S., Ramsey, J. M. (1997). Generation electrospray from microchip devices using electroosmotic pumping. *Anal. Chem.* 69:1174–1178.

Ramsey, J. M., Jacobson, S. C., Knapp, M. R. (1995). Microfabricated chemical measurement systems. *Nat. Med.* 1:1093–1096.

Raymond, D. E., Manz, A., Widmer, H. M. (1994). Continuous sample pretreatment using a free-flow electrophoresis device integrated onto a silicon chip. *Anal. Chem.* 66:2858–2865.

Raymond, D. E., Manz, A., Widmer, H. M. (1996). Continuous separation of high molecular weight compounds using a microliter volume free-flow electrophoresis microstructure. *Anal. Chem.* 68(15):2515–2522.

Rehm, J. E., Shepodd, T. J., Hasselbrink, E. F. (2001). Mobile flow control elements for high-pressure micro-analytical systems fabricated using *in situ* polymerization. Micro Total Analysis Systems 2001. Proceedings μTAS 2001 Symposium. 5th ed. Monterey, CA, Oct. 21–25, 2001, pp. 227–229.

Renaud, P., Lintel, H. V., Heuschkel, M., Guérin, L. (1998). Photo-polymer microchannel technologies and applications. Micro Total Analysis Systems '98. Proceedings μTAS '98 Workshop. Banff, Canada, 13–16 Oct. 1998, pp. 17–22.

Reshni, K. A., Morris, M. D., Johnson, B. N., Burns, M. A. (1998). On-line detection of electrophoretic separations on a microchip by Roman spectroscopy. Micro Total Analysis Systems '98. Proceedings μTAS '98 Workshop. Banff, Canada, 13–16 Oct. 1998, pp. 109–112.

Reyes, D. R., Iossifidis, D., Auroux, P.-A., Manz, A. (2002). Micro total analysis systems. 1. Introduction, theory, and technology. *Anal. Chem.* 74(12):2623–2636.

Ro, K. W., Lim, K., Hahn, J. H. (2001a). PDMS microchip for precolumn reaction and micellar electrokinetic chromatography of biogenic amines. Micro Total Analysis Systems 2001. Proceedings μTAS 2001 Symposium, 5th ed. Monterey, CA, Oct. 21–25, 2001, pp. 561–562.

Ro, K. W., Shim, B. C., Lim, K., Hahn, J. H. (2001b). Integrated light collimating system for extended optical-path-length absorbance detection in microchip-based capillary electrophoresis. Micro Total Analysis Systems 2001. Proceedings μTAS 2001 Symposium. 5th ed. Monterey, CA, Oct. 21–25, 2001, pp. 274–276.

Roberts, M. A., Rossier, J. S., Bercier, P., Girault, H. (1997). UV laser machined polymer substrates for the development of microdiagnostic systems. *Anal. Chem.* 69(11):2035–2042.

Rocklin, R. D., Ramsey, R. S., Ramsey, J. M. (2000). A microfabricated fluidic device for performing two-dimensional liquid-phase separations. *Anal. Chem.* 72(21):5244–5249.

Rohr, T., Yu, C., Davey, M. H., Svec, F., Frechet, J. M. J. (2001). Porous polymer monoliths: simple and efficient mixers prepared by direct polymerization in the channels of microfluidic chips. *Electrophoresis* 22(18):3959–3967.

Rong, W., Kutter, J. P. (2001). On-line chemiluminescence detection of bioprocesses using polymer-based microchips with immobilized enzymes. Micro Total Analysis Systems 2001. Proceedings μTAS 2001 Symposium. 5th ed. Monterey, CA, Oct. 21–25, 2001, pp. 181–182.

Ross, D., Locascio, L. E. (2002). Microfluidic temperature gradient focusing. *Anal. Chem.* 74(11):2556–2564.

Ross, D., Johnson, T. J., Locascio, L. E. (2001). Imaging of electroosmotic flow in plastic microchannels. *Anal. Chem.* 73(11):2509–2515.

Rossier, J. S., Schwarz, A., Reymond, F., Ferrigno, R., Bianchi, F., Girault, H. H. (1999a). Microchannel networks for electrophoretic separations. *Electrophoresis* 20:727–731.

Rossier, J. S., Roberts, M. A., Ferrigno, R., Girault, H. H. (1999b). Electrochemical detection in polymer microchannels. *Anal. Chem.* 71:4294–4299.

Roulet, J., Völkel, R., Herzig, H. P., Verpoorte, E. (2002). Performance of an integrated microoptical system for fluorescence detection in microfluidic systems. *Anal. Chem.* 74:3400–3407.

Rowe, C. A., Scruggs, S. B., Feldstein, M. J., Golden, J. P., Ligler, F. S. (1999). An array immunosensor for simultaneous detection of clinical analytes. *Anal. Chem.* 71(2):433–439.

Russom, A., Ahmadian, A., Andersson, H., Van Der Wijngaart, W., Lundeberg, J., Uhlen, M., Stemme, G., Nilsson, P. (2001). SNP analysis by allele-specific pyrosequencing extension in a micromachined filter-chamber device. Micro Total Analysis Systems 2001. Proceedings μTAS 2001 Symposium. 5th ed. Monterey, CA, Oct. 21–25, 2001, pp. 22–24.

Salas-Solano, O., Schmzlzing, D., Koutny, L., Buonocore, S., Adourian, A., Matsudaira, P., Ehrlich, D. (2000). Optimization of high-performance DNA sequencing on short microfabricated electrophoretic devices. *Anal. Chem.* 72:3129–3137.

Salimi-Moosavi, H., Szarka, R., Andersson, P., Smith, R., Harrision, D. J. (1998). Biology lab-on-a chip for drug screening. Micro Total Analysis Systems '98. Proceedings μTAS '98 Workshop. Banff, Canada, 13–16 Oct. 1998, pp. 69–72.

Salimi-Moosavi, H., Jiang, Y., Lester, L., McKinnon, G., Harrison, D. J. (2000). A multireflection cell for enhanced absorbance detection in microchip-based capillary electrophoresis devices. *Electrophoresis* 21(7):1291–1299.

Sanders, G. H. W., Manz, A. (2000). Chip-based microsystems for genomic and proteomic analysis. *Trends Anal. Chem.* 19:364–378.

Santiago, J. G. (2001). Electroosmotic flows in microchannels with finite inertial and pressure forces. *Anal. Chem.* 73(10):2353–2365.

Sapsford, K. E., Charles, P. T., Patterson, C. H. Jr., Ligler, F. S. (2002). Demonstration of four immunoassay formats using the array biosensor. *Anal. Chem.*, 74(5):1061–1068.

Sato, K., Tokeshi, M., Odake, T., Kimura, H., Ooi, T., Nakao, M., Kitamori, T. (2000). Integration of an immunosorbent assay system: analysis of secretory human immunoglobulin a on polystyrene beads in a microchip. *Anal. Chem.* 72:1144–1147.

Sato, K., Tokeshi, M., Kimura, H., Kitamori, T. (2001a). Determination of carcinoembryonic antigen in human sera by integrated bead-bed immunoasay in a microchip for cancer diagnosis. *Anal. Chem.* 73(6):1213–1218.

Sato, K., Yamanaka, M., Takahashi, H., Uchiyama, K., Tokeshi, M., Katou, H., Kimura, H., Kitamori, T. (2001b). Integrated immunoassay system using multichannel microchip for simultaneous determination. Micro Total Analysis Systems 2001. Proceedings μTAS 2001 Symposium. 5th ed. Monterey, CA, Oct. 21–25, 2001, pp. 511–512.

Schasfoort, R. B. M., Schlautmann, S., Hendrikse, J., Van Den Berg, A. (1999). Field-effect flow control for microfabricated fluidic networks. *Science* 286(5441):942–945.

Schasfoort, R., van Duijn, G., Schlautmann, S., Frank, H., Billiet, H., van Dedem, G., van den Berg, A. (2000). Miniaturized capillary electrophoresis system with integrated conductivity detector. Micro Total Analysis Systems 2000, Proceedings of the μTAS Symposium, 4th ed. Enschede, The Netherlands, May 14–18, 2000, pp. 391–394.

Schasfoort, R. B. M., Luttge, R., Van Den Berg, A. (2001). Magneto-hydrodynamically (MHD) directed flow in microfluidic networks. Micro Total Analysis Systems 2001. Proceedings μTAS 2001 Symposium. 5th ed. Monterey, CA, Oct. 21–25, 2001, pp. 577–578.

Schilling, E. A., Kamholz, A. E., Yager, P. (2001). Cell lysis and protein extraction in a microfluidic device with detection by a fluorogenic enzyme assay. Micro Total Analysis Systems 2001. Proceedings μTAS 2001 Symposium. 5th ed. Monterey, CA, Oct. 21–25, 2001, pp. 265–267.

Schilling, E. A., Kamholz, A. E., Yager, P. (2002). Cell lysis and protein extraction in a microfluidic device with detection by a fluorogenic enzyme assay. *Anal. Chem.* 74(8):1798–1804.

Schmalzing, D., Adourian, A., Koutny, L., Ziaugra, L., Matsudaira, P., Ehrlich, D. (1998). DNA sequencing on microfabricated electrophoresis devices. *Anal. Chem.* 70:2303–2310.

Schmitt, H., Brecht, A., Gauglitz, G. (1996). An integrated system for microscale affinity measurements. Micro Total Analysis Systems '96. Proceedings of 2nd International Symposium on μTAS '96. Basel, 19–22 Nov. 1996, pp. 104–109.

Schoot, B. H., van der Verpoorte, E. M. J., Jeanneret, S., Manz, A., Rooij, N. F. de. (1995). Microsystems for analysis in flowing solutions. Micro Total Anal. Syst. Proc. μTAS '94 Workshop. 1st ed. Meeting Date 1994, pp. 181–190.

Schrum, D. P., Culbertson, C. T., Jacobson, S. C., Ramsey, J. M. (1999). Microchip flow cytometry using electrokinetic focusing. *Anal. Chem.* 71:4173–4177.

Schultz, G. A., Corso, T. N., Prosser, S. J., Zhang, S. (2000). A fully integrated monolithic microchip electrospray device for mass spectrometry. *Anal. Chem.* 72(17):4058–4063.

Schumacher, J., Ranft, M., Wilhelm, T., Dahint, R., Grunze, M. (1998). Chemical analysis based on enviromentally sensitivie hydrogels and optical diffraction. Micro Total Analysis Systems '98. Proceedings μTAS '98 Workshop. Banff, Canada, 13–16 Oct. 1998, pp. 61–64.

Schwarz, A., Rossier, J. S., Bianchi, F., Reymond, F., Ferrigno, R., Girault, H. H. (1998). Micro-TAS on polymer substrates micromachined by laser photoablation. Micro Total Analysis Systems '98. Proceedings μTAS '98 Workshop. Banff, Canada, 13–16 Oct. 1998, pp. 241–244.

Schwarz, M. A., Galliker, B., Fluri, K., Kappes, T., Hauser, P. C. (2001a). A two-electrode configuration for simplified amperometric detection in a microfabricated electrophoretic separation device. *Analyst* 126(2):147–151.

Schwarz, M. A., Hauser, P. C. (2001b). Fast chiral on-chip separations with amperometric detection. Micro Total Analysis Systems 2001. Proceedings μTAS 2001 Symposium. 5th ed. Monterey, CA, Oct. 21–25, 2001, pp. 547–548.

Seiler, K., Harrison, D. J., Manz, A. (1993). Planar glass chips for capillary electrophoresis: repetitive sample injection, quantitation, and separation efficiency. *Anal. Chem.* 65:1481–1488.

Seiler, K., Fan, Z. H., Fluri, K., Harrison, D. J. (1994). Electroosmotic pumping and valveless control of fluid flow within a manifold of capillaries on a glass chip. *Anal. Chem.* 66:3485–3491.

Seki, M., Yamada, M., Ezaki, R., Aoyama, R., Hong, J. W. (2001). Chromatographic separation of proteins on a PDMS-polymer chip by pressure flow. Micro Total Analysis Systems 2001. Proceedings μTAS 2001 Symposium. 5th ed. Monterey, CA, Oct. 21–25, 2001, pp. 48–50.

Seong, G. H., Zhan, W., Crooks, R. M. (2002). Fabrication of microchambers defined by photopolymerized hydrogels and weirs within microfluidic systems: application to DNA hybridization. *Anal. Chem.* 74:3372–3377.

Service, R. F. (1995). The incredible shrinking laboratory. *Science* 268:26–27.

Service, R. F. (1996). Can chip devices keep shrinking? *Science* 274:1834–1836.

Shaw, J., Nudd, R., Naik, B., Turner, C., Rudge, D., Benson, M., Garman, A. (2000). Liquid/liquid extraction systems using microcontractor arrays. Micro Total Analysis Systems 2000. Proceedings of the μTAS Symposium. 4th ed. Enschede, The Netherlands, May 14–18, 2000, pp. 371–374.

Shi, Y., Simpson, P. C., Scherer, J. R., Wexler, D., Skibola, C., Smith, M. T., Mathies, R. A. (1999). Radical capillary array electrophoresis microplate and scanner for high-performance nucleic acid analysis. *Anal. Chem.* 71:5354–5361.

Shoffner, M. A., Cheng, J., Hvichia, G. E., Kricka, L. J., Wilding, P. (1996). Chip PCR. I. surface passivation of microfabricated silicon-glass chips for PCR. *Nucl. Acids Res.*, 24:375–379.

Shoji, S. (1998). Fluids for sensor systems. In: *Microsystem Technology in Chemistry and Life Science*. Berlin, Heidelberg: Springer-Verlag, pp. 163–188.

Shoji, S., Esashi, M. (1992). Micro flow cell for blood gas analysis realizing very small sample volume. *Sensors Actuators B: Chem.* B8(2):205–208.

Shoji, S., Esashi, M. (1995). Bonding and assembling methods for realising a μTAS. Micro Total Analysis Systems '94. Proceedings of the μTAS '94 Workshop. The Netherlands: University of Twente, 21–22 Nov. 1994, pp. 165–179.

Simpson, P. C., Roach, D., Woolley, A. T., Thorsen, T., Johnston, R., Sensabaugh, G. F., Mathies, R. A. (1998). High-throughput genetic analysis using microfabricated 96-sample capillary array electrophoresis microplates. *Proc. Natl. Acad. Sci. USA* 95:2256–2261.

Sirichai, S., de Mello, A. J. (2000). A capillary electrophoresis microchip for the analysis of photographic developer solutions using indirect fluorescence detection. *Analyst* 125(1):133–137.

Sirichai, S., De Mello, A. J. (2001). A capillary electrophoresis chip for the analysis of print and film photographic developing agents in commercial processing solutions using indirect fluorescence detection. *Electrophoresis* 22(2):348–354.

Sjoelander, S., Urbaniczky, C. (1991). Integrated fluid handling system for biomolecular interaction analysis. *Anal. Chem.* 63:2338–2345.

Slentz, B. E., Penner, N. A., Regnier, F. (2002). Sampling BIAS at channel junctions in gated flow injection on chips. *Anal. Chem.* 74(18):4835–4840.

Slyadnev, M. N., Tanaka, Y., Tokeshi, M., Kitamori, T. (2001). Non-contact temperature measurement inside microchannel. Micro Total Analysis Systems 2001. Proceedings μTAS 2001 Symposium. 5th ed. Monterey, CA, Oct. 21–25, 2001, pp. 361–362.

Sobek, D., Young, A. M., Gray, M. L., Senturia, S. D. (1993). A microfabricated flow chamber for optical measurements in fluids. *IEEE* Feb. 7–10:219–224.

Sohn, L. L., Saleh, O. A., Facer, G. R., Beavis, A. J., Allan, R. S., Notterman, D. A. (2000). Capacitance cytometry: measuring biological cells one by one. *Proc. Natl Acad. Sci. USA* 97(20):10687–10690.

Sohn, Y.-S., Goodey, A. P., Anslyn, E. V., McDevitt, J. T., Shear, J. B., Neikirk, D. P. (2001). Development of a micromachined fluidic structure for a biological and chemical sensor array. Micro Total Analysis Systems 2001. Proceedings μTAS 2001 Symposium. 5th ed. Monterey, CA, Oct. 21–25, 2001, pp. 177–178.

Soper, S. A., Murphy, M. C., Mccarley, R. L., Nikitopoulos, D., Liu, X., Vaidya, B., Barrow, J., Bejat, Y., Ford, S. M., Goettert, J. (2001). Fabrication of modular microsystems for analyzing K-ras mutations using LDR. Micro Total Analysis Systems 2001. Proceedings μTAS 2001 Symposium. 5th ed. Monterey, CA, Oct. 21–25, 2001, pp. 459–461.

Stemme, G., Kittilsland, G. (1998). New fluid filter strucure in silicon fabricated using a self-aligning technique. *Appl. Phys. Lett.* 53:1566–1568.

Strömberg, A., Karlsson, A., Ryttsén, F., Davidson, M., Chiu, D. T., Orwar, O. (2001). Microfluidic device for combinatorial fusion of liposomes and cells. *Anal. Chem.* 73:126–130.

Stroock, A. D., Dertinger, S. K., Whitesides, G. M., Ajdari, A. (2002). Patterning flows using grooved surfaces. *Anal. Chem.* 74(20):5306–5312.

Sutton, N., Tracey, M. C., Johnston, I. D. (1997). A novel instrument for studying the flow behaviour of erythrocytes through microchannels simulating human blood capillaries. *Microvasc. Res.* 53:272–281.

Suzuki, S., Shimotsu, N., Honda, S., Arai, A., Nakanishi, H. (2001). Rapid analysis of amino sugars by microchip electrophoresis with laser-induced fluorescence detection. *Electrophoresis* 22(18):4023–4031.

Svasek, P., Jobst, G., Urban, G., Svasek, E. (1996). Dry film resist based fluid handling components for μTAS. Micro Total Analysis Systems '96. Proceedings of 2nd international symposium on μTAS '96. Basel, 19–22 Nov. 1996, pp. 27–29.

Swinney, K., Markov, D., Bornhop, D. J. (2000). Chip-scale universal detection based on backscatter interferometry. *Anal. Chem.* 72(13):2690–2695.

Takamura, Y., Onoda, H., Inokuchi, H., Adachi, S., Oki, A., Horiike, Y. (2001). Low-voltage electroosmosis pump and its application to on-chip linear stepping pneumatic pressure source. Micro Total Analysis Systems 2001. Proceedings µTAS 2001 Symposium. 5th ed. Monterey, CA, Oct. 21–25, 2001, pp. 230–232.

Takao, H., Noda, T., Ashiki, M., Miyamura, K., Sawada, K., Ishida, M. (2001). A silicon microchip for blood hemoglobin measurement using multireflection structure. Micro Total Analysis Systems 2001. Proceedings µTAS 2001 Symposium. 5th ed. Monterey, CA, Oct. 21–25, 2001, pp. 363–364.

Takayama, S., Mcdonald, J. C., Ostuni, E., Liang, M. N., Kenis, P. J. A., Ismagilov, R. F., Whitesides, G. M. (1999). Patterning cells and their environments using multiple laminar fluid flows in capillary networks. *Proc. Natl. Acad. Sci. USA* 96:5545–5548.

Tamaki, E., Sato, K., Tokeshi, M., Sato, K., Aihara, M., Kitamori, T. (2002). Single-cell analysis by a scanning thermal lens microscope with a microchip: direct monitoring of cytochrome c distribution during apoptosis process. *Anal. Chem.* 74:1560–1564.

Tang, T., Ocvirk, G., Harrison, D. J. (1997). ISO-thermal DNA reactions and assays in microfabricated capillary electrophoresis systems. *Transducers* June 16–19:523–526.

Tantra, R., Manz, A. (2000). Integrated potentiometric detector for use in chip-based flow cells. *Anal. Chem.* 72:2875–2878.

Tashiro, K., Ikeda, S., Sekiguchi, T., Shoji, S., Makazu, H., Funatsu, T., Tsukita, S. (2001). A particles and biomolecules sorting micro flow system using thermal gelation of methyl cellulose solution. Micro Total Analysis Systems 2001. Proceedings µTAS 2001 Symposium. 5th ed. Monterey, CA, Oct. 21–25, 2001, pp. 471–473.

Taylor, M. T., Belgrader, P., Joshi, R., Kintz, G. A., Northrup, M. A. (2001). Fully automated sample preparation for pathogen detection performed in a microfluidic cassette. Micro Total Analysis Systems 2001. Proceedings µTAS 2001 Symposium. 5th ed. Monterey, CA, Oct. 21–25, 2001, pp. 670–672.

Terry, S. C., Jerman, J. H., Angell, J. B. (1979). A gas chromatographic air analyzer fabricated on a Silicon wafer. *IEEE Trans. Electron. Devices* 26:1880–1886.

Theemsche, A. V., Deconinck, J., Bossche, Van den B., Bortels, L. (2002). Numerical solution of a multi-ion one-potential model for electroosmotic flow in two-dimensional rectangular microchannels. *Anal. Chem.* 74(19):4919–4926.

Timmer, B. H., Bomer, J. G., Van Delft, K. M., Otjes, R. P., Olthuis, W., Bergveld, P., Van Den Berg, A. (2001). Fluorocarbon coated micromachined gas sampling device. Micro Total Analysis Systems 2001. Proceedings µTAS 2001 Symposium. 5th ed. Monterey, CA, Oct. 21–25, 2001, pp. 381–382.

Tohnson, T. J., Ross, D., Gaitan, M., Locascio, L. E. (2001). Laser modification on channels to reduce band broadening or to increase mixing. Micro Total Analysis Systems 2001. Proceedings µTAS 2001 Symposium. 5th ed. Monterey, CA, Oct. 21–25, 2001, pp. 603–604.

Tokano, H., Sul, J., Mazzanti, M. L., Doyle, R. T., Haydon, P. G., Porter, M. D. (2002). Micropatterned substrates: approach to probing intercellular communication pathways. *Anal. Chem.* 74:4640–4646.

Tokeshi, M., Minagawa, T., Kitamori, T. (2000a). Integration of a microextraction system on a glass chip: ion-pair solvent extraction of Fe(II) with 4,7-diphenyl-1,10-phenanthrolinedisulfonic acid and tri-n-octylmethylammonium Chloride. *Anal. Chem.* 72(7):1711–1714.

Tokeshi, M., Minagawa, T., Kitamori, T. (2000b). Integration of a microextraction system. Solvent extraction of a Co-2-nitroso-5-dimethylaminophenol complex on a microchip. *J. Chromatogr. A* 894(1 + 2):19–23.

Tokeshi, M., Uchida, M., Hibara, A., Sawada, T., Kitamori, T. (2001). Determination of subyoctomole amounts of nonfluorescent molecules using a thermal lens microscope: subsingle-molecule determination. *Anal. Chem.* 73(9): 2112–2116.

Tokeshi, M., Minagawa, T., Uchiyama, K., Hibara, A., Sato, K., Hisamoto, H., Kitamori, T. (2002). Continuous-flow chemical processing on a microchip by combining microunit operations and a multiphase flow network. *Anal. Chem.* 74(7): 1565–1571.

Ueno, Y., Horiuchi, T., Morimoto, T., Niwa, O. (2001). Microfluidic device for airborne BTEX detection. *Anal. Chem.* 73(19):4688–4693.

Ueno, Y., Horiuchi, T., Niwa, O. (2002a). Air-cooled cold trap channel integrated in a microfluidic device for monitoring airborne BTEX with an improved detection limit. *Anal. Chem.* 74(7):1712–1717.

Ueno, Y., Horiuchi, T., Tomita, M., Niwa, O., Zhou, H., Yamada, T., Honma, I. (2002b). Separate detection of BTX mixture gas by a microfluidic device using a function of nanosized pores of mesoporous silica adsorbent. *Anal. Chem.* 74:5257–5262.

Unger, M. A., Chou, H., Thorsen, T., Scherer, A., Quake, S. R. (2000). Monolithic microfabricated valves and pumps by multilayer soft lithography. *Science* 288:113–116.

VanderNoot, V. A., Hux, G., Schoeniger, J., Shepodd, T. (2001). Isoelectric focusing using electrokinetically-generated pressure mobilization. Micro Total Analysis Systems 2001. Proceedings µTAS 2001 Symposium. 5th ed. Monterey, CA, Oct. 21–25, 2001, pp. 127–128.

Verpoorte, E., Manz, A., Luedi, H., Bruno, A. E., Maystre, F., Krattiger, B., Widmer, H. M., Van der Schoot, B. H., De Rooij, N. F. (1992). A silicon flow cell for optical detection in miniaturized total chemical analysis systems. *Sensors Actuators B: Chem.* B6(1–3):66–70.

Verpoorte, E., Manz, A., Widmer, H. M., van der Schoot, B., de Rooij, N. F. (1993). A three-dimensional micro flow system for a multi-step chemical analysis. *Transducer* pp. 939–942.

Vogt, O., Grass, B., Weber, G., Hergenroder, R., Siepe, D., Neyer, A., Pohl, J. P. (2001). Characterization of sputtered thin film electrodes on PMMA microchips with electrochemical impedance spectroscopy and cyclic voltammetry. Micro Total Analysis Systems 2001. Proceedings µTAS 2001 Symposium. 5th ed. Monterey, CA, Oct. 21–25, 2001, pp. 327–328.

Voldman, J., Gray, M. L., Toner, M., Schmidt, M. A. (2002). A microfabrication-based dynamic array cytometer. *Anal. Chem.* 74:3984–3990.

Volkmuth, W. D., Austin, R. H. (1992). DNA electrophoresis in microlithographic arrays. *Nature* 358:600–602.

Volkmuth, W. D., Duke, T., Austin, R. H., Cox, E. C. (1995). Trapping of branched DNA in microfabricated structures. *Proc. Natl. Acad. Sci. USA* 92:6887–6891.

Wabuyele, M. B., Ford, S. M., Stryjewski, W., Barrow, J., Soper, S. A. (2001). Single molecule detection of double-stranded DNA in poly(methylmethacrylate) and polycarbonate microfluidic devices. *Electrophoresis* 22(18): 3939–3948.

Walker, P. A. III., Morris, M. D., Burns, M. A., Johnson, B. N. (1998). Isotachophoretic separations on a microchip. Normal raman spectroscopy detection. *Anal. Chem.* 70(18):3766–3769.

Wallenborg, S. R., Bailey, C. G. (2000). Separation and detection of explosives on a microchip using micellar electrokinetic chromatography and indirect laser-induced fluorescence. *Anal. Chem.* 72:1872–1878.

Wallenborg, S. R., Bailey, C. G., Paul, P. H. (2000). On-chip separation of explosive compounds-divided reservoirs to improve reproducibility and minimize buffer depletion. Micro Total Analysis Systems 2000, Proceedings of the μTAS Symposium. 4th ed. Enschede, The Netherlands, May 14–18, 2000, pp. 355–358.

Wang, J. (2002). Electrochemical detection for microscale analytical systems: a review. *Talanta* 56:223–231.

Wang, S., Morris, M. D. (2000). Plastic microchip electrophoresis with analyte velocity modulation. Application to fluorescence background rejection. *Anal. Chem.* 72:1448–1452.

Wang, J., Pumera, M. (2002). Dual conductivity/amperometric detection system for microchip capillary electrophoresis. *Anal. Chem.* 74(23):5919–5923.

Wang, J., Tian, B., Sahlin, E. (1999a). Integrated electrophoresis chips/amperometric detection with sputtered gold working electrodes. *Anal. Chem.* 71(17):3901–3904.

Wang, J., Tian, B., Sahlin, E. (1999b). Micromachined electrophoresis chips with thick-film electrochemical detectors. *Anal. Chem.* 71(23):5436–5440.

Wang, S.-C., Perso, C. E., Morris, M. D. (2000a). Effects of alkaline hydrolysis and dynamic coating on the electroosmotic flow in polymeric microfabricated channels. *Anal. Chem.* 72(7):1704–1706.

Wang, J., Chatrathi, M. P., Tian, B. (2000b). Micromachined separation chips with a precolumn reactor and end-column electrochemical detector. *Anal. Chem.* 72(23):5774–5778.

Wang, C., Oleschuk, R., Ouchen, F., Li, J., Thibault, P., Harrison, D.J. (2000c). Integration of immobilized trypsin bead beds for protein digestion within a microfluidic chip incorporating capillary electrophoresis separations and an electrospray mass spectrometry interface. *Rapid Commun. Mass Spectrom.* 15:1377–1383.

Wang, J., Chatrathi, M. P., Tian, B., Polsky, R. (2000d). Microfabricated electrophoresis chips for simultaneous bioassays of glucose, uric acid, ascorbic acid, and acetaminophen. *Anal. Chem.* 72(11):2514–2518.

Wang, J., Chatrathi, M. P., Mulchandani, A., Chen, W. (2001a). Capillary electrophoresis microchips for separation and detection of organophosphate nerve agents. *Anal. Chem.* 73(8):1804–1808.

Wang, J., Chatrathi, M. P., Tian, B. (2001b). Microseparation chips for performing multienzymatic dehydrogenase/oxidase assays: simultaneous electrochemical measurement of ethanol and glucose. *Anal. Chem.* 73(6): 1296–1300.

Wang, J., Pumera, M., Collins, G. E., Mulchandani, A. (2002). Measurements of chemical warfare agent degradation products using an electrophoresis microchip with contactless conductivity detector. *Anal. Chem.* 74(23): 6121–6125.

Water, L. C., Jacobson, S. C., Kroutchinina, N., Khandurina, J., Foote, R. S., Ramsey, J. M. (1998a). Multiple sample PCR amplification and electrophoretic analysis on a microchip. *Anal. Chem.* 70:5172–5176.

Waters, L. C., Jacobson, S. C., Kroutchinina, N., Khandurina, J., Foote, R. S., Ramsey, J. M. (1998b). Microchip device for cell lysis, multiplex PCR amplification, and electrophoretic sizing. *Anal. Chem.* 70:158–162.

Weber, G., Johnck, M., Siepe, D., Neyer, A., Hergenroder, R. (2000). Capillary electrophoresis with direct and contactless conductivity detection on a polymer microchip. Micro Total Analysis Systems 2000. Proceedings of the μTAS Symposium. 4th ed. Enschede, The Netherlands, May 14–18, 2000, pp. 383–386.

Webster, J. R., Burns, M. A., Burke, D. T., Mastrangelo, C. H. (1998). An inexpensive plastic technology for microfabricated capillary electrophoresis chips. Micro Total Analysis Systems '98. Proceedings μTAS '98 Workshop. Banff, Canada, 13–16 Oct. 1998, pp. 249–252.

Webster, J. R., Burns, M. A., Burke, D. T., Mastrangelo, C. H. (2001). Monolithic capillary electrophoresis device with integrated fluorescence detector. *Anal. Chem.* 73(7): 1622–1626.

Weigl, B. H., Yager, P. (1999). Microfluidic diffusion-based separation and detection. *Science* 283:346–347.

Weigl, B. H., Holl, M. R., Schutte, D., Brody, J. P., Yager, P. (1996). Diffusion-based optical chemical detection in silicon flow structure. Micro Total Analysis Systems '96. Proceedings of 2nd International Symposium on μTAS '96. Basel, 19–22 Nov. 1996, pp. 174–184.

Weigl, B. H., Kriebel, J., Mayes, K., Yager, P., Wu, C. C., Holl, M., Kenny, M., Zebert, D. (1998). Simultaneous self-referencing analyte determination in complex sample solutions using microfabricated flow structure (T-sensors). Micro Total Analysis Systems '98. Proceedings μTAS '98 Workshop. Banff, Canada, 13–16 Oct. 1998, pp. 81–84.

Weiller, B. H., Ceriotti, L., Shibata, T., Rein, D., Roberts, M. A., Lichtenberg, J., German, J. B., de Rooij, N. F., Verpoorte, E. (2002). Analysis of lipoproteins by capillary zone electrophoresis in microfluidic devices: assay development and surface roughness measurements. *Anal. Chem.* 74(7):1702–1711.

Wensink, H., Elwenspoek, M. C. (2001). New developments in bulk micromachining by powder. Micro Total Analysis Systems 2001. Proceedings μTAS 2001 Symposium. 5th ed. Monterey, CA, Oct. 21–25, 2001, pp. 393–394.

Wheeler, A., Morishima, K., Kirby, B., Leach, A., Zare, R. N. (2001). Cath, a neuron cell analysis on a chip with micellar electrokinetic chromatography. Micro Total Analysis Systems 2001. Proceedings μTAS 2001 Symposium. 5th ed. Monterey, CA, Oct. 21–25, 2001, pp. 311–312.

Wilding, P., Pfahler, J., Bau, H. H., Zemel, J. N., Kricka, L. J. (1994a). Manipulation and flow of biological fluids in straight channels micromachined in silicon. *Clin. Chem.* 40:43–47.

Wilding, P., Shoffner, M. A., Kricka, L. J. (1994b). PCR in a silicon microstructure. *Clin. Chem.* 40:1815–1818.

Wilding, P., Kricka, L. J., Cheng, J., Hvichia, G. (1998). Integrated cell isolation and polymerase chain reaction analysis using silicon microfilter chambers. *Anal. Biochem.* 257:95–100.

Woias. P., Hauser, K., Yacoub-George, E. (2000). An active silicon micromixer for μTAS applications. Micro Total Analysis Systems 2000. Proceedings of the μTAS Symposium. 4th ed. Enschede, The Netherlands, May 14–18, 2000, pp. 277–282.

Wolk, J., Spaid, M., Jensen, M., Macreynolds, R., Steveson, K., Chien, R. (2001). Ultraviolet absorbance spectroscopy in a 3-dimensional microfuidic chip. Micro Total Analysis Systems 2001. Proceedings μTAS 2001 Symposium. 5th ed. Monterey, CA, Oct. 21–25, 2001, pp. 367–368.

Woolley, A. T., Mathies, R. A. (1994). Ultra-high-speed DNA fragment separations using microfabricated capillary array electrophoresis chips. *Proc. Natl. Acad. Sci. USA* 91:11348–11352.

Woolley, A. T., Mathies, R. A. (1995). Ultra-high-speed DNA sequencing using capillary electrophoresis chips. *Anal. Chem.* 67:3676–3680.

Woolley, A. T., Hadley, D., Landre, P., de Mello, A. J., Mathies, R. A., Northrup, M. A. (1996). Functional integration of PCR amplification and capillary electrophoresis in a microfabricated DNA analysis device. *Anal. Chem.* 68:4081–4086.

Woolley, A. T., Sensabaugh, G. F., Mathies, R. A. (1997). High-speed DNA genotyping using microfabricated capillary array electrophoresis chips. *Anal. Chem.* 69:2181–2186.

Woolley, A. T., Lao, K., Glazer, A. N., Mathies, R. A. (1998). Capillary electrophoresis chips with integrated electrochemical detection. *Anal. Chem.* 70:684–688.

Wu, H., Odom, T. W., Whitesides, G. M. (2002). Reduction photolithography using microlens arrays: applications in gray scale photolithography. *Anal. Chem.* 74(14):3267–3273.

Xiang, F., Lin, Y., Wen, J., Matson, D. W., Smith, R. D. (1999). An integrated microfabricated device for dual microdialysis and online ESI-ion trap mass spectrometry for analysis of complex biological samples. *Anal. Chem.* 71(8):1485–1490.

Xu, N., Lin, Y., Hofstadler, S. A., Matson, D., Call, C. J., Smith, R. D. (1998). A microfabricated dialysis device for sample cleanup in electrospray ionization mass spectrometry. *Anal. Chem.* 70:3553–3556.

Xu, J., Locascio, L., Gaitan, M., Lee, C. S. (2000). Room-temperature imprinting method for plastic microchannel fabrication. *Anal. Chem.* 72(8):1930–1933.

Xu, H., Roddy, T. P., Lapos, J. A., Ewing, A. G. (2002). Parallel analysis with optically gated sample introduction on a multi-channel microchip. *Anal. Chem.* 74(21):5517–5522.

Xue, Q., Foret, F., Dunayevskiy, Y. M., Zavracky, P. M., McGruer, N. E., Karger, B. L. (1997). ultichannel microchip electrospray mass spectrometry. *Anal. Chem.* 69:426–430.

Xue, Q., Wainright, A., Gangakhedkar, S., Gibbons, I. (2001). Multiplexed enzyme assays in capillary electrophoretic single-use microfluidic devices. *Electrophoresis* 22(18):4000–4007.

Yager, P., Bell, D., Brody, J. P., Qin, D., Cabrera, C., Kamholz, A., Weigl, B. (1998). Applying microfluidic chemical analytical systems to imperfect samples. Micro Total Analysis Systems '98. Proceedings μTAS '98 Workshop. Banff, Canada, 13–16 Oct. 1998, pp. 207–212.

Yager, P., Cabrera, C., Hatch, A., Hawkins, K., Holl, M., Kamholz, A., Macounova, K., Weigl, B. H. (2000). Analytical devices based on transverse transport in microchannels. Micro Total Analysis Systems 2000. Proceedings of the μTAS Symposium. 4th ed. Enschede, The Netherlands, May 14–18, 2000, pp. 15–18.

Yakovleva, J., Davidsson, R., Lobanova, A., Bengtsson, M., Eremin, S., Laurell, T., Emnéus, J. (2002). Microfluidic enzyme immunoassay using silicon microchip with immobilized antibodies and chemiluminescence detection. *Anal. Chem.* 74(13):2994–3004.

Yang, Z., Goto, H., Matsumoto, M., Yada, T. (1998). Micro mixer incorporated with piezoelctrially driven valveless micropump. Micro Total Analysis Systems '98. Proceedings μTAS '98 Workshop. Banff, Canada, 13–16 Oct. 1998, pp. 177–180.

Yang, T., Jung, S., Mao, H., Cremer, P. S. (2001). Fabrication of phospholipid bilayer-coated microchannels for on-chip immunoassays. *Anal. Chem.* 73:165–169.

Yang, M., Yang, J., Li, C., Zhao, J. (2002a). Genaration of concentration gradient by controlled flow distribution and diffusive mixing in a microfluidic chip. *Lab on a chip*. 2:158–163.

Yang, M., Li, C., Yang, J. (2002b). Cell docking and on-chip monitoring of cellular reactions with a controlled concentration gradient on a microfluidic device. *Anal. Chem.* 74:3991–4001.

Yu, H., Sethu, P., Chan, T., Kroutchinina, N., Blackwell, J., Mastrangelo, C. H., Grodzinski, P. (2000). A miniaturized and integrated plastic thermal chemical reactor for genetic analysis. Micro Total Analysis Systems 2000. Proceedings of the μTAS Symposium. 4th ed. Enschede, The Netherlands, May 14–18, 2000, pp. 545–548.

Zeng, Y., Chen, H., Pang, D., Wang, Z., Cheng, J. (2002). Microchip capillary electrophoresis with electrochemical deetection. *Anal. Chem.* 74:2441–2445.

Zhan, W., Seong, G. H., Crooks, R. M. (2002). Hydrogel-based microreactors as a functional component of microfluidic systems. *Anal. Chem.* 74(18):4647–4652.

Zhang, C., Manz, A. (2001). Narrow sample channel injectors for capillary electrophoresis on microchips. *Anal. Chem.* 73:2656–2662

Zhang, B., Liu, H., Karger, B. L., Foret, F. (1999). Microfabricated device for capillary electrophoresis–electrospray mass spectrometry. *Anal. Chem.* 71:3258–3264.

Zhang, B., Foret, F., Karger, B. L. (2000). A microdevice with integrated liquid junction for facile peptide and protein analysis by capillary electrophoresis/electrospray mass spectrometry. *Anal. Chem.* 72:1015–1022.

Zhao, B., Moore, J. S., Beebe, D. J. (2001). Surface-directed liquid flow inside microchannels. *Science* 291:1023–1026.

21

Biosensor Technology

RALUCA-IOANA STEFAN, JACOBUS FREDEICZZK VAN STADEN
Department of Chemistry, University of Pretoria Pretoria, South Africa

HASSAN Y. ABOUL-ENEIN
Pharmaceutical Analysis and Drug Development Laboratory, Biological and Medical Research Department, King Faisal Specialist Hospital and Research Centre, Riyadh, Saudi Arabia

I. INTRODUCTION

Biosensors are analytical devices that use a biological or biologically derived material immobilized at a transducer to measure one or more analytes (Subrahmanyam et al., 2002). Biosensors couple the sensitivity of the transducer with the selectivity of a reaction. The components of biosensors are: the biological material and the transducer. The biological material can be immobilized physically or chemically on the transducer. The principle of biosensors is simple: the transducer will selectively detect one of the products of the biochemical reaction.

Biosensors are used in clinical, pharmaceutical, food and environmental analyses. The best results, especially in terms of selectivity, are obtained for clinical, pharmaceutical, and food analyses (Aboul-Enein et al., 2000; Stefan et al., 2001). Owing to the complexity of the matrix of the environmental samples, more often only group selectivity is obtained when biosensors are used for environmental analysis. The most popular biosensor is the one developed for glucose assay.

The evolution in the design of biosensors is remarkable, when looking at the materials used as matrices for biosensors' design (plastic, carbon, diamond, etc.) as well as the size of the biosensors (from macro to micro and nano sizes).

II. BIOLOGICAL MATERIALS

The biological materials, also known as molecular recognition elements, can be enzymes, microbes, molecular receptors, cells, nucleic acids, or tissues. The most utilized biological materials in biosensors technology are enzymes. Plant and animal tissues may be used directly for the design of the biosensors (e.g., banana electrode), with minimal preparation (Eggins, 1996).

Molecular receptors are cellular, typically membrane, proteins that bind specific chemicals (ligands) in a manner that results in a conformational change in the protein structure (Subrahmanyam et al., 2002). Four different classes of receptors were identified (Haga, 1995): (1) ion channel receptors; (2) G-protein-linked receptors; (3) receptors with single transmembrane segments; and (4) enzyme-linked receptors.

Nucleic acids are used to detect genetic diseases, cancer, and viral infections. They operate in many cases like antibodies (Subrahmanyam et al., 2002).

The utilization of each of the biological materials has advantages and disadvantages. For example, the main disadvantages of utilization of the enzymes are the short life time of the biosensor and the decrease of activity of the enzyme with time; the main disadvantage of utilizing tissues in the design of the biosensors is the poor selectivity, owing to the multitude of enzymes present in the tissue.

In order to select the best biological material for biosensors design it is necessary to take into account the matrix from where the analyte should be determined to ensure selectivity, and also the sensitivity requested for the biochemical reaction (Aboul-Enein et al., 2000; Stefan et al., 2001). Recently, computational bioanalysis through the evolution of modelling techniques is efficiently solving the selection of the best biological material for biosensors technology (Fischer, 2003).

III. TRANSDUCERS

The transducer must be selected according to the product obtained in the biochemical reaction, and it can be electrochemical, optical, or piezoelectric. The sensitivity of the transducer must not be lower than the sensitivity of the biochemical reaction. Also, it must be selective over the other reaction products as well as over the substrate.

A. Electrochemical Transducers

The electrochemical transducers are classified into potentiometric and amperometric transducers. However, it is advisable to avoid potentiometric transducers, unless absolutely necessary (e.g., the product that should be measured is H^+ or NH_4^+), owing to their low sensitivity.

Amperometric transducers are the most used for the design of biosensors, owing to their high sensitivity. Up till now, the amperometric transducers that gave the best sensitivities were glassy carbon (Freire et al., 2002; Ugo et al., 2002; Wang et al., 2002; Wu et al., 2001), carbon paste (Stefan et al., 2003a–f), and diamond paste electrodes (Stefan and Nejem, 2003; Stefan and Bokretsion, 2003; Stefan et al., 2003g, h). The design of carbon and diamond paste electrodes is shown in Fig. 1.

Diamond paste is a newly developed amperometric transducer obtained by mixing monocrystalline diamond powder with paraffin oil. The advantages of utilizing it as a transducer in the biosensors technology are (Stefan and Nejem, 2003; Stefan and Bokretsion, 2003; Stefan et al., 2003g, h) low background current, wide potential range, lack of adsorption, high signal-to-noise and signal-to-background ratios.

B. Optical Transducers

The main types of optical transducers used in biosensors technology are optical fibres (Andreou and Clonis, 2002; Eggins, 1996; MacCraith, 1998; Marks et al., 2002), fluorescence (Eggins, 1996; Kwakye and Baeumner, 2003; MacCraith, 1998) and chemiluminescence (Li et al., 2002; Premkumar et al., 2002; Zhang et al., 2001; Zhu et al., 2002) transducers, and surface plasmon resonance (SPR) (Eggins, 1996; Subrahmanyam et al., 2002; Naimushin et al., 2002) based transducers. The biosensors designed using an optical transducer have the advantage of being much more sensitive than the ones designed using an amperometric transducer.

Optical fibres transducers are classified into two categories (MacCraith, 1998): (1) direct spectroscopic transducers, and (2) reagent-mediated transducers. In the case of direct spectroscopic transducers, the fibre functions solely as a simple light guide that separates the sensing location from the monitoring instrumentation, whereas in the case of reagent-mediated transducers, the optical fibre is combined with chemical reagents. The features of using optical fibres as transducers in biosensors technology are (MacCraith, 1998): (1) the technology has access to a multiplicity of optical techniques well developed for routine analysis; (2) the low attenuation of optical fibre enables remoter *in situ* monitoring of species in difficult and hazardous locations; (3) the transducers can exploit

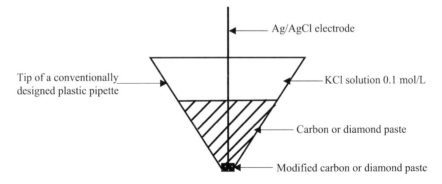

Figure 1 Carbon and diamond paste based electrodes.

the high-quality components (fibres, sources, detectors, connectors, etc.) developed for the more mature fibre optic telecommunications technology; and (4) the geometric flexibility of optical fibres and the feasibility of miniaturization may both be exploited for the development of biosensors.

Fluorescence and chemiluminescence transducers are the most developed within the optical transducer class. Their limits of detection are the lowest that one can obtain using biosensors.

Structural and functional screening for drugs and biological materials are done successfully using the SPR transducers for the design of biosensors. Direct monitoring of receptor–ligand interaction when other transducers are used for the design of the biosensors is difficult owing to the absence of signal amplification present in the case of enzymatic reaction. When SPR is used as the transducer a very high sensitivity is recorded (Subrahmanyam et al., 2002). An advantage of using SPR as the transducer is also the possibility of studying binding events in extracts, as it is not necessary to have highly purified components (Subrahmanyam et al., 2002). Furthermore, owing to the development of instrumentation, by using an Autolab ESPRIT instrument produced by Eco Chemie (Utrecht, The Netherlands) it is possible to perform simultaneous SPR and amperometric measurements with the same electrode. The computer records both signals, and the results can be compared and analyzed at the end of the experiment.

Piezoelectric transducers are very sensitive. Their high sensitivity may have a negative effect on their selectivity. They are used successfully in diagnosis (Zhou et al., 2002).

C. Screen-Printed Electrodes

Screen-printed electrodes (Fig. 2), which are developed as a class of disposable (bio)sensors, are designed mostly for clinical analysis (Pravda et al., 2002; Vasilesou et al., 2003). The reproducibility of the design of screen-printed electrodes is high and assures high reliability of the analytical information.

The working electrode is usually graphite powder-based "ink" printed on to a polyester material. The reference electrode is usually silver/silver chloride ink. Screen-printed electrodes can be modified by incorporation of gold, mercury, chelating agents, or mediators (phthalocyanines, ferrocenes, etc.) into the carbon ink.

IV. IMMOBILIZATION PROCEDURES OF THE BIOLOGICAL MATERIAL

Immobilization procedure is very important especially for the sensitivity and selectivity of the biochemical reaction. The support on which the biological material will be immobilized plays the role of a reaction medium for the biochemical reaction. There are two methods of immobilization of the biological material: (1) physical immobilization and (2) chemical immobilization. A more detailed classification of the methods of immobilization of the biological materials includes (MacCraith, 1998): (1) physical adsorption onto a solid surface; (2) use of cross-linking reagents; (3) entrapment using a gel or polymer; (4) use of membranes to retain the biomolecule close to the electrode surface; (5) covalent binding; and (6) exploitation of biomolecular interactions.

The choice of the immobilization method used depends on the biocomponent to be immobilized, the nature of the solid surface, and the transducing mechanism (MacCraith, 1998). The following factors should be considered in order to decide the type of immobilization(MacCraith, 1998): (1) the activity of the biomolecule must be preserved, and the selectivity of the biochemical reaction must not be modified; (2) the stability of the biomolecule should be preserved or increased; (3) the method of immobilization should be reliable.

Carbon paste (Stefan et al., 2003a–c), diamond paste (Stefan and Nejem, 2003), plastic membranes (Ghosh et al., 2002; Guelce et al., 2002), conducting polymers (Chen et al., 2002; Gerard et al., 2002), sol–gel, (Jia et al., 2002; Jin and Brennan, 2002; Premkumar et al., 2002; Zhu et al., 2002), and nanoparticles (Cai et al., 2003; Gu et al., 2002; Willard, 2003; Ye et al., 2003) are only few of the most utilized supports for the immobilization of biological materials.

A. Physical Immobilization

Physical immobilization is preferred to chemical immobilization as no structural modifications are required for the biological material and support. The most utilized supports for physical immobilization are carbon and diamond pastes.

Figure 2 Screen-printed electrodes.

The physical immobilization of biological materials in these supports assures the highest reliability of the design of the biosensors. The solution containing the biological material is incorporated into the carbon or diamond paste and will form the active part of the electrode. The upper part of the electrode will be filled with carbon or diamond paste free of biological material. Silver wires are recommended to be used for the electrical contact.

B. Chemical Immobilization

A special chemical modification is required for all supports used for covalent immobilization. This step is done in order to favor the binding between the biological material and support.

The most utilized supports for chemical immobilization are carbon paste and nylon (Stefan et al., 2001). For covalent immobilization of an amino acid oxidase into a carbon paste, graphite powder is mixed with a solution of 1-ethyl-3-3(3-dimethylaminopropyl)carbodiimide and heated to 700°C for 60 s in a muffle furnace. After cooling to ambient temperature, polyethylenimine and glutaraldehyde are added. The solution containing the enzyme is added to the pretreated graphite powder. Each graphite enzyme mixture is allowed to react at 4°C for 2 h and then mixed with paraffin oil to produce the modified paste.

Before the enzyme is bound on nylon, the nylon must be specially treated as follows: the disks of nylon are dipped in methanol, rinsed with water, and dried in air stream prior to use. A solution containing glutaraldehyde, bovine serum albumin, and enzyme are added onto the surface of the modified nylon disks. To inactivate the remaining carboxyaldehyde groups, the disks are immersed in certain buffer solutions. The best results for the covalent binding of the enzyme on the plastic membranes are obtain by polymerization under the action of ultraviolet (UV) light and by electropolymerization (Vidal et al., 2002; Xu et al., 2002). These techniques of immobilization assure reliability of the construction of biosensors based on covalent immobilization of an enzyme.

V. DESIGN OF A BIOSENSOR FOR AN EXTENDED USE AND STORAGE LIFE

A layer-after-layer approach can be utilized for the construction of a long-life electrode (Stefan et al., 2001). The enzyme electrode is designed by arranging the three components of the biosensor (electrode material, enzyme, and stabilizer) on a shapable electroconductive (SEC) film surface. First, a thin and compact Pt black layer is prepared onto the SEC film by the heat-press method. Then, an ultrathin layer of enzyme is casted on the platinized SEC film, and after it is dried, a thin gelatin layer is prepared on the enzyme-cast SEC film. Finally, the dried layer assembly is cross-linked by exposing it to a diluted glutaraldehyde solution for a very short time. This design assures a working time of 2 months, 2 years store time—if it is stored in the freezer, and 1 year store time at room temperature.

VI. ARRAY-BASED BIOSENSORS

An amperometric biosensor based on aligned multiwall carbon nanotubes (MWNT) grown on platinum subtrate was proposed (Sotiropoulou and Chaniotakis, 2003). The opening and functionalization by oxidation of the nanotube array allows for the efficient immobilization of the enzyme. The carboxylated open-ends of nanotubes are used for the immobilization of the enzyme, whereas platinum is the transducer. This type of electrode has features in clinical analysis owing to the high reproducibility of its design.

VII. DESIGN OF FLOW INJECTION ANALYSIS/BIOSENSORS AND SEQUENTIAL INJECTION ANALYSIS/BIOSENSORS SYSTEMS

Owing to the high reliability assured by biosensors, they can be successfully used as detectors in flow systems (Stefan et al., 2001). Originally they were used as detectors in flow injection analysis (FIA); and when the sequential injection analysis (SIA) was developed, it was considered that their utilization as detectors in an SIA system is the best alternative for the automatic analysis of a mono- or multicomponent system. The design of such systems is very simple and reliable.

A. Flow Injection/Biosensor(s) System

The electrodes are incorporated into the conduits of a flow injection system (Fig. 3). A Carle microvolume two-position sampling valve (Carle No. 2014) containing two identical sample loops was used. Each loop has a

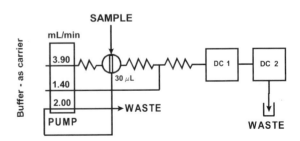

Figure 3 Schematic flow diagram of an FIA system used for the simultaneous determination of two components (HC, holding coil; RC, reaction coil; DC, detection cell).

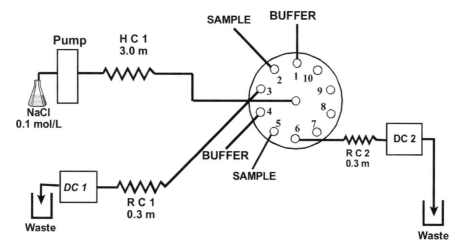

Figure 4 Schematic flow diagram of an SIA system used for the simultaneous determination of two components (HC, holding coil; RC, reaction coil; DC, detection cell).

volume of 30 µL. A Cenco peristaltic pump operating at 10 rev/min supplied the carrier streams to the manifold system. Tygon tubing (0.51 mm i.d.) was used to construct the manifold; coils were wound around suitable lengths of glass tubing (15 mm o.d.). Buffer solution is used as carrier. The sample is injected in the buffer stream.

B. Sequential Injection/Biosensor(s) System

The biosensor(s) is/are incorporated into the conduits of an SIA system (Fig. 4) constructed from a Gilson Minipuls peristaltic pump and a 10-port electrically actuated selection valve (Model ECSD10P, Valco Instruments, Houston, TX). Tygon tubing (0.76 mm i.d. for both holding coils and 0.89 mm i.d. for both mixing coils) is used to construct the manifold; coils are wound around suitable lengths of glass tubing (15 mm o.d.). A 0.1 mol/L NaCl solution is used as carrier.

REFERENCES

Aboul-Enein, H. Y., Stefan, R. I., Baiulescu, G. E. (2000). *Quality and Reliability in Analytical Chemistry*. CRC Press.

Andreou, V. G., Clonis, Y. D. (2002). Novel fibre-optic biosensor based on immobilized glutathione S-transferase and sol-gel entrapped bromcresol green for the determination of atrazine. *Anal. Chim. Acta* 460:151–161.

Cai, H., Cao, X., Jiang, Y., He, P., Fang, Y. (2003). Carbon nanotube-enhanced electrochemical DNA biosensor for DNA hybridization detection. *Anal. Bioanal. Chem.* 375:287–293.

Chen, J., Burrell, A. K., Collis, G. E., Officer, D. L., Swiegers, G. F., Too, C. O., Wallace, G. G. (2002). Preparation, characterization and biosensor application of conducting polymers based on ferrocene-substituted thiophene and terthiophene. *Electrochim. Acta* 47:2715–2724.

Eggins, B. (1996). *Biosensors. An Introduction*. Chichester: Wiley.

Fischer, W. B. (2003). Computational bioanalysis of proteins. *Anal. Bioanal. Chem.* 375:23–25.

Freire, R. S., Duran, N., Kubota, L. T. (2002). Development of a laccase-based flow injection electrochemical biosensor for the determination of phenolic compounds and its application for monitoring remediation of Kraft E1 paper mill effluent. *Anal. Chim. Acta* 463:229–238.

Gerard, M., Chaubey, A., Malhotra, B. D. (2002). Application of conducting polymers to biosensors. *Biosens. Bioelectron.* 17:345–360.

Ghosh, D., Pal, P. S., Sarkar, P. (2002). Glucose biosensor with polymer microencapsulated enzyme. *J. Indian Chem. Soc.* 79:782–783.

Gu, H. Y., Yu, A. M., Yuan, S. S., Chen, H. Y. (2002). Amperometric nitric oxide biosensor based on the immobilization of haemoglobin on a nanometer-sized gold colloid modified gold electrode. *Anal. Lett.* 35:647–661.

Guelce, H., Guelce, A., Kavanoz, M., Coskun, H., Yildiz, A. (2002). A new amperometric enzyme electrode for alcohol determination. *Biosens. Bioelectron.* 17:517–521.

Haga, T. (1995). Receptor biochemistry. In: Meyer, R.A., ed. *Molecular Biology and Biotechnology. A Comprehensive Desk Reference*. New York: VCH.

Jia, J., Wang, B., Wu, A., Cheng, G., Li, Z., Dong, S. (2002). A method to construct a third-generation horseradish peroxidase biosensor: self assembling gold nanoparticles to three-dimensional sol-gel network. *Anal. Chem.* 74:2217–2223.

Jin, W., Brennan, J. D. (2002). Properties and applications of proteins encapsulated within sol-gel derived materials. *Anal. Chim. Acta* 461:1–36.

Kwakye, S., Baeumner, A. (2003). A microfluidic biosensor based on nucleic acid sequence recognition. *Anal. Bioanal. Chem.* 376:1062–1068.

Li, B. X., Zhang, Z. J., Jin, Y. (2002). Plant tissue-based chemiluminescence flow biosensor for determination of unbound dopamine in rabbit blood with on-line microdialysis sampling. *Biosens. Bioelectron.* 17:585–589.

MacCraith, B. D. (1998). In: Diamond, D., ed. *Optical Chemical Sensors, in Principles of Chemical and Biological Sensors*. New York: Wiley.

Marks, R. S., Novoa, A., Thomassey, D. (2002). An innovative strategy or immobilization of receptor proteins on to an optical fibre by use of poly(pyrrole-biotin). *Anal. Bioanal. Chem.* 374:1056–1063.

Naimushin, A. N., Soelberg, S. D., Nguyen, D. K., Dunlap, L., Bartholomew, D., Elkind, J., Melendez, J., Furlong, C. E. (2002). Detection of Staphylococcus aureus enterotoxin B at femtomolar levels with a portable miniature integrated two-channel surface pasmon resonance (SPR) sensor. *Biosens. Bioelectron.* 17:573–584.

Pravda, M., O'Halloran, M. P., Kreuzer, M. P., Guilbault, G. G. (2002). Composite lucose biosensor based on screen-printed electrodes bulk modified with Prussian Blue and glucose oxidase. *Anal. Lett.* 35:959–970.

Premkumar, J. R., Rosen, R., Belkin, S., Lev, O. (2002). Sol-gel luminescence biosensors: encapsulation of recombinant E.coli reporters in thick silicate films. *Anal. Chim. Acta* 462:11–23.

Sotiropoulou, S., Chaniotakis, N. A. (2003). Carbon nanotube array-based biosensor. *Anal. Bioanal. Chem.* 375:103–105.

Stefan, R. I., Bokretsion, R. G. (2003). Determination of creatine and creatinine using diamond paste based electrode. *Instrum. Sci. Technol.* 31:183–188.

Stefan, R. I., Nejem, R. M. (2003). Determination of L- and D-pipecolic acid using diamond paste based amperometric biosensors. *Anal. Lett.* 36:2635–2644.

Stefan, R. I., van Staden, J. F., Aboul-Enein, H. Y. (2001). *Electrochemical Biosensors in Bioanalysis*. New York: Marcel Dekker, Inc.

Stefan, R. I., Bokretsion, R. G., van Staden, J. F., Aboul-Enein, H. Y. (2003a). Determination of L- and D-enantiomers of carnitine using amperometric biosensors. *Anal. Lett.* 36:1089–1100.

Stefan, R. I., Bala, C., Aboul-Enein, H. Y. (2003b). Biosensors for enantioselective analysis of S-captopril. *Sens. Actuators B* 92:228–231.

Stefan, R. I., Bokretsion, R. G., van Staden, J. F., Aboul-Enein, H. Y. (2003c). Biosensors for the determination of ortho-acetyl-L-carnitine. Their utilization as detectors in a sequential injection analysis system. *Prep. Biochem. Biotechnol.* 33:163–171.

Stefan, R. I., Bokretsion, R. G., van Staden, J. F., Aboul-Enein, H. Y. (2003d). Determination of L- and D-enantiomers of methotrexate using amperometric biosensors. *Talanta* 60:983–990.

Stefan, R. I., Nejem, R. M., van Staden, J. F., Aboul-Enein, H. Y. (2003e). Biosensors for the enantioselective analysis of pipecolic acid. *Sens. Actuators B* 94:271–275.

Stefan, R. I., Bokretsion, R. G., van Staden, J. F., Aboul-Enein, H. Y. (2003f). Simultaneous determination of creatine and creatinine using amperometric biosensors. *Talanta* 60:844–847.

Stefan, R. I., Bairu, S. G., van Staden, J. F. (2003g). Diamond paste based electrodes for the determination of iodide in vitamins and table salt. *Anal. Lett.* 36:1493–1500.

Stefan, R. I., Bairu, S. G., van Staden, J. F. (2003h). Diamond paste based electrodes for the determination of Cr(III) in pharmaceutical compounds. *Anal. Bioanal. Chem.* 376:844–847.

Subrahmanyam, S., Piletsky, S. A., Turner, A. P. F. (2002). Application of natural receptors in sensors and assays. *Anal. Chem.* 74:3942–3951.

Ugo, P., Zangrando, V., Moretto, L. M., Brunetti, B. (2002). Ion-exchange voltammetry and electrocatalytic sensing capabilities of cytochrome c at polyestersulfonated ionomer coated glassy carbon electrodes. *Biosens. Bioelectron.* 17:479–487.

Vasilescu, A., Noguer, T., Andreescu, S., Calas-Blanchard, C., Bala, C., Marty, J. L. (2003). Strategies for developing NADH detectors based on Meldola Blue and screen-printed electrodes: a comparative study. *Talanta* 59:751–765.

Vidal, J. C., Garcia, E., Castillo, J. R. (2002). Development of a platinized and ferrocene-mediated cholesterol amperometric biosensor based on electropolymerization of polypyrrole in a flow system. *Anal. Sci.* 18:57–542.

Wang, J., Li, M., Shi, Z., Li, N., Gu, Z. (2002). Direct electrochemistry of cytochrome C at a glassy carbon electrode modified with single-wall carbon nanotubes. *Anal. Chem.* 74:1993–1997.

Willard, D. M. (2003). Nanoparticles in bioanalytics. *Anal. Bioanal. Chem.* 376:284–286.

Wu, Z., Wang, B., Cheng, Z., Yang, X., Dong, S., Wang, E. (2001). A facile approach to immobilize protein for biosensor: self-assembled supported bilayer lipid membranes on glassy carbon electrode. *Biosens. Bioelectron.* 16:47–52.

Xu, J. J., Yu, Z. H., Chen, H. Y. (2002). Glucose biosensors prepared by electropolymerization of p-chlorophenylamine with and without Nafion. *Anal. Chim. Acta* 463:239–247.

Ye, Y. K., Zhao, J. H., Yan, F., Zhu, Y. L., Ju, H. X. (2003). Electrochemical behavior and detection of hepatitis B virus DNA PCR production at gold electrode. *Biosens. Bioelectron.* 18:1501–1508.

Zhang, X., Garcia-Campana, A. M., Baeyens, W. R. G., Stefan, R. I., van Staden, J. F., Aboul-Enein, H. Y. (2001). In: Garcia-Campana, A. M., Baeyens, W. R. G., eds. *Recent Developments in Chemiluminescence Sensors, in Chemiluminescence in Analytical Chemistry*. New York: Marcel Dekker, Inc.

Zhou, X. D., Liu, L. J., Hu, M., Wang, L. L., Hu, J. M. (2002). Detection of hepatitis B virus by piezoelectric sensor. *J. Pharm. Biomed Anal.* 27:341–345.

Zhu, L. D., Li, Y. X., Tian, F. M., Xu, B., Zhu, G. Y. (2002). Electrochemiluminescent determination of glucose with a sol-gel derived ceramic-carbon composite electrode as a renewable optical fiber biosensor. *Sens. Actuators B* 84:265–270.

22

Instrumentation for High-Performance Liquid Chromatography

RAYMOND P. W. SCOTT
7, Great Sanders House, Seddlescombe, 1D East Sussex, UK

I. INTRODUCTION TO THE CHROMATOGRAPHIC PROCESS

To appreciate the relevance of instrument design in the practice of liquid chromatography (LC) it is necessary to have a basic understanding of the chromatographic process. Consequently, the fundamental principles of a chromatographic separation will be first discussed in a general manner and the design of the different parts of the chromatograph will be subsequently explained on the basis of the requirements mandated by the nature of the separation process.

LC is probably the most important analytical technique available to the modern chemist and biotechnologist today. Chromatographic methods are unique in that they possess dual capabilities, the mixture is separated into its components and simultaneously the quantity of each component present is measured. In fact, all forms of chromatography are primarily separation techniques, but by employing detectors with a linear response to monitor the eluent leaving the chromatographic system, the mass of each component present can also be determined. Chromatography was invented in the late 19th century by a Russian botanist called Tswett (1905), who used it to separate some plant pigments. However, although discovered over a century ago, it is only over the last 20 years that LC has been developed into an effective, reliable, and versatile analytical technique. The slow evolution of the technique was partly due to a number of negative reports made by some of Tswett's peers (Willstatter and Stoll, 1913) and partly due to the lack of a suitable detecting system. After the discovery and development of gas chromatography (GC) and after its impressive capabilities had been demonstrated, attention was quickly turned to the development of LC. New and sensitive detectors were invented (Horvath and Lipsky, 1966; Kirkland, 1968) and the technique of LC has now evolved into an analytical method that has proved to be even more useful and versatile than that of GC.

Chromatography has been classically defined (Scott, 1976) as a separation process that is achieved by the distribution of substances between two phases, a stationary phase and a mobile phase. Those solutes distributed preferentially in the mobile phase will move more rapidly through the system than those distributed preferentially in the stationary phase. Thus, the solutes will elute in order of their increasing distribution coefficients with respect to the stationary phase.

A solute is distributed between two phases as a result of the molecular forces (Hirschfelder et al., 1964, p. 835) that exist between the solute molecules and those of the two phases. The stronger the forces between the solute molecules and those of the stationary phase, the greater will be the amount of solute held in that stationary phase under equilibrium conditions. Conversely, the stronger the interactions between the solute molecules and those of the mobile phase, the greater the amount of solute that will be held in the mobile phase. Now, the solute can only move through the chromatographic system while it is in the mobile phase and thus, the speed at which a particular solute will pass through the column will be directly proportional to the concentration of the solute in the mobile phase.

The concentration of the solute in the mobile phase is inversely proportional to the distribution coefficient of the solute *with respect to the stationary phase*, that is,

$$K = \frac{X_s}{X_m} \quad (1)$$

where K is the distribution coefficient of the solute between the two phases, X_m is the concentration of the solute in the mobile phase, and X_s is the concentration of the solute in the stationary phase. Consequently, the solutes will pass through the chromatographic system at speeds that are inversely proportional to their distribution coefficients with respect to the stationary phase. To appreciate the nature of a chromatographic separation, the manner of solute migration through a chromatographic column needs to be understood.

Consider the progress of a solute through a chromatographic column as depicted in diagrammatic form in Fig. 1. The profile of the concentration of a solute in both the mobile and stationary phases has been shown to be Gaussian (Scott, 1992, p. 50) in form. Thus, the flow of mobile phase will slightly displace the concentration profile of the solute in the mobile phase relative to that in the stationary phase; the displacement, grossly exaggerated, is depicted in Fig. 1.

It is seen that, as a result of this displacement, the concentration of solute in the *mobile phase* at the front of the peak *exceeds* the equilibrium concentration with respect to that in the stationary phase. It follows that there is a net transfer of solute from the mobile phase in the front part of the peak to the stationary phase to re-establish equilibrium as the peak progresses along the column. At the rear of the peak, the converse occurs. As the concentration profile moves forward, the concentration of solute in the *stationary phase* at the rear of the peak is now in *excess* of the equilibrium concentration.

Thus, the solute *leaves* the stationary phase and there is a net transfer of solute to the mobile phase in an attempt to re-establish equilibrium. The solute band progresses through the column by a net transfer of solute to the mobile phase at the rear of the peak and a net transfer of solute to the stationary phase at the front of the peak. The mechanism of retention is best explained by means of the plate theory (Scott, 1994). The plate theory provides an equation that describes the elution curve (chromatogram) obtained from a chromatographic column. Consequently, if this expression is differentiated and equated to zero, the position of the peak maximum can be identified and the variables that control its magnitude exposed.

A. The Plate Theory

Consider three consecutive plates in a column, the p − 1, the p and the p+1 plates and let there be a total of n plates in the column. The three plates are depicted in Fig. 2. Let the volumes of mobile phase and stationary phase in each plate be (v_m) and (v_s), respectively, and the concentrations of solute in the mobile and stationary phases in each plate be $X_{m(p-1)}$, $X_{s(p-1)}$, $X_{m(p)}$, $X_{s(p)}$, $X_{m(p+1)}$, and $X_{s(p+1)}$, respectively. Let a volume of mobile phase (dV) pass from plate p−1 into plate p at the same time displacing the same volume of mobile phase from plate p to plate p + 1. As a consequence, there will be a change of mass of solute in plate p that will be equal to the difference in the mass entering plate p from plate p−1 and the mass of solute leaving plate p and entering plate p + 1.

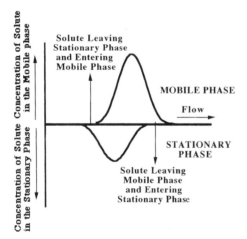

Figure 1 The passage of a solute band along a chromatographic column.

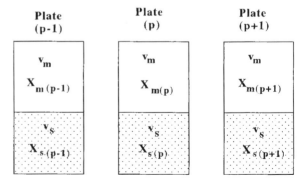

Figure 2 Three consecutive theoretical plates in an LC column.

Thus, bearing in mind that mass is the product of concentration and volume, the change of mass of solute in plate p is:

$$dm = (X_{m(p-1)} - X_{m(p)})dV \quad (2)$$

Now, if equilibrium is to be maintained in the plate p, the mass dm will distribute itself between the two phases, which will result in a change of solute concentration in the mobile phase of $dX_{m(p)}$ and in the stationary phase of $dX_{s(p)}$.

$$dm = v_s dX_{s(p)} + v_m dX_{m(p)} \quad (3)$$

Thus, substituting for $dX_{s(p)}$ from Eq. (1),

$$dm = (v_m + Kv_s)dX_{m(p)} \quad (4)$$

Equating Eqs. (2) and (4) and re-arranging,

$$\frac{dX_{m(p)}}{dV} = \frac{X_{m(p-1)} - X_{(p)}}{v_m + Kv_s} \quad (5)$$

Now, to aid in algebraic manipulation the volume flow of mobile phase will now be measured in units of $v_m + Kv_s$ instead of milliliters. Thus the new variable v can be defined where,

$$v = \frac{V}{v_m + Kv_s} \quad (6)$$

The function $v_m + Kv_s$, has been given the name "plate volume" (Scott, 1992, p. 18) and thus, for the present, the flow of mobile phase through the column will be measured in plate volumes instead of milliliters.

Differentiating Eq. (6),

$$dv = \frac{dV}{v_m + Kv_s} \quad (7)$$

Substituting for dV from Eq. (7) in Eq. (5)

$$\frac{dX_{m(p)}}{dv} = X_{m(p-1)} - X_{(p)} \quad (8)$$

Equation (8) is the basic differential equation that describes the rate of change of concentration of solute in the mobile phase in plate p with the volume flow of mobile phase through it. The integration of Eq. (8) will provide the equation for the elution curve of a solute for any plate in the column.

Integrating Eq. (8),

$$X_{m(n)} = \frac{X_0 e^{-v} v^n}{n!} \quad (9)$$

where $X_{m(n)}$ is the concentration of solute in the mobile phase leaving the nth plate (i.e., the concentration of solute entering the detector), and X_0 is the initial concentration of solute on the first plate of the column.

Equation (9) describes the elution curve obtained from a chromatographic column and is the equation of the curve, or

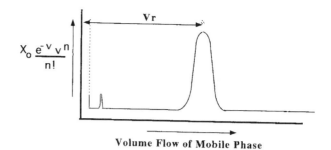

Figure 3 The elution of a single solute.

chromatogram, that is traced by the chart recorder or computer printer. Its pertinence is displayed in Fig. 3.

It can now be seen that an expression for the retention volume (V_r) of a solute, that is the volume that has passed through the column between the injection point and the peak maximum, can be derived by differentiating Eq. (9) and equating to zero.

Differentiating Eq. (9)

$$\frac{dX_{m(n)}}{dv} = X_0 \frac{-e^{-v}v^n + e^{-v}nv^{(n-1)}}{n!}$$

$$= X_0 \frac{-e^{-v}v^{(n-1)}}{n!}(n-v)$$

Equating to zero,

$$n - v = 0$$

or

$$v = n$$

This means that at the peak maximum, n plate volumes of mobile phase have passed through the column. Remembering that the volume flow is measured in plate volumes and not milliliters, the volume passed through the column in milliliters will be obtained by multiplying by the plate volume $v_m + Kv_s$.

Thus, the retention volume (V_r) is given by:

$$V_r = n(v_m + Kv_s)$$
$$= nv_m + nKv_s$$

Now, the total volume of mobile phase in the column V_m will be the volume of mobile phase per plate multiplied by the number of plates (i.e., nv_m). In a similar manner the total volume of stationary phase in the column (V_s) will be the volume of stationary phase per plate multiplied by the total number of plates (i.e., nv_s).

Thus,

$$V_r = V_m + KV_s \quad (10)$$

It is now immediately obvious from Eq. (11) how the separation of two solutes (A) and (B) must be achieved.

For separation,

$V_{r(A)} <> V_{r(B)}$ and $V_{r(A)} \neq V_{r(B)}$

Furthermore, as the retention volumes of each substance must be different, then either

$K_{(A)} <> K_{(B)}$ and $K_{(A)} \neq K_{(B)}$

or

$V_{s(A)} <> V_{s(B)}$ and $V_{s(A)} \neq V_{s(B)}$

Thus, to achieve the required separation,

1. the distribution coefficient (K) of all the solutes must be made to differ, or
2. the amount of stationary phase (V_s), available to each component of the mixture interacts, must be made to differ, and
3. appropriate adjustments to the values of both (K) and V_s must be made.

It is now necessary to identify how K and V_s can be changed to suit the particular sample of interest. Consider, first, the control of the distribution coefficient K. As already stated the magnitude of the distribution coefficient is determined by the nature and magnitude of the molecular forces (Hirschfelder et al., 1964, p. 22) that exist between the solute molecules and those of the two phases. Therefore, the phase system must be selected to provide the appropriate interactive forces that will both retain the solutes and provide the necessary *differential* retention necessary to achieve resolution.

B. Molecular Interactions

Molecular interactions are the direct effect of intermolecular forces that occur between the solute and solvent molecules. There are four basic types of molecular force that can control magnitude of the distribution coefficient of a solute between two phases. Theses forces are *chemical forces*, *ionic forces*, *polar forces*, and *dispersive forces*.

These molecular forces that occur between the solute and the two phases are those that the analyst must modify by choice of the phase system to achieve the necessary separation.

It follows that each type of molecular force enjoins some discussion

1. Chemical Forces

Chemical forces are normally irreversible in nature (at least in chromatography) and thus, the distribution coefficient of the solute with respect to the stationary phase is infinite or close to infinite. Affinity chromatography is an example of the use of chemical forces in a separation process. The stationary phase is formed in such a manner that it will chemically interact with one unique solute present in the sample and, thus, exclusively extract it from the other materials present. The technique of affinity chromatography is, therefore, an extraction process more than a chromatographic separation.

2. Ionic Forces

Ionic forces are electrical in nature and result from the net charge on an atom or molecule caused by ionization. Ionic interactions are exploited in ion chromatography. In the analysis of organic acids, it is the negatively charged acid anions that are separated. Consequently, the stationary phase must contain positively charged cations as counterions to interact with the acid anions, retard them in the column and effect their resolution. Conversely, to separate cations, the stationary phase must contain anions as counterions with which the cations can interact. Ion exchange stationary phases are usually available in the form of cross-linked polymer beads that have been appropriately modified to contain the desired ion exchange groups. Alternatively, ion exchange groups can be chemically bonded to silica by a process similar to the preparation of ordinary bonded phases.

An example of the separation of a series of some inorganic anions on an ion exchange column is shown in Fig. 4. The column is 15 cm long, 4.6 mm in diameter, and packed with a proprietary ion exchange material IonPacAS4A. The mobile phase was an aqueous solution of 1.80 nM sodium carbonate and 1.70 nM sodium bicarbonate at a flow rate of 2.0 mL/min. The volume of charge was 50 µL.

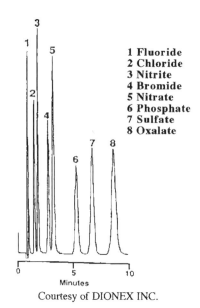

Courtesy of DIONEX INC.

Figure 4 The separation of a series of inorganic anions.

3. Polar Forces

Polar forces also arise from electrical charges on the molecule but in this case from *permanent* or *induced* dipoles. It must be emphasized that there is *no net charge* on a polar molecule unless it is *also* ionic. Examples of substances with permanent dipoles are alcohols, esters, aldehydes, and so on. Examples of polarizable molecules are the aromatic hydrocarbons such as benzene, toluene, and xylene. Some molecules can have permanent dipoles and, at the same time, also be polarizable. An example of such a substance would be phenyl ethyl alcohol.

To separate materials on a basis of their polarity, a *polar* stationary phase must be used. Furthermore, in order to focus the polar forces in the stationary phase, and consequently the selectivity, a relatively *nonpolar* mobile phase would be required.

The use of dissimilar molecular interactions in the two phases, to achieve selectivity, is generally applicable and of fundamental importance (Snyder, 1971).

The interactions chosen to achieve retention must dominate in the stationary phase. It follows, that they should also be as *exclusive as possible* to the stationary phase. It is equally important that in order to maintain the stationary phase selectivity, the interactions in the mobile phase *differ to as great extent as possible* to those in the stationary phase.

To separate some aromatic hydrocarbons, for example, a strongly polar stationary phase would be appropriate as the solutes do not contain permanent dipoles. Under such circumstances, when the aromatic nucleus approaches the strongly polar group on the stationary phase, its nucleus will be polarized and positive and negative charges generated at different positions on the aromatic ring. These charges will then interact with the dipoles of the stationary phase and, as a consequence, the solute molecules will be strongly retained.

In practice, *silica gel* is a highly polar stationary phase, its polar character arising from the dipoles of the surface hydroxyl groups. Consequently, silica gel would be an appropriate stationary phase for the separation of aromatic hydrocarbons. To ensure that polar selectivity *dominates* in the *stationary phase*, and polar interactions in the mobile phase are minimized, a nonpolar or dispersive solvent (e.g., *n*-heptane) could be used as the mobile phase. The separation of some polynuclear aromatic compounds on silica gel is shown in Fig. 5. The separation was carried out on a small bore column 25 cm long and 1 mm i.d. packed with silica gel having a particle diameter of 10 μm. The mobile phase was *n*-hexane saturated with water, and the flow rate, 50 μL/min.

4. Dispersive Forces

Dispersive forces, although electric in nature, result from random charge fluctuations rather than permanent

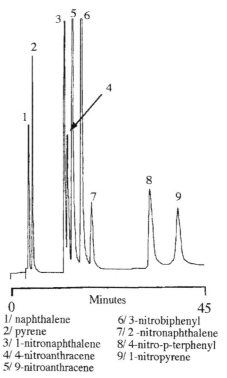

1/ naphthalene
2/ pyrene
3/ 1-nitronaphthalene
4/ 4-nitroanthracene
5/ 9-nitroanthracene
6/ 3-nitrobiphenyl
7/ 2-nitronaphthalene
8/ 4-nitro-p-terphenyl
9/ 1-nitropyrene

Courtesy of Supleco Inc.

Figure 5 The separation of a mixture of aromatic and nitroaromatic hydrocarbons.

electrical charges on the molecule (Glasstone, 1946; Hirschfelder et al., 1964, p. 968). Examples of purely dispersive interactions are the molecular forces that exist between saturated aliphatic hydrocarbon molecules. Saturated aliphatic hydrocarbons are not ionic, have no permanent dipoles, and are not polarizable. Yet molecular forces between hydrocarbons are strong and consequently at normal temperatures and pressures, *n*-heptane is not a gas, but a liquid that boils at 100°C. This is a result of the collective effect of all the dispersive interactions that hold the molecules together as a liquid.

To retain solutes solely by dispersive interactions, the stationary phase must contain no polar or ionic substances, but only hydrocarbon-type materials such as the reverse-bonded phases, now so popular in LC. To ensure that dispersive selectivity dominates in the stationary phase, and dispersive interactions in the mobile phase are minimized, the mobile phase must be strongly polar. Hence the use of methanol–water and acetonitrile–water mixtures as mobile phases in reverse phase chromatography systems. An example of the separation of some antimicrobial agents on Partisil ODS3, particle diameter 5 μm is shown in Fig. 6.

ODS3 is a "bulk type" reverse phase (the meaning of which will be discussed later) which has a fairly high

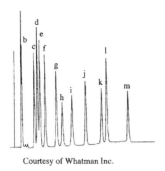

Courtesy of Whatman Inc.

a/ sulfaguanidine, b/ sulfanilamide, c/ sulfadiazine, d/ sufathiazole, e/ sulphapyridine, f/ sulfamerazine, g/ sulfamethazine, h/ sulfachloropyridazine, i/ sulfasoxazole, j/ sulfaethoxypyridazine, k/ sulfadimethoxine, l/ sulfaquinoxine, m/ sulfabromethazine.

Figure 6 The separation of a mixture of antimicrobial agents.

capacity and is reasonably stable to small changes in pH. The column was 25 cm long, 4.6 mm in diameter and the mobile phase a methanol–water mixture containing acetic acid. In this particular separation the solvent mixture was programed, the development procedure of which will also be discussed in a later section.

C. The Thermodynamic Explanation of Retention

Solute retention can also be explained on a thermodynamic basis by considering the change in free energy that occurs when a solute is moved from the environment of one phase to that of the other. The classic thermodynamic expression for the distribution coefficient K of a solute between two phases is given by

$$RT \ln K = -\Delta G_0$$

where R is the gas constant, T is the absolute temperature, and ΔG_0 is the excess free energy.

Now,

$$\Delta G_0 = \Delta H_0 - T\Delta S_0$$

where ΔH_0 is the excess free enthalpy, and ΔS_0 is the excess free entropy.

Thus,

$$\ln K = -\left(\frac{\Delta H_0}{RT} - \frac{\Delta S_0}{R}\right)$$

or,

$$K = e^{-[(\Delta H_0/RT) - (\Delta S_0/R)]} \tag{11}$$

It is seen that, if the *excess free entropy* and *excess free enthalpy* of a solute between two phases can be calculated or predicted, then the magnitude of the distribution coefficient K and consequently, the retention of a solute can also be predicted. Unfortunately, the thermodynamic properties of a distribution system are *bulk* properties, that include the effect of all the different types of molecular interactions in one measurement. As a result it is difficult, if not impossible, to isolate the individual contributions so that the nature of overall distribution can be estimated or controlled. Attempts have been made to allot specific contributions to the total *excess free energy* of a given solute/phase system that represents particular types of interaction. Although the procedure becomes very complicated, this empirical approach has proved to be of some use to the analyst. However, it must be said that thermodynamics has yet to predict a single distribution coefficient from basic physical chemical data. Nevertheless, if sufficient experimental data is available or can be obtained, then empirical equations, similar to those given above, can be used to optimize a particular distribution system once the basic phases have been identified. A range of computer programs, based on this rationale, is available, that purport to carry out optimization for up to three component solvent mixtures. Nevertheless, the appropriate stationary phase is still usually identified from the types of molecular forces that need to be exploited to effect the required separation. Equation (11) can also be used to identify the type of retention mechanism that is taking place in a particular separation by measuring the retention volume of the solute over a range of temperatures. Rearranging Eq. (12),

$$\log K = -\frac{\Delta H_0}{RT} + \frac{\Delta S_0}{R}$$

Bearing in mind,

$$V' = KV_s$$

then

$$\log V' = -\frac{\Delta H_0}{RT} + \frac{\Delta S_0}{R} - \log V_s$$

It is seen that a curve relating $\log V'$ to $1/T$ should give a straight line the slope of which will be proportional to the *enthalpy* change during solute transfer. In a similar way, the intercept will be related to the *entropy* change during solute transfer and thus, the dominant effects in any distribution system can be identified from such curves. Graphs of $\log V'$ against $1/T$ for two different types of distribution systems are given in Fig. 7.

It is seen that distribution system A has a large enthalpy value $(\Delta H_0/RT)_A$ and a low entropy contribution $[-(\Delta S_0/R) - V_s]_A$. In fact, the value of V_s would be common to both systems and thus, the value of the entropy change will be proportional to changes in $[-(\Delta S_0/R) - V_s]_A$. The large value of $(\Delta H_0/RT)_A$ means that the distribution is predominantly controlled by molecular forces. The solute is preferentially distributed in the stationary phase as a result of the interactions of the solute

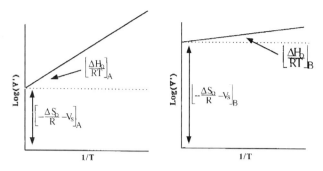

Figure 7 Graph of log corrected retention volume against the reciprocal of the absolute temperature.

molecules with those of the stationary phase being much greater than the interactive forces between the solute molecules and those of the mobile phase.

Because the change in enthalpy is the major contribution to the change in free energy, in thermodynamic terms the distribution is said to be "energy driven". In contrast, it is seen that for distribution system B there is only a small enthalpy change $(\Delta H_0/RT)_B$, but, in this case, a high entropy contribution $[-(\Delta S_0/R) - V_s]_B$. This means that the distribution is *not* predominantly controlled by molecular forces. The entropy change reflects the degree of randomness that a solute molecule experiences in a particular phase. The more random and "more free" the solute molecule is to move in a particular phase, the greater the entropy of the solute. Now, as there is a large entropy change, this means that the solute molecules are more restricted or less random in the stationary phase in system B. This loss of freedom is responsible for the greater distribution of the solute in the stationary phase and thus, its greater retention. Because the change in entropy in system B is the major contribution to the change in free energy, in thermodynamic terms, the distribution is said to be "entropically driven".

Examples of entropically driven systems will be those employing chirally active stationary phases. Returning to the molecular force concept, in any particular distribution system it is rare that only one type of interaction is present and if this occurs, it will certainly be *dispersive* in nature. Polar interactions are always accompanied by dispersive interactions and ionic interactions will, in all probability, be accompanied by both polar *and* dispersive interactions.

However, as shown by Eq. (10), it is not only the *magnitude* of the interacting forces between the solute and the stationary phase that will control the extent of retention, but also the *amount of stationary phase* present in the system and its *accessibility* to the solutes. This leads to the next method of retention control—volume of stationary phase available to the solute.

D. Control of Retention by Stationary Phase Availability

The volume of stationary phase with which the solutes in a mixture can interact [V_s in Eq. (10)] will depend on the physical nature of the stationary phase or support. If the stationary phase is a porous solid, and the sizes of the pores are commensurate with the molecular diameter of the sample components, then the stationary phase becomes size selective. Some solutes (e.g., those of small molecular size) can penetrate and interact with more stationary phase than larger molecules which are partially excluded. Under such circumstances the retention is at least partly controlled by size exclusion and, as one might expect, this type of chromatography is called size exclusion chromatography (SEC). Alternatively, if the stationary phase is *chiral* in character, the interaction of a solute molecule with the surface (and consequently, the amount of stationary phase with which it can interact) will depend on the chirality of the solute molecule and how it fits to the chiral surface. Such separation systems have been informally termed chiral liquid chromatography (CLC).

1. Size Exclusion Chromatography

The term SEC implies that the retention of a solute depends solely on solute molecule size. However, even in SEC, retention may not be *exclusively* controlled by the size of the solute molecule, it will still be controlled by molecular interactions between the solute and the two phases. *Only if the magnitude of the forces between the solute and both phases is the same* will the retention and selectivity of the chromatographic system depend solely on the pore size distribution of the stationary phase. Under such circumstances the larger molecules, being partially or wholly excluded, will elute first and the smaller molecules will elute last. It is interesting to note that, even if the dominant retention mechanism is controlled by molecular forces between the solute and the two phases, if the stationary phase or supporting material has a porosity commensurate with the molecular size of the solutes, *exclusion will still play a part in retention*. The two most common exclusion media employed in LC are silica gel and macroporous polystyrene divinylbenzene resins.

2. Chiral Chromatography

Modern organic chemistry is becoming increasingly involved in methods of asymmetric syntheses (Jung et al., 1994; Stinsen, 1993). This enthusiasm has been fostered by the relatively recent appreciation of the differing physiological activity that has been shown to exist between the geometric isomers of pharmaceutically active compounds. A sad example lies in the drug, thalidomide, which was marketed as a racemic mixture of N-phthalylglutamic acid imide. The desired pharmaceutical activity resides in the R-(+)-isomer and it was not found, until too late, that the corresponding S-enantiomer, was teratogenic and its presence in the racemate caused serious fetal malformations. It follows, that the separation and identification of isomers can be a very important analytical problem and LC can be very effective in the resolution of such mixtures.

Many racemic mixtures can be separated by ordinary reverse phase columns, by adding a suitable chiral reagent to the mobile phase. If the material is adsorbed strongly on the stationary phase then selectivity will reside in the stationary phase, if the reagent is predominantly in the mobile phase then the chiral selectivity will remain in the mobile phase. Examples of some suitable additives are camphor sulphonic acid (Peterson and Schill, 1981) and quinine (Peterson, 1984).

Chiral selectivity can also be achieved by bonding chirally selective compounds to silica in much the same way as a reverse phase. A example of this type of chiral stationary phase is afforded by the cyclodextrins. An example of the separation of the isomers of warfarin is shown in Fig. 8.

The column was 25 cm long and 4.6 mm in diameter packed with 5 μm Cyclobond 1. The mobile phase was approximately 90% v/v acetonitrile, 10% v/v methanol, 0.2% v/v glacial acetic acid, and 0.2% v/v triethylamine. It is seen that an excellent separation has been achieved with the two isomers completely resolved. This separation is an impressive example of an entropically driven distribution system, where the normally random movements of the solute molecules are restricted to different extents depending on the spatial orientation of the substituent groups.

A powerful group of stationary phases with high selectivity for enantiomers are the macrocyclic glycopeptides. One method of preparing such media is to covalently bond vancomycin to the surface of silica gel particles. Vancomycin contains 18 chiral centers surrounding three "pockets" or "cavities" which are bridged by five aromatic rings. Strong polar groups are proximate to the ring structures to offer strong polar interactions with the solutes. This type of stationary phase is stable in mobile phases containing 0–100% organic solvent. The proposed structure of vancomycin is shown in Fig. 9. Vancomycin is a very stable chiral stationary phase, has a relatively high sample capacity, and when covalently bonded to the silica gel, has multiple linkages to the silica gel surface. It can be used with mobile phases with a high water content, as a reversed phase, or with a high solvent content, as a largely polar stationary phase. For example, when used as a reversed phase strongly polar THF–water mixtures are very effective mobile phases. Conversely, when used as a polar stationary phase, n-hexane–ethanol mixtures are appropriate. Vancomycin has a number of ionizing groups and thus can be used over a range of different pH values (pH 4.0–7.0) and exhibit a wide range of retention characteristics and chiral selectivities. Ammonium nitrate, triethylammonium acetate, and sodium citrate buffers have all been used satisfactorily with this stationary phase.

Other than controlling the pH, the effect of the chosen buffer has little or no effect on chiral selectivity. This is verified by the chromatograms shown in Fig. 10. It is

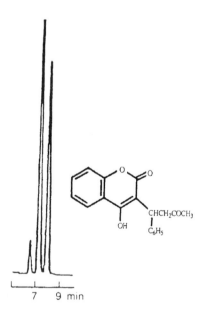

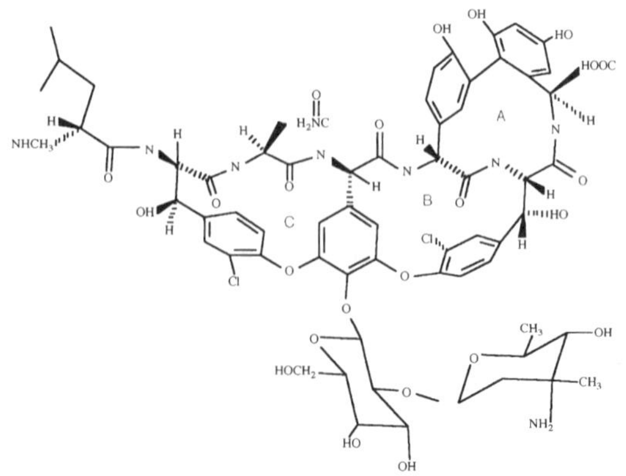

Courtesy of ASTEC Inc.

Figure 8 The separation of warfarin isomers on a Cyclobond® column.

and C are inclusion cavities. Molecular weight 1449. Chiral centers 18. pK's 2.9, 7.2, 8.6, 9.6, 10.4, 11.7. Isoelectric point 7.2

Courtesy of ASTEC Inc.

Figure 9 The proposed structure of vancomycin.

High-Performance Liquid Chromatography

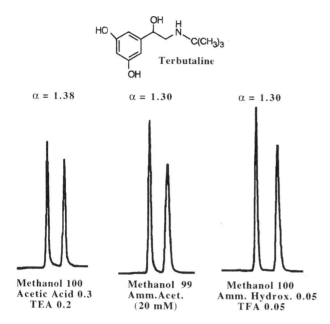

Figure 10 The separations of the isomers of terbutaline employing different buffer solutions.

seen that virtually the same selectivity is obtained from all three buffers irrespective of the actual chemical nature of the buffers themselves. It is interesting to note, however, that, although the difference is exceedingly small, the slightly greater separation ratio obtained from the buffer containing triethylamine might reflect the relatively strong dispersive character of the ethyl groups in the buffer molecule.

The enthalpic contributions to retention can be strongly dispersive, and/or strongly polar, or result from induced dipole interactivity. The aromatic rings will allow induced dipole interactions with the stationary phase and conversely, the strong polar groups on the stationary phase can induce dipole interaction with polarizable groups on the solute. Another macrolytic glycopeptide used in chiral chromatography is the amphoteric glycopeptide teicoplanin which is commercially available under the trade name of Chirobiotic T.

This material can also be bonded to 5 μm silica gel particles by multiple covalent linkages. Teicoplanin contains 20 chiral centers surrounding four molecular "pockets" or "cavities". Neighboring groups are strongly polar and aromatic rings provide ready polarizability. The proposed structure of teicoplanin is shown in Fig. 11.

This stationary phase is claimed to be complementary to the vancomycin phase and can be used with the same types of mobile phase, one often providing chiral selectivity, when the other does not. Teicoplanin can be used in a reversed phase mode using strongly polar mixtures such as acetonitrile–aqueous buffer, 10/90 v/v; THF–aqueous buffer, 10/90 v/v; and ethanol–aqueous buffer, 20/80 v/v.

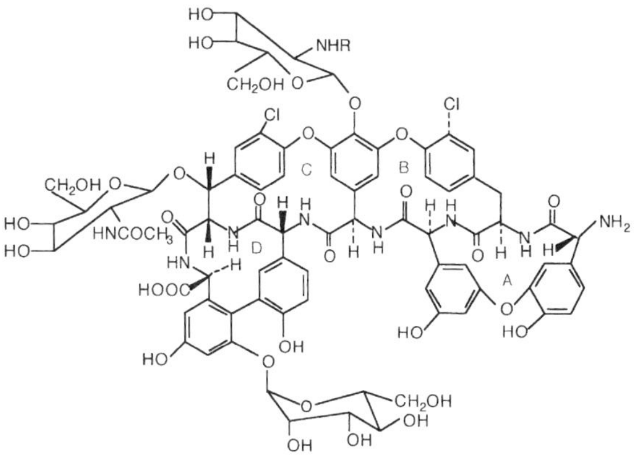

A, B, C and D are inclusion cavities. Molecular weight 1885. Chiral centers 20, Sugar moieties 3, and R is CH3-decanoic acid

Courtesy of ASTEC Inc.

Figure 11 The proposed structure of teicoplanin.

It can also be used as a polar stationary phase using n-hexane–ethanol mixtures as the mobile phase. In some cases it is advisable to control the pH even when the solutes are not ionic, suitable buffers being ammonium nitrate and triethylamine acetate.

One of the most useful and efficient mobile phases for use with the macrocyclic stationary phases, and with Chirobiotic T in particular, consists of 100 parts of methanol with the addition of acid and base (acetic acid and triethylamine) in the range of 1.0–0.1 parts. This system has been termed the *polar/organic mode* of development. The solvent system has been used very successfully for enantiomers that contain two groups that are capable of interacting with the stationary phase. It has also been found that one of these groups must be on or α to the stereogenic center.

The best concentration of the acid and base components appears to depend on the strength of the interaction between the active group and the stationary phase. A graph relating the resolution of the enantiomers and the acid/base composition is shown in Fig. 12. It is seen from Fig. 12 that the acid/base concentration is quite critical. If this type of mobile phase is applicable, there are several advantages. The system offers simplicity and versatility and, in addition, it uses a relatively inexpensive, nontoxic, solvents.

An example of the system in the separation of the propranolol enantiomers is shown in Fig. 13. The separations were carried out on a column 25 cm long, 4.6 mm i.d., packed with Chirobiotic T. The mobile phase was methanol containing acetic acid and triethylamine in the concentrations shown in Fig. 13. The column was operated at room temperature and at a flow rate of 2 mL/min. Teicoplanin is stable over a pH range of 3.8–6.5 although it can

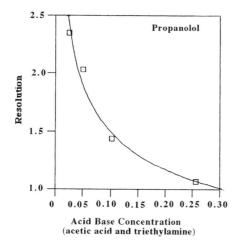

Figure 12 Graph of resolution against acid–base concentration (acid/base ratio 1:1).

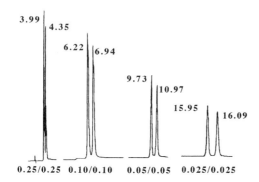

Figure 13 The separation of the enantiomers of propranolol employing different acid/base ratios.

be used for limited periods of time outside this range. In a similar manner to vancomycin, by choice of the mobile phase, teicoplanin can be used in the reversed phase mode, retention and selectivity depending largely on dispersive (hydrophobic) interactions between the solute and the stationary phases.

With a knowledge of the mechanism involved in the control of retention and selectivity, it is now possible to discuss the basic LC system and in particular the rationale behind the design of the instrumentation involved in the solvent supply system.

II. THE BASIC CHROMATOGRAPH

A block diagram of the basic LC system is shown in Fig. 14. It is seen from this figure that the basic LC consists of seven essential parts. The design of the mobile phase supply system (which comprises the solvent reservoirs, the pump, and the solvent programer) will be largely determined by the diverse nature of the solvents that may be

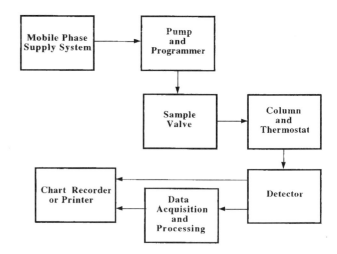

Figure 14 The basic liquid chromatograph.

chosen as the mobile phase. To achieve the necessary (often delicate) interactive balance of molecular forces between the solute molecules and those of the two phases, the solvent system must be designed to accommodate a wide range of diverse solvents and solvent mixtures. The multifarious nature of the solvents and the nature of the sample will affect the design of the sample valve. In contrast, the shape, size, and construction material of connecting tubes, unions, and frits will be determined almost exclusively by the column geometry. The appropriate detector and its specifications will be established from the nature of the sample and the concentration of the components of interest. Finally, the data acquisition and processing equipment and the method of data presentation will be largely determined by the accuracy and precision required from the analysis. It is seen that there are a large number of interacting factors that will determine the design and materials of construction of the different parts of the chromatograph.

A. Solvent Reservoirs

A diagram of a typical solvent reservoir is shown in Fig. 15. Solvent reservoirs were originally made of either glass or stainless steel, but glass is now the preferred material as stainless steel tends to corrode in the presence of certain buffers and dilute chloride solutions. All solvent mixtures that are to be used as a mobile phase should be filtered before placing in the reservoir but, as an extra precaution, it is wise to have a filter placed at the inlet to the tube that leads to the pump. A 10 μm filter is usually adequate. Filters having smaller pores tend to have a relatively high pressure drop across them and, as the mobile phase flows to the pump under atmospheric pressure, the pump can be starved owing to the restricted flow. This must be carefully avoided by ensuring the mobile phase is free

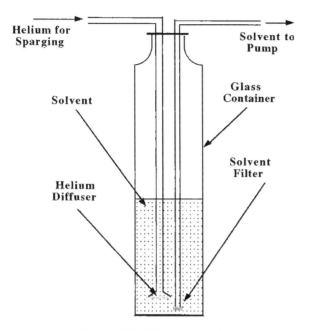

Figure 15 Solvent reservoir.

from particulate matter and by periodically checking that the pump is not starved of solvent. The tubes entering and leaving the reservoir should also be of glass or stainless steel. If stainless steel is used, and the mobile phase is a buffer or contains ionic material, the tubes should be removed and washed after use and not left in the mobile phase overnight. In some designs the base of the reservoir is made conical, thus the minimum amount of solvent is left in the reservoir when empty.

One of the problems met in high-performance liquid chromatography (HPLC) is caused by air dissolved in the mobile phase which can evolve inside the column forming bubbles. These bubbles produce serious and completely unacceptable noise as they pass through the detector. This problem can be particularly serious when using a mobile phase that consists of an aqueous mixture of methanol in which air is considerably soluble. The dissolved air can be removed by bubbling helium through the solvent in the reservoir, a procedure called "sparging". The solubilities of oxygen, nitrogen, and helium in water (Peterson, 1984) are given in Table 1.

It is seen from the table that oxygen is the gas with the highest solubility in water but the solubility of helium is much smaller than either that of nitrogen or that of oxygen. The continuous stream of helium leaches out all the dissolved air and the residual dissolved helium appears to cause no problem in the column or detector. It should be remembered, however, that if the sparging is stopped and the solvent comes again in contact with air, the solvent will rapidly become saturated with oxygen and nitrogen again. Removal of dissolved air by vacuum is an alternative degassing procedure, but is not recommended. Although it initially removes the dissolved air very efficiently, it will quickly become saturated again once the vacuum is released.

There is another type of continuous "flow-through" degassing device that is commercially available, which consists of a semiporous Teflon tube through which the mobile phase passes on the way to the sample valve. The Teflon tube is situated inside a vacuum jacket which extracts any dissolves gasses but the pores are too small to allow the passage of solvent. This is possibly a more compact and efficient form of degasser but another gas will still be required if a flame-proof enclosure is considered necessary.

A minimum of three reservoirs are usually required with a maximum of probably five. The outlet from each reservoir passes to either a programer selector-valve or a pump, depending on whether a high pressure or low pressure solvent programer is used. As it is necessary to accommodate a wide range of solvent types, some solvents will inevitably be highly inflammable. As a consequence it is advisable to place the reservoirs in a flame-proof enclosure. This is fairly easily arranged by appropriate instrument design as the sparging gas helium can be used as an agent to render the enclosure flame proof.

In the past it has been considered advisable to have the reservoir stirred, particularly if mixtures of solvents of extreme density are being used. However, provided there is complete miscibility between the solvents, the mixtures will not separate on standing, and stirring is entirely unnecessary. In some cases, solvent reservoirs have been heated to help degas the solvent. This is a clumsy and only temporary solution to the problem of dissolved gases, as already discussed. Heated solvent reservoirs are also unnecessary and the alternative use of helium sparging is strongly recommended.

B. Solvent Programers

There are two type of solvent programers currently available, both of which provide accurate and reproducible solvent programs, but operate on an entirely different principles. The low pressure programer, as its name suggests, draws solvent from the reservoirs at atmospheric pressure, mixes them in the required proportions, and then passes the mixture to an appropriate high pressure pump that

Table 1 Solubilities of Gases in Water

Gas	Solubility (mL/100 mL of cold water) (15°C)
Helium	0.94
Nitrogen	2.37
Oxygen	4.89

feeds the solvent mixture to the sample valve and then the column. The second type of solvent programer draws the solvent from the reservoir directly with a pump, there is thus a pump for each solvent, and then by controlling the rate of each pump, a solvent mixture is produced at high pressure which is delivered to the sample valve and column. Each method has its advantages and disadvantages but are probably equally popular.

1. The Low Pressure Solvent Programer

A diagram of a low pressure solvent programer is shown in Fig. 16. The low pressure solvent programer consists of three solvent reservoirs (or more if required), each connected to a valve which is controlled by a computer. The outlets of each valve are joined at a manifold and the mixture passes directly to the inlet of a pump. The valves are alternately opened and closed by square waves generated by the computer. The amount of each solvent that is delivered is controlled by the length of the pulse (the "mark-to-space-ratio"), and the total quantity provided from all the reservoirs by the frequency of the pulses.

The valve is quite complex and expensive, although not as expensive as having a pump for each solvent, which is necessary in high pressure solvent programing. A computer can be dedicated to the programer or it can be actuated by the computer that controls the overall operation of the chromatograph. Linear, concave, or complex programs can be generated, as shown in Fig. 17.

The type of program that should be used for a specific separation is impossible to predict and will depend on both the nature of the sample and the solvents being employed. The optimum gradient is almost always identified by experimenting (Berridge, 1985). In practice, the most common gradient form used is linear but there are specific applications that require more complex gradients. The convex gradient is frequently used with methanol–water mixtures in the separation of solute substances of biological origin (e.g., peptides and proteins). Gradient forms can become extremely complex when three or even four solvents are found to be necessary. However, multigradient programs are only necessary for particularly recalcitrant mixtures and, for most separations, dual

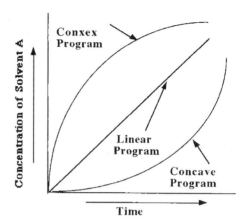

Figure 17 Different forms of solvent gradients.

solvent mixtures usually suffice to provide the necessary resolution. The programer also provides facilities for choosing a specific solvent or solvent mixture for isocratic development. Consequently, the device must ensure that the composition of the solvent mixture is held constant with high precision for long periods of time.

2. The High Pressure Solvent Programers

High pressure solvent programers utilize a pump for each solvent which is independently controlled by the computer. As a consequence, the high pressure programer is somewhat more expensive than the low pressure programer. However, the difference is not as great as one might expect because of the complexity and consequent high cost of the programing valve. A diagram of a high pressure solvent programer is shown in Fig. 18.

Each solvent is delivered to the column by its own high pressure pump which is managed by either a dedicated computer or by the control-computer of the chromatograph. Most pumps are driven by stepping motors and thus the rate of solvent delivery is controlled by adjusting the frequency of the power supply to each pump. The high pressure programer is more bulky than its low pressure counterpart which, together with its higher cost, has made it the less popular of the two. In general, however, owing to the direct flow control of each solvent by its respective

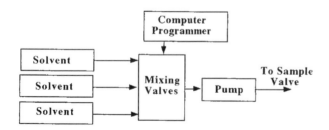

Figure 16 The low pressure solvent programer.

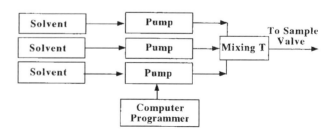

Figure 18 The high pressure solvent programer.

pump, the high pressure programer is considered likely to give the more precise and repeatable programs.

The component of the high pressure solvent programer that requires careful design is the mixing manifold. The solvents must be completely mixed before entering the sample valve and column. This problem is less acute with the low pressure programer as the mixed solvents need to pass through the pump before reaching the valve and column, and the pumping process facilitates mixing. The integrity of the composition of the solvent mixture can be checked by attaching a refractive index detector to the programer outlet and monitoring the solvent over a period of time. Any drift or long term noise would indicate poor control of the solvent composition.

C. Nonreturn Valves

LC columns are operated at exceedingly high pressures of 3000–10,000 p.s.i. and thus depend heavily for their efficient function on well designed nonreturn valves. It is therefore important to understand their action and to recognize problems associated with nonreturn valves. A diagram of a nonreturn valve is shown in Fig. 19. A nonreturn valve consists of a stainless steel body containing a sapphire ball (1–2 mm in diameter) and a sapphire seat. Under some circumstances the valve seat is made of stainless steel but a sapphire seat is preferable. Sapphire is used because it is extremely hard and thus does not scratch easily, which would result in leaks.

There is a hole in the center of the seat in which the ball sits and solvent passes through the hole pushing the ball aside and out through the valve. When the pressure driving the liquid through the valve falls, the ball is forced back onto the seat and stops the flow. In fact, during the return of the ball to the seat a slight flow of liquid does pass back through the aperture before the ball actually seals the valve. However, this flow is minuscule compared with the normal flow from the pump. It is important to appreciate that the ball in the valve is *not* gravity driven, but fluid driven, and thus will work effectively in any position. Nonreturn valves can become contaminated and their action impaired. This can result in a loss of pressure on the column, irregular flow and poor chromatographic reproducibility. It usually arises from particulate matter from contaminated solvents adhering to the sapphire ball or seat and preventing the valve from sealing properly. The material can often be removed by pumping a series of solvents through the valve. An effective series of solvents is, *n*-heptane, methylene dichloride, ethyl acetate, acetone, acetonitrile, 50% v/v aqueous–methanol mixture, and finally a suitable detergent solution. This series of solvents will usually clean the valve but, if they fail, the valve must be taken apart and carefully cleaned mechanically.

D. Chromatography Pumps

There are three critical parts to the LC system, the first and most important is the column in which the separation is carried out. The second is the detector, which monitors the separation and permits a quantitative estimation of each eluted solute. The third is the pump, which forces the flow of mobile phase through the bed of very small particles that are packed in the column to provide the high column efficiency and high resolution. The improved performance realized from HPLC columns is a direct result of the use of small particles as the packing and, therefore, indirectly a result of the use of high pressure pumps. There are certain critical specifications for an LC pump.

1. The pump must be capable of operating continuously at pressures up to 6000 p.s.i. and preferably as high as 10,000 p.s.i.
2. The flow rate range must be appropriate to cover the span of column diameters that are in common use. The column and the effect of its dimensions on instrument design will be discussed later in the chapter but at this time it can be said that the general flow rate ranges necessary for the operation of different types of column are shown in Table 2.

Table 2 Flow Rates for Columns

Column type	Pump flow rate range (mL/min)
Small columns (diameter 0.25–2.0 mm)	0.002–2.00
Normal analytical columns (diameter 1.5–10 mm)	0.2–10.0

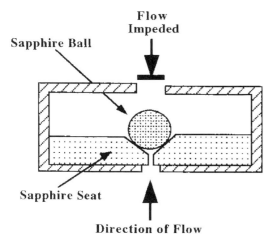

Figure 19 Nonreturn valve.

3. For accurate quantitative and qualitative analysis a flow rate accuracy of $\pm 1\%$ and flow rate precision of $\pm 0.1\%$ are strongly recommended.
4. The flow from the pump should be pulse free. Completely pulse-free flow is almost impossible to achieve with piston or diaphragm type pumps. However, the pump should be designed to minimize such pulses. The column and chromatographic system themselves tend to act as a pulse dampener and reduce pulses, but at the maximum sensitivity, the pulse noise from the pump should be no greater than the total detector noise from all other sources. There are a number of pulse dampening devices available for LC systems but care should be taken in their use. Pulse dampeners inevitably introduce a significant extra volume between the pump and the sample valve, which will distort the concentration profile of any solvent program that may be used. There are a number of different types of pump used in LC systems, each having different attributes; some of these will now be discussed.

1. The Pneumatic Pump

A diagram of a pneumatic pump is shown in Fig. 20. The pneumatic pump has a relatively large flow capacity but, today, it is rarely used in normal LC analysis. It is being used extensively for column packing. It can provide extremely high pressures and is relatively inexpensive, but the high pressure models are a little bulky.

It is seen, in Fig. 20, that the air entering the lower port ($Air_{(2)}$) applies a pressure to the lower piston face that has a diameter y. This total pressure is transferred to the piston controlling the liquid flow that has a diameter x. Because the radii of the pistons differ, there will be a net pressure amplification of y^2/x^2. If, for example, the upper piston is 10 cm in diameter and the lower piston 1 cm in diameter, then the amplification factor will be $(10^2/1^2) = 100$. It is seen that the system can provide very high pressures in a relatively simple manner. The operation of the pump is as follows. Air enters port (2) and applies a pressure to the lower piston that is directly transmitted to the upper piston. If connected to an LC system, or column packing manifold, liquid will flow out of the left-hand side nonreturn valve as shown in the diagram. This will continue until the maximum movement of the piston is reached. At the extreme of movement, a microswitch is activated and the air pressure transferred to port (1). The piston moves downward drawing mobile phase through the right-hand nonreturn valve filling the cylinder with solvent. This occurs extremely rapidly but the pulse resulting from this refill action (and to some extent the accompanying noise) is largely responsible for the unpopularity of this pump for general LC use. This pulse can be reduced with a suitable pulse dampener but the extra volume introduced is unacceptable for the reasons already given. When the cylinder is fully charged with solvent the piston reaches the limit of a second microswitch which activates another valve and returns the air flow to port (2) and the process is repeated.

2. The Single Piston Reciprocating Pump

The single piston pump is one of the most commonly used pumps in LC and although somewhat less popular today, it was once used in over 60% of all LC systems. It is relatively inexpensive and allows the analyst to become involved in LC techniques without excessive investment. A diagram of a single piston pump is shown in Fig. 21. The piston is made of synthetic sapphire; in fact most pistons of modern LC reciprocating pumps are made of sapphire to reduce wear and extend the working life of the pump. The cylinder, which incorporates two nonreturn valves in line with the inlet and outlet connections to the pump, is usually made of stainless steel. The piston is driven by a stainless steel cam which forces the piston into the cylinder expressing the solvent through the exit

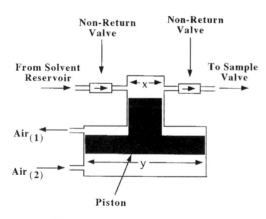

Figure 20 The pneumatic pump.

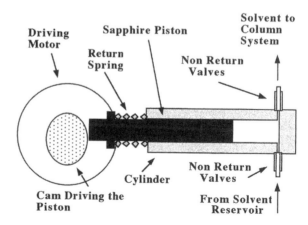

Figure 21 The single piston reciprocating pump.

nonreturn valve. The shape of the cam is cut to provide a linear movement of the piston during expression of the solvent but a sudden return movement on the refill stroke. After reaching the maximum movement, the piston follows the cam and is rapidly returned as a result of the pressure exerted by the return spring. During this movement the cylinder is loaded with more solvent through the inlet nonreturn valve. In this way the pulse effect is reduced. However, the pulses are not completely eliminated and their presence in the flow of mobile phase is probably the most serious disadvantage of the single piston pump. Nevertheless, as a result of its low cost it remains one of the more popular LC pumps. These pumps are manufactured to give a range of flow rates.

Single piston pumps providing flow rates ranging from 0.002 to 2 mL/min are commonly used with small bore (microbore) columns. Flow rate ranges of 0.2–10 mL/min are also available for the more conventional column diameters. In an attempt to reduce the pulses produced by single piston pumps the dual pump was deigned.

3. The Dual Piston Reciprocating Pump

A diagram of the dual piston reciprocating pump is shown in Fig. 22. The actual cylinders and pistons of a two-headed pump are constructed in a very similar manner to the single piston pump with a sapphire piston and a stainless steel cylinder. Each cylinder is fitted with nonreturn valves both at the inlet and outlet. The cams that drive the two pistons are carefully cut to provide an increase in flow from one pump while the other pump is being filled to compensate for the loss of delivery during the refill process and thus, a fall in pressure. There is a common supply of mobile phase from the solvent reservoir or solvent programer to both pumps. The output of each pump also joins and the solvent then passes to the sample valve and then to the column. In the diagram, a single cam drives both pistons, but in some designs, to minimize pressure pulses, each pump has its own cam drive from the motor. The cams must be mechanically connected in an appropriate manner to maintain the correct phase difference and are carefully designed and cut to produce a virtually pulse-free flow. The displacement volume of each pump can vary from 20 or 30 μL to over 1 mL but the usual displacement volume for the Waters pump is about 250 μL. The pump is driven by a stepping motor and thus the delivery depends on the frequency of the supply fed to the motor.

This gives the pump a very wide range of delivery volumes to choose from, extending from a few microliters per minute to over 10 mL/minute. The Waters twin headed pump is shown in Fig. 23. The Waters pump was one of the first twin headed LC pumps to be produced and was probably the major factor in establishing Waters fine reputation in the field of LC instrumentation.

4. The Rapid Refill Pump

In order to avoid the pulses resulting from a single piston refill pump, rapid refilling systems have been devised. These have varied from cleverly designed actuating cams to drive the piston to electronically operated piston movement. An interesting and successful approach to this problem is exemplified by the pump design shown in Fig. 24. The pump consists of two cylinders and a single common piston. The expression of the solvent to the column is depicted in the upper part of Fig. 24. As the piston progresses to the right, the solvent is pumped to the column system but at the same time fresh solvent is being withdrawn into the right chamber from the solvent supply system. At the point where the piston arrives at the extent of its travel, a step in the driving cam is reached and the piston is rapidly reversed. As a result, the contents of the chamber on the right-hand side

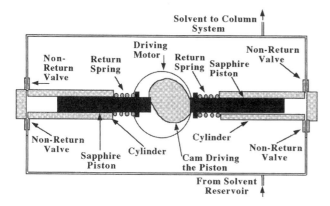

Figure 22 The dual piston reciprocating pump.

Courtesy of Waters Chromatography

Figure 23 An earlier model of the Waters 501 twin headed LC pump.

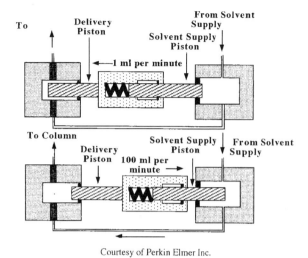

Courtesy of Perkin Elmer Inc.

Figure 24 The rapid refill pump.

are conveyed to the left-hand chamber. This situation is depicted in the lower part of Fig. 24.

The transfer rate of the solvent to the left-hand chamber is 100× as fast as the delivery rate to the column and consequently reduces the pulse on refill very significantly. Furthermore, if a solvent gradient is being used and the right-hand chamber is being filled with a solvent mixture, excellent mixing is achieved during the refill of the left-hand chamber. The rapid refill pump can be easily combined with a low pressure solvent programer to provide a complete and very versatile solvent delivery system.

A diagram of the combined solvent reservoir, pump system, and programer is shown in Fig. 25. The system can be made very compact without compromising either the accuracy and precision of the solvent programer or the versatile performance of the rapid refill pump. The solvent reservoirs are situated above the programers and pump in a flame proof chamber purged with helium.

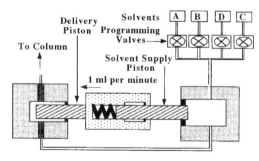

Courtesy of Perkin Elmer Inc.

Figure 25 Rapid refill pump and low pressure solvent programer.

5. The Diaphragm Pump

In an attempt to find an alternative method for reducing pump pulses, the diaphragm pump was developed, which operated with a much smaller stroke, a much wider displacement head, and at a much higher frequency. In the reciprocating diaphragm pump, the actuating piston does not come into direct contact with the mobile phase and thus, the demands on the piston–cylinder seal are not so great. Because the diaphragm has a relatively high surface area, the movement of the diaphragm is very small, allowing the use of fairly high frequencies. The net result is an increase in pulse frequency and a reduction in pulse amplitude. Furthermore, high frequency pulses are more readily damped by the column system. It must be emphasized, however, that diaphragm pumps are not completely pulseless. A diagram of a diaphragm pump, showing its mode of action is depicted in Fig. 26.

The wheel driving the crank is rotating anticlockwise and in stage (1), the diaphragm has been withdrawn and the pumping cavity behind the diaphragm filled with solvent. In stage (2), the piston advances, and when it passes the pumping fluid inlet it starts compressing the diaphragm expressing solvent to the column. In stage (3), the diaphragm has been compressed to its limit and the piston starts to return. In stage (4), the piston moves back withdrawing the diaphragm sucking liquid into the pumping cavity ready for the next thrust. The inlet from the mobile phase supply and the outlet to the column are both fitted with nonreturn valves in the usual manner. Owing to the relatively large volume of the pumping cavity, this type of pump is not often used in analytical chromatographs but is frequently employed for preparative chromatography.

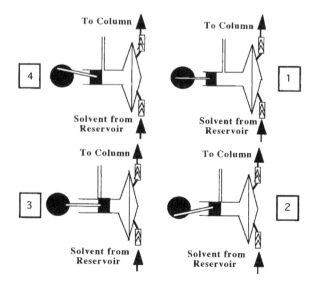

Figure 26 The action of a diaphragm pump.

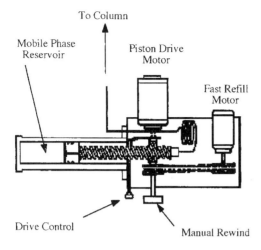

Figure 27 The syringe pump.

6. The Syringe Pump

The syringe pump, as the name implies, is a large, electrically operated simulation of the hypodermic syringe. Although popular in the early days during the renaissance of LC, it is rarely used today as, because of its design, it can provide only a limited pressure and the volume of mobile phase available is restricted to the pump volume. Unless the separation is stopped while the pump is refilled and the development subsequently continued, the pump can only develop a separation to the limit of its own capacity. A diagram of a syringe pump is shown in Fig. 27.

The pump consists of a large metal syringe, the piston being propelled by an electric motor and driven by a worm gear. The speed of the motor determines the pump delivery. Another motor actuates the piston by a different system of gearing to refill the syringe rapidly when required. The solvent is sucked into the cylinder through a hole in the center of the piston, and between the piston and the outlet there is a coil that acts as a dampener. This type of pump is still occasionally used for the mobile phase supply to microbore columns that require small volumes of mobile phase to develop the separation. It is also sometimes used for reagent delivery in postcolumn derivatization as it can be made to deliver a very constant reagent supply at very low flow rates. Its only advantage appears to be that it can be made truly pulseless.

E. Column Design and Performance

The design of the sample valve, connecting tube, and detector cell depends strongly on the characteristics of the column with which they are used. Consequently, before dealing with the design of these components of the chromatograph, the properties of the column will be considered and in particular, how those properties influence instrument design.

The chromatography column has a dichotomy of purpose; firstly, it separates the individual solutes apart by exploiting the different molecular forces that occur between the solutes and the two phases; secondly, it constrains the dispersion or spreading of each solute band so that, having been moved apart from one another, they are eluted discretely. It is the capacity of the column to restrain peak dispersion that determines the design and dimensions of many parts of the chromatograph.

1. Column Efficiency

A chromatographic peak will be close to Gaussian in form, and the peak width at the points of inflexion of the curve (which corresponds to twice the standard deviation of the curve) can be taken as a measure of the ability of the column to restrain dispersion. At the points of inflexion (Scott, 1992, p.45), the second differential of the elution curve equation [Eq. (9)] will equal zero. It follows that,

$$\frac{d_2(X_0(e^{-v}v^n/n!))}{dv^2} = 0$$

Thus,

$$\frac{d_2(X_0(e^{-v}v^n/n!))}{dv^2} = X_0 \frac{e^{-v}v^n - e^{-v}nv^{(n-1)} - e^{-v}nv^{(n-1)} + e^{-v}n(n-1)v^{(n-2)}}{n!}$$

Simplifying and factoring the expression

$$\frac{d_2(X_0(e^{-v}v^n/n!))}{dv^2} = X_0 \frac{e^{-v}v^{(n-2)}(v^2 - 2v + n(n-1))}{n!}$$

Thus, at the points of inflexion,

$$v^2 - 2nv + n(n-1) = 0$$

Then,

$$v = \frac{2n \pm \sqrt{(4n^2 - 4n(n-1))}}{2}$$
$$= \frac{2n \pm \sqrt{4n}}{2}$$
$$= n \pm \sqrt{n}$$

It is seen that the points of inflexion occur after $n - \sqrt{n}$ and $n + \sqrt{n}$ plate volumes of mobile phase has passed through the column. Thus the volume of mobile phase that has passed through the column *between* the inflexion points will be,

$$n + \sqrt{n} - n + \sqrt{n} = 2\sqrt{n} \qquad (12)$$

It follows that the peak width in milliliters of mobile phase will be obtained by multiplying by the *plate volume*, that is,

$$\text{Peak width} = 2\sqrt{n}(v_m + Kv_s) \quad (13)$$

The peak width at the points of inflexion of the elution curve is twice the standard deviation and thus, from Eq. (18) it is seen that the variance (the square of the standard deviation) is equal to n, the total number of plates in the column. Consequently, the variance of the band (σ^2) in milliliters of mobile phase is given by,

$$\sigma^2 = n(v_m + Kv_s)^2$$

Now,

$$V_r = n(v_{m_s} + Kv_s)$$

and

$$\sigma^2 = \frac{V_r^2}{n}$$

It follows, that as the variance of the peak is inversely proportional to the number of theoretical plates in the column, then the larger the number of theoretical plates, the more narrow the peak, and the more efficiently the column has constrained the band dispersion. As a consequence, the number of theoretical plates in a column has been given the term *column efficiency* and is used to describe its quality. The various characteristics of the chromatogram that have so far been considered are shown in Fig. 28. It is important to be able to measure the efficiency of any column and, as can be seen from this figure, this can be carried out in a very simple manner. Let the distance between the injection point and the peak maximum (the retention distance on the chromatogram) be y cm and the peak width at the points of inflexion be x cm, as shown in Fig. 28.

Then, as the retention volume is $n(v_m + Kv_s)$ and twice the peak standard deviation at the points of inflexion is $2\sqrt{n}(v_m + Kv_s)$, then, by simple proportion,

$$\frac{\text{Ret. distance}}{\text{Peak width}} = \frac{y}{x} = \frac{n(v_m + Kv_s)}{2\sqrt{n}(v_m + Kv_s)} = \frac{\sqrt{n}}{2}$$

$$n = 4\left(\frac{y}{x}\right)^2 \quad (14)$$

Equation (14) allows the efficiency of any solute peak, from any column, to be calculated from measurements taken directly from the chromatogram. It is now possible to calculate the maximum sample volume that can be placed on a column and thus determine the volume of the sample valve.

2. The Maximum Sample Volume

Any sample placed on a column will have a finite volume, and the variance of the injected sample will contribute directly to the final peak variance. It follows that the maximum volume of sample that can be placed on the column must be limited, or the column efficiency will be seriously reduced. Consider a volume V_i, injected onto a column. Normal LC injections will start initially as a rectangular distribution and the variance of the eluted peak will be the sum of the variances of the injected sample plus the normal variance of the eluted peak.

Thus,

$$\sigma^2 = \sigma_i^2 + \sigma_c^2$$

where σ^2 is the variance of the eluted peak, σ_i^2 is the variance of the injected sample, and σ_c^2 is the variance due to column dispersion.

It is generally accepted that the maximum increase in band width that can be tolerated owing to extraneous dispersion processes is 5% of the standard deviation (Klinkenberg, 1960) (or 10% of the peak variance). This value, in fact, refers to the total dispersion from all sources that is acceptable for the efficient use of the chromatograph. For convenience, in the following argument it will be assumed that the major contribution to extra column dispersion will arise from the sample volume only, but in practice this is likely to be an ideal and unrealistic case.

Let a volume (V_i) be injected onto a column resulting in a rectangular distribution of sample at the front of the column. According to the principle of the summation of variances, the variance of the final peak will be the sum of the variances of the sample volume plus the normal variance of a peak for a small sample.

Now the variance of the rectangular distribution of a sample, volume V_i, is $V_i^2/12$ and, from the plate theory, $\sigma_v^2 = (\sqrt{n}(v_m + Kv_s))^2$. Then, assuming a 5% increase in peak width due to the sample volume,

$$\frac{V_i^2}{12} + (\sqrt{n}(v_m + Kv_s))^2 = 1.05(\sqrt{n}(v_m + Kv_s))^2$$

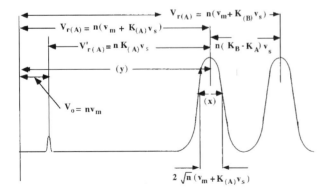

Figure 28 A chromatogram showing the separation of two solutes.

Thus,

$$\frac{V_i^2}{12} = n(v_m + Kv_s)^2(1.05^2 - 1)$$
$$= n(v_m + Kv_s)^2 0.102$$

Consequently,

$$V_i^2 = n(v_m + Kv_s)^2 1.23$$

or,

$$V_i = \sqrt{n}(v_m + Kv_s)1.1$$

Now,

$$V_r = n(v_m + Kv_s)$$

Then,

$$V_i = \frac{1.1 V_r}{\sqrt{n}} \tag{15}$$

It is seen that the that the maximum sample volume that can be tolerated can be calculated from the retention volume of the solute concerned and the efficiency of the column. However, a sample does not consist of a single component, and it is therefore important that the resolution of any solute, irrespective of where it is eluted, is not excessively dispersed by the sample volume.

The solute that will be most affected will be the solute eluted in the smallest volume, which will be the solute eluted close to the dead volume. Consequently, for general use the dead volume should be used in Eq. (15) as opposed to the retention volume of a specific solute.

Thus, Eq. (15) becomes

$$V_i = \frac{1.1 V_c}{\sqrt{n}}$$

Now the total volume (V_C) of a column radius (r) and length (l) will be $\pi r^2 l$. Furthermore, the volume occupied by the mobile phase will be approximately $0.6 V_C$ (60% of the total column volume is occupied by mobile phase). Thus as a general rule the maximum sample volume that can be employed without degrading the resolution of the column is

$$V_i = \frac{0.60 \pi r^2 l}{\sqrt{n}} \tag{16}$$

Bearing in mind that, to a first approximation, $n = 1/2d_p$, this assumes that the contribution to longitudinal diffusion and the resistance to mass transfer in the mobile and stationary phases is small when compared with the multipath contribution to dispersion.

That is,

$$dp \ll \frac{2Dg}{u} + (Cg + Cs)u$$

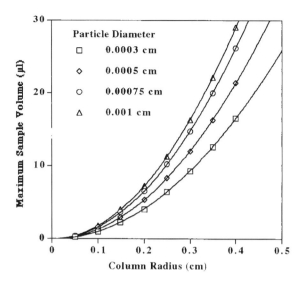

Figure 29 Plot of maximum sample volume against column radius.

which, providing the column is well designed, well packed, and operated at close-to-optimum conditions will be sensibly true,

$$V_i = \frac{0.85 \pi r^2 l}{\sqrt{d_p}} \tag{17}$$

Employing Eq. (17) the maximum sample volume that can be tolerated for columns 5 cm in length packed with different particles is given in Fig. 29. It is seen that for column diameters below 2 mm, the sample volume must be 1 μL or less (Scott and Kucera, 1979a). Such conditions demand a specific type of sample valve, which will be described below. From a practical point of view it is best to use the maximum volume of sample possible (assuming there is no mass overload) as this will allow the detector to be operated at the lowest possible sensitivity and, in doing so, provide the greatest detector stability and, as a consequence, the highest accuracy. The theory given will be referred to when dealing with the design of column connections and detector cell volumes but it is now possible to consider the design of sample valves and their injection volumes.

F. Sample Valves

Originally, when columns were operated at low inlet pressures, samples were placed on the column in much the same way as on a GC column using a hypodermic syringe and a silicone rubber septum. With the introduction of small particle packing and the consequent high inlet pressures, it was found necessary to place the sample on the column with a sample valve. There are two basic types of sample valve, the internal loop sample valve and the external loop

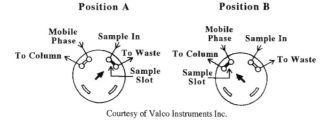

Figure 30 The internal loop sample valve.

sample valve. The internal loop valve can provide the necessary small sample volumes required for small diameter columns whereas the external loop valves can be used for the larger columns and semipreparative and preparative columns.

1. The Internal Loop Sample Valve

A diagram of the internal loop sample valve is shown in Fig. 30. The sample volume is contained in the connecting slot of the valve rotor and can be designed to deliver sample volumes ranging from 0.1 μL to about 1.0 μL to the column. In position A, in Fig. 24, the sample is shown being loaded into the valve. The sample is loaded into the valve by passing the sample solution from an appropriate syringe through the rotor slot to waste. During the procedure the mobile phase passes through the valve directly to the column. On rotation (shown in position B), the valve slot is now imposed between the mobile phase supply and the column and consequently, the sample is swept onto the column by the flow of solvent. This type of valve is used in all cases where peak volumes are small owing to the column design (i.e., the column diameter *or* the length is small).

2. The External Loop Sample Valve

The external loop sample valve is one of the more common sampling systems and operates on exactly the same principle but offers a wide choice of larger sample sizes. A diagram of the external loop sample valve is shown in Fig. 31. In the external loop sample valve, three slots are cut in the rotor so that any adjacent pair of ports can be connected. In the loading position (position A on the left), the mobile phase supply is connected by a rotor slot to port (4) and the column to port (5), thus, allowing mobile phase to flow directly through the column. In this position the sample loop is connected to ports (3) and (6). Sample flows from a syringe into port (1) through the rotor slot to the sample loop at port (6). At the same time the third slot in the rotor connects the exit of the sample loop to waste at port (2). The sampling position is shown in the diagram on the right. On rotating the valve to place the sample on the column, the sample loop is interposed between the column and the mobile phase supply by connecting ports (3) and (4) and ports (5) and (6). Thus, the sample is swept onto the column. In the sampling position, the third rotor slot connects the syringe port to the waste port. After the sample has been placed on the column, the rotor can be returned to the loading position, the system washed with solvent and the sample loop loaded in readiness for the next injection.

The external valve can accommodate sample volumes ranging from about 5 μm to ≥10 mL. In preparative chromatography, however, sample volumes of several hundred milliliters are often used and such volumes are pumped directly onto the column by a separate sample pump. Sample valves are rarely if ever used in large scale preparative chromatography.

A modified form of the external loop sample valve has become very popular for quantitative LC analysis, a diagram of which is shown in Fig. 32. The basic difference between this type of valve and the normal external loop sample valve is the incorporation of an extra port at the front of the valve. This port allows the injection of a sample by a syringe directly into the front of the sample loop. Position (A) shows the inject position. Injection in the front port causes the sample to flow into the sample loop. The tip of the needle passes through the rotor seal and, on injection, is in direct contact with the ceramic

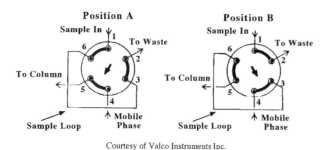

Figure 31 The external loop sample valve.

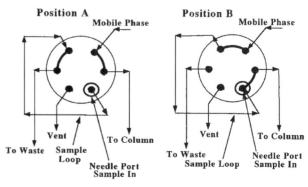

Figure 32 The modified external loop sample valve.

stator face. Note the needle is chosen so that its diameter is too great to enter the hole. After injection, the valve is rotated to position (B) and the mobile phase flushed the sample directly onto the column. The sample is actually forced out of the beginning of the loop so it does not have to flow through the entire length of the loop. This type of injection system is ideally suited for quantitative LC, and is probably by far the most popular injection system in use. Sample valves based on this design is available from a number of manufacturers.

In general, LC sample valves should be able to sustain pressures up to 10,000 p.s.i., although they are likely to operate on a continuous basis, at pressures of 3000 p.s.i. or less. The higher the operating pressure the tighter the valve seating surfaces must be forced together to eliminate any leak. It follows that any abrasive material, however fine, that passes into the valve can cause the valve seating to become seriously scored each time it is rotated. This will ultimately lead to leaks, the sample size will start to vary between samples and eventually the accuracy of the analysis will be seriously affected. It follows that any solid material must be carefully removed from any sample before filling the valve. However, even with stringent precautions, all solid material cannot be completely removed from a sample. Consequently, the operation of the valve and column system at excessively high pressures should be avoided if possible. A good compromise between ensuring satisfactory column efficiency and resolution by the use of adequate pressure and reasonable valve life, is to operate the chromatographic system at or below 3000 p.s.i. A diagram of an LC sample valve is shown in Fig. 33.

As a consequence of the high pressures that must be tolerated, LC sample valves are usually made from stainless steel. The exception to the use of stainless steel will arise in biochemical applications where the materials of construction may need to be biocompatible. In such cases the valves may be made from titanium or some other appropriate biocompatible material having adequate strength. Only those surfaces that actually come in contact with the sample need to be biocompatible and the major parts of the valve can still be manufactured from stainless steel. The actual structure of the valve varies a little from one manufacturer to another but all are modifications of the basic sample valve shown in Fig. 25. The valve consists of a number of parts. There is the control knob or handle that allows the valve selector to be rotated and thus determines the spigot positions. A connecting device communicates the rotary movement of the control knob to the rotor and a valve body contains the different ports necessary to provide connections to the mobile phase supply, the column, the sample loop, the sample injection port, and the connection to waste. The rotor that actually selects the mode of operation of the valve also contains slots that can connect the alternate ports in the valve body to provide loading and sampling functions. Finally there is a preload assembly that furnishes an adequate pressure between the faces of the rotor and the valve body to ensure a leak tight seal.

G. Column Ovens

Both retention and dispersion are temperature dependent, particularly when separations ratios are small (e.g., in chiral separations). Under some circumstances selecting the correct temperature can be critical for a satisfactory separation. As a consequence, the temperature control of the column can be very important and thus, column ovens can be essential.

The effect of temperature on column efficiency can be determined from the Van Deemter equation that describes the variance per unit length (HETP) as a function of the linear velocity.

$$H = 2\lambda d_p + \frac{2\gamma D_m}{u} + \frac{f_1(k')d_p^2}{D_m}u + \frac{f_2(k')d_f^2}{D_s}u \quad (18)$$

where λ and γ are constants, D_m and D_s are the diffusivities of the solute in the mobile and the ionary phases, respectively, d_p is the particle diameter of the packing, d_f is the effective thickness of the film of stationary phase, k' is the capacity ratio of the solute, and u is the linear velocity of the mobile phase.

For a given column, the particle diameter and the film thickness will be constants and k' for a given solute can be assumed to be approximately constant. Furthermore, as the mobile phase and stationary phases are both liquids to a first order of approximation, $D_m = D_s$.

Thus, Eq. (18) can be put in the form,

$$H = A' + \frac{B'D_m}{u} + \frac{C_1 + C_2}{D_m}u \quad (19)$$

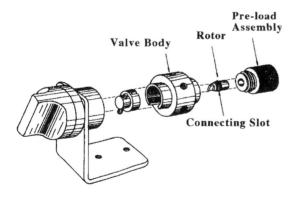

Courtesy of Valco Instruments Inc.

Figure 33 A simple form of the LC sample valve.

Now, as an increase in temperature will increase the value of D_m its effect on H and consequently the column efficiency is clear. If the mobile phase velocity is above the optimum then the function $[(C_1 + C_2)/D_m]u$ will dominate and an increase in temperature will increase D_m, decrease $[(C_1 + C_2)/D_m]u$, and consequently H and thus *increase* the column efficiency. This is the most common situation in LC.

Conversely, if the mobile phase velocity is below the optimum velocity then the function $B'D_m/u$ will dominate and an increase in temperature will increase D_m, increase $B'D_m/u$, and consequently H and thus *decrease* the column efficiency. LC columns are rarely operated below the optimum velocity and thus, this situation is the least likely scenario.

However, from Eq. (3) it is seen that the optimum velocity is directly related to the solute diffusivity and thus, to the temperature. It follows that increasing the temperature, increases the optimum velocity and thus provides a higher efficiency, together with a shorter analysis time. It follows that temperature control can be very important where high efficiencies are in demand (e.g., in size exclusion chromatography) and thus under certain circumstances columns ovens will be required. An example of a commercially available column oven is shown in Fig. 34.

The model shown in Fig. 34 has an operational temperature range from 10°C to 99°C. One of the problems associated with the temperature control of LC ovens is the high thermal capacity of the column and mobile phase. Air-ovens are not very satisfactory owing to the small heat capacity of gases generally. The use of liquid thermostatting media, however, brings its own problems, including evaporation; spillage and column replacement and fitting become difficult. Perkin Elmer avoided this problem by using high thermal conductivity tubing to bring the mobile phase to

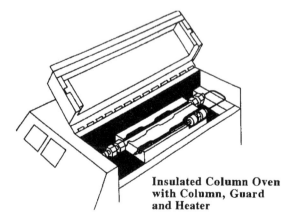

Figure 34 An LC column oven containing column and precolumn.

Figure 35 Heat transfer tube.

thermal equilibrium with the oven temperature prior to entering the column (Katz and Scott, 1983a, 1989). High thermal conductivity tubing is very similar to low dispersion tubing since the conduction of heat across a tube is mathematically equivalent to solute transfer across a tube. The problem of heat transfer across an open tube results from the parabolic velocity profile of the fluid passing through the tube and the relatively low magnitudes of the thermal conductivity of liquids.

To improve both the heat transfer, it is necessary to disrupt the parabolic velocity profile in the tube and introduce radial mixing which will enhance diffusion and thermal conduction. This can be achieved by using a serpentine tube, the form of which is depicted in Fig. 35.

For optimum heat transfer, the ratio of the serpentine amplitude and the tube internal diameter should be less than 4. It functions by the introduction of radial flow as the fluid rapidly changes direction round the serpentine bends as it passes through the tube. The tube is clamped flat to the surface of the oven wall and the heat is transferred from the oven wall to the tube wall and then rapidly to the liquid passing through it. Without this, or some other type of efficient thermal exchange system, air ovens are not to be recommended for LC column temperature control. If the chromatograph can be operated at ambient temperature, provided the solvent mixtures are given time to attain thermal equilibrium with the environment and the apparatus is situated in a thermostatically controlled room, no column oven may be necessary and the LC column can be situated between the injection valve and the detector and exposed to ambient conditions.

H. The LC Column

LC columns have been constructed having widely different lengths and diameters. Analytical columns have ranged from 2 cm in length with a i.d. of 4.6 mm to 15 m in length with a diameter of 1 mm (Scott and Kucera, 1979a). In contrast, preparative columns up to 1 m in diameter have been constructed having lengths of 12–15 ft. The column length and particle diameter of the packing determines the efficiency that is realized, and the column diameter the solvent consumption, sample volume, and mass sensitivity. The type, size, and the kind of analysis determine the optimum dimensions of the column. A very wide range of columns are commercially available and it is unlikely that a special column will need to be packed for a specific application.

The majority of columns are constructed of stainless steel with appropriate fittings at either end. These fittings can also be of stainless steel but more recently they have been constructed from a plastic such as polyetheretherketone (PEEK) and can be hand tightened to provide the necessary leak-free system. The use of hand tightened fittings greatly simplifies columns changing. Columns used for separating labile materials of biological origin need to be made from biocompatible materials to prevent degradation and decomformation of the solutes. In fact, under such circumstances, all materials the solutes come in contact to during their passage through the chromatographic system need to be biocompatible. One of the first biocompatible materials used was titanium and thus the appropriate components were either constructed of titanium or, where possible, fitted with titanium liners. Titanium, however, is a very hard material and difficult to work and so contemporary biocompatible systems usually employ PEEK, or some other biocompatible material for construction, where necessary. Most LC separations can be achieved with a single column but difficult multicomponent mixtures sometimes require the use of a separate column. After separation on one column the eluate is diverted into another column carrying a different stationary phase. Thus the sample is subjected to two LC analyses to achieve the necessary resolution. This technique has been given the pretentious acronym LC/LC. To achieve this double separation a procedure called column switching is used.

1. Column Switching

Column switching can significantly increase the versatility of the chromatograph. The technique employs valves similar to those used for sample injection the procedures. An example of a six-port valve arranged for column switching is given in Fig. 36. The arrangement shown utilizes the external loop valve in a unique manner. Column (1) is connected between ports (5) and (6), and column (2) is connected between ports (2) and (3). Mobile phase from a sample valve, or more usually from another column, enters port (1) and the detector is connected to port (4). In the initial position, the rotor slots connect ports (1) and (6), (2) and (3), and (4) and (5). This results in mobile phase passing from port (1) to port (6), through column (1) to port (5), from port (5) to port (4), and out to the detector. Thus, the separation initially takes place in column (1). The ports connected to column (2) are themselves connected by the third slot and thus isolated. When the valve is rotated [the situation depicted on the right-hand side of Fig. (7)], port (1) is connected to port (2), port (3) connected to port (4), and port (5) connected to port (6). Thus the mobile phase, from either a sample valve or another column, enters port (1) passing to port (2) through column (2) to port (3), then to port (4), and then to the detector. The ports (5) and (6) are connected, isolating column (1).

This arrangement allows either one of two columns to be selected for an analysis or, part of the eluent from another column pass to column (1) for separation and the remainder passed to column (2). This system, although increasing the complexity of the column system renders the chromatographic process far more versatile. The number of LC applications that require such a complex chromatographic arrangement, however, are relatively small but, when necessary, column switching can provide a simple solution to difficult separation problems.

2. Recycling

A method for improving the resolution of a pair of closely eluting solutes (such as enantiomers), is to employ the technique of recycling. After the sample has been placed on the column, the eluent from the column is switched to the pump inlet and thus the mobile phase is continuously circulated round the column. This means that the column is used many times, and each time the sample passes through the column, the resolution is improved. Unfortunately, the resolution is not necessarily proportional to the number of cycles, as significant peak dispersion can occur each time it passes through the pump. Nevertheless, there is a substantial net gain in resolution on each cycle. This procedure can be very time consuming if long retention times are involved, but has the advantage of considerable solvent economy. The recycling procedure, in effect, artificially increases the column length and actually trades in *time* for *solvent economy*.

Recycling can be accomplished using two columns and a simple six-port valve. The valve connections are shown in Fig. 37. With the valve positioned as in the upper diagram, solvent from the sample valve and pump enters port (4) passes to port (5) and then to column A. After

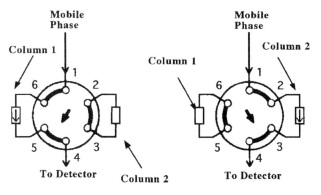

Courtesy of Valco Instruments Inc.

Figure 36 Valve arrangement for column switching.

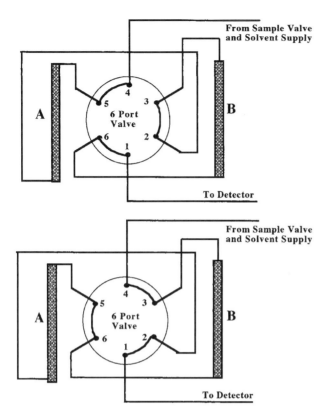

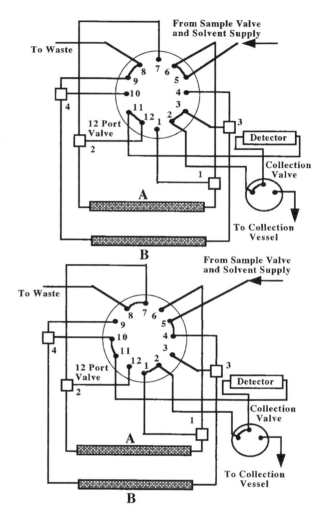

Figure 37 Valve arrangement for recycling.

Figure 38 Valve arrangement for monitored-recycling.

leaving column A the eluent passes to port (3) through to port (4) and then enters the top of column B. After leaving column B the solvent passes to port 6 through to port 1 and out to the detector. When the solutes of interest have left column A and are at the top of column B (usually ascertained by retention time) the valve is rotated to the positions shown in the lower diagram. Solvent now enters the top of column B and elutes the solutes through column B to port (6) through port and then into the top of column A. This procedure is continued for the required number of column passes and the allowed to elute through the last column to the detector where the separation is monitored.

By using a 12-port valve (or two ganged six-port valves) the same procedure can be carried out but the separation achieved after passage through each column can be continuously monitored. The valve connections are shown in Fig. 38. The chromatographic separation starts with the valve in the position shown in the upper diagram. The sample and mobile phases pass to port (5) then to port (6) and to a T piece (1). As port (4) is closed, the flow then passes to column A and the separation commences. When the solutes leave column A, they pass to another T piece (2) and, as port (7) is closed, pass to port (12), then to port (11), and then to the detector. After passing through the detector and the separation monitored, the mobile phase flows to a third T piece (3) and, as port (4) is closed, passes into column B. The mobile phase leaves column B to T piece (4) and, as port (10) is closed, passes to port (9), then to port (8), and then to waste. When the solutes of interest have entered column B (as monitored by the detector) the valve is rotated and now mobile phase enters port (5), passes to port (4), and then to the T piece (3). As port (3) is now closed, the mobile phase passes to column B and continues the separation continues to develop. As the solutes leave column B they pass to T piece (4) and, as port (9) is closed, pass to port (10), then to port (11), and then to the detector where the separation is again monitored. On leaving the detector the mobile phase passes to the collection valve and then to port (2), then port (1), and then to T piece (1). As port (6) is closed, the mobile phase and solutes are directed again into column A. While the solutes were being eluted through column B any remaining solutes of no interest in column A pass to T piece (2) and, as port (12) is closed, passes to port (7), then to (8), and then to

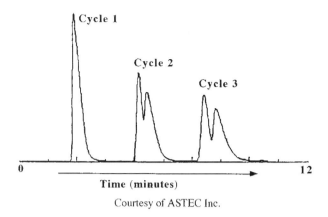

Figure 39 The separation of the warfarin enantiomers by recycling.

waste. When the solutes from column B have re-entered column A, the valve is returned to the initial position (upper diagram) and the whole process repeated. At any time, when the separation is deemed satisfactory, the collection valve can be rotated and the separate solutes collected directly after detection.

An example of the recycling process is demonstrated in the separation of the warfarin enantiomers as shown in Fig. 39. The chromatogram shows the effect of three cycles. In the first cycle there is little or no visible separation. In the second cycle, the two enantiomers are beginning to separate. In the third cycle, the separation is improved further, and is sufficient to allow the collection of significant quantities of the individual isomers at a high purity. It is also seen that the process is fairly rapid as three cycles are completed in less than 12 min.

The separation was carried out on a Cyclobond 12000 column, 25 cm long, 1 in. i.d., using a mobile phase consisting of methanol–acetic acid–triethylamine, 100:0.3:0.2 v/v/v, at a flow rate of 12 mL/min.

3. Column Packing and Stationary Phases

Today there are an immense number of different stationary phases available. The most popular being the bonded phases and, of those, the so-called reversed phases are the most widespread. The bonded phases are produced by reacting an appropriate chloralkylsilane with the hydroxyl groups on the silica surface. Hydrogen chloride is eliminated and the alkane chains are chemically bonded to the surface by a stable silicon–oxygen–silicon link. There are three types of bonded phase. There is the *"brush"* phase (Scott, 1993), which is produced by using monochlorosilanes that add a single alkane chain to each hydroxyl group on the silica. This forms a molecular grass-like surface of alkane chains—hence the term "brush" phase. There is the *oligomeric* phase (Akapo et al., 1989; Odlyha et al., 1991), which is formed by using a dichloroalkylsilane in an alternate reaction sequence of silanization and hydrolysis. This results in the formation of a series alkane groups attached to the same surface silicon atom. Finally, there is the *"bulk"* phase, which is obtained by employing the same alternate synthesis of silanization followed by hydrolysis but with the use of a trichloroalkylsilane as the reagent As two hydroxyl groups are generated on hydrolysis, cross-linking occurs and a polymeric alkane surface is produced. All three different synthesis are completed by capping any unreacted hydroxy groups with either trimethyl chlorosilane or hexamethyldisilazane.

In general, as a result of their reproducibility and potential for somewhat higher efficiencies, the brush phases are the most popular, although the bulk phases exhibit greater retention. The synthesis of the oligomeric phases is involved and for efficient production requires a fluidized-bed reactor (Khong and Simpson, 1986, 1987) and thus such phases are not commercially available at the time of writing this book. Reversed phases are largely employed with aqueous solvent mixtures as the stationary phase.

A contemporary alternative to the bonded phases that are gaining popularity are the macroporous polymers introduced in the early 1960s. The essential technical advance since that time lies in the macroporous nature of the resin packing, which consists of resin particles a few microns in diameter, which in turn comprise a fused mass of polymer microspheres a few angstroms in diameter. The structure is very similar to the silica gel particles, which consist of a fused mass of *primary* particles of polymeric silicic acid. The resin polymer microspheres play the same part as the silica gel primary particles, and confer on the polymer a relatively high surface area as well as high porosity. The more popular resin packings are based on the copolymerization of polystyrene and divinylbenzene. The extent of cross-linking determines its rigidity and the greater the cross-linking the harder the resin becomes until, at extremely high cross-linking, the resin becomes brittle. In order to produce an ion exchange resin, the surface of the polystyrene–divinyl benzene copolymer is finally reacted with appropriate reagents and covered with the required ionogenic interacting groups (e.g., by sulfonation). The high surface area was the key to its efficacy for LC as it provided increased solute retention and selectivity, together with a superior loading capacity, which in turn provided a larger quantitative dynamic range of analysis. The highly cross-linked polystyrene resins, called macroreticular resin, can be produced with almost any desired pore size, ranging from 20 to 5000 Å. They exhibit strong dispersive type interactions with solvents and solutes with some polarizability arising from the aromatic nuclei in the polymer. Consequently, the untreated resin is finding use as an alternative to the C8 and C18 reverse phase columns based on silica. Their use for the separation of peptide and proteins at both high and low pH is well established.

The type of stationary phase employed has little impact on LC instrument design. All stationary phases can be packed in columns of any practical dimensions and do not impose constraints on the material of construction or the type of solvents that can be used.

4. Column Conduits

Column–sample–valve and column–detector connections can be one of the greatest sources of extracolumn dispersion and can completely destroy the resolution of a closely eluted pair of solutes from a high efficiency column. The dispersion that takes place in an open tube (Done, 1976; Katz, 1984; Scott, 1992, p.153; Stephen et al., 1989), measured as the variance per unit length of the tune H, has been shown by Golay (1958) to be

$$H = \frac{2D_m}{u} + \frac{r^2}{24D_m}u$$

where r is the radius of the column, u is the linear velocity of the mobile phase, and D_m is the diffusivity of the solute in the mobile phase.

At the high velocities present in connecting tubes the first term becomes negligible and

$$H = \frac{r^2}{24D_m}u$$

Bearing in mind $Q = \pi r^2 u$, where Q is the flow rate, then,

$$H = \frac{Q}{24\pi D_m}$$

The number of theoretical plates in the tube will be

$$\frac{L}{H} = \frac{24 L \pi D_m}{Q} \quad (20)$$

Taking a value of 0.25 mL/min (0.0.00417 mL/s) for Q, a value for D_m of 3×10^{-5} cm^2/s (Katz and Scott, 1983b) and 0.010 in (0.0127 cm) for the i.d. of the tube, the efficiency of a tube 10 cm long calculated from Eq. (20) will be 5.42.

Now the volume v of the tube is given by

$$v = \pi r^2 L = 0.00127 \, \text{mL}$$

Thus, σ will be given by

$$\frac{v}{\sqrt{n}} = \frac{0.00127}{\sqrt{5.42}} = 0.000546 \, \text{mL} \quad (21)$$

Now, surmise that this tube is used to connect the sample valve to a column or a column to the detector and consider its effect on the performance of a microbore column where extracolumn dispersion can be most detrimental (Kucera, 1984; Scott, 1984a, Scott and Kucera, 1979b).

Consider a small bore column is 15 cm long, 1 mm in diameter packed with particles 5 μm in diameter which will give about 15,000 theoretical plates at the optimum velocity (ca. 0.25 mL/min.). This efficiency should be expected from a commercially available column. Now the column volume (V_c) will be,

$$V_c = \pi R^2 L = 3.14 \times 0.05^2 \times 15 = 0.117 \, \text{mL}$$

where R is the column radius, and L is the column length.

The dead volume (V_0) will be approximately 60% of the column volume and will be given by

$$V_0 = 0.6 V_c = 0.6 \times 0.117 = 0.0707 \, \text{mL}$$

From the plate theory already discussed, the standard deviation (σ_0) of the *dead volume* peak will be

$$\sigma_0 = \frac{V_0}{\sqrt{n}} = \frac{0.0707}{\sqrt{15000}} = 0.000577 \, \text{mL} = 0.577 \, \mu\text{L}$$

Consequently, it is seen that the dispersion in the tube is nearly as great as that which took place in the column. Considering two solutes, the first eluted at a k' of 1.0 and the other 4σ after.

Then, $0.577(1 + k') = 1.154 \, \mu\text{L}$ will be the standard deviation of the dispersion in the column. The elution curves can be constructed [using Eq. (20) and the elution equation] for connecting tubes of different lengths. The curves obtained are shown in Fig. 40.

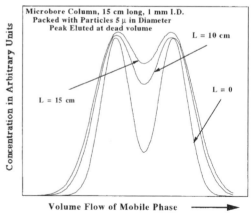

Figure 40 Elution curves for a solute having passed through the column and subsequently through different lengths of connecting tube, 0.010 in i.d.

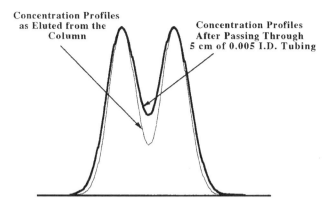

Figure 41 Elution curves for a pair of solutes as eluted from the column and after having passed through a 5 cm of connecting tube, 0.005 in i.d.

It is seen that the dispersion taking place in the connecting tube can have a very serious deleterious effect on the apparent performance of the column. Though 10 cm is not a long length of tubing, it is interesting to repeat the calculations for a tube 5 cm and 0.005 in. i.d. In this case the effect on the resolution of two peaks eluted 4σ apart is calculated and the results are shown depicted in Fig. 41.

It is seen that, even when the diameter of the connecting tube is reduced to 0.005 in. i.d., and the length to 5 cm, the effect on the resolution of the two solutes is still very serious. It is, therefore, extremely important to keep connecting tube lengths to an absolute minimum. In practice, it is difficult to reduce the diameter of connecting tubing below 0.005 in. owing to the likelihood of conduit blockage. The pertinent lengths involve the connection between the sample valve and the column and, even more important, between the column exit and the *sensing cell* or *sensing volume* of the detector. It should be stressed that the length of connecting tube of concern is not that between the column and the detector case or connecting fitting but between the column exit and the *actual* detecting cell. In a badly designed detector there may be several centimeters of connecting tube in the internal conduits of the device itself that will contribute to dispersion (Scott and Kucera, 1971, 1979c).

If significant lengths of connecting tube are necessary (as in the case of recycling systems) then the effect of tube dispersion can be minimized by making the peak volume larger by using wide bore columns. This procedure, will of course require the use of large samples.

I. Detectors

Second to the column that accomplishes the separation, the detector is the most important part of the LC system. In conjunction with a suitable recorder or printer, the detector displays the separation that has been achieved and also allows the components of the mixture to be quantitatively determined. Besides choosing the appropriate detector that will sense the solutes to be separated, however, the analyst must also be concerned with the operating specifications of the detector. The important specifications (Scott, 1975; Atwood and Golay, 1981; Scott and Simpson, 1982) that are pertinent to the satisfactory performance of an LC detector are

detector linearity,
linear dynamic range,
detector noise level,
detector sensitivity or minimum detectable concentration,
pressure sensitivity,
flow sensitivity, and
temperature sensitivity.

1. Detector Linearity and Response Index (α)

True detector linearity is, in fact, a theoretical concept and LC detectors can only *tend* to exhibit this ideal response. The linearity of the detector can be defined and measured in a number of different ways but the following is probably the most useful as it provides a means of correction for any nonlinearity that may occur (Fowliss and Scott, 1963).

$$v = Rc^\alpha$$

where v is the output from the detector, R is a constant, and α is the *response index.*

It is seen that, for a truly linear detector, $\alpha = 1$, and an experimentally determined value of α will indicate the proximity of the response to strict linearity. It is also clear that α could be used to correct for any nonlinearity that might occur and thus, improve the accuracy of an analysis.

Taking the logarithm of the above relationship,

$$\log(v) = \log(R) + \alpha \log(c)$$

It is seen that the value of α can be obtained from the slope of the curve relating the log (detector output) to the log (solute concentration). The value of α can be obtained in practice by measuring the area or height of the peaks for a series of calibration standards. If true linearity is to be assumed, the response index should lie between 0.98 and 1.02. If the response index is not within this range, then it will be necessary to correct for the nonlinearity employing the measured response index.

It is interesting to compare the actual results obtained with the true composition of a mixture for a series analyses that have been carried out using detectors of differing response indices. The actual results that would be obtained from the analysis of a binary mixture containing 10% of one component and 90% of the other, employing detectors with values for α ranging from 0.94 to 1.05, is shown in Table 3.

Table 3 Analysis of a Binary Mixture Employing Detectors with Different Response Indexes

Solute	$\alpha = 0.94$	$\alpha = 0.97$	$\alpha = 1.00$	$\alpha = 1.03$	$\alpha = 1.05$
1	11.25%	10.60%	10.00%	9.42%	9.05%
2	88.75%	89.40%	90.00%	90.58%	90.95%

It is seen that errors in the smaller component can be as great as 12.5% (1.25% absolute) when the response index is 0.94, and when the response index is 1.05 the error is 9.5% (0.95% absolute). As already stated, to obtain accurate results without employing a correction factor, the response index should lie between 0.98 and 1.02. Most LC detectors can be designed to meet this linearity criteria.

2. *Linear Dynamic Range*

The *linear dynamic range* of a detector is that range of solute concentration over which the numerical value of the *response index* falls within a defined range. For example, the linear dynamic range of a detector might be specified as

$$D_L = 3 \times 10^{-8} - 2 \times 10^{-5} \, \text{g/mL} \quad (0.98 < \alpha < 1.02)$$

The *dynamic range* of a detector is that range of solute concentration over which the detector continues to respond. It is *not the same* as the linear dynamic range.

3. *Detector Noise Level*

There are three different types of detector noise (Scott, 1977; Brown et al., 1983), short term noise, long term noise, and drift. These sources of noise combine together to give the composite noise of the detector. The different types of noise are depicted in Fig. 41.

Short term noise consists of base line perturbations that have a frequency significantly higher than the eluted peak. Short term detector noise is not a problem in LC as it can be removed by an appropriate noise filter. Its source is usually electronic, originating from either the detector sensor system or the amplifier.

Long term noise consists of base line perturbations that have a frequency similar to that of the eluted peak. Long term noise is the most serious as it is indiscernible from the small peaks and cannot be removed by electronic filtering without affecting the normal peak profiles. It is seen in Fig. 42 that the peak profile can easily be discerned above the high frequency noise but is lost in the long term noise. The source of long term noise is usually due to changes in either temperature, pressure or flow rate in the sensing cell. Long term noise is largely controlled by detector cell design and ultimately limits the detector *sensitivity* or the *minimum detectable concentration*.

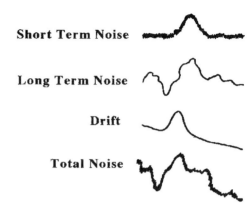

Figure 42 Different types of detector noise.

Drift consists of base line perturbations that have a frequency significantly larger than that of the eluted peak. Drift is almost always due to either changes in ambient temperature, changes in solvent composition, or changes in flow rate. All three sources of noise combine to form the type of trace shown at the bottom of Fig. 42. In general, the sensitivity of the detector should never be set above the level where the combined noise exceeds 2% of the FSD (full scale deflection) of the recorder (if one is used) or appears as more than 2% FSD of the computer simulation of the chromatogram.

4. *Measurement of Detector Noise*

The detector noise is defined as the maximum amplitude of the combined short and long term noise, measured in millivolts, over a period of 15 min.

The value for the detector noise can be obtained by constructing parallel lines embracing the maximum excursions of the recorder trace over the defined time period as shown in Fig. 43 (Scott, 1977; Brown et al., 1983; Scott, 1984b, p. 19; Sternberg, 1966). The distance between the parallel lines measured in millivolts is taken as the noise level.

5. *Detector Sensitivity or the Minimum Detectable Concentration*

Detector sensitivity or the *minimum detectable concentration* has been defined as the minimum concentration of an eluted solute that can be differentiated unambiguously from the noise (Scott, 1977, p. 22). The ratio of the

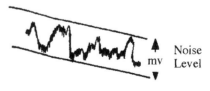

Figure 43 Method for measuring detector noise.

signal to the noise for a peak that is decisively identifiable has been arbitrarily chosen to be two. However, the concentration that will provide a signal equivalent to twice the noise level will usually depend on the physical properties of the solute used for measurement. Consequently, the detector sensitivity, or minimum detectable concentration, must be quoted in conjunction with the solute that is used for measurement.

6. Pressure Sensitivity

The pressure sensitivity of a detector is defined as the change in detector output for unit change in sensor-cell pressure. Pressure sensitivity and flow sensitivity are to some extent interdependent, subject to the manner in which the detector functions. The refractive index detector is the most sensitive to pressure changes.

7. Flow Sensitivity

The flow sensitivity of a detector is taken as the change in detector output for unit change in flow rate through the sensor cell. Again, the refractive index detector is also the most sensitive to flow rate changes.

8. Temperature Sensitivity

Both the sensing device of the LC detector and the associated electronics can be temperature sensitive and cause the detector output to drift as the ambient temperature changes. Consequently, the detecting system should be designed to reduce this drift to a minimum. In practice the drift should be less than 1% of FSD at the maximum sensitivity for 1°C change in ambient temperature.

There are five commonly used LC detectors; the UV detector (in one of its various forms), the fluorescence detector, the electrical conductivity detector, the refractive index detector, and the light scattering detector, with the most popular detector being the UV detector followed by the electrical conductivity detector.

9. The Effect of Detector Sensor Volume on Column Performance

The detector responds to the total amount of solute in the sensor cell of the detector and thus effects the presentation of the separation relative to the true curves for the peaks as eluted from the column. In the extreme case, the detector cell could be large enough to hold two closely eluted peaks and thus give a response that would appear as though only a single solute had been eluted. This extreme condition rarely happens but serious peak distortion and loss of resolution can result, particularly if columns of small diameter are being used. A diagram depicting the integral nature of a detector response is shown in Fig. 44.

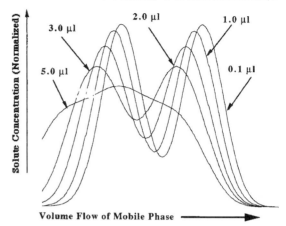

Figure 44 Elution curves of a pair of solutes, 4 s apart, after passing through different lengths of tube.

The column was 15 cm long, 1 mm i.d. packed with particles 5 μm in diameter. The fist peak is eluted close to the dead volume. The connecting tube is 0.010 in. i.d. and 10 cm long. The flow rate through tune and column is 0.25 mL/min.

The example given in Fig. 44, although not the worst case *scenario*, is a condition where the sensing volume of the detector can have a very serious effect on the peak profile and, consequently, the resolution. The column is a small bore column and thus the eluted peaks have a relatively small peak volume which is commensurate with that of the sensing cell. It is seen that even a sensor volume of 1 μL has a significant effect on the peak width and it is clear that if the maximum resolution is to be obtained from the column, then the sensor cell volume should be no greater than 2 μL. It should also be noted that the results from the use of a sensor cell having a volume of 5 μL are virtually useless and that many commercially available detectors do, indeed, have sensor volumes as great, if not greater, than this. It follows that if small bore columns are to be employed such sensor volumes must be avoided.

10. UV Detectors

UV detectors are predominantly "solute property detectors" but if the solvent used for the mobile phase contains a UV chromaphore then there will be a background signal from the mobile phase. As a consequence, the detector will also respond to both the bulk property of the mobile phase and the UV adsorption of the solutes. In practice, the mobile phase is usually chosen to be completely transparent to the wavelength of UV light selected for detection.

This is essential if gradient elution is to be employed, as any adsorption due to one of the solvents will result in a steady and continuous baseline drift throughout the program. There are basically three types of UV detector, the *fixed wavelength detector*, the *variable wavelength detector*, and the *diode array detector*. Each of these detectors, although measuring the amount of UV light adsorbed by the column eluent are quite different in design and therefore will be discussed separately.

The Fixed Wavelength Detector

The fixed wavelength detector operates with a lamp that emits light at a single wavelength (Fig. 45). In fact, although the majority of the light may be emitted at one wavelength, with most single wavelength UV lamps other wavelengths are also always present, although usually at a much lower intensity. The important aspect of the "single wavelength lamp" is that the light emitted at the specific wavelength is at a very high intensity compared with that emitted at the same wavelength by broad spectrum emission lamps. This provides the fixed wavelength UV detector with the capacity for a much higher sensitivity than the other varieties of UV detectors. A diagram of the optical system of a fixed wavelength UV detector is shown in Fig. 45. The light source is a discharge lamp, such as a low pressure mercury vapor lamp, and the light is focused through two adsorption cells onto two photoelectric sensors. The output from the sensors is electronically modified and passed to a recorder or a computer data acquisition system. One adsorption cell carries the column eluent and the other carries the pure mobile phase. The volume of the cell is kept as small as possible to reduce band dispersion and peak distortion. As previously discussed, the sensing volume of a modern LC detector cell should not be greater than 5 μL and preferably less than 2 μL.

The intensity of the UV light adsorbed given as a function of the solute concentration is given by,

$$I_T = I_0 e^{-klc}$$

or

$$\ln I_T = \ln I_0 - klc$$

where I_0 is the intensity of light entering the cell, I_T is the intensity of light transmitted through the cell, l is the path length of the cell, c is the concentration of solute in the cell, and k is the molar adsorption coefficient of the solute at the specific wavelength of the UV light.

Differentiating,

$$\frac{\partial (I_T/I_0)}{\partial c} = -kl \qquad (22)$$

It is seen that the sensitivity of the detector, as measured by the transmitted light, will be directly proportional to the extinction coefficient k and the path length l of the cell. Thus to increase the sensitivity, l must be increased. However, there is a limit to which l can be enlarged as the total volume of the cell must be kept small to ensure there is no loss of resolution by extra column dispersion. It follows, that to increase l, the radius of the cell must also be concurrently reduced to maintain a small cell volume. Reduction in cell radius also reduces the amount of light falling on the photo cell and thus reduces the signal to noise ratio. The two effects oppose one another and consequently the process of increasing detector sensitivity by lengthening the cell is limited. Most modern UV cells have path lengths lying between 1 and 10 mm and internal diameters of 1 mm or less.

From Eq. (22),

$$\log \frac{I_T}{I_0} = k'lc = A$$

where A is the *absorbance*.

ΔA is commonly employed to define the detector sensitivity, where the value of ΔA is the change in absorbance that produces a signal to noise ratio of 2.

Thus,

$$\Delta A = k'l\Delta c$$

where Δc is the detector concentration sensitivity or the "minimum detectable concentration", which is the parameter of importance to chromatographers.

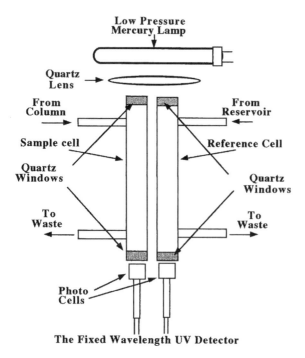

Figure 45 The fixed wavelength UV detector.

High-Performance Liquid Chromatography

Then,

$$\Delta c = \frac{\Delta A}{k'l}$$

It is clear that two detectors having the same sensitivity defined as the minimum detectable change in absorbance will not necessarily have the same sensitivity with respect to solute concentration. Only if the path length of the two cells are identical will they also exhibit the same concentration sensitivity.

This can cause some confusion to the chromatographer who might expect that instruments that are defined as having the same spectroscopic sensitivity to have equivalent chromatographic sensitivity (Fig. 46). To compare the sensitivity of two detectors given in units of absorbance, the path lengths of the cells in each instrument must be taken into account. There are three different UV lamps that are used in fixed wavelength detectors and their spectroscopic specifications are given in Fig. 45. The low pressure mercury lamp (maximum emission at 253.7 nm) is that most commonly used with fixed wavelength LC detectors. The zinc lamp has major emission lines at 213.9 nm which would be ideal for the detection of proteins and polypeptides, but at this time is rarely used. The cadmium lamp has a major emission wavelength at 228.8 which would also be very useful for the detection of protein-like substances. Typical specifications for the fixed wavelength detector employing a low pressure mercury lamp as the light source are given in Table 4.

The Variable Wavelength Detector

The variable wavelength detector employs a lamp that emits light over a wide range of wavelengths and by using a monochromator, light of a particular wavelength can be selected for detection purposes. Alternatively, the flow rate can be arrested, the solute trapped in the detection cell, and by scanning through the different wavelengths an adsorption spectrum of the solute can be obtained. A diagram of a variable wavelength detector is shown in Fig. 47.

Table 4 Typical Specifications for a Fixed Wavelength Detector Employing a Low Pressure Mercury Lamp as the Light Source

Sensitivity (toluene)	5×10^{-8} g/mL
Linear dynamic range	5×10^{-8} to 5×10^{-4} g/mL
Response index	0.98–1.02

The variable wavelength UV detector has several advantages.

1. A UV absorption spectrum can be obtained for any or all the eluted solutes. Such spectra can sometimes help to confirm solute identification. Unfortunately, UV spectra are of limited assistance in solute identification because the majority of substances (excluding those containing aromatic nuclei) exhibit very similar absorption curves.

2. The spectra of a solute can be taken at various points across a peak as it is slowly moved through the cell and if the spectra are consistently the same, the peak is pure. It will be seen later that this can be achieved more elegantly using a diode array detector.

3. The variable wavelength detector can be simply used as a wavelength selector to increases in sensitivity by selecting the optimum wavelengths for detection. This is probably the most popular use of a variable wavelength detector. Typical specifications for a variable wavelength detector are given in Table. 5.

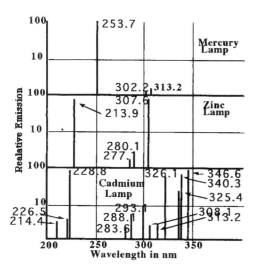

Figure 46 The spectroscopic specifications of three UV lamps.

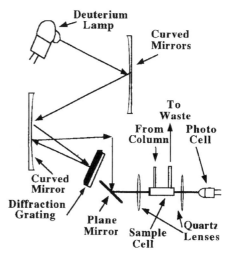

Figure 47 The variable wavelength detector.

Table 5 Typical Specifications of a Variable Wavelength UV Detector

Sensitivity (toluene)	1×10^{-7} g/mL
Linear dynamic range	1×10^{-7} to 5×10^{-4} g/mL
Response index	0.98–1.02

The Diode Array Detector

The diode array detector (Milano et al., 1976; Talmi, 1975a, 1975b; Yates and Kuwana, 1976) also utilizes a deuterium or xenon lamp that emits light over the UV spectrum range. Light from the lamp is focused by means of an achromatic lens through the sample cell and onto a holographic grating. The dispersed light from the grating is arranged to fall on a linear diode array. A diagram of a UV diode array detector is shown in Fig. 48.

The resolution of the detector ($\Delta\lambda$) will depend on the number of diodes (n) in the array, and on the range of wavelengths covered ($\lambda_2 - \lambda_1$).

Thus,

$$\Delta\lambda = \frac{\lambda_2 - \lambda_1}{n}$$

The diode array detector takes a UV spectrum of the eluent continuously throughout the complete development of the chromatogram which can be used to a great advantage. It follows that a chromatogram can be reconstructed by monitoring at a specific wavelength [strictly ($\Delta\lambda$) a narrow group of wavelengths] and thus depict only those substances that adsorb UV light at the chosen wavelength. Consequently, the chromatogram can be arranged to display only those substances that have unique absorbance characteristics.

An interesting example of the use of the diode array detector to confirm the integrity of an eluted peak is afforded by the separation of a mixture of aromatic hydrocarbons. The separation is shown in Fig. 49. This separation was carried out on a column 3 cm long, 4.6 mm in

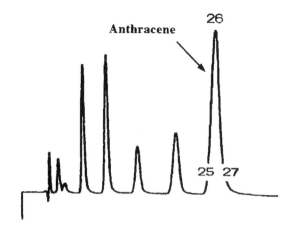

Courtesy of the Perkin Elmer Corporation

Figure 49 The separation of a series of aromatic hydrocarbons.

diameter and packed with a C18 reversed phase on particles 3 μm in diameter. It is seen that the separation appears to be good and all the peaks represent individual solutes. However, by plotting the adsorption ratio, 250 nm/255 nm, for the anthracene peak it became clear that the peak tail contained an impurity.

The absorption ratio peaks are shown in Fig. 50. The ratio peaks in Fig. 48 clearly show the presence of an impurity by the sloping top of the anthracene peak and further work identified the impurity as *t*-butyl benzene at a level of about 5%. The characteristics of both the variable wavelength and the diode array detector are similar and consequently have comparable specifications. The more important properties of the diode array detector are listed in Table 6.

11. The Fluorescence Detector

The fluorescence detector is probably the most sensitive LC detector, but has less versatility than the UV detectors.

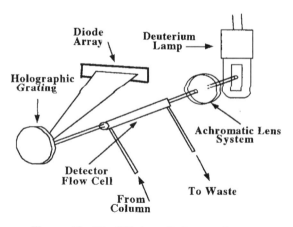

Figure 48 The UV photo diode array detector.

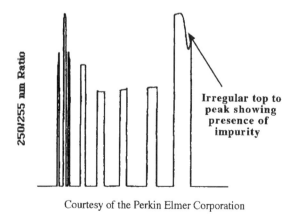

Courtesy of the Perkin Elmer Corporation

Figure 50 Curves relating the adsorption ratio, 250 nm/255 nm, to time.

High-Performance Liquid Chromatography

Table 6 Typical Specifications of a Diode Array Detector

Sensitivity	1.5×10^{-7} g/mL
Linear dynamic range	1.5×10^{-7} to 5×10^{-4} g/mL
Response index	0.97–1.03

Table 7 Typical Specifications for a Fluorescence Detector

Sensitivity (anthracene)	1×10^{-9} g/mL
Linear dynamic range	1×10^{-9} to 5×10^{-6} g/mL
Response index	0.96–1.04

By definition, the fluorescence detector will only detect those materials that will fluoresce or, by appropriate derivatization can be made to fluoresce. Nevertheless, owing to its sensitivity it is one of the most important detectors for use in trace analysis both in environmental and forensic analysis. A diagram of a simple form of fluorescence detector is shown in Fig. 51.

The UV lamp that provides the excitation radiation is often a low pressure mercury lamp as it is comparatively inexpensive and provides relatively high intensity UV light at 253.7 nm. Many substances that fluoresce will, to a lesser or greater extent, be excited by light at this wavelength. The light is focused by a quartz lens through the cell and another lens situated normal to the incident light focuses the fluorescent light onto a photo cell. Typical specifications for a fluorescence detector are as given in Table 7.

A far more sophisticated fluorescence detector is shown in Fig. 52 (Milano et al., 1976). The excitation source that emits UV light over a wide range of wavelengths (usually a deuterium or xenon lamp) is situated at the focal point of an ellipsoidal mirror shown at the top left hand corner of the diagram. The parallel beam of falls on a toroidal mirror that focuses it onto a grating on the left-hand side of the diagram. This grating allows the frequency of the excitation light to be selected or the whole spectrum scanned providing a complete range of excitation wavelengths. The selected wavelength then passes to a spherical mirror and then to a ellipsoidal mirror at the base of the diagram which focuses it on to the sample. Between the spherical mirror and the ellipsoidal mirror, in the center of the diagram, is a beam splitter that reflects a portion of the incident light on to another toroidal mirror.

This toroidal mirror focuses the portion of incident light onto the reference photo cell providing an output that is proportional to the strength of the incident light. Fluorescent light from the cell is focused by an ellipsoidal mirror onto a spherical mirror at the top right-hand side of the diagram. This mirror focuses the light onto a grating situated at about center right of the figure. This grating can select a specific wavelength of the fluorescent light to monitor or scan the fluorescent light and provide a fluorescent spectrum. Light from the grating passes to a photoelectric cell which monitors its intensity. The instrument is extremely complex, however, from the point of view of measuring fluorescence it is extremely versatile.

An example of its use is given in Fig. 53 which depicts the separation of a group of priority pollutants using programed fluorescence detection. The separation was carried out on a column 25 cm long, 4.6 mm in diameter, and packed with a C18 reversed phase. The mobile phase was programed from a 93% acetonitrile, 7% water, to 99% acetonitrile, 1% water over a period of 30 min. The gradient was linear and the flow rate was 1.3 mL/min. All the solutes are separated and the compounds, numbering from the left are given in the table. The separation illustrates the clever use of wavelength programing to obtain the maximum sensitivity. The program used for this particular mixture was rather

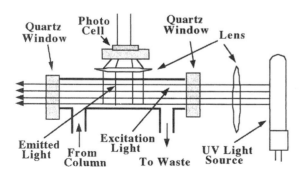

Figure 51 The simple fluorescence detector.

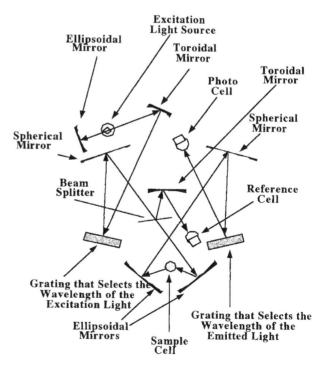

Figure 52 The fluorescence spectrometer detector.

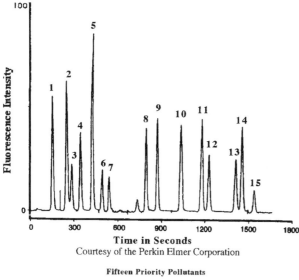

Figure 53 Separation of a series of priority pollutants with programed fluorescence detection.

Fifteen Priority Pollutants

1 Naphthalene
2
3 Fluorine
4 Phenanthrene
5 Anthracene
6 Fluoranthene
7 Pyrene
8 Benz(a)anthracene
9 Chorines
10 Benzo(b)fluoranthene
11 Benz(k)fluoranthene
12 Benzo(a)pyrene
13 Dibenz (a,h)anthracene
14 Benzo(ghi)perylene
15 Indeno(123-cd)pyrene

complex. During the development of the separation, both the wavelength of the excitation light and the wavelength of the emission light that was monitored was changed. This ensured that each solute, as it was eluted, was excited at the most effective wavelength and then, consequently, monitored at the strongest fluorescence wavelength. The system can also provide a fluorescence spectra, should it be required, by arresting the flow of mobile phase when the solute resides in the detecting cell and scanning the fluorescent light as with the variable wavelength detector.

12. The Electrical Conductivity Detector

The electrical conductivity detector is used extensively in ion exchange chromatography. It consists of two electrodes situated in the column eluent, the resistance (or more strictly the impedance) of which is measured by a suitable electronic circuit. A diagram of a simple conductivity sensor is shown in Fig. 54.

It is seen that the basic sensor is, indeed, very simple and can be designed to have an effective sensing volume of a few nanoliters. For these reasons it is often used as a detector in capillary electrophoresis. The out of balance signal that occurs when an ionic solute is present between the electrodes will be proportional to a change in electrical resistance between the electrodes and this will not be

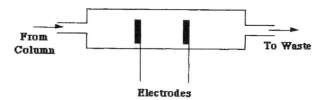

Figure 54 The electrical conductivity detector.

linearly related to the ion concentration. As a consequence if the solutes are eluted over a wide concentration range an appropriate amplifier must be used to provide an output directly proportional to the solute concentration.

The conductivity detector is a bulk property detector and, as such, responds to any electrolytes present in the mobile phase (e.g., buffers etc.) as well as the solutes. It follows that the mobile phase must be arranged to be either nonconducting, or the buffer electrolytes must be removed prior to the detector. This technique of buffer ion removal is called *ion suppression*. The first type of ion suppression involved the use of a second ion exchange column subsequent to the analytical column to remove the appropriate ions. This approach works well provided the ion suppresser (ion remover) does not cause excessive peak dispersion. Another alternative, is to employ an organic buffer, such as methane sulfonic acid and a short length of reversed phase column subsequent to the analytical column. After the separation has been achieved and the individual ions are eluted from the column, the methyl sulfonate that is eluted with them is absorbed almost irreversibly onto the reversed phase leaving the ions of interest to enter the detector in the absence of any buffer ions. It is obvious that this type of suppresser column will have a limited lifetime but can be regenerated by passing acetonitrile through the suppresser volume followed by water. An example of the use of this type of suppression system is given in Fig. 55.

A proprietary ion exchange column, IonPacCS12, was used and the mobile phase consisted of a 20 nM methanesulphonic acid solution in water. A flow rate of 1 mL/min was employed and the sample volume was 25 µL. Although this technique of ion suppression is frequently used in ion exchange chromatography, depending on the nature of the separation, a specific type of ion suppression column or exchange membrane will be required that will be appropriate for the phase system that is chosen. A wide variety of different types of ion suppression columns are available for this purpose. It should be pointed out that any suppresser system introduced between the column and the detector that has a finite volume will cause band dispersion. Consequently, the connecting tubes and suppression column must be very carefully designed to eliminate or reduce this dispersion to an absolute minimum.

High-Performance Liquid Chromatography

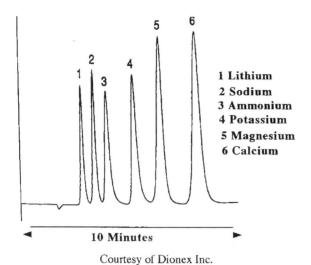

Figure 55 Determination of alkali and alkaline earth cations.

13. The Refractive Index Detector

The refractive index detector [first introduced by Tiselius and Claesson (1942)] is probably the least sensitive of all commercially available and generally useful detectors. It has the disadvantages of being sensitive to changes in ambient temperature, pressure, and flow rate and, as it is a bulk property detector cannot be used for gradient elution. Despite these disadvantages, the refractive index detector has found a niche for monitoring those compounds that are nonionic, do not adsorb in the UV, and do not fluoresce. There are a number of different optical systems used, but the simplest and most common is the differential refractive index detector shown in Fig. 56.

The differential refractometer responds to the deflection of a light beam that is caused by the differing refractive index between the contents of the sample cell and that of the reference cell. A beam of light from an incandescent lamp (usually a tungsten filament lamp) passes through an optical mask that confines the beam to the region of the cell. The lens collimates the light beam, which then passes through both the sample and reference cells to a plane mirror. The mirror reflects the beam back through the sample and reference cells and is focused by a another lens onto a photo cell. The location of the beam, rather than its intensity, is determined by the angular deflection of the beam caused by the refractive index difference between the contents of the two cells. As the beam changes its position of focus on the photo-electric cell, the output changes and the resulting difference signal is electronically modified to provide a signal proportional to the concentration of solute in the sample cell.

In general, the refractive index detector tends to be the "choice of last resort" and is used where, for one reason or another, all other detectors are inappropriate or impractical. Nevertheless, the detector has certain areas of application where it is specially useful, one being for the separation and analysis of polymers. It has been shown that if a polymer contains more than six monomer units, the refractive index (Table 8) becomes directly proportional to the concentration of the polymer and is nearly independent of the molecular weight. Consequently, as there is no need for individual solute response factors, a polymer mixture can be analyzed by peak area normalization, a method that has been previously discussed.

A typical example of the use of the refractive index detector is shown in Fig. 57. Carbohydrates have not UV chromaphores, neither do they fluoresce nor contain ions, and thus unless derivatives are made the only suitable detector that will sense them is the refractive index detector. The column used was Supercosil LC-NH$_2$, an aminopropyl bonded phase that exhibits fairly strong polar interactions in addition to the dispersive interactions from the propyl chains. The support used is a spherical silica having a mean pore size of 100 Å. The column was 25 cm long, 4.6 mm i.d. and the particle size of the packing was 5 μm. The mobile phase consisted of a mixture of 75% v/v acetonitrile and 25% v/v water and the flow rate was 1 mL/min. The sample consisted of 20 mL of a solution containing 10 mg/mL of each analyte in deionized water. It is seen that the carbohydrates are well resolved in little over 16 min. The refractive index

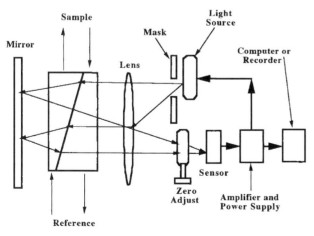

Figure 56 The differential refractometer.

Table 8 Typical Specifications for a Refractive Index Detector

Sensitivity (benzene)	1×10^{-6} g/mL
Linear dynamic range	1×10^{-6} to 1×10^{-4} g/mL
Response index	0.97–1.03

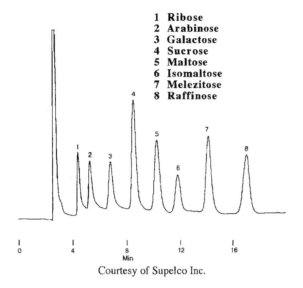

Figure 57 The separation of some carbohydrates.

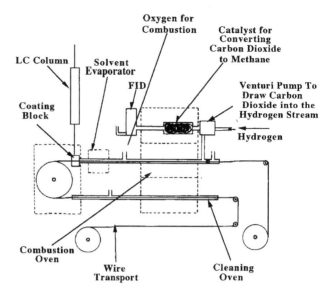

Figure 58 The transport detector with methane converter.

detector is often used in the separation of carbohydrates, aliphatic alcohols, the hydrolysis products of starch, and those natural products that do not have chemical groups that permit other methods of detection. The refractive index detector is, nevertheless, a catholic detector and would be used far more extensively if it could tolerate gradient elution.

14. The Transport Detector

The transport detector was developed in the early 1960s (James et al., 1964; Scott and Lawrence, 1970) and is not commercially available at the time of writing this chapter. However, this device has a number of attributes that seem sufficiently important to have invoked a number of companies to attempt to develop a more compact and more sensitive model. The major attributes of the device are that firstly it is a *universal detector*, in that it detects all involatile materials containing carbon and it secondly, allows *any solvent to be used* as the mobile phase, provided it is relatively volatile. In its original form, eluent from the column passed over a moving wire on which a thin coating of the solvent was deposited. The solvent was then evaporated leaving the solute deposited on the wire. Two forms of detection were originally developed, the first was to pyrolize the solute on the wire and pass the pyrolysis products to an FID or argon detector. The second was to burn the solute to carbon dioxide, and then, by aspirating the oxidation products into a stream of hydrogen and passing it over a heated nickel catalyst, the carbon dioxide is converted to methane and the methane detected with an FID. A diagram of the transport detector with the methane conversion system is shown in Fig. 58.

In its original form the detector was bulky, a little difficult to operate, and had a sensitivity (minimum detectable concentration) little better than the refractive index (i.e., $\sim 1 \times 10^{-6}$ g/mL). It was used (Cropper et al., 1967) and in fact, it still is, for detecting substances that do not have a UV chromaphore and require to be separated under gradient elution conditions (e.g., nonionic detergents, cosmetic products, and carbohydrates). An example of the use of a transport detector for monitoring the separation of a mixture of liver tissue lipids is shown in Fig. 59.

In this separation, a specific series of solvents were employed for gradient elution that clearly illustrate the versatility of the detector for use with a wide range of

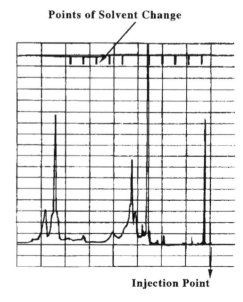

Figure 59 The separation of a sample of liver tissue lipids.

solvent types. The solvents used are as follows: *n*-heptane, carbon tetrachloride, chloroform, ethylene dichloride, 2-nitropropane, propyl acetate, methyl acetate, acetone, ethanol, methanol, and water. It is seen that the detector provides complete freedom for choosing the most appropriate type(s) of solvent for use as the mobile phase.

15. The Electrochemical Detector

The electrochemical detector is one of the most sensitive. Its enthusiasts claim it is even more sensitive than the fluorescence detector. The detector responds to substances that are either oxidizable or reducible, and the electrical output results from an electron flow caused by the reaction that takes place at the surface of the electrodes. If the reaction proceeds to completion, exhausting all the reactant, then the current becomes zero and the total charge that passes will be proportional to the total mass of material that has been reacted. For obvious reasons, this process is called coulometric detection. If on the other hand, the electrode is flowing passed the electrodes the reacting solute will be continuously replaced as the peak passes through the detector. While there is solute present between the electrodes a current will be maintained, albeit, with varying magnitude. This is the most common procedure employed and is called ampiometric detection. Three electrodes are employed: the working electrode (where the oxidation or reduction takes place), the auxiliary electrode, and the reference electrode (which compensates for any changes in the background conductivity of the mobile phase). The reaction is extremely rapid resulting in the layer close to the electrode being completely depleted of reacting material. Thus the output will depend on the rate at which solute reaches the electrode surface and, as this is by diffusion, which is concentration driven, the output is proportional to the concentration of solute in the mobile phase. There are a number of electrode configurations that have been found satisfactory (Poppe, 1978) and a couple of examples of these are shown in Fig. 60. A simplified form of the circuit that is used is shown in Fig. 61.

The auxiliary electrode is held at a fixed potential by the first amplifier, the voltage being selected by the potentiometer P that is connected to a regulated power supply. The current flowing through the working electrode is

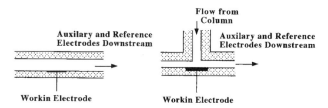

Figure 60 Electrode configurations for the electrochemical detector.

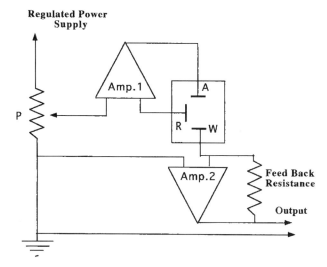

Figure 61 Basic circuit used with the electrochemical detector.

processed by the second amplifier and the output fed to the recorder or data acquisition system.

The electrochemical detector is extremely sensitive but suffers from a number of drawbacks. Firstly, the mobile phase must be extremely pure and in particular free of oxygen and metal ions. A more serious problem arises, however, from the adsorption of the oxidation or reduction products on the surface of the working electrode. The consequent electrode contamination requires that the electrode system must be frequently calibrated to ensure accurate quantitative analysis. Ultimately, the detector must be dissembled and cleaned, usually by mechanical abrasion procedures. Much effort has been put into reducing this contamination problem but, although diminished, the problem has not been eliminated.

Owing to potentially low sensing volume the detector would appear to be very suitable for use with small bore columns. However, it would also appear that the contamination problems are still sufficient to limit its use to a relatively small proportion of LC users.

16. The Evaporative Light Scattering Detector

The detector functions by continuously atomizing the column eluent into small droplets that are allowed to evaporate, leaving the solutes as fine particulate matter suspended in the atomizing gas. The atomizing gas may be air or an inert gas, if so desired. The suspended particulate matter passes through a light beam, and the light scattered by the particles viewed at 45° to the light beam using a pair of optical fibers. The scattered light entering the fibers falls on to a photo multiplier, the output of which is electronically processed and either passed to a computer or to a potentiometric recorder. A diagram of the light scattering detector is shown in Fig. 62.

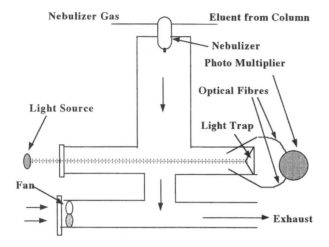

Figure 62 The light scattering detector.

The detector responds to all solutes that are not volatile, and as the light dispersion is largely Raleigh scattering the response is proportion to the mass of solute present. To ensure good linearity the droplet size must be carefully controlled as they, in turn, control the particle size of the dried solutes. The sensitivity of the detector is claimed to be between 10 and 20 ng of solute. However, in these terms it is difficult to compare it with the sensitivities of other detectors. If the solute is assumed to be eluted at a k' of 1 from a column 15 cm long, 4.6 mm in diameter, and packed with 5 μm particles, then this would indicate a sensitivity in terms of concentration of about 2×10^{-7} g/mL, which is equivalent to the response of a fairly sensitive refractive index detector. The great advantage of the detector is its catholic response and that its response is linearly related to the mass of solute present. An example of its use is shown in Fig. 63 (Christie, 1985). Estimation of the minimum detectable concentration from this chromatogram was again made to be about 10 ng of solute. To some extent, this detector provides a replacement for the transport detector as it detects all substances irrespective of their optical or electrical properties.

III. THE MODERN COMPREHENSIVE LC SYSTEM

The complete modern chromatograph can be an exceedingly complex and expensive suite of instruments and, in fact, is very rarely needed in practice. Generally, analytical methods are developed to utilize the minimum, simplest, and least expensive equipment and often comprises a simple pump, sample valve, column, detector, and computer. A diagram showing the layout of a comprehensive chromatograhic system is shown in Fig. 64.

Instrumentation of this caliber, where it occurs, is usually involved in either methods development or

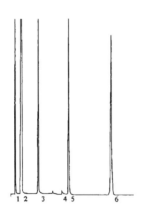

Peak	Compound	Mass (μg)	Retent.Time(min)
1	cholesterol ester	5	0.717
2	triglyceride	18	1.746
3	cholesterol	10	4.687
4	unknown	10	8.860
5	phosphatidyl choline	10	10.028
6	phosphatidyl -ethanolamine		17.390

Figure 63 The separation of some compounds of biochemical interest sensed with a light scattering detector.

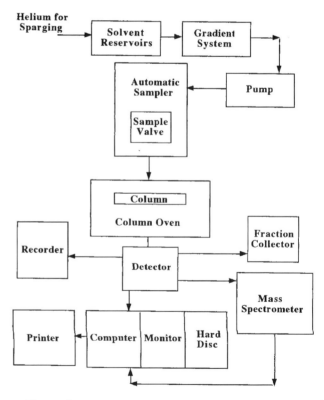

Figure 64 The modern comprehensive chromatograph.

dedicated to very special research or development projects. The chromatograph is usually fitted with at least four (sometimes five) solvent reservoirs, together with a complementary four solvent programer.

An automatic sampler that can hold 100 or more samples injects the sample onto the column and is under the direct control of the chromatograph management software of the computer. A range of columns is usually available with column switching manifolds that allow the column to be selected from a range of options and can also provide multicolumn development, if so desired. A number of detectors can also be provided but these may need to be manually changed as needed. Most sophisticated chromatographs are fitted with a fraction collector also under control of the chromatograph management software of the computer. This allows samples to be collected for subsequent spectroscopic examination or testing for physiological activity.

For *nonroutine* work fraction collection is usually achieved using a multiport valve and a number of collection vessels. For *routine* work, the selector valve should be programable on the basis of time, or be actuated by the detector output signal (preset at the appropriate signal level or signal derivative) or, preferably, a computer interpretation of both. The valve should have at least six ports, or possibly ten. If the system is to be used solely to separate solute pairs (e.g., enantiomer pairs), then a six-port valve would might be adequate. Alternate ports of the valve should be connected to waste to reduce the volume of solvent that must be evaporated during product recovery. In addition, this simplifies solvent recovery for future separations. For small samples a carousel-type fraction collector is very popular, actuated either by the control computer, detector, or timer. Product recovery of large fractions when using normal phase solvents is best carried out in a rotary evaporator under reduced pressure. For reverse phase solvents that have a high water content, recovery can be best achieved by passing the fraction through a reverse phase, C18, column of high capacity. The solute and solvent are adsorbed, and the solute and solvent content of the fraction can be recovered by displacement with another solvent, and the solute recovered by evaporation. For small fractions obtained from a carousel the solvent can be evaporated from the collection vial by means of a warm stream of nitrogen.

The modern trend is to control solvent flows, sampling valves, sample selection, detector operating conditions, and data acquisition and processing from the computer. In fact, if one is used to the manual selection of a chromatograph's operating conditions then, selecting them by way of a computer keyboard, is tedious and time consuming. However, for some reason, the use of the computer for this purpose (for whatever reason) is considered state of the art.

For data acquisition, data processing, and reporting, however, the use of the computer is essential. The software can be very sophisticated with solvent optimization packages, deconvolution routines, together with refined integration methods, peak skimming, and curve fitting procedures. All the reputable packages available today carry out the basic data acquisition and chromatogram evaluation procedures reliably and well, as they have now been established for some time. The special extras such as deconvolution and optimization routines, however, need to be thoroughly examined and demonstrated as some of these are limited in performance and can give poor results when completely novel samples and phase systems are employed.

If expense is no object then the chromatograph may be fitted directly to a mass spectrometer (usually a quadrapole type) with a suitable interface such as the thermal spray or electrospray interfaces. Operating a combined liquid chromatograph/mass spectrometer combination requires specialized knowledge of mass spectrometry as well as chromatography and, consequently, needs well qualified staff if the full value of the combined instrument is to be realized.

Contemporary chromatography instrumentation is reliable and effective but, from an electronic and mechanical engineering point of view, it is still extremely expensive. This becomes particularly obvious when the chromatograph is compared with the design and complexity of a modern automobile of the same price. There are now good and bad buys in chromatographic instrumentation and, as with the purchase of all expensive items, it pays to "shop around". All reputable, manufactures will either lend an instrument with a view to sell, or allow the potential customer to use one in their demonstration laboratories so that it can be tried out on the specific types of sample with which it will be used. It is extremely important to the buyer that advantage is taken of these opportunities.

REFERENCES

Akapo, S., Furst, A., Khong, T. M., Simpson, C. E. (1989). *J. Chromatogr.* 47:283.

Akapo, S., Scott, R. P. W., Simpson, C. E. (1991). *L. Liq. Chromatogr.* 14(2):217.

Atwood, J. G., Golay, M. J. E. (1981). *J. Chromatogr.* 218:97.

Berridge, J. C. (1985). *Techniques for the Automated Optimization of HPLC Separations.* New York: John Wiley and Sons, p. 31.

Brown, A. C. III, Wallace, D. L., Burce, G. L., Mathes, S. (1983). *Liquid Chromatography Detectors.* New York–Basel–Hong Kong, p. 356.

Christie, W. W. (1985). *J. Lipid Res.* 26:507.

Cropper, F. R., Heinberg, D. M., Wedwell, A. (1967). *Analyst* 92:436.

Done, J. N. (1976). In: Simpson, C. F., ed. *Practical High Performance Liquid Chromatography*. Heyden and Son Ltd in Association with The Royal Society of Chemistry, p. 84.
Fowliss, I. A., Scott, R. P. W. (1963). *J. Chromatogr.* 11:1.
Glasstone, S. (1946). *Textbook of Physical Chemistry*. New York: D. Van Nostrand Co., p. 299.
Golay, M. J. E. (1958). In: Desty, D. H., ed. *Gas Chromatography 1958*. London: Butterworths, p. 36.
Hirschfelder, J. O., Curtiss, C. F., Bird, R. B. (1964). *Molecular Theory of Gases and Liquids*. New York: John Wiley and Sons.
Horvath, C. G., Lipsky, S. R. (1966). *Nature* 211:748.
James, A. T., Ravenhill, J. R., Scott, R. P. W. (1964). *Chem. Ind.* 18:746.
Jung, M., Mayer, S., Schurig, V. (1994). *LC–GC* 2(6):458.
Katz, E. D. (1984). In: Scott, R. P. W., ed. *Small Bore Liquid Chromatography Columns*. New York: John Wiley and Sons, p. 23.
Katz, E. K., Scott, R. P. W. (1983a). *J. Chromatogr.* 268:169.
Katz, E. D., Scott, R. P. W. (1983b). *J. Chromatogr.* 270:29.
Katz, E., Scott, R. P. W. (1989). United States Patent No. 4,873,862.
Khong, T. M., Simpson, C. F. (1986). US Patent 8,618,322.
Khong, T. M., Simpson, C. F. (1987). *Chromatographia* 24:385.
Kirkland, J. J. (1968). *Anal. Chem.* 40:391.
Klinkenberg, A. (1960). In: Scott, R. P. W., ed. *Gas Chromatography 1960*. London: Butterworths Scientific Publications, p. 194.
Kucera, P., ed. (1984). *Microcolumn High-Performance Liquid Chromatography*. Amsterdam: Elsevier.
Milano, M. J., Lam, S., Grushka, E. (1976). *J. Chromatogr.* 125:315.
Odlyha, M., Scott, R. P. W., Simpson, C. F. ????.
Peterson, C. (1984). *J. Chromatogr.* 237:553.
Peterson, C., Schill, G. (1981). *J. Chromatogr.* 204:79.
Poppe, H. (1978). In: Huber, J. F. K., ed. *Instrumentaion for High Performance Liquid Chromatography*. Amsterdam: Elsevier.
Rhys Williams, A. T. (1980). *Fluorescence Detection in Liquid Chromatography*. Beaconsfield, UK: Perkin Elmer Corporation.
Scott, R. P. W. (1976). *Contemporary Liquid Chromatography*. New York: John Wiley and Sons, p. 4.
Scott, R. P. W. (1977). *Liquid Chromatography Detection*. Amsterdam: Elsevier.
Scott, R. P. W., ed. (1984a). *Small Bore Liquid Chromatography Columns*. New York: John Wiley and Sons.
Scott, R. P. W. (1984b). *Liquid Chromatography Detectors*. Amsterdam: Elsevier.
Scott, R. P. W. (1992). *Liquid Chromatography Column Theory*. New York: John Wiley and Sons.
Scott, R. P. W. (1993). *Silica Gel and Bonded Phases*. New York: John Wiley and Sons, p. 152.
Scott, R. P. W. (1994). *Liquid Chromatography for the Analyst*. New York: Marcel Dekker Inc., p. 17.
Scott, R. P. W., Lawrence, J. G. (1970). *J. Chromatogr. Sci.* 8(Feb.):65.
Scott, R. P. W., Kucera, P. (1971). *J. Chromatogr. Sci.* 9(Nov.):641.
Scott, R. P. W., Kucera, P. (1979a). *J. Chromatogr.* 169:51.
Scott, R. P. W., Kucera, P. (1979b). *J. Chromatogr.* 185:27.
Scott, R. P. W., Kucera, P. (1979c). *J. Chromatogr.* 186:475.
Scott, R. P. W., Simpson, C. F. J. (1982). *J. Chromatogr. Sci.* 20(Feb.):62.
Snyder, L. R. (1971). In: Kirkland, J. J., ed. *Modern Practice of Chromatography*. New York: John Wiley and Sons, p. 125.
Stephen, G., Weber, G., Carr, P. W. (1989). In: Browen, P. R., Hartwick, R. A., eds. *High Performance Liquid Chromatography*. New York: John Wiley and Sons, p. 33.
Sternberg, J. C. (1966). In: Giddings, J. C., Keller, R. A., eds. *Advances in Chromatography*. New York: Marcel Dekker Inc., p. 205.
Stinsen, S. C. (1993). *Chem. Eng. News*, 7 Sept:38.
Talmi, Y. (1975a). *Anal. Chem.* 47:658A.
Talmi, Y. (1975b). *Anal. Chem.* 47:697A.
Tiselius, A., Claesson, D. (1942). *Ark. Kem. Mineral. Geol.* 15B(18).
Tswett, M. S. (1905). *Tr. Protok. Varshav. Obshch. Estestvoispyt Otd. Biol.* 14.
West, R. C., ed. (1970). *Handbook of Chemistry and Physics*. 51st ed. Cleveland, OH: The Chemical Rubber Company, p. B63.
Willstatter, R., Stoll, A. (1913). *Utersuchungenuber Chlorophy*. Berlin: Springer.
Yates, D. A., Kuwana, T. (1976). *Anal. Chem.* 48:510.

23

Gas Chromatography

MOCHAMMAD YUWONO, GUNAWAN INDRAYANTO
Faculty of Pharmacy, Airlangga University, Surabaya, Indonesia

I. INTRODUCTION

Chromatography covers the separation techniques that involve the repeated distribution or partitioning of the analytes between two different phases, one stationary and another mobile. No other method is as powerful and widely applicable as chromatography. It has recently become the most frequently used analytical method for the separation, identification, and determination of chemical components in various sample matrixes.

There are many kinds of chromatographic techniques that are similar in many ways, such as thin-layer chromatography (TLC), high performance liquid chromatography (HPLC), and gas chromatography (GC). In GC the mobile phase is a gas and called the carrier gas, whereas in HPLC and TLC the mobile phase is a liquid or a mixture of liquids. The separation in GC and HPLC is carried out in columns, whereas TLC is carried out on glass, metal, or plastic sheets coated with silica.

The most commonly used mobile phase in GC is an inert gas, such as helium and nitrogen, which serves as carrier gas alone. Its only function is to carry the components of the sample from the inlet port through the column to the detector. In contrast to most other chromatographic techniques, in GC there is no significant interaction between the mobile phase and the analytes. Thus, the retention time of a sample component depends only on its solubility in the gas and this is directly dependent on its vapor pressure, which is also related to the temperature used and to the intermolecular interaction between the component and the stationary phase. Therefore, the sample to be analyzed by GC must be volatile. The separation of less volatile compounds must be carried out under conditions in which each sample components have a relatively high vapor pressure, for example, at elevated temperature. For nonvolatile compounds a derivatization reaction prior to injection of samples is needed. For instance, GC can successfully separate fatty acids after derivatization reactions to fatty acid methyl esters.

The concept of GC was suggested as early as 1944 by Martin and Synge. In collaboration with James, Martin first published about GC in Biochemical Journal eight years later (James and Martin, 1952). The first commercial GC instrument appeared in the market in 1955, and until 1960s it developed as packed column techniques. The major breakthrough of GC was the introduction of the open tubular column by Golay (1958), and the adoption of fused silica capillary columns by Dandeneau and Zerenner (1979). The great analytical strength of fused silica capillary column lies in its high efficiencies for the separation, so that GC is more favorable. The application of the high-resolution capillary columns with a sensitive detector, for example, GC-mass spectrometry (MS), GC-Fourier transform infra red (FTIR), or GC-FTIR-MS, makes GC to be one of the most important and widely used analytical tools, which can solve various difficult analytical problems, such as isomeric separation, analysis of complex mixtures of natural products and biological samples (Eiceman et al., 2002; Poole and Poole, 1991; Schomburg, 1990).

II. BASIC INSTRUMENTATION

A modern gas chromatographic system is shown schematically in Fig. 1. The basic gas chromatograph consists of the carrier gas supply, the inlet or injector, the column, the detector, and data system. The carrier gas flows through the preheated inlet into which a very small amount of the sample is injected. The vaporized sample is transported by carrier gas into the column, where the separation of the individual components takes place. The column is placed in a thermostatically controlled oven, so that the components remain in vapor form. After separation, the carrier gas and the component bands pass through the detector and are recorded on the recorder or computer data system.

The different parts of the GC instrument are, namely, the injection port, connecting tubes, column, and the detector and the parameters such as the method of injection, flow-rate, temperature, and so on. determine the performance of the gas chromatographic system. If the chromatograph is well designed and the condition of the parameters optimized, the contributions to band broadening from the injection port, detector, and connecting tubes are minimal and constant, so the separation will be efficient and constant.

III. THE SEPARATION SYSTEM

A. Modes of GC

There are two types of GC, gas liquid chromatography (GLC) and gas solid chromatography (GSC). When the capillary column is used for the separation, the system is called capillary GC. High-resolution gas chromatography is termed for the GC using high-resolution fused silica open tubular columns.

1. Gas-Liquid Chromatography

In GLC a liquid substance serves as a stationary phase, which can interact with components of the sample by a multiplicative partition process. For this reason, it is classified as partition chromatography. In practice, GLC is more common type of analytical GC and more widely used than GSC. Therefore, GLC is simplified to GC. GLC can be performed either in packed columns or in capillary columns. In the case of packed column, the liquid stationary phase is nonvolatile and distributed in the form of a thin film on an inert solid support. The most commonly used supports are diatomaceous earths, such as kieselguhr. In the capillary column, the liquid is coated as thin film either on the internal wall of the capillary (wall-coated open tubular, WCOT) or on a fine porous support material as a thin layer attached to the inside column walls (support-coated open tubular, SCOT) (Sandra, 2002).

2. Gas Solid Chromatography

In GSC the stationary phase is an uncoated solid, which may be a simple adsorbent, for example, alumina and silica or a porous solid such as molecular sieve. The separation is based on the adsorption/desorption process of the analytes on the stationary phase, which can occur not only in packed columns but also in capillary porous-layer open tubular (PLOT) columns. Like SCOT columns, the PLOT is prepared by depositing a porous material on the inner column walls. The application of GSC in practice is especially for the analysis of air gases (Ravindranath, 1989; Sandra, 2002).

B. Gas Chromatogram

After being separated in column, the components of the sample are eluted to the detector in the form of characteristic bands. Consequently, the detector generates electric signals, which are amplified and continuously registered by a recorder or displayed in a monitor as a gas chromatogram. The chromatogram is a two-dimensional plot with time as x-axis and the signal of the detector as y-axis. Depending on the detector used, the intensity of the signal is related to the concentration (g/mL) or proportional to the mass flow (g/s) of the components. The chromatogram starts when the sample is presently injected into the inlet port and it finishes when the detector detected the last component of the sample. The chromatogram provides the information of the separated components of the sample, which are known as peaks. The peak begins at the moment when the component comes into the detector. As a result, the detector generates the signal that first

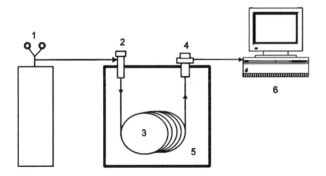

Figure 1 Typical GC System. 1, Carrier gas supply; 2, inlet or injector; 3, column; 4, detector; and 5, oven; and 6, data system.

increases and finally decreases after the maximum value reached. No peak is recorded, when no sample component enters the detector. It means that only pure carrier gas passes through the detector and consequently a straight *baseline* is recorded. The peak in the chromatogram provides useful information, either about the separation of the components or about the qualitative and quantitative data. The number of peaks, which appeared in the chromatogram, indicates the level of complexity of the sample. The shape of the peaks provides the information about the separation and transport processes that occurred in column. The GC system is assumed to yield the symmetrical or Gaussian peak shape. In practice, asymmetrical peaks (tailing or fronting) are obtained when, for example, the GC condition is not optimized. Figure 2 presents a typical gas chromatogram which contains two peaks recorded during elution of the separated components (Schomburg, 1990).

Retention time (R_t) is the time required for a component to migrate from the injection point (column inlet) to the detector (column end). It is usually termed as total or absolute retention time, which is normally represented as minutes. On the chromatogram, it shows also the distance that is measured from the injection point of *x*-axis to the maximum of peak. When a chart recorder is used, the distance should be converted to units of time by considering the chart speed. In the modern GC system where computer data system is used, the retention times of peaks are displayed and printed automatically. The retention times are characteristic of the analytes but not unique. The retention time data of a component being tested and of the reference substance can be used as a feature in construction of an identity profile but is inadequate on its own to confirm identity (USP, 2002). In practice, the total retention time of a given compound frequently varies slightly from run to run due to small alterations in operating parameters. Therefore, the retention time is frequently measured with respect to another peak in the chromatogram, rather than from the origin. Comparisons are normally made in terms of net retention time and relative retention time. The net or corrected retention time (R_t') is the difference between the total retention, R_t, and the dead time, t_0. The dead time is the time required by a nonretained compound, such as methane, to migrate from column inlet to column end without any retardation by the liquid phase. In practice, the use of the relative retention (RR_t) data is preferred and calculated by equation:

$$RR_t = \frac{R_{t2}'}{R_{t1}'}$$

where R_{t1}' and R_{t2}' are the net retention time of the test and the reference substances, respectively, analyzed under identical experimental conditions on the same column. The relative retention also known as separation factor or selectivity factor is useful not only in optimizing chromatographic separation, but also for qualitative analysis by GC. It is independent of carrier gas flow-rate and column construction (quality of packing and column length). If the relative retention of a component being tested and of the reference substance has a same value, the identity of the test substance can be inferred. To obtain more selective results with greater confidence, the same relative retention for the test sample should be achieved from the experiments under two different chromatographic processes (Bailey, 1995). The separation of two components in the sample is also evaluated from the resolution, R, which is determined by the equation:

$$R_s = \frac{2(R_{t2} - R_{t1})}{W_1 + W_2}$$

in which W_1 and W_2 are the widths at the basis of the peaks. Where electronic integrators are used, it may be convenient to determine the resolution by the equation:

$$R_s = \frac{2(R_{t2} - R_{t1})}{1.70(W_{0.5(1)} + W_{0.5(2)})}$$

where the peak widths at the baseline are measured by drawing tangents through the inflection points. The calculation of resolution is useful especially when chromatography is being used to isolate pure compounds, since it gives the chromatographer an indication of where to start and end the collection of the peak in order to achieve the desired purity.

Peak area is area beneath peak, which is proportional to the concentration or mass of the eluted component. Bandwidth, which is also referred as peak width, is the width of the chromatographic band during elution from the column.

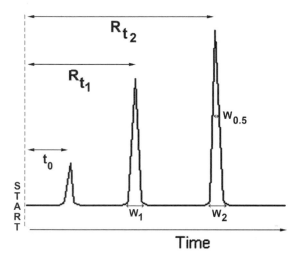

Figure 2 Typical gas chromatogram. R_{t1}, total retention time of peak 1; R_{t2}, total retention time of peak 2, and t_0, dead time.

It is usually measured at the baseline by drawing tangents to the sides of the Gaussian curve representing the peak (Schomburg, 1990).

IV. GAS SUPPLY SYSTEM

In GC a supply of carrier gas is required as the mobile phase to transfer the samples from the injector, through the column, and into the detector. The commonly used carrier gases are helium (He), argon (Ar), nitrogen (N_2), hydrogen (H_2), and carbon dioxide (CO_2). The carrier gas should be inert, dry, and pure.

A. Selection of Carrier Gases

The choice of carrier gas requires consideration of the separation problem to be solved, the detector used, and the purity of the gases. A further consideration in the selection of carrier gas is its availability and its cost. The selection of the best carrier gas is important, because it will determine the column separation processes. In this case, the efficiency of the column separation referred by the height equivalent to a theoretical plate (HETP or H) depends on the solute diffusivity in the carrier. The relation of H and the mobile phase linear velocity (μ) in packed columns was described by van Deemter (Grob, 1997), and developed in capillary columns by Golay (1958). In the van Deemter plot, the minimum of curve (H_{min}) corresponds to the maximum efficiency of the separation. Figure 3 shows the van Deemter plots for the common carrier gases, namely, nitrogen, helium, and hydrogen using conventional column and multicapillary (MC) column.

As shown in Fig. 3, hydrogen and helium offer the minimum H for higher flow-rates over a wider range of mobile phase velocities. It means that the maximum efficiency of the separation and the short analysis times are obtained when hydrogen and helium are used as carrier

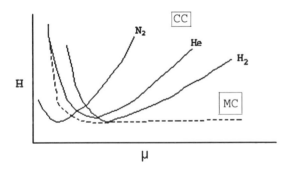

Figure 3 The van Deemter plot. H, HETP, μ, average linear velocity; ------, MC (multicapillary column); ———, CC (conventional column).

gas. On the other hand, due to the relatively high density of nitrogen, the longitudinal diffusional spreading is small and the resistance of component to mass transfer is high, so that the H_{min} value is reached at relatively high flow-rates. The practical consequence is that the flow-rate of the nitrogen as the carrier gas should be set at the slower flow-rate closed to the μ optimal to yield the efficient separation. To yield the faster analysis, it permits the use of shorter columns. In open tubular columns, however, a much higher flow-rate can be used without significant loss of separation power. Figure 3 shows that by using multicapillary column the height of theoretical plate number is smaller compared with the conventional column; and this minimum value is very broad, which allows the use of high flow-rate in order to shorten separation time without sacrificing peak resolution (Pereiro et al., 1998).

In practice, the use of hydrogen as a carrier gas is recommended especially when a thin-film column is used (http://www.gc.discussing.info/gs/f_gas_supply/gas_used_in_GC.html, June 30, 2003). When hydrogen is used, the chromatographer should be aware that it could flow into the oven due to leaks and create a fire or explosion hazard. Therefore, leaks must be tested for all connections, lines, and valves before operating the gas chromatograph. Fortunately, most GC systems today are equipped with hydrogen leak sensors although they are not entirely specific. However, when hydrogen is not favorable because of safety reason, helium should be selected (David and Sandra, 1999). Helium is the preferred gas for capillary column work and the best choice for operating the chromatograph with a mass spectrometer (GC-MS) (Karasek and Clement, 1988). The low operating cost makes also nitrogen the common GC carrier gas, particularly for the separation, in which the speed of analysis is not considered. For the complex separation, the optimum flow-rate should be determined experimentally by constructing the $H-\mu$ plot. The optimum gas velocity is termed as the velocity at which the van Deemter plot becomes linear. The carrier gas flow can be determined by either linear velocity, expressed in cm/s, or volumetric flow-rate, expressed in mL/min. The linear velocity is independent of the column diameter, whereas the flow-rate is dependent on the column diameter. The gas flow can be read electronically on the instrument panel or measured using soap-bubble flow meters at the outlet of the column. The gas is flowing through a bubble meniscus across the tube (http://www.gc.discussing.info/gs/f_gas_supply/pneumatic_system_construction.html, June 30, 2003).

B. Gas Sources and Purity

For laboratory gas chromatograph, the carrier gas is usually obtained from a commercial pressurized (2500 psi or

150–160 atm) gas cylinder. The cylinders should be handled carefully and it is recommended to support them in frames by chains or clamps. When many cylinders are needed for operating several instruments, it is preferable to house them in a separated reinforced room. In using hydrogen as carrier gas the exhaust gases should be removed by an efficient ventilation system. Nitrogen as carrier gas may be supplied either from cryogenic liquid nitrogen tanks or from generators designed specially for chromatography use (http://www.labsolution.net:8080/jsp/info_plaza/Anal/chrom/gc/Basic/CarrieGas-1.htm, April 6, 2003) (Fowlis, 1995).

The gas cylinder is usually equipped with a two-stage regulator for coarse and flow control. In most instruments, provision is also made for secondary fine tuning of pressure and gas flow. This can be provided either with conventional flow/pressure or with electronic pneumatic control (EPC). In the EPC the flow and pressures (inlets, detectors, and auxiliary gas streams) are set at the keyboard, whereas in the non-EPC the inlets use flow controllers and pressure regulator module on any side of GC.

The carrier gas should be pure and free from oxygen, water, and other contaminants especially hydrocarbon. The use of high purity carrier gases becomes more crucial in operating the chromatograph close to its detection limit. The oxygen and water in the carrier gas may give rise to degradation of some column stationary phase at elevated operating temperatures, produce unstable baselines with the electron-capture detector (ECD) and shorten filament lifetime for thermal conductivity detector (TCD). The organic impurities in the carrier, as well as in the hydrogen or air, to the flame ionization detector (FID) may cause additional detector noise and drift; this would reduce the sensitivity of detector. It is recommended that the content of oxygen, water, and hydrocarbon in the carrier gas should be not more than 1–2 ppm each or totally below 10 ppm (http://www.gc.discussing.info/gs/f_gas_supply/gas_used_in_GC.html, June 30, 2003; http://www.gc.discussing.info/gs/f_gas_supply/pneumatic_system_construction.html, June 30, 2003).

Unless the high purity carrier gases of 99.995–99.999% are used, the gas cylinders should be combined with additional chemical and/or catalytic gas purifying devices to minimize contamination. In this case, the carrier gas flow is directed through a sieve trap or through a series of traps to remove the moisture, organic matter, and oxygen, and then through frits to filter off any particulate matter. For this purpose, today some traps or their combinations are available in the market. The oxygen traps usually include a metal equipped with an inert support reagent, which is able to decrease the oxygen up to 15–20 ppb (http://www.gc.discussing.info/gs/f_gas_supply/pneumatic_system_construction.html, June 30, 2003). Traps for removing the moisture can be found in several types of adsorbents that can be refilled, and of the bodies, whether from glass or plastic. In the case of hydrocarbon traps, the adsorbent is usually in the form of active carbon. The traps should be installed in the vertical position on the main line near the gas source with the order starting from moisture, hydrocarbon, and oxygen.

The contamination may also occur from the connecting tubes, flow controllers, and columns, leading to a significant base current response. For that reason, it is suggested to use copper tubing for the connection of the gas cylinder to the gas chromatographs. Plastic tubing should be avoided, since the oxygen and moisture from the atmosphere can permeate the tubing walls and degrade gas purity. The polymer tubing such Teflon, nylon, polyethylene, or polyvinylchloride may also contain contaminants.

For capillary column, make-up gas is added at the column exit to obtain a total gas flow of 30–40 mL/min into the detector, and can be the same gas as the carrier gas or a different gas depending on the type of detector being used (http://www.gc.discussing.info/gs/f_gas_supply/gas_used_in_GC.html, June 30, 2003).

V. GAS CHROMATOGRAPHIC COLUMNS

The column, where the actual chromatographic separation occurs, is described as the heart of the chromatographic system. There are two general types of the gas chromatographic columns most commonly used, namely, packed and open tubular, or capillary. The packed columns were developed first but today the majority of GC has been carried out on capillary columns. The principal difference between the two columns is reflected in their plate numbers. The packed columns are characterized by their relatively low plate number, but they have higher capacities which simplify sample introduction techniques and can accommodate a larger quantity of sample. On the other hand, the capillary columns have very high plate numbers which possess the high efficiencies and separating capabilities. As shown in the van Deemter plot (Fig. 3), the faster gas velocities can be applied on capillary columns, giving rise to shorter analysis. Other important features of capillary columns compared with packed columns are greater inertness, longer life, lower bleed, and more compatible with spectroscopic detectors.

A. Packed Columns

Packed columns consist of metal (stainless steel, copper, or aluminum) or glass tubing filled with solid material either uncoated adsorbents (GSC) or a solid support-coated with a stationary liquid phase (GLC). The selection of tubing material is dictated from the particular analytical use. Glass column can be used at high temperature when the metal tubing would catalyze decomposition of the sample. For this reason, metal columns are undesirable for thermally labile compounds are being analyzed, such as steroids and essential oils. However, metal columns allow the required elevated pressure. If a glass column in GC is used, it is important to integrate a glass line in the injection port, otherwise decomposition of the sample may take place before it reaches the column. Typical packed columns for routine analysis are 2–10 m in length and 2–4 mm in an internal diameter. In order to incorporate them to the oven for thermo-stating, the packed columns are usually formed as coils having diameters of 10–30 cm. It is also available as straight and U-forms for shorter columns (Sandra, 2002).

In the GLC columns, the solid support is needed to hold the liquid stationary phase. The materials should consist of inert uniformly spherical particles having a large surface area per unit volume (particle size between 80 and 120 mesh) in order to minimize the void volume and to supply a specific surface area for interaction with the analytes. In addition, they should be mechanically strong over a wide temperature range. The most frequently used solid supports for GLC are diatomaceous earths, either kieselguhr that is sold under trade names Chromosorb W or G, or crushed firebrick, which has the trade name of Chromosorb P. In GSC the most commonly used adsorbents are activated charcoal, silica gel, alumina, and glass beads.

B. Capillary Columns

Capillary or open tubular columns were introduced by Golay as early as 1957, however, the widespread availability came only in 1979 with the development of fused silica as tubing material. Early columns were constructed of stainless steel, aluminum, copper, plastic, or glass. The fused silica tubing types with a polyimide coating are presently preferred. There are three main types of open tubular columns: WCOT, SCOT, and PLOT.

WCOT capillary columns are the most commonly used in GC. They are prepared by coating the inner column walls with the liquid stationary phase as a thin film. Presently fused silica is the most commonly used column material, which is manufactured from synthetic quartz with very low (less than 1 ppm) metallic impurities. Fused silica has thin walls with the outside diameters about 0.3–0.4 mm and the inside diameters 0.1–0.3 mm.

In the SCOT columns, the liquid is located on porous support as a thin-layer. Compared with WCOT, SCOT columns have a higher sample capacity but less efficiency. PLOT is similar to SCOT columns in that a porous material is deposited on the inner column wall. There

Table 1 Important Properties of GC Columns

Parameters	Packed	WCOT	SCOT
Length (m)	1–6	10–100	10–100
Internal diameter (mm)	2–4	0.20–0.75	0.5
Efficiency (N/m)	500–1000	1000–4000	600–1200
Capacity (ng/peak)	10	10–1000	10–1000

are generally three types of open tubular columns based on the thickness of stationary films: (1) thin film (0.1–0.2 μm), (2) medium film (0.5–1.5 μm), and (3) thick film (3–5 μm). The columns are also available in several standard lengths, including 10, 25, 30, 50, and 60 m. The 25 m or 30 m column is the most frequently used (Ettre, 1981; Kaiser, 1985). Table 1 presents some important properties of each compared with those for packed columns. Figure 4 shows a cross-section of GC columns, and typical chromatogram obtained from packed and capillary coloumn is presented in Fig. 5.

C. Stationary Phases in GC

The stationary phases are usually liquids, which should exhibit thermal stability, chemical inertness, and low volatility to prevent bleeding of the column. The boiling point of the separating liquid should be at least 100°C higher than the required column temperature. In addition, the liquid should demonstrate the selectivity, which produces the partition coefficient of various analytes in a suitable range. The same stationary phases are employed in packed or capillary columns. Although myriad phases were formerly proposed for packed column GC, most analyses today are done on a dozen favored liquid phases. For capillary columns GC, the number of liquids can be still reduced due to their high efficiency characteristics.

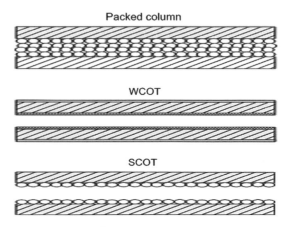

Figure 4 Diagram of the cross-section of the GC columns.

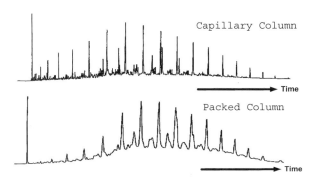

Figure 5 Typical gas chromatograms of light oil obtained from packed and capillary coloumn. GC condition—capillary column CBP1-M25-025: 25 m × 0.2 mm ID, 0.25 μm film thickness, He flow-rate: 1.3 mL/min,; packed column SE-30 5% on AW DMCS chromosorb 80/100:2 m × 3 mm ID, He flow-rate: 40 mL/min; column temperature: 100–250°C, 7°C^{-1} min^{-1}, detector: FID. (Courtesy of Shimadzu Asia Pacific Pte. Ltd., Singapore.)

Table 2 contains a list of some those liquids grouped according to the polarity.

Typical nonpolar phases are hydrocarbon or polydimethyl siloxanes, which are selective for nonpolar analytes, saturated halogenated hydrocarbon, and also alcohols with different enough boiling points. Polar phases have the functional groups –CN, –C=O, OH, or polyester, which are suitable for separating hydrogen-bonded analytes, such as alcohols, organic acids, or amines. On the basis of the structure of poly dimethylsiloxane, some phases with different polarity are produced by replacing the methyl side chain with functional groups, such as phenyl, cyano, trifluoromethyl. Some phenyl derivates have moderately polarity but contain no hydrogen bonding. Cyano and trifluoromethyl derivates are very polar having weak hydrogen bonding characteristic.

Recently Mayer et al. (2002) developed a new stationary phase SOP-75, (a methoxy-terminated, symmetrically substituted 75% diphenyl, 25% dimethyl-polysiloxane), which exhibited alike selectivity to OV-25 or PS126 and have high inertness up to 400–410°C. The authors showed that if the fused silica capillary column was coated with SOP-75 yielded better separation and efficiency compared with OV-25 or PS126.

If the stationary phase consists of binary mixtures, the retention volume of a solute was linearly related to the volume fraction of either one of the two phases, according to following equation:

$$V'_{AB} = \alpha(V'_A - V'_B) + V'_B$$

V'_{AB} is the retention volume of the solute on the binary stationary phase, V'_A and V'_B are the retention volume of the solute in the pure stationary phase A or B, and α is the volume fraction A in the binary mixture (Scott, 2001a).

D. Optimization of the GC Column Parameters

The choice of the stationary phase for particular application in packed columns is more complicated than in open tubular columns. The situation is more difficult by the fact that numerous solvents have been reported as stationary phases. The stationary listed in Table 2 are the most commonly used and it can be considered as a preliminary guide. The important criteria to select the GC Columns for the separation are the polarity of the sample components and the polarity of the stationary phase. This is combined with the general rule "Like dissolves like" for selecting the separating liquid. However, this is not a perfect guide so that trial-and-error tests are also applied to obtain the optimal separation conditions.

Cazes and Scott (2002) derive mathematical equations to calculate the optimum column dimension and operating conditions to achieve a specific separation in the minimum time both for packed and open tubular column. It is shown that small diameter open tubular column is most excellent for separating simple mixture. On the other hand, for difficult separations, wider column is required. This is due to the need of longer column; however, as the inlet pressure is limited, in order to have optimum flow-rate, the column radius should be increased. The optimum column radius is inversely proportional to the separation ratio of the critical pair (see Fig. 6). Figure 7 shows the correlation of the log of minimum column length with the separation ratio of the

Table 2 List of Stationary Phases

Components of interest	Stationary phase	Maximum temperature (°C)	Polarity
Free acids; alcohols	Poly(dicyanoallyldimethyl) siloxane	240	High polarity
Free acids; alcohol, ethers	Polyethylene glycol	250	Polar
Alkyl-substituted Benzenes	Poly(trifluoropropyldimethyl) siloxane	200	Medium polarity
Steriods, pesticides, glycols	Poly(phenylmethyl) siloxane (50% phenyl)	250	Medium polarity
Hydrocarbons, polynuclear aromatics	Polydimethyl siloxane	350	Nonpolar

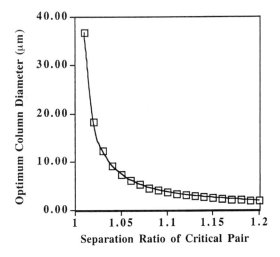

Figure 6 Correlation of optimum column diameter against separation ratio of critical pair. [Reproduced from Cazes and Scott (2002).]

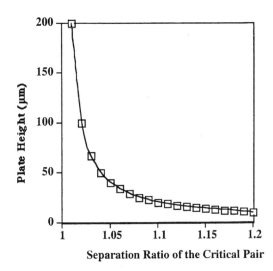

Figure 8 Correlation of the plate height against separation ratio of critical pair. [Reproduced from Cazes and Scott (2002).]

critical pair. Figure 8 describes the correlation between plate height and separation ratio of the critical pair.

1. Column Temperature

Column temperature is the most important variable in GC, since it directly affects the retention and the selectivity of the sample components. Therefore, the optimum column temperature must be found to obtain a good separation in a reasonable analysis time. For simple separations, the analyses are usually performed in isothermal mode, whereby the temperature of the column is held constant throughout the run. In this case, the column temperature is generally held more or less at the average boiling point of the sample. However, problems arise for complex mixtures that contain compounds with widely different boiling points. At too high temperatures, the very volatile components will be eluted quickly but are not fully separated, whereas the high boiling components may be well separated. If the column is operated at low temperature, all volatile components may be separated satisfactorily, however, less volatile components will appear in the chromatogram as flat peaks with a very long retention time, and thus the total analysis time is extended. This problem was overcome by using the mode of *temperature programing* or programed-temperature GC (PTGC). In this method, the temperature of the column is raised during analysis. The technique involves either linear temperature programing, in which the column temperature is continuously increased at a constant rate, or multiple ramping that is a combination of the linear programing at some different rates with isothermal operations. Linear temperature programing is usually performed after an initial isothermal period. Several parameters must be adjusted with respect to the particular separation problems including the initial temperature, initial hold time (i.e., how long the temperature is kept constant after injection); programing rate (usually in degree per minute), final temperature, and final hold time. The initial temperature should be low enough to resolve the low-boiling compounds (k value > 3), whereas the final temperature is selected such that the least volatile compound elutes as rapidly as possible without exceeding the maximum operating limits of the stationary phase. The choice of programing rate is based on the nature of the components of interest

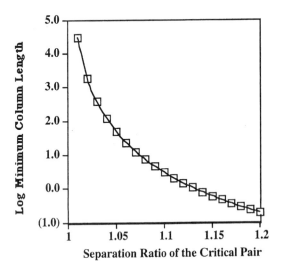

Figure 7 Correlation of logarithm of optimum column length against separation ratio of critical pair. [Reproduced from Cazes and Scott (2002).]

and the complexity of the sample. In general, the programing rates are usually between 0.5 and 10°C/min and set when the components of interest begin to elute. After the components of interest are eluted, higher programing rates are commonly used to remove less volatile but unwanted compounds (Sandra, 2002).

Many separation and analytical problems can be solved so far by using PTGC, such as the analysis of complex mixtures containing components with a wide range of boiling points. One disadvantage with PTGC is that the sample compounds may undergo decomposition as the temperature of column is raised. For those thermally labile compounds, analysis by means of LC is preferable. At elevated temperature, an increasing baseline in the chromatogram is commonly observed due to the column bleed. For GC-MS works, this may lead to the appearance of many extra or background ions in the mass spectra of minor components. In such cases, it is recommended to use WCOT columns of a suitable stationary phase with a reduced liquid film thickness (Karasek and Clement, 1988).

Figure 9 shows the relation between program rate and retention time. In this case the inlet/outlet pressure was assumed to be 2 and the flow-rate of the column exit was 20 mL/min. The curve showed an approximately linear relationship, but for more accurate work a second order polynomial should be used (Scott, 2001b).

2. Carrier Gas Flow

According to the van Deemter equation (see Fig. 3), the theoretical plate numbers (N) of a column rely on the flow of the carrier gas used. As noted earlier, the minimum of curve (minimum H) corresponds to the maximum efficiency of the separation. Increasing the carrier gas flow-rate decreases the elution time of the components and hence shortens analysis time, on the other hand, decreases the theoretical plate and thus decreases the efficiency. For each column, the carrier gas velocity should be practically optimized to find out the flow-rate at which the van Deemter plot becomes linear. Depending on the type of the column and the carrier gas used, a practical compromise must be reached to determine the best flow-rate for minimum H.

In the modern GC, the carrier gas flow during analysis can be changed by gradually increasing the inlet pressure. The advantages of this flow programing are that the analyses can be performed in a reduced separation time and at lower temperatures of the column. The low temperature runs make this method useful for the analysis of thermally labile compounds that may undergo decomposition at high temperatures. In addition, it can minimize column bleed so that the baseline observed is more stable and the column life may be extended (http://www.gc.discussing.info/gs/f_gas_supply/gas_used_in_GC.html, June 30, 2003).

If the flow increased, the inlet pressure will also be increased so the ratio of inlet/outlet pressure will be changed significantly during the flow program. Figure 9 describes the second order relationship between flow-rate and inlet/outlet pressure. The effect of program flow-rate on the retention time is shown in Fig. 10. In this case the column length is 30 m (ID 320 μm), operated at 120°C using nitrogen as carrier gas (Scott, 2001c).

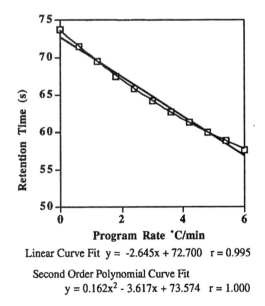

Figure 9 Correlation of the program rate and retention time. [Reproduced from Scott (2001a).]

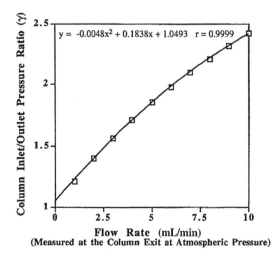

Figure 10 Relationship between flow-rate and inlet/outlet pressure. [Reproduced from Scott (2001b).]

E. Connection Techniques of Capillary Column

The connection of capillary columns to injector and to the detector should be simple to join and to loosen, dead-volume free, and stable at elevated temperature. The most commonly used connection materials include graphite, Vespel, or composites of both materials (Schomburg, 1990).

VI. SAMPLE INLETS

Sample introduction is critical step in GC, especially when WCOT columns are used. Therefore, the inlet system must be well designed to facilitate the injection of the sample onto the head of the column without degradation of the column performance and without discrimination of sample components. In order to minimize the band spreading and to obtain the best resolution, the sample must be injected rapidly as a narrow band and must be of a suitable quantity. Introduction of a large volume or too concentrated sample may give rise to poor resolution and distorted peaks. The sample size depends on the dimensions of the columns and on the sensitivity of the detector. In this case, capillary columns require smaller quantity than do packed columns, and the sample size should not exceed the linear dynamic range of the detector used. For packed column, sample size ranges from a few tenths of a microliter up to 20 μL. Capillary columns need much less sample (0.01–1 μL). For quantitative purpose, the sample introduction must be attained with a high degree of precision and accuracy (Sandra, 1985).

Several techniques have been developed to introduce samples onto the GC column. Unfortunately, no universal inlet system has yet been provided to handle a wide range of samples and different columns. Due to the high sample capacity, the introduction of the sample into packed columns is usually problem-free, so that the design of inlet for packed columns is simple. This is quite different with open tubular columns, in which the sample capacity and carrier gas flowrate characteristics are much lower. In this case, some different techniques have been developed: split, splitless, on-column, and programed-temperature vaporizing (PTV) injection (Grob, 1986, 1987; Grob and Romann, 1981). Split technique is most common, which is used for high-concentration samples. This technique allows injection of samples virtually independent of the selection of solvent, at any column temperature, with a little risk of band broadening or disturbing solvent effect. The splitless technique, on the other hand, is used for trace level analysis. The so-called cold injection technique (on-column, temperature programed vaporization, see Section VI.C.5) has also been developed.

A. Syringe Handling Techniques

The method of injection differs according to the physical state of the sample.

The samples in the form of a liquid are usually injected into GC using a microsyringe through a rubber or silicone septum. For the solid samples, it is convenient to dissolve them in a suitable volatile solvent. For repetitive or periodic injection of a large number of the same or different samples, auto-samplers may be used.

The most commonly used microsyringe is constructed of a syringe needle and a calibrated glass barrel with a close fitting metal plunger. After drawing the sample into the barrel, the injection volume is read and the sample is displaced from the needle. There are several methods applied for sample introduction by means of the syringe: filled needle, cold needle, hot needle, solvent flush, and air flush. The solvent flush mode is the most popular technique. In this technique, the syringe is repeatedly washed with solvent and the solvent is taken up into the syringe. After removing the needle from the solvent, the plunger may be pulled back. The desired volume of sample is then introduced into the syringe and injected to GC in the normal manner. The air barrier may not separate the solvent and sample. The solvent is used to flush the sample out of the syringe and completely occupies the dead space (Grob and Neukom, 1979). The most commonly used silicone-rubber septa may contain impurities that may bleed into the column above a certain temperature, resulting in unsteady baseline and ghost peaks. Recently, variety of septa has become available which can be used at very high temperature (Olsavicky, 1978).

B. Packed Inlet system

Inlet system for packed columns is frequently based on the injection of the liquid sample with a microliter syringe through a self-sealing silicone septum into a vaporizer tube or directly into the column packing. Figure 11 illustrates a typical design of an injection system for packed columns. The vaporizer tube is usually a glass liner, which serves to prevent the sample coming in contact with the heated metal surfaces so that the thermal decomposition can be minimized. Encasing the glass liner is a separate heater, into which the carrier gas continuously flows (Olsavicky, 1978). The temperature of injection port is usually adjusted so as to be approximately 50° higher than of the column. This ensures a process termed flash vaporization of the liquid sample, and the carrier gas sweeps the vaporized sample onto the stationary phase. Injection of the sample into the high temperature of the flash vaporizer, however, may cause chemical changes of the sample components leading to errors. In

Gas Chromatography

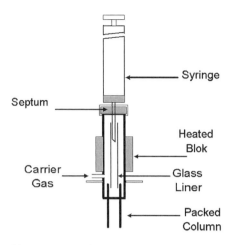

Figure 11 Diagram of the injection system for packed columns.

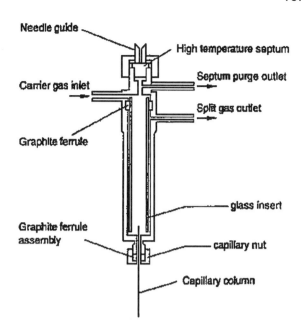

Figure 12 Classical split injector. (Courtesy of Shimadzu Asia Pacific Pte. Ltd., Singapore.)

such cases, the on-column injection technique may be applied. When injection is carried out in the on-column mode, the column is pushed right up so that the column end is close to the septum and the glass wool can be used for packing the injector.

C. Split Injection

Split injection involves injecting a liquid sample into a heated injection port, vaporizing the sample in the injection inlet, and splitting the vaporized samples into two parts so that small fraction of the vaporized sample enters the column and the major portion is vented to waste. Split injection was firstly developed for open tubular columns. It has been one of the most commonly used methods for many applications, since it offers many practical benefits when analyzing concentrated samples with little risk of band broadening. It is easy to automate and it is compatible with both isothermal and temperature-programed operation. The classical split injector is flash vaporization device (see Fig. 12). The device consists of a heated vaporizing chamber, which is usually made from a stainless steel tube lined with a removable glass or quartz liner. The carrier gas at a constant pressure enters behind the glass liner and is therefore preheated. The flow is divided into two different routes, one to purge the septum and the other, and a high flow, to enter the vaporization chamber, where carrier gas is mixed with the vaporized sample. The mixed stream flows by the column inlet, and leaves through split valve exit. The split ratio that is controlled by a needle valve is the ratio of the split flow to the column flow-rate (Grob, 1986).

D. Splitless Injectors

The splitless injection is usually used for analysis of trace compounds. As the name indicates, there is no split in this injection technique, which is achieved by closing the valve in Fig. 12. The design of the splitless injectors differs from the split injectors only in the addition of the solenoid valve downstream from the vent. During the splitless period, most of the sample and solvent enter the column. The residual vapor in the injector, however, may cause peak tailing, especially for the solvent peak. To avoid this problem, the inlet is switched back to the split mode after a nearly complete sample transfer, so that all remaining solvent and vapors are purged out of the split vent. In this case, the splitless period must be optimized for each particular sample matrix since the sample transfer efficiency will be low if the purge is turned on too quickly. This can be determined by a series of injections with increasing length of the period until the analyte concentration plateaus. Factors having an effect on the splitless efficiency include: the starting oven temperature, the solvent boiling point, and the polarity of the solvent (Grob, 1986; Grob and Romann, 1981).

E. Direct Injection

1. Cold On-Column Injectors

The cold on-column injectors allow the injection of liquid sample directly onto the column. During injection, the injection zone is maintained at low temperature to avoid needle discrimination. A major advantage of this injection technique is that the sample is completely transferred without discrimination; and high precision and accuracy of the results can be obtained for samples with a wide range of component volatilities and thermally stabilities. Troubles appear only with samples with a very low volatility since the column inlet temperature cannot be auxiliary reduced even with secondary cooling. Other disadvantage of this injection technique is that the samples enter the column inlet as a liquid plug. The design of a typical cold on-column injector (Fig. 13) is comparatively simple, involving a syringe guide and a stop valve (Grob, 1987).

F. Programed-Temperature Injection

The PTV injection is claimed as a nearly universal injector, since it can be operated in several possibilities, that is, hot or cold split injection, hot or cold splitless injection, cold on-column, and direct injection. In addition, this mode offers the injection of a large sample volume and in multiple parallel capillary columns. In PTV injector, the sample in form of liquid is introduced into a glass liner using an automated or regular syringe. The temperature of the vaporization chamber is kept cold, usually at a temperature below the boiling point of the solvent used. The vaporization chamber is then rapidly heated electrically using preheated compressed air, commonly a few seconds after the withdrawal of the syringe needle. Split or splitless injection can be attained through regulation of the split valve. Dilute samples can be analyzed by introducing samples into cold injector with the split valve open. The split flow is adjusted so that only the solvent evaporates and thus leaves the system through the vent while the solutes remain in the liner. The valve is then closed and the vaporizing chamber is heated.

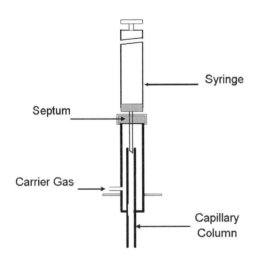

Figure 13 Diagram of a cold on-column injector.

The valve is again opened when the vaporized sample is transferred to the column. This operation is termed as solvent elimination mode, which allows a very dilute sample injected repetitively without the danger of flooding in the column or detector (Tollbäck et al., 2003).

G. Injectors for Gas Samples

For injecting gases and vapors, gas-tight syringes with Teflon-tipped plungers and syringe barrel are available. Many analysts favor using of gas syringes for gas samples; however, the introduction of accurately measured volumes of a gas remains a problem. In the alternative method, gas samples can be introduced onto column using rotary gas switching valves, which can applied in two different modes: the sample is transferred from a gas ampole into the evacuated sample loop, or the loop filled with the sample by flushing it with the sample which must be available under an overpressure (Grecco et al., 2001).

H. Headspace Analysis

The headspace injection permits the analyses of volatile components of complex samples when the matrix is of no interest. A major advantage of this technique is that the sample can be directly analyzed without a complicated extraction of the analytes from the samples. In this case, the sample is transferred in a sealed headspace vial and positioned in a thermo-statted bath, usually at 40–60°C. The volatile components, which have a suitably high vapor pressure above the liquid or solid sample matrix, are in the form of gases distributed in the headspace of a sealed sample vial. The injection of the gaseous sample into the GC column is done by means of a gas-tight syringe or a specially designed valve.

The principles of headspace injection are based on the thermodynamic conditions of the phases. When a sample containing volatile components is placed in an airtight vial, equilibrium is reached between a liquid and its vapor. The vapor phase or headspace can be determined either qualitatively or quantitatively, since the vapor phase has the same composition as the liquid at a given temperature and pressure. The concentration in the vapor phase is a related to the concentration in the original mixture (Scott, 2001d). Headspace injection system can be grouped into two variations of techniques: static headspace sampling and dynamic headspace sampling. Figure 14 shows schematical diagram of static head space. In the static case, the sample is taken only from one phase equilibrium. This technique is useful for analyzing of intermediate volatility components, such as low molecular weight organic acids, alcohols, and ketones. For very volatile compounds, broad initial bands may be resulted due to the much lower sample load ability of

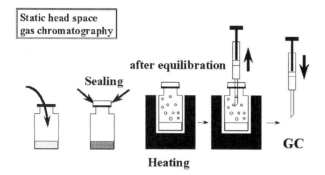

Figure 14 Headspace sampling. (Courtesy of Shimadzu Asia Pacific Pte. Ltd., Singapore.)

capillary columns. In such cases, the split injection mode is commonly performed (Scott, 2001d).

In the dynamic system, the phase equilibrium is transferred continuously over a period of time, in which an inert gas is required to force the headspace out of the vial. The solutes focusing can be completed by cold trap or adsorbents such as Tenax™ and charcoal. Compared with static technique, the dynamic mode is more sensitive (Guimbard et al., 1987; Joffe and Vitenberg, 1984; Scott, 2001d).

I. Solid Phase Microextraction Headspace GC

Arthur and Pawliszyn (1990) and co-workers first introduced a simple extraction technique namely, solid phase microextraction (SPME), which requires no solvents or complicated apparatus. It offers several advantages of sample preparation including elimination of organic solvents, reduced analyte loss, reduced time per sample, and less sample manipulation. In addition, it can be used as a sample introduction method for GC or GC-MS system. The SPME head space unit (Fig. 15) is constructed of a fused silica fiber coated with polydimethylsiloxane and bonded to a stainless steel plunger and holder. It is similar to microsyringe. The coated fiber can be directly immersed in a liquid sample or in the headspace. The headspace technique is favorable especially for the sample which contain nonvolatile or undissolved components (Mills et al., 1999; Pinho et al., 2002; Scarlata and Ebeler, 1999).

J. Purge and Trap GC Systems

The purge and trap have been developed for extracting and concentrating volatile organic compounds (VOCs) that are insoluble and slightly soluble in water samples. The purge and trap coupled with GC procedure comprises purging an inert gas, such as nitrogen or helium, through an aqueous sample placed in a sealed system at ambient temperature, leading to transfer of VOCs to vapor phase. The vapor is

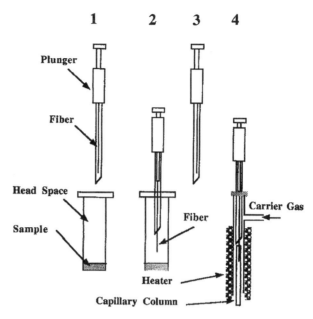

Figure 15 SPME apparatus. [Reproduced from Scott (2001).]

then swept by inert gas flow through a trap containing adsorbent materials (usually containing Tenax™), where it is retained. When the purge process is completed, the carrier gas is directed to the trap by means of a six-way valve. The trap is then rapidly heated to transport them to gas chromatographic column (Lacorte and Rigol, 2002; Wang and Chen, 2001; Wasik et al., 1998).

K. Pyrolysis GC

Pyrolysis is defined as the chemical transformation of a sample when heated at a temperature significantly higher than ambient (Moldoveanu, 1998). In the pyrolysis GC (Py-GC), the pyrolysis procedure is achieved by using a pyrolyzer to load the sample to the gas chromatograph. In this technique, the sample containing very high molecular weight compounds is decomposed in a furnace by controlled heating, usually at the temperature of 400–1000°C. The generated smaller volatile and semivolatile molecules can be easily determined by means of GC, resulting chromatogram that represents a fingerprint of the original sample (Moldoveanu, 2001; Sandra, 1985).

Py-GC is applied frequently for the analysis of large molecules. In this technique, the very high molecular weight compounds are decomposed to small volatile fragments by controlled heating in a furnace, usually at the temperature of 400–1000°C. The small fragments are analyzed by GC resulting chromatogram, which represents a fingerprint of the original sample. The most frequently used analytical pyrolyzers are platinum resistively heated and the Curie point pyrolyzers (Fig. 16).

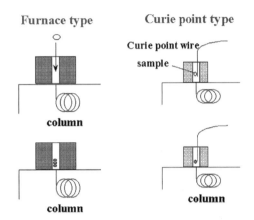

Figure 16 Typical pyrolyzers. (Courtesy of Shimadzu Asia Pacific Pte. Ltd., Singapore.)

Py-GC has been applied for the characterization of polymers, polysaccharides, lignins, soil, sewage sludge, marine sediment, and so on. The analysis is commonly performed using Curie point pyrolyzers. However, the use of filament pyrolyzer is preferred for analyzing a wide range of compounds or when quick sample profile is required. Py-GC was used hyphenated with MS (Py-GC/MS) for the analysis of natural and synthetic polymers, environmental problems, taste of cigarettes and food, and so on (Haken, 1998; Medina-Vera, 1996; Wang and Burleson, 1999). The use of SPME as a sample preparation step for GC-MS was reported by Moldoveanu (1998).

VII. OVEN

A. Conventional GC-Oven

The column is ordinarily housed in a thermostatically controlled oven, which is equipped with fans to ensure a uniform temperature. The column temperature should not be affected by changes in the detector, injector, and ambient temperature. Temperature fluctuations in column ovens can decrease the accuracy of the measured retention times and may also cause the peak splitting effect. For conventional ovens, the oven wall is well insulated using a wire coil of high thermal capacity, which is able to radiate heat into the inner volume of the oven. The characteristics of a more efficient method can control accurately the temperature of column and allow the operator to change the temperature conveniently and rapidly for temperature programing. It is designed by suspending the column in an insulated air oven through which the air circulated at high velocity by means of fans or pumps. Most commercial instruments applied the design and allow for the adjustment and control of temperature between 50°C and 450°C. The sub-ambient-temperature operation would normally require a cryogenic cooling system using liquid nitrogen or carbon dioxide (Ravindranath, 1989; Sandra, 2002; Schomburg, 1990).

B. Flash GC-Oven

Due to the limited heating and cooling rate of conventional air bath oven, a "resistive heating" technique was developed. In this technique, electrical current is employed to heat a conductive material (a metal) located in very close distance from the column. The temperatures can be determined by resistance measurements. The temperature program is converted into a resistance program and the electrical circuits applied power to change the resistance per unit time. The GC using this system is called "flash GC", a conventional GC can be changed into this system by using an upgrade kit (EZ flash). The time of analysis can be reduced by a factor of more than 10 when compared with conventional GC, by using resistive heating and a short column, so the GC method can be used for "fast GC" (Mastovska et al., 2001a).

C. Microwave GC-Oven

By using microwaves to heat only the column, not the surrounding air and oven body, this can minimize heating and cool-down times, thus shortening every analysis time. Heating rate of $10°C\ s^{-1}$ and cooling rate of over $300°C\ min^{-1}$ can be achieved. By true column-only heating can also enable two columns to be programed independently inside an existing column oven. A conventional GC can be changed to microwave GC by installing the dual-oven microwave, so this GC can be upgraded to fast GC (http://www.antekhou.com/products/chrom/oven.htm, July 8, 2003; http://www.antekhou.com/pdf/Literature/3600.pdf, July 21, 2003). When the GC-oven is flushed with nitrogen, the GC column can be heated up to 450°C without loss of strength, eliminating the need for special metal-clad column in high temperature (http://www.scpubs.com/articles/aln/n0207gai.pdf, July 21, 2003).

D. Infrared Heated GC

In this method, the GC-oven is constructed of polyamide foam, which has 5 cm inner diameter and a thickness of 3 cm. Very effective heating rate was performed using a 150 W halogen lamp allows that controlled heating rate up to 1000°C/min. The chamber opened at both sides is cooled down within 60 s assisted by a small fan. A thermocouple was inserted into the steel capillary column positioned inside the oven (http://www.et1.tuharburg.de/downloads_et1/umt/publikationen/irgc.pdf, July 29, 2003).

VIII. DETECTOR

A. Classification and General Properties

The detector in GC senses the differences in the composition of the effluent gases from the column and converts the column's separation process into an electrical signal, which is recorded. There are many detectors that can be used in GC and each detector gives different types of selectivity. An excellent discussion and review on development of GC detectors have been published (Buffington and Wilson, 1987). Detectors may be classified on the basis of its selectivity. A universal detector responds to all compounds in the mobile phase except carrier gas, whereas a selective detector responds only to a related group of substances. Most common GC detectors fall into the selective designation. Examples include FID, ECD, FPD, and flame thermoionic ionization detectors (FTD). The common GC detector that has a truly universal response is the TCD. Mass spectrometer (MS) is other commercial detector with either universal or quasiuniversal response capabilities.

Detectors can also be grouped into concentration-dependent detectors and mass flow dependent detectors. Detectors whose responses are related to the concentration of solute in the detector, and do not destroy the sample are called concentration-dependent detectors, whereas detectors whose responses are related to the rate at which solute molecules enter the detector are called mass flow dependent detectors. Typical concentration-dependent detectors are TCD and FTIR detector (GC-FTIR).

In the concentration-dependent detector, the signal y_i of a sample component i is proportional to the concentration c_i in the carrier gas at any time of the elution (Schomburg, 1990):

$$y_i = a_t c_i = a_t \frac{dQ_i}{dV_i}$$

where a_t is the response factor, Q_i is the total amount of the component i in the volume V of the carrier gas. The peak area A is obtained by the integration between the beginning and the end of y_i.

Important mass flow dependent detectors are the FID, thermoionic detector for N and P (N-, P-FID), FPD for S and P, ECD, and selected ion monitoring (SIM) MS detector. In this type of detectors, the peak area A of the same amount of analyte Q_i is not changed at higher or lower gas flow applied, in this case the signal y_i is dependent only on the properties of the eluted samples.

$$y_i = a_t \frac{dQ_i}{dt}$$

The most important general performances of detectors are detector volume, sensitivity, response time, and (linear) dynamic range (Schomburg, 1990).

In the concentration-dependent detector the effective detector volume should be small in comparison with the volume of carrier gas in which the solute is diluted, otherwise, an increase in peak occurs.

Sensitivity S could be defined as the change of measured detector signal y_i resulting from the change of the concentration of the eluted analyte c_i

$$S = \frac{dy_i}{dc_i}$$

The sensitivity depends on the design and properties of the analyte. The minimum detectability of a detector DL could be calculated from the following equation:

$$DL = K \frac{N}{S}$$

where K is a constant, N is the detector noise expressed as the root-mean-square of the noise signal. The detector noise N can be defined as the maximum amplitude of the combined short- and long-term noise measured over a period of 10 min. It can be calculated by constructing parallel lines embracing the maximum excursion of the recorder trace over a defined time period. The distance between the parallel lines (in millivolt) is taken as the measured noise, v_n, and the noise level, N, is calculated by the following equation:

$$N = v_n AT = \frac{v_n}{B}$$

where v_n is the noise measured in volts from the recorder, AT is the attenuation, and B is alternative amplification factor (Scott, 2001e). For the determination of DL, the authors recommend the method of Funk et al. (1992) by generating linear regression curve of relative low concentration of the analyte and then calculating the test parameter Xp, in this case Xp = DL. DL could be also defined by the amount of the analyte for which the peak height is four times the intrinsic noise height ($(S/N) = 4$) (Sandra, 2002).

For getting a reliable result in quantitative analysis, the value of the detector signal y_i should be proportional to the measured concentration of the analyte. Ideally, there should be a linear relationship between y_i and c_i. The linearity could be proved by various statistical methods [for details the readers could refer to the books by Funk et al. (1992) or Kromidas (1999). It is important to mention that the value of correlation coefficient r alone could not be used anymore to express the linearity of the calibration curve (Analytical Method Committee, 1988; Van Loco et al., 2002).

Another parameter for expressing the detector linearity was the response index (ri) of detector (Scott, 2001f).

$$y_i = K c_i^{ri}$$

If the value of ri = 1, the detector can be defined as truly linear detector. In general if the value of $0.98 < ri < 1.03$, the detector can be assumed to be linear. The value of ri can be estimated by plotting of log(peak height) against solute concentration at peak maximum; ri can be calculated as slope of the curve.

B. Flame Ionization Detector

It is well known that the FID is the most commonly used detector in GC. FID consists of a hydrogen/air flame and a collector electrode plates. The effluent from the column passes through the flame, which oxidizes the organic molecules and produces ions. A collector electrode attracts the negative ions to the electrometer amplifier producing an analog signal, which is connected to the data system. A detailed diagram of an FID detector is shown in Fig. 17 (Jinno, 2002).

FID is sensitive to all compounds which contain C–C or C–H and considerably less sensitive to insensitive to certain functional groups of organic compounds, such as alcohol, amine, carbonyl, and halogen. In addition, the detector is also insensitive toward noncombustible gases such as H_2O, CO_2, SO_2, and NO. This detector has a large dynamic range; its only disadvantage is that FID destroys the samples. The linear dynamic range of the FID detector covers at least four to five orders of magnitude for $0.98 < ri < 1.03$ (Scott, 2001f, g). For operating FID, the hydrogen flow usually ranges between 20 and 30 mL/min, and the airflow is about 120–200 mL/min (for pack column) or approximately the ratio air hydrogen should about 10:1. For capillary columns, the flow-rate may be less than 1 mL/min (Scott, 2001g). To have a maximum sensitivity for FID, it is recommended to use only pure hydrogen. FID is capable for measuring of 10^{-12} g carbon/s. The FID temperature should be maintained hot enough (recommended at least 200°C) so the condensation does not occur in the system.

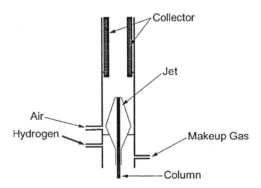

Figure 17 A schematic diagram of FID. [Reproduced from Jinno (2002).]

To enhance the sensitivity of the FID detector toward halogen compounds, the FID detector can be combined with the dry electrolytic conductivity detector (FID/DELCD). The DELCD consists of a small ceramic tube—the DELCD reactor—that is heated to 1000°C. When combined with FID, DELCD is mounted on the FID exhaust. The FID exhaust consists of uncombusted hydrogen, oxygen, nitrogen, water, and carbon dioxide. The reactions between chlorine, bromine, and hydrogen form HCl or HBr, the reactions of chlorine or bromine and oxygen form ClO_2 or BrO_2. DELCD detects the oxidized species of chlorine and bromine. The combination of FID and DELCD is 100 times more sensitive than DELCD operated without FID. Although the FID/DELCD detector is less sensitive than ECD detector, the FID/DELCD detector is more selective for measuring the pesticides (http://www.srigc.com/DELCD.pdf, May 19, 2003; http://www.chromtech.net.au, May 19, 2003).

C. Thermal Conductivity Detector

TCD sometimes referred as *hot-wire detector* or khatarometer (spelled catherometer). TCD is a very early detector for GC, which is based on changes in the thermal conductivity of the gas stream brought by the presence of analyte molecules. Since TCD reacts nonspecifically, it can be used universally for the detection of either organic or inorganic substances. Any component including nitrogen and oxygen, except the gas used for carrier gas, can be detected by TCD. The principle of the detection is using a Wheatstone bridge. Two pairs of TCD are used in GC, one pair is placed in the column effluent to detect the analyte and another is placed before the injector or in separate reference column. There are two basics of TCD design, the "in-line" cell, in which the effluent actually passes directly over the filament, and the "off-line" cell, where the filaments are situated away from the main gases and only reach by diffusion (Fig. 18) (Scott, 2001h).

TCD is a nondestructive detector and capable of measuring 10^{-9} g/mL solute, but it has limited application for narrow-bore capillary column, and can be operated only with light carrier gas such as hydrogen or helium, because their thermal conductivities are 10–15 times greater than that of most organic compounds. TCD detector is flow- and temperature-dependent, and its sensitivity is about 10–100 times less than FID (Karasek and Clement, 1988; Jinno, 2002; Schomburg, 1990).

D. Electron-Capture Detector

In ECD, the column effluent passes over a beta-emitter, such as nickel-63 or tritium. The electrons from emitter

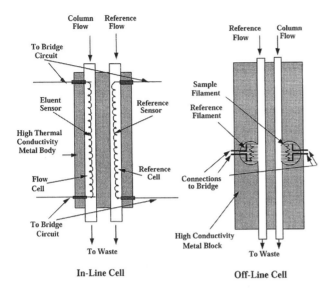

Figure 18 Thermal conductivity detector. [Reproduced from Scott (2001h).]

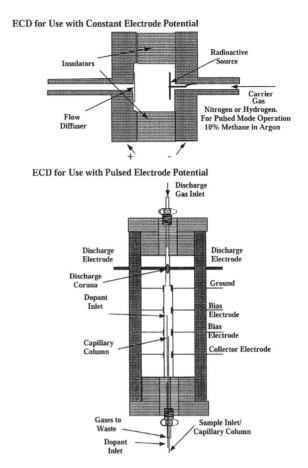

Figure 19 Electron-capture detector. [Reproduced from Scott (2001i).]

bombard the carrier gas (nitrogen), resulting in ions and a burst of electrons. In the absence of an analyte, the ionization process yields a constant standing current. When organic molecules that contain electronegative functional groups, such as halogens and phosphorous groups, pass by the detector, they capture the electrons and reduce the current between the electrodes. The loss of electron stream is related to the quantity of analyte in the carrier gas or make-up gas. The ECD can be classified into a constant potential (DC) mode and a pulsed potential mode. A detailed diagram of the two types of ECD detectors are shown in Fig. 19 (Scott, 2001i).

Sensitivity of ECD increases in the order of F < Cl < Br < I. By chemical derivatization, the applicability of ECD can be expanded for the trace analysis of species that do not or only weakly capture electrons. For these purposes, pentafluorobenzoic acid anhydride and pentafluorobenzoil chloride can be used (for alcohol, phenol, and amine). For using the ECD detector, the carrier and the make-up gas should be very clean and dry. For analysis using packed column, nitrogen, or argon–methane can be used as carrier gas and no make-up gas is needed, whereas in capillary column, helium or hydrogen is preferred as carrier gas, with nitrogen or argon–methane as the make-up gas (25–30 mL/min) (Sandra, 2002).

The applications of ECD illustrate the advantages of a highly sensitive special detector toward molecules that contain electronegative functional groups such as halogens, peroxide, quinines, or nitro groups (Buffington and Wilson, 1987; Pinho et al., 2002).

E. Nitrogen/Phosphorous Detector

The nitrogen phosphorous detector (NPD) is similar in design with FID. This detector is also known as the *thermoionic emission detector* or FTD. An electrically heated (up to 800°C) thermoionic bead is positioned between the jet orifice and the collector. Generally the bead consists of heated silica bead doped with an alkali metal, usually rubidium or cesium salt. Nitrogen and or phosphorous containing molecules will collide with the hot bead and undergo catalytic surface reaction and produce ions, and the ions will be attracted to collector electrode, amplified, and output to the data system. A diagram of the NPD detector is shown in Fig. 20 (Scott, 2001j).

This detector is commonly used for analyzing pesticide. NPD can be combined with DELCD, and this will become a very ideal detector to analyze pesticide. NPD selectively detects organophosphate, whereas DELCD can detect the chlorine species. Comparing with FID, this detector is about 500 times more sensitive for

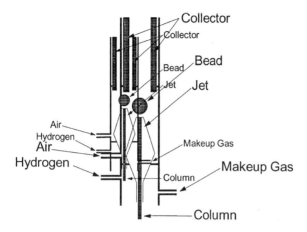

Figure 20 Nitrogen phosphorus detector. [Reproduced from Scott (2001j).]

phosphorous-containing molecules and about 50 times more sensitive for nitrogen-bearing compounds. NPD is a highly sensitive but also specific detector for nitrogen and phosphorous (http://www.srigc.com/catalog/npddetector.html, May 19, 2003; http://www.gc.discussing.info/gs/e_detection/nitrogen-phosphorous-detector.html, May 20, 2003; Jinno, 2002).

F. Flame Photometric Detector

In contrast to the oxygen-rich flame detector FID, FPD uses a hydrogen-rich flame, which is cooler. FPD is selective for S- and P-containing molecules. FPD uses chemiluminescent reaction in a hydrogen/air flame. In this FPD detector, the effluent passed into a low temperature H_2 air flame, which converted phosphorous and sulfur to emitting species. The emitting species for S-compounds is excited S_2, which has λ maximum at about 394 nm, whereas for P-compounds, the emitter is excited HPO, which has λ maximum about 512–526 nm (doublet) (http://www.gc.discussing.info/gs/e_detection/flame-photometric-detector.html, May 20, 2003).

The emitted visible and UV bands are filtered using 526 nm band pass filter (for P-compounds) or 394 nm filter (for S-compounds) and their intensity is recorded photometrically.

Sulfur compounds can be detected to about 200 ppb, whereas phosphorous can be detected down to 20 ppb (http://www.srigc.com/FPD.pdf, May 20, 2003). It should be mentioned that the intensity of light emitted is not linear with the concentration, but approximately proportional to the square of the sulfur atom concentration (Sandra, 2002).

G. Atomic Emission Detection

The recent development of GC detectors is the application of atomic emission detector (AED). The components of AED comprise interface between the incoming capillary column and the microwave induced plasma chamber, the microwave itself, cooling system, a diffraction grating and associated optics to focus then disperse the spectral atomic line, and a adjustable photodiode array interfaced to a computer. The microwave cavity cooling is necessary because most energy focused into the cavity is converted to heat.

The strength of AED is its capability to determine simultaneously the atomic emission of many elements in the samples that elute from the GC effluent. As the solutes are fed into a microwave powered plasma cavity, the compounds are destroyed and their atoms are excited by the energy of the plasma. The light that is emitted by the excited particles is separated into individual lines by a photodiode array. The connected computer then sorts out the individual emission lines and forms a chromatogram made from specific element only (http://www.elchem.kaist.ac.kr/vt/chem-ed/sep/gc/gc-det.htm, April 29, 2003). The advantage of using AED is the possibility of making compound-independent calibration.

Other type of GC-AED systems is using heating quartz tube furnace that is wrapped with nichrome resistance wire and ceramic isolation material, completed with a thermocouple. The furnace can be heated up to 1000°C. The first type comprises a reaction vessel that is connected with U-tube containing GC sorbent. During the reaction, the U-tube is kept in liquid N_2, the reaction products will be trapped in the sorbent, and after the reaction is completed, the liquid N_2 is removed, then the U-tube is heated to separate the analyte(s). In the second type, the effluent of GC is directly connected to the furnace; in this case derivatization reaction is performed prior to the injection of the sample into GC system. For derivatization of several elements (Hg, Ge, Sn, Pb, Se, As, Te, Sb, Bi, and Cd), sodium borohydride ($NaBH_4$) and sodium tetraethylborate ($NaBEt_4$) can be used for hydride formation; the latter reagent is used especially for aqueous derivatization. The Grignard reaction is also widely used (Cai and Zhang, 2001).

Interested reader can refer to an excellent recent review of Anderson (2002) for the application of the AED as GC detector.

H. Gas Chromatography-Mass Spectrometry

To provide GC, the capability for identification of unknown peak(s) and performing the structure elucidation, sensitive and selective quantitation of target compounds in the complex samples, the use of hyphenated techniques such as GC-MS, GC-FTIR, or GC-inductively coupled plasma-MS (GC-ICP-MS) are recommended.

1. GC-MS Interface

In the GC-MS system, the effluent of the column is split and passed directly to mass spectrometer. Due to the high flow of the carrier gas in the packed column of GC, that is, contrary to relatively low vacuum system of the MS, the use of an interface is necessary. The major goal of the interface is to remove the carrier gas without removing the analyte. An ideal interface could transfer the analyte quantitatively, reduce the pressure and flow-rate to a level that mass spectrum can handle. There are four commonly used interfaces for GC-MS system: molecular separator, permeation separator, open split, and capillary direct (http://www.nes.coventry.ac.uk/research/cmbe/306che/306/che58.htm, May 6, 2003). A schematic diagram of the interfaces is shown in Fig. 21.

Molecular Separator

This interface is used in GC-MS system using a packed column. Principle of operation is based on the difference of relative rates of diffusion. Small molecules like carrier gas diffuse more rapidly and will miss the entry jet into the mass spectrum. Solute molecules that are relatively larger molecules will reach the entry jet into the mass spectrum. The rate of diffusion depends on the molecular weight (http://www.nes.coventry.ac.uk/research/cmbe/306che/306/che58.htm, May 6, 2003). An example of this molecular separator is Watson–Bieman separator, which comprises all silanized glass devises, obtainable on extraordinarily inert surface for the interface between GC and MS, thus allowing analysis of highly polar thermal labile molecule (Watson, 1998).

Permeation Interface

In this interface, a semipermeable membrane is placed between the GC effluent and the inlet of the mass spectrum. The membrane selectivity is based on polarity and molecular weight of the solute. This interface is inefficient because only small fraction of analyte can permeate (http://www.nes.coventry.ac.uk/research/cmbe/306che/306/che58.htm, May 6, 2003).

Open-Split Interface

The mass spectrum pulls in about 1 mL/min through a flow resistor. If the flow exceeds this figure, helium from an external source is pulled out. This interface is best for sources with flows close to 1 mL/min, for example, for capillary column. It should be mentioned that split

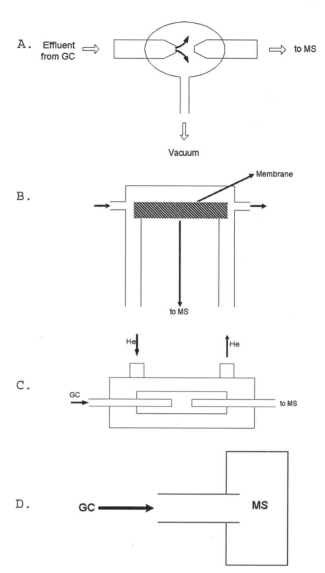

Figure 21 Schematic diagram of GC-MS interfaces. A, molecular separator; B, permeations interface; C, open-split interface; and D, Capillary direct interface.

would change as gas flow changes (http://www.nes.coventry.ac.uk/research/cmbe/306che/306/che58.htm, May 6, 2003).

By using a GC-MS open-split interface device, which has been developed recently by SIS Scientific Instrument Services (Ringoes, NJ, USA), allows the capillary column exit to be maintained at above atmospheric pressure, thus the column performance and retention indices are the same as those when using FID. The device comprises an open-split adaptor, a deactivated fused silica restrictor tubing, and a MS connection (http://www.sisweb.com/ms/sis/opensplt.htm, June 20, 2003).

Capillary Direct Interface

The whole effluent of the capillary column will pass directly into the mass spectrum. This is done by simply inserting the capillary column into the ion source, it has no dead-volume, and so the cost is low.

Dominguez et al. described a universal interface for packed wide-bore and capillary column. They used a length of fused capillary tube permanently (inner diameter 100 μm) connected to the ionization chamber, so no loss of vacuum when the column was changed. The tube is introduced directly inside the capillary column or the wide-bore column. For packed column, a small length of wide-bore tubing is connected at the column outlet, and the connection line to the mass spectrometer (http://www.eurojai.com/jai99/resumenes/css/resumenes_395.htm, June 20, 2003).

A common problem with use of MS as the detector for GC is the amount of time needed to change a capillary column (around 6–12 h). By inserting a GC-MS column changeover device such as EZ No-Vent™ or ms-NoVent™, between capillary exit and MS source, the capillary column can be changed easily without the need to vent the instrument (http://www.restekcorp.com/2003/890-16.pdf, June 20, 2003).

Supersonic Molecular Beam Interface

Recently Amirav et al., 2001 (http://www.tau.ac.il/chemistry/amirav/smbms.html, June 20, 2003) developed a new GC-MS method, namely, "Supersonic GC-MS" that used supersonic molecular beam (SMB) as the interface between GC and MS. A SMB is performed by the expansion of a gas from the GC-outlet (1 atm) through a 0.1 mm nozzle into a vacuum chamber of MS. By this expansion the carrier gas and heavier solute molecules obtain the same final velocity, so the solute molecules are accelerated to the carrier gas velocity, since it is the major gas component. Furthermore, the uniform velocity ensures slow intra-beam relative motion, resulting in the cooling of the internal vibrational degrees of freedom. SMB are characterized by unidirectional motion with controlled hyperthermal kinetic energy (1–20 eV), intermolecular vibrational super cooling, and mass focusing similar to that in molecular separator. The kinetic energy of the solute molecules increases with its molecular weight and nozzle temperature, and is reduced by increasing the carrier gas molecular weight. Due to high gas flowrate requirement in GC-SMB-MS, it enables GC for the analysis of both thermal labile and low-volatility samples. The commercially available GC-MS could be modified to supersonic GC-MS as described by Amirav et al. (2001). The GC-SMB-MS is not yet commercially available.

2. Ionization Methods

In the present time, different methods of ionization are available in GC-MS system, namely electron-impact ionization (EI), chemical ionization (CI), ICP, field ionization (FI), and recently developed hyperthermal surface ionization (HIS).

Electron-Impact Ionization

In EI, the molecules of organic compounds from the GC effluent are bombarded with electron (usually 70 eV) emitted from a rhenium or tungsten filament. Electrons that are generated by a heated filament pass across the ion source to an anode trap. The magnitude of the collection potential may range from 5 to 100 V, depending on the electrode geometry and the molecules being ionized. The sample molecules are introduced in the center of the ion source, and the solute molecules drift, by diffusion, into the path of the electron beam. Collision of the sample molecules with the electrons could produce molecular ions, ionized molecular fragments, radicals, and neutral fragments. The produced ions are driven by a potential applied to the ion-repeller electrode into the accelerating region to the mass spectrometer. A diagram of an EI ion source is shown in Fig. 22 (Scott, 2001k).

The ion fragmentations could be illustrated in the following equations:

$$\begin{aligned} ABCD &\longrightarrow ABCD^{+\bullet}(M^{+\bullet}) + e^- \\ &\longrightarrow AB^{+\bullet} + CD \\ &\longrightarrow AB^+ + CD^\bullet \\ &\longrightarrow ABC^+ + D^\bullet \longrightarrow AB^+ + C + D^\bullet \\ &\longrightarrow A^+ + BC + D^\bullet \end{aligned}$$

Since in the EI method low ion source pressure is employed (10^{-5} torr) and reactions in the source are

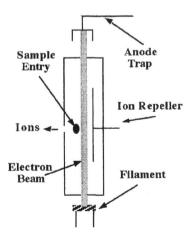

Figure 22 Electron-impact mass spectrometer source. [Reproduced from Scott (2001k).]

unimolecular, association between molecules and fragments does not occur. A disadvantage of the EI method is that complete fragmentation can occur, so sometimes the molecular ion can be observed. Some factors that influence the EI spectrum of the analyte can be summarized as follows: ionization voltage, ion source temperature, sample pressure, and repelled voltage (Anonymous, 1985).

As described earlier, by using SMB as the new interface, the supersonic expansion yields substantial super cooling of the solute molecules at vibration temperature below 70 K, so Amirav et al. (http://www.tau.ac.il/chemistry/amirav/smbms.html, June 20, 2003) named this EI of SMB as *cold EI*. The molecular ion peak of cold EI can be increased three times. With cold EI method deuterium exchange can be employed for NH and OH labeling.

Chemical Ionization

In the CI method, reactions occur between ionized reagent gas (G) and the analyte molecule (M) to produce pseudo-molecular ions. Pseudo-molecular ion positive $[MH]^+$ or negative $[M-H]^-$ can be observed. Unlike the EI method, CI detection occurs in high yield and less fragment ions are observed. For maximizing collision between the ionized gas molecule and the analyte, a tight ion source pressure (0.1–2 torr) is required. In the positive ion mode, the reaction is:

$$[GH]^+ + M \rightarrow [MH]^+ + G$$

whereas in the negative ion (NCI) mode, the reaction is

$$[G-H]^- + M \rightarrow [M-H]^- + G$$

These simple reactions are true gas-phase acid base processes in the Bronsted–Lowrey sense. The reaction process is gentle, and the energy of most reactive ions usually never exceeds 5 eV. Consequently, there is little fragmentation when compared with the EI mode. The main reagent gases used in CI are: methane, ammonia, and isobutane.

In the methane positive ion mode CI, the relevant peaks observed are $[MH]^+$ (meanly peak), $[M+CH_5]^+$, and $[M+C_2H_5]^+$; for ammonia the observed peaks are $[MH]^+$ and $[M+NH_4]^+$; and for isobutane the main peak is $[MH]^+$. In the common negative ion mode the peaks observed are M^- or $[M-H]^-$ if methane is used as reactant. Other reactants that can be used for NCI are CH_2Cl_2, $CHCl_3$, and O_2. Like positive ions, it undergoes proton-transfer reaction, addition, and charge-exchange reactions (Anonymous, 1985).

$$M + Cl^- \rightarrow [M-H]^{-1} + HCl$$

$$M + Cl^- \rightarrow [M+Cl]^{-1}$$

If the instrument has EI/CI common ion source, the electron slit of the CI chamber should be facing the filament.

ICP Source

The ICP ion source is similar to the unit of the ICP atomic emission spectroscopy. The argon plasma initiated by a Tesla coil park, and maintained by high radiofrequency (RF) energy, inductively coupled to the inside of the torch by an external coil, wrapped around the torch system. The plasma is maintained at atmospheric pressure and at temperature about 8000 K. The ICP-MS is based on the ionization of the samples in plasma. From this plasma the ions are transferred by two screens, which are called skimmer and sampler into the evacuated system of the mass spectrometer. A diagram of the ICP ion source is shown in Fig. 23 (Scott, 2001k). The ICP ion source has the advantages that the sample-molecule is in atmospheric pressure, the degree of ionization is relatively uniform for all elements, and singly charged ions are the usually the main products. ICP-MS can be used for wide ranges of elements and isotopes, only H, He, Ne, Ar, and F cannot be directly measured, and very little effect of matrix, water, and acids. By using GC-ICP-MS, a compound independent calibration technique can be applied, so it is not necessary to run calibration for every compounds, quantitation based on elemental (not compound) response, multielement, multilevel calibration can be generated from single injection (http://www.doacs.state.fl.us/~fs-prw/2002/wylie.pdf, June 5, 2003). It should be mentioned that by using ICP-MS, unlike EI and CI, the information of the chemical structure of the analyte is lost.

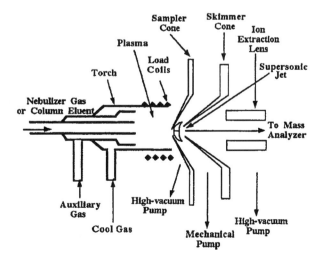

Figure 23 ICP mass spectrometer source. [Reproduced from Scott (2001k).]

Field Ionization

Field ionization (FI) can produce molecular ions for some compounds that do not give molecular ions by EI. Generally FI can produce dominant molecular ion with almost no fragmentation. FI utilizes 10 μ diameter tungsten emitter wires on which carbon whisker, or dendrites, has been grown. A high electrical field gradient 10^8 V/cm at the tips of the whiskers produces ionization by quantum mechanical tunneling of the electrons of the gas phase of the sample-molecules. Compared with CI, FI has less fragmentation, no high-resolution, and less sensitive. For solid samples, this technique calls field desorption (http://www.scs.uiuc.edu/~msweb/ion.fi.html, June 2, 2003). Generally FI yields simple spectrum with intense $[M]^+$ molecular ions, unlike CI, FI produces no adduct molecular ions (e.g., $[M+H]^+$) (http://www.micromass.co.uk).

Hyperthermal Surface Ionization

It is found that unidirectional motion with controlled hyperthemal kinetic energy enables an effective ionization method called HIS. HIS is based on the hyperthermal surface scattering of the sample-molecules from a suitable surface such as rhenium oxide that has high surface work function. HIS is a selective detection method of drugs and polycyclic aromatic hydrocarbon in complex matrix. The interested readers can refer to the original publications by Amirav et al. (http://www.tau.ac.il/chemistry/amirav/smbms.html, June 20, 2003) Amirav and Granot (2000) and Amirav et al. (2001).

3. *Mass Analyzer*

The mass analyzers used in GC-MS system are quadrupole mass spectrometer, magnetic sector instrument mass spectrometer, time-of-flight (TOF), mass spectrometer, and ion trap mass spectrometer. The mass analyzer is used to separate ions according to their mass-to-charge (m/z) ratio. Resolution R of a mass analyzer can be defined as $m/\Delta m$. The SIM method is used to improve sensitivity for quantitative measurements by monitoring only the mass of interest for specific compounds. The use of tandem mass spectrometry (MS/MS) improves the selectivity, by reduction of the background effect, and without loss of information on identification capability.

Magnetic Sector Mass Spectrometer

A magnetic sector will separate ions according to their m/e values. By the fact that the ions leaving the ion source do not have exactly the same energy, therefore do not have same velocity, the resolution will be limited. To achieve a better resolution, it is required to add an electric sector that focuses ions according to their kinetic energy.

If a single charged ion of mass m with velocity v injected to a magnetic field of strength B will follow a circular path of radius r,

$$r = \frac{mv}{Be}$$

where all ions formed in the ion source are accelerated at potential difference V to a kinetic energy T

$$T = eV = \frac{mv^2}{2}$$

$$v = \sqrt{2e\frac{V}{m}}$$

From the Lorentz force of law, the magnetic field applies a force evB which must be equal to the centripetal force as the ion moves in arc through magnetic sector.

$$evB = \frac{mv^2}{r}$$

Substituting for v we get

$$\frac{m}{e} = \frac{B^2 r^2}{2V}$$

The magnetic sector has a fixed radius and typically ions are accelerated through constant V, so by varying B (scanning) a spectrum of m/e can be obtained, an alternative B can hold constant and scan V (http://www.jeol.com/ms/docs/ms_analyzers.html, June 5, 2003) (Ruecker, 1976).

Quadrupole Mass Spectrometer

This is the most common mass analyzer used in GC-MS system, either as single quadrupole or as triple quadrupole, which could provide MS-MS spectra. A diagram of a quadrupole MS is shown in Fig. 24 (Scott, 2001k).

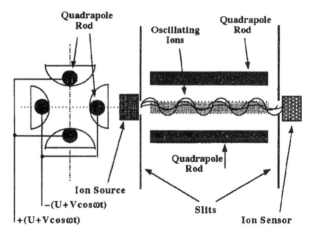

Figure 24 Quadrupole mass spectrometer. [Reproduced from Scott (2001k).]

The instrument consists of four rods that should be precisely straight and parallel, so that the arranged ions beam directly axially between them. A voltage comprising a DC component (U) and RF component ($V\cos \omega t$) is applied to four rods. Ions are accelerated into the center between the rods, by a potential ranging from 10 to 20 V. Ions are vibrating in the x–y plane by the effect of electric field generated by DC and RF. The mass range is scanned by changing U and V, whereas the ratio U/V is held constant. Once the frequency ω of the RF and mass ion are fixed, the vibrating ion becomes stable only when the value of U and V are set in specific region, in this case the ions can pass through in the direction of ion sensor. The stable region differs according to mass of the ion, m. If U and V are scanned along the straight line with U/V as constant, the ions can only pass when U and V are in the stable region for their mass (JEOL, 1993). These parameters of a quadrupole mass spectrometer are described by the Mathieu equation (http://www.webmaster.icp.ms.de, June 4, 2003):

$$\frac{m}{e} = UV\frac{r_0}{\omega}$$

where r_0 is the radius of the quadrupole. As a result, ions having different masses are detected at different periods. The disadvantages of the quadrupole mass analyzer are their moderate resolution ($R < 1000$) and limited mass range that can be analyzed (up to $m/e = 3000$).

The combination of three quadrupole mass spectrometers in series can be constructed to provide MS-MS spectra (Scott, 2001k). In the first quadrupole various ions are separated by usual way, and passed to second quadrupole instead to sensor. In the center quadrupole, the selected ions can be further fragmented by collision ionization, and then the new fragments are passed to the third quadrupole that has function as second analyzer.

TOF Mass Analyzer

The ion source should generate a short (nanosecond) pulse of ions with a fixed and well-defined kinetic energy. The ions can be generated by a pulse ionization method matrix-assisted laser desorption ionization (MALDI), or various kinds of rapid electric field switching are used to release ions from the ion source in a very short time. A detailed discussion on the recent development of MALDI was reported by Franzen et al.(2001). The ions drift along a long flight tube (ca. 1 m in length). Lighter ions travel faster and arrive at the detector first. The kinetic energy of an ion leaving the ion source is (http://www.jeol.com/ms/docs/ms_analyzers.html, June 5, 2003):

$$T = eV = \frac{mv^2}{2}$$

The ion velocity, v, is the length of the flight path, L, divided by the flight time t:

$$v = \frac{L}{t}$$

By substituting v into the kinetic energy relation, the working equation for the TOF mass spectrometer can be derived:

$$\frac{m}{e} = \frac{2Vt^2}{L^2}$$

By rearranging, the equation of TOF is:

$$t = L\sqrt{\frac{m}{e2V}}$$

This equation shows that the mass of the ions is directly proportional to the square of TOF of the sensor. Interested reader could refer to an excellent review on TOF MS by Mamyrin (2001).

Trapped-Ion Mass Analyzer

The two principles of trapped-ion mass analyzer are: three-dimensional quadrupole ion traps (dynamic traps) and ion cyclotron resonance mass spectrometer (static traps). Both operate by storing ions in the trap and manipulating the ions by using DC and RF electrical fields in a series of carefully timed events (http://www.jeol.com/ms/docs/ms_analyzers.html, June 5, 2003).

ION CYCLOTRON RESONANCE. Ions move in a circular path in a magnetic field. The cyclotron frequency, ω, of the ion's circular motion is mass dependent. By measuring the cyclotron frequency, the mass of an ion can be determined. From the Lorentz force of law, the magnetic field applies a force evB which must be equal to the centripetal force as the ion moves in arc through magnetic sector.

$$evB = \frac{mv^2}{r}$$

Solving for the angular frequency, ω, which is equal to v/r:

$$\omega = \frac{v}{r} = \frac{eB}{m}$$

Ions that have the same mass-to-charge ratio will have the same ω, but they will be moving independently and out-of-phase at roughly thermal energy. If an excitation pulse is applied at the cyclotron frequency, the resonance ions will absorb energy and will be brought into phase by the pulse. Since several different masses are present, one should apply an excitation pulse that contains all frequencies of the cyclotron. The image currents induced in the receiver plates will contain all frequencies of the ions.

By using a Fourier transform a time-domain signal (the image currents) could be converted into a frequency-domain spectrum (http://www.jeol.com/ms/docs/ms_analyzers.html, June 5, 2003).

QUADRUPOLE ION TRAPS. The quadrupole ion trap mass analyzer consists of three cylindrically symmetrical electrodes comprising two end caps and rings.

An RF voltage together with an additional DC voltage is applied to the ring, and the end of caps are grounded. The RF and DC potentials can be scanned to eject successive mass-to-charge ratio to from the traps into the detector. This process is called *mass selective* ejection (http://www.jeol.com/ms/docs/ms_analyzers.html, June 5, 2003). Interested readers can refer to the book by March et al. (1989) for the background theory of quadrupole ion trap detector.

4. GC-FTIR Spectroscopy

By using the combination of GC and FTIR spectroscopies (GC-FTIR), IR spectra of the effluent are recorded as they emerge from the column outlet.

The effluent is introduced at a light pipe (LP) where the analyte absorbs radiation at well-defined frequency (Sandra, 2002). The LP interface consists of a flow-through glass tube that is internally coated with a high-reflective gold layer and sealed with IR-transparent alkali halide windows at each end. LP should be heated to prevent condensation that causes band broadening. Typically cell volumes are generally 80–150 μL, cell sizes are 10–20 cm with ca. 1 mm ID (Mossoba et al., 2001).

Due to its low sensitivity, other methods such as direct deposition (DD) or matrix isolation (MI) were preferred. In the MI method, the effluent is sprayed in real-time into the outer rim of slowly rotating gold-plated disk held at 12 K under 10^{-5} torr vacuum, by Joule expansion of compressed helium. Carrier gas usually used is a mixture of helium and argon in the ratio of 98.5:1.5. At 12 K, the helium is pumped away, leaving behind the IR-transparent argon to be frozen with the separated components on the disk. The separated analyte(s) by GC are individually frozen on the outer rims of the moving disk. After the GC separation is completed, each GC-peak is sequentially placed in the path of the IR beam for measurement (Mossoba et al., 2001).

With DD interface, the effluent is directly deposited without argon matrix on a rectangular ZnSe crystal and held under vacuum near 77 K using liquid nitrogen. The ZnSE crystal window is mounted on a computer controlled x–y stage that moves in small increments during the GC separation. Each deposited analyte passes through the IR beam after a brief delay (about 20 s) (Mossoba et al., 2001).

Recently by using a miniaturized radio-frequency plasma, called microplasma, that directly mounted on a top of GC, a hyphenation of a near-infrared (NIR) Echelle spectrometer and GC was developed. An optical fiber with core diameter 400 μm was used to connect the microplasma to the entrance slit of the NIR spectrometer (Cziesla et al., 2001).

5. Hyphenated GC-FTIR/MS and GC-AED/MS

Due to difficulties in isomer-identification using hyphenation GC-MS, it is necessary to add a FTIR in the system to form hyphenated GC-FTIR/MS. By using the combination detectors FTIR and MS, the capability for analysis a complex mixture increases dramatically. Using this hyphenated equipment a pair of closely related compounds geraniol and nerol isomers could be properly identified as well quantified.

Another problem of using GC-MS system is in determining of unknowns which have common structural elements; for example, how many Cl^-- and Br^--containing compounds are present in the sample, and what is the nature. In this case, analyzing with full-scan MS screening is not a good strategy as it is much time-consuming—and peaks of minor constituents partly co-elute with a much larger peak and often go undetected. To solve such a problem, performing two subsequent runs, one with AED and other with MS, could be very helpful. Wilson and Brinkmann (2003) recently published an excellent review article on this hypernation system.

IX. MULTI DIMENSIONAL GC AND COMPREHENSIVE TWO-DIMENSIONAL GC

A. Multidimensional GC

In the multidimensional GC, selected heart-cut transfer of the first column elution zone into the second column was observed, where improved separation occurred. For transferring the bands from the first column, a cryogenic trapping is often used. The cryogenic component is generally located between two columns. In most cases the two columns have different selectivity by asset of their different polarities, some times the second column could be a chiral stationary phase column. The coupling of a narrow to narrow-bore capillaries could enhance the resolution. Selective sample introduction method could be achieved by coupling of wide-bore precolumn to narrow-bore capillary column (Marriott and Kinghorn, 1998).

B. Comprehensive Two-Dimensional GC

The term comprehensive GC (GC × GC) is given to the technique in which two columns are directly serial coupled. The system was termed comprehensive because the entire sample could be analyzed, not just heart-cut fraction. The important factor of the GC × GC is the method of the modulation of the chromatographic band from the first column to the second column. Connection of two columns of different polarity and shorter length than usual enabled rapid and high-resolution separation. This GC × GC technique is very useful for studying complex mixtures such as petroleum fuel and essential oils. The most important instrumentation component of a GC × GC system is the modulator, which is necessary for transferring of the samples from the first column to the second column. It acts as a continuous injector for the secondary column. The interface between two columns should carry out several tasks. First, it must be capable of trapping all the analytes that eluted from the first column; second, it must be able to reintroduce these trapped analytes into the gas stream quantitatively in narrow band, and finally it must start trapping again as soon as possible after an injection into second column. Some important factors that should be mentioned for performing GC × GC are: the second column separation should be rapid in comparison of the solute band with that of the first column, the second column should have a high phase ratio compared with first column, solute(s) from the first column should be focused or compressed before being passed to second column (Bertsch, 1999). Therefore a modulator should be placed between two columns. Different modulation devices will be described in the following section.

1. Thermal Desorption Modulator

This modulator consists of a short piece of capillary tubing that is installed in a cold zone outside the GC-oven. Solutes from the first column are effectively focused on the thick layer of the stationary phase in the modulator tube. Application of electrical pulse to metal-coated tube heats the stationary phase and drives the solutes into gas phase, and then the carrier gas moves the solutes to the secondary column. Two factors should be considered here: the temperature of the thermal desorption modulator should be sufficient to cause rapid heating of the modulation tube, and the duration of the heating period must be long enough to displace the solutes in the modulator with the carrier gas (Bertsch, 2000).

2. Rotating Slotted Heater Modulation

Solutes eluting from the first column were collected in a short segment of a thick capillary column, and then released by application of a heat pulse. The key components of the thermal modulator are a rotating slotted heater and a modulator tube. The modulator tube is used to serially connect the two columns. The heater periodically rotates over the thick film stationary phase section of the modulator tube to desorb the solute and injected it into the second column. The rotation period is usually set to ensure several injections for each first dimension peak. The second analysis dimension may be as rapid as 4 s or faster. All the effluents of the first column will be continuously directed to the second column, so multiple second dimension chromatograms will be observed (Marriott and Kinghorn, 1998; Purch et al., 2002).

3. Longitudinally Modulated Cryogenic System

This modulator regularly traps and releases solutes from the first column by moving a cryogenic trap back and forth along the second column. When the modulator is at top position, a narrow band containing the solutes that eluted from the first column is cold-trapped and focused in the inlet of the second column. When the modulator moves to the bottom, the cold-trapped solutes are rapidly heated by the surrounding oven air and remobilized to second column. The cool trap is about 3–4 cm long, and consists of two concentric stainless steel tubes connected and sealed at the ends to form a chamber between two tubes. The smaller inner tube has an open passage with an inner diameter slightly greater than a capillary column, so the column can pass through this tube. The outer tube has an entry port in which a cryogen gas (usually liquid CO_2) passes, and two-exit port for the expanded gas to pass. Generally the period of the modulator is usually from 2 to 9 s. This longitudinally modulated cryogenic system and rotating slotted heater modulation have a common disadvantage, that is, frequent breakage of the capillary moving parts of the systems (Xu et al., 2003).

4. Jet Cooled/Heated Modulator

This modulator consists of two cold and two hot nitrogen jets that are pulsed to alternately cool and heat two spots at the front end of the second column for focusing and remobilizing the solutes from the first column. Both cold jet tubes lie parallel to each other as do the hot jet tubes, but the cold jet tubes are orthogonal to the hot jet tubes.

Due to the difficulty in getting liquid nitrogen, Beens et al. (2001), developed a simple, nonmoving dual-stage CO_2 jet modulator. This modulator consists of two parts of the capillary that are directly and alternately cooled in order to trap and focus each subsequent fractions, which will remobilize by the heat from the surrounding oven.

5. Diaphragm-Valve Modulation

Fast valve switching could create pulses of samples that are injected into the secondary column. The solutes that are eluting from the first column are split after valve and before the secondary column. Only about 10–20% of the injected samples reach the detector, so the sensitivity of this modulator is significantly lower than the thermal modulator. This valve-based modulator sends only a part of the solutes from the first column to the secondary column, so this method is called "heart-cutting technique". The disadvantage of this modulator is the limited temperature that can be used (not higher than 180–200°C), so it can be used only for the volatile substances (Purch et al., 2002; Xu et al., 2003).

Seely et al. (2001) described a technique called "differential flow modulation GC × GC" that could pass about 90% of the effluent of the primary column into the secondary column, although not 100%, they classify their method as comprehensive. For this purpose they used a high-speed six-port diaphragm valve (maintained at 130°C) fitted with a 20 μL sample loop that was used to collect effluent from the primary column and periodically injected into the secondary column. Secondary injection was performed at a frequency of 1 Hz.

6. Comprehensive Two-Dimensional GC Mass Spectrometric System

The primary difficulty in the integration of the comprehensive two-dimensional GC Mass Spectrometric (GC × GC-MS) system is the slow scan speed of the MS. Full scan spectrum acquisition (45–350 amu) is about 2.4 scan/s compare with GC × GC second dimension peak (about 0.2 s), so the MS scan rate would be insufficient. Simply lengthening the first dimension column can solve this problem (Frysinger and Gaines, 1999).

X. CHROMATOGRAPHY DATA SYSTEM

In the last few years, most of the producers of the GC equipment introduced new 32 bit version programs or have enhanced their existing 16 bit programs. All the programs are mostly designed to work with 32 bit Windows operating system, Windows 95/98/2000 and Windows NT 4.0 or 5.0. Although some GC-producers used other systems such as Macintosh or Unix, their development is very limited. During the 1980s and 1990s the producers produced several softwares simultaneously, perhaps one for GC, another for LC or GC-MS. With the development of the 32 bit programs, many companies are trying to consolidate all the major techniques into one program. In the present time most of the software modules have several functionalities, that is, integrated instrument controls including automatic sampler, data analysis, integrated system automation, software validation tools, custom reports, and user defined security. There is also a trend to include a database with the chromatography software. The new programs can be also used to take control of four chromatographic systems at the same time, either GC or LC.

Lipinski and Stan (1988) developed computer aided pesticide analysis using TURBO PASCAL 3.0. This computer program can be used for evaluation of GC data for residue analysis of foods. Herron and Donnelly (1996) developed software that could help to identify the unknowns by separating their spectra from other background signal; the GC-MS data were first converted into ACII text and imported into personal computer data sheet. A system of pattern recognition for the analysis of complex mixtures using a conventional capillary GC was described by Park et al. (1993), for these purposes they used a multiple reference peak identification method and outlier statistic using on-line data system. Pool et al. (1997) developed an algorithm for automatic processing of GC-MS data in order to quantify individual component. They reported that the method can be used to analyze components with highly similar spectra and severe co-elution.

Various commercial chromatography data system appeared in the marketplace recently. MassTransit^{TM} software could translate GC-MS data and library files between formats used on OEL data systems. This software performs conversion of a wide variety of instrument manufacturer format including Finnigan, Fisons, Variant, Hewlett–Packard, and so on. MassTransit^{TM} has some versions: Window, DOS, and UNIX (http://www.members.optusnet.com.au/~mshivac/mt-tech.htm, June 17, 2003). ChromView^{TM} imports GC and GC-MS data into other Windows application such as MS Word, Excel, or Powerpoint as high-resolution images. ChromView^{TM} is a full OLE server, it can be operated as a stand-alone application with multiple document application interface capabilities, this software could support many GC data file types of different manufacturers (http://www.twcbiosearch.com/DOCS/1/12148.htm, June 17, 2003). ACD/MS Manager^{TM} software allows us to read, analyze, process, a nd store experimental mass spectral data from GC-MS, LC-MS, CE-MS, MS, and MS/MS. This software can extract information from the MS data using sophisticated chemometric algorithms such as CODA and COMPARELCMS (http://www.acdlabs.com/products/spec_lab/exp_spectra/ms/, June 17, 2003). Older GC/LC-MS data systems can be replaced by VECTOR/TWO^{TM} GC/LC-MS data operating system. This software automates nearly all commercial GC/LC-MS instruments, including sample injector, chromatograph, and mass spectrometer, and could be used for library search using probability based matching (http://www.sisweb.com/software/ms/vecror2.ht, June 19, 2003). AMDIS that can be freely downloaded from NIST Mass Spectral Library can help to analyze GC-MS data of complex mixtures, even with strong background and co-eluting peaks. AMDIS analyses steps are: noise analyses, component perception, spectrum deconvolution, and compound identification (http://www.stud.rz.uni-leipzig.de/~che94beq/amdis.htm, June 17, 2003) (Davies, 1998).

XI. QUALITATIVE AND QUANTITATIVE ANALYSES

As other chromatography methods, GC could be used for qualitative and quantitative analyses.

A. Qualitative Analysis

For qualitative analysis using GC some parameters can be used, that is, retention time (R_t), relative retention time (RR_t), and Kovats retention index system (RI). It should be mentioned that if R_t or RR_t of two substances are relatively identical, it does not always mean that the two substances are identical, but if R_t and RR_t are not identical, it means that the two substances are not identical.

For analysis of an unknown substances the combination of GC with other spectroscopic methods such as MS and or FTIR are very helpful to identify the unknown substances. If the analysis was performed using a GC-MS system, the identification of the analyte(s) can be done by using various softwares that have been described earlier.

Besides for analysis of a compound(s), GC can be used for making profile analysis of certain substances such as essential oils, flavors, perfumery, or metabolites of a biological system, and for fatty acids microbial profiling. Park et al. (1993) developed a system of pattern recognition for profile analysis of a complex mixture such as essential oils; the pattern recognition was accomplished by a multiple reference peak identification method and outlier statistics using on-line data system. The MIDI Sherlock™ microbial identification system identifies micro-organisms based on the unique fatty acid patterns using GC analysis (http://www.midi-inc.com/pages/Gcproducts.html, June 25, 2003). AromaTrax™ could be used for aroma profiling using GC, to create an *aromagram* (http://www.mdgc.com/aroma_profiling.htm, June 25, 2003). By using GC-MS the clinicians can determine the urinary steroid profile of the patients for diagnostic purposes (http://www.pupk.unibe.ch/mastering/departemente/durn/abt-nephro/03science/sciencehtml/services/klir, June 25, 2003). Kind and Westwood (2001) reported the application of GC-FID as the primary method for determination of the purity of the organic reference materials of 67 anabolic steroids in sport analysis.

For qualitative analysis of certain analyte(s) by using R_t/RR_t, generally two methods can be used, that is, time window method and time band method, so by defining these parameters first, the GC integrator or software can determine whether the analyte has the identical R_t/RR_t with the authentic standard, and identify it as the substance that has to be analyzed.

The concept of the RI was first published by E. Kovats in 1958. For establishment of any RI system, some requirements are needed, that is, the choice of set of reference compounds, the attribution of standard RI values for those compounds, and the choice of a formula for the calculation of RI values for all other analyte(s). For a GC isothermal condition RI of compound x can be calculated:

$$RI_x = RI_k + [RI_{k+1} - RI_k]\frac{\log R'_{t,x} - \log R'_{t,k}}{\log R'_{t,k+1} - \log R'_{t,k}}$$

where R_t' is corrected R_t, k is n-alkane reference compounds with number of carbon atom k (Zenkevich, 2002).

B. Quantitative Analysis

Methods are available for quantitative analysis using GC, that is, normalization method, external standardization method, internal standardization method, and standard addition method. Before the method(s) can be used routinely, the method(s) should be first validated. Funk et al. (1992), or Kromidas (1999) has published excellent books on the method of validation. Interested readers could refer those books.

1. Normalization Method

In this method, all the analyte(s) present in the sample must elute from the column, with enough resolution and furthermore they have to be detected by the available detector. In this case, the peak area is proportional to the weight of the analyte(s) having passed through the detector. The analytical signal lies within the linearity ranges. Thus, percentage in weight of each analyte can be calculated as follows:

$$c_i = K_i A_i$$

so

$$\% c_i = \frac{A_i}{\Sigma A_i} 100$$

2. External Standardization Method

This method is the most general method for determining the concentration of an analyte in samples. It involves the construction of a calibration plot (area or height vs.

analyte concentrations) by using some concentrations (minimum 3–4 concentrations) of external standard; the concentration of the analyte can be determined by "interpolation". It is not recommended to do "extrapolation". In this case the response factor a_t of the analyte can be calculated as:

$$a_t = \frac{A_i}{c_i}$$

Although this is not recommended by the authors, the concentration of the analyte in the samples can be calculated using by "one-point external standardization" method as follows:

$$c_i = \frac{A_i}{a_t} = \frac{A_i c_{ie}}{A_{ie}}$$

where c_{ie} and A_{ie} are the concentration of the external standard analyte i and its peak area/height, respectively.

3. Internal Standardization Method

The principle of the internal standardization involves the addition of a known quantity of the internal standard to the analyzed sample(s) and to the reference standard(s). This is to compensate errors mostly by variation of the injected amounts. It is not necessary to do "internal standardization" if the GC has an auto-sample injector. The calibration plot is constructed by using the *ratio* of peak area or height of standard(s) and internal standard(s) against concentrations of standard(s) or the ratio of concentrations of standard(s) and internal standard(s) (minimum 3–4 points). The concentration of the analyte(s) can be determined by "interpolation" of the calibration curve(s). In this case the response factor $a_{t/is}$ of the analyte can be calculated as:

$$a_{t/is} = \frac{A_i c_{is}}{c_i A_{is}}$$

where "is" is term for the internal standard. Although this is not recommended by the authors, the concentration of the analyte in the samples can be calculated using by one-point internal standardization method as follows:

$$c_i = \frac{1}{a_{t/is}} \left[\frac{A_i c_{is}}{A_{is}} \right]$$

4. Standard Addition Method

If the concentration of the analyte is below the (linear) range of the calibration curve, the external and internal standardization methods cannot be applied. In this case, one should add certain amount of the analyte in the sample (spiking). It is recommended to add 3–4 concentrations of the analyte into the sample (ca. 30–70% of the original concentration of the analyte in the sample).

A linear regression curve is constructed of the peak area (y-axis) against the added concentrations of the analyte in the sample (x-axis). The concentration of the analyte can be calculated from the intercept of the regression curve with the x-axis.

5. Selected Ion Monitoring

In this SIM method, instead of scanning the whole spectrum, only a few selected specific ions are detected during the GC separation. This can increase the sensitivity to 500-fold. Depending on the analyte, low picogram to nanogram amount can be measured using this SIM-GC-MS system. Quantitation can be performed using peak area or height from the selected ion-plot/chromatogram, as described earlier. It should be mentioned that the SIM-MS data collection rate must be fast enough to give enough points across a peak. It was shown that 7–8 points could recover 99.99% of the peak; this is enough for quantitative purposes (Mastovska and Lehotay, 2003).

XII. RECENT DEVELOPMENTS

A. Fast GC

Some excellent reviews on the fast GC/GC-MS have been published recently by Mastovska and Lehotay (2003), Crammers et al. (1999), Matisova and Dömötöröva (2003). An excellent Ph.D. thesis on fast GC was submitted at Virginia State University, USA by Reed (1999). The definition and classification of faster GC can be summarized in Table 3. Some main approaches toward fast GC are (Matisova and Dömötöröva, 2003):

1. Minimize resolution: using shorter column, higher isothermal separation or faster temperature programing, lower film thickness, lower pressure in the column, optimum carrier gas velocity.
2. Maximize selectivity: using more selective stationary phase, selective detection, 2D-GC, and apply back-flush.

Table 3 Definition and Classification of Fast GC[a]

Type of GC	Analysis time (min)	Peak width[b] (ms)	SEF[c]
Conventional	>10	>1000	1
Fast	1.0–10	200–1000	5–30
Very fast	0.1–1.0	30–200	30–400
Ultra fast	<0.1	5–30	400–4000

[a]Modified from Mastouska and Lehotay (2003) and Matisova and Dömötöröva (2003).
[b]Peak width at half peak maximum.
[c]Speed enhancement factor.

3. Reducing analyzing time at constant resolution: reduce column inner diameter, using hydrogen as carrier gas, applying vacuum-outlet conditions, and turbulent-flow conditions.

As described earlier, by the use of a resistive heating and shorter column (flash GC), the time of analysis can be reduced, so this method can be used for fast GC (Mastovska et al., 2001a).

Another approach for performing fast GC is by using a bundle of MC columns comprising 1000 capillaries, 1 m × 40 μm (ID), WCOT capillaries. By this MC-GC method the HETP value is smaller compared with the conventional capillary column, and these minimum values are very broad (80–280 cm/s) in the Van-Demter (Golay-Giddings) curve, thus allowing the use of high flow-rates to shorten chromatographic separations without sacrificing peak resolution (Pereiro et al., 1998).

The main approaches for applying fast GC-MS are (Mastovska and Lehotay, 2001b, 2003; de Zeeeuw et al., 2000): microbore GC-MS, fast temperature programing GC-MS, SMB GC-MS, and pressure-tunable GC-GC-MS. The mass analyzer often used for fast GC-MS are TOF mass spectrometer, quadrupole mass spectrometer, and ion trap mass spectrometer. The other mass analyzers, such as FT ion cyclotron resonance, ion trap-TOF, TOF-TOF MS, are too expensive to be used for routine application of fast GC-MS.

B. Portable GC

By using a portable instrumental, process analyzer can be moved from the laboratory to the field with on-side analysis using a compact portable instrument such as GC or GC-MS. This portable equipment can be used for forensic science in the field, analysis of compound emitted in fire or chemical disaster, analysis of waste products, monitoring chemical-warfare-related species, and so on. For these purposes, some companies developed compact field-portable miniature GC or GC-MS.

A field equipment (including GC or GC/MS) should have the following requirements (Baykut and Franzen, 1994; McClennen et al., 1994); ease for transporting; rapid system set-up; fast sampling; ability to perform *in-situ* analysis, fast GC, or GC-MS separation, fast quantitative evaluation, minimum dead time between analysis, ability to operate by non-expert personnel; reliable analytical results, and ability to operate while being moved around. The use of portable GC-MS in the present time is increasing in situation where an incident has occurred and rapid identification of chemicals with a high degree of certainty is needed. Usually portable GC-MS is based on linear quadrupoles and the new generation of TOF analyzer (Santos and Galceren, 2003).

C. On-Line LC and GC

On-line coupled LC and GC (LC-GC) is a relative new powerful technique that combines the best feature of LC and GC. It is ideal for the analysis of complex volatile analyte(s), or derivatisation should be possible either before the analysis or on-line. The main advantages of on-line LC-GC are shorter analysis time, better reproducibility, and improved detection limit. An excellent review on this LC-GC combination was recently published by Hyötyläinen and Rieckkola (2003). Ogorka et al. (1992) reported by the identification of unknown impurities in pharmaceutical products using on-line combination of HPLC and GC-MS, whereas Ramsteiner (1987) described the residual pesticide analysis using on-line combination of LC and GC.

ACKNOWLEDGMENTS

The authors gratefully acknowledge the scholarships of the Deutscher Akadamischer Austauschdienst (DAAD), Bonn, Germany in 2002 at the University of Duesseldorf (for G.I.) and Technical University of Braunschweig (for M.Y.); Ms. Ivy Widjaja, M.Sc. (Genom Research Institute, National University of Singapore) for providing many references; Mr. Bambang Riyono Wijaya (PT Ditek Jaya, Jakarta), who helps getting us permission from the Shimadzu for reproducing some figures; and Mr. Deddy Triono (Faculty of Pharmacy Airlangga University) for his technical assistance.

REFERENCES

Amirav, A., Granot, O. (2000). Liquid chromatography mass spectrometry with supersonic molecular beam. *J. Am. Soc. Mass Spectrom.* 11:587–591.

Amirav, A., Gordin, A., Tzanani, N. (2001). Supersonic gas chromatography/mass spectromery. *Rapid Commun. Mass. Spectrom.* 15:811–820.

Analytical Method Committee, (1998). Uses (proper and improper) of correlation coefficient. *Analyst* 113(9):1469–1471.

Anderson, J. T. (2002). Some unique properties of gas chromatography coupled with atomic-emission detection. *Anal. Bioanal. Chem.* 373:344–355.

Anonymous (1985). *Textbook for Mass Spectrometry*, Jeol Ltd., Tokyo.

Arthur, C., Pawliszyn, J. (1990). Solid phase microextraction with thermal desorption fused silica optical fibers, *Anal. Chem.* 62:2145–2148.

Bailey, L. C. (1995). Chromatography, in: Gennaro, A. R. (ed.) Remington Pharmaceutical Sciences: *The Science and Practice of Pharmacy*, Mack Publishing Company, Easton, PA.

Baykut, G., Franzen, J. (1994). Mobile mass spectrometry; a decade of field applications, *Trends Anal. Chem.* 13:267–275.

Beens, J., Adahchour, M., Vreuls, R. J. J., van Atena, K., Brinkman, U. A. Th. (2001). Simple, non-moving modulation interface for comprehensive two-dimensional gas chromatography, *J. Chromatogr. A* 9(919):127–132.

Bertsch, W. (1999). Two-dimensional gas chromatography, concepts, istrumentation, and application—Part 1: fundamentals, conventional two-dimensional gas chromatography, selected application, *J. High. Resol. Chromatogr.* 22:167–181.

Bertsch, W. (2000). Two-dimensional gas chromatography, concepts, istrumentation, and application—Part 2: comprehensive two-dimensional gas chromatography., *J. High. Resol. Chromatogr.*23:167–181.

Buffington, R; Wilson, M. K. (1987). *Detectors for Gas Chromatography—A Practical Primer*, Hewlett–Packard Corporation, Part No. 5958–9433.

Cai, Y., Zhang, W. (2001). Metals and organometallic: gas chromatography for speciation and analysis, in: Cazes, J. (Ed.), *Encyclopedia of Chromatography*, Marcel Dekker, Inc., New York, Basel, pp. 518–521.

Cazes, J., Scott, R. P. W. (2002). *Chromatography Theory*, Marcel Dekker Inc, New York, Basel.

Cramers, C. A., Janssen, H. G., van Deursen, M. M., Leclercq, P. A. (1999). High-Speed gas chromatography: an overview of various concept, *J. Chromatogr. A* 856:315–329.

Cziesla, K., Platzer, B., Okruss, M., Florek, S., Otto, M. (2001). Hyphenation of near-infrared Echelle spectrometer to a microplasma for element-selective detection in gas chromatography, *Frasenius J. Anal. Chem.*, 371:1043–1046.

Dandeneau, R. D., Zerenner, E. H. (1979). *J. High Resol. Chromatogr. Commun.*, 2:351.

David F., Sandra, P. (1999)(September). Use of hydrogen as carrier gas in capillary GC. Application Note. American Laboratory 18–19.

Davies, A. J. (1998). The new Automated Mass Spectrometry Deconvolution and Identification System (AMDIS), Spectroscopy Europe 3:22–25.

Eiceman, G. A., Gardea-Torresdey, J., Overton, E., Carney, K., Dorman, F. (2002). Gas Chromatography. *Anal. Chem.*, 74:2771–2780.

Ettre, L. S. (1981). The evolution of open tubular columns in Jennings W.G, Ed. *Applications of Glass Capillary Gas Chromatography*, Dekker, New York.

Fowlis, I. A. (1995). Gas Chromatography: *Analytical Chemistry by Open Learning*, J. Wiley & Sons, New York.

Franzen, J., Frey, R., Holle, A., Krauter, K. O. (2001). Recent progress in matrix-assisted laser desorption ionization postsource decay, *Int. J. Mass Spectrom.* 206:275–286.

Frysinger, G. S., Gaines, R. B. (1999). Comprehensive two-dimensional gas chromatography with mass spectrometric detection (GCxGC/MS) applied to the analysis of petroleum, *J. High. Resol. Chromatogr.* 22:251–255.

Funk, W., Damman, V., Donnervert, G. (1992). *Qualitätssicherung in der Analytischen Chemie VCH*, Weinheim, pp. 12–36, 161–180.

Golay, M. J. E. (1958). In *D. Detsy. Gas Chromatography*. Butterworths, London, pp. 36–55.

Grecco, S. D., Lopez, A. F., Vasillev, A. N. (2001). An injector for gas sample injections. *J. Sep. Sci.* 24:148–150.

Grob, K. (1986). *Classical Split and Splitless Injection in Capillary GC*, Hüthig Verlag, Heidelberg.

Grob, K. (1987). *On Column Injection in Capillary GC*, Hüthig Verlag, Heidelberg.

Grob, R. L. (1997). Modern Practice of Gas Chromatography, Wiley Interscience, New York.

Grob, K., Neukom, H. P. Jr. (1979). The influence of syringe needle on the precision and accuracy of vaporizing GC injections. *J. High Resol. Chromatogr. Commun.*, 2:15–21.

Grob, K., Romann, A. (1981). Sample transfer in splitless injection in capillary gas chromatography. *J. Chromatogr.* 214:118–121.

Guimbard, J. G., Person, M., Vergnaud, J. P. (1987). Determination of residual solvents in pharmaceutical products by gas chromatography coupled to a headspace injection system and using an external standard. *J. Chromatogr.* 403:109–121.

Haken, J. K. (1998). Pyrolysis gas chromatography of synthetic polymers: a bibliography, *J. Chromatogr. A*, 825(2):171–187.

Herron, N. R., Donnelly, J. R., (1996). Software-based mass spectral enhancement to remove interferences from spectra of unknown, *J. Am. Soc. Mass Spectrom.*, 7:598–604.

Hyötyläinen, T., Rieckkola, M.-L. (2003). On-line coupled liquid chromatography-gas chromatography, *J. Chromatogr. A* 1000:357–384.

James, A. T., Martin, A. J. P. (1952). Gas–liquid partition chromatography: the separation and micro-estimation of volatile fatty acids from formic acid to dodecanoic acid. *Biochem. J.*, 50:679.

JEOL Product Information, (1993). MS116, Jeol Ltd, Tokyo.

Jinno, K. (2002). Detection principle, in: Cazes, J. (Ed.), *Encyclopedia of Chromatography*, On-Line Published, Marcel Dekker, Inc., New York.

Joffe, B. V., Vitenberg, A. G. (1984). Headspace Analysis and Related Methods in Gas Chromatography, John Wiley, New York.

Kaiser, M. A. (1985). *High-Resolution gas chromatography in Grob*, R.L (Ed), Wiley, New York.

Karasek, F. W., Clement, R. E. (1998). *Basic Gas Chromatography-Mass Spectrometry*, Elsevier, Amsterdam, Oxford, New York, Tokyo.

King, B., Westwood, S. (2001). GC-FID as a primary method for establishing the purity of organic CRMs used for drug in sport analysis. *Frasenius J. Anal. Chem.* 370:194–199.

Kromidas, S. (1999). *Validierung in der Analytik*, Wiley-VCH, Weinheim.

Lacorte, S., Rigol, A. (2002). Purge-Backflushing techniques in gas chromatography. in: Cazes, J. (Ed.), *Encyclopedia of Chromatography*, Marcel Dekker, Inc., New York, Basel, pp. 1–7.

Lipinski, J., Stan, H.-J. (1988). CAPA—computer aided pesticide analysis, computer program for the automatic evaluation of chromatographic data for residue analysis of foods, *J. Chromatogr.* 441:213–225.

Mamyrin, B. A. (2001). Time-of-flight mass spectrometry (concepts, achievements, and prospects, *Int. J. Mass Spectrom.* 206:251–266.

March, R. E., Hughes, R. J., Todd, J. F. (1989). Quadrupole storage mass spectrometry, *Chem. Anal.* Vol. 102.

Marriott, P. J., Kinghorn, R. M. (1998). Modulation and manipulation of gas chromatography bands using novel thermal means, *Anal. Sci.*, 14:651–659.

Mastovska, K., Lehotay, S. J. (2003). Practical approaches to fast gas chromatography-mass spectrometry, *J. Chromatogr A* 1000:153–180.

Mastovska, K., Hajslova, J., Godula, M., Krivankova, J., Kocourek, V. (2001a). Fast temperature programming in routine analysis of multiple pesticide residues in food matrice, *J. Chromatogr. A* 907:235–245.

Mastovska, K., Lehotay, S. L., Hajslova, J. (2001b). Optimization and evaluation of low pressure gas chromatography-mass spectrometry for the fast analysis of multiple pesticide residues in a food commodity, *J. Chromatogr. A* 926:291–308.

Matisova, E., Dömötöröva, M. (2003). Fas gas chromatography and its use in trace analysis, *J. Chromatogr. A* 1000:199–221.

Mayer, R. X., Zölner, P., Lorbeer, E., Rauter, W. (2002). A new 75% diphenyl, 25% dimethyl-polysiloxane coated on fused silica capillary column for high temperature gas chromatography, *J. Sep. Sci.*, 25:60–66.

McClennen, W. H., Arnold, N. S., Meuzelaar, H. L. C. (1994). Field portable hyphenated instrumentation: the birth of the tricoder?, *Trends Anal. Chem.* 13:286–293.

Medina-Vera, M. (1996). Pyrolysis-gas chromatography/mass spectrometry used for screening polycyclic aromatic hydrocarbons by desorption from sediment. *J. Anal. Appl. Pyrol.* 36:27–35.

Mills, G. A., Walker, V., Mughal, H. (1999). Quantitative determination of trimethylamine in urine by solid-phase microextraction and gas chromatography mass spectrometry. *J. Chromatogr. B* 723(1–2):281–285.

Moldoveanu, S. C. (1998). *Analytical Pyrolysis of Natural Organic Polymers*, Elsevier, Amsterdam.

Moldoveanu, S. C. (2001). Pyrolysis GC/MS: present and future. *J. Microcolumn Sep.* 13:102–125.

Mossoba, M., McDonald, R. E., Yurawecz, P., Kramer, J. K. G. (2001). Application of on-line GC-FTIR spectroscopy to lipid analysis, *Eur. J. Lipid. Technol.*, 103:826–829.

Ogorka, J., Schwinger, G., Bruat, G. (1992). On-line coupled reversed-phase high-performance liquid chromatography-gas chromatography-mass spectrometry, *J. Chromatogr.* 626:87–96.

Olsavicky, V. M. (1978). A comparison of high temperature septa for gas chromatography. *J. Chromatogr. Sci.* 197–200.

Park, M. K., Cho, J. H., Kim, N. Y., Park, J. H. (1993). Chromatographic pattern recognition for the analysis of complex mixtures, *Anal. Chem. Acta* 284:73–78.

Pereiro, I. R., Wasik, A., Lobinski, R. (1998). Characterization of multicapillary gas chromatograph-microwave induced plasma atomic emission spectrometry for the expeditious analysis for organometallic compounds, *J. Chromatogr. A* 795:359–370.

Pinho, O., Ferreira, I.M.P.L.V.O., Ferreira, M. A. (2002). Solid-phase microextraction in combination with GC/MS for quantification of the major volatile free fatty acids in Ewe Cheese. *Anal. Chem.* 74:5199–5204.

Pool, W. G., Mass, L. R. M., de Leeuw, J. W., van de Graf, B. (1997). Automated processing of GC/MS data: quantitation of the signals of individual components, *J. Mass Spectrom.* 32:1253–1257.

Poole, C. F., Poole, S. K. (1991). *Chromatography Today*, Elsevier, Amsterdam.

Purch, M., Sun, K., Winniford, B., Cortes, H., Weber, A., McCabe, T., Luong, J. (2002). Modulation techniques and applications in comprehensive two-dimensional gas chromatography (GCxGC), *Anal. Bioanal. Chem.* 373:356–367.

Ramsteiner, K. A. (1987). On-line liquid chromatography-gas chromatography in residual analysis, *J. Chromatogr.* 393:123–131.

Ravindranath, B. (1989). *Principles and Practice of Chromatography*, Ellis Horwood Limited, Chichester.

Redd, G. L. (1999). Fast GC: Application and Theoretical Studies, Ph.D thesis, Virginia Polytechnic Institute and State University, USA.

Ruecker, G. (1976). *Spektroskopische Methoden in der Pharmazie*, Wissenschaftliche Verlagsgesselshaft MBH, Stuttgart.

Sandra, P. (1985). *Sample Introduction in Capillary Gas Chromatography*, Hüthig Verlag, Heidelberg.

Sandra, J. F. (2002). Gas chromatography in *Ullmann's Encyclopedia of Industrial Chemistry*, Wiley-VCH Verlag GmbH, Weinheim.

Santos, F. J., Galceren, M. T. (2003). Modern development in gas chromatography-mass spectrometry based on environment analysis, *J. Chromatogr. A* 1000:125–151.

Scarlata, C. J., Ebeler, S. E. (1999). Headspace solid-phase microextraction for the analysis of dimethyl sulfide in beer. *J. Agric. Food Chem.* 47(7):2505–2508.

Schomburg, G. (1990). *Gas Chromatography, A Pratical Course*, VCH Verlagsgesellschaft, Weinheim.

Scott, R. P. W. (2001a). Mixed stationary phase in GC, in: Cazes, J. (Ed.), *Encyclopedia of Chromatography*, Marcel Dekker, Inc., New York, Basel, pp. 523–524.

Scott, R. P. W. (2001b). Programmed temperature gas chromatography in: Cazes, J. (Ed.), *Encyclopedia of Chromatography*, Marcel Dekker, Inc., New York, Basel, pp. 664–666.

Scott, R. P. W. (2001c). Programmed flow gas chromatography in: Cazes, J. (Ed.), *Encyclopedia of Chromatography*, Marcel Dekker, Inc., New York, Basel, pp. 662–66.

Scott, R. P. W. (2001d). Headspace sampling in: Cazes, J. (Ed.), *Encyclopedia of Chromatography*, Marcel Dekker, Inc., New York, Basel, pp. 401–403.

Scott, R. P. W. (2001e). Detector noise, in: Cazes, J. (Ed.), *Encyclopedia of Chromatography*, Marcel Dekker, Inc., New York, Basel, pp. 253–255.

Scott, R. P. W. (2001f). Detector linearity and responses index, in: Cazes, J. (Ed.), *Encyclopedia of Chromatography*, Marcel Dekker, Inc., New York, Basel, pp. 252–253.

Scott, R. P. W. (2001g). Flame ionization detector for GC, in: Cazes, J. (Ed.), *Encyclopedia of Chromatography*, Marcel Dekker, Inc., New York, Basel, pp. 330–333.

Scott, R. P. W. (2001h). Katharomether detector for GC, in: Cazes, J. (Ed.), *Encyclopedia of Chromatography*, Marcel Dekker, Inc., New York, Basel, pp. 465–466.

Scott, R. P. W. (2001i). Electron capture detector, in: Cazes, J. (Ed.), *Encyclopedia of Chromatography*, Marcel Dekker, Inc., New York, Basel, pp. 285–287.

Scott, R. P. W. (2001j). Nitrogen/phosphorous detector, in: Cazes, J. (Ed.), *Encyclopedia of Chromatography*, Marcel Dekker, Inc., New York, Basel, pp. 547–548.

Scott, R. P. W. (2001k). Gas chromatography-mass spectrometry, in: Cazes, J. (Ed.), *Encyclopedia of Chromatography*, Marcel Dekker, Inc., New York, Basel, pp. 366–371.

Seeley, J. V., Kramp, F. J., Sharpe, K. S. (2001). A dual-secondary column comprehensive two-dimensional gas chromatography for the analysis of volatile organic compound mixtures, *J. Sep. Sci.* 24:444–450.

Tollbäck, P., Björklund, C., Östman, C. (2003). Large-volume programmed-temperature vaporiser injection for fast gas chromatography with electron capture and mass spectrometric detection of polybrominated diphenyl ethers. *J. Chromatogr. A* 991:241–153.

USP. (2002). *The United States Pharmacopeia 26, The National Formulary 21*, Asian Edition, United States Pharmacopeial Convention, Inc., Rockville, MD.

Van Loco, J., Elskens, M., Croux, C., Beernaert, H. (2002). Linearity of calibration curve: use and misuse of the correlation coeffcient, *Accredit. Qual. Assur.* 7:281–285.

Wang, F. C. Y., Burleson, A. D. (1999). Development of pyrolysis fast gas chromatography for analysis of synthetic polymers. *J. Chromatogr. A*, 833:111–119.

Wang, J. L., Chen, W. L., (2001). Construction and validation of automated purge-and-trap-gas chromatography for the determination of volatile organic compounds. *J. Chromatogr. A* 927:143–154.

Wasik, A., Pereiro, I. R., Lobinski, R. (1998). Interface for time-resolved introduction of gaseous analyte for atomic spectrometry by purge-and-trap multicapillary gas chromatography. *Spectrochim. Acta B* 53:867–879.

Watson, J. T. (1998). A historical perspective and commentary on pioneering development in gas chromatography/mass spectrometry in MIT, *J. Mass Spectrom.* 33:103–108.

Wilson, I. D., Brinkman, U. A. Th. (2003). Hyphenation and hypernation, the practice and prospects of multiple hypenation, *J. Chromatogr. A* 1000:325–356.

Xu, X., van Stee, L. L. P., Williams, J., Beens, J., Adachour, M., Vreuls, R. J. J., Brinkman, U. A. Th., Lelieveld, J. (2003). Comprehensive two-dimensional gas chromatography (GCxGC) measurements of volatile organic compounds in the atmosphere, *Atmos. Chem. Phys. Discuss.* 3:1139–1181.

de Zeeeuw, J., Peene, J., Jansen, H.-G., Lou, X. (2000). A simple way to speed up separations by GC-MS using short 0.53 column and vacuum outlet conditions, *J. High Resol. Chromatogr.* 23:677–680.

Zenkevich, I. G., (2002). Kovats retention index system, in: Cazes, J. (Ed.), *Encyclopedia of Chromatography*, Marcel Dekker, New York, Basel, pp. 466–470.

24

Supercritical Fluid Chromatography Instrumentation

THOMAS L. CHESTER, J. DAVID PINKSTON
The Procter Gamble Company, Miami Valley Laboratories, Cincinnati, Ohio 45252, USA

I. INTRODUCTION

Although supercritical fluids have been known since the nineteenth century, they have been used as chromatographic mobile phases only since 1962 (Klesper et al., 1962). At first, supercritical fluid chromatography (SFC) (Anton and Berger, 1998; Berger, 1995; Lee and Markides, 1990; Parcher and Chester, 1999; Saito et al., 1994a) was done exclusively with packed columns [e.g., (Gere, 1983; Giddings et al., 1968; Gouw and Jentoft, 1972; Jentoft and Gouw, 1972)]. During the 1980s the focus shifted to wall-coated, open-tubular (OT, or sometimes called capillary) columns as 50 μm inside-diameter (i.d.) fused-silica columns became available, having grown from gas-chromatography (GC) technology (Novotny et al., 1981). In the 1990s, interest and energy returned to packed columns, particularly after the introduction of new instruments with postcolumn pressure control (Anton et al., 1991; Berger, 1992; Verillon et al., 1992), and with business interests in that decade focusing strongly on speed and cost reduction. Currently, packed columns are favored in applications requiring high-speed analyses and with large sample loads, particularly pharmaceutical work. OT-SFC is often preferred for low-temperature or high-molecular-weight, gas-chromatography-like separations requiring flame-ionization or other GC-like detection capabilities, particularly petroleum applications. This chapter will describe the fundamentals of SFC, instrumentation concepts, techniques, and guidelines for using both packed and OT columns.

Many techniques have been named that share some (if not all) of the characteristics of SFC (Parcher and Chester, 1999). These are techniques like near-critical or subcritical fluid chromatography and enhanced-fluidity liquid chromatography (LC) (Cui and Olesik, 1991). Even high-temperature LC and particularly ultra-high-temperature LC share a few of these characteristics (Greibrokk and Andersen, 2003; McGuffin and Evans, 1991; Yan et al., 2000). These do not actually represent a proliferation of separate techniques, but just a proliferation of names describing forms of chromatography with low-viscosity, compressible, and solvating fluids. The key features of these techniques, distinguishing them from GC and conventional LC (or high-performance liquid chromatography, HPLC), are:

1. The mobile-phase viscosity is much lower than in "normal" liquids. This results in much lower pressure drops than in LC and allows longer, more efficient packed columns to be used.
2. Solute diffusion coefficients are higher than in "normal" liquids. This provides faster optimum velocities and shorter analysis times than with LC. Diffusion coefficients in GC are even higher.
3. There are significant mobile phase–solute intermolecular forces contributing to solute partitioning,

just as in LC. (In GC the mobile phase is an inert carrier exerting no significant intermolecular forces on the solutes.)

4. The mobile phase is compressible, and its strength depends on its density. Also, the mobile-phase vapor pressure usually exceeds one atmosphere at the chosen column temperature. Therefore, the column (or detector) outlet cannot simply be vented to atmosphere as in GC and LC; some means must be provided to maintain the pressure and prevent the mobile phase from expanding (or boiling in some techniques) until all the measurements requiring compressed mobile phase are completed.

We will see that it is not always clear when we have a supercritical fluid, so the distinction between the various techniques is often trivial. The information in this chapter is applicable to all of them.

A. Intermolecular Forces in SFC

Retention in chromatography is determined by a balance of intermolecular forces acting on the solutes. Keeping intermolecular forces in mind, and how the functional groups in the solutes may interact with the stationary and mobile phases, is essential in selecting the stationary and mobile-phase components, and in deciding how to change parameter values to improve a separation.

Orientation forces are relatively strong. These result when permanent electrical dipoles approach each other, and are attractive as long as one of the dipoles is free to reorient. (The solute is always free to reorient in chromatography, so these forces are always attractive.) An example is the dipole–dipole attraction between 1-cholorobutane and a cyanopropyl group bound to a stationary phase. *Hydrogen bonding* is a special case of orientation interactions in which the dipole–dipole attraction is further strengthened by the sharing of a hydrogen atom from a donor molecule or functional group with a hydrogen-bond-acceptor (usually a fluorine, oxygen, or nitrogen atom) on the receiving molecule or functional group.

Induction interactions occur when a dipole and a polarizable group near each other. The electric field around the dipole will induce an oppositely aligned dipole in the polarizible group, thus creating an attraction between the dipole and the induced dipole. An example is the interaction of a hydroxyl group (which is dipolar) and a phenyl group (which has polarizable π electrons). The phenyl group may also have a small, permanent dipole moment, and there will be a small dipole–dipole attraction, but the point of this example is that this attraction will be enhanced by the induction effect. Benzene will be retained more strongly by a polar stationary phase than cyclohexane because benzene is more polarizable.

Dispersion or *London* forces result from temporary dipoles that exist in molecules from time to time because of the randomness of the position of electrons. Thus, a temporary dipole randomly created in one molecule may induce and be attracted to temporary dipoles in neighboring molecules. The effect always results in a net attraction between molecules. These are the forces that attract nonpolar molecules to each other.

The order of strength of these interactions is hydrogen bonding > dipole–dipole > dipole–induced-dipole > dispersion. In addition, intermolecular distances, the resulting attractions, and the chromatographic retention will not be the same for an enantiomeric pair of solutes if either the stationary phase or the mobile phase has chiral components. These differences make separations of enantiomers possible. Steric effects may also apply; for example, solutes that are larger than the pore diameter of a porous stationary phase will be unable to interact with stationary phase inside the pores, and will therefore have less retention than would be predicted from the behavior of smaller molecules that can permeate the pore structure.

All of these intermolecular forces listed are attractive. Steric effects are exclusive, not repulsive. In reversed-phase HPLC, ionic attractions and repulsions are also important, but it is not yet clear if isolated ions (as distinguished from ion pairs) can exist in CO_2-based mobile phases in concentrations high enough to influence SFC behavior.

B. Fluids and Phase Behavior

The rules and procedures developed for GC and LC usually cannot be successfully applied to SFC and related techniques with solvating, compressible mobile phases without considering the unique phase behavior of the fluids used. The mobile phase may be a highly pure material, such as SFC-grade CO_2, or it may be a binary or ternary (or more complicated) mixture. CO_2 is usually used as the main component in these mixtures. A second component, usually called the *modifier*, may be added to increase the polarity of the mixture above that of the main component (or, in low concentration, to compete for active sites on the stationary phase). The modifier is often a polar organic solvent like methanol or acetonitrile, although numerous choices are possible. When used, modifier levels are often in the range of a few percent to around 60%. Even higher modifier levels are possible. When the modifier concentration is high enough that it becomes the main component, the technique is sometimes referred to as *enhanced-fluidity LC* (Cui and Olesik, 1991). In this technique, CO_2 is considered a viscosity-lowering agent that also increases diffusion rates making faster analyses possible than when using the pure organic solvent as the mobile phase. There is no discontinuity and no practical distinction between SFC and enhanced-fluidity LC if the two mobile phase components are miscible.

Other components may also be added to mixed mobile phases, usually at 1% or less, to improve solute peak shape, reduce retention, or increase recovery. These components are usually called *additives*.

Some understanding of the phase behavior of fluids is essential for success in SFC and the related techniques.

1. Pure-Substance Phase Behavior

The three common states of matter, solid, liquid, and vapor (or gas), are well known to beginning students. The areas where these phases exist in a two-dimensional pressure–temperature representation for a pure material are depicted in Fig. 1(a). All three phases coexist at just one combination of temperature and pressure, the *triple point*. Liquid (l) and vapor (v) phases can coexist at various combinations of temperature and pressure falling on a line defining the l–v phase boundary, that is, the boiling line. For now, let us consider the properties of the liquid and vapor phases in equilibrium on the l–v phase boundary. If we vary the pressure and temperature exactly as required to move along the l–v line away from the triple point, we would observe that the respective intensive properties [i.e., properties that do not depend on the amount of substance present, such as density, refractive index (RI), dielectric constant, etc.] of the separate liquid and vapor phases approach each other as the pressure and temperature are increased. Eventually, another point is reached, called the *critical point*, where all the respective intensive properties of the liquid and vapor phases merge. Beyond the critical point there are no longer separate liquid and vapor phases, but only a single *supercritical fluid* sharing liquid and vapor properties. The supercritical fluid is compressible and will expand to uniformly fill its container like a vapor, yet it is capable of dissolving other materials like a liquid. The critical point is described by the corresponding pressure and temperature, known as the *critical pressure*, P_c, and *critical temperature*, T_c. There is no phase change separating the supercritical fluid from the liquid and vapor phases. Thus, it is possible to change a fluid from liquid to vapor (or vice versa) without actually going through a phase transition by varying the temperature and pressure to go around the critical point.

Many authors define a pure fluid as *supercritical* when conditions exceed both P_c and T_c. Many literature resources add additional lines and shading to the phase diagram to indicate a "supercritical fluid" region, as shown in Fig. 1(b). However, this is a somewhat arbitrary definition having virtually nothing to do with the properties of the fluid away from the critical point, and can be very misleading. At pressures exceeding P_c, there is no sudden transition from normal liquid to supercritical fluid as the temperature crosses the critical temperature. Similarly, there is no sudden transition from normal gas to supercritical fluid as the pressure is increased beyond the critical pressure at temperatures above T_c. It is more important to realize that the phase diagram is actually as

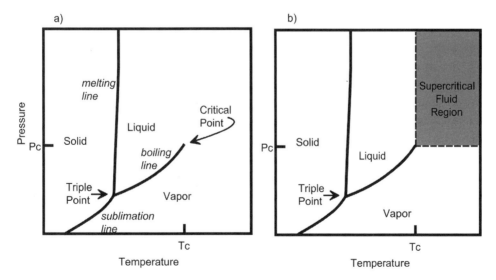

Figure 1 (a) *PT* diagram of CO_2. (b) Many authors define the supercritical fluid region as that where both the critical pressure and temperature are exceeded, as shown by the shaded region. This is a somewhat arbitrary definition because (away from the critical point) there are no abrupt transitions in properties as the dashed lines are crossed. Most simply stated, beyond the critical point there are no separate liquid and vapor phases, but only one fluid phase merging continuously with the liquid and vapor phases and sharing their properties. (a) is the more accurate depiction of phase behavior.

shown in Fig. 1(a), and that the distinction between liquid and vapor simply does not exist beyond the critical point.

The fluid we want to consider, regardless of what we call it, is the state of matter achieved when the pressure and temperature are near or exceed their critical values, the fluid exists in a single phase, and it shares (or at least begins to share) normal liquid and gas properties: the fluid has compressibility significantly higher than normal liquids and often approaches that of vapors, the fluid has significant intermolecular forces and can dissolve solutes like a liquid, and the diffusion rates and viscosity are somewhere between those of "normal" liquids and vapors. We will even extend our treatment to include gases well below their critical temperature but liquefied by application of pressure, and liquids used above their normal boiling points, but still well below their critical temperatures, and kept in the liquid state by the application of pressure. All of these fluids provide properties not found in normal liquids at ambient temperature and pressure.

Furthermore, these properties change continuously with the degree of compression and the resulting density of the fluid, particularly as the temperature approaches or exceeds the critical temperature. Supercritical fluids can span a wide density range, as shown in Fig. 2. Very low-density supercritical fluids (e.g., $T > T_c$, $P \approx P_c$) are much like ordinary gases. The more a supercritical fluid is compressed, the more liquid-like it becomes. Very high-density supercritical fluids ($P \gg P_c$) are very liquid-like. Pressure (or density) programing (Fjeldsted et al., 1983; Jentoft and Gouw, 1970) is often used in OT-SFC, analogously to temperature programing in GC and gradient elution in LC.

So, if we avoid the regions of the l–v transition and the critical point, we can continuously vary the fluid properties between those of liquid and vapor without a phase change. Of course, if we choose to drive the temperature and pressure through the l–v transition, the fluid simply boils or

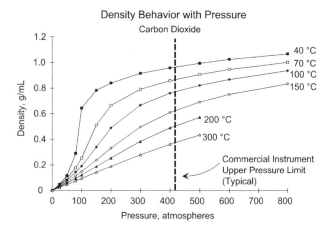

Figure 2 Density–pressure isotherms for CO_2.

condenses, depending on the direction we go. But what happens if we go near the critical point? This is illustrated in Fig. 3. Here we see 0.928 g of CO_2 sealed in a vessel with a 2 mL volume. Two phases form when the vessel is equilibrated at 0°C. As the temperature is raised and re-equilibrated, the respective intensive properties of the two phases approach each other. With the mass and volume used in this example (which correspond to the critical density of CO_2), the experiment will travel up the l–v line and go through the critical point as the temperature is raised. The pressure will be P_c when the temperature is T_c. As the temperature and pressure exceed their critical values, the two phases merge, the phase boundary disappears, and a single supercritical-fluid phase exists as the temperature is raised further. Cooling the vessel reverses the process, and a phase separation occurs upon lowering the temperature through the critical point. Although safety precautions are required, several types of transparent pressure vessels are available

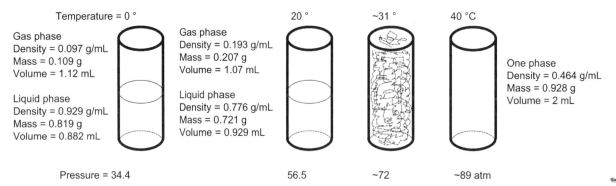

Figure 3 Example of supercritical fluid formation in a sealed vessel. In this example, CO_2 is contained at its critical density, so equilibrium along the entire boiling line can be viewed by equilibrating at any desired temperature.

making experiments like this fairly easy to perform (Berry, 1985; Meyer and Meyer, 1986). These demonstrations give an impressive visual display, called critical opalescence, just below the critical temperature. This results because density and RI change rapidly around the critical point, so small temperature differences within the vessel result in gravity-driven fluid motion that is easy to see.

2. Binary-Mixture Phase Behavior

The previous paragraphs described the behavior of a fluid with only one chemical component such as CO_2. Some basic understanding of the phase behavior of two-component (or binary) mixtures is also essential because we never deal with pure fluids when doing chromatography. Even if the mobile phase has only one component, we must consider what separate phases may be formed when the mobile phase and the sample solvent are mixed, and the consequences the phase behavior may have on the mass transfer of dilute solutes. Six types of binary phase behavior have been identified (van Konyenburg and Scott, 1980). Comparatively little work has been done to characterize ternary and higher mixtures. For simplicity, we will deal only with binary mixtures, and only with the simplest type, type I, in which the two components are completely miscible as liquids.

The complete phase diagram of a binary mixture must contain a third dimension describing the composition of the mixture, x. The boiling lines of the two (neat) components of the mixture exist in the planes at the limits of the composition axis, and every possible mixture has a unique critical point on a curved line connecting the critical points of the neat components. This is called the *lv* (or *vl*) *critical locus*, or simply the *critical-mixture curve*. All of this is illustrated in Fig. 4. However, the two-phase region of a type I binary mixture does not run along a line as for neat liquids, but is a volume in the three-dimensional space of pressure, temperature, and composition coordinates. This is illustrated in Fig. 5. Liquid and vapor phases separate when conditions are within the shaded region of the figure (the two-phase l–v region), and the composition of the resulting phases is given by the intersection of a tie line with the surface of the shaded figure. The tie line intersects the point defining the system parameters and runs parallel to the composition axis. Only a single phase exists in the regions outside this two-phase region, both above it and below. Several isotherms and isopleths are shown in the figure.

Anyone familiar with the work of graphic artist M.C. Escher will fully understand how deceiving two-dimensional representations of three-dimensional objects can be. Projections of the phase information into two dimensions are often easier to comprehend. Projecting the information contained in Fig. 5 into the $P-T$ dimensions

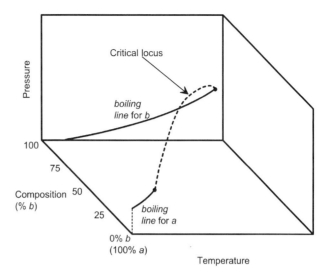

Figure 4 $P-T-x$ (pressure–temperature–composition) plot for a type I binary mixture showing only the boiling lines for the two components (a and b) and the critical locus.

produces plots like Fig. 6. Here, we show the boiling lines of the two pure substances and the critical locus. $T-x$ projections are also possible. However, the $P-T$ projection (Fig. 6) is the one most useful to those doing SFC. In the area above the critical locus only one fluid phase exists for all compositions. Notice that this phase runs continuously between the liquid state of the more-volatile component and the vapor state of the less-volatile component, suggesting the continuous merging of the properties of the two components as the temperature, pressure,

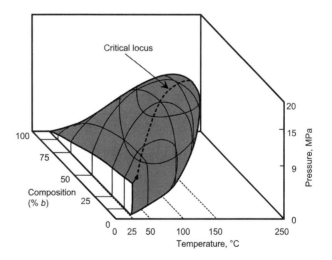

Figure 5 Full $P-T-x$ phase diagram of CO_2–methanol. Shading denotes the two-phase liquid–vapor region. When the overall conditions are at a point in this region, phase separation occurs. The composition of each phase is given by the intersection of a tie line (through the point and parallel to the composition axis) with the surface of the two-phase region.

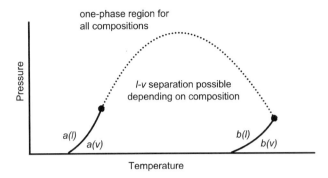

Figure 6 Projection of the pure-component boiling lines (solid) and the critical locus (dashed) of a type I mixture (as in Figs. 4 and 5) into the pressure–temperature plane.

and composition are varied (as long as we remain in the one-phase region). Below the critical locus and between the two pure-fluid boiling lines there exists the possibility of l–v phase separation, depending on the mixture composition.

This kind of plot is useful because in order to avoid phase separation, regardless of composition changes, users need to only operate at temperatures and pressures above the critical locus. Mobile-phase composition programing (gradient elution) is a situation where this is important. Similarly, those who require a phase separation as the composition is widely varied may simply choose T–P points inside the critical locus. In OT-SFC, the direct injection of liquid samples using a retention gap for solvent trapping is a situation where a phase separation is required. Chester (2004) provides a comprehensive collection of critical mixture curves for 23 binary CO_2–solvent systems. Some of the systems described in this reference are not type I, but may be treated as such over much of the temperature range of SFC. Figures 7 and 8 show critical loci for some of the most useful CO_2–solvent systems (Chester, 2004). Descriptions of the behavior of the remaining five types of binary mixtures are available in the literature (van Konyenburg and Scott, 1980; McHugh and Krukonis, 1986; Page et al., 1992).

Ternary (or higher) mixtures involving supercritical fluids are finding increased usage, especially in packed-column SFC; however, practically no information is available on their phase behavior over broad regions of temperature and composition. If the concentration of the third component is low, its influence on phase behavior can usually be ignored, and the binary system of the two most abundant components can be used.

C. Chromatography with Supercritical Fluid Mobile Phases

Because continuity exists between the liquid and vapor states of type I binary mixtures, the distinctions among the various chromatographic techniques performed using such a fluid depend entirely on the temperature, pressure, and composition of the mobile phase; that is to say, the name of the technique depends on where it is performed

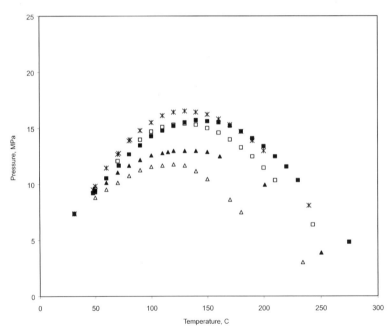

Figure 7 Experimentally determined points on the l–v critical loci of binary solvent systems of CO_2 and: ■ acetonitrile, □ ethanol, ▲ ethyl acetate, △ n-hexane, and * methanol determined by a flow-injection-peak-shape method (Chester, in press).

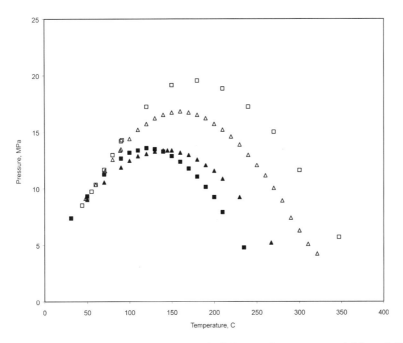

Figure 8 Experimentally determined points on the l–v critical loci of binary solvent systems of CO_2 and: ■ 2-propanol, □ pyridine, ▲ tetrahydrofuran, and △ toluene determined by a flow-injection-peak-shape method (Chester, 2004).

within the phase diagram for the mobile phase. For example, if the temperature is below T_c for the mobile phase, then subcritical-fluid chromatography is being performed. If the mobile-phase vapor pressure is higher than ambient pressure, then the column-outlet pressure must be kept above the vapor pressure to prevent the mobile phase from boiling before it reaches the column outlet (or the detector). The column-inlet pressure must be even higher to produce the necessary flow. Thus, liquid CO_2 can be used as mobile phase, either neat or with modifiers, at temperatures as low as $-50°C$. This would clearly not be SFC as the temperature is far below the critical temperature, but the technique is far from what we consider to be conventional LC because the column outlet must be kept at elevated pressure to prevent the mobile phase from boiling.

Similarly, ordinary LC mobile phases can be used at temperatures significantly higher than their normal boiling points simply by elevating the column-outlet pressure to prevent boiling. This technique is often called ultra-high-temperature LC, but it clearly shares similarities with subcritical-fluid chromatography. The only real difference is if the mobile phase is normally a liquid or a gas at ambient temperature and pressure, which is totally inconsequential to the resulting chromatography except in how the fluids are delivered to the pumps.

SFC, by the strictest definition, results when the temperature exceeds T_c and the pressure exceeds P_c. However, the value of T_c for the mobile phase is not always clear, nor is it constant in a method being operated with gradient elution. Conditions can and often shift from supercritical to subcritical in the course of executing an SFC method. This is totally inconsequential because there is no phase discontinuity as long as the pressure is high enough to prevent l–v phase separation.

This is all part of a *unified* behavior underlying all chromatographic techniques (Parcher and Chester, 1999). The specifically named techniques merely represent various locations in the phase diagram where any particular technique is performed. In this view, conventional GC and LC are simply limiting cases of the larger, unified behavior. GC results when the mobile phase is so weak that it has no ability to contribute to solute partitioning, but is an inert carrier of solute when it is vaporized by virtue of its own vapor pressure. Naturally, hydrogen and helium are best mobile phases for GC because they have the fastest diffusion rates among gases and therefore produce the shortest analysis times possible. They are commonly used with ambient pressure (or detector pressure) at the column outlet by default without regard to the influence of pressure on performance. LC results when the column outlet is held at or near ambient temperature and pressure, and a "normal" liquid is chosen as the mobile phase. Of course, what is normal depends entirely on the ambient temperature and pressure, and vastly different possibilities emerge when these parameters are controlled and "abnormal" fluids are considered.

Although we should always think of all chromatography within the unified chromatography continuum (as

Table 1 Typical Ranges of Density, Viscosity, and Diffusion Coefficients

Fluid	Density (g/mL)	Viscosity (poise)	Diffusion coefficient (cm^2/s)
Gas	10^{-3}	0.5–3.5 ($\times 10^{-4}$)	0.01–1.0
Supercritical fluid	0.2–0.9	0.1–1.0 ($\times 10^{-3}$)	0.1–2.0 ($\times 10^{-3}$)
Liquid	0.8–1.5	0.3–2.4 ($\times 10^{-2}$)	0.5–2.0 ($\times 10^{-5}$)

described by the phase diagram of the mobile phase), there is still instructional value in separating the techniques into classes. We will lump all the techniques requiring elevated outlet pressure together for this purpose, and loosely refer to them as SFC. The density, viscosity, and diffusion coefficient ranges of these three classes of chromatographic mobile phase are summarized in Table 1. The lower viscosity of supercritical (and related) fluids compared with conventional liquids results in relatively low pressure drops and allows longer, more efficient columns to be used than is possible in LC. The mobile phase diffusion coefficient influences the optimum velocity for a particular column. If techniques using different mobile phases were to be compared while using the same column, GC would have the highest optimum velocity because it has the highest diffusion coefficients. LC would be the slowest. The various SFC techniques would be intermediate. But keep in mind that GC is restricted to volatile solutes because the intermolecular forces in the mobile phase are negligible. Supercritical fluids and liquids exert attractive forces on the solutes, enhancing their partitioning into the mobile phase and greatly increasing the range of possible solutes that can be eluted compared with GC.

A list of chromatographically useful supercritical-fluid mobile phases is given in Table 2, along with pertinent physical data for each. Carbon dioxide has been the most popular supercritical-fluid mobile phase. It has a low critical temperature and is compatible with virtually every GC and LC detector. Its compatibility with the flame-ionization detector (FID) is particularly noteworthy. CO_2 is also readily available in high purity, has a low price compared with other solvents, is nontoxic, and is easily and safely disposed by venting.

The stationary phases for packed-column SFC are similar if not identical to LC stationary phases. The surface area of the packing in a typical column is on the order of 100 m^2. The underlying silica surface is often strongly adsorptive. CO_2 is a weak, nonpolar solvent compared with typical organic liquids. These effects combine to produce strong retention for solutes with polar functional groups when neat CO_2 is used as the mobile phase. If solutes are retained by polar or hydrogen-bonding interactions, neat CO_2 may be unable to elute the solutes at all. Therefore, polar modifiers are often used in packed-column SFC mobile phases to provide control of solute retention. Polar modifiers, when used up to a few percent in the mobile phase, are generally believed to primarily compete with solutes for polar or active sites on the stationary phase, thereby lowering solute retention and improving peak shape. At higher levels, modifiers can increase the density and strength of the mobile phase, interact directly with solutes to improve their solubility in the mobile phase, and function as part of the stationary phase (Strubinger et al., 1991).

Like in HPLC, mobile-phase composition programing (gradient elution) is possible in SFC, and is the normal

Table 2 Common Supercritical Fluids (Lee and Markides, 1990)

	T_c (°C)	P_c (atm)	Dipole moment (Debyes)	
CO_2	31.3	72.9	0.00	Safe, cheap, convenient, used almost exclusively
NH_3	132.5	112.5	1.47	Dissolves polyimide (Vespel), corrodes brass, short column life
Pentane	196	33.3	0.00	A good nonpolar solvent, not compatible with FID, leaks can lead to oven explosions
SF_6	45.5	37.1	0.00	Nonpolar, FID compatible but HF combustion product is hazardous and corrosive, rarely used
Xe	16.6	58.4	0.00	Excellent for on-column IR detection, but very expensive

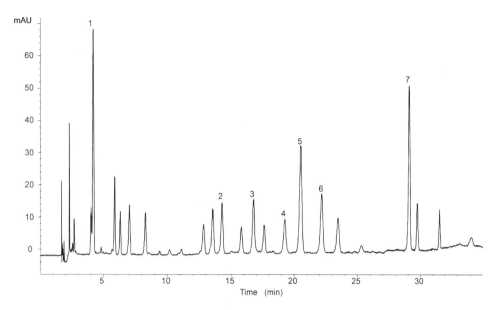

Figure 9 Isothermal gradient-elution SFC chromatogram of paclitaxel and analogs. Conditions: 30°C, 200 bar, 2.0 mL/min, UV detection at 227 nm, Lichrospher diol, 5 μm, 4.6 × 250 mm; CO_2 held with 8% methanol for 8 min, then increased at 0.4%/min to 18%, then at 3%/min to 35%, and held for 4 min. Peaks: (1) taxicin II, (2) baccatin III, (3) cephalomannine, (4) 7-*epi*-10-deacetyltaxol, (5) pacitaxel, (6) 10-deacetylbaccatin III, and (7) 10-deacetyltaxol. (Adapted from Berger Instruments and used with permission.)

programing mode with most packed-column SFC systems. An example of a gradient-elution SFC chromatogram is shown in Fig. 9.

Packed-column SFC provides many speed and efficiency advantages compared with HPLC. In addition to UV, fluorescence, and evaporative light scattering detectors (ELSDs), nitrogen–phosphorus and electron-capture detectors can be used successfully in SFC even with modified mobile phases, within limits (e.g., no acetonitrile modifier with nitrogen detection, no halogenated fluids or additives with electron capture, etc.).

In contrast, OT-SFC resembles GC. The surface area in a wall-coated OT-SFC column is on the order of 10^{-3} m^2 and is highly deactivated. The stationary phase is a relatively thick (0.05–0.5 μm), uniform film of a substituted silicone or other polymer (the same polymers as used in GC in most cases), covering or attached to the surface. The likelihood of solute–active-site interactions is greatly reduced in OT columns compared with packed columns. Thus, OT-SFC works well using neat CO_2 and is rarely done using modifiers or mobile-phase-composition gradients.

Programed elution in OT-SFC is most often done with some form of pressure programing. Typically, the temperature is fixed and the pump delivering the mobile phase is programed to raise the pressure as a function of time. An example of a pressure-programed chromatogram is given in Fig. 10. Linear pressure programing produces nonlinear solvent strength–time profiles because of the curvature of the density–pressure isotherms (Fig. 2), particularly at low temperatures. So, with low-temperature, linear-pressure programing, early eluting peaks belonging to a homologous series are often widely spaced at low pressures where the slope of the density–pressure isotherm is fairly shallow. The next peaks, eluting in the steep part of the density–pressure isotherm, can be crowded as the mobile phase strength changes rapidly. If additional peaks elute as the slope of the isotherm again

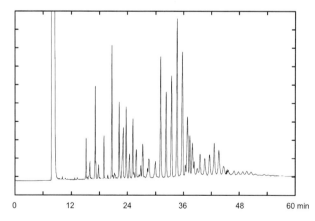

Figure 10 Isothermal pressure programed OT-SFC separation of beeswax. Conditions: 10 m × 50 μm i.d. SB-biphenyl 30 column, direct injection of 0.1 μL with a retention gap, 120°C oven, pressure programed from 115 to 120 atm at 1 atm/min, then increased to 6 atm/min with a hold at 415 atm (at 54 min), and beeswax, 1.91 mg, was dissolved in 2.0 mL of $CHCl_3$.

diminishes at higher pressure, they will be spaced a bit wider.

To avoid these swings in the rate of change of mobile phase strength, commercial instruments may provide the capability of generating density programs (Fjeldsted et al., 1983). Here, the density–time profile desired throughout the chromatogram is specified by the user, and the pump controller continuously computes and updates the required pressure. Linear density programing produces more regularity of elution for homologous series, but the peak spacing tends to slowly diminish with increasing molecular weight, eluting the peaks ever more closely as the series progresses. There is no significant benefit compared with pressure programing.

Temperature can also be positively programed if accompanied by a pressure or density program (van Konyenburg and Scott, 1980). This capability is useful in separating samples containing both high- and low-volatility components, or in tailoring the selectivity during the course of the chromatogram, as shown in Fig. 11. Additional, more specialized programing techniques (e.g., see Pinkston et al., 1991) have been devised, but simple, isothermal pressure programing is usually all that is necessary.

D. Reasons for Considering SFC

With the availability of other, more established chromatographic techniques, why would anyone be interested in using SFC? This can be answered with two considerations: capability and expense.

For packed-column SFC, the speed of methods development and the speed of the resulting methods are important considerations. SFC retains solutes with a different balance of interactions than does HPLC. The selectivity, that is, the ability to distinguish among specific solutes and separate them relative to the other sample components, is therefore different in SFC techniques than in HPLC, even when the same stationary phases are compared in both techniques. Separations that are difficult in HPLC may be much easier in SFC, but not necessarily vice versa.

Solute retention in SFC techniques is highly temperature-dependent, but temperature generally does not affect retention of all solutes to the same extent. This means that selectivity can be adjusted quickly and easily with a small temperature change in SFC. This effect is huge in SFC compared with the temperature dependence of selectivity in HPLC.

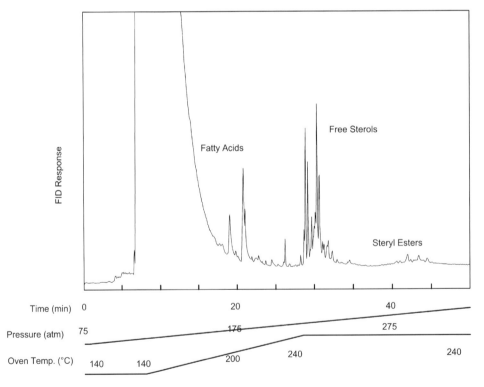

Figure 11 OT-SFC chromatogram of hamster feces extract with simultaneous temperature and pressure programing. This was necessary to get quality peaks of both the relatively volatile fatty acids and the strongly retained steryl esters. Conditions: 10 m × 50 mm i.d., SB methyl 100 column, and time-split injection (Pinkston et al., 1991).

It may appear that this capability, this additional parameter to adjust, would cause more work during methods development. However, method development in packed-column SFC is very fast because selectivity can be so easily fine tuned once a column and mobile phase have been chosen. Fewer stationary phases and mobile phases need to be explored to develop a packed-column SFC method than is usually necessary in HPLC.

Packed-column SFC chromatograms run 3–5× faster than HPLC chromatograms as measured by either theoretical plates or peak capacity per unit time. Fast flow rates and fast gradients can be used. Very little equilibration time is required after making a parameter change or running a gradient, so there is less time required between the end of one analysis and the next injection. This speed advantage can make a tremendous impact in manufacturing situations where product is held pending quality assurance, or during product development when thousands of similar samples simply need to be processed in the shortest time possible.

In addition, operating costs are small in SFC. CO_2 is inexpensive to acquire and may be safely disposed by venting. The total liquid waste generated is small because only the modifier is collected for disposal. Because water is not used in the mobile phases, columns are virtually free of hydrolysis degradation and have very long lifetimes if fouling with sample components is avoided.

OT-SFC is best compared with GC. SFC is not absolutely dependent on "thermal" volatility, and thus has a much larger solute-molecular-weight range than GC. Solutes with molecular weights around 3000 are fairly easy to elute using OT columns and unmodified CO_2 mobile phase. A 10,000–15,000 dalton range is possible for solutes with good solubility in CO_2, even if an FID is required. Prospects for an even higher molecular weight range in OT-SFC, even when using an FID, are good, waiting only on commercial instruments providing more pressure. When modifiers or fluids stronger than CO_2 are used, solutes with molecular weights in the hundreds of thousands can be eluted.

Many workers coming from GC experience are particularly attracted to OT-SFC by its lower temperatures, much higher solute molecular weight range, and selectivity advantages while retaining all the detector flexibility of GC. Packed-column SFC appeals more to those coming from an LC starting point. They are often looking for faster analyses, the ability to use much longer and more efficient columns, the ability to re-equilibrate columns at starting conditions very quickly following gradient elution, additional detection options, reduced solvent waste, and selectivity-tuning capability.

Further guidelines are best made by individuals for their particular situations. SFC is complementary to GC and LC. All three share some overlap, but each also has a unique area of capability. For many problems addressed by analytical chemists, the technique of choice will be a toss-up between two, or even all three of these separation techniques, or between chromatography and some altogether different technique. Only knowledge of the options can tell what choice is best or most cost-effective for a particular problem.

II. INSTRUMENTATION

A. Safety Considerations

Work must be done to compress a supercritical fluid from its volume at ambient pressure to achieve useful densities. This work is stored in the fluid as potential energy, available to expand the fluid rapidly if a high-pressure component or fitting fails. Liquids, with very low compressibility, store practically negligible energy compared with supercritical fluids (or compressed gases) at the same pressure. Saito and Yamauchi (1994) point out that considerably more energy is stored in a familiar device like an LC column when it is pressurized with a supercritical fluid than it would store when containing liquid at the same pressure. Therefore, these authors recommend that higher safety margins are appropriate in SFC than are common in LC: at a minimum, the bursting pressure rating of a component to be used in SFC should equal or exceed 4× the desired maximum operating pressure. Workers must keep in mind at all times that smaller safety factors may have been used to specify the published maximum pressure for LC components, and a reduction in the maximum working pressure may be necessary when these components are used in SFC (Saito and Yamauchi, 1994). Temperature must also be considered. Pressure limits quoted at ambient temperature may need to be reduced as the temperature is elevated. If in doubt, contact the manufacturer of any component in question before pressurizing it.

If a commercial SFC system is ever modified, great care must be taken never to isolate a volume of fluid between two valves without also providing a rupture disk or other pressure relief device in that same volume. This is necessary because if an isolated volume within the system were to be filled with fluid and the temperature should later increase, enormous pressures could result (especially if the fluid is *liquid*).

The use of fused silica in OT-SFC (and sometimes in packed-column SFC) is much different than in GC and deserves special notice. It is necessary on occasion to check fused-silica tubing connections while they are pressurized. There is actually little danger because the volume within the fused-silica tube and the mass of fluid contained are so small; therefore there is very little potential energy in a pressurized fused-silica tube. Furthermore, fused-silica tubing with an i.d. of 100 μm or less is incredibly resistant to pressure uniformly applied to its interior. (However, the probability of failure increases quickly as the i.d. increases. Tubes with i.d. larger than 100 μm should not be used without testing unless a pressure and temperature specification is provided by the supplier.)

Failure of fused-silica columns and tubes are very rare but can occur, especially with old columns. (We have experienced only one column failure in ~40 worker-years of SFC experience using fused-silica tubing.) Most failures are caused by human errors, like dropping a wrench, scratching or abusing the polyimide coating, and so on. The rarity of fused-silica tube failure and the small energy contained within a fused-silica tube may lead workers into a false sense of security and relaxed safety practices. Precautions, including wearing safety glasses and tying down all fused-silica-tube connectors to something sturdy before pressurizing the system, must be routinely followed.

Just as in LC, leaks in an SFC instrument, if explored by hand, can potentially inject matter through skin. Use caution around leaking fittings.

B. General SFC Instrumentation Features

A packed-column SFC instrument resembles a gradient HPLC instrument equipped with a column oven and configured to do high-pressure mixing of the mobile phase from two pumps. The only completely new requirement for packed-column SFC is to control the pressure at the outlet of the column or the detector.

OT-SFC vaguely resembles GC but with a few important differences. The mobile phase must be pumped because the operating pressure is higher than the vapor pressure of the mobile phase. Injection has features of both HPLC and GC techniques, and the flow from the column outlet must be restricted to keep the mobile phase velocity on the column near the desired value and to maintain the pressure and mobile-phase strength over the entire column. The following sections explain these requirements in more detail.

Various combinations of pressure and flow control are encountered in SFC. The most common arrangements are summarized in Fig. 12. The OT-SFC arrangement is simplest because it needs only one pump. The pump actively controls the mobile phase pressure. The pressure drop on an OT-SFC column is very small under typical conditions, so the pressure is essentially constant throughout the column (except for the small pressure drop, usually <1 atm over the column, necessary to produce the mobile phase flow). In this configuration, the mobile-phase flow rate is not actively controlled but is determined by the pressure and by the resistance provided by a flow restrictor attached to the column outlet. Thus, the pump supplies whatever flow rate is required to maintain the pressure at the setpoint. With this arrangement, retention times are reproducible for a given restrictor, and relative retention is reproducible when restrictors are changed or when results on different instruments are compared. But because the flow rate is not adjustable except by changing or altering the restrictor, absolute retention times can vary from one restrictor to another, or can change when a restrictor is damaged or simply ages.

Packed-column SFC was developed and is promoted as a faster, less-expensive alternative to HPLC. To provide the common expectation among HPLC users to be able to reproduce retention times and not just relative retention, both flow control and pressure control are required. The usual arrangement is to control flow at the pumps. This allows volumetric mixing of main fluid from one pump and modifier (with additive) from a second pump. Gradient elution is then possible in a fashion familiar to HPLC users. Pressure must be controlled at the column outlet, as the mobile-phase strength is pressure-dependent, using either a regulator or nozzle under feedback control from a pressure sensor. Thus, the nozzle or regulator will actively change its resistance as necessary to maintain the pressure at the setpoint at the location of the pressure sensor. The pressure at the pumps and the column inlet will be higher and will float at the value required to keep everything else balanced. The arrangement of the pressure regulator and detector(s) will vary depending on the type of detector(s) and interface used. More detail will be given in Section IV.A.3.

In summary, if a passive restrictor is used at the column outlet, then the mobile phase must be delivered with pressure control. However, if active pressure control is used at the column outlet, then the mobile phase is delivered with flow control. It is conceivable that OT-SFC could be performed with upstream flow control and downstream pressure control by adding a make-up flow from a pressure-controlled source at the column outlet, but this would add cost to the instrument and is not offered by manufacturers.

C. Packed-Column SFC Instrumentation

1. Pumps

The pumps used in packed-column SFC [Fig. 12(b) and (c)] resemble HPLC pumps. The modifier pump may be an unaltered HPLC pump, but delivering CO_2 with flow control requires a higher degree of compressibility compensation than is typical in HPLC pumps. This results because the compressibility of liquid CO_2 is not only much larger than for normal liquids, but it also changes with pressure over the range typical of SFC methods. The pump must adjust its rate of volume displacement to maintain the proper CO_2 flow rate when the pressure changes. The CO_2 is delivered to the pump in liquid

Supercritical Fluid Chromatography Instrumentation

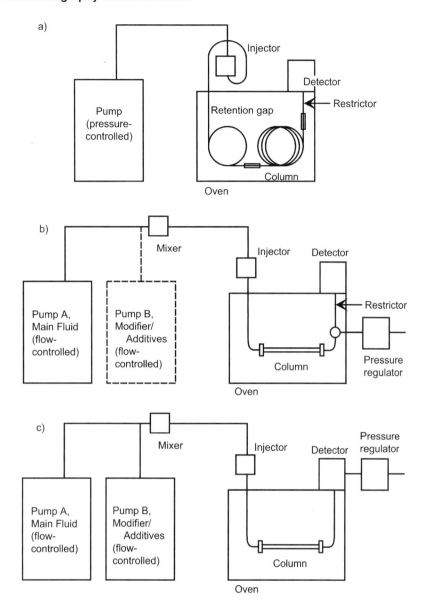

Figure 12 Basic components of supercritical fluid chromatographs. (a) Upstream pressure control with a downstream passive flow restrictor. The retention gap, shown here, is often necessary for direct injection onto OT columns, and is optional with dynamic-flow splitting, time-splitting, and split/splitless injection. (b) Downstream pressure control with upstream flow control used with a low-pressure detector. Either splitting the column effluent, as shown, or adding a pressure-regulated make-up flow at the tee is necessary. (c) Downstream pressure control with upstream flow control used with a column-pressure detector.

form from a cylinder equipped with a dip tube. It is necessary to cool the pump head to prevent the CO_2 from boiling when the local pressure falls during the intake stroke. If the pump cylinder is not completely filled with liquid CO_2, then flow control and compression compensation will not work properly.

It is possible to purchase CO_2 in a cylinder pressurized with a helium headspace (called a helium pad in the trade). This was originally developed to avoid the requirement of cooling the CO_2 pump by delivering the liquid CO_2 at a higher pressure provided by the helium. Unfortunately, helium dissolves in the CO_2 to some extent and weakens the CO_2 as a solvent. To make matters worse, the concentration of helium in the CO_2 changes as the contents of the supply cylinder are consumed and the helium pressure falls, so the strength of the CO_2 is not constant. It is best to avoid CO_2 supplied with helium in analytical SFC.

Because gradients are formed in SFC with high-pressure mixing from separate main and modifier pumps, a high-pressure mixer is necessary to homogenize the mobile phase. Commercial SFC instruments use either an active or a static mixer. An active mixer adds expense to the instrument. Packed-tube static mixers do not mix well enough. We have experienced significant detector baseline noise after mobile phase has been statically mixed, even after this mixture is passed through the analytical column. This noise can be reduced and performance improved by using a 250-μL volume, high-efficiency, vortex-sheer mixer rather than a mixing tube. A high-efficiency mixer has the added benefit of reducing the dwell volume and the gradient delay compared with the alternatives. The pumps may run at slightly higher pressure because of the pressure drop on the mixer, but the pressures at the injector and at all points downstream are unaltered.

2. Injector

Injection in packed-column SFC appears to be identical to HPLC. The same six-port high-pressure injectors and the same basic procedures are used in both. However, there are some important differences that, if ignored, can seriously degrade performance in SFC. A typical specification for peak-area relative standard deviation (RSD) using an autosampler equipped with a typical six-port, external-loop injection valve is 0.5%. Coym and Chester (2003) showed that a typical 10 μL full-loop injection procedure had a peak area RSD of only 0.38% on their particular autosampler when using an HPLC mobile phase, but that the performance was 18× worse when the identical autosampler and procedure were used with an SFC mobile phase.

There were several causes of these problems. First, in HPLC, the injector loop is always left filled with liquid mobile phase after injection. But in SFC, the mobile phase rapidly decompresses and expels most of itself from the loop through the needle port and the waste port when the valve is put in the *load* position, leaving the loop filled with vapor and perhaps a few droplets of modifier. Any gas bubbles that cannot be expelled by loading the next sample will effectively lower the volume of the next sample. If this effect is variable, then so are the peak areas.

This problem was fixed by rinsing the needle port with enough liquid to expel all the gas before loading the next sample. The rinse liquid and the sample solvent were the same solvent used as the modifier so that no unexpected phase separations would occur during sample loading or injection.

Several potential geometry problems that may lead to unwanted siphoning of liquid from the loop were also identified and corrected. With these changes, the peak area RSD with SFC mobile phase was 0.25% or better. The protocol used for a Gilson model 234 autosampler is listed in Table 3. Although the steps are specific for this autosampler and for a 10 μL injection, the same general procedure can be easily adapted to autosamplers from other suppliers. The recommended sample aspiration

Table 3 Recommended Injection Protocol for 10 μL Full-Loop Injections in Packed-Column SFC Using an Autosampler[a]

Step	Task
1[b]	Home the needle
2	Rinse the needle
3	Switch the valve from *inject* to *load*
4	Move the needle to the needle port, and rinse needle port, loop, and waste port with 450 μL of modifier
5	Remove the needle and aspirate a 3 μL air gap into the needle
6[c]	Aspirate 65 μL of sample into the needle and the transfer tube
7[c]	Move the needle to the needle port, and dispense 50 μL of sample into the loop
8	Switch the valve from *load* to *inject*
9	Rinse the needle port and the waste port
10	Remove and rinse the needle
11	Home the needle

[a]This was written specifically for a Gilson model 234 but can be adapted to most others.
[b]The valve is in the *inject* position initially.
[c]The sample aspirate and dispense rate is 0.5 mL/min.

volume is 5 × the loop volume + 15 μL, and the recommended sample dispense volume is 5 × the loop volume. A 3 μL air gap is used to separate the sample from the rinse solvent in the needle and sample delivery tube. The recommended valve and tubing geometry is shown in Fig. 13.

The maximum sample injection volume in packed-column SFC is smaller than in HPLC. This is because, in most cases, the sample must be dissolved in an ordinary liquid that is miscible with the mobile phase, and all possible solvents will be quite strong compared with neat or lightly modified CO_2, as would be appropriate initially in a gradient SFC method. This situation is analogous to dissolving a sample in acetonitrile, then injecting it in reversed-phase HPLC with 90/10 water/acetonitrile mobile phase. You would not expect the solute to be retained well initially if the sample injection volume were very big. If a gradient is applied, some of the solute broadening can be refocused. However, a maximum sample injection volume of ~20 μL is a reasonable limit with 4.6 mm i.d. packed columns.

3. Oven

Packed-column SFC is sometimes performed at temperatures as low as −50°C. (Actually this would be subcritical fluid chromatography if one cares to make the distinction.) Chiral selectivity is often improved at such low temperatures, so this capability is important in many life science applications. Subambient temperature control is offered as

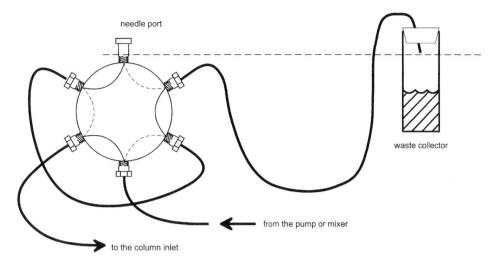

Figure 13 Recommended injector geometry for packed-column SFC (Coym and Chester, 2003). In order to prevent siphoning of the mobile phase or sample solvent through the sample loop, the needle port inlet and waste port outlet should be at the same elevation. In addition, the waste tube should be sufficiently long to contain enough solvent to prevent air from entering the sample loop when the needle is withdrawn. Most of this volume should be in a trap or loop below the needle port and the waste tube outlet. The waste tube outlet should not touch the side of the waste container or the liquid waste being collected. The sample loop will expel its contents when the valve is put in the load position, so the loop must be thoroughly flushed with liquid to displace all the gas before the sample is loaded. The arrangement shown here will prevent gas introduction when the needle is removed after the flushing operation.

an option on some commercial SFC instruments. Without subambient temperature control, it is difficult to operate an instrument reproducibly below ambient temperature plus ~7°C. Some preparative SFC methods are run successfully at ambient temperature without using a column oven by maintaining the temperature of the work area, but we do not recommend this for analytical methods.

The upper practical temperature is set by the stability of the stationary phase or other system components. Many stationary phases are stable at temperatures above 300°C, but other column components may fail at much lower temperatures. Ovens on commercial packed-column SFC instruments typically have a 150 or 200°C specification, but users must be aware of the temperature limits of the components they put in the oven. The requirement for temperature uniformity throughout the oven is similar to GC.

For repetitive analyses, the oven is only required to accurately maintain the set temperature. However, in research applications where many different methods may be executed in a given day, speed of the oven to stabilize at a new temperature is important. The ovens on some commercial packed-column SFC instruments are very slow, perhaps taking 20 min or longer to stabilize at a higher temperature. Others have the ability to rapidly apply heat and stabilize at a higher temperature in only a minute or two, but lack the ability to cool quickly. For a research instrument, a large oven with GC-like performance for both heating and cooling would be ideal, but none is offered on an analytical packed-column SFC at this time.

4. Packed Columns

Porous, spherical packings with particle sizes in the range of 5 μm are widely used as stationary phases. Like in HPLC, larger particles are less efficient and require longer analysis times, but unlike HPLC, smaller particles do not necessarily produce higher efficiency or faster analyses. This is due to a combination of the effects of the longitudinal pressure drop on the local values of plate height, retention factors (k), density, and mobile-phase velocity (Chester, 2002) so that conditions are not necessarily fixed longitudinally in a packed SFC column. The pressure drop becomes large and this complexity of these interacting effects increases rapidly as the particle size drops below ~5 μm in SFC.

Stationary phases for packed-column SFC are similar (and sometimes identical) to many used in HPLC. Common stationary phases are listed in Table 4. Selection of the stationary phase will be discussed in Section III.A.

Columns with i.d. in the range of 2–4.6 mm are considered conventional packed columns. Analytical packed-column SFC generally uses 4.6 mm i.d. columns. These can accommodate reasonable sample volumes and require pumps with maximum flow rates in the range of 10 mL/min. Cartridge-style packed columns may be used in SFC only if they do not contain polymer seals. One must particularly be aware of guard columns in cartridge format. The same warning goes for polymeric compression

Table 4 Stationary Phases Used in SFC

Packed-column SFC	Open-tubular SFC
Silica, bare	Polysiloxanes
Silica, silver-loaded	100% methyl-polysiloxane
Silica-based bonded phases	50% n-Octyl, 50% methylpolysiloxane
Octadecyl	5% Phenyl, 95% methylpolysiloxane
Octyl	50% Phenyl, 50% methylpolysiloxane
Phenyl	30% Biphenyl, 70% methylpolysiloxane
Cyano	25% Cyanopropyl, 75% methylpolysiloxane
Amino	50% Cyanopropyl, 50% methylpolysiloxane
Diol	
2-Ethylpyridine	
Diethylaminopropyl	
SCX (sulfonic acid)	
Other	
Bonded polysiloxane on silica (e.g., DeltaBond)	
Polystyrenedivinylbenzene, porous particles	
Chiral (dozens of phases are available)	

fittings and ferrules. The typical polymeric material used is polyetheretherketone (PEEK), which is fairly inert but swells when contacted with some solvents (e.g., dimethylsulfoxide, tetrahydrofuran, and methylene chloride). The permissible maximum temperature is also greatly reduced when PEEK is present compared with an all-steel system. We recommend that systems containing any PEEK not be used above 75°C in most SFC applications and never higher than 110°C in any circumstance.

Larger-diameter columns handle higher sample mass and volumes without overload, but require moving up to faster, more expensive pumps. Expense rises so rapidly with column diameter that 4.6 mm remains popular for analytical work.

There is little benefit in going to smaller diameters when using conventional concentration detectors (like UV) because the maximum sample volumes and the peak mass drop quickly, and detection limits and detector problems worsen unless very small cell volumes are used. However, 1 mm i.d. columns, called microbore columns, are small enough that the entire effluent flow can be directed to a flame-based detector without splitting. These columns provide over 100× more sample capacity than 50 μm i.d. OT columns, and provide relatively short analysis times. These columns are used extensively for ASTM Method D5186 for the determination of aromatics in diesel and jet fuels and for ASTM Method D6550 for the determination of olefins in gasoline, both using an SFC equipped with an FID. Microbore columns never became popular for general applications because available packings are too retentive for solutes more polar than petroleum components when using neat CO_2 mobile phase. Once the decision is made to use modifier, the detector choice is limited mostly to those that are already compatible with larger flows; hence, conventional columns are chosen.

Packed-capillary columns (Malik et al., 1993; Tong et al., 1994) range in i.d. from ~100 to 300 μm. These columns were popular for a brief period in research literature, primarily because they can be used with very long lengths achieving hundreds of thousands of theoretical plates. This is possible because the pressure drop per unit length (for a given mobile-phase velocity) diminishes as the ratio of the column diameter to the particle size decreases. This results from the packing defects and the lower packing density caused by the increasingly curved tube walls as the diameter decreases. These columns are still interesting but are not widely available.

Even without the benefit of pressure-drop reduction seen with packed-capillary columns, conventional packed columns can also be connected in series to increase the plate number. Because the viscosity of supercritical (or near-critical) fluids is so much lower than ordinary liquids, the column length may be much longer than in conventional HPLC without incurring an excessively large pressure drop over the column set. Impressive efficiency is possible, as demonstrated in Fig. 14 (Berger and Wilson, 1993). When multiple columns are connected in series to increase the peak capacity in a gradient method, or whenever the column length is changed and comparable relative retention is required, it is necessary to change the gradient rate by the ratio of the original total column length to the new total column length. The flow rate should remain unchanged. Thus, doubling the column length will require twice the analysis time, but

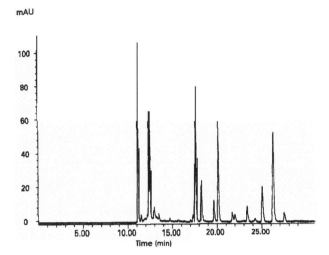

Figure 14 Packed-column SFC chromatogram of Brazilian lemon oil exhibiting >200,000 theoretical plates. This was produced by linking ten 4.6 × 200 mm columns in series. The packing was 5 μm Hypersil silica. Other conditions: 2 mL/min of 5% methanol in CO_2 at 60°C, downstream pressure control at 150 bar, and UV detection at 270 nm. [Reprinted from Berger and Wilson (1993) with permission.]

will keep the relative retention unchanged and will provide all the benefits possible from the extra length.

Some of the apparent efficiency when long packed columns are used may be due to an automatic peak focusing mechanism caused by the pressure drop over the column. The leading edge of a peak is always in weaker mobile phase than the trailing edge, even when there is no programing, thereby automatically reducing the temporal peak width relative to the local mobile phase velocity. This effect with column length works in opposition to the problem mentioned earlier with particles smaller than 5 μm. A more thorough discussion is available for interested readers (Chester, 2002).

Monolithic columns have been used successfully with CO_2–methanol mobile phase at 25°C and 15 MPa outlet pressure (Lesellier et al., 2003), but success at higher temperatures and pressures has not been reported. This is not because of any known problem with the monolithic column material, but because these columns are only available with PEEK jackets. At typical SFC pressures and temperatures, the jacket apparently softens and "balloons" enough to create open channels between the monolith and the jacket wall, thereby ruining any chance of an efficient separation. The monolithic material is promising, especially for applications requiring coupled columns to achieve high plate numbers, but jackets that can withstand significantly higher temperatures and pressures will be necessary to achieve the full benefit of the range of selectivity in SFC.

5. Detectors

Virtually every detector ever devised for any form of chromatography has been used successfully in SFC. There are three basic categories of detector for SFC based on the operating pressure and the interface required: low-pressure (including atmospheric pressure), column pressure, and in-between-pressure detectors. Examples of each are listed in Table 5.

Because mobile-phase strength depends on pressure in SFC techniques, and because many of the mobile phases used are gases at ambient temperature and pressure, low-pressure detectors must be interfaced to supercritical fluid sources using a flow restrictor. Section II.D.5 discusses FIDs and restrictors. Section IV.A discusses interface options to mass spectrometers. These options can be applied to other low- or atmospheric-pressure detectors such as the ELSD.

Detectors equipped with cells that can be operated at column-outlet pressure are generally connected in the flow path between the column outlet and the downstream pressure regulator. Note that common LC detectors cannot be used in this fashion unless they are equipped with high-pressure cells. With UV detectors in SFC, it is also desirable to keep the detector temperature at a subcritical value (i.e., in the liquid range for the mobile phase) in order to reduce the compressibility of the mobile phase and limit the changes in RI that would occur when the pressure is changed. This results in a flatter baseline when performing pressure programing. Lowering the mobile-phase temperature also increases the density and the volumetric concentration of solutes, thus improving the signal-to-noise ratio.

The natural heat loss from a room-temperature transfer line is sufficient to lower the temperature before UV detection as long as the column is used at ambient or slightly elevated temperatures. Some older commercial packed-column SFC instruments were equipped with either a heat exchanger or a heat sink between the column outlet and the detector, or as part of the detector assembly, but this may not be included in the standard configurations of modern instruments. A heat sink may have to be added as an option on one of these instruments to improve detection for high-temperature SFC operation.

The RI detector is called "in between" because the high dependence of RI on pressure and temperature calls for unusual measures. If thermostatted and controlled at a pressure just below the lowest column pressure, the detector may be used even with pressure programing (Hirata and Katoh, 1992). This is accomplished by using a flow restrictor as the column-to-detector interface, and using a downstream pressure regulator at the detector outlet. This arrangement keeps the pressure in the detector constant regardless of the column pressure, and allows

Table 5 Detection Limits of SFC Detectors

SFC detectors	Approximate detection limit, pg (Lee and Markides, 1990)
Low pressure	
Flame-ionization detector (FID)	50
Thermionic detector (TID)	50 (N mode)
	10 (P mode)
Flame photometric detector (FPD)	2,500 (S mode)
	500 (P mode)
Sulfur chemiluminescence detector (SCD)	35
Photo ionization detector (PID)	50 (?)
Electron-capture detector (ECD)	1
Atomic emission spectroscopic detection (AES)	1,000 (?)
Ion-mobility spectrometry detection (IMS)	50 (?)
MS	20
Evaporative light scattering detector (ELS)	$\sim$3,000
Evaporative infrared spectroscopic detection	$\sim$2,000[b]
Column pressure	
Ultraviolet absorption detector (UV)[a]	10
Fluorescence detector[a]	2
On-column infrared detector	10,000
In between	
Refractive index detector (RI)	$\sim$10,000 (?)

[a]With a good chromophore.
[b]Requires signal averaging.

pressure programing with upstream control of the column pressure. The mobile phase flow rate is not regulated, but is determined by the combined flow resistance of the devices beyond the column. This arrangement can also be used with other bulk property detectors, and was actually first reported as a means to remove pressure-related drift from UV absorbance detectors during pressure programing (Hirata and Katoh, 1992). In most cases, however, an ELSD is a much better choice than an RI detector in SFC applications.

Mass spectrometers are such valuable and important detectors in SFC they will be discussed separately in Section IV.

D. OT-Column SFC Instrumentation

1. Pumps

As OT-SFC is usually performed using neat CO_2 mobile phase, only one pump is necessary. Because of the low pressure drop on OT-SFC columns, the pressure–strength relationship of supercritical fluids, and the sensitivity of mass detectors to the flow rate, pumping in OT-SFC must be free of pressure pulses. High-pressure syringe pumps have been preferred by practitioners for many years. Modern instruments use 10 mL syringe pumps equipped with cooling so that they can be rapidly filled with CO_2 from room-temperature sources.

Pressure programing is the predominate method for programed elution in OT-SFC. Therefore, the pumps must not only be able to control pressure at any given setpoint, but also change the setpoints as a function of time. Commercial systems can deliver pressures accurately up to $\sim$400 atm. However, much additional solute molecular weight range and, indirectly, the opportunity to use higher temperatures are available when higher pressures are used with unmodified CO_2, as demonstrated in Fig. 15 (Chester and Innis, 1993). Here, the ability to use higher pressure allowed the use of a more retentive column capable of separating the peaks of interest. This choice would not be viable for systems unable to deliver adequate pressure. It is clear from Fig. 2 that if a temperature above $\sim$100°C is desired, the pressure required to achieve even 90% of the maximum-possible mobile-phase density (with CO_2) is well above 500 atm. At 200°C, a commercial instrument with a 400 atm pressure limit would only deliver $\sim$50% of the mobile-phase density available with higher pressure.

2. Injector

OT-SFC injection is a little more demanding than packed-column SFC because of the small column volume and

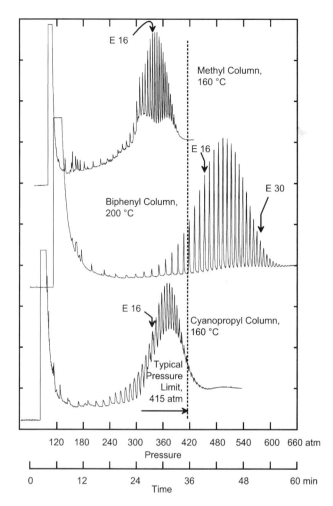

Figure 15 Chromatograms of a surfactant, Brij 78, performed on three different OT columns, each at its optimum temperature. The best separation was provided by the most retentive column, but required pressures far exceeding those available on current commercial instruments (Chester and Innis, 1993).

the likelihood of major flooding in the column inlet. The descriptions in this section may also be of value to users trying to understand mass transfer in packed-column SFC.

Because the dimensions are so small, inlet flooding is particularly bothersome in OT-SFC injections. A 1 μL volume of sample would completely fill about a 50 cm length of 50 μm i.d. column. The inlet flooding resulting from an injection of that volume directly onto a column would distribute solutes over many meters of column before the injection solvent were dissipated. Many solutes, especially the early eluting ones, would be distributed throughout the length of this flooded zone. However, initial solute bandwidths in space must be closer to 1 mm to avoid contributing significantly to the measured peak width at the column exit. Thus, understanding mass transfer, avoiding unrecoverable band spreading, and taking advantage of peak focusing mechanisms are necessary elements of success. OT-SFC can be performed with similar precision as packed-column SFC or HPLC if attention is paid to mass transfer.

Historically, flooding was controlled in OT-SFC by severely limiting the sample volume using either dynamic-flow splitting (Peadon et al., 1982) or time-splitting injection (Richter, 1986), Fig. 16. Both techniques use an ambient temperature, high-pressure, four-port sampling valve, located outside the chromatographic oven. This valve contains an internal sample loop in the range of 0.06–1 μL. With flow splitting, the analytical column is sealed in the bottom fitting of a tee in the oven, with the analytical column extended through a stainless steel tube connecting the tee to the ambient-temperature injection valve. The column inlet is positioned near the valve. Thus, the split point is at ambient temperature (where type I binary CO_2–solvent mixtures exist in a single liquid phase for all possible compositions). The split ratio is adjusted by balancing the resistance of the vent restrictor (attached to the side arm of the tee in the oven) against that of the restrictor at the column outlet.

In the time-splitting technique, the analytical column is attached directly to the injection valve. The valve is cycled very rapidly using a pneumatic actuator powered by compressed helium (used for its low viscosity and the resulting fast valve actuation). The valve is left in the *inject* position for times ranging from ~30 ms to several seconds to transfer a fraction of the sample loop contents, then the valve is switched back to the *load* position. This is a split technique because less than the full content of the sample loop is transferred to the column. The balance is lost through the valve waste port when the valve is flushed with solvent or loaded with the next sample.

The *effective* injection volumes of these techniques range from just a few nanoliters to ~50 nL (for a 10:1 split of a 0.5 μL loop volume). Assuming a nanogram of solute is required on column for accurate and precise *detection*, the corresponding minimum-detectable concentration for each solute in the injection solution is, optimistically, ~20 μg/mL. This concentration must be increased if larger split ratios or smaller valve loops are required to keep the solvent peak from becoming too wide. In addition, internal standards are usually required for quantitation with these injection techniques because there is no guarantee that any two solutions will split with the same ratio. In fact, there is no guarantee that different solutes from the same sample will split with the same ratio, particularly for the flow-splitting technique.

Better performance is available with split/splitless injection (J. Clark, Selerity Technologies, Inc., private communication; Lee et al., 1989; Tuominen et al., 1991). The current commercial implementation has been refined

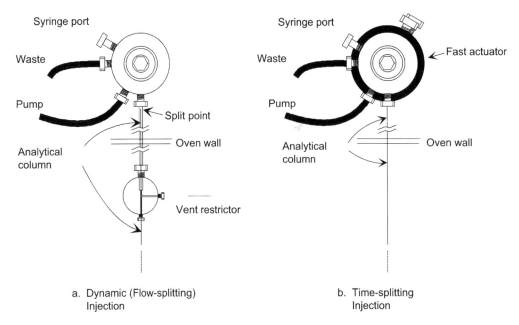

Figure 16 Arrangement for (a) dynamic-flow splitting injection and (b) time-splitting injection for OT-SFC. In both cases the injection valve is outside the oven at room temperature.

and is illustrated in Fig. 17. The same four-port, internal-loop valve is used to inject the sample into a sheath tube containing the column inlet, and a second valve is added to switch the split flow around the column inlet on or off while keeping the flow from the pump constant.

In split mode, all of the mobile-phase flow is directed through the injection valve (the alternate path leading from the tee ahead of the injection valve is blocked), the mobile phase splits continuously around the column inlet, and the injector is operated as described earlier for

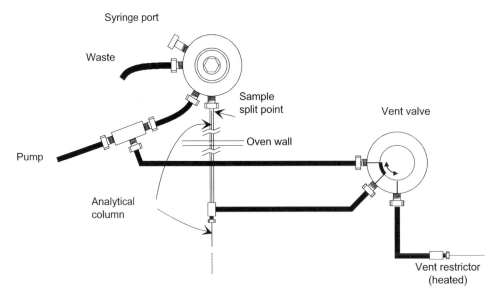

Figure 17 Split/splitless injector for OT-SFC. Note that the flow through the vent is continuous. The position of the vent valve determines one of two states: (1) The flow splits in the tee ahead of the injection valve. This produces a slow flow rate through the internal sample loop that is transferred into the column inlet without splitting. (2) All the flow from the tee ahead of the injection valve is directed through the sample loop at a fast flow rate. This flow is then split around the column inlet. This arrangement keeps the inlet pressure and the flows through the vent and the column constant (at equilibrium) with both positions of the vent valve.

flow splitting. "Splitless" injection starts by changing the split-vent valve so that the mobile phase is split in the tee on the supply side of the injector. This reduces the flow rate through the injection valve, and directs the entire flow out of the injection valve and into the column inlet. Operation requires setting the split-vent valve for splitless operation, equilibrating the pressure, then switching the injection valve from load to inject. If the split flow around the column inlet is left blocked, the entire contents of the injection valve (up to 0.5 μL) can be transferred to the column, but this usually results in some tailing of the solvent peak and perhaps some broadening of some of the solute peaks. The effective injection volume can be varied by switching the split-vent valve to restore the split flow around the column inlet after a prescribed delay period. This allows some flexibility in injections, and is also somewhat analogous to restoring the septum purge in GC. This greatly improves the sample solvent peak shape by rapidly clearing any remaining sample solvent from the injection valve assembly. Nearly splitless transfer with good solvent peak shape results when the split flow around the column inlet is restored ~1 min after sample injection. Performance will also depend on the injection solvent, pressure, oven temperature, and mobile phase velocity. A solute focusing process must be at work behind the scenes, because there can be a significant increase in peak areas comparing a 60 s injection to a 30 s injection, yet well-retained peaks are only ~20 s-wide at the base when they reach the detector. Whenever this is the case, the initial temperature and pressure will have a large influence on the outcome of the method.

If injection of the entire sample-loop contents is the goal, then this is best done by direct injection utilizing a retention gap to recover from the flooding (Chester and Innis, 1995). This technique is simpler to implement and has better precision than those described earlier for OT-SFC, but adds time to the analysis. As there is no possibility of split ratio variations between samples and standards (as there is no splitting), external standards may be used. Flooding is limited to the retention gap and the resulting band broadening in space is handled by taking advantage of the phase behavior of binary mixtures. The hardware is illustrated in Fig. 18.

The steps to perform the direct injection of 0.1 μL onto a 50 μm i.d. column are summarized in Table 6. This volume, though still small by HPLC and packed-column SFC measures, is 2–100× larger than the typical effective injection volumes of the splitting techniques. The direct injection volume can be scaled up to 0.5 μL or more

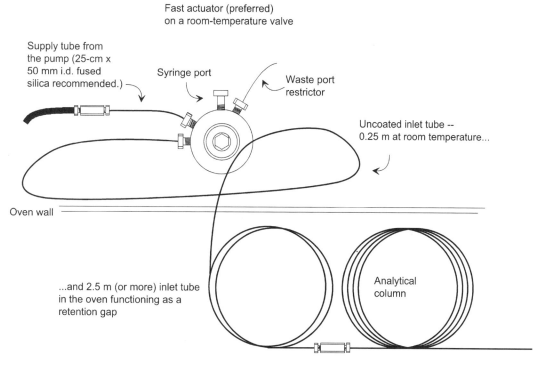

Figure 18 Arrangement for direct injection using a retention gap for OT-SFC (Chester and Innis, 1995). Dimensions for a 0.1 μL injection volume are shown. Larger volumes are possible by scaling up the lengths appropriately. Note, the length of fused-silica inlet tubing outside the oven must be appropriately shielded for safety. Simply encasing it in a loose-fitting plastic tube is sufficient. Also, although shown separately, the retention gap can be wound with the column on a single wire support.

Table 6 Direct Injection of 0.1 μL in OT-SFC

Use a room-temperature inlet tube (0.25 m × 50 μm i.d., continuing into the oven to become a retention gap of an additional 2.5 m length)
Dissolve sample in an appropriate solvent (see Figs. 7 and 8)
Use an oven temperature between 90°C and 150°C, and an initial pressure below 100 atm
Set the injection valve to remain in the *inject* position for 30–60 s after injection, then return it to the *load* position
Hold or program at no more than 1 atm/min for 5 min, then program at 5 atm/min or slower

Note: These are conditions that will work nearly all of the time without any problem. The mass transfer involved is somewhat complicated, but the details are self-working with the recommended parameter settings. Larger volumes are possible with appropriate changes in tube lengths, times, and rates.

with an additional length of inlet tubing and retention gap, and more time for the mass transfer and focusing steps. Here is a brief description for a 0.1 μL direct injection.

The pump is connected to the valve using a section of 50 μm i.d. fused-silica tubing (instead of 1/16 in. o.d. stainless steel) to increase the velocity and reduce the chances of contaminating the supply tube with sample. Sample is loaded into the valve with positive pressure from a syringe or autosampler working against the resistance of a flow restrictor attached to the valve waste port. This ensures that the sample loop is always completely filled with sample. The valve is at room temperature and is connected to a 2.75 m × 50 μm i.d. section of fused-silica tubing. (This tube may need to be longer in some applications.) This tube may be deactivated but is not coated with stationary phase. The first 0.25 m of this tube is kept outside the oven at room temperature (appropriately shielded for safety). This section of the tube is used to receive the sample from the valve into an efficiently swept volume before the liquid front enters the oven and is heated. This helps prevent a compression recoil from propelling a fraction of the sample solution backwards into the mobile phase supply tube. The remaining 2.5 m of this uncoated tube is coiled in the oven where it functions as a retention gap (Chester and Innis, 1989) and is connected to the analytical column.

The key to this technique is to have a single liquid phase at all times outside the oven (to provide good mass transfer from the valve) and to force liquid and vapor phase separation in the oven and on the retention gap. Typical conditions are illustrated on the critical locus in Fig. 19. The liquid is dynamically trapped as a film on the walls of the retention gap where it functions as a temporary stationary phase, retaining the solutes. The injection solvent in liquid form never reaches the analytical column, thus preventing liquid flooding on the stationary phase, premature transfer of solutes down the analytical column, and peak broadening on the column. The liquid film is removed from the retention gap by evaporation and transport as a vapor through the column. Solutes are initially distributed over the length of the flooded zone (i.e., the initial length of the liquid film), which may reach several meters for each 100 nL injected. However, solutes are easily refocused either by solvent focusing on the liquid film as it disappears, or by phase ratio focusing on the head of the analytical column. Refocusing is self-working for programed elution conditions as long as the program rate is not too fast before the refocusing is completed.

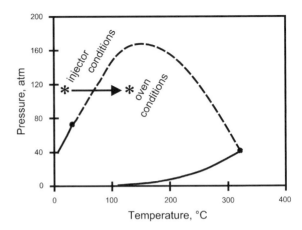

Figure 19 Conditions should be chosen for direct injection so that only one fluid phase exists at all points outside the oven, and that l–v phase separation occurs when the injected sample liquid reaches the retention gap in the oven. The conditions shown here are typical for the injection of samples dissolved in toluene.

Figure 20 illustrates how reproducible this technique can be. For this injection technique, peak area and peak height RSDs are usually in the range of 0.6–1.8% for peaks of a few nanograms or more (Chester and Innis, 1995). Figure 21 illustrates several possible problems and suggests solutions.

3. Oven

OT-SFC ovens resemble GC ovens and have similar programming capabilities and specifications. An upper temperature limit of 400°C is typical. Subambient temperature control is optional.

4. Columns

All OT-SFC is done today using fused-silica columns. The column and stationary phase must be inert and insoluble in the mobile phase. Thus, the biggest single step in the development of OT-SFC was actually the development of nonextractable, crosslinked and surface-bonded GC columns. Stationary phases for OT-SFC are listed in

Supercritical Fluid Chromatography Instrumentation

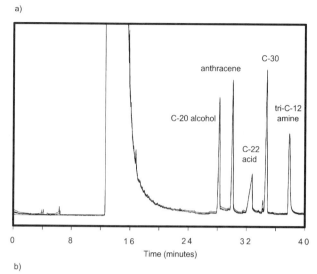

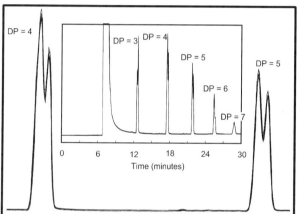

Figure 20 Demonstration of the reproducibility of the direct-injection method in OT-SFC. (a) Five successive 0.5 μL injections of test mix with the resulting chromatograms overlaid. (b) Seven successive 0.1 μL injections of a Polystyrene-400 solution, similarly overlaid. The detail of the seven overlaid chromatograms for the peaks with DP equaling 4 and 5 is enlarged, with the seven complete overlaid chromatograms shown in the inset. The two peaks at each DP result from the presence of diastereomers.

Table 4. Selection of the stationary phase will be addressed in Section III.B.

OT-SFC columns are almost identical to modern GC columns, varying only in i.d. and film thickness. The i.d. is typically 50 μm and the film thickness is typically 0.25 μm. The resulting phase ratio (β, the volume of mobile phase divided by the volume of stationary phase) is lower than in GC columns. This increases the retention of SFC columns relative to GC columns, all else being equal. So if SFC columns were substituted for GC columns, the solute elution temperatures would be considerably higher. However, with the solvating mobile phases in SFC,

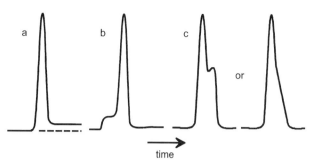

Figure 21 Several problems that may occur with direct injection in OT-SFC. (a) Unequal baselines can result if the supply tube or fitting gets contaminated with sample or sample solvent (from flow recoil after valve actuation) and is slow to clear. Use a 10 cm long, 50 μm i.d. fused-silica supply tube or remake the questionable fitting to remove excess volume. If this does not fix the problem, remove an additional 10–20 cm of the retention gap from the oven to room temperature. (b) Chair-shaped peaks in the direction shown here may result if the retention gap is too short, or the injection volume is too large, and the stationary phase is wet by the liquid sample. Decrease the sample volume or lengthen the retention gap. (c) Split or tailing peaks like those shown here may result if the phase-ratio focusing is not adequate and solutes begin migrating on the column before focusing is complete. Lower the initial pressure, allow more time before beginning the pressure program, or program at a slower rate, at least initially. See Chester and Innis (1995) for more details.

there is no excessive solute retention caused by the phase ratio, and the extra capacity for accommodating solute mass (provided by the smaller phase ratio) is advantageous.

Smaller OT-SFC column diameters are impractical because of injection dead-volume problems. Larger diameters would require significantly longer analysis times. The current 50 μm i.d. columns are a good compromise.

The standard length of commercial OT-SFC columns is 10 m. This is usually a good compromise between providing useful efficiency and keeping analysis times reasonable. Analyses longer than 60 min are rare. If speed is important, shorter columns can provide faster analyses. The typical mobile phase velocity for a 50 μm i.d. OT column is ~1 m/min, so a 20 min analysis would be fairly fast on a 10 m column. However, a 1 m column would only take 2 min with the same relative solute retention. To translate a pressure-programed method between columns of different length and the same diameter requires changing the pressure-program rate by the ratio of the original column length to the new length. The velocity should not be changed.

5. Detectors and Restrictors

All the detectors discussed for packed-column SFC are also applicable to OT-SFC and many GC detectors. The most commonly used detector in OT-SFC is the FID. Thermionic N–P and electron-capture detectors are possible but are rarely used. UV detection is possible but the peak volumes are so small that the detection path length must be kept short; thus, detection limits measured in terms of concentration are not as good as in techniques with larger dimensions.

As we have seen, no column-to-detector interface is required with column-pressure detectors. They are connected directly to the column outlet, and the pressure is reduced downstream from the detector. However, when low-pressure detectors like the FID are used, the pressure must be lowered at some point between the column outlet and the detector inlet. When this occurs, the mobile phase not only loses solvating power, but undergoes adiabatic cooling as it expands. Both effects promote solute condensation. This causes noisy peaks as discrete solute particles enter the detector and each one produces a sudden response. This behavior is generally known as "spiking", named for the appearance of the peaks. In extreme cases, the system can be plugged by condensate. It is the role of the restrictor to interface the column outlet to low-pressure detectors, to maintain the pressure at the column outlet, and to provide a mobile phase flow rate on the column in an adequate range for chromatography as the pressure is programed.

Although straight pieces of narrow-bore tubing can be used as restrictors in situations involving liquid transfer (such as the waste-port restrictor on an injection valve), they usually fail as low-pressure detector interfaces for SFC. Three other types of restrictor are popular: thin-walled, tapered capillaries (Chester, 1984; Chester et al., 1985); short-tapered or integral restrictors (Guthrie and Schwartz, 1986); and porous-frit restrictors (Cortes et al., 1988). They are illustrated in Fig. 22.

The frit restrictor (Cortes et al., 1988) is made by depositing porous ceramic in the end of a fused-silica tube. The ceramic bed may be several centimeters long, providing a vast multiplicity of flow paths. This makes the frit restrictor resistant to plugging. Mass transfer performance is usually adequate as long as the solutes can be eluted below ~400 atm (Pinkston and Hentschel, 1993), so this restrictor is well-matched to the other specifications in commercial OT-SFC instruments. Frit restrictors can be shortened to increase the flow rate. The outside surface of the frit restrictor is coated with polyimide all the way to the outlet making this restrictor very rugged. Frit restrictors are used by most practitioners.

The integral restrictor (Guthrie and Schwartz, 1986) is, in essence, a short, tapered restrictor with thick walls. It is made by closing the end of a tube in a flame, then polishing the end until a small opening forms. The integral restrictor is so named because it may be prepared directly on the end of the column, avoiding a column-to-restrictor union and its possible contribution to peak broadening. The outlet i.d. of an integral restrictor is usually in the range of 0.5 to ~1.5 μm resulting in a very fast outlet velocity and a short pressure transition. Although the polyimide coating is burned off approximately the last millimeter of the tube when the restrictor is made, the walls are so thick that the restrictor is still fairly rugged. Integral restrictors have excellent high-molecular-weight performance if they can be kept from plugging. Plugged integral restrictors can often be opened again by briefly bringing the base of a match flame near the tip while applying 70–100 atm pressure to the inlet (taking appropriate safety measures, of course). This can be done in an SFC–FID instrument without disassembling any high-pressure connections by shutting off the flame gases, cooling the detector, loosening the column nut at the base of the detector (which is not a high-pressure fitting), and extending the restrictor through the detector into the lab air where it can be inspected and treated.

Tapered-capillary restrictors are occasionally worth the effort of making them in the lab. They can be prepared by hand-pulling fused-silica tubes in a hot flame (Chester, 1984), but the best ones are made with some form of automation (Chester et al., 1985). After the pulling operation, the restrictors must be microscopically sized and individually cut to give the desired outlet diameter. The cut must be perfectly square and should be done by scoring and breaking. The taper length is typically 2–4 cm. The i.d. near the outlet is 2–4 μm, giving an outlet velocity much slower than the integral restrictor but much less likely to plug. The mass transfer performance of the tapered-capillary restrictor is excellent, but it allows a large flow rate increase when the column pressure is programed (Pinkston and Hentschel, 1993). This can widen the strongly retained peaks in a programed method. The integral restrictor has a much flatter flow-rate–pressure behavior and often produces narrower high-molecular-weight peaks (Berger, 1989; Pinkston and Hentschel, 1993). Tapered capillary restrictors have no

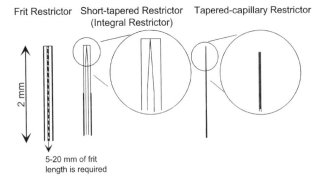

Figure 22 Restrictors for interfacing low-pressure detectors. The last 2 mm is shown for all three restrictors. The insets enlarge the last 0.5 mm.

outside coating over the tapered section, and are more fragile than the previous two, but have long lives once installed in an instrument. The only trick to know in handling tapered restrictors is to avoid scratching the bare silica during installation. This is easily accomplished by inserting the restrictor inlet through the detector outlet, thus imparting any damage to the end that can be easily trimmed once it appears in the oven.

Detector temperature and restrictor position are important considerations for best performance. Berger (1989) recommends keeping the detector temperature low in order to minimize the flow rate increase with pressure programing. The minimum temperature should be $\sim 1.3\times$ the critical temperature (in Kelvin) (Smith et al., 1986). So, with CO_2 mobile phase the minimum restrictor temperature should be $\sim 125°C$. However, the detector heater may not be able to maintain the temperature in the restrictor at the indicated temperature of the detector block, so a higher setting may be necessary. Transfer of high-molecular-weight solutes is often better at high temperatures. Most FIDs can be heated to $400°C$. If mass transfer problems occur (such as spiking or tailing on the late eluting peaks), try increasing the detector temperature, but keep in mind this may worsen the situation if the cause is reaction between solutes and a surface. Also, changing the detector temperature will affect the mobile phase viscosity in the restrictor and will therefore change the amount of restriction and the mobile phase flow rate on the column.

As SFC is often chosen for solutes that are not very volatile, it is important to position the restrictor outlet close to the point of detector signal generation. For example, with flame-based detectors, a jet should be used with an i.d. large enough for the restrictor to pass completely through so that the restrictor outlet can be positioned within 1–2 mm of the base of the flame. It should not extend into the flame.

6. Cutting Fused-Silica Tubing and Making Connections in OT-SFC

OT-SFC tubing must be cleanly cut by scoring and breaking. The diamond-tipped pencils commonly sold by GC supply companies usually do not produce adequately square breaks on the small-diameter tubing commonly used in OT-SFC. The best results are produced using a fiber-optic cleaver with a sapphire cutting element, a sharp, wafer-style ceramic fused-silica tubing cutter, or even the sharp edge of a glass microscope slide. The polyimide coating must be penetrated and the underlying silica scratched in a single stroke, then broken. Every cut should be inspected with a $20\times$ magnifier and redone if not square or if a burr or any raggedness can be detected.

When making connections, avoid actions that may later lead to restrictor plugging. When making a series of connections, it is usually best to work in the direction of the mobile phase flow, starting with the most-upstream connection and working toward the detector. When using fused-silica tubing, place nuts, ferrules, and sleeves (when used) on the tube, then trim the end, inspect with a magnifier to make sure the cut is square and burr-free, then proceed with making the connection.

The most important connection in the system joins the column outlet to the restrictor ahead of a low-pressure detector because a dead volume here introduces unrecoverable broadening. Small imperfections in connectors ahead of the column inlet are not always serious because solutes tend to accumulate and spatially focus on the head of the column when programed elution is used.

Commercial 1/32 in. stainless-steel, zero-dead-volume tube fittings can be used with graphite-polyimide (or other suitable) ferrules to connect fused-silica tubes for high-pressure service. (See, e.g., Technical Note 51, Valco Instruments Co. Inc., Houston, TX.) Traditional-style zero-dead-volume fittings are available with internal bores as small as 0.25 mm (0.010 in.). In addition, steel fittings with even smaller internal volumes (with bores of 0.12 mm) are made by machining the conical ferrule seats and the internal passage in a small steel disk. The disk is then inserted in a steel ring machined to accept the nuts.

In making connections, use ferrules with holes closely matched to the tubing o.d. to avoid the need for a great deal of compression. Using compression to make up for a bad mismatch in ferrule and tube dimensions often results in misalignment or crushing of the ferrule. In an emergency, if a ferrule with a small enough hole is not available, shave ~ 0.1 mm of material off the tip of the ferrule. This will allow the ferrule some room to move against the conical sealing surface and compress sufficiently around the tube before crushing the ferrule tip in the bottom of the fitting. The idea is to have all of the available volume completely filled with the ferrule exactly when sufficient torque has been applied to the nut to hold the tube in place. Of course, shaving too much will result in a large dead volume.

Fittings that butt fused-silica tubes directly together can make true, zero-volume connections, but are very difficult to make correctly with small i.d. fused-silica tubes. A zero-dead-volume fitting with 0.25 mm i.d. internal passage is the best choice for nearly all users.

Tapered quartz or fused-silica couplers into which the ends of the tubes to be joined are inserted and glued in place, are easy to assemble and reliable in use, but have larger mixing volumes than a properly assembled steel fitting.

Connections of fused-silica tubing to (unheated) valves with 1/16 in. fittings are best done with 1/16 in. o.d.

polyimide or PEEK sleeves. Appropriate sleeves can be easily made from PEEK tubing of the appropriate i.d. if the end is cut perfectly square. PEEK sleeves can be used with steel ferrules to grip fused silica. Although the ferrule will be permanently attached to the sleeve, the fused silica can be removed and the sleeve-ferrule combination reused if it is never over-tightened. (PEEK parts should never be used above 110°C, and we do not recommend using PEEK above 75°C in SFC applications.)

E. Testing for Proper Operation

1. Packed-Column SFC

First check the system for leaks. Then pressure stability should be examined at some standard conditions of your choice. We typically use 50°C, 165 bar, 2 mL/min, and with a methanol modifier concentration of 15% with columns 4.6 mm × 150 mm with 5- or 6-μm diameter particles. The inlet pressure should stabilize within the range of 175–185 bar within several minutes of starting the flow. If the pressure does not stabilize, purge the modifier pump and try again. If that does not fix it, then you may have a check valve problem on one of the pumps.

Injecting a test mix without programing will quickly provide performance information about the system. We have found that a mixture of caffeine, 0.20 mg/mL, theophylline, 0.14 mg/mL, and theobromine, 0.060 mg/mL, dissolved in 1.5/98.5 acetone/methanol (v/v) is a good, general test mix for packed-column SFC, and is compatible with a wide variety of stationary phases (but not alkyl-substituted phases like C18). A 1 μL injection of this solution produces unmistakable peaks when detected by UV at 275 nm. The modifier concentration must be adjusted to get the peaks into the range of $1 < k < 5$, which is reasonable for this test. The system passes if the peaks are symmetrical and if the last peak has a reasonable plate height. For well-behaved peaks with k above 5 on a 4.6 mm i.d. column, we generally expect a plate height around 2.5× the particle diameter at 2.5 mL/min. You may get much worse performance, particularly if the stationary phase is not a good choice for these particular solutes. In the absence of an industry-standard test, each lab has to establish its own acceptance criteria.

2. OT-SFC

Again, start by checking for leaks. Grob test mixtures, popular in GC, are not practical in OT-SFC because the components are not retained well enough. The mixture illustrated in Fig. 20(a) is a practical alternative for OT-SFC. A suitable test mix is made by dissolving the test solutes (or lower-molecular-weight analogs) at 0.2 mg/mL each (for direct injection, or 1 mg/mL or higher for splitting injection) using 10:90 methylethyl ketone:toluene (by volume) as the solvent. The methylethyl ketone is required to dissolve the fatty acid. Use a moderate temperature (120°C and pressure program from 120 to 170 bars at 2 bars/min, and then to 270 bars at 5 bars/min. Solvent peaks (when seen by the detector) should rise squarely from the baseline on the ascending side. They are usually not >2–3 min wide in OT-SFC, even with direct injection, and should generally be <2 min wide with split or split/splitless injection. The trailing edge may not be as steep or as square as the ascending side, but should not have excessive tailing. Some tailing on the solvent peak (caused by less-than-perfect connections between the injector and column) is not of concern if the solute peaks are well shaped. Bumps and steps in the solvent peak often occur if the sample solution is capable of forming an azeotrope as it vaporizes, or if hygroscopic sample solutions have taken up some atmospheric water.

Naturally, solute peaks should be narrow and symmetric. Peak splitting often indicates an unwanted phase separation or inadequate refocusing. An elevated baseline following an otherwise well-shaped peak can be caused by the wetting of the mobile phase supply tube by a fraction of the sample. This can happen if the front of the sample plug reaches the oven while some fraction of the sample remains in the injection valve. Chair-shaped peaks (with the "seat" on the leading edge) or split peaks often result with retention gaps when the stationary phase is wet by liquid injection solvent. Milder injection conditions or a longer retention gap will fix it (see Fig. 21).

F. Flow Behavior with Upstream Pressure Control

For those coming to SFC from an LC background, upstream pressure control (rather than flow control) may be difficult to get accustomed to. For example, if a small leak develops in the connections between the pump and injector, there is often no significant consequence. The pump may run a little faster because of the increased flow demand caused by the leak, but often the chromatography is perfectly normal. If a small leak develops in the injector, at the column inlet, or in any other connections between the injector and the column inlet, retention times will still be normal because the pump will simply run faster than usual to maintain the set pressure. However, the leak represents a splitting path for the sample, and peak areas may be reduced significantly while nothing else appears to be wrong. A small leak at the column outlet will result in increased mobile-phase velocity and early eluting peaks combined with smaller-than-normal peak areas. A partial plug anywhere between the pump and the column will always cause delayed peaks, as the flow velocity is reduced and a pressure drop at the plug point will lower the column pressure and increase peak retention. A partial plug at or beyond the column outlet will often cause delayed peaks. But, because the plug lowers column velocity but maintains column pressure, a partial plug on the outlet would not delay the peaks as much as if it were on the inlet side of the column. Partial plugs in the retention gap (if used), column, restrictor, or the intervening connectors will change the split ratio when flow-splitting injectors are used, lowering the mass on column. Partial plugs (or complete plugs) may also occur in the vent of flow-splitting injectors, thus increasing the mass on column, and possibly broadening the peaks beyond usefulness.

III. METHOD DEVELOPMENT

The adjustable parameters in SFC are the choice of the stationary phase (including the column format, that is, a packed or an OT column, the column dimensions, support material, particle size and pore size if applicable, and the bonded phase), the mobile phase (including the identity of the main component and, if applicable, the modifier and additive identities and concentrations), temperature, pressure, and flow rate. Successful SFC methods can be developed completely empirically, but attention to a few important fundamentals can save lots of work.

Understanding intermolecular forces (Section I.B) is necessary to make rational changes in stationary phases, modifiers, and additives. Phase behavior (Section I.C) should always be a primary consideration in setting SFC parameters. A temperature change of a few degrees or a pressure change of a few bars might make the difference between unacceptable peak broadening and self-working peak focusing. Phase separation of mobile phase components in packed-column SFC can greatly degrade separations and introduce noise.

The first task is to choose between packed-column and OT column formats. One common need is to have fast analysis capabilities in the method that will eventually result. If there are convenient detection possibilities for all the solutes, then a packed-column is usually preferred for high-speed analyses. Another common need is to screen samples contatining unknown organic solutes. An FID is excellent for this purpose. If the sample has the required solubility, and if the expected solute molecular weights are in bounds, then OT SFC is the best choice. Table 7 gives more guidance on making the initial choice of column format.

A. Packed-Column SFC Method Development

Forget everything you know about reversed-phase HPLC when developing a packed-column SFC method. Think

Table 7 Choosing Between OT and Packed-Column SFC

Packed-column SFC	OT-SFC–FID
If you need to extend LC capabilities (different selectivity, faster analyses, faster equilibration, longer columns/higher efficiency, and less solvent waste)	If you need to extend GC capabilities (different selectivity, lower temperature, higher molecular weight, and nonvolatile solutes)
If the sample is soluble in methanol or another potential modifier	If the analytes of interest are nonionic and soluble in methanol or a less-polar solvent, or if the sample can be easily derivatized so that the analytes become nonionic and soluble in a suitable solvent. Pyridine is acceptable
If analytes have a convenient detection possibility (UV, N or P content, etc.) or a mass spectrometer is available for detection	If the molecular weight of the largest analyte is below $\sim$3000 (415 atm system) or 10,000 (680 atm or higher system)
	If the analytes have no convenient detection possibility other than their carbon content
	If a chromatogram using an FID to survey a sample with unknown solutes is desired

of packed-column SFC as a normal-phase technique in which the stationary phase is more polar than the mobile phase. The modifier is more polar than the main mobile phase component, so increasing the modifier concentration increases the mobile-phase polarity, strengthening the mobile phase and reducing solute retention. Begin method development by building a polarity window (Berger, 1995): choose a stationary phase that is more polar than the solutes, and a mobile phase that is less polar than the solutes. However, do not become over-reliant on thinking in terms of polarity in SFC methods development. Polarity is just one dimension in solute retention, and does not explain why some solutes elute ahead of others in SFC. Adjustments of other parameters that may not seem to alter polarity can quickly lead to the desired separations. In particular, temperature effects are unpredictable in many cases and are not easily explained from a polarity point of view.

Typical packed-column SFC stationary phases are made from spherical porous silica, usually with $\sim$5 μm diameter and $\sim$6–10 nm (60–100 Å) mean pore diameter. Bare silica may be used, but bonded stationary phases are more typical. Some of the popular ones are listed in Table 4. These are usually monomerically bonded in the typical HPLC fashion, but a few bonded phases (the DeltaBond phases from Thermo-Hypersil-Keystone) are available with a substituted-silicone polymer coating on porous silica (with 30 nm pores in the usual SFC format).

The main mobile-phase component is almost always carbon dioxide. CO_2 is a symmetric, nonpolar molecule, but is weakly polarizable. It may be used alone as a mobile phase in packed-column SFC for low-polarity solutes (like hydrocarbons, petroleum components, silicones, etc.) with low-polarity stationary phases (like C-18), but generally requires the addition of a polar modifier to be useful for separating the more-polar solutes encountered beyond petroleum applications. Other fluids like hydrofluorocarbons, SF_6, and low-molecular-weight alkanes are sometimes used in commercial supercritical-fluid processes, but CO_2 is used almost exclusively in analytical, packed-column SFC, and it is rare that anything else is considered. Nitrous oxide should not be used because of the instability and safety hazard when this strong oxidizer is mixed with an organic sample and pressurized.

Typical modifiers are listed in Table 8. Methanol is by far the most popular choice. Ethanol is favored when toxicity is considered, but, when an alcohol stronger than methanol is required, isopropanol is often the better choice. For the normal alcohols, dispersion and hydrogen-bond-acceptor strength increase and hydrogen-bond-donor strength decreases with increasing molecular weight. The remaining modifiers can be considered when selectivity considerations call for a modifier that cannot donate a hydrogen bond. THF is stronger than acetonitrile in dispersion interactions and weaker in hydrogen-bond-acceptor strength. These choices allow flexibility in selectivity.

Table 8 Modifiers in SFC

Frequently used modifiers in SFC
Methanol
Ethanol
2-Propanol
Tetrahydrofuran
Acetonitrile
n-Hexane
Methyl-*t*-butyl ether
Less frequently used but possible
1-Propanol
Dichloromethane
Propylene carbonate
Formic acid
Water

The additive is the third mobile phase component, not always used, but sometimes required for a particular need. It is added at low levels to the modifier. For example, 0.05–0.5% triethylamine, diethylmethylamine, or isopropylamine can be added to the modifier to improve the peak shape of amines in the sample, and 0.05–0.5% acetic acid, trifluoroacetic acid, and so on, can be used to improve acids. Low levels (1–10 mM) of volatile salts (ammonium acetate, ammonium formate, and ammonium carbonate) can be added to the modifier to elute many polar and ionic analytes (Pinkston et al., 2004). Among all of these possibilities, ammonium acetate is the most universal. It is necessary to make sure that the components of the chromatographic system (modifier pump, column, and detector) are compatible with the additive.

With all this in mind, let us now consider the hypothetical separation of cyclohexane, benzene, toluene, and phenol by packed-column SFC using a cyano column with pure CO_2 mobile phase. Cyclohexane, being of very low polarity, will be retained primarily by dispersion interactions. Benzene is about the same size and is also nonpolar, so dispersion interactions in benzene will be similar to those in cyclohexane. However, benzene is much more polarizable than cyclohexane, and a dipole will be induced when a benzene molecule nears a cyano group on the stationary phase. Toluene is one-carbon larger, so dispersion interactions will be slightly stronger. Toluene is about as polarizable as benzene so the induction effects will be similar. However, toluene also has a small dipole moment. This adds a dipole–dipole interaction absent from the earlier solutes. Finally, phenol is similar to toluene in size and polarizability, but is a much stronger dipole and is also both a hydrogen-bond donor and acceptor. The cyano group on the stationary phase is a hydrogen-bond acceptor. Thus, with pure CO_2 mobile phase, which interacts with solutes primarily by dispersion, it is reasonable to expect the retention times to be ordered cyclohexane < benzene < toluene < phenol.

Now consider adding methanol to the CO_2 mobile phase. The mobile-phase dispersion interactions will be increased, but methanol is polar and undergoes hydrogen bonding. Therefore, we might expect cyclohexane retention to be reduced because of the now-stronger dispersion forces in the mobile phase. Benzene, being about the same size as cyclohexane, would experience approximately the same retention reduction due to the dispersion changes, but the methanol dipole will induce a dipole in the benzene and further reduce its retention relative to the cyclohexane. Toluene will be affected similarly to benzene with regards to dispersion and induction, but toluene also has a small permanent dipole. Therefore, the dipole–dipole interactions possible between toluene and methanol will reduce its retention a bit more than occurs for

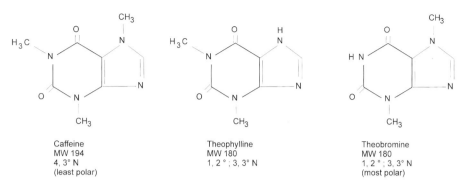

Figure 23 Structures and properties of caffeine, theophylline, and theobromine.

benzene. Phenol would have even stronger dipole–dipole interactions with the methanol than would the toluene, and can interact with methanol through hydrogen bonds. This would tend to reduce its retention more than for the toluene. Therefore, the selectivity will undoubtedly change as methanol is introduced. However, we cannot predict from this simple analysis if the elution order will change at some point as the methanol concentration is increased. This would have to be checked experimentally. The point of this exercise is not to predict elution order or retention times, but to illustrate how selectivity can be altered by making rational changes in the mobile and stationary phases based on the intermolecular forces.

Let us now consider a more subtle situation, the separation of three similar solutes: caffeine, theophylline, and theobromine. The structures and several important parameters are shown in Fig. 23. We examined the retention of these solutes on a variety of columns while holding the temperature and pressure constant at 80°C and 165 bar. The modifier was methanol, and its concentration was adjusted so that k is ~ 1 for the first peak, caffeine. The relative retention on the various columns is shown in Fig. 24.

A C8 column retains solutes primarily by dispersion forces. As these are relatively weak, only 1% methanol is necessary to elute caffeine with $k = 1$. The three solutes are similar in size and structure, so the dispersion interactions are also similar, and the three are not separated on the C8 column. With the remaining columns, interactions with the stationary phase are stronger, and the methanol concentration must be increased to produce the desired k value for caffeine; theophylline and theobromine are retained even stronger and the peaks are separated.

Dispersion interactions are about twice as strong in the C8 phase compared with the other bonded phases, and dispersion was not able to separate the solutes. Other interactions must be responsible for the increased retention and the selectivity when using the other columns. The cyano phase is very polar, but the diol phase has the strongest retention by far, so hydrogen bonding appears to be the most important contributor to retention of these solutes. But why is the elution order of theophylline and theobromine reversed for the amino column?

Hydrogen bonding is known from GC experience to be temperature-dependent, and changing the temperature changes the selectivity for all these polar phases when solutes are retained by hydrogen bonding. The diol phase has similar temperature dependence as the amino phase, meaning the selectivity for theophylline and theobromine changes in the same direction on both phases as the temperature is varied. However, the temperature at which these two solutes coelute is not the same for both phases, and under the conditions used in this example,

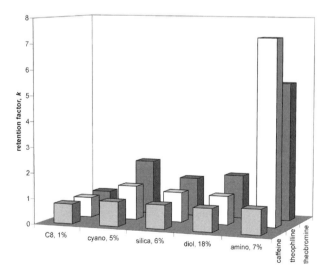

Figure 24 Retention factors of caffeine, theophylline, and theobromine on five stationary phases, as indicated, at 80°C and 165 bar. The modifier was methanol, and its concentration was adjusted in the CO_2 mobile phase so that the first peak had a retention factor of 1. The concentration of methanol is indicated with each stationary phase.

the amino and diol columns give different elution orders. At somewhat lower or higher temperatures they would behave more similarly. Again, the point is not to predict retention, but to suggest how adjustable the selectivity can be. In contrast, the retention and selectivity are much less adjustable with isothermal pressure changes in packed-column SFC once the modifier level is higher than $\sim 10\%$.

In general, to minimize the time required for a separation by packed-column SFC, we need to choose conditions that produce high selectivity for the solutes of interest compared with the other peaks in the chromatogram. High selectivity reduces the retention necessary and allows the successful use of columns with fewer theoretical plates (i.e., shorter columns or higher mobile-phase velocities), thus saving time.

We also need to increase the rate at which we produce the theoretical plates that are necessary. This depends on the rate of diffusion that, in turn, is directly proportional to the temperature, inversely proportional to the viscosity of the mobile phase, and also decreases with increasing pressure and modifier concentration. Because of these interdependencies, increasing the temperature has a compound benefit on analysis speed as long as the mobile phase behaves like a liquid and not like a gas: higher temperatures reduce the amount of modifier required, reduce the viscosity both by the modifier concentration and temperature change, and also improves diffusion on its own. Diffusion rates are proportional to the temperature/viscosity ratio. For neat CO_2 at 200 bar, the temperature/viscosity ratio does not peak until reaching $\sim 200°C$. With higher pressures and with modifiers, even higher temperatures continue to give diffusion benefits. Usually, speed can be tripled by increasing the temperature from $25°C$ to $\sim 100°C$ with neat or lightly modified CO_2.

So, we should prefer to use the lowest modifier concentrations, the lowest pressures, and highest temperatures possible as long as the selectivity is acceptable and the solutes and system components are stable. However, for methods development, we suggest that the pressure initially be set at least 5 bars above the maximum in the critical locus. As the method parameters are refined, pressure can be lowered, if necessary, to help improve column efficiency or allow faster flow rates without a large loss in plate number. Just be careful not to drop the pressure so much that the mobile phase splits into separate liquid and vapor phases. Phase separation would not be directly observable in an SFC instrument, but sudden changes in noise, retention time, and peak shape may occur if the mobile phase separates.

Select the initial column with the polarity window in mind. If in doubt, try a medium-polarity stationary phase. Cyano is a good first choice if you have no better idea where to begin. Select a modifier that dissolves the sample. Methanol is preferred because it is polar enough for most applications and is both a hydrogen-bond donor and acceptor. For most nonionic solutes, it is best to omit the additive, at least at first, and see what happens. If the solutes are very polar or ionic, then 1–5 mM ammonium acetate dissolved in the modifier is a good starting point for the additive.

You may choose to develop an isocratic or a gradient method. Either way, the general procedure is to first get the retention into the right range, then check the selectivity and fix it if necessary, and finally check the peak shape and fix it if necessary.

For methods development in packed-column SFC, it is most convenient to use an initial pressure at least 5 bars higher than the highest pressure in the critical locus for the mobile phase. This allows the composition and temperature to be varied widely with no chance of phase separation of the mobile phase components. If it is necessary or desirable to operate the method at lower pressure, then it is important to pay attention to both the pressure and the temperature to keep the conditions outside the critical locus so that the modifier concentration can be varied over the range required without the occurrence of phase separation.

1. Isocratic Method Development

For an isocratic method, do a scouting run at 40% modifier (with additive, if necessary), $50°C$, 165–200 bar outlet pressure (for methanol modifier, or above the critical locus for others), 2–3 mL/min (for a 150×4.6 mm i.d. column). Make test injections and adjust the modifier concentration in the mobile phase so that k is in the range $1 < k < 10$ (or $0.5 < k < 20$ if you wish) for all the peaks. If the retention is too high with $\sim 60\%$ modifier, then switch to a stronger modifier or a less-polar column. If there is too little retention when the modifier is below $\sim 5\%$, then switch to a more-polar column, or to a weaker modifier. If the peaks are too spread out in their k values, then abandon the isocratic method and try a gradient.

If the retention is reasonable, then try tuning the selectivity next. The goal is to get the peaks as well-spaced as possible. Change the column temperature $\pm 10°C$ and look for improvement trends. Continue in the direction that helps the most. If a temperature change alone is not adequate, then increase the outlet pressure to 250 bar or more and look for trends. Try high temperature and high pressure together. If selectivity is still inadequate, then start over with another column. Choose one *stronger* in the intermolecular forces associated with the differences between the unresolved solutes. For example, theophylline and theobromine are isomers and are not resolved on nonpolar columns, but these solutes differ in their polarity and hydrogen bonding and are easily separated on many more-polar columns. This strategy also works in reverse. For example, if a primary and secondary amine of different molecular weight coelute on a diol column, they may be separated using the same modifier with a column weaker in hydrogen bonding or stronger in dispersion such as a cyano or a phenyl column. Alternatively, if you only have a column that interacts strongly with the common features of the solutes, then try a modifier that interacts strongly with some aspect that is not shared in common among the solutes.

If retention is in the right range and the selectivity is adequate, then examine the peak shapes. If peak shapes are bad, then add 5 mM ammonium acetate to the modifier, or increase the ammonium acetate up to 10 mM, if necessary. If ammonium acetate does not give good peak shapes, then switch to an acidic or basic additive at concentrations up to ~0.5% in the modifier. Use acids to improve acidic solute peak shapes, and bases to improve basic solute peak shapes. If additives do not fix the peak-shape problem, then try a new column or switch to a more retentive stationary phase. If peak shapes are still inadequate, and especially if the solutes are very hydrophilic, then try reversed-phase HPLC.

If there is more resolution than necessary for the worst pair of peaks in the chromatogram, then the analysis speed can be increased. Try increasing the flow rate, increasing the temperature, and lowering the pressure and modifier concentration. If there is much excess resolution, you may consider a shorter column. However, we generally prefer to increase the flow rate before shortening the column. The van Deemter curve is nearly flat in SFC from 2 to 5 mL/min for 4.6 mm i.d. columns, so there is very little efficiency penalty when the flow rate is increased. If the flow rate is as high as practical and there is still excess resolution, then a shorter column is appropriate.

If you have inadequate resolution after finding the conditions giving the best selectivity and peak shape for your problem, the only remaining possibility to increase resolution is to increase the number of theoretical plates. Stationary phase particles smaller than ~5 μm i.d. will not improve efficiency very much in SFC, so flow rate and column length are the only effective possibilities for changes. The flow rate giving the highest plate number (the van Deemter minimum) for a 4.6 mm column packed with 5 μm i.d. particles is ~2 mL/min, but this depends on the temperature and modifier concentration. Consider increasing the column length or putting several columns in series to increase the plate number.

2. Gradient Method Development

Developing a gradient method follows a similar overall strategy. With a 150 × 4.6 mm i.d. column, do a scouting run at 50°C, 165–200 bar (for methanol modifier, or above the critical locus for others), 2–3 mL/min. From a 30 s hold at 1% modifier (+ additive), program to ~50% modifier at 5%/min with a 2–3 min hold at the end. Adjust the initial and final modifier concentrations and the gradient rate to elute the peaks between ~1.5 and 10 min. If the final modifier concentration required is much higher than 60%, then consider a stronger modifier, a weaker column, or both. If the percentage modifier required is very low or the peaks are not adequately retained, use a more polar column, a weaker modifier, or both.

Tune the selectivity next. Change the column temperature ±10°C and look for trends. Continue in the direction that helps. Change the pressure to 250 bar or more and look for trends. Try high temperature and high pressure together. Try another column or another modifier as discussed for isocratic method development.

If peak shapes are bad, add or increase ammonium acetate up to 10 mM in the modifier, or switch to another additive as discussed previously. Try changing the column or try a more retentive column. If none of this is adequate, and especially if the solutes are very hydrophilic, then try reversed-phase HPLC.

To achieve fast analyses, try to use high temperatures and low pressures. If the relative peak spacing is good in a gradient method that you need to go faster, then increase the flow rate and the program rate by the same factor. If the flow rate and program rate are as high as practical and there is still excess resolution, then consider a shorter column. When changing the column length, the program rate must be changed inversely with the column length to maintain the same relative peak spacing.

If you have inadequate resolution after optimizing the flow rate and selectivity, and if an isocratic method is impractical, the only alternative is to increase the column length. Figure 25 is an example of using columns in series to improve the detail in programed chromatograms of a very complicated mixture. When increasing the column length in a gradient method, be sure to change the gradient rate inversely with the column length.

B. OT-SFC Method Development

Dispersion is often the most significant solute–stationary phase force for the OT stationary phases listed in Table 4. Apparently this results even for silicone stationary phases with fairly polar substituents because of the influence of the silicone backbone. Homologs elute in the order of their molecular weights from OT-SFC columns. The "nonpolar" phases, such as methyl and octyl, tend to give an elution order based on relatively "pure" dispersion, minimizing the influence of polar functional groups on the solutes. The phenyl and biphenyl phases are somewhat polar and are polarizible. (Biphenyl is approximately twice

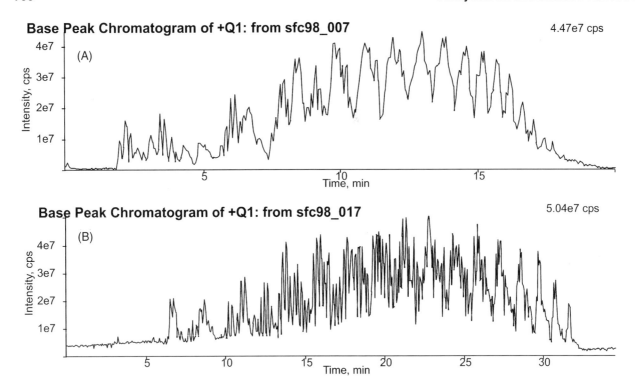

Figure 25 Separations of a low-molecular-weight alkoxylated polymer using column lengths of (a) 25 cm and (b) 75 cm. Outlet pressure (200 bar), column temperature (100°C), and flow rate (2 mL/min) were identical for both runs. The modifier rate was held at 3% methanol for 3 min, then programed at 1.59%/min (a) or 0.5%/min (b) (Pinkston et al., 2002).

as polarizable as phenyl.) So, in addition to their dispersion forces, these phases are capable of providing additional retention for polar solutes. If a nonpolar and a polar solute coelute on a methyl or octyl column, they will likely be separated on a biphenyl column (with the nonpolar solute eluting first). The cyanopropyl and polyether phases are more strongly polar, can have dipole–dipole interactions with polar solutes, and can also induce dipole moments in polarizible solutes. Polar and polarizible solutes tend to be retained quite strongly relative to nonpolar solutes on these stationary phases.

With CO_2 or another nonpolar mobile phase, GC-derived McReynolds constants for similar stationary phases can be used as a rough guide in column selection. This approach is far from perfect, but does offer some idea of the interaction potential of a stationary phase with solute functional groups.

It is best to avoid the extremes of stationary-phase polarity if the solutes contain extremes of polarity within their functionalities. For example, polar lipids can be separated well using a methyl (nonpolar) or a cyanopropyl (polar) stationary phase. However, polar lipids often have poor solubility in both of these stationary phases and tend to overload at low levels. An intermediate-polarity stationary phase, like a biphenyl, gives better results.

With no other clues to follow, an intermediate-polarity stationary phase (like biphenyl) is the first one to try. More applications have been successfully reported with low- and medium-polarity stationary phases than with higher-polarity phases.

The minimum temperature is the one that keeps the mobile phase all in one phase over the range of pressures and mobile phase compositions that may be spanned during the chromatogram. For pure CO_2 when pressure programing is desired, 35°C is a comfortable minimum. However, we have already seen that there is no thermodynamic discontinuity between supercritical-fluid chromatography and near-critical LC. Clearly, subcritical temperatures may be used and the critical temperature or critical pressure may be crossed during programing with absolutely no consequence as long as phase transitions are avoided. An example of subambient-temperature operation will be illustrated in Section VI.

When the solutes are thermally stable, two benefits may be realized by elevating the temperature: diffusion coefficients will improve (thus improving the chromatographic efficiency within a fixed analysis time), and selectivity between dissimilar solutes will change. The extent of the change, and knowledge of whether it improves or worsens the separation, can only be determined by experiment.

From case to case, the upper temperature limit may be set by the stability of solutes, by the stability of the stationary phase, by the swelling or association of the stationary phase with mobile phase and its influence on selectivity, or by the stability or physical limits of the chromatographic system. *Caution*: some stationary phases and other system components may have relatively low temperature limits. Users must be aware of and stay within the temperature and pressure limits of all components in their systems.

In developing isothermal OT-SFC methods it is essential to explore at least two temperatures separated by 20°C or more. With pure CO_2 mobile phase in OT-SFC systems, 100°C is a good initial temperature to explore, followed by a second trial at 120°C or 130°C. If the higher temperature is unsatisfactory, then try 70°C or 80°C. If a temperature-dependent selectivity shift occurs, then additional trials can be added to optimize the temperature.

With OT-SFC, isothermal, linear-pressure programing should be tried first. Here are some rough guidelines for a 10 m × 50 μm i.d. column operated with an initial velocity of ~2 cm/s. If there is no previous experience with similar samples from which to base exploration, then use a biphenyl stationary phase and inject at 100°C and 70 atm. Hold that pressure for ~2 min, then linearly program at 10–20 atm/min to the pressure limit of the system. Hold at the pressure limit for an additional 10–20 min. Subsequent experiments can then be designed to explore: temperature effects or another stationary phase to get adequate selectivity, lowering the program rate before and during the elution of crowded or important parts of the chromatogram to improve resolution, optimizing solute focusing following injection, rapidly purging the system of uninteresting peaks following the last peak of interest, and restoring the system to injection conditions. If speed is more important than initial resolution (such as when screening large numbers of simple samples), then use a 2 m long column, raise the initial program rate to 50 atm/min, and use a final hold time of 10 min.

IV. SUPERCRITICAL FLUID CHROMATOGRAPHY–MASS SPECTROMETRY

Mass spectrometry (MS) is perhaps the single most informative SFC detection means in relatively widespread use. A variety of interfaces have been used to successfully couple these two techniques. Two reviews of the principles and progress in SFC–MS are recommended (Arpino and Haas, 1995; Pinkston and Chester, 1995). The work by Arpino and Haas (1995) is an excellent critical review of progress and a look toward the promising future of packed-column SFC with atmospheric-pressure ionization (API). The review by Pinkston and Chester (1995) is a practical guide to the choices that can make the difference between success and failure in SFC–MS.

A. SFC–MS Interfaces

We will divide SFC–MS interfaces into "low-flow-rate" and "high-flow-rate" interfaces. Low-flow-rate interfaces are compatible with OT and microbore packed-column SFC. This corresponds to SFC systems delivering less than ~15 mL/min of CO_2, measured at atmospheric pressure and 25°C. Some of the high-flow-rate interfaces that will be discussed later are versatile and are compatible with low flow rates as well. Despite the growing popularity of packed-column SFC, which requires the high-flow-rate interface for SFC–MS, we will briefly describe the most common low-flow-rate interface because it is still in use.

1. Low-Flow-Rate Interfaces

The most frequently used interface for low-flow-rate applications has been the direct-fluid-introduction (DFI) interface (Huang et al., 1988; van Leuken et al., 1994; Matsumoto et al., 1986; Pinkston et al., 1988; Ramsey and Raynor, 1996; Reinhold et al., 1988; Smith et al., 1982). The SFC effluent is introduced directly to the electron ionization (EI) or chemical ionization (CI) ion source of a differentially pumped mass spectrometer. The last portion of the OT-SFC column or of the fused-silica transfer line from the column is housed within an interface having two independently heated regions. The heated region farthest from the ion source is generally held at the same temperature as the column oven. The flow restrictor is housed within the heated region closest to the ion source. It is heated from 150°C to 450°C (depending on the particular application) to counteract the Joule–Thompson cooling of the expansion and to provide some volatility to the eluting analytes. The end of the flow restrictor is usually positioned within a few millimeters of the edge of the ion-source ionization volume so that the effluent is introduced directly to the ionization region. The DFI interface is simple and exhibits excellent chromatographic fidelity. However, the mobile phase (most often CO_2) may influence ionization (especially in EI mode) and ion transmission efficiency at the high end of the DFI flow-rate range (Houben et al., 1991; Kalinoski and Hargiss, 1990; Pinkston and Bowling, 1992, 1993).

Commercially available API sources [both electrospray ionization (ESI) and atmospheric pressure chemical ionization (APCI)] are compatible with the low mobile-phase flow rates of OT and micro-packed-column SFC (Arpino et al., 1993; Broadbent et al., 1996; Pinkston and Baker, 1995; Tyrefors et al., 1993). Interfaces incorporating these sources are classified here as high-flow-rate interfaces, which are discussed next.

2. High-Flow-Rate Interfaces

Conventional packed, microbore, and packed-capillary SFC columns operate at flow rates >10–15 mL/min of CO_2 (measured at atmospheric pressure and 25°C). A variety of interfaces have been used for these flow regimes. Some, such as the mobile-phase elimination interfaces [the moving-belt (Berry et al., 1986; Perkins et al., 1991a, b) and the particle-beam (Edlund and Henion, 1989; Jedrzejewski and Taylor, 1995; Sanders et al., 1991) interfaces], the post-expansion-splitting interface (Smith and Udseth, 1987), and the thermospray (TSP) interface (Balsevich et al., 1988; Chapman and Pratt, 1987; Niessen et al., 1989; Saunders et al., 1990; Scalia and Games, 1992; Via and Taylor, 1994), are rarely used today and will not be discussed further. The great majority of the SFC–MS work done today is performed with interfaces incorporating API sources.

API Interfaces

The effluent, or a fraction thereof, is directed to the ionization region, which, as the name implies, is held at or near atmospheric pressure. This allows the chromatograph and mass spectrometer to operate more independently, and facilitates any changes or adjustments to the interface. A portion of the expanded effluent and of the ions produced in the ionization region is sampled by the mass spectrometer. The two most prominent types of API sources are the APCI (Anacleto et al., 1991; Broadbent et al., 1996; Huang et al., 1990; Lazar et al., 1996; Matsumoto et al., 1992; Thomas et al., 1994; Tyrefors et al., 1993) and ESI (Nelieu et al., 1994; Pinkston and Baker, 1995; Sadoun et al., 1993) sources.

The two API techniques are quite different. In the APCI source, a corona discharge is used to produce reagent ions, usually from water in the air or from a protic solvent, which is part of the nebulized effluent. These "primary" reagent ions ionize the analytes in a gas-phase process. The spectra produced by APCI are usually simple, often consisting of the protonated molecule with little fragmentation. In most cases, detection limits are in the low-nanogram to picogram range. Analytes must possess sufficient volatility and thermal stability to be transferred through a heated region, where the solvent is volatilized, to the ionization region without thermal degradation. Also, ions are generally singly charged, so APCI is most commonly used for molecules with molecular mass below 1500–2000 Da.

ESI is quite different from APCI. While many design variations have been introduced, all ESI sources share some features. The effluent is conducted though a small metal or metal-coated tube that is held at high voltage (up to approximately +5 kV in positive-ion mode, and approximately −3.5 kV in negative-ion mode). The outlet of the tube is placed near a conducting surface held at or near ground. An aperture in this conducting surface leads to the high-vacuum portion of the mass spectrometer where mass-to-charge (m/z) analysis is performed. Highly charged droplets are produced at the outlet of the tube. Most experts believe that "pre-formed" ions are desorbed from these evaporating droplets, or are liberated to the gas phase as smaller and smaller droplets are generated via "coulombic explosion" of the evaporating droplets (Fenn et al., 1990; Kebarle and Tang, 1993). "Pure" ESI sources generally can handle only up to a few microliters/minute of fluid flow. However, ESI sources equipped with nebulizing and heating gases can accommodate up to 1 mL/min of fluid flow or more. [Baker and Pinkston (1999) have shown that the addition of supercritical CO_2 to traditional LC/MS effluents can enhance ESI response, presumably because of improved nebulization.]

Multiply charged ions are often generated in ESI, especially of larger molecules with many potential attachment sites for protons or other charged species. This, in effect, extends the mass range of the mass spectrometer. Also, because the ionization process is performed at low temperature from the liquid phase, there is no inherent requirement that the molecule be volatile or thermally stable. Thus ESI can be used for both small and larger molecules, extending to many tens of thousands of daltons in mass for polypeptides and polymers. This can be helpful in polymer analysis by SFC–MS. It is important to note that no ions are observed when ESI is attempted with pure CO_2 mobile phase alone. Droplets are required for ESI, and no droplets are produced during the expansion of pure CO_2 to atmospheric pressure. Therefore, all SFC–ESI–MS interfacing approaches must incorporate the addition of a volatile organic or aqueous–organic solvent when the SFC mobile phase contains little or no organic modifier.

3. Transferring the Mobile Phase to the API Source

A variety of interfacing approaches may be used when coupling SFC to APCI or ESI sources. Each has its advantages and drawbacks. As the analyst must choose which approach to take, we will discuss each here.

Sheath-Flow Interface

For relatively low (15 mL/min CO_2 measured at atmospheric pressure and 25°C) to medium (60 mL/min) flow rates with no active "downstream" (postcolumn) pressure control, the best approach is the sheath-flow interface (McFadden, 1979; Pinkston and Baker, 1995). In this design, the flow restrictor is placed at or near the tip of the ESI sprayer or APCI nebulizer. A concentric flow of a sheath liquid (usually a volatile, protic organic solvent such as methanol or methanol/water, often containing a low millimolar concentration of an acid or of a volatile salt to enhance ionization) is directed about the restrictor and out the sprayer. The sheath flow allows both SFC and mass spectrometer to operate independently. It also provides the solvent droplets necessary for ESI. The sheath flow may also reduce restrictor plugging (Sadoun et al., 1993). Figure 26 shows an annotated diagram of one version of the sheath-flow interface.

Direct Introduction

Similar in some ways to the DFI interface described earlier for low-flow SFC–MS, one may directly introduce the SFC effluent to the API source without a postcolumn pressure regulation device (just as is in traditional LC–MS), but only when three distinct conditions are satisfied. Firstly, for ESI, the SFC mobile phase must contain a significant level of modifier, to ensure formation of droplets (as described earlier) and efficient transfer of ions to the mass spectrometer. While the required minimum level of modifier has not been reported, it is ~10–20%. Secondly, the desired separation must not be sensitive to slight variations in mobile-phase pressure. This condition can often be satisfied at relatively low column temperature (~40°C or lower) and high modifier concentration (~10–20% or higher). Finally, the flow rate, temperature, and post-column restriction must be such that sufficient pressure is maintained in the system to avoid an undesirable phase separation on the column or in the transfer line to the API source. If these three conditions are satisfied, the direct introduction approach can provide excellent results. In one of the most impressive demonstrations of the potential of SFC–MS–MS for rapid analysis, Hoke et al. (2001) used this interfacing approach in the rapid analysis of dextromethorphan from 96-well plates. Figure 27 shows the analysis of a full plate in <10 min. While the cycle time for each separation was only 6 s, the separation was real, with a k for the dextromethorphan peak of ~2. Figure 28 shows a diagram of the direct introduction approach.

Postcolumn Split

Another simple means to SFC–MS interfacing is to perform a postcolumn split of the effluent in the column oven, as shown in Fig. 29. A fraction (10–20%) of the

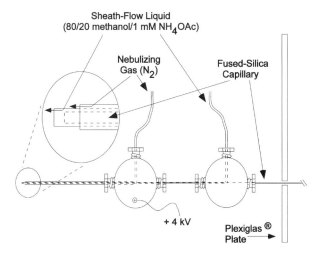

Figure 26 One implementation of the sheath-flow interface. The SFC column effluent proceeds through a fused-silica tube to near the tip of the ESI sprayer. A concentric sheath flow of solvent with good electrospray properties is introduced in the first tee, while a second concentric flow of nebulizing gas is introduced in the second tee. The three independent flows mix at the sprayer outlet. The sprayer is held at a voltage of +1 to +4 kV for positive ion ESI.

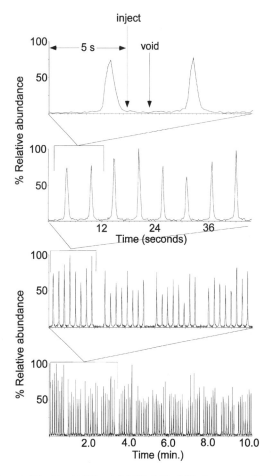

Figure 27 Rapid (10 min) SFC–MS–MS of an entire 96-well plate containing dextromethorphan samples. The time scale of each chromatogram is reduced moving from the bottom to the top chromatogram. The bottom chromatogram shows the analysis of the entire plate, while the uppermost chromatogram shows the analysis of the first two wells of the first row.

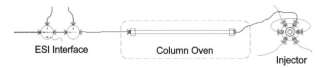

Figure 28 Illustration of the direct introduction approach for packed-column SFC–MS. The downstream pressure is not regulated. While this approach is only successful when certain conditions are met, it can produce very impressive results for high-speed analysis, as described in the text.

effluent is directed to the API source through a narrow-bore transfer line, while the bulk of the effluent moves through the postcolumn pressure-regulation device and on to waste. With ESI, a means of solvent addition, such as the sheath-flow interface described earlier, may be required for efficient ion production when the mobile-phase modifier concentration is low. Generally, no additional solvent is required in this situation for APCI. At least one major vendor and a number of prominent researchers in the field advocate this simple interfacing approach (Morgan et al., 1998; Ventura et al., 1999a, b). While this approach works well in many cases, we have found that the split ratio can change during the course of a pressure, temperature, or mobile-phase composition gradient as the viscosity of the mobile phase changes. This effect can make reproducible quantitation a bit more challenging. However, the effect is sufficiently reproducible that internal or even external standardization can adequately compensate for this variation in the split ratio.

Direct Introduction After Mechanical Pressure Regulation

One way to avoid any variation in a split, as mentioned earlier, and enjoy the benefits of an electronically controlled postcolumn pressure control device, is to simply direct the full flow from the mechanical pressure-regulating device to the API source. This approach is shown in Fig. 30. Pneumatically assisted ESI and APCI sources can easily handle up to 2–8 mL/min of mobile-phase flow, so long as the majority of the mobile phase is carbon dioxide or a similar SFC mobile phase. Some SFC instruments incorporate a pressure transducer in the postcolumn flow path. If the transducer's volume introduces too much band broadening, it must be removed from the flow path (it can still be "tee'ed" into the flow path to provide postcolumn pressure measurement for feedback control).

Also, for efficient ion production in ESI, a low flow (100–200 μL/min) of organic solvent can be introduced into the effluent flow before the mechanical pressure-regulation device. One of the disadvantages of this interfacing approach is that the pressure regulation point (within the SFC instrument) can be 50 cm or more from the API source. The pressure in the transfer line is not regulated, and, under some conditions, phase separation or solute precipitation can occur in the line. This results in "pulsing" at the ESI sprayer or APCI source, a noisy mass spectral signal, and, in the "right" cases, poor mass transfer of analytes through the transfer line. The added organic fluid mentioned earlier improves the mass transfer and reduces the effects of any phase separation.

Pressure-Regulating-Fluid Interface

A "chromatographically simple" interfacing approach is the introduction of a pressure-regulation point

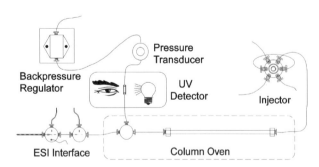

Figure 29 Illustration of the postcolumn split approach for packed-column SFC–MS. While a small portion of the effluent is directed to the API interface, most of the effluent moves through a UV detector (optional), a postcolumn pressure transducer, and a mechanical pressure-regulation device.

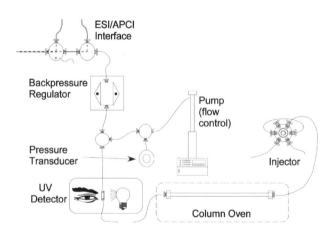

Figure 30 Illustration of the direct introduction of the packed-column SFC effluent into the API interface after mechanical pressure regulation. The postcolumn pressure transducer has been removed from the effluent flow path in this implementation to reduce peak broadening. A low (0.1–0.2 mL/min) flow of solvent sweeps the pressure-transducer cell and is added to the effluent. This eliminates the cell as a source of unswept dead volume.

Supercritical Fluid Chromatography Instrumentation

postcolumn just before the API source (Baker and Pinkston, 1998; Chester and Pinkston, 1998). This pressure-regulation point is provided by a near-zero-dead-volume tee into which is pumped a fluid (such as methanol or methanol/water) under pressure control, as shown in Fig. 31. The fluid is provided by an independent pump. This approach removes the mechanical pressure-regulating device, the pressure transducer, and their associated fittings from the flow path of the effluent, and reduces extra-column band broadening. A short restrictor is installed between the pressure-regulating tee and the API source. The mobile-phase flow and the pressure-regulating-fluid flow vary in inverse proportion as the independent pump maintains the desired pressure within the pressure-regulating tee. The length and internal diameter of the restrictor can be adjusted to accommodate different flow regimes. For example, a 3–4 cm length of 62.5 μm i.d. PEEK tubing provides a pressure-regulating fluid flow of 100–500 μL/min for a mobile-phase flow of 1–3 mL/min. The pressure-regulating fluid acts much like the make-up fluid discussed earlier and improves ESI ionization efficiency at low modifier concentration. This interfacing approach was used to provide the separation shown in Fig. 32 (Pinkston et al., 2002). This chromatogram shows that one can use long columns (1 m here) in SFC to generate high resolution separations of complex, low molecular weight polymers. These separations, combined with MS, provide a powerful tool for characterizing complex mixtures.

While the pressure-regulating-fluid interface provides low extra-column band broadening, it is not as convenient

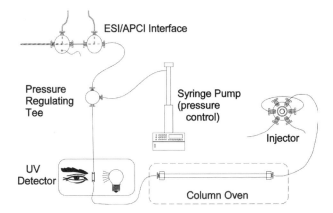

Figure 31 Illustration of the pressure-regulating-fluid approach for postcolumn pressure regulation in packed-column SFC–MS. Here, the mechanical pressure-regulating device is replaced by a low-dead-volume chromatographic tee in which the chromatographic effluent mixes with a fluid pumped under pressure control. The resulting mixture flows to the API interface through a short restrictor, such as a few centimeters of narrow-bore PEEK tubing.

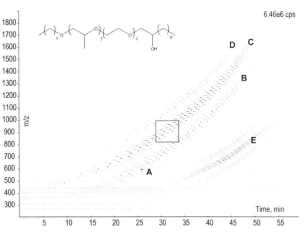

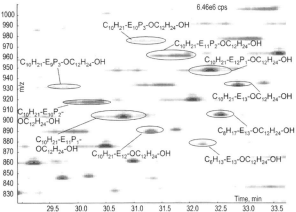

Figure 32 SFC–MS contour plot of a "di-capped" ethoxylate-propoxylate (EO-PO) block copolymer. The x-axis represents retention time, the y-axis m/z, and the intensity is colorized, with red representing the most intense signal. (a) C8E10 (decaoxyethylene octanol, single component) internal standard; (b) singly capped EO-PO chain, lacking the C12-OH capping group; (c) di-capped EO-PO chain, as shown in the structure; (d) di-capped EO-PO chain, but the C12-OH cap is a dimer (C12-O-C12-OH); (e) doubly charged species. In the structure, x has values of 5, 7, and 9, y ranges from 0 to 3, and z from $\sim$3 to at least 34. The enclosed region in the upper contour plot is enlarged in the lower plot. The enlarged plot shows the di-capped EO-PO polymer region about the elution zone of the polymers containing 12 and 13 alkoxylate groups. The C6, C8, and C10 alkyl capping groups and the 0-3 propoxylate groups form a distinctive pattern.

to use, in practice, as the mechanical pressure-regulation system, which is an integral part of the SFC instrument. As the postcolumn pressure is regulated by an independent pump, varying the pressure during a chromatographic separation can be cumbersome. Also, the dimensions of the restrictor can limit the available flow rates. For these reasons, we most often use direct introduction of the effluent after the mechanical pressure-regulating device, as discussed in the previous section.

B. Type of Mass Spectrometer

Modern API sources allow the analyst to use virtually any of the modern types of mass spectrometers. The good performance-to-cost ratio of quadrupole instruments, and the tandem-in-space MS–MS provided by triple quadrupole mass spectrometers, have made these popular choices (Baker and Pinkston, 1998; Dost and Davidson, 2003; Hoke et al., 2001; Wang et al., 2001). However, the advantages of the other types of mass spectrometers have, at one time or another, been explored and exploited. Magnetic sector instruments (Mertens et al., 1996), Paul-type ion traps (Morgan et al., 1998; Pinkston et al., 1992; Todd et al., 1988), and Fourier transform ion cyclotron resonance (FTICR) instruments (Baumeister et al., 1991; Laude et al., 1987; Lee et al., 1987) have been used for SFC–MS. The relatively recent revolution in the data-acquisition speed and high resolution provided by time-of-flight (TOF) instruments has not been lost on the practitioners of SFC–MS (Garzotti et al., 2001; Lazar et al., 1996). Bolanos et al. (2003) argue that the high spectral acquisition rates of modern TOF instruments are best suited to preserve the chromatographic integrity of high speed SFC separations. Baseline peak widths on the order of 1 s require spectral acquisition rates of 10 Hz or greater to accurately reproduce the chromatographic separation.

V. SOLUTE DERIVATIZATION

Modifiers and additives greatly extend SFC capabilities when analyzing polar solutes. However, using modifiers removes some of the detector flexibility. Rather than relying totally on modifiers to deal with strongly retained solutes, it is also possible to derivatize the solutes to add detection options and reduce or sometimes eliminate modifiers.

Derivatization is done in GC specifically to reduce solute polarity, reduce hydrogen bonding, increase volatility, and increase stability in the high-temperature column environment. In SFC, it is still important to reduce solute polarity and hydrogen bonding, but the volatility and thermal stability are no longer important. Instead, we want to maximize solute solubility in the mobile phase, add detection options, or both. Silicones and amorphous fluoropolymers have been identified as "CO_2-philic" (DeSimone et al., 1994). Silyl and fluoroacyl derivatives of polar solutes are also highly soluble in CO_2. Many methods and reagents are available to make these derivatives (Knapp, 1979).

Figure 33 shows the OT-SFC–FID separation of the trimethylsilyl derivative of Maltrin M-100 (Grain Processing Corporation, Muscatine, IA, USA), a maltodextrin made

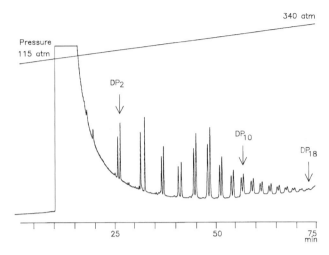

Figure 33 OT-SFC–FID chromatogram of silylated Maltrin M-100. Conditions: 10 m × 50 μm i.d. methyl silicone column with a 0.20 μm film thickness, 0.06 μL direct injection, CO_2 mobile phase at 100°C, pressure programed as indicated. (This particular chromatogram is of historic value with regard to eluting high-molecular-weight derivatives of polar compounds and detecting with an FID. Injection capabilities and solvent peak shapes have improved greatly since this work was originally reported.)

by the acid hydrolysis of corn starch. The degree of polymerization (DP) represents the number of glucose units in each individual species. There are two peaks at each DP corresponding to the α and β anomers. The DP = 18 anomers each have 56 hydroxyl groups before derivatization. After derivatization the molecular weight is 6966. A sample of maltoheptaose (DP = 7), separately derivatized and analyzed by SFC and SFC–MS, showed no traces of incomplete derivatization or degradation during the separation (Chester et al., 1989).

If ionic materials can be made soluble in the mobile phase, then they also can be determined by SFC methods. For example, phosphates can often be silylated, and these derivatives can be eluted using low temperatures, if necessary. Metal ions can also be determined if they can be turned into forms soluble in CO_2 (Carey et al., 1992; Lee and Markides, 1990; Vela and Caruso, 1992; Vela et al., 1993). Many chelants, and particularly N,N-dialkyldithiocarbamates, work well in unmodified CO_2 (Lee and Markides, 1990).

Larger derivatizing groups with better blocking performance can be used in SFC than in GC as long as the requirement for solubility in the mobile phase is met. Cole et al. (1991) systematically studied silylating agents for SFC purposes, including several that add t-butyldimethylsilyl, triisopropylsilyl, and t-butyldiphenylsilyl groups. The value of using a bulky blocking group was also illustrated by Pinkston (1989): *t*-butyldimethylsilyl derivatives of polyacrylic acid oligomers through DP = 28 could easily be separated by OT-SFC while trimethylsilyl derivatives of the same samples could not be separated. Additional research to develop derivatizing agents providing large blocking groups as well as new, fluorinated derivatives would be very useful. Derivatization can be used to impart desirable detection qualities to analytes. For example, Hoffman et al. have shown that alkoxylated low-molecular-weight polymers can be characterized by packed-column SFC with UV detection after derivatization to produce phenyldimethylsislyl derivatives. The derivatives both improve the chromatographic separation and enable a quantitative, simple means of detection.

VI. CHIRAL SFC

Chiral separations are rapidly growing in importance in understanding biological systems and in developing agents that interact with or inhibit these systems. Pharmaceutical and other health-care applications, agriculture, and toxicology all benefit.

A large driving force for the development of chiral separations in the pharmaceutical business is the need to rapidly produce sufficient quantities of a pure enantiomer for efficacy or safety testing early in the development of a drug. It is frequently faster and more cost-effective to synthesize a racemic mixture and then separate the desired enantiomer than to perform a stereoselective synthesis (Toribio et al., 2003). This approach requires the ability to quickly generate the necessary separations and to operate them at the appropriate scale to satisfy the business need.

Impressive resolution is possible by chiral GC when both the stationary phase and the solutes are thermally stable at the required temperature, but drug molecules are frequently polar and not applicable to GC separation at reasonable temperatures. Many chiral stationary phases will interconvert and racemize at high temperatures, as will many solutes. Chiral selectivity often improves at lower temperatures where the stationary phase becomes more highly ordered, but retention in GC soon becomes too high to be practical as the temperature is reduced. Successful separations at lower temperatures require a solvating mobile phase to reduce retention.

Chiral separations of drug candidates are therefore most often performed by HPLC. The stationary phases are rather polar, and separations are generally performed under normal-phase conditions. This is the perfect prerequisite to successfully switch the mobile phase from normal liquids to supercritical or subcritical fluids. Chiral SFC very often has large advantages in both speed and selectivity, and is receiving much attention among researchers, both in analytical and preparative scale. Preparative SFC typically has 3× the production rate of HPLC, measured at the column outlet. In addition, because the SFC mobile phase is very volatile and almost entirely evaporates when depressurized, the desired solute can be collected in concentrated form, avoiding the huge dilution in mobile phase typical in HPLC and the need to evaporate large quantities of liquid solvent. Analytical-scale SFC is extremely useful in scouting and developing conditions for adequate selectivity and resolution before scale-up, and for analyzing the product of preparative chiral separations.

Another significant advantage of SFC compared with HPLC is the possibility of operating at temperatures as low as −50°C. Chiral selectivity frequently continues to improve at these low temperatures, as illustrated in Fig. 34. At these temperatures, liquid mobile phases would become so viscous that slow diffusion and high pressure would make the separation impractical. Of course, a conventional liquid mobile phase will also eventually freeze as the temperature is lowered. However, diffusion and viscosity remain suitable with CO_2-based mobile phases, and any benefits of improved chiral selectivity at such low temperatures can be utilized.

Despite these generalities, selectivity is often different comparing the same column in HPLC mode and SFC mode, just as would be expected comparing HPLC results using two completely different mobile phases. If the chiral selectivity is different in SFC than in HPLC, there is an even chance it will be better. In the majority

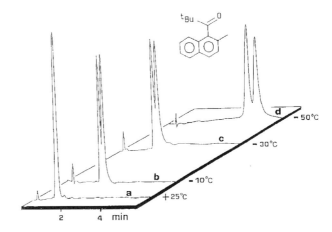

Figure 34 Subambient temperature separation of 2-methyl-1-(2,2-dimethylpropanoyl)-naphthalene enantiomers. Conditions: (*R*,*R*)-DACJ-DNB-modified 5 μm LiChrosorb Si, 100 × 4 mm, eluted with 95:5 CO_2:2-propanol, UV detection at 230 nm, postdetector pressure regulation at 8.0 MPa (79 atm). [Reprinted from Gasparrini et al. (1994) with permission.]

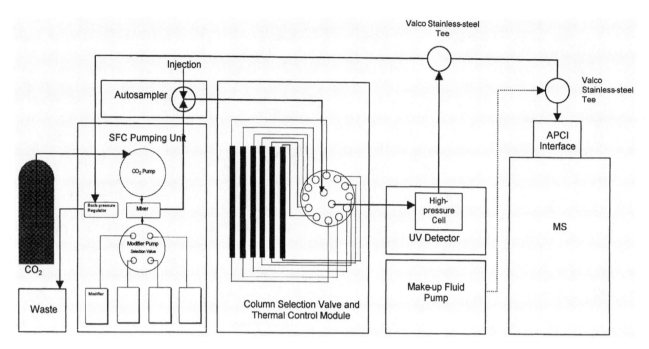

Figure 35 Automated SFC–UV–MS system for method development, as reported by Zhao et al. (2003), used with permission.

of cases reported so far, chiral selectivity in SFC is equal or better than in HPLC using the same stationary phase. And, if the selectivity is only equal, the usual threefold increase in analysis time or production rate can usually be achieved.

Impressive results have been recently reported. For example, del Nozal et al. (2003) separated enantiomers of triadimenol and tridimefon with resolution >3 in under 5 min using a ChiralpakAD column and any of several alcohol modifiers in CO_2 mobile phase. Gyllenhaal and Karlsson (2002) found significant selectivity differences among several alcohol modifiers using L-(+)-tartaric acid as a chiral mobile phase additive in the separation of amino alcohols on a Hypercarb column. Bernal et al. (2002) separated isomers of several cis-2-(2,4-dichlorophenyl)-1,3-dioxolanes and several other materials and found that half the time was required using SFC compared with HPLC, in general. Zhao et al. (2003) reported an automated chiral SFC method development procedure involving screening on several different columns at eight modifier levels and using mass spectrometric detection. An unattended 15 h session produced suitable methods for the problems examined. A schematic diagram of their instrument is shown in Fig. 35. Temperature, pressure, and modifier identity were fixed in this work, but it is easy to envision how such an experiment can be tailored for studies and optimization of numerous parameters for both chiral and conventional separations.

VII. CONCLUSION

Taking pharmaceutical development as an example, SFC is applicable to about the same fraction of problems as is reversed-phase HPLC. Both techniques have about the same probability of solving a randomly selected problem, and when both are applicable, SFC is almost always 3–5× faster (Pinkston et al., 2001). Method development can also be very fast in SFC because of the short analysis times and the ability to easily tune the selectivity. Solvent acquisition and disposal costs are much less expensive in SFC than in HPLC, and particularly when scale-up is contemplated. SFC is not the solution to every problem, especially those involving very hydrophilic solutes that are soluble only in water. However, SFC certainly deserves consideration in the arsenal of modern separation methods.

ACKNOWLEDGMENT

Much credit is due to the past and present colleagues in our Supercritical Fluid Chromatography Group: Claudia Smith, David Innis, Grover Owens, Leisa Burkes, Tom Delaney, Don Bowling, Rosemary Hentschel, Doug Raynie, Brian Haynes, Jil Bos, Darin McDaniel, Rebecca Cunningham, Chris Ott, Michelle Mangels, and Steve Teremi.

REFERENCES

Anacleto, J. F., Ramaley, L., Boyd, R. K., Pleasance, S., Quilliam, M. A., Sim, P. G., Benoit, F. M. (1991). Analysis of Polycyclic Aromatic-Compounds by Supercritical Fluid Chromatography Mass-Spectrometry Using Atmospheric-Pressure Chemical Ionization. *Rapid Commun. Mass Spectrom.* 5:149–155.

Anton, K., Berger, C., eds. (1998). *Supercritical Fluid Chromatography with Packed Columns, Techniques and Applications.* New York: Marcel Dekker, Inc.

Anton, K., Bach, M., Geiser, A. (1991). Supercritical Fluid Chromatography in the Routine Stability Control of Antipruritic Preparations. *J. Chromatogr. A* 553:71–79.

Arpino, P.J., Haas, P. (1995). Recent Developments in Supercritical-Fluid Chromatography Mass-Spectrometry Coupling. *J. Chromatogr. A* 703:479–488.

Arpino, P. J., Sadoun, F., Virelizier., H. (1993). Reviews on Recent Trends in Chromatography Mass-Spectrometry Coupling.4. Reasons Why Supercritical Fluid Chromatography Is Not So Easily Coupled With Mass-Spectrometry As Originally Assessed. *Chromatographia* 36:283–288.

Baker, T. R., Pinkston, J. D. (1998). Development and Application of Packed-Column Supercritical Fluid Chromatography Pneumatically Assisted Electrospray Mass Spectrometry. *J. Am. Soc. Mass Spectrom.* 9:498–509.

Baker, T. R., Pinkston, J. D. (1999). Preliminary Investigation of Supercritical Fluid-Assisted Nebulization for Enhanced Response in Electrospray Mass Spectrometry. *Analusis* 27:701–705.

Balsevich, J., Hogge, L. R., Berry, A. J., Games, D. E., Mylchreest, I. C. (1988). Analysis of Indole Alkaloids from Leaves of Catharanthus-Roseus by Means of Supercritical Fluid Chromatography/Mass Spectrometry. *J. Nat. Prod.* 51:1173–1177.

Baumeister, E. R., West, C. D., Ijames, C. F., Wilkins, C. L. (1991). Probe Interface for Supercritical Fluid Chromatography Fourier-Transform Mass-Spectrometry. *Anal. Chem.* 63:251–255.

Berger, T. A. (1989). Modeling Linear and Tapered Restrictors in Capillary Supercritical Fluid Chromatography. *Anal. Chem.* 61:356–361.

Berger, T. A. (1992). Recent Developments in Packed Column SFC at Hewlett-Packard. The 4th International Symposium on Supercritical Fluid Chromatography and Extraction. Cincinnati, OH, USA, May 20–22, p. 27.

Berger, T. A. (1995). *Packed Column SFC,* Cambridge: The Royal Society of Chemistry.

Berger, T. A., Wilson, W. H. (1993). Packed-Column Supercritical-Fluid Chromatography with 220000 Plates. *Anal. Chem.* 65:1451–1455.

Bernal, J. L., Toribio, L., del Nozal, M. J., Nieto, E. M., Montequi, M. I. (2002). Separation of Antifungal Chiral Drugs by SFC and HPLC: A Comparative Study. *J. Biochem. Biophys. Methods* 54:245–254.

Berry, V. V. (1985). Applications of a Transparent High-Pressure (TrHiP) Container - Supercritical CO_2 and Gradient Generation in Micro-HPLC. *Am. Lab.* 17:33.

Berry, A. J., Games, D. E., Perkins, J. R. (1986). Supercritical Fluid Chromatographic and Supercritical Fluid Chromatographic Mass-Spectrometric Studies of Some Polar Compounds. *J. Chromatogr. A* 363:147–158.

Bolanos, B. J., Ventura, M. C., Greig, M. J. (2003). Preserving the Chromatographic Integrity of High-Speed Supercritical Fluid Chromatography Separations Using Time-Of-Flight Mass Spectrometry. *J. Combinatorial Chem.* 5:451–455.

Broadbent, J. K., Martincigh, B. S., Raynor, M. W., Salter, L. F., Moulder, R., Sjoberg, P., Markides, K. E. (1996). Capillary Supercritical Fluid Chromatography Combined with Atmospheric Pressure Chemical Ionisation Mass Spectrometry for the Investigation of Photoproduct Formation in the Sunscreen Absorber 2-Ethylhexyl-P-Methoxycinnamate. *J. Chromatogr. A* 732:101–110.

Carey, J. M., Vela, N. P, Caruso, J. A. (1992). Multielement Detection for Supercritical Fluid Chromatography by Inductively Coupled Plasma Mass-Spectrometry. *J. Anal. At. Spectrom.* 7:1173–1181.

Chapman, J. R., Pratt, J. A. E. (1987). Improvements in Instrumentation for Thermospray Operation on a Magnetic-Sector Mass-Spectrometer. *J. Chromatogr. A* 394:231–237.

Chester, T. L. (1984). Capillary Supercritical-Fluid Chromatography with Flame-Ionization Detection: Reduction of Detection Artifacts and Extension of Detectable Molecular Weight Range. *J. Chromatogr. A* 299:424–431.

Chester, T. L. (2002). Unified Chromatography: Concepts and Considerations for Multidimensional Chromatography. In: Mondello, L., Bartle, K. D., Lewis, A., eds. *Multidimensional Chromatography.* West Sussex, UK: John Wiley and Sons, pp 151–169.

Chester, T. L. (2004). Determination of Pressure-Temperature Coordinates of Liquid-Vapor Critical Loci by Supercritical Fluid Flow Injection Analysis. *J. Chromatogr. A* 1037:393–403.

Chester, T. L., Innis, D. P. (1989). Sample Loss and Its Control with Internal-Loop Injection Valves in Supercritical Fluid Chromatography. *J. Microcolumn Sep.* 1:230–233.

Chester, T. L., Innis, D. P. (1993). Investigation of Retention and Selectivity in High-Temperature, High-Pressure, Open-Tubular Supercritical-Fluid Chromatography with CO_2 Mobile-Phase. *J. Microcolumn Sep.* 5:441–449.

Chester, T. L., Innis, D. P. (1995). Quantitative Open-Tubular Supercritical-Fluid Chromatography Using Direct-Injection onto a Retention Gap. *Anal. Chem.* 67:3057–3063.

Chester, T. L., Pinkston, J. D. (1998). Pressure-Regulating Fluid Interface and Phase Behavior Considerations in the Coupling of Packed-Column Supercritical Fluid Chromatography with Low-Pressure Detectors. *J. Chromatogr. A* 807:265–273.

Chester, T. L., Innis, D. P., Owens, G. D. (1985). Separation of Sucrose Polyesters by Capillary Supercritical-Fluid Chromatography-Flame Ionization Detection with Robot-Pulled Capillary Restrictors. *Anal. Chem.* 57:2243–2247.

Chester, T. L., Pinkston, J. D., Owens, G. D. (1989). Separation of Malto-Oligosaccharide Derivatives by Capillary Supercritical Fluid Chromatography and Supercritical Fluid Chromatography Mass-Spectrometry. *Carbohydr. Res.* 194:273–279.

Cole, L. A., Dorsey, J. G., Chester, T. L. (1991). Investigation of Derivatizing Agents for Polar Solutes in Supercritical Fluid Chromatography. *Analyst (London)* 116:1287–1291.

Cortes, H. J., Pfeiffer, C. D., Richter, B. E., Stevens, T. S. (1988). U. S. Patent 4793920.

Coym, J. W., Chester, T. L. (2003). Improving Injection Precision in Packed-Column Supercritical Fluid Chromatography. *J. Sep. Sci.* 26:609–613.

Cui, Y., Olesik, S. V. (1991). High-Performance Liquid-Chromatography Using Mobile Phases with Enhanced Fluidity. *Anal. Chem.* 63:1812–1819.

Desimone, J. M., Maury, E. E., Menceloglu, Y. Z., Mcclain, J. B., Romack, T. J., Combes, J. R. (1994). Dispersion Polymerizations in Supercritical Carbon-Dioxide. *Science* 265:356–359.

Dost, K., Davidson, G. (2003). Analysis of Artemisinin by a Packed-Column Supercritical Fluid Chromatography-Atmospheric Pressure Chemical Ionisation Mass Spectrometry Technique. *Analyst* 128:1037–1042.

Edlund, P. O., Henion, J. D. (1989). Packed-Column Supercritical Fluid Chromatography/Mass Spectrometry Via a 2-Stage Momentum Separator. *J. Chromatogr. Sci.* 27:274–282.

Fenn, J. B., Mann, M., Meng, C. K., Wong, S. F., Whitehouse, C. M. (1990). Electrospray Ionization-Principles and Practice. *Mass Spectrom. Rev.* 9: 37–70.

Fjeldsted, J. C., Jackson, W. P., Peaden, P. A., Lee, M. L. (1983). Density Programming in Capillary Supercritical Fluid Chromatography. *J. Chromatogr. Sci.* 21:222–225.

Garzotti, M., Rovatti, L., Hamdan, M. (2001). Coupling of a Supercritical Fluid Chromatography System to a Hybrid (Q-Tof 2) Mass Spectrometer: On-Line Accurate Mass Measurements. *Rapid Commun. Mass Spectrom.* 15:1187–1190.

Gasparrini, F., Maggio, F., Misiti, D., Villani, C., Andreolini, F., Mapelli, G. P. (1994). High-Performance Liquid-Chromatography On The Chiral Stationary-Phase (R,R)-DACH-DNB Using Carbon Dioxide-Based Eluents. *J. High Resolut. Chromatogr./Chromatogr. Commun.* 17:43–45.

Gere, D. R. (1983). Supercritical Fluid Chromatography, *Science* 222:253–259.

Giddings, J. C., Meyers, M. N., McLaren, L., Keller, R. A. (1968). High Pressure Gas Chromatography of Nonvolatile Species, *Science* 162:67–73.

Gouw, T. H., Jentoft, R. E. (1972). Supercritical Fluid Chromatography. *J. Chromatogr. A* 68:303–323.

Greibrokk, T., Andersen, T. (2003). High-Temperature Liquid Chromatography. *J. Chromatogr. A* 1000:743–755.

Guthrie, E. J., Schwartz, H. E. (1986). Integral Pressure Restrictor for Capillary SFC. *J. Chromatogr. Sci.* 24:236–241.

Gyllenhaal, O., Karlsson, A. (2002). Enantiomeric Separations of Amino Alcohols by Packed-Column SFC on Hypercarb with L-(+)-Tartaric Acid as Chiral Selector. *J. Biochem. Biophys. Methods* 54:169–185.

Hirata, Y., Katoh, S. (1992). Temperature and Pressure Controlled UV Detector for Capillary Supercritical Fluid Chromatography. *J. Microcolumn Sep.* 4:503–507.

Hoke, S. H., Tomlinson, J. A., Bolden, R. D., Morand, K. L., Pinkston, J. D., Wehmeyer, K. R. (2001). Increasing Bioanalytical Throughput using pcSFC-MS/MS: 10 minutes per 96-Well Plate. *Anal. Chem.* 73:3083–3088.

Houben, R.J., Leclercq, P.A., Cramers, C.A. (1991). Ionization Mechanisms in Capillary Supercritical Fluid Chromatography Chemical Ionization Mass-Spectrometry. *J. Chromatogr. A* 554:351–358.

Huang, E. C., Jackson, B. J., Markides, K. E., Lee, M. L. (1988). Direct Heated Interface Probe for Capillary Supercritical Fluid Chromatography Double Focusing Mass-Spectrometry. *Anal. Chem.* 60:2715–2719.

Huang, E., Henion, J., Covey T. R. (1990). Packed-Column Supercritical Fluid Chromatography Mass-Spectrometry and Supercritical Fluid Chromatography Tandem Mass-Spectrometry with Ionization at Atmospheric-Pressure. *J. Chromatogr. A* 511:257–270.

Jedrzejewski, P. T., Taylor, L. T. (1995). Packed-Column Supercritical-Fluid Chromatography Mass-Spectrometry With Particle-Beam Interface Aided With Particle Forming Solvent. *J. Chromatogr. A* 703:489–501.

Jentoft, R. E., Gouw, T.H (1970). Pressure-Programmed Supercritical Fluid Chromatography of Wide Molecular Weight Range Mixtures. *J. Chromatogr. Sci.* 8:138–142.

Jentoft, R. E., Gouw, T. H. (1972). Apparatus for Supercritical Fluid Chromatography with Carbon Dioxide as the Mobile Phase. *Anal. Chem.* 44:681–686.

Kalinoski, H. T., Hargiss, L. O. (1990). Supercritical Fluid Chromatography Mass-Spectrometry of Nonionic Surfactant Materials Using Chloride-Attachment Negative-Ion Chemical Ionization. *J. Chromatogr. A* 505:199–213.

Kebarle, P., Tang, L. (1993). From Ions in Solution to Ions in the Gas-Phase - The Mechanism of Electrospray Mass-Spectrometry. *Anal. Chem.* 65:972A–986A.

Klesper, E., Corwin, A. H., Turner, D. A. (1962). High Pressure Gas Chromatography Above Critical Temperatures. *J. Org. Chem.* 27:700–701.

van Konynenburg, P. H., Scott, R. L. (1980). Critical Lines and Phase-Equilibria in Binary Vanderwaals Mixtures. *Phil. Trans. Royal Soc.* 298:495–540.

Knapp, D. R. (1979). *Handbook of Analytical Derivatization Reactions.* New York: John Wiley & Sons.

Laude, D. A., Pentoney, S. L, Griffiths, P. R., Wilkins, C. L. (1987). Supercritical Fluid Chromatography Interface for a Differentially Pumped Dual-Cell Fourier-Transform Mass-Spectrometer. *Anal. Chem.* 59:2283–2288.

Lazar, J. M., Lee, M. L., Lee, E. D. (1996). Design and Optimization of a Corona Discharge Ion Source for Supercritical Fluid Chromatography Time-Of-Flight Mass Spectrometry. *Anal. Chem.* 68:1924–1932.

Lee, M. L., Markides, K. E., eds. (1990). *Analytical Supercritical Fluid Chromatography and Extraction*, Provo, UT: Chromatography Conferences, Inc.

Lee, E. D., Henion, J. D., Cody, R. B., Kinsinger, J. A. Supercritical Fluid Chromatography Fourier-Transform Mass-Spectrometry. *Anal. Chem.* 59:1309–1312.

Lee, M. L., Xu, B., Huang, E. C., Djordjevic, N. M., Chang, H-C. K., Markides, K. E. (1989). Liquid Sample Introduction Methods in Capillary Column Supercritical Fluid Chromatography. *J. Microcolumn Sep.* 1:7–13.

Lesellier, E., West, C., Tchapla, A. (2003). Advantages of the Use of Monolithic Stationary Phases for Modelling the Retention in Sub/Supercritical Chromatography Application to Cis/Trans-Beta-Carotene Separation. *J. Chromatogr. A* 1018:225–232.

van Leuken, R., Mertens, M., Janssen, H. G., Sandra, P., Kwakkenbos, G., Deelder, R. (1994). Optimization Of Capillary SFC-MS for the Determination of Additives in Polymers. *J. High Res. Chromatogr.* 17:573–576.

Malik, A., Li, W. B., Lee, M. L. (1993). Preparation of Long Packed Capillary Columns Using Carbon-Dioxide Slurries. *J. Microcolumn Sep.* 5:361–369.

Matsumoto, K., Tsuge, S., Hirata Y. (1986). Fundamental Conditions in Pressure-Programmed Supercritical Fluid Chromatography-Mass Spectrometry and Some Applications to Vitamin Analysis. *Chromatographia* 21:617–621.

Matsumoto, K., Nagata, S., Hattori, H., Tsuge, H. (1992). Development of Directly Coupled Supercritical Fluid Chromatography with Packed Capillary Column Mass-Spectrometry with Atmospheric-Pressure Chemical Ionization. *J. Chromatogr. A* 605:87–94.

McFadden, W. H. (1979). Interfacing Chromatography and Mass-Spectrometry. *J. Chromatogr. Sci.* 17:2–16.

McGuffin, V. L., Evans, C. E. (1991). Influence of Pressure on Solute Retention in Liquid-Chromatography. *J. Microcolumn Sep.* 3:513–520.

McHugh, M., Krukonis, V. (1986). *Supercritical Fluid Extraction, Principles and Practice.* Chapters 2 and 3. Stoneham, MA: Butterworths,

Mertens, M. A. A., Janssen, H-G. M., Cramers, C. A., Genuit, W. J. L., vanVelzen, G. J., Dirkzwager, H., vanBinsbergen, H. (1996). Development and Evaluation of an Interface for Coupled Capillary Supercritical Fluid Chromatography Magnetic Sector Mass Spectrometry - Application to Thermally Unstable and High Molecular Mass Compounds. *J. High Resolution Chromatogr.* 19:17–22.

Meyer, E. F., Meyer, T. P. (1986). Supercritical Fluid–Liquid, Gas, Both, or Neither—A Different Approach. *J. Chem. Ed.* 63:463–465.

Morgan, D. G., Harbol, K. L., Kitrinos, N. P. (1998). Optimization of a Supercritical Fluid Chromatograph Atmospheric Pressure Chemical Ionization Mass Spectrometer Interface Using an Ion Trap and Two Quadrupole Mass Spectrometers. *J. Chromatogr. A* 800:39–49.

Nelieu, S., Stobiecki, M., Sadoun, F., Virelizier, H., Kerhoas, L., Einhorn J. (1994). Solid-Phase Extraction and LC-MS or SFC-MS for the Analysis of Atrazine Metabolites in Water. *Analusis* 22:70–75.

Niessen, W. M. A., Vanderhoeven, R. A. M., Dekraa, M. A. G., Heeremans, C. E. M, Tjaden, U. R., Vandergreef, J. (1989). Repeller Effects in Discharge Ionization in Liquid and Supercritical-Fluid Chromatography-Mass Spectrometry Using a Thermospray Interface.2. Changes in Some Analyte Spectra. *J. Chromatogr. A* 478:325–338.

Novotny, M., Springston, S. R., Peadon, P. A., Fjeldsted, J. C., Lee, M. L. (1981). Capillary Supercritical Fluid Chromatography, *Anal. Chem.*, 53:407A–414A.

del Nozal, M. J., Toribio, L., Bernal, J. L., Castano, N. (2003). Separation of Triadimefon and Triadimenol Enantiomers and Diastereoisomers by Supercritical Fluid Chromatography. *J. Chromatogr. A* 986:135–141.

Page, S. H., Sumpter, S. R., Lee, M. L. (1992). Fluid Phase-Equilibria in Supercritical Fluid Chromatography with CO_2-Based Mixed Mobile Phases—A Review. *J. Microcolumn Sep.* 4:91–122.

Parcher, J. F., Chester, T. L. (1999). *Unified Chromatography* (ACS Symposium Series 748). Washington, D.C.: American Chemical Society.

Peaden, P. A., Fjeldsted, J. C., Lee, M. L., Springston, S. R., Novotny, M. (1982). Instrumental Aspects of Capillary Supercritical Fluid Chromatography. *Anal. Chem.* 54:1090–1093.

Perkins, J. R., Games, D. E., Startin, J. R., Gilbert, J. (1991a). Analysis of Sulfonamides Using Supercritical Fluid Chromatography and Supercritical Fluid Chromatography Mass-Spectrometry. *J. Chromatogr. A* 540:239–256.

Perkins, J. R., Games, D. E., Startin, J. R., Gilbert, J. (1991b). Analysis of Veterinary Drugs Using Supercritical Fluid Chromatography and Supercritical Fluid Chromatography Mass-Spectrometry. *J. Chromatogr. A* 540:257–270.

Pinkston, J. D. (1989), in Markides, K. E., Lee, M. L., eds. *SFC Applications: 1989 Symposium/Workshop on Supercritical Fluid Chromatography*, Provo, UT: Brigham Young University Press

Pinkston, J. D., Bowling, D. J. (1992). Advances in Capillary SFC-MS. *J. Chromatogr. Libr.* 53:25–46.

Pinkston, J. D., Bowling, D. J. (1993). Investigation of Cryopumping for Enhanced Performance in Supercritical-Fluid Chromatography/Mass Spectrometry. *Anal. Chem.* 65:3534–3539.

Pinkston, J. D., Hentschel, R. T. (1993). Evaluation of flow restrictors for open-tubular supercritical-fluid chromatography at pressures up to 560 atm. *J. High Resolution Chromatogr.* 16:269–274.

Pinkston, J. D., Baker. T. R. (1995). Modified Ionspray Interface for Supercritical-Fluid Chromatography Mass-Spectrometry-Interface Design and Initial Results. *Rapid Commun. Mass Spectrom.* 9:1087–1094.

Pinkston, J. D., Chester, T. L. (1995). Putting opposites together-Guidelines for successful SFC/MS, *Anal. Chem.* 67:650A-656A.

Pinkston, J. D., Owens, G. D., Burkes, L. J., Delaney, T. E., Millington, D. S., Maltby, D. A. (1988). Capillary Supercritical Fluid Chromatography-Mass Spectrometry Using a High Mass Quadrupole and Splitless Injection. *Anal. Chem.* 60:962–966.

Pinkston, J. D., Delaney, T. E., Bowling, D. J., Chester, T. L. (1991). Comparison by Capillary SFC and SFC-MS of Soxhlet and Supercritical Fluid Extraction of Hamster Feces., *J. High Resolution Chromatogr./Chromatogr. Commun.* 14:401–406.

Pinkston, J. D., Delaney, T. E., Morand, K. L., Cooks, R. G. (1992). Supercritical Fluid Chromatography Mass-Spectrometry Using a Quadrupole Mass Filter Quadrupole Ion Trap Hybrid Mass-Spectrometer with External Ion-Source. *Anal. Chem.* 64:1571–1577.

Pinkston, J. D., Wen, D., Morand, K. L., Tirey, D. A., Stanton, D. T. (2001). Screening a Large and Diverse Library of Pharmaceutically Relevant Compounds Using SFC/MS and a Novel Mobile-Phase Additive. The 10th International Symposium on Supercritical Fluid Chromatography, Extraction, and Processing, Myrtle Beach, SC, U.S.A., August 19–22.

Pinkston, J. D., Marapane, S. B., Jordan, G. T., Clair, B. D. (2002). Characterization of Low Molecular Weight Alkoxylated Polymers Using Long Column SFC/MS and an Image Analysis Based Quantitation Approach. *J. Am. Soc. Mass Spectr.* 13:1195–1208.

Pinkston, J. D., Stanton, D. T., Wen, D. (2004). Elution and preliminary structure-retention modeling of polar and ionic substances in supercritical fluid chromatography using volatile ammonium salts as mobile phase additives. *J. Sep. Sci.* 27:115–123.

Ramsey, E. D., Raynor, M. W. (1996). Electron Ionization and Chemical Ionization Sensitivity Studies Involving Capillary Supercritical Fluid Chromatography Combined with Benchtop Mass Spectrometry. *Anal. Commun.* 33:95–97.

Reinhold, V. N., Sheeley, D. M., Kuei, J., Her, G. R. (1988). Analysis of High Molecular-Weight Samples on a Double-Focusing Magnetic-Sector Instrument by Supercritical Fluid Chromatography-Mass Spectrometry. *Anal. Chem.* 60:2719–2722.

Richter, B. E. (1986). Pittsburgh Conference and Exposition. Atlantic City, NJ, March 10–14, Paper No. 514.

Sadoun, F., Virelizier, H., Arpino, P. J. (1993). Packed-Column Supercritical Fluid Chromatography Coupled with Electrospray Ionization Mass Spectrometry. *J. Chromatogr. A,* 647:351–359.

Saito, M., Yamauchi, Y., Okuyama, T. eds. (1994). *Fractionation by Packed-Column SFC and SFE.* New York: VCH Publishers, Inc.

Saito, M., Yamauchi, Y. (1994). Instrumentation. In: Saito, M., Yamauchi, Y., Okuyama, T., eds. *Fractionation by Packed-Column SFC and SFE.* New York: VCH Publishers, Inc., pp. 107–110.

Sanders, P. E., Sheehan, E., Buchner, J., Willoughby, R., Dilts, M., Marecic, T., Dulak, J. (1991). In: Jennings, W. G.; Nikelly, J. G., eds. *Capillary Chromatography.* Heidelberg, Germany: Huethig, pp. 131–153.

Saunders, C. W., Taylor, L. T., Wilkes, J., Vestal, M. (1990). Supercritical Fluid Chromatography Using Microbore Packed-Columns and a Benchtop Thermospray MS. *Am. Lab.* 22:46–53.

Scalia, S., Games, D. E. (1992). Analysis of Conjugated Bile-Acids by Online Supercritical Fluid Chromatography/Thermospray Mass-Spectrometry. *Org. Mass Spectrom.* 27:1266–1270.

Smith, R. M., ed. (1988). *Supercritical Fluid Chromatography.* Cambridge: The Royal Society of Chemistry.

Smith, R. D., Udseth, H. R. (1987). Mass-Spectrometer Interface for Microbore and High Flow-Rate Capillary Supercritical Fluid Chromatography with Splitless Injection. *Anal. Chem.* 59:13–22.

Smith, R. D., Fjeldsted, J. C., Lee, M. L. (1982). Direct Fluid Injection Interface for Capillary Supercritical Fluid Chromatography-Mass Spectrometry. *J. Chromatogr. A* 247:231–243.

Smith, R. D., Fulton, J. L., Petersen, R. C., Kopriva, A. J., Wright, B. W. (1986). Performance of Capillary Restrictors in Supercritical Fluid Chromatography. *Anal. Chem.* 58:2057–2064.

Strubinger, J. R., Song, H. C., Parcher, J. F. (1991). High-Pressure Phase Distribution Isotherms for Supercritical Fluid Chromatographic Systems.2. Binary Isotherms of Carbon-Dioxide and Methanol. *Anal. Chem.* 63:104–108.

Thomas, D., Sim, P. G., Benoit, F. M. (1994). Capillary Column Supercritical-Fluid Chromatography Mass-Spectrometry of Polycyclic Aromatic-Compounds using Atmospheric-Pressure Chemical-Ionization. *Rapid Commun. Mass Spectrom.* 8:105–110.

Todd, J. F. J., Mylchreest, I. C., Berry, A. J., Games, D. E., Smith, R. D. (1988). *Rapid Commun. Mass Spectrom.* 2:55–58.

Tong, D. X., Bartle, K. D., Clifford, A. A. (1994). Preparation and Evaluation of Supercritical Carbon Dioxide-Packed Capillary Columns for HPLC and SFC. *J. Microcolumn Sep.* 6:249–255.

Toribio, L., Nozal, M., Bernal, J. L., Nieto, E. M. (2003). Use of Semipreparative Supercritical Fluid Chromatography to Obtain Small Quantities of the Albendazole Sulfoxide Enantiomers, *J. Chromatogr. A* 1011:155–161.

Tuominen, J. P., Markides, K. E., Lee, M. L. (1991). Optimization of Internal Valve Injection in Open Tubular Column Supercritical Fluid Chromatography, *J. Microcolumn Sep.* 3:229–239.

Tyrefors, L. N., Moulder, R. X., Markides, K. E. (1993). Interface for Open Tubular Column Supercritical Fluid Chromatography/Atmospheric Pressure Chemical Ionization Mass Spectrometry. *Anal. Chem.* 65:2835–2840.5

Vela, N. P., Caruso, J. A. (1992). Determination of Tri-Organotin and Tetra-Organotin Compounds by Supercritical Fluid Chromatography with Inductively Coupled Plasma Mass-Spectrometric Detection. *J. Anal. At. Spectrom.* 7:971–977.

Vela, N. P., Olson, L. K., Caruso, J. A. (1993). Elemental Speciation with Plasma-Mass Spectrometry. *Anal. Chem.* 65: 585A–587A.

Ventura, M. C., Farrell, W. P., Aurigemma, C. M., Greig, M. J. (1999a). Packed Column Supercritical Fluid Chromatography Mass Spectrometry for High-Throughput Analysis. *Anal. Chem.* 71:2410–2416.

Ventura, M. C., Farrell, W. P., Aurigemma, C. M., Greig, M. J. (1999b). Packed Column Supercritical Fluid Chromatography/Mass Spectrometry for High Throughput Analysis. Part 2. *Anal. Chem.* 71:4223–4231.

Verillon, F., Heems, D., Pichon, B., Coleman, K., Robert, J. C. (1992). Supercritical Fluid Chromatography with Independent Programming of Mobile-Phase Pressure, Composition, and Flow-Rate. *Amer. Lab.* 24:45–53.

Via, J., Taylor, L. T. (1994). Packed-Column Supercritical-Fluid Chromatography Chemical-Ionization Mass-Spectrometry of Energetic Material-Extracts Using a Thermospray Interface. *Anal. Chem.* 66:1385–1395.

Wang, T., Barber, M., Hardt, I., Kassel, D. B. (2001). Mass-Directed Fractionation and Isolation of Pharmaceutical Compounds by Packed-Column Supercritical Fluid Chromatography/Mass Spectrometry. *Rapid Commun. Mass Spectrom.* 15:2067–2075.

Yan, B. W., Zhao, J. H., Brown, J. S., Blackwell, J., Carr, P. W. (2000). High-Temperature Ultrafast Liquid Chromatography. *Anal. Chem.* 72:1253–1262.

Zhao, Y.N., Woo, G., Thomas, S., Semin, D., Sandra, P. (2003). Rapid Method Development for Chiral Separation in Drug Discovery Using Sample Pooling and Supercritical Fluid Chromatography-Mass Spectrometry. *J. Chromatogr. A* 1003:157–166.

25

Capillary Electrophoresis

HASSAN Y. ABOUL-ENEIN
*Pharmaceutical Analysis Laboratory, Biological and Medical Research Department (MBC-03),
King Faisal Specialist Hospital and Research Center, Riyadh, Saudi Arabia*

IMRAN ALI
National Institute of Hydrology, Roorkee, India

I. INTRODUCTION

The term electrophoresis refers to the migration of ions or charged particles under the influence of an electric field. Earlier, the electrophoresis technique was performed in a gel or other medium in the form of a bed, slab, rod, and so on, but, due to laborious multistage handling of supporting media and nonreproducibility of the results, the supporting medium was replaced by a capillary and the technique was called capillary electrophoresis (CE). Basically, the root of modern electrophoresis dates back to the experiments of Kohlrausch (1897) dealing with the migration of ions in an electrolyte solution. Later on Tiselius (1930) separated protein mixtures by electrophoresis, which led him to the Nobel prize in 1948. The first CE apparatus was designed by Hjerten (1967), and the modern era of CE is considered to begin with many publications of Jorgenson and Lukacs (1981a, b, 1983) describing instrumentation consisting of a fused silica capillary, an electrode reservoirs, a high-voltage power supply and a detector. Although CE had been a topic of discussion among the scientists, it gained recognition in 1989 with the first International Symposium on High Performance Capillary Electrophoresis, which was held in Boston (Wehr and Zhu, 1993). This meeting attracted over 400 scientists to attend 100 presentations on the theory and practice of CE. In this way, it is a newly developed analytical technique, which is being used for the separation of small ions to biopolymers and even whole cells.

Presently, CE is a versatile technique of high speed, high sensitivity, low limit of detection, and inexpensive running cost, which is a major trend in analytical science, and numerous publications have increased exponentially in separation science using this technique (Foret et al., 1993; Khaledi, 1998; Landers, 1993; Lunn, 2000; Wehr et al., 1998). CE is suitable for samples that may be difficult to separate by liquid chromatography, or at least complements this technique, as the principles of separation are different. It is important to note in this regard that CE is fundamentally a homogenous method of separation, not involving surface adsorption. The lower detection limit of CE leads to the possibility of separating and characterizing very small quantities of material, namely, solution-binding constants of protein–protein or drug–protein interactions can be achieved. Kinetics of this binding process can be directly studied. Moreover, the enzymatic reactions for analytical purposes can be conducted within the capillary column.

II. THEORY OF CE

It is worthy to discuss the basic principle and some other fundamental aspects of CE so that the reader can use this technique in a proper way. The mechanisms of the separation in CE are based on the difference in the electrophoretic mobilities of the analytes. Electro-osmotic flow (EOF) and applied electric potential tend to move analytes from one end of the capillary to other. The different migration velocities of the analytes are due to differences in charge–size ratio. The greater the charge–size ratio, the higher is the mobility, and hence, the lower is the retention time. The different sizes of the analytes are also responsible for their different migrations (steric effect). A schematic representation of the separation in CE is shown in Fig. 1. Basically, the mixture of the analytes moves in the capillary in the form of different zones, and the separation occurs as separate zones during the migration time. Figure 1(a) depicts the loading stage of the mixture, for example, A and B, onto capillary, whereas Fig. 1(b) indicates a partial separation of these two components. Finally, Fig. 1(c) represents the clear separation of analytes A and B, which are detected by the detector and depicted by the recorder in the form of peaks [Fig. 1(d)].

Under the CE conditions, the migration of the molecule is controlled by the sum of intrinsic electrophoretic mobility (μ_{ep}) and electro-osmotic mobility (μ_{eo}), due to the action of EOF. The observed mobility (μ_{obs}) of the analyte is related to μ_{eo} and μ_{ep} by the following equation:

$$\mu_{obs} = E(\mu_{eo} + \mu_{ep}) \quad (1)$$

where E is the applied voltage (kV).

The electrophoretic separations are characterized by retention (k), separation (α), and resolution factors (R_s).

The values of these parameters can be calculated by the following standard equations:

$$k = \frac{t_r - t_0}{t_0} \quad (2)$$

$$\alpha = \frac{k_1}{k_2} \quad (3)$$

$$R_s = \frac{2\Delta t}{w_1 + w_2} \quad (4)$$

where t_r and t_0 are the retention time of the electropheromatogram of the separated analytes and dead time (solvent front) of the capillary in minutes, respectively. Δt, w_1, and w_2 are the difference of the retention times of the two peaks of the separated analyte, and the base widths of peak 1 and peak 2, respectively. If the individual values of α and R_s are >1, the separation is said to be complete. If the individual values of these parameters are <1, the separation is considered to be partial or incomplete.

The simplest way to characterize the separation of two components, resolution factor (R_s), is to divide the difference in the retention times by the average peak width.

The value of separation factor may be correlated with μ_{app} and μ_{ave} by the following equation.

$$R_s = \frac{1}{4}\left(\frac{\Delta\mu_{app}}{\mu_{ave}}\right)N^{1/2} \quad (5)$$

where μ_{app} is the apparent mobility of the two enantiomers and μ_{ave} is the average mobility of the two enantiomers. The utility of Eq. (4) is that it permits independent assessment of the two factors that affect separation, selectivity, and efficiency. The selectivity is reflected in the mobility of the analytes, whereas the efficiency of the separation process is indicated by N. Another expression for N is derived from the following equation.

$$N = 5.54\left(\frac{L}{w_{1/2}}\right)^2 \quad (6)$$

where L and $w_{1/2}$ are capillary length and the peak width at half height, respectively.

It is important to point out that it is misleading to discuss theoretical plates in CE and, simply, it is a carry-over from chromatographic theory. In electrophoresis, the separation is governed by the relative mobilities of the analytes in the applied electric field, which is a function of their charge, mass, and shape. Theoretical plate in CE is merely a convenient concept to describe the analyte peak shapes and to assess the factors that affect the separation. The efficiency of the separation on the column is expressed by N, but it is difficult to assess the factors that affect efficiency. This is because it refers to the behavior of a single component during the separation process and this is not suitable to describe the separation

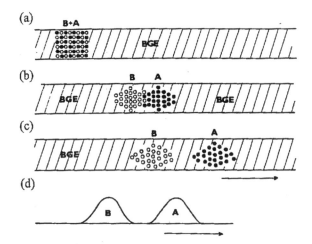

Figure 1 Separation mechanisms in CE.

in CE. However, a more useful parameter is the height equivalent of a theoretical plate (HETP) given as follows:

$$\text{HETP} = \frac{L}{N} = \frac{\sigma_{\text{tot}}^2}{L} \tag{7}$$

HETP may be considered as the function of the length of capillary occupied by the analyte and it is more practical to measure separation efficiency in comparison with N. σ_{tot}^2 is affected not only by diffusion but also by differences in the mobilities, Joule heating of the capillary, and interaction of the analytes with the capillary wall, and hence, σ_{tot}^2 can be represented as follows:

$$\sigma_{\text{tot}}^2 = \sigma_{\text{diff}}^2 + \sigma_{\text{T}}^2 + \sigma_{\text{int}}^2 + \sigma_{\text{wall}}^2 + \sigma_{\text{Electos}}^2 + \sigma_{\text{Electmig}}^2 + \sigma_{\text{Sorp}}^2 + \sigma_{\text{Oth}}^2 \tag{8}$$

where σ_{tot}^2, σ_{diff}^2, σ_{T}^2, σ_{int}^2, σ_{wall}^2, $\sigma_{\text{Electros}}^2$, $\sigma_{\text{Electmig}}^2$, $\sigma_{\text{Sorption}}^2$, and σ_{Oth}^2 are square roots of standard deviations of total, diffusion, Joule heat, injection, wall, electroosmosis, electromigration, sorption, and other phenomena, respectively.

III. MODES OF CE

During the course of time, various modifications have been made in CE and, hence, various forms of CE are currently used. The most important kinds of CE are capillary zone electrophoresis (CZE), capillary isotachphoresis (CITP), capillary gel electrophoresis (CGE), capillary isoelectric focusing (CIEF), affinity capillary electrophoresis (ACE), micellar electrokinetic chromatography (MEKC), capillary electrochromatography (CEC), and capillary electrokinetic chromatography (CEKC). These forms of CE differ in their working principles and can be used for a variety of applications. Among various types of CE, as mentioned earlier, CZE is the most popular due to its wide range of applicability. The analytes in CZE are separated in the form of zones and, hence, they are called capillary zone electrophoresis (CZE); it is often referred to as, simply, capillary electrophoresis (CE).

CITP is a moving boundary capillary electrophoretic technique in which a combination of two buffers is used to create a state that separated zones of all species move at the same velocity. The zones remain sandwiched between so-called leading and terminating electrolytes. The steady state velocity in CITP occurs when the electric field varies in each zone and, hence, very sharp boundaries of the zones appear. CGE refers to the distribution of analytes according to their charges and sizes in a carrier electrolyte to which a gel-forming medium is added. It is the closest technique to traditional slab gel electrophoresis. This technique is applicable for large molecules such as proteins and nucleic acids. In CIEF, a pH gradient is formed within the capillary using compounds that contain both acidic and basic groups and have pI between 3 and 9. A process called focusing, involving basic and acidic solutions at the cathode and anode, respectively, and an applied electric field result in the separation of analytes. This mode of CE is used for the separation of immunoglobulins and hemoglobulins and for the measurement of pIs of proteins. ACE is used for the separation of various biologically related compounds, and the separation in this modality is based on the same types of specific, reversible interactions that are found in biological systems, such as the binding of an enzyme with a substrate or an antibody with antigen. It is also a valuable tool for the separation of halocarbons in the environment and of biomolecules. MEKC permits the separation of uncharged molecules through the clever use of charged micelles as a pseudophase in which partition of the analyte occurs. Sometimes, a surfactant molecule (at a concentration more than its critical micellar concentration) is added for the optimization of the separation in CE, and the separation mechanism is shifted towards a chromatographic mode and, hence, the technique is called micellar electrokinetic chromatography (MEKC). This technique was introduced by Terabe et al. (1984a). The best surfactant for MEKC possesses good solubility in buffer solution and homogeneous micellar solution with compatibility with the detector and with low viscosity. Basically, the micelle and buffer tend to move towards the positive and negative ends, respectively. The movement of the buffer is stronger than the micelle movement and as a result, the micelle and the buffer move towards the negative end. The separation in this mode of CE depends on the distribution of analytes between micellar and aqueous phases. CEC is a hybrid technique between HPLC and CE, and was developed in 1990 (Mayer and Schurig, 1992). It combines the high peak efficiency, which is characteristic of electrically driven separations with the high separation selectivity of HPLC. The chromatographic band broadening mechanisms are quite different in both modes. The chromatographic and electrophoretic mechanisms work simultaneously in CEC, and several combinations are possible. The separation occurs on the mobile/stationary phase interface, and the exchange kinetics between the mobile and the stationary phases are important. Again, in this mode of CE, the separation principle is a mixture of liquid chromatography and CE. CEKC was introduced by Terabe in 1984 (Terabe et al. 1984a). The analytes are separated by electrophoretic migration and is based on chromatographic principles. The most commonly used pseudo phases in CEKC are synthetic and natural micelles and microemulsions, peptides, proteins, charged macromolecules, and charged linear and cyclic oligosaccharides. It is used for the separation of neutral compounds, which is impossible with ordinary CE.

IV. INSTRUMENTATION

The schematic diagram of a CE instrument is shown in Fig. 2. It consists of an injector, a capillary, electrodes, BGE vessels, a high-voltage power supply, a detector, and a recorder. The sample is introduced into the capillary at the injection end, usually by siphoning and/or electromigration. The driving electric current is supplied by a high-voltage power supply. Migrating zones of the sample are detected at the detection end of the capillary by a suitable detector. The simplicity of this experimental setup makes such an apparatus readily accessible to most analytical laboratories. However, for routine practice, more sophisticated, and preferably automated, instrumentation is needed. Various parts of the CE apparatus are discussed in detail in the following sections.

A. Sample Injection

The mode of sample loading is important for reproducible results in CE, and the effects of sample volume on separation efficiency have been reported by several workers (Foret et al., 1988; Hjerten, 1990; Lukacs and Jorgenson, 1985; Terabe et al., 1984a). Therefore, various methods of sample loading have been proposed from time to time. These include a rotary type injector (Tsuda et al., 1981), a miniaturized sampling valve (Mikkers et al., 1979), an electro-osmotic–electromigration technique (Jorgenson and Lukacs, 1981a), an electric sample splitter (Deml et al., 1985), an electrokinetic technique (Jorgenson and Lukacs, 1981a, b, c) and hydrostatic (Terabe et al., 1984; Tsuda et al., 1983a). Among these sample-injection techniques, electrokinetic and hydrostatic techniques are most popular due to their reproducible results and ease of operation and, therefore, are used in all commercially available CE machines.

The electrokinetic sampling technique, the best one for narrow bore capillaries, is based on electrophoretic migration and EOF of electrolyte into the capillary. The sampling end of the capillary and platinum electrode are allowed to dip into the sample vial, and an electric current is switched on for a short period of time, for example, for 10 s. The EOF induced during this sampling procedure moves the sample into the capillary. After the sampling is complete, the end of the capillary is moved back into the buffer reservoir and the analysis proceeds. However, this mode of sampling restricts its use, especially when the sample solution contains more than one component, as the amount of each component introduced into the capillary is selectively determined by its electrophoretic mobility. If the EOF is towards the negative electrode, that is, cathodic flow, then the positively charged solutes are introduced selectively to a greater extent in comparison with neutral and anionic species. Moreover, anions having electrophoretic velocities greater than that of the electro-osmosis do not enter the capillary at all. Therefore, the selection of sampling depends on the type of analytes to be separated.

The sample loading by a hydrostatic mode, also called hydrodynamic or siphoning, is the simplest, most common, and frequently used procedure in most of the commercial CE instruments. In this sampling modality, one end of the capillary is allowed to dip into the sample solution for a specific time interval, for example, between 1 and 30 s. The hydrodynamic flow caused by the hydrostatic pressure introduces a small volume of the sample into the capillary. After this, the end of the capillary is returned back to the buffer vessel and the analysis starts. The hydrostatic method is expected to introduce a real representative aliquot of sample in comparison with the electrokinetic modality. However, reproducibility of sampling is poor if it is done manually and, for this reason, it is convenient to use a fully automated siphon-type sampler.

B. Separation Capillary

The capillary is the most important part of the CE system as the separation depends on the type and nature of the capillary. To obtain good separation, the capillary must meet several requirements, namely, mechanical and chemical stabilities of the capillary are essential along with the good Joule heat releasing capacity for any reproducible analysis. A good transparency to ultraviolet (UV) radiation is most important for on-column optical detection. Nowadays, fused silica capillaries of different lengths and diameters are in use and they meet all of these requirements. Open tubular fused silica capillaries (Polymicro Technologies Inc., Phoenix AZ; Chrompack International B.V. Amsterdam, The Netherlands) are inexpensive and are readily accessible in a wide variety of inner diameters. Capillaries made of other materials,

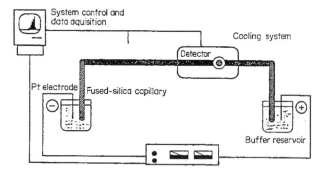

Figure 2 Schematic diagram of CE.

for example, glass or Teflon, are seldom used. Capillaries of rectangular cross-sections, which theoretically possess the best cooling properties and increased path length for absorbance detection, are used only rarely (Thormann et al., 1984; Tsuda et al., 1990a).

Basically, the choice of the diameter and the length of the separation capillary depend on the separation requirements. Generally, longer and narrow capillaries yield the maximum separation efficiency. Normally, capillaries with an inner diameter of 0.2–0.005 mm and a length of 30–100 cm are preferred. The separation efficiencies vary from several thousands to millions of theoretical plates. The narrow bore capillaries are advantageous to use owing to lower Joule heat production. By decreasing the length of the capillary, faster analyses can be carried out, but low volumes of the sample should be loaded to retain a good separation efficiency. The detection window on the capillary can be open by burning off the protection polyimide coating. For the exact and sharp window, two aluminum foil pieces can be used to protect the polyimide layer around the detection window. A detection window of 0.5 mm can easily be prepared by this method. The window can also be prepared by applying a drop of 96–98% sulfuric acid heated to 130°C with the capillary lying between two supporting blocks. The polyimide layer first blackens and then disappears, and the window is then washed with distilled water (Bocek and Chramback, 1991). The window prepared in this manner is not brittle as is the case of the burning method. Moreover, capillary damage risk is low in the acid drop method in comparison with the burning method. Similarly, the window can also be prepared by using a concentrated potassium hydroxide solution. Of course, the window can be prepared by direct scratching of the polyamide layer, but this may damage the capillary and, hence, is not recommended. As discussed earlier, the fused silica capillary has a wide range of applicability; however, coating of the inner surface of capillaries has been reported to meet the requirements for special applications (Landers, 1993). The modified capillaries are hydrophobic, hydrophilic, gel filled, and so on.

C. High-Voltage Power Supply

A high electric field strength is required to facilitate the separation in CE; it ranges from 10 to 100 kV depending on the internal diameter of the capillary, composition of the BGE used, and the nature of analytes to be separated. The resulting driving current rarely exceeds 100 μA; otherwise, capillary overheating would be expected. A reversible polarity, high-voltage power supply is always recommended, with which the electrodes can be grounded, preferably close to the detector. This arrangement generally eliminates any increase in the noise of the detector signal due to high electrostatic tension between the detection end and the detector electronics. A good high-voltage power supply should have a voltage range of 1–20 kV with stability better than 1%, a current range of 0–200 μA, the possibility of working at either constant voltage or current, with reversible polarity, and an interlock for the operator safety.

D. Background Electrolyte

The Background electrolyte (BGE) in CE is used to maintain a high-voltage gradient across the sample containing solution in the capillary. This requires that the conductivity of the electrolyte should be higher than the conductivity of the sample. Therefore, buffers are used as the BGE in most of the CE applications. In addition, the use of buffers is essential to control the pH of the BGE. The electrolyte type and concentration must be chosen carefully for the optimum separation of the analytes. The selection of the BGEs depends on their conductivities and the types of compounds to be analyzed. The relative conductivities of different electrolytes can be estimated from their condosities (defined as the concentration of sodium chloride, which has the same electrical conductance as the substance under study) (Wolf et al., 1987). A wide variety of electrolytes can be used to prepare buffers for CE. Low UV-absorbing components are required for the preparation of the buffers if UV detection is used. In addition, volatile components are required in the case of MS or ICP detection methods. These conditions substantially limit the choice to a moderate number of electrolytes. The pH of the BGE is another factor that determines the choice of the buffers. For low pH buffers, phosphate and citrate have commonly been used, although the latter absorb strongly at wavelengths <260 nm. For basic buffers, borate, Tris, CAPS, and so on, are used as suitable BGEs. A list of useful buffers along with their pHs and working wavelengths are given in Table 1 (Landers, 1993).

E. Detection

Detection is the most important part of any analytical instrument. As with liquid chromatography, CE has been combined ("hyphenated") with many detection devices such as photochemical, electrochemical, fluorescence, conductivity, mass, atomic absorption, inductively coupled plasma, (ICP) and so on. The successful coupling of these detection devices makes CE capable of detecting substances up to 10^{-8}–10^{-9} M levels. Some of the most important detectors used in CE are summarized as follows.

Table 1 Commonly Used Buffers with Their Suitable pHs and Wavelengths for Chiral Separations in CE (Landers, 1993)

Buffers	pH	Wavelength (nm)
Phosphate	1.14–3.14	195
Citrate	3.06–5.40	260
Acetate	3.76–5.76	220
MES	5.15–7.15	230
PIPES	5.80–7.80	215
Phosphate	6.20–8.20	195
HEPES	6.55–8.55	230
Tricine	7.15–9.15	230
Tris	7.30–9.30	220
Borate	8.14–10.14	180
CHES	≥9.50	<190

Note: CHES, 2-(*N*-cyclohexylamino)ethanesulfonic acid; MES, morpholinoethanesulfonic acid; PIPES; piperazine-*N,N′-bis*(2-ethanesulfonic acid); HEPES, *N*-2-hydroxyethyl-piperazine-*N′*-2-ethanesulfonic acid.

1. UV/VIS Absorbance Detectors

The great majority of compounds absorb in the UV and visible regions of spectrum and, hence, more than 50% of analyses have been carried out using photometric detectors in CE. These detectors are based on the absorbance of UV or visible light. The first on-line UV detection in CE was reported by Hjerten (1967). The detection window for light absorbance is established in the capillary as was discussed earlier. Off-line detection systems were also used in CE; as the separated zones leave the capillary, they are pumped to the detection cell of a UV photometer (Hjerten and Zhu, 1987; Tsuda et al., 1983b). The application of off-line detector is limited due to dilution and mutual mixing of previously separated zones.

In these detectors, a narrow beam of radiation from the source passes through the capillary window and is detected by a photosensor (photomultiplier, photodiode). In the case of wide bore capillaries with concentrated solutions of strongly absorbing substances, nonlinearity is important and, in such cases, the recorded peak of a detected zone may not represent the true concentration profile within the zone. The slit width can change the detector response, which is important for stray light rejection and, to increase the linearity of the detection response, the longitudinal height of the slit is important with respect to the efficiency of the separation. Normally, the light source in UV detector is a deuterium (D_2) lamp; However, mercury (254 nm), cadmium (229, 326 nm), zinc (214 nm), and iodine (206, 270 nm) may also be used. A tungsten lamp is generally used for the detection in the visible region. Normally, light intensity reaching the photosensor in CE is slightly reduced in comparison with HPLC detection cells and, hence, small fused silica planoconvex lenses or ball lenses are used to avoid this problem.

Photodiode array detection provides a scanning array detection (Hjerten and Zhu, 1985; Kobayashi et al., 1989; Vindevogel et al., 1990) pattern, which offers several advantages over fixed wavelength detection. In normal photometric detectors, all compounds under analysis are detected at a fixed wavelength; each may have different $\lambda_{\max}$ values and, hence, the detection limits of the detector, for all compounds, cannot be generalized. Conversely, in photodiode array detection, the scanned spectra of all the sample compounds under analysis are recorded by the detector. In a photodiode array detector, radiation from the lamp is focused onto the window of the separation capillary and then it passes to a holographic grating polychromator. The entire wavelength range in the UV and visible regions is scanned and detected by a linear array of photodiodes and, after electronic processing, the signal is evaluated and recorded by a computer. The scan rate, that is, the number of spectra taken during a 1 s interval, depends on the spectral resolution of 2–20 nm.

2. Conductivity Detectors

Conductivity detector (CD) works by measuring the conductance or a voltage drop in the BGE. The CDs are universal and nonselective, that is, they can detect a zone of any substance with an effective mobility different from that of the BGE. The conductivity of a migrating zone differs from that of the BGE and this difference is measured by the detector. If the conductivity of the BGEs ions and the analyte is similar, it is not possible to detect the analyte. The potential gradient detector (PGD) also works on a similar principle; it measures the voltage across two sensing microelectrodes. The net detection signal is proportional to the difference between the electric field strength in the BGE and that in the migrating zone. The surfaces of both CD and PGD cells should be as small as possible to reduce the effects of possible electrode reactions, that is, bubble formation during the analysis. Normally, wires of less than 0.05 mm in diameter of platinum or platinum–iridium alloy are used as electrodes. The construction of detector cell is crucial in CE and is prepared by molding technology to create a detection cell in a block of polyester resin (Foret et al., 1986) or by a laser drilling technique to produce an on-column detection cell directly in a fused silica capillary (Huang et al., 1991). The detection in these detectors depends both on the effective mobility of the analyte and the BGE and, hence, provides useful and complementary information to that provided by a selective detector; thus, they can also be used to identify the zones in CE.

Sometimes, a small potential drop across the sensing electrodes and any current leakage may generate unwanted signals due to the electrochemical reactions resulting in noisy and drifting baseline. In such cases, these signals can be suppressed by the addition of a nonionic detergent into the BGE that forms a protective film on the sensing electrodes. Hydroxypropylcellulose (0.2%) or Triton X-100 can be used for this purpose. Huang et al. (1991) introduced a new conductivity and amperometric detection for CE, which measures the conductivity between the outlet of the separation capillary and the grounded electrode of the power supply. The electrode was made of 50 μm platinum wire fixed in a fused silica capillary. This arrangement avoided the possibility of any electrochemical reaction and improved the stability of the detection signal.

3. Amperometric Detectors

In amperometric detection, the redox potential of an analyte is detected and recorded at the electrode surface. In this technique, the detected ions are chemically oxidized or reduced during their passage through the detection cell. The useful range of potential that can be exploited in practice depends on the solvent and material of the working electrode. The useful range of potential of glassy carbon is from -0.8 to $+1.1$ V vs. Ag/AgCl reference electrode. At a more negative potential, an increase of the background current is obtained, which is due to the reduction of oxygen dissolved in the BGE and the hydrogen overvoltage. The positive potential limit is related to the oxidation of both water and the electrode material. When an electrochemically active solute (oxidizable) is introduced into the electrochemical cell, an electric current is generated, and a characteristic potential ($E_{1/2}$) is obtained, which depends on the nature of the diffusional transport of the solute to the surface of the working electrode. Hence, the further increase of the potential will not increase the detection signal.

In electrophoresis, the separation is achieved by a high-strength electric field and, hence, the placement of amperometric detection cell in the capillary is not possible. Therefore, initially, the separated zones from the outlet of the separation capillary were detected by off-line amperometric detectors (Wallingford and Ewing, 1987a), but, later on, the improved designs for amperometric detection were developed (Enggstrom-Silverman and Ewing, 1991; Gaitonde and Pathak, 1990; Wallingford and Ewing, 1988, 1989) in which the separated zones are transported into the detector connected to the end of a short fused silica detection capillary which is connected to the separation capillary by a specially developed electrically conductive joint made of porous glass. The sensitivity of an amperometric detector is high; it can achieve up to nanomole to attomole detection levels.

4. Thermo-Optical Absorbance Detectors

Thermo-optical absorbance is a laser-based detection technique that is used with CE (Bornhop and Dovichi, 1986; Yu and Dovichi, 1989a). In this detection device, a beam of a high-energy laser is focused onto the separation capillary, and the analytes are excited. Heat is produced in the separated analytes, during nonradiative relaxation of excited states, and the temperature elevation is proportional to the absorbance of the analyte and the laser intensity. The refractive index of the BGE changes with temperature and the heated region acts as a thermal lens. The change in refractive index is measured by another laser beam that is directed across the capillary in the direction perpendicular to the probe beam, and its deflection is monitored by a photodiode. These types of detectors are used only for those substances that strongly absorb the radiation emitted by the laser. Therefore, the analytes should be tagged with a suitable derivatizing agent. The detection limits for these types of detectors can go up to attomole levels.

5. Fluorescence Detectors

In this mode of detection, an analyte molecule in its ground state absorbs a photon and moves to an excited state and then returns back to the ground state while emitting a photon. The excitation radiation from a high-energy source is focused onto the detection window of the separation capillary, and the emitted fluorescence is then measured. The molecules with high absorptivity, good fluorescence quantum yield, and photostability are determined precisely with low limits of detection. In this way, fluorescence detection is more sensitive than light absorbance, and it is more selective. In the simplest arrangements, a high-pressure mercury lamp emitting at 365 nm is used as the excitation source, but in sophisticated detectors, xenon arc lamps, providing a broad spectrum in the entire UV–VIS region, are used as the excitation sources. The fluorescence intensity is directly proportional to the intensity of the excitation radiation and, hence, lasers should be used as the excitation sources. However, laser-based fluorescence detectors have certain drawbacks in comparison with lamp-based detectors. The main disadvantage is that there is no possibility to tune the wavelength of the laser emission to the absorbance maximum of a fluorescent compound and, hence, a limited number of compounds can be detected directly. Therefore, derivatization (pre- or postcolumn) of the compounds, to convert them into fluorescent substances, is required before their detection by this mode.

The most commonly used derivatizing reagents are dansylchloride, naphthalene dialdehyde, fluoresceinisothiocyanate, fluorescamine, phenylthiohydantoin derivatives, 4-chloro-7-nitrobenzofurazan, 2-aminopyridine derivatives, fluorescein derivatives, and 3-(4-carboxybenzoyl)-2-quinolinecarboxyaldehyde (Foret et al., 1993). A good derivatizing reagent should possess a fast reaction rate, good excitation and emission maxima of fluorescence, a good quantum yield, and only one derivative product. In both pre- and postcolumn derivatization, the optimization of the detection is based on a proper selection of the flow rate of the derivatizing reagent; the distance of the detection area from the reaction area, electromigration, electro-osmotic and hydrodynamic flows. The detection limit by these detectors varies from 10^{-8} to 10^{-9} mol/L in CE, but the reported detection limit is 1.2×10^{-20} mol for fluorescein isothiocyanate. However, further improvement of sensitivity in fluorescence detection is expected in the near future with the use of new detection approaches such as two-dimensional charge-coupled devices (Dovichi et al., 1984).

6. Atomic Absorption Spectrometry Detectors

Atomic absorption spectrometry (AAS) was introduced by Walsh in 1950 and, since then, it has become a powerful tool in the quantitative analysis of trace elements. AAS has been coupled, on several occasions, with CE, especially in the application of metal ion analysis and compounds containing metal ions, for example, hemoglobin, cyanocobalamine, and so on. This detection method provides the total concentrations of metal ions and is almost independent of molecular form of the metal ion in the sample. The impurities present in the sample do not affect the detection and, hence, it is very selective. The absorption energy by ground state atoms in the gaseous state forms the basis of AAS. Metal ion atoms are vaporized by a flame in AAS, and atoms in the gaseous form absorb energy from the energy source (the AAS lamp); the absorption of energy is directly proportional to the number of atoms in the separated zone. Various kinds of nebulizers have been developed to couple AAS with CE and this sort of hyphenation approach resulted in the trace analysis of many metal ions (Ergolic et al., 1983; Liu and Lee, 1999; Michalke and Schramel, 1997). The detection limit of this detector has been reported to be from the milligram per liter to the nanogram per liter level.

7. ICP Detectors

ICP spectrometry is a most promising emission spectroscopic technique; commercial ICP systems became available in 1974. In this technique, the carrier gas (argon) is heated to a high temperature (9000–10,000 K), and a plasma is developed in which the excitation of atomic electrons takes place easily and precisely and, hence, ICP is considered to be a better detector than AAS. The metal ion enters into the plasma as an aerosol and the droplets are dried, vaporized and the matrix is decomposed in the plasma. In the high-temperature region of the plasma, atomic and ionic species, in various energy states, are formed. This technique can be coupled with a mass spectrometer (MS); this is thought to be a superior detector with multielemental capabilities and wide linear dynamic range. Quadrupole mass filters are the most common mass analyzers because they are rather inexpensive. Double focusing magnetic/electrostatic sector instruments and time-of-flight mass analyzers are also used (Medina et al., 1993; Timerbaev et al., 1994). Many interfaces have been reported to couple ICP to CE; these couplings resulted in the efficient separations of several metal ions (Magnuson et al., 1992; 1997; Michalke and Schramel, 1997, 1998) with the detection limits ranging from milligram per liter to nanogram per liter.

8. Mass Detectors

In mass detectors, the species under investigation are bombarded with a beam of electrons, which produces ionic fragments from the original species. The resulting assortment of charged particles is then separated according to their masses and charge ratio, and the spectrum produced is called a mass spectrum, which is a record of information regarding various masses produced and their relative abundances. MS has been coupled to CE successfully for many applications.

Despite its immense success, it should be realized that in its standard configuration, that is, equipped with a pneumatic nebulizer for sample introduction and with a quadrupole filter, MS also has a number of important limitations, for example, the occurrence of spectral interferences may hamper accurate trace element determination. The application of a double focusing sector field MS in MS instrumentation offers a higher mass resolution, so that spectral overlap can be avoided to a great extent. Additionally, in a sector field instrument, photons are efficiently eliminated from the ion beam, thereby resulting in very low background intensities, which makes it well suited for extreme trace analysis. Therefore, coupling of an MS to a CE is an important point to be carefully considered, as it represents a new separation dimension. Many papers have been published on this issue, with several designs (Moseley et al., 1990, 1991; Smith and Udseth, 1988; Udseth et al., 1989) and, now, commercial interfaces for coupling of a capillary to a MS are available in the marketplace (Pawliszyn, 1988). Initially, CE–MS coupling was reported by Smith and Udseth (1988), with an electrospray ionization interface for the connection of the separation capillary to a quadrupole MS. The combination of CE–MS can be used for quaternary ammonium salts, azo dyes, vitamins, amino acids, peptides, and even proteins, with good sensitivities (Loo et al., 1989; Smith and Udseth, 1988; Udseth et al., 1989). CE–MS and CE–tandem MS have been reported to have detection limits up to femtomoles for peptides, protein digests, and proteins. The coupling of CE with MS represents one of the most interesting instrumental developments in analytical science. More extensive research is required to explore all the assets of this hyphenated technique in CE.

9. Miscellaneous Detectors

In the early stage of CE development, Lukacs (1983) attempted to use refractive index for detection but, due to the complexity of its signal and its noise and drift, it could not be used successfully. Later on, Chen et al. (1989) and Demana et al. (1990) introduced refractive index detectors for CE. Kaniansky (1981) improved the design of this model of detection. The use of refractive index detectors in CE is not common due to its limited sensitivity.

Another development is the use of radio detectors in which the compounds are detected by measuring their radioactivity and, hence, it is highly selective for detection of only radioactive substances (Pentoney et al., 1989a; Petru et al., 1989). Geiger–Müller and other devices are used for the detection of the separated compounds. Some designs of radio detectors coupled with CE have been described in the literature (Altria et al., 1990; Chen and Morris, 1988; Pentoney et al., 1989b). Radioactive emission is a random process; signal enhancement was achieved by decreasing the driving current at the moment of the detection and, hence, the detection was improved by the prolonged transition time of the detected substances (Kaniansky et al., 1989; Pentoney et al., 1989a, b).

Detection based on Raman spectroscopy relies on the irradiation of a high-energy monochromatic light onto the substances and its measurement after scattering by those substances. Raman bands are a measure of rotational and vibrational bonds in polarizable molecules and, hence, the Raman spectrum is complementary to the classical infrared spectrum. Therefore, the more intense Stokes band at a frequency lower than the incident laser radiation is suitable for the detection. Chen and Morris (1991) designed the first detector for CE based on Raman spectroscopy. Later, Christensen and Yeung (1989) updated this design for multichannel Raman spectroscopic detection. The reported sensitivity was 500 attomoles of methyl orange, corresponding to a concentration of 10^{-7} M.

Chiral optical activity detection is the method of choice in analytical science to determine the optical purities of enantiomers. This detector operates on the principle of rotation of polarized light by enantiomers. The sensitivity of this detector is poor as it works with higher concentrations of the substances and, hence, is seldom used in CE.

10. Indirect Detection

In addition to the direct detectors, indirect detection has also been used in CE, due to its universality. In indirect detection, a specific property of the BGE constituent, such as UV absorbance or fluorescence or electrochemical is measured. The detection signal is achieved owing to a decrease of this background signal because of the migration of the zone for a substance, which replaces the detection active component of the BGE. Initially Hjerten (1967) attempted to use the indirect method of detection in CE but, later on, the use of indirect detection became popular. The application of indirect detection in CE includes the use of UV photometric (Kuhr and Yeung, 1988; Nardi et al., 1990), fluorescence (Garner and Yeung, 1990; Gross and Yeung, 1989; Yeung and Kuhr, 1991), and amperometry (Dovichi et al., 1983; Olefirowicz and Ewing, 1990). Basically, most of the detection methods can be used in the indirect mode. In the indirect mode of detection, the selection of BGE is important; a BGE with co-ion or counter ion should be used. To minimize any zone broadening, sample concentration should be lower than the BGE (about 100-fold). The application range of indirect detection is virtually unlimited, ranging from simple inorganic ions to macromolecules with the detection limit up to femtomole levels (Ali and Aboul-Enein, 2002a, b; Foret et al., 1990). In spite of the universal nature of indirect detection, it suffers from certain disadvantages. For example, the samples with ions having strong direct signal may create detection problem due to the mutual overlap of separated zones; this becomes a serious problem when complex real samples are to be analyzed.

V. DATA INTEGRATION

Data integration is the final important component of a CE system. It provides the direct information about the analysis results. Initially, detectors were coupled to manual integrators or recorders to obtain the electropherograms but, nowadays, computer-based software is available with which the integration of the analysis results can be carried out precisely and accurately. The electropherograms can be obtained at different values of data collection, which provides the accurate information about the analysis. Various manufactures providing the sophisticated data integration units with CE instruments include Waters, USA, Bio-Rad, USA, Lab Alliance, USA, Agilent Technologies, USA, Groton Technology, Inc., USA, Lumex, Russia, Beckman Coulter Ltd., UK, Picometrics, France, and others. As with chromatography, the results of the analysis are reported in terms of retention times, capacity, separation, and resolution.

VI. SAMPLE PREPARATION

If the analysis is to be carried out with unknown samples, for example, of biological or environmental origins, the sample preparation becomes an important issue in CE. In spite of thousands of applications, little attention has been given to sample preparation methodology. The samples containing highly ionic matrices may cause problems in CE, and EOF in the capillary can be altered by the influence of the sample matrix, resulting in poor resolution. Additionally, the detector baseline is usually perturbed when the pH of the sample differs greatly from the pH of the BGE. The samples containing UV-absorbing materials are also problematic in the detection of components from unknown matrices. Owing to these problems, some workers have suggested sample cleanup processes involving various sample extraction procedures (Dabek-Zlotorzynska et al., 2001; Haddad et al., 1999; Malik and Faubel, 2001; Martinez et al., 2000). Real samples often require the application of simple procedures such as filtration, extraction, dilution, and so on. Electromigration sample cleanup suffers from severe matrix dependence effects; even then, it has been used for preconcentration in inorganic analysis. Sample treatment methods have been discussed in several reviews (Haddad et al., 1999; Liu et al., 1999; Pacakova et al., 1999; Valsecchi and Polesello, 1999). In addition, several reports have been published on dialysis and electrodialysis for sample cleanup prior to CE injection (Haddad et al., 1999; Liu et al., 1999; Pacakova et al., 1999; Valsecchi and Polesello, 1999). Several on-line sample preparation techniques have been developed and are used in CE. Whang and Pawliszyn (1998) designed an interface that enables a solid-phase microextraction (SPME) fiber to be inserted directly into the injection end of a CE capillary. They prepared a semi "custom made" polyacrylate fiber to reach the SPME–CE interface (Fig. 3). Similarly, Kuban and Karlberg (1999) have also reported an on-line dialysis/FIA sample cleanup procedure for CE (Fig. 4). Additionally, some reviews have appeared, which deal with sample preparation in CE (Dabek-Zlotorzynska et al., 2001a; Ezzel et al., 1995; Majors, 1995; Martinez et al., 2000).

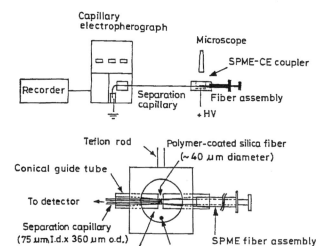

Figure 3 Schematic diagram of the solid-phase micro extraction capillary electrophoresis (SPME-CE) system (Whang and Pawliszyn, 1998).

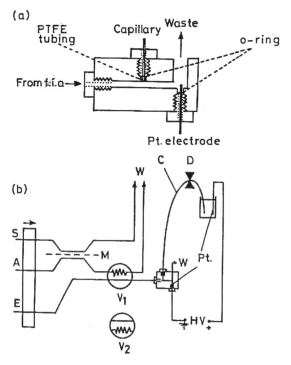

Figure 4 The setup of the on-line dialysis (sample purification) with CE: (a) cross-sectional view of the FIA–CE interface and (b) schematic diagram of the FIA–CE system; A, acceptor stream; C, capillary; HV, high voltage; D, UV detector; E, electrolyte; M, dialysis membrane; Pt, platinum electrode; S, sample, V_1, injection valve in filling position; V_2, injection valve in inject position; and W, waste (Kuban and Karlberg, 1999).

VII. METHOD DEVELOPMENT AND OPTIMIZATION

Method development in CE starts with the selection of suitable BGE, applied voltage, detector, and so on. First of all, the structures of the substances to be analyzed should be explored and a detection mode should be finalized, based on the expected identities of the sample components. If the analytes are UV sensitive, λ_{max} should be determined and used during experiments. The selection of BGE should be made, keeping in mind the wavelength cutoffs of the buffers (Table 1). Initially, 10 kV applied voltage should be used, followed by its optimization as per the requirements of the separation determined experimentally. The separation in CE is optimized by controlling various experimental parameters. These factors may be categorized into two classes, that is, the independent and the dependent. The independent parameters are under the direct control of the operator. These parameters include the choice of the buffer, pH of the buffer, ionic strength of the buffer, applied voltage, temperature of the capillary, dimension of the capillary, and BGE components and additives. In contrast, the dependent parameters are those that are directly affected by the independent parameters and are not under the direct control of the operator. These parameters are field strength (V/m), EOF, Joule heating, BGE viscosity, sample diffusion, sample mobility, sample charge, sample size and shape, sample interaction with capillary, and BGE molar absorptivity. A brief experimental protocol for the separation optimization in CE is given in Sch. 1.

VIII. METHODS VALIDATION

CE is rapidly becoming a preferred fast, efficient, and versatile technique in analytical science, but it still requires improvements in its working conditions to improve reproducibility. To the best of our knowledge, there are only few reports available in the literature dealing with method validation as related to the determination of the compounds by CE (Dabek-Zlotorzynska et al., 2001a; Liu et al., 1999; Macka and Haddad, 1999; Pacakova et al., 1999; Valsecchi and Polesello, 1999). Therefore, at the present time, more emphasis is required in method validation. For routine analysis, it is essential to keep the migration times constant in order to allow automation of peak identification by means of commercial data analysis software. Automatic peak identification and quantification are only possible if the relative standard deviation of migration times is less than 0.5% (Raber and Greschonig, 2000). Several reviews have been published, which describe improvements in the reproducibility of the results (Dabek-Zlotorzynska et al., 1998, 2001b; Dabek-Zlotorzynskaa et al., 1998; Doble and Haddad, 1999; Faller and Engelhardt, 1999; Ikuta et al., 1999; Macka et al., 1999; Timerbaev, 1997). Another approach to qualitative and quantitative reproducibilities is the transformation of the total time x-scale of electrophoretic data into the corresponding effective mobility scale (μ-scale) (Pacakova et al., 1999; Schmitt-Kopplin et al., 2001). The conversion leads to a better interpretation of the obtained electropherograms in terms of separation, and enables better direct comparison of the electropherograms and easier peak tracking when trying to identify single components from complex matrices, especially when UV–VIS signatures of the components are also available (Pacakova et al., 1999). A quantitative improvement was achieved in the μ-scale with significantly better peak area precision, which equates to better precision in quantitative analysis than with the primary time scale integration. However, electrophoretic-based data processing CE software is needed to be able to directly handle the electrophoretic data. Therefore, it may be assumed that the selectivity of different compounds by CE is quite good; however, the reproducibility is still a problem. Many attempts have been made to solve this problem. Organic solvents have been added into the BGE to improve reproducibility. It has been reported that the organic solvents ameliorate the solubility of the hydrophobic complexes, reduce their adsorption onto the capillary wall, regulate the distribution of the substances between aqueous phase and micellar phase, adjust the viscosity of the separation medium, and, accordingly, accomplish the improvement in the reproducibility. The constant pH of the electrolyte solution is very important from the selectivity and reproducibility points of view. pH controls the behavior of the EOF, acid/base dissociation equilibria of complexes, and the states of existing complexes. Therefore, the selectivity and reproducibility can be improved by adjusting and controlling the pH of the BGE.

IX. APPLICATION

CE and related technologies have achieved a status of high separation efficiency, high sensitivity, short analysis time, and low running cost techniques. They have a very broad spectrum of applications, ranging from small ions to macromolecules. CE and related technologies have been used to separate biomolecules, pollutants, and chiral compounds. In biological science, the applications of these include the separation of amino acids, peptides, proteins, enzymes, DNA, carbohydrates, vitamins, β-blockers, profens, antibiotics, toxins, dyes, organic acids, amines, alcohols, and several other drugs and pharmaceuticals, whereas environmental analysis comprises toxic metal ions, pesticides, phenols, plasticizers, polynuclear aromatic

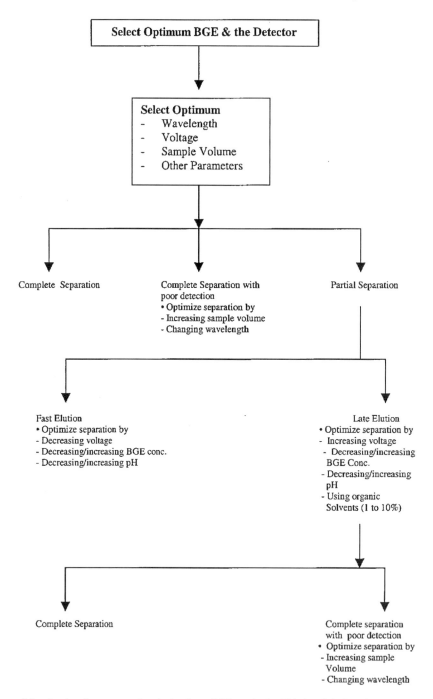

Scheme 1 The protocol for the development and optimization of CE methods. This is a brief outline of the procedure in CE analysis. However, other variations may be carried out.

hydrocarbons (PAHs), and other toxic pollutants. Today, CE has emerged as one of the most important analytical techniques in chiral separation of drugs, pharmaceuticals, pollutants, and other compounds. Unlike chromatography, the chiral selectors in CE are used in the BGE and, hence, called BGE additives. Thousands of applications dealing with the analysis of a variety of compounds using CE and analogous techniques are available in the literature and, hence, many reviews, monographs, and books have been published (Ali et al., 2003; Chankvetadze, 1997; El-Rassi, 2002; El Rassi and Giese, 1997; Foret, 1993; Krull et al., 2000; Landers, 1993; Neubert and Rüttinger, 2003; Rathore, 2003; Wehr et al., 1993). Therefore, it is impossible to cite all of them in this chapter; however, some important applications of CE have been summarized in Table 2.

Table 2 Applications of CE and Related Technologies

Compounds	Sample matrix	Electrolytes	Detection	References
		Capillary Electrophoresis		
Analysis of Biomolecules				
Simple Mixture Analysis				
Amines				
Tetramethyl ammonium bromide, tetramethyl ammonium perchlorate, tetrapropyl ammonium hydroxide, tetrabutyl ammonium hydroxide, and trimethylphenyl ammonium iodide	—	0.1 mM KCl in 50% methanol	MS	Olivers et al. (1987)
Polyamines	—	Formic acid, 0.01% ED, 5% EG	Fluorescence	Tsuda et al. (1988)
Alkylamines	—	0.5 M PB	Fluorescence	Jorgenson and Lukacs (1981a)
Histamine	Wine	0.1 M BB, 0.2 M KCl, pH 9.5	Post column fluorescence	Rose and Jorgenson (1988)
Putrescine, cadaverine, spermidine, and spermine	—	0.005 M BB with 0.1% ED, 2% SDS, and 5% EG	Post column fluorescence	Tsuda et al. (1990b)
Acids				
Formic, acetic, propionic, and butyric	—	0.025 M Na veronal, pH 8.6	Indirect UV 225 nm	Deml et al. (1985)
Malonic, lactic, aspartic, glutamic glucornic, hydrochloric, and phosphoric	—	0.02 M Benzoic acid, histidine, pH 6.2, 0.1% triton	Indirect UV 254 nm	Foret et al. (1989)
Benzoic, benzilic, and naphthoic	—	0.1 M Tris–acetic acid, pH 8.6, 20% dextran	UV 205 nm	Hjerten et al. (1989)
Picric, cinnamic, and sorbic	—	0.01 M KCl, pH 5.6, ME: 0.01 M HCl	UV 254 nm	Bocek et al. (1990)
Hippuric and gibberilic	—	0.1 M Ammonium acetate, MeCN (1:9)	—	Lee et al. (1989)
Carbohydrates				
Neutral	—	0.2 M Boric acid, KOH, pH 5.0	UV 240 nm	Honda et al. (1986)
Oligosaccharides	—	0.01 M Na_2HPO_4, 0.01 M $Na_2B_4O_7$, pH 9.4	Fluorescence	Liu et al. (1991)
Cyclodextrins	—	0.03 M Benzoic acid, Tris, pH 6.2	Indirect UV 254 nm	Nardi et al. (1990)
Amino acids				
All	—	1.0 mM PB, pH 5.31 or 7.12	—	Enggstrom-Silverman and Ewing (1991)
All	—	0.2 mM Na salicylate, 0.04 mM Na_2CO_3, NaOH, pH 9.7	Indirect fluorescence	Kuhr and Yeung (1988)
Debsyl derivatives	Urine	0.02 M PB, pH 7.0, MeCN (1:1)	—	Yu and Dovichi (1989b)
Fluorescamine derivatives	—	10% Propanol, pH 10.16	Fluorescence	Wallingford and Ewing (1987b)

(*continued*)

Table 2 *Continued*

Compounds	Sample matrix	Electrolytes	Detection	References
2,4-Dinitrophenol derivatives	—	LE: 0.02 M HCl, histidine, pH 5.5, 0.1% HEC, TE: 0.01 M MES, histidine, pH 5.3	Visible 405 nm	Kaniansky and Marak (1990)
NDA derivatives	—	0.01 M Boric acid, 0.02 M KCl, pH 9.5	Fluorescence	Nickerson and Jorgenson (1988)
FITC derivatives	—	0.05 M BB, pH 9.0	Fluorescence	Swedler et al. (1991)
	—	0.2 M BB, pH 7.8	Fluorescence	Pentoney et al. (1988)
o-OPA derivatives	—	0.05 M BB, 0.05 M KCL, pH 9.5	Fluorescence	Gassmann et al. (1985)
PTH-derivatives	—	0.033 M Borax, 0.113 M boric acid, 0.1 M SDS, pH 8.6	UV 220 nm	Tehrani and Day (1989)
Dansyl derivatives	—	0.05 M PB, pH 7.0	Fluorescence	Jorgenson and Lukacs (1981a, c, d)
Peptides and proteins				
Dipeptides	—	0.15 M H_3PO_4, pH 1.5	UV 190 nm	McCormick (1988)
Tri- and pentapeptides	—	5 mM Ammonium acetate, NaOH, pH 8	MS	Moseley et al. (1989)ss
Oligopeptides	—	0.05 M BB, pH 9.5	Fluorescence	Gassmann et al. (1985)
β-Lactoglulin	—	0.02 M Citrate buffer, pH 2.5	UV 200 nm	Grossman et al. (1988, 1989)
Ovalbumin and oligopeptide	—	0.0125 M PB, pH 6.86	Fluorescence	Bushey and Jorgenson (1990)
Digest of β-casein	—	0.25 M Ammonium acetate, pH 7.2	UV 210 nm	Tehrani et al. (1991)
Bovine	—	0.1 M Formate buffer, pH 4.0	UV 230 nm	Lukacs (1983) and Jorgenson (1984)
Mycoglobin, cytochrome C, and lysozyme	—	0.03 M KH_2PO_4 buffer, pH 3.8	UV 205 nm	Bruin et al. (1989)
Ribonuclease A, B_1, and B_2	—	0.02 M CAPS buffer, pH 11.0	UV 200 nm	Grossman et al. (1989)
Albumin	Human serum	0.02 M CHES, 0.01 M KCl, KOH, pH 9.0	UV 220 nm	Hecht et al. (1989)
Different proteins	Human serum	0.05 M Na borate, pH 10.0	UV 200 nm	Gordon et al. (1991)
HGH	—	0.1 M PB, pH 2.56	UV 200 nm	Frenz et al. (1989)
Amylase, lipase, and trypsin	—	0.05 M PB, pH 7.0	UV 230 nm	Jorgenson and Lukacs (1985)
Insulin	—	0.09 M Tris–NaH_2PO_4, pH 8.6 + 8.0 M urea, 0.1% SDS	UV	Cohen and Karger (1987)
Different proteins	Human urine	0.1 M Tris–acetic acid, pH 8.6	UV 280 nm	Hjerten (1983)
Nucleic acids				
E. coli t-RNA	E. coli	0.1 M Tris–acetic acid, pH 8.6	UV 280 nm	Hjerten and Hirai (1984)
P(dA)$_{40-60}$	—	0.1 M Tris, 0.25 M borate, 7 M urea, pH 7.6	UV 260 nm	Guttman et al. (1990)
SV40 DNA	—	9 mM Tris–9 mM boric acid, pH 8.0, 2 mM EDTA	Fluorescence	Kasper et al. (1988)
Plasmid DNA	Plasmid	89 mM Tris, 89 mM boric acid, 2 mM EDTA	UV induced fluorescence	Zhu et al. (1989)
DNA, size standards	—	89 mM Tris, 89 mM boric acid, 2.5 mM EDTA	UV 260 nm	Bocek and Chramback (1991)

Drugs				
Profen group		PB, pH 6.1–8.4	UV 215 nm	Wainright (1990)
Sulfa groups		0.03 M, PB, pH 7.0	UV 215 nm	Lux et al. (1990)
Alkaloids		0.05 M PB, pH 7.0	UV 245 nm	Debets et al. (1990)
Toxins		PB, pH 8.7	Fluorescence	Wright et al. (1989)
Antibiotics	Human serum	BB, pH 10.0, 100 mM SDS	—	Kitihashi and Fruta (2001)
Vitamins		0.05 M Tris–borate buffer, pH 8.4, 0.05 M SDS	UV 200 nm	Fujiwara et al. (1988)
β-Blockers		Na$_2$B$_4$O$_7$–H$_3$BO$_3$ (50 mM), pH 9.0	UV	Maguregui et al. (2002)
Histamines		BB, pH 9.0, 100 mM SDS	UV 210 nm	Nishiwaki et al. (2000)
Analgesics	Human urine	PB, 60% EG, pH 7.9	MS	Wey and Thorman (2002)
Anti-allergics		PB	UV	Bernal et al. (1998)
Steroids		PB	UV	Mayer et al. (2002)
Chiral mixture analysis				
Amine drugs		0.1 M BB, pH 9.5, β-CD	UV	Leroy et al. (1995)
Amino acids (AA)		PB with β-CD	UV	Yoshinaga and Tanaka (1994)
AA-dansyl derivatives		PB with β-CD	UV	Schmitt and Engelhardt (1995)
AA-2,4-dinitrophenyl derivatives		PB with 6-amino-6-deoxy-β-CD and its N-hexyl derivatives	UV	Egashira et al. (1996)
AA-AQC derivatives		PB with vancomycin	UV	Grasper et al. (1996)
AA-FMOC derivatives		PB with vancomycin	UV	Grasper et al. (1996)
AA-PTH derivatives		PB with vancomycin	UV	Grasper et al. (1996)
AA-AEOC derivatives		PB with γ-CD	UV	Wan et al. (1996)
β-Blockers		PB with TM-β-CD	UV	Wren (1993)
Profens		PB with β-CD	UV	Lelievre and Gariel (1996)
Dipeptides		PB with (+)-18-crown-6-tetracarboxylic acid	UV	Schmidt and Gübitz (1995)
Antifungals		PB with CDs	UV	Chankvetadze et al. (1995)
Alkaloids		PB with CDs	UV	Swartz (1991)
Benzoin		PB with HSA	UV	Ahmed et al. (1996)
Anti-allergic		PB with CDs	UV	Quang and Khaledi (1995)
Diltiazem		PB with dextran sulfide	UV	Nishi et al. (1995)
Tripeptides		PB with crown ether	UV	Kuhn et al. (1995)
Omeprazole		PB with BSA	UV	Eberle et al. (1997)
Baclofen		PB with β-CDs	UV	Ali and Aboul-Enein (2003)
Thalidomide		PB with β-CDs	UV	Aumatell et al. (1993)
Warfarin		PB with saccharides	UV	D'Hulst and Uerbeke (1992)
Environmental Pollutants Analysis				
Simple mixture analysis				
Alkyl phenols		1.25 mM Na$_2$B$_4$O$_7$, 15 mM NaH$_2$PO$_4$ pH 11.0 with 0.001% HDB	UV 254 nm	Martinez et al. (2000) and Crego and Marina (1997)

(*continued*)

Table 2 Continued

Compounds	Sample matrix	Electrolytes	Detection	References
Chlorophenols		50 mM Na$_2$HPO$_4$/NaH$_2$PO$_4$, pH 6.9	UV 214 nm	Martinez et al. (2000) and Crego and Marina (1997)
Miscellaneous derivatives of phenols		10 mM Na$_2$B$_4$O$_7$/Na$_3$PO$_4$, pH 9.8	UV 210 nm	Crego and Marina (1997) and Dabek-Zlotorzynska (1997)
Pentachlorophenols	Drinking water	40 mM Sodium borate, pH 10	UV	Martinez et al. (2000) and Crego and Marina (1997)
Hexazinone and its metabolite	Ground water	50 mM SDS, 12 mM sodium phosphate, 10 mM sodium borate, and 15% MeOH, pH 9.0	UV 220–247 nm	Malik and Faubel (2001)
PAHs		8 mM Na$_2$B$_4$O$_7$, pH 9.0, 50 mM DOSS,	UV	Dabek-Zlotorzynska (1997)
Methyl, dimethyl, trimethyl, and ethyl amines	Atmospheric aerosols	5 mM DHBP, 6 mM glycine, 2 mM 18-crown-6 ether, pH 6.5	Indirect UV 280 nm	Dabek-Zlotorzynska (1997)
Acetaldehyde, benzaldehyde, formaldehyde, and glyoxal	Rain water	5 mM Na$_3$PO$_4$–10 mM Na$_2$B$_4$O$_7$, pH 8.0, 20% acetonitrile	Laser-induced fluorescence 325 and 442 nm	Dabek-Zlotorzynska (1997)
Synthetic cationic dyes		10 mM Citric acid, pH 3.0, 0.1% PVP	UV 214 nm	Martinez et al. (2000) and Dabek-Zlotorzynska (1997)
Synthetic anionic dyes		10 mM BTP–HCl, pH 6.5, 0.5% PEG,	UV 214 nm	Martinez et al. (2000) and Dabek-Zlotorzynska (1997)
Chiral mixture analysis				
Fenoprop, mecoprop, and dichloroprop.		20 mM Tributyl-β-CD in 50 mM, ammonium acetate, pH 4.6	MS	Otsuka et al. (1998)
Phenoxy acid herbicides		50 mM Sodium acetate, 10 mM dimethyl-β-CD, pH 4.6	UV	Penmetsa et al. (1997)
2-(2-Methyl-4-chloro-phenoxy) propionic acid		0.05 M LiAcO$_2$ with α-CD	UV 200 nm	Nielen (1993)
1,1-Binaphthol, 1,1'-binaphthyl-2-2'-dicarboxylic acid, and 1,1'-binaphthyl-2,2'-dihydrogen phosphate		0.006 M BB (pH 9.0), CDs	UV 254	Nishi (1995)
1,1-Binaphthol, 1,1'-binaphthyl-2,2'-dicarboxylic acid, and 1,1'-binaphthyl-2,2'-dihydrogen phosphate		PB with α-, β-, and γ-CDs	UV 214 nm	Kano et al. (1995)
1,1'-Binaphthyl-2,2-dicarboxylic acid, 1,1'-binaphthyl-2,2-dihydrogen phosphate, and 2,2'-dihydroxy-1,1'-binaphthyl-3,3'-dicarboxylic acid		0.04 M Carbonate buffer (pH 9.0), noncyclic oligosaccharides	UV 215–235 nm	Desiderio et al. (1997)

Analyte	Sample	Buffer	Detection	Reference
Metal ions analysis				
Alkali and alkaline earth metals	Tap and mineral waters	10^{-2} M Imidazole (pH 4.5)	Indirect UV 214 nm	Riviello and Harrold (1993)
Zn and other transition metals	Tap water	10 mM BB, 0.1 mM HQS (pH 9.2)	UV 254 nm	Timerbaev et al. (1993)
Ca, Mg, Ba, Na, K, and Li	Mineral water	3–5 mM Imidazole, pH 4.5	Indirect UV 214 nm	Beck and Engelhardt (1992)
Cu, Ni, Zn, Li, Na, K, Mg, Ca, and Sr	Tap water	6.5 or 8 mM HIBA, pH 4.4	Indirect UV 185 and 214 nm	Weston et al. (1992)
Cu, Ni, Co, Hg, Mn, Fe, Pb, Pd, Zn, Cd, Mg, Sr, Ca, and Ba	River water	2 mM $Na_2B_4O_7$, 2 mM EDTA, pH 4.4	UV 200 nm and 214 nm	Motomizu et al. (1992)
Ba, Ca, K, Mg, Na, and Li	Mineral water	10 mM Imidazole, 1 mM TBABr, pH 4.5, or 10 mM benzylamine, pH 9.0	Indirect UV 204 and 214 nm	Riviello and Harrold (1993)
Cs, K, Li, Ba, Sc, Ca, Mg, and Na	Rain water and cola beverage	500 μM $CsCl_3$, 2 mM 18CR6, 350 μM $CsCl_3$, 150 μM $Cs_2(SO_4)_3$, 2 mM 18C6, 500 μM $Cs_2(SO_4)_3$, 2 mM 18CR6	Fluorescence 251/346 nm	Bachmann et al. (1992)
Organic Pb, Hg, and Se	Tap and sea water	40 mM NaH_2PO_4–$Na_2B_4O_7$, 2.5 mM TTHA, 2 mM SDS (pH 7.5)	UV 200 nm	Liu and Lee (1998a)
Metal ions speciation				
Arsenic species	Drinking water	0.025 mM PB, pH 6.8	UV 190 nm	Lopez-Sanchez et al. (1994)
Cr(IV) and Cr(VI)	Rinse water from chromium platings	1 mM CDTA, 10 mM formate buffer, pH 3.8	UV 214 and 254 nm	Semenova et al. (1996) and Timerbaev et al. (1996)
Fe(II) and Fe(III)	Electroplating water Thermal water	20 mM PB, pH 7.0	UV 214 nm	Aguilar et al. (1989)
Se(IV) and Se(V)		Chromate, 0.5 mM TTAOH, pH 10.5	Indirect UV 254 nm	Gilon and Potin-Gautier (1996)
Co(III) and Co(IV)		NaAc, 0.12 mM5-Br-PAPS, pH 4.9	Indirect visible 550 nm	Motomizu et al. (1994)
Hg(II), CH_3Hg^+, and $CH_3CH_2Hg^+$ TML, TEL, DPhL, and Pb(II)	Tape water	25 mM $Na_2B_4O_7·10H_2O$, pH 9.3 PB	ICP-MS	Silva da Rocha et al. (2000)
Pt(II) and Pt(IV)		0.1 M KSCN, pH 3.0	UV	Liu and Lee (1998b)
Rh species		10 mM KCl–HCl	UV 305	Hamáek and Havel (1999)
			UV 200 nm	Aleksenko et al. (2001)
V(IV) and V(V)	Electroplating bath	20 mM Na_2HPO_4, 5 mM DTPA, pH 8.0 or 8.5	Indirect UV 241 nm	Padarauskas and Schwedt (1997)
Micellar electrokinetic chromatography				
Simple mixture analysis				
Alkyl amines	—	0.01 M Na_2HP_4, 0.006 M $Na_2B_4O_7$, 0.05 SDS, pH 7.0	Fluorescence	Powell and Sepaniak (1992)
Amino acids (AA)	—	0.05 M BB, pH 9.5, MeOH, 2% THF, 0.05 M SDS	Fluorescence	Liu et al. (1988)
AA-PTH derivatives	—	0.1 M BB, 0.05 PB, 0.1 SDS, pH 7, 4.3 M urea	UV 260 nm	Terrabe et al. (1991)
Protein mixtures	—	BB, pH 10.5, 0.8 M SDS	UV 220 nm	Deyl et al. (1989)

(continued)

Table 2 Continued

Compounds	Sample matrix	Electrolytes	Detection	References
Nucleotides	—	6.7 M Tris, 6.7 mM PB	UV 254	Tehrani et al. (1991)
dAMP, dGMP, dTMP, and dCMP	—	0.005 M Na$_2$HP$_4$, 0.02 M Tris, 0.05 SDS	UV 254 nm	Row and Row (1990)
Vitamins	—	0.01 M Na$_2$HP$_4$, 0.006 M Na$_2$B$_4$O$_7$, 0.05 SDS	Fluorescence	Burton et al. (1986)
Antibiotics	—	0.02 M PB–BB, pH 9.0, 0.1 M SDS	UV 220 nm	Nishi et al. (1989)
Analgesics	—	0.02 M PB, pH 7.0, 0.05 M SDS	UV 214 nm	Fujiwara and Honda (1987)
Sulfa drugs	—	0.02 M PB, pH 7.0, 0.05 M SDS	UV 210 nm	Lux et al. (1990)
Alkaloids	—	8.5 mM PB, 8.5 mM borate, 85 mM SDS	UV 210 nm	Weinberger and Lurie (1991)
Phenols	—	PB–BB, pH 7.0, 0.05 M SDS	UV 270	Terabe et al. (1984a)
Herbicides	—	0.01 M Na$_2$HP$_4$, H$_3$PO$_4$, pH 6.75, 0.05 M SDS	UV 254 nm	Rasmussen and McNair (1989)
Chiral mixture analysis				
Amines	—	PB with k,k'-tartaric acid	UV	Dalton et al. (1995)
Amino acids (AA)		PB with Marfey's reagent	UV	Tan et al. (1990)
AA-dansyl derivatives		PB with β-CDs	UV	Terrabe et al. (1993)
AA-PTH derivatives		PB with Na-N-dodecanoyl-L-valinate with SDS and MeOH	UV	Otsuka et al. (1991)
Carboline derivatives		PB with bile salts	UV	Nishi et al. (1990)
Anti-allergic		PB with β-CDs	UV	Ong et al. (1991)
Dipeptides		PB with β-CDs	UV	Wan and Blomberg (1997)
Profens		PB with vancomycin	UV	Rundlet and Armstrong (1995)
β-Blockers		PB with β-CDs	UV	Aumatell and Wells (1994)
Phenoxy acid herbicides		PB and acetate buffers with n-octyl-β-D-maltopyranoside and CDs	UV	Mechref and El Rassi (1996a)
Phenoxy acid herbicides		0.4 m Na borate (pH 10) with N,N-Bis(3D-gluconamidopropyl) deoxycholamide	UV 240 nm	Mechref and El Rassi (1996b)
PCBs		0.09 M CHES, 0.11 M SDS, 2 M urea (pH 10) with β- and γ-CDs	UV 235 nm	Marina et al. (1996)
PCBs	—	CHES with SDS, bile salts, and CDs	—	Crego et al. (2000) and Lin et al. (1999)
1,1'-Bi-2-naphthol		0.016 M NaCl + MeOH (pH 8.1–8.3) with sodium deoxycholate and polyoxyethylene ethers	UV 210 nm	Clothier and Tomellini (1996)
1,1'-Bi-2-naphthol, 1,1'-binaphthyl dicarboxylic acid, and 1,1'-binaphthyl diyl hydrogen phosphate		PB with sodium cholate, sodium deoxycholate, and sodium taurodeoxycholate	UV (Laser etched flow cells and modified detector)	Cole et al. (1990)
Diniconazole and uniconazole		0.1 M SDS, 2 M urea, 0.1 M BB and 5% 2-Me-2-PrOH, CDs	UV 254 nm	Furut and Doi (1994)
1,1'-Bi-2-naphthol, 2,2,2'-trifluoro-1-(9-anthryl)ethanol and 1,1'-binaphthyl-2,		0.05 SDS, 0.02 M PB (pH 9), Na d-camphor sulfonate with γ-CD & β-CD derivatives	UV 220 nm	Nishi et al. (1991)

2'-diyl hydrogen phosphate			
1,1'-Bi-2-naphthol and 1,1'-binaphthyl-2,2'-diyl hydrogen phosphate	0.025 M BB (pH 9.0) γ-CD, poly-(Na N-undecylenyl) D-valinate	UV 280 nm	Wang and Warner (1995)
Metal ions analysis and speciation			
Transition metal ions	0.05 M Na₂HP₄, 0.00125 M Na₂B₄O₇	Visible 500 nm	Saitoh et al. (1989)
As species	10 mM Dodecyltrimethyl-ammonium phosphate, pH 8.0	UV 190 nm	Wuping and Hian (1998a)
Fe(II) and Fe(III)	1 mM Ammoium phosphate buffer, 75 mM SDS, 0.1 mM MPAR, pH 8.0	UV 254 nm	Timerbaev et al. (1994)
Pb(II), triethyl lead(IV), trimethyl lead(IV) and diphenyl lead(IV)	Na₂HPO₄–Na₂B₄O₇, 2.5 mM TTHA, 2.0 mM SDS, pH 7.5	UV 220 nm	Wuping and Hian (1998b)
Se(IV), phenyl-selenium(II), and diphenyl selenium(II)	Na₂HPO₄–Na₂B₄O₇, 2.5 mM TTHA, 2.0 mM SDS, pH 7.5	UV 220 nm	Wuping and Hian (1998b)
Organptin	PB of different pHs and with different concentration of SDS	—	Li et al. (1995)
Capillary electro-chromatography			
Simple mixture analysis			
Alkaloids	MeCN–10 mM Tris, pH 8.3 (80:20)	UV 214 nm	Wei et al. (1998)
Antidepressants	MeCN/50 mM NaH₂PO₄ (pH 2.3)/water (60:20:20)	UV 210 nm	Smith and Evans (1995)
Benzamide	MeCN/10 mM MES, pH 3.0	UV 200 nm	Altria et al. (1998)
Norgestimate drugs	MeCN/THF/25 mM Tris–HCl, (pH 8.0)/water (35:20:25)	UV 225 nm	Wang et al. (1998)
Steroid hormones	MeCN/10 mM borate, pH 8.0 (65:35)	UV 205 nm	Huber et al. (1997)
Parabens	MeCN/25 mM MES, pH 6.0 (80:20)	UV 250 nm	Dittmann and Rozing (1996)
PAHs	MeCN/25 mM Tris, pH 9.0	Fluorescence	Dadoo et al. (1998)
Explosive compounds	MeOH/10 mM MES	UV 254 nm	Bailey and Yan (1998)
Chiral mixture analysis			
Amino acids, dansyl and dinitrophenyl derivatives	MeOH/PB (15:85) and MeOH/15 mM TEA, pH 4.7 (15:85), β-CD	UV	Li and Lloyd (1994)
β-Blockers	MeCN/4 M acetate, pH 3.0 (80:20), imprinted polymers	UV	Schweitz et al. (1997a, b)
Benzoin and temazepam	HAS, 2-PrOH/4 mM PB, pH 7, β-CD, and others	UV and others	Lloyd et al. (1995)
Barbiturates	MeOH/5 mM PB, pH 7.9 (80:20), β-CD	UV	Wistuba et al. (1998)
2-Phenylpropionic acid and warfarin	PB with CDs	UV	Koide and Ueno (2000)
Dichloroprop.	PB with vancomycin	UV	Fanali et al. (2001)

Note: AEOC, 2-(9-anthryl)-ethylchloroformate; AQC, 6-aminopropylhydroxy succinimidyl carbamate; dAMP, 2'-deoxyadenosine-5'-monophosphate; BB, borate buffer; CAPS, 3-[cyclohexylamino]-1-propane-sulfonic acid; CD, cyclodextrin; CHES, 2-(N-cyclohexylaminojethanesulfonic acid; CDTA, cyclohexane-1,2-diaminetetra-acetic acid; dCMP, 2'-deoxycytidine-5'-monophosphate; DHBP, 1,1'-di-n-heptyl-4,4'-bipyridinium hydroxide; DNA, deoxyribonucleic acid; DOSS, sodium dioctyl sulfosuccinate; dTPA, diethylenetriaminepenta-acetic acid; ED, ethylene cyanohydrin, EDTA, ethylenediaminetetra-acetic acid; EG, ethylene glycol; FITC, fluorescein isothiocyanate; FMOC, 9-fluorenylmethyl chloroformate; dGMP, deoxyguanosine-5'-monophosphate; 2'-HDB: hexadimethrine bromide; HEC, hydroxyethyl cellulose; HIBA, α-hydroxyisobutyric acid; HQS, 8-hydroxyquinoline-5-sulfonic acid; HS, human albumin serum; ME, modified electrolyte; MeCN, acetonitrile; MES, morpholinoethanesulfonic acid; MS, mass; NDA, naphthalene-2,3-dicarboxyaldehyde; o-OPA, o-phthaldialdehyde; PAHs, polynuclear aromatic hydrocarbons; PAPS, 3'-phosphoadenosine-5'-phosphosulfate; PB, phosphate buffer; PCBs, polychlorinated biphenyls; PTH, phenylthiohydantion; t-RNA, t-ribonucleic acid; SDS, sodium dodecylsulfate; TBABr, tetrabutylammonium bromide; TE, terminating electrolyte; TEA, triethanolamine; TM-β-CDs, trimethyl-β-cyclodextrins; dTMP, 2'-deoxythymidine-5'-monophosphate; TT-AOH, tetradecyl-trimethylammonium hydroxide; TTHA, triethylenetetraminehexa-acetic acid, and UV, ultraviolet.

X. CONCLUSIONS

Analysis at trace levels is a very important and challenging issue, especially in unknown matrices. CE and related technologies have been used frequently for this purpose but, unfortunately, these have not yet achieved a desirable place in routine analysis due to poor reproducibility. Therefore, many workers have suggested various modifications to establish CE as a method of choice. The detection limit may be improved by using fluorescent and radioactive complexing agents, as the detection by fluorescence and radioactivity detectors is more sensitive and reproducible, with the low limits of detection. To make the CE applications more reproducible, the BGE should be further developed in such a way to ensure that its physical and chemical properties remain unchanged during the experiment. The selection of capillary wall chemistry, pH and ionic strength of the BGE, detectors, and optimization of BGE have been described and suggested in several papers (Horvath and Dolnike, 2001; Liu et al., 1999; Mayer, 2001; Macka and Haddad, 1999; Timerbaev and Buchberger, 1999; Valsecchi and Pdesello, 1999). Apart from the points discussed earlier for the improvement of CE, some other aspects should also be addressed so that CE can be used as the routine method in separation science.

The nonreproducibility of the methods may be due to the heating of BGE after a long CE run. Therefore, to keep the temperature constant through out the experiments, a cooling device should be included in the instrument.

There are only a few reports dealing with method validation. To make the developed method more applicable, the validation of the methodology should be established. A good coupling of a CE system with devices AAS, ICP, mass, and so on should be developed, which may also lead to good reproducibility and low limits of detection. All the capabilities and possibilities of CE as an analytical technique have not been explored until now; research is underway in this direction. However, CE will be realized as a widely recognized method of choice in analytical science. In brief, there is much to be developed for the advancement of CE and, definitely, it will prove itself as one of the best analytical techniques within the next few years.

REFERENCES

Aguilar, M., Huang, X., Zare, R. N. (1989). *J. Chromatogr.* 480:427.
Ahmed, A., Ibrahim, H., Pastore, F., Lloyd, D. K. (1996). *Anal. Chem.* 68:3270.
Aleksenko, S. S., Gumenyuk, A. P., Mushtakova, S. P., Timerbaev, A. R. (2001). *Fresenius J. Anal. Chem.* 370:865.
Ali, I., Aboul-Enein, H. Y. (2002a). *Critical Rev. Anal. Chem.* 32:337.
Ali, I., Aboul-Enein, H. Y. (2002b). *Anal. Lett.* 35:2053.
Ali, I., Aboul-Enein, H. Y. (2003). *Electrophoresis* 24:2064.
Ali, I., Gupta, V. K., Aboul-Enein, H. Y. (2003). *Electrophoresis* 24:1360.
Altria, K. D., Simpson, C. F., Bharij, A. K., Theobald, A. E. (1990). *Electrophoresis* 11:732.
Altria, K. D., Smith, N. W., Turnball, C. H. (1998). *J. Chromatogr. B* 717:341.
Aumatell, A., Wells, R. J. (1994). *J. Chromatogr. A* 688:329.
Aumatell, A., Wells, R. J., Wong, D. K. Y. (1993). *J. Chromatogr. A* 686:293.
Bachmann, K., Boden, J., Haumann, I. (1992). *J. Chromatogr.* 626:259.
Bailey, C. G., Yan, C. (1998). *Anal. Chem.* 70:3275.
Beck, W., Engelhardt, H. (1992). *Chromatographia* 33:313.
Bernal, J. L., del Nozal, M. J., Martin, M. T., Diez-Masa, J. C., Cifuentes, A. (1998). *J. Chromatogr. A* 823:423.
Bocek, P., Chramback, A. (1991). *Electrophoresis* 12:1059.
Bocek, P., Deml, M., Pospichal, J. (1990). *J. Chromatogr.* 500:673.
Bornhop, D. J., Dovichi, N. J. (1986). *Anal. Chem.* 58:504.
Bruin, G. J. M., Chang, J. P., Kuhlman, R. H., Zegers, K., Kraak, J. C., Poppe, H. (1989). *J. Chromatogr.* 471:429.
Burton, D. E., Sepaniak, M. J., Maskarinec, M. P. (1986). *J. Sep. Sci.* 24:347.
Bushey, M. M., Jorgenson, J. W. (1990). *Anal. Chem.* 62:978.
Chankvetadze, B. (1997). *Capillary Electrophoresis in Chiral Analysis*. John Wiley & Sons, p. 353.
Chankvetadze, B., Endresz, G., Blaschke, G. (1995). *J. Chromatogr. A* 700:43.
Chen, C. Y., Morris, M. D. (1988). *J. Appl. Spectrosc.* 42:515.
Chen, C., Morris, M. D. (1991). *J. Chromatogr.* 540:355.
Chen, C., Demana, T., Huang, S., Morris, M. D. (1989). *Anal. Chem.* 61:1590.
Christensen, P., Yeung, E. S. (1989). *Anal. Chem.*, 61:1344.
Clothier, J. G., Tomellini, S. A. (1996). *J. Chromatogr. A* 723:179.
Cohen, A. S., Karger, B. L. (1987). *J. Chromatogr.* 397:409.
Cole, R. O., Sepaniak, M. J., Hinze, W. L. (1990). *J. High Resolut. Chromatogr.* 13:579.
Crego, A. L., Garcia, M. A., Marina, M. L. (2000). *J. Microcol. Sep.* 12:33.
Crego, A. L., Marina, M. L. (1997). *J. Liq. Chrom. Rel. Technol.* 20:1.
Dabek-Zlotorzynska, E. (1997). *Electrophoresis* 18:2453.
Dabek-Zlotorzynska, E., Lai, E. P. C., Timerbaev, A. R. (1998). *Anal. Chim. Acta* 359:1.
Dabek-Zlotorzynska, E., Aranda-Rodriguez, R., Keppel-Jones, K. (2001a). *Electrophoresis* 22:4262.
Dabek-Zlotorzynska, E., Piechowski, M., McGrath, M., Lai, E. P. C. (2001b). *J. Chromatogr. A* 910:331.
Dadoo, R., Zare, R. N., Yan, C., Anex, D. S. (1998). *Anal. Chem.* 70:4787.
Dalton, D. D., Taylor, D. R., Waters, D. G. (1995). *J. Chromatogr. A* 712:365.
Debets, A. J. J., Hupe, K. P., Brinkman, U. A. Th., Kok, W. T. (1990). *Chromatographia* 29:217.

Demana, T., Chen, C., Moris, M. D. (1990). *J. High Resolut. Chromatogr.* 13:587.

Deml, M., Foret, F., Bocek, P. (1985). *J. Chromatogr.* 320:159.

Desiderio, C., Palcaro, C., Fanali, S. (1997). *Electrophoresis* 18:227.

Deyl, Z., Rohlicek, V., Struzinsky, R. (1989). *J. Liq. Chromatogr.* 12:2515.

D'Hulst, A., Uerbeke, N. (1992). *J. Chromatogr.* 608:275.

Dittmann, M. M., Rozing, G. P. (1996). *J. Chromatogr. A* 744:63.

Doble, P., Haddad, P. R. (1999). *J. Chromatogr. A*, 834:189.

Dovichi, N. J., Martin, J. C., Jeff, J. H., Keller, R. A. (1983). *Science* 219:845.

Dovichi, N. J., Nolan, T. G., Weimer, A. W. (1984). *Anal. Chem.* 56:1700.

Eberle, D., Hummel, R. P., Kuhn, R. (1997). *J. Chromatogr. A* 759:85.

Egashira, N., Mutoh, O., Kurauchi, Y., Ogla, K. (1996). *Anal. Sci.* 12:503.

El Rassi, Z. (2002). *CE and CEC Reviews: Advances in the Practice and Application of Capillary Electrophoresis and Capillary Electrochromatography.* New York: John Wiley & Sons.

El Rassi, Z., Giese, R. W., eds. (1997). *Selectivity and Optimization in Capillary Electrophoresis.* Amsterdam: Elsevier.

Enggstrom-Silverman, C. E., Ewing, A. G. (1991). *J. Microcol. Sep.* 3:141.

Ergolic, K. J., Stockton, R. A., Chakarborti, D. (1983). In: Lederer, W. H., Fensterheim, R. J., eds. *Arsenic: Industrial, Biochemical and Environmental Prospectives.* New York: Vnostrand Reinhold.

Ezzel, J. L., Richter, B. E., Felix, W. D., Black, S. R., Meikle, J. E. (1995). *LC–GC* 13:390.

Faller, A., Engelhardt, H. (1999). *J. Chromatogr. A* 853:83.

Fanali, S., Catarcini, P., Blaschke, G., Chankvetadze, B. (2001). *Electrophoresis* 22:3131.

Foret, F., Deml, M., Kahle, V., Bocek, P. (1986). *Electrophoresis* 7:430.

Foret, F., Deml, M., Bocek, P. (1988). *J. Chromatogr.* 452:601.

Foret, F., Fanali, S., Ossicini, L., Bocek, P. (1989). *J. Chromatogr.* 470:299.

Foret, F., Fanali, S., Nardi, A., Bocek, P. (1990). *Electrophoresis* 11:780.

Foret, F., Krivankova, L., Bocek, P. (1993). *Capillary Zone Electrophoresis.* Weinheim: VCH Publ.

Frenz, J., Wu, S. L., Hancock, W. S. (1989). *J. Chromatogr.* 480:379.

Fujiwara, S., Honda, S. (1987) *Anal. Chem.* 59:2773.

Fujiwara, S., Iwase, S., Honda, S. (1988). *J. Chromatogr.* 447:133.

Furut, R., Doi, T. (1994). *J. Chromatogr. A* 676:431.

Gaitonde, C. D., Pathak, P. V. (1990). *J. Chromatogr.* 514:389.

Garner, T. W., Yeung, E. S. (1990). *J. Chromatogr.* 551:639.

Gassmann, E., Kuo, J. E., Zare, R. N. (1985). *Science* 230:813.

Gilon, N., Potin-Gautier, M. (1996). *J. Chromatogr. A* 732:369.

Gordon, M. J., Lee, K. J., Arias, A. A., Zare, R. N. (1991). *Anal. Chem.* 63:69.

Grasper, M. P., Berthod, A., Nair, U. B., Armstrong, D. W. (1996). *Anal. Chem.* 68:2501.

Gross, L., Yeung, E. S. (1989). *J. Chromatogr.* 480:169.

Grossman, P. D., Wilson, K. J., Petrie, G., Lauer, H. H. (1988). *Anal. Biochem.* 173.

Grossman, P. D., Colburn, J. C., Lauer, H. H., Nielsen, R. G., Riggin, R. M., Sittampalam, G. S., Rickard, E. C. (1989). *Anal. Chem.* 61:1186.

Guttman, A., Cohen, A. S., Heiger, D. N., Karger, B. L. (1990). *Anal. Chem.* 62:137.

Haddad, P. R., Doble, P., Macka, M. (1999). *J. Chromatogr. A* 856:145.

Hamáek, J., Havel, J. (1999). *J. Chromatogr. A* 834:321.

Hecht, R. I., Morris, J. C., Stover, F. S., Fossey, L., Demarest, C. (1989). *Prepar. Biochem.* 190:201.

Hjerten, S. (1967). *Chromatogr. Rev.*, 9:122.

Hjerten, S. (1990). *Electrophoresis* 11:665.

Hjerten, S. (1983). *J. Chromatogr. Rev.* 270:1.

Hjerten, S., Hirai, H., eds. (1984). *Electrophoresis.* Berlin: Walter de Gruyter, pp. 71 79.

Hjerten, S., Zhu, M. (1985). *J. Chromatogr.* 327:157.

Hjerten, S., Zhu, M. (1987). In: Rany, B., ed. *Physical Chemistry of Colloids and Macromolecules, IUPAC.* London: Blackwell Scientific Publ., pp. 133–136.

Hjerten, S., Valtcheva, L., Elenbring, K., Eaker, D. (1989). *J. Liq. Chromatogr.* 12:2471.

Honda, S., Iwase, S., Makino, A., Fujiwara, S. (1986). *Anal. Biochem.* 176:72.

Horvath, J., Dolnike, V. (2001). *Electrophoresis* 22:644.

Huang, X., Zare, R. N., Sloss, S., Ewing, A. G. (1991). *Anal. Chem.* 63:189.

Huber, C. G., Chaudhary, G., Horvath, C. (1997). *Anal. Chem.* 69:4429.

Ikuta, N., Yamada, Y., Yoshiyama, T., Hirokawa, T. (1999). *J. Chromatogr. A* 894:11.

Jorgenson, J. W. (1984). *Trends Anal. Chem.* 3:51.

Jorgenson, J. W., Lukacs, K. D. (1981a). *Anal. Chem.* 53:1298.

Jorgenson, J. W., Lukacs, K. D. (1981b). *J. Chromatogr.* 218:209.

Jorgenson, J. W., Lukacs, K. D. (1981c). *Clin. Chem.* 27:1551.

Jorgenson, J. W., Lukacs, K. D. (1981d). *J. High Resolut. Chromatogr.* 4:230.

Jorgenson, J. W., Lukacs, K. D. (1983). *Science* 222:266.

Jorgenson, J. W., Lukacs, K. D. (1985). *J. Chromatogr. Library* 30:121.

Kaniansky, D. (1981). Ph.D. thesis, Komensky University, Bratislava.

Kaniansky, D., Marak, J. (1990). *J. Chromatogr.* 498:191.

Kaniansky, D., Rajec, P., Svec, A., Marak, J., Koval, M., Lucka, M., Franko, S., Sabanos, G. (1989). *J. Radioanal. Nucl. Chem.* 129:305.

Kano, K., Minami, K., Horiguchi, K., Ishimura, T., Kodera, M. (1995). *J. Chromatogr. A* 694:307.

Kasper, T. J., Melera, M., Gozel, P., Brownlee, R. G. (1988). *J. Chromatogr.* 458:303.

Khaledi, M. G., ed. (1998). *High Performance Capillary Electrophoresis: Theory, Techniques and Applications.* New York: John Wiley & Sons.

Kitihashi, T., Fruta, I. (2001). *Clin. Chem. Acta* 312:221.

Kobayashi, S., Ueda, T., Kikumoto, M. (1989). *J. Chromatogr.* 480:179.

Kohlrausch, F. (1897). *Ann. Phys. (Leipzig)* 62:209.
Koide, T., Ueno, K. (2000). *J. High Resolut. Chromatogr.* 23:59.
Krull, I. S., Stevenson, R. L., Mistry, K., Swartz, M. E. (2000). *Capillary Electrochromatography and Pressurized Flow Capillary Electrochromatography.* New York: HNB Publ.
Kuban, P., Karlberg, B. (1999). *Anal. Chem.* 69:1169.
Kuhn, R., Riester, D., Fleckenstein, B., Wiesmüller, K. H. (1995). *J. Chromatogr. A* 916:371.
Kuhr, W. G., Yeung, E. S. (1988). *Anal. Chem.* 60:1832.
Landers, J. P., ed. (1993). *Hand Book of Capillary Electrophoresis.* Boca Raton: CRC Press.
Lee, E. D., Mück, W., Henion, J. D., Covey, T. R. (1989). *Biomed. Environ. Mass Spectrom.* 18:844.
Lelievre, F., Gariel, P. (1996). *J. Chromatogr. A* 735:311.
Leroy, P., Belluci, L., Nicolas, A. (1995). *Chirality* 7:235.
Li, S., Lloyd, D. K. (1994). *J. Chromatogr. A* 666:321.
Li, K., Li, S. F. Y., Lee, H. K. (1995). *J. Liq. Chromatogr.* 18:1325.
Lin, W. C., Chang, C. C., Kuei, C. H. (1999). *J. Microcol. Sep.* 11:231.
Liu, W., Lee, H. K. (1998a). *Anal. Chem.* 70:2666.
Liu, W. P., Lee, H. K. (1998b). *J. Chromatogr. A* 796:385.
Liu, W., Lee, H. K. (1999). *J. Chromatogr. A* 834:45.
Liu, J., Cobb, K. A., Novotny, M. (1988). *J. Chromatogr.* 468:55.
Liu, J., Shirota, O., Wiesler, D., Novotny, M. (1991). *Proc. Natl. Acad. Sci. USA* 88:2302.
Liu, B. F., Liu, B. L., Cheng, J. K. (1999). *J. Chromatogr. A* 834:277.
Lloyd, D. K., Li, S., Ryam, P. (1995). *J. Chromatogr. A* 694:285.
Loo, J. A., Udseth, H. R., Smith, R. D. (1989). *Anal. Biochem.* 179:404.
Lopez-Sanchez, J. F., Amran, M. B., Lakkis, M. D., Lagarde, F., Rauret, G., Leroy, M. J. F. (1994). *Fresenius J. Anal. Chem.* 348:810.
Lukacs, K. D. (1983). Theory, Instrumentation and Application of Capillary Zone Electrophoresis. Ph.D. thesis, University of North Carolina at Chapel Hill, NC, USA.
Lukacs, K. D., Jorgenson, J. W. (1985). *J. High Resolut. Chromatogr.* 8:407.
Lunn, G. (2000). *Capillary Electrophoresis Methods for Pharmaceutical Analysis.* New York: John Wiley & Sons.
Lux, J. A., Yin, H. F., Schomburg, G. (1990). *Chromatographia* 30:7.
Macka, M., Haddad, P. R. (1999). *Electrophoresis* 18:2482.
Macka, M., Johns, C., Doble, P., Haddad, P. R. (1999). *LC–GC* 19:38.
Magnuson, M. L., Creed, J. T., Brockhoff, C. A. (1992). *J. Anal. Atom. Spectrom.* 12:689.
Magnuson, M. L., Creed, J. T., Brockhoff, C. A. (1997). *Analyst* 122:1057.
Maguregui, M. I., Jimenez, R. M., Alonso, R. M., Akesolo, U. (2002). *J. Chromatogr. A* 949:91.
Majors, R. E. (1995). *LC–GC* 3:542.
Malik, A. K., Faubel, W. (2001). *Crit. Rev. Anal. Chem.* 31:223.
Marina, M. L., Benito, I., Diez-Masa, J. C., Gonzalez, M. J. (1996). *J. Chromatogr. A* 752:265.
Martinez, D., Cugat, M. J., Borrull, F., Calull, M. (2000). *J. Chromatogr. A* 902:65.

Mayer, B. X. (2001). *J. Chromatogr. A* 907:21.
Mayer, S., Schurig, V. (1992). *J. High Resolut. Chromatogr.* 15:129.
Mayer, M., Muscate-Magnussen, A., Vogel, H., Ehrat, M., Bruin, G. J. (2002). *Electrophoresis* 23:1255.
McCormick, R. M. (1988). *Anal. Chem.* 60:2322.
Mechref, Y., El Rassi, Z. (1996a). *Chirality* 8:518.
Mechref, Y., El Rassi, Z. (1996b). *J. Chromatogr. A* 724:285.
Medina, I., Rubi, E., Mejuto, M. C., Cela, R. (1993). *Talanta* 40:1631.
Michalke, B., Schramel, P. (1997). *Fresenius J. Anal. Chem.* 357:594.
Michalke, B., Schramel, P. (1998). *Electrophoresis* 19:2220.
Mikkers, F. E. P., Everaerts, F. M., Verheggen, T. P. E. M. (1979). *J. Chromatogr.* 169:11.
Moseley, M. A., Deterding, L. J., Tomer, K. B., Jorgenson, J. W. (1989). *Rapid Commun. Mass Spectrom.* 3:87.
Moseley, M. A., Deterding, L. J., Tomer, K. B., Jorgenson, J. W. (1990). *J. Chromatogr.* 516:167.
Moseley, M. A., Deterding, L. J., Tomer, K. B., Jorgenson, J. W. (1991). *Anal. Chem.* 63:109.
Motomizu, S., Oshima, M., Matsuda, S., Obata, Y., Tanaka, H. (1992). *Anal. Sci.* 8:619.
Motomizu, S., Osima, M., Kuwabara, M. (1994). *Analyst* 119:1787.
Nardi, A., Fanali, S., Foret, F. (1990). *Electrophoresis* 11:774.
Neubert, R., Rüttinger, H. H. (2003). *Affinity Capillary Electrophoresis in Pharmaceutics and Biopharmaceutics.* New York: Marcel Dekker Inc.
Nickerson, B., Jorgenson, J. W. (1988). *J. High Resolut. Chromatogr.* 11:533.
Nielen, M. W. F. (1993). *J. Chromatogr. A* 637:81.
Nishi, H. (1995). *J. High Resolut. Chromatogr.* 18:695.
Nishi, H., Tsumagari, N., Terabbe, S. (1989). *J. Chromatogr.* 477:259.
Nishi, H., Fukuyama, T., Matsu, M., Terrabe, S. (1990). *J. Chromatogr.* 515:233.
Nishi, H., Fukuyarna, T., Terabe, S. (1991). *J. Chromatogr.* 553:503.
Nishi, H., Nakamura, K., Nakai, H., Sato, T. (1995). *Anal. Chem.* 67:2334.
Nishiwaki, F., Kuroda, K., Inoue, Y., Endo, G. (2000). *Biomed. Chromatogr.* 14:184.
Olefirowicz, T. M., Ewing, A. G. (1990). *J. Chromatogr.* 499:713.
Olivers, J. A., Nguyen, N. T., Yonker, C. R., Smith, R. D. (1987). *Anal. Chem.* 59:1232.
Ong, C. P., Ng, C. L., Lee, H. K., Li, S. F. Y. (1991). *J. Chromatogr.* 588:335.
Otsuka, K., Kawahara, J., Takekawa, K., Terabe, S. (1991). *J. Chromatogr.* 559:209.
Otsuka, K., Smith, J. S., Grainger, J., Barr, J. R., Patterson, D. G. Jr., Tanaka, N., Terabe, S. (1998). *J. Chromatogr. A* 817:75.
Pacakova, V., Coufal, P., Stulik, K. (1999). *J. Chromatogr. A* 834:257.
Padarauskas, A., Schwedt, G. (1997). *J. Chromatogr. A* 773:351.
Pawliszyn, J. (1988). *Anal. Chem.* 60:2796.

Penmetsa, K. V., Leidy, R. B., Shea, D. (1997). *J. Chromatogr. A* 790:225.

Pentoney, S. L., Huang, X., Burgi, D. S., Zare, R. N. (1988). *Anal. Chem.* 60:2625.

Pentoney, S. L. Jr., Zare, R. N., Quint, J. F. (1989a). *Anal. Chem.* 61:1642.

Pentoney, S. L. Jr., Zare, R. N., Quint, J. F. (1989b). *J. Chromatogr.* 480:259.

Petru, A., Rajec, P., Cech, R., Kunc, J. (1989). *J. Radioanal. Nucl. Chem.* 129:229.

Powell, A. C., Sepaniak, M. J. (1992). *J. Microcol. Sep.* 2:278.

Quang, C., Khaledi, M. (1995). *J. Chromatogr. A* 692:253.

Raber, G., Greschonig, H. (2000). *J. Chromatogr. A* 890:355.

Rasmussen, H. T., McNair, H. M. (1989). *J. High Resolut. Chromatogr.* 12:635.

Rathore, A. S., Guttman, A. (2003). *Electrokinetic Phenomena.* New York: Marcel Dekker Inc.

Riviello, J. M., Harrold, M. P. (1993). *J. Chromatogr. A* 652:385.

Rose, D. J., Jorgenson, J. W. (1988). *J. Chromatogr.* 447:117.

Row, K. H., Row, J. I. (1990). *Sep. Sci. Technol.* 25:323.

Rundlet, K. L., Armstrong, D. W. (1995). *Anal. Chem.* 67:2088.

Saitoh, T., Nishi, H., Yotsuyanagi, T. (1989). *J. Chromatogr.* 469:175.

Schmidt, M. G., Gübitz, G. (1995). *J. Chromatogr. A* 709:81.

Schmitt, T., Engelhardt, H. (1995). *J. Chromatogr. A* 697:561.

Schmitt-Kopplin, P., Garmash, A. V., Kudryavtsev, A. V., Menzinger, P., Perminova, I. V., Hertkorn, N., Freitag, D., Petrosyan, V. S., Kettrup, A. (2001). *Electrophoresis* 22:77.

Schweitz, L., Andersson, L. I., Nilsson, S. (1997a). *Anal. Chem.* 69:1179.

Schweitz, L., Andersson, L. I., Nilsson, S. (1997b). *J. Chromatogr. A* 792:401.

Semenova, P. P., Timerbaev, A. R., Gagstadter, R., Bonn, G. K. (1996). *J. High. Resolut. Chromatogr.* 19:77.

Silva da Rocha, M., Soldado, A. B., Blanco-Gonzalez, E. B., Sanz-Medel, A. (2000). *Biomed. Chromatogr.* 14:6.

Smith, N. W., Evans, M. B. (1995). *Chromatographia* 41:197.

Smith, R. D., Udseth, H. R. (1988). *Nature* 331:639.

Swartz, M. E. (1991). *J. Liq. Chromatogr.* 14:923.

Sweedler, J. V., Shear, J. B., Fishman, H. A., Zare, R. N., Scheller, R. H. (1991). *Anal. Chem.* 63:496.

Tan, A. D., Blanc, T., Leopold, E. J. (1990). *J. Chromatogr.* 516:241.

Tehrani, J., Day, L. (1989). *Am. Biotechnol. Lab.* Nov./Dec.:32.

Tehrani, J., Macomber, R., Day, L. (1991). *J. High Resolut. Chromatogr.* 14:10.

Terabe, S., Otsuka, K., Ichikawa, A., Ando, T. (1984a). *Anal. Chem.* 56:111.

Terabe, S., Otsuka, K., Ichikawa, K., Tsuchiya, A., Ando, T. (1984b). *Anal. Chem.* 56:113.

Terrabe, S., Ishida, Y., Nishi, H., Fukuyama, T., Otsuka, K. (1991). *J. Chromatogr.* 545:359.

Terrabe, S., Miyashita, Y., Ishihama, Y., Shibata, O. (1993). *J. Chromatogr. A* 636:47.

Thormann, W., Arn, D., Schumacher, E. (1984). *Electrophoresis* 5:323.

Timerbaev, A. R. (1997). *J. Chromatogr. A* 792:495.

Timerbaev, A. R., Buchberger, W. (1999). *J. Chromatogr. A* 834:117.

Timerbaev, A. R., Buchberger, W., Semenova, O. P., Bonn, G. K. (1993). *J. Chromatogr. A* 630:379.

Timerbaev, A. R., Semenova, O. P., Jandik, P., Bonn, G. K. (1994). *J. Chromatogr. A* 671:49.

Timerbaev, A. R., Semanova, O. P., Buchberger, W., Bonn, G. K. (1996). *Fresenius J. Anal. Chem.* 354:414.

Tiselius, A. (1930). The moving boundary method of studying the electrophoresis of proteins. Ph.D. thesis, *Nova acta regiae societatis scientiarum.* Ser. IV. Vol. 17. No. 4. Uppsala, Sweden: Almqvist & Wiksell, pp. 1–107.

Tsuda, T., Mizuno, T., Akiyama, J. (1981). *Anal. Chem.* 59:799.

Tsuda, T., Nomura, K., Nakagawa, G. (1983a). *J. Chromatogr.* 262:385.

Tsuda, T., Nakagawa, G., Sato, M., Yagi, K. (1983b). *J. Appl. Biochem.* 5:330.

Tsuda, T., Kobayashi, Y., Hori, A., Matsumoto, T., Suzuki, O. (1988). *J. Chromatogr.* 456:375.

Tsuda, T., Sweedler, J. V., Zare, R. N. (1990a). *Anal.Chem.* 62:2149.

Tsuda, T., Kobayashi, Y., Hori, A., Matsumoto, T., Suzuki, O. (1990b). *J. Microcol. Sep.* 2:21.

Udseth, H. R., Loo, J. A., Smith, R. D. (1989). *Anal. Chem.* 61:228.

Valsecchi, S. M., Polesello, S. (1999). *J. Chromatogr. A* 834:363.

Vindevogel, J., Sandra, P., Verhagen, L. C. (1990). *J. High Resolut. Chromatogr.* 13:295.

Wainright, A. (1990). *J. Microcol. Sep.* 2:166.

Wallingford, R. A., Ewing, A. G. (1987a). *Anal. Chem.* 59:1762.

Wallingford, R. A., Ewing, A. G. (1987b). *Anal. Chem.* 59:678.

Wallingford, R. A., Ewing, A. G. (1988). *J. Chromatogr.* 441:299.

Wallingford, R. A., Ewing, A. G. (1989). *Anal. Chem.* 61:98.

Wan, H., Blomberg, L. G. (1997). *J. Chromatogr. A* 758:303.

Wan, H., Engstrom, A., Blomberg, L. G. (1996). *J. Chromatogr. A* 731:283.

Wang, J., Warner, I. M. (1995). *J. Chromatogr. A* 711:297.

Wang, J., Schaufelberger, D. E., Guzman, N. A. (1998). *J. Chromatogr. Sci.* 36:155.

Wehr, T., Zhu, M. (1993). In: Landers, J. P., ed. *Capillary Electrophoresis: Historical Perspectives.* Hand Book of Capillary Electrophoresis. Chapter 1. Boca Raton: CRC Press.

Wehr, T., Rodriguez-Diaz, R., Zhu, M. (1998). *Capillary Electrophoresis of Proteins.* Vol. 80. New York: Marcel Dekker, Inc.

Wei, W., Luo, G. A., Hua, G. Y., Yan, C. (1998). *J. Chromatogr. A* 817:65.

Weinberger, R., Lurie, I. S. (1991). *Anal. Chem.* 63:823.

Weston, A., Brown, P. R., Heckenberg, A. L., Jandik, P., Jones, W. R. (1992). *J. Chromatogr.* 602:249.

Wey, A. B., Thorman, W. (2002). *J. Chromatogr. B* 770:191.

Whang, C., Pawliszyn, J. (1998). *Anal. Commun.* 35:353.

Wistuba, D., Czesla, H., Roeder, M., Schurig, V. (1998). *J. Chromatogr. A* 815:183.

Wolf, A. V., Morden, G. B., Phoebe, P. G. (1987). In: Weast, R. C., Astle, M. J., Beyer, W. H., eds. *CRC Hand Book of Chemistry and Physics*. 68th ed. Boca Raton, USA: CRC Press, p. D219.

Wren, S. A. C. (1993). *J. Chromatogr. A* 636:57.

Wright, B. W., Ross, G. A., Smith, R. D. S. (1989). *J. Microcol. Sep.* 1:85.

Wuping, L., Hian, K. L. (1998a) *J. Chromatogr. A* 796:385.

Wuping, L., Hian, K. L. (1998b). *Anal. Chem.* 70:2666.

Yeung, E. S., Kuhr, W. G. (1991). *Anal. Chem.* 63:275A.

Yoshinaga, M., Tanaka, M. (1994). *J. Chromatogr. A* 679:359.

Yu, M., Dovichi, N. J. (1989a). *Appl. Spectrosc.* 43:196.

Yu, M., Dovichi, N. J. (1989b). *Anal. Chem.* 61:37.

Zhu, M., Hansen, D. L., Burd, S., Gannon, F. (1989). *J. Chromatogr.* 480:311.

26

Gel Permeation and Size Exclusion Chromatography

GREGORIO R. MEIRA, JORGE R. VEGA, MARIANA M. YOSSEN
Intec (CONICET and Universidad Nacional del Litoral), Güemes 3450, Santa Fe, Argentina

I. GENERAL CONSIDERATIONS

Size exclusion chromatography (SEC) fractionates polymer molecules according to their sizes in solution (with the larger sizes eluting first and the smaller ones at the end), and then analyzes the different eluting fractions. Ideally, this is a purely entropy-controlled fractionation, and it should not involve any enthalpic interactions. Since its appearance in the mid-60s, the technique has greatly improved and it has been widely investigated (Provder, 1984, 1987; Wu, 1995; Yau et al., 1979). The older terminology, gel permeation chromatography (GPC), stemmed from the fact that the original column packings were soft organic gels. The name gel filtration chromatography (GFC) has been traditionally applied to designate the analysis of hydrophilic biopolymers, and it is also still in use today.

SEC is the main technique for determining the molar mass distribution (MMD) of both synthetic and natural polymers. Other SEC applications include the determination of copolymer chemical composition, of chain branching, of polymer additives, and the pretreatment of biopolymers.

The older equipment only included on-line or instantaneous concentration detectors, such as the differential refractometer (DR) and the single-wavelength UV spectrophotometer. As such, the technique is relative, that is, nonabsolute, in the sense that an independent molar mass calibration is required to transform the chromatogram abscissas (either elution time or elution volume) into molar mass. More recently, the development of "molar mass sensitive detectors", such as on-line specific viscometers (SV) and light-scattering (LS) sensors, has converted SEC into a more absolute technique for determining the MMD, and also makes it possible to indirectly determine other polymer characteristics such as chain branching. Unfortunately, the high cost of these more sophisticated sensors explains their relatively scarce distribution.

The main difficulties of SEC are: (a) even under ideal conditions, the fractionation is by hydrodynamic volume rather than by molar mass, (b) there is an upper molar mass limit of fractionation, above which all molecular species emerge together, and (c) undesirable effects such as secondary (nonentropic) fractionations and band broadening (BB) make the quantification of macrostructural characteristics difficult.

Present trends in SEC include combining the size-controlled fractionation with other mechanisms such as adsorption [for the determination of the chemical composition distribution (CCD)] or eluability (for the determination of the amorphous–crystalline fractions). More recently, true molar mass sensors such as MALDI (matrix-assisted laser desorption ioinization) and mass spectrometers have also been attached to size-exclusion chromatographs (Fei and Murray, 1996), but this instrumentation and data interpretation are still under development (Macha and Limbach, 2002; Murgasova and Hercules, 2003). All of these newer developments are outside the scope of the present review, which is strictly limited to SEC.

II. BASIC CONCEPTS

A. The Chromatographic System

A size-exclusion chromatograph is essentially a high-pressure chromatograph with columns containing a porous packing of pore size in the order of the hydrodynamic volumes of the analyzed polymer molecules. Figure 1 illustrates a typical SEC arrangement. The filtered and degassed solvent flows from the solvent reservoir into a high-pressure isocratic pump. Through intermediate capillaries, the pump forces the mobile phase into the injector, columns, and detectors. The injector introduces the polymeric solution sample into the high-pressure carrier stream. A sample prefilter (with pore diameters between 0.5 and 2 μm) prevents the entrance of particles into the fractionation columns. In the column packing, the porous particle diameters are between 5 and 30 μm. The columns separate the analyte according to hydrodynamic volume, with the biggest molecules emerging first, and the smallest emerging at the end of the chromatogram. Typical operational values are: flow rate, 1 mL/min; injection concentration, 1 mg/mL; and fractionation time, 1 h.

Detectors can be classified into two broad groups: concentration sensitive and molar mass sensitive. In turn, mass sensitive detectors may respond either to the total instantaneous concentration or to the instantaneous concentration of one or more specific chemical groups. The most universal of the total concentration detectors is the DR, whereas (single-wavelength) UV spectrophotometer can be used for detecting the instantaneous concentration of a specific atomic group. The molar-mass sensitive detectors (LS and SV) are both specific to SEC, and their signals must be divided by the instantaneous concentration for estimating the instantaneous molar mass.

A PC is required to operate the chromatographic system and to perform the data treatment. Its operation involves monitoring process variables such as the flow rate, pressures, and temperatures; synchronizing the injection with the acquisition time; digitalizing and storing the detector outputs; and displaying the analysis results.

To produce a molar mass calibration, a set of narrow standards of known molar masses must be analyzed and represented in the format $\log M$ vs. V, where M is the molar mass and V is the elution volume (or elution time). A typical calibration is illustrated in Fig. 2. It exhibits a central linear and two nonlinear regions. The low molar mass nonlinear region may or may not be present. Ideally, the chromatogram should fall inside the linear calibration range. The minimum possible retention volume is the total exclusion limit V_e, where all molecules larger than a certain limiting molar mass emerge together. The maximum retention volume is V_i, where all molecules smaller than a certain molar mass emerge together. The logarithmic nature of Fig. 2 determines that fractionation in SEC is high at the low molar masses, but it loses resolution for increasing molar masses.

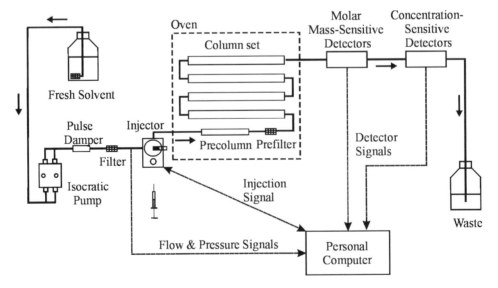

Figure 1 Schematic diagram of a size exclusion chromatograph. The main components are an isocratic pump, a (manual or automatic) sample injector, the fractionation columns, and the detection system. The columns are packed with porous beads that fractionate according to hydrodynamic volume. The sensors may either respond to an instantaneous concentration or to an instantaneous molar mass.

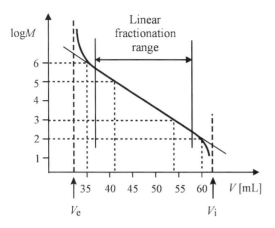

Figure 2 Typical molar mass calibration obtained from a set of narrow standards. The curve exhibits a "linear" behavior in most of its dynamic range. The fractionation limits are determined by the total exclusion and total inclusion volumes V_e and V_i, respectively. Resolution in SEC is highest at low molar masses. In the presented example, this is seen as follows: the same total elution volume of about 6 mL is required to fractionate molecules in the range 100–1000 g/mol (i.e., a molar mass difference of 900), or in the range 10^5–10^6 g/mol (i.e., a molar mass difference of 900,000).

B. Simple and Complex Polymers

A polymer is chromatographically "simple" when, under perfect resolution, all molecules in the detection cell have a common hydrodynamic volume, and there is a one-to-one relationship between hydrodynamic volume and molar mass. Simple polymers are basically linear flexible homopolymers and, in this case, the MMD completely defines their macromolecular structure. In contrast, samples containing a polymer mixture, a copolymer, or a branched homopolymer are chromatographically "complex" in the sense that (for any given hydrodynamic volume) a variety of molar masses coexist in the detector cell. Apart from the MMD, additional distributions are required to characterize complex polymers. For example, copolymers are characterized by a CCD, whereas long-branched homopolymers are characterized by a distribution of the number of branches per molecule (Berek, 2000; Meira, 1991; Radke et al., 1996).

When a simple polymer is analyzed, in ideal SEC, any eluted fraction has a uniform molar mass and, therefore, perfectly accurate MMDs can in principle be obtained. In contrast, when, for example, a randomly branched homopolymer is analyzed under perfect resolution conditions, at any retention volume, a whole variety of molar masses will be present in the detector cell (ranging from linear molecules of a relatively low molar mass to highly branched molecules of a higher molar mass). Thus, complex polymers generate the so-called "local polydispersity phenomenon" in the detection cells, even under conditions of perfect fractionation (Mourey and Balke, 1998; Netopilík, 1998b). In this case, the sensors provide signals that are proportional to some average property instantaneously, such as the chemical composition or the weight-average molar mass.

C. Molar Mass Distribution

The MMD is the most important macromolecular characteristic because (with the exception of the highly cross-linked polymers of infinite molar mass) all synthetic and natural polymers exhibit a distribution of the molar mass.

MMDs are strictly discrete. The number MMD represents the number of moles (or any other variable proportional to it, such as molar fraction, molarity, etc.) vs. the molar mass. The weight MMD represents the mass (or any other variable proportional to it such as weight fraction, mass concentration, etc.) vs. the molar mass. For homopolymers, their molar masses are proportional to the chain lengths and are, therefore, multiples of the repeating unit molar mass. In statistical copolymers, a wide composition distribution is possible at every chain length, thus extraordinarily increasing the number of feasible molar masses (that now are not evenly distributed along the molar mass axis). Clearly, SEC is incapable of separating all possible molar masses and, instead, the detector signals are continuous and smooth, but are sampled at regular intervals of the retention volume. For this reason, only continuous approximations to the true MMD are possible.

Let us first define the continuous cumulative MMD as follows:

$$W(M) = \int_0^M \frac{dw(M)}{dM} dM$$

$$= \text{mass of sample with molar masses} < M \quad (1)$$

where dw/dM vs. M represents the continuous differential MMD exhibiting a linear molar mass axis. Alternatively, the continuous differential MMD exhibiting a logarithmic M axis $dw/d\log M$ vs. $\log M$ is more widely used for being more "characteristic" of the SEC measurement. Note that derivatives are necessary for the ordinates of continuous mass distributions in order to make the area under the curve proportional to a mass (or a concentration).

Two representations of the molar mass calibration are also possible: $M(V)$ and $\log M(V)$ (Fig. 2). For any two pairs (M_1, V_1) and (M_2, V_2) of $M(V)$, the mass of molecules contained in the range $[M_1–M_2]$ (or equivalently, in $[\log M_1–\log M_2]$) must be equal to the mass of molecules eluting in the elution volume range $[V_1–V_2]$; and, therefore:

$$-\int_{M_1}^{M_2} \frac{dw(M)}{dM} dM = -\int_{\log M_1}^{\log M_2} \frac{dw(\log M)}{d, \log M} d\log M$$

$$= \int_{V_1}^{V_2} c(V) dV \quad (2)$$

where $c(V)$ is the continuous (mass concentration) chromatogram. The minus signs in Eq. (2) indicates that positive increments of dM correspond to negative increments of dV. From Eq. (2), the ordinates of the two continuous MMD representations are interrelated through:

$$\frac{dw}{d\log M} = \frac{dM}{d\log M} \frac{dw}{dM} = 2.3026 M \frac{dw}{dM} \quad (3)$$

Assume, now, that the continuous $c(V)$ is sampled at regular ΔV intervals, yielding the discrete chromatogram $c_j(V_j)$, where j (=1, 2, ...) are the sampling instants. We shall assume that the total sample mass is concentrated into the j discretization points. The mass of each point is w_j (=$c_j \Delta V_j$), where ΔV_j is the constant elution volume interval. We shall call the discrete MMD as $w_j(M_j)$. In this distribution, the total sample mass W is distributed among the hypothetical molecular species of molar mass M_j, as given by the discrete chromatogram and the molar mass calibration. Note the following:

$$w_j = c_j \Delta V_j = -\left[\frac{dw}{dM}\right]_j \Delta M_j \quad (4)$$

where $[dw/dM]_j$ is the height (at M_j) of the continuous MMD with a linear M axis dw/dM vs. M. Note that while ΔV_j is constant for all j values, this is not true for the corresponding ΔM_j, due to the logarithmic nature of the molar mass calibration.

A continuous MMD with a linear M axis cannot be directly drawn from the discrete $w_j(M_j)$ by simply joining together the successive w_j values. Consider, now, the height correction that is required for producing a "continuized" MMD with a linear M axis. More precisely, we wish to represent the continuous MMD given by dw/dM vs. M (with a linear M axis) by means of a polygonal that joins together the discrete points of $[dw/dM]_j(M_j)$, with the M_j values taken at irregular ΔM_j intervals according to the discrete chromatogram and the molar mass calibration. Equation (4) yields:

$$\left[\frac{dw(M)}{dM}\right]_j = -\frac{c_j(V_j)}{\Delta M_j/\Delta V_j} \left(\frac{\Delta \log M_j}{\Delta \log M_j}\right)$$
$$= -\frac{0.4343}{M_j(V_j)} \frac{c_j(V_j)}{\Delta \log M_j/\Delta V_j} \quad (5)$$

The first equality of Eq. (5) indicates that each discrete chromatogram height c_j must be divided by the slope of the calibration curve $\Delta M_j/\Delta V_j$. The second equality of Eq. (5) results from the fact that $\Delta \log M_j/\Delta M_j \cong 0.4343/M_j$. Also, note that $\Delta \log M_j/\Delta V_j$ is the slope of the molar mass calibration in the format $\log M_j(V_j)$. When $\log M_j(V_j)$ is linear, then $\Delta \log M_j/\Delta V_j$ is a constant, and the height correction of Eq. (5) basically involves a division by M_j.

To minimize (or even eliminate) the nonlinear height correction required in Eq. (5), the logarithmic relationship of the molar mass calibration suggests that it is preferable to represent MMD as $dw/d\log M$ vs. $\log M$ (i.e., with a logarithmic M axis). To see this, note the following:

$$\left[\frac{dw(\log M)}{d\log M}\right]_j = -\frac{c_j}{\Delta \log M_j/\Delta V_j} \quad (6)$$

In Eq. (6), $\Delta \log M_j/\Delta V_j$ is a constant for a linear calibration and, therefore, $dw_j/d\log M_j$ vs. $\log M_j$ can be directly represented with a polygonal joining the c_j chromatogram heights, without need of any height correction.

Now, let us derive the formulae for calculating the different average molar masses. From Eq. (4), the kth moment of the discrete MMD (λ^k) is:

$$\lambda^k \equiv \sum_j M_j^k w_j = \sum_j M_j^k c_j \Delta V_j$$
$$= -\sum_j M_j^k \left[\frac{dw(M)}{dM}\right]_j \Delta M_j; \quad (k = 1, 2, \ldots) \quad (7)$$

From Eq. (7), and bearing in mind that all ΔV_j values are constant, then the following is derived:

Number-average molar mass:

$$\bar{M}_n = \frac{\lambda_1}{\lambda_0} = \frac{\sum w_j}{\sum w_j/M_j} = \frac{\sum c_j}{\sum c_j/M_j} \quad (8a)$$

Weight-average molar mass:

$$\bar{M}_w = \frac{\lambda_1}{\lambda_0} = \frac{\sum M_j w_j}{\sum w_j} = \frac{\sum M_j c_j}{\sum c_j} \quad (8b)$$

z-average molar mass:

$$\bar{M}_z = \frac{\lambda_2}{\lambda_1} = \frac{\sum M_j^2 w_j}{\sum M_j w_j} = \frac{\sum M_j^2 c_j}{\sum M_j c_j} \quad (8c)$$

Viscosity-average molar mass:

$$\bar{M}_v = \left[\frac{\lambda_\alpha}{\lambda_0}\right]^{1/\alpha} = \left[\frac{\sum M_j^\alpha w_j}{\sum w_j}\right]^{1/\alpha}$$
$$= \left[\frac{\sum M_j^k c_j}{\sum c_j}\right]^{1/\alpha} \quad (8d)$$

The right-hand sides of Eqs. (8a)–(8d) are used to calculate the molar mass averages directly from the

Gel Permeation and Size Exclusion Chromatography

chromatogram heights and the molar mass calibration, without requiring to previously represent the MMD with a linear M axis. This avoids the numerical errors associated with the nonlinear transformation of Eq. (5).

1. Illustration of the Chromatographic Distortion

To illustrate the chromatographic distortion, consider a "synthetic" (or numerical) example that involves the analysis of a most probable (or Schulz–Flory) MMD, defined by:

$$\frac{dw(M)}{dM} = a^2 M(1-a)^{M-1}; \quad a = 2 \times 10^{-6} \qquad (9)$$

In this case, the MMD is normalized and, therefore, its total mass: $W = \int_{0_-}^{\infty} [dw(M)/dM]\,dM \approx 1$. The average molar masses are: $\bar{M}_n = 512{,}380$ g/mol; $\bar{M}_w = 997{,}980$ g/mol; and $\bar{M}_w/\bar{M}_n = 1.948$. Figure 3(c) and (d) represents the continuous MMD with a continuous trace and with a linear M axis. The same continuous MMD, but in the format $dw(\log M)/d\log M$, is obtained from the last equality of Eq. (3) [Fig. 3(b)]. In this example, we adopt a linear molar mass calibration that is represented in Fig. 3(a), and is given by: $M(V) = D_1 \exp(-D_2 V)$, with $D_1 = 1.0176 \times 10^{13}$ and $D_2 = 0.41423$.

From Eq. (2), one obtains the continuous concentration "chromatogram" from: $c(V) = (dw(M)/dM) \times dM/dV$ with $dM/dV = -(D_1 \times D_2)\exp(-D_2 V)$ [Fig. 3(a)]. Then, the discrete concentration chromatogram $c_j(V_j)$ is obtained by sampling the continuous $c(V)$ every $\Delta V = 0.8$ mL [Fig. 3(a)].

Let us now discretize the continuous MMD representation of $dw(\log M)/d\log M$ of Fig. 3(b) at the molar mass values as determined by the discrete chromatogram $c_j(V_j)$ and the calibration. This distribution is then "continuized" in Fig. 3(b) by application of Eq. (6), yielding the polygonal $[dw/d\log M]_j$ vs. $\log M_j$ of Fig. 3(b). Since the molar mass calibration $\log M(V)$ is linear, then the heights $[dw/d\log M]_j$ are proportional to c_j.

The discrete MMD $w_j(M_j)$ with a linear M axis is represented by bars in Fig. 3(c). The bars of $w_j(M_j)$ are unevenly spaced along M, as a consequence of the nonlinear transformation $M_j(V_j)$. Note that large errors with respect to the "true" continuous $dw(M)/dM$ would have been produced if the ordinates of $w_j(M_j)$ had been joined with a polygonal in Fig. 3(c). Instead, Fig. 3(d) illustrates the required height correction of Eq. (5). In this case, the new ordinates $[dw(M)/dM]_j$ are calculated at each sampled M_j, and are seen to fall exactly on the true continuous MMD. However, slight differences are apparent

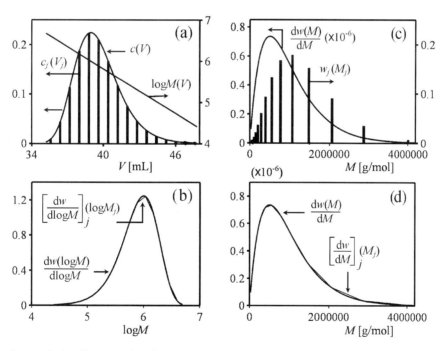

Figure 3 Numerical example that illustrates the effects of the chromatogram discretization and of the distortion produced when a linear molar mass axis is required in the MMD. The raw data are the molar mass calibration $\log M(V)$ in (a), and the continuous MMD presented as $dw(\log M)/d\log M$ in (b), and presented as $dw(M)/dM$ in (c) and (d). The calculation was as follows. First, the continuous and discrete concentration chromatograms $c(V)$ and $c_j(V_j)$ were obtained. From the discrete chromatogram and the calibration, the "continuized" MMD $[dw/d\log M]_j(\log M_j)$ of (b), and the discrete MMD $w_j(M_j)$ of (c) were calculated. Finally, a heights correction was required to transform the discrete $w_j(M_j)$ of (c) into the "continuized" $[dw/dM]_j(M_j)$ of (d).

between the true continuous distribution and its corresponding polygonal [Fig. 3(d)]. Finally, Eqs. (8) were applied to the discrete MMD $w_j(M_j)$ and to the discrete chromatogram $c_j(V_j)$. In either case, the following (identical) results were obtained: $\bar{M}_n = 518{,}010\,\text{g/mol}$, $\bar{M}_w = 997{,}280\,\text{g/mol}$, and $\bar{M}_w/\bar{M}_n = 1.925$. Compared with the "true" original values, negligible errors are observed.

In general, the SEC data treatment involves the processing discrete functions that are, however, represented by continuous polygonals. For simplicity, in the rest of this chapter, the j subindex will not be further included to indicate discretization. Thus, $[dw/d\log M]_j$ vs. $\log M_j$ will be simply written as $dw/d\log M$ vs. $\log M$ to indicate the continuous polygonal MMD obtained by sampling the concentration chromatogram at regular ΔV intervals.

D. Entropy-Controlled Partitioning

The distribution of pore diameters in the packing must match the distribution of hydrodynamic volumes of the analyte molecules. Smaller macromolecules exhibit a higher probability of circulating through most of the system pores, whereas bigger molecules circulate through a reduced fraction of pores. Thus, larger molecules emerge first, whereas smaller emerge at the end of the chromatogram.

In ideal SEC, the fractionation occurs by steric limitations in the molecular movements, and does not involve enthalpic interactions between the analyte and the stationary phase. In ideal SEC, for each hydrodynamic volume, the corresponding retention volume V is obtained from theoretical considerations as follows (Berek, 2000; Janča, 1984; Malawer, 1995):

$$V = V_{\text{int}} + V_p K_{\text{SEC}}; \quad V_p = V_e - V_{\text{int}}; \quad 0 < K_{\text{SEC}} < 1 \tag{10}$$

where V_{int} is the interstitial volume, V_p is the pore volume, V_e is the total exclusion volume (Fig. 2), and K_{SEC} is the SEC distribution or partition coefficient (a function of hydrodynamic volume).

Many theoretical models have been developed to predict the K_{SEC} partition coefficient as a function of the fractionation mechanism, the extracolumn effects, the asymmetric shape of the chromatogram peaks, and so on (Busnel et al., 2001; Cassassa, 1976; Felinger, 1998; Netopilík, 2002; Potschka, 1993; Vander Heyden et al., 2003). These models generally take into consideration processes such as the mass transfer between phases, the molecular dispersion, and the internal diffusion in the column packing. A good theoretical model must predict both the measured chromatograms and the calibration curves. The major difficulty with theoretical models is the estimation of their physico-chemical parameters.

E. Intrinsic Viscosity and Hydrodynamic Volume

In a dilute polymeric solution, the specific viscosity measures the relative increase of the solution viscosity (η) with respect to the solvent viscosity (η_0):

$$\eta_{\text{sp}} = \frac{\eta - \eta_0}{\eta_0} \tag{11}$$

The intrinsic viscosity, or limiting viscosity number (IUPAC), is defined (at the limit of perfect dilution) as follows:

$$[\eta] = \lim_{c \to 0} \frac{\eta_{\text{sp}}}{c} \tag{12}$$

where c is the mass concentration of the eluting polymer, in g/dL. Thus, $[\eta]$ is given in dL/g. The intrinsic viscosity is related to the molar mass through the Mark–Houwink–Sakurada (MHS) equation:

$$[\eta] = K M^\alpha \tag{13}$$

where K and α are functions of the solvent, the polymer, the temperature, and the second virial coefficient, A_2 (Brandrup and Immergut, 1989; Burchard, 1999). The values of K, α, and A_2 are available for many different polymer–solvent–temperature combinations (Brandrup and Immergut, 1989).

In the so-called theta (Θ) condition, the conformation of the dissolved molecules is intermediate between a totally collapsed and a totally swollen conformation. In the (thermodynamically ideal) Θ condition, there are no net interactions between the "poor" solvent and the polymer, and it is simply: $\alpha \cong 0.5$ and $A_2 = 0$ (Berek, 2000; Burchard, 1999). Unfortunately, it is necessary to avoid all possible polymer–polymer interactions; and, for this reason, good solvents that swell and solvate the flexible polymer molecules are required, yielding $0.6 < \alpha < 0.8$ and $A_2 > 0$.

Assume a set of linear and flexible polymer chains of a uniform (or constant) molar mass under Θ conditions. Then, the Flory–Fox equation relates the mean squared radius of gyration R_g^2 with the intrinsic viscosity and the molar mass as follows (Burchard, 1999; Flory and Fox, 1951):

$$R_g^3 = \frac{1}{6^{3/2}} \left(\frac{[\eta] M}{\Phi} \right) \tag{14a}$$

with:

$$\Phi = \Phi_0 (1 - 2.63\alpha' + 2.86\alpha'^2) \tag{14b}$$

$$\alpha' = \frac{2\alpha - 1}{3} \tag{14c}$$

where α' (=0–0.2) is an interaction parameter that takes into account deviations from the Θ condition; and Φ_0(=2.55 × 10^{21}) is a constant for $M > 10^4$ g/mol. For branched flexible molecules, Φ_0 increases with the degree of branching (Burchard, 1999).

The following Einstein equation relates the product $[\eta]M$ with the volume of a hydrodynamically equivalent solid sphere of radius R and volume V_h:

$$[\eta]M = \frac{10}{3}\pi N_A R^3 = 2.5 N_A V_h \quad (15)$$

where V_h is the hydrodynamic volume of the polymer molecule (in dL) and N_A is the Avogadro number. By comparing Eqs. (14a) and (15), it is seen that (under Θ conditions), R_g^3 in Eq. (14a) is another measure of the hydrodynamic volume.

For a given molar mass, a branched molecule exhibits reduced values of the hydrodynamic volume and the intrinsic viscosity than its linear homolog. A measure of this hydrodynamic volume contraction is given by the g branching parameter, which is defined as follows:

$$g = \frac{R_{g,b}^2}{R_{g,l}^2} \leq 1; \quad M_b = M_l \quad (16)$$

where R_g^2 is the mean squared radius of gyration, and the subscripts "b" and "l", respectively, indicate a branched and a linear molecule of the same molar mass. For branched topologies involving tri- and/or tetrafunctional branching points, Zimm and Stockmayer (1949) have developed theoretical expressions that correlate g with the number- or weight-average number of branches per molecule.

A second branching parameter g' is defined from the ratio of hydrodynamic volumes (or equivalently of intrinsic viscosities) as follows:

$$g' = \frac{V_{h,b}}{V_{h,l}} = \frac{[\eta]_b M_b}{[\eta]_l M_l} = \frac{[\eta]_b}{[\eta]_l} \leq 1; \quad M_b = M_l \quad (17)$$

Unlike g, the g' parameter can be easily determined from the ratio of intrinsic viscosities. Unfortunately, however, no theoretical expressions have been developed that correlate g' with the number of branches per molecule; and instead the following empirical expression is proposed for interrelating the two contraction parameters:

$$g' = g^\varepsilon \quad (18)$$

In theory, Eqs. (14a), (16)–(18) suggest that $\varepsilon = 3/2$. However, large deviations with respect to this last value have been reported and, therefore, ε has to be empirically determined. Unfortunately, ε has been estimated for relatively few polymer–solvent systems (Burchard, 1999); and, furthermore, it is now established that ε is a function of the molar mass (Tackx and Tacx, 1998).

F. SEC Resolution and Column Efficiency

Resolution in SEC measures the capacity of a fractionation system for separating two close and narrow peaks. Resolution is improved by either increasing the column efficiency (e.g., using smaller porous particles for a given set of columns) and/or by reducing the calibration slope (e.g., by increasing the column length).

Consider the analysis of two uniform polymers of different molar masses. The column resolution (R_{col}) is defined by (Dawkins, 1978):

$$R_{col} = \frac{2(\bar{V}_1 - \bar{V}_2)}{\Delta w_1 + \Delta w_2} \quad (19)$$

where $\bar{V}_i$ and Δw_i ($i = 1, 2$) are, respectively, the mean retention volume and the mean peak width (as measured at the inflexion points).

The column efficiency is measured by the number of theoretical plates (N). This concept was introduced by Martin and Synge in 1941 for liquid–liquid partition chromatography, neglecting the solvent diffusion between plates (Felinger, 1998; Martin and Synge, 1941; Mori, 1988). The column is divided into a series of consecutive layers (or "plates") and, in each plate, a perfect separation equilibrium is attained. The column efficiency increases for an increasing N. In general, N increases with the temperature and when the following variables are reduced: particle size, molar mass, solvent flow rate, and solvent viscosity. For a uniform polymer, N is given by (Dawkins, 1978):

$$N = 16L\left(\frac{\bar{V}}{\Delta w}\right)^2 \quad (20)$$

where L is the column length.

BB is one possible cause of poor resolution. Due to BB, the chromatogram of a strictly uniform sample is not an impulse, but is, instead, a somewhat skewed peak that depends on the molar mass. The logarithmic nature of the molar mass calibration (Fig. 2) determines that resolution in SEC is high at low molar masses, but poor at high molar masses. Except for the case of some biopolymers, strictly uniform calibration standards are not available and, therefore, the BB function can be only readily determined for pure (nonpolymeric) low molar mass substances.

The column efficiency depends not only on the packing quality and morphology, but also on its aging, eventual presence of channels, and so on. When only the axial dispersion is considered, then the theory indicates that the concentration chromatogram (or band) of a uniform solute is a Gaussian distribution of standard deviation $\sigma = \bar{V}/N$, with N depending on the nature of the sample, its concentration, the injected volume, the column temperature, and the solvent flow rate. Apart from the column itself, other instrumental factors (such as the length and diameter of capillaries, the volume and geometry of the detector cell, etc.) contribute toward a skewed peak broadening. Thus, the efficiencies of different columns are only comparable under identical experimental conditions.

Ideally, the sample employed for determining N must fulfill the following requirements: (a) it must be of a low molar mass, homogeneous, and free of isomers or impurities, (b) it must not present undesirable interactions with the packing material such as adsorption or ionic interchange, and (c) it must provide a good response to DR or UV detectors. In organic SEC with polystyrene (PS)-based packing and tetrahydrofuran (THF) or chloroform as mobile phase, the mentioned conditions are satisfied by benzene, methanol, and acetone. In aqueous SEC, the optimal sample depends on the nature of the packing and on the experimental conditions but, in general, the following are used: ethyleneglycol, sodium p-toluenesulfonate, glucose, and phenylalanine.

III. CONCENTRATION DETECTORS AND MOLAR MASS CALIBRATION

A. Differential Refractometer

The DR is the most "universal" SEC detector, because the refractive index of any polymer solution is generally higher than the refractive index of the pure solvent. For homopolymers, the refractive index difference Δn is strictly proportional to the change in the polymer concentration:

$$\Delta n(V) = \frac{\partial n}{\partial c} c(V) \qquad (21)$$

where $\partial n/\partial c$ is the specific refractive index increment. Each polymer–solvent combination has a characteristic $\partial n/\partial c$ that depends on the temperature and on the wavelength. Unfortunately, $\partial n/\partial c$ also shows a small dependence with the molar mass (it increases with M for oligomers, but it reaches a constant value for $M > 20,000$ g/mol). If this effect is not corrected for, then the low molar mass tail of the concentration chromatogram results in slightly underestimated results, and the global $\bar{M}_n$ is also slightly underestimated. For copolymers, Eq. (21) is only valid when all co-monomeric units exhibit identical specific refractive index increments, and/or when the chemical composition does not change with the molar mass.

Let us call $s_{DR}(V)$ the baseline corrected DR chromatogram. For homopolymers, the DR signal is proportional to the instantaneous mass concentration as follows:

$$s_{DR}(V) = k_{DR} c(V) \qquad (22)$$

where k_{DR} is a calibration constant that depends on the instrument gain and on $\partial n/\partial c$ of the analyzed polymer. Fortunately, k_{DR} is only required when absolute values of $c(V)$ are necessary (e.g., in a signal ratio involving the instantaneous mass concentration). The value of k_{DR} is experimentally obtained by injecting a series of known masses of the analyzed polymer, and measuring the areas under their corresponding chromatograms. Then, the DR gain is obtained from the slope of the resulting linear plot. With this procedure, the $\partial n/\partial c$ value is not required, but the disadvantage is that the resultant detector calibration is only applicable to the analyzed polymer.

Apart from the DR, other concentration sensors are the evaporative LS detector and the density detector (Section VI.E.1).

B. UV Sensor

Many organic solvents are transparent to UV light, whereas some polymers absorb at specific UV wavelengths. For this reason, UV sensors are often employed to detect either the instantaneous total concentration (for homopolymers) or the concentration of specific atomic groups (for copolymers). The absorbance is:

$$A(V, \lambda) = \log\left[\frac{I_0(V,\lambda)}{I_t(V,\lambda)}\right]$$

where I_t and I_0 are the intensities of the transmitted light through the polymer solution and pure solvent cells, respectively. For several absorbing groups, the absorbance is related to their instantaneous concentration through Beer's law:

$$A(V, \lambda) = \log \frac{I_0}{I_t(V,\lambda)} = \sum_i \delta_i(\lambda) l c_i(V) \qquad (23)$$

where δ_i and c_i are the molar absorptivity and the concentration of the different absorbing groups, respectively, and l is the constant flow-cell length. The molar absorptivity depends on the polymer, the solvent, and the wavelength (Snyder et al., 1997).

Most UV sensors do not include a solvent reference cell, and estimate I_0 directly from the sample detector signal when pure solvent is eluting. UV sensors at a fixed wavelength are generally preferred to DR for determining the instantaneous concentration of homopolymers, the reason being that UV sensors are less sensitive to temperature oscillations than are DR. For a given homopolymer–solvent system and at a single-wavelength, the UV chromatogram signal is:

$$s_{UV}(V) = k_{UV} c(V) \qquad (24)$$

where the k_{UV} calibration constant can be estimated following a similar procedure to that previously described for k_{DR}.

An important application of UV sensors is the determination of the instantaneous copolymer composition (see Section III.D). On-line diode array UV–Vis spectrophotometers provide the complete UV–Vis spectrum of the instantaneous eluent, and are particularly useful for determining the chemical nature of copolymers and of low molar mass polymer additives.

C. Molar Mass Calibrations

When the chromatograph is only fit with an instantaneous concentration detector, then a molar mass calibration is required to estimate the MMD. Figure 2 shows a typical calibration obtained from narrow standards. Such calibration depends on the chromatographic system, the nature of the calibration standards, and the operating conditions (solvent flow rate, temperature, etc.). For synthetic polymers, the most common calibration kits contain a set of narrowly distributed PS standards, in turn, synthesized via living anionic polymerizations. For certain polymers, only broad MMD standards are available.

1. Direct Calibration

A direct calibration is normally obtained from a set of narrow standards of the same chemical nature as the analyzed polymer. Each standard of molar mass M_i yields a narrow chromatogram of mean elution volume V_i. Then, the calibration is obtained by interpolating the experimental set of pairs (M_i, V_i) with an analytical expression. The resulting $\log M(V)$ function is normally linear in its central region, but is sometimes interpolated with polynomials of order three or less (higher order polynomials can lead to unrealistic oscillations in the adjusted curve).

Unfortunately, the exact MMD of narrow commercial standards is generally unknown, and only one or more of its absolute average molar masses are provided. For low molar mass standards, $\bar{M}_n$ values are reasonably accurate (as determined, e.g., by vapor pressure osmometry), whereas $\bar{M}_w$ values (obtained by LS) are relatively erroneous. The opposite occurs with high molecular weight standards. The errors in the average molar masses of standards determine that their true polydispersities are unknown.

Commercial standards of almost any molar mass are available and, therefore, it is possible to cover the dynamic range of any column set. Care must be taken to avoid degradation of the ultrahigh standards during their dissolution and fractionation.

A complication of the data processing is how to obtain the experimental set of M_i and V_i values. A widely applied solution is to assign the chromatogram peak at the geometric mean of the average molar masses $(\bar{M}_n \bar{M}_w)^{0.5}$.

This expression is strictly valid for log-normal chromatograms.

2. Universal Calibration

The proportionality between $[\eta]M$ and V_h in Eq. (15) is the basis of the "universal" calibration proposed by Benoit and coworkers (Grubisic et al., 1967). For a chromatographically complex polymer and, even under perfect resolution, a whole distribution of molar masses is instantaneously present in the detector cell. In this case, Hamielec and Ouano (1978) have proven that M in Eq. (15) should be replaced by the instantaneous number-average molar mass $M_n(V)$. Unfortunately, on-line $M_n(V)$ sensors are not commercially available, and also off-line measurements of $\bar{M}_n$ are only possible for molar masses lower than around 1 million Da.

The universal calibration (Grubisic et al., 1967) is required when standards of the analyzed polymer are not available and/or when the nature of the analyzed polymer is unknown. It is applicable, not only for many linear homopolymers, but also for most copolymers, branched polymers, and biopolymers. Furthermore, the universal calibration is required when an unknown polymer is analyzed by on-line viscometry.

Assume that the direct calibration obtained from a set of PS standards $\log M_{PS}(V)$ is available, and that the intrinsic viscosities of such standards $[\eta]_{PS}(V)$ have been determined by capillary viscometry or have been calculated from the MHS constants. The universal calibration is represented by $\log\{[\eta]_{PS} M_{PS}\}$ vs. V. If the fractionation is strictly by hydrodynamic volume, the polymer chains are flexible, and the Θ conditions are assumed, then one can write:

$$\{[\eta]_{PS} M_{PS}\}(V) = \{[\eta]_X M_X\}(V) \qquad (25)$$

where the subscripts PS and X represent the PS standards and the analyzed polymer, respectively. Clearly, this calibration is "universal" only for a given chromatograph under specific operating conditions. The validity and limitations of the universal calibration has been investigated on many occasions (Provder and Kuo, 1999). The shape of $\log\{[\eta]_{PS} M_{PS}\}$ vs. V is similar in shape of the direct calibration presented Fig. 2. In effect, when the direct calibration $\log M_{PS}(V)$ is linear and the MHS coefficients of the calibrants are constant, then $\log\{[\eta]_{PS} M_{PS}\}(V)$ is also linear.

Consider the determination of the MMD of a polymer of unknown chemical nature from the concentration chromatogram and the universal calibration. To transform retention volume into molar mass, either the MHS function $\log[\eta]_X$ vs. $\log M_X$ or (equivalently) the MHS constants of the analyzed polymer are required. In the first case, $M_X(V)$ is obtained by replacing the log–log relationship in Eq. (25) (see the application example of Section VII.B). In the latter case, assume that at the given chromatographic conditions, the MHS constants of the analyzed polymer (K_{PS}, α_{PS}) and of the calibrant (K_X, α_X) are all known. By replacing the MHS equation [Eq. (13)] into Eq. (25), it yields:

$$K_{PS} M_{PS}^{\alpha_{PS}+1}(V) = K_X M_X^{\alpha_X+1}(V) \tag{26}$$

and therefore:

$$M_X(V) = \left[\frac{K_{PS}}{K_X} M_{PS}^{\alpha_{PS}+1}(V)\right]^{\frac{1}{\alpha_X+1}} \tag{27}$$

Equation (27) calculates the molar mass of the analyzed polymer on the basis of the MHS constants and the PS molar mass at the same elution volume. Note that the universal calibration $\log\{[\eta]M\}(V)$ is not explicitly included in Eq. (27). Instead, Eq. (27) is based on the universal calibration concept, but it only requires the $M_{PS}(V)$ values from a direct PS calibration.

D. Chemical Composition

Consider the determination of the copolymer composition. More specifically, we shall determine the instantaneous mass fraction of A of an AB-copolymer [represented by $p_A(V)$], by standard dual-detection, that is, DR + UV at a single-wavelength.

Consider, first, the simplest situation where the UV detector "sees" only repeating unit A, whereas the DR detects the total instantaneous concentration (implying that the $\partial n/\partial c$ of both co-monomers are very similar). In this case, the instantaneous composition is simply obtained from the following signal ratio:

$$p_A(V) = \frac{c_A(V)}{c(V)} = \left[\frac{k_{DR}}{k_{UV,A}}\right]\left[\frac{s_{UV}(V)}{s_{DR}(V)}\right] \tag{28}$$

where c_A is the instantaneous concentration of A, k_{DR} is the DR calibration constant, and $s_{UV}(V)$, $k_{UV,A}$ are, respectively, the UV signal and UV calibration (obtained by injecting known masses of homopolymer A). The propagation of errors in Eq. (28) determines that acceptable estimations of the instantaneous composition are only feasible in the mid-chromatogram region; whereas large errors are produced in the chromatogram tails.

Consider, now, the more general case where the co-monomers exhibit different specific refractive index differences and absorptivities. In this case, the detector equations can be written:

$$s_{DR}(V) = k_{DR,A} p_A(V) c(V) + k_{DR,B}[1 - p_A(V)]c(V) \tag{29a}$$

$$s_{UV}(V) = k_{UV,A} p_A(V) c(V) + k_{UV,B}[1 - p_A(V)]c(V) \tag{29b}$$

In Eqs. (29a) and (29b), $p_A(V) c(V)$ represents the instantaneous concentration of A, while $[1 - p_A(V)] c(V)$ represents the instantaneous concentration of B. Then, from the chromatograms $s_{DR}(V)$ and $s_{UV}(V)$ and the calibrations $k_{DR,A}$, $k_{DR,B}$, $k_{UV,A}$, and $k_{UV,B}$, one can solve for the unknowns $c(V)$ and $p_A(V)$, yielding:

$$c(V) = \frac{k_{UV,A} - k_{UV,B}}{k_{DR,B} k_{UV,A} - k_{DR,A} k_{UV,B}} s_{DR}(V)$$
$$+ \frac{k_{DR,B} - k_{DR,A}}{k_{DR,B} k_{UV,A} - k_{DR,A} k_{UV,B}} s_{UV}(V) \tag{30a}$$

$$p_A(V) = \frac{k_{UV,B} - k_{DR,B}[s_{UV}(V)/s_{DR}(V)]}{(k_{UV,B} - k_{UV,A}) + (k_{DR,A} - k_{DR,B})[s_{UV}(V)/s_{DR}(V)]} \tag{30b}$$

Equation (30a) calculates the instantaneous concentration from a linear combination of the two signals. For this reason, the calculation is numerically "well behaved" in the sense that as $s_{DR}(V)$ and $s_{UV}(V)$ tend to zero, then $c(V)$ also tends to 0. In contrast, Eq. (30b) calculates the instantaneous composition from a nonlinear combination of the signals. For this reason, acceptable estimations of $p_A(V)$ are only feasible in the mid-chromatogram region.

Under special conditions, the CCD represented, for example, by dc/dp_A vs. p_A, can be obtained by simple combination of the individual functions $p_A(V)$ with $c(V)$. For this to be possible, the following (rather hard) conditions must be verified: (a) the instantaneous distribution of the chemical composition in the detection cells is narrow and (b) the chemical composition varies monotonically with the elution volume (Meira and Vega, 2003).

IV. MOLAR MASS SENSITIVE DETECTORS

The LS and SV detectors are particularly useful when the nature of the polymer sample is unknown and/or when a chromatographically complex polymer is analyzed. In either case, a mass concentration sensor (e.g., a DR) is also required to carry out the data treatment.

As a consequence of BB (see Section V.C), of secondary fractionations, and/or of the chromatographically complex nature of many polymers, a variety of molar masses is present in the detector cells. For all these reasons, the ideal LS-DR detector system provides the instantaneous weight-average molar mass $M_w(V)$, whereas the SV-DR combination provides an instantaneous (specific viscosity-based) average molar mass $M_{sv}(V)$.

There are two main problems associated with the dual LS-DR or SV-DR detection. First, a signal ratio must be calculated, and then accurate estimates of the molar mass are only feasible in the mid-chromatogram region. Second, both the LS and SV sensors are quite insensitive to low molar masses.

It should be noted that neither $M_w(V)$ nor $M_{sv}(V)$ should be considered as calibrations, and the reason for this is that fractionation is not strictly by molar mass. In effect, it can be shown that the functions $M_w(V)$ and $M_{sv}(V)$ both depend on the analyzed MMD; and, for this reason, we shall call them *ad-hoc* calibrations. Fortunately, this is not the case for calibrations obtained from narrow PS standards because: (a) the standards are chromatographically simple polymers and (b) by assigning a mean elution volume to some average molar mass, we are somehow avoiding (or correcting for) the BB problem.

A. LS Detector

The LS-DR combination provides an absolute measurement of the instantaneous molar mass detectors and, in this case, an independent molar mass calibration is not required. The LS signal (or signals) is the excess Rayleigh ratio (ΔR_θ), which measures (at one or more detection angles θ) the excess scattering intensity by the polymer solution with respect to the pure solvent scattering. For dilute solutions, the following can be written (Burchard, 1999; Jackson and Barth, 1995; Wyatt, 2003):

$$\frac{K_{LS}c(V)}{\Delta R_\theta(V)} = \frac{1}{M_w(V)P(\theta)} + 2A_2c(V) \tag{31a}$$

with

$$\frac{1}{P(\theta)} \cong 1 + \frac{16\pi^2 n_0^2}{3\lambda_0^2} R_{g,z}^2(V) \sin^2\left(\frac{\theta}{2}\right) \tag{31b}$$

$$K_{LS} = \frac{4\pi^2 n_0^2 (\partial n/\partial c)^2}{\lambda_0^4 N_A} \tag{31c}$$

where K_{LS} is the optical "constant" given by Eq. (31c); $P(\theta)$ is the particle scattering function, which describes the angular variation of the scattered light intensity, and depends on the size and shape of the dissolved polymer molecules; A_2 is the second coefficient of the virial expansion; n_0 is the solvent refractive index; λ_0 is the wavelength; $R_{g,z}^2(V)$ is the instantaneous z-average square radius of gyration; and N_A is the Avogadro constant.

To obtain the instantaneous average molar mass, $M_w(V)$, the double extrapolation $c \to 0$ and $\theta \to 0$ are required. The requirement of $c \approx 0$ is essentially verified at the very low chromatographic concentrations, and this also allows us to neglect the polymer–polymer interaction term ($2A_2c$). To accomplish the $\theta \to 0$ condition, two different sensor types have been developed. In the low-angle LS detectors, measurements are taken at a sufficiently low scattering angle (e.g., between 3° and 7°), where $\theta \approx 0$. In multiangle detectors, the measurements are taken at several fixed angles, and the extrapolation to $\theta = 0$ is required.

The number of measurement angles is fixed for a given equipment, and instruments with 2, 3, 7, and 18 measuring angles are commercially available. From the concentration chromatogram $c(V)$ and the set of LS chromatograms $\Delta R_{\theta,i}(V)$, both $M_w(V)$ and $R_{g,z}^2(V)$ can be obtained from a simplified version of the Zimm-plot procedure (Wyatt, 2003). More specifically, at each elution volume, the experimental function $[K_{LS} c(V)/\Delta R_{\theta,i}(V)]$ vs. $\sin^2(\theta_i/2)$ is fit by a straight line through a least-square regression. Then, at each elution volume, $M_w(V)$ is obtained from the ordinate intercept, whereas $R_{g,z}^2(V)$ is obtained from the slope [see Eq. (31a)].

The representation of $R_{g,z}^2$ vs. $\log M$ is the so-called conformational plot. This plot provides an information on the linearity or nonlinearity of the eluting solution and, in this respect, it is similar to the information provided by the MHS plot $\log[\eta]$ vs. $\log M$. For linear molecules, the slopes of conformational plot vary between 0.5 and 0.6, whereas branched molecules of the same molar mass are more compact and, therefore, exhibit lower slopes. Note that $R_{g,z}^2$ is obtained by interpolation at several measurement angles and, therefore, this variable cannot be obtained from single low-angle LS detectors. Also, note that most LS sensors can also be used in the off-line mode and, in this case, the global averages for the total polymer $\bar{M}_w$ and $\bar{R}_{g,z}^2$ can be directly determined.

The LS sensor calibration consists of estimating the optical constant of Eq. (31c), which unfortunately depends on $\partial n/\partial c$ of the analyzed polymer. Due to the quadratic dependence of K_{LS} on $\partial n/\partial c$, accurate specific refractive index differences are required for quantitative estimations of $M_w(V)$. Some commercially available refractometers (either interferometric or deflection) can be used to measure, off-line, the global $\partial n/\partial c$ values. Values of $\partial n/\partial c$ for many polymer–solvent systems and at several wavelengths are tabulated, but little information is available on $\partial n/\partial c$ values for copolymers.

The following correlation can be used to compensate for possible variations in $\partial n/\partial c$ due to variations in the wavelength (when, e.g., the wavelength employed in the off-line refractometer measurement of $\partial n/\partial c$ differs from that of the LS sensor):

$$\frac{\partial n}{\partial c} = k' + \frac{k''}{\lambda^2} \tag{32}$$

where k' and k'' are, respectively, the intercept and slope of the linear plot $\partial n/\partial c$ vs. $(1/\lambda^2)$. For a given polymer solution, $\partial n/\partial c$ depends not only on the nature of the polymer and concentration, but it also varies (linearly) with the pure solvent refractive index. For this reason, the $\partial n/\partial c$ of a given polymer–solvent system can be estimated by interpolation between known $\partial n/\partial c$ values of a given polymer in different solvents.

For copolymers, the on-line estimation of $M_w(V)$ by SEC-LS is only possible when (Jackson and Barth, 1995; Wyatt, 2003): (1) both co-monomers exhibit the same $\partial n/\partial c$ or (2) the copolymer is homogeneous (i.e., the CCD is very narrow). In such cases, the copolymer can be treated as a pseudo-homopolymer, and Eqs. (31a)–(31c) are applicable. Otherwise, only an apparent instantaneous molar mass $M_W^*(V)$ is determined, which depends on the solvent refractive index. Finally, note the global $\bar{M}_w$ of a copolymer can be estimated from off-line LS measurements in at least three solvents of different refractive indexes (Jackson and Barth, 1995).

1. MMD from LS-DR Detection

For both c and θ tending to 0, and according to Eq. (31a), the (zero-angle) LS sensor signal is:

$$\Delta R_\theta(V) \cong K_{LS} M_w(V) c(V) \tag{33}$$

and the molar mass can be directly calculated from a signal ratio:

$$M_w(V) = \left(\frac{k_{DR}}{K_{LS}}\right)\left[\frac{\Delta R_\theta(V)}{s_{DR}(V)}\right] \tag{34}$$

Finally, the MMD can be obtained from $c(V)$ and $M_w(V)$, yielding: $[dc/d\log M](\log M) \cong [dc/d\log M_w](\log M_w)$.

As mentioned before, Eq. (34) is only applicable in the mid-chromatogram region. To estimate the MMD and to calculate the global averages, a (rather crude) solution is to simply neglect the MMD tails, with the result of a gross underestimation of the global polydispersity. A better solution is to somehow extrapolate the *ad-hoc* calibration $\log M_w(V)$ of the mid-chromatogram region. For example, the following has been applied (Brun et al., 2000): first, estimate a polynomial $\log M_w(V)$ that minimizes the least-square errors between the measured $\Delta R_\theta(V)$ and its prediction according to Eq. (33), and then extrapolate the obtained polynomial toward the chromatogram tails.

As mentioned before, an important limitation of LS is its applicability for copolymers of a varying chemical composition. In this case, $s_{DR}(V)$ is no longer proportional to $c(V)$, and (even worse) $\partial n/\partial c$ changes with elution volume, resulting in a variable LS calibration.

An interesting observation is that the mass chromatogram is not required for calculating the global $\bar{M}_w$. In effect, the following expression obtains the global $\bar{M}_w$ from the injected mass W, the area under the LS chromatogram (at zero degree), and the optical constant:

$$\bar{M}_w = \frac{\int_0^\infty [\Delta R_\theta(V)/K_{LS}]\,dV}{W} \tag{35}$$

This calculation is particularly useful when the higher molar mass fractions remain undetected by the DR sensor.

B. Viscosity Detector

Several viscometer types have been developed for the on-line measurement or estimation of specific viscosity. The specific viscosity is dimensionless [Eq. (11)] and, therefore, no calibration constants are required in this case. Thus, we shall assume that the SV signal is the instantaneous specific viscosity $\eta_{sp}(V)$.

At the (very low) chromatographic concentrations, the instantaneous intrinsic viscosity [Eq. (12)] is directly obtained from a signal ratio (without requiring the extrapolation to zero concentration):

$$[\eta](V) \cong \frac{\eta_{sp}(V)}{c(V)} = k_{DR}\left[\frac{\eta_{sp}(V)}{s_{DR}(V)}\right] \tag{36}$$

The intrinsic viscosity distribution is defined by the function $dc/d[\eta]$ vs. $[\eta]$, and this distribution is readily obtained from $c(V)$ and $[\eta](V)$.

1. MMD from SV-DR Detection

Consider estimating the MMD of a linear homopolymer by combined SV-DR. In this case, the universal calibration $\log\{[\eta]M\}(V)$ [$\equiv \log J(V)$] is required. An instantaneous (specific viscosity based) molar mass $M_{SV}(V)$ results from the universal calibration and the instantaneous intrinsic viscosity measurement as follows:

$$M_{SV}(V) = \frac{J(V)}{[\eta](V)} = J(V)\frac{c(V)}{\eta_{sp}(V)}$$
$$= \frac{J(V)}{k_{DR}} \times \frac{s_{DR}(V)}{\eta_{sp}(V)} \tag{37}$$

Equation (37) is applied for determining the instantaneous molar mass of homopolymers, copolymers, and branched polymers.

Another side application of SV-DR is the determination of the MHS coefficients for a given polymer–solvent–temperature condition. The MHS plot $\log[\eta](V)$ vs. $\log M_{SV}(V)$ is obtained from Eqs. (36) and (37), yielding a straight line for linear homopolymers in good solvents. In this case, the MHS plot can be represented by:

$$\log[\eta] = \log K + \alpha \log M_{SV} \tag{38}$$

and the MHS coefficients K and α can be obtained from the intercept and the slope of the resulting linear plot. Once the

MHS coefficients have been determined for a given polymer–solvent system, then new samples of the same polymer can be directly processed from the concentration signal and the conventional universal calibration procedure, without requiring the SV signal [e.g., by application of Eq. (27)].

Similarly to Eq. (35), the theory indicates that it is possible to estimate the global $\bar{M}_n$ of a complex polymer without requiring its concentration chromatogram. To this end, the following expression is applied, which requires the specific viscosity signal, the total injected mass W, and the universal calibration (Balke et al., 1994):

$$\bar{M}_n = \frac{W}{\int_0^\infty [\eta_{sp}(V)/J(V)]\,dV} \quad (39)$$

2. Degree of Branching by SV-DR

Consider the analysis of a branched homopolymer containing tri- and/or tetrafunctional branching points. The aim here is to determine the instantaneous average number of branches per molecule by combined SV-DR. The following procedure can be applied. First, determine $[\eta](V)$ $\{= [\eta]_b(V)\}$ and $M_{SV}(V)$ $\{= M_b(V)\}$, through Eqs. (36) and (37), respectively. Second, determine $[\eta]_b(M_b)$. Third, calculate the variation of the g branching parameter with the molar mass from the following expression [obtained from Eqs. (13), (17), and (18)]:

$$g(M) = \left[\frac{[\eta]_b(M_b)}{K_l M_l^{\alpha_l}}\right]^{1/\varepsilon}; \quad M_b = M_l \quad (40)$$

where K_l and α_l are the MHS coefficients of the linear homolog of the same molar mass. Fourth, calculate the variation of the instantaneous number- or weight-average number of branching points per molecule with the molar mass $[b_n(M)$ or $b_w(M)$, respectively] through the Zimm–Stockmayer expressions. For trifunctional branching points, these expressions are (Zimm and Stockmayer, 1949):

$$g(M) = \left[\left(1 + \frac{b_n(M)}{7}\right)^{0.5} + \frac{4b_n(M)}{9\pi}\right]^{-0.5} \quad (41a)$$

$$g(M) = \frac{6}{b_w(M)}\left[\frac{1}{2}\left(\frac{2 + b_w(M)}{b_w(M)}\right)^{0.5}\right.$$
$$\left.\times \ln\left(\frac{[2 + b_w(M)]^{0.5} + [b_w(M)]^{0.5}}{[2 + b_w(M)]^{0.5} - [b_w(M)]^{0.5}}\right) - 1\right] \quad (41b)$$

Equivalent expressions to Eqs. (41a) and (41b) have been developed for random long-branched polymers containing tetrafunctional branching points, and for star-type macromolecules (Zimm and Stockmayer, 1949). Note that Eqs. (41a) and (41b) have been developed for polymer solutions of a uniform molar mass under θ conditions, whereas there is an instantaneous distribution of molar masses when branched polymers are analyzed by SEC, and good solvents are also required.

Note, finally, that from the MMD $[dc/d\log M](\log M)$ and from the function $b_n(\log M)$ [or $b_w(\log M)$], it may be possible to represent the degree of branching distribution as $[dc/db_n](b_n)$ [or $dc/db_w(b_w)$]. Necessary conditions for this to be possible are: (a) (most importantly) the instantaneous distribution of number of branches per molecule must be narrow and (b) b_n [or b_w] must increase monotonically with $\log M$. Fortunately, this is almost the case of many free radically produced branched homopolymers, where grafting reactions involve the accumulated polymer chains, and longer chains have a higher grafting probability than shorter chains. Note that, while the instantaneous b_n (or b_w) values are real numbers, the number of branches are strictly integers b, generally lower than 15. For this reason, the branching distribution is a truly discrete distribution that could be represented by $c(b)$.

C. Osmotic Pressure Detector

Osmotic pressure detectors are not yet fully developed, and are not commercially available. This is unfortunate since, in principle, these sensors are highly sensitive to the low molar masses, and this could conveniently compensate the rather low sensitivities of LS and SV sensors to oligomers. Also, as mentioned before, the universal calibration should involve, in theory, the product $\{[\eta]M_n\}$ (Hamielec and Ouano, 1978), and the M_n values of high molar mass standards are generally unavailable.

Several on-line osmotic pressure prototypes have been reported (Yau, 1991; Lehmann et al., 1996), which, in theory, provide an instantaneous $M_n(V)$. The instrument measures the osmotic pressure difference ΔP_{os} between the pure solvent and the polymer solution. At the very low chromatographic concentrations, the following limiting expression can be directly applied:

$$\frac{\Delta P_{os}(V)}{c(V)} \simeq \frac{RT}{M_n(V)} \quad (42)$$

where R is the gas constant and T is the temperature. Clearly, $M_n(V)$ is again obtained from a signal ratio.

V. EXPERIMENTAL DIFFICULTIES

A. The High Molecular Weight Saturation Problem

The total exclusion limit is an important limitation in SEC. In effect, most commercial columns cannot fractionate molar masses that are higher than around 10 million Da. This is a serious limitation for many biopolymers and for many high molar mass synthetic polymers (e.g., produced by emulsion polymerization). Chromatograms with fractions above the total exclusion limit exhibit a high molar mass saturation elbow or peak that is sometimes misinterpreted as a MMD bimodality.

There are two important problems associated with the analysis of samples exhibiting fractions close to the total exclusion limit. First, the resolution is low in that region, and second, the larger molecules can be seriously degraded. The low resolution is the consequence of two combined effects: the logarithmic nature of the molar mass calibration and the BB (i.e., maximum near the total exclusion) (Busnel et al., 2001). Also, the low resolution of high molar mass species generates a high concentration region that further worsens the problem. The molecular degradation by shear stress breaks the large molecules into approximately equal halves, with the effect of seriously underestimating the global $\bar{M}_w$. The degradation problem is particularly serious in columns of small porous beads (e.g., smaller than 7 μm) and in occluded column prefilters. It can be improved by lowering the maximum system pressure, by adding antioxidants such as hydroquinone, and by avoiding the smaller pore and smaller bead columns.

B. Enthalpic Interactions

Ideally, the sought combination of solvent, polymer, and packing must ensure that size exclusion is the only active fractionation mechanism. In other words, no enthalpic interactions must take place between packing and analyte. This is the case of analyzing nonpolar polymers such as PS with polar solvents such as THF in a (nonpolar) PS/polydivinylbenzene (PDVB) crosslinked porous matrix (Berek, 1999). In contrast, the bare silica gel and the porous glass packing are adsorptive, due to the strong interactions that can occur between the packing and the polar or basic groups of the analyte. Several unwanted secondary enthalpic interactions can affect the pure entropic fractionation (Berek, 2000; Berek and Marcinka, 1984). The most important is possibly the polymer-packing adsorption, whereas other secondary mechanisms include partition, ionic effects, and so on.

C. Band Broadening

When analyzing a linear homopolymer, and even in the absence of secondary (or enthalpic) fractionations, the BB determines that (inevitably) a variety of hydrodynamic volumes will be instantaneously present in the detector cell. The main reason for BB is the axial dispersion in the separation columns (Baumgarten et al., 2002; Busnel et al., 2001; Potschka, 1993; Netopilík, 2002; Striegel, 2001). Other reasons are finite injections, finite detection cells, parabolic flow profiles in capillaries, end-column effects, and heterogeneous packing density (Hupe et al., 1984).

BB smoothens and broadens all of the measured chromatograms with a common broadening function. The effect is to overestimate the global polydispersity when the molar masses are calculated from a concentration signal and a calibration, and underestimates the global polydispersity when molar mass sensitive detectors are employed (Jackson and Yau, 1993; Netopilík, 1997).

Fortunately, BB can be generally neglected in the more normal case of smooth and broad chromatograms. But, when quantitative results are required from very narrow or multimodal chromatograms, then all chromatograms should be corrected for BB prior to calculating the molar masses and other polymer characteristics. The standard procedure for BB correction involves the Tung equation (Tung, 1966).

$$s(V) = \int_{-\infty}^{\infty} g(V, V') s^c(V') \, dV' \qquad (43)$$

where $s(V)$ is any measured chromatogram, $s^c(V)$ is the corresponding BB-corrected chromatogram, and $g(V, V')$ is the (in general, nonuniform and asymmetric) broadening function. There are two main problems associated with Eq. (43): (1) the broadening function is difficult to determine except at the low molar mass limit where simple substances can be employed and (2) an ill-posed numerical "deconvolution" is required to calculate $s^c(V)$ from $s(V)$ and $g(V, V')$. Other approximate solutions have been proposed that avoid the numerical deconvolution (Jackson, 1999; Netopilík, 1998a; Prougenes et al., 1998; Yau et al., 1977). Traditionally, only the concentration chromatograms were corrected for BB (Gugliotta et al., 1990; Hamielec, 1984; Meira and Vega, 2001); but, more recently, the molar mass sensitive detectors have spurred the correction of LS and SV chromatograms (Jackson, 1999; Meira and Vega, 2001; Netopilík, 1998a; Prougenes, 1998; Vega and Meira, 2001).

D. Polyelectrolytes

The normal SEC fractionation can be seriously affected by polyelectrolytes, which can be either accelerated or retarded with respect to equivalent neutral molecules. The former situation occurs when an electrostatic repulsion occurs between the polyelectrolyte charges and the residual column packing charges. The latter (and more normal) situation occurs in the presence of attractive forces between solute and packing. Such forces may be the result of electrostatic attractions, hydrogen bonding, or hydrophobic interactions. Hydrophobic interactions can be reduced by addition of surfactants into the mobile phase.

In other chromatographic techniques, the enthalpic interactions are the basis of their fractionation mechanisms. For example, peptides and proteins may be conveniently separated by: (1) ionic exchange chromatography, where the mobile phase exhibits a low ionic force or (2) hydrophobic interaction chromatography, where the mobile phase exhibits a high ionic force.

VI. INSTRUMENTATION

A. High-Temperature Equipment

Most SEC analyses are carried out at ambient temperature. Ambient-temperature equipment is, in general, highly modular, in the sense that most system components are interchangeable. This is not the case of some dedicated high-temperature chromatographs, where the solvent reservoir, the pump, the injector, the columns, and the detectors are all inside a temperature-controlled casing; with the temperature regulated anywhere between 110°C and 150°C. High-temperature systems are specifically designed for analyzing polyolefins like polyethylene and polypropylene. These highly crystalline commodities can be only dissolved at high temperature in chlorinated solvents, and are the most important synthetic polymers from the point of view of world production.

An intermediate possibility between ambient- and high-temperature equipment is to use column ovens and/or detector temperature baths, for the independent temperature control of columns and/or detectors. Column ovens can regulate the temperature between 35°C and 100°C, within ±0.2°C.

When column ovens are applied, the mobile phase suffers an expansion as it enters the columns. This increases the column flow, reduces the solvent viscosity, increases the polymer diffusivity, and increases the column efficiency. This last improvement makes it possible to reduce the total analysis time without affecting resolution. Finally, the column temperature control stabilizes the baseline and improves the system reproducibility.

B. Solvent Delivery System

1. Carrier Solvents

An ideal SEC solvent is characterized by: (a) a high capacity for dissolving many polymers, (b) a low viscosity, (c) a low reactivity, (d) a low inflammability, (e) a capacity for avoiding secondary fractionations, and (f) a good compatibility with the column packing and detectors. The most common solvents are summarized Table 1.

Good solvents must be employed to avoid polymer–polymer interactions, and adsorption between analyte and column packing. For polar packings, it is recommendable to work with highly polar solvents; and eventual enthalpic interactions can be avoided by adding a polar component into the mobile phase that blocks the packing adsorption sites (Glöckner, 1987).

The column packing can be degraded by incompatible solvents. Even though the same packing can be used with a wide range of organic solvents, care must be taken when changing the solvent. The new solvent must be completely miscible with the original solvent, and stabilizers or additives must remain in solution during the solvent transfer.

Some solvents can be viscous at room temperature, and it may be convenient to increase the column temperature (to say, 50–120°C) for improving the mass transfer, reducing the system pressure, and extending the column life.

Independent of its purity, the carrier solvent must be filtered (to eliminate eventual particles and/or bacteria) and degassed, since dissolved gases can generate bubbles in the pump and, therefore, lead to irregular flows. Also, dissolved oxygen can oxidize the analytes and/or alter the UV sensor baseline.

Table 1 Common Solvents in SEC

Solvent	Uses	Temperature
THF	Synthetic polymers and oligomers	Ambient to 40°C
o-Dichlorobenzene	Polyolefines	130–150°C
1,2,4-Trichlorobenzene	Polyolefines	130–150°C
Toluene	Rubbers, elastomers	Ambient to 90°C
m-Cresol	Polyamides, polyesters	100°C
Chloroform	Silicons, epoxi resins	Ambient
Dimethylformamide	Polyurethanes, acrylonitrile	Ambient to 100°C
Trifluoroethanol	Polyamides	Ambient to 40°C
Aqueous solutions of electrolytes and/or buffers	Biopolymer: gelatine, polyacrylates, dextrans, polyglycols, polyacrylamide, polysaccharides, and paper pulp	Ambient to 40°C

The filtering process is usually carried out with membranes of pore size between 0.22 and 0.45 μm. Filtering membranes are placed between the porous glass base of an upper funnel and a lower receiving reservoir, all kept together by means of a metal clamp. The membrane material depends on the selected solvent (Belenkii and Vilenchik, 1983) (Table 2).

The degassing methods include: (a) the reflux technique that requires boiling the mobile phase, (b) the bubbling of an inert gas, and (c) the application of vacuum. This last method is the most commonly used, and it can be combined with the solvent filtration. The batch reflux technique is not applicable to solvent mixtures and is, in general, not recommendable due to possible re-dissolution of the atmospheric gases.

2. Solvent Reservoir, Tubing, and Couplings

The solvent reservoir is made of glass or of a resistant polymer. It contains a cover to minimize evaporation and to prevent the entrance of environmental particles. The reservoir base must be placed above the pump level to ensure a normal pump operation. A porous stainless steel cartridge of porosity between 2 and 10 μm is connected at the solvent inlet tubing to avoid the entrance of particles. Helium gas can be continuously bubbled into the solvent reservoir and, in this case, a control valve is incorporated to ensure a positive gas pressure.

Tubing and capillaries must be chemically inert and must be tightly adjusted. For the high-pressure capillaries, stainless steel tubing is used with external and internal diameters 1/16 in. and 0.2 mm, respectively. The low-pressure tubing may be of polypropylene or Teflon. To minimize extracolumn BB, the capillaries must be kept short, and the interconnectors must contain a minimum of dead volume. Unfortunately, interconnectors are not interchangeable between the various manufacturers. Therefore, one must never use connectors from one instrument to replace a connector in a different one.

Table 2 Filtering Membranes and Solvent Compatibilities

	Solvent		
Membrane	Organic	Aqueous	Aqueous/organic
Regenerated cellulose	R		
Cellulose ester	NR	R	NR
Cellulose nitrate	NR	R	*
Poly(vinylidene fluoride)	R	R	R
Nylon	R	R	R
Teflon	R	NR	NR

Note: R, recommended; NR, not recommended and *, maximum of 30% organic solvent.

3. Positive-Displacement Pumps

In conventional analytical SEC, the maximum pump pressure is around 6000 psi, and the flow rates are controlled between 0.1 and 10 mL/min. In contrast, with microbore columns, flow rates are in the order of μL/min; whereas, in semipreparative and preparative SEC, the flow rates can be 150 mL/min or higher. The pump is made of (both chemically and mechanically) resistant materials such as stainless steel, sapphire, ruby, and Teflon. To preserve the "biological activities" of some biopolymers, the stainless steel can be replaced by titanium (Belenkii and Vilenchik, 1983). Some chromatographs include pressure sensors/switches for interrupting the flow rate when the pressure is either too high (due to obstructions in the filter, etc.) or too low (due to solvent losses or entrance of air bubbles).

An accurate and constant flow rate is necessary for associating the chromatograms abscissas with the retention time (rather than with the retention volume). Small flow fluctuations due to the pump operation are eliminated by a damper placed between the pump and the columns, often just ahead of the sample injection valve. Some slow flow fluctuations may not be due to the pump. For example, on-line viscometers exhibit the so-called Lesec effect, which is produced when the highest molar mass fraction reaches the viscometer (Lesec, 1994; Netopilík et al., 2000). The effect is similar to a shift in the specific viscosity chromatogram, and it can be confused with the interdetector volume effect that is observed when SV and DR sensors are interconnected in series.

The high-frequency pressure noise is normally caused by pump malfunctioning due to retained bubbles and/or leaking pump seals. In contrast, a slow drift in the flow rate (that becomes apparent by a drift in the DR chromatogram baseline) is generally due to density changes in the solvent which are, in turn, produced by temperature drift.

Positive-displacement pumps can be reciprocating, syringe-type, or diaphragm-type. Reciprocating pumps are the most commonly used in high-pressure liquid chromatography (HPLC), and several types of these are available: single-piston, double-pistons, three-pistons, tandem, and piston-and-diaphragm. The flow rate is adjusted by manipulating either the piston speed or the piston chamber volume. Piston chambers are small (between 35 and 400 μL), to allow for quick changes in the flow rate. With saline mobile phases, it is convenient to periodically wash the pump elements with an appropriate solvent, to eliminate possible crystalline deposits that could degrade the pump and/or its operation. Reciprocating "pumps" are, in reality, feedback-controlled constant-delivery systems that maintain a constant speed irrespective of the system counter-pressure.

4. Injectors

The injector is a special valve for introducing the polymer solution into the high-pressure solvent stream without interrupting the flow. Basically, it consists of a fixed body, a sealed rotor, and an external loop with an exchangeable (and fixed-volume) capillary. The polymeric solution is loaded, at ambient pressure, into an injection loop and, then, the loop content is introduced into the solvent stream by rotating the valve rotor. Six- and four-port valves are available from Valco (www.vici.com), Varian (www.varianinc.com), and other manufacturers. The external loops are interchangeable and contain between 5 and 2000 μL. Figure 4 illustrates the operation of a typical six-port valve.

For routine analysis, automatic injectors considerably increase the 24 h throughput of a chromatographic system. The samples must be manually loaded into a set of vials, and the automatic injector injects them one after the other, while the PC stores and processes the emerging chromatograms, without need of human intervention.

C. Sample Preparation and Injection

To inject a polymer sample, the solute must be totally dissolved, and must be free of particles and contaminants. In the simplest case, the dry pure polymer is directly dissolved at ambient temperature in an appropriate solvent. The injection solvent may or may not be of the same nature of the chromatographic carrier solvent (i.e., the mobile phase). This is because, after the fractionation, the injection solvent chromatogram becomes well separated from the polymer chromatogram. Furthermore, a small amount of a different solvent is sometimes purposely added into the injection sample to act as an internal standard. Internal standards reduce the effect of flow rate errors by referring the retention times to the retention time of the internal standard, rather than to the injection time. (The injection time is more distant to the polymer chromatogram than the internal standard peak and, therefore, is more subject to the flow oscillations.)

Crystallinity and high glass transition temperatures can make the polymer dissolution difficult (or even prevent). In general, the speed of dissolution increases with the temperature (which also reduces the solvent viscosity). Also, the polymer solubility decreases with the molar mass, and the use of mechanical agitation or the application of ultrasound must be carefully carried out to prevent polymer degradation. This problem is particularly grave with ultrahigh molar mass samples.

Often, a sample pretreatment is necessary for isolating the analyte from other unwanted components. For example, a solvent extraction procedure is required for isolating a graft copolymer from its corresponding homopolymers (Vega et al., 2001). In the case of biopolymers, the analyte is generally contaminated with other polymers and with low molar mass components, and the following fractionation techniques can be applied: precipitation, centrifugation, lyophilization, evaporation, preparative SEC, and preparative liquid adsorption chromatography.

The injection concentration is an important operational variable, a typical value is 1 mg/mL. When either the injection concentration or the injection volume is too high, a column overload effect is produced which shifts the chromatogram toward higher retention volumes. This effect is enhanced with an increase in flow rate or with

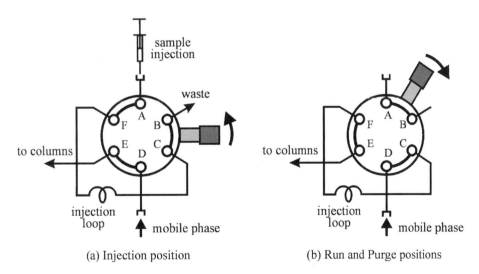

Figure 4 Schematic diagram of six-port injector. Initially, the system is in "purge" (b) position, with the solvent circulating through the injection loop. Then, the valve is turned 60° clockwise, and the sample can be loaded into the loop, "injection" position (a). After the injection, the valve must be returned to the "run" position (b).

an increase in molar mass, and it leads to reduction of the system efficiency and resolution (Mori, 1977, 1984, 2001). Independent of the molar mass, it is preferable if the injection conditions of the analyzed polymer are similar to those of the calibration standards.

Prior to the injection, the polymer solution must be properly filtered to eliminate all suspended particles. This is carried out by means of filter-syringes with membranes of pore size between 0.22 and 0.45 μm. The porous membranes must be solvent-resistant, and should not degrade the analyte. Also, care must be taken to prevent changes in the sample concentration due to solvent evaporation during filtration. Sometimes, a fraction of the analyte may be insoluble, due to the presence of microgels. In this case, it is important to quantify the amount of insolubles, and the filtration itself then becomes an analytical tool.

D. Fractionation Columns

A fractionation column is a stainless steel cylinder which is perfectly packed with porous particles of polymer gels, glass, or silica. The packing process requires a special equipment, and it cannot be carried out by the typical user.

The column selection depends on the analyzed polymer, the solvent, and the range of expected molar masses (Wu, 1999). A broad MMD can be analyzed with a series of columns of varying pore sizes. The smallest pore column must be placed at the beginning of the series, so that the high molar mass fraction may be rapidly diluted in the mobile phase, whereas the smallest molecules are being fractioned. Then, the high molar mass fractions can be separated in the final columns. By increasing the number of columns or the column lengths, the system back pressure and total analysis time are also increased. The optimal column length and type is a compromise among system pressure, analysis time, and resolution.

1. Semi-Rigid Gels

Semirigid (or macroporous) gels are highly crosslinked polymers. They are mechanically stable, exhibit permanent pores and a lower swelling capacity than soft gels. At high temperatures, these packings lose their mechanical resistance, and can be more rapidly degraded. The dimensions of a typical fractionation column are: internal diameter 6–8 mm and length 20–60 cm. Semirigid porous gels can be either nonpolar PS/PDVB gels for organic solvents, or hydrophilic polyhydroximethacrylate- or polyvinylalcohol-modified resin gels for aqueous solvents.

PS/PDVB Columns for Organic Mobile Phase

Porous PS/PDVB is the most widely used SEC packing. Table 3 summarizes many of the available commercial columns. These materials are produced in a wide range of pore sizes; and their main advantage is their low adsorption of eluting polymers in good solvents. The porous beads are produced through a suspension polymerization of a mixture of styrene and divinylbenzene; and the reaction must be controlled to yield spherical particles of uniform channels and narrow particle size distributions (Meehan, 1995). The particles are solvent-swollen, and their pore size is estimated from the analysis of narrow MMD PS standards.

The main criterion for column selection is the molar mass of the analyzed polymer. For highest resolution, the distribution of pore sizes must match the distribution of hydrodynamic volumes of the analyte. Single-bed columns fractionate molar masses over a reduced range, and several of these are required for analyzing a wide distribution sample. Alternatively, mixed-bed or linear columns covering four or five decades of molar mass are also available. Mixed-bed columns are obtained by mixing several uniform gels; and their total exclusion limit is determined by the largest pore size in the mixture (Moody, 1999).

Columns for Aqueous Mobile Phase

Hydrophilic semirigid packings have been developed for aqueous and high-pressure SEC. They permit working over a wide range of pH, and (compared with the soft gels), they exhibit higher adsorption characteristics. Table 4 summarizes the most important commercial columns for aqueous solvents.

Hydrophilic columns are calibrated with standards of poly(ethylene oxide)/poly(ethylene glycol) (PEO/PEG), polysaccharides, or globular proteins. Even for identical hydrodynamic volume ranges, the calibrations may differ considerably with the nature of the injected standard. Thus, columns for aqueous solvents must be carefully selected according to the specific application (Meehan, 1995).

Water-soluble polymers (either synthetic or natural) can be nonionic (neutral) or ionic (polyelectrolytes). Polyelectrolytes are considerably more difficult to quantify, and can be hydrophilic or hydrophobic. If the packing material is not highly hydrophilic, then the hydrophobic interaction between sample and column may be enhanced. Also, the presence of charged sites on the packing may give rise to ionic interactions with polyelectrolytes. In practice, most commercial packings present some degree of hydrophobicity and ionic charge (Meehan, 1995).

The solvent selection depends on the sample type and on the chemical nature of the packing surface. For nonionic polymers, pure water is generally preferred. With ionic polymers, ionic interactions can be suppressed by addition of buffer solutions, whereas hydrophobic interactions can

Table 3 Semi-rigid Commercial Columns: Organic Mobile Phase Columns

Commercial name	Pore size designation	Operational molar mass range (PS in THF)	Particle diameter (µm)	Porosity type	Comments	Supplier	Reference
PLgel	50 Å	0–2,000	5	Unimodal	Synthetic polymers in common solvents: hexane; acetone; toluene; dimethyl acetamide; tetrahydrofurane/dimethyl sulfoxide; chloroform/dimethyl formamide	(1)	Meehan (1999)
	100 Å	0–4,000	10				
	500 Å	500–30,000	20				
	10E3 Å	500–60,000					
	10E4 Å	10,000–600,000					
	10E5 Å	60,000–2,000,000					
	10E6 Å	600,000–10,000,000					
	MIXED-A	1,000–40,000,000	20	Mixed-bed	Common synthetic polymers. Linear calibration curves; analysis of samples with broad MMD and oligomer separation		
	MIXED-B	500–10,000,000	10				
	MIXED-C	200–2,000,000	5				
	MIXED-D	200–4,0000	5				
	MIXED-E	<30,000	3				
PL HFIP gel		500–2,000,000	9	Unimodal	Analysis of polyesters and polyamides with hexafluoroisopropanol (HFIP) as eluant		
SHODEX K series[a]	801 (20 Å)	1.500	6	Unimodal	Synthetic polymers in common solvents	(2)	Suzuki and Mori, (1999)
	802 (60 Å)	5.000	6				
	802.5 (80 Å)	20,000	6				
	803 (100 Å)	70,000	6				
	804 (200 Å)	400,000	7				
	805 (500 Å)	4,000,000	10				
	806 (1000 Å)	(20,000,000)[b]	10				
	807 (>1000 Å)	(200,000,000)[b]	18				
	803L	70,000	6	Mixed-bed	Common synthetic polymers. Linear calibration curves. Analysis of broad MMD samples		
	804L	400,000	7				
	805L	4,000,000	10				
	806L	(20,000,000)[b]	10				
	806M	(20,000,000)[b]	10				
	807L	(200,000,000)[b]	18				
SHODEX HT series[a]	803 (100 Å)	70,000	13	Unimodal	Shodex AT-800, HT-800, and UT-800 are high-temperature columns (for 140°C, 140°C, and 210°C, respectively), useful for analyzing polyolefins; in particular, UT-800 is recommended when the sample contains ultrahigh molar mass components		
	804 (200 Å)	400,000	13				
	805 (500 Å)	4,000,000	13				
	806 (1000 Å)	(20,000,000)[b]	13				
	806-M (1000 Å)	(200,000,000)[b]	13				

(continued)

Table 3 *Continued*

Commercial name	Pore size designation	Operational molar mass range (PS in THF)	Particle diameter (μm)	Porosity type	Comments	Supplier	Reference
SHODEX UT series[a]	802.5 (80 Å)	20,000	30				
	806-M (1,000 Å)	(20,000,000)[b]	30				
	807 (>1,000 Å)	(200,000,000)[b]	30				
SHODEX AT series[a]	806-MS (1000 Å)	20,000,000	12		Analysis of polyesters and polyamides with HFIP as eluent at room temperature		
SHODEX HFIP series[a]	803 (100 Å)	50,000	10				
	804 (200 Å)	150,000	7				
	805 (500 Å)	400,000	10				
	806 (1,000 Å)	(5,000,000)[b]	10	Mixed-bed			
	806-M	(5,000,000)[b]	10	Unimodal			
	807 (>1000 Å)		18				
SHODEX KF-600 series[a]	601 (20 Å)	1,500	3		Downsized columns are packed in 6.0 mm ID × 150 mm length column, which is smaller than the column size of standard type columns; reduce organic solvent consumption, short analysis, time and lower detection limit		
	602 (60 Å)	5,000	3				
	602.5 (80 Å)	20,000	3				
	603 (100 Å)	70,000	3				
	604 (200 Å)	400,000	3				
	605 (500 Å)	4,000,000	3				
	606 (1,000 Å)	(20,000,000)[b]	10				
	606-M	(20,000,000)[b]	10	Mixed-bed			
	607 (>1,000 Å)	(200,000,000)[b]	18	Unimodal			
TSK-Gel H$_6$[a]	G1000	1,000	13	Unimodal	Type H columns: broad separation range; GMH columns: linear calibration curves. Analysis of samples with broad MMD; all the columns contain THF, except GMH-HT that contains dichlorobenzene; HHR columns admit a large variety of solvents and are recommended for temperatures lower than 60°C. GMH-HT columns are used for high-temperature applications (up to 140°C); the small size of Super H columns minimize the solvent consumption and provide a better	(3)	Kato et al. (1999)
	G2000	10,000	13				
	G2500	20,000	13				
	G3000	60,000	13				
	G4000	400,000	13				
	G5000	4,000,000	13				
	G6000	40,000,000	13				
	G7000	400,000,000	13				
	GMH6	400,000,000	13	Mixed-bed			
	GMH6-HT	400,000,000	13				
TSK-Gel H$_8$[a]	G1000	1,000	10	Unimodal			
	G2000	10,000	10				

Gel Permeation and Size Exclusion Chromatography

Series	Model	Exclusion limit		Type	Notes
	G2500	20,000	10		resolution; also, these columns admit a large variety of solvents
	G3000	60,000	10		
	G4000	400,000	10		
TSK-Gel H_{XL}[a]	G1000	1,000	5	Unimodal	
	G2000	10,000	5		
	G2500	20,000	5		
	G3000	60,000	6		
	G4000	400,000	6		
	G5000	4,000,000	9		
	G6000	40,000,000	9		
	G7000	400,000,000	9		
	GMHXL	400,000,000	9	Mixed-bed	
	GMHXL-HT	400,000,000	13		
TSK-Gel H_{HR}[a]	G1000	1,000	5	Unimodal	
	G2000	10,000	5		
	G2500	20,000	5		
	G3000	60,000	5		
	G4000	400,000	5		
	G5000	4,000,000	5		
	G6000	40,000,000	5		
	G7000	400,000,000	5		
	GMHHR-L	4,000,000	5	Mixed-bed	
	GMHHR-N	400,000	5		
	GMHHR-M	4,000,000	5		
	GMHHR-H	400,000,000	5		
TSK-Gel Super H[a]	Super H 1000	1,000	3	Unimodal	
	Super H 2000	10,000	3		
	Super H 2500	20,000	3		
	Super H 3000	60,000	3		
	Super H 4000	400,000	3		
	Super H 5000	4,000,000	3		
	Super H 6000	40,000,000	4		
	Super H 7000	400,000,000	5		
	Super H HM-L	4,000,000	3	Mixed-bed	
	Super H HM-N	400,000	3		
	Super H HM-M	4,000,000	3		
	Super H HM-H	400,000,000	4		

(continued)

Table 3 Continued

Commercial name	Pore size designation	Operational molar mass range (PS in THF)	Particle diameter (μm)	Porosity type	Comments	Supplier	Reference
Styragel HR	HR 0.5	1,000	5	Unimodal	HR columns are of high resolution; used for oligomer analysis, epoxi resins, and additives; mixed-bed columns admit a wide molar mass range, but their resolution is not as good as in unimodal columns; HT columns admit a large variety of solvents, and therefore are recommended for ambient-temperature applications requiring changes of solvents; the large particle size of HMW columns prevents polymer degradation	(4)	Neue (1999)
	HR 1	100–500	5				
	HR 2	500–20,000	5				
	HR 3	500–30,000	5				
	HR 4	5,000–600,000	5				
	HR 5	50,000–4,000,000	5				
	HR 6	200,000–10,000,000	5				
	HR 4E	50–10,000	5	Mixed-bed			
	HR 5E	2,000–4,000,000	5				
Styragel HT	HT 2	100–10,000	10	Unimodal			
	HT 3	500–30,000	10				
	HT 4	5,000–60,000	10				
	HT 5	50,000–4,000,000	10				
	HT 6	200,000–10,000000	10				
	HT 6E	5,000–10,000,000	10	Mixed-bed			
Styragel HMW	HMW2	100–10,000	20	Unimodal			
	HMW7	500,000–100,000,000	20				
	HMW6E	5,000–100,000,000	20	Mixed-bed			
PSS SDB	50, 100, 500, 10^3, 10^4, 10^5, 10^7, linear	<100 to >30,000,000	3, 5, 10, 20	Unimodal and mixed-bed	For use with organic solvents and common synthetic polymers	(5)	Kilz (1999)
PSS PFG	100, 300, 1,000, 4,000, linear	<100 to >30,000,000	7		For use with fluorinated solvents for analyzing polyesters, polyamides, polyacetals, cellulose, and lignines		

[a]Only the exclusion limit is reported.
[b]Estimated value.

Gel Permeation and Size Exclusion Chromatography

Table 4 Semi-rigid Commercial Columns: Aqueous Mobile Phase Columns

Commercial name	Pore size designation	Operational molar mass range	Particle diameter (μm)	Gel chemistry	Comments	Supplier	Reference
PL aquagel-OH	30	100–30,000	8	Hydrophilic macroporous particles with OH functionality; admit pressures up to 140 bar, organic solvent concentrations up to 50%; and pH range 2–10	Particles of size 15 μm for analyzing high molecular weight polymers where degradation can take place	(1)	Meehan (1999)
	40	10,000–200,000	8				
	50	50,000–1,000,000	8				
	60	200,000–10,000,000	8				
	Mixed	100–10,000,000	8				
	40	10,000–200,000	15				
	50	50,000–1,000,000	15				
	60	200,000–10,000,000	15				
Shodex OHpak SB-HQ series[a]	802HQ	4,000;[b] 2,000[c]	8	Hydrophilic poly(hydroxi methacrylate)	For hydrosoluble polymers and proteins; to increase the analyte solubility, protein modifiers (e.g., urea or guanidine-HCl) and surfactants (e.g., SDS or Brij-35) can be added to the mobile phase; for hydrophilic samples, OHpak SB-800HQ columns are more suitable than Asahipak GS/GF	(2)	Suzuki and Mori (1999)
	802.5HQ	10,000;[b] 8,500[c]	6				
	803HQ	100,000;[b] 55,000[c]	6				
	804HQ	1,000,000;[b] 300,000[c]	10				
	805HQ	4,000,000;[b] 2,000,000[c]	13				
	806HQ	2×10^{7};[a,d] 10,000,000[b,d]	13				
	806MHQ	2×10^{7};[a,d] 10,000,000[b,d]	13				
Shodex Asahipak GS/GF series[a]	GS-220HQ	3,000	6	Modified polyvinyl alcohol (PVA) based resin	Recommended for analyzing ionic polymers; admits both organic and aqueous solvents		
	GS-320HQ	40,000	6				
	GS-520HQ	300,000	7				
	GS-620HQ	1,000,000	7				
	GF-310HQ	40,000	5				
	GF-510HQ	300,000	5				
	GF-710HQ	10,000,000	9				
	GF-7MHQ	10,000,000	7				
TSK Gel PW	G1000PW	1,000;[b] <2,000[e]	10	Poly(hydroxi methacrylate)	Separation of polydisperse synthetic macromolecules or biological polymers (polysaccharides); the (mixed-bed) GMPW columns are recommended for samples of wide or unknown MMDs; G-Oligo-PW and G-DNA-PW columns are, respectively, used for oligo-saccharides and DNA or RNA; the small particle size of PWXL ensure higher resolutions than with PW particles; TSK-gel G2500 PW columns are recommended for oligo-nucleotides, peptides, and	(3)	Kato et al. (1999)
	G2000PW	2,000;[b] <5,000[e]	10				
	G2500PW$_{XL}$	3,000;[b] <8,000[e]	6				
	G2500PW	3,000[b]	10				
	G3000PW$_{XL}$	50,000;[b] 60,000;[d] 500–800,000[c]	6				
	G3000PW	50,000[b]	10				
	G4000PW$_{XL}$	2,000–300,000;[b] 1,000–700,000[d]	10				
	G4000PW	10,000–1,500,000[c]	17				
	G5000PW$_{XL}$	4,000–1,000,000;[b] 50,000	10				
	G5000PW	7,000,000;[d] <10000000[c]	17				
	G6000PW$_{XL}$	40,000–8,000,000;[b] 500,000	13				
	G6000PW	50,000,000;[d] <200,000,000[c]	17				

(continued)

Table 4 Continued

Commercial name	Pore size designation	Operational molar mass range	Particle diameter (μm)	Gel chemistry	Comments	Supplier	Reference
	GMPW$_{XL}$	500–8,000,000;[b] <50,000,000[d]	13		anionic compounds, whereas the TSK-gel G-Oligo-PW are recommended for nonionic oligomers		Neue (1999)
	GMPW	<200,000,000[c]	17				
	G-Oligo-PW	3,000;[b] <3,000[c]	6				
	G-DNA-PW	40,000–8,000,000;[b] 200,000,000[c]	10				
Ultrahydrogel	120	5,000	Unavailable information	Poly(hydroxi methacrylate)	For hydrosoluble polymers, columns admit pH values between 2 at 12, organic solvent concentrations up to 50%, salt concentrations up to 0.1 M, temperatures between 10°C and 80°C, and flows between 0.3 and 1.0 mL/min; for long storage periods, replace the mobile phase by a 0.05% solution of sodium azide in distilled water	(4)	
	250	20–80,000					
	500	500–200,000					
	1000	500–1,000,000					
	2000	10,000–20,000,000					
	Linear	20–20,000,000					
PSS HEMA	40, 100, 300, 1,000 linear	100–3,000,000	10	Methacrylic ester copolymer	PSS Suprema columns are used for high-molecular weight PEO, polyacrylamides; and polysaccharides; PSS Suprema can be used with aqueous eluents and PSS HEMA can be used with organic, aqueous or mixtures of solvents	(5)	Kilz (1999)
PSS Suprema	30, 100, 300, 1,000, 3,000, 10,000, 30,000, linear	<100 to >100,000,000	5, 10, 20				

Note: Suppliers—(1), Polymer Laboratories, Shropshire, UK (www.polymerlabs.com); (2), Showa Denko, Tokyo, Japan (www.sdk.co.jp/shodex/english/contents.htm); (3), Tosoh Bioscience, LLC Montgomeryville, PA, USA (www.tosohbiosep.com); (4), Waters Corporation, Milford, Mass. USA (www.waters.com); and (5), PSS Polymer Standard Service, Mainz, Germany (www.polymer.de).
[a] Only the exclusion limit is informed.
[b] Expressed in terms of pullulan.
[c] Expressed in terms of PEG/PEO.
[d] Estimated value.
[e] Expressed in terms of globular proteins in a 0.2 M phosphate buffer, pH 6.8, and flow rate 1 mL/min.
[f] Expressed in dextrans terms in 0.2 M phosphate buffer, pH 6.8, and flow rate 1 mL/min.

Gel Permeation and Size Exclusion Chromatography

be eliminated by addition of organic modifiers into the aqueous solvent (Meehan, 1995).

2. Rigid Packing

Silica-Based Packing

Rigid packings are made of an inorganic porous material that can be either glass or silica. Silica gel is a three-dimensional network of SiO_2 repetitive units with siloxane and silanol terminal units on its surface. Silica gel structures are highly porous and amorphous, and exhibit a large surface area. For protein analysis, high-purity (or metal-free) silica gel is recommended for yielding reproducible retention times without affecting biological activity (Eksteen and Pardue, 1995).

Rigid packings exhibit a narrow and well-defined distributions of pores. They are thermally and mechanically stable, can be easily regenerated, and can be used with either organic or aqueous solvents. Silica particles can be spherical or irregular (Eksteen and Pardue, 1995; Engelhardt and Ahr, 1983; Guiochon and Martin, 1985).

Table 5 presents the main characteristics of the native and derivatized silica columns that are used for analyzing synthetic polymers in organic solvents. Table 6 summarizes the bonded-phase silica columns used for analyzing water-soluble polymers.

In aqueous systems, rigid packings present secondary retentions due to ionic interactions (Unger and Kinkel, 1988). Adsorption can be reduced by adding an organic solvent to the aqueous phase (Montelaro, 1988). Other problems are the dissolution of silica in aqueous systems at pH > 7.5, and the possible degradation of the bonded phase at pH < 2. In general, proteins are more stable at a pH between 7 and 8, but these values are in the upper tolerable limit of dissolution of the silica-based materials (Eksteen and Pardue, 1995). In contrast, the chemical stability of silica is not a concern with organic solvents.

With silica-based diol-bonded columns, the retention volumes of globular proteins vary with their effective charge, and with the pH and ionic strength of the mobile phase (Schmidt et al., 1980). Fortunately, hydrophobic interactions are low because the hydrophobic side chains are predominantly inside the protein molecules. This interaction can be further reduced by adding 5–20% of a non-denaturing solvent such as ethylene glycol, or alternatively by increasing the ionic strength of the mobile phase (Schmidt et al., 1980). In contrast, hydrophobic interactions are important when analyzing polypeptides. In this case, high concentrations of organic solvents are required for ensuring that size exclusion is the main fractionation mechanism (Chang et al., 1976; Gruber et al., 1979; Lau et al., 1984; Pfannkoch et al., 1980; Taneja et al., 1984).

Table 5 Rigid Commercial Columns: Silica-Based Packing for Organic Mobile Phase

Commercial name	Pore size designation	Operation range (PS in THF)	Particle diameter (μm)	Phase chemistry	Comments	Supplier	Reference
Zorbax PSM	PSM 60 (60 Å)	500–10,000	5	Porous spherical native silica and silanized silica (with trimethyl-silane); unimodal and mixed	Silanized packing suppress adsorption effects and are used when the mobile phase contains organic solvents	(1)	Moody (1999)
	PSM 300 (300 Å)	3,000–300,000	5				
	PSM 1000 (1,000 Å)	10,000–1,000,000	5				
	PSM 3000 (3,000 Å)	100,000–9,000,000	5				
	Bimodal (PSM 60 and PSM 100)	500–1,000,000	5				
	Trimodal (PSM 60, PSM 300 and PSM 3000)	500–9,000,000	5				
LiChrospher	Si 60 (60 Å)	50,000	5	Porous spherical silica		(2)	Unger and Kinkel (1988)
	Si 100 (100 Å)	50,000–80,000	5				
	Si 300 (300 Å)	150,000–300,000	10				
	Si 500 (500 Å)	300,000–600,000	10				
	Si 1000 (1,000 Å)	600,000–1,400,000	10				
	Si 4000 (4,000 Å)	2,500,000–8,000,000	10				

Note: Suppliers—(1), Zorbax. MAC-MOD Analytical Inc., Chadds Ford, PA, (www.chem.agilent.com) and (2), Merck, Darmstadt, Germany (www.alltechweb.com).

Table 6 Rigid Commercial Columns: Silica-Based Packing for Aqueous Mobile Phase

Commercial name	Pore size designation	Operation range	Particle diameter (μm)	Phase chemistry	Comments	Supplier	Reference
Zorbax GF	GF-250 (60 Å)	4,000–400,000,000	4	Porous silica stabilized with zirconium and diol-bonded phase	pH range: 3.0–8.5; for estimating protein molar masses, for purifying complex mixtures, and for monitoring chemical reactions and conformational changes	(1)	Moody (1999)
	GF-450 (300 Å)	10,000–1,000,000	6				
TSK Gel SW	Super SW 2000	5,000–150,000,[a]	4	Porous silica with primary alcohols bonded onto the surface. Available in glass and stainless steel columns	Analysis of proteins and peptides; recommended protein concentrations: 1–20 mg/mL, with flow rates 0.5 to 1.0 mL/min	(2)	Kato et al. (1999)
	G2000 SW$_{XL}$	1,000–30,000,[b]	5				
	G2000 SW	500–15,000[c]	10, 13				
	Super SW 3000	10,000–50,000,[a]	4				
	G3000 SW$_{XL}$	2,000–70,000,[b]	5				
	G3000 SW	1,000–35,000[c]	10, 13				
	G4000 SW$_{XL}$	20,000–10,000,000,[a]	8				
		4,000–500,000,[b]					
	G4000 SW	2,000–250,000[c]	13				
Shodex PROTEIN	KW-802.5 (150 Å)	1,000–60,000,[d] 5,000–150,000[a]	5	Poly(hydroxylated silica)	Analysis of proteins and water-soluble polymers; operating conditions: pH 3–7.5, ionic force 0.1–0.3 M, flow rates from 0.5–1.0 mL/min, and protein concentrations up to 10 mg/mL	(3)	Suzuki and Mori (1999)
	KW-803 (300 Å)	2,000–170,000,[d] 5,000–70,000[a]	5				
KW	KW-804 (500 Å)	4,000–500,000,[d] 1,000–1,000,000[a]	7				
SynChropak GPC	GPC PEPTIDE (50 Å)	5,000–35,000,[e] 500–10,000[f]	5	Porous spherical silica bonded with γ-glycidoxy-propylsilane for providing a neutral hydrophilic layer	Analysis of proteins, small peptides, and nucleic acids	(4)	Gooding (1999)
	GPC100 (100 Å)	5,000–160,000,[e] 500–25,000[f]	5				

Gel Permeation and Size Exclusion Chromatography

	GPC300 (300 Å)	10,000–1,000,000,[e]	5			
		2,000–100,000[f]				
	GPC500 (500 Å)	40,000–1,000,000,[e]	7			
		10,000–350,000[f]				
	GPC1000 (1,000 Å)	40,000–10,000,000,[e]	7			
		40,000–1,000,000[f]				
	GPC4000 (4,000 Å)	70,0000–10,000,000[f]	10			
SynChropak CATSEC	CATSEC100 (100 Å)	500–20,000[f]	5	Porous spherical silica bonded with polyamines resulting in a slightly cationic layer	Analysis of cationic polymers such as poly(vinyl pyridines) or PVP that cannot be analyzed with SynChropak GPC columns for the residual negative charge of this packing	
	CATSEC300 (300 Å)	2,000–50,000[f]	5			
	CATSEC1000 (1,000 Å)	10,000–600,000[f]	7			
	CATSEC4000 (4,000 Å)	20,000–1,000,000[f]	10			
Protein Pak	60	1,000–20,000	10	Porous spherical silica derivatized with glycidyl functions	Analysis of proteins with aqueous mobile phases containing salts of concentrations up to 0.1 M (generally, a phosphate buffer at pH 7)	(5) Neue (1999)
	125	20,000–80,000	10			
	300 SW	10,000–500,000	10			

Note: Suppliers—(1), Zorbax. MAC-MOD Analytical Inc., Chadds Ford, PA, USA (www.mac-mod.com; www.chem.agilent.com); (2), Tosoh Bioscience LLC, Montgomeryville, PA, (www.tosohbiosep.com); (3), Showa Denko, Tokyo, Japan (www.sdk.co.jp/shodex/english/contents.htm); (4), SynChropak, Inc., Lafayette, IN, USA (www.cobertassoc.com/hycat.htm#One); and (5), Waters Corporation, Milford, MA, USA (www.waters.com).
[a]Globular proteins: 0.3 M NaCl in phosphate buffer 0.1 M (0.05 M for columns SW$_{XL}$), pH 7.0.
[b]Dextrans.
[c]PEG/ PEO in distilled water.
[d]Pullulans.
[e]DNA and glycidyl-tyrosine in phosphate buffer 0.1 M, pH 7.0.
[f]PVP (60,0000) and cytidine at 0.1% of trifluoroacetic acid and NaCl 0.2 M.

Table 7 Rigid Commercial Columns: PDVB-Based Jordi Columns for Organic or Aqueous Mobile Phase

Commercial name	Pore size designation	Operation range	Particle diameter (μm)	Phase chemistry	Comments	Supplier	Reference
Jordi Gel DVB	100	<100–5,000	5	Native PDVB,[a] neutral charge	Native DVB Jordi columns offer an unparalleled resistance to shrinking and swelling; allowing flexible operating conditions; they admit toluene, THF, methanol, hexane, HFIP, acetone, Freon, or 5% water in THF; the same columns are used at temperatures up to 150°C, pH values from 0 to 14, and with samples containing salts, acids, or bases; useful for separating organic-soluble polymers	(1)	Jordi (1999)
	500	<100–10,000	5				
	10^3	<100–25,000	5				
	10^4	100–2,000,000	5				
	10^5	10,000–10,000,000	5				
	Mixed-bed linear	100 to >10,000,000	5				
Jordi Sulfonated Polar Pac SCX	100	<100–5,000	5	Sulfonated PDVB,[a] neutral charge	The anionic surface of sulfonated-DVB columns separate samples by ion exclusion and size exclusion; for analyzing organic acids, polysaccharides, starches, cellulose, and other water-soluble polymers with a negatively charged surface		
	500	<100–10,000	5				
	10^3	<100–25,000	5				
	10^4	100–2,000,000	5				
	10^5	10,000–10,000,000	5				
	Mixed-Bed linear	100 to >10,000,000	5				
Jordi Polar Pac WAX	100	<100–5,000	5	PDVB[a]–WAX,[b] neutral or positive charge	Polar Pac WAX columns are optimized for separating cationic polymers. Acetic acid mobile phases are good solvents; samples run on this column include chitosan, and water-soluble flocculants		
	500	<100–10,000	5				
	10^3	<100–25,000	5				
	10^4	100–2,000,000	5				
	10^5	10,000–10,000,000	5				
	Mixed-bed linear	100 to >10,000,000	5				
Jordi Gel Glucose	100	<100–5,000	5	PDVB[a]–GBR,[c] neutral charge	Glucose-DVB is a rugged, versatile, highly efficient material for separating polar samples based on hydrodynamic volume; typical samples are polysaccharides, polyacrylic acids, carboxymethyl cellulose, polyacrylamides, carrageenins, starches, hyaluronic acid, and lignins; use buffer salts, strong or weak acids, or organic solvents as mobile phase modifiers		
	500	<100–10,000	5				
	10^3	<100–25,000	5				
	10^4	100–2,000,000	5				
	10^5	10,000–10,000,000	5				
	Mixed-bed linear	100 to >10,000,000	5				

Note: Supplier—(1), Jordi Associates, Bellingham, MA, USA (www.jordiassoc.com).
[a] Poly(divinyl benzene).
[b] Poly(ethylene imine).
[c] Glucose-bounded resin.

Jordi Columns

Jordi Associates (Jordi, 1999) have developed several packings based on native and derivatized PDVB; Table 7 summarizes their main characteristics. Native PDVB columns were developed for conventional synthetic polymers in common organic solvents, and can be either single- or mixed-bed. These gels minimize adsorption of chemical groups such as amines. Several derivatized PDVB columns have been developed for water-soluble polymers. For example, sulfonated- and glucose-bonded resins of PDVD are well suited for analyzing polysaccharides.

3. *Soft Gels*

Soft gels are mainly used in aqueous preparative SEC and very seldom in aqueous analytical SEC. Due to their low mechanical resistance, they cannot withstand the high pressures of analytical SEC. They do not exhibit permanent pores, and their porosity is directly associated with their swelling capacity.

E. Detectors

An ideal SEC detector must be simple to operate, and must exhibit the following characteristics: high sensitivity, high signal-to-noise ratio, wide linearity range, good thermal stability, low time constant, reduced contribution toward BB, and good reliability. The specification sheets of commercially available detectors normally exhibit most of the above-mentioned characteristics together with the typical recommended applications.

In Fig. 1, we have classified the SEC sensors into concentration sensitive and molar mass sensitive. DRs, densitometers, and evaporative LS sensors are all "universal" concentration detectors, in the sense that their signals are *almost* independent of the polymer nature. In contrast, UV–Vis and infrared spectrophotometers are highly "selective" of the eluting chemical groups. The commercially available molar mass detectors are LS sensors and SV. While concentration detectors are also used in adsorptive HPLC (Pasch and Trathnigg, 1999; Snyder et al., 1997), LS and SV detectors are highly specific to SEC (Jackson and Barth, 1995; Pasch and Trathnigg, 1999). Table 8 lists the most important commercially available detectors, together with the manufacturers' information.

1. *Concentration Detectors*

Differential Refractometers

The refraction principle analyzes the change in speed and direction of a monochromatic light when passing from an isotropic medium A into another isotropic medium B. The (dimensionless) relative refractive index n_{BA} is defined by Snell's Law:

$$n_{BA} = \frac{\sin(\theta_i)}{\sin(\theta_r)} \quad (44)$$

where θ_i and θ_r are the incidence and refraction angles, respectively.

The relative refractive index presents a strong (inverse) dependence upon the temperature and, therefore, the detection cell temperature must be appropriately regulated. A baseline drift in the DR signal is often indicative of inadequate temperature control. Also, n_{BA} depends on the wavelength, the pressure, and the polymer concentration. Thus, at constant values of temperature, wavelength, and pressure, the variations of n_{BA} are only due to changes in the sample concentration.

The sensitivity of DR detectors to the polymer concentration is relatively low ($\cong 10^{-3}$ mg/mL). In spite of this, they are preferred to UV sensors when the UV absorptivity is low and/or when the mobile phase strongly absorbs in the UV region of the spectrum. Most DRs exhibit a linear response over a wide range of concentrations that cover two or three orders of magnitude.

The basic components of a DR are: a monochromatic light source, a detection cell, and a photodetector. DR measures the difference in refractive index between the solvent and the polymer solution. To this end, the detection cell is split into a reference cell (for the pure solvent) and a flow cell (for the instantaneously eluting polymer).

Figure 5(a) presents a deflection DR based on model 2414 from Waters. The detection cell is a double-cell prism. The light-emitting diode produces a monochromatic light of a fixed wavelength (690 or 880 nm, according to the model), which is focused onto the cell by means of a lens, a mask, and a mirror. The light path is refracted twice between the cell compartments prior to reaching the two-element photodiode. When pure solvent fills both sides of the cell, then there is no net light refraction, and the two photodiodes receive identical light intensities. A change in the polymer concentration cell causes a deflection in the refracted angle, and the difference between the photodiode signals is proportional to the deflection angle [Fig. 5(a)]. Initially, the reference cell must be flushed with solvent. An electronic auto-zero system compensates for small deviations produced when pure solvent circulates through the sample cell. Figure 5(b) illustrates the normal eluant flow path and the valve system that is required for periodically flushing the reference cell. The heat exchangers contribute to keep a constant eluent temperature.

Apart from the deflection DRs, interferometric DRs are also commercially available. The operation principle of an interferometric refractometer is as follows. A

Table 8 Commercial Detectors for On-Line SEC Applications

Concentration Detectors
Differential Refractive (DR) Index Detectors
  Deflection Type
    Waters[1] 2414. Wavelengths: 690 nm or 880 nm
    Shimadzu[2] RID-10A. For analytical or preparative SEC (up to 150 mL/min). Temperature range: 30–60°C
    Jasco[3] RI-2031. Maximum flow rate: 50 mL/min. Temperature range: 30–60°C
    Knauer[4] WellChrom K-2301 and K-2401. Wavelength: 950 ± 30 nm
    Brookhaven Instruments Corp.[5], BI-DNDC. Estimates $\partial n/\partial c$ in off-line mode at 470, 535, or 620 nm
  Interferometric Type
    Wyatt[6] Optilab DSP. Estimates $\partial n/\partial c$ in off-line mode at 690 nm
    PSS[7] ScanRef. Estimates $\partial n/\partial c$ in on-line and off-line modes, at 632, 544, and 488 nm
Evaporative Light Scattering Detectors (ELSD)
    Waters[1] 2420. Temperature range of heated chamber: ambient to 80°C.
    Shimadzu[2] ELSD-LT. Low temperature: ambient to 80°C
    Alltech[8] ELSD 800 or 2000. With or without the possibility of by-passing part of the flow
    Polymer Lab[9] ELS 1000. Temperature range of heated chamber: ambient to 150°C
    Polymer Lab[9] ELS 2100. Low temperature (for semivolatile compounds)
    ESA[10] ChromaChem. Temperature range of heated chamber: 30–300°C
    Knauer[4] ELSD 800. Temperature range of heated chamber: ambient to 110°C
    Knauer[4] ELSD 2000. Temperature range of heated chamber: ambient to 120°C
Ultraviolet (UV) Detectors
  Fixed wavelength
    Waters[1] 2487. Dual-wavelength absorbance detector. UV range: 190–700 nm
    Gilson[11] 112. UV range: 214–280 nm
    LabAlliance[11] F110. UV range: 214–1000 nm
    Knauer[4] WellChrom K-200. Wavelength: 254 nm (mercury vapor lamp)
  Variable wavelength
    Jasco[3] UV-2070. UV range: 190–900 nm, deuterium and halogen lamps
    Jasco[3] UV-2075. UV range: 190–600 nm, deuterium lamp
    LabAlliance[12] 520. UV range: 195–400 nm, optical range up to NIR (900 nm)
    Merck. LaChrom[13] L-7400. UV range: 190–600 nm, deuterium lamp
    Merck. LaChrom[13] L-7420. UV range: 190–900 nm, deuterium + tungsten lamp
    Shimadzu[2] SPD-10AVP. UV range: 190–600 nm, deuterium lamp
    Shimadzu[2] SPD-10AVVP. UV range: 190–900 nm, deuterium + tungsten lamp
  Photodiode Array Detector
    Waters[1] 2996. UV range: 190–800 nm (± 1 nm). Deuterium lamp
    Waters[1] 996. Variable path length flow cell (0.5 or 3.0 mm). UV range: 190–800 nm (± 1 nm). Deuterium lamp
    Jasco[3] MD-2010. UV range: 195–650 nm. Deuterium lamp
    Jasco[3] MD-2015. UV range: 200–900 nm. Deuterium and tungsten-halogen lamps
    LabAlliance[12] 6000. UV range: 190–800 nm. Deuterium and tungsten-halogen lamps
    Knauer[4] WellChrom K-2800. UV range: 190–740 nm. Deuterium lamp
    Knauer[4] WellChrom K-2800 UV/Vis. UV range: 190–1020 nm. Deuterium and tungsten–halogen lamps
    Shimadzu[2] SPD-M10AVP. UV range: 190–800 nm. Deuterium lamp

Molar Mass Sensitive Detectors
Light Scattering (LS) Detector
    Wyatt[6] miniDAWN Tristar. Three measurement angles: 45°, 90°, and 135°
    Wyatt[6] DAWN EOS. Eighteen measurement angles (10–160°), 690 nm
    Wyatt[6] QELS (quasi-elastic light scattering). Calculates the autocorrelation in miniDAWN Tristar or a DAWN EOS sensors for on-line determination of hydrodynamic radius
    Polymer Lab[9] PD2010. Measurement angles: 90° (small molecules)
    Polymer Lab[9] PD2020. Two measurement angles: 90° and 15°
    Polymer Lab[9] PDDLS (flow-mode dynamic LS). Hydrodynamic radius of molecules: 1.5–1000 nm
    PSS[7] SLD7000. Seven measurement angles: 35°, 50°, 75°, 90°, 105°, 130°, 145°. Laser: 660 nm, vertically polarized
    PD[14] 2010+ and 2020+. Flow-through combined static and dynamic LS. Measurement angles: 90° and 15°. Laser: 680/800 nm

(continued)

Gel Permeation and Size Exclusion Chromatography

Table 8 *Continued*

PD[14] 2040. Flow-through static light scattering for high-temperature (up to 250°C). Measurement angles: 90° and 15°
Intrinsic Viscosity Detector
PSS[7] ETA 1001. For temperatures up to 80°C
PSS[7] ETA 1150. High-temperature viscometer 80–150°C
Multidetector Systems
Viscotek[15] Dual-detector Wheatstone bridge differential viscometer plus LS detector (670 nmm), at 7° (LALS) or 90° (RALS). Both detectors can be independently used
Viscotek[15] TDA™ Triple-detector array. Previous system plus a deflection DR (660 nm) or a UV detector (190–740 nm, deuterium lamp). Any of the three detectors can be independently used

Note: *Suppliers*—(1), Waters Corp., Milford, MA, USA (www.waters.com); (2), Shimadzu, Nakagyo-ku, Kyoto, Japan (www.shimadzu.com); (3), Jasco, Essex, UK (www.jasco.co.uk); (4), Knauer, Berlin, Germany (www.knauer.net); (5), Brookhaven Instruments Corporation, Holtsville, NY, USA (www.bic.com); (6), Wyatt Tech. Corp., Santa Barbara, CA, USA (www.wyatt.com); (7), PSS Polymer Standard Service, Mainz, Germany, (www.polymer.de); (8), Alltech Associates, Inc., Deerfield, IL, USA (www.alltechweb.com); (9), Polymer Laboratories, Shropshire, UK (www.polymerlabs.com); (10), ESA Inc., Chelmsford, MA, USA (www.esainc.com); (11), Gilson, Middleton, WI, USA (www.gilson.com); (12), LabAlliance, State College, PA, USA (www.laballiance.com); (13), Merck, Darmstadt, Germany (www.merck.de); (14), PD Precision Detectors. Bellingham, MA, USA (www.precisiondetectors.com); and (15), Viscotek, Houston, TX, USA (www.viscotek.com; www.viscotek-europe.com).

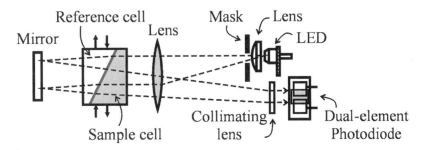

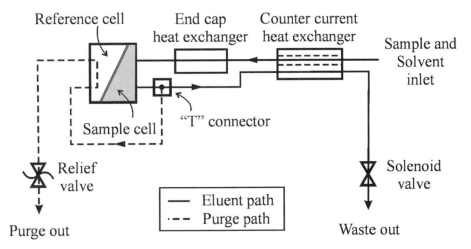

Figure 5 Schematic diagram of a differential refractive index detector (based on Waters Model 2414). When both sides of the measurement cell are full of pure solvent, then the two photodiode elements receive equal amounts of light, and no refractive index difference is detected. When as shown in (a), the polymer starts to flow through the sample cell, then the light beams are bended twice, the photodiodes receive unequal amounts of light, and a signal proportional to the polymer concentration is the output. The solvent cell must be periodically flushed with new solvent. To this effect, the purging system of (b) is employed.

monochromatic light is first polarized, and then split into two parallel beams. One beam is focused onto the reference cell (filled with solvent), while the second beam reaches the sample cell. The difference in their polymer concentrations causes a phase difference between the two beams. The resulting interference modifies the light intensity that reaches the photodetector in a way that is proportional to the polymer concentration in the detection cell. The sensitivity of interferometric detectors can be an order of magnitude higher than that of deflection refractometers.

Some commercial refractometers (either deflection or interferometric) can also be used for off-line measurement of $\partial n/\partial c$ (see Table 8).

Density Sensors

The detection cell of a density sensor is a vibrating U-shaped tube that is rigidly supported at its open ends. The tube is electromagnetically excited to vibrate at its natural resonance frequency, which, in turn, depends on its total mass (or density), which includes the contained eluent. The period of oscillation is measured by means of a reference quartz oscillator (at 5 MHz). After a 4 s measurement time, the sensitivity of the density measurement is around 2.5×10^{-7} g/cm^3 (Trathnigg and Jorde, 1982).

The variation in the period of oscillation (ΔT) is proportional to the change in the sample density ($\Delta \rho$), as follows (Trathnigg and Jorde, 1984):

$$\Delta T = \frac{\Delta \rho}{2AT_0} \quad (45)$$

where A is an instrumental constant and T_0 is the natural period of oscillation when the cell is full of pure solvent. Then, the polymer concentration (c) is estimated from:

$$c(V) = \frac{2AT_0}{1 - \rho_0 \bar{v}^*} \Delta T(V) \quad (46)$$

where ρ_0 is the solvent density and $\bar{v}^*$ is the apparent partial specific volume of the solute.

Density detectors are quite insensitive to the molar mass, except for molar masses lower than 4000 g/mol (Trathnigg and Jorde, 1984). These detectors are particularly useful when the $\partial n/\partial c$ of the polymer–solvent system is low and when there is a significant difference between the solvent and the solute densities (Quivoron, 1984). Density detectors measure the true instantaneous mass (or concentration), whereas DRs can be affected by the chemical compositions of copolymers. The combined density DR detection can provide a measure of the instantaneous copolymer composition. This is particularly useful when UV sensors cannot be used because the analyte's chemical group of interest does not absorb in the UV (Trathnigg, 1991; Trathnigg et al., 1997; Trathnigg, 1995).

UV Sensors

UV absorption detectors are extensively used in SEC. At a fixed wavelength, a UV sensor measures the light absorption of polymers containing specific chromophores. The wavelength range of 180–400 nm is typically covered by a deuterium lamp. When visible wavelengths between 400 and 700 nm are also required, then tungsten halide lamps are employed. In general, UV sensors exhibit a high linear sensitivity over a wide range of concentrations, and are quite insensitive to flow and temperature changes. A necessary condition is that solvents must exhibit a low absorbance at the wavelengths that are being used and, for this reason, aromatic solvents, ketones, DMF, esters, and $CHCl_3$ cannot be used at wavelengths lower than 250 nm. Also, some solvent contaminants (e.g., peroxides in THF) considerably affect the UV signal.

UV detectors operate at a fixed wavelength, at a single selectable wavelength, or at multiple wavelengths (diode arrays). The first two types provide a single signal, whereas diode array detectors typically produce up to 500 simultaneous signals covering the range between 200 and 800 nm.

Fixed-wavelength UV detectors are extremely simple. Typically, a single monochromatic light source (e.g., at 254 nm from a low-pressure mercury lamp), directly falls onto the flow cell; and other (few) wavelengths can be selected through the use of filters (e.g., at 313, 334, and 365 nm). Also, a phosphorous oxide filter allows one to select a wavelength of 280 nm for protein analysis. In contrast, selectable (or variable) wavelength detectors have a moveable diffraction grating, a beam splitter (for compensating instabilities or aging of the light source), and a flow cell placed before the main photodetector [Fig. 6(a)].

In diode array detectors, the flow cell is directly illuminated by a "white" light source, and a fixed diffraction grating (placed after the sample cell) receives the nonabsorbed light and disperses it according to the wavelength [see Fig. 6(b)]. The photodiode signals are proportional to the transmitted light at each given wavelength. The advantage of these detectors is that they do not include moving parts.

Evaporative LS

Figure 7 schematically represents an evaporative light-scattering detector (ELSD). It operates in three steps. In step 1, an inert gas carrier (e.g., nitrogen) is used for nebulizing the eluent into a droplet dispersion or aerosol. The size and uniformity of the droplets determines the system sensitivity and reproducibility. In step 2, the mobile phase is evaporated from the droplets in a heated chamber (drift tube), finally producing a fine mist of dry particles. The heated-chamber temperature is adjusted between ambient

Gel Permeation and Size Exclusion Chromatography

(a) Selectable wavelength UV detector

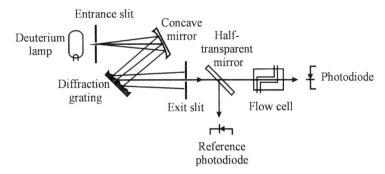

(b) Diode array UV detector

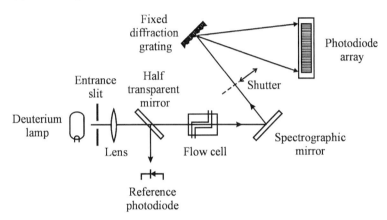

Figure 6 Schematic diagram of two different UV sensors. (a) Selectable wavelength UV detector based on the WellChrom K-2501 spectrophotometer by Knauer. In this typical dispersive design, a (single) wavelength is selectable from the UV range. (b) Photodiode array detector based on Model 2996 by Waters Ass. This equipment measures the instantaneous UV–Vis Spectrum of the eluting solute. It has no moving parts, and the cell is illuminated with white light containing all the wavelengths.

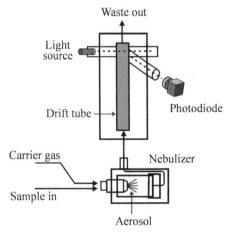

Figure 7 Schematic diagram of an evaporative LS detector for measuring the instantaneous polymer concentration. First, an aerosol is produced from the liquid eluent (in the nebulizer); then, the solvent in the aerosol is evaporated (in the heated drift tube), and finally the solute spray is detected by measuring the scattered light intensity (in the photodiode).

and 120°C, depending on the solute and solvent volatility and on the composition and flow of the mobile phase. In step 3, the dry solute particles are detected (at 90°) as they pass across a laser light beam. The detector signal consists of the scattered light intensity, which is mainly a function of the instantaneous solute mass. Ideally, the operation requires all of the aerosol to travel through the drift tube and reach the optical cell. In some cases, however, it is convenient to split away a fraction of the aerosol droplets, thereby more easily vaporizing the remaining flow. The best results are obtained with an appropriate combination of mobile phase flow rate, analyte volatility, and mobile phase volatility. The scattered light intensity strongly depends on the analyte absorptivity and, therefore, an adequate calibration is required for a quantitative analysis (Pasch and Trathnigg, 1999; Righezza, 1988; Righezza and Guiochon, 1988).

An evaporative detector can sense any substance exhibiting a lower volatility than that of the mobile phase. When the mobile phase contains a buffer or other additives, then these must be volatile enough so as to not contaminate the polymer signal. Their response is practically independent of the analyte optical characteristics and, therefore, these detectors can conveniently replace UV detectors.

ELSDs are almost insensitive to the molar mass; and their baselines are practically unaffected by the drift tube temperature. On the negative side, they exhibit a poor sensitivity to low solute concentrations, and high injection volumes may be then required. A mathematical model has been proposed that simulates the ELSD operation (Trathnigg and Kollroser, 1997; Van der Meeren et al., 1992). The model parameters affecting the ELSD sensitivity are the heated chamber temperature, the flow rates of carrier gas and eluent, and the eluent viscosity. From a phenomenological point of view, the ELSD signal [$s_{ELSD}(V)$] is an exponential function of the instantaneous solute concentration, as follows:

$$s_{ELSD}(V) = a[c(V)]^b \tag{47}$$

where a and b are constants that can sometimes show a small dependence with the molar mass. For hexaethylenglycol and PEG 1000, no significant differences in the a and b parameters were found when the heated-chamber temperature was varied from 30°C to 50°C (Trathnigg and Kollroser, 1997).

2. Molar Mass Sensitive Detectors

Intrinsic Viscosity Detectors

The instantaneous intrinsic viscosity can be determined from individual measurements of the solvent viscosity (η_0), the eluent viscosity (η), and the solution concentration (c) [see Eqs. (11) and (12)]. As an alternative to η_0 and η, the specific viscosity η_{sv} may be directly determined. The very low chromatographic concentrations make it possible to use Eq. (36) for acceptable estimations of the intrinsic viscosity [η]. Slightly improved results are obtained through the following alternative "one-point" expressions:

$$[\eta] = \frac{2^{0.5}}{c}\left[\frac{\eta}{\eta_0} - 1 - \ln\left(\frac{\eta}{\eta_0}\right)\right]^{0.5}$$
$$= \frac{2^{0.5}}{c}\left[\eta_{sp} - \ln(\eta_{sp} + 1)\right]^{0.5} \tag{48}$$

where the first equality is used when c, η, and η_0 are measured, and the second equality when c and η_{sp} are measured. Single-point expressions were developed for the more accurate determination of the intrinsic viscosity from a single concentration measurement, by extrapolating (to $c \to 0$) a low-slope function.

On-line viscometers are all based on Poiseuille's Law, which calculates the pressure drop (ΔP) caused by a polymer solution of flow rate q and viscosity η that circulates through a capillary of radius r_c and length l_c:

$$\Delta P(V) = \frac{8\,l_c}{\pi r_c^4} q \eta(V) \tag{49}$$

Equation (49) indicates that (for a given capillary and a constant flow), the measurement $\Delta P(V)$ is directly proportional to the instantaneous solution viscosity $\eta(V)$.

The more primitive on-line viscometers were single-capillary sensors (Kuo et al., 1990; Lesec et al., 1988; Letot et al., 1980; Lesec and Volet, 1990); but these present: (1) a low sensitivity, as a consequence of the extremely low ΔP signal; (2) a relatively high sensitivity of ΔP to flow fluctuations; and (3) a high sensitivity of both η and η_0 to temperature changes. These problems were considerably overcome in multicapillary viscometers (Lesec, 2001), which aim at either simultaneously measuring η and η_0, or at directly measuring η_{sv}. For measuring of the solvent viscosity while the polymer solution is eluting, hold-up solvent reservoirs (of sufficiently large volumes) are included in the flow path. In these reservoirs, the solution becomes highly diluted and the reservoir output is practically pure solvent. Solvent reservoirs must be frequently purged with solvent to scavenge all polymer traces.

A simple two-capillary viscometer is schematically represented in Fig. 8(a). It consists of two capillaries in series, with an intermediate reservoir. The measurements are the pressure drops along each capillary. In the first capillary, ΔP is proportional to the solution viscosity, whereas, in the second, ΔP_0 is proportional to the pure solvent viscosity. Thus, $\eta_{sp} = (\eta/\eta_0) - 1 = (\Delta P/\Delta P_0) - 1$. In this configuration, the temperature and flow rate oscillations are automatically compensated, since they appear almost simultaneously in both sensors.

The Wheatstone bridge viscometer is schematically represented in Fig. 8(b) (Hanney, 1985a, b). It consists of four capillaries R1–R4 in a Wheatstone bridge arrangement, and a hold-up reservoir in series with capillary R2. In this device, only the pressure drops in the capillaries are significant, while all the other elements do not introduce pressure drops. Initially, the pure solvent circulates through the four capillaries, and $\Delta P \cong 0$ because the mid-points of the bridge branches are perfectly balanced. When the solution reaches the viscometer, it fills the capillaries R1, R3, and R4, but not capillary R2, due to the dilution effect of the solvent reservoir. Thus, a pressure imbalance (ΔP) is produced in the bridge, and it can be shown that the specific viscosity may be directly calculated through (Hanney, 1985b):

$$\eta_{sp}(V) = \frac{4\Delta P(V)}{P_T(V) - 2\Delta P(V)} \tag{50}$$

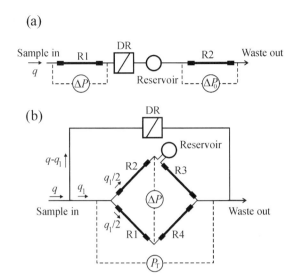

Figure 8 Schematic diagram of two on-line capillary viscometers. (a) Dual-capillary series viscometer. This design independently determines the instantaneous viscosity of the eluent (from the pressure drop across capillary R1), and from the pure solvent (from the pressure drop across capillary R2). (b) Differential refractometer and Wheatstone bridge viscometer, based on the Model 200 by Viscotek. This design direcly estimates the instantaneous specific viscosity, and indirectly the intrinsic viscosity from the viscometer–refractometer signals ratio.

where P_T is the total pressure drop across the sensor. P_T is approximately constant, and considerably larger than ΔP. Thus, η_{sp} is approximately proportional to ΔP. This instrument is a compromise between the higher sensitivity that is provided by the bridge configuration, and the flow imbalances that can be generated in a parallel configuration.

LS Detectors

A typical LS detector consists of: (1) a monochromatic light source (normally, a vertically polarized laser light) that directly falls on the flow cell and (2) one or more photodetectors that collect the light scattered by the molecules in the solution, at each angle θ. Conventionally, $\theta = 0°$ is the direction of the transmitted light. Depending upon their designs, LS detectors are classified into: low-angle-laser LS (LALLS), right-angle-laser LS (RALLS), two-angle-laser LS (TALLS), and multiangle-laser LS (MALLS).

In LALLS, the scattered light is measured at $3 < \theta < 7°$. In this condition, $P(\theta) \cong 1$ in Eq. (31b), and the instantaneous weight-average molar mass can be directly estimated from Eq. (31a): $M_w(V) \cong \Delta R_\theta(V) / [K_{LS} \times c(V)]$. Measurement angles lower than $3°$ cannot be used, due to the contamination by the transmitted light. In practice, LALLS sensors provide strong signals for the high polymer (e.g., $M > 20,000$ g/mol and $R_g > 20$ nm).

In RALLS sensors, the $\theta = 90°$ measurements provide higher signal-to-noise ratios than at low angles. Their signals are mainly intended for measuring oligomers, because (for smaller molecules) it is $P(\theta) \cong 1$ in Eq. (31b) and, again, a simplified LS expression is applicable.

In a TALLS detector, the measurements are taken at two angles (e.g., $90°$ and $15°$) and, therefore, Eq. (31a) can, in principle, be used for simultaneously estimating M_w and R_g through Eqs. (31a)–(31c). Clearly, better estimations of these two quality variables are obtained when measuring at several scattering angles (MALLS) since, in this case, a larger amount of information is available. In practice, however, some of the measurement angles may not provide useful information and, in this case, they must be discarded for data processing.

Figure 9 schematically shows a three-angle LS detector (MiniDawn, from Wyatt Technol. Corp., USA). The light scattered from the flow cell is collected at $45°$, $90°$, and $135°$. The same manufacturer offers a MALLS detector of 18 measurement angles (Table 8).

Combined LS-SV-DR Detection

In recent years, the combined measurement of LS (at $90°$), SV, and DR detectors has been investigated (Brun, 1998; Greene, 2001). The molar mass estimation from either LS-DR or SV-DR involves a signal ratio [see Eqs. (34) and (37)]; and this operation fails at the chromatogram tails. This difficulty is partially overcome with the

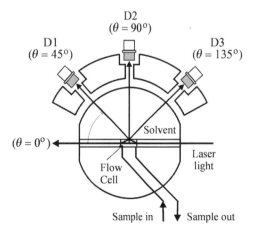

Figure 9 Schematic diagram of a three-angle laser LS photometer (after MiniDawn Tristar, Wyatt Tech. Corp.). This instrument measures the scattered light at $45°$, $90°$, and $135°$. It is a smaller version of the 18-angle Dawn EOS by Wyatt Tech. Corp. In combination with a concentration detector, it is capable of estimating the instantaneous weight-average molar mass and the z-average radius of gyration.

following two configurations. The first configuration is aimed at estimating the high molar masses where the DR signal is too low. In this case, only the LS and SV signals are used, in combination with the universal calibration, with improved estimates of the high molar masses (Brun, 1998). The second configuration uses the three signals simultaneously and, in this case, the universal calibration is not required. The triple sensor has appeared under the name SEC^3 (Viscotek TDA in Table 8).

VII. APPLICATION EXAMPLES

Four application examples are presented. The first two consider the analysis and data processing of two common plastics that are soluble in organic solvents: a linear PS, and a branched poly(vinyl acetate) (PVAc). The third example considers the (rather complex) characterization of amylopectin (a water-soluble ultrahigh molar mass polysaccharide). Finally, some analytical applications of preparative SEC are presented.

A. MMD of a Linear PS

Consider the SEC analysis of a linear, atactic PS, where the MMD completely characterizes its molecular macrostructure. The sample was dissolved in tetrahydrofuran (THF) at room temperature. The injection concentration is not required in the present analysis.

A Waters ALC244 chromatograph was used, fit with a complete set of 6 μ-Styragel columns and a DR. The carrier solvent was THF at 0.9815 mL/min, and the temperature was 25°C. After the injection, 15 min passed before storing the detector signal. Then, the continuous DR signal (in mV) was sampled every 5.4 s during 45 min, thus producing a discrete chromatogram of 500 points. After the baseline correction, the resulting $s_{DR}(V)$ chromatogram contained 138 points, and is represented with a polygonal in Fig. 10(a).

The molar mass calibration was obtained from a set of seven narrow PS standards (Waters). The mean elution volumes were adopted at the peaks of the standard chromatograms Fig. 10(b). Their corresponding molar masses were obtained from the peak molar masses (when available) or, alternatively, were estimated from the geometric mean $(\bar{M}_w \bar{M}_n)^{0.5}$. The resulting experimental points were fitted by the following linear calibration (see Fig. 10b):

$$\log M_{PS}(V) = 12.5531 - 0.1708\,V$$

From the calibration $\log M_{PS}(V)$ and the chromatogram $s_{DR}(V)$, the "continuized" MMD in the format $[dw/d\log M](\log M)$ was directly obtained [see Fig. 10(c)].

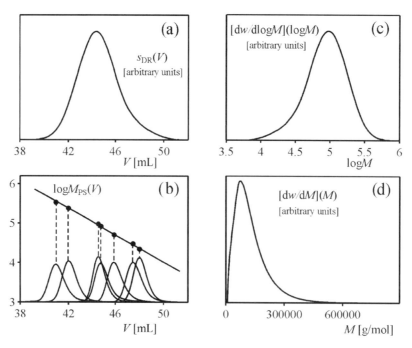

Figure 10 Application example 1: MMD of a linear PS by means of a concentration detector and a direct molar mass calibration obtained from a set of PS standards. (a) Differential refractometer chromatogram, exhibiting 138 points sampled every 0.088 mL. (b) Chromatograms of the calibration standards and resulting (linear) molar mass calibration. (c) MMD in its "chromatographic" format $[dw/d\log M](\log M)$. (d) MMD with a linear molar mass axis $[dw/dM](M)$.

Since the calibration is linear, the shape of $[dw/d\log M]$ ($\log M$) is the mirror image of $s_{DR}(V)$. The average molar masses were calculated from Eqs. 8(a) and (b), yielding: $\bar{M}_n = 79,500$ g/mol, $\bar{M}_w = 128,500$ g/mol, and $\bar{M}_w/\bar{M}_n = 1.62$. The MMD is also represented with a linear molar mass axis in Fig. 10(d) and, to this effect, the heights correction of Eq. (5) was applied.

B. Chain Branching in PVAc

A branched PVAc homopolymer was analyzed by SEC, with the aim of estimating its MMD and the variation of the average number of branches per molecule with the molar mass. The chromatograph was a Waters ALC244 fit with a complete set of 5 Shodex fractionation columns (A 802, 803, 804, 805, 806/S) and a Viscotek™ (Model 200) SV-DR dual-detector (see Fig. 8(b)). The carrier solvent was THF at 0.985 mL/min, and the temperature was 25°C. The polymer was dissolved in THF at room temperature, and the injection concentration was 2.320 mg/mL.

The DR gain was determined as indicated in Section III.A, from known injected masses of PVAc, yielding $k_{DR} = 25.8$ (mV mL/mg). Then, the concentration "chromatogram" was obtained from $c(V) = s_{DR}(V)/k_{DR}$ [Eq. (22)]; and is presented in Fig. 11(a). Also represented in Fig. 11(a) is the specific viscosity chromatogram $\eta_{sp}(V)$. Compared with $c(V)$, note that $\eta_{sp}(V)$ is skewed toward lower retention volumes.

To determine the universal calibration, a set of narrow PS standards (Waters) was used. For each standard, the molar mass of the DR chromatogram peak was known; and the following was determined: (a) the peak elution

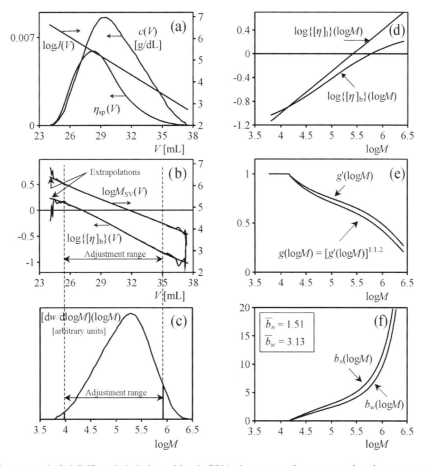

Figure 11 Application example 2: MMD and chain branching in PVAc by means of a concentration detector, a specific viscometer, and a universal calibration. (a) Concentration chromatogram, specific viscosity chromatogram, and universal calibration. (b) Instantaneous intrinsic viscosity and molar mass. Acceptable direct estimations are only feasible in the adjustment range. Outside this range, the estimates were obtained by extrapolation of the smoothened curves. (c) MMD. (d) Intrinsic viscosity vs. $\log M$ for the analyzed polymer, and for its linear homolog. (e) g' and g contraction factors. (f) Number- and weight-average number of trifunctional branching points per molecule vs. $\log M$. Also indicated are the corresponding global averages.

volume of the DR chromatogram (V_p) and (b) the average intrinsic viscosity in THF at 25°C ($[\bar{\eta}]$) by standard glass viscometry (an off-line measurement). The experimental points of $[\bar{\eta}]_{PS} M_{PS}$ vs. V_p were fit with a straight line, yielding the universal calibration:

$$\log J(V) = \log \{[\bar{\eta}]_{PS}(V) M_{PS}(V)\} = 13.60 - 0.2900\, V$$

which is represented in Fig. 11(a).

The intrinsic viscosity of the analyzed branched polymer [Eqs. (12) and (36)] is calculated from the signal ratio: $[\eta]_b(V) \cong \eta_{sp}(V)/c(V)$, and is represented as $\log\{[\eta]_b\}(V)$ in Fig. 11(b). On the basis of Eq. (37), the *ad-hoc* calibration, $\log M_{SV}(V)$ is obtained from: $\log M_{SV}(V) = \log J(V) - \log\{[\eta]_b\}(V)$, and is also represented in Fig. 11(b). Both curves are slightly nonlinear in the mid-chromatogram range, and present important oscillations at the chromatogram tails. With oscillatory values of $\log M_{SV}(V)$, it is impossible to represent a MMD. One crude solution is to limit the range of molar masses to the adjustment range indicated in Fig. 11(b). The resulting (narrower or bounded) MMD is shown in Fig. 11(c) in the format $[dw/d\log M](\log M)$. To this effect, the ordinates of $c(V)$ were slightly corrected to compensate for the very moderate nonlinearities of $\log M_{SV}(V)$. The corresponding average molar masses [directly calculated from $c(V)$ and $\log M_{SV}(V)$ in the Adjustment range] result: $\bar{M}_n = 97{,}600$ g/mol, $\bar{M}_w$ 246,000 g/mol, and $\bar{M}_w/\bar{M}_n = 2.52$.

To extend $\log\{[\eta]_b\}(V)$ into the chromatogram tails, the values in the adjustment range of Fig. 11(b), were first fit with a third-order polynomial, and then this polynomial was extrapolated as shown by the smooth lines in Fig. 11(b). From $J(V)$ and the extrapolated $\log\{[\eta]_b\}(V)$, an extrapolated (and nonoscillatory) *ad-hoc* molar mass calibration $\log M_{SV}(V)$ was obtained [smooth and monotonically varying curve in Fig. 11(b)]. The resulting complete or unbounded MMD is shown in Fig. 11(c); and their average molar masses result: $\bar{M}_n = 76{,}700$ g/mol, $\bar{M}_w = 241{,}500$ g/mol, and $\bar{M}_w/\bar{M}_n = 3.15$. As expected, the global polydispersity has considerably increased with respect to the bounded MMD.

The branching results are presented in Figs. 11(d)–(f). From the smoothened and extrapolated versions $\log\{[\eta]_b\}(V)$ and $\log M_{SV}(V)$, the MHS plot of $\log\{[\eta]_b\}$ vs. $\log M$ was obtained and is shown in Fig. 11(d). Also shown in Fig. 11(d) is the linear MHS plot for the linear PVAc homolog. It is given by: $\log[\eta]_l$ vs. $\log M$, with $K = 1.60 \times 10^{-4}$ dL/g and $\alpha = 0.700$, as taken from the Polymer Handbook (Brandrup and Immergut, 1989). As expected, the slope of $[\eta]_b(\log M)$ slowly drops with $\log M$; and this indicates that branching increases with the molar mass. The (volumetric) contraction index function $g'(\log M)$ was calculated from the ratio $[\eta]_b/[\eta]_l$, and is shown in Fig. 11(e). To this effect, the intrinsic viscosity curves of Fig. 11(d) were employed.

Some recent publications (Grcev et al., 2004) have estimated the branching distribution of PVAc by SEC/MALLS in THF at room temperature. However, intrinsic viscosity measurements require the empirical ε exponent that relates g' with g [Eq. (18)], and very few ε values for PVAc have been reported. In the present example, we have adopted $\varepsilon = 1.2$, as can be calculated from experimental data provided by Park and Graessley (1977). Thus, the g contraction factor was obtained from $(g')^{1/1.2}$, and is also represented in Fig. 11(e). The variations (with the molar mass) of the number- and weight-average number of branches per molecule $b_n(\log M)$ and $b_w(\log M)$ were obtained by injecting $g(\log M)$ into the Zimm–Stockmayer equations [Eqs. (41a) and (41b)], yielding the curves of Fig. 11(f). Finally, the corresponding global number- and weight-averages were obtained from:

$$\bar{b}_n = \frac{\sum_j b_{n,j} \times \{[dw/\log M]_j / M_j\}}{\sum_j [dw/\log M]_j / M_j};$$

$$\bar{b}_w = \frac{\sum_j b_{n,j} \times [dw/\log M]_j}{\sum_j [dw/\log M]_j} \quad (51)$$

and the resulting values are also shown in Fig. 11(f).

C. Characterization of Starch

Starch is a mixture of two biopolymers: amylose and amylopectin, and both of them are based on α,D-glucose monomer (Banks and Greenwood, 1975). Amylose is a mostly linear glucan, containing α,1–4 glycosidic linkages and a limited amount of branching. Its weight-average molar masses are relatively low, between 10^5 and 10^6 g/mol (Roger et al., 1996). In contrast, amylopectin is a highly branched polymer, with short linear chains branched onto longer chains via α,1–6 linkages. The average molar masses of amylopectin are in excess of 10^8 g/mol (Roger et al., 1999). The proportion of amylose to amylopectin in starch varies according to the botanical origin. For example, the weight fraction of amylose ranges from 0% in waxy maize starch to 70–80% in rugosus (wrinkled) pea starch. The amylose content influences both the nutritional and technical properties of starch (e.g., its susceptibility to enzymatic hydrolysis or its gelling and pasting behavior). In this section, the difficulties of the characterization of starch are first considered, and then some of the more recent publications are reviewed.

The determination of molar masses in starch is particularly challenging because: (a) it is extremely difficult to isolate amylopectin from amylose without degrading the polymers, and SEC alone is incapable of disentangling amylose from amylopectin and (b) amylopectin molecules are gigantic, and well outside normal fractionation range of SEC columns. Since SEC is incapable of fractionating amylopectin, then only an average molar mass of this starch component can be measured (but not its MMD). Other problems are the lack of calibration standards, the polymer degradation during SEC analysis, and the determination of the amount of branching. All the recent publications on starch characterization have used on-line LS detectors for estimating the ultrahigh amylopectin molar masses, and have put emphasis on the sample preparation and isolation techniques.

Prior to the SEC measurement, the amylose/amylopectin starch granules must be put into solution; and several temperature and mechanical treatments have been proposed. The inconsistencies in the characterization of starch are partly due to the differences in the sample preparation techniques. Other causes of error are attributable to the SEC analysis, to molecular degradation, and to irreversible retention of the larger molecules in the fractionation columns or frits.

The solubility of starch increases with the amylose content. Also, the solubilized starch chains exhibit a limited stability in neutral aqueous solutions and, for this reason, the structural analysis of starch is difficult in an aqueous medium. In the case of incomplete dissolution, chain aggregation yields molar masses that are too high. Complete dissolution of starch is feasible in alkaline solutions, but these conditions can induce undesirable depolymerization and oxidation reactions. A good solvent for starch is dimethyl sulfoxide (DMSO) but, in this case, it is convenient to add a small amount of water to prevent a too rapid swelling of the starch granules during dissolution. Also, DMSO has been used as mobile phase (Chuang and Sydor, 1987). To minimize degradation, medium- or low-pressure SEC is preferable. Polar packings are undesirable because of possible adsorption between packing and starch. Polymer–polymer, polymer–solvent, and polymer-support interactions can also occur when polar organic solvents are used as mobile phase. These interactions can be eliminated by addition of a low molar mass electrolytes such as sodium nitrate or lithium bromide into the mobile phase (Chuang and Sydor, 1987; Meehan, 1995; Yokoyama et al., 1998).

You and Lim (2000) analyzed a normal corn starch by DR-LALLS using aqueous $NaNO_3$ solution (0.15 M) as a mobile phase and the following columns: TSK G4000 (range: 1000–700,000 g/mol), and 5000 PWXL (range: 50,000–7,000,000 g/mol). The reported values of $\bar{M}_w$ for the amylopectin and amylose fractions were 201×10^6 and 3.3×10^6 g/mol (You and Lim, 2000).

Yoo and Jane (2002) have characterized the amylopectin of different botanical sources with a chromatograph equipped with a DR-MALLS detection system and two Shodex columns: OH pack KB-806 (of exclusion limit 1,000,000 Da) and KB-804 (of exclusion limit 20,000,000 Da). The resulting $\bar{M}_w$ values varied between 7.0×10^7 and 5.7×10^9 g/mol, whereas the corresponding radii of gyration were in the range 191–782 nm.

Han and Lim (2004) analyzed several corn starches with a medium-pressure chromatograph fitted with DR-MALLS and a set Toyoperal HW 65 F columns of range 10,000–1,000,000 Da. The following values of $\bar{M}_w$ were reported: (1) 254×10^6 g/mol for the amylopectin a waxy corn starch (that does not contain amylose), (2) 243×10^6 and 3.13×10^6 g/mol for amylopectin and amylose of a normal corn starch, and (3) 197×10^6 and 1.49×10^6 g/mol for amylopectin and amylose of a high amylose corn starch. Thus, it seems that $\bar{M}_w$ of amylopectin decreases for increasing levels of amylose (Han and Lim, 2004).

D. Sample Fractionation by Preparative SEC

Preparative SEC provides a simple and relatively fast method for fractionating and isolating a wide variety of polymers on the basis of molecular size. Compared with analytical SEC, the column dimensions, particle size, flow rates, and injected samples are all larger in preparative SEC, while the working pressures are lower. Thus, the efficiency of preparative SEC is lower than that of analytical SEC. The SEC operation on a larger scale introduces new problems of cost investment in column materials, of solvent recovery, and so on (Vaughan and Dietz, 1978). Consider some applications of preparative SEC for fractionating polymers for their further analysis by other characterization techniques.

In the preparative isolation of biopolymers, SEC is normally the last stage (or "polishing" stage) of the process. At this stage, the sample volume is already reduced, and the sample has been prepurified. If the component to be separated is physiologically active and will be subsequently used as a medicinal drug, then it is important to not incorporate nonphysiological substances such as detergents or chaotropic reagents into the mobile phase. A better solution is to select an SEC packing that will not interact with the sample components, thus making unnecessary the addition of nonphysiological substances.

Tezuka et al., (2002) fractionated cyclic poly(oxyethylene) (POE) by preparative SEC; and the isolated fractions were later characterized by ^1H NMR, analytical SEC, and time-of-flight mass spectrometry.

Preparative SEC is useful for removing the high and/or low molecular weight fractions of a polymer dispersion (Ekmanis, 1987), for isolating low molar mass additives, and for isolating and purifying proteins (Kato et al., 1987, 1999; Jacob and Britsch, 1999). For example, high molar mass plasma proteins have been isolated by preparative SEC with Fractogel EMD BioSEC (Merck) and Superose 6 (Pharmacie) columns (Josić et al., 1998).

ACKNOWLEDGMENTS

This work was carried out in the frame of the recently approved Project 2003-023-2-G.Meira (IUPAC): "Data Treatment in the Size Exclusion Chromatography of Polymers", http://www.iupac.org/projects/2003/2003-023-2-400.html. We are grateful for the financial support received from the following Argentine institutions: CONICET, Universidad Nacional del Litoral, and SECyT.

REFERENCES

Balke, S., Mourey, T., Harrison, C. (1994). Number-average molecular weight by size exclusion chromatography. *J. Appl. Polym. Sci.* 51:2087–2102.

Banks, W., Greenwood, C. (1975). *Starch and Its Components*. Edinburgh: University Press.

Baumgarten, J., Busnel, J., Meira, G. (2002). Band broadening in size exclusion chromatography of polymers. State of the art and some novel solutions. *J. Liq. Chromatogr. Relat. Technol.* 25(13–15):1967–2001.

Belenkii, B., Vilenchik, L. (1983). *Modern Liquid Chromatography of Macromolecules*. Journal of Chromatography Library. Vol. 25. Amsterdam: Elsevier.

Berek, D. (1999). Interactive properties of polystyrene/ divinylbenzene and divinylbenzene-based commercial size exclusion chromatography columns. In: Wu, Ch., ed. *Column Handbook for Size Exclusion Chromatography*. San Diego, CA: Academic Press, pp. 445–457.

Berek, D. (2000). Coupled liquid chromatography techniques for the separation of complex polymers. *Prog. Polym. Sci.* 25:873–908.

Berek, D., Marcinka, K. (1984). Gel chromatography. In: Deyl, Z., ed. *Separation Methods*. Amsterdam: Elsevier, p. 271.

Brandrup, J., Immergut, E. H., eds. (1989). *Polymer Handbook*, 3rd ed. New York: John Wiley & Sons, Ltd.

Brun, Y. (1998). Data reduction in multidetector size exclusion chromatography. *J. Liq. Chromatogr. Relat. Technol.* 21(13):1979–2015.

Brun, Y., Gorenstein, M., Hay, N. (2000). New approach to model fitting in multi-detector GPC. *J. Liq. Chromatogr. Relat. Technol.* 23(17):2615–2639.

Burchard, W. (1999). Solution properties of branched macromolecules. *Adv. Polym. Sci.* 143:113–194.

Busnel, J. P., Foucault, F., Denis, L., Lee, W., Chang, T. (2001). Investigation and interpretation of band broadening in size exclusion chromatography. *J. Chromatogr. A* 930:61–71.

Cassassa, E. F. (1976). Comments on exclusion of polymer chains from small pores and its relation to gel permeation chromatography. *Macromolecules* 9(1):182–185.

Chang, S., Gooding, K., Regnier, F. (1976). High-performance liquid chromatography of proteins. *J. Chromatogr. A* 125(1):103–114.

Chuang, J., Sydor, R. (1987). High performance size exclusion chromatography of starch with dimethyl sulfoxide as the mobile phase. *J. Appl. Poly. Sci.* 34:1739–1748.

Dawkins, J. (1978). Inorganic packings for GPC. In: Epton, R., ed. *Chromatography of Synthetic and Biological Polymers. Columns Packings, GPC, GF and Gradient Elution*. The Chemical Society Macromolecular Group. Vol. 1. Chichester: Ellis Horwood LTD., pp. 30–56.

Ekmanis, J. (1987). Preparative gel pemeation chromatography. In: Provder, T., ed. *Detection and Data Analysis in Size Exclusion Chromatography*. Washington: American Chemical Society, pp. 47–58.

Eksteen, R., Pardue, K. (1995). Modified silica-based packing materials for size exclusion chromatography. In: Wu, Ch., ed. *Handbook of Size Exclusion Chromatography*. Chromatographic Science Series. Vol. 69. New York: Marcel Dekker, Inc., pp. 47–101.

Engelhardt, H., Ahr, G. (1983). Optimization of efficiency in size-exclusion chromatography. *J. Chromatogr.* 282:385–397.

Fei, X., Murray, K. (1996). On-line coupling of gel permeation chromatography with MALDI mass spectrometry. *Anal. Chem.* 68:3555–3560.

Felinger, A. (1998). *Data Analysis and Signal Processing in Chromatography*. Data Handling in Science and Technology. Vol. 21. Amsterdam: Elsevier.

Flory, P., Fox, T. (1951). Treatment of intrinsic viscosities. *J. Am. Chem. Soc.* 73:1904–1908.

Glöckner, G. (1987). *Polymer Characterization by Liquid Chromatography*. Journal of Chromatography Library. Vol. 34. Amsterdam: Elsevier.

Gooding, K. (1999). SynChropak size exclusion columns. In: Wu, Ch., ed. *Column Handbook for Size Exclusion Chromatography*. San Diego, CA: Academic Press, pp. 305–323.

Grcev, S., Schoenmakers, P., Iedema, P. (2004). Determination of molecular weight and size distribution and branching characteristics of PVAc by means of size exclusion chromatography/multi-angle laser light scattering (SEC/MALLS). *Polymer* 45:39–48.

Greene, S. (2001). SEC with on-line triple detection: light scattering, viscometry, and refractive index. In: Cazes, J., ed. *Encyclopedia of Chromatography*. New York: Marcel Dekker, Inc., pp. 738–742.

Gruber, K., Whitaker, J., Morris, M. (1979). Molecular weight separation of proteins and peptides with a new high-pressure liquid chromatography column. *Anal. Biochem.* 97:176–183.

Grubisic, Z., Rempp, P., Benoit, H. (1967). A universal calibration for gel permeation chromatography. *J. Polym. Sci. B: Polym. Lett.* 5(9):753–759.

Gugliotta, L., Vega, J., Meira, G. (1990). Instrumental broadening correction in size exclusion chromatography. Comparison of several deconvolution techniques. *J. Liq. Chromatogr.* 13:1671–1708.

Guiochon, G., Martin, M. (1985). Theoretical investigation of the optimum particle size for the resolution of proteins by size-exclusion chromatography. *J. Chromatogr. A* 326:3–32.

Hamielec, A. E. (1984). Correction for axial dispersion. In: Janca, J., ed. *Steric Exclusion Liquid Chromatography of Polymers*. Chromatographic Science. Vol. 25. New York: Marcel Dekker, Inc., pp. 117–160.

Hamielec, A., Ouano, A. (1978). Generalized universal molecular weight calibration parameter in GPC. *J. Liq. Chromatogr.* 1:111–120.

Han, J., Lim, S. (2004). Structural changes of corn starches by heating and stirring in DMSO measured by SEC-MALLS-RI system. *Carbohydr. Polym.* 55:265–272.

Hanney, M. (1985). The differential viscometer. I. A new approach to the measurement of specific viscosities of polymer solutions. *J. Appl. Polym. Sci.* 30:3023–3036.

Hanney, M. (1985). The differential viscometer. II. On-line viscosity detector for size-exclusion chromatography. *J. Appl. Polym. Sci.* 30:3037–3049.

Hupe, K., Jonker, R., Rozing, G. (1984). Determination of bandspreading in high-performance liquid chromatographic instruments. *J. Chromatogr.* 285:253–265.

Jackson, C. (1999). Evaluation of the "effective volume shift" method for axial dispersion corrections in multi-detector size exclusion chromatography. *Polymer* 40:3735–3742.

Jackson, C., Yau, W. W. (1993). Computer simulation study of size exclusion chromatography with simultaneous viscometry and light scattering measurements. *J. Chromatogr.* 645:209–217.

Jackson, C., Barth, H. (1995). Molecular weight-sensitive detectors for size exclusion chromatography. In: Wu, Ch., ed. *Handbook of Size Exclusion Chromatography*. Chromatographic Science Series. Vol. 69. New York: Marcel Dekker, Inc., pp. 103–145.

Jacob, L., Britsch, L. (1999). Size exclusion chromatography on fractogel EMD BioSEC. In: Wu, Ch., ed. *Column Handbook for Size Exclusion Chromatography*. San Diego, CA: Academic Press, pp. 219–247.

Janča, J. (1984). Principles of steric exclusion liquid chromatography. In: Janca, J., ed. *Steric Exclusion Liquid Chromatography of Polymers*. Chromatographic Science Series. Vol. 25 New York: Marcel Dekker, Inc., pp. 1–51.

Jordi, H. (1999). Jordi gel columns for size exclusion chromatography. In: Wu, Ch., ed. *Column Handbook for Size Exclusion Chromatography*. San Diego, CA: Academic Press, pp. 367–425.

Josić, D., Horn, H., Schulz, P., Schwinn, H., Britsch, L. (1998). Size-exclusion chromatography of plasma proteins with high molecular masses. *J. Chromatogr. A* 796:289–298.

Kato, Y., O'Donnell, J., Fisher, J. (1999). Size exclusion chromatography using TSK-gel columns and toyopearl resins. In: Wu, Ch., ed. *Column Handbook for Size Exclusion Chromatography*. San Diego, CA: Academic Press, pp. 93–158.

Kato, Y., Yamasaki, Y., Moriyama, H., Tokunaga, K., Hashimoto, T. (1987). New high-performance gel filtration columns for protein separation. *J. Chromatogr. A* 404:333–339.

Kilz, P. (1999). Design, properties, and testing of polymer standards service size exclusion chromatography (SEC) columns and optimization of SEC separations. In: Wu, Ch., ed. *Column Handbook for Size Exclusion Chromatography*. San Diego, CA: Academic Press, pp. 267–303.

Kuo, C., Provder, T., Koehler, M. (1990). Evaluation and application of a commercial single capillary viscometer system for the characterization of molecular weight distribution and polymer chain branching. *J. Liq. Chromatogr.* 13(16):3177–3199.

Lau, S., Taneja, A., Hodges, R. (1984). Effects of high-performance liquid chromatographic solvents and hydrophobic matrices on the secondary and quarternary structure of a model protein: reversed-phase and size-exclusion high-performance liquid chromatography. *J. Chromatogr. A* 317:129–140.

Lehmann, U., Köhler, W., Albrecht, W. (1996). SEC absolute molar mass detection by online membrane osmometry. *Macromolecules* 29:3212–3215.

Lesec, J. (1994). Flow fluctuations in GPC-viscometry. *J. Liq. Chromatogr.* 17(5):1011–1028.

Lesec, J. (2001). Viscosimetric detection in GPC-SEC. In: Cazes, J., ed. *Encyclopedia of Chromatography*. New York: Marcel Dekker, Inc., pp. 872–875.

Lesec, J., Volet, G. (1990). Data treatement in aqueous GPC with on-line viscometer and light scattering detectors. *J. Liq. Chromatogr.* 13(5):831–849.

Lesec, J., Lecacheux, D., Marot, G. (1988). Continuous viscometric detection in size exclusion chromatography. *J. Liq. Chromatogr.* 11(12):2571–2591.

Letot, L., Lesec, J., Quivoron, C. (1980). Performance evaluation of continuous viscometer. *J. Liq. Chromatogr.* 3(3):427–438.

Macha, S., Limbach, P. (2002). Matrix-assisted laser desorption/ionization (MALDI) mass spectrometry of polymers. *Curr. Opin. Solid State Mater. Sci.* 6:213–220.

Malawer, E. (1995). Introduction to size exclusion chromatography. In: Wu, C., ed. *Handbook of Size Exclusion Chromatography*. Chromatographic Science Series. Vol. 69. New York: Marcel Dekker, Inc., pp. 1–24.

Martin, A. J. P., Synge, L. M. (1941). A new form of chromatogram employing two liquid phases. *Biochem. J.* 35:1358–1368.

Meehan, E. (1995). Semirigid polymer gels for size exclusion chromatography. In: Wu, Ch., ed. *Handbook of Size Exclusion Chromatography*. Chromatographic Science Series. Vol. 69. New York: Marcel Dekker, Inc., pp. 25–46.

Meehan, E. (1999). Size exclusion chromatography columns from polymer laboratories. In: Wu, Ch., ed. *Column Handbook for Size Exclusion Chromatography*. San Diego, CA: Academic Press, pp. 349–366.

Meira, G. R. (1991). Data reduction in size exclusion chromatography of polymers. In: Barth, H., Mays, J., eds. *Modern Methods of Polymer Characterization*. New York: J. Wiley & Sons, Inc., pp. 67–101.

Meira, G., Vega, J. (2001). Axial dispersion correction methods in GPC/SEC. In: Cazes, J., ed. *Encyclopedia of Chromatography*. New York: Marcel Dekker, Inc., pp. 71–76.

Meira, G., Vega, J. (2003). Characterization of copolymers by size exclusion chromatography. In: Wu, C., ed. *Handbook of Size Exclusion Chromatography and Related Techniques*. Chromatographic Science Series. 2nd ed. Vol. 91. New York: Marcel Dekker, Inc., pp. 139–156.

Montelaro, R. (1988). Protein chromatography in denaturing and nondenaurig solvents. In: Dubin, P., ed. *Aqueous Size Exclusion Chromatography*. Journal of Chromatography Library. Vol. 40. Amsterdam: Elsevier, pp. 269–296.

Moody, R. (1999). Zorbax porous silica microsphere columns for high-performance size exclusion chromatography. In: Wu, Ch., ed. *Column Handbook for Size Exclusion Chromatography*. San Diego, CA: Academic Press, pp. 75–92.

Mori, S. (1977). High-speed gel permeation chromatography. A study of operational variables. *J. Appl. Polym. Sci.* 21:1921–1932.

Mori, S. (1984). Effect of experimental conditions. In: Janca, J., ed. *Steric Exclusion Liquid Chromatography of Polymers*. Chromatographic Science. Vol. 25. New York: Marcel Dekker, Inc., pp. 161—211.

Mori, S. (1988). Column efficiency. In: Dubin, P., ed. *Aqueous Size-Exclusion Chromatography*. Journal of Chromatography Library. Vol. 40. Amsterdam: Elsevier, pp. 171–190.

Mori, S. (2001). GPC-SEC: effect of experimental conditions. In: Cazes, J., ed. *Encyclopedia of Chromatography*. New York: Marcel Dekker, Inc., pp. 378–380.

Mourey, T. H., Balke, S. T. (1998). Detecting local polydispersity with multidetector SEC from reconstructed DRI chromatograms. *J. Appl. Polym. Sci.* 70(4):831–835.

Murgasova, R., Hercules, D. (2003). MALDI of synthetic polymers. An update. *Int. J. Mass Spectrom.* 226:151–162.

Netopilík, M. (1997). Combined effect of interdetector volume and peak spreading in size exclusion chromatography with dual detection. *Polymer* 38:127–130.

Netopilík, M. (1998a). Effect of interdetector peak broadening and volume in size exclusion chromatography with dual viscometric-concentration detection. *J. Chromatogr. A* 809:1–11.

Netopilík, M. (1998b). Effect of local polydispersity in size exclusion chromatography with dual detection. *J. Chromatogr. A* 793:21–30.

Netopilík, M. (2002). Relations between the separation coefficient, longitudinal displacement and peak broadening in size exclusion chromatography of macromolecules. *J. Chromatogr. A* 978:109–117.

Netopilík, M., Persson, B., Porsch, B., Nilsson, S., Sundelöf, L. (2000). Flow irregularities due to polymer elution in size exclusion chromatography and its effect on viscometric detection. *Int. J. Polym. Anal. Charact.* 5:339–357.

Neue, U. (1999). Waters columns for size exclusion chromatography. In: Wu, Ch., ed. *Column Handbook for Size Exclusion Chromatography*. San Diego, CA: Academic Press, pp. 325–348.

Park, W., Graessley, W. (1977). On-line viscometry combined with gel permeation chromatography. II. Application to branched polymers. *J. Polym. Sci., Polym. Phys. Ed.* 15:85–95.

Pasch, H., Trathnigg, B. (1999). *HPLC of Polymers*. Berlin, Heidelberg: Springer-Verlag.

Pfannkoch, E., Lu, K., Regnier, F., Barth, H. (1980). Characterization of some commercial high performance size-exclusion chromatography columns for water-soluble polymers. *J. Chromatogr. Sci.* 18:430–441.

Potschka, M. (1993). Mechanism of size-exclusion chromatography. I. Role of convection and obstructed diffusion in size-exclusion chromatography. *J. Chromatogr.* 648:41–69.

Prougenes, P., Berek, D., Meira, G. (1998). Size exclusion chromatography of polymers with molar mass detection. Computer simulation study on instrumental broadening biases and proposed correction method. *Polymer* 40:117–124.

Provder, T. ed. (1984). *Size Exclusion Chromatography. Methodology and Characterization of Polymers and Related Material*. ACS Symposium Series 245. Washington: American Chemical Society.

Provder, T., ed. (1987). *Detection and Data Analysis in Size Exclusion Chromatography*. ACS Symposium Series 352. Washington: American Chemical Society.

Provder, T., Kuo, C. (1999). Assessing the validity of the universal calibration concept in THF and DMAC for a variety of polymers and columns supports. *Polym. React. Eng.* 7(4):465–484.

Quivoron, C. (1984). Use for polymer analysis. In: Janca, J., ed. *Steric Exclusion Liquid Chromatography of Polymers*. Chromatographic Science Series. Vol. 25. New York: Marcel Dekker, Inc., pp. 213–280.

Radke, W., Simon, P., Müller, A. (1996). Estimation of number-average molecular weights of copolymers by gel permeation chromatography—light scattering. *Macromolecules* 29:4926–4930.

Righezza, M. (1988). Effects of wavelength of the laser beam on the response of an evaporative light scattering detector. *J. Liq. Chromatogr.* 11(13):2709–2729.

Righezza, M., Guiochon, G. (1988). Effects of the nature of the solvent and solutes on the response of a light scattering detector. *J. Liq. Chromatogr.* 11(9–10):1967–2004.

Roger, P., Bello-Perez, L., Colonna, P. (1999). Contribution of amylose and amylopectin to the light scattering behaviour of starches in aqueous solution. *Polymer* 40:6897–6909.

Roger, P., Tran, V., Lesec, J., Colonna, P. (1996). Isolation and characterisation of single chain amylose. *J. Cereal Sci.* 24:247–262.

Schmidt, D., Giese, R., Conron, D., Karger, B. (1980). High performance liquid chromatography of proteins on a diol-bonded silica gel stationary phase. *Anal. Chem.* 52:177–182.

Snyder, L., Kirkland, J., Glajch, J., Krull, I., Szulc, M. (1997). *Practical HPLC Method Development*. New York: J. Wiley & Sons, Inc.

Striegel, A. (2001). Longitudinal diffusion in size-exclusion chromatography: a stop-flow size-exclusion chromatography study. *J. Chromatogr. A* 932:21–31.

Suzuki, H., Mori, S. (1999). Shodex columns for size exclusion chromatography. In: Wu, Ch., ed. *Column Handbook for Size Exclusion Chromatography*. San Diego, CA: Academic Press, pp. 171–217.

Tackx, P., Tacx, J. (1998). Chain architecture of LDPE as a function of molar mass using size exclusion chromatography and multi-angle laser light scattering (SEC-MALLS). *Polymer* 39(14):3109–3113.

Taneja, A., Lau, S., Hodges, R. (1984). Separation of hydrophobic peptide polymers by size-exclusion and reversed-phase high-performance liquid chromatography. *J. Chromatogr.* 317:1–10.

Tezuka, Y., Mori, K., Oike, H. (2002). Efficient synthesis of cyclic poly(oxyethylene) by electrostatic self-assembly and covalent fixation with telechelic precursor having cyclic ammonium salt groups. *Macromolecules* 35:5707–5711.

Trathnigg, B. (1991). Determination of chemical composition of polymers by size exclusion chromatography with coupled density and refractive index detection. *J. Chromatogr.* 552:507–516.

Trathnigg, B. (1995). Determination of MWD and chemical composition of polymers by chromatographic techniques. *Prog. Polym. Sci.* 20:615–650.

Trathnigg, B., Jorde, C. (1982). Densimetric detection in gel permeation chromatography. VI. An integrating densimetric detector. *J. Chromatogr.* 241:147–151.

Trathnigg, B., Jorde, Ch. (1984). Densimetric detection in SEC. A semi-automated method for calculation of molecular averages. *J. Liq. Chromatogr.* 7(9):1789–1807.

Trathnigg, B., Kollroser, M. (1997). Liquid chromatography of polyethers using universal detectors. V. Quantitative aspects in the analysis of low-molecular-mass poly(ethylene glycol)S and their derivatives by reversed-phase high performance liquid chromatography with an evaporative light scattering detector. *J. Chromatogr. A* 768:223–238.

Trathnigg, B., Feichtenhofer, S., Kollroser, M. (1997). Quantitation in liquid chromatography of polymers: size-exclusion chromatography with dual detection. *J. Chromatogr. A* 786:75–84.

Tung, L. H. (1966). Method of calculating molecular weight distribution function from gel permeation chromatograms. III. Application of the method. *J. Appl. Polym. Sci.* 10(9):1271–1283.

Unger, K., Kinkel, J. (1988). Native and bonded silicas in aqueous SEC. In: Dubin, P., ed. *Aqueous Size-Exclusion Chromatography*. Journal of Chromatography Library. Vol. 40. Amsterdam: Elsevier, pp. 193–234.

Van der Meeren, P., Vanderdeelen, J., Baert, L. (1992). Simulation of the mass response of the evaporative light scattering detector. *Anal. Chem.* 64:1056–1062.

Vander Heyden, Y., Popovici, S., Staal, B., Schoenmakers, P. (2003). Contribution of the polymer standards' polydispersity to the band broadening in size-exclusion chromatography. *J. Chromatogr. A* 986:1–15.

Vaughan, M., Dietz, R. (1978). Preparative gel permeation chromatography (GPC). In: Epton, R., ed. *Column Packings, GPC, GF and Gradient Elution*. Chromatography of Synthetic and Biological Polymers. Vol. 1. Sussex: Ellis Horwood LTD., pp. 199–217.

Vega, J., Meira, G. (2001). SEC of simple polymers with molar mass detection in presence of instrumental broadening. Computer simulation study on the calculation of unbiased molecular weight distribution. *J. Liq. Chromatogr. Relat. Technol.* 24(7):901–919.

Vega, J., Estenoz, D., Oliva, H., Meira, G. (2001). Analysis of a styrene—butadiene graft copolymer by size exclusion chromatography. II. Determination of the branching exponent with the help of a polymerization model. *Int. J. Polym. Anal. Charact.* 6:339–348.

Wu, C., ed. (1995). *Handbook of Size Exclusion Chromatography*. Chromatographic Science Series. Vol. 69. New York: Marcel Dekker, Inc.

Wu, C., ed. (1999). *Column Handbook for Size Exclusion Chromatography*. San Diego, CA: Academic Press.

Wyatt, P. (2003). Light scattering and the solution properties of macromolecules. In: Wu, C., ed. *Handbook of Size Exclusion Chromatography and Related Techniques*. Chromatographic Science Series. 2nd ed. Vol. 91. New York: Marcel Dekker, Inc., pp. 623–655.

Yau, W. (1991). An online osmometer for size exclusion chromatography: preliminary results. *J. Appl. Polym. Sci., Appl. Polym. Symp.* 48:85–94.

Yau, W., Stoklosa, H., Bly, D. (1977). Calibration and molecular weight calculations in GPC using a new practical method for dispersion correction–GPCV2. *J. Appl. Polym. Sci.* 21:1911–1920.

Yau, W., Kirkland, J., Bly, D. (1979). *Modern Size Exclusion Liquid Chromatography. Practice of Gel Permeation and Gel Filtration Chromatography*. New York: John Wiley & Sons, Ltd., pp. 209–211.

Yokoyama, W., Renner-Nantz, J., Shoemaker, C. (1998). Starch molecular mass and size by size-exclusion chromatography in DMSO-LiBr coupled with multiple angle laser light scattering. *Cereal Chem.* 75:530–535.

Yoo, S., Jane, J. (2002). Molecular weights and gyration radii of amylopectins determined by high-performance size-exclusion chromatography equipped with multi-angle laser-light scattering and refractive index detectors. *Carbohydr. Polym.* 49(3):307–314.

You, S., Lim, S. (2000). Molecular characterization of corn starch using an aqueous HPSEC-MALLS-RI system under various dissolution and analytical conditions. *Cereal Chem.* 77(3):303–308.

Zimm, B., Stockmayer, W. (1949). The dimension of chain molecules containing branches and rings. *J. Chem. Phys.* 17:1301–1314.

27

Field-Flow Fractionation

MARTIN E. SCHIMPF
Department of Chemistry, Boise State University, Boise, ID, USA

I. INTRODUCTION

Field-flow fractionation (FFF) is a family of instrumental techniques that separates and characterizes macromolecules, colloids, and particles (macromaterials) on an analytical scale (Colfen and Antonietti, 2000; Schimpf et al., 2000). As illustrated in Fig. 1, the FFF channel has a ribbon-shaped geometry, typically with length 30–50 cm, breadth 1–3 cm, and thickness 0.005–0.025 cm. Because of the high aspect ratio between breadth and thickness, liquid that is pumped through the channel flows in a laminar fashion, with a velocity profile that varies across the thin (x) dimension. A field is applied external to the channel in order to force analyte into the slower flow streams near one wall. The resulting velocity of the analyte through the channel depends on its interaction with the field, and therefore, on physicochemical properties that govern that interaction. Those physicochemical properties vary with the nature of the applied field, but always include the size of the analyte, because size determines the ability of analyte to reach the faster moving flow streams away from the accumulation wall.

The separation of macromaterials by size is a feature that FFF shares with size exclusion chromatography (SEC). However, SEC requires calibration curves to relate retention parameters to size or molar mass. Calibration curves can also be used in FFF, but in many cases analyte size can be calculated directly from retention data (i.e., with little or no calibration) because the dependence of retention on physicochemical properties of the analyte is rigorously defined. For highly complex samples, additional properties can be measured on the separated components with the aid of information-rich detectors or offline instruments.

The FFF instrument is very similar to that used in SEC, with the chromatography column replaced by the FFF channel. A computer is used to store the detector signal and to control the strength of the applied field. Although the nature of the applied field can be varied in order to separate materials by different properties, each type of field requires a separate FFF channel. In sedimentation FFF (SdFFF), for example, the channel is wrapped around a rotor and spun to produce a centrifugal force; SdFFF is used to separates materials larger than about 100 nm in diameter by their size and density. In electrical FFF (ElFFF), a voltage drop across the channel is used to separate colloids larger than about 20 nm by their electrophoretic mobility. In thermal FFF (ThFFF) a temperature gradient is used to separate colloids, particles, and lipophilic polymers by their Soret coefficient, which is related to size and other analyte parameters. Finally, flow FFF (FlFFF) uses a hydrodynamic force to separate a wide range of materials strictly by their size.

The magnitude of the applied field can be varied to optimize a given separation problem, or programmed

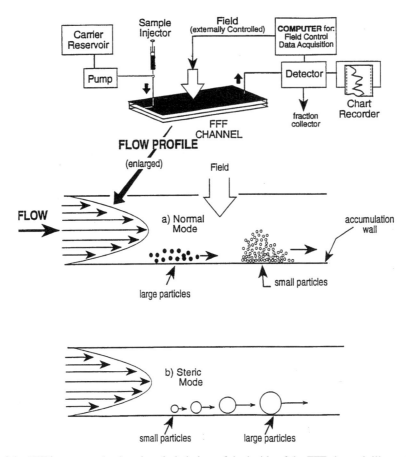

Figure 1 Schematic of the FFF instrument (top) and exploded view of the inside of the FFF channel, illustrating the principles of separation. The field is applied across the thin dimension (0.05–0.25 mm) of a ribbon-shaped channel, which is formed from the cutout of a spacer that is clamped between two plates. The sample is introduced through a hole in one of the clamping plates. (a) In the normal (Brownian) mode of retention, smaller particles elute ahead of larger ones because they diffuse further away from the accumulation wall. (b) In the steric or hyperlayer mode, larger particles elute ahead of smaller ones because they physically protrude into the faster flow streams.

to separate highly polydisperse materials in a single experiment. Varying the field strength is analogous to varying the pore size in SEC packing materials, whereas programing the field is analogous to using a series of SEC columns, or to using a broad distribution of pore sizes in a single SEC column. Thus, FFF is quite versatile, and a complement of the four instruments will separate a wide variety of macromaterials. But this versatility comes with a price. In order to choose the instrument with the most appropriate field, and use that instrument effectively, one must understand the fundamental principles that underlie the mechanism of the separation, and be aware of the limitations of each channel type. This chapter outlines the fundamental theory and operation of FFF, including basic optimization and troubleshooting techniques. For more detailed discussions of FFF and its applications, see Schimpf et al. (2000).

II. PRINCIPLES AND THEORY OF RETENTION

For reasons that will become apparent, we express the level of retention in FFF as the ratio R of the zone velocity v_{zone} to the average velocity of carrier liquid in the channel $\langle v \rangle$:

$$R = \frac{v_{zone}}{\langle v \rangle} = \frac{V^0}{V_r} \tag{1}$$

The retention ratio is also the ratio of the channel void volume V^0 to the retention volume V_r, where V_r is defined as the volume of carrier liquid required to flush a sample component through the channel and V^0 is the volume required to flush a component that is not retained (i.e., a component that does not interact with the field and therefore moves at the average velocity of the carrier liquid). V^0 is called the void volume because it is also the geometric volume of the channel. Thus, each point in the elution profile obtained from an FFF experiment has a unique R value that can be calculated from Eq. (1). For components that do not interact with the field, $v_{\text{zone}} = \langle v \rangle$, so that $V_r = V^0$ and $R = 1$. Note that in the calculation of R, the parameters V_r and V^0 can be replaced by analogous parameters expressed in time units (t_r and t^0). Thus, t_r and t^0 are the times required to elute a retained and nonretained component, respectively.

In many applications of FFF, standards can be used to calibrate the channel under a specific set of conditions. In the simplest of calibration procedures, t_r, V_r, or R is plotted vs. particle diameter (d) or molar mass (M) in logarithmic form. For samples that differ in size or molar mass alone, such plots are valid under a defined set of experimental conditions. However, for samples with a heterogeneous chemical composition, additional parameters that govern the interaction of the analyte with the field must be included in the calibration procedure. In order for the FFF practitioner to incorporate these additional parameters, a basic understanding of retention theory is required. A basic understanding is also necessary for calculating physicochemical parameters from retention data, for understanding the limitations of FFF, and for troubleshooting difficult separations.

A. The Brownian Mode of Retention

Depending on the level of sample compression by the field, one of two principle modes of separation will occur. In the so-called Brownian (or normal) mode, the analyte is small enough to have significant Brownian motion. As a result, the analyte diffuses away from the concentration gradient, or in opposition to the field-induced motion. The result is a zone of material whose concentration decreases exponentially away from the wall. In a mixture of analyte material, dissimilar components form individual zones of different thickness ℓ owing to their different physicochemical properties. It is important to understand that individual particles within a component zone are in continual motion, and therefore sample a range of carrier liquid velocities, although this range is restricted by the field. As a result, particles within a component zone move together at an average velocity defined by the velocity of the carrier liquid at position $x = \ell$ for that zone. Because different components form zones of different thickness, they are carried through the channel at different rates. Thus, small differences in ℓ between analyte components are magnified by the parabolic velocity profile into much larger distances along the channel length, giving rise to the separation.

Zone thickness ℓ is usually expressed in the following dimensionless form:

$$\lambda = \frac{\ell}{w} \tag{2}$$

where w is the thickness of the channel (x dimension) and λ is termed the retention parameter. The dependence of λ on analyte properties differs with the nature of the field. For SdFFF:

$$\lambda = \frac{2D}{d^2|\rho_P - \rho|w^2 r\omega} \quad \text{SdFFF} \tag{3}$$

where D, d, and ρ_P are the diffusion coefficient, hydrodynamic diameter, and density of the analyte, respectively; ρ is the density of the carrier liquid, r is the radius of the rotor, around which the channel is wrapped, and ω is the angular rotation frequency of the rotor ($\omega = \pi\,\text{rpm}/30$ where rpm is the number of revolutions per minute). The analogous equations for other FFF subtechniques are as follows:

$$\lambda = \frac{D}{\mu E w} \quad \text{ElFFF} \tag{4}$$

$$\lambda = \frac{V^0 D}{V_C w^2} \quad \text{FlFFF} \tag{5}$$

$$\lambda = \frac{D}{w D_T (dT/dx)} \cong \frac{D}{D_T \Delta T} \quad \text{ThFFF} \tag{6}$$

The field parameters in Eqs. (3)–(6), which can be programed in the FFF experiment are: rotation frequency ω in SdFFF, electric field E in ElFFF, volumetric rate of cross-flow V_C in FlFFF, and temperature gradient dT/dx (or temperature drop across the channel ΔT) in ThFFF. The sample parameters that govern retention are: particle diameter d in SdFFF, electrophoretic mobility μ in ElFFF, thermal diffusion coefficient D_T in ThFFF, and diffusion coefficient D in all FFF subtechniques.

Although particle diameter is explicitly contained only in Eq. (3), it plays a role in all FFF subtechniques because of its influence on the diffusion coefficient, as expressed by the Stokes–Einstein equation:

$$D = \frac{kT}{3\pi\eta d} \tag{7}$$

Here k is Boltzmann's constant, T is temperature, and η is the viscosity of the carrier liquid. Note that a more compressed zone (smaller λ) is formed by a larger particle (smaller D). The more compressed zones move through the channel at lower velocities (smaller $\lambda \rightarrow$ smaller $R \rightarrow$ smaller v_{zone}). This is characteristic of the Brownian mode of retention (i.e., the elution order is from smaller particles to larger ones).

With the exception of asymmetrical FlFFF (discussed subsequently), the dependence of zone velocity [Eq. (1)]

on zone compression [Eq. (2)] is given by (Schimpf et al., 2000)

$$R = 6\lambda \coth\left(\frac{1}{2\lambda}\right) - 12\lambda^2 \qquad (8a)$$

In practice, a value of V_r (or t_r) associated with a given point in the FFF elution profile is first converted to $R = V^0/V_r$ (or $R = t^0/t_r$), and R is either calibrated to a desired sample parameter (e.g., particle diameter) using standards or converted to the associated λ-value using Eq. (8), followed by the conversion of λ into sample parameters using one of the Eqs. (3)–(6) combined with Eq. (7). The latter method might be used, for example, to calculate the average particle size d directly from an SdFFF elution profile (i.e., without the need for calibration), when the densities of the particle and carrier liquid (ρ_P and ρ, respectively) are known. Note that all other parameters in Eqs. (3) and (7) are either set by the experimenter (w, r, ω) or can be measured independently (η, T).

Because the conversion of R-values into λ-values by Eq. (8a) is not straightforward, several approximate forms have been derived. Three of the more commonly used forms are

$$R = 6\lambda - 12\lambda^2 \qquad (8b)$$

$$\lambda = \frac{R}{6(1-R)^{1/3}} \qquad (8c)$$

$$R = 6\lambda \qquad (8d)$$

Equation (8b) is accurate to within 2% when $R < 0.7$. Equation (8c) is accurate to within 4% for $R < 0.2$, and Eq. (8d) is accurate to within 5% when $R < 0.12$. For the special case of asymmetrical FlFFF, the dependence of R on λ is much more complex, but still follows the simple relationship expressed by Eq. (8d) when $R < 0.12$.

In order to obtain accurate sample parameters for specified points along the FFF elution profile, whether using calibration plots or Eqs. (3)–(8), the FFF practitioner must be aware of the salient features associated with the particular FFF subtechnique involved. In SdFFF, for example, one cannot prepare a calibration curve of log λ or log R vs. log d using particle standards of one density and use it to determine the particle size of an unknown sample with a significantly different density. This is because retention also depends on particle density in SdFFF. On the other hand, if the density of the analyte particle is known, the calibration curve can be adjusted to accommodate the difference in densities between the standard and the analyte, so that the unknown particle sizes can be determined (Giddings et al., 1991a). In ElFFF, one must be aware that the electric field is not simply the voltage difference ΔV divided by the distance between electrodes (w) because of nonlinear voltage drops across electrical double layers at the electrode surfaces (channel walls).

Thus, the effective (working) electric field E_{eff} is less than $\Delta V/w$, and depends on experimental parameters such as ionic strength, pH, and dielectric constant of the carrier liquid, as well as on the chemical nature of the electrode and even the flow rate of the carrier liquid. The double layer around the particle can also affect E_{eff}; therefore, ElFFF should only be used to determine unknown sample parameters when all other parameters are carefully held constant, and the dielectric constant of the calibration standards and the analyte are similar. Still, ElFFF can separate particles by size and charge for subsequent analysis by other techniques with more accuracy and precision than that possible without the separation step.

In FlFFF, the diffusion coefficient can be calculated directly from the retention parameter without complications from other sample parameters. From D, the particle size can be calculated by rearrangement of Eq. (7). For polymer analysis, where the molar mass is often desired, an alternate to Eq. (7) can be used (Rudin and Johnston, 1971):

$$D = \frac{kT}{6\pi\eta}\left(\frac{10\pi N_A}{3[\eta]M_V}\right)^{1/3} \qquad (9)$$

Here M_V is the molar mass (viscosity-average) and $[\eta]$ is the intrinsic viscosity of the polymer-solvent system. Note that if an intrinsic viscosity detector is attached, the retention volume can be converted directly into M_V to obtain a molar mass distribution using Eqs. (1), (5), (8), and (9). Alternatively, FlFFF channels can be calibrated as linear plots of $\log([\eta]M_V)$ vs. $\log V_r$, similar to universal calibration plots in SEC.

In ThFFF there is the additional sample parameter D_T. This parameter can be useful for separating materials by chemical composition because D_T depends strongly on this composition (Gunderson and Giddings, 1986). The utility of ThFFF in the separation of polymers, copolymers, and particles by chemical composition has been widely demonstrated, as discussed subsequently. If, on the other hand, one is interested in characterizing macromaterials by molar mass (or size) alone, then D_T must be known *a priori*. Fortunately, D_T for homopolymers is independent of molar mass, and can be determined using a standard of known D from its ThFFF retention ratio. Once D_T is determined for a given polymer–solvent system, it is just another constant in the calculations involving Eqs. (1), (6), (8), and (9). If calibration curves are used to characterize M, then D_T can be ignored if the calibration standards and unknowns have the same composition. In this case, linear calibration plots are established in the form of log $\lambda\Delta T$ (calculated from experimentally measured values of V_r) vs. log M, as illustrated in Fig. 2. After measuring V_r and calculating R, λ is calculated from Eq. (8), multiplied by ΔT, and the associated M

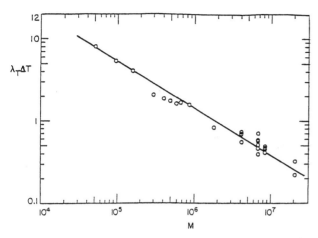

Figure 2 Log–log plot of λD_T vs. M, illustrating the wide range in molecular weights that can be represented in a given polymer-solvent system with a single calibration plot of this type. Such plots are possible because within a given system, M varies predictably with D, while D_T is constant. For universal calibration plots that represent more than one polymer-solvent system, M must be replaced with D/D_T or with $[\eta]M/D_T$. In the latter case, the conformation of the polymers represented must be similar for the calibration points to fall on a single line. [Reprinted with permission from Gao et al., (1985).]

value is obtained from the calibration plot. If the calibration standards and unknowns are not of the same chemical composition, then the calibration curve can be adjusted for the change in D_T, provided the D_T-values are known; this procedure is analogous to the adjustment of calibration curves in SdFFF for density variations. In this case, calibration plots take the form of log $\lambda \Delta T$ vs. log D/D_T (D_T being constant for the calibration standards), and D-values can be calculated and converted into M using Eq. (9). Alternatively, the calibration plot could take the form of log $\lambda \Delta T$ vs. log $[\eta]M/D_T$. In the latter case, an intrinsic viscosity detector would allow the automated conversion of V_r into M using the calibration plot (Gao et al., 1985; Jeon and Schimpf, 1998; Kirkland et al., 1989).

An important note for ThFFF users is that though D_T is independent of molecular weight, it does depend on temperature. Therefore, the retention of a given sample can change when the temperature of the cold wall T_C is changed, even if the applied field (ΔT) is held at a fixed value. The cold wall is typically cooled by heat exchange with water flowing beneath its surface. If either the temperature or the flow rate of that water is unregulated, fluctuations in T_C may occur. Methods have been developed to incorporate T_C into calibration curves (Cao et al., 1999), but it is easier to regulate the temperature and flow rate of the cooling water so that T_C does not fluctuate between calibration and analysis.

B. The Steric/Hyperlayer Mode

The equations derived earlier consider particles as point masses, but as the size of the analyte becomes significant when compared with the channel thickness, it can no longer be treated as such, and we must account for the fact that the particle is excluded from a layer next to the accumulation wall. The thickness of that exclusion layer is, in principle, equal to the radius of the particle, thereby reducing the effective thickness of the channel. Accounting for this exclusion layer results in the following correction in the dependence of R on λ (Giddings, 1978):

$$R_{corr} = 6(\alpha - \alpha^2) + 6\lambda(1 - 2\alpha)\left[\coth\frac{1-2\alpha}{2\lambda} - \frac{2\lambda}{1-2\alpha}\right] \quad (10)$$

Here α is the ratio of the particle radius a to the channel thickness w. Equation (10) should be used as the particle size approaches 1 μm in diameter, or more practically, when calibration plots obtained with the use of Eq. (8) exhibit curvature at high values of d or M.

As the particle size approaches the mean layer thickness predicted from Eq. (2), the concept of an exponentially distributed concentration profile breaks down, and the elution order reverses, that is, larger particles begin to show faster movement through the channel because they physically protrude further from the wall into the faster flow streams (Lee and Giddings, 1988). At this point, the retention volume reaches a limiting value, and Eqs. (2)–(6) and (8) become invalid. Early attempts to model this phenomenon (Giddings, 1978) were not completely successful because they considered particles to extend from the wall, when in reality, they are lifted away from the wall into hyperlayers by hydrodynamic forces that are poorly understood. Although fluid mechanists continue to study the lift forces involved in the formation of hyperlayers, we still do not have robust equations that relate retention to particle size in the hyperlayer mode of retention. Nevertheless, it is still a useful mode, producing highly efficient separations, as illustrated in Fig. 3. Flow rates used in the hyperlayer mode of retention (6.0 mL/min in Fig. 3) tend to be higher than those used in the Brownian mode, which are typically <0.5 mL/min.

The particle size where the elution order reverses is termed the "inversion diameter". The inversion diameter typically lies between 0.5 and 2 μm, but varies with the channel thickness and flow rate of the carrier, as well as the particle shape and its interaction with the field. Polydisperse samples that contain sizes which span the inversion diameter are not completely separated by FFF, as small particles eluting in the Brownian mode coelute with larger particles eluting in the hyperlayer

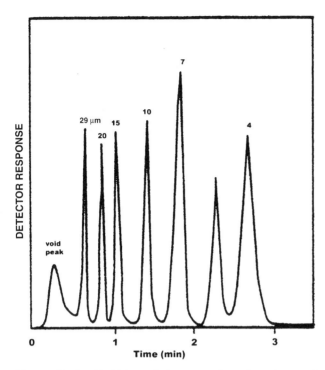

Figure 3 Rapid, high-resolution separation of a mixture of seven latex standards with the indicated diameters by SdFFF run in the hyperlayer mode. The field was 1100 rpm and the flow rate was 6.0 mL/min. [Reprinted with permission from Schimpf et al. (2000).]

mode. Fortunately, the situation is easy to spot as a sharp peak in the elution profile as the limiting retention volume is reached and the elution order reverses, as illustrated in Fig. 4. In some cases, a change in flow rate and/or channel thickness will adjust the inversion diameter enough that a polydisperse sample can be made to elute in one mode or the other. In most cases, however, it is necessary to prefractionate the sample with a filter having a cutoff near the inversion diameter, so that the two resulting fractions can be characterized separately.

III. EXPERIMENTAL PROCEDURES

A. Sample Injection/Relaxation

When the analyte is first injected into the channel, a finite amount of time is required for it to relax into its steady-state concentration profile at the accumulation wall under the influence of the applied field. During this relaxation process, it is not desirable to have the carrier liquid flowing through the channel. First of all, the retention equations assume that analyte is in its steady-state concentration profile from time zero. Second, during the relaxation process the sample will experience a wide range of downstream velocities, which can lead to

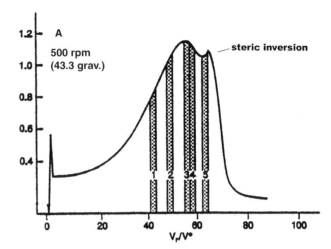

Figure 4 FFF elution profile of a polydisperse PVC sample whose particle sizes span the inversion diameter. The narrow peak fused to the right side of a broader peak indicates foldover into the steric/hyperlayer regime. [Reprinted with permission from Lee and Giddings (1988).]

excessive broadening of the elution profile, and consequently, reduced resolution of components. Several methods have been developed to minimize the impact of sample relaxation. In the simplest method, axial flow is simply routed around the channel, after the sample is injected, for a period of time referred to as the stop-flow time. This allows the sample to relax in the absence of (axial) channel flow. The required stop-flow time is governed by D, λ, and w, and can vary from a few seconds to several minutes. For ThFFF applications, 1 min is generally sufficient for dissolved polymers, and 2–3 min for particle analysis. For FlFFF, the relaxation (stop-flow) time τ is calculated from the rate of cross-flow as follows:

$$\tau = 1.2 \frac{V^0}{V_C} \qquad (11)$$

In applications of SdFFF, particle relaxation can take up to 10 min or more, and is calculated as

$$\tau = \frac{18\eta w}{d^2 \omega^2 r |\rho_P - \rho|} \qquad (12)$$

For new applications, it is best to vary the stop-flow time; if the elution profile does not change when different stop-flow times are used, then the lower value is adequate.

A variety of instrument modifications have been developed to reduce or eliminate the stop-flow time. These include flow splitters, which confine the sample to a region near the accumulation wall during its introduction, as well as special inlets that use hydrodynamic flow to rapidly sweep the sample toward the accumulation wall as it enters the channel. The user should consult

B. Selectivity

Selectivity is a measure of the inherent ability of a technique to separate the components of an analyte mixture. The size-based selectivity is defined as

$$S_d = \left|\frac{d(\log t_r)}{d(\log d)}\right| \quad (13)$$

and the mass-based selectivity as

$$S_m = \left|\frac{d(\log t_r)}{d(\log M)}\right| \quad (14)$$

A high selectivity means that a significant change in elution time occurs over a small interval in particle size or molar mass. Selectivities in FFF differ among the different subtechniques, but generally increase to a plateau value with increasing levels of retention (decreasing R). In SdFFF the plateau value for S_d is 3. This extraordinary selectivity makes SdFFF an extremely powerful technique for separating particles between 100 and 1000 nm in diameter. In fact, the selectivity is so high that the adsorption of macromolecules to the surface of particles can be monitored by shifts in retention (Li and Caldwell 1991).

C. Resolution and Fractionating Power

Resolution R_S is a measure of the separation between two eluting zones; it is defined by

$$R_S = \frac{\Delta z}{2(\sigma_1 + \sigma_2)} \quad (15)$$

where Δz is the gap between the centers of gravity of neighboring zones 1 and 2 and σ is the standard deviation of the zones. The units on Δz and σ can be time, volume, or distance, as long as the same units are used for both terms. For example, if we use units of retention time, resolution is defined as

$$R_S = \frac{\Delta t_r}{2(\sigma_{t,1} + \sigma_{t,2})} \quad (16)$$

where Δt_r is the difference in retention time of two components and $\sigma_{t,1}$ and $\sigma_{t,2}$ are the standard deviation in time units of the elution profiles of components 1 and 2, respectively.

A better parameter for quantifying the resolution of polydisperse materials is the fractionating power (F), which is defined as the relative increment $\delta d/d$ in particle diameter, or the relative increment $\delta M/M$ in molecular weight, that can be separated with unit resolution. Thus

$$F_d = \left(\frac{d}{\delta d}\right)_{R_S=1} \quad (17)$$

$$F_M = \left(\frac{M}{\delta M}\right)_{R_S=1} \quad (18)$$

where the subscripts d and M refer to the diameter-based and mass-based fractionating power, respectively. Fractionating power can be related to selectivity as follows (Schimpf et al., 2000):

$$F_{d,m} = \left[\frac{LD}{384\lambda^3 w^2 \langle v \rangle}\right]^{1/2} S_{d,m} \quad (19)$$

where the subscripts have been generalized to include either mass- or size-based fractionating power. According to Eq. (19), fractionating power increases with channel length L and decreases with increasing velocity of the carrier liquid. Fractionating power also increases dramatically with increased levels of retention (decreasing λ), as well as decreasing channel thickness. In practice, however, thinner channels do not always increase fractionating power because the field strength that can be realized is sometimes diminished as the channel thickness is reduced, thereby placing a lower limit on the value of λ that can be realized.

D. Field Programing

In order to hasten the elution of more strongly retained components in a polydisperse mixture, the applied field can be programmed over the course of the FFF experiment. Field programing in FFF is analogous to temperature programming in gas chromatography and gradient elution in liquid chromatography. Thus, a high field is used initially to retain those components that interact weakly with the field. After a set amount of time, the field is reduced in order to elute the more strongly retained components. Programing also increases the detectability of larger particles in a polydisperse population by reducing the dilution associated with later-eluting peaks.

A simple form of field programing is to decrease the field strength S linearly over time t, but a more effective form decreases the field either exponentially, or according to a power program defined by the following equation (Schimpf et al., 2000):

$$S = S_0 \left(\frac{t_1 - t_a}{t - t_a}\right)^p \quad (20)$$

Here S_0 is the initial field strength, which is held constant for a period of time t_1; t_a is a second time parameter and p is a power greater than zero. The field decay defined by Eq. (20) has the advantage of achieving a fairly uniform fractionating power over wide ranges in particle size.

When the parameters in Eq. (20) are set such that $t_a = -pt_1$, then the retention time of a component is related to the retention parameter as follows:

$$t_r = (p+1)t_1\left[\frac{t^0}{6t_1\lambda_0}\right]^{1/(p+1)} - pt_1 \quad (21)$$

where λ_0 is the retention parameter defined by Eqs. (3)–(6) and the initial field strength.

E. Ancillary Equipment

Like chromatography, the quality and reproducibility of FFF separations depends on ancillary equipment.

1. Pumps

Because the FFF channel is open, pressure drops are not significant. The greatest channel pressures are encountered in FlFFF, and can be as high as 300 psi when low molecular-weight-cutoff membranes are used in combination with high cross-flow rates. Nevertheless, HPLC pumps are commonly used because flow rates can be controlled with high accuracy and precision; they also provide flow-programing capability (Moon et al., 2002). Because most HPLC pumps require a back pressure of at least 100 psi for proper check valve operation, a flow restrictor should be placed between the pump and the FFF channel.

Most analyses require flow rates below 10 mL/min. However, higher flow rates are sometimes desirable in FlFFF, in which case HPLC pumps with a preparatory pump head can be used, or a high-volume, low-pressure pump that pulses at a high rate (to minimize band broadening). Because of their pulsing characteristics, peristaltic pumps are not recommended, except when back pressure from the detector can be kept below 30 psi. Syringe pumps are useful because they give virtually pulse-free flow, which can be important with light scattering detectors, but they have a limited stroke volume before they require refilling. Dual syringe pumps, in which one chamber is filled while the other is emptied, are convenient, but a large pressure disturbance will occur when the direction of the two syringe plungers is reversed.

2. Detectors

The detectors used in FFF are the same as those used in HPLC. However, detector considerations are different in the application of FFF to polymer vs. particle analysis. The response of most detectors varies primarily with sample concentration. Variations in detector response with other sample properties can lead to difficulties in the analysis of the data, especially if those properties are not constant across the elution profile. An example is the application of absorbance detection to a sample whose extinction coefficient changes with elution time. In the analysis of particles, it is important to understand that the detector signal is actually due to the scattering of the light by particles in the detector cell. Because the fraction of forward-scattered light increases with particle size, the fraction of larger particles can be underestimated, particularly when the particle size is similar to the wavelength of the incident light. Therefore, the detection wavelength should be chosen to maximize sensitivity while avoiding such nonlinearities. Methods to correct for the change in sensitivity with particle size have been developed for super-micron-sized particles (Zattoni et al., 2000).

Refractive index (RI) detectors are convenient for polymer analysis when the polymer lacks an accessible UV–Vis absorption band. In this case, the user should be aware that the detector response depends on both concentration and composition of the eluting components, so that for precise quantitative information on polymer mixtures or copolymers, the refractive index increment must be known for each component. RI detector response also varies with particle size in colloidal samples. Therefore, one cannot convert the elution profile of such samples directly into a size distribution.

Evaporative light scattering detectors are extremely sensitive, and therefore especially useful for samples with low absorption coefficients or RI increments. For these detectors it is important to maintain flow through the detector during stop-flow relaxation, in order to avoid unmanageable amounts of noise in the baseline response. Also, the salts and surfactant often used in FFF carrier liquids should be minimized, as they can add excessive noise.

FFF with multi-angle light scattering (MALS) detection is popular for the analysis of macromolecules (Andersson et al., 2001). The combination is also used for colloids in the size range between 20 and 300 nm. In MALS, the angular distribution of the light scattering intensity yields information on molar mass, and if the particle diameter is between 20 and 300 nm, information on size as well. Without the separation step, MALS yields only an average value of the size or molar mass of a sample, with no information on the breadth and modality of the distribution. On the other hand, the determination of size or molar mass by FFF without MALS is complicated by the influence of parameters that govern the interaction of the analyte with the field. By combining the two techniques, size and/or molar mass information is obtained on hundreds of elution slices as they exit the FFF channel, leading to highly precise information on molar mass or particle size distributions.

A variety of other detectors are used with less frequency. As discussed earlier, for example, the intrinsic viscosity detector allows for the determination of polymer molecular weight from measured values of D or from calibration curves (Gao et al., 1985; Jeon and Schimpf, 1998, Kirkland et al., 1989). Laser-induced breakdown has been shown to be more sensitive than light scattering in the characterization of aquatic colloids with $d < 50$ nm (Thang et al., 2000). Flow cytometry is used as a detector for the analysis of red blood cells (Cardot et al., 2002). Atomic absorption and mass spectrometry is becoming popular for the direct detection of a wide range of materials after separation by FFF (Barnes and Siripinyanond, 2002; Blo et al., 1995; Chen and Beckett, 2001), especially environmental materials (see Section IV.E). Each year more FFF detection schemes appear in the literature. In many cases, the low flow rates used in FFF are convenient for designing a simple and efficient interface.

F. Optimization

For characterizing the molar mass or size distribution of macromolecules, the key parameter for maximizing accuracy and precision is the fractionating power. As with any chromatography-like technique, a tradeoff exists between fractionating power and separation time. One cannot, for example, simply increase the flow rate to speed the analysis without expecting to sacrifice fractionating power. However, one can also manipulate the field strength in FFF, and while analysis time increases with field strength, band broadening decreases to a greater extent. Therefore, a simultaneous increase in both the field strength and flow rate can reduce run time while maintaining the fractionating power. Ultimately, however, flow rates are limited by pump specifications or increasing channel pressure (channels begin to leak), whereas field strength is limited by the interaction of samples with the wall when they are compressed too tightly.

If analysis time is not a factor, and one simply wishes to maximize the fractionating power, then the following general equation can be applied (Schimpf et al., 2000):

$$F_{d,m} = S_{d,m} \left(\frac{LD}{384 \lambda^3 w^2 \langle v \rangle} \right)^{1/2} \quad (22)$$

Like SEC, longer channels and slower flow rates yield higher fractionating powers. However, the most important parameter for maximizing the fractionating power is retention parameter λ. Thus, retention levels should be maximized to minimize λ. Higher levels of retention also yield greater selectivities, as discussed earlier. When $R = 0.3$, for example, $S_{d,m}$ is theoretically at 88% of its maximum value. Finally, Eq. (22) predicts that thinner channels will increase fractionating power, but this is not necessarily the case for SdFFF and FlFFF because λ scales inversely with w^2 [see Eqs. (3) and (5)]. As a result, decreasing the channel thickness will only increase fractionating power if retention levels can be maintained by increasing the field strength. Because field strength is also limited, thicker channels are generally used in sedimentation and FlFFF compared to ThFFF and ElFFF.

In situations where sample throughput is a consideration, higher flow rates are used at the expense of fractionating power and consequently, accuracy in the molar mass or size distributions. In order to consider analysis time, we can define a relative fractionating power $F'_{d,m}$ as the fractionating power per unit retention time. Equations similar to Eq. (22) have been derived that relate $F'_{d,m}$ to experimental parameters for each FFF subtechnique. In this case, thinner channels improve $F'_{d,m}$ in all cases, although the relative magnitude of the influence varies with the subtechnique. For a more in-depth discussion of optimization, the reader is referred to Schimpf et al. (2000).

IV. APPLICATIONS

Table 1 summarizes the size and molecular weight range of macromaterials that can be separated by each of the four major FFF subtechniques. This table should be used as a guide only, as parameters other than molar mass or particle size influence the separation. For example, the diameter of particles that can be analyzed by SdFFF decreases as the particle density becomes further removed from that of the carrier liquid.

A search of the chemical literature in July 2003 yielded more than 1500 citations on FFF, with over 250 citations between January 2000 and July 2003. The following discussion categorizes applications into several of the more commonly encountered groups of macromaterials, with representative examples taken primarily from the most recent citations.

A. Polymers

The two FFF subtechniques that are applicable to the characterization of soluble polymers are ThFFF and FlFFF. A general rule for choosing between these two techniques is that polymers dissolved in aqueous solvents should be analyzed by FlFFF, and those dissolved in organic solvents by ThFFF. There are exceptions, however. For example, Kirkland and Yau (1986) demonstrated the ability of ThFFF to separate nonionic water-soluble polymers, as well as the ability of FlFFF to be separate polymers in organic solvents (Kirkland and Dilks, 1992). But in general, the movement of dissolved polymers in a temperature gradient is relatively weak in water (Schimpf and Semenov, 2000), making FlFFF a more efficient technique for separating such materials. On the other hand, most FlFFF channels contain parts that

Table 1 Summary of Macromaterials that are Appropriate for Characterization by FFF Subtechniques

Subtechnique	Carrier liquid	Molecular weight range[a]	Size range (d)[a]
Sedimentation	Aqueous	NA	100 nm–100 μm
Thermal	Organic	>10,000 g/mol	5 nm–100 μm
	Aqueous[b]	>100,000 g/mol	20 nm–100 μm
Flow	Aqueous[c]	>1000 g/mol	2 nm–100 μm
Electrical	Organic	NA	10 nm–100 μm
	Aqueous	NA	20 nm–100 μm

[a]Ranges are approximate for commonly used channel thicknesses.
[b]Aqueous polymer separations by ThFFF are limited to those without charged functional groups.
[c]Specially designed channels have been designed for application to organic carrier liquids, but these channels are not commercially available.

cannot withstand exposure to organic solvents for long periods of time. An exception is the hollow-fiber channel (van Bruijnsvoort et al., 2001a). Although less efficient and not commercially available, the hollow fiber channel is inexpensive to build and can handle organic solvents (Wijnhoven et al., 1995).

ThFFF is applied to a wide variety of lipophilic polymers (Schimpf, 2002). Applications that have appeared recently in the literature include polyvinylpyridines (Lou, 2003), polyolefins (Pasti et al., 2002), styrene–butadiene elastomers (Lewandowski et al., 2000), and natural rubber (Lee et al., 2000). Polyolefins are particularly difficult polymers to characterize because of the high temperatures required for their dissolution, and ThFFF can operate at higher temperatures than SEC. Kassalainen and Williams (2003a) have coupled ThFFF with matrix-assisted laser desorption/ionization time-of-flight mass spectrometry, in order to characterize polymers with high polydispersity.

The lower molecular weight limit for ThFFF applications depends on the chemical composition of both polymer and solvent (Schimpf, 1999; Schimpf and Giddings, 1989). In general, ThFFF begins to lose its advantage in fractionating power over SEC for polymers with molar masses below 10^5 g/mol. Below 10^4 g/mol ThFFF becomes increasingly unreliable, although solvent mixtures have been used to separate polymers with molar masses down to 2500 g/mol (Kassalainen and Williams, 2003b; Rue and Schimpf, 1994).

ThFFF is also used to separate rubbers and gel-containing polymers (Lee et al., 2000; Lewandowski et al., 2000). Sample filtration is not required, so microgels are not lost in the analysis. In these applications, an estimate of the gel content can be obtained by integrating the detector response for the separated gel component.

Because the thermal diffusion coefficient D_T varies strongly with polymer composition, polymer components that are chemically dissimilar can be resolved by ThFFF even when their molar masses are identical (Gunderson and Giddings, 1986; Schimpf and Giddings, 1990). In addition, with the aid of viscometry or light scattering, ThFFF can be used to separate and characterize copolymers according to variations in composition as well as molecular weight (Mes et al., 1999; Schimpf, 1995). An even more powerful technique for characterizing copolymers combines SEC and ThFFF for 2D separations (van Asten et al., 1995). Polymers are first fractionated by size using SEC, then elution slices are cross-fractionated by chemical composition using ThFFF. Figure 5 illustrates a polymer/copolymer mixture that neither SEC nor ThFFF alone could separate. By cross-fractionating the mixture, the three components could be sufficiently resolved to determine both the molar mass and composition of each component (Jeon and Schimpf, 1999, pp. 141–161).

Recent applications of FlFFF to polymer analysis include poly(ethylene oxide) (Benincasa and Caldwell, 2001) glyoxylate copolymers (Glinel et al., 2000), and polyacrylamides (Hecker et al., 2000). FlFFF retention is particularly sensitive to the ionic strength and electrolyte composition of the carrier (Benincasa et al., 2002a), especially with charged polymers, where the size, conformation, and aggregation properties are extremely sensitive to changes in the surrounding environment. Since the seminal report from the Wahlund lab on the aggregation behavior of an amphiphilic graft copolymer (Wittgren et al., 1996), reports on the use of FlFFF with MALS for the study of conformation and aggregation behavior have emerged with increased frequency from a variety of laboratories. For example, FlFFF-MALS has been used to study the conformational shape and degradability of amphiphilic biodegradable copolymers (Glinel et al., 2000). Figure 6 illustrates two FlFFF elution profiles obtained simultaneously on a cellulose sample. One of the profiles was registered by an RI detector, whereas the other was registered from one of the 18 photodiodes in the MALS detector. Note that the response of the MALS detector increases with molar mass, and therefore with retention time, compared to the RI detector. Also superimposed on the elution profile are data points indicating the root-mean-square (rms) radii of the molecules

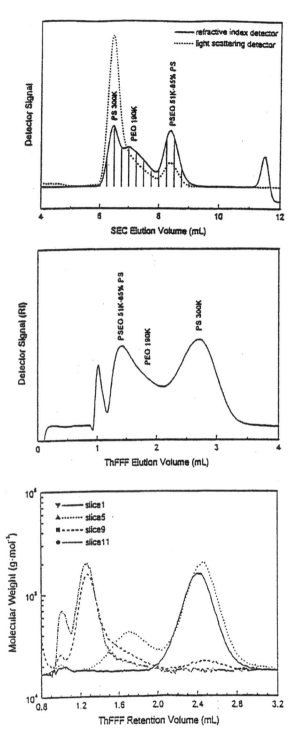

Figure 5 Cross-fractionation of a three-component polymer mixture by SEC and ThFFF. The mixture could not be sufficiently resolved for characterization by either SEC (top) or ThFFF (middle) alone. Cross-fractionation of SEC elution slices (bottom) provided enough resolution to determine the molar mass of each component with an MALS detector. The composition was determined from D and D_T values calculated from SEC and ThFFF retention volumes, respectively. [Reprinted with permission from Schimpf (2002).]

eluting from the channel, as determined by MALS. Note the increase in size with retention time, as expected in the Brownian mode of retention. Also, note the lack of size data above 13 min, which occurs because the calculation of rms radius relies on an RI signal that disappears into the baseline. Figure 6 also contains a log–log plot of the rms radius vs. the molar mass, both measured by the MALS detector. The slope of such plots is indicative of the conformation of the polymer. In this case, the decreasing slope from region A to region B indicates a change in conformation from a random coil to a globular structure with increasing molar mass.

The application of FlFFF-MALS to various polysaccharides has been particularly fruitful. Examples include dextran and pullulan (Duval et al., 2001; Vlebke and Williams, 2000a), carrageenan and xanthan (Viebke and Williams, 2000b), hydroxypropyl methyl cellulose (Wittgren and Wahlund, 2000), ethyl hydroxyethyl cellulose (Andersson et al., 2001), amylopectins (van Bruijnsvoort et al., 2001b; Roger et al., 2001), and other starches and their breakdown products (Chmelik et al., 2001; Wittgren et al., 2002). In one study, FlFFF-MALS was used to monitor conformational ordering in a nondegraded κ-carrageenan, induced at room temperature by switching the solvent from 0.1 M NaCl (coil form) to 0.1 M NaI (helix form) (Wittgren et al., 1998). In another study (Picton et al., 2000) two distinct populations were shown to occur in a complex gum arabic sample. Only with FlFFF-MALS were both populations observed, whereas SEC-MALS gave only partial information. The characterization of larger starch granules is discussed in Section IV.C.

B. Proteins and DNA

The study of proteins and protein complexes is an area of increasing importance in the food and biotechnology industry. For example, conditions that stabilize proteins against denaturation, aggregation, and precipitation are critical to the development of acceptable protein-based pharmaceuticals. It is also widely believed that aggregation plays an important role in the function of certain proteins within the cell (Park et al., 2003). As a result, the separation and characterization of protein complexes is an area of rapid development.

Gel electrophoresis is a relatively easy method for characterizing the approximate size and molecular weight of proteins, but its size range and resolution are limited, and more loosely associated proteins complexes cannot be detected without forming covalent complexes, often through chemical cross-linking. Analytical ultracentrifugation is better, but instrumentation is very expensive and run times are long. SEC is gaining in popularity, especially in concert with MALS detection, but it has

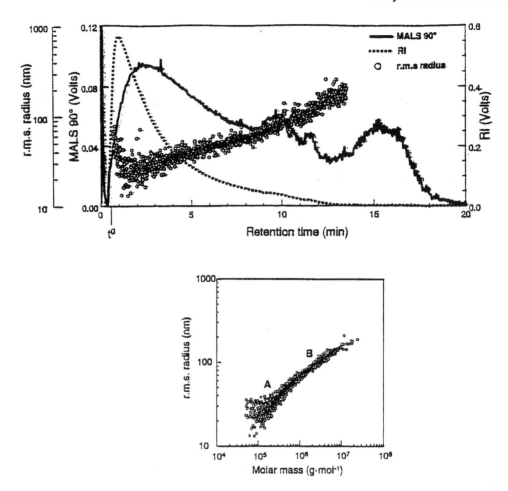

Figure 6 The top figure illustrates the FlFFF elution profiles registered by the RI and MALS detectors, which are hooked in series to the channel outlet. Each circle represents a measurement of the rms radius of eluting material, according to the scale on the far left side of the plot. Above 13 min the recorded rms radii are too imprecise to be reported because of the low RI signal. The bottom figure is a conformation plot. The lines represent a linear regression of the data points in the two molar mass regions A and B. The slope of 0.57 in region A indicates a random coil conformation. The slope of 0.35 in region B indicates a more condensed globular shape. [Reprinted with permission from Andersson et al. (2001).]

limitations as well (Litzén et al., 1993). The more "sticky" protein complexes can be lost through adsorption to the large surface areas available in the SEC column. And although less problematic than gel electrophoresis, the more loosely associated complexes cannot survive the shear forces within the column. Large aggregate populations that do survive the column often elute in the exclusion volume, and therefore, are not fully characterized or, even worse, missed altogether.

FlFFF can be used to analyze DNA and protein complexes over a wide range of sizes, from small molecular weights to insoluble precipitates (Giddings, 1993; Litzén and Wahlund, 1989). Although care must still be taken to avoid the adsorption of material to the channel wall, the available surface area is greatly reduced, allowing a wider range of solvent conditions to be explored. In addition, the range of elution volumes is not as limited as that in SEC, so that a wide range of aggregates can be separated in a single experiment, especially with flow programming (Schimpf et al., 2000). Figure 7 illustrates the separation of a five proteins in under 3 min by asymmetrical FlFFF. Figure 8 illustrates the separation and size analysis of protein aggregates by ElFFF with MALS detection.

FlFFF has proven to be a reliable tool for characterizing the interaction of drugs with human plasma proteins (Madörin et al., 1997). In a more recent application to plasma analysis, the elution profile of lipoprotein samples obtained from patients with coronary artery disease was shown to be distinct from those obtained from healthy persons (Park et al., 2002). In a study of the adsorption of bovine submaxillary gland mucin onto a polystyrene surface, FlFFF was one of several techniques used to investigate the 4–5 nm-thick layer of mucin on the particle surface (Shi and Caldwell, 2000).

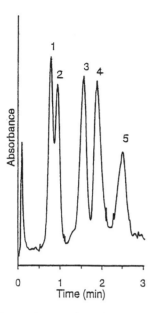

Figure 7 High-speed separation of five proteins of different molecular weights by asymmetrical FlFFF. [1, ovalbumin (43 kDa); 2, bovine serum albumin (67 kDa); 3, monoclonal antibody (156 kDa); 4, ferritin (440 kDa); 5, thyroglobulin (669 kDa)] [Reprinted with permission from Schimpf et al. (2000).]

FlFFF has also been combined with infrared spectroscopy (Wright et al., 2000) and MALS (Tsiami et al., 2000) to characterize gluten proteins. FlFFF was found to be particularly suitable for measuring changes in glutenin size in the very high size range when compared with HPLC and SDS-PAGE (Veraverbeke et al., 2000). The deaggregation of wheat flour proteins by sonication has also been followed using FlFFF (Daqiq et al., 2000; Ueno et al., 2002), as well as the aggregation of self-assembled cationic lipid–DNA complexes (Lee et al., 2001).

C. Colloids (Suspended Particles with $d < 1$ μm)

The most firmly established niche of FFF is certainly the analysis of colloidal-sized particles, defined as suspended material (as opposed to dissolved) with diameters less than 1 μm. For separating these materials, no other analytical technique approaches the resolution that can be achieved with FFF. All four FFF subtechniques can be applied to the colloidal size range. For aqueous suspensions, FlFFF is by far the most versatile, with an applicable size range that spans not only the colloidal size range, but reaches from 2 nm to 100 μm.

ThFFF can also be applied to colloids, but the lower size range for such materials ends at around 20 nm, even though it can separate certain dissolved polymers that are much smaller (see Section IV.A.). ThFFF has the added feature that retention is extremely sensitive to the composition of the particle surface, as well as the nature of the carrier liquid, including additives that affect ionic strength and interfacial forces at the particle surface (Jeon et al., 1997; Mes et al., 2001a; Shiundu and Giddings, 1995). This can be useful in the separation of colloids with chemically modified surfaces, such as silanized silica (Jeon et al., 1997) or core-shell lattices (Ratanathanawongs et al., 1995). Of course, the sensitivity to surface effects, which can be attributed to the role of thermophoresis in ThFFF (Schimpf and Semenov, 2001), can also be an added complication (e.g., if one simply wishes to separate a heterogeneous sample population by size alone). In this case, FlFFF would be the subtechnique of choice, provided the colloid is an aqueous system.

Both FlFFF and ThFFF have a size-based selectivity S_d [Eq. (13)] that plateaus to a value of 1 in the region of high retention, although in ThFFF this value can vary somewhat because D_T for colloids has a slight size dependence (unlike D_T for dissolved polymers). For SdFFF, S_d plateaus at a much higher value of 3, but the size range is limited to diameters above 100 nm and the carrier

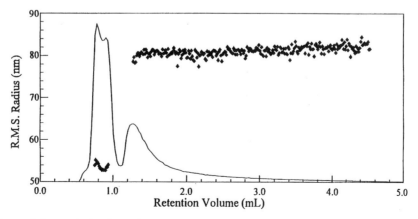

Figure 8 Separation and size analysis of protein aggregates in a pharmaceutical formulation by ElFFF with MALS detection.

liquid must be aqueous. However, for aqueous systems with $d > 100$ nm, SdFFF is the most powerful analytical separation method available. In fact, because of its high size-based selectivity, the SdFFF elution profile can be inordinately broad with highly polydisperse samples, leading to long run times and problems associated with limited detector sensitivities. In such cases, the use of field programing is extremely useful. Figure 9 illustrates the separation of zirconia powder by SdFFF with field programming and the resulting particle size distribution, which was obtained by transformation of the elution profile using procedures outlined in Giddings et al. (1991b).

ElFFF is also used to separate colloidal-sized particles, both in aqueous (Caldwell and Gao, 1993) and nonaqueous (Russell et al., 2000) carrier liquids. In water, electrical double layers are collapsed tightly around the particle surface, as well as at the channel walls. Consequently, the effective field in most of the channel is only a small fraction of the applied field (Palkar and Schure, 1997). Even with special current-carrying additives, the effective field E_{eff} is only a small percent of the applied value (Schimpf and Caldwell, 1995). Unfortunately, the applied voltage is limited to about 2 V, above which the aqueous carrier begins to electrolyze, creating gas bubbles that can reek havoc on the detector output. As a result, ElFFF is not yet suitable to smaller materials, such as dissolved proteins. In principle, the double layers are expanded in organic carrier liquids, increasing E_{eff}, but a systematic study in such carriers is yet to be completed. Work in organic carriers is further complicated by strong electrostatic effects, which can lead to particle adsorption or repulsion from the accumulation wall. Both effects reduce the efficiency and range of applicability. Thus, the potential for ElFFF to become a viable technique remains uncertain, and ElFFF systems are not yet commercially available.

Recent applications of FlFFF to colloid analysis include marine colloids from coastal seawater (Vaillancourt and Balch, 2000), amphiphilic Janus micelles (Erhardt et al., 2003), polyorganosiloxane nanoparticles (Jungmann et al., 2001), and poly(vinyl formamide) (Schuch et al., 2002). Magnetic fluids have also been characterized by coupling FlFFF with photon correlation spectroscopy (PCS) (Anger et al., 2001) and online MALS detection (Rheinlander et al., 2000). Trace metals contained in polishing slurries made of alumina and silica were analyzed by coupling FlFFF with inductively coupled plasma mass spectrometry (ICP-MS) (Siripinyanond and Barnes, 2002). The swelling behavior of core-shell particles with changes in pH and ionic strength was studied by FlFFF with MALS detection (Frankema et al., 2002). In the characterization of carbon nanotubes by FlFFF, it was found that retention is governed principally by the length of the tube (Chen and Selegue, 2002).

SdFFF was coupled with electron microscopy to assess recovery in the characterization of colloidal titanium oxide (Cardot et al., 2001). In a separate study, PCS was combined with SdFFF to characterize colloidal titanium dioxide, and to develop a formula for obtaining well-dispersed particle suspensions with reproducible aggregation properties (Williams et al., 2002).

A simplified form of SdFFF is gravitational FFF, where the earth's gravity is used as the external field. Channel design is simple, consisting of a spacer clamped between two plates that contains holes for sample introduction and egress. Gravitational FFF is more typically applied to micron-sized particles, whose larger mass is more responsive to the gravitational force, which is weak when compared with those achieved with SdFFF and other FFF subtechniques. However, gravitational FFF was recently used to separate a bimodal distribution (310 and 450 nm) of colloidal titanium dioxide (Rasouli et al., 2001).

ThFFF has recently been applied to polystyrene-*co*-butyl methacrylate lattices of various compositions, as well as swellable core-shell particles (Mes et al., 2001b), and to silver, gold, and palladium colloids (Shiundu et al., 2003b).

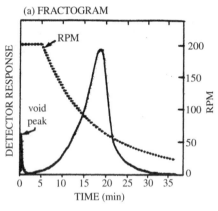

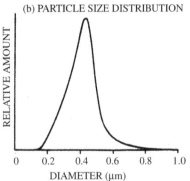

Figure 9 SdFFF fractogram of finely divided zirconia powder (a), and particle size distribution (b). The dotted curve represents the decay of the applied field over time. [Reprinted with permission from Shi and Caldwell (2000).]

D. Starch Granules, Cells, and other Micrometer-Sized Particles ($d > 1$ μm)

With micrometer-sized particles, FFF is operated in the hyperlayer mode, where the presence of lift forces make it impossible to reliably convert elution profiles directly into particle size distributions (i.e., without the use of calibration curves). If calibration curves are used to estimate particle size, the uncertainty of the determination can be reduced by keeping the flow rate and viscosity of the carrier liquid constant between calibration and analysis, as they affect the lift forces acting on the particles. In applications of SdFFF it is also important to compensate for differences in density between the calibration standards and the analyte, as described in Section II and Giddings et al. (1991a). Even so, independent verification of particle size by Coulter counter measurements or microscopy is recommended, and the combination of an FFF separation followed by independent size measurements on collected fractions can be quite powerful. Figure 10 illustrates the fractionation of wheat–starch granules with diameters ranging from 1 to 40 μm, along with optical micrographs taken on several collected fractions. For more examples of the application of FFF to starch granules, see the work reported by Farmakis et al. (2000; 2000a; 2000b).

Any of the FFF subtechniques can be used to separate particles with $d > 1$ μm, but such applications are dominated by FlFFF (Wahlung and Zattoni, 2002; Williams et al., 1991) and SdFFF (Giddings and Williams, 1993). ThFFF and ElFFF can also separate micrometer-sized particles (Liu and Giddings, 1992), but most laboratories with ThFFF are focused on polymer analysis, whereas ElFFF is still under development. FlFFF can in principle be used without a membrane at the accumulation wall (Melucci et al., 2002; Reschiglian et al., 2000) because the underlying porous frit, which supports the membrane, has a pore size <1 μm. However, care must be taken that colloidal impurities are not present because they can eventually plug the frit, requiring an expensive replacement.

The application of FFF to cell analysis dates back to the early 1980s (Caldwell et al., 1984), but the first systematic study of such applications was not reported until 1993 (Barman et al., 1993). In principle, cell separations can be obtained using any of the FFF subtechniques, and though FlFFF has been used to separate cells (Giddings et al., 1991c), the vast majority of reports involve either gravitational FFF (Cardot et al., 1991) or SdFFF (Chianea et al., 2000). The former is popular because of its low cost and simple design, whereas the latter is popular because it can separate cells by both size and density. Barman et al. (1993) demonstrated that FFF can also separate cells by shape and rigidity, and because of its gentle nature, viable, purified cells can be separated and purified in a matter of minutes (Battu et al., 2001,

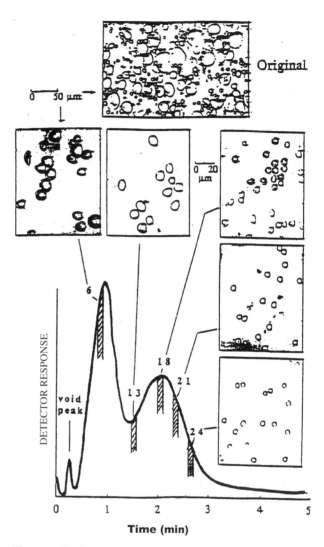

Figure 10 Fractogram and optical micrographs of wheat–starch granules by SdFFF in the hyperlayer mode of retention. The field strength was 400 rpm and the flow rate 6.2 mL/min. The carrier liquid was deionized water containing 0.1% (w/v) FL-70 surfactant and 0.02% (w/v) sodium azide. Note the bimodal distribution and the elution of larger particles ahead of smaller ones. [Reproduced with permission from Schimpf et al. (2000).]

2002). By coupling FFF with additional techniques, information on cells can be obtained with increased efficiency. For example, the use of flow cytometry, which is well known for its analytical power but also has very long sorting times, is benefited by eliminating large groups of inconsequential cells before characterizing subpopulations of particular interest (Cardot et al., 2002). In another example, the separation of bacterial cells with SdFFF, followed by cell counting with epifluorescence microscopy, has proved to be a more efficient method for determining bacterial biomass in sediment samples (Khoshmanesh et al., 2001). FlFFF and gravitational FFF have also been used to sort strains of the same bacterial species on the basis of differences in bacterial membrane characteristics (Reschiglian et al., 2002b). SdFFF and gravitational FFF have been used to distinguish and characterize different varieties of yeast used in wine production (Sanz et al., 2002a, b).

E. Environmental Applications

Environmental scientists deal with materials that vary widely in molecular weight, size, density, reactivity, and other chemical and physical properties. When dealing with such complex samples, the combination of a separation technique with sizing and speciation methods can be a powerful one. Because the separated fractions are more homogeneous in size, and possibly composition, the precision of any subsequent analysis is invariably increased. There are several advantage to using FFF as the separation technique. In addition to its ability to yield size information on an extremely wide range of materials, fractions containing narrow size ranges can be collected and subjected to further, more detailed physical and chemical analysis (Beckett et al., 1992). With an online detector, such analyses can be rapid and efficient (Ranville et al., 1999), but offline analyses can be even more valuable.

Since the initial work by Beckett and coworkers in 1987 (Beckett, 1987; Beckett et al., 1987), the application of FFF to environmental materials has broadened tremendously. Recent work on dissolved organic matter (DOM) include a study of oxidation processes with SdFFF (Win et al., 2000), and a study of the optical properties of DOM from Florida rivers using FlFFF (Zanardi-Lamardo et al., 2002). FlFFF is also used to study the aggregation behavior of DOM under varying conditions (Benincasa et al., 2002b; Schimpf and Petteys, 1997). Figure 11 illustrates the effect of pH on the aggregation of humic acid samples (Schimpf and Wahlund, 1997). A group at Yonsei University uses FFF to characterize fly ash (Kim et al., 2002) and diesel soot particles (Kim et al., 2001a).

FFF is also used to study adsorption behavior in environmental materials. Recent studies include the adsorption of DOM to hematite (Petteys and Schimpf, 1988) and carbon (Schmit and Wells, 2002). An area that currently has a great amount of activity is the adsorption and speciation of metals to environmental colloids. Such studies involve the offline coupling of FlFFF (Benedetti et al., 2002) and gravitational FFF (Chantiwas et al., 2002) with graphite furnace atomic absorption, as well as FlFFF with X-ray fluorescence (Exner et al., 2000). Several reports on the coupling of FFF with ICP-MS for metal speciation have also been published (Al-Ammar et al., 2001; Amarasiriwardena et al., 2001; Anderson et al., 2001c; Chen et al., 2000, 2001; Lyven

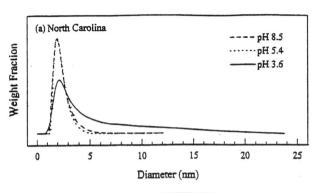

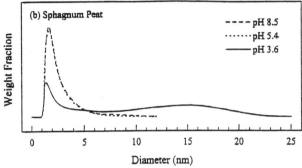

Figure 11 Effect of pH on the aggregation of humic acids. Aggregation is more pronounced in the sample obtained from a peaty soil (b) when compared with that obtained from a more sandy coastal soil (a).

et al., 1997; Schmitt et al., 2002; Siripinyanond et al., 2002). For example, SdFFF has been combined with ICP-MS to study the adsorption of metals onto clay minerals (Schmitt et al., 2002) and sludge-contaminated soil (Chen et al., 2001). Combined with radio-labeling, SdFFF–ICP-MS has been used to study the effect of colloidal surface coatings on the adsorptive behavior of orthophosphate (Chen et al., 2000). FlFFF has been combined with ICP-MS to study the metal complexation by humic acids (Anderson et al., 2001c), dissolved and colloidal organic matter (Amarasiriwardena et al., 2001), and river sediment (Siripinyanond et al., 2002). FlFFF has the advantage that dilute samples can be concentrated online for ultratrace analysis (Al-Ammar et al., 2001; Lyven et al., 1997). A review of FFF–ICP-MS applications is contained in Barnes and Siripinyanond (2002).

The detection of hydrocolloids by thermal lens after FlFFF separation has also been reported (Exner et al., 2001), and the effect of phosphate and silicate on the formation of iron colloids was studied by SdFFF with MALS detection (Magnuson et al., 2001).

Whenever complex mixtures are analyzed, the preferential detection of certain sizes or species over others is a concern, as discussed in Section III.E.2. We must also consider the discrimination of certain sizes or species due to the complete loss of specific components in the analysis. For example, size distributions that span the inversion diameter must be prefiltered to avoid the simultaneous elution of different particle sizes, as discussed in Section II.B. Such prefractionation procedures were considered to be responsible for the loss of submicrometer particles in the analysis of clay by SdFFF (Hassellov et al., 2001). Sample loss can also occur through the adsorption of material to the accumulation wall. In many cases, the adsorption of "sticky" samples can be avoided with the use of additives to the carrier liquid, but finding the appropriate additive can be difficult and time consuming. In the analysis of diesel soot, for example, sample loss through wall adsorption was found to be more problematic with FlFFF compared with SdFFF (Kim et al., 2001b). With FlFFF, low molecular weight components can also be lost by penetration through the accumulation wall.

Sample penetration ultimately limits the size range that can be analyzed by FlFFF, and therefore by FFF in general, as FlFFF has the lowest size range of the FFF subtechniques. Sample loss by penetration can be reduced using additives that lead to charge repulsion at the wall, but calibration curves are likely to be affected (Zanardi-Lamardo et al., 2001). When such a procedure was used to reduce the loss of low molecular weight humic materials, the effect on calibration curves established with sulfonated polystyrene standards was minimized by running samples at multiple concentrations and extrapolating the resulting trends in retention to zero concentration (Manh Thang et al., 2001). If FFF retention theory is used instead to calculate particle sizes, the effect of channel repulsion on the calculation can be accounted for by determining a "particle–wall interaction parameter" (Williams et al., 1997), but this procedure can be extremely time consuming (Du and Schimpf, 2002). Therefore, if one wishes to simply characterize the molecular weight of a low molecular weight sample, it is more efficient to use SEC. But if one wishes to explore the conformational behavior of charged macromolecules (proteins, carbohydrates, humic acids, etc.) under varying conditions, FFF provides greater flexibility, and can yield new information, especially when combined with other analytical instrumentation.

REFERENCES

Al-Ammar, A., Siripinyanond, A., Barnes, R. M. (2001). Simultaneous sample preconcentration and matrix removal using FFF coupled to inductively coupled plasma mass spectrometry. *Spectrochim. Acta B* 56B(10):1951–1962.

Amarasiriwardena, D., Siripinyanond, A., Barnes, R. M. (2001). Trace elemental distribution in soil and compost-derived humic acid molecular fractions and colloidal organic matter in municipal wastewater by FlFFF-ICP-MS. *J. Anal. Atom. Spectrom.* 16:978–986.

Anderson, T., Shifley, L., Amarsiriwardena, D., Siripinyanond, A., Xing, B., Barnes, R. M. (2001c). Characterization of trace metals complexed to humic acids derived from agricultural soils, annelid composts, and sediment by flow FFF inductively coupled plasma-mass spectrometry (flow FFF-ICP-MS). *R. Soc. Chem.* 273:165–177.

Andersson, M., Wittgren, B., Wahlund, K.-G. (2001). Ultrahigh molar mass component detected in ethylhydroxyethyl cellulose by asymmetrical flow FFF coupled to multiangle light scattering. *Anal. Chem.* 73:4852–4861.

Anger, S., Caldwell, K., Muller, R. (2001). Size characterization of magnetic fluids using flow FFF and photon correlation spectroscopy. Proceedings 28th International Symposium on Controlled Release of Bioactive Materials and 4th Consumer & Diversified Products Conference, San Diego, CA, Vol. 2001. No. 1, pp. 349–350.

van Asten, A. C., van Dam, R. J., Kok, W. Th., Tijssen, R., Poppe, H. (1995). Determination of the compositional heterogeneity of polydisperse polymer samples by the coupling of SEC and ThFFF. *J. Chromatogr. A* 703:245–263.

Barman, B. N., Ashwood, E. R., Giddings, J. C. (1993). Separation and size distribution of red blood cells of diverse size, shape, and origin by flow/hyperlayer FFF. *Anal. Biochem.* 212:35–42.

Barnes, R. M., Siripinyanond, A. (2002). FFF-inductively coupled plasma-mass spectrometry. *Adv. Atom. Spectrosc.* 7:179–235.

Battu, S., Elyaman, W., Hugon, J., Cardot, P. J. P. (2001). Cortical cell elution by sedimentation FFF. *Biochim. Biophys. Acta* 1528(2–3):89–96.

Battu, S., Cook-Moreau, J., Cardot, P. J. P. (2002). Sedimentation FFF: methodological basis and applications for cell sorting. *J. Liq. Chromatogr. Rel. Technol.* 25(13–15):2193–2210.

Beckett, R. (1987). The application of FFF techniques to the characterization of complex environmental samples. *Environ. Tech. Lett.* 8:339–354.

Beckett, R., Jue, Z., Giddings, J. C. (1987). Determination of molecular weight distributions of fulvic and humic acids using flow FFF. *Environ. Sci. Technol.* 21:289–295.

Beckett, R., Nicholson, G., Hotchin, D. M., Hart, B. T. (1992). *Hydrobiologia* 235/236:697–710.

Benedetti, M., Ranville, J. F., Ponthieu, M., Pinheiro, J. P. (2002). FFF characterization and binding properties of particulate and colloidal organic matter from the Rio Amazon and Rio Negro. *Org. Geochem.* 33(3):269–279.

Benincasa, M. A., Caldwell, K. D. (2001). Flow FFF of poly(ethylene oxide): effect of carrier ionic strength and composition. *J. Chromatogr. A* 925(1–2):159–169.

Benincasa, M.-A., Caroni, G., Delle Fratte, C. (2002a). FlFFF and characterization of ionic and neutral polysaccharides of vegetable and microbial origin. *J. Chromatogr. A* 967:219–234.

Benincasa, M.-A., Cartoni, G., Imperia, N. (2002b). Effects of ionic strength and electrolyte composition on the aggregation of fractionated humic substances studied by FFF. *J. Sep. Sci.* 25(7):405–415.

Blo, G., Contado, C., Fagioli, F., Bollain Rodriguez, F., Dondi, F. (1995). Analysis of kaolin by sedimentation FFF and electrothermal atomic absorption spectrometry detection. *Chromatographia* 41:715–721.

van Bruijnsvoort, M., Kok, W. Th., Tijssen, R. (2001a). Hollow-fiber FlFFF of synthetic polymers in organic solvents. *Anal. Chem.* 73:4736–4742.

van Bruijnsvoort, M., Wahlund, K.-G., Nilsson, G., Kok, W. T. (2001b). Retention behavior of amylopectins in asymmetrical flow FFF studied by multi-angle light scattering detection. *J. Chromatogr A* 925(1–2):171–182.

Caldwell, K. D., Gao, Y.-S. (1993). ElFFF in particle separations. 1. Monodisperse standards. *Anal. Chem.* 65:1764–1772.

Caldwell, K. D., Cheng, Z. Q., Hradecky, P., Giddings, J. C. (1984). Separation of human and animal cells by steric FFF. *Cell Biophys.* 6:233–251.

Cao, W., Williams, P. S., Myers, M. N., Giddings, J. C. (1999). Thermal FFF universal calibration: extension for consideration of variation of cold wall temperature. *Anal. Chem.* 71:1597–1609.

Cardot, P. J., Gerota, J., Martin, M. (1991). Separation of living red blood cells by gravitational FFF. *J. Chromatogr.* 568:93.

Cardot, P. J. P., Rasouli, S., Blanchart, P. (2001). TiO2 colloidal suspension polydispersity analyzed with sedimentation FFF and electron microscopy. *J. Chromatogr. A* 905(1–2):163–173.

Cardot, P., Battu, S., Simon, A., Delage, C. (2002). Hyphenation of sedimentation of FFF with flow cytometry. *J. Chromatogr. B* 768(2):285–295.

Chantiwas, R., Beckett, R., Jakmunee, J., McKelvie, I. D., Grudpan, K. (2002). Gravitational FFF in combination with flow injection analysis or electrothermal AAS for size based iron speciation of particles. *Talanta* 58(6):1375–1383.

Chen, B., Beckett, R. (2001). Development of SdFFF-ETAAS for characterizing soil and sediment colloids. *Analyst* 126(9):1588–1593.

Chen, B., Selegue, J. P. (2002). Separation and characterization of single-walled and multiwalled carbon nanotubes by using flow FFF. *Anal. Chem.* 74(18):4774–4780.

Chen, B., Hulston, J., Beckett, R. (2000). The effect of surface coatings on the association of orthophosphate with natural colloids. *Sci. Total Environ.* 263(1–3):23–35.

Chen, B., Shand, C. A., Beckett, R. (2001). Determination of total and EDTA extractable metal distributions in the colloidal fraction of contaminated soils using SdFFF-ICP. *J. Environ. Monit.* 3(1):7–14.

Chianea, T., Assidjo, N. E., Cardot, P. J. P. (2000). Sedimentation FFF: emergence of a new cell separation methodology. *Talanta* 51(5):835–847.

Chmelik, J., Krumlova, A., Budinska, M., Kruml, T., Psota, V. Bohacenko, I., Mazal, P., Vydrova, H. (2001). Comparison of size characterization of barley starch granules determined by electron and optical microscopy, low angle laser light scattering and gravitational FFF. *J. Inst. Brew.* 107(1):11–17.

Colfen, H., Antonietti, M. (2000). FFF techniques for polymer and colloid analysis. *Adv. Polym. Sci.* 150:67–187.

Daqiq, L., Larroque, O. R., Stoddard, F. L., Bekes, F. (2000). Extractability and size distribution on wheat proteins using flow FFF. *R. Soc. Chem. Spl. Publ.* 261(Wheat Gluten):149–153.

Du, Q., Schimpf, M. E. (2002). Correction for particle-wall interactions in the separation of colloids by FlFFF. *Anal. Chem.* 74:2478–2485.

Duval, C., Le Cerf, D., Picton, L., Muller, G. (2001). Aggregation of amphiphilic pullulan derivatives evidenced by on-line flow FFF multi-angle laser light scattering. *J. Chromatogr B* 753(1):11–122.

Erhardt, R., Zhang, M., Boeker, A., Zettl, H., Abetz, C., Frederick, P., Krausch, G., Abetz, V., Mueller, A. (2003). Amphiphilic janus micelles with polystyrene and poly(methacrylic acid) hemispheres. *J. Am. Chem. Soc.* 125(11):3260–3267.

Exner, A., Panne, U., Niessner, R. (2001). Characterization of environmental hydrocolloids by asymmetric flow FFF (AF4) and thermal lens spectroscopy. *Anal. Sci.* 17(Spec. Issue):s571–s573.

Exner, A., Theisen, M., Panne, U., Niessner, R. (2000). Combination of asymmetric flow FFF (AF4) and total-reflexion X-ray fluorescence analysis (TXRF) for determination of heavy metals associated with colloidal humic substances. *Fresnius J. Anal. Chem.* 366(3):254–259.

Farmakis, L., Karaiskakis, G., Koliadima, A. (2002a). Investigation of the variation of the mass ratios for large and small starch granules with the pH and ionic strength of the dispersing medium by sedimentation/steric FFF. *J. Liq. Chromatogr.* 25(13–15):2135–2152.

Farmakis, L., Koliadima, A., Karaiskakis, G. (2002b). Study of the influence of ionic strength and pH of the suspending medium on the size of wheat starch granules, measured by sedimentation/steric FFF. *J. Liq. Chromatogr.* 25(2):167–183.

Frankema, W., van Bruijnsvoort, M., Tijssen, R., Kok, W. (2002). Characterisation of core-shell latexes by flow FFF with multi-angle light scattering detection. *J. Chromatogr. A* 943(2):251–261.

Gao, Y. S., Caldwell, K. D., Myers, M. N., Giddings, J. C. (1985). Extension of ThFFF to ultra-high molecular weight polystyrenes. *Macromolecules* 18:1272.

Giddings, J. C. (1978). Displacement and dispersion of particles of finite size in flow channels with lateral forces. FFF and hydrodynamic chromatography. *Sep. Sci. Technol.* 13:241–254.

Giddings, J. C. (1993). FFF: Separation and characterization of macromolecular-colloidal-particulate materials. *Science* 260:1456–1465.

Giddings, J. C., Williams, P. S. (1993). Multifaceted analysis of 0.01 to 100 μm particles by SdFFF. *Am. Lab.* 25:88–95.

Giddings, J. C., Moon, M. H., Williams, P. S., Myers, M. N. (1991a). Particle size distribution by sedimentation/steric FFF: development of a calibration procedure based on density compensation. *Anal. Chem.* 63:1366–1372.

Giddings, J. C., Myers, M. N., Moon, M. H., Barman, B. H. (1991b). Particle separation and size characterization by SdFFF. In: Provder, T., ed. *Particle Size Distribution*. ACS Symp. Series no. 472. Washington, DC: American Chemical Society, pp. 198–216.

Giddings, J. C., Barman, B. N., Liu, M. K. (1991c). Separation of cells by FFF and related methods. In: Kompala, D., Todd, P., eds. *Cell Separation Science and Technology*. ACS Symposium Series no. 464. Chapter 9. Washington, DC: American Chemical Society.

Glinel, K., Vaugelade, C., Muller, G., Bunel, C. (2000). Analysis of new biodegradable amphiphilic water-soluble copolymers with various hydrophobe content by multi-angle light scattering on line with flow FFF. *Int. J. Polym. Anal. Charact.* 6(1–2):89–107.

Gunderson, J. J., Giddings, J. C. (1986). Chemical composition and molecular-size factors in polymer analysis by thermal FFF and SEC. *Macromolecules* 19:2618–2621.

Hassellov, M., Lyven, B., Bengtsson, H., Jansen, R., Turner, D. R., Beckett, R. (2001). Particle size distributions of clay-rich sediments and pure clay minerals: a comparison of grain size analysis with sedimentation FFF. *Aq. Geochem.* 7(2):155–171.

Hecker, R., Fawell, P. D., Jefferson, A., Farrow, J. B. (2000). FlFFF of polyacrylamides. *Sep. Sci. Technol.* 35:593–612.

Jeon, S. J., Nyborg, A., Schimpf, M. E. (1997). Compositional effects in the retention of colloids by ThFFF. *Anal. Chem.* 67:3442–3450.

Jeon, S. J., Schimpf, M. E. (1998). Cross-fractionation of copolymers using SEC and thermal FFF for determination of molecular weight and composition. In: Provder, T., ed. *Chromatography of Polymers: Hyphenated and Multi-Dimensional Techniques*. Washington, DC: American Chemical Society, pp. 182–195.

Jeon, S. J., Schimpf, M. E. (1999). Cross-fractionation of copolymers using SEC and ThFFF for determination of molecular weight and composition. In: Provder, T., ed. *Chromatography of Polymers: Hyphenated and Multi-Dimensional Techniques*. ACS Symposium Series no. 731.

Jungmann, N., Schmidt, M., Maskos, M. (2001). Characterization of polyorganosiloxane nanoparticles in aqueous dispersion by asymmetrical flow FFF. *Macromolecules* 34(23):8347–8353.

Kassalainen, G. E., Williams, K. R. (2003a). Coupling ThFFF with matrix-assisted laser desorption/ionization time-of-flight mass spectrometry for the analysis of synthetic polymers. *Anal. Chem.* 75(8):1887–1894.

Kassalainen, G. E., Williams, S. K. R. (2003b). Lowering the molecular mass limit of thermal field-flow fractionation for polymer separations. *J. Chromatogr. A* 988(2):285–295.

Khoshmanesh, A., Sharma, R., Beckett, R. (2001). Biomass of sediment bacteria by sedimentation FFF. *J. Environ. Eng.* 127(1):19–25.

Kim, W., Lee, D. W., Lee, S. (2002). Size characterization of incinerator fly ash using sedimentation/steric FFF. *Anal. Chem.* 74(4):848–855.

Kim, W., Kim, S. H., Lee, D. W., Lee, S., Lim, C. S., Ryu, J. H. (2001a). Size analysis of automobile soot particles using FFF. *Environ. Sci. Technol.* 35(6):1005–1012.

Kim, W., Young H., Lee, D. W., Lee, S. (2001b). Sample preparation for size analysis of diesel soot particles using FFF. *J. Liq. Chromatogr. Related Technol.* 24(13):1935–1951.

Kirkland, J. J., Yau, W. W. (1986). ThFFF of water-soluble macromolecules. *J. Chromatogr.* 353:95.

Kirkland, J. J., Dilks, C. H. Jr. (1992). FlFFF of polymers in organic solvents. *Anal. Chem.* 64:2836–2840.

Kirkland, J. J., Rementer, S. W., Yau, W. W. (1989). Polymer characterization by thermal FFF with a continuous viscosity detector. *J. Appl. Polym. Sci.* 38:1383–1395.

Lee, S., Giddings, J. C. (1988). Experimental observation of steric transition phenomena in SdFFF. *Anal. Chem.* 60:2328–2333.

Lee, S., Eum, C. H., Plepys, A. (2000). Capability of thermal FFF for analysis of processed natural rubber. *Bull. Kor. Chem. Soc.* 21(1):69–74.

Lee, H., Williams, S. K., Allison, S. D., Anchordoquy, T. J. (2001). Analysis of self-assembled cationic lipid-DNA gene carrier complexes using flow FFF and light scattering. *Anal. Chem.* 73(4):837–843.

Lewandowski, L., Sibbald, M. S., Johnson, E., Mallamaci, M. P. (2000). New emulsion SBR technology: Part I. Raw polymer study. *Rubber Chem. Technol.* 73(4):731–742.

Li, J. T., Caldwell, K. D. (1991). Sedimentation FFF in the determination of adsorbed materials. *Langmuir* 7:2034–2039.

Litzén, A., Wahlund, K.-G. (1989). Improved separation speed and efficiency for proteins and nucleic acids and viruses in asymmetrical FlFFF. *J. Chromatogr.* 476:413–421.

Litzén, A., Walter, J. K., Drischollek, H., Wahlund, K.-G. (1993). Separation and quantitation of monoclonal antibody aggregates by asymmetrical FlFFF. Comparison to gel permeation chromatography. *Anal. Biochem.* 212:469–480.

Liu, G., Giddings, J. C. (1992). Separation of particles in aqueous suspensions by ThFFF. Measurement of thermal diffusion coefficients. *Chromatographia* 34:483–492.

Lou, J. (2003). Salt effect on separation of polyvinylpyridines by thermal FFF. *J. Liq. Chromatogr. Rel. Technol.* 26(2):165–175.

Lyven, B., Hassellv, M., Haraldsson, C., Turner, D. R. (1997). Optimization of on-channel preconcentration in flow FFF for the determination of size distributions of low molecular weight colloidal material in natural waters. *Anal. Chim. Acta* 357:187–196.

Madörin, M., van Hoogevest, P., Hilfker, R., Langwost, B., Kresbach, G. M., Ehrat, M. Leuenberger, H. (1997). Analysis of drug/plasma protein interactions by means of asymmetrical FlFFF. *Pharm. Res.* 14:1706–1712.

Magnuson, M. L., Lytle, D. A., Frietch, C. M., Kelty, C. A. (2001). Characterization of submicrometer aqueous iron (III) colloids formed in the presence of phosphate by sedimentation FFF with multiangle laser light scattering detection. *Anal. Chem.* 73(20):4815–4820.

Manh Thang, N., Geckeis, H., Kim, J. I., Beck, H. P. (2001). Application of the flow FFF to the characterization of aquatic humic colloids: evaluation and optimization of the method. *Colloid. Surf. A* 181(1–3):289–301.

Melucci, D., Zattoni, A., Casolari, S., Reggiani, M., Sanz, R., Reschiglian, P. (2002). Working without accumulation membrane in flow FFF. Effect of sample loading on recovery. *J. Liq. Chromatogr.* 25(13–15):2211–2224.

Mes, E. P. C., Tijssen, R., Kok, W. Th. (1999). Rapid detection of compositional drift of polydisperse copolymers using ThFFF and multi-angle light scattering. *Chromatographia* 50:45–51.

Mes, E. P. C., Tijssen, R., Kok, W. T. (2001a). Influence of the carrier composition on thermal FFF for the characterization of sub-micron polystyrene latex particles. *J. Chromatogr. A* 907(1–2):201–209.

Mes, E. P. C., Kok, W. Th., Tijssen, R. (2001b). Sub-micron particle analysis by thermal FFF and multi-angle light scattering detection. *Chromatographia* 53(11/12):697–703.

Moon, M., Williams, P. S., Kang, D., Hwang, I. (2002). Field and flow programming in frit-inlet asymmetrical flow FFF. *J. Chromatogr. A* 955(2):263–272.

Palkar, S. A., Schure, M. R. (1997). Mechanistic study of ElFFF. I: on the nature of the internal field. *Anal. Chem.* 69:3223–3229.

Park, I., Paeng, K.-J., Yoon, Y., song, J.-H., Moon, M. H. (2002). Separation and selective detection of lipoprotein particles of patients with coronary artery disease by frit-inlet asymmetrical FlFFF. *J. Chromatogr. B* 780:415–422.

Park, H., Wu, S., Dunker, A. K., Kang, C. (2003). Polymerization of calsequestrin: implications for calcium regulation. *J. Biol. Chem.* 278:16176–16182.

Pasti, L., Melucci, D., Contado, C., Dondi, F., Mingozzi, I. (2002). Calibration in thermal FFF with polydisperse standards: application to polyolefin characterization. *J. Sep. Sci.* 25(10–11):691–702.

Petteys, M. P., Schimpf, M. E. (1988). Characterization of hematite and its interaction with humic material using flow FFF. *J. Chromatogr. A* 816:145–158.

Picton, L., Bataille, I., Muller, G. (2000). Analysis of a complex polysaccharide(gum arabic) by multi-angle laser light scattering coupled online to size exclusion chromatography and FFF. *Carbohydr. Polym.* 42(1):23–31.

Ranville, J. F., Chittleborough, D. J., Shanks, F., Morrison, J. S., Harris, T., Doss, F., Beckett, R. (1999). *Anal. Chim. Acta* 381:315–329.

Rasouli, S., Blanchart, P., Cledat, D., Cardot, P. J. P. (2001). Size-and shape-dependent separation of TiO2 colloidal sub-populations with gravitational FFF. *J. Chromatogr. A* 923(1–2):119–126.

Ratanathanawongs, S. K., Shiundu, P. M., Giddings, J. C. (1995). Size and compositional studies of core-shell latexes using flow and thermal FFF. *Coll. Surf. A* 105:243–250.

Reschiglian, P., Melucci, D., Zattoni, A., Mallo, L., Hansen, M., Kummerow, A., Miller, M. (2000). Working without accumulation membrane in flow FFF. *Anal. Chem.* 72(24):5945–5954.

Reschiglian, P., Zattoni, A., Roda, B., Casolari, S., Moon, M. H., Lee, J., Jung, J., Rodmalm, K., Cenacchi, G. (2002b). Bacteria sorting by FFF. Application to whole-cell *Escherichia coli* vaccine strains. *Anal. Chem.* 74(19):4895–4904.

Rheinlander, T., Roessner, D., Weitschies, W., Semmler, W. (2000). Comparison of size-selective techniques for the fractionation of magnetic fluids. *J. Magn. Mat.* 214(3):269–275.

Roger, P., Baud, B., Colonna, P. (2001). Characterization of starch polysaccharides by flow FFF multi-angle laser light scattering-differential refractometer index. *J. Chromatogr. A* 917(1–2):179–185.

Rudin, I., Johnston, H. K. (1971). *Polym. Lett.* 9:55.

Rue, C. A., Schimpf, M. E. (1994). Thermal diffusion in liquid mixtures and its affect on polymer retention in ThFFF. *Anal. Chem.* 66:4054–4062.

Russell, D., Hill, M., Schimpf, M. E. (2000). Characterization of liquid electrophotagraphic toner particles using non-polar ElFFF and MALS. *J. Imag. Sci. Technol.* 44:443–441.

Sanz, R., Cardot, P., Battu, S., Galceran, M. T. (2002a). Steric-Hyperlayer sedimentation FFF and flow cytometry analysis applied to the study of *Saccharomyces cerevisiae*. *Anal. Chem.* 74(17):4496–4504.

Sanz, R., Torsello, B., Reschiglian, P., Puignou, L., Galceran, M. T. (2002b). Improved performance of gravitational FFF for screening wine-making yeast varieties. *J. Chromatogr. A* 966(1–2):135–143.

Schimpf, M. E. (1995). Determination of molecular weight and composition in copolymers using thermal FFF combined with viscometry. In: Provder, T., Barth, H. G., Urban, W., eds. *Chromatographic Characterization of Polymers*. Advances in Chemistry Series 247. Washington, DC: ACS Publications, pp. 183–196.

Schimpf, M. E. (1999). Studies in thermodiffusion by ThFFF. *Entropie* 217:67–71.

Schimpf, M. E. (2002). Polymer analysis by ThFFF. *J. Liq. Chromatogr.* 25:2101–2134.

Schimpf, M. E., Giddings, J. C. (1989). Characterization of thermal diffusion in polymer solutions: dependence on polymer and solvent parameters. *J. Polm. Sci. Polym. Phys. Ed.* 27:1317–1332.

Schimpf, M. E., Giddings, J. C. (1990). Characterization of thermal diffusion of copolymers in solution by ThFFF. *J. Polym. Sci. Polym. Phys. Ed.* 28:2673–2680.

Schimpf, M. E., Caldwell, K. D. (1995). ElFFF for colloid and particle analysis. *Am. Lab.* 27(6):64–68.

Schimpf, M. E., Petteys, M. P. (1997). Characterization of humic materials by flow FFF. *Colloid. Surf. A* 120:87–100.

Schimpf, M. E., Semenov, S. N. (2000). Mechanism of polymer thermophoresis in nonaqueous solvents. *J. Phys. Chem. B* 104:9935–9942.

Schimpf, M. S., Wahlund, K.-G. (1997). Assymetrical flow FFF as a method to study the behavior of humic acids in solution. *J. Microcol. Sep.* 9:535–543.

Schimpf, M. E., Semenov, S. N. (2001). Latex particle thermophoresis in polar solvents. *J. Phys. Chem. B* 105:2285–2290.

Schimpf, M. E., Caldwell, K. D., Giddings, J. C., eds. (2000). *Field-Flow Fractionation Handbook*. New York: Wiley Interscience.

Schmit, K. H., Wells, M. J. (2002). Preferential adsorption of fluorescing fulvic and humic acid components on activated carbon using flow FFF analysis. *J. Environ. Monit.* 4(1):75–84.

Schmitt, D., Taylor, H. E., Aiken, G. R., Roth, D. A., Frimmel, F. H. (2002). Influence of natural organic matter on the adsorption of metal ions onto clay minerals. *Environ. Sci. Technol.* 36(13):2932–2938.

Schuch, H., Frenzel, S., Runge, F. (2002). FFF on poly (vinyl formamide), other polymers and colloids. *Prog. Colloid Polym. Sci.* 121:43–48.

Shi, L., Caldwell, K. D. (2000). Mucin adsorption to hydrophobic surfaces. *J. Colloid Interf. Sci.* 224(2):372–381.

Shiundu, P. M., Giddings, J. C. (1995). Influence of bulk and surface composition on the retention of colloidal particles in ThFFF. *J. Chromatogr.* 715:117–126.

Shiundu, P. M., Munguti, S. M., Williams, S. K. R. (2003a). Practical implications of ionic strength effects on particle retention in thermal FFF. *J. Chromatogr. A* 984(1):67–79.

Shiundu, P. M., Munguti, S., Williams, K. R. (2003b). Retention behavior of metal particle dispersions in aqueous and nonaqueous carriers in thermal FFF. *J. Chromatogr. A* 983(1–2):163–176.

Siripinyanond, A., Barnes, R. M. (2002). Flow FFF inductively coupled plasma mass spectrometry of chemical mechanical polishing slurries. *Spectrochim. Acta B* 57B(12):1885–1896.

Siripinyanond, A., Barnes, R. M., Amarasiriwardena, D. (2002). Flow FFF inductively coupled plasma mass spectrometry for sediment bound trace metal characterization. *J. Anal. Atom. Spectrom.* 17(9):1055–1064.

Thang, N. Knopp, R., Geckeis, H., Kim, J., Beck, H. P. (2000). Detection of nanocolloids with flow FFF and laser-induced breakdown detection. *Anal. Chem.* 72:1–5.

Tsiami, A. A., Every, D., Schofield, J. D. (2000). Redox reactions in dough: effects on molecular weight of glutenin polymers as determined by FFF and MALLS. *R. Soc. Chem.* 261:244–248.

Ueno, T., Stevenson, S. G., Preston, K. R., Nightingale, M. J., Marchylo, B. M. (2002). Simplified dilute acetic acid based extraction procedure for fractionation and analysis of wheat flour protein by size exclusion HPLC and flow FFF. *Cereal Chem.* 79(1):155–161.

Vaillancourt, R., Balch, W. (2000). Size distribution of marine submicron particles determined by FFF. *Limnol. Oceanogr.* 45(2):485–492.

Veraverbeke, W. S., Larroque, O. R., Bekes, F., Delcour, J. A. (2000). In vitro polymerization of wheat glutenin subunits with inorganic oxidizing agents. I. Comparison of single-step and stepwise oxidations of high molecular weight glutenin subunits. *Cereal Chem.* 77(5):582–588.

Viebke, C., Williams, P. A. (2000a). The influence of temperature on the characterization of water-soluble polymers using asymmetric flow FFF coupled to multi-angle laser light scattering. *Anal. Chem.* 72(16):3896–3901.

Viebke, C., Williams, P. A. (2000b). Determination of molecular mass distribution of kappa-carrageenan and xanthan using asymmetrical flow FFF. *Food Hydrocoll.* 14(3):265–270.

Wahlund, K., Zattoni, A. (2002). Size separation of supermicrometer particles in asymmetrical flow FFF. Flow conditions for rapid elution. *Anal. Chem.* 74(21):5621–5628.

Wijnhoven, J. E. G. J., Koorn, J. P., Poppe, H., Kok, W. Th. (1995). Hollow fiber flow FFF of polystyrene sulphonates. *J. Chromatogr.* 699:119–129.

Williams, K. R., Lee, I., Giddings, J. C. (1991). Separation and characterization of 0.01-50 μm particles using FlFFF. In: Provder, T., ed. *Particle Size Distribution II*. ACS Symp. Series no. 472. Chapter 15. Washington, DC: American Chemical Society.

Williams, S. K., Butler-Veytia, B., Lee, H. (2002). Determining the particle size distributions of titanium dioxide using sedimentation field-flow fractionation and photon correlation spectroscopy. *ACS Symp. Ser. (Service Life Prediction)* 805:285–298.

Williams, P. S., Xu, Y., Reschiglian, P., Giddings, J. C. (1997). Colloid characterization by SdFFF: correction for particle-wall interaction. *Anal. Chem.* 69:349–360.

Win, Y. Y., Kumke, M. U., Specht, C. H., Schindelin, A. J., Kolliopoulos, G., Ohlenbusch, G., Kleiser, G., Hesse, S., Frimmel, F. H. (2000). Influence of oxidation of dissolved organic matter (DOM) on subsequent water treatment processes. *Water Res.* 34(7):2098–2104.

Wittgren, B., Wahlund, K. G. (2000). Size characterization of modified celluloses in various solvents using flow FFF-MALS and MB-MALS. *Carbohydrate Polym.* 3(1):63–73.

Wittgren, B., Wahlund, K.-G., Dérand, H., Wesslén, B. (1996). Aggregation behavior of an amphilic graft copolymer in aqueous medium studied by asymmetrical FlFFF. *Macromol.* 29:268–276.

Wittgren, B., Borgström, J., Piculell, L., Wahlund, K.-G. (1998). Conformational change and aggregation of κ-carrageenan studied by FlFFF and MALS. *Biopolymers* 45:85–96.

Wittgren, B., Wahlund, K., Anderson, M., Arfvidsson, C. (2002). Polysaccharide characterization by flow FFF multi-angle light scattering: initial studies of modified starches. *Int. J. Polym. Anal. Charact.* 7(1–2):19–40.

Wright, R. J., Larroque, O. R., Bekes, F., Wellner, N., Tatham, A. S., Shewry, P. R. (2000). Analysis of the gluten proteins in developing spring wheat. *R. Soc. Chem. Spl. Publ.* 261:471–474.

Zanardi-Lamardo, E., Clark, C. D., Zika, R. G. (2001). Frit inlet/frit outlet flow FFF: methodology for colored dissolved organic material in natural waters. *Anal. Chim. Acta* 443(2):171–181.

Zanardi-Lamardo, E., Clark, C. D., Moore, C. A., Zika, R. G. (2002). Comparison of the molecular mass and optical properties of colored dissolved organic material in two rivers and coastal waters by flow FFF. *Environ. Sci. Technol.* 36(13):2806–2814.

Zattoni, A., Torsi, G., Reschiglian, P., Melucci, D. (2000). Quantitative analysis in FFF using UV–Vis. detectors. An experimental design for absolute measurements. *J. Chromatogr. Sci.* 38:122–128.

28

Instrumentation for Countercurrent Chromatography

YOICHIRO ITO

Laboratory of Biophysical Chemistry, National Heart, Lung, and Blood Institute, National Institutes of Health, Bethesda, MD, USA

I. INTRODUCTION

Countercurrent chromatography (CCC) is a liquid–liquid partition chromatography technique that operates without requiring a solid support (Berthod, 2002a; Conway, 1992; Foucault, 1994; Ito, 1981a, 1988a, 1995, 1996a, 2000a, 2001a; Ito and Bowman, 1970a, b; Ito and Menet, 1999). The method has evolved from a unique centrifuge device called the coil planet centrifuge (Ito, 1969; Ito et al., 1966a, b, 1969), which subjects coiled tubes to a planetary motion (i.e., the coil undergoes slow rotation about its own axis along with high-speed rotation around the centrifuge axis). This original coil planet centrifuge was constructed in 1965 where it was aimed at the separation of lymphocytes (Ito et al., 1966b). When the coiled column was filled with nearly equal volumes of two mutually immiscible solvents, one on one side and the other on the opposite side, the planetary motion produced countercurrents of the two phases, which are finally distributed as alternating segments along the coil. When a sample was introduced at the original interface of the two phases, the device separated the solutes according to their partition coefficients, analogous to the countercurrent distribution method (Craig, 1962), but in a continuous fashion (Ito et al., 1969). This success encouraged us to develop a flow-through mechanism so as to perform continuous elution as in liquid chromatography.

Although CCC eliminates various complications arising from the use of solid support, it was necessary to solve the challenging problem of how to permanently retain the stationary phase in the separation column while the mobile phase is continuously flowing. CCC now uses a variety of column geometries, mostly combined with simple rotation or planetary motion, to provide a suitable force field to retain the stationary phase in the column while inducing mixing of the two phases to promote the partition process. All the existing CCC schemes have derived from two basic CCC types, both using a coiled column: the hydrostatic and hydrodynamic equilibrium systems (Ito, 1981a) (Fig. 1).

The basic hydrostatic system uses a stationary coiled tube [Fig. 1(a)]. The coil is first completely filled with the stationary phase, which can be either the lighter or heavier phase. The other phase is then introduced into the inlet of the coil where it percolates through the stationary phase segments on one side of the coil. This process continues until the mobile phase elutes from the outlet of the coil. Solutes introduced locally at the inlet of the coil are subjected to a partition process at every turn of the coil and separated according to their partition coefficient in a manner analogous to that used in liquid chromatography but without requiring a solid support.

In the basic hydrodynamic system the coil is rotated around its horizontal axis [Fig. 1(b)]. This motion creates an Archimedean screw force which drives every

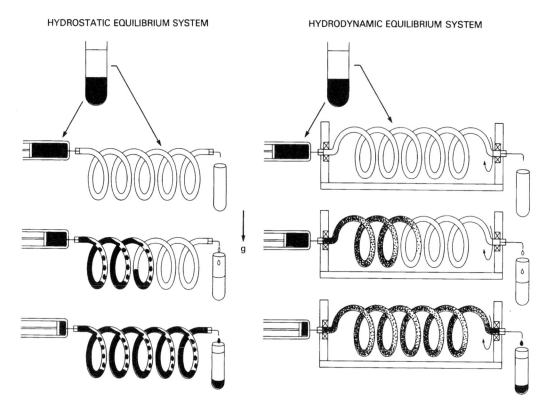

Figure 1 Basic model systems for CCC. [From Ito (1981a).]

object with different densities toward the end of the coil called "the head"; the other end is called "the tail". Under slow rotation of the coil, the mobile phase introduced at the head of the coil is mixed with the stationary phase to establish a hydrodynamic equilibrium where a large amount of the stationary phase is permanently retained in the coil while it is continuously mixed with the mobile phase. This process again continues until the mobile phase reaches the outlet of the coil. Consequently, the solutes introduced locally at the head of the coil are subjected to a continuous partition process between the two phases and separate according to their partition coefficient. Each basic CCC system has its own specific advantages as well as disadvantages, which are summarized in Table 1. In general, the hydrodynamic system yields better separations owing to a lack of dead space in the coil and efficient mixing of the two phases induced by the Archimedean screw effect.

During the 1970s and 1980s many CCC schemes were developed from these two basic CCC systems. In this article, the principle and design of individual CCC schemes are described and their performance evaluated by the separation of a set of dinitrophenyl (DNP) amino

Table 1 Comparison Between Two Basic Countercurrent Chromatographic Systems

	Hydrostatic equilibrium system	Hydrodynamic equilibrium system
Motion of coil	Stationary	Rotation around its own axis
Functional symmetry	Symmetrical	Asymmetrical rotation creates head and tail
Elution	Either direction	Head to tail direction
Retention of stationary phase	Stable and close to 50%	Variable
Partition efficiency	Low	High
Interfacial area	Small	Large
Mixing of phases	Poor	Excellent
Dead space	50% of the column space	None

Source: From Ito (1981a).

acids using a two-phase solvent system composed of chloroform/acetic acid/0.1 M hydrochloric acid at a volume ratio of 2:2:1.

II. HYDROSTATIC EQUILIBRIUM CCC SYSTEMS

A. Development of Hydrostatic CCC Schemes

Because of its simplicity, the basic hydrostatic system was first developed into several efficient CCC schemes for both analytical and preparative separations as shown in Fig. 2.

In the analytical scheme (helix CCC, lower diagram) both helical diameter and internal diameter of the coil are reduced while the coil is subjected to a strong centrifugal force field to retain the stationary phase in each helical turn.

In the preparative scheme (droplet CCC, upper diagram), one side of the coil, which is entirely occupied by the mobile phase and therefore forms an inefficient dead space, is reduced to a narrow bore transfer tube, whereas the other side of the coil where partition is taking place is replaced with a straight tubular column. The mobile phase (heavier in this case) introduced at the upper end of the column forms a train of discrete droplets which travel smoothly through the entire length of the column without coalescence. The necessity of suitable droplet formation limits the choice of the solvent systems in this system. This problem was solved by the locular column which is made by inserting a number of centrally perforated disks into the column at regular intervals to form partition units called locules. In rotation

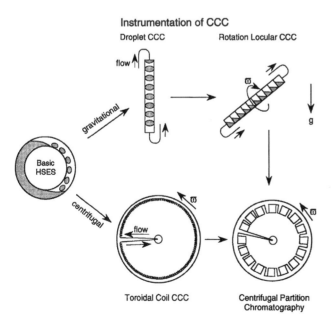

Figure 2 Development of hydrostatic CCC systems.

locular CCC (Fig. 2, upper right), both retention of the stationary phase and interfacial area (where mass transfer takes place) in each locule are optimized by inclining the column. The mixing in each phase is promoted by column rotation as shown in the diagram. This scheme permits universal application of conventional two-phase solvent systems. Centrifugal partition chromatography (CPC) radically improves the efficiency of the scheme by rotating a stack of doughnut-shaped disks containing multiple locules (Fig. 2, lower right).

A detailed description of the individual hydrostatic CCC schemes is given below.

B. Instrumentation of Hydrostatic CCC Schemes

1. Helix CCC (Toroidal Coil CCC)

Original Apparatus

Figure 3(a) shows the essential features of the original apparatus (Ito and Bowman, 1970b). The separation tube is supported at the periphery of the centrifuge head, and is fed from a coaxially rotating syringe. The plunger of a special high-pressure syringe is pushed through as a thrust bearing by a multispeed syringe pusher. Pressure of 16–20 atm was used in these experiments so that it was easier to push a rotating syringe than to make a high pressure rotating seal. Effluent from the helix at approximately atmospheric pressure is conducted from the rotating system through a rotating seal. A model PR-2 centrifuge fitted with a modified Helixtractor head (International Equipment Co.) provides the centrifugal field and holds the helix, syringe, and rotating seals. Centrifugal force is maintained constant by running the centrifuge on DC power from a Lambda LH 122 FM power supply that is regulated to 0.01%.

The separation column was prepared from 0.28 mm i.d. polytetrafluoroethylene (PTFE) tubing (SW 30, Zeus Industrial Products, Raritan, NJ) by pulling it through a plastic die to reduce the i.d. down to 0.20 mm. Two methods of coil preparation were used. In the first, the tubing is wound tightly on a flexible rod support which is subsequently coiled in a number of turns around the inside of the centrifuge head [Fig. 3(a)]. Alternatively, the tubing is folded into, and twisted along, its length to provide a rope-like structure in which the individual strands have the appearance of a stretched helix of small diameter. The coil is wound around a drum support which fits in the centrifuge head. If the centrifugal field, and hence the operating pressure, is high, both coil forms may be encased in a transparent resin for additional support.

Figure 3(b) illustrates the optimal separation achieved for the DNP amino acids using a two-phase solvent

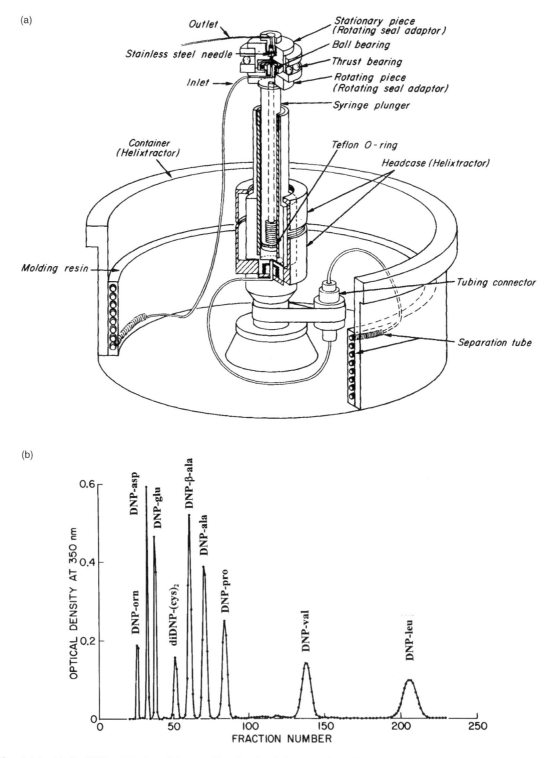

Figure 3 Original helix CCC. (a) Design of the centrifuge head and (b) separation of DNP amino acids. [From Ito and Bowman (1970b).]

system composed of chloroform/acetic acid/0.1 M hydrochloric acid (2:2:1, v/v/v) using the upper aqueous phase as the mobile phase. The separation was performed with a coiled column of 8000 turns with a helical diameter of 0.85 mm i.d. and prepared from a 40 m length of the 0.2 mm i.d. PTFE tubing. The partition efficiencies in this separation calculated according to the conventional equation range from 2500 to 5200 TP (theoretical plates).

Countercurrent Chromatography

Seal-Free Toroidal Coil Centrifuge

The original design of the toroidal coil (helix) CCC apparatus was later improved by adapting a rotary-seal-free flow-through device that was originally developed for apheresis (Ito, 1984; Ito et al., 1975). This seal-less system can tolerate high pressures of several hundred psi, which is required for analytical separation (Ito and Bowman, 1978a).

Figure 4(a) illustrates the principle of the seal-free flow-through system [see also Fig. 8(b) nonplanetary centrifuge system]. A bundle of flow tubes is connected

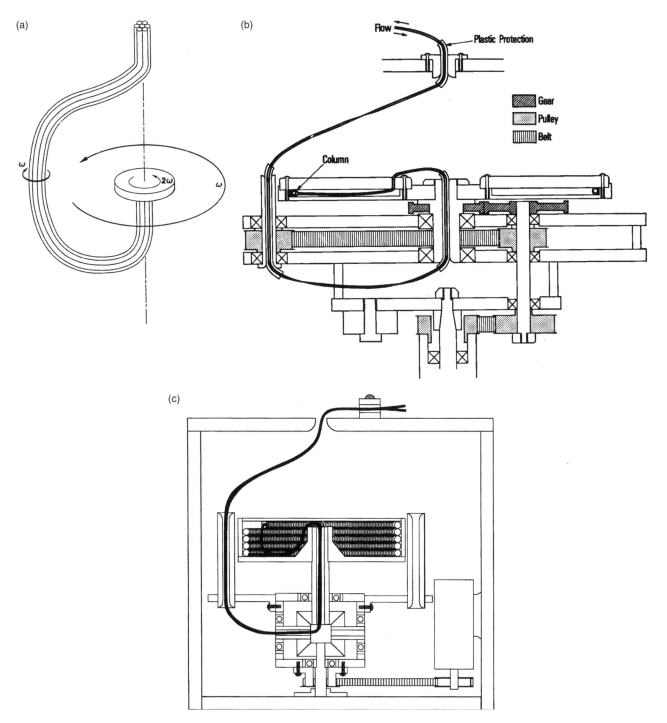

Figure 4 Helix CCC without rotary seals. (a) Principle of twist free system, (b) earlier design of the apparatus, and (c) improved design of the apparatus.

to a bowl at one end while remaining stationary at the other end, forming a loop as shown. When the bowl rotates at an angular velocity of 2ω around the vertical axis and the loop simultaneously revolves around the same axis at ω, the tube bundle remains free from twisting. In so doing, the tube bundle counter-rotates around its own axis at $-\omega$. The design of the centrifuge for performing toroidal coil CCC is illustrated in Fig. 4(b) and (c). The design shown in Fig. 4(b) is an earlier model that counter-rotates the tube supporting pipe as in the first apheresis centrifuge (Ito, 1984; Ito et al., 1975). The frame of the centrifuge head consists of three parallel horizontal plates rigidly linked and driven by the motor shaft as a unit. The frame holds a centrifuge bowl (center), a countershaft (right) and a tube-supporting hollow shaft (left), all mounted through ball bearings. A stationary pulley mounted on the motor housing is coupled through a toothed belt to an identical pulley mounted at the bottom of the countershaft to counter-rotate this shaft with respect to the rotating frame. This motion is further conveyed to the centrifuge bowl by 1:1 gearing between the countershaft and the centrifuge bowl. This arrangement doubles the angular velocity of the centrifuge bowl. To support the counter-rotating flow tubes, the hollow shaft is actively counter-rotated at $-\omega$ by means of a pair of identical toothed pulleys coupled to the hollow shaft and the countershaft.

The column is prepared by winding a narrow-bore PTFE tubing onto a plastic pipe which is in turn coiled around the periphery of the centrifuge bowl. A pair of flow tubes is first passed through the opening of the central shaft and then through the hollow tube supporting pipe, finally exiting the centrifuge at the center of the centrifuge cover where it is tightly supported. These tubes, if protected from direct contact with the metal parts with a lubricated short plastic tube, preserve their integrity for many runs.

The latest model of the toroidal coil centrifuge is shown in Fig. 4(c) where the double-speed rotation of the centrifuge bowl is effected by coupling three identical miter gears at the bottom of the centrifuge while the counter-rotation of the tube-supporting pipe is omitted (Matsuda et al., 1998). Compared with the earlier model this unit is much simpler and the rotor is easier to balance owing to the symmetry of the elements on the rotary frame. However, because the tube-supporting pipe is stationary with respect to the rotating frame, the counter-rotating flow tubes should be thoroughly lubricated to minimize the friction against the inner wall of the tube-supporting pipe.

In the toroidal coil (helix) CCC, the efficiency of separation for a given solvent system may be improved by using a greater length of more tightly wound narrower tubing. However, the practical limit is governed by the centrifugal field and the injection pressure necessary to promote interfacial flow. The minimal pressure P at the inlet of the column, necessary to sustain flow, may be calculated from the following equation:

$$P = R\omega(\rho_l - \rho_u)nd \qquad (1)$$

where R denotes the distance between the center of rotation and the axis of the helix; ω, the angular velocity of rotation; n, the number of turns of the column; d, the helical diameter; and ρ_l and ρ_u, the density of lower and upper phases, respectively.

As mixing of the two phases in this scheme relies entirely on the flow of the mobile phase, the effect of a Coriolis force becomes important, and the direction of the rotation should be chosen so that this force can disperse the mobile phase into the stationary phase rather than forming a stream along the tube wall (Ito, 2001b; Ito and Ma, 1998).

2. Droplet CCC

In this system, the separation is carried out by passing droplets of the mobile phase through a column of the stationary phase [Fig. 5(a)] (Ito and Bowman, 1970b, 1971a; Tanimura and Ito, 1974a, b; Tanimura et al., 1970). The mobile phase may be either heavier or lighter than the stationary phase. When heavier, it is delivered at the top of the column and, when lighter, through the bottom. The column (glass, Teflon, or metal may be used) is mounted perpendicularly and filled with stationary phase. The lighter mobile phase is introduced at the bottom through a tip chosen to give a steady stream of droplets about as large in diameter as the tubing. When a droplet reaches the top of the column it is delivered to the bottom of the next column through narrow-bore Teflon tubing. A small amount of stationary phase may also enter the Teflon tubing initially, but the effect is insignificant.

As the mobile phase moves through the column, turbulence within the droplet promotes efficient partitioning of the solute between the two phases. The expected mixing of the stationary phase along the length of the column is minimized by the advancing series of proper-sized droplets which serves to divide the stationary liquid into distinct segments.

With a given set of phases, generation of droplets of proper size is a function of the column bore, the interfacial tension of the two phases, the flow rate of the mobile phase, and the diameter of the inlet tip. In general, small-bore column (i.e., less than 1.0 mm) produce plug flow in which the entire contents of the tube are displaced. If the bore is too great, large droplets form, which do not allow efficient partitioning. If small droplets are introduced, some coalesce into slower-moving larger droplets which move even more slowly. This can occur throughout

Countercurrent Chromatography

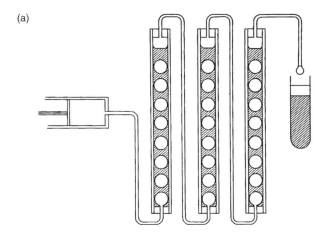

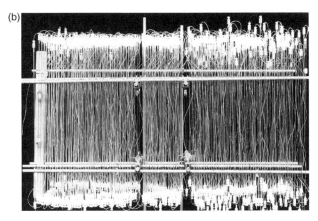

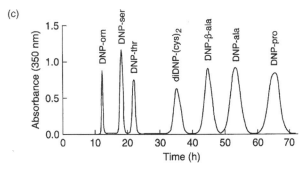

Figure 5 Droplet CCC. (a) Mechanism, (b) original apparatus, and (c) DNP amino acid separation. [From Ito (1988a).]

damping pulsations can be used to deliver a constant flow of droplets. The rate is set at the highest flow compatible with a steady generation of discrete droplets. The highest permissible flow gives droplets which are barely separated.

Figure 5(b) illustrates the original droplet CCC assembly which consists of 300 glass tubes, each 60 cm long and 1.8 mm i.d., with a total capacity of about 540 mL, including about 16% of dead space in the transfer tubing. Using the solvent system composed of chloroform/acetic acid/ 0.1 M hydrochloric acid (2:2:1, v/v/v), 300 mg of DNP amino acids were successfully separated at a partition efficiency of about 900 TP in 70 h [Fig. 5(c)].

In the past, a number of natural products have been purified, mostly using a solvent system composed of chloroform, methanol, and water at various volume ratios. The apparatus is currently available through Tokyo-Rikakikai, Tokyo, Japan.

3. Rotation Locular CCC

This CCC system has been developed to solve various problems inherent to droplet CCC, such as the necessity for suitable droplet formation, longitudinal spreading of the solute band through a column, and the requirement of long separation time.

The column is made by inserting a number of centrally perforated disks into a tubular column at regular intervals so that each compartment (locule) thus formed serves as a partition unit (Ito and Bowman, 1970b, 1971a). In this locular column the retention of the stationary phase and the interfacial area of the two phases are optimized by inclination of the column while the mixing within each phase is accomplished by rotation of the column.

The mechanism of rotation locular CCC is illustrated in Fig. 6(a). The column inclined at angle α from the horizontal plane is filled with the upper stationary phase, and the lower phase is pumped through the first rotating seal connection while the column is rotated around the axis of the column support. The lower phase then displaces the upper stationary phase in each locule down to the level of the hole which leads to the next locule, this process being continued throughout the first column. The mobile phase eluted from the outlet of the first column is then returned to the upper level to enter the second column through a narrow transfer tube accommodated in the hollow column support. A similar process is continued through the second column and the eluted mobile phase is finally led through the second rotating seal at the bottom of the apparatus. In practice, multiple columns can be arranged around the central support to increase the partition efficiency of the system.

In the original apparatus, the column unit was prepared by spacing Teflon rings (2.8 mm, o.d., 0.8 mm i.d., and

the length of the column, sometimes with the result that large segments of the mobile phase ascend the column, which prohibits efficient partitioning. Introduction of the mobile phase at too fast a rate is also deleterious, as the droplets often coalesce into cylinders. The column walls should not have a high affinity for the mobile phase. If the mobile phase is aqueous and the column is glass, the walls must be sylated to prevent the liquid from wetting the walls and moving as a stream. A chromatographic metering pump equipped with a reservoir for

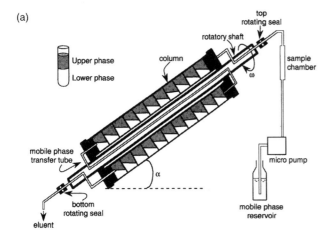

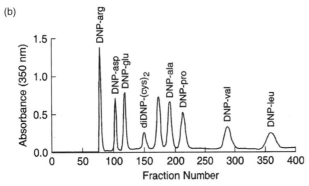

Figure 6 Rotation locular CCC. (a) Mechanism and (b) DNP amino acid separation. [From Ito and Bowman (1970b).]

Although the rotation locular CCC is less efficient than high-speed CCC, in terms of resolution and separation times, it has advantages of a large-scale sample loading capacity and universal application of conventional two-phase solvent systems. Currently, the apparatus is commercially available through Tokyo-Rikakikai, Tokyo, Japan.

4. Centrifugal Partition Chromatography

This CCC system (Foucault, 1994, 2002; Marchal et al., 2002) is considered to be a hybrid between helix CCC and locular CCC. The doughnut-shaped separation column made of inert plastic material contains engraved multiple partition compartments interconnected with fine transfer ducts (Fig. 7). Several units of these disks are stacked to form a column assembly. The whole column assembly is rotated with a flow-through centrifuge system equipped with a pair of rotary seals.

This system was developed by Sanki Engineering, Ltd., Kyoto, Japan (Nunogaki, 1990). A pair of the rotating seals used for the elution can tolerate up to $60 \, kg/cm^2$ (ca. 800 psi), and the whole system is computerized in a fashion similar to commercial liquid chromatography systems. The unit is available from Sanki Engineering, Kyoto, Japan (distributor: EverSeiko Corp., Tokyo, Japan) and Kromaton, Technologies, Villejuif, France (distributor: Richard Scientific Equipment Co. Navato, CA, USA).

1 mm thickness or 4.8 mm o.d., 1.2 mm i.d., and 1.2 mm thickness) into Teflon tubing (3.5 mm o.d., and 2.6 mm i.d. or 5.1 mm o.d., and 4.6 mm i.d., each 47 cm long) at every 3 mm interval to make over 100 locules. In each column, 10–50 such units were mounted side-by-side onto the column holder (1.25 cm diameter, 55 cm long) equipped with a rotating seal at each end. Connections between the column units and between the column units and rotating seals were made of fine Teflon tubing (1 mm o.d., 0.6 mm i.d.). The inclination of the column was adjusted to 25–30° from the horizontal plane. The rotation speed of the column holder was adjustable in several steps up to 300 rpm. The fluid was pumped by either a syringe driver or metering pump at flow rates ranging from 2 to 40 mL/h while the eluate was monitored through an LKB UV monitor and fractionated with an LKB fraction collector. Figure 6(b) shows the separation of a set of DNP amino acid samples with a two-phase solvent system composed of chloroform, acetic acid, and 0.1 M hydrochloric acid (2:2:1, v/v/v) using the original apparatus. Nine DNP amino acids were resolved in 70 h at a partition efficiency of 3000 TP.

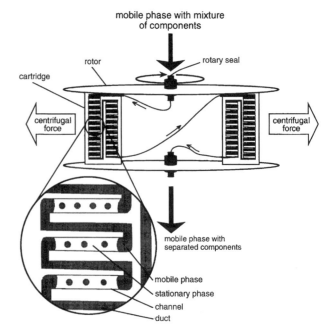

Figure 7 Centrifugal partition chromatograph.

III. HYDRODYNAMIC EQUILIBRIUM SYSTEMS

Solute partitioning in a rotating coil was first demonstrated in an end-closed coiled tube subjected to a planetary motion (Ito et al., 1969). Later, development of the hydrodynamic CCC instruments equipped with a flow-through mechanism was derived from gyration locular CCC (Ito and Bowman, 1970b, 1971a) in which the locular column was held vertical and gyrated around its axis at a high speed [Fig. 8(a)]. This system eliminated the need for the rotating seals and yielded a higher partition efficiency when compared with rotation locular CCC which uses unit gravity for mixing of the two phases. However, application of the high centrifugal force field often deformed the locular columns. This problem led to substituting the locular column with a coiled column. Surprisingly, a simple

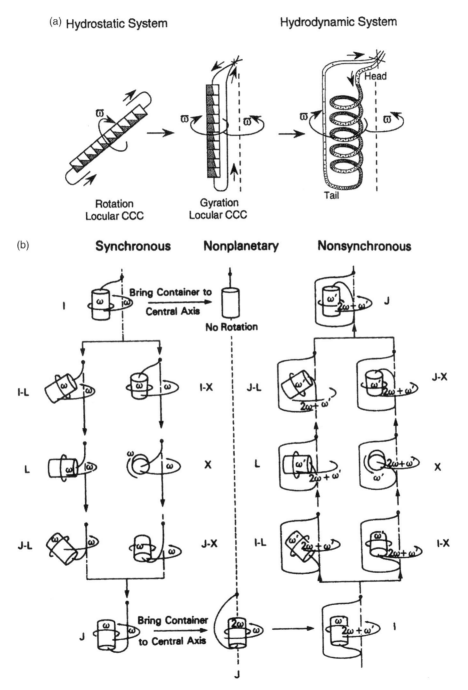

Figure 8 Development of hydrodynamic equilibrium systems. (a) Original flow-through coil planet centrifuge without rotary seals and (b) development of various rotary free flow-through centrifuge systems.

coiled tube mounted around the holder produced substantially higher partition efficiency than the locular column. This finding was an important milestone in the development of the hydrodynamic CCC systems.

A. Rotary-Seal-Free Centrifuge Systems

Development of the hydrodynamic CCC systems was initiated by introduction of various flow-through centrifuge schemes free of rotary seals (Ito, 1981a). These centrifuge schemes are classified into three groups (i.e., synchronous, nonplanetary, and nonsynchronous) according to their mode of planetary motion. In Fig. 8(b) each diagram schematically illustrates rotary-seal-free flow-through centrifuge schemes with a cylindrical coil holder with a bundle of flow tubes, the end of which is tightly supported above the centrifuge.

In the synchronous type I (left, upper), the vertical coil holder revolves around the main axis of the centrifuge and simultaneously rotates about its own axis at the same speed but in the opposite direction. This counter-rotation of the holder steadily unwinds the twist caused in each revolution. Consequently, the system permits continuous elution of mobile phase through the rotating column without the use of rotary-seals which might become potential source of leakage and contamination. This same principle can be equally well applied to other synchronous planetary motions with tilted (types I–L and I–X), horizontal (types L and X), dipping (types J–L and J–X), and even inverted (type J) orientations of the holder.

Although types I and J have similar appearances, there are crucial differences between these two planetary motions. The holder of type I undergoes counter-rotation whereas that of type J rotates in the same direction to the revolution. When the holder of type I is brought to the central axis of the centrifuge it becomes stationary as counter-rotation of the holder cancels out each revolution. On the contrary, similarly shifting the holder of type J synchronous planetary motion toward the center of the centrifuge causes addition of each rotation and revolution of the holder; the result is that the holder rotates at a doubled rate (2ω) around the center of the centrifuge as in the ordinary centrifuge system. However, this shift causes the tube bundle to rotate around the holder at ω as shown in the diagram (nonplanetary, bottom).

This nonplanetary scheme is a transitional form to the nonsynchronous schemes. On the platform of the nonplanetary scheme, the holder is again shifted toward the periphery to undergo a synchronous planetary motion. As the revolution rate of the base is independent of the top planetary motion, the rotation/revolution ratio of the planetary motion becomes adjustable.

Many of these centrifuge schemes have been built and examined for their performance in terms of retention of stationary phase and partition efficiency in CCC. Among them, several synchronous schemes turn out to be most useful. For effective instrumentation of these schemes, it is necessary to understand the acceleration field produced by these planetary motions.

B. Analysis of Acceleration on Synchronous Planetary Motions

1. Schemes I, L, J, and Their Hybrids

We consider a discoid body with radius r and its axis being held at angle Ψ from the central axis of the stationary frame [Fig. 9(a)] (Ito, 1986a). The discoid body undergoes a synchronous planetary motion in such a way that it revolves around the central axis at an angular velocity, ω, with a revolution radius, R, and simultaneously rotates about its own axis at the same angular velocity. Then, we wish to study the motion of an arbitrary point, P, present on the periphery of the discoid body.

Figure 9(b) shows an x–y–z coordinate system where the center of the discoid body initially locates at point Q_0 (R, 0, 0) and the arbitrary point at P_0 ($R - r\cos\Psi$, 0, $r\sin\Psi$). The center of the discoid body circles around point 0 in the x–y plane and at time t moves angle $\theta = \omega t$ to reach point Q ($R\cos\theta$, $R\sin\theta$, 0). By the aid of a diagram illustrated in Fig. 9(c) the location of the arbitrary point on the discoid body is expressed by $P(x, y, z)$ according to the following equations:

$$x = R\cos\theta + r[1 - (1 + \cos\Psi)\cos^2\theta] \tag{2}$$

$$y = R\sin\theta - r(1 + \cos\Psi)\sin\theta\cos\theta \tag{3}$$

$$z = r\sin\Psi\cos\theta \tag{4}$$

where R is the distance OQ; r, the distance QP; and Ψ, the angle formed between the z-axis and the body axis, QA.

The acceleration, α, produced by the planetary motion is then computed from the second derivative of the above equations as

$$\alpha_x = \frac{d^2 x}{dt^2} = -R\omega^2[\cos\theta - 2\beta(1 + \cos\Psi)\cos 2\theta] \tag{5}$$

$$\alpha_y = \frac{d^2 y}{dt^2} = -R\omega^2[\sin\theta - 2\beta(1 + \cos\Psi)\sin 2\theta] \tag{6}$$

$$\alpha_z = \frac{d^2 z}{dt^2} = -r\omega^2 \sin\Psi\cos\theta \tag{7}$$

where $\beta = r/R$ and $R \neq 0$. Using Eqs. (5)–(7), the x, y, z components of the acceleration vector can be computed and displayed separately or in a combined form in the x–y–z coordinate system of the stationary reference frame.

However, the effects of the acceleration on the subjects moving with the discoid body can be more effectively visualized if the acceleration is expressed with respect to the revolving body frame. Figure 9(d) shows the relationship between the original x–y–z coordinate system and the x_b–y_b–z_b body coordinate system in which the z_b-axis coincides with the axis of the discoid body. The x_b–y_b–z_b coordinate system revolves around the z-axis of the original coordinate system in such a way that the radially directed x_b-axis forms angle Ψ against the x–y plane while the y_b-axis is always confined in the same plane. Consequently, the discoid body coaxially rotates around the z_b-axis at angular velocity ω in the x_b–y_b plane.

Transformation of the acceleration from the original x–y–z coordinate system to the x_b–y_b–z_b body coordinate system can be performed by the following equations:

$$\alpha_{x_b} = (\alpha_x \cos\theta + \alpha_y \sin\theta) \cos\Psi - \alpha_z \sin\Psi \quad (8)$$

$$\alpha_{y_b} = -\alpha_x \sin\theta + \alpha_y \cos\theta \quad (9)$$

$$\alpha_{z_b} = (\alpha_x \sin\theta + \alpha_y \cos\theta) \sin\Psi + \alpha_z \cos\Psi \quad (10)$$

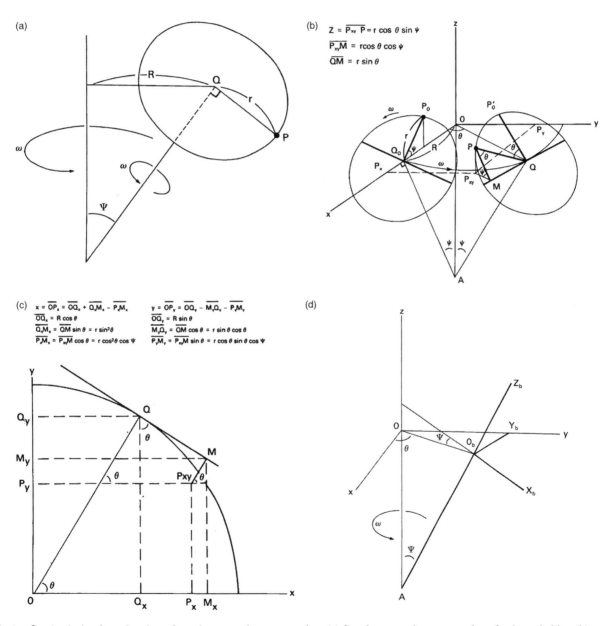

Figure 9 Analysis of acceleration of synchronous planetary motion. (a) Synchronous planetary motion of column holder, (b) x–y–z coordinate system for analysis, (c) analysis of acceleration on x–y plane, (d) reference coordinate system on the moving holder, and (e) results of analyses for various synchronous planetary systems. [From Ito (1986a).]

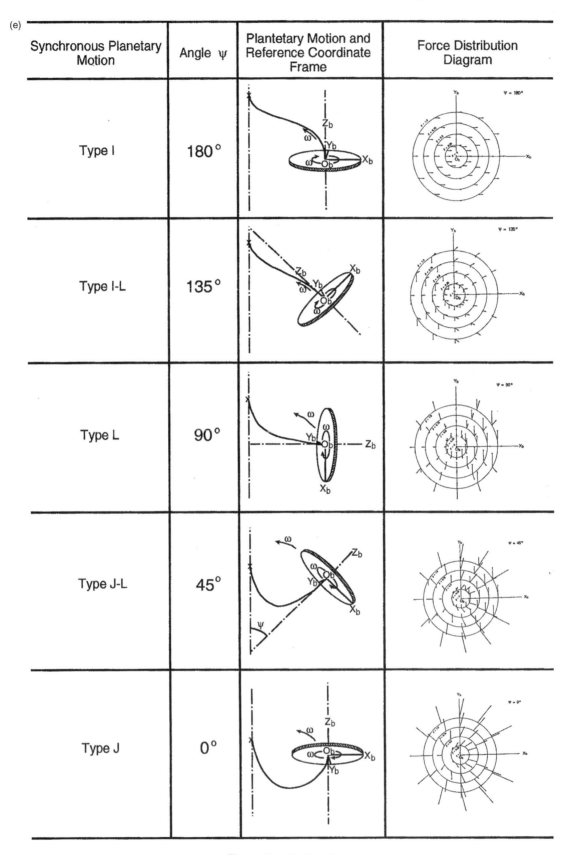

Figure 9 *Continued.*

which become

$$\alpha_{x_b} = -R\omega^2[\cos\Psi - \beta(1+\cos\Psi)^2\cos\theta] \quad (11)$$

$$\alpha_{y_b} = -R\omega^2(1+\cos\Psi)2\beta\sin\theta \quad (12)$$

$$\alpha_{z_b} = -R\omega^2\sin\Psi[1 - \beta(2+\cos\Psi)\cos\theta] \quad (13)$$

For convenience of the diagrammatical expression of acceleration vectors, the first two components, α_{x_b} and α_{y_b}, are combined into a single vector and expressed as an arrow in the x_b–y_b plane, whereas the third vector, α_{z_b}, is separately expressed as a column. Figure 9(e) illustrates a set of the centrifugal force (opposite to the acceleration) distribution diagrams for five types of synchronous planetary motion. The first column (left) gives the types of synchronous planetary motion; the second column, the angle Ψ characteristic of each type of synchronous planetary motion (for the two hybrid types, type I–L and type J–L, Ψ is chosen at 135° and 45°, respectively); the third column, orientation and planetary motion of the discoid body including the x_b–y_b–z_b body coordinate system which serves as the reference frame for the force distribution diagram in the next column; and the fourth column, the distribution diagram of the centrifugal force vectors.

At $\Psi = 180°$, the diagram shows homogeneous 2D distribution of the centrifugal force vectors, each component measuring a unit length and acting in the same direction parallel to the radius of revolution. At $\Psi = 135°$, the force distribution becomes 3D; the first component (arrow) shows somewhat diverged distribution with slight decrease in magnitude, whereas the second component (column) acts strongly downward on the left and slightly upward on the right. When Ψ reaches the transitional angle of 90°, the first force component (arrow) forms symmetrical distribution while always acting outwardly from the circle, whereas the second force components (column) act very strongly upward (or toward the periphery along the radius of revolution) on the right. Here, it is interesting to note that the second force component also acts downward on the left, indicating that the centrifugal force is momentarily directed toward the center of revolution at the very top of the discoid body. This strange phenomenon, which has been reported elsewhere (Ito et al., 1972), is explained on the basis of the Coriolis force produced by the planetary motion. As Ψ is further decreased to 45°, the first force component (arrow) gains strength, especially on the right, whereas the second component (column) maintains a similar profile. Finally, at $\Psi = 0°$, the distribution of the force vectors returns to the 2D pattern as in the first diagram ($\Psi = 180°$), but it consists of a complex arrangement of force vectors diverging from each circle with a great enhancement of magnitude, especially at a remote location from the center of the discoid body.

Comparison between the two extreme synchronous planetary motions, type I and type J, reveals a remarkable difference in both the distribution pattern and magnitude of the centrifugal force vectors under the same revolution radius and speed. The difference is clearly manifested in the hydrodynamic effects on the two immiscible solvent phases in the coiled column. As briefly mentioned earlier, the type I planetary motion produces the basic hydrodynamic equilibrium where the two solvent phases are evenly distributed in the coiled column from the head toward the tail, while any excess of either phase remains at the tail. This hydrodynamic distribution of the two solvent phases can be utilized for performing CCC by introducing the mobile phase, either lighter or heavier phase, through the head of the coiled column. However, in this mode of elution the stationary phase volume retained in the column is substantially less than 50% of the total column capacity. On the other hand, the type J planetary motion distributes the two solvent phases bilaterally (or unilaterally) in the rotating coil coaxially mounted on the holder, where one phase, usually the lighter phase, entirely occupies the head side and the other phase the tail side. This bilateral hydrodynamic distribution is capable of retaining a large column of the stationary phase in the column and is considered to be ideal for performing CCC. The hydrodynamic motion of the two solvent phases produced by this type of planetary motion has been observed under stroboscopic illumination (see Fig. 16). The experiment reveals two distinct zones in each turn of the spiral column: a mixing zone of about a quarter turn located near the center of the centrifuge and a settling zone showing a clear interface of the two phases in the rest of the column. The high partition efficiency characteristic of high-speed CCC, as described later, may be largely attributed to this local mixing unique to the type J synchronous planetary motion.

The force distribution diagrams in Fig. 9(e) also indicate an important information for designing the apparatus: the separation columns of types I and J which show the planar distribution of the force vectors can be balanced simply by mounting a counterweight with the same mass symmetrically on the opposite side of the rotary frame. In contrast, the rest of the schemes, which show 3D distribution of the force vectors, should be balanced by arranging two or more identical columns symmetrically on the rotary frame, all being rotated at the same angular velocity in the same direction.

2. Schemes X, L, and Their Hybrids

The analysis of acceleration on the cross-axis coil planetary centrifuge (types X, L, and their hybrid) was similarly carried out using a 3D coordinate system as shown in Fig. 10(a)–(c) (Ito, 1987a).

906   Analytical Instrumentation Handbook

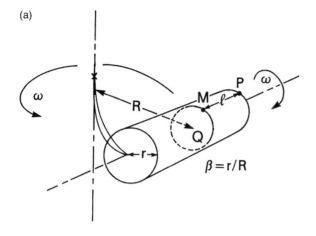

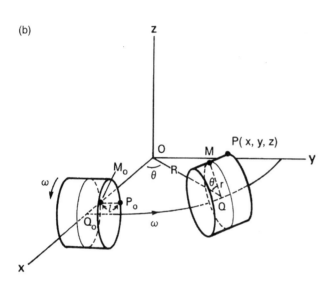

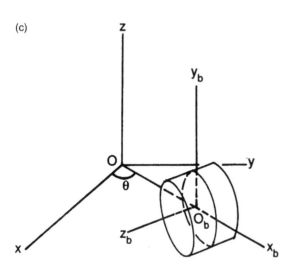

Figure 10 Analysis of acceleration of type X–L synchronous planetary motions. (a) Synchronous planetary motion of column holder, (b) the x–y–z coordinate system for analysis, and (c) reference coordinate system on the moving holder. [From Ito (1987a).]

Figure 10(a) shows planetary motion of a cross-axis CPC where a cylindrical coil holder with radius r revolves around the central axis of the centrifuge system and simultaneously rotates about its own axis at the same angular velocity, ω, in the indicated directions. While doing so, the cylinder maintains the axial orientation perpendicular to, and at a distance R, from the central axis of the centrifuge. An arbitrary point, P, is located at the periphery of the cylinder at a distance l from point M on the central place, as illustrated. Then, we observe the motion of point P as the cylinder undergoes the planetary motion described above.

In the reference x–y–z coordinate system for analysis, illustrated in Fig. 10(b), the cylinder revolves around the z-axis at angular velocity ω and simultaneously rotates about its own axis at the same angular velocity in the indicated directions. The arbitrary point to be analyzed initially locates at point P_0 ($R - r$, l, 0) and, after a lapse of time t, moves to point P (x, y, z) which can be expressed in the following equations:

$$x = R\cos\theta - r\cos^2\theta - l\sin\theta \tag{14}$$

$$y = R\sin\theta - r\sin\theta\cos\theta + l\cos\theta \tag{15}$$

$$z = r\sin\theta \tag{16}$$

The acceleration acting on the arbitrary point is then obtained from the second derivatives of these equations as follows:

$$\frac{d^2x}{dt^2} = -R\omega^2(\cos\theta - 2\beta\cos 2\theta) + l\omega^2\sin\theta \tag{17}$$

$$\frac{d^2y}{dt^2} = -R\omega^2(\sin\theta - 2\beta\sin 2\theta) - l\omega^2\cos\theta \tag{18}$$

$$\frac{d^2z}{dt^2} = -R\omega^2\beta\sin\theta \tag{19}$$

where $\beta = r/R$.

In order to visualize the effects of acceleration on the objects rotating with the cylinder, it is more appropriate to express the acceleration vectors with respect to the body frame or the x_b–y_b–z_b coordinate system illustrated in Fig. 10(c). Transformation of the vectors from the original reference coordinate system to the body coordinate system may be performed according to the following equations:

$$\alpha_{x_b} = \frac{d^2x}{dt^2}\cos\theta + \frac{d^2y}{dt^2}\sin\theta = -R\omega^2(1 - 2\beta\cos\theta) \tag{20}$$

$$\alpha_{y_b} = \frac{d^2z}{dt^2} = -R\omega^2\beta\sin\theta \tag{21}$$

$$\alpha_{z_b} = \frac{d^2x}{dt^2}\sin\theta - \frac{d^2y}{dt^2}\cos\theta = -R\omega^2 2\beta\sin\theta + l\omega^2 \tag{22}$$

where α_{x_b}, α_{y_b}, and α_{z_b} indicate the acceleration vectors acting along the corresponding coordinate axes. Equations (20)–(22), thus obtained, may serve as general formulae of acceleration generated by X, L, and their hybrids by designating two parameters, R and l (e.g., $l/R = \pm 1$ for type X–L, $l = 0$ for type X, and $R = 0$ for type L). From these equations, the centrifugal force vectors at various points on the cylinder are computed for three types of planetary motion, that is, type L ($R = 0$), type X–L ($l/R = -1$), and X ($l = 0$), and diagrammatically illustrated in Fig. 11(a)–(c). In order to express the 3D pattern of the centrifugal force vectors on a 2D diagram, two force vectors, $-\alpha_{x_b}$ and $-\alpha_{y_b}$, are combined into a single arrow forming various angles from the x_b-axis, whereas the third force

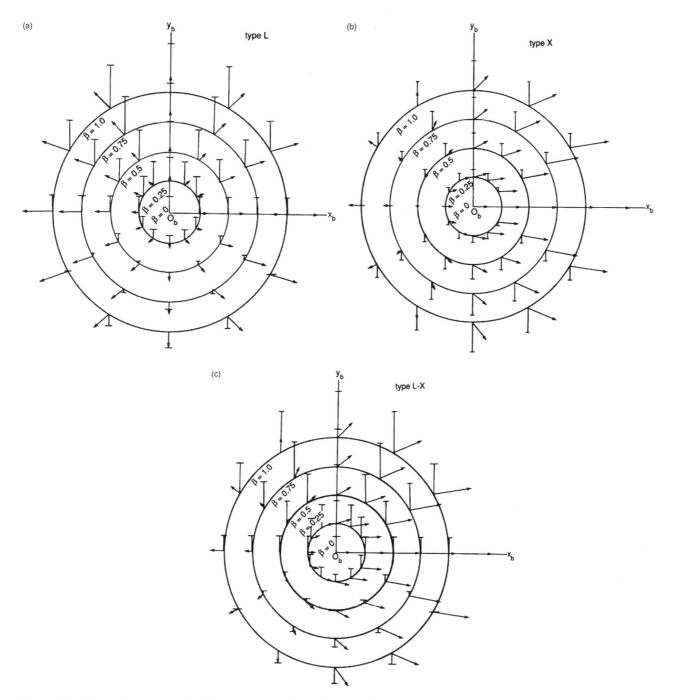

Figure 11 Distribution of centrifugal force vectors on the synchronous planetary motions of L, X, and hybrid systems. Distribution of centrifugal force vectors of (a) type L synchronous planetary motion, (b) type X synchronous planetary motion, and (c) type L–X synchronous planetary motion.

vector, $-\alpha_{z_b}$, which acts perpendicularly to the x_b–y_b plane, is drawn as a vertical column along the y_b-axis. The ascending column indicates the force acting upward ($z_b > 0$) and the descending column, the force acting downward ($z_b < 0$). Several concentric circles around point O_b (the axis of the cylinder) indicate locations on the cylinder corresponding to parameter β or β' ($\beta' = r/l$ for type L planetary motion) indicated in the diagram. The distribution of the centrifugal force vectors in each diagram is fixed to the x_b–y_b–z_b (z_b-axis) in either clockwise or counterclockwise direction as determined by the planetary motion of the cylinder. The centrifugal force distribution diagram of type L planetary motion [Fig. 11(a)] is identical to that shown in Fig. 9(e) (middle), if the x_b–y_b coordinate system of the present diagram is rotated 90° clockwise. The effect of the centrifugal force distribution to the hydrodynamic motion of the two solvent phases in the column may be predicted by comparing the present diagrams to those shown in Fig. 9(e), which also includes the type L force distribution diagram (90°). In these force distribution diagrams, arrows around the circle exert an Archimedean screw force as in types I and J with strong mixing effect while the columns indicate the force acting across the diameter of the coil wound around the cylinder. Therefore, an asymmetrical force distribution of the column seen in type L promotes the separation of the two phases, thus preventing emulsification which may cause carryover and loss of the stationary phase from the coiled column. This stabilizing effect is minimized by the symmetrical force distribution of type X in which the force vibrates across the tubing to enhance the mixing of the two phases [Fig. 11(b)]. Consequently, retention of the stationary phase and mixing of the two phases can be conveniently adjusted by selecting the l/R ratio of the present system such as type X–L [Fig. 11(c)]. In fact, some cross-axis CPCs such as X-1.5L ($l/R = 1.5$) and X-3L ($l/R = 3$) provide satisfactory retention of stationary phase for extremely low-interfacial tension solvent systems including aqueous–aqueous polymer phase systems which are useful for separation of biopolymers.

As mentioned earlier, cross-axis CPCs which display 3D force distribution should be designed by mounting two or more identical columns symmetrically around the rotary frame and rotating all columns in the same direction to achieve perfect balancing of the centrifuge system.

C. Instrumentation

During the past three decades a number of CCC prototypes, mostly based on the hydrodynamic equilibrium system, have been introduced, which include flow-through CPC (type I) (Ito and Bowman, 1971b, 1973a), angle rotor CPC (type I–L) (Ito and Bowman, 1975), elution centrifuge (type L) (Ito and Bowman, 1973b; Ito et al., 1972), new angle rotor CPC (type J–L) (Ito, 1986a, b), horizontal flow-through CPC (type J) (Ito, 1979, 1980a, b, 1986c; Ito and Bowman, 1977, 1978b; Ito and Putterman, 1980; Ito and Oka, 1988), high-speed CCC (type-J) (Ito, 1981b, c, 1987b; Ito and Bhatnagar, 1981; Ito and Chou, 1988; Ito et al., 1982, 1989a, 1990a, b, 1991a; Kitazume et al., 1993; Lee et al., 1988a; Oka et al., 1989a; Nakazawa et al., 1981), dual CCC (type-J) (Bhatnagar and Ito, 1988; Ito, 1985; Lee et al., 1988b; Oka et al., 1989b), cross-axis CPC (types X, X–L) (Bhatnagar et al., 1989; Ito, 1987a, c, d, 1991; Ito and Zhang, 1988a, b; Ito and Menet, 1999; Ito et al., 1989b, 1991b; Menet and Ito, 1993; Menet et al., 1993, 1994; Shibusawa and Ito, 1992; Shinomiya et al., 1993), and nonsynchronous CPC (type I nonsynchronous scheme) (Levine et al., 1984; Ito et al., 1980, 1983). Among these instruments which have not been further developed are not described in this article. Interested readers may refer to the references cited earlier.

1. Horizontal Flow-Through CPC (Type-J)

This model is the first prototype CCC instrument based on type J synchronous planetary motion (Ito and Bowman, 1977, 1978b). As shown in Fig. 12 the planetary motion of the coil holder is effected by coupling a pair of identical gears, one stationary and the other mounted on the axis of the column holder. The column unit was made by winding Teflon tubing, 2.6 mm i.d. onto an aluminum pipe, 48 cm long and 1.25 cm o.d., with 1 mm wall thickness (a vinyl pipe of a similar dimension produced unsatisfactory results owing to its tendency to bend under a centrifugal force field). One column unit was approximately 100 helical turns with a capacity of about 25 mL. A long preparative column was prepared by connecting 10 column units in series with Teflon tubing (0.45 mm i.d.). The use of this narrow tubing appeared to be necessary to prevent pulsatile surging flow of the solvent under the strong centrifugal force field, which causes unnecessary broadening of the sample bands. The column was mounted symmetrically and as close as possible around the rotary shaft while a counterweight was applied on the opposite side of the rotary frame. The flow tubes (0.85 mm i.d. Teflon) of the separation column were first passed through the center hole of the rotary shaft and then through the side-hole of the coupling pipe to lead into the opening of the stationary shaft. The moving portion of the flow tubes was lubricated with grease and protected with a piece of silicone rubber tubing at each supporting point. These tubes, if properly protected, can maintain their integrity over several months of use. This original apparatus has been successfully used for separation of peptides and amino acid derivatives (Ito and Bowman, 1978b).

Later, this original design was improved by replacing the counterweight with an analytical coiled column

Figure 12 Photograph of the original horizontal flow-through coil planet centrifuge. [From Ito and Bowman (1978b).]

which is rotated synchronously in the opposite direction (type I) using a pair of identical pulleys coupled with a toothed belt [Fig. 13(a) and (b)] (Ito, 1979, 1980a, b).

The performance of these horizontal flow-through CPCs was limited by their long separation times, and the application of a high flow rate of the mobile phase caused extensive carryover of the stationary phase from the column. This resulted in loss of resolution. This problem was solved by finding a bilateral hydrodynamic distribution of two solvent phases in a coaxially wound coil on the holder subjected to the type J synchronous planetary motion as described subsequently.

2. High-Speed CCC Centrifuge with Multilayer Coil Assembly (Type J)

Principle

When a tube is directly wound around the holder hub (Fig. 14), type J synchronous planetary motion of the holder distributes the two immiscible solvent phase confined in the tube in such a way that one phase entirely occupies the head side and the other phase the tail side of the coil (Ito, 1981b, c, 1996; Ito et al., 1982). As briefly mentioned earlier, this bilateral phase distribution was found to be useful for performing CCC as illustrated in Fig. 15 (for simplicity all coils are drawn as straight tubes to show overall distribution of the two phases in the coil). An end-closed coil at the top [Fig. 15(a)] shows a bilateral phase distribution where the white phase (head phase) entirely occupies the head side and the black phase (tail phase) the tail side. This hydrodynamic equilibrium of the two phase suggests that the white phase, if introduced at the tail end, would move toward the head, and similarly the black phase introduced at the head end would move toward the tail. This hydrodynamic motion of the two phases can be effectively used for performing CCC as illustrated in Fig. 15(b).

The coil is first entirely filled with the white phase and the black phase is pumped into the head end [Fig. 15(b) upper diagram]. Alternatively, the coil is entirely filled with the black phase followed by pumping the white phase at the tail end. In either case, the introduced mobile phase can quickly move through the coil toward the other end, leaving a large volume of the stationary phase in the column. Consequently, separation can be performed at a high flow rate of the mobile phase with a large amount of the stationary phase retained in the column.

The present system also allows simultaneous introduction of the two phases through the respective terminal of the coil [Fig. 15(c)]. This dual CCC operation requires an additional flow tube at each end of the coil and, if desired, a sample feed tube at the middle portion of the coil. This dual CCC system has been successfully applied to liquid–liquid dual CCC (Lee et al., 1988) and foam CCC (Bhatnagar and Ito, 1988; Ito, 1985; Oka et al., 1989b, c).

The hydrodynamic motion of the two solvent phases in the present system was observed under stroboscopic illumination using a type J planetary centrifuge equipped with a spiral column.

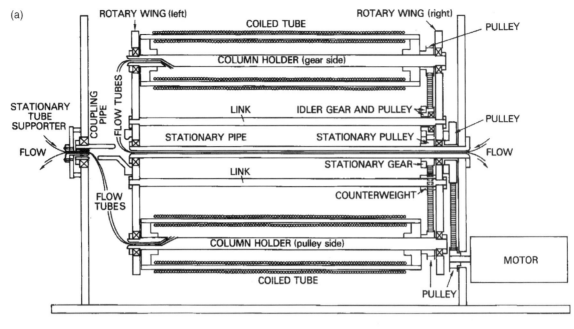

Figure 13 The improved horizontal flow-through coil planet centrifuge for both preparative and analytical separations. (a) Cross-sectional view through the central axis of the apparatus. [From Ito (1979).] (b) Photograph of the apparatus.

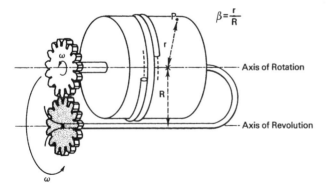

Figure 14 The column geometry of type J coil planet centrifuge for high-speed CCC. [From Ito et al. (1982).]

Figure 16 schematically illustrates the distribution of the two phases in the spiral column where about one-quarter of the area near the center of the centrifuge shows vigorous mixing of the two phases (mixing zone), whereas in the rest of the area the two phases are separated into two layers (settling zone). The bottom diagram shows the motion of the mixing zones through a stretched spiral column. As shown from the top to the bottom, the mixing zone travels through the spiral column at a rate of one round per revolution. This implies that the two solvent phases are subjected to a repetitive partition process of mixing and settling at a very high frequency. At 800 rpm this cyclic partition process is repeated 13× per second.

Countercurrent Chromatography

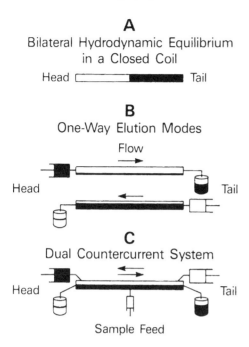

Figure 15 Mechanism of high-speed CCC. For simplicity, coils are drawn as straight tubes to illustrate overall distribution of the two phases. (a) Bilateral hydrodynamic distribution of two phases in a closed coil, (b) one-way elution modes, and (c) dual countercurrent mode. [From Ito (1996a).]

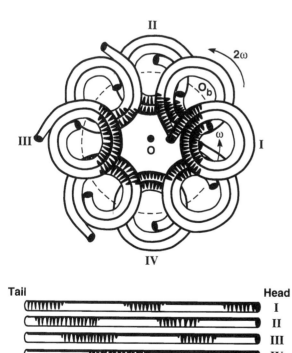

Figure 16 Hydrodynamic distribution of two phases in a spiral tube in type J synchronous planetary motion. [From Ito (1996a).]

Because of this high efficiency and rapid separation, the present system was named high-speed CCC.

Design of the Apparatus

Figure 17(a) shows a photograph of the original high-speed CCC centrifuge (Ito et al., 1982). The multilayer coil separation column was prepared by winding a piece of Teflon tubing (2.6 mm i.d.) making multiple layers of coils between a pair of flanges spaced 2 in. apart. A counterweight was placed on the other side of the rotary frame for balancing the centrifuge system. The mechanical design of the apparatus is illustrated in Fig. 17(b), which schematically shows the cross-sectional view across the central axis of the centrifuge. The motor drives the rotary frame through a pair of toothed pulleys coupled with a toothed belt. The rotary frame holds a column holder and a counterweight through ball bearings in symmetrical positions at a distance 10 cm from the central axis of the centrifuge. The type J planetary motion of the column holder is provided by coupling a pair of identical plastic gears: a stationary gear (dark) mounted around the central hollow shaft (dark), and a planetary gear on the right end of the holder shaft. The column consists of a single piece of Teflon tubing (2.6 mm i.d.) and about 90 m long directly wound around a holder hub (10 cm diameter) forming multiple coiled layers between a pair of flanges spaced 5 cm apart. The total column capacity is about 340 mL. The flow tubes from the coiled column are first passed through the hollow holder shaft and then enter the opening of the central stationary shaft through a hole on the central short support. These flow tubes are tightly supported on the centrifuge wall by a pair of clamps (not shown in the diagram). The β value of the coiled column ranges from 0.5 to 0.75, which produces suitable hydrodynamic behavior of many solvent systems for performing high-speed CCC. This design became the first commercial model of high-speed CCC which was marketed through P.C. Inc. Potomac, MD. Figure 17(c) shows a chromatogram of a DNP amino acid mixture obtained by this apparatus. Although the peak resolution is comparable to that of droplet CCC [Fig. 5(c)], the DNP proline was eluted in only 2.5 h compared with 70 h required by droplet CCC.

This original design of the high-speed CCC apparatus was improved later by eliminating the counterweight and arranging multiple column holders symmetrically around the rotary frame to balance the centrifuge system (Fig. 18) (Ito and Chou, 1988; Ito et al., 1989a, 1990a, b; Oka et al., 1989a). A set of multilayer coils mounted on each holder is interconnected to form a larger-capacity separation column. The interconnection flow tubes are passed through hollow tube supports which are counter-rotated by a 1:1 gear

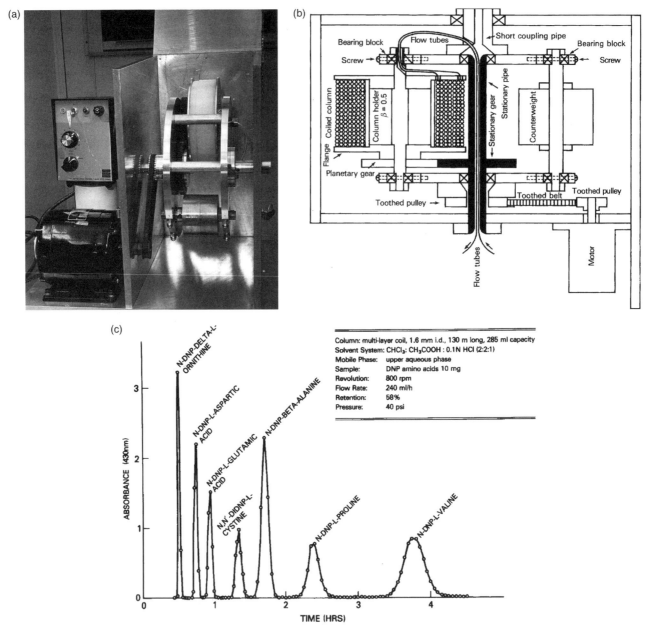

Figure 17 The original type J multilayer coil planet centrifuge for high-speed CCC. (a) Photograph of the apparatus, (b) cross-sectional view through the central axis of the apparatus, and (c) DNP amino acid separation. [From Ito et al. (1982).]

coupling to the neighboring column holders to untwist the flow tubes. The system is perfectly balanced after all columns establish hydrodynamic equilibrium. Figure 19(a) shows the photograph of an analytical high-speed CCC instrument equipped with a pair of multilayer coils connected in series (Ito and Chou, 1988), and Fig. 19(b) the photograph of a semipreparative multilayer coil high-speed CCC instrument equipped with a set of three multilayer coils connected in series providing a total capacity of about 300 mL (Ito et al., 1989a). Large preparative (2.6 mm i.d. and 1.6 L capacity) (Ito et al., 1991a) and analytical (1 mm i.d. and 180 mL capacity) system with three multilayer coils (Oka et al., 1989a) have also been fabricated at our machine shop at the National Institutes of Health, Bethesda, MD.

Currently, multilayer coil high-speed CCC instruments are commercially available from several sources (Berthod, 2002b), including Pharma-Tech Research Corporation, Baltimore, MD; Conway CentriChrom., Buffalo, New York; Renasas Eastern Japan Semiconductor Inc., Tokyo, Japan; AstraZeneca, Macclesfield, Cheshire, UK; and Beijing Institute of New Technology Application, Beijing, China, Tauto Biotech Ltd., Shenzhen, China.

Countercurrent Chromatography

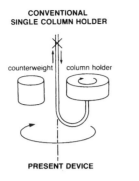

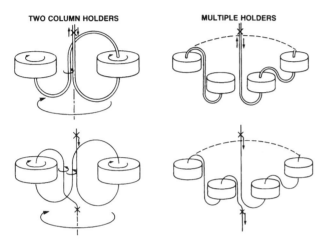

Figure 18 Improved design of the type J high-speed CCC centrifuge. [From Ito (1996a).]

3. High-Speed CCC Centrifuge with Spiral Disk Assembly

The high-speed CCC centrifuges equipped with a multilayer coil described earlier retain the stationary phase in the column using an Archimedean screw force which acts unevenly between the two solvent phases [i.e., one phase (usually the lighter phase) is pushed toward the head and the other phase pushed back toward the tail (Ito, 1992a)]. Although this system works effectively for most of the two-phase solvent systems, it fails to retain aqueous–aqueous polymer phase systems (Albertsson, 1971) which have an extremely low interfacial tension between the two phases. As the polymer phase systems are useful for separation of biopolymers without denaturation of their 3D molecular structure, efforts were made recently to improve the geometry of the separation column for type-J CPC. A new column design called the spiral disk assembly described subsequently allows universal application of solvent systems, including the polymer phase systems, in the type J high-speed CCC system (Ito et al., 2003).

Principle

The spiral column geometry improves the retention of the stationary phase by utilizing a radially acting centrifugal force gradient on the column in addition to an Archimedean Screw force, as the spiral channel helps to distribute the heavier phase in the periphery and the lighter phase in the proximal portion of the column. Naturally, the effect of this radial centrifugal force gradient becomes enhanced by increasing a pitch of the spiral. Although a spiral column can be prepared simply by winding the tubing into a flat spiral configuration, the spiral pitch thus formed is limited to the o.d. of the tubing, and in addition it is difficult to make a connection between the neighboring spiral columns (Shinomiya et al., 2002). These problems can be solved by forming a single or plural spiral channels in a solid disk (Ito et al., 2003).

Design of the Spiral Disk Assembly

In the first model, a single spiral channel is made in a plastic disk which can be serially connected by stacking multiple disks sandwiched with Teflon sheet septa. In the second model, four spiral channels are incorporated in each disk symmetrically around the center so that the spiral pitch is increased by 4× that of the above single spiral channel without losing the column space. Connection between the neighboring spiral channels is made by a short straight channel made radially on the other side of the disk. A set of these disks can be serially connected by spacing a Teflon sheet septum between the neighboring disks so that the column provides a large capacity which is somewhat comparable to that of the multilayer coil. The separation disk with a single spiral serves as a control to evaluate the effect of the pitch on the partition efficiency.

The design of the spiral column assembly is shown in Fig. 20(a) and its column components in Fig. 20(b). It consists of a pair of flanges (one equipped with a gear), nine disks with spiral grooves, and ten Teflon sheet septa. The drawing and dimensions of each component are illustrated in Fig. 21(a)–(e).

The design of the individual doughnut-shaped separation disk (high-density polyethylene) with one spiral groove (17.5 cm o.d. and 3 mm thick with a center hole of 1.25 cm diameter) is shown in Fig. 21(a). The dimensions of the channel (groove) is 3 mm wide, 2 mm deep, with a 1 mm ridge width which gives the spiral pitch of 4 mm. The total channel length is about 4 m, forming 12 spiral turns with a total capacity of ca. 24 mL. The channel starts at the inner inlet (23 mm from the center) and extends to the external outlet (75 mm from the center), which is connected to the radial channel on the other side through a hole of ca. 1 mm diameter. The performance of this single spiral column may be compared to that of the following four channel spiral column using a polar butanol solvent system and polymer phase systems.

Design of a similar separation disk with four separate spiral channels is shown in Fig. 21(b). A plastic disk

Figure 19 Photograph of the improved high-speed CCC apparatus eliminating the counterweight. (a) Analytical high-speed CCC apparatus with a pair of multilayer coils. [From Ito and Chou (1988).] (b) Semipreparative high-speed CCC apparatus with a set of three multilayer coils. [From Ito et al. (1990a).]

(17.5 cm diameter and 4 mm thick with a center hole of 1.25 mm diameter) has four spiral channels (grooves), each 3 mm wide, 2 mm deep, and ca. 1 m long with a capacity of about 6 mL. The ridge between each channel is 1 mm and therefore the pitch of each spiral becomes as large as 16 mm ($4\times$ that of the single spiral channel). Each channel starts at the inner end (I_1, I_2, I_3, and I_4) to reach the outer end (O_1, O_2, O_3, and O_4, respectively). As shown in the diagram, each channel forms 3.25 spiral turns so that the outer end of channel 1 (O_1) radially coincides with the inner end of channel 2 (I_2) and those of channels 2, 3, and 4 ($O_{2,3,4}$) similarly coincide with the inner ends of channel 3, 4, and 1 ($I_{3,4,1}$), respectively. Each end of the channel (except for O_1) has a hole (ca. 1 mm diameter) through the disk to reach the other side where a radial channel (dotted line) leads to the next spiral channel through another hole. The outer end of channel 4 (O_4) is connected to a similar radial channel which leads to the channel 1 (I_1) of the next disk or the column outlet through an opening of the Teflon septum [Fig. 21(c)].

All disks have screw holes (clearing an 8–32 screw) at both inner and outer edges at regular intervals (10° for the outer edge and 45° for the inner edge). Similar holes are also made in both Teflon sheets and the flanges.

The design of the flanges is shown in Fig. 21(d) and (e). They may be made of either stainless steel or aluminum. The upper flange [Fig. 21(d)] is equipped with a plastic gear (9 mm thick, 10 cm pitch diameter) which engages with an identical stationary gear on the high-speed CCC centrifuge. Figure 21(e) shows the lower flange, which has two screw holes (90° apart) for tightly fixing the column assembly against the column holder shaft (0.9 in. diameter). Each flange has an inlet hole which fits to an the hollow central pipe to exit the centrifuge at the center hole of the top plate where it is tightly fixed with a pair of clamps. These tubes are protected with a sheath of Tygon tubing to prevent direct contact with metal parts.

Preliminary studies on this first set of spiral disks revealed that the four spiral channels retain a satisfactory amount of polar low-interfacial tension solvent systems including polymer phase systems. The spiral disk has an advantage over the conventional multilayer coil in allowing modification of the channel geometry to improve the partition efficiency. Figure 22 shows a new spiral disk with four spiral channels having multiple pits (about 1460 each 2.8 mm in diameter and 2 mm deep) connected with narrow short channels (0.8 mm wide, 0.5 mm long, and 2 mm deep). This new disk is called the beads-chain spiral disk, which produced much higher partition efficiency than the original spiral disks.

4. Dual CCC (Type J)

As briefly mentioned earlier (Fig. 15), type J synchronous planetary motion, if combined with a coaxial coil orientation on the holder, permits genuine countercurrent movement of the two solvent phases through a long tubular space. In the past this dual CCC system has been successfully applied to foam separation (Bhatnagar and Ito, 1988; Ito, 1985; Oka et al., 1989b) as well as to liquid–liquid dual CCC (Lee et al., 1988b).

The Original Apparatus with Multilayer Coil

Figure 23(a) schematically illustrates the original column design for dual CCC (Ito, 1985). A coiled column, made of 2.6 mm i.d., 10 m long Teflon tubing with a total capacity of 60 mL, is equipped with three T-junctions, one at each end and one for the sample feed port at the middle portion of the column. In each terminal the feed line made of narrow bore tubing (0.45 mm i.d. Teflon tubing) is extended about 50 cm into the main column so that it prevents the input solvent phase from eluting back from the nearby outlet. In order to accurately regulate the liquid flow, a needle valve is placed on the liquid collection line at the head of the column.

The foam CCC centrifuge is shown in Fig. 23(b). The apparatus holds a pair of column holders on the rotary frame at a distance of 20 cm from the central axis of the centrifuge. The coiled separation column is coaxially mounted on the holder which undergoes type J synchronous planetary motion, while the other holder which gives type I synchronous planetary motion serves as a counterweight. The apparatus was rotated at 500 rpm. The method was applied to the separation of charged samples such as dyes and proteins using a suitable surfactant [i.e., sodium dodecylsulfate for positively charged samples and cetylpyridinium chloride for negatively charged ions (Bhatnagar and Ito, 1988)]. The method

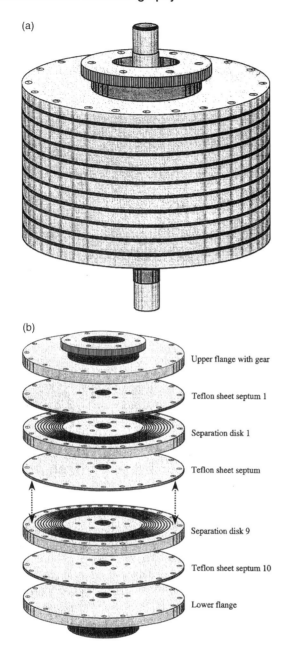

Figure 20 Design of spiral disk assembly for type J high-speed CCC apparatus. (a) Whole view of the spiral disk assembly and (b) Exploded view showing the elements.

adapter with a 1/4–28 screw thread. Direct contact of the solvent with metal parts is prevented by extending the end of the flow tube deep into the hole to reach the Teflon septum. Both upper and lower flanges also have a set of screw holes (clearing an 8–32 screw) around the outer and inner edges as in the separation disks and Teflon septa. The column assembly is mounted on the rotary frame of the multilayer coil centrifuge and counterbalanced with brass blocks. A pair of flow tubes from the column assembly is led through the center hole of the column holder shaft downward and passed through

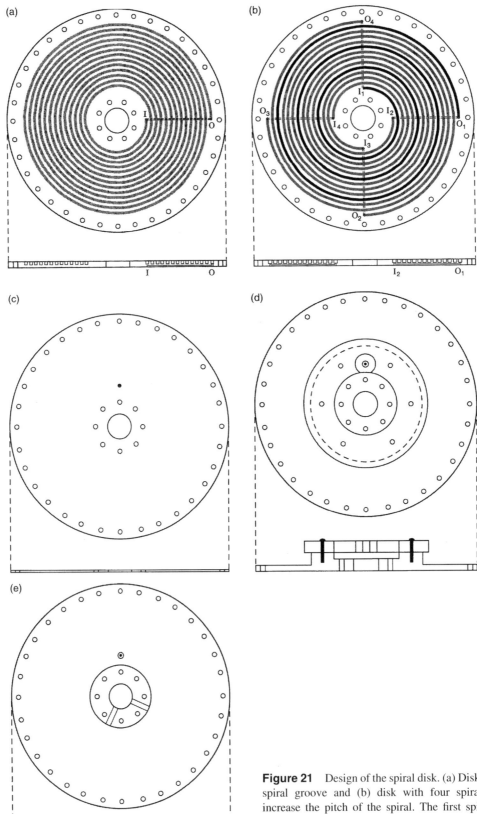

Figure 21 Design of the spiral disk. (a) Disk with a single spiral groove and (b) disk with four spiral grooves to increase the pitch of the spiral. The first spiral groove is highlighted to visualize the increased pitch. (c) Teflon septum, (d) upper flange equipped with a gear, and (e) lower flange. [From Ito et al. (In press).]

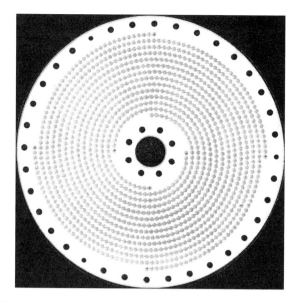

Figure 22 Photograph of the bead-chain spiral disk with four spiral channels having about 1460 pits, each 2.8 mm in diameter and 2 mm deep, connected with short channels of 0.5 mm long and 0.8 mm wide.

was further used for separation of foam producing samples such as bacitracin components using nitrogen gas and water free of surfactant (Oka et al., 1989b).

A similar apparatus has been used for liquid–liquid dual CCC (Lee et al., 1988b). A coiled column prepared from 1.6 mm i.d. Teflon tubing by winding it coaxially onto a holder with a total capacity of 400 mL. The column was rotated at 450 rpm while each phase was fed at a flow rate of 1.5 mL/min from each inlet. The method was successfully used for the preparative separation of steroid mixtures.

Analytical Dual CCC with a Spiral Disk

Recently, an analytical dual CCC model was developed (Ito, 2004a). The column is made of a disk (Kel-F) with a single spiral groove as shown in Fig. 24(a). As in the conventional multilayer column, the spiral channel has five openings, two for each terminal and one at the middle portion. The lighter phase is introduced through the external inlet (I_1) and collected through the internal outlet (O_1), whereas the heavier phase is introduced through the internal inlet (I_2) and collected through the external outlet (O_2). The sample solution is charged through the sample feed (S) by either batch injection or continuous loading. In most cases the column is rotated in such a way that the internal terminal becomes the head. In each opening, the flow tube (0.45 mm i.d.) is tightly connected by first extruding the stretched tubing through the hole into the groove side, cutting it flat with a sharp razor blade, and carefully pulling the tubing back so that the cut end just

rests at the bottom of the groove. In this way the flow tubes preserve the integrity of the shape of the spiral passage. Figure 24(b) shows a cross-sectional view of the analytical dual CCC centrifuge with a revolution radius of 7.5 cm. The spiral disk is placed channel side down onto the column support platform and tightly sealed against a Teflon sheet with a number of screws. The five flow tubes are first passed through the center hole of the column holder shaft whereupon they enter the opening of the central shaft through a side-hole of the central rotary shaft. These tubes are tightly supported on the centrifuge cover with a pair of clamps. This analytical dual CCC instrument is designed for direct interfacing with a mass spectrometer for food analysis.

Retention Time of Solute in Dual CCC

Figure 25 shows a portion of the stretched spiral channel of the dual CCC column. The solute charged through the sample feed (S) is distributed between the two phases according to its partition coefficient, $K = c_U/c_L$, while the two phases move in the opposite direction [i.e., the upper phase moving toward the head (I_1) at a flow rate of v_U (mL/min) and the lower phase toward the tail (I_2) at v_L (mL/min) as indicated in the diagram]. We consider a small portion of the channel (lower diagram) where the cross-sectional area occupied by the upper phase is a_U and that by the lower phase, a_L and the thickness of the sliced channel, Δl. Then, the linear flow rate (cm/min) of the solute in the upper phase is expressed by $(v_U/a_U) c_U a_U \Delta l / (c_U a_U \Delta l + c_L a_L \Delta l)$ or $K v_U / (K a_U + a_L)$, and that in the lower phase, $(v_L/a_L) c_L a_L \Delta l / (c_U a_U \Delta l + c_L a_L \Delta l)$ or $v_L / (K a_U + a_L)$, where $K = c_U/c_L$. The net effect is that the solute moves at a rate and in a direction determined by the difference between these two traveling speeds of the solute. The following three cases are considered:

1. When $K v_U < v_L$, the solute moves through the column space occupied by the lower phase from the sample inlet toward the tail (I_2) at a rate of $v_L/(K a_U + a_L) - v_U K/(K a_U + a_L)$, and reaches the tail at a retention time ($R_{\text{time-L}}$):

$$R_{\text{time-L}} = \frac{l_L}{[v_L/(K a_U + a_L) - K v_U/(K a_U + a_L)]} \quad (23)$$

$$= \frac{V_L + K V_U}{v_L - K v_U} \quad (24)$$

where l_L is a length of the channel from sample port S to I_2 (see the upper diagram), and V_U and V_L are the volumes occupied by the upper and the lower phases in the channel, respectively.

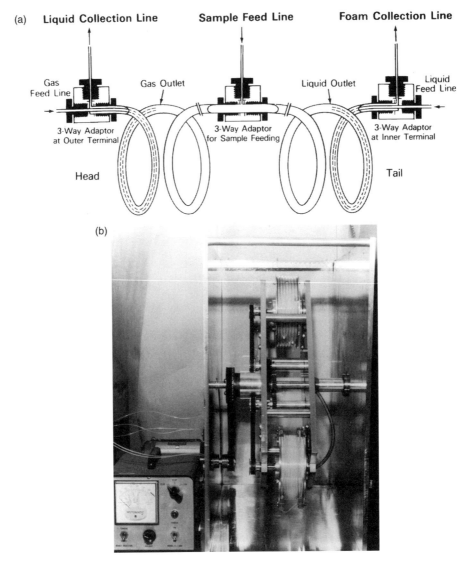

Figure 23 Foam CCC apparatus. (a) Column design. [From Ito (1985).] (b) Photograph of the original apparatus.

2. When $Kv_U = v_L$, $R_{\text{time-L}} \to \infty$, indicating that the solute makes no movement and will be permanently retained in the column, regardless of the volume ratio between the two phases in the channel.

3. When $Kv_U > v_L$, the solute moves toward the head at a rate (cm/min) of $v_L/(Ka_U + a_L) - Kv_U/(Ka_U + a_L)$ through the channel space occupied by the upper phase (V_U) between S and I_1 (see the upper diagram), and reaches the head (I_1) at a retention time ($R_{\text{time-U}}$):

$$R_{\text{time-U}} = \frac{l_U}{[Kv_U/(Ka_U + a_L) - v_L/(Ka_U + a_L)]} \quad (25)$$

$$= \frac{KV_U + V_L}{Kv_U - v_L} \quad (26)$$

The above analysis indicates that the retention time (volume) of the solute can be predicted from the partition coefficient (K) if the volumes occupied by each phase (V_U and V_L) and the flow rates (mL/min) of the two phases are known.

5. Cross-Axis CPC (Types X, L, and Hybrid)

As mentioned earlier, this CCC scheme produces unique planetary motion in which the axis of the holder is perpendicular to the central axis of the centrifuge [see Fig. 8(b)]. As described earlier, the centrifugal force produced by the planetary motion displays 3D fluctuation of the force vectors, one of which acts across the diameter of the tubing to stabilize emulsification of the two phases. Consequently, the system is capable of retaining a large quantity

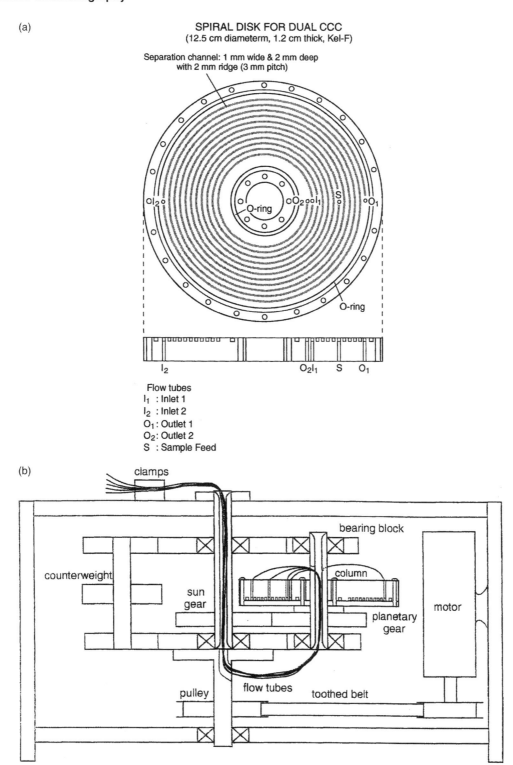

Figure 24 Analytical dual CCC apparatus. (a) Spiral column design and (b) cross-sectional view through the center of the apparatus.

of polar solvent systems including aqueous–aqueous polymer phase systems useful for separation of biopolymers. The present system also produces a bilateral hydrodynamic equilibrium of the two solvent phases, which is characteristic of high-speed CCC.

Design Principle

Figure 26(a) explains designs for various prototypes of the cross-axis CPC using three parameters (i.e., r is the radius of the column holder; R, the distance between

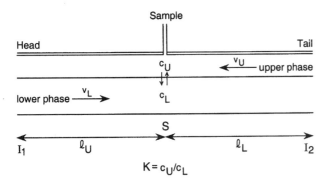

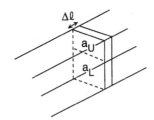

Figure 25 Portion of spiral channel for analysis.

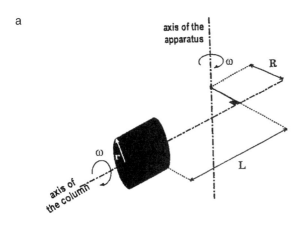

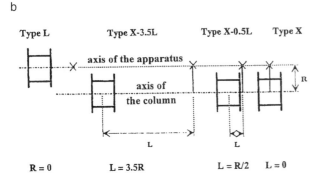

Figure 26 Planetary motion and column position of cross-axis CPC. [From Ito and Menet (1999).]

the two axis; and L, the measure of lateral shift of the column holder along the axis). The nomenclature of various types of the cross-axis CPC is based on the ratio L/R, where $R \neq 0$. Types X and L [also see Fig. 8(b)] represent the limit for the column positions: the first type involves no shift ($L = 0$) of the column holder, whereas the second type corresponds to an infinite shifting ($R = 0$). Figure 26(b) shows some of the examples. Table 2 lists the cross-axis CPCs built at the National Institutes of Health, Bethesda, MD.

Instrumentation

1. X–1.25L CPC. Figure 27 schematically shows the design of X–1.25L cross-axis CPC (Bhatnagar et al., 1989; Ito et al., 1989b) by the side view [Fig. 27(a)] and the cross-sectional view [Fig. 27(b)] through the central axis of the coil holders in a horizontal plane. The motor (1) [Fig. 27(a) bottom] drives the central shaft (2) and the rotary frame around the central axis of the centrifuge. (In the actual design, the motor drives the rotor via a pair of toothed pulleys with a toothed belt.) The rotary frame consists of four side-plates, a pair of inner plates (3) and a pair of outer plates (4), all rigidly bridged together with multiple aluminum links (not shown). A pair of horizontal plates, the upper (5) and the lower (6) plates, connect the pair of inner side-plates to the central shaft. The inner and the outer side-plates horizontally hold a coil holder shaft (7) symmetrically on each side of the rotary frame at a distance of 10 cm from the central axis of the centrifuge. A pair of identical coil holders (8) is mounted around the holder shafts symmetrically, one on each side of the rotary frame, between the inner and the outer side-plates at a location about 12.5 cm from the center of the holder shaft as shown in Fig. 27(b).

The planetary motion of each coil holder is established by the use of a set of miter gears and toothed pulleys as follows. A stationary miter gear (45°) (9) is rigidly mounted on the bottom plate of the centrifuge coaxially around the central shaft (2). This stationary sun gear is coupled to a pair of identical planetary gears (10) attached to the proximal end of a countershaft (11) which radially extends toward the periphery of the rotary frame through the lower portion of the inner side-plate. This gear arrangement produces synchronous rotation of each countershaft on the rotating rotary frame. This motion is further conveyed to each coil holder by coupling a toothed pulley (12) mounted at the peripheral end of the countershaft to the identical pulley (13) on the respective coil holder shaft with a toothed belt (14). Consequently, each coil holder undergoes the desired planetary motion (i.e., revolution around the central axis of the centrifuge and rotation about its own axis at the same angular velocity).

Table 2 Characteristics of Cross-Axis Coil Planet Centrifuges

Year	Reference	Column position	L (cm)	R (cm)	β^a	Column volume (mL)
1987	Ito (1987a)	X	0	10	0.25	15
					0.50	15
					0.75	15
					0.5–0.8	400
					0.19–0.9	25
1988	Ito and Zhang (1988a)	X or X–0.5L	0 or 10	20	0.125	15 or 28
					0.375	15 or 28
					0.625	15 or 28
1988	Ito and Zhang (1988b)				0.375–0.625	800
1989	Ito et al. (1989b)	X–1.25L	12.5	10	0.5	28
					0.75–1.15	750
1991	Ito et al. (1991b)	X–LL	15.2	7.6	0.5	20–30
					1	20–30
					1.6	20–30
					0.25–0.60	280
					0.50–1.00	250
					1.00–1.20	450
1991	Shibusawa and Ito (1992)	X–3.5L	13.5	3.8	0.50–1.30	150
					0.44–1.50	220
1992	Shinomiya et al. (1993)	X–1.5L or L	16.85	10.4 or 0	0.26 or 0.16	20
					0.48 or 0.30	41
					0.26–0.48 or 0.16–0.30	287
					0.10–0.06	18
					0.02–0.01	17

Source: From Ito and Menet (1999).
[a]Calculation of β is as defined in the text ($\beta = r/R$) except for the L position: $\beta = r/L$.

In the actual design of the prototype, each coil holder is made from three portions, that is, a main holder shaft supporting the coil holder and the toothed pulley, an auxiliary shaft placed on the opposite side, and a central pipe connecting these two auxiliary shaft placed on the opposite side, and a central pipe connecting these two shafts where the positions of the main holder shaft and the auxiliary shaft are mutually exchangeable on each side of the rotary frame. Each shaft is removed from the rotary frame by loosening the screws on each side bearing block. Three pairs of spool-shaped coil holders were fabricated, each with hub diameter of 10, 15, or 20 cm and equipped with a pair of flanges 24 cm in diameter spaced 5 cm apart. The coiled column was prepared by winding a piece of Teflon tubing directly around the holder hub, making multiple layers of coil. Each pair of coiled columns on the rotary frame is connected in series to double the column capacity. The layout of the flow tubes (0.85 mm i.d., Teflon) is illustrated in Fig. 27(a) and (b). A pair of flow tube (17a) from the second column [Fig. 27(b), right] first passed through the center hole of the holder shaft and by making a loop (17b), enters the opening of the other holder shaft to reach the first column holder [Fig. 27(b), left], where one flow tube (interconnecting tube) joins one end of the first column. The latter flow tube from the second column exits the holder shaft (17c) and then enter the side-hole (18) [Fig. 27(b)] of the central shaft to reach the stationary tube-guide projecting down from the top plate of the centrifuge. These flow tubes are lubricated with grease and protected with a sheath of Tygon tubing to prevent direct contact with metal parts. Under ordinary conditions, the tubes can maintain their integrity for many months of operation. As in the type X cross-axis CPC [Fig. 8(b)], these tubes are free from twisting and, therefore, serve for the continuous elution through the rotating coils without complications such as leakage and contamination.

As may be visualized from Fig. 27(b), the present design allows mounting another pair of column holders symmetrically on the rotary shafts by either reducing the diameter of the column holders or increasing the revolution radius of the apparatus.

2. Versatile cross-axis CPC. This second design allows a choice of two column positions, X–1.5L (off-center position) or L (central position), simply by switching the column holder unit and the tube support unit

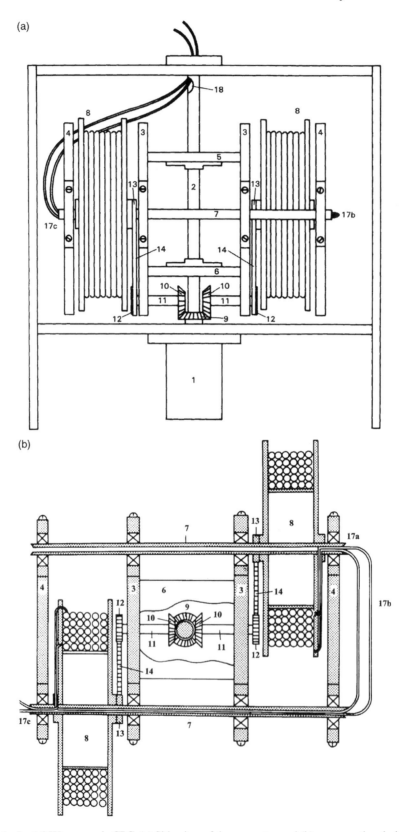

Figure 27 Design of the L–1.25X cross-axis CPC. (a) Side view of the apparatus and (b) cross-sectional view across the center of the column holder. [From Ito et al. (1989b).]

Countercurrent Chromatography

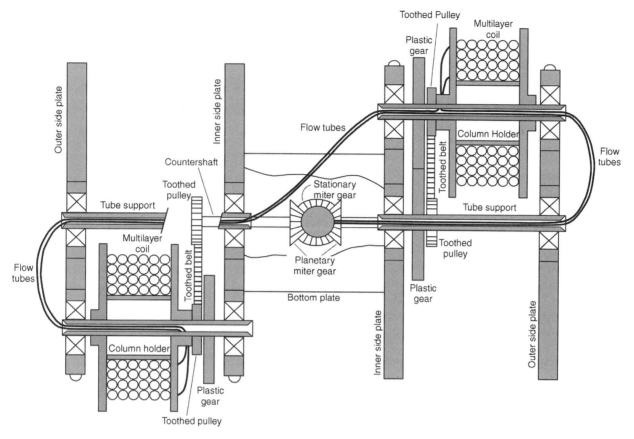

Figure 28 Design of the L-1.5X cross-axis CPC with different flow tube passage. [From Shinomiya et al. (1993).]

(Menet and Ito, 1993; Shinomiya et al., 1993). Figure 28 schematically shows a cross-sectional view across the central axis of the coil holders through a horizontal plane. The motor drives the central shaft and the rotary frame around the central axis of the centrifuge. The rotary frame consists of two pairs of side-plates. A pair of inner side-plates is bridged by a pair of horizontal plates at the upper and the lower edges, rigidly supporting a pair of outer plates with a set of links. Column holders and counter-rotating tube holders are mounted between the inner and outer side-plates in two different positions on each side of the rotary frame [i.e., the off-center position (X−1.5L) (as shown in the diagram) and the central position]. The planetary motion of the column holder is provided by a set of miter gears (45°) and the countershaft where a stationary miter gear (center) is rigidly mounted coaxially around the central shaft at the bottom plate of the centrifuge. This stationary gear is engaged to an identical miter gear affixed at the proximal end of the countershaft radially mounted on each side of the rotary frame. The preceding gearing produces synchronous rotation of the countershaft on the rotating rotary frame. This motion is further conveyed to each column holder by coupling a pair of identical toothed pulleys, one on the distal end of the countershaft and the other on the column holder shaft using a toothed belt. In order to prevent the flow tubes from being twisted, a pair of counter-rotating tube holders is placed one on each side of the rotary frame. The plastic gear (10 cm in pitch diameter) mounted on each tube holder is engaged to an identical gear affixed on the neighboring column holder so that the tube holder synchronously rotates with the column holder in the opposite direction. The positions of the column holder and the tube holder are easily interchanged by loosening the screws on each bearing block from the rotary frame. When the column holder is mounted at the off-center position (as shown in Fig. 28) the tube holder is placed at the central position and vice versa. The layout of the flow tubes connects the pair of separation columns in series and permits continuous elution through the running columns without the use of a rotary seal device. As in other CPCs, the flow tube is lubricated with grease and individually protected with a short sheath of Tygon tubing at each projecting hole to prevent direct contact with metal parts.

The apparatus measures 60 cm wide, 60 cm long, and 34 cm in height. The speed of the apparatus is regulated up to 1000 rpm by a speed controller (Bodine

Electric, Chicago, IL). In the earlier model, a set of metal gears produced considerable noise and vibration. This problem has been largely eliminated by replacing the metal miter gears with plastic gears and restricting the up-and-down movement of the central shaft with shims. Consequently, the noise level of the apparatus became acceptable, provided that the speed is away from the resonance points ranging between 650 and 700 rpm.

Each multilayer coil separation column for preparative separations was prepared by winding a long piece of Teflon tubing (2.6 mm i.d.) coaxially around the holder hub forming a tightly packed coil between the pair of flanges spaced 7.6 cm apart. After completing each coiled layer, the whole layer was wrapped with a piece of adhesive tape and the tubing was directly returned to the original side to start the next layer by winding the tube over the interconnection tube. In order to prevent excessive distortion of the coiled column, interconnection tubes were evenly distributed around the holder with minimum overlapping. Two such columns were serially connected to provide a total capacity of 575 mL.

The analytical column unit was made by winding 0.85 mm i.d. Teflon tubing onto a 7.6 cm long, 5 mm i.d. nylon pipe making a tight left-handed coil. In each column assembly a set of coil units was symmetrically arranged around the holder in parallel to and at the same distance of 4 cm from the holder axis. A pair of the coil assemblies is connected in series to provide a total capacity of 34 mL.

Elution Modes

The partition efficiency and the retention of the stationary phase obtained by the off-center cross-axis CPC are determined by the combination of three elution modes [i.e., head to tail or tail to head elution (as type J CPC), inward or outward elution along the axis of the column holder] and direction of the planetary motion. Although these combined effects are highly complex (Shinomiya et al., 1993), satisfactory results are usually obtained by eluting the lower phase outwardly from the head toward the tail and the upper phase inwardly from the tail toward the head.

The apparatus has been successfully applied to separation of various natural and recombinant proteins using polymer phase systems (Shibusawa, 1996).

6. Nonsynchronous CPC (Nonsynchronous Type I)

We consider the nonsynchronous flow-through CPC is the most versatile scheme in that the rotation and revolution of the coiled column are independently adjustable. Consequently, the separation can be performed in a slowly rotating coil under a strong centrifugal force field. Three different prototypes have been designed: the first model (Ito et al., 1979) with a rotating seal, the second (Ito et al., 1980) and the third (Ito et al., 1983) without rotary seals.

Design of the Apparatus

Figure 29(a) illustrates the design of the most advanced prototype of nonsynchronous flow-through CPC without a rotary seal device, showing a cross-sectional view across the central axis of the apparatus (Ito et al., 1983). The rotor consists of two major rotary structures [i.e., (rotary) frames I and II which are coaxially bridged together with the center piece (dark shade)].

Frame I consists of three plates rigidly linked together and directly driven by motor I. It holds three rotary elements: the centerpiece (center), countershaft I (bottom), and countershaft II (top), all embedded in ball bearings. A pair of long arms extending symmetrically and perpendicularly from the middle plate forms the tube supporting frame which clears over frame II to reach the central shaft on the right side of frame II.

Frame II (light shade) consists of three pairs of arms linked together to rotate around the central shaft. It supports a pair of rotary shafts, one holding a coil holder assembly and the other the counterweight.

There are two motors, motors I and II, to drive the rotor. When motor I drives frame I, the stationary pulley 1 introduces counter-rotation of pulley 2 through a toothed belt and therefore countershaft I rotates at $-\omega_I$ with respect to rotating frame I. This motion is further conveyed to the centerpiece by 1:1 gearing between gears 1 and 2. Thus the centerpiece rotates at $2\omega_I$ or at ω_I with respect to the rotating frame I. The motion of frame II, however, also depends upon the motion of motor II.

If motor II is at rest, pulley 5 becomes stationary as does pulley 1 so that countershaft II counter-rotates at ω_I as does countershaft I. The motion is similarly conveyed to the rotary arms of frame II by 1:1 gearing between gears 3 and 4, resulting in rotation of frame II at $2\omega_I$, or the same angular velocity as that of the centerpiece. Consequently, coupling of pulleys 6 and 7 as well as 8 and 9 with toothed belts produce no additional motion to the rotary shaft which simply revolves with frame II at $2\omega_I$ around the central axis of the apparatus.

When motor II rotates at ω_{II}, idler pulley 4 coupled to pulley 3 on the motor shaft rotates at the same rate. This in turn modifies the rotational rate of pulley 5 on countershaft II. Thus countershaft II now counter-rotates at $\omega_I - \omega_{II}$ on frame I. This motion further alters the rotational rate of frame II through 1:1 gear coupling between gears 3 and 4. Subsequently, frame II rotates at $2\omega_I - \omega_{II}$ with respect to the outside observer or at $-\omega_{II}$ relative to the centerpiece which always rotates at $2\omega_I$.

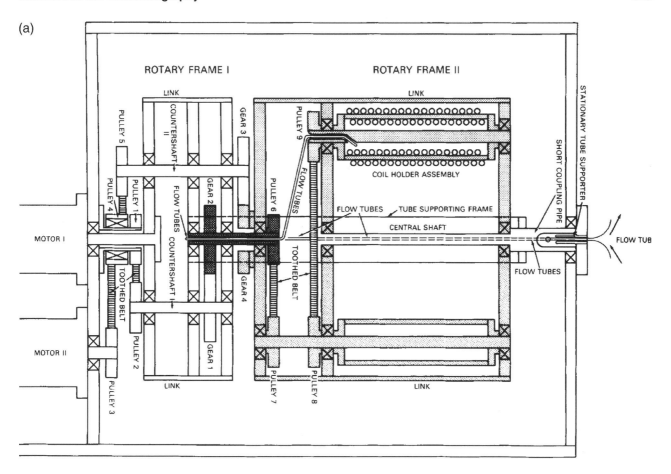

Figure 29 Advanced design of the nonsynchronous flow-through coil planet centrifuge without rotary seals. (a) Cross-sectional view through the central axis of the apparatus and (b) photograph of the apparatus. [From Ito et al. (1991a).]

The difference in rotational rate between frame II and the center piece is conveyed to the rotary shaft through coupling of pulleys 6 to 7 and 8 to 9. Consequently, both rotary shafts rotate at ω_{II} about their own axis while revolving around the central axis of the apparatus at $2\omega_I - \omega_{II}$. This gives the rotation/revolution ratio of the rotary shaft

$$\frac{r}{R} = \frac{\omega_{II}}{2\omega_I - \omega_{II}} \qquad (27)$$

Therefore, any combination between revolution and rotation speeds of the coil holder assembly can be achieved by selecting the proper values for ω_I and ω_{II}.

The coiled column was prepared by winding 1 mm i.d. Teflon tubing continuously onto six units of 0.68 cm o.d. stainless steel pipe to make about 600 helical turns with a total capacity of 15 mL. The column units were symmetrically mounted around the rotary shaft with screws through the hole at each end of the units. The flow tubes connected to the coiled column are first led through the hole of the rotary shaft and then passed through the opening of the centerpiece to exit at the middle portion of frame I. The flow tubes are then led along the tube support frame to clear frame II and then reach the side-hole of the central shaft near the right wall of the centrifuge where they are tightly held by the stationary tube supporter. These tubes are lubricated with grease and protected with a piece of flexible plastic tubing at each supported portion to prevent direct contact with the metal parts.

Figure 29(b) shows the overall picture of the apparatus. The revolution speed of the coil holder assembly is continuously adjustable up to 1000 rpm combined with any given rotational rate between 0 and 50 rpm in either direction.

Applications

Because of the freely adjustable ratio between the rotation and revolution, the method provides satisfactory retention for any kind of two-phase solvent systems including aqueous–aqueous polymer phases systems used for separation of macromolecules (Ito et al., 1980) and cells (Ito et al., 1980; Leive et al., 1984). The method may also be applied to the separation of cells by elutriation using physiological solutions (Ito et al., 1979, 1980; Okada et al., 1996).

7. *Slow Rotary CCC*

Principle

This is essentially the same as the basic hydrodynamic equilibrium system (Fig. 1, right) where a coiled column is rotated around its horizontal axis. Studies on two-phase distribution in the rotating coil under unit gravity revealed that the volume ratio of the two phases alters according to the rotation speed (Ito, 1988b,1992a; Ito and Bhatnagar, 1984). Figure 30 shows the relationship between the rotation speed and the two phase distribution at the head of the end-closed rotating coil with three different helical diameters of 3, 10, and 20 cm as labeled in the diagram.

All three diagrams revealed a common feature of the phase distribution pattern: at the slow rotational speed of 0–30 cm, the two solvent phases are evenly distributed at the head of the coil (stage I). As the rotational speed is increased, the heavier phase starts to occupy more space on the head side of the coil and, at the critical rotational speed between 60 and 100 rpm, the two phases establish bilateral hydrodynamic equilibrium where the heavier phase entirely occupies the head side and the lighter phase, the tail side of the coil (stage II). After this critical range of rpm, the amount of the heavier phase on the head side decreases rapidly, crossing below the 50% line (stage III). Increasing the rpm again yields an even distribution of the two phases in the coil (stage IV). The bilateral phase distribution in stage II is effectively used for CCC if the heavier phase is moved from the tail to head or the lighter phase from the head to tail (Fig. 15) (Du et al., 2000; Ito and Bhatnagar, 1984).

Instrumentation

Figure 31(a) shows a seal-free flow-through rotary device. It holds a cylindrical column holder which is rotated around its own axis at a desired rate ranging from 0 to 100 rpm. A set of gears and pulleys provides a rotation ratio of 2:1 between the column holder and the

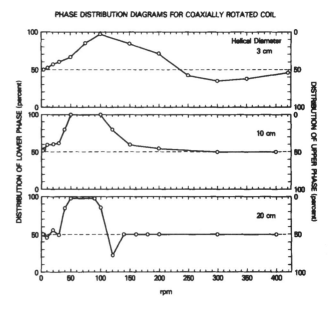

Figure 30 Distribution of two immiscible solvent in a slowly rotating coiled tube. [From Ito and Bhatnagar (1984).]

Countercurrent Chromatography

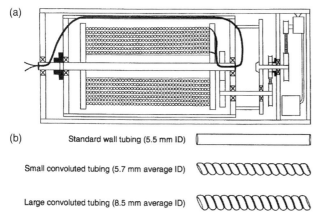

Figure 31 Design of low-speed rotary CCC apparatus. (a) cross-sectional view through the central axis of the apparatus and (b) three different types of Teflon tubing for preparation of the multilayer coil separation column. [From Du et al. (2000).]

rotary frame, which allows the flow tubes to rotate without twisting (Du et al., 2000). Consequently the system permits continuous elution through the rotating column without a conventional rotary seal device that may cause leakage and contamination. Teflon tubing shown in Fig. 31(b) is directly wound around the holder hub forming a multilayer coil separation column. Owing to slow rotation of the column holder, the system can be automated for performing a large-scale separation, which may take several days.

Example of Applications

Preliminary studies showed that the convoluted tubing [Fig. 31(b)] yields better performance in terms of the retention of the stationary phase and partition efficiency than the standard tubing. Figure 32 shows the preparative separation of 150 g of crude tea extract yielding 40 g of epigallocatechin gallate of over 92% purity at a recovery rate of 82.6% (Du et al., 2000).

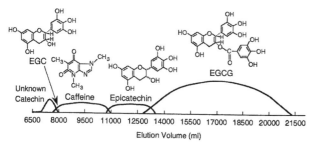

Figure 32 Large-scale preparative separation of tea extract with the low-speed rotary CCC apparatus. [From Du et al. (2000).]

IV. SPECIAL TECHNIQUES

A. pH-Zone-Refining CCC

This powerful preparative CCC technique separates ionized compounds into a train of highly concentrated rectangular peaks with minimum overlapping (Ito, 2000b, 2001c; Ito and Ma, 1994, 1996; Ito et al., 1995).

1. Mechanism

pH-Zone-refining CCC stemmed from an incidental observation that a thyroxine analog produced an unusually sharp elution peak (Ito et al., 1992). The cause was found to be the presence of bromoacetic acid in the sample solution which affected the partition behavior of the solute by protonating the molecule.

The mechanism of this sharp peak formation is illustrated in Fig. 33(a) which shows a portion of the column containing the stationary phase in the upper half and the mobile phase in the lower half. The acid present in the sample solution and/or stationary phase forms a sharp trailing border which moves at a slower rate than the mobile phase. The acidic molecule present at location 1 is protonated and will rapidly distribute to location 2 in the stationary phase. As the sharp trailing border of the acid moves forward, the solute finds itself in location 3 where the stationary phase is in contact with the mobile phase of higher pH. Owing to deprotonation of the molecule, the solute returns back into the mobile phase at location 4, whereupon it rapidly migrates into location 1 to repeat the process. Consequently, the solute elutes with the sharp acid border, forming a sharp elution peak as shown in Fig. 33(b).

When the amount of solute is increased, it is accumulated behind the acid border to form a solute zone at a constant pH (pH-zone) with a trailing border as sharp that of the front border. This second sharp border forces the second solute, with a higher tendency for protonation, to form a second pH-zone, and so forth. Consequently, the solutes are eluted as a train of rectangular peaks with minimum overlap (Ito, 1996b; Ito and Ma, 1994, 1996; Ito et al., 1995; Scher and Ito, 1995; Weisz et al., 1994a). The mechanism of this pH-zone-refining CCC is schematically illustrated in Fig. 34(a) (Ito and Ma, 1996; Ito et al., 1989b) where a portion of the separation column contains the stationary phase containing retainer acid (TFA) in the upper half and the mobile phase containing eluter base (NH_3) in the lower half. The retainer acid, TFA, forms a sharp trailing border which moves at a rate lower than that of the mobile phase. Three acidic solutes—$S_1(R_1COOH)$, $S_2(R_2COOH)$ and $S_3(R_3COOH)$—form their own pH-zones competitively behind the sharp TFA border. Solute S_1, which has the lowest pK_a and hydrophobicity, forms the first pH-zone by protonating other

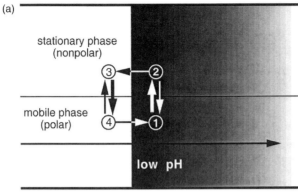

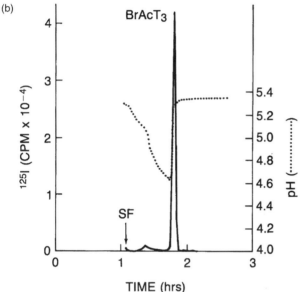

Figure 33 Principle of formation of abnormally sharp peak. (a) Mechanism of sharp peak formation and (b) chromatogram of sharp peak eluted with a sharp acid border. [From Ito et al. (1992).]

components, forcing them into the stationary phase, and delaying their movement while forming a sharp trailing border for the next zone in the same manner as the TFA retainer acid. Competition for the available protons now continues between the other two solutes, and solute S_2 with a lower pK_a and hydrophobicity drives out solute S_3 to establish the second zone. Finally, solute S_3 with the highest pK_a and hydrophobicity occupies the end of the zone train to form its sharp trailing border. As indicated by curved arrows, proton transfer takes place at each zone boundary from the solute in the preceding zone (stationary phase) to that in the following zone (mobile phase), as governed by the difference in pH between these two neighboring zones, while the NH_4^+ counter ions continuously move with the mobile phase. After equilibrium is reached within the column, all solute zones move at the same rate as that of the trailing TFA border

while constantly maintaining their own width and pH levels. The nearly equal height of the flat-topped peaks in Fig. 34(b) is merely a reflection of the nearly equal molar absorptivities and concentrations of the solutes in the effluent.

Although the method bears a close similarity to displacement chromatography such as formation of a train of rectangular peaks, there are some distinct differences between these two methods. Comparison between pH-zone-refining CCC and displacement chromatography is summarized in Table 3 (Ito and Ma, 1996).

2. Selection of Solvent Systems

pH-Zone-refining CCC uses a set of reagents (i.e., a retainer in the stationary phase and eluter in the mobile phase). For separation of acidic solutes, TFA is most often used as the retainer acid and NH_3 as the eluter base. For basic samples, triethylamine is used as the retainer base and HCl as the eluter acid. A two-phase solvent system composed of methyl *tert*-butyl ether/water is effective in many cases: optimization of the solvent system may be started with this binary solvent system. The following method (Ito and Ma, 1996) for an acidic sample represents the solvent condition shown in Fig. 33(a), that is, (1) the left side (basic condition) and (2) right side (acidic condition) of the sharp acid border.

1. Deliver 2 mL each of methyl *tert*-butyl ether and water into a test tube.
2. Dissolve the sample in the solvent.
3. Add NH_3 to bring pH above 9 (basic condition).
4. Equilibrate the contents and measure partition coefficient (K_{base}).
5. If $K_{base} \ll 1$, add TFA so that pH falls below 2 (acidic condition).
6. Equilibrate the contents and measure K_{acid}.
7. If $K_{acid} \gg 1$, the solvent system should be successful.
8. In procedure 5, if K_{base} is not small enough, repeat the earlier procedure using a more hydrophobic solvent system such as hexane/ethyl acetate/methanol/water (1:1:1:1, v/v/v/v).
9. In procedure 7, if K_{acid} is not large enough, repeat the procedure using a more hydrophilic solvent system such as methyl *tert*-butyl ether/acetonitrile/water (4:1:5 or 2:2:3, v/v/v).

For the separation of basic compounds use triethylamine in place of TFA and HCl for NH_3.

The method requires an additional reagent or ligand for particular compounds including catecholamines (Ma et al., 1996), free peptides (Ma and Ito, 1997a), and chiral compounds (Ma and Ito, 1995a, 1996; Ma et al., 1995). For polar catecholamines and zwitter-ions such as free peptides, a ligand, di-(2-ethylhexyl)phosphoric acid

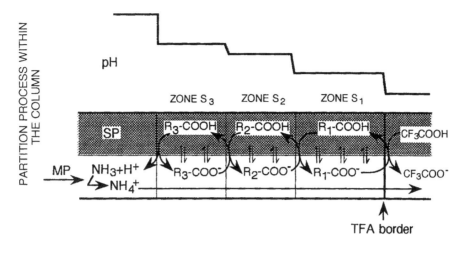

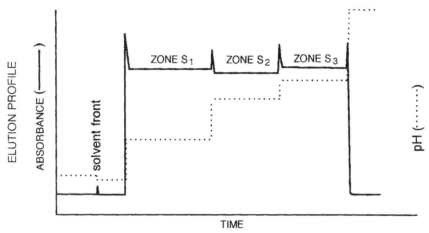

Figure 34 Mechanism of pH-zone-refining CCC. [From Ito (1996b).]

(DEHPA), is added to the stationary phase. For separations of enantiomers, a chiral selector such as N-dodecanoyl-L-proline-3,5-dimethylanilide (DPA) is used in the stationary phase.

The relationship between the zone pH (pH$_{zone}$) and pK_a/hydrophobicity in this method is given by the following equation (Ito and Ma, 1996):

$$\text{pH}_{zone} = pK_a + \log\left\{\frac{K_D}{K} - 1\right\} \quad (28)$$

where K_D and K indicate the distribution coefficient (indicator for hydrophobicity) and distribution ratio (partition coefficient) of the analytes. When the pK_a and K_D of the analyte is known, the zone pH can be computed from the K value. According to a theoretical analysis (Scher and Ito, 1995), the mixture of acids with $\Delta p(K_a/K_D) = 1$ is separated with recoveries of 99% of each acids with over 99% purity.

3. Advantages Over the Standard CCC Method

The pH-zone-refining CCC technique has several advantages over the standard CCC method as itemized below:

1. Sample loading capacity is increased over 10× for a given column.
2. Fractions are highly concentrated.
3. Increase of sample size produces a higher yield of pure fractions.
4. Minor components are concentrated and detected at the boundaries of major peaks.
5. Samples with no chromophore can be effectively monitored by pH.

4. Applications

Figure 35 shows an example of a dye separation by pH-zone-refining CCC (Weisz et al., 1994a). Three chromatograms are obtained from D&C Yellow 5 in different sample sizes of 350, 1, and 5 g as indicated in the

Table 3 Comparison Between pH-Zone-Refining CCC and Displacement Chromatography

	pH-Zone-refining CCC		Displacement chromatography
	Reverse-displacement mode	Normal-displacement mode	
Key reagents	Retainer	Eluter	Displacer
Solute transfer	MP[a] → SP[b]	SP → MP	SP → MP
Acting location	Front of solute bands	Back of solute bands	Back of solute bands
Solute bands			
Form	A train of individual solute bands with minimum overlapping.		
Travelling rate	All move together at the same rate as that of the key reagent.		
K^c value in	Same as and determined by that of the key reagent.		
Impurities	Concentrated at the boundaries of the solute bands.		
Peak profile	A train of rectangular peaks associated with sharp impurity peaks at their boundaries.		
Solute concentration in mobile phase is determined by	Concentration of counterions in aqueous phase	Concentration of counterions in aqueous phase and K value	Solute affinity to stationary phase
Elution order is determined by	Solute pK_a and hydrophobicity	Solute pK_a and hydrophobicity	Solute affinity to stationary phase

Source: From Ito and Ma (1996).
[a]MP, mobile phase.
[b]SP, stationary phase.
[c]K, partition coefficient expressed by solute concentration in the stationary phase divided by that in the mobile phase.

diagram. The shaded areas in each chromatogram indicate the pure fractions. As shown in the diagram, increasing the sample size results in nearly proportional increase of the pure fractions, while the width of the mixing zone remains the same. This indicates an important advantage of pH-zone-refining CCC over standard CCC techniques in that the increase of the sample size improves the yield of pure fractions.

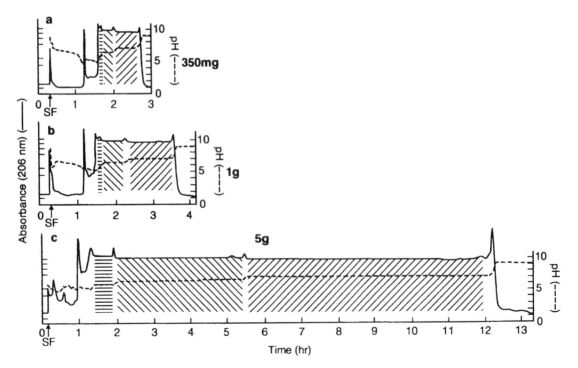

Figure 35 pH-Zone-refining CCC of D&C Yellow No. 5 at three different sample sizes. [From Weisz et al. (1994a).]

In the last decade, pH-zone-refining CCC has been successfully applied to a variety of organic acids and bases, including derivatives of both amino acids (Ito and Ma, 1994; Ma and Ito, 1994) and peptides (Ma and Ito, 1995b), alkaloids (Ma et al., 1994a; Yang et al., 1998), hydroxyxanthine dyes (Shinomiya et al., 1995; Weisz et al., 1994a, b, c, 1995, 1996), anti-human immunodeficiency virus lignans (Ma et al., 1998), curcumin (Patel et al., 2000), indole auxins (Ito and Ma, 1996), and structural and stereoisomers (Denekamp et al., 1994; Dudding et al., 1998; Ma et al., 1994b). Using affinity ligands, the method has been used for separation of catecholamines (Ma et al., 1996), sulfonated dyes (Oka et al., 2002; Weisz et al., 2001, 2002), enantiomers (Ma and Ito, 1995a; Ma et al., 1995), and zwitter-ions such as free peptides (Ma and Ito, 1997a, b). A list of the samples and solvent systems for pH-zone-refining CCC may be found elsewhere (Ito, 1996b; Ito and Ma, 1996; Ito et al., 1995).

The pH-zone-refining CCC technique can be applied to both hydrodynamic and hydrostatic CCC systems provided that the system permits satisfactory retention of the stationary phase in the separation column.

B. Affinity CCC

Analogous to liquid chromatography, CCC permits use of ligands that can be dissolved in the stationary phase to perform affinity separation. According to the nature of the ligand, the method may be classified as ion exchange CCC (tridodecyl amine, di-[2-ethylhexyl] phosphoric acid) (Kitazume, 1996; Kitazume et al., 1990, 1999, 2004; Weisz and Ito, 2000; Weisz et al., 2001), paired ion CCC (tetrabutylammonium hydroxide) (Weisz and Ito, 2000), and chiral CCC using chiral selector such as DPA (Ma and Ito, 1996; Ma et al., 1999a, b). When compared with affinity liquid chromatography, affinity CCC has an advantage in that a large quantity of the ligand can be dissolved into the liquid stationary phase to perform large-scale preparative separations.

C. CCC/MS

An earlier trial (Lee et al., 1988c, 1990a, b) using a thermospray device encountered a problem of high pressure which broke the Teflon tubing used in the multilayer coil separation column of the high-speed CCC apparatus. This necessitated inserting another metering pump at the CCC/MS junction to introduce the CCC effluent into the MS without imposing excess pressure on the high-speed CCC column.

Later, the CCC/MS interfacing method was improved with a frit device (EI, CI, and FAB) which, owing to its low pressure, permits the direct interfacing to MS using a tee splitter illustrated in Fig. 36(a) (Oka, 1991a; 1996). A 0.06 mm i.d. fused silica tube is led to the MS, whereas a 0.5 mm i.d. stainless steel tube is connected to the high-speed CCC column. The other side of the fused silica tube extends deep into the stainless steel tube to receive a small portion of the effluent (ca. 1/40) from the high-speed CCC column while the rest of the effluent is discarded through the 0.1 mm i.d. Teflon tube. The split ratio of the effluent depends on the flow rate of the effluent and the length of the 0.1 mm i.d. Teflon tube. A 2 m length of the 0.1 mm i.d. tube is needed to adjust the split ratio at 1:40. Figure 36(b) shows a flow diagram for high-speed CCC/frit MS using a tee splitter at the junction.

The high-speed CCC/frit MS system including EI, CI, and FAB can be applied to a variety of analytes with a broad range of polarity (Oka, 1991a). For a nonvolatile, thermally labile, and/or polar compound, high-speed CCC/frit FAB is most suitable, whereas both high-speed CCC/frit EI and CI can be effectively used for relatively nonpolar compounds. The high-speed CCC/MS conditions applied to various samples are summarized in Table 4 (Oka, 1996).

V. TWO-PHASE SOLVENT SYSTEMS

A unique feature of CCC over liquid chromatography is that CCC uses a two-phase solvent system, one phase as a stationary phase and the other as a mobile phase. This provides a wide choice of solvent systems, although there are some restrictions such that it should provide stable retention of the stationary phase in the separation column. Also, CCC has a unique advantage that the elution time of solute peaks is predicted from their partition coefficients which can be determined by a simple test tube analysis. Various parameters and their mutual relationship involved in CCC are discussed subsequently.

A. Retention Volume and Partition Coefficient (K)

Figure 37 illustrates the portion of the separation column which shows stationary phase occupying the column space, V_s, in the upper half, and the mobile phase flowing through the column space, V_m, at a flow rate, v_m (mL/min) in the lower half. A solute introduced in the column is distributed between the two phases according to its partition coefficient (K). The amount of the solute distributed in a thin slice of the column (lower diagram) is expressed as $c_s a_s \Delta l$ in the stationary phase and $c_m a_m \Delta l$ in the mobile phase. Then, the moving rate (v) of the solute

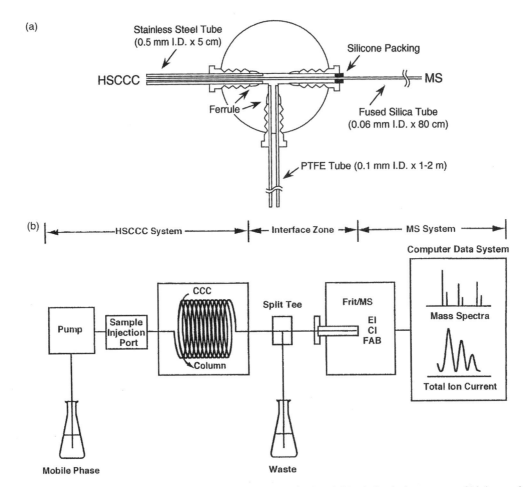

Figure 36 CCC/MS instrumentation. (a) Design of a split tee for CCC/MS and (b) whole elution system of high-speed CCC/frit MS system. [From Oka (1996).]

through the column space occupied by the mobile phase is

$$v = \frac{v_m c_m a_m \Delta l}{c_m a_m \Delta l + c_s a_s \Delta l} = \frac{v_m}{1 + KV_s/V_m} \quad (29)$$

where a_m and a_s are cross-sectional areas of the column occupied by the stationary and the mobile phases, respectively, and Δl, the thickness of the column space. Then the retention volume (R) of the solute in the column is expressed as

$$R = V_m \left(1 + \frac{KV_s}{V_m}\right) = V_m + KV_s \quad (30)$$

Figure 38 illustrates the relationship between these parameters.

From Eq. (30) it is clear that the solute with partition coefficient $K = 1$ would elute at a retention volume equal to the total column capacity. The solute with $K > 1$ elutes at a retention volume less than the total column volume, and the solute with $K < 1$ elutes at a retention volume greater than the total column capacity.

This indicates that for a solute with small K value the elution space is limited between the solvent front (V_m) and the total column volume, ($V_m + V_s$), which is determined by the retention of the stationary phase (V_s). Consequently, the resolution for the solutes with small K values would be improved by increasing the volume of the stationary phase retained in the column.

B. R_s and Retention of the Stationary Phase in CCC

Relationship between the peak resolution (R_s), theoretical plate number (N) and retention volume of the solute peaks (R) is expressed using the following two conventional equations:

$$R_s = \frac{2(R_2 - R_1)}{W_2 + W_1} \quad (31)$$

$$N = 16\left(\frac{R}{W}\right)^2 \quad (32)$$

Table 4 Summary of the Previously Reported High-Speed CCC/MS Conditons

Sample	Column (PTFE tube)	Column capacity (mL)	Revolution speed (rpm)	Solvent system	Mobile phase	Flow-rate (mL/min)	Retention of stationary phase (%)	Ionization
Alkaloids	0.85 mm	38	1500	n-Hexane/ethanol/water (6:5:5)	Lower phase	0.7	—	Thermospray
Triterpenoic Acids	0.85 mm	38	1500	n-Hexane/ethanol/water (6:5:5)	Lower phase	0.7	—	Thermospray
Lignans	0.85 mm	38	1500	n-Hexane/ethanol/water (6:5:5)	Lower phase	0.7	—	Thermospray
Indole auxins	0.3 mm	7	4000	n-Hexane/ethyl acetate/methanol/water (1:1:1:1)	Lower phase	0.2	27.2	Frit-EI
Mycinamicins	0.3 mm	7	4000	n-Hexane/ethyl acetate/methanol/8% ammonia (1:1:1:1)	Lower phase	0.1	40.4	Frit-Cl
Colistins	0.55 mm	6	4000	n-Butanol/0.04 M TFA (1:1)	Lower phase	0.16	34.3	Frit-FAB
Erythromycins	0.85 mm	17	1200	Ethyl acetate/methanol/water (7:4:3)	Lower phase	0.8	—	Electrospray
Didemnins	0.85 mm	17	1200	Ethyl acetate/methanol/water (4:1:4)	Lower phase	0.8	—	Electrospray

Source: From Oka (1996).

where R_1 and R_2 and W_1 and W_2 are the retention volumes and the peak widths of solutes 1 and 2, respectively, and N is the theoretical plate numbers.

Using Eqs. (30)–(32), the peak resolution (R_s) between solutes 1 and 2 with K_1 and K_2 ($K_2 > K_1$, $K_2/K_1 = \alpha$) is expressed according to the following equation (Conway and Ito, 1985):

$$R_s = 0.5\sqrt{N}\frac{\alpha - 1}{(\alpha + 1) + (2/K_1)(1 - S_F)/S_F} \quad (33)$$

where S_F is the fraction of the column volume occupied by the stationary phase or $V_s/(V_m + V_s)$.

Figure 39 illustrates the resolution (R_s) of the pair of solutes 1 and 2 for which the partition coefficient, K_1, of the first eluted peak varies from 0.05 to 20 while α remains constant at a value of 2. Resolution is shown for columns of 100, 250, and 1000 TP. Substances with high partition coefficients (K_1 of 1–20) are best resolved in long columns (1000 TP) with a small stationary phase

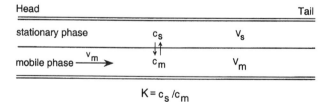

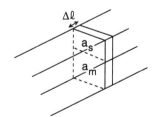

Figure 37 Partition process in the CCC column.

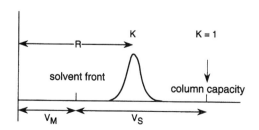

Figure 38 Relationship between various parameters in CCC.

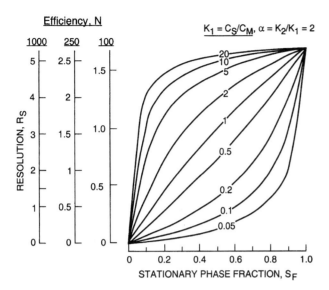

Figure 39 Peak resolution (R_s) vs. retention of stationary phase. [From Conway and Ito (1985).]

volume, S_F about 0.2. As S_F is further increased, resolution of solutes with high K_1 (10–20) improves only slightly, whereas resolution of solutes with intermediate partition coefficients (K_1 of 0.2–2) increases appreciably. When S_F of 0.8 is reached, which is typical for the type J multilayer coil planet centrifuge (high-speed CCC), solutes with partition coefficients almost as low as 0.05 can be resolved on long columns. Solute pairs with K_1 values above about 1 are predicted to be resolved on short columns of about 100 TP if S_F is 0.8.

Short columns having S_F in the region of 0.4, which are typical for the horizontal flow-through coil planet centrifuge and helix CCC centrifuge are predicted to be adequate for resolution of solutes with K_1 values in the range of 5–20.

These estimates are based on a separation factor (α) of 2, and the expected resolution under the same conditions will decrease as α is lowered and will increase with higher α.

The relationship between R_s and percent retention of the stationary phase was experimentally determined using a multilayer coil planet centrifuge (Oka et al., 1992). Three indole auxins were separated with a two-phase solvent system composed of hexane/ethyl acetate/methanol/water (1:1:1:1, v/v/v/v) at various flow rates. When the R_s values were plotted against the retention of the stationary phase ranging 50–85%, the R_s values gradually increased from 3 to 5 until 80% where they sharply rose to 7–11, which is quite comparable with the theoretical presentation in Fig. 39. It is interesting to note that N values similarly plotted against the percent retention of the stationary phase decreased until they reached 75–80%. The results indicate that TP numbers are not always a reliable indicator for evaluating the peak resolution, especially when the separation is performed with low stationary phase retention.

C. Various Parameters of High-Speed CCC Affecting Stationary Phase Retention

In most of the hydrostatic systems and some hydrodynamic systems such as horizontal flow-through CPC and nonsynchronous CPC, retention of the stationary phase is maximized by application of a low flow rate of the mobile phase and/or a high centrifugal force field. However, in type J high-speed CCC systems, retention of the stationary phase is largely affected by many other parameters, some of which have been extensively studied (Ito, 1984b, c; Ito and Conway, 1984).

1. Physical Parameters of the Solvent System

Retention of the stationary phase is affected by a combination of three major physical parameters (i.e., interfacial tension, viscosity, and density difference between the two phases). Our studies revealed that, as expected, a set of three physical parameters show a good correlation to the settling time of each solvent system in a test tube. The solvent system exhibited a short settling time with high interfacial tension, low viscosity, and large difference in density between the two phases.

Table 5 (Ito and Conway, 1984) lists these three parameters for a set of two-phase solvent systems which are classified into three categories (i.e., hydrophobic, intermediate, and hydrophilic) according to their polarity. The hydrophobicity of the solvent systems also shows a good correlation with the settling times: 5–16 s for the hydrophobic system, 15–29 s for intermediate, and 38–59 s for the hydrophilic system. The difference between T (settling time after gentle mixing) and T' (settling time after vigorous mixing) for each solvent system was found to be quite small in both the intermediate and hydrophilic solvent systems. Because of the close correlation to the hydrophobicity of the solvent system, the settling time under unit gravity may serve as a useful parameter for the retention behavior of the two-phase solvent system in the coiled column rotating in a centrifugal force field.

A series of studies using a type J synchronous coil planet centrifuge with a revolution radius of 10 cm, revealed that the helical diameter of the column is another important parameter for retention of the stationary phase, and it is expressed as β ($\beta = r/R$, where r is the helical radius and R is the revolution radius) (Ito, 1984c).

The results indicated that the hydrophobic solvent systems enable high retention of the stationary phase regardless of the β values, if the lower phase is pumped from the head or the upper phase from the tail, whereas

Table 5 Physical Properties of the Two-Phase Solvent System

Solvent group	Two-phase-solvent-system (volume ratio)	Interfacial tension (dyne/cm) $\Delta\gamma$	Viscosity (c.p.) η_U/η_L	$\tilde{\eta}$	Density (g/cm³) ρ_U/ρ_L	$\Delta\rho$	Settling time (s) T	T'
Hydrophobic	Hexane/water	52	0.41/0.95	0.68	0.66/1.00	0.34	<1	8
	Ethyl acetate/water	31	0.47/0.89	0.68	0.92/0.99	0.07	15.5	21
	Chloroform/water	42	0.95/0.57	0.76	1.00/1.50	0.50	3.5	5.5
Intermediate	Hexane/methanol	4	0.50/0.68	0.59	0.67/0.74	0.07	5.5	6
	Ethyl acetate/acetic acid/water (4:1:4, v/v/v)	16	0.76/0.81	0.79	0.94/1.01	0.07	15	16
	Chloroform/acetic acid/water (2:2:1, v/v/v)	12	1.16/0.77	0.97	1.12/1.35	0.24	29	27.5
	n-Butanol/water	3	1.72/1.06	1.40	0.85/0.99	0.14	18	14
	n-Butanol/0.1 M NaCl (1:1, v/v)	4	1.66/1.04	1.35	0.85/0.99	0.15	16	14.5
	n-Butanol/1 M NaCl (1:1, v/v)	5	1.75/1.04	1.40	0.84/1.04	0.20	23.5	21.5
Hydrophilic	n-Butanol/acetic acid/water (4:1:5, v/v/v)	<1	1.63/1.40	1.52	0.90/0.95	0.05	38.5	37.5
	n-Butanol/acetic acid/0.1 M NaCl (4:1:5, v/v/v)	<1	1.68/1.25	1.47	0.89/1.01	0.11	32	30.5
	n-Butanol/acetic acid/1 M NaCl (4:1:5, v/v/v)	1	1.69/1.26	1.48	0.88/1.05	0.16	26.5	24.5
	sec.-Butanol/water	<1	2.7/1.67	2.19	0.87/0.97	0.10	57	58
	sec.-Butanol/0.1 M NaCl (1:1, v/v)	<1	1.96/1.26	1.61	0.86/0.98	0.12	46.5	49.5
	sec.-Butanol/1 M NaCl (1:1, v/v)	3	1.91/1.29	1.60	0.84/1.03	0.19	34	33.5

Note: η_U/η_L = upper phase/lower phase; $\tilde{\eta}$ = mean viscosity of the two phases; ρ_U/ρ_L = upper phase/lower phases; $\Delta\rho$ = density difference between the two phases; T = settling time after gentle mixing; T' = settling time after vigorous shaking.
Source: From Ito and Conway (1984).

the hydrophilic solvent systems generally exhibited a poor retention of the stationary phase in all β values, regardless of the elution mode. The solvent systems with intermediate hydrophobicity show good retention of stationary phase at β values of 0.5 if the lower phase is pumped from the head or the upper phase from the tail, whereas they give a low retention at β value of 0.25 regardless of the elution mode.

On the basis of the earlier findings, the commercial type J high-speed CCC centrifuge is usually equipped with a multilayer coil having a β range between 0.5 and 0.75. However, one must consider the effect of the sample compartment which may alter the settling time, and hence the retention of the stationary phase. Therefore, in a practical application, the settling time of the sample solution should also be measured to ensure retention of the stationary phase.

These findings are only limited to the type J multilayer coil planet centrifuge. A recently developed spiral disk assembly, if mounted on the type J high-speed CCC centrifuge, retains a satisfactory amount of the stationary phase for an extremely hydrophilic solvent systems such as polymer phase systems (Ito et al., 2003). The cross-axis CPC with a relatively high L/X ratio (>1.5) also retains hydrophilic solvent systems, including polymer phase systems, and therefore is suitable for separation of biopolymers.

2. Flow Rate of the Mobile Phase

The flow rate is, as expected, an important parameter to determine the retention of the stationary phase in high-speed CCC. A series of experiments with a set of 15 two-phase solvent systems (hydrophobic to intermediate) revealed that retention of the stationary phase (S_F) is expressed as the following equation (Du et al., 1999):

$$S_F = A - B\sqrt{F_C} \tag{34}$$

where S_F is percent retention of the stationary phase relative to the column capacity; A, relating to the solvent composition; B, relating the volume ratio of the same solvent system; and F_C, the flow rate of the mobile phase (mL/min). The actual A values range from 132.59

for the ethyl acetate/water binary system to 68.91 for the hexane/ethanol/water (6:5:3, v/v/v) system, whereas the B values range from 58.51 to 6.28 for the same solvent pairs. All other solvent systems show both A and B values within the earlier-mentioned ranges. In most cases the A value substantially exceeds 100, suggesting that the linearity of the curve should be limited within some range of S_F, probably between 20% and 80%.

3. Revolution Speed

The effect of revolution speeds on retention of the stationary phase has been extensively studied using a set of two-phase solvent systems with a broad range of hydrophobicity. The type J coil planet centrifuge with three different revolution radii of 5, 10, and 15 cm was employed with three coil positions of $\beta = 0.25-1.9$ (Ito, 1984c).

The results indicated that, with few exceptions, increasing the revolution speed produced a gradual increase in the retention of the stationary phase, and at 800–1000 rpm (maximum) the curve approached a nearly saturating level. This indicates that with a semipreparative high-speed CCC instrument, one should not expect any substantial improvement of the stationary phase retention by applying a revolution speed over 1000 rpm. Using analytical high-speed CCC centrifuge with a revolution radius of 2.5 cm, the maximum revolution speed of 4000 rpm gives a high retention of over 75% for hydrophobic solvent systems at β values ranging from 0.28 to 0.77, if the lower phase is pumped from the head or the upper phase from the tail (Oka et al., 1989a). Among intermediate solvent systems, hexane/ethyl acetate/methanol/water (1:1:1:1, v/v/v/v) produces satisfactory retention, whereas other systems show relatively low retention of the stationary phase.

D. Partition Coefficient of Samples

1. Optimum Range of K Values

The suitable range of the partition coefficient varies according to the CCC instrument used. In the hydrostatic systems where mixing of the two phases is less violent, a higher K value of $K > 1$ will produce better peak resolution, although too large K values require long separation times. On the other hand, for hydrodynamic CCC systems such as high-speed CCC that produce vigorous agitation of the two phases, a lower range of $0.5 < K < 1$ is suitable. Higher K values tend to produce excessive band spreading and dilution of samples.

2. Search for the Suitable Solvent Systems

One of the most important steps in CCC separation is to select a two-phase solvent system which provides an ideal range of the partition coefficient values (K) for the target compound. If a literature search fails to find the separation of similar compounds by CCC, one must rely on a tedious trial and error method to search for a suitable two-phase solvent system. In this situation, one may follow a systematic search which would help reach the desired two-phase solvent system with relatively few trials. Figure 40 illustrates a set of two-phase solvent systems which are arranged according to the hydrophobicity of the organic phase (Ito, 1992b, 1996a; Oka et al., 1991c).

One can start with chloroform methanol water (2:1:1) system [Fig. 39(a)]. If the sample mostly distributes into the lower organic phase ($K \ll 1$, where K is solute concentration in the upper phase divided by that in the lower phase), one should follow the upper arrow to test hexane/ethyl acetate/methanol/water (1:1:1:1). If the solute still mostly distributes into the upper organic phase, one can follow the small arrow upward until reaching the hexane/methanol/water (10:5:5) system at the top of the diagram. If more hydrophobic solvent system is required, one can try hexane/ethanol/water (5:4:1) or hexane/methanol.

When the solute is distributed mostly into the upper aqueous phase in the chloroform solvent system [Fig. 39(a)], one can follow the lower arrow to test ethyl acetate/water binary system, finally reaching n-butanol/water system at the bottom. The more polar solvent system is made modifying this solvent system by adding a modifier such as acetic acid and an inorganic salt such as sodium chloride.

In addition to the hydrophobicity of the solvent system, pH may play an important role in adjusting K values. An addition of acid such as TFA to the solvent system may protonate the acidic compounds to shift their distribution toward the organic phase, and at the same time it often

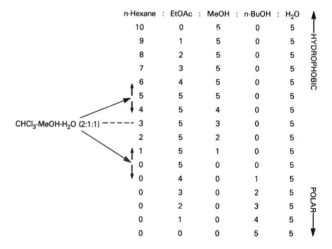

Figure 40 Systematic search for the suitable solvent system. [From Ito (1992b).]

shortens the settling time of the solvent system to yield higher retention of the stationary phase.

In addition to the earlier-mentioned organic–aqueous solvent systems, there are aqueous–aqueous polymer phase systems (Albertsson, 1971) which can be used for the separation of biopolymers such as proteins without denaturation of the molecule. These polymer phase systems are divided into two classes according to their characteristic partition behavior: one is PEG (polyethylene glycol)–inorganic salt including potassium phosphate, ammonium sulfate, and so on, in which most of the low molecular weight compounds are unilaterally distributed into either phase, whereas large molecules such as proteins are rather evenly partitioned between the two phases. The other polymer system is composed of PEG and dextran (MW = 500,000) in dilute buffer solution. In contrast to the PEG–inorganic salt system, the PEG–dextran system distributes small molecules rather evenly between the two phases while large macromolecules such as DNA and RNA are unilaterally distributed in either phase according to the pH of the solvent system. The system can also be used for separation of particles which are mainly partitioned between the upper PEG phase and the interface between the two phases.

3. Measurement of K Values

In CCC, K values can be measured by a simple test tube procedure. If a pure standard is available, equilibrate it with two phases in a test tube, and take an aliquot from each phase to dilute with methanol or other suitable solvent for UV/visible absorbance measurement with a spectrophotometer. For crude samples, after equilibrating with the two phases, analyze an aliquot of each phase with high performance liquid chromatography, and compare the peak height or area of the corresponding peaks. Thin layer chromatography can be useful for a rough estimate of K values.

As mentioned earlier, K values are also computed from the chromatogram, according to the following equation.

$$K = \frac{R - V_m}{V_s} = \frac{R - V_{SF}}{V_c - V_{SF}} \tag{35}$$

where R is the retention volume of the target compound; V_m and V_s, respectively, the volumes of mobile and stationary phases occupying the column space; V_{SF}, the retention volume of solvent front; and V_c, the total column capacity (see also Fig. 38).

VI. FUTURE INSTRUMENTATION

CCC is a versatile separation technique with a high potential for future development. Although the method is more suitable for preparative separations, analytical applications should also be considered.

A. Analytical CCC

Efficient analytical-scale CCC has been reported by one group of workers (Schaufelberger, 1996). Helix CCC and centrifugal partition chromatography may be used for high-resolution analytical separations using a small-bore tubing or narrow path under a high centrifugal force field (Ito and Bowman, 1970a). The limitation in this method is high hydrostatic pressure which may lead to problems of leakage and column damage. Development of a flow tube or separation column tolerating high pressure is necessary.

Analytical high-speed CCC using the type J centrifugal partition chromatography provides less problem of hydrostatic pressure, while the decrease in the column diameter tends to cause a plug flow which may result in loss of the stationary phase (Oka et al., 1991b). This problem might be alleviated by using a narrow Teflon tubing with square or triangular cross-section where the aqueous droplet flows through the center leaving the organic stationary phase at the angular space. Construction of miniature unit of the nonsynchronous CPC may provide efficient analytical separation under low-speed coil rotation in a strong centrifugal force field.

One of the unique applications of CCC is called "dual CCC" where both phases flow against each other. This technique has been applied to foam separation (Bhatnagar and Ito, 1988; Ito, 1985, 1987d; Oka et al., 1989b) and preparative separation (Lee et al., 1988), both using a multilayer coil separation column. Recently, this method has been applied to analytical separations using a spiral channel. Under dual countercurrent of two phases, a small amount of sample injected at the middle portion of the channel is partitioned into the two phases (see dual CCC) so that the major component will move with one phase and quickly elute from the column while a minute amount of the target compound (pollutant) is carried with the other phase in the opposite direction. The method permits CCC/MS for multiple injections at regular intervals without contamination of the major component. The efficiency of the method may be improved by modifying the geometry of the spiral channel (Ito, 2004a US Patent pending).

B. Preparative CCC

Compared with other liquid chromatographic methods, CCC allows scaling up without substantial loss of partition efficiency. In addition, it eliminates the cost of the solid support, consumes much less solvents, and yields highly pure fractions.

Large-scale preparative separations in CCC can be performed simply by using larger and/or longer separation columns. Hydrostatic systems such as helix CCC and centrifugal partition chromatography tend to lose partition efficiency as the column becomes enlarged as mixing of the two phases in these systems entirely depends on the flow of the mobile phase. In hydrodynamic systems, however, mixing the two phases in a large-bore tubing can be efficiently performed by the action of unit gravity or a low centrifugal force field; hence, the loss of the partition efficiency in a large column becomes minimized. There are two promising directions for developing a large-scale preparative CCC instrument.

1. Type J CPC

The multilayer coil planet centrifuge may be further scaled up in both coil diameter and internal diameter of the coil. At the present time our prototype at the National Institutes of Health has a total capacity of 1.6 L which can separate 10–20 g of extracts in a standard operation and around 100 g in pH-zone-refining CCC in several hours. The column can be further enlarged for industrial separations (Ito et al., 1991a).

The maximum sample loading capacity may be achieved using a large multilayer coil rotated around its horizontal axis at a low speed so that the system utilizes both unit gravity as well as the centrifugal force for phase mixing and retention of the stationary phase.

2. Low-Speed Rotary CCC

As described earlier (Fig. 30), a low-speed rotary CCC apparatus produces bilateral distribution of the two phases similar to that by type J high-speed CCC at high revolution rate (Ito, 1988b, 1992a; Ito and Bhatnagar, 1984). Recently, this system has been scaled-up to 10 L, and successfully separated 150 g of tea extract at a relatively high partition efficiency using the standard operation. This suggests that the separation of kilogram quantities by pH-zone-refining CCC (Du et al., 2000) is feasible. Convoluted Teflon tubing used in this separation not only facilitates the preparation of the multilayer coil separation column owing to its high flexibility, but also ensures sufficient stationary phase retention. Because the system uses a low rotation rate at 80–100 rpm, further scaling-up would be less problematical. Although the separation time is much longer than that required for the type J CPC, operation of this simple and inexpensive instrument can be easily automated and left unattended during large-scale industrial separations.

This system may be further improved using a spiral disk assembly (Ito, 2004b US Patent pending; Ito et al., 2003) or a spiral tube support (Ito, 2004c US Patent pending) which utilizes the radial centrifugal force gradient as well as the Archimedean screw effect for enhancing retention of the stationary phase.

ACKNOWLEDGMENT

The author is indebted to Dr. Henry M. Fales for editing the manuscript.

REFERENCES

Albertsson, P. A. (1971). *Partition of Cells and Macromolecules*. New York: Wiley.

Berthod, A. (2002a). In: Barcelo, D., ed. *Countercurrent Chromatography: The Support-Free Liquid Stationary Phase*. Comprehensive Analytical Chemistry. Vol. XXXVIII, Amsterdam: Elsevier Scientific.

Berthod, A. (2002b). Commercially available countercurrent chromatographs. In: Berthod, A., ed. *Countercurrent Chromatography: The Support Free Liquid Stationary Phase*. Vol. XXXVIII. Comprehensive Analytical Chemistry, Appendix, Amsterdam: Elsevier, pp. 379–388.

Bhatnagar, M., Ito, Y. (1988). Foam countercurrent chromatography on various tests samples and the effects of additives on foam affinity. *J. Liq. Chromatogr.* 11(1):21–36.

Bhatnagar, M., Oka, H., Ito, Y. (1989). Improved cross-axis synchronous flow-through coil planet centrifuge for performing countercurrent chromatography. II. Studies on retention of stationary phase in short coils and preparative separation in multilayer coils. *J. Chromatogr.* 463:317–328.

Conway, W. D. (1992). *Countercurrent Chromatography: Principle, Apparatus and Applications*. VCH.

Conway, W. D., Ito, Y. (1985). Resolution in countercurrent chromatography. *J. Liq. Chromatogr.* 8:2281–2291.

Craig, L. C. (1962). *Comprehensive Biochemistry*. Vol. 4. Amsterdam, London, and New York: Elsevier Publishing Co., 1.

Denekamp, C., Mandelbaum, A., Weisz, A., Ito, Y. (1994). Preparative separation of stereoisomeric 1-methyl-4-methoxymethylcyclo-hexanecarboxylic acids by pH-zone-refining countercurrent chromatography. *J. Chromatogr. A* 685:253–257.

Du, Q.-Z., Wu, C.-J., Qian, G.-J., Wu, P.-D., Ito, Y. (1999). Relationship between the flow-rate of the mobile phase and the retention of the stationary phase in countercurrent chromatography. *J. Chromatogr. A* 875:231–235.

Du, Q.-Z., Wu, P.-D., Ito, Y. (2000). Low-speed rotary countercurrent chromatography using convoluted multilayer helical tube for industrial separation. *Anal. Chem.* 72(14):3363–3365.

Dudding, T., Mekonnen, B., Ito, Y., Ziffer, H. (1998). Use of pH-zone-refining countercurrent chromatography to separate 2- and 6-nitro-4-chloro-3-methoxybenzoic acids. *J. Liq. Chromatogr.* 21(1&2):195–201.

Foucault, P. A. (1994). In: Jack Cazes, ed. *Centrifugal Partition Chromatography*. Chromatographic Science Series, Vol. 68. New York: Marcel Dekker Inc.

Foucault, A. P. (2002). Centrifugal partition chromatography: the story of a company. In: Berthod, A., ed. *Countercurrent Chromatography: The Support-Free Liquid Stationary Phase*. In: Barcelo, D., ed. *Comprehensive Analytical Chemistry*. Amsterdam: Elsevier Scientific, Chapter 4, pp. 85–113.

Ito, Y. (1969). Apparatus for Fluid Treatment by Utilizing the Centrifugal Force. US Patent 3,420,436, Jan. 7.

Ito, Y. (1979). Countercurrent chromatography with a new horizontal flow-through coil planet centrifuge. *Anal. Biochem.* 100:271–281.

Ito, Y. (1980a). A new horizontal flow-through coil planet centrifuge for countercurrent chromatography: I. Principle of design and analysis of acceleration. *J. Chromatogr.* 188:33–42.

Ito, Y. (1980b). A new horizontal flow-through coil planet centrifuge for countercurrent chromatography: II. The apparatus and its partition capabilities. *J. Chromatogr.* 188:43–60.

Ito, Y. (1981a). Countercurrent chromatography (minireview). *J. Biochem. Biophys. Methods* 5(2):105–129.

Ito, Y. (1981b). New continuous extraction method with a coil planet centrifuge. *J. Chromatogr.* 207:161–169.

Ito, Y. (1981c). Efficient preparative countercurrent chromatography with a coil planet centrifuge. *J. Chromatogr.* 214:122–125.

Ito, Y. (1984a). Flow-Through Centrifuge Free of Rotating Seals Particularly for Plasmapheresis. US Patent 4,425,112, Jan. 10.

Ito, Y. (1984b). Experimental observations of the hydrodynamic behavior of solvent systems in high-speed countercurrent chromatography: Part I. Hydrodynamic distribution of two solvent phases in a helical column subjected to two types of synchronous planetary motion. *J. Chromatogr.* 301:377–386.

Ito, Y. (1984c). Experimental observations of the hydrodynamic behavior of solvent systems in high-speed countercurrent chromatography: Part II. Phase distribution diagrams for helical and spiral columns. *J. Chromatogr.* 301:387–403.

Ito, Y. (1985). Foam countercurrent chromatography based on dual countercurrent system. *J. Liq. Chromatogr.* 8:2131–2152.

Ito, Y. (1986a). A new angle rotor coil planet centrifuge for countercurrent chromatography: Part I. Analysis of acceleration. *J. Chromatogr.* 358:313–323.

Ito, Y. (1986b). A new angle rotor coil planet centrifuge for counter-current chromatography: Part II. Design of the apparatus and studies on phase retention and partition capability. *J. Chromatogr.* 358:325–336.

Ito, Y. (1986c). High-speed countercurrent chromatography. *CRC Crit. Rev. Anal. Chem.* 17(1):65–143.

Ito, Y. (1987a). Cross-axis synchronous flow-through coil planet centrifuge free of rotary seals for preparative countercurrent chromatography. Part I. Apparatus and analysis of acceleration. *Sep. Sci. Technol.* 22(8&9):1971–1988.

Ito, Y. (1987b). High-speed countercurrent chromatography. *Nature* 326:419–420.

Ito, Y. (1987c). Cross-axis synchronous flow-through coil planet centrifuge free of rotary seals for preparative countercurrent chromatography. Part II. Studies on phase distribution and partition efficiency in coaxial coils. *Sep. Sci. Technol.* 22(8&9):1989–2009.

Ito, Y. (1987d). Foam countercurrent chromatography with the cross-axis synchronous flow-through coil planet centrifuge. *J. Chromatogr.* 403:77–84.

Ito, Y. (1988a). Principle and instrumentation of countercurrent chromatography. In: Mandava, N. B., Ito, Y., eds. *Countercurrent Chromatography: Theory and Practice.* New York: Marcel Dekker Inc., Chapter 3, pp. 79–442.

Ito, Y. (1988b). Studies on hydrodynamic distribution of two immiscible solvent phases in rotating coils. *J. Liq. Chromatogr.* 11(1):1–19.

Ito, Y. (1991). The cross-axis synchronous flow-through coil planet centrifuge (Type XLL) II. Speculation on the hydrodynamic mechanism in stationary phase retention. *J. Chromatogr.* 538:67–80.

Ito, Y. (1992a). Speculation on the mechanism of unilateral hydrodynamic distribution of two immiscible solvent phases in the rotating coil. *J. Liq. Chromatogr.* 15:2639–2675.

Ito, Y. (1992b). Countercurrent chromatography. In: Heftmann, E. ed. *Chromatography, V, J. of Chromatogr. Library, Part A.* Amsterdam: Elsevier Scientific Publishing Co., Chapter 2, pp. A69–A107.

Ito, Y. (1995). Countercurrent chromatography: principles and applications. In: Townshend, A., Fullerlove, G., eds. *Encyclopedia of Analytical Science.* Vol. 2. 1st ed. London: Academic Press, pp. 910–916.

Ito, Y. (1996a). Apparatus and methodology of high-speed countercurrent chromatography. In: Ito, Y. Ito, Conway, W. D., eds. *High-Speed Countercurrent Chromatography.* Vol. 132. Chemical Analysis Series; New York: Wiley Interscience, Chapter 1, pp. 3–43.

Ito, Y. (1996b). pH-peak focusing and pH-zone-refining countercurrent chromatography. In: Ito, Y., Conway, W. D., eds. *High-Speed Countercurrent Chromatography.* Vol. 132. Chemical Analysis Series, New York: Wiley Interscience, Chapter 6, pp. 121–175.

Ito, Y. (2000a). Countercurrent liquid chromatography. In: Wilson, I. D., Adlard, E. R., Cooke, M., Poole, C. F., eds. *Encyclopedia of Separation Science.* Vol. 8 (III). 1st ed. London: Academic Press, pp. 3815–3832.

Ito, Y. (2000b). pH-zone-refining countercurrent chromatography. In: Wilson, I. D., Adlard, E. R., Cooke, M., Poole, C. F., eds. *Encyclopedia of Separation Science*, 1st ed. London: Academic Press, pp. 573–583.

Ito, Y. (2001a). Instrumentation of countercurrent chromatography. In: Cazes, J., ed. *Encyclopedia of Chromatography.* 1st ed. New York: Marcel Dekker Inc., pp. 606–611.

Ito, Y. (2001b). Coriolis force in countercurrent chromatography. In: Cazes, J., ed. *Encyclopedia of Chromatography.* New York: Marcel Dekker Inc., pp. 203–205.

Ito, Y. (2001c). pH-peak-focusing and pH-zone-refining countercurrent chromatography. In: Cazes, J., ed. *Encyclopedia of Chromatography.* 1st ed. New York: Marcel Dekker Inc., pp. 438–440.

Ito, Y. (2004a). Spiral Disk Assembly for Dual Countercurrent Chromatography. US Patent, pending.

Ito, Y. (2004b). Method and Apparatus for Countercurrent Chromatography. US Patent, pending.

Ito, Y. (2004c). Spiral tube support for countercurrent chromatography, US Patent, pending.

Ito, Y., Bowman, R. L. (1970a). Countercurrent chromatography: liquid-liquid partition chromatography without solid support. *Science* 167:281–283.

Ito, Y., Bowman, R. L. (1970b). Countercurrent chromatography: liquid-liquid partition chromatography without solid support. *J. Chromatogr. Sci.* 8:315–323.

Ito, Y., Bowman, R. L. (1971a). Countercurrent chromatography. *Anal. Chem.* 43:69A.

Ito, Y., Bowman, R. L. (1971b). Countercurrent chromatography with flow-through coil planet centrifuge. *Science* 173:420–422.

Ito, Y., Bowman, R. L. (1973a). Countercurrent chromatography with the flow-through coil planet centrifuge. *J. Chromatogr. Sci.* 11:284–291.

Ito, Y., Bowman, R. L. (1973b). Application of the elution centrifuge to separation of polynucleotides with the use of polymer phase systems. *Science* 182:391–393.

Ito, Y., Bowman, R. L. (1977). Horizontal flow-through coil planet centrifuge without rotating seals. *Anal. Biochem.* 82:63–68.

Ito, Y., Bowman, R. L. (1978a). Countercurrent chromatography with flow-through centrifuge without rotating seals. *Anal. Biochem.* 85:614–617.

Ito, Y., Bowman, R. L. (1978b). Preparative countercurrent chromatography with horizontal flow-through coil planet centrifuge. *J. Chromatogr.* 147:221–231.

Ito, Y., Putterman, G. J. (1980). New horizontal flow-through coil planet centrifuge for counter-current chromatography: III. Separation and purification of dinitrophenyl amino acids and peptides. *J. Chromatogr.* 193(1):37–52.

Ito, Y., Bhatnagar, R. (1981). Preparative counter-current chromatography with a rotating coil assembly. *J. Chromatogr.* 207:171–180.

Ito, Y., Bhatnagar, R. (1984). Improved scheme for preparative countercurrent chromatography (CCC) with a rotating coil assembly. *J. Liq. Chromatogr.* 7(2):257–273.

Ito, Y., Conway, W. D. (1984). Experimental observations of the hydrodynamic behavior of solvent systems in high-speed countercurrent chromatography: Part III. Effects of physical properties of the solvent systems and operating temperature on the distribution of two-phase solvent systems. *J. Chromatogr.* 301:405–414.

Ito, Y., Chou, F.-E. (1988). New high-speed countercurrent chromatograph equipped with a pair of separation columns connected in series. *J. Chromatogr.* 454:382–386.

Ito, Y., Oka, H. (1988). Horizontal flow-through coil planet centrifuge equipped with a set of multilayer coils around the holder: countercurrent chromatography of proteins with a polymer phase system. *J. Chromatogr.* 457:393–397.

Ito, Y., Zhang, T.-Y. (1988a). Cross-axis synchronous flow-through coil planet centrifuge for large-scale preparative countercurrent chromatography: Part I. Apparatus and studies on stationary phase retention in short coils. *J. Chromatogr.* 449(1):135–151.

Ito, Y., Zhang, T.-Y. (1988b). Cross-axis synchronous flow-through coil planet centrifuge for large-scale preparative countercurrent chromatography: Part II. Studies on partition efficiency in short coils and preparative separations with multilayer coils. *J. Chromatogr.* 449(1):153–164.

Ito, Y., Ma, Y. (1994). pH-zone-refining countercurrent chromatography: a displacement mode applied to separation of dinitrophenyl amino acids. *J. Chromatogr. A* 672:101–108.

Ito, Y., Ma, Y. (1996). pH-zone-refining countercurrent chromatography. *J. Chromatogr. A* 753:1–36.

Ito, Y., Ma, Y. (1998). Effect of Coriolis force on countercurrent chromatography. *J. Liq. Chromatogr.* 21(1&2):1–17.

Ito, Y., Menet, J.-M. (1999). Coil planet centrifuges for high-speed countercurrent chromatography. In: Menet, J.-M., Thiébaut, D., eds. *Countercurrent Chromatography*. New York: Marcel Dekker Inc., Chapter 3, pp. 87–119.

Ito, Y., Harada, R., Weinstein, M. A., Aoki, I. (1966a). The coil planet centrifuge: principle and application. *Ikakikaigakuzashi* 36(7): (Japanese), 1–4.

Ito, Y., Weinstein, M. A., Aoki, I., Harada, R., Kimura, E., Nunogaki, K. (1966b). The coil planet centrifuge. *Nature* 212:985–987.

Ito, Y., Aoki, I., Kimura, E., Nunogaki, K., Nunogaki, Y. (1969). New micro liquid–liquid partition techniques with the coil planet centrifuge. *Anal. Chem.* 41:1579–1584.

Ito, Y., Bowman, R. L., Noble, F. W. (1972). The elution centrifuge applied to countercurrent chromatography. *Anal. Biochem.* 49:1–8.

Ito, Y., Bowman, R. L. (1975). Angle rotor countercurrent chromatography. *Anal. Biochem.* 65:310–320.

Ito, Y., Suaudeau, J., Bowman, R. L. (1975). New flow-through centrifuge without rotating seals applied to plasmapheresis. *Science* 189:999–1000.

Ito, Y., Carmeci, P., Sutherland, I. A. (1979). Nonsynchronous flow-through coil planet centrifuge applied to cell separation with physiological solution. *Anal. Biochem.* 94:249–252.

Ito, Y., Carmeci, P., Bhatnagar, R., Leighton, S., Seldon, R. (1980). The non-synchronous flow-through coil planet centrifuge without rotating seals applied to cell separation. *Sep. Sci. Technol.* 15(9):1589–1598.

Ito, Y., Sandlin, J. L., Bowers, W. G. (1982). High-speed preparative countercurrent chromatography (CCC) with a coil planet centrifuge. *J. Chromatogr.* 244:247–258.

Ito, Y., Bramblett, G. T., Bhatnagar, R., Huberman, M., Leive, L., Cullinane, L. M., Groves, W. (1983). Improved non-synchronous flow-through coil planet centrifuge without rotating seals. Principle and application. *Sep. Sci. Technol.* 18(1):33–48.

Ito, Y., Oka, H., Slemp, J. L. (1989a). Improved high-speed counter-current chromatograph with three multilayer coils connected in series I. Design of the apparatus and performance of semipreparative columns in DNP amino acid separation. *J. Chromatogr.* 475:219–227.

Ito, Y., Oka, H., Slemp, J. L. (1989b). Improved cross-axis synchronous flow-through coil planet centrifuge for performing countercurrent chromatography. I. Design of the apparatus and analysis of acceleration. *J. Chromatogr.* 463:305–316.

Ito, Y., Oka, H., Lee, Y.-W. (1990a). Improved high-speed countercurrent chromatograph with three multilayer coils connected in series II. Separation of various biological samples with a semipreparative column. *J. Chromatogr.* 498:169–178.

Ito, Y., Oka, H., Kitazume, E., Bhatnagar, M., Lee, Y.-W. (1990b). Improved high-speed countercurrent chromatograph with three multilayer coils connected in series III. Evaluation of semianalytical column with various biological samples and inorganic elements. *J. Liq. Chromatogr.* 13:2329–2349.

Ito, Y., Kitazume, E., Slemp, J. L. (1991a). Improved high-speed countercurrent chromatograph with three multilayer coils connected in series IV. Evaluation of preparative capability with large multilayer coils. *J. Chromatogr.* 538:81–85.

Ito, Y., Kitazume, E., Bhatnagar, M., Trimble, F. D. (1991b). The cross-axis synchronous flow-through coil planet centrifuge (Type XLL) I. Design of the apparatus and studies on retention of stationary phase. *J. Chromatogr.* 538:59–66.

Ito, Y., Shibusawa, Y., Fales, H. M., Cahnmann, H. J. (1992). Studies on an abnormally sharpened elution peak observed in countercurrent chromatography. *J. Chromatogr.* 625:177–181.

Ito, Y., Shinomiya, K., Fales, H. M., Weisz, A., Scher, A. L. (1995). pH-zone-refining countercurrent chromatography: a new technique for preparative separation. In: Conway, W. D., Petroski, R. J., eds. *ACS Monograph on Modern Countercurrent Chromatography*. Chapter 14, pp. 154–183.

Ito, Y., Yang, F.-Q., Fitze, P. E., Sullivan, J. V. (2003). Spiral disk assembly for high-speed countercurrent chromatography. *J. Liq. Chromatogr. Rel. Technol.* 26(9,10):1355–1372.

Kitazume, E. (1996). Separation of rare earth and certain inorganic elements by high-speed countercurrent chromatography. In: Ito, Y., Conway, W. D., eds. *High-Speed Countercurrent Chromatography*. New York: Wiley, Chapter 14, pp. 415–443.

Kitazume, E., Bhatnagar, M., Ito, Y. (1990). Mutual separation of rare earth elements by high-speed countercurrent chromatography. The Proceedings of International Trace Analysis Symposium 103–110, July 23–27.

Kitazume, E., Sato, N., Saito, Y., Ito, Y. (1993). Separation of heavy metals by high-speed countercurrent chromatography. *Anal. Chem.* 65:2225–2228.

Kitazume, E., Kanetomo, M., Tajima, T., Kobayashi, S., Ito, Y. (1999). An on-line micro extraction system of metal traces for their subsequent determination by plasma atomic emission spectrometry using pH-peak-focusing countercurrent chromatography. *Anal. Chem.* 71:5515–5521.

Kitazume, E., Takatsuka, Y., Sato, N., Ito, Y. (2004). Mutual metal separation with enrichment using pH-peak-focusing countercurrent chromatography. *J. Liq. Chromatogr. Rel. Technol.* 27(3):437–449.

Lee, Y.-W., Ito, Y., Fang, Q.-C., Cook, C. E. (1988a). Analytical high speed countercurrent chromatography. *J. Liq. Chromatogr.* 11(1):75–89.

Lee, Y.-W., Cook, C. E., Ito, Y. (1988b). Dual countercurrent chromatography. *J. Liq. Chromatogr.* 11(1):37–53.

Lee, Y.-W., Voyksner, R. D., Fang, Q.-C., Cook, C. E., Ito, Y. (1988c). Application of countercurrent chromatography/thermospray mass spectrometry for the analysis of natural products. *J. Liq. Chromatogr.* 11(1):153–171.

Lee, Y.-W., Voyksner, R. D., Pack, T.-W., Cook, C. E., Fang, Q.-C., Ito, Y. (1990a). Application of countercurrent chromatography/thermo-spray mass spectrometry for the identification of bio-active lignans from plant natural products. *Anal. Chem.* 62:244–248.

Lee, Y.-W., Pack, T. W., Voyksner, R. D., Fang, Q.-C., Ito, Y. (1990b). Application of high speed countercurrent chromatography/thermospray mass spectrometry for the analysis of bio-active triterpenoic acids from *Boswellia carterii*. *J. Liq. Chromatogr.* 13:2389–2398.

Leive, L., Cullinane, L. M., Ito, Y., Bramblett, G. T. (1984). Countercurrent chromatographic separation of bacteria with known differences in surface lipopolysaccharide. *J. Liq. Chromatogr.* 7(2):403–418.

Ma, Y., Ito, Y. (1994). pH-zone-refining countercurrent chromatography: separation of amino acid benzylesters. *J. Chromatogr. A* 678:233–240.

Ma, Y., Ito, Y. (1995a). Chiral separation by high-speed countercurrent chromatography. *Anal. Chem.* 67(17):3069–3074.

Ma, Y., Ito, Y. (1995b). Separation of peptide derivatives by pH-zone-refining countercurrent chromatography. *J. Chromatogr. A* 702:197–206.

Ma, Y., Ito, Y. (1996). Affinity countercurrent chromatography using a ligand in the stationary phase. *Anal. Chem.* 68:1207–1211.

Ma, Y., Ito, Y. (1997a). Peptide separation by pH-zone-refining countercurrent chromatography. *J. Chromatogr. A* 771:81–88.

Ma, Y., Ito, Y. (1997b). Recent advances in peptide countercurrent chromatography. *Anal. Chim. Acta* 352:411–427.

Ma, Y., Ito, Y., Sokoloski, E., Fales, H. M. (1994a). Separation of alkaloids by pH-zone-refining countercurrent chromatography. *J. Chromatogr. A* 685:259–262.

Ma, Y., Ito, Y., Torok, D., Ziffer, H. (1994b). Separation of 2- and 6-nitro-3-acetamido-4-chlorobenozoic acids by pH-zone-refining countercurrent chromatography. *J. Liq. Chromatogr.* 17(16):3507–3517.

Ma, Y., Ito, Y., Foucault, A. (1995). Resolution of gram quantities of racemates by high-speed countercurrent chromatography. *J. Chromatogr. A* 704:75–81.

Ma, Y., Sokoloski, E. A., Ito, Y. (1996). pH-zone-refining countercurrent chromatography of polar catecholamines using di-(2-ethylhexyl)phosphoric acid as a ligand. *J. Chromatogr. A* 724:348–353.

Ma, Y., Qi, L., Gnabre, J. N., Huang, R. C. C., Chou, F.-E., Ito, Y. (1998). Purification of anti-HIV lignans from *Larrea tridentata* by pH-zone-refining countercurrent chromatography. *J. Liq. Chromatogr.* 21(1&2):171–181.

Ma, Y., Ito, Y., Berthod, A. (1999a). A chromatographic method for measuring K_f of enantiomer-chiral selector complexes. *J. Liq. Chromatogr. Rel. Technol.* 22(19):2945–2955.

Ma, Y., Ito, Y., Berthod, A. (1999b). Chiral separation by high-speed countercurrent chromatography. In: Menet, J.-M., Thiébaut, D., eds. *Countercurrent Chromatography*. New York: Marcel Dekker Inc., Chapter 7, pp. 223–248.

Marchal, L., Foucault, A. P., Patissier, G., Rosant, J.-M., Legrand, J. (2002). Centrifugal partition chromatography: an engineering approach. In: Berthod, A., ed. *Countercurrent Chromatography: Support-Free Liquid Stationary Phase*. Amsterdam: Elsevier Scientific, Chapter 5, pp. 115–157.

Matsuda, K., Matsuda, S., Ito, Y. (1998). Toroidal coil countercurrent chromatography: achievement of high resolution by optimizing flow rate, rotation speed, sample volume and tube length. *J. Chromatogr. A* 808(1&2):95–104.

Menet, J.-M., Ito, Y. (1993). Studies on a new cross-axis coil planet centrifuge for performing countercurrent chromatography: II. Path and acceleration of coils and comparison with Type-J coil planet centrifuge. *J. Chromatogr.* 644:231–238.

Menet, J.-M., Shinomiya, K., Ito, Y. (1993). Studies on a new cross-axis coil planet centrifuge for performing countercurrent chromatography: III. Speculations on the hydrodynamic mechanism in stationary phase retention. *J. Chromatogr.* 644:239–252.

Menet, J.-M., Thiébaut, D., Rosset, R. J. (1994). Cross-axis coil planet centrifuge for the separation and purification of polar compounds. *Chromatogr.* 659:3–13.

Nakazawa, H., Andrews, P. A., Bachur, N. R., Ito, Y. (1981). Isolation of daunorubicin derivatives by counter-current chromatography with the horizontal flow-through coil planet centrifuge. *J. Chromatogr.* 205:482–485.

Nunogaki, Y. (1990). Centrifugal Counter-Current Distribution Chromatography. US Patent 4,968,428.

Oka, H. (1996). High-speed countercurrent chromatography/mass spectrometry. In: Ito, Y., Conway, W. D., eds. *High-Speed Countercurrent Chromatography.* New York: Wiley, Chapter 3, pp. 73–91.

Oka, H., Oka, F., Ito, Y. (1989a). Multilayer coil planet centrifuge for analytical high-speed countercurrent chromatography. *J. Chromatogr.* 479:53–60.

Oka, H., Harada, K.-I., Suzuki, M., Nakazawa, H., Ito, Y. (1989b). Foam countercurrent chromatography of bacitracin: Part I. Batch separation with nitrogen and water free of additives. *J. Chromatogr.* 482:197–205.

Oka, H., Harada, K.-I., Suzuki, M., Nakazawa, H., Ito, Y. (1989c). Foam countercurrent chromatography of bacitracin with nitrogen and water free of surfactants and other additives. *Anal. Chem.* 61:1998–2000.

Oka, H., Ikai, Y., Kawamura, N., Yamada, M., Harada, K.-I., Suzuki, M., Chou, F. E., Lee, Y.-W., Ito, Y. (1990). Evaluation of analytical countercurrent chromatographs: high-speed countercurrent chromatograph-400 vs. analytical toroidal coil centrifuge. *J. Liq. Chromatogr.* 13:2309–2328.

Oka, H., Ikai, Y., Kawamura, N., Hayakawa, J., Harada, K.-I., Murata, H., Suzuki, M., Ito, Y. (1991a). Direct interfacing of high speed countercurrent chromatography to frit electron ionization, chemical ionization, and fast atom bombardment mass spectrometry. *Anal. Chem.* 63(24):2861–2865.

Oka, H., Ikai, Y., Kawamura, N., Yamada, M., Hayakawa, J., Harada, K.-I., Nagase, K., Murata, H., Suzuki, M., Ito, Y. (1991b). Optimization of the high-speed countercurrent chromatograph for analytical separations. *J. High Res. Chromatogr. Chromatogr. Commun.* 14(5):306–311.

Oka, F., Oka, H., Ito, Y. (1991c). Systematic search for suitable two-phase solvent systems for high-speed countercurrent chromatography. *J. Chromatogr.* 538:99–108.

Oka, H., Ikai, Y., Hayakawa, J., Harada, K.-I., Nagase, K., Suzuki, M., Nakazawa, H., Ito, Y. (1992). Discrepancy between the theoretical plate number (N) and peak resolution (R_s) for optimizing the flow rate in countercurrent chromatography. *J. Liq. Chromatogr.* 15:2707–2719.

Oka, H., Goto, Y., Ito, Yuko; Matsumoto, H., Harada, K., Suzuki, M., Iwaya, M., Fuji, K., Ito, Y. (2002). Purification of food color Red No. 106 (Acid Red) using pH-zone-refining countercurrent chromatography. *J. Chromatogr. A* 946:157–162.

Okada, T., Metcalfe, D. D., Ito, Y. (1996). Purification of mast cells with an improved nonsynchronous flow-through coil planet centrifuge. *Int. Arch. Allerg. Immunol.* 109:376–382.

Patel, K., Krishna, G., Sokoloski, E. A., Ito, Y. (2000). Large-scale preparative separation curcumin, demethoxycurcumin, and bis-demethoxycurcumin from turmeric powder by pH-zone-refining countercurrent chromatography. *J. Liq. Chromatogr. Rel. Technol.* 23(14):2209–2218.

Schaufelberger, D. E. (1996). Analytical high-speed countercurrent chromatography. In: Ito, Y., Conway, W. D., eds. *High-Speed Countercurrent Chromatography.* Vol. 132. Chemical Analysis Series; Wiley Interscience; Chapter 2, pp. 45–70.

Scher, A. L., Ito, Y. (1995). Equilibrium model for pH-zone-refining countercurrent chromatography. In: Conway, W. D., Petroski, R. J., eds. *ACS Monograph on Modern Countercurrent Chromatography.* Chapter 15, pp. 185–202.

Shibusawa, Y. (1996). Separation of proteins by high-speed countercurrent chromatography. In: Ito, Y., Conway, W. D., eds. *High-Speed Countercurrent Chromatography.* New York: Wiley, Chapter 13, pp. 385–414.

Shibusawa, Y., Ito, Y. (1992). Countercurrent chromatography of proteins with polyethylene glycol-dextran polymer phase systems using type-XLLL cross-axis coil planet centrifuge. *J. Liq. Chromatogr.* 15:2787–2800.

Shinomiya, K., Menet, J.-M., Fales, H. M., Ito, Y. (1993). Studies on a new cross-axis coil planet centrifuge for performing countercurrent chromatography: I. Design of the apparatus, retention of the stationary phase, and efficiency in the separation of proteins with polymer phase systems. *J. Chromatogr.* 644:215–229.

Shinomiya, K., Weisz, A., Ito, Y. (1995). Purification of 4,5,6,7-tetrachlorofluorescein by pH-zone-refining countercurrent chromatography: Effects of sample size, concentration of eluent base, and choice of retainer acids. In: Conway, W. D., Petroski, R. J., eds. *ACS Monograph on Modern Countercurrent Chromatography.* Chapter 17, pp. 218–230.

Shinomiya, K., Kabasawa, Y., Ito, Y. (2002). Protein separation by cross-axis coil planet centrifuge with spiral coil assemblies. *J. Liq. Chromatogr. Rel. Technol.* 25(17):2665–2678.

Tanimura, T., Ito, Y. (1974a). Droplet Countercurrent Chromatography. US Patent 3,784,467, Jan. 8.

Tanimura, T., Ito, Y. (1974b). Droplet Countercurrent Chromatography. US Patent 3,853,765, Dec. 10.

Tanimura, T., Pisano, J. J., Ito, Y., Bowman, R. L. (1970). Droplet countercurrent chromatography. *Science* 169:54–56.

Weisz, A. (1996). Separation and purification of dyes by conventional high-speed countercurrent chromatography and pH-zone-refining countercurrent chromatography. In: Ito, Y., Conway, W. D., eds. *High-Speed Countercurrent Chromatography.* New York: Wiley, Chapter 12, pp. 337–384.

Weisz, A., Ito, Y. (2000). Separation of dyes by high-speed countercurrent chromatography. In: Wilson, I. D., Adlard, E. R., Cooke, M., Poole, C. F., eds. *Encyclopedia of Separation Science.* Vol. 6(III). 1st ed. London: Academic Press, pp. 2822–2829.

Weisz, A., Scher, A. L., Shinomiya, K., Fales, H. M., Ito, Y. (1994a). A new preparative-scale purification technique: pH-zone-refining countercurrent chromatography. *J. Am. Chem. Soc.* 116:704–708.

Weisz, A., Andrzejewski, D., Highet, R. J., Ito, Y. (1994b). Preparative separation of components of the color additive D&C Red No. 3 (erythrosin) by pH-zone-refining countercurrent chromatography. *J. Chromatogr. A* 658: 505–510.

Weisz, A., Andrzejewski, D., Ito, Y. (1994c). Preparative separation of components of the color additive D&C Red No. 28 (Phloxine B) by pH-zone-refining counter-current chromatography. *J. Chromatogr. A* 678:77–84.

Weisz, A., Andrzejewski, D., Shinomiya, K., Ito, Y. (1995). Preparative separation of components of commercial 4,5,6,7-tetrachlorofluorescein by pH-zone-refining countercurrent chromatography. In: Conway, W. D., Petroski, R. J., eds. *ACS Monograph on Modern Countercurrent Chromatography.* Chapter 16, pp. 203–217.

Weisz, A., Scher, A. L., Andrzejewski, D., Ito, Y. (1996). Preparative separation of brominated tetrachlorofluoresceins by pH-zone-refining countercurrent chromatography with computerized scanning uv-vis detection and continuous pH monitoring. *J. Chromatogr. A* 732: 283–290.

Weisz, A., Mazzola, E. U., Matusik, J. E., Ito, Y. (2001). Preparative separation of monosulfonated components of the color additive D&C Yellow No. 10 by pH-zone-refining countercurrent chromatography. *J. Chromatogr. A* 923:87–96.

Weisz, A., Mazzola, E. P., Murphy, C. M., Ito, Y. (2002). Preparative separation of isomeric sulfophthalic acids by conventional and pH-zone-refining countercurrent chromatography. *J. Chromatogr. A* 966:111–118.

Yang, F.-Q., Quan, J., Zhang, T.-Y., Ito, Y. (1998). Preparative separation of alkaloid from the root of *Sophora flavescent* Ait by pH-zone-refining CCC. *J. Chromatogr. A* 822:316–320.

29

HPLC-Hyphenated Techniques

R. A. SHALLIKER, M. J. GRAY
School of Science, Food and Horticulture, University of Western Sydney, NSW, Australia

I. INTRODUCTION

Chromatography, detailed in previous chapters, is one of the most widely employed analytical methods. Chromatography serves to separate species into ideally individual constituents. However, as the technique has evolved and separation performance has increased, scientists are discovering that what they once thought may have been individually resolved isolated components from within mixtures, may now be combinations of coeluting species. 1D chromatography cannot be used to identify species as retention times are not unique. This fact, together with the need to evaluate complex sample matrices, has driven the demand for techniques to be developed that are complementary. As we seek to gain more information regarding the constitution of sample matrices and subsequent quantitation, methods of analysis necessarily become more sophisticated. The increase in separation performance and analysis has followed two paths. There is currently growth in the development of multidimensional methods of analysis namely, coupled chromatographic systems. In particular, multidimensional gas chromatographic methods of analysis have been heavily investigated, but more recently there has been growing interest in multidimensional separations in liquid chromatography. The second path has been in the development of coupled and complementary methods of analysis; that is, chromatography, which is employed for its ability to bring about separation, and other analytical methods of analysis for detection and identification of species. This has developed into what is now referred to as hyphenated methods of analysis.

Hyphenation of liquid chromatographic (LC-hyphenation) methods primarily began with the advent of the diode-array detector (DAD) [or PDA (photodiode array), what it is commonly known as]. The technique is so significantly ingrained in the industry that the process is largely taken for granted. At around the same time, hyphenation with gas chromatographic methods was developed, where mass spectrometry (MS) was coupled with the gas chromatography (GC). Nowadays GC-MS is a routine technique. MS has also been coupled to liquid chromatography, as have Fourier transform infrared (FTIR), nuclear magnetic resonance (NMR), and inductively coupled plasma (ICP) spectroscopies. Also, different types of chromatographies have been coupled, such as, LC-GC, and this raises the question of whether this is a multidimensional separation or a hyphenated technique. Essentially, this is a mute point, because hyphenation can be used as a means to provide separation, largely with regard to selective detection, and multidimensional chromatography can be used for identification based on the low probability of separated species eluting with the same 2D retention time.

This leads us to an important consideration with respect to LC-hyphenation, or for that matter multidimensional chromatography, in general. The primary purpose of an LC separation is to bring about the physical separation of sample constituents; however, as the sample complexity increases this becomes more difficult. Hyphenation and the application of selective detection, for example, can establish the separation of species according to information, but not necessarily in a physical sense. A DAD may be tuned to a particular wavelength such that the desired species are visible but the undesired species are not, or an MS can be tuned to respond to a particular mass ion. So, if the physical collection of sample constituents is a requirement of the separation process, attention must be paid to the chromatography. Furthermore, some hyphenated methods of analysis are best served when good chromatography is applied, such as LC-NMR methods, whereby the interpretation of the NMR information is made easier as the purity of the sample constituent is improved.

This chapter is divided into six sections. First, we cover the rapidly expanding area of LC-MS, which though not yet as popular as GC-MS, will probably develop into the most widely employed hyphenated LC method of analysis. Then, we discuss the technique of LC-FTIR. Although the application of this technique finds a limited audience, this technique follows LC-MS because many of the methods employed for interfacing the LC to the FTIR are very similar to those employed in LC-MS. LC-NMR is covered next, which together with LC-MS primarily forms the two most important methods that allow identification of unknown species. These techniques find important applications in natural product research and studies in environmental analysis. Next, we discuss LC-ICP, which is a new and highly specific hyphenation procedure. We then look at the concept of multiple methods of hyphenation before concluding with a discussion on LC-GC hyphenation. The reader should note that we have largely ignored the hyphenated technique of LC-DAD. We have done so because the technique is so well established that it virtually needs no introduction. The interface is seamless and the operation of the DAD is largely an application of UV/Vis spectroscopy. For further information, refer to the discussion on LC-DAD in a previous chapter. All the instrumental methods of analysis have individually been covered in previous chapters. Our primary interest is in the hyphenation of LC with various methods of analysis, and as a result this chapter is largely focused on the interfacing of such techniques. We have by no means covered every aspect of LC-hyphenation. It becomes difficult to decide as to what constitutes a hyphenated method. Some could argue that LC with chemiluminesence or LC with electrochemical detection are hyphenated methods because they are able to provide selective detection and in some instances identification of at least a class of compounds with some degree of reliability. However, we do not consider that these methods are hyphenated methods, largely because their selectivity and capabilities of identification are insufficient to justify their inclusion.

II. LIQUID CHROMATOGRAPHY—MASS SPECTROSCOPY

The coupling of LC and MS is not a new technique, although academic and commercial interests in the field are intensely competitive. MS is particularly useful in the determination of structural and mass information of molecules. With the use of tandem mass spectrometry (Go et al., 2003; Johnson et al., 1990; Louris et al., 1987; Medzihradszky et al., 2000; Neilsen et al., 2002; Sato et al., 1987) (MS^n) both molecular weight and structural information about a particular molecular species can be determined in a single analysis. MS fails or, at least, becomes more difficult when samples are not pure or there are matrix interferences that reduce ionization efficiencies of the intended sample analyte (Choi et al., 2001; Dijkman et al., 2001; Koch et al., 2003; Liu et al., 2002). Chromatography, on the other hand, is a separation process and provides information about chemical composition and constituent interactions with adsorption surfaces and solvents. Therefore, the unification of these two methods of analysis, where chromatography separates the components of a mixture, while MS provides structural information with regards to each of the eluting molecular species, was an obvious development in analytical sciences.

Long before LC-MS was considered as an analytical tool, GC and MS were unified to form what is now a routine method of analysis of enormous benefit to the analytical chemist. The technique finds widespread application in the pharmaceutical industry, environmental sciences, forensics, and sports, to name just a few. The ease by which MS and GC could be coupled was the driving force behind this technique. The types of carrier gas and flow rates used in GC were compatible with entry into a mass spectrometer ion source and mass analyzer. However, it quickly became apparent that GC-MS was suitable only for the analysis of substances that were relatively nonpolar, low-molecular weight (Whelan, 2000), and of course, thermally volatile. LC, on the other hand, does not have such limitations and its usefulness extends to the analysis of a large variety of compounds with high molecular weights and various other molecular attributes. LC separations can be achieved through the selective choice of stationary phases and mobile phases and seemingly represents a more useful method for separations and hyphenation with MS. However, the liquid mobile phase presents a sizeable challenge as MS primarily operates in the analysis of gas phase ions. The introduction of typical analytical scale LC flow rates (0.5–2.0 mL/min) directly into the mass spectrometer would significantly decrease the sensitivity and working ability of the spectrometer when the liquid is converted to the gas phase. For instance, a flow rate of 1.0 mL/min of GC carrier gas equals 1.0 mL/min of gas entering into the mass spectrometer at atmospheric pressure. In the low-pressure zones (10^{-6} torr) of mass spectrometers the volume becomes equivalent to 12.5 mL/min. In LC, a flow rate of 1.0 mL/min of liquid methanol and water yields 593 and 1240 mL/min of gas at atmospheric pressure, respectively, which is equivalent to 7400 and 15,500 mL/min at a pressure of 10^{-6} torr, respectively (Abian, 1999). These large volumes would severely overload the vacuum equipment of the MS (Arpino et al., 1979).

Interfaces for the hyphenation of LC and MS have been the focus of many scientific groups over the last 30 years. Many excellent reviews (Abian, 1999; Gelpi, 2002; Niessen, 1998, 1999, 2003; Niessen and Tinke, 1995; Vestal, 1984, 2001) have discussed the important features of these interfaces. Following is a discussion of the numerous types and variations of the interfaces for LC and MS.

A. Manual Treatment

The simplest method for coupling LC with MS is one where the LC chromatogram is broken down into fractions or sections. These fractions are generally collected into separate sample vials and blown down under nitrogen or evaporated. These fractions can then be alternately introduced into the mass spectrometer sampling probe and subsequently analyzed. This method is, however, inefficient and also less reproducible because of operator bias. Automated interfaces are therefore the most viable method for LC-MS analysis.

B. Mechanical or Moving Transport Interfaces

Original attempts at interfacing LC with MS treated the hyphenation akin to GC. These processes basically involved transport of the LC eluent to a vaporizer, which subsequently allow transportation of the vapor to the ion source. This method, however, is only viable for volatile samples.

The first attempt to hyphenate LC and MS was illustrated by Scott et al. (1974) employing the so-called moving wire interface. In this interface the chromatographic eluent, at a flow rate of 10 µL/min, is deposited onto a mechanically driven moving wire, which is in turn fed through two differentially pumped chambers (i.e., the pressure is lowered in each consecutive chamber). Small ruby apertures separate these chambers. Helium is passed through a Tee-junction between the chambers to remove contaminants. From here, the wire is fed into the ionization chamber where the sample is heated and vaporized, subsequently allowing for both chemical ionization (CI) and electron impact (EI) ionization. Sensitivity for vitamin A, for example, is ~5×10^{-7} g/mL.

A similar interface designed by McFadden et al. (1976) is that of the classic moving belt interface, which is illustrated (mainly for historical significance) in Fig. 1. This interface increases sample utilization compared with the Scott et al. (1974) design. By using a stainless steel moving belt, an increased amount of sample (flow rate 1 mL/min) is transported to the ionization chamber of the mass spectrometer. Essentially, the eluent from the LC is deposited onto the moving belt, which carries the eluent through vac lock chambers (to aid desolvation) to a chamber connected to the entrance of the MS. In this chamber, the eluent is flash vaporized by thermal and radiant heating, which allows transport of the vapor (approximately 25–40% of the sample) to the ionization chamber. The moving belt method has also been designed to accommodate thermospray and laser ionization interfaces (Hardin et al., 1984). It is noteworthy that although this interface is no longer used for the hyphenation of LC-MS, it has been utilized for the multidimensional chromatographic hyphenation of LC-GC (Yang et al., 1984), the topic of which will be discussed in more detail later in this chapter.

Recently, mechanical interfaces have resurfaced because of the popularity of matrix-assisted laser desorption ionization (MALDI). This ionization mechanism will be discussed in more detail later. Briefly, mobile phase is mixed with a matrix compound [e.g., 2,5-dihydroxybenzoic acid (DBH), 4-hydroxy-3-methoxycinnamic acid, and 33-hydroxypicolinic acid], which is then irradiated using a laser (at ~355 nm) (Ørsnes et al., 2000). The ions produced are then desorbed in a vacuum chamber.

Preisler et al. (1998) developed a rotating wheel approach using MALDI. A diagram of this interface is shown in Fig. 2(a). The eluent from an LC is delivered to a rotating quartz wheel via an interface capillary inside the vacuum region of a time of flight (TOF) mass

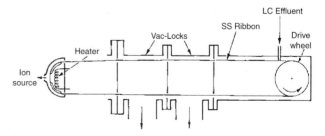

Figure 1 Schematic of LC-MS belt interface. [Reprinted from McFadden, et al. (1976), copyright 1976, with permission from Elsevier.]

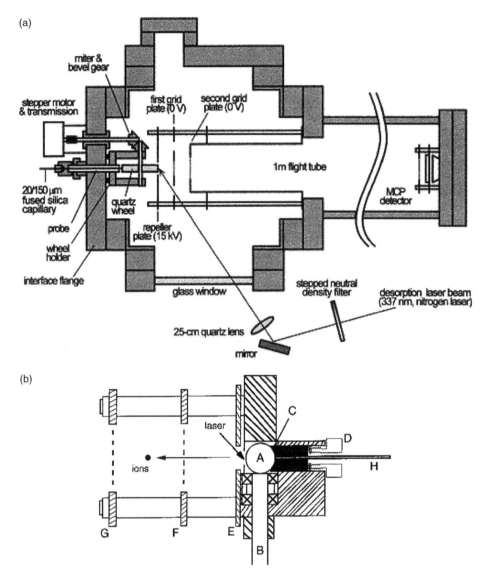

Figure 2 (a) Schematic of the online MALDI-TOF MS instrument (top view) incorporating a rotating quartz wheel. [Reprinted with permission from Preisler et al. (1998a), copyright 1998 American Chemical Society.] (b) Diagram of the online ROBIN-MALDI probe: A, 10 mm in diameter stainless steel ball; B, drive shaft; C, gasket; D, adjustment screw; E, repeller; F, extraction grid; G, ground grid; H, capillary. The ball is rotated through the shaft, which is connected to a gear motor positioned outside the vacuum chamber (not shown). [Reprinted with permission from Ørsnes et al. (2000), copyright 2000 American Chemical Society.]

spectrometer. To prevent freezing of sample solution in the vacuum, the capillary is kept in contact with the rotating wheel. Once inside the vacuum chamber the solvent rapidly evaporates and is subsequently ionized when the sample/matrix is moved past a nitrogen laser set at 337 nm. Following this, the repeller plates accelerate the ionized sample (pulsed) into the TOF analyzer. This construction accommodates flow rates between 100 and 400 nL/min only, which when coupled to an LC, requires a splitting Tee for flow rate reduction. This interface requires cleaning after each analysis, which led Preisler et al. (1998b) to replace the wheel with a disposable Mylar tape. The time available for continuous operation then depends on the length of Mylar tape used and on the speed of the stepping motor, which carries the tape from the capillary probe to the MALDI spot. This method allows for a relatively high degree of reproducibility (standard deviations typically 15% of peak areas during continuous introduction with sensitivities in the nano molar range for peptide mixtures).

Another approach adopted by Ørsnes et al. (2000) uses a similar approach to that shown in Fig. 2(b) for MALDI-TOF, which they called rotating ball inlet or ROBIN. Sample eluent/matrix enters the interface through a capillary (0.06 mm i.d.) and is delivered to the surface of a 10 mm stainless steel rotating ball. The ball is rotated at 13 r.p.m., accommodating a flow rate of up to 2 µL/min, with 5% sample utilization. Similar to the interface developed by Preisler et al. (1998a, b), ions desorbed by MALDI are moved to an extraction grid using a repeller. Ions in the vicinity of the extraction grid are then pulsed down the TOF. The ROBIN-TOF interface can operate for 25 h with no memory effects because the flow capillary is in contact with the rotating ball allowing for a "snow plough" cleaning effect. Flow injection analysis reveals that this interface provides highly reproducible results. Ørsnes et al. (2000) noted that the high pressure (5×10^{-5} torr) in the ion source affects mass resolution because of collisions during acceleration. The ROBIN-TOF system is able to analyze high molecular weight proteins, such as cytochrome c (M_w 12,384) and bovine serum albumin (M_w 66,430). The mass resolution using insulin and bradykinin at 20 fwhm (full width at half maximum) is typically 300.

C. Fast Atom Bombardment Interfaces

There are three variations of the continuous flow fast atom bombardment (cf-FAB) LC-MS interface. These are the frit type, angled tip, and coaxial continuous flow designs. In FAB typically high-speed xenon or argon atoms with a kinetic energy of 2.0–8.0 keV bombard a liquid sample (Arpino and Guiochon, 1982) mixed with a suitable matrix (generally glycerol/water compositions) producing MH^+, and/or MNa^+ ions (among others).

Ito et al. (1985) illustrated the design of the frit-based FAB interface [Fig. 3(a)]. The important feature of the design is the small internal diameter capillary that allows an interface between the HPLC column on the outside of the MS with the internal area of the ion source and frit. In these designs the glycerol matrix can be added prior to HPLC separation or with the use of a mixing Tee after the HPLC separation. The mobile phase/matrix solution passes through the capillary to the frit surface where efficient mixing of the matrix and mobile phase occurs. Generally the solvent evaporates leaving the glycerol and sample. An argon beam bombards the frit surface and produces ions (Ito et al., 1985). However, significant band broadening is observed with this interface, primarily as a result of solute spreading associated with the frit, and this limits sensitivity. Another problem associated with frit-FAB systems is the presence of the glycerol matrix ions in the mass spectra, although this can be overcome by background subtraction. Reproducibility using flow injection is generally excellent.

At around the same time, Caprioli et al. (1986, 1989) developed a cf-FAB interface for MS without the use of a frit. The design consists of a small capillary (0.075 mm i.d.) contained inside a hollow tube with an angled tip. The FAB gun is operated with xenon. These interfaces are generally heated to about 40–60°C to prevent freezing of the eluent in the vacuum ion source chamber (Caprioli et al., 1986). During operation, matrix/eluent is continuously brought to the head of the tip where FAB occurs, producing ions. Optimization of the matrix/eluent flow rate is necessary to prevent evaporation and freezing inside the capillary and, depending on the type of chromatography, flow rates are typically between 1 and 5 µL/min (Caprioli et al., 1989; Davoli et al., 1993), although some report the use of flow rates as high as 10 µL/min (Abian, 1999). Consequently, postcolumn Tee splits for standard LC column dimensions are required. The addition of trifluoroacetic acid (0.3%) is found to enhance the sensitivity of FAB in the analysis of peptides. The use of salts and buffers decreases the sensitivity of the interface limiting the choice of chromatographic modes (Caprioli et al., 1989). The technique is also not suited to gradient elution because the FAB efficiency is mobile phase dependent (Shomo et al., 1985). The cf-FAB interface has a linear response for the peptide substance P over a range of 0.7–135 ng. Flow injection analysis shows that this interface has relatively poor reproducibility (peak height standard deviation $\pm 42\%$, peak areas $\pm 10.7\%$). S/N ratio for the peptide substance P is in the order of 5:1 at 340 pg (Caprioli et al., 1986).

de Wit et al. (1988) and Moseley et al. (1989a, b) produced an enhanced design similar to that of Caprioli et al. (1986, 1989). They constructed a cf-FAB interface that employs a sheath flow of matrix. The interface design is illustrated in Fig. 3(b). In this design, the capillary or open tubular liquid chromatography (OTLC) column carrying the LC eluent is slightly withdrawn compared with the sheath column carrying the FAB matrix at the tip of the interface to prevent premature evaporation inside the OTLC column (Mosely et al., 1989b; de Wit et al., 1988). The ability of the design to independently transport the eluent and the FAB matrix to the interface tip is the important aspect of this construction (de Wit et al., 1988). Transportation can be achieved via a special Tee-piece, as shown in Fig. 3(c). The addition of the matrix prior to the LC separation can cause changes in the selectivity of the chromatography and increase the backpressure of the system (Mosely et al., 1989a). Postcolumn addition at the flow rates of OTLC can cause unnecessary band broadening. Optimizing the matrix composition from 50/50 water/glycerol to 75/25 water/glycerol (Mosely et al., 1989a; de Wit et al., 1988) can visibly reduce band tailing. Limits of detection for certain peptides are in the order of 54 fmol (Mosely et al., 1989a), In some cases, however, the FAB matrix ions dominate the mass spectra.

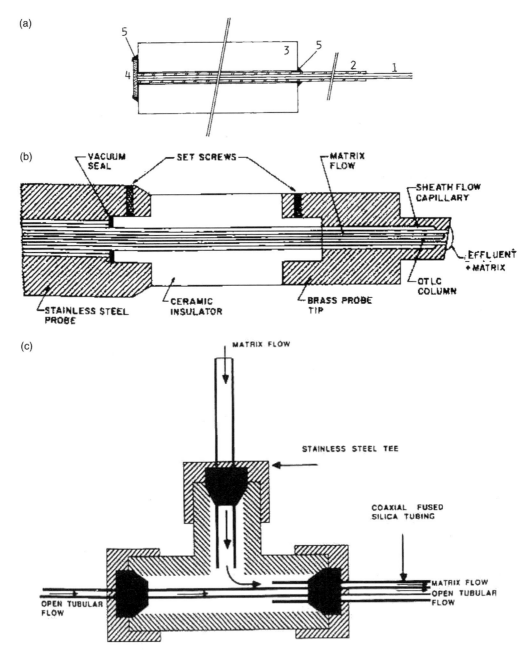

Figure 3 (a) Structure of the frit-HPLC-FABMS interface: 1, Fused silica tubing (40 μm i.d., 0.19 mm o.d.); 2, stainless steel tubing (0.19 mm i.d., 0.41 mm o.d.); 3, glass tubing (0.5 mm i.d., 3 mm o.d.); 4, Stainless steel frit; 5, epoxy-resin adhesive. [Reprinted from Ito et al. (1985), copyright 1985, with permission from Elsevier.] (b) Schematic diagram of the FAB probe tip used for coaxial flow applications. [Reprinted from de Wit et al. (1988), copyright 1988, John Wiley and Sons Limited. Reproduced with permission.] (c) Schematic diagram of the introduction of the matrix flow around the OTLC column. [Reprinted from de Wit et al. (1988), copyright 1988, John Wiley and Sons Limited. Reproduced with permission.] (d) Cross-sectional view of the modified FAB target. The base is screwed on the Vespel support and only the FAB probe tip has to be exchanged. The FAB source box is screwed to the base as normal, holding the probe tip in place. Here, the tip with the drain capillary is shown. [Reprinted from Hau et al. (1993), copyright 1993, John Wiley and Sons Limited. Reproduced with permission.]

More recently Hau et al. (1993a) sought to optimize the cf-FAB and frit-FAB interfaces. Poor reproducibility associated with the cf-FAB interface is improved [as shown in Fig. 3(d)] by the addition of a small-bore drain allowing for efficient drainage from the FAB tip. Hau et al. (1993a) also illustrated the utility of a conical extraction lens in conjunction with quadrupole ion optics, instead of slit extraction plates and standard

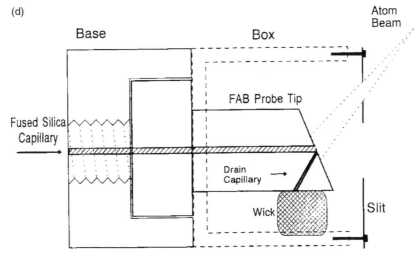

Figure 3 *Continued*

Einzel lenses. This set-up yields better sensitivities and lower background noise because of matrix clusters.

D. Particle Beam and Monodisperse Aerosol Generation Interfaces

The concept of the particle beam (PB) and monodisperse aerosol generation interface for liquid chromatography (MAGIC) has been around for some time. However, intense interest in the development and usage of the interface was generated by the work of Apffel and Perry (1991), Browner and Boorn (1984), and Willoughby and Browner (1984).

It should be noted that the PB and MAGIC interfaces are almost identical with some modifications to the MAGIC to improve analytical performance. The interfaces can generally be divided into three sections (excluding the ion source) which are (1) nebulizer or aerosol generator (2) the desolvation chamber, and (3) the momentum separator.

Approximately 100–500 μL/min of eluent (Willoughby and Browner, 1984) can be accommodated by the interface. A pneumatic nebulizer converts the solvent to an aerosol. Helium is usually used as the nebulizer gas (most probably because of helium's inert nature) (Apffel and Perry, 1991). In the MAGIC interface, the eluent from the HPLC column is pumped through a capillary with a 25 μm diameter orifice resulting in the formation of a liquid jet. The liquid jet then breaks up by Rayleigh interactions with the surrounding gas forming liquid drops of a uniform size. This is called the cylindrical jet break-up process (Willoughby and Browner, 1984; Winkler et al., 1988). Reduction of the orifice to 10 μm diameter produces 20 μm diameter droplets (Winkler et al., 1988). Because of coagulation of droplets immediately following nebulization in the PB interfaces, the variation in droplet diameters is considerably higher than those formed using MAGIC (Willoughby and Browner, 1984). In the development of MAGIC, Willoughby et al. (1984) and Winkler et al. (1988) described a solution for the higher droplet diameter variation problem in PB interfaces in the form of a high velocity helium gas flow perpendicular to the jet spray to prevent coagulation of droplets formed by the cylindrical jet break-up process. This is illustrated in Fig. 4(a).

After the droplet formation, it is imperative that the solvent molecules are removed from the sample analytes. The droplets enter a desolvation chamber that is generally heated to about 40–60°C and held at near-atmospheric pressure to aid in the desolvation process (Apffel and Perry, 1991; Neissen, 1986; Willoughby and Browner, 1984). Heat exchange between droplets and the surrounding gas is more efficient at higher pressures than in a low-pressure chamber (Neissen, 1986).

The droplets then enter into a two-stage *momentum separator*, shown in Fig. 4(a) and (b), held at low pressure ($\sim 10^{-6}$ torr) and evacuated by two pumps. As shown in Fig. 4(b), the enhanced design of the momentum separator consists of a nozzle and two skimmers. The vapor or condensed gas particles enter the momentum separator at high velocities. The heavier particles, which have higher momentum relative to the surrounding vapor and helium molecules, pass through the separator and enter into the MS system. Lighter solvent particles are easily diverted from the sampling orifice and pumped away (Apffel and Perry, 1991). Therefore, the amount of gas entering into the MS is significantly reduced. The separator shown in Fig. 4(b) is enhanced compared with previous designs and allows higher transport efficiency by providing a less disturbed gas flow path through the skimmer region (Winkler et al., 1988). In previous designs, the skimmers

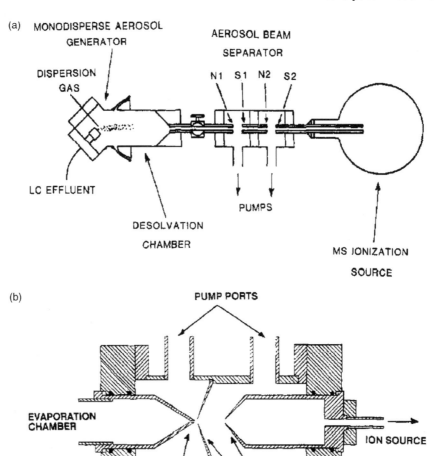

Figure 4 (a) Schematic diagram of MAGIC-LC/MS: N1, nozzle 1; N2, nozzle 2; S1, skimmer 1; S2, skimmer 2. [Reprinted with permission from Willoughby and Browner (1984), copyright 1984, American Chemical Society.] (b) Cross-sectional view of the new MAGIC momentum separator. [Reprinted with permission from Winkler et al. (1988), copyright 1988, American Chemical Society.]

were arranged in a perpendicular fashion to the gas flow path, which created turbulence and lowered particle transport efficiencies. Extrel commercialized a particle beam instrument that used a thermally assisted nebulizer called the Thermabeam (Abian, 1999).

Electron impact (EI), chemical ionization (CI), and electron capture (Giese, 2000) spectra can be produced with this interface. Although the interface provides excellent results in the analysis of polyaromatic hydrocarbons and flavonoids, the interface does have some problems; namely, poor linearity with regards to variation in different concentrations of p-phenylenediamine (Apffel and Perry, 1991) has been observed. However, this has been partially solved by the addition of mobile phase additives (ammonium acetate/oxalate). Also, a nonlinear response has been observed for p-phenylenediamine in the presence of coeluting compounds (Apffel and Perry, 1991). Peak area response also may decrease with a decrease in the volatility of the mobile phase (Willoughby and Browner, 1984). This indicates that the interface may perform poorly during gradient elution chromatography. Therefore, the strength of PB and MAGIC LC-MS lies in isocratic normal-phase separations (Song and Kohlhof, 1999). Nonlinear responses because of coeluting substances are easily resolved through the recent developments of 2D LC (Song and Kohlhof, 1999). With this in mind PB instruments are still used in conjunction with 2D LC in the analysis of vitamin D (Song and Kohlhof, 1999) (using CI and ammonia as the reagent gas). An advantage of this interface (even though it is no longer commercially available) is that spectral information can be applied to current mass-market spectral libraries for structural elucidation.

E. MALDI and Laser-Assisted Interfaces

This section considers the recent advancements made in utilizing laser ionization interfaces, specifically MALDI for LC-MS. Moreover, the pulsed nature of many of the MALDI interfaces makes them extremely suitable for the popular return of many different designs of TOF^n analyzers.

Laser desorption ionization has been used for some years, but this method is a selective technique. For instance, amino acids and dipeptides have strongly different absorption characteristics. Irradiation of both at a particular wavelength and beam strength may allow for absorption of the laser beam by the amino acids and not the dipeptides (Karas et al., 1985). This was illustrated by Wang et al. (1994) in the development of a pulsed sample introduction (PSI) interface using laser induced multiphoton ionization where certain compounds were observed to have different transfer efficiencies, attributed to differences in vaporization efficiencies. Though this attribute can be advantageous in some instances, it can cause problems in the analysis (quantitation) of compounds using LC-MS.

The PSI interface constructed by Wang et al. (1994) is illustrated in Fig. 5(a). This interface was designed to allow the pulsed introduction of a continuous aerosol into a TOF ion source and then to the analyzer. LC eluent is allowed to pass through a heated capillary (200°C) that generates an aerosol/supersaturated vapor.

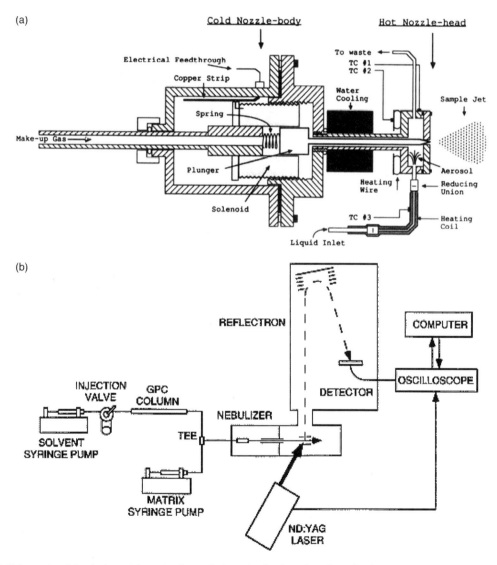

Figure 5 (a) Schematic of the design of the pulsed sample introduction interface. Drawing is not to scale. [Reprinted with permission from Wang et al. (1994), copyright 1994, American Chemical Society.] (b) Aerosol MALDI apparatus configured for online GPC-MS. Column effluent is mixed with matrix solution prior to pneumatic nebulization. Ions are formed from aerosol particles using pulsed 355 nm laser radiation and analyzed in a reflectron TOF mass spectrometer. [Reprinted with permission from Fei and Murray (1996), copyright 1996, American Chemical Society.]

This aerosol expands into the region containing a sample vaporizer contained at ~480°C. A solenoid and spring are used for the movement of a stainless steel needle. When a voltage is applied to the solenoid, this pulls the needle away from the exit orifice allowing a vaporized sample jet to form in the ion source. When the voltage to the solenoid is stopped, the spring causes the needle to move forward to seal the exit aperture. This allows a pulsed sample introduction method to be applied. The sample jet produced is then ionized by a 266 nm ionization laser source operated at 10 Hz. The mass resolution using the TOF system is generally around 300. The interface can accommodate LC flow rates from 0.5 to 1.6 mL/min; however, optimal peak height and width are observed at flow rates ~0.5 mL/min. Many mobile phase combinations can be used with the PSI, except high contents of water decrease detection sensitivity. This has been attributed to an increase in pressure when using water compared with organic mobile phases (from 5×10^{-7} to 5×10^{-6} torr). Nonvolatile buffers can cause clogging of the nebulizer; therefore, they should be avoided.

MALDI interfaces for LC-MS have recently received a lot of attention because of the ability of MALDI to ionize high molecular weight molecules, such as proteins (Karas, 1988; Kaufmann et al., 1993; Overberg et al., 1990). In traditional noncoupled MALDI, a liquid sample is collected, placed on a suitable target, and mixed with a matrix solution (e.g., K-cyano-4-hydroxycinnamic acid) (Laiko et al., 2000; Overberg et al., 1990). The target (with sample/matrix) is then placed inside the source chamber of the MS. Pulsed laser radiation is then absorbed by the matrix, causing both ablation of the surface and ionization of analytes via proton transfer (Murray et al., 1994; Overberg et al., 1990).

Two main MALDI interface types have been recently illustrated, which can be described as (1) an aerosol based MALDI interface (Fei and Murray, 1996; Murray and Russell, 1993; Murray et al., 1994) and (2) continuous flow frit/nonfrit designs (He et al., 1995; Li et al., 1993; Nagra and Li, 1995).

Fei and Murray (1996), Murray and Russell (1993), and Murray et al. (1994) pioneered the aerosol based MALDI technique for LC, which is illustrated in Fig. 5(b). The construction is relatively simple in design but extremely effective for MALDI. Eluent from the LC column is allowed to premix with a soluble matrix solution (e.g., *trans*-4-hydroxy-3-methoxycinnamic acid or 4-nitroaniline), which is usually acidified with trifluoroacetic acid (~5%). This is delivered to a pneumatic nebulizer that sprays the solution into a vacuum chamber (50 μm particles). The aerosol beam is then skimmed through a 25 cm *drying tube* (typically with an internal diameter of 4.0 mm and heated to approximately 300°C to desolvate the particles). The so-formed *dried* aerosols then enter the ion source region in which a 6 in. diffusion pump further lowers the pressure (1×10^{-4} torr) to assist in transport to TOF analyzers. In the ion source the *dried* aerosols cross the path of the 6 ns, 10 Hz pulsed laser source operated at 355 nm and focused to a spot size of 0.5 mm. Ions thus formed enter an extraction region and then enter the TOF, which is further evacuated with a 4 in. diffusion pump (1×10^{-5} torr).

For non-UV absorbing species it is important to include a matrix because without the addition of a matrix, as in the analysis of gramicidin, no signal is observed (Murray and Russell, 1993). An important feature of this method is that the most prominent peak (in most cases) is the MH$^+$ ion. Protonated cluster ions and alkali metal adduct ions are also usually observed (Murray et al., 1994). The utility of aerosol based MALDI was shown in the analysis of bradykinin (M_w 1060), gramicidin (M_w 1142), lysozyme (M_w 14,307), and myoglobin (M_w 16,951) (Murray and Russell, 1993; Murray et al., 1994). Mass resolution is generally in the order of 200–400 (Fei and Murray, 1996) although improvements are expected. LC flow rates between 0.5 and 1.0 mL/min are acceptable.

At around the same time as Fei and Murray (1996), Murray and Russell (1993), and Murray et al. (1994), He et al. (1995), Li et al. (1993), and Nagra and Li (1995) developed an alternative strategy for incorporating MALDI into a hyphenated LC-MS system. Inspired by the work on cf-FAB (Davoli et al., 1993; Hau et al., 1993a; Ito et al., 1985), Li et al. (1993) produced a cf-frit MALDI interface for LC-MS-TOF. As cf-frit MALDI can only accommodate flow rates in the region of 5 μL/min, eluent from an HPLC column, after the combination of mobile phase and a matrix (mostly 3-nitrobenzyl alcohol or 3-NBA), must be split using a splitting Tee (He et al., 1995; Li et al., 1993). The liquid enters the interface region through a silica capillary held in a stainless steel housing. The liquid subsequently passes to the probe tip where a stainless steel frit (1.59 mm o.d. and 0.794 mm thick) is pushed into the exit of the probe tip. A heater is located near the probe tip to prevent freezing of the solution near the exit of the capillary. An Nd:YAG laser (266 nm at 10 Hz to a spot size of 0.5 mm) is aimed at the continuously flowing fluid leaving the frit. Sample molecules are subsequently ionized and exit the source orthogonally via an extraction grid. The system requires a number of pumping stages.

A stable signal was achieved for several hours using this initial design, because the matrix was continuously replenished at the frit, in contrast to static MALDI. However, the signal intensity is usually ~5× less than in the static mode, which is attributed to differences in "local sample concentration" and vacuum pressure differences (Li et al., 1993). This is because of the initial design allowing the sample to diffuse over an area of about

5.0 mm, which is substantially larger than the 0.5 mm laser spot size, leading to inefficient sample utilization.

In an effort to improve mass resolution and signal to noise ratio, a redesigned probe with a Vespel tip and frit has been used. The frit-MALDI probe is locked directly (orthogonal to the TOF) through the inlet ring electrode of an ion trap with the laser opposite. Using the resonant ejection technique of the ion trap, resolution and the S/N were increased by applying an RF frequency voltage on the ion trap endcaps (gradient or step). Background ions are then selectively removed. Detection limits are in the order of 1–5 pmol for certain proteins (He et al., 1995).

Finally in an effort to maximize the ionization efficiency by virtue of a decrease in the sample probe area, the frit-based interface has been replaced by the frit-less design (Nagra and Li, 1995). Signal reproducibility at 10 pmol of cytochrome c is acceptable. No band broadening associated with the interface is observed and signal intensity is good, but mass resolution is generally poor (Nagra and Li, 1995).

Compared with the FAB interface, in which ionization efficiency decreases with increasing molecular weight, cf-MALDI suffers no such affliction (He et al., 1995 and references cited therein). Moreover, the interface can accept many buffers. The flow is continuous in these on-line MALDI interfaces and, as such, delayed extraction methods are typically employed for use in TOF (Juhasz et al., 1996).

F. Atmospheric Pressure Ionization and Nebulizer Based Interfaces

Atmospheric pressure ionization (API) interfaces were introduced for LC in the mid- to late 1970s and early 1980s as a superior alternative to the mechanical and direct liquid introduction (DLI) interfaces of the time. Common to the majority of API interfaces are the initial introduction and nebulization, ionization, and desolvation steps. The initial ionization occurs at atmospheric pressure, which enhances thermal transfer and thus desolvation. API interfaces require one or more intermediate pressure zones, necessitating the use of high power pumps to lower the pressure for final introduction into the low-pressure regions of the mass spectrometer. There are many derivations on the API theme, which include, most importantly, the ionization step and transfer of the liquid sample into the gas phase. These will be discussed separately. Also, other non-API interfaces utilizing a nebulizer will be discussed.

G. Atmospheric Pressure Chemical Ionization Interface

The atmospheric pressure chemical ionization (APCI) interface can be considered one of the original API interface sources for LC-MS. APCI interfaces are excellent in combination with LC as they can easily accommodate the 0.1–2.0 mL/min flow rates used in typical analytical scale HPLC.[1] Chemical ionization in early API interfaces was originally achieved using ^{63}Ni ionization or a corona discharge (Carrol et al., 1975; Proctor and Todd, 1983). Nowadays APCI is dominated by the corona discharge method of chemical ionization because of the improved spatial definition in ion formation (Proctor and Todd, 1983). A benefit of the corona needle arrangement is that it can be positioned to optimize the proportions and amounts of the reagent ions formed (Sunner et al., 1988a).

In earlier designs shown in Fig. 6(a) (Carrol et al., 1975; Henion et al., 1982), eluent from the liquid chromatograph generally enters through a small diameter tube typically heated to around 250–500°C (Draisci et al., 2001; van der Hoeven et al., 1996), which is surrounded by a sheath flow of heated carrier gas (zero grade helium or nitrogen). The heated gas assists desolvation and stabilizes the corona discharge (Carrol et al., 1975). Upon exiting the small diameter tube, the LC eluent is nebulized into a fine spray which, assisted by the flow of heated gas, travels down an *evaporator tube* (typically heated at 250–300°C) to further aid desolvation and evaporation. The mixture of solvent, analytes, and heated carrier gas then enter the ion source, which contains the corona needle (~10 μm tip) (Carral et al., 1975) held between 2 and 5 kV potential (Carrol et al., 1975; Doerge and Bajic, 1992; Draisci et al., 2001; Henion et al., 1982; van der Hoeven et al., 1996; Pfeifer et al., 2001; Proctor and Todd, 1983; Sakairi and Kambara, 1988; Sunner et al., 1988a, b). Here the most abundant molecular species (solvent and heated carrier gas) are ionized from the corona discharge. Analyte molecules become efficiently ionized through charge transfer, dissociate electron attachment, and proton transfer reactions with reagent ions. The most abundant processes are the proton transfer reactions (Doerge and Bajic, 1992; Proctor and Todd, 1983). The region housing the discharge needle is usually kept at a temperature between 50°C and 200°C. The ions formed in the vicinity of the corona needle are then sampled using a small diameter plate aperture. The aperture can be either a small orifice (Henion et al., 1982) or a membrane (Carrol et al., 1975), both of which allow the maintenance of lower pressures in the subsequent MS regions. A potential can also be applied to the sampling orifice to help extract more ions leading to higher ion signal intensity (Proctor and Todd, 1983).

---

[1] http://www.varianinc.com/, *Atmospheric Pressure Chemical Ionization (APCI)*, downloaded: 26/06/2003.

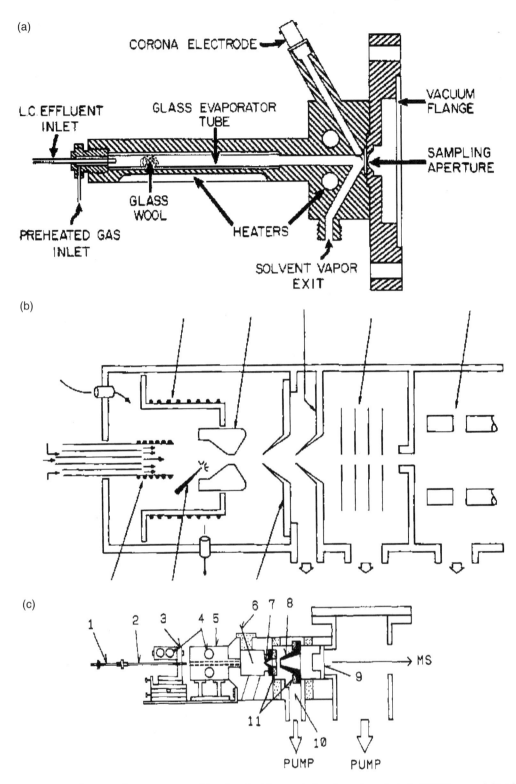

Figure 6 (a) Schematic diagram of corona source. [Reprinted with permission from Carrol et al. (1975), copyright 1975, American Chemical Society.] (b) Schematic diagram of the APCI interface. [Reprinted from Doerge and Bajic (1992), copyright 1992, John Wiley and Sons Limited. Reproduced with permission.] (c) Schematic diagram of interface and API ion source: 1, Teflon tube; 2, stainless steel pipe; 3, copper block; 4, cartridge heater; 5, stainless block; 6, needle electrode; 7, first electrode with aperture; 8, second electrode with aperture; 9, lens; 10, Intermediate region; 11, ceramic heater. [Reprinted with permission from Sakairi and Kambara (1988), copyright 1988, American Chemical Society.]

In some instances, the flow of carrier gas in the direction of nebulization can be used as part of the reagent gas in the ion source region. Zero grade nitrogen can be ionized from the corona discharge (Proctor and Todd, 1983) leading to a number of different ionized species that, through a number of processes (Proctor and Todd, 1983; Sunner et al., 1988a), lead to the formation of reactive hydrate cluster ions. It is then thought that these react through proton transfer processes (Proctor and Todd, 1983). Sunner et al. (1988b) have indicated that the sensitivity of analyte species in APCI can be grouped in terms of gas phase basicity (Sunner et al., 1988b).

A newer APCI interface is illustrated in Fig. 6(b) and (c) (Doerge and Bajic, 1992; Sakairi and Kambara, 1988). Eluent from the LC typically flows through the heated, pneumatically assisted nebulizer where it forms a jet and can enter into either a desolvation chamber or a heated sampling capillary. Upon exiting either the heated sampling capillary or the desolvation chamber, the sample is ionized by the corona discharge (Doerge and Bajic, 1992; Sakairi and Kambara, 1988). These ions move under potential to an intermediate pressure region. The intermediate region is isolated from the ion source and the MS transfer optics (i.e., RF-only quadrupoles) by conical shaped skimmers.

In earlier APCI interfaces, ions formed in the ion source entered into the low-pressure region of the MS, where cluster ions can form and reduce the sensitivity of the analysis (Proctor and Todd, 1983 and references cited therein). The presence of the intermediate region in newer versions serve a number of purposes. A potential can be applied between the sampling cone leading into the intermediate region and the final skimmer. This serves two purposes: (1) to increase the transmission of ions and (2) to break up cluster ions are broken up through collisions with neutral molecules. If a high enough potential is applied in the intermediate region, collisions of ions with gas can result in the breaking of covalent bonds leading to increased fragmentation (Sakairi and Kambara, 1988) (called up-front collision induced dissociation) and can provide a quasi-tandem mass spectrometer. Another analogous intermediate region has been illustrated by Sunner et al. (1988a, b). They used a nitrogen gas flow through the intermediate region (commercially called a gas curtain or nitrogen curtain gas)[2] to prevent solvent droplets and other particles from entering the analyzer region,[2] and also to provide neutral N_2 molecules for declustering cluster ions (Sunner et al., 1988a). This will be discussed later.

Sakairi and Kambara (1988) illustrated the importance of controlling the nebulizer and vaporizer sections independently to obtain optimal ion intensities, van der Hoeven et al. (1996) illustrated the utility of a thermospray adapted APCI source. Although APCI sources can operate in the range of 0.1–2.0 mL/min, it has been shown that the total peak area in total ion chromatograms for diazinon decreased over the flow rate region from 0.5 to 1.25 mL/min. Aqueous mobile phase compositions were also shown to affect the sensitivity of APCI ion responses. No effect was observed when the aqueous content was below 50%; however, at 75% water, a small deviation in response was observed (Doerge and Bajic, 1992). Flow injection analysis has shown that APCI is a very reproducible technique with consecutive responses having a standard deviation less than 3% (Sakairi and Kambara, 1988). Nonvolatile compounds such as certain amines, amino acids, peptides, antibiotics, nucleosides, steroids, and vitamins have been analyzed using this technique. Nucleosides, nucleotides, and alkaloids have recorded detection limits in the range of 10–100 pg (Sakairi and Kambara, 1988).

H. Atmospheric Sampling Glow Discharge Ionization Interface

Atmospheric sampling glow discharge ionization (ASGDI), although not entirely an API interface technique, has recently been shown by Dalton and Glish (2003) to have potential in the analysis of liquid based samples. Because of the similarities with APCI, ASGDI will not be discussed in great detail, but variations will be mentioned as discussed by Dalton and Glish (2003). A schematic of an ES-ASGDI (ES stands for electrospray and will be discussed shortly) is shown in Fig. 7. Briefly,

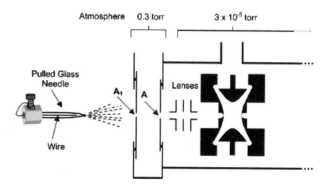

Figure 7 Side view of the custom nanoelectrospray source coupled to a custom atmospheric sampling glow discharge ionization source. Aperture plates A1 and A2 are attached to the flanges with PEEK screws and metal screws, respectively, and each plate is sealed with O-rings. The high voltage is applied to A1 to initiate the glow discharge, which occurs in the region between A1 and A2. [Reprinted with permission from Dalton and Glish (2003), copyright 2003, American Chemical Society.]

---

[2]http://www.appliedbiosystems.com, Document number: 00103840, *API 150 EX*™ *LC/MS System*, downloaded: 22/06/2003.

liquid samples are nebulized by an electrospray needle at atmospheric pressure (Dalton and Glish, 2003). A small aperture (A1) samples the resulting droplets/molecular species. A glow discharge established in the region defined by A1 and A2 (Chambers et al., 1993; McLuckey et al., 1988) results in the formation of ions. Two orthogonal electrodes attached to A1 are used as discharge electrodes. The region between A1 and A2 is under reduced pressure (0.2–0.8 torr) compared with the ionization region in APCI, which is at atmospheric pressure. Another important difference between APCI and ASGDI is that the glow discharge formed in ASGDI is formed at a current of 3–12 mA (Dalton and Glish, 2003; McLuckey et al., 1988), whereas the discharge current is commonly between 2 and 5 μA in the corona discharge in APCI techniques (Sakairi and Kambara, 1988; Sunner et al., 1988a). *Untreated* air is allowed into the ion source chamber, where it is ionized to form reagent ions that can react with solvent and analyte molecules leading to ionization as in APCI.

The types of reagent ions, the degree of fragmentation, and the ion intensity depend on the discharge voltage (Chambers et al., 1993; Dalton and Glish, 2003; McLuckey et al., 1988). Varying the discharge voltage can therefore be used to increase the acquisition of structural knowledge. Increasing fragmentation in this manner is analogous to the potentials applied in the intermediate regions in APCI. Fundamental differences include the operating temperature of the desolvation chamber and spray needle in APCI, which is operated at ambient temperature in ES-ASGDI. As ions are formed under reduced pressure in ASGDI, cluster ion formation is less of a concern compared to APCI (Dalton and Glish, 2003). ASGDI has been used only in the analysis of low molecular weight species such as *n*-butylbenzene, 2,3-butanedione, and other environmental trace organics (Dalton and Glish, 2003; McLuckey et al., 1988; and extensive studies are yet to validate this technique).

I. Thermospray Ionization Interfaces

Thermospray ionization apparently came about through studies in CI technology with the corona needle off (Blackley and Vestal, 1983; Blackley et al., 1980a). A typical thermospray interface for LC is shown in Fig. 8. Eluent from a standard size LC column (0.2–2.0 mL/min) enters the thermospray interface through a heated stainless steel capillary or a vaporizer, which forms a jet of solvent/solute droplets that continue to vaporize in the heated source chamber (Blackley and Vestal, 1983; Niessen et al., 1991). The position of the pumping line opposite the thermosprayer allows for the evacuation and removal of the majority of the nonvaporized solvent droplets (Blackley and Vestal, 1983). Small solvent droplets

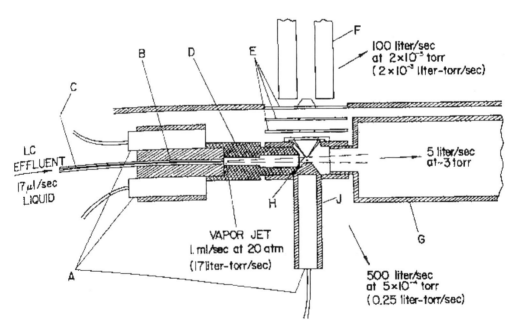

Figure 8 Schematic diagram of thermospray apparatus using electrically heated vaporizer: A, 100 W cartridge heater; B, copper block brazed to stainless steel capillary; C, 1.5 mm o.d. × 0.15 mm i.d. stainless steel capillary; D, thick-walled copper tube; E, ion lenses; F, quadrupole mass filter; G, pump out line to mechanical pump; H, ion exit aperture; J, source heater. [Reprinted with permission from Blackley and Vestal (1983), copyright 1983, American Chemical Society.]

continue to a further heated region consisting of an ion exit aperture and another electrical heater. Desolvated ions exit the source chamber orthogonally under the influence of a repelling device (not shown) (Ashcroft, 1997; Blackley and Vestal, 1983) through to the mass spectrometer ion focussing stages (Hau et al., 1993b). As ions are extracted orthogonally from the nebulizer direction, transport of solvent droplets into the mass spectrometer, which could reduce the sensitivity, is prevented.

Thermospray is known to be an extremely soft ionization technique that produces primarily molecular ions such as $M + H^+$ ions. Generally, in any one instance in certain solutions, statistics dictate the possibility of the presence of equal amounts of positively and negatively charged species. When the solution is rapidly disrupted as in nebulization, the net charge on the resulting droplets is due to statistical variations and could lead to predominantly positively (or negatively) charged droplets. Ions could then be desolvated by complete vaporization or through processes such as Rayleigh subdivision of droplets leading to particles with a single ion (Katta et al., 1991), or through ion evaporation directly from the charged surface of the droplets because of increased surface charge. Gas phase ion molecule reactions can also lead to the ionization of analytes in thermospray (Blackley and Vestal, 1983; Niessen et al., 1991).

Ionization of compounds with ionizable groups are easily analyzed by thermospray; however, polar nonionic compounds can be analyzed by the addition of electrolytes such as ammonium acetate. Ionization of molecules occurs via protonation and cation attachment reactions or vice versa for negative ion spectra. The addition of such electrolytes can decrease sensitivities and in some instances increase fragmentation of molecular ions (Schmelzeisen-Redeker et al., 1989). The sensitivity of the spectrometer depends on the sample properties, solvent composition, and the flow rate that enters the source chamber. Nonpolar and nonionic compounds can be difficult to analyze using thermospray.

Although thermospray can handle flow rates up to 2.0 mL/min, the operation of the interface is more efficient at lower flow rates (Blackley et al., 1980b). Higher flow rates into the source region require higher temperatures (Blackley et al., 1980a). At some flow rates more than 50% of the sample is transported to the mass spectrometer. Although excitement over thermospray was high in the 1980s, it was largely overshadowed by electrospray. In some instances, combination interfaces of electrospray-interfaced thermospray can be found (van der Hoeven et al., 1995). Electrospray can accomplish much of what thermospray can, but without the input of heat. Hence, thermally labile compounds could be analyzed more successfully.

J. Atmospheric Pressure Electrospray/Ionspray Ionization Interfaces

Electrospray as an atmospheric pressure interface has by far received the most attention from the analytical community and this is obvious by way of the numerous improvements made on the interface over its continuing lifespan. This point becomes more obvious when all the major commercial organizations currently focus on variations on electrospray as an ideal interface for liquid samples and thus on HPLC-MS.

Kostiainen and Bruins (1994), Gaskel (1997), and Voress (1994) neatly defined electrospray in three or four distinct steps, from the formation of the liquid jet to the transfer of ions to the gas phase. In the first stage, which can be described as the formation of charged droplets, a positive potential (generally 3–6 kV) is applied between the electrospray capillary and the so-called cylindrical electrode [this can also be a sampling aperture plate or sampling capillary—see Fig. 9(a)–(c)] and this leads to the formation of positive gas phase ions that are subsequently sampled into the mass spectrometer. Formation of negative ions is accomplished by the reversal of voltages applied to the electrospray needle and other electrodes (Chowdhury et al., 1990; Gale and Smith, 1993; Hau and Bovetto, 2001; Schwartz and Jardine, 1992; Smith et al., 1993; 1988a; Sui et al., 1991; Tor et al., 2003; Voress, 1994; Whitehouse et al., 1985a). The high potential field at the tip of the electrospray capillary in effect charges the surface of the emerging liquid (a positive voltage is applied to the electrospray capillary for general discussion), which results in the accumulation of positive ions at and near the surface of the liquid (Gaskel, 1997; Tang and Kebarle, 1993; Whitehouse et al., 1985a). As the surface of the liquid is concentrated with positive ions, the liquid surface is destabilized because of columbic repulsion forces and this leads to the formation of a jet spray commonly known as a Taylor cone (Wilm and Mann, 1994). The droplet size and the formation of the Taylor cone and resulting droplets are highly dependent upon the solvent used in electrospray. This can essentially be viewed in terms of surface tension and the presence of high concentrations of ionics. The electrostatic force formed at the electrospray needle pulls the liquid away from the needle and toward a sampling device (capillary or conical orifice), whereas the surface tension of the liquid effectively prevents this to varying degrees (Sui et al., 1991; Wilm and Mann, 1994). As such, a liquid with a high surface tension, such as highly aqueous phases, are difficult or impossible to electrospray (Bruins, 1998; Smith et al., 1988b; Sui et al., 1991; Voress, 1994). So are low polarity solvents, such as hexane, toluene, and carbon tetrachloride for other reasons (Kostiainen and Bruins, 1994).

Charged droplets [generally 0.5–2.0 μm (Bruins, 1998; Tang and Kebarle, 1993)] formed near the capillary tip in the Taylor cone migrate toward the counter electrode (because of the potential field difference and sometimes also with the assistance of a flow of bath gas in the same direction). Step two of the electrospray process is generally characterized by a reduction in droplet size, which occurs by solvent evaporation. A decrease in the droplet size or radius is accompanied by an increase in the charge density of the positive charges. This leads to increased electrostatic repulsion effects at the surface of the droplets. When the Rayleigh stability limit is reached, which is

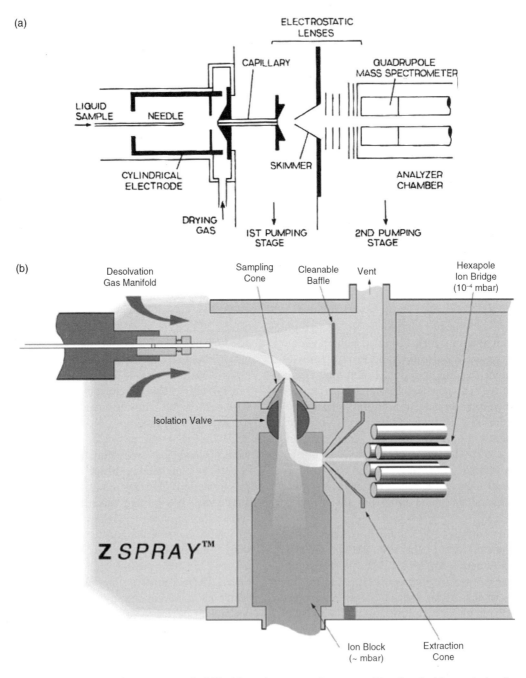

Figure 9 (a) Schematic diagram of the apparatus for MS with an electrospray ion source. [Reprinted with permission from Whitehouse et al. (1985a), copyright 1985, American Chemical Society.] (b) Schematic diagram of the Z spray interface for LC-MS. (c) Front end of the apparatus. Ion-spray capillary movable in three directions. Inner and outer diameters of the capillary are 0.1 and 0.3 mm, respectively (Drawing not to scale.). [Reprinted from Hiraoka et al. (1995), copyright 1995, John Wiley and Sons Limited. Reproduced with permission.]

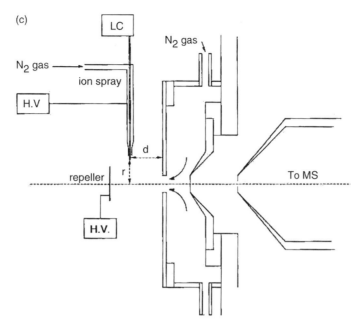

Figure 9 *Continued*

related to the surface tension, droplet disintegration occurs. This process is known to occur a number of times as droplets get smaller and smaller (Kebarle and Ho, 1997; Kostiainen and Bruins, 1994). Step three involves the actual formation of gas phase ions (Kebarle and Ho, 1997), which can be described by two competing mechanisms proposed by Iribarne and Thompson (1976), Thompson and Iribarne (1979), Kostiainen and Bruins (1994), and references cited therein. Röllgen et al. proposed that the continual reduction of droplet size could lead to the formation of a solvated positive ion or a droplet that contains only one ion. This would then be desolvated by evaporation leading to the formation of many gas phase ions (Gaskell, 1997). Alternatively, formation of gas phase ions could occur via ion evaporation, which occurs when the charge on the surface of the droplets becomes sufficient to allow for the direct emission of ions to the gas phase following a number of disintegration processes (Bruins, 1998; Hiraoka, 1992; Kebarle and Ho, 1997; Tang and Kebarle, 1993).

LC-MS systems containing electrospray interfaces are generally all the same, albeit with differences in the number of pumping stages, which depends largely on the type of analyzer. Much of the technology developed using electrospray has involved work on interfacing LC or capillary electrophoretic (CE) systems with MS. It should be noted that interfaces developed for CE-MS could also be utilized as interface designs for LC-MS.

Electrospray can generally accommodate flow rates between 0.5 and 20 µL/min of solvent (Chowdhury et al., 1990; Gale and Smith, 1993; Whitehouse et al., 1985a). An example of an electrospray interface is shown in Fig. 9(a). LC mobile phase is allowed to enter an electrospray capillary. The electrospray capillary is held at a potential of $\sim$3–6 kV relative to another cylindrical electrode, which is often a sampling capillary (Chowdhury et al., 1990; Gale and Smith, 1993; Hau and Bovetto, 2001; Schwartz and Jardine, 1992; Smith et al., 1993; 1988a; Sui et al., 1991; Tor et al., 2003; Voress, 1994; Whitehouse et al., 1985a). Charged droplets produced by the electrospray process in an atmospheric pressure chamber (Chowdhury et al., 1990; Schwartz and Jardine, 1992; Smith et al., 1988a; Whitehouse et al., 1985a) move toward an ion sampling device with the velocity produced from the liquid jet and down the potential gradient. During this transport continual evaporation of the solvent occurs. However, a significant reduction in signal intensity and the S/N ratio, along with increased background noise occurs when solvent droplets, uncharged molecules, and cluster ions reach the mass analyzer (Smith et al., 1988a).[2,3]

This reduction in sensitivity has been addressed in a variety of ways. The simplest solution was found by applying a counter current flow of heated nitrogen that assisted in the desolvation of solvent molecules (Smith et al., 1993; Whitehouse et al., 1985a) while also applying a potential in an intermediate region after the sampling capillary to allow for collision induced dissociation (CID) of cluster

---

[3]http://www.micromass.co.uk/systems/, *LCMS Z Spray*™ *... Presenting The Future of API LC-MS.*, Viewed on 09/04/2003.

ions (Chowdhury et al., 1990; Sui et al., 1991). An alternative involved the application of a heated (40–80°C N$_2$) gas curtain in a chamber prior to the sampling capillary (Bruins, 1991; Bruins et al., 1987; Hiraoka et al., 1995; Lee and Henion, 1992; Loo et al., 1989; Smith et al., 1988a, b). This technology, which is now used in commercialized MDS Sciex instruments,[1] assists in desolvation, prevents neutrals and unwanted neutral solvent particles entering into the analyzer, and finally assists in CID of cluster ions. Another alternative, provided by Micromass, is the so-called Z spray interface, which consists of a two-stage sampling process that consequently allows the introduction of higher flow rates for LC-MS interfaces.[3] This interface is shown in Fig. 9(b). The electrospray aerosol is directed perpendicular to the first sampling aperture, into which ions and gas phase components enter. Droplets and other particles then accumulate in the cleanable baffle, preventing these from entering into the mass analyzer. Meanwhile, in the second stage sensitivity is increased as ions are sampled into the second orthogonal aperture allowing for the introduction of higher flow rates.[3] This can occur as the second sampling aperture in Z spray interfaces can be enlarged when compared with conventional interfaces, thus sampling more ions. In conventional ES interfaces the sampling aperture sizes are limited as direct flow into the apertures increases gas flow in subsequent MS regions and can cause problems in maintaining pressure levels if the sampling apertures are too large.

Other methods of reducing intake of unwanted solvent particles, and so on, into the MS and reducing build-up of residue in the sampling orifice have been squarely aimed at placing the electrospray needle in a number of ways with the aim of increasing the sensitivity of the electrospray system. Off-axis nebulization (Bruins, 1991; Bruins et al., 1987; Sui et al., 1991) is the simplest way that has been achieved. Hiraoka et al. (1995), while studying ionspray (discussed shortly), noted that, although more prominent in ionspray, there is a distribution of droplet sizes from the center of the spray leading to the edges of the spray. Larger droplets are found toward the center of the spray, whereas smaller droplets reside near the edges. This indicates the importance of positioning the electrospray needle to reduce the amount of solvent molecules entering the sampling regions, which, as Bruins (1991) and Bruins et al. (1987) mention, prevents the sampling of large cluster ions. Varian also produces API interfaces that allow for the adjustment of the electrospray needle in the X and Y directions of the atmospheric pressure chamber.[4] In their design, the electrospray needle is offset at 79°. Finally, using ionspray, Hiraoka et al. (1995) illustrated the use of an orthogonal spray design. In this design, the spray needle is positioned parallel to the sampling orifice [Fig. 9(c)]. Ions enter the sampling region with the assistance of a repeller. In this position, increased sensitivity and reproducibility are observed, as are also less fouling or clogging of the sampling orifice. Signal intensities can be optimized by the positional adjustment of the spray needle.

After entering the sampling device, further desolvation occurs, which can be thermally assisted (Chowdhury et al., 1990). Sampled ions then pass through a differentially pumped region consisting of one or more chambers of decreasing pressure (Schwartz and Jardine, 1992). During these stages, ions are generally focused by using electrostatic lenses (Chowdhury et al., 1990; Whitehouse et al., 1985a) or RF-only quadrupoles (Gale and Smith, 1993; Loo et al., 1989). Focused ions then proceed into the low-pressure analyzer region of the MS.

As mentioned earlier, the use of highly aqueous mobile phases and nonvolatile buffers is undesirable with electrospray interfaces as this often leads to a loss in sensitivity and an unstable ion current (Bleicher and Bayer, 1994; Smith et al., 1993; Sui et al., 1991). Furthermore, in contrast to thermospray, a small amount of signal suppression is sometimes observed in the presence of substances such as ammonium acetate (Sui et al., 1991). To overcome such limitations, which are especially important for reversed-phase applications, a number of approaches have been pursued. Gale and Smith (1993), Loo et al. (1989), and Smith et al., (1988b) designed a probe, which includes a sheath flow of liquid methanol or acetonitrile that dilutes and lowers the surface tension of highly aqueous solutions so they can be electrosprayed. Another alternative, by Banks et al. (1994) incorporates an ultrasonically assisted electrospray interface, which utilizes a pair of piezoelectric crystals. The ultrasonically assisted electrospray device has an extremely stable ion count signal even when the mobile phase is changed from 100% methanol to 100% water. Another modification (Lee and Henion, 1992) of the electrospray interface incorporates thermal assistance (150–240°C, which corresponds to a vapor temperature of ~70–125°C). This device is solely aimed at increasing the acceptable flow rate of the electrospray interface to 0.500 mL/min. However, the heat assistance can alter mass spectral information, as shown in the analysis of the hexa-sulfonated azo dye direct red 80 and propylene glycol. The thermally assisted interface is 10× less sensitive than the ionspray interface.

Ionspray (Bruins, 1991; Bruins et al., 1987; Hiraoka et al., 1995; Hopfgartner et al., 1993; Michael et al., 1993) was introduced in the 1980s to overcome some of the limitations of standard electrospray. A diagram of an ionspray interface is shown in Fig. 9(c). Essentially,

---

[4] http://www.varianinc.com/, *Varian 1200L Quadrapole MS/MS*, downloaded: 26/06/2003.

ionspray is a derivation of electrospray with no heat required to form gas phase ions. As shown in Fig. 9(c), the interface consists of an electrospray-style capillary. Charging and formation of droplets are pneumatically assisted by a high velocity flow of nitrogen gas. As a result, the ionspray is able to utilize flow rates up to 1.0 mL/min, although 2.0 mL/min flow rates have been recorded (Bruins et al., 1987; Hopfgartner et al., 1993). These higher flow rates therefore allow ionspray to sample more of the mobile phase than does electrospray. Another advantage of the ionspray compared with electrospray is that ionspray can accommodate aqueous mobile phases as the pneumatic assistance physically acts to break the surface layer (surface tension). Thus, ionspray is a more useful interface than electrospray when gradient analysis is to be performed. Recently, MDS Sciex commercialized the TurboIonSpray®, which increases sensitivity of the ionspray by incorporating a number of ceramic gas heaters, most likely to improve the evaporation processes and enable high flow rates.[5]

A benefit of electrospray type interfaces is that they allow the analysis of high molecular weight compounds such as proteins (Loo et al., 1989). Electrospray is a very soft ionization interface resulting in the formation of molecular ions (MH^+) and multiply charged ions (Chowdhury et al., 1990; Gale and Smith, 1993; Hau and Bovetto, 2001; Lee and Henion, 1992; Loo et al., 1989; Smith et al., 1993; Tor et al., 2003). In this case, mass analyzers with a small mass range (i.e., single quadrupoles) are able to provide mass information on large compounds. Gale and Smith (1993) reported the electrospray analysis of horse heart myoglobin with 23 charges in positive mode analysis. Loo et al. (1989) reported the analysis of conalbumin ($M_w = 77,500$) with 67 positive charges and a bovine albumin dimer ($M_w = 133,000$) with 120 positive charges. In a review, Smith et al. (1993) commented on the analysis of certain polymers whose net charge increased with increasing molecular weight. Most high molecular weight species yielded a "bell-shaped" distribution of charges. A number of articles have discussed methods for the determination of the molecular weight of high molecular weight species from these distributions (Chowdhury et al., 1990; Loo et al., 1989). Therefore, an obvious benefit of electrospray type interfaces is the extension in the mass range that they provide for prospective mass analyzers. Moreover, electrospray devices are extremely useful when coupled to a tandem mass spectrometer where both molecular weight information and structural fragmentation information is obtained. Although, up-front CID and the use of photolysis reactors (Volmer et al., 1996) can also yield more fragmentation information.

TOF mass analyzers have recently gained attention because of short analysis times and other features when used in tandem arrangements. However, TOF^n mass analyzers require a pulsed introduction method. As discussed earlier, aerosol MALDI or laser ionization, has been used in conjunction with pulsed solenoid needle arrangement, which emits pulses of ions into the MS (Juhasz et al., 1996). Though this certainly could be used in electrospray, other methods of pulsed introduction have been illustrated by Michael et al. (1993). They illustrated the use of an ion trap where ions can be ejected using RF field changes. Others have reported on acceleration stages for orthogonal injection into TOF analyzers where only ions in the acceleration region gain any sort of velocity and can subsequently be analyzed. Magnetic sector and double focusing instruments require a highly focused ion beam, which, as mentioned previously, is easily achieved using a series of electrostatic lenses or RF-only octapols.

K. Sonic Spray Interface

Sonic spray (Dams et al., 2002; Hirabayashi et al., 1994; 1995) is a relatively new derivation to the API theme and is shown in Fig. 10. Essentially, liquid at a flow rate of 30 μL/min is passed through a silica capillary that extends ~6 mm into the ion source chamber. Unlike thermospray and electrospray, no heat or potential is applied to the capillary tip to assist in the generation of ions. Sonic spray employs a gas (nitrogen, oxygen, or argon) at ~3.0 L/min, which is passed coaxially in the direction of the flow of liquid in the capillary. At ~3.0 L/min the Mach number for the gas flow is ~1, which is a sonic velocity. Either side of this gas flow, the measured ion current

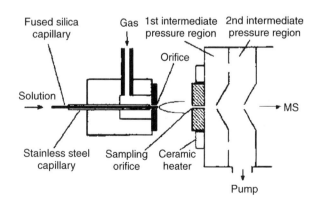

Figure 10 Schematic diagram of the sonic spray interface. [Reprinted with permission from Hirabayashi et al. (1995), copyright 1995, American Chemical Society.]

---

[5]http://www.appliedbiosystems.com, Document number: 00103841, *API 4000 LC/MS/MS System*, downloaded: 22/06/2003.

decreases sharply, for example, in the analysis of gramicidin (Hirabayashi et al., 1994; 1995). Ions produced as a result of the sonic gas velocities are sampled in the first intermediate pressure region, followed by transport to a second pressure region via a number of drift voltages applied to the aperture electrodes in the first and second intermediate regions. It is also important to note that the first and second intermediate pressure regions are differentially pumped (i.e., the second has a lower pressure than the first). Following this, an acceleration potential of 3–8 kV is applied to the second aperture to accelerate ions into the mass analyzer (Hirabayashi et al., 1994; 1995).

A potential is not applied to the capillary and so the relative intensities of mass spectra ion fragments are considerably different to those in ionspray. The ion current is dependent on the type of sonic gas used in the interface. Hirabayashi et al. (1995) speculated that gas phase ions were produced as sonic spray might produce droplet sizes of less than 100 nm. With droplets of this size, Hirabayashi et al. (1995) suggested that the distributions of positive and negative (protonated) ions might be different in particular droplets and when these undergo fission, gas phase ions could be produced following either ejection or complete desolvation.

L. Atmospheric Pressure Spray Ionization

Atmospheric pressure spray (APS) ionization is another variation on the atmospheric pressure source, employing a technique similar to that of thermospray, except that nebulization occurs at atmospheric pressure (Sakairi and Kambara, 1989; Sakairi and Kato, 1998). The interface consists of the usual capillary that facilitates nebulization. The capillary is typically heated to 480°C. A drift voltage of 100 V is applied to the first and second electrodes (sampling apertures) in the intermediate pressure regions (3.1 and 3.0 kV, respectively). From here, the interface is very similar to that already described for the APCI interface.

Although very similar in operation to thermospray, APS yields different mass spectra for certain cationized species than does thermospray. APS has shown utility in the positive ion analysis for cationized or protonated species, which include saccharides, peptides, glycosides, phospholipids, and vitamins. It is also noteworthy that multiply charged species do not display significant intensities in APS mass spectra (Sakairi and Kambara, 1989).

M. Atmospheric Pressure Photoionization Interface

ThermoFinnigan and MSD Sciex have recently introduced another API interface to their repertoire called the atmospheric pressure photoionization (APPI) interface.[6,7] An example of this interface is shown in Fig. 11. The APPI interface utilizes a heated (300–450°C) capillary for nebulization with the assistance of a nebulizer gas (nitrogen). The jet spray passes through a heated desolvation chamber followed by selective ionization by a krypton discharge lamp. Ions are then sampled into a number of intermediate pressure regions. The interface employs a nitrogen gas curtain for purposes mentioned previously (Robb et al., 2000).[6,7]

In APPI, a krypton discharge lamp is utilized, which emits photons in the range of 10–10.6 eV.[7] Krypton discharge lamps are utilized as most reversed-phase solvents have ionization potentials that are more than 10 eV, whereas many organic compounds' ionization potentials are lower than 10 eV. As a consequence, ionization of analytes can be a selective process. However, this can lead to very poor sensitivities. Addition of a dopant (5–15% of mobile phase), such as toluene with an ionization potential of 8.83 eV or acetone, to the flowing phase increases the ionization efficiency markedly. In this instance, the number of dopant molecules is present to a much greater degree than analyte molecules. Dopant molecules at atmospheric pressure then react with analyte or solvent molecules via charge transfer or proton transfer reactions. Solvent molecules then react with analyte molecules along to the same pathways. The use of dopant molecules can increase sensitivity by a factor of 100 (Robb et al., 2000).

APPI interfaces are relatively useful in the ionization of nonpolar low molecular weight analytes. APPI has been shown to achieve significantly higher ionization efficiencies for carbamazepine, acridine, naphthalene, and diphenyl sulfide than does the APCI interface (Robb et al., 2000). In fact APPI gives better signal intensities than APCI and ESI in the analysis of estrone.[7] In the analysis of compounds TBPA and PDP-Ac, APCI was significantly more sensitive in the analysis of PDP-Ac than was APPI. APPI, on the other hand, was more sensitive for the analysis of TBPA. This presented an interesting dilemma and was solved by using a combination of APPI and APCI. That is, the krypton lamp and the corona needle were employed at the same time and yielded ionization efficiencies (thus, signal intensities) significantly greater than when either ionization source was used independently.[7]

---

[6]http://www.appliedbiosystems.com, Document number: 00105894, *Product Bulletin-LCMS-Analyze Additional Compound Classes With The PhotoSpray™ Atmospheric Pressure Photoionization Source*, downloaded: 22/06/2003.

[7]http://www.thermo.com, *APCI/APPI Combination Ion Source*, Document ID: *productPDF_16661*, downloaded: 26/06/2003.

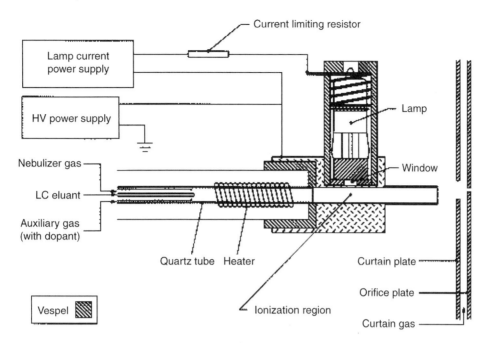

Figure 11 Schematic of the APPI ion source, including the heated nebulizer probe, photoionization lamp, and lamp-mounting bracket. [Reprinted with permission from Robb et al. (2000), copyright 2000, American Chemical Society.]

N. Cold Spray Ionization Interface

Cold spray ionization (CSI) interfaces are essentially identical to an electrospray or ionspray source except that the spray needle and the desolvation chamber is operated at low temperature. A cool nitrogen gas is employed to keep the spray capillary cool and the spray below $-20°C$. The desolvation chamber and ion source block are operated within $-80°C$ to $15°C$ (Yamaguchi, 2003). Although it produces spectra similar to electrospray, CSI finds utility in the analysis of unstable compounds particularly those that are thermally unstable. CSI has been used in the analysis of cage-type platinum complexes with significantly more intense peak profiles for higher charged molecular ions. Many other complex types have been analyzed by the cold spray methods (Yamaguchi, 2003).

O. Interchangeable API Interfaces

The majority of LC-MS manufacturers offer API interfaces that are interchangeable for different compounds. Certainly, this option is advantageous to the consumer as the cost of a specific mass spectrometer is undoubtedly higher than simply purchasing another interface. Previously, the combination of APCI and APPI was discussed as the option for ionization interfaces. Some authors have discussed retrofitting in which API interfaces can be substituted with a different interface. Sakairi and Kato (1998) recently illustrated the use of a multi-API interface that could operate in APCI, ESI, APSI, SSI, and atmospheric pressure electron ionization modes. This certainly gives the analyst more scope in choice for analysis of widely different compound types.

P. High Throughput Strategies in HPLC-MS

Increasing the amount of samples analyzed per unit of time by HPLC-MS is an important commercial strategy, driven by the need to screen many medical samples in a timely fashion. There are a number of different strategies that may be employed, described in two distinct groups: those that focus on the modification of the initial sampling and HPLC steps and those that employ modified interface designs for LC-MS.

Most often the chromatographic separation time limits the number of samples analyzed per unit of time. The simplest strategy involves the implementation of fast chromatographic methodologies. By utilizing smaller columns with smaller stationary phase support particles (Yang et al., 2001a) or monolithic media (Garrido et al., 2003) and higher flow rates, analyses can be achieved in a much shorter frame of time.

Xia et al. (2000) utilized a 10-port two-position switching valve, employing a dual extraction column set-up that preceded an analytical HPLC step followed by MS analysis. Injection of a sample onto one of the extraction columns trapped analyte onto extraction column 1. After the appropriate time period, a valve switch occurred and an alternate mobile phase back-flushed the analyte onto the analytical HPLC column. An extraction would also be performed on extraction column 2 followed by injection onto the analytical column while extraction column 1 was equilibrated with the initial liquid phase. Using this approach less time was wasted with the alternate nature of the first extraction step.

Van Pelt et al. (2001) illustrated the use of a four-column parallel chromatographic system that employed a staggered elution method. The experimental set-up employed a series of custom-built switching valves that connected four parallel identical chromatographic columns. Sections of delay tubing were employed in the gradient elution so that the delay length for column 2 was longer than column 1, and column 3 had a longer delay length than column 2, and so on. Delay times of chromatographic columns 1–4 were 0, 1.65, 3.30, and 4.95 min, respectively, and as such gradient conditions were selected such that the components injected onto each column eluted consecutively.

Another way to increase throughput is to utilize a number of robotic sampling probes using 96-well sample plates. Problems occur because of sampler washing time resulting in increased analysis time. By using at least four robotic sampling probes there is no time delay when performing a multitude of LC-MS analyses (Simpson et al., 1998; Zweigenbaum et al., 1999). Zweigenbaum et al. (1999) illustrated the utility of this procedure and recorded the analysis of 1152 human urine extracts in <12 h.

Employing multiplexed electrospray interfaces is another method for increasing throughput (de Biasi et al., 1999; Yang et al., 2001a, b) (such as the Mux interface commercialized by Waters and Micromass). An example is illustrated in Fig. 12. This interface utilizes either a four- or an eight-channel electrospray interface. The four-channel interface was recently evaluated (de Biasi et al., 1999; Yang et al., 2001b). Sample from the electrospray capillary is alternately introduced into the sampling cone by a rotating sampling aperture driven by a stepping motor. Total cycle times for the interface are generally in the region of 1.2 s (Yang et al., 2001b). A stream of heated nitrogen gas assisted desolvation. Initial tests resulted in the fast analysis of four test compounds (de Biasi et al., 1999). An advantage of the multiplexed interface is decreased analysis times of batch samples by a factor of four for the four-channel device and eight for the eight-channel multiplexed interface. However, the sensitivity of the multiplexed interface was less than that of a single electrospray interface with a signal intensity reduction of a factor of three. This was attributed to the lack of spatial positioning optimization available for the multiplexed interface. Moreover, the amount of spray introduced into the ion source was 4× that of a single sprayer. Another factor responsible for decreased sensitivity may be the reduced dwell time of each of the

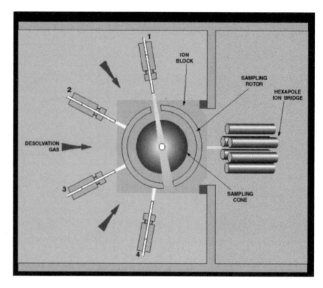

Figure 12 Schematic diagram of the four-channel multiplexed electrospray ion source. [Reprinted from Yang et al. (2001), copyright 2001, with permission from Elsevier.]

sprayers in the multiplexed interface. Crosstalk among the sprayers was also observed. Cross-talk among electrosprays was ~0.01% at 100 ng/mL and 0.08% at 1000 ng/mL. However, the multiplexed interface provides an excellent increase in the achievable throughput, which outweighs its limitations.

III. LIQUID CHROMATOGRAPHY–FOURIER TRANSFORM-INFRARED SPECTROSCOPY

Unlike UV-Vis spectra, the rotational and stretching vibrations resulting from irradiation of chemical species with IR electromagnetic radiation produce relatively narrow spectral lines. Consequently, the spectral information acquired following IR analysis can yield fingerprint identification provided a reference spectrum is available. The coupling of a chromatographic separation step to this form of component identification is potentially a very useful method of analysis, one that is commercially available. Despite the enhanced sensitivity associated with FTIR, the technique is still relatively specialized with only a moderate application base in chromatographic science. The limited sensitivity of the technique, which is primarily due to the opacity of the mobile phase, has lead to the development of somewhat more complicated methods of coupling LC to FTIR compared with UV detection. Essentially two methods of coupling have been developed. These methods are based on either flow cell designs for continuous monitoring or solvent elimination procedures, which can in some instances be effectively described as semicoupled or semicontinuous. The application of LC-FTIR has been extensively reviewed in an excellent publication by Somsen et al. (1999).

A. Flow Cell Methods of Analysis

The first flow cell designs were produced in the mid-1970s (Kizer et al., 1975; Vidrine and Mattson, 1978), but the technique has shown only moderate growth since that time. There are several factors that have contributed to the limited growth. (a) Solvent interference associated with the opaque nature of the mobile phase in the IR region of the spectrum that may overlap the absorption region of the analyte. Consequently, spectral windows may be small. Signal-to-noise ratios become substantial at any region where solvent absorption is appreciable. (b) The spectral windows may be small, limiting the application of the technique for qualitative identification purposes, negating the obvious advantage of the method. (c) Gradient elution procedures are not feasible, as baseline spectral subtraction becomes complicated in the changing solvent environment. (d) The sensitivity of IR detection is limited, due in part, to the necessity to have short path lengths that enable sufficient radiation to reach the detector. Path lengths in organic solvents are generally limited to <1 mm, whereas in aqueous solvents the path length is restricted to <30 μm (Somsen et al., 1999). (e) Signal averaging processes are limited as the technique is carried out in a continuous mode. Consequently, LC-FTIR is generally limited to separations that employ solvents that offer large regions of IR transparency. The strong absorption of IR radiation by water generally makes reversed-phase techniques difficult; solvents such as chloroform or dichloromethane, deuterated solvents (which exhibit less IR absorption), or Freon 113 are all IR compatible. The disadvantage of using deuterated solvents lies in its cost, even if microbore systems are employed. Polar modifiers, such as acetonitrile, can be added to these solvents without a substantial loss in IR transparency. Hence normal phase or size exclusion chromatographies are more frequently coupled than reversed-phase methods or other methods involving aqueous solvents. However, some aqueous methods have been reported that do employ flow cell detection. The area in which the LC-FTIR with flow cell detection is most widespread is in the analysis of polymers. This is largely because of the application of SEC methods, where IR-transparent solvents can generally be found and gradient elution techniques are not necessary.

Despite these inherent problems, two types of flow cells have been developed, which largely find specialized application. The type of chromatography employed (i.e., aqueous or nonaqueous) determines the material from which the flow cell is constructed. Potassium bromide (KBr), for instance, cannot be employed in RPLC because of the solubility of the KBr cell. Calcium fluoride (CaF_2) or zinc selenide (ZnSe) cells are insoluble in aqueous solvents and find more widespread use, even though their IR transparency is less than that of KBr. Transmission flow cells are generally prepared by creating a cavity between two IR transparent plates as illustrated in Fig. 13 (Hellgeth and Taylor, 1987). The LC eluent enters the cavity via capillaries with due care paid to minimize dead volume. The IR beam passes through the cell perpendicular to the flow stream. A study by Vonach et al. (1998) illustrates an unusual application of LC-FTIR that employs flow cell transmission detection (Vonach et al., 1998). What makes this application unusual is that the separation is ion exchange and an aqueous (slightly acidic) mobile phase is employed. The authors used a standard 25 μm CaF_2 transmission flow cell, which was commercially available from Perkin–Elmer. In order to protect the flow cell from the slightly acidic mobile phase, the flow cell was coated with low-density polyethylene. This coating was stable for a period of ~48 h. Spectral information was collected in the region between 900 and 1600 cm^{-1} for the analysis of CO bonding. A typical LC-FTIR chromatogram is illustrated in Fig. 14 (Vonach et al., 1998). The resulting separation showed that for the analysis of the wine samples the technique was both quantitative and qualitative with the organic acids and carbohydrates identified by the characteristic IR spectra. Detection limits for the carbohydrates, alcohols, and organic acids were ~0.2 mg/mL, which suggested that LC-FTIR was more useful than many researchers previously considered.

The cylindrical internal reflectance cell (CIRCLE) is based on the concept of attenuated total reflectance (ATR). The cell consists of a cylindrically shaped IR-transparent rod crystal with cone shaped ends mounted into a stainless steel cell. This cell can be mounted into

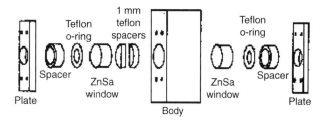

Figure 13 Schematic diagram of a modified Spectra-Tech, Inc. demountable flow cell. Windows are either ZnSE or CaF_2 material. Teflon spacers are 0.5 mm in width. All seals are Teflon O-rings. [Reprinted with permission from Hellgeth and Taylor (1987)].

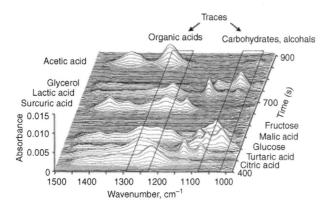

Figure 14 Typical HPLC-FTIR run of a standard solution containing 2 mg/mL of the main components of wine. Only retention times from 400 to 900 s are presented for better visualization (injection volume, 50 μL column temperature, 70°C). [Reprinted from Vonach et al. (1998), copyright 1998, with permission from Elsevier.]

the optical bench allowing the IR beam to be focused onto the incident cone face of the crystal. Mobile phase flows around the crystal, while the IR beam undergoes a series of reflections before exiting the flow cell through the opposite cone. The advantage of this type of flow cell is that the effective path length is increased according to the number of reflections that occur inside the crystal. Though these cells are useful for increasing the sensitivity of the technique, they have limited applications in quantitative work because response is essentially wavelength dependent.

B. Online Removal of Water Prior to Flow Cell Detection

Liquid–liquid partitioning, coupled online with reversed-phase separations and subsequent IR detection, has been employed to improve the detection capabilities, increasing both the sensitivity and the spectral window (Hellgeth and Taylor, 1987; Johnson et al., 1985). This technique eliminates water from the mobile phase by partitioning the solute molecules into an organic flow stream that is directed to the IR detector, while the aqueous eluent is directed to waste. A schematic diagram of this process is given in Fig. 15 (Hellgeth and Taylor, 1987). This process consists of three parts: (1) solvent partitioning of the solute into an immiscible organic solvent, (2) removal of water in a membrane separator, and (3) detection in an IR flow cell. The process of extraction by partitioning is a relatively straightforward task. The removal of water is achieved by passing the eluent through a hydrophobic membrane, allowing only the organic phase to pass. This membrane is illustrated in

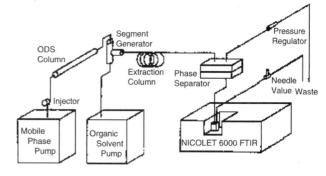

Figure 15 Schematic diagram of RPLC-FTIR instrumentation. [Reprinted with permission from Hellgeth and Taylor (1987).]

Fig. 16. The membrane consists of two stainless steel plates with grooves in each face and three polymeric films made of Gore-Tex material. The total internal volume is 32 μL. The pore size of the inner membrane layer is 0.2 μm, whereas the pore size of the two outer layers is 1 μm. Optimal performance of the membrane was achieved by the careful application of a static back pressure across the membrane. This entails regulation of the organic solvent flow rate by a metering needle valve. The operational lifetime of the membrane is in the order of several months. Using this approach there was minimal disruption to the chromatographic performance with detection limits at the microgram per component level. A similar approach has been employed using a solid-phase extraction procedure coupled online to the chromatographic separation (DiNunzio, 1992). A schematic diagram illustrating this approach is shown in Fig. 17 (DiNunzio, 1992). Here the LC eluent is passed

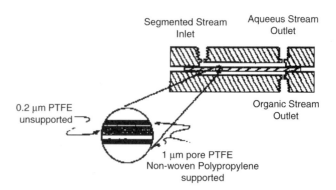

Figure 16 Cross-sectional view of membrane-phase separator with membrane highlighted. Membrane consists of multiple layers of Gore-Tex films: two layers of 0.2 μm pore unsupported PTFE surrounded by single layer of 1 μm pore PTFE with non-woven polypropylene support fabric. Polypropylene support is on the exterior of the membrane. [Reprinted with permission from Hellgeth and Taylor (1987).]

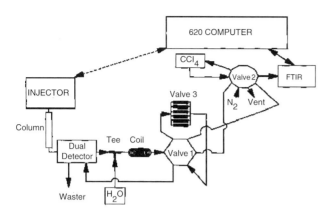

Figure 17 Diagram of the automated solid-phase extraction interface for HPLC-FTIR. [Reprinted from DiNunzio (1992), copyright 1992, with permission from Elsevier.]

through a series of SPE cartridges, after which they are dried with a purge of nitrogen gas. IR compatible solvent (carbon tetrachloride) is then directed through the SPE cartridges, extracting the solutes and subsequently this solvent stream is directed through an IR flow cell (DiNunzio, 1992). Similar designs have been incorporated in flow analysis-FTIR (Daghbouche et al., 1997).

Another process that has been employed for the removal of water incorporates a postcolumn acid catalyzed reaction using 2,2-dimethoxypropane: the reaction products being methanol and acetone. This approach was successful in removing water in mobile phases that contained up to 80% water, in both isocratic and gradient elution (Kalasinsky et al., 1989).

C. Solvent Elimination Methods

Solvent elimination procedures dominate the application of LC-FTIR and have essentially been responsible for the moderate growth of this technique. Without successful desolvation procedures, the technique of LC-FTIR would have floundered and have been confined purely to the realms of academic applications. The basics of solvent elimination techniques can be divided into two parts. The first involves the process of solvent elimination and the second involves the process of sample deposition onto a suitable substrate and subsequent IR analysis. Many of these techniques employ microbore columns in order to minimize the volume of solvent that needs to be eliminated. These techniques offer more versatility than flow cell methods primarily because aqueous eluents may be employed, although with some restrictions, such as limitations in the maximum flow rate because of the low volatility of water. This enables the application of reversed-phase methods, which dominate chromatographic literature. In principle, gradient separations may be employed as the changing solvent environment is no longer a baseline subtraction issue. When operated in this mode, the hyphenated technique (though it can still be dynamic) is usually semicoupled, with the IR analysis taking place subsequent to the chromatographic step. This allows the signal to be enhanced using signal averaging processes, improving the sensitivity.

D. Deposition Substrates

Solutes eluting from the chromatography column are deposited onto suitable IR compatible substrates. There are three basic types of substrates, depending on the mode of IR detection that is to be employed. The first type of substrate is that of IR transparent powders, such as KCl. The second type of substrate is reflective surfaces that is, aluminum mirrors or germanium disks coated with aluminum for enhanced reflectivity, and the third type of substrate is that of IR transparent windows, such as ZnSe or KBr transmission windows. Several different modes of detection can be employed, which to a certain extent is dictated by the type of substrate employed. For example, diffuse reflectance (DRIFT) is often employed when the analyte is deposited on powder, such as KCl or a stainless steel wire mesh (Mottaleb and Kim, 2002). This detection mode is very sensitive, but precision can be variable. Factors such as the homogeneity of the analyte deposition, powder compactness, and solvent evaporation can influence the resulting spectral response. Furthermore, the process of preparing the powder substrate in the collection cups can be tedious and time consuming. Reflection–absorption (R–A) detection is employed when the analyte is deposited onto polished surfaces, such as aluminum plates. More commonly nowadays, the deposition occurs on aluminum-backed germanium disks, because the germanium reduces spectral distortions. An illustration of the R–A detection process is given in Fig. 18 (Schunk et al., 1994). An advantage of the technique is that the IR radiation is passed twice through the sample and increased sensitivity is apparent. When using reflective surfaces for sample deposition, care should be paid to the homogeneity of the deposited sample as well as the width of the deposited sample trace. Inadequate solvent elimination prior to the deposition of the sample analyte can lead to sample spreading on the smooth reflective surface. In general, slower flow rates yield more uniform sample traces with subsequent improved sensitivity (Kok et al., 2002). However, problems such as phosphate buffers, which tend to accumulate on the disk are encountered, when nonvolatile solvent components are employed (Mottaleb and Kim, 2002). Direct transmission detection is employed when the sample is deposited directly onto IR transmission windows, such as ZnSe windows.

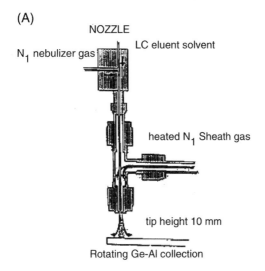

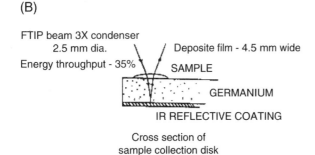

Figure 18 (A) Diagram of the LC-Transform concentric flow nebulizer. (B) Ray diagram of the IR beam path through the sample deposit and rear-surface-aluminized-germanium collection disk. [Reprinted from Schunk et al. (1994), copyright 1994, with permission from Elsevier.]

E. Solvent Elimination Interfaces

A number of solvent elimination interfaces have been employed in LC-FTIR. In many respects the solvent elimination problem that is encountered in LC-FTIR bears similarity to that in LC-MS, albeit for different reasons. Consequently, some solvent elimination processes employed in LC-MS have found favor in LC-FTIR, with some modification. Among the various methods that have been employed are thermospray, pneumatic, concentric flow, ultrasonic, and electrospray interfaces. More details of these solvent elimination procedures are given in the following sections.

F. Thermospray Nebulizer

Thermospray nebulization is achieved through the vaporization of HPLC eluent as it passes through a narrow bore tube with heated walls. The expansion of the mobile phase accelerates the velocity of the solvent and atomizes most of the remaining liquid. A spray is consequently produced, which contains a statistically equal number of positively and negatively charged species (Whitehouse et al., 1985b). The generated spray produces a diffuse deposition onto an IR substrate. Because of this, initial attempts to develop a thermospray deposition nebulizer resulted in generally poor and insensitive detection. However, more recent modifications made by Mottaleb and Kim, (2002) have alleviated some of the early concerns associated with thermospray deposition. In their most recent design, HPLC eluent is heated using a copper-heating block located adjacent to the capillary thermospray tubing. A heated supply of nitrogen gas surrounds the thermospray assembly and aids in the evaporation of the solvent. The desolvated analyte is deposited as individual spots onto a moving stainless steel belt. The spot width is typically around 13 mm with a thickness of 0.025 mm. An additional heat plate is located under the stainless steel belt to assist in the desolation of the analyte and minimize dispersion of the droplet. Variations as little as $\pm 3°C$ in the thermospray capillary temperature can result in deposition problems. More details of thermospray nebulization are covered in the LC-MS section and further discussion is not warranted.

G. Concentric Flow Nebulizer

Concentric flow nebulization was essentially developed by improvements made to the thermospray interface. Griffiths and Lange (1992) and Lange et al. (1991), in a series of publications, illustrated that a concentric flow of gas in a sheath surrounding an inner core of liquid was able to impart enough momentum to break up the liquid stream into a fine stream of droplets. The heated outer sheath of gas surrounding the nebulizer capillary increased the evaporation of the liquid fast enough such that the deposited substrate could be placed within 1 mm of the nebulizer tip. This minimized spreading of the sample analyte. An example of the concentric flow nebulizer together with the thermospray and hydrodynamic nebulizers is shown in Fig. 19. In this instance, the deposition substrate was KCl powder, ZnSe, or a metallic mesh. Measurements were made by DRIFT, or in the case of ZnSe, by direct transmission. An exploded view of the interface is shown in Fig. 19(b), and a close-up of the capillary tip located above the deposition substrate is shown in Fig. 19(c). The concentric flow nebulizer consists of two concentric fused silica tubes. The inner diameter and outer diameter tubes can be varied consistently with the flow of mobile phase. The outer tube is wrapped in a thermo wire, allowing the helium gas flow stream in this tube to be heated by conduction. Helium is used in preference to nitrogen because of the higher thermal conductivity of helium. Hence there is a better transfer of heat

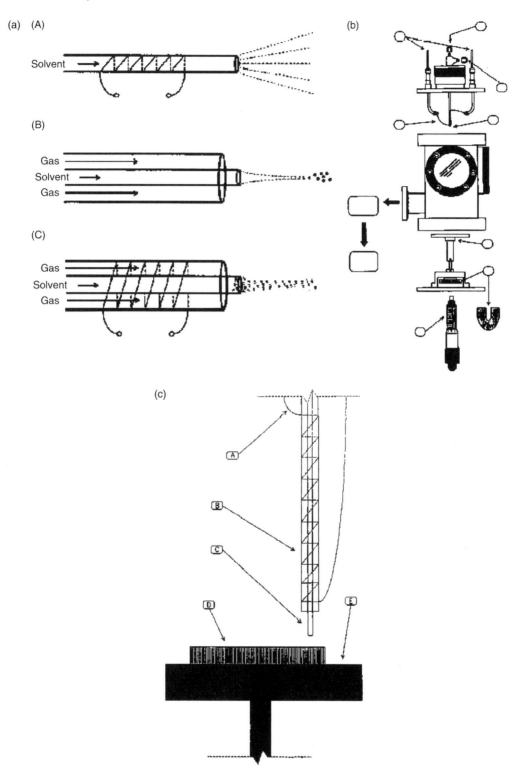

Figure 19 (a) Design elements incorporated into a concentric flow nebulizer: (A) Thermospraying, (B) hydrodynamic focusing, and (C) concentric flow nebulization. [Reprinted with permission from Lange et al. (1991).] (b) Exploded view of HPLC-FTIR interface. (A) Electrical feed-throughs, (B) inlet for chromatographic effluent, (C) inlet for He gas, (D) chromel heating wire, (E) concentric fused-silica tubes, (F) rotatable stage for solute deposition, (G) horseshoe magnet and bar magnet for stage rotation, (H) micrometer for vertical stage positioning. [Reprinted with permission from Lange et al. (1991).] (c) Close-up of stage shown in (b). (A) Chromel heating wire, (B) outer tube carrying He gas, (C) inner tube for carrying chromatographic effluent, (D) zinc selenide plate used as substrate for solute deposition, (E) rotatable stage. [Reprinted with permission from Lange et al. (1991).]

from the outer surface of the outer tube to the mobile phase flowing through the inner tube. Both tubes are connected via a Tee union; gas is directed through one arm and mobile phase through the other. The nebulizer tip is housed inside a vacuum chamber, which contains a rotating stage for solute deposition. Vertical positioning of the stage is achieved using a micrometer. This nebulizer design was found to be useful for mobile phases that contained 100% water, making the design suitable for reversed-phase or ion exchange processes, provided they were carried out in microbore systems. Detection sensitivity is noted to be <1 ng for the test solute methyl violet dye under capillary chromatographic conditions in which the flow rate was 2 μL/min. One critical factor in obtaining the smallest possible spot size on the deposition substrate is the alignment of the inner tube with respect to the outer tube. This must be concentric and becomes particularly important at higher flow rates. A laminar gas flow must also be maintained in order to improve the deposition of a compact sample zone.

H. Pneumatic Nebulizer

In the late 80s, a pneumatic solvent elimination process was developed by Gagel and Biemann (1987). The nebulizer that they developed is illustrated in Fig. 20, and the collection system is shown in Fig. 21 (Gagel and Biemann, 1986). The eluent from a microbore LC was sprayed from this nebulizer continuously onto a rotating reflective disk. The deposited analyte was then detected using R–A IR spectroscopy. In their design the LC eluent is mixed with pressurized nitrogen gas entering the flow stream through a high-pressure Tee-piece and directed to the deposition surface through a syringe needle. A curtain of heated nitrogen gas is directed across the outside of the syringe needle. This heated gas serves to warm the mobile phase eluent and the deposition surface, both of which aid the desolvation process. The heated curtain also serves to envelop the spray of eluent and minimize band dispersion of the deposited analyte. The nebulizer is located 3 mm from the deposition surface at an angle of 45°. Under such conditions the analyte trace is sprayed into a narrow region of 0.5–1.5 mm. Using this system, aqueous mobile phases can be employed allowing reversed-phase systems to utilize FTIR detection. The system also showed compatibility under gradient conditions. However, for gradient conditions, the gas temperature required variation to correspond to changes in solvent volatility throughout the application of the gradient. The nebulizer was tested on a mixture of naphthalenediol isomers in an aqueous reversed-phase system (55% water in methanol). Detection was monitored above 1000 cm^{-1}. Sensitivity of the

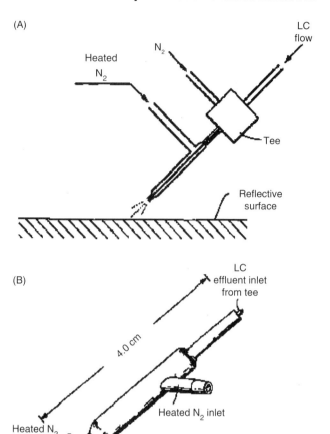

Figure 20 Schematic diagram of a nebulizer: (A) during operation and (B) detailed view of the glass nebulizer design. [Reprinted with permission from Gagel and Biemann (1987).]

technique was ~16 ng; however, for qualitative assessment of the IR spectra, 31 ng was the lower level of sensitivity. Lab Connections commercialized a version of this pneumatic nebulizer design under the name LC-Transform. The LC-Transform comes equipped with optics for R–A analysis. This nebulizer has found use mainly in polymer applications. An illustration of the interface is shown in Fig. 18 (Schunk et al., 1994).

I. Ultrasonic Nebulizer

Solvent elimination using an ultrasonic nebulizer is achieved by allowing the solvent to flow over a transducer that is vibrating at ultrasonic frequencies. This produces a fine spray that is directed to a deposition substrate assisted by a carrier gas (nitrogen). The ultrasonic nebulizer is housed inside a temperature-controlled vacuum chamber

HPLC-Hyphenated Techniques

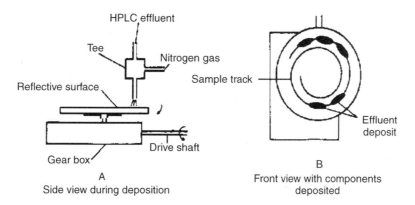

Figure 21 Diagram of the continuous collection device: (A) deposition of HPLC effluent and (B) view of deposited components. [Reprinted with permission from Gagel and Biemann (1986).]

(Torabi et al., 2001). An example of these types of solvent elimination devices is shown in Fig. 22 (Torabi et al., 2001). This design is a modification of the design by Dekmezian and Morioka (1989). The SEC-LC eluent flows directly to an ultrasonic nebulizer located in a vacuum chamber that can be purged with nitrogen. The vacuum chamber can be temperature-controlled. The LC eluent spray is directed onto germanium disks located on a rotating collection wheel. The wheel is 150 mm in diameter and contains 20 equally spaced wells, each holding 13×2 mm^2 polished Ge disks. Following collection of the sample constituents, the sample collection wheel can be removed and relocated in an FTIR instrument with a similar stepper motor.

J. Electrospray Nebulizer

Electrospray nebulization is a process that breaks up a fluid flow stream into a fine mist or spray by the application of an applied electric field (Whitehouse et al., 1985b). This process differs from thermospray in that a charge is used to produce atomization rather than atomization producing charge. In electrospray, a potential of several kilovolts is applied to a tube through which the mobile phase eluent leaves the chromatographic column. A charge is thus imparted onto the liquid as it leaves the tube. The resulting coulombic repulsion forces are sufficient to generate dispersion in a fine mist. Evaporation of the solvent molecules increases the surface charge density, which in turn increases the particle repulsion. Hence the radius of curvature of the liquid leaving the tube increases as the liquid undergoes further solvent evaporation. This phenomenon is referred to as a Taylor cone (Raynor et al., 1992). Stable electrospray conditions can be achieved for a number of solvents, but only at low flow rates and hence the technique is most suited to microcolumn applications. Raynor et al. (1992) illustrated the use of an electrospray interface for the removal of solvent and subsequent deposition of sample analyte onto a ZnSe plate. Mobile phase is transported through a capillary, which has an applied electric field of 3 kV. A sheath of nitrogen gas (1 mL/min) surrounds the capillary and serves to prevent any mobile phase being drawn back up the stainless steel tube by virtue of capillary action. The Taylor cone that results from the electrospray desolvation would appear to be inappropriate for the deposition of solute into a suitably narrow band for IR analysis. However, if the deposition substrate is an earthed semiconductor, such as ZnSe, for example, the divergence of the spray can be limited to a large degree. The optimum distance for the location of the substrate from the outlet tip of the capillary was found to be 1.5 mm. Under these conditions, HPLC-FTIR with an electrospray interface and deposition onto ZnSe plates with subsequent IR-microscopy analysis was successful in the qualitative analysis of caffiene and barbital. However, no information

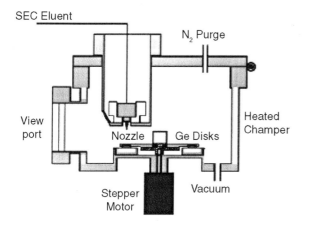

Figure 22 Diagram of the solvent–evaporation interface. [Reprinted from Torabi et al. (2001), copyright 2001, with permission from Elsevier.]

was given regarding quantitation, which was in all probability poor due to the nonuniformity of the deposition of the analyte onto the substrate.

K. Quantitative Considerations

Much of the importance associated with using LC-FTIR as a separation tool is associated with the ability of the FTIR to be used as detector that is capable of identifying and hence additionally resolving species as they elute from the chromatographic column; even if these species coelute, provided they have significantly different IR spectra. Very few studies have illustrated the quantitative aspects of the technique. A study by Kok et al. (2002) detailed the quantitative aspects of FTIR detection employing an LC-transform pneumatic nebulizer for solvent elimination and subsequent deposition of polymer onto a $60 \times 2 \text{ mm}^2$ (diameter × thickness) rear-surface-aluminized germanium substrate. Polymer solutions eluting in a range of mobile phase flow rates in 100% THF were deposited onto the collection substrate. They found that the homogeneity of the deposited sample onto the collection substrate was of the utmost importance if quantitation was important. As the flow rate decreased from 500 to 25 μL/min, the R.S.D. decreased from 26.5% to 13.5%. The chromatographic resolution was dependent on the molecular weight of the polymer to be analyzed. Triplicate depositions of a polystyrene standard mixture yielded a 4.0% variation in response and 0.47% in retention time for high molecular weight polystyrenes, and 5.8% variation in response and 0.21% in retention time for low molecular weight polystyrenes.

IV. LIQUID CHROMATOGRAPHY–NUCLEAR MAGNETIC RESONANCE SPECTROSCOPY

Perhaps, even more useful than MS is NMR, which is arguably the most powerful tool for structural elucidation. It is of little surprise, therefore that the hyphenation of LC and NMR has eventuated, despite the slow development of the technique. This slow development has largely been as a result of a lack of understanding between chromatographers and NMR spectroscopists with regards to their comrades' corresponding methodologies. In a recent review by Albert (1999), he implies that the ownership of knowledge between the respective fields is responsible for the delay in the obvious unification. It is in fact therefore little wonder that LC-NMR is considered to be an expensive technique, one that would gain little following in analytical chemistry. However, the basic requirement for LC-NMR is simply an LC system together with an accessible NMR. The technique is such that a dedicated NMR instrument is not essential for the purpose of hyphenation. Rather, access to a suitable instrument modified appropriately is required only when necessary. As awareness of the virtues of this technique grows, so too does the development of methodologies and instrumentation that support the increasing number of applications.

One of the key limitations that restricted the development of the LC-NMR technique, was the perception that to analyze samples by NMR the sample must be rotated so as to remove magnetic field inhomogeneities. However, with magnetic fields being generated by modern instruments employing cryomagnets, field homogeneity is high and as a consequence the sample need not be rotated. In fact, rotation of the sample leads to distortions in 2D NMR (Albert, 1999). The first reported online coupling of LC and NMR utilized a U-shaped tube as a sample holder located in the NMR probe body. This early design demonstrated that excellent NMR resolution could be obtained without sample rotation.

In comparison with the coupling of LC to MS, the physical unification of LC to NMR is straightforward. But the limitations of the technique lie in two areas. The first is suppression of signals from the commonly employed solvents used in LC, whereas the second is gaining suitable sensitivity and spectral resolution.

A. Solvent Suppression

Typical solvents employed in LC are water, methanol, and acetonitrile, each of which present large ^{1}H NMR signals. In the early days of LC-NMR, deuterated solvents were essential, but the cost of these solvents in potential routine application essentially would have meant the technique would be confined to purely specialized academic laboratories. Nowadays, the modern NMR is able to suppress very efficiently the solvent NMR signal, using techniques such as DRYCLEAN, WATERGATE, Water-PRESS, SLP, and WET (Macnaughtan et al., 2002). It is beyond the scope of this discussion to deal in detail with these methods of signal suppression, except to say, however, that signal suppression in flowing systems becomes less efficient because fresh sample and hence fresh spins are continually being replenished. Furthermore, when gradient systems are employed, solvent suppression becomes less ideal. However, there is one promising solvent suppression process that is worthy of further attention, as the technique may provide improved performance in flowing and gradient systems. This solvent suppression technique employs an NMR difference probe as shown in Fig. 23 (Macnaughtan et al., 2002). This difference probe consists of a dual coil probe that contains two sample

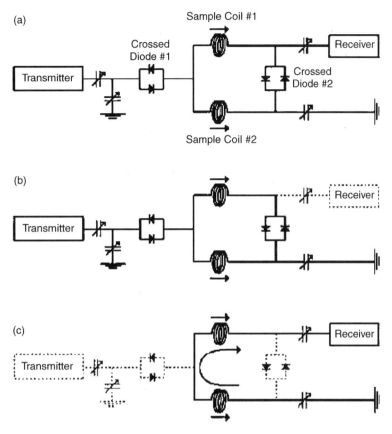

Figure 23 (a) The schematic diagram of the NMR difference probe resonant circuit with two coils and two sets of variable capacitors for tuning and matching each subcircuit. (b) The solid black lines indicate the active part of the circuit during the RF pulse when the crossed-diodes are on. The arrow beside each coil represents the relative phase of the transverse magnetization after the pulse. Because the coils are connected in parallel with respect to the transmitter, the arrows point in the same direction. (c) During acquisition of the NMR signal, the circuit changes. The crossed-diodes are off because the small voltage induced in the coils by the NMR signal is not enough to activate the diodes. The solid black lines indicate the active part of the circuit when the signal is collected at the receiver. The arrow beside the coils represents the phase of the voltages induced in each coil by the excited samples. As the coils are connected in series with respect to the receiver (following the curved arrow), the voltage arrows are opposing and the signals from each coil are 180° out-of-phase. [Reprinted from Macnaughtan et al. (2002), copyright 2002, with permission from Elsevier.]

coils in a resonant circuit that switches between parallel excitation and serial acquisition to cancel common signals, such as solvent and solvent impurities. Essentially, this technique is based on a dual beam background subtraction, where the reference signal and sample signal that are collected simultaneously are subtracted from each other automatically. No software manipulation, pulse sequence modification, or spectrometer alteration is necessary. Hence the technique does not lengthen the pulse sequence but it reduces experimental time. This should yield improved performance in multisolvent systems, such as gradient elution in continuous flow. Background problems are also less significant if nuclei other than ^{1}H or ^{2}H or ^{13}C are employed. For instance, because of their high sensitivity and low impurity, ^{19}F and ^{31}P are regularly employed, especially in the pharmaceutical industry.

B. Sensitivity

The sensitivity of NMR is at present the limiting factor in its application to LC detection. The sensitivity (LOD) of NMR is generally several orders of magnitude higher than both MS and IR. At any given magnetic field strength, the probe sensitivity is primarily determined by the ratio of the sample volume (V_s) to the active coil volume (V_c). That is, S/N as determined according to (Haner et al., 2000):

$$\frac{S}{N} \propto \eta V_c^{1/2} Q^{1/2} = \left(\frac{V_s}{V_c^{1/2}}\right) Q^{1/2}$$

where Q is the quality factor of the resonator circuitry and η is the filling factor ($= V_s/V_c$).

For purposes, largely related to thermal control, the wall thickness of the sample flow cell and the air gap between the outer wall of the flow cell and the detector coil cannot be shrunk below a minimum factor (Haner et al., 2000). The filling factor, therefore, generally decreases as sample tube diameter decreases. Typically continuous NMR flow probes are designed with a volume between 40 and 120 μL, which is much larger than UV detector. Hence, there is conflict between the requirements of chromatographic separation efficiency and NMR detection sensitivity. Even by employing a detector flow cell of this diameter does not ensure adequate sensitivity because dilution of sample constituents following separation and retention in the LC system may be significant. Under such circumstances a stop flow approach is adopted, whereby the flow in the chromatography column is stopped during the NMR analysis of the eluted component. Alternatively, the separated sample constituents can be loaded into an array of sample loops isolated from the NMR and LC via switching valves. These stored sample fractions can be successively loaded into the NMR flow probe for extended analysis until a suitable S/N has been obtained for structural elucidation. When LC-NMR is operated in a stop flow mode, it is essential that the dead volume of the system be accurately known, so as to be certain that the sample located in the flow cell is at the highest possible concentration. That is, the peak maxima are loaded into the flow cell. This has been made more reliable in recent times by the advent of shielded cryomagnets that allow the LC instrument to be located within 30–50 cm of the NMR. A typical instrumental configuration of a Bruker system is shown in Fig. 24 (Albert, 1999).

The diagram in Fig. 25 (Haner et al., 2000) illustrates a 120 μL flow probe designed for automated direct-injection NMR analysis. The region contained within the rectangle depicts the active coil volume, whereas the cross-hatched region depicts the minimum sample volume. The flow probe is compared with the conventional NMR probe tubes. The tapered inlet and outlet regions of the flow probe facilitate sample flow and reduce magnetic field inhomogeneity. The particular flow probe illustrated in Fig. 25 is prepared from thick-walled quartz, which provides a closer match in bulk magnetic properties to water than does thin-walled Pyrex or air. The design of this particular flow probe was such that the filling factor was less than that of a conventional 5 mm NMR tube, but greater than that of a 3 mm conventional NMR tube.

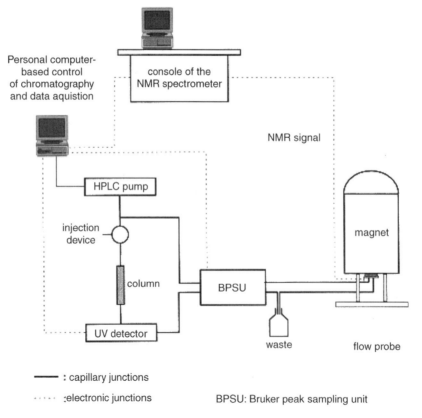

Figure 24 Instrument configuration for analytical LC–NMR experiments. [Reprinted from Albert (1999), copyright 1999, with permission from Elsevier.]

HPLC-Hyphenated Techniques

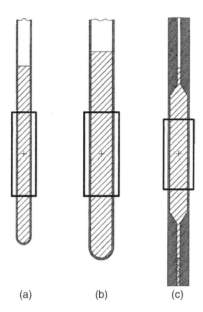

Figure 25 Cross-sectional drawing showing the active volumes and estimated minimum sample volumes for the small volume sample probes. The lengths and diameters of the RF coils are indicated by the rectangles, and the cross-hatched sample regions within the rectangles represent the NMR active volumes. The full cross-hatched regions within the tubes show the approximate minimum sample volumes. The double cross-hatched region at the lower end of the flow cell depicts the approximate location and boundary of the optimal push solvent. Diagrams show (a) the 3 mm probe, (b) the 5 mm probe, and (c) the 3.4 mm i.d. flow probe with 120 μL active volume. [Reprinted from Haner et al. (2000), copyright 2000, with permission from Elsevier.]

In general, the resolution and line shape of the 120 μL flow probe was very similar to that of the conventional 3 mm NMR tube operated under nonspinning conditions. The reproducibility of the flow probe was far superior to that of the conventional 5 mm tubes and the sensitivity of the flow probe compared favorably to that of the conventional 5 and 3 mm NMR tubes. Temperature equilibration (under constant air flow rate) of the 3 mm conventional NMR tube was faster than that of both the 120 μL flow probe and the 5 mm conventional NMR tube, largely as a result of the decrease in surface area. However, because of the rigid mounting of the flow probe, higher rates of temperature-controlled air could be passed through the air gap, which ultimately enabled faster temperature stabilization than the conventional NMR probes.

Mass sensitivity on the other hand, is essentially inversely proportional to the diameter of the RF coil (Lacey et al., 2001). Hence, decreasing the coil diameter leads to an increase in the mass sensitivity. In a recent paper by Lacey et al. (2001) a 3 μm capillary column was linked to a 500 MHz ^{1}H NMR. A capillary HPLC column was coupled to a 2.3-cm long NMR detection capillary. In this particular study the microcoil NMR detection capillary was prepared by wrapping 10 turns of a 2.7 mm long polyurethane-coated flat copper wire around a detection capillary. This created a 1.1 μL coil volume. The microcoil was enclosed by a 10 mL polyethylene bottle fitted with a perfluorinated liquid for magnetic susceptibility matching purposes. A schematic representation of the LC-NMR is given in Fig. 26 (Lacey et al., 2001). Both continuous and stopped flow methods of analysis could be employed and the detection limit of α-pinene was around 37 ng.

C. Trace Enrichment

One method that has been explored as a means to overcome the present limited sensitivity of HPLC-NMR is the incorporation of a solute-trapping process prior to the NMR analysis. Typically, this process involves the separation of species on an analytical column, and the component(s) of interest is (are) cycled through a trapping column. The separation through the analytical column may be repeated

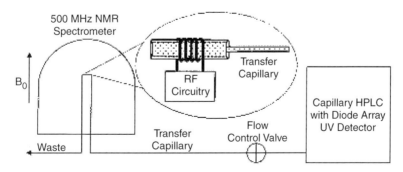

Figure 26 Schematic of the capillary HPLC-NMR set-up in which the capillary HPLC instrument (C_{18} column, 15 cm × 0.32 mm) with a diode array UV detector is located ~4 m from the 11.7 T NMR magnet. The flow control valve is used to halt the flow so that peaks of interest can be positioned in the microcoil NMR detection cell (V_{obs} ~ 1.1 μL) for long data acquisitions. [Reprinted from Lacey et al. (2001), copyright 2001, with permission from Elsevier.]

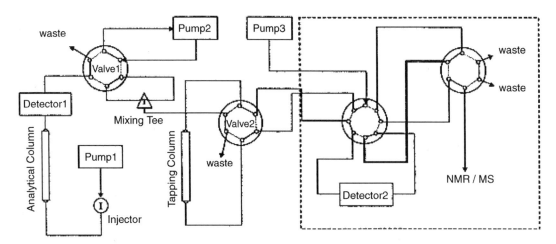

Figure 27 Schematic representation of the column trapping system operated in the back-flush mode, resulting in transfer of the concentrate to NMR probe. The NMR spectrometer and analyte collector (Varian Inc.), surrounded by dashed lines, is in on-flow mode. [Reprinted from Bao et al. (2002), copyright 2002, with permission from Elsevier.]

numerous times with the region of interest subsequently diverted to the trapping column. The solvent/stationary phase conditions in the trapping column are such that the retention factor of the species of interest is very high, hence the solute does not migrate along the trap. Once sufficient sample has been collected on the trap, the flow is reversed and the solute is transported to the NMR for further analysis. An example of this process is given in the diagram illustrated in Fig. 27 (Bao et al., 2002). Using this approach an increase in sensitivity by more than three-fold was observed (Bao et al., 2002) and a four-fold decrease in the stop flow mode of analysis was achieved (de Koning et al., 1998). An extension of this technique has been developed that allows multiple components to be trapped (Sweeney and Shalliker, 2002). Separation of these components is then achieved using a second analytical column (usually of a different selectivity to that of the first).

V. LIQUID CHROMATOGRAPHY-INDUCTIVELY COUPLED PLASMA SPECTROSCOPY

ICP-MS is an element-specific method of analysis that is ideally suited to coupling with LC methods of separation. ICP-MS is a sensitive and selective method of analysis that allows for the detection of metals and nonmetals in a wide number of samples. ICP-MS also offers a wide linear range. The very nature of sample introduction in ICP-MS allows the technique to be almost seamlessly interfaced with LC, albeit with a few restrictions. Furthermore, the wide number of chromatographic modes of separation allow for a great degree of separation potential, hence the coupling of these techniques can yield a very powerful means of speciation. Coupling an LC to an ICP is a simple process of directing the LC flow stream to the nebulizer of the ICP (Klinkenberg et al., 1998). There are a number of interface nebulizers that are employed for LC-ICP, which include the Meinhard cross-flow microconcentric nebulizers (MCN), the direct injection nebulizers (DIN), the high efficiency nebulizer (HEN), the direct-injection high efficiency nebulizer (DIHEN), the ultrasonic nebulizer (USN), and the hydraulic high-pressure nebulizer (Michalke, 2002; McLean et al., 1998). The low nebulization efficiency of the Meinhard and cross-flow nebulizers negates their employment in LC-ICP. However, the DIN, HEN, DIHEN, and USN are more successful in their nebulization efficiency.

The DIN was developed in the mid-1980s (LaFreniere et al., 1987; Lawrence et al. (1984) specifically for LC-ICP and since has received considerable attention, primarily because the DIN is an integral part of the ICP torch. Essentially, 100% of the aerosol enters the plasma (LaFreniere et al., 1987; Lawrence et al., 1984; McLean et al., 1998; Michalke, 2002) and together with the low internal dead volume ($<2\,\mu L$) the DIN displays high precision, fast response times, and low memory effects (McLean et al., 1998). However, the DIN restricts flow rates to less than 0.15 mL/min, and flow splitters are required if conventional analytical scale LC is employed. The technique is therefore more suited to microbore LC. Because of the low flow rates entering into the plasma, little effect on plasma stability is observed and as a consequence detection limits are improved (in general) by $\sim 10\text{--}100\times$ (Michalke, 2002). Although, a study on arsenic speciation by Sun (2003) revealed only a 2.5-fold increase in the detection limit, which was attributed to incomplete vaporization and solvent induced plasma cooling. Wang et al. (1996) also observed similar plasma cooling effects when mobile phases with high organic concentrations were employed. A limitation of the DIN is that it is more expensive and complex compared with the conventional pneumatic nebulizer–spray chamber arrangements, such as the HEN. Also, the DIN requires a high-pressure pump for sample delivery.

Combination of the DIN with the more conventional HEN results in what is referred to as the DIHEN. This nebulizer offers a more cost-effective alternative to the DIN nebulizer by incorporating the direct-injection technique of the DIN to that of the low cost, simplicity, and favourable aerosol characteristics of the HEN. A diagram illustrating the DIHEN is shown in Figs. 28 and 29 (McLean et al., 1998), where formation of the aerosol directly into the plasma can be clearly seen. Detection limits of the DIHEN operated at flow rate of 0.085 mL/min were similar to those obtained using conventional cross-flow pneumatic nebulizers that employed a spray chamber at 1 mL/min. This represented a 12-fold improvement in the absolute detection limits. Sensitivity and precision were generally improved over those of the conventional pneumatic nebulizers. Compared with the DIN, the DIHEN was relatively inexpensive and simple to use and did not require a high-pressure pump for its operation.

The USN offers a high nebulization efficiency with stable signal responses; however, when salt buffer systems are employed quantification can become difficult because of crusting and clogging of transport lines (Michalke, 2002). Furthermore, organic solvents are not suitable when a USN is employed. As little as 2% methanol has been observed to extinguish the plasma torch (Michalke, 2002). MCN interfaces also show promise in application to LC-ICP. This nebulizer is most suited to low flow rates (<0.15 mL/min) and hence is best utilized in microbore LC systems. Nebulization efficiency is $\sim 50\%$ (Sun, 2003) with detection limits similar to the DIN interface.

One of the most important considerations when coupling LC to ICP is maintaining the chromatographic resolution, which may suffer as a consequence of the desolvation process. Passage of the sample analyte through the spray chamber, and so on, can lead to substantial band broadening. In this regard minimizing the dead volume is important in order to maintain resolution. There has been little direct effort given to the optimization of flow chambers (Klinkenberg et al., 1998), although by employing low flow nebulizers, such as the DIN, HIHEN, or MCN interfaces, the chromatographic resolution is better retained (Sun, 2003). In a study by Sun (2003), the chromatographic resolution and band broadening were superior using the DIN interface.

Two of the most significant problems encountered in LC-ICP are the effect of buffer salt and organic modifiers on detection and operation (Michalke, 2002). It goes without saying that buffer precipitation can lead to clogging and hence failure of the system. Nowadays, this problem can, to a certain extent, be addressed by employing new column technologies to achieve separations in new ways using lower or even no buffer content in the mobile phase or 2D arrangements where buffer is excluded in the

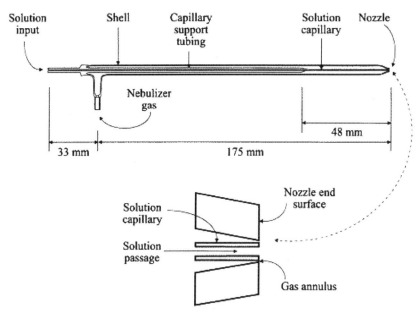

Figure 28 Schematic diagram of the direct-injection high-efficiency nebulizer (DIHEN) and an enlarged section view of the nebulizer tip. [Reprinted with permission from McLean et al. (1998).]

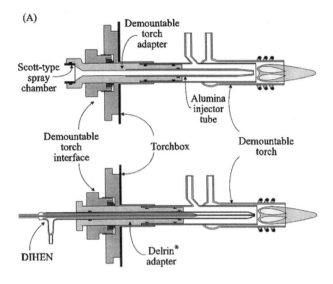

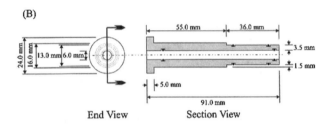

Figure 29 (A) Comparison of the standard demountable torch interface for the Elan 6000 equipped with a cross-flow nebulizer-spray chamber (top) and a DIHEN (bottom). (B) Critical dimensions of the DIHEN Delrin demountable torch adapter. [Reprinted with permission from McLean et al. (1998).]

second dimension. Organic modifiers can be detrimental for several reasons: (a) they can extinguish the plasma; (b) the plasma instability associated with organic solvents leads to difficulties in quantitation because of changes in signal response in gradient applications; and (c) the organic content can lead to a poor burning efficiency. The concentration of organic modifiers can be decreased through dilution by employing a post-column flow stream of water to dilute the mobile phase eluent, but at the expense of diluting the analyte peaks as well. Another method employed to overcome the effect of organic modifiers is to employ a condenser in the transport lines or spray chamber, which reduces the organic modifier content. Under such circumstances methanol concentrations as high as 80% have been successfully employed (Michalke, 2002). Desolvation systems, such as membranes, can also be employed, permitting up to 100% methanol in the mobile phase (see Section III). The drawback of any desolvation step may be, however, that sample could be lost in the process of removing solvent.

VI. MULTIPLE HYPHENATION (OR HYPERNATION) OF LC WITH SPECTROSCOPY

The hyphenation of LC with MS, NMR, and FTIR has been discussed earlier in this chapter and in the case of UV/Vis diode-array in the preceding chapter. Each of these spectrophotometric methods of analysis provides different degrees of information on constituents eluting from the liquid chromatographic system. Often it is necessary to perform a number of analyses, involving different methods of hyphenation in order to provide support for the elucidation of unknown structures. This is a time-consuming process, which can lead to increased running costs and, in the case of small sample quantities, higher consumption of what are very often precious samples.

Multiple hyphenation of spectroscopy, discounting UV diode-array, with LC was first investigated in the mid-1990s with the hyphenation of LC with MS and ^{1}H NMR (Pullen et al., 1995). Recently, Wilson and Brinkman (2003) coined the term hypernation for multiply hyphenated systems. There appears to be two general schemes for multiply hyphenated systems involving LC-NMR-MS systems. The first involves the MS and NMR analyses in series after the LC separation (Dear et al., 1998), and the second involves a parallel run arrangement (Fritsche et al., 2002; Hansen et al., 1999; Holt et al., 1997; Ludlow et al., 1999; Pullen et al., 1995; Sandvoss et al., 2001; Shockor et al., 1996; Taylor et al., 1998; Wilson, 2000; Wilson et al., 1999).

The parallel arrangement of the NMR and MS spectrometers following LC analysis was the first to be investigated and is, in most ways, superior to the serial arrangement. A schematic representation of the parallel arrangement is shown in Fig. 30 (Hansen et al., 1999). Here, the eluent from the liquid chromatograph is split two ways using a standard Tee-split in which the major proportion of the chromatographic flow is diverted to the NMR flow probe and the minor portion to the MS. This is important as MS interfaces, such as electrospray, perform optimally between 2 and 100 μL/min, whereas NMR flow cells generally require large amounts of sample to generate appropriate signals. The length or internal diameter of tubing after the Tee-split is generally adjusted to facilitate the arrival of analyte peaks from the liquid chromatograph to the NMR flow cell coinciding with the MS interface (Wilson et al., 1999). Another important factor in the parallel design is the requirement for data collection to commence simultaneously on the LC-UV detector, NMR computer, and MS controller and is done using an electronic trigger from the autoinjector (Holt et al., 1997; Pullen et al., 1995). In that way, a particular component's spectroscopic signal can be related to

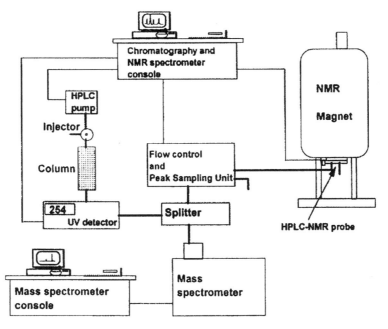

Figure 30 Instrumental set-up of the HPLC-NMR-MS apparatus. [Reprinted with permission Frank Hansen et al. (1999), copyright 1999, American Chemical Society.]

the standard UV/Vis detection response, making component recognition more straightforward.

Dear et al. (1998) illustrated the use of an LC-NMR-MS system in which the LC, NMR, and MS are conducted in series. This is shown in Fig. 31. In this instance, the analyte(s) of interest can be held in storage loops, which are then successively transported to the NMR flow cell for a stopped flow type of analysis. When a sufficient number of scans were recorded for an NMR spectrum, the LC fraction was then transported to a T-junction where a makeup flow of aqueous formic acid facilitated a ^{2}H/^{1}H back exchange resulting in nondeuterated $[M + H]^+$ parent ions in the mass spectral analysis. Alternatively, the deuterium–hydrogen exchange could be negated to produce solely deuterated parent ions (Dear et al., 1998). Parallel designs including ^{2}H/^{1}H exchange systems have also been shown (Sandvoss et al., 2001). A number of reports have indicated problems with the serial arrangement, stemming from the leakage of liquid chromatographic fractions from the NMR flow probe. Attachment of the MS after the NMR apparently yields some backpressure on NMR flow cell, which then causes leaks (Hansen et al., 1999; Ludlow et al., 1999).

A number of analyses have reported on the use of stopped-flow analysis (Dear et al., 1998; Hansen et al., 1999; Sandvoss et al., 2001) or stopped-flow analysis of certain chromatographic peaks (Wilson et al., 1999). This is necessary to generate usable NMR signals for

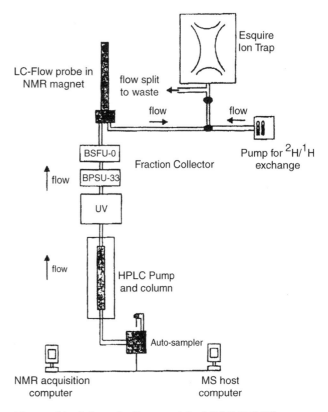

Figure 31 Schematic diagram of the LC/NMR/MS^n system. [Reprinted from Dear et al. (1998), copyright 1998, John Wiley and Sons Limited. Reproduced with permission.]

compounds that have low concentrations. Alternately, some have utilized devices that employ multiple storage loops after the liquid chromatograph. In serial LC-NMR-MS, for instance, when sufficient signal has been achieved in the NMR analysis for an LC fraction, this would then be passed onto the MS for final analysis (Dear et al., 1998). Taylor et al. (1998) reported in their initial investigations into LC-NMR-MS, that in stopped-flow analysis a gradual decrease in signal was observed during the stopped mode of the cycle. This was attributed to a small flow caused by an MS interface, such as ionspray. Switching off the nebulizer gas (Taylor et al., 1998) solved this.

Analysts need to pay attention to the type of solvent used as the mobile phase in the liquid chromatographic step. Most LC steps hyphenated to MS and NMR have been reversed-phase separations using gradients involving acetonitrile and D_2O (Dear et al., 1998; Pullen et al., 1995; Shockor et al., 1996) and/or deuterated acetonitrile (Hyötyläinen et al., 1997). In doing so, the background signal is reduced in the NMR spectra. This has a significant effect on mass spectra. As proton transfer reactions are dominant in most LC-MS interfaces, $[M+H]^+$ parent ions in a positive ion mode, for instance, are replaced by $[M+D]^+$ parent ions (Holt et al., 1997; Wilson et al., 1999). This may cause problems because the mass spectra would undoubtedly be incompatible with mass spectral libraries that have previously been generated. Conversely, the formation of deuterated parent ions can be advantageous as the number of exchangeable hydrogens on a given molecule may be assessed, which yields even more structural information (Wilson et al., 1999). Inorganic buffers can generally be employed in LC-NMR without significantly contributing to any background signal (Ludlow et al., 1999). However, most LC-MS interfaces perform poorly with inorganic buffers. So, in instances where buffers are required to enhance the chromatographic separation, a compromise must be found that suits both hyphenated methods of analysis. Taylor et al. (1998) suggest that buffers such as TFA may be used. However, depending on the sample to be analyzed, TFA can suppress ion formation in interfaces such as electrospray. Formic acid or acetic acid has been suggested as a compromise, though these buffers yield a signal in the NMR spectra (Hansen et al., 1999; Taylor et al., 1998).

The proximity of the mass spectrometer to the NMR magnet can affect the performance of the MS ion optics, such as, ion guides and quadrupoles. The sensitivity and noise of the mass spectrometer were generally seen to decrease by a factor of 100 (Holt et al., 1997; Pullen et al., 1995; Taylor et al., 1998), although this apparently has not hindered the production of quality mass spectra (Pullen et al., 1995). A number of analysts (Holt et al., 1997; Pullen et al., 1995; Taylor et al., 1998) have briefly examined the effect of the MS proximity and orientation to the NMR magnet by placing the MS on a trolley. All agree that the optimal position of the MS, while attempting to minimize band broadening associated with tubing length, is approximately 2 m from the NMR magnet, which occurs at the 5 G line of the magnetic field.

Though it may seem that multiple methods of hyphenation result in an expense that may be difficult to justify for the limited number of samples that a typical laboratory may test, the most obvious cost-cutting can be made by employing portability into the system. That is, the multiple hyphenation system is constructed using "wheel-in" components when the need for such a system exists and can then be justified. Provided the components of the system are located within close proximity, such a method of analysis is cost effective to operate, as each of the spectrometers and liquid chromatograph can be employed in other tasks until coupling is required.

VII. HYPHENATION OF LIQUID CHROMATOGRAPHY WITH GAS CHROMATOGRAPHY

The concept of coupled LC-GC for the analysis of mixtures dates back to the mid-1980s where both heart-cutting (LC-GC) and comprehensive (LC × GC) automated methods were reported. The unification of these two chromatographic techniques relies on the ability to transport solute from the liquid phase to the gas phase without overloading the modern capillary GC column. Transfer of sections >100 μL of LC eluent to an analytical GC capillary column would not yield usable analytical results.

The interfaces in LC-GC systems can be broadly divided into two general groups, with each group containing many subdivisions, within which there are small variations. These two broad classes may be defined in terms of the presence of a vaporizing chamber (most often a programmed temperature vaporizer—PTV) or retention-gap type techniques. In some instances there are combinations that employ both processes and for brevity these will not be repeated for each section.

A. Concurrent Solvent Evaporation Techniques

There are two schemes or modes of operation that may be employed for the hyphenation of LC with GC. These are the fully concurrent (Carlsson and Östman, 1997; Grob and Stoll, 1986; Grimvall et al., 1995; Häkkinen et al., 1988; Hyötyläinen et al., 1997; Jongenotter et al., 1999; Kuosmanen et al., 2001) and partially concurrent (Beens et al., 1997; Biedermann et al., 1989; Blomberg et al.,

1997; Boselli et al., 1998; Hankemeier et al., 1998a; Hyötyläinen et al., 2000); (Schmarr et al., 1989; Shimmo et al., 2002; Trisciani and Munari, 1994)) solvent evaporation techniques. Because of the scope of this text, only the most popular versions of each technique will be discussed.

B. Partially Concurrent Solvent Evaporation

Partially concurrent solvent evaporation techniques are those that are employed for the introduction of LC fractions into a GC system whereby the eluent flow rate entering the GC inlet is adjusted to slightly exceed the evaporation rate of the solvent. A general schematic of a GC utilized for large-volume injection or LC hyphenation is illustrated in Fig. 32. Injection of an LC fraction onto the modified GC system shown in Fig. 32 can be accomplished via a loop-type interface or an on-column injection (Beens et al., 1997; Biedermann et al., 1989; Schmarr et al., 1989; Trisciani and Munari, 1994).

Partially concurrent solvent evaporation techniques are normally conducted using the on-column injection mode and will be the main focus of this discussion. Transfer of the LC eluent via on-column injection generally occurs through modified standard on-column injectors (Shimmo et al., 2002; Trisciani and Munari, 1994). Being on-column, the HPLC pump is used to drive the sample fraction into an uncoated pre-column called the retention gap (Shimmo et al., 2002; Trisciani and Munari, 1994). So that the introduction rate of the liquid exceeds the evaporation rate of the liquid in the uncoated retention gap, the liquid sample fraction is driven some distance into the retention gap. Generally, fraction sizes for partially concurrent solvent evaporation techniques involving retention gaps are ~200–300 μL (Schmarr et al., 1989), although sizes

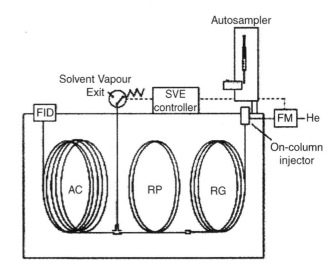

Figure 32 Set-up of the large-volume injection-GC system: AC, analytical column; RP, retaining pre-column; RG, retention gap; FM, flow meter; He, helium; SVE, solvent vapour exit. [Reproduced from Hankemeier et al. (1998), copyright 1998, with permission from Elsevier.]

of 700–1000 μL have been possible (Hyötyläinen et al., 2000), depending on the length and diameter of the retention gap (Schmarr et al., 1989).

Introduction of liquid into the uncoated precolumn or retention gap is shown in Fig. 33. Analytes are trapped by solvent-trapping effects (Boselli et al., 1998; 1999a). The sample is injected into the column as a liquid. The liquid is then driven deeper into the retention gap by the HPLC pump and the carrier gas forming a stable layer that coats the inside of the retention gap. Evaporation initially starts inside the retention gap from both the front and rear boundaries of the liquid plug (Boselli

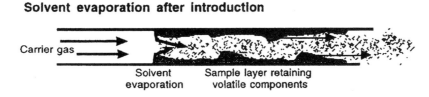

Figure 33 The two steps of solvent evaporation during sample introduction with partially concurrent evaporation according to the classical model. [Reprinted from Boselli et al. (1999), copyright 1999, John Wiley and Sons Limited. Reproduced with permission.]

et al., 1998; Hankemeier et al., 1998a). The amount evaporating from the front is generally considered as insignificant in partially concurrent methods as the solvent evaporates mainly from the rear of the wetted region (Boselli et al., 1998). Evaporation of solvent occurs mainly from the rear of the "flooded zone" because of the increased vapor pressure in the carrier gas (Boselli et al., 1999b; Schmarr et al., 1989). The flooded zone then acts as a temporary stationary phase toward analytes in the liquid. Highly volatile solutes that evaporate into the gas phase are then trapped by the liquid film further down the retention gap. Hankemeier et al. (1998) have shown, however, in the analysis of highly volatile chlorobenzenes, aromatics, anilenes, and phenolic compounds among others, that some analytes can, in some instances, be lost through evaporation from the front of the flooded zone. This was easily solved by the addition of a presolvent ahead of the liquid fraction. As such the presence of the presolvent, in most cases, prevented significant loss in the initial introduction of the liquid fraction (Hankemeier et al., 1998a). This is illustrated in Fig. 34.

An important development in these retention gap techniques is the use of an early solvent vapor exit (SVE). As shown in Fig. 32, the SVE is generally located between a retaining precolumn (discussed subsequently) and an analytical GC column (Blomberg et al., 1997; Schmarr et al., 1989; Trisciani and Munari, 1994). Biedermann et al. (1989) have suggested that the SVE serves two purposes: (1) eluent evaporation rates are enhanced as vapors can be discharged more quickly through the SVE than through a long analytical column (Hankemeier et al., 1998b; Schmarr et al., 1989); and (2) to protect the analytical GC column and, more importantly, the GC detector from excessive amounts of solvent vapor (Boselli et al., 1998; Hankemeier et al., 1998b). Closing the solvent vapor exit toward the end of the evaporation process can prevent loss of analytes through the SVE. Most, however, agree that the best chromatographic performance is attained with closure of the SVE at the end or prior to total solvent evaporation (Hankemeier et al., 1998a).

If the SVE is closed after total solvent evaporation, the retaining precolumn should be used to prevent loss of volatile components through the vapor exit (Biedermann et al., 1989; Schmarr et al., 1989). The retaining precolumn should then have a high retentive power to retain volatile solutes and in most cases consists of a short section, but larger diameter of stationary phase similar to or identical to the analytical column. Using wider bore capillaries for the retaining precolumn would then require a thicker portion of stationary phase (Biedermann et al., 1989). Although, some (Boselli et al., 1999a, b; Schmarr et al., 1989) suggest that the retaining precolumn is not needed as the liquid flooded zone prior to total evaporation concentrates the analytes. In this instance the SVE should be closed prior to total evaporation, although it seems the retaining precolumn is mostly included as a safety-net feature in most systems to prevent incidental loss of highly volatile solutes. In order to determine the correct SVE closure time, the end of solvent evaporation must be determined. The use of a flame ionization detector at the outlet of the SVE allows the analyst to determine the end of the evaporation stage or the evaporation rate of solvent (Biedermann et al., 1989; Hankemeier et al., 1998b; Schmarr et al., 1989), which is useful for determining the optimal SVE closure times. Hankemeier et al. (1998a, b) indicated closure of the SVE could be determined by monitoring pressure and carrier gas flow rates during analysis. In fact Hankemeier et al. (1998) developed a means to automatically close the SVE by

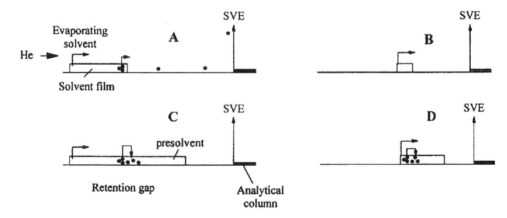

Figure 34 Scenario for analytes (indicated by •) which are not fully trapped by the solvent film during an online SPE-GC transfer. During injection these analytes are deposited (A) at the front of the solvent film if there is no presolvent, or (C) behind a solvent barrier if a presolvent is used. At the end of solvent evaporation, analytes (B) have been lost without presolvent, but (D) are recovered with presolvent. [Reprinted from Hankemeier et al. (1998), copyright 1998, with permission from Elsevier.]

monitoring the first derivative of the carrier gas flow. When evaporation was completed, the carrier gas flow would then increase and with it also would the first derivative of the flow rate. The SVE was closed when the first derivative of the flow exceeded some preset threshold after injection (Hankemeier et al., 1998b).

Boselli et al. (1999) noted that when the SVE is closed prior to total evaporation, secondary expansion of the flooded zone occurs. Closure of the SVE reduces the carrier gas flow through the precolumn, causing the remaining portion of the flooded zone to expand $\sim 2.5\times$. This has significant consequences with regard to the length of the precolumn and volume injected. The SVE must then be closed in the very late stages of the final solvent evaporation to prevent expansion into the analytical GC column (Fig. 32).

C. Fully Concurrent Solvent Evaporation Techniques

Fully concurrent evaporation techniques are those that allow the introduction of analytes in large volumes of LC eluent, typically in the range of 500–2000 µL, to the GC column (although, up to 20 mL has been reported). This interface technique is carried out via the careful adjustment of the inlet pressure, liquid plug delivery, retention-gap temperature, and solvent evaporation rate so that the inlet flow rate is more or less equal to the evaporation rate (Grimvall et al., 1995; Grob and Stoll, 1986; Hyötyläinen and Riekkola, 2003; Hyötyläinen et al., 1997; Mondello et al., 1999).

Fully concurrent evaporation techniques are conducted by on-column injection or the "loop-type" interface. On-column injection simply uses the HPLC pump to deliver the LC fraction to the retention-gap capillary. The limitations and difficulties of this mode of operation are obvious as the LC flow must be delivered at a sufficiently slow rate to compensate for the evaporation rate, while taking into account the carrier gas flow rate. A loop-type interface is illustrated in Fig. 35, where it is shown that the LC flow rate and the introduction rate into the GC retention gap are independent. This is achieved by employing a six-port switching valve where the LC fraction can be stored. Once the storage loop is filled with an LC fraction, a switch of the six-port valve allows the flow-regulated carrier gas to push the plug of solvent into the inlet of the retention gap capillary (Grob and Stoll, 1986; Häkkinen et al., 1988). GC oven temperatures are generally set slightly above the boiling point of the LC eluent (Grob and Stoll, 1986; Hyötyläinen et al., 1997) and in doing so solvent vapor pressure builds and prevents significant movement of the solvent plug. At this stage, in most modern systems, the SVE is kept open to facilitate swift solvent evaporation (Carlsson and Östman, 1997;

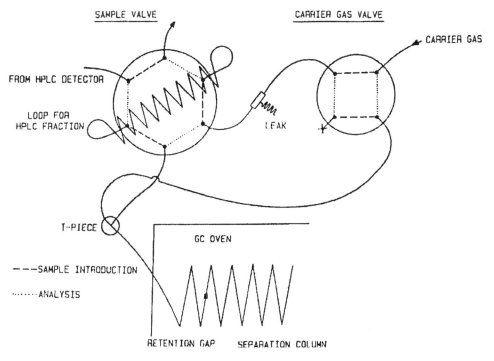

Figure 35 Interface between HPLC and GC. A sample loop with an internal volume corresponding to the volume of the HPLC fraction is built into the six-port switching valve. [Reprinted from Grob and Stoll (1986), copyright 1986, John Wiley and Sons Limited. Reproduced with permission.]

Grimvall et al., 1995). The use of retaining precolumns has been discussed in conjunction with partially concurrent techniques. It is important to note that evaporation occurs from the front of the solvent plug. As a result, a disadvantage of this technique is that highly volatile analytes can be lost with the evaporating solvent, or the first eluting peaks appear broadened or skewed (Barcarolo, 1990). Upon completion of the evaporation of the solvent plug, the SVE is closed to prevent loss of sample (Grimvall et al., 1995). Fully concurrent evaporation is generally utilized for analytes well-above the boiling point of the LC solvent (Kuosmanen et al., 2001), although, this can partially be solved by using a cosolvent with a higher boiling point than the main solvent component. Jongenotter et al. (1999) illustrated the use of a cosolvent in the size-exclusion chromatographic analysis of pesticides. They used a solvent mixture of (54/46 v/v) ethyl acetate/cyclohexane (b.p. 72°C) with nonane as a cosolvent (b.p. 150.8°C). With a transfer temperature of 77°C, the main solvent evaporates. Nonane, the co-solvent, then wets the precolumn ahead of the liquid bulk and acts as a temporary stationary phase and traps volatile components (Hyötyläinen and Riekkola, 2003).

D. Programmed Temperature Vaporizer/ Vaporizer Interfaces

Unlike the partially/concurrent retention-gap techniques discussed earlier, interfaces for LCGC that rely upon vaporizing chambers utilize a permanent stationary phase [packed or as a liner (Staniewski and Rijks, 1993)] inside the vaporizing chamber, rather than the temporary stationary phase provided by solvent trapping in a retention gap. The most common form of interface including a vaporizer is the PTV (Grob, 1990; Señoráns et al., 1995a; Staniewski and Rijks, 1993).

Figure 36(a) and (b) shows a general schematic or approach to the use of PTV as an interface. PTV interfaces can be designed according to whether solvent vapor is discharged or removed with the assistance of a small flow of carrier gas (Mol et al., 1993; Staniewski and Rijks, 1993) or according to vapor overflow (Grob, 1990). In the overflow interface [Fig. 36(a)], LC eluent is generally introduced into the PTV-packed vaporizing chamber, which is held at a temperature below the boiling point of the solvent (Boderius et al., 1995; Engewald et al., 1999; Grob, 1990; 2000; Grob and Biedermann, 1996). The rate of entry should equal the evaporation rate of the solvent (Engewald et al., 1999; Grob, 1990; 2000; Grob and Biedermann, 1996). At the time of liquid transfer, the carrier gas flow is stopped or restricted (Grob, 1990). The liquid sample is introduced at a site opposite to the vapor discharge exit to reduce loss of volatile components as solvent vapor continues to leave through the vapor exit

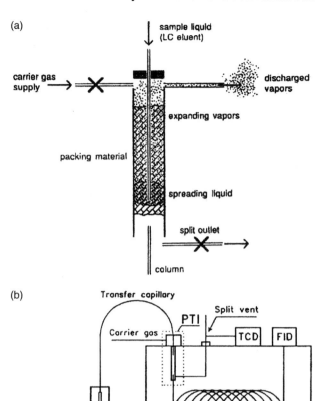

Figure 36 (a) A possible injection device for introducing samples by the vapor overflow technique. Situation during transfer and solvent evaporation. [Reprinted from Grob (1990)], copyright 1990, John Wiley and Sons Limited. Reproduced with permission.] (b) Schematic diagram of large volume introduction instrumentation in capillary GC. [Reprinted from Staniewski and Rijks (1993), copyright 1993, John Wiley and Sons Limited. Reproduced with permission.]

(Grob, 1990). In this interface solvent vapor leaves the chamber because the vapor pressure inside the vaporization chamber is higher than the surrounding environment. Once the liquid sample/fraction has been entirely transferred to the vaporizing chamber, the vapor exit is closed and the temperature of the chamber is heated (in most cases 30–300°C depending upon sample types). Sample is then desorbed, the carrier gas switched on, and the sample transferred to the analytical GC column.

An alternative to the vapor overflow method is to use a small flow of carrier gas to carry the evaporated liquid solvent. This is illustrated in Fig. 36(b). Similar to the vapor overflow PTV methods, sample is introduced into a packed liner from the top of the injector, whereas evaporating solvent is removed with the assistance of carrier gas flow from the bottom of the injector through a split vent or

HPLC-Hyphenated Techniques

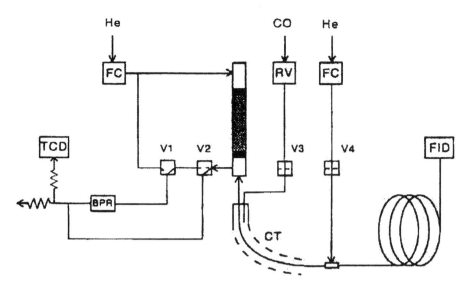

Figure 37 Schematic diagram of PTV-GC system used for direct large volume water introduction: FC, flow controller; BPR, back pressure regulator; RV, reducing valve; V1-V4, valves; TCD, thermal conductivity detector; FID, flame ionization detector; CT, cold trap. [Reprinted from Mol et al. (1993), copyright 1993, John Wiley and Sons Limited. Reproduced with permission.]

vapor exit (Staniewski and Rijks, 1993). Fig. 36(b) illustrates that the evaporation of the liquid solvent can be monitored with the use of a conductivity detector, or alternately an FID. Once solvent evaporation has ceased, the split exit is closed followed by heating of the chamber to desorb analytes (Boderius et al., 1995; Mol et al., 1993; Señoráns et al., 1995b; Staniewski and Rijks, 1993). Alternatively (Barcarolo, 1990), the retention gap and retaining precolumn set-up can be used to vent solvent vapor. Señoráns et al. (1995a, b) on the other hand simply

removed the PTV bottom from the GC column to vent solvent vapors and was subsequently reattached prior to the sample desorption step. Mol et al. (1993) reported a more interesting variation to the theme, which is shown in Fig. 37. Water vapor was pushed through the liner by a purge flow of helium and was allowed to exit through the usual split exit (solid-phase extraction mode). To prevent water from entering the GC column, a helium flow is applied through a helium line and Tee-piece situated 75 cm from the injector. Prior to thermal desorption

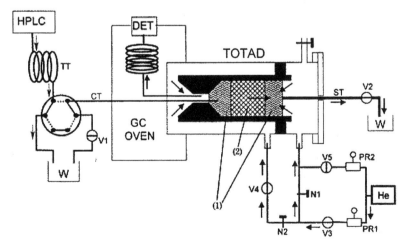

Figure 38 TOTAD interface with valves positioned for both LC separation and interface stabilization and for cleaning steps. Symbols: 1, glass wool; 2, sorbent (Tenax TA); N, needle valve; V, on–off valve; PR, pressure regulator; nondotted arrows, gas flow; dotted arrows, liquid flow; TT, stainless steel tubing to transfer from LC to GC; CT, silica capillary tubing; ST, stainless steel tubing to allow the exit of liquids and gases; W, waste). [Reprinted with permission from Perez et al. (2000), copyright 2000, American Chemical Society.]

of the sample components a flow of carbon dioxide is initiated. Expansion of the carbon dioxide cools the section labeled CT (Fig. 37). This allows for a refocusing effect. When the carbon dioxide was switched off, the trapped components were remobilized following rapid heating of CT.

The vaporizing chamber is filled with chromatographic packing that should have a suitable retention power for the analytes of the sample. Tenax-type packing materials are the most commonly utilized packing for PTV because they have excellent retentive powers and high thermo-desorption efficiencies at high temperature (Señoráns et al., 1995b; Staniewski and Rijks, 1993). Señoráns et al. (1995) have evaluated variables, such as, adsorbent length, Tenax-type and particle size in reversed-phase LC-GC in the analysis of sterols. A problem that can exist in certain instances is that Tenax chamber packing can irreversibly adsorb some compounds. Moreover, degradation of the Tenax can occur leading to the presence of artifacts (Staniewski and Rijks, 1993). Loss of highly volatile components is also known to be a problem in the use of PVT interfaces (Mol et al., 1993). Other common PVT packing materials include glass wool (Señoráns et al., 1995b) and sintered glass (Staniewski and Rijks, 1993). Staniewski and Rijks (1993) mentioned that sintered glass liners allow for the development of a "stable, continuous solvent film in the porous bed". This, however, is not possible with all mobile phase solvents as water, for instance, cannot form such films on the deactivated glass surfaces (Mol et al., 1993). Liquid introduction volumes typically vary from 250 to 1600 μL (Grob, 1990; Señoráns et al., 1995a, b; Staniewski and Rijks, 1993) although the introduction volumes can be changed (i.e., <250 and >1600 μL) by altering the introduction flow rate.

E. Miscellaneous

Perez et al. (2000) have indicated that the use of highly aqueous mobile phases in reversed-phase LC-GC can be problematic. With this in mind they developed an interface titled the through-oven-transfer adsorption/desorption (TOTAD) Interface, shown in Fig. 38. The LC fraction of interest was transferred to the six-port switching valve at 0.1 mL/min through the "CT" tubing to the Tenax TA packed TOTAD chamber. Analytes in the LC eluent were retained in the Tenax packing, whereas the liquid solvent was pushed to waste through the "ST" tubing by a flow of helium. Helium flow was also used to empty the CT tubing of any residual liquid after the fraction of interest was transferred to the packed liner following the switching of the six-port valve back to the waste position. Once the solvent had been removed thermal desorption and transfer to the GC column occurred by employing a reversed flow of helium.

ACKNOWLEDGEMENT

We thank Ms K Mayfield for reading this chapter.

REFERENCES

Abian, J. (1999). The coupling of gas and liquid chromatography with mass spectrometry. *J. Mass Spectrom.*, 34:157.

Albert, K. (1999). Liquid chromatography–nuclear magnetic resonance spectroscopy. *J. Chromatogr. A.*, 856:199.

Apffel, A., Perry, M.L. (1991). Quantitation and linearity for particle-beam liquid-chromatography mass-spectrometry. *J. Chromatogr.*, 554:103.

Arpino, P.J., Guiochon, G. (1982). Optimization of the instrumental parameters of a combined liquid chromatograph-mass spectrometer, coupled by an interface for direct liquid introduction: III. Why the solvent should not be removed in liquid chromatographic-mass spectrometric interfacing methods. *J. Chromatogr.*, 251:153.

Ashcroft, A.E. (1997). In: Barnett, N.'W., ed. *RSC Analytical Spectroscopy Monographs: Ionization Methods in Organic Mass Spectrometry*. Royal Society of Chemistry.

Arpino, P.J., Guiochon, G., Krien, P., Devant, G. (1979). Optimization of the instrumental parameters of a combined liquid chromatograph-mass spectrometer, coupled by an interface for direct liquid introduction: I. Performance of the vacuum equipment. *J. Chromatogr.*, 185:529.

Banks, J.F. Jr., Shen, S., Whitehouse, C.M., Fenn, J.B. (1994). Ultrasonically assisted electrospray ionization for LC/MS determination of nucleosides from a transfer RNA digest. *Anal. Chem.*, 66:406.

Bao, D., Thanabal, V., Pool, W.F. (2002). Determination of tacrine metabolites in microsomal incubate by high performance liquid chromatography–nuclear magnetic resonance/mass spectrometry with a column trapping system. *J. Pharm. Biomed. Anal.*, 28:23.

Barcarolo, R. (1990). Coupled LC–GC—a new method for the online analysis of organochlorine pesticide-residues in fat. *J. High Resolut. Chromatogr.*, 13:465–469.

Beens, J., Tijssen, R. (1997). The characterization and quantitation of sulfur-containing compounds in (heavy) middle distillates by LC-GC-FID-SCD. *J. High Resolut. Chromatogr.*, 20:131–137.

de Biasi, V., Haskins, N., Organ, A., Bateman, R., Giles, K., Jarvis, S. (1999). High throughput liquid chromatography/mass spectrometric analyses using a novel multiplexed electrospray interface. *Rapid Commun. Mass Spectrom.*, 13:1165.

Biedermann, M., Grob, K., Meier, W. (1989). Partially concurrent eluent evaporation with an early vapor exit—detection of food irradiation through coupled LC–GC analysis of the fat. *J. High Resolut. Chromatogr.*, 12:591–598.

Blackley, C.R., Vestal, M.L. (1983). Thermospray interface for liquid chromatography/mass spectrometry. *Anal. Chem.*, 55:750.

Blackley, C.R., Carmody, J.J., Vestal, M.L. (1980). A new soft ionization technique for mass spectrometry of complex molecules. *J. Am. Chem. Soc.*, 102:5931.

Blackley, C.R., Carmody, J.J., Vestal, M.L. (1980). Liquid chromatograph-mass spectrometer for analysis of nonvolatile samples. *Anal. Chem.*, 52:1636.

Bleicher, K., Bayer, E. (1994). Analysis of oligonucleotides using coupled high performance liquid chromatography-electrospray mass spectrometry. *Chromatographia*, 39:405.

Blomberg, J., Mes, E.P.C., Schoenmakers, P.J., van der Does, J.J.B. (1997). Characterization of complex hydrocarbon mixtures using on-line coupling of size-exclusion chromatography and normal-phase liquid chromatography to high-resolution gas chromatography. *J. High Resolut. Chromatogr.*, 20:125–130.

Boderius, U., Grob, K., Biedermann, M. (1995). High-temperature vaporizing chambers for large-volume gc injections and online lc-gc. *J. High Resolut. Chromatogr.*, 18:573–578.

Boselli, E., Grolimund, B., Grob, K., Lercker, G., Amadò, R. (1998). Solvent trapping during large volume injection with an early vapor exit, part 1: Description of the flooding process. *J. High Resolut. Chromatogr.*, 21:355–362.

Boselli, E., Grob, K., Lercker, G. (1999). Solvent trapping during large volume injection with an early vapor exit part—Part 3: The main cause of volatile component loss during partially concurrent evaporation. *J. High Resolut. Chromatogr.*, 22(6):327–334.

Boselli, E., Grob, K., Lercker, G. (1999). Capacity of uncoated 0.53 mm id pre-columns for retaining sample liquid in the presence of a solvent vapor exit. *J. High Resolut. Chromatogr.*, 22(3):149–152.

Browner, R.F., Boorn, A.W. (1984). Sample introduction: the Achilles' heel of atomic spectroscopy? *Anal. Chem.*, 56:786A.

Bruins, A.P. (1991). Liquid-chromatography mass-spectrometry with ionspray and electrospray interfaces in pharmaceutical and biomedical-research. *J. Chromatogr.*, 554:39.

Bruins, A.P. (1998). Mechanistic aspects of electrospray ionisation. *J. Chromatogr. A*, 794:345.

Bruins, A.P., Covey, T.R., Henion, J.D. (1987). Ion spray interface for combined liquid chromatography/atmospheric pressure ionization mass spectrometry. *Anal. Chem.*, 59:2642.

Caprioli, R.M., Fan, T., Cottrell, J. (1986). A continuous-flow sample probe for fast atom bombardment mass spectrometry. *Anal. Chem.*, 58:2949.

Caprioli, R.M., Moore, W.T., Martin, M., DaGue, B.B., Wilson, K., Moring, S. (1989). Coupling capillary zone electrophoresis and continuous-flow fast atom bombardment mass spectrometry for the analysis of peptide mixtures. *J. Chromatogr.*, 480:247.

Carlsson, H., Östman, C. (1997). Clean-up and analysis of carbazole and acridine type polycyclic aromatic nitrogen heterocyclics in complex sample matrices. *J. Chromatogr. A*, 790:73–82.

Carrol, D.I., Dzidic, I., Stillwell, R.N., Haegele, K.D., Horning, E.C. (1975). Atmospheric pressure ionization mass spectrometry. Corona discharge ion source for use in a liquid chromatograph-mass spectrometer-computer analytical system. *Anal. Chem.*, 47:2369.

Chambers, D.M., McLuckey, S.A., Glish, G.L. (1993). Role of gas dynamics in negative ion formation in an atmospheric sampling glow discharge ionization source. *Anal. Chem.*, 65:778.

Choi, B.K., Hercules, D.M., Gusev, A.I. (2001). Effect of liquid chromatography separation of complex matrices on liquid chromatography–tandem mass spectrometry signal suppression. *J. Chromatogr. A*, 907:337.

Chowdhury, S.K., Katta, V., Chait, B.T. (1990). An electrospray-ionization mass-spectrometer with new features. *Rapid Commun. Mass Spectrom.*, 4:81.

Daghbouche, Y., Garrigues, S., Vidal, M.T., de la Guardia, M. (1997). Flow injection Fourier transform infrared determination of caffeine in soft drinks. *Anal Chem.*, 66:1086.

Dalton, C.N., Glish, G.L. (2003). Electrospray-atmospheric sampling glow discharge ionization source for the direct analysis of liquid samples. *Anal. Chem.*, 75:1620.

Dams, R., Benijts, T., Günther, W., Lambert, W., De Leenheer, A. (2002). Sonic spray ionization technology: performance study and application to a LC/MS analysis on a monolithic silica column for heroin impurity profiling. *Anal. Chem.*, 74:3206.

Davoli, E., Fanelli, R., Bagnati, R. (1993). Purification and analysis of drug residues in urine samples by on-line immunoaffinity chromatography/high-performance liquid chromatography/continuous-flow fast-atom-bombardment mass spectrometry. *Anal. Chem.*, 65:2679.

Dear, G.J., Ayrton, J., Plumb, R., Sweatman, B.C., Ismail, I.M., Fraser, I.J., Mutch, P.J. (1998). A rapid and efficient approach to metabolite identification using nuclear magnetic resonance spectroscopy, liquid chromatography mass spectrometry, and liquid chromatography nuclear magnetic resonance spectroscopy sequential mass spectrometry. *Rapid Commun. Mass Spectrom.*, 12:2023–2030.

Dekmezian, A.H., Morioka, T. (1989). Interface to eliminate high-boiling gel permeation chromatographic solvents on-line for polymer composition drift studies. *Anal. Chem.*, 61:458.

Dijkman, E., Mooibroek, D., Hoogerbrugge, R., Hogendoorn, E., Sancho, J.-V., Pozo, O., Hernández, F. (2001). Study of matrix effects on the direct trace analysis of acidic pesticides in water using various liquid chromatographic modes coupled to tandem mass spectrometric detection. *J. Chromatogr. A*, 926:113.

DiNunzio, J.E. (1992). Pharmaceutical applications of high-performance liquid chromatography interfaced with Fourier transform infrared spectroscopy. *J. Chromatogr.*, 626:97.

Doerge, D.R., Bajic, S. (1992). Analysis of pesticides using liquid-chromatography atmospheric-pressure chemical ionization mass-spectrometry. *Rapid Commun. Mass Spectrom.*, 6:663.

Draisci, R., Palleschi, L., Marchiafava, C., Ferretti, E., Quadri, F.D. (2001). Confirmatory analysis of residues of stanozolol and its major metabolite in bovine urine by liquid chromatography–tandem mass spectrometry. *J. Chromatogr. A*, 926:69.

Engewald, W., Teske, J., Efer, J. (1999). Programmed temperature vaporiser-based injection in capillary gas chromatography. *J. Chromatogr. A*, 856:259–278.

Fei, X., Murray, K.K. (1996). On-line coupling of gel permeation chromatography with MALDI mass spectrometry. *Anal. Chem.*, 68:3555.

Fritsche, J., Angoelal, R., Dachtler, M. (2002). On-line liquid-chromatography–nuclear magnetic resonance spectroscopy–mass spectrometry coupling for the separation and characterization of secoisolariciresinol diglucoside isomers in flaxseed. *J. Chromatogr. A*, 972:195–203.

Gagel, J.J., Biemann, K. (1986). Continuous recording of reflection-absorbance Fourier transform infrared spectra of the effluent of a microbore liquid chromatograph. *Anal. Chem.*, 58:2184.

Gagel, J.J., Biemann, K. (1987). Continuous infrared spectroscopic analysis of isocratic and gradient elution reversed-phase liquid chromatography separations. *Anal. Chem.*, 59:1266.

Gale, D.C., Smith, R.D. (1993). Small-volume and low flow-rate electrospray-ionization mass-spectrometry of aqueous samples. *Rapid Commun. Mass Spectrom.*, 7:1017.

Garrido, J.L., Rodríguez, F., Campaña, E., Zapata, M. (2003). Rapid separation of chlorophylls a and b and their demetallated and dephytylated derivatives using a monolithic silica C18 column and a pyridine-containing mobile phase. *J. Chromatogr. A*, 994:85.

Gaskell, S.J. (1997). Electrospray: principles and practice. *J. Mass Spectrom.*, 32:677.

Gelpi, E. (2002). Interfaces for coupled liquid-phase separation/mass spectrometry techniques. An update on recent developments. *J. Mass Spectrom.*, 37:241.

Giese, R.W. (2000). Electron–capture mass spectrometry: recent advances. *J. Chromatogr. A*, 892:329.

Go, A.P., Prenni, J.E., Wei, J., Jones, A., Hall, S.C., Witkowska, H.E., Shen, Z., Siuzdak, G. (2003). Desorption/ionization on silicon time-of-flight/time-of-flight mass spectrometry. *Anal. Chem.*, 75:2504.

Griffiths, P.R., Lange, A.J. (1992). Online use of the concentric flow nebulizer for direct deposition liquid-chromatography Fourier transform-infrared spectrometry. *J. Chromatogr. Sci.*, 30:93.

Grimvall, E., Östmann, C., Nilsson, U. (1995). Determination of polychlorinated biphenyls in human blood plasma by on-line and off-line liquid chromatography—gas chromatography. *J. High Resolut. Chromatogr.*, 18:685–691.

Grob, K. (1990). Ptv vapor overflow—principles of a solvent evaporation technique for introducing large volumes in capillary GC. *J. High Resolut. Chromatogr.*, 13:540–546.

Grob, K. (2000). Efficiency through combining high-performance liquid chromatography and high resolution gas chromatography: progress 1995–1999. *J. Chromatogr. A*, 892:407–420.

Grob, K., Biedermann, M. (1996). Vaporising systems for large volume injection or on-line transfer into gas chromatography: Classification, critical remarks and suggestions. *J. Chromatogr. A*, 750:11–23.

Grob, K., Stoll, J.-M. (1986). Loop-type interface for concurrent solvent evaporation in coupled HPLC-GC. Analysis of raspberry ketone in a raspberry sauce as an example. *J. High Resolut. Chromatogr.*, 9:518–523.

Häkkinen, V.M.A., Virolainen, M.M., Riekkola, M.-L. (1988). New online HPLC-GC coupling system using a 10-port valve interface. *J. High Resolut. Chromatogr.*, 11:214–216.

Haner, R.L., Llanos, W., Mueller, L. (2000). Small volume flow probe for automated direct-injection nmr analysis: design and performance. *J. Mag. Reson.*, 143:69.

Hankemeier, T., van Leeuwen, S.P.J., Vreuls, R.J.J., Brinkman, U.A.Th. (1998). Use of a presolvent to include volatile organic analytes in the application range of on-line solid-phase extraction–gas chromatography–mass spectrometry. *J. Chromatogr. A*, 811:117–133.

Hankemeier, T., Kok, S.J., Vreuls, R.J.J., Brinkman, U.A.Th. (1998). Monitoring the actual carrier gas flow during large-volume on-column injections in gas chromatography as a means to automate closure of the solvent vapour exit. *J. Chromatogr. A*, 811:105–116.

Hansen, S.H., Jensen, A.G., Cornett, C., Björnsdottir, I., Taylor, S., Wright, B., Wilson, I.D. (1999). High-performance liquid chromatography on-line coupled to high-field NMR and mass spectrometry for structure elucidation of constituents of *hypericum perforatum* L. *Anal. Chem.*, 71:5235–5241.

Hardin, E.D., Fan, T.P., Blakley, C.R., Vestal, M.L. (1984). Laser desorption mass spectrometry with thermospray sample deposition for determination of nonvolatile biomolecules. *Anal. Chem.*, 56:2–7.

Hau, J., Bovetto, L. (2001). Characterisation of modified whey protein in milk ingredients by liquid chromatography coupled to electrospray ionisation mass spectrometry. *J. Chromatogr. A*, 926:105.

Hau, J., Schrader, W., Linscheid, M. (1993). Continuous-flow fast-atom-bombardment mass-spectrometry—a concept to improve the sensitivity. *Organic Mass Spectrom.*, 28:216.

Hau, J., Nigge, W., Linscheid, M. (1993). A new ion-source for liquid-chromatography thermospray mass-spectrometry with a magnetic-sector field mass-spectrometer. *Organic Mass Spectrom.*, 28:223.

He, L., Liang, L., Lubman, D.M. (1995). Continuous-flow MALDI mass spectrometry using an ion trap/reflectron time-of-flight detector. *Anal. Chem.*, 67:4127.

Hellgeth, J.W., Taylor, L.T. (1987). Optimization of a flow cell interface for reversed-phase liquid chromatography/Fourier transform infrared spectrometry. *Anal. Chem.* 59:295.

Henion, J.D., Thomson, B.A., Dawson, P.H. (1982). Determination of sulfa drugs in biological fluids by liquid chromatography/mass spectrometry/mass spectrometry. *Anal. Chem.*, 54:451.

Hirabayashi, A., Sakairi, M., Koizumi, H. (1994). Sonic spray ionization method for atmospheric pressure ionization mass spectrometry. *Anal. Chem.*, 66:4557.

Hirabayashi, A., Sakairi, M., Koizumi, H. (1995). Sonic spray mass spectrometry. *Anal. Chem.*, 67:2878.

Hiraoka, K. (1992). How are ions formed from electrosprayed charged liquid droplets. *Rapid Commun. Mass Spectrom.*, 6:463.

Hiraoka, K., Fukasawa, H., Matsushita, F., Aizawa, K. (1995). High-flow liquid-chromatography mass-spectrometry interface using a parallel ion-spray. *Rapid Commun. Mass Spectrom.*, 9:1349.

van der Hoeven, R.A.M., Buscher, B.A.P., Tjaden, U.R., van der Greef, J. (1995). Performance of an electrospray-interfaced thermospray ion source in hyphenated techniques. *J. Chromatogr. A*, 712:211.

van der Hoeven, R.A.M., Tjaden, I.R., van der Greef, J. (1996). Performance of a low cost atmospheric pressure chemical ionization-interfaced thermospray ion source. *Rapid Commun. Mass Spectrom.*, 10:1539.

Holt, R.M., Newman, M.J., Pullen, F.S., Richards, D.S., Swanson, A.G. (1997). High-performance liquid chromatography NMR spectrometry mass spectrometry: further advances in hyphenated technology. *J. Mass Spectrom.*, 32:64–70.

Hopfgartner, G., Wachs, T., Bean, K., Henion, J. (1993). High-flow ion spray liquid chromatography/mass spectrometry. *Anal. Chem.*, 65:439.

Hyötyläinen, T., Riekkola, M.-L. (2003). On-line coupled liquid chromatography–gas chromatography. *J. Chromatogr. A*, 1000:357–384.

Hyötyläinen, T., Keski-Hynnilä, H., Riekkola, M.-L. (1997). Determination of morphine and its analogues in urine by on-line coupled reversed-phase liquid chromatography-gas chromatography with on-line derivatization. *J. Chromatogr. A*, 771:360–365.

Hyötyläinen, T., Andersson, T., Hartonen, K., Kuosmanen, K., Riekkola, M.-L. (2000). Pressurized hot water extraction coupled on-line with LC-GC: determination of polyaromatic hydrocarbons in sediment. *Anal. Chem.*, 72:3070–3076.

Iribarne, J.V., Thompson, B.A. (1976). *J. Chem. Phys.*, 64:2287.

Ito, Y., Takeuchi, T., Ishii, D., Goto, M. (1985). Direct coupling of micro high-performance liquid chromatography with fast atom bombardment mass spectrometry. *J. Chromatogr.*, 346:161.

Johnson, C.C., Hellgeth, J.W., Taylor, L.T. (1985). Reversed-phase liquid chromatography with Fourier infrared spectrometric detection using a flow cell interface. *Anal. Chem.* 57:610.

Johnson, J.V., Yost, R.A., Kelley, P.E., Bradford, D.C. (1990). Tandem-in-space and tandem-in-time mass spectrometry: triple quadrupoles and quadrupole ion traps. *Anal. Chem.*, 62:2162.

Jongenotter, G.A., Kerkhoff, M.A.T., van der Knaap, H.C.M., Vandeginste, B.G.M. (1999). Automated on-line GPC-GC-FPD involving co-solvent trapping and the on-column interface for the determination of organophosphorus pesticides in olive oils. *J. High Resolut. Chromatogr.*, 22(1):17–23.

Juhasz, P., Roskey, M.T., Smirnov, I.P., Haff, L.A., Vestal, M.L., Martin, S.A. (1996). Applications of delayed extraction matrix-assisted laser desorption ionization time-of-flight mass spectrometry to oligonucleotide analysis. *Anal. Chem.*, 68:941.

Kalasinsky, K.S., Smith, J.A.S., Kalasinsky, V.F. (1989). Microbore high-performance liquid chromatography/Fourier transform infrared interface for normal- or reverse-phase liquid chromatography. *Anal. Chem.*, 57:1969.

Karas, M. (1988). Laser desorption ionization of proteins with molecular masses exceeding 10,000 daltons. *Anal. Chem.*, 60:2299–2301.

Karas, M., Bachmann, D., Hillenkamp, F. (1985). Influence of the wavelength in high-irradiance ultraviolet laser desorption mass spectrometry of organic molecules. *Anal. Chem.*, 57:2935.

Katta, V., Rockwood, A.L., Vestal, M.L. (1991). Field limit for ion evaporation from charged thermospray droplets. *Int. J. Mass Spectrom. Ion Processes*, 103:129.

Kaufmann, R., Spengler, B., Lützenkirchen, F. (1993). Mass-spectrometric sequencing of linear peptides by product-ion analysis in a reflectron time-of-flight mass-spectrometer using matrix-assisted laser-desorption ionization. *Rapid Commun. Mass Spectrom.*, 7:902.

Kebarle, P., Ho, Y. (1997). In: Cole, R.B., ed. *Electrospray Ionization Mass Spectrometry: Fundamentals, Instrumentation & Applications.* John Wiley & Sons Inc. (Wiley-Interscience publication, New York, USA).

Kizer, K.L., Mantz, A.W., Bonar, L.C. (1975). *Am. Lab.*, 7(5):85.

Klinkenberg, H., van der Wal, S., de Koster, C., Bart, J. (1998). On the use of inductively coupled plasma mass spectrometry as an element specific detector for liquid chromatography: optimization of an industrial tellurium removal process. *J. Chromatogr. A*, 794:219.

Koch, H.M., Gonzalez-Reche, L.M., Angerer, J. (2003) On-line clean-up by multidimensional liquid chromatography-electrospray ionization tandem mass spectrometry for high throughput quantification of primary and secondary phthalate metabolites in human urine. *J. Chromatogr. B*, 784:169.

Kok, S.J., Arentsen, N.C., Cools, P.J.C.H., Hankemeier, Th., Schoenmakers, P.J. (2002). Fourier transform infrared spectroscopy with a sample deposition interface as a quantitative detector in size-exclusion chromatography. *J. Chromatogr. A*, 948:257.

de Koning, J.A., Hogenboom, A.C., Lacker, T., Strohschein, S., Albert, K., Brinkman, U.A.Th. (1998). On-line trace enrichment in hyphenated liquid chromatography–nuclear magnetic resonance spectroscopy. *J. Chromatogr. A*, 813:55.

Kostiainen, R.K., Bruins, A.P. (1994). Effect of multiple sprayers on dynamic-range and flow-rate limitations in electrospray and ionspray mass-spectrometry. *Rapid Commun. Mass Spectrom.*, 8:549.

Kuosmanen, K., Hyötyläinen, T., Hartonen, K., Riekkola, M.-L. (2001). Pressurised hot water extraction coupled on-line with liquid chromatography-gas chromatography for the determination of brominated flame retardants in sediment sample. *J. Chromatogr. A*, 943:113–122.

Lacey, M.E., Tan, Z.J., Webb, A.G., Sweedler, J.V. (2001). Union of capillary high-performance liquid chromatography and microcoil nuclear magnetic resonance spectroscopy applied to the separation and identification of terpenoids. *J. Chromatogr. A.*, 922:139.

LaFreniere, K.E., Fassel, V.A., Eckels, D.E. (1987). Elemental speciation via high-performance liquid chromatography combined with inductively coupled plasma atomic emission spectroscopic detection: application of a direct injection nebulizer. *Anal. Chem.*, 59:879.

Laiko, V.V., Baldwin, M.A., Burlingame, A.L. (2000). Atmospheric pressure matrix-assisted laser desorption/ionization mass spectrometry. *Anal. Chem.*, 72:652.

Lange, A.J., Griffiths, P.R., Fraser, D.J.J. (1991). Reversed-phase liquid chromatography/Fourier transform infrared spectrometry using concentric flow nebulization. *Anal. Chem.*, 63:782.

Lawrence, K.E., Rice, G.W., Fassel, V.A. (1984). Direct liquid sample introduction for flow injection analysis and liquid chromatography with inductively coupled, argon plasma spectrometric detection. *Anal. Chem.*, 56:289.

Lee, E.D., Henion, J.D. (1992). Thermally assisted electrospray interface for liquid-chromatography mass-spectrometry. *Rapid Commun. Mass Spectrom.*, 6:727.

Li, L., Wang, A.P.L., Coulson, L.D. (1993). Continuous-flow matrix-assisted laser desorption ionization mass spectrometry. *Anal. Chem.*, 65:493.

Liu, H., Berger, S.J., Chakraborty, A.B., Plumb, R.S., Cohen, S.A. (2002). Multidimensional chromatography coupled to electrospray ionization time-of-flight mass spectrometry as an alternative to two-dimensional gels for the identification and analysis of complex mixtures of intact proteins. *J. Chromatogr. B*, 782:267.

Loo, J.A., Udseth, H.R., Smith, R.D. (1989). Peptide and protein analysis by electrospray ionization-mass spectrometry and capillary electrophoresis-mass spectrometry. *Anal. Biochem.*, 179:404.

Louris, J.N., Cooks, R.G., Syka, J.E.P., Kelley, P.E., Stafford, G.C. Jr., Todd, J.F.J. (1987). Instrumentation, applications, and energy deposition in quadrupole ion-trap tandem mass spectrometry. *Anal. Chem.*, 59:1677.

Ludlow, M., Louden, D., Handley, A., Taylor, S., Wright, B., Wilson, I.D. (1999). Size-exclusion chromatography with on-line ultraviolet, proton nuclear magnetic resonance and mass spectrometric detection and on-line collection for off-line Fourier transform infrared spectroscopy. *J. Chromatogr. A*, 857:89–96.

Macnaughtan, M.A., Huo, T., MacNamara, E., Santini, R.E., Raftery, D. (2002). NMR difference probe: A dual-coil probe for NMR difference spectroscopy. *J. Mag. Reson.*, 156:97.

McFadden, W.H., Swartz, H.L., Evans, S. (1976). Direct analysis of liquid chromatographic effluents. *J. Chromatogr.*, 122:389.

McLean, J.A., Zhang, H., Montaser, A. (1998). A direct injection high-efficiency nebulizer for inductively coupled plasma mass spectrometry. *Anal. Chem.*, 70(5):1012.

McLuckey, S.A., Glish, G.L., Asano, K.G., Grant, B.C. (1988). Atmospheric sampling glow discharge ionization source for the determination of trace organic compounds in ambient air. *Anal. Chem.*, 60:2220.

Medzihradszky, K.F., Campbell, J.M., Baldwin, M.A., Falick, A.M., Juhasz, P., Vestal, M.L., Burlingame, A.L. (2000). The characteristics of peptide collision-induced dissociation using a high-performance MALDI-TOF/TOF tandem mass spectrometer. *Anal. Chem.*, 72:552.

Michael, S.M., Chien, B.M., Lubman, D.M. (1993). Detection of electrospray ionization using a quadrupole ion trap storage/reflectron time-of-flight mass spectrometer. *Anal. Chem.*, 65:2614.

Michalke, B. (2002). The coupling of LC to ICP-MS in element speciation: I. General aspects. *TRAC*, 21(2):142.

Mol, H.G.J., Janssen, H.-G.M., Cramers, C.A., Brinkman, U.A.Th., (1993). Use of a temperature programmed injector with a packed liner for direct water analysis and on-line reversed phase LC-GC. *J. High Resolut. Chromatogr.*, 16:459–463.

Mondello, L., Dugo, P., Dugo, G., Lewis, A.C., Bartle, K.D. (1999). High-performance liquid chromatography coupled on-line with high resolution gas chromatography State of the art. *J. Chromatogr. A*, 842:373–390.

Mosely, M.A., Deterding, L.J., de Wit, J.S.M., Tomer, K.B., Kennedy, R.T., Bragg, N., Jorgenson, J.W. (1989). Optimization of a coaxial continuous flow fast atom bombardment interface between capillary liquid chromatography and magnetic sector mass spectrometry for the analysis of biomolecules. *Anal. Chem.*, 61:1577.

Mosely, M.A., Deterding, L.J., Tomer, K.B., Jorgenson, J.W. (1989). Capillary-zone electrophoresis/fast-atom bombardment mass spectrometry: design of an on-line coaxial continuous-flow interface. *Rapid Commun. Mass Spectrom.*, 3:87.

Mottaleb, M.A., Kim (2002). Enhanced desolvation and improved deposition efficiency by modified thermospray for the coupling of HPLC and FT-IR. *Anal. Sci*, 18(5):579.

Murray, K.K., Russell, D.H. (1993). Liquid sample introduction for matrix-assisted laser desorption ionisation. *Anal. Chem.*, 65:2534.

Murray, K.K., Lewis, T.M., Beeson, M.D., Russell, D.H. (1994). Aerosol matrix-assisted laser desorption ionization for liquid chromatography/time-of-flight mass spectrometry. *Anal. Chem.*, 66:1601.

Nagra, D.S., Li, L. (1995). Liquid chromatography-time-of-flight mass spectrometry with continuous-flow matrix-assisted laser desorption ionisation. *J. Chromatogr. A*, 711:235.

Neissen, W.M.A. (1986). A review of direct liquid introduction interfacing for LC/MS part 1: Instrumental aspects. *Chromatographia*, 21:277.

Niessen, W.M.A. (1998). Advances in instrumentation in liquid chromatography–mass spectrometry and related liquid-introduction techniques. *J. Chromatogr. A*, 794:407.

Niessen, W.M.A. (1999). State-of-the-art in liquid chromatography–mass spectrometry. *J. Chromatogr. A*, 856:179.

Niessen, W.M.A. (2003). Progress in liquid chromatography–mass spectrometry instrumentation and its impact on high-throughput screening. *J. Chromatogr. A*, 1000:413.

Niessen, W.M.A., Tinke, A.P. (1995). Liquid chromatography–mass spectrometry: General principles and instrumentation. *J. Chromatogr. A*, 703:37.

Niessen, W.M.A., Tjaden, U.R., van der Greef, J. (1991). Strategies in developing interfaces for coupling liquid chromatography and mass spectrometry. *J. Chromatogr. A*, 554:3.

Neissen, M.L., Bennett, K.L., Larsen, B., Moniatte, M., Mann, M. (2002). Peptide end sequencing by orthogonal maldi tandem mass spectrometry. *J. Proteome Res.*, 1:63.

Ørsnes, H., Graf, T., Degn, H., Murray, K.K. (2000). A rotating ball inlet for on-line MALDI mass spectrometry. *Anal. Chem.*, 72:251.

Overberg, A., Karas, M., Bahr, U., Kaufmann, R., Hillenkamp, F. (1990). Matrix-assisted infrared-laser (2.94 μm) desorption/ionization mass spectrometry of large biomolecules. *Rapid Commun. Mass Spectrom.*, 4:293.

Perez, M., Alario, J., Vazquez, A., Villén, J. (2000). Pesticide residue analysis by off-line SPE and on-line reversed-phase LC-GC using the through-oven-transfer adsorption/desorption interface. *Anal. Chem.*, 72:846–852.

Pfeifer, T., Klaus, U., Hoffmann, R., Spiteller, M. (2001). Characterisation of humic substances using atmospheric pressure chemical ionisation and electrospray ionisation mass spectrometry combined with size-exclusion chromatography. *J.'Chromatogr. A*, 926:151.

Preisler, J., Foret, F., Karger, B.L. (1998). On-line MALDI-TOF MS using a continuous vacuum deposition interface. *Anal. Chem.*, 70:5278.

Preisler, J., Hu, P., Rejtar, T., Karger, B.L. (1998). Capillary electrophoresis-matrix-assisted laser desorption/ionization time-of-flight mass spectrometry using a vacuum deposition interface. *Anal. Chem.*, 72:4785.

Proctor, C.J., Todd, J.F.J. (1983). Atmospheric pressure ionization mass spectrometry. *Org. Mass Spectrom.*, 18:509.

Pullen, F.S., Swanson, A.G., Newman, M.J., Richards, D.S. (1995). "On-line" liquid chromatography/nuclear magnetic resonance mass spectrometry—a powerful spectroscopic tool for the analysis of mixtures of pharmaceutical interest. *Rapid Commun. Mass Spectrom.*, 9:1003–1006.

Raynor, M.W., Bartle, K.D., Cook, B.W. (1992). Electrospray micro liquid-chromatography—Fourier-transform infrared micro spectrometry. *HRC*, 6:361.

Robb, D.B., Covey, T.R., Bruins, A.P. (2000). Atmospheric pressure photoionization: an ionization method for liquid chromatography-mass spectrometry. *Anal. Chem.*, 72:3653.

Sakairi, M., Kambara, H. (1988). Characteristics of a liquid chromatograph/atmospheric pressure ionization mass spectrometer.*Anal. Chem.*, 60:774.

Sakairi, M., Kambara, H. (1989). Atmospheric pressure spray ionization for liquid chromatography/mass spectrometry. *Anal. Chem.*, 61:1159.

Sakairi, M., Kato, Y. (1998). Multi-atmospheric pressure ionisation interface for liquid chromatography–mass spectrometry. *J. Chromatogr. A*, 794:391.

Sandvoss, M., Weltring, A., Preiss, A., Levsen, K., Wuensch, G. (2001). Combination of matrix solid-phase dispersion extraction and direct on-line liquid chromatography–nuclear magnetic resonance spectroscopy–tandem mass spectrometry as a new efficient approach for the rapid screening of natural products: application to the total asterosaponin fraction of the starfish *Asterias rubens. J. Chromatogr. A*, 917:75–86.

Sato, K., Asada, T., Ishihara, M., Kunihiro, F., Kammei, Y., Kubota, E., Costello, C.E., Martin, S.A., Scoble, H.A., Biemann, K. (1987). High-performance tandem mass spectrometry: calibration and performance of linked scans of a four-sector instrument. *Anal. Chem.*, 59:1652.

Schmarr, H.-G., Mosandl, A., Grob, K. (1989). Coupled LC-GC: Evaporation rates for partially concurrent eluent evaporation using an early solvent vapor exit. *J. High Resolut. Chromatogr.*, 12:721–726.

Schmelzeisen-Redeker, G., Bütfering, L., Röllgen, F.W. (1989). Desolvation of ions and molecules in thermospray mass spectrometry. *Int. J. Mass Spectrom. Ion Processes*, 90:139.

Schunk, T.C., Balke, S.T., Cheung, P. (1994). Quantitative polymer composition characterization with a liquid chromatography–Fourier transform infrared spectrometry–solvent-evaporation interface. *J. Chromatogr.*, 661:227.

Schwartz, J.C., Jardine, I. (1992). High resolution parent-ion selection/isolation using a quadrupole ion-trap mass spectrometer. *Rapid Commun. Mass Spectrom.*, 6:313.

Scott, R.P.W., Scott, C.G., Munroe, M., Hess, J. Jr. (1974). Interface for on-line liquid chromatography–mass spectroscopy analysis. *J.'Chromatogr.*, 99:395.

Señoráns, F.J., Herraiz, M., Tabera, J. (1995). On-line reversed-phase liquid chromatography-capillary gas chromatography using a programmed temperature vaporizer as interface. *J. High Resolut. Chromatogr.*, 18:433–438.

Señoráns, F.J., Reglero, G., Herraiz, M. (1995). Use of a programmed temperature injector for on-line reversed-phase liquid chromatography-capillary gas chromatography. *J. Chromatogr. Sci.*, 33:446–450.

Shimmo, M., Adler, H., Hyötyläinen, T., Hartonen, K., Kumala, M., Riekkola, M.-L. (2002). Analysis of particulate polycyclic aromatic hydrocarbons by on-line coupled supercritical fluid extraction-liquid chromatography-gas chromatography-mass spectrometry. *Atmos. Environ.*, 36:2985–2995.

Shockor, J.P., Unger, S.E., Wilson, I.D., Foxall, P.J.D., Nicholson, J.K., Lindon, J.C. (1996). Combined HPLC, NMR spectroscopy, and ion-trap mass spectrometry with application to the detection and characterization of xenobiotic and endogenous metabolites in human urine. *Anal. Chem.*, 68:4431–4435.

Shomo II, R.E., Marshall, A.G., Weisenberger, C.R. (1985). Laser desorption Fourier transform ion cyclotron resonance mass spectrometry vs. Fast atom bombardment magnetic sector mass spectrometry for drug analysis. *Anal. Chem.*, 57:2940.

Simpson, H., Berthemy, A., Buhrman, D., Burton, R., Newton, J., Kealy, M., Wells, D., Wu, D. (1998). High throughput liquid chromatography mass spectrometry bioanalysis using 96-well disk solid phase extraction plate for the sample preparation. *Rapid Commun. Mass Spectrom.*, 12:75.

Smith, R.D., Olivares, J.A., Nguyen, N.T., Udseth, H.R. (1988). Capillary zone electrophoresis-mass spectrometry using an electrospray ionization interface. *Anal. Chem.*, 60:436.

Smith, R.D., Barinaga, C.J., Udseth, H.R. (1988). Improved electrospray ionization interface for capillary zone electrophoresis-mass spectrometry. *Anal. Chem.*, 60:1948.

Smith, R.D., Wahl, J.H., Goodlett, D.R., Hofstadler, S.A. (1993). Capillary electrophoresis/mass spectrometry. *Anal. Chem.*, 65:574A.

Somsen, G.W., Gooijer, C., Brinkman, U.A.Th. (1999). Liquid chromatography-Fourier-transform infrared spectrometry. *J. Chromatogr. A*, 856:213.

Song, D., Kohlhof, K. (1999). Application of two-dimensional high-performance liquid chromatography–mass spectrometry with particle beam interface. *J. Chromatogr. B*, 730:141.

Staniewski, J., Rijks, J.A. (1993). Potential and limitations of differently designed programmed-temperature injector liners for large volume sample introduction in capillary GC. *J. High Resolut. Chromatogr.*, 16:182–187.

Sui, K.W.M., Guevremont, R., Le Blanc, J.C.Y., Gardner, G.J., Berman, S.S. (1991). Electrospray interfacing for the coupling of ion-exchange and ion-pairing chromatography to mass spectrometry. *J. Chromatogr.*, 554:27.

Sun, Y.C. (2003). Comparative study on conventional and low-flow nebulizers for arsenic speciation by means of microbore liquid chromatography with inductively coupled plasma mass spectrometry. *J. Chromatogr. A.*, 1005:207.

Sunner, J., Nicol, G., Kebarle, P. (1988). Factors determining relative sensitivity of analytes in positive mode atmospheric pressure ionization mass spectrometry. *Anal. Chem.*, 60:1300.

Sunner, J., Ikonomou, M.G., Kabarle, P. (1988). Sensitivity enhancements obtained at high temperatures in atmospheric pressure ionization mass spectrometry. *Anal. Chem.*, 60:1308.

Sweeney, A.P., Shalliker, R.A. (2002). Development of a two-dimensional liquid chromatography system with trapping and sample enrichment capabilities. *J. Chromatogr. A.*, 968:41.

Tang, L., Kebarle, P. (1993). Dependence of ion intensity in electrospray mass spectrometry on the concentration of the analytes in the electrosprayed solution. *Anal. Chem.*, 65:3654.

Taylor, S.D., Wright, B., Clayton, E., Wilson, I.D. (1998). Practical aspects of the use of high performance liquid chromatography combined with simultaneous nuclear magnetic resonance and mass spectrometry. *Rapid Commun. Mass Spectrom.*, 12:1732–1736.

Thompson, B.A., Iribarne, J.V. (1979). *J. Chem. Phys.*, 71:4451.

Tor, E.R., Puschner, B., Whitehead, W.E. (2003). Rapid determination of domoic acid in serum and urine by liquid chromatography-electrospray tandem mass spectrometry. *J. Agric. Food Chem.*, 51:1791.

Torabi, K., Karami, A., Balke, S.T., Schunk, T.C. (2001). Quantitative FTIR detection in size-exclusion chromatography. *J.'Chromatogr. A.*, 910:19.

Trisciani, A., Munari, F. (1994). Characterization of fuel samples by on-line LC-GC with automatic group-type separation of hydrocarbons. *J. High Resolut. Chromatogr.*, 17:452–456.

Van Pelt, C.K., Corso, T.N., Schultz, G.A., Lowes, S., Henion, J. (2001). A four-column parallel chromatography system for isocratic or gradient LC/MS analyses. *Anal. Chem.*, 73:582.

Vestal, M.L. (1984). High performance liquid chromatography-mass spectrometry. *Science*, 226:275.

Vestal, M.L. (2001). Methods of ion generation. *Chem. Rev.* 101:361.

Vidrine, D.W., Mattson, D.R. (1978). *Appl. Spectrosc.*, 32:502.

Volmer, D.A., Lay, J.O. Jr., Billedeau, S.M., Vollmer, D.L. (1996). Detection and confirmation of n-nitrosodialkylamines using liquid chromatography-electrospray ionization coupled on-line with a photolysis reactor. *Anal. Chem.*, 68:546.

Vonach, R., Lendl, B., Kellner, R. (1998). High-performance liquid chromatography with real-time Fourier-transform infrared detection for the determination of carbohydrates, alcohols and organic acids in wines. *J. Chromatogr. A.*, 824:159.

Voress, L. (1994). Electrospray MS charging full stream ahead. *Anal. Chem.*, 66:481A.

Wang, A.P.L., Guo, X., Li, L. (1994). Liquid chromatography/time-of-flight mass spectrometry with a pulsed sample introduction interface. *Anal. Chem.*, 66:3664.

Wang, L., May, S.W., Broner, R.F., Pollock, S.H. (1996). Low-flow interface for liquid chromatography inductively coupled plasma mass spectrometry speciation using an oscillating capillary nebulizer. *J. Anal. At. Spectrom.*, 11:1137.

Whelan, T.J. (2000). *The Detection and Quantitation of Impurities in Pharmaceuticals at Trace Levels by Liquid Chromatography Hyphenated Mass Spectrometry*. Masters Thesis, University of Technology Sydney, NSW, Australia.

Whitehouse, C.M., Dreyer, R.N., Yamashita, M., Fenn, J.B. (1985). Electrospray interface for liquid chromatographs and mass spectrometers. *Anal. Chem.*, 57:675.

Whitehouse, C.M., Dreyyer, R.N., Yamashita, M., Fenn, J.B. (1985). Electrospray interface for liquid chromatographs and mass spectrometers. *Anal. Chem.*, 57:675.

Willoughby, R.C., Browner, R.F. (1984). Monodisperse aerosol generation interface for combining liquid chromatography with mass spectroscopy. *Anal. Chem.*, 56:2626.

Wilm, M.S., Mann, M. (1994). Electrospray and Taylor-Cone theory, Dole's beam of macromolecules at last? *Int. J.'Mass Spectrom. Ion Processes*, 136:167.

Wilson, I.D. (2000). Multiple hyphenation of liquid chromatography with nuclear magnetic resonance spectroscopy, mass spectrometry and beyond. *J. Chromatogr. A*, 892:315–327.

Wilson, I.D., Brinkman, U.A.Th. (2003). Hyphenation and hypernation: The practice and prospects of multiple hyphenation. *J. Chromatogr. A*, 1000:325–356.

Wilson, I.D., Morgan, E.D., Lafont, R., Shockor, J.P., Lindon, J.C., Nicholson, J.K., Wright, B. (1999). High performance liquid chromatography coupled to nuclear magnetic resonance spectroscopy and mass spectrometry applied to plant products: Identification of ecdysteroids from silene otites. *Chromatographia*, 49:374–378.

Winkler, P.C., Perkins, D.D., Williams, W.K., Browner, R.F. (1988). Performance of an improved monodisperse aerosol generation interface for liquid chromatography/mass spectrometry. *Anal. Chem.*, 60:489.

de Wit, J.S.M., Deterding, L.J., Moseley, M.A., Tomer, K.B., Jorgenson, J.W. (1988). Design of a coaxial continuous flow fast atom bombardment probe. *Rapid Commun. Mass Spectrom.*, 2:100.

Xia, Y.-Q., Whigan, D.B., Powell, M.L., Jemal, M. (2000). Ternary-column system for high-throughput direct-injection bioanalysis by liquid chromatography/tandem mass spectrometry. *Rapid Commun. Mass Spectrom.*, 14:105.

Yamaguchi, K. (2003). Cold-spray ionization mass spectrometry: principle and applications. *J. Mass Spectrom.*, 38:473.

Yang, L., Fergusson, G.J., Vestal, M.L. (1984). A new transport detector for high-performance liquid chromatography based on thermospray vaporization. *Anal. Chem.*, 56:2632.

Yang, L., Wu, N., Rudewicz, P.J. (2001). Applications of new liquid chromatography–tandem mass spectrometry technologies for drug development support. *J. Chromatogr. A*, 926:43.

Yang, L., Mann, T.D., Little, D., Wu, N., Clement, R.P., Rudewicz, P.J. (2001). Evaluation of a four-channel multiplexed electrospray triple quadrupole mass spectrometer for the simultaneous validation of LC/MS/MS methods in four different preclinical matrixes. *Anal. Chem.*, 73:1740.

Zweigenbaum, J., Heinig, K., Steinborner, S., Wachs, T., Henion, J. (1999). High-throughput bioanalytical LC/MS/MS determination of benzodiazepines in human urine: 1000 samples per 12 hours. *Anal. Chem.*, 71:2294.

30

Thin Layer Chromatography

JOSEPH SHERMA
Department of Chemistry, Lafayette College, Easton, PA, USA

I. INTRODUCTION

Thin layer chromatography (TLC) is a type of liquid chromatography in which the stationary phase is in the form of a layer on a glass, an aluminum, or a plastic support. The term "planar chromatography" is often used for both TLC and paper chromatography (PC) because each employs a planar stationary phase rather than a packed column. PC, which utilizes plain, modified, or impregnated paper (cellulose) as the stationary phase, involves many of the same basic techniques as TLC, but it has not evolved into an efficient, sensitive, quantitative, instrument-based analytical method and has many disadvantages relative to TLC. Consequently, PC will not be covered in this chapter.

As originally developed in 1951 by J.G. Kirchner and colleagues, later standardized by E. Stahl and colleagues, and still widely practiced today (Fried and Sherma, 1999a; Sherma, 2002), classical, capillary-action TLC is an inexpensive, easy technique that requires little instrumentation, which is used for separation of simple mixtures and for qualitative identification or semiquantitative, visual analysis of samples. In contrast, modern TLC [usually termed as high performance thin layer chromatography (HPTLC)], which began around 1975 with the introduction of high efficiency, commercially precoated plates by Merck, is an instrumental technique carried out on efficient, fine-particle layers. Instrumental HPTLC is capable of producing fast, high-resolution separations and qualitative and quantitative results that meet good manufacturing practices (GMP) and good laboratory practices (GLP) standards. The accuracy and precision of data obtained by HPTLC rival those of gas chromatography (GC) and high performance column liquid chromatography (HPLC), and it has many advantages relative to these methods.

TLC is an off-line process in which the various stages (Fig. 1) are carried out independently. The advantages of this arrangement using an open, disposable layer compared with an on-line column process such as HPLC include the possibility of separating up to 70 samples and standards simultaneously on a single plate, leading to high throughput, low cost analyses and the ability to construct calibration curves from standards chromatographed under the same conditions as the samples; analyzing samples with minimum sample preparation without fear of irreversible contamination of the column or carryover of fractions from one sample to another as can occur in HPLC with sequential injections; and analyzing a sample by use of multiple separation steps and static postchromatographic detection procedures with various universal and specific visualization reagents, which is possible because all sample components are "stored" on the layer without chance of loss. TLC is highly selective and flexible because of the great variety of layers that is available commercially. It has proven to be as sensitive

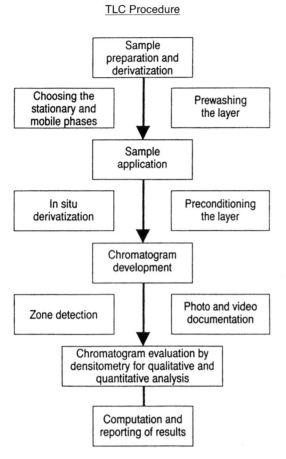

Figure 1 Schematic diagram of the steps in a TLC analysis. (Courtesy of Camag.)

as HPLC in many analyses, and solvent usage per sample is extremely low.

TLC and HPTLC plates are usually developed by capillary flow of the mobile phase without pressure in ascending or horizontal modes, but forced flow methods, in which the mobile phase is driven through the layer by pumping under pressure or by centrifugal force, are also used. In capillary-flow TLC, the migration speed of the mobile phase decreases with the square of the solvent migration distance.

Limited separation efficiency is a disadvantage of TLC. The maximum number of theoretical plates (N) for HPTLC is ca. 5000 compared with ca. 15,000 for HPLC, while the separation number (the number of spots that can be separated over the distance of the run with a resolution of unity) is ca. 15 in TLC compared with 200 in HPLC. This limited efficiency is a result of the capillary-flow mechanism over a restricted migration distance. Under forced flow conditions, TLC separation efficiency is significantly improved. For a discussion on the theory and mechanism of the various modes of TLC, see Fried and Sherma (1999b), Kowalska and Prus (2001), and Kowalska et al. (2003).

An apparent disadvantage of TLC is that although all of the individual steps have been automated and on-line coupling with other chromatographic and spectrometric techniques has been achieved, complete automation of TLC has not been realized. However, changing the offline nature of TLC to an on-line closed system with complete automation would eliminate many of the advantages stated earlier. Several studies have shown that more samples per day can be processed using stepwise-automated TLC compared with a fully automated HPLC system, and with lower cost (Abjean, 1993).

Analytical TLC differs from preparative layer chromatography (PLC) in that larger weights and volumes of samples are applied to thicker (0.5–2 mm) and sometimes larger layers in the latter method, the purpose of which is the isolation of 10–1000 mg of sample for further analysis.

This chapter describes the TLC techniques and instrumentation with different levels of automation that can be used in a contemporary analytical laboratory to produce high-quality analytical results without sacrificing the great flexibility of the method.

II. SAMPLE PREPARATION

Sample preparation (Fried and Sherma, 1999a, c; Sherma, 2003) procedures for TLC are similar to those for GC and HPLC. The solution to be spotted must be sufficiently concentrated so that the analyte can be detected in the applied volume, and pure enough so that it can be separated as a discrete, compact spot or zone. The solvent in which the sample is dissolved must be suitable in terms of viscosity, volatility, ability to wet the layer, and potential for unintended predevelopment during sample application.

Relatively pure samples or their concentrated extracts can often be directly spotted for TLC analysis. If the analyte is present in low concentration in a complex sample, solvent extraction, cleanup (purification), and concentration procedures must precede TLC. Because layers are not reused, it is often possible to apply cruder samples than could be injected into a GC or HPLC column, including samples with irreversibly sorbed impurities. However, impurities that comigrate with the analyte, adversely affect its detection, or distort its zone (i.e., cause streaking or tailing), must be removed prior to TLC. Carryover of material from one sample to another is not a problem as it is in on-line column methods involving sequential injections. Therefore, sample preparation is often simpler for TLC compared with other chromatographic methods.

Common cleanup procedures include liquid–liquid extraction, column chromatography, desalting, and deproteinization. Solid-phase extraction (SPE) using small,

Thin Layer Chromatography

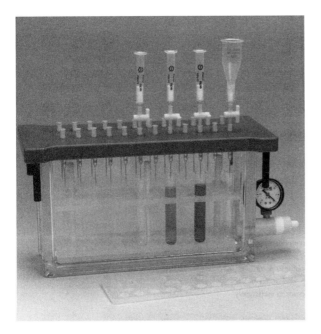

Figure 2 J.T. Baker vacuum processor for 12 or 24 BAKER-BOND SPE or Speedisk columns. (Courtesy of Mallinckrodt Baker Inc., Phillipsburg, NJ.)

disposable columns and a manual vacuum manifold (Fig. 2) or various semiautomated or automated instruments has become widely used for isolation and cleanup of samples prior to TLC analysis, such as pesticide residues in water (Hamada and Wintersteiger, 2002). Supercritical fluid extraction has been used for off-line extraction of analytes from samples (van Beek, 2002), and has been directly coupled to TLC for analysis of solid samples or solutions loaded on glass fiber filters (Esser and Klockow, 1994). A derivative of the analyte can be formed in solution prior to spotting, or *in situ* at the origin by overspotting of a reagent, in order to improve resolution or detection. Special plates with preadsorbent zone serve for sample cleanup by retaining some interfering substances.

III. STATIONARY PHASES

TLC and HPTLC plates are commercially available in the form of precoated layers supported on glass, plastic sheets, or aluminum foil (Fried and Sherma, 1999d; Lepri and Cincinelli, 2001; Rabel, 2003). HPTLC plates are smaller (10×10 or 10×20 cm), have a thinner (0.1–0.2 mm) more uniform layer composed of smaller diameter particles (5–6 μm), and are developed over shorter distances (ca. 3–7 cm) compared with classical 20×20 cm TLC plates, which have a 0.25 mm thick layer of 12–20 μm particle size and are developed for 10–12 cm. Optimal development distances are the points beyond which increased resolution of zones is offset by diffusion effects. In comparison with TLC, HPTLC provides better separation efficiency and lower detection limits.

The choice of the layer and mobile phase is made in relation to the nature of the sample. Normal-phase (NP) or straight-phase adsorption TLC on silica gel with a less polar mobile phase, such as chloroform–methanol, is the most widely used mode. Lipophilic C-18, C-8, and C-2 bonded silica gel phases with a polar aqueous mobile phase, such as methanol–water, are used for reversed-phase (RP) TLC. Other precoated layers include alumina, magnesium silicate (Florisil), polyamide, cellulose, ion exchangers, and chemically bonded phenyl, amino, cyano, and diol layers. The latter three bonded phases can function with multimodal mechanisms, depending on the composition of the mobile phase. Silica gel can be impregnated with various solvents, buffers, and selective reagents to improve separations. Chiral plates composed of a RP layer impregnated with copper acetate and a chiral selector, ($2S,4R,2'RS$)-4-hydroxy-1-(2-hydroxydodecyl)proline, can be used to separate enantiomers through a ligand-exchange mechanism. Preparative layers are thicker than analytical layers to provide higher sample capacity.

Ultra-thin silica layers (Hauck et al., 2002) are the newest type available commercially. Unlike all other TLC and HPTLC layers, these do not consist of particulate material but are characterized by a monolithic silica structure. They are manufactured without a binder, which is usually needed to stabilize the sorbent particles on the support, and have a significantly thinner layer (10 μm), leading to short migration distances, fast development times, and very low solvent consumption.

Important manufacturers of TLC plates include Merck, Whatman, Analtech, and Macherey-Nagel, and literature from these companies should be consulted for details of availability, properties, usage, and applications.

IV. MOBILE PHASES

Unlike GC, in which the mobile phase (carrier gas) is not a factor in the selectivity of the chromatographic system, the mobile phase in liquid chromatography, including TLC, exerts a decisive influence on the separation. In HPLC, the analyte passes through the on-line detector in the presence of the mobile phase. Therefore, solvents must be chosen not only to provide the required resolution, but also to not absorb at the ultraviolet (UV) detection wavelength. Because in TLC the mobile phase is removed (evaporated) before the zones are detected, a wider variety of solvents can be used to prepare mobile phases compared with HPLC.

The mobile phase (Fried and Sherma, 1999e) is usually a mixture of two to five different solvents selected empirically using trial and error guided by prior personal experience and literature reports of similar separations. In addition, various systematic mobile phase optimization approaches (Cimpoiu, 2003; Prus and Kowalska, 2001) involving solvent classification (selectivity) and the elutropic series (strength) patterned after HPLC have been described, most notably, the PRISMA model based on Snyder's solvent classification system as developed by Nyiredy. For NP-TLC (silica gel), the following 10 solvents from Snyder's eight selectivity groups are used for exploratory TLC of the mixture: diethyl ether (group I), isopropanol and ethanol (group II), tetrahydrofuran (group III), acetic acid (group IV), dichloromethane (group V), ethyl acetate and dioxane (group VI), toluene (group VII), and chloroform (group VIII). Hexane (solvent strength = 0) is used to adjust the R_f values within the optimum range (0.2–0.8), if necessary. Between two and five solvents are then selected for construction of the PRISMA model that leads to identification of the optimized mobile phase. A similar procedure is followed for RP-TLC (e.g., C-18 bonded phase layer) using mixtures of methanol, acetonitrile, and/or tetrahydrofuran with water (solvent strength = 0).

TLC is usually carried out with a single mobile phase, rather than a mobile phase gradient as is often used in HPLC. Equilibration between the mobile phase and the layer occurs gradually during TLC development, and the mobile phase composition can change because different constituents migrate through the layer at different rates (solvent demixing). This leads to solvent gradients along the layer during "isocratic" TLC, the formation process of which is very different than the intentional, well-controlled gradients in a fully equilibrated HPLC column. Mobile phase selection for automated multiple development (AMD) is described in Section VI,A,4.

V. APPLICATION OF SAMPLES

Application of small, exactly positioned initial zones of sample and standard solutions (Fried and Sherma, 1999f) having accurate and precise volumes, without damaging the layer surface, is critical for achieving maximum resolution and reliable qualitative and quantitative analysis. The volumes applied and the method of application depend on the type of analysis to be performed (qualitative or quantitative), the layer (TLC or HPTLC), and the detection limit. For the greatest separation efficiency, the solvent in which the sample is dissolved should have high volatility and be as low in solvent strength as possible (nonpolar for NP systems and polar for RP systems) to retard the possibility of "prechromatography" during application.

Application of round spots allows the maximum number of samples to be applied onto a given plate. For TLC, 0.5–5 µL volumes are usually applied manually with a micropipet to produce initial zones with diameters in the 2–4 mm range. For TLC or HPTLC, initial zones in the form of spots can be applied from a disposable 0.5, 1, 2, or 5 µL fixed-volume, selfloading glass capillary pipet, which is held in a rocker-type spotting device (the Nanomat 4, Camag Scientific Inc., Wilmington, NC) that mechanically controls its positioning and brings the capillary tip in gentle and uniform contact with the layer to discharge the solution without damage to the layer. The Nanomat and other instruments for sample application and the later steps in the TLC process have been described by Reich (2003).

The TLS100 (Baron, Reichenau, Germany) is an automated apparatus for applying spots of up to 30 samples and four standards using a 1, 10, or 100 µL motor driven syringe. Locations and volumes can be chosen with a keypad for application onto as many as six HPTLC plates.

Sample application in the form of bands is advantageous for high-resolution separations of complex samples, improved detection limits, and for precise [1% relative standard deviation (RSD)] quantitative scanning densitometry using the aliquot technique (scanning with a slit of one-half to two-thirds the length of the applied band). Narrow, homogeneous sample bands of controlled length [from 1 (spot) to 195 mm] can be applied by use of a spray-on device (the Camag Linomat 5, Fig. 3), in which the plate is mechanically moved right to left in the X-direction beneath a fixed syringe from which

Figure 3 Linomat 5. (Courtesy of Camag.)

0.1–2000 μL of sample is sprayed by an atomizer operating with a controlled nitrogen gas pressure for analytical and preparative applications. The user selects the sample volumes and Y-position via a keypad or by downloading a method from a personal computer, and the instrument exactly positions the initial zones, which facilitates automated scanning after chromatogram development and overspraying samples with a reagent for *in situ* prechromatographic derivatization or with spiking solutions for validation of quantitative analysis by the standard-addition method. With the correct choice of application parameters, less volatile and higher strength sample solvents can be tolerated without forming broadened initial zones. The ability to apply larger volumes to an HPTLC plate without loss of resolution lowers the determination limits with respect to the concentration of the solution, which aids in trace analysis. Complex, impure samples can often be successfully quantified only if bands are applied rather than spots. The Linomat facilitates quantitative analysis by allowing different volumes of the same standard solution to be applied to produce the densitometric calibration curve, rather than the same volume of a series of standards when spots are applied.

The AS30 (Desaga, Weisloch, Germany) is a software controlled, fully automated band or spot applicator that also works according to a spray-on technique, in which a stream of gas carries the sample from the cannula tip onto the plate. The syringe does not have to be manually filled by the user, as with the Linomat. During the filling process, the dosing syringe is positioned over the tray, which collects rinsing and flushing solvent and excess sample. The sample is injected into the body of the syringe through a lateral opening. After the syringe has been filled, a stepping motor moves the piston downwards to dose the fillport. A second stepping motor moves the tower sideways across the plate. The microprocessor controls both motors and the gas valve for accurate and precise application in the form of spots or bands. All parameters for application of up to 30 samples are entered via the keyboard. The user is guided through the clearly structured menu by the two-line LCD display.

Figure 4 shows the Camag Automatic TLC Sampler IV (ATS 4), which is an advanced, fully automated, computer-controlled device for sequential application of up to 66 samples from a rack of vials or 96 samples from well plates through a steel capillary as spots by contact transfer or as bands by the spray-on technique. The speed, volume, and X- and Y-position pattern of application are controllable, and a programmable rinse cycle can eliminate cross-contamination. Low concentration samples can be applied as rectangles, which are focused into narrow bands by predevelopment with a strong mobile phase. An optional heated spray nozzle allows increased application speed, which is important for aqueous solutions. Analyses performed with this

Figure 4 Automatic TLC Sampler 4 (ATS 4). (Courtesy of Camag.)

applicator combined with densitometric chromatogram evaluation controlled by the same computer conform to GMP/GLP standards.

VI. CHROMATOGRAM DEVELOPMENT

TLC is almost always carried out in the elution mode. Methods and applications of displacement TLC have been described (Bariska et al., 2000), but this method is not yet widely used and will not be covered in this chapter. Electro-osmotically driven TLC (Nurok et al., 2002) is a quite new method that has not yet been shown to have a significant number of important practical applications, and it also will not be covered.

Linear development of TLC plates is used almost exclusively today. Therefore, circular (radial) and anticircular development will not be discussed here. TLC development times are typically in the range of 3–60 min, depending on the layer, mobile phase, and development method chosen. However, the development time does not significantly influence the overall analysis time per sample, because many samples and standards can be chromatographed simultaneously. Development modes and chambers have been described in detail elsewhere (Fried and Sherma, 1999g).

A. Capillary-Flow TLC

1. Ascending Development

The results of TLC are strongly dependent upon the environmental conditions during development, such as small changes in mobile phase composition, temperature, humidity, and the size and type of the chamber and its solvent vapor saturation conditions.

In the classical method of linear, ascending development TLC and HPTLC, the developing solvent is contained in a large volume, covered glass tank (N-tank). The spotted plate is inclined against an inside wall of the tank with its lower edge immersed in the developing solvent below the starting line, and the solvent begins to rise immediately through the initial zones due to capillary flow. As the mobile phase ascends, the layer interacts with the vapor phase as well as the mobile phase and the mixture components. The space inside the tank is more or less equilibrated with solvent vapors, depending on the presence or absence of a mobile phase-soaked paper liner, and the period of time the tank is allowed to stand before the plate is inserted. Reproducible results are obtained only if all development conditions are maintained as constant as possible. Solvent consumption is high with classical chambers.

A sandwich chamber (S-chamber) consists of the TLC plate, a spacer of 3 mm thickness, and a cover- or counterplate that is either blank glass or a solvent-soaked TLC plate. These parts are clamped together so that the bottom 2 cm of the layer is uncovered and are placed in a trough containing the mobile phase. Interaction between the layer, dry or wetted, and the gas phase is largely suppressed in an S-chamber, and reproducibility of the separation is improved. Both ascending and horizontal chambers can be operated as S-chambers (Gocan, 2001a).

The twin-trough chamber is an N-chamber modified with an inverted V-shaped ridge on the bottom dividing the tank into two sections, which allow development with only 5–20 mL of solvent, depending on the plate size, on one side, and easy pre-equilibration of the layer with vapors of the mobile phase or another conditioning liquid (e.g., a sulfuric acid–water mixture to control humidity) or volatile reagent on the other side.

2. Horizontal Development

The horizontal developing chamber (Fig. 5) permits simultaneous development from opposite edges to the middle of 72 sample spots on a 20 × 10 cm HPTLC plate, or 36 samples from one end to the other. The developing solvent, held in narrow troughs, is carried to the layer through capillary slits formed between the trough walls and the glass slides. The chamber is covered with a glass plate during pre-equilibration and development and can be operated in N-type or S-type configurations, including humidity control by placing an appropriate sulfuric acid–water mixture in a conditioning tray. Use of the horizontal developing chamber allows separation conditions to be efficiently standardized, and only low amounts of mobile phase are required.

The Camag HPTLC Vario System is a horizontal chamber that facilitates development and optimization of separation parameters. Simultaneous development can be tested with up to six different mobile phases, sandwich or tank configurations, and pre-equilibration conditions in any combination.

3. Continuous Development

The short bed continuous development chamber (Regis, Morton Grove, IL) is used for continuous development, which leads to improved resolution of zones with low migration rates because of the larger effective separation distance. Four glass ridges and the back wall of the chamber support the plate at five different angles of inclination, each of which allows increasing lengths of the layer to protrude out of the top of the chamber. Solvent continually evaporates from the external layer at the top and is replaced by additional solvent drawn from the chamber at the bottom.

4. Gradient Elution TLC Combined with AMD

AMD generally involves 10–30 individual linear ascending developments of an HPTLC plate (usually silica gel) carried out in an AMD instrument (Fig. 6), which includes an N-type chamber equipped with connections for feeding and releasing mobile phase and pumping a gas phase in and out, storage bottles for pure solvents and waste, a gradient mixer, syringes for measuring solvent volumes, and a charge coupled device (CCD) detector for monitoring migration distances. The developments are performed in the same direction with a stepwise mobile phase gradient

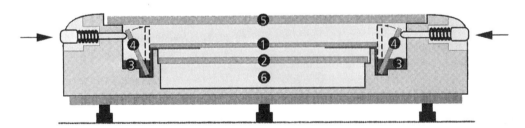

Figure 5 Horizontal development chamber. 1, HPTLC plate with layer facing down; 2, glass plate for sandwich configuration; 3, reservoir for mobile phase; 4, glass strip; 5, cover plate; 6, conditioning tray. (Courtesy of Camag.)

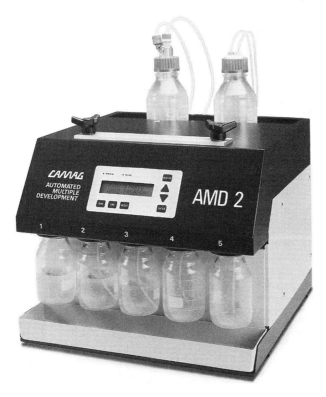

Figure 6 AMD 2 automated multiple development instrument. (Courtesy of Camag Scientific Inc.)

that becomes progressively weaker (i.e., less polar) over distances that increase by 1–5 mm for each stage. The solvent is completely removed from the chamber, and the layer is dried under vacuum applied by a pump for a preselected time after each development. The layer is then preconditioned with the vapor phase of the next batch of fresh solvent, which is fed into the chamber before the following incremental run.

The solvent strength may be changed for each development, or several stages may be carried out with the same solvent before changing its strength. The repeated movement of the solvent front through the chromatographic zones causes them to become compressed into narrow bands during AMD, leading to peak capacities of more than 50 over a separation distance of 80 mm. Typical "universal gradients" for AMD are produced from methanol or acetonitrile (polar); dichloromethane, di-isopropyl ether, or t-butylmethyl ether (medium polarity); and hexane (nonpolar). The central or base solvent and the nonpolar solvent have the greatest effect on selectivity. By superimposing the densitogram of a chromatogram with a matched-scale diagram of the gradient, required modifications of the solvent system can be predicted. Gradient development in TLC has been reviewed (Golkiewicz, 2003).

Complex mixtures containing compounds with widely different polarities can be separated by AMD on one chromatogram, in which sharply focused zones migrate different distances according to their polarities. Zone widths are independent of migration distance and size of the starting zone, leading to high resolution and the ability to resolve relatively large samples for trace analysis. Migration distances of individual components are largely independent of the sample matrix, and detection limits are improved because of the highly concentrated zones (typically 1 mm peak width) that are produced. The densitometer scan shown in Fig. 7 illustrates a high-resolution chromatogram that can be produced by gradient AMD.

5. Two-Dimensional Development

Two-dimensional (2D) TLC involves spotting the sample in one corner of the layer, developing (ascending or horizontal) with the first mobile phase, drying the plate, and developing at a 90° angle with a second mobile phase having a diverse, complementary separation mechanism (selectivity). Computer simulation has been used to optimize 2D separations based on one-dimensional (1D) data. Resolution (spot capacity) in 2D TLC is greatly improved compared with 1D TLC because sample components are resolved over the entire area of the layer, and is usually superior to that obtained by HPLC. Resolution by 2D forced flow planar chromatography (FFPC) (see Section VI.B) exceeds 2D capillary-flow TLC because spot diffusion is smaller. Disadvantages of 2D TLC include a limit of one sample per plate and the time required for two developments and intermediate drying. In addition, quantitative calibration or qualitative identification standards cannot be developed in parallel on the same plate at the same time. Improved video densitometers may allow more reliable quantitative 2D TLC in the future than is possible today with the commonly used slit-scanning densitometers. 2D TLC and other multidimensional methods have been reviewed (Gocan, 2001b).

B. Forced Flow Planar Chromatography

The mobile phase can migrate through the layer by capillary action, as is the case in the procedures described earlier, or under the influence of forced flow. FFPC has theoretical advantages relative to capillary flow, including independent optimization of mobile phase velocity, higher efficiency, lower separation time, and use of solvents that do not wet the layer, but it requires specialized, complex commercial instrumentation. Forced flow is produced by mechanically pumping solvent through a sealed layer [overpressured layer chromatography or optimum performance laminar chromatography (OPLC) (Mincsovics et al., 2003; Rozylo,

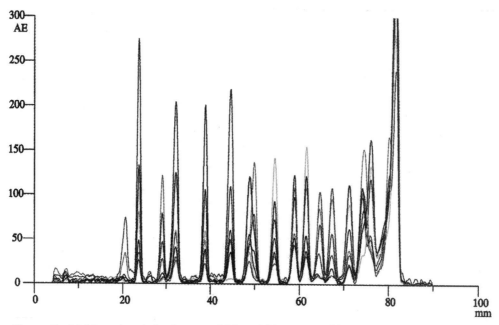

Figure 7 Multiwavelength densitogram of 16 pesticides separated by AMD. (Courtesy of Camag.)

2001)] or by spinning a glass rotor covered with sorbent around a central axis to drive the solvent from the center to the periphery of the layer by centrifugal force [rotation planar chromatography (RPC)]. Separations can be accomplished with a dry layer (off-line FFPC), but the closed system arrangement also allows the separation to be started after the layer is equilibrated with the mobile phase, similar to the situation in HPLC (on-line FFPC).

In RPC, samples are applied to the rotating stationary phase near the center, and centrifugal force along with capillary action drives the mobile phase through the sorbent from the center to the periphery of the plate. Up to 72 samples can be applied for analytical separations, and *in situ* quantification is possible. One circular sample is applied for micropreparative and preparative separations, which can be carried out off- and on-line. Various chambers are used for RPC, which differ mainly in the volume of the vapor space. The major commercial instruments for RPC are the Chromatotron 7924 (Harrison Research, Palo Alto, CA), CLC-5 (Hitachi, Tokyo, Japan), Extrachrom (RIMP, Budakalasz, Hungary), and Rotachrom Model P (Petazon, Zug, Switzerland). Although analytical applications have been proposed, RPC appears to be most useful for preparative applications.

OPLC combines many advantages of classical TLC and HPLC. A special glass- or aluminum-backed layer, sealed at the edges, is covered by a polyethylene or Teflon foil and pressurized by water. Development modes include linear unidirectional, linear bidirectional, circular, on-line, off-line, parallel coupled multilayer, serial coupled multilayer, isocratic, and gradient. The fully automatic, computer controlled personal OPLC BS-50 instrument (OPLC-NIT, Budapest, Hungary) is shown in Fig. 8 (Mincsovics et al., 1999). It consists, in general, of a separation chamber and a liquid delivery system with a

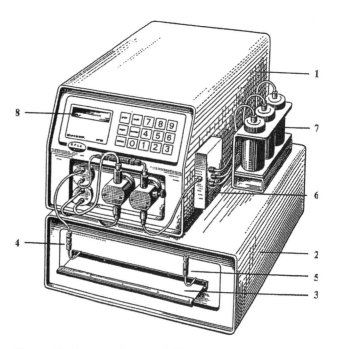

Figure 8 Automated personal OPLC BS-50 system. 1, Liquid delivery system; 2, separation chamber; 3, cassette; 4, mobile phase inlet; 5, mobile phase outlet; 6, mobile phase switching valve; 7, mobile phase reservoirs; 8, liquid crystal display. [Reproduced from Mincsovics et al. (1999) with permission.]

two-in-one hydraulic pump and mobile phase delivery pump. The chamber contains a holding unit, hydraulic unit, tray-like layer cassette, and drain valve. The separation chamber has two mobile phase connections and a 5 MPa maximum external pressure.

VII. ZONE DETECTION

After development with the mobile phase, the plate is dried in a fumehood (with or without heat) to completely evaporate the mobile phase. Separated compounds are detected (or visualized) on the layer by their natural color, natural fluorescence, quenching of fluorescence, or as colored, UV-absorbing, or fluorescent zones after reaction with an appropriate reagent (post-chromatographic derivatization) (Fried and Sherma, 1999h; Sherma, 2001a). Although dependent upon the particular analyte and the detection method chosen, sensitivity values are generally in the low microgram to nanogram range for absorbance and picogram range for fluorescence.

Layers are frequently heated after applying the detection reagent in order to accelerate the reaction upon which detection is based. Heating is carried out with a hair drier in a fume hood, in an oven, or with a TLC plate heater. The plate heater (Fig. 9), which contains a 20×20 cm flat, even heating area, a grid to facilitate proper positioning of TLC and HPTLC plates, programable temperature between 25°C and 200°C, and digital display of the actual temperature, provides the most consistent heating conditions.

Compounds that are naturally colored are viewed directly on the layer in daylight, whereas compounds with native fluorescence are viewed as bright zones on a dark background under UV light. Viewing cabinets incorporating shortwave (254 nm) and longwave (366 nm) UV lamps are available for inspecting chromatograms in an undarkened room.

Compounds that absorb around 254 nm, particularly those with aromatic rings and conjugated double bonds, can be detected on an "F-layer" containing a phosphor or fluorescent indicator (often zinc silicate). When irradiated with 254 nm UV light, absorbing compounds diminish (quench) the uniform layer fluorescence and are detected as dark violet spots on a bright (usually green) background.

Universal or selective chromogenic and fluorogenic liquid detection reagents are applied by spraying or dipping the layer. Various types of aerosol sprayers are available for manual operation, including the Desaga Sprayer SG 1 with a quiet, built-in pump and PTFE spray head (Fig. 10). The ChromaJet DS 20 (Desaga) (Fig. 11) is a spraying instrument that reproducibly applies selectable, accurate amounts of reagent to individual plate tracks under computer control. For safety purposes, spraying is carried out inside a laboratory fumehood or commercial TLC spray cabinet with a blower (fan) and exhaust hose. The most uniform dip application of reagents can be achieved by use of a battery operated chromatogram immersion device (Camag), which provides selectable, consistent vertical immersion and withdrawal speeds between 30 and 50 mm/s and immersion times between 1 and 8 s for plates with 10 or 20 cm heights. This mechanized dipping device can also be used for prewashing TLC plates prior to initial zone application, impregnation of layers with reagents that improve resolution or detection prior to initial zone application and

Figure 9 TLC plate heater. (Courtesy of Camag.)

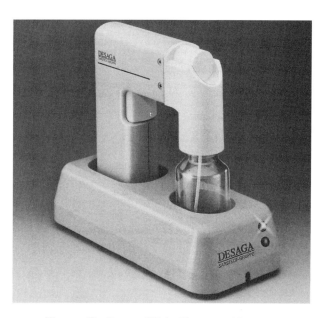

Figure 10 Sprayer SG 1. (Courtesy of Desaga.)

Figure 11 ChromaJet DS 20 automatic spray apparatus. (Courtesy of Desaga.)

development, and for postdevelopment impregnation of chromatograms containing fluorescent zones with a fluorescence enhancer and stabilizer such as paraffin. A few detection reagents (HCl, sulfuryl chloride, iodine) can be transferred uniformly to the layer as vapors in a closed chamber. The preparation, procedure for use, and results for many hundreds of TLC detection reagents have been described (Jork, et al., 1990, 1994; Zweig and Sherma, 1972).

A variety of biological detection methods are available for compound detection. As an example, immunostaining of thin layer chromatograms provides very sensitive detection of glycolipids (Putalan et al., 2001).

VIII. DOCUMENTATION OF CHROMATOGRAMS

TLC plates contain complete chromatograms that provide a great amount of information for sample identification, visual semiquantification, and comparison, especially when different detection methods are used to give multiple images of the same sample.

TLC separations are documented by photography, video recording, or scanning (Fried and Sherma, 1999i; Morlock and Kovar, 2003). Commercial systems for photographic documentation contain a conventional, instant, or digital (Fig. 12) camera and associated lighting accessories for photography of colored, fluorescent, and fluorescence-quenched zones on TLC plates in shortwave, midrange, and longwave UV light and in visible light. Special software is needed for reproducible, GMP/GLP-compliant use of a digital camera.

The Camag video documentation system (VideoStore, Fig. 13) includes visible and UV lighting, CCD camera with zoom lens, monitor, and video color printer. The

Figure 12 Reprostar 3 lighting unit and camera stand with cabinet cover and mounted digital camera. (Courtesy of Camag.)

VideoStore software is GMP/GLP-compliant for capture, editing, annotation, documenting, and archiving of images. Desaga offers a similar color video documentation system (VD 40) with eightfold motor zoom and autofocus CCD camera, pentium powered computer, and ProViDoc GLP-conforming software.

Documentation can also be carried out by scanning the spots on the plate with a flatbed office scanner (Rozylo

Thin Layer Chromatography

Figure 13 Camag VideoStore/VideoScan including Reprostar 3 with cabinet cover, camera bellows, camera support, and 3-CCD camera with zoom objective. (Courtesy of Camag.)

et al., 1997). The equipment required includes a computer, scanner, and monochrome or color printer. Computer scanning can be used only for visible spots, but not those that are fluorescent or quench fluorescence unless the flatbed scanner is modified.

IX. ZONE IDENTIFICATION

The identity of TLC zones (Fried and Sherma, 1999i; Morlock and Kovar, 2003) is obtained in the first instance by comparison of R_f values between samples and reference standards chromatographed on the same plate, where R_f equals the migration distance from the origin to the center of the zone divided by the migration distance of the mobile phase front. Identity is more certain if a selective chromogenic reagent yields the same characteristic color for sample and standard zones, or if an R_f match between samples and standards is obtained in at least two TLC systems with diverse mechanisms, for example, silica gel NP and C-18 bonded silica gel RP. Comparison of standard and sample *in situ* UV-visible absorption or fluorescence emission spectra, obtained by using the spectral mode of a slit-scanning or video densitometer, can also aid identification, but these spectra may contain inadequate structural information for complex mixtures.

The TIDAS TLC 2010 scanner (Flowspek AG, Basel, Switzerland) (Fig. 14) allows rapid scanning of TLC plates with simultaneous acquisition of a complete spectrum for all substances on a layer. Fiber optics technology is combined with diode array detection in this instrument, which has the following specifications: 190–1000 nm wavelength range, 0.8 nm pixel resolution, deuterium and tungsten light sources, <160 μm optical resolution

Figure 14 TLC 2010 diode array scanner. (Courtesy of Flowspek.)

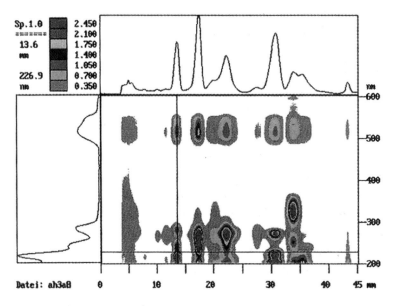

Figure 15 Contour plot, densitogram, and codeine spectrum on one screen shot, with codeine appearing at 13.6 mm. (Courtesy of Flowspek.)

on the layer, 5 mm/s scanning speed, 0.1 mm positioning accuracy, and software with parameters including peak purity, resolution, and identification via spectral library matching. Figures 15 and 16 show the identification and confirmation of codeine in urine with the TLC 2010 diode array scanner.

Zone identity can also be confirmed by the application of combined TLC-spectrometry methods described in Section XI.

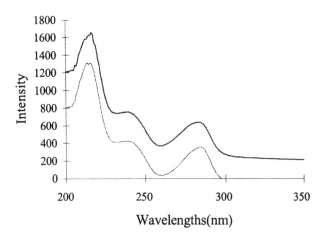

Figure 16 Codeine spectrum (top) and library spectrum (bottom) with a match of 98%. (Courtesy of Flowspek.)

X. QUANTITATIVE ANALYSIS

A. Nondensitometric Methods

Quantification of thin layer chromatograms (Fried and Sherma, 1999j; Prosek and Vovk, 2003) can be performed after manually scraping off the separated zones of samples and standards and elution of the substances from the layer material with a strong, volatile solvent. The eluates are concentrated and analyzed by use of a spectrometry, GC, HPLC, or some other sensitive microanalytical method. This method of quantification is laborious and time consuming, and the difficulty in recovering samples and standards uniformly is a major source of error. Although its importance has declined relative to densitometry, the indirect scraping and elution quantification method is still being rather widely used, for example, for some drug assays according to the US Pharmacopoeia.

Direct TLC semiquantitative analysis can be performed by visual comparison of sample spot intensities with the intensities of reference spots developed simultaneously on the same layer. For this comparison, the bracketing method is used in which standard spots with concentrations equal to, greater than, and less than the expected sample concentration are placed on either side of duplicate sample spots. The concentrations of samples and standards should lie within the linear response range of the detection method. The use of TLC for compliance screening of drug products is an important example of this approach.

B. Densitometric Evaluation

Most modern HPTLC quantitative analyses are performed by *in situ* measurement of the absorbance or fluorescence of the separated zones in the chromatogram tracks using an optical densitometic scanner (Reich, 2003; Sherma, 2001b) operated with a fixed sample light beam in the form of a rectangular slit. The length and width of the slit is selectable for optimized scanning of spot or band-shaped zones with different dimensions. The densitometer measures the difference between the optical signal from a zone-free background area of the plate and that from the calibration standards and sample zones. With automated zone application, precision ranging from 1% to 3% RSD is typical for densitometric analyses.

The plate is mounted on a moveable stage controlled in the *X*- and *Y*-directions by a stepping motor, which allows each chromatogram to be scanned, usually in the direction of development. The single beam, single wavelength scanning mode most often used gives excellent results with high quality plates and a mobile phase that result in compact, well separated zones. A schematic diagram of the light path of a densitometer is shown in Fig. 17. A tungsten-halogen lamp is used as the source for scanning colored zones in the 400–800 nm range (visible absorption) and a deuterium continuum lamp for scanning the absorption of UV light by zones on layers with or without fluorescence indicator in the 190–400 nm range. The monochromator used with these continuous wavelength sources can be a quartz prism or, more often, a grating. The detector is a photomultiplier or a photodiode. For fluorescence scanning, a high intensity xenon or mercury vapor lamp is used as the source, the optimum excitation wavelength is selected by the monochromator, and a cutoff filter is placed between the plate and detector to block the exciting UV radiation and transmit the visible emitted fluorescence. The light beam strikes the plate at a 90° angle, and the photomultiplier for reflectance scanning is at a 30° angle to normal. Part of the light beam is directed to a reference photomultiplier by a beam splitter to compensate for lamp variations and short-term fluctuations. A detector mounted below the stage is used when scanning in the transmission mode (TLC plates or electrophoresis gels). Zig-zag (or meander) scanning with a small spot of light is possible with scanners having two independent stepping motors to move the plate in the *x*- and *y*-axes. Computer algorithms integrate the maximum absorbance measurements from each swing, which

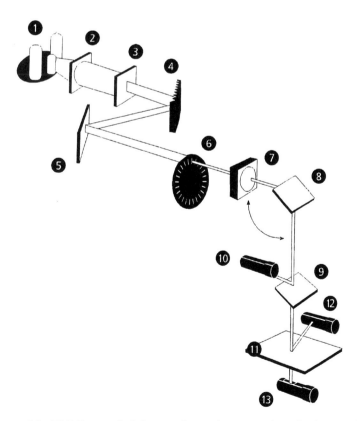

Figure 17 Light path diagram of the TLC Scanner 3. 1, Lamp selector; 2, entrance lens slit; 3, monochromator entry slit; 4, grating monochromator; 5, mirror; 6, slit aperture disk; 7, lens system; 8, mirror; 9, beam splitter; 10, reflectance monochromator; 11, object to be scanned; 12, measuring photomultiplier; 13, photodiode (transmission). [Reproduced from Reich (2003) with permission.]

corresponds to the length of the slit, to produce a distribution profile of zones having any shape. Disadvantages of scanning with a moving light spot include problems with data processing, lower spatial resolution for HPTLC, and unfavorable error propagation upon averaging of readings from different points within the zone.

Most modern scanners have a computer controlled motor driven monochromator that allows automatic recording of *in situ* absorption and fluorescence excitation spectra at multiple wavelengths (Fig. 7). These spectra can aid compound identification by comparison with cochromatographed standards or stored standard spectra, test for identity by superimposition of spectra from different zones on a plate, and check zone purity by superimposition of spectra from different areas of a single zone. The spectral maximum determined from the *in situ* spectrum is usually the optimal wavelength for scanning standard and sample areas for quantitative analysis.

The scanner is connected to a recorder, an integrator, or a computer. A personal computer with software designed specifically for TLC is most common for data processing and automated control of the scanning process in modern instruments. With a fully automated system (e.g., winCATS from Camag), the computer can carry out the following functions: selectable scanning speed up to 100 mm/s; evaluation of 36 tracks with up to 100 substances in sequence; integration with automatic or manual baseline correction; single or multilevel calibration with linear or nonlinear regression using internal or external standards, and statistics such as RSD or confidence interval with full error propagation; subcomponent evaluation to relate unidentified fractions to the main component, as required by various pharmacopoeias; dual-wavelength scan to eliminate matrix effects or quantify incompletely resolved peaks; multiwavelength scan (up to 31 different wavelengths) to obtain the optimum wavelength for quantification of each fraction and achieve maximum analytical selectivity; scanner qualification (automatic tests of mechanical, optical, and electronic functions); track optimization (repeated scanning of each track with small lateral offsets in order to optimize measurements of distorted chromatograms); and spectrum library. For GMP compliance, all conditions and data are automatically recorded and held in a secure format.

Because of light scattering from the sorbent particles, a simple, well-defined mathematical relationship between amount of analyte and the light signal has not been found. Plots relating absorption signal (peak height or area) and concentration or weight of standards on the layer are usually nonlinear, especially at higher concentrations, and do not pass through the origin. Modern integrators and computer software programs can routinely perform linear or polynomial regression of the calibration data, depending upon which is most suitable. Fluorescence calibration curves are generally linear and pass through the origin, and analyses based on fluorescence are more specific and 10–1000 times more sensitive than those employing absorbance. Because of these advantages, compounds that are not naturally fluorescent are often derivatized pre- or postchromatography to allow them to be scanned in the fluorescence mode if an appropriate reagent is available. However, absorbance in the 190–300 nm UV range has been most used for densitometric analyses.

Validation procedures (Fried and Sherma, 1999k) for quantitative analysis are in some aspects very similar to those for HPLC and GC, with additional considerations related to procedural aspects specific to TLC. Protocols for validation of TLC results (Ferenczi-Fodor et al., 2001), especially for pharmaceutical analysis, are available in the literature. Some densitometers include automatic instrument validation programs in their software.

The Camag TLC Scanner 3, Shimadzu (Columbia, MD USA) CS 9000, and Desaga Densitometer CD 60 are examples of modern computer-controlled slit-scanning densitometers.

Video densitometers (image processors) are an alternative to the optical/mechanical slit scanners. Video densitometry is based on electronic point-scanning of a stationary plate using an instrument composed of UV and visible light sources, a CCD camera with zoom capabilities, and a computer with imaging and evaluation software. Video scanners have advantages including rapid data collection and storage, simple design with virtually no moving parts, easy operation, and the ability to quantify 2D chromatograms, but they have not yet been shown to have the required capabilities to replace slit-scanning densitometers. Current video scanners can function only in the visible range to measure colored, fluorescence-quenched, or fluorescent spots. They lack the spectral selectivity and accuracy based on the ability to scan with monochromatic light of selectable wavelength throughout the visible and UV range (190–800 nm) that are inherent in classical densitometry, and they cannot record the *in situ* spectra. The VideoScan software program (Camag) allows quantitative evaluation (videodensitometry) of images at any time after they are captured with the VideoStore documentation system shown in Fig. 13. Integration of analog curves can be performed automatically, quantification performed via peak areas or heights, and single or multilevel calibrations with linear or polynomial regression. The ProResult software package is available for quantification with Desaga's video documentation system.

Special software packages have been used for quantification of zones on chromatogram images produced with a flatbed scanner. Both colored (visible) (Johnson, 2000) and fluorescent (Stroka et al., 2001) zones have been measured for the analyses, the latter after modification of the scanner by adding a black light tube.

XI. TLC COMBINED WITH SPECTROMETRIC METHODS

A. Mass Spectrometry

The identity of TLC zones can be confirmed by mass spectrometry (MS) analysis (Busch, 2003; Rozylo, 2001). The most used ionization modes for TLC/MS include electrospray ionization (ESI), and matrix- and surface-assisted laser desorption ionization (MALDI and SALDI).

ESI is used with solvent extracts of zones scraped from the TLC plate. The liquid is sprayed through a charged needle as an aerosol mist at atmospheric pressure, and the ions created from desolvation and charge distribution processes in the droplets are extracted through skimmer cones into the mass analyzer of the mass spectrometer.

MALDI and SALDI involve ionization directly from the surface layer held under vacuum after addition of an energy-buffering matrix. The ionization occurs as a result of surface irradiation by a laser beam, with mass analysis usually carried with a time of flight (TOF) mass analyzer.

On-line TLC/ESI–MS has been carried out by directly linking an OPLC 50 instrument (Fig. 8) and a Micromass (Beverly, MA USA) Q-TOF mass spectrometer (Chai et al., 2003). Aluminum-backed silica gel layers with a perimeter seal were used. After initial layer preconditioning to reduce background noise, the limit of detection of glycolipids was in the 5–20 pmol range.

Quadrupole, ion trap, and Fourier Transform (FT) mass spectrometers and fast atom bombardment and liquid secondary ion mass spectrometry ionization techniques have also been used in various applications of TLC/MS.

B. Infrared Spectrometry

Zones can also be confirmed by combining TLC with infrared (IR) spectrometry (Morlock and Kovar, 2003). Indirect TLC–IR coupling is carried out by transfer of the sample from the layer to an IR-transparent pellet or powder such as KBr or *in situ* measurement of scraped TLC zones by the diffuse reflectance infrared Fourier transform spectrometry (DRIFT) technique.

TLC has been directly coupled with DRIFT, in which case difficulties occur because conventional stationary phases, for example, silica gel, absorb strongly in the IR region, and the influence of the refractive index on the spectra must be considered. Intense interference bands between 1350 and 1000 cm^{-1} and above 3550 cm^{-1} are superimposed on the DRIFT spectra of the analytes and restrict the available wavenumber range. In addition, the large active surface area of silica gel, with its hydroxyl groups, makes adequate compensation of sample and reference spectra more difficult. It has been found that HPTLC plates with 10 μm particles, small particle size distribution, 0.2 mm layer thickness, and glass backing are best for direct TLC–DRIFT. A Brucker IFS 48 FTIR spectrometer with a special mirror arrangement and MCT (mercury, cadmium, telluride) small band detector has been constructed to enable DRIFT measurement despite the self-absorbance of TLC sorbents (Glauninger et al., 1990).

Although mainly useful for qualitative identification and confirmation of zones, quantification by TLC–IR has been carried out in some applications by evaluation of Kubelka–Munk spectra with integration of their strongest bands, especially for substances lacking chromophores that absorb in the UV-visible range. However, the limit of determination is about 10 times higher than that of densitometry, and precision is poorer.

C. Raman Spectrometry

Surface-enhanced Raman scattering (SERS) spectrometry (Morlock and Kovar, 2003; Somsen et al., 1995) has sensitivity in the nanogram or the picogram range using a Raman spectrometer with argon ion, HeNe, or YAG monochromatic light source and CCD detector. The method is most useful for identification of compounds with groups of atoms that are IR-inactive; quantitative analysis has not been successfully carried out.

For *in situ* analysis, the HPTLC plate, after development and drying, is dipped into or sprayed with a colloidal silver suspension prepared by reduction of silver nitrate with sodium citrate. Alternatively, the silver molecules can be evaporated onto the layer in order to eliminate the zone diffusion and lowering of enhancement caused by dipping or spraying.

Highly Raman-active substances (dyes, optical brighteners) can be determined at low nanogram levels without surface-enhanced scattering on specially modified silica gel plates. EMD chemicals Inc. (Gibbstown, NJ, an affiliate of Merck KGaA) sells silica gel 60 plates with a 0.1 mm layer of 3–5 μm particles on an aluminum support. The spherical silica gel produces a 10-fold increase in Raman spectrometry signal intensity compared with a similar layer made with irregular silica gel particles.

XII. PREPARATIVE LAYER CHROMATOGRAPHY

PLC for isolation of larger amounts of material than normally separated by TLC can be carried out by classical PLC (Fried and Sherma, 1999l) by use of thicker layers (usually 0.5–2 mm) developed in the ascending direction in a large volume chamber with detection of zones by a nondestructive method such as iodine vapor or fluorescence quenching. Fractions are scraped from the layer, and the purified analytes recovered by elution with a solvent for use in other laboratory work or further analysis.

The construction and use of the commercial instruments available for forced flow PLC have been described by Nyiredy (Nyiredy, 2003). These include the older Chrompres 10 and 25 chambers (LABOR Instrument Works, Budapest, Hungary) and Personal OPLC 50 system (Section VI.B, capable of semipreparative separations only) for OPLC and the Chromatotron CLC-5, Rotachrom, and Extachrom for RPC.

An additional commercial instrument for rotation or centrifugal preparative planar chromatography is the CycloGraph II from Analtech (Newark, DE; Fig. 18). Separations typically occur within 20 min without need to scrape off the separated zones. The sample solution is applied using a solvent pump or hand-held syringe inside the adsorbent ring of a precast rotor. The eluent (mobile phase) is pumped through the rotating layer (variable speed control from 100 to 1400 rpm), and as each separated ring reaches the outer rim of the rotor it is spun off of the edge of the glass into a circular trough. The adjustable angle of the trough allows the eluent to settle at the bottom and drip out of the collection port into individual tubes for different fractions. High performance, reusable silica gel rotors (9.5 in diameter) with 1000–8000 μm thickness are available.

XIII. THIN LAYER RADIOCHROMATOGRAPHY

Location and quantification of separated radioisotope-labeled substances on a thin layer requires the use of contact autoradiography, zonal analysis, or direct scanning with a radiation detector. These thin layer radiochromatography (TLRC) methods (Fried and Sherma, 1999m; Hazai and Klebovich, 2003) are especially important in drug and pesticide metabolism studies in plants, animals, and humans and in studies of the fate of labeled chemicals in the environment.

Contact autoradiography involves exposure of X-ray or photographic film to emissions from radioisotope zones to produce an image on the film. After exposure and development of the film, the radioisotopes are visible as dark spots, which can be compared with standards for qualitative identification or quantified by measurement of their optical densities with a scanning densitometer.

Zonal analysis involves scraping radioactive zones from the plate, placing the sorbent in counting vials, adding scintillation fluid or cocktail to elute the radioactive components, and liquid scintillation counting.

Direct measurement of radioactive zones on a layer is most often carried out today by use of a linear analyzer, digital autoradiograph (DAR), or an imaging analyzer. Linear analyzers incorporate a position-sensitive windowless gas-flow proportional counter as the detector, which measures all radioactive zones in a chromatogram (track) simultaneously and then is moved under computer control for measurements at other selected positions across the layer. The fill gas (e.g., argon–methane) is ionized when radioactive emissions from the TLC zones enter the detector, producing electrons. The electron pulses are detected electronically and stored in computer memory to provide a digital image of the distribution of radioactivity on the layer. An example of a TLC linear analyzer is the RITA (Raytest, Wilmington, NC; Fig. 19), which has a gold plated detector with an active length of 200 mm and an active width selectable from 20 to 1 mm by choice of the diaphragm.

Multiwire proportional counters (DAR or microchannel array detector) are 2D detectors based on a measuring principle similar to the linear analyzer, but they can detect all areas of radiation from a 20×20 cm layer simultaneously without moving a detector head. Many research studies have been reported on applications of HPTLC and OPLC coupled with an LB 287 Berthold DAR (EG&G Berthold, Wilbad, Germany) for analysis of tritiated or ^{14}C-labeled metabolites in a variety of

Figure 18 CycloGraph II centrifugal preparative planar chromatography instrument with built-in UV lamp and hinged lid. (Courtesy of Analtech.)

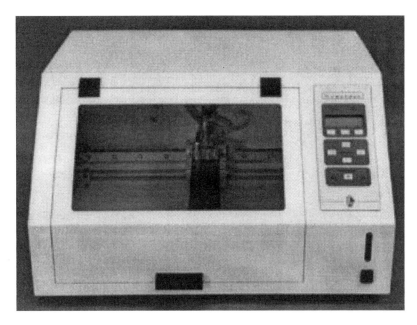

Figure 19 RITA radioactivity intelligent thin layer analyzer. (Courtesy of Raytest.)

biological matrixes. This instrument has a 20×20 cm sensitive area that contains a 600×600 wire grid. Measurements are made using argon–methane (9:1) as counting gas bubbled through methylal at $2.8°C$ and a flow rate of 5 mL/min. The positive high voltage potential used for ^{3}H- and ^{14}C-labeled compounds is 2040 and 1200 V, respectively, and signal analysis is achieved by measuring $5 \times 360{,}000$ detector cells/s.

Bioimaging/phosphor imaging analyzers represent the newest and most advantageous technology for measuring radioactive TLC zones. The layer is exposed to a phosphor imaging plate (IP) that accumulates and stores irradiating radioactive energy from the zones. The plate is then inserted into an image-reading unit and scanned with a fine laser beam. Luminescence is emitted in proportion to the intensity of the recorded radiation, collected by a photomultiplier tube, and converted to electrical energy to produce the instrumental readout. Resolution, linear range, and sensitivity are equal to, or better than, those of the other detection methods. The BAS 5000 (Raytest) (Fig. 20) is an example of a modern phosphor IP TLC scanner. Specifications include 20×25 cm IP size, 25/50 μm pixel size, 5 min (50 μm) reading time, detection limit 0.9 dpm/mm^2/h (^{14}C), and four to five orders of magnitude (16 bits) dynamic range.

XIV. APPLICATIONS OF TLC

TLC can provide rapid, low cost qualitative analyses and screening in order to obtain information such as sample stability, purity, and uniformity and to follow the course of a reaction, whereas instrumental HPTLC can provide accurate and reproducible (1–3% RSD) quantitative results. Samples that are difficult to prepare can be analyzed readily, and detection is especially flexible in the

Figure 20 BAS 5000 phosphor IP scanner. (Courtesy of Raytest.)

Table 1 Sources of Information on the TLC Analysis of Different Compound Classes

	Fried and Sherma (1999n)	Sherma and Fried (2003)	Cazes (2001)
Amino acids	×	×	×
Antibiotics		×	×
Carbohydrates	×	×	×
Carboxylic acids		×	
Ceramides			×
Coumarins			×
Dyes	×	×	×
Enantiomers		×	×
Indoles		×	
Inorganics		×	
Lipids	×	×	×
Nucleic acid derivatives	×	×	
Organometallics		×	
Peptides and proteins		×	
Pesticides		×	×
Pharmaceuticals and drugs		×	×
Phenols		×	×
Pigments	×	×	×
Plant extracts			×
Steroids	×	×	×
Taxoids			×
Terpenoids	×		
Toxins		×	×
Vitamins	×	×	×

Note: × indicates the book contains a chapter on the compound class.

absence of the mobile phase and with a variety of parameters.

TLC has been applied virtually in all areas of analysis, including chemistry, biochemistry, biology, industrial, agricultural, environmental, food, pharmaceutical, clinical, natural products, toxicology, forensics, plant science, bacteriology, parasitology, and entomology. Table 1 lists the compound classes that are covered in three books that contain detailed information on specific applications of TLC analysis, discussion of which is beyond the scope of this chapter. The applications of TLC, as well as theory, techniques, and instrumentation, is regularly updated in a biennial review of planar chromatography (Sherma, 2004) and the Camag Bibliography Service, which is published every March and September and is available in paper and CD-ROM format free of charge.

REFERENCES

Abjean, J. P. (1993). Screening of drug residues in food of animal origin by planar chromatography: application to sulfonamides. *J. Planar Chromatogr.-Mod. TLC* 6(2):147–148.

Bariska, J., Csermely, T., Furst, S., Kalasz, H., Bathori, M. (2000). Displacement thin layer chromatography. *J. Liq. Chromatogr. Relat. Technol.* 23(4):531–549.

van Beek, T. A. (2002). Chemical analysis of *Ginko biloba* leaves and extracts. *J. Chromatogr. A* 967(1):21–55.

Busch, K. L. (2003). Thin layer chromatography coupled with mass spectrometry. In: Sherma, J., Fried, B., eds. *Handbook of Thin Layer Chromatography*. 3rd ed. New York: Marcel Dekker, Inc., pp. 239–275.

Cazes, J., Ed. (2001). *Encyclopedia of Chromatography*. New York: Marcel Dekker, Inc.

Chai, W. G., Leteux, C., Lawson, A. M., Stoll, M. S. (2003). On-line overpressure thin layer chromatographic separation and electrospray mass spectrometric detection of glycolipids. *Anal. Chem.* 75(1):118–125.

Cimpoiu, C. (2003). Optimization. In: Sherma, J., Fried, B., Eds. *Handbook of Thin Layer Chromatography*. 3rd ed. New York: Marcel Dekker, Inc., pp. 81–98.

Esser, G., Klockow, D. (1994). Detection of hydroperoxides in combustion aerosols by supercritical fluid extraction coupled to thin layer chromatography. *Mikrochim. Acta* 113(3–6):373–379.

Ferenczi-Fodor, K., Vegh, Z., Nagy-Turak, A., Renger, B., Zeller, M. (2001). Validation and quality assurance of planar chromatographic procedures in pharmaceutical analysis. *J. AOAC Int.* 84(4):1265–1276.

Fried, B., Sherma, J. (1999a). Introduction and history. *Thin Layer Chromatography*. 4th Ed. New York: Marcel Dekker, Inc., pp. 1–8.

Fried, B., Sherma, J. (1999b). Mechanism and theory. *Thin Layer Chromatography*. 4th Ed. New York: Marcel Dekker, Inc., pp. 9–24.

Fried, B., Sherma, J. (1999c). Obtaining material for TLC and sample preparation. *Thin Layer Chromatography*. 4th Ed. New York: Marcel Dekker, Inc., pp. 51–76.

Fried, B., Sherma, J. (1999d). Sorbents, layers, and precoated plates. *Thin Layer Chromatography*. 4th Ed. New York: Marcel Dekker, Inc., pp. 25–49.

Fried, B., Sherma, J. (1999e). Solvent systems. *Thin Layer Chromatography*. 4th Ed. New York: Marcel Dekker, Inc., pp. 89–107.

Fried, B., Sherma, J. (1999f). Application of samples. *Thin Layer Chromatography*. 4th Ed. New York: Marcel Dekker, Inc., pp. 77–87.

Fried, B., Sherma, J. (1999g). Development techniques. *Thin Layer Chromatography*. 4th Ed. New York: Marcel Dekker, Inc., pp. 109–144.

Fried, B., Sherma, J. (1999h). Detection and visualization. *Thin Layer Chromatography*. 4th Ed. New York: Marcel Dekker, Inc., pp. 145–175.

Fried, B., Sherma, J. (1999i). Qualitative identification and documentation. *Thin Layer Chromatography*. 4th Ed. New York: Marcel Dekker, Inc., pp. 177–196.

Fried, B., Sherma, J. (1999j). Quantification. *Thin Layer Chromatography*. 4th Ed. New York: Marcel Dekker, Inc., pp. 197–222.

Fried, B., Sherma, J. (1999k). Reproducibility and validation of results. *Thin Layer Chromatography*. 4th Ed. New York: Marcel Dekker, Inc., pp. 223–233.

Fried, B., Sherma, J. (1999l). Preparative layer chromatography. *Thin Layer Chromatography*. 4th Ed. New York: Marcel Dekker, Inc., pp. 235–248.

Fried, B., Sherma, J. (1999m). Radiochemical techniques. *Thin Layer Chromatography*. 4th Ed. New York: Marcel Dekker, Inc., pp. 249–267.

Fried, B., Sherma, J. (1999n). *Thin Layer Chromatography*. 4th Ed. New York: Marcel Dekker, Inc.

Glauninger, G., Kovar, K. A., Hoffmann, V. (1990). Possibilities and limits of an online coupling of thin layer chromatography and FTIR spectroscopy. *Fresenius' J. Anal. Chem.* 338(6):710–716.

Gocan, S. (2001a). TLC sandwich chambers. In: Cazes, J., Ed. *Encyclopedia of Chromatography*. New York: Marcel Dekker, Inc., pp. 851–854.

Gocan, S. (2001b). Multidimensional TLC. In: Cazes, J., Ed. *Encyclopedia of Chromatography*. New York: Marcel Dekker, Inc., pp. 533–535.

Golkiewicz, W. (2003). Gradient development in thin layer chromatography. In: Sherma, J., Fried, B., eds. *Handbook of Thin Layer Chromatography*. 3rd Ed. New York: Marcel Dekker, Inc., pp. 153–173.

Hamada, M., Wintersteiger, R. (2002). Determination of phenylurea herbicides in drinking water. *J. Planar Chromatogr.-Mod. TLC* 15(1):11–18.

Hauck, H.-E., Schulz, M., Schaefer, C. (2002). Ultra-thin-layer chromatography (UTLC): the new dimension in planar chromatography. In: Vovk, I., Medja, A., eds. *Proceedings of the International Symposium "Planar Chromatography Today 2002"*. Ljubljana, Slovenia: National Institute of Chemistry, pp. 61–66.

Hazai, I., Klebovich, I. (2003). Thin layer radiochromatography. In: Sherma, J., Fried, B., eds. *Handbook of Thin Layer Chromatography*. 3rd Ed. New York: Marcel Dekker, Inc., pp. 339–360.

Johnson, M. E. (2000). Rapid, simple quantitation in thin layer chromatography using a flatbed scanner. *J. Chem. Educ.* 77(3):368–372.

Jork, H., Funk, W., Fischer, W., Wimmer, H. (1990). *Thin Layer Chromatography, Volume 1a, Physical and Chemical Detection Methods*. Weinheim, Germany: VCH Verlagsgesellschaft mbH.

Jork, H., Funk, W., Fischer, W., Wimmer, H. (1994). *Thin Layer Chromatography, Volume 1b, Reagents and Detection Methods*. Weinheim, Germany: VCH Verlagsgesellschaft mbH.

Kowalska, T., Prus, W. (2001). Theory and mechanism of thin layer chromatography. In: Cazes, J., Ed. *Encyclopedia of Chromatography*. New York: Marcel Dekker, Inc., pp. 821–825.

Kowalska, T., Kaczmarski, K., Prus, W. (2003). Theory and mechanism of thin layer chromatography. In: Sherma, J., Fried, B., eds. *Handbook of Thin Layer Chromatography*. 3rd Ed. New York: Marcel Dekker, Inc., pp. 47–80.

Lepri, L., Cincinelli, A. (2001). TLC sorbents. In: Cazes, J., Ed. *Encyclopedia of Chromatography*. New York: Marcel Dekker, Inc., pp. 854–858.

Mincsovics, E., Garami, M., Kecskes, L., Tapa, B., Katy, G., Tyihak, E. (1999). Personal overpressured layer chromatography (OPLC) basic system 50, flexible tool in analytical and semipreparative work. *J. AOAC Int.* 82(3):587–598.

Mincsovics, E., Ferenczi-Fodor, K., Tyihak, E. (2003). Overpressured layer chromatography. In: Sherma, J., Fried, B., eds. *Handbook of Thin Layer Chromatography*. 3rd Ed. New York: Marcel Dekker, Inc., pp. 175–205.

Morlock, G., Kovar, K.-A. (2003). Detection, identification, and documentation. In: Sherma, J., Fried, B., eds. *Handbook of Thin Layer Chromatography*. 3rd Ed. New York: Marcel Dekker, Inc., pp. 207–238.

Nurok, D., Koers, J. M., Carmichael, M. A., Liao, W.-M., Dzido, T. H. (2002). The performance of planar electrochromatography in a horizontal chamber. *J. Planar Chromatogr.-Mod. TLC* 15(5):320–323.

Nyiredy, Sz. (2003). Preparative layer chromatography. In: Sherma, J., Fried, B., eds. *Handbook of Thin Layer Chromatography*. 3rd Ed. New York: Marcel Dekker, Inc., pp. 307–338.

Prosek, M., Vovk, I. (2003). Basic principles of optical quantification in TLC. In: Sherma, J., Fried, B., eds. *Handbook of Thin Layer Chromatography*. 3rd Ed. New York: Marcel Dekker, Inc., pp. 277–306.

Prus, W., Kowalska, T. (2001). Optimization of thin layer chromatography. In: Cazes, J., Ed. *Encyclopedia of*

Chromatography. New York: Marcel Dekker, Inc., pp. 576–579.

Putalan, W., Tanaka, H., Shoyama, Y. (2001). TLC immunostaining of steroidal alkaloid glycosides. In: Cazes, J., Ed. *Encyclopedia of Chromatography*. New York: Marcel Dekker, Inc., pp. 849–851.

Rabel, F. J. (2003). Sorbents and precoated layers in thin layer chromatography. In: Sherma, J., Fried, B., eds. *Handbook of Thin Layer Chromatography*. 3rd Ed. New York: Marcel Dekker, Inc., pp. 99–134.

Reich, E. (2003). Instrumental thin layer chromatography (planar chromatography). In: Sherma, J., Fried, B., eds. *Handbook of Thin Layer Chromatography*. 3rd Ed. New York: Marcel Dekker, Inc., pp. 135–151.

Rozylo, J. K. (2001). Overpressured layer chromatography. In: Cazes, J., Ed. *Encyclopedia of Chromatography*. New York: Marcel Dekker, Inc., pp. 579–582.

Rozylo, J. K., Siembida, R., Jamrozek-Manko, A. (1997). A new simple method for documentation of TLC plates. *J. Planar Chromatogr.-Mod. TLC* 10(3):225–228.

Rozylo, J. K. (2001). Thin layer chromatography–mass spectrometry. In: Cazes, J., Ed. *Encyclopedia of Chromatography*. New York: Marcel Dekker, Inc., pp. 839–842.

Sherma, J. (2001a). Detection (visualization) of TLC zones. In: Cazes, J., Ed. *Encyclopedia of Chromatography*. New York: Marcel Dekker, Inc., pp. 248–251.

Sherma, J. (2001b). Optical quantification (densitometry) in TLC. In: Cazes, J., Ed. *Encyclopedia of Chromatography*. New York: Marcel Dekker, Inc., pp. 572–576.

Sherma, J. (2002a). Thin layer chromatography. In: Issaq, H. J., Ed. *A Century of Separation Science*. New York: Marcel Dekker, Inc., pp. 49–68.

Sherma, J. (2003). Basic techniques, materials, and apparatus. In: Sherma, J., Fried, B., eds. *Handbook of Thin Layer Chromatography*. 3rd Ed. New York: Marcel Dekker, Inc., pp. 1–46.

Sherma, J., Fried, B., eds. (2003). *Handbook of Thin Layer Chromatography*. 3rd Ed. New York: Marcel Dekker, Inc.

Sherma, J. (2004). Planar chromatography. *Anal. Chem.* 76(12):3251–3262 (Previous biennial reviews of TLC by J. Sherma are contained in issue 12 of this journal each even numbered year dating back to volume *42*. **1970**.).

Somsen, G. W., Morden, W., Wilson, I. D. (1995). Planar chromatography coupled with spectroscopic techniques. *J. Chromatogr. A* 703(1–2):613–665.

Stroka, J., Peschel, T., Tittelbach, G., Weidner, G., van Otterdijk, R., Anklam, E. (2001). Modification of an office scanner for the determination of aflatoxins after TLC separation. *J. Planar Chromatogr.-Mod. TLC* 14(2):109–112.

Zweig, G., Sherma, J. (1972). *Handbook of Chromatography, Volume II*. Boca Raton, FL: CRC Press.

31

Validation of Chromatographic Methods

MARGARET WELLS, MAURICIO DANTUS
Merck & Co. Inc, Rahway, NJ, USA

I. INTRODUCTION

Analytical methods are applied in a number of different industrial sectors (e.g., pharmaceutical, food, water, chemical, environmental, etc.). Analytical methods may be employed for a variety of purposes such as research, development, manufacturing monitoring, or quality control laboratory. Independent of which sector or purpose is being considered, analytical methods are expected to be in a quality control environment and are hence assumed to be accurate and reliable. Analytical method validation is recognized as the means to ensure that these methods are accurate and reliable. The expectations for the method validation process and the laboratories performing the testing may vary from industry to industry, according to recognized standards supported by governmental agencies such as the Food and Drug Administration (FDA), Environmental Protection Agency (EPA), or officially established quality systems (e.g., International Standards Organization ISO 9000 series). The focus of this chapter is to describe the validation of analytical methods in the pharmaceutical industry, as this approach to method validation can be considered the most comprehensive.

In the pharmaceutical industry, analytical method validation is conducted to ensure that the methods used to analyze raw materials, in-process samples from work in-progress, or finished product give consistent results in order to guarantee the safety and purity of the final drug product. Analytical methods may be used to generate data for acceptance, release, stability, or pharmacokinetic studies of the pharmaceutical material (FDA, 1994). These analytical methods ensure that the material produced meets the required specifications. Scientifically valid analytical methods are established using a process called "method validation". The data generated during the method validation is a documented evidence that a method is reliable, accurate, precise, specific, and robust. The regulatory requirements to validate analytical methods for drug products in the USA are included in the following sections in the Code of Federal Regulations (FDA, 1997):

Section 211.165(e): The accuracy, sensitivity, specificity, and reproducibility of test methods employed by the firm shall be established and documented. Such validation and documentation may be accomplished in accordance with 211.194(a)(2).

Section 211.166(a)(3): The written program shall be followed and shall include . . . reliable, meaningful and specific test methods.

Section 211.194(a)(2): A statement of each method used in testing of the sample. The statement shall include the location of data that establish the methods used in the testing of the sample meet proper standards of accuracy and reliability as applied to the product tested. The suitability of all testing methods shall be verified under actual conditions of use.

Hence all analytical procedures such as high performance liquid chromatography (HPLC), gas chromatography (GC), mass spectrometry (MS), and combination techniques (GC-MS, LC-MS) require some degree of method validation.

Analytical method validation guidelines are provided by the United States Pharmacopeia (USP, 2003), British Pharmacopeia, and European Pharmacopeia. These agencies provide analytical method validation guidelines for both regulatory and compendial requirements. In an effort to harmonize the regulatory requirements for the pharmaceutical industry in the USA, Europe, and Japan, the International Conference on Harmonization (ICH) has also issued guidelines for analytical method validation (ICH, 1994, 1996). Analytical methods may vary from accurate and precise measurements to subjective evaluations of physical attributes. Method validation requirements will vary depending on each general type of test method. In addition, requirements may need to be assessed, depending upon the nature of the method and test specifications. Specific procedures and acceptance criteria in analytical method validation are directly related to the purpose of the analytical method.

Starting from the definition of the purpose and scope of the method, a suggested process for validating analytical methods (Shabir, 2003) is shown in Fig. 1. In general terms, the purpose of method validation is to prove that the analytical method is appropriate for its intended use. Typically, analytical methods are grouped according to the following general types of tests.

- *Identification tests*: Identification tests provide data on the identity of a compound in a sample. For example, a chromatographic method used for an identification test would compare the chromatographic behavior of a sample (such as relative retention time) to a known reference standard.
- *Analytical methods for determination of impurities including quantitative assays and limit tests*: Impurity tests provide data on the purity characteristics of a sample. Sometimes, impurity testing may require more than one HPLC method to give an accurate description of the purity of a sample. Depending upon whether the purpose of the method is to provide an actual numeric value (quantitative test) or provide assurance that an impurity is less than a specified amount (limit test), the requirements for method validation vary.
- *Analytical methods for quantitation of major components*: This type of test provides data on the exact quantity of the major compound present in a sample. This value may be used to calculate the potency or purity of the final drug product.

The purpose of the analytical method determines the method validation requirements. Typical method

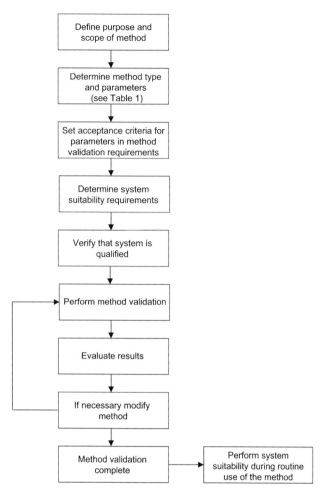

Figure 1 Suggested process for validating analytical methods.

validation parameters for analytical methods are listed as follows.

Precision
Accuracy
Limit of detection (LOD)
Limit of quantitation (LOQ)
Specificity
Range
Linearity
Ruggedness
Robustness
Solution stability

Each of these method validation parameters is discussed in detail in the following sections. The list provided is typical, but not all-inclusive. The requirements for each analytical method and its acceptance criteria should be evaluated on a case-by-case basis. Table 1 illustrates the method validation parameters that are typically required for each general type of test (USP, 2003). The requirements of Table 1 may be applied to any type of analytical

Table 1 Typical Method Validation Requirements Based on USP (2003)

Method validation requirement	Identification test	Impurity quantitative test	Impurity limit test	Major component test
Precision	—	×	—	×
Accuracy	—	×	May be required	×
Limit of detection	—	—	×	—
Limit of quantitation	—	×	—	—
Specificity	×	×	×	×
Range	—	×	May be required	×
Linearity	—	×	—	×
Ruggedness	×	×	×	×
Robustness	×	×	×	×

method. For example, ultraviolet (UV), specific rotation, infrared (IR), and HPLC methods can be classified as identification tests; thin layer chromatography (TLC) and HPLC methods can be classified as impurity limit tests. HPLC methods may be classified as any type of general test depending on the purpose. As HPLC methods are the most common techniques used in pharmaceutical analysis, the discussion of method validation requirements is focused on this analytical technique. The HPLC approach to method validation assumes that the output must be confirmed against a known standard to produce a meaningful result. The methods that do not include a comparison to a standard have a reduced number of validation requirements.

The selection of which performance parameters to include as part of the method validation can be based on the examination of the pertinent literature (Jenke, 1996a). For example, in a survey of 31 references, accuracy and precision were recognized as a universally necessary validation parameter, whereas other parameters such as specificity and linearity were less universally applied (Jenke, 1996a). A similar survey was conducted to determine the use of validation parameters for each type of analytical test (Clarke, 1994).

Each of the following sections describing the method validation parameters contains a list of examples gathered from a reference search of possible procedures and acceptance criteria that, together with any regulatory requirements can be used as guidance. The intent of the list is not to indicate what should be done, but to serve as a starting point for evaluating possible procedures and acceptance criteria in different applications.

Once a method has been validated, it may need to be evaluated on a periodic basis. In cases where the method needs to be modified because of a change in purpose, specifications, or procedure, revalidation of the analytical method should be considered. Method validation may occur more than once during the lifetime of any given method. Depending upon the situation, actual sample analysis may be part of the method revalidation. The type of revalidation that needs to take place will depend upon the type of change proposed (Hokanson, 1994):

Changes in instrumentation: Equipment-specific factors, particularly precision, linearity, range, and limits of detection and qualification may need to be evaluated.

Changes in sample matrix: These changes may require validation of selectivity, accuracy, and precision. If major changes are made in the composition of the sample, the complete validation should be repeated.

Changes in method: If the procedure is modified but the sample and standard preparation process are unchanged, then only equipment-specific parameters (precision, linearity, range, and limits of detection and qualification) may require revalidation. If a change is made in sample preparation but the same concentration range and preparation solvent are used, then accuracy, precision, and ruggedness need to be assessed.

II. PREVALIDATION REQUIREMENTS

There are additional requirements that may need to be met prior to starting a method validation process. Instrument performance contributes to the accuracy of the analytical method. As a result, instrument qualification and system suitability requirements assist in providing assurance of the reliability of the results (FDA, 1994).

A. Instrument Qualification

In the regulated pharmaceutical industry, another part of the validation process is the qualification step. Though qualification is not required *per se* in other industries, a similar concept may be used to ensure that equipment, including analytical instruments, is capable of fulfilling the specified requirements (Rathore et al., 2003). According to the FDA "equipment used in manufacturing, processing ... of a drug product shall be of appropriate design ... and suitably located to facilitate operations for its intended use"

(CFR 211.63) (FDA, 1997). Analytical instrumentation that is used to test raw materials, in-process samples, and final drug product should fully comply with the FDA equipment requirements. These requirements assure appropriate design and adequate capacity for consistent functioning of the instrument (Zanetti and Huber, 1996). Qualification is the process that provides confidence that the equipment is capable of consistently operating within the established design limits. For this reason, analytical instrument qualification is an expected part of analytical method validation. Similar to analytical method validation, qualification provides documented evidence that an instrument is acceptable for its intended use. In the pharmaceutical industry, analytical method validation is assumed to be performed using a qualified instrument. Therefore, instrument qualification is expected to be completed before initiating the analytical method validation process. Qualification is a documented systematic process that confirms that the equipment (analytical instrument) has been installed according to approved design specifications and verifies that the equipment operates according to the approved design.

The qualification process may be divided into three phases. These phases are executed in the following order: installation qualification (IQ), followed by operational qualification (OQ), followed by performance qualification (PQ). IQ is a verification that the actual instrument installed in the laboratory meets the design and installation specifications. OQ is a verification that the instrument operates within the full range established by the design specifications and includes tests for each function of the instrument. PQ is a verification that the ongoing performance of the instrument remains within the designed specifications for its actual use. The intent of qualification is to ensure that the instrument is operating properly, thus ensuring robust and reliable performance. In other words, qualification provides assurance about the validity of the sample results generated by an analytical instrument.

Qualification is accomplished via a documented procedure. The procedure contains the specific instructions and acceptance criteria that need to be executed and met, in order to ensure that the instrument meets the design specifications. The resulting documented evidence verifies that the instrument has been installed correctly and operates properly. The results are evaluated to determine if the acceptance criteria have been satisfied. Then the final executed package is circulated for review and approval. Once approvals of all the appropriate qualification activities have been obtained, the instrument is deemed "qualified" and may be used for analytical method validation.

IQ should include documentation on the following information about the instrument:

Manufacturer
Model and serial numbers
Dimensions (physical description) of the instrument
Materials of construction in contact with mobile or stationary phase
Utility specifications (such as electrical requirements)
Assembly check
Initial calibration records
Spare parts list
Location of supporting documentation (such as operating manuals)

The majority of these items should be specified in the design or by the instrument manufacturer. As part of the IQ, the design specifications are compared with the actual results obtained from the instrument installed in the laboratory.

Ideally, OQ should simulate actual operating conditions. OQ tests should be repeated a sufficient number of times to assure reliable results. "OQ verifies key aspects of instrument performance without the aspects of any contributory effects that could be introduced by a method" (Grisanti and Zachowski, 2002). OQ testing should include:

Verification that all alarms and interlocks are functional. For example, instrument functions only when the top cover is closed.
Test control and functionality of unit operation.
Verification that the instrument operates correctly over the entire specified range.

One purpose of the activities listed earlier is to ensure that the instrument is capable of meeting design specifications for accuracy, linearity, and precision (Grisanti and Zachowski, 2002).

PQ demonstrates that the analytical instrument performs according to specifications appropriate for routine use. PQ may include analyzing known standards and evaluating the results vs. system suitability requirements (see Section II.B. for details). PQ testing should be based on good science and the intended use of the instrument.

Theoretically, there is a distinction between OQ and PQ. OQ verifies that all of the individual units (parts or modules) of an instrument operate properly. On the other hand, PQ verifies that the entire instrument as a system performs properly. In practice, aspects of OQ and PQ may be evaluated together, particularly when the test can be more easily executed at the system level. Tests that may be combined as a joint OQ/PQ are linearity and precision (repeatability) (Grisanti and Zachowski, 2002). For example, in an OQ protocol for an HPLC instrument, the individual modules that might be verified to operate properly are the injector, detector, thermostat, pump, and autosampler. Specifically, key functionalities that might be tested are injection volume, detection wavelength, column temperature, flow rate, and gradient composition. Continuing with this example to illustrate combined OQ/PQ tests, the tests for precision of the injector and pump may be combined with the test for precision of the detector. Here, the precision (reproducibility) of the detector response relative to repeated injections using the same injector volume is evaluated on the system level (Hall and Dolan, 2002).

In conclusion, qualification of analytical instruments should do more than allow companies to comply with regulatory regulations. Qualification should assure scientists of the limits and abilities of their systems and improve their confidence in their analytical results (Grisanti and Zachowski, 2002).

B. System Suitability

The system suitability tests for chromatographic methods are based on the concept that the equipment, electronics, analytical operations, and samples to be analyzed constitute an integral system that can be evaluated as such (ICH, 1996). Through a series of tests and parameters the system suitability ensures that the total system is functioning adequately. If the criteria used to determine the suitability of a system were chosen correctly, the criteria will fail to be met just before the system will generate data that is not acceptable. Hence, appropriate suitability criteria need to be selected during the method validation process. Once the criteria have been determined, the system suitability tests should always be done before starting an analysis.

Depending upon the length of a run or importance of the sample results, system suitability may also be performed during and following the analysis. Typical system suitability parameters for chromatographic methods are listed as follows. Examples were gathered from a reference search of possible procedures and acceptance criteria that, together with any regulatory requirements, can be used as guidance.

Capacity factor (k'):

$$k' = \frac{t_R - t_0}{t_0} \quad (1)$$

where t_R is the retention time of the analyte and t_0 is the elution time of the void volume or nonretained components.

The capacity factor (k') is a measure of where the peak of interest is located with respect to the void volume. It is recommended that the peak should be well resolved from other peaks and the void volume, generally $k' > 2$ (FDA, 1994).

Precision/injection repeatability expressed as the relative standard deviation (RSD):

$$\text{RSD} = \frac{\sigma}{\bar{x}} \times 100 \quad (2)$$

where σ is the standard deviation and $\bar{x}$ is the mean value.

RSD indicates the performance of the system at the time the sample is analyzed. Replicate injections of a standard preparation are compared to ensure that requirements for precision are met (FDA, 1994). One recommendation is to use five replicate injections of the analyte to calculate the RSD if the requirement is $\leq 2.0\%$. If the RSD requirement is >2.0 then six injections are used (USP, 2003).

Relative retention: The relative retention (α) is a measure of the relative location of two peaks:

$$\alpha = \frac{k'_1}{k'_2} \quad (3)$$

This parameter is not essential as long as the resolution is stated (FDA, 1994).

Resolution:

$$R_s = \frac{t_{R2} - t_{R1}}{0.5(t_{W_1} + t_{W_2})} \quad (4)$$

where t_R is the retention time of the analyte and t_W is the peak width measured at baseline of the extrapolated straight sides to baseline.

The resolution (R_s) is a measure of how well two peaks are separated. The resolution should be >2 between the peak of interest and the closest eluting potential interfering peak (FDA, 1994). Though it may not always be possible to achieve a resolution >2, it should be possible to resolve peaks to some degree perhaps by changing method parameters. The acceptable resolution for a method should be evaluated on a case-by-case basis.

Tailing factor (T):

$$T = \frac{W_x}{2f} \quad (5)$$

where W_x is the width of peak determined at either 5% or 10% from the baseline of the peak height and f is the distance between maximum and peak front at W_x

The tailing factor is a measure of peak symmetry and for a perfectly symmetrical peak the value is 1. As peak asymmetry increases, precision becomes less reliable. It is recommended that $T \leq 2$ (FDA, 1994).

Theoretical plates (N):

$$N = 16\left(\frac{t_R}{t_W}\right)^2 \quad (6)$$

The number of theoretical plates is a measure of column efficiency, that is, how many peaks can be located per unit run-time of the chromatogram. The theoretical plate counts depend on elution time, but in general should be >2000 (FDA, 1994).

The number of system suitability tests required for a particular method will depend on the purpose of the test method. The FDA recommends the following minimum suitability tests (FDA, 1994):

For dissolution or release profile tests using external standard method (impurity quantitative test or major component test): capacity factor, tailing factor, and RSD are the minimum recommended system suitability tests.

For acceptance, release, stability, or impurities/degradation methods using external or internal methods (impurity quantitative test or major component test): capacity factor, tailing factor, RSD, and resolution are the minimum recommended system suitability tests.

III. METHOD VALIDATION REQUIREMENTS

A. Precision

1. Definition

The precision parameter demonstrates the ability of an analytical method to produce consistent results. Precision is the degree of reproducibility among individual results, which takes into account internal variation inherent in an analytical method. It is important to understand that the precision parameter does not ensure that a result is the correct value. Rather, precision is the extent of agreement between multiple results. A good example to illustrate the definition of precision uses a bull's eye target. If an archer shoots five arrows at the target, and the arrows become embedded close together, but not near the bull's eye (see Fig. 2), the archer's shots are precise, but not accurate. If the archer shot all five arrows in the center of the target, then the archer is precise and accurate (assuming the bull's eye is the "true" value). Precision may often be divided into the following three categories: repeatability, intermediate precision, and reproducibility. In general,

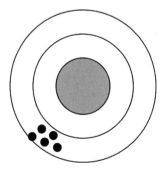

Figure 2 Bull's eye—precision.

these categories are based on where and when the results are collected from a sample that is analyzed by a given analytical method.

Repeatability evaluates data generated using exactly the same operating conditions over a brief timespan. Repeatability is also referred to as intra-assay precision. Repeatability demonstrates that despite inherent variabilities in an analytical method, it is possible to generate multiple data points with close agreement. In theory, a brief time interval is recommended to ensure that each analysis is performed using exactly the same operating conditions. An example of repeatability is when an analyst injects one sample preparation into an HPLC instrument multiple times. In this example, all chromatographic conditions are assumed to remain unchanged throughout the run. Repeatability demonstrates the precision of the instrument itself. In this specific example, the precision of the instrument is the repeatability of the injector.

Intermediate precision evaluates data generated over an extended period of time. Operating conditions are assumed to change when demonstrating intermediate precision. The analysis is assumed to be performed in one laboratory for intermediate precision. However, different analysts, on different instruments or on different days may perform the analysis. Intermediate precision is demonstrated by analyzing multiple samples using a variety of intra-laboratory variations. For example, one analyst may complete repeated analysis of multiple samples using different instruments. Another example is that an analyst may complete repeated analysis of multiple samples on different days. A third example is that two or more analysts may complete repeated analysis of one homogeneous sample. Intermediate precision demonstrates that an analytical method is precise, despite random events affecting operating conditions within the same laboratory.

Reproducibility evaluates the performances of an analytical method in more than one laboratory. Reproducibility is demonstrated by analyzing multiple samples in different laboratories. For example, when a chromatographic method is transferred from a laboratory in a research environment to a quality control laboratory in a manufacturing environment, reproducibility is evaluated. "The evaluation of reproducibility results often focuses more on measuring bias in results than on determining differences in precision alone" (Shabir, 2003). It has been suggested that repeatability is generally one-half to one-third of reproducibility (Jenke, 1996b).

2. Procedures

"The precision of an analytical method is determined by assaying a sufficient number of aliquots of a homogeneous sample to be able to calculate statistically valid estimates of standard deviation or relative standard deviation (coefficient of variation)" (USP, 2003). The number of samples necessary to prove the precision parameter depends on the purpose of the analytical method. For instance, results would need to be much more precise if someone's life depended on the result rather than if only estimates were required (Coleman and Vanatta, 2003). Though precision may vary with sample concentration, the general assumption here is that the precision does not change with sample concentration (Thompson et al., 2002). However, this assumption should be verified if the sample concentration is likely to vary significantly. The list that follows includes results from a reference search on the number of aliquots used to verify precision. Specific procedures and acceptance criteria for precision vary depending on the purpose of the analytical method.

- Minimum of nine determinations covering the specified range of procedures (three concentrations with three replicates of each concentration) (ICH, 1996; Shabir, 2003).
- Minimum of six determinations at 100% target concentration (ICH, 1996; Shabir, 2003).
- Six concentrations analyzed in triplicate (Rathore, 2002).
- Minimum of 10 injections of one sample solution (FDA, 1994; Shabir, 2003).
- Fifteen individual sample preparations covering specified range (five concentration levels with three replicates) (Crowther, 2001).
- Six replicates at three concentrations (top, middle, and bottom) of calibration curve (Buick, 1990).
- Three concentrations (top, middle, and bottom) in duplicate over 5 days (intermediate precision) (Jimidar, 1995).
- For bioanalytical techniques, minimum of five (excluding blank sample) determinations per concentration (Shah et al., 1991, 2000).
- Five replicates of three weighings at 80%, 100%, and 120% of label claim (Qi et al., 2003).
- For release or stability assays, five replicates (Jenke, 1996b).
- At least 6–10 replicates (Jenke, 1996b).
- Same solution, different HPLC system in different laboratories (reproducibility) (Ferretti et al., 1998).
- Duplicate measurements on 10 samples at each of three different analyte levels (Jenke, 1996b).
- Sufficient data should be generated to ensure more than 30 degrees of freedom (Jenke, 1996b).
- Three experiments repeated twice daily for 3 days and then CV calculated (RSD) from mean of three experimental variances estimated for each day using the following equations (Mouatassim-Souali et al., 2003).

$$\text{CV } (\%) = \frac{s_{\text{residual}}}{\bar{X}} \times 100 \quad \text{(for repeatability)}$$
(7)

where s is the square root mean of experimental variance estimated each day.

$$\text{CV } (\%) = \frac{\sqrt{s_{\text{residual}}^2 + s_f^2}}{\bar{X}} \times 100$$

(for intermediate reproducibility) (8)

where $s_f^2 = (s_{\text{interrun}}^2 + s_{\text{residual}}^2)/2$ and s_{interrun}^2 is experimental variance between experiments on different days.

3. Acceptance Criteria

Acceptance criteria as defined by the FDA are "numerical limits, ranges, or other suitable measures for acceptance of the results of analytical procedures" (FDA, 2000). Acceptance criteria should be established prior to initiating method validation activities. Once again, the acceptance criteria should be appropriate for the intended use of the analytical method. The acceptance criteria are most commonly based on calculating the RSD. RSD is chosen because it is usually unrelated to the sample concentration over a logical range of concentrations (Guidelines, 1989). In general the larger the RSD value, the less precise the analytical method. Acceptance criteria for precision may be stated in a number of ways. Some specific suggestions are as follows:

- %RSD not more than 1% (instrument precision) (Shabir, 2003; Ziaková and Brandšteterová, 2003).
- %RSD not more than 2% (intra-assay precision) (Shabir, 2003; Ziaková and Brandšteterová, 2003).
- For impurity assays at LOQ, %RSD not more than 5% (instrument precision) (Shabir, 2003).
- For impurity assays at LOQ, %RSD not more than 10% (intra-assay precision) (Shabir, 2003).
- Assay results in multiple laboratories are statistically equivalent (reproducibility) (Shabir, 2003).
- Assay results in multiple laboratories yield mean results within 2% of value obtained by primary laboratory (reproducibility) (Shabir, 2003).
- For impurity assay results in multiple laboratories yield mean results within 10% of value obtained by primary laboratory (reproducibility) (Shabir, 2003).
- For HPLC assay/purity method, %RSD not more than 3.0% (intermediate precision) (Crowther, 2001).
- Retention times of standards <1% (repeatability) (Ziaková and Brandšteterová, 2003).
- For bioanalytical and pharmacokinetic techniques, precision around mean value should be not more than 15% CV, except for LOQ which is not more than 20% CV (Bressolle et al., 1996; Shah et al., 1991, 2000).
- System precision not more than 1.5% RSD; method precision not more than 2.0% RSD (Jenke, 1996b).

Required discrimination ability must be not more than the quantity 1.96 σ/n (Jenke, 1996b).

For discovery samples, not more than 20% RSD is acceptable. For preclinical and clinical samples, not more than 10% RSD (Jenke, 1996b).

For acceptance ranges of 95–105% and 90–110%, the recommended precision is 2% or 4% RSD, respectively (Jenke, 1996b).

B. Accuracy

1. Definition

Accuracy, or trueness, is the closeness of agreement between the value, which is accepted either as a conventional true value or an accepted reference value, and the value found. Accuracy considers systematic and random errors associated with the analysis. Some of the possible errors include errors associated with the sampling activity, preparation of the sample, and the analysis of the sample. The second bull's eye example illustrates accuracy (see Fig. 3). In this example, the points are scattered around the bull's eye, but not grouped close together. Here, the points are accurate, but not precise.

2. Procedures

There are several methods that can be used for determining accuracy. The most common include:

Analyze a sample of known concentration and compare the measurement to the true value (Ferretti et al., 1998; Qi et al., 2003; Shabir, 2003). In this case, method accuracy is the agreement between the difference in the measured analyte concentration and the known amount of analyte added. That is, the accuracy or %recovered is calculated as:

$$\frac{C_M}{C_T} \times 100 \tag{9}$$

where C_M is the measured concentration and C_T is the theoretical concentration. Accuracy has also been reported as the relative error between the observed and theoretical values (Mouatassim-Souali et al., 2003). An alternative method is to estimate the correlation between true and estimated values with a correlation line slope of one and a zero intercept (Tamisier-Karolak et al., 1995).

Compare the results from the new method with the results of an existing well-characterized method (e.g., compendia and reference) whose accuracy is defined (ICH, 1996; Jimidar et al., 1995; Shabir, 2003; Thompson et al., 2002). Comparison studies can also include methods based on different separation principles. For example, cross-correlation between HPLC and capillary electrophoresis (CE) results has often been used (Fabre and Altria, 2001).

Use of spiking/recovery studies (Ferretti et al., 1998; Izco et al., 2003; Shabir, 2003). The sample is analyzed by the method under validation prior to spiking and it is then analyzed after the addition of a known mass of analyte. The difference between the two results as a proportion of the mass added is called the marginal or surrogate recovery. Methods whose marginal recovery is significantly different than 1, have errors associated with them and are not considered accurate (ICH, 1996).

The list that follows includes results from a reference search on the number of aliquots used to verify accuracy.

The ICH guideline recommends that accuracy should be determined using a minimum of nine determinations over a minimum of three concentration levels covering the specified range (ICH, 1996).

Spiked samples are prepared in triplicate at three levels over a range that covers 50–150% of the target concentration for assay methods and over a range that covers the expected impurity content of a sample for impurity methods (Shabir, 2003).

For bulk drug assay, six determinations over a range that covers 50–150% (Clarke, 1994).

For bulk drug impurity assay, five determinations over a range that covers 50–150% (Clarke, 1994).

For finished product assay, six determinations over a range that covers 75–125% (Clarke, 1994).

Data at least in triplicate at each level (80%, 100%, 120% of label claim) is recommended (FDA, 1994).

Five determinations per concentration (Shah et al., 1991).

Two instruments, six concentrations, three replicates each (Ferretti et al., 1998).

3. Acceptance Criteria

Once again, the acceptance criteria should be appropriate for the intended use of the analytical method. Acceptance

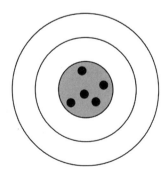

Figure 3 Bull's eye—accuracy.

criteria for accuracy may be stated in a number of ways. Some specific suggestions are as follows:

- For an assay method, mean recovery will be 100% ± 2% at each concentration over the range of 80–120% of the target concentration (Shabir, 2003).
- For an impurity method, the mean recovery will be within 0.1% absolute of the theoretical concentration or 10% relative, whichever is greater for impurities in the range of 0.1–2.5% (v/w) (Shabir, 2003).
- Not more than 15% deviation from the nominal value (Shah et al., 1991).

C. Limit of Detection

1. Definition

LOD is the lowest concentration of analyte in a sample that can be detected but not necessarily quantitated under the experimental conditions of the method. In practice, it is the lowest concentration of analyte that can be distinguished from the blank with a stated degree of confidence (Jenke, 1996b). The LOD is a limit test that evaluates whether the analyte is above or below a given value.

2. Procedure

The rigorous definition of the International Union of Pure and Applied Chemistry (IUPAC) for LOD is based on the statistical theory of hypothesis testing (Castillo and Castells, 2001). In this case, the LOD depends on the blank signal, the standard deviation at the blank level, the risk of detecting the analyte when it is absent, and the risk of not detecting the analyte when it is present (Castillo and Castells, 2001). As the standard deviation of blanks cannot be measured in chromatography, alternative methods need to be applied. The typical methods recommended by ICH are:

- Signal-to-noise ratio: The LOD can be expressed as a concentration at a specified signal-to-noise ratio obtained from samples spiked with analyte. A signal-to-noise ratio between 3:1 and 2:1 is generally considered acceptable (ICH, 1996).
- Standard deviation of the response and the slope of the calibration curve(s) at levels approximating the LOD according to the formula:

$$\text{LOD} = \frac{3.3\sigma}{S} \quad (10)$$

where S is the slope of the calibration curve and σ is the standard deviation. The standard deviation can be determined on the basis of the standard deviation of the blank, the residual standard deviation of the regression line, or the standard deviation of y-intercepts of the regression lines (ICH, 1996).

3. Acceptance Criteria

In general, there are no specific criteria for what value LOD should have, as it is almost never necessary to calculate the actual detection limit. Rather, the detection limit is shown to be sufficiently low by analysis of samples with known concentration of analyte above and below the required detection level (USP, 2003). The LOD values obtained for CE methods are generally higher in concentration than those obtained for HPLC methods (Fabre and Altria, 2001).

D. Limit of Quantitation

1. Definition

LOQ is the lowest concentration of analyte in a sample that can be quantitated. In addition, the LOQ for an analytical procedure is the lowest amount of analyte in a sample that satisfies all other method validation requirements.

2. Procedure

Similar to LOD, the typical methods recommended by ICH to determine the LOQ concentration are:

- Signal-to-noise ratio: The LOQ can be expressed as a concentration at a specified signal-to-noise ratio obtained from samples spiked with analyte. A signal-to-noise ratio of 10:1 is generally considered acceptable (ICH, 1996). This ratio recognized by the ICH (1996) is a general rule. It has been stated that "the determination of LOQ is a compromise between the concentration and the required precision and accuracy. That is, as the LOQ concentration level decreases, the precision increases" (Shabir, 2003).
- Standard deviation of the response and the slope of the calibration curve(s) at levels approximating the LOQ according to the formula:

$$\text{LOQ} = \frac{10\sigma}{S} \quad (11)$$

The standard deviation (σ) can be determined on the basis of the standard deviation of the blank, the residual standard deviation of the regression line, or the standard deviation of y-intercepts of the regression lines (ICH, 1996).

The LOQ can also be determined as the lowest analyte concentration for which duplicate injections result in a %RSD ≤ 2% or by analyzing successively diluted samples until the requisite levels of accuracy and precision are achieved (Jenke, 1996b). In general, there are no specific criteria for what value LOQ should have, as it is almost never necessary to calculate the actual quantitation limit. Rather, the quantitation limit is shown to be sufficiently low by analysis of samples with known concentration of analyte above and below the required quantitation level (USP, 2003).

3. Acceptance Criteria

Some recommended criteria for the LOQ are:

- In pharmaceutical applications, the LOQ is typically set at maximum 0.05% for active pharmaceutical ingredients (Crowther, 2001).
- LOQ defined as the lowest concentration providing a RSD of 5% (Rabouan et al., 2003).
- LOQ should be within the working linear concentration range and the specification limit should be not lower than twice the LOQ (Jenke, 1996b).
- LOQ should be at least 10% of the minimum effective concentration for clinical applications (Jenke, 1996b).

E. Range

1. Definition

Range is the interval between and including the upper and lower limits for which the analytical method has been demonstrated to satisfy all other method validation requirements. The range may be defined as the set of sample concentrations where the sensitivity of the method is essentially constant (Bruce, 1998). The range is established by confirming that the analytical method provides an acceptable linearity, accuracy, and precision at the extremes of the specified range of the method. The range should reflect the purpose of the analytical method. "The range selected for validation should not be unrealistically wide, as this may lead to rejection of a method which is really quite suitable for the intended purpose" (Jenke, 1996b). For this reason, there is a balance between practical restrictions and scientific rigor. The range is determined by verifying that the analytical method meets all other validation parameters at the boundaries and within the range. Practical aspects for performing a method should be considered during the validation. For example, what happens if a result is found to be outside the established range? Is it appropriate to dilute the sample 10-fold and reanalyze the sample? Procedures should be developed to handle these types of situations, specifically when a result is outside of the validated range of the assay (Buick, 1990).

2. Procedure/Acceptance Criteria

The suggested procedures and acceptance criteria for determining range and linearity are intimately linked. Consequently, the methods and criteria are listed in Section III.F.

F. Linearity

1. Definition

Linearity is the ability of an analytical method to obtain results that are directly or indirectly, by a well-defined mathematical transformation, proportional to the concentration of analyte in samples. Analytical procedures may produce output that is an absolute indication of sample property being measured, such as Karl Fisher (KF) or titration instrumentation. On the other hand, instruments may produce output that is mathematically transformed into a result, such as peak areas in HPLC that are related to concentration. This transformation is then used to determine concentration of analyte in a sample. Both direct and transformed output should be verified to be linear. Linearity should be demonstrated within the given range (as discussed in the previous section) and be appropriate for the intended use of the method. Linearity is usually expressed in terms of the square root of the variance (correlation coefficient) around the slope of the regression line. Linearity demonstrates that the relationship between the result and the sample concentration is continuous and reproducible. "A reproducibly nonlinear response curve however can also be acceptable. Nonlinearity is certainly the case with immunological procedures" (Shah et al., 1991). However, more concentrations may be required to define a nonlinear relationship as compared with a linear relationship (Shah et al., 2000).

2. Procedure

Linearity is typically determined by plotting results as a function of analyte concentration. The plot is studied to determine if there is a direct linear correlation between the results and sample concentration or if the data could be mathematically transformed to produce a linear correlation. In the majority of analytical methods, the calibration line (the line that fits best through output signal and corresponding concentration) is expressed by an estimated first-order equation:

$$Y = aX + b \qquad (12)$$

where Y is the measured output signal, X is the concentration of sample, a is the slope, and b is the intercept. Ordinary least squares (OLS) regression is typically used to calculate calibration lines. In order to use OLS regression, the variance of the output signal should be independent from the output signal itself (Crowther, 2001).

Linearity is typically demonstrated by performing duplicate sample preparations on the specified range. As mentioned previously, the specified range should reflect the intended use of the analytical method. Some suggestions for the appropriate range to evaluate are listed as follows:

- Whole range from LOQ to 150% of normal sample concentration (Crowther, 2001).
- Minimum of five concentrations normally being used (ICH, 1996; Shabir, 2003).
- For assay of drug substance or finished product, from 80% to 120% of test concentration (ICH, 1996; Shabir, 2003).
- For impurity assay, from 50% to 120% of specification (ICH, 1996).
- For content uniformity assay, minimum of 70–130% of test concentration unless a wider or more appropriate range on the basis of the nature of the dosage form is justified (ICH, 1996; Shabir, 2003).
- For dissolution testing, ±20% over specified range (ICH, 1996; Shabir, 2003).
- For active ingredient assay in formulation of 50–150%, 80–120%, or 60–140% range of target concentration (Fabre and Altria, 2001).
- Minimum of six calibration standards evenly spread over concentration range and run at least in duplicate in random order (Thompson et al., 2002).
- From 25% to 125% of target range specified (Jenke, 1996b).
- From 25% to 200% of nominal concentration of analyte (Shabir, 2003).
- Range of expected concentrations (Jenke, 1996b).
- From 50% to 150% of expected working range (Jenke, 1996b).
- For degradate assays, 0–2% of active drug level in formation (Jenke, 1996b).
- Minimum of three concentrations representing entire range of calibration curve, one near the LOQ, one near the center, and one near the upper boundary of standard curve (Shah et al., 1991).

3. Acceptance Criteria

As with other validation parameters, the acceptance criteria for linearity are based on appropriate statistical methods. Examples include the correlation coefficient, y-intercept, slope, and RSD [for generated response ratios, the ratio of peak areas of analyte to internal standard vs. concentration is taken (Qi et al., 2003)]. The correlation coefficient is commonly referred to as a test of linearity; however, some literature suggests it should not be used (Jenke, 1996b; Thompson, 2002). The preference is to use a test of significance for the b_2 term in the equation

$$y = b_0 + b_1 X + b_2 X^2 \qquad (13)$$

and residual sum of squares.

Acceptance criteria for the range are embedded in the linearity requirements. Linearity acceptance criteria should be met for the range of results that is suitable for the intended use of the method. Some suggestions for acceptance criteria are as follows:

- Data should be plotted to look for dubious points and to visually establish calibration range (Jenke, 1996b).
- A response factor plot is used to identify concentration where true proportionality is not observed (Jenke, 1996b).
- For HPLC assays of biological or pharmaceuticals, correlation coefficient of >0.995 or >0.997, respectively (Bruce et al., 1998).
- Correlation coefficient of >0.999 and evidence of acceptable fit of data to regression line (Shabir, 2003).
- y-intercept less than a few percent of response obtained for the analyte at target level (Shabir, 2003).
- Goodness-of-fit of data to regression line on the basis of residual sum of squares (Shabir, 2003).
- Assuming regression line is the mean, calculate %RSD for data, %RSD should be less than or equal to 3.0% (Crowther, 2001).
- y-intercept should statistically not differ from zero (Crowther, 2001).
- For HPLC assay/purity method over whole range, y-intercept relative to active ingredient ±10%, RSD response ratio less than or equal to 5.0% and correlation coefficient >0.99 (Crowther, 2001).
- F-test at a 0.05 significance level for lack-of-fit or test of significance of the quadratic regression coefficient obtained on fitting calibration data to a second degree polynomial (Castillo and Castells, 2001).
- The value of n in the equation $y = mx^n + b$ should be between 0.9 and 1.1 and maximum allowable relative error is 1% (Jenke, 1996b).
- Calculated response factor (peak response by CE/concentration), CV of response factor less than or equal to 5% (Izco et al., 2003).

G. Ruggedness

1. Definition

Ruggedness is the degree of reproducibility of an analytical method obtained by analysis of samples under a variety of test conditions. Test conditions should represent parameters expected with normal operating conditions. Test conditions that may be varied include laboratories, analysts, instruments, reagent lots, and elapsed assay times. Ruggedness is a measure of reproducibility of results taking into account normal operating variations. In many respects, ruggedness is an intermediate precision (see Section III.A.) "Generally, a rugged method's reproducibility is 2–3 times greater than the method's repeatability (inherent method precision under "normal" controlled operating conditions)" (Crowther, 2001).

2. Procedure

Ruggedness ensures that an analytical method will perform suitably each time it is used. For example, two analysts may independently perform the assay on at least three sample lots. These analyses may be performed on different days, instruments, standards, and sample solutions to simulate minor variations, which occur under normal operating conditions. Transferring an analytical method from one laboratory (such as research environment) to another laboratory (such as manufacturing environment) may be the definitive ruggedness test (Bruce et al., 1998).

3. Acceptance Criteria

Acceptance criteria that may be applied for ruggedness are those that are listed in Section III.A. Some suggested methods used to determine if an analytical method is rugged are as follows:

- Analysis of aliquots from homogeneous lots in different laboratories, by different analysts using operational and environmental conditions that may differ but are still within the specified parameters of the assay (USP, 2003).
- Four analysts perform one assay per day for 3 days (Jenke, 1996c).
- At least three columns, each one from a different batch produced by recommended manufacturer and at least one column from a different manufacturer (Jenke, 1996c).

H. Robustness

1. Definition

Robustness is a measure of the capacity of an analytical method to remain unaffected by small but deliberate variations in operating parameters. For example, a chromatographic method should not be affected by small variations in different instrument manufacturers, different solvent ratios, or different operating temperatures. Robustness demonstrates that an analytical method is reliable during normal usage. In addition, robustness provides reasonable operating tolerances that may be expected during routine use. The internal variation of the system should be evaluated as part as the OQ/PQ testing. Robustness will encompass at least this variability. The intent of robustness is to determine which parameters need to be tightly controlled and, as appropriate, include these critical variables in the procedural description of the analytical method (Hokanson, 1994). As a result of including robustness as part of a method validation, parameters that can be varied without affecting performance of an analytical method are identified. If a method does not meet acceptable robustness criteria, then optimization of the method should continue until robustness criteria are satisfied.

2. Procedure

Parameters may be varied one at a time and evaluated, or multiple parameters varied as part of factorial experiments (Shabir, 2003). One suggestion is to use a two-level Plackett–Burman statistical approach. In this approach, up to 11 factors can be evaluated in 12 injections (see Table 2).

There are a number of factors that may be varied to evaluate robustness of an analytical method. The acceptable ranges for variation are based on the anticipated

Table 2 Two Factor Plackett–Burman Example

	F1	F2	F3	F4	F5	F6	F7	F8	F9	F10	F11
Inj 1	+	+	−	+	+	+	−	−	−	+	−
Inj 2	+	−	+	+	+	−	−	−	+	−	+
Inj 3	−	+	+	+	−	−	−	+	−	+	+
Inj 4	+	+	+	−	−	−	+	−	+	+	−
Inj 5	+	+	−	−	−	+	−	+	+	−	+
Inj 6	+	−	−	−	+	−	+	+	−	+	+
Inj 7	−	−	−	+	−	+	+	−	+	+	+
Inj 8	−	−	+	−	+	+	−	+	+	+	−
Inj 9	−	+	−	+	+	−	+	+	+	−	−
Inj 10	+	−	+	+	−	+	+	+	−	−	−
Inj 11	−	+	+	−	+	+	+	−	−	−	+
Inj 12	−	−	−	−	−	−	−	−	−	−	−

operating conditions in the laboratories (Virrlichie and Ayache, 1995). Some examples of typical parameters that may need to be studied in an HPLC method are sample preparation, sample concentration, solution stability, extraction time, mobile phase solvent composition, influence of pH variation on mobile phase, buffer pH, buffer ionic strength, different column variations, column temperature, flow rate, injection amount, wavelength of detection, and re-equilibration time. These parameters may impact the peak elution of the analyte. The range used to evaluate these parameters should reflect the potential operating conditions (see example in Section IV).

3. Acceptance Criteria

A suggested approach for the acceptance criteria is to compare the factor effects against pure errors computed for replicates, and determine if errors are statistically significant on the basis of T scores and p values (Lee et al., 2000).

I. Solution Stability

1. Definition

Solution stability should also be evaluated as part of robustness. If a sample decomposes with age, then the precision of the analytical method will also decrease (Szepesi, 1990). This evaluation is important to confirm that the analyte does not degrade over time because of hydrolysis or photolysis or adhesion to glassware (FDA, 1994).

2. Procedure

Sample and standard solution stability may be determined by analyzing aged solutions. One suggestion is to analyze the aging solution every 24 h for up to 3 days against a freshly prepared standard. The aging solutions should be stored in normal laboratory conditions for at least the time necessary to prepare and analyze the sample with a safety factor for instrument delays (Fabre and Altria, 2001). Solution stability may be determined by calculating results at each timepoint and evaluating the absolute difference between results of freshly prepared standards with results obtained at various timepoints. All solutions should be stable for at least the time typically required for analysis.

3. Acceptance Criteria

In general, a standard or system suitability solution is analyzed while varying conditions, and critical system suitability parameters are monitored. Some suggestions for solution stability acceptance criteria are as follows:

> A 2% change in standard or sample response relative to freshly prepared standard (Shabir, 2003).

> Retention time and peak area RSD <2.0% and no significant degradation observed for sample held for one week (Qi et al., 2003).

> For overnight runs, four sample solutions over working concentration range analyzed repetitively over at least 16 h and all concentration values agree to within three times system precisions (Jenke, 1996c).

> For bioanalytical assay, stability should be demonstrated for all aspects of storage and handling with not more than 15–20% deviation from original control value (Needham, 2003).

J. Specificity

Specificity is the ability to assess unequivocally the analyte in the presence of components that may be expected to be present in the sample matrix. For example, in the pharmaceutical industry a sample matrix may include impurities, degradation products, excess raw materials, or excipients. The chromatographic method should be specific and sensitive as required for all known relevant degradation products and/or impurities. Other potential sample components can be generated by exposing the analyte to stress conditions (e.g., temperature, acid, light, base, oxidant, humidity) (Shabir, 2003).

The terms specificity and selectivity are often used interchangeably (Bressolle et al., 1996; Bruce et al., 1998). Compared to the previous definition of specificity, selectivity is the ability of a method to quantify the analyte accurately in the presence of other compounds (Bressolle et al., 1996; Thompson et al., 2002). That is "a method producing a response for only a single analyte can be called specific. A method producing responses for several chemical entities but that is able to distinguish the analyte response from the others can be called selective" (Bruce et al., 1998). In practice, the different specificity and selectivity tests try to approach the same problem, "do we measure what we think we are measuring?" (Bruce et al., 1998).

1. Procedure

There are several tests that can be used to determine whether a method is selective or specific.

> The response of the analyte in a solution containing all impurities, degradation products, and/or excess raw materials is compared with the response of the solution containing only the analyte. This is accomplished by spiking the analyte with appropriate levels of impurities, degradation products, and/or excess raw materials (ICH, 1996; Shabir, 2003).

> If impurity or degradation standards are not available, specificity may be demonstrated by comparing the results of the sample containing the impurities or degradation products with a well-characterized

procedure, such as those found in compendia or other validated method (ICH, 1996).

Peak reanalysis, where the peak is collected and reanalyzed by different chromatographic conditions or with methods sensitive to analyte structure (Jenke, 1996b).

Samples are analyzed using different detection/separation strategies (Jenke, 1996b).

Use of information-rich detectors (e.g., mass spectrometric or multiple wavelength UV) to assess peak purity (Jenke, 1996b).

2. Acceptance Criteria

Acceptance criteria for specificity may be stated in a number of ways. Some specific suggestions are as follows:

Analyte peak will have a baseline chromatographic resolution of at least 1.5 from all other components (for assay methods) (Shabir, 2003).

All impurity peaks that are 0.1% by area will have a baseline chromatographic resolution from the main component (for impurity methods) (Shabir, 2003).

To determine selectivity, at least six blank matrix samples from six different sources are analyzed for possible interferences with the analyte and internal standard (Needham, 2003).

If a peak is collected and reanalyzed on another chromatographic system, it should produce a single response (for peak reanalysis) (Jenke, 1996b).

The relative abundance of all diagnostic ions must be the same in the sample and the standard within a margin of $\pm 10\%$ (EI mode) or $\pm 20\%$ (CI mode) (for low resolution mass spectrometric detection (Jenke, 1996b)).

The peaks should be well separated with a resolution >2 between the peak of interest and the closest eluted peak (Grisanti and Zachowski, 2002).

IV. METHOD VALIDATION EXAMPLE

This section is a hypothetical case study for the validation of an analytical method based on the process outlined in Fig. 1. The intent of this example is to serve as a starting point for evaluating possible procedures and acceptance criteria for an HPLC method validation. The procedures and acceptance criteria to be used should be evaluated on a case-by-case basis.

A. Purpose and Scope

The purpose of the study is to perform an analytical method validation of a stability-indicating HPLC method (compound A).

B. Determine Method Type and Method Requirements

The method is described as a major component method and the method validation requirements to include are provided in Table 3.

C. Set Acceptance Criteria for Method Validation Requirements

The acceptance criteria used to validate the method are given in Table 3. The acceptance criteria were chosen on the basis of the typical expectations from the literature search for analysis of the major component of an active drug substance. If the purpose of the method was to quantify impurity levels, then the acceptance criteria would be different. Specifically, the criteria for accuracy and precision may need to be more rigorous for a lower quantitation level.

D. Determine System Suitability Requirements

For this method, the system suitability parameters were determined to be: tailing factor of main peak <3, RSD for ratio of response of 0.05 mg/mL standard solution for six injections should be $\leq 2.0\%$, capacity factor >2, and the resolution >2. The system suitability parameters

Table 3 Method Validation Requirements for Example

Method validation requirements	Acceptance criteria
Precision	
Assay repeatability	$\leq 1\%$ RSD
Intermediate precision	$\leq 2\%$ RSD
Accuracy	
Mean recovery per concentration	$100.0\% \pm 2.0\%$
Limit of detection	
Signal-to-noise ratio	$\geq 3:1$
Limit of quantitation	
Signal-to-noise ratio	$\geq 10:1$
Linearity/Range	
Correlation coefficient	>0.99
y-Intercept	$\pm 10\%$
Response ratios RSD	$\leq 5.0\%$
Visual	Linear
Ruggedness	
See intermediate precision	
Robustness	
System suitability met	Yes
Solution stability	$\pm 2\%$ change from time zero
Specificity	
Resolution from main peak	>2 min. (retention time)

were chosen on the basis of the minimum requirements of the FDA for stability methods. The acceptance criteria values were based on typical expectations for an active drug substance.

E. Perform/Verify Instrument Qualification

The instrument was qualified by performing IQ, OQ, and PQ through a qualification protocol.

F. Perform Method Validation/Evaluate Results

1. Precision

One analyst prepared a solution at 100% of the target assay concentration and injected it six times on one HPLC instrument to evaluate assay repeatability. Two analysts each prepared a solution at 100% of the target assay concentration and injected it in triplicate on different HPLC instruments on three different days to evaluate intermediate precision. Results are summarized in Tables 4 and 5. The assay repeatability results for peak area and retention time meet the acceptance criteria of $\leq 1\%$ RSD. The intermediate precision results for peak area meet the acceptance criterion of $\leq 2\%$ RSD.

2. Accuracy

Accuracy was evaluated by spiking known amounts of the drug substance (compound A) into a sample, analyzing the sample, and calculating the percent recovered. For the recovery experiments, three solutions were prepared that covered a range of 50–150% of the target concentration and injected in triplicate by one analyst on one HPLC instrument. Results are summarized in Table 6. The accuracy results for mean concentration recovery meet the acceptance criterion of $100.0\% \pm 2.0\%$.

Table 4 Assay Repeatability

Injection number	Peak area (μV s)	Retention time (min)
1	7,977,317	5.52
2	7,976,891	5.55
3	7,984,795	5.57
4	7,976,148	5.70
5	8,003,740	5.51
6	7,989,758	5.53
Average	7,984,775	5.56
Standard deviation	10,739.70	0.0703
%RSD	0.135	1.264

Table 5 Intermediate Precision

Injection number/day	Analyst 1 peak area (area count)	Analyst 2 peak area (area count)
1/1	8,010,482	8,067,462
2/1	8,022,004	8,064,145
3/1	8,028,783	8,041,417
1/2	7,950,075	8,029,446
2/2	8,100,399	7,955,633
3/2	8,050,392	7,973,597
1/3	7,976,669	8,007,109
2/3	7,952,224	8,058,553
3/3	7,977,675	7,986,002
Average	8,007,634	8,020,374
Standard deviation	49,161.93	41,612.51
%RSD	0.614	0.519

3. Linearity/Range

Linearity/range was evaluated over the whole range from the LOQ to 150% of the target assay concentration. One analyst prepared a stock solution of the drug substance and prepared samples by serial dilution until the peak area was no longer detected by the analytical method. The samples were injected in duplicate. From this data, a graph of the average area counts vs. the sample concentration was plotted (see Fig. 4). Using a least squares regression, a linear relationship was observed from 150% to 0.04%. The relevant statistics are slope of 36,536,164, y-intercept of 1877.285, and correlation coefficient of 0.999996. Results are summarized in Table 7. The linearity results meet the acceptance criteria.

4. Robustness

Robustness was evaluated by deliberately varying the experimental parameters outlined in the method and determining the subsequent effect on the system suitability parameters. The parameters investigated include column manufacturer, column temperature (50°C specified), flow rate (1.5 mL/min), detector wavelength (228 nm), and gradient composition (linear gradient from 15% B to 70% B over 20 min). For each variation, the system suitability result was the same as that obtained for an HPLC run conducted using the exact conditions specified in the method. Column manufacturer, column temperature, flow rate, and detector wavelength were determined not to be critical parameters. The gradient composition was determined to be a critical parameter and should be specified as such in the method description. Results are summarized in Table 8.

Solution stability was evaluated by performing and comparing area percent determinations from solutions held at ambient temperature for a 3 day period. One

Table 6 Accuracy

Injection number	Calculated concentration (mg/mL)	Observed concentration (mg/mL)	Recovery (%)
50% of target			
1	0.106	0.104	98.11
2	0.106	0.100	94.34
3	0.106	0.112	105.66
Average		0.105	99.37
100% of target			
1	0.212	0.209	98.58
2	0.212	0.217	102.36
3	0.212	0.198	93.40
Average		0.208	98.11
150% of target			
1	0.318	0.310	97.48
2	0.318	0.316	99.37
3	0.318	0.330	103.77
Average		0.319	100.201

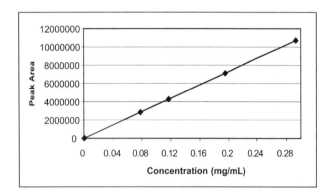

Figure 4 Linearity from 150% to LOQ of target assay concentration.

Table 7 Linearity

Concentration (mg/mL)	Average peak area (area count)	Response ratio
0.293	10,711,428	36,557,773
0.195	7,111,293.6	36,468,172
0.117	4,285,327.8	36,626,733
0.078	2,859,751.4	36,663,479
0.00078	28,511.97	36,553,808
0.000117	3,884.3	33,199,145
0.000078	2,767.5	35,480,769
%RSD		3.548

analyst injected the solution in triplicate every 24 h. The solution stability results meet the acceptance criterion of $\pm 2\%$ change from time zero for all 3 days. Results are summarized in Table 9.

5. Specificity

Specificity was demonstrated by spiking known process impurities into a standard solution and verifying that each impurity is separated from the main peak. Each impurity was well separated from the main peak, with the nearest peaks eluting more than 2 min before and after the main peak; therefore, the acceptance criterion was met. Results are summarized in Table 10.

G. Method Validation Complete

The decision was made not to modify the method. The gradient was determined to be critical and this will be explicitly stated in the method description. As part of the ongoing use of the method, the system suitability tests identified in Section IV.D. will need to be performed on a routine basis. Although LOD and LOQ are not required for the type of analysis presented, as was indicated in Table 1, they are included here to exemplify how each of them is calculated.

1. Limit of Detection

LOD was based on the lowest concentration that gave a minimum signal-to-noise of at least 3:1. One analyst prepared a stock solution at 100% of the target assay

Table 8 Robustness—Method Parameter Variation

Method parameter	System repeatability (RSD ≤ 2.0%)	k'	R_s	Tailing factor of main peak (<3)	System suitability met?
Column manufacturer					
Manufacturer 1	1.8	4.9	2.3	1.7	Yes
Manufacturer 2	1.5	4.6	2.1	1.5	Yes
Column temperature (°C)					
45	0.9	4.8	2.4	1.8	Yes
50	1.8	4.6	2.3	1.7	Yes
55	0.4	5.1	2.5	1.6	Yes
Flow rate (mL/min)					
1.3	0.5	5.2	2.7	1.8	Yes
1.5	1.8	4.6	2.3	1.7	Yes
1.7	0.7	4.3	2.8	1.7	Yes
Detector wavelength (nm)					
220	1.5	5.2	2.4	1.7	Yes
228	1.8	4.6	2.3	1.7	Yes
240	0.8	5.3	2.1	1.7	Yes
Gradient composition (%)					
67	2.8[a]	5.0	2.5	2.7	No
70	1.8[a]	4.6	2.3	1.7	Yes
73	1.4[a]	5.4	2.2	1.7	Yes

[a]End of gradient phase B composition varied.

Table 9 Solution Stability

Time (days)	Area percent (%)	Difference from zero point
0	99.5	—
1	98.9	0.60
2	99.8	0.30
3	99.2	0.30

Table 10 Specificity

Compound	Retention time (min)	Relative retention time
Impurity 1	4.47	0.69
Compound A	6.52	1.00
Impurity 2	8.57	1.31
Impurity 3	11.33	1.74

concentration from which low level standards were prepared by serial dilution; these low level standards were injected in triplicate to determine LOD. At each concentration of less than 1.0% of the target assay concentration, the signal-to-noise ratio was calculated. Results are summarized in Table 11. On the basis of a sample concentration of 0.2 mg/mL, the LOD for this method was determined to be 0.02% by area (~0.00004 mg/mL).

2. Limit of Quantitation

LOQ was based on the lowest concentration that gave a minimum signal-to-noise of at least 10:1. LOQ was determined using the same data generated to calculate the LOD. Results are summarized in Table 11. On the basis of a sample concentration of 0.2 mg/mL, the LOQ for this method was determined to be 0.04% by area (~0.00008 mg/mL).

Table 11 Signal-to-Noise Results

Concentration (mg/mL)	Average peak area (area count)	Standard deviation	%RSD	Signal-to-noise
0.000117	3884	327.8	8.44	17.2
0.000078	2467	79.9	3.24	13.2
0.000039	1144	26.6	2.29	6.6
0.000016	308	73.1	23.75	2.6

V. CONCLUDING REMARKS

Analytical method validation in the pharmaceutical industry is beneficial, not only from a regulatory perspective but also from a scientific point of view, as it ensures that the methods used to analyze the sample are accurate and reliable. The purpose of the analytical method is to determine which method validation requirements are applicable. A discussion of the typical validation requirements is included in this chapter. Each section describing the method validation parameters contains a list of examples gathered from a reference search of possible procedures and acceptance criteria, which together with any regulatory requirements can be used as guidance. The intent of the list is not to indicate what should be done, but to serve as a starting point for evaluating possible procedures and acceptance criteria in different applications. In the end, the requirements for each analytical method and its acceptance criteria should be evaluated on a case-by-case basis.

The benefits of analytical method validation in the pharmaceutical industry can be extended to other industrial sectors that perform some type of analytical testing. Analytical method validation should be viewed as an important developmental tool, not merely a paper-generating activity.

REFERENCES

Bressolle, F., Bromet-Petit, M., Audran, M. (1996). Validation of liquid chromatographic and gas chromatographic methods: application to pharmacokinetics. *J. Chromatogr. B* 686:3–10.

Bruce, P., Minkkinen, P., Riekkola, M. L. (1998). Practical method validation: validation sufficient for an analysis method. *Mikrochim. Acta* 128(1–2):93–106.

Buick, A. R., Doig, M. V., Jeal, S. C., Land, G. S., McDowall, R. D. (1990). Method validation in the bioanalytical laboratory. *J. Pharm. Biomed. Anal.* 8(8–12): 629–637.

Castillo, M. A., Castells, R. C. (2001). Initial evaluation of quantitative performance of chromatographic methods using replicates at multiple concentrations. *J. Chromatogr. A* 921: 121–133.

Clarke, G. S. (1994). The Validation of analytical methods for drug substances and drug products in UK pharmaceutical laboratories. *J. Pharm. Biomed. Anal.* 12(5):643–652.

Coleman, D., Vanatta, L. (2003). Statistics in analytical chemistry—Part 5 calibration: calibration design. *Am. Lab.* 35(12):30–31.

Crowther, J. B. (2001). Validation of pharmaceutical methods. In: Ahuja, S., Scypinski, S., eds. *Handbook of Modern Pharmaceutical Analysis*. San Diego: Academic Press, pp. 415–443.

Fabre, H., Altria, K. D. (2001). Key points for validating CE methods, particularly in pharmaceutical analysis. *LCGC* 19(5):498–505.

FDA (1994). *Reviewer Guidance: Validation of Chromatographic Methods*. Food and Drug Administration: Center for Drug Evaluation and Research.

FDA (1997). *Code of Federal Regulations (21 CFR)*. Washington, DC: Food and Drug Administration.

FDA (2000). *Guidance for Industry: Analytical Procedures and Methods Validation. Chemistry, Manufacturing, and Controls Documentation. Draft Guidance: Validation of Chromatographic Methods*. Food and Drug Administration.

Ferretti, R., Gallinella, B., La Torre, F., Turchetto, L. (1998). Validated chiral high-performance liquid chromatographic method for the determination of trans-(−)-paroxetine and its enantiomer in bulk and pharmaceutical formulations. *J. Chromatogr. B* 710:157–164.

Grisanti, V., Zachowski, E. J. (2002). Operational and performance qualification. *LCGC N. Am.* 20(4):356–362.

IUPAC (1989). Guidelines for collaborative study procedure to validate characteristics of a method of analysis. *J. Assoc. Anal. Chem.* 72(4):694–704.

Hall, G., Dolan, J. W. (2002). Performance qualification of LC systems. *LCGC N. Am.* 20(9):842–848.

Hokanson, G. C. (1994). A life cycle approach to the validation of analytical methods during pharmaceutical product development, Part II: changes and the need for additional validation. *Pharm. Technol.* October:92–100.

ICH (1994). *Text on Validation of Analytical Procedures: Q2A. Recommended for Adoption at Step 4 of the ICH Process*. International Conference of Harmonisation of Technical Requirements for Registration of Pharmaceutical for Human Use.

ICH (1996). *Validation of Analytical Procedures: Methodology Q2B. Recommended for Adoption at Step 4 of the ICH Process*. International Conference of Harmonisation of Technical Requirements for Registration of Pharmaceutical for Human Use.

Izco, J. M., Tormo, M., Harris, A., Tong, P. S. (2003). Optimization and validation of a rapid method to determine citrate and inorganic phosphate in milk by capillary electrophoresis. *J. Dairy Sci.* 86(1):86–95.

Jenke, D. R. (1996a). Chromatographic method validation: a review of current practices and procedures. I. General concepts and guidelines. *J. Liq. Chromatogr. Related Technol.* 19(5):719–736.

Jenke, D. R. (1996b). Chromatographic method validation: a review of current practices and procedures. II. Guidelines for primary validation parameters. *J. Liq. Chromatog. Related Technol.* 19(5):737–757.

Jenke, D. R. (1996). Chromatographic method validation: a review of current practices and procedures. III. Ruggedness, revalidation and system suitability. *J. Liq. Chromatogr. Related Technol.* 19(12):1873–1891.

Jimidar, M., Hartmann, C., Cousement, N., Massart, D. L. (1995). Determination of nitrate and nitrite in vegetables by capillary electrophoresis with indirect detection. *J. Chromatogr. A* 706:479–492.

Lee, K. R., Bongers, J., Gulati, D., Burman, S. (2000). Statistical validation of reproducibility of HPLC peptide mapping for the identity of an investigational drug compound based on principal component analysis. *Drug Dev. Ind. Pharm.* 26(10):1045–1057.

Mouatassim-Souali, A., Tamisier-Karolak, S. L., Perdiz, D., Cargouet, M., Levi, Y. (2003). Validation of a quantitative assay using GC/MS for trace determination of free and conjugated estrogens in environmental water samples. *J. Sep. Sci.* 26:105–111.

Needham, S. (2003). Bioanalytical method validation: example, HPLC/MS/MS bioanalysis of an anti-nerve gas agent drug for homeland security. *Am. Pharm. Rev.* 6(2):86–91.

Qi, M., Chen, R., Cong, R., Wang, P. (2003). A validated method for simultaneous determination of piperacillin sodium and sulbactam sodium in sterile powder for injection by HPLC. *J. Liq. Chromatogr. Related Technol.* 26(4):665–676.

Rabouan, S., Olivier, J. C., Guillemin, H., Barthes, D. (2003). Validation of HPLC analysis of aspartate and glutamate neurotransmitters following o-phthaldialdehyde–mercaptoethanol derivatization. *J. Liq. Chromatogr. Related Technol.* 26(11):1797–1808.

Rathore, A. S., Kennedy, R. M., O'Donnell, J. K., Bemberis, I., Kaltenbrunner, O. (2003). Qualification of a chromatographic column why and how to do it. *Biopharm. Int.* March:30–40.

Rathore, A. S., Kurumbail, R. R., Lasdun, A. M. (2002). Requirements for validating a capillary isoelectric focusing method for impurity quantitation. *LCGC N. Am.* 20(11):1042–1048.

Shabir, G. A. (2003). Validation of high-performance liquid chromatography methods for pharmaceutical analysis: understanding the differences and similarities between validation requirements of the US Food and Drug Administration, the US Pharmacopeia and the International Conference on Harmonization. *J. Chromatogr. A* 987:57–66.

Shah, V. P., Midha, K. K., Dighe, S., McGilveray, I. J., Skelly, J. P., Yacobi, A., Layloff, T., Viswanathan, C. T., Cook, C. E., McDowall, R. D., Pittman, K. A., Spector, S. (1991). Analytical methods validation: bioavailability, bioequivalence and pharmacokinetic studies. *Eur. J. Drug Metab. Pharmacokinet.* 16(4):249–255.

Shah, V. P., Midha, K. K., Findlay, J. W. A., Hill, W. M., Hulse, J. D., McGilveray, I. J., McKay, G., Miller, K. J., Patnaik, R. N., Powell, M. L., Tonelli, A., Viswanathan, C. T., Yacobi, A. (2000). Bioanalytical method validation: a revisit with a decade of progress. *Pharm. Res.* 17(12):1551–1557.

Szepesi, G. (1990). Quantitation in thin-layer chromatography. In: Grinberg, N., ed. *Modern Thin Layer Chromatography*. San Diego: Marcel Dekker, pp. 249–284.

Tamisier-Karolak, S. L., Tod, M., Bonnardel, P., Czok, M., Cardot, P. (1995). Daily validation procedure of chromatographic assay using gaussoexponential modeling. *J. Pharm. Biomed. Anal.* 13:959–970.

Thompson, M., Ellison, S. L., Wood, R. (2002). Harmonized guidelines for single-laboratory validation of methods of analysis (IUPAC Technical Report). *Pure Appl. Chem.* 74(5):835–855.

USP (2003). *United States Pharmacopeia: USP 26. The National Formulary: NF 21*. Washington, DC: United States Pharmacopeial Convention.

Virrlichie, J. L., Ayache, A. (1995). A ruggedness test model and its application for HPLC method validation. *STP Pharma Practiques* 5(1):49–60.

Zanetti, M., Huber, L. (1996). Validating automated analytical instruments to meet regulatory and quality standard regulations. *Mikrochim. Acta* 123(1–4):23–31.

Ziaková, A., Brandšteterová, E. (2003). Validation of HPLC determination of phenolic acids present in some *Lamiaceae* family plants. *J. Liq. Chromatogr. Related Technol.* 26(3):443–453.

Index

Atomic absorption spectrometry, 75
  AFS spectrometers, 118
  Calibration, 119
  Chemical vapor generation, 112
  Flame OES spectrometers, 118
  GFAAS, 101
  High resolution, 92
  History, 76
  Line width, profile, 80
  Spectral interferences, 109
  Spectrometers, 82
  Techniques, 95
Atomic emission spectrometry, 59
  Instruments, 60
  Principles, 59
Atomic spectra, origin, characterization, 79
Auger electron spectroscopy, 399
  Detectors, 410
  Interpretation of spectra, 413
  Quantitative analysis, 418
  Spectrometer design, 404

Beer-Lambert law, 81, 127
Biosensor technology, 681
  Biological materials, 681
  Transducers, 682

Capillary electrophoresis, 803
  Applications, 813
  Data integration, 811
  Instrumentation, 806
  Method development &
    optimization, 813
  Modes, 805
  Sample preparation, 812
  Validation, 813
Chemometrics, 12
Chiroptical spectroscopy, 271
  Circular dichroism (CD), 276

CPL spectroscopy, 278
Fluorescence circular dichroism, 288
Optical rotation (ORD), 272
Polarization of light, 272
Raman optical activity, 285
Vibrational optical activity, 283
Chromatography
  Countercurrent chromatography, 893
  Gas chromatography, 727
  Gel permeation chromatography
    (GPC), 827
  High performance liquid
    chromatography (HPLC), 687
  Ion chromatography, 571
  Size exclusion chromatography, 827
  Supercritical fluid chromatography, 759
  Thin layer chromatography, 995
Computers, lab use of, 3
  A/D conversion, 8
  Analog signals, 7
  Backup, 6
  Cache, 3
  Chemometrics, 12
  Data filtering, digital, 11
  Data storage, 19
  Data transfer/interfaces, 7
  Design considerations, 1
  Disk storage, 3
  Maintenance, 5
  Memory, 3
  Pattern recognition, 16
  Sampling rate, 8
  Signal transmission, 10
  Video, 4
  Virus protection, 6
Conductance, electrolytic, 561
Conductometric titrimetry, 570
Countercurrent chromatography, 893
  Affinity CCC, 931

CCC-MS, 931
  Hydrodynamic equilibrium systems, 901
  Hydrostatic equilibrium systems, 895
  Instrumentation, 908
  Partition coefficients, 936
  pH zone refining CCC, 927
  Solvent systems, two-phase, 931, 935
  Synchronous planetary motion, 902

Electrochemical stripping
    voltammetry, 545
  Applications, 557
  Instrumentation, 552
  Stripping chronopotentiometry, 549
Electrolytic conductance, 561
  Electrodeless measurements, 574
  Instrumentation, 566
  Ion mobility & transport, 564
  Principles, 562
  Strong & weak electrolytes, 563
Electron paramagnetic resonance, 349
  Calibration, 378
  EPR imaging, 388
  Experimental, 350, 381
  Magnet system, 370
  Multiple resonance, 355
  Pulse, 355
  Relaxation time, 360
  Sample processing, 374
  Saturation transfer spectroscopy, 387
  Spectrometer, 363
  Spin–spin coupling, 360
Electrophoresis, capillary, 803
EPR, 349

Field-flow fractionation, 871
  Brownian retention, 873
  Field programming, 877
  Optimization, 879

Field-flow fractionation (*Contd.*)
  Resolution, 877
  Sample injection–relaxation, 876
  Theory, principles, 872
FIR spectroscopy, 209
Flow injection analysis
  Dispersion in, 24
  Gradient techniques, 30
  FIA/SIA, 39
  Instrumentation, 23
  Monographs, 22
  Principles, 23
  Solvent extraction, 39
  Unique features of, 25
Fluorescence, theory, 141

Gas chromatography, 727
  Columns, GC, 731
  Data system, 752
  Detectors, 741
  Direct injection, 737
  Fast GC, 754
  Gas supply system, 730
  GC-MS, 744
  Headspace analysis, 738
  Modes, 728
  Multidimensional GC, 750
  Optimization of parameters, 733
  Oven designs, 740
  Purge & trap, 739
  Pyrolysis GC, 739
  Sample inlets, 736
  Separation system, 728
  Split injection, 737
  Splitless injectors, 737
  Stationary phases, 732
Gel permeation chromatography
    (GPC), 827
  Band broadening, 840
  Chain branching, 863
  Chromatographic system, 828
  Columns, 844
  Concentration detectors, 834
  Enthalpic interactions, 840
  Entropy controlled partitioning, 832
  High MW saturation problem, 839
  Hydrodynamic volume, 832
  Instrumentation, 841
  Intrinsic viscosity, 832
  Molar mass calibration, 835
  Molar mass detectors, 836
  Molar mass distribution, 829
  Resolution & column efficiency, 833
  Sample preparation, 843
  Solvents, 841

High Performance Liquid Chromatography
    (HPLC), 687
  Chromatographic instrumentation, 696

Column design &
    performance, 703, 706
Column ovens, 707
Comprehensive LC system, 724
Control of retention, 693
Detectors, 713
Molecular interactions, 690
Valves, 699, 705
Plate theory, 688
Pumps, 699
Solvent programmers, 697
Thermodynamics of retention, 692
HPLC hyphenated techniques, 945
  LC-FTIR, 966
  LC-GC, 982
  LC-ICP, 978
  LC-MS, 946
  LC-NMR, 974
  Multiple hyphenation, 980
Hyphenated techniques (see HPLC
    hyphenated techniques)

Inductively coupled plasma, 61
  Discharge, 61
  Emission detection, processing, 62, 69
  Nebulizers, 63
  Optics, 66
  Radial ICP, 71
  Spray chambers, 65
Infrared spectroscopy, 163
  Instrumentation, 170
  Fourier transform, 177
  Window materials, 197
Ion chromatography, 571

Lab-on-a-chip (see Microfluidic
    lab-on-a-chip)
Luminescence spectra, 144

Mass spectrometry, 429
  Ionization methods, 430
  Mass analyzers, 434
Microfluidic lab-on-a-chip, 581
  Cellular analysis, 627
  Detection, 617
  DNA analysis, 635
  Microfluidic flow, 594
  Micromachining methods, 582
  Protein analysis, 650
  Sample introduction, 602
  Sample preconcentration, 607
  Separation processes, 610
Molecular fluorescence,
    phosphorescence, 141
  Applications, 153
  Instrumentation, 145
  Theory, 141
  Wavelength calibration, 156

Near infared spectrophotometers, 127
  Instrumentation, 206
Nuclear magnetic resonance, 295
  Data analysis & interpretation, 334
  Experimental methods, 317
  Instrument design, 296
  Internal spin interactions, 311
  Magnetic field gradients, 306
  Multidimensional NMR, 322
  NMR-active nuclei, 309
  Relaxation phenomena, 314
  Theoretical background, 307

Optical emission spectrometry
  Principles, 57

Phosphorescence, theory, 141
Photoacoustic spectroscopy, 257
  Plant science applications, 262
  Principle, 259
  Relevance to plant science, 261
Potentiometry, 509
  Activity coefficients, 510
  Electrochemical cells, 509
  Formal potentials, 512
  Indicator electrodes, 516
  Instrumentation, 523
  Nernst equation, 512
  Reference electrodes, 513

Raman spectroscopy, 163
  Instrumentation, 211

Sequential injection analysis
  Lab-on-valve, 33
  Preconcentration procedures, 36, 44
Size exclusion chromatography (SEC), 827
  See also Gel permeation
    chromatography
Spectrometry
  Atomic absorption, 75
  Atomic emission, 59
  Mass, 429
  NMR, 295
  Optical emission, 59
  X-ray, 239
Spectrophotometers
  Architecture, 131
  Characteristics, 129
  Detectors, 133
  Near infrared, 127
  Signal processing, 136
  Ultraviolet, 127
  Visible, 127
Spectroscopy
  Auger electron, 399
  Chiroptical, 271
  FIR, 209

Index

Spectroscopy (*Contd.*)
  Infrared, 163
  Photoacoustic, 255
  Raman, 163
  Vibrational, 163
  X-ray photoelectron, 399
Stripping voltammetry, 545
Supercritical fluid chromatography, 759
  Chiral SFC, 797
  Flow behavior, 784
  Fluids, 760
  Instrumentation, 769
  Method development, 785
  Mobile phases, 764
  Modifiers for SFC, 786
  OT-column SFC, 776
  Packed-column SFC, 770
  Phase behavior, 760
  SFC–MS, 791
  Testing operation, 784

Thermoanalytical instrumentation, 445
  Calorimetry, 478
  Dielectric thermal analysis, 476
  Differential scanning calorimetry (DSC), 457
  Differential thermal analysis (DTA), 457
  DTA-Rheometry, 494
  Evolved gas analysis (EGA), 467
  Nomenclature, 447
  Optical techniques, 493
  Simultaneous techniques, 491
  Thermogravimetric analysis (TGA), 449
  Thermomechanical methods (TMA), 471
Thin layer chromatography (TLC), 995
  Development of the chromatogram, 999
  Documentation of chromatograms, 1004
  Forced-flow planar chromatography, 1001
  Mobile phases, 997
  Preparative layer chromatography, 1009
  Quantitative analysis, 1006
  Sample application, 998
  Sample preparation, 996
  Stationary phases, 997
  Thin layer radiochromatography, 1010
  TLC-IR, 1009
  TLC-MS, 1009
  Zone detection, 1003
  Zone identification, 1005

Ultraviolet spectrophotometers, 127

Validation of chromatographic methods, 1015
  Accuracy, 1024
  Application validation example, 1028
  Instrument qualification, 1017
  Limit of detection, 1023
  Limit of quantitation, 1023
  Precision, 1020
  Prevalidation requirements, 1017
  Range, 1024
  Robustness, 1026
  Ruggedness, 1025
  Solution stability, 1027
  Specificity, 1027
  System suitability, 1019
Vibrational spectroscopy, 163
  Basic principles, 166

Visible spectrophotometers, 127
Voltammetry, 529
  Cyclic voltammetry, 535
  Faradaic current, 534
  Hydrodynamic voltammetry, 536
  Instrumentation, 529
  IR drop, 531
  Mass transport, 532
  Non-Faradaic current, 534
  Oxidation/reduction, 530
  Polarization, 531
  Polarography, 536
  Waveforms, 538

X-ray photoelectron spectroscopy (XPS), 399
  Angle resolved XPS, 420
  Charge compensation, 406
  Detectors, 410
  Imaging & mapping, 411
  Interpretation of spectra, 413
  Quantitation, 418
  Spectrometer design, 404
X-ray sources, 405
X-ray spectrometry, 239
  Auger effect, 241
  Crystal dispersive elements, 244
  Detectors, 245
  Diffraction, 240
  Energy dispersive, 241
  Instrumentation, 242
  Synchrotron sources, 254
  Total reflection, 252

Zeeman effect, 82